Geological Setting, Palaeoenvironment and Archaeology of the Red Sea

Najeeb M. A. Rasul • Ian C. F. Stewart
Editors

Geological Setting, Palaeoenvironment and Archaeology of the Red Sea

هيئة المساحة الجيولوجية السعودية
SAUDI GEOLOGICAL SURVEY
www.sgs.org.sa

Springer

Editors
Najeeb M. A. Rasul
Center for Marine Geology
Saudi Geological Survey
Jeddah, Saudi Arabia

Ian C. F. Stewart
Stewart Geophysical Consultants Pty. Ltd.
College Park, SA, Australia

ISBN 978-3-319-99407-9   ISBN 978-3-319-99408-6 (eBook)
https://doi.org/10.1007/978-3-319-99408-6

Library of Congress Control Number: 2018952604

This Springer imprint is published by the registered company Springer Nature Switzerland AG
The registered company address is: Gewerbestrasse 11, 6330 Cham, Switzerland

Preface

The Red Sea has a unique tectonic history, environment and biology. It is a young ocean basin that along its length has undergone or is undergoing the transition from a continental rift to true oceanic seafloor spreading, the nature of which is still open to vigorous debate. In addition, due to its semi-enclosed nature and location within an arid region, the environment is affected by high evaporation rates that, together with limited contact with the Indian Ocean, result in high temperatures and salinities. Lower sea levels in the past have also led to extensive evaporite deposition within its basin, while brines and metallic deposits in the axial deeps have been the subject of considerable research. All of this has had a far-reaching impact on the marine and terrestrial life of the region and on its human inhabitants. As a human environment, the Red Sea region is of unusual archaeological and historical interest. It has always been the primary gateway for contact and movement between Africa and Asia, beginning far back in the Quaternary with the earliest expansion of our human ancestors out of Africa, and in later periods becoming a primary conduit for seaborne trade between southern Asia, Arabia, Africa and the Mediterranean.

This is one of a pair of volumes that together represent a successor to an earlier volume published in this series in 2015 under our joint editorship as 'The Red Sea: The Formation, Morphology, Oceanography and Environment of a Young Ocean Basin'. The amount of new information that has become available since then is testament to the range and vigour of new research now being carried out in the region, much of it in Saudi Arabia under the sponsorship of the Saudi Geological Survey, and to the level of international interest. Indeed, so much new research has taken place that we have divided the material into two volumes, this one, which concentrates on geological, environmental and archaeological issues, and a second volume concerned with the oceanography and biology of the Red Sea.

A wide range of topics is examined in this volume, from the geological history of the region to its past and present environments and their effects on prehistoric and historic human activities. The chapters aim to present some of the current thinking and summaries of research in each field of study including useful reference lists for further study.

As with the earlier volume referred to above, which was the outcome of a workshop held in Jeddah, Saudi Arabia, in 2013, most of the chapters in this volume were originally presented at a workshop held in Jeddah, from 15 February to 17 February 2016, under the auspices of the Saudi Geological Survey (SGS), and have been extensively rewritten, independently reviewed and edited for publication.

The support of the Survey in the preparation of this volume is greatly appreciated, and we would like to thank all those who have been involved in its production. We would especially like to thank Dr. Zohair A. Nawab, former President of SGS, and Dr. Abdullah M. Alattas, former Assistant Vice President, as well Eng. Hussain M. Al Otaibi, President of SGS and Mr. Salah A. AlSefry, Assistant Vice President for Technical Affairs. Colleagues at the SGS and the Center for Marine Geology are also thanked for making the workshop a success. Mr. Louiesito Abalos played a substantial part in the preparation of material for publication. We are happy to note our appreciation for the contributions of the technical referees who improved the contents of the chapters as well as the assistance of Femina Joshi Arul Thas, Project Manager, Banu Dhayalan, Project Coordinator, Janet Sterritt-Brunner, Production Books Project Coordinator and Dr. Nabil Khélifi, Senior Editor, of Springer Nature. The assistance and suggestions of Dr. Geoff Bailey, in particular, in preparing some of the chapters greatly helped in the final stages of the reviewing process.

Jeddah, Saudi Arabia — Najeeb M. A. Rasul
College Park, Australia — Ian C. F. Stewart

Contents

Introduction to Geology, Palaeoenvironment and Archaeology of the Red Sea

Najeeb M. A. Rasul, Ian C. F. Stewart, Geoff N. Bailey, and Zohair A. Nawab

1 Introduction

The present volume follows on from an earlier work edited by Rasul and Stewart (2015), in which an extensive introduction (Rasul et al. 2015) outlined the main features of the Red Sea, including an overview of the regional tectonics, geology and geophysics, oceanography and biology. Only one chapter in that volume referred to archaeology, and summarized knowledge then existing about the general Stone Age record with particular emphasis on the role of the Red Sea as a corridor for the earliest expansion of human populations out of Africa during the Pleistocene and the impact of Pleistocene sea-level and climate change on patterns of occupation in the Arabian Peninsula (Bailey 2015). Since that chapter was written, there has been a considerable expansion of archaeological research in Saudi Arabia involving joint Saudi-international teams under the sponsorship of the Saudi Commission for Tourism and National Heritage, along with ongoing geological investigations under the aegis of the Saudi Geological Survey.

Two major projects funded by the European Research Council have got underway in Saudi Arabia to explore in greater detail the early Stone Age history of the region: DISPERSE, focused primarily on the coastal zone of the Red Sea and the submerged landscapes made available to human occupation during periods of low sea level, and PALAEODESERTS, focused mainly on the desert interior and the expansion of early populations made possible by climate change. Since the likely time depth of human occupation extends back to at least 500,000 years ago and most likely much earlier still, geological and climatic changes would have had a major impact on the territory and resources available for human occupation, and of course on the preservation and visibility of archaeological data, including tectonic and volcanic activity, patterns of erosion and sedimentation, and changes in sea-level change, climate and hydrology. There is, therefore, a natural interest in common between geologists and archaeologists, and especially in obtaining improved geological control on changes in the natural environment. Both projects referred to above are multi-national, multi-disciplinary projects including specialists with a range of archaeological, geological and geochronological skills. Their results dominate the archaeological chapters in this volume, and in some cases refer as much to geological issues as to archaeological ones, and the inter-relationship between these two scientific domains.

This volume brings together some of the results of this new phase of research, with 22 chapters primarily on geological themes, and 11 chapters on primarily archaeological ones. In this introductory chapter, we briefly summarise some of the key features of the Red Sea Basin, and provide an introduction to the themes of the ensuing chapters.

2 Location, Bathymetry and Statistics

The Red Sea is a semi-enclosed, elongated warm body of water about 2000 km long with a maximum width of 355 km, a surface area of roughly 458,620 km^2, and a volume of $\sim$250,000 km^3 (Head 1987). The Red Sea is bounded by nine countries, with numerous coastal lagoons

N. M. A. Rasul (✉)
Center for Marine Geology, Saudi Geological Survey, Jeddah, Saudi Arabia
e-mail: najeeb_rasul@hotmail.com; rasul.nm@sgs.org.sa

I. C. F. Stewart
Stewart Geophysical Consultants Pty. Ltd, Adelaide, SA 5069, Australia
e-mail: stewgeop@senet.com.au

G. N. Bailey
Department of Archaeology, University of York, King's Manor, York, YO1 7EP, UK

G. N. Bailey
College of Humanities, Arts and Social Sciences, Flinders University, GPO Box 2100 Adelaide, SA 5000, Australia

Z. A. Nawab
Saudi Geological Survey, Jeddah, Saudi Arabia

N. M. A. Rasul and I. C. F. Stewart (eds.), *Geological Setting, Palaeoenvironment and Archaeology of the Red Sea*,
https://doi.org/10.1007/978-3-319-99408-6_1

and a large number of islands of various dimensions and extensive groups of shoals; it is bifurcated by the Sinai Peninsula into the Gulf of Aqaba and the Gulf of Suez at its northern end (Fig. 1). The sea is connected to the Arabian Sea and Indian Ocean via the Gulf of Aden in the south through the narrow Strait of Bab el Mandab, which has a minimum width of only 30 km, where the main channel is about 310 m deep and 25 km wide at Perim Island (Morcos 1970). Although the Hanish Sill at 13°44'N has a maximum depth of only 137 m, it is likely that the Red Sea has remained connected to the Gulf of Aden for at least the past 400,000 years (Lambeck et al. 2011). However, during the Last Glacial Maximum (LGM), the water depth over the Hanish Sill is estimated to have fallen to only 25–33 m (Biton et al. 2008; Lambeck et al. 2011), with considerable effects on the Red Sea circulation and ecology (Trommer et al. 2011).

The Red Sea has three distinct depth zones; shallow shelves less than 50 m in depth (about 25%), deep shelves depth ranging between 500 and 1000 m, and the central axis with depths between 1000 and 2900 m. The continental slope has an irregular profile, with a series of steps down to about 500 m depth. The 15% of the Red Sea that forms the narrow axial trough is over 1000 m in depth and contains the bathymetric depressions or deeps, some containing hot saline brines (e.g., Hovland et al. 2015) and metalliferous sediments, that were formed by the spreading of the sea; recent data along the axis (Augustin et al. 2014) suggest that the western Suakin Deep, with a depth of 2860 m at 19.6°N is the deepest part of the Red Sea Rift.

The Red Sea is one of the youngest oceanic zones on earth, and together with the Gulf of Aqaba-Dead Sea transform fault it forms the western boundary of the Arabian plate, which is moving in a north-easterly direction. The plate is bounded by the Bitlis Suture and the Zagros fold belt and subduction zone to the north and northeast, and the Gulf of Aden spreading centre and Owen Fracture Zone to the south and south-east (Fig. 2, modified after Stern and Johnson 2010). Along most of its length the Red Sea forms a rift through the Precambrian Arabian-Nubian shield. The Red Sea is of considerable interest, as in the north it is probably undergoing the transition from crustal stretching and thinning to true seafloor spreading, while south of about 24°N the transition has already occurred. The tectonic history of the Red Sea has recently been discussed by Bonatti et al. (2015) and Bosworth (2015). In the Red Sea, the central spreading axis is opening at an average rate of about 1.6 cm per year, with the rate of opening being greater in the south than in the north, as indicated by the width of the Red Sea between the topographic shoulders of the Precambrian basement. These are about 200 km apart in the north at latitude 27°N and 350 km apart in the south at latitude 17°N.

The southern end of the Red Sea joins the Gulf of Aden spreading centre and the northern end of the East African Rift Zone at a triple junction in the Afar region, which may be underlain by an upwelling deep mantle convection plume (e.g., Daradich et al. 2003), resulting in extensive volcanism in the region. In the early stages of rifting, prior to a permanent connection to the Gulf of Aden, thick evaporite deposits accumulated in the Red Sea, and on the shelf and marginal areas these deposits are overlain by Recent sediments. The sea only turned into an open marine environment when the Gulf of Suez in the north and Indian Ocean in the south became connected in the Pliocene.

In the north, the Red Sea bifurcates into the Gulf of Aqaba and Gulf of Suez, where it connects to the Mediterranean Sea via the Suez Canal. The seismically active Gulf of Aqaba is 160–180 km long and 19–25 km wide, narrow in the north and widening to the south with maximum depths of 1850 m toward the east, where shelves and coastal plains are absent. It is part of a left lateral transform fault system moving at about 5–7 mm per year (Ben-Avraham et al. 2008) that forms the north-west boundary of the Arabian Plate and connects the Red Sea, where seafloor spreading occurs, with the Zagros-Taurus zone of continental collision.

The Gulf of Suez is a 300 km long, 25–60 km wide, shallow flat bedded basin with depths ranging between 50 and 75 m. Depths increase toward the south but remain under the 100-m mark at the confluence of the Red Sea and do not exceed 200 m. It is basically a failed continental rift that remains floored by continental crust, with a complex extensional system of blocks that have rotated along low-angle or listric fault planes (Bosworth 1995), with three distinct depocentres for sedimentation. Since the middle Miocene the widening of the Red Sea at its northern end has largely been accommodated by the Gulf of Aqaba-Dead Sea transform, but continuing seismicity in the southern Gulf indicates that there is still some ongoing extensional stress and associated normal faulting.

3 Chapters in this Volume

3.1 Tectonics and Geology

Regional tectonic and geological aspects of the formation of the Red Sea are discussed in the next eight chapters of this volume, followed by two chapters dealing with hydrothermal fluids, brines and salt formation in the Red Sea. Three chapters deal with offshore and onshore structure as well as more the more regional configuration of the crust across Saudi Arabia. A chapter on sediment deposition in the deep Red Sea is followed by two chapters on the fossil reef system in the Gulf of Aqaba. Various aspects of the onshore Cenozoic harrats (lava fields) and dykes are then discussed

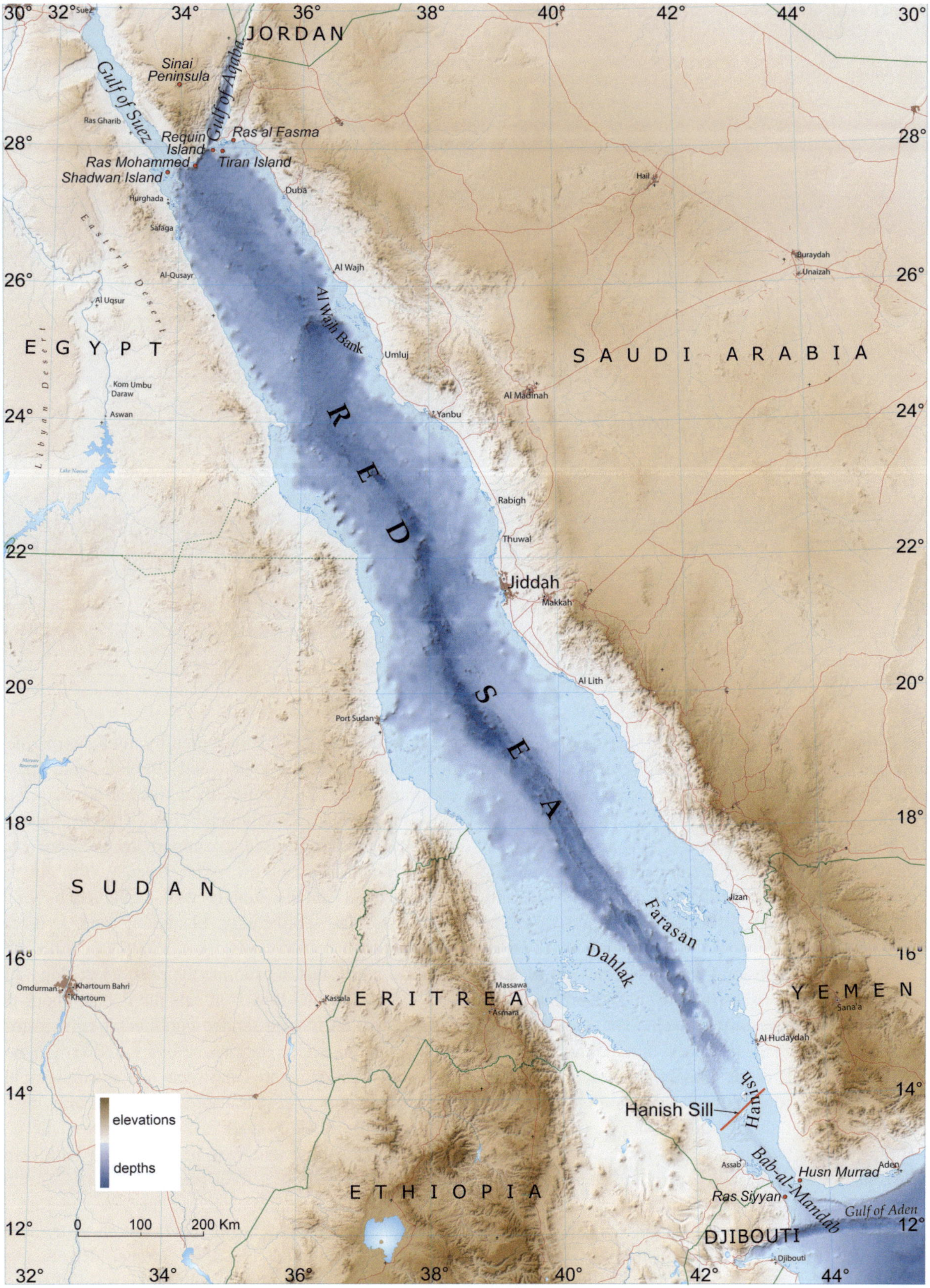

Fig. 1 Geographic map of the Red Sea area, where darker colours indicate greater depths or higher elevations (after Rasul and Stewart 2015)

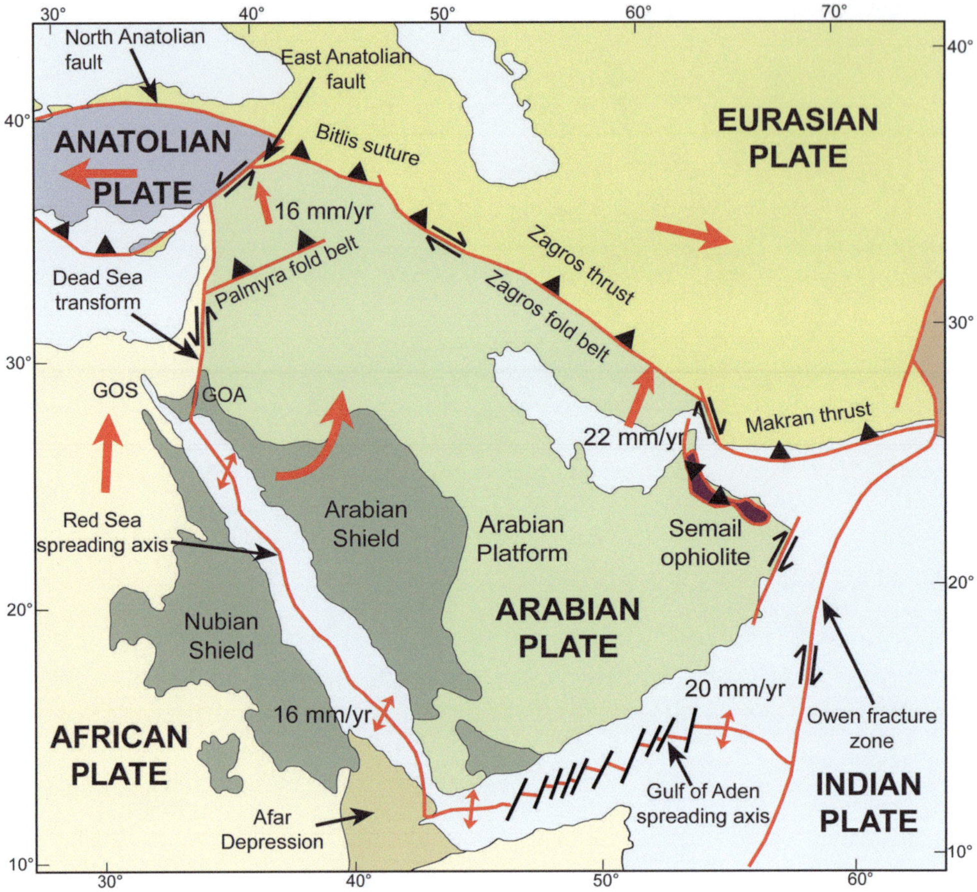

Fig. 2 Main tectonic features of the Arabian Peninsula and surrounding areas (modified from Stern and Johnson 2010)

in the next three chapters. One chapter that describes mineralisation associated with Red Sea rifting is then followed by the last two chapters in the geology section, on recent sedimentation in the Red Sea.

Bosworth et al. discuss the stress regime and ongoing and geologically recent tectonics of the Red Sea and surrounding areas, and how and where rifting and faulting is actually occurring. At the continental scale, the Red Sea is subjected to compression perpendicular to or at a high-angle to its margins. NNE–SSW extension in the southern Gulf of Suez is probably generated by sinistral slip on the similarly oriented Gulf of Aqaba transform margin. The kinematics of the southern Red Sea are complex; not all opening has jumped to the west side of the Danakil horst and significant tectonic activity still occurs along the southernmost Red Sea axis in the vicinity of the Zubair Archipelago. While the most significant neotectonic features of the Red Sea rift system are its southern oceanic spreading centre and the left-lateral Gulf of Aqaba–Levant transform fault in the north, many segments of the rift margins and in particular the Gulf of Suez remain tectonically active.

The process of rifting in the Red Sea, together with its associated volcanism at the spreading centre are described by Augustin and his colleagues. In some parts of the Red Sea the initiation of rifting is obscured by thick submarine flows that blanket the rift valley and obscure what may be a continuous rift axis rather than discrete spreading nodes between regions of continental crust. The geophysical data that was previously used to support the presence of continental crust between the axial basins with outcropping oceanic crust can be equally well explained by processes related to the sedimentary blanketing and hydrothermal alteration.

Stern and Johnson examine the onshore geology adjacent to the Red Sea and how it affects the debate about the tectonic transition from extended continental crust to true seafloor spreading. Correlations across the Red Sea between features such as sutures between tectonostratigraphic terranes, regions of transpressional shortening, and brittle-ductile faults related to Ediacaran orogenic collapse and tectonic escape that vary in orientation with respect to the coastlines require a tight pre-rift fit of the Arabian and Nubian Shields that implies that most of the Red Sea is underlain by oceanic crust. There is clear evidence for oceanic crust along the axis of the southern Red Sea, and the data is suggestive of this for the northern Red Sea.

The voluminous Eocene to Oligocene flood basalts in northern Ethiopia and western Yemen that predate Red Sea extensional faulting and rifting are described by Stockli and Bosworth. Basaltic dike emplacement, syn-rift subsidence and sedimentation, and rapid rift-related fault block exhumation at ~23 Ma along the entire Red Sea-Gulf of Suez rift system then marked the onset of the various stages of Red Sea lithospheric extension and rifting. Middle Miocene onset of left-lateral displacement along the Gulf of Aqaba transform resulted in the tectonic isolation of the Gulf of Suez and a switch from rift-normal to highly oblique extension with the Red Sea that led to the formation of fracture zones, pull-apart basins, and crustal necking, and ultimately local crustal separation and mantle exhumation, prior to Plio-Pleistocene incipient oceanic breakup in the northern Red Sea.

The nature of upper mantle upwelling during initiation of seafloor spreading in the southern Red Sea is studied by Gallacher and his colleagues. They imaged the mantle beneath the southern Red Sea, Afar and the Main Ethiopian rift using Rayleigh-wave tomography to generate a high-resolution 3-dimensional shear-wave velocity model of the upper 250 kilometres that shows the mantle response during the progression from continental rifting to seafloor spreading. The segmented low-velocity anomalies are consistent in scale from the oceanic southern Red Sea rift to the continental Main Ethiopian rift, suggesting that mantle segmentation beneath oceanic rifts initiates early during continental rifting.

Seismic and other geophysical and geological data from the NW Red Sea and the Brothers Islets are used by Ligi et al. to provide constraints on the composition, depth of emplacement and age of early syn-rift magma intrusions into the deep crust. They suggest a stretched and thinned continental crust with few isolated sites of basaltic injections, consistent with a model whereby asthenospheric melt intrusions contribute to weaken the lower crust enabling some decoupling between the upper and lower crust, protracting upper crust extension and delaying crustal breakup. Continental rupture in the northern Red Sea is preceded by intrusion of basaltic melts that cooled forming gabbros at progressively shallower crustal depths as rifting progressed toward continental separation.

Taking a somewhat different approach, Mohriak compares rifting and salt deposition in the Red Sea with those observed in the sedimentary basins on the margins of the South Atlantic. Seismic profiles integrated with gravity and magnetic potential field data suggest alternative models for the birth of oceanic basins that evolve from an earlier phase of intracontinental rifting, salt deposition and continental breakup by mantle exhumation or by development of oceanic spreading centres preceded by igneous intrusions and extrusions in the transition from continental to oceanic crust. Alternative interpretations for syn-rift successions and salt distribution in regional seismic profiles from the Red Sea, together with well data, are compared with similar tectono-stratigraphic settings in the South Atlantic. The data suggest that the Red Sea constitutes a better analogue for the development of the South Atlantic divergent continental margins than the Iberian margin.

Plate motions around the Red Sea since the Early Oligocene are summarised by Schettino and his colleagues. The Red Sea is a very young oceanic basin that formed ~4.6 Ma, although the rifting phase started at ~30 Ma. Two kinematic stages are characterized by distinct directions of extension and different duration; in the first stage, northward motion of the Arabian plate with respect to Africa was accompanied by N–S oriented strike-slip faults and normal faults having E–W strike, with extension that was mainly accommodated by the formation of pull-apart basins. From ~27 Ma (late Oligocene), the extension axes acquired the modern NE–SW pattern, which was conserved until the early Pliocene in the southern Red Sea and is still active in the north.

Hydrothermal fluids in the rift are discussed by van der Zwan et al., with a study of Cl in the assimilation of hydrothermally altered crust at an ultra-slow spreading ridge, due to its saline seawater, the presence of hot brine pools, and the thick evaporite sequences that flank the young rift. Basaltic Cl-excess is spatially closely correlated with evidence of hydrothermal activity, suggesting that deeper assimilation of hydrothermal Cl is the dominant Cl-enrichment process. The basaltic Cl-excess can be used as a tracer together with new bathymetric maps as well as indications of hydrothermal venting (hot brine pools, metalliferous sediments) to predict where hydrothermal venting or now inactive hydrothermal vent fields can be expected.

Salt formation processes during rifting are discussed by Hovland et al. Recent observations of salt flows on the Red Sea floor and huge accumulations of salts in the sub-surface ('Salt Walls' and 'Salt Ridges'), associated with topographical lows (Deeps), suggest that the Red Sea currently produces new volumes of brines and solid salts underground by boiling and supercritical phase separation in forced convection cells (hydrothermal circulation), located above shallow-seated magmatic intrusions along the spreading axis. When reaching the seafloor, the newly formed brines are cooled further, eventually becoming oversaturated in salts, which results in precipitation onto the seafloor, eventually giving rise to salt glaciers, salt walls, salt pinnacles, and 'diapirs' (injectites).

On a more localised topic, the possible origin of the submarine channel north of the Hanish Sill is discussed by Mitchell and Sofianos. Although the currents may help to maintain the upper channel morphology, it is unclear how they would have created the channel, nor can modern currents explain the deeper parts of the channel. The channel is straight and runs parallel with the spreading rift to the north, suggesting that faults may underlie the channel, though a tectonic origin (graben) is not supported by Bouguer gravity anomalies, which reveal no underlying structure. The channel may have originated from massive inflow of Indian Ocean water into the Red Sea following earlier isolation and drawdown of its level.

Onshore, Cenozoic faults, mainly younger than 23 Ma, in NW Saudi Arabia are examined by Roobol and Stewart, and a number of previously unknown structures are identified, including a zone of shattered and intensely faulted rocks 25 km wide that extends inland into Saudi Arabia from the coast of the Gulf of Aqaba. A major Cenozoic fold extends 50 km inland from the coast, where the Precambrian lithologic units, dikes and faults are rotated 90° anticlockwise with the appearance of a drag fold due to the approximately 115 km sinistral offset of the Gulf of Aqaba that probably resulted from the initial displacement of the Arabian Plate from Sinai. Aeromagnetic anomalies show that the traces of major Tertiary gabbro dikes that parallel the Red Sea coast of Saudi Arabia are also curved within the fold, but their curvature of about 45° anticlockwise is less than that of the Precambrian rocks and dikes, which suggests dike emplacement occurred after folding commenced.

In a seismological study, Mai and his colleagues use receiver-functions and surface-wave dispersion curves to determine the crustal and upper-mantle structure of Saudi Arabia, showing first-order differences in crustal thickness between the Arabian Shield in the west and the Arabian Platform in the east. Moho depths generally increase eastward, while crustal thickness varies strongly in the west over the volcanic regions and near the Red Sea. The data refute the hypothesis of a small localized mantle plume as the origin for the volcanic activity in western Saudi Arabia and suggest that the volcanism in western Arabia may be due to the lithospheric mantle being heated from below by lateral flow from the Afar and (possibly) Jordan plumes.

Mitchell et al. examine the deep-water Plio-Pleistocene sediments in the Red Sea, where the sediment distribution does not reflect the pattern of sediment input from the positions of wind gaps through the Red Sea hills and fluvial drainage basin outlets. Near the coast of Egypt, 3D seismic data shows that sediment deposition is unrelated to drainage basins of the adjacent hills, but is strongly affected by halokinetics, with sediment filling evaporite depressions that are elongated sub-parallel with the coast. Profiler data of Pleistocene sedimentation around the Thetis Deep suggests that hemipelagic sedimentation has been almost uniform, and also reveals localized slope failure and sediment flow deposits, as well as tectonic disruptions. The slope failures occurred in the Late Pleistocene after Marine Isotope Stage (MIS) 12 and probably before MIS 6, probably because of seismic ground accelerations. From potential acceleration and earthquake magnitudes, the results suggest that the very low incidence of historical earthquakes in the central Red Sea is not entirely representative of the Late Pleistocene.

The Pleistocene raised marine terraces that occur on both sides of the Gulf of Aqaba are described by Taviani et al. The best developed marine terrace system is reefal, from the last interglacial (Marine Isotope Stage 5e = MIS 5e, ~125 ka BP), although older Pleistocene terraces also occur. All such deposits are very fossiliferous and most carbonates are relatively unaltered, providing suitable material for geochronological purposes. The MIS 5e deposits reflect the structurally-controlled bedrock geology and the Gulf's topography, and the terraces sitting on the crystalline Arabian basement have been tectonically uplifted by up to 26 m above the present mean sea level. The bulk of the marine deposits represent upper fore-reef to beach settings, with back-reef to lagoonal facies only preserved where sufficient accommodation space (wadi valleys) was available during the MIS5e to allow inland marine expansion. Still on the subject of fossil reefs, Angelleti et al. describe the mollusc fauna associated with the Late Pleistocene coral reefs on the Saudi Arabian side of the Gulf of Aqaba, where the MIS 5e deposits document former back reef to fore-reef environments, as well as beach or mangal settings.

Sanfilippo et al. report on a geochemical study of rocks from Harrats Lunayyir and Khaybar, two large lava fields located in the western Arabian Peninsula. The trace element signatures are consistent with alkaline magmas produced by an enriched mantle source, akin to that producing continental flood magmatism in other locations of the Arabian-Nubian plate, with magmatic evolution that occurred in magma chambers located close to the crust-mantle boundary. The results suggest that Cenozoic alkaline volcanism in western Arabia formed mainly by decompression melting of ancient fusible components in the sub-Arabian lithospheric mantle. These were remobilized by lithospheric thinning due to Red Sea rifting and are consistent with progressive thinning of the lithosphere toward the Red Sea and lengthening of the melting column over time. The palaeomagnetism and geochronology of these two harrats are presented in the next chapter by Vigliotti et al. The results imply that the whole rotation of the Arabian Plate took place during the last phase (4–5 Ma) of the opening of the Red Sea, corresponding with true sea floor spreading.

Zanoni et al. study the microstructure and geochemistry of acidic, intermediate, and basic dykes sampled along the Arabian margin. Geochemical results indicate that basanite/basaltic dykes are compatible with a divergent environment such as the Red Sea rifting, whereas andesite dykes are compatible with a convergent setting. The rhyolitic dykes are interpreted as related to the Red Sea rifting as they show geochemical signatures compatible with divergent tectonics and are from a region where rhyolitic dykes were dated around 20 Ma.

Manganese or ferro-manganese mineralization that occurs along the western Red Sea coastal zone and Sinai, Egypt is closely associated with the Red Sea rifting, as described by El Agami. Hydrothermal manganese deposits occur as veins and fracture-filling cutting across the structure of the host rocks, indicating an epigenetic origin, and the mineral forming associations are typical for hydrothermal Mn deposits. The Miocene Mn deposits of the Abu Shaar area are formed by diagenetic replacement of the calcitic materials by Mn-bearing solutions during a marine transgression-regression, and the characteristic mineral-chemical enrichment and geochemical association indicate a marine-diagenetic origin.

One of the offshore reefs known as Shiab Al-Kabeer extends over an area of approximately 6.5 km^2 to the west of Jeddah. Rasul and Qutub have studied the nature and distribution of surface sediments on the shoal and describe sedimentation processes and the likely sources of these sediments, drawing attention to the significant roles played by sea birds, fish, and many invertebrates such as corals in terms of creating sediments. Parrotfish, pufferfish, surgeonfish and shrimp gobies play significant roles in grinding down the dead corals into coral sands, and burrowing organisms such as worms and holothurians also play important roles in recycling sediments. Algae and sponges also bore into the hard calcareous reef structure, leaving burrows and crevices. This study provides a potential model for future sedimentology both here and on other offshore reefs that are attracting the interest of tourism developers. Understanding what has created this species-rich environment will help to ensure that planners do not destroy those very aspects that initially attracted them to these isolated reefs and islands.

Finally, in the geology section, Abuabdullah and Şen study sediment yield rates from drainage basins (wadis) in Saudi Arabia adjacent to the Red Sea. Although wind deposition plays a significant role, the bulk of the sediment yield is due to surface water runoff after each storm rainfall. Drainage basin morphological variables including the basin area, basin slope, main channel slope and the drainage density are used to calculate the sedimentation rates for 3 wadi systems on the Saudi Arabian coast.

3.2 Palaeoenvironment and Archaeology Chapters

The main group of archaeological chapters presents different results from the DISPERSE project, and focuses on the earliest archaeological evidence for human presence along the Red Sea escarpment and the role of coasts and coastal environments in facilitating the expansion of early Stone Age populations out of Africa.

Sinclair et al. present the results of recent archaeological surveys in the south-west provinces of Jizan and Asir, discuss the dating of the important stratified sites of Dhahaban Quarry and Wadi Dabsa, and examine the stone tool assemblages at these locations in relation to their palaeoenvironmental setting. The large number of stone artefacts, and technological and typological differences indicating a wide range of ages, suggest that these were especially attractive locations in the wider landscape, repeatedly visited over long periods of the Pleistocene, most probably because of their association with strategic positions for monitoring the movements of large mammals and facilitating their capture, with nearby sources of basaltic rock for making stone tools, stream channels and water sources. At Wadi Dabsa, lava flows circumscribe an extensive tufa-filled basin, suggesting perennial supplies of slow-moving water that would have made an attractive magnet for animal and human populations. At Dhahaban Quarry, proximity to a palaeo-coastline suggests the additional possibility of exploiting resources along the sea shore.

Inglis et al. report the results of a combined geological and archaeological survey of elevated coral reefs in the same region and on the Farasan Islands, with details of location, elevation, collection of dating samples and archaeological associations. These reefs are of interest both as markers of former high sea level positions, as evidence for late Quaternary tectonic movements, as chronological markers for associated stone tools, and as former coastal environments of potential significance to the Stone Age human populations. On the mainland coastline of Asir, the reef and beachrock elevations indicate a high sea-level stand of about 7 m asl, similar to the evidence of coral reefs reported to the north, and are compatible with an MIS 5e age ($\sim$125 ka), although independent geochronological confirmation is still awaited. These elevations are significantly higher than the MIS 5e sea level predicted by modelling of isostatic adjustment in the Red Sea, indicating either evidence of tectonic uplift over the past 125 kyr, or the need for adjustment of the isostatic model. On the Farasan Islands, reef elevations are more variable, reflecting the effects of salt tectonics.

Bailey et al. look more broadly at the issues of early coastal colonization, examine the arguments for and against early sea crossings at the southern end of the Red Sea when sea level was lower and discuss the significance of now submerged coastal landscapes and submerged palaeoshorelines. They present new information on the mid-Holocene shell mounds of the Farasan Islands and their relationship to geodynamic and ecological changes in shorelines as a case study in analysing the factors that determine the preservation and visibility of archaeological deposits in coastal settings. They refer to some of the results of underwater investigation reported in other chapters (Momber et al., Sakellariou et al.) and highlight the importance of taking into account issues of site formation, preservation and visibility when attempting to interpret the distribution of archaeological sites in terms of palaeo-demography and environmental preferences.

Because sea levels were substantially lower than at present during the late Pleistocene and early Holocene, extensive areas were available for human settlement before being inundated by sea-level rise. Studies in other parts of the world show that coastal archaeological sites such as shell middens can survive inundation and can be discovered by underwater investigation. Momber et al. pursue this line of research and take the archaeological story under water around the Farasan Islands. They describe the range of diving techniques, remote-sensing equipment and other methods used in underwater explorations of the Red Sea and the results of underwater surveys for submerged palaeoshorelines and shell mounds in the Farasan Islands.

All these chapters refer to the results of the R/V Aegaeo survey of the deeper shelf around the Farasan Islands, and the next chapter by Sakellariou et al. gives a more detailed account. This survey was inspired by archaeological questions as part of the DISPERSE project, and though its results provide new data on the geological structure and dynamics of the outer shelf region, they also present new information about the nature of the now-submerged landscapes and topography available to Stone Age hunters and gatherers during the low sea-level episodes of the Last Glacial period. The results give new information on the MIS 3 and MIS 2 submerged coral terraces of the Farasan shelf and evidence of faulting linked to extensional tectonics and mobility of Miocene evaporites. Fault-bounded basins on the shelf associated with evidence of lacustrine sediments, and the presence of narrow valleys that must have been created by sub-aerial erosion by water action, indicate a complex topography with minor barriers, narrow valleys and sources of water that would have created attractive conditions for large animals and their human hunters.

Kübler et al. take up the theme of complex topography and the role of active tectonics and volcanism in creating attractive landscapes for early human evolution and dispersal. They range widely across the Afro-Arabian rift system, drawing on examples of early Stone Age sites and environments in the Kenyan Rift, the Dead Sea Rift, and the Arabian escarpment, the latter including reference to the Wadi Dabsa site described in earlier chapters. They show how modelling of fault motions can aid the reconstruction of an earlier topography, and further demonstrate the importance of tectonic and volcanic features and the edaphic properties of different rock and sediment types as constraints on the seasonal migrations of large mammals that facilitated ambush hunting by early Stone Age hunters.

Finally, in this archaeological grouping, Petraglia et al. summarise the results of the PALAEODESERTS project. They emphasise the important role of episodic climate change during the Pleistocene and 'greening' of the desert interior, which periodically opened up large areas of the Arabian Peninsula to settlement and dispersal following water courses and lakes, and they contrast this model of dispersal through the interior with the models of coastal dispersal examined in the previous chapters. They summarise palaeoclimatic research that demonstrates the time depth of these wetter climatic episodes, which can be identified well back into the Pleistocene, and present evidence for stratified Stone Age sites associated with palaeo-lakes, the earliest in the Nefud region dating back to 211 ka, and a palaeontological site with a large mammal fauna of mixed Eurasian and African affinity dated somewhat earlier and possibly as early as 500 ka.

Dating deposits is a perennial challenge for geologists and archaeologists, and Sanderson and Kinnaird explain the principles and techniques of Optically Stimulated Luminescence (OSL) dating and summarise the growing number of applications of this technique to the dating of palaeoclimatic sequences in the Arabian Peninsula. They also present the result of OSL dating of sediments on land and from underwater cores collected as part of the DISPERSE project. Their results have provided additional constraints on the chronology of some of the Stone Age sediments and sites examined on the Jizan/Asir mainland, and confirmed a Last Glacial Maximum age for lacustrine deposits found in the lower sections of the R/V Aegaeo cores recovered from the Farasan shelf.

The two chapters by Geraga et al. and Hausmann et al. turn the discussion to issues of palaeoclimatic reconstruction. Geraga et al. analyse changes in foraminiferal composition through a radiocarbon-dated sequence of marine sediments in one of the cores recovered from the outer Farasan shelf by the R/V Aegaeo. Their results show interesting changes through the late Pleistocene–Holocene sequence of the core, which demonstrate changes in climate, ecological productivity and oceanographic conditions at that location in the Red Sea, mainly associated with changes in sea-level and the resulting changes in the flow of currents between the Red Sea and the Indian Ocean.

Hausman et al. discuss the application of stable isotope $\delta^{18}O$ and element-ratio variations to the microscopic growth structures of marine mollusc shells recovered from the archaeological shell-mound deposits of the Farasan Islands. These techniques have, in the past, been limited in their application to archaeological shell deposits because of the costs involved and the small sample sizes of measurements that are possible. Hausmann et al., however, show how new techniques are making possible cheaper and more rapid analysis of large samples. These records can provide a very high-resolution climatic record. They can also inform on the season of collection of the molluscs, providing valuable information about the palaeodiet of the human population and patterns of settlement. In discussing the archaeological implications of this technique, the authors return the discussion to its original starting point about the role of coastlines and coastal environments in the chapters that opened this archaeological section.

The last chapter in this archaeological section, by Kotarba-Morley, takes the discussion in a new direction and to a more recent period of Red Sea history and examines the Greco-Roman ports of trade in the Red Sea, which played such a vital role in the trade routes between Asia, southern Arabia, East Africa and the Mediterranean. She focuses on the Egyptian coastal town of *Berenike Troglodytica* and analyses a comprehensive range of variables at the site and its surroundings, including environmental parameters, agricultural productivity, water supplies, sea conditions and socio-political factors, in order to understand the rationale for the choice of location as a port town. Amongst other features of this research is the important point that dynamic changes in coastal environments and geomorphology are not only of relevance to the longer time spans of the Stone Age but play an equally important role and demand an equivalent degree of specialist investigation and analysis in the historical period.

References

Augustin N, Devey CW, van der Zwan FM, Feldens P, Tominaga M, Bantan RA, Kwasnitschka T (2014) The rifting to spreading transition in the Red Sea. Earth Planet Sci Lett 395:217–230. https://doi.org/10.1016/j.epsl.2014.03.047

Bailey GN (2015) The evolution of the red sea as a human habitat during the quaternary period. In: Rasul NMA, Stewart ICF (eds) The Red Sea: the formation, morphology, oceanography and environment of a young ocean basin. Springer Earth System Sciences, Berlin Heidelberg, pp 595–610. https://doi.org/10.1007/978-3-662-45201-1_34

Ben-Avraham Z, Garfunkel Z, Lazar M (2008) Geology and evolution of the southern Dead Sea Fault with emphasis on subsurface structure. Ann Rev Earth Planet Sci 36:357–387

Biton E, Gildor H, Peltier WR (2008) Red Sea during the last glacial maximum: implications for sea level reconstruction. Paleooceanography 23, PA1214. https://doi.org/10.1029/2007pa001431

Bonatti E, Cipriani A, Lupi L (2015) The Red Sea: birth of an ocean. In: Rasul NMA, Stewart ICF (eds) The Red Sea: the formation, morphology, oceanography and environment of a young ocean basin. Springer Earth System Sciences, Berlin Heidelberg, pp 29–44. https://doi.org/10.1007/978-3-662-45201-1_2

Bosworth W (1995) A high-strain rift model for the southern Gulf of Suez (Egypt). Geol Soc London, Spec Publ 80:75–102

Bosworth W (2015) Geological evolution of the Red Sea: Historical background, review, and synthesis. In: Rasul NMA, Stewart ICF (eds) The Red Sea: the formation, morphology, oceanography and environment of a young ocean basin. Springer Earth System Sciences, Berlin Heidelberg, pp 45–78. https://doi.org/10.1007/978-3-662-45201-1_3

Daradich A, Mitrovica JX, Pysklywec RN, Willett SD, Forte AM (2003) Mantle flow, dynamic topography, and rift-flank uplift of Arabia. Geology 31:901–904

Head SM (1987) Red Sea fisheries. In: Edwards AJ, Head SM (eds) Red Sea: Key Environments. Pergamon Press, Oxford, pp 363–382

Hovland M, Rueslåtten H, Johnsen HK (2015) Red Sea salt formations—a result of hydrothermal processes. In: Rasul NMA, Stewart ICF (eds) The Red Sea: the formation, morphology, oceanography and environment of a young ocean basin. Springer Earth System Sciences, Berlin Heidelberg, pp 187–203. https://doi.org/10.1007/978-3-662-45201-1_11

Lambeck K, Purcell A, Flemming NC, Vita-Finzi C, Alsharekh AM, Bailey GN (2011) Sea level and shoreline reconstructions for the Red Sea: isostatic and tectonic considerations and implications for hominin migration out of Africa. Quatern Sci Rev 30:3542–3574

Morcos SA (1970) Physical and chemical oceanography of the Red Sea. Oceanogr Mar Biol Ann Rev 8:73–202

Rasul NMA, Stewart ICF (eds) (2015) The Red Sea: The formation, morphology, oceanography and environment of a young ocean basin. Springer Earth System Sciences, Berlin Heidelberg, 633 pp, ISBN 978-3-662-45200-4, ISBN 978-3-662-45201-1 (eBook). https://doi.org/10.1007/978-3-662-45201-1_1

Rasul NMA, Stewart ICF, Nawab ZA (2015) Introduction to the Red Sea: its origin, structure and environment. In: Rasul NMA, Stewart ICF (eds) The Red Sea: the formation, morphology, oceanography and environment of a young ocean basin. Springer Earth System Sciences, Berlin Heidelberg, pp 1–28. https://doi.org/10.1007/978-3-662-45201-1_1

Stern RJ, Johnson PR (2010) Continental lithosphere of the Arabian Plate: A geologic, petrologic, and geophysical synthesis. Earth-Sci Rev 101:29–67

Trommer G, Siccha M, Rohling EJ, Grant K, van der Meer MTJ, Schouten S, Baranowski U, Kucera M (2011) Sensitivity of Red Sea circulation to sea level and insolation forcing during the last interglacial. Climate of the Past 7:941–955. https://doi.org/10.5194/cp-7-941-2011

Neotectonics of the Red Sea, Gulf of Suez and Gulf of Aqaba

William Bosworth, Marco Taviani, and Najeeb M. A. Rasul

Abstract

The Red Sea, Gulf of Suez, and Gulf of Aqaba comprise the active plate boundaries that separate Africa-Nubia, Arabia and Sinai. This tripartite configuration has been in existence since the Middle Miocene, or about the past 12–14 Ma. We describe the ongoing and geologically recent tectonics of these regions. The Red Sea rift lies east of a broad region of E-W maximum horizontal stress (S_{Hmax}) that covers much of central Africa-Nubia. On its Arabian side, S_{Hmax} is oriented N-S to NE-SW. These far field stresses owe their origins to the spreading centres of the Atlantic Ocean and collision between Arabia and Eurasia along the Bitlis-Zagros suture. At the continental scale, the Red Sea is therefore subjected to compression perpendicular to or at a high-angle to its margins. The realm of shallow crustal stresses conducive to extensional faulting in a Red Sea orientation (rift-normal S_{hmin}) is presently restricted to the Red Sea marine basin itself, and perhaps narrow belts along its shoulders. In the Gulf of Suez there is enough data to show that each of its sub-basins is presently undergoing extension, but in conjunction with differently oriented, sub-regional shallow crustal stress fields. These appear to be spatially related to the original Early Miocene syn-rift basin geometries. NNE-SSW extension in the southern Gulf of Suez is probably generated by sinistral slip on the similarly oriented Gulf of Aqaba transform margin. Large M > 6 earthquakes are generally restricted to the central basin of the Gulf of Aqaba, the southern Gulf of Suez, and the greater Afar region. The geodynamic details responsible for the focusing of these large events are specific to each locale but all are in general associated with the junctions of major plate boundaries. Catalogues of earthquake activity and GPS datasets show that the Sinai micro-plate is still moving away from Africa with a component of left-lateral slip. This results in components of extension perpendicular to both the Gulf of Suez and the Gulf of Aqaba. The kinematics of the southern Red Sea are similarly complex. Not all opening has jumped to the west side of the Danakil horst and significant tectonic activity still occurs along the southernmost Red Sea axis in the vicinity of the Zubair Archipelago. This is the only volcanically active segment of the Red Sea basin that is above sea level. Dike intrusions are ~N-S and not aligned parallel to the rift axis and may indicate that the underlying magmatism is swinging to the west to link with the Afar triple junction. All of the margins of the Red Sea, Gulf of Suez and Gulf of Aqaba underwent tectonically-driven rift shoulder uplift and denudation in the past, particularly during the main phases of continental rifting. However, during the past 125 kyr uplift has been focused along the footwalls of a few, active extensional faults. These include the Hammam Faraun-Tanka fault in the central Gulf of Suez, the Gebel el Zeit-Shadwan Island fault in the southern Gulf of Suez, the Sinai and Arabia coastal boundaries of the Gulf of Aqaba, and faults at Tiran Island at the junction of the Gulf of Aqaba and northern Red Sea. Smaller-scale extensional faulting is also occurring along the Saudi Arabian margin of the northern Red Sea, in the Dahlak and Farasan Archipelagos, and on the volcanically active islands of the Zubair Archipelago in the southernmost

W. Bosworth (✉)
Apache Egypt Companies, 11 Street 281 New Maadi, Cairo, Egypt
e-mail: bill.bosworth@apacheegypt.com; bill.bosworth@apachecorp.com

M. Taviani
Istituto di Scienze Marine (ISMAR-CNR), Via Gobetti 101, 40100 Bologna, Italy

M. Taviani
Biology Department, Woods Hole Oceanographic Institution, 266 Woods Hole Road, Woods Hole, MA 02543, USA

M. Taviani
Stazione Zoologica Anton Dohrn, Villa Comunale, 80121 Naples, Italy

N. M. A. Rasul
Center for Marine Geology, Saudi Geological Survey, Jeddah, Saudi Arabia

N. M. A. Rasul and I. C. F. Stewart (eds.), *Geological Setting, Palaeoenvironment and Archaeology of the Red Sea*,
https://doi.org/10.1007/978-3-319-99408-6_2

Red Sea. On the Farasan and Dahlak islands this is largely related to the movement of underlying salt bodies, similar to effects documented along the coastal plain of Yemen. Though not active at the present time, a broad belt of small-offset, very linear extensional faults dissected the western margin of the central Gulf of Suez during the Plio-Pleistocene. Similar age and style deformation has not been identified in the Suez sub-basins to the north or south. The most significant large-scale neotectonic features of the Red Sea rift system are its southern oceanic spreading centre and the northern linkage to the left-lateral Gulf of Aqaba—Levant transform fault. However, many segments of the rift margins and in particular the Gulf of Suez remain tectonically active. These areas provide stress field and horizontal and vertical displacement data that are relatively inexpensive to acquire and complementary to analyses of the offshore main plate boundaries themselves.

1 Introduction

This paper summarizes neotectonic features of the Red Sea, Gulf of Suez and Gulf of Aqaba (Fig. 1). These three basins, together with Afar, encompass the plate boundaries that now separate Africa-Nubia from Arabia. This configuration has been in place for about the past 12–14 Myr, since the Middle Miocene, when the Gulf of Aqaba—Dead Sea transform plate boundary came into existence. Prior to this, based on regional stratigraphic correlations, the Red Sea and Gulf of Suez were interconnected basins, dominated by open- to marginal-marine sedimentation (Tewfik and Ayyad 1984; Miller and Barakat 1988; Montenat et al. 1988; Hughes and Beydoun 1992; Hughes et al. 1992; Bosworth et al. 1998). With initiation of the transform boundary, extension rates across the Gulf of Suez dropped dramatically but not completely (Joffe and Garfunkel 1987; Steckler et al. 1988; Bosworth et al. 1998). From the Serravallian (late Middle Miocene) onward the north-easternmost boundary of Africa-Nubia involved the additional complexity of the movement of the Sinai micro-plate.

The most tectonically active part of the Red Sea rift system lies in the offshore, along the well-defined spreading centre in the south and its equivalent, though more enigmatic rift axis in the north. A wealth of new data are available for these realms (e.g., Mitchell et al. 2010; Ligi et al. 2012, 2015; Feldens and Mitchell 2015; Ehrhardt and Hübscher 2015). However, localized deformation continues along the rift flanks and as these regions are more accessible this offers a useful complementary picture to the submerged activity. A similar situation exists in the Gulf of Aqaba, where strike-slip faulting is occurring along the offshore basin axis, whereas extensional faulting is dominant in the uplifted margins (Bosworth et al. 2017).

We first review the present understanding of the relative movements between Africa-Nubia, Arabia and Sinai. This is based on geodetic, seismic, and geologic observations. Next we discuss the present-day shallow crustal stress fields that accompany these movements and are responsible for much of the active deformation now seen in outcrop. We then illustrate the most significant neotectonic features we have examined in each segment of these plate boundaries. In some cases, structures can be constrained to be truly active or postdating specific parts of the Pleistocene; in others dating is less constrained and Pliocene activity may be included. The presentation is generally organized from north to south, with youngest or present-day features discussed first.

2 Tectonic Setting

2.1 Plate Motions

Africa and Arabia are presently diverging across the southern Red Sea at a rate of 1.7 ± 0.1 cm/yr in a NE-SW direction (Fig. 1; ArRajehi et al. 2010; 2.4 cm/yr in Reilinger et al. 2015). In the northern Red Sea this reduces to about 0.7 ± 0.1 cm/yr. Plate reconstructions suggest that these opening rates must have been about half these values prior to 11 ± 2 Ma (McQuarrie et al. 2003). This roughly corresponds with geologic evidence for the time of initiation of the Gulf of Aqaba transform boundary (14–12 Ma; Bosworth et al. 2005, and references therein). Reilinger et al. (2015) suggested that the acceleration in the Red Sea opening resulted at least partially from the completion of the oceanic spreading centre along the length of the Gulf of Aden, decoupling the Arabia and Somalia plates. Several lines of evidence therefore point to a significant reorganization of plate kinematics at roughly 14–11 Ma in the greater Red Sea—Gulf of Aden rift system. GPS and geologically derived Arabian plate boundary slip rate determinations suggest that the opening of the Red Sea since the reorganization has, within observational errors, remained constant (Reilinger et al. 2015).

Although most of the northern Red Sea opening now links to the East Anatolia fault via the Gulf of Aqaba—Dead Sea transform, Sinai continues to separate from Africa-Nubia at about 0.15 cm/yr in a NNW direction (Fig. 1; Mahmoud et al. 2005). This equates to about 0.05 cm/yr rift-normal extension and 0.14 cm/yr left-lateral shear parallel to the Gulf of Suez (discussed further below).

Motion on the transform boundary itself is estimated from GPS datasets to be about 0.44 ± 0.03 cm/yr left-lateral movement near the junction with the Red Sea (ArRajehi et al. 2010). There is also about 0.2 cm/yr of opening across the

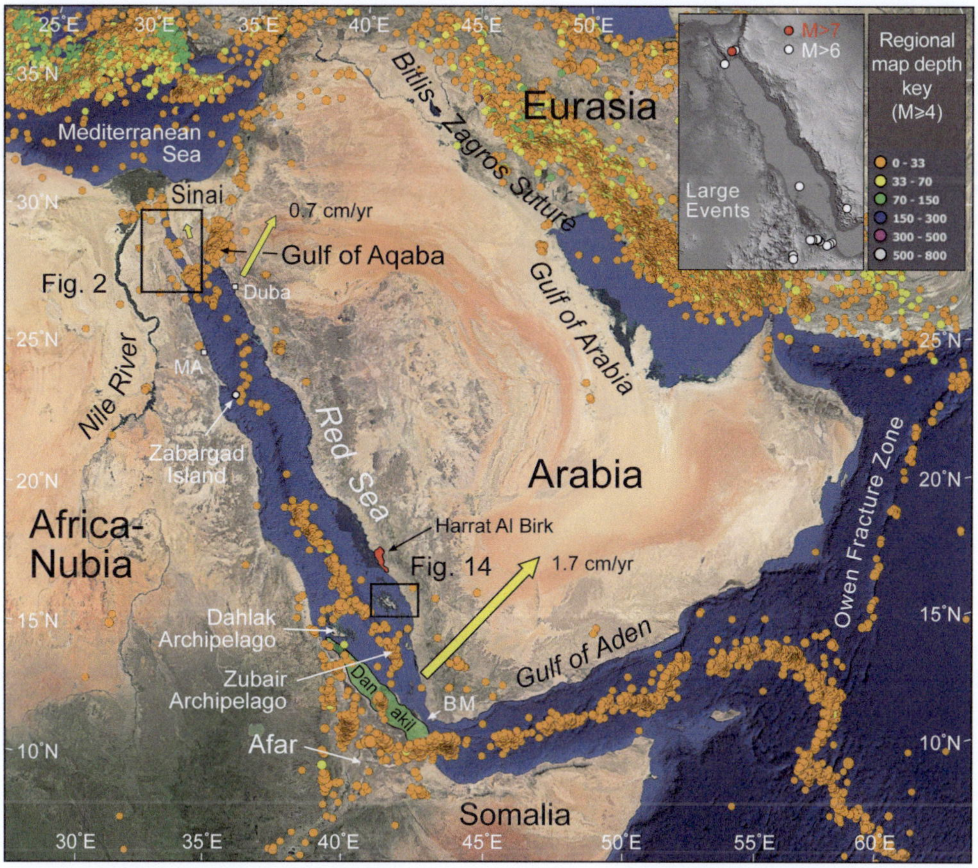

Fig. 1 Regional setting of the Red Sea, Gulf of Suez and Gulf of Aqaba. Instrumentally recorded earthquakes demarcate the plate boundaries of the region (M $\geq$ 4 since 1960-01-01; ANSS 2016). Yellow arrows show the slip rates of Arabia and Sinai relative to Africa (ArRajehi et al. 2010). The small arrow on Sinai equates to 0.15 cm/yr (Mahmoud et al. 2005). Locations of Figs. 2 and 14 are shown by boxes. Volcanic rocks of Harrat Al Birk are shown in red. The continental Danakil Block is shown in green. Base is from Google Earth (Landsat/Copernicus image; data from SIO, NOAA, US Navy, GEBCO). Inset map shows positions of M > 6 earthquakes for the past 50 years for the Red Sea, Gulf of Suez and Gulf of Aqaba (n = 19). BM = Bab el Mandeb, MA = Marsa Alam

southern Gulf of Aqaba (Reilinger et al. 2015) which confirms its early designation as a "leaky transform" boundary (Ben-Avraham et al. 1979). As there are no syn-tectonic magmatic rocks exposed in the Gulf of Aqaba, Ben Avraham et al. used the term leaky to indicate the creation of space perpendicular to the strike-slip boundary, without implications of volcanism. Measurements of offset pre-existing geologic features or pinning points demonstrate that total slip amounts to 107 km (Quennell 1956, 1958). Quennell thought that 62 km of this slip occurred during the Early Miocene to Pliocene, followed by a significant period of quiescence and post-Pliocene to Recent movement of an additional 45 km. The post-Pliocene rate equates to $\sim$1.7 cm/yr. Assuming that the transform did not initiate until the time of the plate reorganization (14–11 Ma) and movement was continuous through time gives long-term averaged slip rates of 0.76–0.97 cm/yr—about twice the present GPS determination or half of Quennell's post-Pliocene rate. These disparencies suggest that the timing of Quennell's two phases of movement probably needs to be adjusted, and perhaps the time gap separating them was not as significant as he thought. It is also possible that slip along the Gulf of Aqaba has been more variable over geologic time and we are presently measuring a relatively slower period of movement. Alternatively the GPS dataset for southern Sinai is simply not robust enough to yet yield accurate instantaneous slip rates.

2.2 Seismicity

Most of the Red Sea axis and the Gulf of Aqaba presently display low to moderate levels of seismic activity (Fig. 1; Ambraseys et al. 1994; Babiker et al. 2015; ANSS 2016). A noticeable gap exists in the central part of the Red Sea at about 21–23°N latitude where no M > 4 earthquakes have been recorded. Overall M > 6 events are rare, with only 19 observed in the past 50 years (Fig. 1 inset). These larger earthquakes are focused in Afar and the central sub-basin of the Gulf of Aqaba, although one occurred in the southern Gulf of Suez and one near the Dahlak Archipelago in the southern Red Sea. Nearly all earthquakes in this region occur in the crust at depths <33 km (ANSS 2016).

Instrumentally recorded earthquakes along the Red Sea axis generally yield normal fault solutions, although strike-slip regime solutions are clustered between the Dahlak and Farasan archipelagos (Delvaux and Barth 2010). Seismicity is locally focused at positions where the axial deep/spreading centre shifts direction (El-Isa and Al Shanti 1989). These are often interpreted as the sites of intersection of transform faults with the ridge axis.

Strike-slip fault solutions predominate for Gulf of Aqaba seismic events (Hofstetter 2003). The largest event ever instrumentally recorded in the Red Sea—Gulf of Suez—Gulf of Aqaba region occurred in the central Gulf of Aqaba on 22 November 1995 (Fig. 2). This earthquake had an interpreted M_S of 7.2 with a left-lateral strike-slip solution on a vertical NNE-SSW striking fault plane (Hofstetter 2003; ANSS 2016). However, normal fault mechanisms are also commonly observed in local network recordings (Alamri et al. 1991; Roobol et al. 1999; Hofstetter 2003). Composite focal mechanisms derived from events that occurred from 1985–1989 with M_L 2.6–3.8 indicated normal dip-slip on approximately north-south striking faults in the northern Aragonese and eastern Elat Deeps (see Fig. 8 for localities; Alamri et al. 1991). Similarly aftershock studies of the large 1995 event show that the total deformation field was complex and involved more than just sinistral strike-slip on the transform boundary fault (Roobol et al. 1999; Hofstetter 2003). Although strike-slip mechanisms predominated in the larger aftershocks, one large normal fault event occurred in the offshore north of the main epicentre. In 1993 an earthquake swarm of over 420 events occurred within and nearby the central Gulf of Aqaba deeps (Hofstetter 2003). The two largest events of the swarm (M_L = 5.8 and 5.6) were principally dip-slip. The epicentres of both of these earthquakes were located on NNE-SSW striking segments of the Arabian coastline. This is the general orientation of the Gulf of Aqaba—Dead Sea transform boundary and such structures would usually be associated with sinistral strike-slip movement. In this case the seismic events document a component of east-west extension in agreement with GPS observations discussed above.

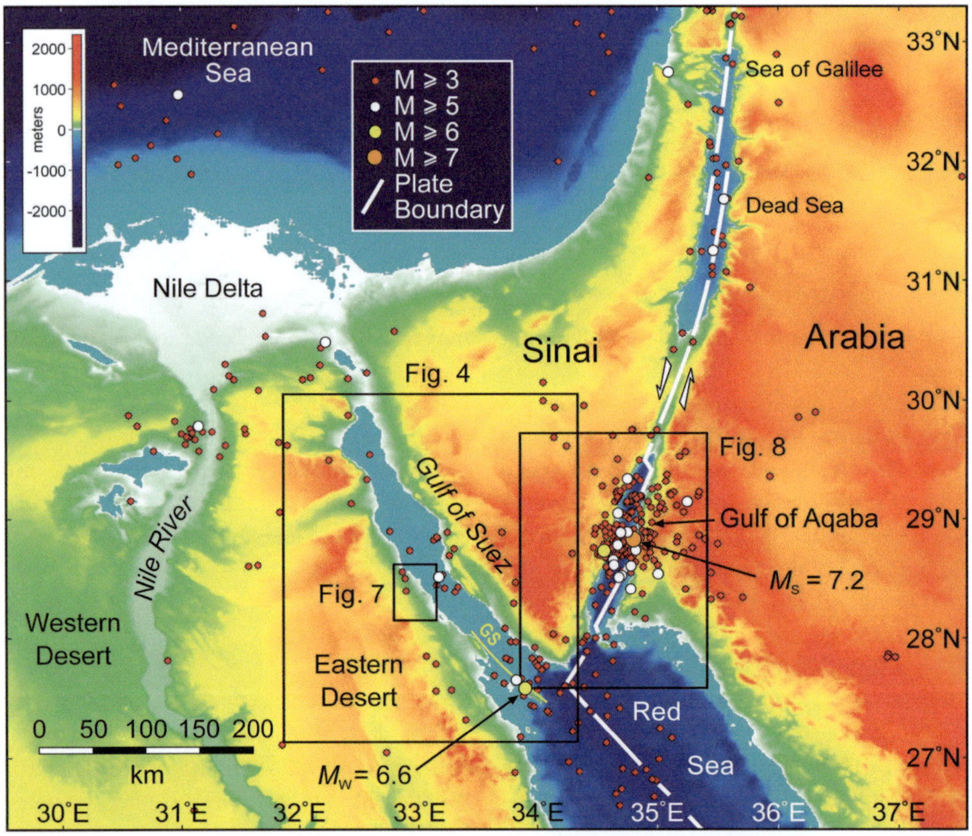

Fig. 2 Tectonic setting of the Gulf of Suez and Gulf of Aqaba and instrumentally recorded seismicity (M $\geq$ 3 since 1960-01-01; ANSS 2016). Base is Etopo merged bathymetry and elevations model (NOAA 2016). Line labeled GS is the Gebel el Zeit-Shadwan Island fault trend where the largest Gulf of Suez events have occurred. Positions of the M_w = 6.6 Gulf of Suez and M_s = 7.2 Gulf of Aqaba earthquakes are indicated. Location of Figs. 4, 7 and 8 are shown by boxes

Despite the fact that the Gulf of Suez is only extending at a very slow rate its seismic activity is locally much higher than most segments of the Red Sea (Figs. 1 and 2). Microseismicity is abundant in the southern Gulf of Suez, particularly for $M = 1 - 3$ (Daggett et al. 1986). Relatively large events were recorded in this area on 31 March 1969 ($M_w = 6.6$) and 28 June 1972 ($M_w = 5.6$) (Huang and Solomon 1987; ANSS 2016). Similarly to the earthquakes recorded in the Red Sea, these were principally normal fault events, although they have been interpreted as including a component of left-lateral shear in a NW-SE direction (Huang and Solomon 1987). This also supports the GPS interpretation that Sinai is moving both away from Africa-Nubia and toward the north.

2.3 Regional Stress Fields

The broad, regional stress field for much of the central part of the African plate displays E-W maximum horizontal stress (S_{Hmax}; Fig. 3; Zoback 1992; Bosworth 2008; Bosworth and Durocher 2017). This is based on earthquake focal mechanism determinations in West Africa (Ayele 2002; Delvaux and Barth 2010), the Sudan (Giardini and Beranzoli 1992; Gaulon et al. 1992; Delvaux and Barth 2010) and southern Egypt (Badawy 2001; Hussein et al. 2013). Industry borehole breakout data from wells in the Mesozoic Sudan rifts similarly show uniform E-W S_{Hmax} (Bosworth et al. 1992). Zoback (1992) attributed this intra-plate compression to the first-order effects of ridge push at the Central and South Atlantic and Indian Ocean spreading centres. This very broad region of relatively uniform E-W S_{Hmax} was termed the Central African Intra-plate (CAIP) stress field (Fig. 3; Bosworth 2008).

Stress field data are not available for most of the northern and central Red Sea basin. The focal mechanism inversion of Delvaux and Barth (2010) indicates NW-SE S_{Hmax}, parallel to the young spreading centre in the southernmost Red Sea (Fig. 3). Offshore Sudan, the mechanisms are normal faults, but as discussed above they become strike-slip offshore Eritrea in the vicinity of the Dahlak archipelago which is probably related to the bend of the plate boundary landward and to the west of the Danakil horst (Fig. 1).

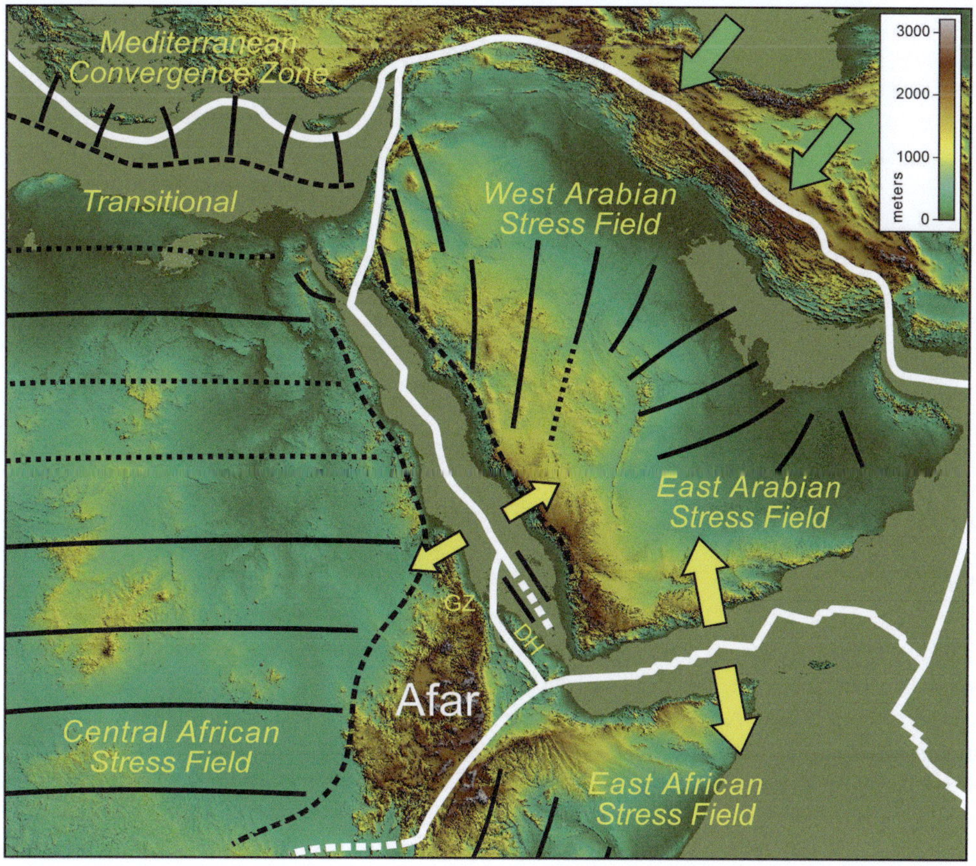

Fig. 3 Plate boundaries (heavy white lines; USGS 2016) of the greater Red Sea region and interpreted orientation of the maximum horizontal stress S_{Hmax} (Bosworth and Durocher 2017). Yellow arrows schematically represent forces from ridge-push at oceanic spreading centres and green arrows the forces associated with convergence between Arabia and Eurasia. Base is Etopo elevation model (NOAA 2016). DH = Danakil Horst; GZ = Gulf of Zula

In the Arabian Gulf region, breakout data define a fanning of S_{Hmax} from NE-SW in Saudi Arabia rotating to approximately N-S in Oman (Fig. 3; Ameen et al. 2010; Ameen 2014). The East Arabia stress field must rotate to N-S in the west as well because young volcanism in this region is being erupted from N-S aligned vents presumably overlying similarly oriented feeder dikes (Camp and Roobol 1992). The geometry of the Arabian regional stress field is strongly influenced by the nature of convergence and compression along the Bitlis-Zagros suture and its continuation into the Arabian Sea (Fig. 3). Along the Gulf of Aqaba transform boundary regional S_{Hmax} rotates further toward the NW-SE in accord with the dominant left-lateral strike-slip movement (Fig. 3; Heidbach et al. 2008). The weakness of the transform fault relative to the adjacent crust of Sinai and Arabia has been interpreted to result in S_{Hmax} swinging into parallelism with the boundary and therefore facilitating extension across the Gulf (Ben-Avraham and Zoback 1992; Bosworth et al. 2017). The stress field in the interior of the Sinai micro-plate is not yet clearly defined as earthquakes are uncommon and no outcrop or subsurface present-day stress field indicators are available.

This brief overview of the stress fields presently enveloping the Red Sea is a simplistic perspective as it does not capture the dynamic nature of this region. Sometime in the Late Pleistocene, after about 500 ka but prior to 125 ka, a significant rotation of the shallow crustal stress field occurred in the Kenya rift valley (~45° clockwise) and the northern Red Sea/Gulf of Suez (~40° counterclockwise) (Strecker et al. 1990; Bosworth et al. 1992; Bosworth and Taviani 1996; Bosworth and Strecker 1997). An adequate explanation of why these stress field rotations occurred has not been presented, nor is it clear if they were truly coeval and/or genetically related. It does suggest that care should be taken in extrapolating stress field datasets to long time intervals in the Red Sea region.

3 Sub-region Details

3.1 Gulf of Suez

Rifting in the Gulf of Suez began at ~23 Ma, at the Oligocene-Miocene transition (reviewed in Bosworth et al. 2005; Bosworth and Stockli 2016). After initiation of the Gulf of Aqaba transform boundary in the Middle Miocene, extension across the Gulf of Suez dropped dramatically to about 0.08–0.16 cm/yr based on a variety of plate circuit (Joffe and Garfunkel 1987) and geologic considerations (Steckler et al. 1988; Bosworth and Taviani 1996). Overall, extension continued to focus through time along the axis of the basin, complicated by continued movement of salt walls and domes (Moretti and Colletta 1987; Perry and Schamel 1990; Bosworth 1995).

The Gulf of Suez is comprised of three along-strike-linked sub-basins, separated by complex accommodation or transfer zones (Fig. 4; Moustafa 1976; Bosworth 1985; Jarrige et al. 1990; Patton et al. 1994; Moustafa 1997). Each of these sub-basins is a mega-half graben, with the large basin-bounding faults switching sides from south to north. The internal structure of each mega-half graben is dominated by nested, rotated fault blocks defined by sets of synthetic faults (Colletta et al. 1988; Perry and Schamel 1990; Patton et al. 1994; Bosworth and McClay 2001).

3.2 Present-Day Stress Fields

Borehole breakout studies in industry exploration wells indicate that the minimum horizontal stress (S_{hmin}) in the southern Gulf of Suez is presently aligned ~NNE-SSW, parallel to the relative slip trajectory of Sinai versus Arabia along the Gulf of Aqaba transform plate boundary (Fig. 4; Bosworth and Taviani 1996; Badawy 2001). A similar orientation is observed in the accommodation zone boundary between the Central and Darag (northern) sub-basins (Bosworth and Durocher 2017). By contrast, breakout and drilling-induced fracture data indicate NE-SW rift-normal extension in the Central sub-basin (Fig. 4).

A variety of models have been proposed regarding Miocene syn-rift stress fields in the Gulf of Suez, but most fault kinematic indicators suggest that NE-SW extension predominated (reviewed in Bosworth and McClay 2001). The subsurface present-day stress data indicate that the central sub-basin is continuing to extend in a manner essentially unchanged from the Miocene period of active continental rifting (Bosworth and Durocher 2017). The situation is very different in the Southern sub-basin, and in the boundary region between the Darag and Central sub-basins. In these areas the NNE-SSW S_{hmin} is compatible with GPS-derived NNW movement of the Sinai micro-plate as discussed above. Bosworth and Taviani (1996) suggested that the NNE-SSW S_{hmin} orientation did not appear in the southern Gulf of Suez until during the late Pleistocene, but before formation of the 125 ka MIS5e terraces discussed below.

Additional stress field data are needed to fully assess the sub-regional stress field variations in the Gulf of Suez. Based on the well data compiled in Fig. 4, it can be concluded that each of the sub-basins possesses a distinct shallow crustal present-day stress field and that the boundaries between these stress domains correspond to major fault geometries inherited from the main phase of Miocene continental rifting (Bosworth and Durocher 2017).

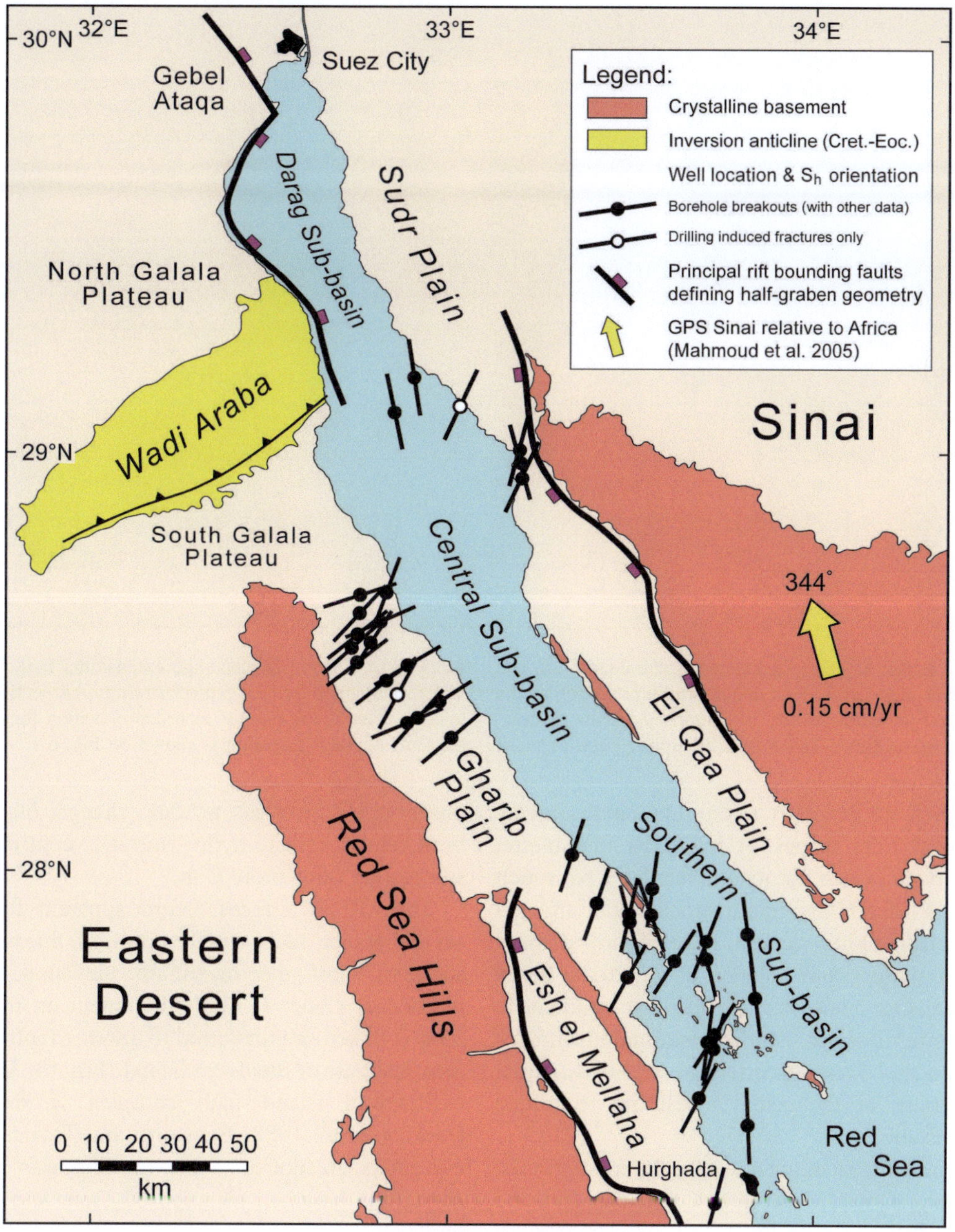

Fig. 4 Borehole breakout and drilling induced fracture data for the Gulf of Suez (Bosworth and Taviani 1996; Bosworth and Durocher 2017). Sticks show average S_{hmin} orientation for each well. The GPS-derived total vector of Sinai movement relative to Africa (Nubia) of ~0.15 cm/yr in the direction of 344° is adapted from Mahmoud et al. (2005). For simplicity only the rift-bounding faults are shown (boxes on downthrown side). The Wadi Araba inversion anticline formed in the Late Cretaceous to Late Eocene

3.3 Coral Terraces

Sea-level fluctuations during the Pleistocene resulted in the formation of several high-stand, interglacial coral terraces in the southern Gulf of Suez, the best preserved being the 125 ka MIS5e ('Eemian' or 'Ipswichian'; Fig. 5) (Andres and Radtke 1988; Reyss et al. 1993; Choukri et al. 1995). A detailed probabilistic and physical modeling approach has suggested that the MIS5e global sea level was 7.2 ± 1.3 m higher than at the present-day (Kopp et al. 2009). Following Kopp et al. we assume in our calculations that the MIS5e eustatic sea level was +7 m. Measurements of terrace elevations should ideally take into account a real-time tidal correction, and consideration of what the actual paleo-water depth for the measured feature was. Live corals are a truer indicator of the low-tide mark rather than average sea level, for example, and their position slightly underestimates the paleo-sea level elevation. But the Gulf of Suez, Gulf of

Fig. 5 MIS5e coral terrace at Gebel Tanka, north-central Gulf of Suez. The Pleistocene corals developed on an unconformity at the top of the Late Eocene Tanka Formation, which is comprised of thin-bedded shallow marine limestone. Gebel Tanka is in the uplifted and exposed footwall of a large Miocene-age extensional fault. The entire syn-rift and a small amount of the pre-rift stratigraphic section are absent at this locality. This is the highest recorded elevation for a MIS5e terrace in the Gulf of Suez. Location is shown on Fig. 6

Aqaba and Red Sea are generally micro-tidal environments today, with typical tidal ranges of less than one metre. Detailed tide charts are often not readily available for much of the coastline. Furthermore, minor erosion has affected many of the terraces. Other workers have discussed these considerations in more detail (Manaa et al. 2016; Inglis et al., this volume). Our analyses of the tectonic significance of the Pleistocene terraces are therefore minimum estimates of paleo-elevation and incur uncertainties of ~1 m (tidal range) beyond those of the actual measuring technique (LiDAR, tape measure, etc.).

A further complexity that might affect the differentiation of tectonic and eustatic signals in the coral terrace dataset is the role of regional isostacy driven by the episodic transfer of mass between the World ocean and continental ice sheets during the Pleistocene. Lambeck et al. (2011) referred to these effects as "glacio-hydro-isotatic contributions". The magnitude of this component of local sea level changes depends on the rheology of the underlying mantle and may be significant even far from the areas that were temporarily glaciated (Lambeck and Chappell 2001). Taking glacio-hydro-isostacy into consideration, Lambeck et al. (2011) estimated that for most of the greater Red Sea region the MIS5e sea level was only 1–2 m above the present-day level.

By surveying numerous MIS5e sites in the southern Gulf of Suez a view of post-125 ka relative subsidence or uplift can be established (Fig. 6; Bosworth and Taviani 1996). Ignoring the effects of isostacy, subtracting 7 m from these observations gives net tectonic change; including the Lambeck et al. glacio-hydro-isostatic contributions requires subtracting only about 2 m.

The MIS5e terraces record apparent footwall uplift of several major faults in the Gulf of Suez (Fig. 6). In the southern Gulf the maximum elevation of ~18.5 m is reached at Gebel el Zeit. Movement on the Gebel el Zeit fault is linked or transferred to another fault to the south that runs just east of Shadwan Island (Fig. 6). The Gebel el Zeit—Shadwan Island fault complex is seismically active (Jackson et al. 1988; Bosworth and Taviani 1996). Terrace elevations are not available for Shadwan Island itself but crystalline basement is exposed there and the crest of the fault block has risen at least relative to the surrounding basin.

MIS5e terrace elevations have not been determined for the Central Gulf of Suez north of Ras Shukeir (Fig. 6). On the east side of the Gulf, basement is exposed at Gebel Araba and patches of MIS5e coral complexes have been reported (Gvirtzman et al. 1992). These are heavily eroded and no terrace elevations were reported, but it is likely that this fault block is also undergoing some footwall uplift. Well-developed coral terraces are present in the northern part of the Central sub-basin at Gebel Tanka (Fig. 5) and less-developed terraces extend north of Hammam Faraun (Fig. 6). The Gebel Tanka terrace sits at +19 m, comparable to the north end of Gebel el Zeit. Unlike the Gebel el Zeit—Shadwan Island fault, the Hammam Faraun—Gebel Tanka

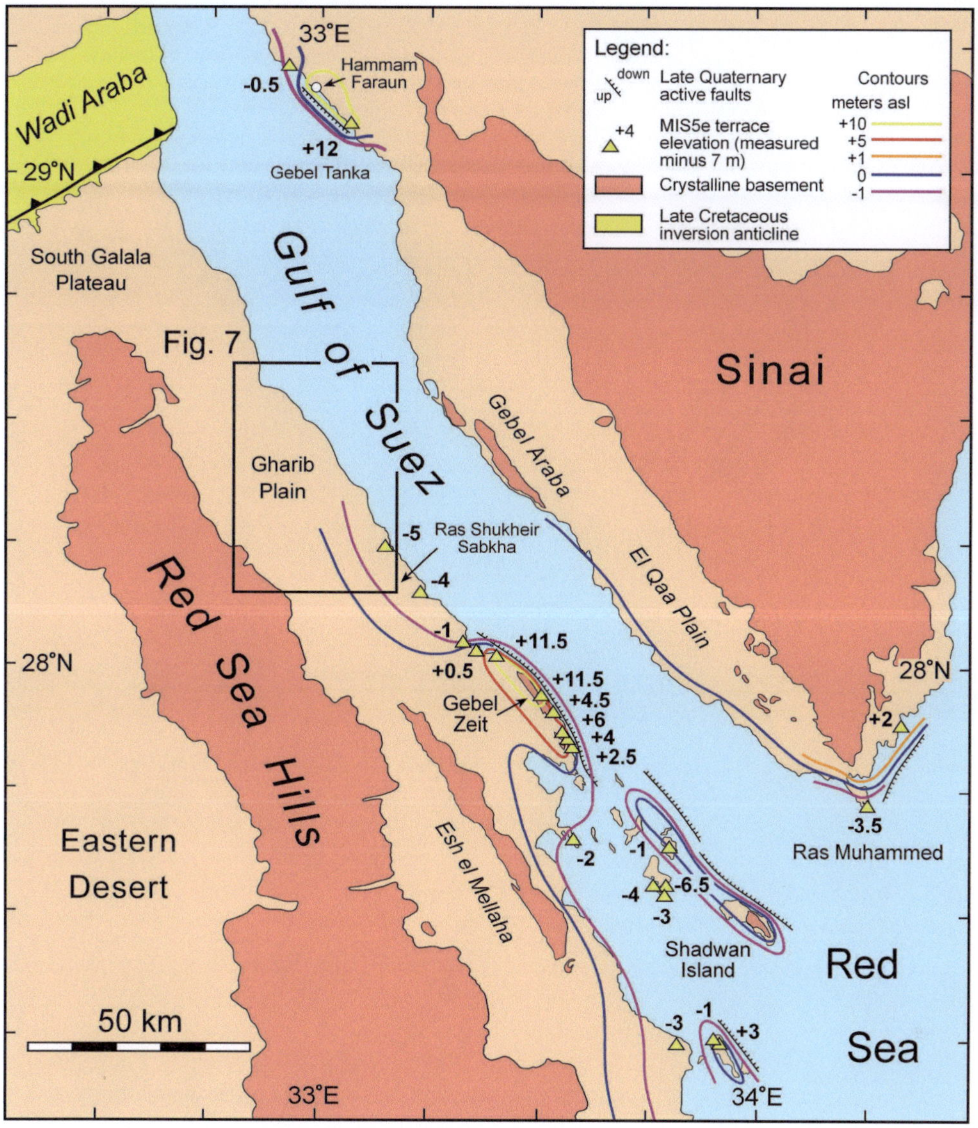

Fig. 6 Net tectonic movement of the MIS5e (Eemian, ~125 ka) coral terraces of the Gulf of Suez. The stated elevations are net tectonic movement after correcting for the +7 m MIS5e paleo-sea level. Glacio-hydro-isostacy considerations suggest that ~5 m should be added to each of the station values (Lambeck et al. 2011) Modified and extended from Bosworth and Taviani 1996. Locations of Fig. 5 (Gebel Tanka) and Fig. 7 (box) are shown

fault has not displayed historical seismicity, but this is certainly a long-term possibility. It is interesting to note that *both* active faults display strong oil and gas seeps at the coastline.

The Gebel el Zeit and Gebel Tanka terraces yield tectonic uplift rates of ~0.01 cm/yr without a glacio-hydro-isostatic correction, or 0.014 cm/yr with a correction.

3.4 Plio-Pleistocene Faulting

Despite the low extension rate (roughly an order of magnitude less than in the northern Red Sea), a broad belt of extensional faults developed along the central western margin of the Gulf of Suez in the Gharib alluvial plain (Fig. 7). Associated surface fault-line scarps are formed in partially lithified sands and gravels. The scarps are largely uneroded except where cut by the active wadi (dry river) systems. Trenching confirmed that the topographic scarps coincide with extensional faults with at least metres of offset. The deformed sands and gravels overlie Late Miocene evaporite beds and are themselves overlain by the MIS5e terraces. Their age is therefore post-Miocene to early-Pleistocene. The deformation appears to have been purely dip-slip although no fault plane kinematic data have been observed. The faults are very straight with an average strike of 145°. The Gharib fault

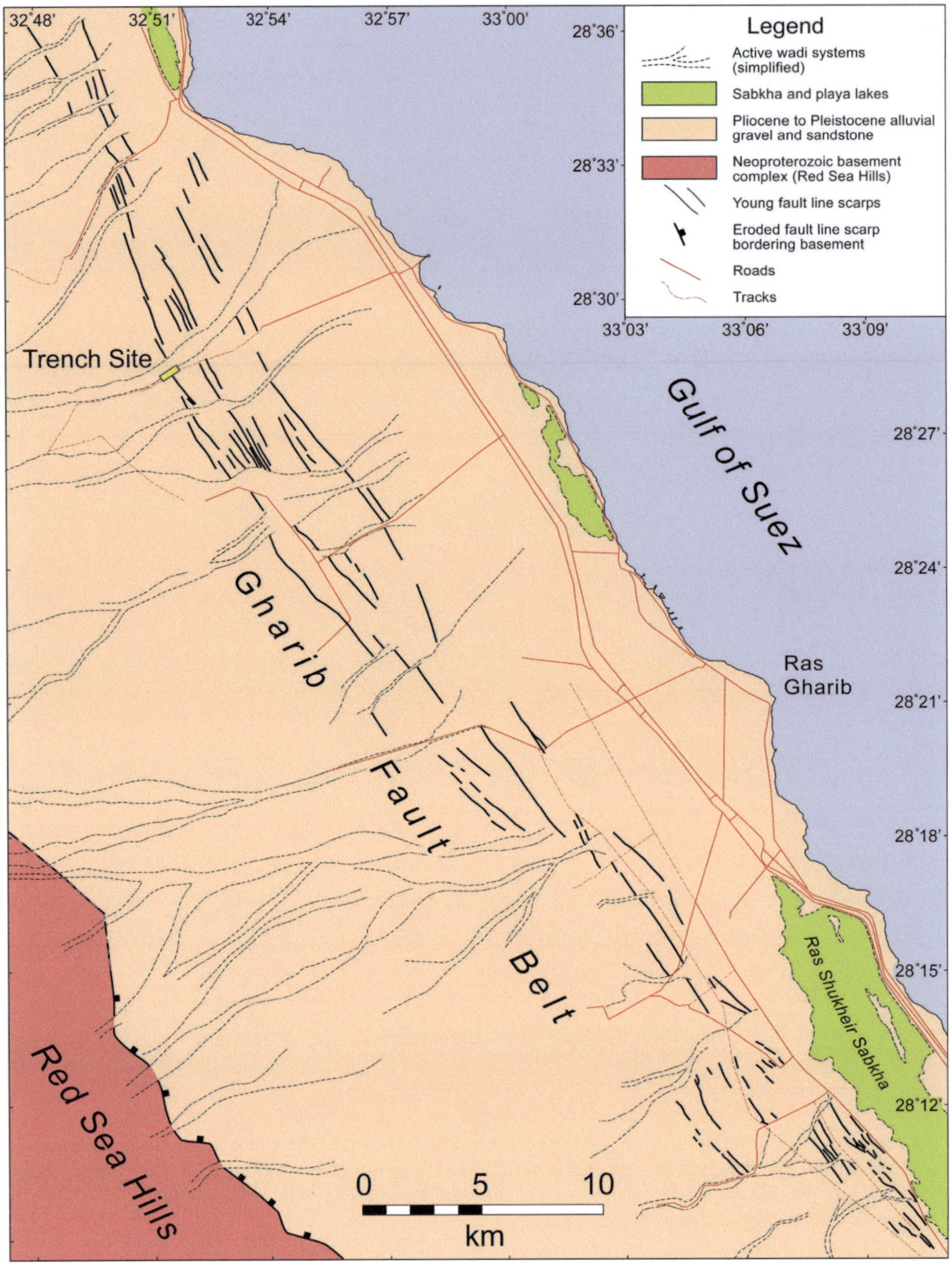

Fig. 7 Young fault-line scarps in Pliocene-Pleistocene alluvium on the west flank of the central Gulf of Suez (Gharib fluvial plain). Faults are straight and most are down-thrown toward the NE. They do not resemble structures that would be associated with underlying salt movement. Labelled yellow bar indicates position of trenching

belt probably developed in a stress field similar to that present today in the Central sub-basin and accommodated rift-normal extension. Because this fault array lies above Late Miocene evaporites it is possible that none of the faults is directly linked to basement structures. However, the straight fault pattern is not suggestive of basin-ward mass wasting or three-dimensional flow above rising salt domes or ridges.

4 Gulf of Aqaba

The Gulf of Aqaba is the southernmost segment of the left-lateral transform boundary that now links the northern Red Sea to the Zagros-Bitlas collision in eastern Turkey (Ben-Avraham et al. 1979; Bartov et al. 1980; Garfunkel

1981; Eyal et al. 1981; Ben-Avraham 1985). The offshore basin consists of three rhombohedral-shaped, left-stepping sub-basins, each with one or more bathymetric deeps (Fig. 8). As discussed above, several of the basin-parallel boundaries of these rhombs experience numerous small and occasional large magnitude earthquakes with strike-slip focal mechanisms, suggestive of traditional pull-apart basin geometries. However, reflection seismic data indicate that left-lateral slip on the west side of the northern (Elat) sub-basin is now being transferred to slip on the east side by a smooth, curvilinear fault, rather than at an orthogonal extensional fault (Ehrhardt et al. 2005). Similar complexities may exist in the central and southern sub-basins.

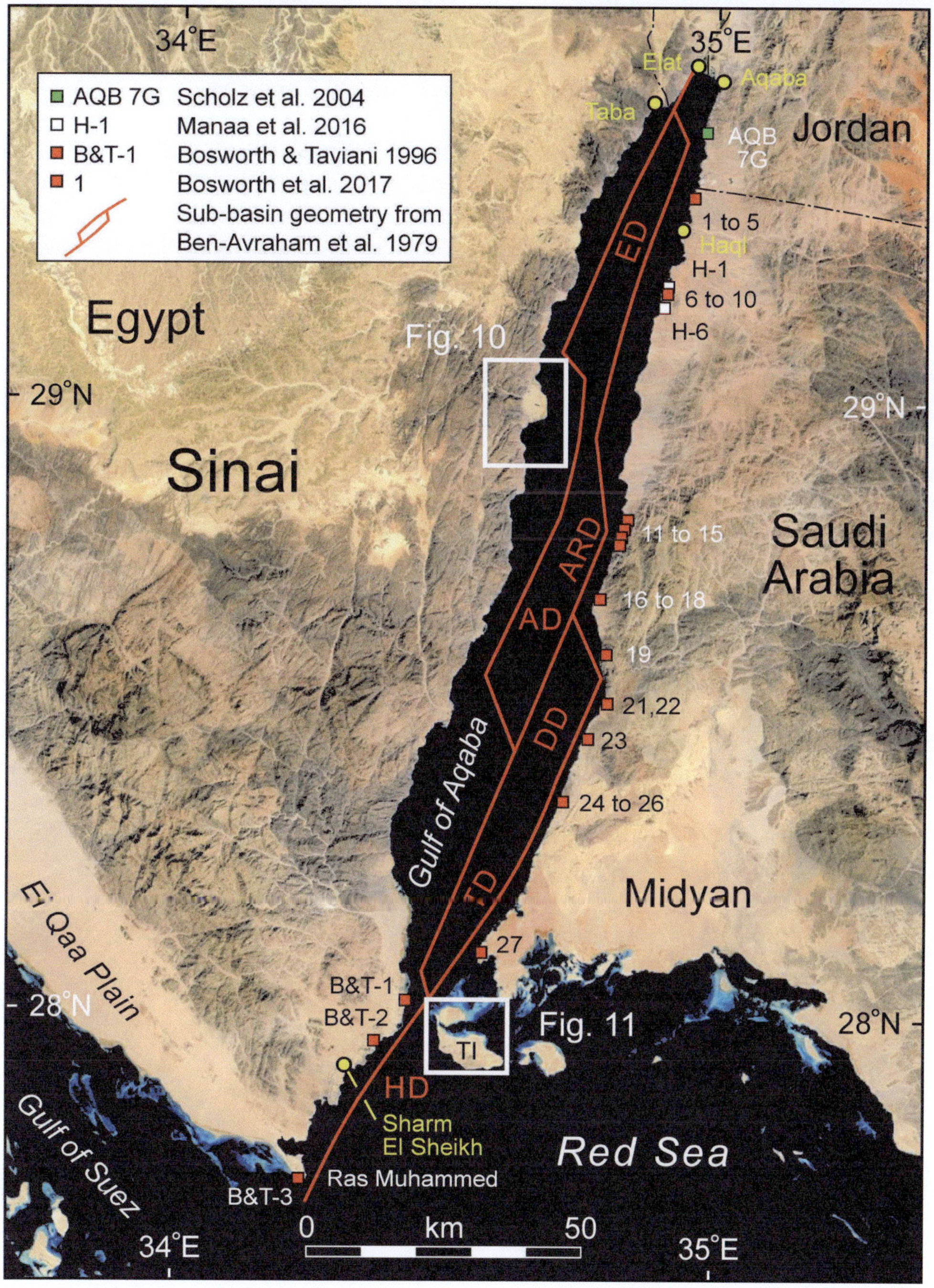

Fig. 8 Gulf of Aqaba MIS5e coral terrace and fault kinematics stations (Bosworth et al. 2017). Locations of Figs. 10 and 11 are shown by boxes. Sub-basin geometry and positions of bathymetric deeps are from Ben-Avraham et al. (1979). HD = Hume Deep; TD = Tiran Deep; DD = Dakar Deep; AD = Arnona Deep; ARD = Aragonese Deep; ED = Elat Deep; TI = Tiran Island. Basemap: Google Earth; Image Landsat

4.1 Present-Day Stress Fields

Industry exploratory wells have not yet been drilled in the Gulf of Aqaba so no borehole breakout data exist. The local present-day stress field can therefore only be interpreted from seismicity and outcrop fault kinematic data. Both large magnitude and small, locally recorded seismic events have been extensively analyzed in the Gulf of Aqaba and its onshore margins. Most $M_L \geq 5$ earthquakes have yielded left-lateral strike-slip focal mechanisms (Salamon et al. 1996; Roobol et al. 1999; Al-Tarazi 2000; Hofstetter 2003). Many of these large strike-slip events are concentrated in the Aragonese-Arnona deep (Fig. 8) and reflect the dominant movement of the Aqaba-Levant plate boundary. The simplest interpretation of these mechanisms suggests a NW-SE S_{Hmax}. However, smaller magnitude normal fault mechanisms are also recorded in local networks and in 1993 $M_L = 5.8$ and 5.6 predominantly normal-slip events occurred as part of a major earthquake swarm (Hofstetter 2003). These large events were located along the Arabian coastline of the Aragonese and Dakar deeps (Fig. 8) and indicated east-west extension. Similarly to the GPS datasets discussed above, seismicity supports the leaky transform model of Ben-Avraham et al. (1979).

East-west extension is not compatible with NW-SE oriented S_{Hmax}. Ben-Avraham and Zoback (1992) analyzed the relationships between the orientations of S_{hmin} and S_{Hmax} and the ratio of the frictional strength of a strike-slip fault versus the shear stress at failure of the adjacent crust. They found that for very weak faults, one or the other of the principal stresses rotates into parallelism with the fault itself. In a convergent setting, S_{hmin} swings into the fault orientation, whereas in a divergent setting it is S_{Hmax} that swings toward the fault orientation. Regional plate geometries and kinematics have produced a divergent setting at the Gulf of Aqaba, causing S_{Hmax} to be aligned roughly N-S along parts of the basin. This facilitates local extension perpendicular to the plate boundary.

4.2 Coral Terraces

Both the western (Sinai) and eastern (Arabia) margins of the Gulf of Aqaba display flights of Pleistocene coral terraces comparable to those in the Gulf of Suez. Pre-MIS5e Pleistocene coral terraces along the Jordanian and Saudi Arabian margin attain maximum elevations of ~100 m above sea level (Dullo 1990; Dullo and Montaggioni 1998). Pleistocene reef complexes are found up to 35 m above sea level in southeastern Sinai, although the upper surfaces of these highest reef terraces have been eroded (Gvirtzman et al. 1992).

Exposures of the MIS5e terrace are very extensive on the eastern margin of the Gulf of Aqaba, from near the city of Aqaba, Jordan, south to Midyan (Fig. 8). Elevations of the terraces are ~10 m in Jordan, rise to ~19–26 m in the central sub-basin margin, and then decline to ~9 m in Midyan (Fig. 9; Scholz et al. 2004; Manaa et al. 2016; Bosworth et al. 2017; Taviani et al., this volume).

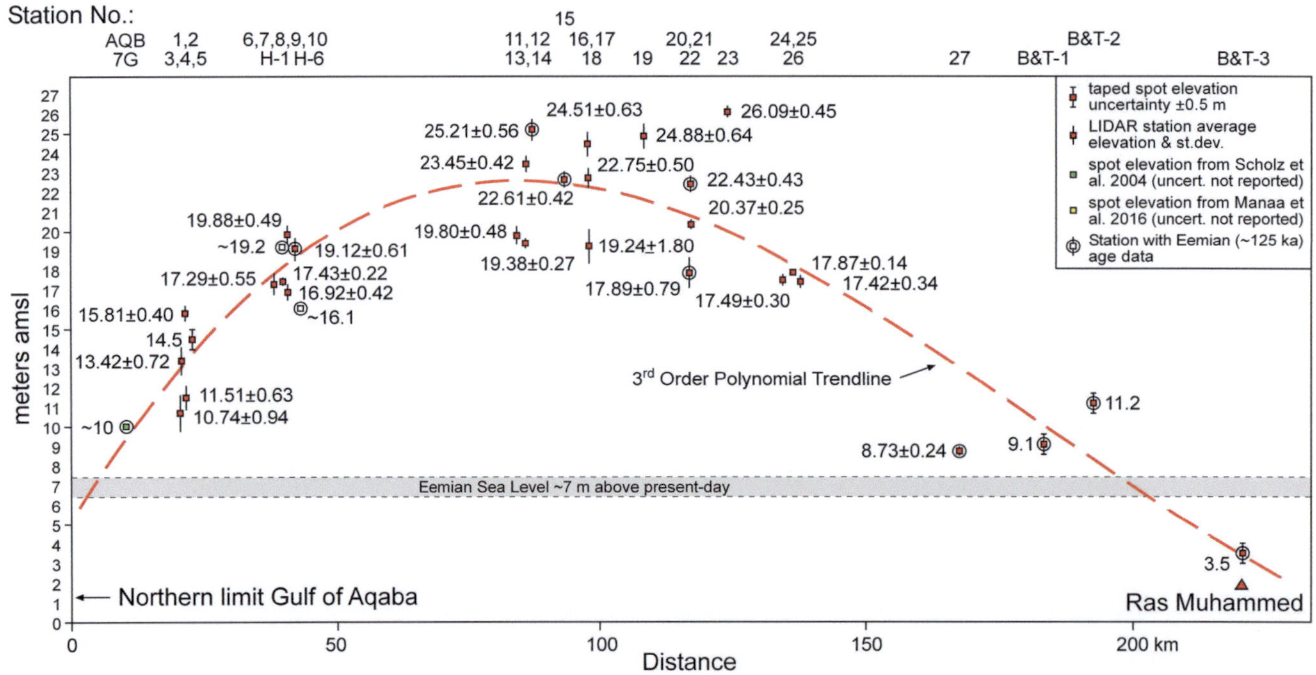

Fig. 9 Gulf of Aqaba MIS5e coral terrace elevations plotted along a north-south section. Station locations are shown in Fig. 8. Elevations are for the tops of the terraces relative to local sea level and without tidal corrections. After Bosworth et al. (2017)

The maximum uplift rate calculated for the central margin is 0.015 cm/yr, somewhat higher than the fastest rising fault blocks in the Gulf of Suez (Bosworth et al. 2017). Incorporating the potential glacio-hydro-isostatic contributions of Lambeck et al. (2011) this maximum uplift rate increases to 0.019 cm/yr. This is the fastest uplift rate in the Gulf of Suez—Gulf of Aqaba—Red Sea rift system where both detailed elevation and radiometric age data are available.

4.3 Plio-Pleistocene to Recent Faulting

The pre-MIS5e partially lithified fluvial sediments near Haql on the Arabian margin (Fig. 8) are cut by numerous extensional faults (Bosworth et al. 2017). The faults are usually down-thrown toward the basin, with offsets of a few metres or less, and strike N-S to NNW-SSE. Smaller-scale extensional faults with similar orientations are also found within the MIS5e terrace carbonate units themselves, although without any kinematic indicators. None of these Plio-Pleistocene faults can be shown to cut crystalline basement. No evaporites are known from the underlying stratigraphic section, but these faults could be generated by large-scale mass-wasting toward the Gulf of Aqaba basin axis.

More spectacular young faults are exposed on the Sinai margin along the coast near Nuweiba (Fig. 10). The footwall of these structures consists of Neoproterozoic granites, and the hanging wall is comprised of partially lithified sandstone and conglomerate that are being actively deposited by large wadi systems. In some cases the faults splay into the alluvial fan deposits themselves. Abundant kinematic indicators show

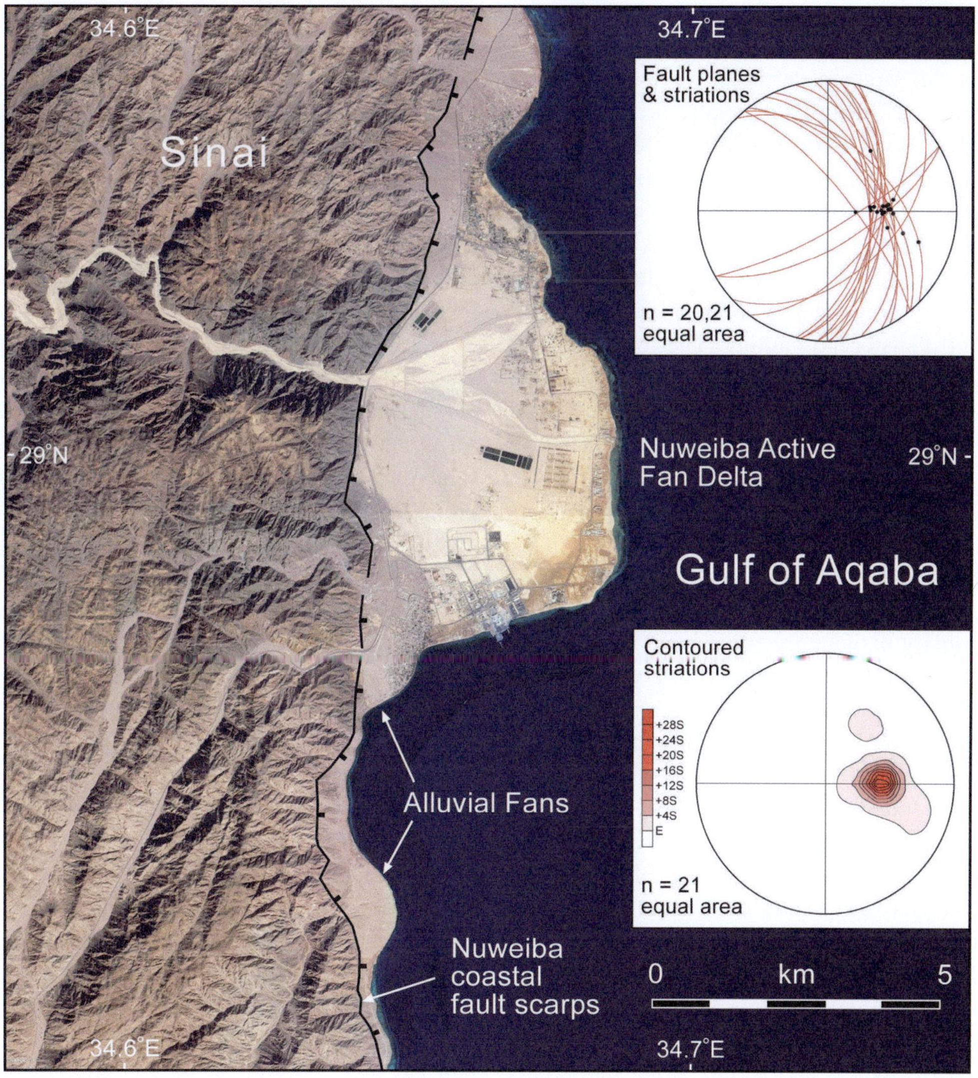

Fig. 10 Extensional fault system bounding the west margin of the central Gulf of Aqaba. All of the observed faults are predominantly dip-slip in nature, down-thrown toward the offshore basin axis. The footwall is comprised of Neoproterozoic igneous rocks and the hanging wall is partially lithified sandstone and conglomerate. Basemap: Google Earth; Image Landsat. After Bosworth et al. (2017)

that these faults are purely dip-slip with an E-W extension direction (Fig. 10; Bosworth et al. 2017). Inversion of the kinematic data from these Nuweiba faults supports the Ben-Avraham and Zoback (1992) model discussed above, wherein S_{hmin} is oriented approximately east-west along the Sinai Gulf of Aqaba margin.

Strike-slip faults are present in both the Sinai and Arabia onshore margins of the Gulf of Aqaba (Roobol and Stewart 2009). However, we are not aware of any that have been shown to cut Pliocene or younger strata. Ongoing deformation in the Gulf of Aqaba is partitioned between onshore extensional faulting and offshore mixed extensional and left-lateral strike-slip faulting.

5 Northern Red Sea

Most of the shorelines and coastal plains of the Red Sea are much less tectonically active than those of the Gulf of Suez or Gulf of Aqaba. MIS5e coral terraces, wave-cut benches and associated beachrock deposits generally lie at about 5–7 m above present-day sea level (El Moursi 1993; Plaziat et al. 1998; Manaa 2011; Inglis et al., this volume). This implies either tectonic stability, or a few metres of uplift when invoking an isostatic correction (Lambeck et al. 2011; Inglis et al., this volume). In contrast, the islands present along both margins display various combinations of uplift, stratal rotation, and Pleistocene faulting. The most interesting and geologically complex locations are Tiran and Zabargad Islands (Fig. 1), but remoteness and limited accessibility have left significant gaps in our understanding of these key localities.

5.1 Present-Day Stress Fields

Several dozen industry exploratory wells have been drilled along the Egyptian and Saudi Arabian northern Red Sea margins, both onshore and offshore (Beydoun 1989; Beydoun and Sikander 1992; Hughes and Beydoun 1992). No borehole stress field interpretations have been published for any of these wells. Modern imaging logs are not available, but borehole breakouts can be identified from the four-arm caliper tools that were run in several of the Egyptian offshore wells. The results from three wells are shown in Table 1. The orientation of S_{hmin} varies between 061° and 103°. This small dataset must be viewed with considerable caution, but generally indicates rift-normal present-day extensional stresses. This contrasts with the NNE-SSW relative plate motion between Africa-Nubia and Arabia along the Aqaba transform fault which results in highly oblique opening of the northern Red Sea.

5.2 Coral Terraces

Well-developed coral terraces and wave-cut notches are present along both the African and Arabian margins of the northern Red Sea. Similarly to the Gulfs of Suez and Aqaba, multiple terrace flights are present in many locales, and the best preserved and most studied are those dated to MIS5e. Near Duba on the Saudi Arabian coastline the terraces lie at 4–5.5 m above sea level (Manaa et al. 2016), and on the conjugate Egyptian margin from the southernmost Gulf of Suez to Marsa Alam they are at about 6 m (El Moursi 1993; El Moursi et al. 1994) (see Fig. 1 for locations). Further south on the Saudi Arabian margin at Harrat Al Birk what are thought to be MIS5e terraces and beach-rocks similarly provide paleo-sea level estimates of 5–7.5 m (Inglis et al., this volume). There are no anomalous, local uplifts in these datasets that would be comparable to the Gulf of Suez extensional fault blocks or the border fault regions of the Gulf of Aqaba. Several islands in the northern Red Sea are, however, more tectonically active.

- **Tiran Island**

The most pronounced offshore uplift in the northern Red Sea is recorded at Tiran Island where the highest Pleistocene (?) carbonate units are now at ~500 m elevation (Fig. 11; Schick 1958; Goldberg and Beyth 1991). A sample dated as ~130 ka and possibly part of an MIS5e terrace was located at ~40 m elevation (Goldberg and Beyth 1991). Goldberg and Beyth calculated an uplift rate of 0.03 cm/yr for this location. Assuming that the actual age of this highstand deposit is 125 ka and removing 7 m of paleo-sea level difference gives a rate of 0.026 cm/yr, still considerably higher than the maximum rates observed in the Gulf of Suez and Gulf of Aqaba. This probably represents the highest Late Pleistocene uplift rates of any location in the greater Red Sea system.

- **Zabargad Island**

Zabargad Island is located 50 km off the southern Egyptian Red Sea coastline (Fig. 1). The geology of the island is extremely complex, but it is most noted for the occurrence of gem-quality peridot within unserpentinized peridotite (Bonatti et al. 1981, 1983). Many models have been proposed relating to the age and uplift history of these peridotites and other rock units exposed on the island. Most pertinent to discussions of neotectonics is the position of the island in a significant kink/offset of the Red Sea axial trough that has been interpreted as a young oceanic fracture zone (Bonatti et al. 1981; Marshak et al. 1992).

Table 1 Borehole breakout measurements from offshore wells along the Egyptian margin of the northern Red Sea

Location – WGS84			Note 1				Breakouts										Method
			KB		TD		Avg. SHmax (deg)	Avg. SHmin (deg)	St. Dev. (deg)	No.	Total Length		Average MD		First	Last	
Well	Latitude	Longitude	(ft)	(m)	(TVD ft ss)	(TVD m ss)					(ft)	(m)	(ft)	(m)	(ft)	(ft)	
RSO T 95-1	27° 27''32.62'' N	34°10′ 30.00″E	46	14.0	6,401	1,951	151	61	2	4	39	12	5,220	1,591	4,865	5,574	4-arm caliper
RSO Z 95-1	26°53′ 38.68″N	34°10′ 56.84″E	46	14.0	11,358	3,645	161	71	6	1	154	47	10,337	3,151	10,260	10,414	4-arm caliper
RSO B 95-1	26°40′ 14.77″N	34°10′ 58.94″E	32	9.8	13,197	4,022	193	103	16	9	218	66	12,645	3,854	12,418	12,872	4-arm caliper
					Average:		168	78	18								

Note 1 KB = Kelly Bushing reference point; MD = measured depth; TVD ss = true vertical depth sub-sea

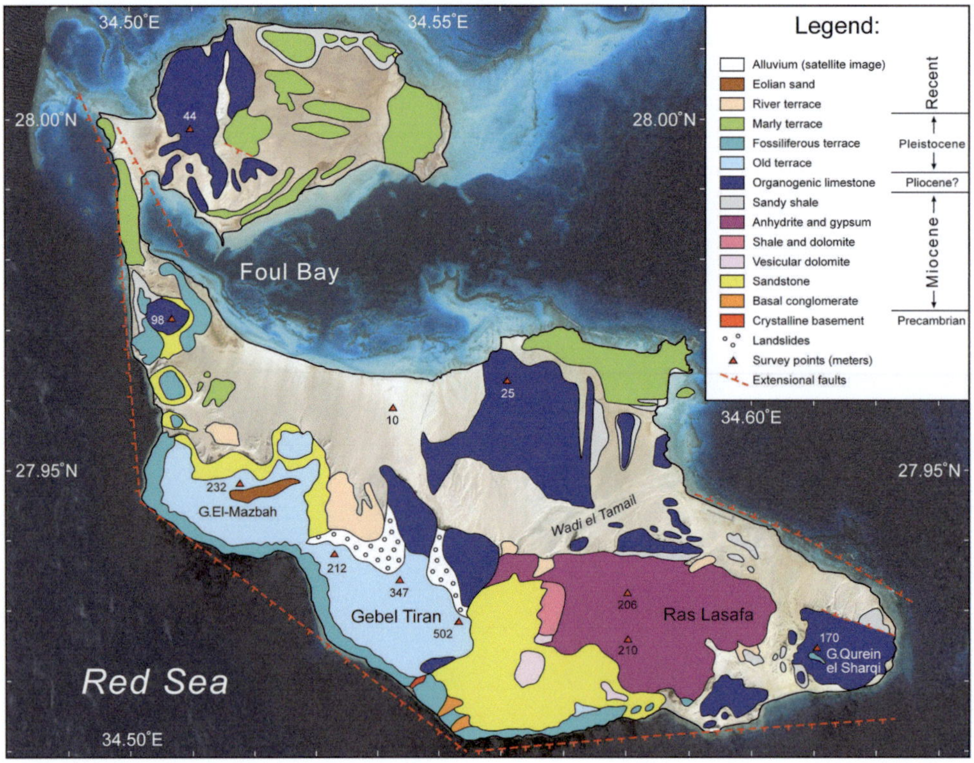

Fig. 11 Geologic map of Tiran Island Modified from Goldberg and Beyth (1991). Adjusted to a high-resolution satellite image base from Digital Globe and Google Earth. The uppermost terrace sits at ~500 m elevation and is the highest recorded limestone terrace in the greater Red Sea basin

Pleistocene coral terraces and geomorphologically less distinct carbonate bodies of several ages are present on Zabargad (Fig. 12a, b). These have been referred to as the Older Coral Limestones (OCL) and the Younger Coral Limestones (YCL) and attain a maximum elevation of ~80 m (Bonatti et al. 1983). Uranium-series dates for some of these units range from ~126 to >345 ka (Hoang and Taviani 1991). Three YCL samples that yielded roughly MIS5e ages (126–138 ka) lie at 6–8 m above sea level, suggesting that the island has behaved similarly to the northern Red Sea onshore margins since the last interglacial. However, one morphologically well-defined terrace on the SE side of the island is rotated ~1.5° to the NE (Fig. 13), and geopetal structures in the OCL often show tilting. This tilted terrace was not dated, but a sample from along-strike and at about the same elevation (10–12 m) produced an age of >300 ka. Re-colonization of older reefs by younger corals is common so caution must be taken in deciphering the succession of these reefs through time.

5.3 Plio-Pleistocene to Recent Faulting

- **Tiran Island**

Goldberg and Beyth (1991) suggested that several of the coastlines of the island are bounded by large-offset extensional faults. The kinematics of these faults are not known but several small-scale strike-slip faults were reported in the older carbonate units ("Organogenic Limestone"; thought to be Messinian to Early Pliocene) near Wadi el Tamail (Fig. 11). Several compressional folds deform these and underlying Miocene strata and the uplift of the island was attributed to the island's collision with south Sinai along faults in the Tiran Straits starting about 2.5–1.7 Ma (Goldberg and Beyth 1991). The high elevations occupied by the numerous Pleistocene coral terraces suggest that much of the uplift is very young.

Crystalline basement is exposed in a small wadi near the south shoreline of Tiran (Fig. 11; Goldberg and Beyth

(a)

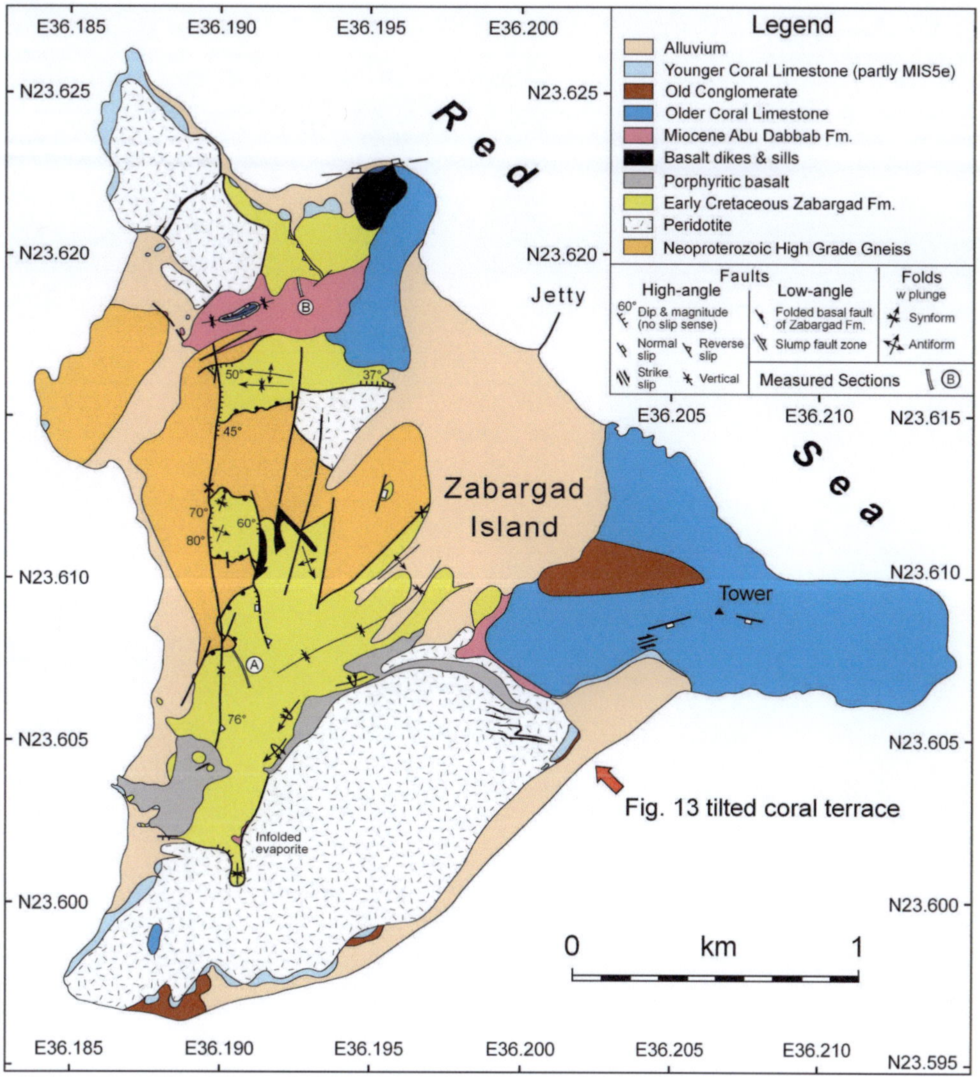

Fig. 12 Geomorphic and geologic features of Zabargad Island. **a** geologic map, modified from Bosworth et al. (1996) and incorporating data from Bonatti et al. (1983) and Marshak et al. (1992); and **b** neotectonic features (Google Earth; Image: Digital Globe). Several generations of small-scale faults cut the Pleistocene carbonate units. The strike-slip faults are preserved only within caves and are therefore thought to be the oldest whereas extensional faults have topographic expression

1991). The highest point on the island is at 502 m elevation (the highest coral terrace), so the total stratigraphic section cannot exceed about 500 m. Goldberg and Beyth (1991) reported that up to ~200 m of this section is composed of Miocene anhydrite and gypsum. Nearby on the mainland at Midyan the thickness of the Miocene evaporite units is much greater, and flowing halite can reach nearly a kilometre in salt walls (Mougenot and Al-Shakhis 1999). Halokinesis may have played a role in uplifting the strata and terraces of Tiran, but given the presence of basement in outcrop and the relatively thin section of evaporites the dominant mechanism appears to have been deeper basement-involved faulting.

- **Zabargad Island**

Numerous small-scale faults have been found that cut the OCL units (Bosworth et al. 1996). These faults are exposed in the large area of OCL outcrop on the eastern peninsula of the island, and between the OCL and the large basalt dike at the north end of the island (Fig. 12a, b). The faults can be separated into two generations, based on their morphologic expression in outcrop and kinematic data.

The faults interpreted as being older are predominantly strike-slip. Most of these are ENE-WSW striking, vertical-to-gently south dipping, and with right-lateral sense of slip. These strike-slip faults do not affect the topography

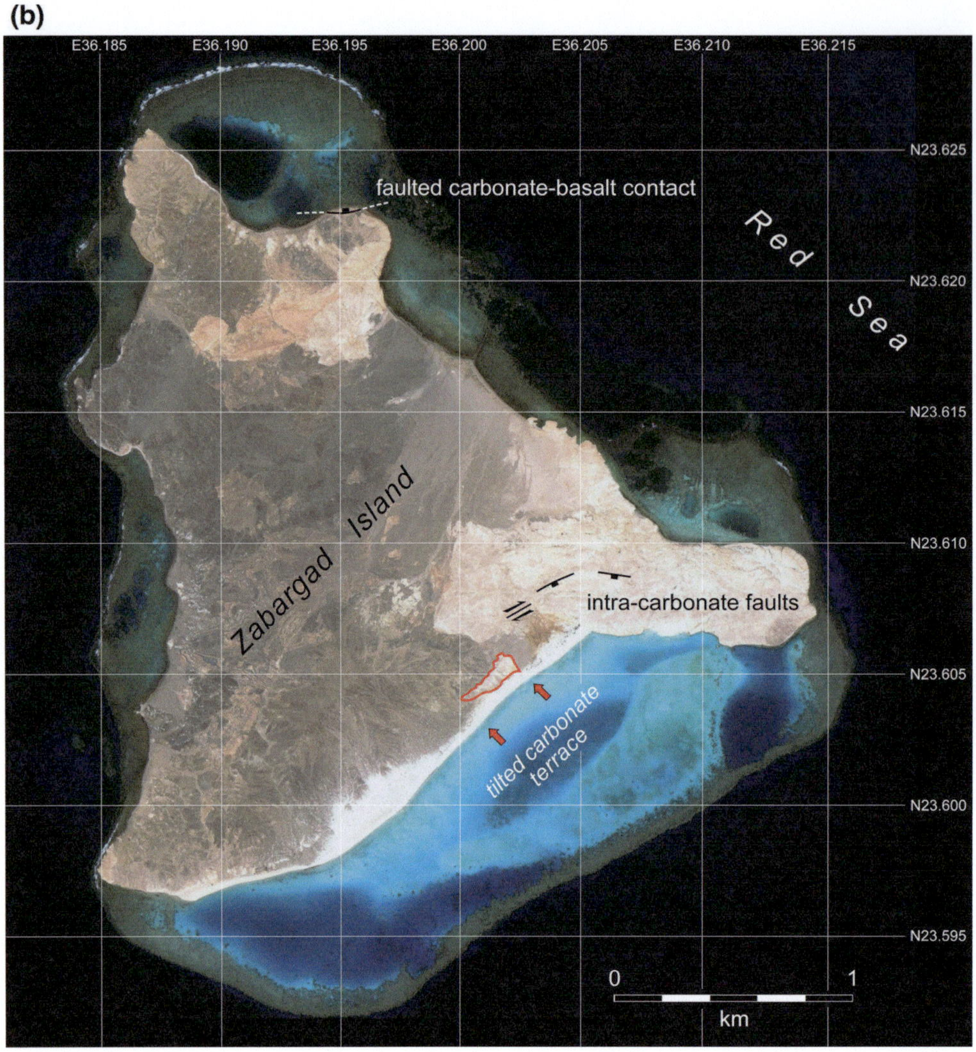

Fig. 12 (continued)

of the OCL, and are found only within caves and grottos in the limestone. Any former topographic expression of these structures must have been removed during subsequent complex Pleistocene eustatic sea-level changes. The amount of offset on the strike-slip faults is not known. The strike-slip faults were interpreted to have formed in a stress field with S_{hmin} oriented NNE-SSW (Bosworth et al. 1996).

The faults cutting the OCL and inferred as being younger are E-W striking, north and south dipping extensional faults. These faults offset terraces within the OCL, help control present-day topography in the southeast of the island, and define part of the northern shoreline of the island (Fig. 12a, b). Most of the normal faults have offsets of a few metres, but the northern shoreline fault, with basalt and OCL in the footwall and OCL in the hanging wall, may have greater displacement. The normal faults formed in a stress field with S_{hmin} oriented approximately N-S (Bosworth et al. 1996).

The multiple generations of faulting that affect the OCL, in conjunction with the tilting of Pleistocene coral terraces, demonstrate that Zabargad Island has experienced relatively young tectonism, perhaps a continuation of the processes that exposed the peridotite at this location within the young fracture zone. It is somewhat perplexing then that the MIS5e terrace appears to have been as "stable" as the Red Sea coastal equivalents.

6 Southern Red Sea

6.1 Present-Day Stress Fields

As discussed in the regional overview, inversion of earthquake focal mechanisms indicates that the southern Red Sea stress field has S_{hmin} aligned perpendicular to the rift axis, with a preponderance of normal fault solutions (Delvaux and

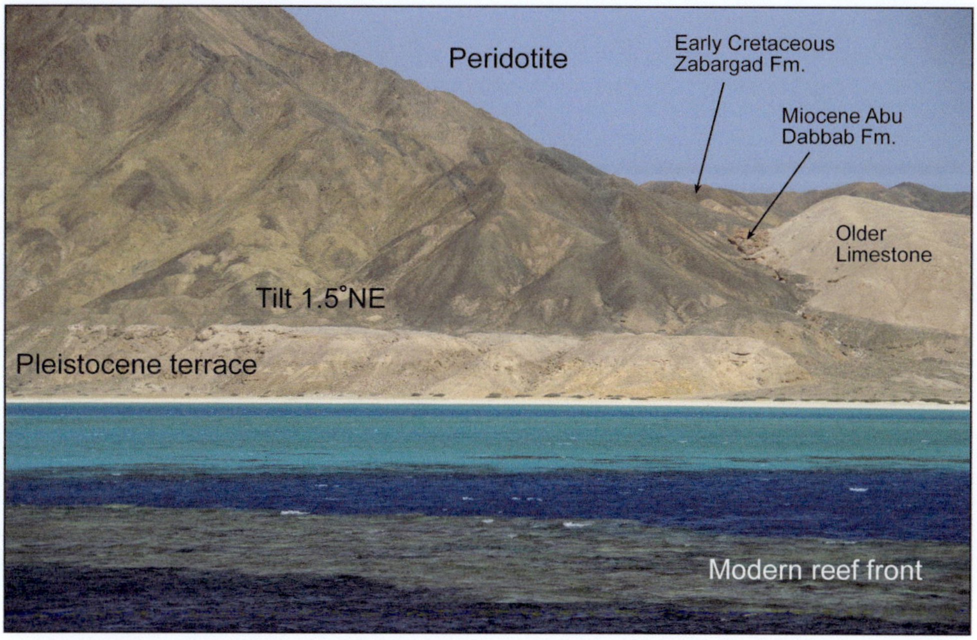

Fig. 13 Zabargad Island tilted Pleistocene coral terrace. South side of island, location given in Fig. 12b

Barth 2010). In this respect, the northern and southern basins may be similar, although the dataset for the north is far from definitive. Beyond this the tectonic state of the two regions is extremely different, as has been long recognized (Girdler 1958; Drake and Girdler 1964; Coleman 1974; Cochran 1983). As is clear from earthquake records (Fig. 1), the principal linkage between extension/oceanic spreading in the southern Red Sea and that in the Gulf of Aden occurs through the Afar depression, west of the Danakil horst via the Gulf of Zula and continuing to the Gulf of Tadjura. This critical region is beyond the scope of our review and many comprehensive studies have previously been published (Manighetti et al. 1998; Eagles et al 2002; Hofstetter and Beyth 2003; Redfield et al. 2003; Wolfenden et al. 2004; Le Gall et al. 2011).

6.2 Plio-Pleistocene to Recent Faulting

The neotectonic signature of the southern Red Sea is significantly different from that of the north. MIS5e coral terraces are present but thus far rarely dated (Inglis et al., this volume). We therefore only discuss some aspects of recent faulting. The tectonically most active regions are observable at the Farasan, Dahlak and Zubair Archipelagos.

- **Farasan and Dahlak Archipelagos**

Offshore from Eritrea the Dahlak Archipelago (Fig. 1) exposes Pleistocene limestones that are cut by active extensional faults (Frazier 1970; Carbone et al. 1998). On the Saudi Arabian mirror margin the Farasan Archipelago (Fig. 14) similarly shows complex active faulting, uplift and block rotation (Dabbagh et al. 1984; Bantan 1999; Inglis et al., this volume). In both areas the dominant effects of flowage of underlying Miocene evaporite units are clear, producing numerous surficial domes and intervening mini-basins (Figs. 14 and 15; Dabbagh et al. 1984; Dullo and Montaggioni 1998). Similar halokinetic effects have been documented along the Red Sea coastal plain in Yemen (Davison et al. 1996; Bosence et al. 1998). The combinations of faulting and salt movement have had pronounced effects on the distribution of Pleistocene carbonate facies. Dahlak lies at the northern intersection of the active Danakil rift with the Red Sea proper and this is also influencing the style and amount of deformation in the archipelago (Carbone et al. 1998).

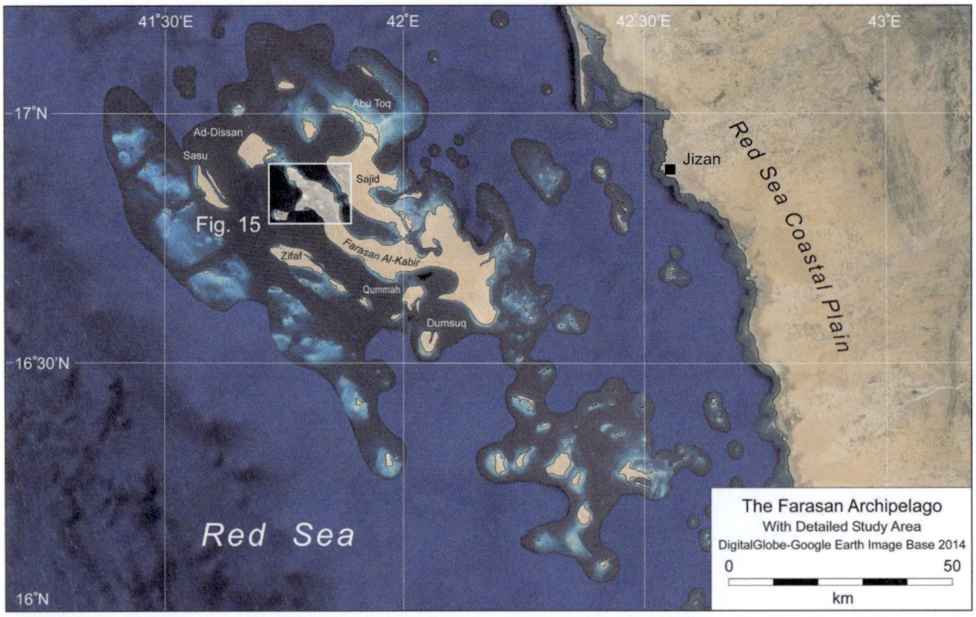

Fig. 14 Farasan Archipelago and location of detailed Fig. 15 structural map. Circular and elliptical dome and bowl structures are evident both onshore and in the shallow waters of the archipelago. The role of salt movement in forming these features is confirmed by exploratory wellbores. Base is from Google Earth; image Digital Globe

Fig. 15 Faulting on Farasan Kebir Island, Farasan Archipelago. Representative field stations were visited throughout the island and in all cases the exposed faults were found to be extensional and pure dip-slip wherever kinematic data were present. The fault pattern however is largely based on satellite image interpretation. See further discussion in Inglis et al. (this volume). Google Earth; image Digital Globe

- **Zubair Archipelago and active volcanism**

The southern Red Sea spreading centre bifurcates at approximately 16.5°N latitude, just north of the Dahlak Archipelago (Fig. 1). The most tectonically active branch abruptly swings toward the African margin via the Gulf of Zula and passes onshore west of the Danakil Horst (Fig. 3). A second branch continues along the axis of the southernmost Red Sea toward the Straits of Bab el Mandab. Seismicity gradually decreases to the south and the straits are not part of the active plate boundary at the present time (Fig. 1).

GPS measurements indicate that at Bab el Mandab the Danakil block is presently moving in-step with Arabia (McClusky et al. 2010; Reilinger et al. 2015). Toward the north, it is separating from both Africa-Nubia and Arabia. The southernmost Red Sea is therefore opening like a 'V' with its tip pinned at Bab el Mandab. This narrow V-shaped region is maintaining a high level of volcanic activity, but unlike other segments of the Red Sea axis the water is relatively shallow and volcanos often break the surface. The Zubair Archipelago and the isolated Jebel at Tair Island just to its north experienced volcanic eruptions in 2007–2008, 2011–2012 and 2013 (Xu et al. 2015; Jónsson and Xu 2015). This is the only subaerial volcanically active region within the confines of the Red Sea.

The southernmost island of the Zubair Archipelago is Center Peak. Xu et al. (2015) performed an InSAR interferometric analysis of satellite passages after the 2011–2012 eruption at the north end of the archipelago. They identified two major N-S faults in the interferogram patterns for Center Peak (Fig. 16). Several other fault scarps are present in the young colluvium on Center Peak, and evidence for active

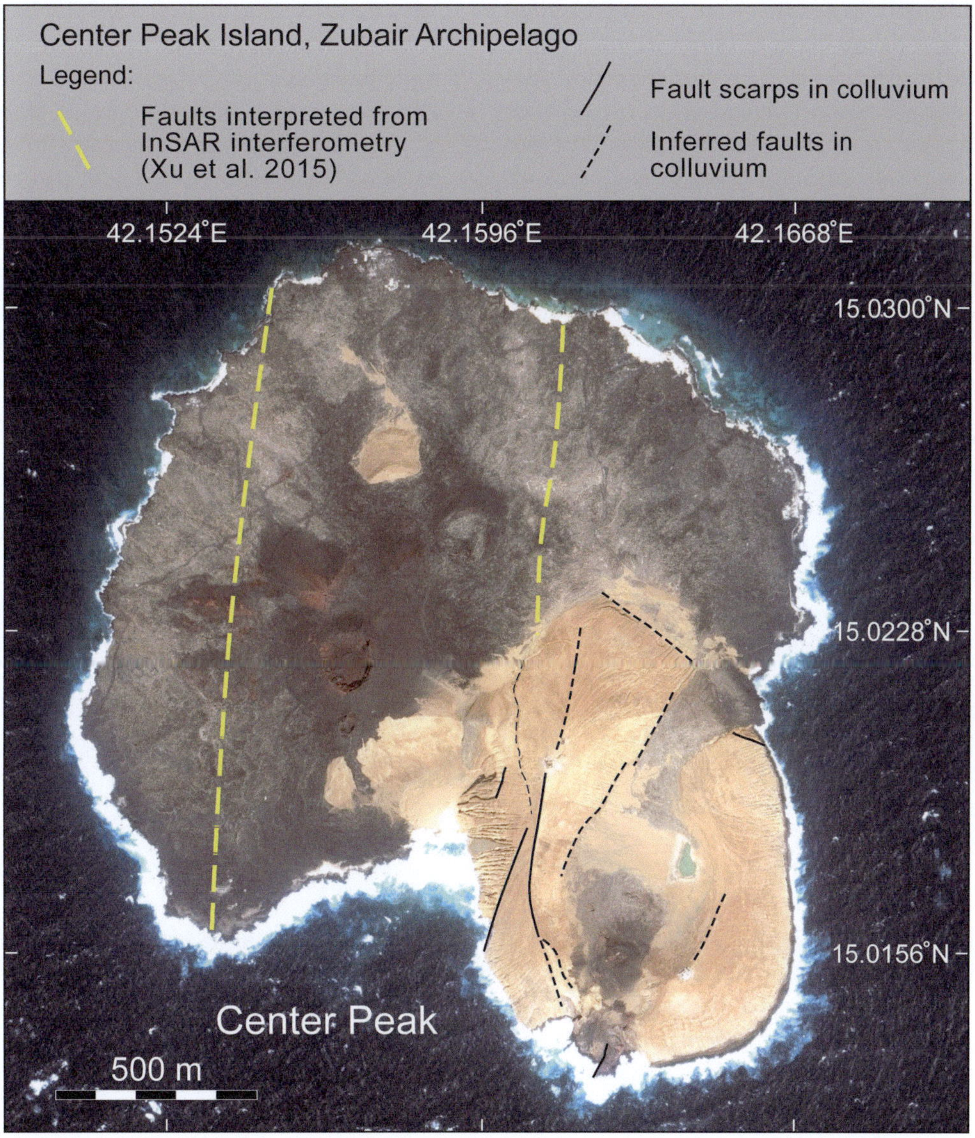

Fig. 16 Faulting on Center Peak Island, Zubair Archipelago. Several large but partly concealed faults were mapped by Xu et al. (2015) using InSAR interferometry. These do not parallel the margins of the southern Red Sea but rather are oriented ~N-S. We have also identified several fault scarps in the young colluvium based solely on satellite image interpretation. Google Earth; image Digital Globe

faulting is common throughout the archipelago. Xu et al. (2015) suggested that the 2011–2012 eruption was fed by a large, N-S oriented dike and the 2013 eruption by a sub-parallel NNW-SSE dike offset about 4 km to the east. Both are rotated clockwise from the overall NW-SE trend of the southern Red Sea rift axis.

7 Concluding Comments

The present-day and recent past tectonic activity of the Red Sea, Gulf of Suez and Gulf of Aqaba are spatially complex and temporally dynamic. Even the regional variations in the present-day stress field are poorly known, let alone how these stress fields may have evolved through time. A more complete understanding of the neotectonics of this complex rift system will depend upon integrating diverse geologic and geophysical datasets in parallel with increased refinements in the absolute dating of these observations. Further studies of exposed Pleistocene to Recent stratigraphic successions, their associated coral terraces, wave-cut benches and beach rocks, and associated tectonic structures are required. This is especially critical in areas where less work has been done in the past, and where human activities are altering coastal exposures. Observations from along the exposed margins and islands of these basins will continue to play a complementary role to new geophysical and geological data gained from the offshore.

8 Summary

- On a plate scale, the Red Sea rift is positioned between E-W S_{Hmax} in Africa-Nubia and N-S to NE-SW S_{Hmax} in Arabia. These far field stresses owe their origin to the spreading centres of the Atlantic Ocean and collision between Arabia and Eurasia along the Bitlis-Zagros suture. Based on limited present-day stress field data and structural observations the realm of rift-normal S_{hmin} is restricted to the Red Sea basin itself, and perhaps narrow belts along its shoulders.
- Within the Gulf of Suez, enough data are available to show that each of its sub-basins is presently experiencing significantly different, sub-regional stress fields that appear to be geographically related to the old Miocene syn-rift basin geometries.
- Large earthquakes ($M > 6$) are generally restricted to the central basin of the Gulf of Aqaba, the southern Gulf of Suez, and the greater Afar region. The detailed geodynamics responsible for this are certainly different but all are associated with the junctions of the major plate boundaries.
- Catalogues of earthquake activity and GPS datasets show that the Sinai micro-plate is still moving away from Africa with a component of left-lateral slip. This results in E-W opening of the Gulf of Aqaba where extensional faults predominate in the onshore realm. Strike-slip faulting is presently largely confined to its offshore sub-basins.
- NNE-SSW S_{hmin} in the southern Gulf of Suez is compatible with the GPS documented relative movement between Africa-Nubia and Sinai, related to the orientation of the Gulf of Aqaba—Dead Sea transform plate boundary. Why extension is rift-normal (NE-SW) in the central Gulf of Suez is more problematic.
- The kinematics of the southern Red Sea are complex. Extension is now occurring both west of the Danakil Horst and along the southernmost Red Sea axis in the vicinity of the Zubair Archipelago. Lack of seismicity suggests that moving south to Bab el Mandeb extension falls to zero. The igneous activity at Zubair may be gradually bending toward the west to link with the Afar triple junction.
- All of the margins of the Red Sea, Gulf of Suez and Gulf of Aqaba underwent tectonically-driven rift shoulder uplift and denudation in the past, particularly during the main phases of continental rifting. However, during the past 125 ka uplift has been much more focused and most pronounced at the footwalls of a few, active extensional faults. These include a few major faults in the Gulf of Suez, the coastal faults of the Gulf of Aqaba, and faults at Tiran Island.
- Smaller-scale extensional faulting is also occurring along the Saudi Arabian margin of the northern Red Sea, in the Dahlak and Farasan archipelagos, and on the volcanically active islands of the Zubair archipelago in the southernmost Red Sea. On the Farasan and Dahlak Islands this is related largely to the movement of underlying Miocene salt bodies, similar to effects documented along the coastal plain of Yemen.
- Though not active at the present time, a broad belt of small-offset, very linear extensional faults dissected the western margin of the central Gulf of Suez during the Plio-Pleistocene. The offshore area immediately to the east maintains a relatively high-level of low magnitude earthquake activity today.

Acknowledgements Daniel Stockli, Scott Durocher, Michele Morsilli, Damian Kelly, Grant Spencer, Miguel Muñoz, Ryan Miller, Douglas Barber and Michael Prior participated in field work in the Gulf of Suez and Egyptian Red Sea islands. Studies in the Gulf of Aqaba and Saudi Arabian Red Sea were promoted and funded by the Saudi Geological Survey in Jeddah and we thank Dr. Zohair Nawab and Dr. Abdullah Alattas for their support and encouragement. We acknowledge all participants of the 2013 and 2014 Saudi Field Parties for field work and post-expedition sample processing. Comments by four anonymous reviewers were very helpful. This paper is part of the PRIN2012 Programme (Project 20125JKANY_002, Principal Investigator Marco Ligi) and is ISMAR-CNR, Bologna, Scientific Contribution no. 1904.

References

Alamri AM, Schult FR, Bufe CG (1991) Seismicity and aeromagnetic features of the Gulf of Aqaba (Elat) region. J Geophys Res 96:20179–20185

Al-Tarazi E (2000) The major Gulf of Aqaba earthquake, 22 November 1995-Maximum Intensity Distribution. Nat Hazards 22:17–27

Ambraseys NN, Melville CP, Adams RD (1994) The seismicity of Egypt, Arabia and the Red Sea: a historical review. Cambridge University Press, Cambridge, p 181

Ameen MS (2014) Fracture and in-situ stress patterns and impact on performance in the Khuff structural prospects, eastern offshore Saudi Arabia. Mar Petrol Geol 50:166–184

Ameen MS, Buhidma IM, Rahim Z (2010) The function of fractures and in-situ stresses in the Khuff reservoir performance, onshore fields, Saudi Arabia. Bull Am Assoc Petrol Geol 94:27–60

Andres W, Radtke U (1988) Quartäre strandterrassen an der Küste des Gebel Zeit (Golf von Suez/Ägypten). Erdkunde 42:7–16

ANSS (2016) Comprehensive earthquake catalog (ComCat). USGS earthquake hazards program. Search Earthquake Catalogue. http://earthquake.usgs.gov/earthquakes/search. Accessed 05 Sept 2016

ArRajehi A, McClusky S, Reilinger R, Daoud M, Alchalbi A, Ergintav S, Gomez F, Sholan J, Bou-Rabee F, Ogubazghi G, Haileab B, Fisseha S, Asfaw L, Mahmoud S, Rayan A, Bendik R, Kogan L (2010) Geodetic constraints on present-day motion of the Arabian plate: implications for Red Sea and Gulf of Aden rifting. Tectonics 29:1–10

Ayele A (2002) Active compressional tectonics in central Africa and implications for plate tectonic models: evidence from fault mechanism studies of the 1998 earthquakes in the Congo Basin. J Afr Earth Sci 35:45–50

Babiker N, Mula AHG, El-Hadidy S (2015) A unified Mw-based earthquake catalogue and seismic source zones for the Red Sea region. J Afr Earth Sci 109:168–176

Badawy A (2001) Status of the crustal stress in Egypt as inferred from earthquake focal mechanisms and borehole breakouts. Tectonophys 343:49–61

Bantan RA (1999) Geology and sedimentary environments of Farasan Bank (Saudi Arabia) southern Red Sea: a combined remote sensing and field study. PhD Thesis, Royal Holloway, University of London, 297 p

Bartov Y, Steinitz G, Eyal M, Eyal Y (1980) Sinistral movement along the Gulf of Aqaba: its age and relation to opening the Red Sea. Nature 285:220–221

Ben-Avraham Z (1985) Structural framework of the Gulf of Elat (Aqaba)—northern Red Sea. J Geophys Res 90:703–726

Ben-Avraham Z, Zoback MD (1992) Transform-normal extension and asymmetric basins: an alternative to pull-apart models. Geology 20:423–426

Ben-Avraham Z, Garfunkel Z, Almagor G, Hall JK (1979) Continental breakup by a leaky transform: the Gulf of Elat (Aqaba). Science 206:214–216

Beydoun ZR (1989) The hydrocarbon prospects of the Red Sea—Gulf of Aden: a review. J Petrol Geol 12:125–144

Beydoun ZR, Sikander AH (1992) The Red Sea-Gulf of Aden: re-assessment of hydrocarbon potential. Marine Petrol Geol 9:474–485

Bonatti E, Hamlyn PB, Ottonello G (1981) The upper mantle beneath a young oceanic rift: peridotites from the Island of Zabargad (Red Sea). Geology 9:474–479

Bonatti E, Clocchiatti R, Colantoni P, Gelmini R, Marinelli G, Ottonello G, Santacroce R, Taviani M, Abdel-Meguid AA, Assaf HS, El Tahir MA (1983) Zabargad (St. John's) Island: an uplifted fragment of sub-Red Sea lithosphere. J Geol Soc Lond 140:677–690

Bosence DWJ, Al-Awah MH, Davison I, Rosen BR, Vita-Finzi C, Whittaker E (1998) Salt domes and their control on basin margin sedimentation: a case study from the Tihama plain, Yemen. In: Purser BH, Bosence DWJ (eds) Sedimentation and tectonics in rift basins: Red Sea—Gulf of Aden. Chapman and Hall, London, pp 448–464

Bosworth W (1985) Geometry of propagating continental rifts. Nature 316:625–627

Bosworth W (1995) A high-strain rift model for the southern Gulf of Suez (Egypt). In: Lambiase JJ (ed) Hydrocarbon habitat in rift basins. Geol Soc London, Special Paper 80:75–112

Bosworth W (2008) North Africa—Mediterranean present-day stress field transition and implications for fractured reservoir production in the Eastern Libyan Basins. In: Salem MJ, Oun KM, Essed AS (eds) Geology of East Libya, vol 4. Earth Science Soc Libya, Tripoli, pp 123–138

Bosworth W, Durocher S (2017) Present-day stress fields of the Gulf of Suez (Egypt) based on exploratory well data: non-uniform regional extension and its relation to inherited structures and local plate motion. J Afr Earth Sci. https://doi.org/10.1016/JjafrearSci2017.04.025

Bosworth W, McClay K (2001) Structural and stratigraphic evolution of the Gulf of Suez rift, Egypt: a synthesis. In: Ziegler PA, Cavazza W, Robertson AHF, Crasquin-Soleau S (eds) Peri-Tethys Memoir 6: Peri-Tethyan Rift/Wrench basins and passive margins. Mémoires du Muséum National d'Histoire Naturelle de Paris 186:567–606

Bosworth W, Stockli DF (2016) Early magmatism in the greater Red Sea rift: timing and significance. Can J Earth Sci 53:1158–1176. https://doi.org/10.1139/cjes-2016-0019

Bosworth W, Strecker MR (1997) Stress field changes in the Afro-Arabian rift system during the miocene to recent period. Tectonophys 278:47–62

Bosworth W, Taviani M (1996) Late Quaternary reorientation of stress field and extension direction in the southern Gulf of Suez, Egypt: evidence from uplifted coral terraces, mesoscopic fault arrays, and borehole breakouts. Tectonics 15:791–802

Bosworth W, Strecker MR, Blisniuk PM (1992) Integration of East African paleo and present-day stress data: implications for continental stress field dynamics. J Geophys Res 97:11851–11865

Bosworth W, Darwish M, Crevello P, Taviani M, Marshak S (1996) Stratigraphic and structural evolution of Zabargad Island (Red Sea, Egypt) since the early cretaceous. In: Youssef El SA (ed) Proceedings of third international conference on geology of the Arab World, vol 1, pp 161–190

Bosworth W, Crevello P, Winn RD, Steinmetz J (1998) Structure, sedimentation, and basin dynamics during rifting of the Gulf of Suez and northwestern Red Sea. In: Purser BH, Bosence DWJ (eds) Sedimentation and tectonics of rift basins: Red Sea-Gulf of Aden. Chapman and Hall, London, pp 77–96

Bosworth W, Huchon P, McClay K (2005) The Red Sea and Gulf of Aden Basins. In: Catuneanu O, Guiraud R, ErikssonP, Thomas B, Shone R, Key R (eds) Phanerozoic evolution of Africa. J African Earth Sci 43:334–378

Bosworth W, Montagna P, Pons-Branchu E, Rasul N, Taviani M (2017) Seismic hazards implications of uplifted Pleistocene coral terraces in the Gulf of Aqaba. Sci Rep 7(1):38, 1–13. https://doi.org/10.1038/s41598-017-00074-2

Camp VE, Roobol MJ (1992) Upwelling asthenosphere beneath western Arabia and its regional implications. J Geophys Res 97 (B11):15255–15271

Carbone F, Matteucci R, Angelucci A (1998) Present-day sedimentation on the carbonate platform of the Dahlak Islands, Eritrea. In: Purser BH, Bosence DWJ (eds) Sedimentation and tectonics of rift basins: Red Sea-Gulf of Aden. Chapman and Hall, London, pp 523–536

Choukri A, Reyss J-L, Plaziat J-C (1995) Datations radiochimiques des hauts niveaux marins de la rive occidentale du Nord de la Mer Rouge au moyen de radioles d'oursin. C. R. Acad. Sci Paris 321, série Iia:25–30

Cochran JR (1983) A model for development of the Red Sea. Bull Am Assoc Petrol Geol 67:41–69

Coleman RG (1974) Geologic background of the Red Sea. In: Whitmarsh RB, Weser OE, Ross DA et al (eds) Initial reports of the deep sea drilling project, vol 23. Government Printing Office, Washington, pp 813–819

Colletta B, Le Quellec P, Letouzey J, Moretti I (1988) Longitudinal evolution of the Suez rift structure (Egypt). Tectonophys 153:221–233

Dabbagh AE, Hoetzl H, Schnier H (1984) South Tihamah and Farasan Islands; Farasan Islands. In: Jado AR, Zoetl JG (eds) Quaternary period in Saudi Arabia, vol 2. Sedimentological, hydrochemical, geomorphological, geochronological and climatological investigations in western Saudi Arabia. Springer, Vienna, pp 212–220

Daggett PH, Morgan P, Boulos FK, Hennin SF, El-Sherif AA, El-Sayed AA, Basta NZ, Melek YS (1986) Seismicity and active tectonics of the Egyptian Red Sea margin and the northern Red Sea. Tectonophys 125:313–324

Davison I, Bosence D, Alsop I, Al-Aawah MH (1996) Deformation and sedimentation around active Miocene salt diapirs on the Tihama Plain, northwest Yemen. In: Alsop I, Blundell D, Davison I (eds) Salt tectonics. Geol Soc London, Special Publication 100:23–39

Delvaux D, Barth A (2010) African stress pattern from formal inversion of focal mechanism data. Tectonophys 482:105–128

Drake CL, Girdler RW (1964) A geophysical study of the Red Sea. Geophys J Roy Astron Soc 8:473–495

Dullo W-C (1990) Facies, fossil record, and age of pleistocene reefs from the Red Sea (Saudi Arabia). Facies 22:1–46

Dullo W-C, Montaggioni L (1998) Modern Red Sea coral reefs: a review of their morphologies and zonation. In: Purser BH, Bosence DWJ (eds) Sedimentation and tectonics of rift basins: Red Sea-Gulf of Aden. Chapman and Hall, London, pp 583–594

Eagles G, Gloaguen R, Ebinger C (2002) Kinematics of the Danakil microplate. Earth Planet Sci Lett 203:607–620

Ehrhardt A, Hübscher C (2015) The northern Red Sea in transition from rifting to drifting- lessons learned from ocean deeps. In: Rasul NMA, Stewart ICF (eds) The Red Sea: the formation, morphology, oceanography and environment of a young ocean basin. Springer Earth System Sciences, Heidelberg, pp 99–121

Ehrhardt A, Hübscher C, Ben-Avraham Z, Gajewski D (2005) Seismic study of pull-apart-induced sedimentation and deformation in the northern Gulf of Aqaba (Elat). Tectonophys 396:59–79

El Moursi MEE (1993) Pleistocene evolution of the reef terraces of the Red Sea coastal plain between Hurghada and Marsa Alam. Egypt J Afr Earth Sci 17:125–127

El Moursi M, Hoang CT, El Fayoumy IF, Hegab O, Faure H (1994) Pleistocene evolution of the Red Sea coastal plain, Egypt: evidence from Uranium-series dating of emerged reef terraces. Quatern Sci Rev 13:345–359

El-Isa ZH, Al Shanti A (1989) Seismicity and tectonics of the Red Sea and western Arabia. Geophys J 97:449–457

Eyal M, Eyal Y, Bartov Y, Steinitz G (1981) The tectonic development of the western margin of the Gulf of Elat (Aqaba) rift. Tectonophys 80:39–66

Feldens P, Mitchells NC (2015) Salt flows in the central Red Sea. In: Rasul NMA, Stewart ICF (eds) The Red Sea: the formation, morphology, oceanography and environment of a young ocean basin. Springer Earth System Sciences, Heidelberg, pp 205–218

Frazier SB (1970) Adjacent structures of Ethiopia, the portion of Red Sea coast including Dahlak Kebir Island and the Gulf of Zula. Phil Trans Roy Soc Lond A267:131–141

Garfunkel Z (1981) Internal structure of the Dead Sea leaky transform (rift) in relation to plate kinematics. Tectonophys 80:81–108

Gaulon R, Chorowicz J, Vidal G, Romanowicz B, Roult G (1992) Regional geodynamic implications of the May–July 1990 earthquake sequence in southern Sudan. Tectonophys 209:87–103

Giardini D, Beranzoli L (1992) Waveform modelling of the May 20, 1990 Sudan earthquake. Tectonophys 209:105–114

Girdler RW (1958) The relationship of the Red Sea to the east African rift system. Quart J Geol Soc London 114:79–105

Gvirtzman G, Kronfeld J, Buchbinder B (1992) Dated coral reefs of southern Sinai (Red Sea) and their implication to late quaternary sea levels. Marine Geol 108:29–37

Goldberg M, Beyth M (1991) Tiran Island: an internal block at the junction of the Red Sea rift and Dead Sea Transform. Tectonophys 198:261–273

Heidbach O, Tingay M, Barth A, Reinecker J, Kurfeß D, Müller B (2008) The world stress map database release 2008. https://doi.org/10.1594/gfz.wsm.rel2008

Hofstetter A (2003) Seismic observations of the 22/11/1995 Gulf of Aqaba earthquake sequence. Tectonophys 369:21–36

Hofstetter A, Beyth M (2003) The Afar depression: interpretation of the 1960–2000 earthquakes. Geophys J Int 155:715–732

Hoang CT, Taviani M (1991) Stratigraphic and tectonic implications of uranium-series-dated coral reefs from uplifted Red Sea Islands. Quat Res 35:264–273

Huang PY, Solomon SC (1987) Centroid depths and mechanisms of mid-ocean ridge earthquakes in the Indian Ocean, Gulf of Aden, and Red Sea. J Geophys Res 92:1361–1382

Hughes GW, Beydoun ZR (1992) The Red Sea-Gulf of Aden: biostratigraphy, lithostratigraphy and palaeoenvironments. J Petrol Geol 15:135–156

Hughes GW, Abdine S, Girgis MH (1992) Miocene biofacies development and geological history of the Gulf of Suez. Egypt Marine Petrol Geol 9:2–28

Hussein HM, Abou Elenean KM, Marzouk IA, Korrat IM, Abu El-Nader IF, Ghazala H, El Gabry MN (2013) Present-day tectonic stress regime in Egypt and surrounding area based on inversion of earthquake focal mechanisms. J Afr Earth Sci 81:1–15

Inglis RH, Bosworth W, Rasul N, Al Saeedi A, Bailey G (this volume) Investigating the palaeoshorelines and coastal archaeology of the southern Red Sea

Jackson JA, White NJ, Garfunkel Z, Anderson H (1988) Relations between normal-fault geometry, tilting and vertical motions in extensional terrains: an example from the southern Gulf of Suez. J Struct Geol 10:155–170

Jarrige J-J, Ott d'Estevou P, Burollet PF, Montenat C, Prat P, Richert J-P, Thiriet J-P (1990) The multistage tectonic evolution of the Gulf of Suez and northern Red Sea continental rift from field observations. Tectonics 9:441–465

Joffe S, Garfunkel Z (1987) Plate kinematics of the circum Red Sea—a re-evaluation. Tectonophys 141:5–22

Jónsson S, Xu W (2015) Volcanic eruptions in the southern Red Sea during 2007–2013. In: Rasul NMA, Stewart ICF (eds) The Red Sea: the formation, morphology, oceanography and environment of a young ocean basin. Springer Earth System Sciences, Heidelberg, pp 175–186

Kopp RE, Simons FJ, Mitrovica JX, Maloof AC, Oppenheimer M (2009) Probabilistic assessment of sea level during the last interglacial stage. Nature 462:863–868

Lambeck K, Chappell J (2001) Sea-level change through the last glacial cycle. Science 292:679–686

Lambeck K, Purcell A, Flemming NC, Vita-Finzi C, Alsharekh AM, Bailey GN (2011) Sea level and shoreline reconstructions for the Red Sea: isostatic and tectonic considerations and implications for hominin migration out of Africa. Quatern Sci Rev 30:3542–3574

Le Gall B, Daoud MA, Rolet J, Egueh NM (2011) Large-scale flexuring and antithetic faulting along a nascent plate boundary in the SE Afar rift. Terra Nova 23:416–420

Ligi M, Bonatti E, Rasul NMA (2015) Seafloor spreading initiation: geophysical and geochemical constraints from the Thetis and Nereus Deeps, central Red Sea. In: Rasul NMA, Stewart ICF (eds) The Red Sea: the formation, morphology, oceanography and environment of a young ocean basin. Springer Earth System Sciences, Heidelberg, pp 79–98

Ligi M, Bonatti E, Bortoluzzi G, Cipriani A, Cocchi L, Caratori Tontini F, Carminati E, Ottolini L, Schettino A (2012) Birth of an ocean in the Red Sea: initial pangs. Geochem Geophys Geosyst 13. 10.1029/2012GC004155

Mahmoud S, Reilinger R, McClusky S, Vernant P, Tealeb A (2005) GPS evidence for northward motion of the Sinai block: implications for E Mediterranean tectonics. Earth Planet Sci Lett 238:217–224

Manaa AA (2011) Late pleistocene raised coral reefs in the eastern Red Sea—Rabigh, Saudi Arabia. M.Sc. Thesis, University of Wollongong. http://ro.uow.edu.au/theses/3501

Manaa AA, Jones BG, McGregor HV, Zhao J, Price DM (2016) Dating quaternary raised coral terraces along the Saudi Arabian Red Sea coast. Mar Geol 374:59–72

Manighetti I, Tapponnier P, Gillot PY, Jacques E, Courtillot V, Armijo R, Ruegg JC, King G (1998) Propagation of rifting along the Arabia-Somalia plate boundary: into Afar. J Geophys Res 103:4947–4974

Marshak S, Bonatti E, Brueckner H, Paulsen T (1992) Fracture-zone tectonics at Zabargad Island, Red Sea (Egypt). Tectonophys 216:379–385

McClusky S, Reilinger R, Ogubazghi G, Amleson A, Haleab B, Vernant P, Sholan J (2010) Kinematics of the southern Red Sea-Afar triple junction and implications for plate dynamics. Geophys Res Lett 37. https://doi.org/10.1029/2009gl041127

McQuarrie N, Stock JM, Verdel C, Wernicke BP (2003) Cenozoic evolution of Neotethys and implications for the causes of plate motions. Geophys Res Lett 30:1–4

Miller PM, Barakat H (1988) Geology of the Safaga concession, northern Red Sea. Egypt Tectonophys 153:123–136

Mitchell NC, Ligi M, Ferrante V, Bonatti E, Rutter E (2010) Submarine salt flows in the central Red Sea. Geol Soc Am Bull 122:701–713

Montenat C, Ott D'estevou P, Purser B, Burollet P, Jarrige J, Sperber F, Philobbos E, Plaziat J-C, Prat P, Richert J, Roussel N, Theiriet J (1988) Tectonic and sedimentary evolution of the Gulf of Suez and the northern western Red Sea. Tectonophys 153:166–177

Moretti I, Colletta B (1987) Spatial and temporal evolution of the Suez rift subsidence. J Geodyn 7:151–168

Mougenot D, Al-Shakhis AA (1999) Depth imaging sub-salt structures: a case study in the Midyan Peninsula (Red Sea). GeoArabia 4:445–463

Moustafa AM (1976) Block faulting of the Gulf of Suez. In: 5th Egyptian general petroleum organization exploration seminar, Cairo, 19 p (unpublished report distributed as paper copies at the conference)

Moustafa AR (1997) Controls on the development and evolution of transfer zones: the influence of basement structure and sedimentary thickness in the Suez rift and Red Sea. J Struct Geol 19:755–768

NOAA (2016) National oceanic and atmospheric administration (NOAA), National Centers for Environmental Information, World Data Service for Geophysics—bathymetry and global relief grid extract. https://www.ngdc.noaa.gov/bathymetry/reliefGridExtract

Patton TL, Moustafa AR, Nelson RA, Abdine SA (1994) Tectonic evolution and structural setting of the Suez Rift. In: Landon SM (ed) Interior rift basins. Am Assoc Petrol Geol Memoir 59:7–55

Perry SK, Schamel S (1990) The role of low-angle normal faulting and isostatic response in the evolution of the Suez rift. Egypt. Tectonophys 174:159–173

Plaziat J-C, Montenat C, Barrier P, Janin M-C, Orszag-Sperber F, Philobbos E (1998) Stratigraphy of the Egyptian syn-rift deposits: Correlations between axial and peripheral sequences of the north-western Red Sea and Gulf of Suez and their relations with tectonics and eustacy. In: Purser BH, Bosence DWJ (eds) Sedimentation and tectonics in rift basins: Red Sea—Gulf of Aden. Chapman and Hall, London, pp 211–222

Quennell AM (1956) Tectonics of the Dead Sea Rift. In: 20th International Geol Congress, Mexico City, 1956, Resumenes de los trabajos presentados, p 407

Quennell AM (1958) The structural and geomorphic evolution of the Dead Sea rift. Quart J Geol Soc London 114:1–24

Redfield TF, Wheeler WH, Often M (2003) A kinematic model for the development of the Afar depression and its paleogeographic implications. Earth Planet Sci Lett 216:383–398

Reilinger R, McClusky S, ArRajehi A (2015) Geodetic constraints on the geodynamic evolution of the Red sea. In: Rasul NMA, Stewart ICF (eds) The Red Sea: the formation, morphology, oceanography and environment of a young ocean basin. Springer Earth System Sciences, Berlin Heidelberg, pp 135–149

Reyss JL, Choukri A, Plaziat JC, Purser BH (1993) Datations radiochimiques des récifs coralliens de la rive occidentale du Nord de la Mer Rouge, premières implications stratigraphiques et tectoniques. C. R. AcadSci Paris 317, série II:487–492

Roobol MJ, Stewart ICF (2009) Cenozoic faults and recent seismicity in northwest Saudi Arabia and the Gulf of Aqaba region. Saudi geological survey technical report SGS-TR-2008-7, 35 p

Roobol MJ, Al-Rehaili M, Arab N, Celebi M, Halawani MA, Janjou D, Kazi A, Martin C, Sahl M, Showail A (1999) The Gulf of Aqaba earthquake of 22 November, 1995: its effects in Saudi Arabia. Saudi Arabian deputy ministry for mineral resources technical report BRGM-TR-99-16, 67 p

Salamon A, Hofstetter A, Garfunkel Z, Ron H (1996) Seismicity of the eastern mediterranean region: perspective from the Sinai subplate. Tectonophys 263:293–305

Schick AP (1958) Marine terraces on Tiran Island, northern Red Sea. Geogr Ann 40:63–66

Scholz D, Mangini A, Felis T (2004) U-series dating of diagenetically altered fossil reef corals. Earth Planet Sci Lett 218:163–178

Steckler MS, Berthelot F, Lyberis N, Le Pichon X (1988) Subsidence in the Gulf of Suez: implications for rifting and plate kinematics. Tectonophys 153:249–270

Strecker MR, Blisniuk PM, Eisbacher GH (1990) Rotation of extension direction in the central Kenya rift. Geology 18:299–302

Taviani M, Montagna P, Rasul NMA, Angeletti L, Bosworth W (this volume) Pleistocene coral reef terraces on the Saudi Arabian side of the Gulf of Aqaba, Red Sea

Tewfik N, Ayyad M (1984) Petroleum exploration in the Red Sea shelf of Egypt. In: Proceeding of 6th exploration seminar, Egyptian general petroleum corporation and Egypt petroleum exploration society, Cairo, March 1982, 1:159–180

USGS (2016) Plate boundaries. http://earthquake.usgs.gov/earthquakes/map

Wolfenden E, Ebinger C, Yirgu G, Deino A, Ayalew D (2004) Evolution of the northern main Ethiopian rift: birth of a triple junction. Earth Planet Sci Lett 224:213–228

Xu W, Ruch J, Jónsson S (2015) Birth of two volcanic islands in the southern Red Sea. Nat Comm 6:7104. https://doi.org/10.1038/ncomms8104

Zoback ML (1992) First- and second-order patterns of stress in the lithosphere: the world stress map project. J Geophys Res 97:11703–11728

A Modern View on the Red Sea Rift: Tectonics, Volcanism and Salt Blankets

Nico Augustin, Colin W. Devey, and Froukje M. van der Zwan

Abstract

Continental rifting and ocean basin formation can be observed at the present day in the Red Sea, which is used as the modern analogue for the formation of mid-ocean ridges. Competing theories for how spreading begins—either by quasi-instantaneous formation of a whole spreading segment or by initiation of spreading at multiple discrete "nodes" separated by thinned continental lithosphere—have been put forward based, until recently, on the observations that many seafloor features and geophysical anomalies (gravity, magnetics) along the axis of the Red Sea appeared anomalous compared to ancient and modern examples of ocean basins in other parts of the world. The latest research shows, however, that most of the differences between the Red Sea Rift (RSR) and other (ultra)slow-spreading mid-ocean ridges can be related to its relatively young age and the presence and movement of giant submarine salt flows that blanket large portions of the rift valley. In addition, the geophysical data that was previously used to support the presence of continental crust between the axial basins with outcropping oceanic crust (formerly named "spreading nodes") can be equally well explained by processes related to the sedimentary blanketing and hydrothermal alteration. The observed spreading nodes are not separated from one another by tectonic boundaries but rather represent "windows" onto a continuous spreading axis which is locally inundated and masked by massive slumping of sediments or evaporites from the rift flanks. Volcanic and tectonic morphologies are comparable to those observed along slow and ultra-slow spreading ridges elsewhere and regional systematics of volcanic occurrences are related to variations in volcanic activity and mantle heat flow. Melt-salt interaction due to salt flows, that locally cover the active spreading segments, and the absence of large detachment faults as a result of the nearby Afar plume are unique features of the RSR. The differences and anomalies seen in the Red Sea still may be applicable to all young oceanic rifts, associated with plumes and/or evaporites, which makes the Red Sea a unique but highly relevant type example for the initiation of slow rifting and seafloor spreading and one of the most interesting targets for future ocean research.

N. Augustin (✉) · C. W. Devey · F. M. van der Zwan
GEOMAR Helmholtz Centre for Ocean Research Kiel, Wischhofstraße 1-3, 24148 Kiel, Germany
e-mail: naugustin@geomar.de

F. M. van der Zwan
Institute of Geosciences, Christian Albrechts University Kiel, Ludewig-Meyn-Straße 10, 24118 Kiel, Germany

N. M. A. Rasul and I. C. F. Stewart (eds.), *Geological Setting, Palaeoenvironment and Archaeology of the Red Sea*,
https://doi.org/10.1007/978-3-319-99408-6_3

1 Introduction

Plate tectonics has been splitting continents and opening new oceans for at least half of all Earth history (Kerrich and Polat 2006). But how the transition from rifting continental lithosphere to spreading new oceanic lithosphere (Courtillot 1982; Whitmarsh et al. 2001) occurs is unclear. Competing hypotheses suggest either geologically instantaneous establishment of a new spreading segment (e.g., Taylor et al. 1995) or the formation of discrete spreading "nodes" within the stretched continent whose activity may continue for several Ma before the continent is finally sundered (Bonatti 1985; Ligi et al. 2012).

Continental rifting and the transition to spreading can be observed at present in the Woodlark Basin (Taylor et al. 1995) and the Red Sea (Bonatti 1985). The point at which rifting changes to spreading in the Woodlark Basin is well defined and migrates westward into the continent by the abrupt initiation and rapid (≈500 mm yr^{-1}) propagation of a new spreading segment within the rifted crust. The change from rifting to spreading in the Woodlark Basin occurs after a relatively constant 200 ± 40 km of overall stretching, consistent with observations from older passive margins such as the northern Atlantic margins, the Eurasian basin and its adjacent margins (e.g., Taylor et al. 1999; White and

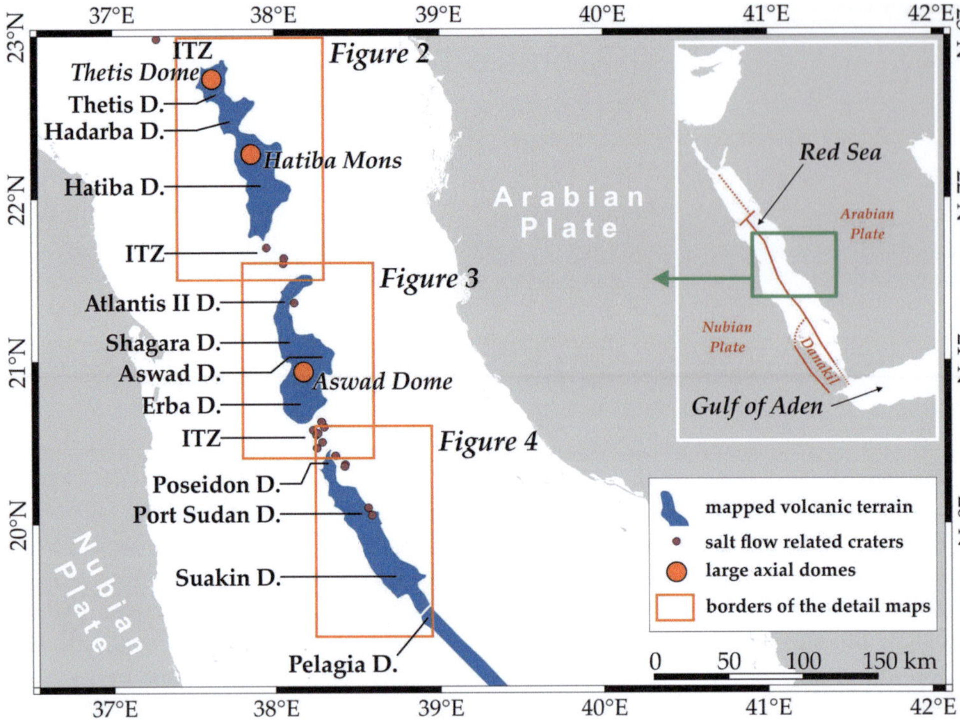

Fig. 1 Overview of the study area with main geographic features, the mapped volcanic terrain (blue areas), large axial dome-shaped volcanoes (orange circles), salt flow related craters (dots) and the borders of the detailed maps shown in Figs. 2, 3, 4. The inset also shows the approximate positions of plate boundaries in the Red Sea (red lines). After Augustin et al. (2016) with permission of Elsevier

McKenzie 1989; Grachev 2003). During its rapid propagation, the new axis is separated from the adjacent previously established axis by an accommodation zone composed of rifting continental crust into which the new axis propagates. When propagation ceases, the new axis develops an overlap-type boundary to the adjacent segment, which can, with continued spreading, evolve into a transform fault (Taylor et al. 1999).

In the Red Sea Rift (RSR), where the axis is uniquely characterized by a line of bathymetric lows known as "Deeps" (Bäcker and Schoell 1972), the transition from rifting to spreading has been proposed to occur concurrently at several "nodes" (marked by the Deeps, see also Fig. 1) scattered over the 670 km-long region between 19.5°N and 24°N (Bonatti 1985). The relative proximity of the Red Sea rift to the Arabian/Nubian plate Euler pole results in significant variations in spreading rate over this distance which in turn implies that, if the "nodes" idea is correct, spreading must have initiated after varying amounts of continental stretching. Propagation of the "nodal" spreading centres into the intervening continental blocks has been proposed to eventually lead to continental break-up (Bonatti 1985; Cochran 2005; Ligi et al. 2012). However, most of this Red Sea tectonic interpretation is based on bathymetric and geophysical data collected prior to 1990, with either large line spacing, relatively low spatial resolution and/or with questionable navigational accuracy.

We present here new high-resolution bathymetric data of the central Red Sea axis combined with backscatter information and ground-truthing. We use this data to study in detail the Red Sea axis, its volcanic and tectonic processes, nature of the inter-trough zones between the deeps and the transition from rifting to spreading in the Red Sea to develop a modern model for the Red Sea. We interpret the data as showing that the flow of submarine salt glaciers (namakiers) strongly influences rift morphology and that the spreading "nodes" merely represent regions of a continuous spreading axis not covered by salt, obviating the need for a special rifting/spreading transition mechanism for the Red Sea.

2 Geological Background

The Red Sea Rift (RSR) developed during the counter-clockwise rotation of Arabia from Nubia (Fig. 1; Girdler and Underwood 1985; Sultan et al. 1993; Cochran 2005). It is an ultraslow-spreading rift with current spreading rates between ≈10 mm yr^{-1} in the northern Red Sea and 15.5 mm yr^{-1} in the central Red Sea (Chu and Gordon 1998; DeMets et al. 2010). In this respect, the RSR is comparable, for example, to the Southwest Indian Ridge (12–16 mm yr^{-1}) or the Gakkel Ridge (11–12 mm yr^{-1}; DeMets et al. 2010).

The oldest seafloor presently known in the axial trough occurs at 17°N and has been assigned an age of 3–5 Ma (Cochran 1983; Gurvich 2006). Older oceanic crust may be present under a blanket of thick sediments on the shelves of the main trough, and it has been proposed that oceanic seafloor spreading may have begun as early as 10–12 Ma (Izzeldin 1987; Augustin et al. 2014, 2016). The sediment cover consists of Miocene evaporites and younger pelagic carbonate ooze, with an estimated overall thickness of up to 7 km (Girdler and Styles 1974; Searle and Ross 1975). Continental break-up, rifting and seafloor spreading caused the kinematic deformation of these evaporites, that in some places have been shown to flow toward the deeper axial rift in the form of salt glaciers, known as submarine namakiers (e.g., Figs. 2, 3, 4, 5, 6; Girdler 1985; Mitchell et al. 2010; Augustin et al. 2014).

In the southern part of the Red Sea (south of 19.5°N) the axial valley is continuous and well developed. It is characterized by the distinct normal-fault-related axis-parallel bathymetric lineations and volcanism seen at other oceanic spreading centres (Gurvich 2006). Between 19°N and 23°N, oceanic volcanics only crop out in the Deeps, which form partially disconnected large basins (Bonatti 1985; Cochran 2005; Ligi et al. 2012). Further north, these Deeps are more widely spaced but at least three of them (Bannock, Mabahiss and Shaban) have been reported to contain basalt (Bonatti et al. 1984; Pautot et al. 1984; Guennoc et al. 1988). All basaltic samples so far recovered from all axial Deeps have been characterized geochemically as typical (for slow spreading) tholeiitic mid-ocean ridge basalts derived from a pure asthenospheric source with no indication of continental input (Altherr et al. 1990; Haase et al. 2000; van der Zwan et al. 2015).

The Deeps are separated by so-called Inter-Trough Zones (ITZ; Figs. 1, 2, 4, 5, 6), which, as well as being shallower, are distinctly different in their geophysical characteristics (mainly gravity and magnetic) compared to the Deeps (Tramontini and Davies 1969; Searle and Ross 1975; Ligi et al. 2012). This led to them being interpreted as thinned continental crust and the widely accepted theory that the Red Sea Deeps are discrete axial spreading "nodes" (Bonatti 1985). Other authors (e.g., Searle and Ross 1975; Girdler and Whitmarsh 1974; Girdler 1985; Girdler and Evans 1977; Izzeldin 1987; Sultan et al. 1992; Augustin et al. 2014, 2016) have discussed the possibility that oceanic crust is present along the entire Red Sea axis but covered in large parts by laterally slumped evaporites and sediments; this alternative interpretation is recently supported by the newly available high resolution multibeam maps and seismic data (e.g., Mitchell et al. 2015; Augustin et al. 2014).

3 General Morphology and Segmentation of the Volcanic Ridge

Prominent first-order features of the Red Sea bathymetry are two major types of seafloor, distinguishable by their average rugosities. These are (I) relatively rugose regions, mainly comprising the Deeps and their immediate surroundings. These areas consist of a series of deep basins (the Red Sea Deeps) cross-cut by volcanic ridges and highs and divided by the ITZs, described in more detail below from N to S, and (II) comparatively smooth regions on the graben flanks and in the ITZs, identified and discussed in detail by Mitchell et al. (2010), Mitchell and Park (2014) and Augustin et al. (2014) as being the result of downslope flows of submarine namakiers.

The regions of the RSR that are not covered by namakiers and that represent the rugose regions show volcanic and tectonic features that are familiar structures on other slow- and ultraslow-spreading ridges (Dick et al. 2003; Cannat et al. 2009; Carbotte et al. 2015). For example, the RSR shows a clear rift valley with numerous volcanic edifices and axial volcanic ridges (AVRs). The partially overlapping AVRs mark the rift axis and split the rift valley into several basins, often bordered by salt flows. In the older low-resolution (>800 m) bathymetric data these basins have been highlighted as the "Deeps" and interpreted as oceanic spreading nodes. The new data shown in Figs. 2, 3, 4, 5, 6 clearly shows that the "Deeps" are analogous to common morphologic features of ultra-slow spreading rifts, related to segmentation by second order (non-transform) offsets. Large axial domes and volcanoes mark the centres of the second order segments, which tend to deepen toward the segment ends and form deep rifts parallel to semi-circular basin structures, typical for slow and ultraslow-spreading ridges (Carbotte et al. 2015). Ultraslow spreading ridges have few, if any, transform faults (they are, for example, absent on the 1900 km-long Gakkel Ridge and along the 1000 km-long segment of SWIR east of the Melville FZ; see also GEBCO or Sandwell et al. (2014) data sets). Except for the Zabargad fracture zone, transform offsets also appear to be absent along the 600 km-long portion of RSR studied here, as characteristic spreading-perpendicular valleys offsetting the rift axis with large-displacement (>30 km) are absent. This is in direct contrast to the large fracture zones or transform faults which have been proposed in the Red Sea by numerous authors (Coleman 1973; Coleman and McGuire 1988; Ghebreab 1998; Almalki et al. 2015; Schettino et al. 2016), that are based on land observations and free-air gravity data but for which, however, no evidence can be found in the bathymetry in form of rift perpendicular graben

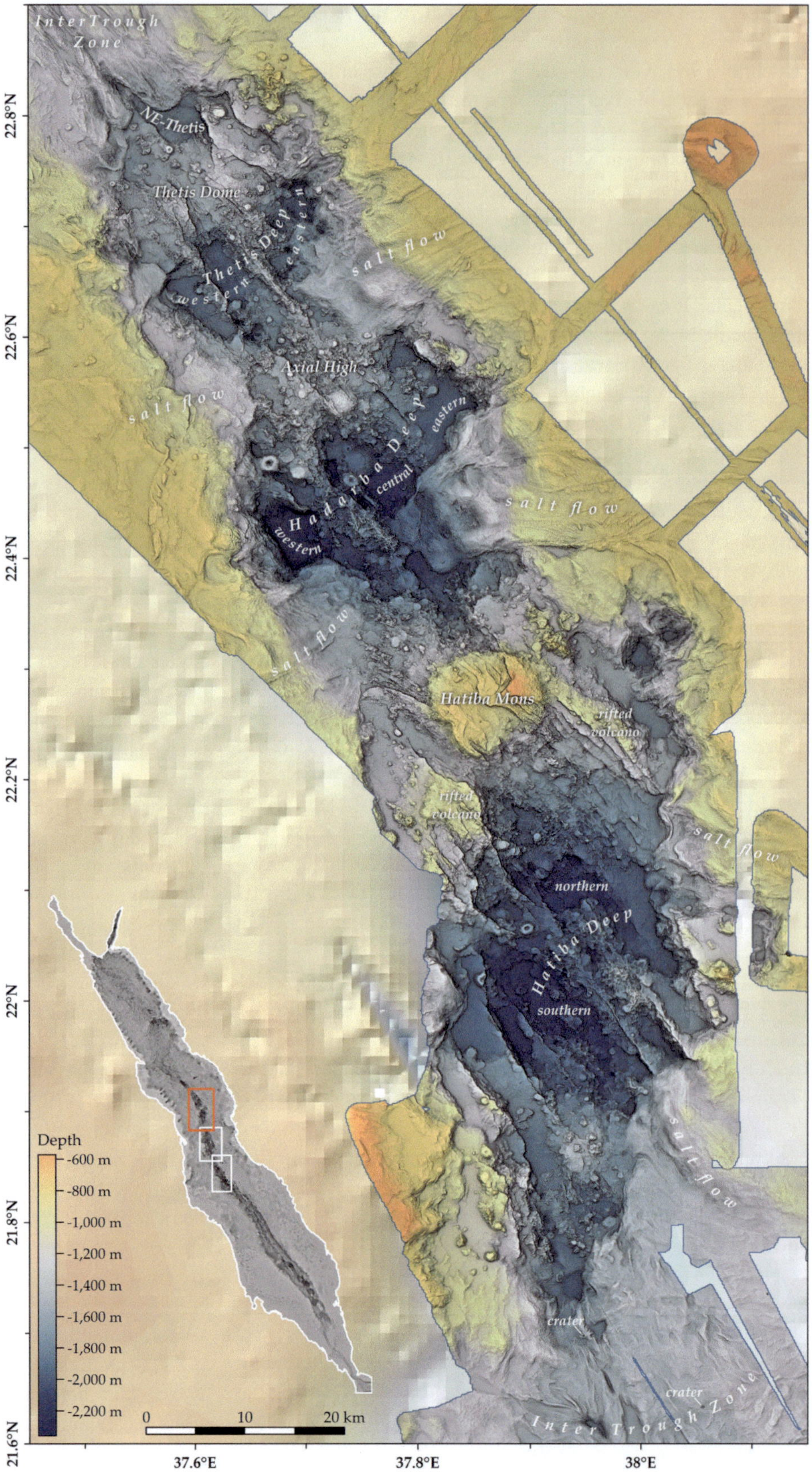

Fig. 2 Bathymetric map of the Thetis–Hadarba–Hatiba Trough, hosting the Thetis Dome and Hatiba Mons volcano. High-resolution data (30 m) from the Urania RS05, Poseidon P408 and Pelagia 64PE-350/351 expeditions. Low-resolution background data: GEBCO. Projection is Geographic WGS84. Reprinted from Augustin et al. (2016) with permission of Elsevier

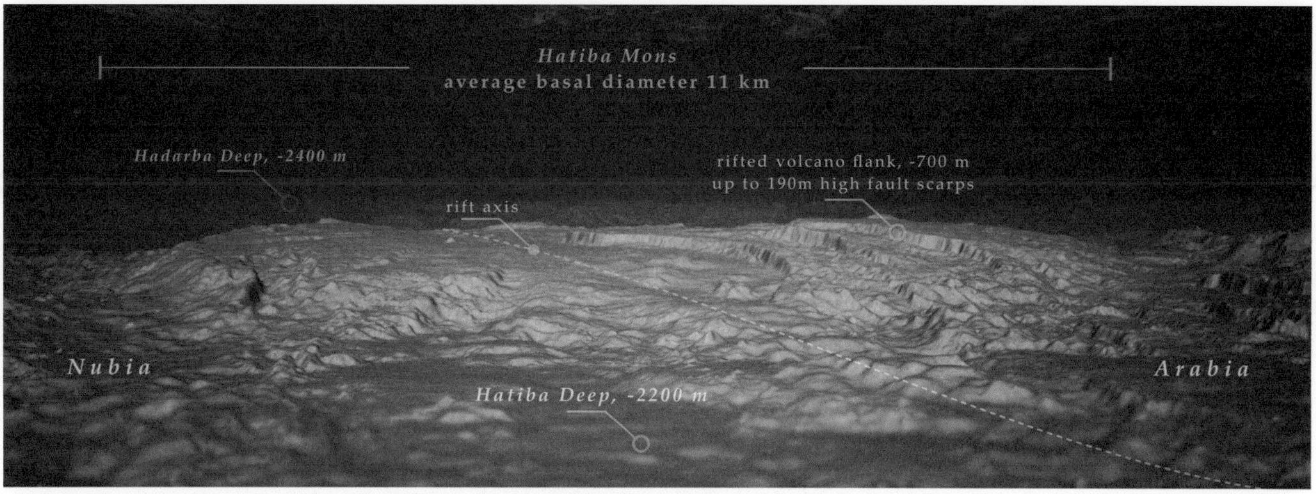

Fig. 3 3D-Visualization (Natural Scene Designer Pro 6 rendering) of Hatiba Mons, the largest axial dome volcano of the central Red Sea Rift. The picture simulates the view of the volcano toward the north. High-resolution data (30 m) from the Urania RS05, Poseidon P408 and Pelagia 64PE-350/351 expeditions

structures that offset the ridge axis, flexure ridges or other morphological features known from transform faults (e.g., Carbotte et al. 2015). The latest bathymetry data revealed that the rift axis in the central RSR instead consists of a sequence of curvilinear second-order segments building one super segment extending 800 km from the Zabargad Fracture zone to the Danakil Triple Junction (see Fig. 1).

4 The Northern Deeps: Thetis–Hadarba–Hatiba Trough

The Thetis–Hadarba–Hatiba Trough consists of four major Deeps and basins along with several smaller basins, bordered by ITZs in the north and south and flanked by submarine namakiers (Mitchell et al. 2010; Augustin et al. 2014). The major Deeps and basins, namely NE-Thetis, Thetis, Hadarba and Hatiba (Figs. 1 and 2), are morphologically separated by three axial highs of which two are characterized by large dome-shaped volcanoes (Thetis Dome, Hatiba Mons). The Thetis Dome is a NW–SE (ridge parallel) faulted axial volcano with an average diameter of 11 km and a height of about 400 m. Despite the faulting that indicates the long history of the volcano, the Thetis Dome shows volcanic activity in backscatter and bathymetry data. Strong backscatter response (indicating hard, non-sedimented seafloor; Augustin et al. 2014), low seafloor ruggedness of these areas and the recovery of unaltered basalts (Bonatti et al. 2005; Ligi et al. 2012) indicate the presence of lava flows that are likely younger than 10 ka. Due to the high average sedimentation rates of 10 cm/ka in the Red Sea (Stoffers and Ross 1974) older lava flows would have been covered by too much sediment to be distinctively visible in the backscatter data (Mitchell 1993; Hewitt et al. 2010). These lava flows run downslope north and south from the central graben into the NE–Thetis basin and the Thetis Deep (Augustin et al. 2016). The Thetis Dome is beset by a few small volcanic edifices and crossed by an axial volcanic ridge (AVR) that runs ridge parallel from the Nereus–Thetis ITZ toward the axial high that separates the Thetis and Hadarba Deeps, thereby splitting the Thetis Deep into two sub-basins (Fig. 2). Both sub-basins have a generally flat seafloor but show high backscatter areas and some volcanic cones and hummocks, and thus were volcanically active in Holocene times (<10 ka). The axial high at 22.5°N is covered with numerous volcanic hummocks, small flat-top volcanoes and domes with diameters of mostly <1 km. This volcanic high is also rift-parallel faulted and shows higher backscatter reflectivity only at the eastern side, suggesting relatively low recent volcanic activity.

The Hadarba Deep, south of the 22.5°N axial high, is 400 m deeper than the Thetis Deep and comprises three sub-basins, separated by AVRs. Volcanic activity, indicated by high backscatter, is concentrated along a prolongation of the NW–SE striking AVR from the Thetis Deep. West of this AVR we see a ridge jump represented by a second ridge-parallel AVR, which fades southward into the axial high topped by the Hatiba Mons volcano (11.4 km base diameter, 530 m high; Fig. 3). The northern end of this ridge is characterized by larger flat-top volcanoes with diameters of up to 3 km (Fig. 2). Along this AVR high backscatter signals and fresh lavas, with little grade of seafloor alteration, are only recovered from the southern end of the Hadarba Deep.

Hatiba Mons is (particularly on its eastern flank) faulted along a NW–SE strike, with minor cross cutting faults

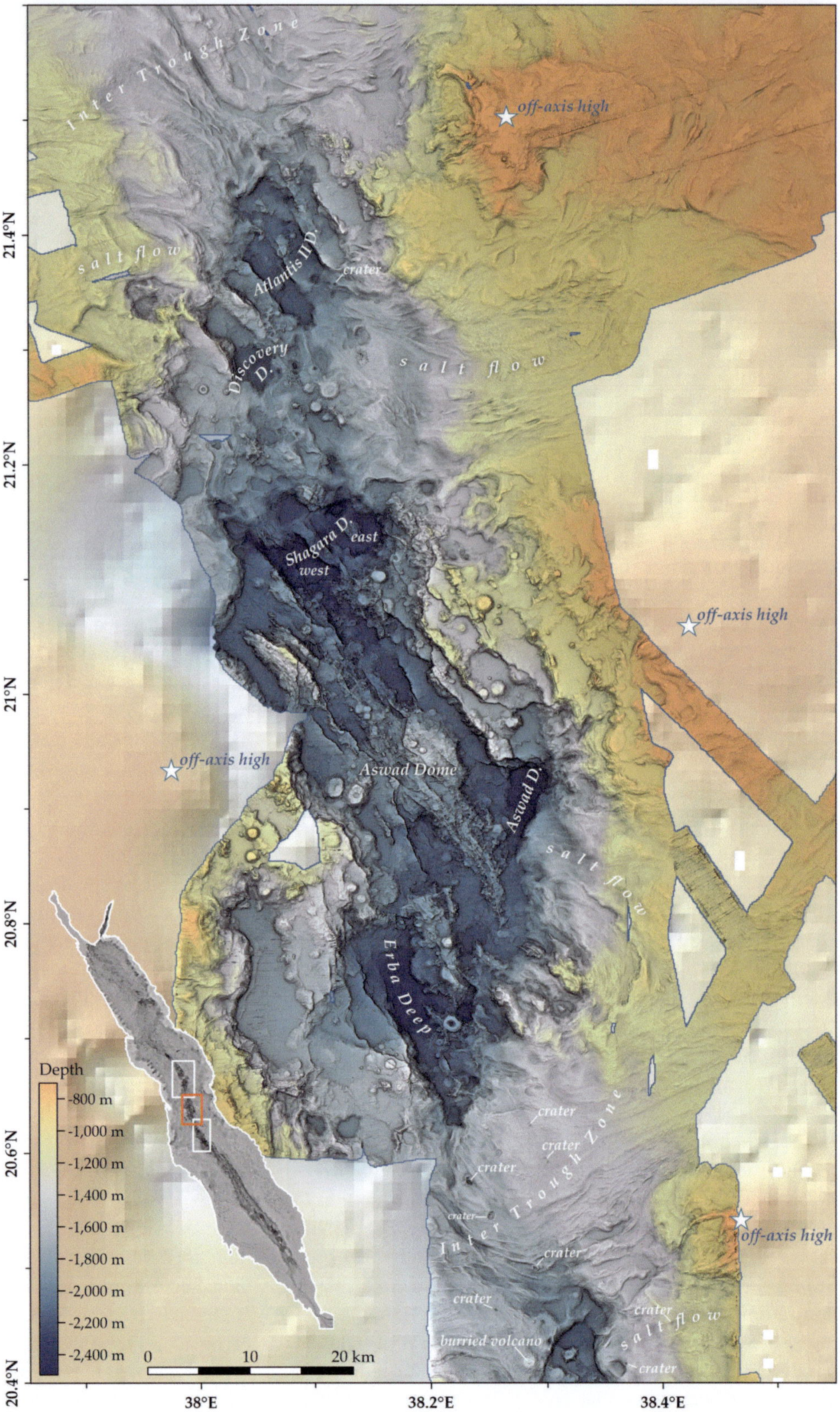

Fig. 4 Bathymetric map of the Red Sea Rift at the Atlantis II Deep area and the Shagara–Erba Trough, hosting the Aswad Dome volcano. High-resolution (30 m) data from the Poseidon P408 and Pelagia 64PE-350/351 expeditions. Stars indicate the positions of off-axis highs that obstruct namakiers corresponding to VGG panel A in Fig. 7. Low-resolution background data: GEBCO. Projection is Geographic WGS84. Reprinted from Augustin et al. (2016) with permission of Elsevier

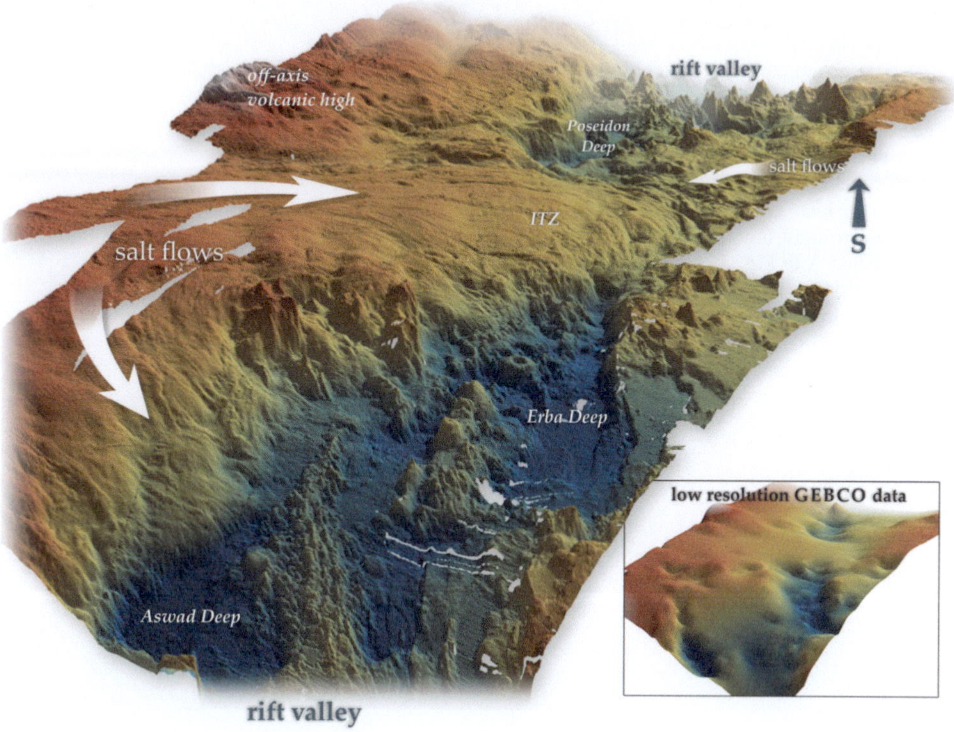

Fig. 5 3D-visualization of the large salt glacier that flows toward the Erba–Poseidon ITZ, creating a significant piedmont-lobe on the rift valley floor. The salt flows are influenced and redirected by volcanic structures. High-resolution data from the Poseidon P408 and Pelagia 64PE-350/351 expeditions. View direction to the south; the low-resolution GEBCO data of the same area is shown for direct comparison. After Mitchell and Augustin (2017) with permission of Elsevier

striking SW–NE. Small satellite cones and hummocky volcanic features are situated at the rim of the main structure. Areas with high backscatter reflectivity along the southern volcano flank, originating from the topographically smooth central part, suggest geologically recent lava flows. This is comparable to the lava flows described from Thetis Dome above. Distinct morphologic highs SW and SE of Hatiba Mons, sharply cut by normal faulting toward the rift valley, suggest a precursor volcanic structure of the recent edifice (marked as "rifted volcano" in Fig. 2).

South of Hatiba Mons lies the 25 km-wide and 60 km-long Hatiba Basin, which is divided by NW–SE striking AVRs and faults into north-eastern and south-western basins. Volcanic hummocks and small flat-top volcanoes are scattered all over the Deep, but are not as abundant as in the Thetis and Hadarba Deeps. High backscatter reflectivity in the northern Deep seems to be directly connected to the lava flow originating from Hatiba Mons. Sampling of high backscatter areas in the deepest area of the southern Deep recovered unaltered basalts. These basalts show sediment filled enclosed pockets as well as irregular, lobate flow patterns indicating that the lavas may have sunk into a soft base. This suggests that the lavas erupted onto a wet sedimented environment. The other parts of the basins as well as adjacent terraces show low backscatter intensities and, based on gravity coring results, are covered by at least 2 m thick sediments (Schmidt et al. 2011). Toward the southwest the Hatiba Deep is flanked by uplifted terraces with scattered volcanic edifices and is bordered to the south by an ITZ.

5 Atlantis II Deep Area and the Shagara–Aswad–Erba Trough

The Atlantis II Deep marks the northernmost end of the Atlantis II–Shagara–Erba Trough section, hosting also numerous sub-deeps, for example, the Discovery and Aswad Deeps (Figs. 1 and 4). The Atlantis II Deep area has been subject to many studies since the 1960s because of the presence of hot brine pools that contain economically valuable metalliferous sediments (Gurvich 2006, and references therein). In general, the area of the Atlantis II Deep appears smoother in bathymetry and less rugged than the basins north and south of it, which is due to massive inflows of salt and sediments from nearly all sides into the Deep (Augustin et al. 2014; Feldens and Mitchell 2015). Clearly visible in the bathymetry are the NW–SE-striking normal faults, tilted blocks and the "central sill" of the Deep described in earlier works (Bäcker and Richter 1973). Only a few prominent volcanoes south of the Atlantis II Deep and west of the Discovery Deep (partially overflowed by namakiers) and little hummocky volcanics are visible in the bathymetry.

The smooth terrain with adjacent namakiers develops southward into the rougher and deeper terrain of the

Fig. 6 Bathymetric map of the Red Sea Rift from the Poseidon Deep to the Pelagia Deep. High-resolution data (30 m) from the Poseidon P408 and Pelagia 64PE-350/351 expeditions. The star indicates the position of an off-axis high that obstruct namakiers corresponding to the southernmost high shown in VGG panel A in Fig. 7. Low-resolution background data: GEBCO. Projection is Geographic WGS84. Reprinted from Augustin et al. (2016) with permission of Elsevier

Shagara–Aswad–Erba Trough. This 65-km long trough is marked by steep very sharp normal faults and is roughly divided by a central 28 km long and 5 km wide AVR into two elongated axis parallel basins (Shagara–Aswad and Erba). This AVR also crosses a strongly faulted central dome volcano that separates the Shagara and Aswad Deeps (Aswad Dome, $\approx$8.4 km in diameter, 470 m high; Fig. 4). Flat-top volcanoes and hummocks are abundantly found on the faulted Aswad Dome and backscatter data indicates Holocene volcanic activity at its axial areas, although less than at Hatiba Mons and the Thetis Dome.

Also in the Shagara Deep, a ridge-parallel AVR separates the Deep into two basins that have a flat seafloor and contain metalliferous sediments. The SE flank of the Shagara Deep shows numerous scattered volcanoes and hummocks on an elevated terrain. Terraces east and west of the Aswad Dome are uplifted to water depths of less than 1200 m and are again scattered with numerous flat-top volcanoes and some hummocky volcanic ridges and edifices. The Aswad Deep (situated east of the Aswad Dome; Pautot et al. 1984; Blanc et al. 1990) is a rather smooth, NE–SW striking basin, but is also volcanically active as indicated by bright backscatter and fresh basaltic glass recovered from its southern part (Schmidt et al. 2013; van der Zwan et al. 2015).

The Erba Deep is situated 15 km south of the Aswad Dome (Gurvich 2006, and references therein; Fig. 2). A NW–SE striking AVR, hosting some flat-top volcanoes and the adjacent flat basin are its main features. Backscatter data as well as the dredging of unaltered basalts revealed volcanic activity in the Erba Deep that might be younger than 10,000 years (van der Zwan et al. 2015). West of the Erba Deep large, $\geq$10 km wide and at least 25 km south trending, sedimented terraces with minor volcanic edifices are dominant. The terraces and the Erba Deep are bordered by namakiers at the westernmost edge of the dataset and in the south by a large piedmont-lobe-like namakier that forms the ITZ toward the southern Deeps (Figs. 4 and 5; Augustin et al. 2014; Feldens and Mitchell 2015).

6 Poseidon, Port Sudan, Suakin and Pelagia Deeps

Between 20.2°N and 20.5°N only a 5–10 km wide and about 30 km long section of the RSR ocean crust outcrops between the salt-covered flanks, consisting of the Poseidon Deep at its northern end (Figs. 1, 5 and 6; Augustin et al. 2016). Backscatter data indicates volcanic activity within the Poseidon Deep and at its eastern flanks. Massive inflows of salt and sediments from both rift flanks and from the north partly cover the AVRs, volcanic edifices and boundary faults (Figs. 1 and 6; Augustin et al. 2014). For example, south of the Poseidon Deep a NW–SE striking ridge of hummocky volcanic edifices, flanked by some flat-top-volcanoes, is crosscut by namakiers in the north and south, while at the north-western edge the shape of an underlying single flat-top volcano can be traced through the evaporite cover. Remarkable is the fact that just north of this ridge recent lava flows seem to originate from the eastern namakiers and run toward the Poseidon Deep as visible in backscatter data (Augustin et al. 2014). A prominent 1.5 km wide and 150 m deep crater structure with a surrounding 20–30 m high ring mound is situated about 5 km southeast of the Poseidon Deep within a namakier (Fig. 6). Numerous comparable but smaller and slightly deformed structures are found within the ITZs.

South of $\sim$20.2°N the RSR axis widens to $\geq$20 km. A straight NW–SE-running AVR separates the Port Sudan Deep in the east from the shallower Volcano Deep in the west (Pautot et al. 1984; Blanc et al. 1990). A 4 km large, smooth, dome-shaped volcano, that is partially covered by namakiers, comprises about two thirds of the Volcano Deep. Recent, <10 ka volcanism in both Deeps is inferred by sampling of unaltered basalt and/or backscatter data although the Port Sudan Deep is also partially filled with hydrothermal sediments (Schmidt et al. 2013; van der Zwan et al. 2015).

South of the Volcano and Port Sudan Deeps the separating hummocky, volcanic ridge continues as a heavily NW–SE faulted discreet axial high at 19.9°N. This high is flanked in the west by inflowing sediments and in the east by a large $\geq$5 km wide and 35 km long NW–SE trending terrace. As an extension of the 19.9°N axial high a 35 km long and up to 10 km wide AVR runs southward to the Suakin Deep, where it continues, less obviously, as a chain of axial hummocks, flat-top and small dome-shaped volcanoes that divide the Suakin Deep into an eastern and a western basin (Fig. 4). The Suakin Deep marks the deepest part of the Red Sea (2860 m), and no signs of recent volcanic activity, either in backscatter data or in rock sampling, could be found. The few recovered basaltic samples show up to cm-thick carbonate and/or manganese encrustations, not visible in the backscatter data and implying an older age (see also Schmidt et al. 2013; van der Zwan et al. 2015). Distinct uplifted terraces with a smooth morphology border the Deep to the east. South of the Suakin Deep at 19.4°N another axial high occurs, which is beset with several flat-top volcanoes and reveals bright backscatter signals, suggesting extensive recent volcanic activity. The axial high is followed by the Pelagia Deep (Augustin et al. 2016), which is a 16-km long basin that marks the geomorphological change from a Deeps-and-Basin morphology to a 300 km-long section of uninterrupted volcanic seafloor with a more constant morphology that is considered as the matured rift (Bonatti 1985).

7 Volcanic Ridges and the Age of the Oceanic Crust

A particular feature of the Red Sea that was not recognised in earlier times is the occurrence of large dome-shaped volcanoes and axial highs. These are a typical feature of (ultra) slow-spreading ridges, where long-term focussed magmatic activity at segment centre highs and domes can be traced over large distances (Cannat et al. 2009; Carbotte et al. 2015). The result is that along (ultra)slow-spreading ridges, spreading-perpendicular volcanic lineaments can be found in bathymetry and gravity data. In the Red Sea, these ridges are partially covered underneath the namakiers, but visible in both the vertical gravity gradient (VGG) and free-air gravity data between 20°N and 23°N. Although this is a simplified interpretation of VGG and free-air gravity data, the large-scale patterns reveal, in addition to ridges adjacent to the three large dome-shaped volcanoes discussed above, at least six other spreading-perpendicular ridges in the central Red Sea (Fig. 7a; Sandwell et al. 2014; Mitchell and Park 2014). The gravity patterns look like those from, for example, the northern MAR (Sandwell et al. 2014; Fig. 7b) but appear less crisp, thus representing a comparable oceanic crust under the thicker Red Sea sediments and salts.

These volcanic ridges appear to span 100–170 km and extend from the ridge axis toward the maximum reach of continental crust as described by Izzeldin (1987), suggesting that the volcanic ridges had a very stable magma production with seafloor spreading over a long interval. As this continental reach is at a distance of 60–180 km from the escarpments, this fits well with a breakup of the continent after total stretching of 200 ± 40 km as given by Taylor et al. (1999), White and McKenzie (1989) or Grachev (2003). Spreading ages (distance from ridge axis times half the spreading rate) of the spreading-perpendicular ridges vary from about 8–9 Myr at the Nereus and Thetis Deeps, up to 12 Myr at the southernmost ridge identified in the VGG (at 20°N, Poseidon Deep area) (Fig. 7a; spreading rates after Chu and Gordon 1998). The calculated spreading ages are related in resolution to the spatial resolution of the used datasets but agree very well with the 8–12 Myr of oceanic spreading discussed by Izzeldin (1987) and Augustin et al. (2014).

The rift-perpendicular ridges disappear south of 20°N, where the character changes from patchy to ridge parallel in VGG-patterns, comparable to other mid-ocean ridges where segmented and unsegmented crust alternate in unequal proportions (Fig. 7a). The absence of spreading-perpendicular ridges in the southern Red Sea was discussed by Mitchell

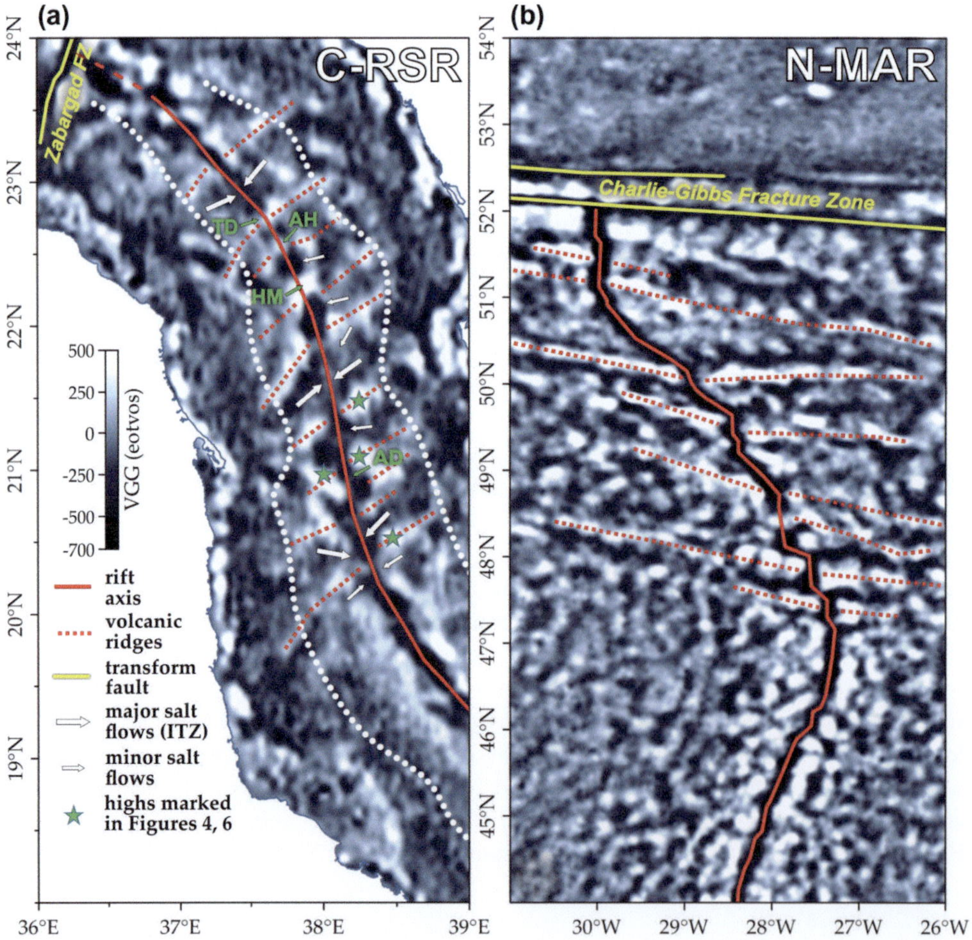

Fig. 7 Vertical gravity gradient (VGG, after Sandwell et al. 2014) of (**a**) the central Red Sea compared to (**b**) the northern Mid-Atlantic Ridge. The comparison shows that the VGG data reveal hidden spreading-perpendicular volcanic ridges (dashed lines) under the sedimentary cover that are comparable to those at the N-MAR. The ridges visible in the Red Sea VGG influence the direction of submarine salt flows (arrows) and end at the maximum extent of continental crust (white dotted line; Izzeldin 1982). Stars mark large off-axis volcanic highs that influence the salt flows and are clearly visible in the bathymetry. TD = Thetis Dome, AH = axial high, HM = Hatiba Mons, AD = Aswad Dome. Projection is Geographic WGS84. After Augustin et al. (2016) with permission of Elsevier

and Park (2014) as being the result of the rising influence of the Afar plume, which is comparable to the Reykjanes Ridge (MAR) where (due to the Icelandic plume) such patterns are absent as well.

8 Inter-trough Zones and Namakiers

In contrast to the volcanic Deeps, the inter-trough zones show significantly lower average rugosities and shallower average depths (1300–1600 m; Figs. 2, 3, 4). The seafloor in the ITZs is, however, not completely smooth, and long-wavelength (relative to the hummocky terrain in the Deeps) linear ridges are visible in many regions. These appear to be flow-structures described by Mitchell et al. (2010) and Augustin et al. (2014) interpreted as resulting from flow in salt, producing salt glaciers (namakiers). The local topographic slope and the orientation of the longitudinal ridges and troughs on the surfaces of the namakiers allow a sense of flow to be determined (Figs. 2 and 3). Where two namakiers collide that flow from opposite sides into the rift, a central cleft or suture is generally visible where the rift axis is covered (Augustin et al. 2014). The bathymetric data also show that namakier flows are affected by the underlying topography as they apparently deviate around volcanoes and volcanic ridges on the rift flanks. The distribution of the large namakier movements, that build up the ITZs, seems also to be strongly influenced by the underlying topography and particularly influenced by the rift-perpendicular volcanic ridges (in VGG), as can be seen by comparing the locations of these ridges to the occurrence and flow directions of namakiers (Fig. 7a; Mitchell and Augustin, in preparation). We observe that the major namakiers which completely blanket the RSR in the ITZs flow toward the rift in the regions between the spreading-perpendicular volcanic ridges. In several cases, higher terrains of volcanic ridges prevent the salt from completely covering the axial rift (marked by stars in Figs. 4, 6, 7a). This explains at least some of the flow directions as seen in bathymetric data (Mitchell et al. 2010; Augustin et al. 2014; Feldens and Mitchell 2015) and why the ITZs and the "Deeps" are where they are.

9 The Nature of the Crust and Volcanism Below Submarine Namakiers

As namakiers blanket the ITZs, conclusive evidence on the subsurface beneath them is difficult to obtain. However, as discussed above, several structures within the namakier give indications that the ITZ subsurface is volcanic and similar to the oceanic rift. At the flow fronts we see linear volcanic ridges apparently being overflowed by the namakiers. These irregular, non-tectonic but sharp contacts suggest an on-lap situation. In some cases, a continuation of the ridge beneath the namakier can be discerned. This is, for example, the case in the Poseidon Deep, where a NNW–SSE striking ridge terminates against the northern namakier but a circular, flat-topped structure, similar in size to an axial flat-top volcano, is visible in the bathymetry of the namakier, probably indicating a covered volcano (Fig. 6). A continuation of volcanic features deep into an ITZ is also apparent where namakiers do not yet completely blanket the rift. For example, small volcanoes can be observed as far as 10 km onto the ITZ south of the Nereus Deep in the still-open ITZ suture (Augustin et al. 2014). These features can be taken as evidence that the oceanic rift is continuous underneath the ITZs and is merely blanketed, rather than being continental crust (Augustin et al. 2014).

Thus, the unique combination of evaporites covering an active rift results in specific features. Within the Nereus–Thetis ITZ and Erba–Poseidon ITZ as well as south-east of the Poseidon Deep we discovered crater structures, often with elevated rims (10–25 m above the surrounding surface; Figs. 1, 2, 4 and 6). The size of the craters varies from 450 m to 1700 m in diameter with a depth of 50 m to 150 m respectively (average aspect ratio is 0.08; Augustin et al. 2016).

An origin of these craters as the formation by karst-like dissolution of salt caused by penetration of seawater, or by dissolution/distortion of salt from the base by escaping gases and/or hydrothermal fluids venting from oceanic crust, can be excluded due to geomorphological misfits and the specific local geological settings of the craters (Augustin et al. 2016). The salt-craters are all located in areas where namakiers are interpreted to overlie the neo-volcanic zone or volcanically active parts near the axial valley (Figs. 1, 2, 4 and 6). This, together with the shape of the craters and the presence of elevated rims around the craters that resemble maar craters on land (Büchel 1993; Geshi et al. 2011), suggests that the craters may be the result of phreatic or phreatomagmatic explosions caused by magma intrusions at or above the basement/salt boundary (Walker 1973; Sato and Taniguchi 1997). Phreatomagmatic eruptions would require that water reaches the base of the namakiers through the pelagic sediment and salt layers. Although little is known about the rheology and permeability of submarine namakiers the penetration of water into and under the salt bodies along faults and basaltic dikes is probable, considering the extensional tectonic setting of the area (e.g., Mitchell et al. 2010; Guennoc et al. 1988; Ross and Schlee 1973). Furthermore, hydrothermal recharge, passing the evaporites, is indicated by evidence for highly saline evaporitic fluids being added to Red Sea magmas (van der Zwan et al. 2015). Therefore, the presence of highly saline pore waters and trapped brines within or at the base of the evaporite sequences is at least locally plausible. Schofield et al. (2014) reported on non-brittle emplacement of magma and magma-salt mingling from the Herfa-Neurode mine (Germany), possibly by melting of salt (melting point of NaCl ≈800 °C, hydrous salts <200 °C; Driesner and Heinrich 2007; Schofield et al. 2014). This indicates that intrusions of about >1000 °C (Moore and Evans 1967) tholeitic magmas (with normal to enriched trace-element compositions, as typical for the Red Sea Rift; Altherr et al. 1988) into a submarine namakier that contains comparable salt assemblages (Manheim et al. 1974; Stoffers and Kühn 1974) most likely melts the surrounding salts and heats up saline fluids significantly. Temperatures of ≥400 °C already vaporize any saline fluids within the salt and trapped below (at pressures of ≈300 bar, e.g., Driesner and Heinrich 2007). Girdler and Whitmarsh (1974) and Girdler (1985) discussed the role of namakiers as thermal insulators. The namakier blanketing could slow down the cooling of magmatic intrusions. The magma is likely heating up and vaporizing significant amounts of the surrounding fluids and salt to build up sufficient steam pressure to blast away overlaying salt and sediments by phreatic eruptions at the seafloor. Thus, the observed craters are likely the product of such eruptions and these craters indicate places of the oceanic crust where volcanism has taken place after these were covered by the salt. Using these craters, we can track the occurrence of oceanic crust also underneath the salt cover.

10 Root Zones and Volumes of Submarine Namakiers

Next to the namakier-covered ITZs we find that the rift flanks are somewhat deeper than the general flank depth, and are characterized by large semi-circular 100–150 m deep embayments (Fig. 8). These depressions east and west of the rift valley likely represent the root zones of the valley-filling namakiers. To test this, Augustin et al. (2014) estimated the volumes, which (I) would be needed to fill up the rift valley and (II) are represented by the missing material from the embayment depressions.

The results of these calculations are shown in Fig. 8 for the ITZs between the Nereus and Thetis Deeps, the Hatiba and Atlantis II Deeps as well as for the namakiers which flow from the eastern flank into the Atlantis II Deep, and between the Erba and Port Sudan Deeps and the piedmont lobe flowing from the western rift-flank toward the Erba–Poseidon ITZ. The calculated loss and fill volumes for postulated namakier root zones and their rift filling are very similar (Fig. 8). Considering the uncertainties involved in estimating these volumes, we consider this close match to be strong supporting evidence for our assumption that the top of basement in the ITZs is a direct continuation of the surrounding ocean crust-namakier contact, also confirming a continuous spreading rift.

We also calculated namakier thicknesses in the ITZ using this top-of-basement estimate (Fig. 8b–h). The Nereus-Thetis Deep ITZ namakier is the thickest, at up to 800 m calculated thickness (Fig. 8b and e), while the namakier flowing from the east toward the Erba–Poseidon Deep ITZ is, with 600 m, somewhat thinner (Fig. 8d and h). These estimates are significantly lower than the 4–5 km salt accumulations that have been estimated for ITZs from earlier seismic studies (Tramontini and Davies 1969; Ligi et al. 2012). The former study (Tramontini and Davies 1969) found a velocity in the Miocene evaporites in the range 4.2–4.6 km s^{-1} based on seismic refraction. This velocity is similar to that reported for massive basaltic flow units from DSDP Hole 504B and IODP Hole 1256D (Little et al. 1985; Swift et al. 2008) and Tramontini and Davies (1969) noted that volcanic and evaporitic layers would most probably be seismically indistinguishable. Far more interesting is that Tramontini and Davies (1969) found what they termed a "lower layer" with a mean seismic velocity between 6.4 and 6.6 km s^{-1} extending beneath both the Deeps and the ITZ at depths of 3–6 km below the sea surface. This velocity is typical of lower oceanic crust (Gilbert and Salisbury 2011; Carlson 2014), which might be expected at this depth but significantly higher than the 5.9–6.4 km s^{-1} typically found in the upper 15 km of the Arabian Shield (e.g., Prodehl and Mechie 1991). This means that the earlier much thicker 4–5 km salt estimations may be the result of the similar salt-basalt velocity and rather represent the combined layer of the two, while the boundary shows the changeover to the lower oceanic crust. We take this as further evidence that the namakiers of the ITZs blanket oceanic crust.

11 Magnetic and Gravity Anomalies and Magnetic Quiet Zones

An important argument that has been previously advanced to support the presence of continental crust beneath the ITZ has been the absence of clear axis-parallel magnetic anomalies, which should be present if these areas were underlain by oceanic lithosphere. However, this argument seems to be negated by magnetic observations from the axial trough of the Red Sea itself. For example, the area between the Hadarba and Nereus Deeps (via the Thetis Deep), despite being unequivocally volcanic, shows an irregular intensity of magnetic anomalies and rock magnetization (Ligi et al. 2012), with only a diffuse and weak boundary (rather than a clear transition as would be expected from an ocean-continent transition) toward the non-magnetic appearance of the ITZ. Areas where Mitchell et al. (2010) and Augustin et al. (2014) described namakiers flowing

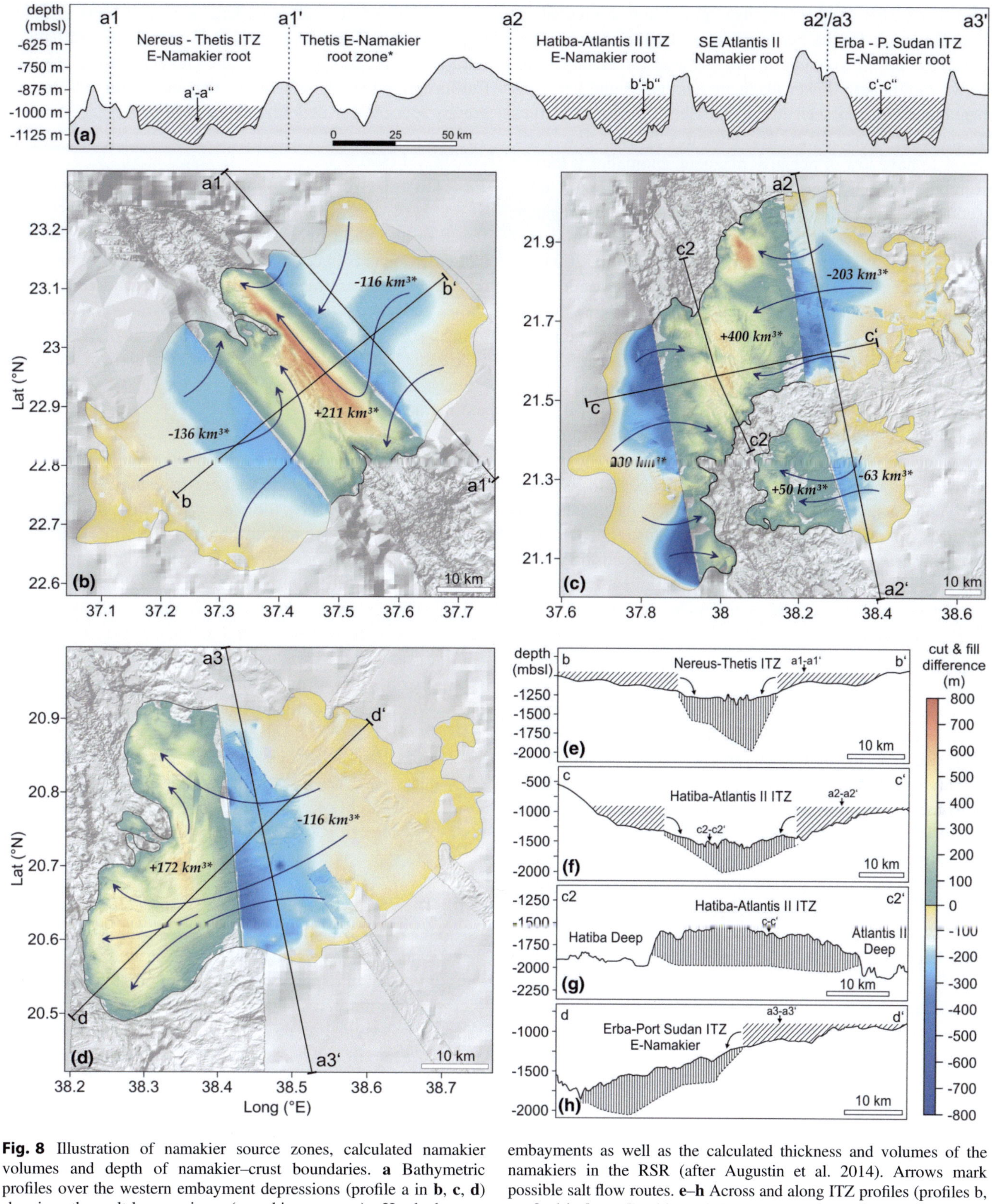

Fig. 8 Illustration of namakier source zones, calculated namakier volumes and depth of namakier–crust boundaries. **a** Bathymetric profiles over the western embayment depressions (profile a in **b**, **c**, **d**) showing the salt-loss regions (namakier sources). Hatched areas indicate loss-thickness. Positions of rift perpendicular profiles b, c, and d are shown by arrows. **b–d** Depths of the namakier source embayments as well as the calculated thickness and volumes of the namakiers in the RSR (after Augustin et al. 2014). Arrows mark possible salt flow routes. **e–h** Across and along ITZ profiles (profiles b, c, c2, d in **b**, **c**, **d**) with modelled namakier–crust boundaries (dashed lines), showing the salt loss and fill areas. After Augustin et al. (2014) with permission of Elsevier

toward the RSR axial valley also appear low in magnetization. The large Hatiba Mons volcano, despite showing evidence for recent volcanic activity (Augustin et al. 2016), shows significantly lower magnetic anomalies and less rock magnetization than seen in parts of the Thetis and Nereus Deeps (Ligi et al. 2012). The tendency for volcanic complexes to have low or negative magnetic anomalies is known from other regions, for instance, the Palinuro Seamount in the Mediterranean Sea shows large regions with strong negative magnetic anomalies where hydrothermal alteration processes have played a relevant role (Caratori Tontini et al. 2009).

Blanketing of the neovolcanic zone by namakiers might significantly reduce the intensity of its magnetization. Magnetic "quiet zones" on sedimented spreading centres are well known from, for example, the Guaymas Basin (Gulf of California), the Escabana Trough (Gorda Ridge) and Middle Valley (Juan de Fuca Ridge) where they have been explained as resulting from enhanced hydrothermal alteration and the breakdown of magnetic minerals under the sedimentary blanket (Levi and Riddihough 1986). Girdler and Whitmarsh (1974) and Girdler (1985) calculated temperatures of up to 100 °C under the salt/sediment covers based on heat-flow measurements and discussed the role of evaporites as thermal isolators, inhibiting a significant magnetization of the oceanic crust under the ITZ.

The interaction between namakiers and the spreading system may be very important for the magnetic structure of the Red Sea Rift. Egloff et al. (1991) showed some magnetic anomalies (adapted from Izzeldin 1987) under the evaporite series west of the Suakin Deep. This Deep belongs morphologically to the continuous southern RSR (which begins roughly with the Port Sudan Deep at 20°N), which shows no interruptions by any ITZs. According to our model this would imply that, either because the spreading rate is too high (spreading velocities south of the Port Sudan Deep increase significantly toward a maximum at 18°N (12.5–16 mm yr^{-1})) or namakier activity too low, the neovolcanic zone does not get covered in salt and only older, cold oceanic crust is buried. Heat flow in this older crust will be significantly lower than at the neovolcanic zone, possibly low enough that the thermal blanketing effect of the salt does not result in loss of magnetization.

The presence of gravity anomalies in, for example, the Thetis–Nereus or Hatiba–Atlantis ITZs (Searle and Ross 1975; Ligi et al. 2012) were also discussed as an argument for the presence of continental crust between the volcanic Deeps and troughs. Nevertheless, Searle and Ross (1975) showed that accumulation of sediments and salt could explain, for example, the 30 mGal difference in Bouguer anomalies observed between the Atlantis II Deep and the ITZ to the north, with no requirement for differences in the underlying crustal structure. This agrees with observations from the eastern Gakkel Ridge, for example, where the active ridge is buried under terrigenous sediments, producing a distinct (free-air) gravity low of −20 to −50 mGal (Drachev et al. 2003) and from the Guaymas Basin (Gulf of California) where a salt body coincides with a (free-air) gravity low (Miller and Lizarralde 2013).

12 Conclusions

Based on the observations that we can make from the new bathymetric maps in combination with the given geophysical data and ground truthing we conclude that the Red Sea Rift shows characteristics that are very comparable to slow- and ultraslow spreading ridges elsewhere, but also some differences that can be attributed to the closeness of the Afar plume and the unique blanketing of the rift valley by large submarine namakiers:

(I) Extensive salt flows (submarine salt glaciers or namakiers) are evident along the entire studied Red Sea Rift. In places, these namakiers flow into the axial region and completely blanket it while the flow directions are strongly influenced by the underlying basement topography. In the namakier covered areas, evidence from bathymetry, volume balance calculations of the salt flows, and geophysical data all seems to support the conclusion that the sub-salt basement is also basaltic in nature and represents oceanic crust.

(II) Where not covered by namakiers, the volcanic seafloor comprises characteristics of a mid-ocean spreading axis. The large geomorphological structures such as, for example, a deep rift valley, steep faults, highly tectonized terrain, rifted volcanoes, bent volcanic ridges with overlapping spreading centres and second order non-transform offsets are typical for slow- and ultraslow spreading ridges elsewhere.

(III) The unique occurrence of evaporites partially covering the active rift leads to distinct volcanic features. We see evidence for volcanic activity under the salt blanket in the form of explosion craters within the ITZ and at the graben flanks, as well as spots that indicate recent lava flows originating through the salt blankets and flowing into the rift valley. Thus, even though large parts of the oceanic crust are buried under layers of salt and sediments, recent volcanism can still be traced.

(IV) Bathymetry and VGG data reveal traces of hidden rift perpendicular ridges under the salt, comparable to the volcanic ridges seen at other slow- and ultraslow spreading ridges. Spreading appears to have begun along the entire Red Sea axis at least as far north as the Nereus Deep at 23°N; these ridges indicate stable magma focusing and oceanic spreading since at least 12 Ma in the Red Sea.

Acknowledgements The authors would like to thank the captains, crews and scientific parties of the expeditions with R/V Poseidon P408 and R/V Pelagia 64PE350/351 as well as the student helpers at GEOMAR, Kiel, who helped in processing the huge amount of data. We kindly thank Marco Ligi for providing the Urania RS05 bathymetry grid. Saudi Geological Survey is thanked for inviting us to the Red Sea Book Workshop. The Jeddah Transect Project between King Abdulaziz University and Helmholtz-Centre for Ocean Research GEOMAR was funded by King Abdulaziz University, Jeddah, Saudi Arabia, under grant no. T-065/430.

References

Almalki KA, Betts PG, Ailleres L (2015) The Red Sea—50 years of geological and geophysical research. Earth-Sci Rev 147:109–140

Altherr R, Henjes-Kunst F, Puchelt H, Baumann A (1988) Volcanic activity in the Red Sea axial trough—evidence for a large mantle diapir? Tectonophysics 150:121–133

Altherr R, Henjeskunst F, Baumann A (1990) Asthenosphere versus lithosphere as possible sources for basaltic magmas erupted during formation of the Red-Sea—constraints from Sr, Pb and Nd isotopes. Earth Planet Sci Lett 96:269–286

Augustin N, Devey CW, van der Zwan FM, Feldens P, Tominaga M, Bantan RA, Kwasnitschka T (2014) The rifting to spreading transition in the Red Sea. Earth Planet Sci Lett 395:217–230

Augustin N, van der Zwan FM, Devey CW, Ligi M, Kwasnitschka T, Feldens P, Bantan RA, Basaham AS (2016) Geomorphology of the central Red Sea Rift: determining spreading processes. Geomorphology 274:162–179

Bäcker H, Richter H (1973) Die rezente hydrothermal-sedimentäre Lagerstätte Atlantis-II-Tief im Roten Meer. Geol Rundsch 62:697–741

Bäcker H, Schoell M (1972) New deeps with brines and metalliferous sediments in Red Sea. Nat Phys Sci 240:153–158

Blanc G, Boulegue J, Charlou JL (1990) Profils d'hydrocarbures légers dans l'eau de mer, les saumures et les eaux intersticielles de la fosse Atlantis II (Mer Rouge). Oceanol Acta 13:187–197

Bonatti E (1985) Punctiform initiation of seafloor spreading in the Red Sea during transition from a continental to an oceanic rift. Nature 316:33–37

Bonatti E, Colantoni P, Vedova BD, Taviani M (1984) Geology of the Red Sea transitional region (22°N-25°N). Oceanol Acta 7:385–398

Bonatti E, Bortoluzzi G, Calafato A, Cipriani A, Ferrante V, Ligi M, Lopez Correa MCM, Redini F, Barabino G, Carminati E, Mitchell N, Sichler B, Schmidt M, Schmitt M, Rasul NRN, Al Nomani S, Bahareth F, Khalil S, Farawati R, Gitto D, Raspagliosi M (2005) Geophysical, geological and oceanographic surveys in the Northern Red Sea. Report on the morphobathymetric, magnetometric, oceanographic, coring and dredging investigations during cruise RS05 aboard R/V Urania. ISMAR Bologna Technical Report 94:1–40 Bologna, Italy

Büchel G (1993) Maars of the Westeifel, Germany. In: Negendank JFW, Zolitschka B (eds) Paleolimnology of European Maar lakes. Springer, Berlin Heidelberg, pp 1–13

Cannat M, Sauter D, Escartín J, Lavier L, Picazo S (2009) Oceanic corrugated surfaces and the strength of the axial lithosphere at slow spreading ridges. Earth Planet Sci Lett 288:174–183

Caratori Tontini F, Cocchi L, Carmisciano C (2009) Rapid 3-D forward model of potential fields with application to the Palinuro Seamount magnetic anomaly (southern Tyrrhenian Sea, Italy). J Geophys Res 114:B02103

Carbotte SM, Smith DK, Cannat M, Klein EM (2015) Tectonic and magmatic segmentation of the Global Ocean Ridge System: a synthesis of observations. In: Wright TJ, Ayele A, Ferguson DJ, Kidane T, Vye-Brown C (eds) Magmatic rifting and active volcanism, vol 420. Geological Society London Special Publications, pp 1–47

Carlson RL (2014) The influence of porosity and crack morphology on seismic velocity and permeability in the upper oceanic crust. Geochem Geophys Geosyst 15. http://dx.doi.org/10.1002/2013GC004965

Chu D, Gordon R (1998) Current plate motions across the Red Sea. Geophys J Int 135:313–328

Cochran JR (1983) A model for the development of the Red Sea. Bull Am Assoc Petrol Geol 67:41–69

Cochran JR (2005) Northern Red Sea: nucleation of an oceanic spreading center within a continental rift. Geochem Geophys Geosyst 6:Q03006. https://doi.org/10.1029/2004GC000826

Coleman RG (1973) Geological map of the Red Sea. U.S. Geological Survey, scale 1:2,000,000

Coleman RG, McGuire AV (1988) Magma systems related to the Red-Sea opening. Tectonophysics 150:77–100

Courtillot V (1982) Propagating rifts and continental breakup. Tectonics 1:239–250

DeMets C, Gordon RG, Argus DF (2010) Geologically current plate motions. Geophys J Int 181:1–80

Dick HJB, Lin J, Schouten H (2003) An ultraslow-spreading class of ocean ridge. Nature 426:405–412

Drachev SS, Kaul N, Beliaev VN (2003) Eurasia spreading basin to Laptev Shelf transition: structural pattern and heat flow. Geophys J R Astron Soc 152:688–698

Driesner T, Heinrich CA (2007) The system H_2O–NaCl. Part I: correlation formulae for phase relations in temperature–pressure–composition space from 0 to 1000 °C, 0 to 5000 bar, and 0 to 1 XNaCl. Geochim Cosmochim Acta 71:4880–4901

Egloff F, Rihm R, Makris J, Izzeldin Y, Bobsien M, Meier K, Junge P, Noman T, Warsi W (1991) Contrasting structural styles of the eastern and western margins of the southern Red-Sea—the 1988 SONNE experiment. Tectonophysics 198:329–353

Feldens P, Mitchell NC (2015) Salt flows in the central Red Sea. In: Rasul NMA, Stewart ICF (eds) The Red Sea: the formation, morphology, oceanography and environment of a young ocean basin. Springer Earth System Sciences, Berlin, pp 205–218

Geshi N, Németh K, Oikawa T (2011) Growth of phreatomagmatic explosion craters: a model inferred from Suoana crater in Miyakejima Volcano, Japan. J Volcanol Geotherm Res 201:30–38

Ghebreab W (1998) Tectonics of the Red Sea region reassessed. Earth-Sci Rev 45:1–44

Gilbert LA, Salisbury MH (2011) Oceanic crustal velocities from laboratory and logging measurements of Integrated Ocean Drilling Program Hole 1256D. Geochem Geophys Geosyst 12:Q09001. https://doi.org/10.1029/2011GC003750

Girdler RW (1985) Problems concerning the evolution of oceanic lithosphere in the northern Red Sea. Tectonophysics 116:109–122

Girdler RW, Evans TR (1977) Red Sea heat flow. Geophys J Int 51:245–251

Girdler RW, Styles P (1974) Two stage Red Sea floor spreading. Nature 247:7–11

Girdler RW, Underwood M (1985) The evolution of early oceanic lithosphere in the southern Red Sea. Tectonophysics 116:95–108

Girdler RW, Whitmarsh RB (1974) Miocene evaporites in Red Sea cores, their relevance to the problem of the width and age of oceanic crust beneath the Red Sea. Init Rep Deep Sea Drill Proj 2:913–921

Grachev AF (2003) The Arctic rift system and the boundary between the Eurasian and North American lithospheric plates: new insight to plate tectonic theory. Rus J Earth Sci 5:307–345

Guennoc P, Pautot G, Coutelle A (1988) Surficial structures of the northern Red Sea axial valley from 23°N to 28°N: time and space evolution of neo-oceanic structures. Tectonophysics 153:1–23

Gurvich EG (2006) Metalliferous sediments of the Red Sea. Metalliferous Sediments of the World Ocean. Springer, Berlin, pp 127–210
Haase KM, Mühe R, Stoffers P (2000) Magmatism during extension of the lithosphere: geochemical constraints from lavas of the Shaban Deep, northern Red Sea. Chem Geol 166:225–239
Hewitt A, Salisbury R, Wilson J (2010) Using multibeam echosounder backscatter to characterize seafloor features. Sea Technol 51:10–13
Izzeldin AY (1987) Seismic, gravity and magnetic surveys in the central part of the Red Sea; their interpretation and implications for the structure and evolution of the Red Sea. Tectonophysics 143:269–306
Kerrich R, Polat A (2006) Archean greenstone-tonalite duality: thermochemical mantle convection models or plate tectonics in the early Earth global dynamics? Tectonophysics 415:141–165
Levi S, Riddihough R (1986) Why are marine magnetic anomalies suppressed over sedimented spreading centers? Geology 14:651–654
Ligi M, Bonatti E, Bortoluzzi G, Cipriani A, Cocchi L, Caratori Tontini F, Carminati E, Ottolini L, Schettino A (2012) Birth of an ocean in the Red Sea: initial pangs. Geochem Geophys Geosyst. https://doi.org/10.1029/2012GC004155
Little SA, Stephen RA, Honnorez J, Adamson AC, Alt JC, Emmermann R, Becker K (1985) Costa Rica Rift borehole seismic experiment, Deep Sea Drilling Project Hole 504B, Leg 92. Init Rep Deep Sea Drill Proj 83:517–528
Manheim FT, Waterman LS, Woo CC, Sayles FL (1974) Interstitial water studies on small core samples, Leg 23 (Red Sea). In: Supko PR, Weser OE (eds) Initial reports of the deep sea drilling program. U.S. Government Printing Office, Washington, pp 955–967
Miller NC, Lizarralde D (2013) Thick evaporites and early rifting in the Guaymas Basin, Gulf of California. Geology 41:283–286
Mitchell NC (1993) A model for attenuation of backscatter due to sediment accumulations and its application to determine sediment thicknesses with Gloria sidescan sonar. J Geophys Res 98:22477–22493
Mitchell NC, Augustin N (2017) Halokinetics and other features of GLORIA long-range sidescan sonar data from the Red Sea. Mar Pet Geol 88:724–738
Mitchell NC, Park Y (2014) Nature of crust in the central Red Sea. Tectonophysics 628:123–139
Mitchell NC, Ligi M, Ferrante V, BonattI E, Rutter E (2010) Submarine salt flows in the central Red Sea. Bull Geol Soc Am 122:701–713
Mitchell NC, Ligi M, Feldens P, Hübscher C (2015) Deformation of a young salt giant: regional topography of the Red Sea Miocene evaporites. Basin Res 29(Suppl 1):352–369
Moore DJG, Evans DBW (1967) The role of olivine in the crystallization of the prehistoric Makaopuhi tholeiitic lava lake, Hawaii. Contrib Mineral Petrol 15:202–223
Pautot G, Guennoc P, Coutelle A, Lyberis N (1984) Discovery of a large brine deep in the northern Red Sea. Nature 310:133–136
Prodehl C, Mechie J (1991) Crustal thinning in relationship to the evolution of the Afro-Arabian rift system—a review of seismic-refraction data. Tectonophysics 198:311–327
Ross DA, Schlee J (1973) Shallow structure and geologic development of the southern Red Sea. Bull Geol Soc Am 84:3827–3848
Sandwell DT, Mueller RD, Smith WHF, Garcia E, Francis R (2014) New global marine gravity model from CryoSat-2 and Jason-1 reveals buried tectonic structure. Science 346:65–67
Sato H, Taniguchi H (1997) Relationship between crater size and ejecta volume of recent magmatic and phreato-magmatic eruptions: implications for energy partitioning. Geophys Res Lett 24:205–208
Schettino A, Macchiavelli C, Pierantoni PP, Zanoni D, Rasul N (2016) Recent kinematics of the tectonic plates surrounding the Red Sea and Gulf of Aden. Geophys J Int 207:457–480
Schmidt M, Devey CW, Eisenhauer A (2011) FS POSEIDON Cruise Report P408. Berichte aus dem Leibniz-Institut für Meereswissenschaften an der Christian-Albrechts-Universität zu Kiel 46
Schmidt M, Al-Farawati R, Al-Aidaroos A, Kürten B (2013) RV PELAGIA Cruise Report 64PE350/64PE351. Berichte aus dem Helmholtz-Zentrum für Ozeanforschung Kiel (GEOMAR) 5
Schofield N, Alsop I, Warren J, Underhill JR, Lehne R, Beer W, Lukas V (2014) Mobilizing salt: magma-salt interactions. Geology 42:599–602
Searle RC, Ross DA (1975) A geophysical study of the Red Sea axial trough between 20.5° and 22°N. Geophys J Int 43:555–572
Stoffers P, Kühn R (1974) Red Sea evaporites: a petrographic and geochemical study. In: Supko PR, Weser OE (eds) initial reports of the deep sea drilling program. U.S. Government Printing Office, Washington, pp 821–847
Stoffers P, Ross DA (1974) Sedimentary history of the Red Sea. In: Supko PR, Weser OE (eds) Initial reports of the deep sea drilling program. U.S. Government Printing Office, Washington, pp 849–865
Sultan M, Becker R, Arvidson RE, Shore P, Stern RJ, Elalfy Z, Guinness EA (1992) Nature of the Red-Sea crust—a controversy revisited. Geology 20:593–596
Sultan M, Becker R, Arvidson RE, Shore P, Stern RJ, Elalfy Z, Attia RI (1993) New constraints on Red-Sea rifting from correlations of Arabian and Nubian Neoproterozoic outcrops. Tectonics 12:1303–1319
Swift S, Reichow M, Tikku A, Tominaga M, Gilbert L (2008) Velocity structure of upper ocean crust at Ocean Drilling Program Site 1256. Geochem Geophys Geosyst 9, Q10O13. http://dx.doi.org/10.1029/2008GC002188
Taylor B, Goodliffe AM, Martinez F, Hey R (1995) Continental rifting and initial sea-floor spreading in the Woodlark basin. Nature 374:534–537
Taylor B, Goodliffe AM, Martinez F (1999) How continents break up: insights from Papua New Guinea. J Geophys Res 104:7497–7512
Tramontini C, Davies D (1969) A seismic refraction survey in the Red Sea. Geophys J Int 17:225–241
van der Zwan FM, Devey CW, Augustin N, Almeev RR, Bantan RA, Basaham A (2015) Hydrothermal activity at the ultraslow- to slow-spreading Red Sea Rift traced by chlorine in basalt. Chem Geol 405:63–81
Walker GPL (1973) Explosive volcanic eruptions—a new classification scheme. Geol Rundsch 62:431–446
White R, McKenzie D (1989) Magmatism at rift zones: the generation of volcanic continental margins. J Geophys Res 94(B6):7685–7729
Whitmarsh RB, Manatschal G, Minshull TA (2001) Evolution of magma-poor continental margins from rifting to seafloor spreading. Nature 413:150–154

Constraining the Opening of the Red Sea: Evidence from the Neoproterozoic Margins and Cenozoic Magmatism for a Volcanic Rifted Margin

Robert J. Stern and Peter R. Johnson

Abstract

As the Earth's best-known example of an active, incipient ocean basin, the Red Sea provides crucial information about continental rifting and the tectonic transition from extended continental crust to seafloor spreading. Study of the Red Sea over the past decades has given many answers, but significant questions remain about how and when it opened because of lacking or ambiguous data and thick salt cover. A key issue is the geometry of the pre-rift join between the Arabian and Nubian Shields that form the basement flanking the Red Sea because this constrains the nature of Red Sea crust. The Neoproterozoic basement rocks flanking the Red Sea contain prominent shears and sutures between amalgamated tectonostratigraphic terranes, regions of transpressional shortening, and brittle-ductile faults related to Ediacaran orogenic collapse and tectonic escape. These structures vary in orientation from orthogonal to oblique with respect to the Red Sea coastlines. Importantly, they correlate across the Red Sea, and provide piercing points for a near coast-to-coast palinspastic reconstruction of the Arabian and Nubian Plates along the entire Red Sea. A tight pre-rift fit of the Arabian and Nubian Shields implies that most of the Red Sea is underlain by oceanic crust. Potential-field data are compelling evidence for oceanic crust along the axis of the Red Sea south of latitude ~22°N, persuasive for the margins of the southern Red Sea, and suggestive for the northern Red Sea. A variety of 20–24 Ma dikes, gabbros, and basaltic flows emplaced during the early stages of Red Sea rifting are consistent with Miocene asthenospheric upwelling, partial melting, and intrusion that would have weakened and facilitated rupture of the ~40-km thick continental crust and thicker mantle lithosphere of the then contiguous Arabian and Nubian Shields. The dikes, gabbros, and basaltic flows emplaced during the early stages of Red Sea rifting are strong evidence furthermore that the Red Sea is an example of a volcanic-rifted margin. Offshore seismic profiling designed to image beneath the salt followed by drilling to basement in the Red Sea are required to test these ideas.

1 Introduction

The Red Sea is a ~300 km wide nascent ocean centred on a well-defined active divergent boundary between the formerly continuous Neoproterozoic rocks of the Arabian Shield and Nubian Shield (ANS) (Fig. 1) that are now parts of the Arabian and Nubian (or African) plates. The Red Sea is Earth's best example of an active, incipient ocean basin that has progressed from continental rifting to seafloor spreading (Bonatti et al. 2015; Bosworth 2015). It is therefore an important natural laboratory in which to test rift-to-drift models of continental rifting and to refine our understanding of the nature of the crust beneath continent-ocean transitions and the formation of passive continental margins. However, although the Red Sea has been studied by geologists, geomorphologists, and geophysicists for more than six decades, so that much is known about the region, many aspects of Red Sea development are unresolved because of geological and logistical problems. Geologic problems include the effect of a thick sediment cover (including salt) that prevents direct observation of most Red Sea crust (Orszag-Sperber et al. 1998); logistical problems for integrated study stem from the Red Sea geographic spread across the jurisdiction of six nations, and the differing approaches and objectives of industrial and academic researchers and terrestrial and marine geoscientists. Most data for models of Red Sea development come from the Gulf of Suez, and the margins of Egypt and Yemen, whereas the rift flanks of Saudi Arabia and Sudan are understudied leaving a large information gap along the central portions of the Red Sea (Szymanski et al. 2007).

R. J. Stern (✉)
Geosciences Department, University of Texas at Dallas, Richardson, TX 75080-0688, USA
e-mail: rjstern@utdallas.edu; bobdadbob@yahoo.com

P. R. Johnson
6016 Haines Street, Portland, OR 97219, USA

N. M. A. Rasul and I. C. F. Stewart (eds.), *Geological Setting, Palaeoenvironment and Archaeology of the Red Sea*,
https://doi.org/10.1007/978-3-319-99408-6_4

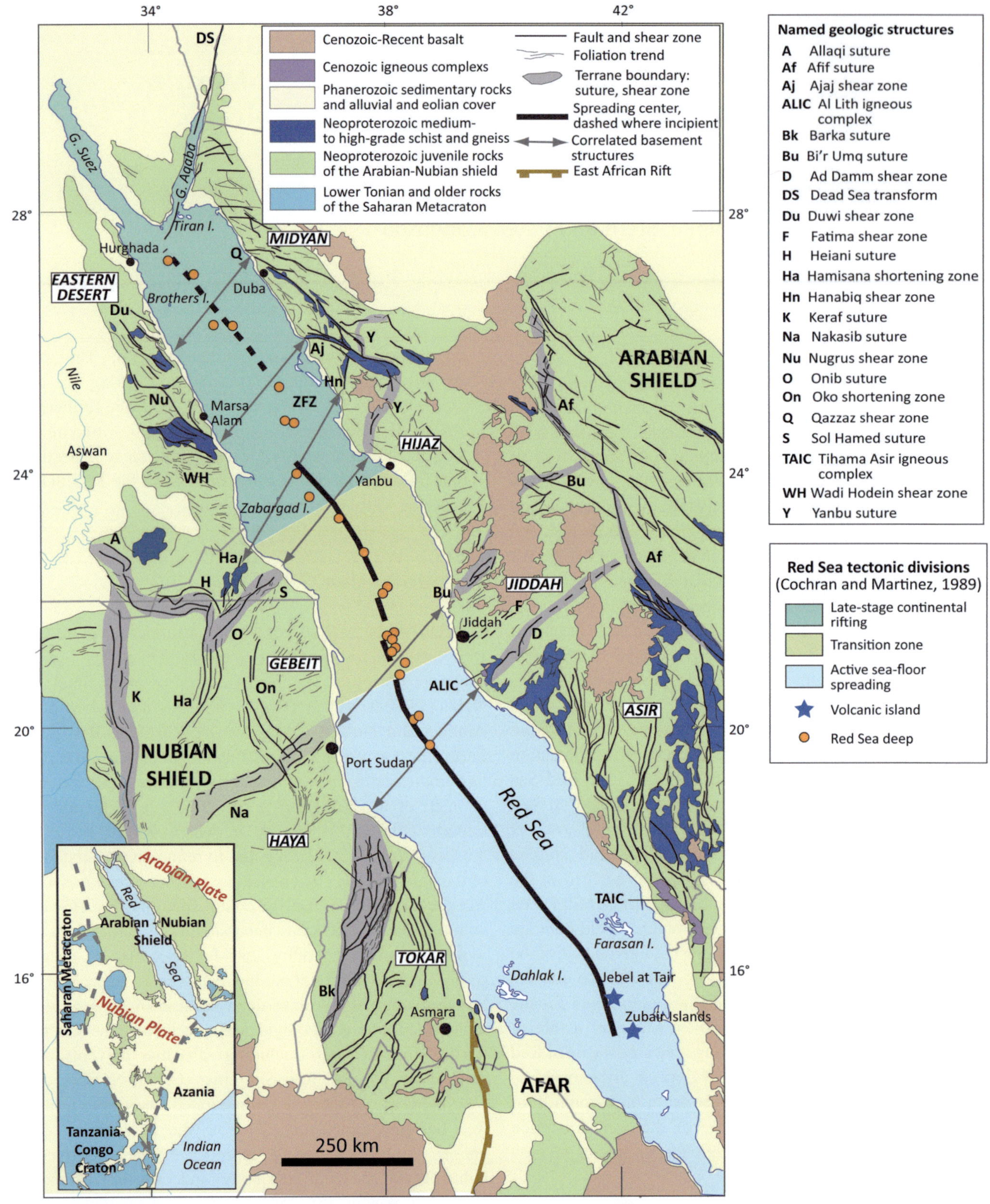

Fig. 1 Major geologic features of the Red Sea (spreading ridges, deeps), surrounding uplifted margins of the Arabian and Nubian Shield (ANS), and distribution of Late Cenozoic lava fields or harrats. These harrats bring fragments of the ANS lower crust and upper mantle to the surface as mafic and ultramafic xenoliths

Nonetheless, an extensive literature about the Red Sea is now available, including the series of papers about Red Sea research between 1970 and 1975 published by the Saudi Arabian Directorate General of Mineral Resources (DGMR 1977) and the recent collection of papers resulting from a 2013 workshop on Red Sea geology, geography, and biology sponsored by the Saudi Geological Survey (Rasul and Stewart 2015). As a consequence, many aspects of Red Sea development are well established; others, however, continue to be debated.

Here we consider two related issues: (1) the geometry of the pre-rift join between the Arabian and Nubian Plates, and (2) the nature of the crust beneath the Red Sea. We do this by reviewing what we know about the nature of Red Sea crust and the magmatic history of the Red Sea basin; by identifying structures in the Neoproterozoic rocks of the ANS that correlate across the Red Sea and provide "piercing-points" for the palinspastic reconstruction of the Arabian and Nubian Plates; and by considering the composition and vertical structure of ANS lithosphere as factors that should affect the strength of the lithosphere and thus how it ruptures. We use this review to argue for a near coast-to-coast palinspastic reconstruction of the Arabian and Nubian Plates, to propose that oceanic crust underlies more of the Red Sea than is generally considered, and to propose a volcanic-rifted margin (VRM) origin for the Red Sea in the light of VRM models used to account for continental-oceanic rifting elsewhere.

A large proprietary geophysical, stratigraphic, and lithologic dataset is held by petroleum exploration companies working on either side of the Red Sea. However, the results are not available for consideration here, and our review is based on published geologic maps and reports, publicly available geophysical information, and journal articles. Consequently, conclusions made here are limited by the level of detail in these sources and, as with all things pertaining to the structure of the Red Sea, would benefit immeasurably from future drilling and geophysical surveys. We are also aware of the difficulty of making pre-Red Sea-spreading reconstructions, as discussed by Coleman (1993), because of assumptions made about the structure of plate boundaries. Kinematic models proposed for the plate motions within the Red Sea area during the last 4–5 million years have met with some success. This is because it is reasonable to assume instantaneous motions of plates with rigid plate boundaries about Euler poles of rotation. However, Coleman (1993) argues that pre-5 Ma reconstruction are problematic because plate boundaries were probably more diffuse as a result of modification to the Red Sea basement prior to the onset of sea-floor spreading by normal faulting, thinning, and mafic intrusion and underplating. Geodetic (GPS), plate tectonic, and geologic observations indicate that the present rate of extension across the Red Sea is 24 ± 1 mm/y in the south and 7 ± 1 mm/y in the north (Reilinger et al. 2015). This rate probably pertained since ~11 Ma, whereas the extension rate prior to that was about half. Recognizing a clear distinction between the onset of rifting and the onset of sea-floor spreading, Coleman (1993) points out that the Gulf of Suez is an example of early extension within the rift zone and that the southern Red Sea underwent considerable extension without the production of oceanic crust. Nonetheless, we are impressed by the overall pattern that emerges from correlating flanking Neoproterozoic structures to argue for a relatively tight pre-Red Sea-spreading juxtaposition of plates. This implies a different history for Red Sea basin evolution than is commonly generally accepted (e.g., Cochran 1983; Bosworth 2015).

2 Red Sea Overview

As recognized by Wegener (1920), the Red Sea is the result of rupture of continental crust and transition into an oceanic basin that has occurred in a complex system of continental extension represented by the Gulf of Aden, East African, and Red Sea rifts. The rifts meet at a triple junction in the Afar region of Eritrea, Djibouti, and Ethiopia, but individually show different stages of development (Bonatti et al. 2015). In this paper, the term "Red Sea" refers to the region covered by present-day seawater, whereas "Red Sea basin" refers to the wider structure comprising the present-day sea and its oceanic crust, Mesozoic-Cenozoic sedimentary rocks deposited on the shelves and coastal plain along the margins of the Red Sea, and coeval Mesozoic-Cenozoic sedimentary rocks preserved in fault basins that reflect pervasive extension across a broader region inland from the Red Sea.

The Red Sea itself is 1900 km long and a maximum of 355 km wide. South of latitude ~21°N (Fig. 1), it has an axial trough 1000–2900 m deep and shallow marine shelves <50 m deep. Between ~21° and 25°N, the bathymetric trough is discontinuous and the axial part of the Red Sea forms a series of deeps that contain basaltic cones, dense, hot brines and deposits of Fe, Cu, and Zn (Schardt 2016). North of latitude 25°N, the Red Sea has a common depth of 1200 m; it is punctuated by deeps but lacks an axial trough. The southern marine shelves and most of the northern Red Sea are underlain by sedimentary rocks that include pre-rift Late Cretaceous to Early Oligocene sandstone, shale and carbonates; syn-rift Early to Middle Miocene Globigerina marl, carbonate, and sandstone; and kilometre-thick sequences of late-rift Middle Miocene to Recent anhydrite, halite, shale, and sandstone (Bosworth 2015).

Although some authors see the Cretaceous-Paleogene embayment that extended south from Tethys Ocean toward Jiddah subparallel to the present day Red Sea axis as a harbinger of the eventual Red Sea rift (d'Almeida 2010), the onset of extension in the Red Sea region is generally considered to be early Miocene, marked by the beginning of deposition of syn-rift sediments (see review by Bosworth 2015). Several stages in development of the overall Gulf of Aden-Red Sea rift system are recognized (Bosworth 2015), including: (1) plume-related basaltic and rhyolitic volcanism at ~31 Ma in the Afar region and SW Yemen that subsequently spread northward into western Saudi Arabia; (2) deposition of marine, syntectonic sedimentary rocks in the central Gulf of Aden between ~29.9 and 28.7 Ma, followed by Early Oligocene rifting in the Gulf of Aden; (3) formation of a small rift basin in the Eritrean Red Sea between ~27.5 and 23.8 Ma, virtually concurrent (~25 Ma) with extension and rifting in the Afar; (4) intrusion of layered gabbro, granophyre, and basaltic dike swarms in southern Arabia at ~24–23 Ma, eruption of early syn-rift basaltic flows (harrats) in southern and northwestern Arabia and in the subsurface in the Cairo region of Egypt, emplacement of basaltic dikes along the margin of, what is now, western Arabia, rift-normal extension, rift-margin uplift (Szymanski 2013), and a deepening of the Red Sea basin marked by a transition to predominantly Globigerina-rich marl and deepwater limestone (Bosworth et al. 2005); (5) penetration of the Dead Sea transform into the continental crust of the northern Arabian Plate at ~14–12 Ma and cessation of rifting in the Gulf of Suez; (6) from ~10 Ma onward, eruption of a younger phase of Oligocene-Recent syn-rift basaltic lava forming the lava fields (harrats) of the western and northern Arabian Plate; and (7) onset of oceanic accretion by ~5 Ma, if not earlier, and the development of organized spreading in the southern Red Sea and, to a debated extent, in the northern Red Sea. As noted by Bosworth et al. (2005), initiation of continental extension marked by the intrusion of mafic dikes and normal faulting, was virtually instantaneous between Eritrea, Egypt, and northwestern Arabia, at the Oligocene-Miocene transition. Extension was initially orthogonal to the trend of the Red Sea, but by 14 Ma, at the time of initiation of the Dead Sea transform, extension became oblique to the Red Sea.

The crux of the debate about Red Sea development hinges on the interpretation of south-to-north morphologic and geophysical variations as approximating a tectonic evolutionary sequence from drifting in the south to rifting in the north (Cochran 1983). The southern axial trough and magnetic anomalies are treated as evidence of active seafloor spreading. In the north, the absence of an axial trough and Vine-Matthews magnetic anomalies and the presence of extensive syn-rift sedimentary rocks are interpreted as evidence that the Red Sea here is a continental rift, which has developed nearly to the point of seafloor spreading (Martinez and Cochran 1988; Lazar et al. 2012; Ligi et al. 2012). Bosworth (2015) concluded that this crust may be highly extended continental material or mixed continental and volcanic material. The central part of the Red Sea between about 21°N and 23°N is regarded as a transitional zone between oceanic and continental rifting (Cochran and Martinez 1988) (Fig. 1) in which the cluster of prominent deeps is interpreted as isolated spreading cells above 'hot points' where upwelling mantle diapirs puncture continental lithosphere (e.g., Bonatti 1985; Bonatti et al. 2015). The Gulf of Suez at the northern end of the Red Sea is an entirely continental rift formed by a complex half graben broken along strike into a series of smaller half grabens. Each half graben consists of nested rotated fault blocks that displace both sedimentary fill and crystalline continental basement (Bosworth 2015). Because its structure is so well known as a result of excellent exposure and a vast amount of drilling and geophysical surveying, the Gulf of Suez is commonly used as a model for the northern Red Sea, although the much deeper seafloor of the northernmost Red Sea requires a different crustal structure.

The widely accepted timing of initial seafloor spreading at ~5 Ma is based on interpreting the age of large-amplitude, short wavelength magnetic anomalies centred on the axial trough of the Red Sea between latitudes ~23.5°N and 15°N. A limited section of these anomalies is shown in Fig. 2a; they are more fully shown in a residual magnetic anomaly map of the Red Sea compiled by Hall et al. (1977). The anomalies trend NW parallel to the Red Sea and are interpreted as Vine-Matthews type magnetic anomalies indicative of newly formed oceanic crust (Girdler and Styles 1974; Roeser 1975; Hall et al. 1977). By comparison with worldwide magnetostratigraphy, the oldest anomalies are estimated to be 5 Ma, the youngest 0 Ma. As shown in Fig. 2a, the linear magnetic anomalies in the axial region of the southern Red Sea are flanked by broader, lower amplitude anomalies that extend as far as the coastlines and even onto part of the coastal plain. The nature of these anomalies is debated, but Girdler and Styles (1974), Hall (1989), and Hall et al. (1977) interpret them as also caused by oceanic crust, although older than the crust in the axial region and reduced in amplitude because of the 4–5 km sedimentary sequence covering them (Rasul et al. 2015). These anomalies are estimated to be either 25–26 Ma or 35–37 Ma, and on this basis two stages of Red Sea sea-floor spreading are proposed (Girdler and Styles 1974). The first stage spanned 41–34 Ma and the second stage occurred from 4–5 Ma to the Present. The southern Red Sea axial trough is further marked by a continuous gravity high (Fig. 2B) and high heat flow, consistent with the emplacement of oceanic crust. Between ~21° and 25°N, where the bathymetric trough is discontinuous, large-amplitude magnetic anomalies are associated with the deeps and basaltic intrusions. North of latitude 25°N, the Red Sea has neither trough nor obvious spreading-related linear magnetic anomalies and the ocean crust is largely considered to be extended continental material analogous to the Gulf of Suez. However, combined

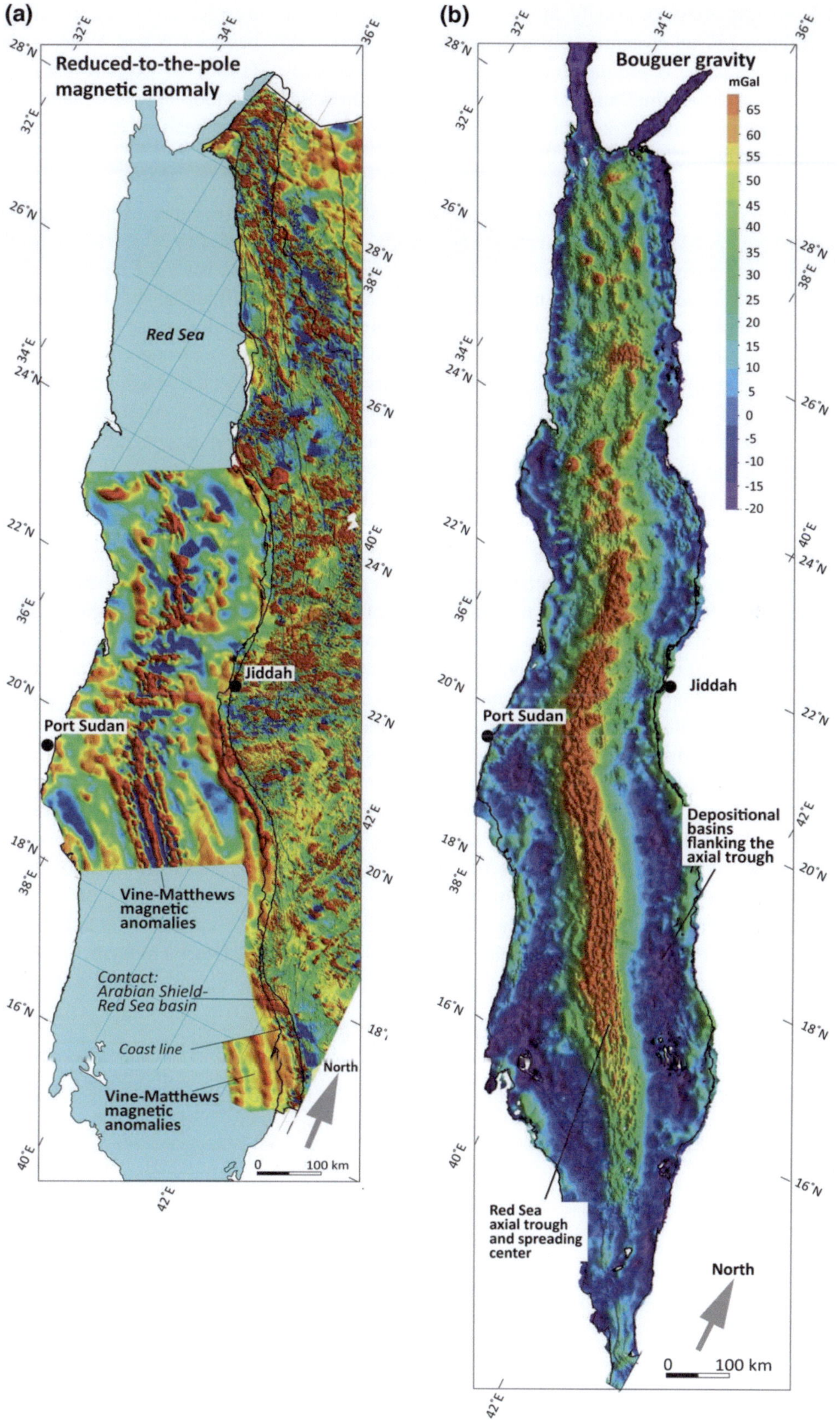

Fig. 2 Potential-field data for the Red Sea. **a** Reduced-to-the-pole magnetic anomaly map for the central Red Sea surveyed in 1976 (after Zahran et al. 2003). The plot uses the conventional color scheme of red for high intensity and blue for low intensity. The image shows magnetic stripes composed of narrow magnetic highs and lows along the Red Sea axial trough and its eastern margin south of 20°N corresponding to Vine-Matthews-type anomalies indicative of seafloor spreading. Discontinuous highs and lows north of 20°N reflect the transition zone of seafloor spreading shown in Fig. 1. **b** Bouguer gravity of the Red Sea generated from TOPEX/Poseidon satellite data (cf., Stewart and Johnson 1994) and presented at a 1-km grid interval. The image shows a virtually continuous gravity high extending from 16°N to 24°N coincident with the axial trough. The gravity high is consistent with the emplacement of mafic material, and suggests active ocean floor spreading at least as far north as 24°N

3-D interpretations of gravity and aeromagnetic data suggest a different crustal structure. Beneath the northern Red Sea, hot upper mantle is present from within a few tens of kilometres of the coastlines and the crust thins from 21 km at the Egyptian coast to 8–12 km thick (excluding the thick sediments) offshore (Saleh et al. 2006).

At the surface, the boundary between the Red Sea basin and the flanking Arabian-Nubian Shield is a depositional-to-faulted contact between sedimentary and locally volcanic rocks of the basin and Neoproterozoic basement rocks. The boundary is reasonably clear in outcrop despite cover by Recent alluvium and colluvium, and well defined in potential-field (magnetic and gravity) (Fig. 2) and seismic-refraction surveys (Fig. 3). A seismic reflection survey of the seaward portion of the coastal plain in southwest Saudi Arabia made in the 1960s (Gillman 1968) indicated that the surface of the Jurassic/basement contact dipped toward the Red Sea at about 10°, deepening from about 2 km deep 20 km inland to nearly 5 km at the coastline near the Mansiyah No. 1 drill hole and the Jizan salt dome (for locations, see Fig. 7). Unfortunately, the Mansiyah drill hole had to be abandoned at 3.9 km depth, short of seismic basement. The reflection survey profiles did not extend northeastward onto the Neoproterozoic basement of the Arabian Shield, and therefore its seismic properties could not be compared with those of basement beneath the coastal plain. This same locality was covered by the 1978 Riyadh-Farasan seismic-refraction survey conducted by the Saudi Arabian Deputy Ministry for Mineral Resources and the U.S. Geological Survey Saudi Arabian Mission (Mooney et al. 1985) (Figs. 3 and 7) which shows an abrupt thinning of continental crust from ~40 km beneath the Arabian Shield to ~20 km beneath the coastal plain and ~10 km beneath the Farasan Islands coincident with an inferred transition from continental to oceanic crust (Mooney et al. 1985; Gettings et al. 1986). The seismically indicated transitional crust is characterized by P-wave velocities between 6 and 8 km/s, and is interpreted by Blank et al. (1986) as oceanic crust produced during early stages of seafloor spreading and strong evidence in support for a hypothesis of essentially shore-to-shore opening in this part of the Red Sea. A recently published study of teleseismic P-wave receiver functions in Egypt (Hosny and Nyblade 2016) indicates thinning of continental crust from 35–39 km inland to 30 km at the northern Red Sea coast. There are no results for the crustal thickness of the Red Sea itself although a figure in the Hosny and Nyblade paper suggests crustal thicknesses of <20 km offshore, but the model illustrated is not precise enough to make any conclusion about the nature of crust beneath the northern Red Sea.

The contact between the ANS basement and the Red Sea basin is generally exposed within 25 km of the shoreline, and the relatively narrow coastal plain between the ANS and the Red Sea coast is underlain by Neogene sedimentary and minor volcanic rocks. In places, however, abrupt changes in the strike of the contact and the development of embayments and fault basins shift the contact more than 100 km inland, testifying to tectonic complexity. The Lisan basin at ~28°N in NW Saudi Arabia, for example, is a broad region of Cenozoic sedimentary rocks and ANS basement horsts in a structurally complex region where the Gulf of Suez, Dead Sea Transform, and Red Sea meet (Fig. 4). Drag along the Dead Sea Transform caused folding and anticlockwise rotation of Cenozoic rocks in the Lisan basin and a 110 km-sinistral slip of the basin rocks from their original position contiguous with the Gulf of Suez. The Gulf of Aqaba remains the most active seismic area in Saudi Arabia and is the site of the last two (1983 and 1995) major earthquakes (Roobol and Stewart 2009).

Other embayments include the Nakheil basin at ~26°N in Egypt (Khalil and McClay 2002; Abd el-Wahed et al. 2010) and Azlam (Aznam) basin at about ~27°N in Saudi

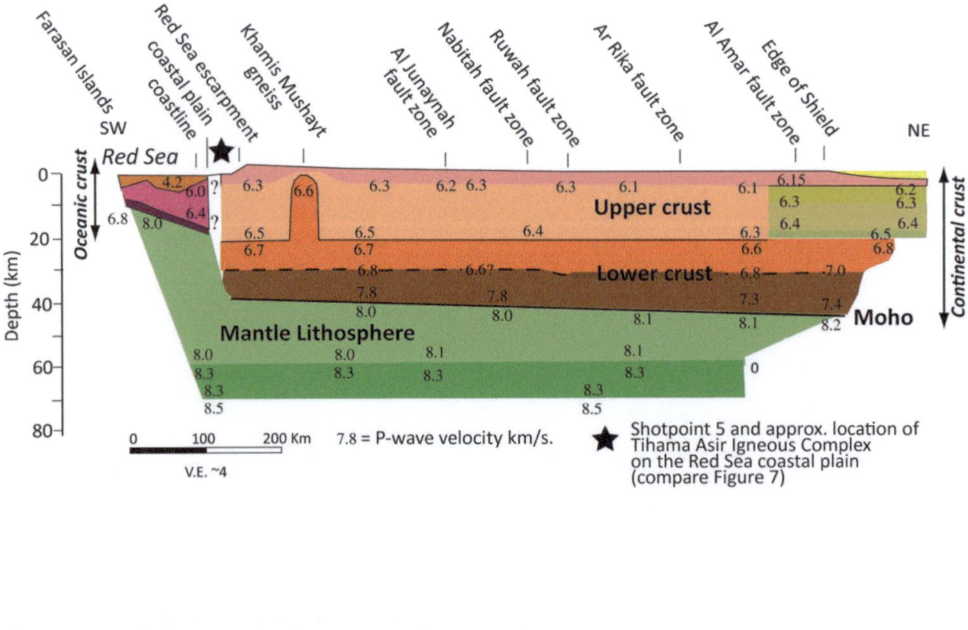

Fig. 3 Interpretive section showing crustal structure based on P-wave velocities (km/s) along the 1978 Riyad-Farasan seismic-refraction profile that extends southwesterly from central Saudi Arabia to the Farasan Islands in the Red Sea (after Mooney et al. 1985)

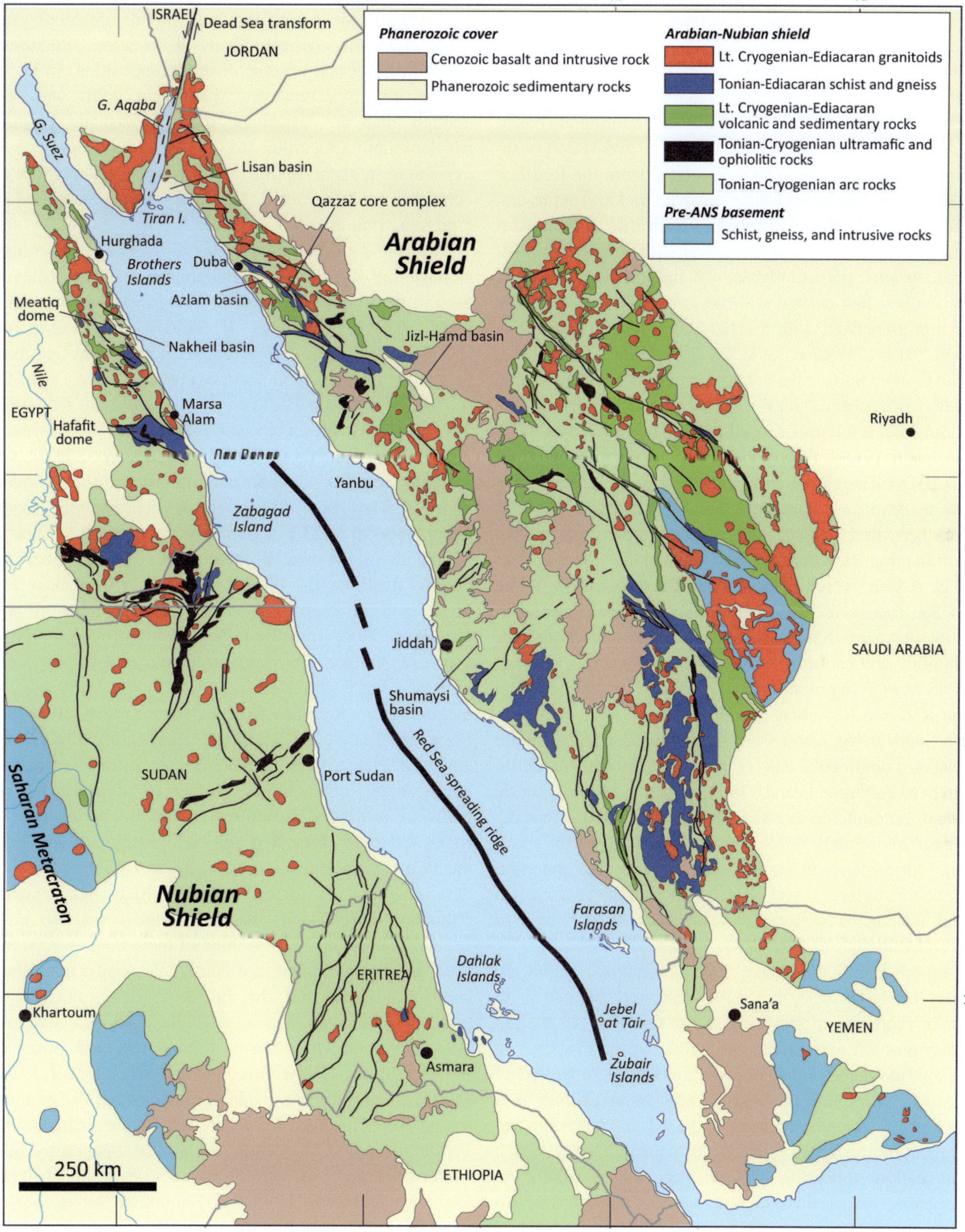

Fig. 4 Simplified geologic map of the Arabian-Nubian Shield

Arabia (Suayah et al. 1991) (Fig. 4). These basins are filled by pre- and syn-rift Mesozoic-Cenozoic sedimentary rocks, and extend into the basement oblique to the general trend of the Red Sea as extensional structures focused along reactivated, preexisting NW-trending Neoproterozoic Najd shear zones. They may have been originally one continuous basin. The Shumaysi basin (E of Jiddah) and the Jizl-Hamd basin (north of Yanbu) are fault basins isolated from the Red Sea by intervening blocks of Neoproterozoic rocks. Apatite and zircon (U-Th)/He thermochronology identifies a prolonged period of diffuse extension associated with the development of these basins, particularly an ~8 million-year period of block faulting that created a ~200 km-wide zone of diffuse extension associated with the Jizl-Hamd half graben (Szymanski 2013; Szymanski et al. 2012, 2016).

The Gulf of Suez at the northern end of the Red Sea lacks evidence of volcanic rocks, intrusive rocks, or sea-floor spreading and is a region of well-defined normal faulting and thinned continental crust (Khalil and McClay 2001; Bosworth 2015). The abrupt change in depth from a few metres below sea level in the southern Gulf of Suez to >1200 m in the northern Red Sea suggests an abrupt change in crustal type at the transition between the Gulf and the Red Sea. The Gulf of Aqaba, which marks the southern extension of the Dead Sea transform, is a narrow, transtensional basin with seafloor down to 1850 m deep. This active plate boundary has faulting and earthquakes but no known igneous activity.

ANS crust is exposed on Tiran Island at the northern end of the Red Sea (Goldberg and Beyth 1991) and is present farther south as Neoproterozoic granitic gneiss and peridotite in a horst of continental crust extending SSE from Ras Benas in Egypt to Zabargad Island (Brueckner et al. 1996). Granitic rocks are encountered elsewhere as much as 20 km from the Egyptian and Saudi Arabian coastlines (Bosworth et al. 2005). Other islands in the Red Sea, however, are underlain by Cenozoic to Recent reef material, such as the Farasan and Dahlak Islands, or by Recent volcanic rock, such as Jebel at Tare and Zubeir Islands (Fig. 4).

It is well known that young rifted continental margins are flanked by large-scale, low-relief plateaux elevated 1 to 2 km or more above sea level as the result of rift-margin uplift or post-rift exhumation and uplift (Japsen et al. 2011). Such uplifted margins may remain exposed for some tens of millions of years after the onset of seafloor spreading, as is the case of crystalline rocks exposed in Norway and Greenland on the flanks of the North Atlantic Ocean, which began seafloor spreading ~54 Ma (Gaina et al. 2009). As the opening ocean matures and widens, the uplifted rift flanks subside and are buried with sediments to become complementary passive continental margins, such as the flanks of the Central Atlantic and Gulf of Mexico, where seafloor spreading began >150 Ma. In these cases, only the largest, most regional basement features can be correlated across the ocean basin, and only broad inferences can be made about the possible effects of basement structures on ocean opening and oceanic crustal composition. In the case of the Red Sea however, rift-margin uplift was recent enough that the basement rocks persist as prominent rift-flank uplifts on either side of the Red Sea, and excellent exposures in these uplifts allow relatively straightforward correlation of basement structures across the Red Sea ("piercing-points").

North-to-south morphologic differences, nevertheless, characterize the dynamic topography resulting from Red Sea rift-flank uplift, mimicking changes seen in the Red Sea itself. Low-temperature (U-Th)/He thermochronology shows that the west-central Saudi Arabian Red Sea margin in an area referred to as the central Arabian rift flank (CARF) was partly exhumed at ~350 Ma but for much of the subsequent Mesozoic was relatively stable (Szymanski 2013; Szymanski et al. 2012, and in prep.). However, a complex Middle Miocene cooling history shows that the onset of Red Sea rifting was associated with footwall exhumation at ~23 Ma during which the CARF was dissected along rift-parallel fault blocks and exhumed from pre-rift flank depths of ~1.5–3.9 km (Szymanski et al. 2012, and in prep.). A comparable exhumation signal elsewhere along the Red Sea Nubian margin and the southern Arabian rift flank in Saudi Arabia and Yemen indicates that Red Sea rifting began along nearly the entire length of the Red Sea-Gulf of Suez system at ~23 Ma (Szymanski et al. 2012, and in prep.). A formerly extensive cover of Phanerozoic sedimentary rocks was stripped off the Arabian-Nubian Shield around the Red Sea but remnants are preserved as outliers unconformable on basement rocks at the top of the Red Sea escarpment and in blocks on the Red Sea coastal plain down-faulted along a W-dipping detachment that formed during Red Sea extension (Bohannon 1989). Rift margins in the south rise in erosional escarpments to elevations of more than ~3000 m. In Arabia, the escarpment is a continuous topographic feature south of Jiddah and the lip of the escarpment is the watershed between drainages to the Red Sea and to the Persian/Arabian Gulf. North of the Jiddah area, in contrast, the escarpment is discontinuous and is no longer a watershed. Drainage from far inland cuts through the mountains inland from the Red Sea in contrast to the sharp drainage divide present in the south. A similar topography occurs west of the Red Sea, with a prominent escarpment flanking the coastal plain in Eritrea and southern Sudan, but only mountains and valleys, not a continuous escarpment, in northern Sudan and Egypt. This suggests that the southern Red Sea rift flanks have been recently rejuvenated but that the northern margins have not.

3 The Red Sea Basin as a Magmatic Province

In addition to its dynamic rift-related topography, development of the Red Sea was accompanied by the emplacement of abundant dikes, sills, and volcanic flows (Bosworth and Stockli 2016) (Fig. 5).

Overall, Red Sea magmatism is decidedly asymmetric, with a preponderance of dike emplacement and basalt eruptions along the western margin of the Arabian Plate (Fig. 5). The earliest magmatism in the region was related to the Afar plume in the form of pre-rift mafic lavas extruded in Ethiopia, Eritrea, and southern Yemen at ~31 Ma. The earliest rift-related Red Sea magmatism comprised swarms of syn-rift mafic tholeiitic dikes, gabbroic intrusions, and volcanic rocks emplaced between ~24 and 20 Ma, and extending over 3000 km from north of Afar into northern Egypt (Bosworth and Stockli 2016) and northern Jordan-southern Syria (Harrat ash Shaam) (Ilani et al. 2001). Particular concentrations of dikes and intrusions make up the Al Lith igneous complex (ALIC) 100 km SE of Jiddah (Figs. 1 and 6) and the Tihama Asir igneous complex (TAIC) 40–100 km east and north of Jizan (Figs. 1 and 7). The dikes formed from magmas produced by partial melting of the mantle (Coleman 1993) and make up swarms of sheeted sequences, as in the igneous complexes, or wide and elongate dikes forming extensive linear structures. The latter do not crop out everywhere but are indicated along most of the Red Sea eastern margin by narrow magnetic lineaments such as those on the RTP aeromagnetic anomaly map of western Saudi Arabia in Fig. 2a. North of latitude 22°N, the magnetic lineaments are virtually continuous, indicating that the dikes extend tens to hundreds of kilometres along strike; south of 22°N, the dikes are less continuous. They yield $^{40}Ar/^{39}Ar$ ages of ~21 Ma (Sebai et al. 1991). Contemporaneous dikes and lava flows (23–22 Ma) mapped in the subsurface in the Cairo area possibly originated from a short-lived "mini-plume" (Bosworth 2015; Bosworth and Stockli 2016).

Following a ~7–9 million-year quiescence, apart from local volcanism in the Afar area and at Harrats Ishara (~17–14.5 Ma; Szymanski 2013) and Harairah between Madinah and Al Wajh (Fig. 5), Middle Miocene resumption of magmatism in the Red Sea basin resulted in new basaltic flows at Harrat ash Shaam (<~13 Ma) and formed the younger harrats of Saudi Arabia (Nawasif/Al Buqum, Kishb, Rahat, Khaybar and Ithnayn, and Uwayrid). Ilani et al. (2001) envisage that this renewal of volcanism was associated with sinistral movement along the Dead Sea transform, and may reflect the emergence of upper-mantle upwelling beneath the western Arabian Plate. Recent volcanic and dike activity is evidenced at Harrats Rahat and Lunayyir in the past 800 years until the Present (Pallister et al. 2010) in northwestern Saudi Arabia and on the coastal plain east of Jizan. As pointed out by Bosworth and Stockli (2016), Red Sea magmatism was episodic and although it gave rise to the prominent fields of basalt that characterize the western Arabian Plate, the exposed erupted and intruded magmatic material does not amount to a large volume when considered in relation to other rifted margins. Nevertheless, the presence of the sheeted dikes and plutons of the Al Lith and Asir Tihama complexes show that oceanic-like rifting, albeit briefly, accompanied the early phase of the Red Sea rift history (Bosworth and Stockli 2016). Furthermore if, as contended in this paper, more of the Red Sea is underlain by oceanic material than conventionally accepted, the overall volume of magmatic material is considerable.

The Al Lith Igneous Complex comprises two sets of dikes and small gabbro and diorite plutons that intrude crystalline rocks of the Arabian Shield (Fig. 6) and volcanic rocks that interdigitate with other epiclastic rocks of the Red Sea basin (Pallister 1987). The volcanic rocks (Sita Formation) are a suite of basalt, dacite, and rhyolite interbedded with volcaniclastic rocks, wacke, and limestone. One set of dikes, referred to as the Ghumayqah complex, consists of widely spaced linear intrusions of gabbro to monzogabbro 10–100 m wide. The Ghumayqah dikes create prominent magnetic anomalies (Fig. 6b). The anomalies and geologic mapping show that two NW-trending dikes shown in Fig. 6a extend more than 50 km SE of the area shown in the figure where they are associated with as many as ten, widely spaced, similar dikes intruded into basement across a width of ~30 km (Pallister 1986). The other set of dikes, referred to as the Damm dike complex, forms swarms of subparallel to parallel, and locally sheeted, dikes and minor sills of alkali basalt and lesser hawaiite, trachyte, comendite, dacite, and rhyolite. This swarm extends over a width of 10 km. Because of greenschist-facies metamorphism, K-Ar dating reported by Pallister (1987) is not as reliable as the $^{40}Ar/^{39}Ar$ dating reported by Sebai et al. (1991). Whole-rock plateau ages range from 26.0 ± 1.3 Ma to 22.3 ± 0.5 Ma and amphibole ages range from 24.0 ± 0.2 Ma to 22.3 ± 0.6 Ma (Sebai et al. 1991). Pallister (1987) interprets the Al Lith Igneous Complex as magmatism in continental crust thinned and extended by horst-graben faulting, and envisages a sheeted mafic dike complex forming the Red Sea oceanic crust farther west (Fig. 6c). The conspicuous magnetic boundary shown on Fig. 6b and regionally shown on Fig. 2a is located where SW-trending anomalies reflecting the structure of the Arabian Shield are truncated by a regional high that extends along the entire eastern margin of the Red Sea south of Jiddah, and is interpreted by us as the boundary between continental and oceanic crust.

The Tihama Asir igneous complex (Fig. 7) comprises swarms of sheeted dikes, bodies of layered gabbro, and

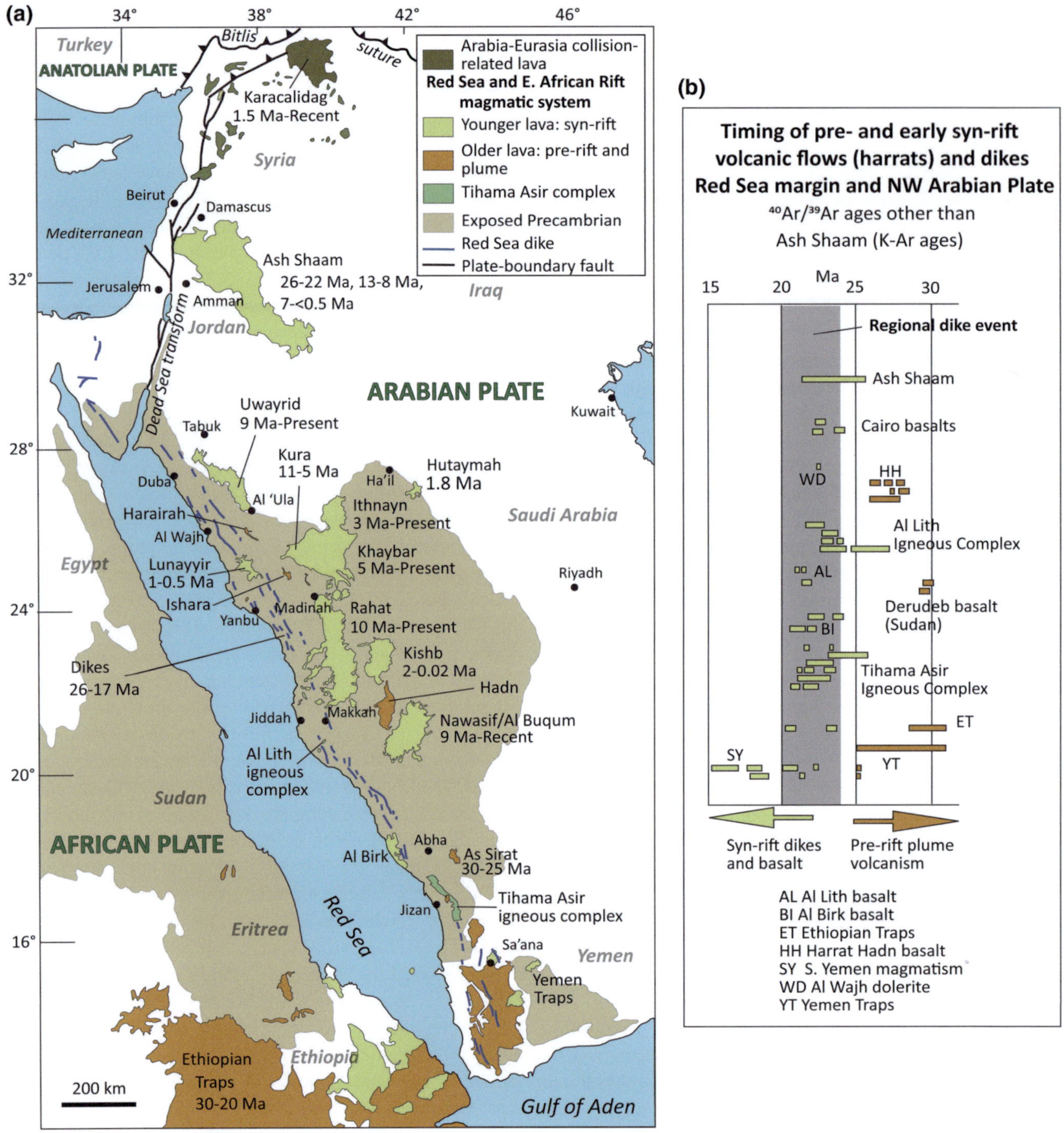

Fig. 5 **a** Cenozoic to Recent volcanic and intrusive rocks of the Red Sea basin and northwestern part of the Arabian Plate. Additional Cenozoic basalt occurs in the Cairo area, west of the western margin of this map (see Bosworth 2015 for details). **b** Compilation of published $^{40}Ar/^{39}Ar$ ages of pre- and syn-rift magmatism in the Red Sea basin (after Bosworth and Stockli 2016) and K-Ar ages in Harrat ash Shaam (after Ilani et al. 2001). Two periods of Oligocene–early Miocene magmatism are distinguishable: (1) initial activity associated with the Afar plume at ~31 Ma, affecting the southern ANS; and (2) syn-rift magmatism along essentially the entire Red Sea rift and northwestern Arabian Plate (Harrat ash Shaam). Widths of boxes indicate published error bars for individual samples except for the Ethiopia and Yemen traps, which are composites of many samples. See Bosworth and Stockli (2016) for sources of data. Lavas in the far north of the Arabian Plate are related to Late Cenozoic collision between the Arabian and Anatolian and Eurasian Plates (Pearce et al. 1990)

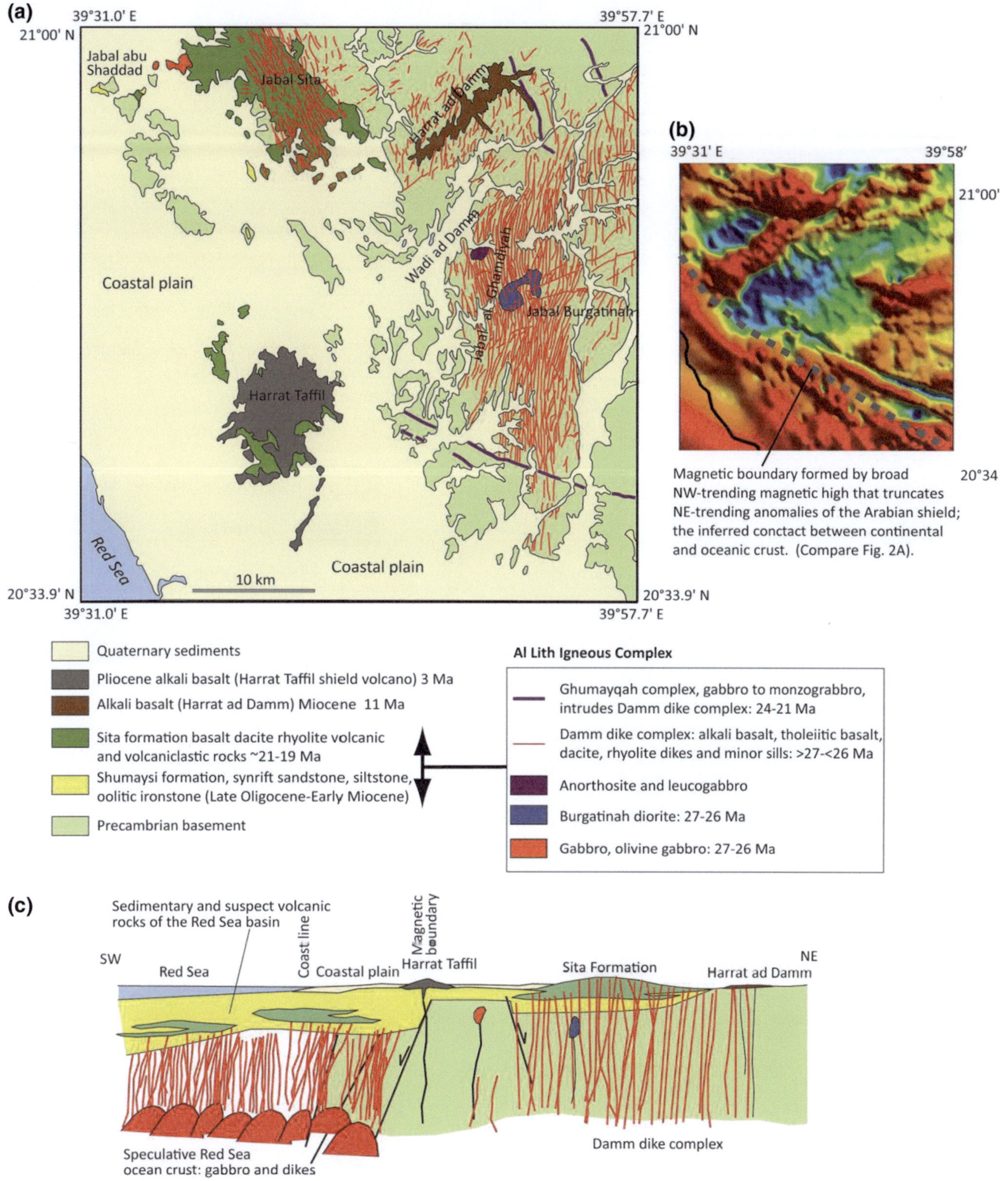

Fig. 6 The Al Lith igneous complex (after Pallister 1987). **a** Map of the complex showing the great concentration of Damm dikes that characterizes the complex. **b** Reduced-to-the-pole aeromagnetic map showing NW-trending linear anomalies coincident with Ghumayqah complex gabbroic dikes and a broad region of high-magnetic intensity in the southwest suggestive of strongly magnetic, possibly oceanic, crust beneath the coastal plain and Red Sea. **c** Schematic paleogeographic reconstruction showing the location of the Al Lith igneous complex in rifted basement of the Arabian Shield at the margin of an inferred swarm of sheeted dikes forming a hypothetical Red Sea crust

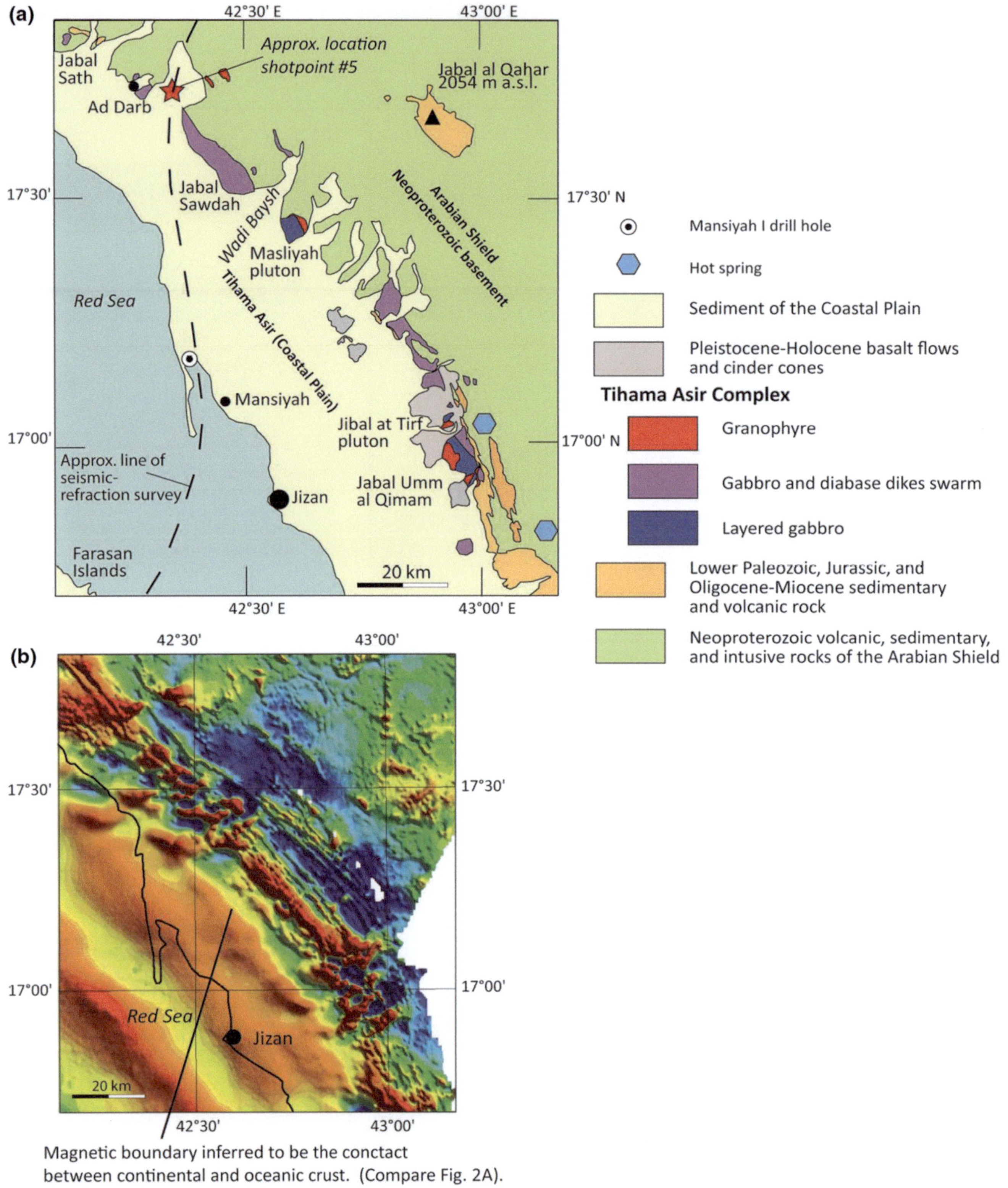

Fig. 7 The Tihama Asir complex (after Coleman et al. 1979 and Blank et al. 1987). **a** Map of the complex showing its strike length of >150 km at the contact between continental crust of the Arabian Shield and Cenozoic sedimentary and volcanic rocks of the Red Sea basin. The complex is partly overlain by Pleistocene-Holocene lava and cinder cones, and details of the contact are obscured by Cenozoic unconsolidated alluvial and eolian sediment. **b** Reduced-to-the-pole aeromagnetic map showing a zone of high-magnetic intensity partly coincident with the Tihama Asir complex and broader zones of high-to-moderate intensity along the coastal plain and Red Sea interpreted here as indicative of oceanic crust

irregular plutons of granophyre exposed inland from Jizan that intruded faulted and extended continental basement at the contact between the Arabian Shield and the Red Sea basin. It has a strike length of >150 km NW–SE parallel to the trend of the Red Sea. Its outcrop width is 5–10 km, but its coincidence with the NW-trending magnetic high shown in Fig. 7b indicates a minimum subsurface or concealed width of at least 20 km. Conspicuous linear, short-wavelength, NW-trending magnetic anomalies farther east suggest the presence of buried Red Sea dikes of the type known elsewhere along the Saudi Arabian margin. The number of dikes and the dike/country-rock volume ratio increase east to west across the shield/coastal plain contact and the western exposures are virtually entirely sheeted dikes, pillow lava, and volcaniclastic rock (Gettings et al. 1986). Individual dikes of the sheeted dike complex are 0.5–19 m wide and vary in density from 100% in areas of dike-on-dike intrusion to 10% or less in areas where screens of country rocks are present. Many dikes intruded along fault planes or joints in the bedrock. Compositionally they are diabase and basalt, with minor gabbro and rhyolite. Layered gabbro, in which variations in plagioclase, clinopyroxene, and olivine contents define subtle, seaward-dipping layers as much as 1 m thick, crops out in plutons as much as 8 km long and 2.5 km wide (Blank et al. 1987). Granophyre occurs as pods and small plutons that intrude the dikes and gabbro. The complex yields K-Ar ages between 24 and 20 Ma for gabbro and granophyre (Coleman et al. 1972), indicating an early Miocene age for the complex, similar to the age estimated for the Al Lith igneous complex. Since the 1950s, the complex was believed to be located at the boundary between the Arabian Shield and Red Sea basin. This conclusion is supported by the coincidence of a region of abrupt thinning of continental crust at the contact between the Arabian Shield and Red Sea basin shown on the seismic-refraction survey profile (Fig. 3) and the NW-trending magnetic boundary shown in Fig. 7b between a region of heterogeneous, shortwave anomalies to the northeast and a region of broad-wavelength anomalies to the southwest.

4 Continental Crust of the Red Sea Margins

The Arabian-Nubian Shield (ANS) extends 3500 km north-south from southern Israel and SW Jordan to Kenya and some 1500 km east-west from central Saudi Arabia to the Nile in Egypt and Sudan, underlying an area of $\sim 2.7 \times 10^6$ km^2 (Fig. 1, inset). This crust formed between $\sim$880 and $\sim$580 Ma as part of the Rodinia-Gondwana supercontinent cycle, during a $\sim$300 m.y. period of Neoproterozoic magmatic and tectonic evolution that yielded a juvenile tract of continental crust, as the ANS was sandwiched between colliding continental blocks of eastern and western Gondwana. Transpressional convergence coupled with tectonic escape and orogenic collapse produced broadly north-trending shear zones and upright tight folds in the southern ANS and northwest-trending strike-slip shear zones of the Najd fault system in the north (Figs. 1 and 4). By early-Cambrian times, the ANS was part of the East Africa–Antarctic Orogen (EAAO; Stern 1994; Fritz et al. 2013), an orogenic belt that stretched southward through the Gondwana supercontinent along the eastern margin of Africa and across Antarctica. Around the Red Sea, ANS rocks crop out in Sinai, Israel, and Jordan in the north; on the Red Sea western flank in the Eastern Desert of Egypt, the Red Sea Hills of Sudan, and the highlands of Eritrea and northern Ethiopia; and on the Red Sea eastern flank in western Saudi Arabia and the highlands of Yemen.

Exposed ANS is composed of subequal proportions of mostly greenschist-facies volcanic and related immature sediments and intrusive plutons (Fig. 8b) that formed during Tonian-Cryogenian and Ediacaran-earliest Cambrian stages. The earlier stage, $\sim$250 m.y. long, corresponds to an episode of accretionary tectonics dominated by formation of volcanic-arc assemblages and accretion of multiple arcs around the Mozambique Ocean to form proto-continental crust. These rocks are dominated by greenschist-facies volcanic and related immature sedimentary rocks and “I-type” granodioritic intrusions that are characterized by similar U-Pb zircon radiometric ages and Nd model ages. In terms of their petrogenesis, they are juvenile additions to the crust from the mantle as indicated by their isotopic compositions (Stern 2002), and are sometimes referred to as the western arc or oceanic terranes of the ANS (Stoeser and Frost 2006). The later, shorter ($\sim$50 m.y.) stage was a period of continental collision, escape tectonics, orogenic collapse, and possible lithospheric delamination; rocks formed during this stage include less-metamorphosed volcanic and sedimentary rocks and widespread granitoid plutons and batholiths (Fig. 4).

Stabilization of the end-Ediacaran ANS crust coincided with shield-wide exhumation, uplift, and the development of a regional angular unconformity (Ram Unconformity of Powell et al. 2014). Uplift possibly resulted from lithospheric delamination coupled with regional compression associated with terminal collision between crustal blocks of eastern and western Gondwana (Avigad and Gvirtzman 2009; Abu-Alam et al. 2011). Subsequent regional subsidence allowed deposition of thick sequences of Cambrian and Ordovician sandstone and siltstone, derived mostly from erosion of the Trans-Gondwana “supermountains” in the south (Squire et al. 2006; Meinhold et al. 2013). Cambro-Ordovician rocks originally extended across the entirety or most of the ANS basement, but have been extensively removed by erosion. Lower Paleozoic sandstone locally crops out on the Saudi Arabian coastal plain as

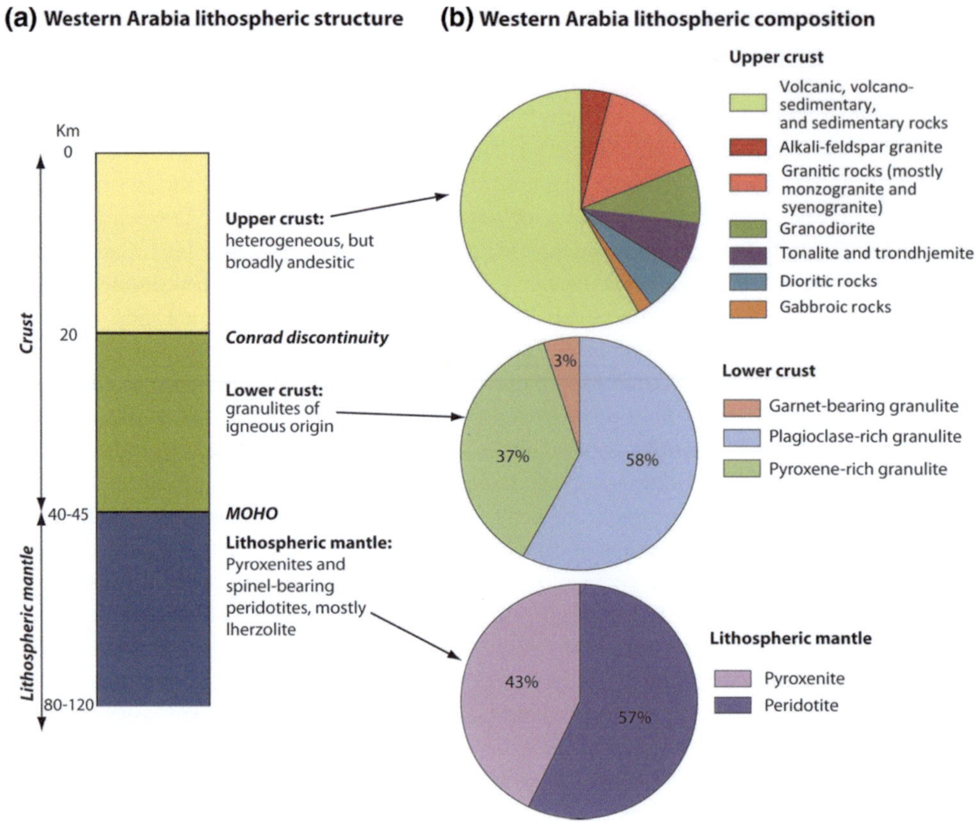

Fig. 8 **a** Structure of western Arabia lithosphere, simplified from Stern and Johnson (2010). **b** Composition of western Arabian lithosphere, as determined from outcrops for upper crust (after Johnson et al. 2011) and reported proportions of lower crust and lithospheric mantle xenoliths brought up in Neogene volcanic eruptions (Harrats; Stern and Johnson 2010). For upper crust, gneiss is calculated in terms of the protolith—thus dioritic gneiss is grouped with diorite, granite gneiss with granite, etc. Ultramafic rocks do not figure either, but amount to about 1%. The Nubian Shield has a similar composition except for a greater proportion of ultramafic rocks

blocks downdropped from their original position at the top of the Red Sea escarpment along seaward-dipping detachment faults (Bohannon 1989), but the main expanse of Phanerozoic rocks east of the Red Sea is 100–700 km inland in central and eastern Arabia. ANS exposures west of the Red Sea extend ~500 km inland in Sudan from the Red Sea coast as far as the Saharan Metacraton (Fig. 1, inset), a region of mainly pre-Neoproterozoic crust that was extensively reworked during the Neoproterozoic (Abdelsalam et al. 2002). To the north, in the Eastern Desert of Egypt, the ANS exposures narrow to less than 100 km.

Tonian-Cryogenian volcanic-arc assemblages constitute tectonostratigraphic terranes that, adjacent to the Red Sea, are referred to as the Eastern Desert, Gebeit, Haya, and Tokar terranes in Egypt, Sudan, and Eritrea, and the Midyan, Hijaz, and Jiddah terranes in Saudi Arabia (Fig. 1). On the basis of isotopic dating and field observations, it is evident that these terranes were once continuous and are now bisected by the Red Sea. The Eastern Desert terrane correlates with the Midyan, the Gebeit terrane with the Hijaz, the Haya terrane with the Jiddah, and the Tokar terrane with the Asir (Fig. 9). The mere fact of correlation, of course, does not itself constrain how close the terranes were juxtaposed prior to Red Sea opening, but the present relationship of the terranes precludes any significant strike-slip displacement of continental crust along the axis of the Red Sea during initiation of Red Sea opening of the type proposed, for example, by Shimron (1989).

4.1 Suture Zones

In addition to the lithostratigraphic terranes that correlate across the Red Sea, there is also general consensus about the correlation of inter-terrane ophiolite-decorated suture zones that resulted from closing the Neoproterozoic ocean basins in which the arc assemblages originated (Azer and Stern 2007; Stern et al. 2004) (Figs. 1 and 9). Of the two ANS sutures that impinge on the Red Sea, the northerly Allaqi-Heiani-Onib-Sol Hamed-Yanbu suture trends E–W to NE–SW across southern Egypt and northern Sudan, and irregularly NE–SW across northwestern Saudi Arabia, forming a composite structure referred to as the YOSHGAH (Yanbu-Onib-Sol Hamed-Gerf-Allaqi-Heiani) suture (Stern et al. 1990). The western part of the suture zone in Egypt and Sudan is a ~400 km-long ophiolite-decorated fold-and-thrust belt composed of imbricate thrust sheets and slices of ophiolite (serpentinite, amphibolite, metagabbro and metabasalt) and island arc metavolcanic/metasedimentary rocks. The Yanbu suture in the Arabian Shield is a similar subvertical to steeply-dipping shear zone containing nappes and fault-bounded lenses of mafic and ultramafic rocks (Pallister et al. 1988). Radiometric dating of ophiolites, metavolcanic rocks, gabbro, quartz diorite and stitching granites suggests that the ophiolites formed ~730 Ma and that accretion between the Gebeit and Eastern Desert terranes and emplacement of the ophiolitic nappes occurred between 730 Ma and 709 Ma (Ali et al. 2010). The timing of convergence along the Yanbu suture is not precisely constrained, but amalgamation probably occurred at ~700 Ma following 808–721 Ma ophiolite formation, and preceding emplacement of 730–690 Ma granodiorite and tonalite (Pallister et al. 1988). The suture zone in Egypt and Sudan extends to the shield/coastal plain boundary; the Yanbu suture stops some 50 km short of the contact with the Red Sea basin because its continuity is interrupted by granite intrusions.

The Bir'Umq-Nakasib suture, one of the best-exposed Neoproterozoic sutures in the world, is a slightly older zone of amalgamation between the Jiddah-Haya and Hijaz-Gebeit terranes in Saudi Arabia and Sudan (Fig. 9). It extends NE across the ANS as a belt of ophiolite nappes and juvenile metavolcanic, metasedimentary, and intrusive rocks 5–65 km wide and over 600 km long. Dating of ophiolitic rocks, volcanic rocks, and pre- and syntectonic plutons indicates that oceanic magmatism in the region was active ~870–830 Ma and suturing occurred ~780–760 Ma (Abdelsalam and Stern 1993; Hargrove 2006).

The Ad Damm-Barka shear zone is a prominent zone of subvertical shearing and folding at the contact between the Jiddah-Haya and Asir-Tokar terranes. The Barka shear zone is a transpressional duplex made up of at least seven structurally distinct units separated by shears (Drury and de Souza Filho 1988). It dips west between 45° and 80° and is as much as 40 km wide. Each unit has large folds; some are internally imbricated by lesser shear zones. The Ad Damm shear zone is a subvertical fault zone 2–4 km wide extending >350 km NE from the Red Sea coastal plain. It is characterized by dextral strike-slip, moderately to steeply NW-plunging stretching lineations, and conspicuous S–C kinematic indicators in greenschist- to amphibolite-facies phyllite and schist developed from volcanic and volcaniclastic rocks as well as mylonitic gneiss developed from granite of the ~620 Ma Numan Complex. Because shearing deforms the Numan Complex, the age of the youngest finite strain of the Ad Damm fault zone is inferred to be Ediacaran, and the structure is considered by some to be a conjugate shear of the otherwise NW-trending Najd fault system (Hamimi et al. 2014). The Ad Damm-Barka shear zone contains little ophiolite and its tectonic significance is debated; it may be a cryptic suture or a young, dextral strike-slip shear zone superimposed on or obliterating an earlier suture. Importantly, however, both the YOSHGAH and Bi'r Umq-Nakasib sutures and the Ad Damm-Barka shear zone impinge, at various oblique angles, on the Red Sea and create a network of structures and piercing-points that must be put back together in any palinspastic model for the Red Sea (Fig. 9).

4.2 North-Trending Shear Zones

Another distinctive structure in the Asir-Tokar and Jeddah-Haya terranes comprises northerly trending belts of shearing, thrusting, and folding coupled with variable amounts of horizontal strike-slip displacements that resulted from terminal transpressional E–W convergence (Abdelsalam and Stern 1996) (Figs. 1 and 9). These shear zones are marked by chlorite-sericite schist, phyllonite, and local serpentinite schist derived by retrograde metamorphism of the greenschist and amphibolite-facies country rocks. They have a marked lithologic and rheologic contrast with the unaltered country rocks, and constitute north-trending zones of potential crustal weakness. The shear zones are subparallel to the Red Sea dikes, the dike swarms of the Al Lith and Tihama Asir igneous complexes, and to the general trend of the southern Red Sea coastline. Consequently, it is likely that such pre-existing Neoproterozoic structures exercised a geometric control on rifting in the southern Red Sea.

Such shear zones are exposed in the Ghedem area of Eritrea, immediately adjacent to the Red Sea coastline (Ghebreab et al. 2005). The basement rocks here include domes of low- to middle-crustal gneiss and schist metamorphosed to amphibolite facies during progressive syn-deformation metamorphism that peaked at ~590 Ma when *P–T* conditions were near 12 kbar and 650 °C. Structurally, the domes are characterized by low-angle, mylonitic ductile shear zones. Subsequent exhumation of the high-grade rocks and extension along E–ESE- and W–WNW-directed low-angle faults during ANS orogenic collapse are denoted by retrogressive metamorphism between 580 and 565 Ma ($^{40}Ar/^{39}Ar$ hornblende and mica ages) (Fritz et al. 2013). The high-grade rocks were further exhumed during Permo-Carboniferous times and were further thinned from a thickness of ~35 to ~14 km during the Cenozoic by low-angle normal faulting. The Cenozoic faults sole-out into detachments among the low-angle Neoproterozoic mylonites, suggesting that the Cenozoic faults reflecting Red Sea extension exploited, or were controlled by Neoproterozoic structures. Unfortunately, $^{40}Ar/^{39}Ar$ cooling ages of hornblende and muscovite fail to constrain the age of detachment, possibly because this occurred below the closing temperature of muscovite (Ghebreab et al. 2005).

The Gebeit terrane contains the Hamisana and Oko shortening zones (Abdelsalam 1994, 2010; Abdelsalam and Stern 1996) (Ha and On in Fig. 1) that formed between ~660–560 Ma (Miller and Dixon 1992; Stern et al. 1989a, b). The Hamisana shortening zone extends north across the YOSHGAH suture, which it displaces by about 50 km (Stern et al. 1990), into the southernmost Eastern Desert terrane and impinges the Red Sea coastline at an acute angle. The correlative Hanabiq shear zone, a belt of highly strained rocks >35 km long and nearly 5 km wide, in the Arabian Shield north of Yanbu (Hn in Fig. 1) (Duncan et al. 1990), contains mylonitic gneiss and schist with dextral kinematic markers and syntectonic granitoids that yield a U-Pb zircon age of 590.5 ± 2.8 Ma (Kennedy et al. 2010) interpreted as the time of syntectonic intrusion. Deformation on the Hanabiq shear zone is estimated to be ~610–585 Ma, overlapping the time of movement on the Hamisana shortening zone. The Hanabiq shear zone is evidently a continuation of the Hamisana shortening zone and the two structures constitute another set of piercing-points for modeling Red Sea closure.

4.3 Najd Shear Zones

The northern ANS terranes lack the N–S shortening structures of the southern ANS. Instead, they are pervasively affected by NW-trending "Najd" shear zones (Sultan et al. 1988) that developed in association with northward-directed tectonic escape of the ANS during terminal collision of E and W Gondwana (Burke and Sengor 1985) and created prominent NW-trending brittle-ductile fault zones (Fig. 1). The Najd system, dating between ~620 and 580 Ma, is one of the largest shear zone systems on Earth (Stern 1985) and constitutes a zone of deformation, sometimes referred to as the Najd fault corridor, that extends across the entire Arabian Shield and northern Nubian Shield.

The composition of mid-level ANS crustal rocks includes upper amphibolite-facies quartzofeldspathic (granitic) gneisses and amphibolites exposed in the Meatiq and Hafafit domes in the Central Eastern Desert of Egypt, the Qazzaz dome in northwestern Saudi Arabia and elsewhere (Fig. 4). These rocks crop out in structural highs surrounded by greenschist-facies ensimatic arc rocks and younger sedimentary basins intruded by dioritic, granodioritic, and granitic plutons, and are interpreted as examples of high-grade and migmatitic infrastructure of the mid-crust ("Tier 1" of Bennett and Mosley 1987) that poke through lower grade superstructure rocks ("Tier 2" of Bennett and Mosley 1987). High-strain mylonitic zones separate the high-grade and low-grade rocks and form a system of crustal dislocations referred to, in the Nubian Shield, as the Eastern Desert Shear Zone (EDSZ) (Andresen et al. 2009, 2010) that is related to ~600 Ma Najd faulting. Similar types of crustal vertical heterogeneity are referred to by Blasband et al. (2000) and Fritz et al. (2002), who describe metamorphic belts in Wadi Kid in Sinai and the Sibai and Meatiq regions of the Central Eastern Desert, Egypt as core complexes. The exhumation of the ANS infracrustal rocks in these localities may reflect interaction of Najd left-lateral strike-slip shearing, extension and partially molten middle crust (Stern 2017). Because of their relatively high metamorphic grades and schistose and gneissic fabric, the ANS infracrustal rocks in Egypt were earlier interpreted as remnants of pre-Neoproterozoic continental crust (e.g., Habib et al. 1985; El-Gaby et al. 1988, 1990). However, the absence of significant geochemical, geochronologic, or isotopic differences between the high- and low-grade rocks strongly supports the view that rocks of the two tiers represent different thermal regimes experienced by different levels of Neoproterozoic ANS crust and are further evidence that the ANS crust in its entire thickness is juvenile.

The Najd system is associated with Neoproterozoic pull-apart volcanosedimentary basins located at bends or jogs in the faults and provided a mechanism for the exhumation of mid-level crustal rocks in gneiss domes or core complexes (Fritz et al. 2002; Abd El-Naby et al. 2008; Meyer et al. 2014). Because the Najd structures are so long and form a network of shears that obliquely intersect the Red Sea coastline, they are particularly useful geologic features for constraining palinspastic reconstruction of the ANS before Red Sea opening. Sultan et al. (1988) (using different names to those used here) correlate specific strands of the Najd system across the Red Sea, linking the Qazzaz and Duwi shears and Ajjaj and Nugrus shears for example. Correlations at the level of such detail may be debated, but the general extension of the Najd fault system from the Arabian Shield into the Nubian Shield is without doubt, and must be accommodated in any model of Red Sea closure. Furthermore, Cenozoic rejuvenation of Najd faults controlled the development of the Late Mesozoic-Cenozoic Nakheil and Azlam (Aznam) basins mentioned above. The faults do not appear to cause any jog in the coastline or affect the location of Red Sea deeps, but they do modify Miocene fault geometry and create the Quseir-Duba accommodation zone, a region across which there is a significant change in Red Sea faulting polarity, from NE-dipping Miocene faults to the north to SW-dipping faults to the south (Bosworth 2015).

5 Vertical Structure of ANS Lithosphere

In all discussions of Red Sea opening, it is accepted that the boundary between the Red Sea basin and the ANS crust approximates a major lithospheric discontinuity that marks where once-continuous continental crust was ruptured. Models of continental rifting necessarily entail an understanding of the overall strength of the lithosphere, a rheologic parameter that is generally considered to be controlled by composition, geothermal gradient or temperature structure (dT/dZ), strain rate, and crustal thickness (Kusznir and Park 1987), along with how much magma was injected (Bialas et al. 2010). Composition affects the rheology of the various layers that make up the lithosphere, and the overall strength of continental lithosphere is considered by some to reflect a relatively strong olivine-rich upper mantle, a weaker granitic to dioritic lower crust, and a stronger upper crust (e.g., Handy and Brun 2004). An alternative view, based on earthquake depth distributions, suggests that the strongest part of the lithosphere is the crust and that the upper mantle is relatively weak (Jackson 2002).

The temperature structure and strain rate of the ANS at the time of rifting is not well established. Limited heat flow data are available from temperatures logged in drill holes at shot points, before explosives were loaded, along the seismic refraction survey conducted in 1974 by the U.S. Geological Survey Saudi Arabian Mission and Saudi Arabian Directorate General of Mineral Resources (USGS/DGMR) from close to Riyadh in the northeast, across the southern Arabian Shield, and finishing at the Farasan Islands in the Red Sea (Fig. 3) (Mooney et al. 1985; Gettings et al. 1986). Together with information from the Mansiyah I deep petroleum exploration drill hole (Fig. 7) (Girdler 1970) and from the Red Sea shelf and axial trough (Girdler and Evans 1977), the data show an apparent increase in heat flow from the Arabian Shield ($\sim$40 mW/m^2) toward the Red Sea margin ($\sim$80–110 mW/m^2). Gettings et al. (1986) account for high heat flow at shot-point 5 close to the Red Sea (Figs. 3 and 7) as the effect of abutting oceanic crust and (or) an enhanced mantle component of heat-flow through the continental crust, which here has been thinned to $\sim$25 km, assuming a temperature regime that has persisted for 10 million years or more.

Comprehensive information about the structure and composition of the ANS lithosphere is given by sources as varied as seismic-refraction surveys, S-wave splitting parameters, lower-crust and upper-mantle xenoliths obtained from Cenozoic lavas (harrats), and structural and lithologic mapping of high-grade rocks exhumed from mid-crustal levels and exposed in gneiss domes or core complexes. One widely accepted model of P-wave arrival times measured during the 1974 Riyadh-Farasan seismic-refraction survey identifies $\sim$40 km thick continental crust divided into an upper $\sim$20 km thick layer of low Vp (<6.5 km/s) of likely felsic composition and a lower $\sim$20 km thick layer of higher Vp (6.7–7.8 km/s) of likely mafic composition (Fig. 3) (Mooney et al. 1985; Gettings et al. 1986). The ANS upper mantle is characterized by high Vp velocities (>8 km/s), suggesting a bulk ultramafic composition. S- and Raleigh-wave functions indicate that the ANS upper (lithospheric) mantle varies in thickness from >40 km at the Saudi Arabian coastline to $\sim$90 km beneath the shield to >170 km just east of the shield (Hansen et al. 2007; Park et al. 2008). Prodehl (1985) modeled the Saudi Arabian crust–mantle boundary at a common depth of 40 km with a maximum of 50 km beneath the southwestern part of the shield and an abrupt thinning to 15 km beneath the Red Sea; a similar result was obtained more recently by Tang et al. (2016). In the Prodehl model, relatively high-velocity material at about 10 km depth in the western shield upper crust is underlain by velocity inversions, and the lower crust with a velocity of about 7 km/s is underlain by a transitional crust-mantle boundary. The upper mantle appears to have a laterally discontinuous lamellar structure with intermixed high-velocity and lower velocity zones.

A limitation of seismic-refraction surveys, of course, is that the results emphasize horizontal structure at the expense of vertical or steep structure. As a consequence, although the Riyadh-Farasan survey clearly shows vertical heterogeneity (layering) in the ANS crust, it provides few instances of conspicuous Vp discontinuities that reflect the many steeply-dipping shear zones and lithologic contacts known to exist in the Arabian Shield (Gettings et al. 1986). The vertical discontinuities that are identified (Mooney et al. 1985) denote a fault in the upper crust $\sim$200 km from the northeastern end of the profile, structures in the upper part of the upper crust southwest of the known Al Amar fault zones, a diapiric structure coincident with the Khamis Mushayt gneiss, and an abrupt thinning of the crust from $\sim$40 km beneath the western edge of the ANS to $\sim$5 km beneath the Red Sea coastal plain (Fig. 3) (Blank et al. 1986; Gettings et al. 1986).

Mafic xenoliths brought to the surface in Cenozoic basalt have been obtained from eight locations in the western part of the Arabian Plate, extending from Harrat Kishb to Harrat ash Shaam (Fig. 5) (see review in Stern and Johnson 2010; Stern et al. 2016). Compositionally, the xenoliths are divided into samples of ANS lower crust and upper mantle on the basis of whether or not the samples contain plagioclase or olivine; plagioclase indicating a lower crustal origin, whereas abundant olivine suggests an upper-mantle origin. Lower crustal xenoliths include mafic granulites, 2-pyroxene gabbro, and rare garnet-bearing granulites. Ten lower crustal xenolith samples from Saudi Arabia yield a mean Nd model age of 0.76 $\pm$ 0.08 Ga (Claesson et al. 1984) indicating that the southern Arabian crustal lithosphere originated during the Neoproterozoic, whereas $\sim$360 Ma and $\sim$560 Ma U-Pb zircon ages of lower crustal samples from Syria and Jordan, respectively, imply that the crust in the northern part of the Arabian Plate is younger (Cadomian and Carboniferous) (Stern et al. 2014, 2016; Golan et al. 2017). Xenolith samples of the upper mantle include spinel lherzolite, dunite, wehrlite, clinopyroxenite, and megacrysts of clinopyroxene and amphibole, representing a suite of peridotite and pyroxenite consistent with the >8 km/s P-wave velocity revealed by the seismic-refraction data.

Another glimpse of mantle structure beneath the Arabian Shield is given by analysis of teleseismic receiver functions and shear-wave splitting (Levin and Park 2000; Hansen et al. 2006) which reveal an upper mantle fabric with anisotropic symmetry axes oriented ~N–S. Levin and Park (2000) suggest that the anisotropy represents shear zones developed during Neoproterozoic continent–continent collision. Hansen et al. (2006, 2007) alternatively argued that neither fossil lithospheric anisotropy nor present-day asthenospheric flow fully explains the observed splitting. They interpreted the splitting as a result of Cenozoic northeast-oriented flow associated with absolute Arabian Plate motion combined with northwest-oriented flow associated with a channelized Afar plume.

6 Discussion

The Red Sea basin is the world's best example of a nascent ocean, juxtaposing well-defined continental and oceanic crusts. It is located between Neoproterozoic crustal blocks of the northeastern Nubian and western Arabian Plates and is the result of rupturing and separation of continental lithospheres.

As described above, the accepted Red Sea model envisages active rifting and emplacement of oceanic crust in the axial region of the southern Red Sea and a transition to a northern region underlain by extended continental crust. In our opinion, however, the geologic, geophysical, and geographic features of the Red Sea strongly support an alternative model in which most of the Red Sea, in the north as well as the south, is underlain by oceanic crust. These features are: (1) the pattern of basement structures that require a virtual coast-to-coast closure of the Red Sea; (2) interpretations of potential field data—gravity and magnetics, that that we find to be compelling in the axial region south of ~22° N, persuasive for the margins of the southern Red Sea, and suggestive of extensive oceanic crust in the northern Red Sea; and (3) the presence of dikes, gabbros, and basalt flows emplaced during the early stages of Red Sea development, which suggest major upwelling of the asthenosphere, partial melting, and intrusion that would have thermally and mechanically weakened the lithosphere and facilitated rupture of the ~40-km thick continental crust and thicker mantle lithosphere that existed prior to Red Sea opening.

Ever since Wegener (1920) there has been a strong opinion that the two plates prior to Red Sea opening were closely juxtaposed, a model referred to by Gettings et al. (1986) as "this venerable concept" and by Coleman (1993), in the context of the reconstruction by Vail (1985), as "the most successful". Wegener (1920) made the proposal because of the close fit of the opposing shorelines, the prominent eastward bend of both shorelines north of Jiddah and Port Sudan, and a westward bend south of Jiddah and Port Sudan (Fig. 1). Palinspastic reconstructions of Precambrian structures across the Red Sea by Abd El-Gawad (1970), Greenwood and Anderson (1977), Vail (1985), and Sultan et al. (1988, 1993) all adopt a close-fit. A variant palinspastic model resulted from a study by the Saudi Geological Survey (SGS) and the Egyptian Mineral Resource Authority (EMRA) designed to test the hypothesis of a virtual coast-to-coast closure in the northern part of the Red Sea (Kozdroj et al. 2011). Although existing geologic maps, rock descriptions, and geochronological results suggested that rock units and structures (piercing-points) on one side of the Red Sea had counterparts on the other side, the investigation revealed that no unique one-to-one correlation of rock units could be made and structural misfits remained. It was concluded that a reconstruction based on tightly fitting the present shorelines was not satisfactory and at least a 15–30 km gap should be left between them, representing an area covered by Cenozoic sediments and underlain by unknown Neoproterozoic rocks (Kozdroj et al. 2011).

The conclusion of Kozdroj et al. (2011) runs counter to the view of the northern Red Sea as being underlain by stretched continental crust, but it is the best way to fit correlative ANS basement rocks and structures on either side of the Red Sea. This fit is the primary reason we also argue for a relatively tight pre-Red Sea juxtaposition of the Arabian and Nubian Plates (Fig. 9). As noted above, the correlative basement structures intersect the Red Sea coastlines at orientations ranging from approximately orthogonal to acute. Structures oblique to the Red Sea provide particularly tight reconstruction constraints because bringing such structures into alignment is very sensitive to minor differences in back rotation. Structures that intersect the Red Sea at high angles are much less sensitive to such small differences in rotation and so are correspondingly less useful in constraining how the Red Sea should be palinspastically closed. Nevertheless, the integrated network of basement structures strongly supports a tight fit of the Arabian and Nubian Plates. Sultan et al. (1993) proposed a similar fit on the basis of juxtaposing satellite imagery (Fig. 10) and our Fig. 9 uses the same pole of rotation to juxtapose the Arabian and Nubian Plates. As in the proposal by Sultan et al. (1988), we envisage that the Qazzaz shear zone has a counterpart in shear zones in the Duwi area; the Ajjaj shear zone continues in the Nugrus shear; the Hamisana shortening zone meets the Hanabiq shear zone; the Sol Hamed sector of the YOSHGAH suture matches up with the Yanbu sector; the Bi'r Umq suture joins with the Nakasib suture, the Ad Damm fault continues in the Barka shear zone, and shear zones in the Asir terrane align with shear zones in the Tokar terrane.

Although Kozdroj et al. (2011) were not able to make one-to-one correlations between ANS rock units on the margins of the northern Red Sea, we note a close

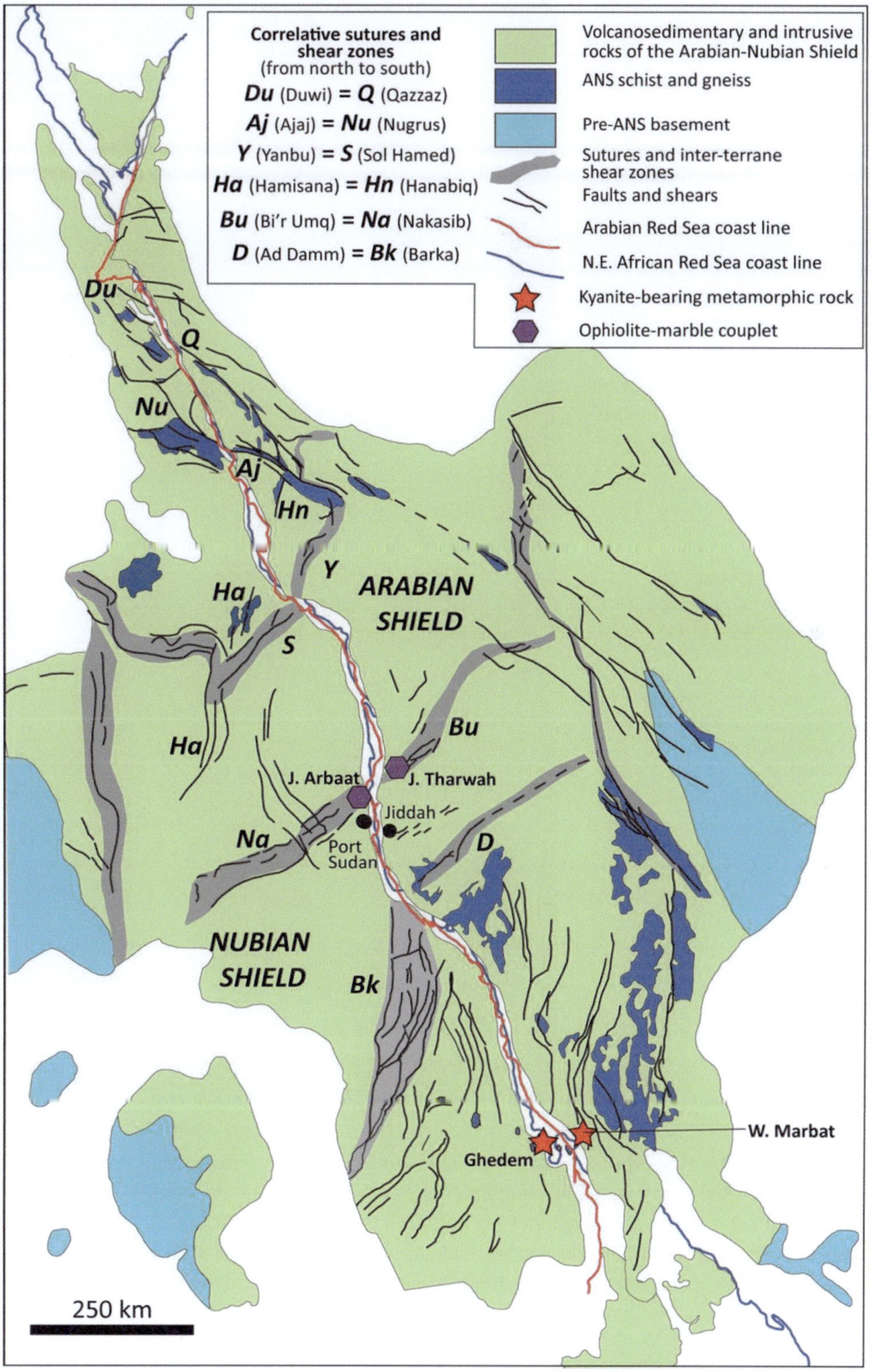

Fig. 9 Preferred palinspastic reconfiguration of the pre-Red Sea rifting relationship of the Arabian and Nubian Shields, showing a near coastline to coastline closure based on the alignment of major structures (sutures and shear zones) in the basement. The configuration is the same as that shown in Fig. 10

juxtaposition of distinctive lithologies in at least two regions on either side of the southern Red Sea, compatible with the reconstruction proposed here. One set of lithologies consists of ophiolite and marble that crop out as components of the Bi'r Umq-Nakasib suture on either side of the Red Sea. The ophiolites are exposed as nappes in Jabal Tharwah in the east and Jabal Arbaat in the west (Fig. 9), underlain in their respective footwalls by thick units of marble. The resulting lithologic couplet—ophiolite and marble—crops out in mountains immediately adjacent to the coastal plain on

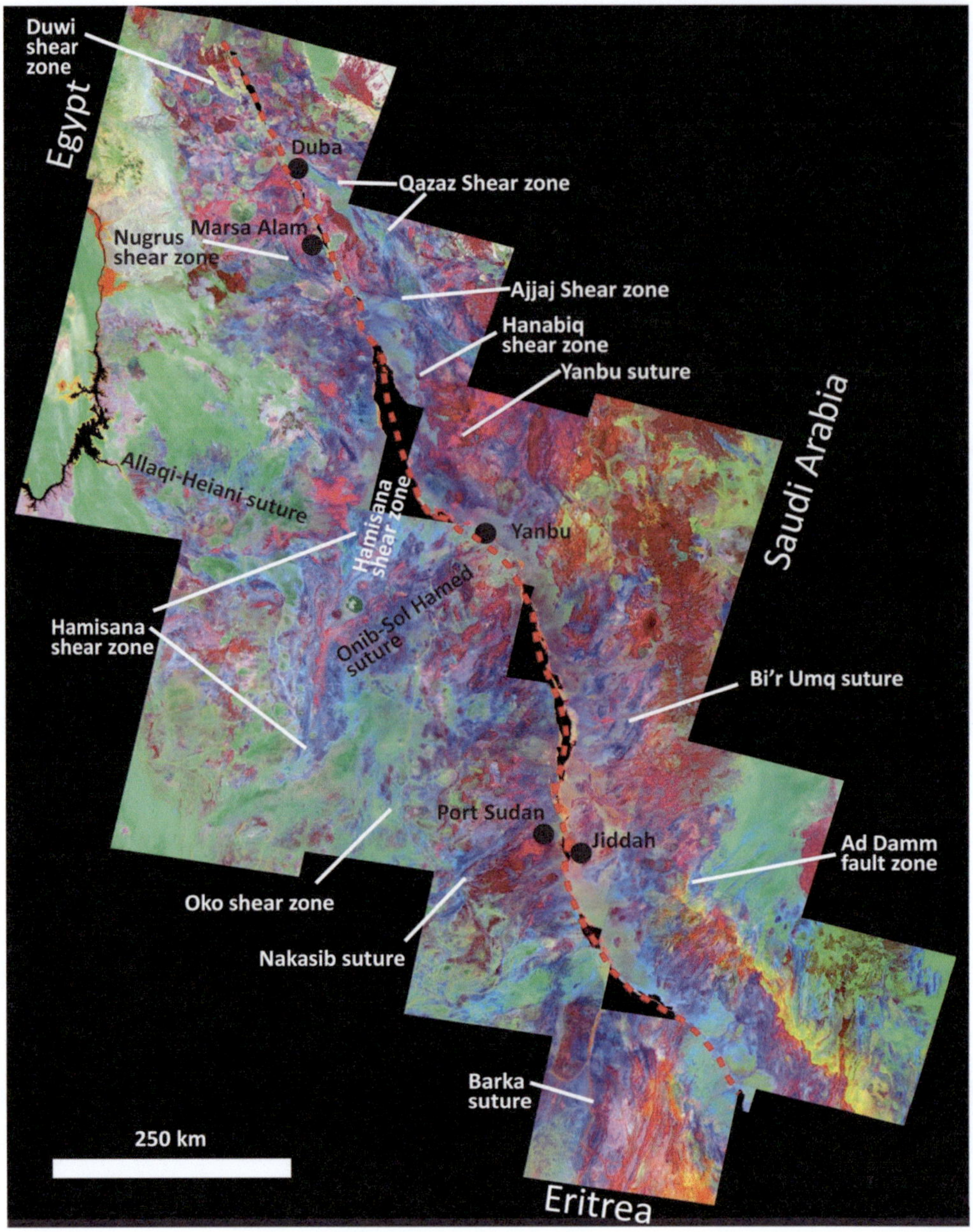

Fig. 10 Colour composite of 23 Landsat thematic mapper scenes (each 185 × 185 km), from Sultan et al. (1993). Dashed red line is preferred closure of Red Sea, near coastline-to-coastline, required to bring structures of the Arabian and Nubian Shields into alignment. The Arabian and Nubian shields were restored to our preferred pre-Red Sea configuration by rotating the Arabian plate around by 6.7° around a pole positioned at 34.6° N, 18.1°E (Note that this is very close to present rotation pole at 32.8°N, 23.8°E (DeMets et al. 2010)). Note that 6.7° of rotation at present rate of 0.38°/Ma would take 17.6 Ma. Restoration is based on matching ANS structures on either side of the Red Sea. TM band ratio product 5/4 × 3/4 (sensitive to the abundance of Fe-bearing aluminosilicates) is assigned blue, band 5/1 (sensitive to the abundance of opaque minerals such as magnetite) is assigned green, and band 5/7 (sensitive to the abundance of hydroxyl- and carbonate-bearing minerals) is assigned red

either side of the Red Sea. A palinspastic reconstruction of the type proposed here would place the lithologic couplets in virtual contact. Farther south, another set of distinctive rock units would be brought within a few tens of kilometres juxtaposition by our preferred reconstruction. The rocks—kyanite-bearing metasediments—crop out in the Ghedem area, Eritrea, and in Wadi Marbat on the Red Sea margin 80 km north of Jizan, Saudi Arabia. In the Ghedem area, the kyanite-bearing rock is paraschist, composed of kyanite, staurolite, almandine garnet, biotite, and quartz derived from a pelitic protolith (Beyth et al. 1977). At Wadi Marbat, the kyanite-bearing rock is kyanite-topaz-lazulite gneiss associated with andalusite-bearing hornfels and quartz-sericite schist that may have originated by high-pressure hydrothermal alteration of pelitic sedimentary rock (Collenette and Grainger 1994). Kyanite-bearing rocks are uncommon in the ANS, being confined to some ten locations in the Arabian Shield (Collenette and Grainger 1994) and in the Nubian Shield to metamorphic core complexes in Egypt and kyanite gneiss in the Duweishat area of northern Sudan. Exposures of kyanite in basement rocks close to both western and eastern contacts of the ANS with the Red Sea basin are therefore noteworthy as a possible piercing-point across the Red Sea.

The satellite imagery shown in Fig. 10 (after Sultan et al. 1993) illustrates the continuity of ANS structures achieved by closing the Red Sea in our preferred ANS reconstruction. In this figure, the Arabian and Nubian Shields were restored to a pre-Red Sea configuration by rotating the Arabian Plate 6.7° clockwise around a pole positioned at 34.6°N, 18.1°E (Sultan et al. 1993). Such a rotation at the present rate for the Arabian Plate of 0.38°/Ma would take 17.6 Ma. This pole is very close to the rotation pole at 32.8°N, 23.8°E more recently determined by DeMets et al. (2010) on the basis of a global assessment of best-fitting angular velocities for the geologically current motions of 25 tectonic plates.

Furthermore, the ANS structures not only help constrain models of Red Sea closure, they also provide insight into Red Sea geometry. As commented earlier, the Red Sea was initiated by propagation of plume-related magmatism northward from the Afar region. Such magmatic extension would have penetrated continental lithosphere, in the region north of Afar, dominated by a N–S structural grain in which zones of shearing and retrogressive metamorphism constituted zones of weakness. We propose that the structural grain of the ANS, at least south of 20°N, was a major factor in controlling the trend of the southern Red Sea by localizing Red Sea extensional faults. It is interesting to note, furthermore, that the upper-mantle anisotropy symmetry axis recognized by Levin and Park (2000) also trends N–S and is explained by Levin and Park as the effect of shear zones developed in the ANS during terminal Neoproterozoic continent–continent collision. This raises the intriguing possibility that there is a causal link between the structure of the upper mantle and trend of the southern Red Sea, as well as between the crust and the Red Sea.

To the north, in the region between latitudes ~20°N and 24°N, the Red Sea axis and coastlines are characterized by sigmoidal bends. This part of the Red Sea is north of the region dominated by north-trending shear zones, but instead is bracketed by the YOSHGAH and Bi'r Umq-Nakasib sutures. As noted earlier, these structures are somewhat orthogonal to the Red Sea trend and therefore would not have directly controlled it. However, they would have constituted zones of lithospheric weakness and they may have indirectly controlled rifting geometry in the manner of tectonic inheritance at a passive margin of the type described, on a larger sale, to account for bends in the eastern North American continental margin during the opening of the Iapetus and Atlantic Oceans (Thomas 2006). The NW-trending Najd structures that dominate the ANS continental crust north of 24°N had little effect on the trend of the notably linear Red Sea and its coastlines. This may be due to the flattening of Najd shears above hot Tier 1 upper crust (Stern 2017), so that they did not penetrate deeply enough into the lithosphere to form significant zones of weakness; alternatively, the Najd shears were annealed by Ediacaran igneous intrusions long before Red Sea opening. The Najd faults, however, helped to localize Cenozoic sedimentary basins that indent the basement exposures in Egypt and Saudi Arabia, and create the Quseir-Duba accommodation zone of faulting in the Red Sea (Bosworth 2015).

The degree of closure accepted for the Red Sea is important because it has a direct bearing on interpretations about the composition of the Red Sea crust, namely that different closures are expected if the Red Sea is underlain by stretched continental crust or Cenozoic oceanic crust. A simple estimate that the Red Sea is underlain by continental crust stretched to $\beta = 2$ implies that basement structures on unrifted crust in NE Africa and Arabia would realign when the Red Sea, presently ~300 km wide, is closed to ~150 km width. In contrast, a Red Sea crust composed of newly emplaced Cenozoic igneous rocks would permit a virtual coastline-to-coastline closure and a tight fit of basement structures.

The alignments of basement structures presented here strongly favour a near coast-to-coast closure of the Red Sea and a model that most if not all of the Red Sea is underlain by oceanic crust. This situation is widely accepted for the southern Red Sea. Short wavelength Vine-Matthews magnetic anomalies along the axial trough south of 22°N provide unequivocal evidence for the emplacement of oceanic crust under the central part of the southern Red Sea, and the presence of broader wavelength, but similarly NW-trending anomalies to the east is permissive of oceanic crust beneath the Red Sea margins as well. Farther inland, the Al Lith and Tihama igneous complexes testify to the emplacement of new igneous material on the margins of the Red Sea, and are consistent with a transition from continental to oceanic crust close to the shoreline in a region where seismic-refraction profiling indicates that the crust changes abruptly in thickness from ~40 km inland to 5 km beneath the Red Sea shelf (Mooney et al. 1985) (Fig. 3) and where regional magnetic boundaries testify to abrupt changes in crustal lithology.

The nature of the northern Red Sea crust is more contentious. As considered by Bosworth (2015), the similarity of Miocene stratigraphy in the northern Red Sea and Gulf of Suez, established by petroleum exploration drilling, is a strong argument in favor of extrapolating models for development of the Gulf of Suez to the northern Red Sea. However, the geophysical interpretations of Hall et al. (1977) and Saleh et al. (2006), for example, strongly favor a conclusion that the northern Red Sea crust contains much more oceanic material than is considered in the conventional interpretation.

If the Red Sea is modeled as largely underlain by oceanic crust in the south and to an unknown but significant degree in the north, an immediate question concerns how the ANS continental crust ruptured and separated to accommodate oceanic crust? As quantitatively modeled by McKenzie (1978), lithospheric extension and rift evolution reflects two phases, extension and subsidence. Rapid stretching thins the lithosphere and allows passive upwelling of hot asthenosphere. This stage is associated with normal faulting and tectonic subsidence. Once extension stops or localizes at a new mid-ocean spreading ridge, the stretched lithosphere cools and re-thickens, accompanied by slow "thermal" subsidence. This model has been applied to rifts all over the world. It is applied to the East African Rift and the Gulf of Suez (e.g., Bosworth 2015; Fig. 11a) and is implicit in models calling for the northern Red Sea to be largely underlain by stretched continental crust (Fig. 11b).

However, one problem arising from applying a stretching model to the Red Sea is the considerable strength of ANS lithosphere. Regardless of the relative importance of a strong or weak upper mantle in controlling lithospheric rheology, the Neoproterozoic-Cadomian age and mafic nature of ANS lower crust would make it cool, strong, and difficult to stretch. A solution to this problem comes from understanding the importance of dike injection as a mechanism that weakens and ultimately allows rupture of otherwise strong lithosphere (Bialas et al. 2010). This has been assessed recently by Ligi et al. (2015) to account for rupture of continental crust and initiation of seafloor spreading in the central part of the Red Sea. Ligi et al. (2015) envisage that the initial phase of continental lithospheric thinning by normal faulting was accompanied by asthenospheric upwelling which resulted in melting at the mantle-crust boundary and underplating of the crust by gabbro. Subsequent weakening of the lithospheric mantle and concomitant injection of gabbroic and basaltic dikes ruptured the lithosphere; mantle partial melting generated mafic magmas that were extracted at the rift axis forming shallow gabbro intrusions, dikes, and basaltic lava. Ligi et al. (2015) see evidence for their model in the swarms of dikes that run the entire length of the eastern Red Sea coast and in the high-pressure gabbro exposed on Zabargad Island and low-pressure gabbroic intrusions and basaltic dikes found on the Brothers Islands off the coast of Egypt. Following modeling by Saleh et al. (2006), we propose that gabbros and dikes representing oceanic crust underlie much or most of the shelf regions of the Red Sea as well as its axial region. Given these thermal and intrusive processes, it is likely that, at its initiation, rifting in the thick, cool, and strong NE-African-Arabian lithosphere resulted in a narrow zone of localized strain of the type described by Kearey et al. (2009), but subsequent thermal weakening of the lithosphere would lead to widening of the zone of strain. The asymmetry of the Red Sea basin shown by more basaltic flows and dikes and greater rift flank elevations in Arabia than NE Africa has been seen as evidence for rifting in terms of simple shear extension on an east dipping, low-angle master fault that penetrates the lithosphere (Dixon et al. 1989) although the model of Ligi et al. (2015), for example, envisages symmetrical extension at the continental to oceanic transition.

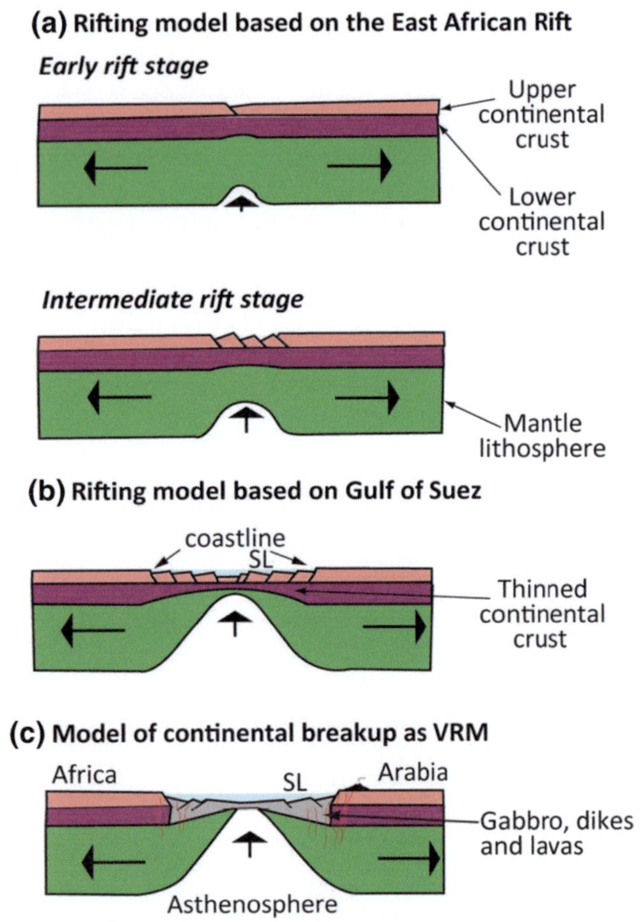

Fig. 11 Models of continental rifting. **a** Continental rifting and breakup based on our understanding of the East African Rift and Gulf of Suez. Early Rift Stage as in the East African Rift; in the initial stage of rifting, continental crust and underlying mantle lithosphere stretches. Intermediate Rift Stage; in this stage continued crustal thinning by normal faulting and lithosphere extended, as experienced by the Gulf of Suez during the Oligocene. **b** Model of a mature rift stage based on the conventionally understood geology of the Gulf of Suez and northern Red Sea, showing thinned continental crust and mantle lithosphere beneath marine basin. **c** Continental breakup with a marine basin underlain by a volcanic rifted margin; our proposed model for the northern Red Sea

On the basis of our interpretation that the Red Sea is largely underlain by oceanic material, is part of a magmatic province, and that dike intrusion was important for weakening and rupturing the lithosphere, we tentatively propose that the Red Sea is a volcanic rifted margin (VRM). The VRM concept derives from modern interpretations of continental breakup that emphasize great magmatic outpourings at the time of rifting. The VRM concept was, in fact, being developed in the early 1980s at about the same time that Red Sea models emphasizing continental stretching were articulated, but for unknown reasons did not get applied to the Red Sea. The VRM concept was triggered by recognition of seaward-dipping layered seismic reflectors in the upper oceanic crust offshore Norway, which Mutter et al. (1982) interpreted as a layered igneous sequence formed at the earliest stage of North Atlantic opening. It is interesting to note that Norwegian coastal regions contain little or no evidence that a VRM exists offshore even if rift-flank crust is exposed; for example, the onshore part of the ~55 Ma Norwegian VRM has no exposed lavas, dikes, or plutons despite the nearby presence of a massive offshore igneous construction. Prior to the work by Mutter et al. (1982), crust marking the transition from continent to ocean—which forms at the time of continental breakup and is buried beneath thick sedimentary sequences of passive continental margins—was thought to be overwhelmingly composed of stretched continental crust. At the present time, the VRM paradigm is the dominant model for continental breakup and ~90% of all passive continental margins are now thought to be VRMs to varying degrees (Menzies et al. 2002).

A model for the crust beneath Red Sea evaporites as a VRM is depicted in Fig. 11c, in contrast to models of Red Sea continental rifting and breakup based on our understanding of the East African Rift and Gulf of Suez. It is interesting to note that thirty years ago, McGuire and Coleman (1986) explicitly compared the Tihama Asir complex to the tholeiitic gabbro and syenite intrusions and dike swarms of East Greenland, which are now accepted as a prime example of a VRM (Voss and Jokat 2007). Emplacement of dikes as well as kilometre-scale thicknesses of basaltic and rhyolitic volcanic formations figure in the VRM model proposed for the onset of rifting in the Afar (Wolfenden et al. 2005). We acknowledge that the volume of magmatic material exposed on the margins of the Red Sea is considerably less that that present in Afar, but if dike intrusion was important for weakening and rupturing the lithosphere, then a VRM interpretation for the Red Sea becomes an attractive interpretation as an alternative to the conventional continent-ocean transition model.

Exceptions to the VRM model occur. The Portugal continental margin is known to lack a VRM (Manatschal 2004). A key factor in support of the Red Sea as a VRM is its obvious association with magmatism. Oceanic magmatism created accreted oceanic crust in the southern part of the present-day axial trough and, depending on interpretation of magnetic striping and the tectonic significance of dike-on-dike intrusions, across the entire southern sea and parts of the coastal plain. Elsewhere on the Red Sea ANS margin continental magmatism created extensive lava fields (harrats). Continental volcanism began in Ethiopia and Yemen ~31 Ma and continued between 24 and 23 Ma along the eastern margin of the Red Sea and in the subsurface near Cairo. A major episode of mafic diking affected Sinai and NW Arabia at 20–24 Ma; a slightly larger range (21–28 Ma) is found for intrusive rocks and dike swarms near 21°N in coastal Arabia.

If a VRM model applies to the Red Sea crust, it probably formed ~20–24 Ma ago, when the basin-wide basaltic volcanism occurred. Early Miocene igneous activity was characterized by tholeiitic basalts, indicating high-degree (10–20%) mantle melts comparable to those that characterize other VRMs. Early Miocene magmas differed in this regard from low-degree (1–5%) mantle melts of basanite and alkali basalt, which erupted more recently in the basalt fields (harrats) to the east. The 20–24 Ma (K-Ar age) sequence of layered gabbro, granophyre, and dike swarms at Tihama Asir in the southern Red Sea coastal region and similar igneous rocks at Al Lith could well be exposed Red Sea VRM. Offshore seismic profiling designed to image beneath the salt followed by drilling to basement in the Red Sea is required to test these ideas.

7 Conclusion

The correlation of Neoproterozoic structures and lithologies across the Red Sea strongly favors a model in which the Red Sea crust is predominantly Cenozoic oceanic material and in which NE Africa and the Arabian Peninsula were originally closely juxtaposed. A model of this type suggests the Red Sea contains minimally stretched continental crust and identifies the Red Sea as a Volcanic Rifted Margins (VRM). It is now recognized that continental breakup often involves great magmatic outpourings and almost 90% of all passive continental margins are now thought to be VRMs (Menzies et al. 2002). By considering the known ages of tholeiitic igneous activity on the margins of the Red Sea basin, we suggest that the Red Sea VRM formed ~20–24 Ma ago. Early Miocene tholeiites around the Red Sea indicate high-degree (10–20%) mantle melts like those of other VRMs. The 20–24 Ma sequence of layered gabbro, granophyre, and dyke swarms at Tihama Asir in the southern Red Sea coastal region (McGuire and Coleman 1986) is likely exposed Red Sea VRM. The emplacement of gabbro and dikes associated with the formation of oceanic crust provides a mechanism for rupturing strong ANS lithosphere by magmatic weakening of the lithosphere and accounts for the magnetic stripes along the Red Sea axial trough as well as under the Red Sea marginal sedimentary basins and coastline. It appears that the northerly trend of the southern Red Sea was controlled by the N–S trending structural fabric of the basement. We concur with previous authors that the bend in the Red Sea between 20°N and 24°N is a consequence of preexisting zones of crustal weakness orthogonal to the Red Sea represented by the YOSHGAH and Bi'r Umq-Nakasib sutures and the Ad Damm-Barka shear zone.

We likewise concur with inferences by other authors that the Najd fault system in the north strongly affected Cenozoic Red Sea faulting.

Acknowledgements We are grateful to Dr. Zohair Nawab and Dr. Najeeb Rasul of the Saudi Geological Survey for organizing the second workshop on the Red Sea in Jeddah in 2016 that gave us the occasion to consolidate our ideas about the initiation of the Red Sea and thank Springer-Verlag for publishing this paper as part of the workshop proceedings. We greatly appreciate the critical comments and suggestions of three anonymous referees. This is UTD Geosciences contribution # 1299.

References

Abd El-Gawad M (1970) Interpretation of satellite photographs of the Red Sea and Gulf of Aden (with discussion). In: A discussion of the structure and evolution of the Red Sea and the nature of the Red Sea, Gulf of Aden and Ethiopia rift junction. Phil Trans R S London Ser A 267:23–40

Abd El-Naby H, Frisch W, Siebel W (2008) Tectono-metamorphic evolution of the Wadi Hafafit Culmination (central Eastern Desert, Egypt). Implication for Neoproterozoic core complex exhumation in NE Africa. Geol Acta 6:293–312

Abd El-Wahed M, Ashmawy M, Tawfik H (2010) Structural setting of Cretaceous pull-part basins and Miocene extensional folds in the Quseir-Umm Gheig region, northwestern Red Sea, Egypt. Lithosphere 2:13–32

Abdelsalam MG (1994) The Oko shear zone: post-accretionary deformations in the Arabian Nubian Shield. J Geol Soc London 151:767–776

Abdelsalam MG (2010) Quantifying 3D post-accretionary tectonic strain in the Arabian-Nubian Shield: superimposition of the Oko Shear Zone on the Nakasib Suture, Red Sea Hills, Sudan. J Afr Earth Sci 56:167–178

Abdelsalam MG, Stern RJ (1993) Tectonic evolution of the Nakasib suture, Red Sea Hills, Sudan: evidence for a late Precambrian Wilson cycle. J Geol Soc London 150:393–404

Abdelsalam MG, Stern RJ (1996) Sutures and shear zones in the Arabian-Nubian Shield. J Afr Earth Sci 23:289–310

Abdelsalam MG, Liégeois JP, Stern RJ (2002) The saharan metacraton. J Afr Earth Sci 34:119–136

Abu-Alam T, Stüwe K, Kadi K (2011) Pan-African exhumation mechanisms. Geophys Res Abstracts, EGU General Assembly 13: EGU2011-10307-1

Ali KA, Azer MK, Gahlan HA, Wilde SA, Samuel MD, Stern RJ (2010) Age constraints on the formation and emplacement of Neoproterozoic ophiolites along the Allaqi-Heiani Suture, South-eastern Desert of Egypt. Gondwana Res 18:583–595

Andresen A, Abu El-Rus MA, Myhre PI, Boghdady GY, Corfu F (2009) U-Pb TIMS age constraints on the evolution of the Neoproterozoic Meatiq Gneiss Dome, Eastern desert, Egypt. Int J Earth Sci 98(3):481–497. https://doi.org/10.1007/s00531-007-0276

Andresen A, Augland LE, Boghdady GY, Lundmark AM, Elnady OM, Hassan MA (2010) Structural constraints on the evolution of the Meatiq gneiss domes (Egypt), East-African Orogen. J Afr Earth Sci 57:413–422

Avigad D, Gvirtzman Z (2009) Late Neoproterozoic rise and fall of the northern Arabian-Nubian Shield: the role of lithospheric mantle delamination and subsequent thermal subsidence. Tectonophysics 477:217–228

Azer M, Stern RJ (2007) Neoproterozoic serpentinites in the Eastern Desert, Egypt: fragments of fore-arc mantle. J Geol 115:457–472

Bennett GD, Mosely P (1987) Tiered tectonics and evolution, Eastern Desert and Sinai, Egypt. In: Matheis G, Schandelmeier H (eds) Current research in African Earth sciences. Balkema, Rotterdam, The Netherlands, pp 79–82

Beyth M, Stern RJ, Matthews A (1977) Significance of high-grade metasediments from the Neoproterozoic basement of Eritrea. Precambrian Res 86:45–58

Bialas RW, Buck WR, Qin R (2010) How much magma is required to rift a continent? Earth Planet Sci Lett 292:68–78

Blank HR, Mooney WD, Healy JH, Gettings MA, Lamson RJ (1986) A seismic refraction interpretation of the eastern margin of the Red Sea depression, Southwest Saudi Arabia. US Geological Survey Open-file Report 86–257, 24 p

Blank HR, Johnson PR, Gettings ME, Simmons GC (1987) Geologic map of the Jizan quadrangle, Sheet 16F, Kingdom of Saudi Arabia. Saudi Arabian Directorate General of Mineral Resources Geoscience Map GM-104

Blasband B, White S, Brooijmans P, De Broorder H, Visser W (2000) Late Proterozoic extensional collapse in the Arabian-Nubian Shield. J Geol Soc London 157:615–628

Bohannon RG (1989) Style of extensional tectonism during rifting, Red Sea and Gulf of Aden. J Afr Earth Sci 8:589–602

Bonatti E (1985) Punctiform initiation of seafloor spreading in the Red Sea during transition from a continental to an oceanic rift. Nature 316:33–37

Bonatti E, Cipriani A, Lupi L (2015) The Red Sea: birth of an ocean. In: Rasul NMA, Stewart ICF (eds) The Red Sea: the formation, morphology, oceanography and environment of a young ocean basin. Springer Earth System Sciences, Berlin, pp 29–44. https://doi.org/10.1007/978-3-662-45201-1_2

Bosworth W (2015) Geological evolution of the Red Sea: historical background, review, and synthesis. In: Rasul NMA, Stewart ICF (eds) The Red Sea: the formation, morphology, oceanography and environment of a young ocean basin. Springer Earth System Sciences, Berlin, pp 45–78. https://doi.org/10.1007/978-3-662-45201-1_3

Bosworth W, Stockli DF (2016) Early magmatism in the greater Red Sea rift: timing and significance. Canadian J Earth Sci. https://doi.org/10.1139/cjes-2016-0019

Bosworth W, Huchon P, McClay K (2005) The Red Sea and Gulf of Aden basins. J Afr Earth Sci 43:334–378

Brueckner HK, Bonatti E, Elhaddad MA, Hamelin B, Kröner A, Reisberg L, Seyler M (1996) A Nd, Sr, Pb and Os study of the gneisses and ultramafic rocks of Zabargad Island, Red Sea: Miocene Moho or Pan African peridotites. J Geophys Res 100(B11):22283–22297

Burke K, Sengor C (1985) Tectonic escape in the evolution of the continental crust. In: Barazangi M, Brown L (eds) Reflection seismology: the continental crust. American geophysical union geodynamics series, vol 14, pp 41–53

Claesson S, Pallister JS, Tatsumoto M (1984) Samarium-neodymium data on two late Proterozoic ophiolites of Saudi Arabia and implications for crustal and mantle evolution. Contrib Mineral Petrol 85:244–252

Cochran JR (1983) A model for development of Red Sea. Bull Am Assoc Petrol Geol 67:41–69

Cochran JR, Martinez F (1988) Evidence from the northern Red Sea on the transition from continental to oceanic rifting. Tectonophysics 153:25–53

Coleman RG (1993) Geologic evolution of the Red Sea. Oxford Monographs on Geology and Geophysics. Oxford University Press, New York, 186 p

Coleman RG, Brown GF, Keith TEC (1972) Layered gabbros in southwest Saudi Arabia. U.S. Geological Survey Professional Paper 800-D:D143–D150

Coleman RG, Hadley DG, Fleck RJ, Hedge CT, Donato MM (1979) The Miocene Tihama Asir ophiolite and its bearing on the opening of the Red Sea. In: Al-Shanti AMS (ed) Evolution and mineralization of the Arabian-Nubian Shield. Pergamon Press, New York, pp 173–187

Collenette P, Grainger DJ (1994) Kyanite and andalusite. In: Collenette P, Grainger DJ (eds) Mineral resources of Saudi Arabia. Saudi Arabian Directorate General of Mineral Resources DGMR Special Publication SP2, pp 149–155

d'Almeida GAF (2010) Structural evolution history of the Red Sea rift. Geotectonics 44:271–282

DeMets C, Gordon RG, Argus DF (2010) Geologically current plate motions. Geophys J Int 181:1–80

DGMR (1977) Red Sea research 1970–1975. In: Saudi Arabian directorate general of mineral resources bulletin, vol 22, 223 p

Dixon TH, Ivins ER, Franklin BJ (1989) Topographic and volcanic asymmetry around the Red Sea: constraints on rift models. Tectonics 8:1193–1216

Drury SA, de Souza Filho CR (1988) Neoproterozoic terrane assemblages in Eritrea: review and prospects. J Afr Earth Sci 27:331–348

Duncan IJ, Rivard B, Arvidson RE, Sultan M (1990) Structural interpretation and tectonic evolution of the Najd Shear Zone (Saudi Arabia) using Landsat thematic-mapper data. Tectonophysics 178:309–315

El-Gaby S, List FK, Tehrani R (1988) Geology, evolution and metallogenesis of the Pan-African Belt in Egypt. In: El Gaby S, Greiling R (eds) The Pan-African Belt of NE Africa and adjacent areas, tectonic evolution and economic aspects. Vieweg, Braunschweig, Wiesbaden, pp 17–68

El-Gaby S, List FK, Tehrani R (1990) The basement complex of the Eastern Desert and Sinai. In: Said R (ed) The geology of Egypt. Balkema, Rotterdam, pp 175–184

Fritz H, Dallmeyer DR, Wallbrecher E, Loizenbauer J, Hoinkes G, Neumayr P, Khudeir AA (2002) Neoproterozoic tectonothermal evolution of the Central Eastern Desert, Egypt: a slow velocity tectonic process of core complex exhumation. J Afr Earth Sci 34:137–155

Fritz H, Abdelsalam M, Ali K, Bingen B, Collins AS, Fowler AR, Ghebreab W, Hauzenberger CA, Johnson P, Kusky T, Macey P, Muhongo S, Stern RJ, Viola G (2013) Orogen styles in the East African Orogens: a review of the Neoproterozoic to Cambrian tectonic evolution. J Afr Earth Sci 86:65–106

Gaina C, Gernigon L, Ball P (2009) Palaeocene-recent plate boundaries in the NE Atlantic and the formation of the Jan Mayen microcontinent. J Geol Soc London 66:606–616

Gettings ME, Blank HR, Mooney WD, Healey JH (1986) Crustal structure of southwestern Saudi Arabia. J Geophys Res 91:6491–6512

Ghebreab W, Talbot CJ, Page L (2005) Time constraints on exhumation of the East African Orogen from field observations and $^{40}Ar/^{39}Ar$ cooling ages of low-angle mylonites in Eritrea, NE Africa. Precamb Res 139:20–41

Gillman M (1968) Preliminary results of a geological and geophysical reconnaissance of the Jizan coastal plain in Saudi Arabia. In: AIME regional tech symposium rep 2d, Dhahran, pp 198–208

Girdler RW (1970) A review of Red Sea heat flow. Phil Trans R Soc Ser A 267:191–203

Girdler RW, Evans TR (1977) Red Sea heat flow. Geophys J R Astron Soc 51:245–252

Girdler RW, Styles P (1974) Two stage seafloor spreading. Nature 247:7–11

Golan, T, Katzir Y, Coble MA (2017) Early Carboniferous anorogenic magmatism in the Levant: implications for rifting in northern Gondwana. Int Geol Rev 60. https://doi.org/10.1080/00206814.2017.1326089

Goldberg M, Beyth M (1991) Tiran Island: an internal block at the junction of the Red Sea and Dead Sea transform. Tectonophysics 198:261–273

Greenwood WR, Anderson RE (1977) Palinspastic map of the Red Sea area prior to Miocene sea-floor spreading. In: Saudi Arabian directorate general of mineral resources bulletin, vol 22, pp Q1–Q6

Habib ME, Ahmed AA, El Nady OM (1985) Tectonic evolution of the Meatiq infrastructure, Central Eastern Desert, Egypt. Tectonics 4:613–627

Hall SA (1989) Magnetic evidence for the nature of the crust beneath the southern Red Sea. J Geophys Res 94:12267–12279

Hall SA, Andresen GE, Girdler RW (1977) Total-intensity magnetic anomaly map of the Red Sea and adjacent coastal area, a description and preliminary interpretation. In: Saudi Arabia directorate general of mineral resources bulletin, vol 22, pp F1–F13

Hamimi Z, El-Sawy EK, El-Fakharani A, Matsah M, Shujoon A, El-Shafei MK (2014) Neoproterozoic structural evolution of the NE-trending Ad-Damm Shear Zone, Arabian Shield, Saudi Arabia. J Afr Earth Sci 99:51–63

Handy MR, Brun J-P (2004) Seismicity, structure and strength of the continental lithosphere. Earth Planet Sci Lett 223:427–441

Hansen S, Schwartz S, Al-Amri A, Rodgers A (2006) Combined plate motion and density-driven flow in the asthenosphere beneath Saudi Arabia: evidence from shear-wave splitting and seismic anisotropy. Geology 34:869–872

Hansen SE, Rodgers AJ, Schwartz SY, Al-Amri AMS (2007) Imaging ruptured lithosphere beneath the Red Sea and Arabian Peninsula. Earth Planet Sci Lett 259:256–265

Hargrove US (2006) Crustal evolution of the Neoproterozoic Bi'r Umq suture zone, Kingdom of Saudi Arabia. Geochronological, isotopic, and geochemical constraints. Unpublished PhD thesis, University of Texas at Dallas, 343 p

Hosny A, Nyblade A (2016) Crustal structure of Egypt from Egyptian National Seismic Network data. Tectonophysics 687:257–267

Ilani S, Harlavan Y, Tarawneh K, Rabba I, Weinberger R, Ibrahim K, Peltz S, Steinitz G (2001) New K-Ar ages of basalts from the Harrat ash Shaam volcanic field in Jordan: implications for the span and duration of the upper-mantle upwelling beneath the western Arabian plate. Geology 29:171–174

Jackson J (2002) Strength of the continental lithosphere: time to abandon the jelly sandwich? GSA Today, September, pp 4–9

Japsen P, Chalmers JA, Green PF, Bonow JM (2011) Elevated, passive continental margins: not rift shoulders, but expressions of episodic, post-rift burial and exhumation. Glob Planet Change. https://doi.org/10.1016/j.gloplacha.2011.05.004

Johnson PR, Andersen A, Collins AS, Fowler AR, Fritz H, Ghebreab W, Kusky T, Stern RJ (2011) Late Cryogenian-Ediacaran history of the Arabian-Nubian Shield: a review of depositional, plutonic, structural, and tectonic events in the closing stages of the northern East African Orogen. J Afr Earth Sci 61:167–232

Kearey P, Klepeis KA, Vine FJ (2009) Global tectonics. Wiley-Blackwell, Chichester, UK, p 482

Kennedy A, Kozdroj W, Kattan FH, Kodzroj MZ, Johnson PR (2010) SHRIMP geochronology in the Arabian Shield (Midyan terrane, Afif terrane, Ad Dawadimi terrane) and Nubian Shield (Central Eastern Desert terrane) Part IV: data acquisition 2008. Saudi Geological Survey Open-File Report SGSOF-2010–10, 101 p

Khalil SM, McClay KR (2001) Tectonic evolution of the NW Red Sea-Gulf of Suez rift system. Geol Soc Spec Publ London 187:453–473

Khalil SM, McClay KR (2002) Extensional fault-related folding, northwestern Red Sea. Egypt. J Struc Geol 24:743–762

Kozdroj W, Kattan FH, Kadi KA, Al Alfy ZFA, Oweiss KA, Mansour MM (2011) SGS-EMRA Project for Trans-Red Sea Correlation between the Central Eastern Desert Terrane (Egypt) and Midyan Terrane (Saudi Arabia). Saudi Geological Survey Open-File Report SGS-TR-2011-5

Kusznir NJ, Park RG (1987) The extensional strength of continental lithosphere: its dependence on geothermal gradient and crustal thickness and composition. Geol Soc Spec Publ London 28:35–52

Lazar M, Ben-Avraham Z, Garfunkel Z (2012) The Red Sea—new insights from recent geophysical studies and the connection to the Dead Sea fault. J Afr Earth Sci 68:96–110

Levin V, Park J (2000) Shear zones in the Proterozoic lithosphere of the Arabian Shield and the nature of the Hales discontinuity. Tectonophysics 323:131–148

Ligi M, Bonatti E, Bortoluzzi G, Cipriani A, Cocchi L, Tontini FC, Carminati E, Ottolini L, Schettino A (2012) Birth of an ocean in the Red Sea: initial pangs. Geochem Geophys Geosyst 13. https://doi.org/10.1029/2012gc004155

Ligi M, Bonatti E, Rasul NMA (2015) Seafloor spreading initiation: geophysical and geochemical constraints from the Thetis and Nereus deeps, central Red Sea. In: Rasul NMA, Stewart ICF (eds) The Red Sea: the formation, morphology, oceanography and environment of a young ocean basin. Springer Earth System Sciences, Berlin, pp 79–98. https://doi.org/10.1007/978-3-662-45201-1_4

Manatschal G (2004) New models for evolution of magma-poor rifted margins based on a review of data and concepts from west Iberia and the Alps. Int J Earth Sci 93:432–466

Martinez F, Cochran JR (1988) Structure and tectonics of the northern Red Sea: catching a continental margin between rifting and drifting. Tectonophysics 150:1–31

McGuire AV, Coleman RG (1986) The Jabal Tirf layered gabbro and associated rocks of the Tihama Asir Complex, SW Saudi Arabia. J Geol 94:651–665

McKenzie D (1978) Some remarks on the development of sedimentary basins. Earth Planet Sci 40:25–32

Meinhold G, Morton AC, Avigad D (2013) New insights into peri-Gondwana paleogeography and the Gondwana super-fan system from detrital zircon U-Pb ages. Gondwana Res 23:661–665

Menzies MA, Klemperer SL, Ebinger CJ, Baker J (2002) Characteristics of volcanic rifted margins. Geol Soc Am Spec Paper 362:1–14

Meyer SE, Passchier C, Abu-Alam T, Stüwe K (2014) A strike-slip core complex from the Najd fault system, Arabian Shield. Terra Nova 26:387–394

Miller MM, Dixon TH (1992) Late Proterozoic evolution of the N part of the Hamisana zone, NE Sudan: constraints on Pan-African accretionary tectonics. J Geol Soc London 149:743–750

Mooney WD, Gettings ME, Blank HR, Healy JH (1985) Saudi Arabian seismic-refraction profile: a traveltime interpretation of crustal and upper mantle structure. Tectonophysics 111:173–246

Mutter JC, Talwani M, Stoffa PL (1982) Origin of seaward-dipping reflectors in oceanic crust off the Norwegian margin by "subaerial sea-floor spreading". Geology 10:353–357

Orszag-Sperber F, Harwood G, Kendall A, Purser BH (1998) A Review of the Evaporites of the Red Sea-Gulf of Suez rift. In: Purser BH, Bosence DWJ (eds) Sedimentation and Tectonics of Rift Basins: Red Sea-Gulf of Aden. Chapman and Hall, London, pp 409–426

Pallister JS (1986) Geologic map of the Al Lith Quadrangle, Sheet 20D, Kingdom of Saudi Arabia. Saudi Arabian Deputy Ministry for Mineral Resources Geoscience Map GM-95

Pallister JS (1987) Magmatic History of Red Sea rifting: perspectives from the central Saudi Arabian coastal plain. Geol Soc Am Bull 98:400–417

Pallister JS, Stacey JS, Fischer LB, Premo WR (1988) Precambrian ophiolites of Arabia: geologic settings, U-Pb geochronology, Pb-isotopes characteristics, and implications for continental accretion. Precamb Res 38:1–54

Pallister JS, McCausland WA, Jónsson S, Lu Z, Zahran HM, Hadidy SE, Aburukbah A, Stewart ICF, Lundgren PR, White RA, Moufti MRH (2010) Broad accommodation of rift-related extension recorded by dyke intrusion in Saudi Arabia. Nat Geosci 3:705–712

Park Y, Nyblade AA, Rodgers AJ, Al-Amri A (2008) S wave velocity structure of the Arabian Shield upper mantle from Rayleigh wave tomography. Geochem Geophys Geosyst 9. https://doi.org/10.1029/2007gc001895

Pearce JA, Bender JF, De Long SE, Kidd WSF, Low PJ, Güner Y, Saroğlu R, Yilmaz Y, Moorbath S, Mitchell JG (1990) Genesis of collision volcanism in Eastern Anatolia, Turkey. J Volcan Geotherm Res 44:189–229

Powell JH, Abed AM, Le Nindre Y-M (2014) Cambrian stratigraphy of Jordan. GeoArabia 19:81–134

Prodehl C (1985) Interpretation of a seismic-refraction survey across the Arabian shield in western Saudi Arabia. Tectonophysics 111:247–282

Rasul NMA, Stewart ICF (2015) The Red Sea: the formation, morphology, oceanography and environment of a young ocean basin. Springer Earth System Sciences, Berlin, 638 p

Rasul NMA, Stewart ICF, Nawab ZA (2015) Introduction to the Red Sea: its origin, structure, and environment. In: Rasul NMA, Stewart ICF (eds) The Red Sea: the formation, morphology, oceanography and environment of a young ocean basin. Springer Earth System Sciences, Berlin, pp 1–28. https://doi.org/10.1007/978-3-662-45201-1_1

Reilinger R, McClusky S, Ar Rajehi A (2015) Geodetic constraints on the geodynamic evolution of the Red Sea. In: Rasul NMA, Stewart ICF (eds) The Red Sea: the formation, morphology, oceanography and environment of a young ocean basin. Springer Earth System Sciences, Berlin, pp 135–149

Roeser HA (1975) A detailed magnetic survey of the southern Red Sea. Geol Jahrb D13:131–153

Roobol MJ, Stewart ICF (2009) Cenozoic faults and recent seismicity in northwest Saudi Arabia and the Gulf of Aqaba region. Saudi geological survey technical report SGS-TR-2008-7, 35 p

Saleh S, Jahr T, Jentzsch G, Saleh A, Abou Ashour NM (2006) Crustal evaluation of the northern Red Sea rift and Gulf of Suez, Egypt from geophysical data: 3-dimensional modeling. J Afr Earth Sci 45:257–278

Schardt C (2016) Hydrothermal fluid migration and brine pool formation in the Red Sea: the Atlantis II Deep. Miner Deposita 51:89–111

Sebai A, Zumbo V, Férand G, Bertrand H, Hussain AG, Giannérini G, Campredon R (1991) 40Ar/39Ar dating of alkaline and tholeiitic magmatism of Saudi Arabia related to the early Red Sea rifting. Earth Planet Sci Lett 104:473–487r

Shimron AE (1989) The Red Sea Line—a Late Proterozoic transcurrent fault. J Afr Earth Sci 11:95–112

Squire RJ, Campbell IH, Allen CM, Wilson CJL (2006) Did the Transgondwanan Supermountain trigger the explosive radiation of animals on Earth? Earth Planet Sci Lett 250:116–133

Stern RJ (1985) The Najd Fault System, Saudi Arabia and Egypt: a late Precambrian rift-related transform system. Tectonics 4:497–511

Stern RJ (1994) Neoproterozoic (900-550 Ma) arc assembly and continental collision in the East African Orogen. Ann Rev Earth Planet Sci 22:319–351

Stern RJ (2002) Crustal evolution in the East African Orogen: a neodymium isotopic perspective. J Afr Earth Sci 34:109–117

Stern RJ (2017) Neoproterozoic formation and evolution of Eastern Desert continental crust—the importance of the infrastructure-superstructure transition. J African Earth Sci. https://doi.org/10.1016/jafreasci.2017.01.001

Stern RJ, Johnson PR (2010) Continental lithosphere of the Arabian Plate: a geologic, petrologic, and geophysical synthesis. Earth-Sci Rev 101:29–67

Stern RJ, Kröner A, Manton WI, Reischmann T, Mansour MM, Hussein IM (1989a) Geochronology of the late Precambrian Hamisana shear zone, Red Sea Hills, Sudan and Egypt. J Geol Soc London 145:1017–1029

Stern RJ, Kröner A, Manton WI, Reischmann T, Mansour M, Hussein IM (1989b) Geochronology of the late Precambrian Hamisana shear zone, Red Sea Hills, Sudan and Egypt. J Geol Soc London 146:1017–1029

Stern RJ, Nielsen KC, Best E, Sultan M, Arvidson RE, Kröner A (1990) Orientation of late Precambrian sutures in the Arabian Nubian shield. Geology 18:1103–1106

Stern RJ, Johnson PJ, Kröner A, Yibas B (2004) Neoproterozoic ophiolites of the Arabian-Nubian Shield. In: Kusky T (ed) Precambrian Ophiolites. Elsevier, pp 95–128

Stern RJ, Ren M, Ali K, Förster H-J, Al Safarjalani A, Nasir S, Whitehouse MJ, Leybourne MI (2014) Early Carboniferous (~357 Ma) crust beneath northern Arabia: tales from Tell Thannoun (southern Syria). Earth Planet Sci Lett 393:83–93

Stern RJ, Ali KA, Ren M, Jarrar GH, Romer RL, Leybourne MI, Whitehouse MJ, Ibrahim KM (2016) Cadomian (~ 560 Ma) crust buried beneath the northern Arabian Peninsula: mineral, chemical, geochronological, and isotopic constraints from NE Jordan xenoliths. Earth Planet Sci Lett 436:31–42

Stewart ICF, Johnson PR (1994) Satellite gravity and Red Sea tectonics. Saudi Arabian Deputy Ministry for Mineral Resources Open-File Report USGS-OF-94-10, 22 p

Stoeser DB, Frost CD (2006) Nd, Pb, Sr, and O isotopic characterization of Saudi Arabian Shield terranes. Chem Geol 226:163–188

Suayah IB, Rogers JJW, Dabbagh ME (1991) High-Ti continental tholeiites from the Aznam trough, northwestern Saudi Arabia: evidence of "abortive" rifting in the "embryonic" stage of Red Sea opening. Tectonophysics 191:75–87

Sultan M, Arvidson RE, Duncan IJ, Stern RJ, El Kaliouby B (1988) Extension of the Najd shear system from Saudi Arabia to the Central Eastern Desert of Egypt based on integrated field and Landsat observations. Tectonics 7:1291–1306

Sultan M, Becker M, Arvidson RE, Shore P, Stern RJ, El Alfy Z, Attia RI (1993) New constraints on Red Sea rifting from correlations of Arabian and Nubian Neoproterozoic outcrops. Tectonics 12:1303–1319

Szymanski E (2013) Timing, kinematics, and spatial distribution of Miocene extension in the central Arabian margin of the Red Sea rift system. PhD dissertation, U Kansas, 430 p

Szymanski E, Stockli DF, Johnson PR (2012) Evidence for an early and sustained mode of diffuse lithospheric extension in the central Arabian flank of the Red Sea rift system: implications for margin structural; kinematics and basin development. In: The American association of petroleum geologists annual convention and exhibition, Long Beach, CA, abstract no. 1235423

Szymanski E, Stockli DF, Johnson PR, Hager C (2016) Thermochronometric evidence for diffuse extension and two-phase rifting within the central Arabian margin of the Red Sea Rift. Tectonics 35. https://doi.org/10.1002/2016tc004336

Szymanski E, Stockli DF, Johnson PR, Kattan FH, Al Shammari A (2007) Observations from fieldwork and (U-Th)/He thermochronologic study of the central Arabian flank of the Red Sea rift system. In: The American geophysical union, fall meeting 2007, abstract #T41A-0379

Tang Z, Julià J, Zahran H, Mai PM (2016) The lithospheric shear-wave velocity structure of Saudi Arabia: young volcanism in an old shield. Tectonophysics 680:8–27

Thomas WA (2006) Tectonic inheritance at a continental margin. GSA Today 16(2):4–11. https://doi.org/10.1130/1052-5173(2006)016%3c4:TIAACM%3e2.0.CO;2

Vail JR (1985) Pan-African (Late Precambrian) tectonic terranes and the reconstruction of the Arabian-Nubian shield. Geology 13:839

Voss M, Jokat W (2007) Continent-ocean transition and voluminous magmatic underplating derived from P-wave velocity modeling of the East Greenland continental margin. Geophys Res Lett 170: 580–604

Wegener A (1920) Die Entstehung der Kontinente und Ozeane. Vieweg, Brunswick, p 135

Wolfenden E, Ebinger C, Yirgu G, Renne P, Kelley S (2005) Evolution of a volcanic rifted margin: Southern Red Sea, Ethiopia. Geol Soc Am Bull 117:846–864

Zahran HM, Stewart ICF, Johnson PR, Basahel MH (2003) Aeromagnetic anomaly maps of central and western Saudi Arabia. Saudi Geological Survey Open-File Report SGS-OF-2002-8

Timing of Extensional Faulting Along the Magma-Poor Central and Northern Red Sea Rift Margin—Transition from Regional Extension to Necking Along a Hyperextended Rifted Margin

Daniel F. Stockli and William Bosworth

Abstract

In light of voluminous Oligocene continental flood basalts in the Afar/Ethiopian region, the Red Sea has often been viewed as a typical volcanic rift, despite evidence for asymmetric extension and hyperextended crust (Zabargad Island). An in-depth analysis of the timing, spatial distribution, and nature of Red Sea volcanism and its relationship to late Cenozoic extensional faulting should shed light on some of the misconceptions. Voluminous Eocene to Oligocene flood basalts in northern Ethiopia and western Yemen at ~31–30 Ma were synchronous with the onset of continental extension in the Gulf of Aden, but demonstrably predate Red Sea extensional faulting and rifting. Basaltic dike emplacement, syn-rift subsidence and sedimentation, and rapid rift-related fault block exhumation at ~23 Ma along the entire Red Sea-Gulf of Suez rift system mark the onset of Red Sea rifting. Early Miocene rifting affected a wide area (~1200 km) around the northern Red Sea with limited strain localization along the main rift axis between ~20 and 14 Ma. While the initiation of lithospheric extension in the northern and central Red Sea and Gulf of Suez was accompanied by only sparse basaltic volcanism and possible underplating, the main phase of rifting in the Miocene Red Sea/Gulf of Suez lacks significant rift-related volcanism. There appears to be no evidence for the formation of SDRs or accretion of a thick proto-oceanic crust. Rift localization and major crustal thinning continued throughout the Early Miocene. Middle Miocene onset of left-lateral displacement along the Gulf of Aqaba transform resulted in the tectonic isolation of the Gulf of Suez and a switch from rift-normal to highly oblique extension with the Red Sea. Oblique extension led to the formation of fracture zones, pull-apart basins, and crustal necking, and ultimately local crustal separation and mantle exhumation, prior to Plio-Pleistocene incipient oceanic breakup in the northern Red Sea. This clearly supports the interpretation of the northern Red Sea as a magma-poor rift system and the importance of the Middle Miocene kinematic reorganization for continental breakup.

1 Introduction

The Tertiary Red Sea-Gulf of Suez rift system (Fig. 1) is one of the premier examples for continental rifting and active continental break-up caught in the transition from rift to drift (e.g., Bosworth et al. 2005). Many fundamental concepts in extensional tectonics have been inspired and influenced by Red Sea studies. The Red Sea system has been used as a poster child or prototype of different modes of continental extension and continental crustal and lithospheric break up. Numerous geodynamic rifting models have relied heavily on interpretations derived from the Red Sea rift system (e.g., Bosworth 1994; Bosence 1998; Bosworth and McClay 2001). Research and hydrocarbon exploration in the Red Sea and Gulf of Suez have greatly influenced basic concepts of extensional faulting and continental break-up for apparent orthogonal rift systems, including interaction of normal faulting and continental and marine sedimentation, magmatism and extension (Fig. 1). Interestingly enough, however, the Red Sea has been used as representative end member models for diametrically opposite processes and concepts (Figs. 2 and 3). It has been cited as an example for both prototypical pure-shear symmetric rift systems with limited crustal stretching and attenuation (e.g., Buck et al. 1988; Hosny and Nyblade 2014; Almalki et al. 2015), and for

D. F. Stockli (✉)
Department of Geological Sciences, Jackson School of Geosciences, The University of Texas at Austin, 1 University, Station C9000, Austin, TX 78712-0254, USA
e-mail: stockli@jsg.utexas.edu

W. Bosworth
Apache Egypt Companies, 11 Street 281, New Maadi, Cairo, Egypt

N. M. A. Rasul and I. C. F. Stewart (eds.), *Geological Setting, Palaeoenvironment and Archaeology of the Red Sea*,
https://doi.org/10.1007/978-3-319-99408-6_5

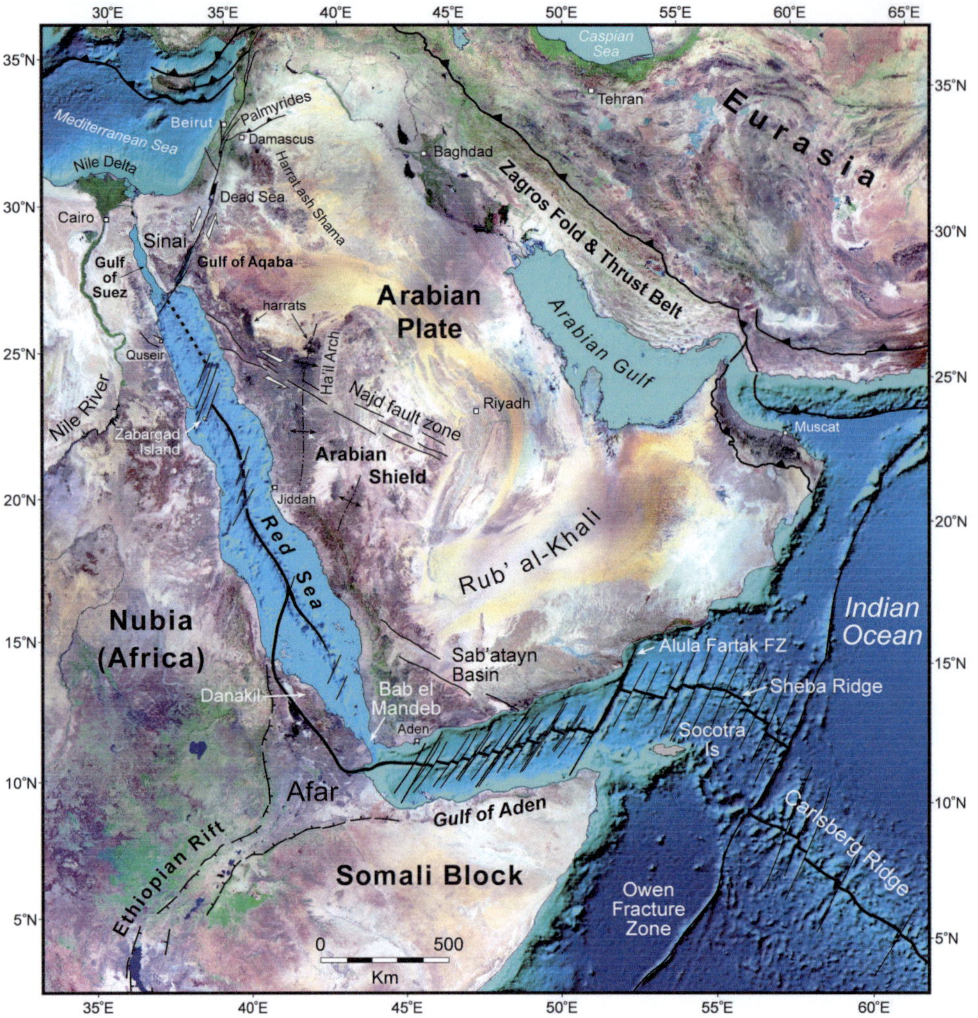

Fig. 1 Overview image (Landsat) of greater Red Sea-Gulf of Suez and Gulf of Aden region, showing divergent, convergent, and transcurrent plate boundaries and tectonic and magmatic elements surrounding the Arabian Plate, including the Zagros Fold and Thrust Belt and the Gulf of Aqaba transform (modified after Bosworth and Stockli 2016)

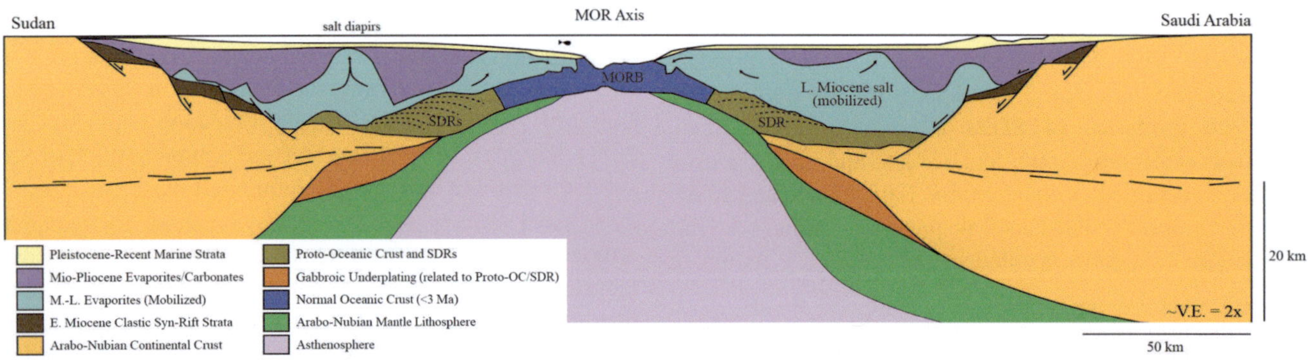

Fig. 2 Red Sea as a magma-rich rift, illustrating conceptual views of a pure-shear dominated symmetrical rift with seaward-dipping normal faults along both margins and limited extent of highly attenuated continental crust. Excess magmatism during continental break-up results in the formation of magmatic seaward dipping reflectors (SDRs) and associated gabbroic underplating and formation of a thick proto-oceanic crust. Note that in this model, proto- and normal oceanic crust stretches from margin to margin and includes an active axial seafloor-spreading centre (partially covered by mobilized salt). The Red Sea, however, lacks documented SDRs, gabbroic underplating, or symmetric syn-extensional rift fault polarities (modified after Mohriak 2014)

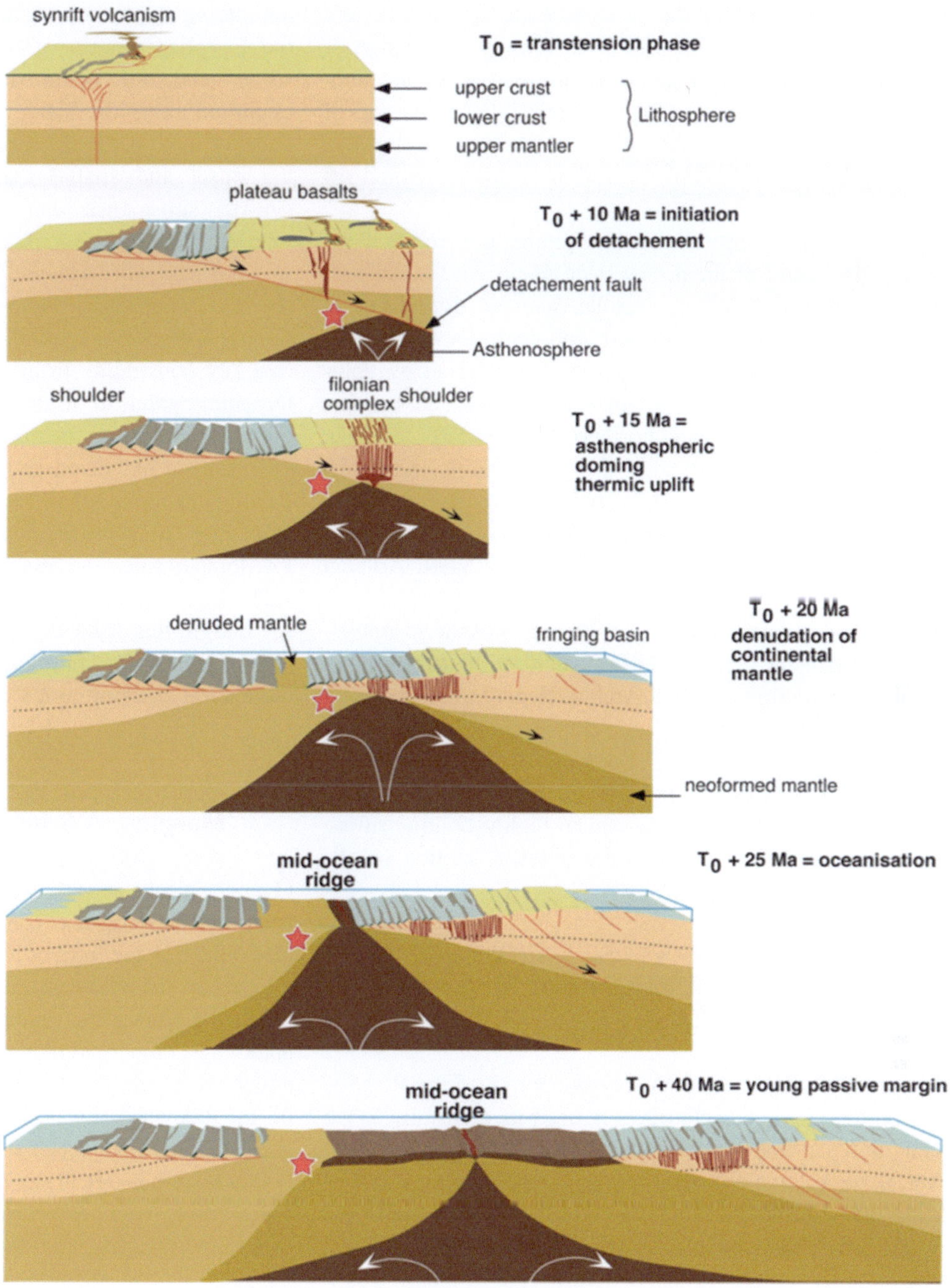

Fig. 3 Schematic rift evolution of asymmetric, magma-poor rifting and break-up of the Red Sea (modified after Stampfli and Marchant 1997), illustrating asymmetric fault-dip polarity across the rift and the future site of crustal separation and seafloor spreading. Asymmetric occurrence of early syn-rift volcanism and dike emplacement is controlled by depth-dependent thinning of the lithospheric mantle with respect to the upper-crustal rift axis. These typical models were inspired by observations in the Red Sea (e.g., Wernicke 1985; Voggenreiter et al. 1988) and asymmetric upper- and lower-plate configurations in other conjugate rifted margins (e.g., Lister et al. 1986)

simple-shear rifting and highly asymmetric rift margins characterized by low-angle normal faults (e.g., Wernicke 1985; Voggenreiter et al. 1988; Bohannon 1989; Tesfaye and Ghebreab 2013). Furthermore, the Red Sea has also been described as a typical active volcanic rift, in part due to the proximity to the Afar hot-spot, and a magma-poor hyperextended rift system (e.g., Mohriak 2014; Colombo et al. 2014; Almalki et al. 2015) (Figs. 2 and 3). Lastly, the Red Sea is commonly viewed as the prototypical orthogonal rift, despite clear evidence for highly oblique rifting after a Middle Miocene kinematic organization linked to the Gulf of Aqaba transform (e.g., Bosworth et al. 2005). Clearly the Red Sea cannot be the prototypical end member for all these contrasting rift scenarios—it cannot be volcanic and magma-poor, orthogonal and oblique, and simple-shear and pure shear at the same time. Such dogmatic abuse of the Red Sea largely stems from the lack of publicly available deep-crustal reflection seismic data, and the ambiguity of rift geometries due to imaging challenges resulting from massive thicknesses of mobile Mio-Pliocene salt are coupled with a lack of reliable information for the timing and amount of rift-related magmatism, and the absence of chronological constraints on the timing of faulting and the temporal and spatial evolution of faulting. While much progress has been made in understanding the plate tectonic framework and modern strain field of the Red Sea, limited knowledge of how extensional strain is geometrically, spatially, and temporally distributed along and across the conjugate margins of the Red Sea has made it difficult to adequately develop and test models for the dynamic evolution of this renowned rift system.

Given the proximity to the Afar plume, the Red Sea has been at the centre of the debate over whether it is the result of active or passive rifting, despite dramatic differences in magmatism between the southern and central/northern Red Sea. While the southern Red Sea is characterized by voluminous Eocene-Oligocene plume- and rift-related basaltic magmatism, the central and northern Red Sea rift system in Saudi Arabia and Egypt/Sudan lacks any significant syn-rift magmatism and appears to be a typical magma-poor continental margin. These data appear to show symmetric, abrupt changes in lithospheric thickness across the Saudi Arabian and Egyptian rift margin consistent with extensional necking and depth-dependent thinning, with asthenospheric upwelling and resultant dynamic surface uplift significantly post-dating rift initiation in Saudi Arabia. Hence, this paper will make the case of the central and northern Red Sea system as a magma-poor continental rift margin resulting from passive rifting and ultimately necking and break-up due to an abrupt kinematic change from orthogonal to highly oblique rifting in the late Miocene, linked to the onset of motion on the Aqaba-Levant transform.

2 The Red Sea—A Volcanic or Magma-Poor Continental Rift

Numerous studies and textbooks have portrayed the Red Sea as a prototypical volcanic rift in light of its proximity to the Afar hotspot and the Ethiopian Rift (e.g., White and McKenzie 1989; Ebinger and Sleep 1998; Mohriak 2014), while other studies have interpreted it as a magma-poor rift system dominated by simple shear (e.g., Wernicke 1985; Voggenreiter et al. 1988; Bosworth and Stockli 2016) (Figs. 2 and 3). Although much progress has been made in understanding the plate tectonic framework and modern strain field of the Red Sea, limited knowledge of how extensional strain is spatially and temporally distributed along the continental margins has made it difficult to adequately evaluate and test models for the evolution of the Red Sea/Gulf of Suez rift system. Traditionally, the timing of growth and exhumation of rift flanks is determined by identifying erosional products within the basin fill. In the Red Sea, however, most of the pre-, syn-, and post- rift sedimentary rocks are either deeply buried within the rift, have been uplifted and eroded away, or are poorly dated due to the scarcity of datable syn-rift Tertiary volcanic rocks, hampering the reconstruction of Red Sea rifting. Although the Egyptian margin of the Red Sea has been studied in some detail, little is known about the timing of Tertiary rift evolution along the Saudi Arabian margin. This study takes a closer look at the rift anatomy, the timing of magmatism, the onset and spatial evolution of extensional faulting, syn-rift subsidence and stratigraphy, and the interplay between faulting, magmatism, and sedimentation in the Gulf of Suez and northern and central Red Sea, to investigate whether the northern Red Sea is indeed an active volcanic or a magma-poor continental rift system, and tries to summarize key constraints from several lines of evidence for the temporal evolution of the northern Red Sea to shed light on the onset of rifting, the nature of rifting, and the progressive evolution of extension and trans-tensional faulting possibly leading to continental break-up and incipient sea-floor spreading. In addition, it also incorporates extensive new low-temperature thermochronometric constraints on the tectonic evolution of the Red Sea rift flank and the proximal rift margin that shed light on the timing, origin, and geometry of extensional faulting and rift flank exhumation along the central and northern Red Sea margin in Saudi Arabia.

2.1 Magma-Poor Versus Magma-Rich Rift Margins: A Brief Review

Over the past decades, fundamental concepts of the understanding of rifting and continental break-up have been revolutionized in the light of new geophysical data, in particular long-offset reflection and refraction seismic data across rifted continental margins, geological studies or exhumed fossil analogues, numerical modeling on the lithospheric scale, and advances in analytical techniques, allowing reconstructions of the temporal and thermal evolution of rifted margins (Fig. 4) (e.g., Manatschal 2004; Lavier and Manatschal 2006; Unternehr et al. 2010). While continental rifts have traditionally been categorized as active and passive rifts, these new insights have led to the depiction of continental margins as end members, based on their magmatic and kinematic evolution—namely as (1) magma-rich, volcanic margins (e.g., SE Greenland, Argentina), (2) magma-poor, hyperextended margins (e.g., Galicia, Newfoundland), and (3) shear margins (e.g., Equatorial Atlantic). Besides the magma-rich versus magma-poor rift controversy, debate also continues on whether crustal and/or lithospheric separation in fact has actually occurred in the northern Red Sea. While there is clear evidence for seafloor spreading in the southern and central parts of the Red Sea since the early Pliocene at $\sim$4.5 Ma, evidence for seafloor spreading in the northern Red Sea is more controversial largely due to very thick salt, obscuring the subsalt geology (e.g., Augustin et al. 2014; Ligi et al. 2018).

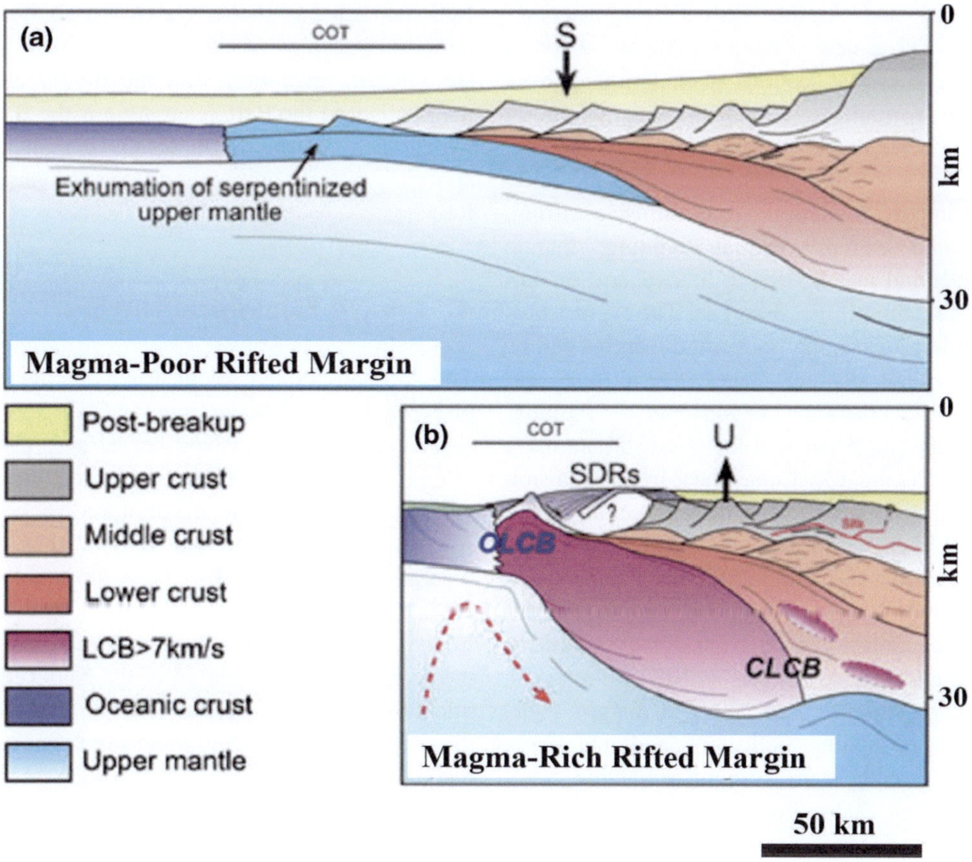

Fig. 4 Schematic depictions of end-member style rifted margins. **a** Magma-poor rifted margins are characterized by a lack of voluminous syn-rift magmatism, resulting in an initial stretching phase, followed by lithospheric necking and hyperextension, crustal separation and lithospheric mantle exhumation during the exhumation phase, and eventual inception of sputtering seafloor spreading. **b** Magma-rich rifted margins are characterized by voluminous magmatism during early stages of rifting and mechanical thinning of the lithosphere, leading to the generation of thick proto-magmatic crust and seaward-dipping reflectors (SDRs) as well as collocated gabbroic underplating. This culminates in an abrupt, magmatically triggered crustal and lithospheric separation and rapid establishment of normal sea-floor spreading. CLCB is continental lower-crustal body, OLCB is oceanic lower-crustal body, S is the locus of post-breakup subsidence, and U is the locus of isostatic uplift, COT is continent-ocean transition. Modified after Gernignon et al. (2005)

2.2 Magma-Rich Margins

Classic volcanic margins of the Atlantic, such as SE Greenland, the Hatton Bank or Pelotas margins, are generally characterized by relatively modest crustal and lithospheric attenuation and hence a narrow continental shelf, an abrupt ocean-continent boundary, voluminous flood basalts (onshore), seaward-dipping reflectors (SDRs) offshore, massive gabbroic underplating, often considered as the possible explanation of "high-velocity bodies", anomalously thick proto-oceanic crust, and a bathymetric step down toward steady-state oceanic crust (Figs. 2 and 4) (e.g., White et al. 1987; White and McKenzie 1989; Gladczenko et al. 1997; Hopper et al. 2003; Stica et al. 2014). Thermal erosion of the mantle lithosphere, diking, and thermal weakening of the lithosphere will lead to relatively rapid continental break-up. The typical progressive rift evolution for volcanic rifted margins can be subdivided into several distinct stages: (1) diffuse and variable-magnitude crustal extensional faulting, possibly associated with regional updoming, (2) effusion of voluminous flood basalts and basaltic underplating, (3) abrupt magma-assisted strain-localization after relatively modest pure-shear lithospheric stretching, (4) magmatically-driven crustal and lithospheric separation, and (5) formation of seaward dipping reflectors parallel to the margin and formation of a thick proto-oceanic crust and transition to steady-state seafloor spreading (e.g., Holbrook and Kelemen 1993; Ebinger and Casey 2001).

2.3 Magma-Poor Margins

The understanding of the evolution of rifting and continental break-up in the absence of copious magmatism has been revolutionized over the past years in light of new concepts developed for magma-poor rift margins such as the

Iberia-Newfoundland and Alpine Tethys margins (e.g., Manatschal 2004). These new concepts integrate observations from mantle and crustal sections at modern and fossil continental margins, formulating a temporal strain evolution for magma-poor hyperextended margins that is compatible with field observations and numerical models that elegantly describes the geometric evolution from diffusive rifting, crustal necking and extreme crustal thinning, mantle exhumation, to eventual magma-driven lithospheric separation to sea floor spreading (Figs. 3 and 4). The typical progressive rift evolution for magma-poor rifted margins can be subdivided into several distinct stages: (1) diffuse and variable-magnitude crustal extensional faulting, (2) strain-localization and upper and lower-crustal coupling resulting in crustal necking, (3) simple-shear hyperextension and extreme crustal thinning seaward of the necking zone, ultimately resulting in crustal separation, (4) exhumation and unroofing of sub-continental mantle after crustal separation, and (5) sputtering initial and transition to steady-state seafloor spreading. Magmatism can interrupt this evolutionary chain at any time and lead to magma-assisted lithospheric rupture or the superposition of a magma-rich margin, characterized by seaward-dipping reflectors and magmatic breakup leading to steady state seafloor spreading (e.g., Lavier and Manatschal 2006; Péron-Pinvidic and Manatschal 2009; Sutra and Manatschal 2012; Tugend et al. 2014).

3 Tectonic Evolution of the N Red Sea

The northern Red Sea margin has undergone a geological and tectonic evolution that reflects interactions between faulting, exhumation, erosion, and volcanism and that manifests itself both in its geomorphologic expression and tectonic anatomy (Fig. 1). The major landscape features of the Red Sea margins comprise a coastal plain that is 0–50 km wide, an erosional escarpment ~50–70 km inland from the present-day coast, and a region of hills between the coastal plain and escarpment that is the remnant of erosional retreat of the escarpment and contains much of the proximal margin extensional structural and stratigraphic record. Structurally, the margin includes (1) the Cenozoic Red Sea basin, which underlies the Red Sea and most of the coastal plain; (2) variably extended and attenuated continental crust composed of Precambrian rocks of the Arabian and Nubian shields; (3) a border fault system that separated highly extended crustal from structurally relatively intact and unfaulted continental crust; (4) extensional half-grabens inboard from the Red Sea border fault system that contain pre-rift sedimentary strata and/or Cenozoic syn-rift sedimentary strata and reflect early distributed crustal extension; (5) Cenozoic mafic dikes that trend parallel to the axis of the Red Sea rift and sparse syn-rift basalt flows. The detailed temporal interplay of extensional faulting, pre- and post-rift stratigraphic evolution, and the timing and nature of rift magmatism in the Red Sea margin is the focus of this study, which seeks to shed light on the overall crustal and lithospheric processes governing extension and break-up in the Red Sea.

4 Pre-rift Geological History

The Arabian-Nubian shield was assembled during the Neoproterozoic East African orogeny (715–530 Ma) through the accretion of multiple island arc terranes (Stoeser and Camp 1985). The crystalline basement is composed of a wide variety of igneous and metamorphic rocks (e.g., Stern and Johnson 2010; Johnson and Kattan 2012). The pre-rift sedimentary sequence that unconformably overlies the Arabian-Nubian basement ranges in age from Cambrian to Eocene. The Triassic to Eocene strata were deposited along the southern passive margin of the Neo-Tethyan, attaining a thickness of >2 km in the Gulf of Suez/Sinai region and thinning to a few hundred metres in the northern Red Sea. In the southern Red Sea, the marine Jurassic is >3 km thick in Ethiopia (Hutchinson and Engels 1970), but pinches out in central Eritrea (Gillman 1968; Beyth 1972) and is overlain by thin Cretaceous to Eocene fluvial-lacustrine strata. In Sudan, little is preserved of the thin pre-rift section that consists of Upper Cretaceous to Paleocene marine marls overlain by non-marine sandstones (Carella and Scarpa 1962).

On the Saudi Arabian side of the Red Sea, the marine and non-marine pre-rift successions have largely been eroded and are only preserved in down-faulted Cenozoic extensional grabens and half-grabens, such as in the Midyan, Azlam, and Jeddah regions (Bohannon 1986; Pallister 1987; Bayer et al. 1988; Bohannon et al. 1989). Minor marginal marine deposits indicate the Arabian coast was near sea level during most of the Late Cretaceous to Oligocene, with thick pre-rift laterite horizons indicating a low-relief topography (e.g., Pallister 1987; Bohannon et al. 1989). The overall picture of the Red Sea prior to the onset of Cenozoic rifting is that of a low relief, low elevation continental to marine realm. Contrary to earlier interpretations (e.g., Gass 1970; Burke and Dewey 1973; Garson and Krs 1976) there appears to be little to no evidence for large scale pre-rift doming or early Cenozoic faulting (Fig. 5). Subsequent uplift related to the rifting in the Red Sea has caused erosion of much of the pre-rift sedimentary strata, which are only preserved in rift structures or as small outliers with Neoproterozoic basement and Paleozoic strata exposed over a broad zone flanking the rift and its inboard areas. However, apparent major thickness variations in Carboniferous strata in the Gulf of Suez and Cretaceous strata in the Azlam and Jeddah areas suggest the

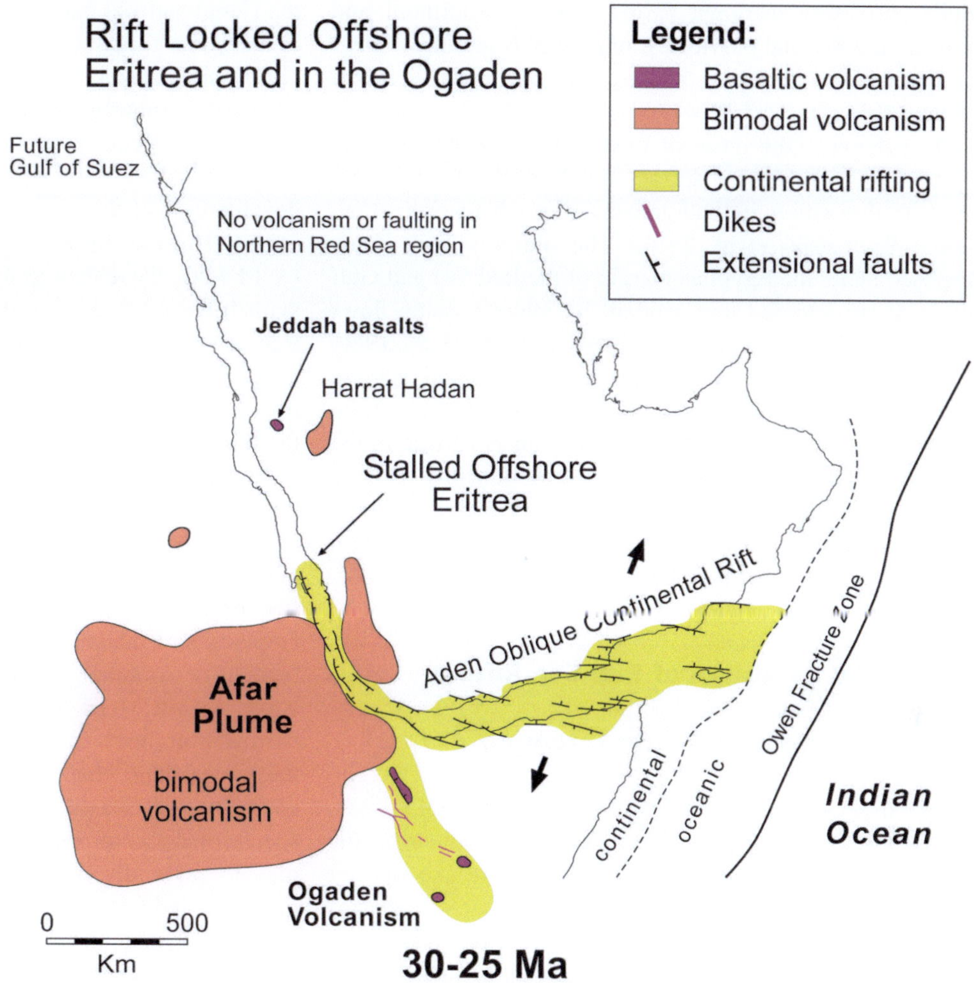

Fig. 5 Tectonic map of the Red Sea-Gulf of Aden rift system in the Oligocene (30–25 Ma), illustrating the location of rifting in the Gulf of Aden and the southernmost Red Sea area and the eruptive footprint of continental flood basalts linked to the Afar Plume. Note the general lack of volcanism or faulting in the future central and northern Red Sea and Gulf of Suez corridor

possible presence of precursor rift structures underlying the Neogene Red Sea rift system, but more work remains to be done to elucidate such structures.

5 Timing of Onset of Red Sea Rifting

In order to elucidate the dynamics of rifting and rupturing of the continental lithosphere in the Red Sea, it is imperative to understand the temporal evolution of rifting, sedimentation, subsidence, and magmatism along both margins of the northern and central Red Sea and the Gulf of Suez (Fig. 1). Knowledge of the temporal and spatial distribution of crustal extension and the timing of flexural amplification and exhumation of a rift flank is critical to evaluating the role of processes such as active versus passive asthenospheric upwelling (Sengör and Burke 1978), secondary convection (Buck 1986), and flexural unloading of the crust (Weissel and Karner 1989), as well as the distribution of sub-crustal lithospheric extension relative to crustal thinning (Lister et al. 1986, 1991). A flexural origin of the uplift predicts uplift coeval with extension, while thermal and magmatic processes predict later and continued uplift.

The timing of growth and exhumation of rift flanks is primarily determined by identification of erosional products within the basin fill and rift-related magmatism (e.g., Evans 1988). In the Red Sea most of the pre-, syn-, and post- rift sedimentary rocks are often deeply buried within the rift, have been uplifted and eroded away, or are poorly dated due to the scarcity of Cenozoic syn-rift volcanic rocks within non-marine strata with poor biostratigraphic control, traditionally hampering attempts to adequately reconstruct the rifting history along the entire margin of the Red Sea. However, the regional tectonic framework of the early stages of rifting is best constrained by the sedimentary record exposed in the Gulf of Suez, which initially formed the northward continuation of the Red Sea until the Middle Miocene and structural isolation by the Aqaba-Levant transform (Tamsett 1984; Steckler and ten Brink 1986). Uplift of marginal basins during rift localization has resulted in widespread exposure of rift sediments. Exploration wells have penetrated much rift section throughout the Gulf of

Suez and other marginal basins along the northern and central Red Sea and provided strong insights into the syn-rift stratigraphic evolution of the Red Sea and Gulf of Suez rift system (Fig. 1).

In summary, the onset of major rifting in the Red Sea system was marked by diffuse initial normal faulting and syn-rift deposition in half-grabens (e.g., Bosworth et al. 2005; Szymanski et al. 2016). The oldest syn-rift strata exposed along the Egyptian margin of the Red Sea and Gulf of Suez are intruded and overlain by minor basaltic flows dated at ~23 Ma (e.g., Steen 1984; Roussel et al. 1986; Bosworth 1995; Bosworth and Stockli 2016). Based on these stratigraphic data, the minimum age for the onset of rifting is ~25 Ma for the latitude of Hurgada and ~23 Ma for the central Gulf of Suez (Figs. 1 and 6). The following sections discuss the structural, stratigraphic, and magmatic evolution in some detail in order to address the aforementioned geodynamic questions surrounding the Red Sea rift evolution.

6 Spatial Evolution of Red Sea Rifting

6.1 Early Miocene Diffuse Extension

At the onset of extensional faulting in the latest Oligocene, the Red Sea rift system exhibited dramatic along-strike differences in terms of rift width (Fig. 6). The rift experienced diffuse stretching, affecting a very wide (~1200 km) rift-perpendicular corridor with active normal faults and dikes from north-central Egypt to north-central Saudi Arabia and Jordan (Fig. 6). Normal faults within the Nile Valley, and extensional faulting and emplacement of basalt dikes and monogenetic cones in the Western Desert of Egypt (Bosworth et al. 2015; Bosworth and Stockli 2016) attest to this Red Sea rift-related ultra-wide domain of diffuse crustal stretching. While extensional strain quickly localized in the Gulf of Suez as illustrated by rapid tectonic subsidence and deposition of marine syn-rift strata (e.g., Bosworth et al. 2005; Bosworth 2015), these diffuse, regionally widespread early Miocene normal faults, affecting the Western Desert, Nile Valley, and Saudi Arabian and Jordanian interior, illustrate the wide regional extent affected by Red Sea rifting. In the Gulf of Suez, older Paleozoic and especially Mesozoic structures, such as Tethyan normal faults, were reactivated during rifting as normal faults, accommodation zones, or regional relay structures (e.g., Bosworth et al. 2005) (Figs. 6 and 7).

In the central Red Sea, Szymanski et al. (2016) documented half-graben formation in the Hamd and Jizl area, north of Madinah, ~180 km inboard from the border fault system near Yanbu along the central Saudi Arabian Red Sea rift margin. There bedrock and detrital apatite (U-Th)/He dating show an initial phase of rapid footwall exhumation in the Hamd and Jizl half-grabens ~22–23 Ma and deposition of the Qattar formation, proximal to the active border fault in both half-grabens. Deposition of the Qattar formation occurred primarily from ~22 to 17 Ma, revealing that syn-rift sedimentary products developed across a broad margin of diffuse continental rifting during the early tectonic development of the nascent Red Sea rift system. Szymanski et al. (2016) also showed that starting in the Middle Miocene (~14 Ma), these half-grabens started to be uplifted and syn-rift basin fill to be eroded during rapid narrowing of the Red Sea rifting. In the Jeddah area, extensional half-grabens are limited to within <80 km of the modern coastline, illustrating the southward narrowing domain of early syn-rift faulting along strike in the Red Sea.

In sharp contrast, in the southern Red Sea (Figs. 1 and 6), the early rift system appears to be much narrower and restricted to well within the modern erosional rift escarpments (<40 km from the modern coastline) with no evidence for any significant normal faulting or syn-rift dike emplacement inboard of the coastal escarpment within the Nubian or Arabian Shields. This pronounced difference in early syn-rift strain distribution from ultra-wide to very narrow from north to south also expresses itself in terms of the geomorphic character of the Red Sea rift flanks. The southern Red Sea (south of Jeddah) is characterized by a spectacular escarpment that represents the erosional retreat of the structurally controlled original rift flank. In contrast, north of Jeddah, this coastal escarpment is much more subdued to non-existent, reflecting the more diffuse strain within the wider to ultra-wide early Red Sea system in the north-central and northern Red Sea rift system (Fig. 1).

This contrast in initial rift-perpendicular rift width can possibly be attributed to several factors, such as pre-rift structural inheritance or thermal structure. The northern Red Sea and Gulf of Suez were previously affected both by late Neoproterozoic Najd transcurrent deformation, Carboniferous block faulting, and early Mesozoic Tethyan rift structures (Fig. 6). These inherited structural grains appear to control early Miocene extensional faulting, as illustrated by dominant hard-linkage of rift normal faults. These inherited structures also control rift accommodation zones and extensional fault polarity domains within the early Gulf of Suez and northern Red Sea rift—the Galala-Abu Zenima, Morgan, and Quesir-Azlam accommodation zones (Fig. 7). South of the Quesir-Azlam accommodation zone, no more early syn-rift accommodation zones are observed in the central Red Sea. However, Voggenreiter and Hotzl (1989) interpreted the southern Saudi Arabian Red Sea margin as a simple-shear, upper-plate margin, requiring an additional accommodation zone south of Jeddah, accommodating this switch in early Miocene syn-rift fault polarity.

In addition to structural inheritance controlling the early rift width, the pre-rift thermal history likely also played an

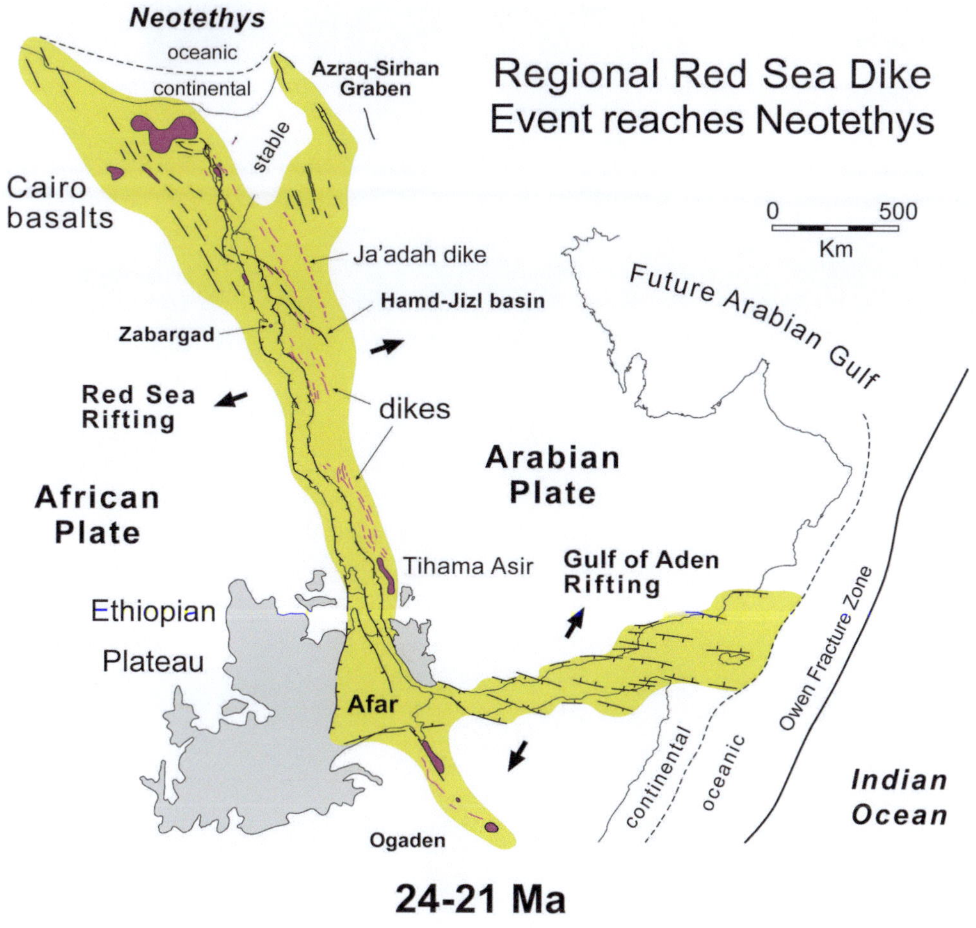

Fig. 6 Tectonic map of the Red Sea-Gulf of Aden rift system in the latest Oligocene and Earliest Miocene, illustrating the location of synchronous extensional faulting along the entire length of the Red Sea accompanied by only minor basaltic volcanism. Volcanism is largely restricted to the NE margin of the Red Sea and very short lived. Extensional faulting occupied an ultra-wide corridor in the Gulf of Suez and northern Red Sea ($\sim$1200 km wide) and narrows southward into the southern Red Sea. Most voluminous magmatism occurred outside the main Red Sea-Gulf of Suez rift in the greater Cairo area and illustrates the overall magma-poor Red Sea rift system

important role. The southern Red Sea experienced significant thermal weakening of the lithosphere and thermal erosion of the sub-lithospheric mantle during Eocene and Oligocene Afar hot-spot activity and continental flood basalt eruption, promoting more localized deformation and a narrow rift. In contrast, the northern Red Sea and the Gulf of Suez did not experience any Afar-related thermal weakening of the lithosphere, promoting distributed rifting within a stronger lithosphere. While speculative, mantle melt extraction during either Neoproterozoic times or during Mesozoic formation of the Neo-Tethys along the NE Africa continental margin could also have led to stronger pre-rift mantle, promoting diffuse stretching in the northern Red Sea domain. In particular, high-flux Neoproterozoic magmatism during both Arabian-Nubian Shield amalgamation (650–720 Ma) and late- to post-Pan-African orogenic collapse (650–560 Ma) likely left the mantle lithosphere significantly melt depleted in the northern Red Sea region (e.g., Johnson and Woldehaimanot 2003; Johnson 2014; Robinson et al. 2014).

6.2 Late Miocene Rift Localization (<14 Ma)

Midway through the Red Sea rift evolution, the Aqaba–Levant transform boundary was initiated, structurally isolating the Gulf of Suez from the rest of the Red Sea rift. This likely occurred at $\sim$14 Ma and was triggered by the hard continent-continent collision of Arabia with Eurasia (e.g., Mouthereau et al. 2012; Koshnaw et al. 2017), resulting in a major change in regional plate kinematics (Le Pichon and Gaulier 1988) and which was responsible for initiating rifting in the western branch of the East African Rift (e.g., Burke 1996) and the onset of displacement along the Aqaba-Levant transform boundary (Fig. 8). The onset of motion along the NNE-trending Aqaba-Levant transform fault resulted in several major transformations in the Red Sea rift: (1) a change in extension direction from rift-normal to oblique extension, (2) tectonic isolation of the Gulf of Suez from the rest of the North Sea basin, (3) strain localization and narrowing of the rift in the Red Sea, and (4) accelerated

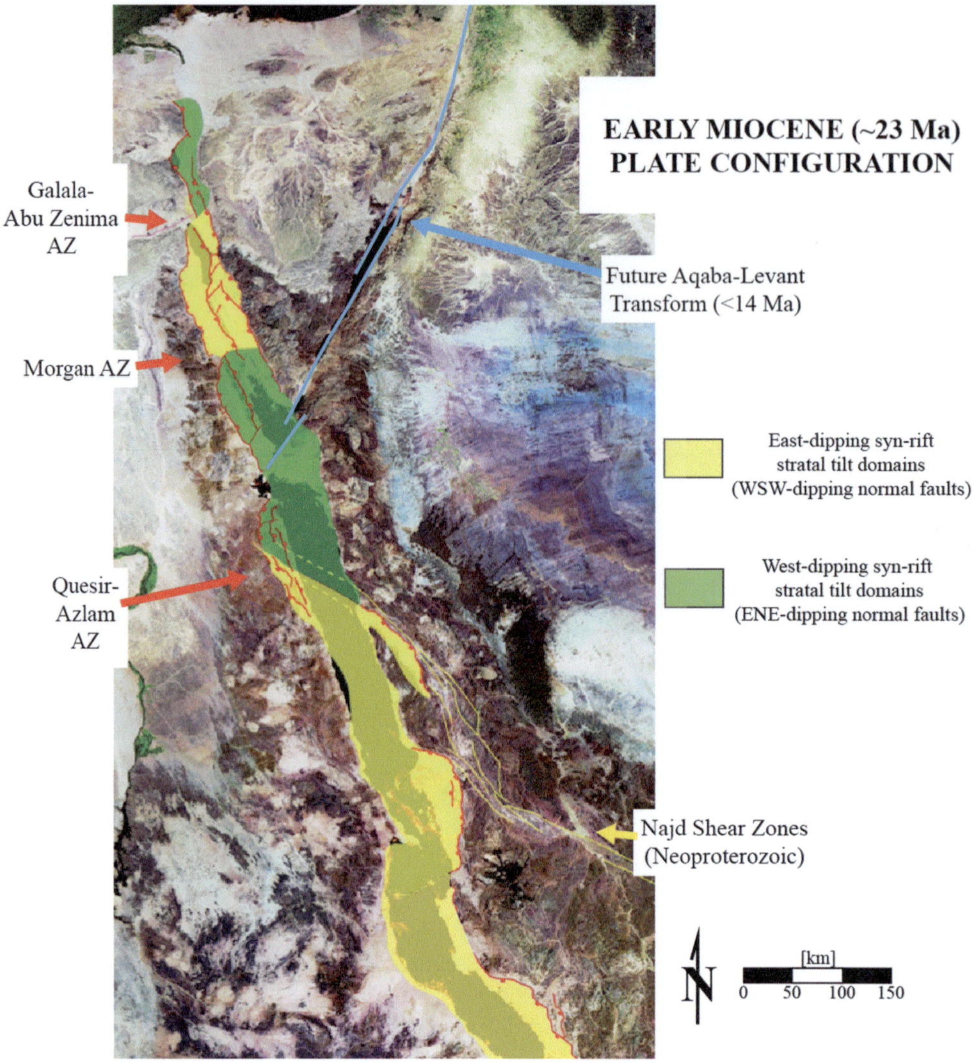

Fig. 7 Early Miocene Gulf of Suez-northern Red Sea rift system restoration shows different asymmetric rift polarity domains and intervening rift-segmenting accommodation zones. Accommodation zones appear to be controlled by both late Neoproterozoic basement shear zones (Najd) and Tethyan extensional structures. Restoration based on ~107 km of left-lateral displacement after ~14 Ma. Note that future Aqaba-Levant transform cuts obliquely across Early Miocene structural grains and might be influenced/controlled by inherited Hercynian basement structures

tectonic subsidence in the northern and central Red Sea and transition to thermal subsidence in the Gulf of Suez.

Early rift dike swarms (24–21 Ma) in NW Saudi Arabia and Sinai and offset basement structures have been used to estimate a total offset of ~107 km along the Aqaba-Levant transform, with ~62 km occurring in the Middle to Late Miocene and ~45 km in the Pliocene to Recent (Quennell 1951, 1958; Bartov et al. 1980; Eyal et al. 1981) (Figs. 7 and 8). Stratigraphic constraints from the Midyan Basin in NW Saudi Arabia suggest that strike-slip movement commenced after Aquitanian times (e.g., Tubbs et al. 2014). Structurally, it has been suggested that faulting likely reactivated a zone of inherited weakness related to formation of the Neo-Tethyan Levant margin or perhaps earlier tectonic structures (De Sitter 1962; Dubertret 1970; Dixon et al. 1987; Guiraud et al. 2001). Along the northern and central Saudi Arabian rift margin multiple ~NNE-trending structures juxtapose blocks with Neoproterozoic-Cambrian cooling signatures against blocks with elevation-invariant Carboniferous zircon (U-Th)/He ages, suggestive of rapid Carboniferous block faulting and exhumation (e.g., Bojar et al. 2002; Szymanski et al. 2016; this paper). These structures seem to be genetically related to ~NNE-trending basement faults, sealed by Permian strata (e.g., Khuff limestone, in the subsurface of central and eastern Saudi Arabia (Ghawar) and Qatar) (e.g., Edgell 1992). Based on these observations it appears plausible that the Aqaba-Levant transform might be reactivating ~NNE-trending Carboniferous crustal-scale faults related to Hercynian block faulting.

With the tectonic isolation of the Gulf of Suez and transfer of motion to the new Aqaba-Levant transform, tectonic extension and subsidence in the Gulf of Suez basin was greatly diminished at the end of deposition of the Kareem Formation in Serravalian times (Bosworth et al. 2005). While the Gulf of Suez entered a phase of thermal subsidence, the northern Red Sea switched from rift-normal to highly oblique extension. This most likely resulted in the abandonment and dismemberment of early Miocene

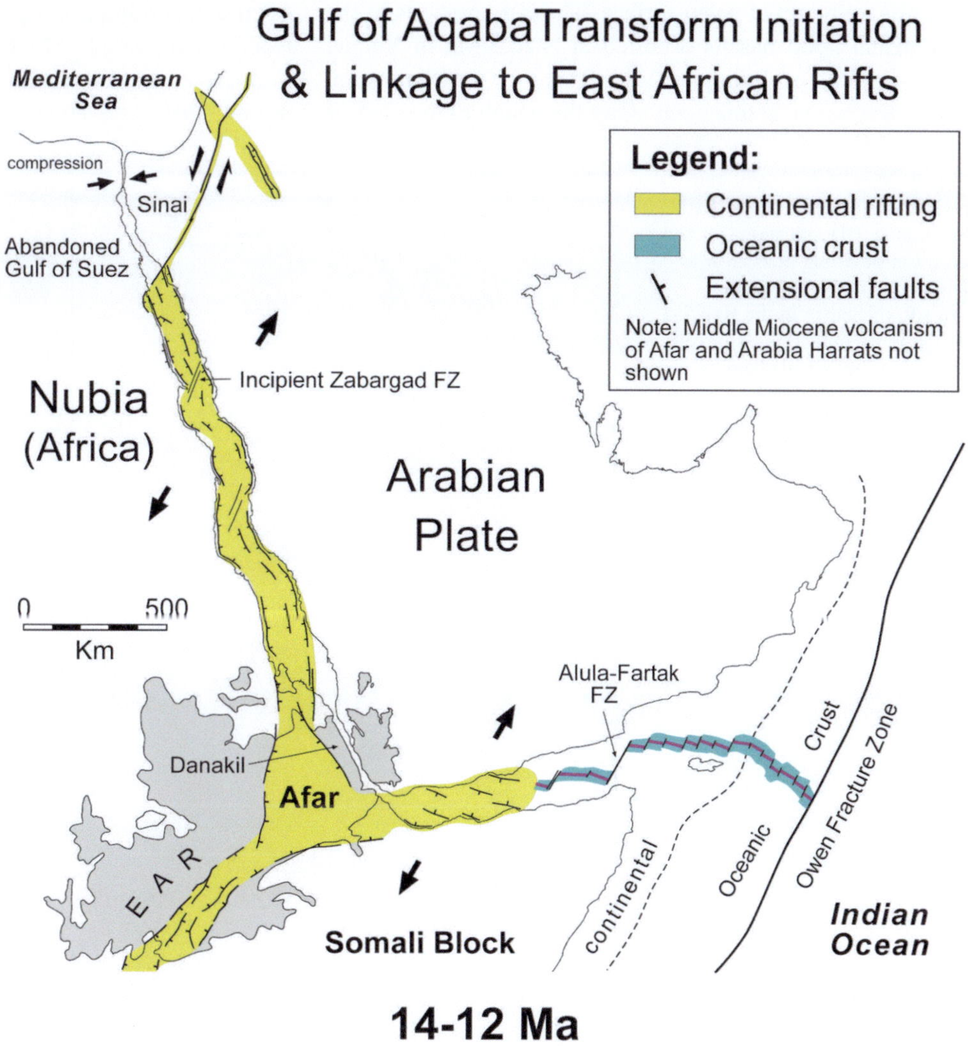

Fig. 8 Tectonic map of the Red Sea-Gulf of Aden rift system in the Middle Miocene (~14 Ma) at the inception of Aqaba-Levant transform motion. The Aqaba-Levant transform structurally isolated the Gulf of Suez, while deformation in the Red Sea underwent a major kinematic shift from roughly orthogonal to highly oblique (parallel to Aqaba-Levant transform). New accommodation and fracture zones (e.g., Zabargad fracture zone) cut obliquely across the Early Miocene extensional grain and accommodation zones. Kinematic reorganization and transition to oblique rifting led to drastic rift localization and lithospheric necking. Despite creation of a composite RRR junction in the Afar, rifting within the Red Sea remains largely magma-poor

accommodation zones in the Red Sea and the establishment of NNE-trending transform zones that still dominate the modern Red Sea rift (e.g., Zabargad fracture zone). These accommodation or transfer zones are roughly parallel to the Aqaba-Levant transform and are plausibly controlled by or reactivating Hercynian crustal-scale fault zones (e.g., Szymanski et al. 2016; this paper).

Along the Red Sea margins, the onset of Aqaba transform motion not only signaled a major kinematic change, but also triggered a dramatic narrowing and strain localization in the rift. This manifests itself as (1) rapid migration of faulting toward the rift axis and abandonment of proximal rift basins (e.g., Hamd-Jizl or Azlam), (2) accelerated tectonic subsidence and creation of massive new accommodation space, (3) uplift of marginal rift blocks capped by Serravalian reef complexes along the entire northern and central Red Sea margins (e.g., Jebel Shar or Dhuba areas), and (4) a second pulse of rapid exhumation along the coastal fault system recorded by low-temperature thermochronometry (e.g., Szymanski et al. 2016).

7 Timing and Nature of Rift-Related Magmatism

In the southern Red Sea regional continental flood volcanism began at ~31 Ma and spread to parts of southeast Ethiopia, southern Sudan, Saudi Arabia and most of western Yemen by 30–29 Ma (Figs. 5 and 9). While this period of flood volcanism was coeval with initial faulting and deposition of syn-rift strata in the Gulf of Aden, no or little extensional faulting occurred along the future Red Sea (Bosworth and Stockli 2016). Hence, flood basalt eruptions are largely considered pre-rift with respect to the Red Sea-Gulf of Suez rifting system. These Oligocene flood basalts unconformably overlie late Mesozoic sedimentary strata and older basement rocks on a low-relief paleo-surface. The last eruptions of the continental flood basalts occurred at ~26.5 Ma in Yemen and ~25 Ma in Ethiopia (Figs. 9 and 10). Continental rifting was active at this point along the length of the Gulf of Aden and reaching into present-day offshore Eritrea. In the Jeddah area, 30–23 Ma basalt flows conformably overlie Cretaceous sandstones of the Usfan Formation, demonstrating that extensional faulting and tilting did not occur

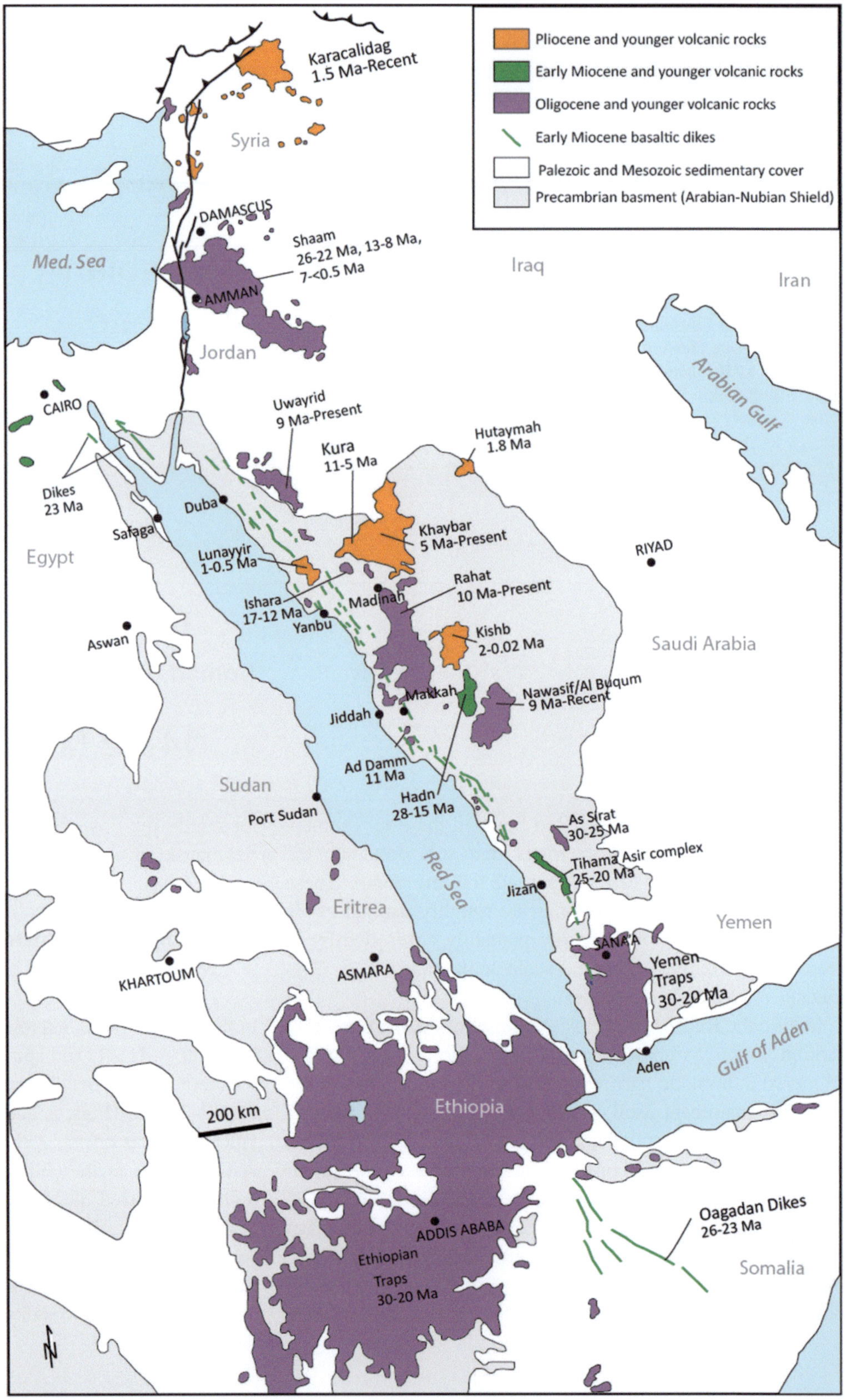

Fig. 9 Map of pre-, syn- and post rift magmatism in the Red Sea region. Classification is based on onset of rifting at ~23 Ma. Distribution of low-volume syn-rift volcanism in the Red Sea is highly asymmetric and largely limited to the NE rift flank in Saudi Arabia and Sinai. Pliocene-Recent volcanism does not appear to be collocated with the Red Sea rift trend, but rather to a more N–S trend, parallel to the Makkah-Madinah-Nafud Line. Modified after Bosworth and Stockli (2016)

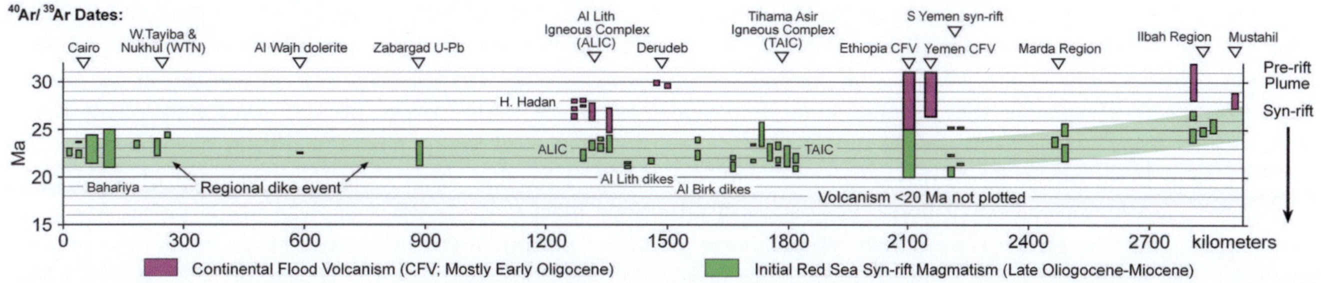

Fig. 10 Compilation of pre- and early syn-rift volcanism in the Red Sea-Gulf of Suez. The data clearly demonstrate the synchronous nature of syn-rift magmatism along strike during the onset of latest Oligocene-earliest Miocene rifting and the very limited to absent pre-rift basaltic continental flood volcanism (CFV) in the central and northern Red Sea

until after 23 Ma and in fact no significant extension or tilting occurred between 30 and 23 Ma (unpublished $^{40}Ar/^{39}Ar$ data).

At ~23 Ma basaltic dikes and igneous complexes formed along the entire length of the Red Sea-Gulf of Suez rift from Yemen through Saudi Arabia into the Sinai (Bosworth and Stockli 2016) (Fig. 10). Coincidentally a basaltic volcanic province developed in northern Egypt and in the Harrat ash Shaam region of Jordan (Bosworth et al. 2015). This phase of volcanism culminated from ~23 Ma and was accompanied by the earliest syn-rift faulting and deposition of continental syn-rift strata, marking the birth of the Red Sea rift (Figs. 9 and 10). The presence of the sheeted dike and associated plutonic complexes at Al Lith and Asir Tihama suggests that oceanic-like rifting might have occurred very early and very briefly in the southernmost Red Sea rift in the region close to the Afar plume. However, only very minor syn-rift volcanism, in the form of some dikes and very few basaltic flows, was recorded in the central and northern Red Sea and the Gulf of Suez.

Eruption of voluminous flood basalts in the southern Red Sea, often cited as evidence for volcanic Red Sea rifting, significantly predate the onset of rifting in the Red Sea by 7–10 m.y. and are not spatially associated with subsequent rifting in the Red Sea. In fact, voluminous syn-rift magmatism is largely absent and the margins remained essentially non-volcanic throughout their evolution. In addition, there is also no published evidence for the presence of offshore SDRs or thick proto-oceanic magmatic crust along the entire Red Sea. While publicly available sub-salt reflection seismic data are of low quality, they do not show any evidence for the formation of syn-rift magmatic crust, although some authors have invoked the presence of prototypical SDRs along the Red Sea, implying the Red Sea formed as a volcanic rift. There is simply no evidence for syn-rift magmatic accretion. However, it is unclear how much gabbroic underplating might have occurred in the early Miocene. Miocene(?) gabbros with MORB-like chemistry in the north-central Red Sea (Brothers Islands) provide interesting clues as they appear to represent either underplated material or intrusions into thinned continental crust (Bonatti et al. 2015). The following section provides a summary of early syn-rift magmatism in the Red Sea-Gulf of Suez rift system (Fig. 9). Geophysical studies of the mantle lithosphere underneath the Arabian Shield have shown the rift axis and the low-velocity mantle are not collocated and do not support an active rifting model for the central and northern Red Sea (e.g., Hansen et al. 2007).

8 Early Syn-rift Magmatism

8.1 Egypt and Egyptian Margins

In Egypt, the earliest Miocene basaltic flows and dikes both within the Gulf of Suez and northern Red Sea occur as basaltic dikes and a single flow on Sinai, as extensive basaltic flows in the greater Cairo area, as monogenetic basaltic cinder cones aligned in a Red Sea parallel trend, as isolated flows west of the Nile near Asyut, and as basaltic flows within the greater Bahariya Oasis (see summary in Bosworth and Stockli 2016). Compositionally, all of these volcanic dikes and flows are sub-alkaline tholeiites (Baldridge et al. 1991; Abdel Aal 1998; Endress et al. 2011; Shallaly et al. 2013) and are temporally and spatially related to NW–SE faults that are present from the Libyan Plateau in the west to Sinai in the east (Bosworth and Stockli 2016) (Fig. 9).

In the Cairo and Faiyum areas, basaltic rocks have been known for over a century (e.g., Beadnell 1902, 1905; Andrew 1937; Rittmann 1954). While surface exposures are sparse, subsurface exploration revealed the occurrence of a 20–80 m thick flood basalt field, extending across a subsurface area of at least 15,000 km^2 (Williams and Small 1984; Bosworth et al. 2015). Bosworth et al. (2015) estimated the original extent prior to uplift and erosion to be ~25,000 km^2. $^{40}Ar/^{39}Ar$ dating of these basalts from Gebel Qatrani, along the northern margin of the Fayoum Depression, yielded ages of ~23.7 Ma. Bosworth et al. (2015) obtained ages of ~22–23 Ma for the same ~22 m thick basalt. A small flow east of Cairo (Qattamia) yielded a $^{40}Ar/^{39}Ar$ age of ~21.4 Ma. In the adjacent Western Desert of Egypt, monogenetic cones west of the Faiyum area, erupted from NW-SE trending dikes parallel to the Red Sea trend, yielded $^{40}Ar/^{39}Ar$ ages between 22 and 24 Ma. In the Bahariya area, Bosworth et al. (2015) obtained $^{40}Ar/^{39}Ar$ ages of ~23 Ma for several basalt flows. Integration of all dates suggests that the greater Cairo basalts erupted over a very brief time period from ~23.7 to 21.5 Ma, contemporaneous with the earliest extensional faulting related to Red Sea rifting.

In the Gulf of Suez, a single ~5–30 m thick basalt lava flow is exposed on the eastern margin of the central Gulf of Suez (Wadi Tayiba, Fig. 11). Bosworth et al. (2015) obtained a $^{40}Ar/^{39}Ar$ age of ~23 Ma for the Wadi Tayiba flow and ~23 Ma for a basaltic dike in Wadi Nukhul. Other isolated basaltic dikes along both margins of the Gulf of Suez yielded K–Ar dates of ~25 Ma at Gebel Ataqa and Gebel Gharamul (Ott d'Estevou et al. 1986; Bosworth and McClay 2001) and a $^{40}Ar/^{39}Ar$ age of ~22 Ma for a large dike exposed on the south side of Wadi Araba (Bosworth et al. 2015). Along the Egyptian Red Sea Margin, basaltic

Fig. 11 Picture of ~23 Ma basaltic lava from the central Gulf of Suez, near Abu Zenima, overlying the pre-rift Tanka Fm., the late Oligocene proto-rift Tayiba Fm., and underlying the continental to marginal marine early syn-rift Nukhul Fm. The Abu Zenima basalt marks the onset of rifting in the central Gulf of Suez and is fed by a ~23 Ma sub-vertical feeder dike, exposed in Wadi Nukhul (Sinai), where it intrudes Eocene Thebes Fm. limestone. From Bosworth et al. (2015)

dikes are rare and often poorly exposed. South of Quseir, thin basalt flows, interbedded with Nakheil Formation sandstone and siltstone, yielded a K–Ar age of ~25 Ma (Roussel et al. 1986).

This exhaustive list of surface and sub-surface basalt dikes and flows from the Red Sea-Gulf of Suez in Egypt illustrates the relative paucity of syn-rift magmatism, with the exception of the "Cairo mini-plume", and the extremely punctuated nature of basaltic magmatism restricted to the very beginning of rifting, and the subsequent apparent lack of syn-rift magmatism (Fig. 10).

8.2 Saudi Arabian Red Sea Margin Dikes and Mafic Igneous Complexes

Along the southernmost Saudi Red Sea margin, in the Jizan area, extensive Red Sea related magmatism has been described in the Tihama Asir igneous complex (Figs. 9 and 10) that includes sheeted dikes, layered gabbros, and granophyre stocks intruded into the Neoproterozoic basement and Paleozoic to Cenozoic sedimentary strata (Coleman et al. 1979; Blank and Gettings 1985). In addition, isolated basalt dikes similar to those present along the rest of the margin are also common. While Tihama Asir sheeted dikes are strongly affected by hydrothermal alteration, K–Ar dates for the gabbros, granophyres and hornfels range in age from ~20 to 24 Ma (Coleman et al. 1979). $^{40}Ar/^{39}Ar$ data for basaltic and rhyolitic dikes and cumulate gabbros yielded ages between 22 and 25 Ma (Sebai et al. 1991). Similarly, a second intrusive complex, located ~70 km NW of Al Lith (Fig. 9), consists of sheeted dikes intruded into Neoproterozoic basement and coeval lava flows (Brown 1972; Pallister 1987). Sebai et al. (1991) analyzed samples from basaltic and felsic dikes and a basalt flow from the Al Lith complex and obtained $^{40}Ar/^{39}Ar$ ages of ~22–24 Ma. While K–Ar ages of all these rocks from the southern Saudi Arabian rift margin exhibit considerable scatter, $^{40}Ar/^{39}Ar$ ages generally tightly cluster between 24.5 and 21 Ma (Fig. 10).

Most impressive, however, in terms of syn-rift magmatism along the Saudi Arabian Red Sea margin is the Late Oligocene-Early Miocene dike system (Figs. 9 and 10). These basaltic dikes occupy a nearly continuous belt that is generally 50–150 km wide and stretches 1700 km from northern Yemen to the Gulf of Aqaba (Bosworth and Stockli 2016). Many individual dikes are over 50 km in length, and the largest—the great Ja'adah dike—is at least 450 km long, based on the interpretation of aeromagnetic data (Zahran et al. 2003). Sebai et al. (1991) obtained $^{40}Ar/^{39}Ar$ dates of 22–24 Ma for six of the Saudi Arabian margin dikes. While spectacular in terms of their lateral continuity along the margin, they do not accommodate any considerable amount of extensional strain, do not appear to heat the supra-crustal rocks, and are temporally extremely limited to the very beginning of rifting.

This magmatic activity associated with early Red Sea rifting is highly asymmetric with almost all of the volcanism north of the Afar occurring on the Saudi Arabian side of the rift (Fig. 9). The age of the oldest volcanism associated with earliest Red Sea rifting is consistently ~22–24 Ma along the entire length of the Red Sea rift system. As discussed in the compilation of data by Bosworth and Stockli (2016), voluminous early Oligocene Afar flood basalt volcanism substantially predates the onset of extensional faulting in the Red Sea. With the exception of the southern Red Sea (Al Lith and Tihama Asir) syn-rift volcanism along the Saudi Arabian Red Sea margin is volumetrically relatively minor.

The highly asymmetric nature of early syn-rift magmatism, structural polarity, and rift topography in the southern Red Sea have led some workers to suggest that the Red Sea may have developed as a result of whole-lithosphere simple shear extension (Wernicke 1985; Voggenreiter et al. 1988). This part of the Red Sea has been used as a potential poster child for simple shear continental rifting and the formation of lower and upper plate margins (Wernicke 1985). However, at present, the relationship of this asymmetry to the tectonic processes and 3-D distribution of extension in the Red Sea remains unclear.

9 Syn-rift Stratigraphy

Despite the resolution that is possible when looking at a geologic system that only started at ~30 Ma, there remain fundamental problems with pegging the timing of Red Sea rift initiation itself. Throughout much of northeast Africa, thin Oligocene strata were deposited in continental, or stressed marginal marine environments. Bosworth et al. (2005; see references therein) compiled data along the length of the Red Sea basin. These immediate pre- or in some cases "proto"-rift strata are all generally of Oligocene age and referred to as the Tayiba, Matiyah, Hamamit, and Dogali Formations along the Gulf of Suez, central Saudi Arabian margin, Sudanese margin, and central Eritrean margin, respectively. The corresponding earliest syn-rift deposits are the Abu Zenima/Nukhul Formations, Al Wajh Formation, lower Maghersum Group, and lower Habab Formation (Fig. 12). All these units are dominated by coarse siliciclastic and frequently redbed sequences. Microfossil assemblages are often poorly preserved and of low diversity, triggering long running debates about the ages of these units and where precisely to place the base-rift unconformity. The following sections attempt to summarize the information with respect to the timing and nature of syn-rift sedimentation in the Red Sea.

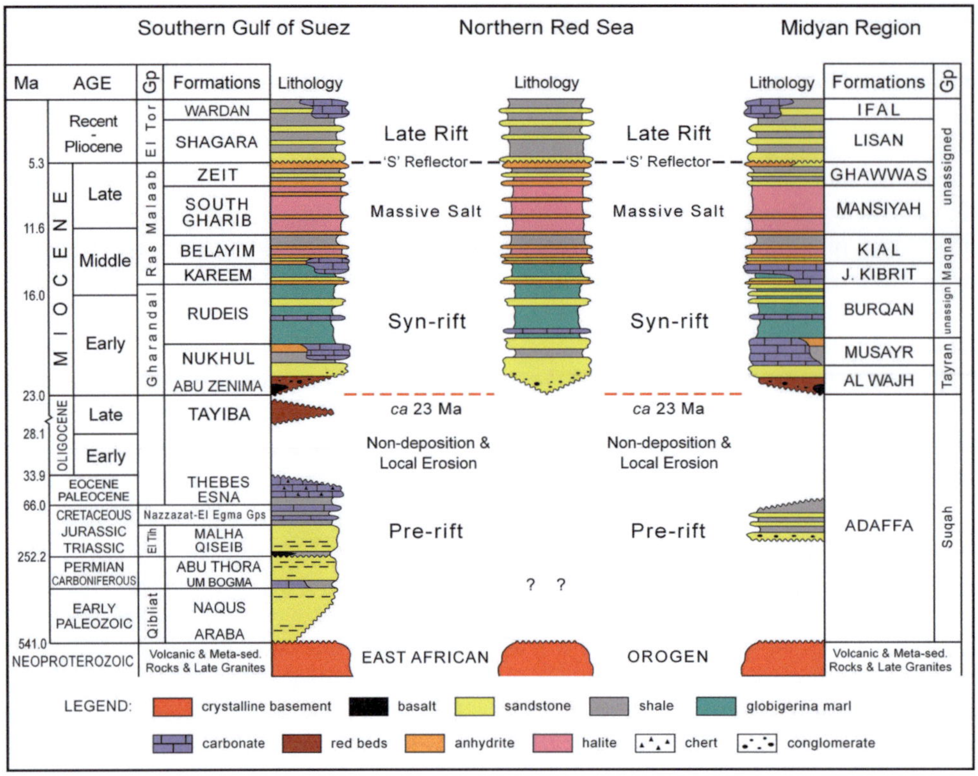

Fig. 12 Summary showing representative stratigraphic sections in the Gulf of Suez, offshore northern Red Sea, and onshore Red Sea in the Midyan region (Saudi Arabia), illustrating the presence of only very limited or absent Oligocene strata and the inception of syn-rift subsidence and deposition of continental and marine sediments after eruption of basal syn-rift basalts at ~23 Ma. The compilation also demonstrates the overall similarities in stratigraphy along the entire length of the Red Sea, arguing against fundamental differences in subsidence history during the syn-rift phase between 23 and 14 Ma. Data from Bosworth (2015) and Hughes et al. (1999)

10 Early Syn-rift Stratigraphy

During the earliest phase of rifting, the entire Red Sea and Gulf of Suez rift system was characterized by deposition of continental siliciclastics and shallow-marine limestones during Aquitanian times (Fig. 12). These basal syn-rift strata were deposited on a low relief surface at or near sea level and appear to lack fault-controlled topography (Garfunkel and Bartov 1977; Sellwood and Netherwood 1984; Bohannon 1986; Bohannon et al. 1989; Coleman 1993). Throughout much of the Red Sea region, they either rest on a pre-Cretaceous peneplanation surface or conformably overlie Cretaceous continental or shallow-marine deposits. This not only argues for little to no pre-Miocene extension, but also for synchronous onset of rifting along the entire length of the Gulf of Suez-Red Sea rift system. There is no evidence for a temporal progression or propagation along the rift axis.

In the Gulf Suez, the earliest definitive syn-rift strata are red beds of the Abu Zenima Formation that are capped by a ~22–23 Ma basalt flow in eastern Sinai (Plaziat et al. 1998; Bosworth et al. 2015; Bosworth and Stockli 2016) and dolomitized carbonates of the Aquitanian Nukhul Formation (Scott and Govean 1985; Bosworth et al. 1998). Initial syn-rift deposition and the stratigraphic response to nucleating and progressively coalescing normal faults and fault block rotation in the Gulf of Suez has been intensely studied and used as an analogue for many other examples of early syn-rift sedimentation around the world (e.g., Gawthorpe et al. 1997; Winn et al. 2001). During the early Miocene subsidence and rift shoulder uplift started to accelerate in the early Burdigalian (~19 Ma) as rift faults continued to coalesce into major rift-block fault and marginal rift basins, such as the Esh el Mellaha Basin in the SW Gulf of Suez, were abandoned during progressive rift localization (Bosworth 1994; Bosworth et al. 1998; Bosworth and McClay 2001).

In the Midyan basin, along the northernmost Saudi Arabian Red Sea margin, rifting initiated in the latest Oligocene to earliest Miocene as recorded by the syn-rift stratigraphic record beginning at ~23 Ma (Fig. 12). These mixed continental, marine, and evaporitic early syn-rift strata of the Tayran Group (Al Wajah Fm.) are characterized by growth faulting and deposition in half-graben wedges and are deposited unconformably on Neoproterozoic basement or

Cretaceous pre-rift strata. Originally dated at Chattian, micro-paleontological and Sr isotopic studies showed an Aquitanian age for these deposits (Hughes et al. 1999). The onset of very rapid syn-rift subsidence is recorded by the marine deposits of the Burqan Fm., marking a regional marine incursion into the Gulf of Suez and Red Sea rift system (Fig. 12). It is followed by sedimentation of the Maqna Group that exhibits progressively more restricted conditions, such as the Mansiyah Fm. evaporites (Hughes and Johnson 2005).

The Azlam Basin (Fig. 1) is an 80 km long and 8–15 km wide half-graben that extends southeast into the Arabian shield from the Red Sea coastal plain (Davies 1985). It is located north of Al Wajah and S of Dhuba and appears to reactivate a latest Neoproterozoic Najd fault system (Fig. 7). The half-graben is filled with up to 2 km of 10–40° NE tilted Cretaceous-Oligocene(?) pre-rift strata, which are juxtaposed against Precambrian basement by normal faults along the SW edge of the basin. Little to no reliable chronostratigraphic work has been carried out on the deposits of the Azlam Basin, but rift-related tilting of the pre-rift strata occurred prior to deposition of only slightly uplifted and tilted Serravalian carbonate platforms south of Dhuba.

In the subsurface of the Al Wajah and Yanbu Basins along the north central Saudi Arabian margin, the early syn-rift Yanbu Fm. of the Tayran Group contains evaporites that have been dated by Sr isotopes at 22–23 Ma (Cocker and Hughes 1993). In the Jizl-Hamd half grabens, inboard from Yanbu, Szymanski et al. (2016) documented that the fault-controlled accumulation of early Micocene syn-rift strata (Qattara formation) started at ~22 Ma, based on detrital thermochronometry, and ceased by ~17 Ma, as evidenced by horizontal, post-faulting basalt flows at Gebel Ishtar.

In the Jeddah-Makkah region along the central Saudi Arabian rift margin, tilted pre-rift and syn-rift strata are exposed in SSE-trending half-grabens that are separated from the main part of the Red Sea basin by a large Precambrian basement horst block. Tilted syn-rift siltstone and sandstone are exposed in several synthetic half-grabens and conformably overlie Cretaceous pre-rift strata (Usfan and Shumaysi Fms.). In the Wadi Shumaysi area, these syn-rift strata overlie pre- and syn-rift basalt flows (~30–23 Ma; Stockli, unpublished ^{40}Ar/^{30}Ar data), making the half-graben fills younger than 23 Ma. These early Miocene (<23 Ma) syn-rift strata within the Jeddah half-grabens dip gently to moderately (up to 30–40°) to the east and are capped by ~5 Ma basalt flows of Harrat Rahat. Just like the Hamd-Jizl half-grabens in the Madinah area, the Jeddah half-grabens only record early Miocene syn-rift sedimentation and were abandoned prior to rift narrowing and localization in the middle to late Miocene.

During the Early Miocene, the Gulf of Suez and Red Sea basins were linked and their structural and stratigraphic evolution was indistinguishable (Steckler and ten Brink 1986; Montenat et al. 1988; Bosworth et al. 1998). From Egypt to Eritrea, early Miocene syn-rift basinal deepwater facies strata can be considered very similar, implying that the principal syn-rift depositional sequences are relatively uniform along strike from the Gulf of Suez into the northern and central Red Sea (Fig. 12). Following the deposition of the Nukhul evaporite and carbonate sequence, sedimentation abruptly shifted to mainly open marine shale, marl, and deepwater limestone of the Rudeis Formation (~1.4 km thick) in Egypt (EGPC 1964; Garfunkel and Bartov 1977), the Burqan Formation (~1.1 km thick) in Saudi Arabia (Hughes and Filatoff 1995), the lower Maghersum Group (~0.8 km thick) in Sudan (Bunter and Abdel Magid 1989), and the Habab Formation (~1.5 km thick) in Eritrea (Hughes and Beydoun 1992; Bunter et al. 1998).

11 Late Miocene Sedimentation (Post-aqaba-Levant Inception)

With the onset of motion along the Aqaba-Levant transform at ~14 Ma, extension shifted from normal to highly oblique relative to the margins and restarted rapid tectonic subsidence and syn-tectonic sedimentation (Fig. 8). In the Midyan area, this resulted in the accumulation of thick (>1.2 km) shallow to marginally marine sandstone and interbedded anhydrite of the late Miocene Ghawwas Fm. (e.g., Tubbs et al. 2014) (Fig. 12). Abrupt thickness changes and stratal geometries in the Ghawwas Fm. indicate this to be a period of active tectonics and salt mobilization (diapirism). In many areas of the Red Sea evaporite deposits were displaced and mobilized down-dip by major detachment faulting or gravitational sliding, believed to be driven by accelerated tectonic subsidence focused in the centre of the rift during abrupt rift localization. Rapid late Miocene subsidence is not observed north of the Aqaba-Levant transform, illustrating the fact that while extension drastically slowed in the Gulf of Suez, it dramatically accelerated in the northern Red Sea. Burial histories from boreholes along the Saudi Arabian margin (Ghassal 2010) show a dramatic increase in sediment accumulation between 12 and 5 Ma (e.g., Ghawwas and Lisan Fms.), suggesting that the kinematic shift led to massive structurally controlled rift localization and tectonic subsidence in the late Miocene in the northern Red Sea.

After inception of motion along the Aqaba-Levant transform, massive halite deposition occurred throughout the Red Sea basin in the early Late Miocene (~12 Ma) (Fig. 12). These evaporite deposits are the South Gharib Fm. in Egypt (Hassan and El-Dashlouty 1970; Fawzy and Abdel Aal 1986), the Mansiyah Fm. in Saudi Arabia (Hughes and Filatoff 1995), the Dungunab Fm. in Sudan (Bunter and

Abdel Magid 1989) and the Amber Fm. in Eritrea (Savoyat et al. 1989; Hughes and Beydoun 1992), and were mostly deposited during Tortonian times during a major eustatic low-stand (e.g., Richardson and Arthur 1988; Wescott et al. 1997).

12 Low-Temperature Thermochronometry

The timing of faulting is commonly assessed by bracketing by either stratigraphic age or intrusive/extrusive units. However, a well-established alternative approach to dating normal fault motion is to determine the low-temperature cooling histories from thermochronometric data from rocks in the footwalls of these faults (e.g., Stockli 2005). Major slip on a normal fault leads to exhumation and cooling of the footwall such that the timing of fault slip can be estimated from the age of this cooling. This approach is limited only by whether exhumation was of sufficient magnitude to expose rocks that resided at temperatures above the closure temperatures of the minerals in question. This approach has been applied for several decades in the Red Sea and Gulf of Suez system to reconstruct the timing of faulting and escarpment retreat.

12.1 Apatite Fission Track Data (Egypt and S Saudi Arabia)

Omar (1985) reported apatite fission track ages and track length data from samples from the basement flanking the Gulf of Suez. Although these samples provided significant constraints on the erosion and differential movement of basement blocks, they were too few to date the initiation of rifting. Data from Cretaceous alkaline ring complexes in southern Egypt indicate that the exhumation is <32 Ma, which is in agreement with earlier data from Sinai (Kohn and Eyal 1981), but do not accurately date its initiation (Omar et al. 1987). Extensive fission track dating of samples from the western margin of the Gulf of Suez indicate that rift flank exhumation started at ~23–21 Ma, approximately coeval with the basin subsidence (Omar et al. 1989). Apatite (U-Th)/He dating on basement rocks and pre-rift strata from the border fault complex and fault blocks (e.g., Gebel Araba) yield ages clustering between 19 and 24 Ma (Pujols 2011).

Apatite FT ages from the southern Eastern Desert of Egypt range from 22 to 23 Ma (Omar and Steckler 1995). Generally, young ages and long mean-track lengths (13–14 μm) are concentrated within ~100 km from the Red Sea coast and progressively increase and decrease inland, respectively, indicating that rift-related exhumation and erosion increase toward the coast. Young apatite samples characterized by long mean track lengths fall on two distinct trends, both arguing for cooling from temperatures >110°C (Omar and Steckler 1995), dating rapid cooling and the timing of erosional denudation following uplift. The older trend attests to commencement of a major cooling and erosion event at ~34 Ma and the younger trend indicates renewed or increased uplift and erosion at ~25 Ma. A third, poorly defined cooling event suggests exhumation during Late Cretaceous-Early Paleocene time. However, the data are limited and the spatial data density not sufficient to determine the geometry and scale of any Late Cretaceous-Early Paleocene tectonic activity. Geological evidence and thermochronometric evidence from other areas make the ~34 Ma cluster enigmatic.

Similarly, Bojar et al. (2002) presented apatite and zircon fission track data from the Eastern Desert of Egypt (Quseir area). They reported a single apatite fission track age of 23 Ma with long confined track length data that was interpreted to constrain the timing of onset of rapid faulting and exhumation in the Duwi basin in the earliest Miocene. Zircon fission track ages mainly yielded Carboniferous cooling ages, ranging from 315 to 366 Ma, and appear to record vertical fault displacement along ~N–S trending faults during Hercynian block faulting.

Bohannon et al. (1989) obtained fission track ages from a large suite of samples from southern Saudi Arabia, yielding a broad range of apparent ages as old as ~568 Ma. The oldest ages come from the Arabian Shield to the east of the rift flank escarpment and imply little erosional denudation. Their youngest sample yielded an age of 14 Ma, indicating that 2.5–5 km of denudation occurred since that time. They suggested that rift flank exhumation may have begun at ~20 Ma, but insufficient age data and lack of mean track length measurements hamper a reliable interpretation of their data.

12.2 Zircon and Apatite (U-Th)/He Data (Central and Northern Saudi Arabia)

A thermochronometric study of the central and northern Saudi Arabian Red Sea rift flank and the proximal rift margin, integrated with structural studies, was conducted with the goal of determining the timing, origin, and geometry of extensional faulting and rift flank exhumation along the central and northern Red Sea margin in Saudi Arabia. These efforts have focused on thermal history reconstruction of the central and northern Saudi Arabian Red Sea rift margin, stretching from the coastal escarpment south of Jeddah to the northern Gulf of Aqaba (Fig. 13). Apatite and zircon (U-Th)/He ages directly date the timing of onset of extension and also quantify the pre- and post-rift erosional denudation of the rift flanks. A critical aspect in understanding the evolution of rifting and rift flank exhumation is to constrain the timing and magnitude of faulting, rift

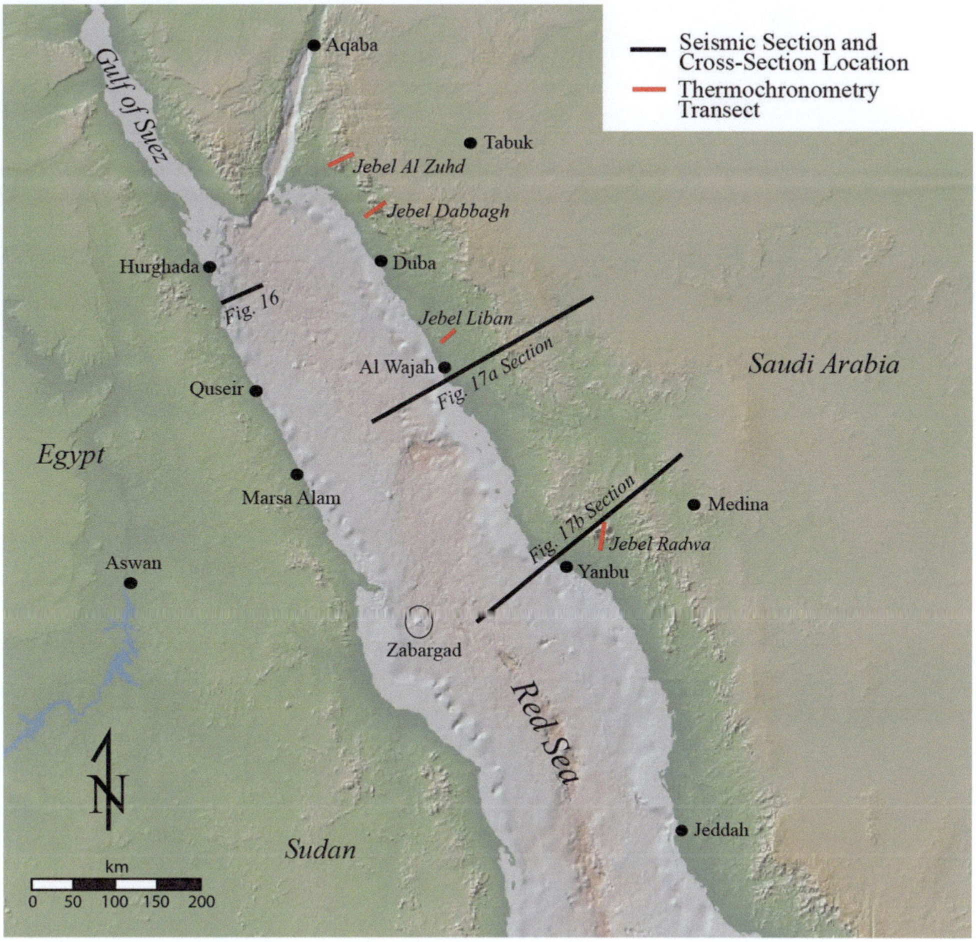

Fig. 13 Digital relief and bathymetry image of the northern Red Sea and Gulf of Suez showing locations of low-temperature thermochronometric transect data (red lines) along and across the Saudi Arabian northern and central Red Sea margin to constrain the timing of faulting and exhumation (Fig. 14). Black lines show locations of seismic line (Fig. 16) and crustal-scale cross-sections of the Saudi Arabian Red Sea margin (Fig. 17). Digital relief base map from GeoMapApp.com

segmentation, and rift localization. The answers to these questions have far-reaching implications for our understanding of rifting dynamics and the interplay between extensional faulting and subsidence during syn-rift sedimentation. More than 400 thermochronometric samples were collected from the central and northern Saudi Arabian Red Sea margin to resolve the timing of extensional faulting and to differentiate between different episodes of cooling/exhumation affecting the Saudi Arabian Red Sea margin. Detailed vertical transects were collected across the exhumed crustal blocks and topographic escarpments to constrain the timing and spatial distribution of extensional faulting and evolution of rift segmentation and localization, as well as rift flank exhumation. Knowledge of the temporal and spatial distribution of crustal attenuation and the timing of flexural amplification and exhumation of the entire width of the rift flank is critical for evaluating the role of processes such as active versus passive asthenospheric upwelling, secondary convection, and flexural unloading of the crust, as well as the distribution of sub-crustal lithospheric extension relative to crustal thinning.

Szymanski et al. (2016) presented apatite and zircon (U-Th)/He data for nearly 100 samples from the central Saudi Arabian Red Sea margin from around the greater Yanbu to Madinah area (Figs. 13 and 14). Thermal modeling of apatite (U-Th)/He and zircon (U-Th)/He thermochronometric data showed a two-stage evolution characterized by Oligocene-Miocene ($\sim$23 Ma) and middle Miocene ($\sim$14) rift pulses and revealed a common three-stage thermo-tectonic evolution for the central Red Sea margin in Saudi Arabia. Initially the pre-rift Arabian-Nubian Shield experienced a Carboniferous cooling event, dominating the ZHe ages (Fig. 14). The main cooling pulse, recorded mainly by AHe ages, relates to major extension $\sim$23 Ma, exhuming a widespread area of the central Saudi Arabian Rift margin (>200 km) and marking the beginning of a phase of diffuse lithospheric stretching. Distributed block faulting, as opposed to regional flank uplift, appears to have been the dominant structural style during early Miocene Red Sea rifting, with rapid early Miocene fault block exhumation and tectonic unroofing of up to 5 km. A second extensional Red Sea rift phase occurred $\sim$14 Ma, focused along the coastal border fault system, marking both the abandonment of proximal rift fault blocks and a dramatic rift narrowing and basinward migration of major extensional faulting during this phase.

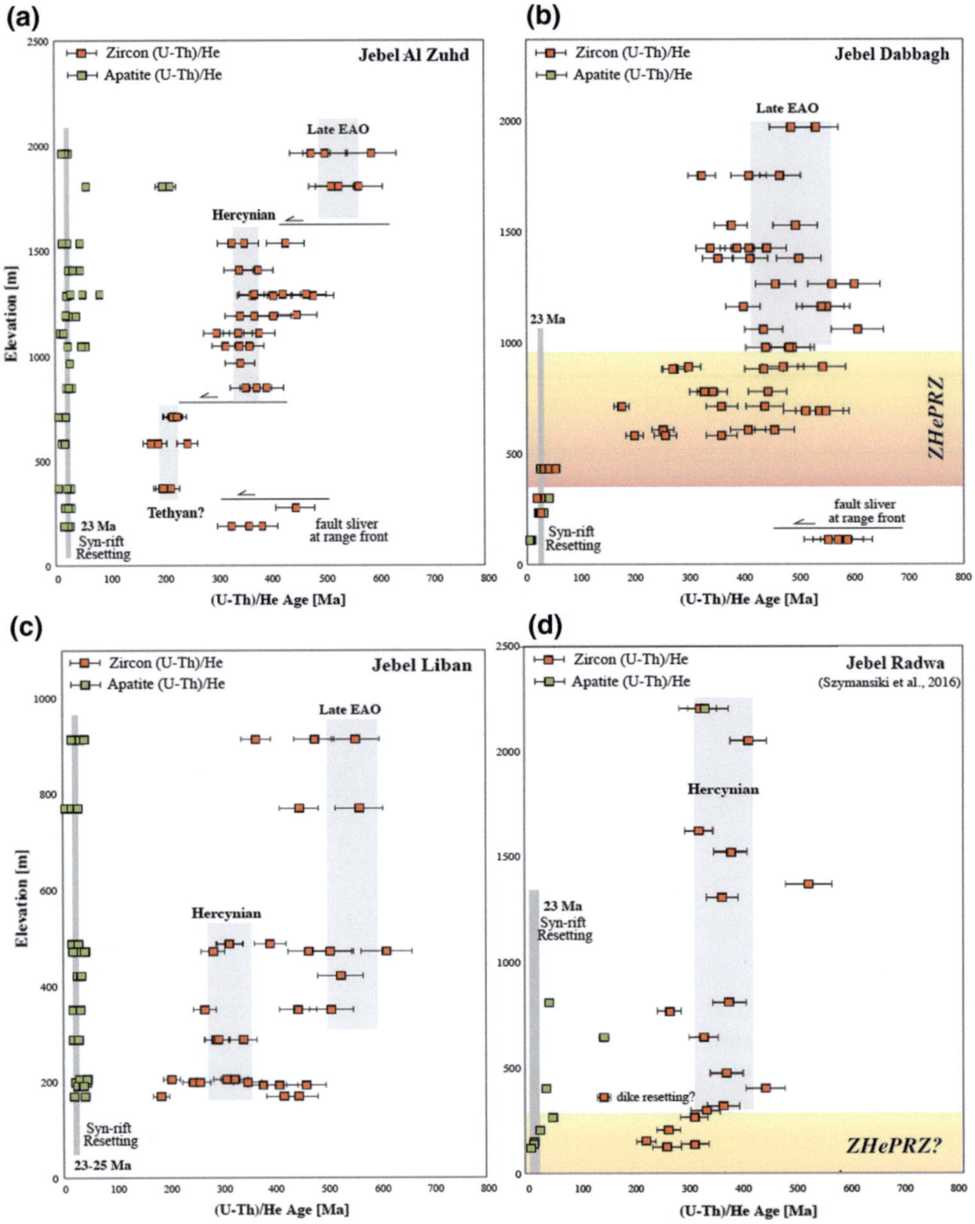

Fig. 14 Apatite and Zircons (U-Th)/He data from the northern Red Sea rift margin of Saudi Arabia (see Fig. 13 for locations). **a** Jebel AL Zhut, located in the footwall of the Midyan Basin. ZHe data show multiple discrete age domains that are juxtaposed by sub-vertical basement fault zones. Late Neoproterozoic against Hercynian ZHe age juxtaposition resembles other similar basement cooling domains in the Red Sea (Szymanski et al. 2016). The lower portion of Jebel Al Zhut exposes ZHe ages that record Triassic Tethyan rifting, similar in age to Triassic deposits (Quseib Fm.) in the Gulf of Suez. The basal portion of the fault block exhibits a Miocene normal fault sliver, duplicating Hercynian ages. AHe data are somewhat scattered, but overall record rapid Early Miocene exhumation, consistent with an Aquitanian age for the earliest syn-rift strata in the hanging wall—the Midyan basin (e.g., Tubbs et al. 2014). **b** Jebel Daggbah, NE of Duba. Elevation-invariant ZHe data show predominantly late Neoproterozoic cooling, a pre-rift partial retention zone (PRZ), and totally reset ZHe ages (~23 Ma), indicating >6 km of vertical throw across the bounding fault at Jebel Daggbah. AHe data were generally of poor quality, yielding reliable data only from the basal section, indicative of rapid Early Miocene exhumation, consistent with the ZHe data. **c** Jebel Liban, located in a horst block NE of Al Wajah and the Azlam half graben. Elevation-invariant ZHe data show both late Neoproterozoic and Early Carboniferous cooling, indicative of crossing a discrete Hercynian fault, with differential accommodation in Early Carboniferous times. AHe data are invariant (~23–25 Ma) across the entire sampled elevation range, recording early Miocene exhumation of the Jebel Liban horst block. **d** Jebel Radwa, located in the footwall of the border fault system NE of Yanbu, Saudi Arabia, shows elevation-invariant ZHe data that record rapid early Carboniferous cooling due to Hercynian (Carboniferous) block faulting, while AHe data along the range front record early Miocene exhumation of the rift flank. Data from Szymanski et al. (2016)

This second pulse appears to be temporally coincident with regional plate reorganization and establishment of the Dead Sea-Gulf of Aqaba transform.

This paper also presents previously unpublished apatite and zircon (U-Th)/He data from the northern Saudi Arabian Red Sea rift margin from vertical transects from Jebel Radwa (NE of Yanbu), Jebel Liban (NE of AL Wajah), Jebel Dabbagh (E of Dhuba), and Jebel Al Zhut (footwall of Midyan basin) (Figs. 13 and 14). In all three vertical transects, apatite (U-Th)/He data show elevation invariant cooling ages clustering around 23–25 Ma, suggesting very rapid exhumation of these fault blocks at the very onset of Red Sea normal faulting. In the Midyan area, the early Miocene cooling ages from Gebel Al Zhut (Fig. 14) are in excellent agreement with the earliest syn-rift strata of Aquitanian age (Tubbs et al. 2014). ZHe data from a >2000 m vertical transect at Jebel Dabbagh (Fig. 14) also record rapid early Miocene cooling. Exhumation of totally reset Miocene ZHe ages along the Red Sea margin is very rare and attests to the >6 km of throw along the border fault system in the Dhuba area. Most ZHe ages, however, from the three transects show discrete domains with contrasting Neoproterozoic, Carboniferous, and early Mesozoic ages. These domains appear to be bounded by $\sim$ N–S trending, sub-vertical faults that are more or less parallel to the Aqaba-Levant transform trend and appear to accommodate Hercynian block faulting. While other thermochronometric studies documented this Hercynian cooling event (e.g., Bojar et al. 2002; Szymanski et al. 2016), these are the first data that show discrete boundaries between crustal blocks/domains with different cooling ages and allow for the evaluation of the tectonic nature of these Hercynian faults. Surprisingly, the lower portion of Jebel Al Zhut in the Midyan area exhibits an early Mesozoic (Triassic) cooling signature, likely related to Mesozoic Tethyan rifting (Fig. 14). While there are Triassic syn-rift strata (Quseib Fm.) in the central Gulf of Suez, restoration of $\sim$107 km of left-lateral displacement along the Aqaba-Levant transform puts the Al Zhut block into the southernmost Gulf of Suez area, making it the southernmost area affected by Mesozoic Tethyan rifting.

In summary, all available thermochronometric data from the Egyptian and Saudi Arabian Red Sea margins suggest a near simultaneous onset of extensional faulting and fault block exhumation along-strike and across-strike during the stretching phase. These thermochronometric constraints are in excellent agreement with constraints from the earliest syn-rift sedimentation and magmatism.

13 Lack of Voluminous Syn-rift Magmatism

Rift-related magmatism in the Red Sea occurred synchronously along strike from the southern Red Sea to the Gulf of Suez, within uncertainty of the thermochronologic data (Bosworth and Stockli 2016). This argues very strongly against any kind of along-strike rift propagation. It is also noteworthy that during the main phase of Red Sea rifting, after 23 Ma and prior to inception of the Aqaba-Levant transform at $\sim$14 Ma, syn-extensional volcanism in the Red Sea is almost completely absent or has not yet been documented (Fig. 9). Furthermore, a significant flare-up in late Miocene and Pliocene volcanism in Saudi Arabia largely does not appear to follow the Red Sea trend, but rather seems to be focused along a N–S trend, the Makkah-Madinah-Nafud line, an inherited lithospheric structure (Hansen et al. 2007).

While the Tihama Asir and Al Lith magmatic complexes could be interpreted as earliest Miocene formation of new magmatic (proto-oceanic) crust in the southern Red Sea, syn-rift magmatism in the central and northern Red Sea is limited to the Red Sea dikes and two early Miocene basalt flows (Jeddah half-graben and Abu Zenima in the Gulf of Suez). There is no evidence from exposed or drilled early Miocene syn-rift strata from the Jeddah, Yanbu, Al Wahaj, and Midyan areas or the Red Sea for any syn-rift basalt outpourings. While there are minor syn-rift volcanic deposits (Szymanski et al. 2016; Ball et al. 2017), overall the remarkable lack of flood basalt volcanism in the proximal or distal continental margin does not seem to be compatible with the formation of magma-rich rift margins in the central and northern Red Sea and the Gulf of Suez. In addition, there does not seem to be any geophysical evidence for the presence of SDRs in the northern and central Red Sea. Some seismic refraction and potential field studies have interpreted the presence of an oceanic domain that stretched nearly coast to coast, but some of these interpretations are highly ambiguous and also conflict with geological interpretations, such as offshore borehole penetrations into continental crust, reflection seismic data showing faulting in sub-salt continental crust, or the subsidence history of the Red Sea basin. Ultimately, only long-offset crustal scale reflection seismic data is likely to resolve the exact geometry and nature of the crustal anatomy under the Red Sea.

14 Large-Magnitude Crustal Extension and Hyperextension

Given the fact that the northern and central Red Sea appear to lack nearly all characteristics of a typical volcanic margin, one should consider the alternatives and answer some very basic questions. If the Red Sea is a volcanic margin, is it a magma-poor margin? Is the northern Red Sea underlain by hyperextended crust or exhumed sub-lithospheric mantle? Has crustal and/or lithospheric separation occurred in the northern Red Sea and if so, when did this happen? These are clearly very fundamental questions and the constraints on the structural, stratigraphic, magmatic, and thermal evolution of the Red Sea should be able to provide some answers despite the lack of direct imaging.

As pointed out above, minor syn-rift basaltic magmatism is temporally limited to the very onset of rifting. Magmatism and faulting are very diffuse and occupy an ultra-wide (>1200 km) rift-perpendicular domain in the Gulf of Suez and northern Red Sea to very wide (>200–300 km) in the central Red Sea. This diffuse stretching of the Arabian and Nubian shield did not last very long. Rifting localized in the Gulf of Suez and the Red Sea still remained relatively diffuse with active half-grabens well inboard from the "border" fault system (e.g., Hamd-Jizil or Jeddah area). Crustal stretching and normal faulting as evidenced by syn-rift stratigraphy, structural reconstructions, and subsidence analysis continued from 23 to 14 Ma, with subsidence in the main rift trend accelerating and establishment of open-marine conditions from 19 Ma onward. Crustal stretching reached beta values of ~2 in the Gulf of Suez and likely the northern Red Sea—or a reduction in crustal thickness of 50%. Given pre-extensional elevations near sea level and isostatically-equilibrated crust (little to no pre-extensional relief), a ~50% crustal reduction would likely result in a crustal thickness of ~10–15 km by the Middle Miocene. This significant crustal thinning should lead to some decompression melting and hence, in the absence of significant volcanism, it appears reasonable to suggest some basaltic underplating during this early rift phase. Fault-bound Cenozoic gabbroic bodies encountered offshore central and southern Egypt, might in fact represent mafic material underplated during crustal stretching (e.g., Ligi et al. 2018).

Numerical modeling and geo-and thermochronometric evidence from several fossil hyper-extended margins has shown that the stretching phase can be associated with significant lower crustal heating and thermal weakening due to depth-dependent thinning of the mantle lithosphere (e.g., Beltrando et al. 2015; Smye and Stockli 2015; Seymour et al. 2016; Hart et al. 2017). Early Miocene U-Pb ages of zircon overgrowth from lower crustal granulites from Zabargad Island suggest high-temperature lower-crustal reheating of thinned Pan-African continental crust during extreme early syn-rift thinning and tectonic or diapiric juxtaposition against sublithospheric mantle (e.g., Oberli et al. 1987; Bosch and Bruguier 1998). The structural juxtaposition of the Zabargad lower-crust against both sub-lithospheric mantle and pre-rift sediments suggests a tectonic position in the hyper-extending distal portion of the continental margin. We interpret the high-temperature heating event to represent a thermal pulse related to depth-dependent thinning at a non-volcanic rift margin.

Thermochronometric data from the Gulf of Suez and the Saudi Arabian Red Sea margin illustrate that rapid fault block exhumation commenced in the earliest Miocene (20–24 Ma) (Fig. 14). However, the data from coastal Saudi Arabia also pointed to a second phase of rapid fault cooling starting at ~15 Ma, roughly coincident with the onset of Gulf of Aqaba-Levant transform motion and the resulting kinematic change and abrupt rift localization. This continued crustal deformation during Serravalian and Tortonian times does not seem to be compatible with an early Miocene transition to sea floor spreading. It appears much more likely that the kinematic change to orthogonal rifting led to an abrupt narrowing and localization of rifting and to crustal and lithospheric necking in the continued absence of voluminous magmatism. This kinematic shift and strain localization resulted in the transition from an early stretching phase to the necking phase of a mostly non-volcanic margin (Fig. 8).

Many questions remain regarding the next phase of rift evolution. For example, what is the duration of this necking phase? Is there evidence for an exhumation phase during which sublithospheric mantle gets exhumed? If and when does oceanic spreading start in the northern Red Sea? These are very controversial questions and the lack of publicly available deep seismic imaging greatly hampers our ability to resolve these questions. Colombo et al. (2014) present a crustal-scale seismic image slice from off the Saudi Arabia coast near Al Wajah with their tectonic interpretation that appears to show highly attenuated and tilted crustal blocks below up to 6 km of mobilized salt (Fig. 15). They interpret these 5–10 km thick crustal blocks as rider blocks at a distal hyper-extended margin and the presence of an exhumed mantle domain below the salt. Given the difficulties and challenges of interpretation of geophysical data in the Red Sea below >5 km of mobilized salt, their interpretation is certainly intriguing, but probably not unequivocal. The image, however, clearly illustrates the presence of distended and attenuated continental crust offshore Saudi Arabia (estimated >50 km). The presence of extended continental crust offshore Quseir (Figs. 13 and 16) or on Zabargad Island (70 km offshore) further corroborates the presence of highly-attenuated continental crust under at least part of the northern and central Red Sea. The presence of exhumed mantle, however, is much more tentative at this point. Importantly, the seismic image by Colombo et al. (2014) also very clearly showed the lack of SDRs in the distal part of the margin and also does not show any oceanic crust, which seriously undermines the interpretation of the northern and central Red Sea as a volcanic rifted margin (Figs. 15 and 17). Given the only tentative nature of the existence of exhumed mantle, it would probably not be prudent to interpret the Red Sea as an Atlantic-style hyperextended margin. More likely, it is a situation similar to that of the opening of the Gulf of California (Sea of Cortez), where orthogonal stretching resulted in high-magnitude crustal stretching, fault block rotation, and even core-complex

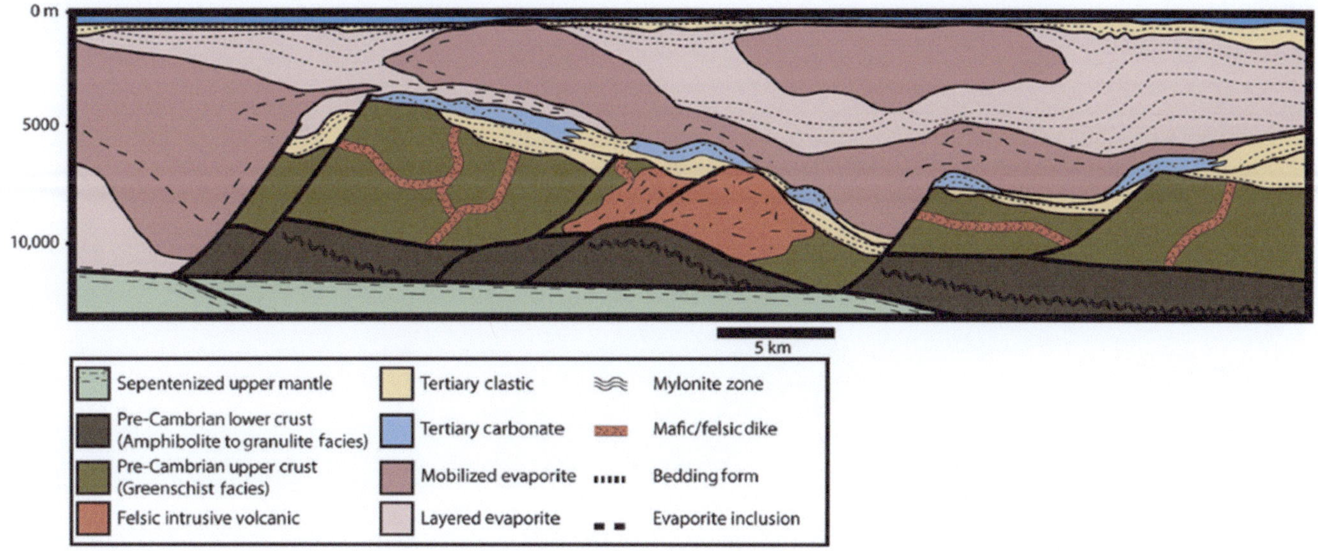

Fig. 15 Geological interpretation of crustal-scale seismic line of a portion of the Saudi Arabian rifted margin offshore Yanbu (modified after Colombo et al. 2014). The section shows >6 km of mobilized salt overlying asymmetrically tilted and attenuated continental crustal fault blocks in the distal margin of the Saudi Arabian central Red Sea. The authors interpreted the salt to lie directly on exhumed mantle SW of the faulted and attenuated crustal blocks of the distal margin. Their data do not appear to show any evidence for massive syn-rift magmatism or the presence of SDRs or basaltic underplating

Fig. 16 Uninterpreted and interpreted reflection seismic sections from the Egyptian portion of the northern Red Sea SE of Hurghada, Egypt. The interpretation of the imaged section is benchmarked by a borehole that penetrated Neoprotereozoic basement below the basal Miocene syn-rift (green) contact (red line) and unequivocally demonstrates the presence of continental basement offshore Egypt. The line also shows multiple NE-dipping low-angle normal faults that root within continental crust in the distal continental margin. Also visible is a flower structure (NE portion of section) related to Aqaba-Levant transform strike-slip faulting

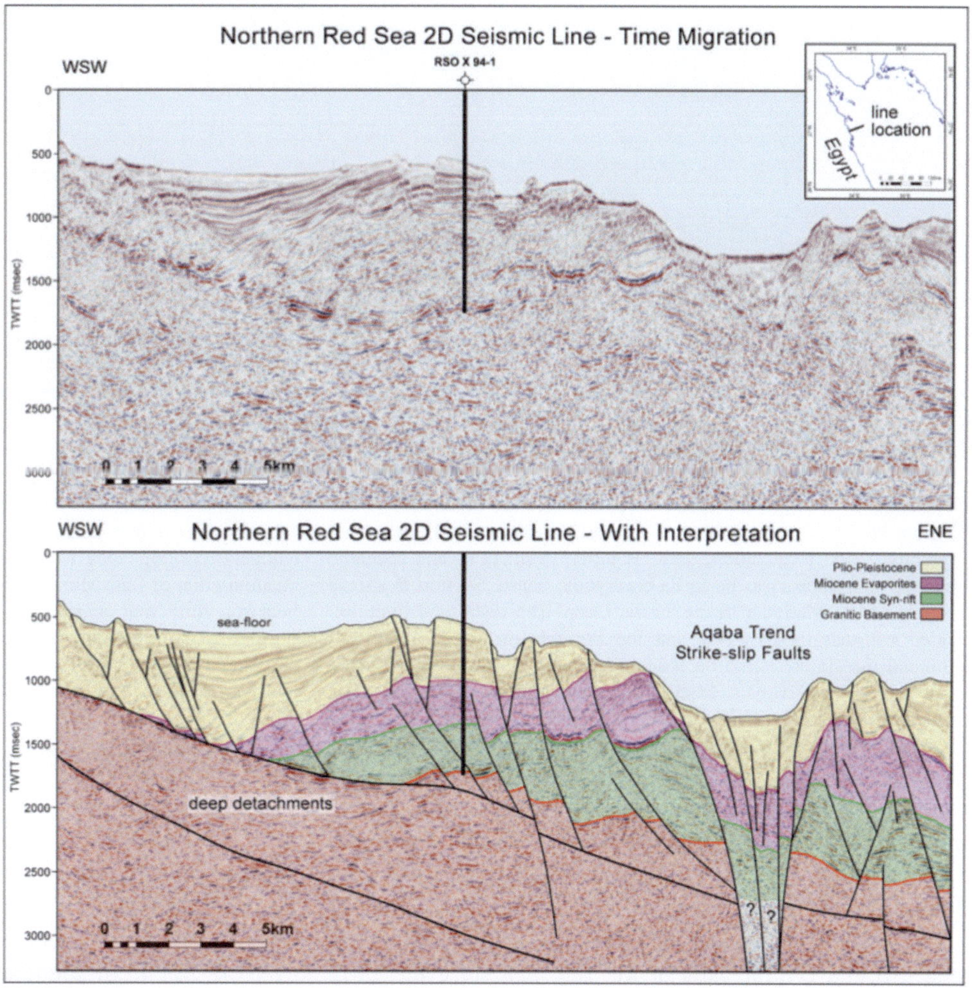

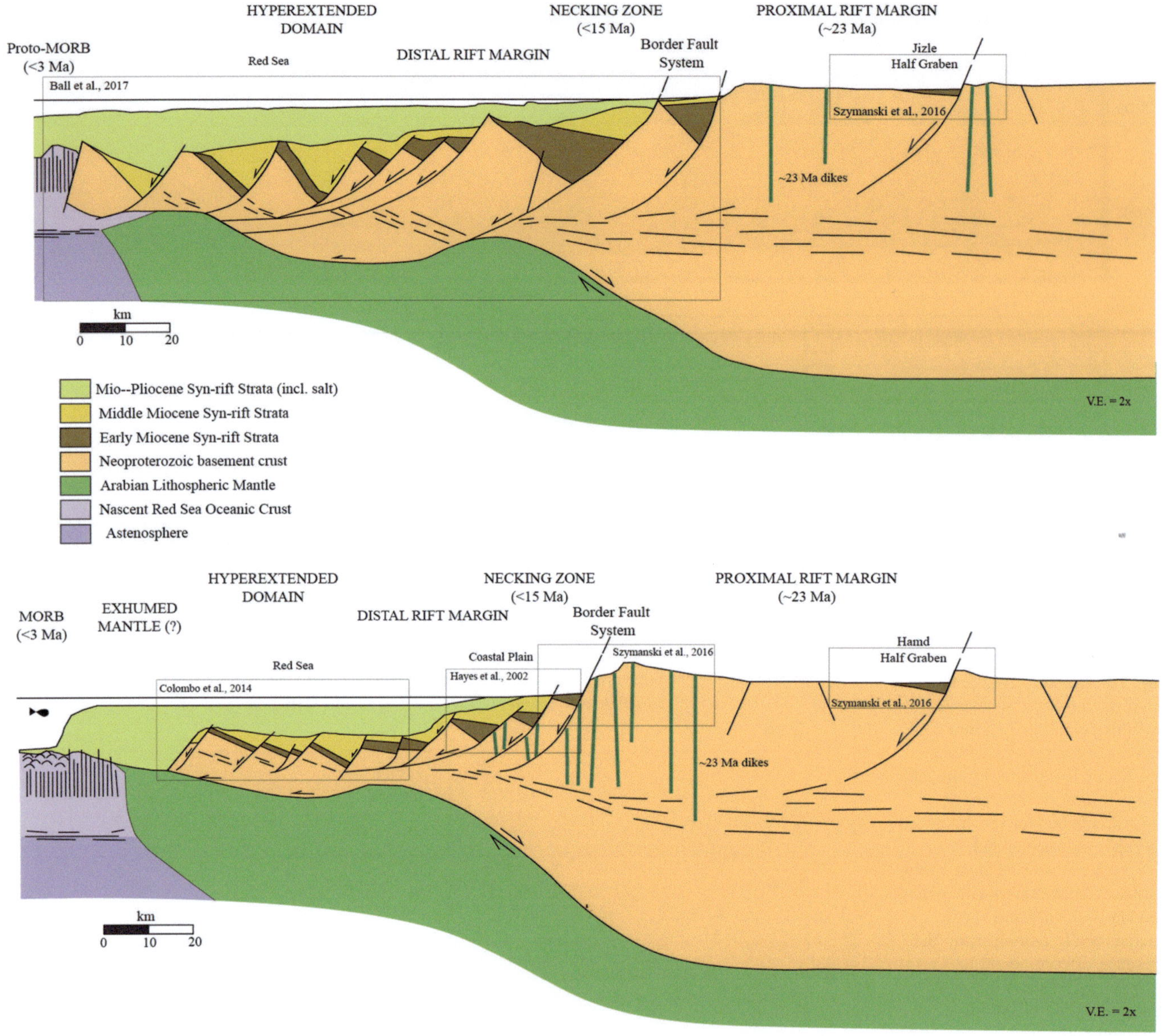

Fig. 17 Crustal-scale cross-sections across the northern and central Saudi Arabian Red Sea rift margins. Section A shows a section across the margin S of Al Wajah based on data from Ball et al. (2017) and Szymanski et al. (2016) that is characterized by Early Miocene stretching, rift localization, and H-block formation and oceanic break-up that appears to be in its embryonic phase. Section B shows a section from Madinah to the Thetis Deep. The distal and hyperextended rift margin geometry and the possible presence of sub-salt exhumed mantle are based on reflection seismic and drilling information, respectively (Hayes et al. 2002; Colombo et al. 2014). The presence of <3 Ma MORB at Thetis rift and the sub-salt extent of oceanic crust is based on gravity data and oceanographic investigations (Augustin et al. 2014, 2016, Ligi et al. 2015). Both sections show Early Miocene rift structures (e.g., Hamd-Jizl half grabens and border fault system), the presence of Early Miocene continental to marine syn-rift strata (brown), and earliest syn-rift basaltic dikes. Middle to Late Miocene cooling ages from near the coast combined with massive accumulation of Late Miocene evaporite and marine strata delineate the necking zone that appears to have formed at the transition from orthogonal to oblique rifting at ~14 Ma with the onset of Aqaba-Levant transform motion. Crustal thicknesses along the Saudi Arabian margin are based on data by Sandvol et al. (1998) and Hansen et al. (2007)

formation with necking and subsequent sea floor spreading only occurring after a kinematic shift to oblique extension and transtension (e.g., Lizarralde et al. 2007; Sutherland et al. 2012).

While there is clear evidence for seafloor spreading in the southern and central parts of the Red Sea since the early Pliocene (~4.5 Ma), evidence for seafloor spreading in the northern Red Sea is more controversial largely due to voluminous salt obscuring the subsalt geology. In the central Red Sea, Ligi et al. (2012) described the Thetis deep, documenting progressive accretion of oceanic crust from ~2.2 to <0.78 Ma from south to north. In the Nereus segment

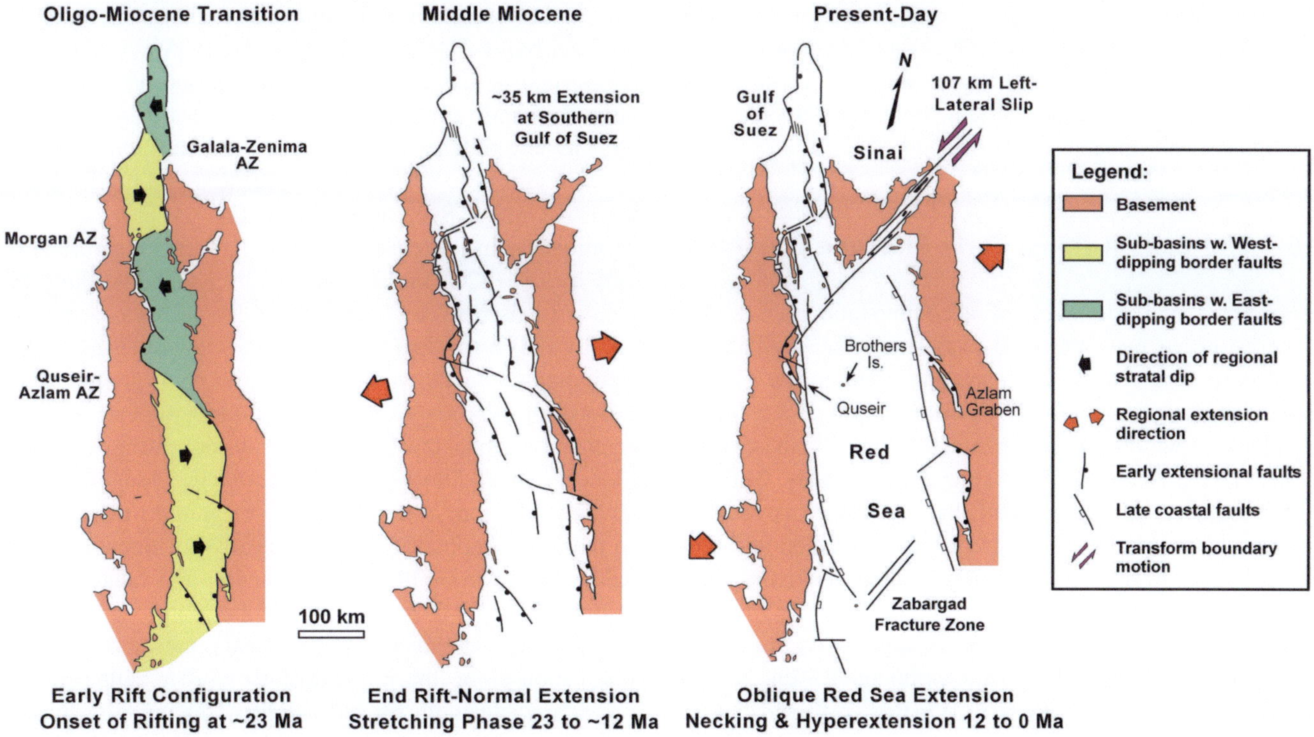

Fig. 18 Summary diagram of the tectonic and kinematic evolution of the Gulf of Suez and northern Red Sea, showing the initial rift configuration and segmentation at the Oligo-Miocene transition, the Early to Middle Miocene phase of orthogonal rifting reaching beta values of ~2, and the transition to oblique rifting since ~14 Ma due to the kinematic reorganization and inception of the Gulf of Aqaba transform, resulting in hyperextension, necking, and mantle exhumation, as well as embryonic mid-oceanic magmatism in Plio-Pleistocene times

oceanic magmatic accretion appears to have started at ~2 Ma (Ligi et al. 2011, 2012, 2015). The zone in between the Thetis and Nereus deeps lacks a central rift valley and discernible magnetic anomalies. Magnetic data from the northern part of the Hadarba Deep immediately south of Thetis, indicate an initial oceanic accretion at ~3 Ma, confirming its south to north progression. According to Ligi et al. (2012) the inter-trough zone between Thetis–Hadarba exposes fragments of continental basement, supporting the notion that much of the central and northern Red Sea is underlain by thinned and hyperextended continental crust. Augustin et al. (2016) beautifully documented the morphology and nature of volcanic activity along the central Red Sea spreading centres, partially covered by collapsing salt, south of the Zabargad Fracture Zone (ZFZ). Vertical gravity gradient analysis appears to reveal the presence and extent of oceanic crust as well as the ocean-continent boundary south of the ZFZ (Augustin et al. 2014, 2016). While the data suggest ~50% oceanic crust in the south-central Red Sea, there appears to be <25% oceanic crust immediately south of the ZFZ, which is dominated by mostly thinned continental crust. This appears to be in good agreement with observations from north of the ZFZ, where there is no clear evidence for the generation of well-organized oceanic crust, with the possible exception of the Mabahiss pull-apart basin. The basin is floored by young basalts at the northern terminus of the ZFZ and scattered deeps with basaltic and hydrothermal activity have been identified, such as the Shaban, Kebrit, and Oceanographer Deeps (Bonatti et al. 2015). These bathymetric deeps north of the ZFZ, with brine pools and metal mineralization along the axial trend of the central and northern Red Sea, appear to be linked to sub-salt magmatic activity, suggesting at least incipient magmatic accretion and seafloor spreading. It is possible, however, that the ultra-thick mobilized salt is retarding the establishment of normal sea floor spreading as the very thick blanket of salt could effectively cool and suppress magma generation and hence the establishment of an incipient oceanic spreading centre.

Overall, the geological evidence points to crustal separation in the Red Sea starting at ~4–5 Ma along the entire length of the Red Sea, including the northern portion. Sinistral slip rates along the Aqaba-Levant transform show a marked increase at ~5 Ma. This temporally coincides with initiation of active seafloor spreading in the southern Red Sea (e.g., Bonatti et al. 2015), arguing for a kinematically linked system. Discrete lithospheric separation appears to migrate northward from ~5 Ma in the southern Red Sea, to

2–3 Ma in the central Red Sea, to incipient in the northern Red Sea. Whether the northern Red Sea has gone through full crustal separation and has developed a full-blown sub-salt seafloor spreading system, however, has yet to be established and imaged.

15 The Case Against Early Miocene Breakup

Subsidence and formation of accommodation space was directly driven by continental extension. The similarity in subsidence and syn-rift sedimentation between the Gulf of Suez and the rest of the Red Sea clearly argues for a similar or identical setting of continental extension and strongly argues against any seafloor spreading in early Miocene times. This is also supported by continued extensional faulting with continental crust in a diffuse fashion and only limited strain localization, strongly arguing against an early rift-to-drift transition. Importantly, during the early Miocene syn-rift phase, basaltic volcanism largely ceased in the Red Sea, rendering this phase of major orthogonal rifting largely non-volcanic from ~19 to ~4 Ma (Bosworth and Stockli 2016).

Structural restorations and subsidence modeling both indicate ~35 km of total extension occurred across the southern Gulf of Suez during the early Miocene (Steckler et al. 1988; Bosworth 1995). This equates to a regional beta factor of ~1.6 and 2.0 in the rift axis—a stretching estimate supported by seismic refraction experiments that documented a ~50% reduction in crustal thickness along the axis of the southern Gulf of Suez (Gaulier et al. 1988).

Considering the estimated ~107 km of left-lateral displacement along the Aqaba-Levant transform, coupled with the similarity of tectonic subsidence and syn-rift accommodation between the Gulf of Suez and the Red Sea, it appears to be a reasonable assumption that the northern and central Red Sea also experienced a similar amount of stretching prior to the onset of Aqaba-Levant transform motion. This is also supported by seismically imaged and drilled continental crust well offshore along the Egyptian and Saudi Arabian Red Sea margins, offshore Quseir and Al Wajah, respectively (Fig. 17). These observations clearly argue for significant crustal extension, but strongly against any early Miocene crustal separation and/or seafloor spreading.

16 Conclusions

A systematic integration of constraints on the spatial distribution of Red Sea extensional faulting, the timing of syn-rift stratigraphy, the timing and distribution of magmatism, and a compilation of extensive regional thermochronometric dating was undertaken to systematically resolve the timing and spatial resolution of extensional faulting, rift localization, and hyper-extension along the Saudi Arabian Red Sea margin and to shed light on different proposed models for the geodynamic evolution of the Red Sea rift system. This large review of data demonstrates a synchronous onset of diffusive crustal extensional in the central and northern Red Sea and Gulf of Suez at ~23 Ma (Fig. 6). Early Miocene rift initiation and extensional faulting affected a surprisingly wide area (~1200 km) before rapidly localizing along the main rift axis between ~20 and 14 Ma and was accompanied by only very sparse basaltic magmatism (Fig. 9). The stratigraphic and thermochronometric data provide constraints for off-axis extension and the progressive strain localization along the border fault system during the early Miocene. In addition to the synchronous along-strike rift initiation and early Miocene large-magnitude exhumation, the thermochronometric data also clearly document a second middle Miocene (~14 Ma) exhumation and faulting event along the entire proximal margin of the central and northern Red Sea, accompanied by large-magnitude faulting and exhumation along the inboard of the border fault complex and large-wavelength rift flank uplift. This middle Miocene phase appears to be kinematically linked to the onset of faulting along the Aqaba transform fault and a switch from rift-normal to highly oblique extension within the Red Sea (Fig. 8). Left-lateral shearing and localized, large-magnitude oblique extension postdate the deposition of early Serravalian-age strata. This rifting phase resulted in major rift flank uplift (>2 km), uplift of Serravalian reef complexes along the crests of proximal extensional fault blocks, and an abrupt change in sedimentation characterized by widespread deposition of up to 800 m of middle to late Miocene evaporites (Fig. 12). We interpret this abrupt localization of large-magnitude extension, rift flank uplift, and major basin subsidence as the onset of crustal necking and crustal hypertension/thinning, triggered by the kinematic reorganization of the Red Sea and transition to oblique rifting. Evidence for extreme crustal attenuation associated with this thinning phase can be observed on Zabargad Island in the central Red Sea, where highly attenuated granulitic lower-crustal rocks are juxtaposed against upper-crustal Cretaceous pre-rift strata above and exhumed sub-lithospheric mantle below along low-angle normal faults at the distal continental margin (Fig. 17). Hyperextension and possibly mantle exhumation were accompanied by evaporite deposition and only sparse mafic magmatic activity during middle and late Miocene times (Fig. 17). It is possible that during the mantle exhumation phase, the thick evaporite blanket in the central and northern Red Sea effectively cooled the upper mantle and inhibited the onset of extension-related decompression melting, effectively retarding the onset of organized seafloor spreading. Seafloor spreading in the southern Red Sea and incipient or sputtering proto-seafloor spreading and basaltic magmatism likely did not start until ~10–12 Ma after Aqaba-Levant transform initiation and necking, signaling a phase of Late Miocene oblique hyperextension and eventual breakup (Fig. 18).

In summary, widespread extensional faulting defines the stretching phases that lasted from ~23 to 14 Ma and was followed by strain localization and low-angle normal faulting in response to crustal necking and high-magnitude crustal thinning (Fig. 18). Dynamic uplift of the rift flank and widespread uplift of early Miocene wave-cut platforms and Serravalian reefs is likely the result of depth-dependent mantle thinning and the establishment of the modern lithosphere-asthenosphere geometry. Apparent tectonic quiescence between middle Miocene necking and the onset of post-Miocene seafloor spreading and related magmatism is not supported by long-term plate reconstruction models and thus does not reflect a hiatus or slow-down in rifting, but rather a time window characterized by migration of deformation and progressive crustal thinning and hyperextension along the distal Red Sea margins, followed by possible exhumation of subcontinental mantle to the ocean floor during the exhumation phase (Fig. 17). While evidence of mantle exhumation is tentative, there is no published evidence for the formation of SDRs or a proto-oceanic crust, suggesting that rifting progressed largely in a non-volcanic, magma-poor fashion. Mantle exhumation in many magma-poor continental margins is commonly accompanied by massive evaporite deposition (e.g., South Atlantic or Gulf of Mexico). These new constraints on the spatial and temporal evolution of extensional faulting of the Red Sea in combination with existing structural, stratigraphic, and geophysical data suggest that the central and northern Red Sea represents a magma-poor hyperextended margin that progressed from early Miocene extension during the stretching phase, to middle Miocene necking and hyperextension during the thinning phase, to likely mantle exhumation during massive evaporite deposition, to eventual incipient organized to sputtering seafloor spreading since the Pliocene (Fig. 17).

Acknowledgements This work was partially supported by a US NSF Margins grant to Stockli (OCE-0305731) and logistical support from the Saudi Geological Survey. We would like to express our sincere gratitude to Dr. Zohair Nawab and Dr. Abdullah Alattas from the Saudi Geological Survey for their support and collaboration. We would like to acknowledge thorough and constructive reviews of the manuscript by P. Ball, M. Masini, G. Manatschal, and anonymous reviewer as well as editorial handling by N. Rasul and I. Stewart. We also would like to thank P. Johnson, F. Katan, K. Kadi, C. Hager, A. Al Shammari for help in the field and Y. Roman for assistance in the laboratory. We also grateful to G. Karner for stimulating discussions on continental margin tectonics and his persistent support for the Red Sea NSF Margins program.

References

Abdel Aal AY (1998) Mineral and chemical composition of basalts in the neighbourhood of Giza, Egypt. J Afr Earth Sci 26:101–117

Almalki KA, Betts PG, Ailleres L (2015) The Red Sea—50 years of geological and geophysical research. Earth Sci Rev 147:109–140

Andrew G (1937) The late tertiary igneous rocks of Egypt (field relations). Bull Fac Sci, Cairo University (10):61

Augustin N, Devey CW, van der Zwan FM, Feldens P, Tominaga M, Bantan RA, Kwasnitschka T (2014) The rifting to spreading transition in the Red Sea. Earth Planet Sci Lett 395:217–230

Augustin N, van der Zwan FM, Devey CW, Ligi M, Kwasnitschka T, Feldens P, Bantan RA, Basaham AS (2016) Geomorphology of the central Red Sea rift: determining spreading processes. Geomorphology 274:162–179. https://doi.org/10.1016/j.geomorph.2016.08.028

Baldridge WS, Eyal Y, Bartov Y, Steinitz G, Eyal M (1991) Miocene magmatism of Sinai related to the opening of the Red Sea. Tectonophysics 197:181–201

Ball P, Incerpi N, Birkle P, Lacsamana E, Manatschal G, Agar S, Zhang S, Borsato R (2017) Comparison of hydrothermal activity between the Adriatic and the Red Sea rift margins. EGU General Assembly Research Abstracts 19:5949–5951

Bartov Y, Steinitz G, Eyal M, Eyal U (1980) Sinistral movement along the Gulf of Aqaba—its age and relation to the opening of the Red Sea. Nature 285:220–221

Bayer H-J, Hotzl H, Jado AR, Roscher B, Voggenreiter W (1988) Sedimentary and structural evolution of the northwest Arabian Red Sea margin. Tectonophysics 153:137–151

Beadnell HJL (1902) The Cretaceous region of Abu Roash, near the Pyramids of Giza. Egypt, Egyptian Survey Dept, p 48

Beadnell HJL (1905) The topography and geology of the Fayum province of Egypt. Egyptian Survey Dept, 101 pp

Beltrando M, Stockli DF, Decarlis A, Manatschal G (2015) A crustal-scale view at rift localization along the fossil Adriatic margin of the Alpine Tethys preserved in NW Italy. Tectonics 34 (9):1927–1951

Beyth M (1972) Paleozoic-Mesozoic sedimentary basin of Mekele outlier, northern Ethiopia. Am Assoc Petrol Geol Bull 56:2426–2439

Blank HR, Gettings ME (1985) Reconnaissance geologic map of the Jizan Quadrangle, sheet 16/42B, Kingdom of Saudi Arabia (No. 85-724). US Geological Survey and Saudi Arabian Deputy Ministry for Mineral Resources, Open-File Report 85-724

Bohannon RG (1986) Tectonic configuration of the western Arabian continental margin, southern Red Sea. Tectonics 5(4):477–499

Bohannon RG, Naeser CW, Schmidt DL, Zimmermann RA (1989) The timing of uplift, volcanism, and rifting peripheral to the Red Sea: a case for passive rifting? J Geophys Res Solid Earth 94(B2):1683–1701

Bohannon RG (1989) Style of extensional tectonism during rifting, Red Sea and Gulf of Aden. J Afr Earth Sci (and Middle East) 8(2–4):589–602

Bojar AV, Fritz H, Kargl S, Unzog W (2002) Phanerozoic tectonothermal history of the Arabian-Nubian shield in the Eastern Desert of Egypt: evidence from fission track and paleostress data. J Afr Earth Sci 34(3):191–202

Bonatti E, Cipriani A, Lupi L (2015) The Red Sea: birth of an ocean. In: Rasul NMA, Stewart ICF (eds) The Red Sea: the formation, morphology, oceanography and environment of a young ocean basin. Springer Earth System Sciences, Berlin, pp 29–44

Bosch D, Bruguier O (1998) An early Miocene age for a high-temperature event in gneisses from Zabargad Island (Red Sea, Egypt): Mantle diapirism? Terra Nova 10(5):274–279

Bosence DWJ (1998) Stratigraphic and sedimentological models of rift basins. In: Purser BH, Bosence DWJ (eds) Sedimentation and tectonics of rift basins: Red Sea-Gulf of Aden. Chapman and Hall, London, pp 9–25

Bosworth W (1994) A model for the three-dimensional evolution of continental rift basins, north-east Africa. Geologisch Rundschau 83:671–688

Bosworth W (1995) A high-strain rift model for the southern Gulf of Suez (Egypt). In: Lambiase JJ (ed) Hydrocarbon Habitat in Rift Basins. Geol Soc London, Spec Publ 80, pp 75–112

Bosworth W (2015) Geological evolution of the Red Sea: historical background, review, and synthesis. In: Rasul NMA, Stewart ICF (eds) The Red Sea: the formation, morphology, oceanography and environment of a Young Ocean Basin. Springer Earth System Sciences, Berlin, pp 45–78

Bosworth W, Huchon P, McClay K (2005) The Red Sea and Gulf of Aden Basins. J Afr Earth Sci 43(1–3):334–378

Bosworth W, Crevello P, Winn RD, Steinmetz J (1998) Structure, sedimentation, and basin dynamics during rifting of the Gulf of Suez and northwestern Red Sea. In: Purser BH, Bosence DWJ (eds) Sedimentation and tectonics of rift basins: Red Sea-Gulf of Aden. Chapman and Hall, London, pp 77–96

Bosworth W, Stockli DF, Helgeson DE (2015) Integrated outcrop, 3D seismic, and geochronologic interpretation of Red Sea dike-related deformation in the Western Desert, Egypt—the role of the 23 Ma Cairo "mini-plume". J Afr Earth Sci 109:107–119

Bosworth W, McClay K (2001) Structural and stratigraphic evolution of the Gulf of Suez rift, Egypt: a synthesis. In: Ziegler PA, Cavazza W, Robertson AHF, Crasquin-Soleau S (eds) Peri-Tethys Memoir 6: Peri-Tethyan Rift/Wrench Basins and Passive Margins. Memoires du Museum National d'Histoire Naturelle de Paris 186, pp 567–606

Bosworth W, Stockli DF (2016) Early magmatism in the greater Red Sea rift: timing and significance. Can J Earth Sci 53(11):1158–1176

Brown GF (1972) Tectonic map of the Arabian Peninsula. US Geological Survey, scale 1:4,000,000

Buck WR (1986) Small-scale convection induced by passive rifting: the cause for uplift of rift shoulders. Earth Planet Sci Lett 77:362–372

Buck WR, Martinez F, Steckler MS, Cochran JR (1988) Thermal consequences of lithospheric extension: pure and simple. Tectonics 7:213–234

Bunter MAG, Abdel Magid AEM (1989) The Sudanese Red Sea: 1. New developments in stratigraphy and petroleum-geological evolution. J Petrol Geol 12:145–166

Bunter MAG, Debretsion T, Woldegiorgis L (1998) New developments in the pre-rift prospectivity of the Eritrean Red Sea. J Petrol Geol 21 (4):373–400

Burke K (1996) The African plate. South Afr J Geol 99:341–409

Burke K, Dewey JF (1973) Plume-generated triple junctions: key indicators in applying plate tectonics to old rocks. J Geology 81 (4):406–433

Carella R, Scarpa N (1962) Geological results of exploration in Sudan by AGIP Mineraria. Proc Fourth Arab Petrol Cong, Beirut, p 23

Cocker JD, Hughes GW (1993) Seawater strontium stratigraphy applied to anhydrites; tertiary basins of the Saudi Arabian Red Sea. Geol Soc London meeting, May 1993

Coleman RG (1993) Geologic evolution of the Red Sea. Oxford University Press

Coleman RG, Hadley DG, Fleck RG, Hedge CT, Donato MM (1979) The Miocene Tihama Asir Ophiolite and its bearing on the opening of the Red Sea. In: Al-Shanti AMS (ed) Evolution and mineralization of the Arabian-Nubian shield. Inst Applied Geology Bull 3(1). Pergamon Press, Oxford, pp 173–186

Colombo D, McNeice G, Raterman N, Zinger M, Rovetta D, Curiel ES (2014) Exploration beyond seismic: the role of electromagnetics and gravity gradiometry in deep water subsalt plays of the Red Sea. Interpretation 2(3):SH33–SH53

Davies FB (1985) Geologic map of the Al Wajh quadrangle, sheet 26B, Kingdom of Saudi Arabia. Saudi Arabian Deputy Ministry for Mineral Resources Geoscience Map GM 83, scale 1:250,000, 27 pp

De Sitter LU (1962) Structural development of the Arabian Shield in Palestine. Geol Mijnbouw 41:116–124

Dixon TH, Stern RJ, Hussein IM (1987) Control of Red Sea rift geometry by Precambrian structures. Tectonics 6:551–571

Dubertret L (1970) Review of structural geology of the Red Sea and surrounding area. Phil Trans Roy Soc London A 267:9–20

Ebinger CJ, Sleep NH (1998) Cenozoic magmatism throughout east Africa resulting from impact of a single plume. Nature 395 (6704):788

Ebinger CJ, Casey M (2001) Continental breakup in magmatic provinces: an Ethiopian example. Geology 29:527–530

Edgell HS (1992) Basement tectonics of Saudi Arabia as related to oil field structures. Basement Tectonics 9:169–193

EGPC (1964) Oligocene and miocene rock-stratigraphy of the Gulf of Suez region. Consultative Stratigraphic Committee of the Egyptian General Petroleum Corporation, Cairo, Report E.R. 575, pp 1–142

Endress C, Furman T, Abu El-Rus MA, Hanan BB (2011) Geochemistry of 24 Ma basalts from NE Egypt: Source components and fractionation history. In: Van Hinsbergen DJJ, Buiter SJH, Torsvik TH, Gaina C, Webb SJ (eds) The formation and evolution of Africa: a synopsis of 3.8 Ga of Earth History. Geol Soc London, Spec Publ 357, pp 265–283

Evans AL (1988) Neogene tectonic and stratigraphic events in the Gulf of Suez rift area, Egypt. Tectonophysics 153:235–247

Eyal M, Eyal Y, Bartov Y, Steinitz G (1981) The tectonic development of the western margin of the Gulf of Elat (Aqaba) rift. Tectonophysics 80(1–4):39–66

Fawzy H, Abdel Aal A (1986) Regional study of Miocene evaporites and Pliocene-Recent sediments in the Gulf of Suez. In: Proceedings of 7th Exploration seminar, March 1984. Egyptian General Petroleum Corp and Egypt Petroleum Exploration Soc, Cairo, pp 49–74

Garfunkel Z, Bartov Y (1977) The tectonics of the Suez rift. Geol Survey Israel Bull 71(44):1–48

Garson MS, Krs M (1976) Geophysical and geological evidence of the relationship of Red Sea transverse tectonics to ancient fractures. Geol Soc Am Bull 87:169–181

Gass IG (1970) The evolution of volcanism in the junction area of the Red Sea, Gulf of Aden and Ethiopian rifts. Phil Trans Roy Soc London 267A:369–381

Gaulier JM, Le Pichon X, Lyberis N, Avedik F, Geli L, Moretti I, Deschamps A, Hafez S (1988) Seismic study of the crust of the northern Red Sea and Gulf of Suez. Tectonophysics 153:55–88

Gawthorpe RL, Sharp I, Underhill JR, Gupta S (1997) Linked sequence stratigraphic and structural evolution of propagating normal faults. Geology 25:795–798

Gernignon L, Planke S, Ringenbach JC, Le Gall B (2005) Tectonic and deep crustal structures along the Norwegian volcanic margin: implications for the "Mantle Plume or Not?" debate. www.MantlePlumes.org

Ghassal BI (2010) Petroleum geochemistry and geology of the Midyan and Al Wajah basins, northern Red Sea, Saudi Arabia. Unpublished MSc thesis, University of Utah, Salt Lake City

Gillman M (1968) Preliminary results of a geological and geophysical reconnaissance of the Jizan coastal plain in Saudi Arabia. In: Second Reg Tech Symp Proceedings, AIME, Saudi Arabian Sect. Dhahran, Saudi Arabia, pp 189-208

Gladczenko TP, Hinz K, Eldholm O, Meyer H, Neben S, Skogseid J (1997) South Atlantic volcanic margins. J Geol Soc London 154 (3):465–470

Guiraud R, Issawi B, Bosworth W (2001) Phanerozic history of Egypt and surrounding areas. In: Ziegler PA, Cavazza W, Robertson AHF, Crasquin-Soleau S (eds) Peri-Tethys Memoir 6: Peri- Tethyan Rift/Wrench Basins and Passive Margins. Mémoires du Muséum National d'Histoire Naturelle de Paris 186, pp 469–509

Hassan F, El-Dashlouty S (1970) Miocene evaporites of Gulf of Suez region and their significance. Bull Am Assoc Petrol Geol 54:1686–1696

Hansen SE, Rodgers AJ, Schwartz SY, Al-Amri AM (2007) Imaging ruptured lithosphere beneath the Red Sea and Arabian Peninsula. Earth Planet Sci Lett 259(3):256–265

Hart NR, Stockli DF, Lavier LL, Hayman NW (2017) Thermal evolution of a hyperextended rift basin, Mauléon Basin, western Pyrenees. Tectonics 36. https://doi.org/10.1002/2016tc004365

Hayes TS, Sutley SJ, Kadi KA, Balkhiyour, MB, Siddiqui, AA, Beshir Z, Hashem, HI (2002) Jabal Dhaylan zinc-lead deposits, geological setting, genesis, and 1996–2000 exploration programs. Saudi Geological Survey Open File Report SGS-OF-2001-5, 71 pp

Holbrook WS, Kelemen PB (1993) Large igneous province on the US Atlantic margin and implications for magmatism during continental breakup. Nature 364(6436):433–436

Hopper JR, Dahl-Jensen T, Holbrook WS, Larsen HC, Lizarralde D, Korenaga J, Kent GM, Kelemen PB (2003) Structure of the SE Greenland margin from seismic reflection and refraction data: Implications for nascent spreading center subsidence and asymmetric crustal accretion during North Atlantic opening. J Geophys Res Solid Earth 108(B5):2269–2291

Hosny A, Nyblade A (2014) Crustal structure in southeastern Egypt: Symmetric thinning of the northern Red Sea rifted margins. Geology 42:219–222

Hughes GW, Beydoun ZR (1992) The Red Sea-Gulf of Aden: Biostratigraphy, lithostratigraphy and palaeoenvironments. J Petrol Geol 15:135–156

Hughes GW, Perincek D, Grainger DJ, Abu-Bshait A-J, Jarad A-RM (1999) Lithostratigraphy and depositional history of part of the Midyan region, northwestern Saudi Arabia. GeoArabia 4:503–541

Hughes GW, Filatoff J (1995) New biostratigraphic constraints on Saudi Arabian Red Sea pre- and syn-rift sequences. In: Al-Husseini MI (ed) Middle east petroleum geosciences, Geo-94, vol 2. Gulf PetroLink, Bahrain, pp 517–528

Hughes GW, Johnson RS (2005) Lithostratigraphy of the Red Sea region. GeoArabia 10:49–126

Hutchinson RW, Engels GG (1970) Tectonic evolution in the southern Red Sea and its possible significance to older rifted continental margins. Geol Soc Am Bull 83:2989–3002

Johnson PR (2014) An expanding Arabian-Nubian Shield geochronologic and isotopic dataset: Defining limits and confirming the tectonic setting of a Neoproterozoic accretionary orogen. Open Geol J 8(1):3–33. https://doi.org/10.2174/1874262901408010003

Johnson PR, Woldehaimanot B (2003) Development of the Arabian-Nubian Shield: perspectives on accretion and deformation in the northern East African Orogeny and the assembly of Gondwana. Geol Soc, London, Spec Publ 206(1):289–325

Johnson P, Kattan F (2012) The geology of the Saudi Arabian Shield. Saudi Geological Survey, Jeddah, Saudi Arabia, p 479

Kohn BP, Eyal M (1981) History of uplift of the crystalline basement of Sinai and its relation to opening of the Red Sea as revealed by fission track dating of apatites. Earth Planet Sci Lett 52:129–141

Koshnaw RI, Horton BK, Stockli DF, Barber DE, Tamar-Agha MY, Kendall JJ (2017) Neogene shortening and exhumation of the Zagros fold-thrust belt and foreland basin in the Kurdistan region of northern Iraq. Tectonophysics 694:332–355

Lavier LL, Manatschal G (2006) A mechanism to thin the continental lithosphere at magma-poor margins. Nature 440(7082):324–328

Le Pichon X, Gaulier JM (1988) The rotation of Arabia and the Levant fault system. Tectonophysics 153:209–220

Ligi M, Bonatti E, Tontini FC, Cipriani A, Cocchi L, Schettino A, Bortoluzzi G, Ferrante V, Khalil S, Mitchell NC, Rasul N (2011) Initial burst of oceanic crust accretion in the Red Sea due to edge-driven mantle convection. Geology 39(11):1019–1022

Ligi M, Bonatti E, Bortoluzzi G, Cipriani A, Cocchi L, Caratori Tontini F, Carminati E, Ottolini L, Schettino A (2012) Birth of an ocean in the Red Sea: initial pangs. Geochem Geophys Geosyst 13 (8):Q08009. https://doi.org/10.1029/2012GC004155

Ligi M, Bonatti E, Rasul NMA (2015) Seafloor spreading initiation: Geophysical and geochemical constraints from the Thetis and Nereus Deeps, central Red Sea. In: Rasul NMA, Stewart ICF (eds) The Red Sea: the formation, morphology, oceanography and environment of a Young Ocean Basin. Springer Earth System Sciences, Berlin, pp 79–98

Ligi M, Bonatti E, Bosworth W, Cai Y, Cipriani A, Pamiotto C, Ronca S, Seyler M (2018) Birth of an ocean in the Red Sea: Oceanic-type basaltic melt intrusions precede continental rupture. Gondwana Res 54:150–160

Lister GS, Etheridge MA, Symonds PA (1986) Detachment faulting and the evolution of passive continental margins. Geology 14:246–250

Lister GS, Etheridge MA, Symonds P (1991) Detachment models for the formation of passive continental margins. Tectonics 10:1038–1064

Lizarralde D, Axen GJ, Brown HE, Fletcher JM, González-Fernández A, Harding AJ, Holbrook WS, Kent GM, Paramo P, Sutherland F, Umhoefer PJ (2007) Variation in styles of rifting in the Gulf of California. Nature 448(7152):466

Manatschal G (2004) New models for evolution of magma-poor rifted margins based on a review of data and concepts from West Iberia and the Alps. Int J Earth Sci 93:432–466

Mohriak W (2014) Birth and development of continental margin basins: Analogies from the South Atlantic, North Atlantic, and the Red Sea. Am Assoc Petrol Geol Search Discovery, Article 41502

Montenat C, D'Estevou PO, Purser B, Burollet PF, Jarrige JJ, Orszag-Sperber F, Philobbos E, Plaziat JC, Prat P, Richert JP, Roussel N (1988) Tectonic and sedimentary evolution of the Gulf of Suez and the northwestern Red Sea. Tectonophysics 153(1–4):161–177

Mouthereau F, Lacombe O, Vergés J (2012) Building the Zagros collisional orogen: timing, strain distribution and the dynamics of Arabia/Eurasia plate convergence. Tectonophysics 532:27–60

Oberli F, Ntaflos T, Meier M, Kurat G (1987) Emplacement age of the peridotites from Zabargad Island (Red Sea): A zircon U-Pb isotope study. Terra Cognita 7:334

Omar GI (1985) Phanerozoic tectono-thermal history of the Nubian massif, Eastern Desert, Egypt and its relationship to the opening of the Red Sea as revealed by fission track studies. PhD thesis, University of Pennsylvania, Philadelphia, 156 pp

Omar GI, Steckler MS, Buck WR, Kohn BP (1989) Fission-track analysis of basement apatites at the western margin of the Gulf of Suez rift, Egypt: Evidence for synchronicity of uplift and subsidence. Earth Planet Sci Lett 94:316–328

Omar GI, Kohn BP, Lutz TM, Faul H (1987) The cooling history of Silurian to Cretaceous alkaline ring complexes, south Eastern Desert, Egypt, as revealed by fission-track analysis. Earth Planet Sci Lett 83:94–108

Omar GI, Steckler MS (1995) Fission-track evidence on the initial opening of the Red Sea: two pulses, no propagation. Science 270:1341–1344

Ott d'Estevou P, Bolze J, Montenat C (1986) Etude géologique de la marge occidentale du Golfe de Suez: le Massif des Gharamul e le Gebel Dara. Documents et Travaux, Institut Géologique Albert de Lapparent 10:19–44
Pallister JS (1987) Magmatic history of Red Sea rifting: Perspective from the central Saudi Arabian coastal plain. Geol Soc Am Bull 98:400–417
Péron-Pinvidic G, Manatschal G (2009) The final rifting evolution at deep magma-poor passive margins from Iberia-Newfoundland: A new point of view. Int J Earth Sci 98(7):1581–1597
Plaziat J-C, Montenat C, Barrier P, Janin M-C, Orszag-Sperber F, Philobbos E (1998) Stratigraphy of the Egyptian syn-rift deposits: Correlations between axial and peripheral sequences of the north-western Red Sea and Gulf of Suez and their relations with tectonics and eustacy. In: Purser BH, Bosence DWJ (eds) Sedimentation and Tectonics of Rift Basins: Red Sea-Gulf of Aden. Chapman and Hall, London, pp 211–222
Pujols EJ (2011) Structural evolution and timing of extensional faulting in the central Gulf of Suez. Unpubl MSc thesis, University of Kansas, Lawrence, Kansas, 145 pp
Quennell AM (1951) The geology and mineral resources of (former). In: Jordan (Trans) HM Stationery Office
Quennell AM (1958) The structural and geomorphic evolution of the Dead Sea rift. Q J Geol Soc 114(1–4):1–24
Richardson M, Arthur MA (1988) The Gulf of Suez—northern Red Sea Neogene rift: A quantitative basin analysis. Marine Petrol Geol 5:247–270
Rittmann A (1954) Remarks on the eruptive mechanism of the Tertiary volcanoes of Egypt. Bull Volcanologique 15:109–117
Robinson FA, Foden JD, Collins AS, Payne JL (2014) Arabian Shield magmatic cycles and their relationship with Gondwana assembly: insights from zircon U-Pb and Hf isotopes. Earth Planet Sci Lett 408:207–225
Roussel N, Purser BH, Orszag-Sperber F, Plaziat J-C, Soliman M, El Haddad AA (1986) Géologie de la région de Qusier. Documents et Travaux, Institut Géologique Albert de Lapparent, Paris 10:129–144
Sandvol E, Seber D, Barazangi M, Vernon F, Mellors R, Al-Amri A (1998) Lithospheric seismic velocity discontinuities beneath the Arabian Shield. Geophys Res Lett 25:2873–2876
Savoyat E, Shiferaw A, Balcha T (1989) Petroleum exploration in the Ethiopian Red Sea. J Petrol Geol 12:187–204
Scott RW, Govean FM (1985) Early depositional history of a rift basin: Miocene in the western Sinai. Palaeogeog Palaeoclimat Palaeoecol 52:143–158
Sebai A, Zumbo V, Féraud G, Bertrand H, Hussain AG, Giannerini G, Campredon R (1991) $^{40}Ar/^{39}Ar$ dating of alkaline and tholeiitic magmatism of Saudi Arabia related to the early Red Sea rifting. Earth Planet Sci Lett 104:473–487
Sellwood BW, Netherwood RE (1984) Facies evolution in the Gulf of Suez area: sedimentation history as an indicator of rift initiation and development. Mod Geol 9:43–69
Sengör AHC, Burke K (1978) Relative timing of rifting and volcanism on earth and its tectonic implications. Geophys Res Lett 5:419–421
Seymour NM, Stockli DF, Beltrando M, Smye AJ (2016) Tracing the thermal evolution of the corsican lower crust during tethyan rifting. Tectonics 35. https://doi.org/10.1002/2016tc004178
Shallaly NA, Beier C, Haase KM, Hammed MS (2013) Petrology and geochemistry of the Tertiary Suez rift volcanism, Sinai, Egypt. J Volcanol Geoth Res 267:119–137
Smye A, Stockli DF (2015) Rutile U-Pb depth profiling: a continuous record of lithospheric thermal evolution. Earth Planet Sci Lett 408:171–182
Stampfli GM, Marchant RH (1997) Geodynamic evolution of the Tethyan margins of the Western Alps. In: Pfiffner OA, Lehner P, Heitzmann P, Mueller S, Steck A (eds) Deep structure of the Swiss Alps—results from NRP 20. Birkhaüser, pp 223–239
Steckler MS, ten Brink US (1986) Lithospheric strength variations as a control on new plate boundaries: Examples from the northern Red Sea region. Earth Planet Sci Lett 79:120–132
Steckler MS, Berthelot F, Lyberis N, Le Pichon X (1988) Subsidence in the Gulf of Suez: implications for rifting and plate kinematics. Tectonophysics 153:249–270
Stern RJ, Johnson PR (2010) Continental lithosphere of the Arabian Plate: a geologic, petrologic, and geophysical synthesis. Earth-Sci Rev 101(1):29–67
Steen G (1984) Radiometric age dating of some Gulf of Suez igneous rocks. In: Proceedings of the 6th Exploration conference, Egyptian General Petroleum Corporation, Cairo
Stica JM, Zalán PV, Ferrari AL (2014) The evolution of rifting on the volcanic margin of the Pelotas Basin and the contextualization of the Paraná-Etendeka LIP in the separation of Gondwana in the South Atlantic. Marine Petrol Geol 50:1–21
Stockli DF (2005) Application of low-temperature thermochronometry to extensional tectonic settings. In: Reiners P, Ehlers T (eds) Thermochronometry. Rev Mineral Geochem 58:420–461
Stoeser DB, Camp VE (1985) Pan-African microplate accretion of the Arabian shield. Geol Soc Am Bull 96:817–826
Sutherland FH, Kent GM, Harding AJ, Umhoefer PJ, Driscoll NW, Lizarralde D, Fletcher JM, Axen GJ, Holbrook WS, González-Fernández A, Lonsdale P (2012) Middle Miocene to early Pliocene oblique extension in the southern Gulf of California. Geosphere 8 (4):752–770
Sutra E, Manatschal G (2012) How does the continental crust thin in a hyperextended rifted margin? Insights from the Iberia margin. Geology 40(2):139–142
Szymanski E, Stockli DF, Johnson PR, Hager C (2016) Thermochronometric evidence for diffuse extension and two-phase rifting within the central Arabian margin of the Red Sea rift. Tectonics 35(12):2863–2895
Tamsett D (1984) Comments on the development of rifts and transform faults during continental breakup; examples from the Gulf of Aden and northern Red Sea. Tectonophysics 104:35–46
Tesfaye S, Ghebreab W (2013) Simple shear detachment fault system and marginal grabens in the southernmost Red Sea rift. Tectonophysics 608:1268–1279
Tubbs RE, Fouda H, Afifi A, Raterman N, Hughes G, Fadolalkarem Y (2014) Midyan Peninsula, northern Red Sea, Saudi Arabia: Seismic imaging and regional interpretation. GeoArabia 19(3):165–184
Tugend J, Manatschal G, Kusznir NJ, Masini E, Mohn G, Thinon I (2014) Formation and deformation of hyperextended rift systems: Insights from rift domain mapping in the Bay of Biscay-Pyrenees. Tectonics 33(7):1239–1276
Unternehr P, Péron-Pinvidic G, Manatschal G, Sutra E (2010) Hyper-extended crust in the South Atlantic: in search of a model. Petrol Geosci 16(3):207–215
Voggenreiter W, Hotzl H (1989) Kinematic evolution of the southwestern Arabian continental margin: Implications for the origin of the Red Sea. J Afr Earth Sci 8:541–564
Voggenreiter W, Hötzl H, Mechie J (1988) Low-angle detachment origin for the development of the Red Sea? Tectonics 150:51–75
Weissel JK, Karner GD (1989) Flexural uplift of rift flanks due to mechanical unloading of the lithosphere during extension. J Geophys Res 94:13919–13950
Wernicke B (1985) Uniform-sense normal simple shear of the continental lithosphere. Can J Earth Sci 22:108–125
Wescott WA, Krebs WN, Dolson JC, Ramzy M, Karamat SA, Moustafa T (1997) Chronostratigraphy, sedimentary facies, and architecture of tectono-stratigraphic sequences: An integrated approach to rift basin exploration, Gulf of Suez, Egypt. In: Shanley KW, Perkins BF (eds) Gulf Coast Section SEPM Foundation 18th Annual Research Conference, Shallow Marine and Nonmarine Reservoirs, 7–10 Dec 1997, pp 377–399

White RS, Westbrook GK, Fowler SR, Spence GD, Barton PJ, Joppen M, Morgan J, Bowen AN, Prestcott C, Bott MHP (1987) Hatton Bank (northwest UK) continental margin structure. Geophys J Int 89(1):265–272

White R, McKenzie D (1989) Magmatism at rift zones: The generation of volcanic continental margins and flood basalts. J Geophys Res Solid Earth 94(B6):7685–7729

Williams G, Small J (1984) A study of the Oligo-Miocene basalts in the Western Desert. In: Proceedings of the 7th Petroleum exploration seminar. Egyptian General Petroleum Corporation, Cairo, pp 252–268

Winn RD, Crevello PD, Bosworth W (2001) Lower Miocene Nukhul Formation of Gebel el Zeit, Egypt: Sedimentation and structural movement during early Gulf of Suez rifting. Am Assoc Petrol Geol Bull 85:1871–1890

Zahran HM, Stewart ICF, Johnson PR, Basahel MH (2003) Aeromagnetic-anomaly maps of central and western Saudi Arabia. Scale 1:2 million. Saudi Geological Survey Open-File Report SGS-OF-2002-8, 6 pp, 4 plates

The Nature of Upper Mantle Upwelling During Initiation of Seafloor Spreading in the Southern Red Sea

Ryan Gallacher, Derek Keir, and Nicholas Harmon

Abstract

The response of the mantle and specifically melt generation during the progression from continental to oceanic rifting is poorly understood due to an absence of direct observations. Specifically, it is unclear when during the rifting process the segmented nature of magma supply typical of seafloor spreading initiates. We have imaged the mantle beneath the southern Red Sea, Afar and the Main Ethiopian rift using Rayleigh-wave tomography. From this we have generated a high-resolution absolute 3-dimensional shear-wave velocity model of the upper 250 kilometres imaging the mantle response during the progression from continental rifting to seafloor spreading. We interpret segmented partial melt beneath the rifts based on observations of low seismic velocities and rift parallel anisotropy. The low velocities are likely formed by focusing of partial melt along the base of the lithosphere in addition to buoyancy driven active upwelling. The segmented anomalies are consistent in scale from the oceanic southern Red Sea rift to the continental Main Ethiopian, suggesting that mantle segmentation beneath oceanic rifts initiates early during continental rifting. The similarity between the observed mantle segmentation and mantle segmentation observed at oceanic rifts worldwide indicates that these interpretations may apply globally.

R. Gallacher (✉)
Department of Earth and Environmental Sciences,
Tulane University, New Orleans, LA, USA
e-mail: rgallacher@tulane.edu

R. Gallacher · D. Keir · N. Harmon
National Oceanography Centre Southampton,
University of Southampton, Southampton, UK

D. Keir
Dipartimento di Scienze della Terra, Universita Degli Studi di Firenze, Florence, Italy

N. M. A. Rasul and I. C. F. Stewart (eds.), *Geological Setting, Palaeoenvironment and Archaeology of the Red Sea*,
https://doi.org/10.1007/978-3-319-99408-6_6

1 Introduction

Explanations for melt production during rifting have largely been developed from modeling mantle conditions using the geochemistry of volcanic rocks and also from the seismic velocities of igneous crustal intrusion at ancient rifted passive continental margins (White et al. 2008; Shillington et al. 2009; Armitage et al. 2010). Due to the lack of direct observations, however, the relative contributions of mantle potential temperature (White and McKenzie 1989), mantle composition (Shillington et al. 2009), melt focusing along the lithosphere-asthenosphere boundary (LAB) (Holtzman and Kendall 2010) or prior rift history (Armitage et al. 2010) in determining the locus and volume of melt production are debated. In addition, the importance of buoyancy driven active upwelling in controlling the 3-dimensional geometry of melt production during rifting remains unclear (e.g., Wang et al. 2009).

Rifting in Afar and along the Red Sea rift shows a progression from continental rifting to incipient seafloor spreading, providing a unique opportunity to directly constrain melting processes. Previous geophysical, geological, geochemical and modelling studies of the Red Sea rift have suggested that melt production and crustal intrusion vary in intensity along the rift (e.g., Bonatti 1985; Bonatti and Seyler 1987; Corti et al. 2003). This has been proposed to be due to segmented active upwelling in the mantle, a consequence of diapiric rise caused by Rayleigh–Taylor instability resulting from density and viscosity inversions (Bonatti 1985). An alternative explanation at other continental and oceanic rifts worldwide relies on variations in lithospheric structure during the initiation of continental breakup (e.g., Madge and Sparks 1997). We test and develop these hypotheses using seismic imaging from Gallacher et al. (2016) in the East African, Afar and the southern Red Sea rifts, and provide direct evidence that segmented active upwelling in the mantle occurs during continental breakup and persists through to oceanic rifting. We here further develop the ideas presented in Gallacher et al. (2016) in light of previous constraints on mantle dynamics in the Red Sea, and also using new constraints from rifts elsewhere.

2 The Breakup of Continents and Initiation of Seafloor Spreading

The breakup of continents and subsequent formation of new ocean basins is a fundamental component of plate tectonics that has shaped the geological record and distribution of natural resources along continental margins such as the North Atlantic. Continental rifts are initially characterized by relatively broad zones of mechanical extension in which faulting, stretching and heating of the tectonic plate accommodates strain and defines the primary architecture of the rift zone (e.g., McKenzie 1978; Braun and Beaumont 1989; Weissel and Karner 1989). Ultimately, however, the locus of strain must shift toward a narrowing zone that becomes the newly formed seafloor-spreading centre. It is here that magma formed from decompression melting of the mantle intrudes the plate and creates new ocean floor (e.g., Schouten et al. 1985; Delaney et al. 1998). Despite the importance of continental breakup in plate tectonic theory, it remains unclear how and when the transition from mechanical to magmatic extension of the plate occurs (e.g., Hayward and Ebinger 1996; Hopper et al. 2004). It also remains ambiguous how important the thermal, chemical, and physical structure of the mantle are in aiding the passively upwelling mantle to melt, and in so doing supplying magma that is intruded into the plate and/or erupted to the surface (e.g., Lizarralde et al. 2007; Shillington et al. 2009).

As lithosphere thins during rifting, the underlying asthenosphere upwells and may melt as it decompresses (McKenzie 1978; Bown and White 1995). It is now well established from seismic imaging of continental margins, as well as from numerical simulations, that a higher rate of extension increases the supply of melt beneath a rift (Bown and White 1995). However, increased temperature and anomalous chemistry of the mantle can induce asthenospheric melting with or without extension of the overlying plate (e.g., King and Anderson 1995; White et al. 2008; Shillington et al. 2009). Furthermore, steep gradients on the lithosphere-asthenosphere boundary (LAB) beneath the rift valley flanks provide a means to locally enhance melting (e.g., King and Anderson 1998; Holtzman et al. 2003), and also function to guide and focus melts riftward (e.g., Ebinger and Sleep 1998; Holtzman et al. 2003).

Establishing detailed 3D constraints on the locus, depth, geometry and elastic properties of the upwelling asthenosphere beneath a continent currently breaking apart will enable discrimination between these mechanisms for generating spatial and temporal variations in magma supply. In addition, it is important to determine at what stage during the development of a new ocean spreading centre does the mantle flow field organize into discrete, passive upwellings beneath individual segments of the rift thought to operate at the more mature ocean basins (e.g., Wang et al. 2009).

3 Horn of Africa Rifting Summary

These problems are addressed by focusing on high resolution absolute shear wave velocity tomographic imaging beneath the southern Red Sea and northern East African rifts, showing how magma is produced beneath an evolving continental rift and young mid ocean ridge (Fig. 1). This region is the best natural laboratory on Earth for the study of continental breakup because all stages of the breakup process are exposed within a relatively small area, with continental rifting transitioning to full seafloor spreading. In addition, we interpret our images in light of both past and present geodynamical and geochemical constraints on melting conditions beneath the region.

Continental rifting is ongoing in the Main Ethiopian rift (MER), the northernmost sector of the East African rift. Here the rift valley is relatively narrow, with extension occurring on both large, steep border faults, and also on rift axial fault networks (e.g., Hayward and Ebinger 1996; Wolfenden et al. 2004; Corti 2008) (Fig. 1). Further north near the Afar triple junction, rifting is more evolved with the later stages of continental breakup exposed subaerially (Fig. 1). The geology in Afar shows that extensional deformation near the surface is currently localised to ~15-km-wide, ~60-km-long faulted volcanic ranges with aligned chains of basaltic cones and fissural flows (Fig. 1; Barberi and Varet 1977). These rift segments (e.g., Dabbahu segment) are similar in size, morphology, and spacing to those observed along much of the globe's slow-spreading mid-ocean ridge system (Hayward and Ebinger 1996; Manighetti et al. 1998). Modeling of gravity data suggests the crust beneath rift segments is intruded by dense, likely mafic rock, and that the crust is thinner (e.g., Tiberi et al. 2005; Lewi et al. 2016). Some regions of the Afar Depression have already subsided below sea level, suggesting strongly that seafloor spreading is imminent in Afar (Fig. 1). The final stage of the breakup process, the formation of a young seafloor-spreading centre, is now occurring in the southern Red Sea and western Gulf of Aden (GoA) where spreading rates are relatively slow at ~2 cm yr^{-1} (Fig. 1; e.g., Bonatti 1985; Manighetti et al. 1998; Leroy et al. 2010).

Studying the style of plate deformation and concurrent dynamics of the upwelling mantle during continental breakup has added significance in the Horn of Africa. Other regions commonly regarded as ideal candidates for studying breakup, such as the Woodlark Basin and Taupo Volcanic

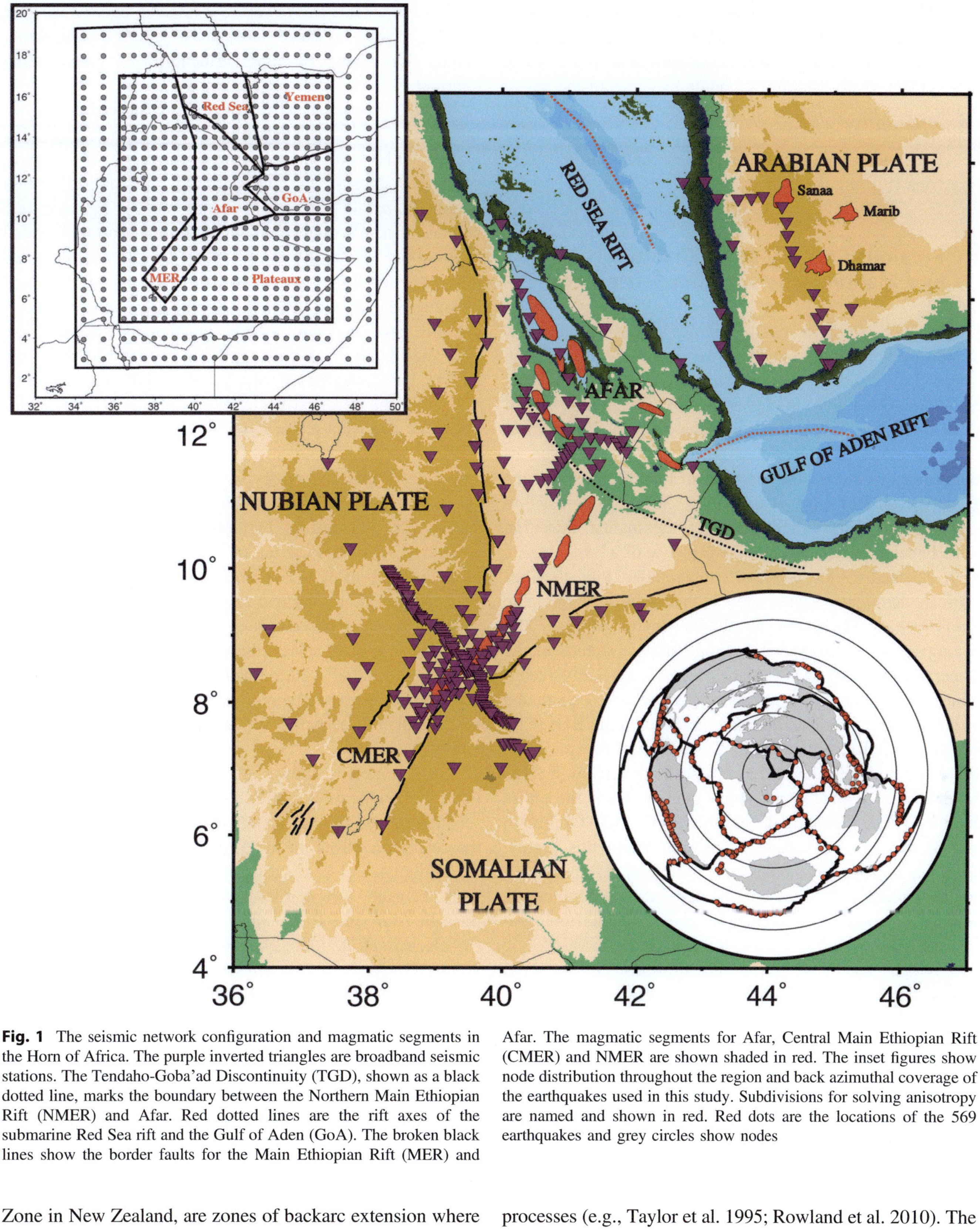

Fig. 1 The seismic network configuration and magmatic segments in the Horn of Africa. The purple inverted triangles are broadband seismic stations. The Tendaho-Goba'ad Discontinuity (TGD), shown as a black dotted line, marks the boundary between the Northern Main Ethiopian Rift (NMER) and Afar. Red dotted lines are the rift axes of the submarine Red Sea rift and the Gulf of Aden (GoA). The broken black lines show the border faults for the Main Ethiopian Rift (MER) and Afar. The magmatic segments for Afar, Central Main Ethiopian Rift (CMER) and NMER are shown shaded in red. The inset figures show node distribution throughout the region and back azimuthal coverage of the earthquakes used in this study. Subdivisions for solving anisotropy are named and shown in red. Red dots are the locations of the 569 earthquakes and grey circles show nodes

Zone in New Zealand, are zones of backarc extension where Earth structure, as well as mantle dynamics and chemistry, are significantly influenced by ongoing subduction zone processes (e.g., Taylor et al. 1995; Rowland et al. 2010). The Gulf of California also offers an opportunity to study breakup process, but a tectonic history of fairly recent

(<15 My ago) and nearby plate subduction beneath the rift also complicates the interpretation of key components required to rift a continent (e.g., Lizarralde et al. 2007; Wang et al. 2009). Even when the complex tectonics of the above regions are ignored, they represent the division of small continental fragments rather than the continental scale rifting of Africa and Arabia that is currently occurring in the Horn of Africa (Fig. 1).

4 Previous Constraints on Crustal and Mantle Structure from Seismology

Global scale mantle tomography shows a large region of low velocities stretching from the lower mantle at the African Large Low Shear Velocity Province (LLSVP) beneath Southern Africa up to the upper mantle beneath the Horn of Africa. This feature is known as the African Superplume, with its elevated temperature having been interpreted as being responsible for the magmatism found throughout the East African Rift (e.g., Hansen et al. 2012). Regional P–S receiver function studies analyzing the mantle transition zone beneath Africa interpret the presence of a broad region with elevated mantle potential temperatures along the East African Rift, consistent with global scale tomographic studies (e.g., Owens et al. 2000; Huerta et al. 2009; Thompson et al. 2015). Global tomographic studies of the upper mantle (P, S and surface wave) consistently record some of the lowest mantle seismic velocities beneath the Horn of Africa, comparable with the East Pacific Rise (e.g., Shapiro and Ritzwoller 2002; Debayle and Ricard 2012; Schaeffer and Lebedev 2013). These models show a large low velocity region in the upper mantle across the entire Horn of Africa including the Red Sea, GoA and Afar, but do not, however, show detailed structure within this region. Regional tomography studies have been used to improve our understanding of the smaller scale upper mantle structure by improving upon the resolution of global seismic tomography. A regional surface wave study imaged a broad mantle low velocity zone beneath Ethiopia; however, this study did not greatly improve upon the resolution of previous studies (e.g., Sicilia et al. 2008). Regional relative arrival time body wave tomography studies improve on this and identify continuous low velocities focused mainly beneath the MER and Afar, in addition to beneath the southern Red Sea, interpreted as being due to low mantle potential temperatures and the presence of partial melt (e.g., Park et al. 2007; Bastow et al. 2008; Hammond et al. 2013; Civiero et al. 2015). Pn tomography potentially shows segmentation of low velocities, interpreted as partial melt, directly beneath the magmatic segments in Afar (e.g., Stork et al. 2013; Corbeau et al. 2014). However, it remains unclear whether the Pn phase in these studies is sampling the upper mantle lithosphere, or upper asthenosphere, complicating interpretation.

Multiple methods have constrained anisotropy in the lithosphere throughout the Horn of Africa, including shear wave splitting and surface wave tomography. Shear wave splitting studies in the Main Ethiopia rift and Afar find strong rift parallel anisotropy which reduces in magnitude and becomes less uniform in direction further north (Kendall et al. 2005; Hammond et al. 2014). This is interpreted as being due to aligned partial melt beneath the MER, which becomes less aligned beneath Afar due to complicated stress patterns associated with a triple junction. For the southern Red Sea region shear wave splitting is found to be strong and sub parallel to the rift, interpreted as being due to mantle flow from the northward progression of the African Superplume or the Afar plume (Hansen et al. 2006; Gao et al. 2010). Azimuthal anisotropy constrained by surface wave tomography broadly agrees with the shear wave splitting results across both regions (Sicilia et al. 2008). Despite these studies, the relationship between anisotropy and velocity structure, and the location of partial melt in the mantle beneath the Horn of Africa remains unclear. We improve on both the lateral and vertical resolution of previous methods by using an array based surface wave method incorporating all of the temporary seismic networks deployed in the region from 1999 to 2013. Additionally, we obtain absolute shear velocities throughout the region, which have not previously been available, allowing for interpretation of partial melt volumes.

Rifting since ~30 Myr ago has impacted crustal structure (Wolfenden et al. 2004). Passive and controlled source seismology, along with inversion of gravity data show that the crust varies from 40 to 45 km and 35 km thick beneath the Ethiopian and Southeast plateaux respectively, to 26 km thick beneath the northernmost MER, 20–26 km thick beneath Afar, and 16 km thick beneath northernmost Afar (MacKenzie et al. 2005; Hammond et al. 2011; Lewi et al. 2016). The Vp/Vs ratio varies from ~1.8 beneath the plateaux not covered by flood basalts to 1.9 beneath the flood basalt provinces and above 1.95 beneath the rift (Hammond et al. 2011). The data was interpreted to show that extension occurs by a combination of mechanical and magmatic extension, with magma intrusion mostly focused within the rift (Hammond et al. 2011). In our study, we focus on understanding the feedbacks between extension and mantle processes. We interpret the temporal evolution of the process by imaging beneath several stages of rift evolution, from early continental rifting in the MER, to early seafloor spreading in the Red Sea.

5 Station and Event Data

We used Rayleigh wave vertical component records for earthquakes recorded at 290 broadband stations from 12 temporary seismic deployments during 1999–2013, and 6 permanent stations; AAUS, ANKE, ATD, DAMY, FURI, RAYN (Fig. 1). The temporary seismic deployments are: the Reseau Large Bande Mobile Horn of Africa experiment (RLBM: June 1999–December 2002) (Sebai et al. 2006), Ethiopia Kenya Broadband Seismic Experiment (Ethiopia/Kenya: January 2000–December 2002) (Benoit et al. 2006), Ethiopia Afar Geoscientific Lithospheric Experiment (EAGLE: October 2001–February 2003) (Bastow et al. 2008), Boina Broadband Network (Boina: October 2005–October 2007) (Keir et al. 2009), SEARIFT array (SEARIFT: March 2007–October 2009), UK–US Afar Consortium Experiment (Afar Consortium: October 2007–October 2009) (Belachew et al. 2011; Hammond et al. 2011; Keir et al. 2011), Danakil Depression array (Afar Depression: October 2009–February 2013), Eritrea Seismic Project (Eritrea: June 2011–October 2012) (Hammond et al. 2013) and Young Conjugate Margins Laboratory in the GoA (YOCMAL: March 2009-February 2010) (Ahmed et al. 2013). This resulted in a network consisting of 162 6T seismometers (sensor response: 0.01–30 s), 59 3ESP seismometers (sensor response: 0.01–120 s), 56 3T seismometers (sensor response: 0.01–120 s), 52 40T seismometers (sensor response: 0.02–50 s), 11 STS1 seismometers (sensor response: 0.02–120 s) and 1 STS2 seismometer (sensor response: 0.02–120 s). The total network is approximately 1100 km × 1100 km in extent with average station spacing of 25–50 km and maximum station spacing of <100 km. Some of the stations were not used either due to poor data quality or the time period of operation.

6 Pre-processing

A full description of the methodology used in this study can be found in Gallacher et al. (2016). Initial processing removes the mean and trend from each signal; additionally, the instrument response is removed for each station. The signals were then filtered with a zero-phase four-pole Butterworth filter with corner frequencies of 0.004 and 0.1 Hz. Zero-phase implementation was used to ensure the phase information of the signal remained unaltered and was achieved by running the filter both forwards and backwards in time.

Quality control was performed manually by inspecting the Signal-to-Noise ratio (SNR), group velocity dispersion curve and wave energy for the vertical component at every station for each event. For an event to be included, the seismograms required a SNR of greater than 3, continuous group velocity dispersion curve and peak energy between the approximate frequencies of interest, 0.01 to 0.07 Hz. A further requirement was for more than 7 good seismograms per event as this is the minimum station coverage resulting in a stable answer. In total, 569 teleseismic events with Ms >5.5 and epicentral distances of 25–150° were used, with the epicentral distance range ensuring that the energy of the Rayleigh wave is clearly distinguishable from that of the body waves and excluding the possibility of the Rayleigh wave arriving at the station from opposite directions at the same time.

Due to the dispersive nature of Rayleigh waves it was necessary to separate seismic signal information into 28 period results from 10 to 200 s. This was achieved by filtering the amplitude values at ±0.002 Hz around frequencies from 0.005 to 0.015 Hz, at 0.001 Hz increments, and ±0.005 Hz around frequencies from 0.02 to 0.1 Hz, at 0.005 Hz increments. The negative spectral information for each was removed using a Hilbert transform thus leaving the time analytic signal, which represents the envelope of the wave. Thus, the signals contain information about the group velocity for each period of interest. The signal power at each station for each period is found by interpolating the signal amplitude with the predicted group velocity, based on the station event distance and the travel time. For each event the signal power for each good station was summed and from the resulting plot of group velocity against period the group velocity dispersion curve was picked manually along the path of peak power. The period and velocity information from the dispersion curve was then interpolated, thus obtaining velocity information at a subset of 18 periods of interest, from 20 to 167 s. This period range was chosen as it encapsulated the majority of the signal power. This information was used to extract an idealized time varied filtered signal for each event, station and period (Landisman et al. 1969). The amplitude and phase information from each signal was obtained using a Fast Fourier transform and saved for use in the 1D inversion.

7 1D Phase Velocity Inversion

To solve for the phase velocity, we invert amplitude and phase information extracted from the individual seismograms by minimising the misfit in both the real and the imaginary components of the bandpass filtered seismograms. We use an array method that accounts for distortions in the incoming wavefield using a two-plane wave approximation (Forsyth and Li 2005) and accounts for first order scattering using 2-dimensional Born approximation sensitivity kernels for amplitude and phase (Zhou et al. 2004; Yang and Forsyth

2006). The inversion is completed in two stages, with the first stage utilising a simulated annealing method (Press et al. 1992) and the second stage utilising a standard linearised inversion technique (Tarantola and Valette 1982). The inversion simultaneously solves for the phase velocity, azimuthal anisotropy and wave parameters for each event. The first stage is necessary due to the periodic non-linearity of the problem, in addition to the solution being ambiguous where the two plane waves have similar azimuths. By performing the simulated annealing inversion first we ensure a global minimum is found in the second stage. We determine the average dispersion curve for the region using a 1-dimensional version of the two-plane wave inversion method, which solves for the 6 wave parameters (3 for each plane wave; amplitude, phase and wave number) for each event and a single-phase velocity for the region. We try a range of starting phase velocities for the model ranging from 3.00 to 4.00 km s^{-1}, and typically the models converge on a single phase velocity for each period. Thus, we use an average of the output from these starting velocities as our starting model. After an initial set of inversions for the period range of interest, we quality control the data and assess the quality of fit, prior to a final set of inversions for the final 1-D dispersion curve. The quality control consists of identifying and removing events that cannot be fitted with a plane wave model, either due to incoherent propagation across the array or complications in the sources.

8 2D Phase Velocity Inversion

The node structure for the 2-dimensional phase velocity inversion was 25 × 29 nodes covering the region of interest (Fig. 1). The node spacing is 0.5° × 0.5° with the outermost two rows and columns spaced at 1° to absorb velocity heterogeneities outside the region of interest. The variation of velocities allowed at the outside edges is increased by a factor of 10 to allow for this. For the 2-dimensional phase velocity map inversions we start with a uniform velocity and no anisotropy at each point across the map, where the velocity value is taken from our 1-dimensional dispersion curve for each period of interest. Any event >0.05 out of phase is removed and the inversion is repeated once using the output phase velocity maps as input for the following iteration. The culling of this data is necessary as it removes waveforms with complicated source radiation patterns and other effects not accounted for in the inversion. At short periods, a single average phase velocity may not adequately capture rapid variations in crustal structure across the region, especially given the transitions from continental to oceanic crust. To account for this, we repeated the 2-dimensional phase velocity inversion procedure, as described above, using the predicted phase velocity structure from the CRUST 1.0 model (Laske et al. 2013) between 20 and 40 s and a homogeneous mantle velocity structure derived from our best-fit 1-dimensional shear velocity model between 40 and 167 s as our starting model. The inversion solves for the phase velocity and cos2θ and sin2θ azimuthal anisotropy terms for each node. Higher order azimuthal anisotropy terms are ignored as they are assumed to be negligible for Rayleigh waves. The values of azimuthal anisotropy at each individual node are not well resolved, thus we subdivided the nodes into 6 well-resolved regions and averaged the anisotropy within each region. This provides 6 values for anisotropy across the entire study area, one per region. The regions are labelled as MER, Afar, Nubian and Somalian Plateaux, Yemen, Red Sea, GoA, and the external nodes are unlabelled (Fig. 1).

9 Shear Velocity Inversion

For the shear velocity inversion we used the Occam inversion (Constable et al. 1987), which linearizes the nonlinear forward problem for a starting model, in our case a combination of CRUST1.0 from 0 to 40 km depth and ak135 (Kennett et al. 1995) from 40 to 250 km. This procedure has been designed to produce a smooth output model for shear velocity, which minimises the potential for spurious features to be over-interpreted. The smoothness is achieved by parameterising the model for a roughness factor taken as the integrated square of the first derivative with respect to depth for the shear velocity model. The linearized problem is then solved for the desired model, rather than for a model correction, where the roughness factor is smallest while still fitting the data. This method is combined with the solutions for the Jacobian matrixes found by DISPER80 (Saito 1988) to output a 1-dimensional shear velocity profile at a given point accounting for all periods. Performing the above procedure for each individual point across all of the 2-dimensional phase velocity maps generates the 3-dimensional shear velocity model. The model is shown as a depth averaged slice from 30 to 112 km in Fig. 2 as this represents a minimum structure view of the model and includes the likely zone of melt generation throughout much of the region. Figure 2 also includes 3 transects taken through the model which intersect all of the rifts in the region.

10 Resolution Tests

We assess the lateral resolution of our model using a measure based on the formal resolution matrix of the linearized inversion for the 2-dimensional phase velocity maps. We present the values of the diagonal resolution matrix, which provides an indication of the independence of each velocity

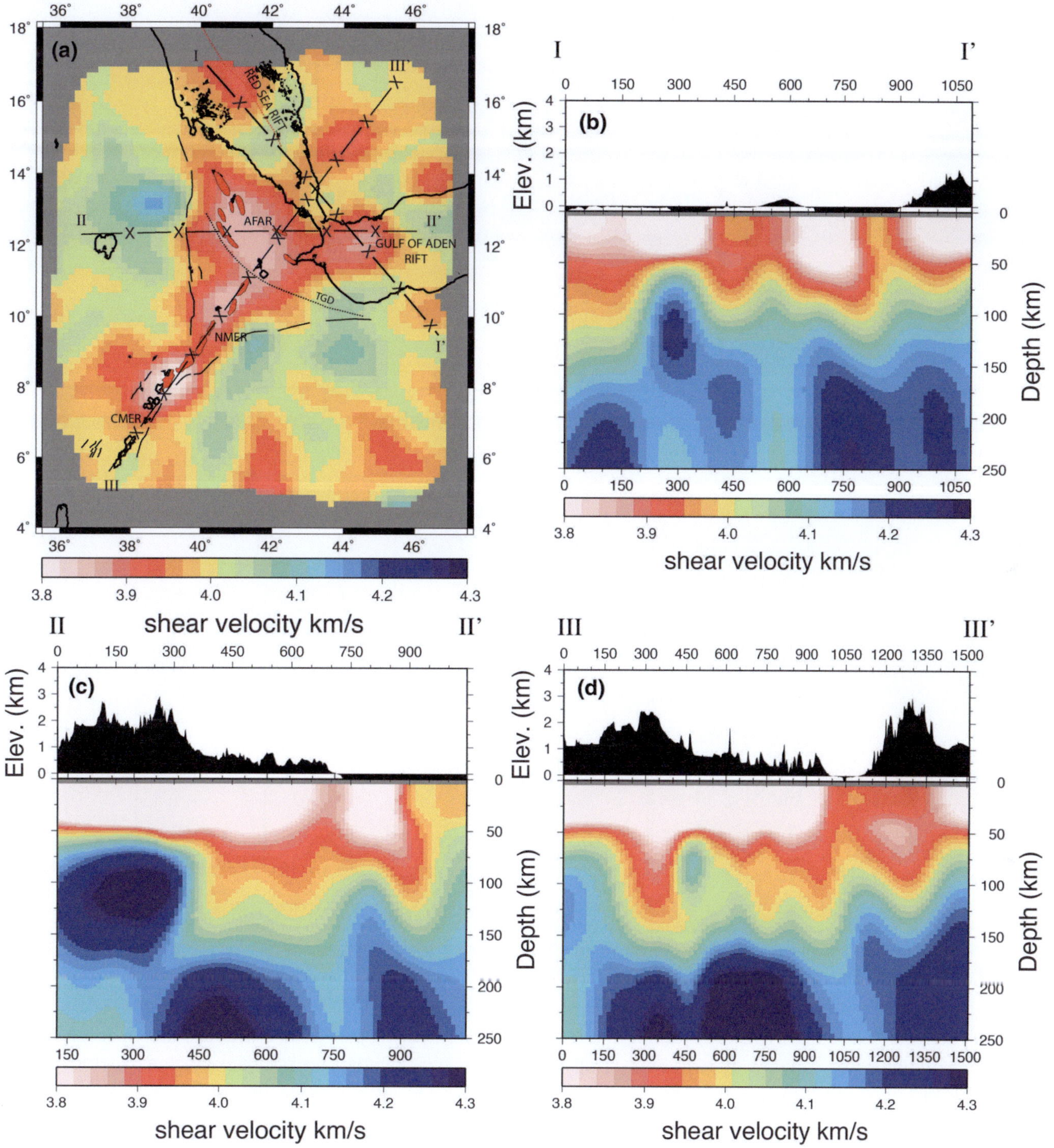

Fig. 2 A depth averaged 3-dimensional model slice and transects. **a** The 30–112 km averaged depth slice shows that the Gulf of Aden (GoA) and Afar form a continuous low velocity (<3.90 km s^{-1}) region. Higher velocities (>3.85 km s^{-1}) separate low velocities between the NMER (3.80 km s^{-1}) and a focused low velocity (3.80–3.85 km s^{-1}) along the CMER. Transects are labelled I-I', II-II' and III-III'. X's mark intervals of 150 km on the transects. The Tendaho-Goba'ad Discontinuity (TGD), shown as a black dotted line, marks the boundary between the Northern Main Ethiopian Rift (NMER) and Afar. Red dotted lines are the rift axes of the submarine Red Sea rift and the GoA. The broken black lines show the border faults for the Main Ethiopian Rift (MER) and Afar. The magmatic segments for Afar, Central Main Ethiopian Rift (CMER) and NMER are shown shaded in red. **b** The I-I' transect shows three separate mantle low velocities beneath the Red Sea and Gulf of Aden (<3.90 km s^{-1}) **c** The II-II' transect shows low velocities beneath Afar (3.80–4.00 km s^{-1}) and beneath the GoA (3.80–4.00 km s^{-1}). **D** The III-III' transect shows an approximately 200 km long low velocity (3.80–4.00 km s^{-1}) beneath the CMER and an approximately 300 km long low velocity (3.90–4.00 km s^{-1}) beneath the NMER and Afar. The spacing between the low velocities is 150 km. Crustal velocities <3.8 km s^{-1} are saturated at 3.8 km s^{-1}

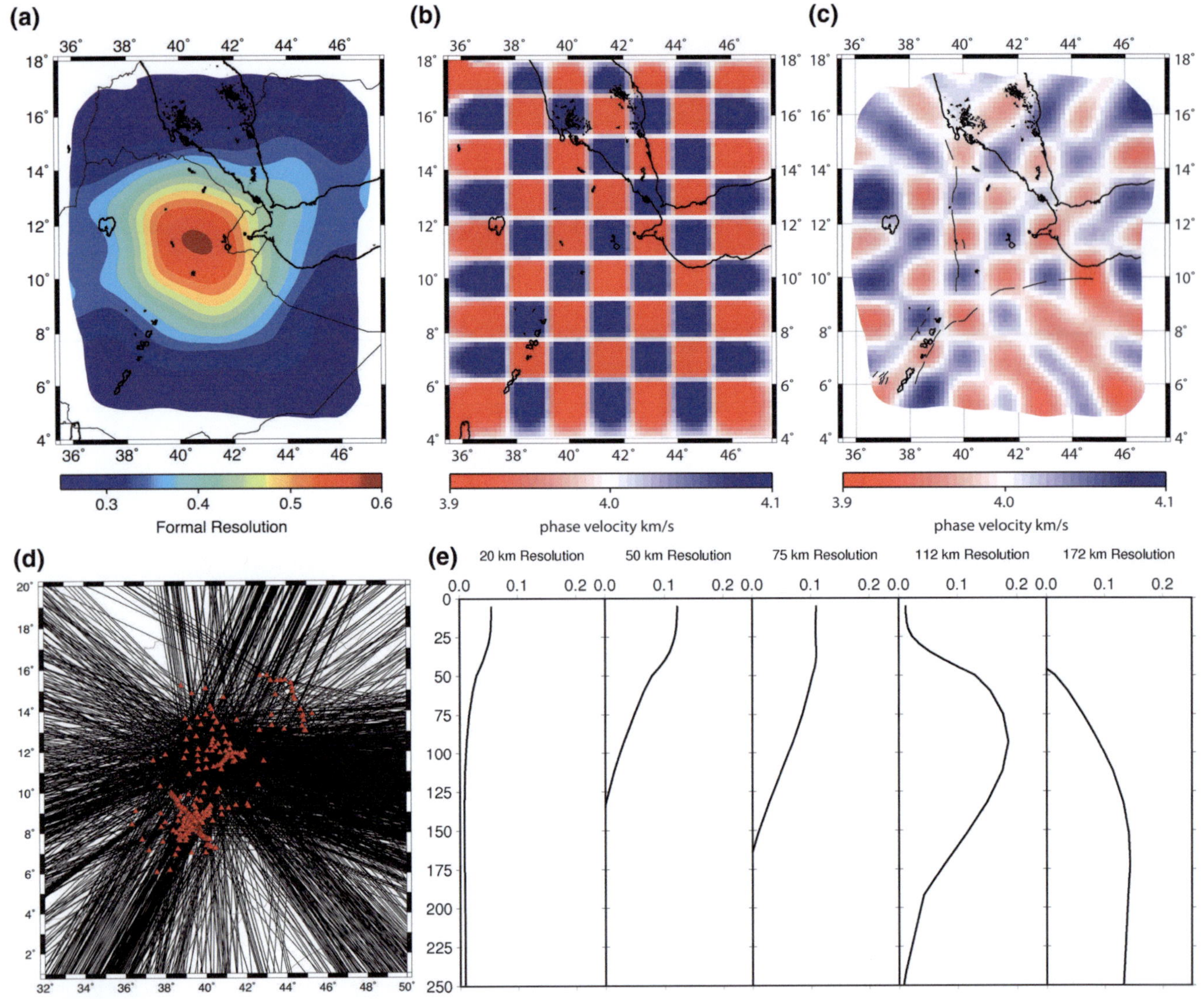

Fig. 3 Horizontal and vertical resolution tests. **a** Checkerboard test, 150 km scale input. **b** the 40-s period checkerboard test, 150 km scale output. **c** Node independence from formal resolution matrix for the 40-s period. **d** Raypath map for the 40 s period. Coverage of crossing raypaths is good within the centre of the study region. **e** Backus–Gilbert resolution kernels are shown for depths of 20, 50, 75, 112 and 172 km. Smearing ranges from ±25 at 20 km depth to ±55 km at 112 km depth

node (0 indicating not resolved, 1 meaning perfectly resolved, with numbers in between indicating fractional interdependence with adjacent parameters). This diagonal resolution plot is shown as a 40-s 2-dimensional phase velocity map (Fig. 3a). The map shows independence values of 0.3–0.6 within the interpreted regions. This means that a minimum of approximately three nodes is required to have a well-resolved model feature. As the node spacing is 50 km we can thus interpret features of 150 km or greater.

To further test the resolution of our model we conduct "checkerboard" tests where we create synthetic 2-dimensional phase velocity maps for the 40-s period with the velocity blocks varying between 3.9 and 4.1 km s^{-1} at a length scale of 150 km (Fig. 3b). Figure 3c shows the checkerboard output for 40-s with 150 km input checkers. It can be seen that the checkerboard structure is retrieved with a slight underestimation of the anomalies ($\sim$0.05 km s^{-1}), which matches well with the result from formal resolution. Figure 3d shows a raypath map for the 40 s period with the majority of the rays crossing within the array.

To test the vertical resolution, we present Backus–Gilbert resolving kernels (Backus and Gilbert 1968) for a range of depths in our model (Fig. 3e). These kernels show the recovery of a delta function at the target depth and are based on the formal resolution matrix. This shows that shear velocities can be smeared over ±25 km at 20 km depth, ±45 km at 75 km depth and ±55 km at 112 km depth. Where these values are obtained over the

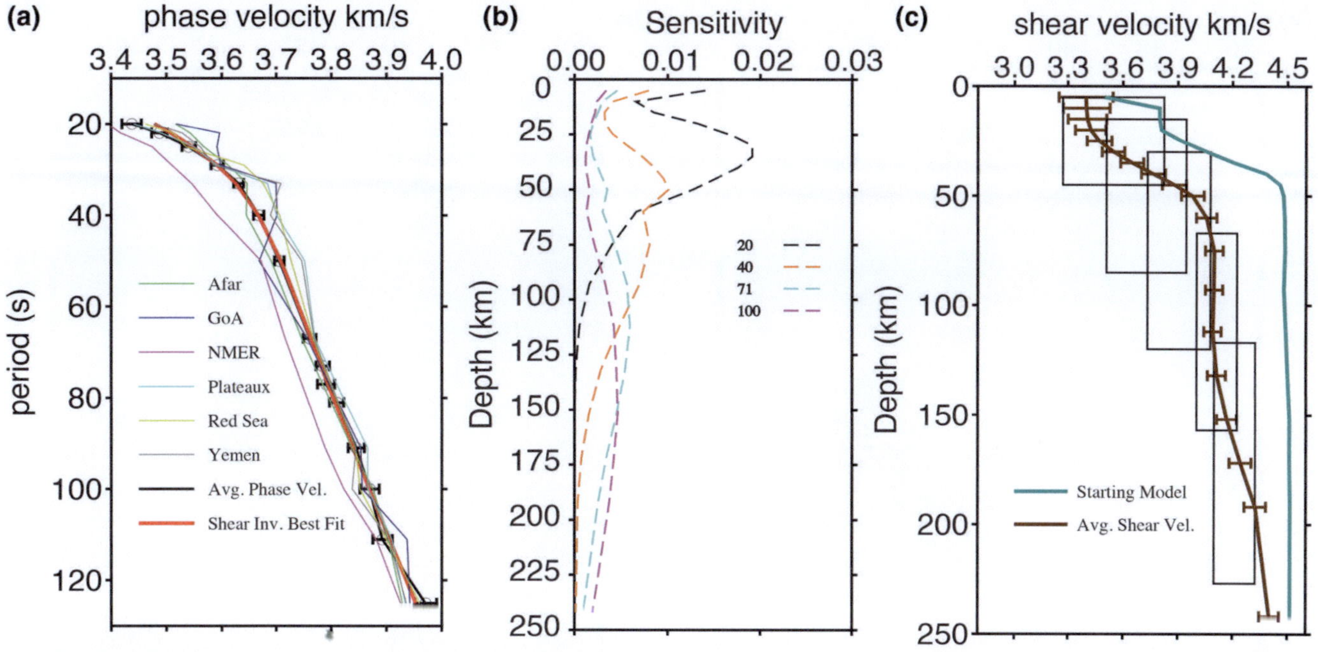

Fig. 4 One-dimensional phase velocity dispersion curves, sensitivity kernels and 1-dimensional shear velocity profile. **a** The predicted phase velocity dispersion curve (red) obtained from the average 1-dimensional shear velocity profile for the region fits well with the average phase velocity dispersion curve (black) for the whole region. Black horizontal bars show 3× standard deviation. NMER is Northern Main Ethiopian Rift and GoA is Gulf of Aden. Regional phase velocity dispersion curves are determined using a subset of the nodes used for the whole region. **b** The sensitivity kernels (dashed) calculated from DISPER80 (Saito 1988) are shown for representative periods between 20 and 100 s. **c** The average regional 1-dimensional shear velocity model (brown) is up to 11% lower than the starting model (turquoise), which combines CRUST 1.0 (0–40 km) and ak135 (40–250 km), at depths >50 km. Black rectangles show the average error in shear velocity over well resolved depth ranges. Errors are obtained from a Monte-Carlo estimate (100,000 random perturbations from our best fit model) showing the range of possible solutions to the inversion that fit the dispersion curve within error. Resolution at 25, 50, 75, 112 and 167 km is shown as vertical black lines

well-resolved part of the kernel, which is defined as greater than 0.025 for 20 km input depth, 0.075 for 75 km input depth and 0.1 for 112 km input depth.

11 Regional Phase Velocity Results

We obtain an average phase velocity dispersion curve for periods of 20–125 s and 2-dimensional phase velocity maps at each period including regionally averaged azimuthal anisotropy. We also obtain an average phase velocity dispersion curve for each of the subdivided regions (MER, Afar, Nubian and Somalian Plateaux, Yemen, Red Sea, GoA) previously defined for averaging anisotropy (Fig. 4a). Regionalized phase velocity curves show that overall the entire study region is anomalously slow for a continental setting, with the average phase velocities at all periods being <4.0 km s^{-1}. Relative to the average dispersion and all other dispersion curves across the entire region, the MER has the lowest phase velocities at all periods. These suggest that the source of low velocities at the MER is distributed throughout a larger range of depths compared with the other regions given the depth sensitivity of Rayleigh waves in this period range (Fig. 4b). The Red Sea phase velocities are on average 0.04 km/s slower than the GoA phase velocities from 20 to 30 s and are within 0.01 km s^{-1} from 30 to 40 s. The GoA phase velocities are 0.03 km s^{-1} slower than those for the Red Sea from 50 to 70 s. This suggests that the source of low velocities is shallower along the Red Sea and deeper along the GoA. Afar has phase velocities that are very similar to the average phase velocity at all depths, being faster than the Red Sea but slower than the GoA from 20 to 30 s, slower than the Red Sea from 30 to 80 s and slower than the GoA from 30 to 50 s. The Plateaux and Yemen are faster than the average dispersion for the entire region from 20 to 60 s periods by up to 0.05 km s^{-1}, indicating that the crust and uppermost mantle is faster in these regions given the depth sensitivity at these periods (Fig. 4). The differences in velocities are an order of magnitude greater than the calculated errors for the phase velocities.

12 Regional Anisotropy

The regional anisotropy is shown in Fig. 5 for the 40 s (Fig. 5a) and 100 s (Fig. 5b) 2-dimensional phase velocity maps. Figure 5c shows anisotropy results with well resolved

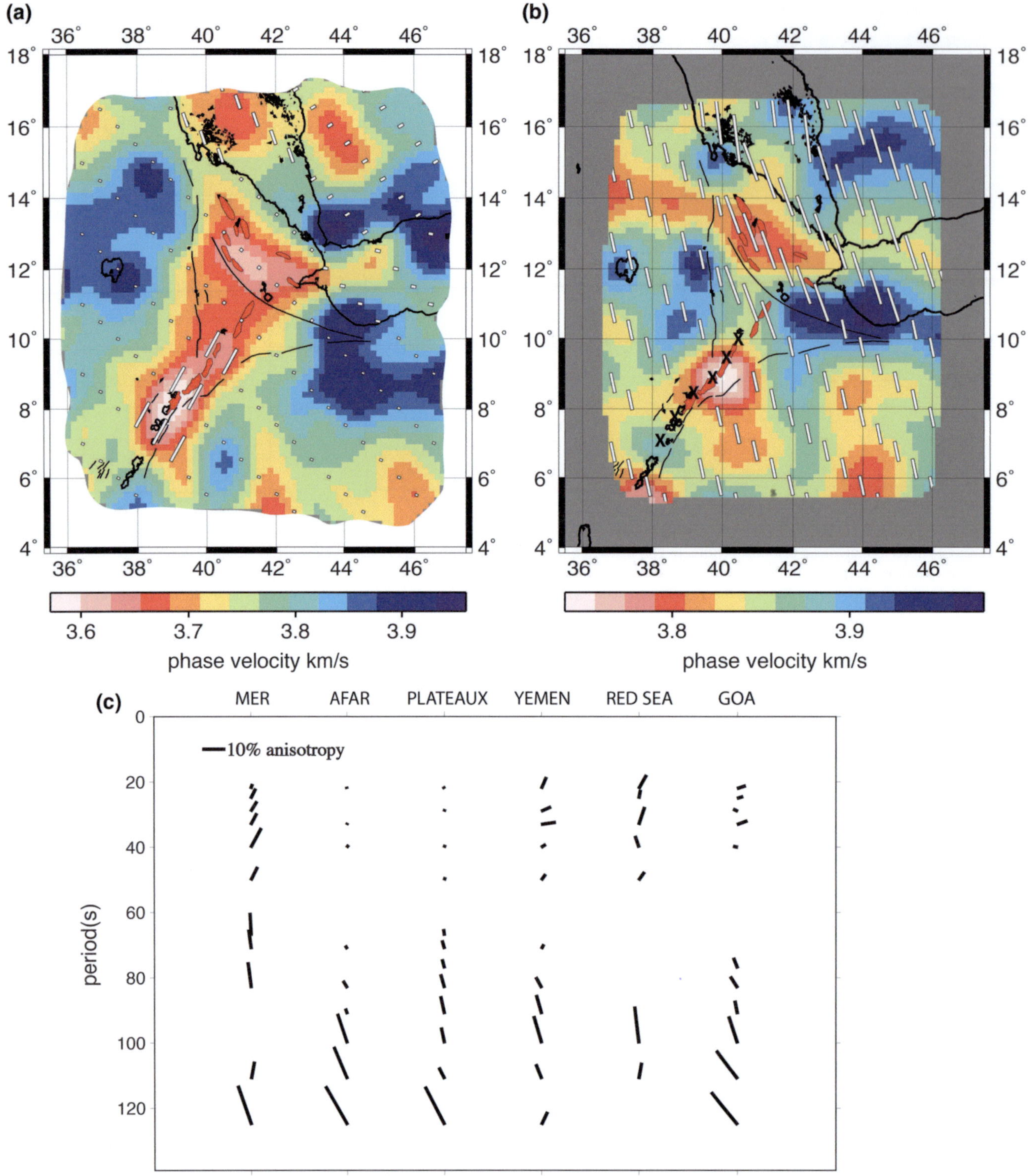

Fig. 5 Two-dimensional phase velocity maps at **a** 40 s and **b** 100 s with the magnitude and azimuth of regional anisotropy plotted on both maps as white rectangles. The result for each region is plotted repetitively. X's are shown when the region does not have a well resolved result for anisotropy. **c** The magnitude and azimuth (black rectangles) of anisotropy for well-resolved periods across all regions are shown. The study area was divided into six separate regions each solved separately for anisotropy

values greater than the error values for all regions and periods. Based on these plots the anisotropy can be divided into shallow (20–50 s) and deep (67–125 s), where the shallow is sensitive to 0–100 km and the deep is sensitive to 75–200 km. Shallow azimuthal anisotropy has a consistent direction of NE–SW beneath the MER with the strength of anisotropy varying from 2.5 to 10%. Strong, shallow azimuthal anisotropy is also evident beneath the Red Sea from 3 to 8% with directions approximately N–S sub-parallel to the rift. The GoA has rift parallel, E–W, azimuthal anisotropy of 2.5–4% magnitude with similar magnitude anisotropy in NE–SW directions beneath Yemen. Shallow azimuthal anisotropy beneath the Afar and the Plateaux regions is negligible. Deep azimuthal anisotropy trending NW-SE is strikingly consistent in direction across all regions while being distinct from any observed shallow anisotropy. The magnitude generally increases with depth from 2.5 to 20%.

13 Shear Velocity Results

The average 1-dimensional shear velocity model has velocities up to 11% lower than ak135 at depths of 40–150 km, indicating that the mantle shear velocities in the study area are significantly lower than the global average (Fig. 4c). At mantle depths (>40 km), shear wave velocities of less than 4.00 km s^{-1} underlie the rifts, which are lower than those observed beneath the adjacent flanks (>4.05 km s^{-1}) (Fig. 2). These lower wave-speeds are not homogeneous; instead, shear wave velocities of less than 3.95 km s^{-1} are segmented beneath the rifts with each lobe typically 200–300 km long and a regular spacing between lobes of 100–150 km (Fig. 2). These lowest velocities can be divided into 5 regions: Red Sea (3.80 km s^{-1}), GoA (3.9 km s^{-1}), Yemen (4.0 km s^{-1}), MER (3.85 km s^{-1}) and Afar (3.9 km s^{-1}). The maximum depth extent of the low velocity anomalies varies along the rift from 120 ± 55 km beneath the less mature Central Main Ethiopian Rift (CMER) to 40 ± 25 km beneath the more mature Red Sea. Additionally, the mantle low velocity anomalies generally directly underlie the surface volcanic segments beneath the Red Sea rift. Xu et al. (2014) report submarine eruptions which created new volcanic islands in the Zubair archipelago which is located above the lowest shear velocities along the Southern Red Sea in our model. An exception to this pattern is the low velocity anomaly beneath the volcanic fields of Sana'a in western Yemen which last erupted in 500 AD (Ambraseys et al. 1994). The mantle beneath the border faults, rift flanks and surrounding plateaux is characterized by higher velocity mantle (>4.1 km s^{-1}).

14 Interpretation of Azimuthal Anisotropy

The sharp change in anisotropy between shorter (shallow) and longer (deep) periods, in addition to the observed variation in shallow anisotropy and consistency of deeper anisotropy suggests that the processes which create anisotropy change with depth. This sharp change in the nature of anisotropy occurs across the entire region, suggesting it is most likely associated with a distinct change in the mantle. Given the depth sensitivity of each period range a plausible interpretation is that the change in anisotropy occurs across the lithosphere-asthenosphere boundary. The interpreted lithospheric azimuthal anisotropy along both the MER and GoA is generally rift parallel and of similar magnitude. This suggests that a similar process during rifting is responsible for the observed anisotropy along both rifts. Previous studies of anisotropy along the rifts, using shear wave splitting, have also found strong rift parallel anisotropy (Kendall et al. 2005; Hansen et al. 2006; Hammond et al. 2014). These studies have interpreted the anisotropy as being due to either aligned melt or mantle flow from the African Superplume. They were unable, however, to distinguish between these hypotheses, as anisotropy interpreted from shear wave splitting has no depth dependence. Hammond et al. (2014), however, were able to invert for multiple layers of anisotropy, finding a two-layered model for anisotropy throughout the region. This result, while consistent with the interpretation of anisotropy in our study, does not constrain the depths at which the layers are present. Using the depth dependency of anisotropy from our surface wave study and the observation that the anisotropy is always rift parallel irrespective of rift orientation, it can be interpreted that rift parallel anisotropy is within the lithosphere, supporting the aligned melt hypothesis. Anisotropy from the African Superplume would be expected to be approximately unidirectional and primarily within the asthenosphere. This hypothesis is further supported by the change in the direction of anisotropy between the MER and Red Sea rift, which would not necessarily be expected if the anisotropy resulted from the African Superplume.

The interpreted lithosphere anisotropy for the GoA, E–W, also agrees with shear wave splitting from the Horn of Africa array station YAF which was located in southwestern Yemen near the ocean-continent transition (Hammond et al. 2014). Data from this station was included in this study. Due to the limited extent of sampling within the GoA in our surface wave model we cannot conclusively state that rift parallel anisotropy is due to aligned melt as has been suggested for the MER and Red Sea rift, however, this remains the most plausible explanation. The negligible lithospheric anisotropy observed in our model for Afar and the Somali and Nubian Plateaux does not agree well with previous shear wave splitting studies within the region (Hammond et al. 2014). These studies show weak anisotropy with a greater range of azimuths than within the rifts; however, the anisotropy is significantly stronger than observed in our study. One possible explanation of this is that the lithosphere in both regions is either relatively isotropic or lithospheric anisotropy is sufficiently heterogeneous to result in no observable anisotropy. Shear wave splitting within these regions would thus be sampling only asthenospheric anisotropy and the lithosphere would have negligible anisotropy as suggested by our study. Alternatively, the greater spatial sampling of the surface waves may average out heterogeneity which was imaged as weak anisotropy by SKS studies. The lithospheric anisotropy within Yemen does not agree with either previous shear wave splitting studies (Wolfe et al. 1999; Hansen et al. 2006), the lithospheric anisotropy interpreted for the Red Sea rift or the lithospheric anisotropy interpreted for the GoA. We suggest that the observed lithospheric anisotropy may result from a combination of an averaging of the bordering rifts lithospheric anisotropy, due to the subdivision of nodes inadvertently including Red Sea rift and GoA influenced nodes within the Yemen region. Other potential explanations include strong anisotropy associated with partial melt at the Sana'a, Marib and Dhamar Volcanic Fields being averaged across the rest of Yemen, which has negligible lithospheric anisotropy or a potential bias in the direction of sampled raypaths leading to artificial anisotropy.

While the shallower anisotropy can be best explained by regionally varying lithospheric processes, the consistency of the deeper anisotropy indicates that a single process is responsible for the asthenospheric anisotropy throughout the entire study region. The deeper anisotropy does not parallel any of the current plate motion directions within the region, and further, these directions are not consistent with each other, thus it is unlikely that the observed deeper anisotropy is due to mantle flow from plate motion. One potential explanation for this result is that mantle flow across the entire region is influenced at depth by the northward progression of the African Superplume. While appealing, this hypothesis does not explain the NW–SE direction of interpreted flow as previous studies have suggested that mantle flow within the plume-affected region is N–S. A NW–SE mantle flow, which could create the observed deeper anisotropy, is predicted by some models of mantle circulation accounting for driving forces induced by mantle upwelling and slab pull. The slab pull of the Aegean trench rollback could led to enhanced NW–SE mantle flow from all regions with the exception of Yemen, as is observed in our results (Facenna et al. 2013).

15 Interpretation of Seismic Velocities

We estimate the maximum velocity increase from attenuation by inverting 1-dimensional profiles while accounting for attenuation (Levshin et al. 1989), with a shear Q factor of 60 (Dalton et al. 2008) at periods of 20–125 s. We find that high attenuation, Q factor $\sim$60, can decrease shear velocities at all depths by up to $\sim$0.20 km s^{-1}. Additionally, a mantle potential temperature of 1450 °C has been estimated for the region from modelling REE compositions of recent basaltic lava flows (Rooney et al. 2012; Ferguson et al. 2013; Armitage et al. 2015). This, coupled with the effect of attenuation can account for the difference between the un-rifted plateaux and ak135 and has been suggested as being due to the presence of a broad mantle plume encompassing the whole region (Benoit et al. 2006). Percentages of partial melt in Afar, the GoA and the Red Sea are calculated using the experimental relationship of 1% partial melt lowering shear velocities by 7.9% (Hammond and Humphreys 2000). For the CMER a different relationship between partial melt and shear velocities (11% shear velocity reduction for 1% partial melt) (Blackman and Kendall 1997) is used assuming the presence of vertically aligned melt inclusions. The shear velocities beneath the CMER, Afar and GoA can be explained by less than 0.5% partial melt after similarly accounting for a mantle potential temperature of 1450°C and a Q factor (attenuation) of 60. For the Red Sea rift, the low shear velocities can be explained by 1% partial melt using the same assumptions.

For the CMER, Afar and the Red Sea rift the shallow maximum depths of 120 ± 55 km, 80 ± 45 km and 40 ± 25 km respectively for the slowest mantle anomalies strongly suggest that decompression melting in response to extension is the dominant source of partial melt. The segmented nature of upwelling, however, is more indicative of active upwelling, possibly due to buoyancy from enhanced melt retention. Focused elevated melt fractions beneath the CMER and Northern Afar occur where there is a known step in crustal thicknesses (e.g., Makris and Ginzburg 1987; Maguire et al. 2006) and therefore inferred step in lithospheric thickness. We therefore interpret that elevated melt fraction may be due to 3-D melt focusing at steep LAB topography. The interpretation of the significantly greater volumes of partial melt beneath the Red Sea rift, 0.6%, compared with Afar and the CMER is likely due to greater decompression and a larger melting zone in the older and more developed rift.

The mantle low velocities beneath Yemen require the presence of 0.2% partial melt, using the same assumptions in temperature and attenuation as for the rifts. The proposed mechanism for melt generation within the rifts,

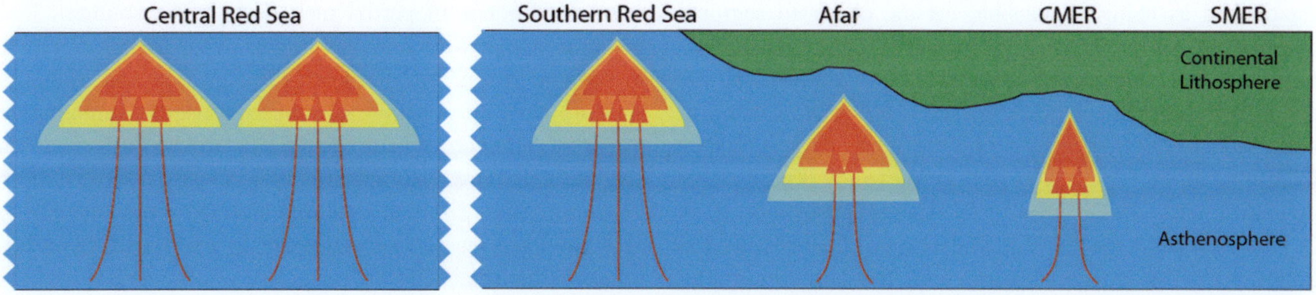

Fig. 6 An illustrative representation of the interpretation of segmented mantle upwelling and partial melt distributed along the Main Ethiopian Rift, Afar and the southern Red Sea rift. As the age of rifting increases there is a progression from deeper more focused melting to shallower melting occurring throughout a larger volume of the upper mantle. The schematic for the central Red Sea rift is altered after Ligi et al. (2012). SMER is Southern Main Ethiopian Rift and NMER is Northern Main Ethiopian Rift

decompression melting resulting from extension related thinning, does not seem an obvious explanation in a region of un-rifted continental lithosphere. This region of low velocities does, however, correspond with the location of volcanic fields (Sana'a, Marib and Dhamar volcanic fields) (Davison et al. 1994) (Fig. 1). Additionally, previous seismic tomography studies have imaged the anomaly and interpreted it as an off-axis diapiric mantle upwelling (Korostelev et al. 2014). Alternatively, lithospheric thinning could indeed be the result of ongoing extension beneath the rift margin (Ebinger and Belachew 2010).

16 Discussion

The interpretations for low seismic velocities in the mantle beneath the rifts and lithospheric rift parallel anisotropy are both consistent with the hypothesis of segmented pockets of focused and aligned partial melt. Segmented mantle upwellings have previously been proposed to occur within the Red Sea rift. Bonatti (1985) suggested that oceanic crust forms along the Red Sea rift in different segments, which start at "hot points" in the crust. This is taken to imply that there are focused regions of mantle upwelling beneath each "hot point". Similarly to this study, Bonatti (1985) argues that the interpreted active upwelling would lead to enhanced melt generation which contributes to the creation of new oceanic crust. Bonatti (1985) argues that the source of the mantle upwelling is due to density inversions from the presence of partial melt leading to Raleigh-Taylor instabilities generating regularly spaced upwellings. It is noted, however, that lithospheric structure and thickness can influence the spacing of the upwellings. Further, it is suggested that the mantle upwelling may be occurring beneath continental lithosphere along the northern Red Sea.

Bonatti and Seyler (1987) perform petrological analysis of samples obtained at sub-aerially exposed crustal sections from islands in the Red Sea. There is broad agreement between Bonatti (1985) and Bonatti and Seyler (1987). The latter study provides estimates for the depths at which magmatic addition to the crust occurs, finding depths generally >30 km. This is interpreted as suggesting that magmatic addition to the crust and hence the mantle upwelling responsible for melt generation precedes rifting of the crust. Further, there is strong evidence that the magmatic intrusions underwent retrograde metamorphism events, which require that significant decompression occurred after emplacement, likely due to the crustal thinning associated with the progression from continental to oceanic rifting.

Further investigation of the crustal structure within the Red Sea rift by Ligi et al. (2011, 2012) led to conclusions similar to the previous studies, that the Red Sea rift segments originate at regularly spaced intervals above regions of mantle upwelling. These studies suggest that there was a higher degree of melting, possibly due to increased upwelling, just after continental crust had been replaced by oceanic rifting. The increased upwelling is suggested to result from small scale convection driven by a combination of warm mantle upwelling beneath the rift and cooler mantle down-welling when in contact with the continental lithosphere. It can thus be seen that the results presented in this study are a continuation of the studies conducted in the Red Sea rift. In our study the maximum depth of slow wave-speeds is beneath the youngest and least mature rift in our system, consistent with increased and deeper partial melting. Shallower melt and reduced volumes of partial melting are interpreted to be beneath the older and more mature section of the continental rift. We have thus imaged features that are best explained by segmented partial melt present within the oceanic southern Red Sea, continuing into the transitional regional of Afar and finally in the fully continental MER. All studies agree that the source of additional melt generation is likely increased mantle upwelling.

Our results further show the influence of continental lithosphere on the mantle upwelling which has been established as being responsible for the observed segmentation along the Red Sea. We see increased mantle upwelling, and thus increased partial melting, where there are changes in lithospheric thickness and this is interpreted as being due to melt focusing along the LAB. Increased melt retention can be expected to lead to buoyancy driven active upwelling and thus increased partial melting (Rychert et al. 2012). This suggests that the lithosphere influences the initiation of mantle segmentation and that Raleigh-Taylor instabilities that subsequently develop due to enhanced partial melting were caused by the initial melt focusing of the lithosphere (Fig. 6).

Segmentation of mantle melting in the rifting process has been imaged using seismic tomography at other oceanic rifts and continental rifts. Wang et al. (2009) performed surface wave tomography on the Gulf of California and found segmented mantle low velocities interpreted as segmented partial melt with a similar spacing, $\sim$100 km, to those observed in this study. Accardo et al. (2017) interpreted low seismic velocities from ambient noise and Rayleigh wave tomography within the Malawi rift as the onset of segmented melt supply. Multiple studies of mid ocean ridges have found along rift segmentation of similar scales to those observed in this study and along the Red Sea rift (e.g., Langmuir et al. 1986; Sempere et al. 1993; Niu et al. 2001). The similarity in segmentation lengths and interpreted mantle segmentation between the Gulf of California, Mid Atlantic ridge, Red Sea rift, East Pacific Rise, Malawi rift and the MER suggests that mantle segmentation may have initiated during continental rifting at other oceanic rifts.

17 Conclusions

We have used Rayleigh-wave tomography to construct a high-resolution absolute 3-dimensional shear-wave velocity model of the mantle beneath the Afar triple junction, imaging the mantle response during progressive continental breakup from the Red Sea rift to the MER. Low seismic velocities, <4.00 km s^{-1}, are imaged within the rifts as well as rift parallel anisotropy at short periods. The low velocities are interpreted as regions of focused partial melt with the anisotropy explained by vertical alignment of this partial melt. A region-wide sharp change in anisotropy at periods >50 s suggests that a minimum of two layers of anisotropy is present. The change in anisotropy is interpreted to occur at the LAB based on the depth sensitivity kernels for each period. Partial melt beneath the rifts is attributed to decompression melting. We attribute the focused and segmented zones of partial melt to a combination of melt focusing along the LAB and enhanced melt generation due to buoyancy driven active upwelling. The interpretation of segmented zones of partial melt beneath the southern Red Sea agrees well with previous studies of the mantle response to oceanic rifting both beneath the Red Sea and globally. This study is able to show continuity between segmentation of partial melt beneath continental rifts and beneath an ocean spreading centre, suggesting that mantle segmentation beneath oceanic rifts initiates early during continental rifting and the location of each segment is controlled by lithospheric structure.

Acknowledgements Funding was provided by NERC grants NE/E007414/1, NE/D008611/1, NE/J012297/1 and NE/I020342/1, NE/K500926/1, NE/L013932/1 and BHP-Billiton. We thank the Saudi Geological Survey for their invitation to participate in the Saudi Geological Survey Red Sea Workshop. We also thank the three anonymous reviewers who helped improve this chapter.

References

Accardo NJ, Gaherty JB, Shillington DJ, Ebinger CJ, Nyblade AA, Mbogoni GJ, Chindandali PRN, Ferdinand RW, Mulibo GD, Kamihanda G, Keir D, Scholz C, Selway K, O'Donnell JP, Tepp G, Gallacher RJ, Mtelela K, Salima J, Mruma A (2017) Surface wave imaging of the weakly extended Malawi Rift from ambient-noise and teleseismic Rayleigh waves from onshore and lake-bottom seismometers. Geophys J Int 209:1892–1905

Ahmed A, Tiberi C, Leroy S, Stuart GW, Keir D, Sholan J, Khanbari K, Al-Ganad I, Basuyau C (2013) Crustal structure of the rifted volcanic margins and uplifted plateau of Western Yemen from receiver function analysis. Geophys J Int 193:1673–1690

Ambraseys NN, Melvile CP, Adams RD (1994) The seismicity of Egypt, Arabia and the Red Sea. Cambridge University Press, Cambridge, UK, p 201

Armitage JJ, Collier JS, Minshull TA (2010) The importance of rift history for volcanic margin formation. Nature 465:913–917

Armitage JJ, Ferguson DJ, Goes S, Hammond JOS, Calais E, Rychert CA, Harmon N (2015) Upper mantle temperature and the onset of extension and break-up in Afar, Africa. Earth Planet Sci Lett 418:78–90

Barberi F, Varet J (1977) Volcanism of Afar: Small-scale plate tectonics implications. Bull Geol Soc Am 88:1251–1266

Backus G, Gilbert F (1968) The resolving power of gross earth data. Geophys J Int 16:169–205

Bastow ID, Nyblade AA, Stuart GW, Rooney TO, Benoit MH (2008) Upper mantle seismic structure beneath the Ethiopian hot spot: rifting at the edge of the African low-velocity anomaly. Geochem Geophys Geosyst 9:Q12022

Belachew M, Ebinger CJ, Coté D, Keir D, Rowland JV, Hammond JOS, Ayele A (2011) Comparison of dike intrusions in an incipient seafloor-spreading segment in Afar, Ethiopia: seismicity perspectives. J Geophys Res 116:B06405

Benoit MH, Nyblade AA, VanDecar JC (2006) Upper mantle P-wave speed variations beneath Ethiopia and the origin of the Afar hotspot. Geology 34:329–332

Blackman DK, Kendall JM (1997) Sensitivity of teleseismic body waves to mineral texture and melt in the mantle beneath a mid–ocean ridge. Phil Trans R Soc Lond 355:217–231

Bonatti E (1985) Punctiform initiation of seafloor spreading in the Red Sea during transition from a continental to an oceanic rift. Nature 316:33–37

Bonatti E, Seyler M (1987) Crustal underplating and evolution in the Red Sea rift: Uplifted gabbro/gneiss crustal complexes on Zabargad and Brothers Islands. J Geophys Res 92:12803–12821
Bown JW, White RS (1995) Effect of finite extension rate on melt generation at rifted continental margins. J Geophys Res 100:18011–18029
Braun J, Beaumont C (1989) A physical explanation of the relation between flank uplifts and the breakup unconformity at rifted continental margins. Geology 17:760–764
Constable SC, Parker RL, Constable CG (1987) Occam's inversion: A practical algorithm for generating smooth models from electromagnetic sounding data. Geophysics 52:289–300
Civiero C, Hammond JOS, Goes S, Fishwick S, Ahmed A, Ayele A, Doubre C, Goitom B, Keir D, Kendall J-M, Leroy S (2015) Multiple mantle upwellings in the transition zone beneath the northern East-African Rift system from relative P-wave travel-time tomography. Geochem Geophys Geosyst 16:2949–2968
Corbeau J, Rolandone F, Leroy S, Al-Lazki A, Stork AL, Keir D, Stuart GW, Hammond JOS, Doubre C, Vergne J, Ahmed A (2014) Uppermost mantle velocity from Pn tomography in the Gulf of Aden. Geosphere 10:958–968
Corti G (2008) Control of rift obliquity on the evolution and segmentation of the main Ethiopian rift. Nat Geo 1:258–262
Corti G, Van Wijk J, Bonini M, Sokoutis D, Cloetingh S, Innocenti F, Manetti P (2003) Transition from continental break-up to punctiform seafloor spreading: how fast, symmetric and magmatic. Geophys Res Lett 30:1604
Dalton CA, Ekström G, Dziewoński AM (2008) The global attenuation structure of the upper mantle. J Geophys Res 113:B09303
Davison I, Al-Kadasi M, Al-Khirbash S, Al-Subbary AK, Baker J, Blakey S, Bosence D, Dart C, Heaton R, McClay K, Menzies M (1994) Geological evolution of the southeastern Red Sea Rift margin, Republic of Yemen. Bull Geol Soc Am 106:1474–1493
Debayle E, Ricard Y (2012) A global shear velocity model of the upper mantle from fundamental and higher Rayleigh mode measurements. J Geophys Res 117:B10308
Delaney JR, Kelley DS, Lilley MD, Butterfield DA, Baross JA, Wilcock WSD, Embley RW, Summit M (1998) The quantum event of oceanic crustal accretion: impacts of diking at mid-ocean ridges. Science 281:222–230
Ebinger CJ, Sleep NH (1998) Cenozoic magmatism throughout east Africa resulting from impact of a single plume. Nature 395:788–791
Ebinger CJ, Belachew M (2010) Geodynamics: Active passive margins. Nature Geosci 3:670
Facenna C, Becker TW, Jolivet L, Keskin M (2013) Mantle convection in the Middle East: reconciling Afar upwelling, Arabia indentation and Aegean trench rollback. Earth Planet Sci Lett 375:254–269
Ferguson DJ, Maclennan J, Bastow ID, Pyle DM, Jones SM, Keir D, Blundy JD, Plank T, Yirgu G (2013) Melting during late-stage rifting in Afar is hot and deep. Nature 499:70–73
Forsyth DW, Li A (2005) Array analysis of two-dimensional variations in surface wave phase velocity and azimuthal anisotropy in the presence of multipathing interference. In: Levander A, Nolet G (eds) Seismic earth: Array analysis of broadband seismograms. Am Geophys Union Monograph 157, pp 81–97
Gao SS, Liu KH, Abdelsalam MG (2010) Seismic anisotropy beneath the Afar depression and adjacent areas: implications for mantle flow. J Geophys Res 115:B12330
Gallacher RJ, Keir D, Harmon N, Stuart GW, Leroy S, Hammond JOS, Kendall JM, Ayele A, Goitom B, Ogubazghi G, Ahmed A (2016) The initiation of segmented buoyancy-driven melting during continental breakup. Nature Commun 7:13110
Hammond WC, Humphreys ED (2000) Upper mantle seismic wave velocity: effects of realistic partial melt geometries. J Geophys Res 105:10975–10986
Hammond JOS, Kendall J-M, Stuart GW, Keir D, Ebinger CJ, Ayele A, Belachew M (2011) The nature of the crust beneath the Afar triple junction: evidence from receiver functions. Geochem Geophys Geosyst 12:Q12004
Hammond JOS, Kendall JM, Stuart GW, Ebinger CJ, Bastow ID, Keir D, Ayele A, Belachew M, Goitom B, Ogubazghi G, Wright TJ (2013) Mantle upwelling and initiation of rift segmentation beneath the Afar depression. Geology 41:635–638
Hammond JOS, Kendall JM, Wookey J, Stuart GW, Keir D, Ayele A (2014) Differentiating flow, melt, or fossil seismic anisotropy beneath Ethiopia. Geochem Geophys Geosyst 15:1878–1894
Hansen S, Schwartz S, Al-Amri A, Rodgers A (2006) Combined plate motion and density-driven flow in the asthenosphere beneath Saudi Arabia: Evidence from shear-wave splitting and seismic anisotropy. Geology 34:869–872
Hansen SE, Nyblade AA, Benoit MH (2012) Mantle structure beneath Africa and Arabia from adaptively parameterized P-wave tomography: implications for the origin of Cenozoic Afro-Arabian tectonism. Earth Planet Sci Lett 319:23–34
Hayward NJ, Ebinger CJ (1996) Variations in the along-axis segmentation of the Afar Rift system. Tectonics 15:244–257
Holtzman BK, Kohlstedt DL, Zimmerman ME, Heidelbach F, Hiraga T, Hustoft J (2003) Melt segregation and strain partitioning: Implications for seismic anisotropy and mantle flow. Science 301:1227–1230
Holtzman BK, Kendall JM (2010) Organized melt, seismic anisotropy, and plate boundary lubrication. Geochem Geophys Geosyst 11: Q0AB06
Hopper JR, Funck T, Tucholke BE, Larsen HC, Holbrook WS, Louden KE, Shillington D, Lau H (2004) Continental breakup and the onset of ultraslow seafloor spreading off Flemish Cap on the Newfoundland rifted margin. Geology 32:93–96
Huerta AD, Nyblade AA, Reusch AM (2009) Mantle transition zone structure beneath Kenya and Tanzania: More evidence for a deep-seated thermal upwelling in the mantle. Geophys J Int 177:1249–1255
Keir D, Hamling IJ, Ayele A, Calais E, Ebinger CJ, Wright TJ, Jacques E, Mohamed K, Hammond JOS, Belachew M, Baker E (2009) Evidence for focused magmatic accretion at segment centers from lateral dike injections captured beneath the Red Sea rift in Afar. Geology 37:59–62
Keir D, Belachew M, Ebinger CJ, Kendall JM, Hammond JOS, Stuart GW, Ayele A, Rowland JV (2011) Mapping the evolving strain field during continental breakup from crustal anisotropy in the Afar Depression. Nature Commun 2:285
Kendall JM, Stuart GW, Ebinger CJ, Bastow ID, Keir D (2005) Magma-assisted rifting in Ethiopia. Nature 433:146–148
Kennett BLN, Engdahl ER, Buland R (1995) Constraints on seismic velocities in the Earth from travel times. Geophys J Int 122:108–124
King SD, Anderson DL (1995) An alternative mechanism of flood basalt formation. Earth Planet Sci Lett 136:269–279
King SD, Anderson DL (1998) Edge-driven convection. Earth Planet Sci Lett 160:289–296
Korostelev F, Basuyau C, Leroy S, Tiberi C, Ahmed A, Stuart GW, Keir D, Rolandone F, Ganad I, Khanbari K, Boschi L (2014) Crustal and upper mantle structure beneath south-western margin of the Arabian Peninsula from teleseismic tomography. Geochem Geophys Geosyst 15:2850–2864
Landisman M, Dziewonski A, Sato Y (1969) Recent improvements in the analysis of surface wave observations. Geophys J R Astr Soc 17:369–403
Langmuir CH, Bender JF, Batiza R (1986) Petrological and tectonic segmentation of the East Pacific Rise, 5°30′–14°30′ N. Nature 322:422–429

Laske G, Masters G, Ma Z, Pasyanos M (2013) Update on CRUST1.0—a 1-degree global model of Earth's crust. EGU General Assembly 2013, Geophys Res Abstr 15, EGU2013–2658

Leroy S, Lucazeau F, d'Acremont E, Watremez L, Autin J, Rouzo S, Bellahsen N, Tiberi C, Ebinger C, Beslier MO, Perrot J (2010) Contrasted styles of rifting in the eastern Gulf of Aden: A combined wide-angle, multichannel seismic, and heat flow survey. Geochem Geophys Geosyst 11:Q07004

Levshin AL, Yanovskaya TB, Lander AV, Bukchin BG, Barmin MP, Ratnikova LI, Its EN (1989) Recording identification and measurements of surface wave parameters. In: Keilis-Borok VI (ed) Seismic surface waves in laterally inhomogeneous earth. Kluwer Academic Publishers, Dordrecht, pp 131–182

Lewi E, Keir D, Birhanu Y, Blundy J, Stuart G, Wright T, Calais E (2016) Use of a high-precision gravity survey to understand the formation of oceanic crust and the role of melt at the southern Red Sea rift in Afar, Ethiopia. Geol Soc Spec Publ 420:165–180

Ligi M, Bonatti E, Tontini FC, Cipriani A, Cocchi L, Schettino A, Bortoluzzi G, Ferrante V, Khalil S, Mitchell NC, Rasul N (2011) Initial burst of oceanic crust accretion in the Red Sea due to edge-driven mantle convection. Geology 39:1019–1022

Ligi M, Bonatti E, Bortoluzzi G, Cipriani A, Cocchi L, Caratori Tontini F, Carminati E, Ottolini L, Schettino A (2012) Birth of an ocean in the Red Sea: initial pangs. Geochem Geophys Geosyst 13: Q08009

Lizarralde D, Axen GJ, Brown HE, Fletcher JM, González-Fernández A, Harding AJ, Holbrook WS, Kent GM, Paramo P, Sutherland F, Umhoefer PJ (2007) Variation in styles of rifting in the Gulf of California. Nature 448:466–469

Madge LS, Sparks DW (1997) Three-dimensional mantle upwelling, melt generation, and melt migration beneath segment slow spreading ridges. J Geophys Res 102:20571–20583

Mackenzie GD, Thybo H, Maguire PK (2005) Crustal velocity structure across the Main Ethiopian Rift: Results from two-dimensional wide-angle seismic modeling. Geophys J Int 162:994–1006

Maguire PKH, Keller GR, Klemperer SL, Mackenzie GD, Keranen K, Harder S, O'Reilly B, Thybo H, Asfaw L, Khan MA, Amha M (2006) Crustal structure of the northern Main Ethiopian Rift from the EAGLE controlled-source survey; a snapshot of incipient lithospheric break-up. Geol Soc Spec Publ 259:269–292

Makris J, Ginzburg A (1987) The Afar depression: transition between continental rifting and sea-floor spreading. Tectonophysics 141:199–214

Manighetti I, Tapponnier P, Gillot PY, Jacques E, Courtillot V, Armijo R, Ruegg JC, King G (1998) Propagation of rifting along the Arabia-Somalia plate boundary: Into Afar. J Geophys Res 103:4947–4974

McKenzie D (1978) Some remarks on the development of sedimentary basins. Earth Planet Sci Lett 40:25–32

Niu Y, Bideau D, Hekinian R, Batiza R (2001) Mantle composition control on the extent of mantle melting, crust production, gravity anomaly, ridge morphology, and ridge segmentation: a case study at the Mid-Atlantic Ridge 33–35°N. Earth Planet Sci Lett 186:383–399

Owens TJ, Nyblade AA, Gurrola H, Langston CA (2000) Mantle transition zone structure beneath Tanzania, East Africa. Geophys Res Lett 27:827–830

Park Y, Nyblade AA, Rodgers AJ, Al-Amri A (2007) Upper mantle structure beneath the Arabian Peninsula and northern Red Sea from teleseismic body wave tomography: implications for the origin of Cenozoic uplift and volcanism in the Arabian Shield. Geochem Geophys Geosyst 8:Q06021

Press WH, Teukolsky SA, Vetterling WT, Flannery BP (1992) Simulated annealing methods. In: Numerical recipes in C: the art of scientific computing (2nd edn). Cambridge University Press, pp 444–455

Rooney TO, Herzberg C, Bastow ID (2012) Elevated mantle temperature beneath East Africa. Geology 40:27–30

Rowland JV, Wilson CJ, Gravley DM (2010) Spatial and temporal variations in magma-assisted rifting, Taupo Volcanic Zone, New Zealand. J Volc Geotherm Res 190:89–108

Rychert CA, Hammond JOS, Harmon N, Kendall JM, Keir D, Ebinger C, Bastow ID, Ayele A, Belachew M, Stuart G (2012) Volcanism in the Afar rift sustained by decompression melting with minimal plume influence. Nature Geosci 5:406–409

Saito M (1988) DISPER80: A subroutine package for the calculation of seismic normal-mode solutions. In: Doornbos DJ (ed) Seismological algorithms: computational methods and computer programs. Academic Press, New York, pp 293–319

Schaeffer AJ, Lebedev S (2013) Global shear speed structure of the upper mantle and transition zone. Geophys J Int 194:417–449

Schouten H, Klitgord KD, Whitehead JA (1985) Segmentation of mid-ocean ridges. Nature 317:225–229

Sebai A, Stutzmann E, Montagner J, Sicilia D, Beucler E (2006) Anisotropic structure of the African upper mantle from Rayleigh and Love wave tomography. Phys Earth Planet Int 155:48–62

Sempere JC, Lin J, Brown HS, Schouten H, Purdy GM (1993) Segmentation and morphotectonic variations along a slow-spreading center: The Mid-Atlantic Ridge (24°00′ N–30°40′ N). Mar Geophys Res 15:153–200

Shapiro NM, Ritzwoller MH (2002) Monte-Carlo inversion for a global shear-velocity model of the crust and upper mantle. Geophys J Int 151:88–105

Shillington DJ, Scott CL, Minshull TA, Edwards RA, Brown PJ, White N (2009) Abrupt transition from magma-starved to magma-rich rifting in the eastern Black Sea. Geology 37:7–10

Sicilia D, Montagner JP, Cara M, Stutzmann E, Debayle E, Lépine JC, Lévêque JJ, Beucler E, Sebai A, Roult G, Ayele A (2008) Upper mantle structure of shear-waves velocities and stratification of anisotropy in the Afar Hotspot region. Tectonophysics 462: 164–177

Stork AL, Stuart GW, Henderson CM, Keir D, Hammond JOS (2013) Uppermost mantle (Pn) velocity model for the Afar region, Ethiopia: an insight into rifting processes. Geophys J Int 193:321–328

Tarantola A, Valette B (1982) Generalized nonlinear inverse problems solved using the least squares criterion. Rev Geophys 20:219–232

Taylor B, Goodliffe A, Martinez F, Hey R (1995) Continental rifting and initial sea-floor spreading in the Woodlark basin. Nature 374:534–537

Thompson DA, Hammond JOS, Kendall J-M, Stuart GW, Helffrich GR, Keir D, Ayele A, Goitom B (2015) Hydrous upwelling across the mantle transition zone beneath the Afar Triple junction. Geochem Geophys Geosyst 16:834–846

Tiberi C, Ebinger CJ, Ballu V, Stuart G, Oluma B (2005) Inverse models of gravity data from the Red Sea-Aden-East African rifts triple junction zone. Geophys J Int 163:775–787

Wang Y, Forsyth DW, Savage B (2009) Convective upwelling in the mantle beneath the Gulf of California. Nature 462:499–501

Weissel JK, Karner GD (1989) Flexural uplift of rift flanks due to mechanical unloading of the lithosphere during extension. J Geophys Res 94:13919–13950

White RS, McKenzie D (1989) Magmatism at rift zones: The generation of volcanic continental margins and flood basalts. J Geophys Res 94:7685–7729

White RS, Smith LK, Roberts AW, Christie PAF, Kusznir NJ (2008) Lower-crustal intrusion on the North Atlantic continental margin. Nature 452:460–464

Wolfe CJ, Vernon FJ, Al-Amri A (1999) Shear-wave splitting across western Saudi Arabia: The pattern of upper mantle anisotropy at a Proterozoic shield. Geophys Res Lett 26:779–782

Wolfenden E, Ebinger CJ, Yirgu G, Deino A, Ayalew D (2004) Evolution of the northern Main Ethiopian rift: birth of a triple junction. Earth Planet Sci Lett 224:213–228

Xu W, Ruch J, Jónsson S (2014) Birth of two volcanic islands in the southern Red Sea. Nature Commun 6:7104

Yang Y, Forsyth DW (2006) Regional tomographic inversion of the amplitude and phase of Rayleigh waves with 2-D sensitivity kernels. Geophys J Int 166:1148–1160

Zhou Y, Dahlen FA, Nolet G (2004) Three-dimensional sensitivity kernels for surface wave observables. Geophys J Int 158:142–168

Oceanization Starts at Depth During Continental Rupturing in the Northern Red Sea

Marco Ligi, Enrico Bonatti, William Bosworth, and Sara Ronca

Abstract

We present here 3D seismic reflection and gravity data obtained from an off-axis area of the NW Red Sea, as well as results of a study of gabbroic rocks recovered in the same area both from an oil well below a thick evaporitic-sedimentary sequence, and from a layered mafic complex exposed on the Brothers Islets. These new data provide constraints on the composition, depth of emplacement and age of early syn-rift magma intrusions into the deep crust. The Brothers are part of a series of sub-parallel NW-striking topographic highs associated with SW-dipping extensional fault blocks with significant footwall uplift during rifting that brought early syn-rift deep crustal rocks up to the seafloor. Assuming an important role played by magmatism in the evolution of narrow rifts helps to solve the controversy on the nature of the crust in the northern/central Red Sea (i.e., the crust outside the axial oceanic cells is either oceanic or it consists of melt-intruded extended continental crust). Gabbros show petrologic and geochemical signatures similar to those of MORB-type gabbroic cumulates and are compatible with their having been emplaced either in a continental or in an oceanic context. We explored the different hypotheses proposed to explain the lack of magnetic anomalies in the presence of oceanic crust in the northern Red Sea. Our results, combined with a review of all the geophysical and geological data in the area, suggest a stretched and thinned continental crust with few isolated sites of basaltic injections, in line with a model whereby asthenospheric melt intrusions contribute to weaken the lower crust enabling some decoupling between upper and lower crust, protracting upper crust extension and delaying crustal breakup. Our findings show that continental rupture in the northern Red Sea is preceded by intrusion of basaltic melts with MORB-type elemental and isotopic signature, that cooled forming gabbros at progressively shallower crustal depths as rifting progressed toward continental separation.

1 Introduction

The Red Sea formed in response to divergence and anti-clockwise rotation of the Arabian plate with respect to Nubia due to extensional forces consistent with the far field stress related to the Zagros subduction (Bellahsen et al. 2003; Bosworth 2015; Kendall and Lithgow-Bertelloni 2016; Demets and Merkouriev 2016). The Red Sea preserves the entire tectonic and sedimentary record from continental rifting to break-up stages, plus ~30 Ma of magmatic activity prevalent on the western side of the Arabian plate.

Large amounts of geophysical and geological data have been collected along the Red Sea during the past 50 years (Almalki et al. 2015, and reference therein); however, they provide partial viewpoints on Red Sea crustal structure and basin architecture, often extrapolated along the entire length of the basin to infer general tectonic models. Due to this, several controversies about the evolution of the Red Sea are still open, such as: (a) the role of mantle upwelling and mantle plume versus far field stresses in localizing and propagating rifting and sea-floor spreading (Bellahsen et al. 2003; Lee et al. 2011; Chang and Van der Lee 2011), and (b) whether continental rifting was asymmetric (simple shear) (Voggenreiter and Hötzl 1989), symmetrical (pure

M. Ligi (✉) · E. Bonatti
Istituto di Scienze Marine, CNR, Bologna, Italy
e-mail: marco.ligi@bo.ismar.cnr.it

E. Bonatti
Lamont Doherty Earth Observatory, Columbia University, Palisades, NY, USA

W. Bosworth
Apache Egypt Companies, 11 Street 281, New Maadi, Cairo, Egypt

S. Ronca
Dipartimento di Scienze della Terra, Sapienza Università di Roma, Rome, Italy

N. M. A. Rasul and I. C. F. Stewart (eds.), *Geological Setting, Palaeoenvironment and Archaeology of the Red Sea*,
https://doi.org/10.1007/978-3-319-99408-6_7

shear) (Bohannon and Eittreim 1991), or strike-slip (pull-apart) (Makris and Rhim 1991). Another important unresolved question concerns the extent of oceanic crust within the Red Sea basin, particularly in the northern Red Sea, with some authors preferring an oceanic regime and others the prevalence of attenuated continental crust (cf. Girdler 1985, 1991; Sultan et al. 1992 versus Mitchell and Park 2014; Almalki et al. 2016). No borehole data are available as to the nature of the crust beneath the main trough. Gravity and magnetic data can be interpreted as due to either continental or oceanic sources, and the thick and high-velocity evaporitic layer, reducing penetration of multichannel seismic reflection profiles, masks the tectonic style of the basement, hiding data potentially useful for clarifying its origin (e.g., Mitchell et al. 2017).

The lack of organised magnetic anomalies north of the Zabargad Island, where the coasts are straight, the high-density crustal layers just beneath evaporites, the presence of low seismic- refraction velocity gradients, and the presence of Precambrian shield rocks in boreholes near the coasts (Fig. 1), all support the idea that the northern Red Sea is carpeted by stretched and thinned continental crust, with a few isolated sites of basaltic injection (Cochran 1983, 2005; Bonatti 1985; Cochran and Martinez 1988; Guennoc et al. 1988; Bosworth 1993; Cochran and Karner 2007; Stern and Johnson 2010; Mitchell and Park 2014; Almalki et al. 2016). Other authors suggest that, given that in the northern Red Sea the plate separation rates are small (of the order of 6–10 mm/yr), the absence of organised magnetic anomalies might be due to the blanketing effect of large thicknesses of salt which flow faster than the opening rate (LaBrecque and Zitellini 1985; Gaulier et al. 1988; Girdler 1985, 1991; Sultan et al. 1992; Dyment et al. 2013). The magnetic anomalies are subdued due to the high heat flow and slow cooling of the intruding rocks (Dyment et al. 2013; Tapponnier et al. 2013).

In this paper, we present seismic reflection and gravity data from an area adjacent to two small islets (Brothers Islands) located about 40 km west of the axis of the northern Red Sea, as well as geochemical data from gabbroic rocks recovered from the islets and from the base of an offshore well located southwest of Brothers ~50 km from the axis of the basin (Fig. 2). We will test the hypotheses that these gabbros either represent former basaltic melt intrusions in thinned continental crust in a pre-oceanic rift setting, or they were emplaced in an oceanic context due to seafloor spreading. Clarifying the nature of the northern Red Sea crust outside the axial oceanic cells will help our understanding of Red Sea rift evolution and generally of processes that drive continental rupture.

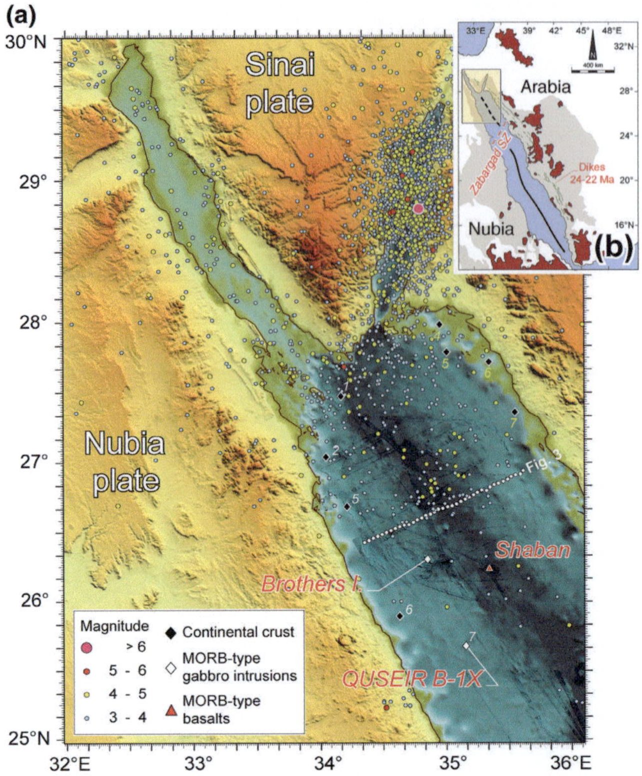

Fig. 1 **a** Seismicity of the northern Red Sea. Colour filled circles are earthquake epicentres with colour indicating magnitudes. Epicentres recorded during 1973–2017 from International Seismological Centre (ISC), Thatcham, UK (*OnLine Bulletin*: http://www.isc.ac.uk). Black filled diamonds show offshore wells where Nubian shield basement rocks have been encountered at total depth; white numbers from Bosworth (1993) and yellow numbers from Almalki et al. (2015). Brothers and QUSEIR MORB-gabbros locations are also indicated. White filled circles indicate heat flow stations of Martinez and Cochran (1989) shown in Fig. 3. **b** Simplified plate tectonic framework of the Red Sea/Gulf of Aden area. Thick black and red dashed lines indicate the Red Sea axis and the Zabargad Shear Zone, respectively. Dark red areas indicate intraplate Cenozoic lava fields and green lines, dyke swarm (emplaced at 24–22 Ma) running along the entire coast of the Arabia plate (Bosworth and Stockli 2016)

2 The Northern Red Sea

Modern geodetic constraints on plate spreading extrapolated over time suggest that rifting in the Red Sea started 22 ± 3 Ma in the late Oligocene/early Miocene, assuming that the initial rate of extension across the rift was roughly half the present-day rate (Reilinger et al. 2015). Seafloor spreading anomalies from surrounding ocean basins constrain opening rates to at least 12 Ma, with large ambiguities prior to that (DeMets and Merkouriev 2016). At 11 ± 2 Ma, when rifting in the northern Red Sea shifted from the Gulf of Suez to

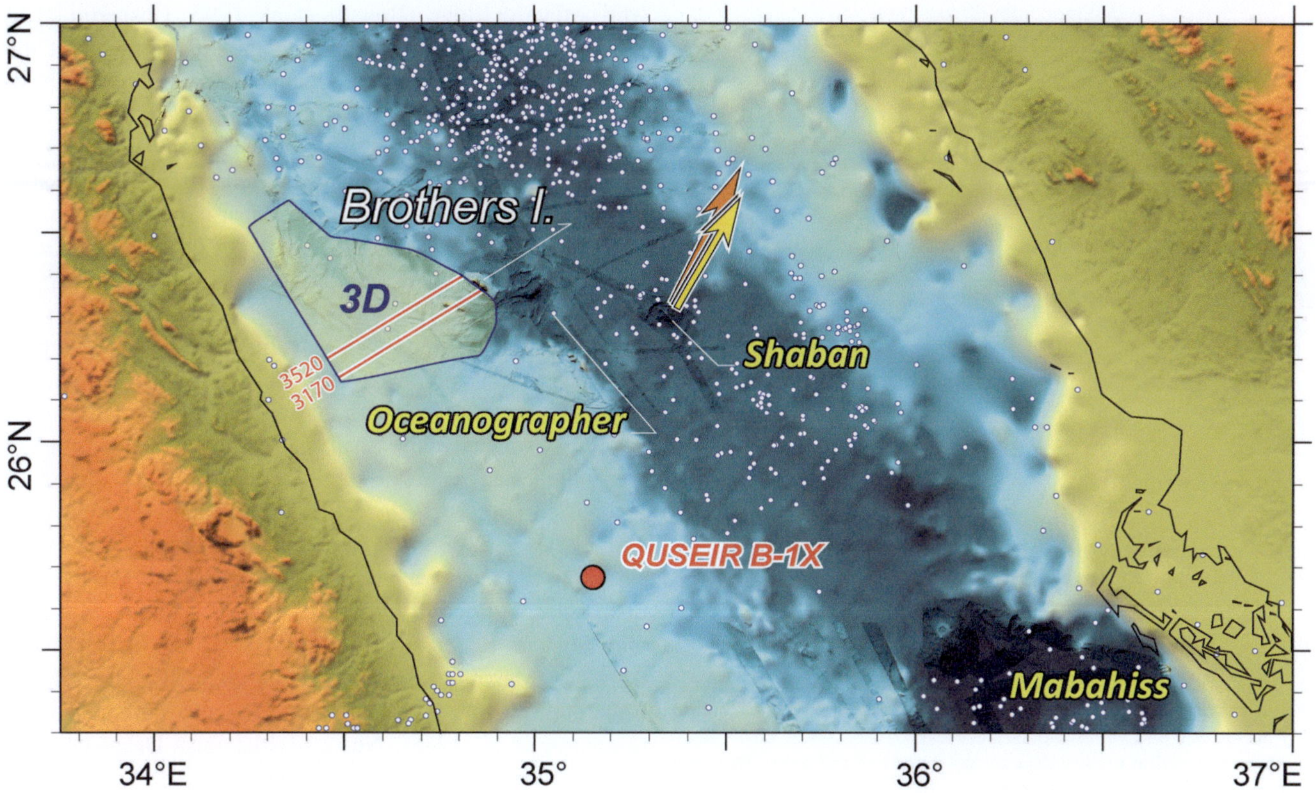

Fig. 2 Shaded relief map showing the major morphotectonic features of the northern Red Sea. White dots, earthquake epicentres since 1973 from ISC. The blue polygon marks location of the 3D seismic survey. The red circle indicates the location of the Phillips oil well QUSEIR B-1X and the red line marks the location of seismic cross-lines 3520 and 3170 shown in Fig. 7. Orange, grey and yellow arrows indicate velocity vectors of Arabia relative to the Nubia plate at the latitude of Shaban Deep (10.47, 8.81 and 8.82 mm/yr) from Chu and Gordon (1998), DeMets and Merkouriev (2016) and ArRajehi et al. (2010), respectively. Plate separation rates of Arabia from Nubia range from 15–18 mm/yr in the southern Red Sea to 8–10 mm/yr at Brothers (Chu and Gordon 1998; ArRajehi et al. 2010; DeMets and Merkouriev 2016)

the Dead Sea/Gulf of Aqaba fault system, plate separation rate doubled to the present rate (7 ± 1 mm/yr at ~27°N, Reilinger et al. 2015). Bosworth and Stockli (2016), based on geochronology of early rift magmatism and on stratigraphic reconstructions, reach similar conclusions, dating the initial phase of extension at ~24–23 Ma for most of the Red Sea. Alternatively, Wolfenden et al. (2005) date eruptive volcanic centres along large offset border fault systems at ~28 Ma, suggesting that rift onset was coincident with flood magmatism in the southern Red Sea rift, an interpretation compatible with the hypothesis that continental rifting began first in the southern Red Sea (offshore Eritrea) during the Late Oligocene and then stalled for few million years (Bosworth et al. 2005; Bosworth 2015).

The crustal structure across the northern Red Sea based on P-wave receiver functions in southeastern Egypt and other estimates of crustal thickness in the northern Red Sea region (Hosny and Nyblade 2014, 2016; Tang et al. 2016), reveal a symmetric pattern of crustal thickness beneath the conjugate margins. Crustal thickness along the rifted margins of the Red Sea, Gulf of Suez and Gulf of Aqaba ranges from 25 to 30 km, whereas beneath northern and central Egypt and northern Saudi Arabia crustal thickness ranges from 32 to 38 km (Hosny and Nyblade 2016; Tang et al. 2016). STEFAN E project refraction profiles (Voggenreiter et al. 1988) show a good agreement with these results, with a thickness of ~20 km near the coast decreasing toward the rift axis down to ~10 km. Thus, the 35–40 km pre-rift crustal thickness implies a 5–10 km crustal thinning during rifting beneath the rifted margins near the coast to take into account the 25–30 km thick crust observed there (Hosny and Nyblade 2016).

Seismicity in the northern Red Sea, south of the Aqaba-Dead Sea transform, is scattered across the basin. It is dominated by a large number of earthquakes with low and moderate magnitudes on the Richter scale (Figs. 1 and 2), in contrast with the central and southern Red Sea where several moderate to high magnitude events are focused along the rift axis (Al-Ahmadi et al. 2014; Bosworth et al. 2018). The low level of seismic activity in the northern compared with the southern Red Sea and the Gulf of Aqaba may be due to: (1) elastic energy due to extension being released through a large number of low-magnitude earthquakes; and/or (2) lithospheric deformations in this region being

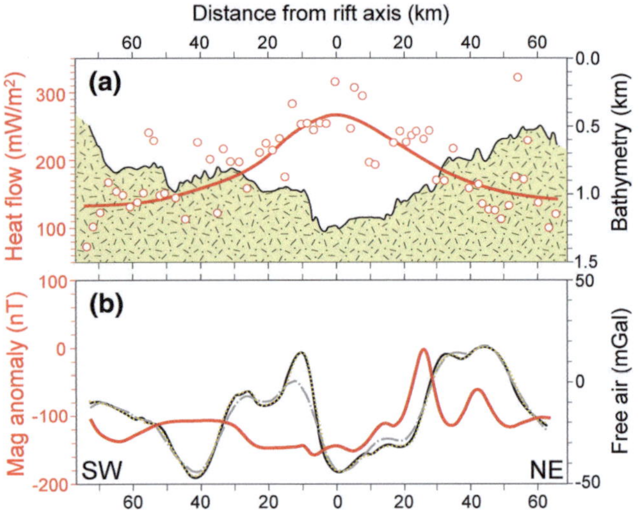

Fig. 3 Heat flow, bathymetry, magnetic anomaly and free air gravity along a profile projected at N60°E across the northern Red Sea north of the Brothers Islets from Martinez and Cochran (1989). **a** Heat flow stations: red circles, and general trend fitting observations: red line. Topography: black line. **b** Total intensity magnetic anomaly: red line; and shipborne, satellite derived and combined satellite and ship-borne free air gravity anomaly interpolated along the ship-track: black solid, grey dot-dashed and yellow dotted line, respectively. Large errors near the shore of satellite gravity data are not observed and the differences between shipborne and satellite derived data are mostly due to the high-frequency content of the ship-gravity data. Standard deviation of ship-satellite free air differences along the line is 3.83 mGal

accommodated over a wide area across the basin, offshore and inland. The majority of earthquakes in the northern Red Sea are clustered in a ~40 km-wide strip centred along the main trough, suggesting that the rift axis is located in the deepest part of the basin. This is confirmed by the heat flow distribution. Figure 3 shows heat flow measurements from Martinez and Cochran (1989) along a profile perpendicular to the rift axis just north of Brothers Islands (see Fig. 1 for location).

Initial accretion of oceanic crust accompanied by Vine-Matthews magnetic anomalies started roughly 5 Ma in the southern Red Sea, and 3 to 1 Ma in discrete axial cells within the central Red Sea in a pattern suggesting northward propagation of the nascent oceanic rift (Girdler and Styles 1974; Cochran 1983; Bonatti 1985; Ghebreab 1998; Bosworth et al. 2005; Ligi et al. 2012, Gallacher et al. 2018). The Red Sea northward-propagating oceanic rift impacts against the Zabargad Shear Zone (Bonatti 1985), a major morphotectonic feature striking almost N-S that intersects and offsets the Red Sea axis northward by ~100 km, and marks the southern limit of the northern Red Sea (Fig. 1). We suggest the Zabargad Shear Zone (SZ) may be a "proto-transform fault" that, if the Red Sea were to continue opening, might develop into an "initial" major oceanic transform, similar to those offsetting today the equatorial Mid Atlantic Ridge (Bonatti et al. 1996; Gasperini et al. 2001). Zabargad Island lies along the southern end of this feature, while a probably extensional (pull apart) basin (Mabahiss Deep) lies at its NE end (Figs. 1 and 2).

Zabargad Island represents an emerged sliver of sub-Red Sea lithosphere, probably uplifted by transpressive/transtensive tectonics along the Zabargad SZ (Crane and Bonatti 1987); it provides a sample of continent-ocean transitional lithosphere (Bonatti et al. 1981, 1983, 1986; Nicolas et al. 1987). Zabargad exposes mantle-derived peridotites of sub-continental affinity in faulted contact with an interlayered amphibolite-gneiss unit containing relicts of pyroxenites and gabbros recrystallized into mafic granulites, intruded by basaltic dykes. The igneous relicts were suggested to be originally part of a mafic-ultramafic layered complex which crystallized at relatively high pressure in the lower continental crust (Bonatti and Seyler 1987).

Along the main trough of the northern Red Sea, the only deeps clearly floored by MORB-type basalts are Mabahiss and Shaban (Guennoc et al. 1988; Haase et al. 2000) (Fig. 2). Mabahiss Deep lies at the NE end of the Zabargad SZ and has been interpreted as a pull-apart basin (Guennoc et al. 1988). North of Mabahiss Deep, the only area where recent continental rupture can be observed is at Shaban Deep (Fig. 2), where an elongated NW-SE 6 km-long volcanic ridge rises to a level of 900 m from a maximum depth of 1600 m (Ehrhardt and Hubscher 2015). Basaltic glasses from the axial ridge (Haase et al. 2000) indicate derivation from a depleted mantle source (MORB source) with no contamination by continental lithosphere, similarly to basaltic glasses sampled along the axial zone of Thetis and Nereus (Ligi et al. 2012, 2015) in the northern sector of the central Red Sea.

The Brothers are two islets located about 40 km west of the Shaban Deep at about 26°10'N (Fig. 2). They rise steeply from the surrounding sea floor by about 700 m on the western side and 1000 m on the eastern side. They are both elongated on a NW-SE direction. The northern islet is about 600 m long; the southern islet is 300 m long; they are flat topped, reaching about 10 m and 6 m above sea level, respectively, and they are covered by a late Pleistocene carbonate reef unit (Taviani and Rabbi 1984; Hoang and Taviani 1991). The structural lineations (fractures, fissures, etc.) observed on the northern island trend prevalently in an E–W direction. This system offsets a N–S system but is offset by a Dead Sea-Gulf of Aqaba trending system (NE–SW), that appears to be the youngest structures affecting the island (Taviani et al. 1986). Reconnaissance work has shown that below the carbonate caps the islets expose gabbroic rocks cut by a few doleritic dykes striking NW-SE, parallel to the Red Sea rift trend (Shukri 1944; Taviani and Rabbi 1984; Taviani et al. 1986; Seyler and Bonatti 1988; Bosworth and Stockli 2016).

Gabbros were also sampled beneath a ~4 km thick sedimentary sequence, dominated by evaporites, at the base of the QUSEIR B-1X drill hole, ~80 km south of the Brothers (Fig. 2). Samples of these rocks are the object of this study.

3 Datasets and Methods

Multibeam bathymetry: Bathymetry was acquired during the 2005 R/V Urania (RS05) cruise with a RESON SeaBat 8160 multibeam, DGPS positioning, and TSS MAHRS MRU and gyrocompass. Data processed by the Kongsberg Neptune package produced Digital Terrain Models with up to 25 m of grid resolution. Topography of the northern Red Sea was obtained from a synthesis of our own multibeam and single-beam data (Bonatti et al. 1984; Mitchell et al. 2010; Ligi et al. 2011) acquired in 1979 (R/V Salernum, MR79) and 1983 (R/V Bannock, MR83), multibeam data (Cochran and Martinez 1988; Guennoc et al. 1988; Haase et al. 2000; Augustin et al. 2014) from cruises MEROU (R/V Charcot, 1978), RC2507 (R/V Conrad, 1984), M31L2 (R/V Meteor, 1995) and 64PE350/351 (R/V Pelagia, 2012), plus water depth data obtained by converting to depth the two-way travel time of the 3D seismic seafloor reflector using a constant velocity of 1525 m/s, bathymetric grids from GEBCO and NGDC databases (https://www.bodc.ac.uk/data/; http://www.ngdc.noaa.gov/mgg/), and elevation data from the Shuttle Radar Topography Mission (SRTM) database (http://srtm.usgs.gov/). Spatial analysis and mapping were performed using the PLOTMAP package (Ligi and Bortoluzzi 1989).

Gravity: Shipboard gravity measurements from the NGDC database were corrected with the techniques outlined in Ligi et al. (2012) and combined with satellite-derived free-air gravity data (Sandwell et al. 2014, version 24.1) in order to add the ship-gravity high-frequency content to the full coverage of satellite data (Fig. 3b). A total of 27796 shipboard gravity measurements were used from the NGDC database (cruises: RC2507, SHA1079 and 83005911). The overall statistics at 1687 cross-over points give a mean and standard deviation of −0.36 and 5.02 mGal, respectively. In order to integrate the different data sets, the ship gravity for each line was adjusted in a least-square sense by applying a series of corrections aimed at reducing the errors at cross-over points and at matching the satellite-derived marine gravity field (Ligi et al. 2012). After adjustment, the overall cross-track discrepancies are reduced to zero mean and standard deviation of 4.26 mGal, equivalent to an accuracy of 3.0 mGal. The adjusted ship gravity is then blended with satellite-derived gravity by the input-output system theory (IOST) method (Fig. 3b), with power spectral densities estimated directly from the data (Li and Sideris 1997; Tziavos et al. 1998). We assume gravity anomalies are affected by a gaussian random noise with zero mean and variance of 18.15 and 25.0 $mGal^2$ for ship-board and satellite-derived gravity data, respectively. Bouguer anomalies were obtained by subtracting from the free air anomalies the attraction of seafloor topography and sediments. The contribution of topography and sediments to the local gravity field has been computed from grids of bathymetry, post-evaporitic unconsolidated sediments and Miocene evaporites by a FFT technique, that uses a series expansion of the Fourier transformed powers of the base of each layer to represent the Fourier transform of the gravity anomaly (Parker 1973). The first nine terms of the series expansion were retained in our calculations to account for the non-linear gravitational attraction of the large topographic relief. The Bouguer correction was obtained by replacing with a layer of crustal material (density of 2670 kg/m^3) the water layer (density of 1040 kg/m^3), unconsolidated sediments (density of 1790 kg/m^3) and evaporites (density 2160 kg/m^3). Grids of bathymetry and sediment thickness were produced at 0.1 km spatial resolution in order to perform upward continuation also in the shallowest portion of the northern Red Sea. Each grid was mirrored to avoid edge effects introduced by the implicit periodic assumption of the FFT routine. The predicted gravity contribution of the crustal layers, computed at bathymetry grid points, was interpolated on to combined ship-satellite gravity grid points and then subtracted from the corresponding free air anomalies (Figs. 4 and 5). The zero level of the Bouguer anomalies is arbitrary and corresponds to the centre of the range in anomaly amplitudes.

Seismic reflection: We interpreted a commercial 3D seismic survey acquired by BG in 1999 offshore the Egyptian coast between Safaga and Quseir and extending seaward up to 40 km west of the rift axis (Gordon et al. 2010). The time migrated reflection profiles from the 3D survey were included in the seismic and geological interpretation software IHS *Kingdom*® to map the top and the base of the major tectonic sequences, such as the top of evaporites, the base of evaporites, the rift-onset unconformity and the top of igneous basement. The thicknesses of hemipelagic sediment cover and of the evaporites were estimated from the 3D seismic survey and from single-channel seismic reflection data (cruises MR79, MR83 and RC0911), adopting average interval seismic velocities (Tramontini and Davis 1969) of 2 km/s and 3.75 km/s, respectively. The resulting high-resolution grids of bathymetry, and top and base of the evaporites were included in the calculation of the Bouguer anomaly (Fig. 5).

Elemental chemistry: Gabbroic rocks from the northern Red Sea recovered on the Brothers Islets and from the bottom of oil well QUSEIR B-1X (QUSEIR) were studied in order to unravel their role in the opening of the northern Red

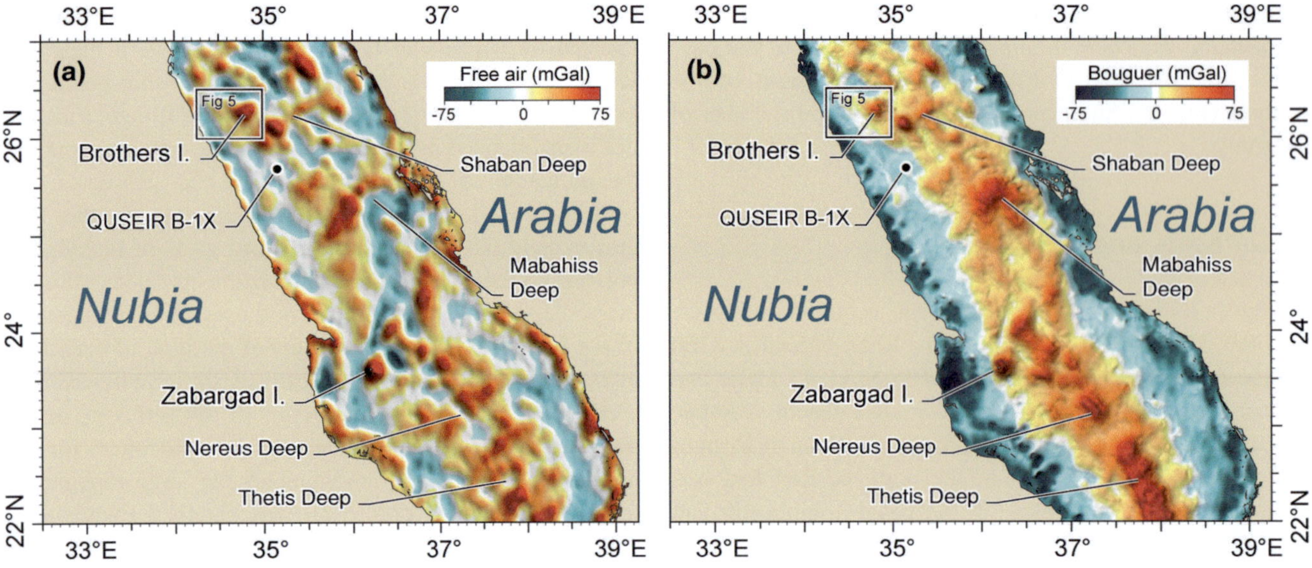

Fig. 4 Gravity anomalies of the northern Red Sea. **a** free air anomalies from satellite-derived gravity data (Sandwell et al. 2014) (version 24.1). **b** Bouguer anomalies from free air data corrected for seabed topography using a density contrast of 1630 kg/m^3. The black box refers to the area displayed in Fig. 5

Sea. Whole rock major and trace element compositions were obtained by lithium metaborate/tetraborate fusion ICP-AES and ICP–MS, at Activation Laboratories Inc. (Ontario, Canada) according to the Code 4Lithoresearch package and by ICP-OES (Agilent 720) at Lamont-Doherty Earth Observatory of Columbia University (LDEO) on solutions prepared with lithium metaborate fusion. FeO was determined by redox titration ($KMnO_4$ titration). Loss On Ignition (LOI) was measured according to standard gravimetric procedures and corrected for Fe oxidation. Mineral compositions were determined using an automated CAMECA SX50 microprobe operating in full WDS mode at the Istituto di Geologia Ambientale e Geoingegneria (IGAG-CNR), Sapienza University of Rome. An acceleration potential of 15 kV with a sample current of 15 nA (measured on brass) was applied. The beam size was varied as a function of the analyzed phase. Natural and synthetic oxides and silicates were used as standards. Samples and standards were carbon coated. On-line corrections for drift, dead-time and background were applied to the raw data. In situ trace elements have been measured at the Centro Interdipartimentale Grandi Strumenti (CIGS) of the Università of Modena and Reggio Emilia using a Nd:YAG deep UV (213 nm) New Wave Research UP-213 laser ablation system (LA) coupled to a Thermo Fisher Scientific X-Series II Induced Coupled Plasma Mass Spectrometer (ICP-MS). Instrumental drift correction was computed by linear correction of measured intensities among repeated measurements of the NIST 610, 612 and 614 glasses with ^{44}Ca as the internal standard. The analytical routine includes 100 µm pre-ablation scan (dwelling time: 2 s, 5 Hz laser fire, laser fluency 18÷20 J/cm^2) followed by 80 µm ablation scan (dwelling time: 30 s, 20 Hz laser fire, laser fluency 18÷20 J/cm^2). Data reduction was performed with Plasma Lab® software by Thermo Scientific. Precision and accuracy, both better than 10% for concentrations at the ppm level, were assessed from repeated analyses of NIST 610, 612 and 614 standards.

Geothermometry and geobarometry: Pressure conditions during crystallisation were estimated using the cpx geobarometer of Nimis and Ulmer (1998), and Nimis (1999) based on the cpx structural parameters calculated from major-element composition. Equilibration temperatures of the cumulate rocks were estimated by means of various geothermometers; an olivine-augite geothermometer based on Fe–Mg exchange (Loucks 1996), applicable to mineral assemblages including ol + pl + cpx ± opx or pigeonite, a two-pyroxene geothermometer (Frost and Lindsley 1992) in the QUILF software (Andersen et al. 1993), a pyroxene thermometer for cpx +opx intercumulus assemblages (Brey and Koehler 1990), projection of the pyroxene compositions onto Lindsley's isotherms (Lindsley 1983) providing minimum equilibration temperature for cpx, and a Ti-in-amphibole geothermometer (Otten 1984; Féménias et al. 2006) in order to estimate crystallization temperatures of interstitial magmatic, late-magmatic and subsolidus amphiboles.

Isotope chemistry: Nd, Sr and Pb isotope ratios of QUSEIR and Brothers gabbros were measured on whole rock powders and mineral separates. Mineral separates were carefully checked under a binocular microscope to avoid any inclusion or alteration. Minerals and powders were strongly

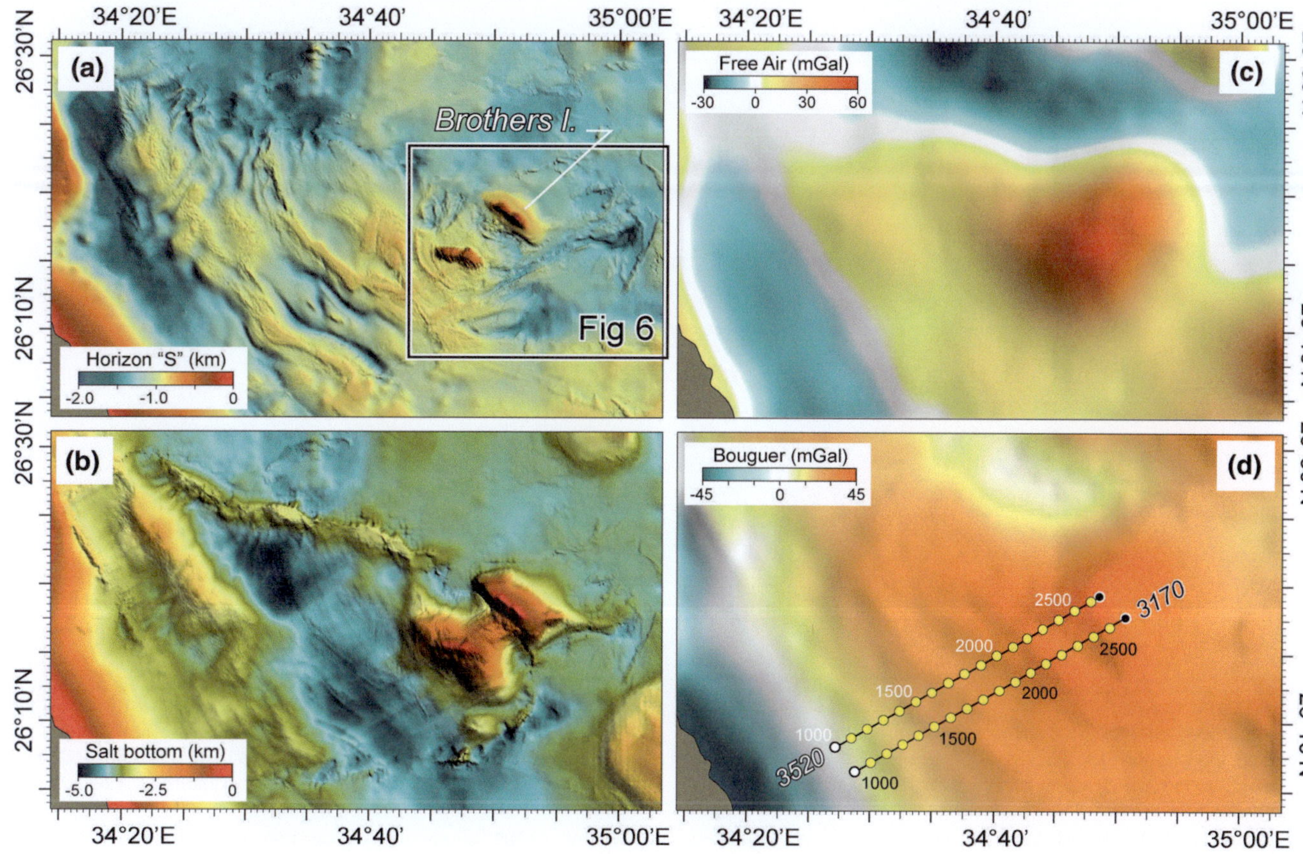

Fig. 5 Gravity imagery in the Brothers Islets region. Details on the gravity data reduction method are in the text. **a** Depth below sea level of "S" reflector (top of Miocene evaporites). Time to depth conversion of two-way travel times was obtained assuming a P-wave velocity of 2 km/s in the unconsolidated sediments. The black box refers to the area displayed in Fig. 6. **b** Depth below sea level of reflector marking the base of Miocene evaporites. Time to depth conversion obtained assuming an average P-wave velocity of 3.75 km/s in the evaporites. The structural high bounding the deep salt basin to the north in the central part of the displayed region is probably an artefact due to migration edge effects in the 3D seismic survey. **c** Combined ship-satellite free air gravity anomaly. **d** Bouguer anomalies obtained by subtracting from the free air anomalies the attraction of seafloor topography, unconsolidated sediments and evaporites. Locations of seismic cross-lines 3520 and 3170 shown in Fig. 7 are also indicated

leached with a solution of 6.2 N HCl and 5% HF. Pb was separated using AG1-X8 anion resin, Sr was separated using Eichrom Sr resin and Nd was separated in a two-column procedure using Eichrom TRU-spec resin to separate the REE, followed by alfa-hydroxy isobutyric acid. The Sr, Nd and Pb isotopic compositions were measured on a VG Sector 54 multicollector thermal ionization mass spectrometer at LDEO. Standardization and normalization details can be found in Mazzucchelli et al. (2016) and Ligi et al. (2018).

4 Results

4.1 3D Seismic Interpretation

A 3D seismic reflection survey of a limited but complex area from the shoreline to the Brothers Islets in the western African side of the northern Red Sea (Gordon et al. 2010) allows us an evaluation of the structural setting of the margin. The top and the base of major tectonic sequences have been mapped, such as the top of the evaporites, the base of the evaporites, the rift-onset unconformity and the top of the igneous basement. Mapping of these features allows detailed structural interpretation and geological models. The resulting high-resolution grids of bathymetry (Fig. 6), as well as the top and base of the evaporites, were included in the calculation of the Bouguer anomaly of the Brothers-QUSEIR region allowing us to infer crustal density variations (Fig. 5).

Nicolas et al. (1987) proposed that the Brothers Islets represent a "subsidence-resistant" block within the stretching regime, whereas, according to Taviani et al. (1986) they are culminations of a sliver of lower crust tectonically uplifted in response to localized compression related to a presumed transform fault (Brothers Fracture Zone of Crane and Bonatti 1987), in a setting similar to that proposed for Zabargad (Crane and Bonatti 1987). Seismic reflection profiles

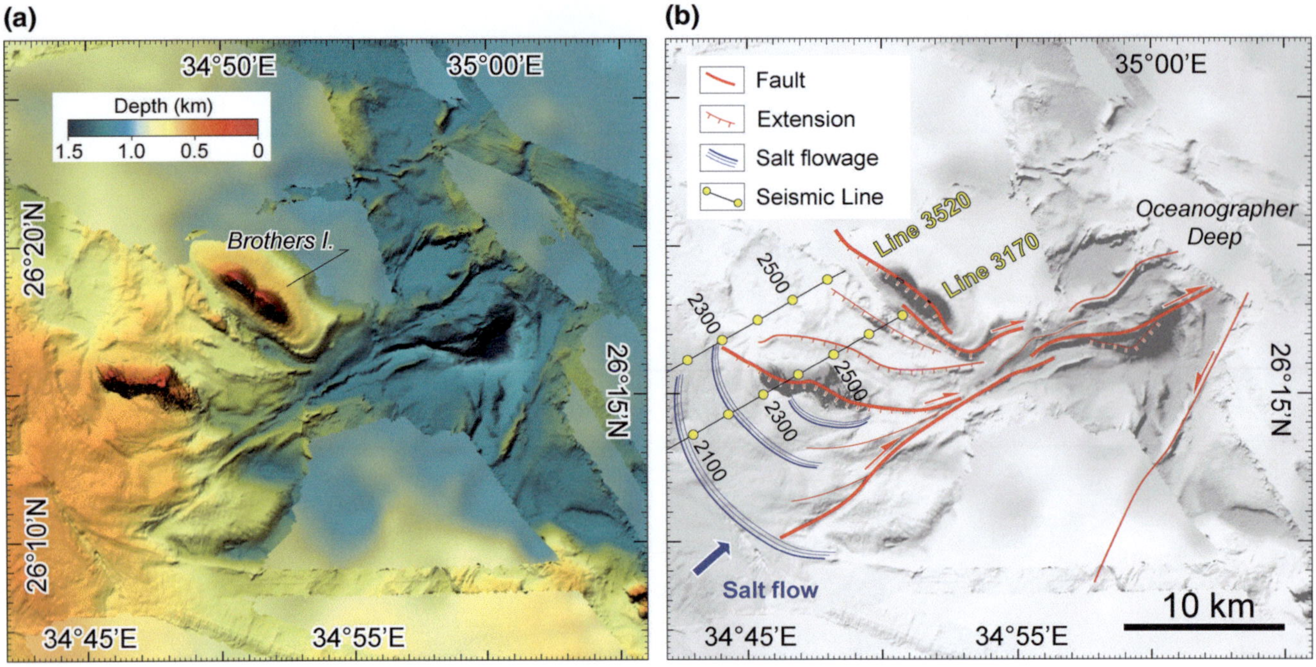

Fig. 6 Morphotectonic setting of Brothers Islets region. Bathymetry and morphology show the transtensional origin of the Oceanographer Deep and the ongoing tectonics causing uplift of the crustal blocks culminating in the two Brothers Islets and in the structural high located to southwest of the islets. The Oceanographer Deep is located east of the two crustal blocks at the intersection of a left stepping transcurrent fault with Dead Sea Transform direction and a NE dextral transcurrent fault oriented normal to the present-day rift axis. The recent uplift of the two crustal blocks is due to extension associated with splay faults at the southern end of the NE dextral transcurrent fault. **a** and **b** Shaded relief images from a compilation of single beam and multibeam bathymetry together with water depth obtained by converting to depth the two-way travel time of the 3D seismic seafloor reflector using a constant water velocity of 1.525 km/s. Source of light from NE, grid resolution 25 m

extending across the western Red Sea close to the Brothers Islets provide some insight as to the mechanisms that allowed uplift of the gabbroid body and its emplacement at shallow crustal levels (Fig. 7). The Brothers Islets constitute the emerged summit of a NW-SE-elongated extensional crustal block. Another uplifted crustal block was detected further to the west, nearly parallel to the Brothers (Fig. 6). Thus, the Brothers are part of a series of sub-parallel topographic highs associated with right lateral transtension, including to the southwest a prominent NW-striking SW-dipping extensional fault block (Fig. 6). A few kilometres east of the Brothers, a WSW-ENE elongated transtensional depression (Oceanographer Deep) lies at the intersection of a left-stepping transcurrent fault parallel to the Dead Sea transform and a NE right-stepping transcurrent fault (Brothers Fracture Zone of Crane and Bonatti 1987) normal to the present rift axis (Fig. 6).

The strong reflector we mapped beneath the Miocene evaporites can be interpreted as the top of the crystalline basement (Fig. 7). Bouguer gravity data suggest that it marks the top of the same gabbro unit sampled in the QUSEIR drill hole and on the Brothers (Figs. 4 and 5). This implies that the Brothers and QUSEIR gabbros intruded into sub-rift crust and then were exhumed by footwall-uplift during extension and block rotation, in a mechanism similar to that proposed for the basement cores of fault blocks exposed in the Gulf of Suez (Bosworth et al. 2005).

5 Brothers and QUSEIR Gabbroic Rocks

The rocks outcropping on the Brothers (BRG) include layered gabbros, ranging from leucotroctolites to olivine gabbros, coarse-grained gabbros and anorthositic gabbros intruded by amphibole-bearing olivine microgabbronorite dykes (Fig. 8). They generally preserve primary igneous textures and mineral assemblages; late-magmatic alteration processes affect mostly the coarse-grained gabbros and with replacement of pyroxenes by brown amphiboles followed by lower temperature alteration crystallizing green amphibole; olivine is sporadically altered to talc or colourless amphibole. Moreover, discontinuous veins of green-colourless amphibole plus albite cut all lithologies.

Medium-grained leucotroctolite/troctolitic leucogabbros to medium/coarse-grained olivine gabbros consist of 1–2 cm thick layers, displaying modal and grain-size micro-rhythmic layering, locally accompanied by a weak lamination, due to the alignment of plagioclase laths and elongated olivine aggregates. The samples show adcumulate to mesocumulate textures. The main cumulus phases

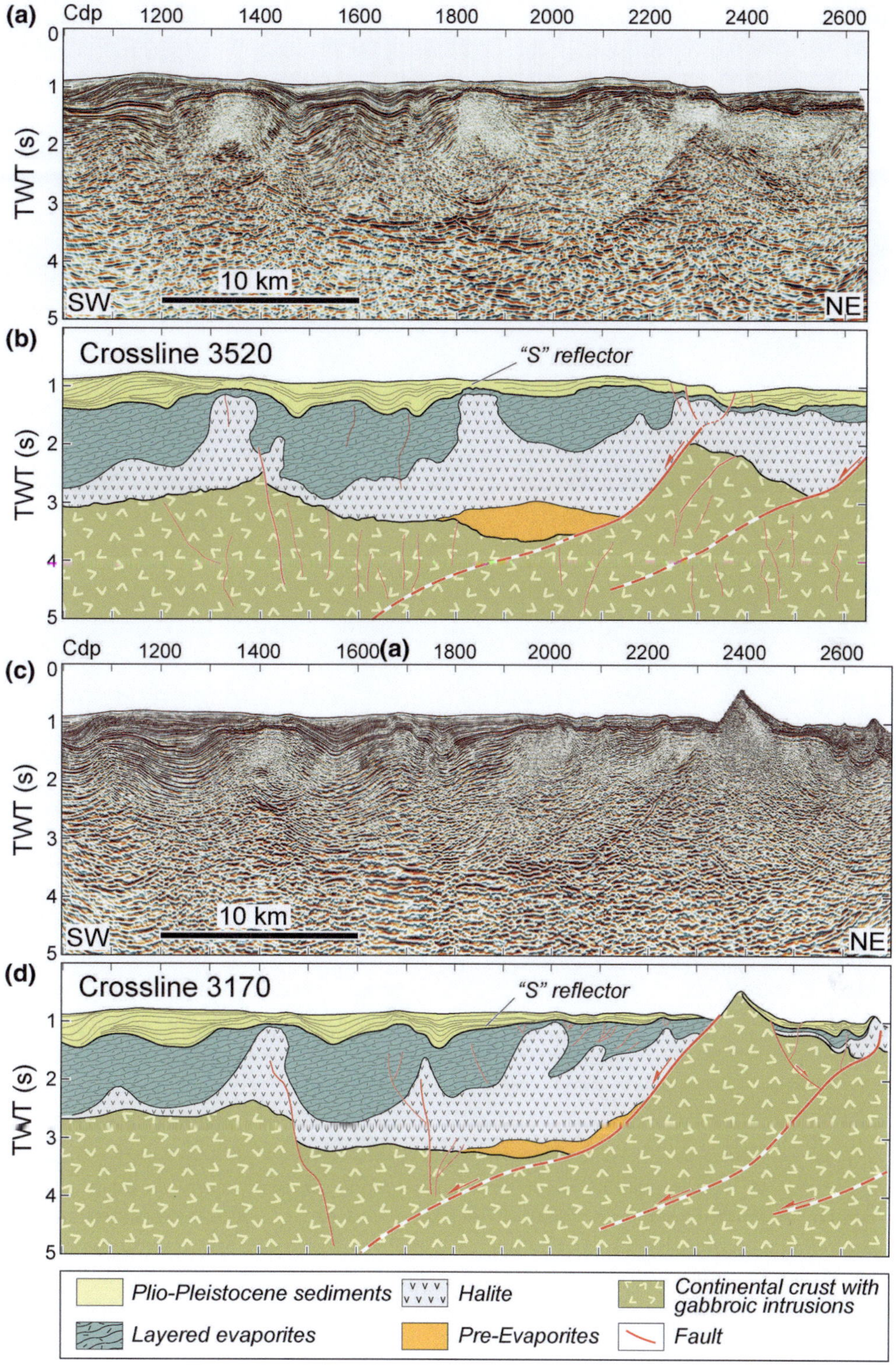

Fig. 7 Time migrated seismic cross-lines 3520 and 3170 extracted from the 3D seismic survey and running perpendicular to the axis of the northern Red Sea and to the crustal block lying to the southwest of Brothers Islets. Full profile locations are indicated in Fig. 5d. **a** and **c** Processed data only. **b** and **d** Interpretation of the processed data. The Brothers Islets reached the present level before 125 ka and since then they have been tectonically quiescent (Hoang and Taviani 1991). However, a Plio-Quaternary uplift of the south-western crustal block is shown in seismic crossline 3170. In fact, the Miocene evaporites are absent on the NE flank of the tilted block and the Plio-Quaternary sequence conformably follows the tilting of the basement

are olivine and plagioclase, whereas clinopyroxene (cpx) is both a cumulus and intercumulus phase in the olivine gabbros.

Leucotroctolite/troctolitic leucogabbro layers: These are granular, medium-grained (grain size ≤ 2–3 mm), and consist of cumulus plagioclase (modal content 70–75 vol.%; $An_{61.4-57.4}Ab_{38.2-42.1}Or_{0.4-0.6}$), cumulus to intercumulus olivine (25 vol.%; $Fo_{72.9-71.9}$) and intercumulus cpx ($En_{51.6-44.1}Fs_{11.1-16.8}Wo_{31.6-45.3}$; $Mg^{\#}$ = 86.3–79) and orthopyroxene (opx)—(0–5 vol.%) with minor brown-green pargasite ($Mg^{\#}$ = 74) and opaques. Thin rims of orthopyroxene (opx) surround olivine and, more rarely, cpx crystals.

Fig. 8 Thin-section micrographs (crossed polar view) of selected samples of the Brothers and QUSEIR rocks. **a** Medium to coarse grained olivine gabbro and **b** coarse grained leucograbbro from Brothers. **c** QUSEIR medium to coarse-grained olivine leucogabbro. **d** Amphibole-bearing olivine micro-gabbronorite from Brothers

Olivine gabbro layers: These are medium (grain-size 0.5–2 mm) to coarse-grained (2–3 up to 10 mm) with granular to heterogranular to poikilitic textures, and are composed of olivine (15–20 vol.%; $Fo_{74.6-70.6}$), plagioclase (45–55 vol.%; $An_{71.3-60.9}Ab_{28.6-41}Or_{0.1-0.6}$) cpx (30–35 vol.%; $En_{51.61-42}Fs_{6.1-17.3.}Wo_{31.6-46.9}$; $Mg^{\#}$ = 91.8–77) opx (3–5 vol.%; $En_{75.4-72.2}Fs_{22.6.-27.1}Wo_{0.7-1.8}$; $Mg^{\#}$ = 79–75.1), and minor brown pargasitic amphibole ($Mg^{\#}$ = 75–70) and iron oxides (Fig. 8a). The medium-grained layers have commonly a higher modal content of plagioclase but their modal olivine/cpx ratios may vary. In the coarse-grained layers, larger grains of cpx occur; locally, they are poikilitic, enclosing olivine and plagioclase. Opx rims the cumulus phases (olivine, cpx and plagioclase) in the olivine gabbros, more commonly than in the troctolitic layers. Fe–Ti oxides commonly occur as inclusions in cpx.

Coarse-grained gabbroic rocks with grain size between 0.5 and 20 mm are mostly isotropic, generally inequigranular, sometimes poikilitic with adcumulate to mesocumulate textures and include dominantly olivine-poor (leuco-) gabbros and minor olivine (leuco-) gabbros (Fig. 8b). The modal olivine content is up to 10–14%, but generally does not exceed 5%. The modal proportion of cumulus phases is slightly variable but plagioclase and cpx are generally dominant. Olivine ($Fo_{75.9-68.7}$) crystallizes as large strained grains or more often as small anhedral crystals interstitial to plagioclase and cpx or in clusters with cpx; it is generally rimmed by opx or symplectite coronas of opx + opaques.

Plagioclase (50–77 vol.%; $An_{82.6-53.1}Ab_{17.3-46.1}Or_{0.1-0.9}$) occurs as large euhedral-subeuhedral laths or as smaller polygonal crystals. Cpx (10–45 vol.%; $En_{48.1-41.9}Fs_{6.1-17.3}Wo_{31.6-46.9}$; $Mg^{\#}$ = 88.9–76.6) is present as large to small crystals, euhedral to subhedral, more rarely as oikocrysts. Opx (tr-6 vol.% $En_{80.2-70.6}Fs_{19.6-27.5}Wo_{0.1-2}$; $Mg^{\#}$ = 85.6–73.1), titanian pargasitic to edenitic amphibole (tr-1 vol.%; $Mg^{\#}$ = 72–77) and Fe–Ti oxides (tr-3 vol.%) are interstitial phases. Opx is present as an intercumulus phase and as strings between plagioclase and olivine or cpx, at times forming symplectites with Fe–Ti oxides around corroded olivine. Brown amphibole partially replaces cpx grains, forming patchy blebs and lamellae in cpx, and coronas around Fe–Ti oxide and olivine.

Anorthositic gabbros have heterogranular texture with centimetre-size (>2 cm) euhedral-subhedral plagioclase grains (80–85 vol.%) and intercumulus/intergranular patches consisting mainly of serpentine, chlorite, actinolite and minor brown-green Ca-amphibole wholly replacing magmatic mafic minerals. Iron oxides occur in minor amounts (1–2 vol.%). The large primary labradoritic plagioclase laths showing deformation twinning are surrounded by andesine-oligoclase grains of smaller size (0.5–1 mm) with common polygonal shapes and 120° triple junctions of grain boundaries, suggesting high-temperature recrystallisation or crystal-plastic deformation.

Olivine microgabbronorites are aphyric or subaphyric fine-grained rocks with intergranular to subophytic textures, although some very fine-grained (grain size < 0.1 mm) allothiomorphic varieties also occur (Fig. 8d). They consist of rare olivine ($Fo_{79.4-70.9}$) and plagioclase phenocrysts (>7–8 mm; $An_{71.7-59.1}Ab_{28.1.3-48.7}Or_{0.2-0.8}$) set in a groundmass (grain-size 0.2–1 mm) of rounded equigranular, polygonal assemblages of olivine, plagioclase ($An_{73.28-56.4}Ab_{26.5-43.4}Or_{0-0.5}$), cpx ($En_{45.4-41.9}Fs_{9.6.-11.3}Wo_{43.5-47.4}$; $Mg^{\#}$ = 93.8−80.3), subordinate opx ($En_{75.9-72.8}Fs_{22.5-25.2}Wo_{1.4-2.1}$; $Mg^{\#}$ = 81.5−75.5), and brown amphibole (titanian pargasite, $Mg^{\#}$ = 75−72) associated with opaques (ilmenite and magnetite). Modal abundances of opx and brown amphibole are variable.

In most of the BRG, brown pargasitic (often Ti-rich) amphibole occurs interstitially or as overgrowth rims of the mafic minerals such as cpx and olivine. Moreover, brown amphibole also crystallizes as patchy blebs within the cpx, as a replacement phase. Along with opx, very low modal brown amphibole derives probably from late stage crystallisation of interstitial melts (Ross and Elthon 1997; Tribuzio et al. 1999; Charlier et al. 2005; Bernstein 2006; Borghini and Rampone 2007). The lack of zoning in the cumulus phases supports exchanges between interstitial liquids and fresh MORB-type melt in open system percolation.

The QUSEIR borehole sample (QG) is a medium to coarse-grained olivine leucogabbro (Fig. 8c). It shows mesocumulate texture with dominant plagioclase (70% vol) and minor olivine (5–10% vol) as cumulus phases, ophitically enclosed by cpx. Some cpx oikocrysts enclose small plagioclase laths. Small euhedral spinel crystals occur as an accessory phase. No foliations and igneous laminations, and no magmatic layering were observed. Primary phases are variably affected by alteration under a range of temperatures from amphibolite facies (near-solidus) to low temperature conditions. Olivine is altered to serpentine, tremolite-actinolite, carbonates and iron oxides; cpx is partially replaced by brown to green amphiboles with titanian edenite to edenite composition ($Mg^{\#}$ = 77.3−74.8). Plagioclase is mostly fresh, except for minor alteration to sericite, and occurs as euhedral to subhedral laths up to 10–12 mm in length, sometime containing euhedral olivine. The largest grains show optical zonation, with composition ranging from $An_{84-78}Ab_{16-22}Or_{0-0.2}$ (cores) to $An_{68-40}Ab_{31-58}Or_{0.2-0.5}$ (rims). The small plagioclase crystals have bytownite ($An_{81-70}Ab_{19-30}Or_{0.1-0.2}$) composition. Sodic plagioclase ($An_{13}Ab_{82}Or_5$) forms intergranular patches and narrow plaques replacing primary plagioclase at grain margins and along internal cracks ($An_{5-7}Ab_{94-92}Or_{0-1}$). A trend of increasing FeO with decreasing An% suggests Fe-enrichment in the melt during plagioclase crystallization. Cpxs have ferroan diopside to magnesium-rich augite compositions ($Wo_{43.1-47.1}En_{43.6-48.6}Fs_{5.9-13.3}$), with $Mg^{\#}$ ranging from 77.1 to 89.4.

Compared with the Brothers cumulate gabbros, QUSEIR olivine gabbros are more primitive, with Mg-richer cpx and more anorthitic plagioclases. Overall, the mineral composition variations are similar to those observed in oceanic cumulate gabbros.

6 Brothers and QUSEIR Rocks Geochemistry

Representative major and trace-element whole-rock compositions for the Brothers Gabbros (BRG) and QUSEIR Gabbros (QG) are shown in Figs. 9 and 10 and given in Table 1 of Ligi et al. (2018). Gabbros are silica-saturated (Ol-Hy normative) basic rocks (SiO_2 = 47.0−50.5) and have $Mg^{\#}$ ($100 * Mg/Mg + 0.9Fe^{2+}_{tot}$ molar ratio) in the range 70.4–78.6. These values are higher than those of primitive basaltic liquids (Kempton and Harmon 1992) consistent with a process of olivine + pyroxene accumulation. In addition, the rather low and variable SiO_2/Al_2O_3 values at constant Mg# (Fig. 11) indicate that the whole-rock composition is also controlled by plagioclase accumulation. On the AFM (Na_2O + K2O −FeOtot-MgO) diagram (Fig. 12), the BRG and QG fall between the origin of the Skaergaard trend and the MgO apex, in the field of gabbroic cumulates from mid-oceanic ridges. The total FeO content is low (4–7 wt%) reflecting the minimum modal abundance of Fe–Ti-oxides in the cumulates. The

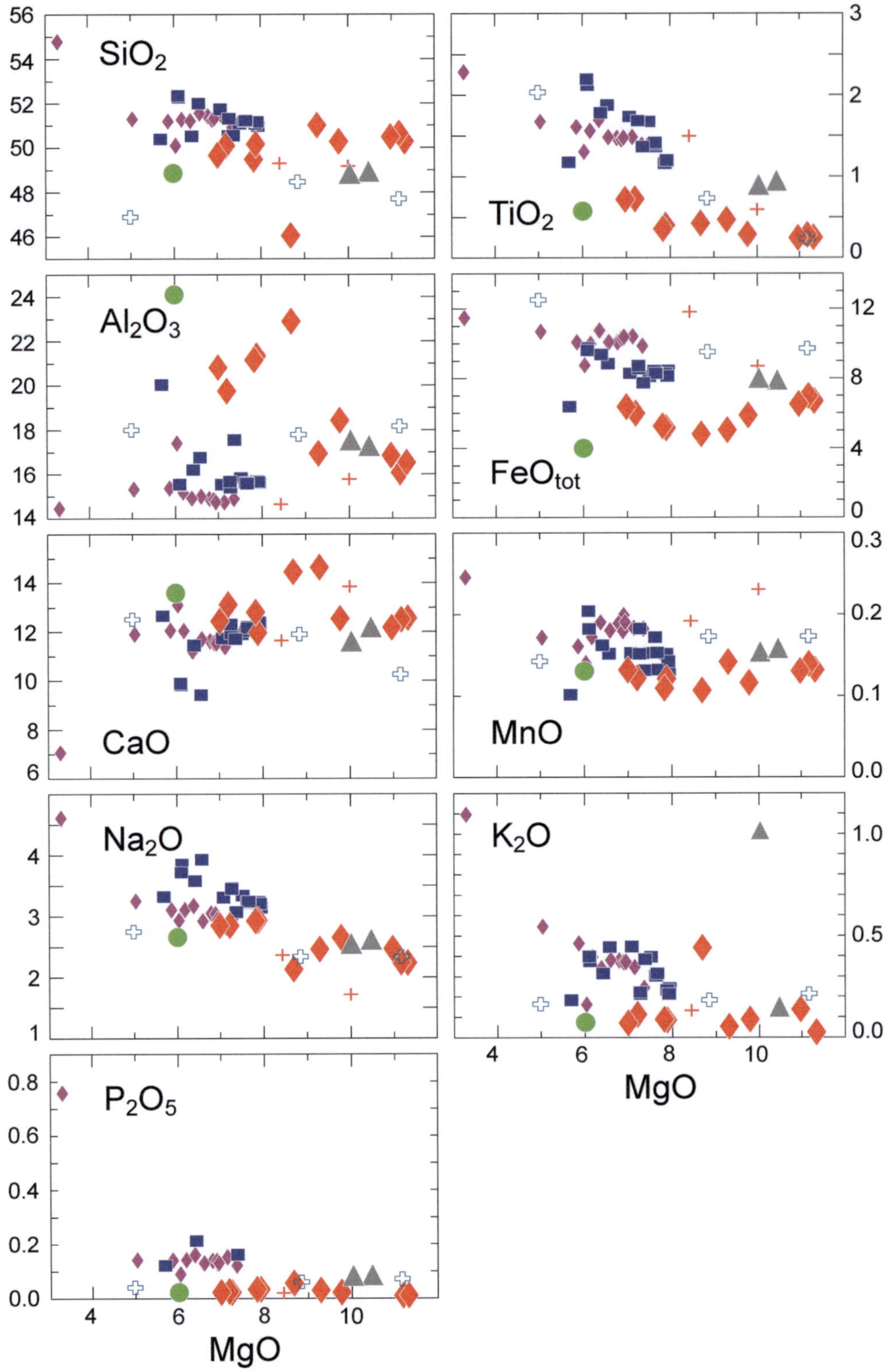

Fig. 9 Major element compositions versus MgO of Brothers and QUSEIR gabbros. Brothers gabbros (red diamond) and dolerites (gray triangle), QUSEIR gabbros (green circle) and gabbroic rocks from Jabal Tirf (McGuire and Coleman 1986) (open blue cross). Whole rock compositions of basalts from Shaban (filled blue square), Mabahiss (filled purple diamond) and Nereus (red plus) deeps are included for comparison (Haase et al. 2000; Altherr et al. 1988)

concentrations of TiO_2 (0.24–0.72 wt%), K_2O (0–0.13 wt%) and P_2O_5 (0–0.03 wt%) and some incompatible trace elements such as Zr (<29 ppm), and Y (<12 ppm) are low, as expected for troctolitic and olivine gabbro cumulates (Fig. 9).

The compatible trace elements such as Cr and Ni are relatively low (122–260, 92–132 ppm, respectively) in the BRG suggesting a depletion of these elements in the parental magma, whereas Cr content in QG is higher (503 ppm) due

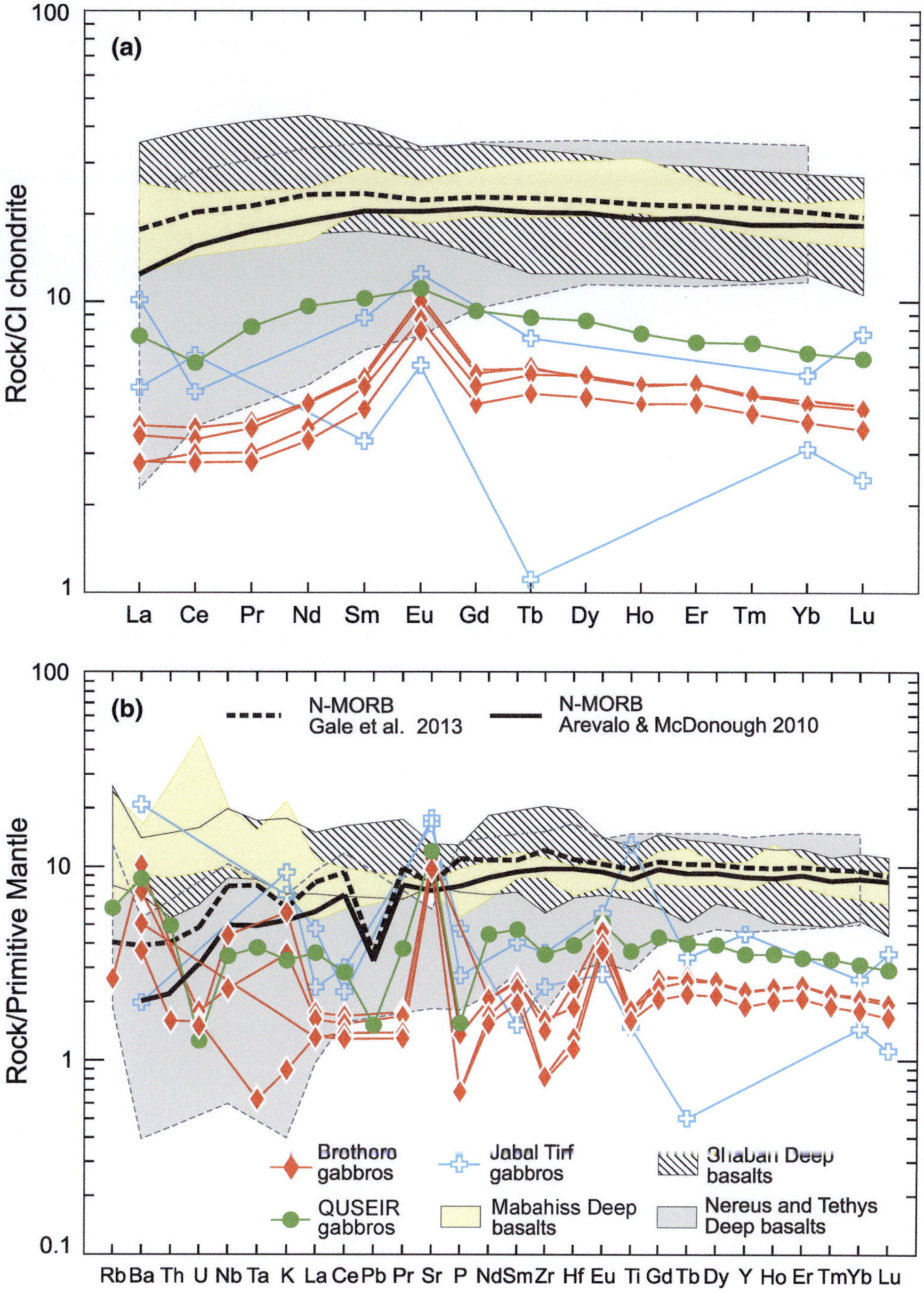

Fig. 10 Trace element composition for Brothers and QUSEIR gabbros. Whole rock compositions of Jabal Tirf layered gabbros (Coleman et al. 1977, 1979; McGuire and Coleman 1986), and of Thetis, Nereus, Shaban and Mabahiss basaltic glasses (Altherr et al. 1988; Haase et al. 2000; Ligi et al. 2012, 2015) are also shown for comparison. Thick black dashed and solid lines indicate average N-MORB patterns from Gale et al. (2013) and Arevalo and McDonough (2010), respectively. **a** REE patterns. Normalization to chondritic values is from McDonough and Sun (1995). **b** Primitive mantle-normalized trace element patterns. Primitive mantle composition is from Lyubetskaya and Korenaga (2007)

to the occurrence of Cr-spinel in the rock. Compared with samples BRO1457b and BRO1458, samples BRO1457a and BRO1457 show higher Cr content, in agreement with the higher modal abundance of cpx of sample BRO1457a (also as a cumulus phase) and the slightly Cr-richer cpx composition.

The primitive mantle-normalized trace element patterns (Fig. 10b) of BRG and QG are LREE and HFSE depleted compared to N-MORB compositions, with prominent positive spikes of Ba, Sr and Eu and negative P, Zr and Ti anomalies. The QG has trace element abundances higher than those observed in BRG, with less pronounced anomalies.

The chondrite-normalized REE patterns for BRG are flat $[(La/Yb)_n = 0.6–0.8]$ for QG with very low REE enrichment (La ≈ 2–3 × chondrite), depletion of LREE $[(La/Sm)_n = 0.55–0.67]$ yielding a concave-upward LREE pattern, low HREE fractionation $[(Tb/Yb)_n = 1.2–1.3)$ and positive Eu anomalies (Eu/Eu* = 1.7–1.8). QG displays higher REE abundances (La = ≈ 8 × chondrite), a concave-downward LREE pattern with a minor depletion of LREE $[(La/Yb)_n = 1.1]$ and a weak

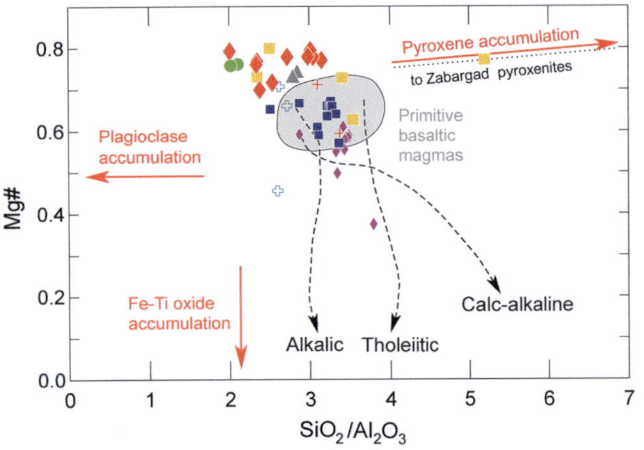

Fig. 11 Mg# (Mg/(Mg + 0.85Fe_{tot}) versus SiO_2/Al_2O_3 plot. Symbols as in Fig. 8. Whole rock compositions of Zabargad pyroxenites and gabbros (Seyler and Bonatti 1988) (filled orange square) are also shown for comparison. Gray filled area indicate compositions of primitive basaltic magmas (Kempton and Harmon 1992). Dashed black arrows denote average differentiation trends for alkalic, tholeiitic and calc-alkaline melts. Red arrows show generalized direction of compositional change with accumulation of indicated mineral phase (Kempton et al. 1997)

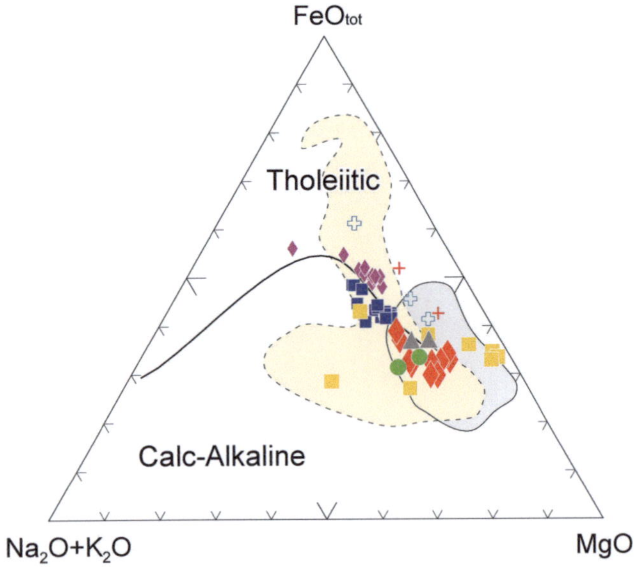

Fig. 12 A ($Na_2O + K_2O$)–M (MgO)–F (FeO_{tot}) plot. Brothers gabbros (red diamond) and dolerites (gray triangle), QUSEIR gabbros (green circle), Zabargad pyroxenites and gabbros (yellow square; Petrini et al. 1988; Seyler and Bonatti 1988) and gabbroic rocks from Jabal Tirf (open blue cross; Coleman et al. 1977, 1979; McGuire and Coleman 1986). Whole rock compositions of basalts from Shaban (filled blue square), Mabahiss (filled purple diamond) and Nereus (red plus) deeps are included for comparison (Altherr et al. 1988; Haase et al. 2000). Yellow and gray filled areas indicate compositions of gabbroic rocks from ODP hole 1109A (Casey et al. 2007) and of gabbroic cumulates from ophiolites (Beccaluva et al. 1977; Phelps and Avé Lallemant 1980; Jaques and Chappell 1980) and from Mid-Atlantic (Tiezzi and Scott 1980) and South West Indian ridges (Meyer et al. 1989), respectively

positive anomaly Eu (Eu/Eu* = 1.1). Such REE patterns and low abundances do not resemble liquid composition; rather they are largely controlled by the partitioning of REE between the dominant cumulus phases (olivine + pl ± cpx) and basaltic liquids (Fig. 10a).

The cpxs from the BRG and QG show chondrite-normalized REE patterns with LREE depletion, almost flat MREE and HREE in the range 6÷15 × chondrite and a variable positive Eu anomaly (Fig. 3 of Ligi et al. 2018). Overall the REE patterns are similar to the calculated cpx composition in equilibrium with N-MORB.

7 Isotope Geochemistry

The Sr, Nd and Pb isotopic compositions of bulk rock samples and cpx separates from BRG and QG are presented in Table 2 of Ligi et al. (2018). Our data together with data from the Red Sea region are plotted in Figs. 13 and 14. Nd isotope ratios of the BRG and QG and their mineral separates plot within the field of Red Sea "true" oceanic crust and global MORB (Fig. 13).

The BRG show a significant $^{87}Sr/^{86}Sr$ and $^{143}Nd/^{144}Nd$ co-variation with ranges from 0.70298 to 0.70354 and from 0.51297 to 0.51308 respectively. The four whole rock samples (BRO14–5 series) have basically identical Nd isotopic compositions (0.513041 on average). They plot within the field occupied by the Red Sea "true" oceanic crust and next to the Shaban lavas. The cpxs plot into two distinct groups; one group has about the same Nd isotopic composition of whole rock samples but higher $^{87}Sr/^{86}Sr$, the second

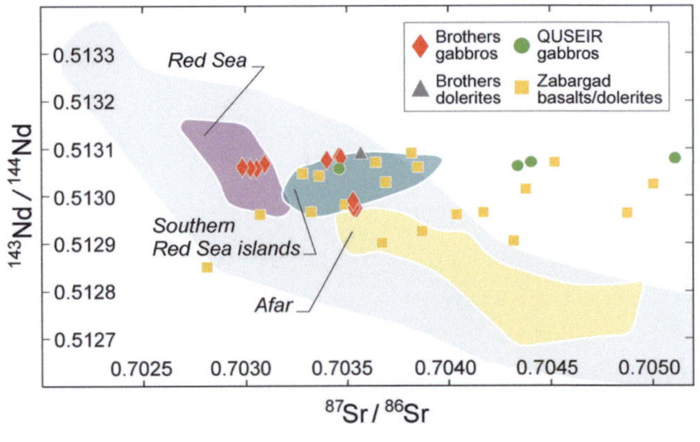

Fig. 13 Sr-Nd isotopic composition of Brothers and QUSEIR gabbros superimposed on global MORB data (gray field), on basalt data from the Red Sea axis (purple field), from the Red Sea southern islands (blue field) and from the Afar (yellow field). Brothers gabbros (red diamond) and dolerites (gray triangle), QUSEIR gabbros (green circle), Zabargad pyroxenites and gabbros (yellow square; Seyler and Bonatti 1988; Petrini et al. 1988) and gabbroic rocks from Jabal Tirf (open blue cross; Coleman et al. 1977, 1979; McGuire and Coleman 1986)

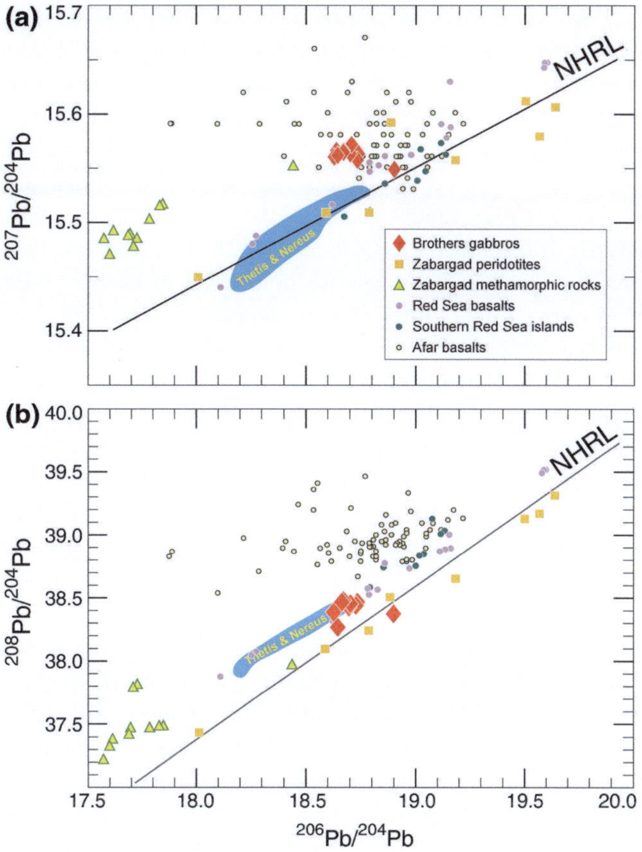

Fig. 14 Pb isotopic composition for the Brothers gabbros. Fields of Thetis and Nereus basaltic glasses (blue filled area, A. Cipriani, pers. comm.) and data from Red Sea MORB (Haase et al. 2000; Volker and McCulloch 1993; Altherr et al. 1988; Schilling et al. 1992), southern Red Sea islands, Zabargad peridotites and metamorphic rocks (Brueckner et al. 1995), and Afar basalts (Deniel et al. 1994), are plotted for comparison

group has both lower $^{143}Nd/^{144}Nd$ and higher $^{87}Sr/^{86}Sr$ than whole rock samples. The first group plots within the field defined by the southern Red Sea islands, the latter plots within the Afar plume field as defined by Deniel et al. (1994). A diabase from Brothers measured by Brueckner et al. (1995) plots close to the cpx group with the most radiogenic Nd isotopic composition but has even higher $^{87}Sr/^{86}Sr$.

If weathering introduced the high $^{87}Sr/^{86}Sr$ component in some of our BRG samples we would expect the whole rock to have higher Sr isotopic composition than the cpx separates, as it is more difficult to remove a radiogenic Sr component from whole rock than from pure mineral separates. However, a high $^{87}Sr/^{86}Sr$ signature is common in the Red sea region and could therefore point to an enriched source signal. Unfortunately, whole rock and mineral separates belong to different samples and we cannot rule out secondary alteration for some of our samples. QG samples show high Sr isotope ratios at relatively constant Nd isotope ratios probably resulting from hydrothermal alteration (Fig. 13). Cpx, plg and bulk data of QG define a pseudo-isochron of about 20 Ma. Given the cpx alteration, this age could indicate a metamorphic age of the gabbros due to reaction with seawater.

The Pb isotopic composition of BRG is quite uniform (on average $^{206}Pb/^{204}Pb = 18.6795$; $^{207}Pb/^{204}Pb = 15.564$; $^{208}Pb/^{204}Pb = 38.4231$). Only one cpx sample is shifted toward slightly higher $^{206}Pb/^{204}Pb$ and lower $^{208}Pb/^{204}Pb$ isotopic ratios. They have slightly more "enriched" isotopic compositions than average MORB, possibly reflecting the influence of the Afar plume (Deniel et al. 1994) (Fig. 14). In a plot of $^{207}Pb/^{204}Pb$ versus $^{206}Pb/^{204}Pb$, most Brothers samples fall next to the Afar plume field, slightly above the NHRL, but close to the values for the Red Sea region (Volker and McCulloch 1993) (Fig. 14a). In a $^{208}Pb/^{204}Pb$ versus $^{206}Pb/^{204}Pb$ plot the samples lie within a trend explained by binary mixing between a depleted and an enriched mantle component (Fig. 14b). The presence of an ancient incompatible element enriched component in the source of the Brothers gabbros supports the idea of a long-term influence of the Afar plume in the Red Sea region (Volker and McCulloch 1993).

8 Geothermometry and Geobarometry

We applied a number of geothermometers (Loucks 1996; Frost and Lindsley 1992; Lindsley and Frost 1992) to estimate the temperature of crystallization of BRG and QG. Olivine-augite geothermometer (Loucks 1996) yielded temperature estimates for the layered BRG ranging from 1080–1000°C in the olivine gabbros to 1130–1020°C in the troctolitic layers. The calculated crystallization temperatures of the olivine gabbros BRO1457b and BRO1457a range from 1000° to 1100°C (average 1061 ± 30.67°C) and from 1054° to 1122°C (average 1095 ± 22.6°C), respectively. Minimum equilibrium temperatures obtained from two-pyroxene thermometry (Frost and Lindsley 1992; Lindsley and Frost 1992; Andersen et al. 1993) for coexisting intercumulus opx and cpx are 910–940°C for intercumulus cpx (T_{cpx}), and 840–890°C for opx (T_{opx}), in agreement with studies (Mitra et al. 1999; Drüppel et al. 2001) suggesting variable blocking temperature for different mineral phases during slow cooling (i.e., $T_{cpx} > T_{opx}$). Minimum crystallization temperatures for intercumulus cpx also agree with those derived from the two-pyroxene geothermometer of Brey and Koehler (1990) (T = 926–933°C). The lack of fresh olivine and coexisting cpx-opx in the mineral assemblage hampered the application of the olivine-augite and two-pyroxene thermometers to QG.

Projection onto Lindsley's (1983) isotherms at P = 5 kbar of all pyroxene (cumulus to intercumulus) compositions indicates for BRG minimum crystallization temperatures from about 1150 down to 750 °C for cpx and less than 900 °C for opx. The graphic thermometer indicates for QG equilibration temperatures from 1140 to 700 °C, although most of the cpx data fall in between 1040 and 1000 °C. Furthermore, pyroxenes from Zabargad gabbros yielded a minimum temperature of equilibration in the range 1200–950°C for primary cpx magmatic relics and 900–700 °C for the granulitic cpx.

Temperatures of crystallization of the BRG and QG amphiboles were estimated with the Ti-in-amphibole geothermometer of Helz (1973) and Ernst and Liu (1998), calibrated by Otten (1984) and Féménias et al. (2006). In the BRG, post-cumulus titanian pargasites yield temperatures in the range 940–840 °C, while vein-filling green amphibole temperatures range between 557 and 665 °C. However, brown edenite locally crystallising along the veins yielded higher T (~798 °C) suggesting that brittle deformation started at a relatively high temperature (about 800 °C) just after the latest stage of crystallization of the magma. In the QG, the brown amphiboles replacing cpx yield temperatures in the range 865–750 °C; lower temperatures (<700 °C) were obtained on green amphiboles.

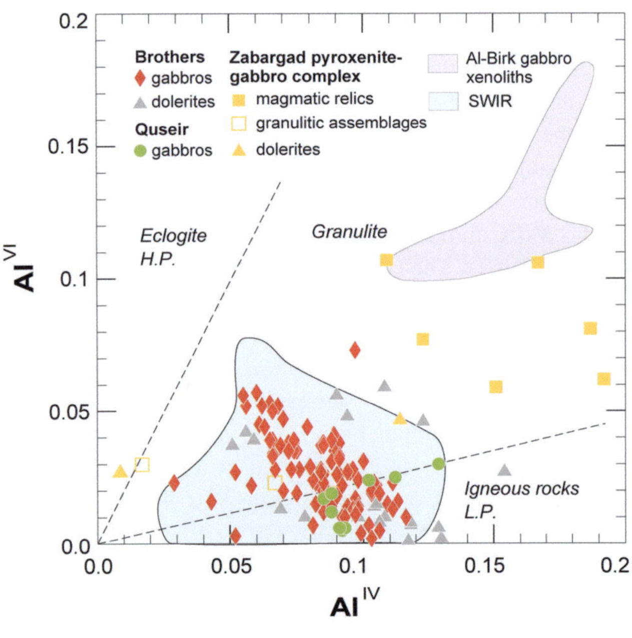

Fig. 15 Al^{VI} versus Al^{IV} for cpx (structural formulas were calculated on the basis of six oxygens and four cations) from Brothers gabbros and dolerites (red diamond and gray triangle, respectively) and QUSEIR gabbros (green circle). The light blue and the pink areas refer to cpx from SW Indian Ridge gabbros (Hebert et al. 1991; Stakes et al. 1991) and from Al Birk gabbro xenolites (Ghent et al. 1980). Cpx compositions from Zabargad gabbro-pyroxenite complex, i.e., magmatic relicts (filled orange square), granulitic assemblage (hollow orange squares) and dolerites (orange triangle) are displayed for comparison. Dashed lines indicate pressure domains (Aoki and Kushiro 1968)

Pressure conditions during crystallization can be estimated with the cpx geobarometer of Nimis and Ulmer (1998) and Nimis (1999) assuming tholeiitic bulk rock composition (TH formulation). Cpx crystallization temperatures used for geobarometric calculations were derived from various geothermometers (Lindsley 1983; Frost and Lindsley 1992; Lindsley and Frost 1992; Loucks 1996) (BRG: 1000–1130 °C; QG: 950–1120 °C). The resulting pressures are relatively low for both BRG (0.14–0.44 GPa, with most of data falling in the interval 0.2–0.36 GPa), and for QG (0.23–0.37 GPa average 0.30 ± 0.04 GPa for n = 11). The same geobarometer (MA-formulation) applied to the Zabargad gabbro pyroxenes yields pressures as high as 0.8–0.9 GPa, (corresponding to roughly 30–35 km depth). The results should be viewed with caution due to uncertainties in the temperature estimates (poorly constrained crystallization temperature) and to the possible re-equilibration of the primary cpx composition under granulite-facies conditions (Seyler and Bonatti 1988). However, the pressure estimates, considering the widespread occurrence of pyroxenite cumulate in the Zabargad complex, are consistent with experimental studies (Villiger et al. 2004, 2007) showing that the crystallization of tholeiitic liquids at 0.7–1 GPa, lacking plagioclase as early liquid phase, led to the production of large volume of wehrlite and pyroxenite cumulates. The $^{VI}Al/^{IV}Al$ ratios of cpx (Fig. 15) may also provide indications on crystallization pressure (Aoki and Shiba 1973). They confirm a shallow level of intrusion for Brothers and Quseir gabbros, comparable to that of the SW Indian Ridge gabbros (Hebert et al. 1991), but contrasting with the values for Zabargad gabbros that imply significantly higher pressure.

9 Geochronology

A fragment of QUSEIR gabbro sample was crushed and sieved for $^{40}Ar/^{39}Ar$ age determination (Ligi et al. 2018). Argon data were corrected for backgrounds and for nuclear interferences (Dalrymple et al. 1981). The amount of Ar released was low, and the ages derived from all the heating steps fall within error of each other and yield a plateau age of 25 ± 6 Ma (n = 10). These steps fall on an isochron with a younger age of 16 ± 7 Ma and an initial $^{40}Ar/^{36}Ar$ value of 305 ± 4 that is similar to that of air. The two ages overlap within error (Ligi et al. 2018).

10 Discussion

Petrographic features (cumulate textures as well as modal and grain-size rhythmic layering) and bulk-rock major and trace element compositions of the Brothers and QUSEIR gabbroic rocks indicate that they formed as ol + plagioclase ± cpx cumulates. Their $Mg^{\#}$ value, which is higher than those of primitive basaltic liquids, reflects mafic phases accumulation. Their positive Sr and Ba spikes in the primitive mantle-normalized trace elements patterns, and positive Eu anomalies in chondrite normalized REE patterns, are consistent with the presence of cumulus plagioclase. The depletion of Ti, Zr, Hf and P correlates with the low modal abundances of Fe–Ti-oxides and the lack of other accessory phases such as zircon and apatite. The low contents of Zr and Hf as well as of Ta, Nb, and U, along with the low concentration of K_2O and P_2O_5, considering their highly incompatible behaviour during crystallization of olivine, plagioclase and cpx from basaltic melts (GERM database, www.earthref.org), are consistent with the cumulitic nature of the Brothers gabbroic rocks (i.e., accumulation of plagioclase and olivine, and/or cpx) and support a low proportion of interstitial melts, in agreement with the adcumulate and mesocumulate textures observed in the BRG. The preferred orientation of plagioclase and olivine occurring locally in the layered gabbros, together with deformation features such as pointed, sometime curved, twinning and shadowy extinction in plagioclase may indicate compaction as a possible mechanism of expulsion of the interstitial liquids (McBirney and Nicolas 1997). Evidence of brittle deformation is given by the alteration veins cutting the gabbroic complex.

The Brothers and QUSEIR gabbros REE patterns and concentrations are similar to those of oceanic gabbros (Fig. 10), as well as of MORB-derived gabbroic cumulates emplaced at the ocean-continent transition in the western Iberia margin (Seifert et al. 1996, 1997; Cornen et al. 1999) and similar intrusions from ophiolitic complexes (Montanini et al. 2008). Trace element and REE patterns are also comparable to those of the Jabal Tirf layered gabbro, a Miocene intrusion exposed along the Red Sea coastal plain in southwest Saudi Arabia (Tihama Asir complex; McGuire and Coleman 1986; Coleman and McGuire 1988). In comparison with the tholeiitic lavas from the Red Sea Shaban, Mabahiss, Nereus and Tethys Deeps, the Brothers gabbros have a significantly lower REE total abundance; nevertheless, their pattern for HREE is similar. The low content of REE further supports the very limited amount of entrapped residual melt in the cumulate rocks.

Reconstructing the nature of the parental melt of the Brothers and QUSEIR gabbros is hindered by the absence in our samples of chilled margins, non-cumulate rocks or comagmatic dykes. However, phase compositions and the data give some qualitative constraints. Mineral chemistry indicates a non-orogenic affinity (mid-ocean ridge affinity). In a Fo (mol%) of olivine versus An (mol%) of plagioclase diagram (Beard 1986), coexisting olivine and plagioclase plot within the fields of cumulates from mid-ocean ridges, oceanic islands and tholeiitic layered intrusions.

The percentage of tetrahedral Al (%AlTsite) versus TiO_2 in cpx may provide a further indication of the nature of the crystallization magma (Le Bas 1962; Loucks 1990). Cpx from the Brothers and QUSEIR gabbros show Al and Ti contents comparable to those of cpx from Mid Atlantic Ridge plutonic cumulates, following, together with cpx from Zabargad mafic complex, the low Al/Ti ratio trend defined by rift-related cumulates. A subalkaline composition of the parent magma is also suggested by the composition of magmatic amphiboles.

The olivine composition (Fo content) may be used to infer the MgO/FeO ratio of the parental magma. However, estimates obtained from cumulates must be regarded with caution as in a cumulitic environment, olivine may be modified by intercumulus liquids (Barnes 1986; McCarter et al. 2006) forming orthopyroxene from the peritectic reaction of Ol + melt. However, olivine found in the Brothers gabbros is commonly unzoned with no clear signs of resorption, and is often free of opx rims. Moreover, opx forms as an intercumulus phase not only around olivine but also cpx and plagioclase. These features suggest that olivine maintained its primary composition, being unaffected by magmatic or secondary modifications.

Fe–Mg exchange between olivine and melt indicates that the Brothers gabbros derived from evolved melts. In fact, according to experimentally established basalt phase equilibria (Roeder and Emslie 1970; Bender et al. 1978; Walker et al. 1979; Langmuir and Hanson 1981; Nielsen and Dungan 1983; Weaver and Langmuir 1990), the most fosteritic olivine of our gabbros must have crystallized from a basaltic melt with $Mg^{\#}$ in the range 43.2–48.1 (for $Kd = 0.27–0.33$) at a liquidus temperature of 1137 °C. Although the Fe/Mg exchange partition coefficient between cpx and basaltic liquid is less well defined, being sensitive also to changes in redox conditions (Botcharnikov et al. 2008), the calculated $Mg^{\#}$ values of the liquids in equilibrium with cpx, using cpx/melt partition coefficients of 0.23 for FeO/MgO (Toplis and Carroll 1995), also support for BRG a crystallization from variably evolved basic melts ($Mg^{\#}_{melt}$ 44–63), whereas they indicate for QG a derivation from less differentiated basaltic liquids ($Mg^{\#}_{melt}$ 49–77),

The olivine liquidus temperature derived from the experimental basalt phase equilibria is comparable to the higher temperatures estimated from thermobarometry. These temperatures are compatible with the cotectic crystallization of ol + pl + cpx and the observed modal abundances are in good agreement with the cotectic proportion obtained in experimental studies (Toplis and Carroll 1995; Feig et al. 2006).

The crystallization sequence observed in gabbroic rocks (ol + pl-ol + pl + cpx) with plagioclase that crystallizes together with olivine before cpx, well in agreement with the pressure estimates (0.2–0.3 Gpa), indicates that the Brothers gabbros formed at low pressure from a relatively dry magma (Niu and O'Hara 2003; Feig et al. 2006). Water poor and low oxygen fugacity conditions are also suggested by the near absence of hydrous mineral phases and by the low modal abundances of iron oxides.

Mineral phases and bulk rock chemistry of BRG and QG are consistent with the evolved character of the basaltic magma that generated the gabbros, implying extensive early fractionation of a MORB-type primary magma with the consequent formation of cumulates. The limited exposure of the Brothers mafic complex has not allowed us to identify ultramafic cumulates (primitive cumulates) attesting that fractional crystallization of the parent magma developed mostly at low pressure. On the other hand, even in other areas close to the Red Sea where layered gabbros are more widely exposed (Tihama Asir complex) primitive cumulates are also lacking, providing evidence for polybaric differentiation. This idea is supported by the occurrence of high-pressure mafic intrusions at Zabargad. Therefore, BRG and QG could represent pods of differentiated (evolved) magma that, fractionated previously at the base of the crust (in the lower crust), would have risen and consolidated in the thinned upper crust.

Mineral chemistry as well as whole rock major and trace element composition of the QUSEIR and Brothers gabbros are compatible with the hypothesis that the gabbroic complex crystallized from near-MORB magma composition. They may have been emplaced either in a continental or in an oceanic context. This finding renews the controversy about northern/central Red Sea crust outside the axial oceanic cells: is this crust oceanic or does it consist of melt-intruded extended continental crust as observed along the East African rift zone from Kenya to Eritrea (e.g., Hammond et al. 2011, 2013; Ebinger and Scholz 2012)? Resolving this controversy can nevertheless be reached from available data in the northern Red Sea.

11 Northern Red Sea: Oceanic Crust or Melt Intruded Extended Continental Crust?

If we assume that the main trough in the northern Red Sea is floored largely by oceanic crust and that seafloor spreading initiated 16 Ma (LaBrecque and Zitellini 1985), given that Brothers and QUSEIR gabbros are located 40 to 50 km from the present-day axis and that the plate separation rate has been similar to the present-day rate (8–9 mm/yr at Brothers Islets) since 11 ± 2 Ma (Reilinger et al. 2015), the gabbros should have an age of ~12–10 Ma. Although the initiation of the salinity crisis in the Red Sea is not dated precisely, evaporites were probably deposited during Mid to Late Miocene (Bosworth et al. 2005; Almalki et al. 2015; Mitchell et al. 2015). According to this scenario, seafloor spreading in the Red Sea started during evaporite deposition. It has been suggested that the absence of magnetic lineations in the northern Red Sea, in contrast with the southern Red Sea, may be due to a thick salt layer and to the slow spreading rate. There are two possible explanation for this: (1) salt flowage exceeds the spreading rate and eruptions at the base of salt would form no extrusives but only intrusive rocks (Girdler 1985; Dyment et al. 2013; Tapponnier et al. 2013); (2) pervasive hydrothermal alteration in basalt would destroy the titanomagnetite minerals and erase the associated magnetic anomalies (Rona 1978; Levi and Riddihough 1986; Gaulier et al. 1988; Augustin et al. 2014).

According to the first model, the oceanic crust is intruded beneath the salt where temperatures are high. Melt crystallizing below the salt will not form pillow lavas but will cool slowly to produce sills and dykes with larger grain size, inhibiting the acquisition of strong magnetization. In order to explore this hypothesis, we simulated Mid to Late Miocene Red Sea thermal conditions at the base of the evaporitic layer to assess if temperatures exceeded the Curie temperature. We carried out numerical modelling of heat flow, revisiting a former work by Hutchison (1985), to obtain an estimate of the perturbed temperatures at the top and within the basement through time during sedimentation, including the effects of compaction and seawater temperature changes. We simulated salt deposition using both constant and temperature dependent thermal properties for halite, assuming a constant sedimentation rate of 0.5 mm/yr, a constant temperature of 15 °C at the seafloor and a constant heat flux of 250 mW/m^2 at the base of the model (depth of 150 km). Numerical results are shown in Fig. 16. Given the high conductivity of halite, temperatures at the base of the salt and within the shallow basement (sub-basement depth <2 km) remain well below the Curie point of magnetite (550–580 °C).

These temperatures are relatively low if compared with temperatures predicted during sedimentation with a similar accumulation rate of other sediment types with lower conductivity (Fig. 16b). Levi (1979) reported magnetic properties comparable to those of extrusive pillow basalts in dyke samples cored at a sub-basement depth of more than 500 m at DSDP sites 417D and 418A, and given by Bleil et al. (1982) on rock samples from a 3–4 km thick vertical volcanic section in eastern Iceland with magnetization of the subaerially extruded flows. The time required for magma to cool below a mid-ocean ridge segment (i.e., intrusive versus extrusive rocks) depends on several factors such as volume of the magmatic body, melt replenishment rate and how the heat is transferred efficiently toward the surface (host rock conductivity, thermal gradient and hydrothermal circulation). To be as conservative as possible, we assume that below a non-sedimented mid-ocean ridge segment, gabbros

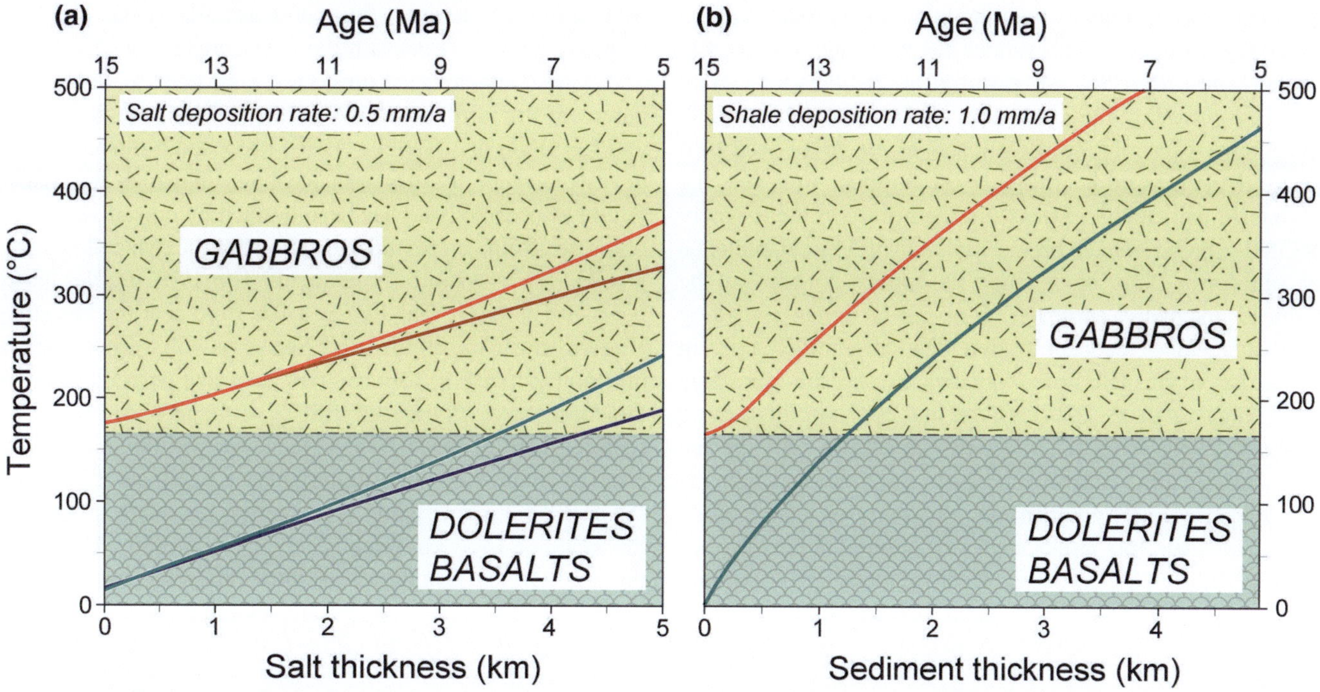

Fig. 16 Predicted temperatures below a depositing sedimentary layer during the rift to drift transition assuming a constant heat-flux from below of 250 mW/m^2. We assume that beneath a mature oceanic ridge segment, basalts and dolerites form at the surface and at shallow depths beneath the top of oceanic crust, whereas gabbros form at a depth >2 km. **a** Predicted temperature 2 km below (red and brown solid lines) and at the top (cyan and blue solid lines) of the basement during deposition of halite. Brown and blue lines indicate temperature variations using constant values for thermal conductivity and diffusivity of halite (conductivity: 5.86 Wm^{-1} K^{-1}, diffusivity: 3.18 • 10^{-6} m^2/s). Red and cyan lines are predicted values, including thermal property changes with temperature of halite determined experimentally by Urquhart and Bauer (2015). We assume a constant rate of 0.5 km/Ma in order to fit the ~4–5 km thick Mid to Late Miocene evaporitic layer in the northern Red Sea. **b** Predicted temperature at the top of the basement (cyan line) and 2 km below the sediment-basement interface (red line) during shale-like sedimentation at a rate of 1 km/Ma including the effects of compaction

form at a sub-basement depth > 2 km, that is, at a temperature >170 °C with a constant heat flow from below of 250 mW/m^2. Accordingly, we would expect that gabbros be emplaced in the Red Sea when the salt layer exceeds a thickness of ~3–4 km, with >170 °C temperatures at the top of basement. Thus, our model predicts that in the early phases of seafloor spreading in the northern Red Sea (from ~15 to ~10–7 Ma), when salt thickness was lower and with limited possibility for salt flowage toward the axis, we would observe well developed magnetic lineations if normal oceanic crust formed beneath the evaporites. However, magnetic anomalies parallel to the rift axis are not observed in the northern Red Sea, implying an absence of oceanic crust.

In addition, Gaulier et al. (1988), referring to drill hole QUSEIR B-1X, reported on a 35 m-thick metavolcanoclastic sequence resting above the QUSEIR gabbros and below the evaporites. The Precambrian basement complex outcropping onshore in the Quseir area consists of highly deformed volcanics and volcanogenic sediment, pervasively metamorphosed to lower greenschist facies and intruded by large granitic plutons. The volcanic rocks are calc-alkaline, predominatly andesitic, although basaltic, dacitic and rhyolitic rocks are present (Greene 1984). The ^{40}Ar-^{39}Ar age of 25 ± 6 Ma (Ligi et al. 2018) suggests that QUSEIR MORB gabbros intruded continental lithosphere during the very early Red Sea rift. In addition, the isotopic composition (Nd and Sm) of these gabbros indicates a pseudo-isochron for exposure at seafloor of ~20 Ma. Without any additional constraints on the nature of the metavolcanic material overlying the QUSEIR gabbros, we speculate that it may represent sediments deposited above mid-crust intruded gabbros exhumed during an early tectonic rift phase about 20 Ma, before the deposition of evaporites.

The second model to explain the lack of organized magnetic stripes parallel to the rift axis in the northern Red Sea includes pervasive hydrothermal alteration of the oceanic crust due to a close hydrothermal system below extensively sedimented, relatively-young spreading centres (Levi and Riddihough 1986). The sedimentary cover acts as a cap trapping the hydrothermal fluids with long residence time in the basalt that will cool more slowly while fluids will thermally and chemically equilibrate with host rocks and the overlying sediments. This model has been applied recently in the central Red Sea (Augustin et al. 2014), where giant salt flows invading the axial valley and covering the oceanic crust have been observed (Mitchell et al. 2010; Augustin et al. 2014; Ligi et al. 2015). In the northern Red Sea, the Mabahiss and Shaban Deeps are the only two areas where MORB-type lavas have been sampled, with limited evidence of salt flows. This model may be applied to explain the reduced magnetic anomaly observed in some spreading segments along the central Red Sea present-day axis, in particular where salt flow velocities exceed the spreading rate (Augustin et al. 2014). Given that evaporite deposition stopped at the end of Miocene, salt flows tapping the neo-volcanic zone may create closed hydrothermal systems allowing intense hydrothermal alteration of the basaltic crust. However, this model cannot be applied in the Red Sea during Mid to Late Miocene, when seafloor spreading and evaporite deposition acted contemporaneously, because pervasive hydrothermal alteration was unlikely due to a lack of fluids (seawater) beneath the evaporitic cover.

Crustal seismic velocities may also help to define the nature of the northern Red Sea crust. Gaulier et al. (1988) and LePichon and Gaulier (1988) proposed a model for the development of the northern Red Sea in which the distribution of oceanic crust has been defined by crustal velocities obtained from an extensive two-ships expanded-spread-profile (ESP) seismic experiment. The ESP profiles have been shot parallel to the Red Sea axis from near the Egyptian coast to the axis of the Red Sea (Fig. 17) with common mid-points (CMP's) forming two transects perpendicular to the coast (northern transect: ESP 3, 4, 5 and 6, and southern transect: ESP 7, 8, 9, 10 and 11) and one parallel to the coast running along the main trough (ESP 13, 6, 14 and 15). All of the profiles show very thin crust underlying the evaporites in the range of 14–6 km thick (Gaulier et al. 1988). However, the northern and southern sets show different crustal velocities, with relatively low crustal velocities (5–6.4 km/s) to

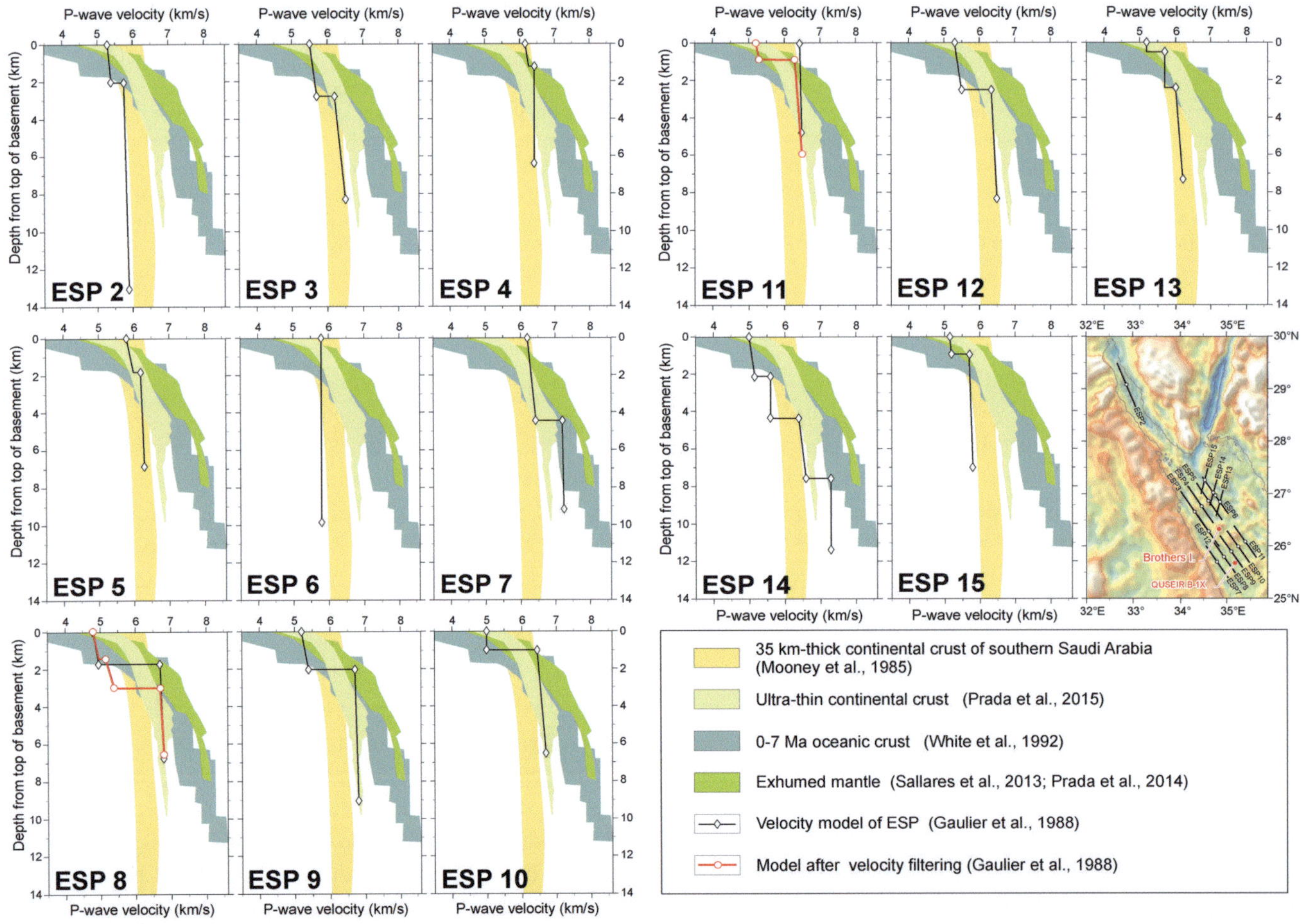

Fig. 17 ESP vertical velocity profiles of Gaulier et al. (1988) compared with P-wave velocity-depth envelopes from different domains. Velocity structures for Arabian 35 km-thick continental crust (Mooney et al. 1985), for young oceanic crust (White et al. 1992), for thinned continental crust (Prada et al. 2015) and for exhumed mantle (Sallarès et al. 2013; Prada et al. 2015), are indicated as coloured curves marking the lateral variability of velocity within the given region

the north and higher velocities to the south (5–6.8 km/s). Gaulier et al. (1988) interpreted the higher velocities in the southern transect as being diagnostic of oceanic crust. LePichon and Gaulier (1988) proposed an abrupt limit related to the Dead Sea-Aqaba transform boundary from oceanic crust in the south to very thin continental crust in the north that extends from roughly near the Brothers Islets to the axis of the sea and then north to reach the Arabian coast near 28°N.

Crustal velocities below evaporites from ESP profiles in the northern Red Sea (Fig. 17) of Gaulier et al. (1988) are compared with compilations of velocity-depth profiles corresponding to: ~35 km-thick continental crust from southern Saudi Arabia (Mooney et al. 1985), thinned continental crust from the Salton Trough (Parsons and McCarthy 1996), Afar (Hammond et al. 2011), Nova Scotia rifted margin, Grand Banks of Newfoundland and Moroccan margin (Prada et al. 2015), 0–7 Ma-old oceanic crust (White et al. 1992), and exhumed mantle rocks from the central Tyrrhenian Sea (Prada et al. 2014) and the Gulf of Cadiz (Sallarès et al. 2013). Overall, the vertical velocity crustal structure inferred from the northern transect (ESP profiles 3, 4, 5, and 6; 13, 14 and 15) matches well with continental crust compilation from the Arabian plate, while the southern transect (ESP profiles 7, 8, 9, 10 and 11) and ESP 12 show lower-crust velocities, slower than those expected for young oceanic crust and rather similar to velocities from thinned continental crust. The vertical profile ESP 7 shows the highest velocity gradient and can be classified as either continental or oceanic. High lower-crust seismic velocities (7.0–7.6 km/s) recorded in poorly evolved rifts such as Baikal and southern Kenya rifts have been interpreted due to magmatic intrusions that have compensated the amount of crustal thinning by the addition of new material (Birt et al. 1997; Thybo and Nielsen 2009). However, ESP 7 is the nearest to the coast, suggesting that the difference in the velocity structure between the northern and the southern transects in the northern Red Sea may also correspond to a north-south change in the nature of the basement rocks observed onshore in Egypt (Stern et al. 1984; Stern and Hedge 1985; Greiling et al. 1988; Cochran 2005). The northern Egyptian basement is mainly composed of granite, granodiorite and weakly deformed plutonic rocks. In contrast, the southern basement consists of mafic metavolcanics, gabbros, ultramafic rocks and associated metasedimentary rocks, with the boundary between the two provinces located between Safaga and Quseir (Cochran 2005). These observations suggest that the north-to-south change in the basement velocity structure in the northern Red Sea, interpreted by LePichon and Gaulier (1988) as a change from continental to oceanic crust, is actually a change between two very different, although both continental, types of pre-rift basement (Cochran 2005).

Diffuse seismicity in the northern Red Sea contrasts with the axial concentrated seismicity of the southern Red Sea and in general of mid-ocean ridges. In addition, there is no evidence for sea dipping reflectors (SDRs) in the northern Red Sea, implying a magma-poor rift zone (Stockly and Bosworth 2018). Exposed peridotites in magma poor rifts such as Iberia-Newfoundland or the Alps show a complex interaction between melt and lithospheric mantle with widespread occurrence of refertilized plagioclase peridotites (Müntener et al. 2004, 2010; Le Roux et al. 2007). Basaltic glasses from the northern Red Sea axial ridge (Haase et al. 2000) are geochemically similar to basaltic glasses sampled along the axial zone of Thetis and Nereus (Ligi et al. 2012, 2015) in the northern sector of the central Red Sea, that is, they derive from a depleted mantle source (MORB source) with no contamination by continental lithosphere.

A rifting model for the central and northern Red Sea that fits the observations, requires: (1) a mechanism able to protract rifting for over 23–24 Ma, and (2) an upper crust that during the rift to drift transition rests above an asthenospheric mantle. In fact, emplacement of MORB with no lithospheric contamination in the axial part of the central and northern Red Sea during the very early stages of seafloor spreading implies replacement of lower crust and mantle lithosphere by active upwelling asthenosphere before breakup, as proposed for the southernmost Red Sea region in Afar (Rychert et al. 2012).

Continental rifting induced stretching and thinning of the lithosphere in the northern Red Sea with formation of extensional shear zones along the margins and with upwelling of the asthenosphere and partial melting beneath the axis. Melt percolation and storage into the crust generated low-P intrusions exemplified by the QUSEIR and Brothers gabbros. When magma volumes and rates of extension are low and the melt does not repeatedly intrude the same axial zone, the compositional effect of dyke intrusions dominates over heating, leading to an increase in crustal strength (Beutel et al. 2010; Daniels et al. 2014). Thus, during the initial phases of melt intrusion, episodic dyke injections accommodate extension in the upper crust, maintaining its thickness and eventually increasing its strength, thus delaying upper crust rupture (Daniels et al. 2014). Moreover, a thick evaporite blanket (given the high conductivity of halite) may contribute to increasing the strength of the upper crust compared to other highly clastic-sedimented margins. Extension in the ductile melt-rich lower crust is accommodated continuously by a combination of magmatic addition and viscous flow, favouring the rise at shallow levels of the asthenosphere (Bastow and Keir 2011; Wright et al. 2012). Brothers and QUSEIR gabbros represent thinned-continental lower-crust intrusions of asthenospheric melt that may be later exposed at the seafloor during rifting. $^{40}Ar/^{39}Ar$ ages and isotopic composition of QUSEIR gabbros indicate that they may be related to the early syn-rift intense magmatic event at ~22 Ma, that originated the dyke swarm running along the entire Arabian Red Sea coast (Bosworth et al. 2005; Bosworth and Stockli 2016; Stern and Johnson 2018).

Magmatic intrusions since the early stages of rifting represented by Zabargad, Brothers and QUSEIR mafic complexes suggest that the northern Red Sea is a non-volcanic, but not a magma-poor rift, as generally believed. Magmatism may have contributed to weakening the lower crust during the early phase of rifting favouring decoupling between the upper and lower crust, protracting upper crustal extension and delaying crustal breakup.

12 Conclusions

Gabbros drilled offshore Egypt in the northern Red Sea and mafic complexes exposed on the Brothers and Zabargad Islands suggest that continental breakup in the northern Red Sea, a relatively non-volcanic narrow rift, is preceded by intrusions of basaltic melt that crystallize at progressively shallower crustal depths as rifting progresses toward continental break-up. A seismic reflection profile running across the central part of the southern Thetis basin shows a ~5 km wide reflector that marks the roof of a magma chamber located ~3.5 km below the seafloor (Ligi et al. 2018). The presence of a few kilometres deep sub-rift magma chamber soon after the initiation of oceanic spreading implies crystallization of lower oceanic crust intrusives as a last step in a sequence of basaltic melt intrusion from pre-oceanic continental rifting to oceanic spreading. This allows us to describe the transition from continental rift to oceanic seafloor spreading in the central and northern Red Sea as resulting from the interactions between structural and magmatic processes.

Tethys ocean subduction and mantle dynamics induced stretching and thinning of the lithosphere beneath the Red Sea with formation of crustal-scale fault zones along the margins. Lithospheric thinning was accompanied by passive asthenospheric upwelling beneath the rift axis. This enabled deep partial melting of the rising asthenosphere with melt intruding the continental lithosphere. Melt solidification generated high-pressure gabbroic rocks similar to those exposed on Zabargad Island (Bonatti and Seyler 1987). Increased temperatures deepened the solidus enabling melt migration toward the rift axis and enhanced short-lived small-scale convection by locally decreasing viscosity and increasing buoyancy of the upwelling asthenosphere (Ligi et al. 2011, 2015). Melt percolation and storage generated low-P oceanic-type gabbroic intrusions, exemplified by the drilled gabbros on the western side of the northern Red Sea and by those exposed on the Brothers Islets, probably feeding the swarm dykes now exposed along the entire coastal plain of the Arabian Peninsula (24–22 Ma; Bosworth et al. 2005). Active upwelling of the deeper asthenosphere forced the final rupture of the upper continental lithosphere. Mantle partial melting generated aggregated N-MORB melts that migrated and were extracted at the rift axis, forming shallow gabbroic intrusions and basaltic lava flows. Thus, oceanic crust accretion in the Red Sea rift starts at depth before continental rupturing, emplacement of oceanic basalt at the sea floor, and development of Vine-Matthews magnetic anomalies.

Acknowledgements The research was sponsored by the PRIN2012 Programme (Project 20125JKANY_002). The work was supported by the Saudi Geological Survey and the Italian Consiglio Nazionale Ricerche. Fruitful discussions during SGS workshop held in Jeddah on February 14–17, 2016 improved this work. We thank O. R. Berg for providing the gabbro sample from the QUSEIR B-1X drill hole. We are grateful to Y. Cai, A. Cipriani, C. Palmiotto, M. Seyler, G. Barabino and G. Traversa for carrying out part of the analytical work. We thank Dr. Z. A. Nawab, SGS President and Dr A. M. Al Attas, SGS Assistant President, and Dr N. Rasul for their support during this work. We particularly thank P. Betts, C. Ebinger and two anonymous reviewers for their helpful and constructive comments.

References

Al-Ahmadi K, Al-Amri A, See L (2014) A spatial statistical analysis of the occurrence of earthquakes along the Red Sea floor spreading: clusters of seismicity. Arab J Geosci 7:2893–2904

Almalki KA, Betts PG, Ailleres L (2015) The Red Sea—50 years of geological and geophysical research. Earth Sci Rev 147:109–140

Almalki KA, Betts PG, Ailleres L (2016) Incipient seafloor spreading segments: insights from the Red Sea. Geophys Res Lett 43:2709–2715

Altherr R, Henjes-Kunst F, Puchelt H, Baumann A (1988) Volcanic activity in the Red Sea axial trough: Evidence for a large mantle diapir? Tectonophysics 150:121–133

Andersen DJ, Lindsley D, Davidson PM (1993) QUILF: A Pascal program to assess equilibria among Fe–Mg–Mn–Ti-oxides, pyroxenes, olivine, and quartz. Comp Geosci 19:1333–1350

Aoki K, Kushiro I (1968) Some clinopyroxenes from ultramafic inclusions in Dreiser Weiher, Eifel. Contrib Mineral Petrol 18:326–337

Aoki K, Shiba I (1973) Pyroxene from lherzolite inclusions of Itinomegata, Japan. Lithos 6:41–51

Arevalo R, McDonough WF (2010) Chemical variations and regional diversity observed in MORB. Chem Geol 271:70–85

ArRajehi A, McClusky S, Reilinger R, Daoud M, Alchalbi A, Ergintav S, Gomez F, Sholan J, Bou-Rabee F, Ogubazghi G, Haileab B, Fisseha S, Asfaw L, Mahmoud S, Rayan A, Bendik R, Kogan L (2010) Geodetic constraints on present-day motion of the Arabian Plate: implications for Red Sea and Gulf of Aden rifting. Tectonics 29:TC3011. https://doi.org/10.1029/2009tc002482

Augustin N, Devey CW, van der Zwan FM, Feldens P, Tominaga M, Bantan R, Kwasnitschka T (2014) The transition from rifting to spreading in the Red Sea. Earth Planet Sci Lett 395:217–230

Barnes SJ (1986) The distribution of chromium among orthopyroxene, spinel and silicate liquid at atmospheric pressure. Geochim Cosmochim Acta 50:1889–1909

Bastow ID, Keir D (2011) The protracted development of the continent-ocean transition in Afar. Nature Geosci 4:248–250

Beard JS (1986) Characteristic mineralogy of arc-related cumulate gabbros: Implications for the tectonic setting of gabbroic plutons and for andesite genesis. Geology 14:848–851

Beccaluva L, Ohnenstetter D, Ohnenstetter M, Venturelli G (1977) The trace element geochemistry of Corsican ophiolites. Contrib Mineral Petrol 64:11–31

Bellahsen N, Faccenna C, Funiciello F, Daniel JM, Jolivet L (2003) Why did Arabia separate from Africa? Insights from 3-D laboratory experiments. Earth Planet Sci Lett 216:365–381

Bender JF, Hodges FN, Bence AE (1978) Petrogenesis of basalts from the project Famous area: Experimental study from 0 to 15 kbars. Earth Planet Sci Lett 41:277–302

Bernstein S (2006) In situ fractional crystallization of a mafic pluton: Microanalytical study of a Palaeogene gabbronorite plug in East Greenland. Lithos 92:222–237

Beutel E, van Wijk J, Ebinger C, Keir D, Agostini A (2010) Formation and stability of magmatic segments in the Main Ethiopian and Afar rifts. Earth Planet Sci Lett 293:225–235

Birt CS, Maguire PKH, Khan MA, Thybo H, Keller GR, Patel J (1997) The influence of pre-existing structures on the evolution of the southern Kenya Rift Valley – evidence from seismic and gravity studies. Tectonophysics 278:211–242

Bleil U, Hall JH, Johnson HP, Levi S, Schonharting G (1982) The natural magnetization of a 3-kilometer section of Icelandic crust. J Geophys Res 87:6569–6589

Bohannon RG, Eittreim SL (1991) Tectonic development of passive continental margins of the southern and central Red Sea with a comparison to Wilkes Land, Antarctica. Tectonophysics 198:129–154

Bonatti E (1985) Punctiform initiation of seafloor spreading in the Red Sea during transition from continental to an oceanic rift. Nature 316:33–37

Bonatti E, Clocchiatti R, Colantoni P, Gelmini R, Marinelli G, Ottonello G, Santacroce R, Taviani M, Abdel-Meguid AA, Assaf HS, El Tahir MA (1983) Zabargad (St. John) Island: An uplifted fragment of sub-Red Sea lithosphere. J Geol Soc London 14D:667–690

Bonatti E, Colantoni P, Della Vedova B, Taviani M (1984) Geology of the Red Sea transitional region (22°-25°N). Ocean Acta 7:385–398

Bonatti E, Hamlyn P, Ottonello G (1981) Upper mantle beneath a young oceanic rift - peridotites from the island of Zabargad (Red-Sea). Geology 9:474–479

Bonatti E, Ligi M, Carrara G, Gasperini L, Turko N, Perfiliev A, Peyve A, Sciuto PF (1996) Diffuse impact of the Mid Atlantic Ridge with the Romanche transform: An Ultracold Ridge/Transform Intersection. J Geophys Res 101:8043–8054

Bonatti E, Ottonello G, Hamlyn PR (1986) Peridotites from the island of Zabargad (Red Sea). J Geophys Res 91:599–631

Bonatti E, Seyler M (1987) Crustal underplating and evolution in the Red Sea rift. J Geophys Res 92:12083–12821

Borghini G, Rampone E (2007) Postcumulus processes in oceanic-type olivine-rich cumulates: The role of trapped melt crystallization versus melt/rock interaction. Contrib Mineral Petrol 154:619–633

Bosworth W (1993) Nature of the Red Sea crust. A controversy revisited: Comment and reply. Geology 21:574–575

Bosworth W (2015) Geological evolution of the Red Sea: Historical background, review, and syntesis. In: Rasul NMA, Stewart ICF (eds) The Red Sea: The Formation, Morphology, Oceanography and Environment of a Young Ocean Basin. Springer Earth System Sciences, Berlin Heidelberg, pp 45–78

Bosworth W, Huchon P, McClay K (2005) The Red Sea and Gulf of Aden basins. J African Earth Sci 43:334–378

Bosworth W, Stockli D (2016) Early magmatism in the greater Red Sea rift: Timing and significance. Canadian J Earth Sci 53(11):1158–1176. https://doi.org/10.1139/cjes-2016-0019

Bosworth W, Taviani M, Rasul N (2018) Neotectonics of the Red Sea, Gulf of Suez and Gulf of Aqaba. In: Rasul NMA, Stewart ICF (eds) The Red Sea. Springer Earth System Sciences, Berlin Heidelberg, this issue

Botcharnikov RE, Almeev RR, Koepke J, Holtz F (2008) Phase relations and liquid lines of descent in hydrous ferrobasalt – implications for the Skaergaard intrusion and Columbia River flood basalts. J Petrol 29:1687–1727

Brey GP, Koehler T (1990) Geothermobarometry in four-phase lherzolites II. New thermobarometers, and practical assessment of existing thermobarometers. J Petrology 31:1353–1378

Brueckner HK, Elhaddad MA, Hamelin B, Hemming S, Kröner A, Reisberg L, Seyler M (1995) A Pan-African origin and uplift for gneisses and peridotites of Zabargad Island, Red Sea: A Nd, Sr, Pb and Os isotope study. J Geophys Res 100:22283–22297

Casey JF, Banerji D, Zarian P (2007) Geochemical composition of gabbroic rocks from ODP Hole 179–1105A, southwest Indian Ridge. In: Casey JF, Miller DJ (eds) Proceedings of the Ocean Drilling Program, Scientific Results, vol. 179. Ocean Drilling Program, College Station, TX, pp 1-125

Chang SJ, Van der Lee S (2011) Mantle plumes and associated flow beneath Arabia and East Africa. Earth Planetary Sci Lett 302:448–454

Charlier BLA, Wilson CJN, Lowenstern JB, Blake S, Van Calsteren PW, Davidson JP (2005) Magma generation at a large, hyperactive silicic volcano (Taupo, New Zealand) revealed by U-Th and U–Pb systematics in zircons. J Petrology 46:3–32

Chu D, Gordon RG (1998) Current plate motions across the Red Sea. Geophys J Int 135:313–328

Cochran JR (1983) A model for the development of the Red Sea. Am Assoc Petrol Geol Bull 67:41–69

Cochran JR (2005) Northern Red Sea: Nucleation of an oceanic spreading center within a continental rift. Geochem Geophys Geosyst 6:Q03006

Cochran JR, Karner GD (2007) Constraints on the deformation and rupturing of continental lithosphere of the Red Sea: The transition from rifting to drifting. In: Karner GD, Manatschal G, Pinheiro LM (eds) Imaging, mapping and modeling continental lithosphere extension and breakup. Geol Soc London, Spec Publ, vol. 282, pp 265–289

Cochran JR, Martinez F (1988) Evidence from the northern Red Sea on the transition from continental to oceanic rifting. Tectonophysics 153:25–53

Coleman RG, Fleck RJ, Hedge CE, Ghent ED (1977) The volcanic rocks of southwest Saudi Arabia and the opening of the Red Sea. In: Red Sea Research 1970–1975. Saudi Arabian Directorate General of Mineral Resources Bull 22:Dl-D30

Coleman RG, Hadley DG, Fleck RG, Hedge CT, Donato MM (1979) The Miocene Tihama Asir ophiolite and its bearing on the opening of the Red Sea. In: Al-Shanti AM (ed) Evolution and Mineralization of the Arabian Shield. Pergamon Press Ltd, Oxford, pp 173–186

Coleman RG, McGuire AV (1988) Magma systems related to the Red Sea opening. Tectonophysics 150:77–100

Cornen G, Girardeau J, Monnier C (1999) Basalts, underplated gabbros and pyroxenites record the rifting process of the West Iberian margin. Mineral Petrol 67:111–142

Crane K, Bonatti E (1987) The role of fracture zones during early Red Sea rifting: Structural analysis using Space Shuttle radar and LANDSAT imagery. J Geol Soc London 144:407–420

Dalrymple GB, Alexander EC, Lanphere MA, Kraker GP (1981) Irradiation of samples for $^{40}Ar/^{39}Ar$ dating using the Geological Survey TRIGA reactor. USGS Professional Papers 1176, U.S. Geological Survey, Reston, VA, USA, p 29

Daniels KA, Bastow ID, Keir D, Sparks RSJ, Menard T (2014) Thermal models of dyke intrusion during the development of continent-ocean transition. Earth Planet Sci Lett 285:145–153

DeMets C, Merkouriev S (2016) High-resolution estimates of Nubia-Somalia plate motion since 20 Ma from reconstructions of the Southwest Indian Ridge, Red Sea and Gulf of Aden. Geophys J Int 207:317–332

Deniel C, Vidal P, Coulon C, Vellutini P-J, Piguet P (1994) Temporal evolution of mantle sources during continental rifting: The volcanism of Djibouti (Afar). J Geophys Res 99:2853–2869

Drüppel K, von Seckendorff V, Okrusch M (2001) Subsolidus reaction textures in anorthosites of the Kunene Intrusive Complex, NW Namibia. European J Mineralogy 13:289–309

Dyment J, Tapponnier P, Afifi AM, Zinger MA, Franken D, Muzaiyen EA (2013) New seafloor spreading model of the Red Sea: Magnetic anomalies and plate kinematics. AGU Fall Meeting 2013, San Francisco, T21A-2512

Ebinger CE, Scholz CA (2012) Continental rift basins: The East African perspective. In: Busby C, Azor A (eds) Tectonics of Sedimentary Basins: Recent Advances. Wiley-Blackwell, Oxford, pp 185–208

Ehrhardt A, Hubscher C (2015) The northern Red Sea in transition from rifting to drifting—lessons learned from ocean deeps. In: Rasul NMA, Stewart ICF (eds) The Red Sea: the formation, morphology, oceanography and environment of a Young Ocean Basin. Springer Earth System Sciences, Berlin, pp 121–135

Ernst WG, Liu J (1998) Experimental phase-equilibrium study of Al- and Ti-contents of calcic amphibole in MORB-A semiquantitative thermobarometer. Am Mineral 83:952–969

Feig ST, Koepke J, Snow JE (2006) Effect of water on tholeiitic basalt phase equilibria: an experimental study under oxidizing conditions. Contrib Mineral Pet 152:611–638

Féménias O, Mercier JCC, Nkono C, Diot H, Berza T, Tatu M, Demaiffe D (2006) Calcic amphibole growth and compositions in calc-alkaline magmas: evidence from the Motru dike swarm (Southern Carpathians, Romania). Am Mineral 91:73–81

Frost BR, Lindsley DH (1992) Equilibria among Fe–Ti oxides, pyroxenes, olivine, and quartz: part II application. Am Mineral 77:1004–1020

Gale A, Dalton CA, Langmuir CH, Su Y, Schilling J-G (2013) The mean composition of ocean ridge basalts. Geochem Geophys Geosyst 14:489–518

Gallacher R, Keir D, Harmon N (2018) The nature of upper mantle upwelling during initiation of seafloor spreading in the southern Red Sea. In: Rasul NMA, Stewart ICF (eds) The Red Sea. Springer Earth System Sciences, Heidelberg (this issue)

Gasperini L, Bernoulli D, Bonatti E, Borsetti AM, Ligi M, Negri A, Sartori R, von Salis K (2001) Lower cretaceous to Eocene sedimentary transverse ridge at the Romanche fracture zone and the opening of the equatorial Atlantic. Mar Geol 176:101–119

Gaulier J-M, Le Pichon X, Lyberis N, Avedik F, Geli L, Moretti I, Deschamps A, Hafez S (1988) Seismic study of the crustal thickness, northern Red Sea and Gulf of Suez. Tectonophysics 153:55–88

Ghebreab W (1998) Tectonics of the Red Sea region reassessed. Earth Sci Rev 45:1–44

Girdler RW (1985) Problems concerning the evolution of oceanic lithosphere in the northern Red Sea. Tectonophysics 116:109–122

Girdler RW (1991) The Afro-Arabian rift system; an overview. Tectonophysics 197:139–153

Girdler RW, Styles P (1974) Two stage seafloor spreading. Nature 247:7–11

Ghent ED, Coleman RG, Hadley DG (1980) Ultramafic inclusions and host alkali olivine basalts of the southern coastal plain of the Red Sea. Am J Sci 280A:499–527

Gordon G, Hansen B, Scott J, Hirst C, Graham R, Grow T, Spedding A, Fairhead S, Fullarton L, Griffin D (2010) The hydrocarbon prospectivity of the Egyptian North Red Sea basin. In: Vining BA, Pickering SC (eds) Petroleum geology: from mature basins to new frontiers. Proceedings of the 7th Petroleum geology conference, Geol Soc London, pp 783–789

Greene DC (1984) Structural geology of the Quseir area, Red Sea coast, Egypt. Dept Geology and Geography, University of Massachusetts, Amherst, MA (contribution no. 52, pp 158)

Greiling RO, El Ramly MF, El Arhal H, Stern RJ (1988) Tectonic evolution of the northwestern Red Sea margin as related to basement structure. Tectonophysics 153:179–191

Guennoc P, Pautot G, Coutelle (1988) Surficial structures of the northern Red Sea axial valley from 23°N to 28°N: Time and space evolution of neo-oceanic structures. Tectonophysics 153:1–23

Haase KM, Muhe R, Stoffers P (2000) Magmatism during extension of the lithosphere: geochemical constraints from lavas of the Shaban Deep, northern Red Sea. Chem Geol 166:225–239

Hammond JOS, Kendall J-M, Stuart GW, Keir D, Ebinger C, Ayele A, Belachew M (2011) The nature of the crust beneath the Afar triple junction: evidence from receiver functions. Geochem Geophys Geosyst 12:Q12004

Hammond JOS, Kendall J-M, Stuart GW, Ebinger CJ, Bastow ID, Keir D, Ayele A, Belachew M, Goitom B, Ogubazghi G, Wright TJ (2013) Mantle upwelling and initiation of rift segmentation beneath the Afar Depression. Geology 41:635–639

Hebert R, Constantin M, Robinson PT (1991) Primary mineralogy of Leg 118 gabbroic rocks and their place in the spectrum of oceanic mafic igneous rocks. In: Von Herzen RP, Robinson PT et al (eds) Proceedings of ocean drilling program, Scientific Results 118. Ocean Drilling Program, College Station, TX, pp 3–20

Helz RT (1973) Phase relations of basalts in their melting range at P (sub H_2O) = 5 kb as a function of oxygen fugacity; part I, mafic phases. J Pet 14:249–302

Hoang CT, Taviani M (1991) Stratigraphic and tectonic implications of uranium-series dated coral reefs from uplifted Red Sea islands. Quat Res 35:264–273

Hosny A, Nyblade A (2014) Crustal structure in southeastern Egypt: Symmetric thinning of the northern Red Sea rifted margins. Geology 42:219–222

Hosny A, Nyblade A (2016) The crustal structure of Egypt and the northern Red Sea region. Tectonophysics 687:257–267

Hutchison I (1985) The effects of sedimentation and compaction on oceanic heat flow. Geophys J Int 82:439–459

Jaques AL, Chappell BW (1980) Petrology and trace element geochemistry of the Papuan ultramafic belt. Contrib Mineral Petrol 75:55–70

Kempton PD, Downes H, Embey-Istzin A (1997) Mafic granulite xenoliths in Neogene alkali basalts from the western Pannonian basin: Insights into the lower crust of a collapsed orogen. J Petrol 38:941–970

Kempton PD, Harmon RS (1992) Oxygen isotope evidence for large-scale hybridization of the lower crust during magmatic underplating. Geochim Cosmochim Acta 56:971–986

Kendall JM, Lithgow-Bertelloni C (2016) Why is Africa rifting? In: Wright TJ, Ayele A, Ferguson DJ, Kidane T, Vye-Brown C (eds) Magmatic rifting and active volcanism. Geol Soc London, Spec Publ 420, pp 11–30

LaBrecque JL, Zitellini N (1985) Continuous sea-floor spreading in Red Sea: an alternative interpretation of magnetic anomaly patterns. Bull Am Assoc Petrol Geol 69:513–524

Langmuir CH, Hanson GN (1981) Calculating mineral-melt equilibria with stoichiometry, mass balance and single component distribution coefficients. In: Newton RC, Navrotsky A, Wood BJ (eds) Advances in physical geochemistry 1: thermodynamics of minerals and melts. Springer, Berlin, New York, pp 247–271

Le Bas MJ (1962) The role of aluminium in igneous clinopyroxenes with relation to their parentage. Am J Sci 260:267–288

Lee CTA, Luffi P, Chin EJ (2011) Building and destroying continental mantle. Ann Rev Earth Planet Sci 39:59–90

LePichon X, Gaulier J-M (1988) The rotation of Arabia and the Levant fault system. Tectonophysics 153:271–294

Le Roux V, Bodinier JL, Tommasi A, Alard O, Dautria JM, Vauchez A, Riches AJV (2007) The Lherz spinel lherzolite: refertilized rather than pristine mantle. Earth Planet Sci Lett 259:599–612

Levi S (1979) Paleomagnetism and some magnetic properties of basalts from the Bermuda Triangle. In: Initial reports of the Deep Sea Drilling Project, vol. 52. U.S. Government Printing Office, Washington, DC, pp 1363–1377

Levi S, Riddihough R (1986) Why are marine magnetic anomalies suppressed over sedimented spreading centers. Geology 14:651–654

Li J, Sideris MG (1997) Marine gravity and geoid determination by optimal combination of satellite altimetry and shipborne gravimetry data. J Geodesy 71:209–216

Ligi M, Bonatti E, Bortoluzzi G, Cipriani A, Cocchi L, Caratori Tontini F, Carminati E, Ottolini L, Schettino A (2012) Birth of an ocean in the Red Sea: Initial pangs. Geochem Geophys Geosyst 13: Q08009

Ligi M, Bonatti E, Bosworth W, Cai Y, Cipriani A, Palmiotto C, Ronca S, Seyler M (2018) Birth of an ocean in the Red Sea: oceanic-type basaltic melt intrusions precede continental rupture. Gondwana Res 54:150–160

Ligi M, Bonatti E, Caratori Tontini F, Cipriani A, Cocchi L, Schettino A, Bortoluzzi G, Ferrante V, Khalil SM, Mitchell NC, Rasul N (2011) Initial burst of oceanic crust accretion in the Red Sea due to edge-driven mantle convection. Geology 39:1019–1022

Ligi M, Bonatti E, Rasul NMA (2015) Seafloor spreading initiation: geophysical and geochemical constraints from the Thetis and Nereus Deeps, Central Red Sea. In: Rasul NMA, Stewart ICF (eds) The Red Sea: the formation, morphology, oceanography and environment of a Young Ocean Basin. Springer Earth System Sciences, Berlin, pp 79–98

Ligi M, Bortoluzzi G (1989) PLOTMAP: Geophysical and geological applications of good standard quality cartographic software. Comput Geosci 15:519–585

Lindsley DH (1983) Pyroxene thermometry. Am Mineral 68:477–493

Lindsley DH, Frost BR (1992) Equilibria among Fe-Ti oxides, pyroxenes, olivine, and quartz: part I theory. Am Mineral 77:987–1003

Loucks RR (1990) Discrimination of ophiolitic from non-ophiolitic ultramafic-mafic allochthon sinorogenic belts by the Al/Ti ratio in clinopyroxenes. Geology 18:346–349

Loucks RR (1996) A precise olivine-augite Mg-Fe-exchange geothermometer. Contrib Mineral Petrol 125:140–150

Lyubetskaya T, Korenaga J (2007) Chemical composition of Earth's primitive mantle and its variance: 1 method and results. J Geophys Res 112:B03211

Makris MJ, Rhim R (1991) Shear-controlled evolution of the Red Sea: Pull-apart model. Tectonophysics 198:441–466

Martinez F, Cochran JR (1989) Geothermal measurements in the northern Red Sea: Implications for lithospheric thermal structure and mode of extension during continental rifting. J Geophys Res 94:12239–12265

Mazzucchelli M, Cipriani A, Hemond C, Zanetti A, Bertotto GW, Cingolani CA (2016) Origin of the DUPAL anomaly in mantle xenoliths of Patagonia (Argentina) and geodynamic consequences. Lithos 248–251:257–271

McBirney AR, Nicolas A (1997) The skaergaard layered series part ii: magmatic flow and dynamic layering. J Petrol 38:569–580

McCarter RL, Fodor RV, Trusdell F (2006) Perspectives on basaltic magma crystallization and differentiation: lava-lake blocks erupted at Mauna Loa Volcano summit, Hawaii. Lithos 90:187–213

McDonough WF, Sun S-S (1995) Composition of the earth. Chem Geol 120:223–253

McGuire AV, Coleman RG (1986) The Jabal Tirf layered gabbro and associated rocks of the Tihama Asir complex, SW Saudi Arabia. J Geol 94:651–665

Meyer PS, Dick HJB, Thompson G (1989) Cumulate gabbros from the Southwest Indian Ridge, 54°S–7°16'E: Implications for magmatic processes at a slow spreading ridge. Contrib Mineral Petrol 103:44–63

Mitchell NC, Ligi M, Ferrante V, Bonatti E, Rutter E (2010) Submarine salt flows in the central Red Sea. Geol Soc Am Bull 122:701–713

Mitchell NC, Ligi M, Feldens P, Hubscher C (2017) Deformation of a young salt giant: Regional topography of the Red Sea Miocene evaporites. Basin Res 29:352–369

Mitchell NC, Ligi M, Rohling EJ (2015) Red Sea isolation history by Plio-Pleistocene seismic reflection sequences. Earth Planet Sci Lett 430:387–397

Mitchell NC, Park Y (2014) Nature of crust in the central Red Sea. Tectonophysics 628:123–139

Mitra S, Princivalle F, Samanta AK, Moon H-S (1999) Geothermometry and mineralogy of two-pyroxene granulites: Evaluation from Mossbauer and X-ray single crystal cation partitioning of Ca-Poor and Ca-Rich pyroxenes. J Geol Soc India 53:537–548

Montanini A, Tribuzio R, Vernia L (2008) Petrogenesis of basalts and gabbros from an ancient continent–ocean transition (External Liguride ophiolites, northern Italy). Lithos 101:453–479

Mooney WD, Gettings ME, Blank HR, Healy JH (1985) Saudi Arabian seismic refraction profile: A traveltime interpretation of crustal and upper mantle structure. Tectonophysics 111:173–246

Müntener O, Manatschal G, Desmurs L, Pettke T (2010) Plagioclase peridotites in ocean–continent transitions: Refertilized mantle domains generated by melt stagnation in the shallow mantle lithosphere. J Petrol 51:255–294

Müntener O, Pettke T, Desmurs L, Meier M, Schaltegger U (2004) Refertilization of mantle peridotite in embryonic ocean basins: trace element and Nd-isotope evidence and implications for crust-mantle relationships. Earth Planet Sci Lett 221:293–308

Nicolas A, Boudier F, Montigny R (1987) Structure of Zabargad Island and early rifting of the Red Sea. J Geophys Res 92:461–474

Nielsen RL, Dungan MA (1983) Low pressure mineral-melt equilibria in natural anhydrous mafic systems. Contrib Mineral Petrol 84:310–326

Nimis P, Ulmer P (1998) Clinopyroxene geobarometry of magmatic rocks, part 1: an expanded structural geobarometer for anhydrous and hydrous, basic and ultrabasic systems. Contrib Mineral Petrol 133:314–327

Nimis P (1999) Clinopyroxene geobarometry of magmatic rocks: Part 2. Structural geobarometers for basic to acid, tholeiitic and mildly alkaline magmatic systems. Contrib Mineral Petrol 135:62–74

Niu Y, O'Hara MJ (2003) Origin of ocean island basalts: A new perspective from petrology, geochemistry, and mineral physics considerations. J Geophys Res 108:ECV5-1–ECV5-19

Otten MP (1984) The origin of brown hornblende in the Artfjället gabbro and dolerites. Contrib Mineral Petrol 86:189–199

Parker RL (1973) The rapid calculation of potential anomalies. Geophys J R Astr Soc 31:447–455

Parsons T, McCarthy J (1996) Crustal and upper mantle velocity structure of the Salton Trough, southeast California. Tectonics 15:456–471

Petrini R, Joron JL, Ottonello G, Bonatti E, Seyler M (1988) Basaltic dykes from Zabargad Island, Red Sea: petrology and geochemistry. Tectonophysics 150:229–248

Phelps D, Avé Lallemant HG (1980) The Sparta ophiolite complex, northeast Oregon: a plutonic equivalent to low K_2O island-arc volcanism. Am J Sci 280A:345–358

Prada M, Sallares V, Ranero CR, Vendrell MG, Grevemeyer I, Zitellini N, de Franco R (2014) Seismic structure of the Central Tyrrhenian basin: geophysical constraints on the nature of the main crustal domains. J Geophys Res 119:52–70

Prada M, Sallares V, Ranero CR, Vendrell MG, Grevemeyer I, Zitellini N, de Franco R (2015) The complex 3-D transition from continental crust to backarc magmatism and exhumed mantle in the Central Tyrrhenian basin. Geophys J Int 203:63–78

Reilinger R, McClusky S, ArRajehi A (2015) Geodetic constraints on the geodynamic evolution of the Red Sea. In: Rasul NMA, Stewart ICF (eds) The Red Sea: the formation, morphology, oceanography and environment of a Young Ocean Basin. Springer Earth System Sciences, Berlin, pp 121–135

Rychert CA, Hammond JOS, Harmon N, Kendall JM, Keir D, Ebinger C, Bastow ID, Ayele A, Belachew M, Stuart G (2012) Volcanism in the Afar Rift sustained by decompression melting with minimal plume influence. Nature Geosci 5:406–409

Roeder PL, Emslie RF (1970) Olivine-liquid equilibrium. Contrib Mineral Petrol 29:275–289

Rona PA (1978) Magnetic signatures of hydrothermal alteration and volcanogenic mineral deposits in oceanic crust. J Volcanology Geothermal Res 3:219–225

Ross K, Elthon D (1997). Cumulus and postcumulus crystallization in the oceanic crust: Major and trace element geochemistry of Leg 153 gabbroic rocks. In: Karson JA, Cannat M, Miller DJ, Elthon D (eds) Proceedings of the ocean drilling program, scientific results, vol. 153. Ocean Drilling Program, College Station, TX, pp 333–350

Sallarès V, Martínez-Loriente S, Prada M, Gràcia E, Ranero CR, Gutscher MA, Bartolome R, Gailler A, Dañobeitia JJ, Zitellini N (2013) Seismic evidence of exhumed mantle rock basement at the Gorringe Bank and the adjacent Horseshoe and Tagus abyssal plains (SW Iberia). Earth Planet Sci Lett 365:120–131

Sandwell DT, Muller RD, Smith WHF, Garcia E, Francis R (2014) New global marine gravity model from CryoSat-2 and Jason-1 reveals buried tectonic structure. Science 346:65–67

Schilling JG, Kingsley RH, Hanan BB, McCully BL (1992) Nd–Sr–Pb isotopic variations along the Gulf of Aden; evidence for Afar mantle plume–continental lithosphere interaction. J Geophys Res 97:10927–10966

Seifert KE, Chang C-W, Brunotte DA (1997) Evidence from Ocean Drilling Program Leg 149 mafic igneous rocks for ocean crust in the Iberia Abyssal Plain ocean–continent transition. J Geophys Res 102:7915–7928

Seifert KE, Gibson I, Weis D, Brunotte D (1996) Geochemistry of metamorphosed cumulate gabbros from Hole 900A, Iberia Abyssal Plain. Proc Ocean Drill Program Sci Results 149:471–485

Seyler M, Bonatti E (1988) Petrology of the gneiss/amphibolite metamorphic unit from Zabargad Island, Red Sea. Tectonophysics 150:177–207

Shukri NM (1944) On the geology of the Brothers Islets–northern Red Sea. Bull Fac Sci Cairo Univ 25:175–196

Stakes D, Mével C, Cannat M, Chaput T (1991) Metamorphic stratigraphy of Hole 735B. In: Von Herzen RP, Robinson PT (eds) Proceedings of the ocean drilling program, scientific results, vol. 118. Ocean Drilling Program, College Station, TX, pp 153–180

Stern RJ, Gottfried D, Hedge CE (1984) Late Precambrian rifting and crustal evolution in the Northeastern Desert of Egypt. Geology 12:168–172

Stern RJ, Hedge CE (1985) Geochronologic and isotopic constraints on Late Precambrian crustal evolution in the Eastern Desert of Egypt. Am J Sci 285:97–127

Stern RJ, Johnson P (2010) Continental lithosphere of the Arabian Plate: a geologic, petrologic, and geophysical synthesis. Earth Sci Rev 101:29–67

Stern RJ, Johnson P (2018) Constraining the opening of the Red Sea: evidence from the Neoproterozoic margins and Cenozoic magmatism for a Volcanic Rifted Margin. In: Rasul NMA, Stewart ICF (eds) The Red Sea. Springer Earth System Sciences, Berlin (this issue)

Stockly D, Bosworth W (2018) In: Rasul NMA, Stewart ICF (eds) The Red Sea. Springer Earth System Sciences, Heidelberg (this issue)

Sultan M, Becker R, Arvidson RE, Sore P, Stern RJ, El-Alfy Z, Guinnes EA (1992) Nature of the Red Sea crust, a controversy revisited. Geology 20:593–596

Tang Z, Julià J, Zahran H, Mai PM (2016) The lithospheric shear-wave velocity structure of Saudi Arabia: young volcanism in an old shield. Tectonophysics 680:8–27

Tapponnier P, Dyment J, Zinger MA, Franken D, Afifi AM, Wyllie A, Ali HG, Hanbal I (2013) Revisiting seafloor-spreading in the Red Sea: basement nature, transforms and ocean-continent boundary. In: AGU Fall Meeting 2013, San Francisco, T12B-04

Taviani M, Bonatti E, Colantoni P, Rossi PL (1986) Tectonically uplifted crustal blocks in the northern Red Sea: data from the Brothers Islets. Mem Soc Geol It 27:47–50

Taviani M, Rabbi E (1984) Marine botryoidal aragonite in Pleistocene reef limestones of Red Sea offshore islands (Northern Brother and Rocky Island). Miner Petrogr Acta 28:49–58

Thybo H, Nielsen CA (2009) Magma-compensated crustal thinning in continental rift zones. Nature 457:873–876

Tiezzi LJ, Scott RB (1980) Crystal fractionation in a cumulate gabbro, Mid-Atlantic Ridge, 26°N. J Geophys Res 85:5438–5454

Toplis MJ, Carroll MR (1995) An experimental study of the influence of oxygen fugacity on Fe-Ti oxide stability, phase relations, and mineral-melt equilibria in ferro-basaltic systems. J Petrol 36:1137–1170

Tramontini C, Davis D (1969) A seismic refraction survey in the Red Sea. Geophys J R Astr Soc 17:225–241

Tribuzio R, Tiepolo M, Vannucci R, Bottazzi P (1999) Trace element distribution within olivine-bearing gabbros from the northern Apennine ophiolites (Italy): Evidence for post-cumulus crystallization in MOR-type gabbroic rocks. Contrib Mineral Petrol 134:123–133

Tziavos IN, Sideris MG, Forsberg R (1998) Combined satellite altimetry and shipborne gravimetry data processing. Mar Geodesy 21:299–317

Urquhart A, Bauer S (2015) Experimental determination of single-crystal halite thermal conductivity, diffusivity and specific heat from −75 to 300 °C. Int J Rock Mech Min Sci 78:350–352

Villiger S, Ulmer P, Müntener O, Thompson AB (2004) The liquid line of descent of anhydrous, mantle-derived, tholeiitic liquids by fractional and equilibrium crystallization—An experimental study at 1.0 GPa. J Petrol 45:2369–2388

Villiger S, Ulmer P, Müntener O (2007) Equilibrium and fractional crystallization experiments at 0.7 GPa. The effect of pressure on phase relations and liquid compositions of tholeiitic magmas. J Petrol 48:159–184

Voggenreiter W, Hötzl H (1989) Kinematic evolution of the southwestern Arabian continental margin; implications for the origin of the Red Sea. J Afr Earth Sci 8:541–564

Voggenreiter W, Hötzl H, Mechie J (1988) Low-angle detachment origin for the Red Sea Rift System? Tectonophysics 150:51–75

Volker F, McCulloch MT (1993) Submarine basalts from the Red Sea: New Pb, Sr, and Nd isotopic data. Geophys Res Lett 20:927–930

Walker D, Shibata T, Delong SE (1979) Abyssal tholeiites from the Oceanographer fracture zone. II, Phase equilibria and mixing. Contrib Mineral Petrol 70:111–125

Weaver JS, Langmuir CH (1990) Calculation of phase equilibrium in mineral-melt systems. Comput Geosci 16:1–19

White RS, McKenzie D, O'Nions RK (1992) Oceanic crustal thickness from seismic measurements and rare earth element inversions. J Geophys Res 97:19683–19715

Wolfenden E, Ebinger C, Yirgu G, Renne PR, Kelley SP (2005) Evolution of a volcanic rifted margin: Southern Red Sea, Ethiopia. Bull Geol Soc Am 117:846–864

Wright T, Sigmundsson F, Ayele A, Belachew M, Brandsdottir B, Calais E, Ebinger C, Einarsson P, Hamling I, Keir D, Lewi E, Pagli C, Pedersen R (2012) Geophysical constraints on the dynamics of spreading centres from rifting episodes on land. Nature Geosci 5:242–249

Rifting and Salt Deposition on Continental Margins: Differences and Similarities Between the Red Sea and the South Atlantic Sedimentary Basins

Webster Mohriak

Abstract

The results of regional deep seismic acquisition in the South Atlantic continental margins provide new constraints on the birth and development of sedimentary basins formed during the Gondwana breakup. The interpretation of these seismic profiles integrated with gravity and magnetic potential field data suggest alternative models for the birth of oceanic basins that evolve from an earlier phase of intracontinental rift, salt deposition and continental breakup by mantle exhumation or by development of oceanic spreading centres preceded by igneous intrusions and extrusions in the transition from continental to oceanic crust. The analysis of regional deep-penetrating seismic profiles in the South Atlantic and Red Sea, integrated with potential field methods and plate reconstructions, provides a template for the interpretation of the tectono-sedimentary features that are characterized from the proximal rifts onshore and in the platform. Basinward, more elusive features are characterized toward the transitional and oceanic crust in divergent margins. This work discusses alternative interpretations for syn-rift successions and salt distribution in regional seismic profiles from the Red Sea, which have been integrated with results of wells that penetrated the stratigraphic section below the evaporites in a few exploratory wells along the Arabian and African conjugate margins. These interpretations can be compared with similar tectono-stratigraphic settings in the South Atlantic, which are constrained by several exploratory wells that penetrated the syn-rift sequence in both shallow and deep waters. The temporal development of syn-rift structures, magmatism, salt deposition, oceanic propagators and development of the divergent margins suggest that the Red Sea constitutes a better analogue for the development of the South Atlantic divergent continental margins than the Iberian margin.

W. Mohriak (✉)
UERJ—State University of Rio de Janeiro, Rio de Janeiro, Brazil
e-mail: webmohr@gmail.com

N. M. A. Rasul and I. C. F. Stewart (eds.), *Geological Setting, Palaeoenvironment and Archaeology of the Red Sea*,
https://doi.org/10.1007/978-3-319-99408-6_8

1 Introduction

The Red Sea is a natural laboratory for studying continental breakup processes at an embryonic stage, whereas the Central and South Atlantic Oceans correspond to more developed divergent margins with tectonic plates in advanced drift stages. Classical works on plate tectonic concepts (as for example, in the pioneering publications by Wegener (1912) and Wilson (1966)) have pointed out the Red Sea as a present-day paradigm for the early stages of development of continental margins. The Red Sea has also been used by geoscientists in basin analysis studies as a modern analog for sedimentary environments, structural geology, and salt tectonics, particularly in comparisons of rift development, magmatism and halokinesis (see for example, Purser and Bosence 1988; Mohriak and Leroy 2013; Bosworth 2015; Rasul et al. 2015).

Several exploratory plays in the South Atlantic have analogs in sedimentary basins across conjugate margins (Brazil—West Africa). These pre-salt and post-salt reservoirs can be analyzed in an earlier stage of development in the Red Sea, which offers a perspective of an early post-rift stage in some areas where breakup has occurred (Mohriak 2014). Risk analysis of petroleum exploration plays in the deep-water setting of the South Atlantic includes the assessment of rift architecture and distribution of source rocks and reservoirs, as well as salt tectonics, migration pathways and trap development. The Red Sea and Gulf of Aden provide important constraints on tectonic models for divergent continental margins and their petroleum systems in the early phases of plate divergence.

This work will briefly summarize some petroleum exploration highlights from the South Atlantic's largest oil fields, and thoroughly discuss the tectonic analogies between the South Atlantic, the North Atlantic and the Red Sea, aiming to provide clues on how divergent margins form and develop through time. Based on a regional integration of the central Red Sea potential field (gravity and magnetic) and seismic datasets, fundamental aspects of rifting, evaporite deposition, continental breakup and early stages of margin divergence are analyzed. The interpretation is based on the structural and stratigraphic analysis of the regional geoseismic profiles extending across the proximal and distal rift basin near the Mabahiss Deep in the north-central Red Sea, and by discussing seismic profiles reaching the main trough in the Nereus Deep or crossing the axial trough spreading centre in the Thetis Deep. Seismic interpretation of selected structures along transects in this segment of the Red Sea, from the platform toward the embryonic oceanic crust, provides a template to compare with the South Atlantic and other rifted continental margins.

2 Petroleum Exploration and Giant Oil Fields in the South Atlantic

The South Atlantic sedimentary basins that occur from onshore to offshore along the continental margins of Brazil, Uruguay, and Argentina (Fig. 1) are characterized by Mesozoic rifts associated with the Gondwana breakup and

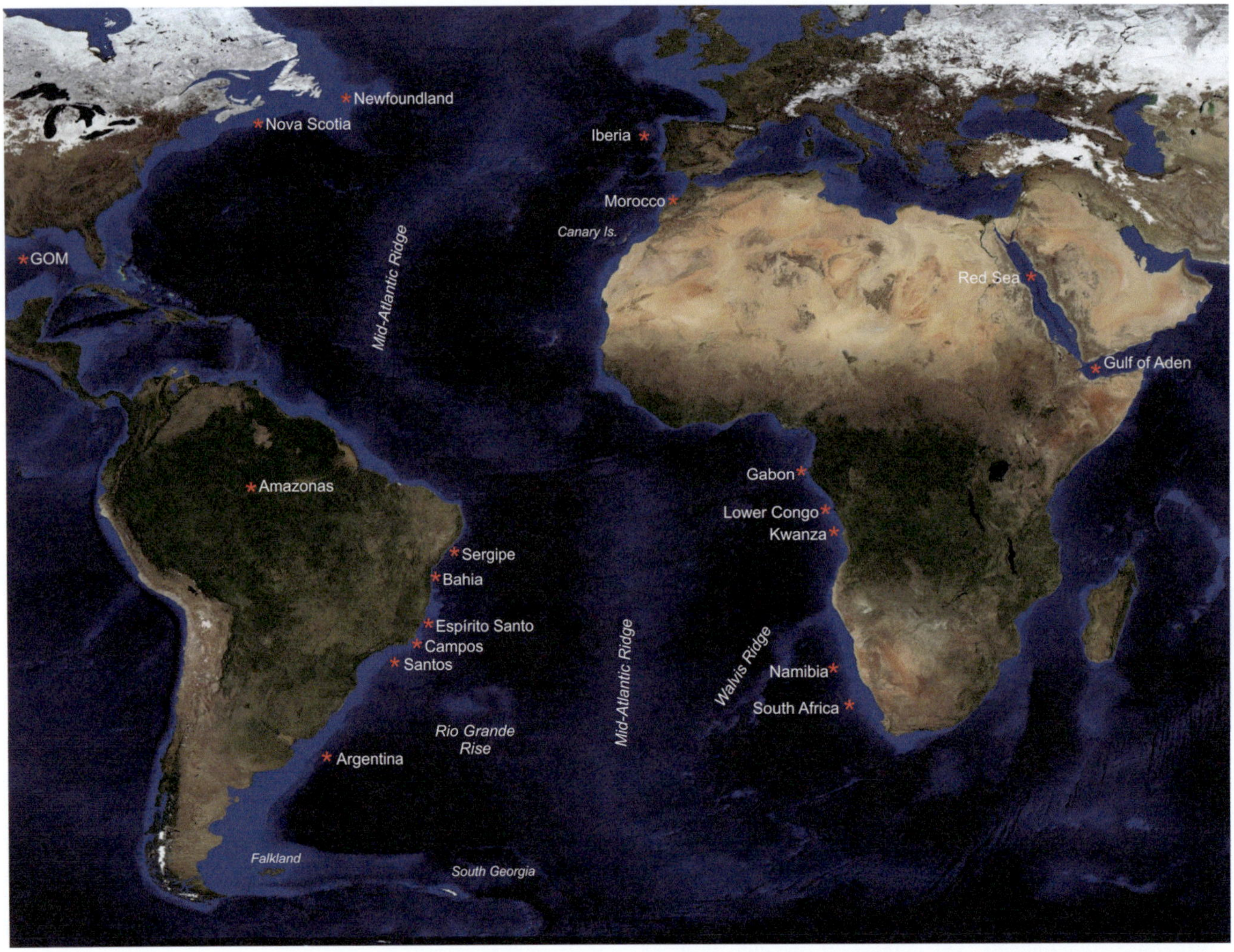

Fig. 1 World satellite map showing archetypal divergent margin sedimentary basins in the Atlantic Ocean, associated with an active Mid-Atlantic Ridge. Newfoundland and Iberia are conjugate margins characterized by magma-poor sedimentary basins with mantle exhumation during continental breakup. In the Brazilian margin, the Santos, Campos and Espírito Santo basins are prolific hydrocarbon provinces with rich synrift lacustrine source rocks overlain by thick salt layers. The conjugate margin basins in West Africa (Kwanza, Lower Congo and Gabon) are also characterized by major salt structures developed after deposition of the evaporites in the Late Aptian. The Argentina and Namibia/South Africa basins are volcanic margins where salt is absent in the transitional sequence, which is characterized by wedges of seaward-dipping reflectors. The Red Sea and the Gulf of Aden are divergent margins in the early stages of plate tectonics. The Middle to Late Miocene salt layer in the Red Sea extends from the Gulf of Suez toward the triple junction in the Afar region, where the Red Sea rift joins the East African and the Gulf of Aden rifts

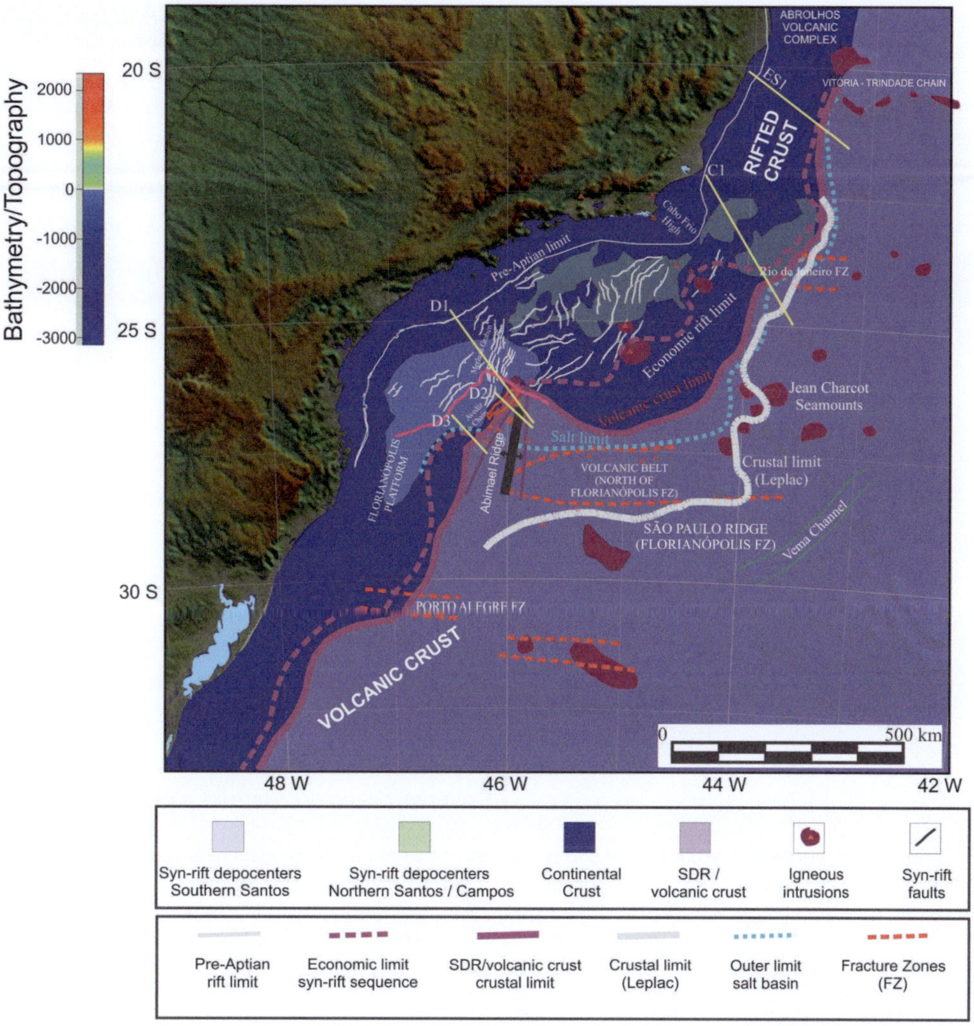

Modified from Mohriak, 2001.

Fig. 2 Simplified tectonic map of the Pelotas, Santos, Campos and Espírito Santo basins offshore SE Brazil. This map shows a comparison of different interpretations of the continent-ocean boundary (COB) based on analysis of potential field data. The Leplac limit by Gomes (1992) and the volcanic crust limit by Mohriak (2001), are similar in the Campos and Espírito Basins, but very different in the southern Santos Basin. The distribution of syn-rift depocentres is based on seismic interpretation and exploratory well control. Basinward of the dashed purple line (economic rift limit) the syn-rift succession is not clearly visible in the seismic profiles along the Campos and Santos Basins, suggesting a volcanic basement below the salt layer. In the southern Santos Basin, the thick salt diapir province extends basinward of the economic rift limit, reaching the possible oceanic crust that is indicated by the solid purple line (volcanic crust limit). Observe that the Abimael Ridge (triangular shaped feature westward of the volcanic belt located north of the Florianópolis Fracture Zone) propagated into the salt basin of the southern Santos Basin. The location of the regional profiles ES1 (Espírito Santo Basin), C1 (Campos Basin) and D1, D2 and D3 (Santos Basin) are indicated by yellow lines

development of divergent margins that are conjugate to the West African basins offshore South Africa, Namibia, Angola, Gabon and Rio Muni. The Santos, Campos and Espírito Santo basins (Fig. 2) are the most prolific basins of offshore Brazil, with reservoirs ranging in age from Neogene (Miocene) to Early Cretaceous (Rangel and Martins 1998; Davison 1999; Beglinger et al. 2012). These basins are characterized by a thick Late Aptian salt layer that is also observed in the conjugate margin basins of offshore Angola and Gabon (Davison 2007; Mohriak and Fainstein 2012). Regional deep seismic profiles extending from the continental to oceanic crust along the SE Brazilian margin (Fig. 2) will be presented and discussed in the analysis of the rift architecture and salt tectonics.

Petroleum exploration in the South Atlantic has been dominated in the early years of the 21st century by giant discoveries in the deep-water region of the Brazilian and West African continental margins. The largest oil fields discovered in the world's southern hemisphere in the past 15 years (2000–2015) are characterized by carbonate rocks as the main reservoir, with thick evaporite layers forming the seal for the traps (Mohriak 2015). The Tupi discovery in the

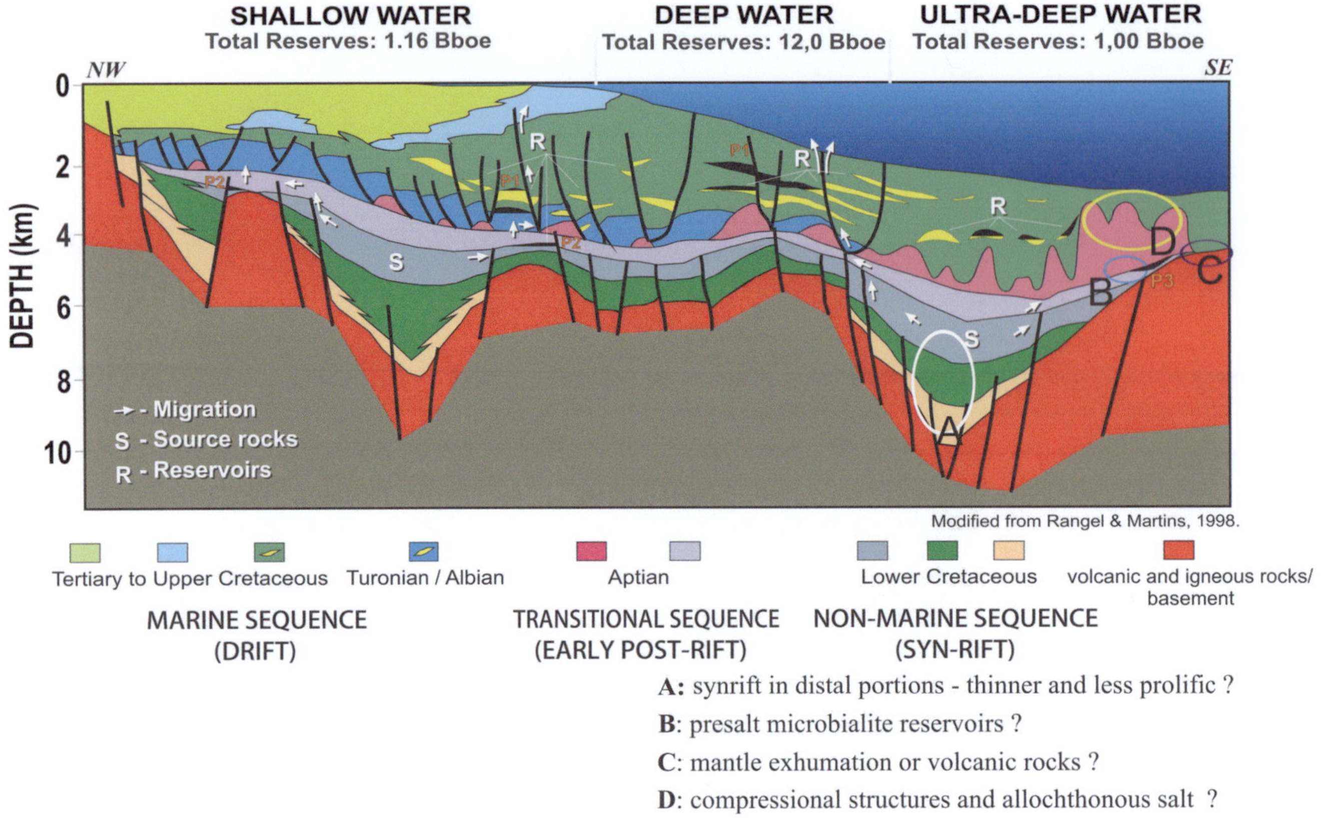

Fig. 3 Schematic (but insightful) geological section of the Campos Basin, from the platform toward the oceanic crust. Many changes in geological interpretation and petroleum assessment have been observed in the past decade. New models have been proposed by applying modern geophysical acquisition and processing of long cable regional 2D deep seismic data and 3D surveys covering large areas in the deep-water salt wall province. Interpretations challenging old paradigms in the ultra-deep-water province are shown by ellipses and marked as A, B, C and D. Conceptual geodynamic models still remain to be tested in the transition from continental to oceanic crust. P1, P2 and P3 (orange coloured letters) are the main exploratory plays and petroleum systems in the Campos Basin. P1: post-salt carbonate and siliciclastic (turbidite) reservoirs; P2: pre-salt source rocks and reservoirs (coquinas); P3: pre-salt carbonates (microbialites and coquinas) constitute a deep-water new play in the distal margin. Total petroleum reserves are now estimated to exceed 50 Bboe with the discovery of the presalt carbonate play in the ultradeep-water region

Santos Basin, which heralded a new era of Brazilian offshore oil production, is characterized by microbialites sealed by Late Aptian evaporites (Berman 2008; Carminatti et al. 2009; Mohriak et al. 2012). Salt deposition along the incipient continental margins is thus considered an essential element for this type of exploratory play.

This play type is very similar to the Kashagan Field, which was discovered in 2000 in shallow waters of the North Caspian Sea, and corresponds to the largest discovery worldwide in 21st century (Mohriak et al. 2012). In addition, the largest hydrocarbon discoveries in the Eastern Mediterranean Sea, including the Zohr gas field discovered in 2015, are also associated with carbonate build-ups sealed by evaporites (Skiple et al. 2012; Esestime et al. 2016).

The interpretation of a schematic geological section in the Campos Basin (Fig. 3), one of the most prolific basins in the South Atlantic (Guardado et al. 1989; Mohriak et al. 1990b; Rangel and Martins 1998) shows three main tectono-stratigraphic sequences: The syn-rift (non-marine, Lower Cretaceous), transitional (evaporitic, Aptian) and the drift (marine, Albian to Tertiary) sequences (Winter et al. 2007). The petroleum systems in the basin are associated with the Barremian syn-rift lacustrine source rocks, which occur below the Late Aptian evaporites (Mello et al. 1994; Beglinger et al. 2012). The giant oil fields in deep waters are associated with siliciclastic and carbonate reservoirs in the post- and pre-salt stratigraphic successions respectively (Fig. 3). In the mid- to late-1970's, the first hydrocarbon accumulations in the Campos Basin were found in the post-salt Albian carbonate rocks and Late Tertiary to Late Cretaceous turbidite sandstone reservoirs (Play P1 in Fig. 3), and also in the pre-salt lacustrine carbonate reservoirs (coquinas) in the proximal basin (Play P2). More recently, giant accumulations have been discovered in the distal part of the basin, where new plays are characterized by carbonate rocks (microbialites and coquinas) as the main reservoir rocks with stratified evaporites forming the top seal (Play P3).

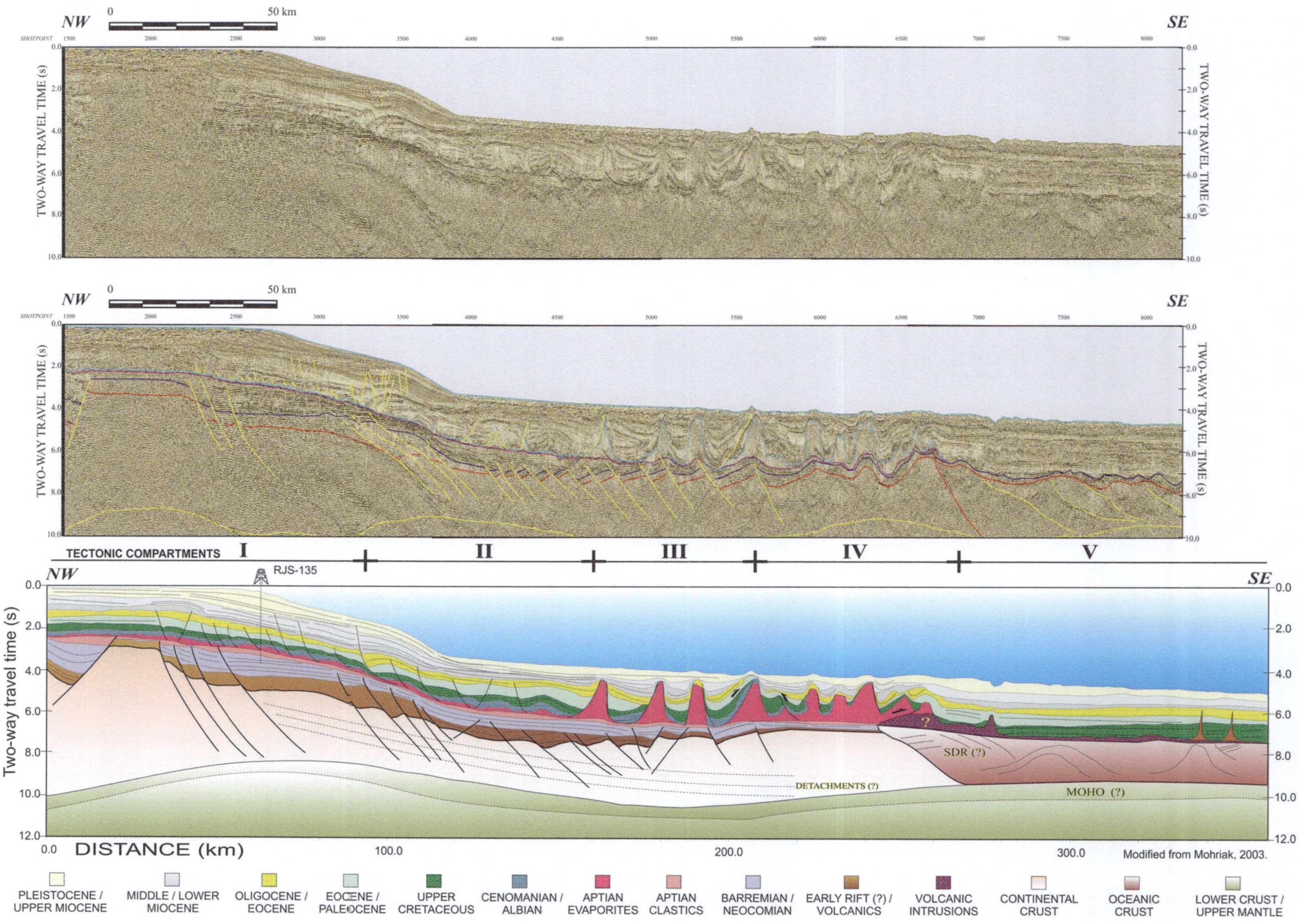

Fig. 4 Regional deep seismic profile in the Campos Basin (profile C1 in Fig. 6) with uninterpreted seismic (top), a simplified seismic interpretation (middle), and a schematic geoseismic interpretation (bottom) based on integration with potential field data and other regional seismic profiles. The salt tectonics compartments (Domains I–IV) are illustrated in the geoseismic profile, from the platform toward the oceanic crust. The shallow water extensional domain (I) is characterized by small listric faults resulting in the collapse of the Albian carbonate platform; the deep-water extensional domain is marked by large roll-overs (II), the compressional domain is associated with tall salt diapirs (III), the strongly compressional domain is marked by allochthonous salt tongues advancing above transitional crust basement or volcanic highs (IV); and the oceanic crust domain is characterized by lack of salt structures (V). Assuming this interpretation, the syn-rift sediments pinch-out toward the salt wall province. The transition to oceanic crust is characterized by volcanic intrusions, magmatic accretion and possibly by wedges of seaward-dipping reflectors

Table 1 Interpretation concepts in the Campos Basin

Previous interpretation	Alternative interpretation	Observations
Thickest syn-rift troughs occur over the deep-water province[a,b]	Syn-rift sediments show a reduced thickness below the massive salt walls in deep waters[e,f]	Figure 3 illustrates the interpretation of a thick syn-rift trough in the ultradeep-water region; Fig. 4 illustrates the interpretation of a reduced syn-rift thickness in the ultradeep-water province, underlying the salt walls
Continent-ocean boundary is related to pre-rift basement structural highs (with pinch-out of syn-rift sediments)[a,b]	Outermost highs interpreted as post-rift structures[e,f]	New seismic data indicate that the outermost highs were formed after the syn-rift deposits and are overlain by allochthonous salt structures; locally SDR wedges and volcanic highs can be observed in the transition from continental to oceanic crust
Autochthonous salt walls and diapirs occur over the continent-ocean boundary[a,b,c]	Allochthonous salt tongues occur in the continent-ocean boundary[d,e,f]	Allochthonous salt is clearly indicated by modern 3D seismic acquisitions; Tertiary reservoirs below Aptian salt have been drilled both in Brazil and West Africa
Minor amounts of hydrocarbons occur over the transitional crust[a,b,c]	Giant fields have been identified in the ultradeep-water province associated with pre-salt carbonate rocks[g,h,i]	Figure 3 showed an estimated 1 Bbbl of oil in ultradeep-water settings in the late 1990's; nowadays the pre-salt discoveries in microbialite reservoirs may exceed 50 Bbbl of oil according to mid-2010's estimates

Key references
[a]Guardado et al. (1989), [b]Rangel and Martins (1998), [c]Guardado et al. (2000)
[d]Demercian et al. (1993), [e]Mohriak et al. (2008), [f]Mohriak et al. (2012)
[g]Berman (2008); [h]Carminatti et al. (2009); [i]Jones and Chaves (2015)

The tectonic evolution of the Campos Basin is associated with lithospheric extension and crustal thinning that resulted in development of an Early Cretaceous rift phase with lacustrine sedimentation (Guardado et al. 1989; Mohriak et al. 1990a; Winter et al. 2007; Karner 2000). This is followed by a late syn-rift to early drift phase of thermal subsidence that resulted in the deposition of Late Aptian evaporites and marine Early Albian carbonates. A regional deep seismic profile in the central part of the Campos Basin (Fig. 4, modified from Cainelli and Mohriak (1998) and Mohriak (2003)) shows the interpretation of the crustal architecture and the distribution of the rift depocentres. The geoseismic section interpretation (Fig. 4-bottom) illustrates the interpretation of the Moho uplift in the platform and a thinned continental crust at the transition to oceanic basement at the distal end of the massive salt diapir province. This interpretation assumes a pure shear mechanism for lithospheric extension and a transition to oceanic crust characterized by increasing magmatic activity (Mohriak et al. 1990a, 2008; Meisling et al. 2001; Mohriak 2003; Blaich et al. 2011).

With the advance of petroleum exploration from deep to ultradeep waters, several important changes in the geological interpretation of the Brazilian divergent margin basins have been proposed in recent years by different authors. These interpretations are based on integration of potential field (gravity and magnetics) and modern seismic acquisition, including 2D deep seismic profiles extending from the platform to the oceanic crust, and also huge 3D acquisition in large exploratory areas, allowing a better imaging of syn-rift structures and salt tectonics. For example, alternative interpretations that have changed or challenged previous paradigms for the Campos Basin are implicit in the geoseismic profiles presented in Figs. 3 and 4. Table 1 presents some of these concepts and also includes a few key-references for the alternative interpretations:

In the late 1990's and early 2000's a large number of tectonic models for continental rifting and breakup forming divergent margins have increasingly assimilated the results from the Ocean Drilling Program (ODP), one of the most significant international scientific endeavours ever undertaken in the study of ocean basins. In the North Atlantic, seminal projects sampled peridotites from many drilling sites of offshore Iberia (Shipboard Scientific Party 1985, 1998). These models assume that mantle exhumation precedes the inception of oceanic crust in magma-poor margins, as indicated by the presence of peridotite ridges near the continental-oceanic crust boundary (Boillot et al. 1980, 1987; Whitmarsh and Wallace 1998; Manatschal 2004; Péron-Pinvidic et al. 2013).

Some authors (e.g., Unternehr et al. 2010; Zalán et al. 2011) have proposed a direct application of the Iberia mantle exhumation model for the South Atlantic, using deep seismic profiles integrated with potential field data (e.g., Figure 5). Based on these interpretations, the salt diapir province in the deep-water region of the Campos Basin overlies sag basin sediments deposited directly on exhumed mantle rocks, with a top-basement detachment fault controlling a block of continental crust that separates the proximal from the distal segments of the margin (Lavier and Manatschal 2006). However, the extensive application of the Iberian model for the South Atlantic is highly controversial, given the striking differences of tectono-magmatic evolution between these divergent margins, and the general lack of seismic evidence for a Moho reflector rising up to intersect the base Aptian salt in the Campos and Kwanza basins (e.g., Mohriak et al. 2008; Lentini et al. 2010; Mohriak and Leroy 2013).

This work presents an overview of the rift architecture of the Red Sea sedimentary basins, from the onshore area toward the offshore distal margin and oceanic crust, comparing their characteristics with the South Atlantic and Central Atlantic margins, where regional deep seismic profiles have been available since the 1980's (Mohriak et al. 1990a, 2008). The line interpretations of regional seismic profiles in the Red Sea are included to show the geological and geophysical characteristics of the syn-rift succession and salt deposition in the onshore and offshore segments of the continental margin. The incipient spreading centre in the central Red Sea is also discussed using seismic and potential field data. This aims at providing clues on how the continents are split during the rift phase, how salt basins are formed at the conjugate margins, and the relative ages of rifting, salt deposition, magmatism and continental breakup. Understanding the differences and similarities between the South Atlantic and the Red Sea sedimentary basins and their tectonic controls has important implications for tectonic and exploratory models, particularly in the evaluation of active petroleum systems in the early stages of development of rifted continental margins.

3 Tectonic Models for the Formation of the Atlantic Continental Margins

The tectonic models for the development of divergent margins are most often based on industry seismic data acquisition and potential field data integrated with results of exploratory wells. A recent review of these geological and geophysical interpretations for the Central and South Atlantic suggests that these continental margins might be compared to the Iberian margin, where a predominantly simple shear model of lithospheric extension associated with detachment faults resulted in mantle exhumation (Perón-Pivindic et al. 2013). Regional 2D and 3D seismic data acquired in the Iberian margin have been used to investigate fundamental basin-forming processes, and several Ocean Drilling Program (ODP) scientific wells have confirmed that serpentinized peridotites outcrop along ridges near the continent-ocean boundary (Perón-Pinvidic and Manatschal 2009).

Based on results of ODP scientific wells, numerous authors have proposed a tectonic model for the Atlantic continental margins (such as Newfoundland and Iberia) involving mantle exhumation associated with detachment faults before the emplacement of oceanic crust (Boillot et al. 1980;

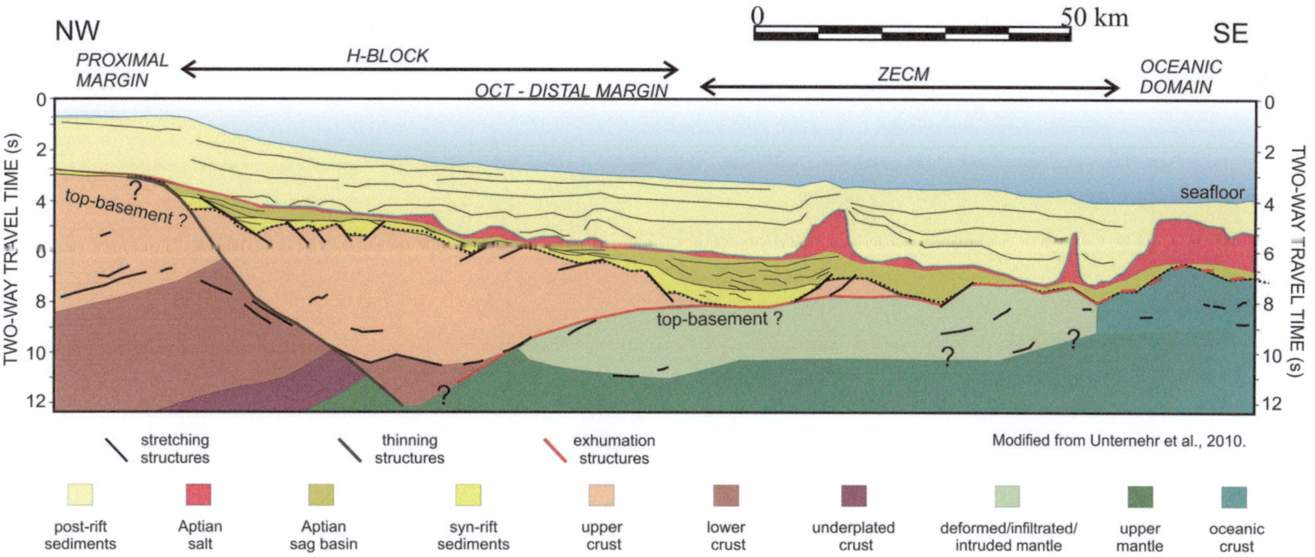

Fig. 5 Regional seismic profile in the Campos Basin assuming the exhumed mantle model by Unternehr et al. (2010). The proximal margin is characterized by reduced thickness of the syn-rift sediments and a residual salt layer overlies the syn-rift and sag basin sediments in the platform and deep waters. An H-Block is interpreted between the shelf-break and the deep-water basin, with increasing thickness of the Aptian sag basin sediments. The salt diapir province overlies the sag basin in the distal margin (where the continental crust is abruptly thinned by a top-basement detachment fault), whereas the syn-rift sediments pinch-out basinward and are not present in the zone of extended continental mantle (ZECM). The compressional salt wall province is interpreted to overlie the ZECM with Late Aptian sediments and evaporites deposited directly on the upper mantle peridotites

Perón-Pinvidic and Manatschal 2009). Reconstructions of regional deep seismic profiles in magma-poor basins along the conjugate margins of Canada and Iberia indicate that lithospheric stretching is associated with crustal thinning and mantle exhumation by detachment faults in the deep-water region before emplacement of oceanic crust (Pérez-Gussinyé 2012).

The Iberian model of mantle exhumation involving simple shear detachment faults has recently been extensively applied to the interpretation of several basins in the Eastern Brazilian and West African conjugate margins. Some authors have re-interpreted the Eastern Brazilian and West African margins as magma-poor (Unternehr et al. 2010; Zalán et al. 2011; Péron-Pinvidic et al. 2013), contradicting the results of the exploration wells drilled in the deep-water region of these prolific basins, which indicate an abundance of volcanic rocks underlying and within the syn-rift sequences (Mohriak et al. 2008). A recent work by Alvarenga et al. (2016) has convincingly interpreted the presence of hydrothermal vents in the deep-water region of the Campos Basin. These features, imaged within the Early Cretaceous rift sequence, commonly occur in volcanically active areas worldwide, and have been characterized in present-day settings such as the East African Rift System as well as in ancient rift basins, as for example the Vøring Basin, offshore Norway (Planke et al. 2005).

A direct, but questionable, application of the Iberian model for the South Atlantic margin (Fig. 6) assumes that the sedimentary basins in southeast Brazil and West Africa are magma-poor margins where the outermost high at the eastern limit of the salt diapir province corresponds to serpentinized peridotite rocks (Zalán et al. 2011). In this model, the exhumed mantle in the outer high has been divided into

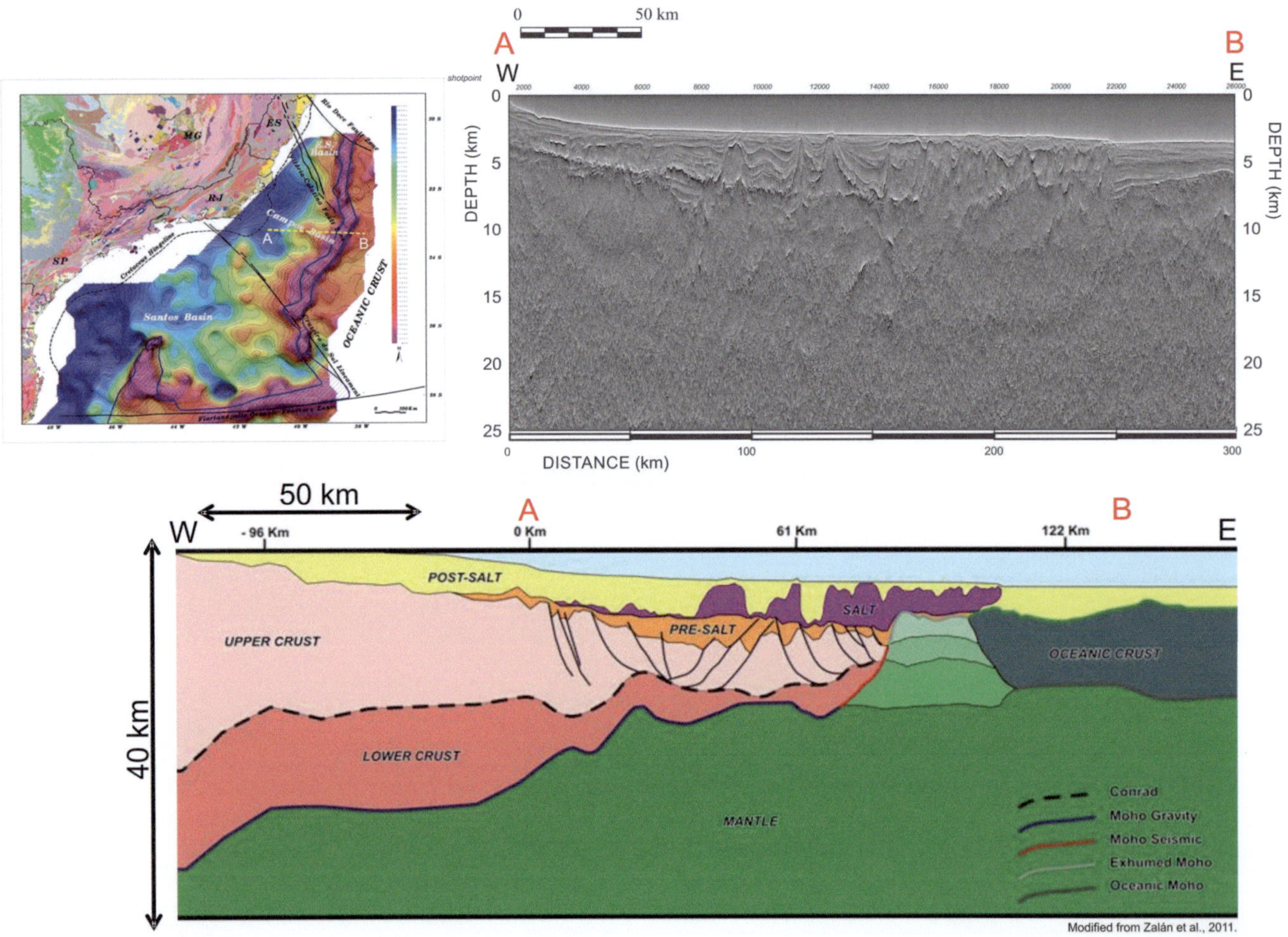

Fig. 6 Top left: Structural map showing the interpreted top of the Moho Discontinuity in depth (blue colours: structural lows; red to purple: structural highs), with a blue line outlining the regions with interpretation of peridotite ridges or exhumed mantle (modified from Zalán et al. 2011). A regional deep seismic profile extending along a W-E direction (top right, line A-B on the map) shows a reduced amount of salt in the Cabo Frio Platform and massive salt diapirs in the ultradeep-water region, where an allochthonous salt tongue is imaged advancing toward the oceanic crust. The gravity model of a regional profile based on the interpretation of the seismic profile A-B (bottom) shows the exhumed mantle rocks between the continental and oceanic crusts. According to this interpretation, serpentinized peridotites occur below the salt tongue that has advanced basinward of a conspicuous structural high that previously formed a barrier for salt deposition

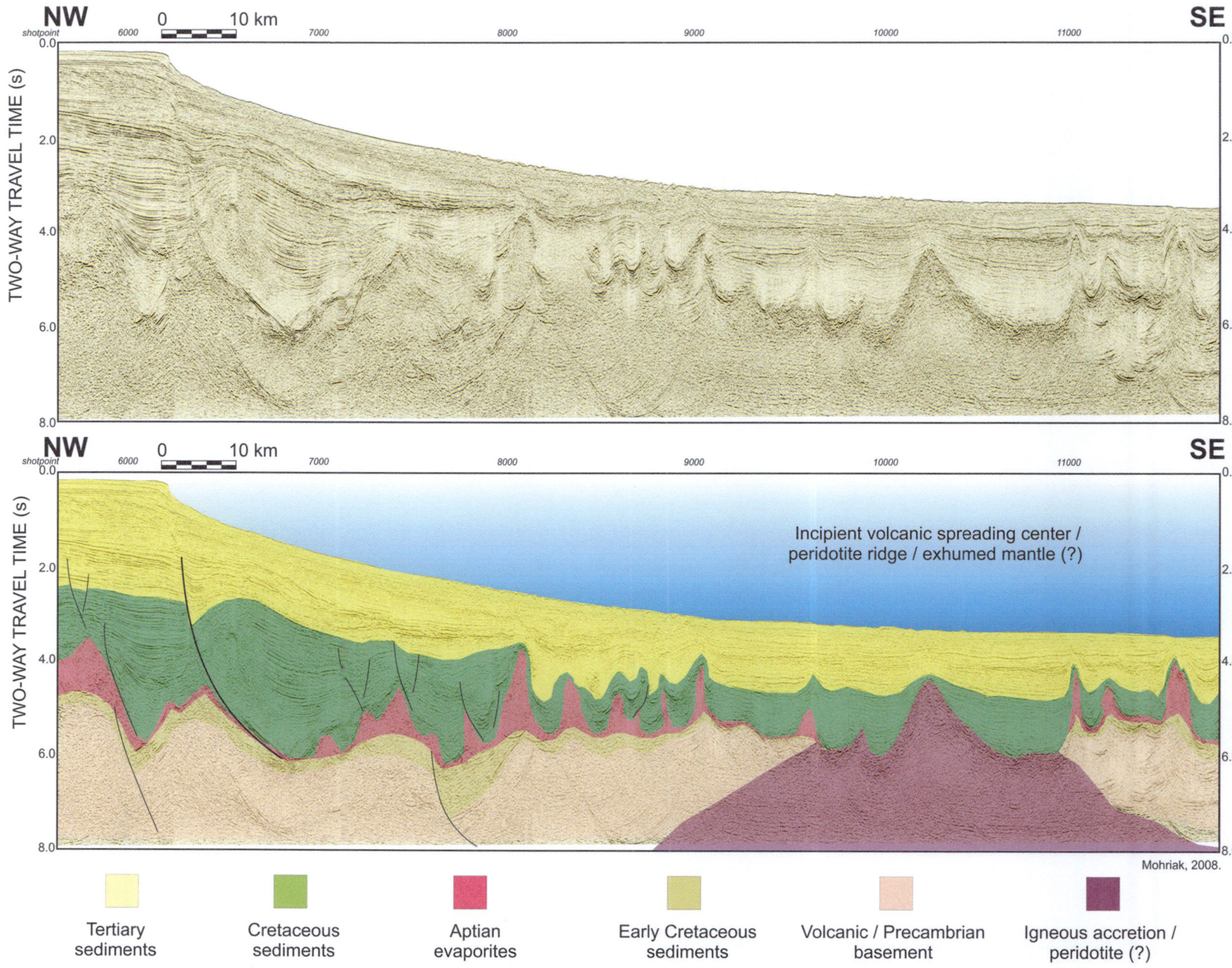

Fig. 7 Seismic Profile D1 (uninterpreted on top, interpretation below), from the platform toward the deep-water region of the Santos Basin. It shows two salt basins split by a triangular-shaped structure characterized by a strong positive gravity anomaly, which is associated with seaward-dipping reflectors toward the Pelotas Basin. This feature probably corresponds to an igneous intrusion associated with an oceanic propagator that split the main salt basin in the southern Santos Basin and was aborted after Late Aptian times (Mohriak 2001). The remnant distal salt basin located at the SE extremity of the profile is limited by a volcanic belt north of the Florianópolis Fracture Zone

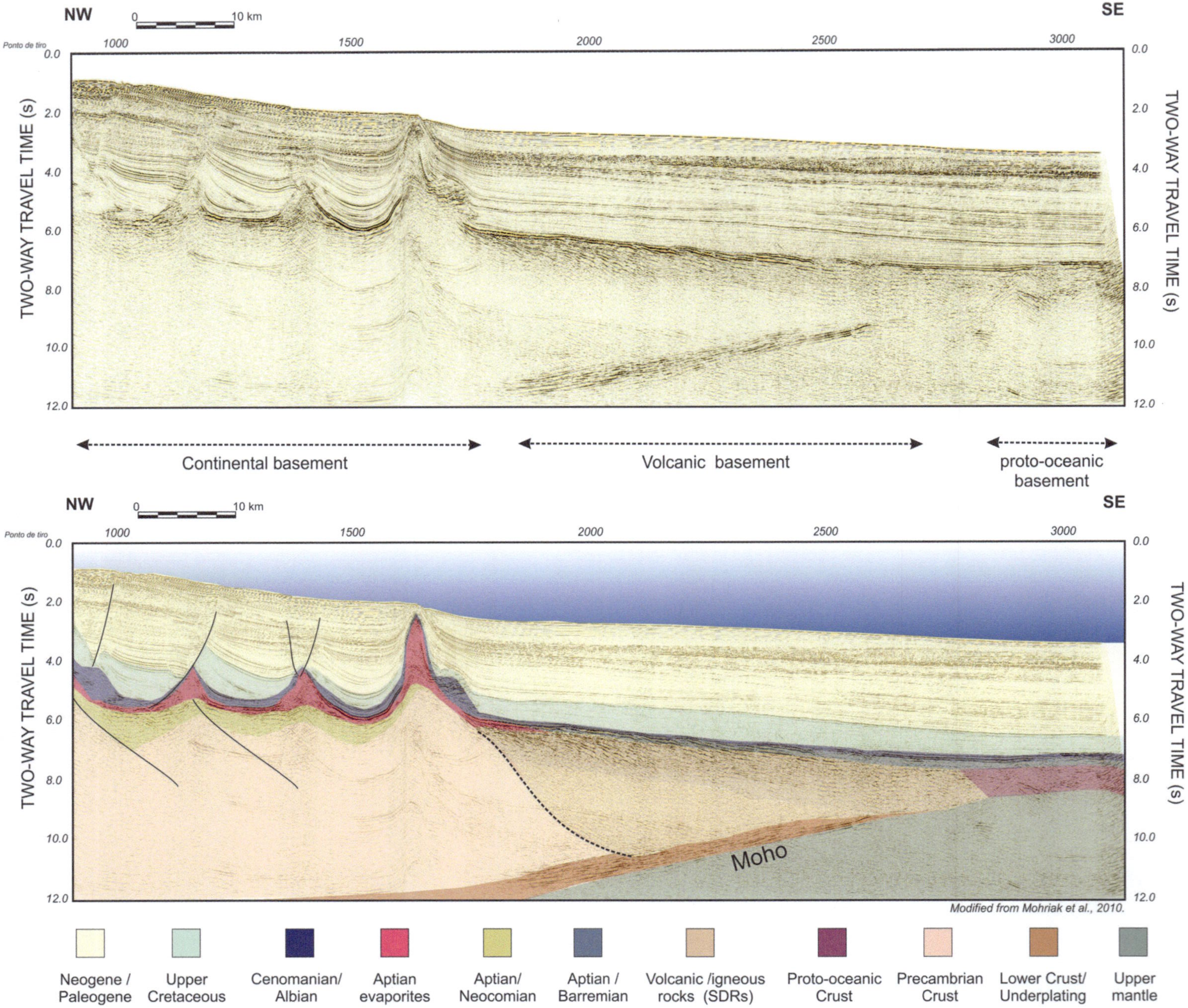

Fig. 8 Seismic Profile D3 (uninterpreted on top, interpretation below), showing the southern limit of the salt basin limited by a volcanic basement characterized by seaward-dipping reflectors that are not as well developed as in the Pelotas Basin. Basinward of the salt basin there is a strong reflector package at the base of the crust which abruptly rises from 10 to 9 s TWT. The proto-oceanic crust of the Abimael propagator is characterized by a very shallow Moho discontinuity between 9 and 8 s TWT, but with no clear indication of mantle exhumation

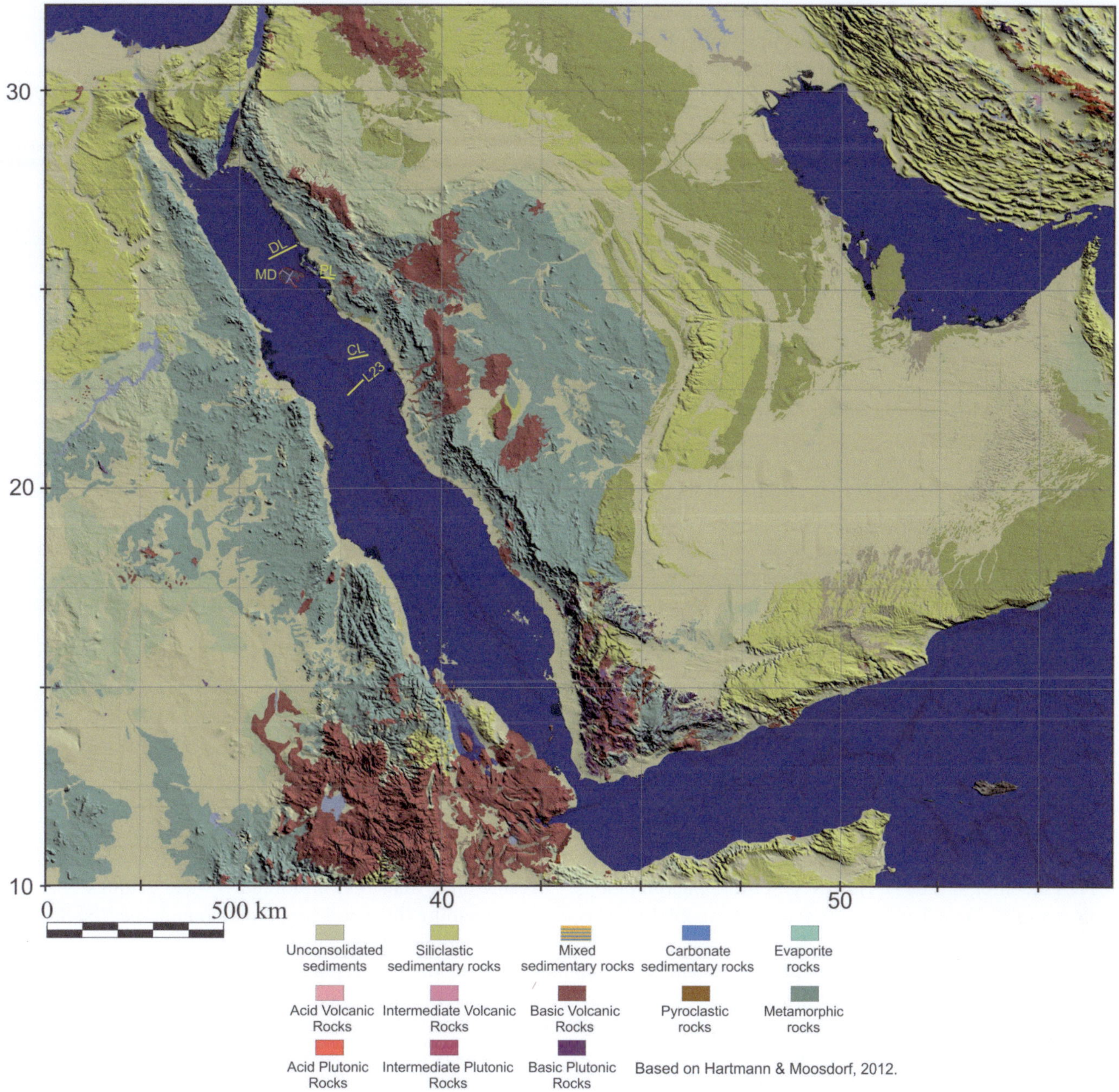

Fig. 9 Regional geological map of the African and Arabian plates with main lithological units draped on topobathymetric data. The location of the proximal and distal profiles (PL and DL) are located near the Mabahiss Deep (MD) in the north-central Red Sea. The seismic lines published by Colombo et al. (2014) and Mitchell et al. (2010) are indicated as CL and L23. Extensive lava fields (harrats) are observed on the Arabian plate, and these basalts range in age from about 30 Ma to Recent. Volcanic islands have been formed in the southern axial trough, particularly offshore Yemen. The volcanic rocks in the Afar area are associated with the triple junction between the Red Sea and Gulf of Aden oceanic ridges and the northern branch of the continental East African Rift System

zones of different densities (related to variable degrees of serpentinization) to fit the measured and modeled gravity data. However, other authors (e.g., Blaich et al. 2011) have conducted similar gravity modelling using regional seismic profiles and obtained very different results for equivalent geoseismic transects across the Brazilian margin. The exhumed mantle interpretation for the Campos and Santos basins (Unternehr et al. 2010; Zalán et al. 2011; Gomes et al. 2012) suggests that the outer high and the Florianópolis Fracture Zone (FFZ) are characterized by serpentinites outcropping at the bottom of the ocean. Assuming this model, even the oceanic propagator (Abimael Ridge, compare Fig. 2 with Fig. 6) that has been postulated to advance from the northern Pelotas Basin toward the southern Santos

Basin (Mohriak 2001) should correspond with peridotites formed in a magma-poor setting. In this work, we present seismic data suggesting that the propagator in the Santos Basin is locally associated with wedges of seaward-dipping reflectors (SDRs), and thus might be related to igneous intrusions and extrusions derived from mantle melting. In addition, recently acquired wide-angle seismic profiles acquired in the distal part of the Santos Basin suggest velocities compatible with thinned continental crust until normal oceanic crust is reached south of the Florianópolis Fracture Zone (Evain et al. 2015).

The model of propagating rifts as proposed by Courtillot (1982) assumes that continental breakup starts with a phase of continental rifting and continues with the propagation of rifts toward locked zones, where the crust is deformed by extension whereas oceanic crust is generated in the wake of the V-shaped rift zone. The conspicuous Abimael Ridge has been interpreted as a failed oceanic propagator in the Santos Basin (Mohriak 2001; Mohriak et al. 2008; Carminatti et al. 2009; Gomes et al. 2012; Kumar et al. 2012; Dehler et al. 2016) and is characterized by a strongly positive gravity anomaly indicating denser rocks in the lower crust, and the peculiarly negative magnetic anomalies suggest that the feature is associated with a volcanic belt north of the Florianópolis Fracture Zone (Fig. 2). The positive Bouguer gravity anomaly associated with this feature has had several interpretations in the geological literature, including volcanic rocks on continental crust (Karner 2000), an aborted spreading centre with volcanics on salt (Meisling et al. 2001), a volcanic wedge of proto-oceanic crust (Gomes et al. 2012; Evain et al. 2015), and exhumed mantle (Zalán et al. 2011; Pindell et al. 2015). However, seismic interpretations from regional 2D seismic profiles indicate it might correspond to an igneous feature (Fig. 7), locally associated with seaward-dipping reflectors at the transition from the southern Santos to the northern Pelotas basins (Mohriak et al. 2008, 2010; Quirk et al. 2013). The volcanic basement interpretation is supported by deep seismic profiles that show the Moho discontinuity rising abruptly basinward of the salt basin (Fig. 8), and this discontinuity appears to occur below volcanic rocks and igneous crust characterized as a proto-oceanic basement oceanward of the salt basin. Although this proto-oceanic basement adjacent to the Abimael Ridge has not been drilled in the Santos Basin, the potential field data and seismic interpretation suggest a relationship with the volcanic belt north of the Florianópolis Fracture Zone (FFZ, see Fig. 2).

Apparently, the region where the magmatic accretion is observed (and interpreted as proto-oceanic basement oceanward of the incipient SDRs in Fig. 8) acted as an oceanic propagator that advanced from the Pelotas Basin, where well-developed wedges of SDRs are clearly visible in seismic profiles (e.g., Koopmann et al. 2014). The propagator penetrated the salt basin in the southern Santos Basin, westward of the Florianópolis Fracture Zone (FFZ), during the last stages of the Gondwana breakup. The salt layer was incipiently split and the embryonic spreading centre was aborted, resulting in a ridge jump that transferred the mid-ocean ridge to an eastward position (Mohriak 2001). It is remarkable that the largest transform fault zone in the Santos Basin, the FFZ, is associated with the tip of the Abimael propagator. Similar relationships between propagators and fracture zones have been observed and discussed in other rifted continental margins as well (e.g., Courtillot 1982; Taylor et al. 2009; Franke 2013).

4 Tectonic Models for the Formation of the Red Sea Continental Margins

The Red Sea (Fig. 9) is one of the largest salt basins in the world, comparable in size with the South Atlantic salt basins offshore Brazil and West Africa, and much larger than the salt basins in the Central Atlantic (NE North America—NW Africa conjugate margins). Regional deep seismic profiles have been extensively acquired in the Central and South Atlantic continental margins in the past decade, particularly along the Brazilian and West African margins (Kumar et al. 2012). Refraction transects have been obtained in the Red Sea and Gulf of Aden (e.g., Mooney et al. 1985; Egloff et al. 1991; Watremez et al. 2011), but these regions are not as well sampled by long cable deep seismic reflection profiles as the South Atlantic, where the petroleum industry has extended the exploration frontiers beyond the continental-oceanic crust transition in ultradeep water settings.

Concerning the petroleum exploration in the Red Sea, the tectonic interpretations have important consequences for the assessment of the deep-water frontier regions. Decades of activity in the conjugate margins offshore Egypt, Sudan, Eritrea, Saudi Arabia and Yemen resulted in acquisition of an extremely large dataset from the platform toward the axial trough. A huge thickness of more than 5 km of salt has been suggested by many researchers based on seismic interpretation of the evaporite layer in the deep-water region (Lowell and Genik 1972; Guennoc et al. 1990). These evaporites have been proved in the southern, central and northern Red Sea, and sampled by exploratory wells drilled both onshore and offshore, not only in the platform but also in the deep-water region (Beydoun 1989; Tubbs et al. 2014; Rowan 2014; Hadad and Abdullah 2015). These wells indicate that the heat flow is considerably higher in the main trough, reaching more than 100 mW m^{-2} off the Sudanese coast, for example (Hadad et al. 2016), thus making it a gas-prone basin. More importantly, the majority of wells that were drilled offshore the Red Sea and penetrated through the Miocene salt have not shown any significantly rich pre-salt

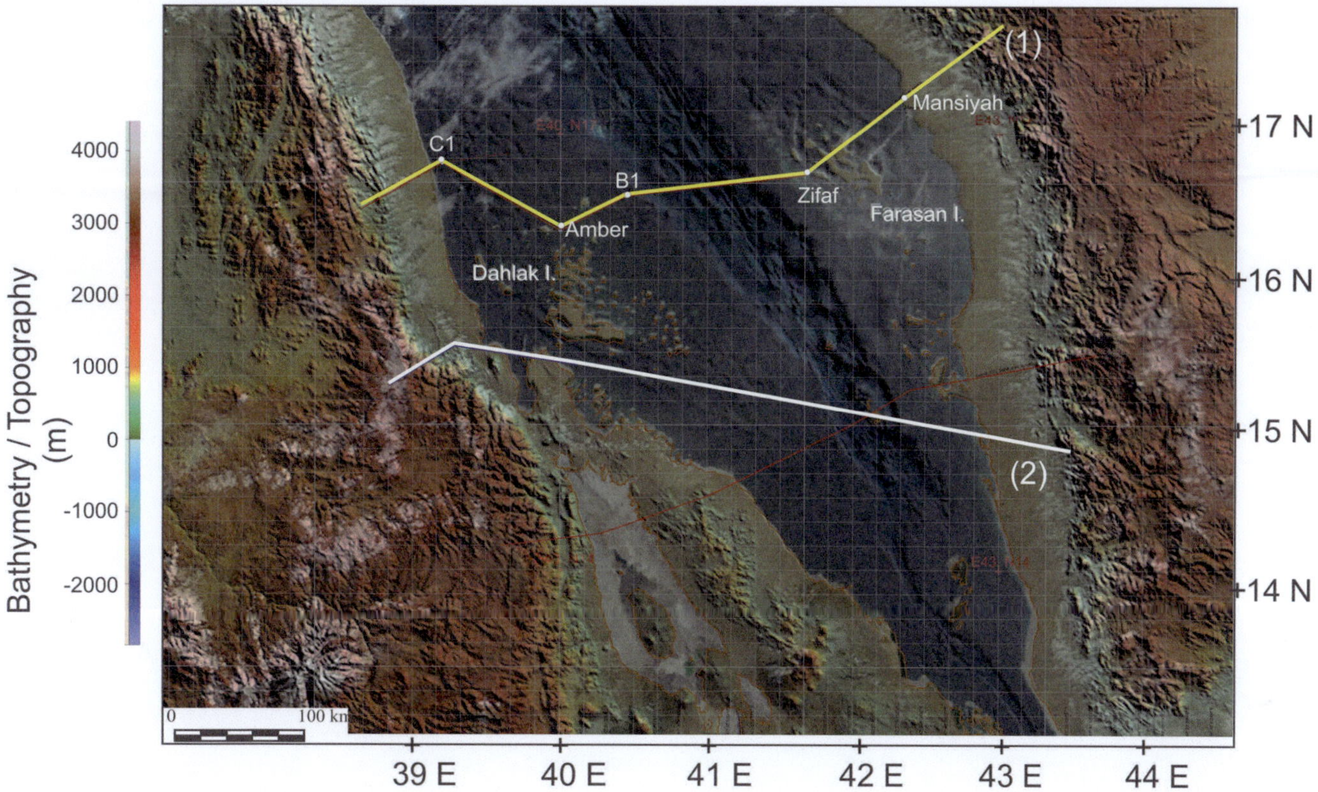

Fig. 10 Regional transects across the southern Red Sea (north and south of Dahlak and Farasan Islands) with two models for the development of the continental margins. Transect (1), north of the islands, is constrained by exploratory wells drilled in the platform of Eritrea and Saudi Arabia. Transect (2), south of the islands, extends from the onshore region north of the Danakil Depression in Ethiopia and reaches the Tihama Plain and the Yemen Escarpment in the conjugate margin

(pre-rift or syn-rift) oil-prone source beds. Although the syn-rift sediments in the Gulf of Suez and Red Sea have similar quality kerogen and potential for hydrocarbon generation, they are genetically different in terms of biomarkers (Alsharhan and Salah 1997) and the maturity is higher in the Red Sea than in the Gulf of Suez (Hadad and Abdullah 2015). The imaging below the thick salt layer in the Red Sea is much poorer than the resolution presently available with the modern seismic technologies applied to the interpretation of the South Atlantic prolific basins, and only a few wells have been drilled by the national oil companies in the deep-water region, resulting in small gas discoveries along the northern Arabian coast (Mohriak 2015).

The nature of the crust underlying the massive salt layer that is observed in the Red Sea axial trough has been debated for more than five decades (Almalki et al. 2015a). Several authors have gathered data that suggest that the oceanic basement is widespread throughout the Red Sea margins, extending almost from coastline to coastline (e.g., McKenzie et al. 1970; Hall 1979; Sultan et al. 1992). On the other hand, several geoscientists have proposed that most of the Red Sea is continental crust and oceanic spreading is only observed in the central to southern Red Sea with the oldest oceanic crust dated as Late Miocene/Pliocene ($\sim$5 Ma) and confined to the axial trough (Bosworth et al. 2005; Cochran and Karner 2007; Lazar et al. 2012; Schettino et al. 2016).

Since the development of the early plate tectonics concepts, the Red Sea and Gulf of Aden regions have been considered as paradigms for the evolution of continental rift basins that evolved into a gulf stage with incipient divergent margins. However, up to the present there is no deep seismic reflection profile extending across the Red Sea and imaging the conjugate margins from Africa to Arabia to corroborate either the pure shear or the simple shear hypothesis. Nonetheless, a large number of tectono-sedimentary studies integrating geological and geophysical datasets of the Red Sea and Gulf of Aden have been published in the geological literature (see for example, Leroy et al. 2004, 2012; Almalki et al. 2015a) discussing the pros and cons of each hypothesis. Some of these studies (e.g., Ligi et al. 2012; Augustin et al. 2014) indicate that in some segments of the central and southern Red Sea there is an active spreading centre in the axial trough; moreover, volcanic rocks with mid-ocean ridge affinity have been recovered by dredging of the seafloor, or are indicated by indirect methods (seismic and potential field data) in areas partially covered by salt (Guennoc et al. 1990).

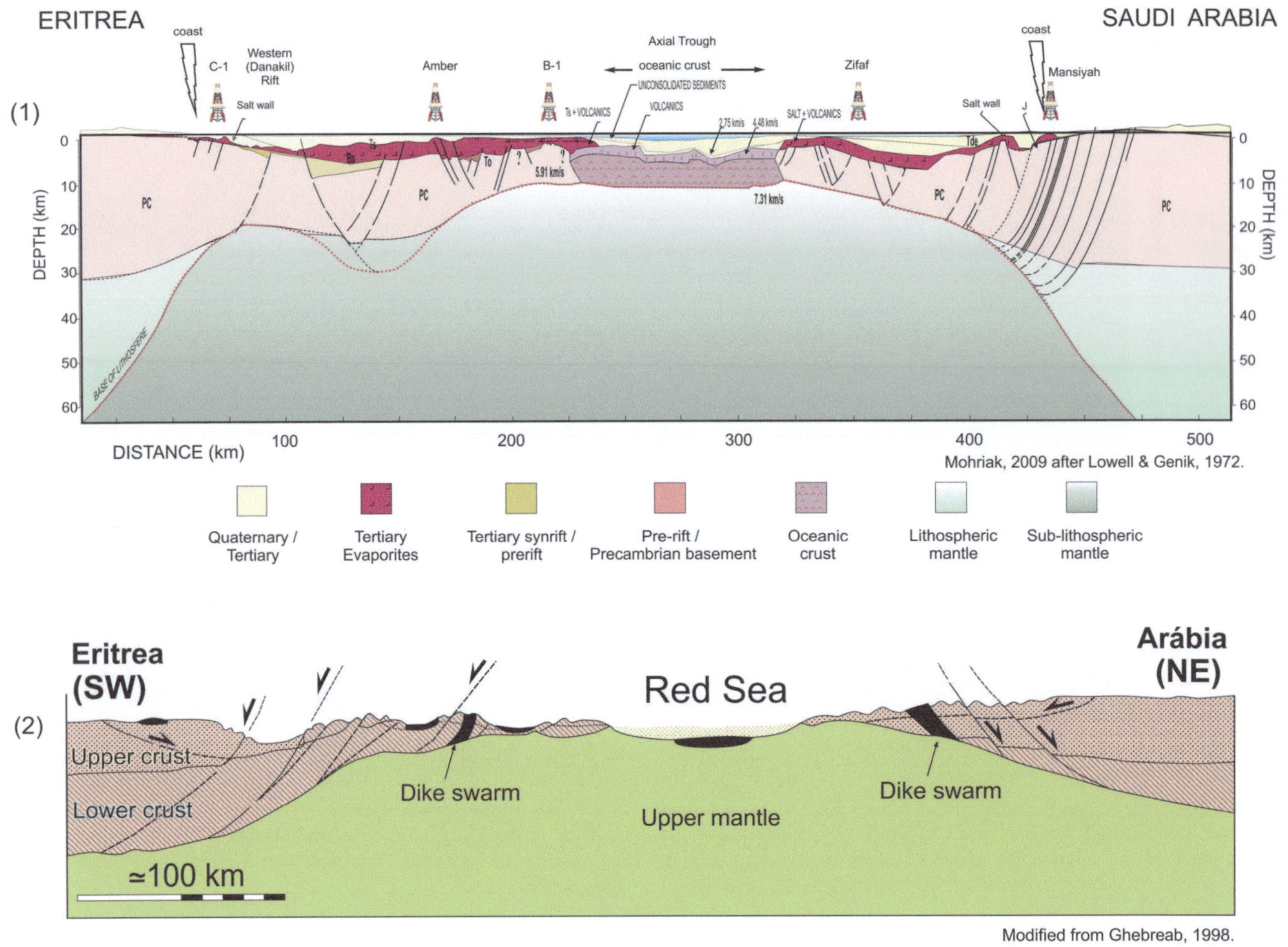

Fig. 11 Two end-member models for the development of the Red Sea continental margins are presented for the two transects shown in Fig. 10. Transect (1) assumes a volcanic basement in the axial trough and is constrained by exploratory wells drilled in the platform of Eritrea and Saudi Arabia. The well Amber-1 offshore Eritrea penetrated a thick salt sequence and bottomed at pre-rift metamorphic rocks. The well Mansiyah-1 onshore Saudi Arabia penetrated a thick sequence of evaporites and bottomed at volcanic rocks. Transect (2) extends from the onshore region north of the Danakil Depression in Ethiopia and reaches the Tihama Plain and the Yemen Escarpment. This model assumes mantle exhumation controlled by detachment faults that expose the upper mantle peridotites at the Red Sea axial trough

This work will analyze the crustal architecture of the central to northern Red Sea by integrating potential field datasets with seismic profiles from the proximal to the distal margin (Fig. 9).

Several authors have proposed different geodynamic models for the development of the Red Sea (e.g., Bonatti 1985; Voggenreiter et al. 1988; Bosworth et al. 2005). Mohriak (2014) has critically analyzed the pure shear and simple shear extensional models that have been proposed for the axial trough in the Red Sea (e.g., Lowell and Genik 1972; Ghebreab 1998). A comparison of these end-member models can be summarized by analyzing two geological cross sections in the southern Red Sea (Fig. 10), one north of the Dahlak Islands (Lowell and Genik 1972) and one south of these islands (Ghebreab 1998), in the offshore region just north of the Danakil Depression.

The geological transects in the southern Red Sea (Fig. 11) show two end-member models of continental breakup and formation of oceanic crust. Model 1 (Fig. 11-top, corresponding to the northern transect in Fig. 10) suggests continental breakup by pure shear extension and formation of oceanic crust in the axial trough, which might be filled with Late Tertiary sediments (Lowell and Genik 1972). This model is supported by the interpretation of high angle normal faults affecting the continental crust, and assumes that salt was deposited on continental crust, but is presently advancing toward oceanic crust in the main trough, and the axial trough might be covered by Neogene siliciclastic sediments overlying volcanic rocks (Ligi et al. 2012; Augustin et al. 2014).

The southern transect (Fig. 11-bottom, Model 2 in Fig. 10) assumes continental breakup by simple shear

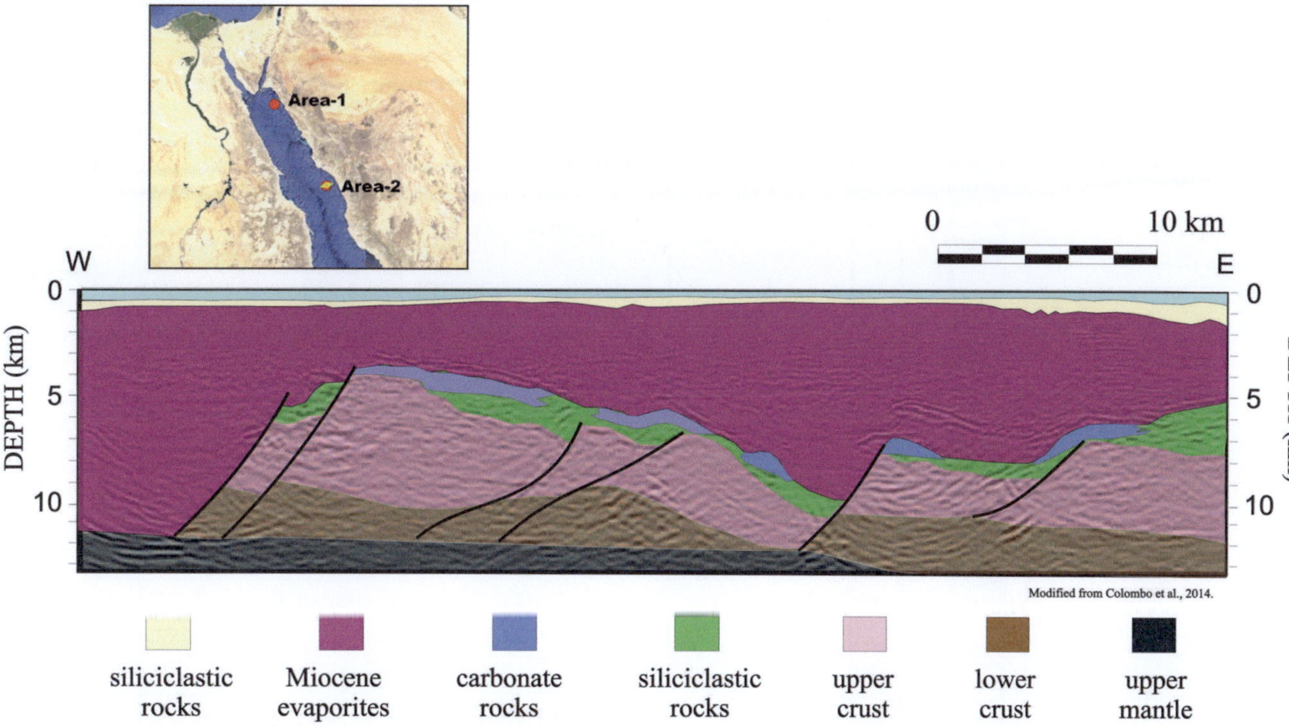

Fig. 12 3D seismic profile in the Area 2 of the eastern central Red Sea, between the Mabahiss Deep and the Nereus Deep, eastward of Zabargad Island (about 24° N). The proposed interpretation (Colombo et al. 2014) shows a thick salt layer on hyperextended continental crust and a massive salt mass (more than 10 km of stratified evaporites) overlying upper mantle rocks in the main trough

extension, with the axial trough located adjacent to portions of the lithosphere where mantle rocks are exhumed by extensional faults that detach at the lower crust—upper mantle boundary (Ghebreab 1998; Voggenreiter et al. 1988; Voggenreiter and Hotzl 1989). This model implies low angle normal faulting affecting continental crust, with detachment faults soling on the upper mantle peridotites. Consequently, hyper-extended continental crust would be associated with rotated fault blocks, serpentinized upper mantle rocks might be exposed in the main trough, and dike swarms would be asymmetric along the continental margins. This model implies that salt may overlie continental crust and exhumed mantle, thus the salt layer might be associated with a syn-thinning desiccation process that resulted in massive evaporite deposition preceding the onset of oceanic spreading (Rowan 2014).

The mantle exhumation model has also been tentatively applied to the central Red Sea, assuming simple shear stretching and salt deposition directly on exhumed mantle rocks in the deep-water province (e.g., Colombo et al. 2014; Rowan 2014). A recent application of the exhumed mantle model, based on the interpretation of the large 2D and 3D seismic dataset available for the central Red Sea, is shown in Fig. 12. The seismic profile in Area 2 corresponds to a 3D seismic profile extending across the eastern central Red Sea main trough near the Nereus Deep (Colombo et al. 2014). This model assumes that the Miocene evaporites cover hyper-extended continental crust in the platform main trough, whereas the axial trough is characterized by more than 10 km of salt directly overlying mantle rocks. Similar models have also been applied to several other basins worldwide, such as the Gulf of Mexico, as proposed by Pindell et al. (2015). Following the mantle exhumation model, it was assumed that in the early development of the Gulf of Mexico active detachment faults resulted in outer marginal collapse by the Early Oxfordian, and the spreading ridge that was active until the Middle Jurassic separated two salt basins offshore the USA and Mexico, with the distal parts of the slumped salt covering subcontinental mantle peridotites.

However, integration of potential field and seismic data, and exploratory wells in the Red Sea suggest that a pure-shear model associated with lithospheric stretching and development of magmatic centres in the axial trough may be considered as the model that better fits the observations of syn-rift structures, magmatism and salt deposition in the conjugate margins of Saudi Arabia-Yemen and Egypt-Sudan-Eritrea (Mohriak and Leroy 2013). The simple shear model of mantle exhumation would imply very asymmetric basins, whereas the gravity and magnetic anomalies, as well as P-wave receiver function modelling, are indicative of symmetric structures relative to the axial trough (see for example, Hosny

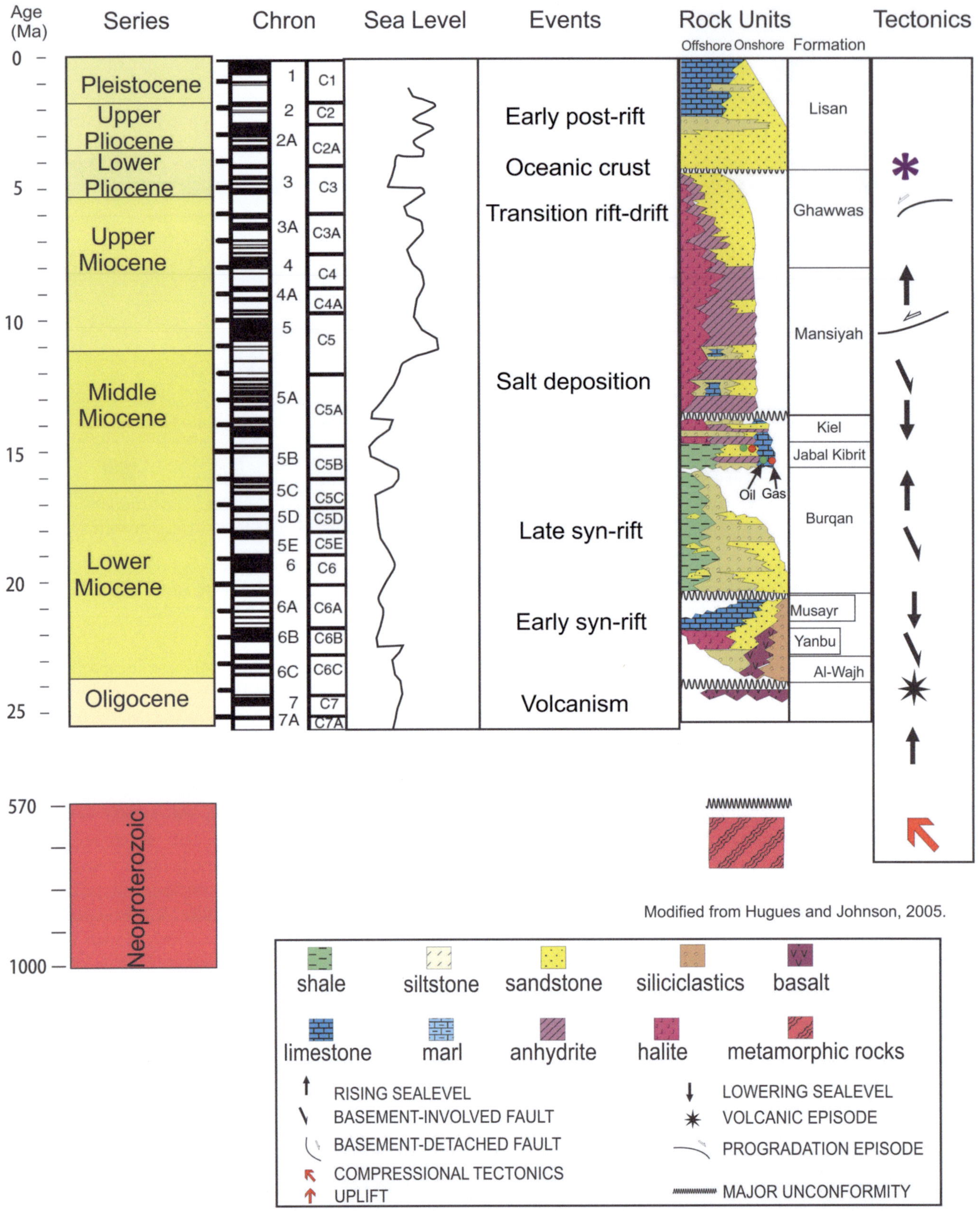

Fig. 13 Schematic stratigraphic section of the Red Sea with magnetic geochrons, sea-level variations and main tectonic events. The Lower Miocene stratigraphic successions (Al-Wajh and Burqan Fms.) are interpreted as syn-rift, and the early Middle Miocene Kiel and Jabal Kibrit formations are interpreted as deposited in a late syn-rift stage or sag basin. The Middle to Upper Miocene Mansiyah Fm. evaporites are interpreted as deposited in the rift-drift transition, although some authors assume these evaporites are late syn-rift. The Pliocene-Pleistocene post-salt successions include the Ghawwas Fm. and the Lisan Gp., which are separated by a regional and widespread unconformity in the Red Sea. There is a possible correlation of this regional unconformity with the Messinian desiccation event, which some authors associate with the breakup unconformity, marking the inception of the oceanic crust (~5 Mybp)

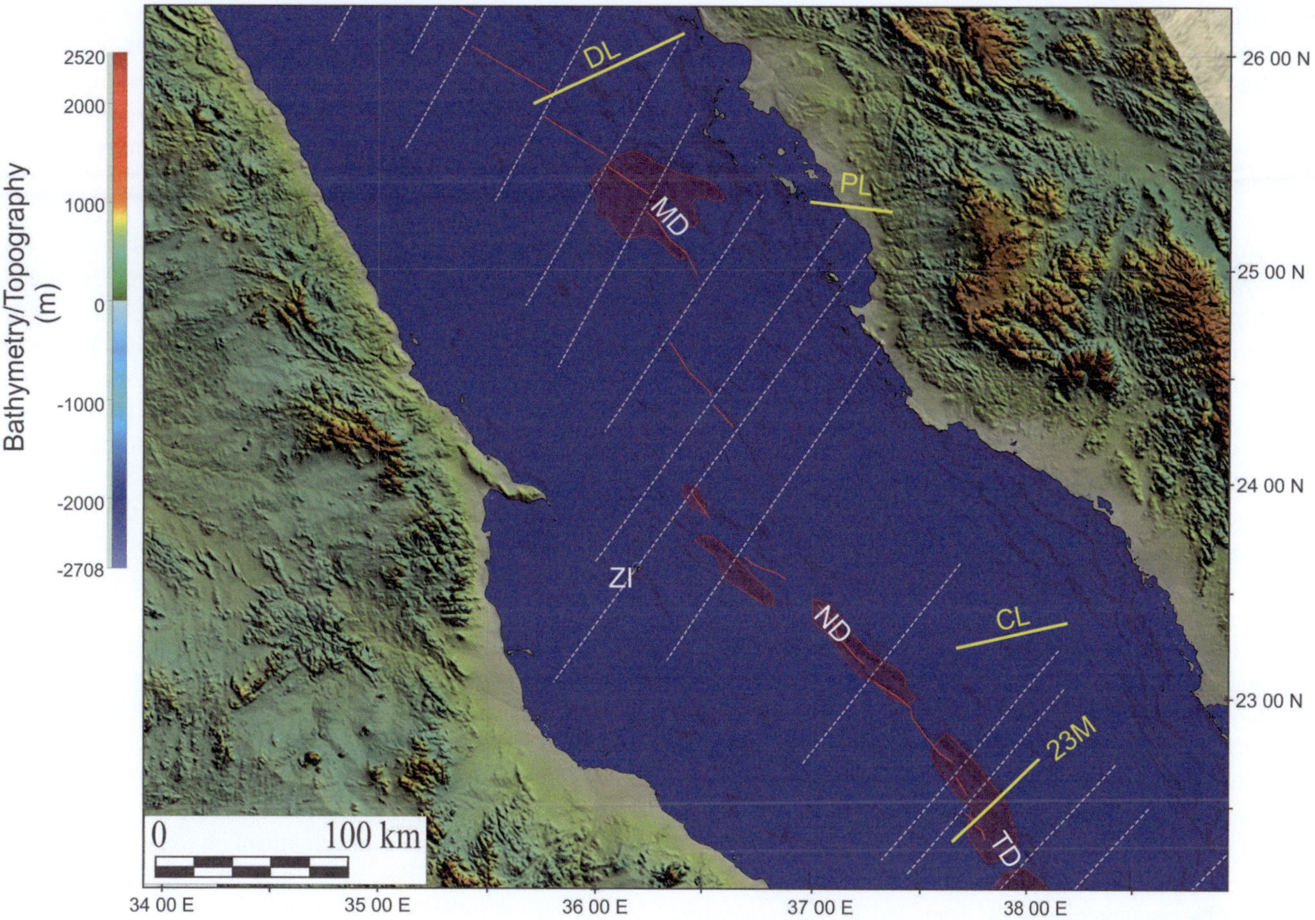

Fig. 14 Topobathymetric map of the central Red Sea with the locations of regional seismic profiles both onshore and offshore shown by yellow lines. The central Red Sea is characterized by an axial trough with depths that may reach about 2000 m in the Thetis, Nereus and Mabahiss Deeps (TD, ND and MD). The volcanic basement in the Red Sea axial trough is indicated by a purple colour adjacent to the embryonic oceanic ridges. Transform faults are indicated by dashed white lines. Several seismic profiles have been acquired in the Thetis Deep by academic institutions; Seismic Line 23 M, which crosses the axial trough spreading centre, is discussed here. The schematic geoseismic sections (proximal and distal profiles, PL and DL) present geological models for the onshore-offshore rift architecture and the alternative interpretations for the deep-water rift basins, with important implications for the syn-rift sediments and salt distribution. The approximate position of the seismic profile interpreted by Colombo et al. (2014) is also shown as CL, extending from the platform toward the Nereus Deep, located eastward of Zabargad Island (ZI) offshore Egypt

and Nyblade 2014). The bathymetry and development of basin depocentres and salt distribution are more or less symmetric across the margins (Mohriak and Leroy 2013). With the exception of the Mabahiss Deep, which is clearly offset toward the Arabian side, most of the Red Sea deeps are located in the centre of the axial trough, resulting in approximately symmetric margins. However, geological maps indicate a much larger area of Tertiary to Quaternary pre- and syn-rift volcanic rocks on Arabia's continental crust when compared to the African crust (Fig. 9). The harrats (lava fields) onshore Saudi Arabia extend for more than 200 km in length along a N–NW direction, and geological maps (e.g., Bosworth et al. 2005; Hartmann and Moosdorf 2012), indicate that the African side of the Red Sea margin lacks extensive lava fields, except toward the Afar province in the south, where large volcanoes such as Erta Ale are presently outpouring basaltic lavas in the Danakil Depression. Davison et al. (1994) emphasized that the major episodes of lava flows in Yemen span the interval of Oligo-Miocene (30–20 Ma), but volcanism is also registered in the Afar area onshore Ethiopia from Middle to Late Miocene to the present-day, concomitant with the syn-rift and early drift phases that affected the Red Sea region (Corti et al. 2015). Volcanic structures are also characterized below the salt in the Red Sea main axial trough, and in modern times a number of new-born volcanic islands have grown above sea-level offshore Yemen (e.g., Gass et al. 1973; Xu et al. 2015).

5 Rift Development, Stratigraphic Successions and Salt Tectonics in the Red Sea Continental Margins

Similar to the South Atlantic, the stratigraphy of the Red Sea continental margins can be divided into three main tectonic phases (Fig. 13); the syn-rift, the transitional, and the post-rift or early drift phases. The syn-rift phase, dated as Early Miocene, post-dates the volcanic rocks that have been dated by Ar–Ar at $\sim$ 30 Ma both in the Afar region in the southern Red Sea, as well as in the onshore region of Yemen and Saudi Arabia, where the youngest flood basalts show Ar–Ar isotopic ages of about 26 Ma (Baker et al. 1996). These volcanic rocks probably constitute the pre-rift volcanics that have been sampled by several wells in the Red Sea, such as the onshore Jizan wells at the border between Saudi Arabia and Yemen (Hughes and Johnson 2005).

The late syn-rift phase or sag basin is less affected by basement-involved faults, which locally penetrate the Middle to Late Miocene Mansiyah Fm., which corresponds to the main evaporite layer in the Saudi Arabian Red Sea (Bosworth et al. 2005; Hughes and Johnson 2005; Tubbs et al. 2014). Above the salt layer there is a thick sequence of Late Miocene to Pliocene sediments (Ghawwas Fm.) locally associated with salt mobilization. Finally, the Pleistocene to Recent siliciclastic and carbonate rocks of the Lisan Gp. cover the previous stratigraphic successions. These stratigraphic sequences will be discussed in the geoseismic profiles that extend from the border fault in the onshore region (proximal profile, PL in Fig. 9) toward the Red Sea axial trough (distal profile, DL in Fig. 9). A detailed map of the central to northern Red Sea (Fig. 14) shows the location of these profiles and the interpreted outline of the Mabahiss Deep, with the inferred position of the axial trough oceanic ridges and transform faults (Bonatti and Seyler 1987; Guennoc et al. 1990).

The age of oceanic crust formation in the Red Sea and Gulf of Aden has been discussed by many authors. Bosworth et al. (2005) advocated that seafloor spreading started in the Gulf of Aden at about 19–18 Ma and propagated westward toward the Afar plume, whereas in the Red Sea organized spreading commenced only at about 5 Ma. The model is based on stretching of continental crust simultaneously with magmatic activity (dyke intrusions and lava extrusions), and involves only a limited zone of oceanic crust in the axial trough of the Red Sea and Gulf of Aden, while allowing for the rapid thinning of the continental crust close to the coastline as indicated by geophysical studies in Saudi Arabia and Yemen (Mooney et al. 1985).

Several explorationists and researchers working in the Red Sea continental margins have interpreted that the Middle to Upper Miocene evaporites were deposited during the syn-rift (Bosworth et al. 2005; Allen and Beaumont 2015) or syn-thinning phase preceding mantle exhumation (Rowan 2014). The regional and widespread unconformity that is observed above the salt layer might correspond to the rift-drift transition or to the breakup unconformity that followed the rift-drift transition in the Upper Miocene—Lower Pliocene, and is often associated with the inception of oceanic crust in the central to southern Red Sea (Fig. 13). The characterization of major unconformities in the transition from syn-rift to post-rift strata in continental margins worldwide has long been debated in the geological literature, and its relationship with uplift of continental blocks or to the development of oceanic crust is controversial (Benson and Doyle 1988; Braun and Beaumont 1989). However, there is geological and geophysical evidence that some of these unconformities are related to a change in the locus of extension, from the proximal to the distal margin. There is clear seismic indication of oceanward younging of fault activity until the locus of extension is focused on an embryonic spreading centre. The syn-kinematic sediments young basinward and an unconformity develops when extension abandons one area in favour of new faults forming oceanward. It is also noteworthy to observe that these unconformities are developed regionally, and may constitute the register of events formed elsewhere in the margin, as a result of the diachronous development of oceanic ridges, and multiple unconformities may span an interval of several million years, rather than correspond to an instantaneous or short-term event preceding the sea-floor spreading phase.

In order to elucidate the controversial issues related to where and when oceanic crust might have formed in the Red Sea, this work analyses four regional profiles in the Arabian margin, interpreted from seismic lines acquired both onshore and offshore by academic institutions and the petroleum industry (Figs. 9 and 14). Two of these profiles are shown as schematic line-interpretation diagrams (Mohriak 2014) based on seismic data acquired from the onshore rift basin border toward the platform (proximal line, PL), and from the platform to the deep-water region (distal line, DL). Alternative hypotheses for the rift architecture of the Red Sea will be discussed based on interpretations of the distal profile.

The seismic profile east of the Nereus Deep (seismic line CL in Fig. 14) is based on a 3D seismic program (Area 2, see Fig. 12) acquired in the deep-water region of the Red Sea, as indicated by Colombo et al. (2014). Its location in Fig. 14 is only approximately indicated, and is based on the bathymetric data and length of the published seismic line (about 55 km). Finally, this work discusses a 2D seismic line that extends across the axial trough spreading centre (Line 23 M from Mitchell et al. (2010) and Ligi et al. (2012)). The integration of these profiles provides a unique geoseismic transect from the basin limit near the Precambrian outcrops

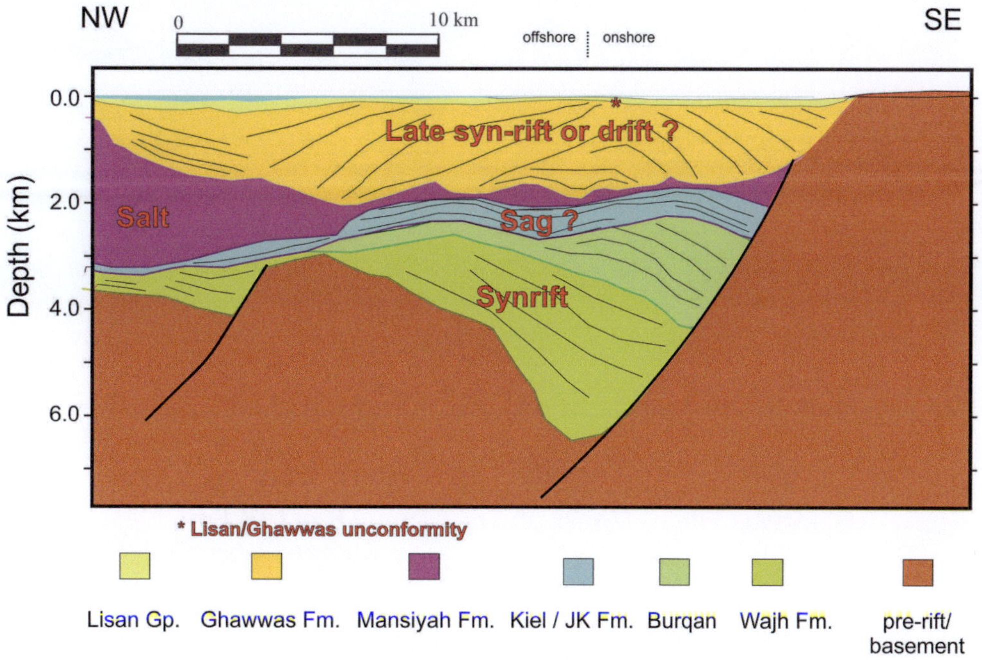

Fig. 15 Schematic geological section showing the onshore and the continental platform rift (proximal profile, PL in Fig. 14). The syn-rift stratigraphic successions (Al-Wajh and Burqan Fms.) are overlain by the late rift or sag basin deposits of the Kiel and Jabal Kibrit Formations. The Middle to Late Miocene evaporites (Mansiyah Fm.) have been mobilized from the onshore region to inflate the large salt diapirs in the offshore region. The Pliocene-Pleistocene post-salt successions (Ghawwas Fm. and Lisan Gp.) form a major roll-over structure associated with salt evacuation. The Lisan/Ghawwas unconformity (*) has a regional occurrence in the Red Sea and is possibly associated with the Messinian desiccation event. Several researchers also suggest a possible correlation with the breakup unconformity, marking the inception of the oceanic crust (~5 Mybp)

toward the oceanic crust spreading centre, as will be discussed subsequently.

Figure 15 shows the schematic seismic line interpretation of the proximal profile at the transition from the onshore to the offshore region in the central Red Sea. The main structure is characterized by the high-angle rift border fault near the eastern extremity of the seismic section, where the Precambrian basement rocks outcrop at the surface. The rift unit is overlain by salt and siliciclastic sediments in the proximal basin, and salt evacuation resulted in development of large turtle structures and salt diapirs in the continental platform.

The Precambrian basement may be interpreted at depths exceeding 6 km in the transition from the onshore to the offshore region. The normal fault that controls the syn-rift depocentre offsets the basement and rotates the sedimentary packages (Al-Wajh Fm. and Burqan Fm.). The Early Miocene syn-rift sequence is associated with the Al-Wajh coarse clastic rocks, that are locally covered by evaporites (Yanbu Fm.) and reefal carbonates (Musayr Fm.). The Burqan Fm., which is typically characterized by fine-grained sediments, and the Kiel and Jebal Kibrit Fm. successions, associated with Early to Middle Miocene carbonate and evaporite rocks, might correspond to a reduced phase of tectonic activity comparable to the sag basin in the South Atlantic (Mohriak 2014). These formations may be related to a deep marine environment, particularly the Burqan Fm. (Hughes and Johnson 2005), and the shallow marine carbonate of the Kiel and Jabal Kibrit formations may constitute reservoirs, as observed in the Midyan Basin (Tubbs et al. 2014). Some faults affect these layers and some even penetrate the base of the thick and widespread evaporite layer (Mansiyah Fm., Middle to Late Miocene).

The Middle to Late Miocene Mansiyah Fm. evaporites show indications of basinward salt flow forming large diapirs in the platform and deep water (Fig. 15). The post-salt stratigraphic succession is characterized by a thick halokinetic growth section controlled by listric basinward-dipping faults that form a major roll-over structure. This was produced by salt evacuation and migration to inflate the large salt diapirs in the platform, in a structural style that is similar to that observed in the Midyan Basin onshore, near the Gulf of Aqaba (Tubbs et al. 2014). Above the evaporites the stratigraphic sequences include the Ghawwas Fm., characterized by halokinetic growth sequences, and the Lisan Gp., which onshore is characterized by a thin veneer of sediments sealing the halokinetic structures.

The angular unconformity that corresponds to the boundary between the Ghawwas Fm. (Late Miocene) and the Lisan Gp. (Pliocene to Pleistocene) might be positioned around 5 Ma in the stratigraphic chart (Fig. 13). The

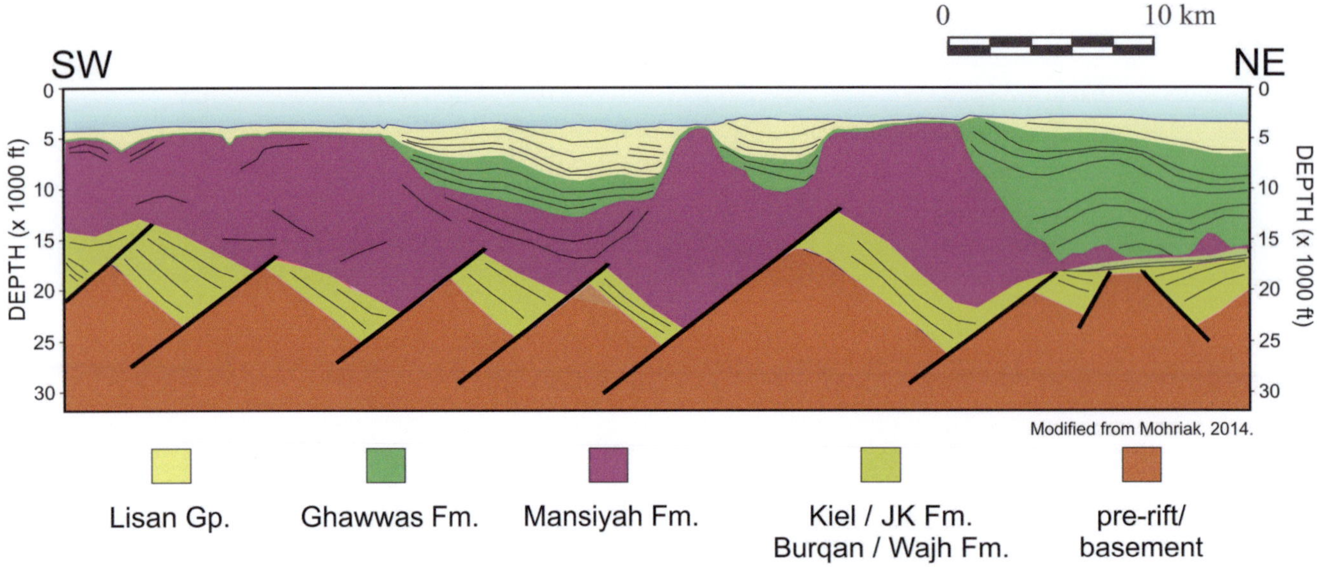

Fig. 16 Hypothesis 1 for the interpretation of the Red Sea distal profile (DL in Fig. 14), assuming a magma-poor margin and salt deposited during a late syn-rift stage, before the continental breakup. No oceanic crust is present in the basin centre, as interpreted for the northern Red Sea and Gulf of Suez. Maximum salt thickness overlying the rift sequence is estimated at about 15000 ft ($\sim$5000 m) of stratified evaporites

interpretations of the magnetic anomalies in the central to southern Red Sea have been correlated with the Chron C3n.2n ($\sim$4.6 Ma; Schettino et al. 2016), at about the same age of the transition from Late Miocene to Pliocene, thus corresponding with the unconformity between the Ghawwas Fm. and the Lisan Gp., which some authors also associate with the widespread Messinian Unconformity that is characterized in the Mediterranean but may also affect the Red Sea (Bosworth et al. 2005; Afifi et al. 2014). Many researchers and explorationists have envisaged a possible correlation of this regional unconformity with the breakup unconformity, which may herald the inception of oceanic crust as suggested by many authors (e.g., Falvey 1974; Esedo et al. 2012; Franke 2013). However, several authors consider there is no proven link between the inception of oceanic crust and a phase of uplift affecting the rifted continental crust prior to the breakup, and this unconformity might in fact correspond with a mantle upwelling developing a lithospheric breakup surface during the rift-drift transition (Duarte et al. 2012). The assumption that the unconformity observed in the Miocene/Pliocene boundary might correspond with the breakup would thus point to the generally accepted age of inception of the oceanic crust in the Red Sea around 5 Ma (Bayer et al. 1989; Bosworth et al. 2005; Cochran and Karner 2007).

This interpretation has important consequences for the Red Sea rift architecture in the distal margin, because if the breakup unconformity in the central Red Sea is the Messinian Unconformity, then there is oceanic crust only in the axial trough where the anomaly C3 is clearly identified, and most of the area covered by thick salt masses might correspond to extended continental crust with syn-rift sediments, as suggested by several researchers (e.g., Cochran and Karner 2007) and also by explorationists working in the area (Tubbs et al. 2014). This possibility is illustrated as hypothesis 1 in Fig. 16, with tilted rift blocks covered by thick salt in the deep-water region (Mohriak 2014).

On the other hand, if the oceanic crust or volcanic basement extends landward of the axial trough, then the breakup unconformity should be older than the Chron C3. Comparing the proximal geoseismic profile (Fig. 15) with equivalent geoseismic sections in the South Atlantic, Mohriak (2014) suggested that the breakup unconformity might be positioned below the salt, and not above it. A number of authors have used the magnetic anomaly pattern in the Red Sea to suggest that the formation of the initial oceanic crust might be positioned closer in time to the development of the syn-rift sequence, in Early Miocene times (Labrecque and Zittelini 1985). This interpretation advocates that the main trough might be constituted by a thickened oceanic crust with effusion of large quantities of stratoid basalts during the initial phases of continental breakup. The axial trough would then correspond with a more advanced stage of focusing the magma production in the spreading centre developed in the past few million years.

The distal profile in the central Red Sea extends from the platform toward the main trough and axial trough north of the Mabahiss Deep (Fig. 14), which does not show bathymetric depressions similar to the depressions in the southern Red Sea, where the axial trough is clearly characterized by a

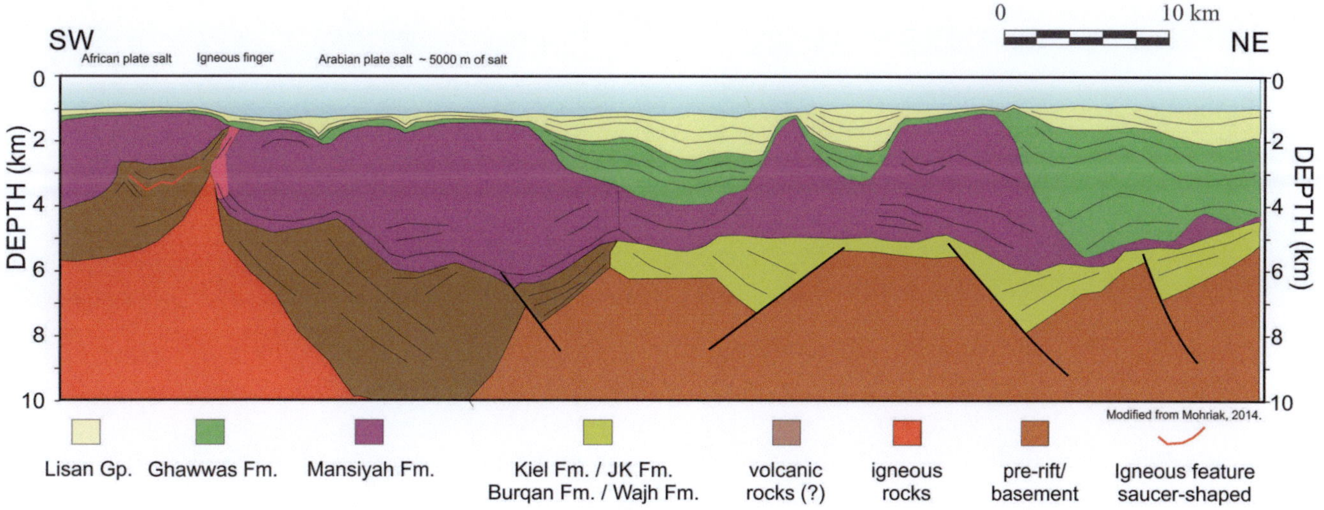

Fig. 17 Hypothesis 2 for the interpretation of the Red Sea distal profile (DL in Fig. 14). Assuming this interpretation, salt was deposited above the breakup unconformity, at the transition from late synrift (sag basin) to an early drift phase. Salt deposition was concomitant with igneous extrusions and intrusions in the volcanic basement. Igneous fingers penetrate the salt layer in the axial trough, and volcanic intrusions have started to split the evaporite basin into two conjugate margin salt bodies. Eventually one of these salt masses will be attached to the African plate (Egypt/Sudan) and the other to the Arabian plate (Saudi Arabia/Yemen). The salt mass in the Arabian side may reach more than 5000 m of evaporites overlying late syn-rift sediments and volcanic basement rocks

magnetic anomaly zebra pattern associated with organized spreading in the oceanic crust (Hall 1979; Zahran et al. 2003). Two hypotheses have been analyzed for the interpretation of this profile. Hypothesis 1 (Fig. 16) assumes that the thick salt masses in the axial trough cover tilted rift blocks on the continental crust, which may be affected by extensional faults with large offsets. Following this interpretation, the Mansiyah Fm. evaporites were deposited during a late syn-rift phase before continental breakup and no oceanic crust has been developed in the basin centre. This is similar to the interpretations of diffuse extension still active in the northern Red Sea and Gulf of Suez (Cochran 1983; Bonatti 1985; Bosworth and McClay 2001; Bonatti et al. 2015). This hypothesis has been favored by many explorationists working in the area, particularly along the northern Arabian Red Sea, which is considered much less volcanic than the southern Red Sea. The intrasalt reflectors might then be interpreted as a suture of sedimentary rock trapped below amalgamated salt bodies.

Hypothesis 2 (Fig. 17) suggests an alternative interpretation for the distal profile, assuming that the thick salt masses are advancing toward a volcanic basement associated with incipient oceanization and protracted magmatic activity that resulted in the formation of large volcanoes below the salt, and intrusive bodies might occur within the salt layer (Mohriak 2014). Assuming this interpretation, the continental crust has already been affected by igneous intrusions that will eventually split the salt basins apart by embryonic spreading centres. The Middle to Late Miocene evaporites were deposited during the transition from an early post-rift to an early drift phase, overlying volcanic basement rocks in the distal basin. The model envisages that the salt masses in the margins are still amalgamated in the central to northern Red Sea, forming a single salt basin in the main and axial troughs. However, locally they are starting to be separated by embryonic spreading centres that have formed in the past few million years, with intrusive bodies or igneous fingers invading the crust and also penetrating the salt layer, which is mainly allochthonous at this portion of the basin. In other areas, the process is more advanced and the salt masses are totally separated by oceanic ridges and newly formed volcanic basement, as in the Thetis Deep and more incipiently in the Mabahiss Deep (Mohriak 2014).

These hypotheses might be tentatively tested by integration of the geoseismic interpretation with potential fields (gravity and magnetic datasets) and by analysis of seismic profiles in the axial trough, which might also be relevant to the interpretation of the simple shear and pure shear mechanisms of margin development (as discussed in the transects presented in Figs. 10 and 11). In addition, the analysis of the Mabahiss Deep, located about 50 km south of the Distal Line, might throw some lights on the tectono-magmatic processes that are presently occurring in the Red Sea axial trough.

Figure 18 shows (at the same scale) the potential field regional maps for the central Red Sea, based on worldwide datasets (Sandwell and Smith 2009; Maus et al. 2009; Balmino et al. 2012). These maps illustrate the topobathymetry (Fig. 18a), the free-air gravity anomaly field (Fig. 18b), the Bouguer gravity anomaly field (Fig. 18c) and the total field magnetic anomaly (Fig. 18d). The locations of the proximal

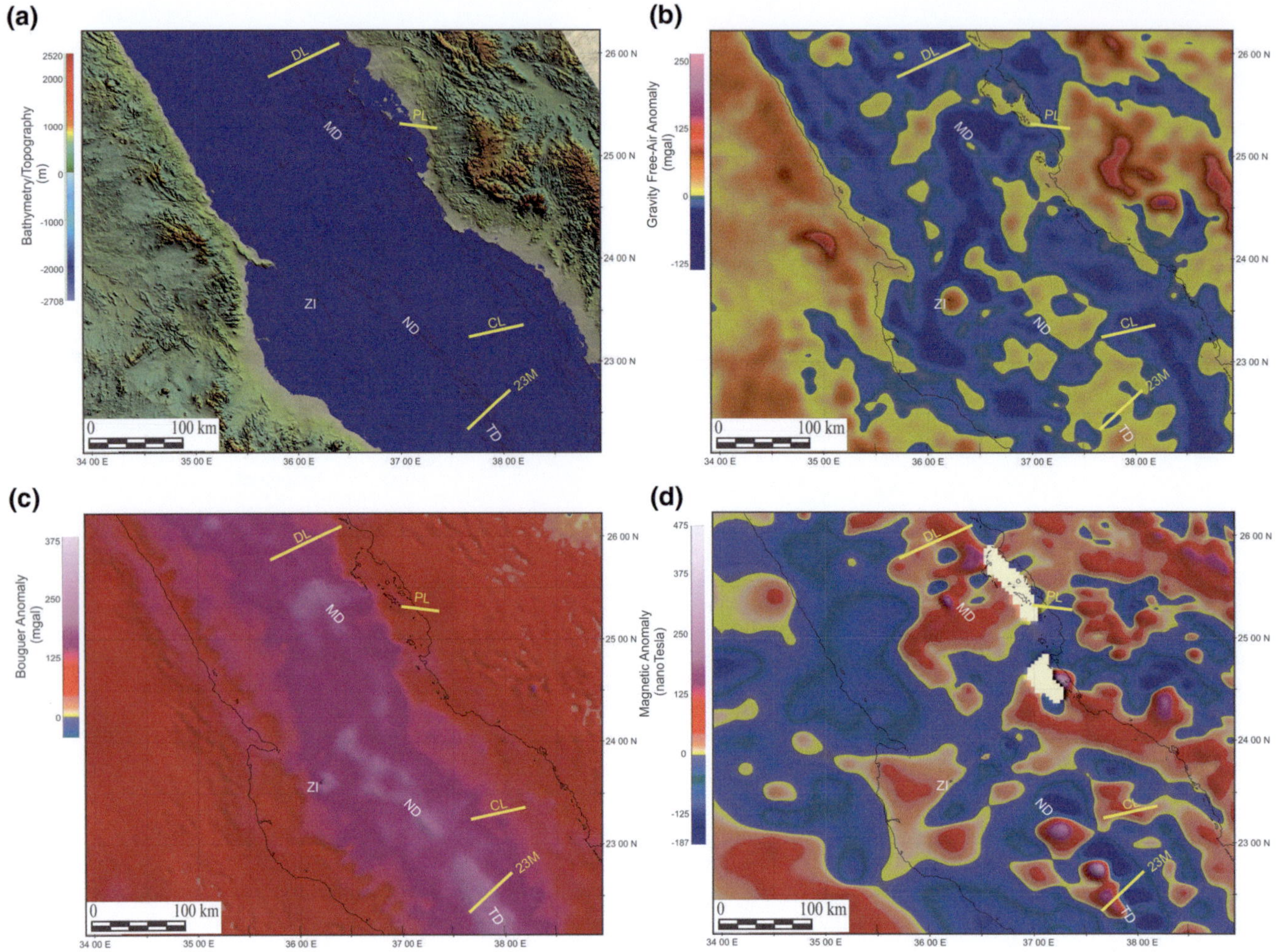

Fig. 18 Regional maps of the central Red Sea with the location of the proximal (~25°N) and distal (~26°N) geoseismic lines. Simplified tectonic elements such as the interpreted oceanic spreading ridge (red line) and transform faults (dashed lines) are included in the maps and shown in more detail in Fig. 14. The central to northern Red Sea deeps (Mabahiss, Nereus and Thetis) are located in the axial trough and indicated as MD, ND and TD, respectively. The Colombo et al. (2014) and Mitchell et al. (2010) seismic profiles (CL and 23 M), located between the Nereus and the Thetis Deeps, are also located in the potential field maps, as well as Zabargad Island (ZI) offshore Egypt. **a** Topobathymetric map of the central to southern Red Sea; **b** Free-Air gravity anomaly map; **c** Bouguer gravity anomaly map; **d** Magnetic (total field) anomaly map

and distal geoseismic profiles (as well as the industry 3D seismic profile CL and the academic 2D seismic line 23 M), previously shown in Fig. 14, are also plotted on these maps. The seismic profile 23 M is particularly of interest for the interpretation, because it is located in deep to ultradeep waters, with bathymetry around 800 m in the NE end of the profile but reaching more than 2000 m in the axial trough at the Thetis Deep, where it crosses the embryonic spreading centre (Mitchell et al. 2010; Ligi et al. 2012).

The free-air gravity anomaly map (Fig. 18b) is characterized by N–NW positive anomaly trends onshore in both Egypt and Saudi Arabia. In the offshore region, the NW trends are also observed, but local NE trends are also conspicuous, and these might be associated with transfer zones or transform faults, as for example in the Zabargad Shear Zone (Bonatti et al. 1984; Ligi et al. 2012). Zabargad Island is associated with a major positive gravity anomaly in the African plate, which is associated with the lower crust and upper mantle rocks that outcrop at this island (Bonatti et al. 1981; Bonatti and Seyler 1987). The proximal line is characterized by a positive anomaly near the basement outcrop and by a negative anomaly toward the coastline. The distal line is characterized by a gravity anomaly that decreases from the platform toward the distal portion of the line, and rises again near the transform fault at the interpreted embryonic spreading centre.

A cross-plot of the potential field anomalies with the interpretation discussed for hypothesis 2 for the distal transect is shown in Fig. 19. The distal profile shows the bathymetry reaching a maximum of about 1400 m in the

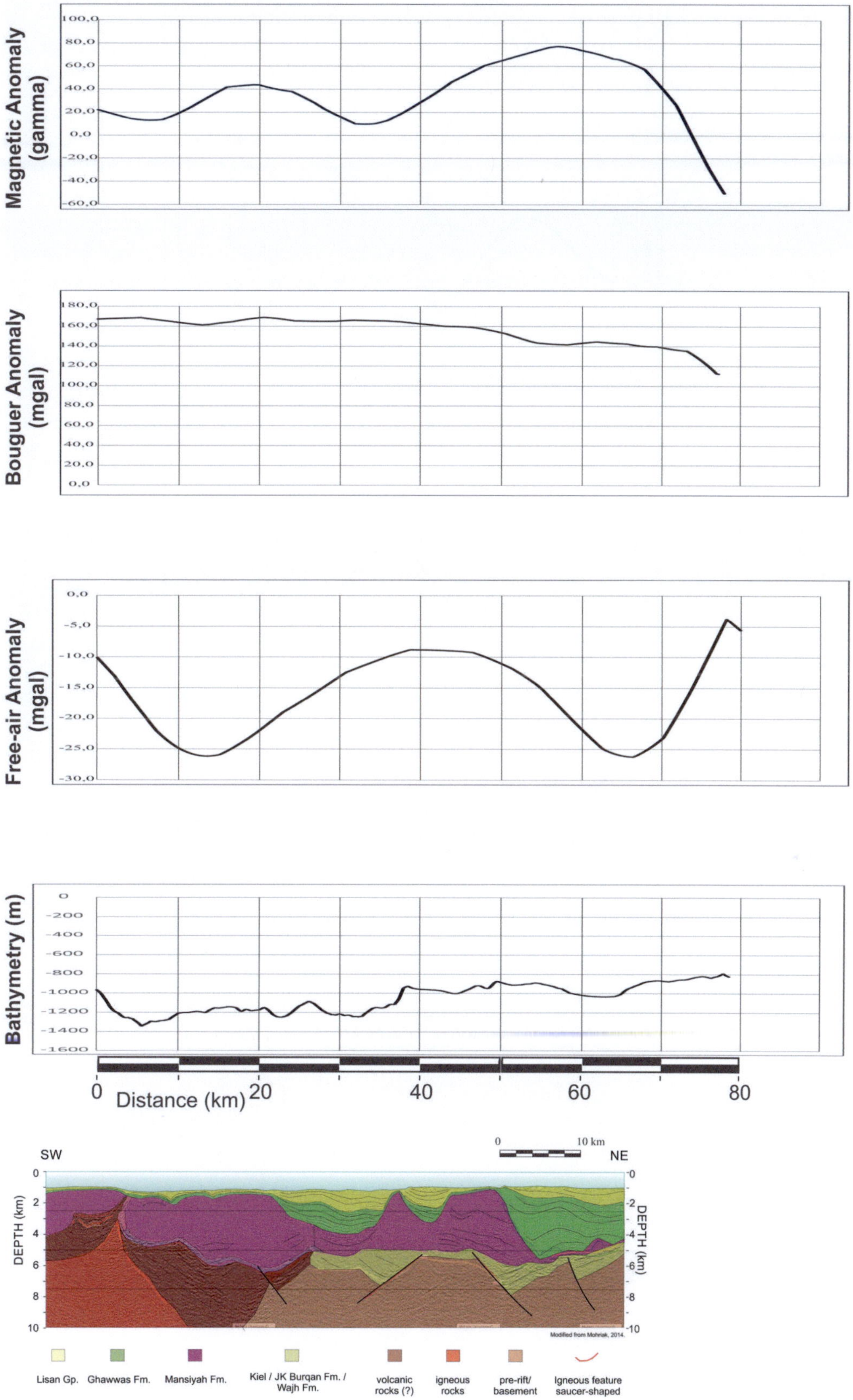

Fig. 19 Bathymetry, Free-Air, Bouguer and Magnetic anomaly profiles for the distal profile based on industry high-resolution and world global geophysical datasets. The correspondence between the potential field data and the geoseismic profile is only approximate, based on the same horizontal scale and the bathymetry variation along the seismic line position. The high-resolution gravity and magnetic anomaly datasets used to create the profiles do not extend toward the SW extremity of the seismic profile. On the other hand, the potential field datasets extend landward of the NE extremity of the seismic line

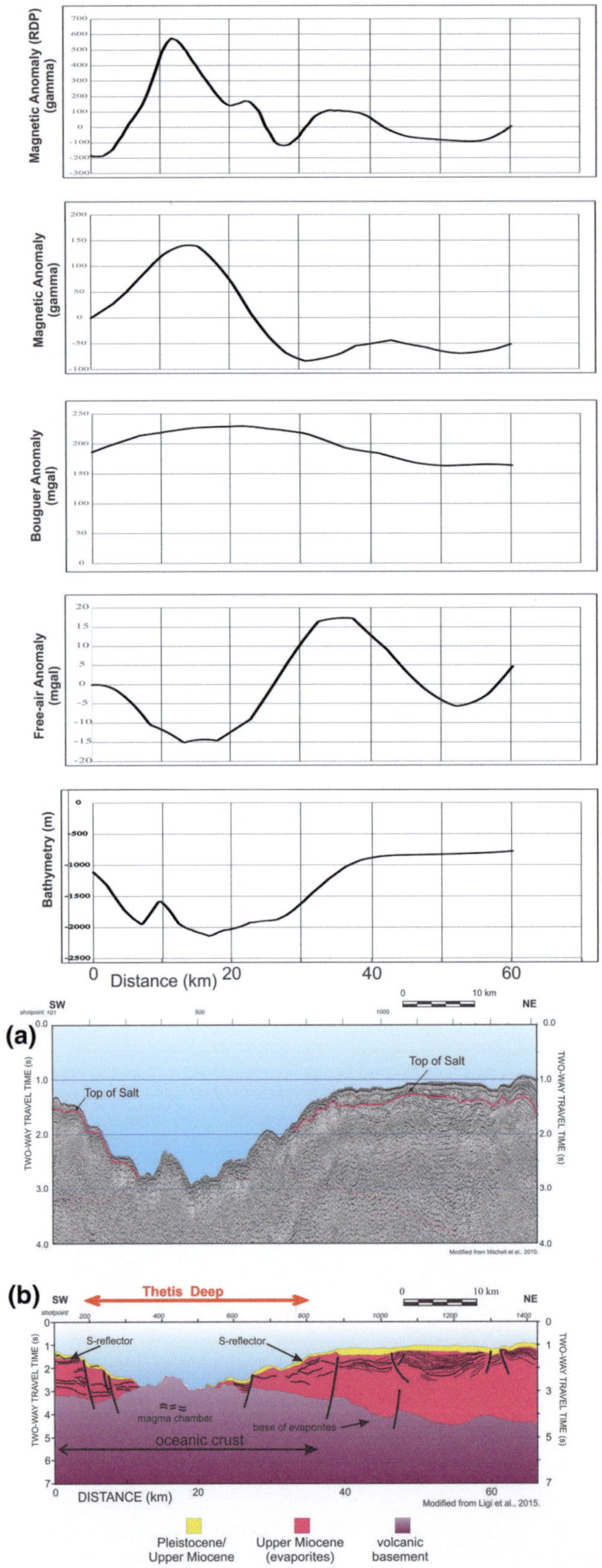

Fig. 20 Bouguer, total field magnetic anomaly and reduced to the pole magnetic anomaly profiles for seismic line 23 M across the Thetis deep (Line 23 M in Figs. 14 and 18), in the central Red Sea, based on industry and world global geophysical datasets. The correspondence between the potential field data and the seismic/geoseismic profiles is only approximate, based on the same horizontal scale and the bathymetry variation along the seismic line position. The interpretation of the seismic profile shows an incipient spreading centre separating two salt bodies that are attached to the Egyptian and Saudi Arabian conjugate margins. No sediments are observed above the active spreading ridge, but the allochthonous salt masses seem to be advancing toward the bathymetric abyss

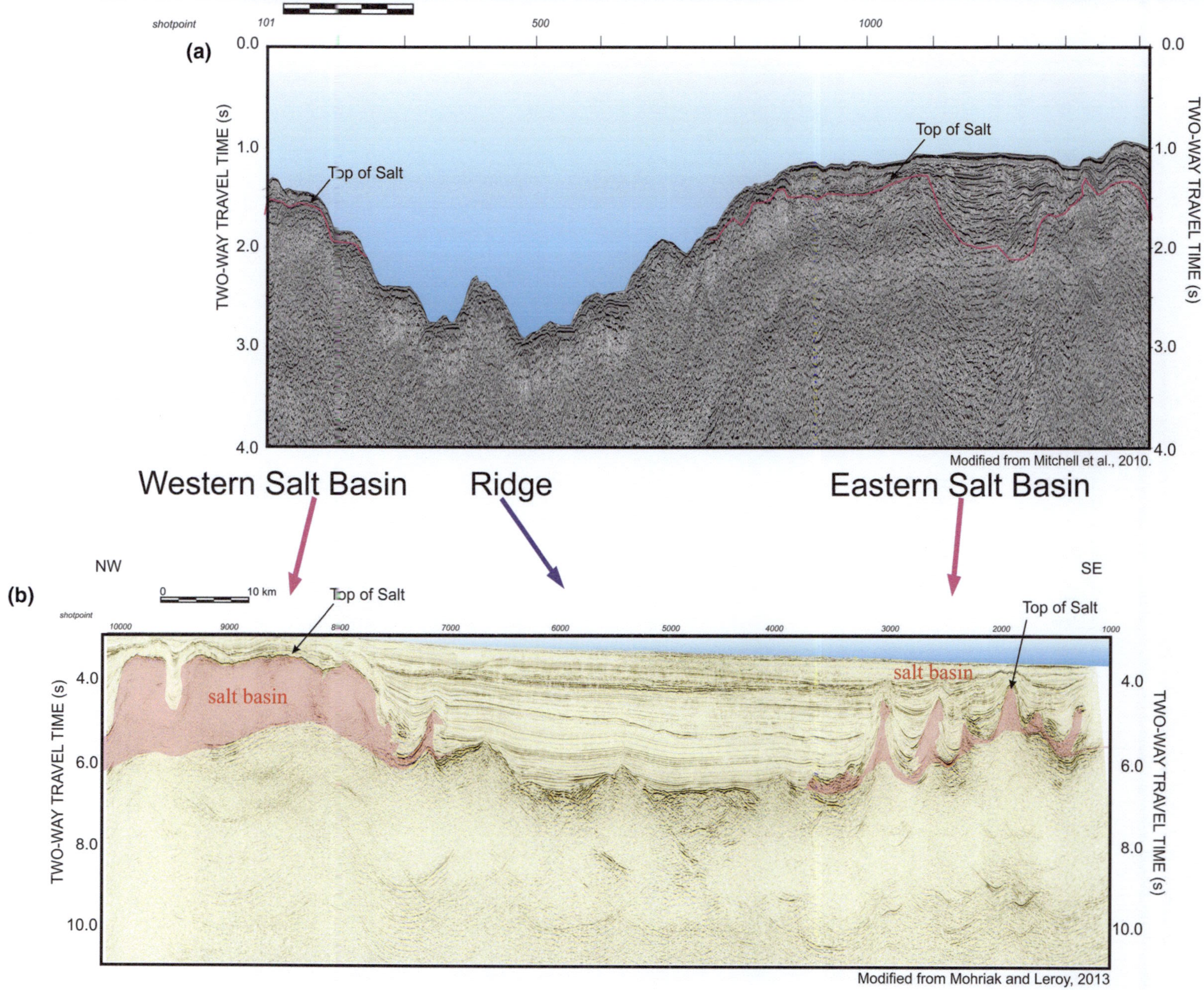

Fig. 21 Comparison of the Red Sea spreading centre in the Thetis Deep (**a**), corresponding to Line 23 M in Fig. 14, with a possible aborted spreading centre in the southern Santos Basin (**b**), corresponding to the profile D2 in Fig. 2. The two salt basins in the central Red Sea, between the African and Arabian conjugate margins (**a**), are separated by incipient spreading centres in the axial troughs with no evaporites and no sediments above the structural high which may be associated with an embryonic spreading ridge developed in the last 2–3 Ma. Salt was deposited about 12–14 Mabp in the platform and deep waters, and is now advancing as an allochthonous salt tongue toward the bathymetric abyss. This picture illustrates the development of a divergent continental margin 10 Ma after salt deposition. In the southern Santos Basin offshore Brazil (**b**) the two salt masses are split by a proto-oceanic crust (volcanic basement) associated with the Abimael Ridge, and the embryonic axial trough has been filled with sediments deposited in the last 115 Ma. Only a residual salt basin is observed eastward of the ridge, and the southern salt limit is associated with a volcanic belt north of the Florianópolis Fracture Zone (Fig. 2)

axial trough, at the SW extremity of the profile, and the free-air gravity anomaly is negative in this region, which is characterized by a thick salt mass. The axial trough located south of 24°N (Nereus Deep) is characterized by a slightly positive free-air anomaly, with local gravity lows, as in the position of the interpreted spreading ridge. The seismic line 23 M when cross-plotted against the potential field data (Fig. 19) is characterized by a positive free-air anomaly, not only at the axial trough with the spreading centre, but also in the adjacent regions.

The Bouguer anomaly map (Fig. 18c) clearly illustrates the segmented character of the axial trough by the transform faults. The axial trough south of 24°N is characterized by a positive anomaly that deflects toward the NW north of 23°N. Zabargad Island is associated with a major transfer fault zone that offsets the Bouguer anomaly pattern toward the Arabian side (Ligi et al. 2012). Compression due to shearing along the transfer zone might be responsible for uplifting the lower crust and upper mantle rocks that outcrop at this island (Bonatti et al. 1983, 2015). The proximal line is characterized by a positive anomaly that increases from the onshore to the offshore region. The distal line is characterized by a much higher gravity anomaly that gradually increases from the platform toward the distal portion of the line (Fig. 19). The 3D seismic profile published by Colombo et al. (2014) extends from the platform to the main trough, where the Bouguer anomalies are considerably higher. The seismic profile 23 M crosses the interpreted spreading centre of the Red Sea at 23°N (Thetis Deep), and is marked by a strongly positive Bouguer gravity anomaly centred at the bathymetric abyss (Fig. 18c).

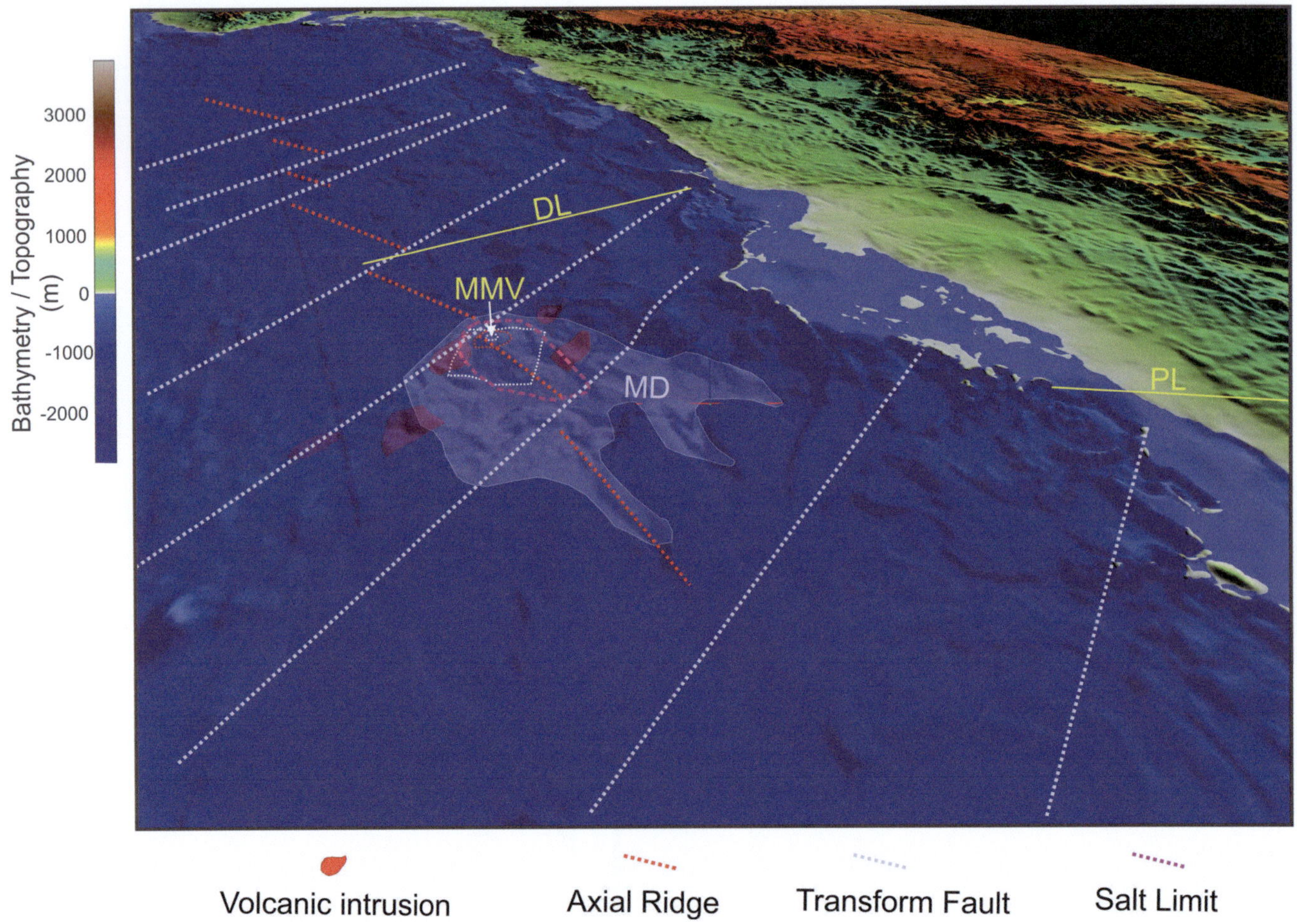

Fig. 22 Three-dimensional oblique visualization of the topobathymetric map of the central Red Sea with location of the Mabahiss Deep (MD), which is characterized by a volcanic basement with no evaporites in the NW part, indicated by a dashed magenta colour outline enveloping the axial ridge, where basement is exposed and salt is presently advancing as allochthonous masses flowing toward the bathymetric depression. The area with detailed bathymetric survey in the Mabahiss Deep (rectangle with dashed white line) is presented with a 3D horizon visualization in Fig. 23. The Mabahiss Seamount (Mabahiss Mons Volcano, MMV) is located at the northwest corner of the white square, near the inferred spreading centre or axial ridge. Several igneous plugs have also been identified in the region by Guennoc et al. (1990) and are marked by red ellipses at the northern part of the Mabahiss Deep. The distal and proximal geoseismic profiles are also indicated by yellow lines (DL and PL)

The magnetic anomaly map of the central Red Sea (Fig. 18d) is more difficult to interpret due to low resolution of the world dataset and the structural and lithologic complexity of the basement rocks that occur both onshore and offshore (Blank 1977; Blank and Andreasen 1991). The axial trough south of 24°N is characterized by a positive-negative anomaly pair with a NE trend. Zabargad Island is associated with a major transform fault that offsets the magnetic anomaly along a NE direction. The proximal line is characterized by a negative anomaly onshore that slightly increases toward the coastline. The distal line is characterized by a much higher magnetic anomaly that shows a typical dipole at the axial trough. The 23 M seismic profile crosses the interpreted spreading centre at the Thetis Deep, and the strong magnetic anomaly dipole is indicative of igneous rocks along a NW trend, with NE trends associated with the interpreted transform faults.

The potential field dataset is an important background to regionally assess the seismic interpretation of the proximal and distal profiles. Several researchers have used the gravity anomaly datasets acquired both onshore and offshore of the Red Sea to constrain the crustal architecture and propose tectonic models for the lithospheric extension that resulted in continental breakup in some parts of the axial trough (e.g., Saleh et al. 2006). Mitchell and Park (2014) have integrated several geophysical methods and pointed out that refraction data collected east of Thetis Deep might indicate the presence of gabbros underlying the salt layer for a distance of about 65 km outside of the axis in the deep, thus leaving little space for rifted continental crust in the distal margin.

The seismic profile 23 M across the axial trough in the Thetis Deep (Fig. 20) shows that the salt basin only occurs in the elevated regions (with bathymetry less than 750 m), whereas in the bathymetric depression there is an abyss deeper than 1500 m, and there is no evidence of salt accumulation on the newly-formed volcanic basement. The reduced to the pole magnetic anomaly crossplot (Fig. 20-top profile) shows a large anomaly centred at the centre of the axial trough. The protuberant structural high in the internal zone of the abyss is interpreted to correspond to an active spreading ridge (Mitchell et al. 2010; Mohriak and Leroy 2013) that was formed about 2 Mabp (Ligi et al. 2012).

One important question is whether this type of structure, characterized as an embryonic spreading centre within an axial trough, might exist in other places in the South Atlantic salt basin, with implications for timing of the breakup, tectonics, sedimentary facies and petroleum systems in the Brazilian and West African conjugate margins.

Mohriak and Leroy (2013) suggested that the mid-ocean spreading centre in the central Red Sea (Thetis Deep) might

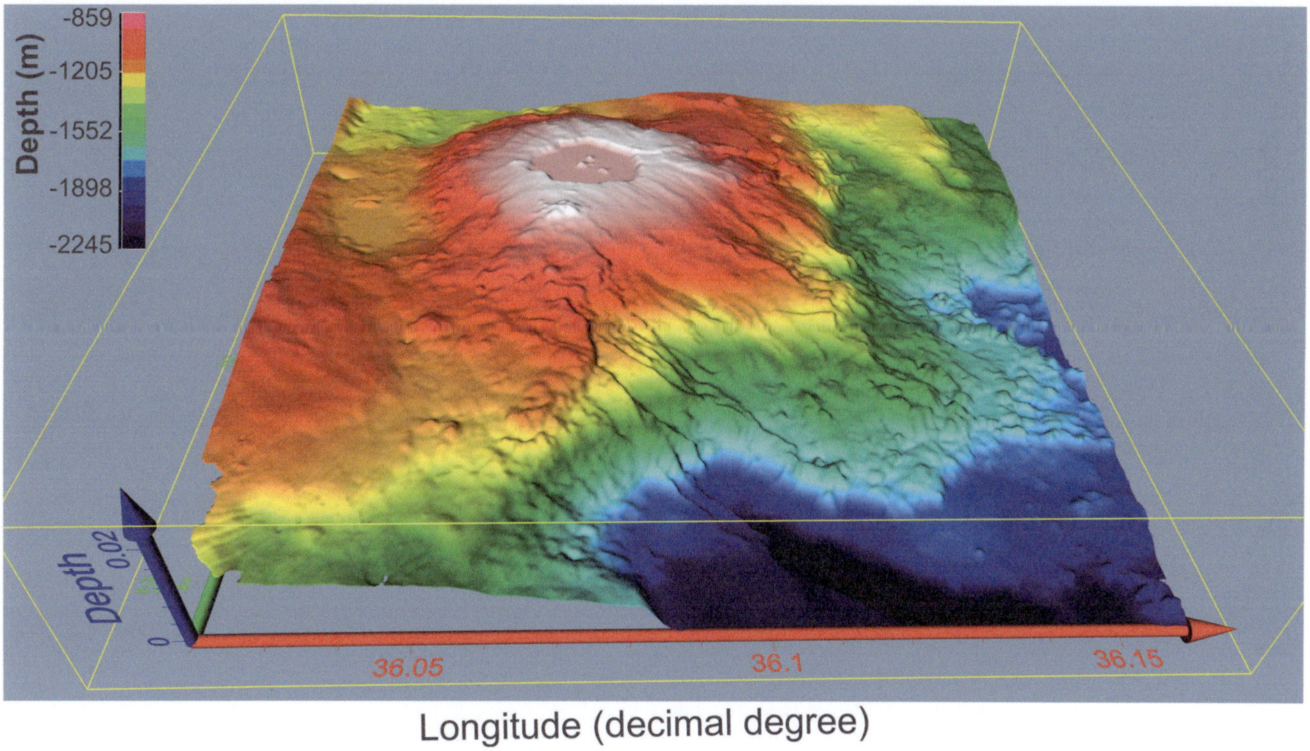

Fig. 23 Three-dimensional view of the seafloor surface horizon at the northern part of the Mabahiss Deep in the central Red Sea (see location in Fig. 22). This remarkable structure, known as Mabahiss Seamount (or Mabahiss Mons Volcano) is located at the inferred Red Sea spreading centre (longitude 36.1°E, latitude 25.5°N), south of the distal profile DL (Fig. 14)

be compared with the embryonic spreading centre in the southern Santos Basin (Fig. 21), which is associated with the Abimael propagator in the northern Pelotas Basin (Mohriak 2001). This feature may be interpreted as an igneous intrusion advancing northward, and it was active during and following salt deposition and initial breakup. After impinging on the salt basin, the propagator was aborted when the active spreading centre shifted eastward (Mohriak et al. 2008). The comparison of the seismic profile in the Thetis Deep (Fig. 21a) with the seismic profile in the Santos Basin (Fig. 21b) indicates that in both cases the salt masses pinch-out toward the abyss where the intrusive body is emplaced. In the Santos Basin, the pre-salt rift blocks are highly eroded below the main salt mass, and in the abyss only post-salt sediments are observed covering the volcanic rocks. The Red Sea profile might correspond to an instant photograph of the Santos Basin 10 Ma after salt deposition, with sediments filling up the initial abyss in the proto-oceanic axial trough (Mohriak 2014). If the propagator had not failed to propagate northward, then the residual salt bodies observed in the SE extremity of the Santos Basin profile would belong to the African plate, and oceanic crust would be observed between the two salt basins.

There are some important constraints on the early post-salt sedimentary environments and relative timing of salt deposition and breakup when the seismic data in the Red Sea is analyzed and compared with the South Atlantic salt basins. There is a marked unconformity below a thin veneer of sediments that covers the top of the salt mass (S reflector) that is observed in the Red Sea profile (Fig. 21a), and locally we observe undulating features that suggest mini-basins or internal deformation of the stratified evaporites, indicating early halokinesis. The salt masses in the central Red Sea are separated by newly-formed oceanic crust associated with an active spreading ridge, with possible magma chambers at depth (Ligi et al. 2015). In large areas covered by salt, huge volcanoes might be present, even in regions with subdued gravity and magnetic anomalies, such as in the Mabahiss Deep, which is located west of the proximal profile and south of the distal profile (see location in Fig. 14).

Figure 22 shows a three-dimensional visualization of the topography and bathymetry of the central Red Sea, with the outline of the Mabahiss Deep (light purple colour) and some tectonic elements such as the location of the inferred axial ridges and transform fault zones (Bonatti and Seyler 1987; Guennoc et al. 1990). Guennoc et al. (1990) suggest that the Mabahiss Deep might be associated with a young oceanic

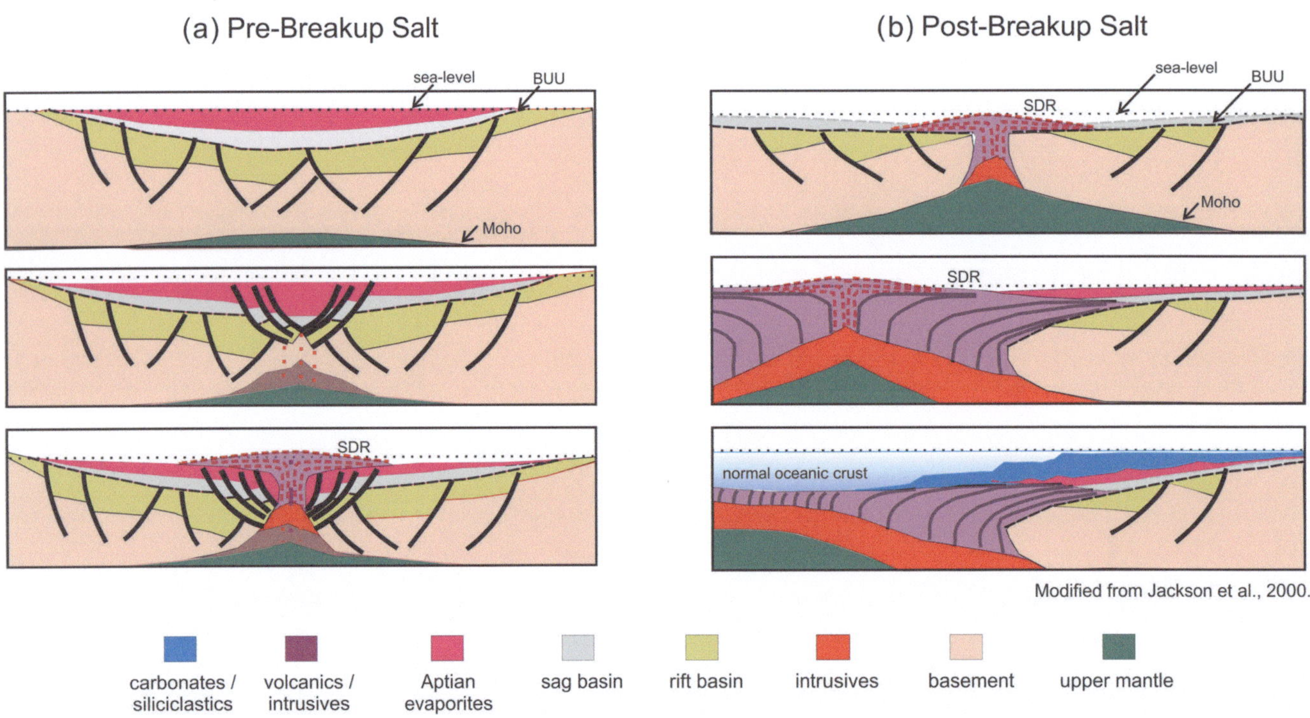

Fig. 24 Schematic diagrams showing conceptual models for salt deposition in the South Atlantic with two end-member hypotheses: (**a**) pre-breakup salt, with a single salt basin in the marine gulf between the Brazilian and West African margins; extensional tectonics and magmatism resulted in development of seaward-dipping reflectors (SDRs) that post-date the evaporites. Salt extends toward the active spreading ridges. **b** post-breakup salt, with magmatism and SDR following the rift phase but preceding the salt deposition; the SDRs and volcanic buildups constitute a barrier for the salt layer that develops only in one side of the margin, covering rifted continental crust and only part of the volcanic wedges feather edge. Assuming this model, salt does not advance toward normal oceanic crust

rift with an embryonic spreading centre, whereas Ligi et al. (2012) assume this bathymetric depression may be associated with a pull-apart basin floored by basalts, corresponding to the northernmost occurrence of oceanic crust in the Red Sea, formed in a tectonic setting of very slow spreading centres (about 0.5 cm/year).

The Mabahiss Deep is characterized by a large circular edifice at its northern segment, interpreted as a seamount with a pancake dome geomorphology (Fig. 23). This feature is known as the Mabahiss Seamount (or Mabahiss Mons Volcano) and represents one of the largest individual volcanoes in the Afro-Arabian rift system. It has a summit at the depth of about −860 m (Guennoc et al. 1990) and detailed bathymetric surveys indicate a sub-circular crater with small cones inside (Augustin et al. 2016). This structure displays a radius of about 4 km and covers an area of about 34 km^2, with an average height of about 260 m above the surrounding seafloor and a conspicuous crater at depths less than 1000 m below sea level. Dredging of the seafloor on the volcano flanks at depths around 1000–1500 m recovered basalts with geochemical affinity with mid-ocean ridges.

The Red Sea distal profile in the central to northern Red Sea (Figs. 14 and 22) shows an incipient stage of plate separation and is characterized by amalgamation of salt masses that are only beginning to split apart between Egypt and Saudi Arabia, north of the Mabahiss Deep. There are some interesting analogies between the Mabahiss Deep with the alternative interpretations of the Red Sea distal profile discussed previously. The first interpretation (hypothesis 1) assumes rifted continental crust throughout the basin (Fig. 16) and that the reflectors within the distal profile salt layer might correspond to sediments trapped in the suture between two colliding salt sheets. Hypothesis 2 (Fig. 17) assumes that the breakup processes might have already started in the central to northern areas by magmatic intrusions or igneous fingers piercing through the salt layer. This possibility is enhanced by (1) the occurrence of saucer-shaped structures or sills at the SW extremity of the distal profile; (2) the neo-formed volcanic basement in the Mabahiss Deep; (3) the separation of salt masses by active oceanic spreading in the Mabahiss Deep axial trough; and (4) by the large Mabahiss Seamount interpreted as a volcanic structure associated with a mid-ocean ridge at the northern end of the Mabahiss Deep (Guennoc et al. 1990; Augustin et al. 2016).

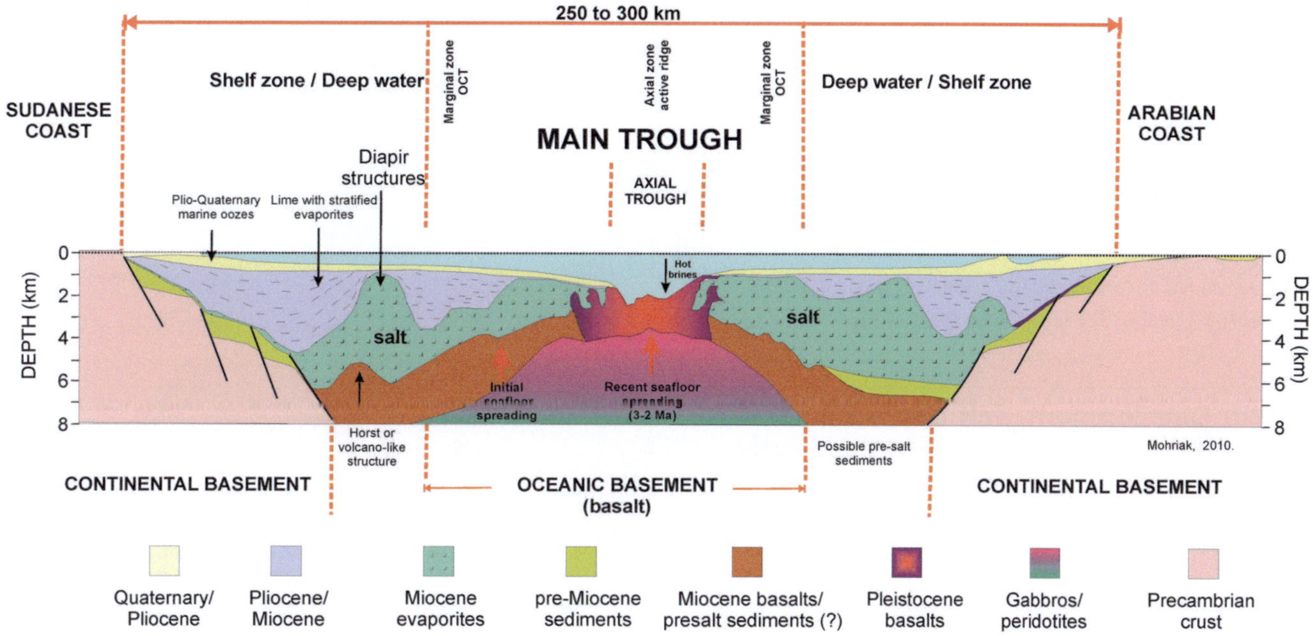

Fig. 25 Schematic geological model for the central to southern Red Sea architecture, with syn-rift sediments extending from the coastline toward the platform and deep-water province, covered by a thick salt mass that is allochthonous at the transition from the continental to the oceanic crust. The axial trough is characterized by an embryonic spreading centre with allochthonous salt advancing toward the bathymetric abyss. Below the massive salt in deep water the basement is probably constituted by volcanic rocks. Large volcanoes can be identified in areas with significant salt thickness, and recent volcanic episodes at the main trough also indicate igneous activity associated with the incipient divergence of the Arabian and African plates. The model also suggests the temporal development of structures from the platform, where the pre-salt sequences (syn-rift sediments) are observed, to the deep-water province, where volcanic basement and seaward-dipping reflectors might be present in the continent-ocean transition. In the distal salt basin, igneous fingers are intruding into the salt mass and eventually an embryonic spreading centre will form and separate the salt basins. Subsequently the spreading ridge will subside and allochthonous salt will advance toward the oceanic crust basement

The salt thickness near the axial trough in the area north of the Mabahiss deep is extremely large, about 5–6 km, with the salt being thicker on the Arabian side (Fig. 17). Eventually, the inception of an incipient spreading axis in the central to northern Red Sea will result in allochthonous salt bodies overlying the volcanic basement, as presently observed in the Mabahiss and Thetis deeps (Guennoc et al. 1990; Ligi et al. 2012).

The central to southern Red Sea, south of the Zabargad Fracture Zone (Fig. 14) shows evidence of a more advanced stage in the tectonic evolution, with an active spreading ridge separating the salt masses by newly-formed oceanic crust in the Thetis Deep. The seismic profile across the axial trough (Fig. 21) corresponds to a snapshot 10 Ma after salt deposition, with allochthonous salt bodies split by igneous intrusions and an incipient proto-oceanic crust starting to develop, associated with an embryonic spreading centre formed less than 2 Mybp (Ligi et al. 2012).

Several igneous rocks dredged at the sea floor of the Red Sea deeps indicate volcanic rocks with a tholeiitic composition. Antonini et al. (1998) analyzed the glass and whole rock compositions of basalt samples dredged from the Nereus Deep (Fig. 14) and confirmed their affinity with mid-ocean ridge basalts. Volcanic structures as observed in the Red Sea deeps are not present in magma-poor margins such as the Iberian margin but are commonly associated with magmatic centres in active rift zones such as the East African Rift System, the Afar province and in the axial trough of the southern Red Sea. In 2011–2013, volcanic eruptions related to recent magmatic activity resulted in the birth of two volcanic islands in the Zubair Archipelago offshore Yemen, westward of the coastal village of As-Salif (Xu et al. 2015). In the conjugate margin setting, large volcanoes in the Danakil Depression (as for example the Erta Ale), are presently active, outpouring basaltic lavas through their calderas.

6 Rift Development, Salt Deposition and Continental Breakup in the South Atlantic: Analogies with the Red Sea

The interpretation of the salt basin development in the South Atlantic has been discussed by several authors over the past decades. Based on the analysis of geological and geophysical datasets along the West African and Brazilian continental margins, Jackson et al. (2000) in a classical paper that addressed the relationship between rift tectonics, salt deposition and magmatism in the South Atlantic, discussed two end-members models for the evaporites (Fig. 24), whose relative age of deposition might be constrained by the occurrence of seaward-dipping reflectors (SDRs), which were interpreted in the transition from continental to oceanic

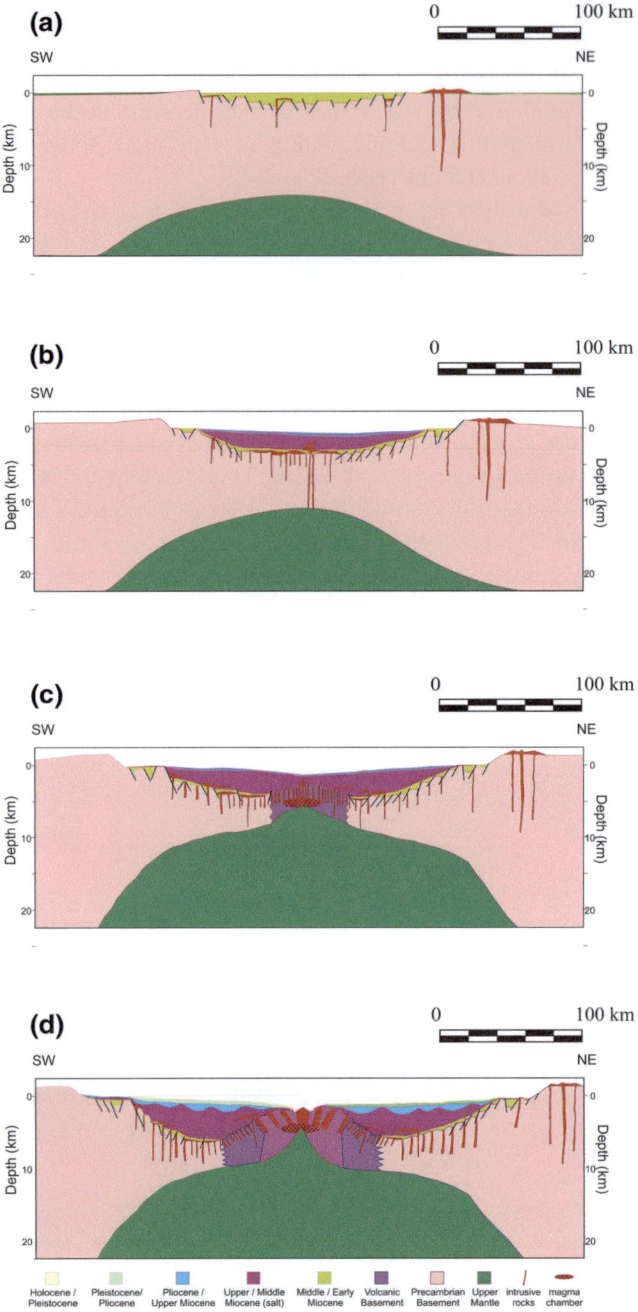

Fig. 26 Schematic geological model for the Red Sea evolution based on observations in the area of the Mabahiss Deep (Guennoc et al. 1990). **a** syn-rift sediments (Early to Middle Miocene) deposited on Precambrian basement, pre-rift sediments or volcanic rocks extruded from about 30 Mabp; **b** Magmatic activity started to focus on the incipient spreading centre by Middle to Upper Miocene, when the basin was filled with evaporites, which were initially pierced by igneous fingers; **c** Increased magmatic activity associated with continued extension and thinning of the continental crust formed igneous intrusions and volcanic edifices in the salt basin by Upper Miocene; **d** finally, an embryonic spreading centre started organized spreading by Upper Miocene/Lower Pliocene, resulting in the development of oceanic crust that eventually separated the salt bodies in the conjugate margins. Allochthonous salt has migrated oceanward over volcanic basement highs and advanced toward the abyss where the active spreading centre has developed in the last 5 My

crust and also in the proximal basins along the West African margin.

There are two fundamental currents of interpretation, based on development of salt basins either before or after the breakup that split the conjugate margins (Jackson et al. 2000). The pre-breakup hypothesis (Fig. 24a) assumes that the salt basins were formed almost simultaneously, forming a continuous layer from the Eastern Brazilian to the West African margins. The post-breakup hypothesis (Fig. 24b) assumes these salt basins might have slightly different ages (e.g., Davison 2007; Karner and Gambôa 2007) or were formed completely separated by a spreading centre along the incipient gulf between the conjugate margins (Jackson et al. 2000).

The following themes were thoroughly discussed to test each hypothesis (Jackson et al. 2000): (1) tectono-stratigraphy of salt; (2) tectono-stratigraphy of SDRs; (3) seismic characterization of the distal margin of salt basins; and (4) the map pattern of salt basins reconstructed for geological ages before the continental breakup and plate separation. These themes were focused on geological information mainly obtained from the African margins, and considering all lines of evidence that contradicted the pre-breakup hypothesis, strongly favouring a post-breakup origin for the Aptian salt basins of the South Atlantic margins, the authors advocated that the African and Brazilian salt basins must have always been separated by the mid-oceanic ridge, and a single salt basin could not have existed.

However, the analysis of the same themes for the Brazilian margin allows the following conclusions for the salt basins: (1) the evaporite layers post-date the main syn-rift phase and extend at least toward the proximal edge of the SDRs at the continental-oceanic crust transition (see for example, Fig. 8);

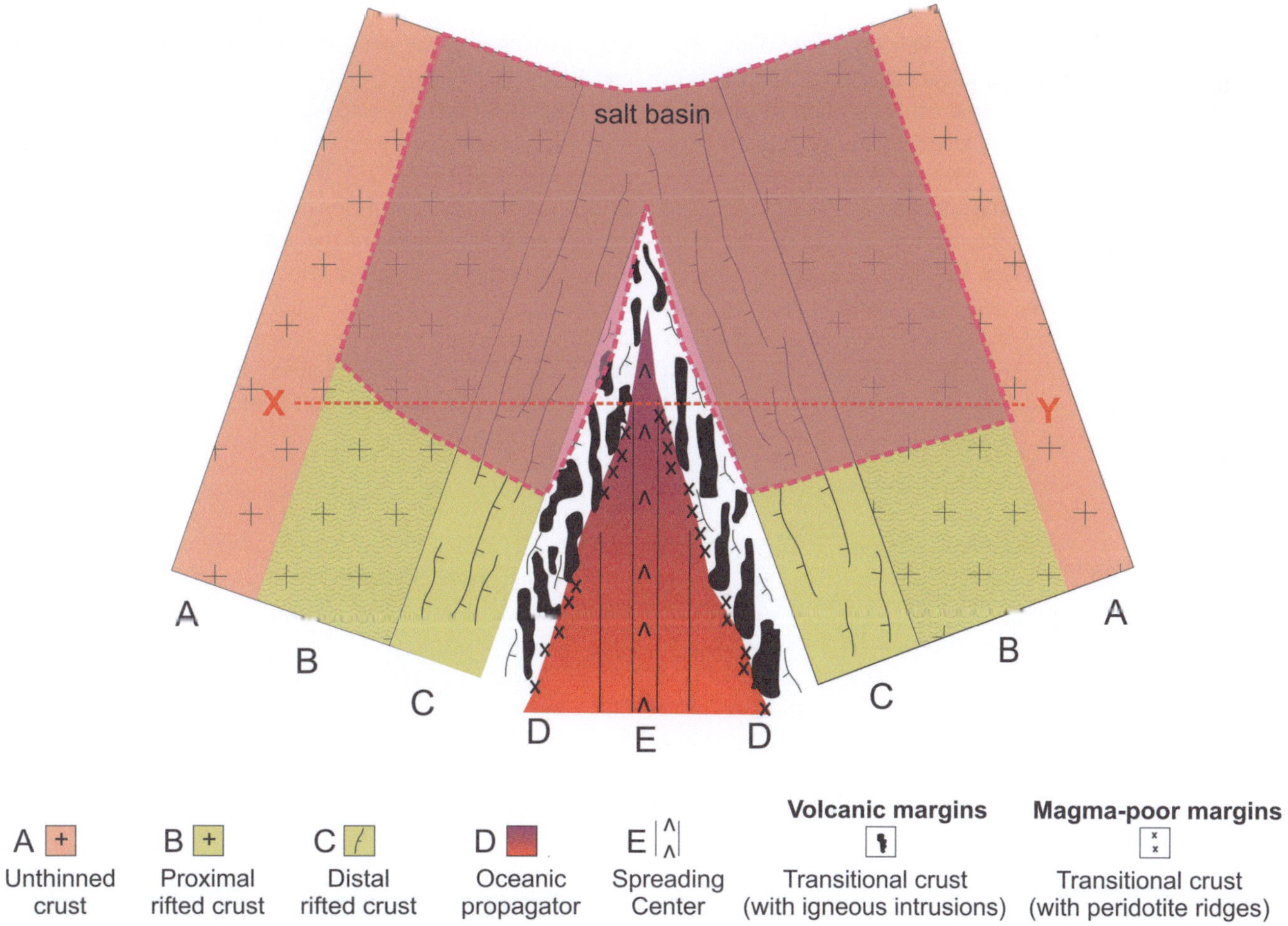

Fig. 27 Propagator model with four types of crust representing different stages of breakup in the same ocean. The sectors A, B and C correspond to continental crust affected by increasing extensional factors associated with the rifting process, which in the South Atlantic and Red Sea is followed by salt deposition. A transitional crust between C and D might involve igneous intrusions and extrusions in volcanic margins, or mantle exhumation in magma-poor margins. Salt was deposited in a single basin across the rifted conjugate margins. Allochthonous salt masses may advance toward the neo-formed volcanic basement region. The oceanic propagators (D–E) are associated with embryonic mid-ocean ridges or spreading centres that will develop oceanic crust splitting the salt basin along the conjugate divergent margins, which in the South Atlantic might correspond to the latest Aptian/earliest Albian

(2) some of the salt is allochthonous at the outermost highs that are located near the continent-ocean boundary (as, for example, the salt that occurs oceanward of the rifted crust in the Campos Basin, Fig. 4); (3) reconstructions of the plates for geological ages older than Aptian indicate that the salt masses may overlap in the distal portions of the conjugate margin basins (Heine et al. 2013), indicating that some of the salt may have advanced toward a volcanic basement, as suggested for the southern Santos Basin (see Fig. 2, with the salt limit oceanward of the volcanic basement limit); and (4) if we assume that the two salt basins imaged in the area of the Abimael propagator were once continuous and split apart by an embryonic spreading centre (Fig. 21b), then the final plate separation actually post-dates both the rift-drift unconformity and the evaporite deposition.

The Red Sea might be an interesting analogue for the interpretation of pre-breakup salt. There are many similarities with the South Atlantic in terms of tectonic events; the syn-rift phase is interpreted below the massive salt deposition, salt extends toward the volcanic basement, and allochthonous salt is presently advancing toward the abyss in the axial trough (Fig. 21). Based on integration of potential field, seismic refraction and reflection data (e.g., Egloff et al. 1991), and constrained by geological information from exploratory wells (Hughes and Johnson 2005), Mohriak and Leroy (2013) proposed a schematic model for the architecture of the central to southern Red Sea (Fig. 25).

This model suggests that the shelf zone and the deep-water provinces are characterized by large salt diapir structures, and the salt layer advances toward the main trough as allochthonous salt masses. Rifted continental crust with syn-rift sediments is interpreted below the salt at least in the proximal region and in the platform, and basinward, a volcanic basement may be present below the thick salt mass, as indicated by recent interpretations in the southern Red Sea (e.g., Almalki et al. 2014, 2015b). The axial trough may reach bathymetries exceeding 2000 m in the abyss and the salt masses are beginning to be separated by embryonic or active spreading centres associated with an incipient oceanic crust. This model assumes that the igneous intrusions in the axial trough correspond to a present-day spreading ridge formed about 2 Ma, and the volcanic and igneous crust may be about 7 km in thickness. The axial trough is a starved basin, with only a veneer of siliciclastic sediments covering the volcanic basement. Landward of the axial trough the volcanic basement may correspond to highly intruded continental crust or proto-oceanic crust with seaward-dipping reflectors, thus giving an age for the breakup unconformity older than the Miocene evaporites, with thick salt masses overlying the continental-oceanic crust transition in the marginal zone main trough (Mohriak and Leroy 2013).

These elements can also be envisaged in the numerical modelling conducted by Allen and Beaumont (2015). In the conceptual model (Fig. 25), the Red Sea is characterized by an onshore rift basin with residual salt covering pre-rift and syn-rift sediments; most of the salt has been mobilized basinward forming large salt diapirs in the platform and in the deep-water region. The deep-water region is characterized by salt layers covering a volcanic basement that rises toward the spreading centre, which is characterized by an abyss in the axial trough. Allochthonous salt is moving toward the abyss

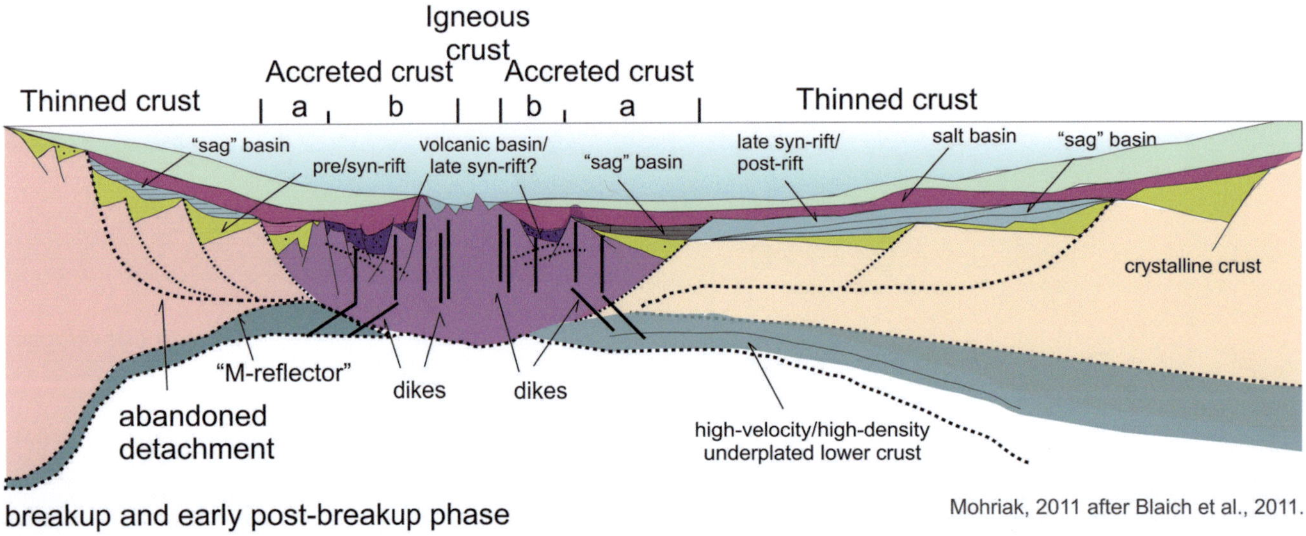

Fig. 28 Schematic geological transect (line X–Y in Fig. 27) across the South Atlantic Ocean showing an incipient oceanic basin that is similar to the Red Sea present-day stage. The salt basin was formed on sectors B and C. Salt is not deposited on the spreading ridge (sector E), but as observed in the Red Sea, allochthonous salt may advance toward the abyss where the active spreading centre is observed, and evaporites may occur on the incipiently accreted igneous crust (sectors C and D)

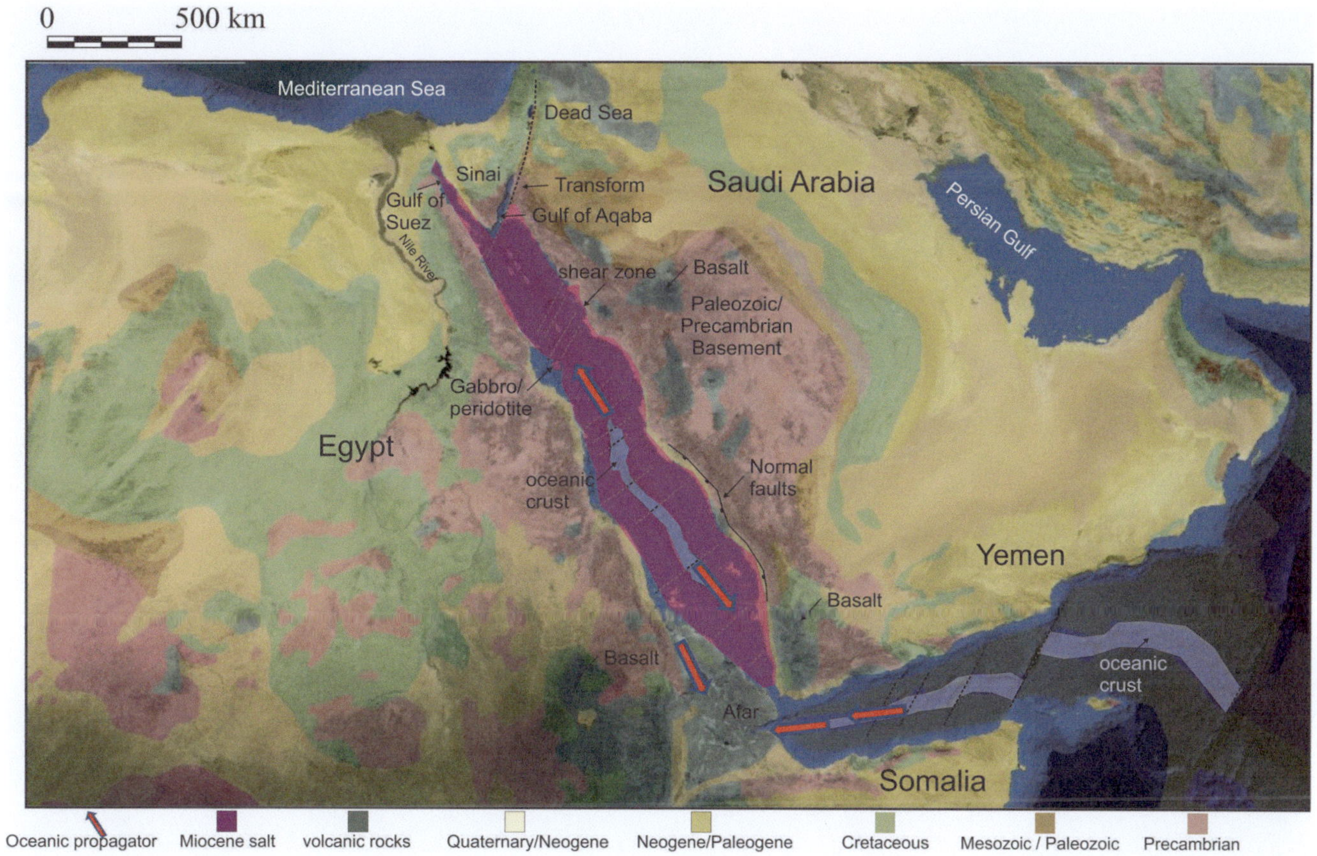

Fig. 29 Satellite image of the African (Nubian) and Arabian plates with a simplified geological map and schematic distribution of salt in the Red Sea, which is registered along the Egypt, Sudan and Eritrea margins in the western plate, with equivalent evaporites along Saudi Arabia and Yemen. Red arrows indicate the propagation of oceanic rifting both in the Gulf of Aden (with well-developed oceanic crust covered by siliciclastic rocks) and in the southern Red Sea (with embryonic spreading centres and a younger breakup unconformity, with the volcanic basement locally covered by evaporites). A single salt basin is observed in the northern Red Sea, whereas in the southern Red Sea the salt basins have been split apart by the active spreading centre. The salt basin between Africa and Arabia extends for about 2000 km along strike and is limited by the Afar hotspot in the south and by the transform fault at the Gulf of Aqaba in the north. The Gulf of Aden is characterized by older oceanic crust when compared to the Red Sea, and there are no Miocene evaporites in the rift basins along the Somalia–Yemen conjugate margins. North of the Zabargad shear zone there is a limited area of no evaporites in the northern part of the Mabahiss Deep, suggesting an incipient spreading centre with volcanic basement separating the salt masses

and is advancing above a volcanic basement adjacent to the present-day spreading centre. In the numerical model (Allen and Beaumont 2015), salt was deposited in a late syn-rift phase, and the salt mobilization resulted in salt flow toward the spreading centre, with allochthonous salt extending over previously formed oceanic crust.

The Mabahiss Deep has been characterized as the northermost segment of neo-formed oceanic basement in the central to northern Red Sea (Bonatti and Seyler 1987; Guennoc et al. 1990). The distal profile (Fig. 17) suggested initial piercing of the salt basin by igneous structures forming northward of the Mabahiss Deep. The sequence of events presented as schematic diagrams for the Red Sea evolution (Fig. 26) illustrates a possible mechanism of piercing the salt layer by igneous intrusions and subsequently splitting the margins as an embryonic spreading centre formed southward of the Mabahiss Seamount.

The conceptual model for the breakup process assuming the propagator model (Fig. 27) suggests that as we see today in the Red Sea, a single salt basin might have existed in the South Atlantic by Late Aptian times. The salt basin might have been locally split by oceanic propagators as we observe in the southern Santos Basin, where portions of the salt are separated by newly-formed volcanic basement (Mohriak 2014). The breakup unconformity, which could arguably be associated with the first inception of oceanic crust, predates the salt deposition, thus it was initially formed in other parts of the elongated gulf that extended from offshore southern Brazil to Argentina, where the rifting process was in a more advanced stage of evolution, as indicated by older magnetic

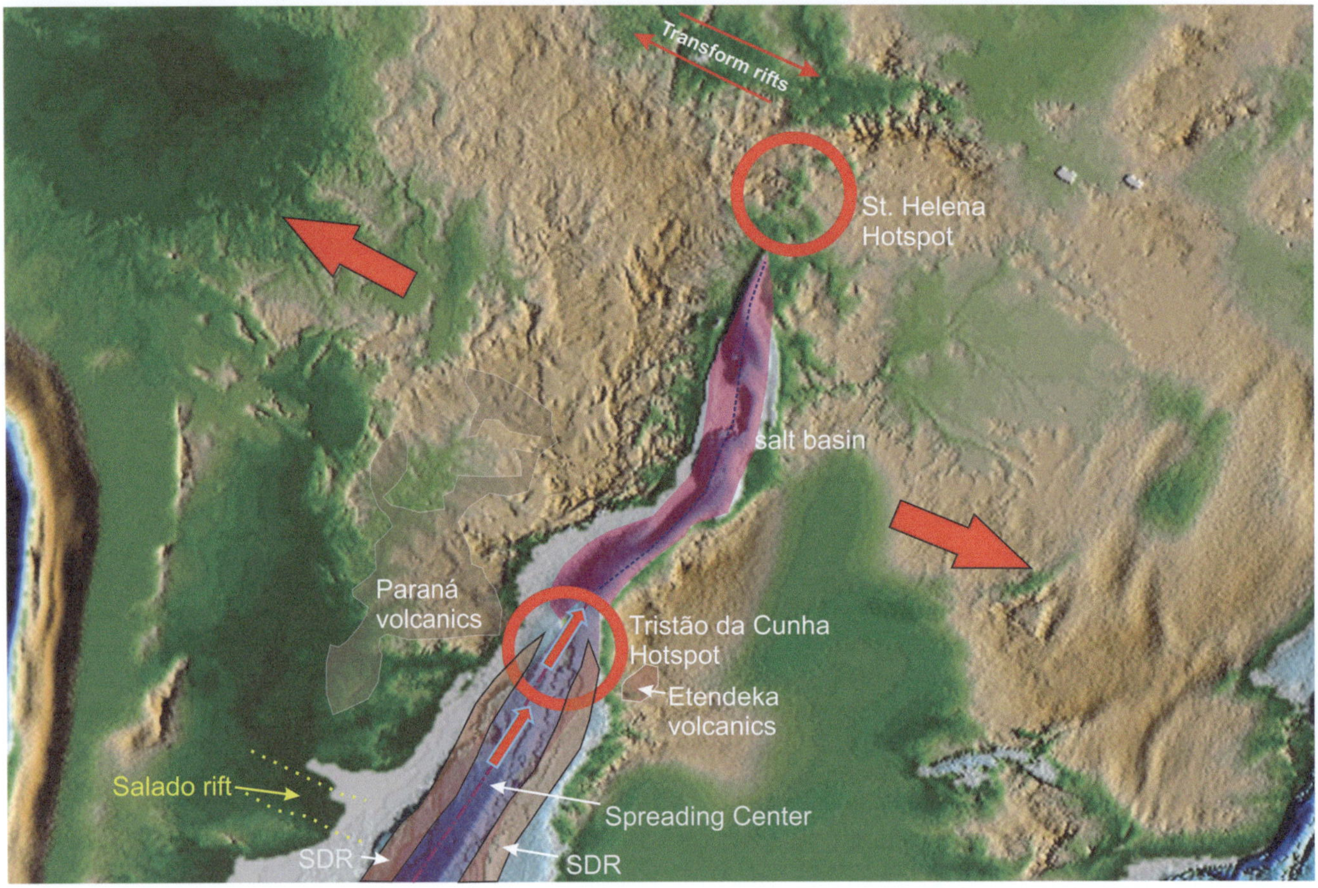

Fig. 30 Palinspastic reconstruction of the South Atlantic about 120 Ma (Scotese 2002), or probably at the transition from Aptian to Albian about 112 Ma, with schematic distribution of the salt basins and volcanic margins between Brazil and West Africa. Large arrows in South America and West Africa indicate the extensional stresses in the continental crust affecting the Santos, Campos, Espírito Santo and Bahia State Basins in Brazil, and the Namibia, Angola and Gabon conjugate divergent margins. The northern South Atlantic was affected by the Central Atlantic transtensional shear zone that evolved to the transform continental margins after continental breakup in the Albian, with no widespread evaporite deposition in the Late Aptian. The northern limit of the Late Aptian salt basin (magenta colour) is characterized by the transform rifts north of the St. Helena hotspot. The salt basin extends for about 2000 km along strike and is limited to the south by the Tristão da Cunha hotspot and the Florianópolis fracture zone at the southern Santos Basin. The southern South Atlantic is characterized by Early Cretaceous tholeiitic lava flows onshore (Paraná Basin volcanics of the Serra Geral Fm. and Etendeka basalts in Namibia). Offshore these volcanic margins are characterized by seaward-dipping wedges and active oceanic spreading centres that resulted in older oceanic crust and no evaporite deposition in the conjugate margins. Red arrows in the marine gulf between South America and Africa indicate the propagation of oceanic rifting from Argentina toward the Brazilian margin

anomalies and tectonic reconstructions (e.g., Moulin et al. 2010).

The diachronous nature of the breakup might be envisaged as a propagator advancing from areas with oceanic crust to areas with continental crust still in the rifting process (Fig. 27). The areal distribution of the rocks in this model suggests four types of crust: In the basement region adjacent to the rifted margins the crust has not been thinned (sector A in Fig. 27); the rifted continental crust in the proximal basins might be characterized by shallow lakes and continental sediments (sector B); the distal rifted continental crust might be associated with a gulf that was periodically affected by marine incursions, and basin desiccation resulted in deposition of evaporites overlying syn-rift lacustrine sediments (sectors B to C); and the distal margin might be associated with mantle exhumation in magma-poor margins such as in Iberia or to a volcanic basement in a transitional crust in magma-rich margins such as in the South Atlantic (sector D). Spreading centres would be formed in different episodes during the rifting process that led to the continental breakup and formation of oceanic crust. The oceanic ridges might start with igneous fingers that pierced through the pre-breakup rift and salt sequence, and subsequently split the salt basin, forming conjugate divergent margins, as observed in the central Red Sea.

Figure 28 schematically shows a cross-section for the South Atlantic at the transition from the gulf to the open ocean stage of the Wilson cycle, showing the distribution of

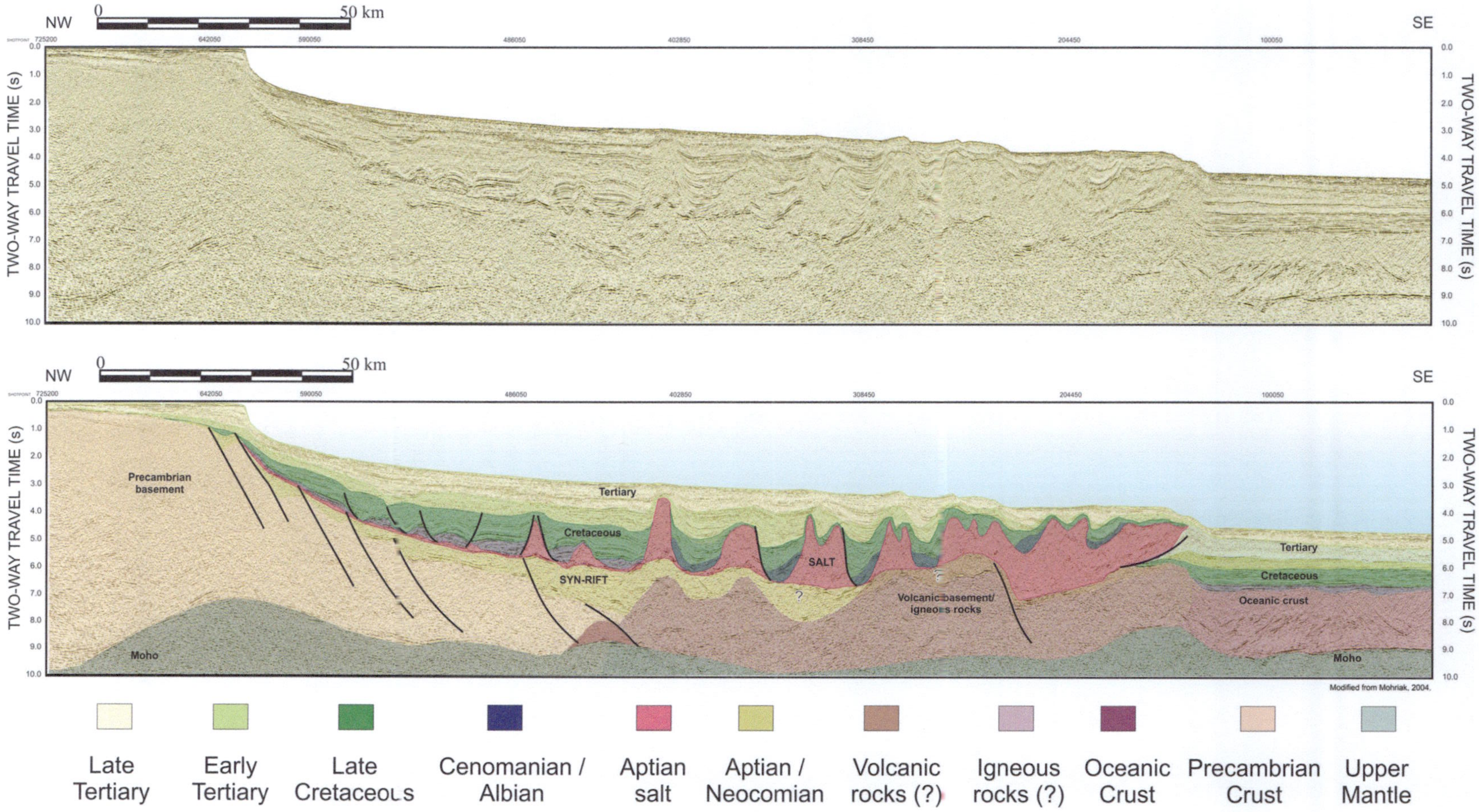

Fig. 31 Regional deep seismic profile in the Espírito Santo basin, offshore Brazil (seismic line ES1, Fig. 2), located south of the Abrolhos Volcanic Complex. Top: Uninterpreted seismic line; bottom: Geoseismic profile with a simplified interpretation (based on Mohriak 2003 and Mohriak et al. 2008). A residual salt layer is present in the narrow platform and in the proximal deep-water region. Salt mobilization resulted in large diapirs in the ultradeep-water province, and allochthonous salt is observed overriding a structural high, advancing toward the oceanic crust, forming a bathymetric escarpment. The massive salt walls basinward of the structural high are here interpreted to overlie a volcanic basement, associated with transitional crust highly intruded by igneous rocks, although other interpretations (e.g., Zalán et al. 2011) suggest it might correspond to exhumed mantle rocks (serpentinized peridotites). The oceanic crust at the SE extremity of the profile is clearly marked by a strong Moho reflector at 9.0 s TWT

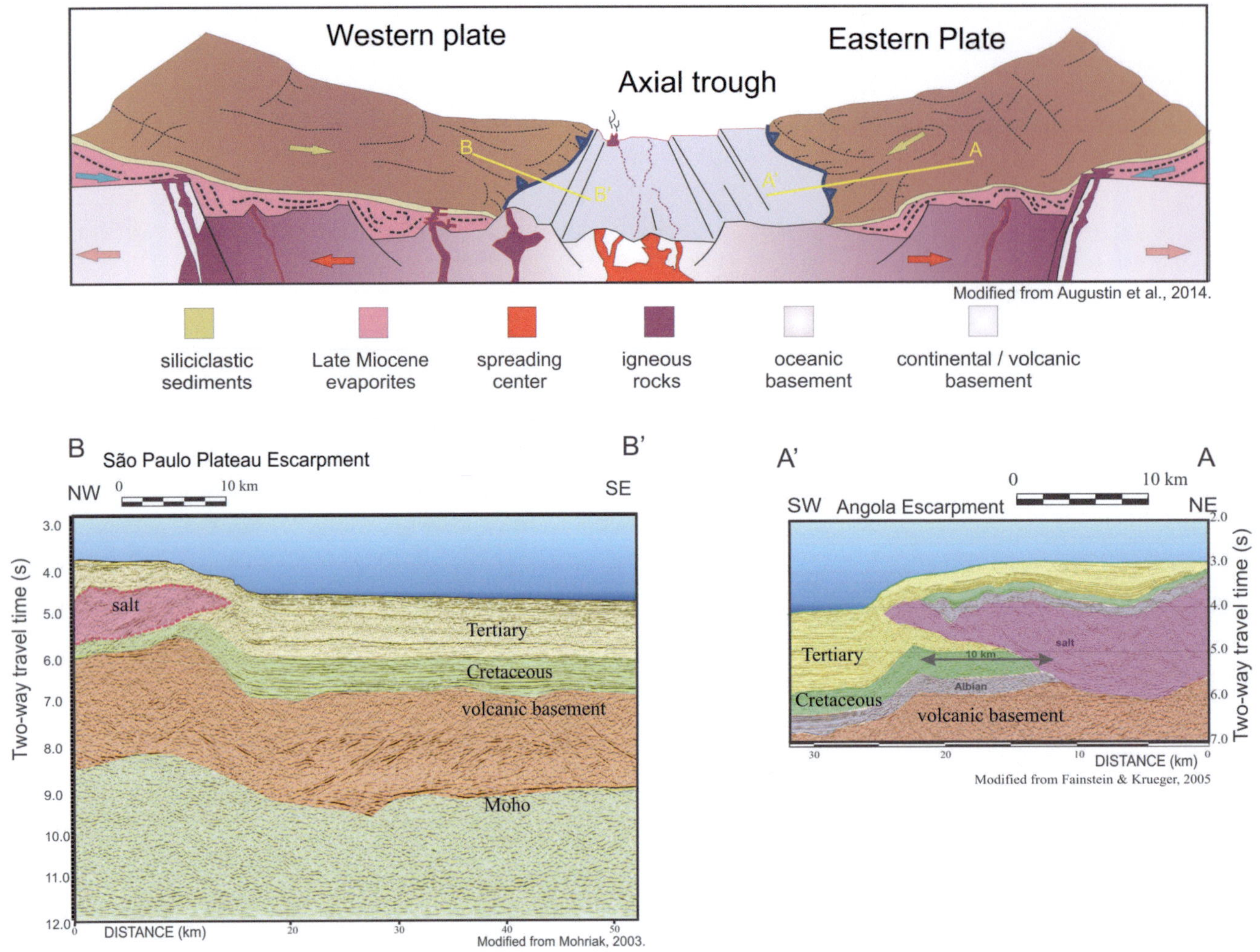

Fig. 32 Top: Schematic model of the Red Sea salt basins with allochthonous masses advancing toward the abyss in the axial trough along thrust faults, overriding volcanic basement highs (Augustin et al. 2014). The Western Plate (Nubian plate in the Red Sea, or South American plate in the South Atlantic reconstruction) may be characterized by an active spreading centre separating the conjugate margin salt basin formed on the Eastern Plate (Arabian plate, or African plate in the South Atlantic reconstruction). Sections B–B' and A'–A in the upper diagram correspond to the approximate location of seismic profiles along the Brazilian and West African margins. Bottom: Seismic profiles showing a present-day snapshot of the allochthonous salt tongues along the Espírito Santo Basin in Brazil (line B–B', corresponding to distal extremity of profile ES1, see location in Fig. 2 and regional profile in Fig. 31) and Kwanza Basin in West Africa (line A–A'). The distal salt tongue formed bathymetric escarpments in the São Paulo Plateau (left) and offshore Angola (right)

the evaporites and the rift structures in the conjugate margins that were separated by incipient spreading centres and volcanic crust around Aptian-Albian times. The two salt basins are separated by an incipient oceanic basin developed after the Early Cretaceous rifting that resulted in proximal and distal basins (Fig. 28, sectors B and C). Similar to the Red Sea, the model suggests that the evaporites in the distal margin might be advancing toward the embryonic spreading centre and covering a substratum associated with a late syn-rift sedimentation or volcanic basement characterized by SDRs, as recently suggested by Norton et al. (2016). Thus, the distal salt basin might overlie a transitional crust with igneous rocks accreted during the breakup process by Late Aptian times (Fig. 28, sectors C and D). The schematic transect, based on gravity modelling and seismic interpretation (Blaich et al. 2011; Mohriak and Leroy 2013), indicates that the salt basin extends from sectors B to C and may advance toward sector D as allochthonous salt masses (Fig. 28, sectors C and D), overlying volcanic rocks associated with inception of the active spreading centre that eventually separated the salt basins in the conjugate margins (Fig. 28, sector E, cf. Figure 27).

The salt distribution in the Red Sea is thus characterized by a single salt basin in the north, probably overlying a transitional crust highly intruded by igneous rocks, and by two salt basins in areas where the breakup processes resulted in splitting the salt masses apart (as in the southern Red Sea and more incipiently in the Mabahiss Deep). In these regions, the axial trough is an abyss with no evaporites, and newly-formed volcanic basement is observed in the regions adjacent to the spreading ridges (Fig. 29). The oceanic propagators that are well developed in regions with oceanic crust might advance toward triple junctions (e.g., Afar) or toward fracture zones (e.g., Zabargad Shear Zone).

In the South Atlantic, the salt distribution is characterized by tectonic limits similar to the Red Sea, involving hotspots and transform fault zones (Mohriak 2014). Several researchers assume that the Tristan da Cunha plume may be associated with the Early Cretaceous flood basalts (Serra Geral Fm. in the Paraná Basin, and Etendeka volcanics in Namibia). Tholeiitic basalts have also been drilled in most wells that penetrated the rift sequences offshore SE Brazil and Namibia, and the Ar–Ar ages for these igneous rocks indicate ages similar to the Paraná—Etendeka lava flows, around 130 ± 3 Ma (Almeida et al. 2013), thus they pre-date the development of the rift basins in the continental margins (Almeida et al. 2013).

The plate tectonic reconstruction of the South Atlantic at the gulf stage (Fig. 30) shows the location of the volcanic margins south of the Tristan da Cunha hotspot, which are characterized by seaward-dipping reflector wedges in the transition zone from continental to oceanic crust (Mello et al. 2013). The southern segment of the South Atlantic (Austral Segment) is interpreted to have opened between anomalies M13 and M4 (139.5 and 130 Ma, respectively) whereas the oceanic spreading in the central segment is believed to have started more than 18 Ma later, at the end of evaporite deposition (Moulin et al. 2010). The southern segment is also characterized by onshore-offshore rift basins in Argentina, dated as Jurassic to Early Cretaceous, with axes that trend oblique to the continental margin (e.g., Salado Basin in Fig. 30). There was no significant deposition of evaporites in the volcanic margins or in the onshore rifts south of the Tristan da Cunha hotspot, which may have acted as a volcanic barrier isolating the Santos Basin from the southern volcanic margin basins (Hinz et al. 1999; Mohriak 2001). A similar role can be envisaged for the Afar plume, which separates the Red Sea from the Gulf of Aden. The salt basins between the eastern Brazilian and West African margins overlie the Early Cretaceous syn-rift succession (Neocomian to Barremian) and also the Aptian sag basin, and might be dated as Late Aptian (around 112–115 Ma), although some authors suggest slightly different ages for the evaporites in Brazil and Africa (e.g., Davison 2007; Karner and Gambôa 2007).

The ages for the main tectonic events presented here differ slightly from other reconstructions of the South Atlantic, as for example Scotese (2002) assumes salt deposition at 120 Ma, and Heine et al. (2013) assume that by 115 Ma the salt basins were already separated by an active spreading centre. In this work, we envisage the final separation only after salt deposition (which probably ended with the marine incursions in the earliest Albian, probably slightly younger than 112 Ma). Most explorationists working in the Brazilian margins envisage the breakup unconformity at the top of the syn-rift tilted blocks, overlain by late syn-rift or sag basin sediments, thus predating the sag and salt deposition (e.g., Winter et al. 2007). This rift-drift unconformity in the Brazilian margin might be related to the inception of oceanic crust in the Austral Segment of the continental margin, considering that the age of the first oceanic crust varied progressively from south to north as indicated by magnetic anomalies, flow lines and tectonic reconstructions (Moulin et al. 2010). However, the final separation between the plates with the salt basins occurred only after salt deposition, as indicated by the aborted spreading centre in the southern Santos Basin, pointing to an age of Late Aptian-Early Albian for final splitting of the Santos-Campos and Kwanza conjugate basins. This corroborates the assumption that the breakup is diachronous and may occur in some parts of the basin coeval with active rifting processes in other parts of the continental margin.

Figure 31 shows a regional deep seismic profile in the Espírito Santo Basin, south of the Abrolhos Volcanic Complex (see location in Fig. 2). The seismic line extends from the continental crust in the platform toward the oceanic crust beyond the salt limit. A conspicuous structural high is observed in the ultradeep-water region, and although there are interpretations of exhumed mantle at this position, the integration with potential field data suggests it corresponds to a volcanic basement high, probably formed around the breakup time, after the rift phase (Mohriak et al. 2008).

In the Espírito Santo profile (Fig. 31), the tectonic processes that led to the development of conjugate divergent margins are much more advanced than in the Red Sea. Figure 32 shows the schematic model of the distal part of the Red Sea salt basin (Augustin et al. 2014) compared with seismic profiles in the distal parts of the South Atlantic salt basins. The Red Sea model corresponds to an analogue of the South Atlantic salt basins at the transition from Aptian to Albian about 112 Ma, and the conjugate margins at present are characterized by allochthonous salt tongues advancing toward oceanic crust (Mohriak 2014). In the South Atlantic, the allochthonous salt bodies have advanced basinward of the basement or volcanic high and are now forming salt tongues that are migrating toward the oceanic crust, resulting in a bathymetric escarpment that is rather similar to the Sigsbee Escarpment in the Gulf of Mexico (Fainstein and Krueger 2005). The salt body at the SE extremity of the Espírito Santo Basin profile (Fig. 32, bottom left) has a thickness of about 3 km. The oceanic crust beyond the salt tongue has a thickness of about 7 km and overlies the oceanic Moho, which is clearly imaged at a depth of about 14 km. This seismic profile corresponds to a snapshot of a salt basin 115 Ma after salt deposition, with allochthonous salt advancing toward the oldest oceanic crust. The salt tongue in the Angolan margin (Fig. 32, bottom right) might transport Cretaceous strata in a piggy-back style, with about 10 km of Albian to Maastrichtian sequences overlapping above the volcanic basement.

These analogies provide important clues on how salt basins in the South Atlantic might have evolved through time. At present, the allochthonous salt front in Brazil is more than 4000 km from its conjugate margin salt front in the Kwanza Basin, which suggests that salt advanced toward an oceanic crust basement that was formed by the Late Aptian—Early Albian (Mohriak et al. 2008). The tectonic processes are much more advanced in the South Atlantic than in the Red Sea, where only a portion of the salt masses have been split by incipient oceanic spreading in the last 2–3 Ma (Ligi et al. 2012).

7 Conclusions

Several changes in the interpretation of the South Atlantic salt basins have been discussed in the last 20 years as a result of intensive exploratory work in the distal margins of these prolific basins. These changes in paradigms and search of new interpretations based on analogies with magma-poor margins include: (1) the lack of thick syn-rift depocentres in ultradeep-water distal sedimentary provinces; (2) the presence of important carbonate (microbialite) reservoirs in the pre-salt sequence; (3) the questionable proposal of exhumed mantle in the transition from continental to oceanic crust; and (4) the characterization of different styles of salt tectonics associated with extensional and compressional events, including the presence of allochthonous salt tongues in the transition from continental to oceanic crust.

The Iberian model of mantle exhumation, which recently has been extensively applied for the Brazilian and West African basins (Péron-Pinvidic et al. 2013), shows many important differences in tectonic context and stratigraphy from the South Atlantic margins (see Mohriak and Leroy 2013—Table 1). The Iberian model assumes that the separation of extending continental crust is achieved by detachment faults that expose large stretches of mantle peridotites before an oceanic spreading centre is formed. This occurs in areas where volcanic activity is negligible both before the rift and during breakup. This is not the case of the Red Sea during the Neogene and Quaternary periods, which are marked by magmatic activity during the rift and early drift phases. Volcanic rocks related to the rifting process are observed both onshore (for example, in Saudi Arabia and Yemen, and in the Afar province) and offshore (for example, in volcanic islands recently formed near the southern Red Sea axial trough, or as volcanic structures partially covered by salt in the central to northern Red Sea, for example, in the Mabahiss Deep). Volcanic structures such as wedges of seaward-dipping reflectors, volcanoes and hydrothermal vents have also been characterized in the South Atlantic during the Early Cretaceous rift phase and in the early drift phase.

The Red Sea basins can be considered better analogues for the continental breakup that occurred in the South Atlantic by the Late Aptian—Early Albian, after the magmatic episode that formed the Paraná-Etendeka large igneous province (LIP) and the extensional stresses that formed continental lakes in the rifted margins. Similar to the Gulf of Aden–Red Sea system, the South Atlantic is marked by two major sedimentary provinces, which are separated by the Tristan da Cunha mantle plume. The volcanic rocks of the Paraná-Etendeka LIP may correspond to the initial impact of a mantle plume in the continental crust, and thus are similar to the pre-rift volcanism in the Afar region, between the Red Sea and the Gulf of Aden. The region north of the Tristan da Cunha hotspot is characterized by major salt basins that are extremely prolific in hydrocarbons, and the basins south of the hotspot (Pelotas Basin in southern Brazil, and the basins offshore Uruguay and Argentina), are characterized by large igneous provinces both onshore and offshore, with development of large wedges of seaward-dipping reflectors, with minor oil production both in South America and in Africa. Syn-rift magmatic activity resulting in formation of volcanic basement and large volcanoes is registered both in the Red Sea (as in the Mabahiss Deep) and in the South Atlantic salt basins. Several exploratory wells in the Campos and Santos basins have sampled volcanic rocks in the syn-rift and pre-rift sequences, and seismic interpretation suggests the presence of volcanic layers (even forming incipient wedges of SDRs) or magmatic structures such as saucer-shaped sills and hydrothermal vents.

The analysis of industry and academic seismic profiles in the Red Sea indicates that the known distributions of volcanic structures onshore and offshore are associated with development of an incipient divergent margin, with an axial trough locally characterized by embryonic active spreading centres that correspond with volcanic rocks (tholeiitic basalts). These features have been imaged in the central Red Sea and can be compared to the aborted spreading centres in the South Atlantic salt basins, particularly to the Abimael Ridge in the southern Santos Basin, which apparently propagated from the Pelotas Basin toward the southern Santos Basin, but failed to develop an active and successful spreading ridge that would split the South American and African plates at the western tip of the Florianópolis Fracture Zone. This feature penetrated the southern Santos salt basin and constitutes an important element to calibrate the relative ages of rifting, salt deposition and continental breakup. It is interpreted as an igneous intrusion associated with seaward-dipping reflector wedges in the Pelotas Basin. The

Moho rises rapidly basinward of the salt limit in the deep-water region of the southern Santos Basin, but there is no clear mantle exhumation or salt deposition in the region of the oceanic propagator.

Two end-member hypotheses for salt deposition in the South Atlantic have been discussed; one (A) that assumes that salt was deposited before, and (B) the other after continental breakup. Jackson et al. (2000) discarded hypothesis (A) and favored hypothesis (B), suggesting that the salt basins in the South Atlantic have always been separated by a spreading centre formed before salt deposition, and the salt layer does not extend toward the normal oceanic crust. Using the Red Sea as a present-day analogue, this work suggests that hypothesis (A) may be observed in some parts of the central Red Sea, with spreading ridges splitting a single salt basin and forming conjugate margins with evaporites separated by newly-formed oceanic crust. The breakup unconformity, however, may be interpreted as older than the evaporites, assuming a diachronous propagation of the oceanic rifting. This interpretation would imply that the evaporites advanced over a volcanic basement developed before the formation of the active spreading centres that are splitting the salt layer in regions such as the Mabahiss Deep or the Thetis Deep.

Assuming the model based on the Red Sea observations, the South Atlantic salt basin was split after salt deposition by igneous intrusions and embryonic spreading centres that propagated from regions with oceanic crust already formed (as for example, offshore Argentina). Salt migrated toward the newly formed oceanic crust axial troughs after initial halokinesis, forming allochthonous salt tongues overlying volcanic basement. Oceanic propagators in the South Atlantic and in the Red Sea advanced from oceanic crust toward a mantle plume thermal anomaly and transform fault zones, and the breakup is diachronous along the length of the continental margins. These tectonic models have important implications for understanding the petroleum systems that may be active in the early stages of basin development, particularly for the evaluation of pre-salt stratigraphic successions, source rock and reservoir distribution.

Acknowledgements The author wishes to thank several geoscientists at PETROBRAS—Petroleo Brasileiro S. A. for enlightening discussions and participation in previous projects conducted in the 2000's, which focused on the geology of the Red Sea and the analogies with petroleum systems and exploratory plays in the South Atlantic. I am also grateful to many colleagues and students at UERJ—State University of Rio de Janeiro for providing suggestions and improvements for this work, particularly S. Wischer who revised the text of an earlier draft. The Red Sea Team at Saudi Aramco provided a unique opportunity to participate in a field trip in the Midyan Basin in 2013, and I enjoyed working with the large regional dataset offshore Saudi Arabia.The AAPG Distinguished Lecturer Program is also thanked for their invitation to participate in the 2014 program and present the summary of the chapter included in the book "Conjugate Divergent Margins", published by the Geological Society of London in 2013. I am grateful to the Saudi Geological Survey staff for the organization of the Jeddah meeting in early 2016, which was very well organized and integrated several branches of science in the search for a better understanding of the geological, biological and anthropogenic backgrounds of the Red Sea. Dr. Najeeb Rasul and Dr. Ian Stewart had key roles in the conference logistics and also helped in the preparation of the book chapters. Special thanks are owed to N. Augustin for kindly providing high-resolution bathymetric data on the Mabahiss Deep and to M. Ligi for technical discussions on the Red Sea geology. The final review of this contribution was provided by Ian Davison, Michael R. Hudec and Nickolas Raterman. I greatly appreciated their enlightening comments on key concepts and constructive suggestions that helped to substantially improve the scientific content and clarity of the text.

References

Afifi AM, Tapponnier P, Raterman NS (2014) The Messinian unconformity in the Red Sea: evidence for widespread dessication? AAPG search and discovery article 90188, GEO-2014. In: 11th Middle east geosciences conference and exhibition, March 2014, Manama, Bahrain, pp 10–12

Allen J, Beaumont C (2015) Continental margin syn-rift salt tectonics at intermediate width margins. Basin Res 27:1–36. https://doi.org/10.1111/bre.12123

Almalki KA, Betts PG, Ailleres L (2014) Episodic sea-floor spreading in the southern Red Sea. Tectonophysics 617:140–149. https://doi.org/10.1016/j.tecto.2014.01.030

Almalki KA, Betts PG, Ailleres L (2015a) The Red Sea—50 years of geological and geophysical research. Earth-Sci Rev 147:109–140

Almalki KA, Ailleres L, Betts PG, Bantan RA (2015b) Evidence for and relationship between recent distributed extension and halokinesis in the Farasan Islands, southern Red Sea, Saudi Arabia. Arabian J Geosci 8:8753–8766

Almeida J, Dios F, Mohriak WU, Valeriano CM, Heilbron M, Eirado LG, Tomazzoli E (2013) Pre-rift tectonic scenario of the eo-cretaceous Gondwana break-up along SE Brazil–SW Africa: insights from tholeiitic mafic dyke swarms. In: Mohriak WU, Danforth A, Post PJ, Brown DE, Tari GC, Nemcok M, Sinha ST (eds) Conjugate divergent margins, vol 369. Geological Society of London, Special Publications, pp 11–40

Alsharhan AS, Salah MG (1997) A common source rock for Egyptian and Saudi hydrocarbons in the Red Sea. Am Assoc Petrol Geol Bull 81(10):1640–1659

Alvarenga RS, Iacopini D, Kuchle J, Scherer CMS, Goldberg K (2016) Seismic characteristics and distribution of hydrothermal vent complexes in the cretaceous offshore rift section of the Campos Basin, offshore Brazil. Marine Petrol Geol 74:12–25

Antonini P, Petrini R, Contin G (1998) A segment of sea-floor spreading in the central Red Sea: Basalts from the Nereus deep (23°00'–23°20'N). J African Earth Sci 27:107–114

Augustin N, Devey CW, van der Zwan FM, Feldens P, Tominaga M, Bantan RA, Kwasnitschka T (2014) The rifting to spreading transition in the Red Sea. Earth Planet Sci Lett 395:217–230. https://doi.org/10.1016/j.epsl.2014.03.047

Augustin N, Zwan FM, Devey CW, Ligi M, Kwasnitschka T, Feldens P, Bantan RA, Basaham AS (2016) Geomorphology of the central Red Sea Rift: determining spreading processes. Geomorphology 274:162–179. https://doi.org/10.1016/j.geomorph.2016.08.028

Baker J, Snee L, Menzies M (1996) A brief Oligocene period of flood volcanism in Yemen: implications for the duration and rate of continental flood volcanism at the Afro-Arabian triple junction. Earth Planet Sci Lett 138:39–55

Balmino G, Vales N, Bonvalot S, Briais A (2012) Spherical harmonic modeling to ultra-high degree of Bouguer and isostatic anomalies. J Geodesy 86:499–520. https://doi.org/10.1007/s00190-011-0533-4

Bayer HJ, El-Isa Z, Hotzl H, Mechie J, Prodehl C, Saffarini G (1989) Large tectonic and lithospheric structures of the Red Sea region. J African Earth Sci 8:565–587

Beglinger SE, Doust H, Cloetingh S (2012) Relating petroleum system and play development to basin evolution: Brazilian South Atlantic margin. Petrol Geosci 18:315–336. https://doi.org/10.1144/1354-079311-022

Benson RN, Doyle RG (1988) Early Mesozoic rift basins and the development of the United States middle Atlantic continental margin. In: Manspeizer W (ed) Triassic-Jurassic rifting—continental breakup and the origin of the Atlantic ocean and passive margins, vol 22. Development in Geotectonics. Elsevier Science Publishers BV, Chapter 5, pp 99–127

Berman A (2008) Three super-giant fields discovered in Brazil's Santos Basin. World Oil 23–24

Beydoun ZR (1989) The hydrocarbon prospects of the Red Sea—Gulf of Aden: a review. J Petrol Geol 12:125–144

Blaich OA, Faleide JI, Tsikalas F (2011) Crustal breakup and continent-ocean transition at South Atlantic conjugate margins. J Geophys Res 116:B01402. https://doi.org/10.1029/2010JB007686

Blank HR, Andreasen GE (1991) Compilation and interpretation of aeromagnetic data for the Precambrian Arabian shield, Kingdom of Saudi Arabia. Saudi Arabian Directorate General of Mineral Resources Open-File Report USGS-OF-10-8, 54 p

Blank HR (1977) Aeromagnetic and geological study of Tertiary dikes and related structures on the Arabian margin of the Red Sea. In: Saudi Arabian directorate general of mineral resources, Red Sea research 1970–1975, Bulletin 22:G1–18

Boillot G, Grimaud S, Mauffret A, Mougenot D, Kornprobst J, Mergoil-Daniel J, Torrent G (1980) Ocean-continent boundary off the Iberian margin: a serpentinite diapir west of the Galicia bank. Earth Planet Sci Lett 48:23–34

Boillot G, Winterer EL, Meyer AW et al (1987) Proc ODP, initial reports 103. College station, TX (Ocean drilling program). https://doi.org/10.2973/odp.proc.ir.103.1987

Bonatti E (1985) Punctiform initiation of seafloor spreading in the Red Sea during transition from a continental to an oceanic rift. Nature 316:33–37

Bonatti E, Seyler M (1987) Crustal underplating and evolution in the Red Sea rift: uplifted gabbro/gneiss crustal complexes on Zabargad and Brothers Islands. J Geophys Res 92:12803–12821

Bonatti E, Cipriani A, Lupi L (2015) The Red Sea: birth of an Ocean. In: Rasul NMA, Stewart ICF (eds) The Red Sea: the formation, morphology, oceanography and environment of a Young Ocean Basin. Springer Earth System Sciences. Springer, Heidelberg, pp 29–44. https://doi.org/10.1007/978-3-662-45201-1

Bonatti E, Clocchiatti R, Colantoni P, Gelmini R, Marinelli G, Ottonello G, Santacroce R, Taviani M, Abdel-Meguid AA, Assaf HS, Tahir MA (1983) Zabargad (St. John's) Island: an uplifted fragment of sub-Red Sea lithosphere. J Geol Soc London 140:677–690

Bonatti E, Colantoni P, Vedova BD, Taviani M (1984) Geology of the Red Sea transitional region (22°N–25°N). Oceanol Acta 7:385–398

Bonattti E, Hamlyn P, Ottonello G (1981) Upper mantle beneath a young oceanic rift: peridotites from the Island of Zabargad (Red Sea). Geology 9:474–479

Bosworth W (2015) Geological evolution of the Red Sea: historical background, review, and synthesis. In: Rasul NMA, Stewart ICF (eds) The Red Sea: the formation, morphology, oceanography and environment of a young ocean basin. Springer Earth System Sciences. Springer, Heidelberg, pp 45–78. https://doi.org/10.1007/978-3-662-45201-1_3

Bosworth W, McClay K (2001) Structural and stratigraphic evolution of the Gulf of Suez Rift, Egypt: a synthesis. In: Ziegler PA, Cavazza W, Robertson AHF, Crasquin-Soleau S (eds) Peri-Tethys Memoir 6: Peri-Tethyan Rift/Wrench basins and passive margins, vol 186. Memoires du Museum National d'Histoire Naturelle, Paris, pp 567–606

Bosworth W, Huchon P, McClay K (2005) The Red Sea and Gulf of Aden basins. J African Earth Sci 43:334–378

Braun J, Beaumont C (1989) A physical explanation of the relation between flank uplifts and the breakup unconformity at rifted continental margins. Geology 17:760–764

Cainelli C, Mohriak WU (1998) Geology of Atlantic Eastern Brazilian basins. In: 1998 AAPG international conference & exhibition short course—Brazilian geology Part II, 8–11 November 1998, Rio de Janeiro, Brazil, 67 p

Carminatti M, Dias JL, Wolf B (2009) From turbidites to carbonates: breaking paradigms in deep waters. OTC, Abstract Volume, OTC-20124, 7 p

Cochran JR (1983) A model for development of Red Sea. Am Assoc Petrol Geol Bull 67:41–69

Cochran JR, Karner GD (2007) Constraints on the deformation and rupturing of continental lithosphere of the Red Sea: the transition from rifting to drifting. In: Karner GD, Manatschal G, Pinheiro LM (eds) Imaging, mapping and modelling continental lithosphere extension and breakup, vol 282. Geological Society of London, Special Publication, pp 265–289

Colombo D, McNeice G, Raterman N, Zinger M, Rovetta D, Curiel ES (2014) Exploration beyond seismic: the role of electromagnetics and gravity gradiometry in deep water subsalt plays of the Red Sea. Interpretation 2:SH33–SH53

Corti G, Bastow ID, Keir D, Pagli C, Baker E (2015) Rift-related morphology of the Afar depression. In: Billi P (ed) Landscapes and landforms of Ethiopia. World Geomorphological Landscapes Series. Springer, Chapter 15, pp 251–274. https://doi.org/10.1007/978-94-017-8026-1_15

Courtillot VE (1982) Propagating rifts and continental breakup. Tectonics 1:239–250

Davison I (1999) Tectonics and hydrocarbon distribution along the Brazilian South Atlantic margin. In: Cameron NR, Bate RH, Clure VS (eds) The oil and gas habitats of the South Atlantic, vol 153. Geological Society of London, Special Publications, pp 133–151

Davison I (2007) Geology and tectonics of the South Atlantic Brazilian salt basins. In: Ries AC, Butler RWH, Graham, RH (eds) Deformation of the continental crust: the legacy of Mike Coward, vol 272. Geological Society of London, Special Publications, pp 345–359

Davison I, Al-Kadasi M, Al-Khirbash S, Al-Subbary A, Baker J, Blakey S, Bosence D, Dart C, Heaton R, McClay K, Menzies M, Nichols G, Owen L, Yelland A (1994) Geological evolution of the southeastern Red Sea rift margin: Republic of Yemen. Geol Soc Am Bull 106:1474–1493

Dehler NM, Magnavita LP, Gomes LC, Rigoti CA, Oliveira JAB, Sant'Anna M, Costa FGD (2016) The 'Helmut' geophysical anomaly: a regional left-lateral transtensional shear zone system connecting Santos and Campos basins, southeastern Brazil. Marine Petrol Geol 72:412–422

Demercian LS, Szatmari P, Cobbold PR (1993) Style and pattern of salt diapirs due to thin-skinned gravitational gliding, Campos and Santos basins, offshore Brazil. Tectonophysics 228:393–433

Duarte MS, Alves TM, Terrinha P (2012) The breakup sequence and associated lithospheric breakup surface: their significance in the context of rifted continental margins (West Iberia and Newfoundland margins, North Atlantic). Earth Planet Sci Lett 355–356:311–326

Egloff F, Rihm R, Makris J, Izzeldin YA, Bobsien M, Meir K, Junge P, Noman T, Warsi W (1991) Contrasting structural styles of the

eastern and western margins of the southern Red Sea: the 1988 SONNE experiment. Tectonophysics 198:329–353
Esedo R, van Wijk J, Coblentz D, Meyer R (2012) Uplift prior to continental breakup: indication for removal of mantle lithosphere? Geosphere 8:1078–1085. https://doi.org/10.1130/GES00748.1
Esestime P, Hewitt A, Hodgson N (2016) Zohr—a newborn carbonate play in the Levantine Basin, East-Mediterranean. First Break 34:87–93
Evain M, Afilhado A, Rigoti C, Loureiro A, Alves D, Klingelhoefer F, Schnurle P, Feld A, Fuck R, Soares J, Lima MV, Corela C, Matias L, Benabdellouahed M, Baltzer A, Rabineau M, Viana A, Moulin M, Aslanian D (2015) Deep structure of the Santos Basin-São Paulo Plateau system, SE Brazil. J Geophys Res Solid Earth 120:5401–5431. https://doi.org/10.1002/2014JB011561
Fainstein R, Krueger A (2005) Salt tectonics comparisons near the continent-ocean boundary escarpments. In: Post PJ, Rosen NC, Olson DL, Palmes SL, Lyons KT, Newton GB (eds) GCSSEPM 25th Annual Bob F. Perkins research conference: petroleum systems of divergent continental margin basins, Abstracts CD, pp 510–540
Falvey DA (1974) The development of continental margins in plate tectonic theory. APEA J 14(1):95–106
Franke D (2013) Rifting, lithosphere breakup and volcanism: comparison of magma-poor and volcanic rifted margins. Marine Petrol Geol 43:63–87
Gass IG, Mallick DJ, Cox KG (1973) Volcanic islands of the Red Sea. J Geol Soc London 29:275–310
Ghebreab W (1998) Tectonics of the Red Sea region reassessed. Earth-Sci Rev 45:1–44
Gomes BS (1992) Integração preliminar dos dados gravimetricos marítimos da Petrobras e do Projeto Leplac: Bacias de Campos, Santos e Pelotas. Anais do XXXVII Congresso Brasileiro de Geologia, São Paulo, SP 1:559–560
Gomes PO, Kilsdonk W, Grow T, Minken J, Barragan R (2012) Tectonic evolution of the outer high of Santos Basin, southern São Paulo Plateau, Brazil, and implications for hydrocarbon exploration. In: Gao D (ed) Tectonics and sedimentation: implications for petroleum systems. Am Assoc Petrol Geol Memoir 100:125–142
Guardado LR, Gamboa LAP, Luchesi CF (1989) Petroleum geology of the Campos basin, a model for a producing Atlantic-type basin. In: Edwards JD, Santogrossi PA (eds) Divergent/Passive margin basins. Am Assoc Petrol Geol Memoir 48:3–79
Guardado LR, Spadini AR, Brandão JSL, Mello MR (2000) Petroleum system of the Campos basin, Brazil. In: Mello MR, Katz BJ (eds) Petroleum systems of South Atlantic margins. Am Assoc Petrol Geol Memoir 73:317– 324
Guennoc P, Pautot G, Leqentrec MF, Coutelle A (1990) Structure of an early oceanic rift in the northern Red Sea. Oceanol Acta 13: 145–157
Hadad YT, Abdullah WH (2015) Hydrocarbon source rock generative potential of the Sudanese Red Sea basin. Marine Petrol Geol 65:269–289
Hadad YT, Hakimi MH, Abdullah WH, Makeen YM (2016) Thermal maturity history reconstruction and hydrocarbon generation/ expulsion modeling of the syn-rift Rudeis and Kareem source rocks in the Red Sea rift basin, Sudan. Arabian J Geosci 9:442, 17 p. https://doi.org/10.1007/s12517-016-2458-y
Hall SA (1979) A total intensity magnetic anomaly map of the Red Sea and its interpretation. U.S. geological survey Saudi Arabian project report 275, Open File 80-131, 260 p
Hartmann J, Moosdorf N (2012) The new global lithological map database GLiM: a representation of rock properties at the earth surface. Geochem Geophys Geosyst 13:Q12004. https://doi.org/10.1029/2012GC004370
Heine C, Zoethourt J, Muller RD (2013) Kinematics of the South Atlantic rift. Solid Earth 4:215–253. https://doi.org/10.5194/se-4-215-2013
Hinz K, Neben S, Schreckenberger B, Roeser HA, Block M, Souza KG, Meyer H (1999) The Argentine continental margin north of 48° S: sedimentary successions, volcanic activity during breakup. Marine Petrol Geol 16(1):1–25
Hosny A, Nyblade A (2014) Crustal structure in southeastern Egypt: symmetric thinning of the northern Red Sea rifted margins. Geology 42:219–222. https://doi.org/10.1130/G34726.1
Hughes G, Johnson RS (2005) Lithostratigraphy of the Red Sea region. GeoArabia 10(3):49–126
Jackson MPA, Cramez C, Fonck JM (2000) Role of subaerial volcanic rocks and mantle plumes in creation of South Atlantic margins: implications for salt tectonics and source rocks. Marine Petrol Geol 17:477–498
Jones CM, Chaves HAF (2015) Re-Assessment of yet-to-find oil in the pre-salt province of Brazil. In: 14th International congress of the Brazilian geophysical society, Rio de Janeiro, Brazil, August 3–6, 2015, 6 p
Karner GD (2000) Rifts of the Campos and Santos basin, southeastern Brazil: distribution and timing. In: Mello MR, Katz BJ (eds) Petroleum systems of South Atlantic margins. Am Assoc Petrol Geol Memoir 73:301–315
Karner GD, Gambôa LAP (2007) Timing and origin of the South Atlantic pre-salt sag basins and their capping evaporites. In: Schreiber BC, Lugli S, Babel M (eds) Evaporites through space and time, vol 285. Geological Society of London, Special Publication, pp. 15–35
Koopmann H, Schreckenberger B, Franke D, Becker K, Schnabel M (2014) The late rifting phase and continental break-up of the southern South Atlantic: the mode and timing of volcanic rifting and formation of earliest oceanic crust. In: Wright TJ, Ayele A, Ferguson DJ, Kidane T, Vye-Brown C (eds) Magmatic rifting and active volcanism, vol 420. Geological Society of London, Special Publication, pp 315–340. http://dx.doi.org/10.1144/SP420.2
Kumar N, Danforth A, Nuttall P, Helwig J, Bird DE, Venkatraman S (2012) From oceanic crust to exhumed mantle: a 40 year (1970–2010) perspective on the nature of crust under the Santos basin, SE Brazil. In: Mohriak WU, Danforth A, Post PJ, Brown DE, Tari GC, Nemcok M, Sinha ST (eds) Conjugate divergent margins, vol 369. Geological Society of London, Special Publication, pp 147–165. http://dx.doi.org/10.1144/SP369.16
Labrecque JL, Zitellini N (1985) Continuous sea-floor spreading in Red Sea: an alternative interpretation of magnetic anomaly pattern. Am Assoc Petrol Geol Bull 69:513–524
Lavier L, Manatschal G (2006) A mechanism to thin the continental lithosphere at magma-poor margins. Nature 440:324–328
Lazar M, Ben-Avraham Z, Garfunkel Z (2012) The Red Sea—new insights from recent geophysical studies and the connection to the Dead Sea fault. J African Earth Sci 68:96–110
Lentini MR, Fraser SI, Sumner HS, Davies RJ (2010) Geodynamics of the central South Atlantic conjugate margins: implications for hydrocarbon potential. Petrol Geosci 16:217–229
Leroy S, Gente P, Fournier M, d'Acremont E, Patriat P, Beslier MO, Bellahsen N, Maia M, Blais A, Perrot J, Al-Kathiri A, Merkouriev S, Fleury JM, Ruellan PY, Lepvrier C, Huchon P (2004) From rifting to spreading in the eastern Gulf of Aden: a geophysical survey of a young oceanic basin from margin to margin. Terra Nova 16:185–192. https://doi.org/10.1111/j.1365-3121.2004.00550.x
Leroy S, Razin P, Autin J, Bache F, Acremont E, Watremez L, Robinet J, Baurion C, Denèle Y, Bellahsen N, Lucazeau F, Rolandone F, Rouzo S, Kiel JS, Robin C, Guillocheau F, Tiberi C, Basuyau C, Beslier MO, Ebinger C, Stuart G, Ahmed A, Khanbari K, Ganad I, Clarens P, Unterneher P, Toubi K, Lazki AA (2012) From rifting to oceanic spreading in the Gulf of Aden: a synthesis. Arabian J Geosci 5:859–901. https://doi.org/10.1007/s12517-011-0475-4

Ligi M, Bonatti E, Bortoluzzi G, Cipriani A, Cocchi L, Tontini FC, EL Carminati Ottolini, Schettino A (2012) Birth of an ocean in the Red Sea: initial pangs. Geochem Geophys Geosyst 13:Q08009. https://doi.org/10.1029/2012GC004155

Ligi M, Bonatti E, Rasul NMA (2015) Seafloor spreading initiation: Geophysical and geochemical constraints from the Thetis and Nereus Deeps, central Red Sea. In: Rasul NMA, Stewart ICF (eds) The Red Sea: the formation, morphology, oceanography and environment of a young ocean basin. Springer Earth System Sciences. Springer, Heidelberg, pp 79–98.6 https://doi.org/10.1007/978-3-662-45201-1_4

Lowell JD, Genik GJ (1972) Sea-floor spreading and structural evolution of the southern Red Sea. Am Assoc Petrol Geol Bull 56:247–259

Manatschal G (2004) New models for evolution of magma-poor rifted margins based on a review of data and concepts from West Iberia and the Alps. Intl J Earth Sci 93:432–466

Maus S, Barckhausen U, Berkenbosch H, Bournas N, Brozena J, Childers V, Dostaler F, Fairhead JD, Finn C, Frese RRB, Gaina C, Golynsky S, Kucks R, Luhr H, Milligan P, Muller RD, Olesen O, Pilkington M, Saltus R, Schreckenberger B, Thebault E, Tontini FC (2009) EMAG2: A 2-arc-minute resolution Earth Magnetic Anomaly Grid compiled from satellite, airborne, and marine magnetic measurements. Geochem Geophys Geosyst 10:1–12. https://doi.org/10.1029/2009GC002471

McKenzie DP, Davies D, Molnar P (1970) Plate tectonics of the Red Sea and East Africa. Nature 226:242–248

Meisling KE, Cobbold PR, Mount VS (2001) Segmentation of an obliquely rifted margin, Campos and Santos basins, southeastern Brazil. Am Assoc Petrol Geol Bull 11:1903–1924

Mello MR, Mohriak WU, Koutsoukos EAM, Bacoccoli G (1994) Selected petroleum systems in Brazil. In: Magoon LB, Dow WG (eds) The petroleum system—from source to trap. Am Assoc Petrol Geol Memoir 60:499–512

Mello MR, Azambuja Filho NC, Bender AA, Barbanti SM, Mohriak W, Schmitt P, Jesus CLC (2013) The Namibian and Brazilian southern South Atlantic petroleum systems: are they comparable analogues? In: Mohriak WU, Danforth A, Post PJ, Brown DE, Tari GC, Nemcok M, Sinha ST (eds) Conjugate Divergent Margins, vol 369. Geological Society of London, Special Publication, pp 249–266

Mitchell NC, Ligi M, Ferrante V, Bonatti E, Rutter E (2010) Submarine salt flows in the central Red Sea. Geol Soc Am Bull 122:701–713

Mitchell NC, Park Y (2014) Nature of crust in the central Red Sea. Tectonophysics 628:123–139

Mohriak WU (2001) Salt tectonics, volcanic centers, fracture zones and their relationship with the origin and evolution of the South Atlantic ocean: geophysical evidence in the Brazilian and West African margins. In: 7th International congress of the Brazilian Geophys Soc, Salvador—Bahia, Brazil, October 28–31, 2001, Expanded Abstract, pp 1594–1597

Mohriak WU (2003) Bacias sedimentares da margem continental Brasileira. In: Bizzi LA, Schobbenhaus C, Vidotti RM, Gonçalves JH (eds) Geologia. Tectônica e Recursos Minerais do Brasil, CPRM, Brasília, Capítulo III, pp 87–165

Mohriak WU (2014) Birth and development of continental margin basins: Analogies from the South Atlantic, North Atlantic, and the Red Sea. Am Assoc Petrol Geol Search and Discovery, Article 41502 (posted December 22, 2014)

Mohriak WU (2015) Rift basins in the Red Sea and Gulf of Aden: Analogies with the southern Atlantic. In: Post PJ, Coleman J, Rosen NC, Brown DE, Roberts-Ashby T, Kahn P, Rowan M (eds) 34th annual GCSSEPM foundation Perkins-Rosen research conference, petroleum systems in "rift" basins, Houston, December 13–15, Expanded Abstracts, pp 789–826

Mohriak WU, Fainstein R (2012) Phanerozoic regional geology of the eastern Brazilian margin. In: Roberts D, Bally A (eds) Phanerozoic passive margins, cratonic basins and global tectonic maps. Elsevier BV, Chap 7, pp 223–282. https://doi.org/10.1016/b978-0-444-56357-6.00006-8

Mohriak WU, Leroy S (2013) Architecture of rifted continental margins and break-up evolution: insights from the South Atlantic, North Atlantic and Red Sea–Gulf of Aden conjugate margins. In: Mohriak WU, Danforth A, Post PJ, Brown DE, Tari GC, Nemcok M, Sinha ST (eds) Conjugate divergent margins, vol 369. Geological Society of London, Special Publication, pp 497–535. http://dx.doi.org/10.1144/SP369.17

Mohriak WU, Hobbs R, Dewey JF (1990a) Basin-forming processes and the deep structure of the Campos Basin, offshore Brazil. Marine Petrol Geol 7:94–122

Mohriak W, Mello MR, Dewey JF, Maxwell JR (1990b) Petroleum geology of the Campos Basin, offshore Brazil. In: Brooks J (ed) Classic petroleum provinces, vol 50. Geological Society of London, Special Publication, pp 119–141

Mohriak WU, Nemcok M, Enciso G (2008) South Atlantic divergent margin evolution: rift-border uplift and salt tectonics in the basins of SE Brazil. In: Pankhurst RJ, Trouw RAJ, Brito Neves BB, de Wit MJ (eds) West Gondwana pre-Cenozoic correlations across the South Atlantic region, vol 294. Geological Society of London, Special Publication, pp 365–398

Mohriak WU, Nóbrega M, Odegard ME, Gomes BS, Dickson WG (2010) Geological and geophysical interpretation of the Rio Grande Rise, south-eastern Brazilian margin: extensional tectonics and rifting of continental and oceanic crusts. Petrol Geosci 16:231–245

Mohriak WU, Szatmari P, Anjos S (2012) Salt: Geology and tectonics of selected Brazilian basins in their global context. In: Alsop GI, Archer SG, Hartley AJ, Grant NT, Hodgkinson R (eds) Salt tectonics, sediments and prospectivity, vol 363. Geological Society of London. Special Publication, pp 131–158. https://doi.org/10.1144/sp363.7

Mooney WD, Gettings ME, Blank HR, Healy JH (1985) Saudi Arabian seismic-refraction profile: a traveltime interpretation of crustal and upper mantle structure. Tectonophysics 111:173–246

Moulin M, Aslanian D, Unternehr P (2010) A new starting point for the south and equatorial Atlantic ocean. Earth-Sci Rev 98:1–37

Norton IO, Carruthers DT, Hudec MR (2016) Rift to drift transition in the South Atlantic salt basins: A new flavor of oceanic crust. Geology 44:55–58

Pérez-Gussinyé M (2012) A tectonic model for hyperextension at magma-poor rifted margins: An example from the West Iberia–Newfoundland conjugate margins. In: Mohriak WU, Danforth A, Post PJ, Brown DE, Tari GC, Nemcok M, Sinha ST (eds) Conjugate divergent margins, vol 369. Geological Society of London. Special Publication, pp 403–427. http://dx.doi.org/10.1144/SP369.19

Péron-Pinvidic G, Manatschal G (2009) The final rifting evolution at deep magma-poor passive margins from Iberia-newfoundland: a new point of view. Intl J Earth Sci (Geol Rundsch) 98:1581–1597. https://doi.org/10.1007/s00531-008-0337-9

Péron-Pinvidic G, Manatschal G, Osmundsen PT (2013) Structural comparison of archetypal Atlantic rifted margins: a review of observations and concepts. Marine Petrol Geol 43:21–47

Pindell J, Graham R, Bellingham P, McDermott K, Kaminski M, Paton D, Horn B (2015) Continental margin formation and creation of "lateral tectonic accommodation space" for salt deposition, Campos and Santos basins, São Paulo Plateau, Brazil. In: Post PJ, Coleman J, Rosen NC, Brown DE, Roberts-Ashby T, Kahn P, Rowan M (eds) 34th Annual GCSSEPM foundation Perkins-Rosen research conference, petroleum systems in "rift" basins, Expanded Abstracts, Houston, 13–15 December, pp 787–788

Planke S, Rasmussen T, Rey SS, Myklebust R (2005) Seismic characteristics and distribution of volcanic intrusions and hydrothermal vent complexes in the Vøring and Møre basins. In: Doré AG, Vining BA (eds) Petroleum geology: North-West Europe and global perspectives. Proceedings of the 6th petroleum geology conference, pp 833–844

Purser BH, Bosence DWJ (1988) Sedimentation and tectonics in rift basins: Red Sea—Gulf of Aden. Chapman & Hall, London, p 663

Quirk DG, Hertle M, Jeppesen JW, Raven M, Mohriak WU, Kann DJ, Norgaard M, Howe MJ, Hsu D, Coffey B, Mendes MP (2013) Rifting, subsidence and continental break-up above a mantle plume in the central South Atlantic. In: Mohriak WU, Danforth A, Post PJ, Brown DE, Tari GC, Nemcok M, Sinha ST (eds) Conjugate divergent margins, vol 369. Geological Society of London, Special Publication, pp 185–214. http://dx.doi.org/10.1144/SP369.20

Rangel HD, Martins CC (1998) Principais compartimentos exploratórios, Bacia de Campos. In: Schlumberger, Search November 1998, Searching for Oil and Gas in the Land of Giants, Cenário geológico nas bacias sedimentares no Brasil, pp 16–40

Rasul NMA, Stewart ICF, Nawab ZA (2015) Introduction to the Red Sea: its origin, structure, and environment. In: Rasul NMA, Stewart ICF (eds) The Red Sea: the formation, morphology, oceanography and environment of a Young Ocean Basin. Springer Earth System Sciences. Springer, Heidelberg, pp 1–28. https://doi.org/10.1007/978-3-662-45201-1

Rowan MG (2014) Passive-margin salt basins: hyperextension, evaporite deposition, and salt tectonics. Basin Res 26:154–182. https://doi.org/10.1111/bre12043

Saleh S, Jahr T, Jentzsch G, Saleh A, Abou Ashour NM (2006) Crustal evaluation of the northern Red Sea rift and Gulf of Suez, Egypt from geophysical data: 3-dimensional modeling. J African Earth Sci 45:257–278

Sandwell DT, Smith WHF (2009) Global marine gravity from retracked Geosat and ERS-1 altimetry: ridge segmentation versus spreading rate. J Geophys Res 114:B01411. https://doi.org/10.1029/2008JB006008

Schettino A, Macchiavelli C, Pierantoni PP, Zanoni D, Rasul N (2016) Recent kinematics of the tectonic plates surrounding the Red Sea and Gulf of Aden. Geophys J International 207(1):457–480. https://doi.org/10.1093/gji/ggw280

Shipboard Scientific Party (1985) Introduction, objectives, and principal results: ocean drilling program Leg 103, West Galicia Margin. In: Boillot G et al (eds) Proceedings of the ocean drilling program, vol 103, Part A—Initial Report, Galicia margin, pp 3–17

Shipboard Scientific Party (1998) Leg 173 introduction. In: Whitmarsh RB, Beslier M-O, Wallace PJ, et al. (eds) Proceedings of the ocean drilling program, Initial reports, vol 173, pp 7–23

Scotese CR (2002) Paleomap project. http://www.scotese.com

Skiple C, Anderson E, Furstenau J (2012) Seismic interpretation and attribute analysis of the Herodotus and the Levantine Basin, offshore Cyprus and Lebanon. Petrol Geosci 18:433–442. https://doi.org/10.1144/petgeo2011-072

Sultan M, Becker R, Arvidson RE, Shore P, Stern RJ, Alfy ZE, Guinness EA (1992) Nature of the Red Sea crust: a controversy revisited. Geology 20:593–596

Taylor B, Goodliffe A, Martinez F (2009) Initiation of transform faults at rifted continental margins. CR Geosci 341:428–438

Tubbs RE, Fouda HGA, Afifi AM, Raterman NS, Hughes GW, Fadolalkarem YK (2014) Midyan Peninsula, northern Red Sea, Saudi Arabia: seismic imaging and regional interpretation. GeoArabia 19:165–184

Unternehr P, Peron-Pinvidic G, Manatschal G, Sutra E (2010) Hyper-extended crust in the South Atlantic: in search of a model. Petrol Geosci 16:207–215

Voggenreiter W, Hotzl H (1989) Kinematic evolution of the southwestern Arabian continental margin: implications for the origin of the Red Sea. J African Earth Sci 8:541–564

Voggenreiter W, Hotzl H, Mechie J (1988) Low-angle detachment origin for the Red Sea rift system? Tectonophysics 150:51–75

Watremez L, Leroy S, Rouzo S, d'Acremont E, Unternehr P, Ebinger C, Lucazeau F, Al Lazki A (2011) The crustal structure of the northeastern Gulf of Aden continental margin: insights from wide-angle seismic data. Geophys J Intl 184:575–594. https://doi.org/10.1111/j.1365-246X.2010.04881.x

Wegener A (1912) The origins of the continents. In: Jacoby WR (2001) Translation of Die Entstehung der Kontinente, Dr Alfred Wegener, Petermanns Geographische Mitteilungen, 58 I, 185–195, 253–256, 305–309. J Geodynamics 32:29–63

Whitmarsh RB, Wallace PJ (1998) The rift-to-drift development of the West Iberia nonvolcanic continental margin: A summary and review of the contribution of the ocean drilling program Leg 173. In: Beslier MO, Whitmarsh RB, Wallace PJ, Girardeau J (eds) Proceedings of the ocean drilling program, scientific results, vol 173, pp 1–36

Wilson JT (1966) Did the Atlantic close and then re-open? Nature 211:676–681. https://doi.org/10.1038/211676a0

Winter WR, Jahnert RJ, França AB (2007) Bacia de Campos. Boletim de Geociências da Petrobras 15(2):511–529

Xu W, Ruch J, Jónsson S (2015) Birth of two volcanic islands in the southern Red Sea. Nat Comm 6:7104. https://doi.org/10.1038/ncomms8104

Zahran HM, Stewart ICF, Johnson PR, Basahel MH (2003) Aeromagnetic-anomaly maps of central and western Saudi Arabia. Saudi Geological Survey Open-File Report SGS-OF-2002-8, 6 p, 1 fig., 1 table, 4 plates

Zalán PV, Severino MCG, Rigoti CA, Magnavita LP, Oliveira JAB, Vianna AR (2011) An entirely new 3D-view of the crustal and mantle structure of a South Atlantic passive margin—Santos, Campos and Espírito Santo Basins, Brazil. Am Assoc Petrol Geol Search and Discovery, article 30177, 12 p

Plate Motions Around the Red Sea Since the Early Oligocene

Antonio Schettino, Chiara Macchiavelli, and Najeeb M. A. Rasul

Abstract

The Red Sea represents a very young oceanic basin that formed between Nubia and Arabia since chron C3 (~4.6 Ma). The rifting phase started at ~30 Ma (early Oligocene) and can be represented by two kinematic stages, characterized by distinct directions of extension and different duration. Deformation associated with rifting was accommodated through the reactivation of the inherited Proterozoic structures. We show that the first stage was characterized by the northward motion of the Arabian plate with respect to Africa, accompanied by a pattern of deformation that included N–S oriented strike–slip faults and normal faults having E–W strike. During this stage, extension was mainly accommodated by the formation of pull–apart basins. Starting from ~27 Ma (late Oligocene), the extension axes changed dramatically and acquired the modern NE–SW pattern, which was conserved until the early Pliocene in the southern Red Sea and is still active in the northern region. In this time interval, an inherited system of NW–SE structures was reactivated as normal faults accommodating NE–SW extension, while NE–SW Proterozoic structures were reactivated as transfer strike–slip faults. Although no changes in the directions of extension are observed during this interval, two significant tectonic events occurred around 14 Ma and at 1.77 Ma. During the Langhian, two intervening microplates formed between Nubia and Arabia; the Danakil and Sinai microplates, whose motion determined the formation of the Afar Depression and the Gulf of Aqaba, respectively. Finally, starting from the Pleistocene, ongoing collision of Arabia with Eurasia along the Zagros mountains resulted into a dramatic slowdown in the Red Sea opening rates.

1 Introduction

Determining the kinematics of the tectonic plates surrounding the Red Sea and the Gulf of Aden (Fig. 1) represents a fundamental step toward a comprehension of the fundamental geodynamic processes driving the formation of new tectonic plates. The rifting and spreading history of the Gulf of Aden has been described accurately during the past few years by the analysis of marine magnetic anomalies, fracture zones, and seismic profiles (e.g., d'Acremont et al. 2005; Fournier et al. 2010; Leroy et al. 2010). In the case of the Red Sea, a similar approach has recently led to the characterization of the spreading kinematics during the last ~4.6 Ma (Schettino et al. 2016). However, a quantitative description of the Nubia–Arabia kinematics during the rifting phase is still controversial and several alternative solutions have been proposed so far, with regard to the age of initiation of rifting, the age of rift–drift transition, the width of the thinned continental margins, and the amount and directions of extension.

McKenzie et al. (1970) were the first to propose a reconstruction of the relative positions of Arabia and Nubia during the Miocene, based on a best–fitting procedure of coastline segments north of 15°N and the assumption that most of the Red Sea was floored by oceanic crust. These authors mentioned some evidence that the Euler pole of rotation between these plates had remained constant during the opening of the Red Sea and inferred an age of 15 Ma for the initiation of this process, assuming that a steady spreading rate of 10 mm yr^{-1} was maintained and that neither Africa nor Arabia deformed significantly during the

A. Schettino (✉)
School of Science and Technology – Geology Division, University of Camerino, Via Gentile III Da Varano, 62032 Camerino, MC, Italy
e-mail: antonio.schettino@unicam.it

C. Macchiavelli
Institute of Earth Sciences Jaume Almera, ICTJA-CSIC, Lluis Sole I Sabaris S/N, 08028 Barcelona, Spain

N. M. A. Rasul
Center for Marine Geology, Saudi Geological Survey, Jeddah, Saudi Arabia

N. M. A. Rasul and I. C. F. Stewart (eds.), *Geological Setting, Palaeoenvironment and Archaeology of the Red Sea*,
https://doi.org/10.1007/978-3-319-99408-6_9

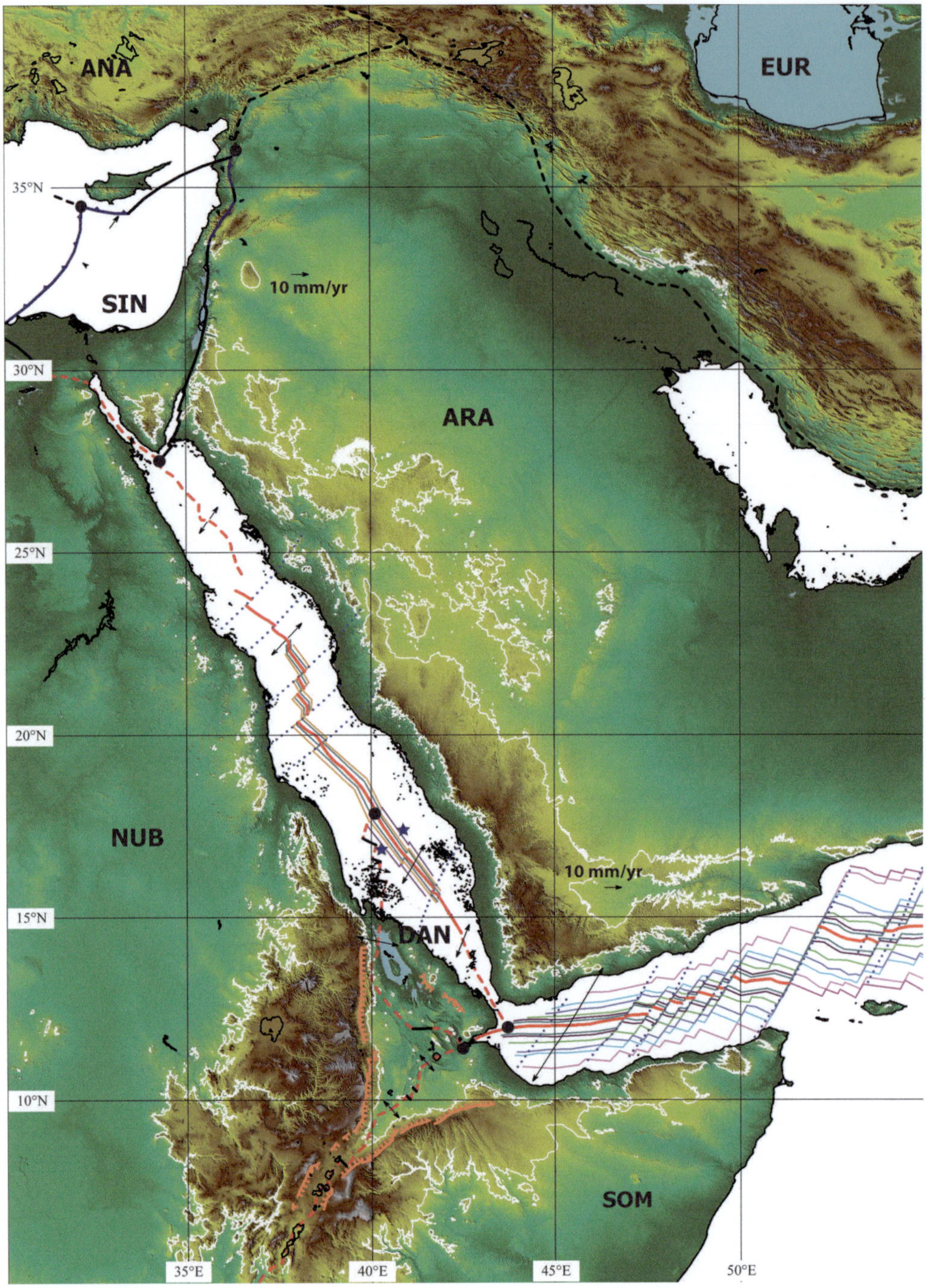

Fig. 1 Location map, showing present-day plate boundaries, velocity fields, triple junctions, magnetic isochrons, and fracture zones around the Red Sea and Gulf of Aden (from Schettino et al. 2016 and Fournier et al. 2010). Red solid lines: Mid-ocean ridges; Red dashed lines: Rift axes; Blue dotted lines: Fracture zones; Black solid lines: Strike-slip faults; Blue lines with barbs: Convergent boundaries; Black dashed lines: Plate boundaries outside the study area; White line: 1000 m topography contour; Orange lines with barbs: Main rift shoulders; Black dots: Triple junctions; Stars in the southern Red Sea indicate the locations where the oldest oceanic crust (4.62 Ma, early Pliocene) has been identified; Magnetic isochrons 2, 2A, and 3 in the Red Sea are shown in green, ochre, and purple, respectively; ANA = Anatolia, EUR = Eurasia, SIN = Sinai, ARA = Arabia, NUB = Nubia, DAN = Danakil, SOM = Somalia. Areas in blue are continental inland below sea level. Fracture zones in the Gulf of Aden are from Leroy et al. (2012). Magnetic isochrons in the Gulf of Aden are: 2Ay, 2Ao, 3A, 4A, 5, 5C, 5D, and 6 (after Fournier et al. 2010)

rifting stage (i.e., the amount of extension of the margins was assumed to be nearly zero, so that $\beta \approx 1$). A thorough discussion of the discrepancies between such a rigid plates approximation and geological observations can be found in Le Pichon and Francheteau (1978). In addition, these authors showed that models based on the assumption that the whole Red Sea is floored by Oligocene to recent oceanic crust extending from coast to coast are not compatible with simple geodynamic considerations. Some years later, Cochran (1983) provided the first estimate of stretching for the conjugate continental margins of Arabia and Nubia in the northern Red Sea, about 160 km at 25°N ($\beta = 1.3$). However, a much higher amount of extension, with β ranging from 3.45 to 2.6, was proposed by Joffe and Garfunkel (1987) for the area north of ~17°N.

Regarding the age of initiation of sea floor spreading, while there is now general agreement that it was Chron C6n (~20.1 Ma, early Burdigalian) in the eastern Gulf of Aden (Fournier et al. 2010), the timing is still controversial in the case of the Red Sea. In fact, it depends on the different interpretations given to the nature of the crust below the up to 5 km thick layer of Miocene evaporites overlying the basement. For example, Le Pichon and Gaulier (1988) assumed that the Red Sea crust is oceanic even in the northernmost area, up to 26.3°N, from the axial zone up to a distance of about 35 km from the coastlines. Accordingly, they estimated that sea floor spreading would have initiated at ~13 Ma if the spreading rate remained constant. In general, in the model of Le Pichon and Gaulier (1988) rifting started at ~30 Ma and proceeded as a slow rotation (~0.06°/Myr) about a constant Euler pole up to ~13 Ma. Then, starting from this age the rotation of Arabia relative to Nubia was accompanied by sea floor spreading at a rate of ~0.42°/Myr about the same Euler pole. More recent GPS estimates have suggested an early Miocene (~24 Ma) age of initiation of rifting in both the Red Sea and the Gulf of Aden (ArRajehi et al. 2010), with a 70% increase in the angular velocity of separation between Nubia and Arabia at 13 Ma without changes in the stage pole location. This kinematic model is partially retained in the very recent work of DeMets and Merkouriev (2016), although these authors propose a three-stages tectonic evolution of the Red Sea, such that two early stages before Chron C5C (~16 Ma) have stage poles slightly different from the pole of rotation of ArRajehi et al. (2010).

In this paper, we estimate the amount of syn-rift extension around the Red Sea on the basis of a palinspastic restoration of the stretched margins to their pre-rift width. To this purpose, we will compile a Moho grid for the Red Sea region, which will be combined with Aster GDEM topography to analyse a series of crustal cross-sections across the Red Sea using the method illustrated in Schettino (2014). Then, taking into account that no change in the directions of extension has been observed along the conjugate margins (e.g., Schettino et al. 2016), the resulting finite strain can be used to determine the angular syn-rift separation between Arabia and Nubia.

2 Recent Kinematics of the Red Sea

A quantitative determination of the sea-floor spreading history of an oceanic basin requires the identification of magnetic anomaly crossings, transform faults, and fracture zone trends. The first comprehensive analysis of sea-floor spreading anomalies in the Red Sea was performed by Chu and Gordon (1998), who did not include transform fault offsets in their study because of the short offset of these features, which never exceeds 5 km in the southern Red Sea. More recently, we have shown that a set of transverse structures exists in the Red Sea region, whose formation is associated with the progressive oceanization of the conjugate margins. Schettino et al. (2016) argued that these features have a strike that is representative of the directions of relative motion during the latest rifting stage and onset of sea floor spreading. Therefore, these authors combined both directional data and sea-floor spreading anomalies to build a refined kinematic model since 4.6 Ma, which is the age of the oldest oceanic crust identified in the southern Red Sea around 17.1°N (Fig. 1). The magnetic isochrons of this new model are shown in Fig. 1. The model describes the five-plates system formed by Nubia, Arabia, Somalia, and the Sinai and Danakil microplates. The predicted flow lines of relative motion are illustrated in Fig. 2. These lines determine the local azimuth of relative motion between any plate pair. The trends in Fig. 2 show that Danakil is separating from Arabia by anti-clockwise rotation about a pole located in the Gulf of Aden, whereby the velocity of separation between the two plates decreases southward. Consequently, rifting of Danakil from Arabia occurred with strain rates higher in the northern part of the southern Red Sea. Actually, the southern part of this region from 14.8°N up to the Bab al Mandeb area is still in the rifting stage. Figure 2 also shows that Danakil is rifting from Nubia with extension directions that vary from E–W in southern Afar to NW–SE close to the triple junction. Figure 2 also illustrates the pattern of relative motions in the northernmost Red Sea. In the model of Schettino et al. (2016), Sinai is an independent microplate that is separating from Nubia through a trans-tensional boundary in the Gulf of Suez while sliding apart with respect to Arabia through the transcurrent Dead Sea Fault Zone (DSFZ). Finally, as shown in Figs. 1 and 2, the northern Red Sea at latitudes higher than 24°N is still in the rifting stage. A very young spreading ridge, which is not yet flanked by magnetic isochrons, can be found north of 22.6°N. This ridge is formed by three spreading segments separated by two transform faults, whose landward continuations form two major active transverse structures.

3 Plate Reconstructions for the Late Langhian—Early Pliocene Time Interval

A set of plate reconstructions at anomalies 3n.2r (4.62 Ma), 3n (4.18 Ma), 2A (2.58 Ma), and 2 (1.77 Ma), illustrating the plate tectonic evolution of the Red Sea since the early Pliocene, has been presented by Schettino et al. (2016). In this study, the starting point is represented by the oldest of these reconstructions, namely by the early Pliocene configuration (4.62 Ma), which will be extrapolated backward to reconstruct the most recent stage of rifting in the Red Sea. As mentioned above, the early Pliocene represents the age of the oldest oceanic crust identified in the Red Sea. Figure 3a and b show plate boundaries and velocity fields during this time interval. There is strong geological and structural evidence that this configuration of the plate boundaries is representative of a tectonic phase that started in the middle Miocene

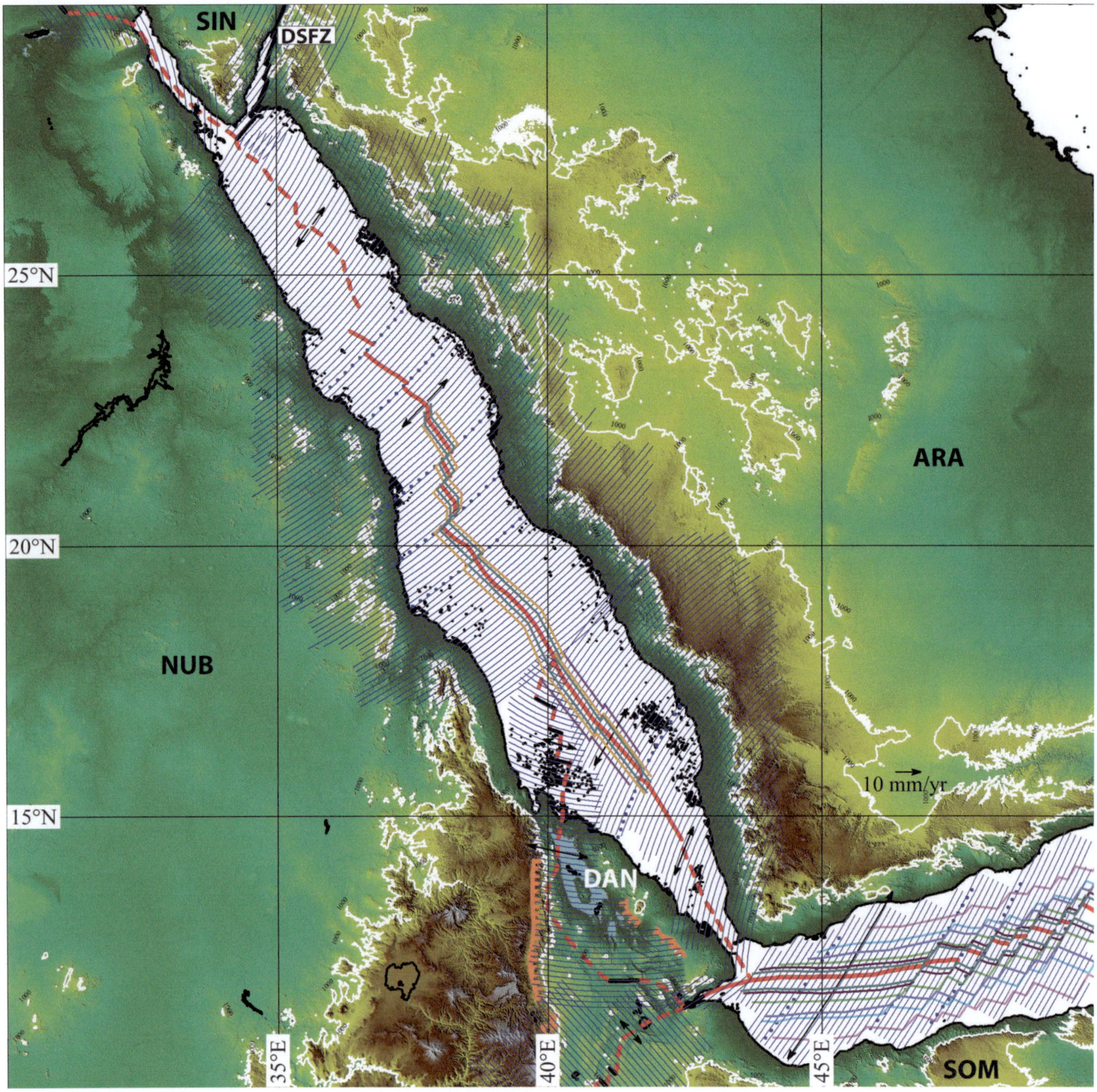

Fig. 2 Flow lines and relative velocity field of current plate motions in the Red Sea and in the Gulf of Aden. Local relative velocity between two plates along a boundary region or a deformation zone is always tangent to these lines. DSFZ = Dead Sea Fault Zone. (From Schettino et al. 2016, their Figs. 13 and 14)

(between 17 and 14 Ma; Bosworth et al. 2005; Nuriel et al. 2017). In addition, geological fieldwork conducted during three successive research expeditions along the Saudi Arabian margin by the authors (in 2015–2016) shows that the recent pole of opening of the Red Sea has remained invariant at least since the late Oligocene (~ 27 Ma).

The observed stability of the stage pole of rifting during the late Oligocene to recent time interval allowed us to determine the total angle of rotation Ω through balancing of 23 crustal cross-sections across the Red Sea using the method illustrated in Schettino and Turco (2006) and Schettino (2014). In this approach, crustal profiles are generated and restored palinspastically to the pre-rift configuration. The trace of these profiles is chosen to coincide with flow lines of relative motion about the stage pole, while the profiles are built combining Moho and topography at each

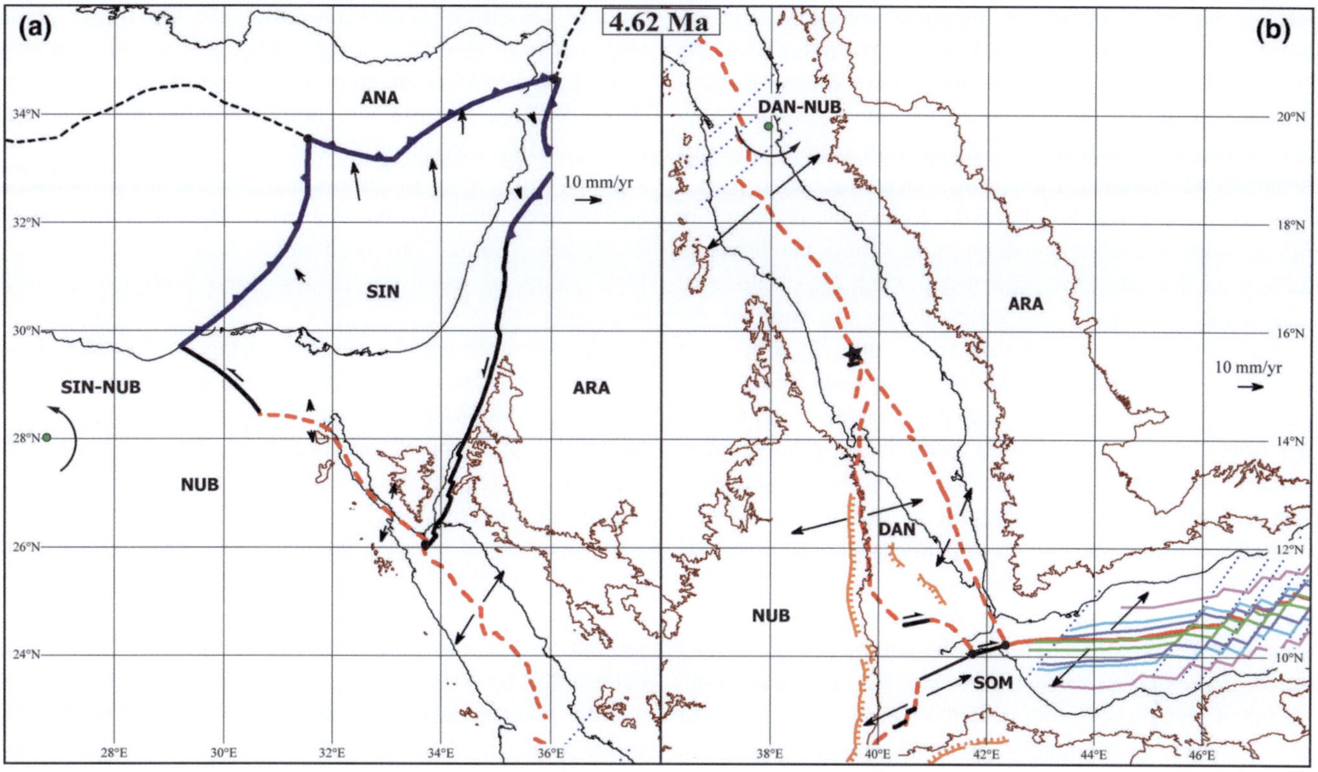

Fig. 3 Plate reconstructions at chron C3n.2ry (4.62 Ma) for the northern Red Sea and eastern Mediterranean (**a**) and for the southern Red Sea, Afar, and western Gulf of Aden (**b**), showing past plate boundaries and velocity vectors of relative motion. Also shown are instantaneous Euler poles of Sinai–Nubia and Danakil–Nubia plate pairs. The star in (**b**) indicates the reconstructed location of the first oceanic crust formed in the southern Red Sea, which coincides with the Danakil–Nubia–Arabia triple junction. Black dots are triple junctions. Red solid lines: Mid–ocean ridges; Red dashed lines: Rift axes; Black dotted lines: Fracture zones and transform faults; Black solid lines: Strike–slip faults; Blue lines with triangular barbs: Convergent boundaries; Orange lines with squared barbs: Major rift structures; Brown lines: Reconstructed modern 1000 m topographic contour. After Schettino et al. (2016)

location along the traces. This method is based on two simplifying assumptions: (1) Crustal mass is conserved along cross-sections during rifting (no transversal flow), and (2) Sediment flow is intra-basinal. The first assumption implies that for any point in the rift valley the ductile flow in the middle–lower crust does not have a component orthogonal to the local velocity vector of relative motion, while the second assumption guarantees that all the balanced mass was originally in the rift region. In addition, the method requires preliminary removal of the oceanic crust in the case of profiles that cross the area of active sea floor spreading. Computation of the pre-rift size of the continental margins at some distance ζ from the Euler pole, $L_0(\zeta)$, and the average stretching factor, $\beta(\zeta)$, requires first the determination of the stretched thickness function $H = H(x,\zeta)$ along the corresponding crustal cross-section, x being the distance along profile, and the initial unstretched crustal thickness $H_0(\zeta)$. Then, if $L(\zeta)$ is the present day size of the rift valley at distance ζ from the Euler pole, the pre-rift restored size, L_0, and the stretching factor, β, will be given by:

$$L_0(\zeta) = \frac{1}{H_0(\zeta)} \int_0^{L(\zeta)} H(x,\zeta)dx \quad (1)$$

$$\beta(\zeta) = L(\zeta)/L_0(\zeta) = \frac{L(\zeta)H_0(\zeta)}{\int_0^{L(\zeta)} H(x,\zeta)dx} \quad (2)$$

In order to determine the function $H(x,\zeta)$ and the quantity $H_0(\zeta)$ for each profile, we compiled a Moho grid for the Red Sea region starting from published data. This is illustrated in Fig. 4 and includes data from several sources (Salem et al. 2013; Hansen et al. 2007; Al–Damegh et al. 2005), as well as oceanic Moho data determined on the basis of the isochron maps of Schettino et al. (2016) and Fournier et al. (2010), respectively for the Red Sea and the Gulf of Aden. To generate the crustal profiles, we combined Moho profiles along the selected traces with Aster GDEM topography. As illustrated in Fig. 5, a total of 23 crustal profiles were created along the central and northern Red Sea, with the objective of determining the variability of the stretching factor β in the region of separation between Nubia and Arabia and the corresponding finite strain ε. This quantity can be used in turn to estimate the total angle of rotation between Nubia and Arabia, Ω, which restores the pre-rift configuration of the conjugate margins. A plot of the stretching factor, β, as a function of the distance ζ from the Euler pole is shown in Fig. 6, while the corresponding finite strains are listed in Table 1. In this table, the present-day width $L(\zeta)$ at distance ζ from the Euler pole was estimated on the basis of the function $H(x,\zeta)$, which stabilizes to a value $H_0(\zeta)$ at some distance from the rift axis. We note that both β and ε are essentially independent from ζ, despite that we could expect that both rise for increasing ζ, because the relative plate velocity increases with the distance from the Euler pole. However, numerical modelling shows that rifting velocity exerts a control on the strain localization width, so that wide basins have the tendency to form under conditions of slow extension and vice versa (Van Wijk and Cloetingh 2002; Salerno et al. 2016). Therefore, although the rifting velocity $v(\zeta)$ decreases when we move toward the Euler pole, both the initial and present widths of the rift increase, and their ratio remains approximately invariant with respect to ζ. Consequently, by Eq. (2) the quantity β does not change significantly. The average stretching factor resulting from the data listed in Table 1 is: $\beta = 1.47 \pm 0.02$. The corresponding finite strain is: $\varepsilon = \ln(\beta) = 0.39 \pm 0.02$. To assess the sensitivity of these values to errors in crustal thickness H or rift width L, we note that by Eqs. (1) and (2) a very large 10% error in either H or L translates directly into a 10% error in β.

The stretching parameters obtained above can be used to achieve a preliminary estimate of the total angle of rotation that restores the pre-rift configuration. This turned out to be $\Omega = 14.31°$. However, it would not be correct to use this result in a kinematic model of the Red Sea rifting, because there is strong field evidence that a short initial phase of extension occurred during the Oligocene, possibly between 30 and 27 Ma, characterized by N–S extension (i.e., N–S trending strike-slip faults and E–W normal faults) as initially suggested by Makris and Rihm (1991) and Ghebreab (1998). Such an early phase of N–S extension led to the formation of left-lateral pull-apart basins along the rifted margins of Arabia and Nubia. Therefore, a small fraction of the total extension that can be estimated through crustal balancing is not associated with the late Oligocene to recent phase of NE–SW extension but it is related to an earlier stage of separation between Arabia and Nubia. On the basis of kinematic considerations, we suggest that a better estimate for the total angle of rotation during the NE–SW phase of extension is $\Omega = 12.15°$. In this instance, the average angular velocity of separation during the last rifting phase turns out to be $\omega = \Omega/27 = 0.45°/\text{Myr}$. The reduced rotation angle $\Omega = 12.15°$ is the angle that brings the N–S trending Great Yemeni Escarpment 1000 m contour to match the corresponding contour line along the Afar Escarpment (see Fig. 8), while more to the north this rotation still shows a gap between the NW–SE oriented 1000 m contour lines of Nubia and Arabia. This observation suggests that an early N–S phase of extension is associated with the residual stretching.

The estimated angular velocity during the NE–SW phase of extension allows us to determine the relative positions of Arabia and Nubia at any time during the late Oligocene—early Pliocene interval, because a unique stage pole characterized this tectonic phase. Similarly, the relative positions of Danakil and Arabia during rifting in the southern Red Sea were determined considering that there is no evidence of changes in the Euler pole of separation between these plates and assuming a constant angular velocity. Finally, the motion of the Sinai microplate with respect to Arabia was determined by extrapolation of the current stage pole and angular velocity (see Schettino et al. 2016). An interesting consequence of this approach is that at 14 Ma (late Langhian) the Danakil microplate displays a tight fit against Nubia, determining complete closure of the Afar Depression (Fig. 7). We interpret this feature as evidence that this microplate started separating from Nubia during the late Langhian, whereby it must be considered as part of Nubia for times older than ~14 Ma. It is also interesting to note that at the same time the western coastline and the southern tip of the Sinai microplate are perfectly aligned with the Arabian Peninsula coastline (Fig. 7). This feature and other geological evidence discussed by Bosworth et al. (2005) support the assumption that the starting time of left-lateral strike-slip motion along the DSFZ is 14 Ma, thereby coinciding with the time of initiation of the main phase of rifting in Afar but more probably related to the time of final collision between Arabia and Eurasia along the Zagros Mountains (Agard et al. 2005; Mouthereau et al. 2007). Although very recent data provide evidence that the formation of the DSFZ occurred earlier by northward propagation in the time interval between 21 and 17 Ma (Nuriel et al. 2017), it is likely that the development of a true plate boundary linked to the global kinematic circuit required more time, therefore we prefer to retain 14 Ma as the representative age of formation of the DSFZ as a plate boundary. We also note that according to Bosworth et al. (2005) extension and rifting initiated in Afar at ~25 Ma. Therefore, our result is much more compatible with geological studies that argue a later time of rifting between Danakil and Nubia and an early rift in the southern Red Sea (e.g., Wolfenden et al. 2004).

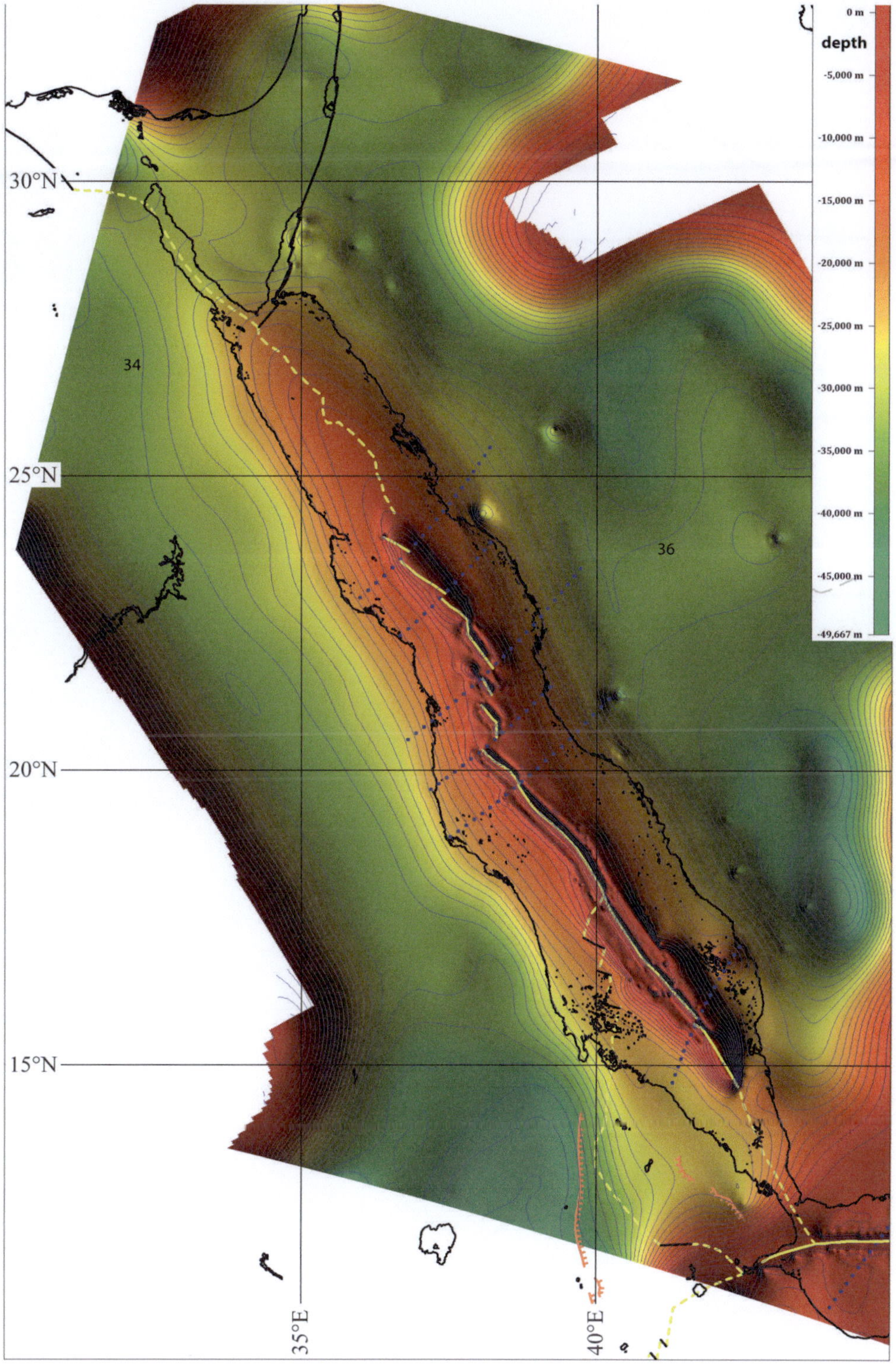

Fig. 4 A Moho map for the Red Sea region. Contour lines are spaced 2 km. Numbers are depths in km

A late Langhian reconstruction of the plate boundaries around the Red Sea is illustrated in Fig. 7 in a fixed Nubian frame of reference. This reconstruction is representative of the time of initiation of rifting between Danakil and Nubia and shows a tight fit of Danakil against the main Afar Escarpment (1000 m contour). Although an earlier phase of extension in Afar is not excluded, it strongly supports kinematic models in which most of the deformation is concentrated in the area to the east of Danakil in the time interval between the late Oligocene and the middle Miocene. Consequently, our preferred scenario assumes that separation of Danakil from Arabia in the southern Red Sea and

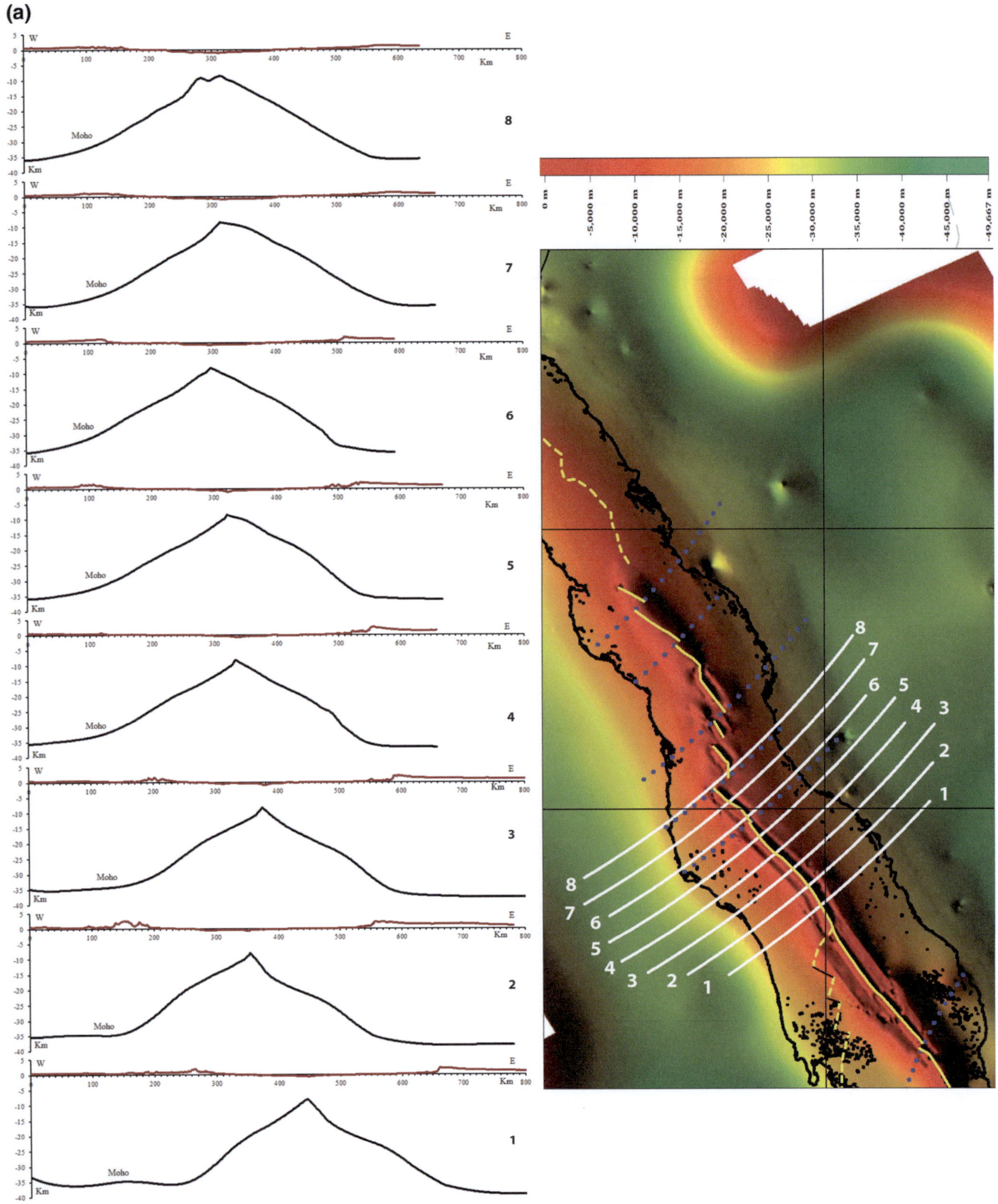

Fig. 5 Crustal profiles across the Red Sea and conjugate thinned continental margins, with the central stripe of oceanic crust. The profiles show the lateral extent of the rift area and follow the flow lines of relative motion between Nubia and Arabia. The map to the right shows Moho depths (see Fig. 4)

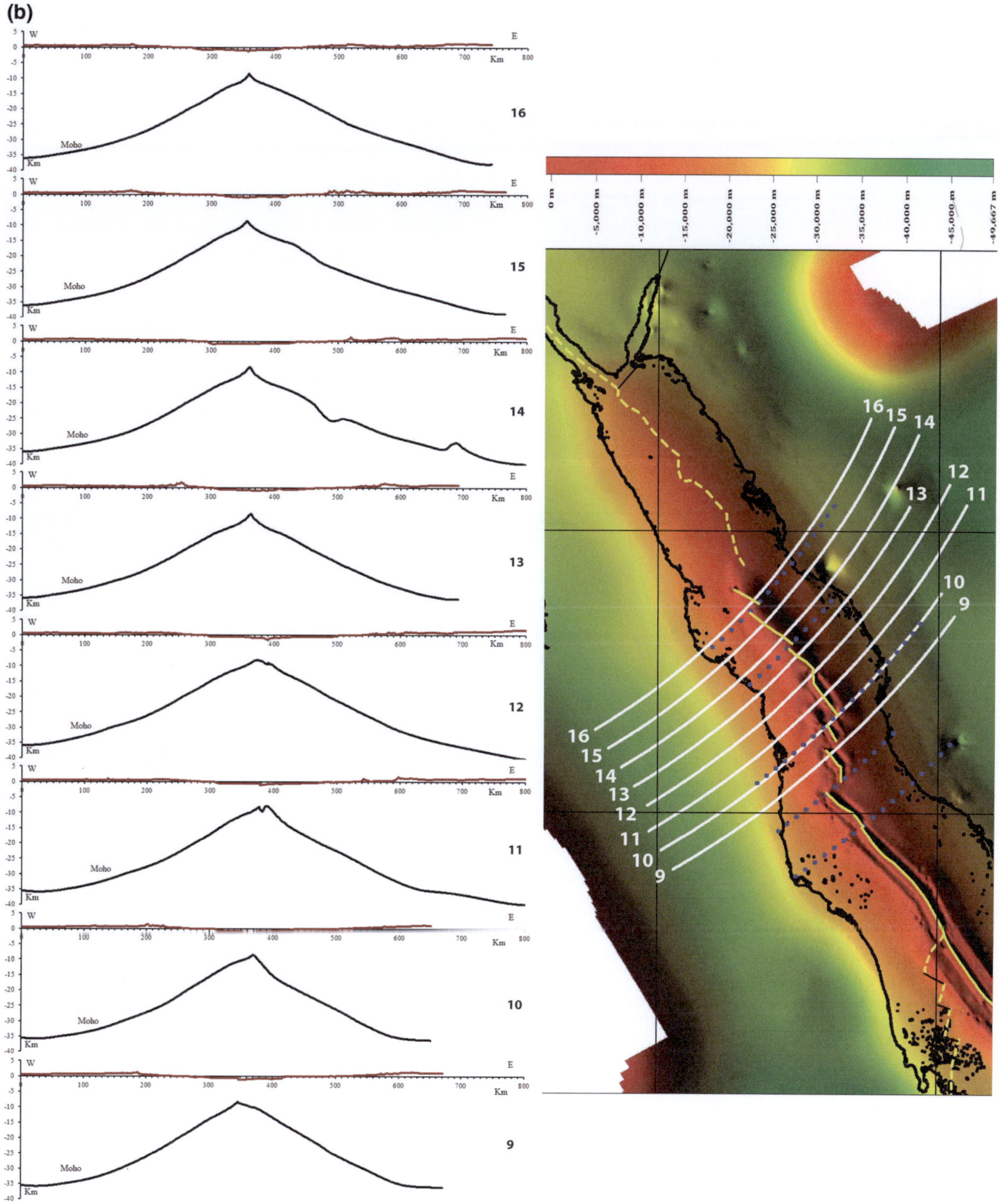

Fig. 5 (continued)

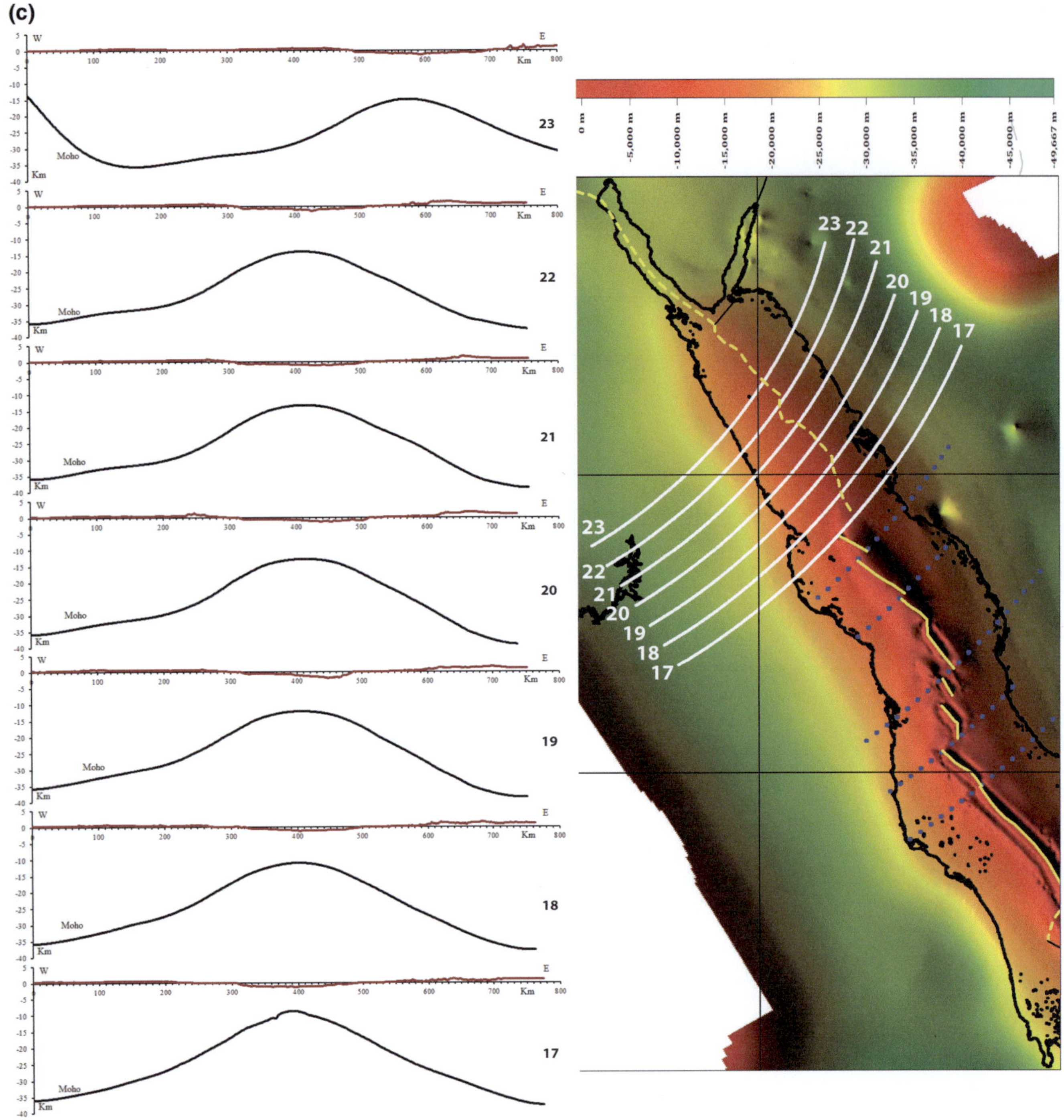

Fig. 5 (continued)

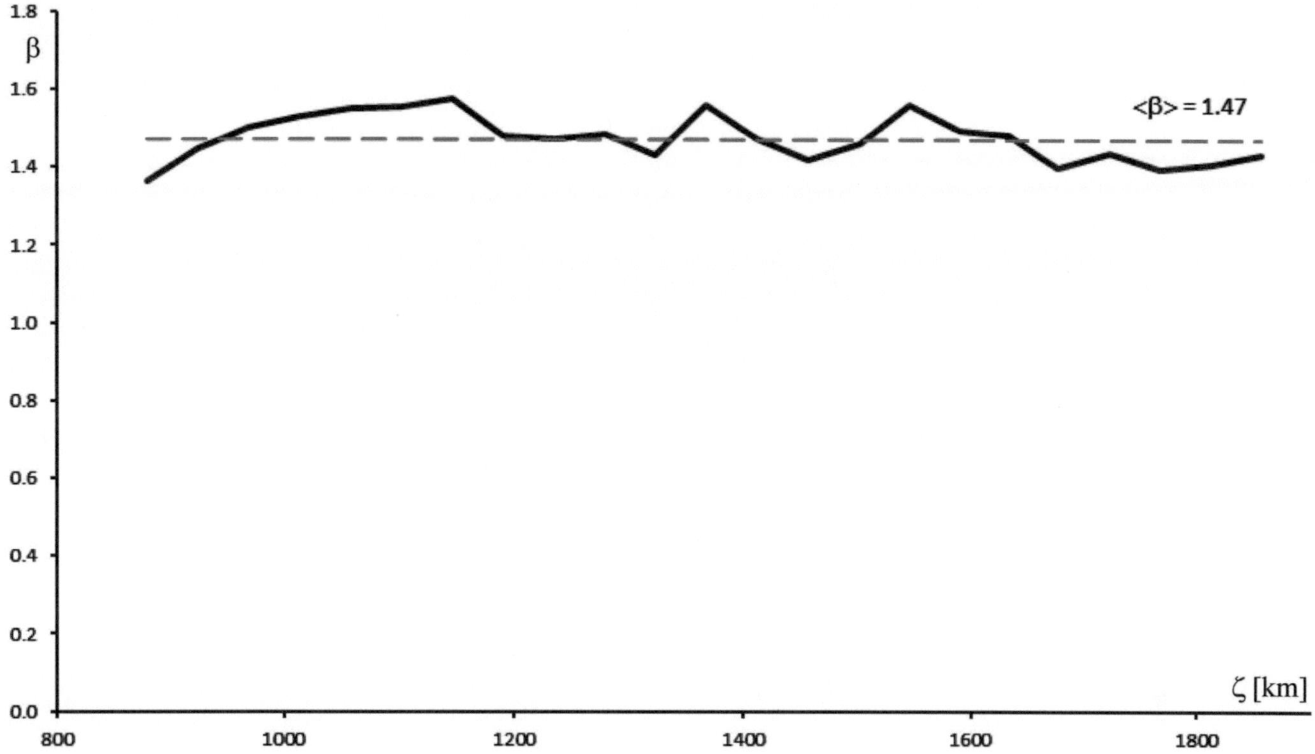

Fig. 6 Estimated beta factor, β, as a function of the distance ζ from the Euler pole of opening of the Red Sea

rifting between Nubia and Arabia in the central and northern Red Sea started simultaneously in the late Oligocene, while partial westward transfer of extension from the southern Red Sea to Afar occurred at a later time during the middle Miocene.

4 Rifting Kinematics During the Late Oligocene—Middle Miocene

The arguments discussed in the previous section imply that Danakil and Sinai were part of the Nubian and Arabian plates, respectively, before the middle Miocene. Consequently, in order to obtain the late Oligocene (27 Ma) configuration, we must keep the Danakil microplate fixed to Nubia and the Sinai microplate attached to Arabia as in the 14 Ma reconstruction. The position of Arabia with respect to Nubia can be obtained applying a clockwise rotation of Ω = 12.15° from its present-day position about the stage pole of Nubia–Arabia rifting, which is located at 30.32°N, 27.18°E (Schettino et al. 2016). More problematic is determining the relative position of Nubia and Somalia, because a large amount of uncertainty exists about timing and kinematics of rifting events along the East African Rift (EAR). In a recent paper, Fournier et al. (2010) have shown that the oldest pair of magnetic anomalies in the Gulf of Aden has age C6no (20.13 Ma in the geomagnetic polarity time scale of Cande and Kent 1995). Some authors argue that the oldest time of extensional processes along the Main Ethiopian Rift is younger than 11 Ma (e.g., Wolfenden et al. 2004). In this instance, the existence of a precise independent determination of the relative motion between Somalia and Arabia through marine magnetic anomalies would force plate motions in the Red Sea to coincide with plate motions in the Gulf of Aden for times older than 11 Ma, because Nubia and Somalia would form a unique African plate before this time. However, it can be shown that Euler poles from the Gulf of Aden cannot account for the rifting history of the Red Sea during the time interval between 11 Ma and the late Oligocene. Therefore, the formation of the EAR must be older than 11 Ma. Here we assume that Chron C6n (~20.13 Ma) is the initial time of rifting between Nubia and Somalia. This hypothesis does not contrast with the timing determined by marine magnetic anomalies in the Gulf of Aden and is supported by (U–Th)/He thermochronometry observations (Pik et al. 2008).

Figure 8 illustrates the resulting configuration of the plate boundaries and the reconstructed 1000 m topographic contours at 27 Ma. The latter are particularly interesting. If we move northward starting from the Afar zone, we note that the Arabian and Nubian contours display a perfect match between 10° and 15°N, where they have a straight N–S trend. Further north, the two lines both assume a NW–SE trend, but are now separated by a ~40 km gap in N–S

Table 1 Finite strains across the central and northern Red Sea

N	ζ	λ	L	L_0	β	ε'	ε
1	1855	18.29	704	493	1.4284	0.4284	0.3565
2	1811	18.62	694	493	1.4068	0.4068	0.3413
3	1767	18.92	707	507	1.3943	0.3943	0.3324
4	1722	19.21	658	459	1.4325	0.4325	0.3594
5	1677	19.50	664	475	1.3976	0.3976	0.3347
6	1633	19.86	590	398	1.4811	0.4811	0.3928
7	1589	20.12	609	408	1.4942	0.4942	0.4016
8	1545	20.65	602	386	1.5588	0.5588	0.4439
9	1501	21.06	635	436	1.4572	0.4572	0.3765
10	1456	21.57	626	442	1.4162	0.4162	0.3480
11	1412	22.08	795	541	1.4700	0.4700	0.3853
12	1367	22.50	811	520	1.5589	0.5589	0.4440
13	1323	22.78	678	474	1.4291	0.4291	0.3571
14	1279	23.13	795	536	1.4838	0.4838	0.3946
15	1235	23.35	760	516	1.4721	0.4721	0.3867
16	1190	23.76	735	497	1.4787	0.4787	0.3912
17	1145	23.99	772	490	1.5756	0.5756	0.4547
18	1102	24.69	753	484	1.5561	0.5561	0.4422
19	1057	25.20	746	481	1.5521	0.5521	0.4396
20	1013	25.58	736	481	1.5287	0.5287	0.4244
21	968	25.86	750	500	1.5009	0.5009	0.4061
22	924	26.17	751	520	1.4446	0.4446	0.3678
23	880	26.58	804	589	1.3648	0.3648	0.3110

N = Profile number
ζ = Distance from the Euler pole [km]
λ = Latitude of Red Sea axis intersection
β = β factor
L = Rift width [km]
L_0 = Initial width of the rifting area
ε' = Observed engineering strain (= $\beta - 1$)
ε = Observed true strain (= $\ln(1 + \varepsilon')$)

direction up to 15.8°N. This is indicative of an early phase of N–S extension, which could have led to the formation of left-lateral pull-apart basins as suggested by Makris and Rihm (1991) and Ghebreab (1998). Field evidence along the Arabian margin also supports this scenario. Starting from the configuration illustrated in Fig. 8, we can reconstruct the initial shape of the Ethiopian–Yemeni Plateau at ~30 Ma before the onset of rifting and the subsequent break-up of the Pan-African assembly. To this purpose, it is sufficient to rotate southward the Arabian plate about an equatorial pole by ~40 km, as required by the size of the gap. The resulting reassembly of the Ethiopian–Yemeni Plateau is shown in Fig. 9, while the rotation parameters that reconstruct the tectonic history of the plates surrounding the Red Sea are listed in Table 2.

5 Discussion

The kinematic model (and associated tectonic history) of the Red Sea, illustrated in the previous sections, results from a combination of marine geophysical data acquired in the oceanic areas of the Red Sea and the Gulf of Aden (Schettino et al. 2016; Fournier et al. 2010), geological and geophysical observations from the conjugate margins of Nubia, Arabia and Somalia (e.g., Schettino et al. 2016; d'Acremont et al. 2005; Leroy et al. 2010), and balancing of 23 crustal cross-sections across the Red Sea. With respect to other recent kinematic models (ArRajehi et al. 2010; DeMets and Merkouriev 2016), the scenario illustrated above includes and integrates different sources of data, resulting from both

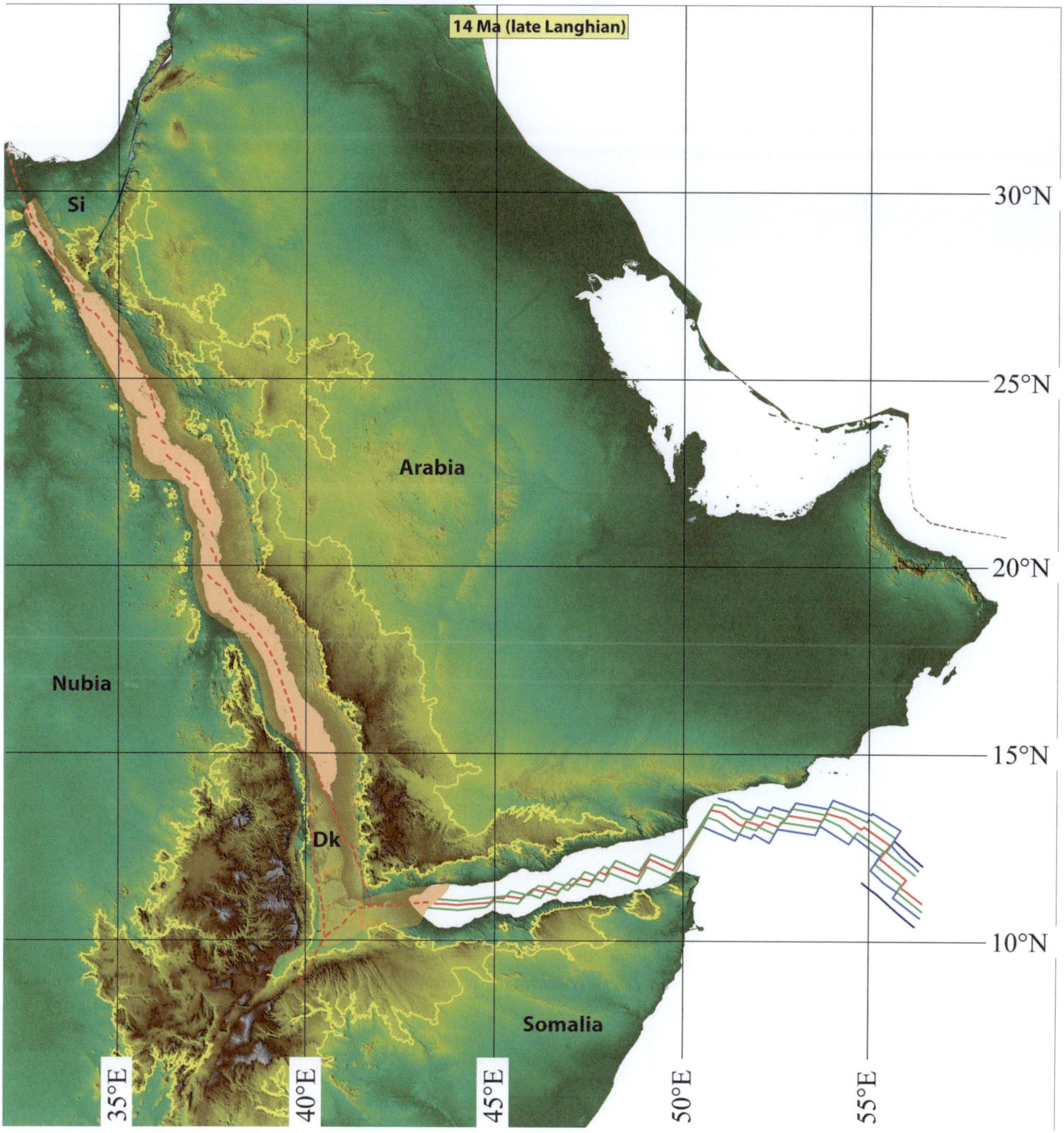

Fig. 7 Reconstruction of the 1000 m topography contour (yellow lines) at 14 Ma (late Langhian) in a fixed Nubian frame of reference. Dk = Danakil microplate; Si = Sinai microplate. The orange area indicates areas of active rifting. Oceanic crust in the Gulf of Aden is indicated by magnetic isochrons 5C (green lines, 16 Ma) and 5D (blue lines, 17.5 Ma) (after Fournier et al. 2010) and by the active spreading ridge (red line)

geophysical and geological observations. A key point in our approach was the determination of the closure angle associated with the pre-rift configuration starting from a palinspastic restoration of the initial size of the rift valley. Although this procedure provided a closure angle $\Omega = 14.31°$, using this parameter would have caused a relevant overlap of the N–S trending 1000 m contour lines of the Great Yemeni Escarpment against the Afar Escarpment. Therefore, we assumed that a small fraction of the total stretching occurred during an early short phase of N–S

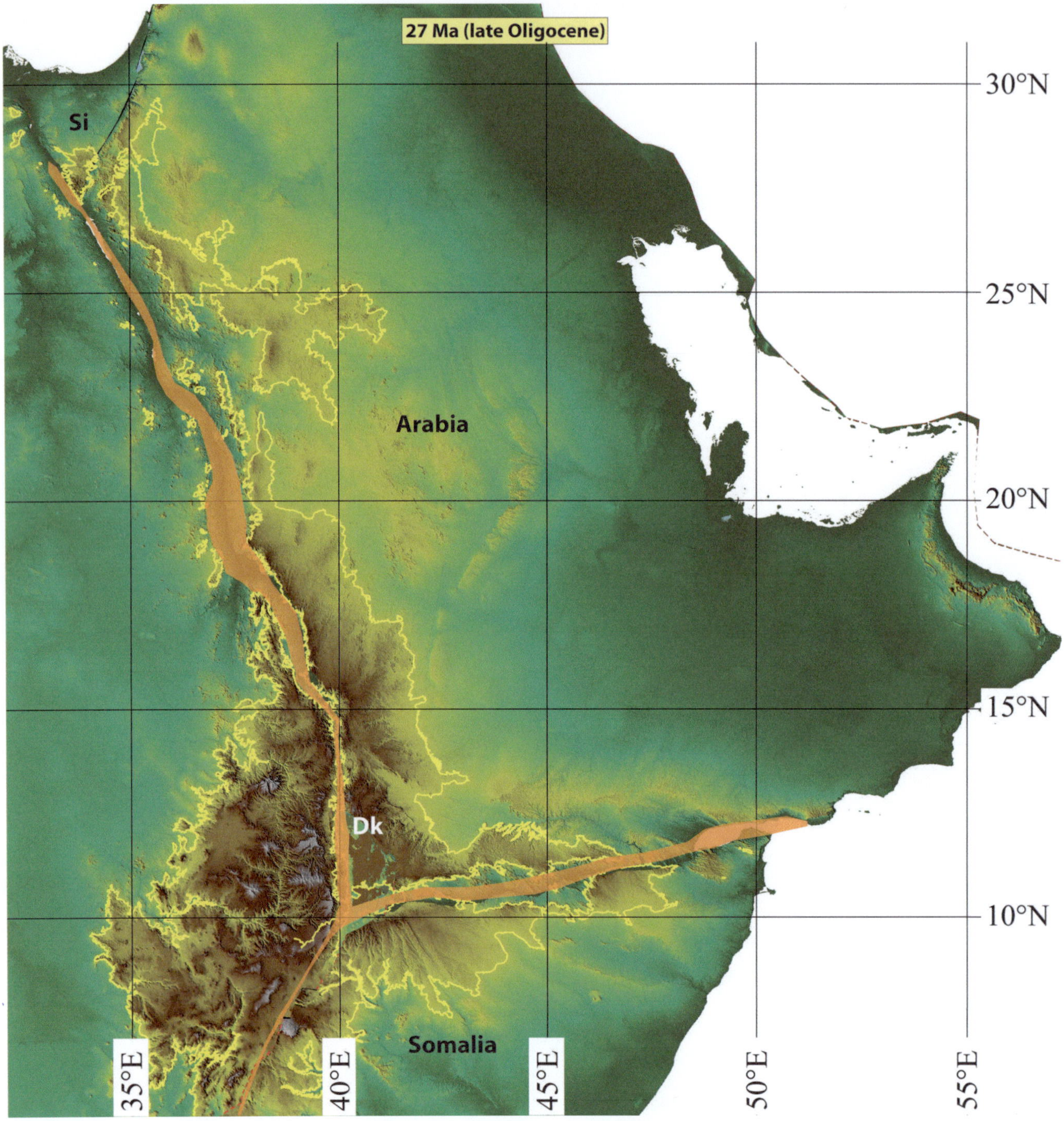

Fig. 8 Reconstruction of the 1000 m topography contour (yellow lines) at 27 Ma (late Oligocene) in a fixed Nubian frame of reference. Dk = Danakil microplate; Si = Sinai microplate. The orange zone indicates areas of active rifting

extension, characterized by the formation of pull-apart basins between Nubia and Arabia, as originally suggested by Makris and Rihm (1991) and Ghebreab (1998). Geological fieldwork conducted during three successive research expeditions along the Saudi Arabian margin by the authors (in 2015–2016) also supports this interpretation. It could be argued that ~40 km of N–S misfit are probably below the resolution of plate kinematics. However, the possibility that the initial stage of formation of the Red Sea was accommodated by N–S strike-slip faults and pull-apart basins does not contrast with some fundamental geodynamic considerations. In fact, the far-field system of forces that drove the

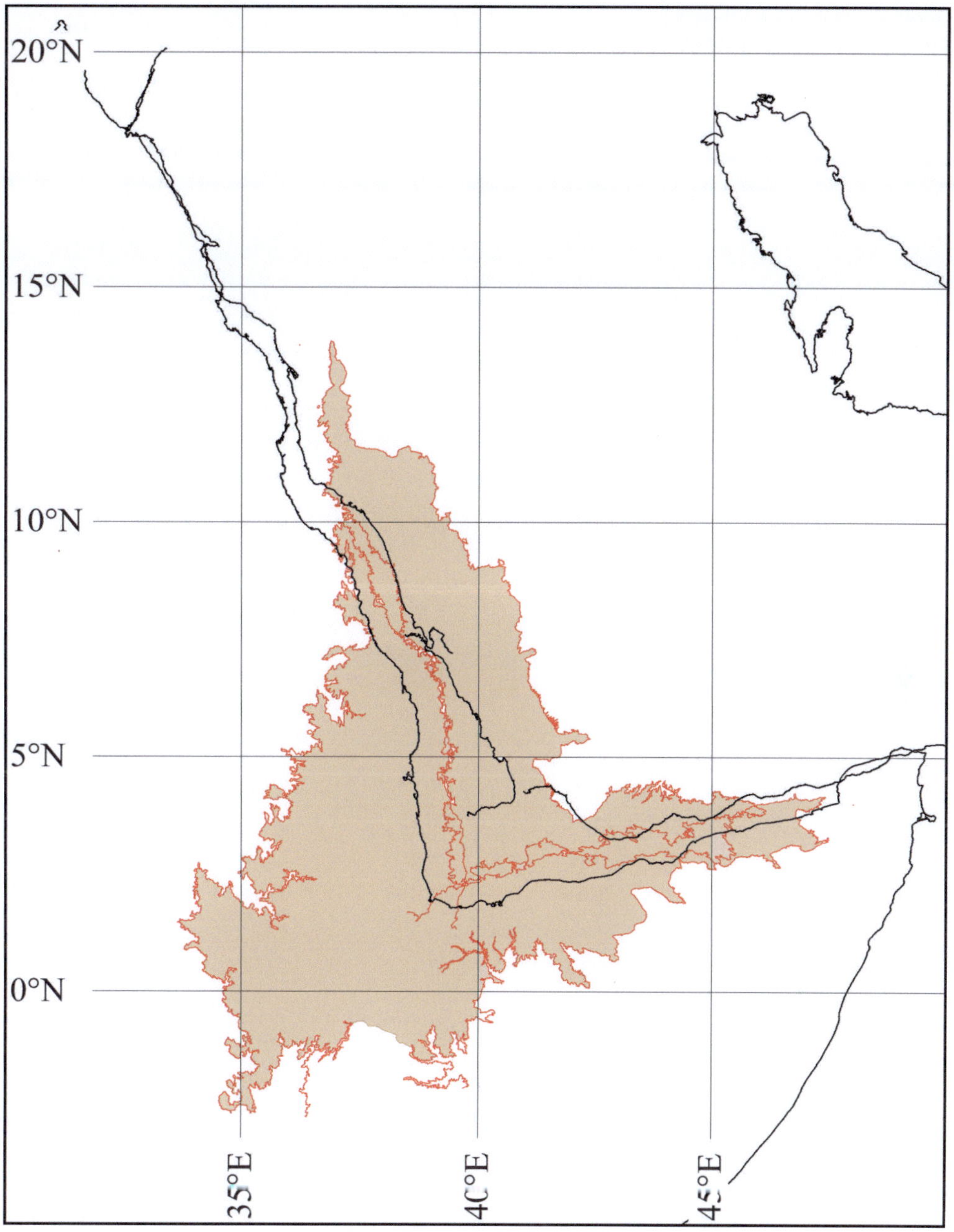

Fig. 9 Reconstruction of the Ethiopian–Yemeni Plateau at ~30 Ma in the palaeomagnetic reference frame of Schettino and Scotese (2005)

initial rifting and rupture between Arabia and Nubia was most probably associated with NE–SW directed slab pull exerted by subducting Neo–Tethys attached to the northeastern margin of Arabia (Schettino and Turco 2011). In this case, the formation of N–S structures that accommodated the initial phase of extension is perfectly compatible with the orientation of the stress field at that time.

The reconstructions presented in this paper are incompatible with tectonic scenarios that envisage a Red Sea floored entirely or for most of its extent by oceanic crust (e.g., McKenzie et al. 1970; Sultan et al. 1992, 1993). In fact, our starting point was the kinematic model of Schettino et al. (2016) and their isochron map of the Red Sea, which puts a strong constraint on the distribution of oceanic crust in this region. Our reconstructions show that the opening history of the Red Sea can be described by the relative motion of two large plates, Nubia and Arabia, and two intervening microplates, Sinai and Danakil, which formed some time during the Langhian respectively at the northern and southern ends of the rift valley. In the case of Sinai, it was part of the NE-moving Arabian plate from the late Oligocene to the Langhian (~14–16 Ma), in which during this time interval the Red Sea rift continued northward determining the formation of the Gulf of Suez. The details of this process are described extensively in Bosworth et al. (2005). Starting from the Langhian, the onset of left-lateral strike slip along

Table 2 Finite reconstruction poles for the Red Sea and Gulf of Aden regions

Age	Lat	Lon	Angle	References
Arabia–Nubia				
1.77	+30.32	27.18	–0.86	Schettino et al. (2016)
2.58	+30.32	27.18	–1.59	Schettino et al. (2016)
4.62	+30.32	27.18	–3.43	Schettino et al. (2016)
27.00	+30.32	27.18	–12.15	This paper
30.00	+29.87	25.13	–12.25	This paper
Danaki–Arabia				
1.77	+11.68	49.74	–1.26	Schettino et al. (2016)
2.58	+11.68	49.74	–2.00	Schettino et al. (2016)
4.18	+11.68	49.74	–3.38	Schettino et al. (2016)
4.62	+11.68	49.74	–3.76	Schettino et al. (2016)
14.00	+11.68	49.74	–11.86	This paper
Danakil–Nubia				
14.00	–18.14	221.24	+18.42	This paper
Somalia–Arabia				
1.00	+23.67	22.21	+0.52	Schettino et al. (2016)
2.58	+23.67	22.21	+0.94	Fournier et al. (2010)
3.58	+21.28	28.50	+1.62	Fournier et al. (2010)
5.89	+25.46	25.41	+2.40	Fournier et al. (2010)
8.70	+22.56	27.71	+3.99	Fournier et al. (2010)
10.95	+23.88	26.66	+4.74	Fournier et al. (2010)
16.01	+25.85	25.40	+6.85	Fournier et al. (2010)
17.62	+26.10	22.98	+7.28	Fournier et al. (2010)
20.13	+26.46	21.66	+7.83	Fournier et al. (2010)
Somalia–Nubia				
20.13	–41.18	237.32	+1.89	This paper
Sinai–Arabia				
4.62	32.37	27.02	1.28	Schettino et al. (2016)
14.00	32.37	27.02	3.92	This paper

the DSFZ strongly reduced the magnitude of relative motion between Sinai and Nubia and the rift rates in the Gulf of Suez, while the Gulf of Aqaba formed as a pull-apart basin during the development of the DSFZ. Plate kinematics around Danakil is a little bit more complex, because this microplate formed by strain partitioning during the rift between Nubia and Arabia. In the time interval between the late Oligocene and the Langhian ($\sim$14 Ma), it was part of the Nubian plate. Consequently, rifting must have started to the east of the Danakil Horst in an ENE direction ($\sim$N58–60E, according to the rotation model of Table 2). As mentioned above, such a scenario of late initiation of extension in Afar has been proposed by several authors (Wolfenden et al. 2004; Bonini et al. 2005; Corti 2009) and implies that basaltic magmatism in this area occurred long before a true plate boundary developed between Nubia and Danakil. Starting from the Langhian, the newly formed Danakil microplate moved with respect to both Nubia and Arabia, determining the formation of the Afar Depression. An important feature of the Red Sea rift, which is shared by the Gulf of Aden, is represented by its along-strike segmentation in half-graben sub-basins that are separated by transverse accommodation zones (Bosworth et al. 2005; Leroy et al. 2012). The latter are strongly influenced by pre-existing basement structures, for example by the Najd fault system (Younes and McClay 2002).

A comparison between the relatively complex kinematic model discussed above and more simple plate motion models based exclusively on extrapolation of GPS data can be performed considering the displacement or velocity history of a point along the Nubia–Arabia boundary. This is illustrated in Fig. 10. We note that for a point presently at 9.8°N, 38.6°E both ArRajehi et al. (2010) and Reilinger and McClusky (2011) predict a constant rate between 12 and

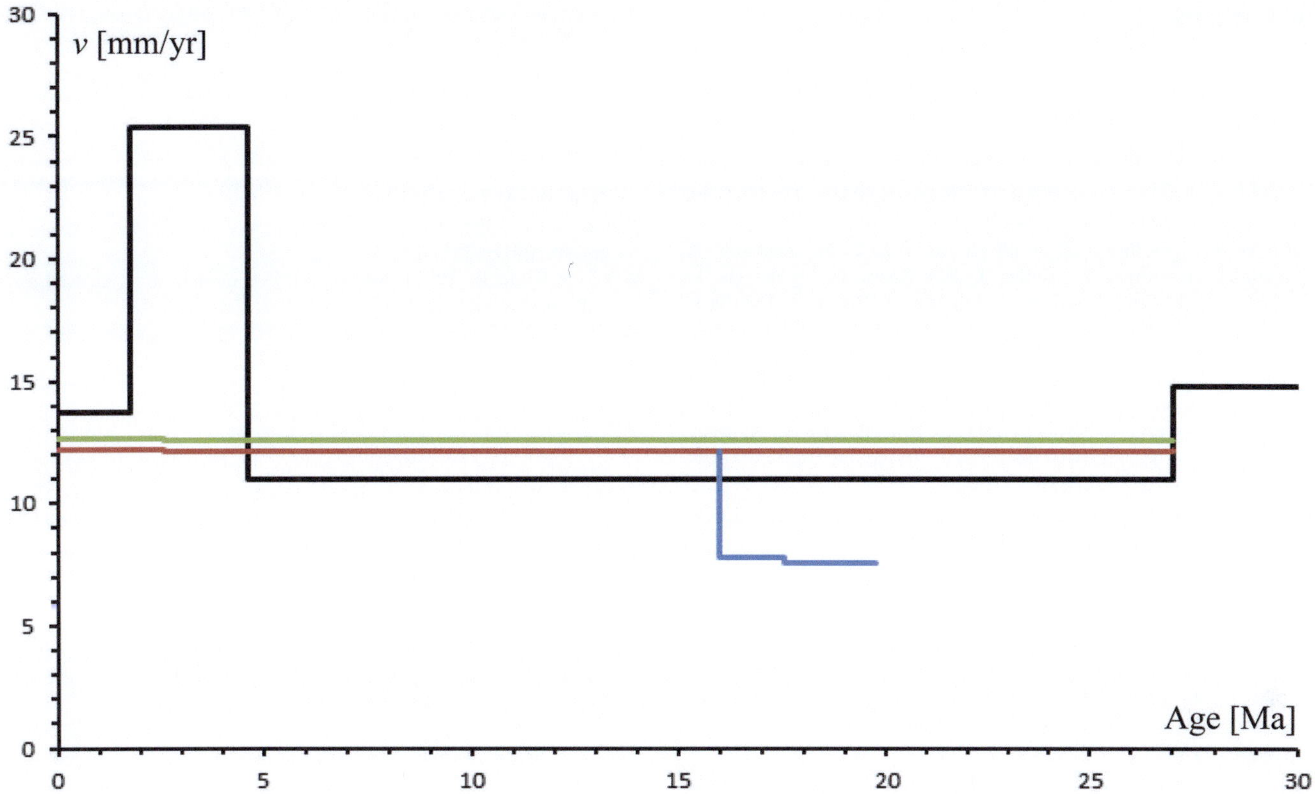

Fig. 10 Predicted spreading/rifting velocities for a point along the Nubia–Arabia boundary, presently located at 19.8°N, 38.6°E. Black line = this paper (and Schettino et al. 2016, in particular the spreading peak between 4.6 and 1.77 Ma); Green line = ArRajehi et al. (2010); Brown line = Reilinger and McClusky (2011); Blue line = DeMets and Merkouriev (2016)

13 mm/yr since the Oligocene, independently from the fact that important geological changes occurred during this long time interval. For example, it is now widely accepted that the rift-drift transition is accompanied by important kinematic changes (Brune et al. 2016). In addition, the end of Neo–Tethys oceanic subduction along the northeastern margin of Arabia (Schettino and Turco 2011) and the subsequent collision with Eurasia, with underthrusting of the Arabian crust beneath Iran starting from ~10 Ma (Mouthereau et al. 2007), cannot be considered uninfluential for plate kinematics. In our model, these two events are associated with an increase of velocity at 4.62 Ma and a subsequent slowdown at 1.77 Ma respectively (Fig. 10). Therefore, the model provides a more realistic view of the tectonic history of the Red Sea region.

6 Conclusion

The kinematic model illustrated above is mainly based on the analysis of marine magnetic data integrated with kinematic indicators observed along the Arabian margin during two research expeditions performed in 2015 and 2016. This model illustrates additional detail on the plate tectonic evolution of the Red Sea since the early Oligocene (~30 Ma). Although a rigorous determination of plate motions in the Red Sea for times older than the late Pliocene is impeded by the absence of magnetic anomalies older than Anomaly 3, structural data show that the Euler poles of relative motion remained stable during most of the rifting stage. Consequently, plate reconstructions of the pre-drift configurations can be obtained by assuming invariant stage poles and performing a palinspastic balance of crustal profiles across the Red Sea. A reconstruction at 27 Ma, which is representative of the onset of NE–SW extension in the Red Sea, supports the idea that a short initial phase of N–S strike-slip between Arabia and Nubia occurred before the development of a system of NW–SE normal faults associated with the main phase of extension.

Acknowledgements This work was funded by the Italian Ministry of University and Scientific Research, PRIN prot. 20125JKANY, and by the Saudi Geological Survey (SGS). The authors are grateful to the SGS personnel who helped them in surveying the area and to the SGS drivers who showed great professionality in their difficult work. Finally, the authors thank four anonymous reviewers for their accurate reviews and useful suggestions that considerably improved this paper.

References

Agard P, Omrani J, Jolivet L, Mouthereau F (2005) Convergence history across Zagros (Iran): constraints from collisional and earlier deformation. Int J Earth Sci 94(3):401–419

Al-Damegh K, Sandvol E, Barazangi M (2005) Crustal structure of the Arabian plate: new constraints from the analysis of teleseismic receiver functions. Earth Planet Sci Lett 231(3):177–196

ArRajehi A, McClusky S, Reilinger R, Daoud M, Alchalbi A, Ergintav S, Gomez F, Sholan J, Bou-Rabee F, Ogubazghi G, Haileab B, Fisseha S, Asfaw L, Mahmoud S, Rayan A, Bendik R, Kogan L (2010) Geodetic constraints on present-day motion of the Arabian Plate: implications for Red Sea and Gulf of Aden rifting. Tectonics 29:TC3011. https://doi.org/10.1029/2009tc002482

Bonini M, Corti G, Innocenti F, Manetti P, Mazzarini F, Abebe T, Pecskay Z (2005) Evolution of the Main Ethiopian Rift in the frame of Afar and Kenya rifts propagation. Tectonics 24(1):TC1007. https://doi.org/10.1029/2004tc001680

Bosworth W, Huchon P, McClay K (2005) The Red Sea and Gulf of Aden basins. J Afr Earth Sci 43(1):334–378

Brune S, Williams SE, Butterworth NP, Müller RD (2016) Abrupt plate accelerations shape rifted continental margins. Nature 536 (7615):201–204

Cande SC, Kent DV (1995) Revised calibration of the geomagnetic polarity timescale for the late Cretaceous and Cenozoic. J Geophys Res 100(B4):6093–6095

Chu D, Gordon G (1998) Current plate motions across the Red Sea. Geophys J Int 135:313–328

Cochran JR (1983) A model for development of Red Sea. Am Assoc Petrol Geol Bull 67(1):41–69

Corti G (2009) Continental rift evolution: from rift initiation to incipient break-up in the Main Ethiopian Rift, East Africa. Earth Sci Rev 96(1):1–53

d'Acremont E, Leroy S, Beslier M-O, Bellahsen N, Fournier M, Robin C, Maia M, Gente P (2005) Structure and evolution of the eastern Gulf of Aden conjugate margins from seismic reflection data. Geophys J Int 160:869–890. https://doi.org/10.1111/j.1365-246X.2005.02524.x

DeMets C, Merkouriev S (2016) High-resolution estimates of Nubia-Somalia plate motion since 20 Ma from reconstructions of the Southwest Indian Ridge, Red Sea and Gulf of Aden. Geophys J Int 207(1):317–332. https://doi.org/10.1093/gji/ggw276

Fournier M, Chamot-Rooke N, Petit C, Huchon P, Al-Kathiri A, Audin L, Beslier M-O, d'Acremont E, Fabbri O, Fleury J-M, Khanbari K, Lepvrier C, Leroy S, Maillot B, Merkouriev S (2010) Arabia-Somalia plate kinematics, evolution of the Aden-Owen-Carlsberg triple junction, and opening of the Gulf of Aden. J Geophys Res 115:B04102. https://doi.org/10.1029/2008JB006257

Ghebreab W (1998) Tectonics of the Red Sea region reassessed. Earth-Sci Rev 45(1):1–44

Hansen SE, Rodgers AJ, Schwartz SY, Al-Amri AM (2007) Imaging ruptured lithosphere beneath the Red Sea and Arabian Peninsula. Earth Planet Sci Lett 259(3):256–265

Joffe S, Garfunkel Z (1987) Plate kinematics of the circum Red Sea—a re-evaluation. Tectonophysics 141(1):5–22

Le Pichon X, Francheteau J (1978) A plate-tectonic analysis of the Red Sea-Gulf of Aden area. Tectonophysics 46:369–406

Le Pichon XT, Gaulier JM (1988) The rotation of Arabia and the Levant fault system. Tectonophysics 153(1):271–294

Leroy S, Lucazeau F, D'Acremont E, Watremez L, Autin J, Rouzo S, Bellahsen N, Tiberi C, Ebinger C, Beslier M-O, Perrot J, Razin P, Rolandone F, Sloan H, Stuart G, Lazki AA, Al-Toubi K, Bache F, Bonneville A, Goutorbe B, Huchon P, Unternehr P, Khanbari K (2010) Contrasted styles of rifting in the eastern Gulf of Aden: a combined wide-angle, multichannel seismic, and heat flow survey. Geochem Geophys Geosys 11(7):Q07004. https://doi.org/10.1029/2009GC002963

Leroy S, Razin P, Autin J, Bache F, d'Acremont E, Watremez L, Robinet J, Baurion C, Denèle Y, Bellahsen N, Lucazeau F, Rolandone F, Rouzo S, Serra Kiel J, Robin C, Guillocheau F, Tiberi C, Basuyau C, Beslier M-O, Ebinger C, Stuart G, Abdulhakim A, Khanbari K, Al Ganad I, de Clarens P, Unternehr P, Al Toubi K, Al Lazki K (2012) From rifting to oceanic spreading in the Gulf of Aden: a synthesis. Arab J Geosci 5:859–901. https://doi.org/10.1007/s12517-011-0475-4

Makris J, Rihm R (1991) Shear-controlled evolution of the Red Sea: pull apart model. Tectonophysics 198(2):441–466

McKenzie DP, Davies D, Molnar P (1970) Plate tectonics of the Red Sea and East Africa. Nature 226(5242):243–248

Mouthereau F, Tensi J, Bellahsen N, Lacombe O, de Boisgrollier T, Kargar S (2007) Tertiary sequence of deformation in a thin-skinned/thick-skinned collision belt: the Zagros Folded Belt (Fars, Iran). Tectonics 26:TC5006. https://doi.org/10.1029/2007tc002098

Nuriel P, Weinberger R, Kylander-Clark ARC, Hacker BR, Craddock JP (2017) The onset of the Dead Sea transform based on calcite age-strain analyses. Geology 45(7):587–590

Pik R, Marty B, Carignan J, Yirgu G, Ayalew T (2008) Timing of East African Rift development in southern Ethiopia: implication for mantle plume activity and evolution of topography. Geology 36(2):167–170

Reilinger R, McClusky S (2011) Nubia-Arabia-Eurasia plate motions and the dynamics of Mediterranean and Middle East tectonics. Geophys J Int 186:971–979

Salem A, Green C, Campbell S, Fairhead JD, Cascone L, Moorhead L (2013) Moho depth and sediment thickness estimation beneath the Red Sea derived from satellite and terrestrial gravity data. Geophysics 78(5):G89–G101

Salerno VM, Capitanio FA, Farrington RJ, Riel N (2016) The role of long-term rifting history on modes of continental lithosphere extension. J Geophys Res 121. https://doi.org/10.1002/2016jb013005

Schettino A (2014) Quantitative plate tectonics. Springer, Berlin, 403 p. ISBN 978-3-319-09134-1

Schettino A, Scotese CR (2005) Apparent polar wander paths for the major continents (200 Ma—present day): a paleomagnetic reference frame for global plate tectonic reconstructions. Geophys J Int 163(2):727–759

Schettino A, Turco E (2006) Plate kinematics of the Western Mediterranean region during the Oligocene and early Miocene. Geophys J Int 166(3):1398–1423

Schettino A, Turco E (2011) Tectonic history of the western Tethys since the late Triassic. Geol Soc Am Bull 123(1/2):89–105

Schettino A, Macchiavelli C, Pierantoni PP, Zanoni D, Rasul N (2016) Recent kinematics of the tectonic plates surrounding the Red Sea and Gulf of Aden. Geophys J Int 207:457–480. https://doi.org/10.1093/gji/ggw280

Sultan M, Becker R, Arvidson RE, Shore P, Stern RJ, El Alfy Z, Guinness EA (1992) Nature of the Red Sea crust: a controversy revisited. Geology 20(7):593–596

Sultan M, Becker R, Arvidson RE, Shore P, Stern RJ, El Alfy Z, Attia RI (1993) New constraints on Red Sea rifting from correlations of Arabian and Nubian Neoproterozoic outcrops. Tectonics 12(6):1303–1319

van Wijk JW, Cloetingh SAPL (2002) Basin migration caused by slow lithospheric extension. Earth Planet Sci Lett 198:275–288

Wolfenden E, Ebinger C, Yirgu G, Deino A, Ayalew D (2004) Evolution of the northern Main Ethiopian rift: birth of a triple junction. Earth Planet Sci Lett 224(1):213–228

Younes AI, McClay KR (2002) Development of accommodation zones in the Gulf of Suez-Red Sea rift, Egypt. Am Assoc Petrol Geol Bull 86:1003–1026

Hydrothermal Prospection in the Red Sea Rift: Geochemical Messages from Basalts

Froukje M. van der Zwan, Colin W. Devey, and Nico Augustin

Abstract

Hydrothermal circulation at mid-ocean ridges and assimilation of hydrothermally altered crust or hydrothermal fluids by rising magma can be traced by measuring chlorine (Cl) excess in erupted lavas. The Red Sea Rift provides a unique opportunity to study assimilation of hydrothermally altered crust at an ultra-slow spreading ridge (maximum 1.6 cm yr^{-1} full spreading rate) by Cl, due to its saline seawater (40–42‰, cf. 35‰ in open ocean water), the presence of (hot) brine pools (up to 270‰ salinity and 68 °C) and the thick evaporite sequences that flank the young rift. Absolute chlorine concentrations (up to 1300 ppm) and Cl concentrations relative to minor or trace elements of similar mantle incompatibility (e.g., K, Nb) are much higher in Red Sea basalts than in basalts from average slow spreading ridges. Mantle Cl/Nb concentrations can be used to calculate the Cl-excess, above the magmatic Cl, that is present in the samples. Homogeneous within-sample Cl concentrations, high Cl/H_2O, the decoupling of Cl-excess from other trace elements and its independence of the presence of highly saline seafloor brines at the site of eruption indicate that Cl is not enriched at the seafloor. Instead we find basaltic Cl-excess to be spatially closely correlated with evidence of hydrothermal activity, suggesting that deeper assimilation of hydrothermal Cl is the dominant Cl-enrichment process. A proximity of samples to both evaporite outcrops and bathymetric signs of volcanism on the seafloor enhance Cl-excess in basalts. The basaltic Cl-excess can be used as a tracer together with new bathymetric maps as well as indications of hydrothermal venting (hot brine pools, metalliferous sediments) to predict where hydrothermal venting or now inactive hydrothermal vent fields can be expected. Sites of particular interest for future hydrothermal research are the Mabahiss Deep, the Thetis-Hadarba-Hatiba Deeps and Shagara Aswad Erba Deeps (especially their large axial domes), and Poseidon Deep. Older hydrothermal vent fields may be present at the Nereus and Suakin Deeps. These sites significantly increase the potential of hydrothermal vent field prospection in the Red Sea.

1 Introduction

1.1 Hydrothermal Circulation at (Ultra) Slow-Spreading Ridges

Hydrothermal circulation of seawater, driven by the magmatic heat of young oceanic lithosphere is an important process at all mid-ocean ridges (MORs). There, cold seawater that penetrates the crust provides an effective way to cool the newly formed oceanic crust. The fluids that get heated rise again to the seafloor, where hydrothermal activity can be expressed by diffuse (low temperature) or focussed (high temperature) venting in the form of (black or white) smoker fields. Interaction of the circulating seawater with the hot magma and rocks of the oceanic crust leads to a metal enrichment of the fluids. Upon cooling of these fluids at the surface this metal enrichment can build economically interesting seafloor massive sulphide (SMS) deposits (Rona et al. 1986; Hannington et al. 2011). In addition, the heat and specific chemistry of the venting fluids create habitats for chemosynthetic communities at the seafloor.

Hydrothermal activity occurs at all MORs, but the along-axis frequency of high temperature hydrothermal venting increases with spreading rate; calculations for (ultra) slow-spreading ridges, like the Red Sea Rift (<10–16 mm yr^{-1}; Chu and Gordon 1998) plot on the lower end and

F. M. van der Zwan (✉) · C. W. Devey · N. Augustin
GEOMAR Helmholtz Centre for Ocean Research Kiel, Wischhofstraße 1-3, 24148 Kiel, Germany
e-mail: fzwan@geomar.de

F. M. van der Zwan
Institute of Geosciences, Christian Albrechts University Kiel, Ludewig-Meyn-Straße 10, 24118 Kiel, Germany

N. M. A. Rasul and I. C. F. Stewart (eds.), *Geological Setting, Palaeoenvironment and Archaeology of the Red Sea*,
https://doi.org/10.1007/978-3-319-99408-6_10

indicate at least one active vent field every 100 km of ridge axis (e.g., Baker and German 2004; Hannington et al. 2011; Beaulieu et al. 2015). Moreover, ultraslow-spreading ridges such as the Gakkel Ridge and the Southwest Indian Ridge that were believed to be hydrothermally less active show—wherever investigated in more detail—more sites of active venting to be present than expected (e.g., Edmonds et al. 2003; German et al. 2016). Despite a lower along-axis frequency, the size of hydrothermal deposits at (ultra) slow-spreading ridges is generally larger compared to fast-spreading ridges (Fouquet 1997) and they have a higher Cu and Au potential, which makes them thus most attractive from an economic point of view (German et al. 2016). The largest known SMS deposits, like TAG and Semenov 4, have been discovered at the slow-spreading Mid-Atlantic Ridge (Petersen et al. 2000; Pertsev et al. 2012).

1.2 Current Inferred Sites of Hydrothermal Activity in the Red Sea

Hydrothermal venting in the Red Sea Rift (RSR) is inferred to occur in some of the Red Sea Deeps that contain highly saline (up to 270‰ NaCl) brine pools (e.g., Swallow and Crease 1965; Fig. 1). Although the brine pools are not of hydrothermal origin (Pierret et al. 2010; van der Zwan et al. 2015), the immobility of the dense pools gives them the capability to preserve and accumulate temperature and chemical attributes from hydrothermal fluids that are vented into a brine, in contrast to ambient seawater, where those attributes are quickly dispersed by the convecting seawater and can only be traced close to the source. The brine pools in the Red Sea vary in temperature from being similar to Red Sea bottom water (about 22 °C) up to 68 °C (in the Atlantis II Deep) and can have a characteristic chemical composition, for example up to 5 orders of magnitude higher trace element and gas concentrations compared to ambient Red Sea bottom water (e.g., Schmidt et al. 2015). High brine temperatures and specific chemistries are often associated with the occurrence of metalliferous sediments (Gurvich 2006, and references therein). In the Atlantis II Deep, the metalliferous sediments and their sulphur isotope compositions, together with the temperature and composition of the brine pool, suggest the strongest hydrothermal venting in the Red Sea into a brine (Bäcker and Richter 1973; Pierret et al. 2001; Laurila et al. 2014; Schmidt et al. 2015). With the large amount of metalliferous sediments, Atlantis II Deep is hosting the only known currently forming sedimentary exhalative ore (SEDEX) deposit in the world, underlining its hydrothermal origin. Besides Atlantis II Deep, the adjacent Discovery Deep and, further south, the Port Sudan Deep show metalliferous sediments and seafloor brine pools with elevated temperatures, indicating hydrothermal venting into

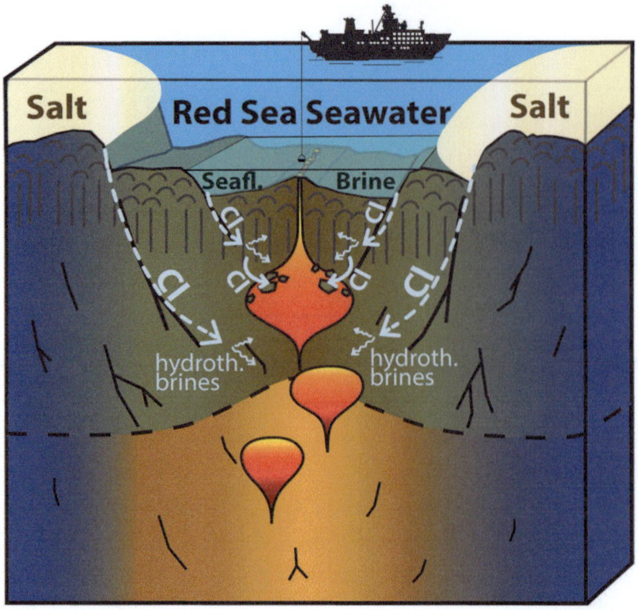

Fig. 1 Conceptual sketch showing the increase of Cl in the crust by hydrothermal alteration. The magmatic assimilation of hydrothermal fluids or hydrothermally altered crust causes a Cl gain in the magma leading to basalts with Cl-excess that can be dredged. High Cl in hydrothermal fluids is caused by phase separation (small arrows). Hydrothermal fluids interacting with evaporites or seafloor brines may be even more highly charged in Cl

the brine pools there as well (Pierret et al. 2001; Gurvich 2006; Schmidt et al. 2015). In addition, the Nereus Deep brine also shows elevated temperatures. Metalliferous sediments are not exclusively formed in Deeps with brines, but are always associated with hydrothermal activity (e.g. Thetis Deep; Pierret et al. 2010). Metalliferous sediments in some of these Deeps without brine pools or in Deeps with brine pools that have ambient temperatures (e.g., Shagara, Suakin Deeps) may suggest past occurrences of hydrothermal venting (Gurvich 2006; Pierret et al. 2010). Nevertheless, no active hydrothermal vents (black smokers) have been confirmed so far in the Red Sea by direct observation, due to the limited accessibility to the brine pools by modern camera sampling systems. Only some (extinct) black smokers and sulphide fragments have been reported from Kebrit Deep and from the 18°N RSR respectively (Monin et al. 1982; Blum and Puchelt 1991), but without any further indications of recent activity. From other areas on the RSR that do not host a brine pool, no information is known on hydrothermal venting due to a lack of detailed investigations.

This information implies that, even when counting the 3–4 inferred active vents in brines that are inaccessible for state-of-the-art, high-resolution visualisation equipment (underwater robots and autonomous vehicles), the number of active hydrothermal vent fields is significantly less than expected over the >1300 km length of the RSR. At this spreading rate, we would expect a minimum of 13 active hydrothermal systems (based on the correlation between spreading rate and the statistical occurance of hydrothermal vents per 1000 km rift axis; Hannington et al. 2010, 2011; Beaulieu et al. 2015), which implies that there are more vent fields to be discovered in the Red Sea. The Red Sea may have an even further enhanced capacity for hydrothermal activity as the ocean crust has a significantly higher heat flow than other ultraslow-spreading ridges (Girdler and Evans 1977; Augustin et al. 2016). Each of these undiscovered fields is expected to contain considerable amounts of SMS due to the ultraslow-spreading rate of the RSR (cf. Fouquet 1997; Hannington et al. 2010, 2011). The unique high intrinsic seawater salinity (40–42‰ compared to 34.5‰ for average ocean water) of the Red Sea, the (hot) saline brine pools (Pierret et al. 2001), and thick evaporite sequences flanking the RSR (e.g., Whitmarsh et al. 1974; Mitchell et al. 2010; Augustin et al. 2014a; Fig. 1) likely result in highly saline hydrothermal fluids that may enhance fluid-rock chemistry exchange. The large tonnage of 90 Mt of metalliferous sediments (with up to 2.06% Zn, 0.46% Cu, 41 g/t Ag, and 3 g/t Au) in Atlantis II Deep shows the resource potential of a single RSR hydrothermal vent area (Guney et al. 1988; Laurila et al. 2014).

1.3 Hydrothermal Vent Field Prospection

Active high-temperature vent fields can be found by detecting their effluent in the overlying water column. This task can be performed by MAPR casts and Tow-Yos across the areas of interests (Klinkhammer et al. 1977; Baker and Massoth 1987; Edmonds et al. 2003; Devey et al. 2010) or by CTD-systems equipped with video-systems, water samplers and additional sensors (i.e., CH_4, CO_2, pH, Eh) that support visual ground-truthing of venting features and provide the opportunity for confirming, mapping and quantifying hydrothermal fluid input into the water column (Schmidt et al. 2013a; Linke et al. 2015). In addition, if a rough target area is known, vent fields can be detected using high-resolution sensors (Eh, turbidity, temperature, side-scan, magnetometer) from deep towed instrument platforms (TOBI) or autonomous underwater vehicles (AUV) that can provide measurements of water anomalies, magnetic anomalies and sidescan reflections over small areas. For detailed investigations those instruments can also directly image hydrothermal vent fields associated with hydrothermal plume signals (e.g., Blondel 2010; Szitkar et al. 2015). However, the search for hydrothermal fields using these methods is a difficult and time-consuming process, particularly due to the scarcity of detailed maps of the ocean floor. The discovery of extinct hydrothermal fields that are economically interesting and give information on hydrothermal activity over a longer time span, is at best extremely challenging.

As all hydrothermal venting is the seafloor expression of deep hydrothermal fluid circulation and alteration of the crust, an alternative prospection method for finding active and ancient, now inactive hydrothermal occurrences is to search for geochemical traces of deeper hydrothermal alteration of the crust. Evidence from fast-spreading ridges shows that hydrothermal activity and high-temperature alteration of the oceanic crust can indirectly be traced by the chemistry of submarine erupted mid-ocean ridge basalts (MORB) for which chlorine (Cl) is specifically indicative. This is because the Cl contents of seawater and magma are vastly different (1.9 wt% and generally <500 ppm respectively; Michael and Schilling 1989; Michael and Cornell 1998). Seawater infiltration and interaction with the oceanic crust will hence increase the Cl content of the crust by hydrothermal alteration (Barnes and Cisneros 2012; Fig. 1). Assimilation of this hydrothermally altered crust or of hydrothermal fluids (inclusions) by rising magma can subsequently increase the Cl content of this magma that later can be recovered as basalt from the seafloor (Fig. 1; e.g., Michael and Schilling 1989; Coogan et al. 2002; Gillis et al. 2003). Therefore, Cl in erupted lavas is a sensitive tracer for hydrothermal activity. Chlorine addition due to interaction of basalts with hydrothermal fluids was previously identified at fast-spreading mid-ocean ridges (e.g., Michael and Schilling 1989; Gillis et al. 2003; le Roux et al. 2006; France et al. 2009, 2010; Kendrick et al. 2013), in ophiolites (Coogan et al. 2002; Coogan 2003), at back-arc basins (Kent et al. 2002; Sun et al. 2007), as well as in ocean island basalts (Kent et al. 1999a, b; Kendrick et al. 2017).

The Cl concentrations in basalts from (ultra) slow-spreading ridges like the RSR are relatively low (Cl $\sim$50–200 ppm), compared to fast-spreading ridges (Cl <1300 ppm; e.g., Michael and Cornell 1998). Nevertheless, using a high-precision method to measure Cl in basalts (van der Zwan et al. 2012) reveals that at slow-spreading ridges variations in Cl due to the assimilation of hydrothermal Cl can also be observed (van der Zwan 2014; van der Zwan et al. 2015, 2017). Furthermore, the highly saline environment of the Red Sea has the potential to create circulating hydrothermal fluids with high Cl contents (Fig. 1); therefore, the Cl contamination in basalts by hydrothermal circulation can be much higher than on average for (ultra)slow-spreading ridges. By obtaining Cl data (ranging from 50 to 1400 ppm) from Red Sea basalts by our high precision Cl measurement routine ($\pm$1–2 ppm SD) with low detection limits ($\sim$10 ppm; van der Zwan et al. 2012), we can use Cl enrichment in basalt as an excellent magmatic marker of present and past hydrothermal circulation that can be used to trace hydrothermalism in the RSR.

To encircle the potential locations of hydrothermal venting further, the geomorphological features along mid-ocean ridges can be analysed, in order to predict potential locations of hydrothermal venting (Yoshikawa et al. 2012). At (ultra)slow-spreading ridges, hydrothermal vent fields have often been found at two specific types of rift morphologies. All closely investigated oceanic core complexes revealed ultramafic hosted vent fields (www.inter-ridge.org; Früh-Green et al. 2003; Petersen et al. 2009; Pertsev et al. 2012; Escartín et al. 2017), but core complexes have not been identified in the RSR so far (Augustin et al. 2016). Many basalt-hosted hydrothermal systems at (ultra) slow-spreading ridges are related to large axial volcanoes, highs and volcanic ridges (Lein et al. 2010; Fontaine et al. 2014; Anderson et al. 2016). For example, along the Mid-Atlantic Ridge at 13–33 °S all large axial volcanoes show active hydrothermal venting (Devey et al. 2013). These axial volcanoes have a relatively smooth surface, indicating recent volcanic activity, and are remarkably similar in their morphological appearance to the axial domes found in the RSR (Augustin et al. 2016). Here the use of the latest high-resolution bathymetry data of the RSR (Augustin et al. 2014a, b, 2016) together with targets derived from hydrothermal Cl-excess in basalt and additional indicators for the occurrence of hydrothermal activity in the RSR (hot brine pools, metalliferous sediments) enable us to define the most promising areas for currently active and older, inactive hydrothermal vent fields and to evaluate hydrothermal activity in the Red Sea.

2 Chlorine Excess in Red Sea Basalts

The samples are derived from 25.5°N to 16.5°N on the RSR (van der Zwan et al. 2015) and are dredged basalts that come in the form of pillow lavas, lobate and sheet lavas or rock fragments with glassy surfaces (Fig. 2). The freshest (and therefore likely youngest) samples (with no visible alteration) were collected from Erba Deep (Fig. 2a), while some samples from Atlantis II (Fig. 2b) and all samples from

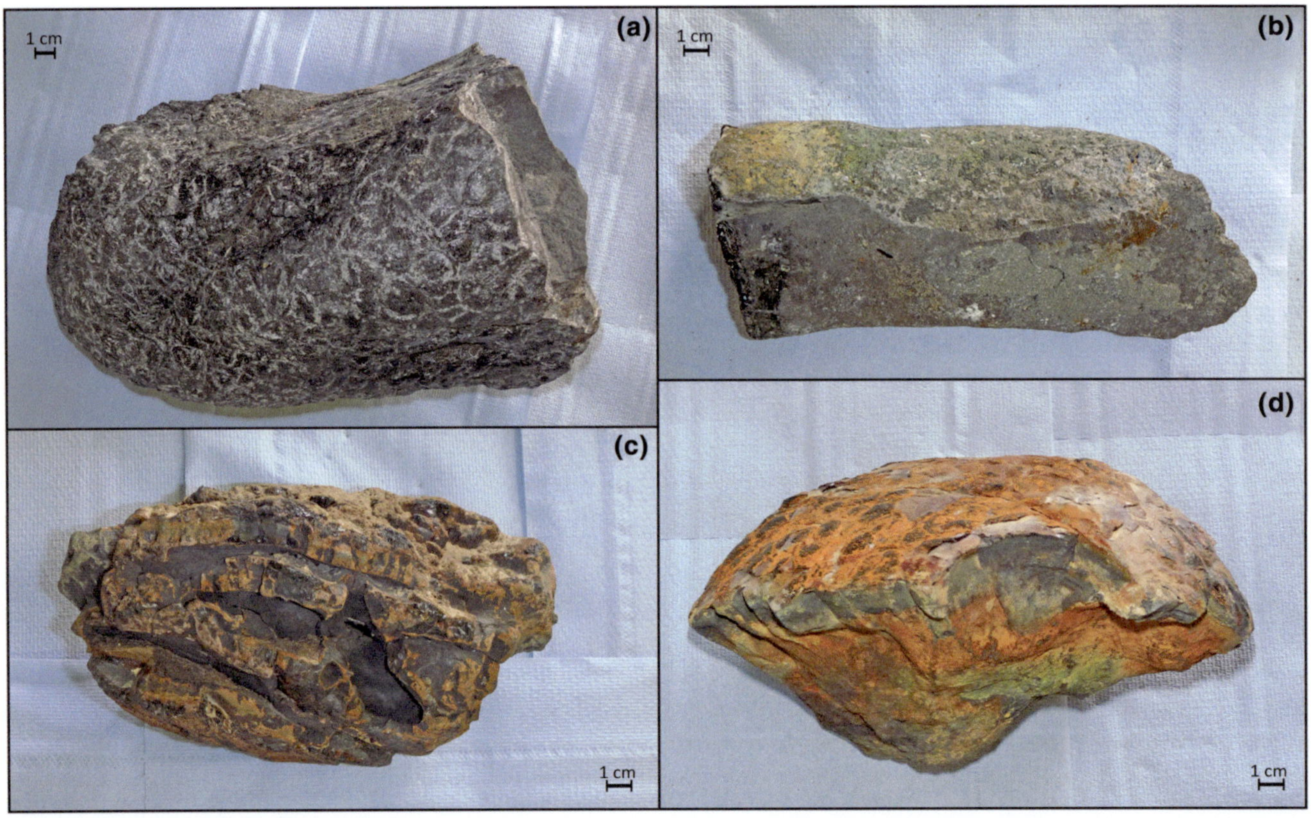

Fig. 2 Examples of basalt samples from the Red Sea. **a** Freshest recovered pillow basalt from Erba Deep (64PE350-43DR). The salt encrustations (whitish network on the rock) show that the basalt was recovered from a seafloor brine. **b** Blocky basalt from within Atlantis II Deep (64PE350-33DR) shows most alteration, but nevertheless has fresh glass on the top (left in the picture). **c** Basalt from Mabahiss Deep (64PE351-14DR) shows sheet flows with fresh glass under a thin palagonite layer. **d** Well-formed pillow basalt from Hadarba Deep (64PE350-23DR) with a fine crystalline interior but fresh glasses on top under a thin palagonite layer

Shaban Deep show the strongest alteration, although they still contain fresh glass. The samples from Erba and Northern Atlantis II Deeps were recovered out of the seafloor brine pools (Schmidt 2013b). Fresh glasses from the rocks have been selected and analysed for Cl, major and trace elements by electron microprobe and laser-induced-coupled-plasma-spectrometer (for details see van der Zwan et al. 2015). The analytical precision (2 standard deviations) for Cl is generally <3 ppm, for major elements <2.5%, but up to 5% for Na and MgO and ~30% for Mn and P and typically <2–5% (one standard deviation) for trace elements. The reproducibility is 1–2 ppm for internal standards (van der Zwan et al. 2012).

Red Sea glasses are both on grain and sample scale very homogeneous in their Cl, major and trace element concentrations, without any gradients. The samples have Cl values that range from 58.7 to 1399 ppm (van der Zwan et al. 2015) that are elevated compared to the majority of other slow-spreading ridges (<200 ppm; Michael and Cornell 1998). To determine the amount of Cl enrichment by hydrothermal processes, without disturbing this signal by variations in magmatic Cl, it is important to estimate the amount of magmatic Cl in each of the samples. As magmatic Cl concentrations vary with melting or crystallisation processes, we can use elements that behave similarly during these processes, that is, with a similar mineral-melt distribution coefficient, such as K or Nb to estimate the magmatic Cl concentrations (Sun et al. 2007). As Nb is the closest to Cl in incompatibility and less mobile than K, we prefer to use the Cl/Nb ratio from the mantle (that will not be changed during magmatic processes as the elements behave similarly) to calculate magmatic Cl from Nb. The global Cl/Nb mantle ratio is <30 (e.g., McDonough and Sun 1995; Saal et al. 2002; le Roux et al. 2006), but varies with region. For the Red Sea, the Cl/Nb ratio can be refined based on the lowest measured Cl/Nb values of the samples, since neither magmatic processes, nor interaction with seawater can explain a lower Cl/Nb than the mantle Cl/Nb ratio (Fig. 3a). This mantle ratio seems to be homogeneous in the Red Sea, as Cl/Nb is not correlated with any trace elements or trace element ratios that indicate source processes (van der Zwan et al. 2015). The lowest Cl/Nb samples of the Red Sea fall over a range of compositions with Nb from 1 to 35 ppm on a line of 18.3 ± 2.5 that can be taken as the maximum Cl/Nb mantle value (Fig. 3a), and which falls within the range of Cl/Nb mantle values in the literature (see discussion in van der Zwan et al. 2017). The samples that coincide with this line do not show Cl-excess; for all other samples, the magmatic Cl can be calculated and from that the amount of Cl-excess, by extracting the magmatic Cl from the measured Cl using the formula: $Cl_{excess} = Cl_{measured} - (Cl/Nb_{mantleRedSea} \times Nb_{measured})$. The result is shown in Fig. 3b and shows that in contrast to Cl/Nb ratios, the Cl-excess is independent from the Nb content of the samples (and hence their magmatic Cl content), thereby excluding any magmatic effects. The Cl-excess in the Red Sea basalts is visible in 84% of the samples and ranges from 15 ppm up to 1340 ppm (Fig. 3b). Cl-excess is present in samples from the entire RSR apart from the RSR at 17°N, but is more evident at the Hadarba, Hatiba, Atlantis II, Discovery, Aswad and Port Sudan Deeps (Fig. 4).

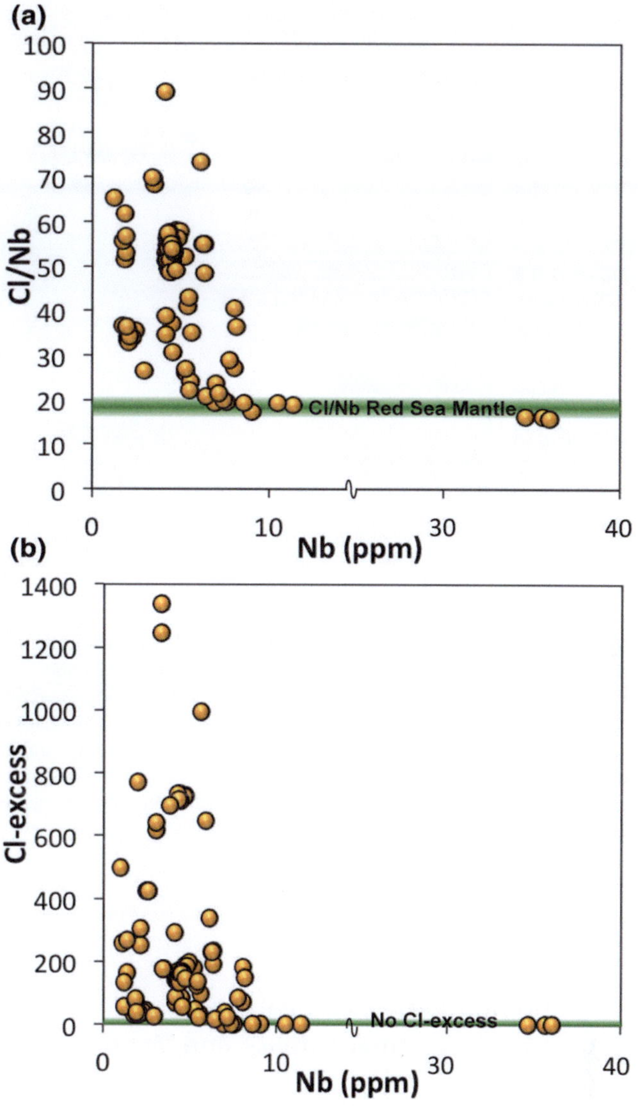

Fig. 3 Chlorine chemistry of the Red Sea basalts. **a** Cl/Nb versus Nb of the samples compared to the Cl/Nb of the Red Sea mantle shows strongly elevated Cl values for most samples. **b** Calculated Cl-excess against Nb reveals that most samples have Cl excess which is independent of the trace element chemistry (Nb-content) of the samples

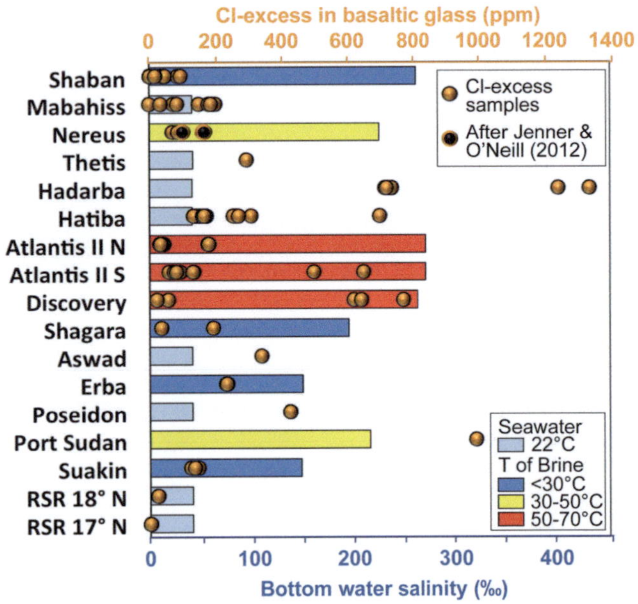

Fig. 4 Cl-excess of basalts (dots; this study used data from van der Zwan et al. 2015 and additionally Cl-excess after data of Jenner and O'Neill 2012 is shown) grouped by geographic location along the RSR, compared to local bottom water salinity (bars) and temperatures (indicated by the colour of the bars). Brines with high temperatures consistently contain some basalts with Cl-excess (100–800 ppm), indicating that basaltic Cl-excess shows hydrothermal activity in the crust

3 Discussion

3.1 Chlorine Excess and its Relation to Hydrothermal Activity and Venting

By using Cl-excess above the magmatic Cl contents, we are able to exclude variations in mantle Cl contents and the effects of crustal magmatic processes on the Cl contents. Therefore, any Cl-excess should occur due to the interaction of magma with seawater (cf. Jambon et al. 1995). Interaction with seawater would be possible at the seafloor, either post-eruptive, by direct low temperature alteration (e.g., Hart et al. 1974), or syn-eruptive, by incorporation of vaporized seawater in sheet lavas during emplacement (Perfit et al. 2003; Soule et al. 2006). However, for the Red Sea basalts Cl addition at the seafloor is excluded based on: (I) The fresh appearance of the glass chips and lack of pervasive alteration (minerals); (II) Some of the optically most altered samples (Shaban, Atlantis II Deeps) show the lowest Cl-excess, while the freshest samples from Erba Deep show significant Cl-excess (Figs. 2 and 4); (III) Homogeneous Cl-excess in the samples is not consistent with alteration, where gradients would be expected; (IV) The lack of any correlation of Cl to chemical indicators of surficial alteration (e.g., U, Ba, Pb, Rb; Alt et al. 1986; Alt and Teagle 2003; Schramm et al. 2005; Augustin et al. 2008); (V) Samples that display Cl-excess have lower H_2O contents than expected for pure seawater alteration (cf. Ito et al. 1983; Staudigel et al. 1996; Bach et al. 2003; le Roux et al. 2006). To generate the observed Cl-excess in basalts without significantly elevating the H_2O contents of the samples requires a salty contaminant with Cl/H_2O ratios of >0.25, which is higher than seawater ($Cl/H_2O_{seawater}$ ~0.02) or the seafloor brines ($Cl/H_2O_{seafloor\ brine}$ <0.22) (for details see van der Zwan et al. 2015). In addition, we see no systematics in Cl-excess for samples that are recovered from brine pools or from ambient seawater (Fig. 4).

Since Cl addition due to interaction with seawater at the seafloor can be excluded, Cl must have been added deeper in the crust. This is consistent with homogenous Cl contents within each of the samples that suggest mixing and homogenisation of magma after Cl incorporation as shown by van der Zwan et al. (2015). The Cl contaminant, which must have a relatively high Cl/H_2O ratio, cannot be simply heated seawater. Neither does bulk hydrothermally altered crust have a high enough Cl/H_2O ratio to produce the observed Cl-excess when assimilated (<0.036; Ito et al. 1983; Bach et al. 2003; Barnes and Cisneros 2012; van der Zwan et al. 2015). Fluids with high Cl/H_2O ratios are found in the crust in the form of hydrothermal brines with up to 50% salinity (not to be confused with the seafloor brines) that are formed by phase separation of heated seawater (Bischoff and Rosenbauer 1987; Fournier 1987; Kelley and Delaney 1987; Berndt and Seyfried 1990; Von Damm et al. 2003; Fig. 1). Chlorine from these hydrothermal brines can be added to a magma in diverse ways. They can be directly incorporated as a liquid into a magma (Kendrick et al. 2013), or by assimilation of hydrothermally altered crust, if brine fluids are present in the porosity of the rocks (Coogan et al. 2003; Gillis et al. 2003). Also in high-Cl hydrothermally formed minerals such as amphibole (Barnes and Cisneros 2012), brines can be present in the form of brine inclusions or in the crystal structure, which provides an alternative mechanism to add Cl to a melt if this altered crust is assimilated (Michael and Schilling 1989; Kent et al. 1999a). Assimilation of only partial melts of hydrothermally altered crust, which are particularly formed by these secondary Cl and/or brine-rich minerals that have the lowest solidus temperatures, will also result in high Cl/H_2O liquids as the melt takes up more Cl than H_2O (Kent et al. 1999a; France et al. 2010; Wanless et al. 2010). In all cases, even though the exact mechanism to incorporate Cl that was introduced by hydrothermal fluids in the sub-surface to a magma is disputed, the resulting erupted basalts with Cl-excess indicate hydrothermal activity.

This is confirmed by a comparison of Cl-excess in basalts to sites with other, independent evidence for hydrothermal activity. The Red Sea data show that Cl-excess related to hydrothermal activity is widespread all along the RSR, demonstrating that hydrothermal circulation is occurring everywhere along the ridge to a certain degree. However, we see a clear variation in Cl-excess between samples from different locations. The high variation in Cl-excess between samples, also within a Deep, reflects the high compositional variety with heterogeneous amounts of hydrothermal alteration of slow-spreading oceanic crust and the relatively small scale of processes acting there (e.g., Dick et al. 2000). As hydrothermal circulation is not pervasive throughout the crust and magmas may assimilate different sections of this crust, the lack of Cl-excess does not preclude hydrothermal circulation, but samples with Cl-excess always point to interaction with hydrothermal brines. The regions in the Red Sea with evidence for focussed former or present-day seafloor extrusion of hydrothermal fluids in the form of metalliferous seafloor precipitates (e.g., Thetis, Shagara and Suakin Deeps) and/or hot brine pools (Nereus, Atlantis II, Discovery and Port Sudan Deeps; red and yellow bars in Fig. 4) always yielded basalts with the highest Cl-excess of >100 ppm (Cl/Nb >50; van der Zwan et al. 2015). Within the Atlantis II Deep samples with the strongest Cl-excesses are found close to the SW basin and on the western flank of Discovery Deep (van der Zwan et al. 2015), areas where most probably current hydrothermal venting into the brine is occurring, based ontrace element studies (Brewer and Spencer 1969; Bäcker and Schoell 1972; Schoell and Hartmann 1973; Monin et al. 1981). This suggests that on the scale of a single Deep, high Cl-excess in basalt can also indicate the location of hydrothermal venting. In conclusion, Cl-excess in basalts can be used as a firm indicator of the strongest (past and present) hydrothermal circulation and the amount of Cl-excess can be applied to indicate potential areas affected by hydrothermal venting (cf. van der Zwan et al. 2015, 2017).

3.2 Red Sea Morphology and Cl-Excess in Basalts

Chlorine excess due to hydrothermal activity in Red Sea basalts is strongly linked to the morphological features of the Red Sea Rift, particularly to volcanic activity and the proximity to evaporites on the rift flanks. Samples with the highest Cl-excess (>250 ppm; Cl/Nb >100; van der Zwan et al. 2015) are always found within 5 km distance of evaporites, indicating that close to evaporites the inflowing water becomes highly charged with Cl as a result of evaporite dissolution and 5 km may be indicative of the scale of the recharge zone. Beyond 5 km, Cl-excess is still obvious but less strong, representing hydrothermal alteration of the crust by circulation of ambient Red Sea bottom water. When evaluating the areas for the best hydrothermal prospection, it needs to be kept in mind that some of the extreme values may be a result of evaporite presence and therefore not directly indicative of the strongest hydrothermal activity or the location of vent fields. More indicative for hydrothermal vent field prospection is the relation between Cl-excess and volcanism.

It was shown that magmatic intensity based on the percentage area of volcanic edifices and ridges, discernible in seafloor morphology, has a positive relation with Cl-excess in Red Sea basalts (van der Zwan et al. 2015). In addition, we note that in the Thetis-Hadarba-Hatiba area, which was determined by Augustin et al. (2016) to be the most volcanically active area in the central Red Sea, some of the highest Cl-excess in basalts is found (Fig. 4). The fact that the southern RSR, which is volcanically even more active, does not show the same Cl-excess has likely to do with the absence of evaporites close by (van der Zwan et al. 2015). In the central RSR faults are not related to Cl-excess and are probably ubiquitously present so that the pathways for hydrothermal activity are always available (van der Zwan et al. 2015). As shown by the relation between volcanism and Cl-excess, here the presence or absence of a heat source (volcanic activity) is essential for hydrothermalism to take place. Strong and robust magmatism that has been active since several Ma can be found in the Red Sea in the form of large axial dome volcanoes (Mabahiss Mons, Thetis Dome, Hatiba Mons, Aswad Dome) and associated spreading-perpendicular volcanic ridges (Metz et al. 2013; Augustin et al. 2014b, 2016). These domes are very similar to large dome volcanoes at other (ultra)slow-spreading ridges that host active hydrothermal vent systems, such as at the southern (Devey et al. 2013) and northern Mid-Atlantic Ridge (Marcon et al. 2013; Escartín et al. 2014). Indeed, the domes in the Red Sea show recent activity in the form of lava flows that are visible in multibeam backscatter data (Augustin et al. 2016) and all hosted basalts display Cl-excess (Fig. 4). This demonstrates that geomorphological evidence for high volcanic activity can be used together with the Cl-excess to point to the most promising areas where hydrothermal venting is presently taking place.

3.3 Hydrothermal Vent Field Potential of the Red Sea Rift

With the new advances in the geomorphology of the RSR and the geochemistry of its basalts, we are able to make a contemporary evaluation of the present (and past) areas influenced by hydrothermal activity along the RSR (Fig. 5). For this we used previously known indications for hydrothermal activity such as the temperatures of the brines and occurrences of metalliferous sediments in the deeps, together with new information of basaltic Cl-excess and the

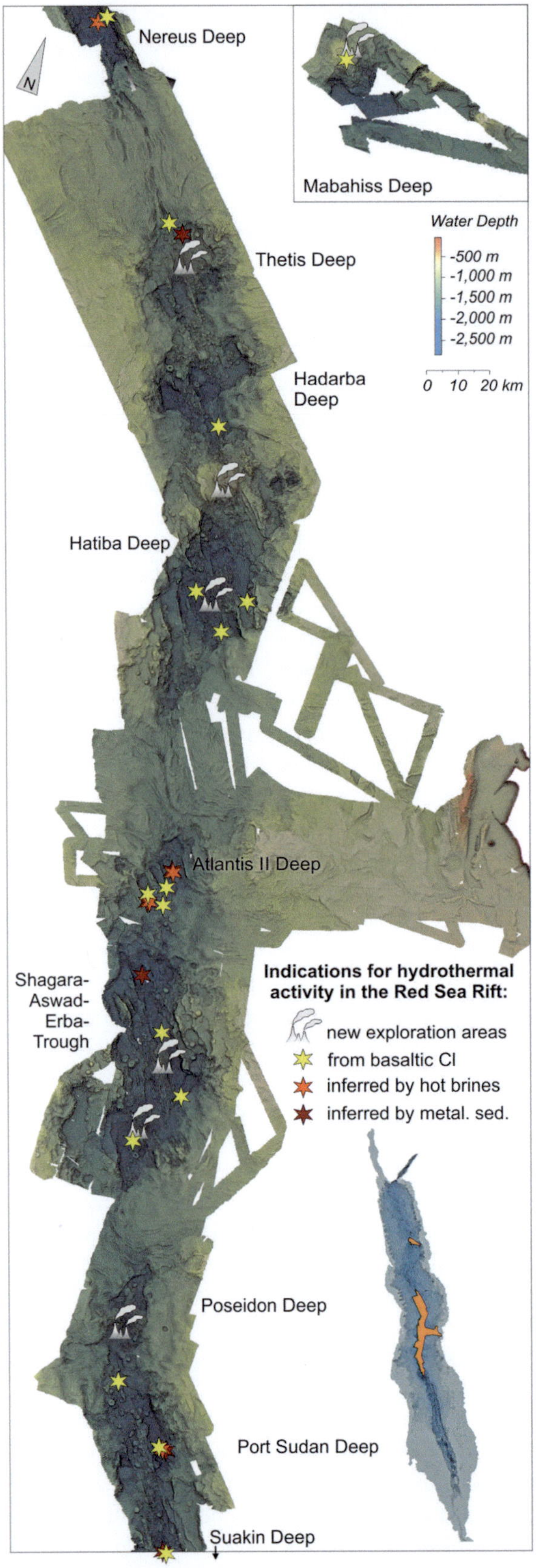

Fig. 5 Bathymetric map of the central RSR with inferred sites of hydrothermal activity based on basaltic Cl-excess (yellow stars), brine temperatures (orange stars) and metalliferous sediments (red stars). New exploration areas based on Cl-excess and rift morphology are also indicated (smokers)

latest bathymetric models of the RSR. While the metalliferous sediments and Cl-excess give information on the possible occurrence of massive sulphide deposits formed by hydrothermal activity at any possible time, the temperatures of the brines and indications from morphology for recent or past volcanism (providing a heat source to drive hydrothermalism) give insights as to whether hydrothermal venting at these locations may be current or extinct.

In the previous paragraphs, we have shown that all sites that have high temperature seafloor brine pools (Atlantis II, Discovery and Port Sudan, Nereus Deeps; red and yellow bars in Fig. 4) and hence must have hydrothermal venting into the brine, also have high basaltic Cl-excess. However, we see differences between the locations, with Nereus Deep showing clearly lower Cl-excess in its rock samples and having the lowest elevated brine temperature (30°; Schmidt et al. 2015). Bathymetry from Nereus Deep shows little evidence of active volcanism, and a greater depth of the Deep (Fig. 5). Also, basalt samples collected from there were embedded in carbonates and thus are older. This implies that hydrothermal venting is presently limited in Nereus Deep, due to the lack of recent volcanism. In the other high-temperature brines, venting seems very active, also shown by the abundant metalliferous sediments there and increasing brine temperatures of the Atlantis II Deep brines during the last decades (Schmidt et al. 2015).

At Shagara, Erba and Suakin Deeps hydrothermal activity is indicated by Cl-excess and in some cases metalliferous sediments, but the seafloor brines are colder (within 5 °C of ambient temperatures; Fig. 4). Thus, although hydrothermal activity is indicated, this is either not presently active, or venting is not taking place within the brine pool. The first is most likely the case for Suakin Deep, which is at present the deepest area of the whole RSR and characterised by extensive tectonic and little volcanic activity (Fig. 5); only 5% of the area shows high multibeam backscatter indicating recent volcanism (Augustin et al. 2014a, b). Also, samples recovered from here indicate older ages as they were covered by thick manganese encrustations, indicating former volcanism and hydrothermal venting. Samples from Shagara and Erba Deeps, in contrast, are fresh glasses with Cl-excess coupled with higher volcanic activity and recent lava flows seen in backscatter data (Augustin et al. 2014b, 2016; Fig. 5). Here hydrothermal venting is most likely presently occurring outside of the brine pools (and therefore any heat released through focussed venting is released to ambient seawater and not trapped in the brine pools), perhaps not surprisingly in view of their relatively small volume (≤ 11 km^2) compared to their large basins (>800 km^2). In these Deeps, recent hydrothermal venting seems likely and they represent good targets for hydrothermal prospection for non-brine-covered vent fields. Shaban Deep, which also hosts a cool brine pool, does not show any evidence of hydrothermal activity, other than low Cl-excess, which probably merely represents diffuse background hydrothermal circulation and does not represent the presence of focussed hydrothermal venting, leading to black smokers.

Chlorine-excess as an indicator of hydrothermal activity is most valuable for the areas in the RSR that do not host brine pools, as little other evidence is present there due to the lack of brines to store signals of hydrothermal venting and the few investigations of these areas. The strongest indications for hydrothermal activity from Cl-excess come from basalts of the Thetis-Hadarba-Hatiba Deeps area, which is hence a primary target for hydrothermal exploration (Figs. 4 and 5). This area also displays a comparably shallow rift axis and discernible volcanic edifices cover 44% of the seafloor (average of the RSR is ~33%), with 20% displaying fresh lava flows (Augustin et al. 2016). This implies that sufficient heat must have been available to drive hydrothermal circulation and that hydrothermal activity is most likely recent. In more detail, the highest Cl-excess is found in Hadarba deep close to the largest dome volcano, Hatiba Mons (Fig. 5), confirming the relation between hydrothermal activity and dome volcanoes. Moreover, the other large volcanoes on the RSR (Thetis Dome, Aswad Dome and Mabahiss Mons) are interpreted to host hydrothermal venting, due to their strong association with Cl-excess and additionally the occurrence of metalliferous sediments in the Thetis Deep. As volcanism seems to be active at all domes based on recent lava flows and fresh samples (Metz et al. 2013; Augustin et al. 2016), hydrothermal venting is also likely to be presently active there. Another volcanically active area that hosts basalts with strong Cl-excess is 20 km north of Port Sudan Deep (Poseidon Deep), indicating possible current hydrothermal venting (Fig. 5).

In the southern Red Sea, Cl-excess is lower, due to the lack of evaporites and also high trace element concentrations (and thus high magmatic Cl), which makes it more difficult to detect Cl-excess. Nevertheless, Cl-excess is visible in some basalts at 18°N, consistent with fragments of sulphide chimneys, indicating hydrothermal vents there (Monin et al. 1982). Although bathymetric data show considerable volcanism, hydrothermal venting may have been present only for a limited time, due to the higher magmatic activity of the crust that may make it less permeable, comparable to variations observed at 5–11 °S MAR (Devey et al. 2010). With the current data, we therefore cannot determine whether or not this hydrothermal venting is occurring now.

4 Conclusions

Chlorine-excess in basalts shows that the RSR is hydrothermally active all along the axis. However, the temporal and spatial intensity of this activity is varying and focussed hydrothermal venting is only indicated by higher basaltic Cl-excess. The target areas derived from Cl-excess in basalt, together with the good bathymetric data of the Red Sea Rift and additional indicators for the occurrence of hydrothermal activity in the RSR (hot brine pools, metalliferous sediments), enable us to define the most promising locations for hydrothermal vent fields and associated seafloor massive sulphide deposits and vent communities. The data shows that hydrothermal venting is probably currently occurring at Mabahiss Mons, in the Thetis-Hadarba-Hatiba Trough and the Shagara-Aswad-Erba Trough (particularly at the large axial volcanic Thetis Dome, Hatiba Mons and Aswad Dome) and in Poseidon Deep, in addition to previously known locations in the Atlantis II, Discovery and Port Sudan Deeps. Older hydrothermal vent fields may be present at Nereus and Suakin Deeps, while at the RSR at 18°N it is unclear if hydrothermal venting is still taking place. Our new evidence for hydrothermal activity significantly increases the potential of hydrothermal vent field prospection in the Red Sea and these new sites will be of particular interest for future hydrothermal research in the Red Sea.

Acknowledgements We are grateful for the help of the captains, crews and scientific shipboard parties of RV Poseidon and RV Pelagia expeditions P408 and PE350/351. We gratefully thank Jan Fietzke for the help with the Cl measurements and Mario Thöner, Matthias Frische, Dagmar Rau (all at GEOMAR) and Renat Almeev (University of Hannover) for technical support with the EMP, LA-ICP-MS and FTIR measurements, respectively. Antoine Bézos (University of Nantes) and Anna Krätschell are thanked for providing additional (sub)samples of the Red Sea. Reviews by three anonymous reviewers are greatly appreciated. We thank the Saudi Geological Survey for accommodating the workshop in preparation of this volume. We would like to acknowledge generous financial support from the Jeddah Transect Project, between King Abdulaziz University and Helmholtz-Centre for Ocean Research GEOMAR, that was funded by King Abdulaziz University (KAU), Jeddah, Saudi Arabia, under grant no. T-065/430.

References

Alt JC, Teagle DA (2003) Hydrothermal alteration of upper oceanic crust formed at a fast-spreading ridge: mineral, chemical, and isotopic evidence from ODP Site 801. Chem Geol 201:191–211

Alt JC, Honnorez J, Laverne C, Emmermann R (1986) Hydrothermal alteration of a 1 km section through the upper oceanic crust, Deep Sea Drilling Project Hole 504B: mineralogy, chemistry and evolution of seawater-basalt interactions. J Geophys Res Solid Earth 91:10309–10335

Anderson MO, Hannington MD, Haase K, Schwarz-Schampera U, Augustin N, McConachy TF, Allen K (2016) Tectonic focusing of voluminous basaltic eruptions in magma-deficient backarc rifts. Earth Planet Sci Lett 440:43–55

Augustin N, Lackschewitz K, Kuhn T, Devey CW (2008) Mineralogical and chemical mass changes in mafic and ultramafic rocks from the Logatchev hydrothermal field (MAR 15 N). Mar Geol 256:18–29

Augustin N, Devey CW, van der Zwan FM, Feldens P, Bantan RA, Kwasnitschka T (2014a) The rifting to spreading transition in the Red Sea. Earth Planet Sci Lett 395:217–230

Augustin N, Schmidt M, Devey CW, Al-Aidaroos A, Kürten B, Eisenhauer A, Brückmann W, Dengler M, van der Zwan FM, Feldens P (2014b) The Jeddah Transect Project: extensive mapping of the Red Sea Rift. InterRidge News 22:68–73

Augustin N, van der Zwan FM, Devey CW, Ligi M, Kwasnitschka T, Feldens P, Bantan RA, Basaham AS (2016) Geomorphology of the central Red Sea Rift: determining spreading processes. Geomorph 274:162–179

Bach W, Peucker-Ehrenbrink B, Hart SR, Blusztajn JS (2003) Geochemistry of hydrothermally altered oceanic crust: DSDP/ODP Hole 504B—implications for seawater-crust exchange budgets and Sr-and Pb-isotopic evolution of the mantle. Geochem Geophys Geosyst 4(3). https://doi.org/10.1029/2002gc000419

Bäcker H, Richter H (1973) Die rezente hydrothermal-sedimentäre Lagerstätte Atlantis-II-Tief im Roten Meer. Geol Rundsch 62:697–737

Bäcker H, Schoell M (1972) New deeps with brines and metalliferous sediments in the Red Sea. Nat Phys Sci 240:153–158

Baker ET, German CR (2004) On the global distribution of hydrothermal vent fields. In: German CR, Lin J, Parson LM (eds) Mid-ocean ridges: hydrothermal interactions between the lithosphere and oceans, vol 148. American Geophysical Union, Geophys Monograph, pp 245–266

Baker ET, Massoth GJ (1987) Characteristics of hydrothermal plumes from two vent fields on the Juan de Fuca Ridge, northeast Pacific Ocean. Earth Planet Sci Lett 85(1):59–73

Barnes JD, Cisneros M (2012) Mineralogical control on the chlorine isotope composition of altered oceanic crust. Chem Geol 326–327:51–60

Beaulieu SE, Baker ET, German CR (2015) Where are the undiscovered hydrothermal vents on oceanic spreading ridges? Deep Sea Res Part II 121:202–212

Berndt ME, Seyfried WE (1990) Boron, bromine, and other trace elements as clues to the fate of chlorine in mid-ocean ridge vent fluids. Geochim Cosmochim Acta 54:2235–2245

Bischoff JL, Rosenbauer RJ (1987) Phase separation in seafloor geothermal systems; an experimental study of the effects on metal transport. Am J Sci 287:953–978

Blondel P (2010) The handbook of sidescan sonar. Springer Science & Business Media

Blum N, Puchelt H (1991) Sedimentary-hosted polymetallic massive sulfide deposits of the Kebrit and Shaban Deeps, Red Sea. Mineral Depos 26:217–227

Brewer PG, Spencer DW (1969) A note on the chemical composition of the Red Sea brines. In: Degens ET, Ross DA (eds) Hot brines and recent heavy metal deposits in the Red Sea. Springer, Berlin, pp 174–179

Chu D, Gordon RG (1998) Current plate motions across the Red Sea. Geophys J Int 135:313–328

Coogan LA (2003) Contaminating the lower crust in the Oman ophiolite. Geology 31:1065–1068

Coogan LA, Mitchell NC, O'Hara MJ (2002) Roof assimilation at fast spreading ridges: an investigation combining geophysical, geochemical, and field evidence. J Geophys Res 108:2002

Coogan LA, Mitchell NC, O'Hara MJ (2003) Roof assimilation at fast spreading ridges: an investigation combining geophysical, geochemical, and field evidence. J Geophys Res Solid Earth 108:ECV 2-1–ECV 2-14

Devey CW, Scientific Shipboard Party (2013) SoMARTherm: the Mid-Atlantic ridge 13–33 °S. Cruise No. MSM25. Leitstelle Deutsche Forschungsschiffe, Universität Hamburg, 113 p

Devey CW, German C, Haase K, Lackschewitz K, Melchert B, Connelly D (2010) The relationships between volcanism, tectonism, and hydrothermal activity on the southern equatorial Mid-Atlantic Ridge. In: Rona PA, Devey CW, Dyment J, Murton BJ (eds) Diversity of hydrothermal systems on slow spreading ocean ridges, vol 188. American Geophysical Union, Geophys Monograph, pp 133–152

Dick HJB, Natland JH, Alt JC, Bach W, Bideau D, Gee JS, Haggas S, Hertogen JGH, Hirth G, Holm PM, Ildefonse B, Iturrino GJ, John BE, Kelley DS, Kikawa E, Kingdon A, LeRoux PJ, Maeda J, Meyer PS, Miller DJ, Naslund HR, Niu Y-L, Robinson PT, Snow J, Stephen RA, Trimby PW, Worm H-U, Yoshinobu A (2000) A long in situ section of the lower ocean crust: results of ODP Leg 176 drilling at the Southwest Indian Ridge. Earth Planet Sci Lett 179:31–51

Edmonds HN, Michael PJ, Baker ET, Connelly DP, Snow JE, Langmuir CH, Dick HJB, Muhe R, German CR, Graham DW (2003) Discovery of abundant hydrothermal venting on the ultraslow-spreading Gakkel ridge in the Arctic Ocean. Nature 421:252–256

Escartín J, Soule SA, Cannat M, Fornari DJ, Düşünür D, Garcia R (2014) Lucky strike seamount: Implications for the emplacement and rifting of segment-centered volcanoes at slow spreading mid-ocean ridges. Geochem Geophys Geosyst 15:4157–4179

Escartín J, Mével C, Petersen S, Bonnemains D, Cannat M, Andreani M, Augustin N, Bezos A, Chavagnac V, Choi Y et al (2017) Tectonic structure, evolution, and the nature of oceanic core complexes and their detachment fault zones (13°20'N and 13°30'N, Mid-Atlantic Ridge). Geochem Geophys Geosyst 18(4):1451–1482. https://doi.org/10.1002/2016GC006775

Fontaine FJ, Cannat M, Escartín J, Crawford WC (2014) Along-axis hydrothermal flow at the axis of slow spreading Mid-Ocean Ridges: insights from numerical models of the Lucky Strike vent field (MAR). Geochem Geophys Geosyst 15:2918–2931

Fouquet Y (1997) Where are the large hydrothermal sulphide deposits in the oceans? Phil Trans Roy Soc Math Phys Eng Sci 355:427–441

Fournier R (1987) Conceptual models of brine evolution in magmatic-hydrothermal systems. US Geol Surv Prof Pap 1350:1487–1506

France L, Ildefonse B, Koepke J (2009) Interactions between magma and hydrothermal system in Oman ophiolite and in IODP Hole 1256D: fossilization of a dynamic melt lens at fast spreading ridges. Geochem Geophys Geosyst 10:Q10O19. https://doi.org/10.1029/2009gc002652

France L, Koepke J, Ildefonse B, Cichy SB, Deschamps F (2010) Hydrous partial melting in the sheeted dike complex at fast spreading ridges: experimental and natural observations. Contrib Mineral Petrol 160:683–704

Früh-Green GL, Kelley DS, Bernasconi SM, Karson JA, Ludwig KA, Butterfield DA, Boschi C, Proskurowski G (2003) 30,000 years of hydrothermal activity at the Lost City vent field. Science 301:495–498

German CR, Petersen S, Hannington MD (2016) Hydrothermal exploration of mid-ocean ridges: where might the largest sulfide deposits be forming? Chem Geol 420:114–126

Gillis KM, Coogan LA, Chaussidon M (2003) Volatile element (B, Cl, F) behaviour in the roof of an axial magma chamber from the East Pacific Rise. Earth Planet Sci Lett 213:447–462

Girdler RW, Evans TR (1977) Red Sea heat flow. Geophys J Int 51:245–251

Guney M, Al-Marhoun MA, Nawab ZA (1988) Metalliferous sub-marine sediments of the Atlantis-II-Deep, Red Sea. Can Min Metall Bull 81:33–39

Gurvich EG (2006) Metalliferous sediments of the Red Sea. Metalliferous sediments of the World Ocean: fundamental theory of Deep-Sea hydrothermal sedimentation. Springer, Berlin, pp 127–210

Hannington M, Jamieson J, Monecke T, Petersen S (2010) Modern sea-floor massive sulfides and base metal resources: toward an estimate of global sea-floor massive sulfide potential. Soc Econ Geol Spec Pub 15:317–338

Hannington M, Jamieson J, Monecke T, Petersen S, Beaulieu S (2011) The abundance of seafloor massive sulfide deposits. Geology 39:1155–1158

Hart S, Erlank A, Kable E (1974) Sea floor basalt alteration: some chemical and Sr isotopic effects. Contrib Mineral Petrol 44:219–230

Ito E, Harris DM, Anderson AT (1983) Alteration of oceanic crust and geologic cycling of chlorine and water. Geochim Cosmochim Acta 47:1613–1624

Jambon A, Déruelle B, Dreibus G, Pineau F (1995) Chlorine and bromine abundance in MORB: the contrasting behaviour of the Mid-Atlantic Ridge and East Pacific Rise and implications for chlorine geodynamic cycle. Chem Geol 126:101–117

Jenner FE, O'Neill HSC (2012) Analysis of 60 elements in 616 ocean floor basaltic glasses. Geochem Geophys Geosyst 13(2):Q02005

Kelley DS, Delaney JR (1987) Two-phase separation and fracturing in mid-ocean ridge gabbros at temperatures greater than 700 °C. Earth Planet Sci Lett 83:53–66

Kendrick MA, Arculus R, Burnard P, Honda M (2013) Quantifying brine assimilation by submarine magmas: examples from the Galápagos Spreading Centre and Lau Basin. Geochim Cosmochim Acta 123:150–165

Kendrick MA, Hemond C, Kamenetsky VS, Danyushevsky L, Devey CW, Rodemann T, Jackson MG, Perfit MR (2017) Seawater cycled throughout Earth's mantle in partially serpentinized lithosphere. Nat Geosci 10:222–228. https://doi.org/10.1038/NGEO2902

Kent AJR, Norman MD, Hutcheon ID, Stolper EM (1999a) Assimilation of seawater-derived components in an oceanic volcano: evidence from matrix glasses and glass inclusions from Loihi seamount, Hawaii. Chem Geol 156:299–319

Kent AJR, Clague DA, Honda M, Stolper EM, Hutcheon ID, Norman MD (1999b) Widespread assimilation of a seawater-derived component at Loihi Seamount, Hawaii. Geochim Cosmochim Acta 63:2749–2761

Kent AJR, Peate DW, Newman S, Stolper EM, Pearce JA (2002) Chlorine in submarine glasses from the Lau Basin: seawater contamination and constraints on the composition of slab-derived fluids. Earth Planet Sci Lett 202:361–377

Klinkhammer G, Bender M, Weiss RF (1977) Hydrothermal manganese in the Galapagos Rift. Nature 269:319–320

Laurila TE, Hannington MD, Petersen S, Garbe-Schönberg D (2014) Early depositional history of metalliferous sediments in the Atlantis II Deep of the Red Sea: evidence from rare earth element geochemistry. Geochim Cosmochim Acta 126:146–168

le Roux PJ, Shirey SB, Hauri EH, Perfit MR, Bender JF (2006) The effects of variable sources, processes and contaminants on the composition of northern EPR MORB (8–10°N and 12–14°N): evidence from volatiles (H_2O, CO_2, S) and halogens (F, Cl). Earth Planet Sci Lett 251:209–231

Lein AY, Bogdanov YA, Maslennikov V, Li S, Ulyanova N, Maslennikova S, Ulyanov A (2010) Sulfide minerals in the Menez Gwen nonmetallic hydrothermal field (Mid-Atlantic Ridge). Lithol Mineral Resour 45:305–323

Linke P, Schmidt M, Rohleder M, Al-Barakati A, Al-Farawati R (2015) Novel online digital video and high-speed data broadcasting via standard coaxial cable onboard marine operating vessels. Mar Technol Soc Bull 49:7–18

Marcon Y, Sahling H, Borowski C, dos Santos FC, Thal J, Bohrmann G (2013) Megafaunal distribution and assessment of total methane and sulfide consumption by mussel beds at Menez Gwen hydrothermal vent, based on geo-referenced photomosaics. Deep Sea Res Part I 75:93–109

McDonough WF, Sun SS (1995) The composition of the Earth. Chem Geol 120:223–253

Metz D, Augustin N, van der Zwan FM, Bantan RA, Al-Aidaroos AM (2013) Mabahiss Mons, 25.5 °N Red Sea Rift: tectonics and volcanism of a large submarine dome volcano. In: EGU general assembly conference abstracts, vol 15, p 10487

Michael PJ, Cornell WC (1998) Influence of spreading rate and magma supply on crystallization and assimilation beneath mid-ocean ridges: evidence from chlorine and major element chemistry of mid-ocean ridge basalts. J Geophys Res 103:18325–18356

Michael PJ, Schilling J-G (1989) Chlorine in mid-ocean ridge magmas: evidence for assimilation of seawater-influenced components. Geochim Cosmochim Acta 53:3131–3143

Mitchell NC, Ligi M, Ferrante V, Bonatti E, Rutter E (2010) Submarine salt flows in the central Red Sea. Geol Soc Am Bull 122:701–713

Monin A, Plakhin E, Podrazhansky A, Sagalevich A, Sorokhtin O (1981) Visual observations of the Red Sea hot brines. Nature 291:222–225

Monin A, Litvin V, Podrazhansky A, Sagalevich A, Sorokhtin O, Voitov V, Yastrebov V, Zonenshain L (1982) Red sea submersible research expedition. Deep Sea Res Part A Oceanographic Res Papers 29(3):361–373

Perfit MR, Cann JR, Fornari DJ, Engels J, Smith DK, Ian Ridley W, Edwards MH (2003) Interaction of sea water and lava during submarine eruptions at mid-ocean ridges. Nature 426:62–65

Pertsev A, Bortnikov N, Vlasov E, Beltenev V, Dobretsova I, Ageeva O (2012) Recent massive sulfide deposits of the Semenov ore district, Mid-Atlantic Ridge, 13°31′ N: associated rocks of the oceanic core complex and their hydrothermal alteration. Geol Ore Deps 54:334–346

Petersen S, Herzig P, Hannington MD (2000) Third dimension of a presently forming VMS deposit: TAG hydrothermal mound, Mid-Atlantic Ridge, 26 °N. Mineral Deposita 35:233–259

Petersen S, Kuhn K, Kuhn T, Augustin N, Hékinian R, Franz L, Borowski C (2009) The geological setting of the ultramafic-hosted Logatchev hydrothermal field (14°45′ N, Mid-Atlantic Ridge) and its influence on massive sulfide formation. Lithos 112:40–56

Pierret MC, Clauer N, Bosch D, Blanc G, France-Lanord C (2001) Chemical and isotopic (87Sr/86Sr, δ18O, δD) constraints to the formation processes of Red-Sea brines. Geochim Cosmochim Acta 65:1259–1275

Pierret M, Clauer N, Bosch D, Blanc G (2010) Formation of Thetis Deep metal-rich sediments in the absence of brines, Red Sea. J Geochem Explor 104:12–26

Rona PA, Klinkhammer G, Nelsen TA, Trefry JH, Elderfield H (1986) Black smokers, massive sulphides and vent biota at the Mid-Atlantic Ridge. Nature 321:33–37

Saal AE, Hauri EH, Langmuir CH, Perfit MR (2002) Vapour under saturation in primitive mid-ocean-ridge basalt and the volatile content of Earth's upper mantle. Nature 419:451–455

Schmidt M, Linke P, Esser D (2013a) Recent development in IR sensor technology for monitoring subsea methane discharge. Mar Technol Soc Bull 47:27–36

Schmidt M, Al-Farawati R, Al-Aidaroos A, Kürten B (eds) (2013b) RV PELAGIA cruise report 64PE350/64PE351. Berichte aus dem Helmholtz-Zentrum für Ozeanforschung Kiel (GEOMAR) 5

Schmidt M, Al-Farawati R, Botz R (2015) Geochemical classification of brine-filled Red Sea deeps. In: Rasul NMA, Stewart ICF (eds) The Red Sea: the formation, morphology, oceanography and environment of a young ocean basin. Springer Earth System Sciences, Berlin, pp 219–233

Schoell M, Hartmann M (1973) Detailed temperature structure of the hot brines in the Atlantis II Deep area (Red Sea). Mar Geol 14:1–14

Schramm B, Devey CW, Gillis KM, Lackschewitz K (2005) Quantitative assessment of chemical and mineralogical changes due to progressive low-temperature alteration of East Pacific Rise basalts from 0 to 9 Ma. Chem Geol 218:281–313

Soule SA, Fornari DJ, Perfit MR, Ridley WI, Reed MH, Cann JR (2006) Incorporation of seawater into mid-ocean ridge lava flows during emplacement. Earth Planet Sci Lett 252:289–307

Staudigel H, Plank T, White B, Schmincke HU (1996) Geochemical fluxes during seafloor alteration of the basaltic upper oceanic Crust: DSDP sites 417 and 418. In: Bebout GE, Scholl DW, Kirby SH, Platt JP (eds) Subduction top to bottom. American Geophysical Union, Washington, DC, pp 19–38

Sun WD, Binns RA, Fan AC, Kamenetsky VS, Wysoczanski R, Wei GJ, Hu YH, Arculus RJ (2007) Chlorine in submarine volcanic glasses from the eastern Manus Basin. Geochim Cosmochim Acta 71:1542–1552

Swallow JC, Crease J (1965) Hot salty water at the bottom of the Red Sea. Nature 205:165–166

Szitkar F, Petersen S, Caratori Tontini F, Cocchi L (2015) High-resolution magnetics reveal the deep structure of a volcanic-arc-related basalt-hosted hydrothermal site (Palinuro, Tyrrhenian Sea). Geochem Geophys Geosyst 16:1950–1961

van der Zwan FM (2014) Hydrothermal activity at slow-spreading mid-ocean ridges: evidence from chlorine in basalt. PhD thesis, Christian-Albrechts-Universität Kiel

van der Zwan FM, Fietzke J, Devey CW (2012) Precise measurement of low (<100 ppm) chlorine concentrations in submarine basaltic glass by electron microprobe. J Analyt At Spectrom 27:1966–1974

van der Zwan FM, Devey CW, Augustin N, Almeev RR, Bantan RA, Basaham A (2015) Hydrothermal activity at the ultraslow- to slow-spreading Red Sea Rift traced by chlorine in basalt. Chem Geol 405:63–81

van der Zwan FM, Devey CW, Hansteen TH, Almeev RR, Augustin N, Frische M, Haase KM, Basaham A, Snow JE (2017) Lower crustal hydrothermal circulation at slow-spreading ridges: Evidence from chlorine in Arctic and South Atlantic basalt glasses and melt inclusions. Contrib Mineral Petrol 172(11–12):97. https://doi.org/10.1007/s00410-017-1418-1

Von Damm K, Lilley M, Shanks W III, Brockington M, Bray A, O'Grady K, Olson E, Graham A, Proskurowski G (2003) Extraordinary phase separation and segregation in vent fluids from the southern East Pacific Rise. Earth Planet Sci Lett 206:365–378

Wanless V, Perfit M, Ridley W, Klein E (2010) Dacite petrogenesis on mid-ocean ridges: evidence for oceanic crustal melting and assimilation. J Petrol 51:2377–2410

Whitmarsh RB, Weser OE, Ross DA (1974) Initial reports of the deep sea drilling project, vol 23. U.S. Government Printing Office, Washington, DC, pp 35–56

Yoshikawa S, Okino K, Asada M (2012) Geomorphological variations at hydrothermal sites in the southern Mariana Trough: relationship between hydrothermal activity and topographic characteristics. Mar Geol 303:172–182

Salt Formation, Accumulation, and Expulsion Processes During Ocean Rifting—New Insight Gained from the Red Sea

Martin Hovland, Håkon Rueslåtten, and Hans Konrad Johnsen

Abstract

Recent observations of thick carpets of mobile salt slurries on the Red Sea floor (Salt Flows) and huge accumulations of salts in the sub-surface ('Salt Walls' and 'Salt Ridges'), associated with topographical lows (Deeps), suggest that the Red Sea currently produces new volumes of brines and solid salts underground. The salt producing zone is focused around the central rifting axis and represents about 15% of the entire Red Sea area. The brines and solid salts are formed by boiling and supercritical phase separation in forced convection cells (hydrothermal circulation), located above shallow-seated magmatic intrusions along the spreading axis. The descending water of the convection cells attains increasing pressure and temperature, resulting in supercritical water conditions, giving rise to phase separation. Salts are therefore deposited underground and accumulate in the heavily fractured country rocks in the rift zone. Dense brines also migrate further down and concentrate beyond saturation. Conversely, the ascending limbs of the hydrothermal cells consist of low salinity vapor which condenses upon cooling, hence dissolving previously deposited salts. The different solubilities of sea salts lead to a refining of the salt types. When reaching the seafloor, the newly formed brines are cooled further, eventually becoming oversaturated in salts, which results in precipitation onto the seafloor. The dense brine layers also protect seafloor salts from re-dissolution by normal seawater. In the continued process, brines will migrate through salt deposits (as they build up) eventually giving rise to salt glaciers, salt walls, salt pinnacles, and 'diapirs' (injectites). Ores, hydrocarbons and clays are often associated with salts, being part of such a robust, self-organizing and self-sustaining hydrothermal system. Although this model arose from geological and geophysical observations performed in the Red Sea area, it may also be applicable to present and past rift zones worldwide, especially those with low spreading rates, in their early rifting stage. In addition to solar evaporation of seawater, our model also considers the significance of temporal variations in mass- and heat-flow and its impact on hydrothermal flow that governs underground salt formation, accumulation and mobility.

1 Introduction

The Red Sea is the youngest actively rifting marine basin in the world and it is also one of the few giant salt basins that are still evolving, documented by a plethora of recent discoveries of seafloor and sub-seafloor features, some of which do not fit into the conventional salt tectonic model of evaporites (Warren 2010). Exactly how the thick layers of salt (evaporites) were formed in the Red Sea is not well understood. The two main problems, according to Stoffers and Kühn (1974): "*...is whether the evaporites were precipitated under shallow- or deep-water conditions; the second is the question of the depth of the basin in which the evaporites were formed*".

Currently, the short conventional version of the salt deposition history of the Red Sea, in modern literature, runs as follows: "*Evaporite deposition took place throughout the entire Red Sea basin starting at around 14 Ma in the Middle-Miocene* (*Serravallian*) *and reaching a thickness of thousands of meters* (*Mitchell* et al. 2010; *Bosworth* 2015). *Halite and anhydrite-gypsum were the main mineral phases; they are locally interbedded with sandstones and shales as well as with carbonate platform deposits, indicating intervals of normal marine deposition*" (Bonatti et al. 2015). Another short description of the salt formation history is: "*The sediment cover consists of Miocene evaporites and*

M. Hovland (✉)
4055 Sola, Norway
e-mail: mthovland@gmail.com

H. Rueslåtten · H. K. Johnsen
7000 Trondheim, Norway

N. M. A. Rasul and I. C. F. Stewart (eds.), *Geological Setting, Palaeoenvironment and Archaeology of the Red Sea*,
https://doi.org/10.1007/978-3-319-99408-6_11

younger pelagic carbonate ooze, with large thicknesses of up to 5 km (e.g. Girdler and Whitmarsh 1974*; Searle and Ross* 1975*; Girdler* 1984*). Continental break-up and sea-floor spreading caused the disruption of these evaporites, which in some places have been shown to flow toward the deeper axial rift in the form of salt glaciers, known as submarine namakiers* (*Girdler* 1985*; Mitchell* et al. 2010)" (Augustin et al. 2014).

In other words, the formation mechanisms of salt deposition are basically unknown, and may be summed up as was done by Stoffers and Kühn (1974): "*Considering the idea of a repeated refilling of the basin during Miocene time, it seems that a desiccated deep-basin hypothesis is not in contrast to the deep water—deep basin model postulated by Schmalz* (1969). *The desiccated deep-basin model may be regarded as a specific stage in a cycle of repeated refilling and evaporation. Therefore, it is concluded that the origin of the Mediterranean and Red Sea evaporites could be accounted for by Schmalz's "Deep-Water Model" modified by periods of desiccation caused by occasional isolation of the Mediterranean from the Atlantic.*"

Because there are currently no known tectonic processes that can open and shut ocean straits rhythmically to produce many subsequent desiccation episodes, we believe that the knowledge about how giant salt basins form is currently inadequate and that new understanding of the process is overdue. This inadequacy has also been discussed in a recent paper by Lugli et al. (2015), entitled: "*The deep record of the Messinian salinity crisis: Evidence of a non-desiccated Mediterranean*", an article that actually refutes the generally accepted conventional 'desiccation' model by Hsü et al. (1977).

Some of the main salt-related features of the Red Sea which we discuss in this article are: Brine pools (Deeps); salt ridges and walls; salt flows (carpets of moving salt); fluid flow through salt (permeable salt); and salt stocks ('diapirs') as conduits. These features are difficult to explain by the conventional 'evaporite' model, because that model mainly relies on solar evaporation and desiccation. These features may therefore, rather be described by a different paradigm, namely that of hydrothermal salt formation (Hovland et al. 2006a, b; 2015; Driesner 2007; Coumou et al. 2009; Lecumberri-Sanchez et al. 2015). Our objective is to investigate and interpret some pertinent new and old observations in the Red Sea and elsewhere, by bringing in advances in thermo- and geodynamic modelling. We, thus, attempt to assemble these salt related features into an integrated, comprehensive conceptual model, which may help to explain the sequence of stages and events in rift-related salt basin developments.

In addition to refining salt by hydrothermal phase separation, hydrothermal systems are also involved in serpentinization processes, which are important in the rifting process, including that of the Red Sea. During this process, where seawater is in contact with mantle rocks (peridotites), there is brine formation as only pure water (H_2O) enters into the serpentinization process. In fact, there may, thus, be a significant salt production, depending on the volume and degree of active serpentinization, as briefly explained by Scribano and Viccaro (2014) and Scribano et al. (2017): "*Since the serpentinization process requires pure H_2O, the involved seawater-derived serpentinizing fluid will undergo total out-salting. In particular, about 10.5 kg of salts can be deposited per m^3 of serpentinized peridotite.*" However, in this article, because of lack of pertinent data from the Red Sea, we refrain from addressing this process in more detail.

Salt formation is also to be expected during subduction processes, where seawater is transported inside crustal rocks to great depth. The subducting slab encounters pressures and temperatures where both serpentinization and hydrothermal outsalting occur (Hovland et al. 2016). Neither of these processes is, however, the subject of this paper.

2 A Proposed New Model with Supporting Observations

2.1 Model Description

Hydrothermal systems set up by shallow-seated igneous heat sources lead to dissolution and subsequent precipitation of many minerals including common sea salts that represent a separate subgroup. One of the prerequisites for the formation of solid salt and observable salt accumulations associated with hydrothermal systems is the supply of seawater.

In this section, we describe how rift-related hydrothermal systems may produce, store, and move brines and salts. Our scientific basis is molecular and numerical modelling (Hovland et al. 2006a), thermodynamic modelling (Driesner 2007; Coumou et al. 2009), new and old observations from the Red Sea (Rasul et al. 2015; Feldens and Mitchell 2015; Ehrhardt and Hübscher 2015; Anschutz 2015; Augustin et al. 2014, 2016; Hovland et al. 2015), and observations and geodynamic modelling from other regions (Miller et al. 2009; Gruen et al. 2014; Lecumberri-Sanchez et al. 2015; Norton et al. 2015). In addition, it has to be mentioned that the P/T-conditioned permeability of halite (Holness and Lewis 1997) is also an important aspect of our model.

Continental rifting always starts out on land, without access to sea water. During the subaerial rifting, several hydrothermal processes take place, involving meteoric water. However, this phase of the rifting story is not part of our model. Eventually, the sea invades the rift and starts to interact with rocks heated by the hot igneous intrusion. Here is where our model description starts. We have identified 5 pertinent steps in the evolution of salt-related rift basins, whereby salt is naturally processed in some way, to finally end up forming giant salt basins containing different kinds of salts in different structures and accumulations, including 'diapirs' and layered deposits.

The pertinent steps are:

(1) Seawater meets hot rocks at moderate depths in the fault zones. Salt precipitation takes place due to boiling. Upon cooling of this system, salts are dissolved and most of them are returned to the sea.
(2) As fracture volume expands and hydrothermal brines move further down toward the hot igneous body, high-pressure 'supercritical' phase separation occurs, resulting in salt precipitation and the formation of heavy brines that migrate further down and concentrate beyond saturation.
(3) The circulating fluids cool the system further, and the upward migrating low salinity vapour is condensing, and starts to dissolve previously deposited salts. The different solubilities of sea salts lead to a refining of the salt types, whereby the most soluble are predominantly dissolved first. These brines are cooled further during the upward migration, and some salts are re-precipitated before the brine reaches the seafloor.
(4) Upon reaching the seafloor, the brines are cooled further, and will easily become supersaturated with respect to some of the salt types. Salts will then precipitate from these heavy brines on the seafloor (ponding/mini-basin formation). This process includes further refining of the salts.
(5) In the continued process, brines will migrate through salt deposits (as they build up). This brine migration also includes, (a) dissolution and re-precipitation, (b) pressure build-up due to salt-clogging and release, and (c) extrusion of salts/salt slurries and brines (e.g., the formation of salt glaciers, salt walls, salt pinnacles, and 'diapirs').

The steps are not to be understood as completely isolated or sequential, as several of the steps may occur simultaneously, although one step may at a certain time be more dominant than the other. For more details about the prerequisites of the hydrothermal salt model, see also Hovland et al. (2018a, b).

2.2 Evidence for Model Steps 1 and 2

For the most extreme cases of igneous conditions, with high temperatures, vigorous fluid movements, large temperature gradients and a short distance to the sea floor, the salt formed may be returned directly to the sea, without having accumulated significant quantities of solid salt in the subsurface (Bischoff and Rosenbauer 1989; Von Damm 1995; Von Damm et al. 1995; Lowell and Germanovich 1997; Lilley et al. 2003; Fontaine and Wilcock 2006; Singh et al. 2013; Lowell et al. 2014).

According to Lowell et al. (2014), the broad range of permeability values estimated from field and modelling studies clearly indicate that permeability in hydrothermal regions is fracture-controlled: "*Fracture-controlled permeability is typically heterogeneous and possibly anisotropic. Fracture concentration and orientation may result from tectonic stresses as well as processes related to magma emplacement and volcanic eruptions and thermal stresses. Recharge zones are more problematical and are typically hard to identify, but faulting could be important there as well*".

Hence, one may conclude that the prerequisite for salt accumulation in the igneous rocks of a rift system is the development of fractures, that is, systems that are large enough to accommodate large quantities of solid salts or brines. At a later stage, the phase separation takes place at sufficient depth to produce brines that are not easily swept out of the system. These brines will remain deep in the rocks and concentrate further upon later phase separation at different P/T conditions, as illustrated by Bischoff and Rosenbauer (1989). It represents a time step in the development of a salt basin comparable to our step 2 above (Fig. 1).

Coumou et al. (2009) made numerical simulations coupling thermodynamics and fluid dynamics for several hydrothermal systems. They confirmed the observations made by Bischoff and Rosenbauer (1989). In their Table 1 we see that normal seawater at 3.2 wt% salt enters the hydrothermal systems, but only water with less salinity comes out of the hydrothermal chimneys on the seafloor. This means that salt has been retained in the underground in all of their cases (Coumou et al. 2009) which are simulated only with water and NaCl, without any other salts.

To understand all the stages in the development of a salt basin, one has to move fluids around in the water/salt phase diagram in the time, space, salinity, pressure and temperature domains (Fig. 2). The temperature is changing over time for a given depth above the igneous intrusion as it cools down. This moves the salt forming processes (systematically) around and allows for condensation of both salt-laden vapors and fresh vapors as they move upward.

As rising vapors and light liquids bring thermal energy with them to shallower levels, a typical change for rising fluid systems is a decrease in pressure while maintaining most of the temperature. Eventually, such fluids may enter the boiling zone on their way up and even drop out all the dissolved salts in them. A saline vapor may undergo cooling at almost constant pressure and thereby enter its supercritical domain, thereby increasing its ability to hold more salt and start to dissolve precipitated salt.

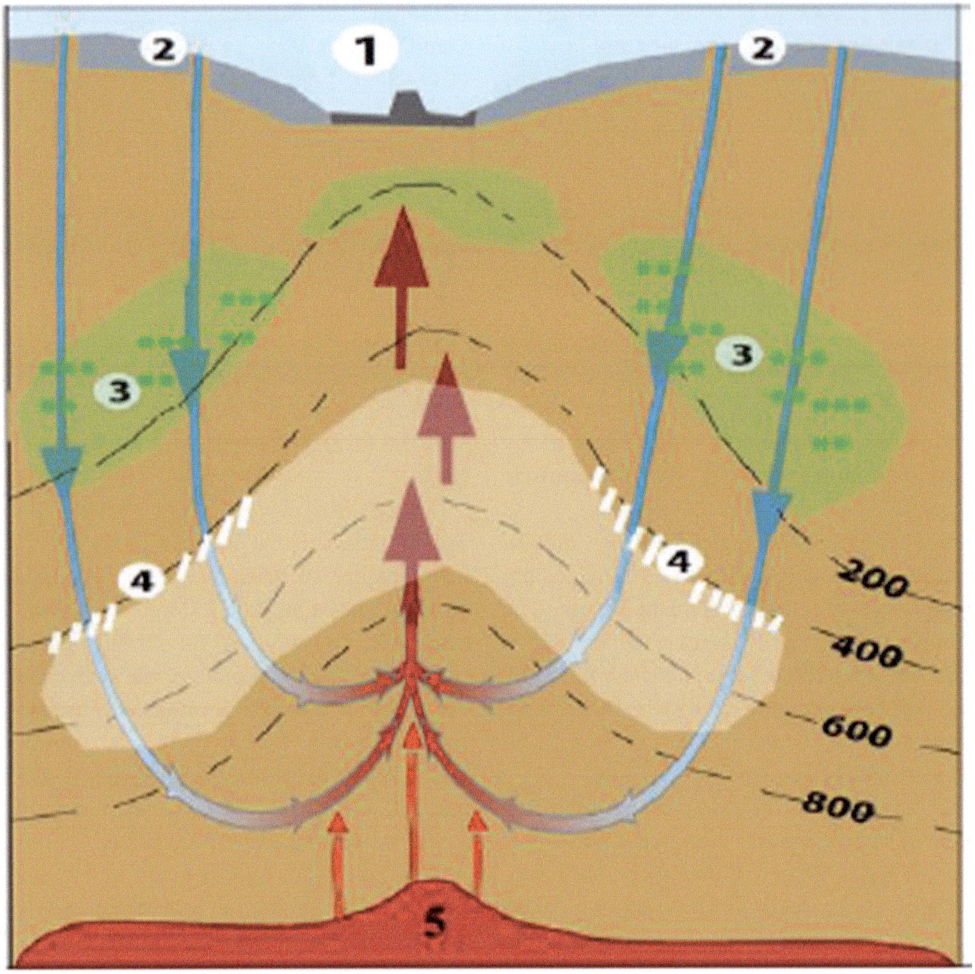

Fig. 1 Sketch of the inferred salt-producing and salt retaining hydrothermal system (Case 2). (1) Rising vapor from phase separation of interstitial fluids (bold red arrows = reflux zone); (2) Recharging seawater, 'sucked' into the hydrothermal system through permeable seafloor sediments and fractures; (3) Green area with stars: precipitation of anhydrite as recharging seawater is heated above ~120–130 °C; (4) Whitened area is zone where phase separation of seawater causes formation of dense brines and solid halite. This is equivalent to the 'Brine Cell' of Bischoff and Rosenbauer (1989). (5) The 'magma chamber' or igneous intrustion (T ~ 1200 °C). Note that halite melts at ~800 °C and this liquid may occupy some of the available void space near 5. The red vertical arrows immediately above 5 indicate magmatic (saline) fluids that may migrate out of the igneous body (according to Lecamberri-Sanchez et al. 2015). Inferred isotherms are shown as broken curves with black temperature labels (°C)

Table 1 From Coumou et al. (2009), this clearly shows how seawater with a salinity of 3.2 wt% NaCl percolates into the hydrothermal system (given by simulation numbers in the first column), and lower salinity water (from 1.4 to 2.1 wt% salt) exits out of the system. In practical terms, this means that salt has been retained (and become 'immobilized') underground, as clearly demonstrated by numerous real observations at hydrothermal vents

	Model setup			Venting			
	P_{top} (Mpa)	$Q^{max}_{bot.}$ (Wm^{-2})	k (m^2)	T (°C)	X_{av} (wt%)	Variation period	Irregular outliers
1	15	60	10^{-14}	345.8 + 3.0	1.4 + 1.0	~years	up to 12 wt%
2	15	10	10^{-15}	345.0 + 2.0	1.5 + 1.0	~decades	up to 12 wt%
3	25	60	10^{-14}	380.0 + 8.0	1.9 + 0.5	~decades	no
4	25	10	10^{-15}	383.7 + 0.7	1.7 + 0.1	~centuries	no
5	35	60	10^{-14}	407.8 + 7.0	2.1 + 0.1	~decades	no
6	35	10	10^{-15}	399.0 + 2.0	1.8 + 0.2	~centuries	no

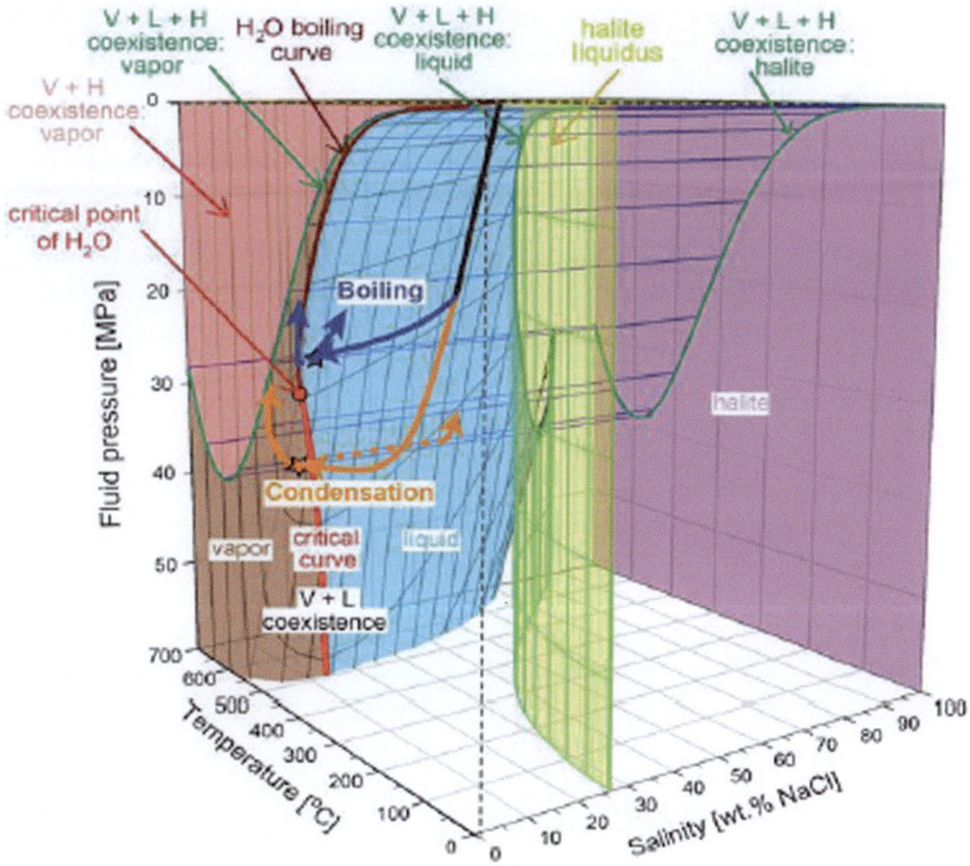

Fig. 2 Phase diagram of the system H_2O-NaCl from Scott et al. (2017) illustrating the effect of intrusion depth on the style of phase separation of an initial seawater-salinity fluid. The blue path represents circulation from the surface to ~2.5 km depth, followed by upflow and boiling, while the orange path represents circulation to ~4.5 km depth, followed by upflow and condensation. (Reproduced with permission from Scott et al. 2017)

Brines that are initially formed during phase separation, will, once they exceed a certain density, not be carried away with the rising vapor from the phase separation. As discussed by Bischoff and Rosenbauer (1989), such brines will become trapped near the igneous intrusion (e.g., form a 'brine cell') (Fig. 1) and undergo repeated phase changes (due to intermittent, cyclical or irregular changes in P and T) whereupon they become more and more concentrated (Bischoff and Rosenbauer 1989; Lilley et al. 2003). Many more scenarios can be envisioned by varying the conditions in a typical flowing system; however, this requires coupled thermodynamic and fluid dynamic models to keep track of all the variables and changes.

Various types of salt have different solubilities, and their solubilities are also varying with temperature. Calcium and magnesium salts in particular, have high solubilities at high temperatures, and tend to leave the system early in the process. Chlorides of magnesium and calcium are known to bind crystal water even as solid salts. Thus, they are likely to capture whatever water is left in the system to complete their molecular structure. This will enhance halite precipitation from mixed dense brines. It may be less known that the same Mg and Ca salts in solid form will become liquid brines by heating them above 120–123 °C. They dissolve in their own crystal water. Hence, these salts will stay as brines at elevated temperatures and solidify upon cooling. Later heating may dissolve them again and enable them to flow to a different location. This ability is, therefore, of great importance as a refining process for seawater salts. Halite is the least soluble chloride salt at high temperatures, and will therefore be refined in the subsurface to higher purity during all processes involving phase separation and condensation.

In both low-pressure (boiling) and high-pressure phase separation, salt is accumulating in the porous and fractured rock systems, and the condensation vapor (e.g., 'fresh' water) is led out of the hydrothermal system, driven by its low density. As mentioned earlier, super-critical undersaturated vapors may pick up salt on their way up and become saturated. On the other hand, a hot, saturated brine may also deposit salt along a cooling gradient.

Solubility of halite in water is reduced gradually with the lowering of temperature (Warren 2006). Whereas water may hold up to 80% halite at 600 °C it may only hold 25% at ambient seawater temperature (25 °C). A concentrated NaCl-brine can, therefore, deposit solid salt along the ascending flow path, over long distances (or high columns). It follows that the ability to deposit salt upward in the subsurface will be greater if the flow path is long and the temperature gradient is restricted (low).

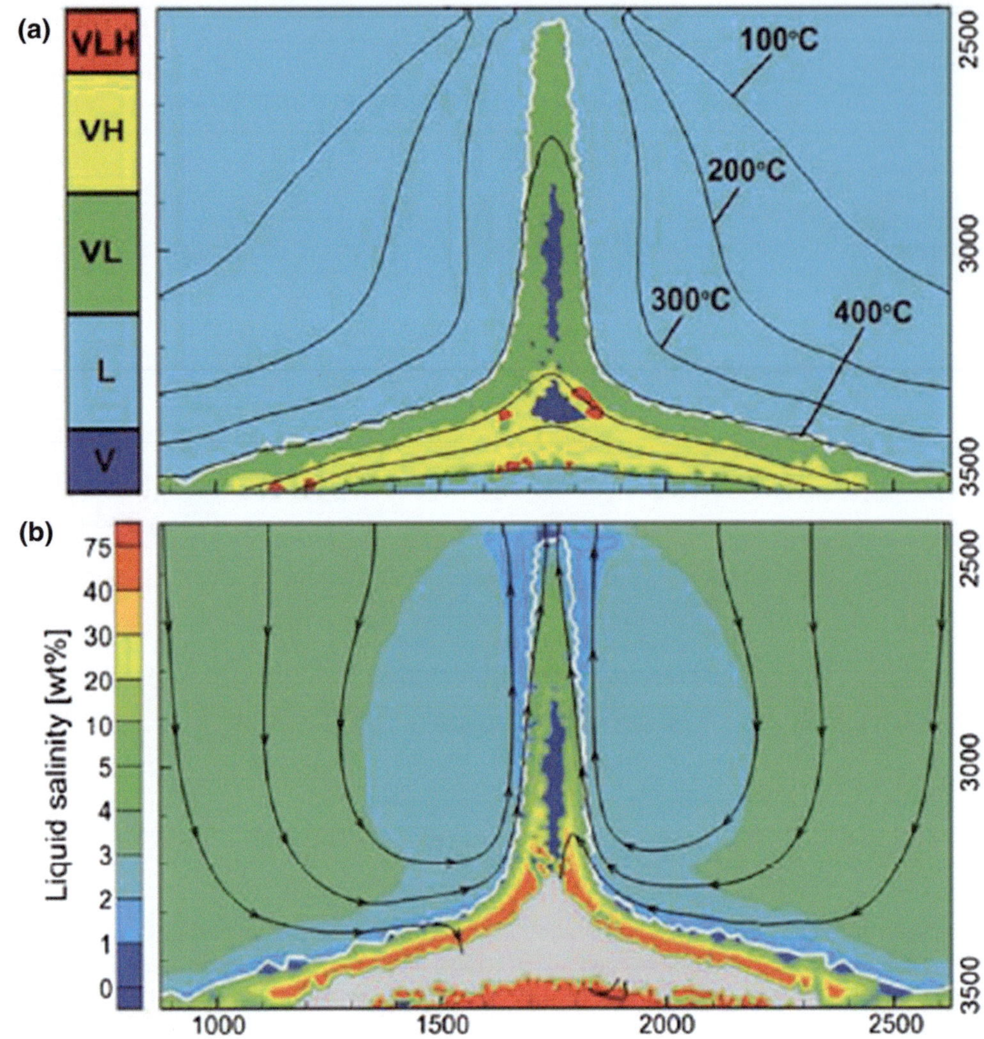

Fig. 3 Modified from Coumou et al. (2009), Simulation #6 (Table 1). Their figure text: "*Results from simulation 4 (Table 1) after (e.g.) 4000 years.* **a** *Phase state in color and isotherms (black lines) every 100 °C. LH, liquid-halite; VH, vapor-halite; VL, vapor-liquid; L, liquid; and V, vapor.* **b** *Liquid salinity and liquid flow lines. The white line in all plots marks the boundary between single-phase (L) and other phase regions. Grey areas in Figure b, indicate that the phase does not exist there. Depth is given in meters below sea level.*"

Once the hydrothermal system is dominated by liquids and solids, rather than vapours, the next step in the process becomes more important. As cooling goes on, the deposited salts are exposed to lower salinity liquids and vapors from below. We infer that transport of salts from the 'brine cell'-zone above the igneous intrusion and up to the seabed surface is mostly done by flow of precipitating brines.

The solid ('immobile') salt precipitation zones, according to Coumou et al. (2009) are shown in red and yellow colours, in their thermodynamic simulations in Fig. 3. In Fig. 4, we can follow fluid flow trajectories in the phase diagram (Fig. 2) and see how halite saturation leads to halite precipitation further up in the hydrothermal system as the fluids cool (right side of Fig. 4).

2.3 Evidence for Model Steps 3 and 4

Norton et al. (2015) have attempted a numerical modelling of events leading up to salt emplacement during the Atlantic rifting, which occurred more than a hundred million years ago (Fig. 5). They have done this by reconstructing the events by modern geodynamic modeling, based on careful seismic interpretation of sediment layers on both sides of the Atlantic, from the Santos Basin in the west across the spreading centre to the Kwanza Basin in the east. The result is a modern time-lapse view of the rifting, including the heat source and the salt emplacement over time (Fig. 5).

Thick salts were present in some of the early rift basins of the Atlantic Ocean as demonstrated by Norton et al. (2015).

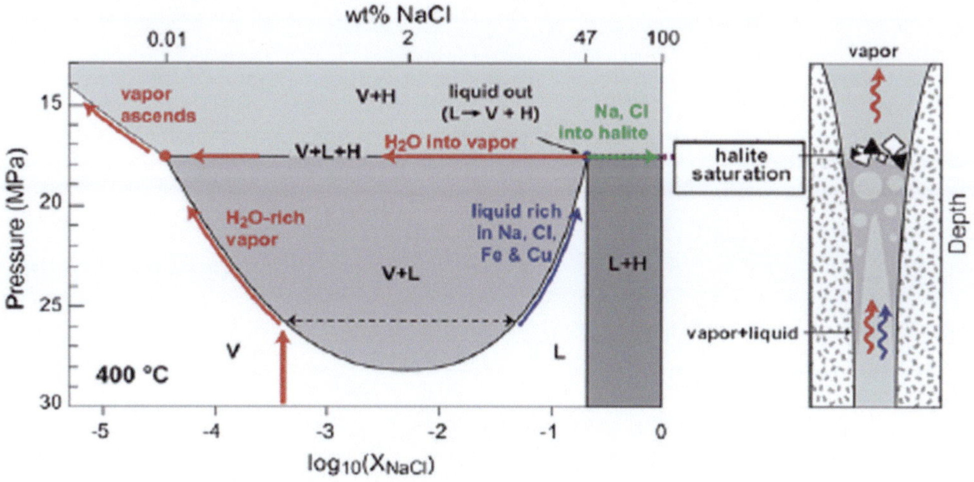

Fig. 4 Modified from Lecamberri-Chanchez et al. (2015). Their figure text: "*Left: Pressure-salinity projection of the H_2O-NaCl phase diagram at 400 °C (Driesner and Heinrich,* 2007) *showing a potential mechanism for copper sulfide mineralization via halite (H) saturation. Destruction of the liquid (L) phase results in partitioning H_2O to the vapor (V), and Cu and Fe to the solid phase. Right side shows the same process schematically.*"

According to their geophysical data, salt was deposited over a short time interval. However, to Norton et al. (2015), it seems that the salt was deposited due to solar evaporation, as they state the following: "*Salt was therefore deposited shortly after the end of sag basin deposition and extrusion of the lava flows contained in the SDRs* ('*Seaward Dipping Reflectors*'). *We thus conclude that salt was deposited during the transition from rifting to seafloor spreading, after SDR formation but before significant generation of oceanic crust*". And, according to Mohriak et al. (2012) who also mention SDRs: "*The region between the continental rifted margin and the active spreading centre is characterized in the South Atlantic and probably in the Red Sea by magmatic features and locally by seaward-dipping wedges* (*SDRs*) *that may underlie rift sediments and evaporites* (*Mohriak* 2001; *Mohriak* et al. 2010)". Their conceptual figure, therefore, has strong similarities with that of Norton et al. (2015) (e.g. Fig. 5), in that the evaporites were deposited after the formation of the SDRs.

The occurrence of thick salt deposits associated with the onset of SDR formation is interesting. According to common knowledge, SDRs are composed of varying mixtures of subaerial volcanic flows and clastics and also submarine basaltic flows. Thus, they probably belong to a general geology, which is not unlike the current geology of the Afar depression, between the Erta Ale volcano and the Dallol salt volcano (Talbot 2007). In this area, there is even an active volcano 'Gada Ale', which has an associated 'salt dome', partly covered in volcanic lava, described by Barberi and Varet (1971): "*West of Gada Ale volcano, a salt dome, mainly constituted by halite, has pushed up by as much as 100 m the whole lava series over an area approximately 2 km in diameter. It forms a N-S elongated ellipse and is linked with the Gade Ale volcano by a fault system forming a small horst. The Gada Ale volcano itself seems to be located on a domed* (*salt diapir controlled*) *structure*".

This process is also in accordance with the "*Volcanogenic salt model*", by Momenzadeh (1990), where he terms the precipitated salts 'saturites', rather than 'evaporites'. In our view, this demonstrates the chronology of rifting, SDRs, and salt deposition and the subsequent generation of oceanic crust. Consequently, some of the answers to the fundamental questions regarding the formation and accumulation of salts in rifted basins may be found in the Afar–Red Sea areas.

In their discussion of the rift to drift transition in the South Atlantic salt basins, Norton et al. (2015) pose some relevant questions that are equally important for the Red Sea: "*…however, several questions remain: Why do the inner* (*seaward facing*) *and outer* (*landward facing*) *basement ramps exist and what is the relationship between the inner ramp and LOC* (*'Limit of Oceanic Crust'*)? *Why is the oceanic crust between the two ramps so rugose? At what point in this evolution was salt deposited, and what role did it play in the development of these structures*?" The last question especially is relevant for the Red Sea. As Miller et al. (1966) clearly show in a simple sketch of the sea, more-or-less the whole cross section of the Red Sea became covered by a thick layer of salts at an early rifting stage.

A common problem with time lapse modeling of rift-related salt basins is that the salt volume seems to exceed the available basin volume at the time of salt deposition. The fact that salts occur to completely cover the intrusive oceanic crust from Stage C to D, in Fig. 5, points in the direction of a

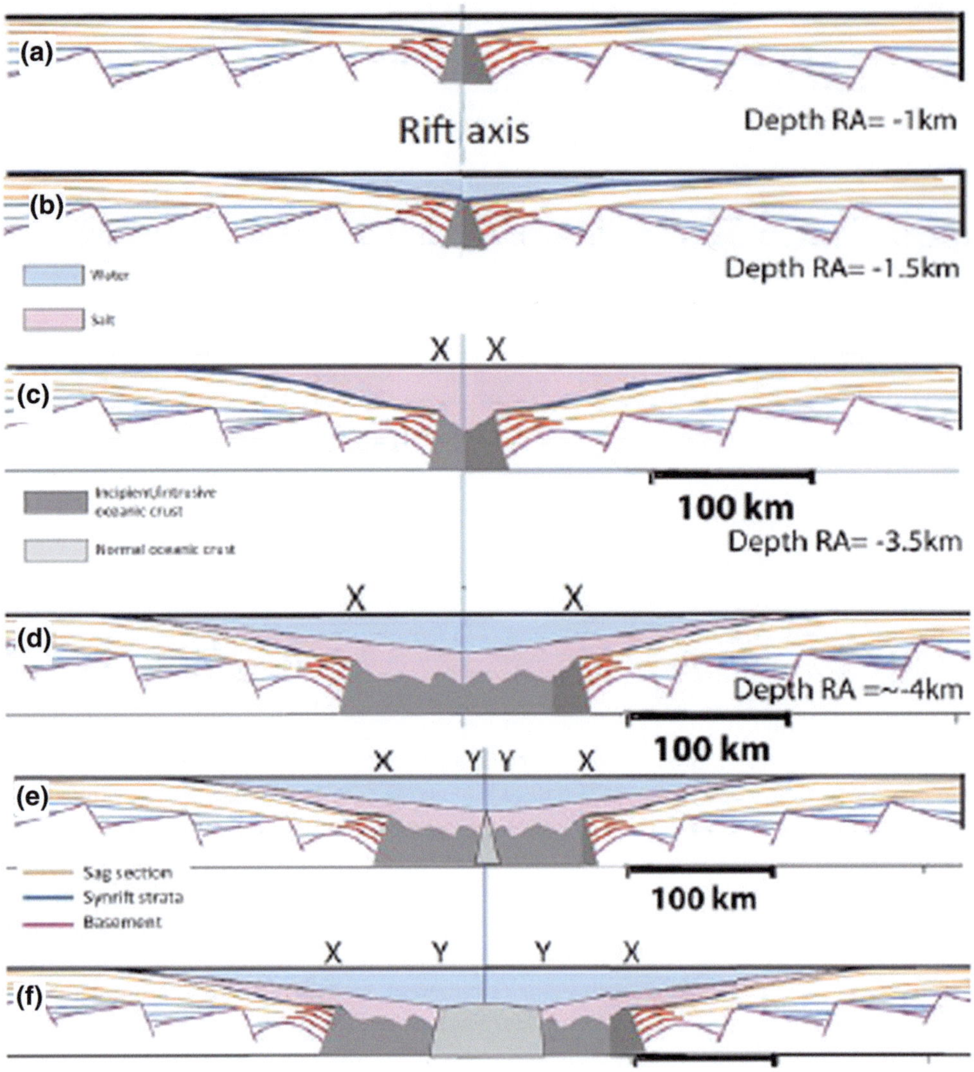

Fig. 5 These sketches, modified from Norton et al. (2015), represent a modern view on how and when salt is deposited in a marine spreading centre (in this case the Central Atlantic, Campos and Kwanza basins). RA is the depth of the Rift Axis below sea level. The figure text of Norton et al. (2015) is as follows: *"The case of seaward-dipping reflector (SDR) margins like part of the Campos Basin is illustrated. Very similar geometries of the salt basin are expected for non-volcanic simple rift to drift margins like Santos and Kwanza. A: Rift basin is subaerial and transitioning to oceanic crust formation. Rift axis is ~1 km below sea level. B: Same time as in A, but now seawater has flooded the basin. Rift axis is ~1500 m below sea level. C: Salt has filled the basin to sea level; rift axis is ~3500 m below sea level. D: 3–4 m.y. after C, with ~100 km total motion since C. Intrusive oceanic crust is being created under salt. E: Shortly after D, with volcanism breaking through the salt, forming the breakthrough volcanoes. F: A few million years after E, with normal oceanic spreading."* Note that SDRs occur as red 'fans' in the sketches

basin filled with dense precipitating brines, rather than a layer of solid salt.

Norton et al. (2015) actually suggest that the salt formation, in scenario C, of Fig. 5, is more-or-less 'instantaneous': "*Estimates of duration of salt deposition are poorly constrained, ranging from 600 k.y.* (*Dias* 2005) *to 5 m.y.* (*Davison* et al. 2012). *Given Aptian plate spreading rates of 50–60* km/m.y. (*Heine* et al. 2013), *longer intervals of salt deposition would mean that the conjugate salt basins separated before the end of salt deposition; this is implausible. We thus prefer a short period of salt deposition. We also assume that isostatic loading, occurring as it does over thousands of years as opposed to hundreds of thousands of years for salt deposition* (*Van den Belt and de Boer* 2007), *was instantaneous*". However, to us, this seems very improbable, especially when we know that isostatic forces will not be powerful enough to provide the needed subsidence, as argued by Van den Belt and de Boer (2007) in 'Shallow basin model for 'saline giants' based on isostacy-driven subsidence'.

Filling the basin with salts having a relatively low density will not be able to isostatically push the deeper, heavy rocks down to such an extent that the entire salt accumulation stays below sea level. We thus doubt that their model based on isostasy alone is viable. In our view, the salt basin was filled during periods of high heat-flow, for example, SDR-formation with accumulated salts (saturated brines) from below, while simultaneously sinking due to compaction of the rocks beneath the seabed as the salts moved out. In other words, the lowering of the basin floor was held back by salts and brines in the sub surface, until the transport stage became dominant and moved the salts up. Even during the period of salt transport, more salt/brine was produced and the added volume from this may have totally flooded the basin before the brines dried out to produce both layered and 'amorphous' solid salts, as also seen in the Red Sea.

We infer that a similar process is currently taking place in the northern Red Sea and Afar regions and is, thus, demonstrating the close relationship between hydrothermal salt production and volcanism, as well as brine migration and refining of different salts on the surface in a dry climate or on the seafloor, as high-density brine pools. However, the most interesting aspect of the conclusions of Norton et al. (2015) is their transition from Stage B to Stage C (Fig. 5), which occurs 'instantaneous'. This is indeed parallel to what seems to have occurred also in the Mediterranean Sea (e.g., the Messinian Salinity Crisis) and is currently occurring in the Red Sea.

2.4 Evidence for Model Step 5

Some of the observations from the Afar (Danakil) subaerial volcanic-hydrothermal region may also be relevant for our model. Talbot (2007), describes and discusses the widespread occurrence of hydrothermally produced salt and hot brines at Dallol salt "*mountain*" and on the adjacent salt flats of Lake Asele, about 18 km north of the very active Erta Ale volcano in the northern Afar region of Ethiopia/Eritrea. Talbot (2007) describes some of the structures found on the SE flank of Dallol, as: "*Mounds or towers of hydrothermal salt precipitated from hot springs…*" (Fig. 6).

But, perhaps, the most intriguing feature he discusses is the so-called "skating rink", located only ~5 km north of Erta Ale, on the Lake Asele salt flats. According to Holwerda and Hutchinson's (1968) pioneering descriptions of this feature, the 17 m high and 100 m wide skating rink has "*a core of pure salt surrounded by an upturned ring of impure salt*". The boiling brine bubbling up in the skating rink holds a temperature of 130 °C and consists of $MgCl_2$. Both the high-magnesium salts "*Bischofite* ($MgCl_2.6H_2O$) *and Carnellite* ($KMgCl_3.6H_2O$) *crystallize around this spring but the* $MgCl_2$ *deliquesces in atmospheric-derived moisture and the solutions drain away leaving sylvite*" (Talbot 2007). These high-solubility salts are the last of the salts to precipitate out in hydrothermal systems, after all other (lower-solubility) salts have precipitated. These (Mg) salts are therefore likely to leave the hydrothermal system entirely as brines and may end up in the sea or in separate brine pools on the seafloor (e.g., 'ponding').

Talbot (2007) also assembled a very interesting interpretational map of "upwelling features" on the Lake Asele salt flats, which renders a megapolygonal pattern, typical for convective cells: "*The cell centres are spaced about 1.7 km and 1.2 km apart along and across the depression, respectively. This polygonal pattern is more consistent with the salt having undergone solid-state thermal convection above a hot basement than being influenced by igneous centres like those to the south that are aligned along potential spreading ridges. Talbot (1978) calculated a near-critical Rayleigh number for the salt flats using a value of thermal expansivity that was two orders of magnitude too high but a thermal gradient that may have been even lower, given that the thermal gradient along the axial zone of the Red Sea is near 120* °C/km" (Talbot 2007).

Talbot (2007) also mentions other boiling brine pools in the area, such as "*Round-mountain*", "*Horseshoe-mountain*" and "*Flat-mountain*" that occur on the Asele salt flat, SE of Dallol. Furthermore, he mentions that native sulphur occurs mixed with halite in places, which clearly demonstrates how closely the salt production is associated with volcanic heat and general hydrothermal processes.

In step F of Fig. 5 (Norton et al. 2015), the volcanic process must start all over again. But now there is a dense salt layer covering large parts of the basin floor, and the

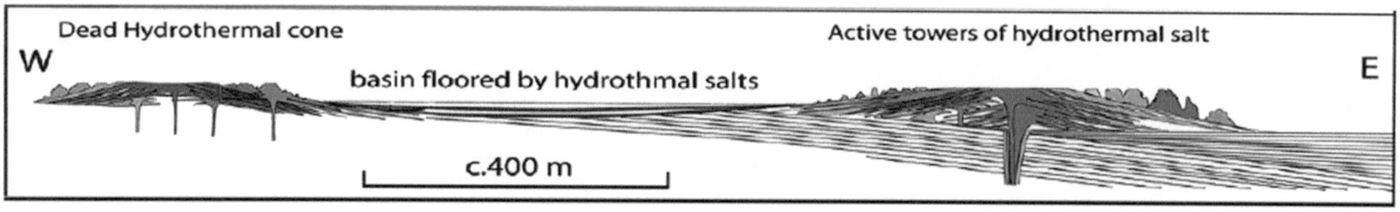

Fig. 6 A modified drawing of the Dallol hydrothermally produced salt volcano and stratified subaerial salt basin (Lake Asele, Afar Depression), based on Talbot (2007), whose figure text is: "*Dogleg geological profile across Dallol 'mountain' from NNE shore to summit and then W–E across the 'crater'.*"

igneous rocks that initially were characterized by a large fracture system are now clogged by a variety of precipitates. The access of seawater to the fracture system of the magma is therefore much reduced. The accumulation of salt therefore decreases, and the magma body develops into a conventional oceanic crust, without accumulated salt.

Salt may also accumulate in deep igneous systems following the same thermodynamic process, as in a subduction setting. Subsequent rifting in such environments may encounter the appearance of accumulated salt associated with ascending plutons (Lecumberri-Sanchez et al. 2015). The flow of brines and precipitation of salts from the now inactive subduction zone may be analogous to the salt accumulation processes associated with rifting. Perhaps the most remarkable discoveries of brine flow up to the seafloor are the warm brine lakes occurring north of the Mid Mediterranean Ridge, directly above a marine subduction zone (Wallmann et al. 1997). According to the MEDRIFF consortium (1995): "*The lakes have a distinctive appearance on side-scan sonar images with no backscattering from the lake surface and enhanced backscattering at their edges.*" Furthermore: "*The chemistry of the brines is different in all three lakes and shows that the brines are derived from selective solution of different parts of the Messinian evaporites that lie beneath the sea floor*". From this brief description, there can be no doubt that large portions of the buried salt must have very permeable vertical conduits that have opened up and are non-sealing for the warm brines to ascend to the surface.

Salt diapirs are traditionally regarded as being completely sealed and constitute sealing bodies for the migration of water and petroleum. Hence, they were considered to be ideal repositories for the storage of radioactive waste, etc. This changed totally when Lewis and Holness (1996) published their findings about the permeabilities of halite (Fig. 7).

According to Lewis and Holness (1996) and Holness and Lewis (1997), the PT-conditioned halite permeability may lead to vertical salt mobilization, that is, diapirism: "*Our results explain recent observations of major salt-fluid interactions at depth, and suggest that deep-rooted salt diapirs may act as conduits for basinal formation water*" (Lewis and Holness 1996). The permeability is caused by P/T-dependent alteration of the dihedral angles of the salt crystals (Fig. 7). Consequently, saturated hot brines and other fluids (including condensed water vapor) can easily be transported through salt formations at depths greater than ~3 km, provided sufficiently high temperatures (Schoenherr et al. 2007a, b; Ghanbarzadeh et al. 2015). Salts may then be re-dissolved and re-deposited repeatedly from upwards migrating brines throughout the Holness Zone, according to the changing temperatures and pressures. Above the Holness Zone, hydraulic fracturing of the

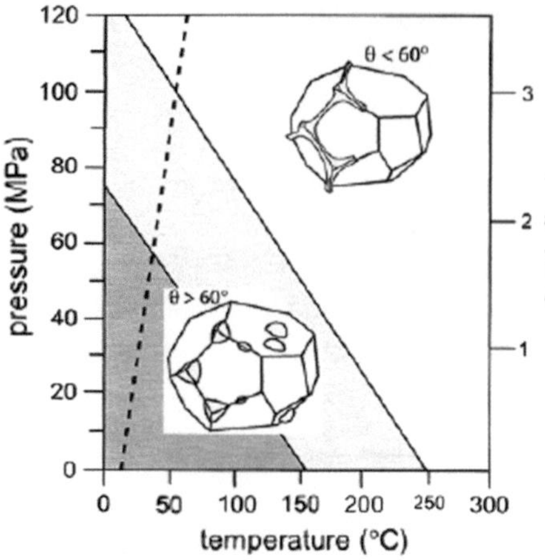

Fig. 7 This cartoon shows that the equilibrium halite water dihedral angle (θ) is a function of P and T. In the white area, θ is less than 60°, resulting in the formation of continuous pore-fluid connectivity in the halite-water system. The light grey area defines the P-T region in which the permeability of halite may be significant. The dark grey area (θ > 60°) represents the region in which permeability of halite is low or non-existent. The dashed line indicates a normal geothermal temperature pressure gradient. Drawings of halite crystals with high connectivity and non-connectivity are shown (Modified from Lewis and Holness 1996)

impermeable salt may create transportation routes for the migrating brines and slurries of brines and solid salt particles. These brines and slurries are then venting out at the top of the salt structures (and faults above), and may flow downwards on the seafloor like 'salt glaciers' (namakiers), as is seen in the Thetis Deep in the Red Sea (see 'Salt flows' and Fig. 15).

The discovery of a large field of seafloor pockmark craters at ~400 m water depth in the eastern Red Sea shows that even salt thicknesses of several hundred metres to a few kilometres do not hinder fluids from moving through the salt. These results were reported by Feldens et al. (2016). The pockmark field, containing hundreds of pockmark craters that have modal diameters of 140–150 m, was mapped in an area of about 1,000 km^2, located at ~20 °N, 80 km east of the central Red Sea axis. Analysis of sediment cores from the pockmarks showed porewaters of higher salinity than seawater and adsorbed light hydrocarbons that have an isotopic signature of thermogenic petroleum. The researchers conclude that: "*The most likely explanation for the increased permeability of the evaporite sequence is brittle deformation triggered by extensive local tectonic movements and supported by high fluid overpressure within the evaporite sequence*".

These observations show clearly that brines are transporting pre-salt hydrocarbons via fluid flow 'chimneys' to the surface where pockmark craters form due to local fluid flow induced erosion (Judd and Hovland 2007; Hovland and Judd 1988). Worldwide, however, it is not uncommon to find fluid flow chimneys, pockmarks and even mud volcanoes in areas with significant thicknesses of layered and 'amorphous' salt and salt diapirs. Regions where such features have been mapped are the northern Gulf of Mexico, the eastern Mediterranean, and especially the outer Nile Delta Fan, all of which contain substantial amounts of sub-surface salts. In all of these cases, the model of conditioned halite permeability by Holness and Lewis (1997) plays an important role.

Fluids within a hydrothermal system may exert large overpressures on their surroundings. The maximum possible overpressure may be estimated as the difference between the lithostatic pressure created by a column of rock/solid salt, minus the hydrostatic pressure from a saturated brine/vapour of the same column height. If the overpressure is great enough due to column height, the flow may go all the way to the surface/seabed. Due to adiabatic temperature loss in the brine column, it will precipitate salt all the way along its flow path once it has reached saturation at some depth/temperature. This will gradually reduce the ascending brine density and increase the ability to create overpressure per column height.

The ultimate source of the overpressure in the hydrothermal system is the heat source and the ability of water to dissolve and precipitate matter and to form low-density vapour/steam. This leads to an intermittent system where pressure, temperature and mass-flow vary over time as a result of combined heating, dissolution, clogging, pressure build up and fracturing. During certain periods, such a system may produce maximum pressures as described, before dropping back to hydrostatic again. Hydrothermal systems therefore cannot be modelled or understood on the basis of averaged parameters.

In addition to the brittle deformation triggered by extensive local tectonic movements and sub-surface pressure build-up, suggested by Feldens et al. (2016), the fact that halite is conditionally permeable (Lewis and Holness 1996; Hovland et al. 2006a, 2012, 2015; Schoenherr et al. 2007a; Ghanbarzadeh et al. 2015), is, in our view, a more plausible explanation for the pockmark formation. The reason why we prefer this mechanism is that it can actually start off at depth, by fluid flow through the salt matrix itself, according to Lewis and Holness (1996). The fluids migrate upward into shallower regions where the halite turns impermeable and brittle. From here on, the flow follows fractures that are 'spearheaded' by hydraulic fracturing due to fluid pressure build-up. According to Ghanbarzadeh et al. (2015), the permeability of natural rock salt may not be negligible: "*Brine-filled pore networks in rock salt approach textural equilibrium due to fast reaction kinetics of salt dissolution and reprecipitation*". "*We confirm the static pore-scale theory in undrained laboratory experiments on synthetic salt samples that have been imaged with non-destructive x-ray microtomography*".

In support of the conditioned halite permeability model (Holness and Lewis 1997), it is known that fluids have been observed both to flow inside salt stocks and to exit at the surface after having flowed through buried, deep-rooted salt stocks (Aharon et al. 1992; Sassen et al.1994; Gay et al. 2011). Some examples are reviewed below.

3 Other Relevant Observations

3.1 'Five Islands' Salt Domes, Louisiania

Unexpected expulsions of gases, brines and salt from production faces inside salt stock mines are commonly known as 'outbursts'. Outbursts can release over one million cubic feet of methane, brines and fractured salt, and are responsible for the deaths of numerous miners (Ehgartner et al. 1998). In southern Louisiana, USA, buried salt domes occurring with accompanying hills on the surface are called 'Islands'. Inside 'Winnfield Island' there is evidence of fluid flow: "*When new excavations are made, both brine and CO_2 come up through the floor, out through the walls, and down through the ceiling. The flow of brine and gas is largest immediately after an excavation is formed. It gradually tapers off over a period of several years as if the salt had been drained—or possibly the fractures sealed by precipitating salt. High-pressure brine and gas inclusions are common throughout the salt, particularly within anhydrite bands* (*Hoy* et al. 1962; *Kyle* et al. 1987; *Loof* 1999)" (Ehgartner et al. 1998). Peculiar features of mineral growth are also seen, including vertically oriented anhydrite 'pencils', 4–5 inches in cross section and 12–18 inches in length (Hoy et al. 1962).

Over 'Weeks Island' the land surface is characterised by an elongated 50 m high ridge, the 'Devil's Backbone', expected to be associated with an internal shear zone exposed in the Weeks Island salt mine (Kupfer 1976). Later studies by Kupfer et al. (1998) revealed that these shear zones are better described as 'anomalous zones', as there are: "*…inclusions of gases* (*hydrocarbon gases, carbon dioxide and nitrogen*), *liquids* (*brine, liquid hydrocarbons*), *and clastic sediments* (*clay, marl, sand*)" (Kupfer et al. 1998). Similar observations are made inside 'Avery Island' salt dome where a vertical anomalous zone contains: "*…considerable water in local pockets, most of which eventually drains and dries up over a period of months*" (Kupfer 1976). On the land surface, directly above this shear zone there is a series of saline ponds, indicating direct leakage of brines upward through the apex of the salt dome. Similar 'anomalous zones' occur inside all of the Five-Island Louisianian mined salt domes (Kupfer et al. 1998).

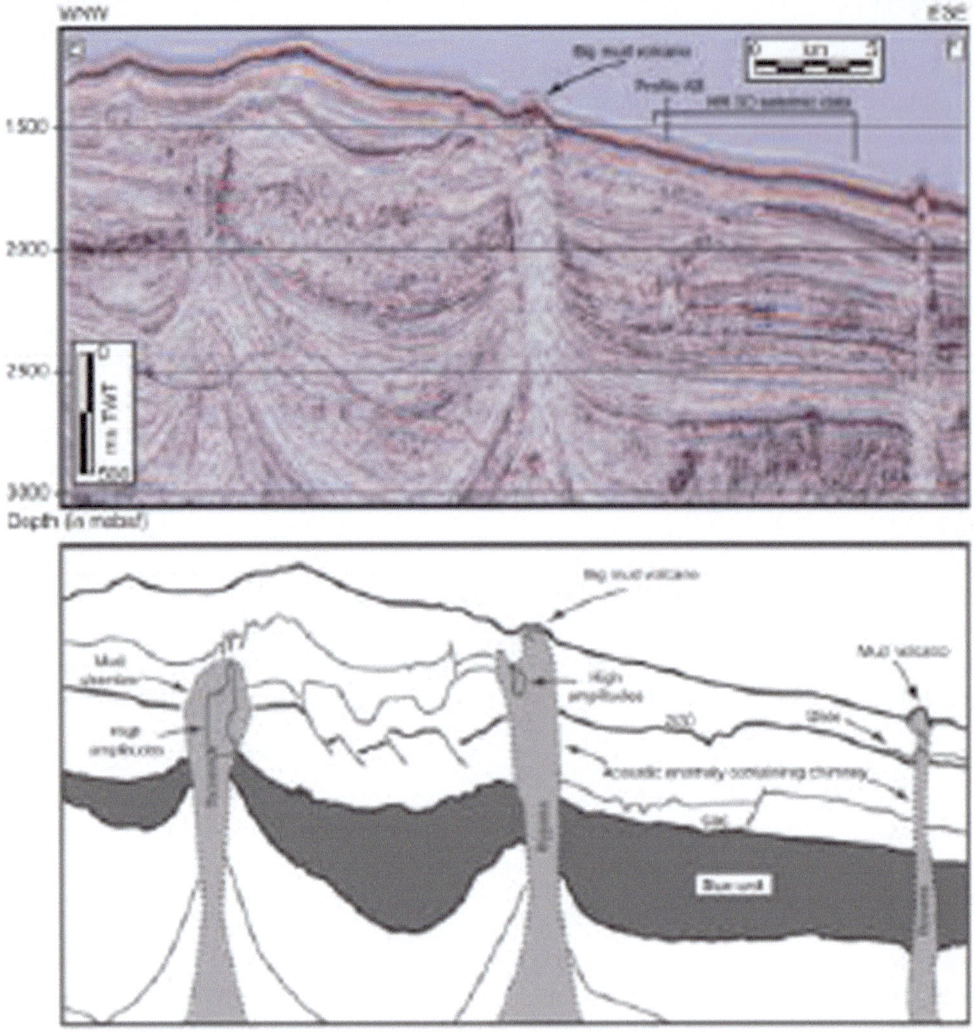

Fig. 8 Upper image: A 2D-seismic section across the salt-dome-related mud volcano at the Ursa field (Gulf of Mexico). It is named 'Big Mud Volcano' by Gay et al. (2011). Beneath the mud volcano, the salt reflector is seen to form an extrusive pinnacle (or conduit). Lower image: An interpretation of the same seismic section, by Gay et al. (2011). This is a clear example of a large salt dome ('diapir') having one or several active central conduits (Images courtesy of Gay et al. 2011)

3.2 The Ursa Salt Dome, Gulf of Mexico

Another, perhaps more relevant example for Red Sea analogies is a relatively large mud volcano located on top of the central salt dome of the Ursa gas and oil field in the Gulf of Mexico (Ruppel et al. 2005; Gay et al. 2011). During the development of the field there was an uncontrolled 'shallow water flow' episode near the salt dome. A steel template with 6 drill slots was lost due to over-pressured water flow through the seafloor. There have been discussions over the causes for this flow, and the scientific IODP (Integrated Ocean Drilling Program (now called the 'International Ocean Discovery Program'), Expedition # 308) drilled some geotechnical holes to attempt finding out what happened. However, these holes were too shallow and no final conclusions were made (Flemings et al. 2005).

Gay et al. (2011) published some data from the Ursa field, showing that the mud volcano is located immediately above the summit of the main salt dome at Ursa, above a slender salt structure called a 'bypass chimney' by Gay et al. (Figure 8). Of great interest, here, are the observations of oil, free gas bubbles, and temperature in this Ursa mud volcano. Measurements show anomalously high temperatures of up to 23 °C (Ruppel et al. 2005). These observations clearly suggest that there is an active conduit underneath the volcano, which is obviously fed by deep-sourced fluids through the main salt body. Three sediment cores (PC-25, −26, and −27) were taken across the mud volcano ('mound'), as described by Ruppel et al. (2005): "*On the mound* (mud volcano) *crest, PC-25 has a pronounced concave down salinity gradient, with maximum values reaching* >4 *times seawater salinity and a decrease in salinity below 1.25 m depth. The concave down gradient is less pronounced at PC-26, which is located just inside the zone of acoustic transparency. PC-27, collected in reference sediments 1 km from the mound, has nearly seawater salinity throughout". "Simple numerical models yield estimated upward flow rates of 0.15 m yr*$^{-1}$ *at PC-25 if the decrease in salinity below*

1.25 m is neglected and 0.05 m yr^{-1} *at PC-26. Both hydrate and oil were recovered on the mound crest (PC-25 and PC-30), and oil was found in PC-26*" (Ruppel et al. 2005).

Although Gay et al. (2011) do not discuss or suggest fluid flow conduits through the salt dome at Ursa, the seismic, topographic, geochemical data, and the very existence of the mud volcano demonstrate that there must be such flow. According to these researchers, overpressure in the 'lower compartment' leads to episodic and focused fluid migration through deep conduits that bypass the upper 'compartment', thus forming mud volcanoes at the seabed.

The above observations seem to confirm that salt accumulations may be subjected to internal pressures in encapsulated brine/gas systems. The pressure buildup in these fluids may be the result of a combined effect of increased pressures in the vapor/steam phases, tectonic forces, or due to the salt itself causing accommodation problems by clogging the fluid flow channels. A column of isolated brine/gas subjected to such pressure will exert large forces on the salt located above, due to the large difference in density between salt and brine. Such forces may open up fractures in the salt and let the brine leak out. However, in some situations it may also lift out large "pistons" of impermeable/impure salt on top of the brine channel. A sealed off brine column within salt of roughly 550 metres height, will in theory be capable of lifting out a 250 meters high pinnacle of impure salt above it, by hydraulic forces, as also inferred by Hoy et al. (1962) and Loof (1999).

3.3 Red Sea Salt Structures and Features

In the following we will apply the proposed model on relevant (anomalous) observations in the Red Sea. Thus, we will attempt to explain these structures and features, one by one, and attempt to identify where they belong in our 'Model Step' system. We start with the 'Deeps', continue with 'Salt Flows', thereafter, 'Salt Ridges' and 'Salt Walls', and finally discuss the conditioned permeability of thick salt deposits.

3.4 Brine Pools (Deeps), Red Sea

Over several decades, the Red Sea Deeps have been studied with geophysical, oceanographic, biological, and geochemical methods and instruments. The 25 or so Deeps form along the axial trough where there is active spreading and likely manifestations of shallow-seated high-temperature igneous intrusions that host hydrothermal circulation cells, as suggested by Mitchell et al. (2010), Augustin et al. (2014), and Ehrhardt and Hübscher (2015). The discovery of the dense, high-temperature brines in the Deeps came as a surprise to the early investigators of the central Red Sea. According to Lowell et al. (1995), the first evidence of hydrothermal activity at a submerged seafloor spreading centre appeared as an anomaly in temperature and salinity in hydrographic data. Subsequent sampling of the water column and seafloor also revealed that horizontal mid-water acoustic reflections were produced by the impedance contrast caused by hydrothermal brines containing suspended metal sulphides (Lowell and Germanovich 1997). Most of the deepest troughs contain hot brines with these characteristic horizontal seismic reflections (Ehrhardt and Hübscher 2015). But there is one notable exception, the Thetis Deep, which will be discussed at a later stage.

As to the origin of the hot brines in the Deeps of the Red Sea, there was at first great controversy. Although Craig (1969) suggests that "*the probable source lies about 800 km to the south near the strait of Bab el Mandeb…* ", other contemporary researchers had different opinions and ideas. Thus Brooks et al. (1969) stated: "*On the basis of trace metal enrichment that is higher in brine and interstitial water samples from the Atlantis II Deep than from the Discovery Deep, it is suggested that the source of the brine is derived from beneath the floor of the Atlantis II Deep*". "*It is impossible to determine from our data if there is one or several vents and where they are located…, but since the brine spreads over the bottom of the basin more data are needed to determine the exact location of the vents*".

However, Miller et al. (1966) were the first to recognize that salts were precipitated ('leached') by hydrothermal activity in the Red Sea: "*The Red Sea brines may be related to oilfield brines in mechanism and formation, or may be produced through subsurface circulation and leaching of evaporites by normal Red Sea water, or formation waters of near-normal oceanic salinity. The proximity of an intrusive igneous body also may be effective as a source of heat and possibly of volatile materials. In the Imperial Valley of Southern California, an area similar in geologic characteristics to the Red Sea, a deep geothermal well tapped a brine considerably higher in temperature (*~*300* °C*) but of similarly high salinity. A magmatic origin for the water was proposed*". "*Three principal hypotheses can be advanced to account for the temperature and salinity of the brines:(1) Basinward flow of brines concentrated by local surface evaporation and solar heating. (2) Basinwide evaporation during a former isolated stage of the Red Sea. (3) Submarine discharge of brine*" (Miller et al. 1966).

The active emplacement by hydrothermal systems was also suggested by Erickson and Simmons (1969), based on thermal measurements: "*The heat balance in the 56* °C *water in the Atlantis II Deep requires thermal transfer mechanisms other than molecular conduction for the transport of heat through the sediment. The existence of distinct, isothermal water layers throughout the entire Atlantis II Deep suggests that heat transfer through the bottom of the 56* °C *water is more uniform than would be expected if transfer occurs by hydrothermal discharge through a few localized vents. Uniform convection of the brine through the porous sediment is the most likely mechanism at the present time, although it was probably preceded by periods of hydrothermal discharge*" (Erickson and Simmons 1969).

The actual heat-flow values of the Red Sea Deeps were investigated by Anschutz et al. (1999), and were calculated to be somewhere between 250 and 600 mW m^{-2}. Furthermore, they noted that the range of heat-flow values corresponds to bottom heat-flow values that have been reported earlier for the axial zone of the Red Sea. Whereas other studied brine pools indicate stable (steady-state) temperature and salinity conditions, it was found that the Valdivia Deep brine increased by 4.1 °C and 10‰, respectively, between 1972 and 1992. This suggests that the Deeps in the Red Sea undergo stages of heating (e.g., with hydrothermal brine discharge, Anschutz et al. 1999) and stages of cooling. These processes of heating and cooling are, in essence, the most important aspect of our model. Thus, we expect that the transition from the continental rifting stage to the oceanic crustal stage is equivalent to what Norton et al. (2015) term 'incipient/intrusive oceanic crust', for example, the sudden transition from Stage B to Stage C, in Fig. 5, where salts move up from the underground storage reservoirs to the surface, inducing dramatic subsidence and seafloor collapse, as a consequence.

3.5 Atlantis II Deep

Three pools of hot brines are found within the Atlantis II Deep (Degens and Ross 1969), and sediments underlying these brines are enriched in heavy metals, which are actively accumulating through the discharge of hydrothermal solutions (Pirajno 2009). The Atlantis II Deep is defined by the 2000 m isobath. It is 14 km long and about 5 km wide. The temperatures of the hydrothermal fluids that enter the Deep are variously estimated to range from approximately 210 °C to as high as ~450 °C. These hot brines have high concentrations of Fe, Mn, Zn, Cu, Pb, Co, Ba, Li, and Si, which precipitate as layers of sulphides and silicates (Pirajno 2009). This indicates seawater contact with the mantle (peridotite), also suggesting that some of the circulating seawater may participate in serpentinization reactions.

According to Anschutz (2015): "*The Atlantis II Deep is one of the most remarkable seafloor settings of the planet. It is the largest known present-day submarine hydrothermal environment, consisting of a 65 km^2 topographic depression located in the axial trough of the Red Sea at 2,000 m depth. The depression traps hot and dense brines fed by hydrothermal fluids, the source of which have not yet been sampled*".

The Deep Sea Drilling Project (DSDP, Leg 23) Site 226 was located at -2169 m in the Atlantis II Deep: "*This was a high risk hole in that it was not clear if enough sediment was present to satisfactorily bury the drill string, and indeed, drilling attempts at three different offset positions failed to penetrate a shallow basalt. Only a total of 14 m was penetrated before part of the bottom-hole assembly was lost. The two cores we obtained contained sediments similar to those previously obtained from the area—a Late Quaternary mixed montmorillonite, anhydrite, and goethite-hematite facies in the upper 5 m with fresh basalt fragments in the lower 9 m. The basalts, which show no indications of chemical reaction with the hot brines, are similar to previously collected ridge basalts*" (Whitmarsh et al. 1973). To us, this suggests that the seafloor inside the Atlantis II Deep is very complex, and that it is likely composed of alternating layers of salts (evaporites), sediments (including hydrothermal-associated clays), and volcanic sills. The modern swath maps also demonstrate that the topography is rugged, with local highs and troughs (Augustin et al. 2014). Because this initial drilling more-or-less failed because of the problematic nature of the drilling conditions on the seafloor, the next hole was moved out of the axial trough to DSDP Site 225, which was located on the eastern flank, 16 km to the east of the Atlantis II Deep, at -1228 m. It penetrated 230 m of sediments and salts, and terminated 54 m into the Late Miocene 'evaporite' sequence.

According to our own model, steps 1–3, large masses of salt have been precipitated in the buried hydrothermal systems over time-scales of a few hundred thousand years. We also predict that the salts are formed sub-surface, near the heat-source, but that there is a semi-continuous or intermittent flow of hot brines and particulate salts to the seafloor surface (e.g., model steps 4–5). Other hydrothermally associated minerals are also transported by these brines, such as metalliferous products, Fe, Mn, etc.

Although very limited visual observation has been achieved inside and adjacent to the Atlantis II Deep, there was a survey conducted in 2008, with the possibility of visual observation (Bower 2009). This cruise was run with the vessel R/V Oceanus, and it discovered seepage of brines (named the 'Oceanus Seep') along the western wall of the central rift valley. Here, at -1840 m, the ambient seawater temperature is around 23 °C. The temperature of the Oceanus Seep water is about 33 °C (Bower 2009). Adjacent to the Oceanus Seep, they also discovered some diffusive seeps: "*At several positions,..., a thick layer of slightly warm, saline water was observed at the bottom. Current measurements indicate that these thin, dense layers were flowing downslope. This suggests that dense briny water is seeping out of the walls of the rift valley and flowing down the walls*" (Bower 2009).

At deeper elevations (e.g., deeper than -2020 m), inside the Atlantis II Deep, brine has a temperature of ~68 °C. Here, they reported the following observations: "*Samples from near the top of the brine layer were yellow/orange, whereas those from near the bottom were clear,* (*showing*) *possible evidence of precipitation of iron particles at the brine interface due to the presence of dissolved oxygen in the overlying water*" (Bower, 2009). Both of these pertinent observations support our notion that there may be extensive seepage of hot both under-saturated and saturated brines flowing out of the flanks of the central rift valley. This indicates a late stage hydrothermal system, where brines are flushed upward (e.g., our steps 3–4).

Anschutz (2015) mentions that: "*The hydrothermal end member* (*of the brines*) *is highly depleted in dissolved magnesium and sulphate and enriched in calcium and dissolved metals. The isotopic ratio of dissolved strontium is lower than that of sea water and close to that of ocean crust basalt.*" Also with respect to helium and argon isotopes, there is evidence of a deep origin of the brines: "*Helium and argon isotopes measured in the brine have a mantle-derived origin* (*Lupton* et al. 1977*; Winckler* et al. 2001), *which confirms that fluids that feed the Atlantis II Deep have interacted with newly injected fresh basaltic material*" (Anschutz 2015).

And, finally, he discusses the interesting saturation state of the brine: "*The saturation indices reveal that the brine is under-saturated with respect to halite, gypsum, and anhydrite at* in situ *temperature* (*Monnin and Ramboz* 1996*; Anschutz* et al. 2000), *suggesting that the major salt concentrations are not controlled by* in situ *equilibrium of the brine pool with respect to salt minerals. The brine is close to equilibrium with halite at a temperature close to that of Red Sea bottom water* (22 °C), *which supports the hypothesis that the salinity of the incoming brine has been reached through equilibrium with evaporite minerals at seawater temperature before heating of the solution through fractured Red Sea basaltic crust*" (Anschutz 2015). However, he does not discuss any of the other high-solubility (high-magnesium) salts, such as bischofite. According to our model they may escape through the reflux zone.

3.6 Other Deeps

The northernmost active Deeps in the Red Sea are the Conrad, Kebrit and Shaban Deeps, located up to 500 km north of the Atlantis II Deep (Blum and Puchelt 1991; Ehrhardt and Hübscher 2015). According to Bonatti (1985), this part of the Red Sea consists predominantly of stretched and thinned continental crust with a few tectonically formed isolated Deeps that mark initial stages of seafloor spreading with local emplacement of oceanic crust (e.g., Stages B and C in Fig. 5 (Norton et al. 2015)). Recent bathymetry and shallow seismic surveys of these three deeps have revealed some interesting features regarding precipitation and dissolution of salt. Ehrhardt and Hübscher (2015) suggest that these data provide information about how thick layers of salts may dissolve, deform, flow, re-precipitate, collapse, and sediments become 'altered'. This is mainly due to hydrothermal circulation, causing 'subrosion' (under-ground salt dissolution) and re-precipitation of salts.

According to our model, we offer other interpretations to some of these findings, and question:

(1) why do the Deeps form in the first place?
(2) what is the nature of the "*sediment alteration*"?

The answer to question (1) can simply be that the Deeps are local topographical lows in the seafloor that form as a consequence of a combination of special physical, thermodynamic, and lithologic conditions. Ehrhardt and Hübscher (2015) term it "*subrosion*". However, their more descriptive term, "*collapse structure formation*" suggests that these are lows caused by the collapse of the seafloor, due partly to local removal of mass by salt dissolution and/or subrosion. We suggest that where the Deeps occur, it is rather a more complex combination of causes. To us, the Deeps may signify actively cooling hydrothermal cells, which most probably only occur immediately above the shallowest portion of the igneous intrusions, where the upward-directed reflux zone not only acts as a conduit for hot, phase separated condensation water, but also a rich mixture of minerals, including clays, volcanic fragments and precipitating brines. We infer that a stockwork conduit-system, crack/fissure/brecchia structure, experiences pressure build-up in periods. In other words, salt and other minerals are moving out of the deep brine reservoirs of the hydrothermal systems to the seafloor (model steps 4–5). This process actually causes a form of seafloor collapse, because the ocean floor subsides, when salt moves from immediately below the surface to above it.

With respect to question (2), it is inferred that "*sediment alteration*" means that sediment particles, such as feldspars, have been dissolved (by hot hydrothermal fluids) and re-crystallized to form clay minerals. It is believed that such processes occur fast, driven by superheated water. New minerals, including clays, may also form by interaction between such hot water and gabbro or peridotite in serpentinization reactions in deeper parts of the hydrothermal circulation cell (Fryer and Fryer 1987; Hübner et al. 2003).

Thus, there is no longer any doubt that the high heat-flow and the hydrothermal processes in the Red Sea Deeps contribute significantly to the production of saturated brines on the sea floor. The question is if these brines have any relevance to the formation of solid salt accumulations.

According to model steps 3 and 4, it is envisioned that when the hydrothermally formed brines cool, solid salt will precipitate, for example, equivalent to processes known to occur in the Dead Sea (Warren 2010). In this way, it could be said that the ongoing 'collapse' or subrosion actually is caused by the displacement of salt, via solutes (brines) up from depth and re-precipitation upon cooling on the ocean floor, perhaps in the shape of particulate salt flows, as demonstrated by Mitchell et al. (2011) in the Thetis Deep and elsewhere.

4 Salt Walls and Salt Ridges

4.1 Northern Red Sea

In the northern Red Sea, Mart and Ross (1987) described numerous 'diapirs' occurring in water depths ranging from 700 to 1300 m (Fig. 9). They are typically elongated, attaining lengths exceeding 40 km and widths of 4 km, and they are either exposed or covered by up to one km

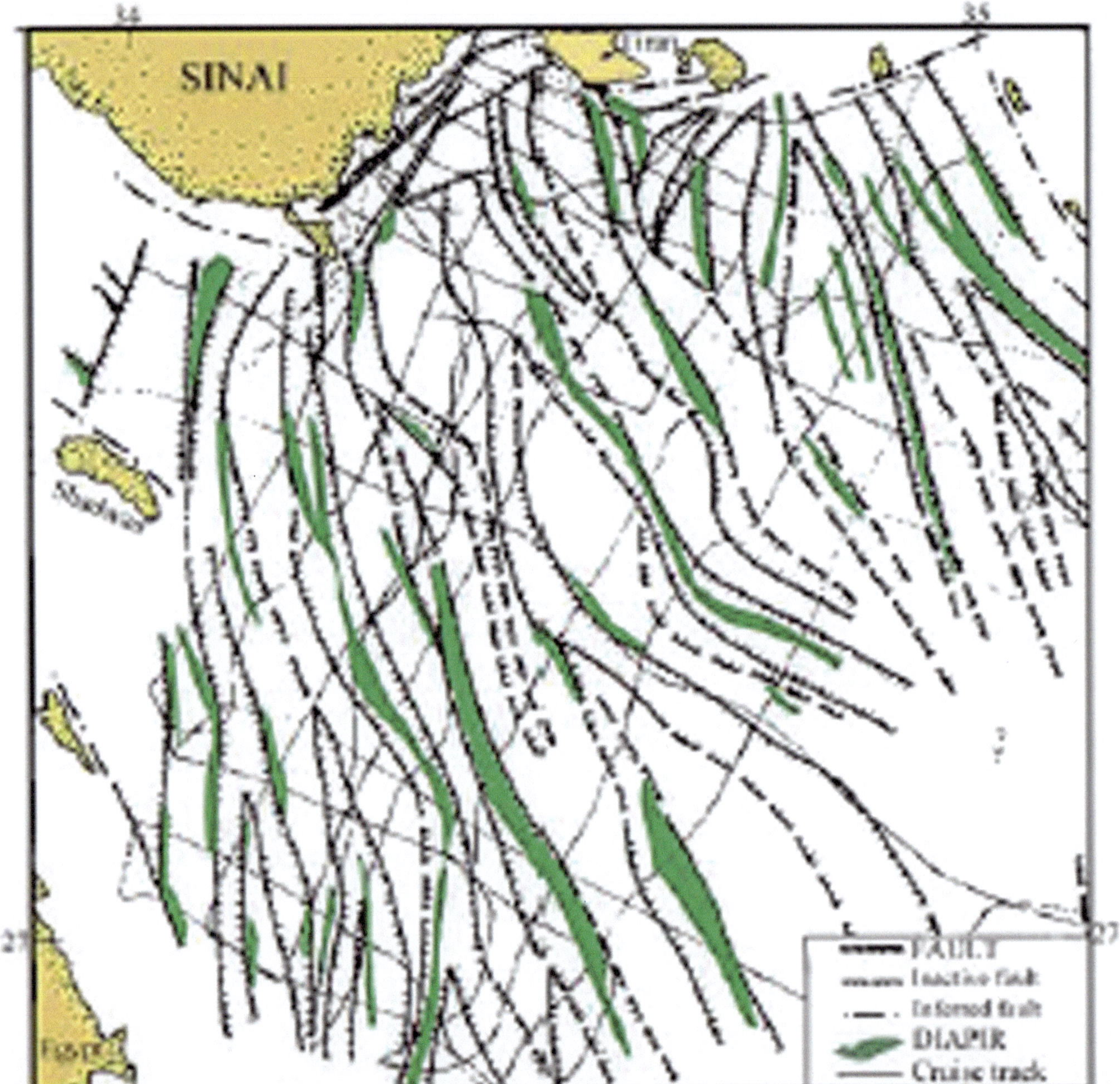

Fig. 9 Map of up to 40 km long exposed salt-cored piercement ridges (in green) in the northern Red Sea (from Mart and Ross 1987). Note that the ridges are generally parallel with the spreading axis of the northern Red Sea. This suggests that they form along deep faults associated with the tectonic extension process

overburden (Fig. 9). These 'diapirs' commonly have formed either along the boundary faults of extensional rift half-grabens or inside the rift floors. One of the most pronounced positive features is about 500 m wide and rises at least 250 m above the sediment-covered seabed. Some 'diapirs' are also topped by negative 'collapse' structures, one of which measures about 2 km across and 75 m in depth. Mart and Ross (1987) interpreted these to represent salt dissolution pits.

Because of their anomalous appearance, Mart and Ross (1987) question if these structures represent normal salt diapirs or not (Figs. 10 and 11). They are very difficult to understand using the conventional halokinetic explanations and interpretations, particularly because they seem to have been 'squeezed' up into the water column without being severely leached. Another possibility is that they represent viscous pressure ridges ('salt injectites') with surface liquid outflow and corresponding crater formation (Hovland et al. 2015). But these normally only occur in compressional settings, involving sedimentation onto salt layers (Ings and Beaumont 2010).

However, as a consequence of the hydrothermal salt theory, we suggest they represent ductile and over-pressured salt masses originating underground, which have been pushed upward from the subsurface formation zones along active deep faults associated with tectonic extension. Thus, we interpret these structures as salt injectites (extrudites) caused by high fluid pressure and the lack of accommodation space in the subsurface, where solid salt is suspected to be produced hydrothermally (Hovland et al. 2015).

The exposed portions of these salt piercement structures are probably protected against re-dissolution in seawater by a cap of anhydrite/gypsum and/or sediments. The locally observed depressions in their summit either represent dissolution craters, where the capping sediments or anhydrites have been breached (Mart and Ross 1987), or more probable expulsion craters, where fluids emanate from a conduit and locally erode the outlet areas (Hovland et al. 2015).

In our view, these structures are inferred to represent hydrothermally associated salt ridges, some of which, or parts of which, may be acting as fluid conduits in a high heat-flow tectonic setting. We have already described the formation of these piercement structures in our model step 5, with analogies to other salt piercement structures elsewhere. As these piercement features have, to our knowledge, not been explained in this manner previously, we infer that their formation is highly relevant for understanding underground accumulation of hydrothermal solid salts in the Red Sea and elsewhere, where subsurface rift-related salt masses have been documented by seismic and gravity investigations (Miller et al. 2009; Hovland et al. 2015).

4.2 Ridges and Walls in the Deeps

Although there are numerous salt-cored ridges along the flanks of the northern Red Sea, similar ridges of apparently massive salt were also discovered crossing over one of the hot spreading centres (Deeps) of the Red Sea. According to Ehrhardt et al. (2005) and Ehrhardt and Hübscher (2015), the Conrad Deep in the northern Red Sea is a poorly understood basin (Deep), associated with rifting and shear accommodation. It has a series of anomalous features, some of which Ehrhardt et al. (2005) and Ehrhardt and Hübscher (2015) suspected to be salt diapirism, by applying the conventional evaporite model.

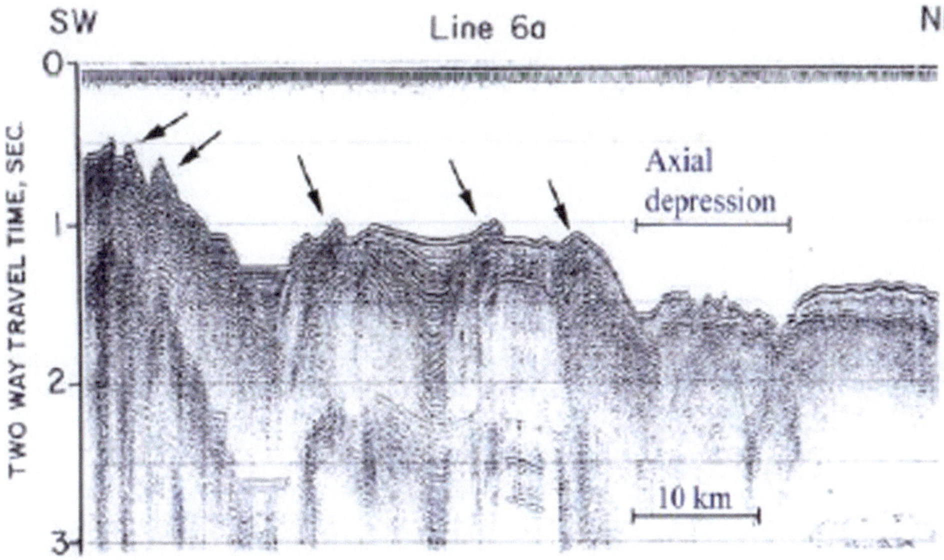

Fig. 10 Reflection seismic line no 6a, modified from Cochran (2005), showing 'salt diapirs' or piercement salt ridges (arrowed) in the northern Red Sea. They correspond to the salt-covered piercement ridges shown in Fig. 9

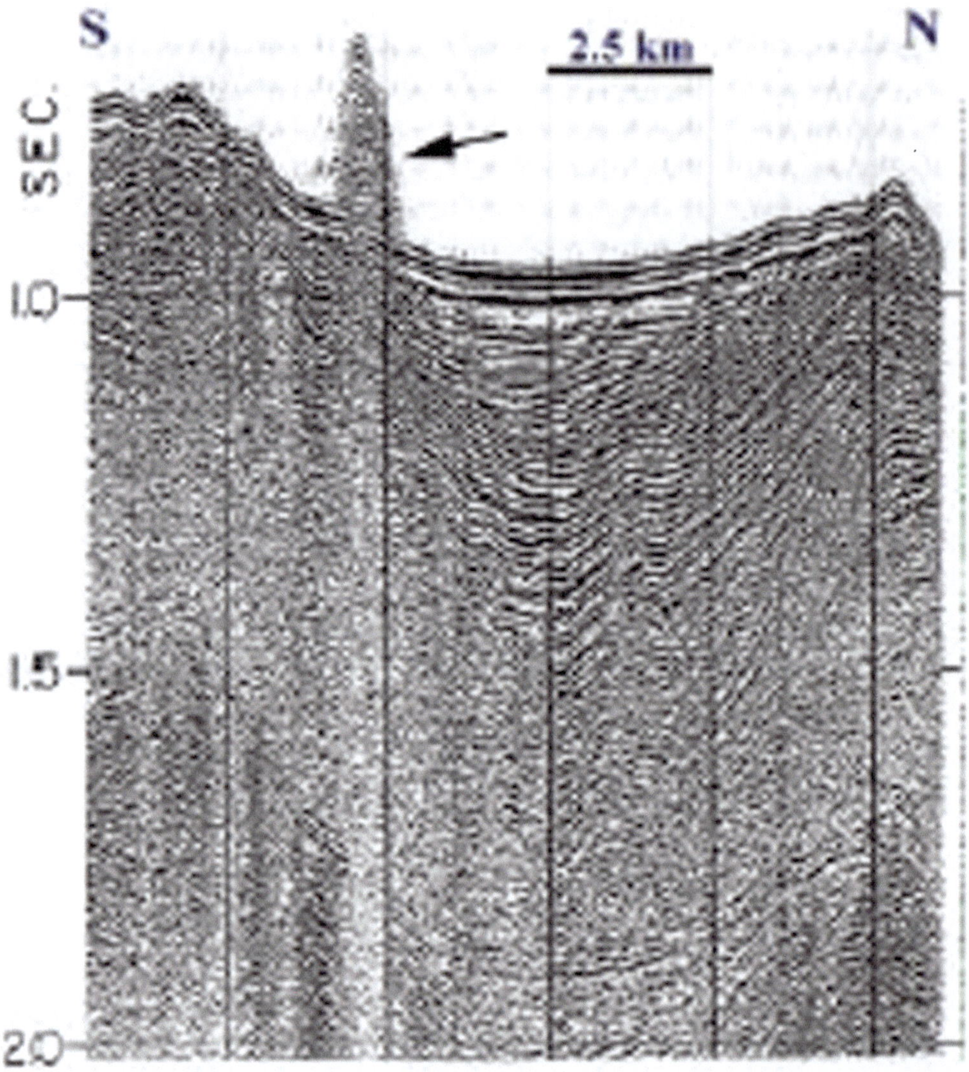

Fig. 11 A 2D-seismic section from the northern Red Sea (based on Mart and Ross 1987). The image crosses over one of the many Ridges, consisting of salt (probably halite with an outer layer of sediments). The Ridge rises about 250 m above the surrounding seafloor and its base is about 500 m wide

On 2D-seismic data over this area, there is a total salt (evaporite) deposit of 3000 m thickness (Cochran et al. 1986). The deposit is characterized by having a high amplitude low frequency reflection, located 200 to 300 ms TWT (two-way-time) below the seafloor. Depending on the facies at the upper part of the Miocene salts (halite, anhydrite, or gypsum) the internal reflectivity of this sediment body alternates from seismically transparent (amorphous) to a layered pattern (Ehrhardt et al. 2005; Ehrhardt and Hübscher 2015; Ligi et al. 2015).

Ehrhardt and Hübscher (2015) suggest that the lateral change of evaporite facies (e.g., from amorphous/transparent to stratified) is due to halite folding (diapirism) and deposition of other salts, for example gypsum, carbonate, and anhydrite (stratified). We agree that the halite bodies are acoustically transparent, but suggest that the seismic transparency of halite may be caused by both depositional (e.g., hydrothermal processes) and tectonic processes (Hovland et al. 2015).

The Conrad Deep is characterized by steep slopes on both flanks (9–18°), and is composed of at least two sub-basins, of which the south-western is shallower and separated from the deeper north-eastern basin. A brine layer was detected by Ehrhardt et al. (2005) in both these basins. Salt diapirs or volcanic extrusions are located at the southern flank of the Deep (Fig. 12). They are 150 to 190 m higher than the surrounding seafloor of the main trough. The elongated rise that strikes parallel to the Deep is called the '*SE-Rise*' by Ehrhardt et al. (2005) and is concluded to be an uplift related to the emplacement of a magmatic body at the flanks of the Conrad Deep axial depression.

Another feature of the Conrad Deep is suspected to represent the development of an elongated diapiric wall, called the '*Salt Wall*' (Ehrhardt et al. 2005). It runs obliquely to the strike of the Conrad Deep's SE-shoulder (Figs. 12 and 13) and strikes N34°E along its SW-extension into the Deep and acts partly as the SE slope of the Deep (Fig. 13). This rise must have occurred most recently and is still active. Thus, it

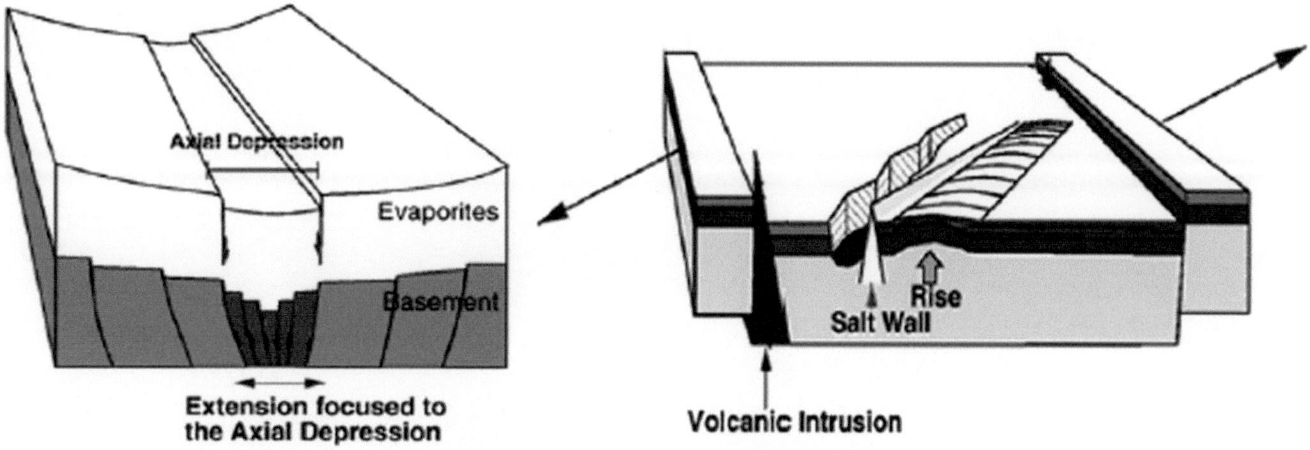

Fig. 12 Conceptual sketch of the Conrad Deep, modified from Ehrhardt et al. (2005). On the left is a general drawing of the axial depression showing that it basically is a graben structure in the thick evaporites over the downfaulted basinal axial depression. On the right is a conceptual close-up sketch of the Conrad Deep features (Ehrhardt et al. 2005). See also text for further interpretations

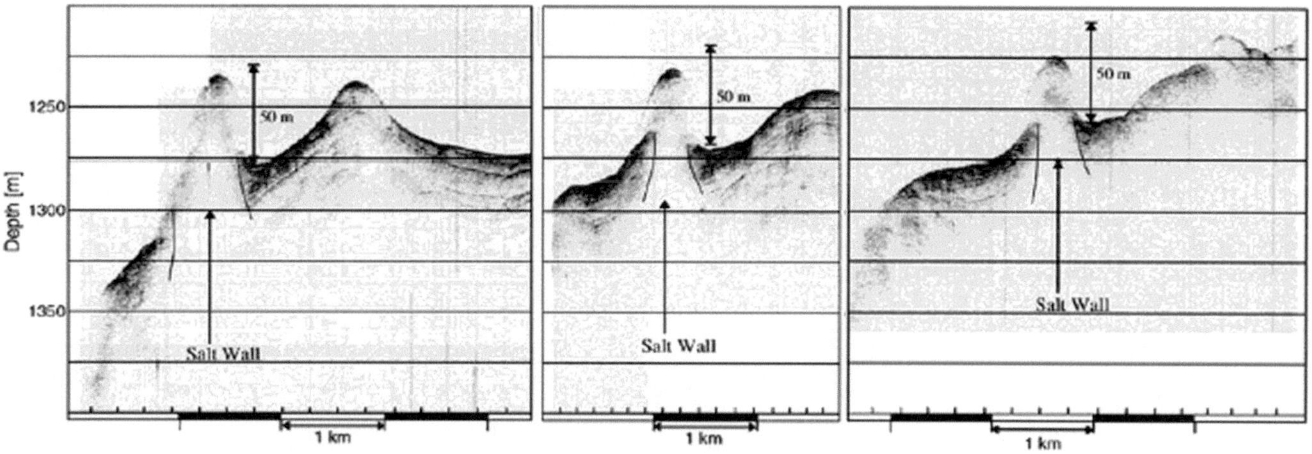

Fig. 13 Three consecutive high-resolution shallow seismic (parasound) reflection records across the salt wall and salt rise described by Ehrhardt et al. (2005). Each of the lines was run in a NW–SE direction, and they are shown from south to north (left to right, the southernmost line seen on the left) (Modified from Fig. 12 of Ehrhardt et al. 2005). See also previous figure, which shows the relation of this salt wall to the Conrad Deep

is inferred by Ehrhardt et al. (2005) and Ehrhardt and Hübscher (2015) that 'salt diapirism' is still active on the flanks of the Conrad Deep (Figs. 12 and 13). However, in our interpretation, this is evidence of on-going underground subrosion combined with hydrothermal salt production and storage in the deep 'brine cell' (Fig. 1), caused by phase separation in model steps 1 and 2. However, at steps 3–4, the whole system starts to cool, resulting in deep saturated brines precipitating solid salts and, therefore, causing subsurface accommodation problems and pressure build-up. The pressure builds up so much that the rocks fail along an inferred oblique weak shear-stress zone, and conduits form. This facilitates accommodation space and a vertical salt flow conduit, where overpressured salt masses can escape in the form of an extrudite. In our view, this is how the salt walls and ridges may form. Thus, the surface expression of underground salt accommodation is rock salt extrusion or salt injection. In our model, the combination of salt precipitation from an ascending brine causes clogging and overpressures that lead to upward migration of solid salts above (model steps 3–5). We thus infer that the anomalies documented by Ehrhardt et al. (2005) and Ehrhardt and Hübscher (2015) are mainly the results of locally produced hydrothermal salt, underground. The '*SE-Rise*' and '*Salt Wall*' structures could, therefore, probably more aptly be termed salt injectites or extrusives (Fig. 14).

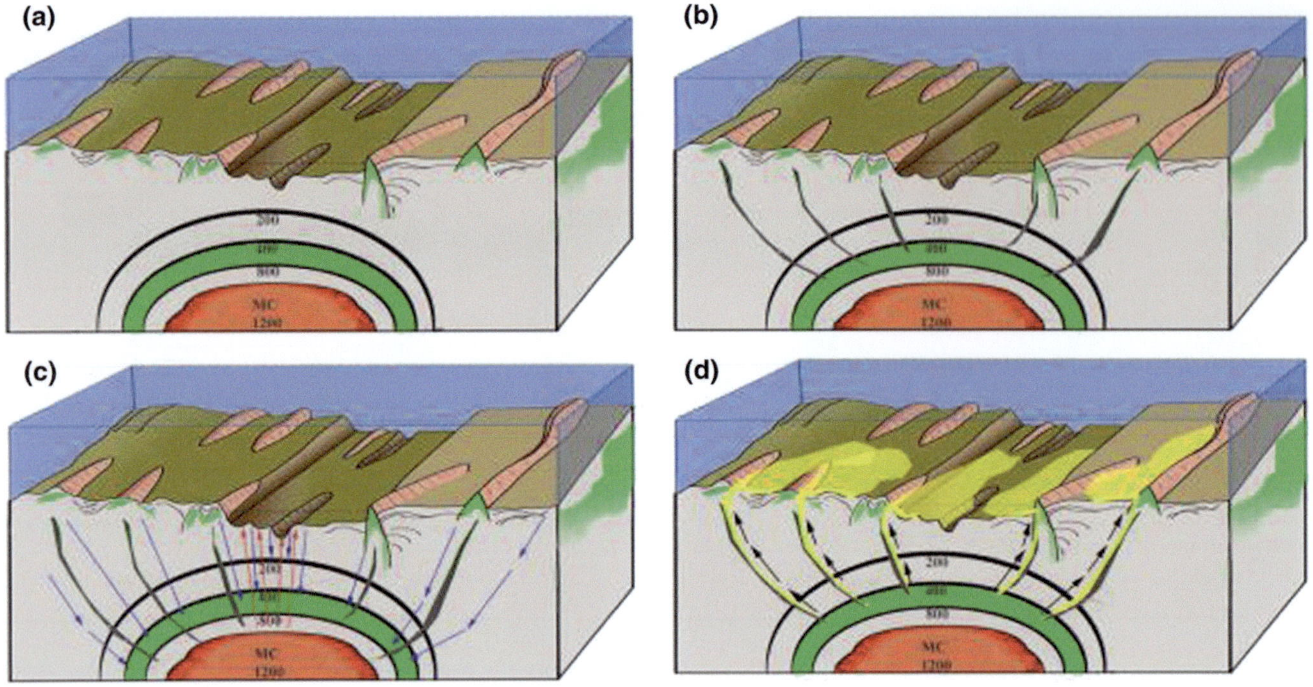

Fig. 14 Conceptual sketches based on new knowledge from Deeps in the Red Sea (e.g., Ehrhardt et al. 2005; Cochran 2005; Hovland et al. 2015; Augustin et al. 2016). It shows **a** inferred isotherms 200 °C, 400 °C, and 800 °C, and a shallow magma chamber 'MC' (e.g., igneous intrusion) with an inferred temperature of 1200 °C (model steps 1–3). **b** Inferred normal faults or weakness zones, caused by tectonic extension. **c** Inferred hydrothermal water circulation. Blue arrows: recharge; Red arrows: reflux (model steps 3–5). **d** Inferred under-ground transport pathways (black arrows) for precipitated salt particles (yellow) from the hydrothermal 'out-salting'/'Brine cell' zone (green) upward along normal faults (model steps 4–5). The surfacing salt slurry is then expected to move by gravitational forces down-slope toward the central graben and Deeps (yellow: Salt Flows)

4.3 Salt Flows

Mitchell et al. (2010) discovered submarine salt flows in the Thetis Deep of the central Red Sea. Using swathe multi beam echo-sounder mapping, they interpreted the flow-structured surface features on the seafloor as mobile carpets of salt. This interpretation has become more detailed recently, as it has been found that the salt carpets may flow at a velocity of more than 1 m per year (Feldens and Mitchell 2015). These surprisingly large flow-like features, estimated to have reliefs of 200–500 m, were also identified on topographic data (Augustin et al. 2014, 2016). Augustin et al. (2014), calculated their greatest thickness at 800 m in the Nereus-Thetis Deep Inter-Trough-Zone (ITZ). Because the regional upper hemipelagic sediment layer is much thinner than 200 m, the only possibility for understanding these features is that they represent 'fluidized evaporites' or 'salt flows' as named by Mitchell et al. (2010, 2011).

The enormous extent of the moving salt flows was mapped by Augustin et al. (2014, 2016). They infer that the triggering mechanism is a combination of dissolution by seawater and tectonic extension. Furthermore, they term the salt flows as 'submarine namakiers', named after plastic subaerial movement of glacier-like rock salt flows down the 'Kuh-i-Namak' and other salt domes in the Dashti region of Iran (Lees 1931; Talbot 1998; Augustin et al. 2014, 2016).

Elsewhere along the axial zone of the Red Sea, Augustin et al. (2014) find similar salt flows especially well developed in the ITZs, where: *"...long-wavelength (relative to the hummocky terrain in the Deeps) linear ridges are visible in many regions. These appear to be similar to flow-structures described by Mitchell* et al. (2010) *from the Thetis Deep region, which they interpreted as resulting from flow in salt, producing salt glaciers (namakiers). The local topographic slope and the orientation of the longitudinal ridges and troughs on the surface of the namakiers allow a sense of flow to be determined. Where two namakiers flowing from opposite sides of the rift meet, a central cleft or suture is generally visible. The bathymetric data also show namakier flow ridges apparently deviating around volcanoes and volcanic ridges on the rift flanks, suggesting that these are influencing namakier flow*" (Augustin et al. 2014).

Our model steps 4 and 5 predict that large quantities of hydrothermally produced particulate salt will precipitate underground in the central sections of the Red Sea. Therefore, the observation of salt flows is taken as another indication that this model is realistic. However, it was rather unsuspected that the flows are so extensive and that they seem to emanate from the sides of the central graben, rather than from point sources inside the graben. The longest distance from the central axis of the Red Sea Rift to their source area is up to 30 km (Augustin et al. 2014, 2016). This suggests that the flows are both composed of newly precipitated surfacing salt slurries together with older salt layers and sediments that have been disrupted and mobilized, as suggested by Mitchell et al. (2010), Feldens and Mitchell (2015), and Augustin et al. (2014, 2016).

According to Feldens and Mitchell (2015), the flowing material is suspected to consist of a dispersion of particulate solid salt mixed with clastic sediment particles in concentrated brine, for example a salt slurry. We would add the possibility of a concentrated cooling and precipitating brine. Mitchell et al. (2010) and Feldens and Mitchell (2015) interpret these salt flows to occur as a consequence of Miocene evaporite dissolution by hydrothermal activity. A similar view is also that of Augustin et al. (2014, 2016), although, here, hydrothermal processes are not inferred to have a central role.

Some of these salt flows occur as over 30 km long drifts along the central valley of the Red Sea (graben trough). From their morphology, they can be described as flow deposits, blankets, or drifts. The flows are recognized by their smooth surface morphology which contrasts with the rugged morphology typical of exposed volcanic ridges and hydrothermal craters and cones on the floor of the Deeps (Fig. 15). One of the salt flows that apparently issues out of the eastern wall of the Thetis Deep terminates inside that Deep, and is named 'Flow B' (Fig. 15; Mitchell et al. 2010).

However, there is an apparent anomaly associated with the Thetis Deep: Despite it being a locally enclosed Deep, totally surrounded by higher ground, there are no dense brines trapped inside it (Pierret et al. 2010; Mitchell et al. 2011). This is strange for two reasons: a) there are metal-rich sediments at the bottom of the Deep, suspected of having been deposited by concentrated, hydrothermal brines (Pierret et al. 2010), and there are salt flows that terminate inside the Deep (Mitchell et al. 2011), without the enclosed body of still water being affected by an increase in salinity: "*Ultimately, the lack of a brine lake here may therefore have a structural origin, if tectonic movements on boundary wall faults are too small and/or infrequent significantly to offset the hemipelagic and evaporite beds to allow dissolution, while evaporite flowage and seabed processes acting on surface hemipelagic sediments are continually healing such offsets to prevent water circulation*" (Mitchell et al. 2011).

In contrast to this conclusion, we offer a totally different explanation, one that is a consequence of active hydrothermal activity inside (or beneath) the Thetis Deep. In the hydrothermal salt model, there is a requirement for the release and loss of relatively large quantities of condensation water from the supercritical or boiling zone of the hydrothermal system. Condensation (distilled) water vapor is hot and devoid of solutes. Therefore, it is, besides any gases produced, actually the lightest substance occurring in the hydrothermal system. It is therefore suspected that it escapes upward in the 'reflux zone', which is located immediately above the hottest zone of the system, for example, through the deepest portion of the Deeps (see Fig. 14c). This is also supported by Lecumberri-Sanchez et al. (2015), who write: "*The coexisting vapor phase, which ascends more rapidly than the liquid, would transport sulfur and gold upward*". "*Salt precipitation appears to be a ubiquitous feature in hydrous volcanic and geothermal environments, invoking dynamic behavior due to rapid multiphase reactions involving liquids, vapors, and solids.*"

There is no doubt that this 'fresh', hot water is highly corrosive to any nearby body of salt, including salt flows. In

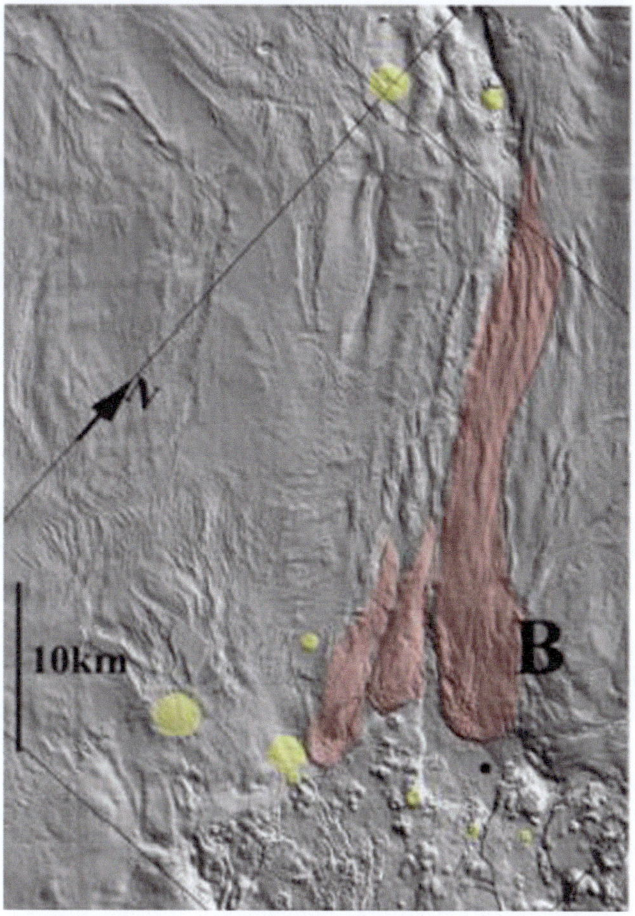

Fig. 15 Shaded relief image of bathymetric data of the central graben of the Thetis Deep (modified from Mitchell et al. 2010). The image shows evidence of massive Salt Flows. The Thetis Deep is seen as the flat-floored, rugged depression in the lower right. Three of the large Salt Flow lobes (according to Mitchell et al. 2010) have been highlighted here (in pink). Note the Salt Flow (named 'B' by Mitchell et al. 2010) flowing down the slope on the right, foreground. The salt fronts are lobate and relatively steep, where they terminate in the deepest portions of the Thetis Deep (from Hovland et al. 2015). The circular yellow structures are inferred volcanic domes or hydrothermal vents

our view, the escape of condensation water through some of the Deeps, including the Thetis Deep, may hinder the formation of a concentrated brine layer, and may also rapidly dissolve salt in salt flows (such as Flow B) that happen to intercept their upward flow path. This dynamic and relatively rapid process may also explain the steep lobate forms of the terminations of the salt flows (e.g., their 'fronts'). In the description by Mitchell et al. (2011) of the Thetis Deep, they infer that there is active hydrothermal activity there: "*Perhaps because any volcanic and/or hydrothermal activity is fortuitously located away from the evaporite beds (cones to the SW and SE) and because of the hemipelagic cover, relatively little of the halite has been dissolving and contributing to a brine layer in the Thetis Deep in the geologically recent past*".

Further support for the suggestion that the mid-central graben of the Red Sea may contain an upward directed corrosive condensation water flux is gained from the numerical modelling performed by Coumou et al. (2008). Thus, as also shown in Fig. 3, it is indeed seen that the relatively high-temperature and low-salinity water escapes above the centre of the subsurface heat source. Their most surprising result, however, is that the upflow zones are associated with closely spaced recharging cells: "*Using 3D numerical simulations, we demonstrated that convection cells self-organize into pipelike upflow zones surrounded by narrow zones of focused and relatively warm downflow. This configuration ensures optimal heat transfer and efficient metal leaching for ore-deposit formation*" (Coumou et al. 2008).

Another mechanism that may disrupt thick layers of salts is the movement and emplacement of volcanic material, including sills and dykes. Salt flowage triggered by sills was therefore studied by Underhill (2009), in the greater Silverpit area of the North Sea. He concluded that igneous dyke intrusion below the salt could explain the onset of halokinesis. This could be done by inducing possible salt melting, aided by supercritical waters, which caused the viscosity of the salt to be lowered: "*...explained by the presence of a gas- or fluid-release pipe above deeply buried dykes and sills, similar to intrusions mapped in other areas. Furthermore, the occurrence of an intrusion or pipe would provide a simple explanation for the occurrence of the so-called 'central uplift' affecting the Jurassic stratigraphic levels beneath the Silverpit Crater syncline and explain the small radial faults emanating from it too*" (Underhill 2009). In our view, we do not need to include molten salt (halite melts at around 800 °C) to explain the salt flows, although we cannot rule out that it may also represent a (minor) factor.

5 Conclusions

Our main conclusion is that the Red Sea represents a giant natural 'salt-factory', which constantly produces new volumes of brines and solid salts underground, and which currently manages to accumulate and protect giant amounts of salt formations in the sea from re-dissolution in seawater. The process is driven by the high heat-flow caused by shallow igneous intrusions in the active oceanic rifting process. The salt producing zone is focused around the central rifting axis and represents about 15% of the entire Red Sea. The brines and solid salts are produced by boiling and supercritical phase change in the deep forced convection cells (hydrothermal systems) located above shallow igneous intrusions in the Deeps along the spreading axis. It is a robust self-organizing and self-sustaining system, which functions as a consequence of local high heat-flow and gravitationally stable water stratification.

To our knowledge, the Red Sea is currently the only place on Earth where highly concentrated brines and solid salt can migrate and form salt bodies inside the sediments, below the high-temperature grabens and protected by dense, warm brines and sediments of the Deeps. Thus, in the Red Sea axial trough, dense brines recharge into the hydrothermal systems and become refined for each new circulation, until solid salt precipitates, both underground and on the seafloor as the saturated brines cool down.

Acknowledgements We would like to thank Saudi Geological Survey, for their initiative and invitation to participate in the Red Sea book project, Charlotte B. Schreiber for keeping us updated on the Messinian Salinity Crisis debate, and, not least, Vittorio Scribano, Fabio Manuella and an anonymous reviewer for constructive review comments and suggestions. Najeeb Rasul is thanked for his efforts in compiling the Red Sea book.

References

Aharon P, Roberts HH, Snelling R (1992) Submarine venting of brines in the deep Gulf of Mexico: observations and geochemistry. Geology 20:483–486

Anschutz P, Blanc G, Chatin F, Geiller M, Pierret M-C (1999) Hydrographic changes during 20 years in the brine-filled basins of the Red Sea. Deep Sea Res Part I: Oceanographic Res Papers 46 (10):1779–1792

Anschutz P (2015) Hydrothermal activity and paleoenvironments of the Atlantis II Deep. In: Rasul NMA, Stewart ICF (eds) The Red Sea: The formation, morphology, oceanography and environment of a young ocean basin. Springer Earth System Sciences, Berlin Heidelberg, pp 235–249. https://doi.org/10.1007/978-3-662-45201-1_14

Anschutz P, Blanc G, Monnin C, Boulègue J (2000) Geochemical dynamics of the Atlantis II Deep (Red Sea). II: pore water composition of metalliferous sediments. Geochim Cosmochim Acta 64:3995–4006

Augustin N, Devey CW, van der Zwan FM, Feldens P, Tominaga M, Bantan RA, Kwasnitschka T (2014) The rifting to spreading transition in the Red Sea. Earth Planet Sci Lett. https://doi.org/10.1016/j.epsl.2014.03.047

Augustin N, van der Zwan FM, Devey CW, Ligi M, Kwasnitschka T, Feldens P, Bantan RA, Basaham AS (2016) Geomorphology of the Central Red Sea Rift: Determining spreading processes. J Geomorphology. https://doi.org/10.1016/j.geomorph.2016.08.028

Barberi F, Varet J (1971) The Erta Ale volcanic range (Danakil depression, northern Afar, Ethiopia). Bull Volcanology 34(4):848–917. https://doi.org/10.1007/BF02596805

Bischoff JL, Rosenbauer RJ (1989) Salinity variations in submarine hydrothermal systems by layered double-diffusive convection. J Geol 97:613–623

Blum N, Puchelt H (1991) Sedimentary-hosted polymetallic massive sulfide deposits of the Kebrit and Shaban Deeps, Red Sea. Miner Deposita 26:217–227

Bonatti E (1985) Punctiform initiation of seafloor spreading in the Red Sea during transition from a continent to an oceanic rift. Nature 316:33–37

Bonatti E, Cipriani A, Lupi L (2015) The Red Sea: Birth of an ocean. In: Rasul NMA, Stewart ICF (eds) The Red Sea: The formation, morphology, oceanography and environment of a young ocean basin. Springer Earth System Sciences, Berlin Heidelberg, pp 29–44

Bosworth W (2015) Geological evolution of the Red Sea: Historical background, review, and synthesis. In: Rasul NMA, Stewart ICF (eds) The Red Sea: The formation, morphology, oceanography and environment of a young ocean basin. Springer Earth System Sciences, Berlin Heidelberg, pp 45–78

Bower A (2009) Field report. R/V Oceanus Voyage 449–6 Red Sea. WHOI-KAUST-CTR 2009-1, 40 pp

Brooks RR, Kaplan IR, Peterson MNA (1969) Trace element composition of Red Sea geothermal brine and interstitial water. In: Degens ET, Ross DA (eds) Hot brines and recent heavy metal deposits in the Red Sea. Springer, pp 180–203

Cochran JR (2005) Northern Red Sea: Nucleation of an oceanic spreading center within a continental rift. Geochem Geophys Geosys, AGU https://doi.org/10.1029/2004GC000826, Q03006

Cochran JR, Martinez F, Steckler MS, Hobart MA (1986) Conrad Deep: A new Northern Red Sea Deep. Origin and implications for continental rifting. Earth Planet Sci Lett 78:18–32

Coumou D, Driesner T, Heinrich CA (2008) The structure and dynamics of mid-ocean ridge hydrothermal systems. Science 321:1825–1828

Coumou D, Driesner T, Weis P, Heinrich CA (2009) Phase separation, brine formation, and salinity variation at Black Smoker hydrothermal systems. J Geophys Res 114:B03212

Craig H (1969) Geochemistry and origin of the Red Sea brines. In: Degens ET, Ross DA (eds) Hot brines and recent heavy metal deposits in the Red Sea. Springer, pp 208–242

Davison I, Anderson L, Nuttall P (2012) Salt deposition, loading and gravity drainage in the Campos and Santos salt basins. In: Alsop GI, Archer SG, Hartley AJ, Grant NT, Hodgkinson R (eds) Salt tectonics, sediments and prospectivity. Geol Soc London Spec Publ 363, pp 159–174. doi:10.1144 /SP363.8

Degens ET, Ross DA (1969) Hot brines and recent heavy metal deposits in the Red Sea. Springer Verlag, New York, pp 535–541

Dias JL (2005) Tectônica, estratigrafia e sedimentação no Andar Aptiano da margem leste Brasileira: Boletim de Geociencias Petrobras 13, pp 7–25

Driesner T (2007) The system H2O–NaCl. Part II: Correlations for molar volume, enthalpy, and isobaric heat capacity from 0 to 1000 & #xB0;C, 1 to 5000 bar, and 0 to 1 XNaCl. Geochim Cosmochim Acta 71:4902–4919

Driesner T, Heinrich CA (2007) The system NaCl-H2O. I. Correlation formulae for phase relations in temperature-pressure-composition space from 0 to 1000 & #xB0;C, 0 to 5000 bar, and 0 to 1 XNaCl. Geochim Cosmochim Acta 71:4880–4901

Ehrhardt A, Hübscher C (2015) The northern Red Sea in transition from rifting to drifting—Lessons learned from ocean deeps. In: Rasul NMA, Stewart ICF (eds) The Red Sea: The formation, morphology, oceanography and environment of a young ocean basin. Springer Earth System Sciences, Berlin Heidelberg, pp 99–121

Ehrhardt A, Hübscher C, Gajewski D (2005) Conrad Deep, northern Red Sea: Development of an early stage ocean deep within the axial depression. Tectonophysics 411:19–40

Ehgartner B, Neal J, Hinkebein T (1998) Gas release from salt. Nec Researchindex, www.citeseer.nj.nec.com

Erickson AJ, Simmons G (1969) Thermal measurements in the Red Sea hot brine pools. In: Degens ET, Ross DA (eds) Hot brines and recent heavy metal deposits in the Red Sea. Springer, pp 114–121

Feldens P, Mitchell NC (2015) Salt flows in the Central Red Sea. In: Rasul NMA, Stewart ICF (eds) The Red Sea: The formation, morphology, oceanography and environment of a young ocean basin. Springer Earth System Sciences, pp 205–218. https://doi.org/10.1007/978-3-662-45201-1_14

Feldens P, Schmidt M, Mücke I, Augustin N, Al-Farawati R, Faber E (2016) Expelled subsalt fluids form a pockmark field in the eastern Red Sea. Geo-Marine Lett 36(5):339–352. https://doi.org/10.1007/s00367016-0451-9

Flemings PB, Behrmann I, John C (2005) Gulf of Mexico hydrogeology-overpressure and slope stability, seeps, and shallow-water flow. IODP Scientific Prospectus 308. http://iodp.tamu.edu/publications/SP/308SP/308SP.PDF

Fontaine FJ, Wilcock WSD (2006) Dynamics and storage of brine in mid-ocean ridge hydrothermal systems. J Geophys Res 111: B06102. https://doi.org/10.1029/2005JB003866

Fryer P, Fryer GJ (1987) Origins of nonvolcanic seamounts in a forearc environment. In: Keating B, Fryer P, Batiza R (eds) Seamounts, Islands, and Atolls. Am Geophys Union, Geophys Monograph 43, pp 61–69

Gay A, Takano Y, Gilhooly WP III, Berndt C, Heeschen K, Suzuki N, Saegusa S, Nakagawa F, Tsunogai U, Jiang SY, Lopez M (2011) Geophysical and geochemical evidence of large scale fluid flow within shallow sediments in the eastern Gulf of Mexico, offshore Louisiana. Geofluids 11:34–47

Ghanbarzadeh S, Hesse MA, Prodanovic M, Gardner JE (2015) Deformation-assisted fluid percolation in rock salt. Science 350 (6264):1069–1072. https://doi.org/10.1126/science.aac8747

Girdler RW (1984) The evolution of the Gulf of Aden and Red Sea in space and time. Deep-Sea Res 31:747–762

Girdler RW (1985) Problems concerning the evolution of oceanic lithosphere in the northern Red Sea. Tectonophysics 116:7–11

Girdler RW, Whitmarsh RB (1974) Miocene evaporites in Red Sea cores, their relevance to the problem of the width and age of oceanic crust beneath the Red Sea. In: Whitmarsh RB, Weser PE, Ross DA (eds) Initial reports of the Deep Sea Drilling Project 23. US Government Printing Office, Washington, pp 913–921

Gruen G, Weis P, Driesner T, Heinrich CA, de Ronde CE (2014) Hydrodynamic modeling of magmatic–hydrothermal activity at submarine arc volcanoes, with implications for ore formation. Earth Planet Sci Lett 404:307–318. https://doi.org/10.1016/j.epsl.2014.07.041

Heine C, Zoethout J, Müller RD (2013) Kinematics of the South Atlantic rift. Solid Earth :215–253. https://doi.org/10.5194/se-4-215-2013

Holness MB, Lewis S (1997) The structure of halite-brine interface inferred from pressure and temperature variations of equilibrium dihedral angles in the halite-H_2O-CO_2 system. Geochim Cosmochim Acta 61(4):795–804

Holwerda JG, Hutchinson RW (1968) Potash-bearing evaporites in the Danakil area. Ethiopia. Econ Geol 63(2):124–150

Hovland M, Jensen S, Fichler C (2012) Methane and minor oil macro-seep systems — Their complexity and environmental significance. Marine Geol 332–334:163–173. https://doi.org/10.1016/j.margeo.2012.02.014

Hovland M, Judd AG (1988) Seabed pockmarks and seepages—impact on geology, biology and the marine environment. Graham & Trotman Ltd, London, p 293

Hovland M, Kutznetsova T, Rueslåtten H, Kvamme B, Johnsen HK, Fladmark GE, Hebach A (2006a) Sub-surface precipitation of salts in supercritical seawater. Basin Res 18(2):221–230

Hovland M, Rueslåtten H, Johnsen HK, Kvamme B, Kutznetsova T (2006b) Salt formation associated with sub-surface boiling and supercritical water. Marine Petrol Geol 23:855–869

Hovland M, Rueslåtten H, Johnsen HK (2015) Red Sea salt formations—A result of hydrothermal processes. In: Rasul NMA, Stewart ICF (eds) The Red Sea: The formation, morphology, oceanography and environment of a young ocean basin. Springer Earth System Sciences, Berlin Heidelberg, pp 187–203

Hovland M, Rueslåtten H, Johnsen HK, Manuella F (2016) Possible role of salt accumulations in Wilson cycles (abstract). Proc Arthur Holmes Meeting, London, May 23–25, 2016. Geol Soc London

Hovland M, Rueslåtten H, Johnsen HK (2018a) Large salt accumulations as a consequence of hydrothermal processes associated with 'Wilson cycles'. A review Part 1: Towards a new understanding. Mar Pet Geol 92:987–1009. https://doi.org/10.1016/j.marpetgeo.2017.12.029

Hovland M, Rueslåtten H, Johnsen HK (2018b) Large salt accumulations as a consequence of hydrothermal processes associated with 'Wilson cycles'. A review Part 2: Application of a new salt-forming model on selected cases. Mar Pet Geol 92:128–148. https://doi.org/10.1016/j.marpetgeo.2018.02.015

Hoy RB, Foose RM, O'Neill BJ (1962) Structure of Winnfield salt dome, Winn Parish, Louisiana. Am Assoc Petrol Geol Bull 46:1444–1459

Hsü KJ, Montadert L, Bernoulli D, Cita MB, Erickson A, Garrison RE, Kidd RB, Mèlierés F, Müller C, Wright R (1977) History of the Mediterranean salinity crisis. Nature 267:399–403

Hübner A, De Lange GJ, Dittmer J, Halbach P (2003) Geochemistry of an exotic sediment layer above a sapropel S-1: Mud expulsion from the Urania Basin, eastern Mediterranean? Marine Geol 197:49–61

Ings SJ, Beaumont C (2010) Shortening viscous pressure ridges, a solution to the enigma of initiating salt 'withdrawal' minibasins. Geology 38(4):339–342

Judd A, Hovland M (2007) Seabed fluid flow, impact on geology, biology and the marine environment. Cambridge University Press, Cambridge, UK, p 300

Kupfer DH (1976) Shear zones inside Gulf Coast salt stocks help to delineate spines of movement. Am Assoc Petrol Geol Bull 60 (9):1434–1447

Kupfer DH, Lock BE, Schank PR (1998) Anomalous zones within the salt at Weeks Island, Louisiana. Gulf Coast Association of Geological Societies Trans XLVIII:181–191

Kyle RJ, Ulrich MR, Gose WA (1987) Textural and paleomagnetic evidence for the mechanism and timing of anhydrite cap rock formation, Winnfield salt dome, Louisiana. In: Lerche I, O'Brien JJ (eds) Dynamical geology of salt and related structures. Academic Press, New York, pp 497–543

Lecumberri-Sanchez P, Steele-MacInnis M, Weis P, Driesner T, Bodnar RJ (2015) Salt precipitation in magmatic-hydrothermal systems associated with upper crustal plutons. Geology 43 (12):1063–1066. https://doi.org/10.1130/G37163.1

Lees GM (1931) Salt—some depositional and deformational problems. Inst Petrol Technologists J 17(91):259–280

Lewis S, Holness M (1996) Equilibrium halite-H_2O dihedral angles: High rock-salt permeability in the shallow crust? Geology 24:432–434

Ligi M, Bonatti E, Rasul NMA (2015) Seafloor spreading initiation: Geophysical and geochemical constraints from the Thetis and Nereus Deeps, Central Red Sea. In: Rasul NMA, Stewart ICF (eds) The Red Sea: The formation, morphology, oceanography and environment of a young ocean basin. Springer Earth System Sciences, Berlin Heidelberg, pp 79–98

Lilley MD, Butterfield DA, Lupton JE, Olson EJ (2003) Magmatic events can produce rapid changes in hydrothermal vent chemistry. Nature 422:878–881

Lonsdale P (1977) Clustering of suspension-feeding macrobenthos near abyssal hydrothermal vents at oceanic spreading centers. Deep Sea Res 24, 857IN3859–3858IN4863

Loof KM (1999) Microstructural and structural analysis of three salt structures of different tectonic regimes. Unpublished Ph.D. dissertation, Texas A&M University, College Station, Texas, 283 pp (excerpts on the Internet)

Lowell RP, Germanovich LN (1997) Evolution of a brine-saturated layer at the base of a ridge-crest hydrothermal system. J Geophys Res 102:10245–10255

Lowell RP, Rona PA, Kolandaivelu K (2014) Hydrothermal activity. Elsevier Reference module, Earth Systems and Environmental Sciences. https://doi.org/10.1016/B978-0-12-4095489.09132-6

Lugli S, Manzi V, Roveri M, Schreiber BC (2015) The deep record of the Messinian salinity crisis: Evidence of a non-desiccated Mediterranean Sea. Palaeogeog Palaeoclimat Palaeoecol 433:201–218

Lupton JE, Weiss RF, Craig H (1977) Mantle helium in the Red Sea brines. Nature 266:244–246

Mart Y, Ross DA (1987) Post-Miocene rifting and diapirism in the northern Red Sea. Marine Geol 74:173–190

Miller AR, Densmore CD, Degens ET, Hathaway JC, Manheim FT, McFarlin PF, Pocklington R, Jokela A (1966) Hot brines and recent iron deposits in deeps of the Red Sea. Geochim Cosmochim Acta 30:341–369

Miller NC, Lizarralde D, Harding A, Kent G (2009) Constraints on early Gulf of California rifting from seismic images across the eastern margin of Guaymas Basin. AGU Fall Meeting, Abstract, Poster T31A-1783

Mitchell NC, Ligi M, Ferrante V, Bonatti E, Rutter E (2010) Submarine salt flows in the central Red Sea. Geol Soc Am Bull 122(5/6):701–713

Mitchell NC, Schmidt M, Ligi M (2011) Comment on "Formation of Thetis Deep metal-rich sediments in the absence of brines, Red Sea" by Pierret et al. (2010). J Geochem Explor 108:112–113

Mohriak WU (2001) Salt tectonics, volcanic centers, fracture zones and their relationship with the origin and evolution of the South Atlantic Ocean: Geophysical evidence in the Brazilian and West African margins. 7th International Congress of the Brazilian Geophysical Society, Salvador, Expanded Abstracts, p 1594

Mohriak WU, Nóbrega M, Odegard ME, Gomes BS, Dickson WG (2010) Geological and geophysical interpretation of the Rio Grande Rise, southeastern Brazilian margin: Extensional tectonics and rifting of continental and oceanic crusts. Petrol Geosci 16:231–245

Mohriak WU, Szatmari P, Anjos S (2012) Salt: Geology and tectonics of selected Brazilian basins in their global context. In: Alsop GI, Archer SG, Hartley AJ, Grant NT, Hodgkinson (eds) Salt Tectonics, Sediments and Prospectivity. Geol Soc London Spec Publ 363, pp 131–158

Monnin C, Ramboz C (1996) The anhydrite saturation index of the Red Sea ponded brines and sediment pore waters of the Red Sea deeps. Chem Geol 127:141–159

Momenzadeh M (1990) Saline deposits and alkaline magmatism: A genetic model. J Petrol Geol 13(3):341–356

Norton IO, Carruthers DT, Hudec MR (2015) Rift to drift transition in the South Atlantic salt basins: A new flavor of oceanic crust. Geology 44(1):55–58. https://doi.org/10.1130/G37265.1

Pierret MC, Clauer N, Bosch D, Blanc G (2010) Formation of Thetis Deep metal-rich sediments in the absence of brines, Red Sea. J Geochem Explor 104:12–26

Pirajno F (2009) Hydrothermal processes and mineral systems. Springer, Berlin, p 1250

Rasul NMA, Stewart ICF, Nawab ZA (2015) Introduction to the Red Sea: Its origin, structure, and environment. In: Rasul NMA, Stewart ICF (eds) The Red Sea: The formation, morphology, oceanography and environment of a young ocean basin. Springer Earth System Sciences, Berlin Heidelberg, pp 1–28

Ruppel C, Dickens GR, Castellini DG, Gilhooly W, Lizarralde D (2005) Heat and salt inhibition of gas hydrate formation in the northern Gulf of Mexico. Geophys Res Lett 32:L04605

Schoenherr J, Littke R, Urai JL, Kukla PA, Rawahi Z (2007a) Polyphase thermal evolution in the infra-Cambrian Ara Group (South Oman Salt Basin) as deduced by maturity of solid reservoir bitumen. Organic Geochem 38:1293–1318

Schoenherr J, Urai JL, Kukla PA, Littke R, Schléder Z, Larroque J-M, Newall MJ, Al-Abry N, Al-Siyabi H, Rawahi Z (2007b) Limits to the sealing capacity of rock salt: A case study of the infra-Cambrian Ara salt from the south Oman salt basin. Am Assoc Petrol Geol Bull 91(11):1541–1557

Schmalz RF (1969) Deep-water evaporite deposition: A genetic model. Am Assoc Petrol Geol Bull 53(4):798–823

Scott S, Driesner T, Weis P (2017) Boiling and condensation of saline geothermal fluids above magmatic intrusions. Geophys Res Lett 44 (4):1696–1705. https://doi.org/10.1002/2016GL071891

Scribano V, Viccaro M (2014) En-route formation of highly silica-undersaturated melts through interaction between ascending basalt and serpentinite-related saline brines: Inference from Hyblean cenozoic nephelinites, Sicily. Nicolosi (Catania) 29 Miscellanea INGV ISSN 2039–6651, vol 25, p 107

Scribano V, Carbone S, Manuella FC, Hovland M, Rueslåtten H, Johnsen H-K (2017) Origin of salt giants in abyssal serpentinite systems. Int J Earth Sci (Geol Rundsch). https://doi.org/10.1007/s00531-017-1448-y

Searle RC, Ross DA (1975) A geophysical study of the Red Sea axial trough between 2.5° and 22°N. Geophys J R Astron Soc 43:555–572

Singh S, Lowell RP, Lewis KC (2013) Numerical modeling of phase separation at the Main Endeavour Field, Juan de Fuca ridge. Geochem Geophys Geosyst 14(10):4021–4034. https://doi.org/10.1002/ggge.20249

Stoffers P, Kühn R (1974) Red Sea evaporites: A petrographic and geochemical study. In: Whitmarsh RB, Weser OE, Ross DA et al (eds) Initial Reports of the Deep Sea Drilling Project 23. US Govt Printing Office, Washington, DC, pp 821–847

Talbot C (1978) Halokinesis and thermal convection. Nature 273:739–741

Talbot CJ (1998) Extrusions of hormuz salt in Iran Geol Soc Lond Special Publ 143:315–334

Talbot CJ (2007) Hydrothermal salt—but how much?. Marine Petrol Geol, Discussion. https://doi.org/10.1016/J.Marpetgeo.2007.05.005

Underhill JR (2009) Role of intrusion-induced salt mobility in controlling the formation of the enigmatic 'Silverpit Crater', UK southern North Sea. Petrol Geosci 15:197–216

Van den Belt FJ, de Boer PL (2007) Shallow-basin model for 'saline giants' based on isostacy-driven subsidence. In: Smith ND, Rogers J (eds) Fluvial sedimentology VI. Internat Assoc Sedimentologists Spec Publ 38, pp 241–252

Von Damm KL (1995) Controls on the chemistry and temporal variability of seafloor hydrothermal fluids. In: Humphris SE, Zierenberg RA, Mullineaux LS, Thompson RE (eds) Seafloor hydrothermal systems: Physical, chemical, and biological interactions. Am Geophys Union, Washington, DC, Geophys Monograph Series 91, pp 222–247

Von Damm KL, Oosting SE, Kozlowski R, Buttermore LG, Colodner D, Edmonds HN, Edmond JM, Grebmeir JM (1995) Evolution of East Pacific Rise hydrothermal vent fluids following a volcanic eruption. Nature 375:47–50

Wallmann K, Suess E, Westbrook GH, Winckler G, Cita M (1997) Salty brines on the Mediterranean Sea floor. Nature 387:31–32

Warren JK (2006) Evaporites: Sediments, resources and hydrocarbons. Springer, 1023 pp

Warren JK (2010) Evaporites through time: Tectonic, climatic and eustatic controls in marine and nonmarine deposits. Earth-Sci Rev 98:217–268

Whitmarsh RB, Weser PE, Ross DA (1973) Initial reports of the Deep Sea Drilling Project, vol 23. U.S. Govt Printing Office, Washington, pp 821–847

Winckler G, Aeschbach-Hertig W, Kipfer R, Botz R, Rübel AP, Bayer R, Stoffers P (2001) Constraints on origin and evolution of Red Sea brines from helium and argon isotopes. Earth Planet Sci Lett 184:671–683

Origin of Submarine Channel North of Hanish Sill, Red Sea

Neil C. Mitchell and Sarantis S. Sofianos

Abstract

Submarine channels several kilometres wide can be found near sills between major basins but they typically lie on the sides of the sills where dense bottom waters passing between the basins form gravity currents. In the southernmost Red Sea, in contrast, an 8 km-wide and 100- to 250-m deep channel lies on the north side of Hanish Sill, in an area where the strongest bottom currents flow southward, associated with winter expulsion of dense saline Red Sea deep water. Current meter data collected 10 m or more above the seabed over the sill reveal speeds occasionally exceeding 1 m s^{-1}, which are sufficient to mobilize very coarse sand and have likely prevented deposition of finer sediments in the channel, particularly for parts of the channel affected by Red Sea Outflow Water (RSOW). However, the channel extends below 200 m depths, where Red Sea Deep Water is more sluggish (typically <1 cm s^{-1}). Although the stronger currents may help to maintain the upper channel morphology, it is unclear how they would have created the channel, nor can modern currents explain the deeper parts of the channel. The channel is straight and runs parallel with the spreading rift to the north, suggesting that faults may underlie the channel, though a tectonic origin (graben) is not supported by Bouguer gravity anomalies, which reveal no underlying structure. Alternatively, the channel may have originated much earlier, from massive inflow of Indian Ocean water into the Red Sea following earlier isolation and drawdown of its level. These and other possible origins of the channel are discussed in the light of limited public data from the area.

N. C. Mitchell (✉)
School of Earth and Environmental Sciences, University of Manchester, Williamson Building, Oxford Road, Manchester, M13 9PL, UK
e-mail: neil.mitchell@manchester.ac.uk

S. S. Sofianos
Department of Physics, University of Athens, University Campus Buildings Phys-5, 15784 Athens, Greece

N. M. A. Rasul and I. C. F. Stewart (eds.), *Geological Setting, Palaeoenvironment and Archaeology of the Red Sea*,
https://doi.org/10.1007/978-3-319-99408-6_12

1 Introduction

The regional bathymetry of the southernmost Red Sea (Fig. 1) reveals submarine channels emanating from Hanish Sill (the shallowest point between the Red Sea and Gulf of Aden) both northward into the Red Sea and southward toward the Perim Narrows and Gulf of Aden. The strongest bottom water flow across Hanish Sill occurs in winter, comprising Red Sea deep waters made saline and dense by evaporation in the desert climate (Maillard and Soliman 1986; Pratt et al. 1999; Smeed 2004). In winter, those waters are expelled in a bottom layer while overlying less saline Indian Ocean water enters the Red Sea in a shallow surface layer to balance the outflow and evaporation losses. During summer, a three-layer water structure can develop, with Gulf of Aden Intermediate Water intruding over the sill between outflowing surface water and RSOW. The channel continuing south of Hanish Sill is associated with a density current, which is created as the dense Red Sea deep waters enter the Gulf of Aden (Chang et al. 2008; Ilicak et al. 2008; Peters and Johns 2006), forming bottom surges over about 20% of the winter period (Bower et al. 2005). In such channels, the shear stress imposed by the flow on the bed prevents deposition of finer-grade sediments, while sediments are able to accumulate on the areas outside the channel where bottom currents are weaker, leading to aggradation there. The channel is therefore a product of gravity currents as well as constraining them. Similar channels associated with density currents have been found in the 'downstream' sides of sills between other major basins, such as in the Black Sea north of the Bosphorus (Parsons et al. 2010) and Atlantic west of the Strait of Gibraltar (Garcîa et al. 2008).

In contrast, the Bosphorus shelf 'upstream' (south) of the Sea of Marmara–Black Sea sill has a broad funnel-shaped depression narrowing toward the Bosphorus (Eriş et al. 2007; Hiscott et al. 2002). Although that morphology has been complicated by fluvial erosion and delta emplacement during and following the Last Glacial Maximum (LGM) (the

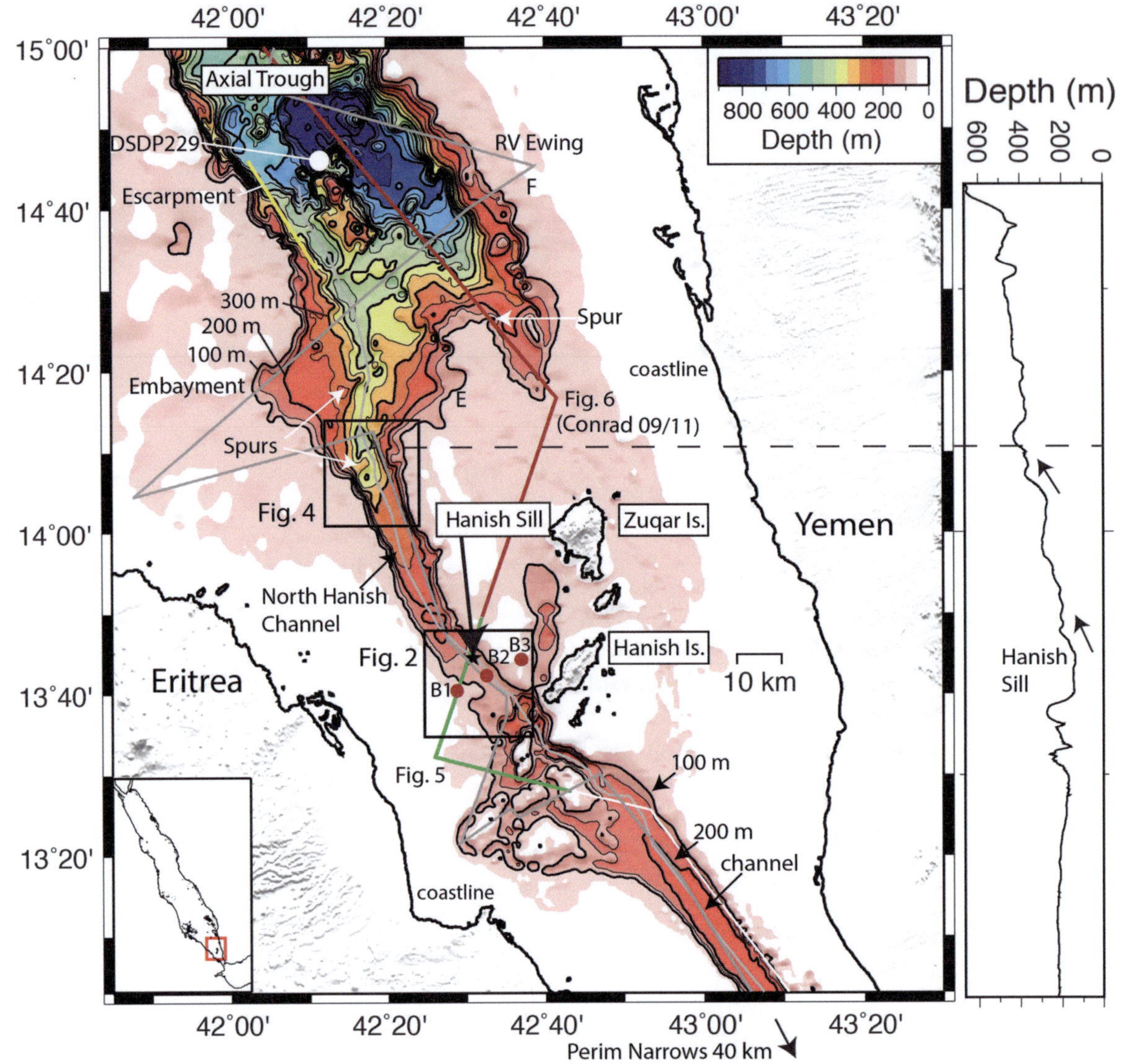

Fig. 1 Topography of the Hanish Sill area derived from the Global Multi-Resolution Topography Synthesis (Ryan et al. 2009) with color changes every 50 m (inset lower-left locates the map relative to the coastlines). Black contours are shown from 100 m to maximum depth every 50 m, with every 100 m highlighted. Black star marked "Hanish Sill" locates minimum sill depth according hydrographic charts reviewed by Lambeck et al. (2011). Solid white circle locates DSDP Site 229. White/green/red line is path of RV *Conrad* cruise rc0911 with extent of seismic reflection data shown in Figs. 5 and 6 marked in green and dark red. Darker grey line is path of 2001 RV *Ewing* cruise (Sofianos and Johns 2007). Three solid red circles annotated B1 to B3 locate current meter stations (Sofianos et al. 2002) where the data in Fig. 3 were acquired. Annotation: E, embayment; F, line F of Sofianos and Johns (2007). Graph to right shows the depth profile of the channel (along lighter grey line in main figure) plotted versus latitude at the same distance scale as the map. Arrows mark two segments of steeper gradients discussed in the main text

shelf lies above the LGM relative sea level), the pattern of later Holocene sediment thickness (Eriş et al. 2007) does not obviously show a tendency toward developing a narrow linear channel like the north channel in Fig. 1. Additionally, no surface channel can be observed in multibeam bathymetry data collected upstream (on the east or Mediterranean side) of the Camarinal Sill (Gibraltar Strait; Luján et al. 2010). The channel north of Hanish Sill (Fig. 1) therefore seems unusual.

The data available to address the North Hanish Channel origin are unfortunately very limited. The area lies in sea-lanes and is an area of political instability, making planning of further marine data acquisition in this region problematic. Nevertheless, we make best use of the information in the limited data available to debate the different possible explanations for this channel, which we hope will help to identify areas where new data (e.g., sampling by drilling and/or seismic reflection surveys) may address this issue more effectively in the future when better conditions prevail. We suggest that this debate is worthwhile as one of the potential origins is a massive inflow of Indian Ocean waters. If such inflow occurred, it may have been a dramatic geological event, similar to an event suggested to explain a buried channel potentially representing the Mediterranean re-flooding following the Messinian Salinity Crisis (MSC) (Garcia-Castellanos et al. 2009). As with the Mediterranean re-flooding, the event would have potential implications for how unconformities around the basin were created, how isostatic adjustment occurred as a result of water loading of the basin, how abruptly water exchanges with the Indian Ocean changed (with paleoceanographic effects) and how abruptly the regional climate changed.

2 Datasets and Methods

Current meter records were collected over an approximately 18-month period with instruments installed close to the seabed at sites B1–B3 (Figs. 1 and 2) (Sofianos et al. 2002). Data were collected at the deepest site B2 with an acoustic Doppler current profiler (ADCP) whereas conventional Aanderaa impeller meters were used at the other sites. Summaries of the data collected are shown along with information on the measurements in Fig. 3.

The bathymetry and elevation data shown in Fig. 1 are from the Global Multi-Resolution Topography Synthesis (Ryan et al. 2009), with the bathymetry constructed from track-line data (mainly passage soundings) and depths between ship tracks interpolated using the marine gravity field derived from satellite altimetry data (Sandwell and Smith 1997; Smith and Sandwell 1997). Features interpreted from this grid are constrained by numerous passage soundings of vessels passing through the Bab al Mandab Strait. The data are shown contoured every 50 m below 100 m to highlight the channel and Axial Trough morphology (for comparison, the 113-m depth level corresponds with the local water level over the sill at the time of the LGM (Lambeck et al. 2011)).

Echo-soundings were collected around the Hanish Sill in 1971 and 1972 by Werner and Lange (1975). Figure 2 shows a map constructed by digitizing the 10-m depth contours shown by Werner and Lange (1975) and interpolating using the "surface" software module (Smith and Wessel 1990). The 1971 surveys (white lines in Fig. 2b) were carried out with visual bearings to Hanish Island, whereas the 1972 survey (black lines in Fig. 2b) was aided by radar reflectors placed on two navigation buoys located in Fig. 2b. The large open circle in Fig. 2a highlights a small inner channel observable immediately north of Hanish Sill. Of the profiles shown in Fig. 2a, 5022 runs along the lip of the sill, where the shallowest point of the channel reaches 137 m depth in the Werner and Lange (1975) survey, and 5025 crosses the channel on the Red Sea side of the sill. In Fig. 2c, magnetic anomalies collected along the RV *Jean Charcot* track crossing the sill are shown projected against the vessel's trackline.

Figure 4 shows the bathymetry of the lower part of the channel immediately before it reaches the Axial Trough. This map was obtained by gridding the available track-line sounding data from the National Geophysical Data Center (NGDC). In Fig. 4, bathymetry along four tracks are also shown in red (profiles projected perpendicular to tracks after removing an offset of 300 m).

Seismic reflection data were collected during RV *Conrad* cruise rc0911 along the line marked in Fig. 1. According to information from the NGDC, these data were collected with a 25 cubic inch airgun. Despite their age, the records shown in Figs. 5 and 6 are of sufficient quality to reveal sub-surface reflections beneath the slope NW of Hanish Sill and the area NW of Hanish Island. Their large vertical exaggeration ($\sim$15 X) needs to be considered when attempting to interpret dips of features in the data.

Structure beneath Hanish Sill, the channel and surrounding areas can also be assessed from gravity data. The marine free-air gravity field shown in Fig. 7a was created using data from satellite altimeters; in open ocean areas these gravity data have comparable precision to shipboard gravity data (Sandwell et al. 2014). However, a study by Mitchell (2015) comparing shipboard gravity measurements in the northern and central Red Sea with the earlier version 18 of the altimeter-based gravity field found long-wavelength mismatches of up to 10–20 mGal that are potentially artifacts of ties with terrestrial gravity measurements. Hence, gravity anomalies along the coasts, such as that along the Yemen coast in Fig. 7a, should be interpreted cautiously.

To reduce the effects on the gravity anomalies of the seabed topography, a map of Bouguer gravity anomalies was calculated and is shown in Fig. 7b. This was computed below sea level using a density contrast of 1.77–1.03 g cm^{-3}, where 1.77 g cm^{-3} is the mean density of the upper 200 m of sediment from physical property measurements made on DSDP Site 229 samples (Manheim et al. 1974; Whitmarsh et al. 1974) (Fig. 1) and 1.03 g cm^{-3} is seawater density. Above sea level, a 2.3 g cm^{-3} density that is typical for the upper 1 km of volcanic rocks on Hawaii and in the rift zone of Iceland was used based on sources compiled by Rubin (1992).

3 Information on Seabed Sediments

Sediment grain size and density are important for inferring particle threshold of motion and for assessing how the geomorphology may have formed from sediment erosion or deposition. Unfortunately, there is very little information on the seabed sediments around the sill. Werner and Lange

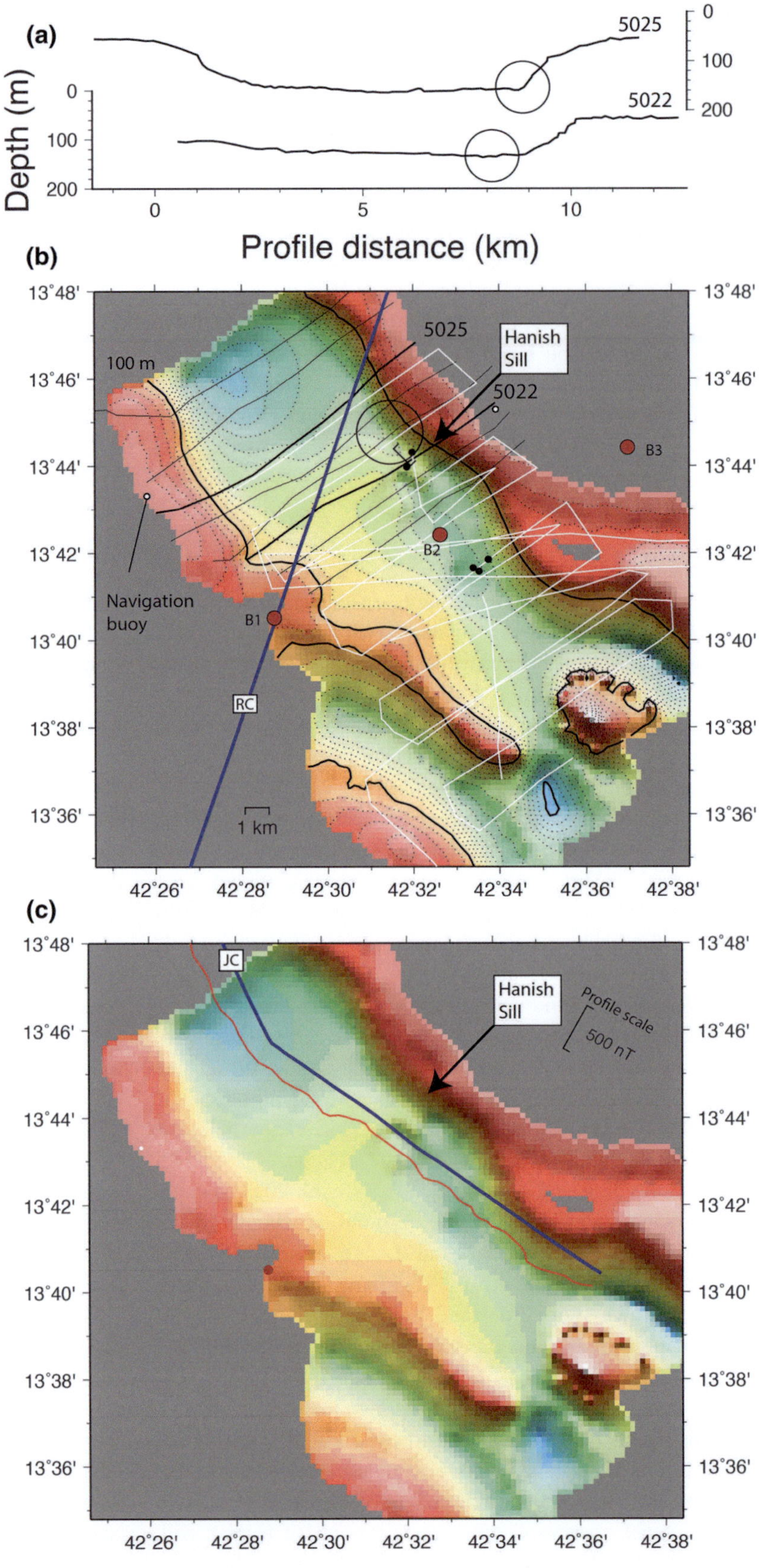

Fig. 2 Bathymetry of the sill based on data of Werner and Lange (1975). **a** Two bathymetry profiles located with bold lines in **b**. Circles mark a channel in profile 5025 and the minimum elevation at the sill in 5022. **b** Map of bathymetry constructed from the 10-m depth contours in the map of Werner and Lange (1975), with 100 m contour in bold and 10 m contours dotted. Open circles: navigation buoys; closed circles: bottom sample stations; white lines: 1971 surveys; solid straight lines: 1972 survey; dark red symbols, current meter sites as Fig. 1; blue line, path of RV *Conrad* (RC). **c** Red line with scale to upper right shows residual magnetic anomalies collected along RV *Jean Charcot* track (blue line)

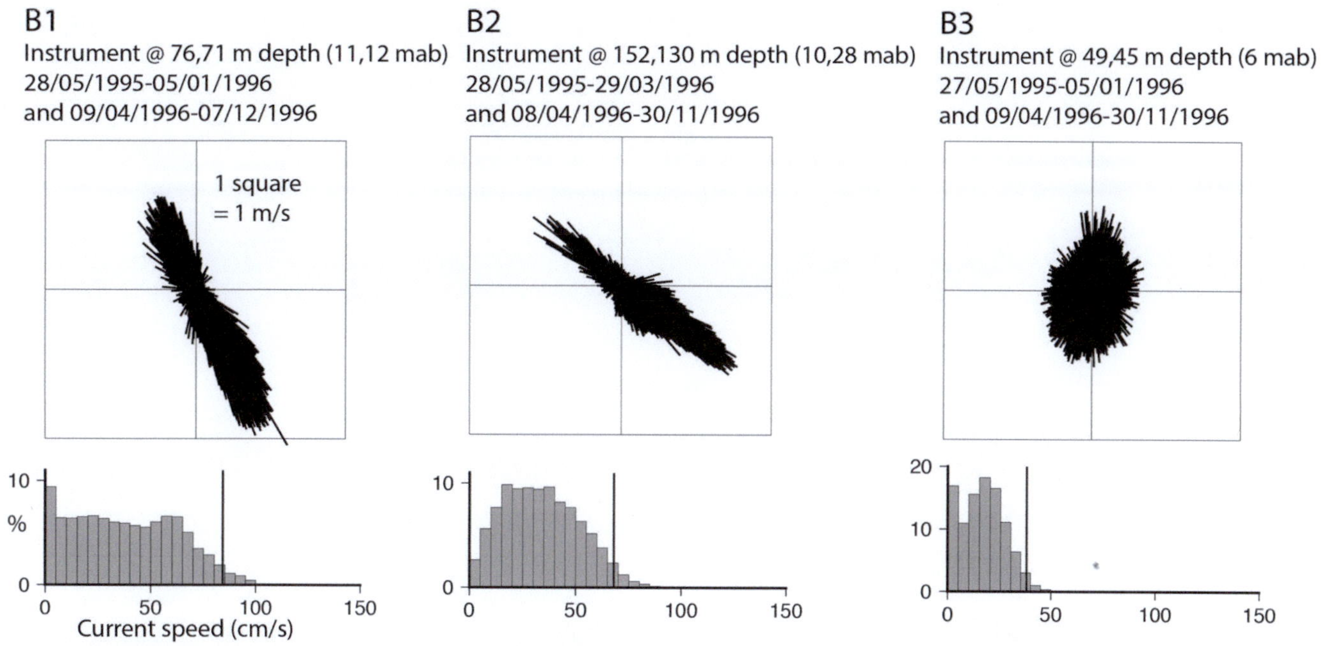

Fig. 3 Summaries of data from current meters deployed near the seabed at sites B1–B3 shown in Fig. 1 (Sofianos et al. 2002). Data in the vector plots (upper graphs) are drawn at a scale such that a distance across one square is equivalent to a current speed of 1 m s^{-1}. In the histograms of current speeds, vertical bars locate the 95-percentile of each distribution

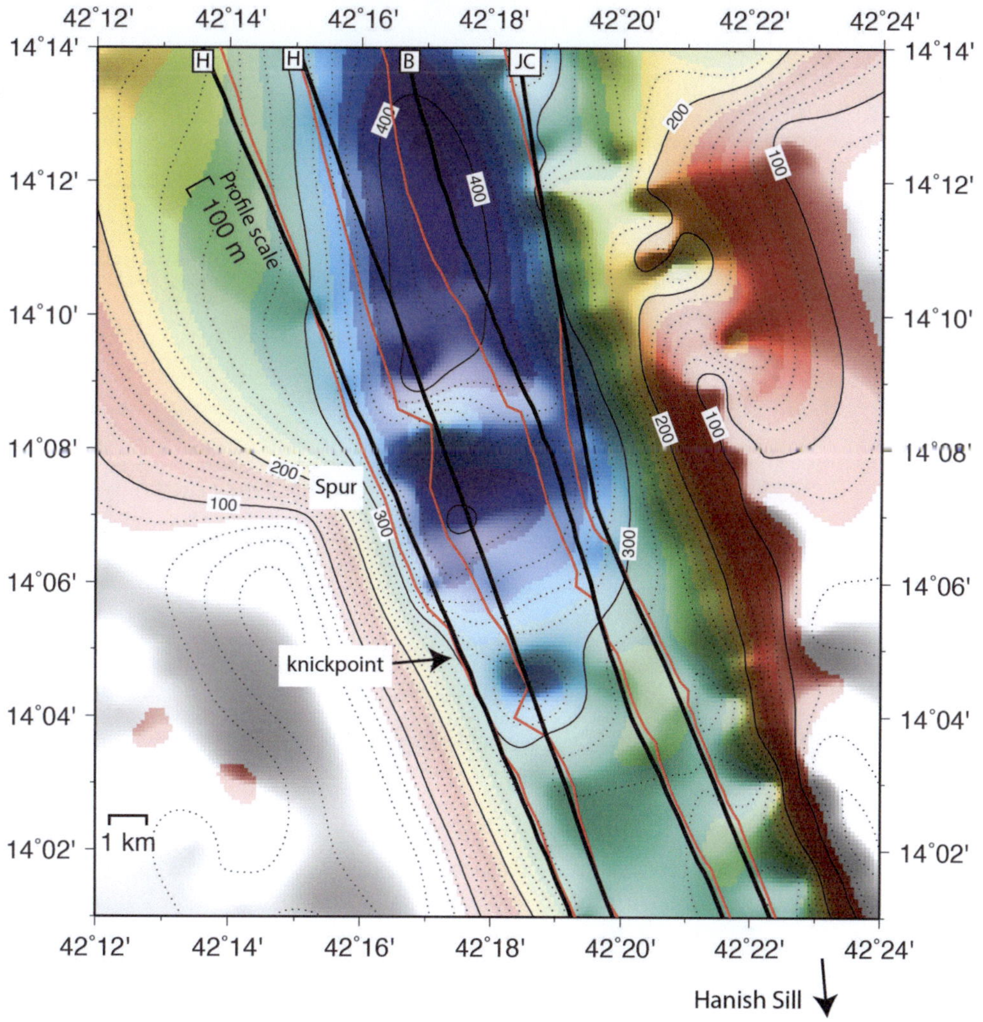

Fig. 4 Detail of the lower channel knickpoint. Bathymetry in background was generated by gridding the available track-line data and contoured every 20 m (dotted lines) with 100 m intervals annotated (bold lines). Also shown are tracks with depth soundings plotted (red lines) toward N060°E with profile scale in lower left for (H) two passages of HMS *Herald* (1996), (B) one passage of HMS *Beagle* (1995) and (JC) one passage of RV *Jean Charcot* (1983). Depths have been offset by 300 m before projecting perpendicular to vessel tracks. (The bathymetry interpolation was carried out with the minimum curvature option of the "surface" module of the GMT software, which creates artifacts in areas lacking data (Smith and Wessel 1990), hence the hill under annotation "knickpoint" in the figure is an artifact)

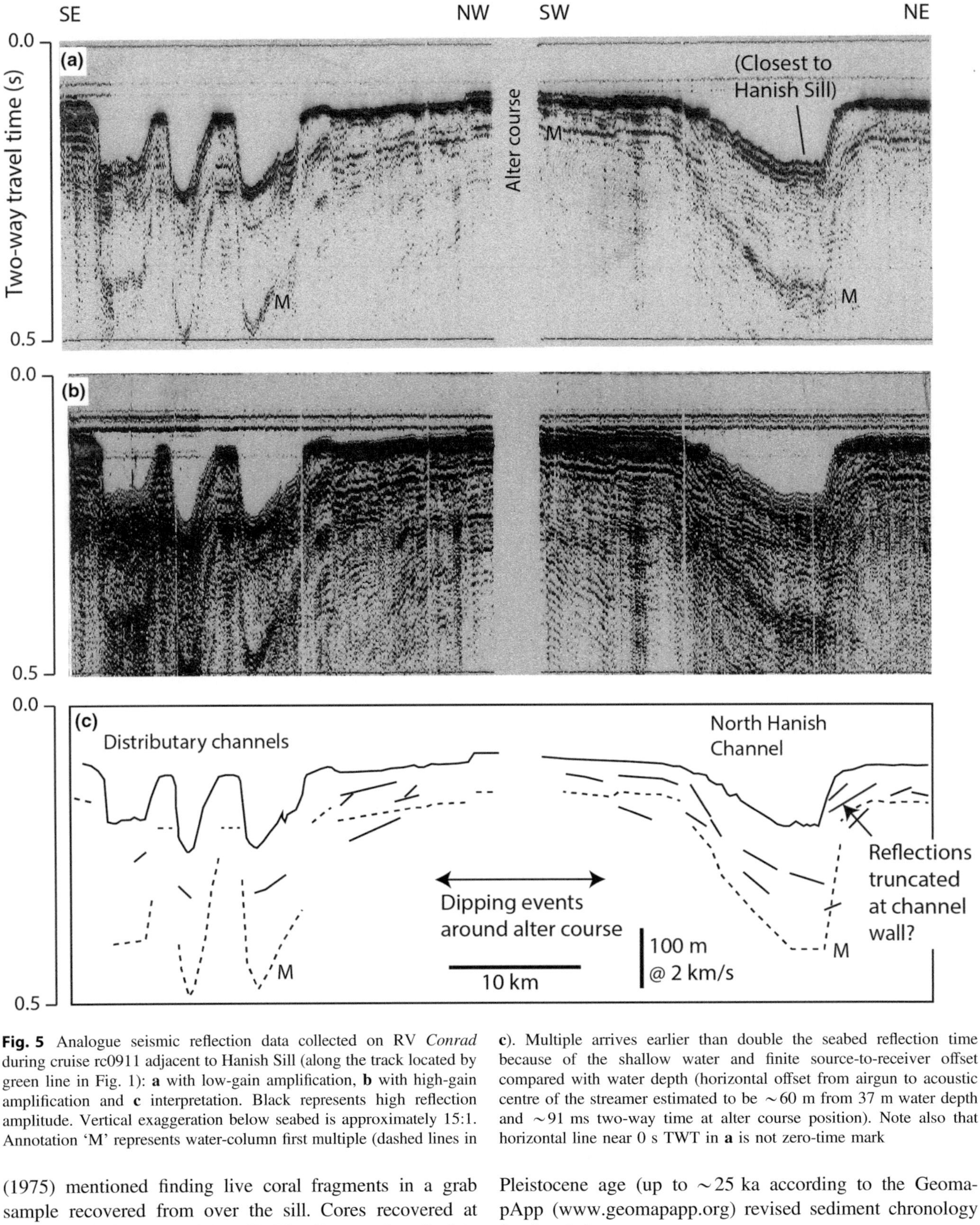

Fig. 5 Analogue seismic reflection data collected on RV *Conrad* during cruise rc0911 adjacent to Hanish Sill (along the track located by green line in Fig. 1): **a** with low-gain amplification, **b** with high-gain amplification and **c** interpretation. Black represents high reflection amplitude. Vertical exaggeration below seabed is approximately 15:1. Annotation 'M' represents water-column first multiple (dashed lines in **c**). Multiple arrives earlier than double the seabed reflection time because of the shallow water and finite source-to-receiver offset compared with water depth (horizontal offset from airgun to acoustic centre of the streamer estimated to be ~60 m from 37 m water depth and ~91 ms two-way time at alter course position). Note also that horizontal line near 0 s TWT in **a** is not zero-time mark

(1975) mentioned finding live coral fragments in a grab sample recovered from over the sill. Cores recovered at DSDP Site 229 north of the sill (Fig. 1) were described in the drilling reports (Whitmarsh et al. 1974) as containing 212 m of clay-rich nanofossil carbonate ooze of Late Pleistocene age (up to ~25 ka according to the GeomapApp (www.geomapapp.org) revised sediment chronology for the site).

Einsele and Werner (1972) described the textures and mineralogy of seabed sediment recovered at 12 sites south of

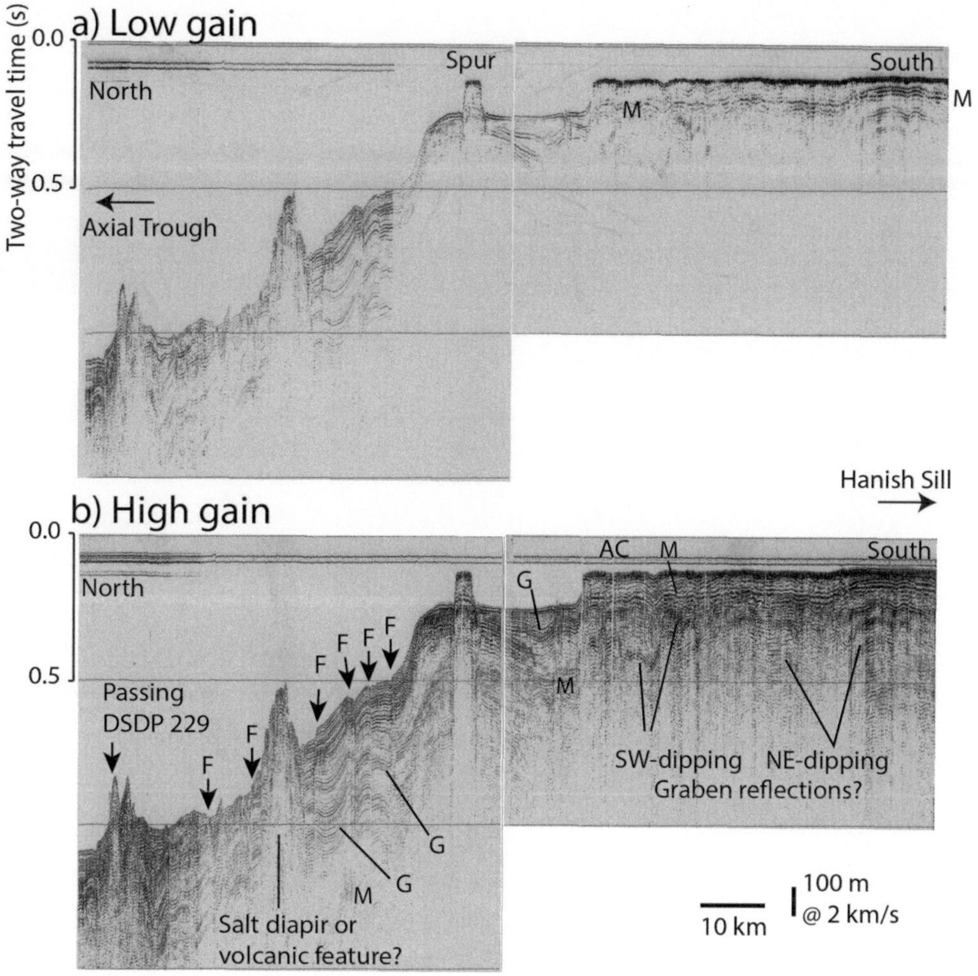

Fig. 6 Analogue seismic reflection data collected on RV *Conrad* during cruise rc0911 north of Hanish (located by red line in Fig. 1): **a** low-gain amplification and **b** high-gain amplification (black is high reflection amplitude), as in Fig. 5. Annotation: 'M', water-column first multiple; 'F', shallow fault; 'G', tectonic growth structure; "AC", alter course

the sill in an area centred on the Perim Narrows. They found that the shelf and channel sediments comprise dominantly biogenic particles derived from the adjacent nearshore areas and from in situ production. On the shelf immediately north of the Perim Narrows, bryozoan fragments dominated, with lesser benthic foraminifera, coral fragments and septula. Some sampling sites within the narrows contain hardgrounds with pebbles and shell fragments. At one site further north within the Axial Trough at ~14° 50'N, a sample was recovered containing dominantly pelagic foraminifera and gastrapoda, and lesser scaphapoda and pteropoda.

Based on these studies, the particles entering the area of the channel north of Hanish Sill are likely mostly biogenic and have a wide range of grain sizes. Grain size is not conservative as biological particles can be easily broken into smaller particles during transport by currents. The authors of the report for DSDP Site 229 (Whitmarsh et al. 1974) suggested that the fine particles there had been swept from the shelves by seasonal currents. Given the dominant biogenic nature of these sediments, their bulk grain densities are likely lower than siliclastic particles such as quartz because of pore spaces commonly within them.

4 Observations

4.1 Bathymetry

4.1.1 Overview

Hanish Sill (Fig. 1) lies in the area of a southward-narrowing sea between the coasts of Arabia and Africa, though it is not at the narrowest point (which lies further south at Perim Narrows). Volcanically active islands (Hanish and Zuqar) lie east of the sill (Gass 1970). Bathymetry of the area is generally shallow, forming broad shelves of <100 m depth. This shallow bathymetry is interrupted north of 14°20'N by the southernmost end of the Axial Trough, which contains the active spreading centre (Garfunkel et al. 1987). The seabed within the trough deepens below 500 m to the north in Fig. 1. South of Hanish Sill, the bathymetry contains a series of closed-contour highs, which likely formed islands during the LGM (Lambeck et al. 2011). Two major channels can be observed in Fig. 1. One runs to the SE, whereas a second (the focus of this article and referred to as "North Hanish Channel") runs NNW of the sill, on its Red Sea side.

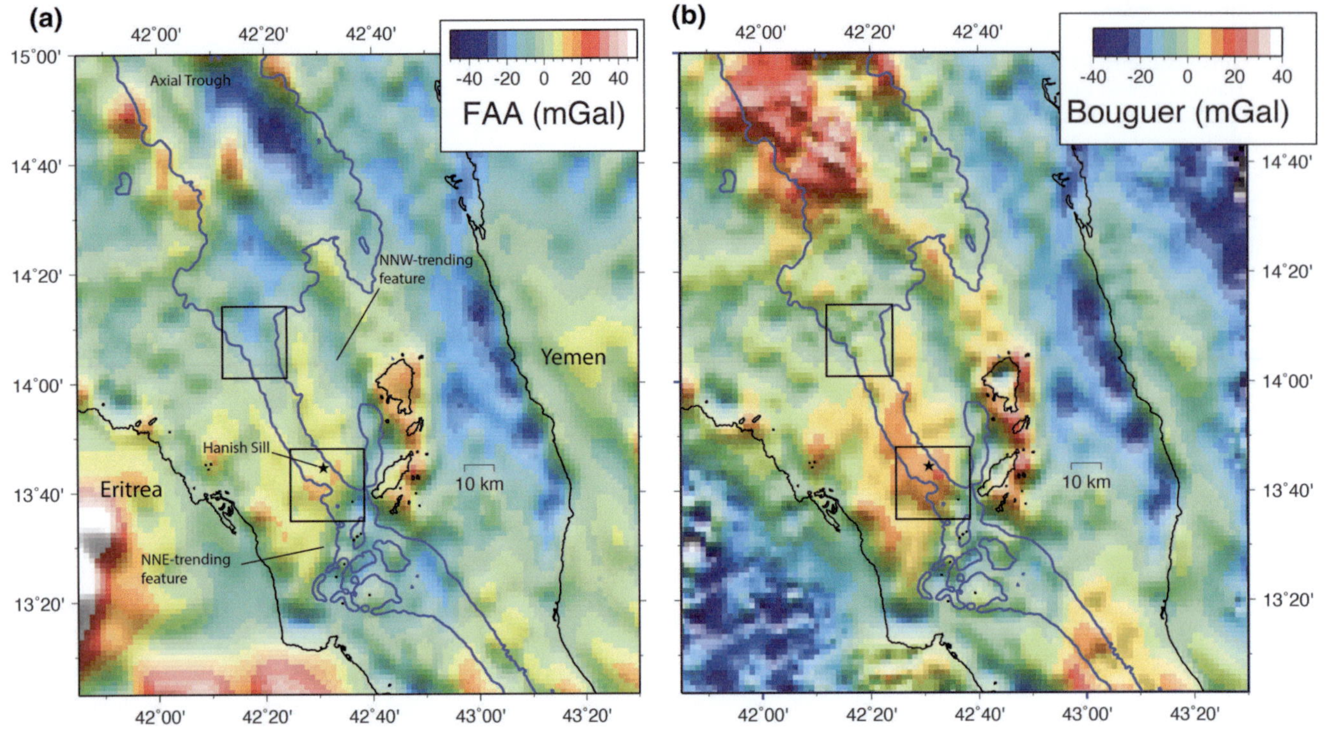

Fig. 7 **a** Free-air gravity anomalies derived from satellite altimetry (marine areas) from Sandwell et al. (2014) (their version 22). **b** Bouguer gravity anomalies derived using the gravity slab formula by correcting the free-air anomalies for a 2.3 g cm^{-3} density above sea level and for a (1.77–1.03 g cm^{-3}) seabed density contrast below sea level. Blue contour is 100 m depth level. Other annotation as Fig. 1, which covers the same area.

North Hanish Channel is 8-km wide. In cross-section, it has a vertical relief of 100–250 m and its base is mostly flat. The mean gradient from the sill to the edge of the Axial Trough is ~0.9% (0.5°). Steep sections of the along-channel profile (marked by arrows in the graph in the right panel of Fig. 1) occur just below the edge of the sill and about 10 km south of where the channel enters the Axial Trough. Two spurs of 50–100 m elevation extending into the channel are marked in Fig. 1.

4.1.2 Hanish Sill

In Fig. 2b, the sill lies within a broad depression, about 8 km across, on the east side of a broad NE-SW-oriented ridge crossing the depression. Immediately north of the sill, a small ~10-m deep channel runs NW (circled in Fig. 2b). A further 3 km down the channel, profile 5025 shows a corresponding small depression (circled in Fig. 2a). Profile 5022 runs along the sill and reveals a mainly low-relief seabed, though with a small depression of perhaps a few metres (circled) where the sill minimum elevation was proposed by Werner and Lange (1975) to be located at 137 m depth. Running southeast of the site of current meter B1, a ridge of >50 m elevation causes a modest narrowing of the central depression to the SE and a closed-contour high occurs east of its tip (13°38'N, 42°36'E). The *Jean Charcot* magnetic anomalies (Fig. 2c) vary modestly with amplitudes of ~100 nT and rise only by 100–200 nT on crossing the sill.

4.1.3 Lower Channel

Figure 4 shows enlarged bathymetry where the channel is unusually steep (14°04'–14°07'N). In river channels, such morphological features are described as "knickpoints", so we use this term here. The four lines of passage soundings plotted as profiles (red lines) in Fig. 4 also reveal this morphology. One of the spurs mentioned earlier can be observed in the central two lines of passage soundings.

4.2 Currents

The current meter data summaries (Fig. 3) reveal the stronger outflows (to SE) for sites B1 and B2 relative to inflows, while currents at the shallower site B3 are smaller and less asymmetric. The 95-percentile current speeds (marked by vertical bars in the lower graphs of Fig. 3) are 84, 68 and 38 cm s^{-1} for B1, B2 and B3. The constriction provided by the ridge (Fig. 2b) may explain the greater speeds at B1 compared with B2, much like how headland-attached sand banks can lead to accelerated currents passing over them.

4.3 Gravity Anomalies

The free-air anomalies (Fig. 7a) reveal two gravity depressions, one running NNE just west of Hanish Island and another running NNW roughly orthogonal to the long axis of Hanish Island. These two trends also occur in the computed Bouguer anomalies (Fig. 7b), so they most likely represent underlying linear bodies of lower density than their surrounding rocks. The NNW feature forms a Bouguer anomaly trough of about 10–20 mGal. The source of the anomaly can be very roughly evaluated using the gravity slab formula (Telford et al. 1976). Thus, a 20 mGal anomaly could potentially be created by a density contrast of 1.0 g cm^{-3} between superficial sediments and a denser basement with 500 m of relief. Hanish Sill is associated with somewhat elevated Bouguer anomalies of about 15–20 mGal.

In contrast, the linear feature in the free air anomaly following the channel north of Hanish Sill does not appear in the Bouguer anomaly; the free air feature is explained simply by the effect of seabed topography. Bouguer anomalies computed using an alternative seabed density contrast of 2.0–1.03 g cm^{-3}, where 2.0 g cm^{-3} is a typical density of shelf sediments (Hamilton and Bachman 1982), produced a similar result.

4.4 Seismic Reflection Data

Although water-column multiples obscure some features, the seismic data reveal dipping reflections beneath the slope immediately north of Hanish Sill (SW to NE section in Fig. 5). The large vertical exaggeration of the data make these shallow dipping features appear steep; the west bank of the North Hanish Channel marked in Fig. 5c has an along-profile gradient of less than 1°. Beneath the North Hanish Channel, sub-bottom reflections all appear to dip with the same sense as the channel floor. Reflections can be observed up to 150 ms two-way travel time below the seabed reflection ($\sim$150 m sub-bottom depth if the shallow sediments have 2 km s^{-1} seismic velocity). However, beneath the northeast wall of North Hanish Channel, one or two reflections dip to the southwest and appear to be truncated at the seabed (channel wall).

Further sub-surface reflections can be observed in the data beneath the distributary channels (left side of Fig. 5). Around the change of vessel course in Fig. 5, a series of reflections can be observed. They appear to dip to the SE before the course change and to the NE after the course change (dip components along the sections). These may represent strata or other layered features dipping to the east beneath the west side of Hanish Sill.

To the south in Fig. 6, some dipping reflections underlie the area of the NNW-trending Bouguer gravity anomaly trough described earlier and thus may represent a small graben. Immediately northwest of there, the *Conrad* track ran over the southeasterly embayment of the Axial Trough (Fig. 1) and the corresponding seismic record in Fig. 6 reveals locally thickened intervals between reflections suggesting a tectonic growth structure ("G" in Fig. 6). The track then crosses a spur of elevated seabed with character in the seismic data much like those collected over the shelves.

Further northwest of there, where *Conrad* crossed the southeast submarine slope of the Axial Trough, the record reveals a series of abrupt disruptions of reflections suggesting the presence of shallow faults ("F" in Fig. 6). These faults and further growth structures most likely represent the effects of slumping of superficial sediments into the Axial Trough. The slope also contains an isolated peak without internal structure, which we speculate is a salt diapir or volcanic edifice. The seismic data collected adjacent to DSDP Site 229 reveal some seabed layering, which may represent the same hemipelagic sediments recovered at that site (Whitmarsh et al. 1974).

5 Discussion

5.1 Currents Affecting Channel Morphology

The $\sim$1 m s^{-1} peak currents recorded at sites B1 and B2 are likely to prevent deposition of fine and silt grade particles and many sand grade particles. For example, a 1 m s^{-1} threshold velocity at 1 m above the bed would mobilize quartz sand up to 4 mm in diameter according to Miller et al. (1977). Threshold speeds for biogenic clasts can be smaller because of the lower net densities of porous clasts. In settling experiments, densities of foraminifera tests have been found to be 1.162 g cm^{-3} (Berthois and Le Calvez 1960) and 1.500 g cm^{-3} (Berger and Piper 1972). Their densities relative to seawater ($\sim$1.03 g cm^{-3} in the Red Sea) are thus small compared with quartz. Southard et al. (1971) measured the thresholds of motion of carbonate ooze comprising mainly foraminiferal tests and coccolithophorid remains, which Miller and Komar (1977) converted to velocity at 1 m above the bed. Those values vary from only 9–13 cm s^{-1} with no compaction to $\sim$20 cm s^{-1} after some compaction. Thresholds of motion of larger particles are difficult to predict given the strongly varied effect of grain shape of biogenic particles (Paphitis et al. 2002). Nevertheless, threshold velocities should be lower than for quartz clasts. The 95-percentile velocities recorded at current meter sites B1, B2 and B3 imply mobilization of quartz grains up to diameters of 2, 1.4 and 0.2 mm according to the data of Miller et al. (1977), which are therefore minima. Hence, at least very coarse sand would be mobilized in the channel and fine sand would be mobilized at Site B3. These currents maintain the geomorphology of the sill by preventing the finer sediment grades from depositing within the central depression (Fig. 2).

However, going northwest from the sill and into deeper water, bottom currents are likely to be weaker. This is partly a consequence of the funnel shape of the southernmost Red Sea. Furthermore, potential densities in the southernmost Red Sea increase with depth (e.g., Jarosz et al. 2005) and oppose any significant upward movement of deep waters. RSOW with a potential density of ~1028 kg m^{-3} overlies Red Sea Deep Water (RSDW) with a potential density of ~1028.6 kg m^{-3}, with a boundary between the two at around 200 m (Sofianos and Johns 2003). Consequently, RSOW probably originates from intermediate depths in the Red Sea (Cember 1988; Neumann and McGill 1962; Sofianos and Johns 2003). Although collected in summer when the outflow is less intense, ADCP data collected along line F in Fig. 1 (Sofianos and Johns 2007) revealed negligible current speeds (<1 cm s^{-1}) below 200 m. Thus, the northern part of the channel, where it lies deeper than 200 m, is likely to contain less mobile RSDW. The prevalence of slope failure features in Fig. 6 further supports this view, as such features typically form in muds (deposited under weak currents) rather than coarser grade unconsolidated clastic particles. Deposition of muds is more likely to fill in the lower channel than create its morphology.

5.2 Slope Failure Origin

The faults and tectonic growth stratigraphy in the southern slope to the Axial Trough in Fig. 6 are interpreted here as due to slope failure. As similar sediments most likely underlie the southerly slope of the Axial Trough (area annotated "Embayment" and "E" in Fig. 1), slope instability may have also occurred there. As mentioned earlier, the mean channel gradient is only ~0.5°. While retrogressive failure cannot be ruled out as having created the channel morphology, such a gradient would be small for failure in carbonate shelf sediments. McAdoo et al. (2000) evaluated evidence for slope failure in extensive multibeam bathymetry data around the USA. They found slope failure had occurred in areas of <1° seabed gradient, though they occurred in the clay-rich muddy sediments of the Gulf of Mexico and may have been caused or affected by underlying weak halite deposits. Additionally, the regular, straight morphology of the North Hanish Channel does not match the morphologies of failed slope deposits documented by McAdoo et al. (2000), so it is unlikely to have been created simply by slope failure.

However, the embayments to either side of the channel where it enters the trough (Fig. 1) may have been produced by slope instability. Failure in the lower part of the channel where it enters the Axial Trough may also potentially explain the knickpoint morphology.

5.3 Tectonic Origin

The present-day stability of the area is suggested by GPS measurements, which show little horizontal movement between the Arabian and African coasts on either side of Hanish Sill (McClusky et al. 2003; Reilinger et al. 2015). The on-going separation of the Nubian and Arabian tectonic plates occurs instead to the west of the Danakil Block in the Danakil Depression. How far back in time this limited movement persisted is difficult to assess from plate reconstruction data, though the data also suggest limited movement. Roeser (1975) interpreted seafloor spreading magnetic anomalies along the Red Sea spreading axis to anomaly 3 (4–5 Ma; Cande and Kent 1995) at 16°N, but the anomalies become confused south of 15°30'N and die out entirely south of 15°N (Chu and Gordon 1998), as would be expected if the Danakil Block had continued rotating while effectively joined to Arabia in the south after anomaly 3. The rotation pole for Danakil relative to Arabia of Chu and Gordon (1998) lies to the SE with a 95% error ellipse almost reaching the Bab al Mandab Strait. Plate tectonic reconstructions of the Danakil Block (Eagles et al. 2002) based on the Chu and Gordon (1998) rotation poles show only modest (a few tens of km at most) separation of the southern part of that block relative to Arabia since 5 Ma.

The success in detecting the NNW-trending feature in the Bouguer anomalies (Fig. 7b) suggests that sediment-filled features of a few hundred metres in relief should be observable in the data. The lack of any such feature over the North Hanish Channel suggests that any fault throw if it exists is likely to be only modest (<100 m). The Bouguer data rather suggest that two tectonic features occur with significant relief, the NNW- and NNE-trending features. Hanish Island and its superimposed cones are also aligned with the latter orientation. Hanish Sill appears to be an isolated block, possibly underlain by denser material, encompassed on its NE and SE sides by faults or fault zones. The sill may have been stable or only gradually moving vertically, as modelling of sedimentary $\delta^{18}O$ suggests that the sill has been uplifting during the last 500 ky by ~0.02 m kyr^{-1} (Rohling et al. 2009). This interpretation might be supported if the two trends marked on Fig. 7 are normal faults or transcurrent faults that have had normal components of slip, so that Hanish Sill lies within their footwalls and the faults have been modestly active. Overall, the gravity data do not obviously support a tectonic origin for the north channel.

5.4 Massive Red Sea Inflow Origin

5.4.1 Morphological Evidence

We compare the North Hanish Channel with the buried channel east of the Camarinal Sill that has been suggested to have been excavated by a massive flow during re-flooding of the Mediterranean (Garcia-Castellanos et al. 2009). That channel has a steeper gradient of $\sim 2°$. It has a similar 8-km width to the North Hanish Channel, although the Hanish channel may have originally been wider and has since narrowed by progradation of sediments shed from the shelves, such as is suggested by some dipping stratigraphy in Fig. 5. Unfortunately, the seismic data in Fig. 5 do not reveal many details of the reflections, although arguably reflections beneath the east slope of Hanish Sill are truncated at the seabed, suggesting erosion.

We highlighted the knickpoint (Fig. 4) in our observations section. Steep sections of river bedrock channels and submarine channels form either by strata that are locally resistant to erosion (Miller 1991) or from migration of a steepened gradient because that gradient leads to an intensification of flow and enhanced erosion (Gardner 1983; Mitchell 2006; Seidl and Dietrich 1992). Rapid inflow of Gulf of Aden waters could potentially explain the occurrence of the knickpoint in the North Hanish Channel (Fig. 4). Accordingly, during inflow, the acceleration of the inflowing waters as they entered the Axial Trough could have produced a wave of erosion.

When we describe river or submarine channels as "tectonically controlled", in practice we mean that fault or fold topography has guided the flows, leading to greater erosion along their paths, or that breccias created by fault movements left the rocks more susceptible to erosion. Although there is no Bouguer anomaly associated with the channel, less significant fault displacemet (~ 100 m) is not ruled out by the gravity data and smaller faults aligned with the Axial Trough would explain the straight morphology of the channel in plan view.

5.4.2 Feasibility of Channel Excavation

We cannot accurately know the discharge of any influx without knowing the geometry of any structure that was breached. However, we can say that it would have been larger than the net evaporation loss of the Red Sea in order to re-flood the sea. In the following section, we explore the implications of such a discharge on the shear stresses within a channel 8-km wide and with a gradient of 1% like the modern North Hanish Channel. We calculate the potential shear stresses imposed by such a flow on its bed. Although we do not have geotechnical information on the eroded materials, we can say that the shear stresses are generally large compared with shear stresses that occur in bedrock rivers in flood, where erosion is known to occur.

The modern net evaporation (evaporation minus precipitation) is about 30000 $m^3 s^{-1}$ (Maillard and Soliman 1986; Sofianos et al. 2002). If the influx occurred following a Pleistocene sea level drawdown, the net evaporation could have been around half this value because sea level lowered below the shelves would have left the sea covering roughly half of its modern surface area. The LGM (and likely other Pleistocene glacial maxima) evaporation rates were probably comparable to present rates because greater wind speed and lowered atmospheric humidity compensated partly for lowered temperature (Biton et al. 2008). If the event occurred as early as the Late Miocene or early Pliocene, evaporation rates are less well known. The Red Sea basin was narrower than it is today because it has widened subsequently with plate tectonic spreading (by about 60 km on average based on a 12 mm yr^{-1} spreading rate in the basin center (Mitchell and Park 2014), or about 20% of the present area). Common occurrences of fluvial clastic deposits in the upper Miocene Zeit Formation suggest that the climate was wetter in the Miocene (Griffin 1999), further suggesting that the evaporative loss was much less than the present-day 30000 $m^3 s^{-1}$.

After accelerating from Hanish Sill onto the channel gradient, the flow would have approached equilibrium. The velocity u (m s^{-1}) of that spillway flow can be estimated using standard hydraulic relationships. Manning's Formula for depth-averaged flow speed (Webber 1971) suggests at equilibrium:

$$u = \frac{R^{2/3} S^{1/2}}{n} \quad (1)$$

where R is the hydraulic radius (m), S is the dimensionless channel gradient and n is a constant representing friction of the channel boundaries. For the following estimates, values of $n = 0.03$ and 0.07 were used, appropriate for channel roughness corresponding with gravels or boulders (Webber 1971). For wide channels such as the channel here, the hydraulic radius is approximately equal to the flow depth d.

Discharge $Q = uA$, where A is the flow cross-sectional area $A = bd$ and b is the depth-averaged width. Substituting u from Eq. (1) in $Q = ubd$ and rearranging:

$$d = \left(\frac{nQ}{S^{1/2} b}\right)^{3/5} \quad (2)$$

Various properties of a hypothetical spillway flow were computed using these relationships, with the modern spillway average gradient $S = 0.01$ (0.5°) and with discharge Q varied up to 45000 $m^3 s^{-1}$. The shear stress for equilibrium flow shown in Fig. 8 was estimated from flow weight resolved parallel to the bed (i.e., $\rho_w g d S$).

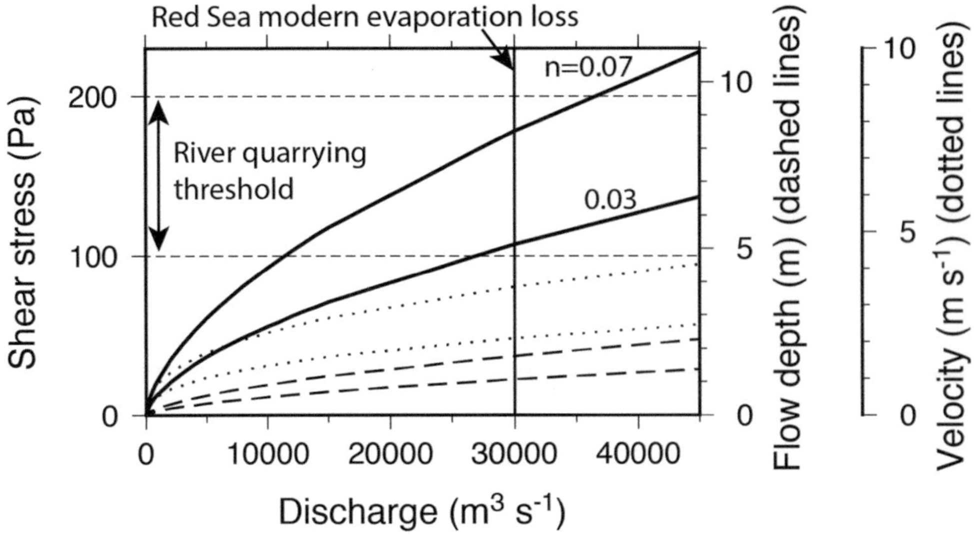

Fig. 8 Bed shear stress and other hydraulic properties of an open channel equilibrium flow of 8 km width, 1% gradient and with varied discharge. Values were calculated separately with Mannings' constant *n* of 0.03 and 0.07 (in the case of velocity, the lower dotted line was computed with the higher *n*-value)

The bed shear thresholds for quarrying in Fig. 8 are based on observations of visible movement of bedrock blocks at 100–200 Pa in a river in flood by Snyder et al. (2003). That this threshold might apply more generally to rivers is suggested by compiled flood shear stresses from various tectonic and climatic settings, which have values limited mostly to the range 100–1000 Pa (Mitchell 2014; Mitchell et al. 2012). If the substrate eroded along the channel comprised jointed indurated materials, this threshold could be appropriate. If the materials were less indurated, a lower threshold would be more appropriate.

The graphs in Fig. 8 suggest that inflow discharge at half the modern evaporation loss would produce stresses mostly just below the threshold for bedrock erosion, though probably within the range of erosion of less indurated materials. On the other hand, if the inflow occurred at greater discharge because of abrupt failure of an obstruction, the stresses could have been sufficient to erode bedrock. These calculations are not intended to test the likelihood of an inflow event, though they demonstrate that channel excavation with reasonable discharge values is a viable explanation for the channel.

5.4.3 Paleoceanography

The possibility of a deep drawdown of the Red Sea and subsequent re-flooding is not expected from reading the literature of the Plio-Pleistocene. During the Late Pleistocene glacial stages (to ~500 ka, constrained by continuous sediment cores), the sea appears not to have become completely isolated and hypersaline because δ^{18}O values in Red Sea pelagic sediments remained comparable with LGM values that represent continuous though restricted exchange with the Gulf of Aden (Biton et al. 2008; Rohling et al. 2009; Siddall et al. 2003, 2004). Furthermore, some benthic foraminifera and pteropods survived through those stages, suggesting a lack of hypersalinity (Almogi-Labin et al. 1998; Fenton et al. 2000).

In the Miocene, the solutes that formed the Red Sea evaporites were probably supplied by seawater from the Mediterranean (Bosworth et al. 2005; Coleman 1974; Hughes et al. 1992; Orszag-Sperber et al. 1998). Evidence for this includes calcareous nannofossils identified by Boudreaux (1974) within the intercalated black shales of the evaporites drilled at DSDP Sites 225 and 227 in the central Red Sea, which have affinities with the Mediterranean (Orszag-Sperber et al. 1998), and Mediterranean fauna in marginal commercial wells (e.g., Bunter and Abdel Magid 1989; Hughes 2014).

At some uncertain date, a transition occurred between Mediterranean exchange to Indian Ocean exchange with the opening of the Bab el Mandab Strait (Hughes et al. 1992; Orszag-Sperber et al. 1998; Richardson and Arthur 1988) (Fig. 1). By the early Pliocene, an uplifted sill separated the Red Sea from the Mediterranean and an Indian Ocean fauna invaded the Red Sea (Orszag-Sperber et al. 1998; Purser and Hötzl 1988; Swartz and Arden 1960). Seismic reflection data bear on the issue of isolation. Those data from the central Red Sea show a reflective upper Pleistocene, which can be explained by layers of preferentially more rigid, and hence higher seismic velocity, aragonite-rich sediment produced when the sea was more saline during sea-level lowstands (Almogi-Labin et al. 1996; Milliman et al. 1969; Mitchell et al. 2015). In contrast, the lower Plio-Pleistocene is typically more transparent in seismic reflection data from the central and southern Red Sea (summarized by Mitchell et al. 2015). Such transparency would be consistent with a lack of seismic impedance variation over sediment intervals equal to or greater than a quarter of a seismic wavelength (a few metres for these data), implying a lack of extreme density and velocity produced by aragonite layers, and hence ultimately that the sea was continuously exchanging water with the global oceans (Mitchell et al. 2015, 2017).

According to the literature, the only period in which a rapid inflow event is expected to have occurred was at the end of the evaporite stage at the Miocene-Pliocene boundary. Widespread unconformities have been detected at this stratigraphic level in commercial wells (Bunter and Abdel Magid 1989; Hughes et al. 1992) and are observed in seismic data of the S reflection representing the top of the Miocene (Izzeldin 1987; Mitchell et al. 2017). Although coastal erosion during a rising sea could explain these features (Mitchell et al. 2017), rapid sea currents may also help to explain them.

An origin of the channel at the Miocene-Pliocene boundary would seem to be an extreme interpretation given that the boundary is 5.3 million years old. Nevertheless, the currents are maintaining at least the shallower parts of the channel morphology by preventing some sediment deposition. GPS and plate-reconstruction data indicate that modest tectonic movements have occurred across the sill for the last few million years. As the tectonic and slope failure explanations seem weak or at least equivocal, we suggest this more extreme explanation of rapid inflow at the end of the Miocene is worthy of further consideration. Clearly more data (seismic, coring and in situ oceanographic measurements) are required around the North Hanish Channel and the section where it enters the Axial Trough to test this idea.

6 Conclusions

Current meter measurements show large outflows of $\sim 1 \text{ m s}^{-1}$ over Hanish Sill at present, which are likely maintaining the morphology of the shallower parts of the channel around Hanish Sill. However, currents are anticipated to be much weaker below 200 m within Red Sea Deep Water, which lies below Red Sea Outflow Water at intermediate depths, so the lower channel is not obviously explained by the currents. Evidence for slope failure was found in seismic data of the southern slope of the Axial Trough and may have contributed to the lower channel morphology, although the 0.5° gradient of the channel would be low for slope failure in carbonate shelf sediments. Although the available data do not rule out the existence of faults parallel with the Axial Trough, which may have guided inflows and thus explain the straight morphology of the channel, Bouguer gravity anomalies do not reveal any major graben structure under the North Hanish Channel. Therefore, the channel most likely does not have a tectonic origin. By a process of elimination, therefore, the North Hanish Channel may have formed by a massive inflow event, perhaps as long ago as 5.3 Ma at the end of the Miocene evaporite phase.

Acknowledgements Rose Anne Weissel is thanked for help in locating and scanning the RV *Conrad* data used in this study. All other data besides the current meter records used here were obtained from public sources (http://www.ngdc.noaa.gov/, http://www.geomapapp.org/, http://topex.ucsd.edu/). Andrew Goodwillie clarified the origin of data contributing to the GMRT grid in this area. Eelco Rohling explained why deep drawdown was unlikely to have occurred in the late Pleistocene. We would like to thank the Saudi Geological Survey for organizing the publication of this book on the Red Sea. Thanks also to Graeme Eagles for reviewing an earlier version of this chapter and to the three anonymous reviewers of the present chapter for helpful comments. Figures in this article were created with the "GMT" software system (Wessel and Smith 1991).

References

Almogi-Labin A, Hemleben C, Meischner D, Erlenkeuser H (1996) Response of Red Sea deep-water agglutinated foraminifera to water-mass changes during the Late Quaternary. Mar Micropal 28:283–297

Almogi-Labin A, Hemleben C, Meischner D (1998) Carbonate preservation and climate changes in the central Red Sea during the last 380 kyr as recorded by pteropods. Mar Micropal 33:87–107

Berger WH, Piper DJW (1972) Planktonic foraminifera: differential settling, dissolution and redeposition. Limn Oceanogr 17:275–287

Berthois L, Le Calvez Y (1960) Etude de la vitesse de chute des coquilles de foraminiferes planctoniques dans un fluide comparativement a celle de grains de quartz. Inst Peches Mar 24:293–301

Biton E, Hildor H, Peltier WR (2008) Red Sea during the last glacial maximum: implications for sea level reconstruction. Paleocean 23:paper PA1214. https://doi.org/10.1029/2007pa001431

Bosworth W, Huchon P, McClay K (2005) The Red Sea and Gulf of Aden basins. J African Earth Sci 43:334–378

Boudreaux JE (1974) Calcareous nannoplankton ranges, Deep Sea Drilling Project Leg 23. In: Whitmarsh RB, Wesser DE, Ross DA (eds) Initial reports of the Deep Sea Drilling Project. US Government Printing Office, Washington, DC, pp 1073–1090

Bower AS, Johns WE, Fratantoni DM, Peters H (2005) Equilibration and circulation of Red Sea outflow water in the western Gulf of Aden. J Phys Ocean 35:1963–1985

Bunter MAG, Abdel Magid AEM (1989) The Sudanese Red Sea: 1. New developments in stratigraphy and petroleum-geological evolution. J Petrol Geol 12:145–166

Cande SC, Kent DV (1995) Revised calibration of the geomagnetic polarity timescale for the late Cretaceous and Cenozoic. J Geophys Res 100:6093–6095

Cember RP (1988) On the sources, formation, and circulation of Red Sea deep water. J Geophys Res 93:8175–8191

Chang YS, Özgökmen TM, Peters H, Xu X (2008) Numerical simulation of the Red Sea outflow using HYCOM and comparison with REDSOX observations. J Phys Ocean 38:337–358

Chu D, Gordon RG (1998) Current plate motions across the Red Sea. Geophys J Int 135:313–328

Coleman RG (1974) Geologic background of the Red Sea. In: Whitmarsh RB, Weser OE, Ross DA (eds) Initial reports of the Deep Sea Drilling Project, vol 23. U.S. Government Printing Office, Washington, DC, pp 813–819

Eagles G, Gloaguen R, Ebinger C (2002) Kinematics of the Danakil microplate. Earth Planet Sci Lett 203:607–620

Einsele G, Werner F (1972) Sedimentary processes at the entrance Gulf of Aden/Red Sea. In: Seibold E, Closs H (eds) "Meteor" Forschungsergebnisse, Reihe C. Bornträger, Berlin, pp 39–62

Eriş KK, Ryan WBF, Çağatay MN, Sancar U, Lericolais G, Ménot G, Bard E (2007) The timing and evolution of the post-glacial transgression across the Sea of Marmara shelf south of İstanbul. Mar Geol 243:57–76

Fenton M, Geiselhart S, Rohling EJ, Hemleben C (2000) Aplanktonic zones in the Red Sea. Mar Micropal 40:277–294

Garcîa M, Hernández-Molina FJ, Llave E, Stow DAV, Léon R, Fernández-Puga MC, Río V, Somoza L (2008) Contourite erosive features caused by the Mediterranean outflow water in the Gulf of Cadiz: quaternary tectonic and oceanographic implications. Mar Geol 257:24–40

Garcia-Castellanos D, Estrada F, Jiménez-Munt I, Gorini C, Fernàndez M, Vergés J, De Vicente R (2009) Catastrophic flood of the Mediterranean after the Messinian salinity crisis. Nature 462:778–782

Gardner TW (1983) Experimental study of knickpoint migration and longitudinal profile evolution in cohesive homogenous material. Geol Soc Am Bull 94:664–672

Garfunkel Z, Ginzburg A, Searle RC (1987) Fault pattern and mechanism of crustal separation along the axis of the Red Sea from side scan sonar (GLORIA) data. Ann Geophys 5B:187–200

Gass IG (1970) The evolution of volcanism in the junction area of the Red Sea, Gulf of Aden and Ethiopian Rift. Phil Trans Roy Soc Lond A267:369–382

Griffin DL (1999) The late Miocene climate of northeast Africa: unravelling the signals in the sedimentary succession. J Geol Soc Lond 156:817–826

Hamilton EL, Bachman RT (1982) Sound velocity and related properties of marine sediments. J Acoust Soc Am 72:1891–1904

Hiscott RN, Aksu AE, Yaşar D, Kaminski MA, Mudie PJ, Kostylev VE, MacDonalda JC, Işler FI, Lord AR (2002) Deltas south of the Bosphorus Strait record persistent Black Sea outflow to the Marmara Sea since ~10 ka. Mar Geol 190:95–118

Hughes GW (2014) Micropalaeontology and palaeoenvironments of the Miocene Wadi Waqb carbonate of the northern Saudi Arabian Red Sea. GeoArabia 19:59–108

Hughes GW, Abdine S, Girgis MH (1992) Miocene biofacies development and geological history of the Gulf of Suez. Egypt. Mar Petrol Geol 9:2–28

Ilicak M, Özgökmen TM, Peters H, Baumert HZ, Iskandarani M (2008) Performance of two-equation turbulence closures in three-dimensional simulations of the Red Sea overflow. Ocean Model 34:122–139

Izzeldin AY (1987) Seismic, gravity and magnetic surveys in the central part of the Red Sea: their interpretation and implications for the structure and evolution of the Red Sea. Tectonophysics 143:269–306

Jarosz E, Murray SP, Inoue M (2005) Observations on the characteristics of tides in the Bab el Mandab Strait. J Geophys Res 110:paper C03015. https://doi.org/10.01029/02004jc002299

Lambeck K, Purcell A, Flemming NC, Vita-Finzi C, Alsharekh AM, Bailey GN (2011) Sea level and shoreline reconstructions for the Red Sea: isostatic and tectonic considerations and implications for hominin migration out of Africa. Quat Sci Rev 30:3542–3574

Luján M, Crespo-Blanc A, Comas M (2010) Morphology and structure of the Camarinal Sill from high-resolution bathymetry: evidence of fault zones in the Gibraltar Strait. Geo-Mar Lett 31:163–174

Maillard C, Soliman G (1986) Hydrography of the Red Sea and exchanges with the Indian Ocean in summer. Oceanol Acta 9: 249–269

Manheim FT, Dwight L, Belastock RA (1974) Porosity, density, grain density, and related physical properties of sediments from the Red Sea drill cores. In: Whitmarsh RB, Weser OE, Ross DA (eds) Initial reports of the Deep Sea Drilling Project, vol 23. U.S. Government Printing Office, Washington, DC, pp 887–907

McAdoo BG, Pratson LF, Orange DL (2000) Submarine landslide geomorphology, US continental slope. Mar Geol 169:103–136

McClusky S, Reilinger R, Mahmoud S, Ben Sari D, Tealeb A (2003) GPS constraints on Africa (Nubia) and Arabia plate motions. Geophys J Int 155:126–138

Miller JR (1991) The influence of bedrock geology on knickpoint development and channel-bed degradation along downcutting streams in south-central Indiana. J Geol 99:591–605

Miller MC, Komar PD (1977) The development of sediment threshold curves for unusual environments (Mars) and for inadequately studied materials (foram sands). Sedimentology 24:709–721

Miller MC, McCave IN, Komar PD (1977) Threshold of sediment motion under unidirectional currents. Sedimentology 24:507–528

Milliman JD, Ross DA, Ku T-L (1969) Precipitation and lithification of deep-sea carbonates in the Red Sea. J Sed Petrol 39:724–736

Mitchell NC (2006) Morphologies of knickpoints in submarine canyons. Geol Soc Am Bull 118:589–605

Mitchell NC (2014) Bedrock erosion by sedimentary flows in submarine canyons. Geosphere 10. https://doi.org/10.1130/ges01008.01001

Mitchell NC (2015) Lineaments in gravity data of the Red Sea. In: Rasul NMA, Stewart ICF (eds) The Red Sea: the formation, morphology, oceanography and environment of a young ocean basin. Springer Earth System Sciences, Berlin, pp 123–133

Mitchell NC, Park Y (2014) Nature of crust in the central Red Sea. Tectonophys 628:123–139

Mitchell NC, Huthnance JM, Schmitt T, Todd B (2012) Threshold of erosion of submarine bedrock landscapes by tidal currents. Earth Surf Proc Landf 38:627–639

Mitchell NC, Ligi M, Rohling EJ (2015) Red Sea isolation history suggested by Plio-Pleistocene seismic reflection sequences. Earth Planet Sci Lett 430:387–397

Mitchell NC, Ligi M, Feldens P, Hübscher C (2017) Deformation of a young salt giant: Regional topography of the Red Sea Miocene evaporites. Basin Res 29:352–369

Neumann AC, McGill DA (1962) Circulation of the Red Sea in early summer. Deep Sea Res 8:223–235

Orszag-Sperber F, Harwood G, Kendall A, Purser BH (1998) Review of the evaporites of the Red Sea-Gulf of Suez rift. In: Purser BH, Bosence DWJ (eds) Sedimentation and tectonics of rift basins: Red Sea-Gulf of Aden. Chapman & Hall, London, pp 409–426

Paphitis D, Collins MB, Nash LA, Wallbridge S (2002) Settling velocities and entrainment thresholds of biogenic sands (shell fragments) under unidirectional flow. Sedimentology 49:211–225

Parsons DR, Peakall J, Aksu AE, Flood RD, Hiscott RN, Besiktepe S, Mouland D (2010) Gravity-driven flow in a submarine channel bend: direct field evidence of helical flow reversal. Geology 38:1063–1066

Peters H, Johns WE (2006) Bottom layer turbulence in the Red Sea outflow plume. J Phys Ocean 36:1764–1785

Pratt LJ, Johns W, Murray SP, Katsumata K (1999) Hydraulic interpretation of direct velocity measurements in the Bab el Mandab. J Phys Ocean 29:2769–2784

Purser BH, Hötzl H (1988) The sedimentary evolution of the Red Sea rift: A comparison of the northwest (Egyptian) and northeast (Saudi Arabian) margins. Tectonophys 153:193–208

Reilinger R, McClusky S, ArRajehi A (2015) Geodetic constraints on the geodynamic evolution of the Red Sea. In: Rasul NMA, Stewart ICF (eds) The Red Sea: the formation, morphology, oceanography and environment of a young ocean basin. Springer Earth System Sciences, Berlin, pp 135–150

Richardson M, Arthur MA (1988) The Gulf of Suez—northern Red Sea Neogene rift: a quantitative basin analysis. Mar Petrol Geol 5:247–270

Roeser HA (1975) A detailed magnetic survey of the southern Red Sea. Geol Jahrb 13:131–153

Rohling EJ, Grant K, Bolshaw M, Roberts AP, Siddall M, Hemleben C, Kucera M (2009) Antarctic temperature and global sea level closely coupled over the past five glacial cycles. Nature Geosci 2:500–504

Rubin AM (1992) Dike-induced faulting and graben subsidence in volcanic rift zones. J Geophys Res 97:1839–1858
Ryan WBF, Carbotte SM, Coplan JO, O'Hara S, Melkonian A, Arko R, Wiessel RA, Ferrini V, Goodwillie A, Nitsche F, Bonczkowski J, Zemsky R (2009) Global multi-resolution topography synthesis. Geochem Geophys Geosys 10:Paper Q03014
Sandwell DT, Smith WHF (1997) Marine gravity anomaly from Geosat and ERS-1 satellite altimetry. J Geophys Res 102:10039–10054
Sandwell D, Müller RD, Smith WHF, Garcia E, Francis R (2014) New global marine gravity model from CryoSat-2 and Jason-1 reveals buried tectonic structure. Science 346:65–67
Seidl M, Dietrich WE (1992) The problem of channel erosion into bedrock. In: Schmidt KH, de Ploey J (eds) Functional geomorphology: landform analysis and models. Catena Supplement, vol 23, pp 101–124
Siddall M, Rohling EJ, Almogi-Labin A, Hemleben C, Meischner D, Schmeizer I, Smeed DA (2003) Sea-level fluctuations during the last glacial cycle. Nature 423:853–858
Siddall M, Smeed DA, Hemleben C, Rohling EJ, Schmeizer I, Peltier WR (2004) Understanding the Red Sea response to sea level. Earth Planet Sci Lett 225:421–434
Smeed DA (2004) Exchange through the Bab el Mandab. Deep-Sea Res II 51:455–474
Smith WHF, Sandwell DT (1997) Global sea floor topography from satellite altimetry and ship soundings. Science 277:1956–1962
Smith WHF, Wessel P (1990) Gridding with continuous curvature splines in tension. Geophysics 55(3):293–305
Snyder NP, Whipple KX, Tucker GE, Merritts DJ (2003) Importance of stochastic distribution of floods and erosion thresholds in the bedrock river incision problem. J Geophys Res 108:ETG 17-11–ETG 17-15. https://doi.org/10.1029/2001jb001655
Sofianos SS, Johns EW (2003) An oceanic general circulation model (OGCM) investigation of the Red Sea circulation, 2. Three-dimensional circulation in the Red Sea. J Geophys Res.107:Paper 3066. https://doi.org/10.1029/2001jc001185
Sofianos SS, Johns EW (2007) Observations of the summer Red Sea circulation. J Geophys Res 112:Paper C06025. https://doi.org/10.01029/02006jc003886
Sofianos SS, Johns EW, Murray SP (2002) Heat and freshwater budgets in the Red Sea from direct observations at Bab el Mandeb. Deep-Sea Res II 49:1323–1340
Southard JB, Young RA, Hollister CD (1971) Experimental erosion of calcareous ooze. J Geophys Res 76:5903–5909
Swartz DH, Arden DD (1960) Geologic history of the Red Sea area. Bull Am Assoc Petrol Geol 44:1621–1637
Telford WM, Geldart LP, Sheriff RE, Keys DA (1976) Applied geophysics. Cambridge University Press, New York, p 860
Webber NB (1971) Fluid mechanics for civil engineers. Chapman and Hall, New York, p 340
Werner G, Lange K (1975) A bathymetric survey of the sill area between the Red Sea and the Gulf of Aden. Geologisches Jarbuch D 13:125–130
Wessel P, Smith WHF (1991) Free software helps map and display data. EOS Trans Am Geophys Union 72:441
Whitmarsh RB, Weser OE, Ross DA (1974) Initial reports of the Deep Sea Drilling Project, 23B. U. S. Government Printing Office, Washington, DC

Cenozoic Faults and Seismicity in Northwest Saudi Arabia and the Gulf of Aqaba Region

M. John Roobol and Ian C. F. Stewart

Abstract

Following the November 1995 earthquake in the Gulf of Aqaba that affected countries in the area including northwest Saudi Arabia, a study was made to identify the youngest geological faults in the area. These faults have been termed Cenozoic faults, although most are Neogene in age, being younger than 23 Ma. A number of previously unknown structures have been identified. These include a zone of shattered and intensely faulted rocks 25 km wide that extends inland into Saudi Arabia from the coast of the Gulf of Aqaba. In this zone, shattered granite intrusions are eroded in a similar manner to sandstone and produce deep, vertical walled, narrow gorges, such as Tayyib al Ism. Faults parallel to the Gulf of Aqaba occur as deep wadis where the geological contacts are offset by sinistral displacements of up to 9 km, and show systematic shoulder uplift away from the gulf. Another major Cenozoic structure contained within the Saudi Arabian side of the Gulf of Aqaba is a major fold that extends 50 km inland from the coast. In this fold the Precambrian lithologic units, dikes and faults are rotated 90 degrees anticlockwise with the appearance of a drag fold due to the approximately 115 km sinistral offset of the Gulf of Aqaba. This structure is not repeated in Sinai and occurs only on the Arabian side of the Gulf. It is interpreted not as a drag fold but as a fold that probably resulted from the initial displacement along the Arabian Plate from Sinai. Later faults isolated the structure within Saudi Arabia. Reactivated Precambrian faults within the structure are also curved. Aeromagnetic anomalies show that the traces of major Tertiary gabbro dikes that parallel the Red Sea coast of Saudi Arabia are also curved within the fold, but their curvature of about 45 degrees anticlockwise is less than that of the Precambrian rocks and dikes, which suggests dike emplacement occurred after folding commenced. Maps that cover an area 550 km east-west by 360 km north-south and include the Gulf of Aqaba and adjacent areas show the whole area to be cut by thousands of Cenozoic faults (about two and a half thousand of the larger faults are shown here). The oldest faults occur in a 250-km wide belt of pre-rift structures that include graben structures that lie to the east of the Gulf of Aqaba fault system. These faults are oriented north-northwest in the north and curve to the northwest in the south of the study area, and are named Ribbon faults because they divide the country rocks into long thin strips from 100 to 500 m wide. These faults mainly occur within Paleozoic sandstone where they are preserved as walls cemented by iron oxides/hydroxides. Where the Ribbon faults occur within Precambrian Shield and Tertiary sediments the oxide/hydroxide fills are absent. The Ribbon faults result from crustal extension as they form horst-graben structures with topography of a few metres. In some areas open fissures occur without apparent displacement. The aeromagnetic maps show that some of the major graben structures are underlain by large Tertiary gabbro dikes. The system is mainly seismically inactive today with a few local exceptions, such as a cluster of earthquake activity about 75 km southeast of Tabuk city. Most active are the north-south to north-northeast-trending Gulf of Aqaba system of faults that include those of the floor of the Gulf of Aqaba. Other notable fault systems in the area are a prominent set of pre-rift east–west faults, some of which contain traces of Tertiary gabbro dikes and some of which become curved toward the Gulf of Aqaba like the reactivated Precambrian faults. The map area preserves a record of about 25 million years of faulting that commenced in pre-Red Sea rift time, and also records the initiation of separation of the Arabian Plate from Sinai by fold deformation prior to

M. J. Roobol (✉)
The Anchorage, Sandy Haven, St. Ishmaels, Haverfordwest, Pembrokeshire, SA62 3DN, UK
e-mail: roobolinc2000@hotmail.com

I. C. F. Stewart
Stewart Geophysical Consultants Pty. Ltd, Adelaide, SA 5069, Australia
e-mail: stewgeop@senet.com.au

N. M. A. Rasul and I. C. F. Stewart (eds.), *Geological Setting, Palaeoenvironment and Archaeology of the Red Sea*,
https://doi.org/10.1007/978-3-319-99408-6_13

faulting. Some minor folding occurs from drag and compression of a belt of Tertiary sedimentary rocks originally continuous with the Gulf of Suez which were transported 115 km into the Gulf of Aqaba. The Gulf of Aqaba region remains the most active seismic area in Saudi Arabia and is the site of three large earthquakes in 1983, 1993 and 1995.

1 Introduction

Several of the most powerful earthquakes experienced in Saudi Arabia during recent decades were centred in the Gulf of Aqaba and affected the inhabited areas of northwest Saudi Arabia, especially the town of Haql (Figs. 1 and 2). These include the 1983 swarm, with a maximum magnitude M_l of 5 (El Isa et al. 1984), a swarm in 1993 with a maximum M_w of 6.1, and the November 1995 M_w 7.3 event (Smith and Bokhari 1984; Bazzari et al. 1990; Klinger et al. 1999; Roobol et al. 1999). Following the latter magnitude 7.3 event a geological investigation was initiated by the Saudi Geological Survey to identify the youngest fault systems responsible for these earthquakes. The results of this study for the northwest of Saudi Arabia and adjacent regions are presented here in a revised version of that given by Roobol and Stewart (2009), including more recent seismicity data.

The study area forms the westernmost part of the Arabian lithospheric plate at the junctions of the Red Sea and Gulf of Aqaba. Today, the Red Sea is a young ocean basin with a central spreading axis opening at a mean rate of about 16 mm per year, with a present-day rate of 24 ± 1 mm/year in the south (at 12°N) and 7 ± 1 mm/year in the north (Reilinger et al. 2015). The opening of the Red Sea is considered to have taken place in two stages that correspond to two distinct volcanic episodes along the western margin of the Arabian Plate (Camp and Roobol 1992). An earlier stage of rifting occurred between about 30 to 20 Ma, followed by a later stage of rifting during at least the past 5 million years and possibly 10 million years (Girdler and Southren 1987; Bayer et al. 1988; McGuire and Bohannon 1989; Bohannon et al. 1989; Johnson 1998). The deeply-eroded fault boundary between the Cenozoic coast plain sediments and the Precambrian shield has been named the "Master Listric Coastal Fault" by Roobol and Kadi (2008).

The Stage 1 opening of the Red Sea corresponded to the eruption of the older Tertiary harrats (lava fields) of western Saudi Arabia between about 30 and 14 Ma. The second stage of opening resulted in the formation of the 200-km long Gulf of Aqaba and its 1,000 km extension as the Dead Sea Rift Zone. This is a left lateral transform fault system that forms the northwest boundary of the Arabian Plate. It is essentially a number of subparallel sinistral transform faults with associated pull-apart basins. The Stage 2 opening corresponds to the formation of the younger harrats of the western Arabian Plate from about 10 Ma to the present time. Ben-Avraham et al. (2008) estimate the sinistral motion on the Gulf of Aqaba and Dead Sea Rift to be approximately 5–7 mm/year. Reilinger et al. (2015) suggest that at the southern end of the Gulf of Aqaba the left-lateral motion of Arabia with respect to Sinai is 4.5 mm/year with 2.2 mm/year extension. From GPS measurements, the present slip rate on the Wadi Araba Fault section of the Rift immediately to the north of the Gulf of Aqaba is given by al Tarazi et al. (2011) as 4.9 ± 0.4 mm/year, while Sadeh et al. (2012) obtained a slip rate of 4.7–5.4 mm/year. Reilinger et al. (2015) indicate that there is almost pure left lateral strike slip with a geodetic slip rate along the southern Dead Sea Fault of 4.4 ± 0.3 mm/year where they quote a total post-Late Miocene (11 Ma) offset on the Fault of 45 ± 5 km. The fault zone is described as a transform and not a transcurrent fault because it separates two tectonic plates. Ben-Avraham (1985) describes it as "a plate boundary of the transform type that connects the Red Sea, where seafloor spreading occurs, with the Zagros-Taurus zone of continental collision". He also notes that it "is of particular interest because it is one of the two places in the world (the Gulf of California-Imperial Valley is the other), where a mid-ocean ridge system changes into a transform system and runs into a continent".

The fault-bounded Maqna Block of Tertiary sedimentary units measures 50 km by 50 km (Fig. 3) and constitutes the extreme western point of Saudi Arabia. Internally the sedimentary units are folded into an anticlinorium in the west (with a core of Precambrian rocks) and a synclinorium in the east. Following Stage 1 opening of the Red Sea–Gulf of Suez, this large block of Tertiary sediments was part of the Tertiary sedimentary belts that line the margins of the Gulf of Suez and the early Red Sea (Coleman 1977). The Maqna Block can be regarded as part of the Gulf of Suez Tertiary sedimentary belt that has been displaced 115 km into the present Gulf of Aqaba by sinistral movement along the Aqaba-Levant fault system. The folding of the Tertiary sediments into an anticlinorium with a core of Precambrian basement and associated synclinorium is believed to result from lateral compression of the Maqna Block as it was transported into the Aqaba-Levant fault system.

A first attempt to fit the data shown here to a Landsat base failed because detailed structures could not be seen or positioned accurately. However, when the Shuttle Radar Topography Mission (SRTM) topographic grid data became available in 2000, it was found to be ideally suited to the deserts of Saudi Arabia, where the lack of vegetation enables the detailed SRTM information to be a good approximation

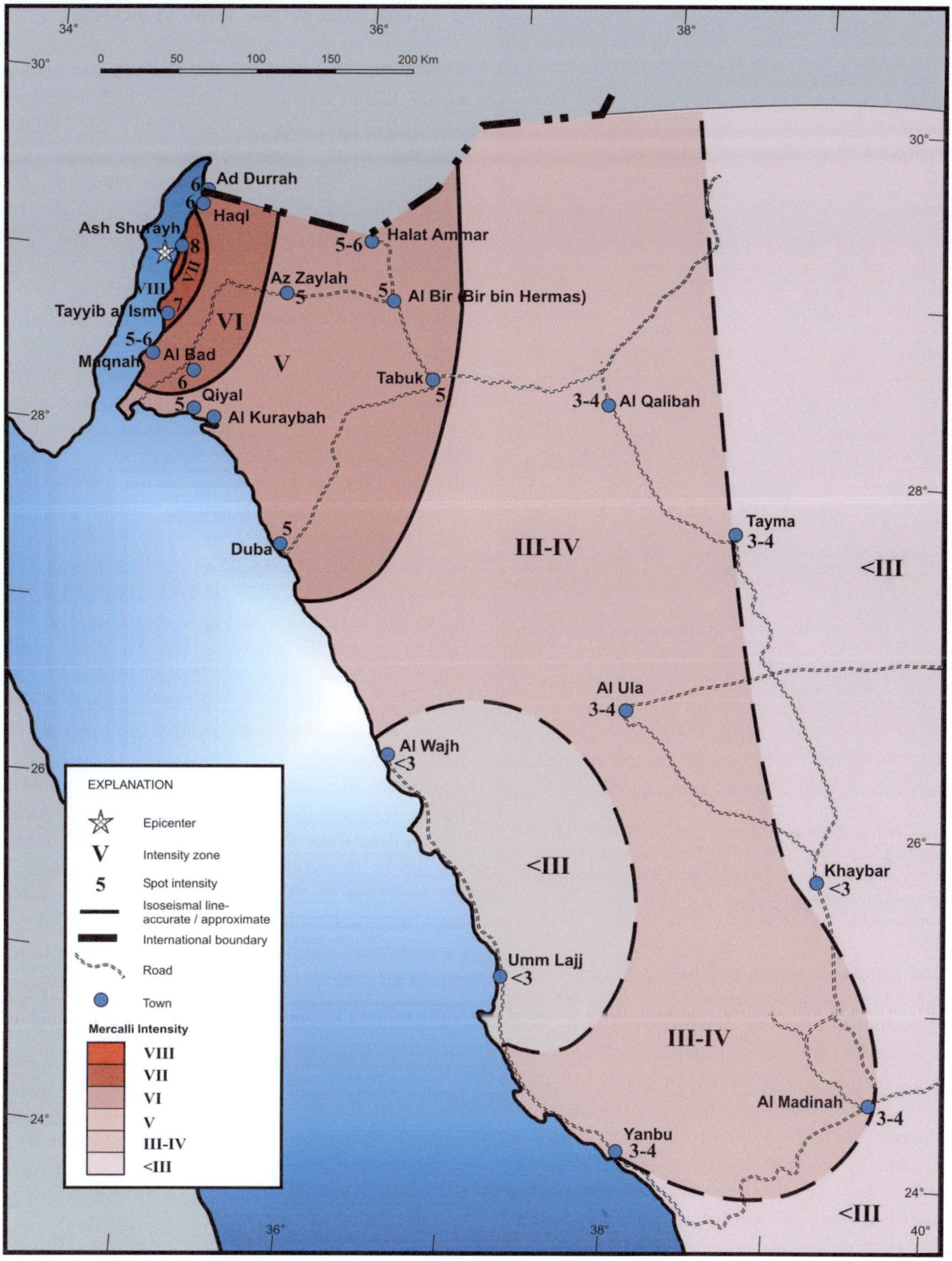

Fig. 1 Isoseismal map of the effects of the Gulf of Aqaba earthquake of November 22, 1995 in Saudi Arabia, showing zones of Mercalli intensity (after Roobol et al. 1999)

Fig. 2 One of the fault scarps of November 22, 1995 in the zone of epicentral rupture. The scarp is 40 cm high and the fault shows 10 cm of opening. It is situated at the foot of a prehistoric fault scarp 12 m high in unconsolidated Quaternary/Tertiary fanglomerate. The location is latitude 29°02'44.97" N, longitude 34°51'36.84" E

to the actual surface elevation. A 200-m grid of the SRTM data on a UTM projection is shown in Fig. 4, where the colour interval is 100 m and the maximum height is 2608 m above sea level. The first vertical derivative (FVD) for the northern half of this area is shown in Fig. 5 in order to accentuate local variations used here for the analysis of structural trends and faults. In this FVD image locally high values are white or light and low values are dark.

Many of the main structural trends can also be observed on images of aeromagnetic data, which were only available to us for that part of the study area within Saudi Arabia. The magnetic grid used here was compiled from a number of airborne surveys carried out over the Precambrian shield and adjacent Phanerozoic rocks, as described by Zahran et al. (2003). The reduced-to-the-pole (RTP) magnetic grid is shown in Fig. 6, where high values are red and low values are blue or purple.

2 Structure of the Gulf of Aqaba

The 200-km long Gulf of Aqaba with its 1,000-km extension as the Aqaba-Levant fault system, which includes the Dead Sea Rift Zone, has a long history of seismic activity (Al-Amri et al. 1991; Ben-Avraham et al. 1979, 2008; Hamiel et al. 2009). A paleoseismic study of the Lebanese section by Daëron et al. (2007) indicated that there have been 10–12 major earthquakes in the past 12,000 years. A similar study of the Jordanian section by Klinger et al. (2015) indicated 9 major earthquakes in the past 5,000 years. The November 1995 earthquake had a moment magnitude M_w of 7.3 with an epicentre just off the Saudi Arabian coastline. A 20-km long section of uninhabited coastline south of Haql showed epicentral rupture (Fig. 2; Williams et al. 2001), and aftershocks continued into 1996. Modelling of InSAR data by Klinger et al. (2000a) indicates that this event was due to mainly left-lateral slip of 2.1 m along a 56 km-long fault striking N195°E between the Elat (Eilat) and Aragonese Deeps.

The structure of the Gulf of Aqaba has been described by Ben-Avraham et al. (1979) and Ben-Avraham (1985) and the fault data is included in Fig. 7 (based on the FVD of the SRTM topography image in Fig. 5), which also includes data taken from Bender (1975) and Eyal et al. (1980). The major faults visible on the SRTM base maps are included in Fig. 7, but there are many smaller ones that are not shown here. In its simplest form, the Gulf of Aqaba can be regarded as a series of sinistral transform fault controlled pull-apart basins or rhomb-shaped grabens that are in places more than 10 km deep (Ben-Avraham et al. 2008). However, as pointed out by Ben-Avraham (1985) this is oversimplified and the structures are more complex. The steep sea cliffs along the northeastern shore of the Gulf of Aqaba are fault zones of shattered rock that plunge into the deep waters of the pull-apart basins of the Gulf (Fig. 8). There is an anticlockwise rotation of the Arabian Plate of 0.37° per Myr relative to Nubia (Reilinger et al. 2006; ArRajehi et al. 2010), as indicated by the width of the Red Sea between the shoulders of Precambrian basement. These are about 200 km apart in the north at latitude 27 degrees north and 350 km apart in the south at latitude 17 degrees north. The rotation requires a progressive opening of the Gulf of Aqaba so the basins become wider with time.

3 Asymmetric Shoulder Uplift Along the Red Sea

The Precambrian shield on both sides of the Red Sea, like that on either side of the Gulfs of Suez and Aqaba, shows topographic shoulders. The phenomenon is clearly expressed

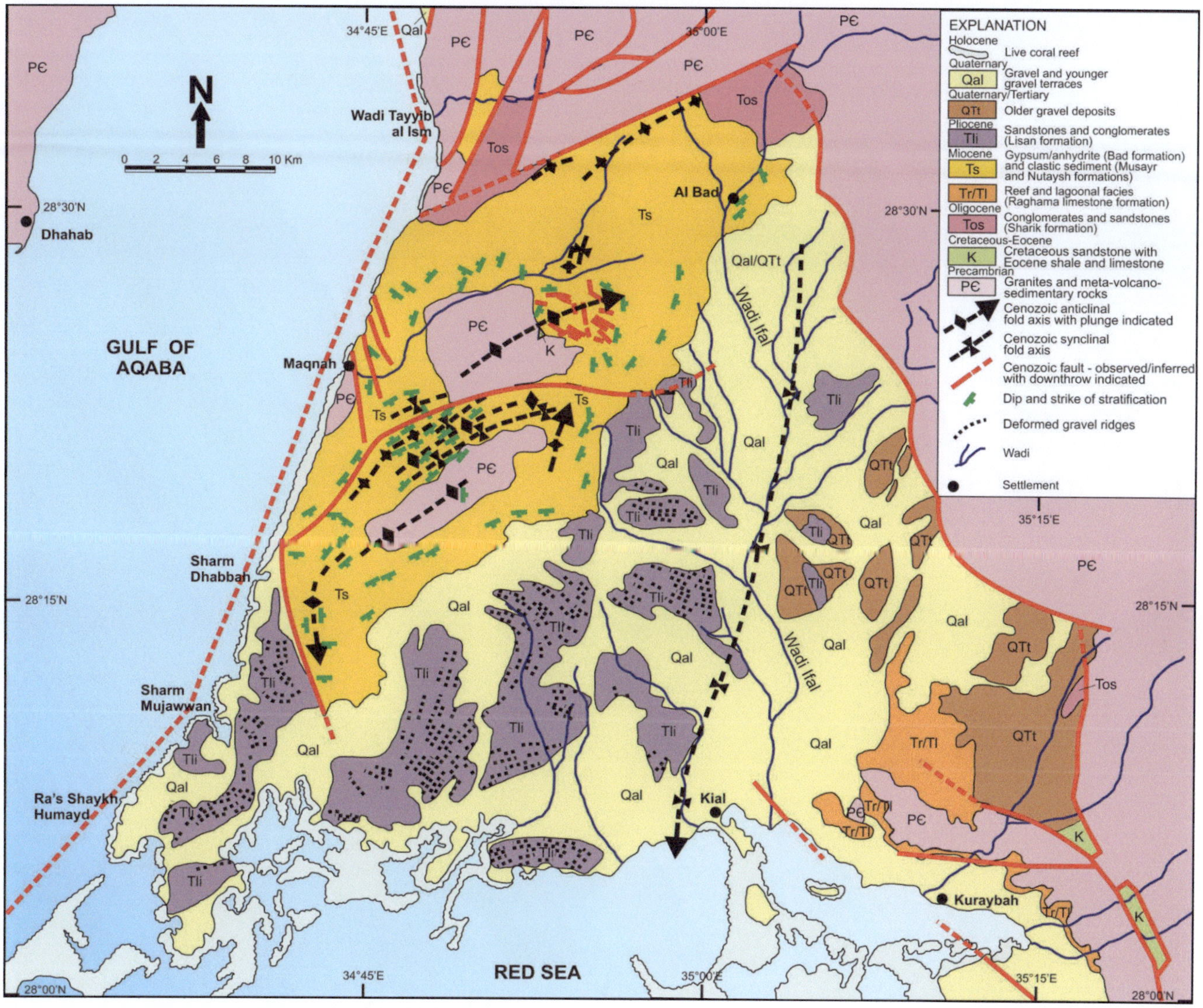

Fig. 3 Cenozoic folding and sedimentary units in the Maqna Block (modified after Motti et al. 1982; Bayer et al. 1988)

in Saudi Arabia along the road leading from the Red Sea port of Duba to the inland city of Tabuk. The road crosses the Red Sea coastal plain and then climbs the Red Sea escarpment to come out on the exhumed peneplain on top of the Precambrian shield that is overlain by erosional outliers of Cambrian/Ordovician Saq Sandstone. Looking northwest toward the Gulf of Aqaba, the exhumed peneplain with its cap of reddish sandstone can be seen to rise systematically toward the Gulf. Along the Saudi Arabian side of the Gulf of Aqaba considerable shoulder uplift has occurred of the Precambrian rocks which rise to a maximum of 2,580 m on Jabal Laws. Similarly, along the length of the Red Sea there are distinctive shoulders along the escarpments, as shown in Fig. 4. Fission track ages of apatite (Bohannon et al. 1989) have dated the uplift to between 13.8 and 5 Ma, and these dates suggest that 2.5–4 km of uplift occurred along the Arabian escarpment during the past 13.8 million years.

The escarpment crest on the eastern side of the Red Sea is higher than that on the west. The presence of topographic shoulders, but not the asymmetry, was regarded as due to isostatic rebound, possibly because of erosion of the cover. A mechanism proposed by Doglioni et al. (2003) that accounts for the asymmetry as a worldwide phenomenon associated with ocean ridges and rifts requires that depleted and lighter asthenosphere generated below an ocean ridge is shifted eastward relative to the lithosphere, resulting in a density deficit below the eastern flank so that the lithosphere rises or is buoyed up. Their model suggests that the westward drift of the lithosphere relative to the underlying mantle might be a global phenomenon.

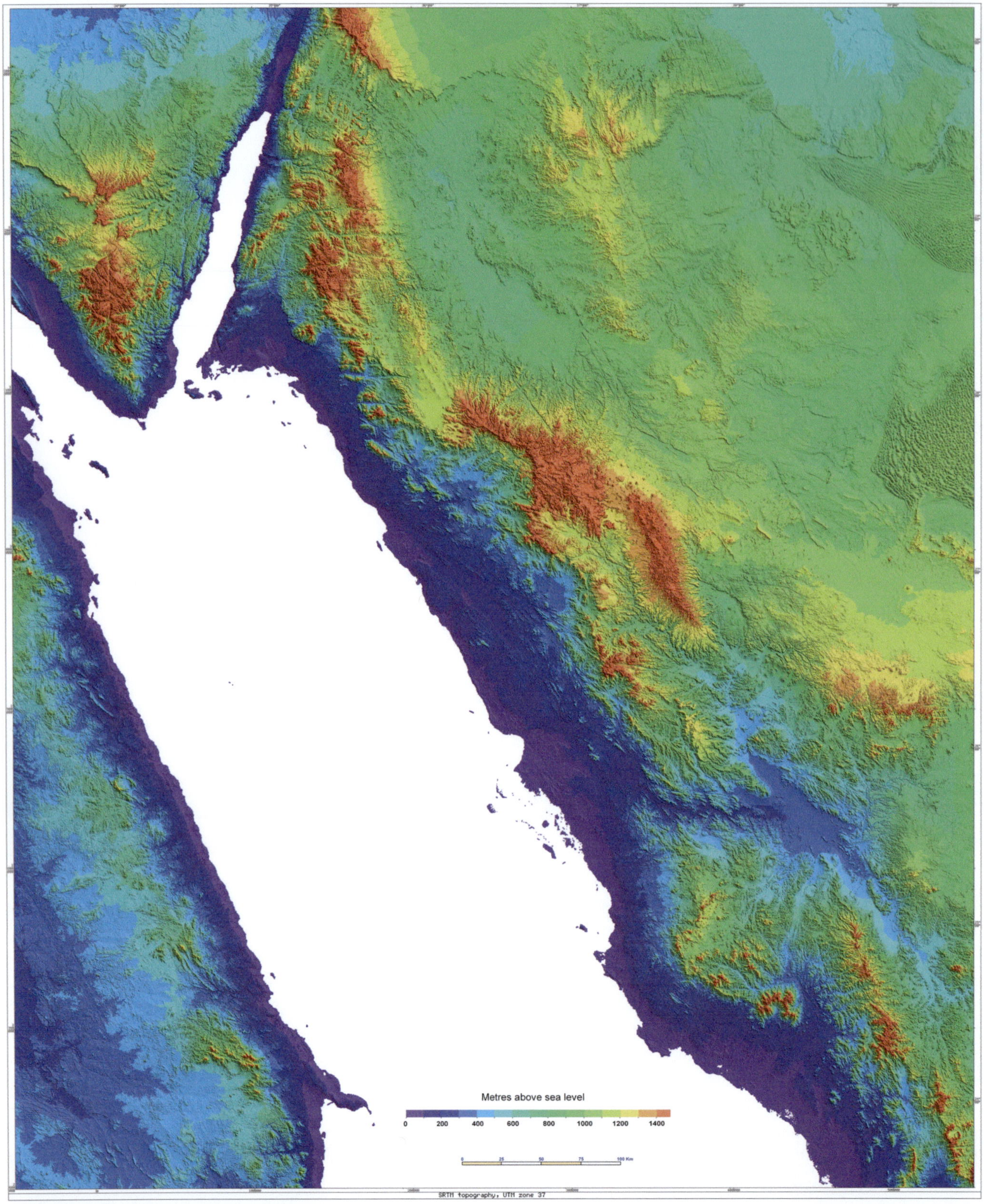

Fig. 4 Shuttle Radar Topography Mission (SRTM) surface elevation, around northern Red Sea and northwest Saudi Arabia, showing uplifted shoulders on both sides of the Red Sea including southern Sinai. The uplift is higher on the eastern side of the Red Sea. In northwest Saudi Arabia the Cenozoic lava field of Harrats Ar Raha and Uwayrid lies on top of the Red Sea coastal escarpment

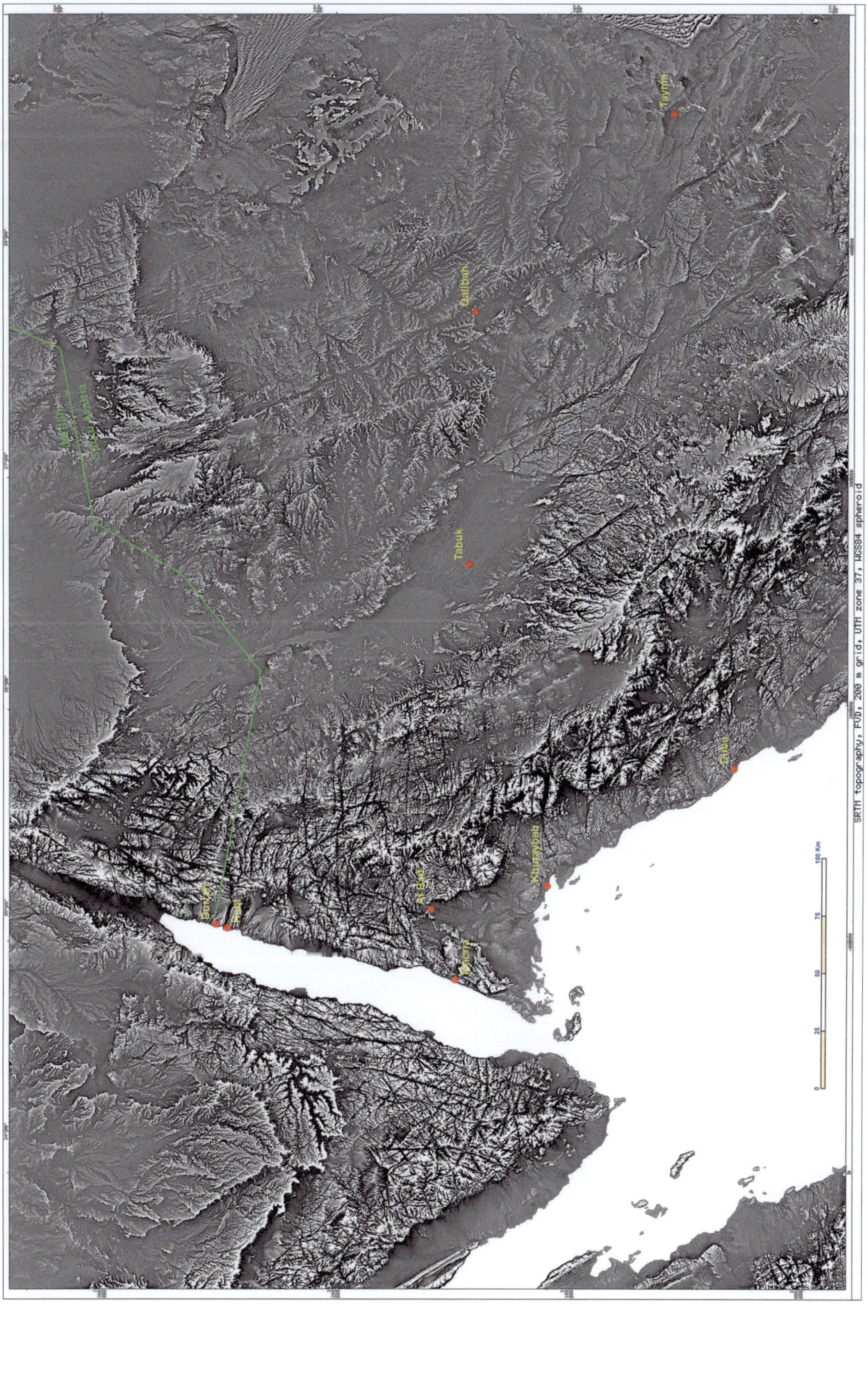

Fig. 5 SRTM surface elevation, first vertical derivative, northwest Saudi Arabia and adjacent areas

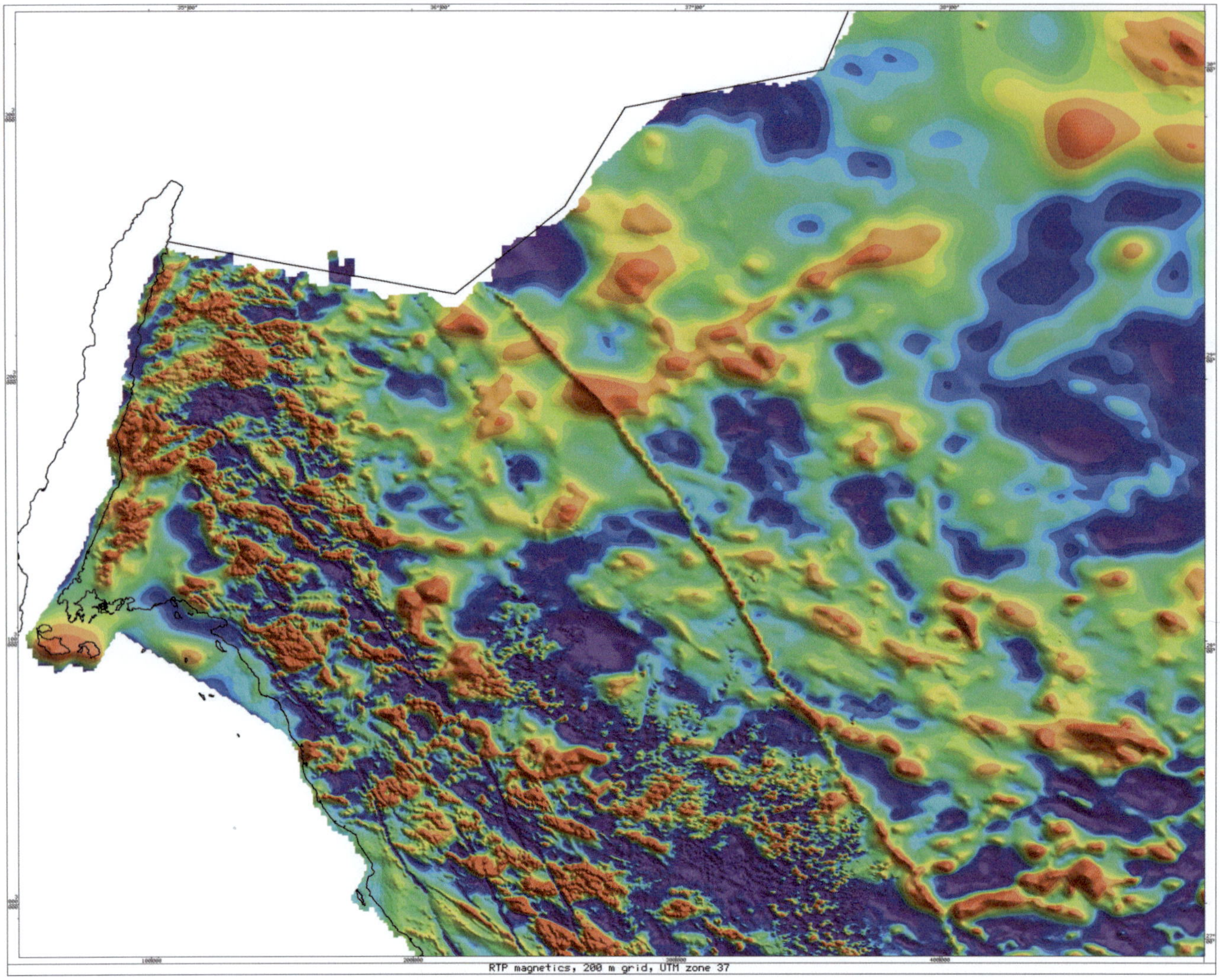

Fig. 6 Reduced to the pole aeromagnetic map of northwest Saudi Arabia

4 Recognition of Cenozoic Faults

In the course of this work a number of criteria evolved to separate Cenozoic from Precambrian faults. Cenozoic faults were readily identified because they offset the coastal Tertiary sediment such as the yellow-weathering green siltstone and sandstone. Another valuable tool was the Quaternary raised reefs, composed of resistant reef-facies limestone and more friable lagoonal facies calcarenites. Along the margins of the Red Sea, the Lower Miocene Raghama limestone provided a good reference rock that is cut by Cenozoic faults. The Raghama paleoreef is remarkably well preserved along the northernmost 600 km of Red Sea coastline where it coats the deeply-eroded Master Listric Fault that separates the coast plain Cenozoic sediments from the Precambrian shield. The internal structure of the limestone is exposed in gorges eroded through it, and commonly the limestone shows steep fore-reef bedding in the reef front facies which is often successively overlain by flat bedded limestones that overlap the foreset bedding. The top of the reef is flat and approximates the Lower Miocene sea level. This is also a valuable reference surface to use to identify Cenozoic faults. Along the front of the Raghama limestone reef are patches of black limestone where hot springs deposited Pb-Zn-Ag sulfide minerals now weathered to secondary minerals at the surface. The best known of these localities is the Jabal Dhaylan base-metal prospect (Hayes et al. 2002). Some of these sites were mined by the Nabatean people whose tombs, like those of Petra in Jordan and Madain Saleh in Saudi Arabia are found at Al Bad within the Maqna Block. Many of these ancient mining areas are now fenced off as archeological sites.

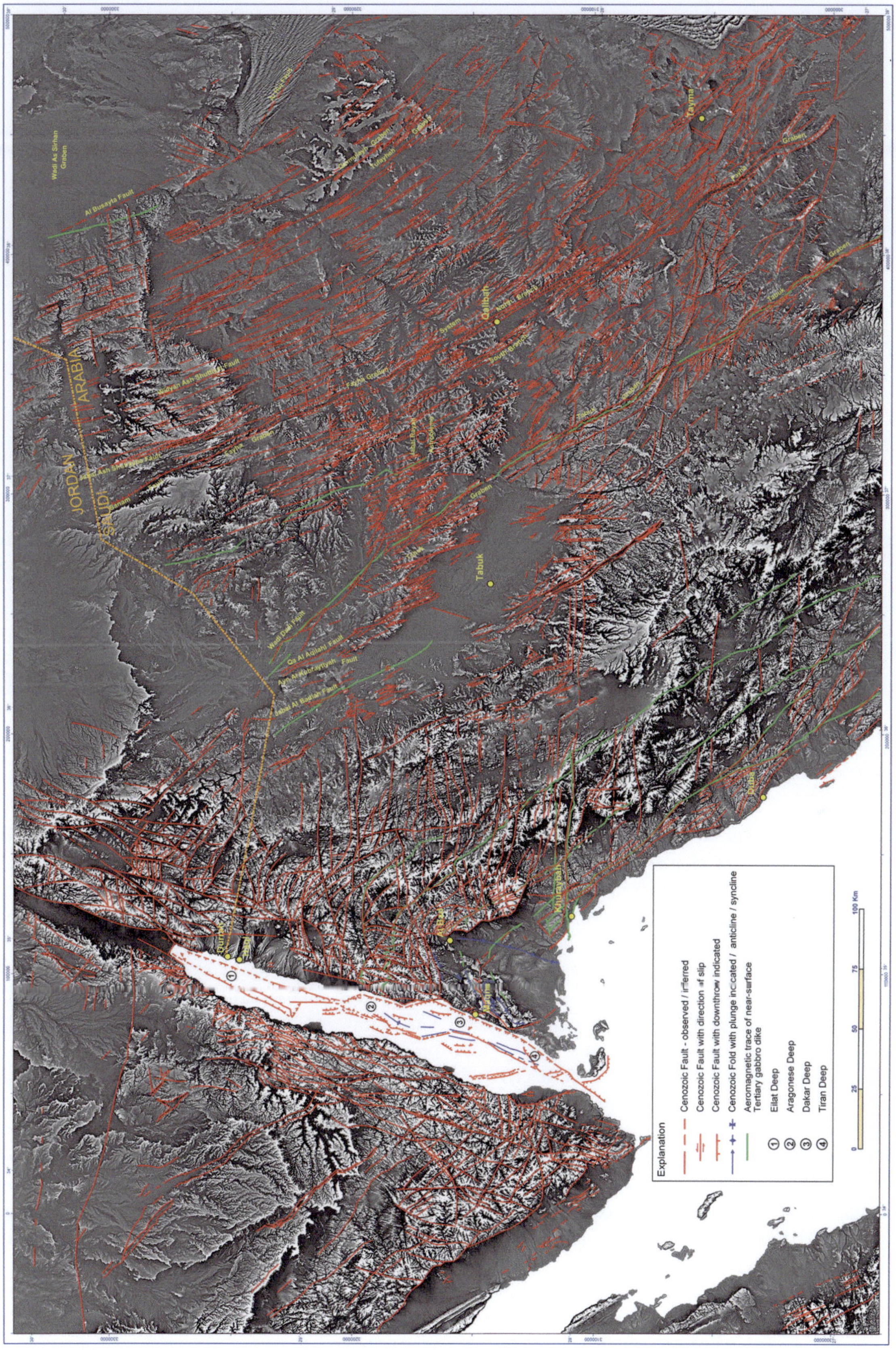

Fig. 7 SRTM surface elevation, first vertical derivative showing Cenozoic faults and folds of northwest Saudi Arabia and adjacent areas. Fault locations in the Gulf of Aqaba from Ben-Avraham (1985)

Fig. 8 Walls of shattered granite oriented parallel to the Gulf of Aqaba. The walls are separated by cataclastic zones of finely crushed rock that erode more rapidly. The coast line here is a fault zone that plunges to water depths of 1,100 m in the Aragonese Deep. The location is latitude 28°35'34.09"N, longitude 34°47'32.95"E

Another valuable indicator of Cenozoic faulting is found immediately behind or inland of the Raghama Limestone where the flat upper limestone surface continues inland as a broad series of marine terraces cut into the Precambrian shield. Directly north of Duba this marine peneplain is capped by erosional remnants of flat-bedded lagoonal facies Raghama Limestone. The marine peneplain is up to 30 km wide between Kuraybah and Al Wajh and although dissected by erosion can be seen to rise from the flat top of the Raghama (usually around 110 m above sea level (a.s.l)) in a series of steps to over 300 m a.s.l. and is terminated abruptly in high raised sea cliffs (Fig. 9). Some of the latter are spectacular where they have cut into granite plutons of the Arabian Shield. The marine peneplain provides a datum surface in which uplift of slices of the Precambrian shield bounded by the Red Sea fault systems (sub parallel to the Red Sea coastline) is obvious.

Having recognized horst blocks bounded by Cenozoic faults (relative to the marine peneplain) another criterion of Cenozoic fault recognition emerged. Within areas of Precambrian shield where there may or may not be a preserved marine peneplain, the Cenozoic fault scarps can be recognized because they produce broad fluvial debris aprons that have been mapped as piedmont fan deposits. The many terrace deposits on the Red Sea coastal plain are all coarse grained and dark in colour with a patina of desert varnish. However, the piedmont fan deposits that radiate away from the Cenozoic fault scarps in fans up to 10 km long are very fine grained and lack desert varnish. These light coloured fans and their adjacent linear fault scarps can be recognised on satellite and Google Earth images.

Since there are many sets of Cenozoic faults as well as Precambrian faults in the area, down-dropped blocks of pre-rift Cretaceous and Tertiary sedimentary rocks occur adjacent to the Red Sea within the Precambrian shield. The largest of these blocks is the Wadi Azlam graben immediately south of the coastal town of Duba on the Red Sea coastal plain at latitude 27°N. The northwest oriented graben is 85 km long in Saudi Arabia and the other half is found in Egypt where it is known as the Quseir Basin (Alabouvette et al. 1979). Other smaller down-dropped blocks within the Precambrian basement occur near the village of Khuraybah and one of these has yielded the first dinosaur bones found in Saudi Arabia (Hughes and Filatoff 1994; Hughes and

Fig. 9 The foreground shows a broad marine platform cut into the Precambrian shield after Stage 1 opening of the Red Sea. The platform exists along the northwest coast of Saudi Arabia from Duba northward. In the distance are high former sea cliffs cut into late granite intrusions. The location is latitude 27°34'31.76"N, longitude 35°36'04.85"E

Johnson 2005). Two small blocks occur east of the Gulf of Aqaba. One is a small block of Cretaceous-Eocene sediment present on the southern flanks of Jabal Hazma that was described by Motti et al. (1982). The second block is a larger graben of Cambrian/Ordovician Saq Sandstone that has been down-dropped about 2000 m into the basement. The graben is tilted so the Saq Sandstone remains only on the northeast corner and the unconformity overlying the Precambrian surface is exposed (Fig. 10). The block is contained between two E-W faults.

Our interpretation of Cenozoic faults of northwest Saudi Arabia and adjacent areas is shown on a simplified geological map in Fig. 11. The geology is based on maps by Bender (1975), Eyal et al. (1980), and 1:250,000 quadrangle maps for the Kingdom of Saudi Arabia, and data has also been taken from Ben-Avraham et al. (1979) and Ben-Avraham (1985). Inconsistencies in the previous mapping of the Cretaceous and Tertiary across the Jordan-Saudi Arabia border were resolved using unpublished data from a pterosaur fossil collected recently in the region by Dr. Thomas Rich (pers. comm.). These paleontological data showed that some areas mapped earlier as Tertiary in Saudi Arabia are in fact obviously Cretaceous in age. Faults in the Gulf of Aqaba are compiled from Ben-Avraham (1985).

5 The North-South to North-Northeast-Trending Gulf of Aqaba Fault System

Numerous north-south to north-northeast-trending faults up to 50 km long are parallel and subparallel to both shores of the Gulf of Aqaba (Fig. 7). These faults are best developed on the Sinai side in a belt up to 40 km wide, and are less common but still conspicuous on the Arabian side of the

Fig. 10 Down-dropped Saq sandstone in Precambrian basement. This graben is exposed on both sides of the asphalt road between Haql and Al Bad. The location is latitude 29°02'22.25"N, longitude 35°10'41.69"E

Gulf. The faults all show sinistral displacement that can be measured by the horizontal offset of distinctive rock units (Precambrian markers) as described by Eyal et al. (1981), who record sinistral displacements totalling 18–20 km on both sides of the Gulf of Aqaba. The amount of sinistral displacement on the faults increases toward the Gulf of Aqaba. The three largest of these faults in Sinai have a total sinistral offset of about 22 km and the two largest sinistral displacements in Saudi Arabia total 8 km of offset. The surface expressions of these major faults are linear open wadis, usually without any outcrop of the faults, the presence of which is demonstrated by dissimilar geology on either side of each wadi (Fig. 12). A minimum figure for the vertical component of shoulder uplift is obtained from the maximum topographic height of the uplifted fault-bounded Precambrian block. Precambrian peaks inland of the coast have been uplifted higher than the Cambrian peneplain on which the Cambro-Ordovician Saq sandstone of the Cover Rocks lies. The greatest preserved shoulder uplift is 2,580 m on Jabal Lawz.

Two fault plane solutions for the November 22, 1995 magnitude 7.3 earthquake calculated by Harvard University and the National Earthquake Information Center of the USA both indicate a north-northeast fault plane and a sinistral displacement. The zone of epicentral rupture on the Saudi Arabian shore showed the shoreline unchanged, but open fault planes where the land was shoulder uplifted in steps with an accumulated total of 90 cm of uplift (Roobol et al. 1999).

The age of the Aqaba fault system is younger Cenozoic or Neogene, being younger than 20 Ma and resulted from the second stage of opening of the Red Sea. Stage one opening is preserved in the Gulf of Suez but an additional 115 km of sinistral displacement has widened the northern part of the Red Sea trough.

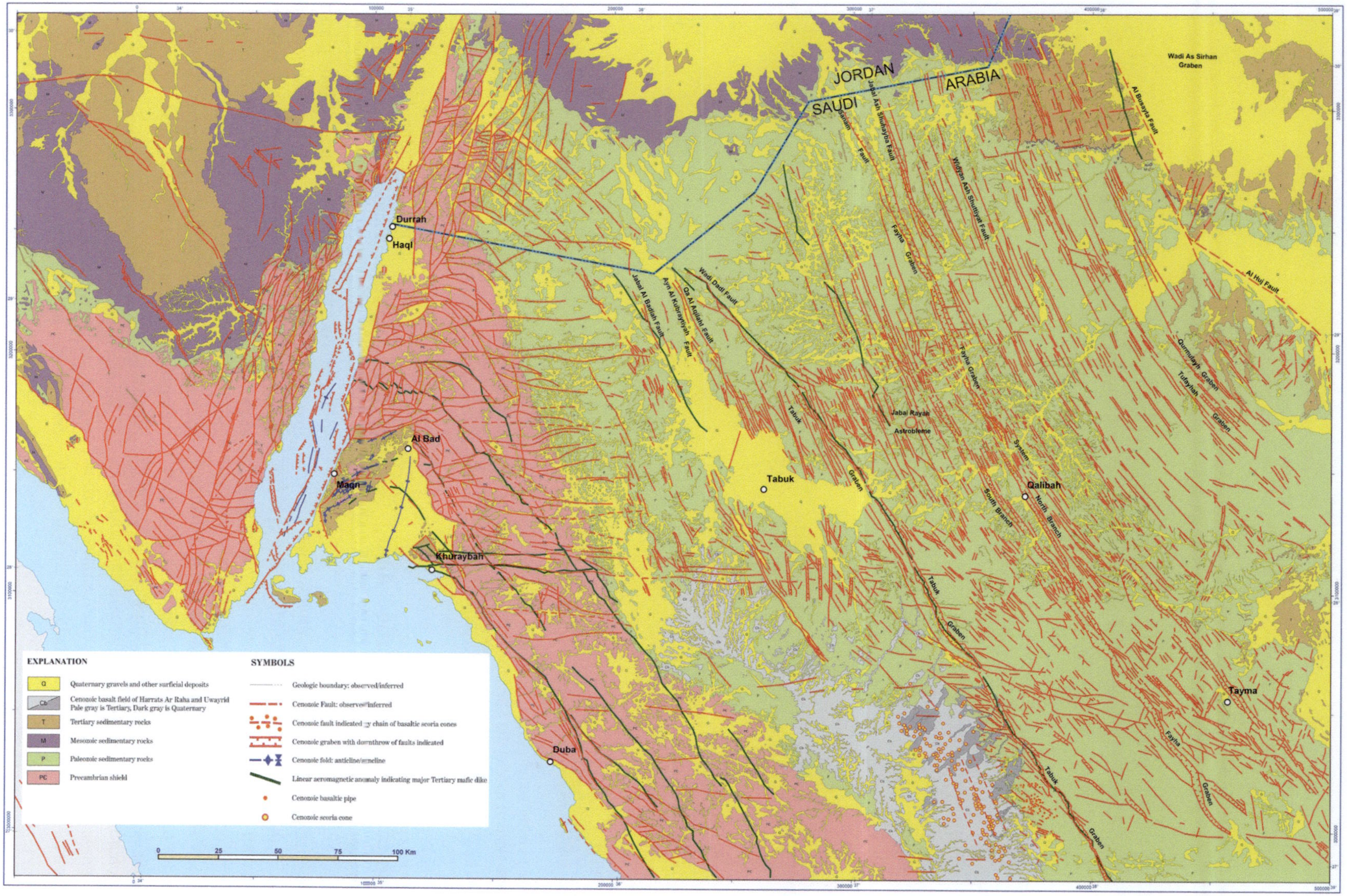

Fig. 11 Cenozoic faults and folds of northwest Saudi Arabia and adjacent areas shown on a geological base map simplified after Clark (1987), Davies and Grainger (1985), Grainger and Hanif (1989), Janjou et al. (1996, 1997), Rowaihy (1985), Vaslet et al. (1994), and Wallace et al. (2000a, 2000b, 2001)

Fig. 12 View looking south along a prominent north-northeast fault (Gulf of Aqaba system) cutting the Cambro-Ordovician Saq sandstone on the Haql to Tabuk road at 28°51.990'N, 35°33.463'E. The block on the right (west) is downthrown at least 70 m whereas Precambrian basement is exposed under the sandstone of the other block

6 A Zone of Intense Deformation Along the Eastern Margin of the Gulf of Aqaba

A major feature that was previously undescribed on the eastern side of the Gulf of Aqaba in Saudi Arabia is a belt extending for up to 25 km inland where many granite plutons are shattered with cataclastic zones of finely crushed rock. In Wadi Mabrak (along the Jordan border) at a distance of 18 km from the coast the shattered zone is 20 km wide. The shattering is best noticed in the granite plutons which have an uncharacteristic very crumbly type of erosion in which rock pinnacles rise from scree covered slopes. Erosion produced vertical gorges through the shattered granite which erodes similarly to sandstone. The best example of such a gorge in shattered granite is that of Tayyib al Ism (Fig. 13). The belt of shattered granite plutons and their debris curtains can be recognized on Landsat images and differs from the normal blocky weathering of granite plutons further east and throughout the Arabian shield.

During fieldwork over two seasons, it was found that in some areas of steep cliffs along the coastline parts of the road maintained by the Saudi Arabian Border Patrol were incomplete along the section of shattered granites immediately adjacent to the Aragonese Deep. Here, the high granite cliffs plunge down directly into the Deep, which has a maximum depth of 1800 m. Along this coastal section, cliff collapses are frequent and the road is constantly being renewed.

Whereas the granites in a belt 25 km wide along the eastern shore of the Gulf of Aqaba are shattered, the associated rocks are intensely faulted. The latter comprise Tertiary sedimentary rocks of the Maqna block along the southern shoreline of the Gulf of Aqaba and the metavolcanic-sedimentary rocks of the Precambrian shield along the northern shoreline. Both types of rock are intensely

Fig. 13 The gorge of Tayyib al Ism. This vertical walled gorge is cut into shattered granite with cataclastic zones of crushed rock adjacent to the Gulf of Aqaba (note the Toyota Landcruiser for scale, circled). The location is latitude 28°33'35.74"N, longitude 34°48'09.98"E

faulted and the belt of intense faulting extends beyond the belt of shattered granites for a distance of 50 km from the coast. Immediately adjacent to the coastline, the Tertiary sedimentary rocks are cut by numerous small faults with spacings of only a few metres and vertical displacements of centimetres to metres (Figs. 14 and 15). In the Precambrian shield along the northern shores some indication of this coastal intense faulting can be obtained from the 1:100,000 WSA aerial photomosaics where the larger faults have been traced, and a section of this fault map is recorded in Fig. 16.

Fig. 14 The Oligocene Sharik formation sediment in Wadi Sukk showing numerous small faults typical of the coastal belt along the Gulf of Aqaba. Two of the main faults and their sense of motion are indicated by arrows. Looking south with the Gulf to right at latitude 28°31'15.7"N, longitude 34°49'21.7"E

7 A Major Fold Structure to the East of the Gulf of Aqaba

Within a 50-km wide belt adjacent to the Gulf of Aqaba, which includes the 25-km wide zone of coastal shattered granite, the rock units of the Precambrian shield form a broad fold having the appearance of a drag fold consistent with the sinistral displacement of the Gulf of Aqaba (Figs. 6 and 7). This fold affects the Precambrian rock units of northwest Saudi Arabia as well as the many Precambrian dikes contained within them. When the structure is traced from east to west toward the Gulf of Aqaba, inland northwest trends become east-west oriented and then turn to northeast-southwest adjacent to the Gulf. This structure can be seen on Landsat and Google Earth images but no similar structure can be seen on the Sinai side of the Gulf. The fold structure was first recognized by Clark (1987) when he mapped the Al Bad 1:250,000 quadrangle, but no comment was made on the age of the structure. A clear case can be made that this fold, including the rotation of Precambrian rock units, Precambrian dikes and Tertiary dikes, is of Cenozoic age corresponding to the second stage of opening of the Red Sea. Although the structure resembles a drag fold resulting from the sinistral displacement along the Gulf of Aqaba, the absence of a similar structure on the Sinai side does not support this interpretation. Rather a different origin is indicated, namely that the 50-km wide fold structure formed before the main transform faulting and pull-apart basin formation along the Gulf of Aqaba. That is, the second stage of opening of the Red sea was accommodated at first by folding, then by the production of a shatter zone and belt of intense local faulting and finally by pull-apart basins related to large movements on a small number of faults. Thus, rather than being a simple drag fold, the structure records the earliest sinistral displacement of the Arabian Plate from Sinai. Later, a belt of north-northeast faults of the Gulf of Aqaba system formed over a broad zone with individual sinistral displacements of up to 9 km. Finally, major displacement on a few of these faults led to the formation of pull-apart basins. This later displacement resulted in separation of the entire fold structure which is now preserved intact within the Arabian Plate in Saudi Arabia.

Fig. 15 Miocene Nutaysh formation sediment on the coast immediately south of Maqna showing numerous small faults typical of the coastal belt along the Gulf of Aqaba. The location is latitude 28°23'27.93"N, longitude 34°44'13.67"E

Another indication of the age of the fold structure is obtained from the fact that the Precambrian lithologies and dikes were rotated through 90 degrees anticlockwise to form the fold. However, the aeromagnetic traces of the major gabbro dikes that parallel the margins of the Red Sea are rotated through only about 45 degrees in an anticlockwise direction. This suggests that they were emplaced well after the fold structure started to form. Although poorly exposed, these dikes have a range in ages of 24 to 22 Ma (cf. Table 1 in Bosworth 2015) which dates the latter part of the folding.

8 Arcuate Faults and Offset of Tertiary Gabbro Dikes

The 50-km wide fold structure adjacent to the Gulf of Aqaba within Saudi Arabia shows a number of arcuate faults that parallel the deformed Precambrian lithologic units and include Precambrian dikes (Figs. 6 and 7). These structures are Precambrian faults that have also been deformed by the major fold structure. Originally the Precambrian faults were planar with linear surface traces but they were reactivated in Neogene time as part of the fold structure and became arcuate. This can be demonstrated because of the presence of thick gabbro dikes that occupy Tertiary pre-rift faults that parallel the entire length of the Red Sea coast. The dikes were first described by Blank (1977; Plate 3 therein), Blank and Andreasen (1991) and Johnson and Vranas (1992), and are particularly prominent on the aeromagnetic map of Saudi Arabia (Zahran et al. 2003). These dikes are rarely exposed at the surface (Fig. 17), where they are usually 200 to 300 m thick and nearly vertical. 40Ar/39Ar ages show the dikes to have formed mainly from 24 to 22 Ma (Bosworth 2015). The northwesterly-trending Tabuk graben, a youthful morphologic structure, can be traced right across the study area (Fig. 7) and aeromagnetic data show it to be underlain by a Tertiary gabbro dike (which is a prominent linear feature across the centre of Fig. 6) that has been named the Great Ja'adah Dike (Phoenix Corporation 1985). Conversely the aeromagnetic data also show that the parallel Fayha graben and another unnamed graben further to the east are not underlain by gabbro dikes. As the aeromagnetic traces of the dikes are followed into the 50-km wide fold structure alongside the Gulf, they can also be seen to be rotated. One

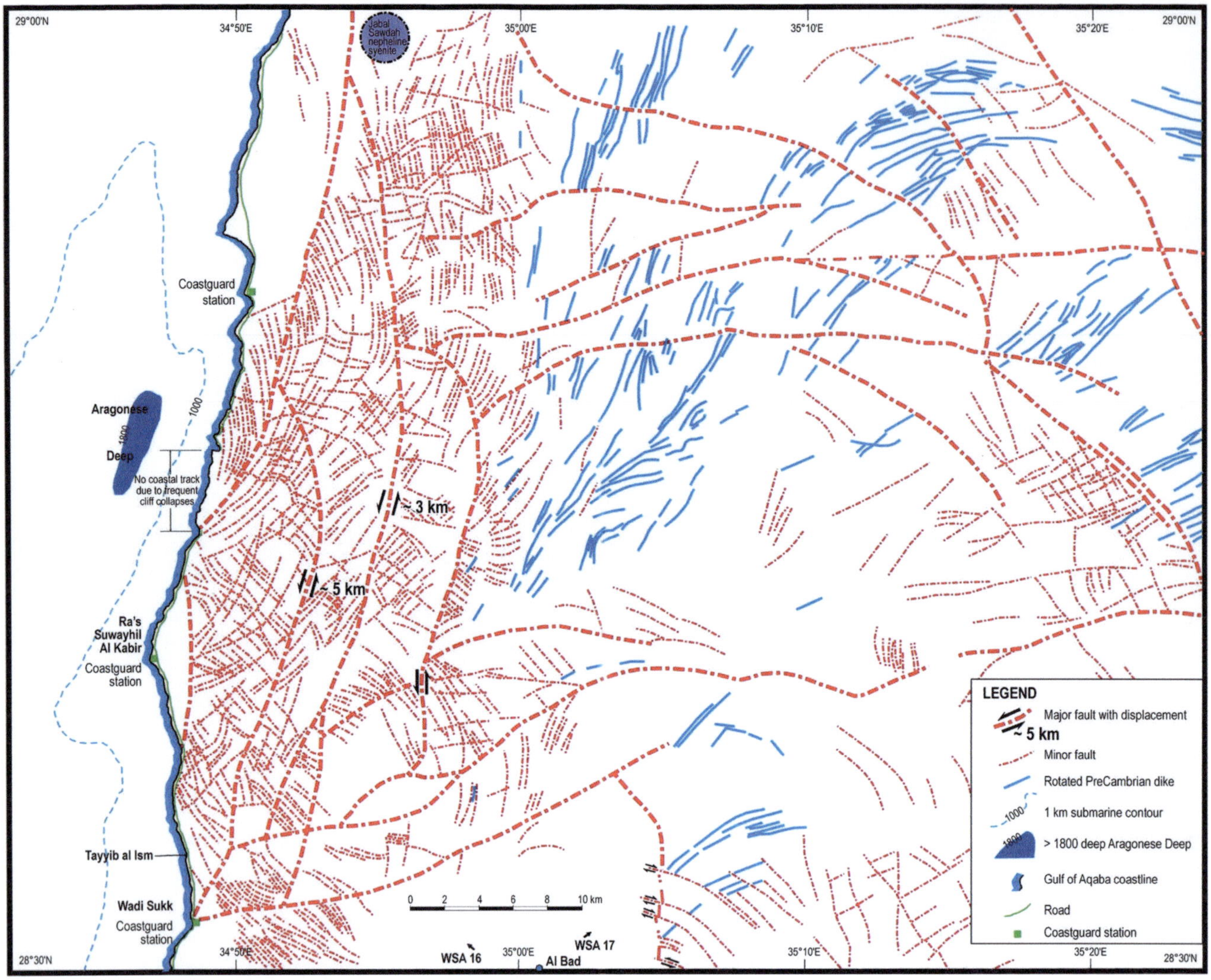

Fig. 16 Fracture pattern and rotated Precambrian dikes on the east side of the Gulf of Aqaba

of them can be traced in detail through the zone of intense faulting (Fig. 18). More detail of the fault-dike relationship is shown in Fig. 7. Thus, the age of the major fold structure, and the arcuate nature of the Precambrian faults is demonstrated to be Neogene in age or in part younger than the gabbro dikes at about 24–20 Ma, which corresponds to the second stage of opening of the Red Sea. It should be noted that the curvature of the trace of the gabbro dikes is less arcuate than that of the structural trends in the Precambrian shield, because the Tertiary dikes probably were emplaced after the folding had started.

We have been unable to locate aeromagnetic data for the Sinai Peninsula, but youthful graben structures parallel to those in northwest Saudi Arabia exist there and some are underlain by Tertiary gabbro dikes as described by Eyal et al. (1981).

9 Cenozoic Reactivation of a Precambrian East-West Fault System

East-west faults form a well-developed Precambrian fault system that is present throughout the Arabian shield. In the Gulf of Aqaba region faults of this system were reactivated during the Cenozoic to form a belt between the Aqaba fault system and the Ribbon fault system. Some of these faults can be traced to the west where they become arcuate faults within the major fold structure. Others cut the Phanerozoic sedimentary rocks and some cut and down-drop the Phanerozoic Cover Rocks of northwest Saudi Arabia as blocks and grabens into the Precambrian Shield. Figure 19 shows one of these deeply-eroded faults (near the village of Shigri, after which it is here named) that can be traced for 150 km (cf. Fig. 7). At its

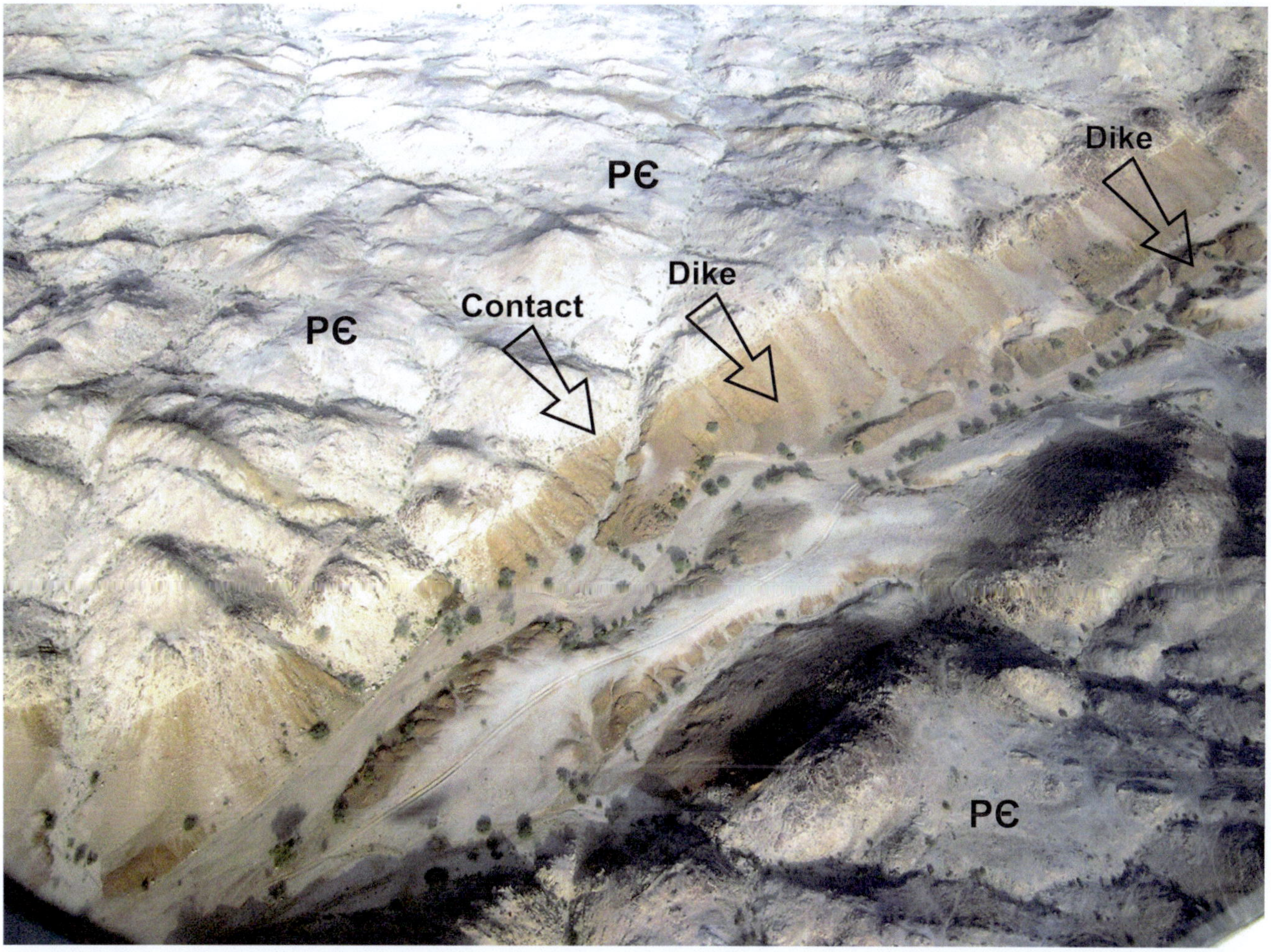

Fig. 17 Tertiary gabbro dike exposed near coastline near Duba. These dikes are seldom exposed at the surface but have very strong aeromagnetic signatures. The dike forms the dark-coloured walls of the wadi and has a sharp contact with the country rock. The location is latitude 27°28'29.16"N, longitude 35°45'46.68"E

western end on the Red Sea coast near the village of Khuraybah (Fig. 7) the fault scarp is mantled in Lower Miocene Raghama reef facies limestone. In its central part to the west of the Duba-Tabuk road the fault is represented by a deep erosional canyon cutting through the Phanerozoic Cover Rocks (Saq Sandstone) and deep into the Precambrian basement. The Saq Sandstone is offset across this fault (Fig. 19). At its eastern end the fault appears to cut Tertiary basalts of Harrat Ar Raha (Fig. 7). Much of this fault is underlain by an aeromagnetic anomaly indicating the subsurface presence of a major gabbro dike. Another prominent east-west fault in the Khuraybah area had its westernmost end reactivated in Cenozoic times because it truncates and down-drops a block of Lower Miocene Raghama limestone that can be seen adjacent to the coast road.

Some of the east-west faults are accompanied by aeromagnetic anomalies that are thought to represent Tertiary gabbro dikes. East-west aeromagnetic anomalies occur on Harrat Lunayyir where they are associated with Quaternary chains of basaltic scoria cones. These relations suggest that the east-west faults were reactivated in both pre-and syn-Red Sea rift times.

10 The North-Northwest Red Sea Coastal Fault System

The broad coastal plain that borders the Red Sea obscures the Precambrian shield but seismic refraction profiles show that the 40-km thick Precambrian Shield (Gettings et al. 1986) thins to about 15 km beneath the coastal plain (Makris et al. 1983), as is also indicated by gravity data. The thinning of the Precambrian shield to around 15 km is believed (from mapping the coast plain elsewhere, e.g. Roobol and Kadi 2008), to be achieved by normal faulting. The faults strike parallel to the Red Sea and comprise the north-northwest Red Sea coastal fault system. Roobol and Kadi (2008) were able to subdivide the fault system into two distinct ages corresponding with pre-and syn-rift stages

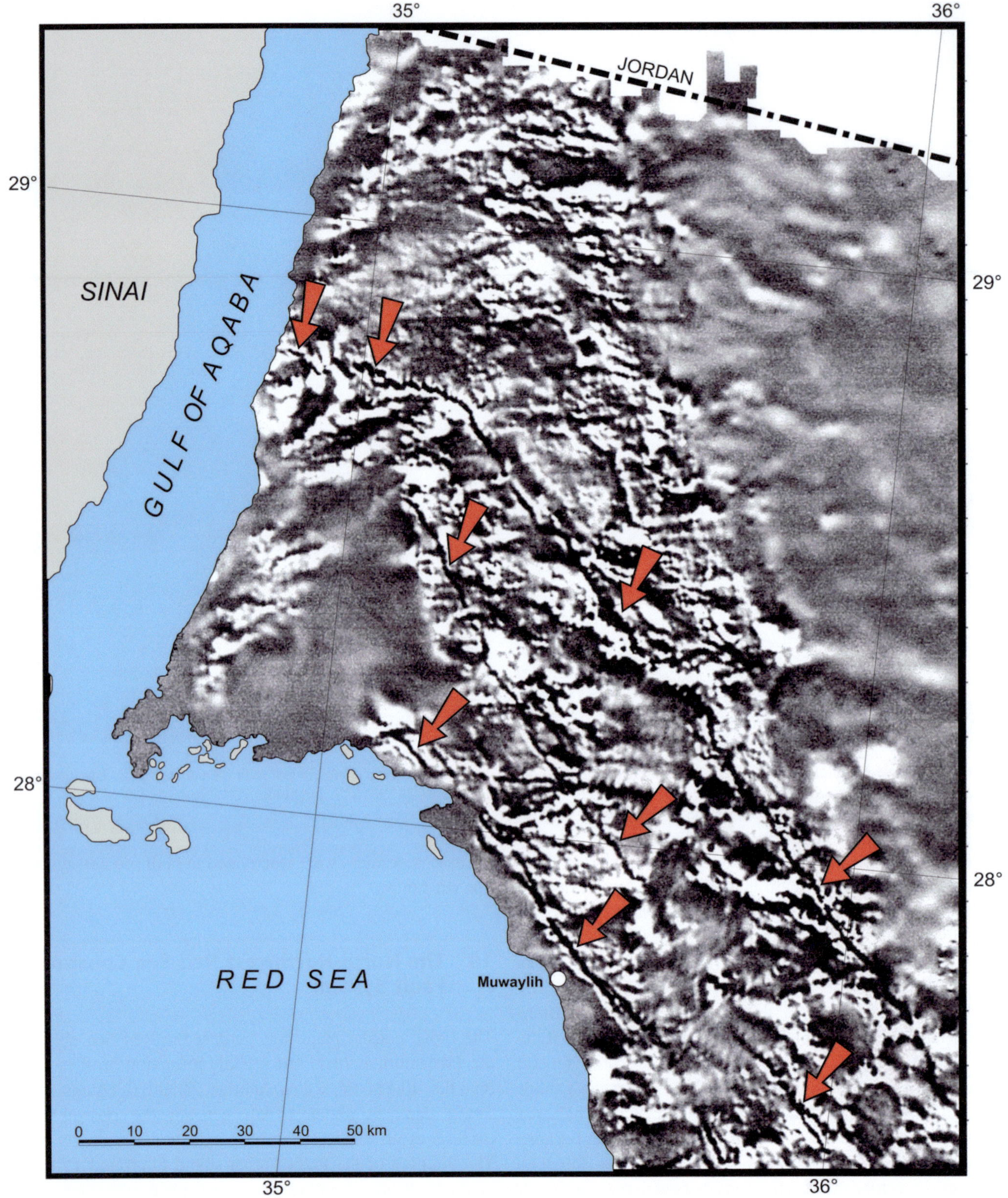

Fig. 18 Detail showing curved aeromagnetic traces of hidden gabbro dikes (indicated by red arrows) through the zone of intense faulting immediately adjacent to the Gulf of Aqaba. The relationships of the most prominent dikes to the Cenozoic faults are shown in Fig. 7

Fig. 19 Part of the 150 km long east-west fault eroded into a deep canyon near the Duba-Tabuk road north of Shigri village at latitude 28°03'34.6"N, longitude 35°55'58.8"E. There is a small downthrow of the Cambro-Ordovician cover to the north (right)

of the opening of the Red Sea. First the Afro-Arabian shield was attenuated in pre-rift times to form a "Pre-rift Foothills Normal Fault System". These faults are inclined steeply both toward and away from the Red Sea and create a horst and graben topography that has not yet been removed by erosion. This fault set is distinct because many of the faults contain major Tertiary gabbro dikes with distinctive aeromagnetic signatures. A later set formed as a result of the Stage 1 opening of the Red Sea. This later set, also parallel to the Red Sea coast, was named "The Red Sea Coast Plain Fault System". The faults are syn-rift rotational normal faults (see cross section in Roobol and Kadi 2008) and are believed to result from gravity collapse of the Precambrian shield along the margins of the newly-opened Red Sea. In northwestern Saudi Arabia only one of these younger faults is visible as the Master Listric Fault behind the Red Sea coastal plain. This is a prominent but deeply-eroded fault scarp separating the coast plain from the Precambrian shield.

The north-northwest Red Sea Coastal fault system can be traced along the entire length of the Red Sea and locally shows Tertiary base metal mineralization described by Hayes et al. (2002). A similar fault system exists on the African side of the Red Sea and can be seen on Landsat and Google Earth images.

11 The Ribbon Fault System

East of longitude 36°E, both the Gulf of Aqaba and the east-west fault systems die out and are replaced by a different fault system named here the Ribbon fault system. This is a belt of normal faults 250 km wide that trends north-northwest in the north and arcs around to northwest in the south of the map area (Fig. 7). The eastern limit of this fault system has been documented at longitude 38°E in the Sirhan-Turayf Basin 1:250,000 sheet (Johnson, unpublished 2007). On the map (Fig. 7) approximately 2,400 of these

Fig. 20 Google Earth image showing ribbon faults in the Marthoum area. Chains of elliptical columnar-jointed basalt pipes have been emplaced along the faults. Horst blocks are marked "h" and grabens "g". The width of the field of view is approximately 9.4 km

faults are shown and these can all be seen on the SRTM image. However, in the field, the faults are much more numerous and only about a third (the longest and most prominent ones) are shown in Fig. 7. Minor faults can be seen on Google Earth images and they are particularly prominent in the field where they divide the Phanerozoic sandstones into long narrow horst-graben strips with topographic differences of up to 4 m. These faults are a swarm and are named "Ribbon faults" because they cut up the rocks into long ribbon-like strips. They are of small displacement with fault scarps seldom more than 3 m high. In many places the faults are distinct as they are filled with iron oxide/hydroxide that resists erosion to form ridges. In some cases there are open fissures with little apparent vertical displacement. In all cases, however, these faults and fissures appear to result from crustal extension.

An example of ribbon faulting in the Marthoum area of the Tayma quadrangle (Vaslet et al. 1994) is shown in a Google Earth image in Fig. 20 at latitude 27°15'N, longitude 37°50'E. The ribbon faults are apparent as sand and gravel filled grabens (marked as "g" in the diagrams) separated by hummocky sandstone horsts (marked as "h") with a light colour of wind-blown sand. The faults are marked by chains of Saq Sandstone hills that have joints filled with limonite ironstone (due to weathering of mafic minerals in the sandstone) that resists erosion (Fig. 21). In this area between the Tabuk graben to the east and Harrat Uwayrid to the west there are 100 Tertiary basalt pipes (Liddicoat 1982) that are intruded along the ribbon faults and occur in chains. Individual pipes are lenticular along the host ribbon fault (Fig. 20).

Ribbon faults are particularly striking in an area of Devonian Tawil Sandstone in the Thaniyat-Turayf

Fig. 21 Line of limonitized Wajid sandstone hills along a Cenozoic ribbon fault in the Marthoum area at latitude 27°14'36.75"N, longitude 37°48'42.29"E

Fig. 22 Google Earth image of ribbon faults in the Devonian Tawil formation sandstone at latitude 29°30'N, longitude 37°45'E. The faults are represented by linear chains of limonitized sandstone caused by weathering of the rocks and migration of limonite into the fault zone. The width of the field of view is approximately 7.8 km

Fig. 23 Line of limonitized Devonian Tawil sandstone hills along a Cenozoic ribbon fault. Behind the line of hills, the sandstone is exposed in a low horst block. In the foreground the sandstone has subsided in a graben now largely filled with dark gravel of ferruginated sandstone. The location is latitude 29°06'14.15"N, longitude 37°45'44.87"E

Fig. 24 NASA World Wind pseudocolour image of the Thaniyat-Turayf escarpment. The Devonian Tawil sandstone formation is exposed in the southern half of the image and includes the dark rocks (indicated by the arrows) along the base of the escarpment. In the north the sandstone is overlain by a thin layer of Cretaceous Aruma Group. The white coloured sediments in the top half of the image are the Tertiary Turayf Group, and all can be seen to be cut by the ribbon faults which are therefore Cenozoic in age. The width of the field of view is approximately 47.7 km

quadrangle (Wallace et al. 2000a). Here long thin strips of rock bounded by ribbon faults can be seen where light coloured sandstone horst blocks stand above dark-coloured gravel-filled grabens. Some of the ribbon faults are exposed as fault scarps up to 3 m high along which yellow wind-blown sand is banked.

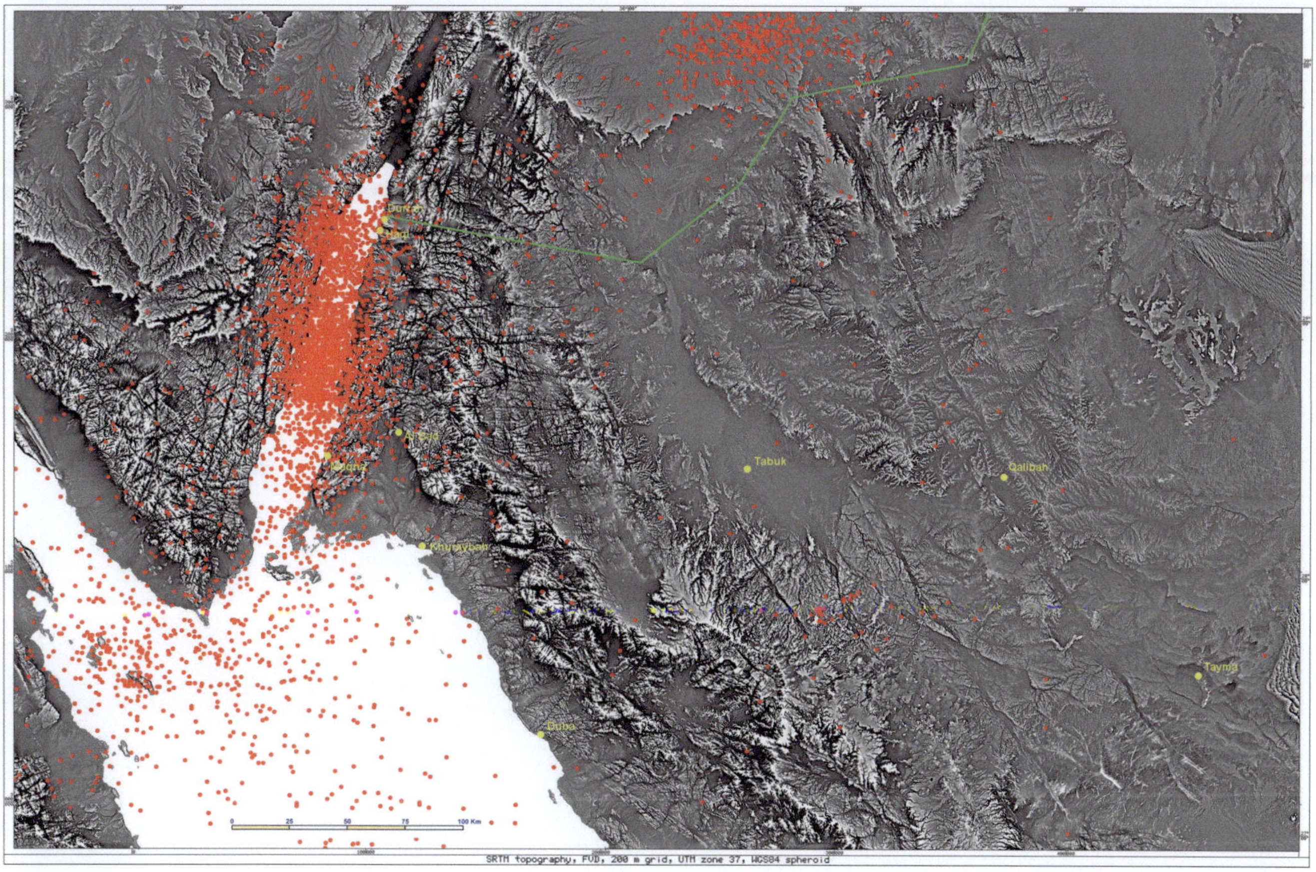

Fig. 25 SRTM surface elevation, First Vertical Derivative, showing earthquake epicentres greater than magnitude 3 for northwest Saudi Arabia and adjacent areas

To the north of the location in Fig. 20 in an area of spectacularly beautiful desert scenery are areas of ribbon faulting that appear truly remarkable on Google Earth images. The faults themselves are marked by straight lines of sandstone hills bounded by joints filled with limonite. An example at about latitude 29°30'N, longitude 37°45'E is shown in Fig. 22. A field photograph taken in a traverse through the area is shown in Fig. 23. The desert tracks mainly follow the grabens because they are infilled with brown limonitized fragments derived by erosion of the limonitized ironstone fill of the ribbon faults. In contrast, the sandstone horsts form light coloured hummocky ground where yellow wind-blown sand has gathered.

Evidence for the Cenozoic age of the ribbon faults is found in an area of Mesozoic and Tertiary sedimentary rocks exposed in the Thaniyat-Turayf quadrangle (Wallace et al. 2000a). The area is well known because of the prominent Thaniyat-Turayf escarpment showing the Devonian Tawil sandstone formation in the lower part, unconformably overlain by Cretaceous Aruma Group, and in turn unconformably overlain by the Tertiary Turayf Group. The latter is well known for the two phosphorite members and a USGS adit into the lower part of the Thaniyat phosphorite member. Wallace et al. (2000a) are non-committal on the age of the faults but the Geocover 2000 images from NASA World Wind (Fig. 24) and a field visit clearly indicate that some ribbon faults cut all members especially the Tertiary sediments, indicating that they are Cenozoic faults. A small number of faults in the area may be Pre-Cretaceous in age as Wallace et al. (2000a) show one such fault on their map.

12 Recent Seismicity

Earthquake epicentres from 1950 up to 2015 taken from the Saudi Geological Survey database were plotted on the FVD images of the SRTM topography for possible correlation with mapped fault trends. The seismic hazard for the Red Sea, the Gulf of Aqaba and western Saudi Arabia is discussed by Stewart (2007) and Zahran et al. (2015, 2016), and hazard estimates for the Sinai area are also given by Sawires et al. (2016). The epicentres are plotted as red dots on the same base map as shown in Fig. 5, with 5007 events above magnitude 3 shown in Fig. 25, and 986 events greater than magnitude 4 shown in Fig. 26. Earthquakes larger than magnitude 4 are presumably reasonably well located

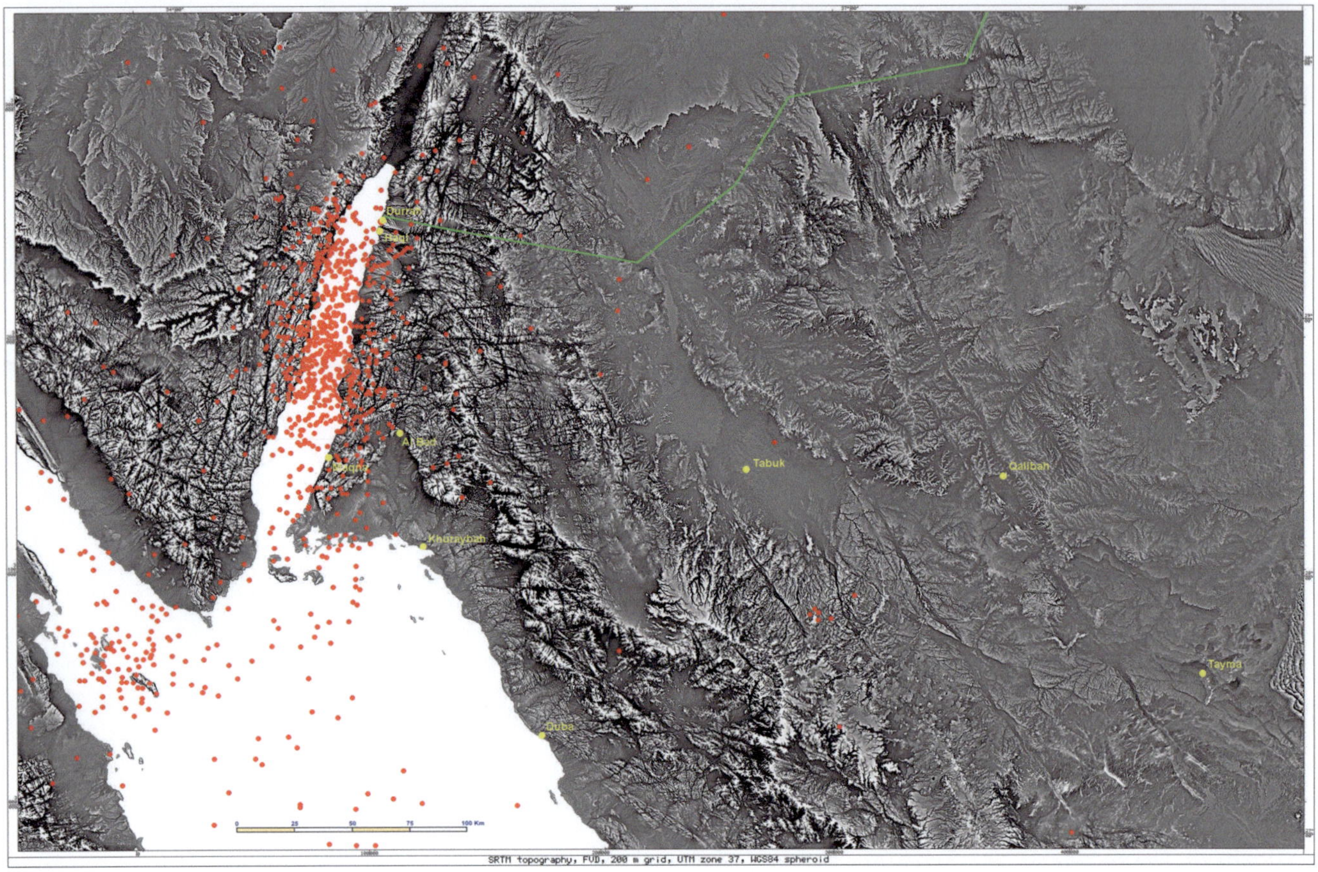

Fig. 26 SRTM surface elevation, First Vertical Derivative, showing earthquake epicentres greater than magnitude 4 for northwest Saudi Arabia and adjacent areas

(probably to within 10 km) and should exclude most events of cultural origin such as quarry blasts. The large cluster of events near the centre of the upper margin of Fig. 25, between 36°E and 37°E in fact probably represents activity related to phosphate mining rather than naturally occurring earthquakes, and is absent in Fig. 26. The earthquake activity extends onshore on both sides of the Gulf, and the seismicity in the Sinai Peninsula is discussed by Abdel-Rahman et al. (2009). A more complete epicentre plot, especially for the region of the Dead Sea Transform to the north of the Gulf of Aqaba, has recently been presented by Wetzler and Kurzon (2016), and the seismicity of the northern Red Sea in particular has also recently been discussed by Mitchell and Stewart (2018).

The spatial variation in earthquake activity can also be shown by the rate of release of seismic moment, which for an earthquake is the product of fault plane area, the fault slip or displacement and the shear modulus of the rock, and is probably the most physically realistic measure of the size of an earthquake. Using earthquakes greater than magnitude 4 since 1950, the logarithm of the cumulative seismic moment in 5 km cells (in units of N m) is shown in Fig. 27.

The Araba (Arava) Fault section of the Dead Sea Rift from about 29.5°N northward (between the Gulf of Aqaba and the Dead Sea) shows relatively low seismicity compared to adjacent sections of the rift zone (Ben-Menahem and Aboodi 1981), and the southern end of the Gulf of Aqaba currently also has relatively low activity. These two zones may only be temporarily quiescent, and could in fact represent areas with relatively high earthquake risk, illustrating the problem in estimating hazard from incomplete data sets. Klinger et al. (2000b) also suggest that historically there have been fewer large earthquakes than expected along the Arava Valley in Jordan, although much of the observed rate of movement could be accommodated by an M_w 7.6 event about every 6000 years. Paleoseismic studies along the southern Dead Sea fault show that the return intervals of large earthquakes over the past 5000 years are very variable, with long periods of relative quiescence (Klinger et al. 2015), which illustrates the difficulty in determining the long-term seismicity and risk from instrumental data recorded over only a few tens of years. Presumably the average left lateral rate of movement in the southern Gulf of Aqaba and near the Tiran Strait is similar to that further north (about

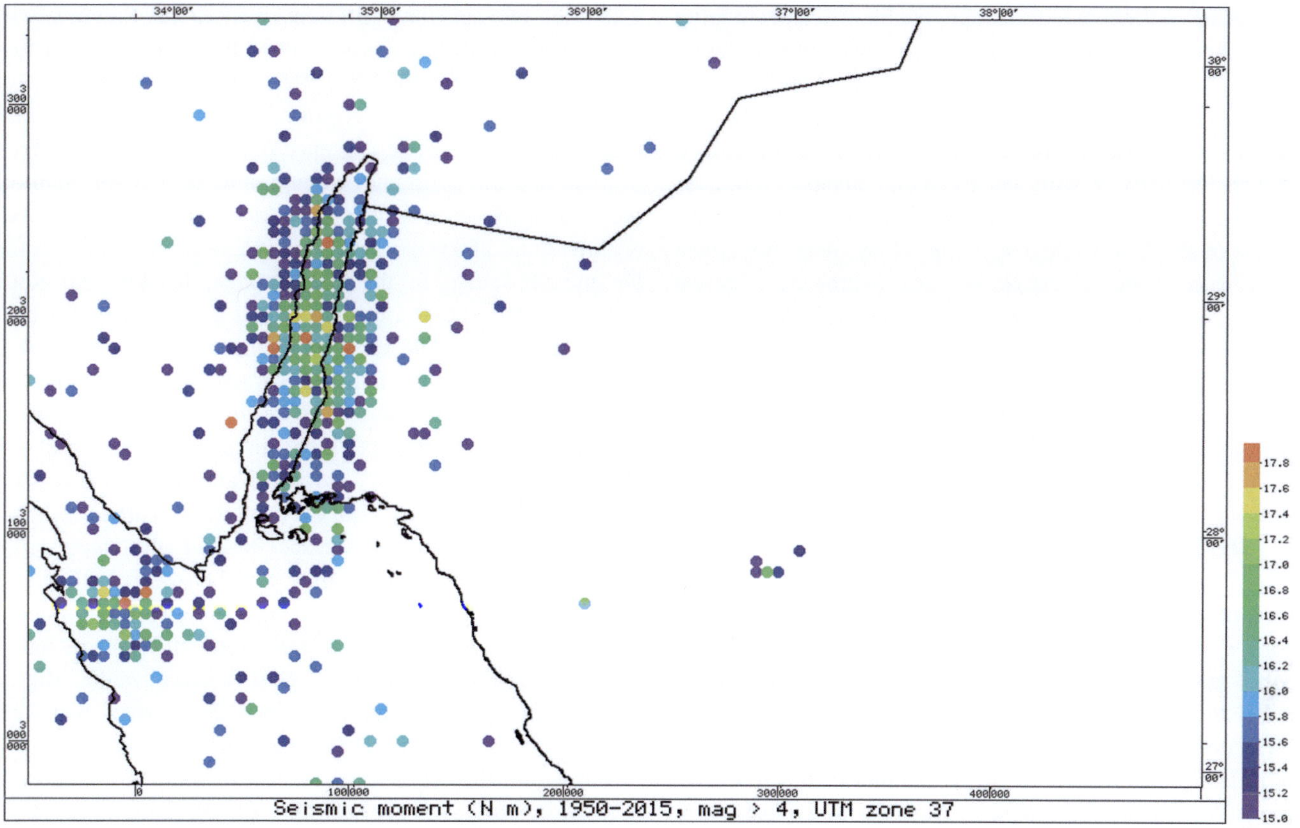

Fig. 27 Earthquake seismic moment (a measure of energy release, in N m units) summed over 5 km cells using seismicity data from 1950 to 2010 and events greater than magnitude 4

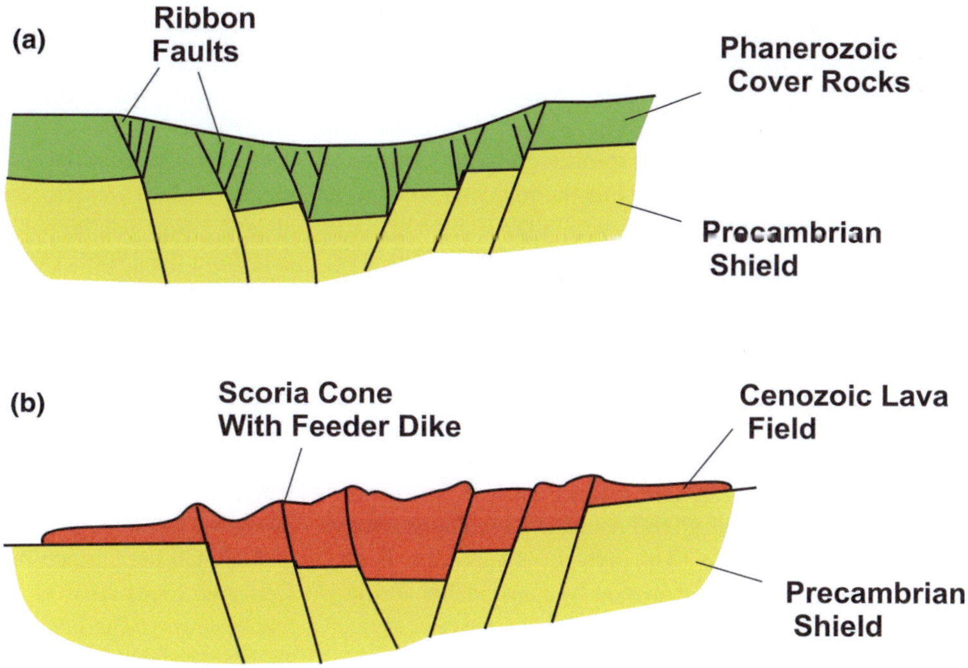

Fig. 28 Diagram that illustrates a possible explanation for the two types of crustal extension along the western side of the Arabian Plate. In the north in the unmetamorphosed sedimentary cover rocks the faults feather to form a wide belt of ribbon faults (**a**). At depth the crystalline basement of the Precambrian shield forms a simple graben as found along the Makkah-Madinah-Nafud volcanic line (**b**)

5 mm/year) so there may be an appreciable risk of damaging earthquakes that could affect any planned infrastructure despite the current moderate seismicity in this area.

The epicentres plotted on the SRTM images are likely to be in error by at least several km, and hence it is difficult to correlate the activity with any particular mapped fault. As expected, most of the seismic activity lies within and on the margins of the Gulf of Aqaba, along which most movement occurs. Even if the epicentres and focal depths were known precisely, correlation of events with known faults may be difficult except for very shallow earthquakes since most faults are not vertical. However, the seismicity does show the areas that are currently undergoing movement, but this does not show the longer-term changes over a geologic timeframe.

13 Discussion

The thousands of Cenozoic faults in NW Saudi Arabia record a tectonic history that extends to at least 25 Ma. Roobol and Kadi (2008) were able to subdivide the faults parallel to the Red Sea coast into two distinct age groups, pre-rift and syn-rift relative to the opening of the Red Sea. The oldest pre-rift faults are those parallel to the Red Sea, some of which the aeromagnetic maps show to host major gabbro dikes. Most prominent of these is the Tabuk graben which is underlain by the Great Ja'adah Dike. These dikes (which rarely outcrop at the surface) have recently had their age range revised to 24–22 Ma by 40Ar/39Ar dating (Bosworth 2015). The syn-rift group are normal rotational faults resulting from the gravity collapse of the 40-km thick Arabian shield into the Red Sea rift. In most places along the Red Sea coast of Saudi Arabia the two fault systems are parallel and because both systems can dip either to the east or west it is sometimes difficult to distinguish between such structures present in the field as long narrow wadis bounded by high hills or mountains. However, in the Rabigh area the pre-rift faults are truncated by one of the syn-rift faults. This was named the "Master Listric Fault" system by Roobol and Kadi (2008) and it can be traced along the length of the Red Sea where it separates the Precambrian shield from the Cenozoic sediment of the Red Sea coastal plain.

The pre-rift faults are most common in the foothills in Precambrian rocks that border the Red Sea coastal plain and decrease in numbers eastward toward the Red Sea escarpment. Roobol and Kadi (2008) named this fault system the Foothills Fault System. The syn-rift faults are poorly exposed in the gravel and in poorly consolidated Tertiary sediments of the Red Sea coastal plain but have been proved by drilling. Roobol and Kadi (2008) named this group of faults the "Red Sea Coastal Plain Faults". Roobol and Kadi (2008) regard the Rabigh area as the best locality to distinguish the two fault systems because it is the only place where the two sets have different orientations. Previously in areas where they are parallel they have not been distinguished and were named the "Red Sea Fault System".

Previous workers have regarded all these faults as normal rotational faults, but the presence of the major gabbro dikes in them and the accuracy of being able to position them on the published aeromagnetic map of western Saudi Arabia (Zahran et al. 2003) have resulted in a new interpretation for the pre-rift faults. In the Rabigh area, Roobol and Kadi (2008) used 1:50,000 topographic and aeromagnetic images, where it was possible to identify the signatures of hidden dikes that followed the faults and sometimes stepped across from one fault to another. Because the aeromagnetic signatures of the gabbro dikes (which run along the length of the Red Sea, cf. Blank 1977) are sharp and distinct, it can be argued that the pre-rift dikes cannot be curved normal rotational faults near the surface. Instead, the sharp aeromagnetic traces indicate that they must be steeply inclined, at least for the upper several kilometres. The two fault systems are considered to be physically different—the pre-rift faults are steeply inclined planes often hosting major dikes, whereas the syn-rift faults of the Red Sea coastal plain are normal rotational faults (Roobol and Kadi 2008).

Although the numerous northwest-southeast faults in northwest Saudi Arabia (including ribbon faults in the Turayf Group) began forming in pre-rift times (and some may be reactivated pre-Cretaceous faults), geologic and seismic evidence supports the proposition that minor activity continues on them up to the present day. This activity might be regarded as syn-rift reactivation of pre-rift structures. Vaslet et al. (1994) describe the Shagab quadrangle where a tongue of basalt lava that extends northeast from Harrat Uwayrid is down-faulted into a 100 m deep northwest-southeast graben (at 27°49'06"N, 37°21'18"E). The uppermost basalt flow is down faulted only 10 m. A K/Ar age for this lava flow yielded 7.1 ± 0.9 Ma. Thus, the graben fault was active after 7.1 Ma. Additional information comes from seismic data from the 1960's onward. Although mainly clustered in the Gulf of Aqaba, there are a number of scattered epicentres well to the east of the Gulf of Aqaba in the area of northwest-southeast Cenozoic faults. In April 2008 an earthquake swarm (maximum magnitude 4.1 at 27.914°N, 37.055°E) occurred in an uninhabited area 75 km southeast of Tabuk and 25 km north of Harrat Uwayrid. This is an area where seismic activity has occurred since instrumental recording began and some epicentres are shown in Figs. 25 and 26. On the surface there are many northwest-southeast Cenozoic faults as well as some oriented both north-south and east-west. The epicentres cannot be correlated with any specific surface fault but correspond to an area of crustal extension with abundant Cenozoic faults.

Comparison of the seismic data layers with the fault map clearly indicates that since instrumental earthquake recording began, activity in northwest Arabia occurs mainly in the Gulf of Aqaba and represents the Gulf of Aqaba fault system. The seismicity (Figs. 25, 26 and 27) gives little indication of present day activity on either the syn-rift arcuate reactivated Precambrian faults or the prominent east-west fault system. There is some evidence of seismic activity both on the Red Sea coastal plain and just offshore, suggesting some of the syn-rift Red Sea coast plain faults may be still active, or alternatively the seismicity is due to salt tectonics from the Miocene evaporate deposits as seen in the southern coastal city of Jizan, where a salt dome emerges onto the surface.

The seismicity (Figs. 25 and 26) also shows that the ribbon fault system is today mainly inactive. However, some geologic evidence supports syn-rift activity in this system. Many of the horst-graben strips have morphological expressions of a few metres suggesting displacement in the not too distant past. There are also areas of open fissures in some of the Paleozoic sandstone units near Tabuk that indicate crustal extension in the past few thousand years.

The ribbon fault system, as well the pre-rift grabens (some with gabbro dikes), indicate northeast to southwest crustal extension in northwest Saudi Arabia in the form of thousands of small horst and graben blocks. This style of crustal extension in the NW of Saudi Arabia contrasts greatly with the system of crustal extension seen on the west-central part of the Arabian shield, where a well-defined graben system underlies the Cenozoic basaltic lava fields of the Makkah-Madinah-Nafud volcanic line (MMN line). The difference in the physical expression of the crustal extension on the back or passive margin of the Arabian tectonic plate may reflect mainly the physical nature of the host rocks. In the northwest, layered sedimentary rocks of the Phanerozoic cover rocks dominate, whereas in west central Saudi Arabia the Precambrian shield rocks dominate. The different styles of extension may reflect only lithological differences. Such an interpretation is shown in Fig. 28. The extension in the underlying crystalline Precambrian shield is expressed as a simple graben system but in the overlying less competent and unmetamorphosed Phanerozoic cover rocks, the basement faults are shown as feathering or splaying into a fault swarm of many small faults as they penetrate the sedimentary cover rocks.

14 Conclusion

Today the most seismically active area of Saudi Arabia is centred on the Gulf of Aqaba and extends to the shores on both sides. The distribution of the epicentres corresponds to the pattern of the Gulf of Aqaba fault system. Geological work in the area reveals thousands of young Cenozoic faults, mainly Neogene (younger than 23 Ma), that extend over an area up to 300 km east of the Gulf of Aqaba. In this area to the east of the Gulf there are many fault systems of different ages. Some faults predate rifting of the Red Sea, such as the Tabuk graben that aeromagnetic data shows to be underlain by the Great Ja'adah Gabbro Dike. This dike is actually about 30–40 km to the northeast of a much more subdued magnetic trend that passes close to the city of Tabuk. The Ja'adah dike (or Tabuk graben) extends northwest out of Saudi Arabia up to the Dead Sea Rift Zone (cf. Hatcher et al. 1981), as is also indicated by the fault trends shown in the geological map of Jordan (Bender 1975). Some of the northwest-southeast faults of this system show seismic activity and can be regarded as syn-rift reactivation of older pre-rift faults. The current earthquake record contains little evidence for the thousands of Cenozoic faults in the 300-km wide area east of the Gulf of Aqaba.

At the present time there is considerable development of infrastructure and new towns and cities throughout Saudi Arabia. From Haql and the Saudi Arabian side of the Gulf of Aqaba, the high-rise hotels of Elat and Sharm el Shaik can be seen, and there is the possibility of developing hotels and a tourist industry on the Saudi Arabian side of the Gulf. However, the work of Ben-Avraham et al. (1979) and Ben-Avraham (1985) clearly shows that active faults underlying the Gulf of Aqaba pass close to the Saudi Arabian side of the Gulf. Roobol et al. (1999) describe an area of epicentral rupture of the ground along the Saudi Arabian coastline between Haql and Maqna. The fault map (Fig. 7) shows the distribution of the active Gulf of Aqaba fault system and its proximity to the Saudi Arabian coastline. Hence any future development along the shoreline of the Saudi Arabian side of the Gulf of Aqaba must take into consideration the close proximity of major faults belonging to the Gulf of Aqaba system to the Saudi Arabian shore. The danger of development in a coastal area of potential epicentral rupture, sand volcano formation (due to P wave expulsion of groundwater) and small tsunamis (as was observed to happen in November 1995) must be considered.

Acknowledgements The work was initiated in 2003 by the geohazard section of the Saudi Geological Survey, when raised coral reefs were surveyed along the coast of the Gulf of Aqaba by a team comprising John Roobol, Mohammed Al Rehaili, Ghassan Al-Sulaimani, Faisal Al-Ammawi, and A.A. Al-Gafour Turkostani and was completed by the National Center for Earthquakes and Volcanoes, partly as compilation work and partly as a field traverse accompanied by Mr. Khalid A. Kadi. The earthquake database was edited by Dr. S.Y. El-Hadidy. The data is released with the permission of President of the Saudi Geological Survey, His Excellency Dr. Z.A. Nawab. The SRTM and magnetic data processing and images in the report employ software from Stewart Geophysical Consultants Pty. Ltd. (stewgeop@senet.com.au). Drafting of the diagrams was done by Mr. Diosado G. Quero, Mr. George Certeza and Mr. Mazen Abuabdullah.

References

Abdel-Rahman K, Al-Amri AMS, Abdel-Moneim E (2009) Seismicity of the Sinai Peninsula, Egypt. Arabian J Geosci 2:103–118

Alabouvette B, Motti E, Redmond CP, Villemur JR (1979) The Wadi Azlam prospect—results of the 1974 program. Saudi Arabian Directorate General of Mineral Resources, BRGM Saudi Arabian Mission Report 97-JED-15, 38 p, 7 figs, 21 app

Al Amri AM, Schult FR, Bufe CG (1991) Seismicity and aeromagnetic features of the Gulf of Aqaba (Elat) region. J Geophys Res 96 (B12):179–185

Al Tarazi E, Abu Rajab J, Gomez F, Cochran W, Jaafar R, Ferry M (2011) GPS measurements of nearfield deformation along the southern Dead Sea Fault System. Geophys Geochem Geosyst (G3) 12(12). https://doi.org/10.1029/2011gc003736, 17 p

ArRajehi A, McClusky S, Reilinger R, Daoud M, Alchalbi A, Ergintav S, Gomez F, Sholan J, Bou-Rabee F, Ogubazghi G, Haileab B, Fisseha S, Asfaw L, Mahmoud S, Rayan A, Bendik R, Kogan L (2010) Geodetic constraints on present-day motion of the Arabian Plate: Implications for Red Sea and Gulf of Aden rifting. Tectonics 29, TC3011. https://doi.org/10.1029/2009tc002482, 10 p

Bayer H-J, Hotzl H, Jado AR, Roscher B, Voggenreiter W (1988) Sedimentary and structural evolution of the northwest Arabian Red Sea margin. Tectonophysics 153:137–151

Bazzari M, Merghelani H, Badawi H (1990) Seismicity of the Haql Region, Gulf of Aqaba, Kingdom of Saudi Arabia. Saudi Arabian Directorate General of Mineral Resources Open-File Report USGS-OF-10-9 (IR 825), 39 p

Bender F (1975) Geologic map of Jordan. Scale 1:500,000. U.S. Geological Survey, Reston, Virginia, Professional paper 560-I, Plate 1

Ben-Avraham Z (1985) Structural framework of the Gulf of Elat (Aqaba), northern Red Sea. J Geophys Res 90(B1):703–716

Ben-Avraham Z, Garfunkel Z, Almagor G, Hall JH (1979) Continental breakup by a leaky transform: the Gulf of Elat (Aqaba). Science 206:214–216

Ben-Avraham Z, Garfunkel Z, Lazar M (2008) Geology and evolution of the Southern Dead Sea Fault with emphasis on subsurface structure. Ann Rev Earth Planet Sic 36:357–387

Ben-Menahem A, Aboodi E (1981) Micro- and macro seismicity of the Dead Sea Rift and off-coast eastern Mediterranean. Tectonophysics 80:199–233

Blank HR (1977) Aeromagnetic and geological study of Tertiary dikes and related structures on the Arabian margin of the Red Sea. Saudi Arabian Directorate General of Mineral Resources, Red Sea Research 1970-1975, Bulletin no. 22, pp G1-18

Blank HR, Andresen GE (1991) Compilation and interpretation of aeromagnetic data for the Precambrian Arabian shield, Kingdom of Saudi Arabia. Saudi Arabian Directorate General of Mineral Resources Open-File Report USGS-OF-10-8, 54 p

Bohannon RG, Nasser CW, Schmidt DL, Zimmermann RA (1989) The timing of uplift, volcanism, and rifting peripheral to the Red Sea: a case for passive rifting? J Geophys Res 94:1683–1701

Bosworth W (2015) Geological evolution of the Red Sea: Historical background, review and synthesis. In: Rasul NMA, Stewart ICF (eds) The Red Sea: the formation, morphology, oceanography and environment of a young ocean basin. Springer Earth System Sciences, Berlin Heidelberg, pp 45–78. https://doi.org/10.1007/978-3-662-45201-1_3

Camp VE, Roobol MJ (1992) Upwelling asthenosphere beneath western Arabia and its regional implications. J Geophysics Res 97 (B11):15255–15271

Clark MD (1987) Geological map of the Al Bad Quadrangle, sheet 28A: Kingdom of Saudi Arabia. Saudi Arabian Directorate General of Mineral Resources Geoscience Map GM-81C, scale 1:250,000, with text, 46 p

Coleman RG (1977) Geologic background of the Red Sea. In: Red Sea Research (anon. ed.) Saudi Arabian Directorate General of Mineral Resources Bulletin no. 22:C1-C9

Daëron M, Klinger Y, Tapponier P, Elias A, Jacques E, Sursock A (2007) 12,000-year-long record of 10 to 13 paleoearthquakes on the Yammouneh Fault, Levant Fault System, Lebanon. Bull Geol Soc Am 97(3):749–771

Davies FB, Grainger DJ (1985) Geologic map of the Al Muwaylih quadrangle, sheet 27A, Kingdom of Saudi Arabia. Saudi Arabian Deputy Ministry for Mineral Resources Geoscience Map GM-82A, scale 1:250,000, with text 32 p

Doglioni C, Carminati E, Bonatti E (2003) Rift asymmetry and continental uplift. Tectonics 22(3):1024. https://doi.org/10.1029/2002tc001459

El-Isa ZH, Merghelani HM, Bazzari MA (1984) The Gulf of Aqaba earthquake swarm of 1983 January-April. Geophys J Roy Astron Soc 78:711–722

Eyal M, Bartov Y, Shimron AE, Bentor YK (1980) Sinai geological map. Geol Surv Isr, Scale 1:500,000

Eyal M, Eyal Y, Bartov Y, Steinitz G (1981) The tectonic development of the western margin of the Gulf of Elat (Aqaba) rift. Tectonophysics 80:39–66

Gettings ME, Blank HR, Mooney WD, Healy JH (1986) Crustal structure of southwestern Saudi Arabia. J Geophys Res 91:6491–6512

Girdler RW, Southren TC (1987) Structure and evolution of the northern Red Sea. Nature 330:716–721

Grainger DJ, Hanif MR (1989) Geologic map of the Shagab quadrangle, sheet 27B, Kingdom of Saudi Arabia. Saudi Arabian Directorate General of Mineral Resources Geoscience Map GM-109C, scale 1:250,000, with text 31 p

Hamiel Y, Amit R, Begin ZB, Marco S, Katz O, Salamon A, Zilberman E, Porat N (2009) The seismicity along the Dead Sea fault during the last 60,000 years. Bull Seismol Soc Am 99:2020–2026

Hatcher RD, Zietz I, Regan RD, Abu-Ajamieh M (1981) Sinistral strike slip motion on the Dead Sea Rift: Confirmation from new magnetic data. Geology 9:458–462

Hayes TS, Sutley SJ, Kadi KA, Balkhiyour MB, Siddiqui AA, Beshir Z, Hashem HI (2002) Jabal Dhaylan zinc-lead deposits, geological setting, genesis and 1996–2000 exploration progress. Saudi Geological Survey Open-File Report SGS-OF-2001-5; vol.1—Text, App 1 and Plates; vol. 2—Apps 2, 3, 4 and 5; text 71 p

Hughes GW, Filatoff J (1994) New biostratigraphic constraints on Saudi Arabian Red Sea pre- and syn-rift sequences. In: Al-Husseini MI (ed) Middle East Petroleum Geosciences Conference, Geo'94, April 25–27, 1994, Bahrain. Selected Paper, Gulf Petrolink, Manama, Bahrain, vol. II: 517–528

Hughes GW, Johnson RS (2005) Lithostratigraphy of the Saudi Arabian Red Sea. GeoArabia 10(3):49–126

Janjou D, Halawani MA, Al-Muallem MS, Robelin C, Brosse J-M, Courbouleix S, Dagain J, Genna A, Razin T, Roobol MJ, Shorbaji H, Wyns R (1996) Geologic map of the al-Qalibah quadrangle, sheet 28C, Kingdom of Saudi Arabia. Saudi Arabian Directorate General of Mineral Resources Geoscience Map GM-135, scale 1:250,000, with text 44 p

Janjou D, Halawani MA, Brosse J-M, Al-Muallem MS, Becq-Giraudon J-S, Dagain J, Genna A, Razin T, Roobol MJ, Shorbaji H, Wyns R (1997) Geologic map of the Tabuk quadrangle, sheet 28B, Kingdom of Saudi Arabia. Saudi Arabian Deputy Ministry for Mineral Resources Geoscience Map GM-137, scale 1:250,000, with text 49 p

Johnson PR (1998) Tectonic map of Saudi Arabia and adjacent areas. Saudi Arabian Deputy Ministry for Mineral Resources Technical Report USGS-TR-98-3 (IR 948)

Johnson PR (2007) Geologic map of the Sirhan-Turayf Basin compiled by P.R. Johnson after mapping by Chester A. Wallace, Salah M. Dini, Anwar A. al-Farasani, Ahmed S. Banaker, Abdullah F. al-Khattabi and Mohammed H. al-Kaff. Saudi Geological Survey, unpublished map, scale 1:250,000

Johnson PR, Vranas GJ (1992) Qualitative interpretation of aeromagnetic data for the Arabian shield. Saudi Arabian Directorate General of Mineral Resources Open-File Report USGS-OF-92-1, 34 p

Klinger Y, Rivera L, Haessler H, Maurin J-C (1999) Active faulting in the Gulf of Aqaba: New knowledge from the Mw 7.3 earthquake of 22 November 1995. Bull Seismol Soc Am 89:1025–1036

Klinger Y, Michel R, Avouac J-P (2000a) Co-seismic deformation during the Mw 7.3 Aqaba earthquake (1995) from ERS-SAR interferometry. Geophys Res Lett 27(22):3651–3654

Klinger Y, Avouac JP, Dorbath L, Abou Karaki N, Tisnerat N (2000b) Seismic behavior of the Dead Sea fault along Araba valley, Jordan. Geophys J Int 142:769–782

Klinger Y, Le Béon M, Al-Qaryouti M (2015) 5000 yr of paleoseismicity along the southern Dead Sea fault. Geophys J Int 202:313–327

Liddicoat WK (1982) Cenozoic volcanic pipes in the Marythoum area (27/37D). Saudi Arabian Deputy Ministry for Mineral resources, Jeddah, Open-File Report DGMR-OF-03-11, 47 p

Makris J, Allam A, Mokhtar T, Basahel A, Dehghani GA, Bazari M (1983) Crustal structure in the northwestern region of the Arabian shield and its transition to the Red Sea. Bull Faculty Earth Sci, King Abdulaziz University, 6:435–447

McGuire AV, Bohannon RG (1989) Timing of mantle upwelling: Evidence for a passive origin for the Red Sea rift. J Geophys Res 94:1677–1682

Mitchell NC, Stewart ICF (2018) The modest seismicity of the northern Red Sea rift. Geophys J Int 214:1507–1523

Motti E, Teixido L, Vazquez-Lopez R, Vial A (1982) Maqna Massif area—Geology and Mineralization. Saudi Arabian Deputy Ministry for Mineral Resources Open-File Report BRGM-OF-02-16, 44 p

Phoenix Corporation (1985) The interpretation of an airborne geophysical survey of the Cover Rocks region. Kingdom of Saudi Arabia, Saudi Arabian Deputy Ministry for Mineral Resources, 76 p

Reilinger R, McClusky S, Vernant P, Lawrence S, Ergintav S, Cakmak R, Ozener H, Kadirov F, Guliev I, Stepanyan R, Nadariya M, Hahubia G, Mahmoud S, Sakr K, ArRajehi A, Paradissis D, Al-Aydrus A, Prilepin M, Guseva T, Evren E, Dmitrotsa A, Filikov SV, Gomez F, Al-Ghazzi R, Karam G (2006) GPS constraints on continental deformation in the Africa-Arabia-Eurasia continental collision zone and implications for the dynamics of plate interactions. J Geophys Res 111, B05411. https://doi.org/10.1029/2005jb004051, 26 p

Reilinger R, McClusky S, ArRajehi A (2015) Geodetic constraints on the geodynamic evolution of the Red Sea. In: Rasul NMA, Stewart ICF (eds) The Red Sea: The formation, morphology, oceanography and environment of a young ocean basin. Springer Earth System Sciences, Berlin Heidelberg, pp 135–149. https://doi.org/10.1007/978-3-662-45201-1_7

Roobol MJ, Al-Rehaili M, Arab N, Celebi M, Halawani MA, Janjou D, Kazi A, Martin C, Sahl M, Showail A (1999) The Gulf of Aqaba Earthquake of 22 November, 1995: Its effects in Saudi Arabia. Saudi Arabian Deputy Ministry for Mineral Resources Technical Report BRGM-TR-99-16, 67 pp, 49 figs, 4 tables, 1 app

Roobol MJ, Kadi KA (2008) Cenozoic faulting in the Rabigh area, central west Saudi Arabia (including the sites of King Abdullah Economic City and King Abdullah University for Science and Technology). Saudi Geological Survey Technical Report SGS-TR-2008-6, 1:250,000 map, with text 12 p

Roobol MJ, Stewart ICF (2009) Cenozoic faults and recent seismicity in northwest Saudi Arabia and the Gulf of Aqaba region. Saudi Geological Survey Technical Report SGS-TR-2008-7, 35 pp, 35 figs, 2 apps, 10 plates

Rowaihy MN (1985) Geological map of the Haql quadrangle, sheet 29A, Kingdom of Saudi Arabia. Saudi Arabian Directorate General of Mineral Resources Geoscience Map GM-80C, scale 1:250,000, with text 15 p

Sadeh M, Hamiel Y, Ziv A, Bock Y, Fang P, Wdowinski S (2012) Crustal deformation along the Dead Sea Transform and the Carmel Fault inferred from 12 years of GPS measurements. J Geophys Res 117, B08410. https://doi.org/10.1029/2012jb009241, 14 p

Sawires R, Peláez JA, Fat-Helbary RE, Ibrahim HA (2016) Updated probabilistic seismic-hazard values for Egypt. Bull Seismol Soc Am 106:1788–1801

Smith J, Bokhari M (1984) The Haql Earthquake; preliminary report for the period 6–30 Rabi al Thani, 1403/ January-February, 1983. Saudi Arabian Directorate General of Mineral Resources Open-File Report DGMR-OF-03-29, 12 p

Stewart ICF (2007) Earthquake risk in western Saudi Arabia and the Red Sea from seismic moment. Saudi Geological Survey Technical Report SGS-TR-2007-4, 41 pp, 32 figs

Vaslet D, Janjou D, Robelin C, Al-Muallem MS, Halawani MA, Brosse J-M, Breton J-P, Courboulix S, Roobol MJ, Dagain J (1994) Geologic map of the Tayma Quadrangle, Sheet 27C, Kingdom of Saudi Arabia. Saudi Arabian Directorate General of Mineral Resources Geoscience Map GM-134, scale 1:250,000, with explanatory text 51 p

Wallace CA, Dini SM, Al-Farasani AA (2000a) Geologic map of the Thaniyat Turayf quadrangle, sheet 29C, Kingdom of Saudi Arabia. Saudi Geological Survey Geoscience Map GM-129C, scale 1:250,000 with text 31 p

Wallace CA, Dini SM, Al-Farasani AA (2000b) Geologic map of the Wadi as Sirhan quadrangle, sheet 30C, Kingdom of Saudi Arabia. Saudi Geological Survey Geoscience Map GM-127C, scale 1:250,000 with text 27 p

Wallace CA, Dini SM, Al-Farasani AA (2001) Geologic map of the Jibal at Tubayq quadrangle, sheet 29B, Kingdom of Saudi Arabia. Saudi Geological Survey Geoscience Map GM-133C, scale 1:250,000 with text 36 p

Wetzler N, Kurzon I (2016) The earthquake activity of Israel: Revisiting 30 years of local and regional seismic records along the Dead Sea Transform. Seismol Res Lett 87(1):47–58

Williams VS, Al-Rehaili MH, Al-Sulaimani G, Showail AA, Bahloul S, Zahrani M (2001) Preliminary report on investigations of neotectonics and paleoseismicity on the east coast of the Gulf of Aqaba. Saudi Geological Survey Open-File Report SGS-OF-2001-2, 11 p

Zahran HM, Stewart ICF, Johnson PR, Basahel MH (2003) Aeromagnetic-anomaly maps of central and western Saudi Arabia. Saudi Geological Survey Open-File Report SGS-OF-2002-8, 6 pp, 1 fig, 1 table, 4 sheets (scale 1:2,000,000)

Zahran HM, Sokolov V, El-Hadidy Youssef S, Alraddadi WW (2015) Preliminary probabilistic seismic hazard assessment for the Kingdom of Saudi Arabia based on combined areal source model: Monte Carlo approach and sensitivity analyses. Soil Dynamics Earthquake Eng 77:453–468

Zahran HM, Sokolov V, Roobol MJ, Stewart ICF, El-Hadidy Youssef S, El-Hadidy M (2016) On the development of a seismic source zonation model for seismic hazard assessment in western Saudi Arabia. J Seismology. https://doi.org/10.1007/s10950-016-9555-y, 23 p

Crustal and Upper-Mantle Structure Beneath Saudi Arabia from Receiver Functions and Surface Wave Analysis

P. Martin Mai, Jordi Julià, and Zheng Tang

Abstract

Using receiver-functions and surface-wave dispersion curves, we study the crustal and upper-mantle structure of Saudi Arabia. Our results reveal first-order differences in crustal thickness between the Arabian Shield in the west and the Arabian Platform in the east. Moho depths generally increase eastward, while crustal thickness varies strongly in the west over the volcanic regions and near the Red Sea. Localized zones of increased P-wave speed in the west may indicate solidified magmatic intrusions within the area of recent volcanism. Our receiver-function analysis for deep converted phases reveals that the transition zone thickness between the 410 km and the 660 km discontinuities is not anomalously thinned, refuting the hypothesis of a small localized mantle plume as the origin for the volcanic activity in western Saudi Arabia. Our results suggest that the volcanism in western Arabia may be due to the lithospheric mantle being heated from below by lateral flow from the Afar and (possibly) Jordan plumes. This triggers localized melts that ascend adiabatically through the lithosphere as magma diapirs. Recent xenolith measurements that provide information on temperatures and depths of melting are overall consistent with this hypothesis. However, further dedicated localized tomographic studies are needed to decipher the details of the origin of the volcanism and its relation to the overall geodynamics of the region.

1 Introduction

The Arabian Plate consists of the western Arabian Shield and the eastern Arabian Platform (Fig. 1). While the Arabian Platform is predominantly composed of Paleozoic, Mesozoic and Cenozoic sedimentary rocks with eastwardly increasing thickness (Brown 1972; Stoeser and Camp 1985), the Arabian Shield is a complex assemblage of Proterozoic terrains, separated by major ophiolite-bearing suture zones (Schmidt et al. 1979). Several previous studies investigated the large-scale lithospheric structure of Saudi Arabia, but only sparse datasets were available at that time (e.g., Sandvol et al. 1998; Kumar et al. 2002; Julià et al. 2003; Al-Damegh et al. 2005; Tkalčić et al. 2006; Hansen et al. 2007). These studies revealed that Moho depth generally increases eastward, but also indicated that lateral Moho-depth variations exist in the west. However, due to the complex geology and tectonic history of the Arabian plate, these spatially coarse measurements of crustal thickness and lithospheric structure do not allow more detailed interpretation on the correlations between deep Earth structure and surface geology.

Moreover, the geology and recent tectonic history of western Saudi Arabia is strongly affected by the opening of the Red Sea that started when Arabia began to separate from Africa ~30 Ma ago (Camp and Roobol 1992; Bosworth et al. 2005; Garfunkel and Beyth 2006). The Red Sea opening is thought to be governed by periods of both passive and active rifting. Rifting initiated passively, and was followed by a period of active rifting (Camp and Roobol 1992; Ebinger and Sleep 1998; Daradich et al. 2003). Rifting is also non-uniform along the Red Sea. The southern Red Sea (south of ~21°N) already shows seafloor spreading and volcanic activity (Ebinger and Sleep 1998; Daradich et al. 2003; Chang et al. 2011; Xu and Jónsson 2014), while the northern Red Sea (north of ~21°N) is interpreted as an active rift at the end of its continental stage that transitions into oceanic seafloor spreading (Cochran and Martinez 1988).

P. Martin Mai (✉) · Z. Tang
Division of Physical Sciences and Engineering, King Abdullah University of Science and Technology, Thuwal, 23955-6900, Saudi Arabia
e-mail: martin.mai@kaust.edu.sa

J. Julià
Universidade Federal Do Rio Grande Do Norte, Natal, Brazil

N. M. A. Rasul and I. C. F. Stewart (eds.), *Geological Setting, Palaeoenvironment and Archaeology of the Red Sea*,
https://doi.org/10.1007/978-3-319-99408-6_14

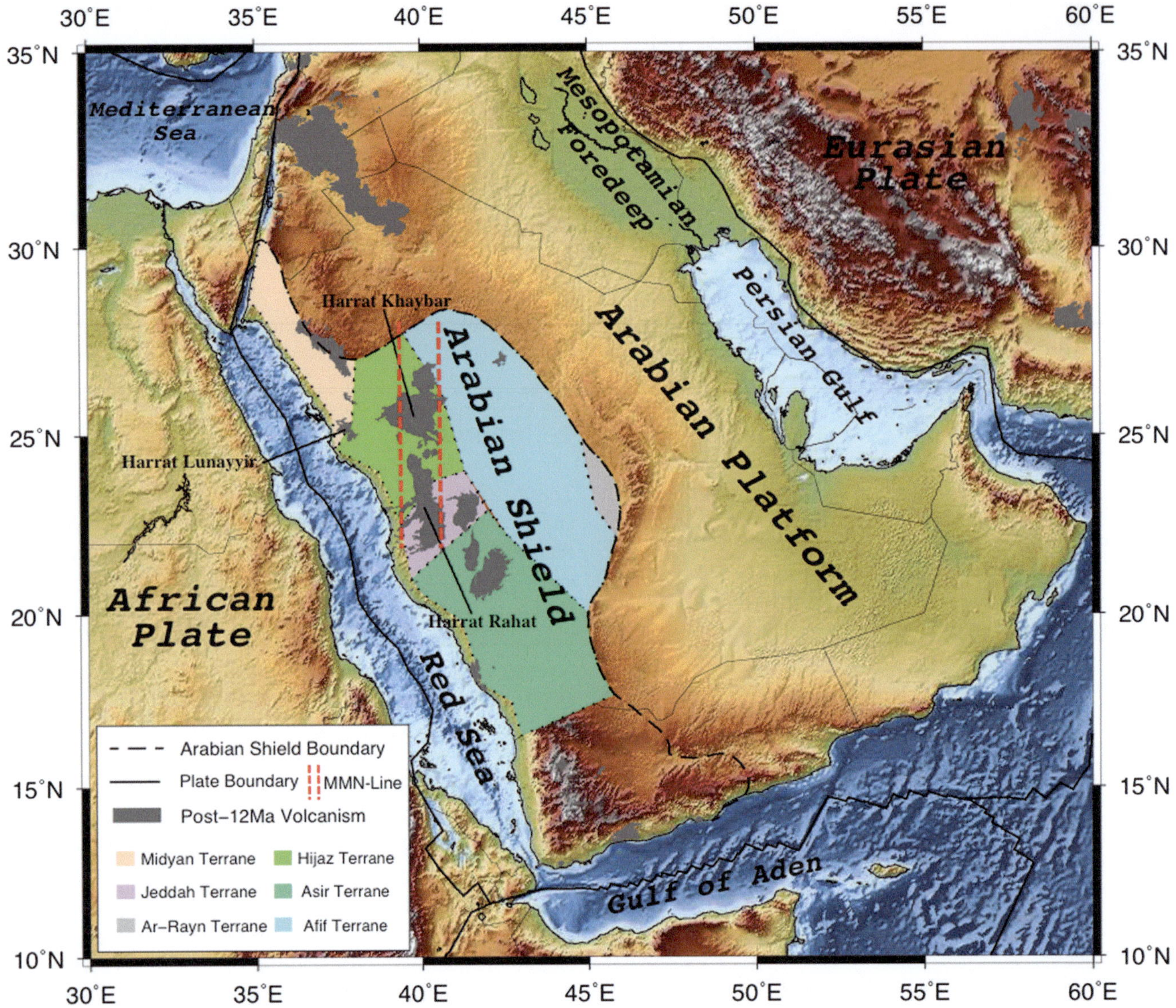

Fig. 1 Geological map of the Arabian plate and neighbouring tectonic plates. Major plate boundaries are shown by black lines. The bright-coloured regions in the Arabian Shield outline six accreted terranes (Midyan, Hijiaz, Jeddah, Asir, Afif and Ar-Rayn terrane). Areas of recent volcanic activity (past 12 million years) are shown in dark-gray (modified after Tang et al. 2016)

Western Saudi Arabia is dotted with prominent Cenozoic volcanic regions (so-called harrats, Figs. 1 and 2). The detailed crustal and upper mantle structure underneath these lava fields is still poorly known, although recent studies investigated the velocity structure underneath Harrat Lunayyir (Hansen et al. 2013; Koulakov et al. 2015), where a stalled dike intrusion caused significant seismic activity in 2009 (Pallister et al. 2010). In fact, the exact origin of these lava fields is not well understood. They are found to become progressively younger northward along the north-south oriented Makkah–Madinah–Nafud (MMN) line (Fig. 1). They also show a bimodal chemical character: older lavas (30–20 Ma) are tholeiitic to transitional in composition while younger volcanics (12 Ma to recent) are transitional to strongly-alkalic. The genesis of the older harrats, approximately oriented parallel to the Red Sea, is commonly associated with the early rifting stage of the Red Sea. However, it remains elusive as to how the harrats and their recent magmatic activities are related to the current rift-axis of the Red Sea, which is located 200–400 km farther to the west. A local mantle plume beneath the Arabian shield was suggested by Camp and Roobol (1992), while Chang and van der Lee (2011) proposed that the Cenozoic volcanism may be due to either lateral mantle flow from the Afar and, perhaps, the Jordan hotspot. A region of deep ($\sim$150 km) low shear velocities is mapped under the southern Red Sea and the western Arabian plate that may be considered as an indication for lateral mantle flow from the Afar region (Chang et al. 2011). The global isotropic and

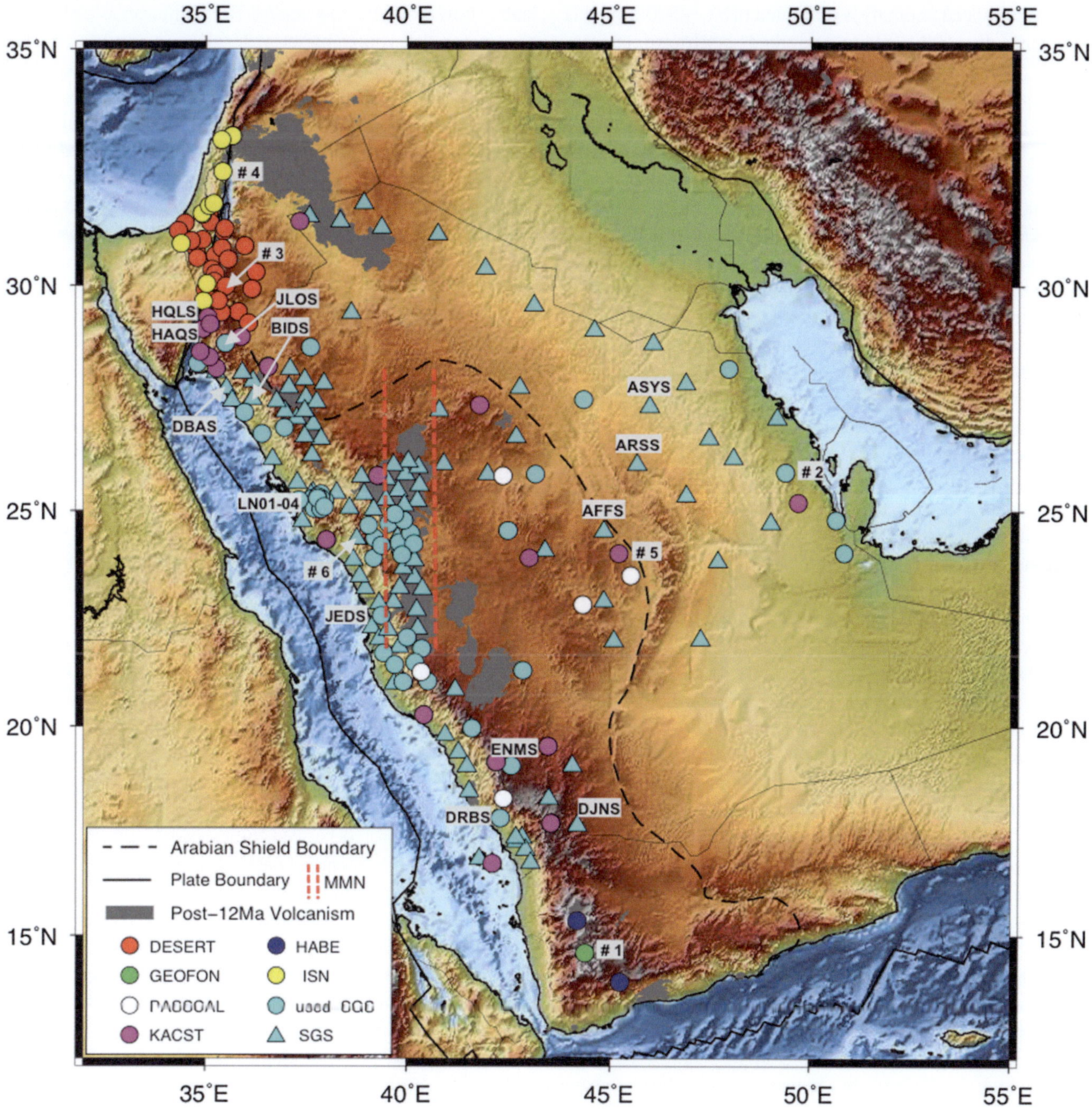

Fig. 2 Topographic map of the Arabian Plate and surrounding regions, showing the broadband seismic stations operated by the SGS (cyan circles and triangles), and locations of several other seismic deployments. Note the increased station density in the western part of Saudi Arabia, particularly in regions of recent volcanic activity. Stations marked by circles (including DESERT, GEOFON, PASSCAL, HABE, ISN, KACST and 56 SGS stations) are utilized for estimating the transition zone thickness. 56 SGS stations (filled cyan circles) are used for joint inversion to investigate the crustal and upper mantle structures of Saudi Arabia (see Tang et al. 2016). For sites with station names, we show receiver-function examples in Fig. 4, for the stations numbered 1–6, in Fig. 9

radially anistropic mantle-structure model SGLOBE-rani (Chang et al. 2015) finds a broad lower mantle low-velocity region beneath southern Africa ("Africa superplume") that connects to the upper-mantle low-velocity zone beneath the Afar, but the relationship of the low-velocity region underneath western Arabia to this large-scale low-velocity zone remains unclear. Recent regional tomography studies confirm the ambiguous spatial relation between the location of the Cenozoic volcanism in western Saudi Arabia, the rift axis of the Red Sea (and an expected zone of low shear-wave speed underneath it), and a more deeply rooted low-shear-wave speed region of unknown origin (e.g., Chang et al. 2011). This shear-wave speed anomaly is estimated to be $\sim$500 km wide at a depth of $\sim$100–300 km, and trends northward, approximately parallel to the MMN-line.

To understand the geodynamics and present-day geology specifically of western Saudi Arabia, the origin and activity of the harrat volcanism needs to be investigated. Is the volcanism controlled by a localized small plume underneath western Saudi Arabia, or by lateral flow of mantle material from Afar? What does the deeper lithospheric structure look like? Is there any indication in the thickness of the upper mantle transition zone that may hint at the presence of a deep heat source, potentially related to a localized mantle plume? Can we reconcile the volcanism and related seismological observations/parameters with previous petrological work on xenoliths from various harrats that provide point-wise estimates of crustal temperature-pressure conditions (e.g., McGuire 1988)?

Here, we combine two sets of receiver-function studies that help to illuminate the crustal structure and the upper-mantle transition zone, and then relate our findings to information gleaned from xenolith data on pressure- and temperature-dependent melting. First, we summarize our recent work focused on jointly inverting teleseismic P-receiver functions and Rayleigh wave dispersion curves at 56 broadband stations across Saudi Arabia. The joint inversion reduces the non-uniqueness of the receiver-function inversion and its starting-model dependence. We then describe our observed lateral variations in crustal thickness and shear-wave speeds in the crust and uppermost mantle. These measurements allow us to estimate the temperature of the mantle lid, which displays significant lateral variations that we reconcile with thermobarometric data from xenolith analyses. The picture emerging from our work supports localized small-scale melt ascent from the mantle into the crust. Second, we analyze a dataset of over 4,500 teleseismic P-waveforms recorded at about 100 broadband stations in the region. This allows us to test the hypothesis of a deep mantle plume under Arabia and/or Jordan, based on estimates of the thickness of the upper-mantle transition zone, which is defined by temperature-dependent phase transformations in the olivine mineral system (nominally at 410 and 660 km depth). Variations in transition-zone thickness may serve as a proxy for transition-zone temperature, such that a positive temperature anomaly due to hot mantle upwellings leads to a thin transition zone. Our analysis demonstrates that the transition zone is thermally unperturbed under the Arabian shield.

Finally, we discuss our combined findings, related to Moho depth and its lateral variations, inferred mantle-lid temperatures, previous xenolith studies in the region, and our results on transition-zone thickness. These results are incompatible with the presence of a mantle plume below western Arabia. However, our analyses are consistent with lateral flow from the Afar and/or Jordan plume as a plausible source for the young harrat volcanism in the region. The precise origin of the individual harrats cannot be determined with the available data, calling for dedicated high-resolution experiments (small-scale seismic networks and local tomographic studies). However, the presence of localized zones with high Vp/Vs-ratios, indicative of partial melts, as well as localized high P-wave anomalies, potentially due to solidified melts, suggest episodic melt ascent into the crust.

2 Data and Data Processing

Since 2006, the Saudi Geological Survey (SGS) has operated the permanent Saudi National Seismic Network (SNSN), which consists of modern broadband seismic sensors. The number of stations has increased over time to more than 200 (as of December 2016). Stations are installed at an average spacing of about 70 km in western Saudi Arabia, but at much larger and irregular spacing in the Eastern Province (Fig. 2). Locally, the network is densified in areas of special interest, such as regions of increased seismicity due to ongoing deep-seated volcanic processes. The SNSN stations are equipped with either T40 or T120 Nanometrics sensors, whose instrument response is flat up to 40 s and 120 s, respectively. At several SNSN sites STS-2 Streckeisen instruments are operated. The recorded data are sent via VSAT to the National Center for Earthquakes and Volcanoes (NCEV), where routine earthquake detection and location is conducted, and the data is archived. We utilize the seismic dataset described subsequently to estimate the crustal and upper-mantle shear-wave speed (i.e., S-wave velocity structure) underneath Saudi Arabia.

Applying rigorous data-quality criteria, we select a subset of 56 SNSN stations installed until late 2012 (Fig. 2). Since the modern SGS broadband seismic array began operating only in 2006, our database is limited to 157 earthquakes at teleseismic distances (epicentral distances $\Delta = 30°$ to $90°$, no depth constraints) with magnitude $m_b > 5.5$. Most events arrive from back-azimuths between 25° and 110°, and epicentral distances between 60° and 90° (hence, their ray-parameters fall between 0.04 and 0.06 s/km), corresponding to predominant source locations in the subduction zones of the western Pacific and Eurasian plates. For this dataset, we calculate receiver functions (RF) from the teleseismic P-waveforms, while station-specific surface-wave dispersion (SWD) curves were extracted from the independent tomographic study of Pasyanos (2005). We then jointly invert the RFs and SWD curves to estimate the depth-dependent S-wave speed underneath each station (Tang et al. 2016).

RFs are time series calculated by deconvolving the vertical component of motion from the corresponding

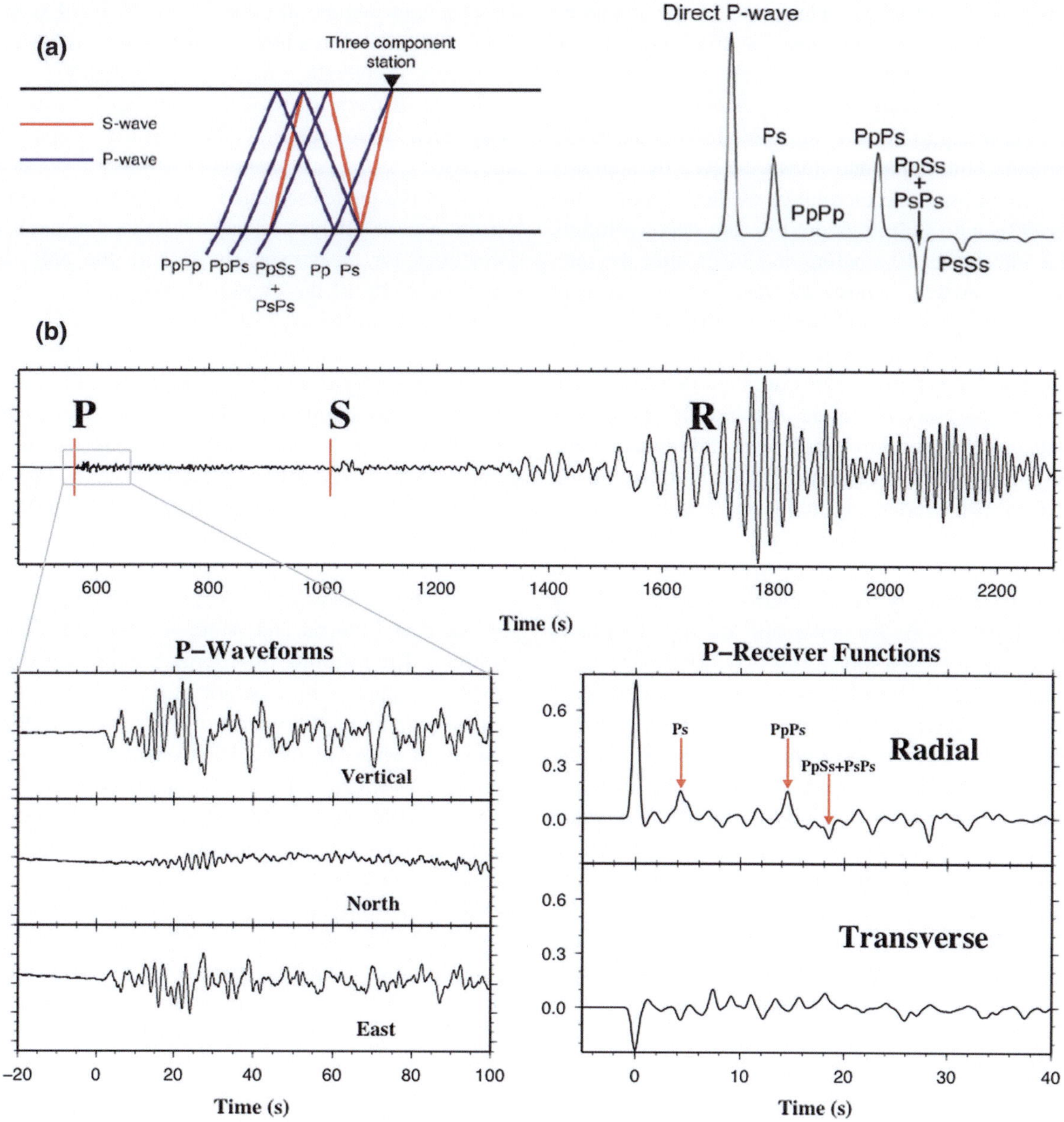

Fig. 3 **a** (left) Sketch illustrating how P-waves (blue) and S-waves (red) interact with the subsurface before being recorded at a seismic station located at the Earth surface. (right) Receiver function example that documents the multiple seismic-wave interactions with the layer interface. **b** Graphical summary of teleseismic waveform characteristics and receiver function processing. (top) At large distances from an earthquake, the first arriving P-wave and the secondary S-wave are well separated; the S-wave is followed by large-amplitude surface waves (here: Rayleigh waves). (bottom left) The complex wavetrain of the P-waves contains information about the earthquake source as well as the Earth structure through which the seismic waves travel, recorded on three mutually perpendicular components. (bottom right) The receiver-function (RF) processing procedure of the two horizontal components returns a radial and transverse RF; the peaks in the radial RF indicate interaction of the P-wave with the layered structure of the Earth

horizontal components (Fig. 3). The deconvolution removes the signature of the instrument response and the source time functions, leaving the near-receiver propagation effects in the deconvolved waveforms (Langston 1979). For a simple layer over half-space model, the RF time series contains the direct P-wave, the P-to-S conversion (Ps) at the major internal discontinuity, and two reverberations (PpPs and PsPs + PpSs) between the free surface and the discontinuity (Fig. 3a). The largest amplitudes in the RF waveforms are generally associated with the crust-mantle boundary (i.e., the Mohorovicic discontinuity, “Moho”). Therefore, modeling the S-P travel times associated with those amplitudes can be used to constrain crustal thickness and the bulk Vp/Vs ratio of the crust (e.g., Zhu and Kanamori 2000). More complex structures involving multiple seismic discontinuities complicate the RF waveforms by adding further Ps conversions and reverberations due to the interaction of the incoming P-wavefront with each discontinuity. Detailed modeling of RF waveforms therefore allows the estimation of depth-dependent variations in the S-velocity below the recording station (e.g., Owens et al. 1984; Julià et al. 1998).

We compute both radial and transverse RFs by applying an iterative time-domain deconvolution method (Ligorría and Ammon 1999), with 500 iterations. Before deconvolution, we cut the waveforms 10 s before and 110 s after the teleseismic P-wave arrival, remove the mean, de-trend, taper with a 5% cosine window, and band-pass filter the seismograms between 0.05 and 4 Hz to remove low- and high-frequency noise. Seismograms are then down-sampled to 10 samples per second before rotating the horizontal components into the great-circle path to obtain the radial and tangential component seismograms. Furthermore, we compute the RFs at two overlapping frequency ranges with fc < 1.25 Hz and fc < 0.5 Hz, using an acausal Gaussian filter with width factors of a = 2.5 and a = 1.0, which helps discriminating rapid velocity variations from gradual velocity transitions (Julià 2007). An automatic quality control is applied to the deconvolved waveforms to remove RFs that do not reproduce at least 85% of the original horizontal waveforms when convolved back with the corresponding vertical traces. Finally, both radial and transverse RFs are visually inspected to remove unstable and/or significantly distorted RFs from the dataset. Figure 4 displays selected RF averages for several stations (computed with a Gaussian width of 2.5). For most stations, the radial RFs display clear Ps conversions and multiples due to the Moho discontinuity. Also, the transverse RFs reveal comparatively small amplitudes, inferring a nearly isotropic and laterally homogenous underlying medium. The results of the RF-analysis for crustal thickness and upper-mantle lid properties are described in detail in the section on crustal structure from RF and surface-wave inversion.

RFs are also utilized to map variations in transition zone thickness under the Arabian Peninsula. The selection criteria for the seismic sources are the same as for the crustal study: magnitude m_b 5.5 or larger, epicentral distances between 30° and 90°, and no restrictions on source depth. To extend our sampling further north, we include additional seismic stations in Israel and Jordan, considering broadband stations from the Israeli Broadband Seismic Network (GII-Net) and from the DEad SEa Rift Transect (DESERT) temporary deployment (see e.g., Mohsen et al. 2005) for which data are publically accessible. GII-Net stations are equipped with Streckeisen STS-2 sensors and Quanterra Q380 dataloggers, and they continuously sample at 20, 40 or 80 Hz, depending on the station. The DESERT stations include a variety of sensors—either Streckeisen STS-2, Guralp CMG-3T, and Guralp CMG-40T—connected to RefTek 72A digitizers; these stations sample at either 20 or 50 Hz. Except for the CMG-40T, all sensors have a flat response in velocity down to 120 s. Overall, more than 20,000 teleseismic P-waveforms recorded at 120 broadband stations across the study area were selected for transition zone thickness analysis.

Processing the teleseismic P-waveforms to obtain RFs for transition zone analysis closely follows that applied to the waveforms for the crustal study, with the only difference being the width of the chosen Gaussian filter. The original P-waveforms are windowed 10 s before and 120 s after the P-wave arrival, demeaned, detrended, and tapered with a 5% cosine taper before applying a high-pass filter at 20 s to eliminate long-period noise. The data are also decimated to 10 samples per second, and are then rotated into the great-circle path and low-pass filtered with a Gaussian filter of width of 0.5 (f_c < 0.2 Hz) before deconvolution. We then apply a time-domain deconvolution of the vertical component from the corresponding radial and transverse components of the teleseismic P-waveforms using the iterative procedure of Ligorría and Ammon (1999), again with 500 iterations. The resulting time series are then convolved with a Gaussian pulse with a width of 0.5. As with the crustal RFs, we conduct quality control by convolving the vertical component with the radial RF to assess the percentage of recovery of the original radial component, and those records with a recovery under 85% are automatically rejected. The transverse RFs are then visually inspected, and those displaying amplitudes comparable to the radial component are excluded from further analysis. Finally, the remaining radial RFs are superimposed to visually identify and eliminate possible outliers. In total, 7010 RFs pass our quality control and are utilized in the mapping of the transition zone thickness. An additional step in the quality control is performing single-station stacks with the selected receiver functions to assess the recovery of the P410s and P660s converted phases. The single-station stacks are produced after migrating the RFs into the depth domain and then averaging all migrated RFs for a given station. Migration is achieved by mapping S-P travel times in the RFs to P-to-S conversion depths after ray-tracing through the global velocity model ak135 (Kennett et al. 1995). Notice that our velocity models do not reach transition zone depths, and hence cannot be utilized for ray-tracing in this case. The mapping is performed at 10 km depth intervals for depths between 0 and 800 km using the TauP Toolkit (Crotwell et al. 1999). Our single-station migrations show an excellent recovery of the P660 s conversion and generally good recovery of the P410s conversion. Further details and results focused on using RFs to more accurately map the transition zone thickness are described in the section on transition zone thickness from deep converted waves.

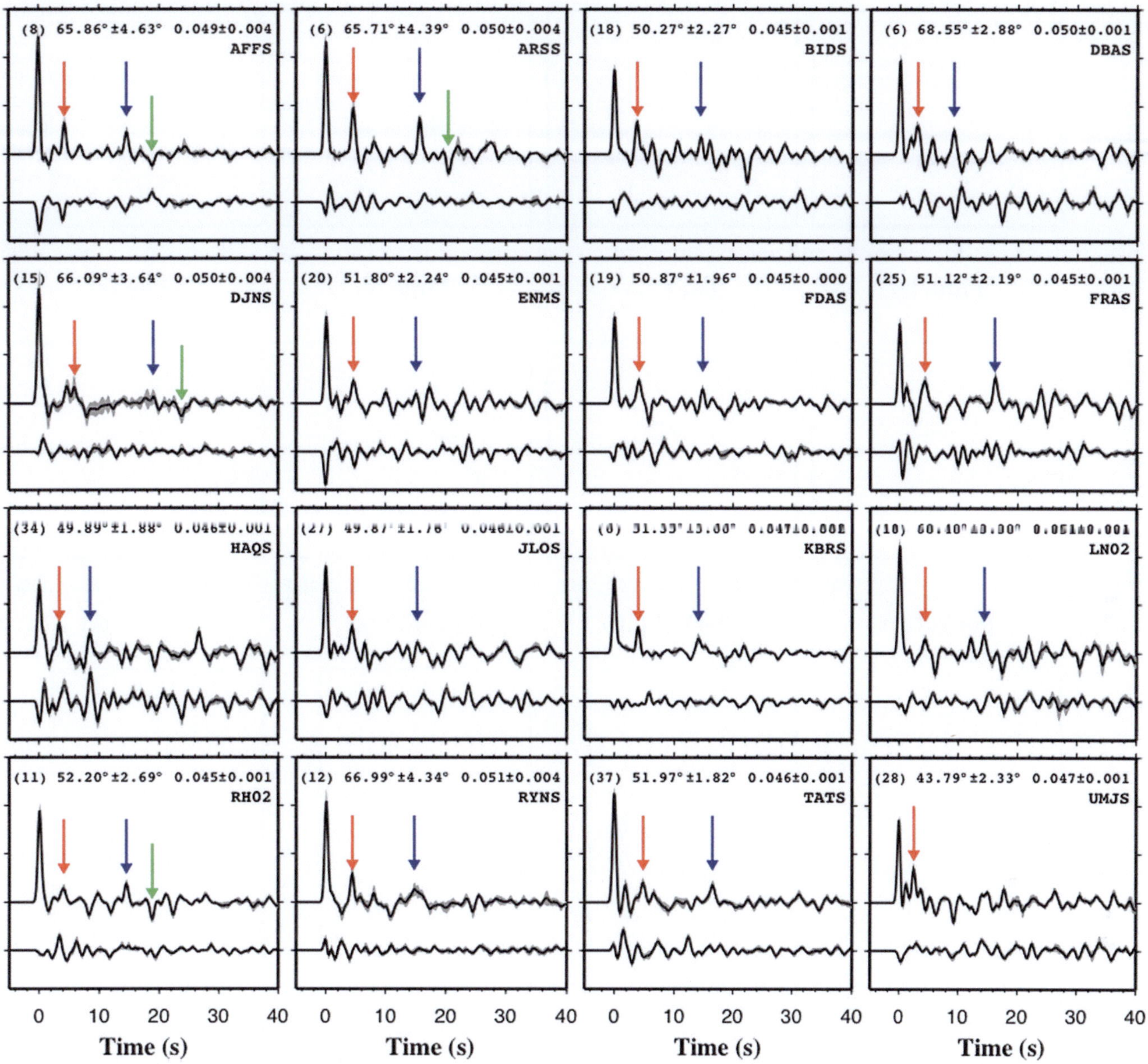

Fig. 4 Examples of receiver functions for selected stations, whose names are indicated in the upper right corner. In each panel, the average radial (top) and the transverse (bottom) RFs are shown, with a Gaussian width of 2.5. The small numbers indicate the number of earthquakes used to obtain the displayed RFs, their average back-azimuth (and its variation, in degrees), as well as the ray parameter (and its variation, in s/km). The gray-shading beneath each average RF (plotted in black) illustrates the confidence bounds. The red, blue and green arrows provide visual guidance (for a few cases) to identify the Ps conversion, first multiple (PpPs) and second multiple (PpSs + PsPs) described in Fig. 3

3 Crustal Structure from Receiver Function and Surface-Wave Inversion

We apply two standard techniques to our RF dataset to extract information about crustal structure under each recording station. First, we use the H-κ stacking procedure of Zhu and Kanamori (2000) to estimate the crustal thickness (H) and the bulk Vp/Vs (κ) ratio. The "H-κ stacking" approach constructs a stacking surface by adding RF amplitudes along S-P travel-time curves for a range of crustal models that are assumed to consist of a single layer over a half-space. The amplitudes are thus stacked along phase move-out curves for the Ps, PpPs and PpSs + PsPs phases for each possible pair of H and κ within pre-defined ranges and at regular intervals. If the true model under the given station can be well approximated by a layer-over-half space model, then the stacking surface will display a single

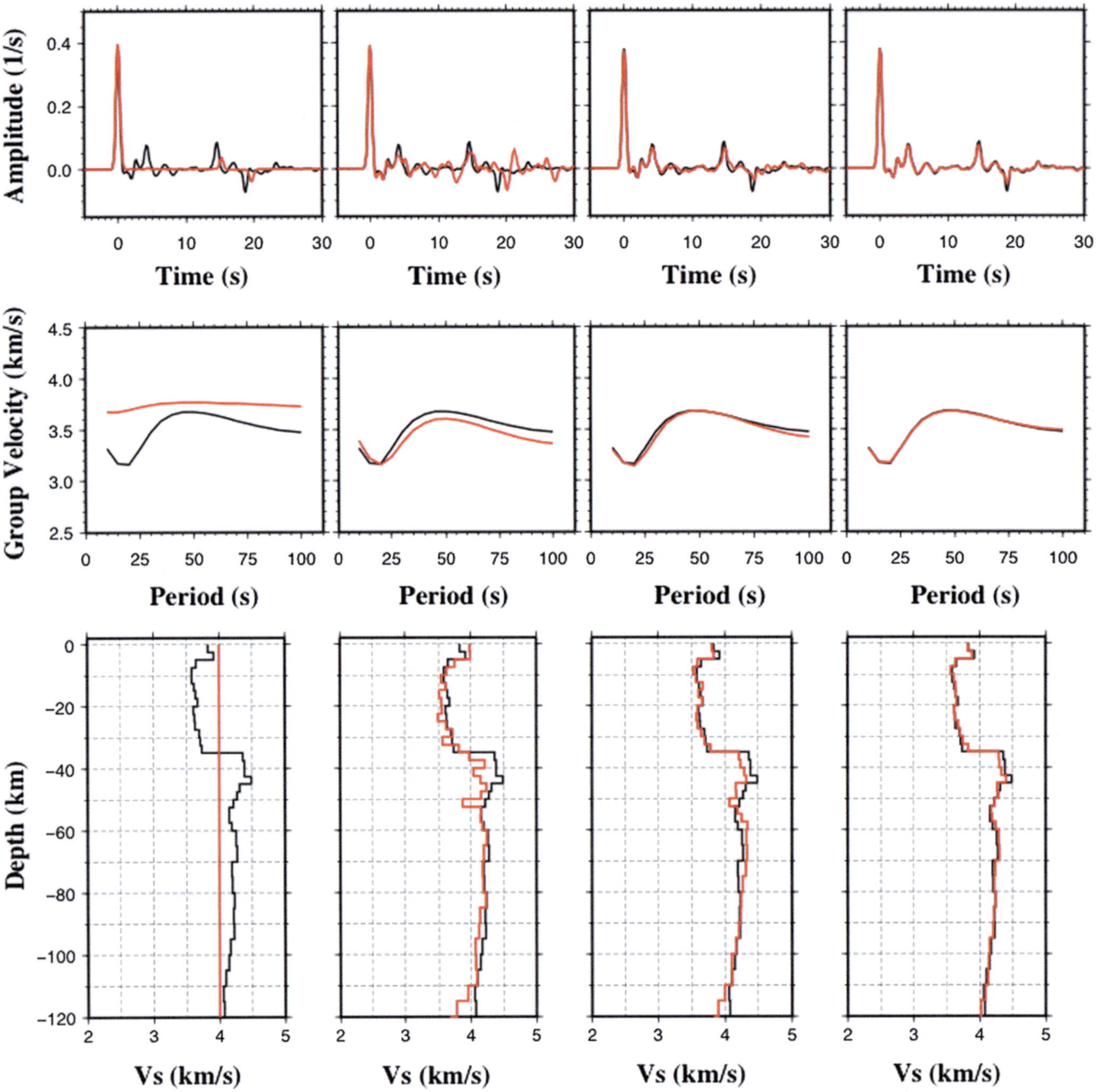

Fig. 5 Illustration of the joint inversion approach, using receiver functions (top row) and surface-wave dispersion curves (center row), to determine the one-dimensional shear-wave velocity structure underneath station AFFS, located on the Arabian Shield. RF and SWD observations are shown in black, the modeled RF and SWD are shown in red. Iteration steps proceed from left to right, showing how the initial model (red) converges to the final solution (black)

best-fitting maximum for the values of H and κ that coincide with the actual values under the station. For more complex structures, the H-κ stacking surface may display multiple maxima corresponding to other intra-crustal discontinuities or to artifacts from combining P-to-S conversions from different discontinuities. Thus, this method alone provides only a first-order estimation for H and κ for the crust.

Therefore, we subsequently invert the radial RFs jointly with fundamental-mode Rayleigh-wave group velocities to develop 1D velocity-depth profiles that constrain the detailed S-velocity variation with depth (Fig. 5), following the approach described by Julià et al. (2000, 2003). Both RFs and SWDs are mainly sensitive to S-wave velocity, but they help to constrain different features of the velocity-depth profiles. RFs constrain the velocity contrast across the discontinuities, and the S-P travel times between the discontinuity and the surface, while SWD measurements control absolute S-velocity averages within frequency-dependent depth-ranges. In the ideal case of noise-free observations, jointly inverting RFs and SWDs uniquely allows for detailed estimation of depth-dependent S-velocity variations (Julià et al. 2000, 2003), thus reducing the dependence on the starting model that characterizes the RF-only inversion (Ammon et al. 1990).

Our starting model for the joint inversion is parameterized as a stack of thin layers with variable S-velocity. Layer thicknesses are 2.5 km at crustal and uppermost mantle levels, and then progressively increase to 5 km at lithospheric mantle levels, and 10 km at larger depths. The P-velocity is calculated from the S-velocity by assuming an a priori Vp/Vs ratio for each layer, while the density is obtained from the P-velocity using an empirical relationship for crustal rocks (Berteussen 1977). The starting model consists of a 40 km thick crust with a gradual increase in S-wave speed from 3.4 to 4.0 km/s, overlying an uppermost mantle with velocities around 4.5 km/s down to ~180 km depth. The bottom portion of the starting model follows a flattened PREM (Dziewonski and Anderson 1981), and is parameterized to a depth of ~400 km.

Our results are described in detail in Tang et al. (2016) and are summarized here (Fig. 6). We find significant lateral variations in crustal thickness, shear-wave velocity, and bulk Vp/Vs ratio. The Arabian shield has a 27.5 to 45.0 km thick crust with crustal shear-velocities of 3.48 to 3.95 km/s and an average bulk Vp/Vs ratio of 1.73 ± 0.07. In the densely instrumented volcanic region of Harrat Lunayyir, the crustal thickness ranges between 35.0 and 37.5 km, while the crustal S-velocity ranges between 3.85 and 3.94 km/s, with average Vp/Vs of 1.79 ± 0.11. In the northern Harrat Rahat, the crustal thickness varies between 32.5 and 37.5 km, with crustal S-velocities between 3.76 and 3.88 km/s and an average bulk Vp/Vs ratio of 1.77 ± 0.07. Thinner crust (25.0–32.5 km thick) with strong lateral variations is present along the Saudi Arabian coast of the Red Sea. Beneath the northwestern part of the Arabian platform, we find roughly 35 km thick crust, with crustal thickness increasing to 40.0 to 45.0 km under the eastern platform. The Arabian platform has crustal Vs of 3.44–3.68 km/s and an average Vp/Vs ratio of 1.77 ± 0.09. These results reveal a rapid transition from oceanic to continental crust in the Red Sea margin and the western Arabian shield, as well as remarkable crustal thickening from the Arabian shield to the platform. In addition, we observe that crustal thickness tends to increase from the northern to the southern Arabian shield, parallel to the Red Sea (Fig. 6).

To further investigate the spatial pattern of depth-dependent S-wave speeds and relate it to independent geophysical/geologic data, we estimate the upper mantle lid temperature beneath each station utilizing the first-order Taylor expansion of the pressure-temperature dependence of the shear-wave speed. We assume that the dominant rock type for the upper mantle lid is peridotite, with laboratory measurements for the reference velocity (i.e., S-velocity at pressure 0 kbar and temperature 0 °C) and partial derivatives

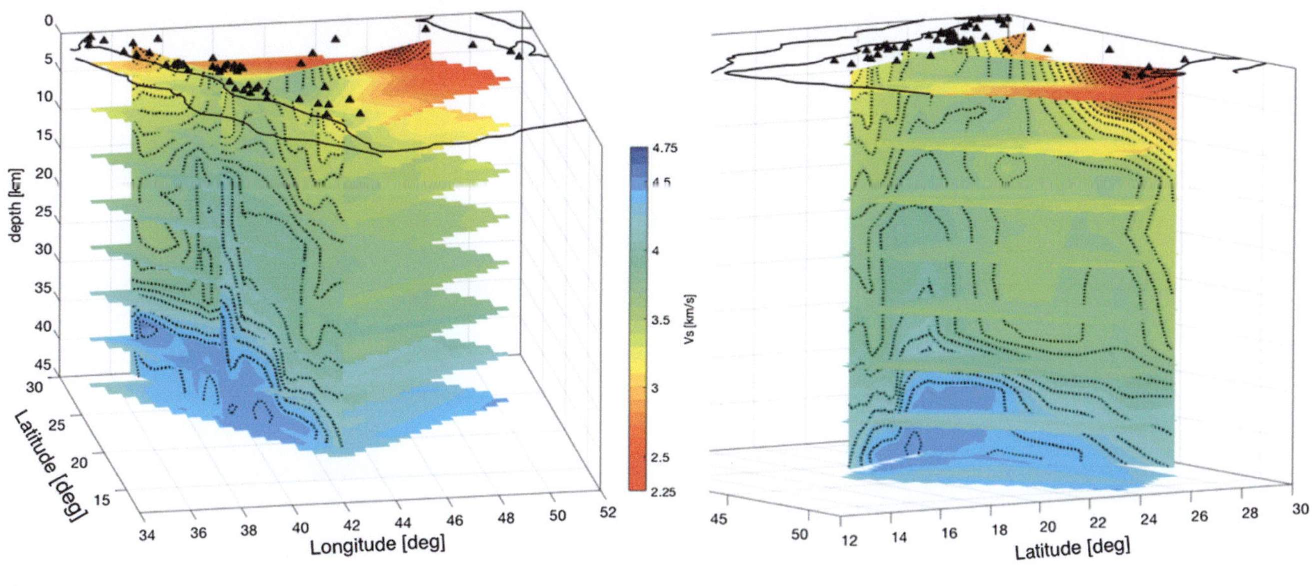

Fig. 6 Three-dimensional rendering of the inferred S-wave velocity structure underneath Saudi Arabia. Seismic stations used for the inversion are plotted as triangles, the coastlines of the Arabian Peninsula and east Africa are shown by black lines. Horizontal slices are shown at 4 km depth intervals, the vertical slices are contoured (black-dotted line) at 0.2 km/s intervals. (left) View from south-west. (right) View from south-east. Note the lateral variations of S-wave speed associated with variations in Moho depth, and the crustal thickening from the Arabian shield to the platform

(i.e., pressure and temperature derivatives) taken from Kern and Richter (1981). Pressure is estimated as the crustal overburden, taking into account the densities and layer thicknesses from the inverted models of S-wave speed, while the S-velocity of the lid is estimated as a 3-layer average within the mantle lid portion of our velocity-depth profiles. The resulting mantle lid temperature and its corresponding uncertainty under each station are displayed in Fig. 7; the uncertainties are estimated by error propagation of the inferred S-wave speed, and are generally in the range 200–300 °C.

The most important geophysical signatures we find are the patterns of lateral shear velocity and temperature variation in the upper-mantle lid beneath the Arabian shield (Fig. 7) as well as the anomalously high bulk Vp/Vs ratios in the Cenozoic volcanic area of Harrat Lunayyir (Tang et al. 2016). The patterns indicate that the southern (and northern) portion of the shield is characterized by lower shear

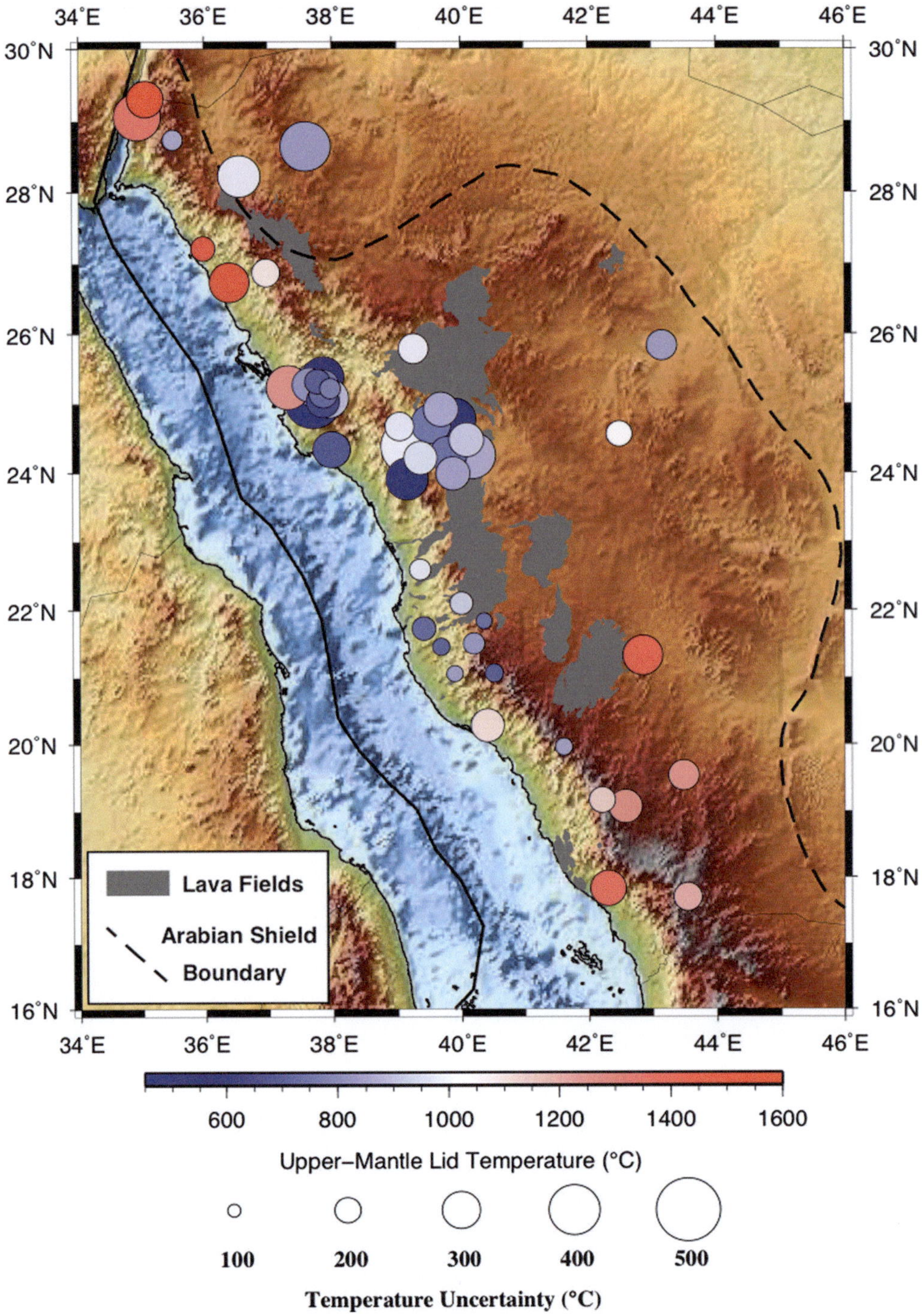

Fig. 7 Estimated upper-mantle lid temperatures across the Arabian Shield. The circle size scales with the uncertainty in the temperature estimation. The dark-gray-shaded regions mark volcanic provinces (modified after Tang et al. 2016)

velocities and higher temperatures across the upper-mantle lid. We interpret these spatial patterns in terms of lateral heating by thermal conduction from a mantle upwelling below Afar (and, perhaps, Jordan) toward the central shield. The presence of high Vp/Vs ratios and relatively large, lower crustal S-velocities suggests a compositional anomaly (i.e., solidified magma intrusions from old magmatic episodes), rather than the presence of partial melt under Harrat Lunayyir. Our inferred lower-crustal relatively higher Vs at Harrat Lunayyir is difficult to reconcile with partial melts, but is consistent with a middle-crustal high Vp anomaly observed by Hansen et al. (2013). Thus, the magma sources feeding the recent volcanic eruptions must be preferentially located in the lithospheric mantle, which is consistent with the model of lateral flow from the Afar and (possibly) Jordan plumes. We speculate the lateral flow heats the bottom of the shield's lithosphere, triggering localized melts in the lower lithosphere that buoyantly rise to the surface on a local scale (perhaps through zones of crustal weakness). Because we did not find high temperatures in the uppermost mantle lid under those lava fields, we argue that the volcanism in western Arabia may result from small-scale adiabatic ascent of magma diapirs.

4 Transition Zone Thickness from Deep Converted Waves

In this section, we analyze RF measurements to map the thickness of the transition zone between the 410 km and the 660 km discontinuity in the Earth's mantle. Figure 8 conceptually displays the pressure-temperature-compositional conditions for these discontinuities and how the exact depths change for either colder or hotter than average regimes. The pressure-temperature dependence *dP/dT* can be written as:

$$\frac{dP}{dT} = \frac{L}{T(\rho_2 - \rho_{21})}$$

where P is pressure, T is temperature, $\rho_2 - \rho_1$ represents density change, and L is the latent heat associated with the phase transformation. For $L > 0$, an exothermic process occurs (at the 410 km discontinuity); for $L < 0$, the process is endothermic (at the 660 km discontinuity). Because elevated or depressed discontinuities lead to changes in the arrival times of the corresponding converted seismic phases, the RF properties will also change. This is illustrated for a set of single-station migrated RFs in Fig. 9, showing excellent recovery of the P660 s conversion and a generally good recovery of the P410s conversion. Some stations display a disrupted P410s conversion (e.g., station #2, on the eastern platform, see Fig. 2), while others seem to have a very depressed and thin transition zone (e.g., station #3, in Jordan).

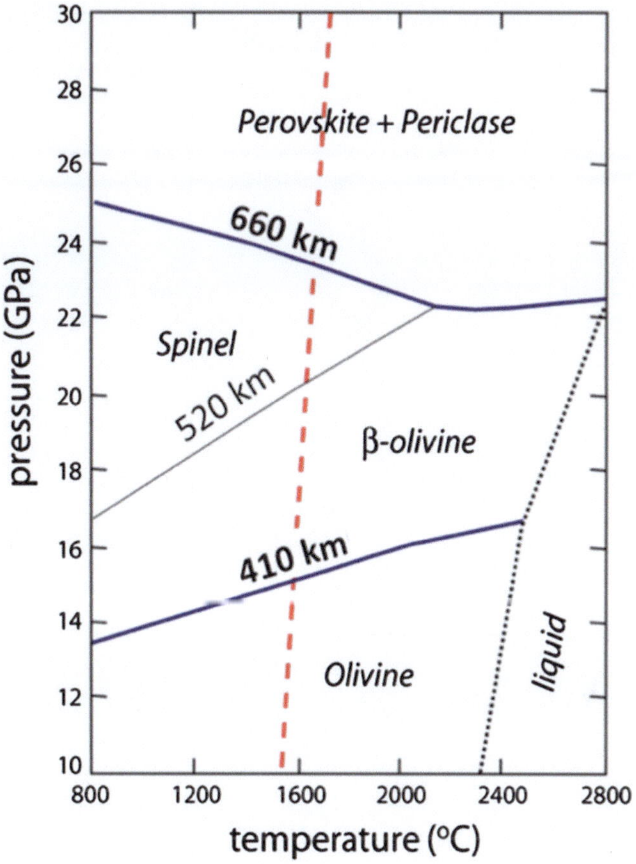

Fig. 8 Conceptual diagram for the thickness of the transition zone between the 410 km and the 660 km discontinuities in the Earth's mantle. The pressure-temperature (P-T) diagram shows the phase-equilibrium boundaries for pure Mg_2SiO_4 (calculated using Fei et al. 1992). Depending on the temperature gradient, the thickness of the transition changes. A thinned transition zone may be indicative of deep, hot material that signifies a potential mantle plume

To map lateral variations in transition-zone thickness under the Arabian Peninsula, we migrate and stack low-frequency (Gaussian width 0.5) P-wave RFs following the variable bin radius stacking scheme (Owens et al. 2000). First, the locations of piercing points (i.e., geographical locations of Ps conversions at a given depth) are determined for all RFs, considering conversion depths between 0 and 800 km with 10 km depth intervals and ray-tracing through the global Earth model ak135 (Kennett et al. 1995). Second, a grid of nodes (spaced at 0.50°) is defined over the study area sampled by the piercing points. The piercing points are then grouped into bins with radii varying between 0.25° and 2.00° around the nodes, with bin radii being set to the smallest value that contained a minimum of 30 piercing points from at least four different stations. In practice, the minimum radius is found by initially setting its value to 0.25° and progressively increasing it with 0.25° intervals until the two criteria are met. The location of the piercing points at 410 km and 660 km depth are displayed in

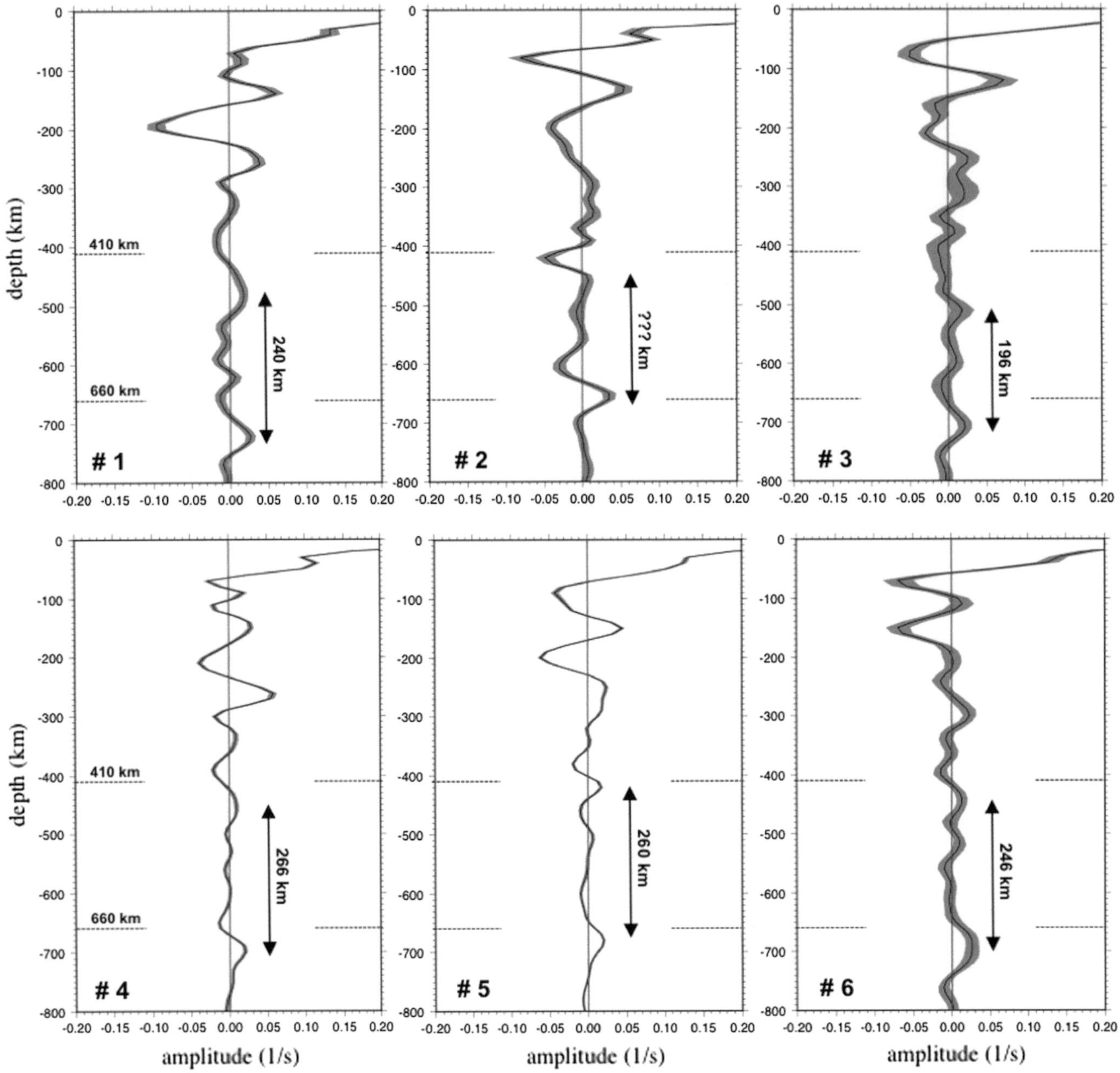

Fig. 9 Examples of six RFs to estimate the transition zones thickness between the 410 km and 660 km discontinuities across the Arabian Plate. The gray band around each RF-waveform indicates the confidence interval. Station #3 shows some indication of a decrease in transition zone thickness; all other stations reveal little evidence for transition-zone variations. At station #2, the transition zone cannot be estimated unambiguously due to a disrupted 410 km discontinuity. See Fig. 2 for station locations

Fig. 10a, estimated transition-zone thicknesses are shown in Fig. 10b, and the map of bin radii is displayed in Fig. 11. After the bins are defined for each node and at each depth, the amplitudes associated with the piercing points within each bin are averaged (stacked) to develop a bin-averaged RF function for each node.

We estimate transition zone thickness by applying an automated "picker" through each of the bin-averaged RFs to determine the migrated depths for the P410s and P660s phase conversions. The automated procedure searches for the largest peak within prescribed depth ranges of 350–500 km for the P410s and 600–750 km for the P660s, and calculates transition zone thickness through the difference of the inferred depth values. The results are displayed in Fig. 10b for nodes with bin radii smaller than or equal to 1.0°. The figure demonstrates that transition zone thickness under the Arabian Shield is close to the nominal value of 250 km shown in global Earth models, suggesting that the

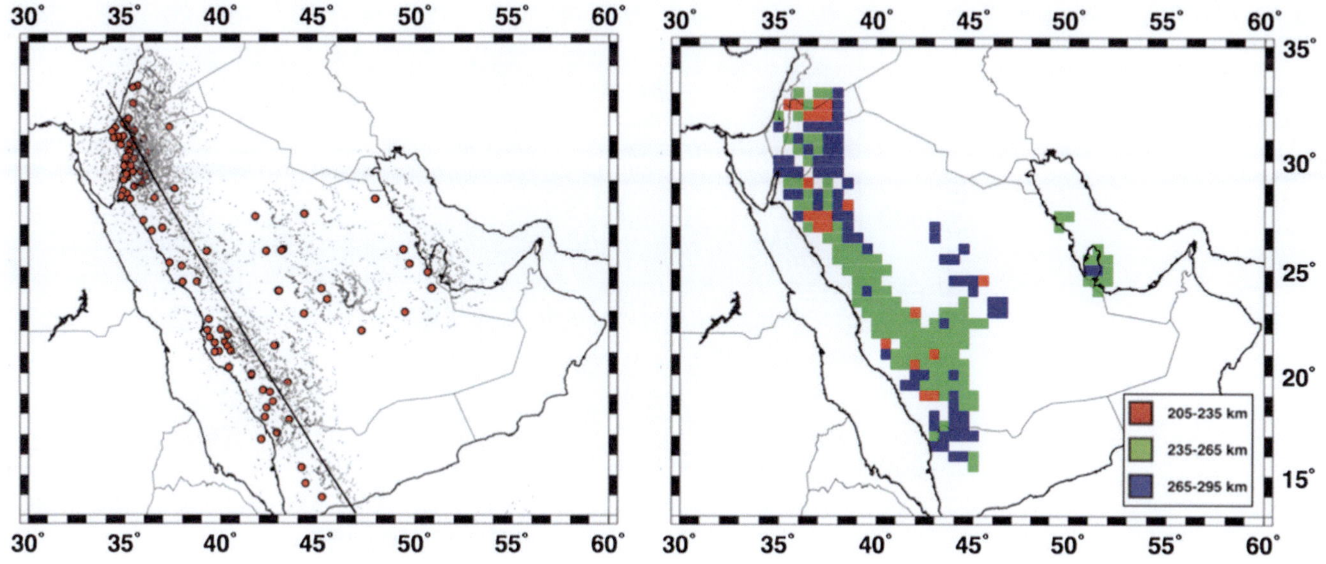

Fig. 10 Transition-zone thickness across the Arabian Plate. (left) Stations (red circles) and piercing points (gray dots); the NW–SE-trending line marks the profile shown in Fig. 12. The study area was binned into 0.5° × 0.5° bins to estimate the transition-zone thickness. (right) The Arabian Shield shows largely a normal transition-zone thickness of around 250 km

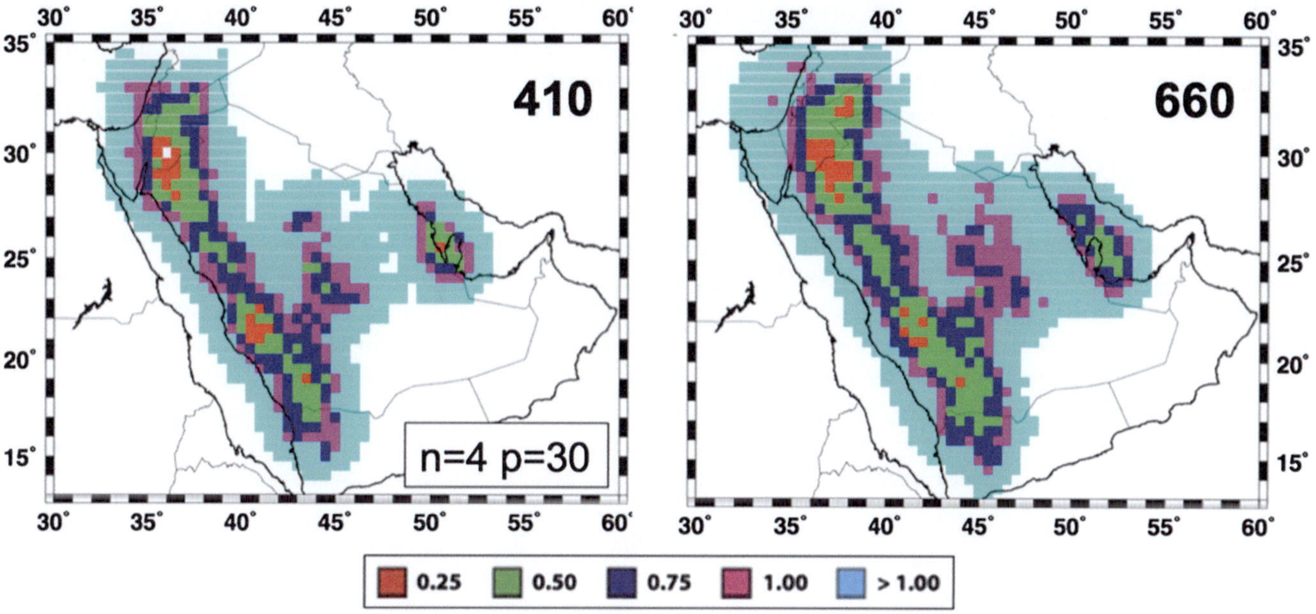

Fig. 11 Bin radii at depths of 410 km and 660 km for bins containing at least 30 piercing points from at least four different stations (see text for details). The bin size has been colour-coded as specified in the legend. Note that the smallest radii (0.25°) are achieved in regions with high piercing-point density. Bins with radii over 1° (or low piercing point density) were not considered in the development of the transition zone map displayed in Fig. 10b

transition zone is not thermally perturbed under the central parts of the Arabian Shield. This rules out the presence of a mantle plume, as proposed by Camp and Roobol (1992). This is further illustrated in Fig. 12, where a NW–SE RFs cross-section oriented parallel to the Red Sea coast is displayed. The cross-section shows that the 410 and 660 km discontinuities are generally well imaged by the RFs, although statically shifted to larger depths. The shift is due to the assumption of a global Earth model with an upper mantle faster than that imaged in tomographic studies for the region (e.g., Chang et al. 2011). This static shift mostly prevents us from recovering absolute depths for the discontinuities bounding the transition zone, but has a minimal effect on transition zone thickness (Owens et al. 2000). The

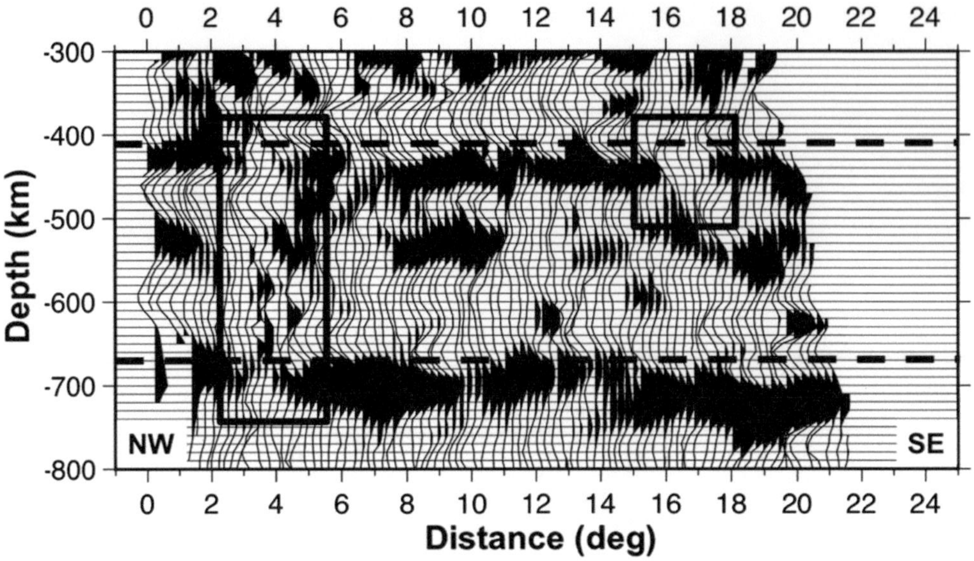

Fig. 12 NW–SE cross-section of RFs along the profile shown in Fig. 10. Black rectangles mark regions of limited to no RF-signal amplitudes at the 410 km or 660 km discontinuities (dotted lines), suggesting that at these locations, these discontinuities are poorly developed

cross-section also shows that the 410 and 660 km discontinuities are disrupted at 300–400 km from the NW end, which we attribute to a "leaking" effect, deriving from complex wave propagation in the vicinity of the Dead Sea Transform. The cross-section also shows that the 410 km discontinuity is not imaged at distances 1650–1800 km away from the NW end, further emphasizing the need for 3D travel-time corrections to fully interpret the topography of the transition zone discontinuities.

Based on our spatially expanded dataset for transition-zone thickness estimation (Fig. 10b), we also observe that under northern Jordan, immediately east of the Dead Sea Transform, the transition zone is anomalously thin (Fig. 10b). Such thinning is further demonstrated in Fig. 9 through the single-station stack for station #3, suggesting that the thinning could be on the order of $\sim$50 km or, equivalently, that the associated thermal anomaly could be $\sim$300°K. Interestingly, this anomalous region broadly coincides with the location of the postulated Jordan plume (Chang and Van der Lee 2011).

5 Discussion and Conclusions

The picture emerging from our results outlined in the section on crustal structure from RF and surface-wave inversion is that the Arabian lithospheric mantle is heated from below by lateral flow from the Afar and (possibly) Jordan plumes. This triggers localized melts that ascend adiabatically through the lithosphere as magma diapirs. This interpretation, as discussed in detail by Tang et al. (2016), is based on our observations of mantle lid temperatures increasing toward the location of the Afar plume and the lack of correlation between high lid temperatures and the surface location of the Cenozoic harrats (Fig. 7). The possibility of direct heating by a mantle plume under the central shield, as proposed by Camp and Roobol (1992), is ruled out by the lack of transition zone thinning under the shield (Fig. 10b) and seismic images of the Arabian upper mantle (e.g., Chang and van der Lee 2011).

The precise relationship between heating at the base of the lithosphere and surface volcanism is still unknown. Is laterally-transported plume material from Afar melted and entrained under the Arabian Shield to finally make its way up to the surface? Or is this transported plume material just providing the heat to trigger melting that will eventually ascend through the lithosphere? To address these questions, mantle xenolith data may provide relevant information. Xenoliths are inclusions of rock fragments that become entrained in igneous rocks during magmatic eruptions (or emplacement), and they provide temperature, pressure, and compositional information about the Earth's mantle. Konrad et al. (2016) compiled Helium isotope ratios from lava flows and mantle xenoliths from several harrats in Saudi Arabia and Yemen, and demonstrated a bimodal distribution: For Harrat Rahat, along the MMN line, $^{3}He/^{4}He$ ratios are in the 9.3–11.8 Ra range (where Ra is the atmospheric ratio), suggesting a "weak but detectable plume helium signal"; for all other harrats in the Arabian Shield, ratios are in the 6.8–8.2 Ra range, where the lower bound is representative of the ambient Proterozoic lithospheric mantle, and the upper bound is representative of mixing with the underlying asthenosphere. Konrad et al. (2016) propose that the lithosphere under the MMN line may have been thinned by lateral flow from the Afar plume and, as thermal erosion progressed, convective instabilities may have mixed asthenospheric material with overlying lithospheric rocks. Plume material would have then become entrained in the mix for lavas found at Harrat Rahat (directly above the MMN line), but not for the low $^{3}He/^{4}He$ harrat basalts found elsewhere. These findings are, to a first order, consistent with our interpretation of the Afar and Jordan upwellings being diverted laterally and flooding the base of the Arabian lithosphere. Why entrainment of plume material may only happen beneath Harrat Rahat remains, nonetheless, an open question that requires detailed mapping of the lithosphere-asthenosphere boundary and modeling of lateral asthenospheric flow under the Arabian Shield (e.g., Faccenna et al. 2013).

The bimodal character of Cenozoic volcanics in the Arabian Shield was already reported by Camp and Roobol (1992), who divided its continental magmatism into two distinct phases: An older phase (30–20 Ma) that produced tholeiitic-to-transitional lavas emplaced along NW trends, and a younger phase (12 Ma to Recent) that produced transitional-to-strongly-alkalic lavas emplaced along NS trends. The older phase was related to passive rifting along the Red Sea basin, while the younger one was related to active upwelling from a mantle plume. The bimodal character was further emphasized by petrological evidence showing that the younger harrats were derived by greater degrees of partial melting at shallower depths than those harrats of older age. The $^3He/^4He$ ratios reported in Konrad et al. (2016) further emphasize this bimodal character and provide additional evidence for entrainment of plume material or lack thereof in the two phases of magmatism, respectively.

The adiabatic ascent of magma diapirs on a local scale as the feeding mechanism for harrat volcanism, as proposed by the interpretation of our seismic results, is also supported by mantle xenoliths. Kaliwoda et al. (2007) investigated chemical zonation patterns in xenoliths found in Harrat Uwayrid lavas, in the northwest of the Arabian Shield, and identified a first high-T stage (also recorded in other volcanic fields) related to basal flooding of the lithosphere by the Afar plume material, and a renewed heating event of a more local nature, probably caused by ascending and stagnating magmas. The regional high-T stage was first identified from geothermobarometry on Harrat Al-Kishb samples and geothermal modeling by McGuire and Bohannon (1989). The renewed high-T stage is consistent with our interpretation of an "adiabatic" ascent of magma diapirs, as the local nature of the heating event would not produce a thermal anomaly on a scale large enough to be detected seismically. Interestingly, the regional heating event is postulated to have pervaded the Arabian lithosphere and caused widespread partial melting. We remark that widespread partial melting at the base of the lithosphere would remove the necessity for lateral flow of the lithosphere-asthenosphere mix proposed by Konrad et al. (2016) away from the MMN line.

Finally, the role of a possible Jordan plume in the generation of the melts that fed the northern harrats needs further study. On one hand, our velocity models do not sample the mantle lid close enough to the postulated location of the Jordan plume to observe a marked lid temperature increase close to the proposed plume. Geothermobarometry on xenoliths found at Harrat Ash Saam in eastern Jordan, as summarized by Shaw et al. (2007), reveal temperatures of 900–1030 °C at 11–18 kbar (35–60 km). These temperatures are in the upper (hot) limit of the temperature range inferred for the central shield from seismic velocities and are also similar to temperatures inferred under the two northeastern shield stations (Fig. 7). Relatively low $^3He/^4He$ ratios determined for lava flows from Harrat Ash Saam in eastern Jordan suggest a shallow lithosphere-asthenosphere source for the melts and, therefore, no entrainment of a putative Jordan plume.

On the other hand, the only evidence for a Jordan plume is found in seismic studies, such as Chang and van der Lee (2011) —who show a vertical seismic cross-section that depicts a low-velocity anomaly under Jordan—and the current work, where thinning of the transition zone under eastern Jordan is reported (Fig. 10b). Lateral flow and entrainment of Jordan plume material under the Arabian lithosphere, however, is yet to be assessed. Therefore, to unravel the exact source of the volcanism on the Arabian plate and to understand its occurrence in the context of the geodynamic processes associated with the Red Sea opening and the Afar plume, dedicated seismic experiments are needed to create higher-resolution images of the lithosphere in the region.

Acknowledgements We thank the Saudi Geological Survey (SGS) for sharing their seismic data with us, in particular Hani Zahran and Mahmoud Salam. Laura Parisi helped to create Fig. 6. Many thanks to Najeeb Rasul (SGS, Jeddah) for organizing the workshop on "The Geological Setting, Oceanography and Environment of the Red Sea" in February 2016 in Jeddah, which inspired this chapter. We are grateful to anonymous reviewers for their constructive criticism and careful reviews that helped to improve this study. The research reported in this publication was supported by funding from King Abdullah University of Science and Technology (KAUST), grant number BAS/1/1339-01-01. We also acknowledge support by Universidade Federal do Rio Grande do Norte for hosting Z.T. during two research visits.

References

Al-Damegh K, Sandvol E, Barazangi M (2005) Crustal structure of the Arabian plate: new constraints from the analysis of teleseismic receiver functions. Earth Planet Sci Lett 231(3):177–196

Ammon CJ, Randall GE, Zandt G (1990) On the nonuniqueness of receiver function inversions. J Geophys Res Solid Earth 95 (B10):15303–15318

Berteussen KA (1977) Moho depth determinations based on spectral-ratio analysis of NORSAR long-period P waves. Phys Earth Planet Inter 15(1):13–27

Bosworth W, Huchon P, McClay K (2005) The Red Sea and Gulf of Aden basins. J Afr Earth Sci 43(1):334–378

Brown GF (1972) Tectonic map of the Arabian Peninsula (No. 72-52). US Geological Survey

Camp VE, Roobol MJ (1992) Upwelling asthenosphere beneath western Arabia and its regional implications. J Geophys Res 97 (B11):15255–15271

Chang SJ, Van der Lee S (2011) Mantle plumes and associated flow beneath Arabia and East Africa. Earth Planet Sci Lett 302(3):448–454

Chang SJ, Merino M, Van der Lee S, Stein S, Stein CA (2011) Mantle flow beneath Arabia offset from the opening Red Sea. Geophys Res Lett 38(4):L04301

Chang SJ, Ferreira AMG, Ritsema J, van Heijst HJ, Woodhouse JH (2015) Joint inversion for global isotropic and radially anisotropic mantle structure including crustal thickness perturbations. J Geophys Res Solid Earth 120:4278–4300. https://doi.org/10.1002/2014JB011824

Cochran JR, Martinez F (1988) Evidence from the northern Red Sea on the transition from continental to oceanic rifting. Tectonophysics 153(1–4):25–53

Crotwell HP, Owens TJ, Ritsema J (1999) The TauP Toolkit: flexible seismic travel-time and ray-path utilities. Seismol Res Lett 70 (2):154–160

Daradich A, Mitrovica JX, Pysklywec RN, Willett SD, Forte AM (2003) Mantle flow, dynamic topography, and rift-flank uplift of Arabia. Geology 31(10):901–904

Dziewonski AM, Anderson DL (1981) Preliminary reference Earth model. Phys Earth Planet Inter 25(4):297–356

Ebinger CJ, Sleep NH (1998) Cenozoic magmatism throughout east Africa resulting from impact of a single plume. Nature 395 (6704):788–791

Faccenna C, Becker T, Jolivet L, Keskin M (2013) Mantle convection in the Middle East: reconciling Afar upwelling, Arabia indentation and Aegean trench rollback. Earth Planet Sci Lett 375:254–269. https://doi.org/10.1016/j.epsl.2013.05.043

Fei Y, Mao H, Shu J, Parthasarathy G, Bassett WA, Ko J (1992) Simultaneous high-P, high-T X ray diffraction study of β-(Mg, $Fe)_2SiO_4$ to 26 GPa and 900 K. J Geophys Res 97(B4):4489–4495. https://doi.org/10.1029/92JB00076

Garfunkel Z, Beyth M (2006) Constraints on the structural development of Afar imposed by the kinematics of the major surrounding plates. Geol Soc Lond Spec Publ 259(1):23–42

Hansen SE, Rodgers AJ, Schwartz SY, Al-Amri AM (2007) Imaging ruptured lithosphere beneath the Red Sea and Arabian Peninsula. Earth Planet Sci Lett 259(3):256–265

Hansen SE, DeShon HR, Moore-Driskell MM, Al-Amri AMS (2013) Investigating the P wave velocity structure beneath Harrat Lunayyir, northwestern Saudi Arabia, using double-difference tomography and earthquakes from the 2009 seismic swarm. J Geophys Res Solid Earth 118(9):4814–4826

Julià J (2007) Constraining velocity and density contrasts across the crust-mantle boundary with receiver function amplitudes. Geophys J Int 171(1):286–301

Julià J, Vila J, Macià R (1998) The receiver structure beneath the Ebro basin, Iberian Peninsula. Bull Seismol Soc Am 88(6):1538–1547

Julià J, Ammon CJ, Herrmann RB, Correig AM (2000) Joint inversion of receiver function and surface wave dispersion observations. Geophys J Int 143(1):99–112

Julià J, Ammon CJ, Herrmann RB (2003) Lithospheric structure of the Arabian Shield from the joint inversion of receiver functions and surface-wave group velocities. Tectonophysics 371(1):1–21

Kaliwoda M, Altherr R, Meyer HP (2007) Composition and thermal evolution of the lithospheric mantle beneath the Harrat Uwayrid, eastern flank of the Red Sea rift (Saudi Arabia). Lithos 99(1):105–120

Kennett BLN, Engdahl ER, Buland R (1995) Constraints on seismic velocities in the Earth from traveltimes. Geophys J Int 122(1):108–124

Kern H, Richter A (1981) Temperature derivatives of compressional and shear-wave velocities in crustal and mantle rocks at 6 kbar confining pressure. J Geophys 49(1):47–56

Konrad K, Graham DW, Thornber CR, Duncan RA, Kent AJ, Al-Amri AM (2016) Asthenosphere–lithosphere interactions in Western Saudi Arabia: Inferences from $^3He/^4He$ in xenoliths and lava flows from Harrat Hutaymah. Lithos 248:339–352

Koulakov I, El Khrepy S, Al-Arifi N, Kuznetsov P, Kasatkina E (2015) Structural cause of a missed eruption in the Harrat Lunayyir basaltic field (Saudi Arabia) in 2009. Geology 43(5):395–398

Kumar MR, Ramesh DS, Saul J, Sarkar D, Kind R (2002) Crustal structure and upper mantle stratigraphy of the Arabian shield. Geophys Res Lett 29(8):1301–1304

Langston CA (1979) Structure under Mount Rainier, Washington, inferred from teleseismic body waves. J Geophys Res Solid Earth 84(B9):4749–4762

Ligorría JP, Ammon CJ (1999) Iterative deconvolution and receiver-function estimation. Bull Seismol Soc Am 89(5):1395–1400

McGuire AV (1988) The mantle beneath the Red Sea margin: xenoliths from western Saudi Arabia. Tectonophysics 150(1):101–119

McGuire AV, Bohannon RG (1989) Timing of mantle upwelling: evidence for a passive origin for the Red Sea rift. J Geophys Res 94:1677–1682. https://doi.org/10.1029/88JB03905

Mohsen A, Hofstetter R, Bock G, Kind R, Weber M, Wylegalla K, DESERT Group (2005) A receiver function study across the Dead Sea Transform. Geophys J Int 160(3):948–960

Owens TJ, Zandt G, Taylor SR (1984) Seismic evidence for an ancient rift beneath the Cumberland Plateau, Tennessee: a detailed analysis of broadband teleseismic P waveforms. J Geophys Res Solid Earth 89(B9):7783–7795

Owens TJ, Nyblade AA, Gurrola H, Langston CA (2000) Mantle transition zone structure beneath Tanzania, East Africa. Geophys Res Lett 27(6):827–830

Pallister JS, McCausland WA, Jónsson S, Lu Z, Zahran HM, El Hadidy S, Aburukbah A, Stewart ICF, Lundgren PR, White RA, Moufti MRH (2010) Broad accommodation of rift-related extension recorded by dike intrusion in Saudi Arabia. Nat Geosci 3(10):705–712

Pasyanos ME (2005) A variable resolution surface wave dispersion study of Eurasia, North Africa, and surrounding regions. J Geophys Res Solid Earth 110(B12):B12301. https://doi.org/10.1029/2005JB003749

Sandvol E, Seber D, Calvert A, Barazangi M (1998) Grid search modeling of receiver functions: implications for crustal structure in the Middle East and North Africa. J Geophys Res Solid Earth 103 (B11):26899–26917

Schmidt DL, Hadley DG, Stoeser DB (1979) Late Proterozoic crustal history of the Arabian Shield, southern Najd Province, Kingdom of Saudi Arabia. In: Tarhouns A (ed) Evolution and mineralization of the Arabian-Nubian Shield, vol 2. Institute for Applied Geology, Jeddah, pp 41–58

Shaw JE, Baker JA, Kent AJR, Ibrahim KM, Menzies MA (2007) The geochemistry of the Arabian lithospheric mantle—a source for intraplate volcanism? J Petrology 48(8):1495–1512

Stoeser DB, Camp VE (1985) Pan-African microplate accretion of the Arabian Shield. Geol Soc Am Bull 96(7):817–826

Tang Z, Julià J, Zahran H, Mai PM (2016) The lithospheric shear-wave velocity structure of Saudi Arabia: young volcanism in an old shield. Tectonophysics 680:8–27

Tkalčić H, Pasyanos ME, Rodgers AJ, Gök R, Walter WR, Al-Amri A (2006) A multistep approach for joint modeling of surface wave dispersion and teleseismic receiver functions: Implications for lithospheric structure of the Arabian Peninsula. J Geophys Res Solid Earth 111(B11):B11311. https://doi.org/10.1029/2005JB004130

Xu W, Jónsson S (2014) The 2007–8 volcanic eruption on Jebel at Tair Island (Red Sea) observed by satellite radar and optical images. Bull Volcanol 76(2):1–14

Zhu L, Kanamori H (2000) Moho depth variation in southern California from teleseismic receiver functions. J Geophys Res Solid Earth 105(B2):2969–2980

Variations in Plio-Pleistocene Deposition in the Red Sea

Neil C. Mitchell, Marco Ligi, and Najeeb M. A. Rasul

Abstract

The thickness of deep-water Plio-Pleistocene (PP) sediments in the Red Sea varies somewhat, as expected from increased biogenic pelagic production rates in the south and with input of aeolian and fluvial sediments through the Tokar Gap in the Sudanese hills. Otherwise, however, the sediment distribution does not obviously reflect the likely pattern of sediment input (from the positions of wind gaps through the Red Sea hills and fluvial drainage basin outlets). We use localized seismic surveys to investigate sediment distribution of two areas in more detail. The first, located near the coast of Egypt, utilized 3D seismic data collected for oil and gas exploration. The data reveal a pattern of sediment deposition that is unrelated to drainage basins of the adjacent hills. Instead, deposition here has been strongly affected by halokinetics, with sediment filling evaporite depressions that are elongated sub-parallel with the coast. For the second, Chirp sediment profiler data allow study of finer scale Pleistocene sedimentation around Thetis Deep in the central Red Sea. The data contain a common sequence of reflections, which suggests that hemipelagic sedimentation has been almost uniform about the deep. The seismic time interval between the seabed and one reflection at ~20–30 ms sub-bottom was mapped out and varies little either side of the deep, but does reveal a systematic thickening of the interval with increasing water depth. The data also reveal structures indicating localized slope failure and sediment flow deposits, as well as tectonic disruptions. From correlations of reflections with sea level curves and sediment core data, we suggest that the slope failures occurred in the Late Pleistocene after Marine Isotope Stage (MIS) 12 and probably before MIS 6. We suggest that these slopes likely failed because of seismic ground accelerations. Applying a pseudo-static slope stability model and assuming shear strengths of comparable carbonate-rich sediments, we estimate the potential acceleration and earthquake magnitude. The results suggest that the very low incidence of historical earthquakes in the central Red Sea is not entirely representative of the Late Pleistocene.

1 Introduction

The Red Sea Plio-Pleistocene (PP) sediments are important as they provide information on the regional climate. For example, magnetic remanence measurements have usefully recorded fluctuations in haematite content in sediment cores that reflect past aridity and wind conditions (Roberts et al. 2011; Rohling et al. 2008). They also provide important information on global climate, as the Red Sea deep-water sediments contain an amplified glacial-interglacial oxygen isotope variation because of evaporation in the desert climate and because the narrow and shallow sill between the sea and the Indian Ocean further limited exchange during times of lowered sea level (Hemleben et al. 1996). Modeling these effects has yielded a relative sea level record for the late Pleistocene that matches global records when adjusted for hydro-isostatic effects (Rohling et al. 2009). Studies such as those carried out in the present chapter can help paleoclimate work, as analyses may suggest how much the input of particles has varied spatially, which may be important for depositional fluxes, and assist in optimal siting of new cores.

The early sedimentary history of extensional continental margins is commonly obscured by burial under kilometres of

N. C. Mitchell (✉)
School of Earth and Environmental Sciences, University of Manchester, Williamson Building, Oxford Road, Manchester, M13 9PL, UK
e-mail: neil.mitchell@manchester.ac.uk

M. Ligi
Istituto di Scienze Marine, Consiglio Nazionale delle Richerche, Via Gobetti 101, 40129 Bologna, Italy

N. M. A. Rasul
Center for Marine Geology, Saudi Geological Survey, Jeddah, Saudi Arabia

N. M. A. Rasul and I. C. F. Stewart (eds.), *Geological Setting, Palaeoenvironment and Archaeology of the Red Sea*,
https://doi.org/10.1007/978-3-319-99408-6_15

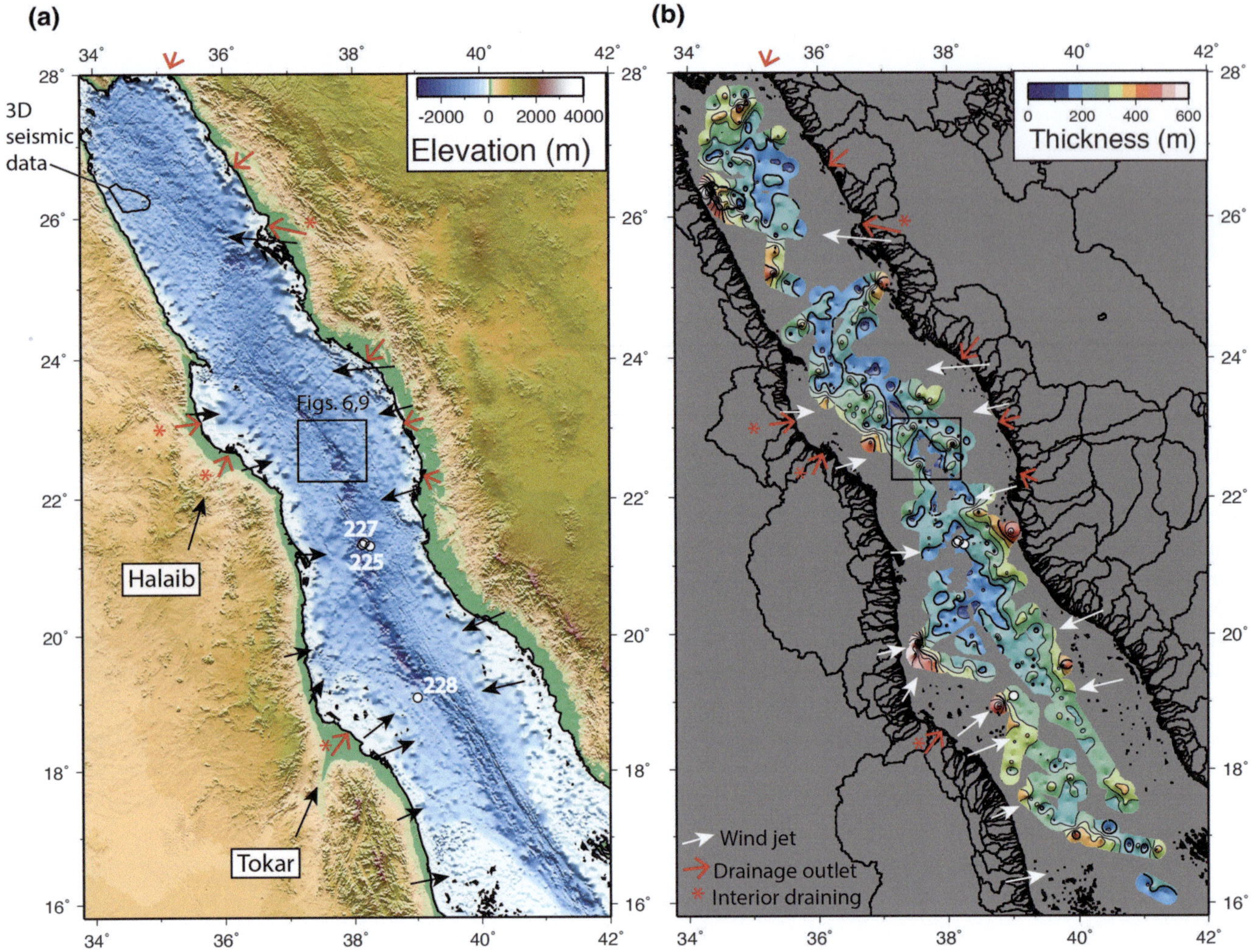

Fig. 1 Overviews of **a** the topography of the Red Sea region and **b** thickness of Plio-Pleistocene sediments overlying the evaporites according to Mitchell et al. (2017) updated using 3D seismic data. Elevations are from Becker et al. (2009). Open circles annotated 225–228 in **a** locate Deep Sea Drilling Project sites (Whitmarsh et al. 1974). Black arrows in **a** and corresponding white arrows in **b** locate significant wind jets predicted with atmospheric models for 12 July 2008 and 14 January 2009 (Jiang et al. 2009) (easterly and westerly wind jets, respectively). Watersheds in **b** were derived from the Shuttle Radar Topography Mission (Lehner et al. 2008). Red arrows locate drainage basins of greater than ~ 5000 km^2. Basins breaching the Red Sea hills are marked with asterisks (including "Tokar" and "Halaib")

later terrigenous sediments and, where thick salt deposits occur, by halokinetics. The Red Sea is a young ocean basin formed by extension of the African-Arabian shield and the patterns of sedimentation found there may provide clues to help us to understand those older margins or guide the collection of samples or other data that would usefully investigate processes. For example, as in the Red Sea, seismic reflection data collected offshore Brazil show abundant salt diapirs and walls (Davison et al. 2012). The reasons why salt diapirs and walls nucleate at some locations and not others are still not fully understood. Working out the pattern of sediment deposition in the Red Sea may ultimately help in assessing the role that the sediment loading has in initiating halokinetics and whether it relates to the distribution of primary sediment inputs.

Modern sedimentation in the central trough of the Red Sea is likely to be mostly pelagic, as the present-day deep waters move only sluggishly, renewed slowly by subsiding saline waters in the north (Maillard and Soliman 1986). Sofianos and Johns (2007) reported essentially unresolvable velocities ($\ll$20 cm s^{-1}) in shipboard acoustic Doppler current meter data below 250 m depth in the central Red Sea. Simulated eddies based on wind stresses also only reach to 250 m depth (Nicholls et al. 2015), which is shallower than typical water depths of the central trough of the Red Sea (Fig. 1a). The presence of aplanktonic zones implying >49 ppt salinity during glacial periods throughout much of the Red Sea (Fenton et al. 2000) suggests that only limited renewal of deep waters occurred during glacial lowstand times over the period 500 ka to present, also consistent with sluggish circulation.

However, it is presently unclear if such sluggish conditions existed in the Red Sea before 500 ka. For example, the Hanish Sill may not have been as shallow and therefore capable of retarding loss of saline deep water from the Red Sea into the Indian Ocean as it is today. If deep waters were less saline, the sea would have been less stratified and thus more prone to over-turning circulation. Hanish Sill lies adjacent to the volcanically active Hanish Island (Gass 1970) and the evolution of the sill's morphology is unknown for this earlier period. The lower PP section in seismic reflection data from the Red Sea is commonly transparent, which Mitchell et al. (2017) argued may indirectly indicate that extreme salinity did not develop as it has in the Late Pleistocene. Furthermore, tectonic subsidence and flowage of underlying evaporite beds have depressed the seabed throughout much of the basin, which was likely closer to sea level in the earlier Pliocene, as evidenced by unconformities around the Miocene-Pliocene boundary (Izzeldin 1987; Mitchell et al. 2017) and reworked anhydrite clasts (Stoffers and Kühn 1974; Stoffers and Ross 1974). Some of the older PP sediments were therefore potentially influenced by surface currents and waves during deposition.

In this chapter, after updating earlier work on the regional pattern of PP sediment thickness derived from shallow seismic data, we introduce two datasets that illustrate the variations in sediment deposition at a finer scale. The first, located in the northwestern Red Sea, comprises industry 3D seismic reflection data, revealing how sediments have been deposited in depressions created by halokinetic deformation. The second, located in a more axial location in the central Red Sea, comprises high-resolution Chirp seismic data. In the latter, a characteristic sequence of reflections occurs throughout much of the area surveyed. These were suggested by Mitchell et al. (2015) to originate from rigid aragonite-rich sediment layers deposited during glacial lowstand conditions (Gevirtz and Friedman 1966; Milliman et al. 1969; Stoffers and Ross 1974) and provide an excellent relative chronology with which to characterize patterns of sediment deposition.

2 Previous Work

Stoffers and Ross (1974) summarized the PP sediments recovered during the DSDP expeditions at the sites shown in Fig. 1a. Textures were described as uniformly silty clay. Terrigenous components comprise a variety of minerals likely sourced from the surrounding continents (quartz, feldspars, micas and clays). Crossley et al. (1992) and Hughes and Beydoun (1992) interpreted sedimentary environments from various industry well and seismic reflection data. The reconstructions of Crossley et al. (1992) show sand deposits overlying the Miocene evaporites in elongate depressions oriented parallel to the coast. Other studies that have addressed aspects of the evaporite and overlying PP sediment geometry have been more localized (Cochran and Karner 2007; Davison et al. 1994; Ehrhardt and Hübscher 2015; Heaton et al. 1995; Hutchinson and Engels 1972; Izzeldin 1987; Mart and Ross 1987; Martinez and Cochran 1988; Miller and Barakat 1988; Richter et al. 1991; Savoyat et al. 1989).

Mitchell et al. (1992) mapped out the PP sediment thickness overlying the Miocene evaporites from seismic data as part of a hydrocarbon evaluation project. According to them, the original seismic reflection data were of poor quality. Because their thicknesses are much greater than those found here and are incompatible with more recent seismic results (Heaton et al. 1995), it is suspected that the digitized horizon corresponds with the top of massive halite rather than the top of the evaporite, as the massive halite surface likely generates a stronger reflection than the top of the Miocene in seismic surveys designed for deeper imaging. Hence, their PP thickness probably includes some or all of the upper Miocene layered evaporites that overlie the massive halite (Zeit Formation (Hughes and Beydoun 1992)). In the south, the mapped top of the Miocene is incompatible with known outcrops of volcanic rock of the spreading centres, illustrating the problem of data quality. Despite confusion over the interval that it represents, their results nevertheless reveal >3 km thick sediments immediately seaward of two major gaps in the westerly Red Sea hills (Halaib and Tokar, Fig. 1a), as expected if fluvial and aeolian sediment sources were important. Thicknesses are generally greater south of 23°N than north of 23°N, reflecting the greater relief of the hills surrounding the Red Sea, and hence sediment released by erosion of those hills.

3 Background—Regional Variations in Plio-Pleistocene Sediment Thickness

A regional map of the thickness of PP sediments was derived by Mitchell et al. (2017) from shallow seismic reflection data, exploiting a prominent reflection "S" marking the top of the Miocene evaporites (Ross and Schlee 1973). Figure 1b shows this map updated with industry 3D seismic data from offshore Egypt (Gordon et al. 2010) introduced later. Mitchell et al. (2017) had previously described some potential sources of sediments that may partly explain variations in Fig. 1b. Biological productivity is enhanced where inflowing Indian Ocean waters mix with Red Sea waters in the south, so thicker sediments south of 19°N could be due to greater biogenic sediment as well as aeolian and fluvial particles supplied via the Tokar gap. Climatic conditions were likely wetter during the earlier Pliocene (Macgregor 2012) and terrigenous content increases in the DSDP cores into the Pliocene (Stoffers and Ross 1974) also suggesting greater fluvial input. Progressive deposition from wind and water flows of aeolian and fluvial origin may explain why sediment thickness generally declines going away from the coast toward the spreading axis of the Red Sea.

We elaborate on those comments here with the aid of the catchments surrounding the Red Sea in Fig. 1b derived from Shuttle Radar Topography Mission data (Lehner et al. 2008). Although the topography of the hills has undoubtedly been modified during escarpment retreat (Steckler and Omar 1994), the pattern of drainage basins has probably not changed dramatically over the last 5.3 my of the Plio-Pleistocene, a small interval compared with the >20 my since the Red Sea hills first developed (Bosworth et al. 2005). Also shown (long arrows) are the positions of prominent wind jets for dates representing summer westerly and winter easterly winds from an atmospheric model (Jiang et al. 2009).

The main drainage divide of the Red Sea hills is asymmetric, lying roughly double the distance from the coast on the Arabian side compared with the African side. This suggests that more fluvial sediment volume was likely transported toward the Red Sea from terrigenous erosion on the east side. On the other hand, four drainage basins extend beyond the Red Sea hills into the continental interiors, including Halaib and Tokar, implying substantial localized sources of sediments where their outlets meet the Red Sea. The modeled wind jets (Jiang et al. 2009) unsurprisingly also exploit gaps in the hills, suggesting that aeolian input has varied along the Red Sea in a somewhat similar way to the fluvial input. Summer westerly winds are strongest over the Tokar delta, whereas winter wind jets (easterlies) occur at various points along the Red Sea, including two either side of our study area of Thetis Deep (rectangle in Fig. 1).

Whereas sediment supplied via the Tokar delta may have contributed to the thicker sediments south of 19°N, there is less evidence in the mainly deep-water sediment thicknesses in Fig. 1b for any relationship with the other major catchments or with the above-mentioned asymmetry between the two coasts. Most coarser grade sediment was probably deposited close to the coasts as shown in seismic data (Egloff et al. 1991; Heaton et al. 1995; Izzeldin 1987; Mitchell et al. 1992), leaving little to reach deep water.

We suggest that the general lack of correlation between the deep-water thickness in Fig. 1b and the along-rift distribution of localized sediment sources can be explained if currents in the Red Sea have dispersed the finer sediments derived from them. Winds create four gyres in the surface waters of the Red Sea, each ~200 km in extent, with geostrophic current speeds of up to 0.6 m s^{-1} (Quadfasel and Baudner 1993). As the winds vary seasonally, the pattern of surface currents varies, with reversed current directions possible (Chen et al. 2014; Clifford et al. 1997; Yao et al. 2014a, b; Zhai and Bower 2013). Shipboard ADCP measurements of Sofianos and Johns (2007) revealed surface currents in places exceeding 1 m s^{-1}, diminishing with depth, though influencing the upper 200 m. Boundary currents of up to tens of cm s^{-1} flowing parallel to the axis of the Red Sea (Bower and Farrar 2015) are present in numerical simulations (Chen et al. 2014; Clifford et al. 1997; Sofianos and Johns 2003; Yao et al. 2014a, b). These velocities are large compared with the settling rates of silt and clay grade sediments (McCave 2005).

4 Data and Methods

4.1 3D Seismic Data

The block of 3D seismic reflection data from offshore Egypt is located in Fig. 1a. Little detail of acquisition and processing methods is available to us. According to Gordon et al. (2010), the data were originally acquired in 1999 for BG plc. Time migrated reflection lines, extracted from the 3D data block, were imported into the seismic and geological interpretation software IHS *Kingdom*® in order to map the top and the base of major sedimentary sequences. As elsewhere in the Red Sea, the top of the evaporites creates a strong reflection (S), as marked on examples of the data in Fig. 2. Depths to S and the seabed reflection were mapped out and are shown in Fig. 3. Time to depth conversion was obtained by adopting average interval velocities of 1.525 and 2 km s^{-1} for the water layer and the hemipelagic sediment cover, respectively. Figure 4 shows the PP sediment thickness derived from them and Fig. 5 is a structural interpretation.

The two-way time to reflection S below the seabed is typically <0.5 s in Fig. 2. For comparison, Heaton et al. (1995) presented industry data collected off the Red Sea coast of Yemen in which the main (pre-mobilized) salt layer is up to 2 s (two-way time, TWT) below the seabed. The salt there has been more greatly mobilized than for the data presented here and forms canopies. Industry data collected off the coast of Saudi Arabia show salt depth varying up to 2 s TWT below seabed and with bodies roughly aligned with the coast (Richter et al. 1991), similar to those presented in Figs. 3, 4 and 5. Therefore, the data we present here are somewhat representative of marginal settings in the Red Sea, though without the more extreme mobilization that results from 2 s TWT of sediment overburden.

4.2 Chirp Data Around Thetis Deep

Sediment profiler data were collected with a Benthos CAP-6600 CHIRP-II around Thetis Deep while mapping the bathymetry in Fig. 6 with a RESON SeaBat 8160 multibeam system on RV *Urania*. Chirp systems produce high-resolution

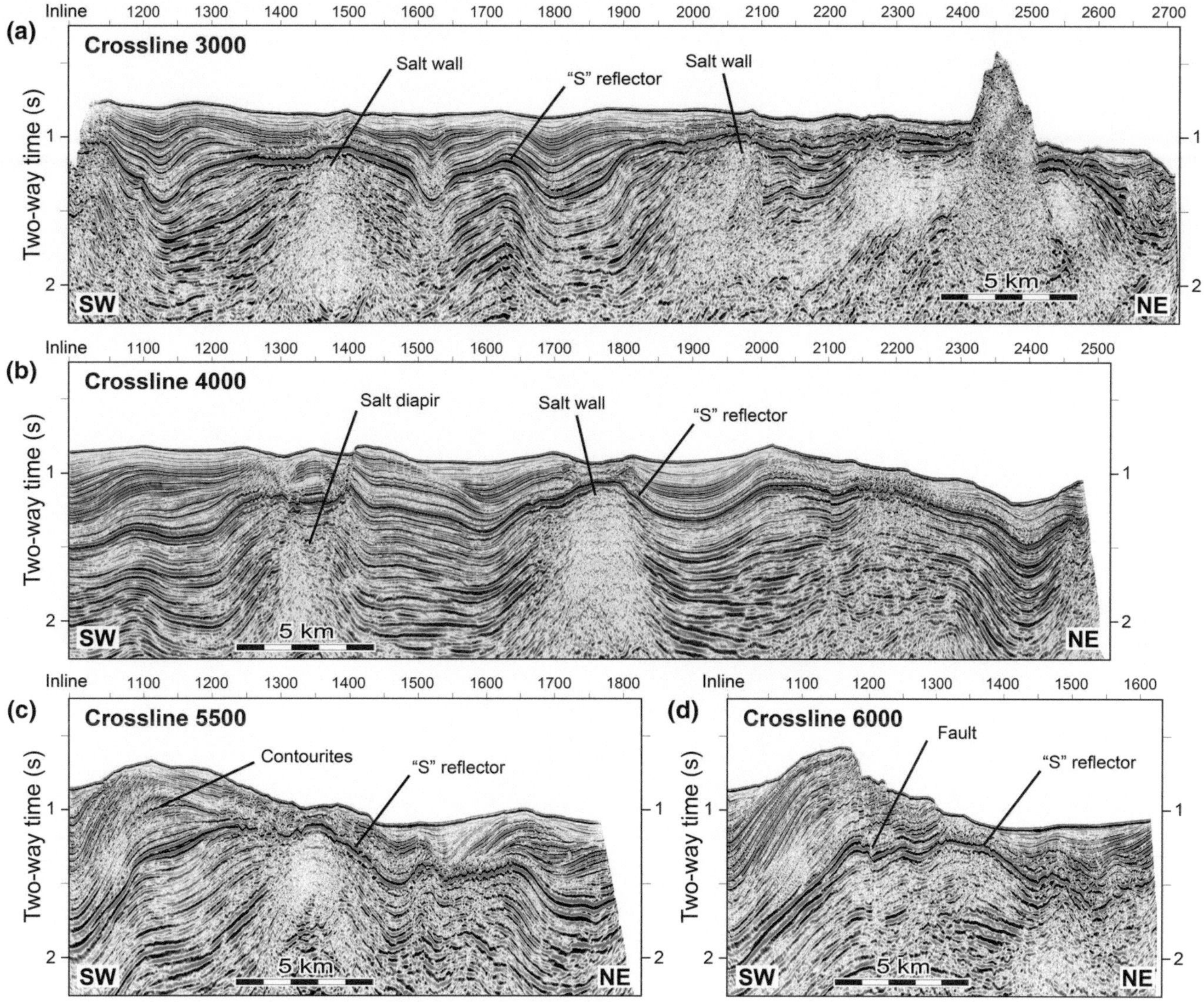

Fig. 2 Selected cross-lines through the 3D seismic block located in Fig. 3. "S" reflector marks the top of the Miocene evaporites

images of sediments by transmitting long pulses with swept frequency and compressing the returned signal. The theoretical resolution of the data shown here is ~0.1 ms two-way time (Mitchell et al. 2015) although noise and sea conditions have degraded this somewhat. Parts of the reflection sequence R1–R5 identified by Mitchell et al. (2015) are marked on the example sections in Fig. 7a–d, which are located in Fig. 6. These sections were deliberately chosen for showing sedimentary mass-movement features, though a simple draping morphology of the reflections, typical of pelagic sediments (Tominaga et al. 2011), is more common away from faults and away from the floor of the deep. Note the reflectivity in the water column, including the ghost of the seabed (e.g., Figure 7b), is partly an artifact of processing.

Figure 8 shows averaged amplitudes of six traces of the data with reflections R1–R5 marked. The amplitudes are shown relative to the seabed reflection amplitude in decibels, a logarithmic scale that ensures that any exponential decrease in amplitudes with distance and time due to seismic attenuation appears linear if the sediments have a uniform attenuation coefficient (Mitchell, 2016). Also shown is the tentative chronology put forward by Mitchell et al. (2015), whereby reflections were interpreted to arise mainly from the seismic impedance contrasts on the upper surfaces of the aragonite-rich layers deposited during sea level lowstands. Reflections hence are generally associated with sea level rise.

The reflection sequence starts with extended seabed reflectivity R1. Although not visible in the grey-scale figures, in detail this comprises two close reflections separated by ~1 m (Fig. 8), which were suggested to originate from the seabed and the aragonite-rich Last Glacial Maximum (LGM) layer underlying the Holocene sediments.

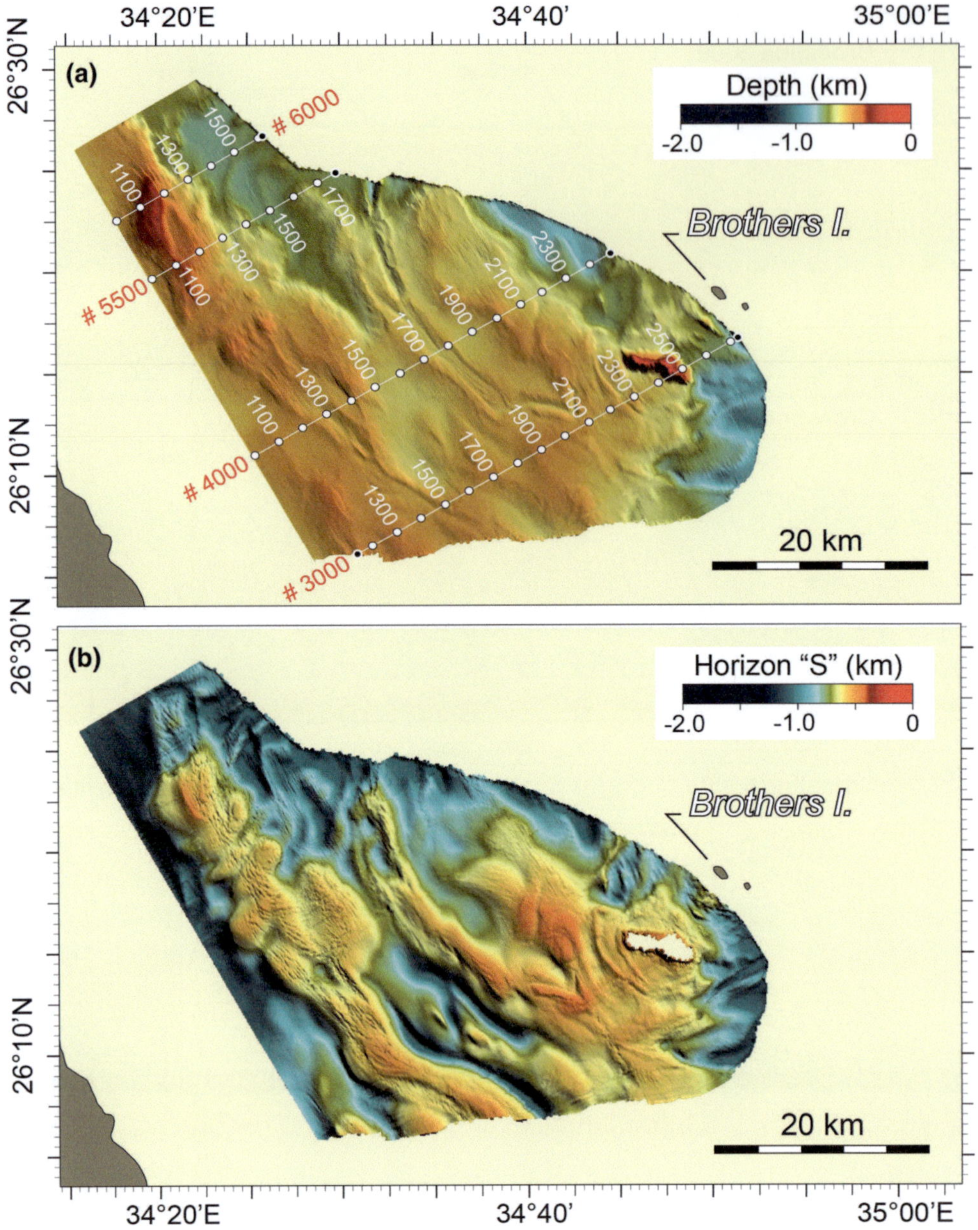

Fig. 3 Depth below sea level to the **a** seabed and **b** reflection S. Lines highlighted are shown in Fig. 2

Corroborating this, a 3.3-m gravity core collected at site RS05-01 (Fig. 6) contained carbonate rich hemipelagic sediment of typically sandy clayey silt texture. XRD analysis revealed the presence of aragonite in the upper 20 cm. Although not containing aragonite, stiff carbonate was found at 2–3 m depth. This core is not dated at present, but the LGM layer was identified in core MD92-1039 just SW of the survey area (Fig. 6), where it lies at 75–142 cm depth (Mitchell et al. 2015).

Below R1 and prominent reflection R2, reflectivity steadily declines with two-way time. A strong reflection R3 is underlain by a weak reflection R4, and after typically ~ 10 ms, a modest reflection R5. R4 was suggested to arise from the impedance contrast at the base of the aragonite-rich layer deposited during Marine Isotope Stage (MIS) 12, as the sea level of Rohling et al. (2009) sharply declines into MIS 12, in contrast with a more gradual fall into the other isotope stages. As there are no long cores collected exactly over the Chirp lines, this chronology is tentative, although it is consistent with the pelagic sequence in other hemipelagic Red Sea cores.

In many parts of the dataset, this sequence is well developed (e.g., NW end of Fig. 7a). In some other areas, one or two prominent reflection can be present, but the full sequence is less clear, either because the data are noisier (poor sea conditions) or because of sediment mass flows or erosion. An example is marked in the SE side of Fig. 7a where it is annotated "R3?" In some of these places, the reflections can still be identified from their context, for

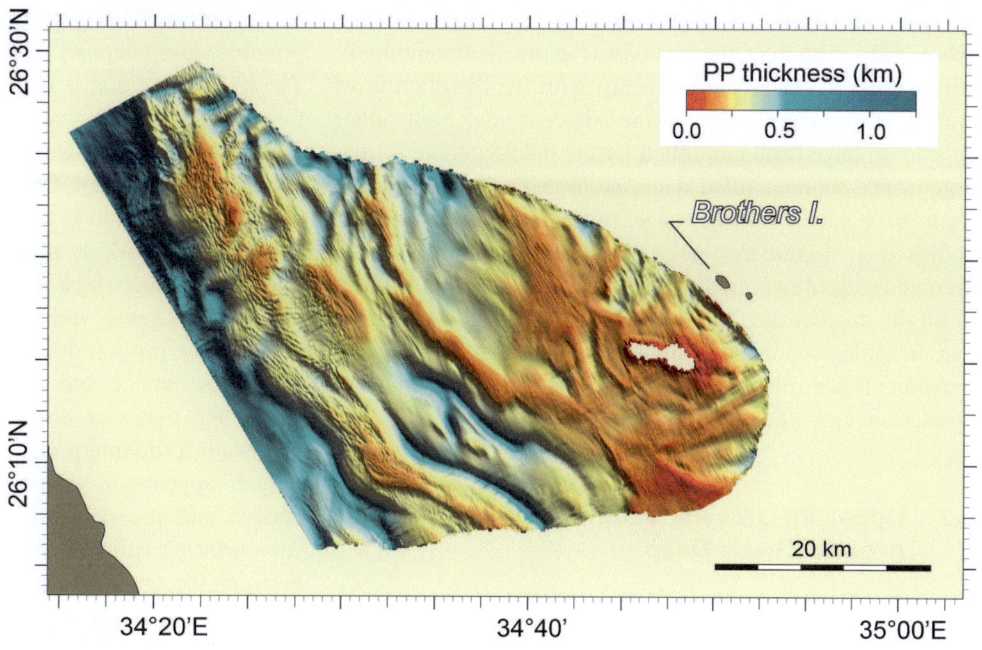

Fig. 4 Total thickness of the Plio-Pleistocene strata obtained by differencing the two surfaces shown in Fig. 3

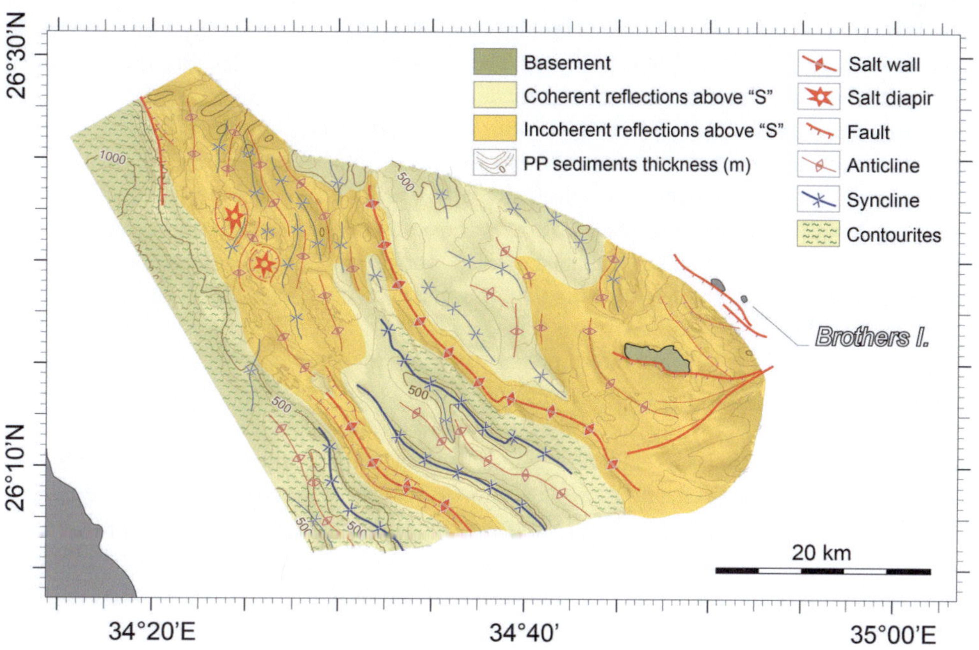

Fig. 5 Structural interpretation of the 3D seismic volume

example, a reflection can be traced from where it can be identified within a clearer sequence. R3 was thus classified into three groups as indicated along ship tracks in Fig. 6: Red for well-identified, green for weakly developed but still identifiable and blue for weakly identifiable and not confirmed by context. The two-way time of R3 to the seabed reflection was mapped out (Fig. 9) and is shown versus geographical location and depth in Fig. 10 to help reveal potential causes.

5 Observations and Interpretations

5.1 3D Seismic Data

The topography of S is dominated by salt diapirs and walls with a typical relief of up to roughly a kilometre (Figs. 3b and 5). As the seabed is more subdued with a local relief of typically <100 m (Fig. 3a), the PP sediment thickness contains elongate structures of also kilometre amplitude aligned sub-parallel with the adjacent coast (Fig. 4). Sediment thickness bears no apparent relationship with the distribution of outlets of drainage basins on the adjacent coast, but rather appears to have been controlled by the halokinetic deformation, with sediment filled depressions between diapirs and walls. In detail, the seismic images reveal some contourite-like bodies (e.g., Figure 2c). Therefore, sediment transport has involved some down-gradient movement of sediment in order to fill the depressions, but also some transport along the Red Sea in order to create contourites. Continual halokinetic movements have disrupted the Plio-Pleistocene strata in many areas, such as at "Fault" in Fig. 2d.

5.2 Upper Pleistocene Sediments in Chirp Data Around Thetis Deep

The absence of parts of the R1–R5 sequence ("*" in Fig. 7) helps us to identify features produced by sedimentary mass movements. For example, in Fig. 7a where marked "Thin R3-seabed", this interval is several ms thinner than where it is annotated on the far NW end of the section. The stratigraphic location of failure surfaces or mass movement deposits relative to such reflections also provides a relative chronology of the failure events. In none of the areas where this reflection sequence is fully developed are observed any current-generated bedforms (e.g., mud-waves, furrows or elongated sediment drifts (Mitchell and Huthnance 2013)), consistent with deep-water currents being very sluggish. In some parts of the reflection data, the record is more transparent, such as where marked "T" in Fig. 7. We interpret many of these parts as due to debrites. Where they contain some reflectivity (e.g., "DL" at the NW end of Fig. 7c and at the SE end of Fig. 7a), reflections tend to be less regular and not parallel with underlying or overlying reflections, so they are suspected to represent layering within single debrite beds, although multiple debrites cannot be ruled out. Where marked "DS" in Fig. 7a, a debrite layer and overlying reflections are stepped and diffraction hyperbolae can be observed. However, there are hints of the reflection sequence fading into some transparent areas, such as T1 in Fig. 7a. This suggests that some transparent areas are due to seismic attenuation in overlying deposits.

Where marked by an asterisk in Fig. 7b, some of the sequence above R3 appears to be missing. Two diffraction hyperbolae can be observed, the upper of which marks the top of a seabed escarpment ("Headwall"), whereas the lower marks a reflector termination either at the base of the escarpment or within the subsurface. Immediately SE of the escarpment, the seabed reflectivity is extended as it is for R1 in unfailed areas. It is also more extended than R2 immediately NW of the escarpment (where it is marked "Sharp R2"). To the SE in Fig. 7b, transparent deposits (T4) and a possible slump deposit lie down-slope of the thinned section. The missing section is likely to have been removed by slope failure and the displaced material has either formed a localized slump deposit or disaggregated to form a debrite (transparent deposit). The extended seabed reflection in the slope failure scar in Fig. 7b suggests that this occurred here prior to the LGM, so that both LGM aragonites and a thin Holocene layer (which is typically 1 m thick in other cores) have left extreme variations in seismic impedance at the seabed. The proposed slide surface is underlain by a similar R1–R2 sequence but lacks the reflectivity immediately below R2 that can be observed immediately NW of the headwall. If the interpretation of R2 to the SE end of Fig. 7b (which appears to continue beneath the slide surface) is correct and the deeper diffraction hyperbolae indicates a sub-surface termination of a reflector, the failure occurred between the ages corresponding with R2 and R3 (MIS 6 and 12, Fig. 8) creating the transparent deposit. Sediment corresponding with R2 and R1 was subsequently deposited leaving the laterally continuous reflection pair. Alternatively, it is younger and created the marked slump deposit, but we consider this less likely as that deposit has a much smaller cross-sectional area than the slide scar.

Figure 11 shows a selection of traces of the data of Fig. 7b, confirming the double reflection nature of the near-seabed reflectivity over the proposed slide scar. Furthermore, in Fig. 11a the second (deeper) diffraction hyperbola can clearly be seen terminating ~2–3 ms below the TWT of R2. The landslide event therefore probably occurred a little before the time of R2, or MIS 6 if the chronology in Fig. 8 is correct.

A similar configuration of transparent deposits and missing section can be observed for the slide marked in Fig. 7a. Although neither R2 nor R3 can be traced continuously to the transparent deposit T3 to the SE, a case can be made that these reflections encompass that deposit. The debrite step ("DS" in Fig. 7a) implies that halokinetic deformation has continued after emplacement of the debrite.

Above the other sections of transparent data, parts of the R1–R5 sequence are also observed. Above T2 and T5 in Fig. 7a and c at least one prominent sub-bottom reflection is observed, which is perhaps R2, suggesting that failure occurred before the time of R2 (MIS 6?; Figure 8). It is unclear if R2 is present over the thinned section in Fig. 7c, but the uppermost sub-bottom reflection overlying the adjacent transparent deposit T6 to the SE does appear to merge with a reflection in the slope sediments of the thinned section, so slope failure probably occurred before the age of that reflection, hence it is also probably between R2 and R3 in age. This line almost intersects that of Fig. 7b. The failure probably occurred as one event crossing both these lines and leaving one or more debrites deposited in the oblique depression marked in Fig. 6. Line 18 m is ~5 km to the

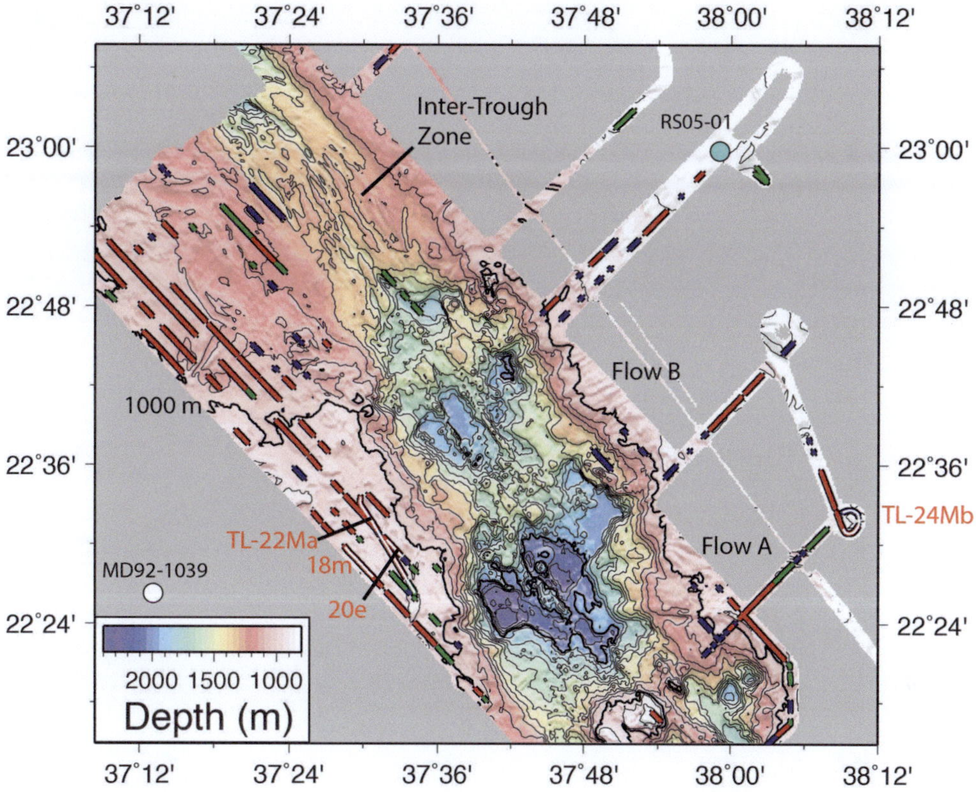

Fig. 6 Bathymetry map created from *Urania* multibeam sonar overlain with classification of the Chirp data where the R3 reflection was identified confidently (red), moderately confidently (green) and weakly (blue). The four segments of white tracks and associated red lettering locate lines 18 m, 20e, TL-22 Ma and TL-24 Mb shown in Fig. 7. Open circle annotated MD92-1039 locates a long piston core site. Turquoise circle marked RS05-01 is site of 3.3 m gravity core collected on *Urania*. Depth contours are every 100 m with 1000 m contours bold

southeast, though the thinned section and debrite may also have formed in the same event.

The data in Fig. 7d were collected during a gradual change of ship heading on a track that intersected the edge of a hill. Reflection sequence R1–R5 can be observed to the south of the hill between "Headwall or sidewall" and "RS" (rugged seabed, extended echo with numerous overlapping diffraction hyperbolae). The transparent data T7 north of the hill is speculatively underlain by R3 and R5. Above T7 are around 15 ms, or 10 m, of a well-developed reflection sequence. The slope failure event leading to T7 is therefore probably also post-R3. Immediately adjacent to the hill on both sides are RS, suggesting localized mass-flow deposits that are younger than the transparent deposits. South of the seabed step marked "Headwall or sidewall" is another area of rugged seabed, with a thin R3-seabed interval. This is interpreted to be the result of slope failure, which occurred after the age of R3.

5.3 Sediment Thickness Between R3 and Seabed

A colour table was chosen for the thicknesses in Fig. 9 to highlight areas where the R3-seabed interval is anomalously thick (red) or thin (blue) compared with more typical hemipelagic values (white, 25 ms). The interval is anomalously thick in some elongate depressions are marked with arrows ("Aligned features" in Fig. 9). Many of these depressions contain debrites like those shown in Fig. 7. Localized ridges and valleys (which formed by movements in the underlying evaporites (Mitchell et al. 2010)) are commonly associated with erosion and deposition, respectively, as observed in Fig. 7.

The R3-seabed interval plotted versus longitude (Fig. 10a) reveals many localized deviations from the typical ~25–29 ms. The better classified sequences (red symbols), however, show similar thicknesses on the east side of the deep (38°03′E) as on the west side. Similarly, there is little systematic trend going north (Fig. 10b). Nevertheless, the thickness is found to increase systematically with water depth; the robust regression computed with the software of Wessel and Smith (1991) and shown by the solid line in Fig. 10c has a gradient of 0.014 ms m^{-1} or ~11 m per km depth change (assuming 1500 m s^{-1} seismic velocity for these shallow sediments). The 6.4 ms change over the regression shown (763–1211 m) is 25% of the average thickness.

6 Discussion

6.1 Locally, Sedimentation Rates Tend to Increase with Depth

The sediments accumulate preferentially in valleys or depressions, as illustrated by the R3-seabed interval

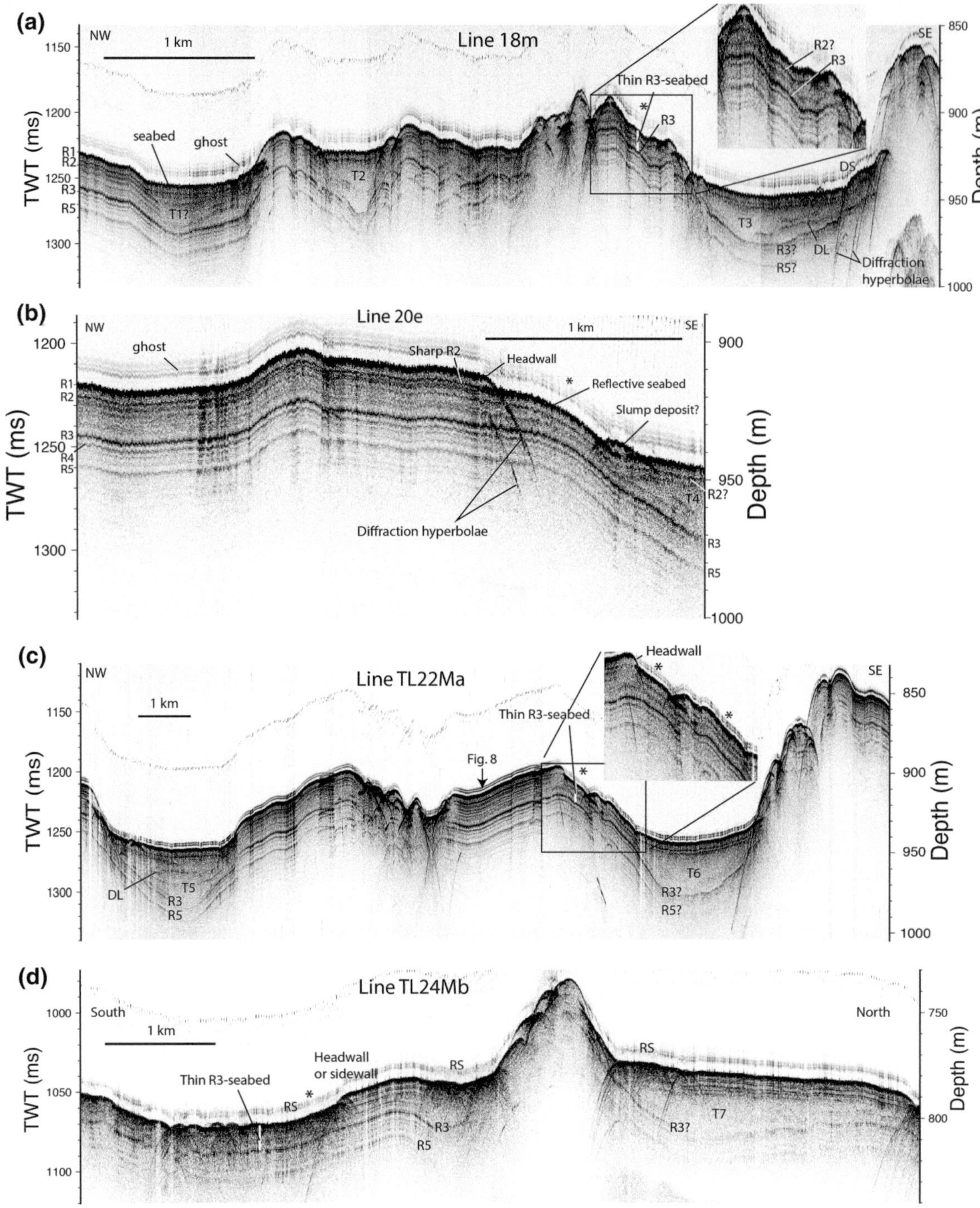

Fig. 7 Sections of Chirp data located in Fig. 6; parts of lines **a** 18 m, **b** 20e, **c** TL22 Ma and **d** TL24 Mb. Asterisks locate slope failure features. RS: rough seabed, T: transparent sediments, DL: debrite layering, DS: debrite step

increasing with depth (Fig. 10c). This general tendency is also shown by the seismic record in Fig. 2, in which the relief of the seabed is smaller than that of S. Other examples can be seen in published seismic datasets (Cochran 2005; Ehrhardt and Hübscher 2015; Mitchell et al. 2015, 2017).

Preferential deposition in valleys could occur because currents become more sluggish as they expand over deeper water (Flood 1988), but this explanation seems unsatisfactory here given that bottom water is generally slow moving. A more likely explanation is that sediments re-suspended by organisms interacting with the bottom and occasional slope failure move down-slope under gravity until the suspension has re-deposited. Repeated sediment re-suspensions by bottom interactions by organisms revealed in abyssal

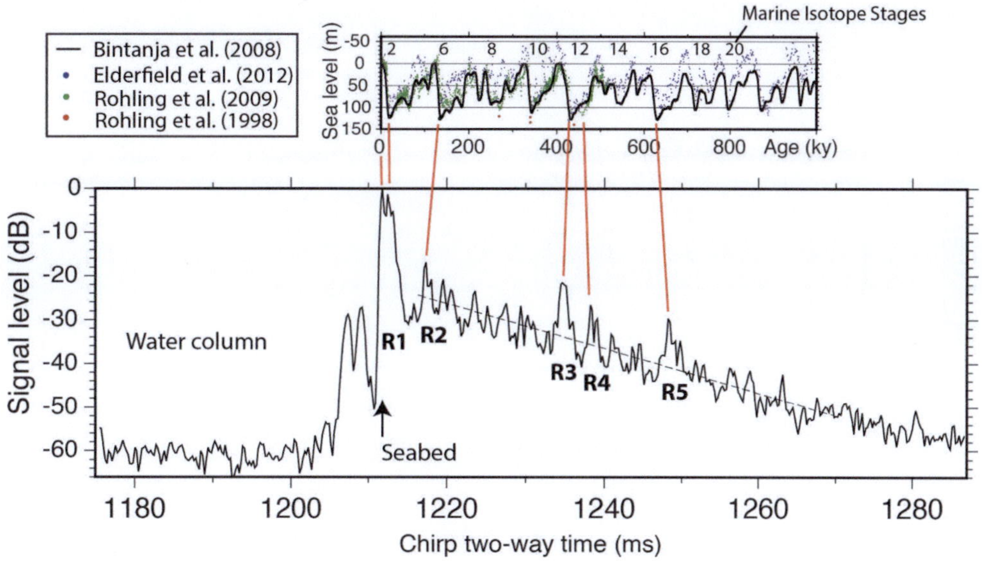

Fig. 8 Main (lower) panel shows amplitude versus two-way time (average of 6 traces) of Chirp sediment profiler data collected during the 2005 cruise of RV *Urania* (from Mitchell et al. 2015). The traces were extracted where marked in Fig. 7c. The data are shown as signal level in decibels relative to the seabed reflection (i.e., $20\log_{10}(A/A_0)$, where A is the data amplitude and A_0 is the seabed reflection amplitude). Straight dashed line is a regression line fitted to the data shown and implies an attenuation coefficient of 0.34 dB m^{-1}, assuming sediment velocity of 1500 m s^{-1}. This coefficient estimate is an upper bound because progressive compaction with depth tends to reduce density and velocity contrasts (reducing reflection amplitudes with depth), though 0.34 dB m^{-1} is compatible with other attenuation data on similar sediments for this 2–7 kHz frequency band (Mitchell 1993). R1–R5 reflections are revealed here as deviations from the regression line. Top panel shows estimates of sea level variation of the late Pleistocene (Bintanja and van de Wal 2008; Elderfield et al. 2012; Rohling et al. 1998, 2009). Red lines link times of rapid changes in sea level generally following lowstands to reflections in the Chirp data caused by aragonite-rich layers deposited during sea level lowstands (Mitchell et al. 2015). R4 was interpreted as associated with the base of the MIS 12 layer

Fig. 9 Thickness of sediment between the seabed and reflection R3 overlain on a grey-shaded version of the bathymetry in Fig. 6. Thickness is shown in seismic two-way time (TWT) in milliseconds (1 ms = 0.75 m at 1500 m s^{-1})

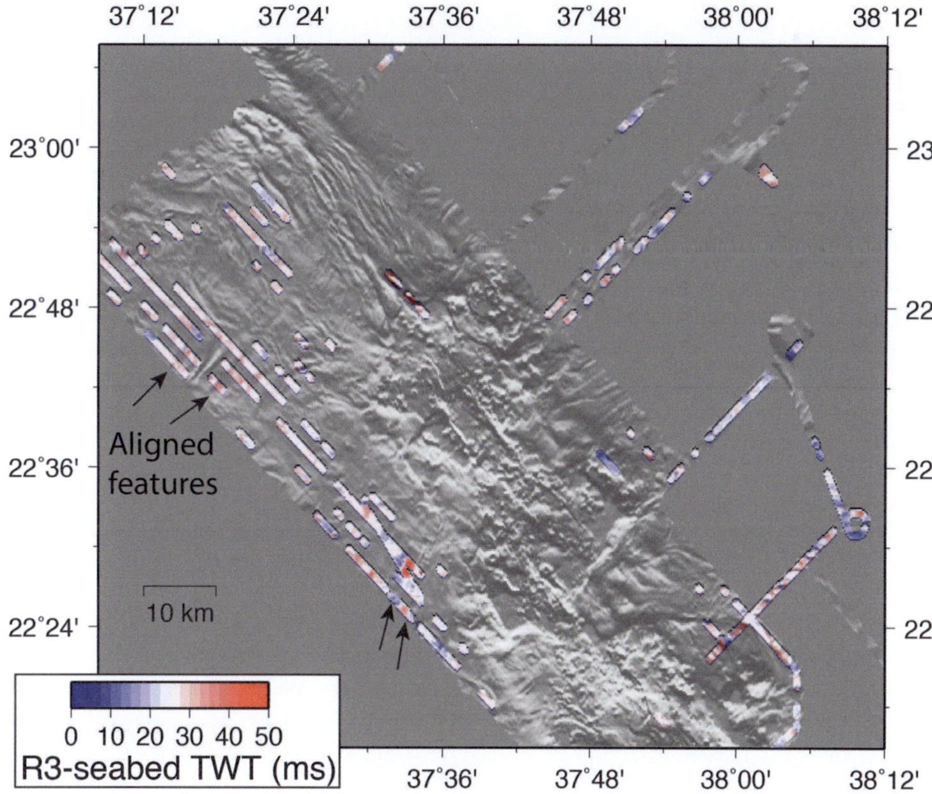

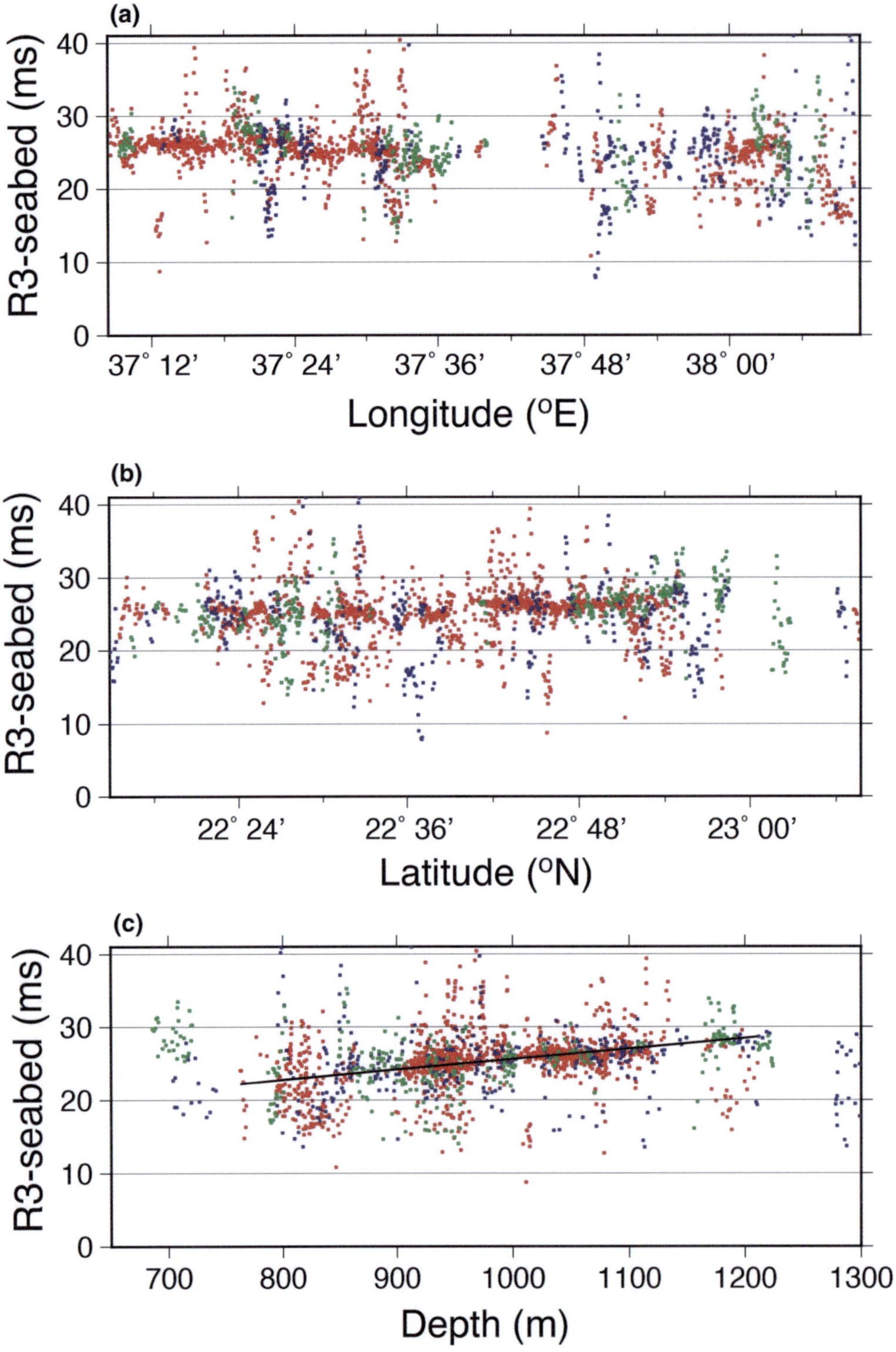

Fig. 10 Variation in the seabed to R3 two-way time with **a** longitude, **b** latitude and **c** water depth. Colour coding as in Fig. 6

photographs (Arrhenius 1963; Seibold and Futterer 1982) have been speculated to lead to fall-out under the influence of gravity in the form of small sedimentary flows (Arrhenius 1963; Luyendyk 1970; Marks 1981). These in turn may explain the typically curved morphology of pelagic deposits on the Mid-Atlantic Ridge (Mitchell 1995) and a tendency for surface grain size to decrease systematically with depth there (Marks 1981). Outside of the deep, the seabed deepens in the west side of Fig. 6 over a distance of ~20 km while the R3-seabed interval thickens. Thus, sediment movements

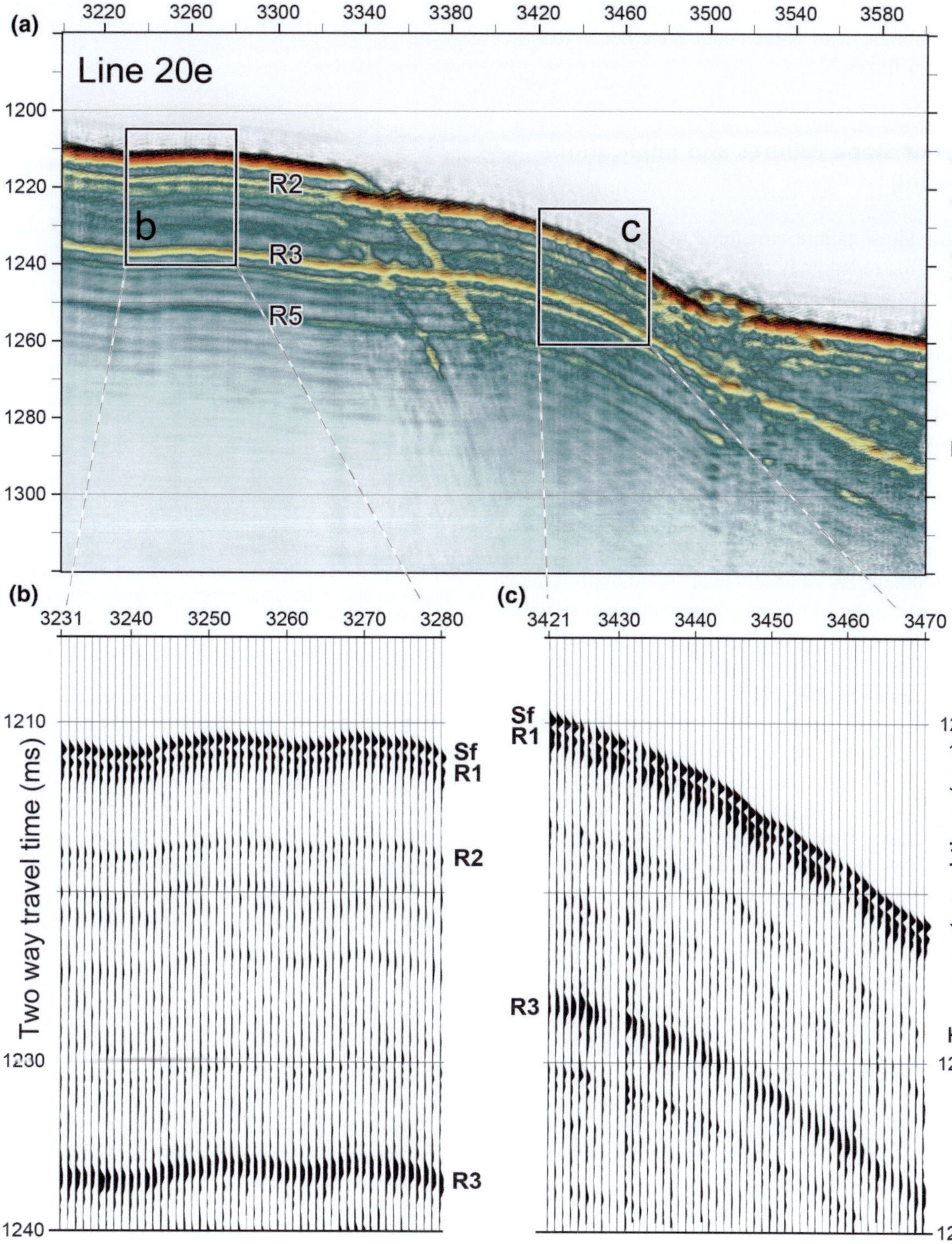

Fig. 11 Colour-coded amplitude (**a**) and trace displays (**b**) and (**c**) of the Chirp data in Fig. 7b (Sf: seafloor; other reflections as in Fig. 8)

by this re-suspension mechanism may have occurred over distances of ~20 km.

Alternatively, the variation in R3-seabed thickness with depth in Fig. 10c arises instead from a spatial variation in aeolian input. There are two areas of thicker sediment in Fig. 9, one in the northwest of the survey area and one to the far southeast. The more reliable Chirp interpretations in the northwest area may lie closer to a westerly wind jet from the African coast (Fig. 1a) (Jiang et al. 2009). Collecting cores here could resolve this issue, from the relative aeolian and biogenic composition. More generally, extensive mapping with Chirp combined with cores could do a better job of resolving aeolian fluxes.

6.2 Origin of Slope Failures and Implications for Coring

In none of the slope failure structures identified does the basal surface of the failed body cross strata. Rather these are all likely strata-parallel failures, probably because of varied geotechnical properties between the layers as inferred for failed sediments elsewhere (Schwab et al. 1988). Although the failure events are not easily dated, those for the westerly sites shown in Fig. 7 all seem to have occurred at about the same time and between the dates represented by the R3 reflection and the LGM (given the extended seabed in each case) and probably prior to R2 (given the reflections overlying the debrites). Steepening because of the on-going halokinetics (Mitchell et al. 2010) is possible, although steepness is unlikely an isolated factor as similarly steep deposits have not failed. Overpressure is unlikely to be a cause in such shallow slowly depositing carbonates. We suggest that seismically triggered failure would explain their isolated occurrence and similar age.

To assess the shear strengths of the sediments implied by their failure, we have used the pseudo-static model of Morgenstern (1967). The approach is unsophisticated, but can be useful if we use it to estimate extremes of properties (Mitchell and Searle 1998). Here, ignoring effects of cyclic loading implies that we under-estimate sediment undrained shear strength (S_u, kPa). Using a wet density of 1.80 g cm^{-3} (an average value derived from a similarly hemipelagic sediment core KL11 (Mitchell et al. 2015)) and the depth and gradients of the landslide scars, we obtained the variation of potential S_u with peak horizontal ground acceleration (k, fraction of g). Where the scars vary in gradient, we used maximum and minimum values. The results are shown in Fig. 12.

The northern Red Sea experiences remarkably low rates of seismicity (Al-Almadi et al. 2014; Al-Amri 1995; El-Isa and Al. Shanti 1989; Fairhead and Girdler 1970; Mitchell and Stewart 2018; Zahran et al. 2016). Consequently, the peak ground acceleration was estimated from historical seismicity by Stewart (2007) to be only $\sim$10 cm s^{-2} in the area of Thetis Deep. This value is plotted on Fig. 12. If this acceleration were representative of the longer term over which these failures occurred, it would imply S_u< 2 kPa.

Such low values are more compatible with weak clay-rich sediments than carbonates such as here. For example, Johnson et al. (1977) found S_u= 5–25 kPa for calcareous ooze from the Ontong Java Plateau. Kenter and Schlager (1989) reviewed more than 2000 measurements of S_u from

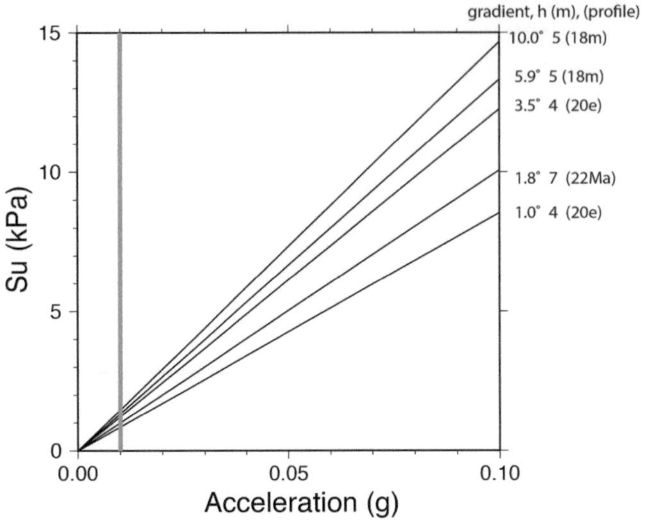

Fig. 12 Assessment of sediment undrained shear strength (S_u) and seismic horizontal ground acceleration implied by slope failure geometries using a pseudo-static analysis. Grey bar is peak ground acceleration for 950-year return period for this area according to historical seismicity analysis of Stewart (2007). Gradient and thickness (h) values to right correspond with slope failures marked ("*") in the Chirp profiles in Fig. 7

the literature and their own values and found typically S_u> 15 kPa below 1.5 m depth in carbonates (%$CaCO_3$ > 60%) and S_u> 10 kPa below 3.0 m depth in marls (30% < $CaCO_3$ < 60%). They found that S_u correlated strongly with carbonate content. Given that sediment cores from the Red Sea deep-waters typically contain inorganically precipitated aragonite cemented layers (Milliman et al. 1969) and are generally high in carbonate (Stoffers and Ross 1974), the failed sediments are more likely to be strong, such as these other carbonates.

We therefore suggest that S_u> 5.0 kPa and that these slope failures imply a horizontal ground acceleration of >0.05 g. Strong motion data compiled by Westaway and Smith (1989) for normal fault earthquakes suggest that earthquakes of M = 4.0–4.5 can generate >0.05 g peak horizontal accelerations within a few km of the source, but accelerations are ≪0.05 g at 100 km distance from source. The localized slope failures here therefore imply the occurrence of an isolated M = 4.0–4.5 event, possibly involving slip on the westerly border fault of the axial rift beneath the areas of Chirp profiles TL-22 Ma, 18 m and 20e. More generally, the acquisition of more extensive geophysical data along with geotechnical data from cores could potentially be used to carry out a regional paleoseismological analysis of the Red Sea.

The presence of slope failures has implications for coring and paleoceanographic studies based on them. These slope failure features occur in a minority of the Chirp data. Mitchell et al. (2015) summarized the stratigraphy of seven of the long piston cores collected in the Red Sea, which contain all of the isotope stages and no evidence for removed intervals. Nevertheless, the possibility of encountering missing stratigraphy while coring suggests that, for sediment dating by oxygen isotopes to be rigorous, it needs to be corroborated with other methods. Removal of layers by slope failure also opens up the possibility of obtaining samples extending beyond 500 ka by long coring in such areas.

7 Conclusions

There appears little correspondence between the thickness of Plio-Pleistocene sediments overlying the Miocene evaporites in deep-water areas of the Red Sea with the pattern of likely sedimentary terrigenous inputs (wind jets and drainage basin outlets). As the spatial distributions of drainage basins and wind gaps has likely not changed much over the past 5 my, a short interval compared with the 20 my history of continental rifting here, we suggest that wind-driven surface currents and boundary currents have caused significant along-rift transport of the finer particles before deposition, while coarser sediment has deposited closer to the coasts.

The 3D seismic reflection data from offshore Egypt reveal sediments deposited in depressions that have developed by halokinetics. Those depressions are aligned with salt walls that lie sub-parallel with the coast and do not obviously relate to the distribution of drainage basins on the adjacent African coast. Within the deeper parts of the sediments, the data suggest the presence of contourites, implying that significant along-basin redistribution of particles occurred in the earlier Pliocene.

High-resolution Chirp data of the later Pleistocene sediments show a trend of increasing sedimentation rates with depth (by 25% on average from 763 to 1211 m). Given the normally sluggish deep waters and lack of any current-generated bedforms in the data, this increase may have been caused by repeated re-suspension of bottom sediments and their re-deposition under gravity. Alternatively, the variation could reflect enhanced input of sediment to the northwest of the survey area, such as aeolian input via a wind gap in the Sudanese Red Sea hills.

Slope failure structures are uncommon in these data, but reflection sequences around slope failure scars and debrites in three Chirp profiles to the west of the axial deep suggest that they occurred at roughly the same time. Based on reflections encompassing the slide scars and deposits, they occurred after the sediment ages corresponding to R3 but before that of R2 (suggesting between MIS 6 and 12, if the chronology of Mitchell et al. (2015) is correct). Using a pseudo-static slope stability model, failure could have involved very weak sediments ($S_u < 2$ kPa) if ground accelerations derived from seismicity of 0.01 g here (Stewart 2007) are representative. However, such low shear strengths are not commonly found in carbonate-rich sediments such as these, so we suggest that these failed deposits instead imply the occurrence of one or more $M > 4.0$ earthquakes under the westerly side of the deep, perhaps caused by movement on an underlying rift border normal fault. The analysis suggests that more extensive high-resolution geophysical data combined with geotechnical data from cores could be used for paleoseismological analysis of the Red Sea.

Acknowledgements Thanks to Rose Anne Weissel for help in locating and scanning the RV *Conrad* data used in this study. Permissions of the governments of Egypt, Sudan and Saudi Arabia to carry out the surveys on RVs *Urania, Poseidon* and *Pelagia* contributing to this study are gratefully acknowledged. The *Urania* cruise was funded by the Consiglio Nazionale delle Ricerche under project LEC-EMA21F of the European Science Foundation programme EUROMARGINS (contract ERAS-CT-2003-980409 of the European Commission, DG Research FP6). The *Poseidon* and *Pelagia* cruises were part of the Jeddah Transect Project, a collaboration between King Abdulaziz University and Helmholtz-Center for Ocean Research GEOMAR Kiel funded by King Abdulaziz University (KAU) Jeddah, Saudi Arabia, under grant No. T-065/430-DSR. Houshuo Jiang is thanked for supplying a copy of his model output used for Fig. 1. Rickbir Bahia is thanked for help in extracting the catchment boundaries in Fig. 1b. We also thank Bill Bosworth for allowing access to the 3D seismic data used in generating Figs. 2, 3, 4 and 5. Figures in this article were created with the "GMT" and "PLOTMAP" software systems (Wessel and Smith 1991; Ligi and Bortoluzzi 1989). We thank four anonymous reviewers for comments that led to a significant improvement in this chapter and the editors and Saudi Geological Survey for organizing the publication of this book on the Red Sea.

References

Al-Almadi K, Al-Amri A, See L (2014) A spatial statistical analysis of the occurrence of earthquakes along the Red Sea floor spreading: clusters of seismicity. Arab J Geosci 7:2893–2904

Al-Amri AMS (1995) Recent seismic activity in the northern Red Sea. J Geodyn 20:243–253

Arrhenius GA (1963) Pelagic sediments. In: Hill MN (ed) The Sea. Wiley-Interscience, New York, pp 655–727

Becker JJ, Sandwell DT, Smith WHF, Braud J, Binder B, Depner J, Fabre D, Factor J, Ingalls S, Kim S-H, Ladner R, Marks K, Nelson S, Pharaoh A, Trimmer R, Von Rosenberg J, Wallace G, Weatherall P (2009) Global bathymetry and elevation data at 30 arc seconds resolution: SRTM30_PLUS. Mar Geod 32:355–371

Bintanja R, van de Wal RSW (2008) North American ice-sheet dynamics and the onset of 100,000-year glacial cycles. Nature 454:869–872

Bosworth W, Huchon P, McClay K (2005) The Red Sea and Gulf of Aden basins. J Afr Earth Sci 43:334–378

Bower AS, Farrar JT (2015) Air-sea interaction and horizontal circulation in the Red Sea. In: Rasul NMA, Stewart ICF (eds) The Red Sea: the formation, morphology, oceanography and environment of a young ocean basin. Springer Earth System Sciences, Heidelberg, pp 329–342

Chen C, Li R, Pratt L, Limeburner R, Beardsley RC, Bower A, Jiang H, Abualnaja Y, Xu Q, Lin H, Liu X, Lan J, Kim T (2014) Process

modeling studies of physical mechanisms of the formation of an anticyclonic eddy in the central Red Sea. J Geophys Res 119:1445–1464. https://doi.org/10.1002/2013JC009351
Clifford M, Horton C, Schmitz J, Kantha LH (1997) An oceanographic nowcast/forecast system for the Red Sea. J Geophys Res 102:25101–25122
Cochran JR (2005) Northern Red Sea: nucleation of an oceanic spreading center within a continental rift. Geochem Geophys Geosyst 6. Paper Q03006, https://doi.org/10.01029/02004GC000826
Cochran JR, Karner GD (2007) Constraints on the deformation and rupturing of continental lithosphere of the Red Sea: the transition from rifting to drifting. In: Karner GD, Manatschal G, Pinheiro LM (eds) Imaging, mapping and modelling continental lithosphere extension and breakup. Geol Soc London, Spec Publ 282, pp 265–289
Crossley R, Watkins C, Raven M, Cripps D, Carnell A, Williams D (1992) The sedimentary evolution of the Red Sea and Gulf of Aden. J Petrol Geol 15:157–172
Davison I, Al-Kadashi M, Al-Khirbash S, Al-Subbary AK, Baker J, Blakey S, Bosence D, Dart C, Heaton R, McClay K, Menzies M, Nichols G, Owen L, Yellend A (1994) Geological evolution of the southeastern Red Sea Rift margin, Republic of Yemen. Geol Soc Am Bull 106:1474–1493
Davison I, Anderson L, Nuttall P (2012) Salt deposition, loading and gravity drainage in the Campos and Santos salt basins. In: Alsop GI, Archer SG, Hartley AJ, Grant NT, Hodgkinson R (eds) Salt tectonics, sediments and prospectivity. Geol Soc London, Spec Publ 363, pp 159–173
Egloff F, Rihm R, Makris J, Izzeldin YA, Bobsien M, Meier K, Junge P, Noman T, Warsi W (1991) Contrasting structural styles of the eastern and western margins of the southern Red Sea: the 1988 SONNE experiment. Tectonophys 198:329–353
Ehrhardt A, Hübscher C (2015) The northern Red Sea in transition from rifting to drifting—lessons learned from ocean deeps. In: Rasul NMA, Stewart ICF (eds) The Red Sea: the formation, morphology, oceanography and environment of a young ocean basin. Springer Earth System Sciences, Heidelberg, pp 99–121
El-Isa ZH, Al Shanti A (1989) Seismicity and tectonics of the Red Sea and western Arabia. Geophys J 97:449–457
Elderfield H, Ferretti P, Greaves M, Crowhurst S, McCave IN, Hodell D, Piotrowski AM (2012) Evolution of ocean temperature and ice volume through the mid-Pleistocene climate transition. Science 337:704–709
Fairhead JD, Girdler RW (1970) The seismicity of the Red Sea, Gulf of Aden and Afar triangle. Phil Trans Royal Soc Lond A267:49–74
Fenton M, Geiselhart S, Rohling EJ, Hemleben C (2000) Aplanktonic zones in the Red Sea. Mar Micropal 40:277–294
Flood RD (1988) A lee wave model for deep-sea mudwave activity. Deep-Sea Res 35:973–983
Gass IG (1970) The evolution of volcanism in the junction area of the Red Sea, Gulf of Aden and Ethiopian rift. Phil Trans Roy Soc Lond A267:369–382
Gevirtz JL, Friedman GM (1966) Deep-sea carbonate sediments of the Red Sea and their implications on marine lithification. J Sediment Petrol 36:143–151
Gordon G, Hansen B, Scott J, Hirst C, Graham R, Grow T, Spedding A, Fairhead S, Fullarton L, Griffin D (2010) The hydrocarbon prospectivity of the Egyptian North Red Sea basin. In: Vining BA, Pickering SC (eds) Petroleum geology: from mature basins to new frontiers. Proceedings of the 7th Petroleum Geology Conference. Geol Soc London, pp 783–789. https://doi.org/10.1144/0070783
Heaton RC, Jackson MPA, Bamahmoud M, Nani ASO (1995) Superimposed Neogene extension, contraction, and salt canopy emplacement in the Yemeni Red Sea. In: Jackson MPA, Roberts DG, Snelson S (eds) Salt tectonics: a global perspective. Am Assoc Petrol Geol, pp 333–351
Hemleben C, Meischner D, Zahn R, Almogi-Labin A, Erlenkeuser H, Hiller B (1996) Three hundred eighty thousand year long stable isotope and faunal records from the Red Sea: influence of global sea level change on hydrography. Paleoceanography 11:147–156
Hughes GW, Beydoun ZR (1992) The Red Sea—Gulf of Aden: biostratigraphy, lithostratigraphy and palaeoenvironments. J Petrol Geol 15:135–156
Hutchinson RW, Engels GG (1972) Tectonic evolution in the southern Red Sea and its possible significance to older rifted continental margins. Geol Soc Am Bull 83:2989–3002
Izzeldin AY (1987) Seismic, gravity and magnetic surveys in the central part of the Red Sea: their interpretation and implications for the structure and evolution of the Red Sea. Tectonophysics 143:269–306
Jiang H, Farrar JT, Beardsley RC, Chen R, Chen C (2009) Zonal surface wind jets across the Red Sea due to mountain gap forcing along both sides of the Red Sea. Geophys Res Lett 36. Article L19605. https://doi.org/10.11029/12009GL040008
Johnson TC, Hamilton EL, Berger WH (1977) Physical properties of calcareous ooze: control by dissolution at depth. Mar Geol 24:259–277
Kenter JAM, Schlager W (1989) A comparison of shear strength in calcareous and siliclastic marine sediments. Mar Geol 88:145–152
Lehner B, Verdin K, Jarvis A (2008) New global hydrography derived from spaceborne elevation data. EOS Trans Am Geophys Union 89:93–94
Ligi M, Bortoluzzi G (1989) PLOTMAP: geohysical and geological applications of good standard quality cartographic software. Comput Geosc 15:519–585
Luyendyk BP (1970) Origin and history of abyssal hills in the northeast Pacific Ocean. Geol Soc Am Bull 81:2237–2260
Macgregor DS (2012) The development of the Nile drainage system: integration of onshore and offshore evidence. Petrol Geosci 18:417–431
Maillard C, Soliman G (1986) Hydrography of the Red Sea and exchanges with the Indian Ocean in summer. Oceanol Acta 9:249–269
Marks NS (1981) Sedimentation on New Ocean crust: the Mid-Atlantic Ridge at 37°N. Mar Geol 43:65–82
Mart Y, Ross DA (1987) Post-Miocene rifting and diapirism in the northern Red Sea. Mar Geol 74:173–190
Martinez F, Cochran JR (1988) Structure and tectonics in the northern Red Sea: catching a continental margin between rifting and drifting. Tectonophys 150:1–32
McCave IN (2005) Deposition from suspension. In: Selley RC, Cocks LRM, Malone MJ (eds) Encyclopedia of geology. Elsevier, Oxford, pp 8–17
Miller PM, Barakat H (1988) Geology of the safaga concession, northern Red Sea. Egypt Tectonophy 153:123–136
Milliman JD, Ross DA, Ku T-L (1969) Precipitation and lithification of deep-sea carbonates in the Red Sea. J Sed Petrol 39:724–736
Mitchell DJW, Allen RB, Salama W, Abouzakm A (1992) Tectonostratigraphic framework and hydrocarbon potential of the Red Sea. J Petrol Geol 15:187–210
Mitchell NC (1993) A model for attenuation of backscatter due to sediment accumulations and its application to determine sediment thickness with GLORIA sidescan sonar. J Geophys Res 98:22477–22493
Mitchell NC (1995) Diffusion transport model for pelagic sediments on the Mid-Atlantic Ridge. J Geophys Res 100(B10):19,991–920,009
Mitchell NC (2016) Comment on: "The spatial extent of the Deep Western Boundary Current into the Bounty Trough: new evidence

from parasound sub-bottom profiling" by Horn and Uenzelmann Neben. Marine Geophysical Research 37:371–374

Mitchell NC, Huthnance JM (2013) Geomorphological and geochemical evidence (^{230}Th anomalies) for cross-equatorial currents in the central Pacific. Deep-Sea Res I 78:24–41

Mitchell NC, Ligi M, Farrante V, Bonatti E, Rutter E (2010) Submarine salt flows in the central Red Sea. Geol Soc Am Bull 122:701–713

Mitchell NC, Ligi M, Feldens P, Hübscher C (2017) Deformation of a young salt giant: regional topography of the Red Sea Miocene evaporites. Basin Res 29:352–369

Mitchell NC, Ligi M, Rohling EJ (2015) Red Sea isolation history suggested by Plio-Pleistocene seismic reflection sequences. Earth Planet Sci Lett 430:387–397

Mitchell NC, Searle RC (1998) Fault scarp statistics at the Galapagos spreading centre from deep tow data. Mar Geophys Res 20:183–193

Mitchell NC, Stewart ICF (2018) The modest seismicity of the northern Red Sea rift. Geophys J Int 214(3):1507–1523

Morgenstern NR (1967) Submarine slumping and the initiation of turbidity currents. In: Richards AF (ed) Marine Geotechnique. University of Illinois Press, Urbana, pp 189–210

Nicholls JF, Toumi R, Stenchikov G (2015) Effects of unsteady mountain-gap winds on eddies in the Red Sea. Atm Sci Lett 16:279–284

Quadfasel D, Baudner H (1993) Gyre-scale circulation cells in the Red Sea. Oceanol Acta 16:221–229

Richter H, Makris J, Rihm R (1991) Geophysical observations offshore Saudi Arabia: seismic and magnetic observations. Tectonophysics 198:297–310

Roberts AP, Rohling EJ, Grant KM, Larrasoaña JC, Liu Q (2011) Atmospheric dust variability from Arabia and China over the last 500,000 years. Quat Sci Rev 30:3537–3541

Rohling, Grant K, Hemleben C, Kucera M, Roberts AP, Schmeltzer I, Schulz H, Siccha M, Siddall M, Trommer G (2008) New constraints on the timing of sea level fluctuations during early to middle marine isotope stage 3. Paleocean 23. Article PA3219. https://doi.org/10.1029/2008PA001617

Rohling EJ, Fenton M, Jorissen FJ, Bertrand P, Ganssen G, Caulet JP (1998) Magnitudes of sea-level lowstands of the past 500,000 years. Nature 394:162–165

Rohling EJ, Grant K, Bolshaw M, Roberts AP, Siddall M, Hemleben C, Kucera M (2009) Antarctic temperature and global sea level closely coupled over the past five glacial cycles. Nat Geosci 2:500–504

Ross DA, Schlee J (1973) Shallow structure and geologic development of the southern Red Sea. Geol Soc Am Bull 84:3827–3848

Savoyat E, Shiferaw A, Balcha T (1989) Petroleum exploration in the Ethiopian Red Sea. J Petrol Geol 12:187–204

Schwab WC, Lee HJ, Kayen RE, Quinterno PJ, Tate GB (1988) Erosion and slope instability on Horizon Guyot, Mid-Pacific mountains. Geo-Mar Lett 8:1–10

Seibold E, Futterer D (1982) Sediment dynamics on the northwest African continental margin. In: Scrutton RA, Talwani M (eds) The ocean floor. John Wiley, New York, pp 147–163

Sofianos SS, Johns EW (2003) An oceanic general circulation model (OGCM) investigation of the Red Sea circulation, 2. Three-dimensional circulation in the Red Sea. J Geophys Res 107. Paper 3066. https://doi.org/10.1029/2001JC001185

Sofianos SS, Johns EW (2007) Observations of the summer Red Sea circulation. J Geophys Res 112. Paper C06025. https://doi.org/10.01029/02006JC003886

Steckler MS, Omar GI (1994) Controls on erosional retreat of the uplifted rift flanks at the Gulf of Suez and northern Red Sea. J Geophys Res 99:12159–12173

Stewart ICF (2007) Earthquake risk in western Saudi Arabia and the Red Sea from seismic moment. Saudi Geological Survey, Jeddah, Technical report SGS-TR-2007-4, 41 p

Stoffers P, Kühn R (1974) Red Sea evaporites: a petrographic and geochemical study. In: Whitmarsh RB, Weser OE, Ross DA et al. (eds) Initial reports of the deep sea drilling project, vol 23. US Govt Printing Office, Washington, DC, pp 821–847

Stoffers P, Ross DA (1974) Sedimentary history of the Red Sea. In: Whitmarsh RB, Weser OE, Ross DA et al (eds) Initial reports of the deep sea drilling project, vol 23. US Govt Printing Office, Washington, DC, pp 849–865

Tominaga M, Lyle M, Mitchell NC (2011) Seismic interpretation of pelagic sedimentation regimes in the 18–53 Ma eastern equatorial Pacific: basin-scale sedimentation and infilling of abyssal valleys. Geochem Geophys Geosyst 12. Paper Q03004. https://doi.org/10.01029/02010GC003347

Wessel P, Smith WHF (1991) Free software helps map and display data. EOS Trans Am Geophys Union 72:441

Westaway R, Smith RB (1989) Strong ground motion in normal-faulting earthquakes. Geophys J 96:529–559

Whitmarsh RB, Weser OE, Ross DA (1974) Initial reports of the deep sea drilling project, 23B. US Govt Printing Office, Washington, DC

Yao F, Hoteit I, Pratt LJ, Bower AS, Zhai P, Köhl A, Gopalakrishnan G (2014a) Seasonal overturning circulation in the Red Sea: 1. Model validation and summer circulation. J Geophys Res 119:2238–2262. https://doi.org/10.1002/2013JC009004

Yao F, Hoteit I, Pratt LJ, Bower AS, Köhl A, Gopalakrishnan G, Rivas D (2014b) Seasonal overturning circulation in the Red Sea: 2. Winter circulation. J Geophys Res 119:2263–2289. https://doi.org/10.1002/2013JC009331

Zahran HM, Sokolov V, Roobol MJ, Stewart ICF, El-Hadidy Youssef S, Hadidy M E (2016) On the development of a seismic source zonation model for seismic hazard assessment in western Saudi Arabia. J Seismol 20:747–769

Zhai P, Bower A (2013) The response of the Red Sea to a strong wind jet near the Tokar Gap in summer. J Geophys Res 118:422–434. https://doi.org/10.1029/2012JC008444

Pleistocene Coral Reef Terraces on the Saudi Arabian Side of the Gulf of Aqaba, Red Sea

Marco Taviani, Paolo Montagna, Najeeb M. A. Rasul, Lorenzo Angeletti, and William Bosworth

Abstract

A major geomorphic feature of the coastal Red Sea region is represented by Pleistocene raised marine terraces that occur on both sides of the Gulf of Aqaba. Those bordering the Saudi Arabian sector have received little attention thus far, and are comparatively less known than their counterparts in the Sinai sector of the Gulf and in the Red Sea. As is the rule in the Red Sea region, the best developed marine terrace system is reefal and pertains to the last interglacial (Marine Isotope Stage 5e = MIS5e, ~125 ka BP), although older Pleistocene terraces also occur. All such deposits are very fossiliferous and most carbonates are relatively unaltered, providing suitable material for geochronological purposes. Syndepositional marine botryoidal aragonite cements have been identified infilling vugs in the host bedrock at some sites. In some respect, the MIS5e deposits are unique, reflecting the structurally-controlled bedrock geology and the Gulf's topography. The Gulf of Aqaba is rather narrow and characterized by steep and precipitous topography along its flanks. Coastal marine deposits commonly plaster the crystalline Arabian basement which faces the present seashore, extending from the Jordan border to almost two thirds of the coastal strip. Terraces sitting on this basement have been tectonically uplifted to considerable altitudes (up to 26 m) over the present mean sea level (m.s.l.). The bulk of the marine deposits represent upper fore-reef to beach settings, with better developed back-reef to lagoonal facies only preserved in those favourable conditions (wadi valleys) where sufficient accommodation space was available during the MIS5e to allow inland marine expansion. This is observed in the north at Al Wasel, and ~14 km south of Ra's Suwayhil as Saghir. The terraces further to the south lie instead over a more recent bedrock, including Miocene sedimentary strata. Here MIS5e deposits are found close to standard altitudes between ~4–8 m above present m.s.l., and preserve shallow reefal habitats, as seen at Ash Shaykh Humayd. A rare example of a putative MIS5e salina-mangal complex has been identified in the area of Ra's Suwayhil as Saghir at ~23 m above m.s.l.

M. Taviani · P. Montagna · L. Angeletti (✉)
Istituto di Scienze Marine (ISMAR-CNR), Via Gobetti 101, 40129 Bologna, Italy
e-mail: lorenzo.angeletti@bo.ismar.cnr.it

M. Taviani
e-mail: marco.taviani@bo.ismar.cnr.it

M. Taviani
Biology Department, Woods Hole Oceanographic Institution, 266 Woods Hole Road, Woods Hole, MA 02543, USA

M. Taviani
Stazione Zoologica Anton Dohrn, Villa Comunale, 80121 Naples, Italy

P. Montagna
Lamont-Doherty Earth Observatory, Columbia University, 61 Route 9W, Palisades, NY 10964, USA

N. M. A. Rasul
Center for Marine Geology, Saudi Geological Survey, Jeddah, Saudi Arabia

W. Bosworth
Apache Egypt Companies, 11 Street 281, New Maadi, Cairo, Egypt

1 Introduction

Raised terraces representing prevalent former coral reef systems stretch discontinuously the entire length of the Red Sea, including islands, and the Gulf Aden (e.g., Hume and Little 1928; Dreyfuss 1931; Sandford and Arkell 1928; Faure et al. 1973; Faure 1975; Hoang et al. 1974, 1996; Taviani et al. 1986; Dullo 1986, 1990; Plaziat et al. 1989, 1995, 1998, 2008; Hoang and Taviani 1991; Bosworth and Taviani 1996; El-Asmar 1997; El-Sorogy 1997; Taviani 1998a; Dawood et al. 2013; Bantan et al. 2015; Hamed 2015; Mansour and Madkour 2015; Hamed et al. 2016). An extensive flight of Pleistocene marine terraces equally borders both sides of the narrow and deep Gulf of

N. M. A. Rasul and I. C. F. Stewart (eds.), *Geological Setting, Palaeoenvironment and Archaeology of the Red Sea*,
https://doi.org/10.1007/978-3-319-99408-6_16

Aqaba, along the Sinai Peninsula and Jordan-Saudi Arabian coastline, and also occurs on Tiran Island (Hume 1906; Schick 1958; Goldberg and Yaron 1978; Gvirtzman et al. 1977, 1992; Gvirtzman and Buchbinder 1978; Al-Sayari et al. 1984; Dullo 1984, 1990; Al-Rifaiy and Cherif 1988; Strasser et al. 1992; Gvirtzman 1994; Bosworth and Taviani 1996; Strasser and Strohmenger 1997; Alhejoj et al. 2016; Manaa et al. 2016; Bosworth et al. 2017, and this volume). Furthermore, still-submerged Quaternary terraced reefs have been identified in the Gulf of Aqaba (Gvirtzman 1994; Hartman et al. 2015).

This type of coastal feature testifies to sea-level fluctuations during the Pleistocene (e.g., Aharon and Chappell 1986; Montaggioni and Braithwaite 2009; Lambeck et al. 2011; Camoin and Webster 2015; Dutton et al. 2015), but their current altitude with respect to the present mean sea level (m.s.l.) also responds to tectonic movements, a case often documented in the Red Sea (Faure 1975; Goldberg and Yaron 1978; Faure et al. 1980; Dullo 1990; Hoang and Taviani, 1991; Bosworth and Taviani 1996; El-Asmar 1997; Bosence et al. 1998; Bosworth et al. 2017, and this volume).

The raised Pleistocene coral terraces of the Red Sea region are known to pertain to various interglacial times, as supported by information based on dating techniques (e.g., Berry et al. 1966; Veeh and Giegengack 1970; Conforto et al. 1976; Goldberg and Yaron 1978; Andres and Radtke 1988; Dullo 1990; Gvirtzman et al. 1992; Choukri et al. 1995; Plaziat et al. 1998; Taviani 1998a; Dawood et al. 2013: but see Plaziat et al. 2008 warning about potential age errors and caveats linked to coral diagenesis). Most of them formed during the Marine Isotope Stage 5e (MIS5e) at about 125 ka BP (Dabbagh et al. 1984; Dullo 1990; Hoang and Taviani 1991; Gvirtzman et al. 1992; El-Moursi et al. 1994; Bosworth and Taviani 1996; Hoang et al. 1996; El-Asmar 1997; Plaziat et al. 1998, 2008; Walter et al. 2000; Scholz et al. 2004; Manaa et al. 2016; Bosworth et al. 2017; Casazza 2017; Al-Mikhlafi et al. 2018).

The Red Sea fossil coral reef systems host a remarkable paleobiological legacy which is useful to reconstruct former environments, to recognize past biodiversity, to unravel biogeographic connections and disclose paleoclimatic events (e.g., Newton 1900; Hall and Standen 1907; Cox 1929; Brighton 1931; Nardini 1937; Selli 1973; Dollfus and Roman 1981; Borri et al. 1982; Dullo 1990; Marchesan and Taviani 1994; Taviani 1994, 1998b; Ragani 1997; Plaziat et al. 1998; Mewis and Kiessling 2013; El-Sorogy et al. 2014; Bantan et al. 2015; Hamed 2015; Alexandroff et al. 2016; Angeletti et al., this volume). The marine terraces also provide pristine biogenic carbonates that are the prime material for dating purposes (Ivanovich and Harmon 1992) and are critical material for geochemically-based paleoclimate reconstructions (e.g., El-Asmar 1997; Felis et al. 2000, 2004; McCulloch and Esat 2000; Rimbu et al. 2001).

As reported above, MIS5e and older coral terraces have previously been studied and dated in the Gulf of Aqaba, but observations were not sufficient to provide a detailed assessment of Pleistocene vertical movements, nor were the coral terrace measurements combined with other types of structural data. Therefore, an expedition was conducted in winter 2013 to study the coastal terraces located on the Saudi Arabian part of the Gulf of Aqaba. This survey covered more than four fifths of the entire length of the Gulf's eastern coastal strip. One of the key objectives of this expedition was to obtain enough control points to determine where uplift is occurring along the Saudi Arabian margin of the Gulf of Aqaba, what the rate of that uplift (or subsidence) is, and what faulting style might be producing these vertical changes. The tectonic interpretation of the specific coral elevations calibrated by geological and radiometric data is discussed in Bosworth et al. (2017, and this volume), whilst preliminary results on the superb MIS5e fauna is reported by Angeletti et al. (this volume).

The goal of the present chapter is set at providing introductory information on the coastal terraces outcropping along the Saudi Arabian Gulf of Aqaba. These Pleistocene features surely deserve to be considered in the future for better stratigraphic assessment and accomplishing accurate facies reconstructions, inclusive of the pre-MIS5e carbonates, as well as for their paleoclimatic potential.

2 Materials and Methods

The winter 2013 expedition (December 6–15, 2013) was conducted by an international geological team under the leadership of the Saudi Geological Survey (SGS) in Jeddah, which supplied vehicles, full field work logistics and technical assistance (Bosworth et al. 2013). In particular, the SGS Lidar survey group has provided geomorphologic/topographic data for most stations that were surveyed. We explored the Saudi Arabian coastal strip of the Gulf of Aqaba from the border with the Kingdom of Jordan in the north to the Straits of Tiran in the south, a distance of about 140 km. Cursory observations were recorded at 27 principal stations along the coastline (Fig. 1, Table 1). In order to obtain a better appreciation of the Pleistocene reef systems, an exploratory SCUBA diving survey was also conducted for comparative purposes on a modern fringing reef 2.3 km south of Al Wasel (in front of St. 16).

Qualitative geo-paleontological sampling was conducted at most sites and the obtained material later shipped to ISMAR-CNR Bologna to be examined in the laboratory. Corals and other carbonate material (e.g., botryoidal aragonite) suitable for dating purposes by the U-series method were collected and elevations measured in most cases by Lidar survey. Uranium and thorium isotopes were analyzed

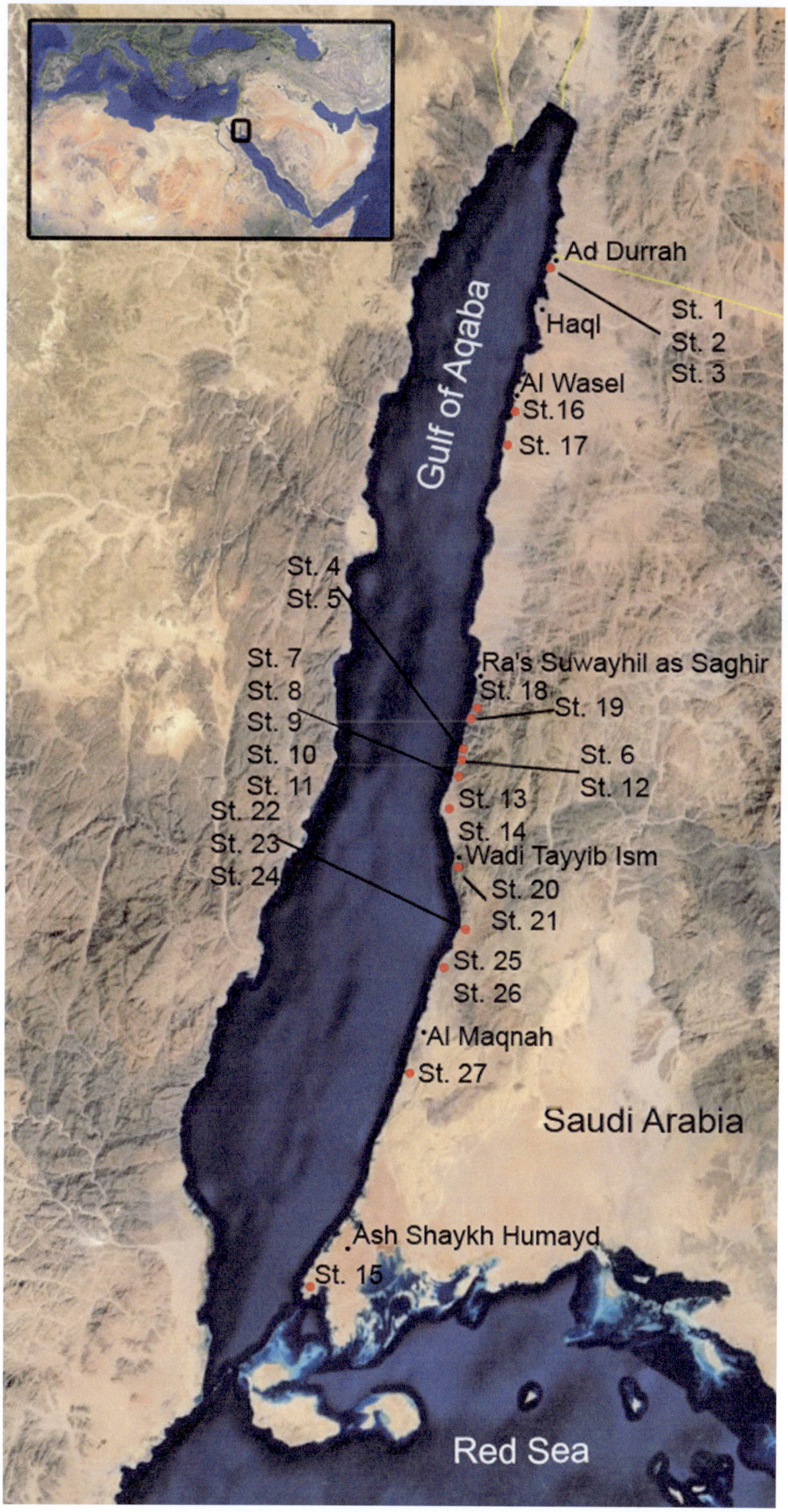

Fig. 1 Map of the Gulf of Aqaba showing the location of the stations discussed in the text (inset shows the regional geographic position of the Gulf of Aqaba)

Table 1 Station data of sites discussed in the text, indicating latitude and longitude, and terrace elevations above present m.s.l.; (*) marks elevation of terrace crests measured by Lidar. Note that station positions and altitudes correspond to those reported by Angeletti et al. (this volume); however, station numbers have been coded differently in Bosworth et al. (2017, and this volume), although from the same 2013 expedition

Station name	Latitude N	Longitude E	Elevation (m)
1*	29°21′19.60″	34°57′26.57″	11–13
2*	29°21′02.13″	34°57′23.09″	11
3*	29°20′19.18″	34°56′57.21″	14
4*	28°47′39.27″	34°49′33.02″	20
5*	28°46′46.89″	34°49′10.33″	19–23
6	28°45′29.59″	34°48′46.18″	23–25
7	28°42′36.15″	34°47′49.91″	22
8	28°42′30.28″	34°47′46.67″	22
9	28°42′24.91″	34°47′44.67″	22
10	28°42′26.44″	34°47′48.29″	22
11*	28°42′45.59″	34°47′52.91″	23
12*	28°46′06.37″	34°49′00.37″	25
13*	28°40′41.83″	34°46′52.21″	24
14*	28°40′32.74″	34°46′46.30″	19
15*	28°05′47.43″	34°34′21.16″	9
16*	29°11′22.95″	34°54′01.72″	19
17	29°11′09.33″	34°54′03.97″	19
18	28°48′36.63″	34°49′52.80″	19–26
19	28°48′32.05″	34°49′52.11″	19–27
20*	28°35′06.83″	34°47′35.14″	25
21	28°35′19.96″	34°47′32.39″	25
22*	28°30′23.94″	34°47′44.18″	18
23	28°30′17.31″	34°47′37.17″	14–22
24*	28°30′13.85″	34°47′39.01″	22
25*	28°26′58.80″	34°45′36.60″	26
26	28°26′23.32″	34°45′20.61″	26
27*	28°20′41.92″	34°43′09.55″	17

at the Laboratoire des Sciences du Climat et de l'Environnement (Gif-sur-Yvette, France) following laboratory procedures reported in Pons-Branchu et al. (2014). The U-series analysis was conducted only on pristine fossil material, identified based on a detailed mineralogical and petrographic screening of the entire coral collection. The U-series ages are presented in Bosworth et al. (2017) and are also reported in Table 2.

3 Geo-Structural Setting

The basement complexes of the Gulf of Aqaba region were accreted during the complex Neoproterozoic Pan-African orogenies that spanned from ~1000 to 540 Ma (Gass 1977; Kröner 1979; Stern 1994; Stern and Johnson 2010; Johnson and Kattan 2012). Granites and lesser metavolcanic rock

Table 2 U/Th dated MIS5e corals and botryoidal aragonite from the Saudi Arabian Gulf of Aqaba

Station name	Age (ka)	Note
6	125.56 ± 1.48	Fossil coral
8	118.58 ± 1.45	Botryoidal aragonite
15A	121.73 ± 0.96	Fossil coral
15D	120.14 ± 1.11	Fossil coral
16	121.54 ± 0.84	Fossil coral
22	122.76 ± 1.96	Fossil coral
24	117.42 ± 1.23	Fossil coral

types predominated in this area. The region was subsequently covered by a veneer of sedimentary rocks during the Paleozoic and Mesozoic, dominated by coarse siliciclastic continental deposits (Tawadros 2001; Guiraud et al. 2005). Much of this stratigraphic section was subsequently eroded, particularly during the Cenozoic Red Sea rifting event which initiated in this area at ~23 Ma (Coleman 1993; Bosworth et al. 2005). The Red Sea opened northward into modern day Egypt, stalling in the Neotethyan/Mediterranean margin (Steckler and ten Brink 1986). This finished creation of the Arabian plate, which at that time included Sinai. During the Middle Miocene, the Gulf of Aqaba—Dead Sea transform margin formed, and the Gulf of Suez was largely abandoned as an active continental rift (Bayer et al. 1988; Bosworth and McClay 2001). The Gulf of Aqaba evolved as series of pull-apart basins along this sinistral strike-slip plate boundary that now separates the Sinai micro-plate from the Arabian plate (Ben-Avraham et al. 1979; Bartov et al. 1980). The style of deformation within the basin is therefore dominated by strike-slip faulting which is attested to by the observed seismicity (Alamri et al. 1991; Roobol et al. 1999; Hofstetter 2003).

This strike-slip faulting will not, in and of itself, produce the significant uplift of the crystalline basement seen along its margins. This uplift could be formed at either extensional or convergent steps along the plate boundary, as in classical pull-apart basins or restraining bends. However, as noted by Ben-Avraham and Zoback (1992), the basin-margin (footwall) uplift is distributed along much of the Gulf of Aqaba rather than simply focused at the long-recognized intra-gulf left-stepping faults. The Gulf of Aqaba is experiencing a significant amount of opening perpendicular to its length, resulting in the descriptor of a "leaky" transform (Ben-Avraham et al. 1979). A large percentage of the small to moderate magnitude earthquakes in the Gulf are actually dip-slip normal fault events with ~ WNW-ESE tension axes (Alamri et al. 1991; Roobol et al. 1999; Hofstetter 2003). This component of extensional faulting along the length of the Gulf of Aqaba can be driving the uplift and exhumation of the basement complexes along the basin margins (Bosworth et al. 2017, and this volume).

In the study area, faulting was observed in both the strata and crystalline basement rocks beneath the coral terraces and within the carbonate units of the terraces themselves. Where kinematic data were present these faults are essentially all dip-slip and extensional in character; a few strike-slip faults of unknown age were observed in the basement granites (Bosworth et al., this volume).

In addition to general uplift of the Gulf of Aqaba margins the extensional faulting is also causing minor local tilting of the terrace units. We did not observe any significant horizontal repetition of terrace outcrops that could be attributed to extensional faulting.

4 The Coral Reef Terraces

Not surprisingly, when considering that a great percentage of the coastal area in the northern-central part of the narrow and deep Gulf of Aqaba (Figs. 1 and 2a) is rimmed by granitoid hills intruded by dikes of the Arabian basement complex reaching up to the present coastline, the modern coral reefs in the Gulf of Aqaba are narrow fringing reefs attached to the bedrock (Fig. 2b, c) with little if any development of back-reef lagoonal habitats (Mergner and Schuhmacher 1974; Bouchon et al. 1981; Gabrié and Montaggioni 1982; Dullo 1990; Dullo and Montaggioni 1998; Al-Horani et al. 2006). A few exceptions in the Saudi Arabian sector of the Gulf are offered by coasts facing wadi valleys and the lowlands to the south in the Midyan area (Fig. 2d, e). These accommodation space constraints hold valid also for their Pleistocene counterparts. This common physiography of limited beach development, absence of lagoons and extremely narrow fringing reefs is well exemplified at 2.3 km south of Al Wasel. This site has, therefore, been examined also for its modern reef facies, from shore to fore-reef, to complement the Pleistocene analysis and improve the facies interpretation (Figs. 3 and 4).

The coastal terraces rimming the Arabian side of the Gulf of Aqaba belong to more than one generation of reefal growth (Dullo 1984, 1990; Manaa 2011; Alhejoj et al. 2016; our own observations). Although the large majority of the terraces pertain to the last interglacial (MIS5e), older reefal limestone, possibly belonging to MIS7 (Manaa 2011) and older (Dullo 1990), occurs at many sites at places serving as substrate to the later development of MIS5e reefs. The degree of diagenesis, including more pronounced lithification and advanced dissolution of carbonate allochems (Dullo 1984, 1990), easily permits distinguishing older coral limestones from their last interglacial counterparts (Fig. 5). In the Red Sea region, most corals predating MIS5e reef growth suffered advanced to complete alteration and replacement of original aragonite into calcite, and the same holds true for aragonitic molluscs and cements alike (Gvirtzman and Friedman 1977; Dullo 1984, 1986). The difficulty of finding un-recrystallized corals in the pre-MIS5e is at present a major obstacle to establishing a sound chronological frame for all coral terraces in this region. We have concentrated on the better preserved and datable MIS5e systems.

The MIS5e coral reefs grew over older Pleistocene reefal limestones, Arabian basement crystalline rocks, and Miocene sedimentary rocks (Fig. 6). They attain present altitudes from about 5 m up to about 26 m above mean sea level (m.s.l.), thus indicating that this sector of the Saudi Arabian coastline underwent remarkable but also differential uplift since that time. Here we do not comment further on such neotectonic aspects which are instead fully discussed in Bosworth et al. (2017, and this volume). Because of the substantial uplift, 'deeper' parts of the former coral reef system, seldom exposed in outcrop, are visible.

Fig. 2 Physiographic and geological constraints controlling coral reef development along the Saudi Arabian coast; **a** general view of the coast with alluvial plain, Pleistocene remnant of former reefs (arrow), modern sandy beach and fringing reef. Note in the background the mountain ranges of Sinai documenting the narrowness of the Gulf of Aqaba; **b** note the modest extension of the fringing reef hanging already at a few metres from the coastline in the deep waters of the Gulf of Aqaba, about 3 km north of Wadi Tayyib Ism; **c** the crystalline rocks of the Arabian basement reach up to the seashore along a considerable portion of the Saudi Arabian coast, about 4 km south of Ra's Suwayhil as Saghir; **d** lagoonal facies could in principle only develop on lowlands such as this wadi valley, prone to be invaded by the sea at times of relative high sea level, south of Ra's Suwayhil as Saghir; **e** general view of the southernmost tip of the Gulf, a lowland typified by Pleistocene reef growth in absence of uplifted bedrock, Ash Shaykh Humayd

5 Paleobiological Importance of the Pleistocene Coral Terraces

The Pleistocene terraces under scrutiny hold a remarkable fossil content of fundamental importance for reconstructing the former depositional and ecological environments and similarly for dating purposes. The pre-MIS5e carbonates have been by and large affected by diagenetic processes intense enough to dissolve in most cases any original aragonitic mineralogy. Calcitic fossils are often preserved (e.g., foraminifers, echinoids, pectinids). The MIS5e terraces are on the contrary little affected by diagenesis, leaving the aragonite record intact in many cases. This material is critical for identifying proper dating material, and, in perspective, may serve for paleoclimatic-oriented geochemical analyses.

The MIS5e macrofauna of these coral terraces is highly diverse and comprises scleractinian corals (*Acropora*,

Fig. 3 The modern fringing reef at Al Wasel explored by SCUBA diving on 11 December 2013: **a** general view of the site, the tabular relief on the right is St. 16; **b** tabular and shallow (50 cm) reef flat with predominant abraded, almost barren carbonate bedrock with algal coverage; **c** rippled sandy bottom with scant algal coverage and partly unburied carbonate substrate, 1 m depth (bar = 10 cm); **d** reef-edge and beginning of the reef slope showing intense coral growth, characterized by hydroids *Millepora platyphylla* (foreground) and *Millepora dichotoma* (isoriented fans), and scleractinians (*Porites* and *Acropora*, in the centre), about 1.5 m depth (bar = 10 cm); **e** reef slope, about 15 m depth, characterized by patchy coral colonies and sandy sediment; **f** outer reef at about 5 m, with scleractinian colonies (*Stylophora*, *Acropora*, faviids) and *Millepora* growing on dead coral limestone bedrock. Reef-edge and upper reef slope analog facies are represented in the visible part of the MIS5e terrace to the right. The MIS5e St. 16 includes back reef sandy habitats not occurring today at the diving spot. Underwater pictures by Abdulnasser Al-Qutub

Fig. 4 Details of modern habitats at Al Wasel: **a** eroded reefal bedrock colonized by intertidal molluscs, such the chiton *Acanthopleura vaillantii* (bar = 1 cm); **b** the large gastropod *Lambis truncata sebae* camouflaged among algae on the reef flat; note dead but fresh *Tridacna* shell on the left (arrow) (bar = 10 cm); **c** gaping specimen of the large clam *Tridacna maxima* on the reef flat (bar = 10 cm); **d** reef edge at −1 m with eroded bedrock remnant and in the foreground a colony of *Millepora platyphylla* embedding various specimens of the giant vermetid gastropod *Ceraesignum maximum* (arrows) (bar = 5 cm); **e** reef edge with carbonate bedrock settled by scattered "faviid" corals and vermetids (arrow) (bar = 5 cm); **f** coarse sandy skeletal hash (arrow) along the reef slope, about −10 m. Underwater pictures by Abdulnasser Al-Qutub

Fungia, *Pocillopora*, *Stylophora*, *Porites*, *Pavona*, *Leptoseris*, *Lobophyllia*, *Caulastraea*, *Favites*, *Goniastrea*, *Platygyra*, *Cyphastrea*, *Echinopora*, *Dipsastraea*, *Phymastrea*, *Galaxea*, *Leptastrea*, etc.), octocorals, hydroids (*Millepora*), molluscs, decapods, barnacles, echinoids, serpulids, and calcareous algae. Scleractinian corals are the main framebuilders (work in progress), but molluscs are the most diverse group (>400 species; Angeletti et al. this volume). Additionally, a variety of benthic microfossils occur in these deposits, mainly foraminifera (*Quinqueloculina* and other miliolids, *Sorites*, *Peneroplis*, *Heterostegina*, *Amphisorus*, *Amphistegina*, *Acervulina*, etc.) and ostracods. Noticeably, epipelagic foraminifera (e.g., *Globigerinoides ruber*, *Orbulina universa*), thecosomatous pteropods (e.g., *Creseis clava*, *Limacina bulimoides*, *Heliconoides inflatus*), abundant species of the Gulf of Aqaba holoplankton (Reiss and Hottinger 1984), and calcified meroplankton are a significant component of the fossil assemblages identified in the MIS5e deposits (e.g., St. 8 and 12), documenting the relevance of a pelagic input which is amplified and facilitated by the reduced size of fringing reef systems directly hanging over the open sea.

The in-depth study of this rich Pleistocene fauna will hopefully contribute to various aspects of the biogeography of the Red Sea-Indian Ocean region and the evolution of their biota. It will hopefully impact the debate about the supposedly complete sterilization of the Red Sea biota in response to the establishment of highly-saline basinal conditions during the last glaciation (Reiss et al. 1980; Locke and Thunell 1988; Taviani 1998b). Such faunal annihilation was followed by a complete post-glacial renovation, including coral reefs (Gvirtzman et al. 1977; Taviani 1998c), but this hypothesis is somewhat challenged by genetic data on endemic fish (Dibattista et al. 2016). More generally, appreciation of any difference in the coral reef assemblages versus the modern counterparts also bears on the global

Fig. 5 Pre-MIS5e Pleistocene carbonates: **a** Panoramic view of St. 2, Ad Durrah exposing pre-MIS5e carbonates; **b** Spur consisting of pre-MIS5e carbonates, with modest pockets of MIS5e sediments, St. 3, Ad Durrah; **c** Limestone block showing the severity of carbonate dissolution in pre-MIS5e reefal rocks, St. 14, 17.9 km south of Ra's Suwayhil as Saghir; **d** degraded and recrystallized "faviid" coral head at St. 1, Ad Durrah

understanding of coral reef ecosystemic stability *versus* the paroxysmal ice ages of the Pleistocene (e.g., Taviani 1998c; Camoin and Webster 2015).

6 Site Description

Sites visited during the winter 2013 field trip are shown in Fig. 1 and relevant data are reported in Table 1. Here we briefly describe the main sites from the north (close to the border with the Kingdom of Jordan) to the southernmost site at the very end of the Gulf. A thorough study of these terraces was not possible during this exploratory survey but all sites presented here have been georeferenced and photographed to permit their easier identification in the field by other researchers. This very basic and preliminary information is intended to help in orienting further scientific research on such coastal features. The sites are described from north to south.

The first sites are three isolated outcrops, whose crests are between 10–14 m above m.s.l., found close to the Jordan border as distinct almost adjacent geomorphic expressions of Pleistocene coral carbonates; they occur as spurs in small coastal indentations at the locality of Ad Durrah.

Station 1: is a coastal spur made up by a composite carbonate body (Fig. 6a); the basal coral carbonate is of unknown Pleistocene age, and is plastered by younger but undated carbonates above; the carbonate bedrock hosts recrystallized coral heads (Fig. 5d) and is bored by *Lithophaga* mussels and settled by *Spondylus* bivalves; large coral heads occur on the upper part of the section; no MIS5e deposits have been clearly recognized here.

Station 2: is close to the south in the same small embayment and displays a rather similar geological situation (Fig. 5a); it is noteworthy that it exposes coastal marine sands enriched in pectinids and irregular echinoids (*Clypeaster* and *Laganum*) whose tests are aligned along the bedding.

Fig. 6 Typology of contacts of MIS5e reefs with respect to substrate in the Saudi Arabian Gulf of Aqaba: **a** Young Pleistocene reef (white arrow) growing over preceding older Pleistocene reefs rocks (yellow arrow), St. 1, Ad Durrah; **b** remnants of MIS5e reefs (arrow) attached to Arabian basement rocks (granitoids crossed by basic dykes) at St. 12, 7.1 km south of Ra's Suwayhil as Saghir, dotted line marks the upper part of the reefal carbonates; **c** Reefal growth over large bedrock boulders (dotted line) at St. 6, 8.3 km south of Ra's Suwayhil as Saghir; **d** Reef growing over dipping Miocene sandstones (boundary between reefal carbonates and substrate marked by dotted line), St. 27, 6.75 km south of Al Maqnah

Station 3: is a small rocky spur immediately before the entrance fence of a coastguard site (Fig. 5b); besides older Pleistocene carbonates as at the previous two stations, last interglacial (MIS5e) reefal carbonates are exposed as documented by classic index-fossils, that is, the molluscs *Diodora impedimentum* and *Euplica turturina* (Taviani 1998a, 1998c).

Station 16: 2.3 km south of Al Wasel (Fig. 3a). This station is one of the best exposures of MIS5e coral terraces in the entire Gulf of Aqaba area coral reef systems, culminating at ~19–20 m above m.s.l. The preservation state of the host fossil content here is exquisite; noticeably fossiliferous, this site contains a *Pinna* bed with articulated shells amidst echinoids, other bivalves and gastropods. Corals are represented by a number of scleractinians (e.g., *Fungia*, *Acropora*, *Porites*, "faviids"), organ-pipe alcyonarians (*Tubipora*), and hydroids (*Millepora platyphylla*). The mollusc fauna of this terrace is one of the most diverse of the study area (Angeletti et al., this volume), representing various habitats including shore (the gastropods *Nerita orbignyana* and *N. sanguinolenta*), backreef (strombids, *Mammilla*, *Rhinoclavis*), reef-flat and edge (*Tectus dentatus*, *Trapezium oblongum*), and fore-reef (*Pinctada margaritifera*, *Spondylus* spp.).

Station 17: 2.65 km south of Al Wasel. The site exposes a poorly preserved reefal terrace, with degraded corals, *Tridacna* fragments, partly dissolved and replaced molluscs, including *Conus* spp., pectinids, and *Lithophaga* mostly as moulds; vestiges of ancient Pleistocene marine coral reef deposits are found at ~44 m above m.s.l.

Station 18: 2.4 km south of Ra's Suwayhil as Saghir. The site is represented by a granitic bedrock with carbonate veins and Neptunian dikes; some carbonate served as hardground substrate to encrusting serpulids.

Station 19: 2.45 km south of Ra's Suwayhil as Saghir. A spur-like outcrop of undated Pleistocene coral carbonates, macrofossiliferous, and at places cemented by botryoidal aragonite or bored by *Lithophaga*. Mixed carbonate-arkosic breccias (from the dismantling of the Arabian basement) also show cementation by aragonitic botryoids (Fig. 15c); the most recent reefal carbonates (MIS5e) contain molluscs and corals.

Station 4: 4.1 km south of Ra's Suwayhil as Saghir. MIS5e reefal terraces partly covered by alluvium along the coastal road at an elevation of ~20 m above m.s.l., are rich in well preserved frame-building scleractinian corals and molluscs; botryoidal aragonite cement occurs in the host rock (Fig. 15d).

Station 5: 5.8 km south of Ra's Suwayhil as Saghir. A fresh road cut exposes pre-MIS5e carbonate bedrock between 19–23 m above m.s.l.; next to the outcrop, loose but fresh *Porites* erratic blocks are found bored by *Lithophaga* mussels. This site was not examined in detail.

Station 12: 7.1 km south of Ra's Suwayhil as Saghir. Hanging Pleistocene coral limestones and semilithified carbonates plaster as relics the crystalline bedrock at 25 m above m.s.l. (Figs. 6b and 7d); this relic feature is extremely interesting since it documents a rare example of a late Pleistocene exposed slope, under pelagic input (pteropods, globigerinids) with-reef talus deposition. Abundant environmentally-valuable fossil material is found at this site, such as *Spondylus*, large barnacles and oysters on limestone bedrock, micromolluscs, melobesian fragments, decapod claws, rare corals, benthic foraminifers (including *Amphistegina*), ostracods, and echinoids in the loose or firm sediment.

Station 6: 8.3 km south of Ra's Suwayhil as Saghir. This outcrop is noticeable since it exposes large boulders overgrown by reefal carbonates attributable to MIS5e (Fig. 6c). This last interglacial fauna includes a variety of well-preserved scleractinian corals and molluscs, such as articulated (*Trapezium oblongum*) and disarticulated bivalves (*Spondylus* spp., *Trachycardium* sp.), and gastropods (e.g., Trochidae spp. and Cypraeidae spp.); large mamelon-like aragonite botryoids have been found here cementing Pleistocene reefal carbonates.

Station 7: 13.8 km south of Ra's Suwayhil as Saghir. A well-preserved flat-topped MIS5e coral terrace is exposed reaching up to ~22 m above m.s.l. (Fig. 7a–c). Fresh carbonates and scleractinian corals are exposed on the flanks along the wadi cut, revealing very large and articulated shells of the giant clam *Tridacna maxima*. The fossil system comprises a paleoshoreline, testified by *Cellana radiata* shells. The preservation of reef-building scleractinian corals is excellent, providing suitable material for geochronological dating and geochemical analyses.

Station 8: 14 km south of Ra's Suwayhil as Saghir. There is a large exposure of MIS5e reefal terraces at the same altitude as the previous station, with intervening bioturbated arkosic coarse-sandy deposits; the rich mollusc assemblage includes index-fossils such as *Diodora impedimentum* and *Euplica turturina*. This site is characterized by reef edge and upper slope abundant scleractinian corals including *Acropora*, *Porites*, *Caulastrea*, *Fungia* and "faviids" among others (Fig. 8d). The back-reef coarse coral sands host numerous articulated *Pinna* bivalves, and many other molluscs (especially common is *Architectonica trochlearis*). Finally, a MIS5e former shoreline is here documented by the limpet *Cellana radiata*.

Station 9: this site is immediately south of St. 8 and displays MIS5e tabular coral terrace remnants, some of which collapsed, and whose facies are equivalent to the previous station.

Station 10: this site is represented by a series of well-preserved coral terraces dissected by wadis at ~20 m above m.s.l. that is a conspicuous geomorphological aspect of the coastal plain at this location situated between stations 8 and 9 (Fig. 8a–c). Terraces consist of reefal facies enriched in scleractinian corals such as *Lobophyllia*, *Platygyra* (Fig. 8b), and molluscs as trochids, *Chicoreus ramosus, Spondylus* spp. and *Tridacna* sp. (Fig. 8c). This site is relevant for reconstructing the paleogeography and assessing post-depositional vertical movements since it keeps vestiges of the former last interglacial shoreline, also evidenced by outwashed worn shells of intertidal gastropods (*Cellana radiata*, *Nerita* sp.), other shallow marine molluscs (*Tridacna* sp., *Lambis truncata sebae*), and corals.

Station 13: 17.6 km south of Ra's Suwayhil as Saghir. The coral terraces reaching up to 25 m above m.s.l. (Figure 9a) display *Platygyra* colonies on top (Fig. 9b), and contain terebrid gastropods and sands with in situ *Tridacna* shells. The topmost coral rich carbonates cap a melobesian-rich whitish siliciclastic-carbonate sand, with scattered coral heads in situ (Fig. 9e). In this area we have identified a concentration of broken and bleached *Lambis* shells on the ground, possibly representing a sub-recent shell midden related to human activity (Fig. 9f).

Station 14: 17.9 km south of Ra's Suwayhil as Saghir. A complex coral reefal situation occurs at this site exposing un-dated Pleistocene coral bedrock affected by diagenetic processes, as documented by more or less completely dissolved corals and molluscs (Fig. 5c), and younger deposits with such features such as *Tridacna* beds, whitish melobesian sand, and frame-building corals.

Station 20: 3 km north of Wadi Tayyib Ism. This outcrop exposes the interfingering of MIS5e reefal carbonates with conglomeratic layers sitting on Arabian basement granitoids (Fig. 9c). Abundant scleractinian heads are observed in the

Fig. 7 **a** The MIS5e coral terrace 13.8 km south of Ra's Suwayhil as Saghir (St. 7); **b** an exposure of upper fore-reef coral-rich facies with numerous branching scleractinian corals; **c** detail showing large coral accumulation with *Porites*, "faviids" and other stony corals; **d** hanging Pleistocene coral limestones and semilithified carbonates plaster as relics (marked by dotted line) the crystalline bedrock, 7.1 km south of Ra's Suwayhil as Saghir (St. 12)

middle part of the section and are more common on the top carbonates, including *Porites*, *Acropora*, *Fungia* and "faviids". The accompanying benthic fauna includes the index-fossil *Diodora impedimentum*, and other gastropods (e.g., *Turbo radiatus, Ceraesignum maximum*), cemented (*Chama* spp., *Spondylus* spp.), epifaunal and infaunal bivalves (*Tridacna maxima, Trachycardium* sp.).

Station 21: 3.4 km north of Wadi Tayyib Ism, and before Al Maqnah. Here a presumed MIS5e coral growth at ~25

Fig. 8 **a** Well-preserved coral terraces at St. 10; **b** detail of an upper fore-reef facies typified by stony corals such as *Lobophyllia*, *Porites*, *Acropora* and "faviids"; **c** giant valve of *Tridacna maxima*; **d** underroof of a MIS5e tabular coral terrace remnants immediately to the south of St. 8 (St. 9) revealing various frame-building scleractinians such as *Galaxea*, *Lobophyllia*, *Porites* and *Acropora*

m above m.s.l. caps the Pliocene/Miocene bedrock; this outcrop has not been analyzed in detail.

Station 22: 5.9 km south of Wadi Tayyib Ism. This site consists of a cluster of three disconnected main outcrops culminating at altitudes between ~14–22 m above m.s.l., a few tens of metres apart separated by gullies on a wide coastal lowland. The reefal terraces (Fig. 10a–f) expose coral-rich facies, back-reef lagoonal red sands with marine

Fig. 9 **a** Coral terraces 17.6 km south of Ra's Suwayhil as Saghir (St. 13); **b** large *Platygyra* coral on the top of the terrace (St. 13); **c** small *Porites* coral heads (arrows) within siliciclastic-carbonate sand (St. 14); **d** broken *Lambis* shells scattered on the ground likely testify to a subrecent anthropogenic midden (St. 14); **e** coral terrace at St. 14; **f** general view of the sites showing relative positions of St. 13 and 14; **g** outcrop 3 km north of Wadi Tayyib Ism (St. 20) exposing the reefal carbonates, interfingering; **h** with basal conglomeratic deposits sitting on Arabian basement granitoids

Fig. 10 **a–c** General views of the site 5.9 km south of Wadi Tayyib Ism (St. 22) consisting of three disconnected main MIS5e outcrops, a few tens of metres apart separated by gullies emerging on a wide coastal lowland, arrows pointing out the top of the terraces; **d** massive *Platygyra* heads on the top of the terrace; **e** intensely bioturbated (*Thalassinoides*) pebbly sand below reefal carbonates; **f** coarse reefal-sediment accumulation enriched by displaced skeletal remains, including echinoids (red arrow), bivalves (*Codakia tigerina*, black arrow) and gastropods (*Turbo radiatus*, white arrow)

molluscs (*Codakia tigerina*, *Lambis truncata sebae*), at places intensely bioturbated (*Thalassinoides*: Fig. 10e), and arkosic gravels. The raised reef deposit at ~12 m above m. s.l. hosts a dense coral growth mostly represented by well-preserved and large *Porites* heads (Fig. 10d), *Fungia*, "faviids", *Millepora* and other corals, which are draped by a veneer of coral debris which contains also highly degraded branching hydroids (*Millepora*) and molluscs.

Station 23: 6.1 km south of Wadi Tayyib Ism. The site is a MIS5e coral terrace whose dismantling sheds well preserved fossils in the talus.

Station 24: 6.2 km south of Wadi Tayyib Ism. The area is characterized by prominent coral terraces (Fig. 11a) and displays large coral heads on top but has not been analyzed in detail (Fig. 11b).

The Al Maqnah (or Maqna) area is described and commented in detail for its Pleistocene marine terraces by Dullo (1990).

Station 25: 5.6 km north of Al Maqnah. Here the faulted Miocene substrate is plastered by Pleistocene reefal deposits at 26 m above m.s.l. (Fig. 11c, d). The scant paleontological legacy includes clusters of the intertidal barnacle *Tetraclita*.

Station 26: 4.5 km north of Al Maqnah. This site exposes a flat-topped coral terrace flight at the same elevation as the previous station over Miocene sandstones (Fig. 12a), and is characterized by a rich coral content with *Lobophyllia*, *Acropora*, *Leptoseris*, *Platygyra*, *Stylophora*, *Porites*, *Fungia*, *Favites*, documenting a fore-reef setting (Fig. 12c, d); spectacular giant coral heads (*Galaxea fascicularis*: Fig. 12 b) are found here.

Station 27: 6.75 km south of Al Maqnah. A series of terraces around ~17 m above m.s.l. separated by gullies,

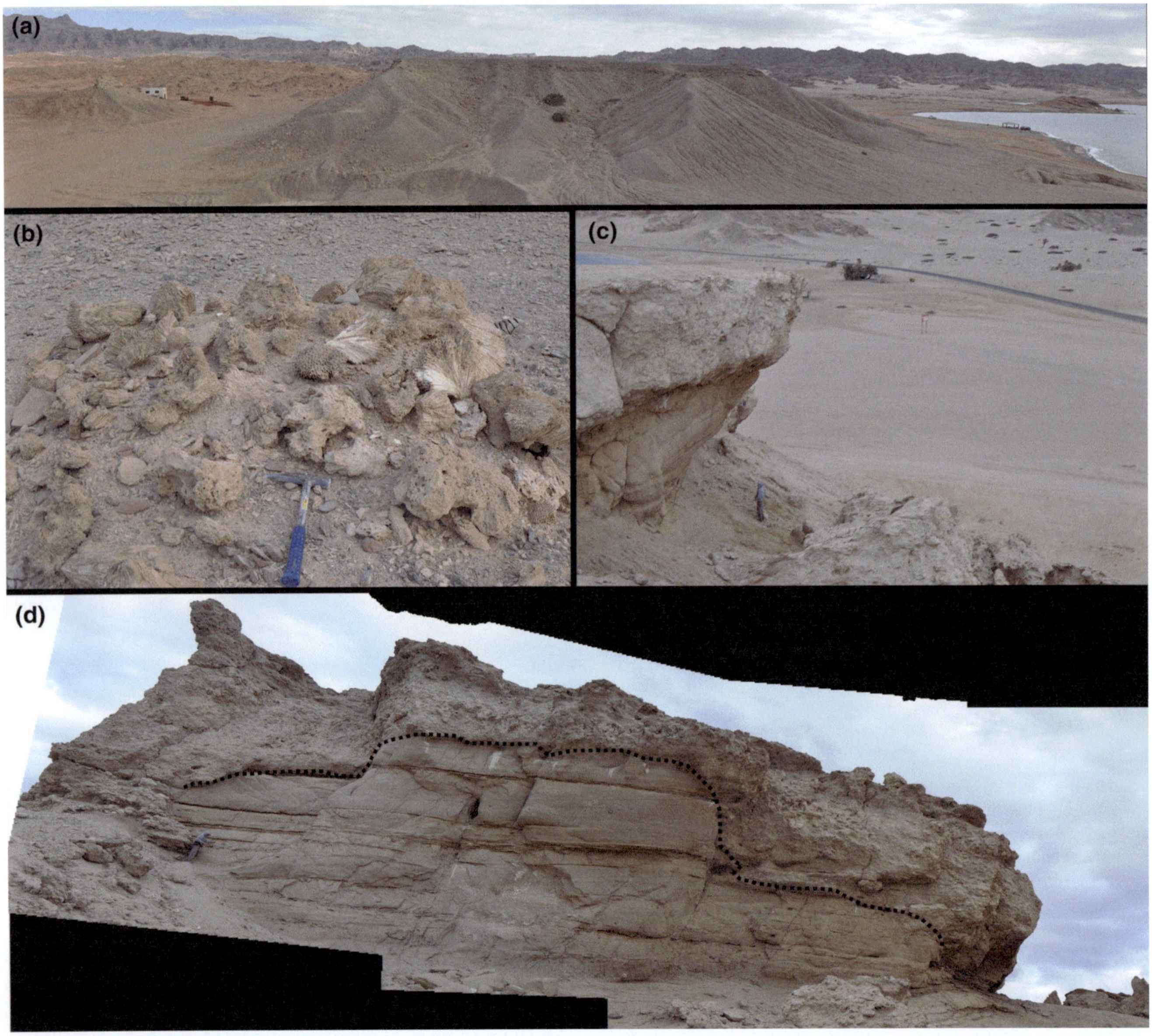

Fig. 11 **a** General view of the area 6.2 km south of Wadi Tayyib Ism (St. 24), characterized by prominent coral terraces; **b** detail of a coral terrace top showing large coral heads, prevalent *Platygyra* and "faviids"; **c**, **d** the faulted Miocene substrate 5.6 km north of Al Maqnah (St. 25), plastered by Pleistocene reefal deposits, dotted line marks the boundary between the MIS5e reef and the Miocene substrate, person for scale

Fig. 12 **a** Flat-topped coral terrace capping Miocene sandstones, 4.5 km north of Al Maqnah (St. 26); **b** spectacular *Galaxea fascicularis* giant heads (>2 m in height) occur here; **c** outer-reef facies with *Lobophyllia hemprichii*, *Porites*, "faviids" and other stony corals; **d** detail of the contact between the MIS5e reef and Miocene sandstones showing large *Porites* colonies in growth-position

typifies the coastal setting with MIS5e coral terraces growing onto dipping Miocene strata (Fig. 6d). *Porites*-dominated reefs and intervening sandy sediments host abundant and diverse associated fauna which includes large bivalves (e.g., *Tridacna* spp, *Codakia tigerina*) and gastropods.

Station 15: Ash Shaykh Humayd. At the extreme south of the Gulf of Aqaba facing the Straits of Tiran and unconformably overlying Miocene sandstones (Fig. 13), there is a noticeable exposure of MIS5e coral reef terraces reaching a maximum elevation of ~8.7 m above m.s.l., which also host typical index fossils (*Diodora impedimentum*, *Euplica turturina*). The terraces expose reef-edge and back-reef lagoonal deposits, the latter culminating at ~4–6 m above m.s.l. This site is extremely rich in well preserved and diverse macrofossils (Fig. 13), including various stony corals (*Porites*, *Acropora*, *Montastraea*, *Favia*, *Favites*, *Fungia*, etc.), a number of molluscs (the bivalves *Tridacna* spp., Tellinidae spp., *Circe* spp., the gastropods *Conomurex fasciatus*, Cypraeidae spp., *Conus arenatus, Bulla ampulla*, and many others; Fig. 13f), echinoids (clypeasterids and *Heterocentrotus*), *Tetraclita* barnacles, melobesians, and much more. Botryoidal aragonite has been also found here.

7 Pleistocene Salina-Mangal Terrace

A unique situation for the Saudi Arabian Gulf of Aqaba to date is found 13.5 km south of Ra's Suwayhil as Saghir (St. 11) where a hyperhaline (evaporitic) and mangal complex crops out (Fig. 14). The sub-horizontal terrace reaches up to ~23 m above m.s.l. (Fig. 14a). Stratigraphically, the sequence exposes at its base a few metres of clayey-sandy lagoonal sediments containing the diagnostic mangrove-associated brackish gastropod *Terebralia palustris* (see Angeletti et al., this volume), numerous shells of the euryhaline gastropod *Pirenella conica*, articulated shallow-infaunal bivalves (*Gafrarium* sp., *Anodontia* sp.: Fig. 14b), and disarticulated intertidal bivalve shells (*Atactodea striata*). This basal lagoonal salina-mangal sequence is followed upward by a lenticular chaotic deposit (flash-flood?) and then by sediments documenting a confined saline microbasin (Fig. 14c). The evaporitic sequence consists of about 1 m of laminated gypsified evaporites, entrapping occasional gastropod marine shells (*Nerita* spp. and *Turbo radiatus*, possibly transported by hermit crabs), then by 40 cm of selenitic gypsum cumulates with macrocrystals up to 4 cm in size (Fig. 14d, e); this unit is affected upward by a second chaotic unit, and the sequence ends up with a second package of laminated evaporites (Fig. 14c).

Although the lack of corals in this site prevents precise U-series dating, we tentatively attribute this complex to MIS5e because of its altitude that is comparable to coral terraces, absence of evident diagenesis, and sub-horizontality. Analogous situations of this age are part of the coastal stratigraphy of the Red Sea proper, and have been treated in detail by Plaziat et al. (1998, 2008) and Orszag-Sperber et al. (2001).

8 Botryoidal Aragonite

Large aragonite fans and botryoids up to 8 cm in diameter have been discovered on terraces south of Ra's Suwayhil as Saghir (St. 4, 8 and 19), and at Ash Shaykh Humayd in the extreme south of the Gulf, as filling vugs into MIS5e and pre-MIS5e reefal carbonate bedrock, at places also cementing mixed carbonate-arkosic breccias (Fig. 15). Botryoidal aragonite is a distinct form of calcium carbonate micro- to macrofabric characterized by a radial-fibrous growth often achieving a mamelon-like final shape (Fig. 15b). Botryoids form in a variety of contrasting genetic scenarios; subaerial caves (Ostermann et al. 2007; Fairchild and Baker 2012), marine hydrocarbon seeps (e.g., Roberts et al. 1993; Savard et al. 1996; Taviani 2001; Feng et al. 2008; Taviani et al. 2016), or shallow-marine settings (Ginsburg and James 1976; Aissaoui 1985). Regarding the latter, spectacularly large aggregates of botryoidal aragonite are found as epigenetic marine cements filling vugs in Quaternary shallow-water carbonates, preferentially reefal host-rocks, from various tropical localities including New Caledonia, Bahamas, Belize and the Red Sea (Ginsburg and James 1976; Taviani and Rabbi 1984; Aissaoui 1985; Grammer et al. 1993; Dill et al. 1998). In the Red Sea basin, botryoidal aragonite has been reported from Pleistocene coral reef terraces along the Egyptian mainland and islands, and proved to represent synsedimentary marine cements based on field and stable isotope data (Taviani and Rabbi 1984; Aissaoui 1985; Taviani et al. 1987). The Zabargad Island's aragonite includes the largest known such crystals and mineralizes not only carbonate rocks but also fissures in basic intrusives and mixed breccias (Moon 1923; El-Shazly et al. 1974; Bonatti et al. 1983; Taviani et al. 1987). Botryoidal fabric is known in the geological record since the Paleozoic (Mazzullo and Cys 1978; Aissaoui 1985; Lake 2004; Brachert et al. 2007), more often as calcite or partially-dolomitized pseudomorphs of the original aragonite.

The aragonite botryoids from the Saudi Arabian fossil reefs are interpreted as synsedimentary fast-growing marine cements, similarly to their Gulf of Mexico analogues (Grammer et al. 1993), or Red Sea equivalents (Aissaoui 1985; Taviani et al. 1987). Regarding the time of their genesis, botryoidal aragonite from a MIS5e terrace south of Ra's Suwayhil as Saghir (St. 8) provided a U/Th age of about 119 ka (Bosworth et al. 2017; Table 2). This figure is consistent with a process of marine carbonate precipitation

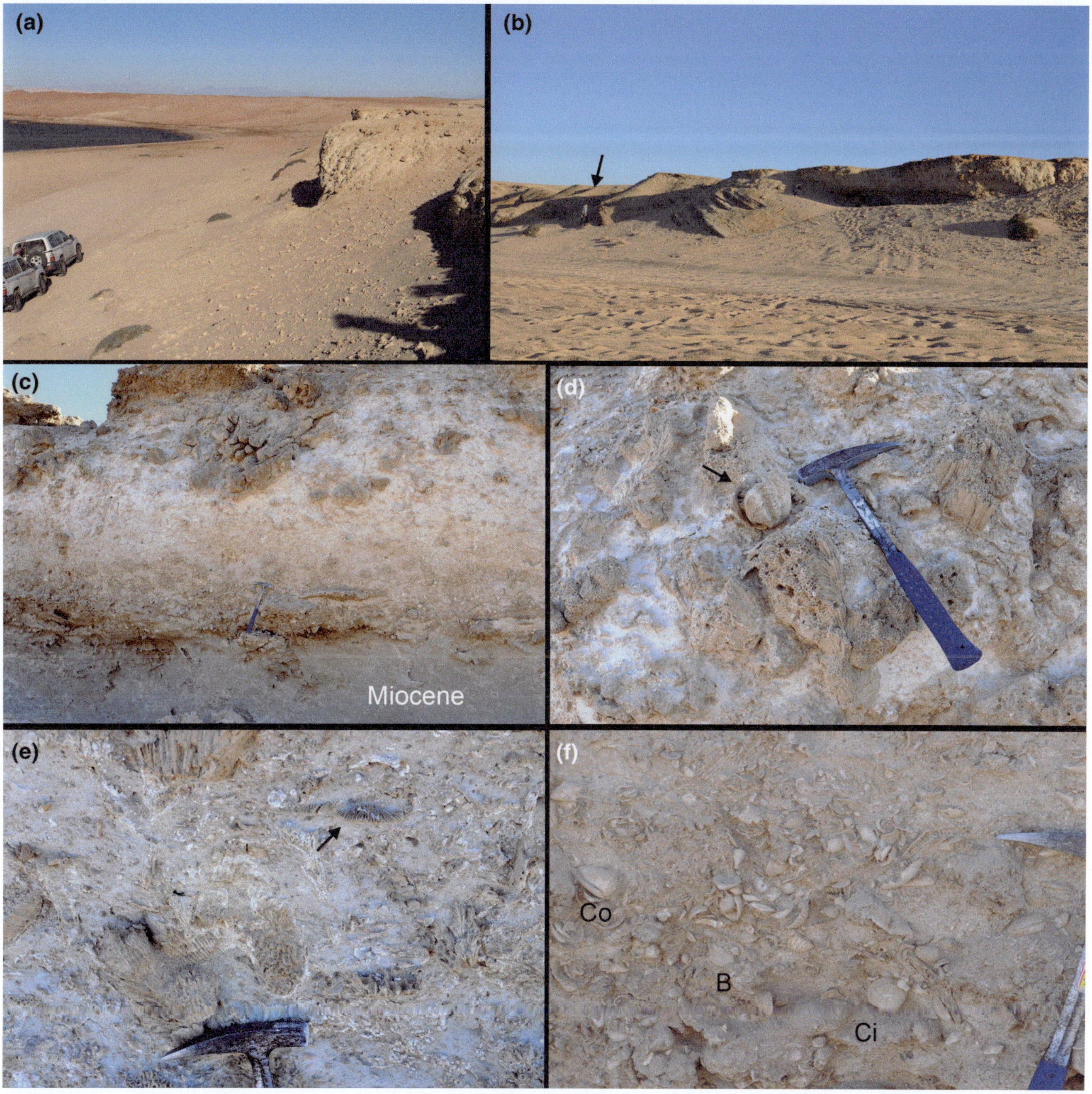

Fig. 13 **a** The MIS5e terrace of Ash Shaykh Humayd here culminating at about +6 m above m.s.l.; **b** overview showing the same MIS5e reefal terraces, person for scale (arrow); **c** section documenting that the MIS5e overlies unconformably the inclined Miocene strata, more shelly at the base, and upward grading into a reef edge facies with *Platygyra*, *Porites*, *Acropora* and "faviids", more abundant on the top, as well as *Heterocentrotus* spines; **d** detail of a reef-edge facies with in-situ *Favites* and other stony coral colonies, and articulated *Tridacna* (arrow); **e** melobesian and stony coral-rich facies with colonial scleractinians such as "faviids" and the solitary *Fungia scutaria* (arrow), and *Heterocentrotus* spines; **f** detail of a chaotic shell-rich deposit sourced from back-reef sandy habitat; the highly diverse paleontological content is dominated by molluscs, including the lagoonal gastropods *Conus arenatus* (Co), *Bulla ampulla* (B) and the bivalve *Circe* sp. (Ci), and subordinately, "faviids" and irregular echinoids

Fig. 14 The lagoonal salina-mangal complex 13.5 km south of Ra's Suwayhil as Saghir (St. 11), **a** general view of the outcrop, with muddy-sandy lagoonal and mangal (mangrove derived) sediment at the base, followed upward by evaporitic deposits; asterisk marks the best stratigraphic exposure; **b** detail of the lagoonal salina-mangal sediments showing articulated *Anodontia* and *Gafrarium* bivalves still in situ (arrow); **c** the stratigraphy of the uppermost part of the evaporitic section (asterisk in a), documenting the top of the first laminite unit partly masked by talus (1), the selenitic cumulate unit (2), the second chaotic unit (3) and the second laminite unit (4); **d** detail of the selenitic gypsum cumulates, observe inverse grading; **e** hand sample of large selenitic gypsum crystals

Fig. 15 Aragonite cements: **a** large fans within Pleistocene coral carbonates, 2.45 km south of Ra's Suwayhil as Saghir (St. 19); **b** mamelon-botryoid cementing Pleistocene reefal carbonates (St. 8); **c** mixed carbonate-arkosic breccia cemented by aragonite botryoids (St. 19); **d** polished slab of carbonate reefal bedrock showing large botryoidal aragonite cement, with clear accretionary bands (arrow) (St. 4); **e** drusy cements within vuggy carbonate rock (St. 19). Bars = 1 cm

during MIS5e, although this does not necessarily imply that all such aragonite cements observed in the terraces examined here share this age.

9 Summary and Conclusions

The Saudi Arabian coast of the Gulf of Aqaba boasts a superb set of marine coral terraces all along its length, connecting to the north with their equivalent in Jordanian territory and terminating to the south at the tip of the Arabian Peninsula and junction with the Red Sea. These terraces are Pleistocene in age, with the majority belonging to coral reef complexes of the last interglacial MIS5e. Besides coral reef terraces, this Pleistocene record also preserves a presumably coeval salina-mangal setting as well as spectacular examples of syndepositional aragonitic cement.

The altitude of the MIS5e terraces is variable, responding to tectonics, in most cases documenting considerable uplift. For future developments, it is important to note that the rich fossiliferous content provides a means to identify former environments with confidence, supplying ideal material for dating and geochemical paleoclimatology, and contributing taxonomic information valuable for evolutionary and biogeography purposes.

The exceptional uplift of MIS5e terraces makes these Saudi Arabian settings unique in the frame of the Red Sea region as a whole.

Acknowledgements We thank the Saudi Geological Survey for the invitation to present this study at the '2nd Red Sea Book Workshop' held in Jeddah on February 2016. The expedition in the Gulf of Aqaba has been promoted and funded by the Saudi Geological Survey in Jeddah. We acknowledge all participants to the Winter 2013 Field Party SGS personnel for field work (Ali Saeedi, Abdulnasser Al-Qutub, Salem Al-Nomani, Nawaf Widinly, Mostafa Khorshid, Thamer Bakarman) and post-expedition sample processing (Adel Jarees, Ali Al-Tharowi, Mohamed M. Bin Himd, Hassan Khamis Ali). Abdulnasser Al-Qutub was also part of the diving team and responsible for underwater photography. We thank Francesca Bosellini for helping with coral identification. Michele Morsilli and two anonymous referees provided valuable critical comments. This paper is part of PRIN2012 Programme (Project 20125JKANY_002, Principal Investigator Marco Ligi) and is Ismar-CNR, Bologna, scientific contribution no. 1905.

References

Aharon P, Chappell J (1986) Oxygen isotopes, sea-level changes and the temperature history of a coral reef environment in New Guinea over the last 105 years. Palaeogeogr Palaeocl 56:337–379

Aissaoui DM (1985) Botryoidal aragonite and its diagenesis. Sedimentology 32:345–361

Alamri AM, Schult FR, Bufe CG (1991) Seismicity and aeromagnetic features of the Gulf of Aqaba (Elat) region. J Geophys Res 96:20179–20185

Alexandroff SJ, Zuschin M, Kroh A (2016) Quantitative comparison of Pleistocene and Recent coral reef habitats in the northern Red Sea (El Quseir, Egypt). Facies 62:15. https://doi.org/10.1007/s10347-016-0468-6

Alhejoj I, Bandel K, Salameh E (2016) The fossil beach and reef terraces of the Gulf of Aqaba coast, Jordan—its environment, formation, and relation to mountain uplifting mechanism. Arab J Geosci 9:275. https://doi.org/10.1007/s12517-015-2293-6

Al-Horani FA, Al-Rousan SA, Al-Zibdeh M, Khalaf MA (2006) The status of coral reefs on the Jordanian coast of the Gulf of Aqaba, Red Sea. Zool Middle East 38:99–110

Al-Mikhlafi AS, Edwards LR, Cheng I (2018) Sea-level history and tectonic uplift during the last-interglacial period (LIG): Inferred from the Bab al-Mandab coral reef terraces, Southern Red Sea. J Afr Earth Sci 138:133–148

Al-Rifaiy IA, Cherif OH (1988) The fossil coral reefs of Al-Aqaba, Jordan. Facies 18:219–230

Al-Sayari SS, Dullo W-C, Hoetzl H, Jado AR, Zotl JG (1984) The Quaternary along the coast of the Gulf of Aqaba. In: Jado AR, Zotl JG (eds) Quaternary period in Saudi Arabia 2. Springer, Wien, pp 32–47

Andres W, Radtke U (1988) Quartäre strandterrassen an der küste des Gebel Zeit (Golf von Suez/Ägypten). Erdkunde 42:7–16

Angeletti L, Rasul N, Taviani M (this volume) Mollusc fauna associated with Late Pleistocene coral reef systems of the Saudi Arabian side of the Gulf of Aqaba

Bantan RA, Abu-Zied RH, Haredy RA (2015) Lithology, fauna and environmental conditions of the Late Pleistocene raised reefal limestone of the Jeddah coastal plain, Saudi Arabia. Arab J Geosci 8:9887–9904

Bartov Y, Steinitz G, Eyal M, Eyal Y (1980) Sinistral movement along the Gulf of Aqaba: Its age and relation to opening the Red Sea. Nature 285:220–221

Bayer H-J, Hötzl H, Jado AR, Roscher B, Voggenreiter W (1988) Sedimentary and structural evolution of the northwest Arabian Red Sea margin. Tectonophysics 153:137–151

Ben-Avraham Z, Zoback MD (1992) Transform-normal extension and asymmetric basins: An alternative to pull-apart models. Geology 20:423–426

Ben-Avraham Z, Garfunkel Z, Almagor G, Hall JK (1979) Continental breakup by a leaky transform: The Gulf of Elat (Aqaba). Science 206:214–216

Berry L, Whiteman AJ, Bell SV (1966) Some radiocarbon dates and their geomorphological significance, emerged reef complex of the Sudan. Z Geomorphol 10:119–143

Bonatti E, Clocchiatti R, Colantoni P, Gelmini R, Marinelli G, Ottonello G, Santacroce R, Taviani M, Abdel-Meguid AA, Assaf HS, El Tahir MA (1983) Zabargad (St. John's) Island: An uplifted fragment of sub-Red Sea lithosphere. J Geol Soc 140:677–690

Borri M, Sabelli B, Taviani M (1982) *Cypraea* (*Talparia*) *exusta* Sowerby, 1832, from the central Red Sea: New data. Lavori Soc Malacol Ital 103–107

Bosence DWJ, Al-Aawah MH, Davidson I, Rosen B, Vita-Finzi C, Whittaker L (1998) Salt domes and their control on basin margin sedimentation: a case study from Tihama Plain, Yemen. In: Purser BH, Bosence DWJ (eds) Sedimentation and tectonics of rift basins: Red Sea-Gulf of Aden. Chapman and Hall, London, pp 450–467

Bosworth W, McClay K (2001) Structural and stratigraphic evolution of the Gulf of Suez rift, Egypt: A synthesis. In: Ziegler PA, Cavazza W, Robertson AHF, Crasquin-Soleau S (eds) Peri-Tethys Mem 6: Peri-Tethyan Rift/Wrench Basins and Passive Margins. Mém Muséum Nat Hist Natur Paris 186, pp 567–606

Bosworth W, Taviani M (1996) Late Quaternary reorientation of stress field and extension direction in the southern Gulf of Suez, Egypt: Evidence from uplifted coral terraces, mesoscopic fault arrays, and borehole breakouts. Tectonics 15:791–802

Bosworth W, Huchon P, McClay K (2005) The Red Sea and Gulf of Aden Basins. In: Catuneanu O, Guiraud R, Eriksson P, Thomas B, Shone R, Key R (eds) Phanerozoic evolution of Africa. J Afr Earth Sci 43:334–378

Bosworth W, Taviani M, Rasul, N, Angeletti L, Montagna P, and Team (2013) Pleistocene Coral Terrace Project—Gulf of Aqaba, Saudi Arabia, Phase 1: December 6–15, 2013. SGS Report, 61 pp

Bosworth W, Montagna P, Pons-Branchu E, Rasul N, Taviani M (2017) Seismic hazards implications of uplifted Pleistocene coral terraces in the Gulf of Aqaba. Sci Repts 7(38):1–13. https://doi.org/10.1038/s41598-017-00074-2

Bosworth W, Rasul N, Taviani M (this volume) Neotectonics of the Red Sea, Gulf of Suez and Gulf of Aqaba

Bouchon C, Jaubert J, Montaggioni L, Pichon M (1981) Morphology and evolution of the coral reefs of the Jordan coast of the Gulf of Aqaba (Red Sea). Proc Fourth Intern Coral Reef Symp, Manila 1:559–565

Brachert TC, Vescogni A, Bosellini F, Reuter M, Mertz-Krauss M (2007) High salinity variability during the early Messinian revealed by stable isotope signatures from vermetid and *Halimeda* reefs of the Mediterranean region. Geol Romana 40:51–66

Brighton AG (1931) The geology of the Farasan Islands, Gizan and Kamaran Island, Red Sea. Part 3. Echinoidea. Geol Mag 68:323–333

Camoin G, Webster JM (2015) Coral reef response to Quaternary sea-level and environmental changes: State of the science. Sedimentology 62:401–428

Casazza LR (2017) Pleistocene reefs of the Egyptian Red Sea: Environmental change and community persistence. PeerJ 5:e3504. https://doi.org/10.7717/peerj.3504

Choukri A, Reyss JL, Plaziat JC (1995) Datations radiochimiques des hauts niveaux marins de la rive occidentale du Nord de la Mer Rouge au moyen de radioles d'oursins. CR Acad Sci Paris, (Sér 2, Sci Terre) 321:25–30

Coleman RG (1993) Geologic evolution of the Red Sea. Oxford Monographs on Geology and Geophysics, vol 24. Oxford University Press, Oxford, p 186

Conforto L, Delitala MC, Taddeucci (1976) Datazioni col 230Th di alcune formazioni coralligene delle Isola Dahlak (Mar Rosso). Rend Soc Ital Miner Petrol 32:153–158

Cox LR (1929) Notes on the post-Miocene Ostreidae and Pectinidae of the Red Sea region, with remarks on the geological significance of their distribution. Proc Malacol Soc London 18:165–209

Dabbagh A, Hoetzl H, Schnier H (1984) Farasan Islands. In: Jado AR, Zotl JG (eds) Quaternary period in Saudi Arabia 2. Springer, Wien, pp 212–220

Dawood YH, Aref MA, Mandurah MH, Hakami A, Gameil M (2013) Isotope geochemistry of the Miocene and Quaternary carbonate rocks in Rabigh area, Red Sea coast, Saudi Arabia. J Asian Earth Sci 77:151–162

Dibattista JD, Choat JH, Gaither MR, Hobbs J-PA, Lozano-Cortés DF, Myers RF, Paulay G, Rocha LA, Toonen RJ, Westneat MW, Berumen ML (2016) On the origin of endemic species in the Red Sea. J Biogeogr 43:13–30

Dill RF, Land LS, Mack LE, Schwarcz HP (1998) A submerged stalactite from Belize: Petrography, geochemistry, and geochronology of massive marine cementation. Carbon Evapor 13:189–197

Dollfus RP, Roman J (1981) Les echinides de la Mer Rouge. Monographie zoologique et paléontologique, Ministères des Universités, Comité des travaux historiques et scientifiques, Bibliothèque Nationale de Paris, 145 pp

Dreyfuss M (1931) Etudes de géologie et de géographie physique sur la cote francaise des Somalis. Rev Géogr Phys Géol Dyn 4:311–327

Dullo W-C (1984) Progressive diagenetic sequence of aragonite structures: Pleistocene coral reefs and their modern counterparts on the eastern Red Sea coast, Saudi Arabia. Palaeontogr Am 54:254–260

Dullo W-C (1986) Variation in diagenetic sequences: an example from Pleistocene coral reefs, Red Sea, Saudi Arabia. In: Schroeder JH, Purser BH (eds) Reef Diagenesis. Springer, Berlin, pp 77–90

Dullo W-C (1990) Facies, fossil record, and age of Pleistocene reefs from the Red Sea (Saudi Arabia). Facies 22:1–46

Dullo W-C, Montaggioni L (1998) Modern Red Sea coral reefs: a review of their morphologies and zonation. In: Purser BH, Bosence DWJ (eds) Sedimentation and tectonics of rift basins: Red Sea-Gulf of Aden. Chapman and Hall, London, pp 583–594

Dutton A, Carlson AE, Long AJ, Milne GA, Clark PU, DeConto R, Horton BP, Rahmstorf S, Raymo ME (2015) Sea level rise due to polar ice-sheet mass loss during past warm periods. Science 349 (6244). https://doi.org/10.1126/science.aaa4019

El-Asmar HM (1997) Quaternary isotope stratigraphy and paleoclimate of coral reef terraces, Gulf of Aqaba, South Sinai. Egypt. Quatern Sci Rev 16:911–924

El-Moursi M, Hoang CT, El Fayoumy IF, Hegab O, Faure H (1994) Pleistocene evolution of the Red Sea coastal plain, Egypt: Evidence from Uranium-series dating of emerged reef terraces. Quatern Sci Rev 13:345–359

El-Shazly EM, Saleeb GB, Zaki N (1974) Quaternary basalt in St. John's Island, Red Sea. Egypt. Egypt J Geol 18:137–148

El-Sorogy A (1997) Pleistocene coral reefs of southern Sinai, Egypt: Fossil record, facies analysis and diagenetic alterations. Middle East Research Center, Ain Shams University, Earth Sci Series 11:17–36

El-Sorogy A, Youssef M, Al-Sabrouty M, Al-Otaiby N (2014) Facies pattern and molluscan fauna of the Late Pleistocene raised coral reef of Rabigh area, Red Sea coast, Saudi Arabia. Indian J Geo-Mar Sci 43:1571–1580

Fairchild IJ, Baker A (2012) Speleothem science: from process to past environments. Wiley-Blackwell, 450 pp

Faure H (1975) Recent crustal movements along the Red Sea and Gulf of Aden coasts in Afar (Ethiopia and T.F.A.I.). Tectonophysics 29:479–486

Faure H, Hoang CT, Lalou C (1973) Structure et géochronologie (230Th/234U) des récifs coralliens soulevés à l'ouest du Golf d'Aden (T.F.A.I). Rev Géogr Phys Géol Dynam 15:393–403

Faure H, Hoang CT, Lalou C (1980) Datations 234U/230Th des calcaires coralliens et mouvements verticaux à Djibouti. Bull Soc Géol France 22:959–962

Felis T, Lohmann G, Kuhnert H, Lorenz SJ, Scholz D, Pätzold J, Al-Rousan SA, Al-Moghrabi SM (2004) Increased seasonality in Middle East temperatures during the last interglacial period. Nature 429:164–168

Felis T, Pätzold J, Loya Y, Moaz F, Nawar AH, Wefer G (2000) A coral oxygen isotope record from the northern Red Sea documenting NAO, ENSO, and North Pacific teleconnections on Middle East climate variability since the year 1750. Paleoceanography 15:679–694

Feng D, Cheng D, Roberts HH (2008) Sedimentary fabrics in the authigenic carbonates from Bush Hill: Implication for seabed fluid flow and its dynamic signature. Geofluids 8:301–310

Gabrié C, Montaggioni L (1982) Sedimentary facies from the modern coral reefs, Jordan Gulf of Aqaba, Red Sea. Coral Reefs 1:115–124

Gass IG (1977) The evolution of the Pan African crystalline basement in NE Africa and Arabia. J Geol Soc London 134:129–138

Ginsburg RN, James NP (1976) Submarine botryoidal aragonite in Holocene reef limestones, Belize. Geology 4:431–436

Goldberg M, Yaron F (1978) Th 230/U 234 ages of raised Pleistocene marine terraces on the Island of Tiran and the southeastern coast of Sinai with some tectonic implications. IAS 10th International Congress on Sedimentology. Jerusalem, Abs 1:258–259

Grammer GM, Ginsburg RN, Swart PK, McNeill DF, Jull AJT, Prezbindowski DR (1993) Rapid growth rates of syndepositional marine aragonite cements in steep marginal slope deposits, Bahamas and Belize. J Sedimen Petrol 63:983–989

Guiraud R, Bosworth W, Thierry J, Delplanque A (2005) Phanerozoic geological evolution of Northern and Central Africa: An overview. In: Catuneanu O, Guiraud R, Eriksson P, Thomas B, Shone R, Key R (eds) Phanerozoic evolution of Africa. J Afr Earth Sci 43:83–143

Gvirtzman G (1994) Fluctuations of sea-level during the past 400,000 years: The record of Sinai, Egypt (northern Red Sea). Coral Reefs 13:203–214

Gvirtzman G, Buchbinder B (1978) Recent and Pleistocene coral reefs and coastal sediments of the Gulf of Eilat. Tenth International Congress of Sedimentology, Jerusalem, Post-Congress Excursion Y4:163–191

Gvirtzman G, Buchbinder B, Sneh A, Nir Y, Friedman GM (1977) Morphology of the Red Sea fringing reefs: A result of the erosional pattern of the last-glacial-low-stand sea level and the following Holocene recolonization. 2e Symposium International sur les Coraux et Récifs Coralliens Fossils. Mem Bur Rech Min 89:480–491

Gvirtzman G, Friedman GM (1977) Sequence of progressive diagenesis in coral reefs. Am Assoc Petrol Geol Stud Geol 4:357–380

Gvirtzman G, Kronfeld J, Buchbinder B (1992) Dated coral reefs of southern Sinai (Red Sea) and their implication to the Late Quaternary sea-levels. Mar Geol 108:29–37

Hall WJ, Standen R (1907) On the Mollusca of a raised coral reef on the Red Sea coast. J Conch 12:65–68

Hamed B (2015) Pleistocene reefs of the Red Sea coast, Sudan: Depositional environments, fossil coral, age dating and diagenesis. Ph.D. Thesis, Technische Universität Berlin, 131 pp. https://doi.org/10.14279/depositonce-4414

Hamed B, Bussert R, Dominik W (2016) Stratigraphy and evolution of emerged Pleistocene reefs at the Red Sea coast of Sudan. J Afr Earth Sci 114:133–142

Hartman G, Niemi TN, Ben-Avraham Z, Tibor G, Al-Zoubi A, Sade RA, Hall JK, Akawi E, Abueladas A, Al-Ruzouq R, Makovsky Y (2015) Distinct relict fringing reefs in the northern shelf of the Gulf of Elat/Aqaba: Markers of Quaternary eustatic and climatic episodes. Sedimentology 62:516–540

Hoang CT, Dalongeville R, Sanlaville P (1996) Stratigraphy, tectonics and palaeoclimatic implications of uranium-series-dated coral reefs from the Sudanese coast of the Red Sea. Quatern Int 31:47–51

Hoang CT, Lalou C, Faure H (1974) Les récifs soulevés à l'ouest du golfe d'Aden (TFAI) et les hauts niveaux de coraux de la dépression de l'Afar (Ethiopie), géochronologie et paléoclimats interglaciaires. In: Les méthodes quantitatives d'étude des variations du climat au cours du Pléistocène. Colloq Internation Centre Nation Rech Scient, Paris, 219:103–114

Hoang CT, Taviani M (1991) Stratigraphic and tectonic implications of uranium-series-dated coral reefs from uplifted Red Sea Islands. Quatern Res 35:264–273

Hofstetter A (2003) Seismic observations of the 22/11/1995 Gulf of Aqaba earthquake sequence. Tectonophysics 369:21–36

Hume WF (1906) The topography and geology of the peninsula of Sinai (southern portion). Survey Dept Egypt, Cairo, p 280

Hume WF, Little WH (1928) Raised beaches and terraces of Egypt. First Rep Comm on Pliocene and Pleistocene Terraces, Paris, pp 9–15

Ivanovich M, Harmon RS (1992) Uranium-series Disequilibrium: Applications to Earth, Marine, and Environmental Sciences. Oxford University Press, Oxford, p 911

Johnson PR, Kattan FH (2012) The Geology of the Saudi Arabian Shield. Saudi Geological Survey, Jeddah, p 479

Kröner A (1979) Pan African plate tectonics and its repercussions on the crust of northeast Africa. Geol Rundsch 68:565–583

Lake JH (2004) Early marine cementation in Upper Devonian (Duperow Formation) carbonates in southwestern Saskatchewan. In: Summary of Investigations 2004, 1. Saskatchewan Geological Survey, Sask Industry Resources, Misc Rep 2004–4.1, CD-ROM, Paper A-6, 3 pp

Lambeck K, Purcell A, Flemming NC, Vita-Finzi C, Alsharekh AM, Bailey GN (2011) Sea level and shoreline reconstructions for the Red Sea: Isostatic and tectonic considerations and implications for homin migration out of Africa. Quatern Sci Rev 30:3542–3574

Locke S, Thunell RC (1988) Paleoceanographic record of the last glacial/interglacial cycle in the Red Sea and Gulf of Aden. Palaeogeogr Paleoclimatol Palaeocol 64:163–187

Manaa AA (2011) Late Pleistocene raised coral reefs in the eastern Red Sea—Rabigh, Saudi Arabia. M.Sc. Thesis, School of Earth and Environmental Sciences, University of Wollongong, Australia. http://ro.uow.edu.au/theses/

Manaa AA, Jones BG, McGregor HV, Zhao J, Price DM (2016) Dating Quaternary raised coral terraces along the Saudi Arabian Red Sea coast. Mar Geol 374:59–72

Mansour AM, Madkour HA (2015) Raised coral reefs and sediments in the coastal area of the Red Sea. In: Rasul NMA, Stewart ICF (eds) The Red Sea: the formation, morphology, oceanography and environment of a young ocean Basin. Springer Earth System Sciences, Berlin Heidelberg, pp 379–393

Marchesan M, Taviani M (1994) Biological modification of coral reef molluscan assemblages during a glacial-interglacial cycle (Red Sea, Egypt). Mem Descr Carta Geol It 52:57–58

Mazzullo SJ, Cys JM (1978) In-situ formation of botryoidal aragonite on Permian seafloor. Am Assoc Petrol Geol Bull 63:493

McCulloch MT, Esat T (2000) The coral record of last interglacial sea levels and sea surface temperatures. Chem Geol 169:107–129

Mergner H, Schuhmacher H (1974) Morphologie, okologie und zonierung von korallenriffen bei Aqaba (Gulf von Aqaba, Rotes Meer). Helgoländer Wissensc Meeresunt 26:238–258

Mewis H, Kiessling W (2013) Environmentally controlled succession in a late Pleistocene coral reef (Sinai, Egypt). Coral Reefs 32:49–58

Montaggioni LF, Braithwaite CJR (2009) Quaternary Coral Reef Systems, History, development processes and controlling factors. Developments in Marine Geology 5, Elsevier, 550 pp

Moon FW (1923) Preliminary geological report on St. John's Island (Red Sea). Geol Surv Egypt Government Press, Cairo, Egypt, p 36

Nardini S (1937) Molluschi delle spiagge emerse del Mar Rosso e dell'Oceano Indiano. Parte II (Lamellibranchi). Palaeontogr It 37:225–247

Newton RB (1900) Pleistocene shells from the raised beach deposits of the Red Sea. Geol Mag 7(500–514):544–560

Orszag-Sperber F, Plaziat F, Baltzer F, Purser BH (2001) Gypsum salina-coral reef relationships during the Last Interglacial (MIS 5e) on the Egyptian Red Sea coast: A Quaternary analogue for Neogene marginal evaporites? Sediment Geol 140:65–85

Ostermann M, Sanders D, Prager C, Kramers J (2007) Aragonite and calcite cementation in "boulder-controlled" meteoric environments on the Fern Pass rockslide (Austria): Implications for radiometric age dating of catastrophic mass movements. Facies 53:189–208

Plaziat J-C, Purser BH, Soliman M (1989) Localisaton et organisation interne de récifs coralliens immatures sur un cone alluvial du Quaternarie ancien de la Mer Rouge (Sud de l'Egypt). Géol Médit 16:41–59

Plaziat J-C, Baltzer F, Choukri A, Conchon O, Freytet P, Orszag-Sperber F, Purser BH, Raguideau A, Reyss J-L (1995)

Quaternary changes in the Egyptian shoreline of the northwestern Red Sea and Gulf of Suez. Quatern Int 29–30:11–22

Plaziat J-C, Baltzer F, Choukri A, Conchon O, Freytet P, Orszag-Sperber F, Raguideau A, Reyss JL (1998) Quaternary marine and continental sedimentation in the northern Red Sea and Gulf of Suez (Egyptian coast): Influences of rift tectonics, climatic changes and sea-level fluctuations. In: Purser BH, Bosence DWJ (eds) Sedimentation and tectonics of rift basins: Red Sea-Gulf of Aden. Chapman and Hall, London, pp 537–573

Plaziat J-C, Reyss J-L, Choukri A, Cazala C (2008) Diagenetic rejuvenation of raised coral reefs and precision of dating. The contribution of the Red Sea reefs to the question of reliability of the Uranium-series datings of middle to late Pleistocene key reef-terraces of the world. Carn Géol 4:1–40

Pons-Branchu E, Douville E, Roy-Barman M, Dumont E, Branchu P, Thil F, Frank N, Bordier L, Borst W (2014) A geochemical perspective on Parisian urban history based on U-Th dating, laminae counting and yttrium and REE concentrations of recent carbonates in underground aqueducts. Quatern Geochronol 24:44–53

Ragani L (1997) Echinoid fauna from the Quaternary fringing reef of Aqaba (Jordan). Palaeontogr It 84:1–19

Reiss Z, Hottinger L (1984) The Gulf of Aqaba: Ecological Micropaleontology. Ecological Studies 50, Springer Verlag, Berlin Heidelberg New York Tokyo, 355 pp

Reiss Z, Luz B, Almogi-Labin A, Halicz E, Winter A, Wolf M, Ross DA (1980) Late Quaternary paleoceanography of the Gulf of Aqaba (Eilat), Red Sea. Quatern Res 14:294–308

Rimbu N, Lohmann G, Felis T, Pätzold J (2001) Arctic Oscillation signature in a Red Sea coral. Geophys Res Lett 28(15):2959–2962

Roberts HH, Aharon P, Walsh MM (1993) Cold-seep carbonates of the Louisiana continental slope-to-basin floor. In: Rezak R, Lavoie DL (eds) Carbonate Microfabrics. Springer, Berlin, pp 95–104

Roobol MJ, Al-Rehaili M, Arab N, Celebi M, Halawani MA, Janjou D, Kazi A, Martin C, Sahl M, Showail A (1999) The Gulf of Aqaba Earthquake of 22 November, 1995: Its effects in Saudi Arabia. Saudi Arabian Deputy Ministry for Mineral Resources Technical Report BRGM-TR-99-16, 67 pp

Sandford KS, Arkell WJ (1928) Paleolithic Man and the Nile Valley in lower Egypt with some notes upon a part of the Red Sea littoral. A study of the regions during Pliocene and Pleistocene times. The University of Chicago Press, pp 1–189

Savard MM, Beauchamp B, Veizer J (1996) Significance of aragonite cements around Cretaceous marine methane seeps. J Sediment Res 66(3):430–438

Schick AP (1958) Tiran: The straits, the island, and its terraces. Israel Explor J 8(3):189–196

Scholz D, Mangini A, Felis T (2004) U-series dating of diagenetically altered fossil reef corals. Earth Planet Sci Lett 218:163–178

Selli R (1973) Molluschi quaternari di Massaua e di Gibuti. In: Missione Geologica A.G.I.P. nella Dancalia meridionale e sugli altopiani Hararini (1936–1938), Docum Paleont Accad Naz Lincei Roma 4:151-444

Steckler MS, ten Brink US (1986) Lithospheric strength variations as a control on new plate boundaries; examples from the northern Red Sea region. Earth Planet Sci Lett 79:12159–12173

Stern RJ (1994) Arc assembly and continental collision in the Neoproterozoic East African Orogen: implications for consolidation of Gondwanaland. Ann Rev Earth Planet Sci 22:319–351

Stern RJ, Johnson P (2010) Continental lithosphere of the Arabian Plate: a geologic, petrologic, and geophysical synthesis. Earth-Sci Rev 101:29–67

Strasser A, Strohmenger C (1997) Early diagenesis in Pleistocene coral reefs, southern Sinai, Egypt: Response to tectonics, sea-level and climate. Sedimentology 44:537–558

Strasser A, Strohmenger C, Davaud E, Bach A (1992) Sequential evolution and diagenesis of Pleistocene coral reefs (South Sinai; Egypt). Sediment Geol 78:59–79

Taviani M (1994) The ever changing climate: Late Quaternary paleoclimatic modifications of the Red Sea region as deduced from coastal and deep-sea geological data. In: Proc Egyptian-Italian Seminar on "Geosciences and Archeology in the Mediterranean Countries". Geol Survey of Egypt, Spec Publ 70:193–200

Taviani M (1998a) Post-Miocene reef faunas of the Red Sea: Glacio-eustatic controls. In: Purser BH, Bosence DWJ (eds) Sedimentation and tectonics of rift basins: Red Sea-Gulf of Aden. Chapman and Hall, London, pp 574–582

Taviani M (1998b) Axial sedimentation of the Red Sea Transitional Region (22–25 N): Pelagic, gravity flow and sapropel deposition during the late Quaternary. In: Purser BH, Bosence DWJ (eds) Sedimentation and tectonics of rift basins: Red Sea-Gulf of Aden. Chapman and Hall, London, pp 467–478

Taviani M (1998c) Stable tropics not so stable: Climatically driven extinctions of reef-associated molluscan assemblages (Red Sea and western Indian Ocean; last interglaciation to present). In: Camoin GF, Davis G (eds) Reefs and Carbonate platforms in the Pacific and Indian oceans. Spec Publ Intl Assoc Sedimentologists, pp 69–76

Taviani M (2001) Fluid venting and associated processes. In: Martini P (ed) Vai GB. The Apennines and adjacent Mediterranean Basin. Kluwer Academic Publishers, Anatomy of an Orogen, pp 351–366

Taviani M, Franchi F, Angeletti L, Correggiari A, Lopez Correa M, Maselli V, Mazzoli C, Peckmann J (2016) Biodetrital carbonates on the Adriatic continental shelf imprinted by oxidation of seeping hydrocarbons. Mar Petrol Geol 66:511–531

Taviani M, Hoang C-T, Pini A, Rabbi E (1987) Massive aragonite submarine lithification of reefal and non-reefal host rocks: Examples from the Red Sea. 8th IAS Regional Meeting of Sedimentology, Tunis (Tunisia), 1–3 April 1987, Abstracts, pp 534–535

Taviani M, Rabbi E (1984) Marine botryoidal aragonite in Pleistocene reef limestones of Red Sea offshore islands (Northern Brother and Rocky Island). Min Petrogr Acta 27:49–56

Taviani M, Bonatti E, Colantoni P, Rossi PL (1986) Tectonically uplifted crustal blocks in the northern Red Sea: Data from the Brothers islets. Mem Soc Geol Ital 27(1984):47–50

Tawadros E (2001) Geology of Egypt and Libya. AA Balkema, Rotterdam, p 468

Veeh HH, Giegengack R (1970) Uranium-series ages of corals from the Red Sea. Nature 226:155–156

Walter RC, Buffler RT, Bruggemann JH, Guillaume MMM, Berhe SM, Negassi B, Libsekal Y, Cheng H, Edwards RL, von Cosel R, Néeraudeau D, Gagnon M (2000) Early human occupation of the Red Sea coast of Eritrea during the last interglacial. Nature 405(6782):65

Mollusc Fauna Associated with Late Pleistocene Coral Reef Systems of the Saudi Arabian Side of the Gulf of Aqaba

Lorenzo Angeletti, Najeeb M. A. Rasul, and Marco Taviani

Abstract

The Saudi Arabian eastern side of the Gulf of Aqaba has been recently explored all along its length, and the main Pleistocene coral reef outcrops identified and sampled for macrofossils. Terraces ascribable to the last interglacial, the Marine Isotope Stage 5e (MIS5e, ~125kyr BP), predominate and contain a well preserved fossil legacy of sea life at the time. Molluscs proved to represent the most diverse component in the macrobenthos, overwhelming at species up to family levels other benthic groups. As many as 124 families of marine molluscs have been identified in this fossil system, 86 of which pertain to Gastropoda, 31 to Bivalvia, 5 to Polyplacophora and 2 to Scaphopoda. No less than 277 genera are represented in the fossil mollusc fauna, and the diversity is well in excess of 400 species. Molluscs are a fundamental clue for classifying with relative precision the former habitats preserved in these MIS5e coral systems. By and large, the MIS5e deposits document former back reef to fore-reef environments and, to a minor degree, beach or mangal settings. The fauna includes also some index-fossils of ecostratigraphic significance for recognizing MIS5e deposits in the Red Sea. The MIS5e fauna of the Gulf of Aqaba is an astonishing example of a tropical marine biodiversity hotspot whose number of species and genera is not rivalled globally by any known coeval mollusc fauna.

L. Angeletti · M. Taviani (✉)
Istituto di Scienze Marine (ISMAR-CNR), Via Gobetti, 10140129 Bologna, Italy
e-mail: marco.taviani@bo.ismar.cnr.it

L. Angeletti
e-mail: lorenzo.angeletti@bo.ismar.cnr.it

N. M. A. Rasul
Center for Marine Geology, Saudi Geological Survey, Jeddah, Saudi Arabia

M. Taviani
Biology Department, Woods Hole Oceanographic Institution, 266 Woods Hole Road, Woods Hole, MA 02543, USA

M. Taviani
Stazione Zoologica Anton Dohrn, Villa Comunale, 80121 Naples, Italy

N. M. A. Rasul and I. C. F. Stewart (eds.), *Geological Setting, Palaeoenvironment and Archaeology of the Red Sea*,
https://doi.org/10.1007/978-3-319-99408-6_17

1 Introduction

An extensive flight of Pleistocene marine terraces rims, albeit discontinuously, the entire eastern coast of the Gulf of Aqaba (Taviani et al., this volume). Terraces formed in response to Pleistocene sea-level fluctuations and mostly represent former fringing coral reef systems that are largely ascribable to the Marine Isotope Stage 5e (MIS5e) dated at about 125 kyr BP (Dullo 1990; Alhejoj et al. 2016a; Manaa et al. 2016; Bosworth et al. 2017; Taviani et al., this volume). Pleistocene marine terraces also occur on the opposite side of the Gulf of Aqaba, outcropping along the Sinai Peninsula (Gvirtzman et al. 1992; Strasser et al. 1992; Gvirtzman 1994; Bosworth and Taviani 1996; Mewis and Kiessling 2013; Taviani et al., this volume), or are still submerged (Hartman et al. 2015). Outside the Gulf of Aqaba, MIS5e terraces stretch the entire length of the Red Sea and Gulf of Aden (e.g., Berry et al. 1966; Hoang and Taviani 1991; Reyss et al. 1993; Bosworth and Taviani 1996; El-Asmar 1997; El-Sorogy 1997; Plaziat et al. 1998, 2008; Taviani 1998a; Mansour and Madkour 2015; Alexandroff et al. 2016; Hamed et al. 2016; Al-Mikhlafi et al. 2017; Casazza 2017).

All such Red Sea Pleistocene marine terraces are very fossiliferous (Issel 1869, 1873; Newton 1900; Bullen 1901; Hall and Standen 1907; Arkell 1928; Jousseaume 1930; Macfayden 1930; Nardini 1937; Selli 1944, 1973; Leigh 1968; Goreau et al. 1976; Borri et al. 1982; Sabelli and Taviani 1984; Dullo 1990; Lorenz 1992; Ragani 1997; Plaziat et al. 1998; Taviani 1998a; Walter et al. 2000; Bantan and Abu-Zied 2014; El-Sorogy et al. 2014; Al-Mikhlafi et al. 2017; Casazza 2017; Abu-Zied and Bantan 2018), providing important hints about climate ecosystem turnovers in the semi-enclosed Red Sea basin (Dullo 1990; Marchesan and Taviani 1994; Taviani 1994, 1998a, b; Plaziat et al. 1998; Mewis 2016).

Since little is known on the paleontological content of Pleistocene terraces in the key area of the Gulf of Aqaba besides scleractinian corals, various sites in the Saudi Arabian section of the Gulf have been surveyed and sampled for macrofossils during an international mission in 2013 (Fig. 1 and Table 1).

Here we report on molluscs from MIS5e outcrops, with a preliminary discussion of their diversity and providing hints about their paleoecological potential. The present chapter is intended as an introduction to this important fossil fauna, certainly meritorious to be considered in the future for more detailed taxonomic and paleoenvironmentally-oriented studies.

2 Materials and Methods

Details about the 2013 survey of the marine terraces along the Saudi Arabian section of the Gulf of Aqaba are reported in Bosworth et al. (2013, this volume) and Taviani et al. (this volume), and are not repeated here. Qualitative paleontological sampling in the field was carried out through collecting bulk sediment for minute species and hand picking for larger ones (Fig. 2). A substantial contribution of smaller shells was also derived by washing the inside of large shells over a 0.1 mm mesh. Sample sieving was carried out in the Saudi Geological Survey (SGS) headquarters in Jeddah immediately after the expedition and fractions were washed in freshwater and dried at room temperature or in the oven at $T < 40$ °C. Taxonomic identification was carried out by referring mostly to Oliver (1992), Dekker and Orlin (2000), Zuschin and Oliver (2003), Zuschin et al. (2009), and Janssen et al. (2011). Classification follows the World Register of Marine Species (WoRMS: http://www.marinespecies.org/), in agreement with Bouchet and Rocroi (2005, 2010), amended in the case of Nassariidae (Galindo et al. 2016) and Conidae (Puillandre et al. 2014, 2015). Although *Aqabarella* appears in the MIS5e fauna, the family Aqabarellidae recently proposed from the modern coral reef environment in the Gulf of Aqaba (Alhejoj et al. 2016b) is not computed in our study since it has not been validated to date by WoRMS.

3 Comparison with Pleistocene and Recent Mollusc Fauna

Molluscs prove to be the most abundant and diverse component in the fossil legacy of such terraces, overwhelming, in terms of number of species, co-occurring scleractinians (main frame-building corals), octocorals, hydroids, echinoids, crustaceans, polychaetes, bryozoans, brachiopods, foraminifera, ostracods, fishes, and calcareous algae. The malacological material consists of many thousands of shells whose taxonomic assessment at the species level is underway and the present study is mainly oriented at the supraspecific level with a focus on family and genus level (Appendix 1).

We have identified our material at family level, recognizing 124 families thus far, grouping not less than 277 genera (Table 2) containing more than 400 species (work in progress). The most diverse group is Gastropoda (86 families), followed by Bivalvia (31), Polyplacophora (5), and Scaphopoda (2). These figures must be taken as conservative and set to grow since not all the material is yet scrutinized, many fragmented shells and juveniles are unaccounted for, and some species-rich groups (e.g., Cerithiopsidae, Triphoridae and Pyramidellidae) are still under-studied. Those numbers are significantly higher than the taxa recorded by Selli (1973) in the rich Quaternary deposits (in all likeness MIS5e) of Massawa and Djibouti in the proximity of the Bab el-Mandeb Strait, accounting for 80 families, 134 genera and 224 species.

If we analyse our data set at a family level (with the exclusion of Nudibranchia that leave no obvious fossils for comparison) in the light of the compilation by Bouchet and Rocroi (2005, 2010), it appears that the MIS5e fauna contains as much as 36 and 30% of all extant marine gastropod and bivalve families, respectively. This is indeed remarkable, making this late Pleistocene fauna a key reference for historical biology. The high number of taxa is amplified by the occurrence in these fossil strata of many micromolluscs, whose shells represent a substantial share of the Indo-Pacific coral reef mollusc fauna especially among gastropods and, furthermore, provide rare hints of secluded habitats of the reef system (cf. Bouchet et al. 2002; Sasaki 2008).

In the fossil fauna under scrutiny, Class Polyplacophora accounts for 2 superfamilies, 5 families and 9 genera. Bivalvia is represented by 21 superfamilies, 31 families and 79 genera. Scaphopoda account for 2 families and 2 genera. Gastropoda is represented by 44 superfamilies, 86 families and 187 genera (Table 2). The most recent checklist of the Red Sea mollusc fauna was compiled by Dekker and Orlin (2000). With respect to that list (updated to the present taxonomic acceptance, and exclusive of deep-water taxa added by Janssen and Taviani 2014), the Gulf of Aqaba MIS5e fauna accounts for 86 out of 102 Gastropoda families (excluding Nudibranchia and the other shell-less molluscs), 31 out of 51 Bivalvia and 2 out of 3 among Scaphopoda and 5 out of 6 among Polyplacophora. Lastly, the comparison of the number of genera present in the families shared by the modern whole Red Sea and the MIS5e fauna underlines further the remarkable richness of the fossil fauna of the Saudi Arabian Gulf of Aqaba (Fig. 3).

Fig. 1 Location of last interglacial MIS5e terraces along the Saudi Arabian side of the Gulf of Aqaba discussed in the text

Table 1 Location of last interglacial MIS5e terraces along the Saudi Arabian side of the Gulf of Aqaba discussed in the text

Station name	Latitude N	Longitude E	Notes
Station 04	28°47′ 39.27″	34°49′ 33.02″	4.1 km south of Ra's Suwayhil as Saghir
Station 05	28°46′ 46.89″	34°49′ 10.33″	5.8 km south of Ra's Suwayhil as Saghir
Station 07	28°42′ 36.15'″	34°47′ 49.91″	13.8 km south of Ra's Suwayhil as Saghir
Station 08	28°42′ 30.28″	34°47′ 46.67″	14 km south of Ra's Suwayhil as Saghir
Station 09	28°42′ 30.28″	34°47′ 46.67″	14.2 km south of Ra's Suwayhil as Saghir
Station 10	28°42′ 30.28″	34°47′ 46.67″	14.5 km south of Ra's Suwayhil as Saghir
Station 11	28°42′ 45.59″	34°47′ 52.91″	13.5 km south of Ra's Suwayhil as Saghir
Station 15	28°05′ 47.94″	34°34′ 18.74″	Ash Shaykh Humayd
Station 16	29°11′ 22.95″	34°54′ 01.72″	2.3 km south of Al Wasel
Station 19	28°48′ 32.05″	34°49′ 52.12″	2.5 km south of Ra's Suwayhil as Saghir
Station 20	28°35′ 06.83″	34°47′ 35.15″	3 km north of Wadi Tayyib Ism
Station 23	28°30′ 17.31″	34°47′ 37.17″	6.1 km south of Wadi Tayyib Ism
Station 25	28°26′ 58.80″	34°45′ 36.60″	5.6 km north of Maqna
Station 26	28°26′ 23.32″	34°45′ 20.61″	4.5 km north of Maqna

Noticeably, the MIS5e mollusc fauna under scrutiny includes classic index fossils (Fig. 4) such as the extinct gastropod *Diodora impedimentum* (Cooke, 1885) or other gastropods such as *Euplica turturina* (Lamarck, 1822) that seem excluded from the present Red Sea fauna (Newton 1900; Mienis 1981; Bosworth and Taviani 1996; Taviani 1998a, 1998b). Additionally, our fossil fauna contains the gastropods *Monetaria moneta* (Linnaeus, 1758), *Oliva inflata* Lamarck, 1811 (= *O. bulbosa* (Röding, 1798)), and *Corbula taitensis* Lamarck, 1818 (here synonymized with *C. acutangula* Issel 1869: see also Selli 1973) that, albeit still occurring at present in the Red Sea (Dekker and Orlin 2000), are consistently more abundant and widespread in the last interglacial deposits of this basin than at present (Taviani 1982, 1994, 1998a; Taviani et al. 1995).

Altogether these molluscs have an ecostratigraphic meaning in a way comparable to that of the Senegalese guests in the Mediterranean MIS5e (Sabelli and Taviani 2014; Taviani 2015, with references therein). They are, in fact, recurrent and often abundant in Red Sea MIS5e deposits and their chronostratigraphic value has been confirmed by U/Th ages of some such outcrops (e.g., Hoang and Taviani 1991; Bosworth and Taviani 1996; Bosworth et al. 2017). *D. impedimentum* and *E. turturina* are mainly found in reef edge and fore-reef former habits while *M. moneta*, *O. inflata* and *C. taitensis* are mostly from back-reef lagoonal sandy habitats. The recently described gastropod *Turbo moolenbeecki* Dekkers and Dekker (2016), suggested to be a Red Sea endemic (Pleistocene-Recent) belonging to the Indo-Pacific *T. petholatus*-group, occurs in the MIS5e fauna of the Saudi Arabian Gulf of Aqaba.

4 Paleoecological Aspects

Today, as for the recent past, the physiography of the study area prevents a significant seaward aggradation of a fringing reef system since a substantial length of the cliffy coastline hangs over the rather precipitous walls of the Gulf down to bathyal depths (Fig. 5). Back-reef to lagoonal settings only developed at propitious sites characterized by flatter topography, namely wadi systems and coastal plains (Taviani et al., this volume). Whenever permitted by the (paleo)topography the higher-than-present MIS5e sea-level forced an expansion of marine conditions inland, so that back-reef and lagoonal systems are better represented in the last interglacial than today (see Taviani et al., this volume). On the other hand, the impressive uplift of part of the MIS5e system in the Gulf of Aqaba, reaching up to 26 m above present sea-level (Bosworth et al. 2017, this volume; Taviani et al., this volume), offers a rare view on deeper fore-reef habitats, seldom to be seen in the Red Sea geological legacy.

Macrofossils are a fundamental clue to classifying with relative precision the former habitats (Figs. 6, 7, 8 and 9). Those preserved in these MIS5e terraces document former fore-reef, reef edge and back-reef environments and, to a minor degree, beach, lagoonal or mangal settings (Taviani et al., this volume). Among reef macrobenthic invertebrates, molluscs stand as one of the most diverse and ecologically-significant group (Sorokin 1995). The MIS5e fauna considered here contains a variety of extant molluscs, whose preservation is generally excellent since the MIS5e terraces underwent little diagenesis with respect to their older

Fig. 2 Field examples of macrofossil occurrences: **a** In situ articulated but gaping shell of the giant clam *Tridacna maxima* Roeding, 1798 (St. 14); **b** bleached large shells belonging to *Lambis truncata sebae* (Kiener, 1843), *Tricornis tricornis* (Lightfoot, 1786), *Tectus dentatus* (Forsskål in Niebuhr, 1775) or *T. maxima* (photo: St. 1) are not necessarily MIS5e fauna but may at times represent anthropogenic deposits (e.g., food remains) (bar = 15 cm); **c** dense accumulation of molluscan shells (e.g., venerid and cardiid clams and *Conus*) and other skeletal remains in sandy matrix (St. 15, near Ash Shaykh Humayd); **d** *Pinna*-layer from St. 16, with articulated *Pinna muricata* Linnaeus, 1758 shells amidst other bivalves and gastropods and echinoid tests; **e** partially dissolved shell of *Turbo radiatus* Gmelin, 1791, embedded in evaporitic sediment (St. 11); **f** partially exfoliated shell of the large strombid *L. truncata sebae* emerging from host coarse-skeletal sand (St. 13, bar = 10 cm)

Table 2 Superfamilies, families and genera recorded in the MIS5e mollusc fauna of the Saudi Arabian Gulf of Aqaba

	Superfamily	Family	Genera
Polyplacophora	2	5	9
Bivalvia	21	31	79
Scaphopoda	–	2	2
Gastropoda	44	86	187
Total	67	124	277

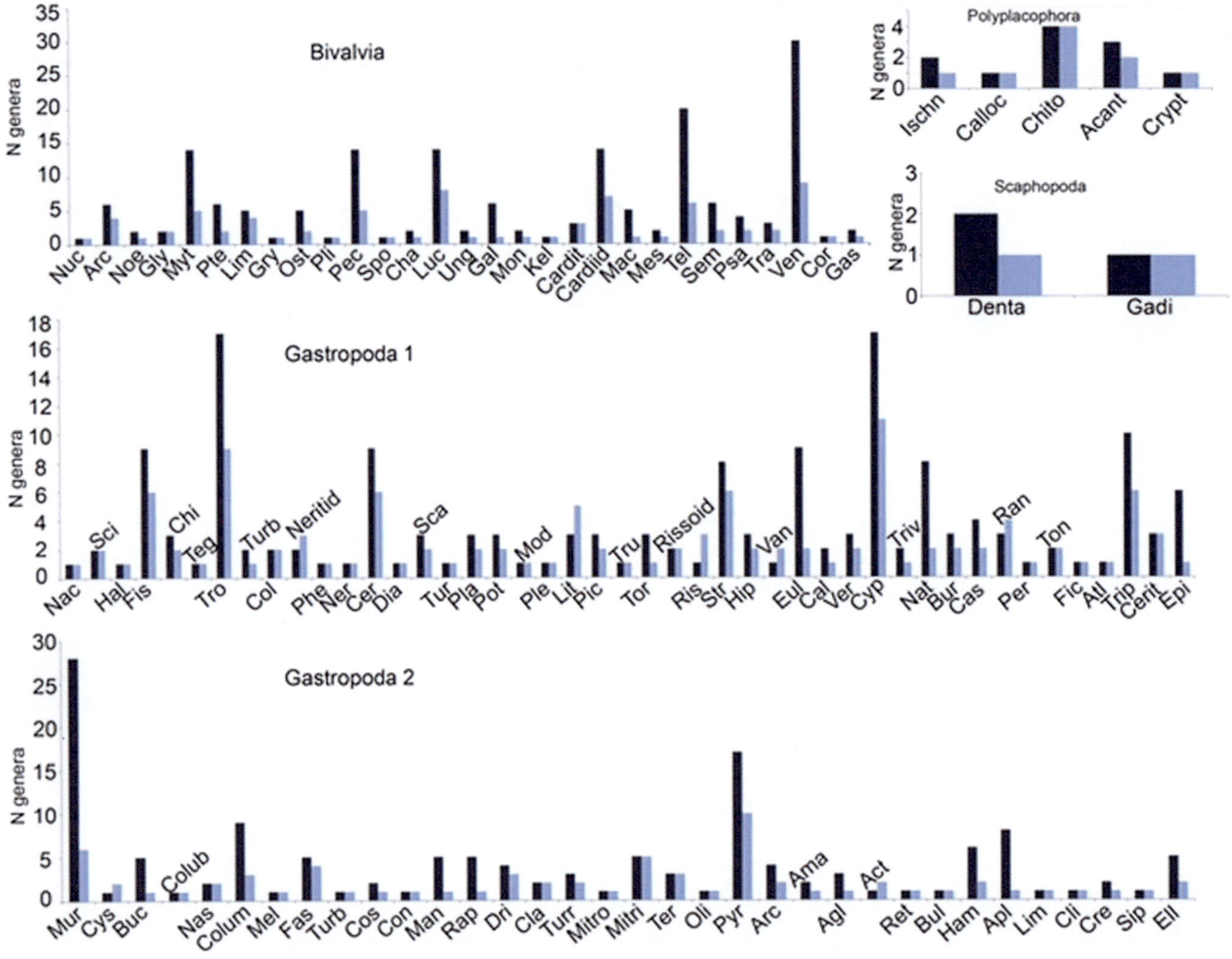

Fig. 3 Number of genera identified for the 124 families found in the Saudi Arabian Gulf of Aqaba MIS5e marine terraces (light blue) with respect to those listed by Dekker and Orlin (2000) for the modern Red Sea (dark blue). Gastropoda 1 refers to families and genera belonging to Patellogastropoda, Vetigastropoda, Neritimorpha and Caenogastropoda; Gastropoda 2 refers to families and genera pertaining to Caenogastropoda, Neogastropoda, Heterobranchia, Opistobranchia and Pulmonata

Fig. 4 Index-fossils from MIS5e marine terraces: The cowry *Monetaria moneta* (Linnaeus, 1758) (**a** St. 16, bar = 1 cm), the keyhole limpet *Diodora impedimentum* (Cooke, 1885) (**b** St. 15, bar = 5 mm), the columbellid *Euplica turturina* (Lamarck, 1822) (**c** St. 16, bar = 5 mm) and the olivid *Oliva inflata* Lamarck, 1811(**d** St. 15, bar = 1 cm)

Fig. 5 Panoramic view of a MIS5e marine terrace about 14 km south of Ra's Suwayhil as Saghir (St. 7). The fossil reef edge is indicated by the black arrow, with the modern fringing reef equivalent shown by the white arrow; the modern wadi termination is visible in the foreground while the Panafrican crystalline basement is in the background

counterparts, resulting in a general absence of lithification of the encasing sediment and little shell degradation. Occasional exceptions are some seaside cliffs exposed to salty spray, some hanging reefs stacked on crystalline bedrock, and some sandy deposits, where incipient lithification or partial shell dissolution was observed. The exquisite preservation of a good share of the MIS5e fossil fauna is further confirmed by the fact that some shells still retain part of the original colour patterns, especially those characterized by brownish to reddish original hues.

Fig. 6 Examples of gastropods, bivalves and polyplacophorans diagnostic of different habitats. Among molluscs indicative of intertidal rocky shore settings, thus attesting to former shorelines, are *Nerita* cf. *orbignyana* Récluz, 1841 (**a** St. 16, bar = 5 mm), *Acanthopleura vaillantii* Rochebrune, 1882 (**b** St. 16, bar = 5 mm), *Cellana rota* (Gmelin, 1791) (**c** St. 8, bar = 1 cm), *Siphonaria* sp. (**d** St. 15, bar = 1 mm), *Angiola punctostriata* (E.A. Smith, 1872) (**e** St. 16, bar = 2 mm), and *Modiolus auriculatus* (Krauss, 1848) (**g** St. 16, bar = 5 mm); the tiny neritid *Smaragdia* sp. is a seagrass indicator (**f** St. 15, bar = 1 mm); the potamidid *Terebralia palustris* (Linnaeus, 1767) (**h** St. 11, bar = 1 cm) is exclusive of mangrove habitats

It is important to call attention to the paleoecological value provided by the Gulf of Aqaba MIS5e mollusc fauna that represents a wide array of life habits in terms of trophic and substrate requirements (Figs. 4, 5 and 6). Gastropoda are mostly mobile and vagrant epifauna and infauna, inclusive of macro and microcarnivore predators (i.e., Naticidae, Ranellidae, Cassidae, Tonnidae, Muricidae, Olividae, Conidae, Turridae, Raphitomidae, Mitridae, Costellariidae, Bullidae), scavengers (e.g., Nassariidae), herbivores (e.g., Haliotidae, Trochidae, Tegulidae, Neritidae, Cerithiidae) and detritivores (e.g., Strombidae); specialized corallivorous species (e.g., Epitoniidae, Coralliophiinae, Architectonicidae), spongivorous (e.g., Triphoridae), and ectoparasites (e.g., Eulimidae) also occur. Bivalvia include primarily sediment infaunal/semi-infaunal (e.g., Arcidae, Glycymeridae, Pinnidae, Cardiidae, Tellinidae, Veneridae), epifaunal (e.g., Pectinidae), or cemented (e.g., Ostreidae, Spondylidae, Plicatulidae), filter feeders but also deposit feeders, commensals, borers (e.g., Mytilidae), and chemosymbiotic representatives (e.g., Lucinidae), plus the classic symbiont-bearing taxa such as *Tridacna* spp.

Many diagnostic species are found that testify to a former habitat in the MIS5e terraces. Emblematic examples occurring in the fossil fauna under scrutiny are gastropods such as *Cellana rota* (Gmelin, 1791), *Nerita* spp., *Peasiella isseli* (Issel 1869), *Angiola punctostriata* (E.A. Smith, 1872), *Siphonaria* sp., the mytilid bivalve *Modiolus auriculatus* (Krauss, 1848), or the polyplacophoran *Acanthopleura vaillantii* Rochebrune, 1882, which are typical intertidal inhabitants of rocky shores (Safriel and Lipkin 1964; Reid 1989; Zuschin and Oliver 2003; Anseeuw and Terryn 2004; Zuschin et al. 2009; Janssen et al. 2011). The bivalve *Atactodea striata* (Gmelin, 1791) is a strictly beach-intertidal bivalve burrowed in coarse sand. The presence of these molluscs in the fossil assemblages can be taken as a reliable indication of former shorelines as documented, as an example, for the raised terraces south of Al Wasel (St. 16) or south of Ra's Suwayhil as Saghir (St. 7, 8 and 13). Amphibious Ellobiidae gastropods (*Allochroa* sp.) testify to supratidal conditions, and the large gastropod *Terebralia palustris* (Linnaeus, 1767), now extinct in the Red Sea region and identified south of Ra's Suwayhil as Saghir at St. 11, to mangrove settings (mangals: Taviani 1994; Plaziat et al. 1998). Gastropods like the large tegulid *Tectus dentatus* (Forsskål in Niebuhr, 1775) and giant vermetid *Ceraesignum maximum* (G.B. Sowerby I, 1825), are consistent with reef flat to reef edge situations (Kappner et al. 2000; Zuschin et al. 2009; Mahmoud et al. 2012) as seen in the terraces south of Al Wasel (St. 16) and south of Ra's Suwayhil as Saghir (St. 6 and 8). Strombids like *Conomurex*

Fig. 7 Examples of molluscs of paleoecological significance: the tegulid *Tectus dentatus* (Forsskål in Niebuhr, 1775) (**a** St. 16, bar = 1 cm) and the large vermetid *Ceraesignum maximum* (G.B. Sowerby I, 1825) (**c** St. 22, bar = 1 cm) are preferential dwellers of the reef flat and reef edge; the herbivore gastropod *Turbo radiatus* Gmelin, 1791 is a common component of shallow-marine back-reef and reef flat habitats (**e** St. 8, bar = 1 cm); *Spondylus* is a bivalve typically cemented to calcareous substrates, including scleractinian corals (**d** St. 16, bar = 1 cm), while *Trapezium oblongum* (Linnaeus, 1758) is consistently found in coral crevices from the reef edge down to the slope (**b** St. 27, bar = 1 cm); the occurrence of the shallow-water branching scleractinian *Pocillopora* is substantiated by the presence of the corallivorous muricid coralliophilinin *Coralliophila monodonta* (Blainville, 1832) (**g** St. 16, bar = 5 mm); the chemosymbiotic lucinid *Anodontia* sp. testifies at dysaerobic sediment, probably an internal bay or lagoon (**f** St. 16, bar = 1 cm)

fasciatus (Born, 1778) and *Tricornis tricornis* ([Lightfoot], 1786) or cerithiids such as *Rhinoclavus fasciata* (Bruguière, 1792), document sandy habitats, often vegetated, commonly back-reef lagoons but are also found in the fore-reef (Janssen et al. 2011). Examples occur south of Al Wasel (St. 16) and near Ash Shaykh Humayd (St. 15). The presence of seagrass habitats is 'indirectly' documented by diagnostic herbivores such as *Smaragdia* (Reich et al. 2015), and aggregations of *Pinna muricata* Linnaeus, 1758 (Zuschin and Oliver 2003), while the presence of sacoglossan prasinid gastropods, such as *Julia* (near Ash Shaykh Humayd, St. 15) testifies to the past occurrence of *Caulerpa* seaweeds (see Le Renard et al. 1996). Noticeable is the rare occurrence in these MIS5e terraces of holoplanktonic gastropods (thecosomatous pteropods and heteropods) and meroplanktonic mollusc larvae (Fig. 10), whose relative abundance in fore-reef facies reflects very narrow reefal systems facing open and deep marine conditions. The MIS5e holoplanktonic fauna contains taxa known from the late Pleistocene to Recent Red Sea fauna (Almogi-Labin et al. 1991; Janssen 2007).

Many more past habitats are documented in the Pleistocene terraces of the Gulf of Aqaba but, as mentioned, the in-depth paleoecological analysis is not the scope of this introductory chapter on the MIS5e mollusc fauna. Surely, the recent contributions on the ecology and taphonomy of modern Red Sea mollusc communities provide an excellent

Fig. 8 Examples of molluscs and their preferred habitats: *Bulla ampulla* Linnaeus, 1758 (**a** St. 16, bar = 1 cm) is a common inhabitant of protected and vegetated sandy bays; **b** the venomous snail *Conus arenatus* Hwass in Bruguière, 1792 (St. 16, bar = 5 mm) and the strombid *Conomurex fasciatus* (Born, 1778) (**e** St. 8, bar = 5 mm) likely point out former shallow-water sandy back-reef settings; *Tricornis tricornis* ([Lightfoot], 1786) is a large strombid mainly encountered in subtidal sandy settings as deep back-reef lagoons (**c** St. 15, bar = 1 cm), *Indomodulus tectum* (Gmelin, 1791) is not exclusive of a specific habitat but preferably inhabits shallow depths in the reef-flat or back-reef lagoons (**g** St. 19C, bar = 5 mm), while the cerithiid *Rhinoclavis fasciata* (Bruguière, 1792) is an eurybathic sand dweller (**d** St. 16, bar = 1 cm); the sacoglossan *Julia* cf. *exquisita* Gould, 1862, documents the past existence of caulerpid green algae (**f** St. 15, bar = 500 μm)

baseline for a comparison of Pleistocene *versus* modern counterparts (e.g., Zuschin and Oliver 2003, 2005; Zuschin and Stachowitsch 2007; Zuschin et al. 2000, 2001, 2009; Janssen et al. 2011; Bantan and Abu-Zied 2014; Zuschin and Ebner 2015a, 2015b; Zauner and Zuschin 2016).

5 The Gulf of Aqaba Case in the Global Context of MIS5e Mollusc Fauna

A considerable bibliography exists reporting on MIS5e marine molluscs from deposits around the world, although often limited to incomplete lists or notes on some taxa (e.g., Taylor 1978; Crame 1986; Brigham-Grette et al. 2001; Muhs et al. 2002; Charò et al. 2014; Montesinos et al. 2014; Ávila et al. 2015; Martínez et al. 2016). Regarding biodiversity (number of species), this literature permits us to identify the Red Sea—Gulf of Aden, the Mediterranean and the Azores as the prime regions for which a substantial number of mollusc taxa of this age has been listed. The deposits from the southern Red Sea and Gulf of Aden described by Selli (1973), in all likelihood coeval to ours, are relatively rich and diverse, but still fewer taxa are represented there, although encompassing a wider coastal region. In fact, Selli (1973) recorded 224 species (vs. >400 in this study) distributed into 134 (vs. 275 in this study) genera and 81 (vs. 124 in this study) families.

The classic last interglacial mollusc fauna of the temperate Mediterranean basin (Amorosi et al. 2014; Sabelli and Taviani 2014), although reasonably rich in molluscs (see references in Taviani 2015), is not so diverse as the Gulf of Aqaba at any taxonomic level below Class. The same reasoning applies to the Azores (Ávila et al. 2015).

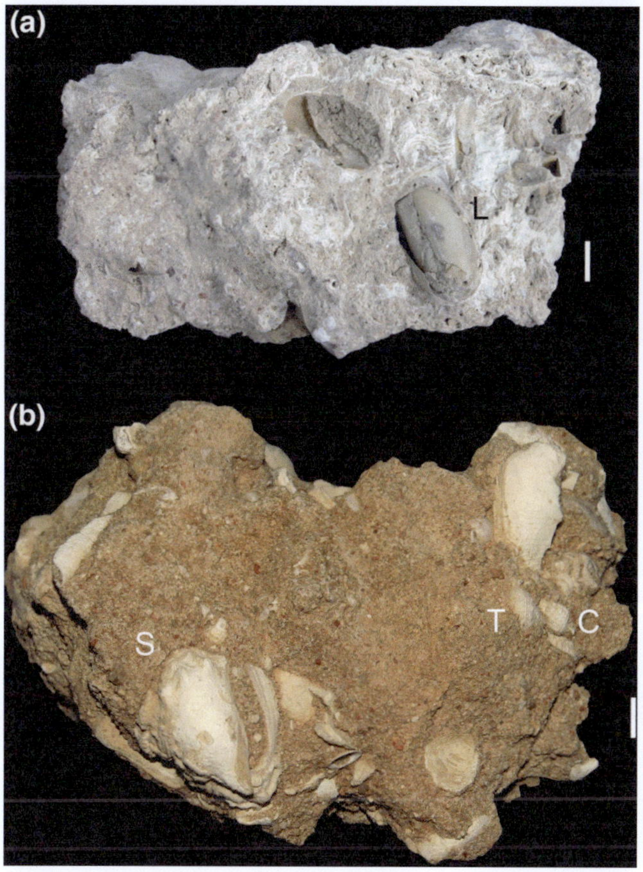

Fig. 9 **a** Massive coral rock is often home to the boring mytilid bivalve *Lithophaga* (L, St. 19, bar = 1 cm); **b** one of the best preserved MIS5e coral reef terrace is located south of Al Wasel (St. 16) and exposes also back-reef lagoonal facies as testified by loose sands and sandstones containing gastropods as strombids (S = *Conomurex fasciatus*), cerithiids (C), trochids (T = *Priotrochus obscurus* (W. Wood, 1828), naticids (*Mammilla*), conids and infaunal bivalves (*Timoclea*, etc.)

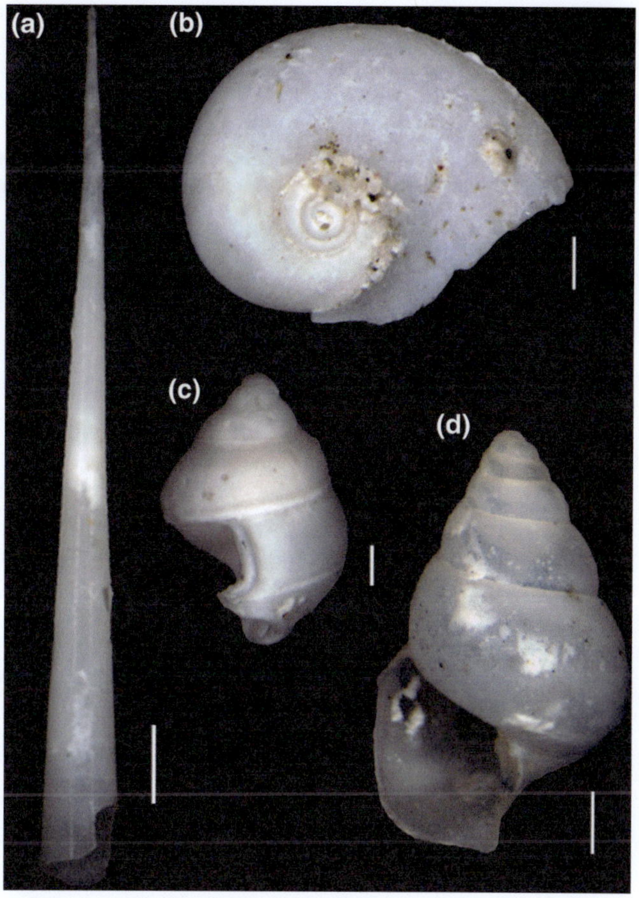

Fig. 10 Pelagic input in the MIS5e former coral environments is testified by holoplanktonic gastropods such as the Thecosomata pteropods *Creseis clava* (Rang, 1828) (**a** St. 15, bar = 500 μm) and *Limacina bulimoides* (d'Orbigny, 1834) (**d** St. 15, bar = 200 μm) and heteropods (**b** *Atlanta* sp., St. 8, bar = 200 μm), and by meroplanktic gastropod larvae (**c** St. 15, bar = 250 μm)

The remarkable richness of the Pleistocene fauna of the Gulf of Aqaba offers a rare window on the 'magnitude of biodiversity' (Bouchet 2006) in the recent past, contributing to a better temporal appreciation of biogeographic and evolutionary issues in the Indo-west Pacific coral reef systems.

6 Conclusions

The Last Interglacial (MIS5e) terraces along the Saudi Arabian coast in the Gulf of Aqaba, representing former coral reef complexes, inclusive of beach, lagoonal and fore-reef settings, host a rich molluscan component. Our reconnaissance survey documents no less than 124 families and 277 genera, grouping a number of species in excess of 400. The fossil fauna by and large matches the modern Red Sea fauna in terms of families and genera. The mollusc biodiversity of these terraces is the most pronounced of this age known thus far on a global scale.

Acknowledgements This paper is dedicated to the memory of geologist Marcello Facundo Quarantini (1972-2018). We thank the Saudi Geological Survey for the invitation to present this study at the '2nd Red Sea Book Workshop' held in Jeddah on February 2016. The expedition in the Gulf of Aqaba was promoted and funded by the Saudi Geological Survey in Jeddah. We acknowledge all participants in the winter 2013 field party and the SGS personnel for field work and post-expedition sample processing. In particular, the raised coral reefs considered here were surveyed together with Bill Bosworth and Paolo Montagna. We are very grateful to Martin Zuschin and two anonymous referees for their critical comments. This paper is part of the PRIN2012 Programme (Project 20125JKANY_002, Principal Investigator Marco Ligi) and is Ismar-CNR, Bologna, Scientific contribution no. 1903.

Appendix

Preliminary list of mollusc from marine terrace of the Marine Isotope Stage 5e from the Gulf of Aqaba.

Phylum	Classis	Subclassis	Ordo	Superfamilia	Familia	Genus
Mollusca	**Polyplacophora**					
		Neoloricata	Chitonida	Chitonoidea	Callochitonidae	*Callochiton*
					Chitonidae	*Chiton*
						Rhyssoplax
						Lucilina
						Acanthopleura
					Ischnochitonidae	*Ischnochiton*
				Cryptoplacoidea	Acanthochitonidae	*Acanthochitona*
						Choneplax
					Cryptoplacidae	*Cryptoplax*
	Bivalvia					
		Protobranchia	Solemyoidea	Manzanelloidea	Nucinellidae	*Huxleia*
		Pteriomorpha	Mytiloida	Mytiloidea	Mytilidae	*Brachidontes*
						Modiolus
						Musculus
						Lihophaga
						Septifer
			Arcoida	Arcoidea	Arcidae	*Acar*
						Anadara
						Arca
						Barbatia
					Noetidae	*Arcopsis*
					Glycymeridae	*Glycymeris*
						Tucetona
			Pterioida	Pterioidea	Pteriidae	*Pinctada*
						Isognomon
				Pinnoidea	Pinnidae	*Pinna*
			Ostreoida	Ostreoidea	Ostreidae	*Saccostrea*
						sp.
					Gryphaeidae	*Hyotissa*
			Pectinoida	Pectinoidea	Pectinidae	*Mimachlamys*

(continued)

Phylum	Classis	Subclassis	Ordo	Superfamilia	Familia	Genus
						Scaeochlamys
						Semipallium
						Laevichlamys
						Decatopecten
					Spondylidae	*Spondylus*
				Plicatuloidea	Plicatulidae	*Plicatula*
			Limoida	Limoidea	Limidae	*Ctenoides*
						Lima
						Limaria
						Limatula
		Heterodonta	Lucinoida	Lucinoidea	Lucinidae	*Anodontia*
						Codakia
						Ctena
						Lamellolucina
						Cardiolucina
						Divalinga
						Pillucina
						Wallucina
			Carditoida	Carditoidea	Carditidae	*Beguina*
						Cardita
						Cardites
			Veneroida	Arcticoidea	Trapeziidae	*Trapezium*
						Coralliophaga
				Cardioidea	Cardiidae	*Acrosterigma*
						Fragum
						Tridacna
						Ctenocardia
						Vasticardium
						Lunulicardia
						Parvicardium
				Chamoidea	Chamidae	*Chama*
				Galeommatoidea	Galeommatidae	sp.
					Montacutidae	*Curvemysella*
					Kelliidae	*Kellia*
				Mactroidea	Mactridae	*Mactra*
				Tellinoidea	Semelidae	*Ervilia*
						Iacra
					Tellinidae	*Tellina*
						Tellinella
						Phylloda
						Scutarcopagia
						Semelangulus
						Pinguitellina
					Psammobiidae	*Gari*
						Asaphis

(continued)

Phylum	Classis	Subclassis	Ordo	Superfamilia	Familia	Genus
					Mesodesmatidae	*Atactodea*
				Ungulinoidea	Ungulinidae	*Diplodonta*
				Veneroidea	Veneridae	*Callista*
						Dosinia
						Gafrarium
						Lioconcha
						Periglypta
						Sunetta
						Timoclea
						Tapes
						Circe
				Gastrochaenoidea	Gastrochaenidae	*Gastrochaena*
			Myoida	Myoidea	Corbulidae	*Corbula*
					Hiatellidae	*Hiatella*
	Scaphopoda					
			Dentaliida		Dentaliidae	*Dentalium*
			Gadilida		Gadilidae	*Cadulus*
	Gastropoda					
		Patellogastropoda		Patelloidae	Patellidae	*Cellana*
				Eoacmaeoidae	Eoacmaeidae	*Eoacmea*
		Vetigastropoda		Fissurelloidea	Fissurellidae	*Diodora*
						Emarginella
						Emarginula
						Scutus
						Hemitoma
						Zeidora
				Haliotoidea	Haliotidae	*Haliotis*
				Scissurelloidea	Scissurellidae	*Scissurella*
						Sukashitrochus
				Seguenzioidea	Chilodontidae	*Perrinia*
						Vaceucheus
				Trochoidea	Tegulidae	*Tectus*
					Trochidae	*Pagodatrochus*
						Priotrochus
						Fossarina
						Stomatella
						Stomatia
						Clanculus
						Rubritrochus
						Trochus
						Ethminolia
					Turbinidae	*Turbo*
				Phasianelloidea	Colloniidae	*Collonista*

(continued)

Phylum	Classis	Subclassis	Ordo	Superfamilia	Familia	Genus
						Homalopoma
		Neritimorpha	Cycloneritimorpha	Neritoidea	Neritidae	*Nerita*
						Smaragdia
						Pisulina
					Phenacolepadidae	*Plesiotyreus*
				Neritopsoidea	Neritopsidae	*Neritopsis*
			Architaenioglossa	Cyclophoroidea	Neocyclotidae	*Daronia*
		Caenogastropoda	Sorbeoconcha	Cerithioidea	Cerithiidae	*Cerithidium*
						Cerithium
						Rhinoclavis
						Clypeomorus
						Bittium
						Selia
					Dialidae	*Diala*
					Modulidae	*Indomodulus*
					Planaxidae	*Hinea*
						Planaxis
					Potamididae	*Terebralia*
						Pirenella
					Scaliolidae	*Finella*
						Scaliola
					Turritellidae	*Turritella*
				Campaniloidea	Plesiotrochidae	*Pleiotrochus*
			Littorinimorpha	Calyptraeiodea	Calyptraeidae	sp.
				Cypraeioidea	Cypraeaidae	*Cypraea*
						Erosaria
						Erronea
						Luria
						Lyncina
						Mauritia
						Monetaria
						Nucleolaria
						Pustularia
						Staphylaea
						Talparia
				Ficoidea	Ficidae	*Ficus*
				Littorinoidea	Littorinidae	*Bembicium*
						Echinolittorina
						Littorina
						Nodilittorina
						Peasiella
					Pickworthiidae	*Clatrosansonia*
						Sansonia

(continued)

Phylum	Classis	Subclassis	Ordo	Superfamilia	Familia	Genus
				Naticoidea	Naticidae	*Notococlis*
						Mammilla
				Pterotracheoidea	Atlantidae	*Atlanta*
				Rissooidea	Rissoidae	*Voorwindia*
						Parasceila
					Rissoinidae	*Rissoina*
						Schwartziella
						Stosicia
				Truncatelloidea	Caecidae	*Caecum*
					Tornidae	*Circulus*
					Truncatellidae	*Truncatella*
				Stromboidea	Strombidae	*Canarium*
						Conomurex
						Gibberulus
						Lambis
						Terestrombus
						Tricornis
					Seraphisidae	*Terebellum*
				Tonnoidea	Tonnidae	*Malea*
						Tonna
					Cassidae	*Casmaria*
						Cassis
					Bursidae	*Bursa*
						Tutufa
					Personidae	*Distorsio*
					Ranellidae	*Charonia*
						Monoplex
						Ranularia
						Septa
				Vanikoroidea	Vanikoridae	*Vanikoro*
						Macromphalus
					Hipponicidae	*Sabia*
						Cheilea
				Velutinoidea	Triviidae	*Trivirostra*
				Vermetoidea	Vermetidae	*Cerasignum*
						Serpulorbis
			Ptenoglossa	Epitonioidea	Epitoniidae	*Epitonium*
					Eulimidae	*Pyramidelloides*
						Eulima
				Triphoroidea	Cerithiopsidae	*Cerithiopsis*
						Joculator
						Seila
					Triphoridae	*Metaxia*
						Viriola
						Inella

(continued)

Phylum	Classis	Subclassis	Ordo	Superfamilia	Familia	Genus
						Mastonia
						Obesula
						Triphora
			Neogastropoda	Buccinoidea	Buccinidae	*Engina*
					Colubrariidae	*Colubraria*
					Columbellidae	*Euplica*
						Graphicomassa
						Zafra
					Fasciolariidae	*Pleuroploca*
						Fusinus
						Peristernia
						Turrilatirus
					Nassariidae	*Nassarius*
						Phos
					Melongenidae	*Volema*
				Muricoidea	Muricidae	*Coralliophila*
						Drupella
						Chicoreus
						Murex
						Nassa
						Ergalatax
					Costellariidae	*Vexillum*
					Cystiscidae	*Gibberula*
						Granulina
					Mitridae	*Mitra*
						Domiporta
						Nebularia
						Strigatella
						Scabricola
					Turbinellidae	*Vasum*
				Olivoidea	Olividae	*Oliva*
				Conoidea	Conidae	*Conus*
					Clathurellidae	*Etrema*
						Lienardia
					Drillidae	*Drillia*
						Clavus
						Kernia
					Mangeliidae	*Eucithara*
					Mitromorphidae	*Anarithma*
					Raphitomidae	*Microdaphne*
					Terebridae	*Myurella*
						Oxymeris
						Terebra
					Turridae	*Iotyrris*
						Pseudoraphitoma

(continued)

Phylum	Classis	Subclassis	Ordo	Superfamilia	Familia	Genus
		Heterobranchia	Incertae sedis	Actonoidea	Acteonidae	sp.
				Architectonicoidea	Architectonicidae	*Architectonica*
						Heliacus
				Pyramidelloidea	Pyramidellidae	*Otopleura*
						Pyramidella
						Odostomia
						Pyrgulina
						Cingulina
						Costabieta
						Turbonilla
						Puposyrnola
						Angiola
					Amathinidae	*Leucotina*
		Opistobranchia	Cephalapsidea	Bulloidea	Acteocinidae	*Pupa*
						Acteocina
					Bullidae	*Bulla*
					Retusidae	*Retusa*
				Diaphanoidea	Cylichnidae	*Cylichna*
				Haminoeoidea	Haminoeidae	*Atys*
						Diniathys
				Philinoidea	Aglajidae	*Chelidonura*
					Colpodaspidae	*Colpodaspis*
			Anapsidea	Aplysioidea	Aplysiidae	*Petalifera*
			Thecosomata	Cavolinioidea	Cliidae	*Clio*
					Creseidae	*Creseis*
				Limacinoidea	Limacinidae	*Limacina*
		Pulmonata	Incertae sedis	Siphonarioidea	Siphonariidae	*Siphonaria*
				Ellobioidea	Ellobidae	*Allochroa*
						sp.

References

Abu-Zied RH, Bantan RA (2018) Late Pleistocene gastropods from the raised reefal limestone of Jeddah, Saudi Arabia: taxonomic and palaeoenvironmental implications. Paleontol Z 92:65–86

Alexandroff SJ, Zuschin KA (2016) Quantitative comparison of Pleistocene and recent coral reef habitats in the northern Red Sea (El Quseir, Egypt). Facies 62:15. https://doi.org/10.1007/s10347-016-0468-6

Alhejoj I, Bandel K, Al-Najjar T (2016a) Pickworthiidae and Aqabarellidae new family (Caenogastropoda, Mollusca) of Aqaba, Jordan: their larval shells and remarks about their evolution and relation. Natural Sci 8:403–430

Alhejoj I, Bandel K, Salameh E (2016b) The fossil beach and reef terraces of the Gulf of Aqaba coast, Jordan—its environment, formation, and relation to mountain uplifting mechanism. Arab J Geosci 9:275

Al-Mikhlafi AS, Edwards LR, Cheng I (2017) Sea-level history and tectonic uplift during the last-interglacial period (LIG): Inferred from the Bab al-Mandab coral reef terraces, southern Red Sea. J Afr Earth Sci. https://doi.org/10.1016/j.jafrearsci.2017.10.023

Almogi-Labin A, Hemleben C, Meischner D,ErlenkeusH (1991) Paleoenvironmental events during the last 13,000 years in the Central Red Sea as recorded by Pteropoda. Paleoceanogr 6:83–98

Amorosi A, Antonioli F, Bertini A, Marabini S, Mastronuzzi G, Montagna P, Negri A, Rossi V, Scarponi D, Taviani M, Angeletti L, Piva A, Vai GB (2014) The Middle-Upper Pleistocene Fronte Section (Taranto, Italy): An exceptionally preserved marine record of the Last Interglacial. Glob Planet Change 119:23–38

Anseeuw B, Terryn Y (2004) Intertidal chitons (Mollusca: Polyplacophora) from the coast of Jordan, Red Sea, with the description of a new species of *Parachiton* Thiele, 1909. Boll Malacol, Suppl 5:1–24, 2003:1–24

Arkell WJ (1928) Aspects of the ecology of certain fossil coral reefs. J Ecol 16:134–149

Ávila SP, Melo C, Silva L, Ramalho RS, Quartau R, Hipólito A, Cordeiro R, Rebelo AC, Madeira P, Rovere A, Hearty PJ, Henriques D, Marques da Silva C, de Frias Martins AM, Zazo C (2015) A review of the MIS 5e highstand deposits from Santa Maria

Island (Azores, NE Atlantic): Palaeobiodiversity, palaeoecology and palaeobiogeography. Quat Sci Rev 114:126–148

Bantan RA, Abu-Zied RH (2014) Sediment characteristics and molluscan fossils of the Farasan Islands shorelines, southern Red Sea, Saudi Arabia. Arab J Geosc 7:773–787

Berry L, Whiteman AJ, Bell SV (1966) Some radiocarbon dates and their geomorphological significance, emerged reef complex of the Sudan. Z Geomorphol 10:119–143

Borri M, Sabelli B, Taviani M (1982) *Cypraea* (*Talparia*) *exusta* Sowerby, 1832, from the central Red Sea: new data. Lavori Soc Malacol Ital, pp 103–107

Bosworth W, Taviani M, Rasul N, Angeletti L, Montagna P, and Team (2013) Pleistocene Coral Terrace Project—Gulf of Aqaba, Saudi Arabia, Phase 1: December 6–15, 2013. Saudi Geological Survey Technical Report (unpublished), 61 pp

Bosworth W, Montagna P, Pons-Branchu E, Rasul N, Taviani M (2017) Seismic hazards implications of uplifted Pleistocene coral terraces in the Gulf of Aqaba. Sci Repts 7, 38, 13 pp. https://doi.org/10.1038/s41598-017-00074-2

Bosworth W, Rasul NMA, Taviani M (this volume) Neotectonics of the Red Sea, Gulf of Suez and Gulf of Aqaba

Bosworth W, Taviani M (1996) Late Quaternary reorientation of stress field and extension direction in the southern Gulf of Suez, Egypt: Evidence from uplifted coral terraces, mesoscopic fault arrays, and borehole breakouts. Tectonics 15:791–802

Bouchet P (2006) The magnitude of marine biodiversity. In: Duarte CM (ed) The Exploration of marine biodiversity: scientific and technological challenges. Fundación BBVA, Bilbao, Spain, Braby, pp 31–62

Bouchet P, Lozouet P, Maestrati P, Heros V (2002) Assessing the magnitude of species richness in tropical marine environments: Exceptionally high numbers of molluscs at a New Caledonia site. Biol J Linnean Soc 75:421–436

Bouchet P, Rocroi J-P (2005) Classification and nomenclator of gastropod families (with a classification of gastropod families by J Frýda, B Hausdorf, W Ponder, A Valdés, A Warén). Malacologia 47:1–397

Bouchet P, Rocroi J-P (2010) Nomenclator of bivalve families (with a classification of bivalve families by R Bieler, JG Carter, EV Coan). Malacologia 52:1–184

Brigham-Grette J, Hopkins DM, Ivanov VF, Basilyan AE, Benson SL, Heiser PA, Pushkar VS (2001) Last Interglacial (isotope stage 5) glacial and sea-level history of coastal Chukotka Peninsula and St. Lawrence Island. Western Beringia. Quat Sci Rev 20:419–436

Bullen RA (1901) Pleistocene Mollusca from the raised beach deposits of Perim Island. Proc Malacol Soc Lon 4.254–255

Casazza LR (2017) Pleistocene reefs of the Egyptian Red Sea: Environmental change and community persistence. Peer J 5:e3504. https://doi.org/10.7717/peerj.3504

Charó MP, Gordillo S, Fucks EE, Giaconi LM (2014) Late Quaternary molluscs from the northern San Matías Gulf (Northern Patagonia, Argentina), Southwestern Atlantic: Faunistic changes and paleoenvironmental interpretation. Quatern Intern 352:26–47

Crame JA (1986) Late Pleistocene molluscan assemblages from the coral reefs of the Kenya coast. Coral Reefs 4:183–196

Dekker H, Orlin Z (2000) Check-list of Red Sea of Red Sea Mollusca. Spirula 47(suppl):1–46

Dekkers AM, Dekker H (2016) A new species of *Turbo* from the Red Sea. Basteria 80:31–38

Dullo W-C (1990) Facies, fossil record, and age of Pleistocene reefs from the Red Sea (Saudi Arabia). Facies 22:1–46

El-Asmar HM (1997) Quaternary isotope stratigraphy and paleoclimate of coral reef terraces, Gulf of Aqaba, South Sinai. Egypt. Quatern Sci Rev 16:911–924

El-Sorogy A (1997) Pleistocene coral reefs of southern Sinai, Egypt: Fossil record, facies analysis and diagenetic alterations. Middle East Research Center, Ain Shams University, Earth Science Series 11:17–36

El-Sorogy A, Youssef M, Al-Sabrouty M, Al-Otaiby N (2014) Facies pattern and molluscan fauna of the Late Pleistocene raised coral reef of Rabigh area, Red Sea coast, Saudi Arabia. Indian J Geo-Mar Sci 43:1571–1580

Galindo LA, Puillandre N, Utge J, Lozouet P, Bouchet P (2016) The phylogeny and systematics of the Nassariidae revisited (Gastropoda, Buccinoidea). Mol Phylogenet Evol 99:337–353

Goreau TF, Goreau NI, Goreau TJ, Carter JG (1976) *Fungiacava eilatensis* burrows in fossil *Fungia* (Pleistocene) from the Sinai Peninsula. Proc R Soc B: Biol Sci 193:245–252

Gvirtzman G (1994) Fluctuations of sea-level during the past 400,000 years: The record of Sinai, Egypt (northern Red Sea). Coral Reefs 13:203–214

Gvirtzman G, Kronfeld J, Buchbinder B (1992) Dated coral reefs of southern Sinai (Red Sea) and their implication to the Late Quaternary sea-levels. Mar Geol 108:29–37

Hall WJ, Standen R (1907) On the Mollusca of a raised coral reef on the Red Sea coast. J Conch 12:65–68

Hamed B, Bussert R, Dominik W (2016) Stratigraphy and evolution of emerged Pleistocene reefs at the Red Sea coast of Sudan. J Afr Earth Sci 114:133–142

Hartman G, Niemi TM, Ben-Avraham Z, Tibor G, Al-Zoubi A, Sade RA, Hall JK, Akawi E, Abueladas A, Al-Ruzouq R, Makowsky Y (2015) Distinct relict fringing reefs in the northern shelf of the Gulf of Elat/Aqaba: Markers of Quaternary eustatic and climatic episodes. Sedimentology 62:516–540

Hoang CT, Taviani M (1991) Stratigraphic and tectonic implications of Uranium-series-dated coral reefs from uplifted Red Sea islands. Quatern Res 35:264–273

Issel A (1869) Malacologia del Mar Rosso, ricerche zoologiche e paleontologiche. Volume I della Biblioteca Malacologica, Pisa, 387 pp

Issel A (1873) Iconografia di alcune conchiglie raccolte nel Golfo di Suez e sulle spiagge emerse del Mar Rosso. Tipografia Sordomuti, Genova

Janssen A (2007) Holoplanktonic Mollusca (Gastropoda) from the Gulf of Aqaba, Red Sea and Gulf of Aden (Late Holocene-Recent). The Veliger 49:140–195

Janssen R, Taviani M (2014) Taxonomic, ecological and historical considerations on the deep-water benthic mollusc fauna of the Red Sea. In: Rasul NMA, Stewart ICF (eds) The Red Sea: the formation, Morphology, Oceanography and environment of a Young Ocean Basin. Springer Earth System Sciences, Berlin Heidelberg, pp 511–528

Janssen R, Zuschin M, Baal C (2011) Gastropods and their habitats from the northern Red Sea (Egypt: Safaga) Part 2: Caenogastropoda: Sorbeochonca and Littorinimorpha. Ann Naturhist Mus Wien 113:373–509

Jousseaume F (1930) Cerithiidae de la Mer Rouge. Journ Conchyl 74:270–296

Kappner I, Al-Moghrabi SM, Richter C (2000) Mucus-net feeding by the vermetid gastropod *Dendropoma maxima* in coral reefs. Mar Ecol Prog Ser 204:309–313

Leigh EG (1968) Modern and Fossil Marine Mollusca of the Red Sea Littoral at Mersa Alam. In: Butzer KW, Hansen CL (eds) Desert and River in Nubia. Geomorphology and Prehistoric Environments at te Aswan Reservoir. The University of Wisconsin Press, pp 499-508

Lorenz F (1992) Pleistocene Cypraeacea from the vicinity of Hurghada, Egypt. Schr Malakozool 5, 19 pp

Le Renard J, Sabelli B, Taviani M (1996) On *Candinia* (Sacoglossa: Juliidae), a new fossil genus of bivalved gastropods. J Paleont 70:230–235

Macfadyen WA (1930) The Geology of the Farsan Islands, Gizan and Kamaran Island, Red Sea. Geol Mag 67:310–315

Marchesan M, Taviani M (1994) Biological modification of coral reed molluscan assemblages during a glacial-interglacial cycle (Red Sea, Egypt). Mem Descr Carta Geol It 52:57–58

Mahmoud MAM, Yassien MH, Abdel-Razek FA (2012) Growth rate of the top shell *Tectus dentatus* (Forskål, 1775) under laboratory conditions, 1- Effect of density. IJESE 2:75–82

Manaa AA, Jones BG, McGregor HV, Zhao J, Price DM (2016) Dating Quaternary raised coral terraces along the Saudi Arabian Red Sea coast. Mar Geol 374:59–72

Mansour AM, Madkour HA (2015) Raised coral reefs and sediments in the coastal area of the Red Sea. In: Rasul NMA, Stewart ICF (eds) The Red Sea: The Formation, Morphology, Oceanography and Environment of a Young Ocean Basin. Springer Earth System Sciences, Berlin Heidelberg, pp 379–393

Martínez S, del Río CJ, Rojas A (2016) A Pleistocene (MIS 5e) mollusk assemblage from Ezeiza (Buenos Aires Province, Argentina). J South Am Earth Sci 70:174–187

Mewis H (2016) Ecological stability of Indo-Pacific coral reefs during Quaternary climatic fluctuations: Case studies from Vanuatu and Egypt. Ph.D. Thesis (Dr. rer. nat.), Humboldt-Universität Berlin, 328 pp. https://doi.org/10.18452/17456

Mewis H, Kiessling W (2013) Environmentally controlled succession in a late Pleistocene coral reef (Sinai, Egypt). Coral Reefs 32:49–58

Mienis HK (1981) Note on the identity of *Fissurella impedimentum* Cooke, 1885 (Prosobranchia: Fissurellidae). J Conch 30:303–304

Montesinos M, Ramos AGJ, Lomoschitz A, Coca J, Redondo A, Betancort JF, Meco J (2014) Extralimital Senegalese species during Marine Isotope Stages 5.5 and 11 in the Canary Islands (29° N): Sea surface temperature estimates. Palaeogeogr Palaeoclimat Palaeoecol 410:153–163

Muhs DR, Simmons KR, Kennedy V, Rockwell TK (2002) The last interglacial period on the Pacific Coast of North America: timing and paleoclimate. Geol Soc Am Bull Bull 114:569–592

Nardini S (1937) Molluschi delle spiagge emerse del Mar Rosso e dell'Oceano Indiano. Parte II (Lamellibranchi). Palaeontogr It 37:225–247

Newton RB (1900) Pleistocene shells from the raised beach deposits of the Red Sea. Geol Mag 7(500–514):544–560

Oliver PG (1992) Bivalved Seashells of the Red Sea. Verlag Christa Hemmen & National Museum of Wales, Wiesbaden & Cardiff, p 330

Plaziat J-C, Baltzer F, Choukri A, Conchon O, Freytet P, Orszag-Sperber F, Raguideau A, Reyss JL (1998) Quaternary marine and continental sedimentation in the northern Red Sea and Gulf of Suez (Egyptian coast): Influences of rift tectonics, climatic changes and sea-level fluctuations. In: Purser BH, Bosence DWJ (eds) Sedimentation and tectonics of rift basins: Red Sea-Gulf of Aden. Chapman and Hall, London, pp 537–573

Plaziat J-C, Reyss J-L, Choukri A, Cazala C (2008) Diagenetic rejuvenation of raised coral reefs and precision of dating: The contribution of the Red Sea reefs to the question of reliability of the Uranium-series dating's of middle to late Pleistocene key reef-terraces of the world. Notebooks on Geology 4:1–35

Pulliandre N, Duda TF, Mayer C, Olivera BM, Bouchet P (2015) One, four or 100 genera? A new classification of the cone snails. J Mollus Stud 81:1–23

Pulliandre N, Bouchet P, Duda TF, Kauferstein S, Kohn AJ, Olivera BM, Watkins M, Meyer C (2014) Molecular phylogeny and evolution of the cone snails (Gastropoda, Conoidea). Mol Phylogenet Evol 78:290–303

Ragani L (1997) Echinoid fauna from the Quaternary fringing reef of Aqaba (Jordan). Palaeontogr It 84:1–19

Reich S, Di Martino E, Todd JA, Wesselingh FP, Renema W (2015) Indirect paleo-seagrass indicators (IPSIs): A review. Earth-Sci Rev 143:161–186

Reid DG (1989) Systematic revision of the Recent species of *Peasiella* Nevill, 1885 (Gastropoda: Littorinidae), with notes on the fossil species. The Nautilus 103:43–69

Reyss J-L, Choukri A, Plaziat J-C, Purser BH (1993) Datations radiochimiques des récifs coralliens de la rive occidentale du Nord de la Mer Rouge, premières implications stratigraphiques et tectoniques. C R Acad Sci Paris 317:487–492

Sabelli B, Taviani M (1984) The genus *Nodilittorina* Von Martens, 1897, in the Red Sea: New data. Boll Malacol 20:95–100

Sabelli B, Taviani M (2014) The Making of the Mediterranean molluscan biodiversity. In: Goffredo S, Dubinsky Z (eds) The Mediterranean Sea: Its history and present challenges. Springer Science and Business Media, Dordrecht, pp 285–306

Safriel U, Lipkin Y (1964) Note on the intertidal zonation of the rocky shores at Eilat (Red Sea, Israel). Isr J Zool 13:187–190

Sasaki T (2008) Micromolluscs in Japan: Taxonomic composition, habitats, and future topics. Zoosymp 1:147–232

Selli R (1944) I caratteri e le affinità delle malacofaune quaternarie del Mar Rosso. Giorn Geol 17:5–22

Selli R(1973) Molluschi quaternari di Massaua e di Gibuti. In: Missione Geologica A.G.I.P. nella Dancalia meridionale e sugli altopiani Hararini (1936–1938), Docum Paleont Accad Naz Lincei Roma 4:151–444

Sorokin YI (1995) Coral reef ecology. Ecological Studies 102, Springer, 466 pp

Strasser A, Strohmenger C, Davaud E, Bach A (1992) Sequential evolution and diagenesis of Pleistocene coral reefs (South Sinai; Egypt). Sed Geol 78:59–79

Taylor JD (1978) Faunal response to the instability of reef habitats: Pleistocene molluscan assemblages of Aldabra Atoll. Paleontology 21(1):1–30

Taviani M (1982) Paleontological markers in Late Pleistocene coral reefs in the Red Sea. XI INQUA Congress, Moscow 1982, Abstract vol 1:307

Taviani M (1994) The ever-changing climate: Late Quaternary paleoclimatic modifications of the Red Sea region as deduced from coastal and deep-sea geological data. In: Proceedings Egyptian-Italian seminar "On geosciences and archaeology in the Mediterranean countries. Geological Survey of Egypt, Special Publication 70:193–200

Taviani M, Marchesan M, Quarantini M, Salmi M (1995) Reef-associated mollusk fauna of the Gulf of Suez (Egypt): last interglacial to present. Twelfith International Malacological Congress, Vigo, Spain, 3-8 September 1995, Abstract vol:333–334

Taviani M (1998a) Post-Miocene reef faunas of the Red Sea: glacio-eustatic controls. In: Purser BH, Bosence DWJ (eds) Sedimentation and tectonics of rift basins: Red Sea-Gulf of Aden. Chapman and Hall, London, pp 574–582

Taviani M (1998b) Stable tropics not so stable: Climatically driven extinctions of reef-associated molluscan assemblages (Red Sea and western Indian Ocean; last interglaciation to present). In: Camoin GF, Davis G (eds) Reefs and carbonate platforms in the Pacific and Indian Oceans. Spec Pub Int Assoc Sedimentol 25:69–76

Taviani M (2015) Unpersisting *Persististrombus*: A Mediterranean story. Viaraea 42:9–18

Taviani M, Montagna P, Rasul NMA, Bosworth W, Angeletti L (this volume). Pleistocene coral reef terraces on the Saudi Arabian side of the Gulf of Aqaba, Red Sea

Walter RC, Buffler RT, Bruggemann JH, Guillaume MMM, Berhe SM, Negassi B, Libsekal Y, Cheng H, Edwards RL, von Cosel R, Néraudeau D, Gagnon M (2000) Early human occupation of the Red Sea coast of Eritrea during the last interglacial. Nature 405:65–69

WoRMS Editorial Board (2017) World Register of Marine Species. Available from http://www.marinespecies.org at VLIZ. Accessed 2017-03-05. https://doi.org/10.14284/170

Zauner S, Zuschin M (2016) Diversity, habitats and size-frequency distribution of the gastropod genus *Conus* at Dahab in the Gulf of Aqaba, Northern Red Sea. Zool Middle East 62:125–136

Zuschin M, Ebner C (2015a) Compositional fidelity of death assemblages from a coral reef-associated tidal-flat and shallow lagoon in the northern Red Sea. Palaios 30:181–191

Zuschin M, Ebner C (2015b) Actuopaleontological characterization and molluscan biodiversity of a protected tidal flat and shallow subtidal at the northern Red Sea. Facies 61:5. https://doi.org/10.1007/s10347-015-0428-6

Zuschin M, Oliver PG (2003) Bivalves and bivalve habitats in the northern Red Sea: The northern Bay of Safaga (Red Sea, Egypt): An actuopalaeontological approach. VI. Bivalvia, Naturhistorisches Museum Wien, p 304

Zuschin M, Oliver PG (2005) Diversity patterns of bivalves in a coral dominated shallow-water bay in the northern Red Sea–high species richness on a local scale. Mar Biol Res 1:396–410

Zuschin M, Stachowitsch M (2007) The distribution of molluscan assemblages and their postmortem fate on coral reefs in the Gulf of Aqaba (northern Red Sea). Mar Biol 151:2217–2230

Zuschin M, Hohenegger J, Steininger FF (2000) A comparison of living and dead molluscs on coral reef associated hard substrata in the northern Red Sea—implications for the fossil record: Palaeogeog Palaeoclimat Palaeoecol 159:167–190

Zuschin M, Hohenegger J, Steininger FF (2001) Molluscan assemblages on coral reefs and associated hard substrata in the northern Red Sea. Coral Reef 20:107–116

Zuschin M, Janssen R, Baal C (2009) Gastropods and their habitats from the northern Red Sea (Egypt: Safaga) Part 1: Patellogastropoda, Vetigastropoda and Cycloneritimorpha. Ann Naturhist Mus Wien 111A:73–158

Geochemistry of the Lunayyir and Khaybar Volcanic Fields (Saudi Arabia): Insights into the Origin of Cenozoic Arabian Volcanism

Alessio Sanfilippo, (Merry) Yue Cai, Ana Paula Gouveia Jácome, and Marco Ligi

Abstract

This paper reports on a detailed geochemical study of rocks from Harrats Lunayyir and Khaybar, two large lava fields located in the central portion of the western Arabian Peninsula. Lavas from young flows north of Al Birk were also considered. Sample composition ranges from basanite to basalts with transitional to alkaline affinity. Their incompatible trace element signatures are consistent with alkaline magmas produced by an enriched mantle source, akin to that producing continental flood magmatism in other locations of the Arabian-Nubian plate. Large variations in major (Al_2O_3, CaO, NaO, TiO_2) and trace (e.g., Ni, Cr, Nb, Sr, Zr, Ti, Y and REE) element compositions at a given Mg/(Mg+Fe) indicate that magmatic evolution occurred in magma chambers located at or close to the crust-mantle boundary, constrained by fractionation of olivine, clinopyroxene, plagioclase and Ti–Fe oxides. Their Ba/Nb and K/La ratios (7–10 and 200–300, respectively) are similar to those of ocean island basalts (OIBs) and with no evidence of crustal assimilation. Fractionations between incompatible trace elements are used to investigate differences in mantle composition and melting conditions in the studied localities and in other lava fields in Arabia, Yemen and Syria. Variable La/Yb and Dy/Yb fractionations of the lavas can be reproduced by mixing different proportions of partial melts produced within the garnet and the spinel stability fields. Lavas from Harrats Lunayyir, Khaybar and Al-Birk show relative depletion in Rb, Th and U compared to Ba and Nb, and negative K anomalies. These compositions are consistent with the local occurrence of an amphibole-bearing source, located most likely in the subcontinental lithospheric mantle. This model agrees with the idea that the sub-continental mantle in this region may have formed in a supra-subduction environment with residual amphiboles that preferentially withheld Ba and Nb during fluid-fluxed melting. Our results, when analyzed together with existing data from the region, suggest that Cenzoic alkaline volcanism in western Arabia formed mainly by decompression melting of ancient fusible components in the sub-Arabian lithospheric mantle, that were remobilized by lithospheric thinning due to Red Sea rifting. Additionally, our data are consistent with progressive thinning of the lithosphere toward the Red Sea and lengthening of the melting column over time.

1 Introduction

The Cenozoic volcanic fields on the western Arabian Peninsula (Fig. 1) form one of the largest alkali basalt provinces in the world (area $\sim 180{,}000$ km^2; Coleman et al. 1983). The oldest volcanism initiated in Yemen ~ 30 Ma, and has been related with the activity of the Afar mantle plume (Stern and Johnson 2010). This early phase predated the onset of volcanism in the central/northern part of the Arabian Peninsula, which was roughly synchronous with the Red Sea rifting and uplift of the Arabian shield (McGuire and Bohannon 1989; Camp and Robool 1991). Magmatic activity in the central/northern region of the Arabian Peninsula initiated as early as 27–25 Ma (Bohannon et al. 1989; Bosworth and Stockli 2016) and gave rise to voluminous lava fields (called *harrats*) that spread over 1400 km from north to south in Syria, Israel, Jordan and Saudi Arabia. Although this volcanism is widespread in the entire western

A. Sanfilippo (✉)
Dipartimento di Scienze della Terra, Università di Pavia, Pavia, Italy
e-mail: alessio.sanfilippo@unipv.it

(Merry) Y. Cai
Lamont Doherty Earth Observatory, Columbia University, Palisades, NY, USA

A. P. G. Jácome
Universidade Federal de Ouro Preto, Ouro Preto, MG, Brazil

M. Ligi
Istituto di Scienze Marine, CNR, Bologna, Italy

N. M. A. Rasul and I. C. F. Stewart (eds.), *Geological Setting, Palaeoenvironment and Archaeology of the Red Sea*,
https://doi.org/10.1007/978-3-319-99408-6_18

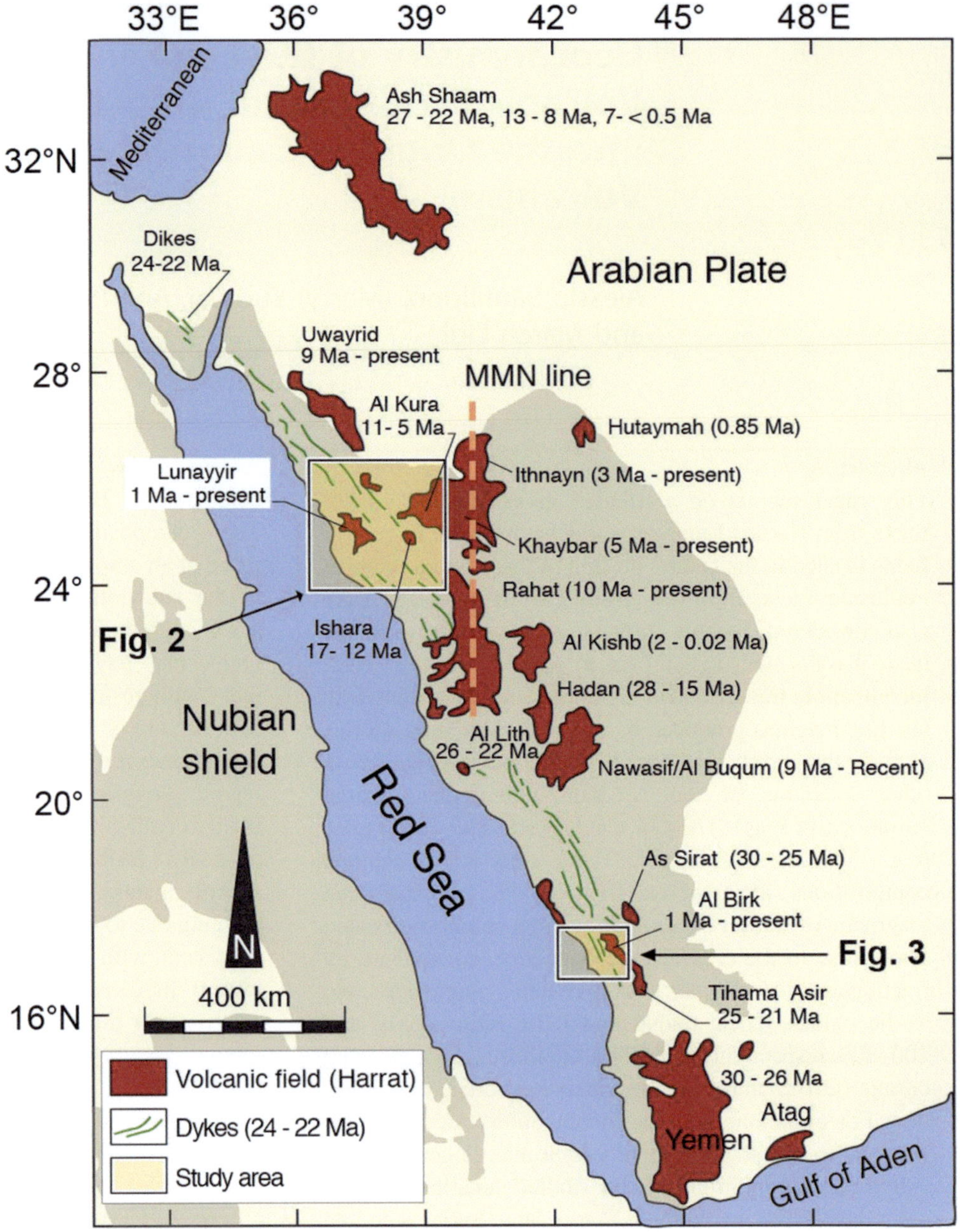

Fig. 1 Distribution and ages of Cenozoic Arabian volcanism (dark red areas), modified from Stern and Johnson (2010) and Bosworth and Stockli (2016). Gray filled area shows the extent of the Nubian and Arabian shield. Black boxes refer to the study areas displayed in Figs. 2 and 3

Arabian plate (Coleman et al. 1983), uplift or volcanism of similar scale is missing in the corresponding area of the African plate, to the west of the Red Sea. For this reason, many authors have questioned the link between this volcanism and the rifting event that generated the Red Sea (Camp and Robool 1992; Duncan and Al-Amri 2013; Moufti et al. 2013; Duncan et al. 2016). In addition, the widespread HIMU-like (high U/Pb ratio, μ) isotopic signature of most basalts in the region (Stein and Hofmann 1992; Baker et al. 1996; Bertrand et al. 2003; Shaw et al 2003; Rooney et al. 2014; Sgualdo et al. 2015; Natali et al. 2016) and geophysical evidence for a N-S low velocity zone under the Arabian shield (Krienitz et al. 2009; Chang and Lee 2011) suggest that the Afar plume or a different mantle plume triggered this large continental alkaline volcanic province.

Based on these observations, several scenarios for the origin of the western Arabian volcanism have been proposed, including: (i) Mobilized fossil plume material beneath the sub-continental lithosphere (Stein and Hofmann 1992); (ii) progressive lithospheric thinning tapped lithospheric to asthenospheric sources (Bertrand et al. 2003; Shaw et al. 2003); (iii) anomalously hot asthenospheric mantle triggering melting of the overlying hydrous lithospheric mantle (Stein et al. 1997); (iv) a separate mantle plume existing beneath northern Arabia (Chang and Van der Lee 2011);

(v) north-westward channeling of hot Afar plume material (Camp and Roobol 1992; Krienitz et al. 2009). However, the uncertainty in the origin of the western Arabian volcanism is linked to the limited knowledge of the different volcanic features and products in this large region. Pioneering studies carried out in the late twentieth century (Camp and Roobol 1989, 1991, 1992) provided a preliminary characterization of the age and stratigraphy of most lava fields. However, trace elements and isotopic compositions available only from a few harrats in Saudi Arabia (Duncan and Al-Amri 2013; Moufti et al. 2013; Duncan et al. 2016). Here, we report a detailed geochemical study of basalts from three areas poorly known in the literature, Lunayyir, Khaybar and Al Birk (Figs. 1, 2, 3 and 4). Whole rock major and trace element compositions are used to evaluate the magmatic evolution of these lavas and to obtain information on the composition of their mantle source. Comparing our new data with published data from other volcanic fields in the central/northern region of the Western Arabian plate, we discuss the melting conditions of the mantle source, that provide new insights into the formation of this large alkaline volcanic province.

2 Cenozoic Volcanic Fields in Saudi Arabia

The main volcanic fields in the Red Sea-rift region are located in western Saudi Arabia. These basalts were emplaced within 300 km of the NW-trending eastern margin of the Red Sea on an elevated and rugged terrain along the rift shoulders (Bohannon et al. 1989). Compared with tholeiitic basalts from the Red Sea rift (Altherr et al. 1990; Volker et al. 1993; Antonini et al. 1998; Haase et al. 2000; Ligi et al. 2011, 2012, 2015), Cenozoic lavas from the Arabian volcanic province are predominantly alkali olivine–basalt, basanite and hawaiite, locally associated with evolved products (Camp and Roobol 1989, 1992; Camp et al. 1991). The Harrat volcanisms is roughly contemporaneous with the Trap Series basalts in Yemen and Ethiopia ($\sim$30 Ma; Camp and Roobol 1992; Mohr and Zanettin 1988) and continued until $\sim$20 Ma. This early magmatism occurred as dikes and eruptive centers parallel to the NW–SE Red Sea margin and consists of volcanics with tholeiitic to transitional compositions, well documented in Harrats Uwayrid, Kura and Isahara (Pallister 1987; Coleman and McGuire 1988). Volcanism then evolved toward more alkaline compositions, which characterize the majority of lavas in the studied region. During this late stage, volcanic centres extend along N–S alignments, that probably reflect a major change in the overall regional stress regime (Ashwal and Burke 1989). The most voluminous harrats associated with the late stage volcanism are aligned along the Makkah–Madinah–Nafud (MMN) line and consist of Harrats Rahat, Khaybar and Ithnayn. More recent, less voluminous harrats formed west (Harrat Lunayyir) and east (Harrat Kishb and Harrat Hutaymah) of the MMN line (Fig. 1).

2.1 Harrat Lunayyir

Harrat Lunayyir is one of the smaller and younger Cenozoic volcanic fields in western Arabia. It covers an area of

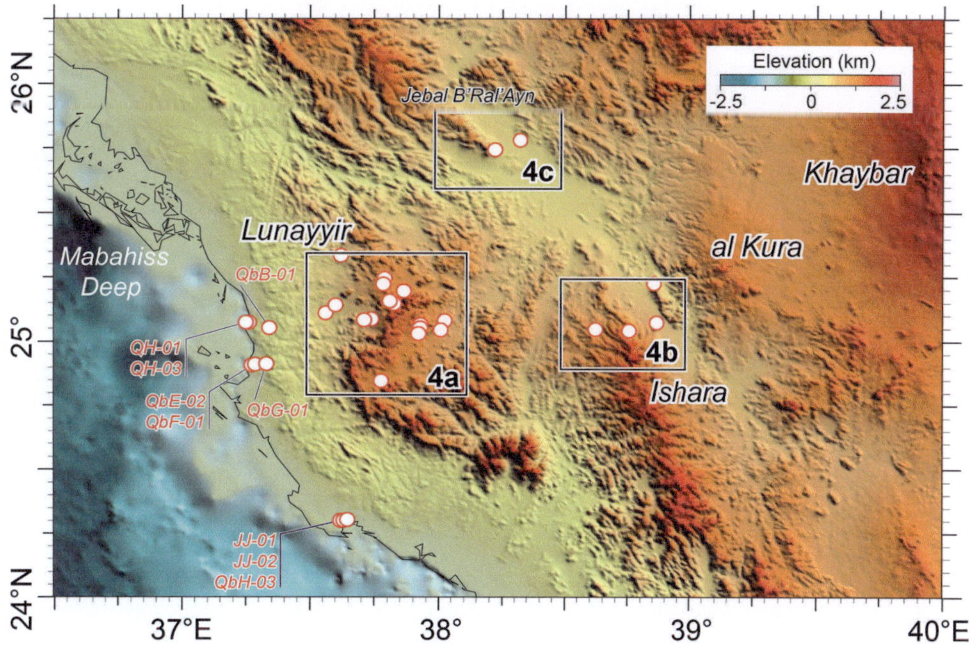

Fig. 2 Topography of Harrats Lunayyir, Ishara and al Kura-Khaybar in the north-western part of the Arabian shield and bathymetry of the northern Mabahiss Deep in the northern Red Sea. Bathymetric grids from GEBCO (https://www.bodc.ac.uk/data/) and NGDC databases (http://www.ngdc.noaa.gov/mgg/), and elevation data from the Shuttle Radar Topography Mission (SRTM) database (http://srtm.usgs.gov/). Circles mark locations of sample sites

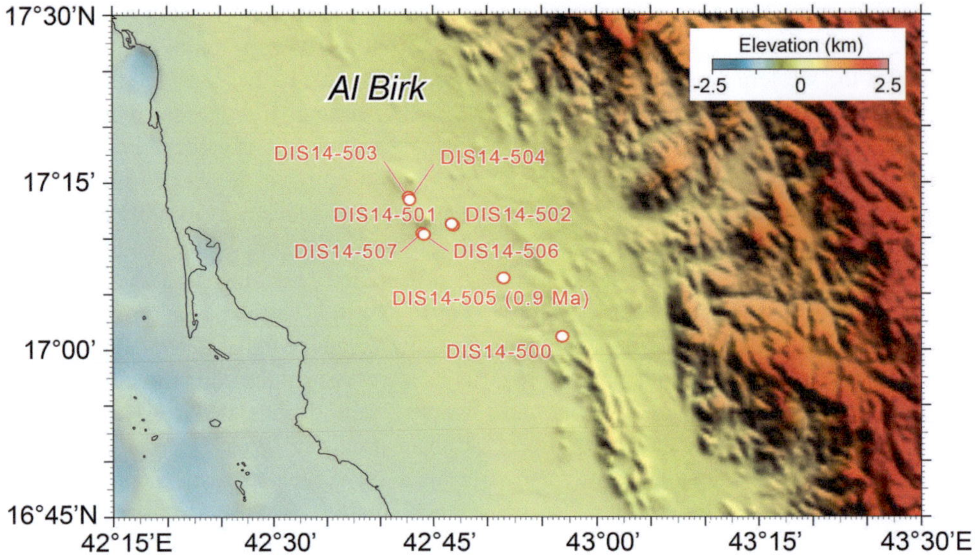

Fig. 3 Topography of the Al Birk region, south-western Saudi Arabia. Red circles mark locations of sample sites

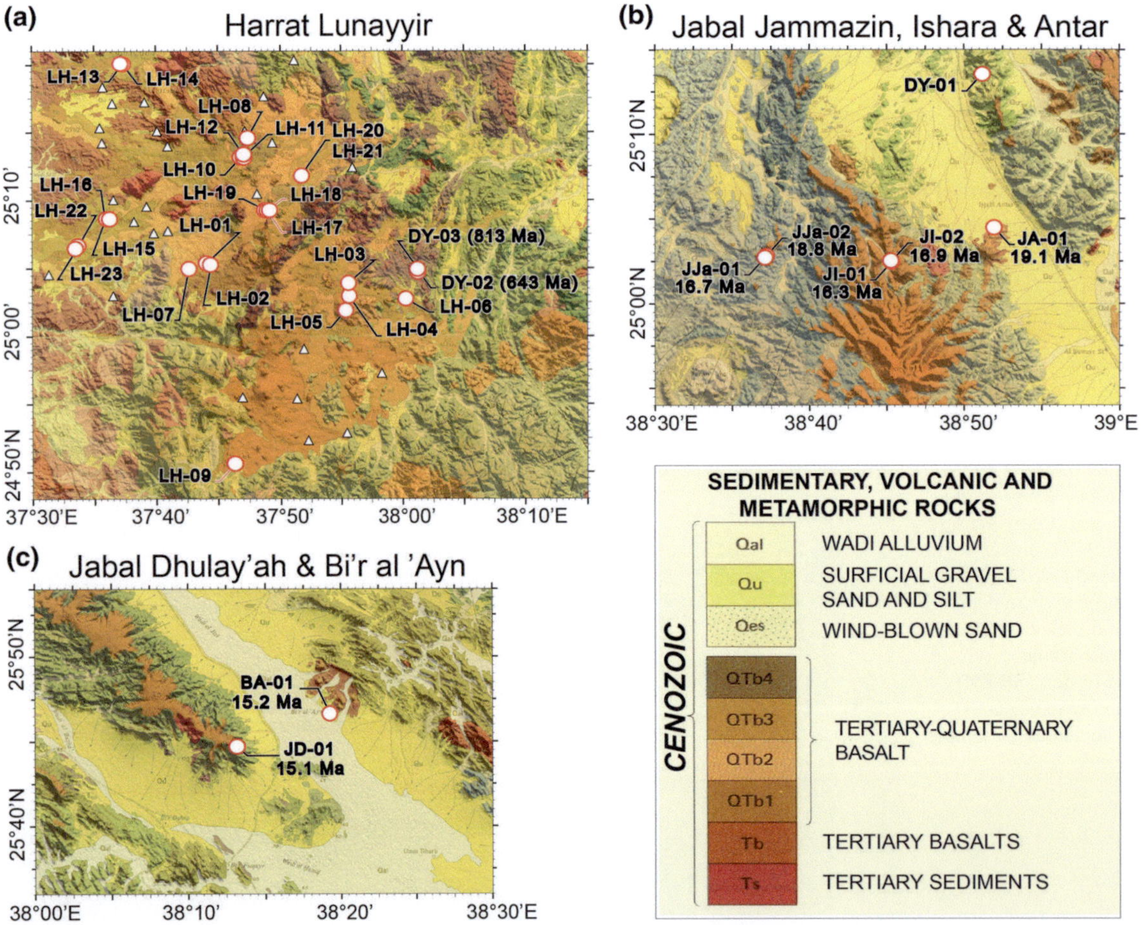

Fig. 4 Sample locations. Geological map from Kemp (1981) is superimposed on the topography of the Lunayyir (**a**), Ishara (**b**) and Jabal Dhulay'ah (**c**) regions

~3575 km^2, ~100 km east of the Red Sea margin, and north of the city of Yanbu. Harrat Lunayyir was formed from a series of volcanic eruptions that occurred along N–S and NW–SE trending fissures forming about 50 cinder cones. Lavas are typically a'a type with thicknesses ranging from 3 to 5 m. Scoria cones are commonly well-preserved from erosion and typically 400–500 m in diameter and 100–200 m in height. Based on combined field observations, geomorphologic data and satellite imaging, Al-Amri et al. (2012) and Duncan and Al-Amri (2013) identified six volcanic units, each with different degrees of erosion and saturation with wind-blown dust. They are subdivided into one Tertiary unit (T) separated by a thin, intermittent soil horizon from five units of Quaternary age (Q5–Q1). However, $^{40}Ar/^{39}Ar$ age determinations constrain the volcanic activity within the last 600 kyr (Al-Amri et al. 2012; Duncan and Al-Amri 2013). Based on temporal variation of middle to heavy REE ratios, Duncan and Al-Amri (2013) suggested that the depth of melt production decreased through time as a result of gradual thinning of the lithosphere under Harrat Lunayyir in the Quaternary.

2.2 Harrat Khaybar, Harrat Al Kura and Harrat Ishara

Harrat Khaybar forms one of the largest lava fields in Saudi Arabia, with an age spanning from 5 Ma to present. This lava field is characterized by an elevated central region of a linear vent system containing pyroclastic deposits, white felsic domes and tuff cones (Camp et al. 1991). On its western margin, we have the much smaller Harrat Al Kura, that forms a sloping highland area deeply incised on its western margin and degrading towards Harrat Khaybar towards the east. The oldest basalts of Harrat Al Kura are along the eastern margin, with an age of about 11 Ma, whereas the oldest basalts of Harrat Khaybar that erupted in the study area have an estimated age of 6 Ma (Kemp 1981; this study). The volcanic products of Harrats Khaybar and Kura mostly range from alkaline olivine basalts to hawaiites and basanites.

Jabal Ishara extends over a highland area along the southwestern margin of Harrat Kura with a lava flow thickness from 30 to 400 m (Coleman et al. 1983). Jabal Ishara is widely eroded and the basal sections are well exposed in many wadis allowing reconstruction of the entire sequence of events. The lavas of the Ishara basalt consist mainly of alkali olivine basalt with rare occurrences of clinopyroxene and plagioclase phenocrysts. The age of the Ishara basalts has been estimated to be ~20 Ma, among the oldest products of the Cenozoic volcanism in the region (Coleman et al. 1975).

2.3 Al Birk Lava Flows

Five basalt samples were collected from a small lava field ~100 km south of the city of Al Birk. The lava flows are composed mainly of olivine-pyric alkaline basalts frequently containing mantle xenoliths. The lavas are in place associated with less voluminous pyroclastic deposits, mainly tufaceous. One sample furnished an Ar–Ar age of 0.9 Ma (Vigliotti et al., this volume).

3 Samples Selection and Petrographic Description

We conducted field work and sampling on a representative subset of harrats to the NW and N of Jeddah, over 150 km east the shoreline (Fig. 2). Within the study area, the volcanic activity is represented by 2 principal groups: (i) lavas exposed on the deeply dissected plateau east of the Red Sea rift shoulders, extending from Harrat Khaybar (including Jabal Isahara in the east and Harrat al Kura in the west) to the southern and eastern margins of the survey area; and (ii) Tertiary and Quaternary lava flows from Harrat Lunayyir. In addition, we analyzed basalts from young lava flows along the Red Sea coastline, north of Al Birk (Fig. 3).

Samples from Lunayyir were collected from different volcanic units (T to Q1–Q5 of Duncan and Al-Amri 2013) with sampling sites specifically selected to supplement the previous sampling effort of Duncan and Al-Amri (2013; see Fig. 4a). Lava flows close to the Red Sea coast, west of the main volcanic field were also collected (Umlujj and Qalib areas, west of Lunayyir). Although they are not physically connected to the main volcanic field, geological mapping and field observations suggest that these lava flows are part of a late period of activity of Harrat Lunayyir. Samples from Harrats Khaybar, Al Kura and Ishara were collected mainly from the top of well-exposed basaltic flows. We selected several locations along the eastern (i.e., Harrat Khaybar and Harrat al Kura) and southern (i.e., Jabal Jammazin, Jabal Ishara and Jabal Antar) edges of the study area (Fig. 4b and c). These samples have ages (Vigliotti et al. 2017, this volume) ranging from ~16 Ma (Jabal Isahara) to ~13 Ma (Jabal Dhulay'ah). For simplicity, they will be referred to be part of Harrat Khaybar.

Selected samples are mostly olivine-basalts and hawaiites. They show two distinctive textures: (i) fine to medium grained subophitic texture, characterizing mainly basalts from the Khaybar and Tertiary Lunayyir flows. They are formed by euhedral to subhedral olivine and plagioclase locally included within granular to interstitial clinopyroxene grains (Fig. 5a). Locally, large, euhedral olivine grains also occur; (ii) Porphyric to sparsely phyric basalts containing olivine and plagioclase phenocrysts. The phenocrysts are commonly included within an intergranular microcrystalline to cryptocrystalline groundmass of plagioclase laths (Fig. 5 b). Olivine and plagioclase are often reabsorbed and show compositional zoning (Fig. 5c). The porphyric texture is typical of the Quaternary lava flows from Lunayyir and Al Birk. The Al Birk samples contain large proportions of olivine phenocrysts (up to 20 vol.%) and local occurrences of olivine megacrysts or mantle xenoliths (Fig. 5d).

4 Whole Rock Major and Trace Elements Compositions

A subset of 47 samples (33 from Lunayyir, 9 from Khaybar and 5 from Al Birk) were selected for geochemical analyses, based on stratigraphy and texture. To minimize the effect of alteration, we selected samples from the innermost portion of the basaltic flows with minimal amounts of vesicles. Details of sample locations and textural characteristics are summarized in Table 1. Selected portions of each sample were crushed and pulverized in a tungsten carbide swing mill and measured for major and trace element concentrations using inductively coupled plasma optical emission spectroscopy (ICP-OES) and ICP mass spectrometry (ICP-MS) at Activation Laboratories Ltd. (Ancaster, Ontario) and are reported in Table 2. Total uncertainties of the element analyses are generally within 10%. Details of the analytical techniques and detection limits are available from the company website (www.actlabs.com).

SiO_2 content of the lavas ranges within 45–51 wt% and do not show any correlation with $Na_2O + K_2O$ in a total alkali versus silica (TAS) diagram (Fig. 6a). The samples range from alkali to transitional basalts with a wide range of K_2O concentrations (0.5–2 wt%) (Fig. 6b). In particular, the Lunayyir and Al Birk samples are mostly alkali basalts, whereas the Khaybar samples are mostly transitional basalts and rare subalkaline. Basalts from the lava flows west of

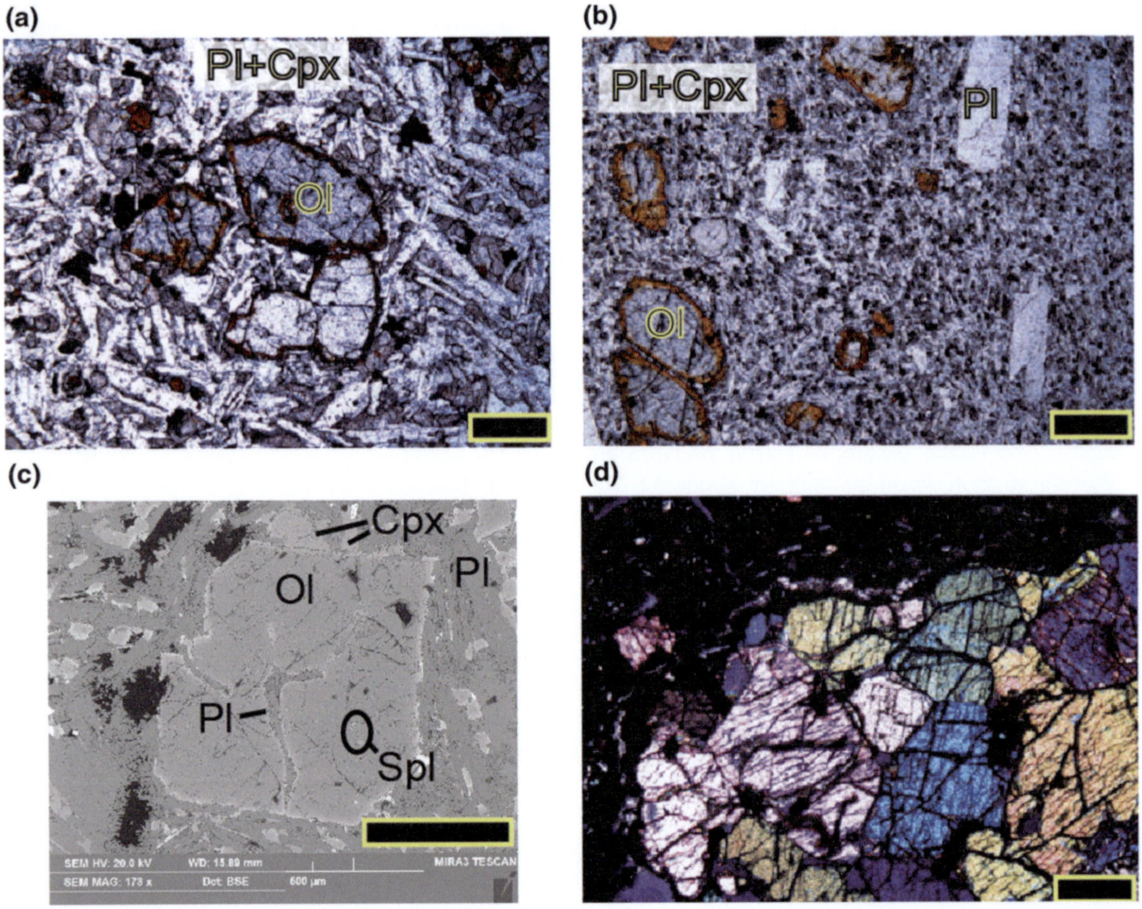

Fig. 5 Microphotographs showing the main textural features of the studied samples. Scale bar 0.5 mm. **a** euhedral olivine (Ol) phenocryst in a medium grained groundmass. Plagioclase (Pl) and clinopyroxene (Cpx) in the groundmass have a subophitic texture; **b** olivine and plagioclase phenocryst within a microcrystalline groundmass; **c** Back-scattered electron image of a olivine phenocryst having irregular grain-boundaries and embayments filled by plagioclase. Small spinel grains are locally included within the olivine; **d** small-sized (~10 mm) mantle xenolith in a basanite from the Al-Birk lava field

Table 1 Location and main petrographical characteristics of selected samples from the Khaybar, Lunayyir and Al-Birk areas

Area	Sample name	Locality	Weathering	Lithology	Ave grain size	Texture	Vesicles		Plagioclase		Olivine		Cpx		Magmatic Oxides	
							%	Ave. Size	%	Size (mm)	%	Size (mm)	%	Size (mm)	%	Size (mm)
Khaybar	BA-01/Tb-1	Bi'r al ' Ayn	Light	Ol-basalt	F	Sparsely phyric	2	1			15	2				
	JA-01/Tb-1	Jabal Antar	Light	Basalt	F	Aphyric	0									
	JD-01/Tb-1	Jabal Dhulay'ah	Medium/Heavy	Ol-Pl	M	Phyric	5	1	10	1	20–20.5	2			5	1
	JI-01/Tb-1	Jabal Isahara	Light	Basalt	F	Aphyric	0									
	JI-02/Tb-1	Jabal Isahara	Light	Ol-basalt	F	Sparsely phyric	0				3	3				
	JJ-01/1	Jabal Jarbul	Medium	Ol-Pl	F	Phyric	5	1	15	2	10	2			2	1
	JJ-02/Tb-1	Jabal Jarbul	Medium	Ol-PL	F	Highly phyric	5	1	20	2	30	3			5	1
	JJa-01/Tb-1	Jabal Jammazın	Light	Ol-basalt	F	Phyric					20	3			2	1
	JJa-02/Tb-2	Jabal Jammazin	Light	Ol-basalt	F	Sparsely phyric	0				10	1				
Lunayyir	LH-01/Tb-3	Lunayyr Harrat	Light	Ol-basalt	M	Highly phyric	0				15	3	5	2		
	LH-02/Qtb-1	Lunayyr Harrat	Light	Basalt	F	Aphyric	10	2–10								
	LH-03/Tb-1	Lunayyr Harrat	Light	Basalt	F	Sparsely phyric	5	1								
	LH-04/Qtb-1	Lunayyr Harrat	Light	Pl-Cpx	F	Sparsely phyric	10	4	5	2	1	1	3	1	0	
	LH-05/Tb-1	Lunayyr Harrat	Light	Pl-basalt	F	Sparsely phyric	20	2	5	2						
	LH-06/Qtb-2	Lunayyr Harrat	Medium	Ol-basalt	F	Sparsely phyric	5	2			5	1				
	LH-07/Qtb-2	Lunayyr Harrat	Light	Basalt	F	Aphyric										
	LH-08/Tb-1	Lunayyr Harrat	Light	Pl-basalt	F	Sparsely phyric			20	2						
	LH-09/b1-1	Lunayyr Harrat	Light	Ol-basalt	F	Phyric					20	1				
	LH-10/Qtb-2	Lunayyr Harrat	Light	Basalt	F	Aphyric	20	3								
	LH-11/Tb-1	Lunayyr Harrat	Light	Pl-Ol	F	Sparsely phyric	5	2	20	2	5	2	0		0	
	LH-12/Qtb-1	Lunayyr Harrat	Light	Ol-basalt	F	Phyric	5	1			10	1				
	LH-13/Tb-2	Lunayyr Harrat	Light	Ol-Cpx	F	Sparsely phyric	20	2–3			1	2	5	2	0	
	LH-14/Tb-2	Lunayyr Harrat	Light	Pl-basalt	F	Sparsely phyric	5	2	10	2						
	LH-15/Tb-1	Lunayyr Harrat	Light	Pl-basalt	F	Sparsely phyric	3	2	10	1	2	2			5	1
	LH-17/Qtb-1	Lunayyr Harrat	Light	Pl-basalt	F	Sparsely phyric	5	1	10	3	0					
	LH-18/Qtb-1	Lunayyr Harrat	Light	Ol-basalt	F	Sparsely phyric	2	1			2	2	2	2	2	1
	LH-19/Qtb-1		Light	Pl-basalt	F		0		5	4						

(continued)

Table 1 (continued)

Area	Sample name	Locality	Weathering	Lithology	Ave grain size	Texture	Vesicles		Plagioclase		Olivine		Cpx		Magmatic Oxides	
							%	Ave. Size	%	Size (mm)	%	Size (mm)	%	Size (mm)	%	Size (mm)
		Lunayyr Harrat				Sparsely phyric										
	LH-20/Qtb-1	Lunayyr Harrat	Light	Pl-basalt	F	Sparsely phyric	2	1	5	2–5	2	1	0		0	
	LH-21/Qtb-1b	Lunayyr Harrat	Light	Pl-basalt	F	Sparsely phyric	5	1	3	4						
	LH-22/Qtb-1	Lunayyr Harrat	Light	Ol-basalt	F	Sparsely phyric	10	2			5	3	2	2		
	LH-23/Qtb-1	Lunayyr Harrat	Light	Basalt	F	Aphyric			1	1						
	QbB-01/1	Umlujj	Light	Ol-basalt	F	Sparsely phyric	0				10	2			2	1
	QbE-02/1	Umlujj	Light	Ol-basalt	M	Sparsely phyric	0				10	2			5	1
	Qbf-01/1	Umlujj	Medium	Ol-basalt	F	Highly phyric	0				20	1.5	0		5	1
	Qbf-01/2	Umlujj	Medium	Ol-basalt	F	Highly phyric	0				20	2			2	1
	QbG-01/2	Umlujj	Light	Pl-basalt	F	Sparsely phyric	5	2	4	1	1	1				
	QbH/03-1	Umlujj	Medium	Ol-Pl	F	Phyric	0		20	2	20	1				
	QH-01/1	Qalib Harrat	Light	Pl-Ol	F	Highly phyric	0		15	3	10	1				
	QH-03/1	Qalib Harrat	Light	Pl-Ol	F	Sparsely phyric	0		20	2	20	1				
Al-Birk	DIS14-500	Al-Birk	Light	Ol-basalt	F	Sparsely phyric	1	2	2	1	5	2	1	2		
	DIS14-503	Al-Birk	Light	Pl-basalt	F	Sparsely phyric	40	4	5	2						
	DIS14-504	Al-Birk	Medium	Ol-basalt	F	Sparsely phyric	10	2			2	1				
	DIS14-506	Al-Birk	Light	Ol-basalt	F	Sparsely phyric	0		2	1	2	2	3	2		
	DIS14-507	Al-Birk	Light	Ol-basalt	F	Phyric	0		2	1	5	1	10	2	0	

Harrat Lunayyir have lower alkalis and more generally primitive compositions.

Basalts from Lunayyir and Khaybar have wider ranges of MgO (3.2–10.4 wt%), CaO (2.3–14.1 wt%), Al_2O_3 (14.0–16.9 wt%) and FeO_{TOT} (7.9–14.2 wt%) (Fig. 6). Similarly, their incompatible elements show larger variations in TiO_2, Na_2O and K_2O, ranging from 0.66 to 3.09, 0.93 to 5.12 and 0.43 to 1.91 wt%, respectively. Their Mg# [= Mg/(Mg+Fe^{2+}) mol%] ranges from 31 to 70 and correlates with CaO and CaO/Al_2O_3 ratios (Fig. 7). The Al_2O_3 contents first increase with decreasing Mg# until Mg# reaches ~45, and then decrease in the most evolved samples. Similarly, FeO_{TOT} and TiO_2 increase initially during fractionation, until Mg# reaches ~40, then gradually decrease until Mg# reaches ~30. At a given Mg#, the Al Birk samples show higher TiO_2, K_2O and Na_2O than our other samples.

Trace elements also show a large compositional range. Compatible trace elements such as Ni, Cr and Co decrease with decreasing Mg# (Fig. 8). The more primitive samples have Ni up to 300 ppm and Cr up to 450 ppm. Al Birk samples have the lowest Co contents, although Ni is similar to the most primitive samples from Lunayyir and Khaybar. The abundances of REE, Y (15–49 ppm), Zr (24–403 ppm) and Hf (0.70–7.80 ppm) increase with decreasing Mg#, delineating good correlations with the samples from Lunayyir and Khaybar (Fig. 8). Notably, highly incompatible elements such as Rb (5.0–58.0 ppm), Ba (131–690 ppm), Th (0.06–9.65 ppm), U (0.03–2.55 ppm), Nb

Table 2 Major and trace element compositions of selected samples from the Khaybar, Lunayyir and Al-Birk areas

Area	Khaybar								
Sample name	BA-01/Tb-1	JA-01/Tb-1	JD-01/Tb-1	JI-01/Tb-1	JI-02/Tb-1	JJa-02/Tb-2	JJa-01/Tb-1	JJ-01/1	JJ-02/Tb-1
Locality	Bi'r al 'Ayn	Jabal Antar	Jabal Dhulay'ar	Jabal Isahara	Jabal Isahara	Jabal Jammazin	Jabal Jammazin	Jabal Jarbul	Jabal Jarbul
Rock type	Basalt	Basalt	Basalt	Basalt	Basalt	Basalt	Basanite	Basalt	Basalt
(wt%)									
SiO_2	46.65	46.84	49.45	46.33	48.91	45.28	46.25	45.05	46.88
Al_2O_3	15.18	16.04	15.97	16.09	15.44	15.86	15.55	14.36	15.26
FeO(T)	11.10	9.35	10.25	10.32	10.19	10.18	10.37	10.54	10.87
MnO	0.17	0.16	0.16	0.16	0.16	0.18	0.15	0.16	0.17
MgO	9.05	9.88	4.18	7.39	6.77	6.00	7.55	9.14	9.01
CaO	10.08	8.96	11.13	9.94	8.28	9.85	9.95	10.94	9.73
Na_2O	3.12	2.70	3.13	2.68	3.02	3.70	3.92	2.45	3.29
K_2O	0.76	0.43	0.88	0.46	0.53	0.88	1.46	0.85	0.92
TiO_2	1.57	1.34	1.99	2.20	2.24	2.23	2.60	1.80	1.87
P_2O_5	0.22	0.25	0.31	0.30	0.31	0.46	0.34	0.36	0.40
LOI	0.69	2.32	2.07	2.16	3.13	2.68	1.31	2.82	0.81
Total	99.85	99.32	100.70	99.19	100.10	98.43	100.60	99.62	100.40
Mg# (ppm)	59.23	65.32	42.10	56.07	54.21	51.24	56.49	60.73	59.64
Sc	28.0	29.0	27.0	30.0	22.0	21.0	23.0	25.0	26.0
V	215.0	164.0	243.0	223.0	182.0	162.0	202.0	206.0	211.0
Cr	340.0	370.0	150.0	90.0	170.0	190.0	310.0	330.0	330.0
Co	57.0	57.0	44.0	45.0	43.0	41.0	50.0	58.0	56.0
Ni	230.0	230.0	60.0	70.0	110.0	110.0	170.0	250.0	220.0
Cu	80.0	70.0	60.0	60.0	50.0	50.0	60.0	80.0	80.0
Zn	80.0	70.0	90.0	70.0	80.0	70.0	70.0	80.0	70.0
Rb	11.0	6.0	15.0	6.0	7.0	13.0	25.0	10.0	14.0
Sr	400.0	667.0	443.0	415.0	420.0	483.0	582.0	539.0	495.0
Y	20.90	25.80	22.80	26.70	27.60	30.90	23.50	22.50	22.70
Zr	128.0	129.0	136.0	168.0	203.0	264.0	202.0	145.0	171.0
Nb	13.30	12.60	15.30	17.90	16.60	28.80	38.00	25.50	29.70
Ba	180.0	189.0	352.0	131.0	225.0	205.0	296.0	268.0	207.0
La	11.70	11.60	1[illegible].70	12.40	13.10	19.30	18.70	20.70	22.40
Ce	26.90	27.20	27.10	28.90	31.80	44.70	40.60	43.00	45.80
Pr	3.45	3.57	3.64	3.97	4.28	5.76	5.10	5.07	5.33
Nd	15.40	17.00	16.40	17.90	20.00	27.20	22.10	20.80	21.80
Sm	3.88	4.36	4.34	4.86	5.01	6.21	5.16	4.58	4.62

(continued)

Table 2 (continued)

Area	Khaybar								
Sample name	BA-01/Tb-1	JA-01/Tb-1	JD-01/Tb-1	JI-01/Tb-1	JI-02/Tb-1	JJa-02/Tb-2	JJa-01/Tb-1	JJ-01/1	JJ-02/Tb-1
Locality	Bi'r al 'Ayn	Jabal Antar	Jabal Dhulay'ar	Jabal Isahara	Jabal Isahara	Jabal Jammazin	Jabal Jammazin	Jabal Jarbul	Jabal Jarbul
Rock type	Basalt	Basalt	Basalt	Basalt	Basalt	Basalt	Basanite	Basalt	Basalt
(wt%)									
Eu	1.43	1.61	1.59	1.85	1.80	2.19	1.87	1.64	1.62
Gd	4.21	4.77	4.58	5.32	5.57	6.50	5.24	4.71	4.52
Tb	0.69	0.78	0.72	0.86	0.90	1.01	0.80	0.73	0.72
Dy	4.05	4.87	4.37	5.28	5.23	5.86	4.59	4.34	4.26
Ho	0.77	0.95	0.86	0.99	1.02	1.11	0.88	0.87	0.84
Er	2.30	2.67	2.44	2.87	2.81	3.21	2.41	2.36	2.30
Tm	0.33	0.39	0.33	0.42	0.40	0.45	0.33	0.33	0.31
Yb	2.13	2.48	2.12	2.50	2.46	2.82	2.05	1.95	1.99
Lu	0.30	0.38	0.32	0.37	0.39	0.40	0.32	0.29	0.31
Hf	3.10	3.00	3.10	3.80	4.20	5.60	4.60	3.30	3.60
Ta	1.12	0.90	1.16	1.20	1.14	2.06	2.84	1.93	2.17
Th	1.22	0.98	0.98	1.10	1.09	1.91	1.99	2.07	2.33
U	0.36	0.27	0.30	0.29	0.33	0.54	0.55	0.34	0.67

Area	Al-Birk					Lunayyir							
Sample name	DIS14-503	DIS14-507	DIS14-500	DIS14-506	DIS14-504	QH-01/1	QH-03/1	QbH-03/1	QbF-01/1	QbF-01/2	QbB-01/1	QbG-01/2	QbE-02/1
Locality						Qalib Harrat	Qalib Harrat	Umluji	Umluji	Umluji	Umluji	Umluji	Umluji
Rock type	Basalt	Basalt	Basalt	Basanite	Basanite	Basalt	Basalt	Basalt	Basalt	Basalt	Basalt	Basalt	Basanite
(wt%)													
SiO_2	43.80	43.70	45.58	44.54	46.08	47.03	47.38	44.50	45.53	45.93	47.62	46.94	48.14
Al_2O_3	14.79	14.73	16.78	14.98	15.77	15.02	15.37	14.01	14.69	14.76	15.55	16.02	16.29
FeO(T)	7.91	8.13	8.95	8.38	8.59	9.47	9.72	10.97	10.04	10.28	10.14	10.53	10.75
MnO	0.15	0.15	0.16	0.16	0.16	0.16	0.16	0.16	0.17	0.17	0.17	0.16	0.16
MgO	10.40	9.64	7.36	10.13	9.27	9.43	9.51	10.03	9.68	9.16	9.74	6.93	6.90
CaO	10.48	9.94	10.30	10.03	9.62	10.44	9.80	11.07	10.28	10.11	9.85	8.56	8.35
Na_2O	2.73	3.32	3.41	3.38	4.00	2.86	2.97	2.22	2.91	3.01	3.10	3.66	4.11
K_2O	1.50	1.19	1.29	1.49	1.91	0.73	0.75	0.74	0.74	0.81	0.75	1.15	1.13
TiO_2	2.69	2.71	2.95	2.68	2.76	1.24	1.29	1.75	1.34	1.46	1.44	1.99	2.03
P_2O_5	0.73	0.71	0.69	0.70	0.80	0.25	0.25	0.27	0.28	0.33	0.30	0.60	0.62
LOI	3.13	3.26	1.84	2.81	0.48	2.29	1.33	3.58	3.36	3.16	0.69	0.95	0.71
Total	99.19	98.38	100.30	100.20	100.40	99.97	99.63	100.50	100.10	100.30	100.50	98.66	100.40
Mg# (ppm)	70.10	67.90	59.44	68.31	65.79	63.98	63.56	61.98	63.21	61.38	63.13	53.99	53.36

(continued)

Table 2 (continued)

Area	Al-Birk					Lunayyir							
Sample name	DIS14-503	DIS14-507	DIS14-500	DIS14-506	DIS14-504	QH-01/1	QH-03/1	QbH-03/1	QbF-01/1	QbF-01/2	QbB-01/1	QbG-01/2	QbE-02/1
Locality						Qalib Harrat	Qalib Harrat	Umluji	Umluji	Umluji	Umluji	Umluji	Umluji
Rock type	Basalt	Basalt	Basalt	Basanite	Basanite	Basalt	Basalt	Basalt	Basalt	Basalt	Basalt	Basalt	Basanite
Sc	25.0	26.0	28.0	26.0	25.0	27.0	27.0	26.0	25.0	27.0	29.0	20.0	20.0
V	220.0	237.0	264.0	230.0	229.0	190.0	194.0	214.0	178.0	195.0	206.0	155.0	160.0
Cr	310.0	320.0	210.0	340.0	240.0	400.0	430.0	380.0	370.0	350.0	390.0	210.0	210.0
Co	43.0	42.0	43.0	42.0	41.0	52.0	54.0	60.0	53.0	53.0	52.0	44.0	44.0
Ni	250.0	240.0	170.0	240.0	200.0	250.0	240.0	310.0	240.0	220.0	230.0	110.0	100.0
Cu	40.0	50.0	40.0	50.0	40.0	90.0	80.0	80.0	70.0	70.0	80.0	60.0	60.0
Zn	60.0	60.0	50.0	60.0	60.0	80.0	70.0	70.0	70.0	70.0	70.0	70.0	70.0
Rb	31.0	38.0	31.0	45.0	58.0	10.0	10.0	7.0	13.0	17.0	13.0	19.0	21.0
Sr	1189.0	755.0	833.0	749.0	840.0	487.0	385.0	746.0	474.0	495.0	640.0	750.0	714.0
Y	29.40	28.80	28.90	29.20	31.30	21.10	21.40	19.50	20.40	22.30	22.30	25.40	25.50
Zr	241.0	233.0	285.0	237.0	292.0	141.0	145.0	133.0	150.0	162.0	154.0	230.0	277.0
Nb	80.90	81.40	83.20	81.60	96.60	18.10	18.50	20.30	23.00	26.20	20.30	31.70	38.70
Ba	642.0	614.0	644.0	520.0	690.0	174.0	174.0	181.0	197.0	201.0	227.0	276.0	277.0
La	66.10	62.50	56.70	62.30	72.20	14.60	15.20	14.70	18.30	20.30	16.40	30.50	29.20
Ce	122.00	117.00	108.00	116.00	133.00	30.40	31.80	31.20	37.00	41.10	34.00	64.10	61.30
Pr	13.00	12.40	11.80	12.30	14.00	3.65	3.82	3.83	4.43	4.79	4.06	7.70	7.27
Nd	48.40	45.80	43.10	45.80	51.80	15.60	16.10	16.00	17.30	19.60	17.10	30.30	29.20
Sm	8.68	8.05	8.12	8.14	9.20	3.57	3.69	3.87	3.85	4.30	4.06	6.62	6.48
Eu	2.83	2.61	2.57	2.67	2.93	1.29	1.38	1.38	1.36	1.40	1.34	2.25	2.21
Gd	7.42	7.25	7.65	7.19	7.67	3.77	3.98	3.99	3.88	4.26	4.31	6.20	5.86
Tb	1.07	1.07	1.05	1.05	1.10	0.62	0.67	0.60	0.62	0.70	0.66	0.93	0.88
Dy	6.00	5.99	5.76	5.63	6.27	3.77	4.02	3.63	3.61	4.12	4.03	5.28	4.99
Ho	1.10	1.05	1.07	1.06	1.15	0.73	0.77	0.72	0.74	0.81	0.83	0.98	0.94
Er	2.91	3.03	2.92	2.92	3.24	2.28	2.31	1.93	2.13	2.29	2.32	2.47	2.51
Tm	0.40	0.42	0.42	0.41	0.47	0.31	0.35	0.28	0.30	0.33	0.32	0.33	0.32
Yb	2.49	2.53	2.78	2.59	2.95	1.95	2.09	1.91	1.99	2.18	2.26	2.13	2.02
Lu	0.37	0.39	0.40	0.40	0.42	0.32	0.33	0.28	0.30	0.34	0.36	0.31	0.32
Hf	5.10	5.00	5.10	4.90	6.40	3.10	3.20	2.80	3.20	3.50	3.30	4.40	5.30
Ta	6.28	6.33	5.24	6.22	7.08	1.24	1.35	1.44	1.70	1.92	1.44	2.30	2.63
Th	8.85	8.27	6.94	8.40	9.65	1.86	1.98	1.40	2.06	2.35	1.98	3.22	3.11
U	2.29	1.70	1.80	1.83	2.55	0.39	0.58	0.27	0.43	0.67	0.38	0.63	0.90

Table 2 (continued)

Area	Lunayyir											
Sample name	LH-01/Tb-3	LH-13/Tb-2	LH-15/Tb-1	LH-05/Tb-1	LH-11/Tb-1	LH-03/Tb-1	LH-14/Tb-2	LH-08/Tb-1	LH-09/Tb1-1	LH21/Qtb-1	LH-19/Qtb-1	LH-04/Qtb-1
Locality	Lunayyir Harrat	Lunayyir Harrat	Lunayyir Harrat	Lunayyir Harrat	Lunayyir Harrat	Lunayyir Harrat	Lunayyir Harrat	Lunayyir Harrat	Lunayyir Harrat	Lunayyir Harrat	Lunayyir Harrat	Lunayyir Harrat
Rock type	Basalt	Basalt	Basalt	Basanite	Basanite	Basanite	Basanite	Basanite	Basanite	Basanite	Basanite	Basalt
(wt%)												
SiO_2	48.81	47.28	47.65	48.13	48.19	48.41	47.36	47.67	47.84	48.47	46.26	46.91
Al_2O_3	14.91	15.19	16.80	15.27	15.79	15.77	16.41	15.67	15.92	15.79	16.60	15.66
FeO(T)	10.30	11.20	9.61	10.53	13.24	10.59	13.76	13.17	11.09	13.61	12.99	12.02
MnO	0.16	0.18	0.17	0.18	0.21	0.18	0.22	0.21	0.18	0.22	0.21	0.19
MgO	9.64	8.41	7.06	6.60	4.80	7.01	4.86	4.53	6.61	4.92	4.91	6.91
CaO	9.67	9.55	10.92	8.12	8.03	8.27	7.49	6.95	7.89	7.54	7.71	9.09
Na_2O	3.10	3.10	3.35	4.06	4.28	4.05	4.40	4.41	4.51	4.24	4.21	3.58
K_2O	0.60	0.63	1.00	1.20	1.10	1.40	1.24	1.25	1.24	1.20	1.11	0.78
TiO_2	1.71	1.87	2.00	1.87	2.75	1.76	3.09	2.79	2.26	3.04	2.93	2.32
P_2O_5	0.32	0.40	0.45	0.58	0.69	0.61	0.77	0.81	0.68	0.76	0.68	0.46
LOI	-0.18	0.39	0.32	1.01	0.27	0.87	-0.66	-0.57	0.51	-0.72	-0.76	0.45
Total	100.20	99.45	100.40	98.71	100.80	100.10	100.50	98.37	99.98	100.60	98.30	99.71
Mg# (ppm)	62.52	57.23	56.70	52.78	39.26	54.13	38.64	38.00	51.51	39.18	40.25	50.61
Sc	28.0	26.0	32.0	20.0	21.0	21.0	21.0	18.0	18.0	22.0	21.0	23.0
V	219.0	209.0	243.0	158.0	204.0	161.0	206.0	157.0	153.0	209.0	209.0	205.0
Cr	380.0	220.0	140.0	190.0	< 20	260.0	< 20	< 20	110.0	30.0	< 20	110.0
Co	51.0	51.0	40.0	39.0	42.0	45.0	43.0	40.0	41.0	44.0	43.0	51.0
Ni	230.0	160.0	60.0	100.0	< 20	130.0	< 20	< 20	90.0	< 20	20.0	80.0
Cu	80.0	70.0	60.0	50.0	40.0	60.0	40.0	30.0	60.0	40.0	40.0	60.0
Zn	70.0	80.0	60.0	60.0	110.0	70.0	110.0	110.0	70.0	110.0	100.0	90.0
Rb	9.0	8.0	15.0	23.0	15.0	29.0	19.0	17.0	19.0	18.0	17.0	11.0
Sr	454.0	432.0	576.0	632.0	532.0	627.0	541.0	544.0	656.0	518.0	539.0	525.0
Y	22.10	27.30	27.80	27.20	40.10	29.30	40.40	43.50	28.80	41.80	37.00	30.70
Zr	161.0	179.0	184.0	229.0	301.0	267.0	323.0	354.0	277.0	311.0	242.0	198.0
Nb	20.40	19.50	32.20	38.70	32.90	47.10	39.50	37.30	42.30	37.30	28.90	21.70
Ba	172.0	177.0	255.0	319.0	292.0	330.0	314.0	301.0	295.0	307.0	276.0	208.0
La	15.90	17.00	25.20	34.20	29.00	40.00	33.50	32.90	33.00	32.20	28.50	21.90
Ce	34.70	37.60	50.40	68.40	65.60	78.90	73.40	74.50	68.20	71.50	63.50	48.10
Pr	4.20	4.88	5.81	7.75	8.39	8.60	9.10	9.36	7.93	8.92	7.95	5.98
Nd	17.30	20.50	24.10	30.50	35.90	34.60	38.80	39.80	32.40	37.80	33.10	25.90
Sm	3.88	5.01	5.28	5.88	7.90	6.56	8.30	8.62	6.41	8.24	7.52	5.79

(continued)

Table 2 (continued)

Area	Lunayyir											
Sample name	LH-01/Tb-3	LH-13/Tb-2	LH-15/Tb-1	LH-05/Tb-1	LH-11/Tb-1	LH-03/Tb-1	LH-14/Tb-2	LH-08/Tb-1	LH-09/Tb1-1	LH21/Qtb-1	LH-19/Qtb-1	LH-04/Qtb-1
Locality	Lunayyir Harrat	Lunayyir Harrat	Lunayyir Harrat	Lunayyir Harrat	Lunayyir Harrat	Lunayyir Harrat	Lunayyir Harrat	Lunayyir Harrat	Lunayyir Harrat	Lunayyir Harrat	Lunayyir Harrat	Lunayyir Harrat
Rock type	Basalt	Basalt	Basalt	Basanite	Basanite	Basanite	Basanite	Basanite	Basanite	Basanite	Basanite	Basalt
(wt%)												
Eu	1.43	1.71	1.89	1.99	2.60	2.24	2.82	2.80	2.25	2.65	2.47	2.02
Gd	4.42	5.12	5.43	5.54	7.86	6.19	8.00	9.05	5.90	8.39	7.30	5.98
Tb	0.68	0.87	0.86	0.85	1.25	0.96	1.26	1.42	0.96	1.34	1.16	0.96
Dy	4.10	5.09	5.15	5.12	7.43	5.54	7.44	8.22	5.54	7.86	6.82	5.76
Ho	0.81	0.99	1.05	1.00	1.44	1.04	1.53	1.62	1.07	1.54	1.35	1.12
Er	2.39	2.92	2.93	2.83	4.26	2.93	4.39	4.61	2.91	4.39	3.87	3.24
Tm	0.34	0.42	0.40	0.41	0.59	0.43	0.60	0.65	0.42	0.63	0.56	0.47
Yb	2.14	2.64	2.62	2.65	3.79	2.81	3.89	4.14	2.63	4.01	3.71	3.08
Lu	0.35	0.38	0.42	0.39	0.62	0.42	0.61	0.63	0.39	0.62	0.55	0.45
Hf	3.30	3.80	4.10	4.60	6.20	5.60	6.40	7.30	5.50	6.20	4.40	4.00
Ta	1.43	1.35	2.28	2.94	2.35	3.83	2.75	2.70	2.87	2.66	2.07	1.70
Th	1.65	1.61	3.02	4.12	2.50	5.09	3.14	3.14	3.60	3.02	2.53	1.99
U	0.41	0.56	1.15	1.17	0.82	1.43	0.95	1.01	1.10	0.92	0.81	0.60

Area	Lunayyir										
Sample name	LH-06/Qtb-2	LH-18/Qtb-1	LH-02/Qtb-1	LH-10/Qtb-2	LH-17/Qtb-1	LH-23/Qtb-1	LH-12/Qtb-1	LH-22/Qtb-1	LH-16/Qtb-1	LH-20/Qtb-1	LH-07/Qtb-2
Locality	Lunayyir Harrat	Lunayyir Harrat	Lunayyir Harrat	Lunayyir Harrat	Lunayyir Harrat	Lunayyir Harrat	Lunayyir Harrat	Lunayyir Harrat	Lunayyir Harrat	Lunayyir Harrat	Lunayyir Harrat
Rock type	Basalt	Basalt	Basalt	Basanite	Basanite	Basanite	Basanite	Basalt	Basanite	Basanite	Basanite
(wt%)											
SiO_2	47.24	46.69	46.16	47.68	47.50	46.72	47.69	48.04	48.33	48.17	50.99
Al_2O_3	15.89	15.97	15.95	15.56	16.72	15.80	15.28	15.61	15.44	16.87	15.73
FeO(T)	11.93	12.98	12.52	12.79	12.95	14.22	13.53	11.58	13.34	12.82	12.38
MnO	0.19	0.21	0.20	0.21	0.20	0.22	0.23	0.19	0.22	0.21	0.23
MgO	7.16	7.04	6.52	4.90	4.76	4.81	4.51	8.36	4.27	3.89	3.17
CaO	8.64	8.86	8.42	7.79	7.75	7.53	7.17	9.48	6.98	7.06	6.05
Na_2O	3.72	3.66	3.78	4.18	4.15	4.25	4.31	3.54	4.49	4.70	5.12
K_2O	0.70	0.79	0.97	1.05	1.09	1.20	1.34	0.69	1.41	1.23	1.62
TiO_2	2.26	2.58	2.67	2.68	2.88	3.03	2.98	1.86	2.92	2.69	1.89
P_2O_5	0.40	0.53	0.55	0.65	0.69	0.79	0.82	0.42	0.95	0.82	1.09
LOI	0.03	-0.27	-0.73	-0.02	-0.23	-0.46	-0.83	-0.43	-0.37	-0.46	0.19
Total	99.49	100.40	98.41	98.89	99.91	99.69	98.53	100.60	99.47	99.43	99.84
Mg# (ppm)	51.69	49.17	48.15	40.59	39.59	37.62	37.27	56.27	36.32	35.10	31.34

(continued)

Table 2 (continued)

Area	Lunayyir										
Sample name	LH-06/Qtb-2	LH-18/Qtb-1	LH-02/Qtb-1	LH-10/Qtb-2	LH-17/Qtb-1	LH-23/Qtb-1	LH-12/Qtb-1	LH-22/Qtb-1	LH-16/Qtb-1	LH-20/Qtb-1	LH-07/Qtb-2
Locality	Lunayyir Harrat	Lunayyir Harrat	Lunayyir Harrat	Lunayyir Harrat	Lunayyir Harrat	Lunayyir Harrat	Lunayyir Harrat	Lunayyir Harrat	Lunayyir Harrat	Lunayyir Harrat	Lunayyir Harrat
Rock type	Basalt	Basalt	Basalt	Basanite	Basanite	Basanite	Basanite	Basalt	Basanite	Basanite	Basanite
(wt%)											
Sc	24.0	25.0	25.0	20.0	21.0	20.0	22.0	27.0	19.0	16.0	11.0
V	222.0	224.0	227.0	200.0	208.0	196.0	181.0	211.0	139.0	126.0	61.0
Cr	160.0	140.0	90.0	< 20	< 20	< 20	< 20	240.0	< 20	< 20	20.0
Co	51.0	52.0	48.0	43.0	42.0	44.0	41.0	50.0	39.0	36.0	26.0
Ni	80.0	90.0	60.0	< 20	< 20	< 20	< 20	130.0	< 20	< 20	< 20
Cu	60.0	70.0	60.0	40.0	40.0	40.0	40.0	80.0	40.0	30.0	40.0
Zn	80.0	90.0	90.0	100.0	100.0	120.0	120.0	80.0	130.0	100.0	100.0
Rb	8.0	11.0	14.0	15.0	16.0	16.0	19.0	9.0	21.0	18.0	23.0
Sr	533.0	474.0	490.0	529.0	539.0	534.0	487.0	412.0	534.0	586.0	632.0
Y	28.40	34.70	34.40	39.00	37.30	43.20	48.10	27.60	49.00	42.30	41.50
Zr	200.0	239.0	225.0	297.0	279.0	331.0	320.0	204.0	379.0	344.0	403.0
Nb	21.10	24.70	28.20	33.00	35.00	36.30	35.10	23.00	43.90	38.70	53.90
Ba	151.0	188.0	237.0	285.0	274.0	310.0	358.0	156.0	390.0	294.0	409.0
La	18.80	21.40	26.00	28.80	29.20	32.70	35.70	19.20	39.70	34.00	48.20
Ce	42.50	49.30	56.80	64.10	63.70	73.40	78.90	41.70	86.90	76.60	103.00
Pr	5.36	6.28	7.09	8.12	7.95	9.19	10.00	5.14	11.00	9.38	12.20
Nd	23.40	27.20	30.40	34.30	33.50	39.60	44.10	21.70	47.10	39.90	48.40
Sm	5.09	6.25	6.60	7.56	7.59	8.56	9.53	5.19	10.20	8.76	9.14
Eu	1.88	2.15	2.23	2.55	2.42	2.95	3.06	1.76	3.19	2.72	2.94
Gd	5.85	6.59	6.61	7.84	7.23	9.02	9.66	5.21	10.10	8.65	8.56
Tb	0.91	1.03	1.03	1.21	1.11	1.37	1.51	0.85	1.52	1.40	1.32
Dy	5.25	6.28	6.15	7.14	6.70	7.96	9.04	5.12	9.00	8.26	7.88
Ho	1.03	1.23	1.21	1.39	1.34	1.57	1.75	1.00	1.76	1.54	1.48
Er	3.06	3.59	3.64	4.02	4.06	4.41	5.03	2.83	4.99	4.41	4.26
Tm	0.44	0.53	0.52	0.56	0.57	0.64	0.73	0.41	0.70	0.62	0.60
Yb	2.80	3.58	3.27	3.58	3.58	4.07	4.78	2.74	4.60	4.02	3.91
Lu	0.43	0.53	0.52	0.55	0.57	0.63	0.73	0.40	0.70	0.63	0.64
Hf	4.40	4.80	4.40	6.30	5.80	6.70	6.10	4.20	7.60	6.90	7.80
Ta	1.63	1.68	2.19	2.29	2.39	2.50	2.59	1.68	2.98	2.72	3.87
Th	1.66	1.80	2.46	2.49	2.58	2.76	3.10	1.93	3.41	3.04	4.45
U	0.54	0.60	0.76	0.80	0.84	0.89	0.96	0.49	1.06	0.90	1.24

Mg# = 100*Mg/(Mg+Fe) mol%

Fig. 6 **a** Total alkalis and **b** K_2O versus SiO_2 diagrams showing the chemical classification of selected rocks

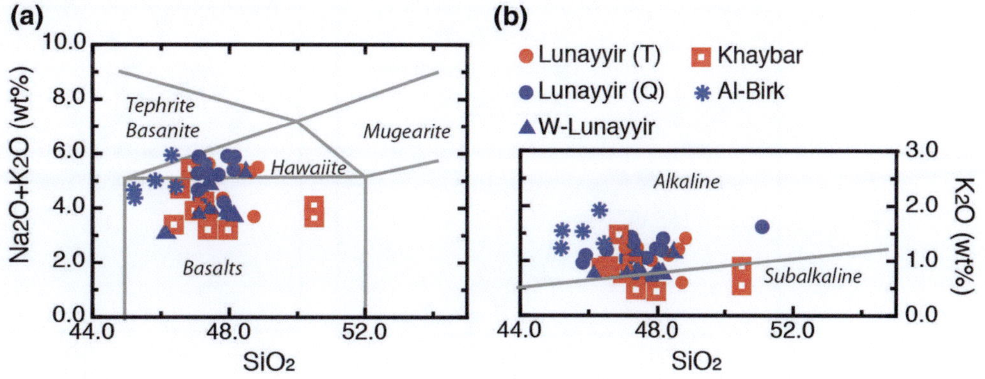

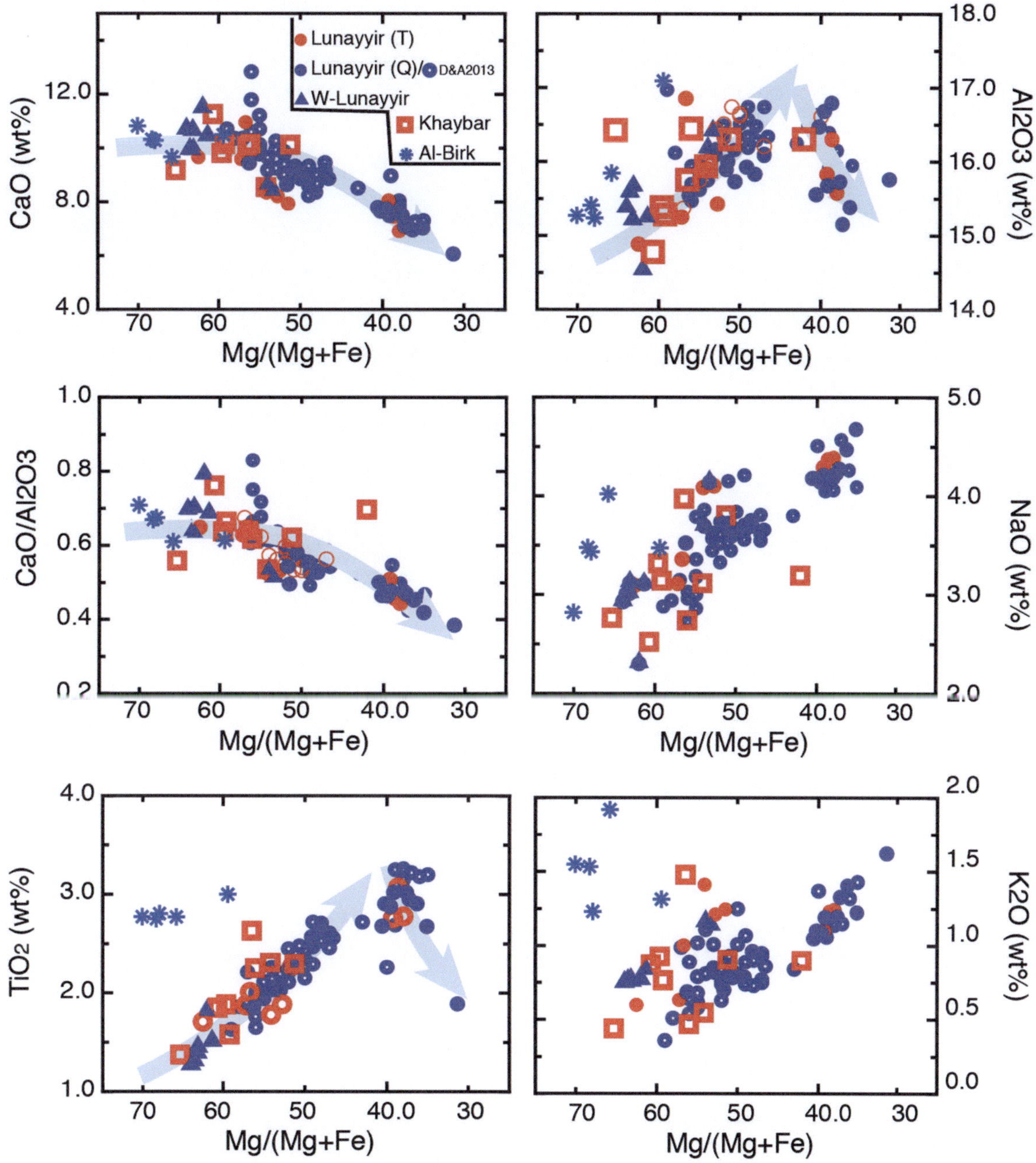

Fig. 7 Variations in Mg/(Mg+Fe) (mol%) versus CaO wt%, Al_2O_3 wt%, CaO/Al_2O_3 ratios, Na_2O wt%, TiO_2 wt% and Na_2O wt% of lavas from the Lunayyir, Khaybar and Al Birk areas. Compositions of lavas from Lunayyir from Duncan and Al-Amri (2013) are also shown

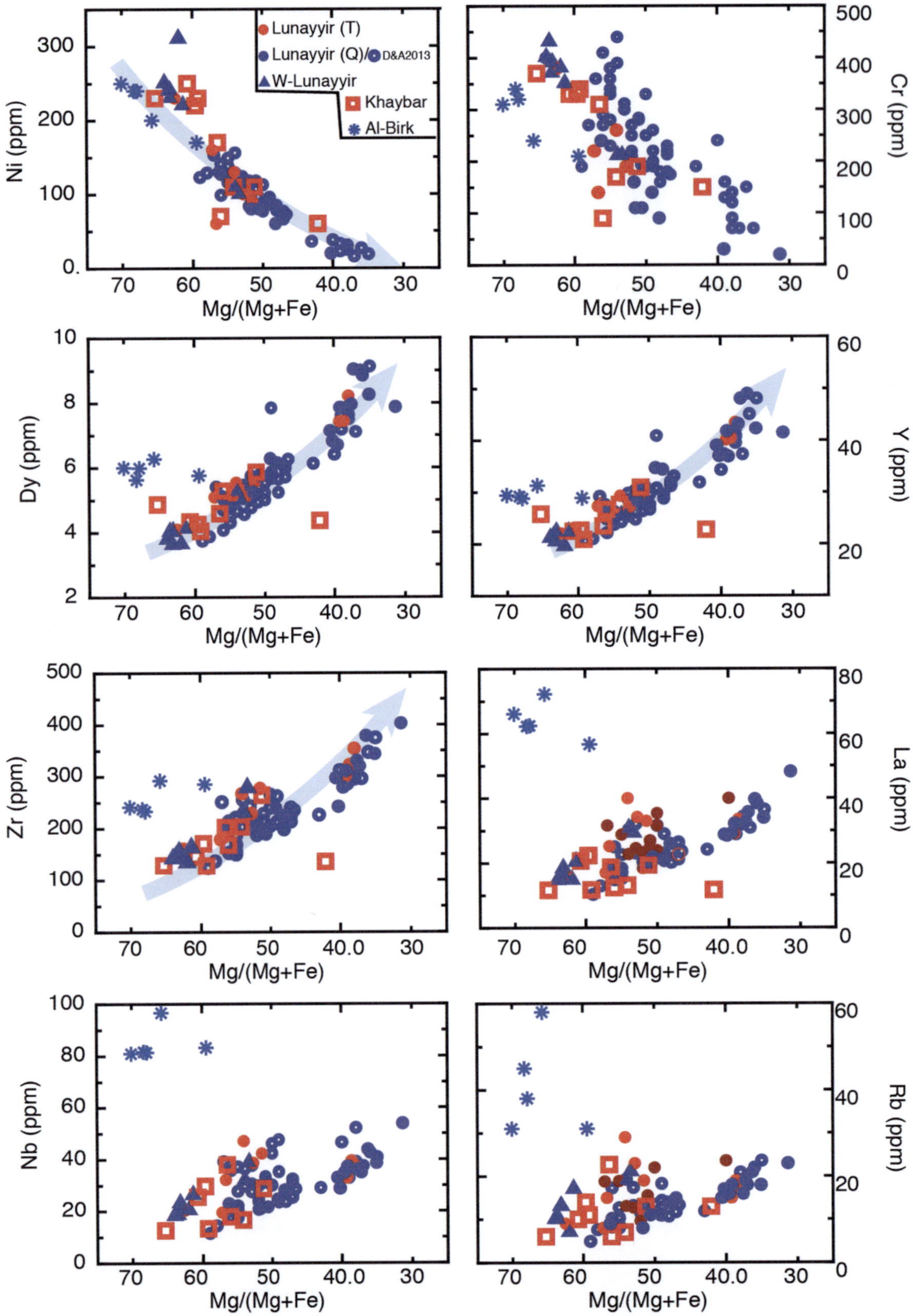

Fig. 8 Variations in Mg/(Mg+Fe) (mol%) versus some trace elements (Ni, Cr, Dy, Y, Zr, La, Nb, Rb, ppm) of lavas from the Lunayyir, Khaybar and Al-Birk areas. Compositions of lavas from Lunayyir from Duncan and Al-Amri (2013) are also shown

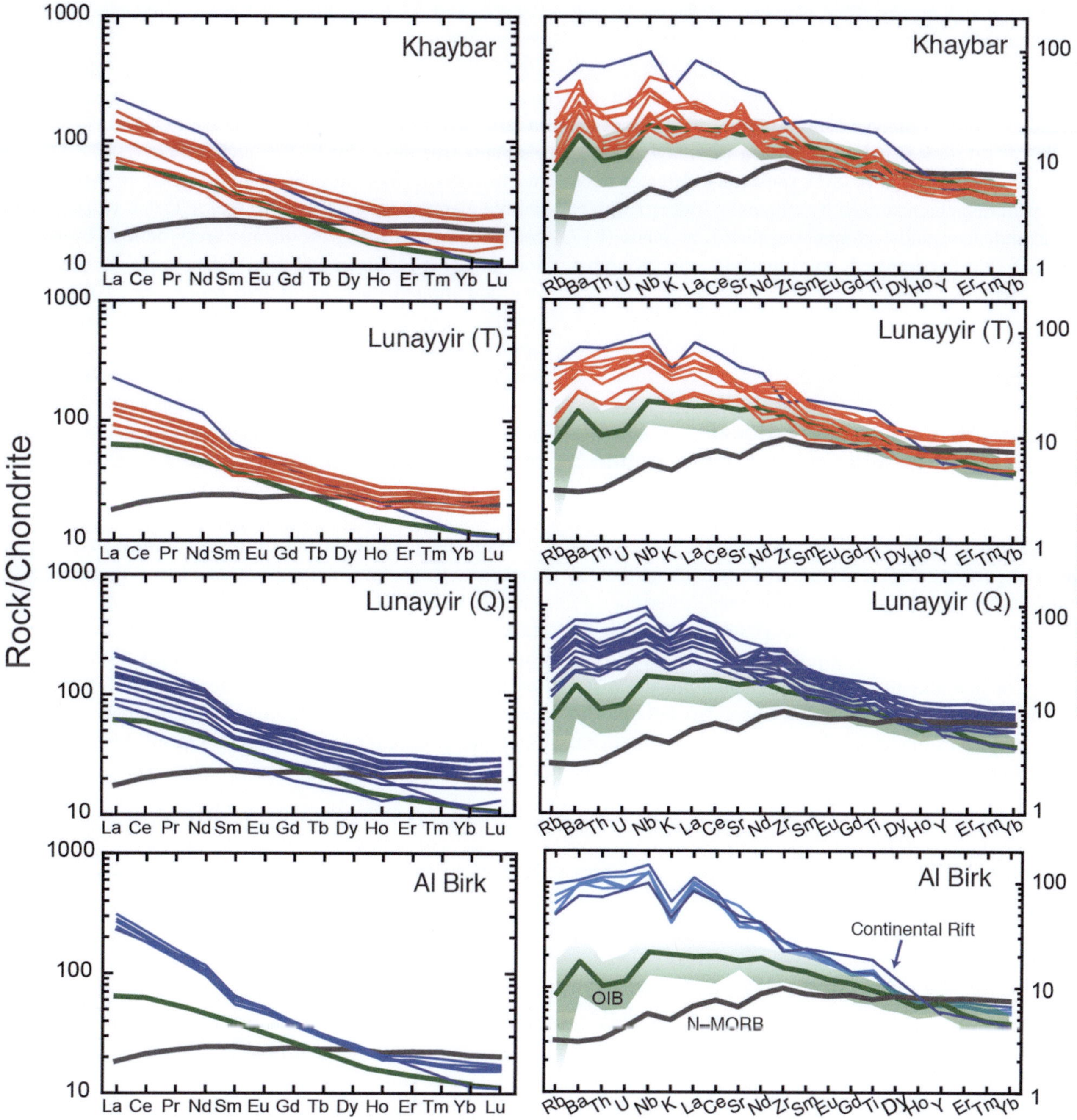

Fig. 9 Chondrite-normalized and primitive mantle-normalized (Sun and McDonough 1989) diagrams of samples from the Lunayyir, Khaybar and Al Birk areas. Representative compositions of N-MORB and E-MORB (Gale et al. 2013); Hawaii OIB (Hofmann and Jochum 1996) and Kenya continental rift (Ewart et al. 2004) are also reported for comparison

(0.80–96.6 ppm) and K do not display clear correlations with Mg#. Samples from Al Birk are distinct from the other groups with overall higher contents of incompatible trace elements for a given Mg# (Fig. 8).

All selected samples have variably fractionated chondrite-normalized REE patterns (Fig. 9), typical for alkali basalts from this region. Lunayyir samples show wide variations in REE abundances (MREE ranging between 10 and 40 times that of chondrite), and variable REE fractionations (La_N/Yb_N 4–10) (Fig. 9). Basalts from Harrat Khaybar show limited variability in absolute REE abundances but variable REE fractionations (La_N/Yb_N = 3–7). Al Birk samples are highly enriched in LREE compared to the other samples and display strong L/MREE and M/HREE fractionations (La_N/Sm_N ~5; Dy_N/Yb_N = 1.3–1.6). Europium anomalies are absent in all studied samples. Primitive mantle-normalized incompatible elements of Lunayyir and Khaybar samples show steep trends and a characteristic "OIB-like" geochemical signature (Fig. 9). They have high Nb content (20–146 times CI), and show troughs in Th and U. Samples from Lunayyir show more pronounced K anomalies (K_N/La_N = 0.7–1.1) compared to Khaybar lavas (K_N/La_N = 0.9–1.7). Al Birk lavas show very different trace element patterns, characterized by extremely higher Rb, Nb, and La abundances, and a stronger negative K anomaly (K_N/La_N = 0.42–0.59) compared to Lunayyir and Khaybar samples.

5 Discussion

5.1 Magmatic Evolution: Fractional Crystallization and Crustal Contamination Processes

The variability of Mg# (31–70) in the studied lavas suggests that they experienced some degree of fractionations before eruption, in agreement with the widespread occurrence of olivine and plagioclase phenocrysts. Therefore, the magma most likely resided in the crust prior to eruption raising the possibility of interaction with the continental crust. Crustal assimilation has been shown to be minor for basalts and basanites erupted in the Lunayyir, based on their non-crustal trace element pattern (e.g., low and constant Th/Ta ratios; Duncan and Al-Amri 2013) and non radiogenic Sr isotope compositions ($^{87}Sr/^{86}Sr$ = 0.7029–0.7031; Altherr et al. 1990; Bertrand et al. 2003). Pearce (1983) inferred that assimilation of continental crust should lead to enrichments in large ion lithophile elements relative to high field strength elements (see also Abu El-Rus and Rooney 2017). Most of our samples do not show anomalous spikes in their incompatible element patterns (Fig. 9) and a Ta/Yb versus Th/Yb correlation lying along the mantle array defined by MORB and OIB (Fig. 10a). In addition, most of our samples show low and limited range of Ba/Nb ratios (Fig. 10b). A few samples from Khaybar show anomalously high Ba/Nb ratios (up to 23). These samples have relatively high L.O.I (~2–3 wt%), high K (K_N/La_N = 1.4–1.7) and low Rb and Sr contents, which suggest that low temperature alteration rather than crustal assimilation have caused their anomalous composition. Taken as a whole, crustal assimilation did not play a substantial role in defining the composition of our selected basalts.

Magmatic evolution can be inferred from co-variations in major and trace element compositions. Lavas from Lunayyir and Khaybar follow similar trends in Mg# versus major and trace element compositions, suggesting that they experienced similar fractionation processes (Figs. 7 and 8). For instance, the decrease in Ni with decreasing Mg# indicates fractionation of olivine, whereas the gradual decrease in Cr, CaO and CaO/Al_2O_3 ratios with decreasing Mg# is consistent with fractionation of clinopyroxene and/or spinel. In

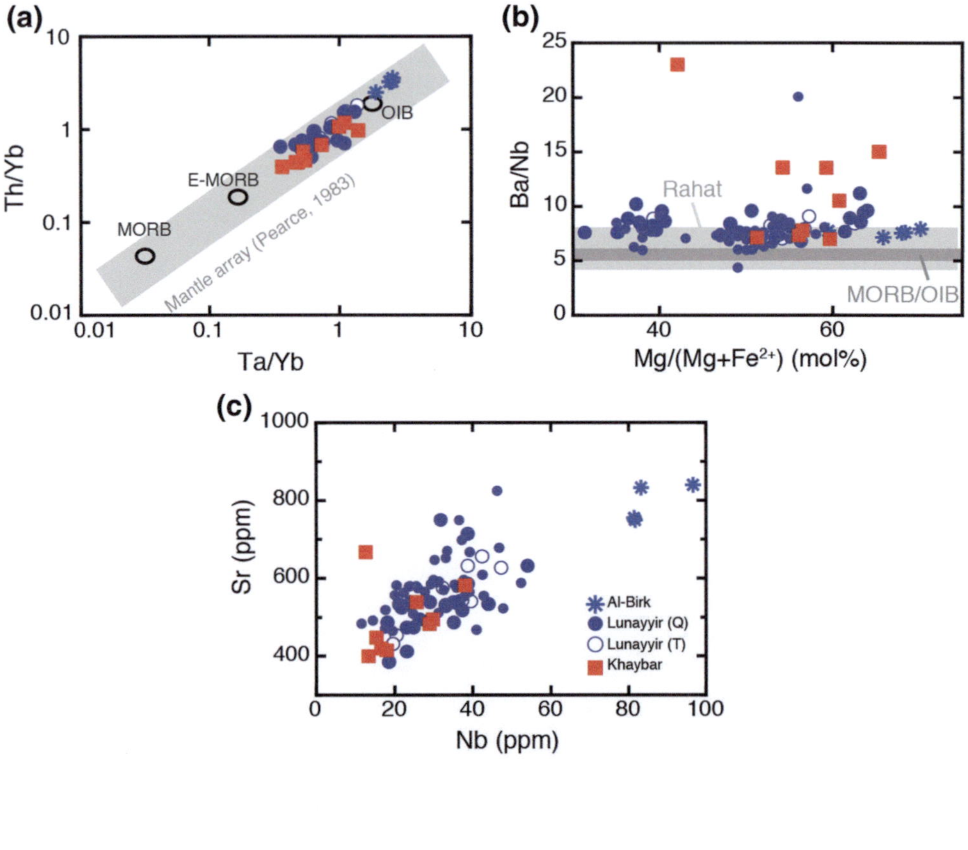

Fig. 10 **a** Ta/Yb versus Th/Yb variation diagram after Pearce (1983) showing the mantle array (grey field) relating N-MORB, E-NORB and OIB (data Sun and McDonough 1989). **b** Variations in MgO (wt%) versus Ba/Nb ratios (**a**) and Sr (ppm) versus Nb (ppm) (**b**) of lavas from the Lunayyir, Khaybar and Al Birk areas. The ranges of Rahat mafic lavas (MgO >6 wt%) (Moufti et al. 2012) and MORB-OIB (Hofmann 1988) are also shown

particular the increase in CaO/Al_2O_3 ratios and the lack of a substantial Eu anomaly in the REE pattern of most samples indicate that plagioclase did not played a major role into the petrological evolution of these melts. This is in agreement with the positive correlation between Sr and Nb (Fig. 10c), that also indicates minimal Sr removal during the early evolution of the primary melts. On the other hand, the gradual increase in Al_2O_3 up to Mg# of ~50, followed by rather constant Al_2O_3 contents in the more evolved samples, and the local occurrence of plagioclase phenocrysts in the more evolved samples from the Lunayyir, suggest that fractionation of plagioclase occurred in the late stage of evolution. Fractional crystallization experiments and thermodynamic calculations show that at high pressure, clinopyroxene crystallizes before plagioclase (Presnall et al. 1978; Tormey et al. 1987; Grove et al. 1992; Yang et al. 1996). The same crystallization order was inferred for lavas from Harrats Rahat and Hutaymah (Moufti et al. 2012; Duncan et al. 2016), where covariations between major elements also suggest an early crystallization of clinopyroxene relative to plagioclase. Hence, although substantial differences between the liquid lines of descent of different lava fields may exist locally (see for instance Duncan et al. 2016) the early occurrence of clinopyroxene as a solid phase appears to be a common feature in alkali basalts of the Western Arabian province, suggesting that fractionation occurred within magma chambers located at relatively high depths, either within the upper mantle or at the crust-mantle boundary.

As expected, Mg# correlates well with moderately incompatible elements such as MREE/HREE, Zr, Y and Ti, forming negative trends (Fig. 8). On the other hand, well-defined correlations are less obvious between Mg# and highly-incompatible elements like Rb, Ba, Th, U, Nb, K and La (Fig. 8). At a given Mg#, lavas from each volcanic field show different highly-incompatible elements contents, resulting in variable highly-incompatible elements/moderately-incompatible elements ratios (e.g., Ce/Y, Nb/Zr; Fig. 11). This is especially the case for the Al Birk basalts, that have high Mg# approaching those of undifferentiated melts, but are highly enriched in incompatible elements compared to Lunayyir and Khaybar samples at a given Mg# (Figs. 7 and 8). Below, we will discuss the chemical variability of these lavas in terms of differences in melting conditions (either melting pressure or melting degrees) or derivation from a heterogeneous source.

5.2 Constraints on the Mantle Source

The Nb/U ratios of the studied samples (35–60) generally fall in the same range as MORB and OIB (47 ± 10; Hofmann 1988). The lack of LILE enrichments relative to HFSE and REE and the OIB-like enriched incompatible element

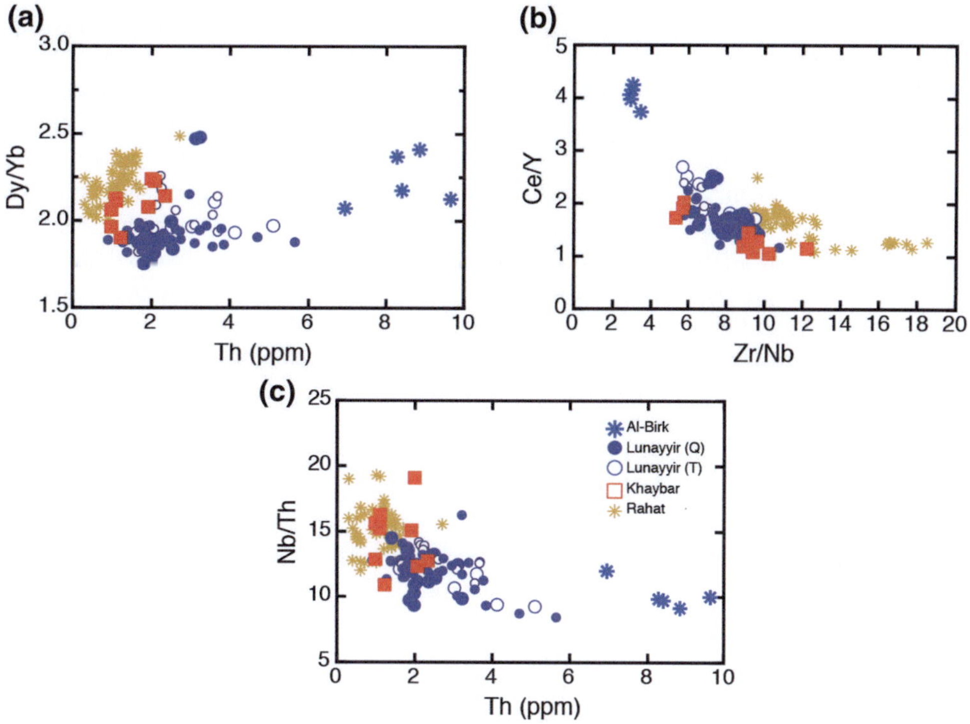

Fig. 11 Variations in Th (ppm) versus Dy/Yb (**a**); Zr/Nb versus Ce/Y (**b**) and Th(ppm) versus Nb/Th (**c**) for lavas from the Lunayyir, Khaybar and Al Birk areas. Mafic lavas (MgO >6 wt%) from Rahat (Moufti et al. 2012) are also shown for comparison

patterns (Fig. 9) suggest that their mantle source was enriched compared to a the depleted mantle (DM) (see also Duncan and Al-Amri 2013). Previous isotopic studies on the harrats indicated concordantly that Cenozoic magmatism in western Arabia was produced mainly by an enriched mantle source compared to the asthenospheric source that produced MORB-type magmatism along the Red Sea axis. In particular, nearly all the alkaline lavas from the Arabian plate have low $^{143}Nd/^{144}Nd$ and $^{87}Sr/^{86}Sr$, and high $^{206}Pb/^{204}Pb$ ratios compared to Red Sea MORBs (Altherr et al. 1990; Stein and Hofmann 1992; Bertrand et al. 2003; Shaw et al. 2003; Moufti et al. 2013; Rooney et al. 2014). These isotopic signatures point to the involvement of a source with HIMU signature (Zindler and Hart 1986; Hofmann 1997). The origin of this HIMU signature in the source of Arabian lavas has been discussed widely and three main scenarios have been proposed: (i) deep-stored material transported and emplaced below the Arabian lithosphere by a fossil plume (e.g., Stein and Hoffman 1992); (ii) enriched material related to the northward channelization of the Afar plume along a N-S lithospheric channel (Camp and Robool 1992; Krienitz et al. 2009; Chang and Van der Lee 2011; Duncan and Al-Amri 2013; Duncan et al. 2016); and (iii) chromatographic metasomatism of the Arabian lithosphere during a Neoproterozoic subduction related to the Pan-African orogeny (Bertrand et al. 2003; Shaw et al. 2003, 2007; Moufti et al. 2013; Rooney et al. 2014). A discussion on the origin of the HIMU signature in the alkaline basalts from the western Arabian Peninsula is beyond the scope of the present study. However, based on trace element compositions we can discuss the mineralogy of the source of the Lunayyir and Khaybar lavas, and their relationships with basalts erupted at other localities.

In the previous section we demonstrated that most of the samples considered in this study suffered some extent of fractionation and do not represent primary melts. Abu El-Rus and Rooney (2017) applied a successful method for the calculation of primary melts of alkaline and subalkaline basalts from middle Egypt, adding a liquidus phase crystals back into the measured compositions to attain the composition of the primary melt. These authors showed that the compositions of fractionation-corrected melts do not show any difference from the measured ratios between incompatible trace elements, that are nearly constant during the fractionation of these liquids within magma chambers. Hence, the fractionation of olivine, clinopyroxene and plagioclase do not lead to substantial modifications of the incompatible trace elements patterns of the calculated melts, leaving constant incompatible trace elements ratios. Based on this, we can infer the effect of source mineralogy on the chemistry of lavas from incompatible elements ratio versus contents of highly incompatible elements (Hofmann et al. 1984). This would allow us to infer the effect of residual minerals that may control bulk-solid/melt partition coefficients (Yang et al. 2003). The compatibility of Yb into garnet (Green et al. 2000) allows the use of Dy/Yb ratio to infer the possible occurrence of residual garnet in the mantle source, which in turn can be used as proxy for the pressure of melting. Different Dy/Yb ratios of the lavas in this study suggest that garnet had variable effects on mantle melting in the three localities (Fig. 11a). With the exception of two samples with unusually high Dy/Yb ratios ($\sim$2.5), most Lunayyir lavas display a positive correlation between Dy/Yb and Th. Although the Th horizontal scatter for each lava flow can be partly related to fractional crystallization, the relatively low Dy/Yb ratios of the Quaternary basalts from Lunayyir compared to the Tertiary lavas suggest a higher average melting pressure (or higher proportion of partial melt generated in the garnet stability field). Likewise, the Tertiary samples show overall higher Ce/Y ratios than Quaternary Lunayyir lavas, which are instead indicative of higher degrees of melting. These variations are consistent with decreasing average pressure of melting associated with progressive thinning of the lithospheric lid and increasing degrees of melting over time. The temporal evolution of the source is in agreement with the geochemical variability of the basalts from the different Quaternary units of the Lunayyir as previously documented by Duncan and Al-Amri (2013). Samples from Al Birk show higher Dy/Yb than Lunayyir and Khaybar lavas, chemical features also ascribable to higher average pressure of melting. However, the high Th and Dy/Yb ratios together with a high Ce/Y cannot be easily explained by different melting pressures alone, but require melts produced at different melting conditions and/or from different mantle sources. In particular, high Th and high Dy/Yb ratios may result from low degree of melting mainly in the garnet stability field, whereas melts with low Th contents and little fractionation in Dy/Yb ratios may result from higher degrees of partial melting mainly in the spinel stability field. Alternatively, high Th contents and Dy/Yb ratios can also be related to a high contribution of an incompatible element enriched fusible component in the source of these melts, such as enriched peridotites or pyroxenite veins, which start to melt at higher pressure. Low Th contents and low Dy/Yb ratios may reflect a lower contribution of this enriched component, that may have been exhausted at lower pressures. Radiogenic isotope data will be necessary to identify clearly the presence of different components in the source of these lavas. However, hereafter we will show that based on elemental data alone, we favor the hypothesis of an heterogeneous source.

The primitive mantle (PM) normalized trace element patterns of the Lunayyir, Khaybar and Al Birk lavas display positive anomalies in Ba, Nb, Zr and Ti and negative anomalies of Rb, Th and U (Fig. 9). Since there is no residual phase that can preferentially retain Rb, Th and U in normal mantle peridotite, these anomalous chemical features may reflect high Ba/Rb and Nb/Th-U ratios of the mantle source. These chemical characteristics resemble those of the metasomatized mantle wedge above a subduction front proposed by Stein et al. (1997). These authors suggested that subduction during the Pan-African orogeny might have created a mantle column whose lower portion contains abundant amphibole. Successive dehydration and/or flux melting of this mantle would have preferentially removed mobile trace elements that are not compatible with residual amphibole (Rb, Th and U) creating excess in the immobile elements preferentially retained in the amphibole phases, such as Nb and Ba, (LaTourette et al. 1995). In addition, the occurrence of amphibole in the mantle source is in agreement with the negative K anomaly of most samples in this study (Fig. 9), that requires a K-bearing phase (amphibole or phlogopite) in their mantle source. Magmas like those considered here, having relatively high NaO/K_2O ratio (2–5), cannot be produced by a phlogopite, but by a source containing pargasitic amphibole (e.g., Rosenthal et al. 2009; Moufti et al. 2013; Abu El-Rus and Rooney 2017). Given the stability of pargasitic amphibole in peridotite at a maximum temperature as low as 1170 °C (i.e., Green 1973; Niida and Green 1999), this amphibole-bearing source may be located within the subcontinental lithosphere (Späth et al. 1996; Rooney et al. 2014). This metasomatic mantle column could have accreted below the Arabian mantle lithosphere in the Phanerozoic being isolated from the convecting upper mantle (Stein et al. 1997). Decompression melting due to lithospheric thinning would easily melt out the residual amphibole form these fertile metasomes and generate lavas with negative K-anomaly but high Ba/Rb, Nb/Th-U ratios and HIMU-type isotopic signature (see also Stein et al. 1997; Rooney et al. 2014; Weiss et al. 2016; Abu El-Rus and Rooney 2017). The idea of a metasomatized lower mantle is supported by the occurrence of amphibole megacrysts and veins within mantle xenoliths from Egypt, Jordan and western Arabia (Ionov and Hofmann 1995; Shaw et al. 2007; Lucassen et al. 2008; Sgualdo et al. 2015). These amphiboles are characterized by enrichment in Nb and Ba relative to Th, U and La, which may account for the positive Ba, Nb, Ti and Zr anomalies observed in the alkaline lavas.

Together with the previous observations on the variability in Dy/Yb ratios, we conclude that the studied basalts formed by mixing melts produced by a garnet-bearing peridotite and an amphibole-bearing spinel peridotite. Mixing melts produced at different pressures and melting degrees have been proposed by Baker et al. (1996) on the basis of the variable incompatible element ratios. These authors argued that the variable Zr/Nb and Ce/Y ratios of Quaternary basalts in Yemen can be attributed to mixing different melt batches derived from an amphibole-bearing spinel peridotite and a garnet-bearing anhydrous peridotite. Similarly, Lunayyir, Khaybar and Al Birk (in this study) and Harrat Rahat (Moufti et al. 2013) show highly variable Zr/Nb ratios, which decrease with increasing Ce/Y (Fig. 11b). This correlation is also consistent with mixing melts produced at different melting conditions, which is a common process proposed for Cenozoic alkaline volcanism in the Arabian Peninsula. In summary, we can infer that the distinctive trace element patterns of lavas from Lunayyir, Khaybar and Al-Birk may derive from a mixed source made up of a garnet-bearing peridotite and an amphibole-bearing peridotite melted in the spinel-stability field.

5.3 Quantitative Melting Models and Implications for Volcanism in Western Arabia

Our qualitative analyses suggested that most of the chemical variability shown by Lunayyir, Khaybar and Al Birk lavas can be ascribed to polybaric melting of a garnet-bearing peridotite and an amphibole-bearing mantle. Below we will demonstrate with semi-quantitative modeling that lavas from each locality can be modelled as variable mixtures between partial melts generated in the garnet versus spinel stability field. In order to reproduce the compositions of the lavas from this study, we applied the non-modal fractional melting model from Hellebrand et al. (2002), based on critical melting equations in Sobolev and Shimizu (1993). This model takes into account a *critical melt porosity* (minimum melting degree at which the instantaneous melt is extracted from the mantle source), that for simplicity is fixed here at 1%. In addition, the model considers the progressive change in mineralogy of the source material, and the bulk partition coefficient *(bulk D)* is adjusted at each step of the process (see Fig. 11 and Table 3 for details). Based on the generally enriched isotope compositions of basalts in the region (e.g., Altherr et al. 1990; Bertrand et al. 2003), and the probable presence of amphibole in the mantle, it is unlikely that the source of the western Arabia volcanism was a mantle with a DM-like composition. In addition, DM-like mantle source (La/Yb = 0.52; Dy/Yb = 1.38, Yb = 0.35; Salters and Stracke 2004) cannot produce melts with La/Yb ratios >5 (Baker et al. 1996), while the La/Yb ratios of our samples are general higher. We thereby used as initial source composition those defined by Baker et al. (1996) and Shaw et al. (2003) as sources of the Arabian alkaline volcanism. This composition corresponds to an upper mantle peridotite enriched in LREE (La/Yb ~1.6; Dy/Yb ~1.5; Yb = 0.37) with an amphibole-bearing upper mantle peridotite produced by metasomatism during a Proterozoic subduction event (Stein et al. 1997).

Fractional melting curves for enriched garnet and spinel lherzolites are shown in Fig. 12 (see Table 3 for starting compositions, mineral modes and partition coefficients). The lavas from this study (Lunayyir, Khaybar and Al Birk) are compared with those from Harrats Rahat and Hutaymah. In addition, the field of Harrat Ash Shaam in Syria (Shaw et al. 2003) and Sana'a in Yemen (Baker et al. 1997) are also plotted for comparison. Because Yb is highly compatible in garnet (with Kd of 4–9; e.g., Salters and Longhi 1999), partial melts generated in the presence of garnet should have elevated La/Yb and Dy/Yb ratios. In contrast, partial melt generated in the spinel stability field should only have a slightly elevated La/Yb ratio, but Dy/Yb ratios roughly similar to those of the peridotite source (Fig. 12). The lavas from different harrats define distinct fields, most of which follow linear trends. As in the lavas from Yemen and Syria (see Baker et al. 1996; Shaw et al. 2003), lavas from Harrats Lunayyir, Rahat and Khaybar show positive La/Yb versus Dy/Yb correlations. To the first order, these linear trends can be reproduced by mixing different melt batches produced in the garnet and in the spinel stability fields (cf., Thirlwall et al. 1994). Using more depleted or more enriched peridotite compositions (see figure caption for details) would shift the melting curves toward more fractionated or less fractionated patterns, respectively. This would eventually change the quantification of the garnet versus spinel component required to reproduce the trends of the natural samples, but it will not affect the variability shown by lavas from each harrat.

Starting from an enriched mantle source, our model suggests that nearly all the lavas in the region require mixing between small melt fractions (from <1 to 2%) produced in the garnet stability field with variable extents of partial melts (1–10%) produced in the spinel stability field (Fig. 11). In particular, the Khaybar and Rahat lavas follow similar correlation lines, consistent with ~5% of partial melt generated in the garnet stability field mixed with up to 10% partial melt generated in the spinel stability field. Interestingly, most of these samples require mixing with melts produced by a garnet-bearing source equivalent to 50–70% of the total melt mass. Lunayyir lavas show distinctly lower degrees of melting and a lower contribution of melts produced in the garnet (Grt) stability field compared to Khaybar lavas. There are also clear differences between the Quaternary and the Tertiary lava flows in the Lunayyir area, with the latter shifted toward higher La/Yb and Dy/Yb ratios. This difference requires a higher proportion of melts produced in the Grt-stability field in the Tertiary lavas (>50% produced by a Grt-peridotite) compared to the Quaternary lavas (10–50%). Such temporal evolution of Harrat Lunayyir has been suggested by Duncan and Al-Amri (2013) and attributed to progressive thinning of the lithospheric lid under Lunayyir associated with regional extension. The samples from Al Birk are characterized by very high La/Yb ratios, which cannot be produced by melting a spinel peridotite with La/Yb ratios of an enriched mantle source (La/Yb ~1.2). However, these high La/Yb ratios are coupled with high Dy/Yb (2–2.4), which require mixing with melts derived from a garnet-bearing source. In particular, mixing between very low degrees of melting (~1%) in the garnet stability field and melts produced at a few percent degrees of melting of a spinel peridotite can reproduce the compositions of the basanites from Al Birk very well. Similar melting conditions have been suggested to explain the highly fractionated patterns of some Quaternary Sana'a lavas in Yemen (Baker et al. 1996). Based on trace elements alone we cannot exclude that Al Birk lavas were produced in a more enriched mantle domain. Figure 12 also shows lavas from Harrat Hutaymah (Duncan et al. 2016). Compared to Khaybar, Rahat and Lunayyir, Hutaymah lavas are characterized by extremely high and nearly constant Dy/Yb values but variable La/Yb ratios, which is consistent with partial melts generated mainly in the presence of garnet from an enriched peridotite source. This is in agreement with the location of Harrat Hutaymah ~600 km east of the Red Sea axis, above a thick continental lithosphere.

In conclusion, the variable melting conditions between the different harrats indicate a change in the melting depth associated with the extension of the Arabian plate. Our data are consistent with the model of a melting process governed by lithospheric thinning and rising of the asthenosphere with time and toward the Red Sea axis (Altherr et al. 1990; Bertrand et al. 2003; Shaw et al. 2003; Moufti et al. 2012). In particular, the melting region below the western Arabia shield becomes shallower from east to west, that is, from nearly unmodified cratonic lithosphere underneath Harrat Hutaymah, to the thinned lithosphere below Lunayyir. In addition, our melting model also agrees with previous inferences that the degrees of melting under Harrat Khaybar and Harrat Rahat (aligned along the MMN line) are amongst the highest in the region (Camp and Roobol 1992; Moufti et al. 2013). This observation has been used to support the idea that a radial flow of hot mantle material from the Afar mantle plume (Camp and Roobol 1992; Krienitz et al. 2009) or the presence of a separate mantle plume (Chang and Van der Lee 2011) causes extensive lithospheric melting below the MMN line (see also Duncan and Al-Amri 2013; Duncan et al. 2016). However, isotopic studies on basalts (Altherr et al. 1990; Bertrand et al. 2003; Moufti et al. 2012) and associated mantle xenoliths (Konrad et al. 2016) do not show evidence for involvement of Afar mantle plume material in the harrat volcanism. Rooney et al. (2014) and Abu El-Rus and Rooney (2017) suggested that highly fusible metasomes (amphibole-bearing peridotitic/pyroxenitic

Table 3 Melt compositions and paramentes of the critica. porosity melting model

Melting conditions		Grt-Peri	Ol	Opx	Cpx	Grt		
		Source Mode (X̄)	0.57	0.21	0.13	0.09		
		Melt Mode	0.15	−0.4	1.05	0.2		
		Spl-Peri	Ol	Opx	Cpx	Amp		
		Source Mode (X̄)	0.54	0.28	0.13	0.05		
		Melt Mode	0.05	0.25	0.5	0.2		
Garnet stability field								
		La	Sm	Dy	Yb			
	Source composition	0.6	0.239	0.55	0.374			
	Bulk D	0.003138	0.045144	0.172289	0.5557			
	P (D*Melt Mode)	0.014365	0.19661	0.452905	1.3335			
F	F%	La	Sm	Dy	Yb	La/Sm	La/Yb	Dy/Yb
0.001	0.1	91.33692784	5.138246396	3.179993756	0.673458095	17.77589489	135.6237731	4.72188809
0.01	1	55.30625891	5.00816652	3.412059386	0.744434332	11.04321485	74.29299872	4.583425613
0.02	2	38.32414759	4.429778085	3.205533368	0.712837435	8.651482502	53.76281563	4.496864517
0.03	3	28.78871092	4.046323464	3.105160946	0.7037329	7.114782389	40.90857615	4.412414067
0.04	4	22.68063079	3.728014091	3.031444866	0.700286192	6.083837196	32.38765956	4.328865683
0.05	5	18.52879815	3.447865569	2.968487853	0.699132072	5.373990887	26.50257208	4.245961486
0.06	6	15.57954711	3.19606039	2.910992726	0.699150307	4.87460974	22.2835447	4.163615031
0.07	7	13.40462128	2.96771418	2.856690989	0.69986221	4.516816805	19.15322913	4.081790596
0.08	8	11.74761098	2.759817387	2.80444999	0.701029797	4.256662428	16.75764859	4.000471883
0.09	9	10.44909409	2.570215975	2.753641977	0.702522172	4.065453719	14.87368586	3.919651346
0.1	10	9.406636724	2.397203807	2.70389173	0.704262109	3.924003748	13.356727	3.839325863
0.15	15	6.271103456	1.61216134	1.923242248	0.519562657	3.889873365	12.06996571	3.701656039
0.2	20	4.703327628	1.210321764	1.511418802	0.429161981	3.886014256	10.95932966	3.521791
Garnet stability field								
		La	Sm	Dy	Yb			
	Source composition	0.6	0.239	0.55	0.374			
	Bulk D	0.007538	0.045064	0.0998	0.1026			
	P (D*Melt Mode)	0.0258	0.154555	0.3005	0.29			
F	F%	La	Sm	Dy	Yb	La/Sm	La/Yb	Dy/Yb
0.001	0.1	73.00584337	5.13957355	6.775004719	4.122337933	14.20464999	17.70981529	1.643486009
0.01	1	47.63853627	4.994460229	6.95496133	4.266352813	9.538275227	11.16610331	1.630188978

(continued)

Table 3 (continued)

0.02	2	34.12850014	4.410789225	6.354208968	3.9177666	7.737504197	8.711213205	1.621895742
0.03	3	26.2198031	4.023901269	5.9891406	3.711098304	6.516015514	7.065240787	1.613845851
0.04	4	20.95923331	3.703665673	5.690176827	3.54322393	5.659051102	5.915300225	1.605932038
0.05	5	17.27556092	3.422812391	5.423237868	3.393465698	5.047183118	5.090831159	1.598141355
0.06	6	14.60109201	3.171271484	5.176826797	3.254886393	4.604175986	4.485899122	1.590478491
0.07	7	12.59881956	2.943935059	4.945898572	3.124474249	4.279584743	4.032300655	1.582953859
0.08	8	11.0582516	2.737596981	4.727827268	3.000689161	4.039400861	3.685237295	1.57558048
0.09	9	9.843604469	2.549932095	4.521061823	2.882644731	3.860339845	3.414782391	1.568372881
0.1	10	8.864971029	2.379087267	4.324588314	2.769781081	3.726206748	3.200603502	1.561346614
0.15	15	5.910035094	1.601330222	2.971764041	1.912979103	3.690703524	3.089440488	1.553474387
0.2	20	4.432526645	1.202803552	2.257446022	1.459032404	3.685162582	3.037990542	1.547221306

Garnet-peridotite melting modes and amphibole-spinel peridotite melting modes are from Longhi (2002) and Baker et al. (1997). Partition coefficents from Salters and Longhi (1999) and Adam and Green (2006)

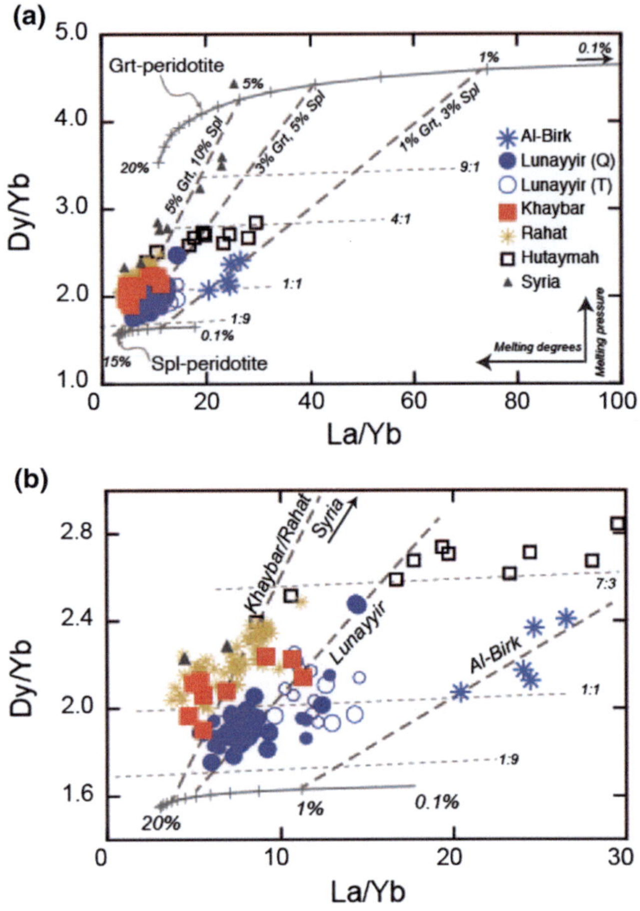

Fig. 12 **a–b** Calculated melting curves using the critical porosity melting model of Hellebrand et al. (2002) for garnet-peridotite and spinel-peridotite sources. Starting composition is an enriched peridotite with La/Yb ~1.6; Dy/Yb ~1.5 and Yb = 0.37 ppm (see also Baker et al. 1996; Shaw et al. 2003). Melting modes for garnet (Ol 0.15; Opx −0.4; Cpx 1.05; Grt 0.2) and spinel (Ol 0.05; Opx 0.25; Cpx 0.27; Spl 0.13; Amp 0.27) peridotites are from Longhi (2002) and Baker et al. 1997. Partition coefficients from Salters and Longhi (1999) and Adam and Green (2006). Dashed horizontal lines depict mixing values between garnet- and spinel-peridotite derived melts as indicated by italic numbers. Composition of lavas from Rahat (Moufti et al. 2012), Hutaymah (Duncan et al. 2016), Yemen (Baker et al. 1996) and Syria (Shaw et al. 2003) are also shown

domains) are widespread within the Arabian-Nubian lithospheric mantle, where they remained physically isolated from the depleted upper mantle (see also Stein and Goldstein 1996; Stein et al. 1997). During lithospheric thinning as a consequence of Red Sea rifting, these highly fusible components may undergo extensive melting at depth, without the need for elevated mantle potential temperatures. Thus, the higher melting degrees documented by the lavas from Harrats Khaybar and Rahat compared to Lunayyir could be attributed to the localization of fertile metasomes mainly along the MMN line, probably an inherited suture originated during the Pan-African orogeny.

6 Conclusions

Lavas from the Lunayyir, Khaybar and Al Birk lava fields are used to define the geochemical variability of the volcanic activity in three poorly known areas of Cenozoic volcanism exposed along the western margin of the Arabian Peninsula. Selected samples range from basalts to basanites retaining a transitional to alkaline affinity. The lavas reveal some degree of fractionation within magma chambers located at or close to the crust-mantle boundary, with minimal interaction with the continental crust.

Like other primary lavas in the region, the incompatible trace element signature is similar to that of Oceanic Island basalts (OIB), and generally produced by an enriched mantle source. Variation in incompatible trace element ratios (e.g., Dy/Yb, Nb/Zr and Ce/Y) indicate that the lavas formed by mixing melts batches produced at different melting conditions and/or different mantle sources. In addition, high Ba/Rb, Nb/Th-U ratios and a negative K anomaly suggest the involvement of an amphibole-bearing, metasomatized mantle. This idea is consistent with the HIMU isotopic signature typically shown by the Cenozoic volcanics in the Arabian-Nubian plate (e.g., Altherr et al. 1990; Stein and Hofmann 1992; Bertrand et al. 2003; Shaw et al. 2003; Moufti et al. 2013), which was also attributed to the occurrence of a suprasubduction metasomatic mantle source (Stein et al. 1997; Rooney et al. 2014; Weiss et al. 2016).

In comparison with the lava fields in central Arabia, Yemen and Syria, our data from Lunayyir, Khaybar and Al Birk are consistent with progressive thinning of the lithosphere toward the Red Sea axis. In particular, semi-quantitative geochemical modeling suggests that the volcanism in the Lunayyir area (i.e., close to the Red Sea axis) is mainly produced by a low degree of mantle melting ($\sim 5\%$) mostly occurring in the spinel stability field, while magmas erupted in more internal regions (Harrats Khaybar, Rahat and Hutaymah, and Syria) show more contribution of melts produced in the presence of garnet. This observation is consistent with a melting process governed by progressive lithospheric thinning and rising of the asthenosphere. Evidence that degrees of melting are higher ($\sim 10\%$) at Harrats Khaybar and Rahat (aligned along the MMN line) compared to lavas from Lunayyir can be explained by the heterogeneous distribution of fertile metasomes in the deepest portion of the Arabian continental mantle. We conclude that the Cenozoic alkaline volcanism in western Arabia was produced mainly by decompression melting of a heterogeneous mantle caused by progressive thinning of the lithosphere as a consequence of Red Sea rifting.

Acknowledgements This work is the result of a joint effort of the Saudi Geological Survey (SGS), and the Istituto di Scienze Marine, CNR of Bologna (ISMAR-CNR). We thank Dr. Z. A. Nawab, SGS President and Dr. A. M. AlAttas, SGS Assistant President for Technical Support. We particularly thank Dr. Najeeb Rasul and SGS team (A. O. Saeedi, A. Zahrani, Z. A. Otaibi, H. H. Subahi, M. M. Khorsheed and A. M. Jarees), Captain P. Dimala and helicopter assistants F. Abdulhadi and A. Al-Harbi for their collaboration during the field work. E. Billotta, Pavia University is thanked for petrographic analyses. The first version of this work benefit of useful comments by three anonymous reviewers. The work was supported by the SGS, the Italian Consiglio Nazionale Ricerche and the US National Science Foundation. We thank Dr. G. Bailey for providing basalt samples from the Al Birk region (Saudi Arabia). The research was sponsored by the PRIN2012 Programme (Project 20125JKANY_002).

References

Abu El Rus MMA, Rooney AO (2017) Insights into the lithosphere to asthenosphere melting transition in northeast Africa: evidence from the Tertiary volcanism in middle Egypt. Chem Geol 455:282–303

Adam J, Green T (2006) Trace element partitioning between mica- and amphibole-bearing garnet lherzolite and hydrous basanitic melt: 1. Experimental results and the investigation of controls on partitioning behaviour. Contrib Mineral Petrol 152:1–17

Al-Amri AM, Fnais MS, Abdel-Rahman K, Mogren S, Al-Dabbagh M (2012) Geochronological dating and stratigraphic sequences of Harrat Lunayyir, NW Saudi. Int J Phys Sci 7:2791–2805

Altherr R, Henjes-Kunst F, Baumann A (1990) Asthenosphere versus lithosphere as possible sources for basaltic magmas erupted during formation of the Red Sea: constraints from Sr, Pb and Nd isotopes. Earth Planet Sci Lett 96:269–286

Antonini P, Petrini R, Contin G (1998) A segment of sea-floor spreading in the central Red Sea: basalts from the Nereus Deep (23° 00′–23°20′N). J Afr Earth Sci 27:107–114

Ashwal LD, Burke K (1989) African lithospheric structure, volcanism, and topography. Earth Planet Sci Lett 96:8–14

Baker JA, Menzies MA, Thirlwall MF, MacPherson CG (1997) Petrogenesis of Quaternary intraplate volcanism, Sana'a, Yemen; implications for plume–lithosphere interaction and polybaric melt hybridization. J Petrol 38:1359–1390

Baker JA, Thirwall MF, Menzies M (1996) Sr–Nd–Pb isotopic and trace element evidence for crustal contamination of plume-derived flood basalts: oligocene flood volcanism in western Yemen. Geochim Cosmochim Acta 60:2559–2581

Bertrand H, Chazot G, Blichert-Toft J, Thoral S (2003) Implications of widespread high-m volcanism on the Arabian Plate for Afar mantle plume and lithosphere composition. Chem Geology 198:47–61

Bohannon RG, Naeser CW, Schmidt DL, Zimmerman RA (1989) The timing of uplift, volcanism, and rifting peripheral to the Red Sea—a case for passive rifting? J Geophys Res 94:1683–1701

Bosworth W, Stockli DF (2016) Early magmatism in the greater Red Sea rift: timing and significance. Can J Earth Sci 53(11):1158–1176. https://doi.org/10.1139/cjes-2016-0019

Camp VE, Roobol MJ (1989) The Arabian continental alkali basalt province: part I. Evolution of Harrat Rahat, Kingdom of Saudi Arabia. Geol Soc Am Bull 101:71–95

Camp VE, Roobol MJ (1991) Geologic map of the Cenozoic lava field of Harrats Khaybar, Ithnayn, and Kura, Kingdom of Saudi Arabia. Saudi Arabian Directorate General of Mineral Resources Geosciences Map GM-131, scale 1:250,000 with text, 40 p

Camp VE, Roobol MJ, Hooper HP (1991) The Arabian continental alkali basalt province: part II. Evolution of Harrats Khaybar,

Ithnayn, and Kura, Kingdom of Saudi Arabia. Geol Soc Am Bull 103:363–391

Camp VE, Roobol MJ (1992) Upwelling asthenosphere beneath western Arabia and its regional implications. J Geophys Res 90:15255–15271

Chang S-J, Van der Lee S (2011) Mantle plumes and associated flow beneath Arabia and East Africa. Earth Planet Sci Lett 302:448–454

Coleman RG, McGuire AV (1988) Magma systems related to the Red Sea opening. Tectonophysics 150:77–100

Coleman RG, Blank HR, Hadley DG, Fleck RG (1975) A Miocene ophiolite on the Red Sea coastal plain. Am Geophys Union Trans 56:1080

Coleman RG, Gregory RT, Brown GF (1983) Cenozoic volcanic rocks of Saudi Arabia. Saudi Arabian Deputy Minister of Mineral Resources, Open File Report USGS-OF93, 82 p

Duncan RA, Al-Amri AM (2013) Timing and composition of volcanic activity at Harrat Lunayyir, western Saudi Arabia. J Volcanol Geotherm Res 260:103–116

Duncan RA, Kent AJR, Thornber CR, Schlieder TD, Al-Amri AM (2016) Timing and composition of continental volcanism at Harrat Hutaymah, western Saudi Arabia. J Volcanol Geotherm Res 313: 1–14

Ewart A, Marsh JS, Milner SC, Duncan RA, Kamber BS, Armstrong RA (2004) Petrology and geochemistry of early Cretaceous bimodal continental flood volcanism of the NW Etendeka, Namibia. Part 1: introduction, mafic lavas and re-evaluation of mantle source components. J Petrol 45:59–105

Gale A, Dalton CA, Langmuir CH, Su Y, Schilling JG (2013) The mean composition of ocean ridge basalts. Geochem Geophys Geosyst 14:489–518

Green DH (1973) Experimental studies on a model upper mantle composition at high pressure under water-undersaturated and water-saturated conditions. Earth Planet Sci Lett 19:37–53

Green T, Blundy J, Adam J, Yaxley G (2000) SIMS determinations of trace element partition coefficients between garnet, clinopyroxenes and hydrous basaltic liquids at 2–7.5 GPa and 1080–1200 °C. Lithos 53:165–187

Grove TL, Kinzler RJ, Bryan WB (1992) Fractionation of mid-ocean ridge basalt. In: Morgan JP, Blackman DK, Sinton JM (eds) Mantle flow and melt generation at mid-ocean ridges, vol 71. American Geophysical Union, pp 281–311

Haase KM, Muhe R, Stoffers P (2000) Magmatism during extension of the lithosphere: geochemical constraints from lavas of the Shaban Deep, northern Red Sea. Chem Geology 166:225–239

Hellebrand E, Snow JE, Hoppe P, Hofmann AW (2002) Garnet-field melting and late-stage refertilization in 'residual' abyssal peridotites from the Central Indian Ridge. J Petrol 43:2305–2338

Hofmann AW (1988) Chemical differentiation of the Earth: the relationship between mantle, continental crust, and oceanic crust. Earth Planet Sci Lett 90:297–314

Hofmann AW (1997) Mantle geochemistry: the message from oceanic volcanism. Nature 385:219–229

Hofmann AW, Jochum KP (1996) Source characteristics derived from very incompatible trace elements in Mauna Loa and Mauna Kea basalts, Hawaii Scientific Drilling Project. J Geophys Res 101:11831–11839

Hofmann AW, Feigenson MD, Raczek I (1984) Case studies on the origin of basalt. III. Petrogenesis of the Mauna Ulu eruption Kilauea, 1969–1971. Contrib Mineral Petrol 88:24–35

Ionov DA, Hofmann AW (1995) Nb–Ta-rich mantle amphiboles and micas: implications for subduction-related metasomatic trace element fractionation. Earth Planet Sci Lett 131:341–356

Kemp J (1981) Explanatory notes to the geological map of the Wadi al'Ays quadrangle, sheet 25C, Kingdom of Saudi Arabia. Saudi Arabian Ministry of Petroleum and Mineral Resources, 39 p

Konrad K, Graham DW, Thornber CR, Duncan RA, Kent AJR, Al-Amri AM (2016) Asthenosphere-lithosphere interactions in Western Saudi Arabia: inferences from 3He/4He in xenoliths and lava flows from Harrat Hutaymah. Lithos 248:339–352

Krienitz MS, Haase KM, Mezger K, van den Bogaard P, Thiemann V, Shaikh-Mashail MA (2009) Tectonic events, continental intraplate volcanism, and mantle plume activity in northern Arabia: constraints from geochemistry and Ar–Ar dating of Syrian lavas. Geochem Geophys Geosyst 10. https://doi.org/10.1029/2008gc002254

LaTourette T, Hervig RL, Holloway JR (1995) Trace element partitioning between amphibole, phlogopite, and basanite melt. Earth Planet Sci Lett 135:13–30

Ligi M, Bonatti E, Rasul NMA (2015) Seafloor spreading initiation: geophysical and geochemical constraints from the Thetis and Nereus Deeps, central Red Sea. In: Rasul NMA, Stewart ICF (eds) The Red Sea- The formation, morphology, oceanography and environment of a young ocean basin. Springer Earth Sciences Series, pp 79–98. Springer, Heidelberg

Ligi M, Bonatti E, Bortoluzzi G, Cipriani A, Cocchi L, Caratori Tontini F, Carminati E, Ottolini L, Schettino A (2012) Birth of an ocean in the Red Sea: initial pangs. Geochem Geophys Geosystems 13:Q08009. https://doi.org/10.1029/2012GC004155

Ligi M, Bonatti E, Caratori Tontini F, Cipriani A, Cocchi L, Schettino A, Bortoluzzi G, Ferrante V, Khalil SM, Mitchell NC, Rasul N (2011) initial burst of oceanic crust accretion in the Red Sea due to edge-driven mantle convection. Geology 39:1019–1022

Longhi J (2002) Some phase equilibrium systematics of lherzolite melting: I. Geochem Geophys Geosyst 3:1020

Lucassen F, Franz G, Romer RL, Pudlo D, Dulski P (2008) Nd, Pb, and Sr isotope composition of Late Mesozoic to Quaternary intraplate magmatism in NE-Africa (Sudan, Egypt): high-μ signatures from the mantle lithosphere: the propagating tip of the western branch of the East African Rift. Earth Planet Sci Lett 284:236–248

McGuire AV, Bohannon RG (1989) Timing of mantle upwelling: evidence for a passive origin for the Red Sea rift. J Geophys Res 94:1677–1682

Mohr P, Zanettin B (1988) The Ethiopian flood basalt province. In: Macdougall JD (ed) Continental flood basalts. Kluwer Academic, Dordrecht, pp 63–110

Moufti MR, Moghazi AM, Ali KA (2012) Geochemistry and Sr–Nd–Pb isotopic composition of the Harrat Al-Madinah volcanic field, Saudi Arabia. Gondwana Res 21:670–689

Moufti MR, Moghazi AM, Ali KA (2013) $^{40}Ar/^{39}Ar$ geochronology of the Neogene-Quaternary Harrat Al-Madinah intercontinental volcanic field, Saudi Arabia: implication for duration and migration of volcanic activity. J Asian Earth Sci 62:253–268

Natali C, Beccaluva L, Bianchini G, Ellam RM, Savo A, Siena F, Stuart FM (2016) High-MgO lavas associated to CFB as indicators of plume-related thermochemical effects: the case of ultra-titaniferous picrite–basalt from the Northern Ethiopian-Yemeni Plateau. Gondwana Res 34:29–48

Niida K, Green DH (1999) Stability and chemical composition of pargasitic amphibole in MORB pyrolite under upper mantle conditions. Contrib Mineral Petrol 135:18–40

Pallister JS (1987) Magmatic history of Red Sea rifting: perspective from the central Saudi Arabian coastal plain. Geol Soc Am Bull 98:400–417

Pearce JA (1983) The role of subcontinental lithosphere in magma genesis at destructive plate margins. In: Hawkesworth CJ, Norry HJ (eds) Continental basalt and mantle xenoliths. Shiva, Nantwich, pp 230–249

Presnall DC, Dixon SA, Dixon JR, O'Donnell TH, Brenner NL, Schrock RL, Dycus DW (1978) Liquidus phase relations on the join diopside-forsterite-anorthite from 1 atm to 20 kbar: their bearing on the generation and crystallization of basaltic magma. Contrib Mineral Petrol 66:203–220

Rosenthal A, Foley SF, Pearson DG, Nowell GM, Tappe S (2009) Petrogenesis of strongly alkaline primitive volcanic rocks at the propagating tip of the western branch of the East African rift. Earth Planet Sci Lett 284:236–248

Rooney TO, Nelson WR, Dosso L, Furman T, Hanan B (2014) The role of continental lithosphere metasomes in the production of HIMU-like magmatism on the northeast African and Arabian plates. Geology 42(5):419–422. https://doi.org/10.1130/G35216.1

Salters VJM, Longhi J (1999) Trace element partitioning during the initial stages of melting beneath mid-ocean ridges. Earth Planet Sci Lett 166:15–30

Salters VJM, Stracke A (2004) Composition of the depleted mantle. Geochem Geophys Geosyst 5:Q05B07. https://doi.org/10.1029/2003gc000597

Sgualdo P, Aviado K, Beccaluva L, Bianchini G, Blichert-Toft J, Bryce JG, Graham DW, Natali C, Siena F (2015) Lithospheric mantle evolution in the Afro-Arabian domain: insights from Bir Ali mantle xenoliths (Yemen). Tectonophysics 650:3–17

Shaw JE, Baker JA, Kent AJR, Ibrahim KM, Menzies MA (2007) The geochemistry of the Arabian lithospheric mantle source for intraplate volcanism. J Petrol 47:1–18

Shaw JE, Baker JA, Menzies MA, Thirlwall MF, Ibrahim KM (2003) Petrogenesis of the largest intraplate volcanic field on the Arabian Plate (Jordan): a mixed lithosphere–asthenosphere source activated by lithospheric extension. J Petrol 44:1657–1679

Späth A, le Roex AP, Duncan RA (1996) The geochemistry of lavas from the Comores Archipelago, western Indian Ocean: petrogenesis and mantle source region characterstics. J Petrol 37:961–991

Sobolev AV, Shimizu N (1993) Ultra-depleted primary melt included in an olivine from the Mid-Atlantic Ridge. Nature 363:151–154

Stein M, Hofmann AW (1992) Fossil plume head beneath the Arabian lithosphere? Earth Planet Sci Lett 114:193–209

Stein M, Goldstein SL (1996) From plume head to continental lithosphere in the Arabian-Nubian Shield. Nature 382:773–778

Stein M, Navon O, Kessel R (1997) Chromatographic metasomatism of the Arabian-Nubian lithosphere. Earth Planet Sci Lett 152:75–91

Stern RJ, Johnson PR (2010) Continental lithosphere of the Arabian Plate: a geologic, petrologic, and geophysical synthesis. Earth Sci Rev 101:29–67

Sun SS, McDonough WF (1989) Chemical and isotopic systematics of oceanic basalts: implications for mantle composition and processes. In: Saunders AD, Norry MJ (eds) Magmatism in ocean basins, vol 42. Geological Society of London, Special Publication, pp 313–345

Thirlwall MF, Upton BGJ, Jenkins C (1994) Interaction between continental lithosphere and the Iceland plume—Sr–Nd–Pb isotope geochemistry of Tertiary basalts, NE Greenland. J Petrol 35:839–879

Tormey DR, Grove TL, Bryan WB (1987) Experimental petrology of normal MORB near the Kane Fracture Zone: 22°–25°N, Mid-Atlantic Ridge. Contrib Mineral Petrol 96:121–139

Vigliotti L, Cai Y, Rasul NMA, Hemming SR (2017) Paleomagnetism and geochronology of the Lunayyir and Khaybar lava fields, Saudi Arabia (this volume)

Volker F, McCulloch MT, Altherr R (1993) Submarine basalts from the Red Sea: new Pb, Sr, and Nd isotopic data. Geophys Res Lett 20:927–930

Weiss Y, Class C, Goldstein SL, Hanyu T (2016) Key new pieces of the HIMU puzzle from olivines and diamond inclusions. Nature 537:666–670

Yang HJ, Frey FA, Clague DA (2003) Constraints on the source components of lavas forming the Hawaiian North Arch and Honolulu volcanics. J Petrol 44:603–627

Yang HJ, Kinzlerand RJ, Grove TL (1996) Experiments and models of anhydrous, basaltic olivine-plagioclase-augite saturated melts from 0.001 to 10 kbar. Contrib Mineral Petrol 124:1–18

Zindler A, Hart S (1986) Chemical geodynamics. Ann Rev Earth Planet Sci 14:493–571

Palaeomagnetism and Geochronology of the Harrats Lunayyir and Khaybar Lava Fields, Saudi Arabia

Luigi Vigliotti, (Merry) Yue Cai, Najeeb M. A. Rasul, and Salem M. S. Al-Nomani

Abstract

To better constrain the tectonic history of the Arabian craton in the Tertiary, we carried out a combined paleomagnetic and $^{40}Ar/^{39}Ar$ geochronological study on volcanic rocks from the Khaybar and Lunayyir Harrats plus a site of sediments deposited below the Miocene rocks in the former area. Progressive thermal or alternating field demagnetization successfully isolated stable characteristic magnetizations (ChRM) that are consistent with a primary magnetization only in the Late Quaternary lava flows from Harrat Lunayyir. The Harrat Lunayyir paleomagnetic data set of 11 flow-mean directions (D = 0.31°, I = 36.9°, $\alpha 95$ = 10.5) is statistically indistinguishable from the present field and the virtual geomagnetic poles (VGP: 214.1°E, 85.1°N; A95 = 12.3°) which indicate a negligible rotation (R = −1.98 ± 10.49°) with respect to the coeval African pole position. The paleomagnetic signal of the Miocene lava flows from the Harrat Khaybar area appear to be contaminated by the effect of lightning and weathering and consequently no tectonic/plate movement significance may be attributed to the large CCW rotation shown from 2 sites with antipodal directions. The direction of the high coercivity chemical remanent magnetization (CRM) isolated after thermal cleaning from the Pre-Miocene siltstones (D = 169.6°, I = −44.8°; $\alpha 95$ = 5.4°) is consistent with the few existing paleomagnetic results from Arabia. The associated VGP (314.4°E, 80.6°N) is close to the Pliocene VGP of the Arabian Plate and CCW rotated (R = 14.86 ± 6.38°) with respect to the Oligocene African VGP. The results imply that the whole rotation of the Arabian Plate took place during the last phase (4–5 Ma) of the opening of the Red Sea, corresponding with the true sea floor spreading as already noted in the past by other authors. Whole rock $^{39}Ar/^{40}Ar$ step-heating analyses yield whole-rock plateau ages of 12.8 to 16.3 Ma for the alkaline lava flows from the Khaybar area, which is consistent with the estimated age range of the region-wide late Cenozoic alkaline volcanism in Saudi Arabia.

L. Vigliotti (✉)
Istituto di Scienze Marine, CNR Bologna, Bologna, Italy
e-mail: luigi.vigliotti@bo.ismar.cnr.it

(Merry)Y. Cai
Lamont Doherty Earth Observatory, Columbia University, Palisades, NY, USA

N. M. A. Rasul · S. M. S. Al-Nomani
Center for Marine Geology, Saudi Geological Survey, Jeddah, Saudi Arabia

1 Introduction

The Red Sea has been considered the closest modern example of initiation of sea floor spreading between continental blocks (Wegener 1929; Bayer et al. 1989). The Red Sea was also one of the first oceanic basins to be interpreted in the context of plate tectonics theory (e.g., McKenzie et al. 1970) and to be included in a tectonic reconstruction based on paleomagnetic data (Irving and Tarling 1961). The Arabian Plate originated ~30 Ma by rifting of NE Africa to form the Gulf of Aden and the Red Sea. It contains extensive Cenozoic to Recent volcanic fields (harrats) dominated by alkali olivine basalts and hawaiite representing one of the largest alkali basalt provinces in the world (area 180,000 km^2). The Cenozoic basaltic lava fields of western Saudi Arabia were emplaced directly on the stable upper Proterozoic metavolcanics and granitic plutons of the Precambrian Arabian shield (Camp et al. 1991). The Arabian harrats are not associated with a well-developed continental rift system, such as the coeval lava fields of east Africa which are typical of the eastern margin of the Red Sea, but largely missing in the western margin on the African Plate. They occur on the uplifted eastern flank (rift shoulder) of the Red Sea depression in a tectonic environment similar to that of the continental basalt fields of eastern Australia (Johnson et al. 1989). In addition, their extension into Jordan, Syria, and Turkey, well beyond the northern latitude of the Red Sea

N. M. A. Rasul and I. C. F. Stewart (eds.), *Geological Setting, Palaeoenvironment and Archaeology of the Red Sea*,
https://doi.org/10.1007/978-3-319-99408-6_19

suggests that they may be not strictly related to the rifting of Gulf of Aden/Red Sea (Camp et al. 1991).

Volcanic activity began during the Oligocene (30–31 Ma ago) with continental flood volcanism in northern Ethiopia, Eritrea and western Yemen which was coeval with the onset of rifting in the Gulf of Aden (Bosworth and Stockli 2016). Volcanism shifted from an initial tholeiitic and transitional composition to a more alkalic composition during the Miocene (<12 Ma ago) (Camp and Roobol 1992). It is generally agreed that the Oligocene alkali basalts around the Red Sea are associated with the divergence between the African and Arabian Plates (Brown 1970; Jones and Rex 1974; Coleman et al. 1977, 1979). However, it is unclear whether the rifting was passive or active and debates exist regarding the mechanism that drives the spreading and the development of the oceanic crust (e.g., Burke and Dewey 1973; Tapponnier and Francheteau 1978; Collins 2003). Stern and Johnson (this volume) propose that the Red Sea is a volcanic rifted margin (VRM) formed 20–24 Ma ago.

The exact timing of rifting and continental separation is unclear. Girdler and Styles (1974) suggested that an initial period of rifting and spreading occurred during late Eocene and Early Oligocene time (41–34 Ma ago) followed by a period of quiescence that ended as recently as 4–5 Ma, after which a new period of spreading began. Hall et al. (1976) and Hall (1979) pointed out that the onset of the initial spreading is not well-constrained by the interpretation of the magnetic anomalies, which can be indicative of two distinct periods either between 40 and 34 Ma or between 29 and 24 Ma. Cochran (1981) suggested that the rotation of Arabia away from Africa took place in late Oligocene to early Miocene with true seafloor spreading starting only about 4–5 Ma, when the Red Sea axial trough formed. McKenzie et al. (1970) calculated that about 7° of total rotation has taken place about a pole west of Greece (36°N, 17°E). Le Pichon and Francheteau (1978) suggested that this model only applies for the region of the Red Sea north of 18°N, while in the region south of 18°N the seafloor data indicate a spreading pole at 7°N, 50°E, south of the Red Sea, with about 3° of opening over the past 5 Ma about this pole.

Synthesizing available literature magnetic data, Hall (1979) suggested that the formation of the Red Sea can be best explained by 8.8° of counter-clockwise rotation of Arabia with respect to Africa associated with the early (main) period of opening, followed by 1.2° of rotation during the late (post 5 Ma) period of opening. He also suggested that the poles of rotation for the two periods are significantly displaced from each other (29°N, 27°E for early phase spreading and 45°N, 10°E for the later spreading), indicating that initially Arabia moved away from Africa more toward the north than the more recent E–W movement. The mean pole of rotation for an angle of opening of 9.9° is located at 31°N, 25°E. According to Roeser (1975), the axial trough is underlain by oceanic crust and well-developed magnetic anomalies can be reconstructed back to anomaly 3 (about 5 Ma) from 16°N to 23°N, whereas Girdler (1983, 1985) suggests two or three phases for the opening of the Red Sea with an initial stage around 25 Ma ago and a third phase in the Plio-Pleistocene. The absence of magnetic anomalies older than Anomaly 3 hampers a detailed reconstruction of the plate motions before the Pliocene; however, Schettino et al. (this volume) suggests that the Euler poles of relative motions between the plates remained stable during most of the rifting stage.

Reilinger et al. (2015), based on GPS data, suggested that rifting in the Red Sea started at 22 ± 3 Ma, with an initial rate of extension across the rift at half of the present-day rate. At 11 ± 2 Ma, when rifting in the northern Red Sea shifted from the Gulf of Suez to the Red Sea/Aqaba fault system, the rate doubled to the current rate. On the base of geological data Bosworth and Stockli (2016) reach similar conclusions, dating the initial phase of extension at $\sim$23 Ma.

The Arabian Plate was one of the first whose motion was investigated using paleomagnetic data (Irving and Tarling 1961). However, very few additional results have appeared in the literature since then. Kellogg and Reynolds (1983) published a few questionable data from Oligocene alkali basalts from As Sarat in the southern part of Arabia. Based on these data and new results from the younger Pliocene lava flows collected from Harrat Rahat, Hussain and Bakor (1989) defined a Tertiary apparent polar wander path (APWP) for Arabia characterized by the same clockwise loop detected in the African APWP during the Miocene to Holocene, with the best fit obtained by closing the Red Sea through a clockwise rotation of Arabia by an angle of 9° around the same pole indicated by Hall (1979) at 29.3°N, 27.1°E. An improvement in this fit included closing the Gulf of Aden through the shifting of the Arabian Plate by 120 km in a SSE direction, parallel to the Red Sea axial trough. The authors recognized that available data were few and far from satisfactorily accounting for the tectonics of the Red Sea. However, after more than a quarter of a century there are still no additional paleomagnetic data from this area that could better constrain the history of the Arabian Plate. The existing data are quite puzzling which make it difficult to constrain the timing of Red Sea opening and rotation. Thus, one of our goals was to estimate the extent of rotation of the Arabian plate since the beginning of the Red Sea opening and to constrain the timing of the rotation in recent times (late Quaternary).

In this work, we carried out a geochronological and palaeomagnetic study on 86 oriented hand samples collected from 20 sites that represent Cenozoic volcanic rocks from the lava fields in Harrats Lunayyir, Ishara, Al Kura and Khaybar located in the north-western region of Saudi Arabia in order to improve the resolution of the rotational history of

Arabia relative to Nubia during the past 30 Ma (Fig. 1). Previous attempts to determine the ages of Ishara, Al Kura and Khaybar rocks have been based on the K–Ar total fusion method with ages ranging from zero to 27 Ma (Camp and Roobol 1992). Duncan and Al-Amri (2013) conducted $^{40}Ar/^{39}Ar$ step-heating experiments on lavas from Lunayyir and found that most of the volcanism there occurred within the past 600 Ka. Here, we report new $^{40}Ar/^{39}Ar$ age determinations of Jabal Ishara, Jabal Anthar, Bi'r al' Ayn, Jabal Dhulay'ah and Jabal Jammazin on the same samples that have been used for paleomagnetic analysis.

2 Geological Framework and Sampling

We conducted field work in volcanic fields lying along the 25°N parallel, from the eastern Red Sea margin to ~150 km inland (Fig. 1a). Several recent lava flows perpendicular to the Red Sea margin dissect the Arabian coastal plain. Moving toward the east, one of the smaller and younger volcanic fields (Harrat Lunayyir) crops out, composed of relatively older flows in the south-western part and very young pyroclastic cones and flows that formed along related fissures. Some ~50 cinder cones form N–S and NW–SE trends and cover an area of about 3600 km^2 (Duncan and Al-Amri 2013) (Fig. 1b). The Lunayyir samples are generally fine-grained alkali olivine basalts and trachy–basalts with aphyric to only moderately phyric textures. Olivine and plagioclase phenocrysts are present within the groundmass that consists mainly of micro- to cryptocrystalline plagioclase and augite with minor olivine (Sanfilippo et al., this volume). To the east of Lunayyir, olivine basalts of Harrat Ishara occupy a highland area extending along the south-western margin of Wadi al Jiz'l and Wadi Hamd (Fig. 1c). The volcanic field is deeply eroded with exposed basal units implying a relatively old age for the basalt preserved on the plateau relics. Along the eastern edge of the study area, to the east of Wadi al' Ays, basaltic lava flows form the most westerly part of the large Harrat Khaybar and of the smaller Harrat al Kura (Fig. 1c).

According to the geological maps (Kemp 1981) the collected rocks belong to Tertiary/Quaternary volcanic units (Tb; Qb) that characterize the Wadi al' Ays region, except for one site sampled at Jabal Bi'r al' Ayn that represents continental deposits that consist of red quartz-arenite and claystone of Eocene/Oligocene age underlying the volcanic units (Fig. 1d). These sediments offer the possibility of applying a tectonic correction to the paleomagnetic data that usually is impossible with volcanic rocks.

Hand samples were collected in the field and oriented using only a magnetic compass, which does not rule out the possibility of a bias in the magnetic direction due to large local magnetic anomalies from strongly magnetized rock units. Deflection of the compass needle was observed in a few outcrops and a back-sighting technique was used to check the reliability of the orientation of the fiducial arrow marked on the rock. In a couple of outcrops, the local field was so strong that it was impossible to collect any oriented samples. Usually 4–6 hand samples were collected at each site, representing a typical paleomagnetic site, but at a few sites only 1–3 samples were included in the collection. It would have been desirable to get more samples from each lava flow, but the limited time for sampling and the roughness of the area hampered the possibility of improving on this number.

3 Methods

Between 1 and 5 standard paleomagnetic cylinders of 2.54 × 2.2 cm were drilled in the laboratory from each rock sample, for a total of 218 specimens. The natural remanent magnetization (NRM) was measured using a Molspin spinner magnetometer at the Paleomagnetic Laboratory of ISMAR-CNR at Bologna (Italy). Both alternating field (AF) (up to 80 mT) and thermal demagnetization (up to 600 °C) were carried out on pilot samples to check the reliability and the magnetic behaviour during the cleaning. Stepwise AF demagnetization was applied to the volcanics with an AGICO LDA-3A demagnetizer, while a MMTD1 furnace was used for thermal demagnetization of the sediments characterized by the presence of high coercitivity minerals. The results were analysed by using the DAIE program (Sagnotti 2013) based on Fisher statistics (Fisher 1953) and principal component analysis (Kirshvink 1980). The mean directions were used to calculate the virtual geomagnetic pole (VGP) by using the PaleoMac program developed by Cogné (2003). Finally, the rotation with respect to Africa was evaluated following the procedure suggested by Demarest (1983) using coeval reference paleopoles from Torsvik et al. (2008). The site mean directions are given in Table 1. Magnetic susceptibility (κ) was measured for all but a few samples using a Bartington MS2 m. Hysteresis properties were measured at the INGV paleomagnetic laboratory in Rome by using a Micromag alternating gradient magnetometer (AGM-2900) with a maximum field of 1T. The ratio of the remanent magnetization to the induced magnetization produced by the Earth's field (29.86 A/m) was used to calculate the Koenigsberger ratio (Q). The results of these measurements are reported in Tables 2 and 3 for Khaybar and Lunayyir respectively.

$^{40}Ar/^{39}Ar$ age determinations were carried out at the AGES-laboratory at Lamont Doherty Earth Observatory of Columbia University using a Micromass VG 5400. The argon-isotopic ratios were determined in static mode using automated data collection software developed by Alan Deino

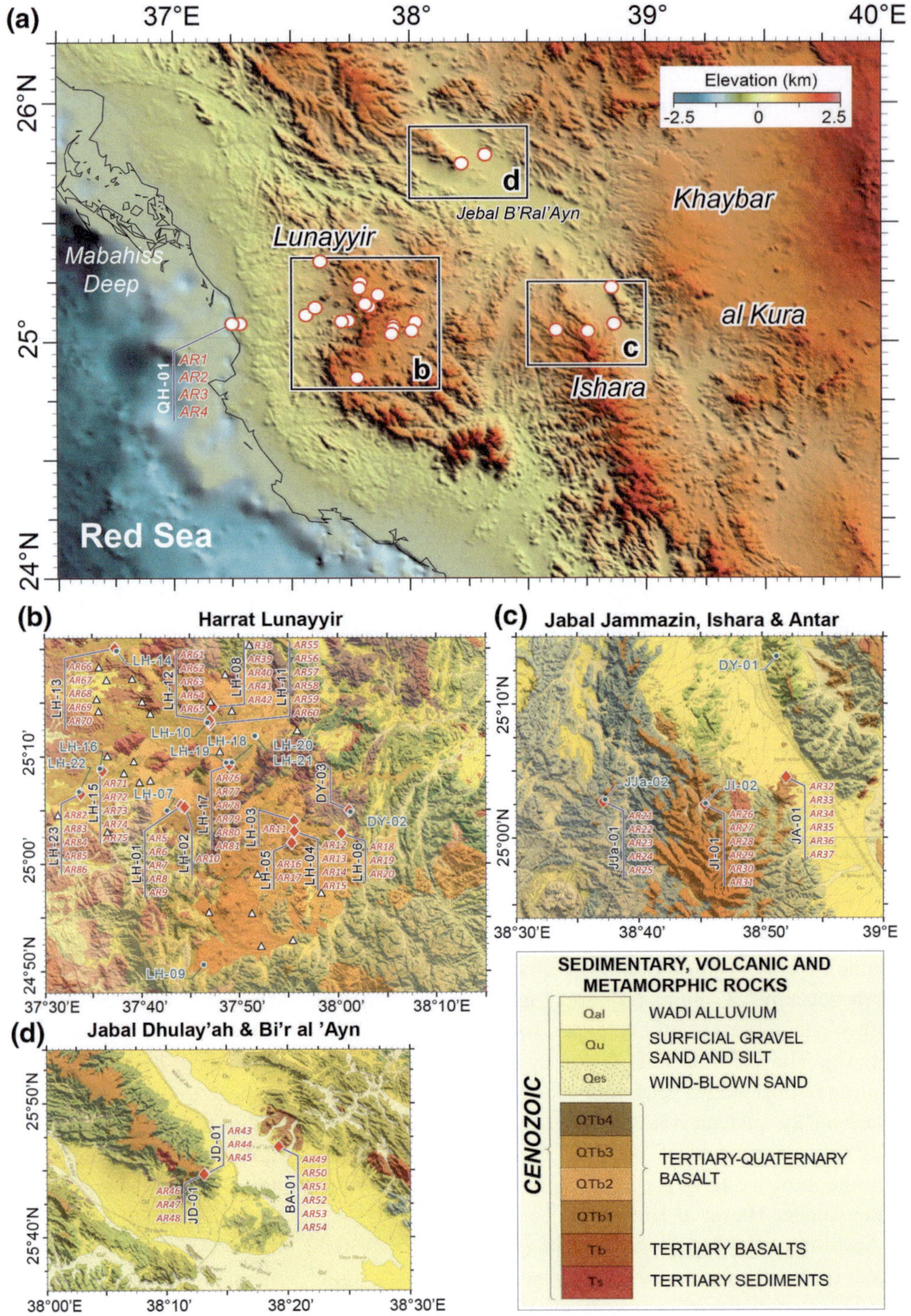

Fig. 1 Shaded relief map showing the major morphotectonic features of the northern Red Sea and adjacent volcanic lava fields on the western Arabian Plate with the locations of the sampling sites (**b**, **c**, **d**). Age assignments are based on morphologic data (relative extents of erosion, alteration, soil cover; Kemp 1981). Paleomagnetic samples are marked in red while blue colours indicate the samples used for $^{40}Ar/^{39}Ar$ age determinations. White triangles in Harrat Lunayyir refer to the samples studied by Duncan and Al-Amri (2013)

of the Berkeley Geochronology Center. Alteration of primary phases to secondary minerals is relatively limited in this very arid region. Thus, the volcanic samples were simply crushed and rinsed with deionized water. 0.5–1 mm representative rock chips without visible signs of alteration were selected under the microscope and co-irradiated with Fish Canyon sanidine at the USGS TRIGA reactor in Denver, CO, U.S.A for 8 h. The irradiated samples were placed in Ta tubes, and evacuated in a chamber with a glass window for incremental heating with a diode laser (PhotonMachines). Stepwise heating of the samples was conducted with energies of 1–18 W, in increments designed to spread the ^{39}Ar out as much as possible. Released argon was cleaned of active gases by interaction with Zr-Al getters set at 2 amps, and then allowed into the Micromass VG5400 noble gas mass spectrometer for analyses using peak-hopping on an

Table 1 Paleomagnetic mean directions, Fisher statistics (K and α95/A95 parameters) and Paleopoles for the studied sites. N/N_0 = number of samples used for the mean/number of measured samples. Qb, Tb in the age column refer to the stratigraphic age given in the geologic map (see Fig. 1)

Site	N/N_0	AGE		Decl (°)	Incl (°)	K	α95-A95 (°)
QH-01	7/12	Qb		345.3	42.5	417.0	2.96
LH-01	9/11	Tb		352.2	34.0	170.0	3.96
LH-02	10/10	Qb		334.1	39.6	52.2	6.75
LH-04	10/11	Qb+Tb		353.2	40.5	108.1	4.70
LH-05	9/9	Tb		33.1	14.9	303.8	2.96
LH-06	6/7	Qb		344.8	19.9	23.5	14.10
LH-08	9/9	Tb		354.9	34.9	268.0	3.20
LH-11	12/12	Tb		14.7	40.3	72.5	5.10
LH-12	9/9	Qb		3.6	41.7	41.4	8.10
LH-15	12/12	Tb		355.6	51.8	130.5	3.80
LH-17	10/12	Qb		25.4	30.3	33.3	8.50
ALL	N = 11/11	Quaternary		0.31	36.9	19.8	10.5
			VGP	214.1 E	85.1 N	P_{lat} = 20.6	12.3
Selected	N = 9/11 (excl LH05-LH06)			358	40.4	36.6	8.6
			VGP	259.9 E	87.3 N	P_{lat} = 23.1	10.4
BA-01	14/14	Oligocene?	Before TC	204.3	−38.2	3.5	24.8
BA-01	10/14		Selected	182.2	−30.7	43.7	7.4
BA-01	10/14		After TC	169.6	−44.8	82.3	5.40
				314.4 E	80.6 N	P_{lat} = 26.4	6.8
JJA-01	11/12	14.5±0.2		306.9	7.9	106	4.50
JD-01	9/9	13.2±0.6		343.0	64.0	163	4.00
JD-01	7/9	13.2±0.6		206.5	−18.1	627	2.40
JA-01	13/13	16.3±0.3		125.2	6.3	20.3	9.40
JI-01	12	14.6±1.7	Scattered	–	–	–	–

analogue multiplier. Argon data were corrected for backgrounds and discrimination using frequently measured blanks and air pipettes. Corrections were made for nuclear interferences using the information from Dalrymple et al. (1981) for the USGS reactor. The J-values were calculated from the average ages of Fish Canyon (FC) sanidines that were co-irradiated with the sample, for a given irradiation level assuming that the true age of FC is 28.201 ± 0.046 Ma (Kuiper et al. 2008).

4 Results

4.1 Ar/Ar Datings

Age determinations of samples selected from the top of the lava flow sequences and from the contact between volcanics and the Precambrian basement at three major peaks of Harrat Ishara, now a deeply dissected plateau, range from 16.3 to 14.5 Ma (Table 4), in agreement with previously estimated ages ranging from 17 to 12 Ma (Bosworth and Stockli 2016). In particular, samples JJa-01/Tb-1 and JJa-02/Tb-2, representing the top and base of the ~500 m-thick volcanic sequence of Jabal Jammazin (Fig. 1c), after 9 and 7 heating steps, respectively, released plateau ages of 14.5 ± 0.2 Ma and 15.6±0.8 Ma, respectively (Fig. 2). Although the two ages overlap within errors, they are consistent with their stratigraphic relationships. Sample JI-01/Tb-1 from the top of the ~300 m-thick sequence of Jabal Ishara (Fig. 1c), run in 8 heating steps, yielded a plateau age of 14.6 ± 1.7 Ma (Fig. 2c, d). The ~350 m-thick lava flow sequence of Jabal Antar (Fig. 1c) lies above a 50 m-thick sedimentary sequence resting above the Precambrian basement and includes boulder conglomerate and several upward-fining

Table 2 Natural remanent magnetization (NRM), magnetic susceptibility (κ) and Koenigsberger ratio (Q) of the volcanic rocks from the Khaybar area. NRM values in A/m, κ values in 10^{-5} SI units. Q was calculated by using a local magnetic field of 29.86 A/m

Samples	NRM	κ(10^{-5} SI)	Q
AR-21aI	0.41	503	2.73
AR-22bII	0.64	487	4.41
AR-23aI	0.54	506	3.57
AR-24cII	0.41	559	2.48
AR-24bII	0.42	536	2.63
AR-25bI	0.53	657	2.72
AR-25dI	0.63	640	3.3
Site JJa-01mean	0.51	555.43	3.12
AR-26bI	434	1288	1128.45
AR-26bII	349	996	1173.48
AR-27aII	73.2	977	250.92
AR-27b	72.7	895	272.03
AR-28	102	694	492.21
AR-29	478	613	2611.43
AR-30	25.9	790	109.8
AR-31aI	24.5	571	143.69
AR-31aII	16.4	458	119.92
AR-31b	26.6	486	183.3
SiteJI-01 mean	180.57	833.73	672.89
AR-32a	0.3	486	2.05
AR-32bII	0.26	435	2.04
AR-33aI	0.3	428	2.35
AR-33aII	0.28	364	2.59
AR-33bI	0.32	385	2.77
AR-34bI	0.24	581	1.38
AR-34bII	0.25	536	1.54
AR-35	0.23	215	3.55
AR-36aI	0.16	250	2.21
AR-36aII	0.22	289	2.49
AR-36bII	0.2	257	2.65
AR-37a	10.7	454	78.93
AR-37c	11	477	77.23
SiteJA-01 mean	1.88	396.69	13.98
AR-43a	1.77	610	9.72
AR-43b	1.75	597	9.82
AR-44aI	1.84	593	10.39
AR-44aII	1.92	605	10.63
AR-44cI	1.78	571	10.44
AR-44cII	1.85	612	10.12
AR-45a	102	672	508.32
AR-45b	70.5	615	383.91

(continued)

Table 2 (continued)

Samples	NRM	κ(10^{-5} SI)	Q
SiteJD-01A mean	22.93	609.38	119.17
AR-46aII	2.23	720	10.37
AR-46cI	1.99	725	9.19
AR-46cII	1.72	724	7.96
AR-47aI	4.71	761	20.73
AR-47bII	3.07	771	13.34
AR-47cI	3.9	690	18.93
AR-48aI	3.94	773	17.07
AR-48bII	2.81	773	12.17
AR-48bIII	3.59	740	16.25
SiteJD-01B mean	3.11	741.89	14

sequences of coarse-grained sandstone to red siltstone with gypsiferous bases (Kemp 1981). Sample JA-01/Tb-1 from the base of this volcanic sequence, after 8 heating steps, yielded a plateau age of 16.3±0.3 Ma (Fig. 2a, b), refining previously estimated ages of 20 Ma by Coleman et al. (1975) and 9 Ma by Dhellemmes and Delfour (1979).

Samples JD-01/Tb-1 and BA-01/Tb-1 from the northern sector of the study area are considered to be part of the Ishara volcanic field (Kemp 1981). Sample JD-01/Tb-1, recovered at the base of the volcanic sequence of Jabal Dhulay'ah (Fig. 1d), which is part of the Jabal Khirsat plateau, run in 9 heating steps, yielded a plateau age of 13.2 ± 0.6 Ma (Fig. 2g, h). Sample BA-01/Tb-1 (Fig. 1d) is from the top of a volcanic sequence that lies above a Cenozoic sedimentary sequence of poorly indurated argillo-arenaceous sediments (Pellaton and Dhellemmes 1978), and after 7 heating steps yielded a plateau age of 12.8 ± 0.13 Ma (Fig. 2e, f). This new age is younger than those estimated for the Ishara plateau basalts (~20 Ma, Coleman et al. 1975), which suggests that the Khirsat plateau basalts may be generated by a distinct volcanic event.

5 Paleomagnetism

5.1 Sediments

The sediments collected in the Bi'r al' Ayn (Site BA-01; Fig. 1d) region exhibit differences in the intensity of magnetization and magnetic coercivity reflecting changes in the magnetic mineralogy even in specimens drilled from the same hand-sample. Usually the NRM intensity is within 10^{-2} A/m, but significant variations up to one order of magnitude were observed even in specimens from the same

Table 3 Natural remanent magnetization (NRM), magnetic susceptibility (κ) and Koenigsberger ratio (Q) of the volcanic rocks from Harrat Lunayyir. NRM values in A/m, κ values in 10^{-5} SI units. Q was calculated by using a local magnetic field of 29.86 A/m

Samples	NRM	κ(10^{-5} SI)	Q	Samples	NRM	K (10^{-5} SI)	Q	Samples	NRM	κ(10^{-5} SI)	Q
AR-1a	1.57	488	10.77	AR-5aI	17	1284	44.34	AR-76aI	15.7	2590	20.3
AR-1b	1.58	501	10.56	AR-5bI	17.5	1368	42.84	AR-76bI	17.8	3315	17.98
AR-2aI	2.14	679	10.55	AR-6aI	7.79	2936	8.89	AR-77a	24.1	3032	26.62
AR-2aII	2.14	641	11.18	AR-6cII	7.41	2262	10.97	AR-77b	22.6	2956	25.6
AR-2bII	2.15	631	11.41	AR-7aII	2.76	1180	7.83	AR-78a	26.8	3789	23.69
AR-3aI	2.1	672	10.47	AR-7c	2.4	1095	7.34	AR-79a	13.6	3611	12.61
AR-3b	2.24	703	10.67	AR-8a	2.14	1580	4.54	AR-79b	7.73	4022	6.44
AR-4aI	1.68	541	10.4	AR-8b	1.96	1540	4.26	AR-80aI	21	6297	11.17
AR-4bII	1.71	542	10.57	AR-9aI	1.56	1240	4.21	AR-80aII	19.1	6159	10.39
AR-4aII	1.28	521	8.23	AR-9bII	2.15	1417	5.08	AR-81a	5.98	3755	5.33
AR-4c	1.34	540	8.31	AR-9cI	1.9	1349	4.72	AR-81b	6.03	3743	5.4
Site-QN mean	1.81	587.18	10.28	LH-01 mean	5.87	1568.27	13.18	LH-17 mean	16.4	3933.55	15.05
AR-55a	20.6	890	77.52	AR-71aI	4.55	1185	12.86	AR-66a	154	2500	206.3
AR-55b	19.5	655	99.7	AR-71bIII	4.26	1255	11.37	AR-66b	120	2405	167.1
AR-57a	6.63	3976	5.58	AR-72aI	7.68	602	42.72	AR-67aI	209	2075	337.32
AR-57b	6.09	3894	5.24	AR-73aI	4.09	1176	11.65	AR-67bI	233	2281	342.09
AR-57c	7.42	4167	5.96	AR-73bII	6.16	757	27.25	AR-68aII	268	1951	460.03
AR-58a	7.84	2488	10.55	AR-73aII	5.35	912	19.65	AR-68bI	185	1975	313.7
AR-58bII	7.91	2401	11.03	AR-73bI	5.32	897	19.86	AR-69aII	6.5	493	44.15
AR-58c	7.26	2296	10.59	AR-74bII	8.96	695	43.18	AR-69bI	6.16	463	44.56
AR-59a	7.09	3150	7.54	AR-74c	8.14	625	43.62	AR-69c	6.65	453	49.16
AR-59b	6.99	2701	8.67	AR-75aII	5.77	796	24.28	AR-70aII	24	518	155.16
AR-60b	12.6	4594	9.19	AR-75bI	6.99	898	26.07	AR-70bI	25.3	750	112.97
LH-11 mean	9.99	2837.45	22.87	LH-15 mean	6.12	890.73	25.68	LH-13 mean	112.51	1442.18	202.96
AR-38aI	26.5	6410	13.85	AR-11aII	2.46	1047	7.87	AR-61aII	6.31	4447	4.75
AR-38bII	23.7	7066	11.23	AR-11bI	9.34	1914	16.34	AR-61aIII	8.64	4281	6.76
AR-39aI	11.9	5499	7.25	AR-12a	1.83	405	15.13	AR-62aII	9.88	4008	8.26
AR-39bII	10.2	5317	6.42	AR-12b	2.23	523	14.28	AR-62b	7.11	4592	5.19
AR-40aII	9.32	6000	5.2	AR-13bI	2	433	15.47	AR-63aI	18.5	5321	11.64
AR-40bI	9.02	4917	6.14	AR-14aI	2.87	1117	8.6	AR-63bI	30.6	4879	21
AR40cII	10.3	4902	7.04	AR-14b	3.1	1137	9.13	AR-64a	47.2	4774	33.11

(continued)

Table 3 (continued)

Samples	NRM	κ(10^{-5} SI)	Q	Samples	NRM	K (10^{-5} SI)	Q	Samples	NRM	κ(10^{-5} SI)	Q
AR-42a	10.6	5193	6.84	AR-15aI	3.13	331	31.67	AR-64b	32.5	5113	21.29
AR-42c	13.9	5121	9.09	AR-15aII	3.37	354	31.88	AR-65BI	34.5	5891	19.61
LH-08 mean	13.94	5602.78	8.12	LH03-04 mean	3.37	806.78	16.71	LH-12 mean	21.69	4811.78	14.62
AR-82bI	12.2	2558	15.97	AR-16aIII	3.87	1882	6.89	AR-18a	39.6	2065	64.22
AR-83aI	6.29	862	24.44	AR-16bI	3.73	1997	6.26	AR-18bI	32.7	1851	59.16
AR-83b	6.21	845	24.61	AR-16bII	4.43	1882	7.88	AR-18bII	32.5	2451	44.41
AR-84aII	9.17	3116	9.86	AR-16cI	3.73	1600	7.81	AR-19b	37.9	3605	35.21
AR-84bI	3.37	3213	3.51	AR-17aII	4.15	1119	12.42	AR-19c	40.9	3709	36.93
AR-84cI	6.53	2947	7.42	AR-17aIII	3.97	1355	9.81				
AR-85aI	9.9	2764	12	AR-17bI	3.62	1220	9.94				
AR-85aIII	5.18	1194	14.53	AR-17bII	5.15	1737	9.93				
AR-86aI	18.3	4067	15.07	AR-17cI	4.61	1300	11.88				
AR-86aII	16.3	4140	13.19								
LH-23 mean	9.34	2570.6	14.06	LH-05	4.14	1565.78	9.2	LH-06 mean	36.72	2736.2	47.99

Table 4 $^{40}Ar/^{39}Ar$ ages of the volcanics from the Khaybar area. Three types of ages are reported with the preferred ages shown in bold: (1) integrated ages are obtained by summing all of the argon fractions extracted during a step-heating experiment; (2) isochron ages are calculated from regression analysis of $^{36}Ar/^{40}Ar$ vs. $^{39}Ar/^{40}Ar$ data; (3) plateau ages are calculated based on trapped initial $^{36}Ar/^{40}Ar$ values calculated from the isochrons and are defined as the weighted mean age of three or more contiguous heating steps that correspond to a minimum of 50% of the ^{39}Ar released. Column N lists the number of heating steps used to calculate the plateau/the total number of heating steps

40Ar-39Ar age determination										
Sample	Integrated age (Ma)	2σ error (Ma)	Plateau age (Ma)	2σ error (Ma)	N	MSWD	Isochron age (Ma)	2σ error (Ma)	$^{40}Ar/^{36}Ar$ initial	2σ error
JA01-TB-1	16.3	1.3	**16.3**	0.3	7/7	0.4	16.3	0.4	301	3
JI01/Tb-1	14	2	**14.6**	1.7	6/8	0.09	14.7	1.8	293	10
BA01/Tb-1	25.9	1.4	**12.8**	0.13	6/9	0.33	12.8	0.2	304	2
JD01/Tb-1	13.1	1.6	**13.2**	0.6	6/9	0.03	13.2	0.8	297	7
JJa-01/Tb-1	14.3	0.4	**14.5**	0.2	8/9	0.13	14.5	0.3	299	4
JJa02/Tb-2	15	3	**15.6**	0.8	4/7	0.24	15.5	1.2	301	3

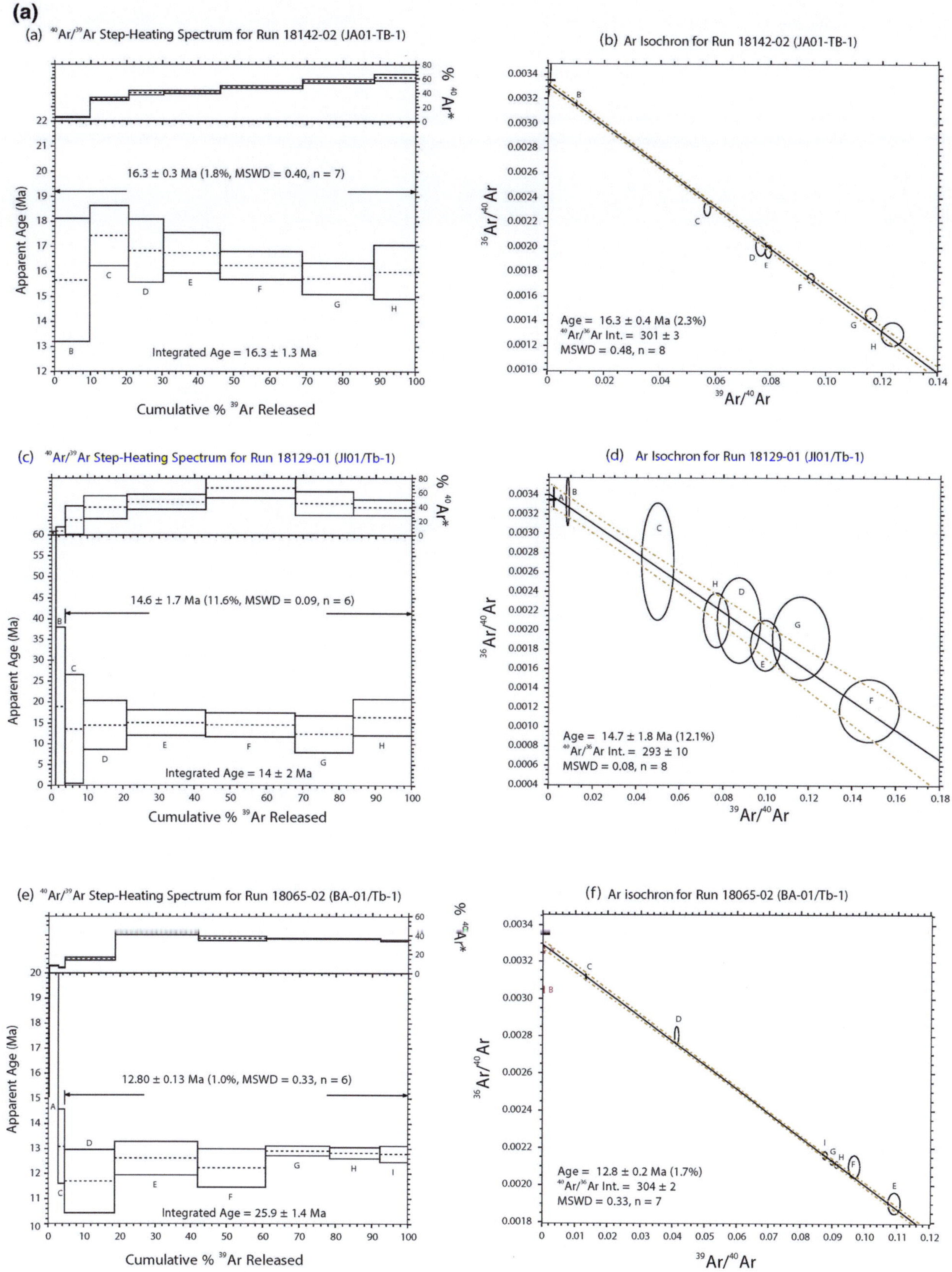

Fig. 2 Radiometric dating of the studied samples (2a and 2b). Step-heating spectra (diagrams a-c-e) and isochrons (diagrams b-d-f) of the whole-rock $^{40}Ar/^{39}Ar$ analyses. The apparent ages reported on the step-heating spectra are calculated using trapped initial Ar compositions that are delineated by the isochron of the same sample. The associated raw data are reported in Table 2. 1SD errors are shown as boxes and ellipses about the mean values. The first two heating steps of BA-01/Tb-01 in panels **e** and **f** show anomalously old ages at very low percent radiogenic $^{40}Ar^*$. These steps likely reflect secondary effects and are therefore excluded from the calculations for the plateau and isochron ages

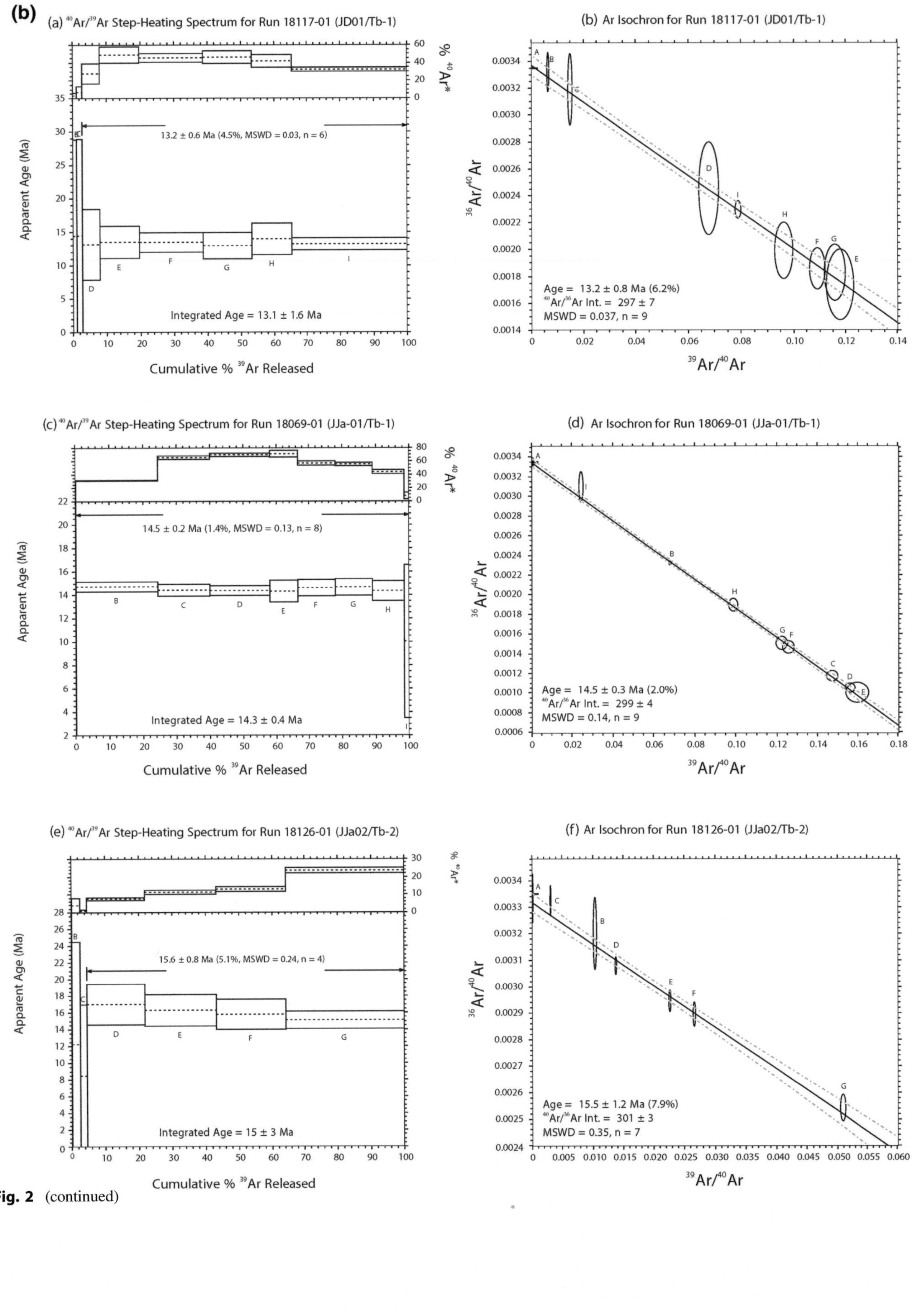

Fig. 2 (continued)

sample. The stepwise thermal demagnetization (Fig. 3a) result shows that a significant amount of magnetization survives at temperatures of 570–600° C (i.e., AR-54bIII, AR-49aII; Fig. 3b, c), which suggests that hematite is the carrier of the remanence as also indicated by the reddish colour of the rock. However, some samples lost part of their magnetization at lower temperatures (i.e., AR-49aII; Fig. 3a, c) and even AF treatment (AR-52bI; Fig. 3d) appears able to remove part of the magnetization, indicating that the hematite content is not constant. This is confirmed by the differences in the hysteresis properties, with coercivity of the remanence (B_{cr}) ranging from 32.4 mT in sample AR-51 (Fig. 3e) to 158.9 mT in sample AR-52 (Fig. 3f). The differences in the magnetic mineralogy appear to play a role in the paleomagnetic record, with stable and clear paleomagnetic directions isolated only from samples characterized by hard coercivity. All the samples exhibit negative inclinations, which indicate that the present field is not responsible for the observed directions. After the tectonic corrections, 10 out of 14 specimens give means of D = 169.6°, I = −44.8° with an α_{95} = 5.4° (Fig. 4). The interpretation of the paleomagnetic data should take into account that the sampling was restricted to a small outcrop and the results confined to the samples collected from a thin layer that possibly represents a post-depositional chemical remanent magnetization (CRM). It cannot be excluded that this CRM could have been acquired during a time interval that was insufficient to fully average the secular variations. The grouping of the paleomagnetic directions appears to become better after the tectonic correction (Fig. 4) which indicates that the age of the magnetization predates the tilting of the rock. The associated VGP is Long: 316.2°E, Lat: 80.6°N, K = 4.3°, A95 = 6.8°, Plat = 26.4°N.

5.2 Miocene Lava Flows

Scattered results mostly with shallow and negative inclinations characterize the paleomagnetic directions measured in the lava flows of Miocene age. The magnetic content of the rock appears very variable between and even within each site, with differences well represented by magnetic behaviour during thermal (Fig. 5a) and AF demagnetization (Fig. 5b). Almost certainly, weathering plays an important role in the magnetization of these rocks. Most of the rocks exhibit a secondary overprint, probably of viscous origin. This secondary overprint is removed after AF cleaning at 5–10 mT, after which the directions stabilize indicating the presence of a stable vector representing the Characteristic Remanent Magnetization (ChRM) (Fig. 5c). As shown in the plot of the induced magnetization (Ji) versus NRM (Fig. 6), the lavas from Jabal Jammazin (JJa-01) and Jabal Anthar (JA-01) yield a quite weak NRM intensity (J < 1 A/m) with low Koenigsberger ratio (Q < 5) whereas scattered and very high NRM values (up to 10^2 A/m), with low coercivity (median destructive field, MDF <5 mT) and high Koenigsberger ratio (Q > 100), occur in the rocks collected at Jabal Ishara (JI-01) (Table 2). Probably lightning strikes are responsible for the magnetization measured at this site, and the scattered directions (Fig. 7) should be considered unreliable for any paleomagnetic discussion. More constant values of both NRM and Ji characterize the lavas in the northern area of Jabal Dhulay'ah (JD-01) (Fig. 6), with two close lava flows showing a different polarity (Fig. 7) with a normal/reverse (N/R) stratigraphic sequence. If no significant time gap occurred between the two flows the Ar/Ar dating of the rock (13.2 ± 0.6 Ma) indicates that they can be correlated with one of the few N/R sequences occurring within the chron C5A (5AB-5AA) that characterizes the first half of the Serravalian (Gradstein et al. 2012). The mean values exhibit conflicting directions in terms of both declination and inclination (D = 343°, I = 64, α95 = 4°; D = 206.5°, I = 18.1°, α95 = 2.4°) (Fig. 7; Table 1), but are still in the range of the secular variation. Antipodal directions characterize the paleomagnetic directions measured in the rocks from Jabal Jammazin (JJa-01, D = 306.9°, I = 7.9°) and Jabal Anthar (JA-01, D = 125.1°, I = 6.3°) (Fig. 7; Table 1), with shallow inclinations that, coupled with the already discussed weak intensity of magnetization, make suspicious the source of the magnetization. The scatter of the results does not allow the calculation of a mean value that can yield a VGP representative of the Miocene direction for the Arabian Plate.

5.3 Quaternary Lava Flows

The lava flows collected from Harrat Lunayyir display a stable behaviour with only a minor viscous secondary overprint, which could be removed in the early steps (<10 mT) of AF demagnetization (Fig. 8) that were applied to these volcanics as a cleaning procedure. Most of the magnetization shows a spectrum of coercitivity restricted within 20–25 mT (Fig. 8) with a MDF <10 mT reflecting a predominance of MD grain size. The plot of Ji versus NRM shown in Fig. 6 attests to a better grouping of the Lunayyir samples with respect to the Miocene volcanics. The Q ratio is mainly between 5 and 70, with higher values only measured in a few samples from site LH-13 likely affected by lightning (Table 3, Fig. 6). After cleaning, the paleomagnetic directions show a normal polarity except for the above-mentioned samples from site LH-13 (Fig. 9). The results of this site have not been included in the calculations of the Fisher's statistics. A poor grouping characterized also Site LH-06 (α_{95} = 19.9°) and a few samples from other sites (i.e., LH-17) that lie far from the mean, although they did not

Fig. 3 Normalized intensity of magnetization (**a**), orthogonal vector diagrams (tilt corrected) during thermal (**b**, **c**) and AF (**d**) cleaning and hysteresis loops (**e**, **f**) for representative samples of pre-Miocene sediments. Red (blue) circles represent projection onto the horizontal (vertical) plane

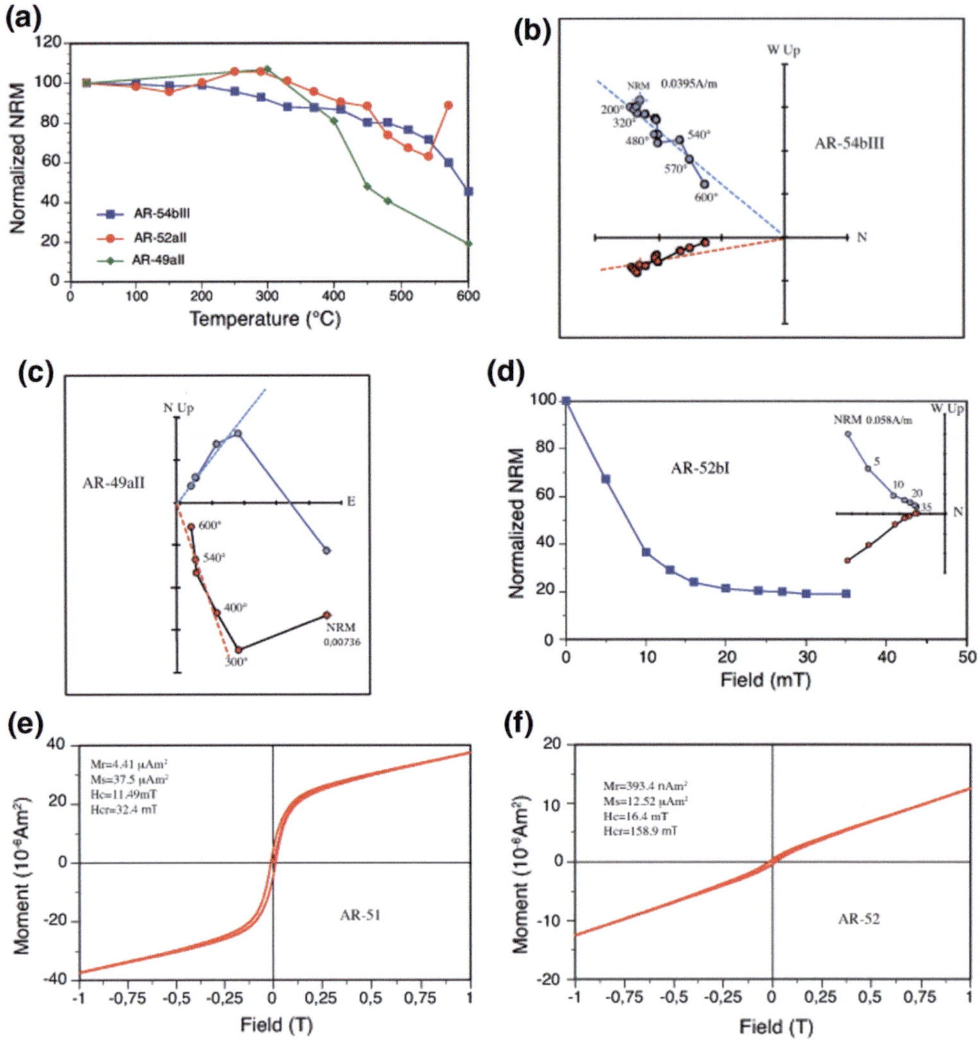

Fig. 4 Equal area projection of the paleomagnetic data from the pre-Miocene sediments collected at Bi'r al 'Ayn before and after tectonic correction. Open (closed) symbols refer to negative (positive) inclinations. Open circles in blue represent the MAD of each sample and the α95 for the mean values (red square)

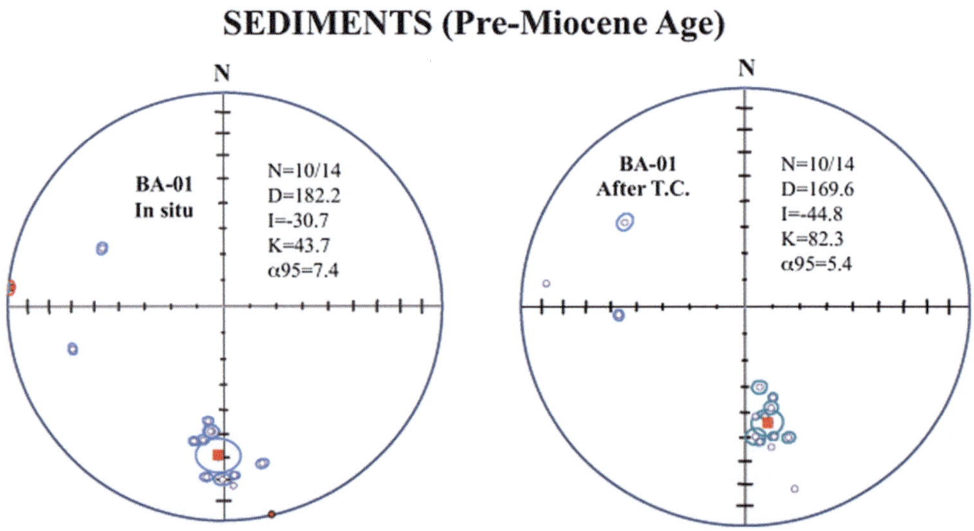

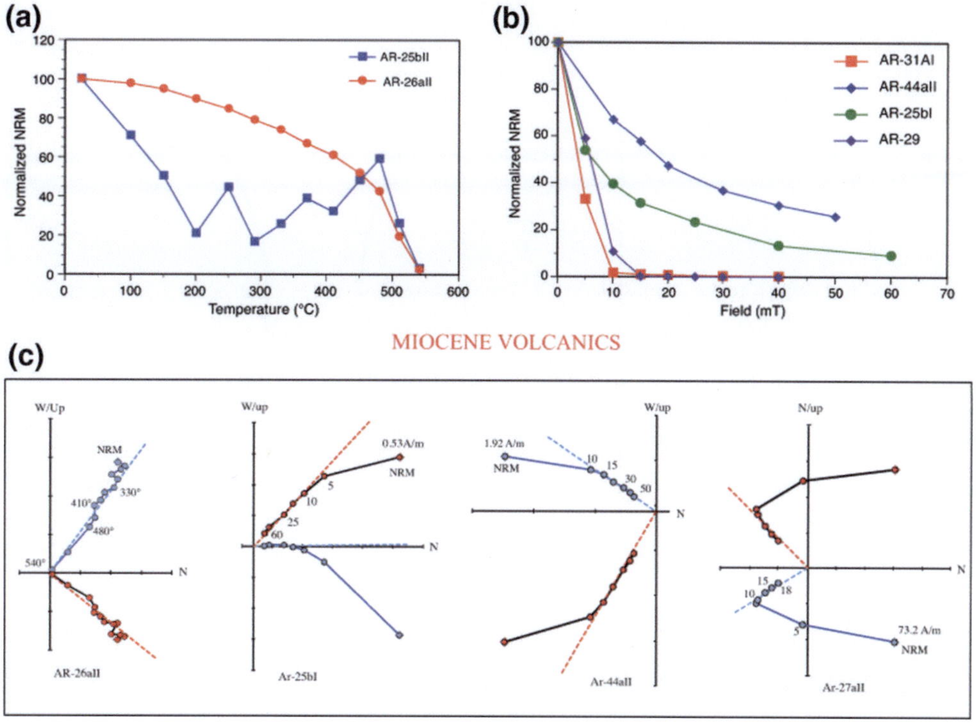

Fig. 5 Normalized intensity of magnetization during thermal (**a**) and AF (**b**) cleaning, and orthogonal vector diagrams (**c**) for representative samples of volcanic lava flows of Miocene age. Red (blue) circles represent projection onto the horizontal (vertical) plane

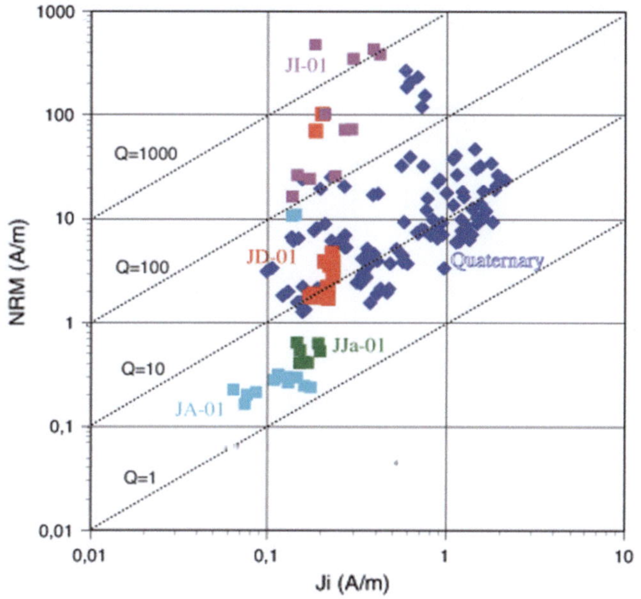

Fig. 6 Induced magnetization (Ji) versus NRM for the Miocene volcanics from Harrat Khaybar (Sites JA-01; JJa-01; JD-01; Ji-01) and the Quaternary lavas from Lunayyir

exhibit directional change during AF demagnetization. The nature of the outcrops restricted the sampling to the surface of the lava flows so it is possible that these samples may have been displaced during the acquisition of their thermoremanent magnetization (TRM). They have been excluded in the calculations of the means.

The results from the 11 sites with reliable magnetization measurements clustered around a mean value of D = 0.31°, I = 36.9° with statistical parameters K = 19.8 and α_{95} = 10.5° (Fig. 10; Table 1). An improvement of the clustering (D = 358°, I = 40.4°; K = 36.6; α_{95} = 8.6°) with an inclination that is closer to the expected value for the latitude of the sites (43°) can be obtained by excluding the mean values of sites LH05 and LH06 with shallower inclinations (Table 1). However, considering that the results are restricted to 11 sites we prefer to use the whole data set for a better average of the secular variation. The mean direction for the Lunayyir volcanics corresponds to a pole position at 85.1°N Lat., 214.1°E Long (Fig. 11; Table 1).

6 Discussion and Conclusions

As discussed previously the ChRM of the rocks was isolated from most of the studied sites, but only for the Quaternary volcanics can it for certain be considered representative of a primary magnetization. The CRM carried by the high coercitivity minerals present in the red siltstone from the Bi'r al' Ayn region could be younger than the Oligocene age of the rock but the reverse polarity indicates that it is certainly older than the Brunhes (0.78 Ma). The mean declination value (169.6°) is compatible with the counter-clockwise rotation of the Arabian Plate established by the existing paleomagnetic results. Nevertheless, the inclination (44.8°)

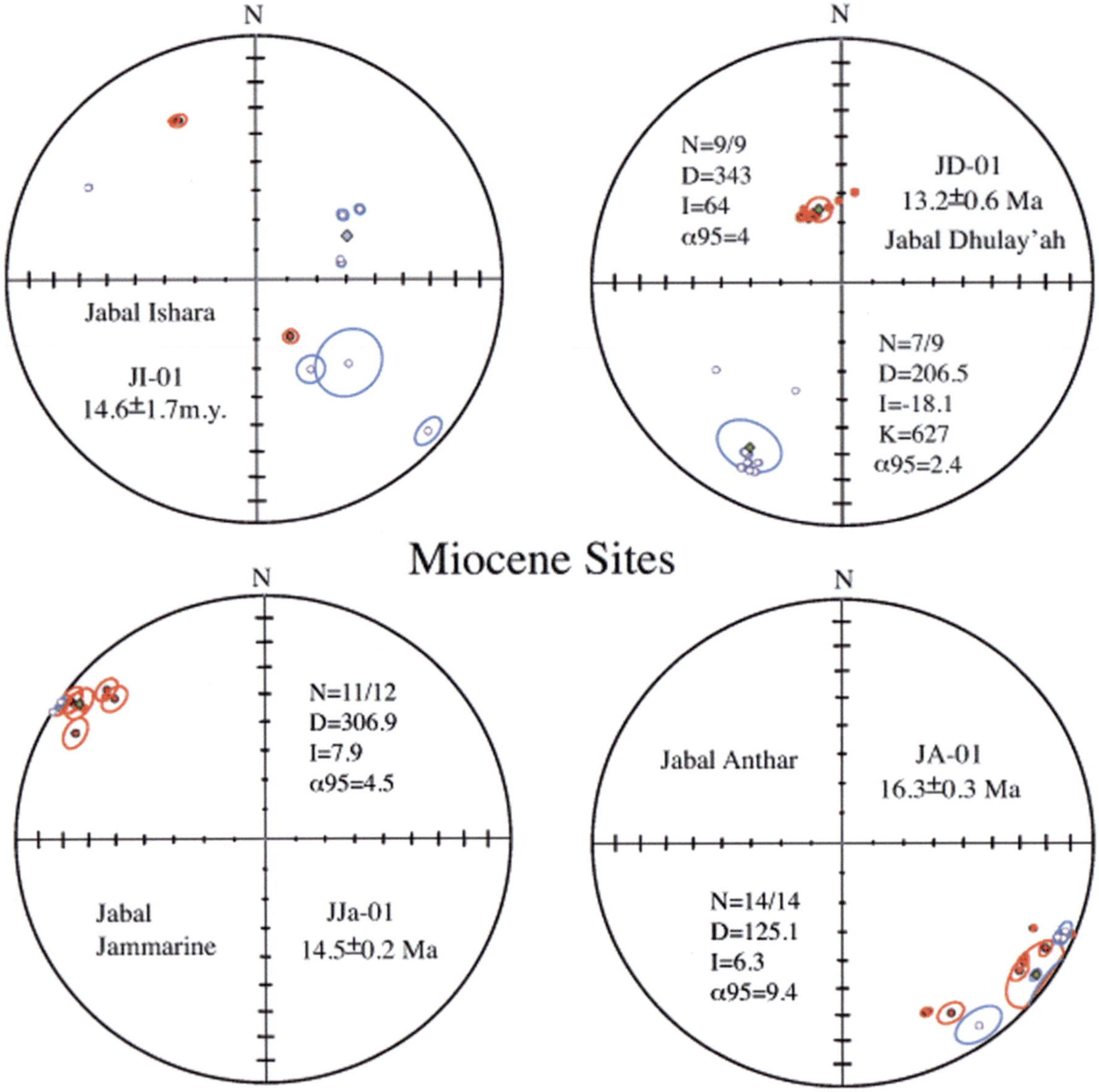

Fig. 7 Equal area projection and mean values of the paleomagnetic data from the Miocene volcanics. Closed (open) symbols refer to positive (negative) inclinations. Open circles represent the MAD of each sample and the α95 for the mean value (green square)

implies a paleolatitude (26.4° N) that is in contrast with the northern drift of the Arabian Plate during the Tertiary.

A comparison of the VGP (316.2°E, 80.6°N) with the reference paleopoles from Africa (SAFR; Torsvik et al. 2008) was evaluated using the approach suggested by Demarest (1983). The rotation and flattening values with respect to the Oligocene (30 Ma) paleopole give values that indicate a significant paleomagnetic rotation ($R = 14.86 \pm 6.38°$) and differences in latitudes ($F = -10.76 \pm 5.38°$). The comparison of the same African pole with the VGP from the As Sarat basaltic flows of Oligocene age (29–24 Ma) studied by Kellogg and Reynolds (1983) point to a smaller counter clockwise rotation (CCW) rotation ($R = -9.65 \pm 4.05°$) with a significant and contrasting difference in the flattening ($F = 9.13° \pm 4.88°$). The uncertainty in both the age of the rock and magnetization in the Bi'r al' Ayn siltstones could explain the observed differences. Additionally, the As Sarat data show inconsistent antipodal directions that cannot be easily interpreted. The flows with normal ($D = 354.8$; $I = 19$) and reverse polarity ($D = 186.3$; $I = -10.2$) represent two distinct populations that were interpreted by the authors as characteristic of the Earth's magnetic field model described by Wilson (1971). However, as shown by Mc Elhinny (2004), there is no evidence that the normal and reverse mean fields differ from one another. Hence it is more likely that these results reflect partially uncleaned magnetic directions, especially for the lava flows with reverse polarity.

The critical constraint on the timing of the Arabian rotation could be provided by the paleomagnetic data of the Miocene

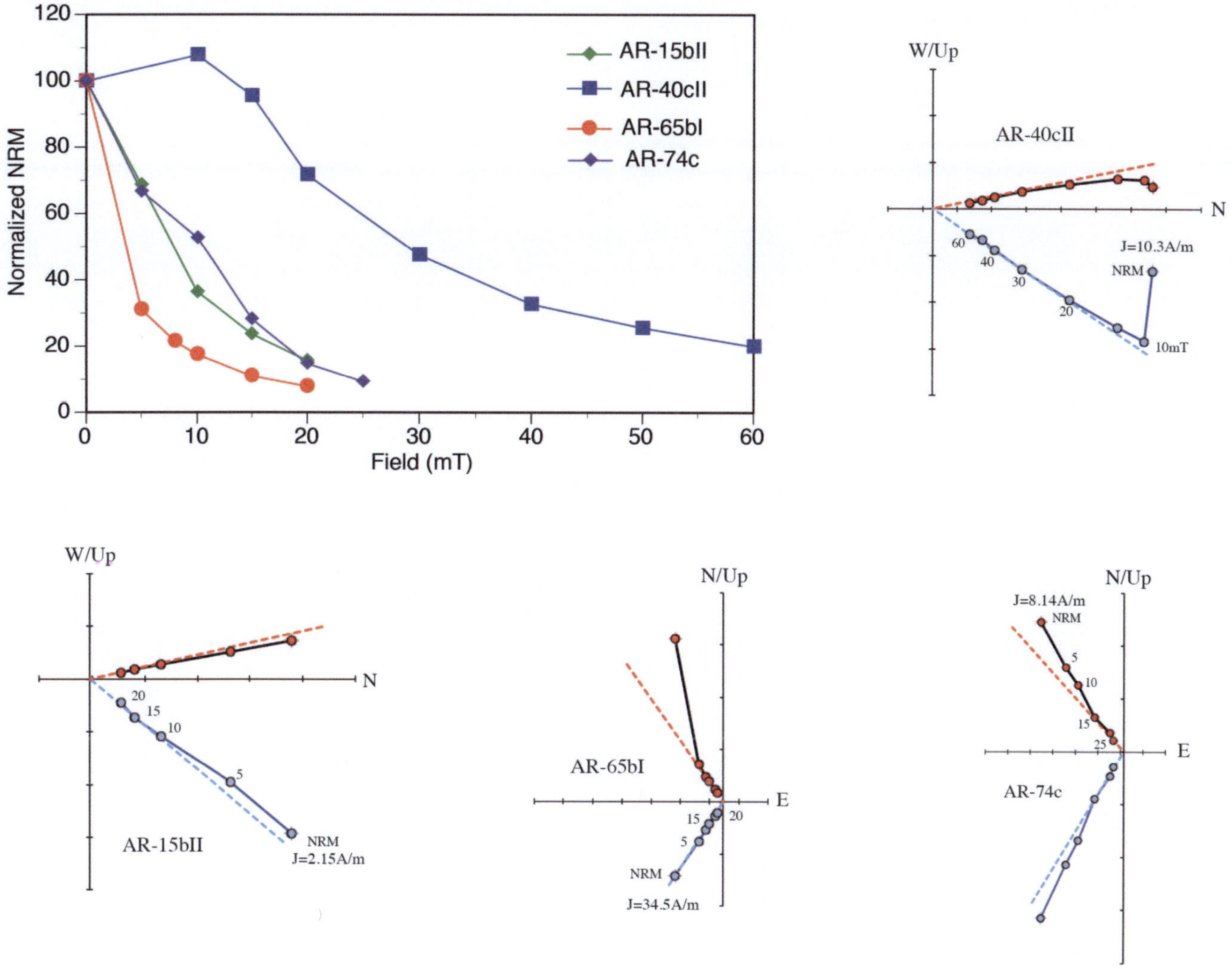

Fig. 8 Normalized intensity of magnetization and orthogonal vector diagrams during AF cleaning of selected samples of Quaternary lava flows from Harrat Lunayyir. Red (blue) circles represent projection onto the horizontal (vertical) plane

rocks and this was the target of our sampling of the Khaybar lava flows. Unfortunately, the quality of the paleomagnetic results from these lava flows is quite poor and certainly inadequate to calculate a mean value representative of a Miocene paleopole for Arabia. The effect of lightning, an artefact that has been observed in several rocks from Arabia (Tarling 1970; Kellog and Reynolds 1983) hampered the possibility of isolating a stable component of magnetization from Jabal Ishara. Meanwhile, contrasting directions from two lava flows collected at Jabal Dhulay'ah could represent only spots of the magnetic field considering that the directions are compatible with secular variations. The results from Jabal Jammazin (D = 306.9°; I = 7.9°) and Jabal Anthar (D = 125.1°; I = 6.3°) (Fig. 7) show shallow and almost antipodal directions that are consistently (57/58°) CCW deviated with respect to the early Miocene (20 Ma) paleopole for Africa (SAFR; Torsvik et al. 2008). This result cannot be justified by the rotation of the Arabian Plate since that time. The sub-horizontal bedding of the lava flows excludes tectonic tilt as the main cause for this apparent rotation and additional explanations are necessary. The nature of the outcrops limited the sampling only to the exposed part of the lava flows where the weathering is certainly more pronounced. The low susceptibility and NRM intensity (Fig. 5; Table 2) observed in these rocks is a typical signature of a post-depositional alteration that could be responsible for the anomalous directions. Furthermore, in several outcrops we detected the presence of a strong and deviating local magnetic field caused by the morphology of the cooled parts of the lava. The paleomagnetic record can be significantly affected by this effect if it was already present during the emplacement and cooling of lava flows (Baag et al. 1995; Valet and Soler 1999).

Although some scattering characterizes our late Quaternary volcanics from Lunayyir, the mean value (D = 0.31°; I = 36.9; α95 = 10.4°) is indistinguishable from the direction

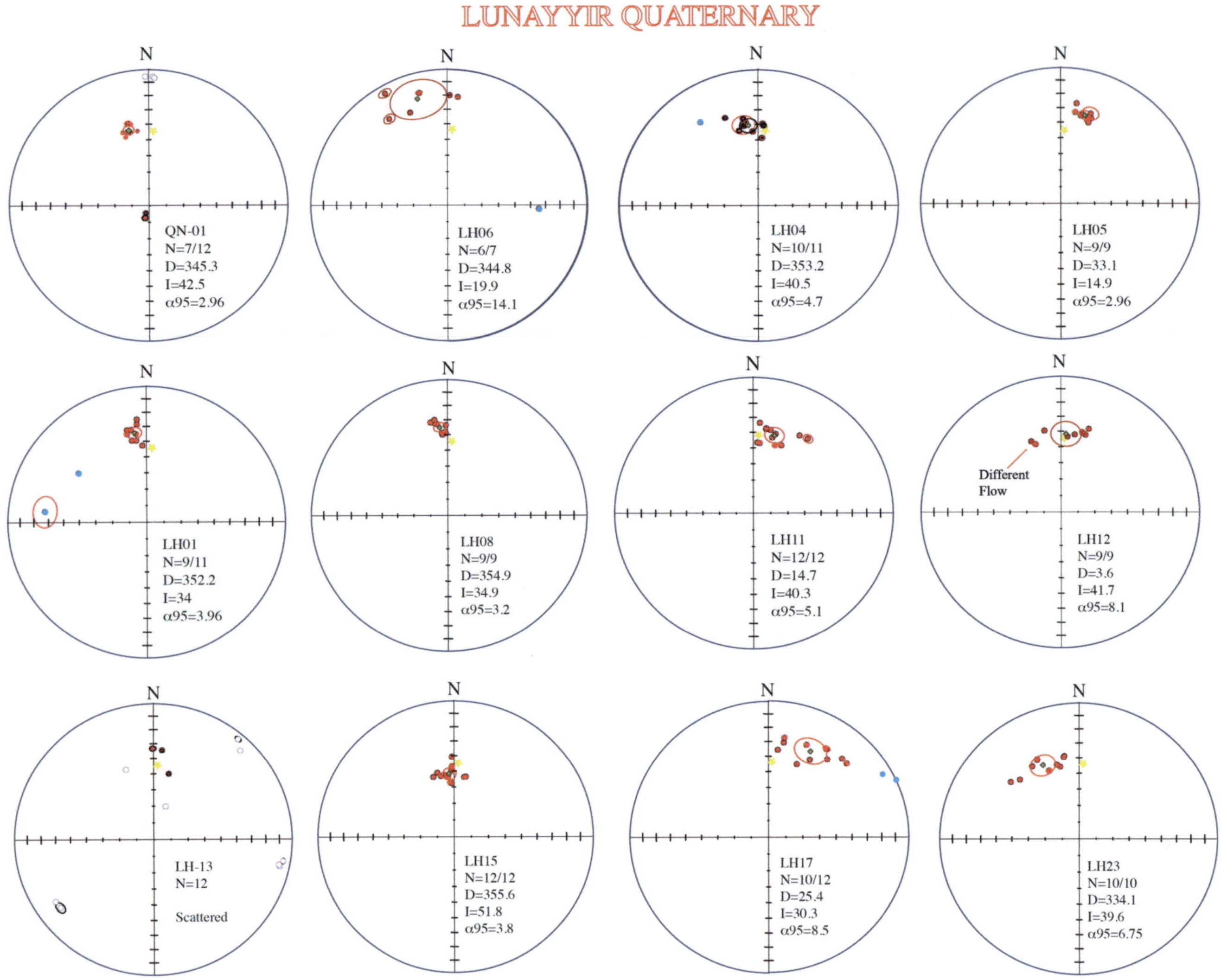

Fig. 9 Equal area projection of the mean paleomagnetic directions and statistical parameter α95 for the sites studied in Harrat Lunayyir. Closed (open) symbols refer to positive (negative) inclinations. Symbols in blue refer to samples excluded by the mean. Open circles represent the MAD of each sample and the α95 for the mean value (green square). Yellow star represents the present magnetic field

of the present field, indicating that no rotation occurred since their emplacement. The observed normal polarity suggests that these flows were emplaced during the Brunhes Chron (0–0.78 Ma), in agreement with the ages (<0.57 Ma) reported by Duncan and Al-Amri (2013) for the rocks of Lunayyir. Even though paleomagnetic data have limited use in constraining small rotations, especially if uncertainties are not very small, a comparison with the coeval African paleopole (SAFR; Torsvik et al. 2008) indicates a negligible rotation between the two plates in recent times (R = −1.98 ± 10.49° and F = 5.29 ± 8.68°).

There are few Tertiary paleomagnetic results from the Arabian Peninsula that can be compared with our data. The middle Pliocene volcanics (age 3.3–3.73 Ma) from Harrat Rahat (Hussain and Bakor 1989) and the early Pliocene volcanics (5 Ma) from Aden (Irving and Tarling 1961) indicate that about 11° and 7° of CCW rotation occurred since their formation, respectively (Fig. 11a). It should be noted that the amount of rotation of the younger volcanics (middle Pliocene age) is a little bit larger than those of the early Pliocene from Aden, even though statistically they are indistinguishable. These data are in agreement with the Oligocene data from As Sarat (Kellogg and Reynolds 1983) and they seem consistent with our results from the sediments of Bi'r al 'Ayn (Fig. 11a). The existing data seem to indicate that the Arabian Plate undertook a CCW rotation of about 10° with respect to the magnetic pole and about 15° with respect to stable Africa. However, the existing paleomagnetic directions of magnetization and associated pole positions show an agreement from the Oligocene to the Pliocene (Fig. 10), which can be interpreted only by assuming that the entire rotation occurred during the last 4–5 Ma as suggested

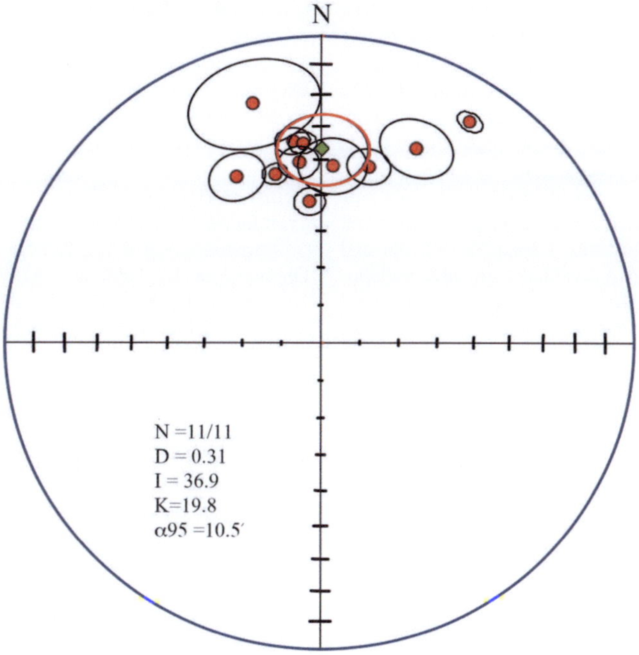

Fig. 10 Equal area projection of the paleomagnetic directions and statistical parameters for the Quaternary volcanics from Harrat Lunayyir. Green square refers to the mean value with α95 in red

Fig. 11 Comparison of the paleomagnetic results from this study with the existing paleomagnetic data (**a**) and relative VGP position (**b**) for Arabia during the Tertiary. Circles represent uncertainties (α95 for the paleomagnetic directions and A95 for the poles)

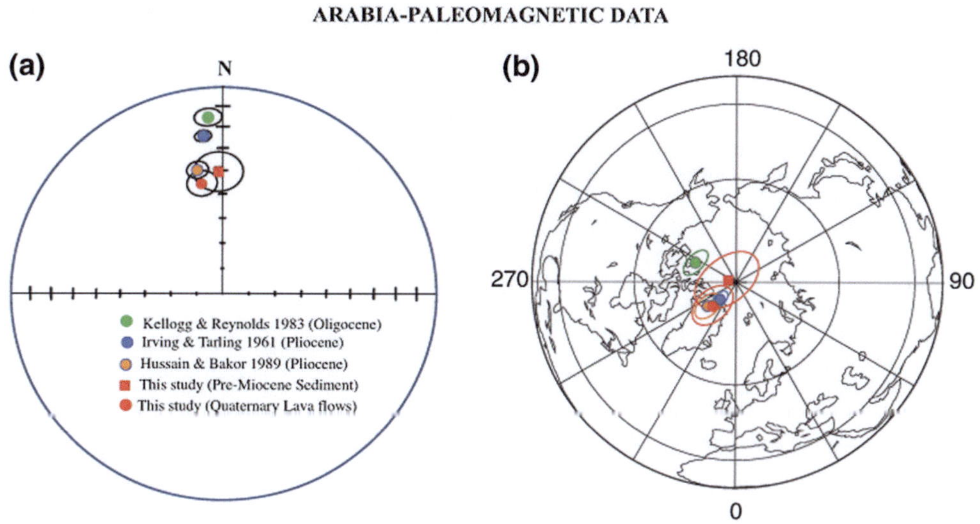

by other authors (Tarling 1970; Kellog and Reynolds 1983). This would imply that the whole rotation took place during the last phase of the opening that corresponds to true sea-floor spreading. This is in contrast with the hypothesis that suggested different phases of rotation or a constant opening of the Red Sea since the beginning of the rifting (Hall 1979; Girdler 1991; Labreque and Zitellini 1985; Almalki et al. 2015). However, the existing paleomagnetic data are inadequate to enlighten the early rotational history of the Arabian Plate.

The Koenigsberger ratio (Q) (Tables 2 and 3; Fig. 6) is above 1 in all the rocks indicating that the remanence predominates over the induced magnetization and it should be taken into account when interpreting magnetic anomalies. Considering that in the Lunayyir area these data can be integrated with geochemical and petrological data (Duncan and Al-Amri 2013; Sanfilippo et al., this volume) there is a significant data set that could be used to test the reliability of the aeromagnetic data interpretations.

Acknowledgements The research was sponsored by the PRIN2012 Programme (Project 20125JKANY_002) and supported by the Italian Consiglio Nazionale Ricerche and the US National Science Foundation. The authors are grateful to Dr. Aldo Winkler (INGV-Rome) for the measurements of the hysteresis properties and Dr. Marco Ligi for help during the sampling and improvements of the manuscript. This work is the result of a joint effort of the Saudi Geological Survey (SGS) and the Istituto di Scienze Marine, CNR of Bologna (ISMAR-CNR). We particularly thank the SGS team: A.O. Saeedi, A. Zahrani, Z.A. Otaibi, H. H. Subahi, M.M. Khorshid and A.M. Jarees, Captain P. Dimala and helicopter assistants F. Abdulhadi and A. Al-Harbi for their collaboration during the field work. We thank Dr. Z.A. Nawab, SGS President and Dr. A.M. AlAttas, SGS Assistant President for Technical Support.

References

Almalki KA, Ailleres L, Betts PG, Bantan RA (2015) Evidence for and relationship between recent distributed extension and halokinesis in the Farasan Islands, southern Red Sea, Saudi Arabia. Arab J Geosci 8:8753–8766. https://doi.org/10.1007/s12517-015-1792-9

Baag C, Helsley CE, Xu SZ, Lienert RB (1995) Deflection of paleomagnetic directions due to magnetization of the underlying terrain. J Geophys Res 100(B7):10013–10027

Bayer HJ, El-Isa Z, Hotzl H, Mechie J, Prodehl C, Saffarini G (1989) Large tectonic and lithospheric structures of the Red Sea region. J African Earth Sci (and the Middle East) 8(2):565–587. https://doi.org/10.1016/S0899-5362(89)80045-4

Bosworth W, Stockli DF (2016) Early magmatism in the greater Red Sea rift: timing and significance. Can J Earth Sci 53(11):1158–1176. https://doi.org/10.1139/cjes-2016-0019

Brown GF (1970) Eastern margin of the Red Sea and the coastal structures in Saudi Arabia. Phil Trans Roy Soc London A 267:75–87

Burke K, Dewey JF (1973) Plume-generated triple junctions: key indicators in applying plate tectonics to old rocks. J Geology 81:406–433

Camp VE, Roobol MJ (1992) Upwelling asthenosphere beneath western Arabia and its regional implications. J Geophys Res 97:15,255–15,271

Camp VE, Roobol MJ, Hooper PR (1991) The Arabian continental alkali basalt province: part II. Evolution of Harrats Khaybar, Ithnayn, and Kura, Kingdom of Saudi Arabia. Bull Geol Soc Am 103:363–391

Cochran JR (1981) The Gulf of Aden: structure and evolution of a very young ocean basin and continental margin. J Geophys Res 86:263–287

Cogné JP (2003) PaleoMac: a MacintoshTM application for treating paleomagnetic data and making plate reconstructions. Geochem Geophys Geosyst 4(1):1007. https://doi.org/10.1029/2001GC000227

Coleman RG, Hadley DG, Fleck RG, Hedge CT, Donato MM (1979) The Miocene Tihama Asir ophiolite and its bearing on the opening of the Red Sea. In: Al-Shanti AM (ed) Evolution and Mineralization of the Arabian Shield, vol 1. Pergamon Press, Oxford, pp 173–186

Coleman RG, Fleck RJ, Hedge CE, Ghent ED (1977) The volcanic rocks of southwest Saudi Arabia and the opening of the Red Sea. Mineral Resources Bull 22D, Red Sea Research. Jeddah, Saudi Arabia, pp D1–D30

Coleman RG, Fleck RJ, Hedge CE, Ghent ED (1975) The volcanic rocks of southwestern Saudi Arabia and the opening of the Red Sea. U.S. Geol Survey Saudi Arabian Project Report 194

Collins WJ (2003) Slab pull, mantle convection, and Pangaean assembly and dispersal. Earth Planet Sci Lett 205:225–237. https://doi.org/10.1016/S0012-821X(02)01043-9

Dalrymple GB, Alexander EC, Lanphere MA, Kraker GP (1981) Irradiation of samples for 40Ar/39Ar dating using the Geological Survey TRIGA reactor. In: USGS Professional Papers, U.S. Geol Surv, Reston, VA, 1176:29

Demarest HH (1983) Error analysis for the determination of tectonic rotation from paleomagnetic data. J Geophys Res 88:4321–4328. https://doi.org/10.1029/JB088iB05p04321

Dhellemmes R, Delfour J (1979) Geologic map of the Khaybar quadrangle, sheet 25 D, Kingdom of Saudi Arabia (with text). Saudi Arabian Directorate General of Mineral Resources Geologic Map GM-50A, scale 1:250000

Duncan RA, Al-Amri AM (2013) Timing and composition of volcanic activity at Harrat Lunayyir, western Saudi Arabia. J Volcanology Geothermal Res 260:103–116

Fisher RA (1953) Dispersion on a sphere. Proc Roy Soc London A 217:295–305

Girdler RW, Styles P (1974) Two stage Red Sea floor spreading. Nature 247:7–11

Girdler RW (1983) The evolution of the Gulf of Aden and Red Sea in space and time. In: Angel MY (ed) Marine science of the northwest Indian Ocean and adjacent waters. Pergamon Press, New York, p 747

Girdler RW (1985) Problems concerning the evolution of oceanic lithosphere in the northern Red Sea. Tectonophysics 116:109–122

Girdler RW (1991) The Afro-Arabian rift system—an overview. Tectonophysics 197:139–153

Gradstein FM, Ogg JG, Schmitz M, Ogg G (eds) (2012) The geologic time scale 2012. Elsevier, 1144 pp

Hall SA, Andreasen GE, Girdler RW (1976) Total intensity magnetic anomaly map of the Red Sea and adjacent coastal areas, a description and preliminary interpretation. U.S. Geol Surv Saudi Arabian Project Rep 206, 36 pp

Hall SA (1979) A total intensity magnetic anomaly map of the Red Sea and its interpretation. U.S. Geol Surv Saudi Arabian Project Rep 275, 260 pp

Hussain AG, Bakor AR (1989) Petrography and palaeomagnetism of the basalts, southwest Harrat Rahat, Saudi Arabia. Geophys J Int 99:687–698. https://doi.org/10.1111/j.1365-246x.1989.tb02051.x

Irving E, Tarling DH (1961) The palaeomagnetism of the Aden volcanics. J Geophys Res 66:549–555

Johnson RW, Knutson J, Taylor SR (eds) (1989) Intraplate volcanism in eastern Australia and New Zealand. Cambridge University Press, Cambridge, UK, p 408

Jones PW, Rex DC (1974) New dates from the Ethiopian plateau volcanic. Nature 252:218–219

Kellog KS, Reynolds RC (1983) Opening of the Red Sea: constraints from a paleomagnetic study of the As Sarat volcanic field, south-western Saudi Arabia. Geophys J R Astron Soc 74:649–665

Kemp J (1981) Explanatory notes to the geological map of the Wadi al' Ays quadrangle, sheet 25C, Kingdom of Saudi Arabia. Ministry of Petroleum and Mineral Resources, Saudi Arabia, 39 pp

Kirschvink JL (1980) The least-squares line and plane and the analysis of palaeomagnetic data. Geophys J Int 62:699–718

Kuiper KF, Deino A, Hilgen FJ, Krijgsman W, Renne PR, Wijbrans JR (2008) Synchronizing rock clocks of Earth history. Science 320:500–504. https://doi.org/10.1126/science.1154339

Labrecque JL, Zitellini N (1985) Continuous seafloor spreading in Red Sea: an alternative interpretation of magnetic anomaly pattern. Bull Am Assoc Pet Geol 69:513–524

Le Pichon X, Franchetcau J (1978) A plate tectonic analysis of the Red Sea-Gulf of Aden area. Tectonophysics 46:369–406

McElhinny M (2004) Geocentric axial dipole hypothesis: a least squares perspective. In: Channell JET, Kent DV, Lowrie W, Meert JG (eds) Timescales of the paleomagnetic field. Am Geophys Union, Washington, DC. https://doi.org/10.1029/145gm01

McKenzie D, Davies D, Molnar P (1970) Plate tectonics of the Red Sea and East Africa. Nature 226:243–248

Pellaton C, Dhellemes R (1978) Geology and mineral exploration of the Jabal Dhulay'ah quadrangle, 25/38A. Bureau de Recherches Geologiques et Minieres Technical Record 78-JED-3, 16 p, 1 fig., 2 maps, 1 app

Reilinger R, McClusky S, Ar Rajehi A (2015) Geodetic constraints on the geodynamic evolution of the Red Sea. In: Rasul NMA, Stewart ICF (eds) The Red Sea: the formation, morphology, oceanography and environment of a young ocean basin. Springer Earth System Sciences, Berlin Heidelberg, pp 135–149

Roeser HA (1975) A detailed magnetic survey of the southern Red Sea. Geol Jahrb 13:131–153

Sagnotti L (2013) Demagnetization analysis in Excel (DAIE)—An open source workbook in Excel for viewing and analyzing demagnetization data from paleomagnetic discrete samples and u-channels. Annals Geophys 56(1):D0114. https://doi.org/10.4401/ag-6282

Sanfilippo A, Cai Y, Gouveia Jácome AP, Ligi M (this volume) Geochemistry of the Lunayyir and Khaybar volcanic fields (Saudi Arabia): Insights into the origin of Cenozoic Arabian volcanism

Schettino A, Macchiavelli C, Rasul N (this volume) Plate motions around the Red Sea since the Early Oligocene

Stern RJ, Johnson PR (this volume) Constraining the opening of the Red Sea: evidence from the Neoproterozoic margins and Cenozoic magmatism for a Volcanic Rifted Margin

Tarling DH (1970) Palaeomagnetism and the origin of the Red Sea and the Gulf of Aden. Phil Trans Roy Soc A 267:219–226

Tapponnier P, Francheteau J (1978) Necking of the lithosphere and the mechanics of slowly accreting plate boundaries. J Geophys Res 83:3955–3970. https://doi.org/10.1029/JB083iB08p03955

Torsvik TH, Müller RD, Van der Voo R, Steinberger B, Gaina C (2008) Global plate motion frames: toward a unified model. Rev Geophys 46:RG3004. https://doi.org/10.1029/2007rg000227

Valet JP, Soler V (1999) Magnetic anomalies of lava fields in the Canary Islands: possible consequences for paleomagnetic records. Phys Earth Planet Int 115:109–118

Wegener A (1929) Die Entstehung der Kontinente und Ozeane, 4th edn. Friedrich Vieweg and Sohn, Braunschweig

Wilson RL (1971) Dipole offset—the time average palaeomagnetic field over the past 25 million years. Geophys J R Astr Soc 22:491–504

Microstructure and Geochemistry of Magmatic Dykes from the Arabian Margin, Red Sea

Davide Zanoni, Najeeb M. A. Rasul, Antonio Langone, and Moustafa Khorshid

Abstract

The Red Sea rifting nucleated within the Neoproterozoic Arabian-Nubian shield, which formed during the Pan-African orogeny over a time period of about 300 million years. The Red Sea rifting started about at 30 Ma and was assisted by much diffused magmatism that lasted until recent times. The majority of magmatic rocks consist of basalts that constitute volcanic plateaux (harrats), which represent one of the largest Cenozoic volcanic provinces in the world. Volcanic rocks are distributed all along the Arabian plate margin of the Red Sea, from Yemen to Jordan. Some of the oldest magmatic rocks are acidic in composition, especially along the southern part of the Arabian margin. In this chapter microstructure and geochemistry of acidic, intermediate, and basic dykes sampled along the Arabian margin are described. Acidic dykes consist of granitoids and porphyritic rhyolites. Intermediate and basic dykes consist of andesite and basanite/basalt, respectively. Granitoid dykes show equal-granular coarse-grained texture and mostly consist of euhedral crystals. Other than local displays of crystal-plastic deformation in quartz, these dykes have primary magmatic textures. Dykes consisting of rhyolites contain euhedral K-feldspar phenocrysts with frequent perthitic intergrowth of albite. The rock matrix consists of quartz, K-feldspar, and albite. Basanite/basaltic dykes consist of plagioclase, pyroxene, and amphibole phenocrysts. Plagioclase is also abundant in the groundmass where glass is also preserved. Andesite dykes are characterised by a pervasive alteration that in some instances prevents the identification of original phenocrysts. Where identifiable, phenocrysts consist of plagioclase, amphibole, and pyroxene. Locally, in the groundmass interstitial quartz shows crystal-plastic deformation. Fluidal magmatic structures are recorded locally in basanitic and basaltic dykes and only weakly in rhyolitic dykes. The fine-grained texture of the rock groundmass and vesicular structures indicate that dyke emplacement is quite shallow (hypabyssal conditions). However, the Al-content in amphibole phenocrysts of basanite/basalt dykes is consistent with phenocryst crystallisation depths of 15–20 km. The first U-Pb tests on granitoids reveal that they contain zircon grains from Cryogenian to Ediacaran (middle to late Neoproterozoic) age. Geochemical results indicate that basanite/basaltic dykes are compatible with a divergent environment such as the Red Sea rifting, whereas andesite dykes are compatible with a convergent setting. The rhyolitic dykes are interpreted as related to the Red Sea rifting as they show geochemical signatures compatible with divergent tectonics and are from a region where rhyolitic dykes were dated around 20 Ma.

1 Introduction

Microstructural analysis is a fundamental tool that contributes to the investigation of rock forming processes, deformation, and transformation, giving the rock history throughout different petrogenetic environments. Such a rock history is reconstructed by analysing relict microstructural sites saved because of the heterogeneous partitioning of deformation and mineral transformation, which commonly affects crystalline rocks, and individuating overprinting relationships of micro-structures and mineral assemblages (e.g., Gosso et al. 2015, and references therein). This chapter discusses microstructures and geochemistry of five acidic, three intermediate, and two basic dykes occurring along the

D. Zanoni (✉)
Dipartimento di Scienze della Terra "A. Desio", Università degli Studi di Milano, Via Mangiagalli 34, 20133 Milano, Italy
e-mail: davide.zanoni@unimi.it

N. M. A. Rasul · M. Khorshid
Center for Marine Geology, Saudi Geological Survey, Jeddah, Saudi Arabia

A. Langone
CNR-IGG U.O.S. Pavia, Via Ferrata 1, 27100 Pavia, Italy

N. M. A. Rasul and I. C. F. Stewart (eds.), *Geological Setting, Palaeoenvironment and Archaeology of the Red Sea*,
https://doi.org/10.1007/978-3-319-99408-6_20

Arabian side of the Red Sea (Fig. 1), between the Al Lith and Tabuk regions that are located on the Red Sea coast approximately 150 km south of Jeddah, and 100 km south of the Jordan-Saudi Arabian border about 100 km east of the Red Sea coast, respectively. Acidic dykes consist of granitoid and rhyolite, intermediate dykes consist of andesite, and basic dykes consist of basalt and basanite. This chapter aims to provide information on the mineral and bulk rock chemical composition, microstructures, and alteration state recorded by the dykes. Also, we report the first test results on zircon U-Pb geochronology of three acidic dykes. On the basis of these data, the tectonic setting of dyke emplacement is inferred, along with estimations of the depth and temperature of dyke emplacement.

2 Geological Setting of the Red Sea

The western and southern margins of the Arabian plate are bounded by two rift basins, namely the Red Sea and Gulf of Aden, which nucleated within Precambrian rocks of the Pan-African Nubian-Arabian shield. These two rift basins are connected to the Afar zone where a large mantle plume is documented (Rooney et al. 2012). In particular, the rifting propagated southward and westward in the southern Red Sea and Gulf of Aden, respectively (Cochran 1981; Courtillot et al. 1987; Audin et al. 2004; Schettino et al. 2016). However, continental rifting and oceanisation are older in the Gulf of Aden than in the Red Sea. In the Gulf of Aden rifting started at 35 Ma and predated the formation of the first oceanic spreading that occurred at 20 Ma in the eastern part (Roger et al. 1989; Fournier et al. 2010).

Lying between Africa and Arabia, the Red Sea is a narrow basin (about 2000 km long and up to 355 km wide) floored by oceanic, transitional, or thinned continental crust. According to most workers, the Red Sea rifting started in the late Oligocene and it is still ongoing north of 24 °N, where there is still a thinned continental crust, although there are many small isolated deeps characterised by mixed oceanic and continental crust. South of 20 °N a linear spreading ridge formed in the Pliocene with production of MORB oceanic crust at 4.6 Ma (Cochran et al. 1986; Gaulier et al. 1988; Bayer et al. 1989; Bohannon et al. 1989; Khattab 1992; Bosworth et al. 2005; Ehrhardt et al. 2005; Almalki et al. 2014; Schettino et al. 2016). Between 19°30'N and 23°30'N a transitional zone is characterised by an attenuated continental crust injected by basaltic dykes, where ocean floor cells have nucleated but are not yet linked into a continuous axis (Cochran 1983; Pautot 1983; Bonatti et al. 1984; Bonatti 1985; Bicknell et al. 1986).

The Red Sea rifting was preceded by conspicuous volcanism in Ethiopia and southern Yemen over a 1.5 Ma time period, during which most of the basalt volcanics were produced (Mohr 1983; Baker et al. 1996; Hofmann et al. 1997; Coulié et al. 2003; Audin et al. 2004). Starting from the Oligocene, in the Arabian margin of the Red Sea rift, Precambrian rocks experienced uplift and magmatism (intrusive and effusive) as an effect, at first, of a pre-rift mantle plume, and afterward of the asthenosphere decompression assisted by lithospheric thinning during the Red Sea opening (Pallister 1987; Coleman and McGuire 1988; Bayer et al. 1989; Camp and Roobol 1992; Daradich et al. 2003; Bosworth et al. 2005; Wilson et al. 2014). The lithospheric thinning of the Arabian-Nubian shield leading to the Red Sea opening possibly took place along inherited Pan-African structurally weak zones that are both steep and slightly oblique (N-trending) to the Red Sea axis and flat-lying shear zones (Garson and Krs 1976; Dixon et al. 1987; Makris and Rihm 1991; Ghebreab and Talbot 2000; Bosworth et al. 2005). On the other hand, the rift nucleation may have been controlled also by the interplay between plate motion driven by pulling of the Arabian slab, subducting beneath the Zagros thrust-fold belt, and the Afar plume operating a thermal weakening of the lithosphere, thus allowing localisation of deformation along the future Red Sea zone (ArRajehi et al. 2010). The activation of the Dead Sea fault zone and the westward extrusion of the Anatolia wedge during the middle Miocene and Pliocene, respectively, facilitated the northward motion of the Arabia plate that could move faster than Africa, thus aiding the rifting process (Hempton 1987; Makris and Rihm 1991; Lazar et al. 2012) that induced a renewed production of basalt in the northern Arabian plate (Ilani et al. 2001).

The onset of the lithospheric thinning heralding the Red Sea opening interrupted a tectonic and magmatic quiescence that basically lasted since the end of the Neoproterozoic. Indeed, in the region that nowadays is the Arabian shield, the Pan-African cycle ended at about 580 Ma and was followed by widespread crustal thinning and magmatism until about 530 Ma (Ibrahim and McCourt 1995; Genna et al. 2002; Johnson et al. 2011; Fritz et al. 2013; Veeningen et al. 2015).

2.1 The Arabian Shield

The Arabian shield is located in the northern part of the East African orogen that formed during the Pan-African orogeny in the late Neoproterozoic. The Pan-African orogeny was derived by suturing East and West Gondwana to form the Gondwana super continent, as a consequence of the complete closure of the Mozambique ocean (Stern 1994; Johnson et al. 2011; Fritz et al. 2013). To the north and east, Phanerozoic sedimentary cover unconformably overlies the crystalline rocks of the Arabian shield. The shield consists of different terranes (e.g., Kröner 1985; Quick 1991) that along the Arabian shoulder of the Red Sea rift, from north to south, are named the Midyan, Hijaz, Jiddah, and Asir terranes (Fig. 1). The Arabian shield terranes, which consist of island

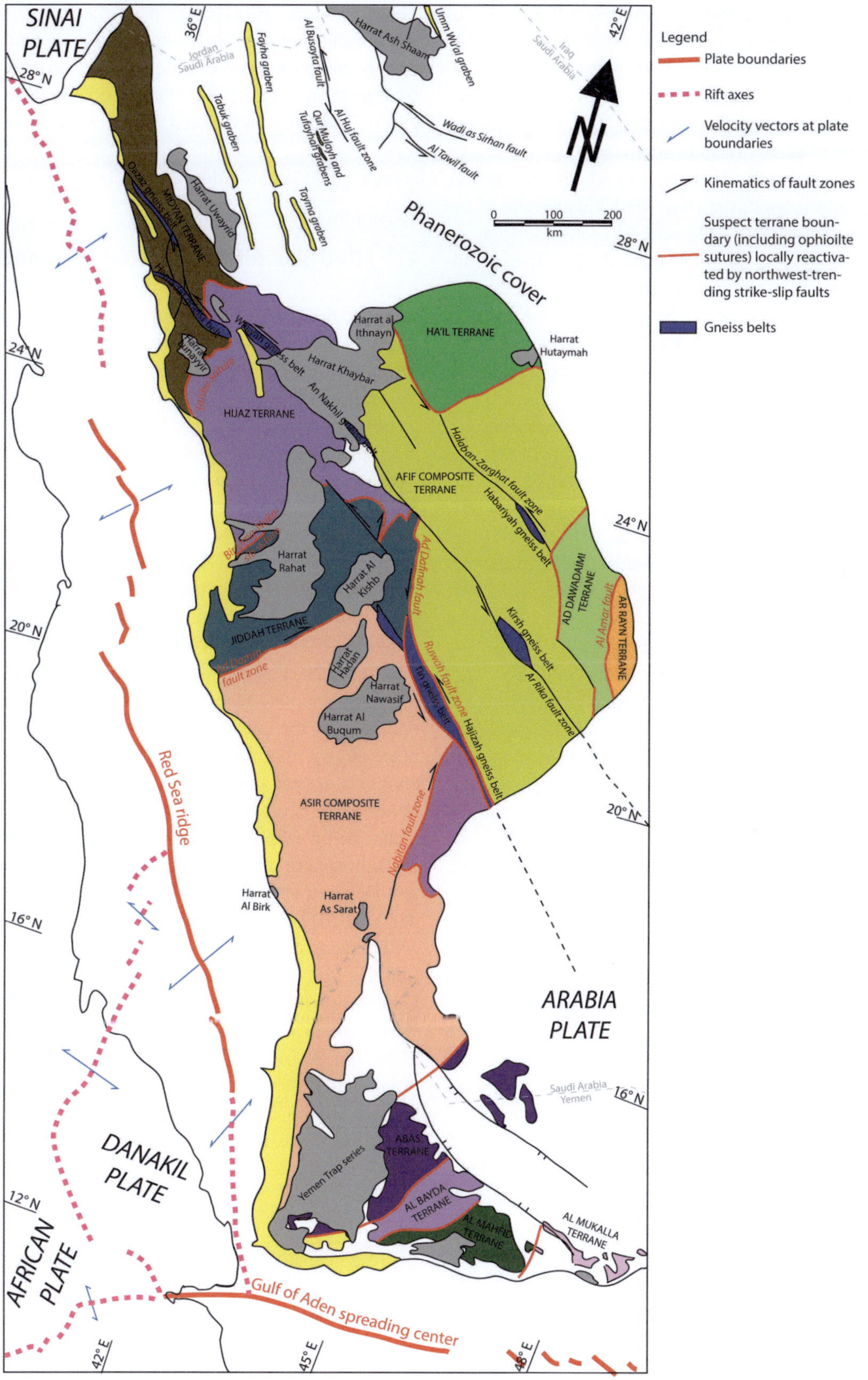

Fig. 1 Tectonic sketch map of the Arabian shield showing the Neoproterozoic terranes, the major suture belts and fault zones (Modified from Johnson 1998), and the Cenozoic harrats or volcanic fields (grey). The Neoproterozoic structures reactivated during the Red Sea rifting are highlighted in red. Plate boundaries, rift axes, and velocity vectors at plate boundaries are from Schettino et al. (2016)

arcs and micro-continents, were assembled over a period of time of about 300 million years between 890 and 580 Ma and are formed by different types of intrusive and metamorphic rock assemblages representing juvenile Neoproterozoic crust (Stoeser and Camp 1985; Vail 1985; Kröner et al. 1991; Fritz et al. 2013). In general, there is no evidence of the pre-Neoproterozoic geologic record in the Arabian shield, with local exceptions such as the Khida terrane of the Afif composite terrane, the Abas and Al Mahfid terranes in western Yemen, and the Sa'al complex in the Sinai peninsula (Windley et al. 1996; Stoeser et al. 2001; Stoeser and Frost 2006; Be'eri-Shlevin et al. 2012). The Arabian shield is also characterised by gneiss belts elongated into the main fault zones that are parallel to the Red Sea axis (Johnson et al. 2011, and references therein).

The different terranes of the Arabian shield are separated by ophiolitic suture zones (Fig. 1) that formed between 780 and 600 Ma and testify to a progressive accretion of the terranes via different subduction/collision events. Ophiolite rocks involved in the suture zones span between 870 and 630 Ma and are associated with volcanic arc rock assemblages (Shanti and Roobol 1979; Pallister et al. 1987; Al-Saleh and Boyle 2001; Nehlig et al. 2002; Dilek and Ahmed 2003; Johnson et al. 2004; Fritz et al. 2013). The oldest ophiolites outcrop along the Yanbu and Bir Umq suture zones that are interposed between the Midyan and Hijaz terranes and Hijaz and Jiddah terranes, respectively (Fig. 1). These two suture zones are intersected at a high angle by the Red Sea rifting and thus outcrop also in the Nubian shoulder of the Red Sea rift, running up to southern Egypt-northern Sudan and central Sudan, where they are named the Heiani-Allaqi and Nakasib suture zones, respectively (Sultan et al. 1993; Abdelsalam and Stern 1996; Ali et al. 2010).

The intrusive rocks of the Arabian shield can be divided into two major groups with respect to the Pan-African orogenic history (Fig. 2). The early calc-alkaline granitoid rocks of the shield are thought to be generated in a subduction-related magmatic arc setting, whereas the youngest alkaline granitoid rocks are interpreted as being formed during the collapse of the orogen under an extensional tectonic regime (Nehlig et al. 2001; Genna et al. 2002; Hargrove et al. 2006; Farahat et al. 2007; Be'eri-Shlevin et al. 2009; Johnson et al. 2011; Fritz et al. 2013; Beyth et al. 2014; El-Bialy and Omar 2015). However, four distinct periods of magmatism have been identified in the western Arabian shield, as follows: Island arc about 845 Ma, syn-collisional about 736–700 Ma (mainly characterised by calc-alkaline magmatism), post-tectonic about 620 Ma, and anorogenic until about 525 Ma (mainly characterised by alkaline magmatism) (Robinson et al. 2014). The magmatism coeval with crustal extension, postdating the Pan-African collision in the Eastern Egyptian Desert and Midyan terrane, is responsible for the formation of much of the present day crust in contrast with the rest of the shield, where the crust was produced under convergent tectonic settings (Stern et al. 1984; Clark 1985; Ibrahim and McCourt 1995). This extensional stage was also characterised by the formation of a basin filled with terrigenous sediments and connected to oceanic domains (Johnson et al. 2011).

2.2 Red Sea Rifting-Related Magmatism

All over the Arabian shoulder of the Red Sea rift, Cenozoic volcanic plateaux (harrats) rest mostly on the Precambrian crystalline rocks and represent one of the largest Cenozoic volcanic fields in the world. Indeed, the Arabian harrats form an almost continuous roughly N–S linear volcanic belt about 3000 km long, from Yemen and Afar to the Bitlis suture in Turkey. In the central part of the Arabian shield, Harrats Rahat, Kishb, Khaybar, and Ithnayn (Fig. 3) are N–S aligned to form the Makkah-Madinah-Nafud volcanic line. The linear distribution of this Cenozoic volcanism has been associated with a N–S trending regional lithospheric flexure that may induce a linear weak zone in the crust (Camp and Roobol 1992). Alternatively this linear distribution may be due to the N–S alignment of several mantle plumes beneath the Arabian plate (Pavoni 1992), or to a localised thermal anomaly in the mantle (Park et al. 2007). Recently it has been shown that from the Afar plume, horizontal mantle channels spread radially and one of them flows northward beneath the southern Red Sea and Arabian plate, thus most likely being responsible for thermal erosion of the base of the lithosphere and for such a linear distribution of volcanism (Hansen et al. 2007; Chang et al. 2011; Chang and Van der Lee 2011).

The age of the Cenozoic magmatic rocks of the Arabian margin lies between late Oligocene (about 30 Ma) and Quaternary (about 5 ka) (Fig. 4) (Camp and Roobol 1992; Bosworth et al. 2005; Al-Amri et al. 2012; Runge et al. 2016). In the Arabian margin a few historical eruptions are documented (Moufti et al. 2013) and present day magmatic activity is the cause of earthquakes linked to extensional tectonics (Pallister et al. 2010; Koulakov et al. 2014).

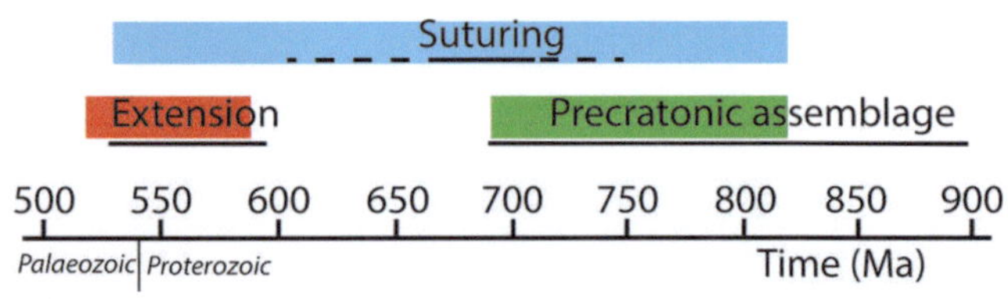

Fig. 2 Summary of the tectonic and magmatic events recorded in the Arabian shield Modified after Genna et al. (2002) and Fritz et al. (2013). Red = post-orogenic magmatism; green = oceanic magmatism, pale blue = arc magmatism. Time scale in Ma (million years)

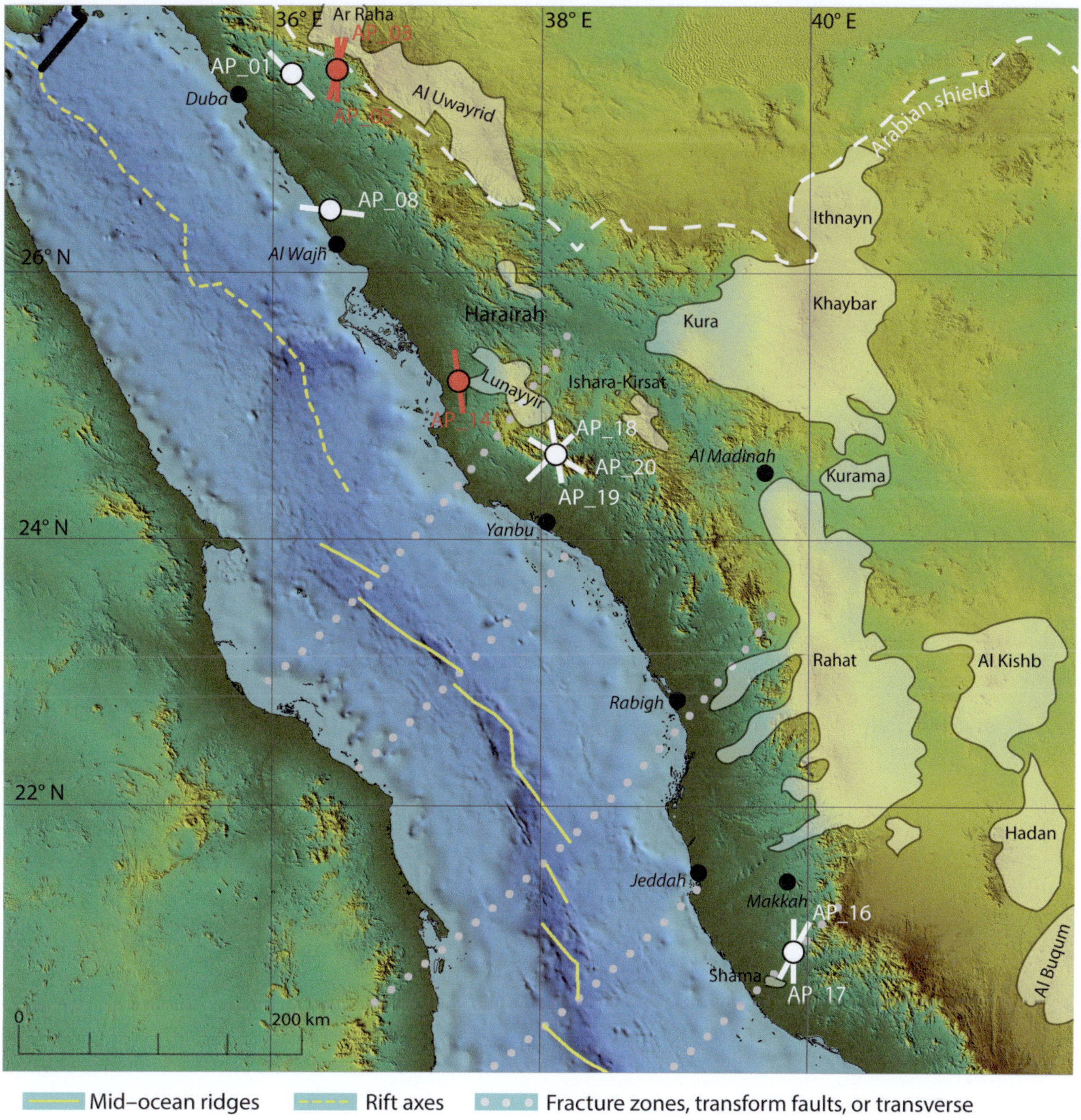

Fig. 3 Location map of studied dykes showing present–day plate boundaries and transverse structures around the Red Sea (after Schettino et al. 2016). White shaded polygons: Cenozoic volcanic harrats redrawn after Coleman et al. (1983) and Roobol and Al-Rehaili (1998). White (not dated) and red (dated) segments indicate the strikes of the studied dykes (not to scale)

Magmatism associated with seismicity has also been well documented in the Afar region (Bendick et al. 2006; Keir et al. 2006; Wright et al. 2006; Ebinger et al. 2008, 2010; Belachew et al. 2013; Desissa et al. 2013).

The development of the Cenozoic continental magmatism in Arabia is partitioned into two major stages between 30 and 20 Ma and from 12 Ma to the present, respectively (Camp and Roobol 1992). Until 25 Ma, magmatism developed without significant crustal extension, whereas starting from 24 Ma magmatism was coeval with strong crustal extension (Bosworth et al. 2005). Between 30 and 20 Ma magmatic rocks consist of subalkaline basalts intercalated with rhyolites, which are concentrated in the southern part of the Arabian margin, and after 12 Ma alkaline basalts are

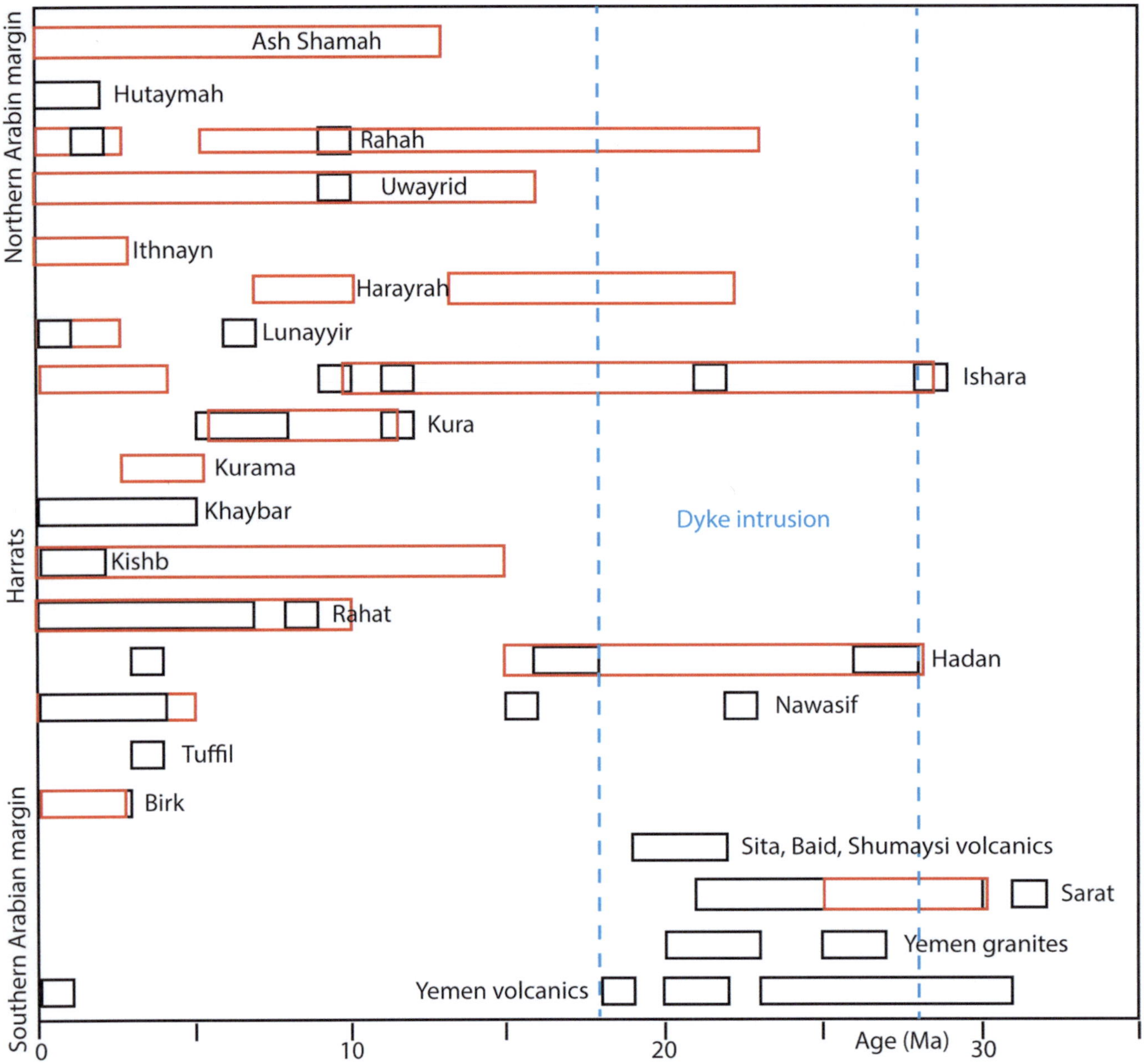

Fig. 4 Age compilation of Cenozoic volcanic harrats in the Arabian shield. Black from Camp and Roobol (1992) and red from Runge et al. (2016) compilations, respectively

dominant and are scattered all over the Arabian margin (Schmidt et al. 1983; Manetti et al. 1991; Baker et al. 1996).

The southern Arabian margin, northern Yemen, is characterised by a first magmatic cycle between 30 and 26 Ma with flood basalts intercalated with ignimbrites until 22 and 20 Ma (Riisager et al. 2005). These volcanics are coeval with basaltic to granitic dykes and alkali granite emplacement between 25 and 16 Ma, and 26 and 20 Ma (Bosworth and Stockli 2016). A second and third magmatic cycle mostly consist of alkali basalts with rare alkali rhyolites and took place at about 10 Ma and after 6.5 Ma, respectively (Manetti et al. 1991).

Geochemical evidence indicates that the Yemen magmatism may result from extensive crustal contamination and fractionation processes (Manetti et al. 1991) and was fed by a plume source (Davison et al. 1994; Baker et al. 1996), which is consistent with active rifting because volcanism predates extension of the upper crust (Ghebreab 1998). In particular, Miocene to Quaternary basalts from Yemen display a higher radiogenic component with respect to basalts from further north along the Arabian margin and this may suggest that the Afar plume is responsible for the partial melting of the lithosphere beneath Yemen (Bertrand et al. 2003). Although basaltic rocks are more abundant than felsic rocks, further north, in the central coastal plain of the Arabian margin, the Ad Damm dyke suite intruded the Precambrian basement between 22 and 28 Ma and is characterised by bimodal composition varying between alkali basalt and rhyolite (Pallister 1986, 1987; Coleman and McGuire 1988; Sebai et al. 1991). Moreover, in this region Harrat Tufail consists of Miocene and Pliocene basalt and rhyolite lava flows that are blanketed by Quaternary alkali basalts (Bakhsh 2015). To explain rhyolites among the Ad Damm dykes, partial crustal melting has been proposed (Pallister 1987). In general, fractionation and assimilation are used to explain the intrusion of rhyolitic dykes in the Jizan region, which is located south of Wadi ad Damm along the Arabian margin (Coleman and McGuire 1988). More to the north, Harrat Lunayyir is characterised by basanite, alkali olivine basalt, and trachy-basalt that are fed by a N–S trending fissure. In this case isotopic data show that there is no evidence of crustal melting and magmatic source varies throughout time with the asthenosphere rising (Duncan and Al-Amri 2013). Harrat Uwayrid results from two magmatic cycles; the first during Miocene and Pliocene is dominated by alkali basalt flows; the second, Quaternary in age, is characterised by basanite and tephrites (Kaliwoda et al. 2007). At the very northern end of the volcanic belt, in Harrat Ash Shaam in Jordan, basaltic dykes are aged between 23 and 1.7 Ma. The dykes show a change in composition through time from tholeiitic, to pyroxene-olivine basalt, to olivine basalt, to basanite, marking an increase of the alkaline character (Ibrahim et al. 2003). Volcanics from Harrat Ash Shaam consist of mafic alkali basalt and basanite distributed in time over three cycles from 26 and 0.5 Ma. The Harrat Ash Shaam magmatism is interpreted as being derived from partial melting of lithospheric mantle as a consequence of lithospheric extension (Ilani et al. 2001; Shaw et al. 2003). In general, the data discussed show that Cenozoic magmatism along the western Arabian margin becomes younger northward with younger magmatic rocks being more alkaline than older magmatic rocks (Figs. 1, 3, and 4).

3 Field Outlines of the Studied Dykes

Among the ten dyke samples, whose location is shown in Fig. 3, three samples consist of granitoid dykes (AP-03, AP-05, and AP-14) and two of basic dykes (AP-01 and AP-08), all intruded in the Midyan terrane rocks (Figs. 1 and 3). Other samples represent three andesite dykes intruded in the Hijaz terrane (AP-18, AP-19, and AP-20), and two rhyolite dykes that are emplaced in the Jiddah terrane, close to the tectonic contact (Ad Damm fault zone) with Asir composite terrane (AP-16 and AP-17).

Sample AP-03 and AP-05 were selected from two subvertical dykes that trend 0 °N and 195 °N, respectively. These dykes intersect Neoproterozoic Qaraqir alkali granite of the Midyan suite, close to the tectonic contact with the Atiyah monzogranite (Grainger and Hanif 1989). The age of the Qaraqir formation lies between 630 Ma and 609 Ma (Al-Husseini 2015). Sample AP-14 is from Wadi al Hayil, which corresponds to a dextral fault. This subvertical dyke trends 352 °N and intrudes gabbroic rocks not far from the boundary with an amphibole-bearing tonalite, both belonging to the Neoproterozoic Nabt complex (Pellaton 1982). The Nabt complex was intruded by the Salajah tonalite at 725 ± 12 Ma (Pallister et al. 1987).

Sample AP-01 is located in the same area as samples AP-03 and AP-05, near Jabal Masah, and it is part of a subvertical dyke swarm trending 317 °N (Fig. 5a). The dyke swarm intruded the Neoproterozoic alkali granite of the Massah complex, which is dated at 629 ± 12 Ma (Grainger and Hanif 1989). The locality is in the proximity of a 120 °N trending fault that put the alkali granites in contact with Neoproterozoic volcanoclastic rocks of the Bayda complex (Grainger and Hanif 1989). Sample AP-08 represents a subvertical dyke trending 277 °N to the north of Wadi Marabit and Wadi ar Ramah, only 6.5 km from the Red Sea shore. This dyke is intruded in the Neoproterozoic monzogranite of the Liban complex, which is part of the Marabit suite (Davies 1985). The Liban monzogranite is dated at 621 ± 7 Ma or 638 ± 10 Ma (Hedge, in Al-Husseini 2015). This dyke shows a marked morphology that shapes a hill in the middle of the wadi plain filled with Quaternary sediments (Fig. 5b).

Samples AP-18, AP-19, and AP-20 are from Wadi Jari, 5 km west from the intersection with Wadi Rakhu. These are subvertical dykes (Fig. 5c, d) trending 255 °N, 170 °N, and 302 °N, respectively and are intruded into a Neoproterozoic granitoid complex consisting of alkali granite, monzogranite, and granodiorite (Pellaton 1979).

Samples AP-16 and AP-17 are from 1.5 km southwest of Bir ad Damm village. These represent two subvertical dykes trending 28 °N and 0 °N intruded into the Neoproterozoic meta-basalts of the Baish group, whose protoliths are older than 804 Ma. Here the dykes and country rocks are affected by the Ad Damm fault zone that put them in contact with the Neoproterozoic granite gneiss of the Namam complex (Pallister 1986). These two dykes show a good morphological expression and are responsible for tracing the hill ridges that are continuous for about 10 km (Fig. 5e, f).

Fig. 5 **a** Granitoid rocks intruded by basaltic dyke. AP-01 was sampled from this dyke. A. Schettino for scale; **b** metre-thick basaltic dyke intruded in granitoid rocks. The dyke is responsible for shaping the hill that dominates the middle of the wadi. AP-08 was sampled from this dyke; **c** decimetre-thick basaltic dyke intruded in granitoid rocks. AP-19 was sampled from this dyke. P.P. Pierantoni for scale; **d** 1-metre-thick basaltic dyke intruded in granitoid rocks. AP-20 was sampled from this dyke; **e** metre-thick rhyolitic dyke intruded in metavolcanic rocks. AP-16 was sampled from this dyke; **f** metre-thick rhyolitic dyke intruded in metavolcanic rocks. Clearly the dyke is responsible for shaping the ridge over a kilometric distance. AP-17 was sampled from this dyke

4 Microstructural and Petrographical Analysis

Microstructural and petrographical analysis has been applied to the dyke samples in order to integrate the field data collected in 2015. Besides the optical microscope, samples have been analysed in part also by means of a scanning electron microscope (SEM). In the following, granitoid, basic, andesitic, and rhyolitic dykes are described separately.

4.1 Granitoid Dykes

The coarse-grained granitoid dykes consist of granite (AP-03), monzogranite (AP-05), and granodiorite (AP-14). The main forming minerals are quartz, K-feldspar, plagioclase, and biotite. Accessory minerals consist of medium to fine-grained opaque minerals, zircon, apatite, and rutile. These minerals constitute the magmatic phases that are altered by post-magmatic mineral phases such as white mica and chlorite. Rocks show holocrystalline hypidiomorphic granular texture. According to modal analysis, K-feldspar (30–40%) is more abundant than plagioclase (10–20%) with the exception of sample AP-14 where plagioclase (35%) is more abundant than K-feldspar (10%). K-feldspar mostly forms euhedral to subhedral crystals and shows perthite and minor flame perthite exsolution of albite in sample AP-03 and microcline twinning in sample AP-05 (Fig. 6a). At the margins of K-feldspar phenocrysts graphic intergrowths of K-feldspar and quartz occur (AP-03); the graphic intergrowth may also consist of sub-rounded quartz crystals that testify that, after complete crystallisation, graphic structure recrystallized (cf., Vernon 2004) (Fig. 6b). Plagioclase forms medium- to coarse-grained subhedral crystals and locally shows polysynthetic twinning. In sample AP-05 plagioclase forms aggregates with K-feldspar or is enclosed in coarse-grained K-feldspar phenocrysts (Fig. 6c). Often both plagioclase and K-feldspar are altered by a very fine-grained aggregate of white mica and minor chlorite and epidote (Fig. 6c). In sample AP-03 such alteration is so intense that locally it makes optical distinction of plagioclase and K-feldspar difficult. Locally the outer rims of plagioclase survived the prominent replacement by white mica, chlorite, and epidote (Fig. 6d), suggesting that the plagioclase rims may be poorer in Ca. Quartz forms mostly medium- to coarse-grained sub-rounded anhedral crystals that show homogeneous extinction or only a weak undulose extinction. In sample AP-14 quartz shows a more evident undulose and subgranular extinction than in the other granitoid rock samples. Although in general it records a gentle deformation, quartz is the only completely pristine major forming magmatic mineral. Locally quartz forms rounded crystals surrounded by interstitial K-feldspar (Fig. 6e). Biotite occurs in all three samples and forms medium-grained crystals that are often pervasively replaced by chlorite, white mica, and fine-grained opaque minerals. The minerals replacing biotite grew mimetically along cleavages and locally overgrow entirely single grains. Locally chlorite constitutes layers of randomly oriented crystals (e.g., AP-03), which may be due to fluid circulations during the late stage of crystallisation or after the complete crystallisation. In sample AP-14 coarse-grained green poikilitic hornblende crystals occur (Fig. 6f). These crystals enclose rounded quartz grains and fine-grained biotite and are associated with coarse-grained biotite aggregates and zircon grains. In these aggregates biotite shows slight undulose extinction. Apatite forms fine-grained crystals enclosed in K-feldspar. Among accessory minerals titanite occurs in AP-03 and forms very fine-grained sub-rounded crystals enclosed in quartz.

4.2 Basic Dykes

Samples AP-01 and AP-08 consist of basaltic dykes that show intergranular and porphyritic texture, respectively. Both dykes contain euhedral (AP-08) and subhedral (AP-01) plagioclase phenocrysts that are less than 1 mm and up to 5 mm in size in AP-01 and AP-08, respectively (Fig. 7a). Plagioclase also forms interstitial grains (AP-01) and euhedral microlites in the groundmass (AP-08). In both samples plagioclase shows growth polysynthetic twinning and zoning. In sample AP-01 plagioclase phenocrysts are replaced by fine-grained white mica, which forms randomly oriented aggregates affecting especially the crystal core and leaving the very outer rim unaltered (Fig. 7b). Groundmass plagioclase survived white mica alteration. In contrast, in sample AP-08 plagioclase is very pristine (Fig. 7a). The mafic mineral phases consist of brown amphibole, pyroxene, olivine, and rare biotite. Amphibole is more abundant in AP-01 than in AP-08. In sample AP-01 amphibole forms euhedral grains up to 0.1 mm in size, just bigger than the groundmass (Fig. 7b). Locally, amphibole encloses acicular apatite crystals. Rarely amphibole is rimmed by calcite, Fe–Mg-carbonates (determined by the SEM analysis), and chlorite. In AP-08 amphibole forms up to 0.5 mm-sized euhedral phenocrysts rimmed by coronitic aggregates of very fine-grained Fe-oxide (Fig. 7c), which replace almost entirely amphibole. This coronitic aggregate may result from disequilibrium between amphibole and liquid either because of liquid changing composition during crystallisation or amphibole being a xenocryst. Clinopyroxene forms elongated crystals in sample AP-01 (Fig. 7d) that locally are partially enclosed in amphibole. In AP-08 pyroxene consists of more abundant clinopyroxene than orthopyroxene.

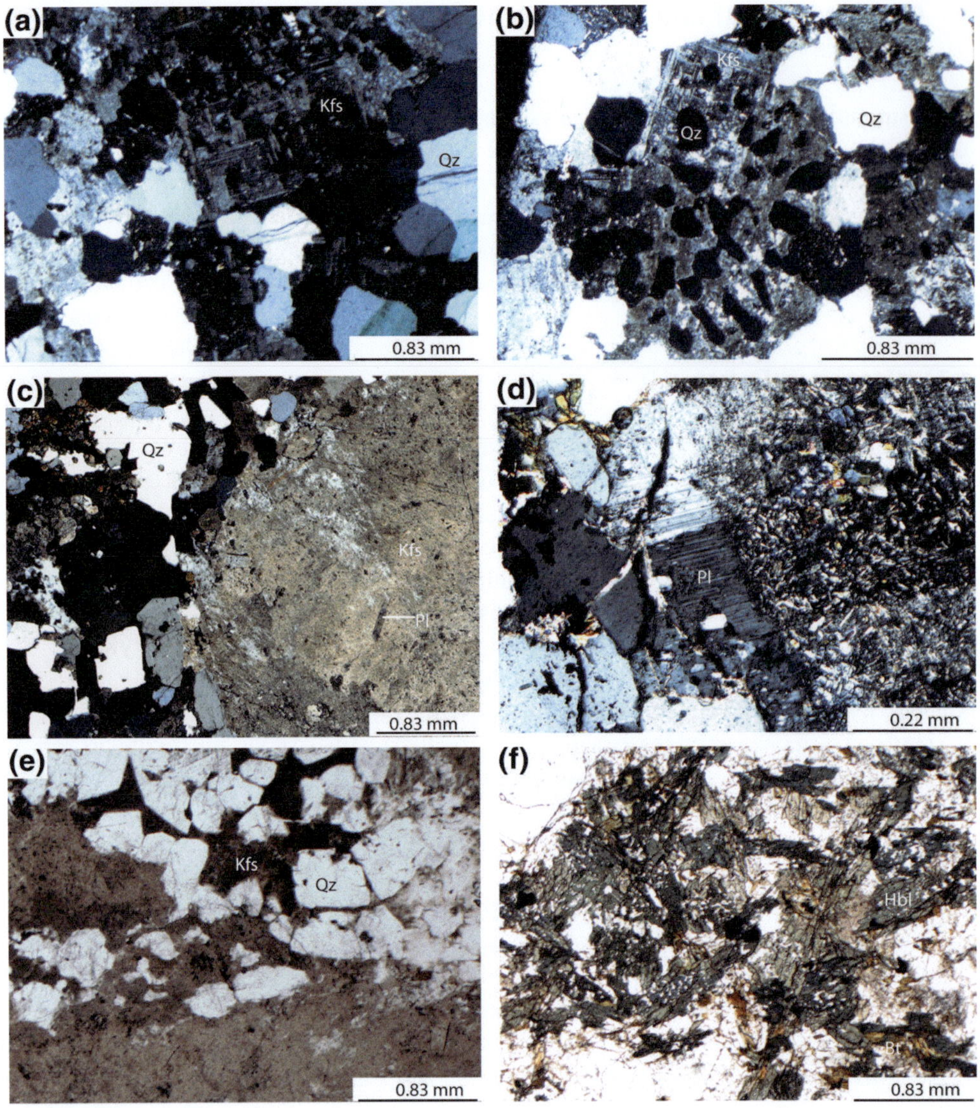

Fig. 6 **a** Tartan twinning in microcline (Kfs) (sample AP-05); crossed polars; **b** graphic intergrowth between microcline (Kfs) and quartz (Qz) (sample AP-05). Quartz shows subrounded faces against K-feldspar that suggest recrystallization; **c** very altered K-feldspar (Kfs) phenocryst that encloses fine-grained euhedral plagioclase (Pl) (sample AP-03); crossed polars; **d** plagioclase (Pl) rim free of replacing mineral phases such as white mica, chlorite, and epidote (sample AP-14). Note that locally epidote in the replacing aggregate is in well-formed crystals; crossed polars; **e** rounded quartz (Qz) crystals surrounded by interstitial K-feldspar (Kfs), close to a K-feldspar phenocryst (sample AP-03). K-feldspar is completely altered by very fine-grained aggregate mostly consisting of white mica. Plane polarised light; **f** massive aggregate of green hornblende (Hbl) and minor biotite (Bt) (sample AP-14). Plane polarised light

Clinopyroxene forms euhedral medium-grained crystals scattered in the groundmass and up to 4 mm sized crystals (Fig. 7e). Medium-grained crystals contain fine-grained ilmenite and apatite. Orthopyroxene occurs in rounded crystals scattered in the groundmass, enclosed in plagioclase phenocrysts (Fig. 7a), and enclosed in and aggregate with the large clinopyroxene crystals (Fig. 7e). Olivine occurs only in sample AP-01 as rounded crystals scattered in the rock matrix. Normally olivine grains are up to 0.5 mm in size with the exception of few mm-sized grains. Olivine shows a zoned rim that is partially replaced by serpentine and talc. In sample AP-08 mm-sized phenocrysts that are totally replaced by talc/serpentine and carbonates may represent former olivine. This kind of replacement occurs also within coarse-grained aggregates of clinopyroxene. Biotite occurs only in AP-01 in the form of a few μm-sized crystals in contact with olivine or its rimming reaction products. Sample AP-01 is characterised by a few mm-sized whitish aggregates consisting of very large-grained euhedral calcite and analcime (Fig. 7f). Analcime contains rare fine-grained K-feldspar, epidote, chlorite, and quartz. Glass is still preserved in sample AP-08 and in plane polarised light it looks yellowish. It is enclosed in coarse-grained plagioclase phenocrysts, both in the cores and rims (Fig. 7a), and is interstitial within plagioclase microlites of the groundmass. Medium-grained internal strain-free plagioclase laths show shape-preferred orientation (Fig. 7c, e). This evidence is consistent with a magmatic fluidal structure that wraps around plagioclase, clinopyroxene, and olivine phenocrysts. In AP-08 the fluidal structure is defined also by shape preferred orientation of clinopyroxene, and minor amphibole (Fig. 7c).

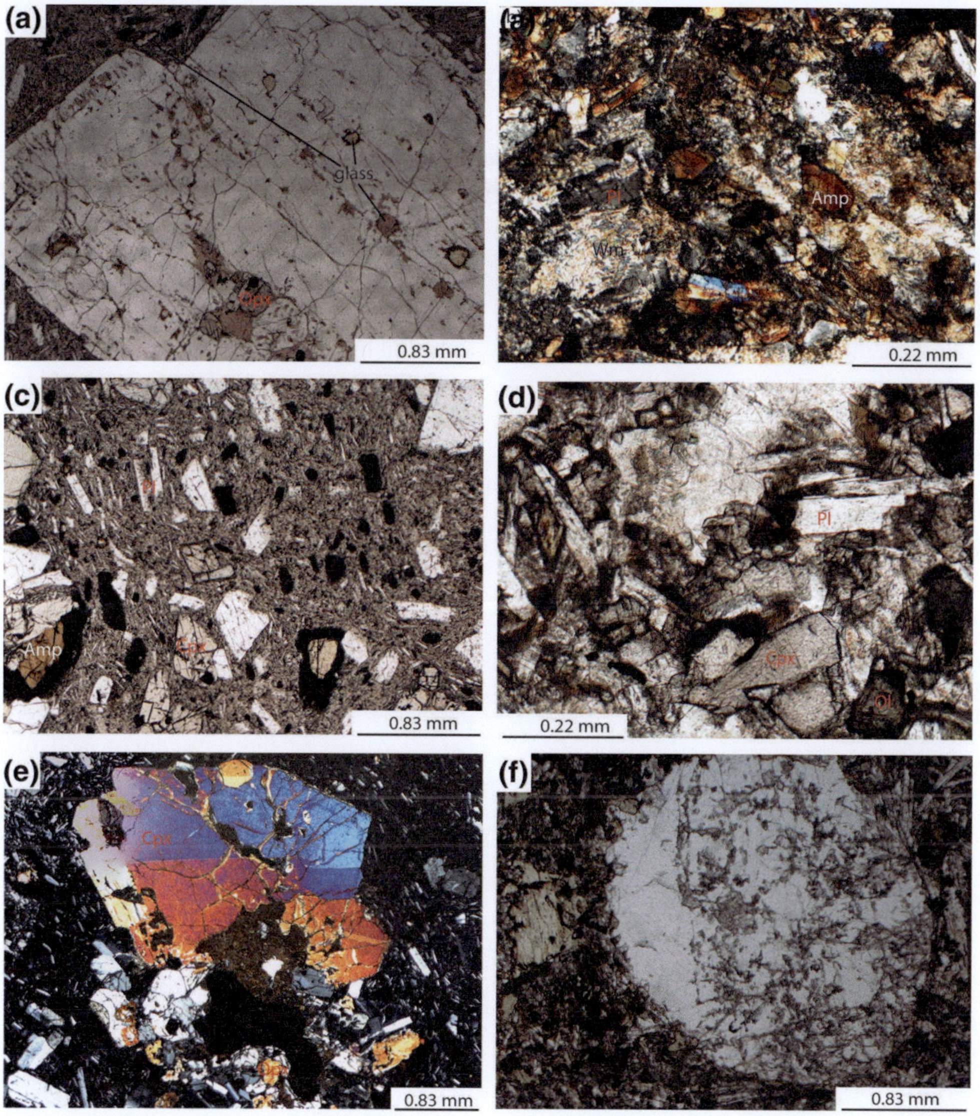

Fig. 7 **a** Euhedral plagioclase phenocryst that contains glass (yellowish patches) in the core and in the rim (sample AP-08). Phenocryst contains also rounded orthopyroxene (Opx); plane polarized light; **b** plagioclase (Pl) phenocrysts widely overgrown by white mica (Wm) (sample AP-01). Euhedral to sub-euhedral brown amphibole (Amp) grains are surrounded by the rock matrix; crossed polars; **c** brown amphibole (Amp) phenocrysts with a thick rim of opaque minerals (sample AP-08). Amphibole, plagioclase (Pl), and clinopyroxene (Cpx) show shape-preferred orientation defining the fluidal structure; plane polarized light; **d** euhedral plagioclase (Pl) laths and subhedral clinopyroxene (Cpx) (sample AP-01); plagioclase also forms interstitial grains; plane polarized light; **e** very coarse-grained clinopyroxene (Cpx) surrounded by the groundmass in which shape preferred orientation of fine-grained plagioclase defines a fluidal structure (sample AP-08). Finer-grained orthopyroxene (Opx) crystals occur at margins of and are contained in clinopyroxene; crossed polars (photo merging of photomicrographs, which results into an apparent plane within clinopyroxene); **f** rounded analcime crystals enclosing K-feldspar and epidote (sample AP-01); plane polarized light

4.3 Andesitic Dykes

AP-18, AP-19, and AP-20 dykes of intermediate composition are strongly altered and often recognising the original magmatic minerals and textures is difficult. These dykes are characterised by a porphyritic texture that is more evident in AP-18 and AP-20 and is defined by very altered plagioclase phenocrysts, and amphibole (Fig. 8a). Some of the mineral phases are recognizable by their shape and alteration products. Plagioclase phenocrysts are generally totally replaced by fine-grained aggregates of white mica and minor epidote. Amphibole (hornblende) phenocrysts are replaced by green amphibole (actinolite) and chlorite. Plagioclase also occurs as fine-grained laths that locally (AP-20) show shape-preferred orientation defining a fluidal structure. In the latter case, plagioclase shows polysynthetic growth twinning. Generally, the lath-shaped plagioclase crystals are only partially replaced by fine-grained white mica and epidote (Fig. 8b). Locally zoned interstitial grains of plagioclase are intergrown with albite and minor K-feldspar (Fig. 8c). In particular, K-feldspar occupies grain boundaries or forms skeletal microlites. Besides plagioclase, clinopyroxene is another surviving magmatic mineral (e.g., AP-18), along with minor euhedral and lozenge-shaped titanite crystals (e.g., AP-20). Clinopyroxene shows fractures filled by chlorite (Fig. 8d). Quartz and magnetite are pristine and form interstitial and euhedral crystals, respectively. Quartz shows undulose extinction that testifies to the effect of crystal plastic strain. In AP-18 fine-grained aggregates of white mica, epidote, and albite and green amphibole (actinolite), chlorite, and calcite totally replaced large phenocrysts. In AP-19 the phenocrysts are totally replaced by fine-grained chlorite aggregates with a rim of white mica. Fine-grained chlorite and epidote are diffused all over the rock matrix, likely as replacement of the original groundmass minerals. Epidote also forms rounded coarse-grained aggregates. Late-stage micro-veins, filled by calcite and minor chlorite, intersect the rock groundmass.

4.4 Rhyolitic Dykes

AP-16 and AP-17 are inequigranular porphyritic rhyolitic dykes containing K-feldspar phenocrysts less than 1 mm in size. These euhedral phenocrysts locally are grouped into aggregates (Fig. 9a) and show Carlsbad twinning and ubiquitously are rich in perthitic exsolution of albite (Fig. 9b). Perthitic albite forms blebs that are slightly elongated. The groundmass consists of up to 0.1 mm long crystals of K-feldspar, albite, quartz, and minor magnetite. Euhedral to subhedral K-feldspar in the rock matrix contains perthitic albite and displays a weak shape preferred orientation, especially in sample AP-16. Albite and quartz occur as interstitial grains in the rock matrix and are of roughly the same size as K-feldspar in the matrix. Magnetite forms euhedral to subhedral crystals diffused in the rock matrix (Fig. 9b). Magnetite that also grew in patches occurs within K-feldspar phenocrysts. In the latter case magnetite forms very fine-grained crystals (Fig. 9b) giving to K-feldspar a mottled appearance. However, K-feldspar can still be recognized locally by its distinct shape. In AP-17, along with magnetite, rare calcite patches also overgrew K-feldspar. Quartz occurs in the groundmass and forms subhedral or

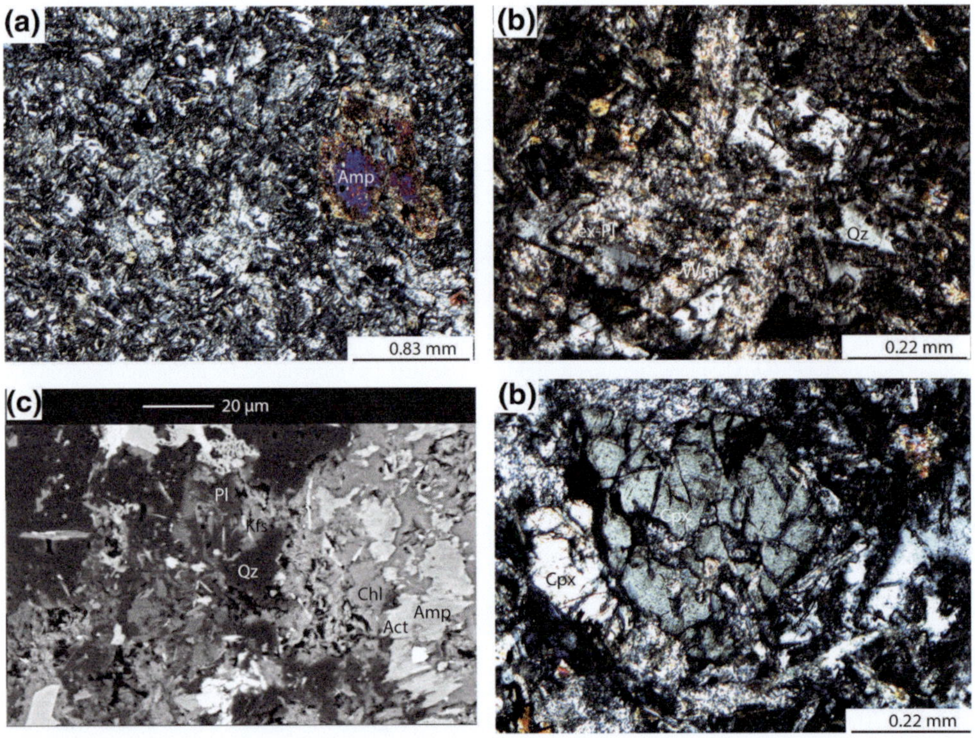

Fig. 8 **a** Relict amphibole (Amp) phenocryst that is completely replaced mainly by actinolite (sample AP-18). The original magmatic minerals in the rock matrix are hardly recognizable; crossed polars; **b** close up of Fig. 8a; in the rock matrix plagioclase (ex-Pl) laths widely replaced by white mica (Wm) are recognizable as well as fine grained amphibole; quartz (Qz) forms pristine interstitial grains; crossed polars; **c** relict amphibole (Amp) in the rock matrix replaced by actinolite (Act) and chlorite (Chl) (sample AP-18); K-feldspar (Kfs) marks the boundaries between plagioclase and quartz; backscattered electron image; **d** pristine clinopyroxene (Cpx) phenocryst intersected by chlorite-bearing fractures; crossed polars

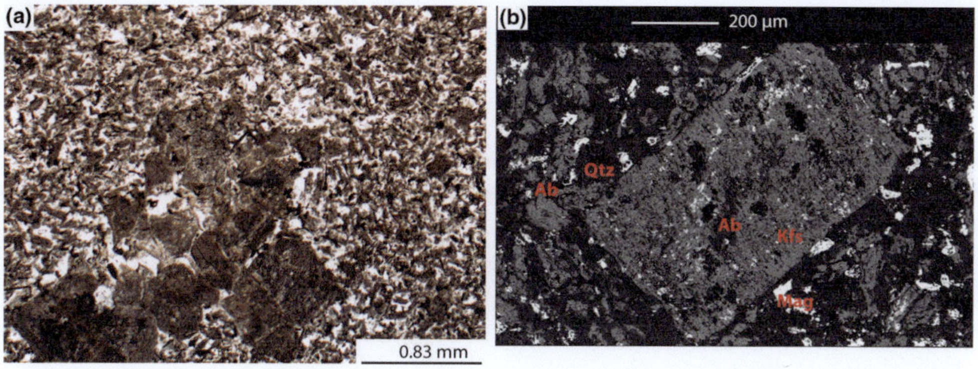

Fig. 9 **a** Aggregate of K-feldspar phenocrysts (mottled) with interstitial quartz (white patches) (sample AP-17); quartz is very diffused in the ground mass; plane polarized light; **b** euhedral K-feldspar (Kfs) phenocryst with perthitic albite (Ab) (sample AP-16); magnetite (Mag) forms very-fine grained crystal aggregates on K-feldspar phenocrysts and subhedral to euhedral crystals in the groundmass; backscattered electron image

interstitial crystals (Fig. 9b) that show very weak undulose extinction.

5 U–Pb Geochronology

5.1 Analytical Methods

We acquired preliminary U-Pb data in order to test the age of three granitoid dykes (AP-03, AP-05, and AP-14) that were selected because they contain zircon grains larger than 25 µm. U–Pb geochronology was carried out on zircon grains in polished thin section at the CNR–IGG UOS of Pavia (Italy) with Laser Ablation (LA)–ICP–MS (Table 1). The system couples an ArF excimer laser microprobe (type GeoLas102 from MicroLas) with a sector field ICP–MS (type Element from ThermoFinnigan).

The signals of ^{202}Hg, ^{204}(Pb + Hg), ^{206}Pb, ^{207}Pb, ^{208}Pb, ^{232}Th and ^{238}U masses were acquired. ^{202}Hg is acquired to correct the isobaric interference of ^{204}Hg on ^{204}Pb, so that the presence of common Pb in the sample can be monitored. However, the background of mass 204 is relatively high, due to traces of Hg in the He gas, and does not allow us to determine small amounts of common Pb. Remarkably, in the investigated samples the signal of ^{204}(Pb + Hg) was always indistinguishable from the background. The ^{235}U signal is calculated from ^{238}U on the basis of the ratio ^{238}U/^{235}U = 137.818 (Hiess et al. 2012).

U–Pb fractionation effects in zircon were simultaneously corrected using a matrix matched external standard and considering the same integration intervals on the standard and the unknowns. Analyses were carried out using a spot size of 25 µm, a frequency of 5 Hz and a laser fluence of 8 J cm^{-2}. The reference zircon GJ–1 (Jackson et al. 2004) was adopted as the external standard and the reference zircon 02123 (Ketchum et al. 2001) was selected as the validation standard (Table 2). Data reduction was carried out with the software package GLITTER® (Van Achterbergh et al. 2001). Individual uncertainties given by the software for the isotope ratios were propagated relative to the respective reproducibility of the standard. This procedure was carried out for each analytical run as reported in Horstwood et al. (2003). The ISOPLOT/Ex 3.00 software package by Ludwig (2003) was used for U–Pb apparent age calculations and representations.

5.2 LA-ICP-MS U-Pb Dating Results

Before U-Pb dating, zircon grains were localised at the SEM by acquiring BSE images of the textural site (Fig. 10). Zircon commonly occurs as euhedral to subhedral grains; inclusions are common and fractures, with or without fragment displacement, are also frequent. In the BSE images, zircon grains revealed zoning features, that locally are oscillatory (Fig. 10b–f). These features are coherent with the magmatic origin of the zircon grains. U-Pb isotopic data and calculated ages of samples and standards are reported in Tables 1 and 2, respectively. Eighteen analyses on seventeen zircon grains were collected for sample AP-05, and eleven of them yielded concordant ages in the range 602 ± 15 Ma to 650 ± 16 Ma (2σ error, Fig. 11). Nineteen analyses on fourteen zircon grains were collected for sample AP-03; U-Pb results were mainly discordant with only six concordant data at 604 ± 22 Ma, 689 ± 24 Ma, 732 ± 25 Ma, 735 ± 28 Ma, 748 ± 27 Ma, and 805 ± 30 Ma (Fig. 11). Only one zircon grain was analysed for the sample AP-14 and the two analysis results were concordant at 724 ± 18 Ma and 792 ± 21 Ma (Fig. 11).

Since this preliminary geochronological investigation revealed ages concerning the Arabian shield history, we did not proceed with a quantitative data acquisition on zircon

Table 1 U-Pb isotopic data and calculated ages for samples AP-03, AP-05, and AP-14

24th September 2015, IGG-CNR U.O.S. of Pavia					Data for Wetherill plot[3]					Ages[3]							Concordant ages	
Sample	Identifier	Zrc#	$^{207}Pb/^{206}Pb$	1 s abs	$^{207}Pb/^{235}U$	1 s abs	$^{206}Pb/^{238}U$	1 s abs	Rho	$^{207}Pb/^{206}Pb$	1 s abs	$^{207}Pb/^{235}U$	1 s abs	$^{206}Pb/^{238}U$	1 s abs	% U-Pb disc[4]		2 s abs
AP-05	Se24a006	1	0.06286	0.00117	0.88477	0.01827	0.10206	0.00154	0.7	703	13	644	13	626	9	2.7		
AP-05	Se24a007	2	0.06051	0.00094	0.88527	0.01608	0.10606	0.00158	0.8	622	10	644	12	650	10	−0.9	646	17
AP-05	Se24a008	3	0.06151	0.00093	0.89734	0.01573	0.10587	0.00149	0.8	657	10	650	11	649	9	0.2	650	16
AP-05	Se24a009	4	0.06065	0.00093	0.87581	0.01552	0.10475	0.00150	0.8	627	10	639	11	642	9	−0.5	640	16
AP-05	Se24a010	5	0.09040	0.00172	1.28106	0.02658	0.10276	0.00164	0.8	1434	27	837	17	631	10	24.7		
AP-05	Se24a011	6	0.05977	0.00095	0.81905	0.01506	0.09935	0.00146	0.8	595	9	608	11	611	9	−0.5	609	16
AP-05	Se24a012	7	0.06009	0.00092	0.82213	0.01461	0.09927	0.00144	0.8	607	9	609	11	610	9	−0.1	610	16
AP-05	Se24a013	8	0.05478	0.00084	0.77893	0.01401	0.10311	0.00151	0.8	403	6	585	11	633	9	−8.2		
AP-05	Se24a014	9	0.07576	0.00114	0.57120	0.01017	0.05465	0.00082	0.8	1089	16	459	8	343	5	25.2		
AP-05	Se24a015	10	0.06333	0.00128	0.86362	0.01910	0.09883	0.00152	0.7	719	15	632	14	608	9	3.9		
AP-05	Se24a016	11	0.05910	0.00088	0.80404	0.01389	0.09870	0.00138	0.8	571	8	599	10	607	8	−1.3	602	15
AP-05	Se24a017	12a	0.06037	0.00092	0.86481	0.01527	0.10395	0.00146	0.8	617	9	633	11	638	9	-0.8	635	16
AP-05	Se24a018	12b	0.06105	0.00097	0.85751	0.01562	0.10190	0.00144	0.8	641	10	629	11	626	9	0.5	627	16
AP-05	Se24a019	13	0.09201	0.00158	0.77158	0.01508	0.06079	0.00093	0.8	1468	25	581	11	380	6	34.5		
AP-05	Se24a020	14	0.07059	0.00099	0.71568	0.01193	0.07353	0.00103	0.8	946	13	548	9	457	6	16.6		
AP-05	Se24a021	15	0.05981	0.00109	0.82738	0.01696	0.10030	0.00149	0.7	597	11	612	13	616	9	−0.7	615	17
AP-05	Se24a022	16	0.06016	0.00099	0.84603	0.01572	0.10208	0.00144	0.8	609	10	622	12	627	9	-0.7	625	16
AP-05	Se24a023	17	0.06079	0.00105	0.85211	0.01643	0.10175	0.00144	0.7	632	11	626	12	625	9	0.2	625	16
AP-03	Se24b006	1	0.06468	0.00143	1.06700	0.03232	0.11996	0.00222	0.6	764	17	737	22	730	14	0.9	732	25
AP-03	Se24b007	2	0.12895	0.00239	2.34674	0.06497	0.13210	0.00245	0.7	2084	39	1227	34	800	15	34.8		
AP-03	Se24b008	3	0.24758	0.00433	5.11647	0.13800	0.15049	0.00271	0.7	3169	55	1839	50	904	16	50.9		
AP-03	Se24b009	4	0.23574	0.00443	4.28480	0.11801	0.13205	0.00236	0.6	3091	58	1690	47	800	14	52.7		
AP-03	Se24b010	5a	0.09653	0.00280	1.57762	0.05501	0.11846	0.00220	0.5	1558	45	961	34	722	13	24.9		
AP-03	Se24b011	5b	0.05984	0.00168	0.81068	0.02782	0.09828	0.00191	0.6	598	17	603	21	604	12	−0.2	604	22
AP-03	Se24b012	6	0.18133	0.00597	2.87623	0.10701	0.11514	0.00253	0.6	2665	88	1376	51	703	15	48.9		
AP-03	Se24b015	9	0.06608	0.00242	1.21282	0.05018	0.13303	0.00264	0.5	809	30	806	33	805	16	0.2	805	30
AP-03	Se24b017	11	0.12814	0.00227	2.14304	0.05853	0.12147	0.00221	0.7	2073	37	1163	32	739	13	36.4		
AP-03	Se24b020	14	0.11664	0.00224	2.25379	0.06346	0.14008	0.00253	0.6	1905	37	1198	34	845	15	29.5		
AP-03	Se24b022	16a	0.06414	0.00290	1.06791	0.05216	0.12080	0.00245	0.4	746	34	738	36	735	15	0.3	735	28
AP-03	Se24b023	16b	0.06974	0.00190	1.12459	0.03796	0.11688	0.00213	0.5	921	25	765	26	713	13	6.9		
AP-03	Se24b024	17	0.06496	0.00128	1.03918	0.02965	0.11599	0.00205	0.6	773	15	724	21	707	13	2.2		

(continued)

Table 1 (continued)

24th September 2015, IGG-CNR U.O.S. of Pavia						Data for Wetherill plot[3]				Ages[3]							Concordant ages	
Sample	Identifier	Zrc#	$^{207}Pb/^{206}Pb$	1 s abs	$^{207}Pb/^{235}U$	1 s abs	$^{206}Pb/^{238}U$	1 s abs	Rho	$^{207}Pb/^{206}Pb$	1 s abs	$^{207}Pb/^{235}U$	1 s abs	$^{206}Pb/^{238}U$	1 s abs	% U-Pb disc[4]		2 s abs
AP-03	Se24b025	18a	0.06442	0.00228	1.09203	0.04425	0.12290	0.00237	0.5	755	27	749	30	747	14	0.3	748	27
AP-03	Se24b026	18b	0.07580	0.00175	1.30455	0.03988	0.12478	0.00220	0.6	1090	25	848	26	758	13	10.6		
AP-03	Se24b027	18c	0.07620	0.00418	1.34725	0.07694	0.12806	0.00291	0.4	1100	60	866	49	777	18	10.3		
AP-03	Se24b028	19a	0.14380	0.00255	2.38235	0.06504	0.11992	0.00216	0.7	2273	40	1237	34	730	13	41.0		
AP-03	Se24b029	19b	0.06224	0.00201	0.96873	0.03666	0.11290	0.00211	0.5	682	22	688	26	690	13	−0.3	689	24
AP-03	Se24b030	20	0.25056	0.00418	5.74509	0.15170	0.16628	0.00289	0.7	3188	53	1938	51	992	17	48.8		
AP-14	Se24c008	1a	0.06374	0.00167	1.04461	0.02924	0.11886	0.00159	0.5	733	19	726	20	724	10	0.3	724	18
AP-14	Se24c009	1b	0.06527	0.00174	1.17803	0.03399	0.13085	0.00191	0.5	783	21	790	23	793	12	−0.3	792	21

Table 2 Standards used for U-Pb geochronology

24 September 2015, IGG-CNR U.O.S. of Pavia			$^{207}Pb/^{206}Pb$			$^{207}Pb/^{235}U$			$^{206}Pb/^{238}U$			Ages					
Identifier	Comments	Position within the run	Ratio	1 s %	1 s % (Prop)	Ratio	1 s %	1 s % (Prop)	Ratio	1 s %	1 s % (Prop)	$^{207}Pb/^{206}Pb$	1 s abs	$^{207}Pb/^{235}U$	1 s abs	$^{206}Pb/^{238}U$	1 s abs
Se24a001	GJ-1	Begin	0.05996	0.06%	0.08%	0.7994	0.92%	1.31%	0.0967	0.12%	0.13%	602	8	596	10	595	8
Se24a002	GJ-1	Begin	0.06007	0.06%	0.09%	0.8147	0.96%	1.35%	0.0984	0.12%	0.14%	606	9	605	10	605	8
Se24a003	GJ-1	Begin	0.06041	0.06%	0.09%	0.8186	0.98%	1.37%	0.0983	0.12%	0.14%	618	9	607	10	604	8
Se24a004	GJ-1	Begin	0.06024	0.06%	0.09%	0.8111	0.96%	1.35%	0.0977	0.12%	0.14%	612	9	603	10	601	8
Se24a026	GJ-1	End	0.06080	0.07%	0.09%	0.814	1.06%	1.43%	0.0971	0.12%	0.14%	632	10	605	11	598	8
Se24a027	GJ-1	End	0.05897	0.07%	0.09%	0.7941	1.06%	1.40%	0.0977	0.12%	0.14%	566	9	593	10	601	8
Se24a028	GJ-1	End	0.06055	0.07%	0.09%	0.8175	1.08%	1.44%	0.0979	0.12%	0.14%	623	10	607	11	602	8
Se24b002	GJ-1	Begin	0.06088	0.07%	0.10%	0.8315	1.06%	2.18%	0.099	0.12%	0.17%	635	11	614	16	609	10
Se24b003	GJ-1	Begin	0.05891	0.07%	0.10%	0.7843	1.03%	2.07%	0.0965	0.12%	0.17%	564	9	588	16	594	10
Se24b004	GJ-1	Begin	0.06061	0.07%	0.10%	0.8167	1.08%	2.16%	0.0977	0.13%	0.17%	625	10	606	16	601	10
Se24b032	GJ-1	End	0.06036	0.07%	0.10%	0.8208	1.13%	2.20%	0.0986	0.13%	0.17%	617	11	608	16	606	10
Se24b034	GJ-1	End	0.06047	0.07%	0.10%	0.8204	1.13%	2.19%	0.0984	0.13%	0.17%	620	11	608	16	605	10
Se24b035	GJ-1	End	0.05957	0.07%	0.10%	0.7914	1.08%	2.11%	0.0963	0.12%	0.17%	588	10	592	16	593	10
Se24c004	GJ-1	Begin	0.05991	0.08%	0.16%	0.8122	1.17%	2.36%	0.0983	0.13%	0.14%	600	16	604	18	605	9
Se24c005	GJ-1	Begin	0.06022	0.08%	0.16%	0.8044	1.15%	2.34%	0.0969	0.12%	0.14%	611	17	599	17	596	8
Se24c014	GJ-1	End	0.06210	0.08%	0.17%	0.8371	1.24%	2.46%	0.0978	0.12%	0.14%	678	19	618	18	601	8
Se24c015	GJ-1	End	0.05858	0.08%	0.16%	0.7878	1.18%	2.32%	0.0976	0.12%	0.14%	552	15	590	17	600	8
Se24a024	02123	End	0.05241		0.21%	0.35045		1.41%	0.04849		0.08%	303	12	305	12	305	5
Se24b005	02123	Begin	0.05529		0.10%	0.35516		0.97%	0.04656		0.08%	424	8	309	8	293	5
Se24c006	02123	Begin	0.05184		0.20%	0.33359		1.30%	0.04674		0.08%	278	11	292	11	294	5

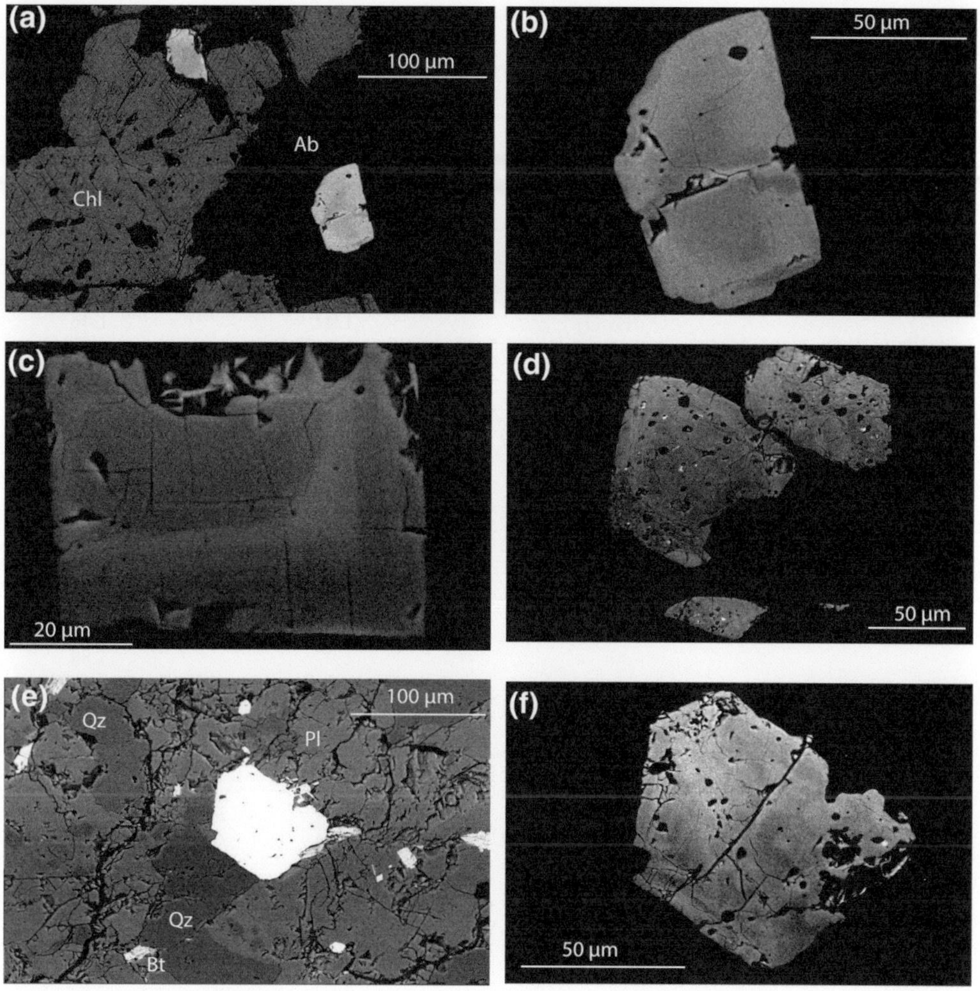

Fig. 10 Backscattered electron images of zircon grains; **a** zircon grain in a groundmass of albite (Ab) and chlorite (Chl) (sample AP-03); **b** close-up of the zircon grain that shows zoning features (sample AP-03); **c** zircon grain with zoning features (sample AP-05); **d** zircon grains containing quartz-inclusions (sample AP-14); **e** euhedral zircon grain surrounded by quartz (Qz) and plagioclase (Pl) crystals, with minor biotite (Bt = small white grains) (sample AP-14); **f** close-up of the image "e" showing zoning features in zircon grain

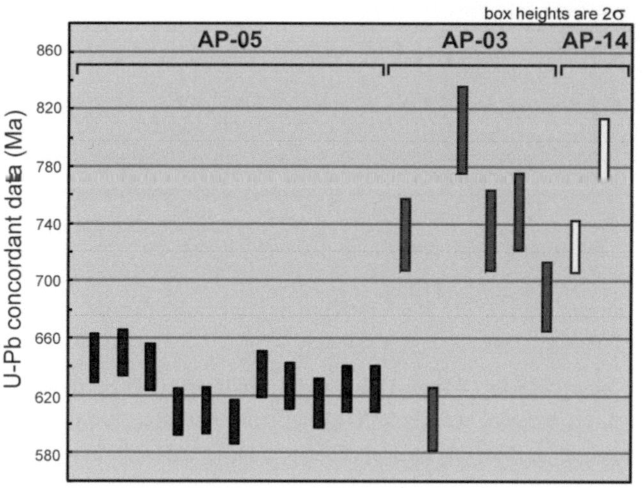

Fig. 11 Distribution diagram of U-Pb concordant age data for samples AP-03, AP-05, and AP-14

separates, as usually it is done for geochronological investigation of intrusive rocks. Even though the emplacement ages cannot be constrained from the collected data, U-Pb results suggest a Neoproterozoic origin for the investigated samples. Tentatively, we can envisage an intrusive age between late Cryogenian and early Ediacaran for the AP-05 dyke, whereas the scattered concordant U-Pb data obtained for samples AP-03 and AP-14 might be due to the inheritance signature of the country rocks related to the complex tectonic history of the Arabian shield.

6 Mineral Chemistry

Mineral chemical compositions were acquired by energy dispersive X-ray spectroscopy (EDS) analysis performed using the Cambridge Stereoscan 360 scanning electron microscope (SEM) of CNR Milano (Italy), operating at the Earth Science Department of Milano University. Six dyke samples were analysed: two basaltic (AP-01 and AP-08), two andesitic (AP-18 and AP-19), and two rhyolitic (AP-16 and AP-17) dykes. Mineral stoichiometric formulae were recalculated from the weight percent of element oxides on the basis of the following numbers of oxygen atoms: 4 for

Table 3 Representative analyses of amphibole. Sample label: first 5 digits are sample number, last three digits are point analyses

Amphibole			
Sample label	AP_01_13	AP_08_57	AP_18_20
SiO_2	40.07	42.25	51.02
TiO_2	5.37	3.67	0.74
Al_2O_3	12.11	13.17	4.19
Cr_2O_3	0.00	0.27	0.03
MnO	0.33	0.30	0.48
FeO	16.05	9.56	10.84
Fe_2O_3	1.92	1.44	4.39
NiO	0.02	0.13	0.00
MgO	8.79	13.56	14.36
CaO	11.80	12.04	12.51
Na_2O	0.94	1.90	0.27
K_2O	0.94	0.53	0.19
Total	98.33	98.83	99.02
T site			
Si	6.03	6.12	7.32
Al	1.97	1.88	0.68
C site			
Ti	0.61	0.40	0.08
Al	0.18	0.36	0.03
Cr	0.00	0.03	0.00
Fe^{3+}	0.22	0.16	0.48
Ni	0.00	0.02	0.00
Mn^{2+}	0.00	0.00	0.04
Fe^{2+}	2.02	1.11	1.30
Mg	1.97	2.93	3.07
B site			
Mn^{2+}	0.04	0.04	0.02
Fe^{2+}	0.00	0.05	0.00
Ca	1.90	1.87	1.92
Na	0.05	0.05	0.05
A site			
Ca	0.00	0.00	0.00
Na	0.22	0.49	0.02
K	0.18	0.10	0.03

olivine and spinel, 6 for pyroxene, 7 for serpentine, 8 for feldspars, 12.5 for epidote, 22 for biotite, white mica, and talc, 23 for amphibole, 28 for chlorite. Fe^{3+} was recalculated for epidote, pyroxene, amphibole, and spinel on the basis of charge balance. In particular, for pyroxene, amphibole, and spinel, the calculation schemes of Morimoto (1988), Locock (2014), and Droop (1987) were followed, respectively. Hereafter the compositional variation of each mineral phase is described in comparison with the different samples and in relation to the microstructural site.

Amphibole (Table 3) has been analysed in two basaltic dykes (samples AP-01 and AP-08) and in one andesitic dyke (sample AP-18). Amphibole in sample AP-08 shows the highest content in Na that is variable between 0.95 and 1.06 atoms per formula unit (a.p.f.u.) and the lowest in Ca that varies between 1.82 and 1.85 a.p.f.u. In contrast, amphibole in sample AP-18 shows the highest Ca (1.50–2.00 a.p.f.u.), and the lowest Na (0.11 and 0.68 a.p.f.u.) concentration. Amphibole in AP-01 shows intermediate values of Na (0.53–0.56 a.p.f.u.) and Ca (1.89–197 a.p.f.u.) with respect to amphibole in samples AP-08 and AP-18 (Fig. 12a). However, amphibole in AP-01 reaches the highest content in Al (1.86–2.72 a.p.f.u.) and Ti (0.42–0.62 a.p.f.u.), with maximum values recorded in the core of the grains. Al (0.46–1.47 a.p.f.u.) and Ti (0.01–0.44 a.p.f.u.) contents show the lowest values in amphibole from sample AP-18 (Fig. 12b). Amphibole in sample AP-08 has an intermediate composition of Al (2.07–2.23 a.p.f.u.) and Ti (0.40–0.46 a.p.f.u.) with respect to amphibole in samples AP-01 and AP-18 (Fig. 12b). In sample AP-18 amphiboles forming both phenocrysts and matrix crystals are mostly actinolite. Since the sample is affected by a strong alteration, it is suggested that the actinolite derives from the alteration of primary amphibole (hornblende). Amphibole from samples AP-01 and AP-08 shows Si content between 5.67 and 6.33 a.p.f.u. and between 6.11 and 6.12 a.p.f.u. and Mg content between 1.87 and 2.48 a.p.f.u. and 2.93 and 3.13 a.p.f.u., respectively.

Clinopyroxene and orthopyroxene (Table 4) have been analysed in samples AP-01, AP-08, and AP-18 and consist of augite and diopside, and enstatite, respectively (Fig. 12c). Clinopyroxene in sample AP-01 is the richest in Ca (0.92–0.94 a.p.f.u.) and consists of diopside. The Ca in clinopyroxene in samples AP-08 (0.74–0.89 a.p.f.u.) and AP-18 (0.79–0.83 a.p.f.u.) has similar concentrations and is consistent with pyroxene having an augite composition. In sample AP-08 only one core of a crystal defining the fluidal structures has diopside composition. In the same sample, finer-grained augite grains defining the fluidal structure show the lowest content in X_{Mg} ($Mg/(Mg + Fe^{2+})$) and Ca, whereas coarse-grained augite grains record the highest X_{Mg} (Fig. 12d). X_{Mg} is lower and similar in clinopyroxene from samples AP-01 and AP-18 and it is higher in the grain core in both samples (Fig. 12d). Orthopyroxene (sample AP08) shows the highest X_{Mg} ratio (0.78–0.82) in grains associated with the coarse grained clinopyroxene, whereas crystals in the rock matrix and enclosed in plagioclase phenocrysts show X_{Mg} ranging between 0.68 and 0.77 (Fig. 12d).

Feldspar (Table 5) was analysed in all samples and consists of plagioclase and alkali feldspar (Fig. 12e). Coarser-grained plagioclase from sample AP-01 reaches the highest content in anorthite (An) and plots in the compositional field of bytwonite (An = 72–78%). These crystals show a rim of oligoclase. In the same sample, finer-grained euhedral laths show a labradoritic composition (An = 57%), whereas interstitial plagioclase mostly consists of oligoclase and minor albite (An = 2–14%). In sample AP-08 the composition of plagioclase varies from labradorite to oligoclase. Most of the plagioclase crystals in this sample show an andesinic composition, such as the coarse-grained phenocrysts (An = 40–43%). Finer-grained crystals show a similar composition with anorthite varying from 40 and 42%. The highest anorthite content is reached by plagioclase crystals enclosed in amphibole (An = 58%) and coarse-grained pyroxene (An = 46%). The plagioclase microlites preserved in micro-glass pockets show an anorthite content between 20 and 32%. In sample AP-18 plagioclase shows andesinic and albitic composition. Andesinic plagioclase forms fine-grained crystals in the rock matrix and shows an increase of anorthite from core (33%) to rim (47%). Albitic plagioclase forms interstitial crystals and shows content in anorthite variable between 1 and 10%. In this sample K-feldspar microlites show content in orthoclase between 93 and 97%. Sample AP-19 (andesite) shows plagioclase with composition varying from labradorite to albite. A few dozen micron-sized laths show anorthite variable between 58 and 65% with a few exceptions of 29% anorthite. These laths show a rim of albite with an anorthite content of 10%. Samples AP-16 and AP-17 (rhyolite) contain only alkali feldspar that consists of orthoclase and albite (Fig. 12e). In sample AP-17 K-feldspar phenocrysts and microlites are 100% orthoclase, whereas in sample AP-16 K-feldspar phenocrysts and microlites show content of orthoclase ranging between 96 and 97%. Both in K-feldspar phenocrysts and laths, perthitic exsolution shows albite contents varying from 95 and 99%, with the exception of a few cases in which perthite consists of 78% of albite and 22% of orthoclase. This last example may suggest that locally the subsolidus unmixing is not complete. Fine-grained albite in the rock matrix consists of 98 and 99% of albite.

Olivine (Table 6) occurs only in sample AP-01 (basalt) and shows an X_{Mg} variable between 0.69 and 0.81, with lower values detected from crystal rims. Higher amount of iron in olivine rims than cores suggest that olivine crystallised during changes in residual melt composition. Biotite (Table 6) was analysed in sample AP-01 as well and is characterised by X_{Mg} between 0.40 and 0.51, with lower values characterising biotite in fine-grained acicular grains and higher values in biotite facing the coronitic reaction product of olivine destabilisation.

Spinel consists of magnetite with an X_{Fe}^{3+} ($X_{Fe}^{3+}/(X_{Fe}^{3+} + Cr + Al)$) ratio varying between 0.91 and 0.99 (Fig. 13 a). In basaltic and andesitic dyke samples (AP-08 and AP-19, respectively) magnetite shows significant Ti content of 0.28 and 0.19 a.p.f.u., respectively (Fig. 13b). In samples AP-16 and AP-17 (rhyolite dykes) magnetite forming euhedral crystals in the rock matrix shows the highest values of X_{Fe}^{3+} compared to very fine-grained magnetite occurring in patches within K-feldspar phenocrysts. These magnetite patches contain higher X_{Cr} ($Cr/(Cr + X_{Fe}^{3+} + Al)$) (0.06) than euhedral crystals (0.00–0.03).

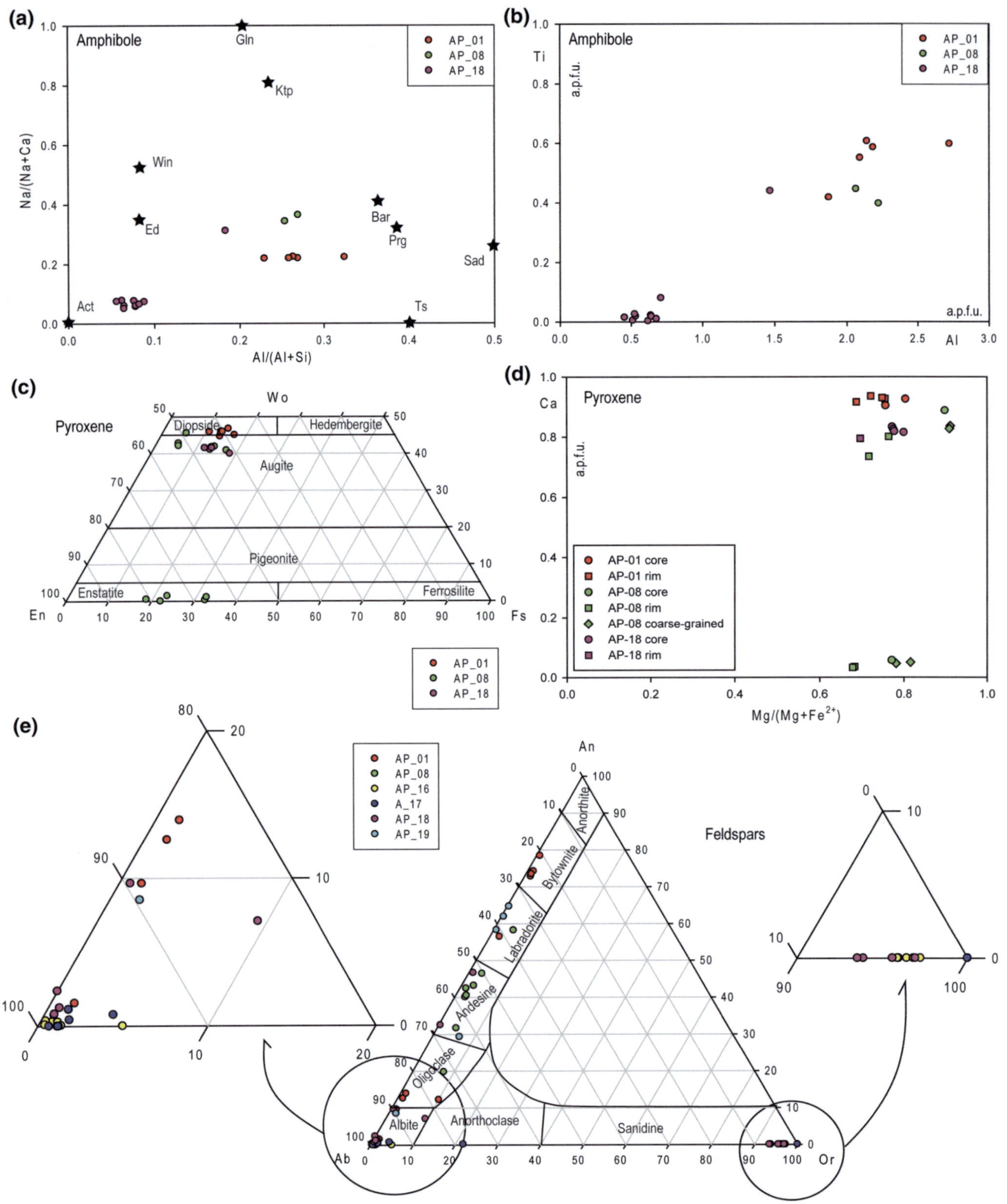

Fig. 12 **a** Binary diagram for amphibole showing Na/(Na + Ca) versus Al/(Al + Si). Black stars locate the end-member compositions according to Leake et al. (1997): Act = actinolite, Ts = tschermakite, Sad = sadanagaite, Prg = pargasite, Bar = barroisite, Ktp = katophorite, Gl = glaucophane, Win = winchite, Ed = edenite; **b** binary diagram showing the variation of Ti versus Al in amphibole; **c** classification ternary diagram Wo (wollastonite) versus En (enstatite) versus Fs (ferrosilite) in pyroxene (Morimoto 1988); **d** binary diagram showing the variation of Ca versus Mg/Mg + Fe^{2+} in pyroxene; **e** classification ternary diagram of feldspar; An = anorthite, Ab = albite, Or = orthoclase

Table 4 Representative analyses of pyroxene and epidote. Sample label: first 5 digits are sample number, last three digits are point analyses

Piroxene					Epidote	
Sample label	AP_01_20	AP_08_54	AP_08_58	AP_18_29	AP_18_83	AP_19_98
K_2O	0.03	0.00	0.01	0.00	0.05	0.09
CaO	23.44	0.82	21.18	20.78	23.72	23.05
TiO_2	1.69	0.16	1.13	0.55	0.00	0.10
Cr_2O_3	0.00	0.05	0.28	0.07	0.03	0.00
MnO	0.07	0.48	0.23	0.16	0.27	0.00
FeO	8.75	20.08	2.70	7.69	0.00	0.00
Fe_2O_3	0.00	0.00	4.44	2.05	15.89	13.95
NiO	0.01	0.03	0.31	0.20	0.00	0.00
Na_2O	0.00	0.15	0.70	0.25	0.02	0.00
SiO_2	50.42	54.08	50.15	51.37	38.16	36.70
Al_2O_3	3.49	0.74	4.71	1.85	22.01	21.88
MgO	12.73	23.67	15.33	14.64	0.01	0.11
TOTAL	100.61	100.27	100.74	99.40	100.16	95.87
Si	1.88	1.99	1.83	1.92	3.00	2.99
Al^{IV}	0.12	0.01	0.17	0.08	0.00	0.01
Al^{VI}	0.03	0.02	0.03	0.00	2.04	2.10
Ti	0.05	0.00	0.03	0.02	0.00	0.01
Cr	0.00	0.00	0.01	0.00	0.00	0.00
Fe^{3+}	0.00	0.00	0.12	0.06	0.94	0.86
Fe^{2+}	0.27	0.62	0.08	0.24	0.00	0.00
Mg	0.71	1.30	0.83	0.82	0.00	0.01
Ni	0.00	0.00	0.01	0.01	0.00	0.00
Mn	0.00	0.02	0.01	0.01	0.02	0.00
Ca	0.94	0.03	0.83	0.83	2.00	2.02
Na	0.00	0.01	0.05	0.02	0.00	0.00
K	0.00	0.00	0.00	0.00	0.00	0.01

Chlorite (Table 6) plots in the compositional field of pycnochlorite with minor compositions in the field of diabanite in sample AP-01 (Fig. 13c). In samples AP-18 and AP-19 (andesitic dyke) X_{Fe} varies between 0.43 and 0.49, with the highest values recorded in lamellae that overgrew amphibole phenocrysts and lower values in fine grained aggregates that completely replaced former phenocrysts. Si comprises between 5.76 and 6.24 a.p.f.u. and characterises chlorite associated with epidote aggregates and that overgrew the rock matrix, respectively. Chlorite in sample AP-01 shows lower X_{Fe} and much higher Si than in samples AP-18 and AP-19 (Fig. 13c).

Epidote (Fig. 13d, Table 4) was analysed in samples AP-18 and AP-19 (andesitic dyke) and shows a composition characterised by an X^{3+}_{Fe} ($Fe^{3+}/(Fe^{3+} + Al^{VI} + Cr + Ti)$) varying between 0.28 and 032, with higher values obtained from sample AP-18. The only exception is a crystal core in which iron is very low ($X^{3+}_{Fe} = 0.02$).

White mica (Table 6) has Si content between 6.11 and 6.58 a.p.f.u., Al between 4.64 and 5.73 a.p.f.u., and Fe + Mg between 0.15 and 1.08 a.p.f.u. (Fig. 13e). The highest and lowest values are recorded in sample AP-19 (andesitic dyke) in grains associated with fine-grained epidote. White mica with the lowest Si value shows the lowest Fe + Mg and highest Al values. Sample AP-18 records the lowest Al value in white mica forming fine-grained aggregates with epidote diffused in the rock matrix. The highest Fe + Mg value is recorded in white mica replacing coarse-grained plagioclase phenocrysts in sample AP-01, although, in this sample, white mica shows values as low as 0.33 and 0.66 a.p.f.u. Paragonite content is mostly zero and is detected in white mica from sample AP-18 ($X_{Pg} = 0.02$–0.03) in grains replacing plagioclase. The content in muscovite is higher than that of paragonite, according to X_{Ms} (Green and Usdansky 1986) that varies between 0.62 and 0.96 (Fig. 13f). The lowest X_{Ms} values are recorded in white mica

Table 5 Representative analyses of feldspars. Sample label: first 5 digits are sample number, last three digits are point analyses

Feldspar							
Sample label	AP_01_6	AP_01_24	AP_08_43	AP_08_81	AP_16_65	AP_16_70	AP_19_91
K_2O	0.20	0.39	0.28	1.39	15.82	0.05	0.10
CaO	14.74	11.30	9.30	4.30	0.00	0.07	11.88
TiO_2	0.02	0.14	0.07	0.08	0.12	0.00	0.13
Cr_2O_3	0.00	0.00	0.16	0.00	0.00	0.00	0.00
MnO	0.08	0.00	0.00	0.00	0.04	0.03	0.09
FeO	0.59	0.28	0.30	0.48	0.14	0.03	1.06
NiO	0.07	0.10	0.04	0.00	0.19	0.00	0.00
Na_2O	2.74	4.55	6.79	8.73	0.43	11.58	3.97
SiO_2	50.48	55.66	56.54	65.42	64.62	67.97	52.84
Al_2O_3	30.81	28.54	26.71	22.11	18.35	18.82	28.47
MgO	0.03	0.00	0.00	0.03	0.00	0.00	0.06
TOTAL	99.75	100.96	100.19	102.54	99.72	98.55	98.60
Si	2.31	2.49	2.55	2.84	2.99	3.01	2.43
Al	1.66	1.50	1.42	1.13	1.00	0.98	1.54
Ti	0.00	0.00	0.00	0.00	0.00	0.00	0.00
Cr	0.00	0.00	0.01	0.00	0.00	0.00	0.00
Fe^{2+}	0.02	0.01	0.01	0.02	0.01	0.00	0.04
Mn	0.00	0.00	0.00	0.00	0.00	0.00	0.00
Mg	0.00	0.00	0.00	0.00	0.00	0.00	0.00
Ni	0.00	0.00	0.00	0.00	0.01	0.00	0.00
Ca	0.72	0.54	0.45	0.20	0.00	0.00	0.59
Na	0.24	0.39	0.59	0.73	0.04	0.99	0.35
K	0.01	0.02	0.02	0.08	0.94	0.00	0.01
X_{An}	74.84	57.85	43.08	21.40	0.00	0.33	62.32
X_{Or}	1.19	2.32	1.52	7.61	96.03	0.28	0.62
X_{Ab}	24.86	41.17	56.05	72.62	3.97	99.39	37.45

replacing plagioclase in sample AP-01 and the highest values in white mica associated with epidote in sample AP-19.

Serpentine and talc form the coronitic reaction products that rim olivine in sample AP-01. Serpentine shows an X_{Mg} ratio of 0.85 and in talc the X_{Mg} ratio varies between 0.64 and 0.76. In sample AP-08 talc is associated with carbonates, forms the reaction product of coarse-grained completely replaced phenocrysts, and domains enclosed in coarse-grained clinopyroxene probably representing former olivine. In this sample talc is characterised by an X_{Mg} value varying between 0.65 and 0.74.

Glass (Table 7) occurs only in sample AP-08 (basaltic dyke) and shows a variable composition in different microstructural sites. Glass trapped within both core and rim of plagioclase phenocrysts shows the highest SiO_2, Na_2O, and K_2O contents, although plagioclase phenocrysts also enclose glass with 68 weight percent of SiO_2 and 0.5 weight percent of $Na_2O + K_2O$. Glass with similar composition occurs also in the groundmass of nearby plagioclase phenocrysts. Glass with high SiO_2 also occurs in the groundmass next to amphibole and clinopyroxene.

7 Whole Rock Geochemistry

Seven samples were sent for whole rock geochemistry analysis to Activation Laboratories in Ancaster, Ontario, Canada. Major and trace elements (Table 8) were analysed via lithium metaborate/tetraborate fusion with ICP and ICP/MS (inductively coupled plasma mass spectrometry).

The TAS (total alkali versus silica) diagram reveals that basic dyke samples AP-01 and AP-08 consist of basanite and basalt, respectively (Fig. 14). Intermediate dyke samples AP-18, AP-19, and AP-20 have andesitic composition and in particular sample AP-19 is basaltic andesite, sample AP-20 is trachyandesite, and sample AP-18 is andesite.

Table 6 Representative analyses of olivine, biotite, chlorite, and white mica. Sample label: first 5 digits are sample number, last three digits are point analyses

	Olivine	Biotite	Chlorite				White mica	
Sample label	AP_01_35	AP_01_10	AP_01_41	AP_18_84	AP_18_93	AP_19_90	AP_01_1	AP_19_83
SiO_2	37.88	36.00	35.17	30.22	25.76	30.45	47.71	48.04
TiO_2	0.00	1.85	0.13	0.10	0.00	0.17	0.06	0.10
Al_2O_3	0.08	18.79	13.44	18.72	17.06	18.57	29.07	29.80
FeO	28.04	20.46	19.67	23.58	21.76	22.69	4.50	3.24
MnO	0.46	0.24	0.19	0.54	0.52	0.23	0.00	0.00
MgO	34.76	7.58	17.68	17.85	14.30	15.40	2.83	1.31
CaO	0.45	0.29	0.22	0.21	0.24	0.39	2.06	0.00
Na_2O	0.00	0.00	0.00	0.54	0.53	0.00	0.00	0.00
K_2O	0.05	10.30	0.43	0.01	0.01	0.08	8.64	10.73
SO_3	0.14	0.00	0.14	0.02	0.00	0.00	0.00	0.10
NiO	0.06	0.04	0.09	0.14	0.00	0.00	0.00	0.04
Sum	101.92	95.55	87.16	91.93	80.18	87.98	94.87	93.36
Si	0.99	5.52	7.13	5.98	5.89	6.24	6.46	6.58
Ti	0.00	0.21	0.02	0.01	0.00	0.03	0.01	0.01
Al	0.00	3.39	3.21	4.37	4.60	4.49	4.64	4.81
Fe^{2+}	0.62	2.62	3.33	3.90	4.16	3.89	0.51	0.37
Mn	0.01	0.03	0.03	0.09	0.10	0.04	0.00	0.00
Mg	1.36	1.73	5.34	5.27	4.88	4.71	0.57	0.27
Ca	0.01	0.05	0.05	0.04	0.06	0.09	0.30	0.00
Na	0.00	0.00	0.00	0.21	0.24	0.00	0.00	0.00
K	0.00	2.01	0.11	0.00	0.00	0.02	1.49	1.87
S	0.01	0.00	0.04	0.01	0.00	0.00	0.00	0.02
Ni	0.00	0.00	0.01	0.02	0.00	0.00	0.00	0.00

Acidic dyke samples AP-16 and AP-17 are subalkaline rhyolite (Fig. 14). In the TAS diagram, data from Cenozoic magmatic rocks from different localities of the Arabian plate are compared with the data presented in this chapter (Fig. 14). Samples AP-01 and AP-08 plot in the alkaline field and show a similar composition to volcanic rocks from Harrat Lunayyir (Duncan and Al-Amri 2013), the Jordanian part of the Harrat Ash Shaam, which are younger than 8 Ma (Shaw et al. 2003), and from a region 100 km south of Aleppo in Syria (Krienitz et al. 2009). Andesitic samples show a composition quite close to that of magmatic rocks from the central coastal plain of Arabia. Samples AP-18 and AP-19 have a composition quite close to that of dykes of the Gumayqah complex, which are dated between 21 and 24 Ma (Pallister 1987), and sample AP-20 shows a composition quite close to that of Burgatinah diorite (Pallister 1987). Samples AP-16 and AP-17 are from dykes close to the Ad Damm fault zone and they display a similar composition to other dykes and volcanic rocks of the Sita formation. At Jabal Sita, which is located about 25 km west from sampling sites of samples AP-16 and AP-17, rhyolites have been dated between 19 and 20 Ma (Pallister 1987).

In the classification diagram involving Zr, TiO_2, Y, and Nb (Winchester and Floyd 1977), AP-01 and AP-08 samples plot in the field of alkaline basalt, samples AP-18 and AP-20 plot in the field of andesite, and sample AP-19 is in the field of andesite basalt (Fig. 15a). These plots are consistent with the data plotted in the TAS diagram. The sole exception concerns sample AP-20 that plots in the field of andesite, whereas in the TAS diagram it plots in the field of trachyandesite. Basic and intermediate dykes are also plotted in a diagram that relates Th and Nb (Saccani 2015), which are normalised with respect to the normal MORB (mid ocean ridge basalt) composition (Sun and McDonough 1989). Samples AP-01 and AP-08 are classifiable as alkaline ocean-island basalt, although sample AP-08 falls also in the compositional field of plume-type MORB and enriched-type MORB. Samples AP-18 and AP-19 plot in the compositional field of calc-alkaline basalts, whereas sample AP-20 is classifiable as back arc basin basalt or low-Ti island arc

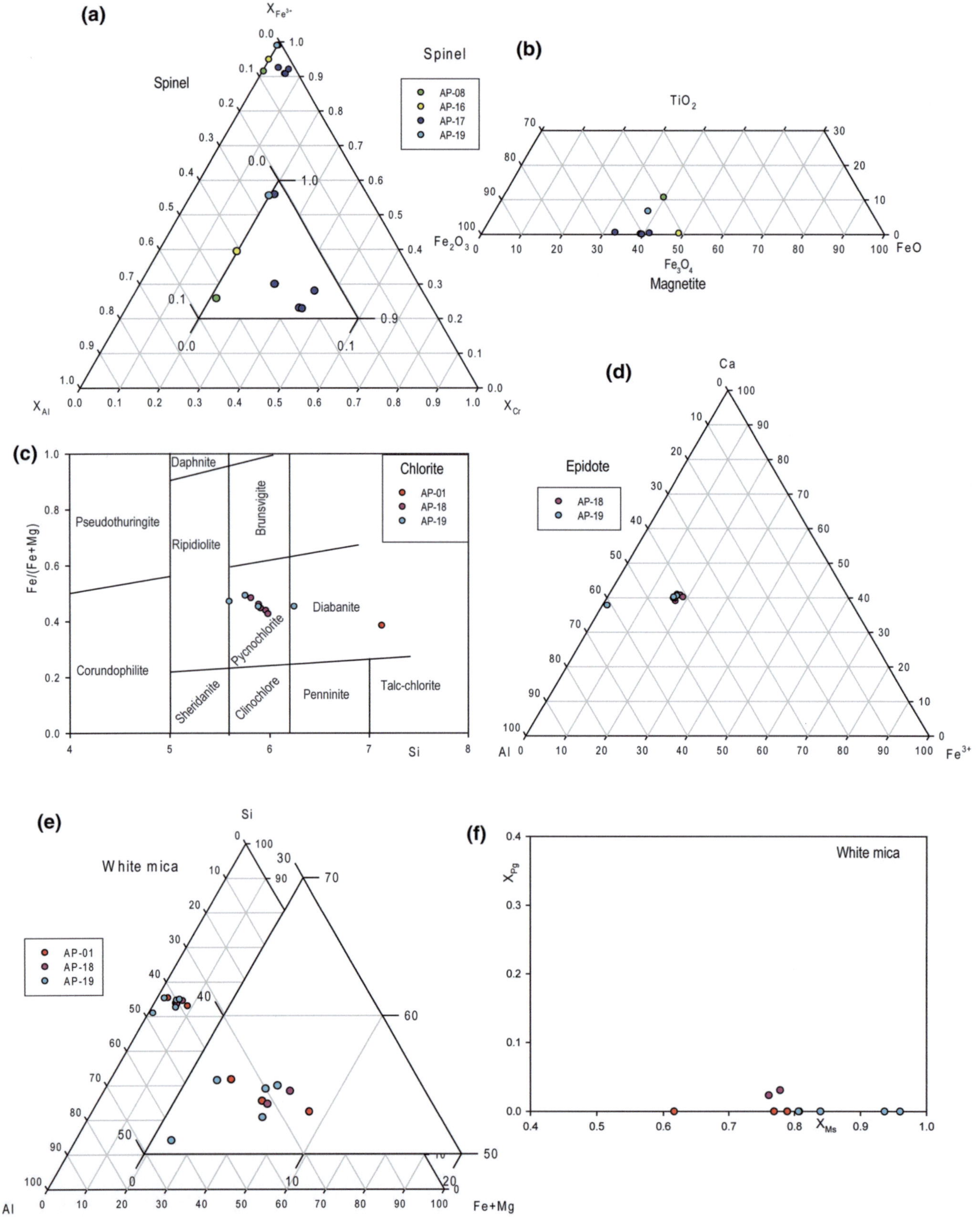

Fig. 13 **a** Ternary diagram showing the variation of X_{Al} (Al/(Al + Cr + Fe^{3+})) versus X_{Cr} (Cr/(Al + Cr + Fe^{3+})) versus X_{Fe}^{3+} (Fe^{3+}/(Al + Cr + Fe^{3+})) in spinel; **b** classification ternary diagram for spinel that shows the variation of Fe_2O_3 versus FeO versus TiO_2 (Buddington and Lindsley 1964); **c** classification binary diagram for chlorite (Hey 1954); **d** ternary diagram showing the variation of Al versus Fe^{3+} versus Ca in epidote; **e** ternary diagram showing the variation of Al versus Fe + Mg versus Si in white mica; **f** X_{Pg} (molar fraction of paragonite) versus X_{Ms} (molar fraction of muscovite) binary diagram; X_{Ms} is according to Green and Usdansky (1986)

Table 7 Analyses of glass. The microstructural position is indicated for any point analysis. Sample label: first 5 digits are sample number, last three digits are point analyses

Sample label	AP_08_32	AP_08_34	AP_08_36	AP_08_38	AP_08_44	AP_08_53	AP_08_79	AP_08_80
	Next pl	In pl	In pl	In pl	Groundmass	Next amp	Next cpx	Next cpx
Na_2O	0.21	1.96	3.01	0.12	0.60	2.59	2.11	0.76
MgO	4.86	0.07	0.16	6.53	0.17	0.09	0.12	0.11
Al_2O_3	5.56	11.64	12.12	6.00	8.19	12.87	11.25	10.16
SiO_2	60.82	72.25	74.56	68.21	58.60	73.65	72.21	67.78
SO_3	0.01	0.11	0.00	0.08	0.08	0.11	0.00	0.12
K_2O	0.37	5.24	5.90	0.43	2.38	4.69	5.14	1.83
CaO	2.36	0.43	0.34	1.92	3.04	0.54	0.40	3.10
TiO_2	0.19	0.68	0.59	0.13	0.20	0.76	0.52	0.49
Cr_2O_3	0.08	0.00	0.00	0.12	0.05	0.06	0.05	0.07
MnO	0.07	0.10	0.07	0.00	0.08	0.08	0.05	0.03
FeO	3.93	1.90	1.61	4.52	0.29	1.41	1.66	0.86
NiO	0.14	0.00	0.06	0.00	0.09	0.19	0.00	0.17
Sum	78.59	94.40	98.42	88.06	73.79	97.06	93.52	85.50

tholeiite (Fig. 15b). According to the classification diagram of Saccani (2015), the composition of samples AP-01 and AP-08 is consistent with a divergent tectonic setting, whereas the composition of samples AP-18, AP-19, and AP-20 is consistent with a convergent tectonic setting.

The rhyolitic dykes AP-16 and AP-17 in the diagram that relate both $FeO^{tot}/(FeO^{tot} + MgO)$ and $Na_2O + K_2O - CaO$ with SiO_2 (Frost et al. 2001) are classifiable as alkali and alkali-calcic ferroan A-type granites (Fig. 16a and 16b). Moreover, on the binary diagram Rb/Nb versus Y/Nb (Eby 1992) the samples AP-16 and AP-17 plot in the field of A-type granites characteristic of continental rifting or intraplate magmatism (Fig. 16c).

Finally, in the AFM diagram (Fig. 17) most of the samples plot in the calcalkaline field. Only a basic (AP-01) and intermediate (AP-19) dyke samples plot over the divide of the calcalkaline and tholeiitic fields. The sole sample that definitely plots in the tholeiitic field is AP-08. The dyke samples of basic and intermediate composition that plot in the calcalkaline field may have recorded contamination with crustal material.

8 Discussion and Conclusions

Microstructural and geochemical features of dykes that have been sampled along the Arabian Red Sea margin, between the Al Lith and Tabuk regions (Fig. 3) are here summarised. Microstructural analysis shows that generally the texture developed during the magmatic emplacement of the dykes and that post-magmatic crystal-plastic deformation is quite scarce. According to the slight undulose extinction, the mineral phase that records the highest amount of post-magmatic internal strain is quartz, mostly in granitoid but also in andesitic dykes. Feldspars, contain growth twinning and are deformation twinning-free. The sole microstructure in alkali feldspar imputable to post-magmatic tectonics is the albite flame perthite that occurs rarely in granitoid dykes (sample AP-14) and testifies to unmixing assisted by differential stress (Pryer and Robin 1996). Since quartz may be affected by crystal plasticity under low temperatures (e.g., Stipp et al. 2002) and feldspar does not show clear evidence of internal strain, the weak crystal plastic deformation in quartz should have been recorded at temperatures below 400 °C (e.g., Passchier and Trouw 2005). The highest amount of internal strain in quartz is recorded in the granitoid dykes, in particular in the sample AP-14, whereas in andesitic dykes, quartz shows gentle undulose extinction. In granitoid dykes graphic intergrowth between K-feldspar and quartz testifies that the two minerals crystallised together under cotectic conditions (e.g., Vernon 2004). Similarly, the mineral composition of rhyolitic dykes, which basically consists of K-feldspar, quartz, and albite, is consistent with minimum-temperature composition of a much-differentiated granitic system.

The dykes show either massive or oriented texture. The granitoid dykes are coarse-grained holocrystalline rocks and record massive texture, whereas oriented texture is evident in some of the mafic dykes and it is poorly defined in the rhyolitic dykes. Generally, shape-preferred orientation of euhedral and strain-free crystals (mostly plagioclase) defines the oriented texture. These characteristics are consistent with

Table 8 Major and trace elements for two basic dykes (AP-01 and AP-08), three intermediate dykes (AP-18, AP-19, and AP-20), and two rhyolitic dykes (AP-16 and AP-17). Note that in Figs. 14, 15, and 16, major elements were normalised to 100% without considering loss on ignition (LOI)

Weight %	**SiO_2**	**Al_2O_3**	**Fe_2O_3**	**MnO**	**MgO**	**CaO**	**Na_2O**	**K_2O**	**TiO_2**	**P_2O_5**	**LOI**	**Total**
AP_01	40.78	14.57	11.49	0.168	7.7	10.29	2.35	1.82	2.099	0.38	8.81	100.4
AP_08	41.12	14.11	15.12	0.167	5.08	7.76	2.17	1.22	2.982	0.55	9.34	99.61
AP_16	68.14	13.4	5.39	0.132	0.17	0.38	4.17	4.68	0.344	0.16	2.4	99.36
AP_17	71.58	12.17	4.18	0.073	0.24	0.62	4.1	4.44	0.249	0.02	1.54	99.21
AP_18	57.84	16.23	8.01	0.166	3.85	5.27	3.46	1.88	0.915	0.2	2.91	100.7
AP_19	51.65	17.09	10.41	0.146	5.51	5.51	2.21	1.42	1.018	0.35	4.95	100.3
AP-20	54.08	16.7	8.6	0.164	3.22	3.48	6.19	1.03	0.989	0.24	4.18	98.88
ppm	**Sc**	**Be**	**V**	**Cr**	**Co**	**Ni**	**Cu**	**Zn**	**Ga**	**Ge**	**As**	**Rb**
AP_01	26	2	242	180	45	110	70	70	15	1.4	<5	50
AP_08	21	2	211	30	44	50	30	110	18	1.3	<5	38
AP_16	1	6	9	<20	<1	<20	<10	150	28	1.2	<5	93
AP_17	2	7	11	40	2	<20	10	170	31	2	<5	112
AP_18	21	1	166	30	22	<20	10	140	15	1.3	20	47
AP_19	28	<1	229	60	31	30	50	90	16	1.1	23	29
AP-20	12	1	152	<20	21	20	<10	150	16	2.6	5	24
ppm	**Sr**	**Y**	**Zr**	**Nb**	**Mo**	**Ag**	**In**	**Sn**	**Sb**	**Cs**	**Ba**	**La**
AP_01	466	23.5	169	38.8	<2	<0.5	<0.1	1	<0.2	7.3	330	28.8
AP_08	188	31.4	216	22.2	<2	<0.5	<0.1	1	0.3	4	169	19.2
AP_16	241	99.5	806	59	<2	1.5	0.2	7	<0.2	0.2	117	45.9
AP_17	26	134	1118	71	2	6.4	0.2	8	<0.2	0.2	63	72.4
AP_18	435	24.9	154	4.2	<2	<0.5	<0.1	1	3.4	1	633	12
AP_19	462	17.5	75	2	<2	<0.5	<0.1	<1	1.9	1.5	781	9.33
AP-20	369	17.2	105	2.1	2	<0.5	0.1	<1	3.7	0.7	213	10
ppm	**Ce**	**Pr**	**Nd**	**Sm**	**Eu**	**Gd**	**Tb**	**Dy**	**Ho**	**Er**	**Tm**	**Yb**
AP_01	59.5	6.98	28.2	6.02	1.92	5.35	0.8	4.6	0.83	2.26	0.308	2.06
AP_08	46.6	6.21	28.4	6.78	2.35	6.98	1.05	5.87	1.11	3.19	0.434	2.73
AP_16	127	13.8	54.7	14	1.86	14.6	2.63	16.6	3.38	10.1	1.55	10.5
AP_17	162	18.8	75.3	17.9	1.56	18.7	3.5	22.8	4.6	14.1	2.14	14.4
AP_18	29.2	3.89	17.4	4.34	1.16	4.36	0.72	4.33	0.85	2.47	0.358	2.44
AP_19	22.7	3.32	15.5	3.8	1.26	3.69	0.55	3.11	0.62	1.71	0.258	1.67
AP-20	24.7	3.44	16.3	4.21	1.3	3.78	0.57	3.17	0.61	1.61	0.23	1.47
ppm	**Lu**	**Hf**	**Ta**	**W**	**Tl**	**Pb**	**Bi**	**Th**	**U**	Fe_2O_3 =total iron		
AP_01	0.311	3.7	2.68	<0.5	0.09	<5	<0.1	2.99	0.89			
AP_08	0.438	4.4	1.52	<0.5	0.11	<5	<0.1	1.48	0.64			
AP_16	1.62	15.8	4.28	<0.5	0.13	20	<0.1	8.54	2.01			
AP_17	2.2	25.3	6.5	<0.5	0.1	20	<0.1	13.2	3.8			
AP_18	0.38	3.8	0.27	<0.5	0.26	<5	<0.1	1.71	1.02			
AP_19	0.244	1.9	0.07	<0.5	<0.05	<5	<0.1	0.14	0.24			
AP-20	0.235	2.4	0.06	<0.5	0.16	7	0.3	0.58	0.43			

this texture being interpreted as a magmatic fluidal structure (Paterson et al. 1989). In mafic dykes, the fluidal structure is defined in general by plagioclase crystals and in rhyolitic dykes by K-feldspar. In particular, in sample AP-08 the fluidal structure is defined by clinopyroxene and amphibole medium-grained phenocrysts and plagioclase microlites that

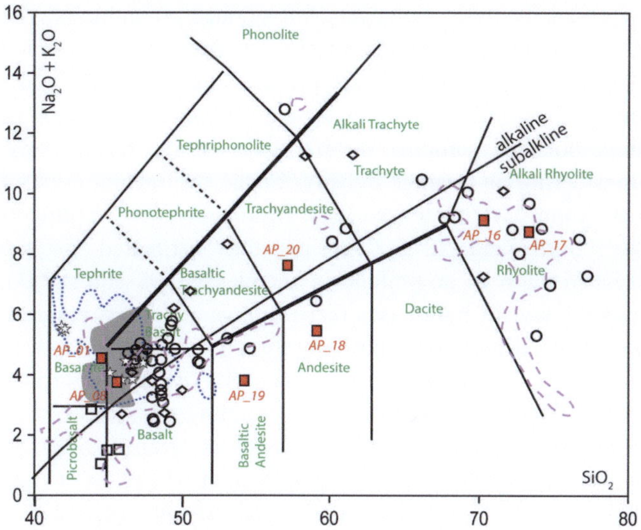

Fig. 14 TAS (total alkali versus silica) classification diagram. In red data analysed in this chapter are indicated and are compared with the composition of Cenozoic magmatic rocks in the Arabian plate from: Al Lith area in open circles (Pallister 1987), Harrat as Sirat, Harrat Khaybar, and Harrat al Birk in open diamonds (Coleman et al. 1983), Harrat Lunayyir in grey shading (Duncan and Al-Amri 2013), Harrat Al Birk in open squares (Ghent et al. 1980), Harrat Ash Shaam, northeast Jordan in blue dotted line (Ibrahim et al. 2003; Shaw et al. 2003; Ibrahim and Al-Malabeh 2006), Yemen in pink dashed line (Manetti et al. 1991), and Syria in white stars (Krienitz et al. 2009). Alkaline–subalkaline divide is from Irvine and Baragar (1971)

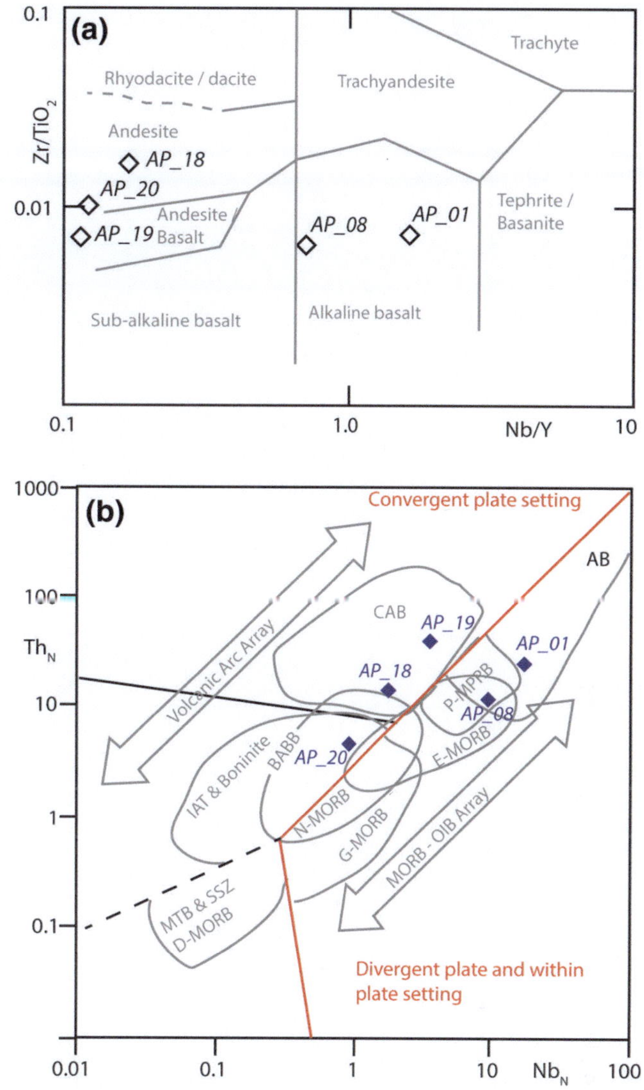

Fig. 15 **a** Zr/TiO_2 versus Nb/Y classification diagram (Winchester and Floyd 1977); **b** Th_N versus Nb_N classification diagram (Saccani 2015). Th (Th_N) and Nb (Nb_N) are normalised with respect to the normal MORB (Sun and McDonough 1989). MORB = mid-ocean ridge basalt; G-MORB = garnet-influenced MORB; N-MORB = normal-type MORB; E-MORB = enriched-type MORB; p-MORB = plume-type MORB; AB = alkaline ocean-island basalt; IAT = low-Ti, island arc tholeiite; BON = very low-Ti, boninitic basalt; CAB = calc-alkaline basalt; MTB = medium-Ti basalt; D-MORB = depleted-type MORB; BABB = back arc basin basalt

wrap around plagioclase and clinopyroxene coarse-grained phenocrysts. This evidence suggests that coarse-grained phenocrysts may have started growing earlier than the other crystals. This interpretation is also consistent with coarse-grained plagioclase phenocrysts having a higher content in anorthite with respect to groundmass plagioclase microlites and with a higher X_{Mg} ratio in coarse- and medium-grained clinopyroxene than in groundmass clinopyroxene. On the other hand, in this sample, coarse-grained clinopyroxene shows an X_{Mg} ratio and Ca content similar to clinopyroxene of websterite xenoliths contained in alkali basalts from Harrat Al Birk and Harrat Ash Shaam (Ghent et al. 1980; Shaw et al. 2007). On the basis of this evidence we can argue that coarse-grained clinopyroxene and enclosed orthopyroxene may represent millimetre-sized websterite xenoliths.

Post-magmatic alteration affects the dykes with different intensities. White mica replaces feldspars and it is associated with chlorite and epidote if it replaces plagioclase. In intermediate dykes AP-18, AP-19, and AP-20 alteration is widespread and much diffused. Phenocrysts are nearly completely replaced and locally their forming mineral phases are not recognizable. The groundmass is mostly affected by overgrowth of chlorite and epidote, which also forms granular aggregates. In rhyolitic samples (AP-16 and AP-17), K-feldspar is only affected by the overgrowth of very fine-grained magnetite. In basaltic/basanitic samples alteration is more diffused in sample AP-01 than in sample AP-08, which is quite pristine. In sample AP-01 coarse-grained plagioclase is replaced by abundant white mica, whereas finer-grained plagioclase laths are pristine. Olivine is partially replaced by rims of talc and serpentine.

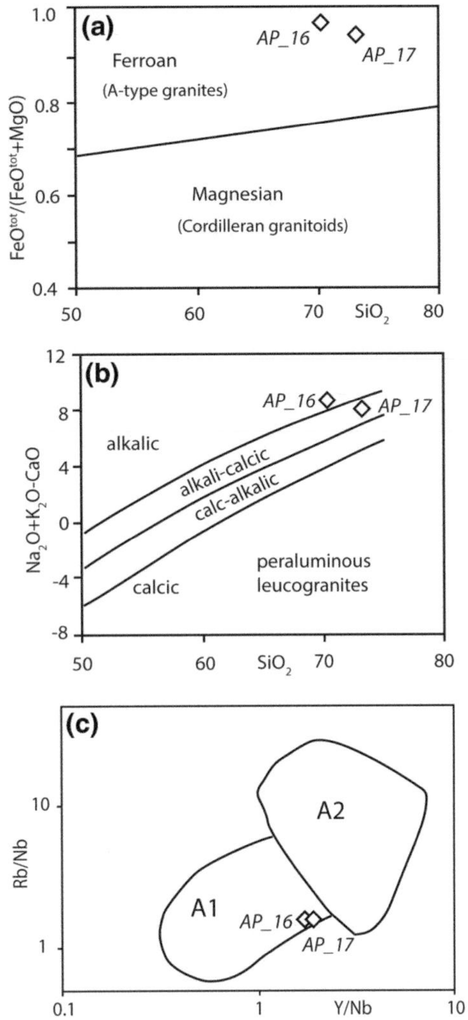

Fig. 16 **a** $FeO^{tot}/(FeO^{tot} + MgO)$ versus SiO_2 classification diagram (Frost et al. 2001); **b** $Na_2O + K_2O - CaO$ versus SiO_2 classification diagram (Frost et al. 2001); **c** classification diagram Rb/Nb versus Y/Nb (Eby 1992); A1 = magmas emplaced during continental rifting or intraplate; A2 = continent/continent collision or island arc magmatism

In sample AP-08 talc and carbonate replaced millimetre-sized phenocrysts and domains in coarse-grained clinopyroxene. Since samples AP-01 and AP-08 have a similar composition (Figs. 14 and 15), talc and carbonates may represent the replacement of olivine. In sample AP-01 degassing bubbles are filled by mm-sized calcite and analcime, which contains minor K-feldspar, epidote, chlorite, and quartz. This mineral assemblage may result from late hydrothermal alteration (e.g., Rona et al. 1980). Also, degassing bubbles suggest that the final emplacement depth was as shallow as a few kilometres (e.g., Westrich et al. 1988; Aubry et al. 2013). Shallow levels of emplacement can be envisaged also for rhyolitic samples and sample AP-08 on the basis of the very fine-grained groundmass. The Al content in amphibole is correlated to pressure (Ridolfi and Renzulli 2012). On this basis, pressures of 0.39 ± 0.13 GPa and 0.55 ± 0.02 GPa are estimated for the crystallisation of amphibole in samples AP-01 and AP-08, respectively. Considering an average density for the continental crust of 2.83 g cm^{-3} (Christensen and Mooney 1995) depths of 14 ± 5 km and 20 ± 1 km are thus estimated for the crystallisation of amphibole in samples AP-01 and AP-08, respectively. In both cases these estimates do not represent the emplacement depth but only the crystallisation depth of amphibole, which may have been carried toward the surface by the magma flow into the dyke as solid grains. According to the geothermobarometer of Ridolfi and Renzulli (2012), amphibole crystallised at temperatures of 842 ± 58 °C and 983 ± 25 °C in samples AP-01 and AP-08, respectively. According to the thermometer based on X_{Fe} and Al content in chlorite (Cathelineau 1988; Jowett 1991), a temperature of 274 ± 31 °C can be estimated for the development of the alteration in andesitic samples AP-18 and AP-19.

We started with U-Pb testing on samples containing zircon grains at least as large as 25 μm. These samples (AP-03, AP-05, and AP-14) consist of three granitoid dykes enclosing zircon grains that turned out to be Neoproterozoic in age. In particular, for sample AP-05, a late Cryogenian and early Ediacaran age of dyke emplacement can be tentatively proposed and ascribable to suturing events during the assemblage of the Arabian shield (see Fig. 2 and Table 9). Zircon grains from samples AP-03 and AP-14 have a strong inherited imprint, which most likely is related to the tectonic history predating the Arabian shield assemblage (see Fig. 2).

The mineral and geochemical composition reveals that the other seven dykes have a variable composition spanning from basanites to andesite and rhyolite (Fig. 14). On the ground of whole rock content in Nb and Th, samples AP-01 and AP-08 are compatible with extensional tectonics, whereas samples AP-18, AP-19, and AP-20 with convergent tectonics (see Fig. 15b and Table 9). Therefore, basanitic/basaltic samples AP-01 and AP-08 may be related to the Red Sea rifting and andesitic samples AP-18, AP-19, and AP-20 may be related to magmatism during the Pan-African orogeny. In support of this interpretation there are plenty of known Neoproterozoic mafic dykes in the Arabian shield, among which andesite and trachyandesite rocks occur (e.g., Kessel et al. 1998; Jarrar 2001), and that AP-18, AP-19, and AP-20 dykes were sampled about 50 km southeast from the Neoproterozoic Yanbu suture zone (Figs. 1 and 3). Indeed, andesitic magmas are generally common in subduction/collision contexts. Moreover, these

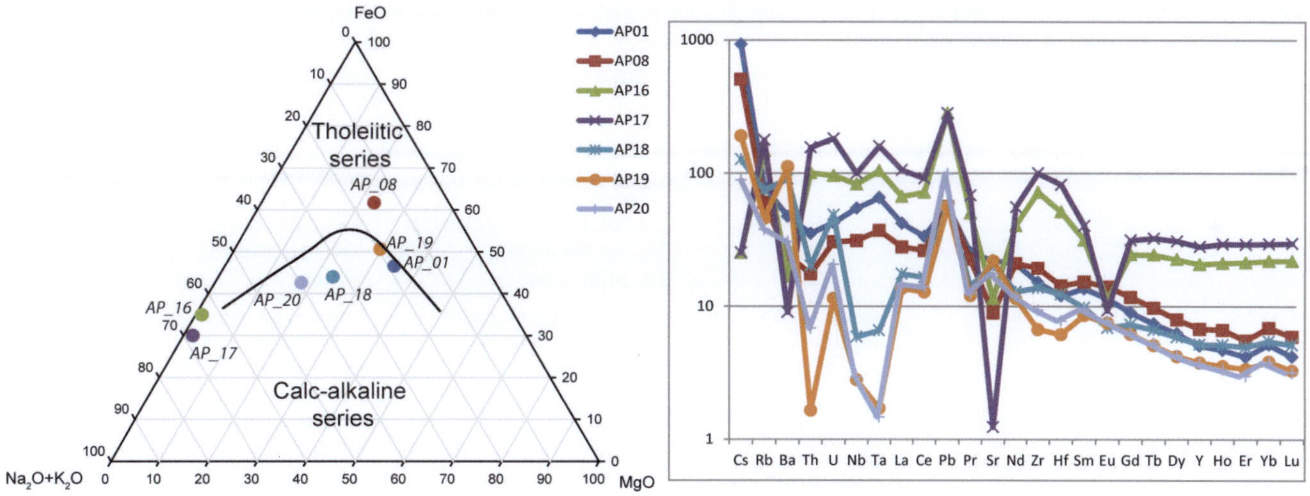

Fig. 17 AFM diagram (A = K_2O Na_2O; F = FeO; M = MgO) and multi-element diagram. In the AFM diagram, FeO is the total iron as recommended by Rollinson (1993) and the black line is the divide of tholeiitic and calcalkaline series (Irvine and Baragar 1971). In the multi-element diagram, elements are normalised with respect to the normal MORB (Sun and McDonough 1989)

three dykes are strongly altered and in the TAS diagram their composition is generally poorly comparable with the geochemical composition of magmatic rocks of Cenozoic harrats (Fig. 14). In contrast, the AP-01 and AP-08 dykes (especially AP-08) are less altered than andesitic dykes and have a composition more similar to the Cenozoic magmatic rocks from Harrat Ash Shaam (Shaw et al. 2003) and central Syria (Krienitz et al. 2009) (see Fig. 14). Nevertheless, it should be noted that in the northern Arabian plate, late Neoproterozoic mafic dykes, which however show higher SiO_2 and lower alkali content than samples AP-01 and AP-08, are interpreted as related to extensional tectonics (Kessel et al. 1998; Jarrar 2001). The two rhyolitic dykes in the TAS diagram (Fig. 14) show a composition very similar to part of the Ad Damm dyke suite in the central coastal plain of the Red Sea Arabian margin. As these dykes are ferroan rocks, most likely an origin in a convergent context (i.e., Pan-African orogeny) could be ruled out, since Cordilleran-type granitoid rocks usually show a marked magnesian signature (Frost et al. 2001, and references therein). Also, these dykes have been sampled in the region where rhyolitic dykes have been dated at around 20 Ma (Pallister 1987) and their geochemical features can be typical of rhyolites emplaced during continental rifting or intraplate magmatism (Fig. 16). The rhyolitic dykes may have originated during partial melting of the continental crust due to the positive perturbation of thermal structure of the lithosphere during the Red Sea continental rifting. Alternatively, rhyolitic dykes may have been derived from differentiation of basaltic magma (see also discussions in Pallister 1987), as is also suggested by the ferroan chemical signature (Frost et al. 2001). In magma derived from the partial melting of the continental crust, minerals such as white mica or garnet usually testify to its peraluminous chemistry. Since the rhyolitic dykes do not contain such minerals, the second option might be more reliable. Finally, the strong morphological evidence of rhyolite (samples AP-16 and AP-17) and basaltic (sample AP-08) dykes (Fig. 5b, e, f) is also consistent with these dykes being recent (Table 9).

Table 9 Summary of the dyke features for each sample with possible tectonic interpretation. Asterisk indicates the dyke that turned out to be Neaproterozoic in age after radiometric age testing

Sample #	Dyke type	Lithology	Coordinates	Host rocks	Inferences and Remarks	Possible tectonic setting
AP-01	Basaltic	Basanite	Latitude: 27.504511° N Longitude: 36.121809° E	Alkali granite of the Massah complex 629 ± 12 Ma (Grainger and Hanif 1989) (Midyan terrane)	Subvolcanic (based on fine-grained crystallinity) dyke emplaced in divergent tectonic environment	Red Sea rifting
AP-03*	Granitoid	Granite	Latitude: 27.535205° N Longitude: 36.457635° E	Qaraqir alkali granite (Midyan terrane)	Intrusive (as suggested by coarse-grained crystallinity) dyke probably emplaced due to suturing event	Arabian shield accretion
AP-05*	Granitoid	Monzogranite	Latitude: 27.534905° N Longitude: 36.458497° E	Qaraqir alkali granite (Midyan terrane)	Intrusive (as suggested by coarse-grained crystallinity) dyke probably emplaced due to suturing event	Arabian shield accretion
AP-08	Basaltic	Basalt	Latitude: 26.488274° N Longitude: 36.389255° E	Liban monzogranite 621 ± 7 Ma or 638 ± 10 Ma (Hedge, in Al-Husseini 2015) (Midyan terrane)	Subvolcanic (based on fine-grained crystallinity) dyke emplaced in divergent tectonic environment	Red Sea rifting
AP-14*	Granitoid	Granodiorite	Latitude: 25.188390° N Longitude: 37.416037° E	Neoproterozoic gabbroic rocks (Midyan terrane)	Intrusive (as suggested by coarse-grained crystallinity) dyke probably emplaced due to suturing event	Arabian shield accretion
AP-16	Rhyolite	Rhyolite	Latitude:20.914031° N Longitude: 39.901878° E	meta-basalts of the Baish group protoliths older than 804 Ma (Asir composite terrane)	Subvolcanic (based on fine-grained crystallinity) dyke emplaced during continental rifting	Red Sea rifting
AP-17	Rhyolite	Rhyolite	Latitude:20.896851° N Longitude:39.891866° E	meta-basalts of the Baish group protoliths older than 804 Ma (Asir composite terrane)	Subvolcanic (based on fine-grained crystallinity) dyke emplaced during continental rifting	Red Sea rifting
AP-18	Andesitic	Andesite	Latitude:24.6262148° N Longitude: 38.0937260° E	Neoproterozoic granitoid complex (Hijaz terrane)	Subvolcanic (based on fine-grained crystallinity) of convergent tectonic environment	Arabian shield accretion
AP-19	Andesitic	Basaltic andesite	Latitude:24.6249047° N Longitude: 38.0948120° E	Neoproterozoic granitoid complex (Hijaz terrane)	Subvolcanic (based on fine-grained crystallinity) of convergent tectonic environment	Arabian shield accretion
AP-20	Andesitic	Trachyandesite	Latitude: 24.6371164 ° N Longitude:38.0697803° E	Neoproterozoic granitoid complex (Hijaz terrane)	Subvolcanic (based on fine-grained crystallinity) of convergent tectonic environment	Arabian shield accretion

Acknowledgements This work was funded by the MIUR (Ministero dell'Istruzione, dell'Università e della Ricerca)—PRIN (Progetti di Ricerca di Interesse Nazionale) protocol 20125JKANY "Rift-to-Drift Transition in the Red Sea". A. Schettino and P. P. Pierantoni carried out the fieldwork and sampled the dykes in 2015 with the financial support and logistical assistance of the Saudi Geological Survey. A. Schettino read the first draft of the chapter, A. Sanfilippo advised on whole rock geochemistry, and A. Rizzi assisted the work at the SEM. Four anonymous reviewers greatly helped with valuable advice.

References

Abdelsalam MG, Stern RJ (1996) Sutures and shear zones in the Arabian-Nubian Shield. J African Earth Sci 23:289–310. https://doi.org/10.1016/S0899-5362(97)00003-1

Al-Amri AM, Fnais MS, Abdel-Rahman K, Mogren S, Al-Dabbagh M (2012) Geochronological dating and stratigraphic sequences of Harrat Lunayyir, NW Saudi Arabia. Int J Phys Sci 7:2791–2805. https://doi.org/10.5897/IJPS12.178

Al-Husseini M (2015) Ediacaran-Cambrian Middle East geologic time scale 2015: spatio-temporal position of the Ediacaran Thalbah Basin in the Najd Fault System, Arabian Shield. GeoArabia 20:17–44

Ali KA, Azer MK, Gahlan HA, Wilde SA, Samuel MD, Stern RJ (2010) Age constraints on the formation and emplacement of Neoproterozoic ophiolites along the Allaqi-Heiani Suture, South Eastern Desert of Egypt. Gondwana Res 18:583–595. https://doi.org/10.1016/j.gr.2010.03.002

Almalki KA, Betts PG, Ailleres L (2014) Episodic sea-floor spreading in the southern Red Sea. Tectonophysics 617:140–149. https://doi.org/10.1016/j.tecto.2014.01.030

Al-Saleh AM, Boyle AP (2001) Structural rejuvenation of the eastern Arabian Shield during continental collision: $^{40}Ar/^{39}Ar$ evidence from the Ar Ridayniyah ophiolitic mélange. J African Earth Sci 33:135–141. https://doi.org/10.1016/S0899-5362(01)90094-6

ArRajehi A, McClusky S, Reilinger R, Daoud M, Alchalbi A, Ergintav S, Gomez F, Sholan J, Bou-Rabee F, Ogubazghi G, Haileab B, Fisseha S, Asfaw L, Mahmoud S, Rayan A, Bendik R, Kogan L (2010) Geodetic constraints on present-day motion of the Arabian Plate: implications for Red Sea and Gulf of Aden rifting. Tectonics 29:TC3011. https://doi.org/10.1029/2009tc002482

Aubry GJ, Sator N, Guillot B (2013) Vesicularity, bubble formation and noble gas fractionation during MORB degassing. Chem Geol 343:85–98. https://doi.org/10.1016/j.chemgeo.2013.02.006

Audin L, Quidelleur X, Coulié E, Courtillot V, Gilder S, Manighetti I, Gillot P-Y, Tapponnier P, Kidane T (2004) Palaeomagnetism and K-Ar and $^{40}Ar/^{39}Ar$ ages in the Ali Sabieh area (Republic of Djibouti and Ethiopia): constraints on the mechanism of Aden ridge propagation into southeastern Afar during the last 10 Myr. Geophys J Int 158:327–345. https://doi.org/10.1111/j.1365-246X.2004.02286.x

Baker J, Snee L, Menzies M (1996) A brief Oligocene period of flood volcanism in Yemen: implications for the duration and rate of continental flood volcanism at the Afro-Arabian triple junction. Earth Planet Sci Lett 138:39–55. https://doi.org/10.1016/0012-821X(95)00229-6

Bakhsh RA (2015) Pliocene-Quaternary basalts from the Harrat Tufail, western Saudi Arabia: recycling of ancient oceanic slabs and generation of alkaline intra-plate magma. J African Earth Sci 112:37–54. https://doi.org/10.1016/j.jafrearsci.2015.08.010

Bayer H-J, El-Isa Z, Hotzl H, Mechie J, Prodehl C, Saffarini G (1989) Large tectonic and lithospheric structures of the Red Sea region. J African Earth Sci (and Middle East) 8:565–587. https://doi.org/10.1016/S0899-5362(89)80045-4

Be'eri-Shlevin Y, Eyal M, Eyal Y, Whitehouse MJ, Litvinovsky B (2012) The Sa'al volcano-sedimentary complex (Sinai, Egypt): a latest Mesoproterozoic volcanic arc in the northern Arabian Nubian Shield. Geology 40:403–406. https://doi.org/10.1130/g32788.1

Be'eri-Shlevin Y, Katzir Y, Whitehouse M (2009) Post-collisional tectonomagmatic evolution in the northern Arabian-Nubian Shield: time constraints from ion-probe U-Pb dating of zircon. J Geol Soc London 166:71–85. https://doi.org/10.1144/0016-76492007-169

Belachew M, Ebinger C, Cote D (2013) Source mechanisms of dike-induced earthquakes in the Dabbahu-Manda Hararo rift segment in Afar, Ethiopia: implications for faulting above dikes. Geophys J Int 192:907–917. https://doi.org/10.1093/gji/ggs076

Bendick R, McClusky S, Bilham R, Asfaw L, Klemperer S (2006) Distributed Nubia-Somalia relative motion and dike intrusion in the Main Ethiopian Rift. Geophys J Int 165:303–310. https://doi.org/10.1111/j.1365-246X.2006.02904.x

Bertrand H, Chazot G, Blichert-Toft J, Thoral S (2003) Implications of widespread high-μ volcanism on the Arabian Plate for Afar mantle plume and lithosphere composition. Chem Geol 198:47–61. https://doi.org/10.1016/S0009-2541(02)00418-7

Beyth M, Eyal Y, Garfunkel Z (2014) The geology of the northern tip of the Arabian-Nubian Shield. J African Earth Sci 99:332–341. https://doi.org/10.1016/j.jafrearsci.2014.03.028

Bicknell JD, MacDonald KC, Miller SP, Lonsdale PF, Becker K (1986) Tectonics of the Nereus Deep, Red Sea: a deep-tow investigation of a site of initial rifting. Mar Geophys Res 8:131–148. https://doi.org/10.1007/BF00338225

Bohannon RG, Naeser CW, Schmidt DL, Zimmermann RA (1989) The timing of uplift, volcanism, and rifting peripheral to the Red Sea: a case for passive rifting? J Geophys Res 94:1683–1701. https://doi.org/10.1029/JB094iB02p01683

Bonatti E (1985) Punctiform initiation of seafloor spreading in the Red Sea during transition from a continental to an oceanic rift. Nature 316:33–37. https://doi.org/10.1038/316033a0

Bonatti E, Colantoni P, Della Vedova B, Taviani M (1984) Geology of the Red Sea transitional region (22 N-25 N). Oceanol Acta 7:385–398

Bosworth W, Huchon P, McClay K (2005) The Red Sea and Gulf of Aden Basins. J African Earth Sci 43:334–378

Bosworth W, Stockli DF (2016) Early magmatism in the greater Red Sea rift: timing and significance. Can J Earth Sci 53:1158–1176. https://doi.org/10.1139/cjes-2016-0019

Buddington AF, Lindsley DH (1964) Iron-titanium oxide minerals and synthetic equivalents. J Petrol 5:310–357. https://doi.org/10.1093/petrology/5.2.310

Camp VE, Roobol MJ (1992) Upwelling asthenosphere beneath western Arabia and its regional implications. J Geophys Res 97:15255–15271. https://doi.org/10.1029/92JB00943

Cathelineau M (1988) Cation site occupancy in chlorites and illites as a function of temperature. Clay Miner 23:471–485

Chang S-J, Merino M, Van der Lee S, Stein S, Stein CA (2011) Mantle flow beneath Arabia offset from the opening Red Sea. Geophys Res Lett 38:L04301. https://doi.org/10.1029/2010GL045852

Chang S-J, Van der Lee S (2011) Mantle plumes and associated flow beneath Arabia and East Africa. Earth Planet Sci Lett 302:448–454. https://doi.org/10.1016/j.epsl.2010.12.050

Christensen NI, Mooney WD (1995) Seismic velocity structure and composition of the continental crust: a global view. J Geophys Res Solid Earth 100:9761–9788. https://doi.org/10.1029/95JB00259

Clark MD (1985) Late Proterozoic crustal evolution of the Midyan region, northwestern Saudi Arabia. Geology 13:611–615. https://doi.org/10.1130/0091-7613

Cochran JR (1981) The Gulf of Aden: structure and evolution of a young ocean basin and continental margin. J Geophys Res 86:263. https://doi.org/10.1029/JB086iB01p00263

Cochran JR (1983) A model for development of Red Sea. Am Assoc Pet Geol Bull 67:41–69

Cochran JR, Martinez F, Steckler MS, Hobart MA (1986) Conrad Deep: a new northern Red Sea deep. Earth Planet Sci Lett 78:18–32. https://doi.org/10.1016/0012-821X(86)90169-X

Coleman RG, Gregory RT, Brown GF (1983) Cenozoic volcanic rocks of Saudi Arabia. US Geol Surv Open File Rep USGS-OF-03-93, 82 pp. https://doi.org/10.3133/ofr83788

Coleman RG, McGuire AV (1988) Magma systems related to the Red Sea opening. Tectonophysics 150:77–100. https://doi.org/10.1016/0040-1951(88)90296-X

Coulié E, Quidelleur X, Gillot P-Y, Courtillot V, Lefèvre J-C, Chiesa S (2003) Comparative K-Ar and Ar/Ar dating of Ethiopian and Yemenite Oligocene volcanism: implications for timing and duration of the Ethiopian traps. Earth Planet Sci Lett 206:477–492. https://doi.org/10.1016/S0012-821X(02)01089-0

Courtillot V, Armijo R, Tapponnier P (1987) Kinematics of the Sinai triple junction and a two-phase model of Arabia-Africa rifting. Geol Soc London, Spec Publ 28:559–573. https://doi.org/10.1144/GSL.SP.1987.028.01.37

Daradich A, Mitrovica JX, Pysklywec RN, Willett SD, Forte AM (2003) Mantle flow, dynamic topography, and rift-flank uplift of Arabia. Geology 31:901–904. https://doi.org/10.1130/G19661.1

Davies FB (1985) Geological map of the Al Wahj Quadrangle, Sheet 26B, Kingdom of Saudi Arabia. Saudi Arabian Ministry of Petroleum and Mineral Resources, Directorate General of Mineral Resources

Davison I, Al-Kadasi M, Al-Khirbash S, Al-Subbary AK, Baker J, Blakey S, Bosence D, Dart C, Heaton R, McClay K, Menzies M, Nichols G, Owen L, Yelland A (1994) Geological evolution of the southeastern Red Sea Rift margin, Republic of Yemen. Geol Soc Am Bull 106:1474–1493. https://doi.org/10.1130/0016-7606

Desissa M, Johnson NE, Whaler KA, Hautot S, Fisseha S, Dawes GJK (2013) A mantle magma reservoir beneath an incipient mid-ocean ridge in Afar, Ethiopia. Nature Geosci 6:861–865. https://doi.org/10.1038/ngeo1925

Dilek Y, Ahmed Z (2003) Proterozoic ophiolites of the Arabian Shield and their significance in Precambrian tectonics. Geol Soc London, Spec Publ 218:685–700. https://doi.org/10.1144/GSL.SP.2003.218.01.33

Dixon TH, Stern RJ, Hussein IM (1987) Control of Red Sea rift geometry by Precambrian structures. Tectonics 6:551–571. https://doi.org/10.1029/TC006i005p00551

Droop GTR (1987) A general equation for estimating Fe^{3+} concentrations in ferromagnesian silicates and oxides from microprobe analyses, using stoichiometric criteria. Mineral Mag 51:431–435

Duncan RA, Al-Amri AM (2013) Timing and composition of volcanic activity at Harrat Lunayyir, western Saudi Arabia. J Volcanol Geotherm Res 260:103–116. https://doi.org/10.1016/j.jvolgeores.2013.05.006

Ebinger CJ, Ayele A, Keir D, Rowland J, Yirgu G, Wright T, Belachew M, Hamling I (2010) Length and timescales of rift faulting and magma intrusion: the Afar rifting cycle from 2005 to Present. Ann Rev Earth Planet Sci 38:439–466. https://doi.org/10.1146/annurev-earth-040809-152333

Ebinger CJ, Keir D, Ayele A, Calais E, Wright TJ, Belachew M, Hammond JOS, Campbell E, Buck WR (2008) Capturing magma intrusion and faulting processes during continental rupture: seismicity of the Dabbahu (Afar) rift. Geophys J Int 174:1138–1152. https://doi.org/10.1111/j.1365-246X.2008.03877.x

Eby GN (1992) Chemical subdivision of the A-type granitoids: petrogenetic and tectonic implications. Geology 20:661–664

Ehrhardt A, Hübscher C, Gajewski D (2005) Conrad Deep, northern Red Sea: development of an early stage ocean deep within the axial depression. Tectonophysics 411:19–40. https://doi.org/10.1016/j.tecto.2005.08.011

El-Bialy MZ, Omar MM (2015) Spatial association of Neoproterozoic continental arc I-type and post-collision A-type granitoids in the Arabian-Nubian Shield: the Wadi Al-Baroud Older and Younger Granites, North Eastern Desert. Egypt. J African Earth Sci 103:1–29. https://doi.org/10.1016/j.jafrearsci.2014.11.013

Farahat ES, Mohamed HA, Ahmed AF, El Mahallawi MM (2007) Origin of I- and A-type granitoids from the Eastern Desert of Egypt: implications for crustal growth in the northern Arabian-Nubian Shield. J African Earth Sci 49:43–58. https://doi.org/10.1016/j.jafrearsci.2007.07.002

Fournier M, Chamot-Rooke N, Petit C, Huchon P, Al-Kathiri A, Audin L, Beslier M-O, d'Acremont E, Fabbri O, Fleury J-M, Khanbari K, Lepvrier C, Leroy S, Maillot B, Merkouriev S (2010) Arabia-Somalia plate kinematics, evolution of the Aden-Owen-Carlsberg triple junction, and opening of the Gulf of Aden. J Geophys Res 115:B04102. https://doi.org/10.1029/2008JB006257

Fritz H, Abdelsalam M, Ali KA, Bingen B, Collins AS, Fowler AR, Ghebreab W, Hauzenberger CA, Johnson PR, Kusky TM, Macey P, Muhongo S, Stern RJ, Viola G (2013) Orogen styles in the East African Orogen: a review of the Neoproterozoic to Cambrian tectonic evolution. J African Earth Sci 86:65–106. https://doi.org/10.1016/j.jafrearsci.2013.06.004

Frost BR, Barnes CG, Collins WJ, Arculus RJ, Ellis DJ, Frost CD (2001) A geochemical classification for granitic rocks. J Petrol 42:2033–2048. https://doi.org/10.1093/petrology/42.11.2033

Garson MS, Krs M (1976) Geophysical and geological evidence of the relationship of Red Sea transverse tectonics to ancient fractures. Geol Soc Am Bull 87:169–181. https://doi.org/10.1130/0016-7606

Gaulier JM, Le Pichon X, Lyberis N, Avedik F, Geli L, Moretti I, Deschamps A, Hafez S (1988) Seismic study of the crust of the northern Red Sea and Gulf of Suez. Tectonophysics 153:55–88. https://doi.org/10.1016/0040-1951(88)90007-8

Genna A, Nehlig P, Le Goff E, Guerrot C, Shanti M (2002) Proterozoic tectonism of the Arabian Shield. Precambrian Res 117:21–40

Ghebreab W (1998) Tectonics of the Red Sea region reassessed. Earth-Sci Rev 45:1–44. https://doi.org/10.1016/S0012-8252(98)00036-1

Ghebreab W, Talbot CJ (2000) Red Sea extension influenced by Pan-African tectonic grain in eastern Eritrea. J Struct Geol 22:931–946. https://doi.org/10.1016/S0191-8141(00)00022-5

Ghent ED, Coleman RG, Hadley DG (1980) Ultramafic inclusions and host alkali olivine basalts of the southern coastal plain of the Red Sea, Saudi Arabia. Am J Sci 280A:499–527

Gosso G, Rebay G, Roda M, Spalla MI, Tarallo M, Zanoni D, Zucali M (2015) Taking advantage of petrostructural heterogeneities in subduction-collisional orogens, and effect on the scale of analysis. Period Mineral 84(3B):779–825. https://doi.org/10.2451/2015PM0452

Grainger DJ, Hanif R (1989) Geologic map of the Shaghab Quadrangle, Sheet 27 B, Kingdom of Saudi Arabia. Saudi Arabian Ministry of Petroleum and Mineral Resources, Directorate General of Mineral Resources

Green NL, Usdansky SI (1986) Toward a practical plagioclase-muscovite thermometer. Am Mineral 71:1109–1117

Hansen SE, Rodgers AJ, Schwartz SY, Al-Amri AMS (2007) Imaging ruptured lithosphere beneath the Red Sea and Arabian Peninsula. Earth Planet Sci Lett 259:256–265. https://doi.org/10.1016/j.epsl.2007.04.035

Hargrove US, Stern RJ, Kimura J-I, Manton WI, Johnson PR (2006) How juvenile is the Arabian-Nubian Shield? Evidence from Nd isotopes and pre-Neoproterozoic inherited zircon in the Bi'r Umq suture zone, Saudi Arabia. Earth Planet Sci Lett 252:308–326. https://doi.org/10.1016/j.epsl.2006.10.002

Hempton MR (1987) Constraints on Arabian Plate motion and extensional history of the Red Sea. Tectonics 6:687–705. https://doi.org/10.1029/TC006i006p00687

Hey MH (1954) A new review of the chlorites. Mineral Mag 30:277–292

Hiess J, Condon DJ, McLean N, Noble SR (2012) 238U/235U systematics in terrestrial uranium-bearing minerals. Geology 335:1610–1614. https://doi.org/10.1126/science.1215507

Hofmann C, Courtillot V, Féraud G, Rochette P, Yirgu G, Ketefo E, Pik R (1997) Timing of the Ethiopian flood basalt event and implications for plume birth and global change. Nature 389:838–841. https://doi.org/10.1038/39853

Horstwood MSA, Foster GL, Parrish RR, Noble SR, Nowell GM (2003) Common-Pb corrected in situ U-Pb accessory mineral geochronology by LA-MC-ICP-MS. J Analyt At Spectrom 18:837–846. https://doi.org/10.1039/B304365G

Ibrahim KM, Al-Malabeh A (2006) Geochemistry and volcanic features of Harrat El Fahda: a young volcanic field in northwest Arabia, Jordan. J Asian Earth Sci 27:147–154. https://doi.org/10.1016/j.jseaes.2005.01.009

Ibrahim KM, McCourt WJ (1995) Neoproterozoic granitic magmatism and tectonic evolution of the northern Arabian Shield: evidence from southwest Jordan. J African Earth Sci 20:103–118. https://doi.org/10.1016/0899-5362(95)00037-T

Ibrahim KM, Tarawneh K, Rabba' I (2003) Phases of activity and geochemistry of basaltic dike systems in northeast Jordan parallel to the Red Sea. J Asian Earth Sci 21:467–472. https://doi.org/10.1016/s1367-9120(02)00075-5

Ilani S, Harlavan Y, Tarawneh K, Rabba I, Weinberger R, Ibrahim K, Peltz S, Steinitz G (2001) New K-Ar ages of basalts from the Harrat Ash Shaam volcanic field in Jordan: implications for the span and duration of the upper-mantle upwelling beneath the western Arabian plate. Geology 29:171–174. https://doi.org/10.1130/0091-7613

Irvine TN, Baragar WRA (1971) A guide to the chemical classification of the common volcanic rocks. Can J Earth Sci 8:523–548. https://doi.org/10.1139/e71-055

Jackson SE, Pearson NJ, Griffin WL, Belousova EA (2004) The application of laser ablation-inductively coupled plasma-mass spectrometry to in situ U-Pb zircon geochronology. Chem Geol 211:47–69. https://doi.org/10.1016/j.chemgeo.2004.06.017

Jarrar G (2001) The youngest Neoproterozoic mafic dyke suite in the Arabian Shield: mildly alkaline dolerites from South Jordan—their geochemistry and petrogenesis. Geol Mag 138:309–323. https://doi.org/10.1017/S0016756801005295

Johnson PR (1998) Tectonic map of Saudi Arabia and adjacent areas. Saudi Arabian Ministry of Petroleum and Mineral Resources, Deputy Ministry for Mineral Resources Technical Report USGS-TR-98-3 (IR 948)

Johnson PR, Andresen A, Collins AS, Fowler AR, Fritz H, Ghebreab W, Kusky T, Stern RJ (2011) Late Cryogenian-Ediacaran history of the Arabian-Nubian Shield: a review of depositional, plutonic, structural, and tectonic events in the closing stages of the northern East African Orogen. J African Earth Sci 61:167–232. https://doi.org/10.1016/j.jafrearsci.2011.07.003

Johnson PR, Kattan FH, Al-Saleh AM (2004) Neoproterozoic ophiolites in the Arabian Shield: field relations and structure. Dev Precambrian Geol 13:129–162. https://doi.org/10.1016/S0166-2635(04)13004-1

Jowett EC (1991) Fitting iron and magnesium into the hydrothermal chlorite geothermometer. In: Geol Assoc Canada/Mineral Assoc Canada/Soc Econ Geol Joint Annual Meeting, abstracts, p 62

Kaliwoda M, Altherr R, Meyer H-P (2007) Composition and thermal evolution of the lithospheric mantle beneath the Harrat Uwayrid, eastern flank of the Red Sea rift (Saudi Arabia). Lithos 99:105–120. https://doi.org/10.1016/j.lithos.2007.06.013

Keir D, Ebinger CJ, Stuart GW, Daly E, Ayele A (2006) Strain accommodation by magmatism and faulting as rifting proceeds to breakup: seismicity of the northern Ethiopian rift. J Geophys Res Solid Earth 111:B05314. https://doi.org/10.1029/2005JB003748

Kessel R, Stein M, Navon O (1998) Petrogenesis of late Neoproterozoic dikes in the northern Arabian-Nubian Shield: implications for the origin of A-type granites. Precambrian Res 92:195–213. https://doi.org/10.1016/S0301-9268(98)00075-8

Ketchum JWF, Jackson SE, Culshaw NG, Barr SM (2001) Depositional and tectonic setting of the Paleoproterozoic Lower Aillik Group, Makkovik Province, Canada: evolution of a passive margin-foredeep sequence based on petrochemistry and U-Pb (TIMS and LAM-ICP-MS) geochronology. Precambrian Res 105:331–356. https://doi.org/10.1016/S0301-9268(00)00118-2

Khattab MM (1992) The extent of oceanization of the Egyptian northern Red Sea crust indicated by gravity and magnetic data. J African Earth Sci (and Middle East) 14:493–498. https://doi.org/10.1016/0899-5362(92)90081-M

Koulakov I, El Khrepy S, Al-Arifi N, Sychev I, Kuznetsov P (2014) Evidence of magma activation beneath the Lunayyir basaltic field (Saudi Arabia) from attenuation tomography. Solid Earth Discuss 6:1401–1421. https://doi.org/10.5194/sed-6-1401-2014

Krienitz M-S, Haase KM, Mezger K, van den Bogaard P, Thiemann V, Shaikh-Mashail MA (2009) Tectonic events, continental intraplate volcanism, and mantle plume activity in northern Arabia: constraints from geochemistry and Ar-Ar dating of Syrian lavas. Geochem Geophys Geosyst 10:Q04008. https://doi.org/10.1029/2008gc002254

Kröner A (1985) Ophiolites and the evolution of tectonic boundaries in the late proterozoic Arabian-Nubian shield of northeast Africa and Arabia. Precambrian Res 27:277–300. https://doi.org/10.1016/0301-9268(85)90016-6

Kröner A, Linnebacher P, Stern RJ, Reischmann T, Manton W, Hussein IM (1991) Evolution of Pan-African island arc assemblages in the southern Red Sea Hills, Sudan, and in southwestern Arabia as exemplified by geochemistry and geochronology. Precambrian Res 53:99–118. https://doi.org/10.1016/0301-9268(91)90007-W

Lazar M, Ben-Avraham Z, Garfunkel Z (2012) The Red Sea—new insights from recent geophysical studies and the connection to the Dead Sea fault. J African Earth Sci 68:96–110. https://doi.org/10.1016/j.jafrearsci.2012.04.001

Leake BE, Woolley AR, Arps CES, Birch WD, Gilbert MC, Grice JD, Hawthorne FC, Kato A, Kisch HJ, Krivovichev VG, Linthout K, Laird J, Mandarino JA, Maresch WV, Nickel EH, Rock NMS, Schumacher JC, Smith DC, Stephenson NCN, Ungaretti L, Whittaker EJW, Guo Y (1997) Nomenclature of amphiboles; report of the subcommittee on amphiboles of the international mineralogical association, commission on new minerals and mineral names. Am Mineral 82:1019–1037

Locock AJ (2014) An Excel spreadsheet to classify chemical analyses of amphiboles following the IMA 2012 recommendations. Comput Geosci 62:1–11. https://doi.org/10.1016/j.cageo.2013.09.011

Ludwig KR (2003) Isoplot/Ex version 3.0: a geochronological toolkit for Microsoft Excel. In: Berkeley Geochronology Center Special Publication 4, Berkeley, 70 pp

Makris J, Rihm R (1991) Shear-controlled evolution of the Red Sea: pull apart model. Tectonophysics 198:441–466. https://doi.org/10.1016/0040-1951(91)90166-P

Manetti P, Capaldi G, Chiesa S, Civetta L, Conticelli S, Gasparon M, La Volpe L, Orsi G (1991) Magmatism of the eastern Red Sea margin in the northern part of Yemen from Oligocene to present. Tectonophysics 198:181–202. https://doi.org/10.1016/0040-1951(91)90150-Q

Mohr P (1983) Ethiopian flood basalt province. Nature 303:577–584. https://doi.org/10.1038/303577a0

Morimoto N (1988) Nomenclature of pyroxenes. Mineral Mag 52:535–550

Moufti MR, Moghazi AM, Ali KA (2013) $^{40}Ar/^{39}Ar$ geochronology of the Neogene-Quaternary Harrat Al-Madinah intercontinental volcanic field, Saudi Arabia: implications for duration and migration of volcanic activity. J Asian Earth Sci 62:253–268. https://doi.org/10.1016/j.jseaes.2012.09.027

Nehlig P, Asfirane F, Genna A, Guerrot C, Nicol N, Salpeteur I, Shanti M, Thiéblemont D (2001) Aeromagnetic map constrains cratonization of the Arabian Shield. Terra Nova 13:347–353. https://doi.org/10.1046/j.1365-3121.2001.00370.x

Nehlig P, Genna A, Asfirane F, Dubreuil N, Guerrot C, Eberlé JM, Kluyver HM, Lasserre JM, Le Goff E, Nicol N, Salpeteur N, Shanti M, Thiéblemont D, Truffert C (2002) A review of the Pan-African evolution of the Arabian Shield. GeoArabia 7:103–124

Pallister JS (1987) Magmatic history of Red Sea rifting: perspective from the central Saudi Arabian coastal plain. Geol Soc Am Bull 98:400–417. https://doi.org/10.1130/0016-7606

Pallister JS (1986) Geologic map of the Al Lith Quadrangle, Sheet 20D, Kingdom of Saudi Arabia. Ministry of Petroleum and Mineral Resources, Directorate General of Mineral Resources

Pallister JS, McCausland WA, Jónsson S, Lu Z, Zahran HM, El Hadidy S, Aburukbah A, Stewart ICF, Lundgren PR, White RA, Moufti MRH (2010) Broad accommodation of rift-related extension recorded by dyke intrusion in Saudi Arabia. Nat Geosci 3:705–712. https://doi.org/10.1038/ngeo966

Pallister JS, Stacey JS, Fischer LB, Premo WR (1987) Arabian Shield ophiolites and Late Proterozoic microplate accretion. Geology 15:320–323. https://doi.org/10.1130/0091-7613

Park Y, Nyblade AA, Rodgers AJ, Al-Amri A (2007) Upper mantle structure beneath the Arabian Peninsula and northern Red Sea from teleseismic body wave tomography: implications for the origin of Cenozoic uplift and volcanism in the Arabian Shield. Geochem Geophys Geosyst 8:Q06021. https://doi.org/10.1029/2006GC001566

Passchier CW, Trouw RAJ (2005) Microtectonics. Springer, Berlin, Heidelberg, New York

Paterson SR, Vernon RH, Tobisch OT (1989) A review of criteria for the identification of magmatic and tectonic foliation in granitoids. J Struct Geol 11:349–363

Pautot G (1983) Les fosses de la Mer Rouge: approche géomorphologique d'un stade initial d'ouverture océanique réalisée à l'aide du Seabeam. Oceanol Acta 6:235–244

Pavoni N (1992) Rifting of Africa and pattern of mantle convection beneath the African plate. Tectonophysics 215:35–53. https://doi.org/10.1016/0040-1951(92)90073-F

Pellaton C (1982) Geological map of the Umm Laji Quadrangle, Sheet 25 B, Kingdom of Saudi Arabia. Saudi Arabian Ministry of Petroleum and Mineral Resources, Directorate General of Mineral Resources

Pellaton C (1979) Geological map the Yanbu al Bahr Quadrangle, Sheet 24C, Kingdom of Saudi Arabia. Saudi Arabian Ministry of Petroleum and Mineral Resources, Directorate General of Mineral Resources

Pryer LL, Robin P-YF (1996) Differential stress control on the growth and orientation of flame perthite: a palaeostress direction indicator. J Struct Geol 18:1151–1166

Quick JE (1991) Late Proterozoic transpression on the Nabitah fault system - implications for the assembly of the Arabian Shield. Precambrian Res 53:119–147. https://doi.org/10.1016/0301-9268(91)90008-X

Ridolfi F, Renzulli A (2012) Calcic amphiboles in calc-alkaline and alkaline magmas: thermobarometric and chemometric empirical equations valid up to 1,130 C and 2.2 GPa. Contrib Mineral Petrol 163:877–895. https://doi.org/10.1007/s00410-011-0704-6

Riisager P, Knight KB, Baker JA, Ukstins Peate I, Al-Kadasi M, Al-Subbary A, Renne PR (2005) Paleomagnetism and $^{40}Ar/^{39}Ar$ Geochronology of Yemeni Oligocene volcanics: implications for timing and duration of Afro-Arabian traps and geometry of the Oligocene paleomagnetic field. Earth Planet Sci Lett 237:647–672. https://doi.org/10.1016/j.epsl.2005.06.016

Robinson FA, Foden JD, Collins AS, Payne JL (2014) Arabian Shield magmatic cycles and their relationship with Gondwana assembly: insights from zircon U-Pb and Hf isotopes. Earth Planet Sci Lett 408:207–225. https://doi.org/10.1016/j.epsl.2014.10.010

Roger J, Platel JP, Cavelier C, Bourdillon-de-Grissac C (1989) Données nouvelles sur la stratigraphie et l'histoire géologique du Dhofar (sultanat d'Oman). Bull Société Géologique Fr 8:265–277. https://doi.org/10.2113/gssgfbull.v.2.265

Rollinson H (1993) Using geochemical data: evaluation, presentation, interpretation. Longman, Essex

Rona PA, Boström K, Epstein S (1980) Hydrothermal quartz vug from the Mid-Atlantic Ridge. Geology 8:569–572. https://doi.org/10.1130/0091-7613

Roobol MJ, Al-Rehaili M (1998) Geohazards along the Makkah-Madinah-Nafud (MMN) volcanic line. Saudi Arabian Deputy Ministry for Mineral Resources Tech Rep BRGM-TR-98, 20 pp

Rooney TO, Hanan BB, Graham DW, Furman T, Blichert-Toft J, Schilling J-G (2012) Upper mantle pollution during Afar plume-continental rift interaction. J Petrol 53:365–389. https://doi.org/10.1093/petrology/egr065

Runge MG, Bebbington MS, Cronin SJ, Lindsay JM, Moufti MR (2016) Integrating geological and geophysical data to improve probabilistic hazard forecasting of Arabian Shield volcanism. J Volcanol Geotherm Res 311:41–59. https://doi.org/10.1016/j.jvolgeores.2016.01.007

Saccani E (2015) A new method of discriminating different types of post-Archean ophiolitic basalts and their tectonic significance using Th-Nb and Ce-Dy-Yb systematics. Geosci Front 6:481–501. https://doi.org/10.1016/j.gsf.2014.03.006

Schettino A, Macchiavelli C, Pierantoni PP, Zanoni D, Rasul N (2016) Recent kinematics of the tectonic plates surrounding the Red Sea and Gulf of Aden. Geophys J Int 207:457–480. https://doi.org/10.1093/gji/ggw280

Schmidt DL, Hadley DG, Brown GF (1983) Middle Tertiary continental rift and evolution of the Red Sea in southwestern Saudi Arabia. Saudi Arabian Deputy Ministry for Mineral Resources, Open-File Report, USGS-OF-03-06, 56 pp

Sebai A, Zumbo V, Féraud G, Bertrand H, Hussain AG, Giannerini G, Campredon R (1991) 40Ar/39Ar dating of alkaline and tholeiitic magmatism of Saudi Arabia related to the early Red Sea Rifting. Earth Planet Sci Lett 104:473–487. https://doi.org/10.1016/0012-821X(91)90223-5

Shanti M, Roobol MJ (1979) A late Proterozoic ophiolite complex at Jabal Ess in northern Saudi Arabia. Nature 279:488–491. https://doi.org/10.1038/279488a0

Shaw JE, Baker JA, Kent AJR, Ibrahim KM, Menzies MA (2007) The geochemistry of the Arabian lithospheric mantle—a source for intraplate volcanism? J Petrol 48:1495–1512. https://doi.org/10.1093/petrology/egm027

Shaw JE, Baker JA, Menzies MA, Thirwall MF, Ibrahim KM (2003) Petrogenesis of the largest intraplate volcanic field on the Arabian plate (Jordan): a mixed lithosphere-asthenosphere source activated by lithospheric extension. J Petrol 44:1657–1679. https://doi.org/10.1093/petrology/egg052

Stern RJ (1994) Neoproterozoic (900-550 Ma) arc assembly and continental collision in the East African orogen: implications for the consolidation of Gondwanaland. Ann Rev Earth Planet Sci 22:319–351

Stern RJ, Gottfried D, Hedge CE (1984) Late Precambrian rifting and crustal evolution in the northeastern desert of Egypt. Geology 12:168–172. https://doi.org/10.1130/0091-7613

Stipp M, Stünitz H, Heilbronner R, Schmid SM (2002) The eastern Tonale fault zone: a "natural laboratory" for crystal plastic deformation of quartz over a temperature range from 250 to 700 C. J Struct Geol 24:1861–1884. https://doi.org/10.1016/S0191-8141(02)00035-4

Stoeser DB, Camp VE (1985) Pan-African microplate accretion of the Arabian Shield. Geol Soc Am Bull 96:817–826. https://doi.org/10.1130/0016-7606

Stoeser DB, Frost CD (2006) Nd, Pb, Sr, and O isotopic characterization of Saudi Arabian Shield terranes. Chem Geol 226:163–188. https://doi.org/10.1016/j.chemgeo.2005.09.019

Stoeser DB, Whitehouse MJ, Stacey JS (2001) The Khida terrane? Geology of Paleoproterozoic rocks in the Muhayil area, eastern Arabian shield, Saudi Arabia. Gondwana Res 4:192–194. https://doi.org/10.1016/S1342-937X(05)70691-2

Sultan M, Becker R, Arvidson RE, Shore P, Stern RJ, El Alfy Z, Attia RI (1993) New constraints on Red Sea rifting from correlations of Arabian and Nubian Neoproterozoic outcrops. Tectonics 12:1303–1319. https://doi.org/10.1029/93TC00819

Sun S-S, McDonough WF (1989) Chemical and isotopic systematics of oceanic basalts: implications for mantle composition and processes. Geol Soc London, Spec Publ 42:313–345. https://doi.org/10.1144/GSL.SP.1989.042.01.19

Vail JR (1985) Pan-African (late Precambrian) tectonic terrains and the reconstruction of the Arabian-Nubian Shield. Geology 13:839–842. https://doi.org/10.1130/0091-7613

Van Achterbergh E, Ryan CG, Jackson SE, Griffin WL (2001) Data reduction software for LAICPMS. In: Sylvester P (ed) Laser ablation ICPMS in the earth sciences: Principles and applications. Mineralogical Assoc Canada Short Course Ser 29, pp 239–243

Veeningen R, Rice AHN, Schneider DA, Grasemann B (2015) Thermochronology and geochemistry of the Pan-African basement below the Sab'atayn Basin, Yemen. J African Earth Sci 102:131–148. https://doi.org/10.1016/j.jafrearsci.2014.11.009

Vernon RH (2004) A practical guide to rock microstructure. Cambridge University Press

Westrich HR, Stockman HW, Eichelberger JC (1988) Degassing of rhyolitic magma during ascent and emplacement. J Geophys Res 93:6503–6511. https://doi.org/10.1029/JB093iB06p06503

Wilson JWP, Roberts GG, Hoggard MJ, White NJ (2014) Cenozoic epeirogeny of the Arabian Peninsula from drainage modeling. Geochem Geophys Geosyst 15:3723–3761. https://doi.org/10.1002/2014GC005283

Winchester JA, Floyd PA (1977) Geochemical discrimination of different magma series and their differentiation products using immobile elements. Chem Geol 20:325–343. https://doi.org/10.1016/0009-2541(77)90057-2

Windley BF, Whitehouse MJ, Ba-Bttat MAO (1996) Early Precambrian gneiss terranes and Pan-African island arcs in Yemen: crustal accretion of the eastern Arabian Shield. Geology 24:131–134. https://doi.org/10.1130/0091-7613

Wright TJ, Ebinger C, Biggs J, Ayele A, Yirgu G, Keir D, Stork A (2006) Magma-maintained rift segmentation at continental rupture in the 2005 Afar dyking episode. Nature 442:291–294. https://doi.org/10.1038/nature04978

Manganese Mineralization Related to the Red Sea Rift System: Examples from the Red Sea Coast and Sinai, Egypt

Nasser L. El Agami

Abstract

Manganese or ferro-manganese mineralizations occur along the western Red Sea coastal zone and Sinai, Egypt. Their occurrence and distribution patterns are closely associated with the Red Sea rifting. The Halaib area and Wadi Araba in the Eastern Desert, and Sharm El Sheikh and Um Zariq areas in South Sinai represent the hydrothermal manganese deposits along the Red Sea coast. These deposits occur as veins and fracture filling cutting across the structure of the host rocks (Paleozoic, Miocene and Plio-Pleistocene sedimentary rocks, Precambrian granite and volcanic rocks), indicating the epigenetic origin. The mineral forming associations are typical for hydrothermal Mn deposits. This mineralization is characterized by enrichment in Tl, Mo, Zn, Cu, Sr, Ba, Pb and U. The REE patterns are characterized by negative to no Ce-and negative Eu-anomalies. The high Tl and Mo contents, U/Th ratios and REE patterns provide a firm evidence for a hydrothermal origin, related to Red Sea rifting. The Miocene Mn deposits of the Abu Shaar area on the Red Sea coast are formed by diagenetic replacement of the calcitic materials by Mn-bearing solutions during a marine transgression-regression. Based on petrographic and texture evidence (i.e., graded pisoliths and ooliths accumulation), characteristic mineral-chemical enrichment and geochemical association (low Tl, Zn, U, Th, low REE budget and positive Ce-anomaly) indicate a marine-diagenetic origin.

N. L. El Agami (✉)
Applied Geology Division, Department of Environmental Geology, Saudi Geological Survey, Jeddah, Saudi Arabia
e-mail: ellagami.nl@sgs.org.sa

N. L. El Agami
Nuclear Materials Authority, Cairo, Egypt

1 Introduction

Oceanic and paleo-environmental studies, especially over the last half century, have contributed much to the understanding of the nature of manganese deposit formation (Rona 1984; Frakes and Bolton 1992). Marine Mn-bearing deposits are generally classified as three genetic types, hydrogenetic, diagenetic and hydrothermal (Bolton et al. 1988; Hein et al. 1997). The first type is represented by ferromanganese crusts, which slowly precipitate from seawater at the seafloor substrate, while the diagenetic deposits, which are Mn-carbonate or oxides, result from direct precipitation of Mn during early diagenesis either at the seafloor or within the soft sediments due to changing redox conditions (Coleman 1985; Hein and Koski 1987; Polgari et al. 1991). The manganese hydrothermal deposits belonging to the third type precipitate from marine low-to medium-temperature hydrothermal fluids when mixed with cold seawater. These three types of Mn deposits differ in their mineralogy, geochemistry, and mode of occurrence. However, there are Mn deposits that originate from a combination of processes, for example, most Mn nodules that form on abyssal plains derive Mn from seawater (hydrogenetic) as well as from pore waters (diagenetic; Dymond et al. 1984). In addition, some Fe–Mn oxy-hydroxide crusts precipitate from mixed hydrothermal fluids and seawater (Glasby et al. 1997; Hein et al. 1997). Thus, differentiation among the genetic types can be ambiguous, especially in fossil deposits.

Egypt lies at the northeastern corner of the African Plate and the exposure of the Precambrian basement rocks in the Eastern Desert, southern Western Desert and Sinai constitute the extreme northwestern part of the Arabian–Nubian Shield, above which the Phanerozoic strata were deposited. The basement rocks have undergone various crustal disturbances and deformational events that were terminated by the late-Proterozoic Pan-African tectono-thermal episode. The Phanerozoic intra-plate deformation and associated processes of erosion and sedimentation were generally controlled by reactivated fault systems and global eustatic

N. M. A. Rasul and I. C. F. Stewart (eds.), *Geological Setting, Palaeoenvironment and Archaeology of the Red Sea*,
https://doi.org/10.1007/978-3-319-99408-6_21

changes throughout this Era (Schandelmeir et al. 1987; Meshref 1990; El Aref 1994). Polymetallic mineralizations occur along the western Red Sea coastal zone and Sinai Peninsula; these are iron, manganese, barite, lead, zinc, bauxite, copper, and uranium (Hussein and El Sharkawi 1990; El Aref 1994; Zaghloul et al. 1995; El Agami 1996; Abd El Monem et al. 1997; El Agami et al. 1999, 2000; Hassan 2002; Fahim et al. 2013; Baiyoumi et al. 2014). Miller et al. (1966) mentioned that deposits of hydrothermal Fe, Mn and Ba-bearing brines occur in the Red Sea. Hassan (2002) concluded that the Ba–Fe–Mn and U–Pb–Zn mineralizations occurring along the western Red Sea coastal zone of Egypt are of hydrothermal origin genetically related to Oligo-Miocene volcanic activity associated with the Red Sea rifting. On the contrary, El Aref (1994) mentioned that Neogene rift-related Pb, Zn, Mn, S, Ba, and Sr stratiform and stratabound deposits coincided with the Red Sea syn-rift coastal facies, uplift and erosions. He furthermore added that it is unlikely that these Neogene ore deposits with their host lithologies and facies can be attributed to rift-related sea-floor spreading and hot brines. It is worth mentioning that the northern part of the Red Sea rift is in an embryonic stage, and sea-floor spreading with hot brines is only present in the Red Sea trough. In addition, neither deep-water sediments nor evidence from hot brine pools have been recorded in the syn-rift Neogene succession of Egypt.

Several manganese or ferro-manganese mineralizations occur along the western Red Sea coastal zone and Sinai. Their occurrence and distribution patterns are closely associated with Carboniferous and Tethyan oceans as well as the Red Sea rifting, and therefore the metallogenic provinces of manganese in Egypt are overprinted with its major tectono-stratigraphic units. These Mn deposits range in age from Carboniferous to Plio-Pleistocene. The genesis of the Um-Bogma Mn deposits of southwestern Sinai is a Carboniferous sedimentary origin and not related to Red Sea rifting (El Agami et al. 2000; El Agami 2006); therefore, it is excluded from this study.

The principle objective of this paper is to review, describe the occurrence and distribution, and clarify the tectonic setting and host-rock lithologies as well as depositional environment of the Mn deposits along the Red Sea coast in Egypt. This paper also presents data on the rare-earth element geochemistry of Mn deposits in Egypt.

2 Samples and Analytical Techniques (Methodology)

Eleven Mn–Fe samples from the studied occurrences were analyzed chemically. The samples were collected from trenches and outcrops at the different localities, from Sinai and along the western coastal zone of the Red Sea. There are three samples from the Sharm el Sheikh (Shr) area, two samples from the Um Zariq (Umz) area on the Gulf of Aqaba, Sinai, and three samples from Wadi Araba (Arb) on the western Red Sea coast. One sample is from the Halaib (Hlb) area in the extreme southern part of the Eastern Desert (in addition to the results of major and trace elements analysis of another five samples of the Halaib area from El Agami et al. (1999) that are plotted in Figs. 8 and 9). Two samples are from the Abu Shaar (Ash) Mn deposits on the Red Sea coastal zone.

Mineral identification was carried out using a combination of optical microscopy, X-ray diffraction (XRD) and scanning electron microscope (SEM) supported by an energy dispersive system (EDS). A Philips PW1350 instrument equipped with Co-k radiation and filter at the Geological Survey of Canada (GSC, Ottawa) made the XRD analysis. Phase relations and morphology were examined by the SEM using secondary electron (SE) and backscattered electron (BSE) detectors at the GSC and Nuclear Materials Authority (NMA) in Cairo.

The major elements were determined by automated X-ray fluorescence (XRF) spectrometry using a Philips PW1400 technique, with a relative accuracy of 1% for major elements. Inductively Coupled-Plasma-Mass Spectrometry (ICP-MS) was used for analyzing the trace elements and REE. Analyses were performed at the laboratories of GSC. The chemical analysis data and some elemental ratios are given in Table 1 and for REE in Table 2.

3 Manganese Deposits

Mn–Fe mineralization occurs in the Red Sea coast at the Halaib area in the extreme southern part of the Eastern Desert and Wadi Araba on the Gulf of Suez, as well as at the Sharm el Sheikh Area and Wadi Um Zariq on the Gulf of Aqaba in south Sinai. Moreover, a Mn deposit is located at Gebel Abu Shaar, in the southern part of the Esh el Mellaha range, on the Red Sea coastal zone. The locations of these deposits are shown in Fig. 1.

4 Mode of Occurrence

4.1 Halaib Area

At the Halaib area, vein type Mn deposits are localized along the NW–NS trending tensional fractures within the Precambrian granite, acid volcanic rocks and Miocene sediments. The marginal parts of the Qash Amer granite pluton are characterized by manganese fracture filling (2–50 cm) following a N–S direction (Fig. 2a); Mn veins (0.5–1 m thick) extend for more than 100 m, tending NNE and dipping 5°–

Table 1 Elemental concentrations (major and trace elements) for the MnFe deposits related to the Red Sea

Oxides/trace	Mn deposits from Sinai & Red Sea Coast									Abu Shaar Mn	
S. No.	Umz1	Umz2	Hl5	Arb 1	Arb2	Arb3	Shr1	Shr2	Shr 3	Ab.sh1	Ab.sh2
SiO_2	75.6	63	9.6	31.5	23	43.4	42.4	59.5	8.1	2	1.1
TiO_2	0.19	0.21	0.09	0.14	0.16	0.09	0.29	0.22	0.06	0.08	0.85
Al_2O_3	12.6	8.1	1.4	2.5	1.3	1.8	4.2	1.5	0.4	1.2	0.95
Fe_2O_3	1.4	1.04	1.2	0.7	15	39.1	2.9	6.9	41.9	2.6	2.1
MnO_2	1.14	15.6	68.67	49.47	37.24	0.22	27.92	17.37	23.14	18.2	20.3
MgO	0.15	0.35	1.3	0.45	0.96	0.25	0.78	0.53	0.36	16.1	16.6
CaO	0.24	0.65	1.02	1.41	2.04	0.66	1.34	1.86	6.16	20.9	20.3
Na_2O	3.73	1.85	0.4	1.1	3	0.4	1.9	0.5	0.7	1.4	1.1
K_2O	5	3.87	4.03	1.25	1.94	0.16	1.93	0.84	0.26	0.4	0.3
CO_2T	0.2	0.2	0.2	0.7	0.3	1.73	0.4	0.4	0.3	36.3	35.5
P_2O_5	0.02	0.03	0.15	0.02	0.04	0.03	0.17	0.63	0.15	0.05	0.06
S	0.00	0.14	0.07	0.19	0.81	0.24	0.19	0.06	0.14	0.09	0.1
LOI	0.4	2.6	7.2	8.1	9.7	13.7	10.3	7.2	13.8	0.5	0.9
Total	100.67	97.64	95.38	97.53	95.49	101.78	94.72	97.51	95.47	99.82	100.16
Ag	0.00	0.00	0.1	0.1	0.4	0.2	0.5	0.7	0.2	0.1	0.1
Ba	1510	17300	22000	17000	19000	250	41000	18000	32000	82.2	255
Be	19	2.2	30	3.2	4.1	1.3	12	7.4	26	0.15	0.17
Cd	1.7	0.6	7.5	0.00	3	0.00	2.7	2.7	3.6	na	na
Co	0.00	0.00	72	80	73	45	7	57	68	240	215
Cr	34	31	132	107	82	90	66	60	42	na	na
Cu	16	0.00	50	36	124	201	83	298	2290	60	50
Ga	22	16	31	30	27	8.4	21	12	14	na	na
Hf	5.4	4.6	1	2.2	5.4	0.82	8.5	6.5	6.5	na	na
Mo	51	655	455	32	22	3.7	1155	578	1533	98	70
Nb	10	8.8	7	2.6	5.6	3.4	14	5.4	4	na	na
Ni	0.00	0.00	17	101	34	134	29	24	12	95	110
Pb	808	236	336	3	22	41	18	61	32	na	na
Rb	81	130	42	8.7	13	5.8	60	19	8.7	na	na
Sb	0.2	0.6	0.5	0.00	0.4	0.2	0.4	0.3	0.00	na	na
Sc	1.2	3.8	1.6	2.8	1.9	1.9	2.9	2.2	0.8	na	na
Sn	1.9	4.2	0.00	0.7	0.5	1	2.4	1	0.00	na	na

(continued)

Table 1 (continued)

Oxides/trace	Mn deposits from Sinai & Red Sea Coast									Abu Shaar Mn	
S. No.	Umz1	Umz2	Hlb	Arb 1	Arb2	Arb3	Shr1	Shr2	Shr 3	Ab.sh1	Ab.sh2
Sr	66	336	5500	1000	1200	39	1700	1700	1400	97.7	120
Ta	1.2	0.99	0.68	0.26	0.37	0.23	1.2	0.44	0.52	na	na
Th	17	19	6.9	7.5	10	5.2	11	5.9	18	0.15	0.2
Tl	1.8	3.8	870	190	210	0.94	89	76	170	0.05	0.05
U	16	200	22	2.2	14	3.1	37	26	99	1.74	2
V	9	65	78	48	238	13	41	39	11	2.95	28
Zn	56	593	2260	52	273	196	227	228	889	360	290
Zr	172	199	23	210	250	29	340	280	240	na	na
Y	13	27	110	14	22	17	82	35	310	3.24	5

Note na = not analyzed

30° ESE, intruded into the acidic volcanics. In addition, the Elba alkaline granites are traversed by Mn–oxide veins (1–2 m thick) formed along steeply dipping NNW faults (El Agami et al. 1999). Mn deposits also occur in more or less parallel NW-trending veins that follow steeply dipping faults and fissures cutting across middle Miocene rocks, mainly sandstone conglomerate, limestone and clays. The Mn veins range in length from 200 m to 1 km and in apparent thickness from 0.5 to 3 m. Wall rock alteration at the site of the vein deposits in different rock types is represented by the formation of black calcite. Sometimes both Mn ores and barite occur in the same vein, or barite occurs as separate veins parallel to a Mn vein (El Agami et al. 1999).

4.2 Wadi Araba

At W. Araba (8–9 km SW of the Zafarana lighthouse) Mn deposits occur in the form of lenticular bodies, elongated pockets and bed-like lenses in coarse-grained sandstone of late Paleozoic and lower Cretaceous age (Fig. 2b). Moreover, goethite and limonite cementing in the sandstone are also recorded.

4.3 Sharm El Sheikh

At the Sharm el Sheikh Area, vein type Mn mineralization is localized along N–S trending faults in younger granites (Shr. 3), or within N–S and NW trending faults, along the contact between granite and Oligocene sandstone and conglomerate (Shr.1). Mn ore is interbedded with Oligocene red beds (Shr. 2); the thickness of Mn beds ranges from 30 cm to 60 cm. In addition, Mn occurs as cementing materials for Oligocene and Plio-Pleistocene conglomerate and tectonic breccias developed along the NW-trending faults (Figs. 2c, d).

4.4 Um Zariq Area

The Mn mineralization is concentrated in fractures along brecciated fault planes (trending NE and N–S). These faults and shear zones are developed along the margin of the Gulf Aqaba (5 km from Nuweiba City) because of Miocene to Recent tectonic activity related to strike slip fault and shearing along the Gulf of Aqaba. Some of the Mn lenses are sheared, which points to continuous tectonic activity after the emplacement of Mn mineralization.

4.5 Abu Shaar

At Gebel Abu Shaar el Qebli, and Wadi Beli the Mn ores occur in the syn-rift, Middle Miocene carbonates of the Um Mahara

Table 2 Rare Earth Elements (REE) concentration and ratios for the Mn–Fe deposits related to the Red Sea

REE (ppm)		Mn deposits								Abu Shaar Mn	
S. No.	Umz1	Umz2	Hlb	Arb 1	Arb2	Arb3	Shr1	Shr2	Shr 3	Ab.sh1	Ab.sh2
La	74	54	184	21	13	8.5	28	11	17	1.5	4.59
Ce	71	110	290	66	41	22	51	23	52	3.5	10.6
Pr	14	12	60	6.6	4	2.8	6.6	2.7	4	0.3	0.95
Nd	48	38	240	26	17	11	27	11	18	1.3	4.02
Sm	7.1	5.3	63	6.6	4	2.8	5.9	2.6	6	0.4	1.07
Eu	0.35	0.61	2.9	0.75	0.72	0.44	0.2	0.2	1.1	0.1	0.23
Gd	5.3	3.3	63	4.4	3.8	2.6	7.40	3.1	19	0.3	0.99
Tb	0.76	0.41	11	0.59	0.6	0.36	1.5	0.59	5.4	0.05	0.16
Dy	3.9	2	54	3.2	3.4	2.1	11	4.3	46	0.34	1.02
Ho	0.79	0.39	8.7	0.56	0.72	0.39	2.5	1	11	0.07	0.24
Er	2.1	1	18	1.4	1.8	1	6.7	2.9	32	0.21	0.63

Note Eu/Eu*; Eu- anomaly value calculated as $Eu/Eu^* = Eu_N/(Sm_N \times Gd_N)^{0.5}$ & $Ce/Ce^* = Ce_N/(La_N)$ 2/3 $(Nd_N)^{1/3,}$ Umz = Um Zariq; Hlb = Halaib; Arb = Araba; Shr = Sharm El Sheikh; Ab.sh = Abu Shaar

Formation. The rocks units exposed at G. Abu Shaar el Qebli (southern part of the Esh el Mellaha range) are, from the base, the Precambrian basement rocks (mainly the Dokhan volcanics), the Cenomanian Nubia sandstone, and Miocene rock (basal conglomerate, sandy coralline limestone, dolomitic limestone and gypsum). The plateau is affected by a complex of faults, which can be grouped into three sets of NW–SE, NW–SW and nearly E–W trends (Hassan 2002).

The Mn deposits of the Abu Shaar el Qebli form a bed in the middle Miocene carbonate, and foraminiferal dolomitic limestone with a maximum thickness of about 80 cm, and with sharp contacts between the Mn bed and the overlying and underlying limestone (Fig. 3). This bed extends for 2–3 km. Oxidic manganese ooliths and pisoliths exhibit several cycles of normal grading, and the size of the grading pisoliths often ranges between 2–5 mm over intervals of 20–50 cm (Fig. 3). Abdel Moteleb (1996) divided the Mn ore bed into 3 zones. A lower, red zone up to 20 cm thick consists mainly of iron rich pisoliths and ooliths, red clays, carbonate fragment and calcareous mud. A middle black zone, up to 50 cm thick, consists entirely of Mn-pisoliths and ooliths embedded in a manganiferous and calcareous matrix. An upper red zone, that reaches up to 10 cm in thickness consists of iron ooliths and/or pisoliths and laminated red clay.

5 Mineralogy

5.1 Halaib Area

The microscopic investigations, XRD and SEM with EDX analyses revealed that the Halaib manganese ores are composed of pyrolusite, hollandite, cryptomelane, psilomelane coronadite, manjiroite, coronadite, chalcophanite and Mn-silicate, with minor amounts of goethite and hematite. The main gangue minerals are quartz, calcite, barite and clay minerals.

The paragenetic sequence of the Mn minerals started with the deposition of hollandite, psilomelane, and high barium cryptomelane, which was corroded and replaced by quartz grains, then coronadite, manjorite and finally Mn silicate and pyrolusite. Pyrolusite was formed at the expense of hollandite and high Ba-cryptomelane as in Fig. 4a, b and c, where zones of feathery pyrolusite developed on hollandite and cryptomelane as well as quartz grains (Fig. 4d, e and f). Pyrolusite exhibits needle-like forms at its outer margin. These pyrolusite needles are enclosed in hollandite, cryptomelane and even the quartz grains (Fig. 2e and f). Fragment of polycrystalline pyrolusite show well developed cleavage traces, and typical shrinkage cracks. The pyrolusite phase is usually associated with Mn-silicate, which is sometimes enclosed within pyrolusite (Fig. 4f). The presence of Mn-silicate in association with Mn-oxides provides firm evidence for a hydrothermal origin.

5.2 Wadi Araba

The principal Mn minerals at W. Araba are pyrolusite, cryptomelane, hollandite, and psilomelane, with hematite, goethite, and limonite. Quartz constitutes the main gangue mineral. Pyrolusite is formed at the expense of other Mn-minerals. Colloform texture with concentric bands of Fe and Mn oxides are occasionally observed.

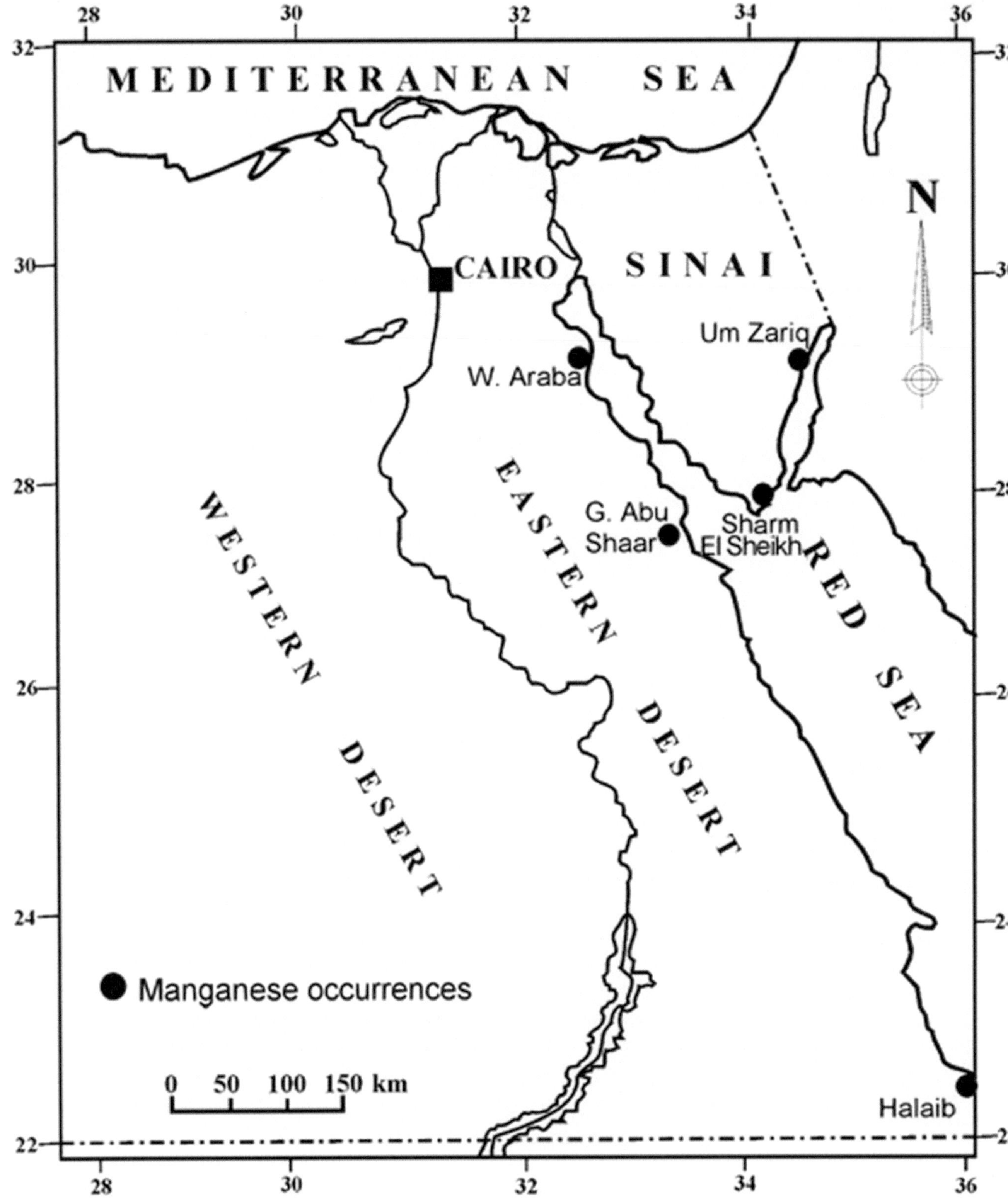

Fig. 1 Location map for Mn occurrences along the Red Sea coast

Fig. 2 **a** Field photograph, Qash Amir granitic pluton, Halaib area. The marginal part of the pluton is characterized by manganese fracture filling (2–50 cm in width) following the main structural trends (N–S, NW and E–W fracture planes). **b** Field photograph showing W. Araba Mn ore that occurs in the form of lenticular bodies, elongated pockets and bed-like lenses in the coarse-grained sandstone of late Paleozoic and lower Cretaceous age. The Mn–Fe ore lenses are intimately connected with fractures, fissures and fault lines (NW fault trend). **c** and D) Field photographs showing that Sharm el Sheikh vein type Mn mineralization is localized along the N–S fault within the granite (Shr.3; 2C), or within N–S faults along the contact between granite and Oligocene sandstone and conglomerate (Shr.1; 2D)

5.3 Sharm El Sheikh Area

The principal manganese minerals at the Sharm el Sheikh area are pyrolusite, cryptomelane, psilomelane, and rare chalcophanite. Goethite usually occurs with Mn-oxides, and gangue minerals including quartz, fluorite and clay. The Mn ore is mainly concretionary and pisolitic; this is in the form of botryoidally kidney-shaped masses, ranging from 5 to 20 cm. The concretions have concentric structures, with iron oxide and Mn–Fe oxide at the outer cortex (70 µm wide laminae) and Mn at the core. This ore is characterized by a colloform texture (Fig. 5). Sometimes a banding of Fe and Mn–Fe oxides exhibits the so-called liesegang rings. However, the liesegang rings in gels may indicate a rhythmical precipitation of Fe and Mn. These liesegang bands and colloform texture reflect that Fe and Mn were transported as gel and colloid, which precipitate, solidify and harden to rounded and reniform masses, known as colloform. The outer cortex is composed of thin laminae of iron and Mn–Fe oxide, and then the core is composed of hollandite and pyrolusite (Fig. 5). The contact between the two phases is sharp and shows no evidence of one mineral replacing the other, suggesting that both phases grew alternately by direct precipitation from solution.

5.4 Um Zariq Area

The main Mn minerals at the Um Zariq area are hollandite, pyrolusite, coronadite and psilomelane, with hematite, quartz and calcite constituting the main gangue minerals. Hollandite and psilomelane were deposited first and then replaced by pyrolusite. Mn may show colloform texture or occurs as cement for granitic fragments.

Fig. 3 Field photographs showing Abu Shaar Mn deposits that consist of Mn- and Fe-rich ooliths and pisoliths embedded in a calcareous and manganiferous matrix, forming a bed in the Middle Miocene carbonate

5.5 Abu Shaar

The mineralogy of the Mn deposits at G. Abu Shaar is pyrolusite, psilomelane, cryptomelane, some manganite and mangano-calcite. Gangue minerals include dolomite, calcite, quartz and limonite. The ooliths and pisoliths are concentrically laminated, with alternating laminae of psilomelane and mangano-calcite or dolomite, and they range from 0.2 mm to 1.2 mm in diameter; both spherical and oval shaped ooliths and pisoliths are encountered (Fig. 6) and are embedded in a calcareous (mangano-calcite) and manganifrous matrix. Foraminifera 1–2.5 mm in diameter, superficial ooliths and cracked pellets (up to 0.5 mm) and several algae were recorded. The chambers of the foraminifera are filled with wad, todorokite (Ca, Mg–Mn oxide), psilomelane and pyrolusite. The mangano-calcite and dolomite occasionally form the barrier between the foraminifer chambers and the groundmass (Fig. 7). Both ooliths and pisoliths have undergone different degrees of deformation and cracking, which may have happened during transportation or diagenesis. A microbial origin of the Mn-mineralized crusts is supported by the presence of encrusted foraminifera, which were also found in the Fe-rich layer. Recently Gradzinski et al. (2004) suggested the existence of a trophic relationship between bacteria and encrusting foraminifera, the former being nutrients for the latter.

6 Geochemistry

The MnO_2 content in mineralized veins of the Halaib area reach up to 68.57%, while the Fe_2O_3 content is 1.2%. The Mn/Fe ratio is 63.4. The MnO_2 content in the sedimentary rocks of Wadi Araba (Arb) ranges from 0.22 to 49.47% with an average of 22.3%, while the Fe_2O_3 content ranges from 0.7 to 39.1% and the Mn/Fe ratio ranges from 0.007 to 78.18. The MnO_2 content of the Sharm el Sheikh Mn deposits ranges from 17.37 to 27.9%, while the Fe_2O_3 content ranges from 2.9% to 41.9%, and the Mn/Fe ratio ranges from 0.6 to 11. The MnO_2 content of the Um Zariq mineralization ranges from 1.14% to 15.6%, while Fe_2O_3 ranges from 1.04 to 1.4% (Table 1). Mn is strongly fractionated from Fe within these deposits.

The Mn contents of the Abu Shaar el Qebli Mn deposits range from 18–20%, with an average of 19.3%, while Fe_2O_3 contents range from 2 to 2.6% with an average of 2.35%. Mn/Fe ratios range from 7.8 to 10.7, so Mn is strongly fractionated from Fe within the main Mn bed at Gebel Abu Shaar.

Generally, the hydrothermal Mn deposits are characterized by depletion in Co, Ni and Cu (Co+Ni+Cu $<$ 0.01%) (Bolton et al. 1988; Usui et al. 1997). Moreover, Nicholson (1992) postulated that hydrothermal Mn deposits are characterized by enrichment in As, Ba, Cu, Pb, Sb, Sr, Li, Cd, Mo, V and Zn. The studied Mn deposits show these features, as seen on a ternary diagram of Co+Ni+Cu versus Fe and Mn (Fig. 8). A diagram of Co+Ni wt. percent against As+Cu+Mo+Pb+V+Zn wt. percent (after Nicholson 1992) is given in Fig. 9, which shows that the studied Mn samples of Abu Shaar area lie in the marine field and all other studied samples in the hydrothermal field. Co/Zn versus Cu+Ni+Co content (from Tooth 1980) is shown in Fig. 10, where the studied samples plot in the field of Mn deposits of hydrothermal origin.

Barium is among the elements considered to be derived from hydrothermal solutions, and is found to be markedly enriched and positively correlated with Mn in recent and ancient hydrothermal deposits (Bonatti et al. 1972; Rossi et al. 1980). In this context, all of the studied hydrothermal Mn deposits are enriched in Ba. The Ba content ranges from 250 to 41000 ppm in the studied locations, except Mn samples of Abu Shaar that show a lower content of Ba that

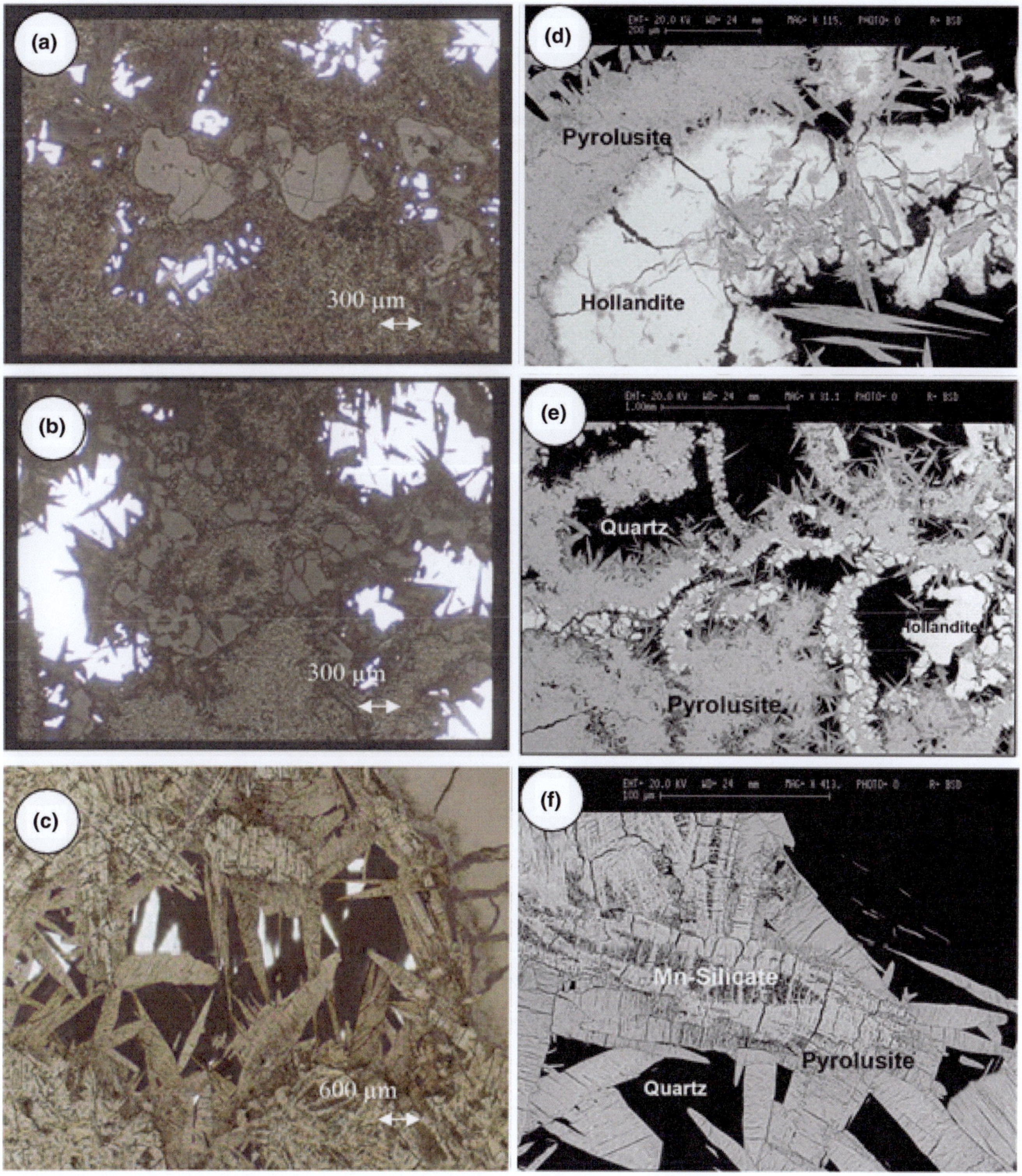

Fig. 4 Halaib Mn deposits. Reflected light photomicrographs (**a–c**), and SEM, BSI (d-f) showing feathery pyrolusite developed on hollandite, pyrolusite needles enclosed in hollandite and quartz grains (b and e), pyrolusite showing a well-developed cleavage traces and shrinkage cracks, and Mn-silicates associated with pyrolusite (**c** and **f**)

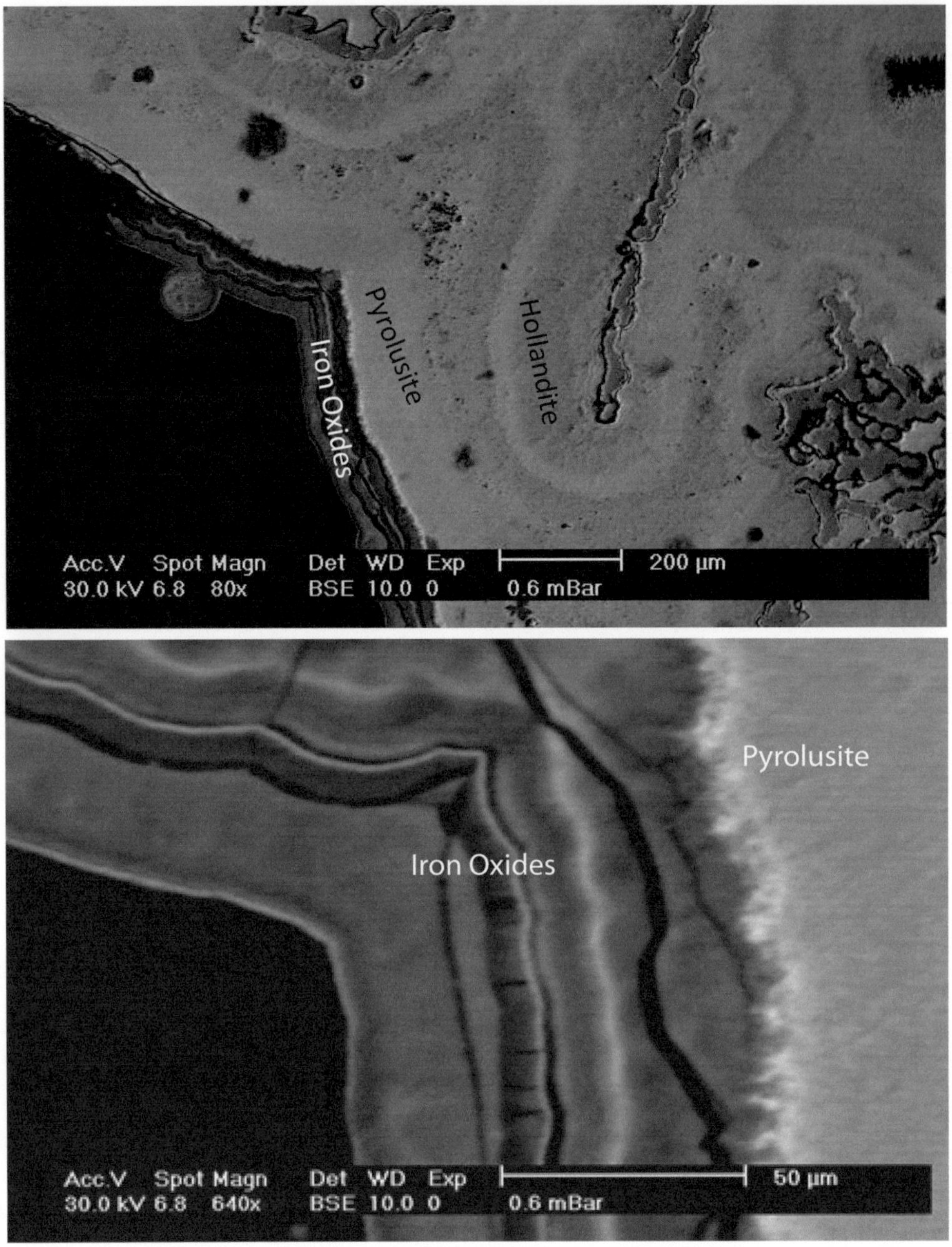

Fig. 5 Sharm el Sheikh Mn deposits. Colloform texture with rhythmic bands of iron and Mn–Fe oxides, the outer cortex is mainly of iron oxides and the core of hollandite and pyrolusite (SEM, BSI)

ranges from 82 to 250 ppm. Similarly, strong enrichment of Mo and Cd was found in hydrothermal Mn crust from the Galapagos Mounds Field (Moorby and Cronan 1983) and the Tonga-Kermedec Ridge (Moorby et al. 1984), which was considered to be of hydrothermal origin. Significant Mo enrichments of hydrothermal origin are also known from the Explorer Ridge Mn crust (Grill et al. 1981). In this context, most of the studied Mn deposits are enriched in Mo, especially the Sharm el Sheikh Mn deposits in which the Mo content reaches up to 1533 ppm.

Hewett et al. (1963) and Roy (1981) mentioned that the high Tl (thallium) content in Mn oxide indicates that it is hypogene in origin rather than supergene. Most of the studied Mn oxide samples are enriched in Tl (ranging from 1.8 to 870 ppm), and the high Tl content provides a firm evidence for a hydrothermal origin; only Abu Shaar Mn

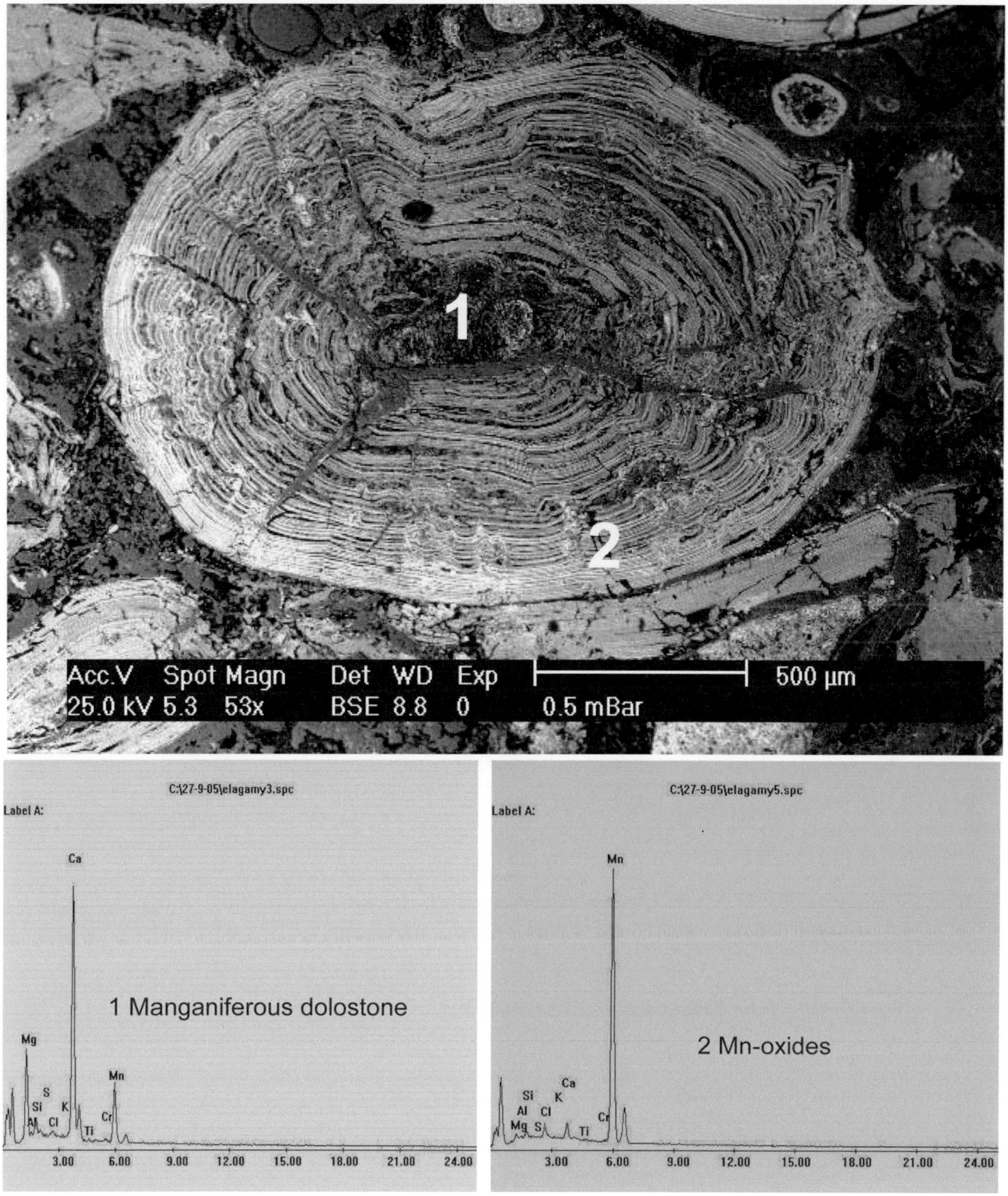

Fig. 6 Abu Shaar Mn deposits. SEM, BSI and EDX analysis for concentric oolitic grainstone, with nucleus of mangano-dolomite and cortex of Mn oxides

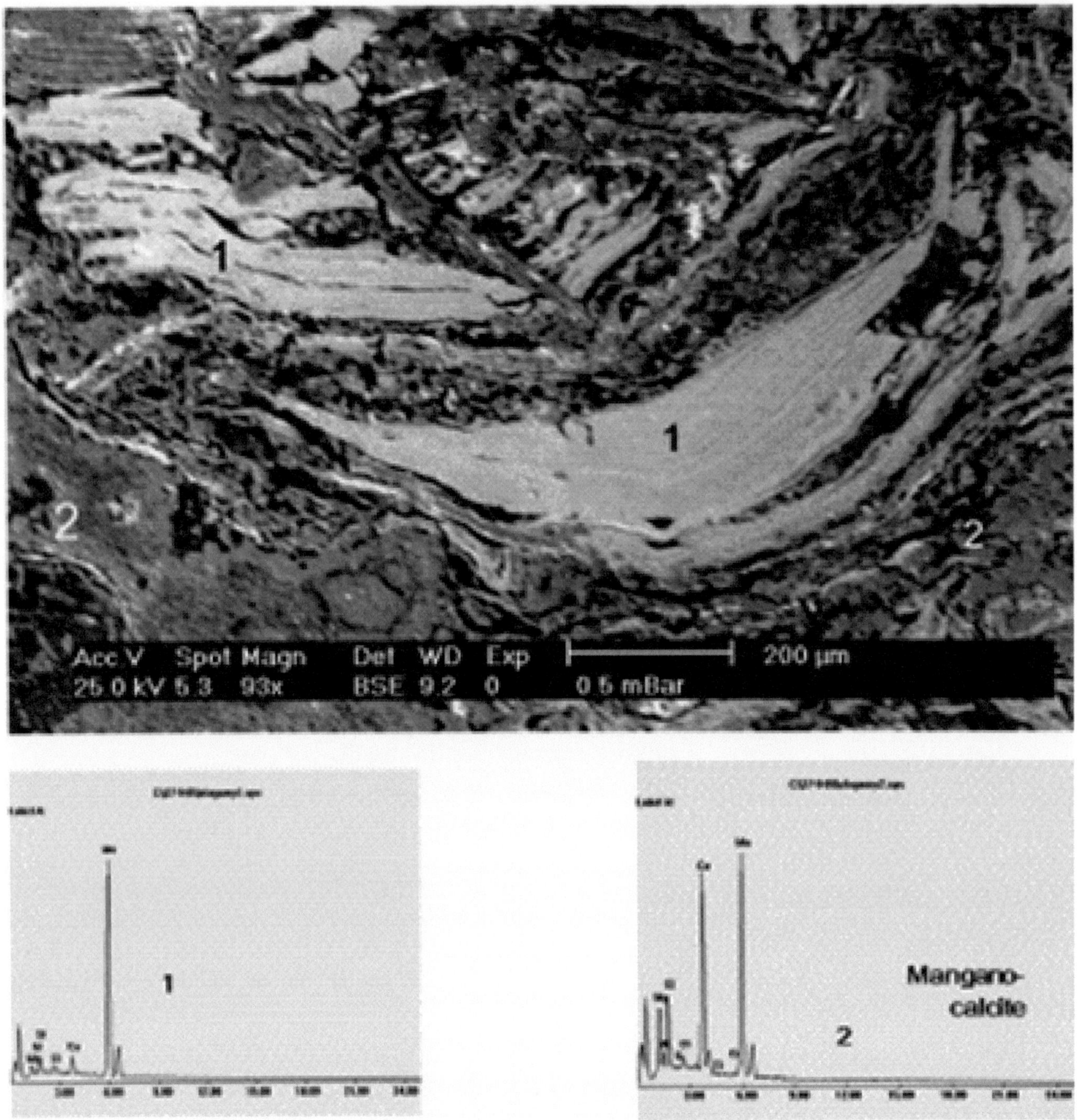

Fig. 7 Abu Shaar Mn deposits. SEM, BSI and EDX analysis for the chambers of foraminifera that are filled with manganese minerals; mainly fine crystals of the chambers are replaced by mangano-calcite

samples show a very low Tl content (0.05 ppm), which indicates a hydrogenous origin. The U/Th ratio may similarly be used to discriminate hydrothermal deposits from sediments and normal pelagic sediments (Bonatti 1975). The U/Th ratios for the studied Mn ores are high (1.4 to 11.6), several orders of magnitude greater than typical marine sediment, indicating a hydrothermal origin for these deposits.

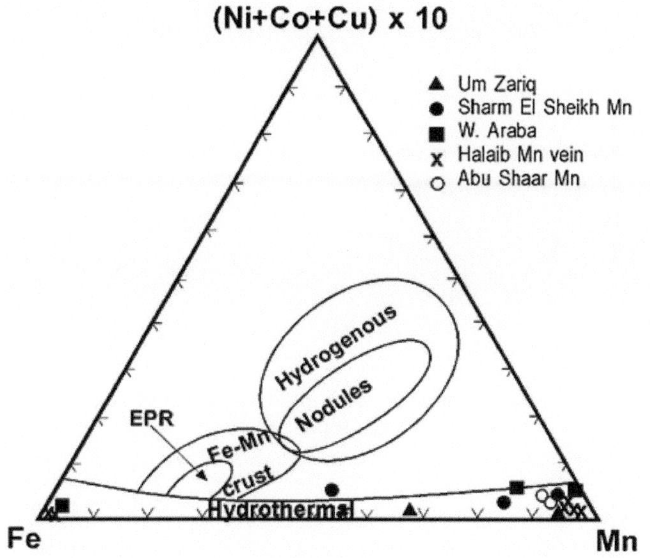

Fig. 8 Ternary diagram showing (Co+Ni+Cu) × 10 versus Fe versus Mn, comparing data from different studied Mn areas with the generalized field for marine and hydrothermal Mn deposits (modified from Bonatti et al. 1972)

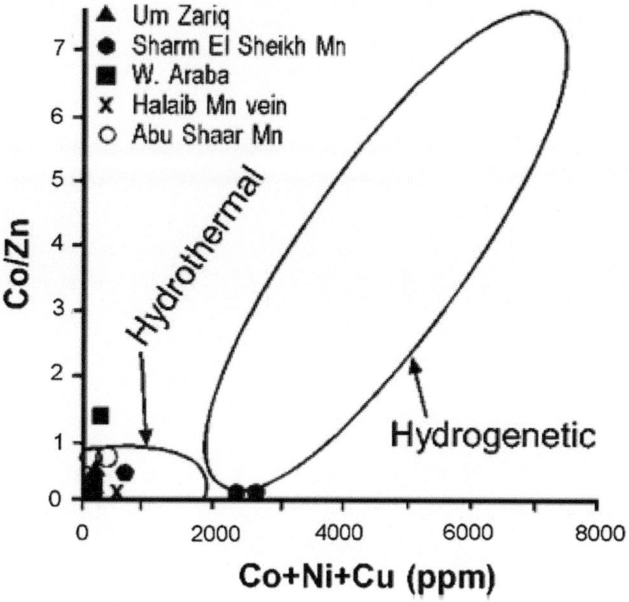

Fig. 10 Co/Zn versus Co+Ni+Cu content (modified from Tooth 1980) for Mn-mineralization at different localities

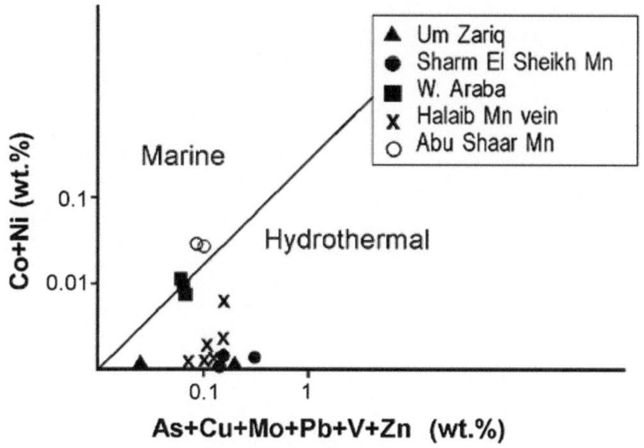

Fig. 9 Diagram of Co+Ni wt. percent against As+Cu+Mo+Pb+V+Zn wt. percent. Diagnostic plot to differentiate hydrothermal manganese oxide and sedimentary marine (cf. Nicholson 1992)

Generally, the hydrogenous Mn deposits are characterized by enrichment in Co, Ni and Cu (Co+Ni+Cu > 0.01%), which reflects the very slow accumulation from oxidizing sea water (Hein et al. 1997). The studied Mn deposits of Abu Shaar show these features, as seen on a bivariant plot of Co+Ni wt. percent versus As+Cu+Mo+Pb+V+Zn wt. percent (after Nicholson 1992), where the studied samples plot in the area of Mn deposits of hydrogenetic origin (Fig. 9). Moreover, on a ternary diagram of Co+Ni+Cu versus Fe and Mn of Bonatti et al. (1972) (Fig. 8), and on the Co/Zn versus Cu+Ni+Co content (from Toth 1980; Fig. 10), the studied samples plot in the area of Mn deposits of hydrothermal and diagenetic origin.

The elemental concentration in Abu Shaar Mn ore is considered to have the lowest content of trace and REE elements among all other types of Mn-deposits in Egypt. This may be attributed to diagenetic processes to which this ore was subjected, or perhaps it was deposited in a protected lagoon or bay which had not enough input from these elements as it was formed with lagoonal carbonate. The most enriched elements in Abu Shaar Mn ore are Ni (95–110 ppm), Co (215–240 ppm), Zn (290–360), Cu (50–60 ppm), Sr (98–120 ppm), V (2–28), but it is depleted in Tl (0.05 ppm), U (1.7–2 ppm), and Th (0.15–0.2).

7 REE (Rare Earth Elements) Geochemistry

The abundance and distribution of rare earth elements (REE) in the studied Mn deposits are given in Table 2. Total REE contents range from 8.4 to 249 ppm, with exception of one sample from the Halaib Mn-oxide that has up to 1009 ppm total REE. These values are small compared with hydrogenetic Mn–Fe crusts, where the concentration of total REE exceeds 1400 ppm (Hein et al. 1997). On the other hand, low contents of REE (about 100 ppm) are characteristic of hydrothermal Mn deposits (Usui et al. 1997). Light Rare Earth Elements (LREE) enrichment relative to Heavy Rare Earth Elements (HREE), which is seen in

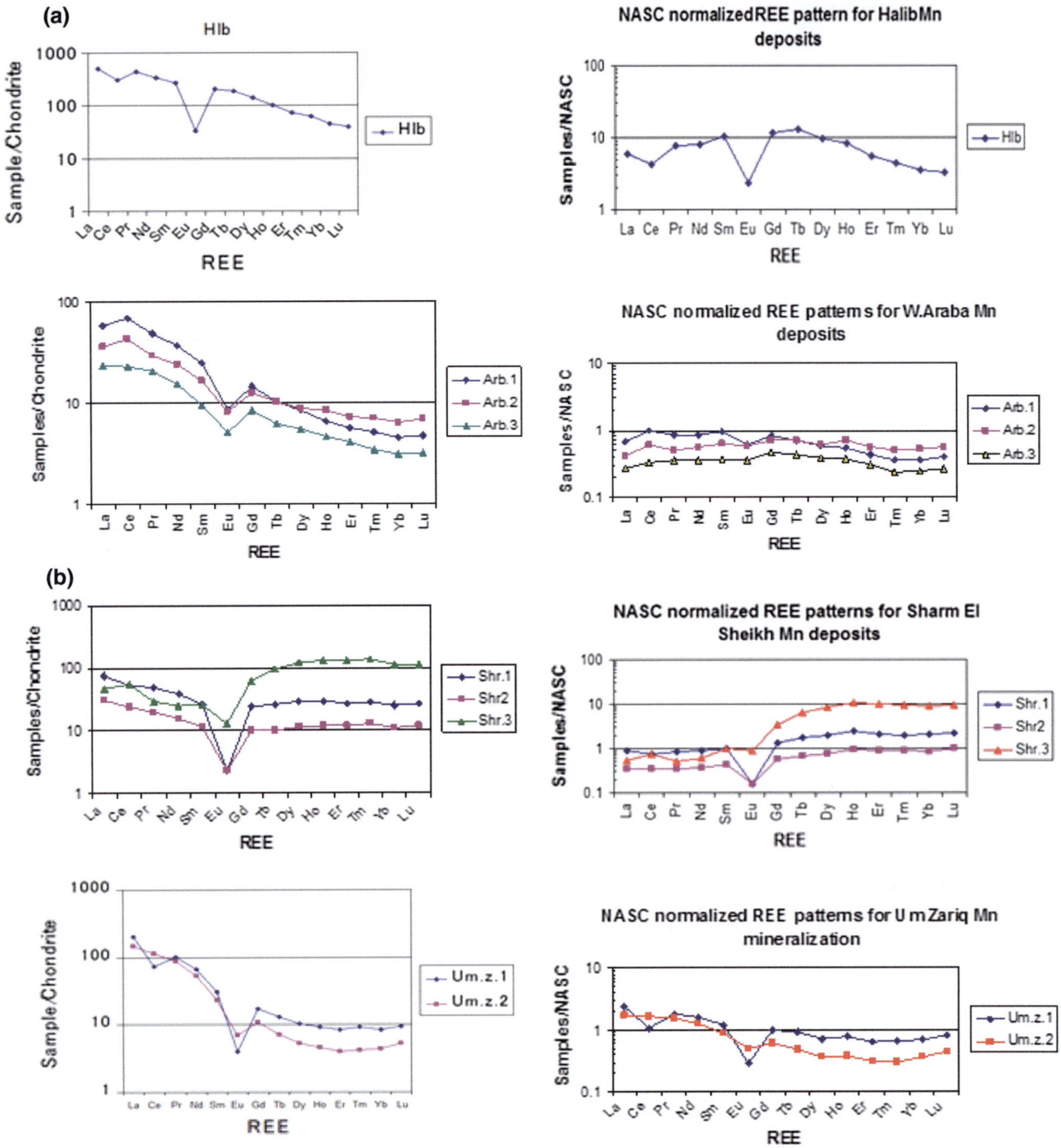

Fig. 11 **a** and **b** Chondrite and NASC normalized REE patterns for hydrothermal Mn mineralization along the Red Sea coast

the studied samples, is also an indicator of hydrothermal origin (Mills and Elderfield 1995).

Chondrite-normalized REE distributions in Mn ore exhibit LREE enrichment (~ 10–$100 \times$ chondrite; only one sample of Halaib ore reaches up to $900 \times$ chondrite) (Figs. 11a, b). LREE/HREE ranges from 0.652 to 25.7. Sample Shr3 of the Sharm el Sheikh Mn deposits, which occur in granite, is enriched in HREE (LREE/HREE 0.65; Table 2). Chondrite-normalized REE patterns for Mn ores exhibit a high negative Eu–anomaly (Eu/Eu* 0.1 to 0.56)

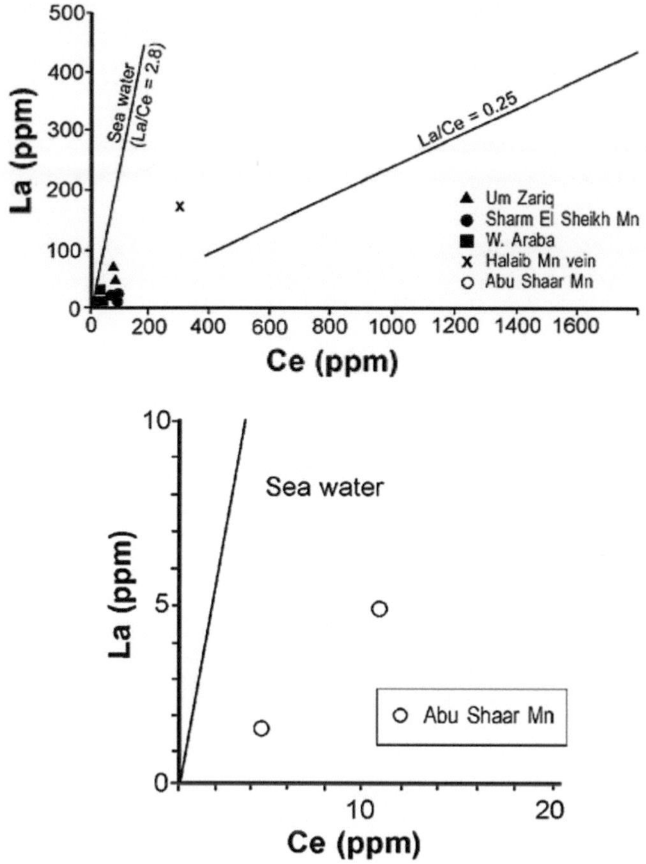

Fig. 12 **a** and **b** La versus Ce concentrations. Hydrothermal crust La/Ce ratios are similar to seawater. All other deposits are enriched in Ce relative to seawater, approaching an apparent lower limit of La/Ce ~0.25

and −ΔCe (Ce/Ce* 0.53–0.88); only the Mn ore of W. Araba and Mn ore in granite at Sharm el Sheikh show a positive Ce anomaly (Ce/Ce* 1.14 to 1.43).

The NASC (North American Shale Composite) normalized REE patterns show that Mn ores have nearly flat patterns and exhibit high −ΔEu and a negative to positive Ce anomaly for Um Zariq and Wadi Araba Mn ores. While Sharm el Sheikh Mn deposits show a depleted pattern (HREE enrichment), the NASC-normalized pattern of Halaib Mn ore is convex compared to those of other localities and exhibits a strong −ΔEu (Eu/Eu* 0.14) and negative Ce anomaly (Ce/Ce* 0.69), which is typical for a hydrothermal pattern (Fig. 11a).

Both the Halaib and Um Zariq patterns show negative Ce, and positive to negative Eu-anomaly. In contrast, W. Araba Mn samples and Mn ore in granite at Sharm el Sheikh show a positive or very weak negative Ce-anomaly, and a negative Eu-anomaly. The highest positive Ce-anomaly (Ce/Ce* 1.4) was detected in Mn ore in granite from the Sharm el Sheikh area, as well as in Wadi Araba Mn ore.

The negative Eu-anomaly (−0.1 to −0.56) may reflect the influence of seawater on the formation of the studied Mn ore (Hein et al. 2000), but the presence of an Eu-anomaly largely depends on the composition of the rocks that the hydrothermal fluids leached (Michard and Albarede 1986; Hein et al. 1997; Glasby et al. 1997). Thus, the Eu anomaly does not always provide an unequivocal indicator of the origin of a deposit, and it is likely that the hydrothermal solutions responsible for the formation of Mn deposits were in contact with and leached a thick continental crust, composed mainly of highly differentiated granite (Eu–depleted).

Most of the studied Mn deposits show negative to no La_{NASC} anomalies as illustrated by a $(La/Pr)_{NASC}$ ratio below unity. The La_{NASC} anomalies range from 0.69 to 1.04, which points to deposition in an area with hydrothermal activity. Moreover, the vast majority of the studied Mn mineralizations show no to negative Gd_{NASC} anomalies, as quantified by Bau et al. (1996), where $Gd/Gd^*_{NASC} = Gd/Sm_{NASC} \times 0.33 + Tb_{NASC} \times 0.67$. Gd/Gd^*_{NASC} for the studied ore deposits are 1.02, 0.89, 1.06 and 1.1 for the Halaib, Sharm el Sheikh, Um Zariq and Araba Mn deposits, respectively. A plot between La and Ce (Figs. 12a, b) illustrates the behaviour of Ce in the studied Mn deposits. La/Ce ratios of the studied Mn ores are low (values <1, with the exception of sample Umz1 which has value equal to or more than 1) and lie between the seawater ratio and the line defining hydrothermal Mn–Fe rich crust, and hydrogenous crusts.

The chondrite normalized REE patterns for Abu Shaar Mn ore show that it is enriched in LREE over HREE (LREE/RREE ranges from 3.7 to 5.7) and exhibits −ΔEu (Eu/Eu* 0.68–0.69) and +ΔCe (Ce/Ce* 1.15). The total REE budget is very low, ranging from 8.5 to 25 ppm; these values are small compared with hydrogenetic Mn–Fe crusts, where the concentration of total REE exceeds 1400 ppm (Hein et al. 1997). On the other hand, low contents of REE (about 100 ppm) are characteristic of hydrothermal and diagenetic Mn deposits (Usui et al. 1997). These REE comparisons indicate that the deposits studied may have formed by diagenetic processes, with a hydrogenetic source as proved by a positive Ce anomaly (Fig. 13).

The Mn deposits of Abu Shaar, are characterized by positive Gd_{NASC} anomalies of 1.1; these are a general feature of hydrogenetic Fe–Mn crusts. The positive Gd_{NASC} anomalies may be inherited from seawater or may be generated during REE sorption. The Abu Shaar Mn deposits also exhibit positive La_{SN} anomalies $(La/Pr)_{SN}$ ratios above unity; this is characteristic of hydrogenetic Fe–Mn crust and reflects the size of the La_{SN} anomaly in seawater. Moreover, the Abu Shaar Mn deposits have positive Y_{SN} anomalies (Y/Ho = 41), as shown by Y/Ho ratios above 28. The positive Y_{SN} anomalies are consistent with the positive Y_{SN} anomaly

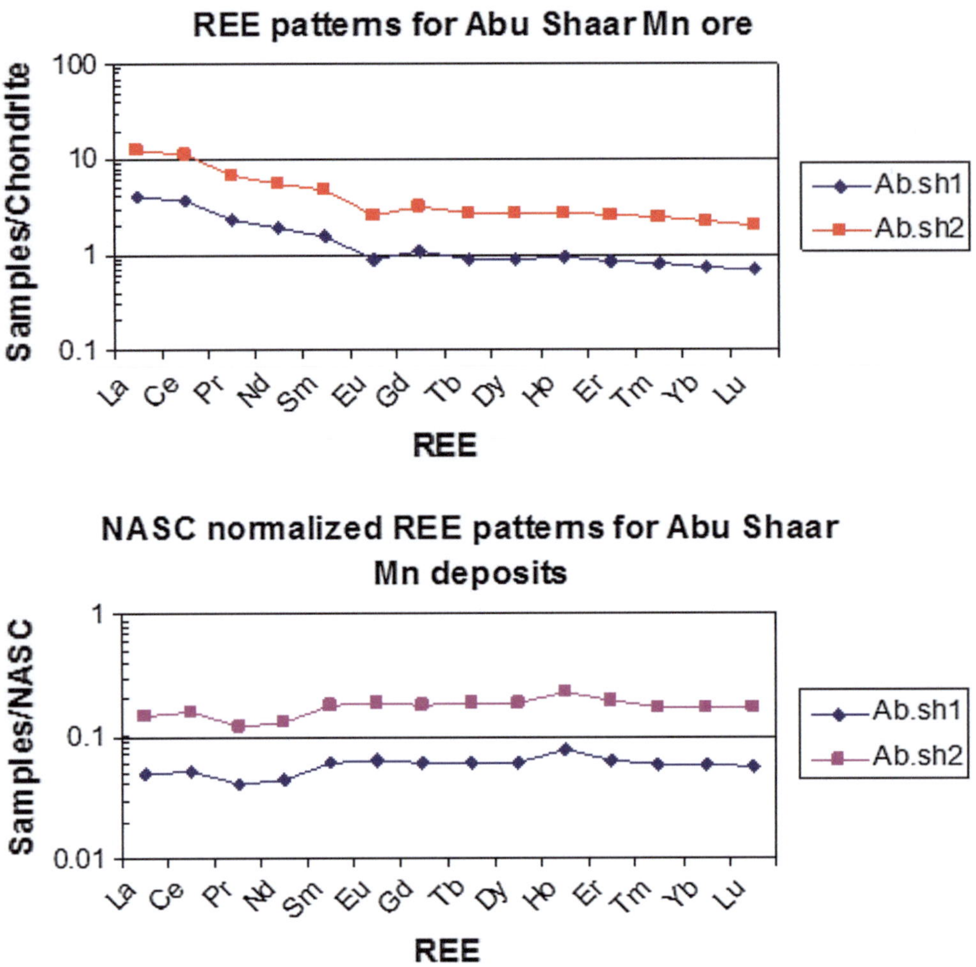

Fig. 13 Chondrite and NASC normalized REE patterns for Abu Shaar Mn deposits

of seawater (Bau et al. 1995). In the Abu Shaar Mn deposits, the La/Ce ratio varies from 0.42 to 0.43. In the light of the above and the conclusions of Dubinin and Volkov (1986), the La/Ce ratio is an expression for the rate of Ce depletion in sediments, and a terrigenous source of REE can be inferred for the Abu Shaar Mn deposits.

8 Discussion

The liesegang bands and colloform texture in the Halaib Mn deposits reflect that iron and Mn were transported as gel and colloid precipitates that then hardened to rounded and reniform masses referred to as colloform. However, the liesegang rings in gels may indicate a rhythmical precipitation of iron and Mn.

Sample Shr3 of the Sharm el Sheikh Mn deposits, which occur in granite, is enriched in HREE (LREE/HREE 0.65; Table 2); this may be attributed to Mn ore associated with fluorite, reflects that the hydrothermal solution is fluorine rich, and therefore has an affinity for HREE rather than LREE.

Negative Ce and positive to negative Eu-anomalies (Mills and Elderfield 1995; Hein et al. 1997) usually characterize hydrothermal Fe–Mn deposits. Both the Halaib and Um Zariq patterns are typically similar to hydrothermal Mn ore. The highest positive Ce-anomaly (Ce/Ce* 1.4) was detected in Mn ore in granite from the Sharm el Sheikh area, as well as in Wadi Araba Mn ore. This may result from a relatively low rate of precipitation and oxygenated conditions during formation of this layer.

The negative Eu-anomaly "between 0.1 to 0.69" can reflect the influence of seawater on the formation of the studied Mn ore (Hein et al. 2000), but the presence of an Eu-anomaly largely depends on the composition of the rocks that the hydrothermal fluids leached (Michard and Albarede 1986; Hein et al. 1997; Glasby et al. 1997). Thus, the Eu anomaly does not always provide an unequivocal indicator of the origin of a deposit, and it is likely that the hydrothermal solutions responsible for the formation of Mn deposits were in contact with and leached a thick continental crust, composed mainly of highly differentiated granite (Eu-depleted).

The La_{NASC} anomalies in the studied Mn deposits point to deposition in an area with hydrothermal activity (Hein et al. 1997). The weak or negative Gd_{NASC} anomalies point to a hydrothermal contribution, because high positive Gd_{NASC} anomalies are a general feature of hydrogenetic Fe–Mn crust (Bau et al. 1996).

The origin and occurrence of the Mn–Fe ore deposits at G. Abu Shaar and surrounding areas were discussed by Burek (1969), Abu Bakr (1979), Hassan and Sabet (1979), and Hassan (2002), in which they suggested a hydrothermal origin. Ghobrial (1963) proposed a diagenetic replacement of calcareous ooids, and El Aref (1994) and Abdel Moteleb (1996) suggested a stratiform sedimentary origin.

Normal graded bedding of Mn-oxide pisoliths and ooliths has been interpreted as due to precipitation of Mn during peaks of transgression (Bolton and Frakes 1985). The spatial distribution of ore types at Abu Shaar, found in more sheltered areas, suggests such deposits may have accumulated in lagoons or sheltered bays and this is in agreement with the conclusions of El Aref (1994) and Abdel Moteleb (1996). That this deposit is indeed shallow marine is attested by the primarily pisolitic-oolitic character of the ore. Frakes and Bolton (1992) demonstrated that sea level changes (transgression, regression) lead to a variation in a number of physico-chemical parameters, such as energy level (e.g., transport of precipitates and detrital materials) and oxygen concentration (e.g., mixing with surface water, re-oxygenation), that in turn can produce conditions that favor precipitation. The hydrogenous Mn deposits of Abu Shaar are characterized by enrichment in Co, Ni, and Cu, which reflects the very slow accumulation from oxidizing seawater, which is indicated by the low contents of Tl, U and Th.

9 Summary and Conclusions

The hydrothermal manganese deposits include the Halaib area and Wadi Araba on the western Red Sea coastal zone, and the Sharm el Sheikh and Um Zariq areas in south Sinai. The characteristic features are suggestive of a hydrothermal origin for Mn ore in the studied localities, and most of these deposits occur as veins or fracture filling cutting across the structure of the host rocks (Miocene and Plio-Pleistocene sedimentary rocks, Precambrian granite and volcanic rocks), indicating an epigenetic origin. In addition, the presence of barite, fluorite and carbonates, and the presence of elements such as Tl, Mo, As, U, and REE distributions confirm such an identification. In addition, the following features may indicate a hydrothermal origin: The separation of Fe from Mn (high Mn/Fe ratios), high Ba, low Ni, Co, silicification, a low REE budget, and high LREE/HREE ratios. Moreover, characteristic mineral-chemical enrichment, and geochemical association signatures, in conjunction with diagnostic plots indicate a hydrothermal origin. This mineralization is coeval with other metallic mineralization occurrences along the Red Sea system (both Gulf of Suez and Gulf of Aqaba) which are also connected with Miocene to Recent tectonic activity. Deposits of hydrothermal iron, Mn and barium-bearing brines occur in the Red Sea (Miller et al. 1966).

The Middle Miocene Mn deposits of the Abu Shaar ore occur as oxidic manganese ooliths and pisoliths with normal grading. The ooliths and pisoliths are concentrically laminated, with alternating lamina of psilomelane and mangano-calcite or dolomite. Normal graded bedding of Mn oxide is interpreted as due to precipitation of Mn during peak transgression. The character of the ore and hosting rock reflects that the Abu Shaar Mn deposits are hydrogenetic-diagenetic and may have accumulated in lagoons or sheltered bays. Low Tl, U, Th and REE budget and high Co and Ni, and positive Ce and Gd anomalies indicate the marine origin.

References

Abd El Monem A, El Aassy I, Hegab O, El Fayoumy I, El Agami N (1997) Gibbsite, uranium and copper mineralization, Um Bogma area, southwestern Sinai. Egypt. Sedimentology of Egypt 5:47–132

Abdel Moteleb A (1996) Geological and mineralogical studies of some manganese occurrence of Egypt. Ph.D. Thesis, Cairo University, 304 pp

Abu Bakr M (1979) Geology and geochemistry of manganese ore deposits in Esh Mellaha Range. Thesis, Ain Shams University, Egypt, M.Sc, p 194

Bau M, Koschinsky A, Dulski P, Rein I (1996) Comparison of the partitioning behavior of yttrium, rare earth elements and titanium between hydrogenetic marine ferromanganese crusts and sea water. Geochim Cosmochim Acta 60(10):1725–1709

Bau M, Dulski P, Moller P (1995) Yttrium and holmium in South Pacific seawater: vertical distribution and possible fractionation mechanisms. Chem Erde 55:1–15

Baiyoumi H, Ahmed H, Kheder M (2014) A mixed hydrogenous and hydrothermal origin of the Bahariya iron ores, Egypt: evidences from the trace and rare earth element geochemistry. J Geochem Explor 146:149–162

Bolton B, Frakes L (1985) Geology and genesis of manganese oolite, Chiatura, Georgia, U.SS.R. Geol Soc Am Bull 96:1398–1406

Bolton B, Both R, Exon N, Hamilton T, Ostwal J, Smith J (1988) Geochemistry and mineralogy of sea floor hydrothermal and hydrogenetic Mn-oxide deposits from the Manus basin and Bismarck Archipelago region of the southwest Pacific Ocean. Mar Geol 85:65–87

Bonatti E (1975) Metallogenesis at oceanic spreading centers. Ann Rev Earth Planet Sci 3:401–431

Bonatti E, Kraemer T, Rydell H (1972) Classification and genesis of submarine iron-manganese deposits. In: Horn D

(ed) Ferromanganese deposit in the ocean floor. National Science Foundation, Washington DC, pp 149–165
Burek P (1969) Structural effects of sea-floor spreading in the Gulf of Aden and the Red Sea on the Arabian Shield. In: Degens E, Ross D (eds) Hot brines and recent heavy metal deposits in the Red Sea. Springer, New York, pp 59–70
Coleman M (1985) Geochemistry of diagenetic non-silicate minerals, kinetic considerations. Phil Trans Royal Soc London A 315:39–56
Dymond J, Mitchell L, Finney B, Piper D, Murphy K, Conrad R, Pisias N (1984) Ferromanganese nodules of mineralogical and chemical composition by multiple accretionary processes. Geochim Cosmochim Acta 48:931–949
Dubinin A, Volkov I (1986) Rare earth elements in metalliferous sediments of thc East Pacific Rise. Geokhimija 5:645–662
El Agami N (1996) Geology and radioactivity studies on the Paleozoic rock units in Sinai Peninsula, Egypt. Ph.D. Thesis, Faculty of Science, Mansoura University, Egypt, 302 pp
El Agami N, Abu Baker M, Ibrahim M, Rashed M (1999) Mineralogical and geochemical studies on some mineralization in Halaib area, South Eastern Desert, Egypt. Egyptian J Geology 43:27–38
El Agami NL, Ibrahim E, Odah H (2000) Sedimentary origin of the Mn–Fe ore of Urn Bogma, southwest Sinai: geochemical and paleomagnetic evidence. Econ Geol 95:607–620
El Agami NL (2006) Mode of occurrence, mineralogical and geochemical investigations for Mn–Fe ore deposits in Egypt. Application to Genesis of the Ore Deposits. The Ninth Arab Conference for Mineral Resources, KSA, Jeddah, 30 Oct-1 Nov, 2006
El Aref MM (1994) Phanerozoic stratiform and stratabound deposits of Egypt; their stratigraphic, paleo-geographic, topographic and environmental controls. Proc International Conference on Geology of the Arab World, Cairo University 1:97–124
Fahim M, El Faramawy H, Ahmed A, Ghali S, Kandil H (2013) Characterization of Egyptian manganese ores for production of high carbon ferromanganese. J Minerals Materials Characterization Engineering 1:68–74
Frakes L, Bolton BR (1992) Effects of ocean chemistry, sea level and climate on the formation of primary sedimentary manganese ore deposit. Econ Geol 87:1207–1218
Glasby G, Stuben D, Jeschke G, Stoffers P, Garbe Schonberg C (1997) A model for the formation of hydrothermal manganese crusts from the Pitcairn Island hotspot. Geochim Cosmochim Acta 61:4583–4597
Ghobrial M (1963) Some new occurrences of manganese deposits in the Eastern Desert of Egypt. Egyptian Geol Surv, Cairo, Paper 20, 25 pp
Gradzinski M, Tyszka J, Uchman A, Jach R (2004) Large microbial - foraminifera oncoids from condensed lower-Middle Jurassic deposits: a case study from the Tatra Mountains, Poland. Paleogeog Palaeoclimat Palaoecol 213:133–151
Grill E, Chase R, MacDonald R, Murray J (1981) A hydrothermal deposit from Explorer Ridge in the northeast Pacific Ocean. Earth Planet Sci Lett 52:142–150
Hassan M (2002) Genesis of Miocene polymetal mineralization, western Red Sea coastal zone. Egypt. Sedimentology of Egypt 10:1–16
Hassan M, Sabet A (1979) Geochemical and mineralogical studies on Esh El Mellaha manganese deposits, Eastern Desert. Egypt. First Mid East Geol Congress, Mineral Res Explor Inst, Ankara, Turkey, pp 143–178
Hein J, Koski R (1987) Bacterially mediated diagenetic origin for chert-hosted manganese deposits in the Franciscan Complex, California coast Range. Geology 15:722–726
Hein J, Stamatakis M, Dowling J (2000) Trace metal-rich Quaternary hydrothermal manganese oxide and barite deposits, Milos Island, Greece. Trans Inst Mining Metallurgy, Section B, Applied Earth Sci 109:67–76
Hein J, Koschinsky A, Halbach P, Manheim FT, Bau M, Kang J-K, Lubick N (1997) Iron and manganese oxide mineralization in the Pacific. In: Nicholson K, Hein J, Buhn B, Dasgupta S (eds) Manganese mineralization: Geochemistry and mineralogy of terrestrial and marine deposits. Geol Soc Spec Publ 119:123–138
Hewett D, Fleischer M, Conklin N (1963) Deposits of the manganese oxides. Econ Geol 58:1–51
Hussein A, El Sharkawi M (1990) Mineral deposits. In: Said R (ed) The Geology of Egypt. Balkema, Rotterdam, Broookfield, pp 511–566
Meshref W (1990) Tectonic framework. In: Said R (ed) The Geology of Egypt. Balkema, Rotterdam, Brookfield, pp 113-156
Michard A, Albarede F (1986) The REE content of some hydrothermal fluids. Chem Geol 55:51–60
Miller A, Desmore E, Degens J, Hathaway F, Manheim P, McFarlin R, Pocklington JA (1966) Hot brines and recent iron deposits in deeps of the Red Sea. Geochim Cosmochim Acta 30:341–359
Mills R, Elderfield H (1995) Hydrothermal activity and the geochemistry of metalliferous sediment. In: Humphris S, Zierenberg R, Mallineaux L, Thomson R (eds) Seafloor hydrothermal systems. Am Geophys Union, Washington DC, Monograph 91, pp 392–407
Moorby SA, Cronan D (1983) The geochemistry of hydrothermal and pelagic sediments from the Galapagos hydrothermal mounds field, DSDP Leg 70. Mineral Mag 47:291–300
Moorby SA, Cronan D, Glasby G (1984) Geochemistry of hydrothermal Mn-oxide deposits from the S.W. Pacific Island Arc. Geochim Cosmochim Acta 148:433–441
Nicholson K (1992) Contrasting mineralogical-geochemical signature of manganese oxides: guides to metallogensis. Econ Geol 87:1253–1264
Polgari M, Okita PM, Hein JR (1991) Stable isotope evidence for the origin of the Urkut manganese ore deposits, Hungary. J Sediment Petrology 61:384–393
Rona PA (1984) Hydrothermal mineralization at sea floor spreading centers. Earth Sci Rev 20:1-104
Rossi PL, Bocchi G, Lucchini F (1980) A manganese deposit from the south Tyrrhenian region. Oceanol Acta 3:107–113
Roy S (1981) Manganese deposits. Academic press, New York, p 456
Schandelmeir A, Klitzsch E, Hendriks F, Wycisk P (1987) Structural development of northeast Africa since Precambrian times. Berliner Geowiss Abh 75:5–24
Tooth JR (1980) Deposition of submarine crusts rich in manganese and iron. Geol Soc Am Bull 91:44–54
Usui A, Ban M, Yamazaki T (1997) Manganese micro-chimneys buried in the central Pacific Pelagic sediments: evidence of intraplate water circulation. Marine Geol 141:258–269
Zaghloul ZM, El Shahat A, Kora M, Abu Shabana M (1995) Modes of occurrence, geochemistry and origin of the Mn–Fe deposits of Um Bogma, west central Sinai. Egypt. J Environ Sci Egypt 9:29–59

The Spatial Distribution Pattern of Surficial Sediment in Shiab Al-Kabeer, a Shoal in the Red Sea of Saudi Arabia

Najeeb M. A. Rasul and Abdulnasser S. Al-Qutub

Abstract

A unique feature commonly referred to as a '*shoal*', called Shiab Al-Kabeer, is located in the Red Sea offshore from Saudi Arabia. It is made up mainly of live and dead corals with over 90% calcium carbonate material largely contributed by the corals. Aeolian quartz from the adjacent deserts and fluvial material derived from numerous wadis during flood discharge make up a small part of the sediment veneer. The shoal is inhabited by various fauna and flora exclusive to the Red Sea. Bio-eroders such as grazers, borers and burrowers, especially some species of fish like parrotfish and shrimp-gobies, create cavities and holes in the shoal area, and generate new sediment by excavating the cavities, forming mounds on the sea bed. Apart from being an excellent ground for spawning, the shoal is also used for overnight dwelling by picnickers, as an anchorage for small boats, as a nesting ground for seabirds, and is an attractive site for marine tourism (SCUBA diving). The shoal undergoes a degree of stress either through human activities or natural processes. However, the clear and uncontaminated water keeps the sub-environment of the shoal healthy, although minor irreversible processes caused by contamination might be of some concern. Some coral debris, molluscs, and foraminiferal sand are stained black, because of precipitation of free pyrite in a reducing environment caused by weak currents and quiescent conditions, in the presence of plant and organic debris, mainly algae and sea-grass. The dynamics and shape of the shoal are controlled primarily by water currents generated either by wind or by tidal cycles, breaking waves and physical factors related to human activities.

N. M. A. Rasul (✉) · A. S. Al-Qutub
Center for Marine Geology, Saudi Geological Survey, Jeddah, Saudi Arabia
e-mail: najeeb_rasul@hotmail.com; rasul.nm@sgs.org.sa

N. M. A. Rasul and I. C. F. Stewart (eds.), *Geological Setting, Palaeoenvironment and Archaeology of the Red Sea*,
https://doi.org/10.1007/978-3-319-99408-6_22

1 Introduction

The Red Sea is an elongated body of water about 2000 km long with a maximum width of 355 km. It contains over a thousand islands, most of which have been named but remain unexplored due to inaccessibility. In addition, a large number of shoals with sizes ranging from a few metres across to several kilometres in length form part of the unique Red Sea environment. The Red Sea includes 3.8% of the world's coral reefs (Sea Around Us 2007). The major Red Sea barrier reef is 400 km long and several kilometres wide, and located between 10 and 40 km off the coast of Saudi Arabia (Rasul et al. 2015). Most of the bays and lagoons along the Saudi Arabian side of the Red Sea are shallow and bordered by shallow fringing reefs. Many steep-sided patch reefs sit on the seabed platform, which starts at a depth of 30–60 m. Because of the absence of rivers and permanent streams, most terrigenous material inputs to the Red Sea take place only during heavy rainfall and flash floods through the wadis (seasonal streams). In arid coastal regions where riverine sediments are scarce or absent, aeolian material and carbonate materials form the major contribution to the marine sediments.

Islands and shoals associated with coral reefs are formed by various processes related to volcanic and tectonic activities. It is known that some islands are formed by hydrodynamic processes (e.g., barrier islands) and terrigenous sources (input) in deltaic and estuarine environments (Maul 1996). In tropical zones many islands are formed by coral reefs—a very common feature in the Red Sea. The Red Sea is regarded as one of the world's richest seas in terms of its coral reef population. The reefs are more common in the north because of the favourable conditions for their growth as compared to the south. Dead corals that act as substrate (carbonate crust) are the building blocks for coral reefs. The cycle of building and destruction continues, resulting in the shoals made up of coral reefs. These contribute substantial amounts of carbonate debris to the Red Sea, especially by the breaking down of the reef system either by human activities or naturally, supplemented by biological productivity. The bases of the islands are associated mostly with shoal-like features that on the Saudi Arabian side of the Red Sea occur with relatively large dimensions (e.g., the Al-Wajh Bank area and the Al-Khuraybah lagoon). In the central Red Sea, about 24 km offshore from Jeddah, is the Shiab Al-Kabeer, where the shoal emerges during low tide. These shoals form excellent grounds for spawning and fishing, and are environmentally sensitive ecosystems with high carbonate content that is supplied to the area in the form of sediment by the reef system.

Detailed work on the shoals and islands is lacking, particularly with regard to the processes responsible for the transportation, distribution and dispersal pattern of sediments. Such data is not available, or if present, it is poorly integrated or too broad in scope to fully describe the system. The objectives of this study are to investigate the contemporary sedimentation processes, understand the sediment distribution pattern and determine the source and sediment transport mechanism in this carbonate-rich sub environment.

There has been limited research interest in the shoals and islands of the Red Sea because of the rocky nature and shallow water surrounding them making access to them very difficult. This is why most of these features have long remained unexplored, one of them being the Shiab Al-Kabeer. However, a few studies on the environmental aspects of the islands and shoals of the Red Sea (e.g., Ras Al-Qasabah, Al Wajh, and Yanbu areas in the north, and the Farasan Banks and Farasan Islands in the south), have been conducted by the Khaled bin Sultan Living Oceans Foundation (Bruckner et al. 2011, 2012). In 1997 and 1998, a semi-quantitative study related to the health of the coral reefs and water quality assessment at 79 sites in the Farasan Islands' Marine Protected Area was carried out by DeVantier and Pilcher (2000) and De Vantier et al. (2000). Detailed descriptions of the geology, physical environment, climate, hydrology, oceanography and habitats including the coral reefs of the Red Sea and wider Arabian Region have been presented by Fishelson (1971), Mergner (1971), Ormond et al. (1984), Behairy and Jaubert (1984), Edwards (1987), Crossland (1987), UNEP/IUCN (1988), Sheppard and Sheppard (1991), Behairy et al. (1992), Sheppard et al. (1992), Rowlands et al. (2014), Klaus (2015), Rasul (2015), Bruckner and Dempsey (2015), and Rowlands and Purkis (2015).

2 Study Area-Geographical Setting and Morphology

The Shiab Al-Kabeer is one of the many shoals in the Red Sea, and lies at latitude 21°42′15″N and longitude 38°50′15″ E. It is ~24 km west of Sharm Obhur, toward the central axis of the Red Sea dividing the Red Sea into Saudi Arabian and African waters (Fig. 1).

It is also known as "*Abu Tair*", referring to the marine birds such as the Brown Booby (*Sula leucogaster*) and white eyed gull (*Larus leucophthalmus*) that colonise the shoal. The surface area of the shoal is ~6.5 km^2, with a width of 2.8 km, while the length of the coral patch facing the open sea toward the west is ~4 km, trending NNW–SSE.

The Shiab Al-Kabeer has an average water depth of about 10 m and is surrounded by deep water, especially on the western side. The variation in the water depths within the sub environment is not as great as expected because of the numerous small shoals detached from the main shoal. However, the shoal facing the open sea to the west shows a greater and more abrupt fall where the direct actions of waves and currents create its elongated and solid appearance. The bathymetric map in Fig. 1 was produced in ArcGIS in conjunction with spot depth recordings during sediment sampling. The multi-spectral signature is based on reflectance from the seabed and geomorphological features extracted from satellite images.

The geomorphological and seabed features associated with the shoals show the variations in the sub environment (Fig. 2a–l). Reef walls facing the open sea or drop-offs are steep and near vertical, with coral cover ranging from 10 to 70%. All coral forms are present, but mainly massive and foliose growth forms on exposed points. The coral community consists mainly of *Pocillopora* spp., *Favia* spp., and *Porites* spp. The space between coral colonies can be bare or occupied by macro-algae, coralline algae, and other species of soft coral. The area has isolated rubble accumulation along with sand at the foot of the numerous small shoals. The rubble and sand consolidate into a substrate and are colonized by stony and soft corals in a calm environment, whereas live coral cover is generally low due to the continual movement of sediment by bottom currents. Dead corals are covered with algae.

3 Materials and Methods

A total of fifty (50) surficial sediment samples were collected from the uppermost 15–30 cm of the seabed using a Van Veen grab sampler (Fig. 2). The samples were subjected to detailed grain-size analysis following the procedures outlined by Folk (1980). A Munsell Colour Chart was used to assign colour to the sediment veneer. During sediment sampling, turbidity in the water column was also determined by a standard Secchi disk, lowered from the shadow side of the boat to limit the effect of direct sunlight. A multi-parameter Hydrolab MS5 probe was used for collecting temperature, salinity and pH data from all the stations. Aanderaa RCM-9 and Valeport-106 current meters were

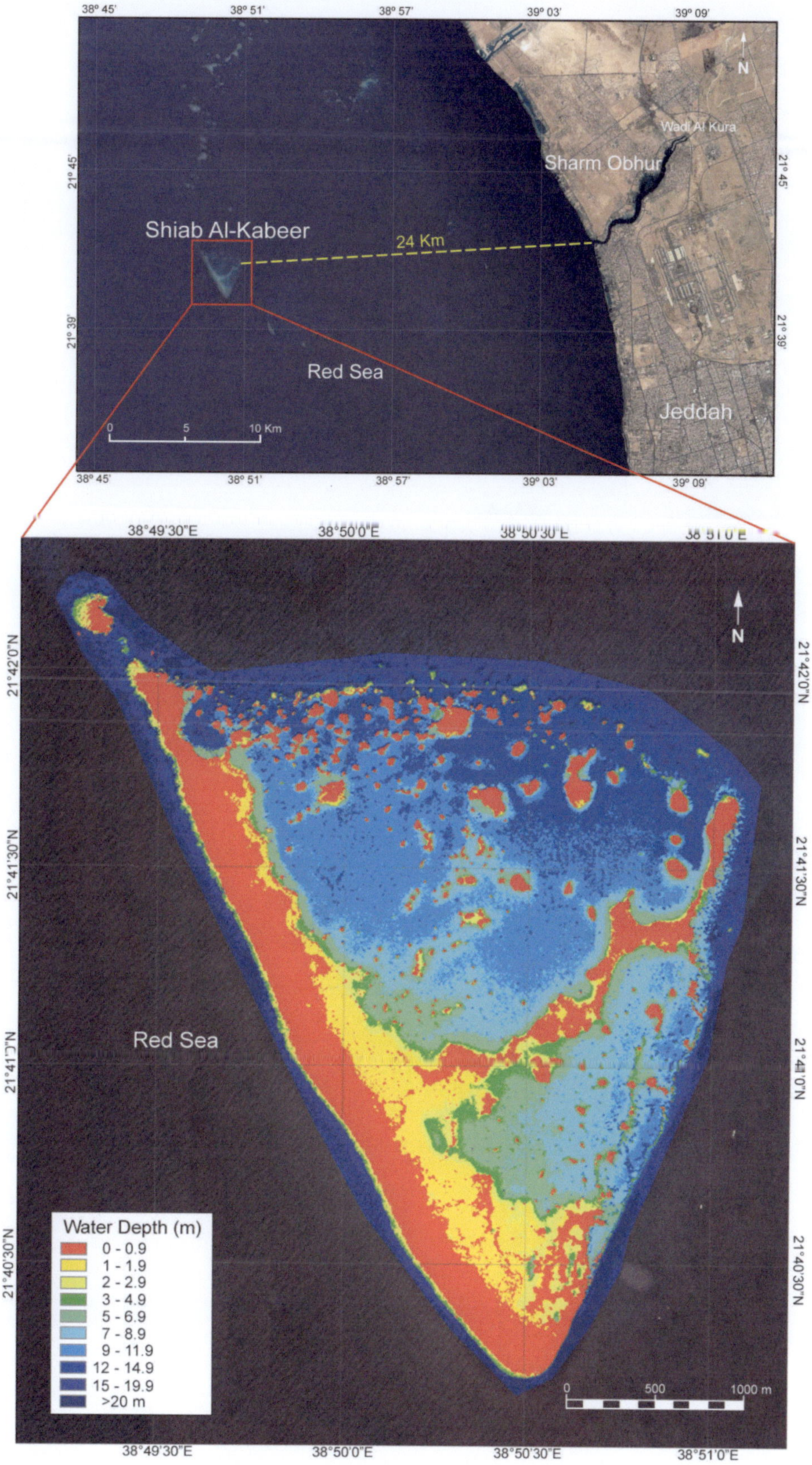

Fig. 1 Location map of the Shiab Al-Kabeer 24 km due west from the mouth of Sharm Obhur. GIS based bathymetric map was produced by applying different wavelengths of the multispectral WV-2 bands

(a)
(b)
(c)
(d)
(e)
(f)
(g)
(h)
(i)
(j)
(k)
(l)
Red Sea
500
1000 m

◀ **Fig. 2** **a** A colony of the stony coral *Goniastrea* sp. growing on top of dead coral substrate. Soft coral *Xenia* sp. and *Acropora* spp. are also common; **b** Shallow water (12 m) sand flat. In general, fine sediment is flushed by current and wave activity, allowing coral colonization, leading to a sparse coral community composed mainly of *Favia* spp., *Fungia* spp., *Acropora* spp., and *Xenia* spp. (soft coral); **c** exposed reef, commonly seen during low tide. Coral heads can be seen protruding from the water; **d** fish that prefers to live in burrows is seen peeping out from its home; **e** low-relief grooves run sporadically down the vertical reef face; **f** reef wall on the southern tip; **g** water is extremely transparent because of lack of sediment supply; **h** garbage dumping by seafarers; **i** shoal during high tide. The rock like feature is exposed at all times; **j** carbonate crust and reef flats typically exposed during low tide, characterized by 1 cm thin veneer of sediment and macro-algae, *Pocillopora* spp.; **k** reef flat community at the back reef near the cemented base buoy used for navigational purposes; and **l** water clarity is reduced in shallow waters especially during spring tide and ebbing currents

deployed at various depths to obtain current velocities and direction. SCUBA diving was carried out to understand the flora and fauna inhabiting the area. A LEICA MZ 16A stereo-microscope with an attached high-resolution LEICA DFC 320 camera was used for studying the general composition, grain roundness and angularity of the coarse fractions. A TESCAN-VEGA 8 scanning electron microscope (SEM) was used to identify and image individual grains. Individual sand grains were stuck on conductive carbon tape stubs (standard sample holders). Each sample was gold-palladium coated using a Quorum coating machine to make the sample conductive. A Shimadzu 6000 X-ray diffractometer (XRD) with CuKa radiation, a voltage of 30 kV, and current of 10 mA was used for bulk mineralogy on finely powdered sediments at a scanning speed of 1°/min with an angle ranging from 2 to 60°. The relative abundance of individual minerals was determined by comparing their peak heights on 16 representative samples. An AC-280 Automated Calcimeter was used to determine calcium carbonate by treating the ground sediment samples with hydrocholoric acid. The total organic carbon (TOC) was determined from sediment samples by the wet oxidation method adopted by Le Corre (1983), involving a chromic acid digestion, which was then back titrated with ferrous ammonium sulphate using diphenylamine as an indicator.

4 Results and Discussion

4.1 Surficial Sediment Characteristics

The sediment texture ranges between muddy sand and sandy gravel, dominated by gravelly sand with a few patches of other fractions (Table 1 and Fig. 3). Isolated and sheltered areas have coarse textured material along with a smaller proportion of biogenic carbonate mud. Gravel and sand, which are the product of coral disintegration by intruders, are the most common. The sediment veneer is comprised of 1–51% gravel with an average of 14%. Sand dominates the veneer ranging from 47 to 93% (average = 76%), whereas mud ranges from 1 to 46% (average = 10%).

Numerous attempts have been made in the past to identify sedimentary environments from grain-size parameters but the results have been debatable (e.g., Mason and Folk 1958; Shepard and Young 1961; Schlee et al. 1964; Solohub and Klovan 1970; Rasul et al. 2017). The use of bivariate plots (scattergrams) is based on the fact that the processes of (sand) transportation and deposition tend to produce sediments with a characteristic range of grain sizes, so the size parameters that define the sediment distribution are therefore often diagnostic of the environment. The inter-relationship of various statistical data from the study area did not help in discriminating the environment of deposition because of the carbonate-rich environment with a wide range of grain sizes. No distinct environments could be identified but a broad conclusion could be drawn in relation to the mixed source of sediments and erratic tidal and wind generated currents. However, the lack of terrigenous material and the abundant carbonate material produced by coral erosion and disintegration settle at the foot of the reefal system and therefore cannot be part of the sediment transport mechanism as perceived by many sedimentologists.

4.2 Sediment Colour and Staining

The sediment colour is important in understanding the history and process of sedimentation because it points to the sediment source(s), dispersal and distribution patterns, including any geochemical reactions that may have taken place in situ. Based on the Munsell colour chart the sediment colour ranged between 5Y 8/1/1 white and 10 YR 8/2/2 very pale brown. In general, the sediment colour has a lighter hue compared to other areas in the Red Sea (see Table 1 and Fig. 4). However, in places where the environment is less dynamic two coloured sediments are observed, darker in the bottom layer and lighter on the top (Fig. 5). The dark coloured sediment is due to the reducing conditions on the sea bottom instigated by the decay of organic debris.

Most of the area is covered by coarse grained, light coloured sediment derived from the corals inhabiting the shoal. However, in some fine sediments varying shades of gray are observed that could be related to sluggish conditions, and to anthropogenic input in the form of plant, seagrass and algal debris whose decay enhances the process of

Table 1 Sediment texture data based on various size fractions

Station	Water depth (m)	SDDD (m)	Gravel (%)	Sand (%)	Mud (%)	Sediment texture	Munsell colour
1	8.0	8.0	28	69	3	Gravelly sand	2.5Y 8/2/2 pale yellow
2	3.0	3.0	3	58	39	Muddy sand	5Y 8/2/2 pale yellow
3	8.3	8.3	2	85	13	Muddy sand	5Y 8/1/1 white
4	10.0	10.0	8	86	6	Muddy gravelly sand	2.5Y 8/2/2 pale yellow
5	7.9	7.9	3	70	27	Muddy sand	5Y 8/2/2 pale yellow
6	7.0	7.0	1	53	46	Muddy sand	5Y 8/2/2 pale yellow
7	7.4	7.4	30	67	3	Gravelly sand	5Y 8/2/2 pale yellow
8	6.6	6.6	4	74	22	Muddy sand	5Y 8/2/2 pale yellow
9	7.1	7.1	27	70	3	Gravelly sand	5Y 8/2/2 pale yellow
10	5.1	5.1	18	80	2	Gravelly sand	2.5Y 8/2/2 pale yellow
11	1.7	1.7	16	82	2	Gravelly sand	5Y 8/2/2 pale yellow
12	1.5	1.5	17	80	3	Gravelly sand	5Y 8/2/2 pale yellow
13	2.0	2.0	20	79	1	Gravelly sand	5Y 8/2/2 pale yellow
14	2.0	2.0	1	91	8	Muddy sand	5Y 8/2/2 pale yellow
15	7.0	7.0	40	57	3	Gravelly sand	5Y 8/2/2 pale yellow
16	12.5	11.0	1	85	14	Muddy sand	5Y 8/2/2 pale yellow
17	8.0	7.0	1	88	11	Muddy sand	5Y 8/2/2 pale yellow
18	6.6	6.6	10	78	12	Muddy gravelly sand	5Y 8/2/2 pale yellow
19	7.9	7.9	12	78	10	Muddy gravelly sand	5Y 8/2/2 pale yellow
20	8.5	7.0	11	85	4	Gravelly sand	5Y 8/1/1 white
21	5.6	5.6	7	82	11	Gravelly muddy sand	5Y 8/2/2 pale yellow
22	5.0	5.0	9	80	11	Gravelly muddy sand	5Y 8/2/2 pale yellow
23	1.5	1.5	10	78	12	Muddy gravelly sand	5Y 8/2/2 pale yellow
24	1.2	1.2	7	65	28	Muddy gravelly sand	5Y 8/2/2 pale yellow
25	2.8	2.8	28	64	8	Muddy gravelly sand	5Y 8/2/2 pale yellow
26	4.1	4.1	1	80	19	Muddy sand	5Y 8/2/2 pale yellow
27	5.8	5.8	11	84	5	Muddy gravelly sand	5Y 8/2/2 pale yellow
28	8.8	8.8	7	84	9	Gravelly muddy sand	5Y 8/2/2 pale yellow
29	6.2	6.2	1	75	24	Muddy sand	5Y 8/2/2 pale yellow
30	12.0	12.0	1	86	13	Muddy sand	2.5Y 8/2/2 pale yellow
31	14.5	14.5	51	47	2	Sandy gravel	5Y 8/2/2 pale yellow
32	13.0	13.0	7	90	3	Gravelly sand	2.5Y 8/2/2 pale yellow
33	18.7	18.7	6	93	1	Gravelly sand	2.5Y 8/2/2 pale yellow
34	25.0	25.0	2	92	6	Muddy sand	10YR 8/2/2 very pale brown
35	26.0	26.0	32	67	1	Gravelly sand	10YR 8/2/2 very pale brown
36	30.0	30.0	36	62	2	Gravelly sand	10YR 8/2/2 very pale brown
37	32.0	32.0	14	85	1	Gravelly sand	10YR 8/2/2 very pale brown
38	8.5	8.5	4	52	44	Muddy sand	2.5Y 8/2/2 pale yellow
39	14.8	14.8	26	70	4	Gravelly sand	2.5Y 8/2/2 pale yellow
40	17.2	17.2	16	80	4	Gravelly sand	2.5Y 8/2/2 pale yellow
41	16.8	16.8	12	80	8	Muddy gravelly sand	2.5Y 8/2/2 pale yellow
42	11.0	11.0	5	74	21	Gravelly muddy sand	5Y 8/1/1 white
43	11.5	9.0	18	80	2	Gravelly sand	5Y 8/2/2 pale yellow
44	11.0	9.0	36	61	3	Gravelly sand	5Y 8/2/2 pale yellow

(continued)

Table 1 (continued)

Station	Water depth (m)	SDDD (m)	Gravel (%)	Sand (%)	Mud (%)	Sediment texture	Munsell colour
45	11.0	8.5	26	72	2	Gravelly sand	5Y 8/2/2 pale yellow
46	8.3	7.0	17	80	3	Gravelly sand	5Y 8/2/2 pale yellow
47	10.0	9.0	3	66	31	Muddy gravelly sand	5Y 8/2/2 pale yellow
48	8.4	8.4	6	90	4	Gravelly sand	5Y 8/2/2 pale yellow
49	10.7	9.5	9	88	3	Gravelly sand	5Y 8/2/2 pale yellow
50	7.6	7.6	14	83	3	Gravelly sand	5Y 8/2/2 pale yellow
Min	1.2	1.2	1	47	1		
Max	32.0	32.0	51	93	46		
Avg	9.7	9.5	14	76	10		

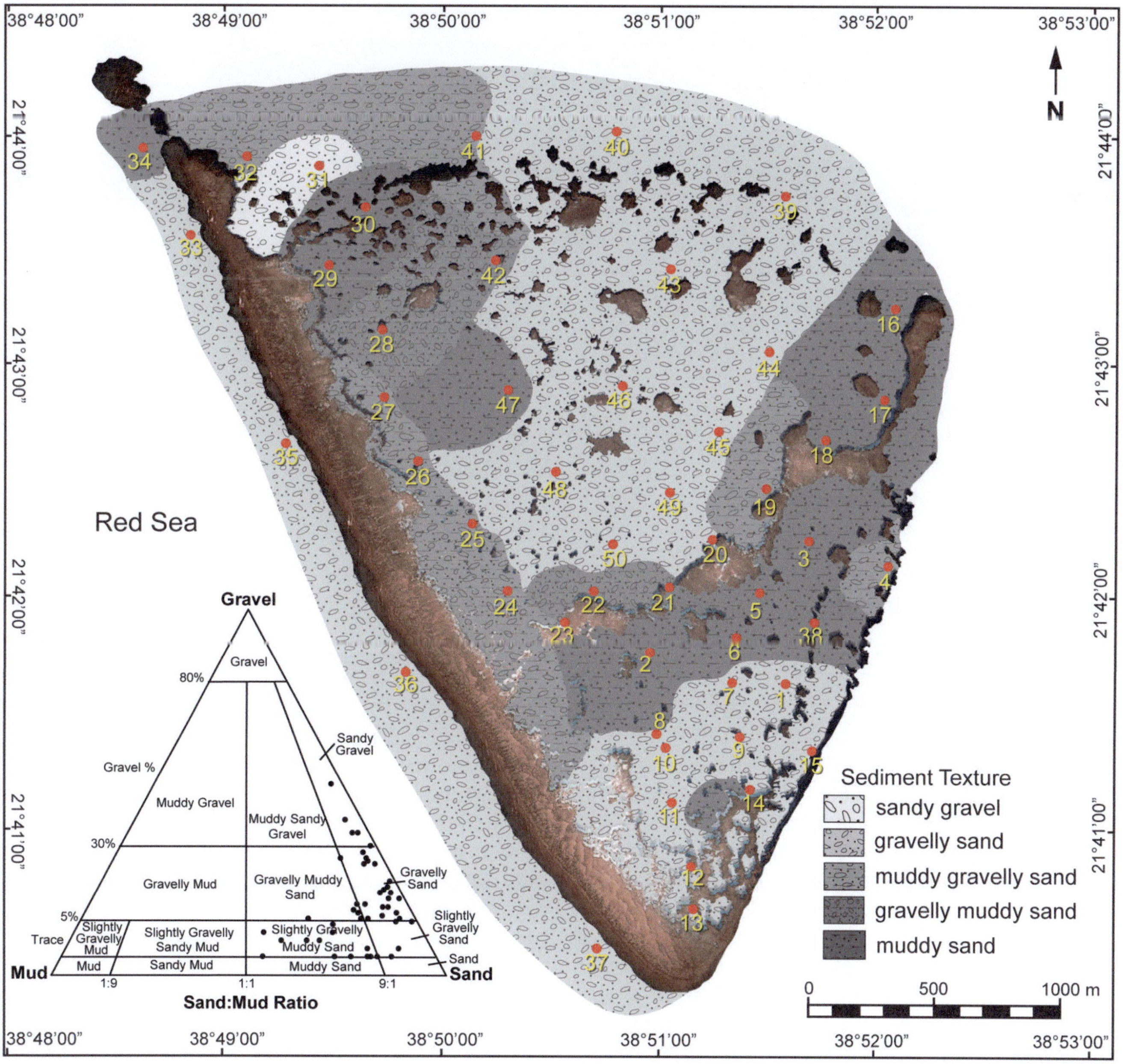

Fig. 3 Sediment distribution map based on the textural classification scheme proposed by Folk (1980)

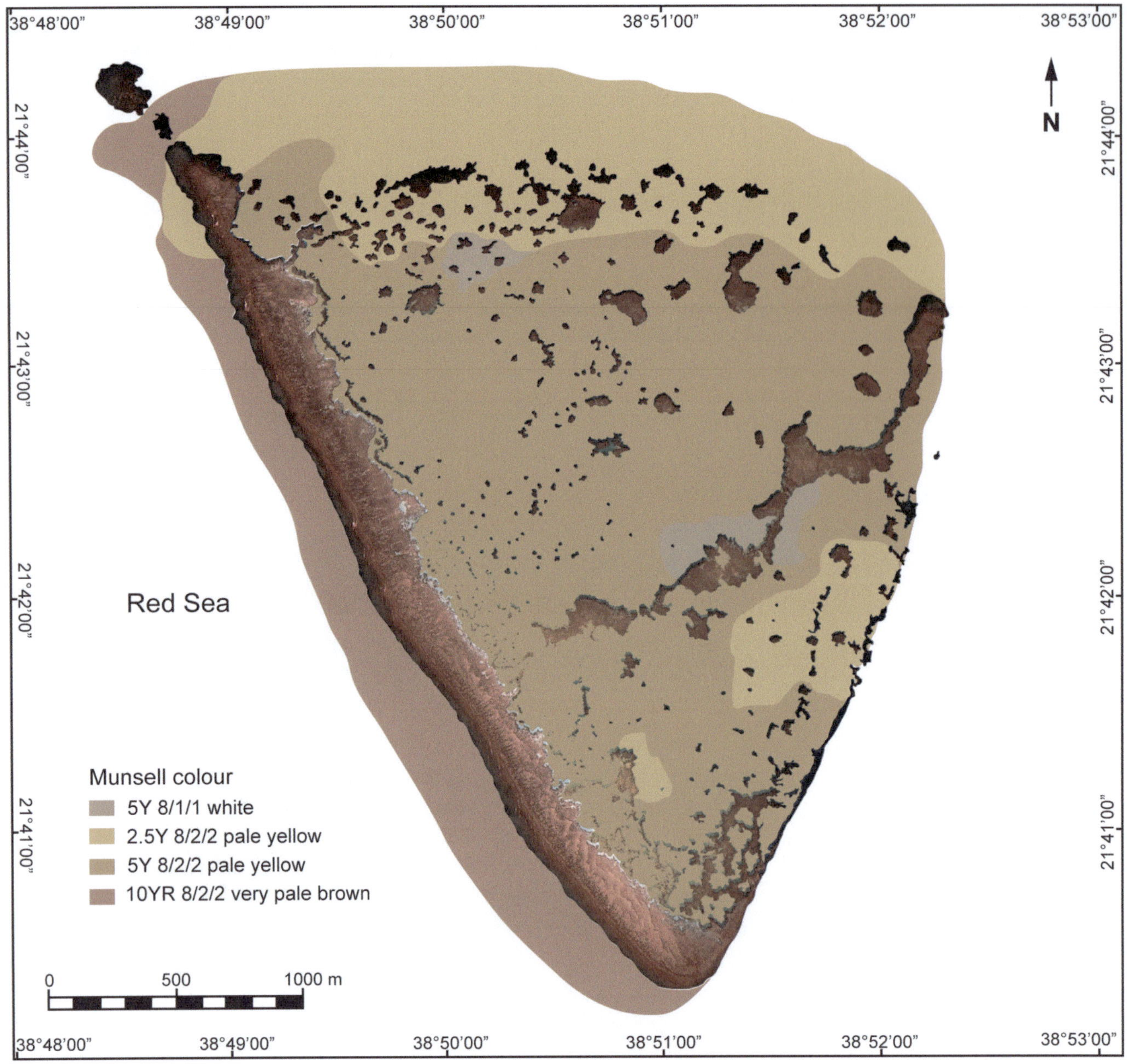

Fig. 4 Sediment colour (top 30 cm) defined using the Munsell colour chart

reduction. The resulting loss of oxygen promotes the staining of the sediments (Rasul et al. 2017). It is worth noting that in the area where stagnant conditions prevail and seabirds are present, the air has a disagreeable odour, mostly because of the tons of bird droppings. The relatively darker sediment in the shallow water, especially in the sheltered area on the southeastern side of the shoal, is where calmer conditions prevail because of the many small shoals that act as barriers to currents and water movement. The area of stained gray sediment could therefore be referred to as an environment of deposition where the effects of tide- wind- and wave-generated currents are limited, thus restricting sediment movement and promoting the staining of the coral debris, shell fragments and forams.

Relatively older biogenic fragments are stained gray or are worn on the outer surface and sometimes on the inside as well. Similar staining is present in the many semi-enclosed bays and lagoons along the Red Sea as well as in many coastal and marine environments, especially where stagnation prevails (Rasul 2015; Rasul et al. 2017). Both ancient and recent calcareous marine sediments stained gray or black with finely divided pyrite have been reported in pellety debris of carbonate sediments from the Middle East, Britain and other areas (Rasul 2015; Sugden 1966). Once the

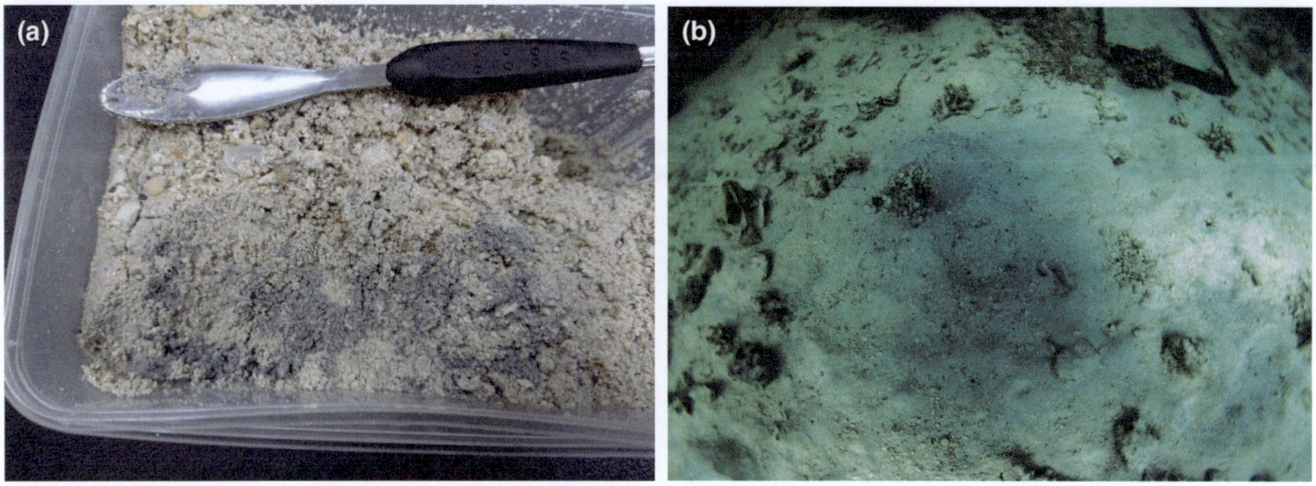

Fig. 5 **a** The staining shows a reducing environment; top layer is lighter than the underlying layer. Although the composition of the sediment is the same, the colour is different because of a reducing environment, and **b** a process that takes place in the presence of organic debris, staining under a thin veneer of recently contributed organic material in sheltered areas

sediment is buried under the pile of either terrigenous or anthropogenic material in the form of organic debris, a reducing environment develops because of the consumption of oxygen and organic matter by bacteria during decay. Aerobic bacterial activity, utilizing dissolved oxygen, penetrates a surficial sediment layer from a few cm to tens of cm in thickness (Sugden 1963). Some organic matter is oxidized and destroyed and some gets buried to depths where free oxygen decreases and is then subjected to anaerobic bacterial decay that produces hydrogen sulphide. However, the black colour on the shell fragments can change to brown when sediment from the reducing environment is exposed to an oxidizing environment by burrowing organisms, erosion or currents (Maiklem 1967). The blackening or staining of shells may be due to the presence of pyrite in sediment from the coastal plain that contains iron-rich clays and sands (Jado and Hötzl 1984; Jado et al. 1989; Abou-Ouf and El-Shater 1993) and not from organic matter. The location of the shoal in an open environment of the Red Sea does not always promote a reducing environment because of the constant movement and mixing of water during tidal cycles and wind generated currents, but in the micro-environment within the shoal stagnant conditions prevail resulting in a reducing environment in places.

4.3 Water Transparency

The dispersion of light in the water column and water transparency play an important role in understanding the effect of the local currents on the sediment veneer, including sediment colour, source, type and composition. Apart from the water current, water depths also play an important role in the resuspension of bottom sediment (Rasul et al. 2017).

The Secchi Disk Disappearance Depth (SDDD) in the study area ranged between 1.20 and 32 m with an average of 9.45 m (see Table 1 and Fig. 6). The water clarity is exceptionally good and it seems that the local currents do not have an impact on the turbidity level because of a lack of terrigenous input even in shallow waters. SDDD and water depth are equal, indicating that the current in general fails to resuspend the bottom sediment. In addition, sediment supply to the area is negligible and therefore the clarity is good. The light colour of the carbonate material on the bottom also enhances the light reflection. Since the sediment veneer is important in defining the water clarity and turbidity level, the water clarity is directly related to the character of the seabed. In general, the SDDD is the same as the spot depths except at some stations, although the difference between the SDDD and spot depth does not exceed 1.5 m. The data show the probable influence of bottom currents on the resuspension of bottom sediment. However, the very poor sorting values of the bottom sediment show that the water currents are ineffective in disturbing the sediment veneer, resulting in an increase in water clarity that allows the fauna to colonise the area. The clarity of the water seems to be excellent because of the coarser texture of the seabed and the lack of input from the wadis. Since the seabed is coarser in nature and very little mud is present, the water is less turbid, allowing sunlight to penetrate through the water column much more effectively. However, where the bottom current is strong the calcareous fine materials are resuspended, resulting in a decrease in water clarity.

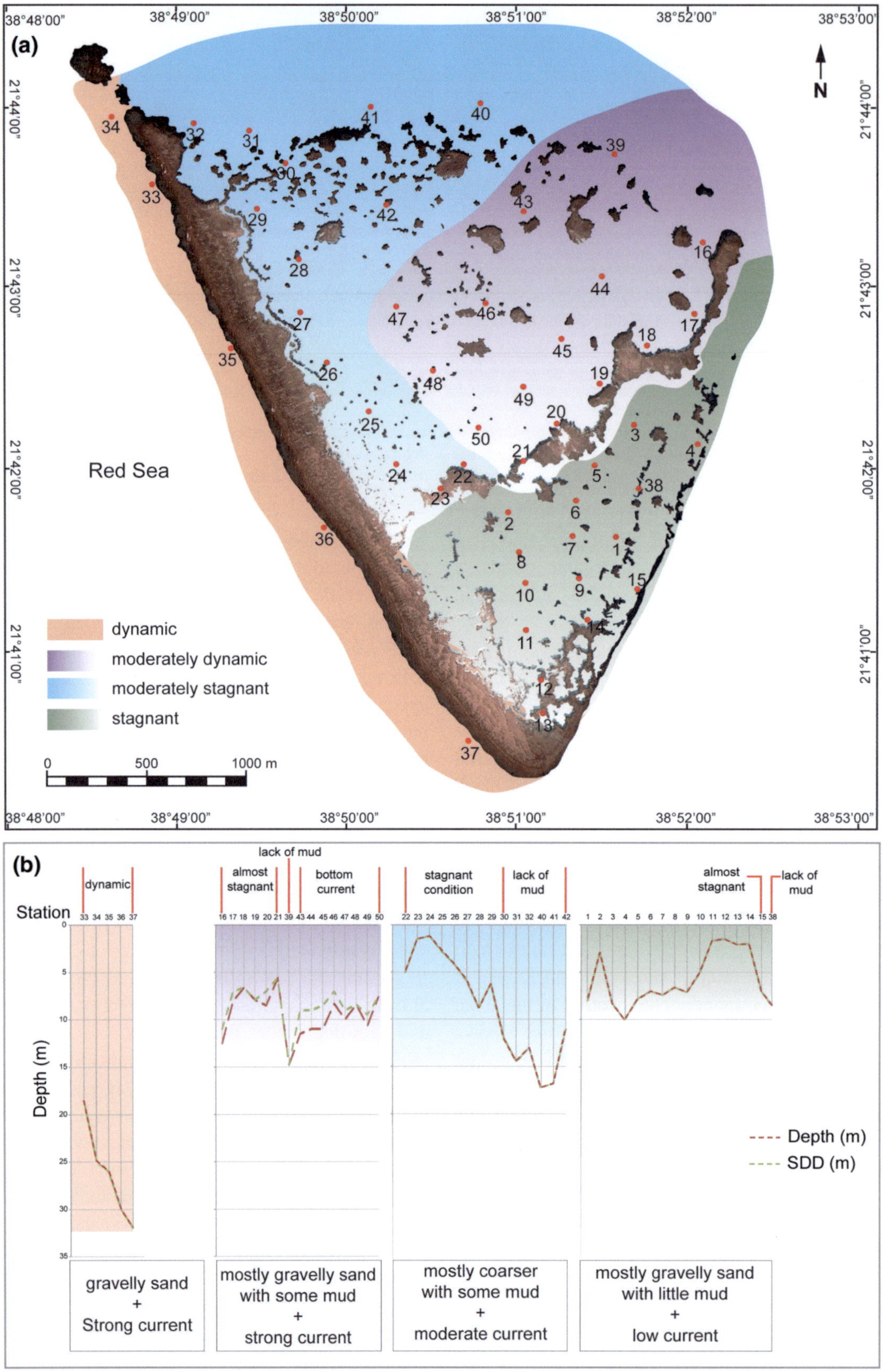

Fig. 6 **a** Station locations and water transparency, and **b** the relationship between water depth, Secchi disk disappearance depth (SDDD) and sediment type

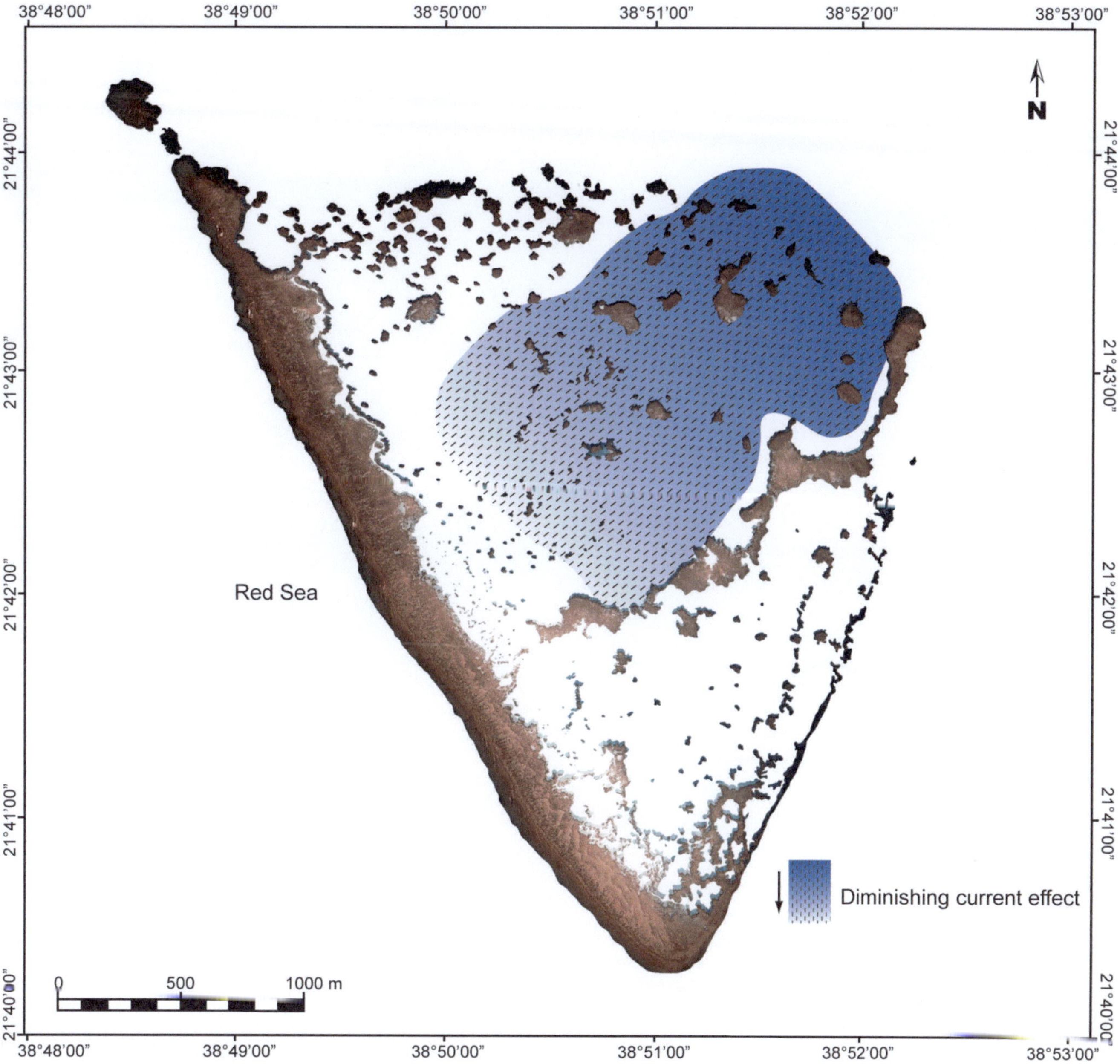

Fig. 7 The diminishing effect of the current during ebbing. The energy of the current is checked and subsides after encountering the numerous shoals

4.4 Water Current Measurement

Water currents are important in understanding the sediment transport pathways and patterns. Where the water depths were 7 and 36 m, the current meters were deployed at 5 and 18 m from the water surface respectively. The current velocities ranged between 3 and 47 cm s^{-1} with an average of 21 cm s^{-1}, mostly in the ESE direction and occasionally in the WSW. The current movement is mostly related to tidal cycles and to some extent is generated by the westerly wind. Weak, moderate to strong current velocities have been observed, at times reaching a maximum of about 1 m s^{-1} during flood tide on the eastern side of the shoal. In general, the water current is not strong in the area, and becomes weaker and gradually fades until the influence of currents ceases on reaching the numerous small shoals that act as barriers to sediment movement (Fig. 7).

Breaking waves and tidal currents are responsible for the erosion of the small-scale underwater features (coral reefs) and shoals. The coarser material in the area facing the open sea is attributed to the removal of fine sediment by the sifting action of the strong tide, wind and wave-generated currents

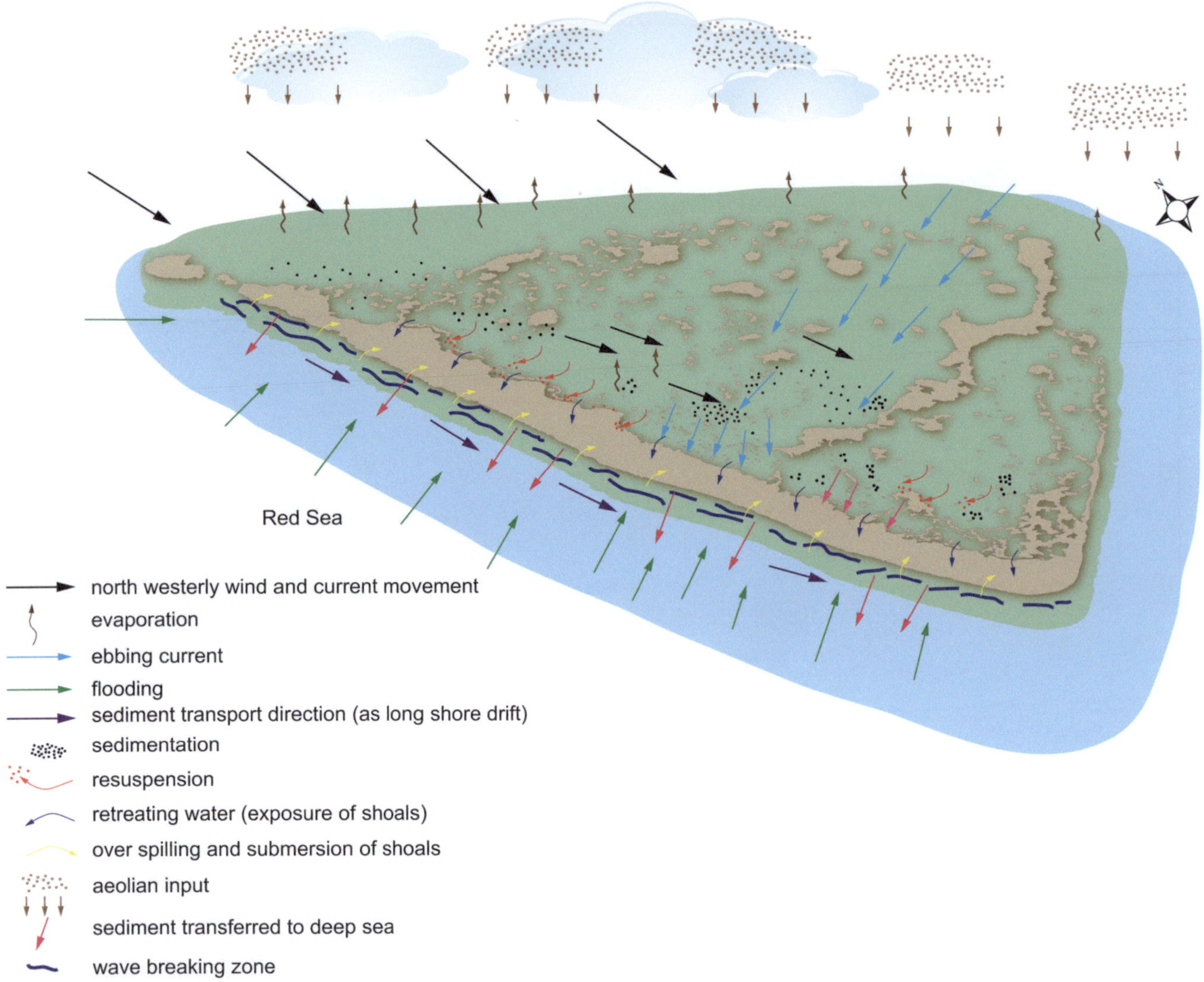

Fig. 8 A 3D representation of Shiab Al-Kabeer depicting the sediment and water movement during tidal cycles and the effect of wave- and wind-generated currents in the movement of locally produced sediments

Table 2 The degree of roundness of sedimentary grains

Station	Sediment texture	Sample type	Roundness/Angularity
1	Gravelly sand	Bimodal, very poorly sorted	Sub-angular
2	Muddy sand	Unimodal, poorly sorted	Angular
3	Muddy sand	Unimodal, moderately well sorted	Sub-angular
4	Muddy gravelly sand	Unimodal, moderately sorted	Sub-angular
5	Muddy sand	Unimodal, poorly sorted	Sub-angular
6	Muddy sand	Unimodal, poorly sorted	Sub-angular
7	Gravelly sand	Bimodal, very poorly sorted	Sub-angular
8	Muddy sand	Unimodal, moderately sorted	Very angular
9	Gravelly sand	Bimodal, very poorly sorted	Sub-angular
10	Gravelly sand	Bimodal, very poorly sorted	Angular
11	Gravelly sand	Bimodal, very poorly sorted	Angular
12	Gravelly sand	Bimodal, very poorly sorted	Angular
13	Gravelly sand	Bimodal, very poorly sorted	Angular
14	Muddy sand	Unimodal, well sorted	Sub-angular

(continued)

Table 2 (continued)

Station	Sediment texture	Sample type	Roundness/Angularity
15	Gravelly sand	Bimodal, very poorly sorted	Sub-angular
16	Muddy sand	Unimodal, moderately well sorted	Rounded
17	Muddy sand	Unimodal, moderately well sorted	Rounded
18	Muddy gravelly sand	Unimodal, poorly sorted	Rounded
19	Muddy gravelly sand	Bimodal, poorly sorted	Rounded
20	Gravelly sand	Unimodal, moderately sorted	Rounded
21	Gravelly muddy sand	Unimodal, poorly sorted	Rounded
22	Gravelly muddy sand	Unimodal, poorly sorted	Sub-angular
23	Muddy gravelly sand	Unimodal, Poorly Sorted	Sub-angular
24	Muddy gravelly sand	Unimodal, poorly sorted	Sub-angular
25	Muddy gravelly sand	Bimodal, very poorly sorted	Sub-angular
26	Muddy sand	Unimodal, moderately sorted	Sub-angular
27	Muddy gravelly sand	Unimodal, moderately sorted	Sub-angular
28	Gravelly muddy sand	Unimodal, poorly sorted	Sub-angular
29	Muddy sand	Unimodal, moderately sorted	Sub-angular
30	Muddy sand	Unimodal, moderately well sorted	Sub-angular
31	Sandy gravel	Bimodal, very poorly sorted	Sub-angular
32	Gravelly sand	Unimodal, moderately sorted	Sub-angular
33	Gravelly sand	Unimodal, moderately sorted	Very angular
34	Muddy sand	Unimodal, very well sorted	Very angular
35	Gravelly sand	Bimodal, very poorly sorted	Very angular
36	Gravelly sand	Bimodal, very poorly sorted	Very angular
37	Gravelly sand	Bimodal, moderately sorted	Very angular
38	Muddy sand	Unimodal, poorly sorted	Sub-angular
39	Gravelly sand	Bimodal, very poorly sorted	Rounded
40	Gravelly sand	Bimodal, very poorly sorted	Sub-angular
41	Muddy gravelly sand	Bimodal, poorly sorted	Sub-angular
42	Gravelly muddy sand	Unimodal, poorly sorted	Sub-angular
43	Gravelly sand	Bimodal, very poorly sorted	Sub-rounded
44	Gravelly sand	Bimodal, very poorly sorted	Rounded
45	Gravelly sand	Bimodal, very poorly sorted	Rounded
46	Gravelly sand	Bimodal, very poorly sorted	Sub-rounded
47	Muddy gravelly sand	Unimodal, poorly sorted	Sub-rounded
48	Gravelly sand	Unimodal, moderately sorted	Sub-angular
49	Gravelly sand	Unimodal, moderately sorted	Rounded
50	Gravelly sand	Bimodal, moderately sorted	Sub-angular

(Fig. 8). The maximum movement of sediment due to wave action takes place just seaward of where waves break, as seen at the edge of the windward (west) side, where the sediment is coarse grained because fine sediments are removed and transported to deeper waters.

At high tide (as high as 30 cm), small waves may pass over the edge or top of the shoal, causing minor disturbance to bottom sediments, whereas waves or disturbance generated during low tide (ebbing) are effective in resuspension because of stronger velocities than during flood tide. Therefore, the movement of sediment is governed primarily by the tidal cycle and, depending on the strength of the tide, either the sediments settle down within the shoal area or are transported offshore during ebbing. Granules may not be transported because of the weak energy, and remain as lag deposits, while mud is resuspended and transported to deeper water but may

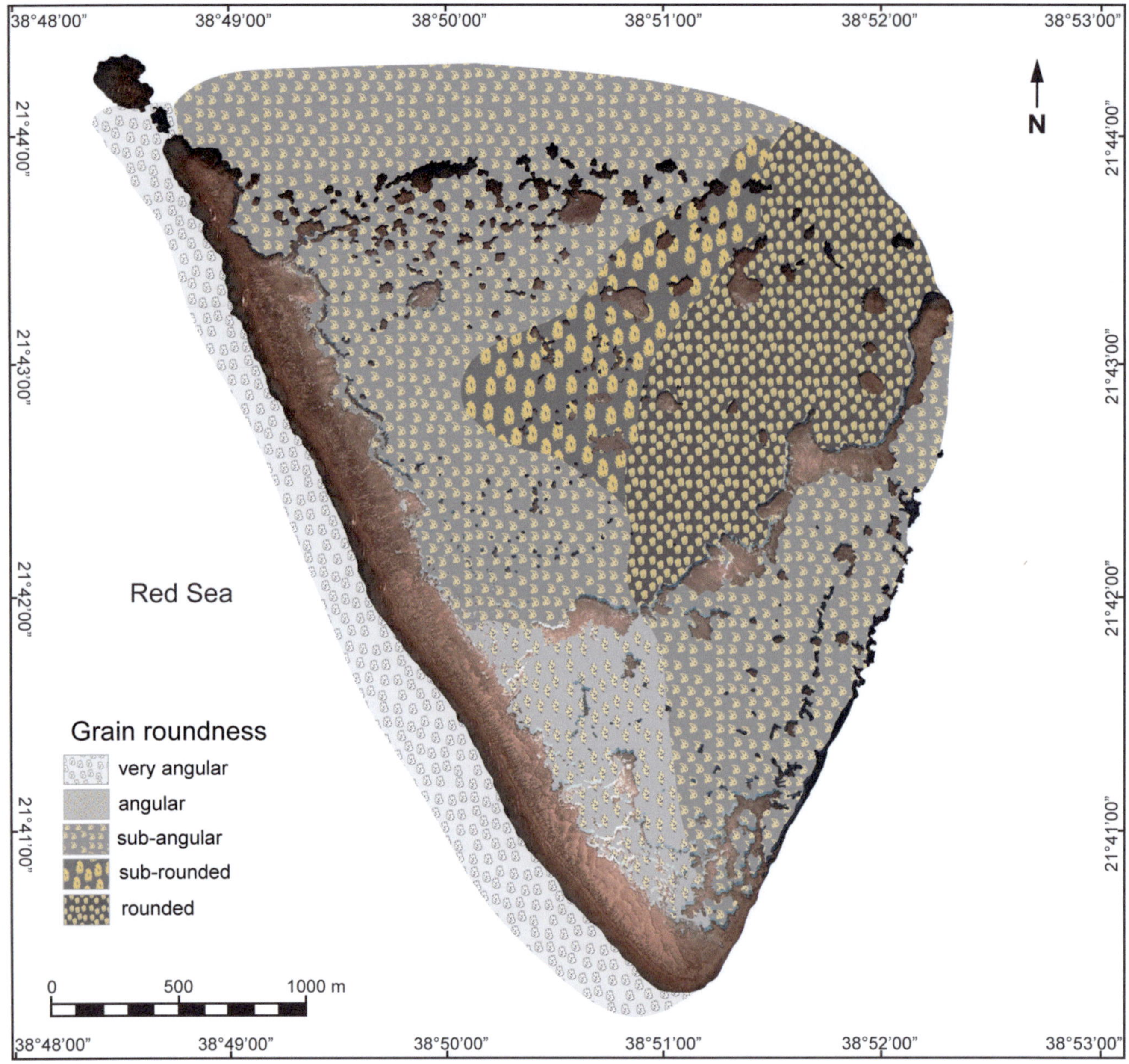

Fig. 9 The degree of roundness and angularity of grains in the study area

also settle down under low energy within the shoal. However, trapped granules could be transported to the deeper waters only when the water currents get stronger.

4.5 Grain Morphology

Grain roundness and angularity are used in sedimentology for deciphering the severity of the mode of transportation, and the impact the grains undergo during their transit (Folk 1980). Roundness also relates to the distance and history of travel. The roundness of sediments on the shoal ranged between angular to rounded. Well-rounded grains are absent, indicating either a low-energy environment or a short history of transportation (Table 2 and Fig. 9). The data presented are related mostly to biogenic material to investigate the effect of physical parameters, rather than source, in order to identify the cycle and history of transportation. Under normal circumstances, changes in grain size, angularity, sphericity and roundness are dependent on the environmental dynamic, the distance the grain has travelled, and the composition and strength of the individual grain.

An admixture of angular, sub-angular and sub-rounded grains is observed. The low degree of roundness of the grain is because of their being constantly broken up by strong currents and the boring of the substrate by the marine organisms that weaken and break the shells and corals, giving the grains their angularity. The whole shells and shell fragments are recent and fresh in appearance. The freshness and polished surface of shell fragments are from the constant agitation of water. The number of whole shells (intact) increases because of the restricted water agitation, especially because of the shoals present in the area. The weak water current does not have much impact on the breaking-up of biogenic grains nor is it capable of polishing the grain surface. This is one of the many reasons the biogenic materials are intact and dull brown in appearance. It is observed that the roundness and sorting decreases with diminishing current velocity. Recently broken corals forming the debris are angular, whereas ones with dull lustre are sub-angular, indicating that with time the roundness increases because of the to and fro movement of grains by water currents.

4.6 Sediment Contributors

Sediment is contributed to the shoal by different means and sources, derived either locally or transported by wind or by wadis only during flash floods. The sediment can be biogenic or lithogenic in composition and the mechanism of sediment production can differ significantly for each type.

Fig. 10 **a** Typical shallow reef-flat habitat that has ~1 cm thick cover of algae, and minor coral coverage; **b** Example of shallow lagoonal sand and mud; the clean fine sediment occasionally contains rubble and small broken coral heads; sand may be colonized by small patches of green–blue algae called Brown slime (*Schizothrix sigalcicola*), is a sign of stagnant conditions

Fig. 11 Typical types of bio-erosion grazers; **a** Parrotfish nibble on hard corals, and **b** Surgeonfish (Acanthurids) scrape the coral surface with their sharp teeth

Biological processes involving the flora and fauna inhabiting the shoal are the principal contributors to the carbonate-rich environment where dead corals dominate. The contribution is from the bio-erosion of the coral reefs where marine organisms disintegrate the reefal system by chemical and physical processes. This bio-erosion plays two key roles; it creates new cavities in the reef, and produces new sediment by excavating the holes that are created by various marine fauna, especially parrotfish, shrimp-gobies and others capable of grinding the corals into small particles. The bio-eroders have been classified as grazers, borers, and burrowers by Dudley (2003), who defined the roles of each in supplying sediments, especially carbonate material. Grazers are organisms that feed and nibble on live corals, encrusting coralline algae, and on tufted, filamentous and endolithic algae growing on dead reef substrate (Fig. 10). The most dramatic of the grazers are ones that attack the coral directly. Live corals are consumed by various species of worms, gastropods, sponges, nudibranchs, crabs, starfish and fish. Pufferfish (Tetradontids) are known to bite off and swallow the soft tips of growing coral, whereas surgeonfish (Acanthurids) scrape the coral surface with their teeth. The most important contributor is the parrotfish (Scarids) that feed mainly on soft algae growing on dead coral substrates and seldom on live corals (Fig. 11). When feeding on the encrusting layer on dead coral they remove part of the substrate beneath it, generating new sediment. Some species have been known to swallow mostly sand and recycle old reef sediment by ingesting. The pharyngeal mill of parrotfish, like the gizzard of surgeonfish, is known to grind fragments of carbonate material into tiny sizes. It is likely that some fine carbonate sediment is excreted in its original

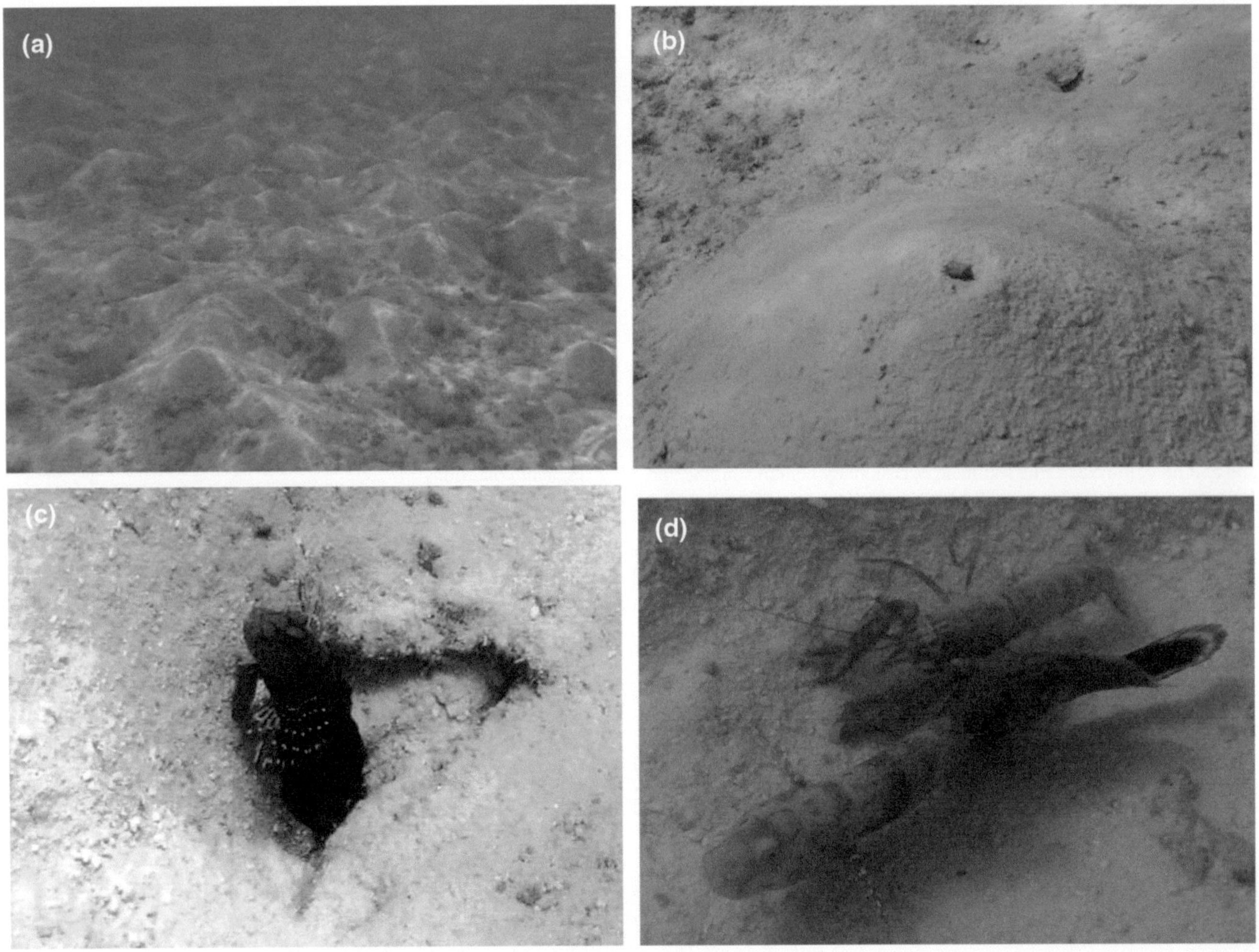

Fig. 12 **a** Cluster of burrows in a shallow lagoonal sand and mud environment, some areas covered with small patches of algae indicating a stagnant environment; **b** fine to coarse sand burrow construction known as a Symmetrical Mound, resulting from *Alpheus bellulus* shrimp boring activity; **c** fish known as Luther's shrimp-goby (*Cyptocentrus lutheri*) inhabits expanses of fine to coarse sand in protected lagoons, and lives in burrows with the shrimps. It bores into the carbonate sediment with the help of *Alpheus bellulus* shrimp, as shown in (**d**)

form whereas some particles dissolve while passing through the gut. Parrotfish can therefore be regarded as an important bio-eroder by generating new sediment as a result of nibbling on coral substrates, and by reducing the size of pre-existing sediments they ingest (Dudley 2003).

Many different plants and animals produce holes ranging in size from a few microns (μm) to several centimetres by boring on the seabed (Rasul et al. 2017). However, holes between 1 and 5 μm in diameter are known to be produced by bacteria and fungi. Boring activity of endolithic algae is well known. Green, red, and blue–green algae (cyanophyta bacteria) are also known to be capable of eroding carbonate substrates. The blue–green algae possess special boring filaments which produce cavities between 5 and 15 μm in diameter. The micro-boring activity of algae, bacteria and fungi is believed to be accomplished through biochemical dissolution of the carbonate (Dudley 2003). Many examples have been observed in the marine environment, especially where the area is not dynamic.

Fish species such as Luther's shrimp-goby (*Cyptocentrus lutheri*) inhabit areas rich in fine to coarse sand in protected and sheltered environments like lagoons and bays, and live in burrows with the shrimp *Alpheus bellulus*. They make burrows by boring into the carbonate-rich sediment veneer with the assistance of shrimps. Mounds are created by the excavated sediments and are abandoned with time for unknown reasons. Once evacuated, the mounds become somewhat cemented and covered with algae (Fig. 12). Bioturbation seems to be quite active in areas where the bottom currents are not strong, making the calmer environment the best place to inhabit and to produce sediment. In order for the shrimp-gobies to house themselves or avoid being preyed upon, these two species make burrows in either hard ground or on loose sediment veneer. The dynamics of burrows are well known and documented. Karplus (1987) divided the concept of burrows into three; their structure, physical construction and their dynamics. All three are dependent on the species and the type of sediment (Cummins 1979; Karplus et al. 1974; Yanagisawa 1984). Some burrows may be shallow and branching while others could be short and deep. The morphologies of tunnel-like features, commonly referred to as burrows, are relatively consistent in all shrimp, and are used as homes to inhabit and to hide from predators. The floors are sandy as the shrimp excavates sand and the tops are rocky or filled in with coral debris to prevent collapse (Yanagisawa 1984). Farrow (1971) concluded that the physical characteristic of the sediment is critical in determining the structure and size of the burrows. When the substrate is hard and rocky, the holes are irregular, but when the substrate is muddy and soft, they are regular and dichotomously branching.

The process of bioturbation is quite common in the benthic marine environment. Apart from fish and shrimps, burrowing organisms such as worms and holothurians (sea cucumbers) take in fresh sediment at one end and excrete the processed sediment at the other. As the sediment passes through the gut not only is the organic matter removed but in some cases the sediment grains are also reduced in size due

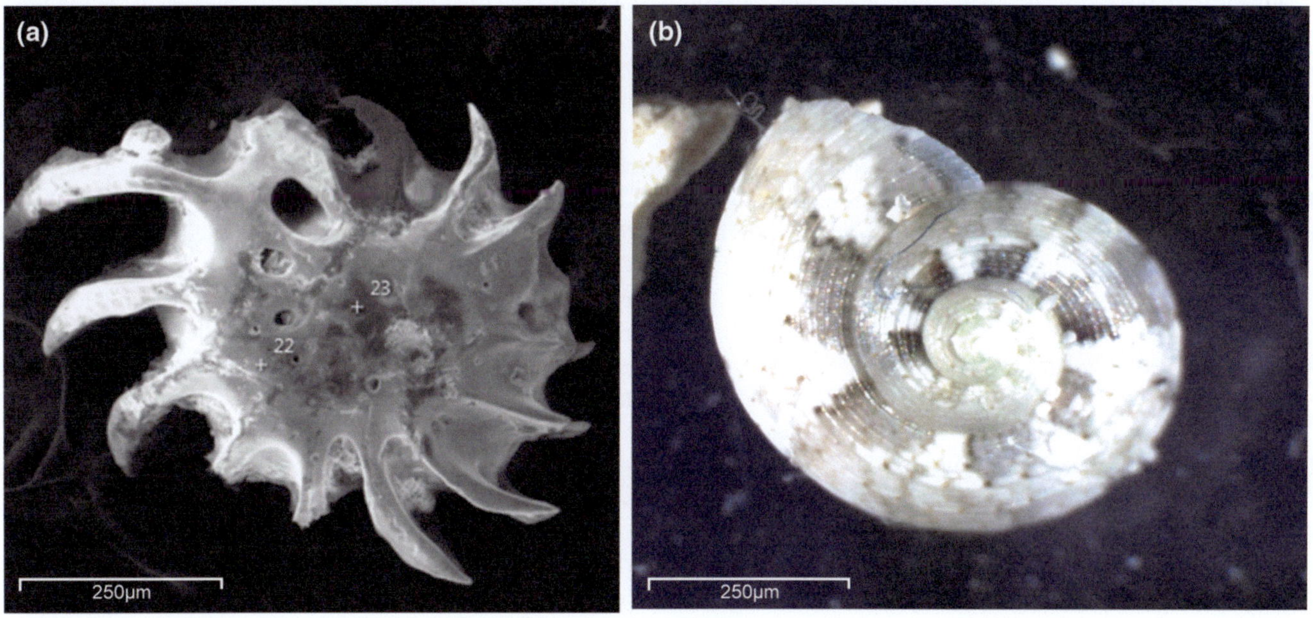

Fig. 13 **a** Broken benthic foraminifera (250 μm medium sand), also altered by taphonomic processes such as breakage by currents, micro-borings by endolithic microorganisms and fungi, and dissolution by low pH; **b** microgastropod (250 μm medium sand) also altered and microbored by endolithic microorganisms and fungi, and dissolution by low pH

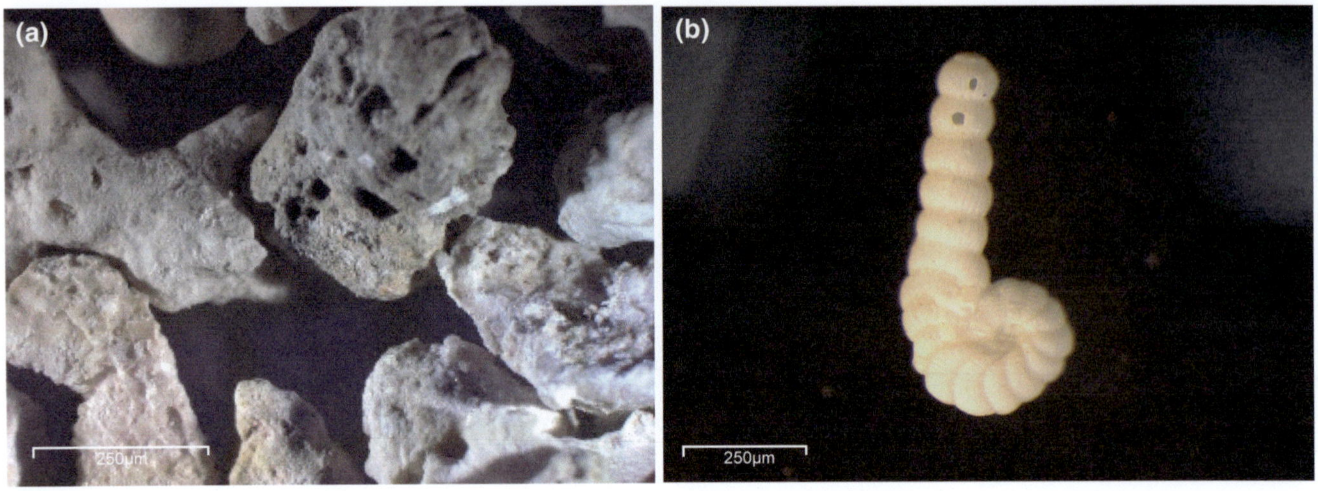

Fig. 14 **a** Pitted carbonate grain (250 µm medium sand), an indication of stagnant conditions, and **b** micro-borings by endolithic micro-organisms

to chemical dissolution or abrasion. Some organisms segregate sediments according to size, moving fine-grained material toward the surface (Dudley 2003). They are responsible for recycling the seabed and contribute sediments in different forms to the shoal in addition to the other sources.

Pitting observed on the surface of shells and forams could be related to algal growth (Rasul 1992, 2015; Rasul et al. 2017). The algae bore into the shell surface or wall and with time the algal filaments die and decay, leaving the trace of the boring. Algal boring is also observed on dead corals. However, the green and blue–green algae colonies within the pits could not be observed because of the hydrogen peroxide treatment to remove organic matter from the grains, leaving the boring visible. Shells may break because of strong currents and when the forams die the outer surface becomes weaker; therefore, they could be altered by taphonomic processes such as breakage by currents, microborings by endolithic micro-organisms, fungi and dissolution (Fig. 13). It is observed that where sediment input is absent or minimal boring or pitting takes place especially in calmer environments (Fig. 14).

4.7 Sediment Composition

The shoal is veneered mostly with gastropods, forams (Fig. 15) and coral-reef debris (Fig. 16a) of biogenic origin (carbonate), followed by limited amounts of detrital material (Fig. 16b and c). Since no major wadis (seasonal streams) drain into the Red Sea, terrigenous input from the adjoining land by wadis is negligible and carbonate material dominates the sediment veneer. The distribution of detrital material is limited and controlled mostly by the hydrodynamics and by the atmospheric dust brought in by the shamal (wind). The relative abundance is also governed by the particle size, and the environment of deposition, but the adjoining deserts contribute stained quartz to the area (Fig. 16d).

The destruction of reefs and shoals contributes most, if not all of the carbonate material to the study area. The supply and breakdown of reefs are either by wave and current actions, human activities and natural causes (weakening and breaking of the corals due to natural deterioration), and by boring marine organisms, including coralline algae, sponges and bryzoans that are related to high productivity in the Red Sea. The forams, gastropods and other shelly materials are the product of the typical marine environment of the Red Sea. A microscopic examination showed that most of the detrital materials consists of quartz, feldspars, and occasionally mica (biotite). The feldspars are fresh in appearance, while angular to sub-angular quartz grains, mostly of aeolian origin, are appreciably frosted. Relatively limited terrigenous material is supplied to the shoal, probably by Wadi Al-Kura or by wadis located north of Sharm Obhur. Since the veneer is made up of quartz, feldspars and mica, it is suggested that the probable source of these minerals, although in limited amounts, is from Wadi-Al Kura during the times when it was directly connected to the sharm, that is, before the construction of a bridge and barriers erected to control debris and garbage transport to the sharm. However, it is also possible that mica has drifted from the north (Rasul 2015; Rasul et al. 2009).

4.8 Bulk Mineralogy

A total of sixteen representative samples were analysed and the following minerals were determined in order of abundance: aragonite, High-Mg calcite, quartz, feldspars, pyrite, dolomite and traces of halite and gypsum (Table 3 and Fig. 17). A few peaks were not taken into account because

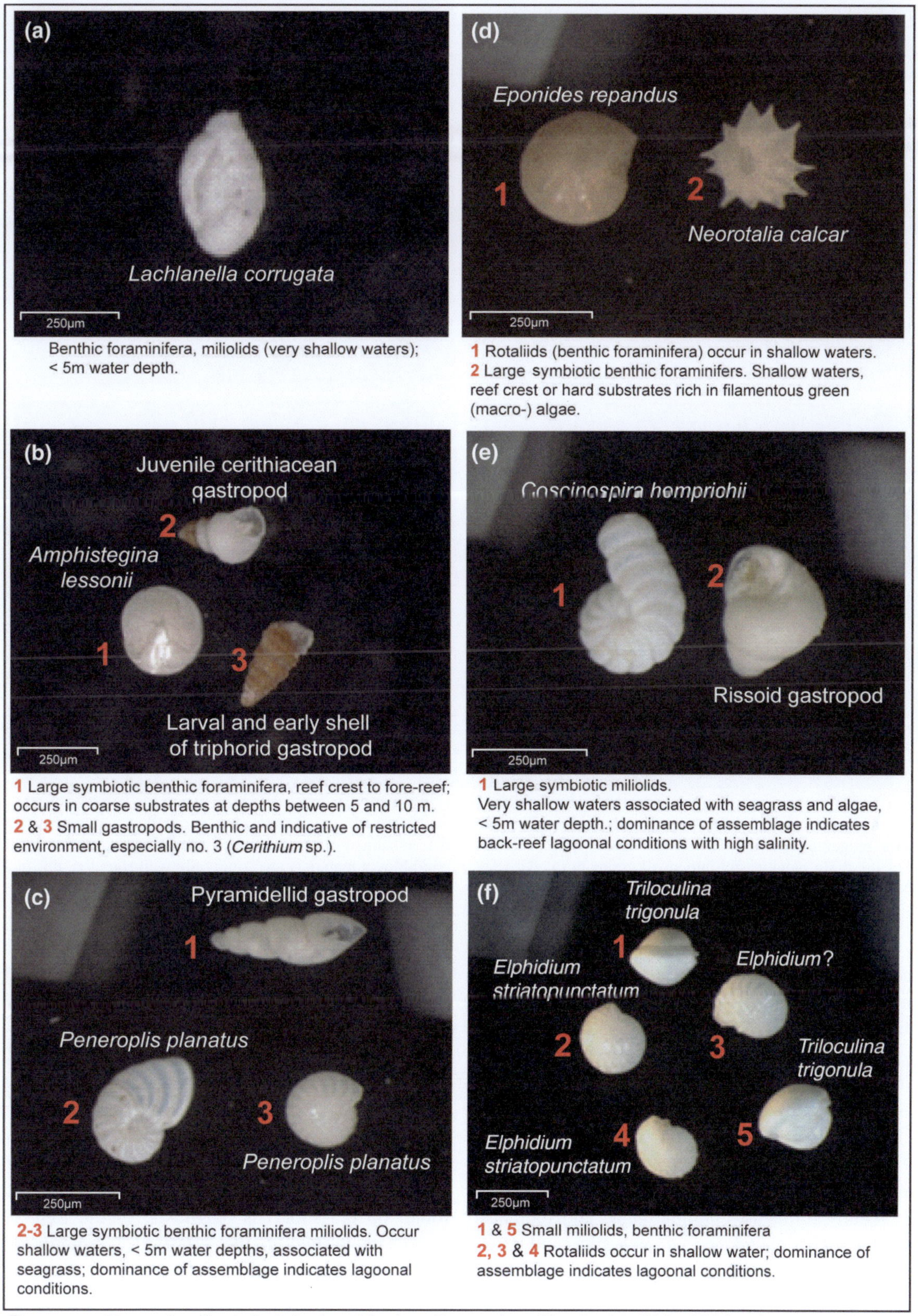

These assemblages altogether indicate back-reef (back-reef lagoonal conditions) or intertidal-subtidal environments with water depths of < 5 m, and enriched in sea-grasses and algae dominated area.

Fig. 15 The assemblages indicate back-reef lagoonal conditions or inter-tidal to sub-tidal environments at <5 m water depth. They flourish in areas where seagrass and algae dominate

Fig. 16 **a** Typical shoal sample. Dark grains rich in carbon are associated with biogenic material. The brittle nature of the carbon material makes the grain break easily into small grains but it does not have a pre-determined source or association. **b** Mixture of quartz and feldspars. Some quartz is aeolian (wind blown), some may be wadi-derived. **c** Biotite in a shallow quiescent environment. **d** Iron-stained quartz grain, a confirmation of aeolian input. Although wind-blown materials are commonly well-rounded, this quartz grain is sub-angular, indicating a short history of transportation

of the faint signal. Environmental conditions and the proportions of lithogenic and biological materials control the relative abundance of these minerals. Traces of halite (crystals developing from high evaporation rates) and gypsum were observed but the faint peak on the XRD diffractogram failed to show a distinct distribution pattern. Aragonite is produced originally from coral fragments, for example, *Helmida* sp., including inorganic carbonate precipitates, whereas High-Mg calcite and aragonite are produced from reefal sediments, coralline algae and mollusks (El-Sayed 1984). The abundance of aragonite and High Mg-calcite is probably derived from the coral fragments as observed by Friedman (1968) and Behairy and El-Sayed (1984). However, it is also suggested that the water flooding the sabkhas (supra-tidal deposits) is rapidly evaporated to an extent that allows the precipitation of High Mg-calcite, dolomite and gypsum (Behairy et al. 1991), although dolomite was not present or identified. According to Levy (1977), dolomite is present in some sabkhas, including traces in the sabkha sediments of Wadi Al-Kura, where it forms diagenetically through the interaction between previously precipitated calcium carbonate and concentrated marine brines rich in magnesium. The dominance of aragonite and High-Mg calcite and limited pure calcite, and dolomite (not found on the shoal) is diagnostic of the Red Sea coastal sediments located in the subtropical region (Rasul et al. 2010; Bahafzalah and El-Askary 1981; Rao and Behairy 1986; Al-Washmi and Rasul 2003). The presence of some crystals of halite is the result of high evaporation and the hypersaline water.

Table 3 Bulk minerals identified by X-ray diffratogram (XRD)

Station no.	Abundant	Moderate	Traces
4	Aragonite, High Mg–Calcite		Plagioclene, quartz
6	Aragonite, High Mg–Calcite		Clay minerals, mica, quartz, plagioclene
8	Aragonite, High Mg–Calcite		
14	Aragonite, High Mg–Calcite, quartz		
16	High Mg–Calcite, aragonite		Plagioclene, quartz
19	Aragonite, High Mg–Calcite		Plagioclene, feldspar
22	Aragonite, High Mg–Calcite		Plagioclene
26	Aragonite, High Mg–Calcite		Plagioclene, quartz
28	Aragonite, High Mg–Calcite		Quartz
29	High Mg–Calcite, aragonite		Low Mg–Calcite, quartz, gypsum
30	Aragonite, High Mg–Calcite		
34	High Mg–Calcite, aragonite		Clay minerals, mica?
36	Aragonite, High Mg–Calcite	Low Mg–Calcite	
38	High Mg–Calcite, aragonite	High Mg–Calcite, quartz, mica?	Gypsum, clay minerals
41	High Mg–Calcite, aragonite		
47	Aragonite, High Mg–Calcite		

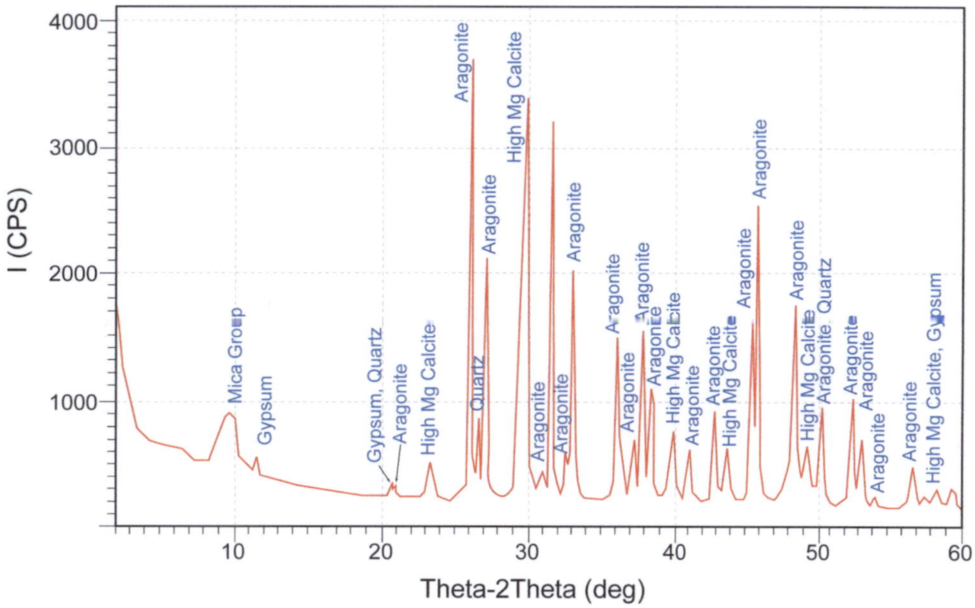

Fig. 17 X-ray diffractogram of Red Sea sediment showing various minerals identified by measuring the XRD peaks

4.9 Calcium Carbonate ($CaCO_3$)

The calcium carbonate content ranges from 66% at station 33 to 99% at station 19, with an average of 83% (Table 4 and Fig. 18). The higher content is because carbonate is contributed mostly by the shoal and from biogenic materials such as gastropods and foraminiferal sand. Basins in reef-bound coasts without any significant runoff and high aridity are the most probable source of carbonate material in the Red Sea. Therefore, the particle size and their characteristics are controlled primarily by the in situ production of carbonate material. In the Red Sea, the corals and presence

Table 4 Calcium carbonate data obtained from an automated calcimeter

Station	Water depth (m)	Textural class (Folk 1980)	$CaCO_3$ (%)
1	8.00	Gravelly sand	91
2	3.00	Muddy sand	92
3	8.30	Muddy sand	88
4	10.00	Muddy gravelly sand	99
5	7.90	Muddy sand	88
6	7.00	Muddy sand	96
7	7.40	Gravelly sand	93
8	6.60	Muddy sand	96
9	7.10	Gravelly sand	91
10	5.10	Gravelly sand	90
11	1.70	Gravelly sand	74
12	1.50	Gravelly sand	75
13	2.00	Gravelly sand	92
14	2.00	Muddy sand	98
15	7.00	Gravelly sand	92
16	12.50	Muddy sand	95
17	8.00	Muddy sand	85
18	6.60	Muddy gravelly sand	87
19	7.90	Muddy gravelly sand	99
20	8.50	Gravelly sand	71
21	5.60	Gravelly muddy sand	88
22	5.00	Gravelly muddy sand	87
23	1.50	Muddy gravelly sand	70
24	1.20	Muddy gravelly sand	71
25	2.80	Muddy gravelly sand	72
26	4.10	Muddy sand	90
27	5.80	Muddy gravelly sand	92
28	8.80	Gravelly muddy sand	95
29	6.20	Muddy sand	91
30	12.00	Muddy sand	95
31	14.50	Sandy gravel	70
32	13.00	Gravelly sand	73
33	18.70	Gravelly sand	66
34	25.00	Muddy sand	68
35	26.00	Gravelly sand	72
36	30.00	Gravelly sand	75
37	32.00	Gravelly sand	75
38	8.50	Muddy sand	90
39	14.80	Gravelly sand	71
40	17.20	Gravelly sand	68
41	16.80	Muddy gravelly sand	67
42	11.00	Gravelly muddy sand	81
43	11.50	Gravelly sand	73
44	11.00	Gravelly sand	70

(continued)

Table 4 (continued)

Station	Water depth (m)	Textural class (Folk 1980)	$CaCO_3$ (%)
45	11.00	Gravelly sand	86
46	8.30	Gravelly sand	86
47	10.00	Muddy gravelly sand	80
48	8.40	Gravelly sand	87
49	10.70	Gravelly sand	77
50	7.60	Gravelly sand	88
Min	1.2		66.0
Max	32.0		99.0
Avg	9.7		83.3

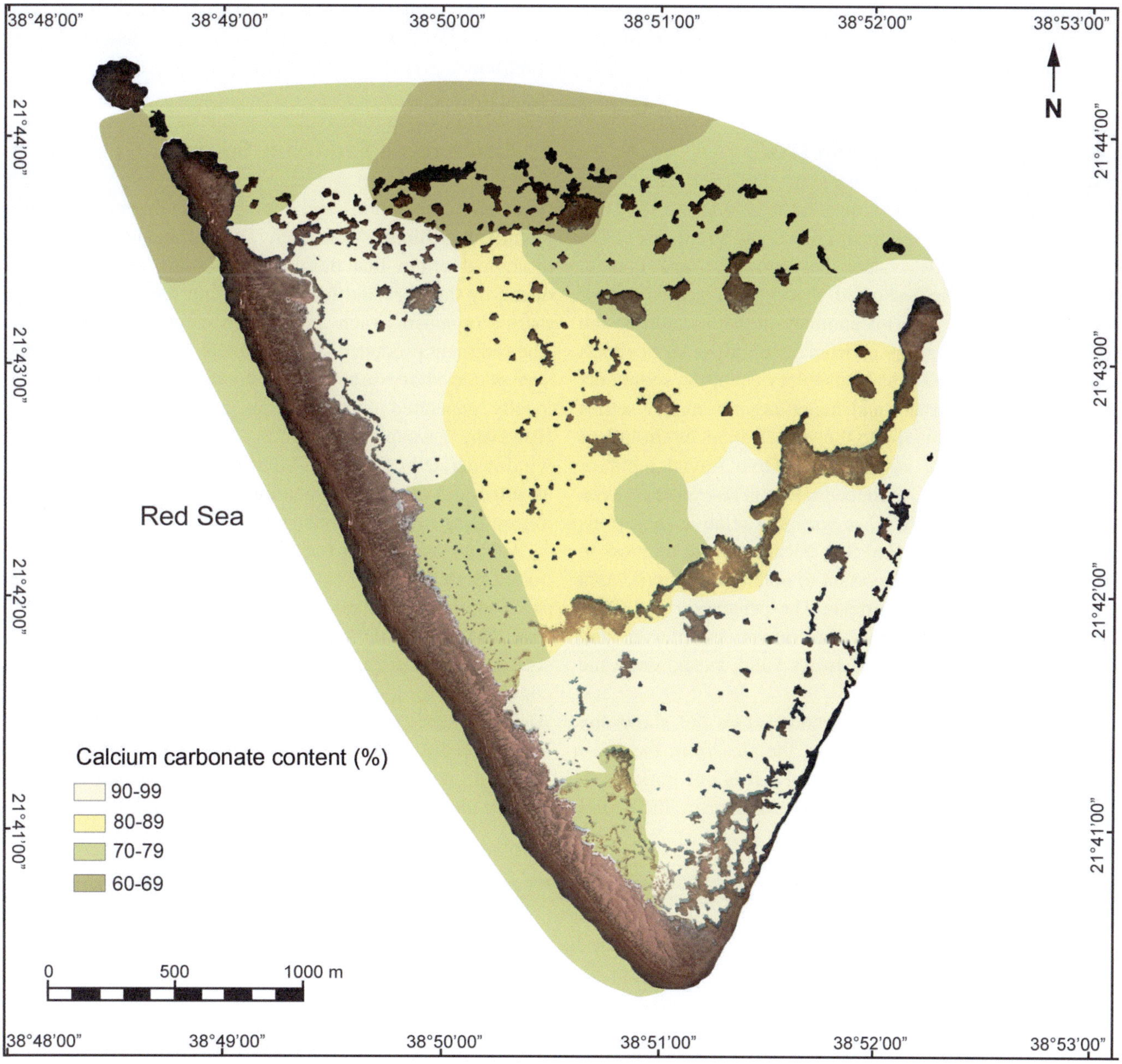

Fig. 18 Distribution of calcium carbonate content in the surficial sediment from the study area

of fringing and barrier reefs play a significant role in the high production of reefal sediments. At present, the restricted flow of Wadi Al-Kura into Sharm Obhur (a creek) and numerous wadis to the north of the sharm mean that the neighbouring areas are not influenced by flash floods, and are rich in carbonate content. As expected the carbonate material over-shadows the limited terrigenous input from the past. The carbonate is supplied to the area by the erosion of reefs by wave and current action, break-up of reefs by fishermen and nibbling of the corals by fish. The percentage of calcium carbonate content contributed by each is undetermined but the source of the carbonate is undoubtedly the shoal itself. However, the biogenic components are further fragmented by human activity and pounded by wave action, especially that part of the shoal facing the open sea, causing the reef to break and thus initiating the production of fine calcareous material.

4.10 Total Organic Carbon (TOC)

Of the 50 samples recovered, only six were selected for TOC analysis based on the sediment colour and grain size. Five showed minimal TOC values between 0.01 and 0.16%, while only one sample showed a very high percentage of TOC (3.12%; Table 5). This anomaly of 3.1% could be the product of bird droppings or from green algae or seagrass found in the area. This area is also very calm, the sediments are not greatly disturbed, and hence organic matter settles down and is incorporated into the veneer. In an environment where there is hardly any influence from fluvial input or organic sources, the TOC concentration is restricted or even absent. In general, the TOC is very low relative to various sub-environments in the Red Sea, because of the richness in carbonate material. It is important to note that the entire area contains light-coloured sediment. The perception that TOC is higher in fine sediments and lower in coarser and carbonate-rich material is not always valid, as shown in the preceding sections.

The distribution of TOC in general does not show any trend because of the remoteness of the area from the coast and lack of organic sources of supply. It is not possible to obtain a trend based on only 6 samples. However, the values of TOC are not compatible with values reported by other workers in the Red Sea (El-Sayed and Hosny 1980; Behairy et al. 1983; Al-Washmi and Rasul 2003), nor are they present in fine sediments and related to sediment texture. The abundance in coarse-grained carbonate material is somewhat unusual but not abnormal. The higher values of TOC are related to weak currents, and to a sufficient supply of organic material from seagrass and algae, and including the droppings contributed by seabirds. The tendency of shells and forams to adsorb and absorb the organic material effectively has been observed in the shoal area. The degree of absorption is higher in species of forams and gastropods that have porous surfaces that result in the penetration of the nutrient-rich material, giving rise to algal growth (green–red algae) thus enhancing the process of reduction and contributing to TOC increase over time under stagnant conditions.

5 Discernible Sediment Sources

Different compositions and sources of sediments have been identified in Shiab Al-Kabeer. The carbonate sediments of biological origin that dominate the shoal are locally produced by the flora and fauna inhabiting the shoal, including marine organisms such as algae growing on the reef itself. The sediment produced is deposited at the foot of the reef or transported short distances within the sub-environment by locally generated currents, especially during tidal cycles. However, water turbulence generated mostly by wind-generated currents and during spring tidal cycles plays an important role in resuspending the bottom sediments and transporting them to deeper waters.

Aeolian material in the form of stained quartz is transported from the desert by strong winds and deposited on the shoal during dust storms that pass over the Red Sea (Fig. 19e). Although other terrigenous components are scarce, drifted mica, fluvial quartz and feldspars are also contributed to the area either by the once-active Wadi Al-Kura located at the head of the sharm, or numerous wadis located to the north of Sharm Obhur. The numerous small wadis are not very active and therefore the input to the sea may be negligible and would be noticeable in the short term

Table 5 Distribution of TOC in selected areas of the shoal

Station	Sediment texture	TOC (%)	Mud (%)
8	Muddy sand	3.12	39
28	Gravelly muddy sand	0.14	9
38	Muddy sand	0.12	44
40	Gravelly sand	0.03	4
42	Gravelly muddy sand	0.16	21
45	Gravelly sand	0.01	2

Fig. 19 Probable modes of sediment input to the shoal. **a** Floaters (in yellow) deployed to restrict the effect of sediment plumes on the sharm, although the plume is eventually dispersed to the open sea by the surface current; **b** Dredging to increase the water depth for deep-keel boats to manoeuvre easily. The dredged material is dumped into the deeper water regardless of the environmental hazard or impact; **c** sediment plume from Sharm Obhur moving in a westerly direction. This event is after rainfall; **d** inadequately treated sewage and ground water being discharged into the sea at the Corniche area. The murky water indicates the huge area under the threat of contamination; and **e** dust storm over the Red Sea area between Sharm Obhur and Shiab Al-Kabeer. The storm might influence the sediment distribution over the shoal and adjoining areas. Although dust storms are frequent over the Red Sea, the chance of finding aeolian quartz is low because of the constant carbonate input from the shoal

only. The groundwater that is discharged into the Red Sea, and the sediment plume that is generated during rainfall by the seasonal streams are possible contributors of sediments to the shoal (Fig. 19). However, because of the lack of terrigenous input it is very difficult to determine a single source for the veneer composition in the Red Sea and Shiab al-Kabeer.

6 Conclusion

The surface veneer rich in biogenic material in the form of shells, forams and abundant coral debris is relatively coarser, dark in colour and poorly sorted with little fine sediment, which if present is in minor amounts and is calcareous in nature. This is related to weak currents where the environment is relatively stagnant, but coarse debris is contributed mostly by human activities and fine sediment by bio-erosion. Many fish species inhabiting the coral reefs and algae covering the dead corals in quiescent areas are exclusively responsible for the break-up of the carbonate crust or carbonate-rich rubble that constitute part of the sediment veneer of biogenic origin in the shoal area. Quartz, mica and feldspars make up the terrigenous input.

The area has mostly weak to moderate currents because the shoals have a significant role in controlling the movement of water and sediment. Staining, boring and pitting of sediments are found in less dynamic areas. The sediment colour ranges from light yellow white to very pale brown, where darker sediments are related to the reducing environment and limited supply of terrigenous material. In general, the sediment on the shoal is contemporary and its distribution is controlled primarily by biological processes, physical erosion and how dynamic the area is in terms of wind-wave-generated currents, including the tidal current responsible for sediment transport.

Additional environmental and geological studies are warranted since investments related to tourism in many islands and shoals in the Saudi Arabian Red Sea are now being promoted.

Acknowledgements We wish to thank the Saudi Geological Survey for giving us the opportunity to use the data that were generated under the technical work program, and data used in a Master's thesis. We would also like to express our thanks to Dr. Marco Taviani and Dr. Ramadan Abu-Zeid in helping us in the identification of biogenic components. We also wish to extend our thanks to Drs. G. Owen, J. Coakley, N. Hashimi, G. Bailey, A. Khalik and A. Rushdi for reviewing the chapter. Many discussions with Dr. I. Stewart and his comments helped us tremendously in compiling this research. Louiesito Abalos is thanked for helping us in formatting the chapter. Special thanks to all the members of the Center for Marine Geology and colleagues at Saudi Geological Survey for helping us in the field and laboratory work.

References

Abou-Ouf MA, El Shater A (1993) Black benthic foraminifera in carbonate facies of a coastal sabkha, Saudi Arabian Red Sea coast. J Fac Mar Sci King Abdulaziz Univ 4:133–141

Al-Washmi HA, Rasul NM (2003) Sedimentary processes of the Quaternary sediments of Al-Kharrar Lagoon, Red Sea, Saudi Arabia. J Environ Sci Univ Mansoura Egypt 26(1):273–296

Bahafzalah AAK, El Askary MA (1981) Sedimentological and micropaleontological investigations of the beach sands around Jeddah, Saudi Arabia. Bull Fac Earth Sci KAU 4:25–42

Behairy AKA, El-Sayed MKh (1984) Carbonate cements in modern Red Sea reef, north of Jeddah, Saudi Arabia. Mar Geol 58:443–450

Behairy AKA, Jaubert J (eds) (1984) Ecology of a coral reef complex and of an inshore lagoon near Sharm Obhur, Red Sea, Saudi Arabia. Final Report, KAU. FMS/UN 163 p (unpublished)

Behairy AKA, El-Rayis OA, Ibrahim AM (1983) Preliminary investigations of some heavy metals in water, sediments and plankton in Obhur Creek (eastern Red Sea). J Fac Mar Sci King Abdulaziz Univ 3:129–139

Behairy AKA, Rao DNVN, El-Shater A (1991) A siliciclastic coastal sabkha, Red Sea Coast, Saudi Arabia. J Fac Mar Sci King Abdulaziz Univ 2:65–77

Behairy AKA, Sheppard CRC, El-Sayed MK (1992) A review of the geology of coral reefs in the Red Sea. UNEP Regional Seas Reports and Studies No. 152, 39 p

Bruckner A, Dempsey CA (2015) The status, threats, and resilience of reef-building corals of the Saudi Arabian Red Sea. In: Rasul NMA, Stewart ICF (eds) The Red Sea: the formation, morphology, oceanography and environment of a young ocean basin. Springer Earth System Sciences, Berlin, pp 471–486

Bruckner A, Rowlands G, Riegl B, Purkis S, Williams A, Renaud P (2011) Khaled bin Sultan Living Oceans Foundation atlas of Saudi Arabian Red Sea marine habitats. Panoramic Press, Phoenix, Arizona, 274 p

Bruckner A, Rowlands G, Riegl B, Williams A, Renaud P (2012) Khaled bin sultan living oceans foundation atlas of Saudi Arabian Red Sea marine habitats. Panoramic Press, 262 p

Crossland CJ (1987) In situ release of mucus and DOC-lipid from the corals *Acropora variabilis* and *Stylophora pistillata* in different light regimes. Coral Reefs 6:35–42

Cummins RA (1979) Ecology of gobiid fishes associated with alpheid shrimps at one tree reef. PhD thesis, University of Sydney, Sydney, Australia, 252 p

De Vantier L, Pilcher N (2000) The status of coral reefs in Saudi Arabia. In: Global Coral Reef Monitoring Network (GCRMN)

De Vantier L, Turak E, Al-Shaikh K, De'ath G (2000) Coral communities of the central-northern Saudi Arabian Red Sea. Fauna Arab 18:23–66

Dudley WC (2003) Coral reef sedimentology. Kalakua Marine Education Center, University of Hawaii, Hilo, Hawaii. Marine Science Course 461

Edwards FJ (1987) Climate and oceanography. In: Edwards AJ, Head SM (eds) The Red Sea. Pergamon Press, Oxford, pp 45–68

El-Sayed SZ (1984) Productivity of the Antarctic waters—a reappraisal. In: Holm-Hansen O, Bolis L, Gllles R (eds) Marine phytoplankton and productivity. Springer, Berlin, pp 19–34

El-Sayed MK, Hosny CF (1980) Sediments of the intertidal zone off Ghardaqa, Red Sea, Egypt. In: Proceedings of the symposium on coastal and marine environments of the Red Sea, Khartoum

Farrow GE (1971) Back-reef and lagoonal environments of Aldabra atoll distinguished by their crustacean burrows. Symp Zool Soc Lond 28:455–500

Fishelson I (1971) Ecology and distribution of the benthic fauna in the shallow waters of the Red Sea. Mar Biol 10:113–133

Folk RL (1980) Petrology of sedimentary rocks. Hemphill's, Austin, Texas, 182 p

Friedman G (1968) Geology and geochemistry of reefs carbonate sediments, and waters, Gulf of Aqaba (Elat), Red Sea. J Sed Petrol 38(2):895–919

Jado AR, Hötzl H (1984) Quaternary period in Saudi Arabia. Springer, 361 p

Jado AR, Hötzl, H Boscher B (1989) Development of sedimentation along the Saudi Arabian Red Sea coast. J Earth Sci King Abdulaziz Univ 3:863–888

Karplus I (1987) The association between gobiid fishes and burrowing alpheid shrimps. Oceanogr Mar Biol Ann Rev 25:507–562

Karplus I, Szlep R, Tsurnamal M (1974) The burrows of alpheid shrimp associated with gobiid fish in the northern Red Sea. Mar Biol 24:259–268

Klaus R (2015) Coral reefs and communities of the central and southern Red Sea (Sudan, Eritrea, Djibouti, and Yemen). In: Rasul NMA, Stewart ICF (eds) The Red Sea: the formation, morphology, oceanography and environment of a young ocean basin. Springer Earth System Sciences, Berlin, pp 409–451

Le Corre (1983) Dosage du carbone organique paticulaire. In: Aminot A, Chaussepeid M (eds) Manuel des analyses chimiques en milieu marin. CNEXO, France, pp 203–208

Levy Y (1977) The origin and evolution of brine in coastal sabkhas, northern Sinai. J Sediment Petrol 47(1):451–462

Maiklem WR (1967) Black and brown speckled foraminiferal sand from the southern part of the Great Barrier Reef. J Sediment Petrol 37:1023–1030

Mason C, Folk RL (1958) Differentiation of beach dune and aeolian flat environment by size analysis, Mustang Island, Texas. J Sediment Petrol 28:211–226

Maul GA (1996) Small islands: marine science and sustainable development. Coastal and Estuarine Studies, American Geophysical Union, Washington, DC, 467 p

Mergner H (1971) Structure, ecology and zonation of Red Sea reefs. Symp Zool Soc Lond 28:141–161

Ormond RFG, Shepherd AR, Price ARG, Pitts JR (1984) Report on the distribution of habitats and species in the Saudi Arabian Red Sea. No. 1. Kingdom of Saudi Arabia. IUCN/MEPA

Rao DNVN, Behairy AKA (1986) Nature and composition of shore-zone sediments between Jeddah and Yanbu, eastern Red Sea. Mar Geol 70:287–305

Rasul NMA (1992) Late Quaternary to present day coarse-grained sedimentation of the Indus fluvial-marine system. PhD thesis (unpublished), University of Wales, 248 p

Rasul NM (2015) Lagoon sediments of the eastern Red Sea: distribution processes, pathways and patterns. In: Rasul NMA, Stewart ICF (eds) The Red Sea: the formation, morphology, oceanography and environment of a young ocean basin. Springer Earth System Sciences, Berlin, pp 281–316

Rasul N, Al-Nomani S, Al-Hazmi O, Widinly N, Qutub A, Bantan R (2009) Bathymetric survey in Sharm Obhur. Saudi Geological Survey Technical Report SGS-TR-2008-11, 15 p

Rasul NMA, Abushosha M, Al-Hazmi O, Al-Nomani S, Bahareth F, Widinly N, Akbar A, Qutub AS, Farawati R (2010) Bathymetric survey and ground-truthing of the Al-Kharrar lagoon, Red Sea coast of Saudi Arabia. Saudi Geological Survey Technical Report SGS-TR-2008-8, 77 p

Rasul NMA, Stewart ICF, Nawab ZA (2015) Introduction to the Red Sea: its origin, structure, and environment. In: Rasul NMA, Stewart ICF (eds) The Red Sea: the formation, morphology, oceanography and environment of a young ocean basin. Springer Earth System Sciences, Berlin, pp 1–28

Rasul NMA, Al-Kotob AN, Al-Nomani, S (2017) Geological and oceanographical studies of the coastal waters of Rabigh and Masturah regions, SGS Technical Report (submitted)

Rowlands G, Purkis SJ (2015) The geomorphology of shallow water coral reef environments in the Red Sea. In: Rasul NMA, Stewart ICF (eds) The Red Sea: the formation, morphology, oceanography and environment of a young ocean basin. Springer Earth System Sciences, Berlin, pp 395–408

Rowlands G, Purkis SJ, Bruckner A (2014) Diversity in the geomorphology of shallow-water carbonate depositional systems in the Saudi Arabian Red Sea. Geomorphology. https://doi.org/10.1016/j.geomorph.2014.03.014

Schlee J, Uchupi E, Trumbull JVA (1964) Statistical parameters of Cape Cod beach and eolian sands. US Geological Survey Professional Paper 501-D, pp D118–D122

Sea Around Us (2007) A global database on marine fisheries and ecosystems. Fisheries Centre, University of British Columbia, Vancouver, Canada. www.seaaroundus.org/lme/SummaryInfo.aspx?LME=33

Shepard FP, Young R (1961) Distinguishing between beach and dune sands. J Sediment Petrol 31(2):196–214

Sheppard CRC, Sheppard ALS (1991) Corals and coral communities of Arabia. Fauna of Saudi Arabia 12:3–170

Sheppard C, Price A, Roberts C (1992) Marine ecology of the Arabian region: patterns and processes in the extreme tropical environments. Academic Press, The Netherlands

Solohub JT, Klovan JE (1970) Evaluation of grain-size parameters in lacustrine environments. J Sediment Petrol 40:81–101

Sugden W (1963) Some aspects of sedimentation in the Persian Gulf. J Sediment Petrol 33:355–364

Sugden W (1966) Pyrite staining of pellety debris in carbonate sediments from the Middle East and elsewhere. Geol Mag 103:250–256

UNEP/IUCN (1988) Coral reefs of the world. UNEP Regional Sea Directories and Bibliographies. IUCN, Gland, Switzerland and Cambridge, UK; UNEP, Nairobi, Kenya

Yanagisawa Y (1984) Studies on the interspecific relationship between gobiid fish and snapping shrimp. 2. Life history and pair formation of snapping shrimp *Alpheus bellulus*. Publ Seto Mar Biol Lab 29:93–116

Sediment Yield Calculation Along the Red Sea Coastal Drainage Basins

Mazen Abuabdullah and Zekâi Şen

Abstract

Sediment yield rate calculation formulations are lacking in arid regions, because almost all the studies are concerned with humid environments. Although drainage basins (wadis) in arid regions are studied from different points of view, unfortunately there has been insufficient work on the possible sediment yield and transportation. Sediment yield within a drainage basin is a combined consequence of upland, gully, and channel erosion, transportation, and depositional processes. In arid zones, although wind deposition plays a significant role, occasional floods and flash floods are the main causes along the main channel. In fact, the bulk of the sediment yield is due to surface water runoff after each storm rainfall. This paper provides an efficient application of some available sediment yield formulations for arid regions by considering drainage basin morphology, and runoff discharge calculated for a 100-year return period. Morphological variables include the drainage basin area, drainage basin slope, main channel slope and the drainage density. This paper, after the presentation of brief information about the sediment yield process in arid regions, suggests the application of some formulas developed for arid region sediment calculation to three watersheds along the Red Sea coastal area within the Kingdom of Saudi Arabia. Two different sediment yield formulations are employed, where the first includes the drainage area, slope and the flood discharge, and the other includes additionally the drainage density. These wadis (drainage basins) from the north to the south are Al-Amud, Masturah and Yabah. According to the second formulation it is calculated that for a 100-year return period sediment rates are 0.00048 m^3/s, 0.02165 m^3/s and 0.03064 m^3/s, respectively.

M. Abuabdullah (✉) · Z. Şen
Department of Applied Geology, Saudi Geological Survey, Jeddah, Saudi Arabia
e-mail: Abuabdullah.mm@sgs.org.sa

Z. Şen
Turkish Water Foundation, Libadiye Caddesi Doğanay Sokak No: 10 Kat: 4, Üsküdar, İstanbul, Turkey

N. M. A. Rasul and I. C. F. Stewart (eds.), *Geological Setting, Palaeoenvironment and Archaeology of the Red Sea*,
https://doi.org/10.1007/978-3-319-99408-6_23

1 Introduction

Watershed sediment yield systems are, in general, nonlinear and time-variant. Although their parameters vary in time and space, they are assumed to be constant in many modeling studies. For instance, the errors in a sediment yield model may arise due to parameter uncertainty, inadequacy of the model itself, and inadequate understanding of the watershed sediment yield process due to randomness. In general, the error in the sediment yield prediction is due to the uncertainty caused by the physical process. Although the model and the input data can be reduced by use of an appropriate sediment yield error, unfortunately data is very scarce or almost non-existent in arid regions and therefore it is not possible to use efficient time series models such as Kalman filtering (Lee and Singh 1999).

The quantitative prediction of sediment transport across areas of net deposition is an essential part of assessing the offsite effects of soil erosion on hillslopes. An understanding of sediment delivery from the hillslopes to the stream requires information on rock weatherability, soil erosion and sediment deposition. Although a vast number of studies concerning different aspects of soil erosion processes at various spatial and temporal scales exist, there are very few detailed studies on sediment transport by overland flow through net deposition zones.

Despite extensive research efforts, knowledge of erosion and sediment transport still remains incomplete, and there is no generally accepted formula to be used for an accurate solution of the sediment transport rate and watershed sediment yield (Maidment 1993). Especially in arid environments, erosion and sediment yield are closely related to geology through weathering of the bedrock exposed to the flow of water, wind speed and indirectly through the character of the parent material whose properties are similar to the bedrock. The geological conditions of the area and soil properties affect rock weatherability and soil erodibility respectively, and thereby also the intensity of erosion and sedimentation processes. Erosion and sediment yield are also closely related to soil properties through the susceptibility of soil to both detachment and transport. This susceptibility defines soil erodibility, which varies with soil texture, aggregate stability, shear strength, infiltration capacity and organic as well as chemical contents.

The earliest work on the erodibility indices of soil was presented by Middleton (1930). He analyzed samples of several erodible and non-erodible soils and found that some physical properties of soil such as dispersion and erosion ratios could be used to differentiate between erodible and non-erodible soils. These two indices are defined such that the dispersion ratio is equal to suspension percentage divided by ultimate silt plus clay, and the erosion ratio is equal to dispersion ratio divided by ratio of colloid percentage to moisture equivalent. From a study of erosion in southern California, Anderson (1951) supported the use of Middleton's dispersion ratio and later he introduced a new index of soil erodibility. It is the surface-aggregation ratio, which is defined as the surface area requiring binding, divided by the aggregated silt and clay content. He observed that this index is well correlated with suspended-sediment discharge from 33 watersheds in western Oregon. Andre and Anderson (1961) and Scott and Williams (1978) concluded that both dispersion and surface-aggregation ratios are significantly related to soil-geologic rock type and such indices provide practical tools in evaluating erosion and sediment yields. Semi-arid and arid environments are distinctive in terms of factors affecting erosion and sediment yield (Jansson 1982; Hadley 1986; Pearce 1986; Walling 1986; Walling and Webb 1986). Al-Suba'i (1991) presented a detailed evaluation of erosion and sedimentation features in arid regions.

Dimensional analysis has been the concern of many engineers and physicists throughout the ages. A brief statement of the development and concepts of this operational tool has been given by Staicu (1982). Examples outside engineering practices and pure physics are applications in structural geology (Hubbert 1937), geophysics (Ramberg 1967), geomorphology (Strahler 1958), and biology (Thompson 1959). Numerous applications of the basic ideas are also found in hydraulics, hydrology, geography, economics, business, sociology and physiography (Backer 1976). Dimensional analysis (DA) has its concept in physics in that physical laws do not depend on the choice of measurement scales and units. Always any physical equation has homogeneous dimensions. This means that the right hand and left hand sides should have the same physical dimensions. DA is an efficient and powerful analytical tool in almost all branches of physics as well as in disciplines such as engineering and chemistry (Bridgman 1922; Barenblatt 1979, 1987). The method is used from simple error-checking of physical equations, to obtaining explicit functional relations among variables relevant to a phenomenon being investigated.

In this paper, first a general account of rainfall and runoff phenomena contributions to the sediment yield mechanism is presented by considering dry seasonal climatic environments. Subsequently, rock weatherability and soil erodibility accounts are presented from a sediment yield point of view. Applications are presented for three wadis (drainage basins) from western Saudi Arabia along the Red Sea coast.

2 Sedimentation Agents

Sedimentation takes place as a result of complex interactions of rainfall, runoff, geomorphological features, wind power (especially in arid regions), rock weatherability and soil erodibility processes.

2.1 Rainfall and Sedimentation

Sediment is closely related to rainfall, partly through the detaching and splash power of raindrops striking the surface materials and partly through the contribution of rain to runoff. The rainfall potential to cause erosion is dependent upon its kinetic energy, which is determined by fall velocity and rain-drop-diameter and the total number of raindrops per unit time, which is represented by rainfall intensity (Onstad 1986). Since rainfall consists of a series of different intensity events, it is a major factor that explains temporal variations in erosion and sediment yield. Generally, the most important rainfall features of the arid and semi-arid environments can be summarized as follows:

1. Rainfall can be very varied and erratic, spatially as well as temporally.
2. Individual storm-totals can be very high, where in many cases a single storm rainfall exceeds the mean annual rainfall.
3. Rainfall intensities can be very high. Sediment yield can increase greatly due to high rainfall intensity.
4. The amount of runoff is increased by the scaling effects of rainfall impact. Rainfall impact on runoff increases its transport capacity.
5. Due to the schematically represented monthly patterns of the rainfall, vegetation and erosion, the most significant period for erosion and sediment yield is the early part of the wet season when the rainfall is high, but the vegetation has not grown sufficiently to protect the surface.

Nevertheless, questions arise as to how much rain is required to induce significant erosion and sediment yield, which events are to be employed, moderate or extreme, and what is the most suitable expression of the rainfall erosivity.

Rainfall threshold values may vary, however, with the variation of other factors controlling erosion and sediment yield. For instance, an intensity of 25 mm/h is discussed in the literature as a threshold of erosive rains (Hudson 1981). Quoting from research in Zimbabwe, he stated that 50% of the annual soil loss occurred in only two storms in one year, and 75% of the erosion took place in 10 min. The opinion that most erosion and sediment take place during events of moderate frequency and magnitude exemplified by rainstorms yielding 30–60 mm with a frequency of 1–10 times per year is generally accepted by geomorphologists (Morgan 1986). Roose (1970) stated that rainfall of extreme intensity, for example, storms with a 10–100-year recurrence interval, may accomplish a large part of the erosion, transportation and mass-movement work in some climatic settings, such as in the arid and semi-arid ones of the Sahara and Mediterranean zones. Wischmeier and Smith (1978) concluded that the best representation is the product of the total storm kinetic energy, E and the maximum 30-min intensity, I_{30}. They further stated that the expression EI_{30} is widely used in USA as the basic erosivity expression for the Universal Soil Loss Equation. Lal (1976) in Nigeria reported a better correlation between the product of total rainfall, A_r, and the maximum 30-min intensity, I_{30}, than with EI_{30}. Foster and Meyer (1975) indicated that rainfall amounts and maximum 30-min intensity are the two most important general measures of rainfall erosivity. Table 1 shows the coefficient of determination for some erosivity expressions applied to three locations in the USA, and one in each of Brazil and Kenya (Onstad 1986), and R0 indicates rainfall erosivity factor.

Values in Table 1 show wide variations in coefficients of determination, which indicate that expressions for erosivity need to be independently examined for various regions. As can be seen from this table, the A_r-I_{30} index shows about the same amount of variability in the data as the EI_{30} index but the former can be computed more sensitively depending on the data size.

2.2 Runoff and Sedimentation

Runoff indicates sediment yield by detachment of the sediment from the surface and by the subsequent accumulation and transportation of eroded sediment. It occurs when the rainfall intensity is higher than the infiltration capacity and when surface depression storage has been filled and the soil is saturated. Osborn and Lane (1969) have established the minimum amount of rainfall required to initiate runoff in semi-arid environment small catchments as 0.81 cm. High intensity rain, as in semi-arid zones, seals the surface of bare soil very quickly, and consequently only a shallow depth of soil moisture may be achieved before pounding and surface runoff start. When runoff is concentrated in rills or when there is a deep sheet of flow on a slope, erosion and transportation by flow are important. The water exerts a drag force on the surface particles and with increasing depth and gradient, the velocity and hence the shear stress increase and the water has a greater ability both to erode and transport materials. Quite a lot of evidence suggests that the rate of detachment by flowing water decreases as the sediment load increases (Willis 1971; Meyer and Monke 1965) or that the rate of deposition is directly proportional to the difference between the sediment concentration in the flow and the equilibrium concentration for that flow condition (Einstein 1968). Foster and Meyer (1975) expressed this concept by the relation:

$$\frac{\text{Detachment rate by flow}}{\text{Detachment capacity of flow}} + \frac{\text{Se diment load of flow}}{\text{Transport capacity of flow}} = 1 \tag{1}$$

This equation indicates that rilling will be greatest when the sediment load is low, but that little or no rilling will occur when the transport capacity of flow is satisfied with the sediment produced by raindrop-impact erosion. It also shows that deposition may occur when transport capacity of sediment-load flow is decreased due to reduced flow gradient

Table 1 Coefficient of determination for various erosivity expressions

Erosivity factor	Watkinsville Georgia, USA	Morris Minnnesota, USA	Presque Isle Maine, USA	Campinas Brazil	Katumani Kenya
A_r	0.46	0.11	0.01	0.45	0.66
R0	0.64	0.71	0.18	0.62	0.71
E	–	–	–	–	0.64
EI_{30}	0.70	0.67	0.35	0.58	0.69
A_r-I_{30}	0.62	0.56	0.23	–	0.72
EI_{30}+R0	0.74	0.84	0.38	0.70	0.74

or increased hydraulic roughness. Many researchers have correlated runoff with sediment yield in several regions. For instance, in the USA, Langbein and Schumm (1958) and Judson and Ritter (1964) showed that in very large basins with considerable human activities, the sediment yield per square kilometre is a maximum at about 10–20 mm of runoff, decreasing sharply on both sides of the maximum. A curve by Douglas (1967) based on data from the main Asian rivers exhibits two maxima, one at about 50 mm runoff and another at about 1145 mm runoff. Under natural vegetation in eastern Australia, he indicated that the sediment yield per square kilometre is very low even at high runoff.

Due to the technical difficulties in measuring bed load and the relative paucity of observations on solute concentrations, most sediment yield estimation studies tend to relate water discharge to suspended sediments. This simple relationship has been questioned by Walling (1977). He stated that an error of ±50% or more may be associated with many rating curve estimates of sediment load.

In his study on discharge frequency compared to long-term sediment yield, Neff (1967) pointed out that there is a direct relationship between the variability of annual peak discharge and the amount of sediment moved by less frequent flows. He stated that in arid and semi-arid areas, which have the greatest variability in annual peak discharge, only 40% of the long-term sediment load is moved by flows having a frequency of less than 10 years.

Obviously, generalizations about the influence of arid and semi-arid environment runoff on sediment yield are difficult to make but the following features are important:

1. There is a great difference between the theoretical limits of the basin, the divide or watershed, and the effective area which yields runoff and sediment.
2. Irregular patterns of flow events make collection and interpretation of discharge data difficult or rather impossible in the case of very extreme conditions. This is not peculiar to events only but within the event itself there is irregularity, where the flash flood events vary in time and space with very large variations in discharge over a short period of time during the rising stage (Campbell 1977; Al-Khafif 1986).
3. Extreme events of high magnitude and low frequency may be associated with sediment yield exceeding the mean annual value by 50 times or more.
4. There are several types of sediment yield pattern associated with hydrograph rise (Schick and Sharon 1974). The rising stage is normally associated with a turbulent front wave of high energy carrying a large amount of eroded materials. It would be practically impossible to carry out accurate sampling of sediment concentration or precise measurement of flow parameters during this stage. Hence, it will be equally impossible to correlate the sediment discharge to the flow. This sharp rise is followed by a relatively sharp crest before the relatively smooth recession occurs. The sediment discharge rate varies markedly between these stages. Hence, computation of erosion or sediment yield based on such data becomes highly speculative.
5. In rock gullies, torrential runoff can cause movement of all sizes of material so that tributary channels are at times swept clean of debris. At gully mouths, there is a hazard of small fan and delta development in the main channel which, in turn, may lead to bank erosion, the formation of gravel sheets, lobs, bars and alterations of bed elevation.
6. Deposits from high magnitude tributary flows may produce temporary terraces that are steadily removed by smaller magnitude flows or even swept clean by a further major spate.
7. As with rainfall, the most vulnerable time for runoff to contribute to sediment yield is the early part of the wet season when the rainfall is high but vegetation has not grown sufficiently to protect the surface.

2.3 Erosion Assessment in Dry Environments

There are erosion assessment studies in the southwestern part of the Kingdom of Saudi Arabia. The Tihamat Asir region began to develop during the early Tertiary period and is influenced by the geodynamic processes that resulted in the Red Sea opening and its current widening. The Red Sea rifting and the uplifting of the graben shoulder along the western margin of the Arabian Shield (which is indicated by the lateritic surfaces and Oligocene basalts near the summit of the Red Sea escarpment) to heights up to 3000 m above m.s.l. are the principle events responsible for the main physiographic features of the region. The tectonic forces accountable for these events are still active today (Whitney 1983; El-Isa and Al-Shanti 1989) and are continuing to cause strong denudation and fast sediment deposition in the Red Sea basin.

The catchments of the region are characterized by terrains with an abrupt mountainous nature (elevation in the range 100–3000 m above m.s.l.), and narrow streams with steep gradients (slopes in the range 5–80%). The stream patterns are sub parallel to dendritic-type drainage systems and the first order streams flow generally perpendicular to the axis of the mountain chain. The coastal plain is narrow in the north (less than 40 km), and increases to the south until it becomes about 150 km; this is due to several reasons, the most important of which is the increase in the amount of sediment received by the Red Sea from the southern wadis compared with the northern wadis (Bird 2010; Rasul et al. 2015).

The hydrological condition of the region is characterized by high rainfall intensity and flash floods of short duration and sizeable magnitude. Floods with discharges of about 4000–10000 $m^3\ s^{-1}$ have been estimated for wadis in the region. Such floods usually wash away particles of all sizes or even sweep clean the stream channel materials.

Sheet, rill and gully erosion all seem to be active in the region. Sediment production by water erosion appears to be more pronounced at the upper and middle elevations during both normal and major storms. The downslope movement of dry sediment (by gravitational force) as individual particles is locally an important erosional process in the region. This process supplies detritus to stream channels at the base of steep hillslopes, where it may await a major flood before further transportation can occur. In some cases, talus accumulations are associated with either rock slides or debris slides. Wind is a relatively minor eroding force in the study catchments compared to the leaching and cutting forces of water.

The assessment of erosion is a special form of sediment yield evaluation, the objective of which is to identify the sediment-source and critical erosion areas. Such an assessment aims at dividing a land area into regions, similar in their degree of weathering and erosion intensity, which can be used as a basis for sediment yield evaluation. This technique is, generally, used for a quick and regional assessment of erosion in the case of absence of erosion and sedimentation measurements or in which erosion and sedimentation act selectively. Two indices of erosion intensity, namely rock weatherability index W_i (the opposite of rock durability), and soil erodibility index, e_i have been used herein. The values of these two indices for the study region together with geological and physiographic maps of the area are used primarily to develop rock weatherability and soil erodibility maps that expose the potential of erosion in the study area. The study by Al-Suba'i (1991) shows that the rock weatherability and soil erodibility range from very low to very high and zones of low rock weatherability and medium soil erodibility are rather dominant in the region. He considered only the catchments of the central part of the Tihamat Asir region, extending over the Red Sea littoral zone between latitudes 16°52'N and 18°05'N and stretching eastward from the coast at longitude 42°35'E to the summit of the Red Sea escarpment at longitude 43°30'E. He used topography and general geological maps of the region.

3 Geomorphological and Rainfall Pattern Characteristics

Semi-arid and arid regions are distinctive in terms of erosion and sedimentation yield (Jansson 1982; Hadley 1986; Walling 1986; Walling and Webb 1986). The most important agents are:

(i) Erosion and transport characteristics.
(ii) Basin morphological features.
(iii) Geological layout and soil coverage.
(iv) Vegetal cover factors.
(v) Land-use and management practices.

In arid regions, hill slopes are the primary source of erosion by occasional runoff with channel banks and flood plains as secondary sources. The sediment yield phenomenon can be divided into two broad categories as the upland phase, and the lowland stream or in-channel phase.

The upland phase emphasizes the erosion process of detachment and transport in rill and inter-rill areas where the mechanics of the rainfall event and surface flow are the major agents. Major variables influencing the yield in this phase are rock units, soil type, environmental conditions and moisture content at the start of the event, in addition to the slope, vegetation, litter cover, rainfall amount, intensity and duration. In the in-channel phase, sediment transport and deposition processes predominate and consequently channel transport capacity becomes more important. Pertinent variables in this phase are the velocity, depth of flow, channel slope, wash-load, water temperature, grain size distribution and median size of bed material. Although not all of the sub-processes occur on all source areas, each has its part in the total sediment yield process.

In a refined assessment, a drainage area is divided into sub-areas, similar in their degree of weathering and erosion intensity, which can be used as a basis for sediment yield evaluation. This technique is used for a quick regional assessment of erosion in the absence of measurements, as it is in southwestern Saudi Arabia. Accurate prediction of sediment yield rate in arid and semi-arid regions is a difficult task that is not handled well by most models.

The main purpose here is to develop simple models for sediment yield rate estimations in arid regions utilizing dimensional analysis. The relevant variables considered for sediment yield rate formulation are the runoff discharge, drainage area, mean channel slope and drainage density. These variables are explained in any standard handbook on hydrology (Maidment 1993).

4 Study Area

The arid region sediment yield formulations developed by Şen (2014) are applied for three wadis, namely Wadis Al-Amud, Masturah and Yabah, along the Red Sea coastal area (Fig. 1). These have various geomorphological features, different rainfall regimes and are also equipped with few rainfall stations. The choice of each one is based on the consideration that there are negligible urban areas, agricultural activities and land use developments, and hence natural

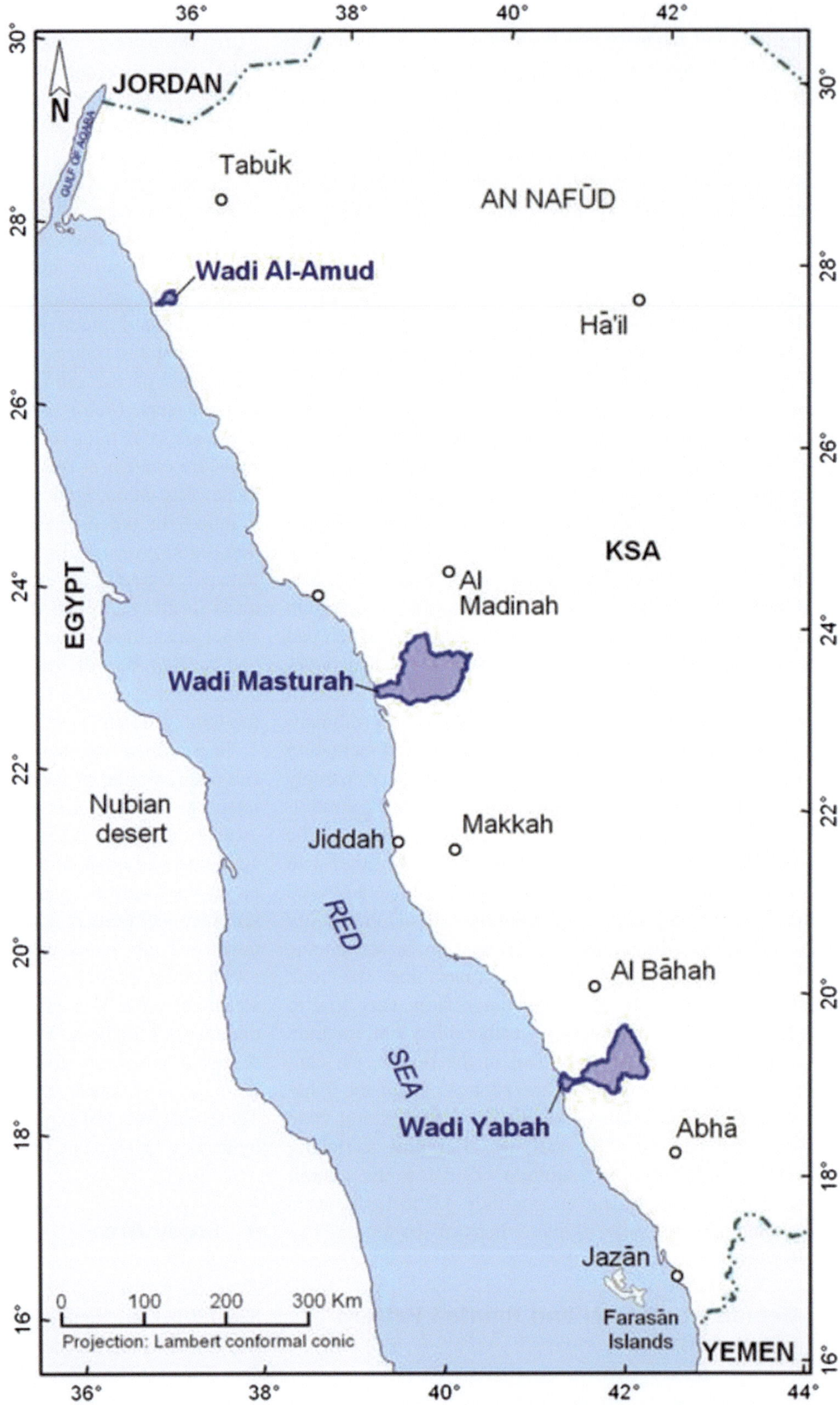

Fig. 1 Sediment yield calculation wadis along the Red Sea coastal region

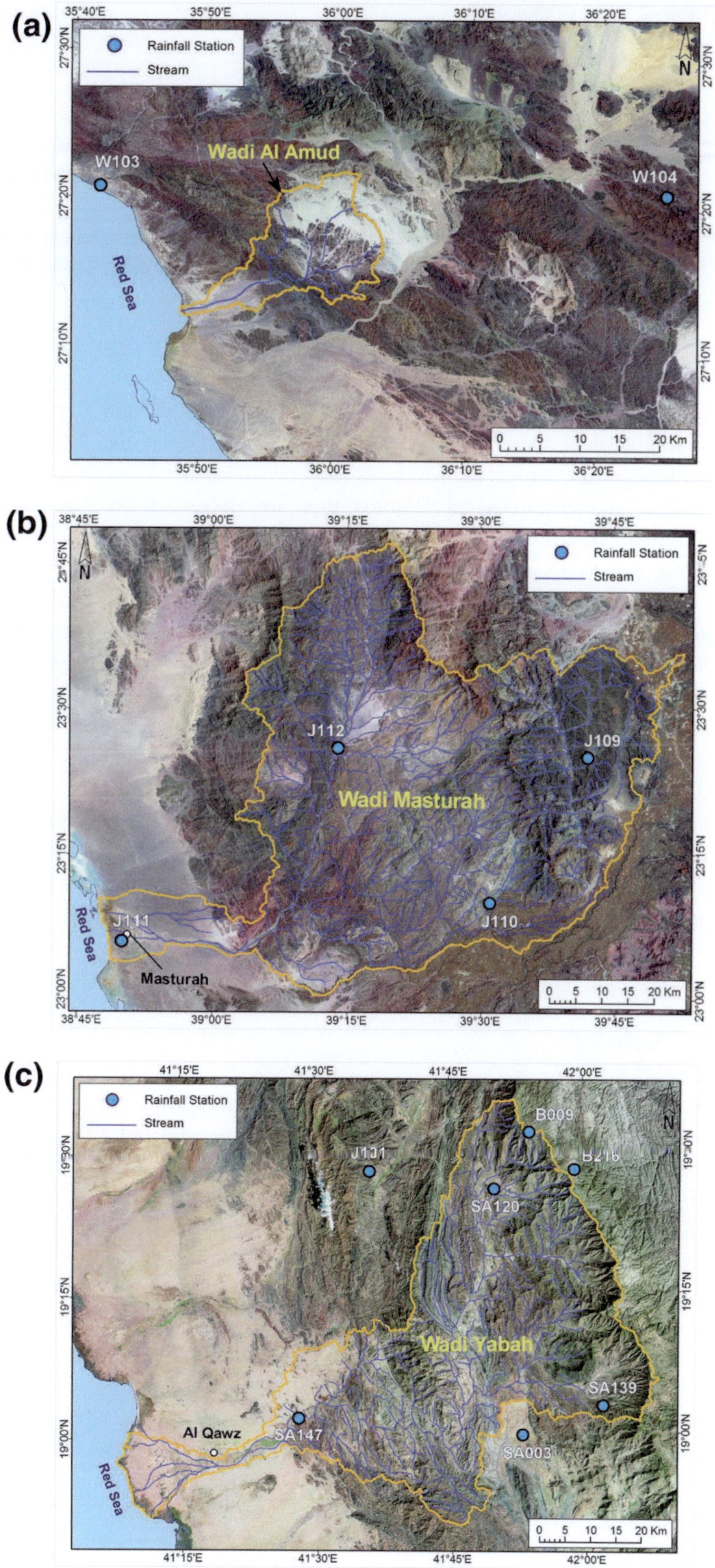

Fig. 2 Drainage areas (orange outlines): **a** Al-Amud, **b** Masturah, **c** Yabah

Table 2 Geomorphological characteristics

Name	Area (km^2)	Length (km)	MFD[a] (km)	B.Slope (m/m)	MFD slope (m/m)	Drainage density[b] (1/m)
Wadi Al-Amud	214.4	27.8	37.2	0.056	0.0188	0.00077
Wadi Masturah	4694.0	121.5	188.4	0.150	0.0053	0.00074
Wadi Yabah	3151.5	105.7	169.9	0.199	0.0135	0.00070

[a]MFD: Max flow distance (Basin length along main channel from outlet to upstream boundary)
[b]Drainage density was calculated by selecting the threshold value (1 km^2)

conditions are maintained. In order to achieve detailed calculations, the sub-basins in each one are also taken into consideration with their environmental, geological and meteorological conditions. The separate features of each one of these wadis are given in Fig. 2.

The geomorphological features that are necessary for sediment yield calculation later in this study are given in Table 2, based on SRTM elevation data (90-m resolution). Land use and soil type information were extracted from geological maps (Saudi Geological Survey, GM Series, Scale 1:250000; GM82 and GM109 for Wadi Al Amud, GM87 and GM49 for Wadi Masturah, GM70, GM74 and GM94 for Wadi Yabah) and modern satellite images, and the main geological zones are given in Fig. 3. The curve number (CN) for each basin was estimated based on these maps, for later use within the hydrological model, as follows: Wadi Al-Amud, 76.73; Wadi Masturah, 79.42; and Wadi Yabah, 80.22. The CN (also known as the runoff curve number) is an empirical parameter used in the SCS method for predicting direct runoff from rainfall excess (USDA-SCS 1985).

5 Rainfall

Rainfall data for the three wadis were collected from the rainfall stations of MOW available within or around these wadis (Fig. 2). Table 3 gives the name, position, the available record years, rainfall depth of a 100-year return period and the type of distribution for each station, in addition to weighted rainfall depth of a 100-year return period for each wadi, which were later used in the hydrological model.

For the peak discharge estimate of 100-year return period in three wadis, the SCS curve number method (integral to HEC-HMS) is applied using the geomorphological features given in Table 2, weighted rainfall depth given in Table 3 and the CN values as mentioned above. The results of the hydrological model of the three wadis are obtained as: Wadi Al-Amud, 138.7 m^3/s; Wadi Masturah, 2,597 m^3/s; and Wadi Yabah, 2,971 m^3/s, and the hydrograph of each wadi is given in Fig. 4. These hydrographs show the change in the outflow of each wadi resulting from the rainstorm used in the hydrological model (100-year return period), peak discharge and time to peak, as well as precipitation and losses during a storm. Their comparison shows that Wadi Al-Amud had more of a flash discharge according to the values of time to peak ($\sim$13 h), due to the small area of Wadi Al-Amud, whereas wadis Masturah and Yabah are $\sim$20 and $\sim$17 h, respectively. The catchment size difference with the difference in rainfall depth also led to differences in peak discharges, in addition to the differences in the geomorphological characteristics for each wadi.

6 Methodology

Şen (2014) proposed sediment yield formulations S_y based on the drainage basin characteristics and discharge, two of which are as follows:

$$S_y = k_{f1}\frac{Q}{A}S \tag{2}$$

and

$$S_y = k_{f2}D_dQ\sqrt{\frac{S}{A}} \tag{3}$$

where A (catchment area), D_d (drainage density), Q (runoff discharge) and S (main channel slope) as the independent variables in the sediment yield formulation. In these equations k_{f1} and k_{f2} are the model constants that must be calculated from actual field measurements. An important point is that both models provide sediment yield amount per unit area per time. Hence, their units are in $m^3/(m^{2\ s})$, that is, m sec^{-1}. The first formulation reflects a rather linear form of approach, which may not be very desirable in practical applications, because as has been explained in the previous sections the sedimentation process has a non-linear dynamism. Along this line the second model in its non-linear form provides comparatively better results than the first one.

Fig. 3 Land use and soil type units: **a** Al-Amud, **b** Masturah, **c** Yabah

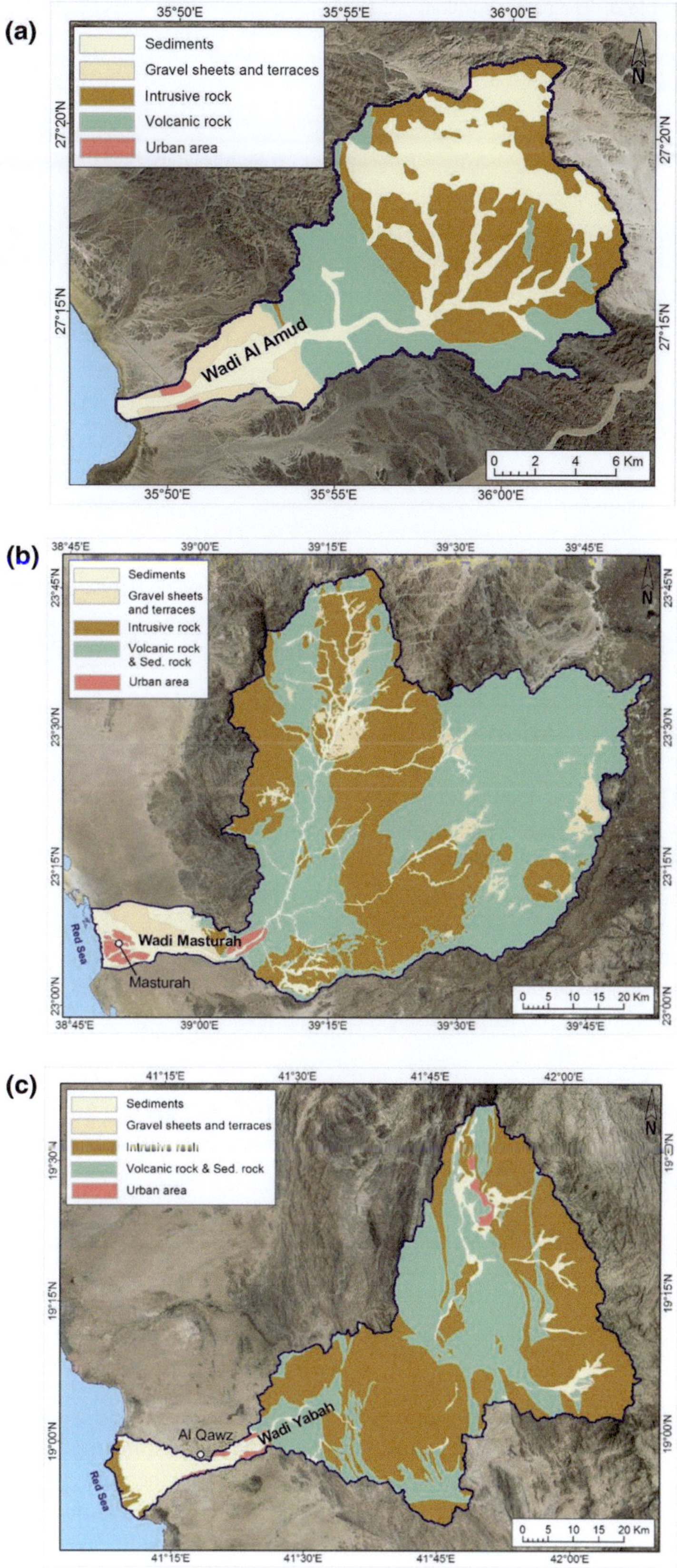

Table 3 The name, position, available years, 100-year rainfall depth and type of distribution for each station

ID	Wadi	Name	Long	Lat	Years	100 y	Distribution
1	Al Amud	W103	35.66667	27.35000	31	61.1	Gamma
2		W104	36.41667	27.35000	37	83.8	Gamma
Weighted rainfall depth (Average of two values)						*72.45*	
3	Wadi Masturah	J109	39.70000	23.41667	49	80.3	GEV
4		J110	39.51667	23.16667	48	97.8	Gamma
5		J111	38.83333	23.10000	38	76	Gamma
6		J112	39.23333	23.43333	39	91.2	GEV
Weighted rainfall depth (Using isohyetal method)						*86.86*	
7	Wadi Yabah	B009	41.90000	19.53333	25	140.4	LogNormal
8		B216	41.98333	19.46667	31	98	GEV
9		J131	41.60000	19.46667	39	135.6	LogNormal
10		SA003	41.88333	19.00000	38	99.5	LogNormal
11		SA120	41.83333	19.43333	42	105.8	Gumbel
12		SA139	42.03333	19.05000	43	104.6	LogNormal
13		SA147	41.46667	19.03333	24	71.4	Gamma
Weighted rainfall depth (Using isohyetal method)						*101.98*	

Hydrological model

7 Application and Discussion

The proposed methodology was applied for the three wadis, Al-Amud, Masturah and Yabah. Based on Al-Suba'i (1991), for Saudi Arabia the model constants are given by Şen (2014) as $k_{f1} = 10^{-4}$ and $k_{f2} = 1.74 \times 10^{-6}$ using the data of Wadi Jizan ($A = 1{,}430$ km^2, $D_d = 0.48$ m^{-1}, $Q = 3{,}362$ m^3/s, $S = 0.095$ and $S_y = 2.239 \times 10^{-11}$ m s^{-1}). Since the current study is based on the use of digital elevation models (DEM90 m), the model constants are recalculated using the geomorphological features as given in Table 4.

Substitution of the values in Table 4 into Eqs. (2) and (3) gives model constants as $k_{f1} = 3.04 \times 10^{-4}$ and $k_{f2} = 2.26$ 10^{-6}, and hence, the same expressions take the following form:

$$S_y = 0.000304 \frac{Q}{A} S \tag{4}$$

and

$$S_y = 0.00000226\, D_d Q \sqrt{\frac{S}{A}} \tag{5}$$

These predictive models provide tools for estimating sediment yield rates for planning and design in the study region, and also for other dams. By taking into consideration the geomorphological parameters from Table 2 with the calculated flood discharge values for a 100-year return period, the calculations from the linear model (Eq. 4) are presented in Table 5.

On the other hand, with the import of the drainage density factor the non-linear sediment yield formulation in Eq. (5) yields the results in Table 6.

The sediment yield amounts are given in the last two columns of these tables are per area per time and volume per time, respectively. Comparison of the columns between two tables indicates that the non-linear case provides lower sediment yield amounts than the linear model for small wadis (Wadi Al-Amud), but the opposite for the big wadis (Wadi Masturah and Wadi Yabah). In comparison with the results of Wadi Jazan (Al-Suba'i 1991), it is observed that the values resulting from the non-linear model are more logical than the linear model, and this is consistent with the idea that sedimentation process has a non-linear dynamism.

8 Conclusion

In arid regions, the study of sediment yield has proven to be very complex and difficult because of the number of variables involved, the inconsistency or interrelation among those variables, and the lack of an appropriate method to quantify some of those variables. In spite of these difficulties, the present work is an attempt to calculate sediment

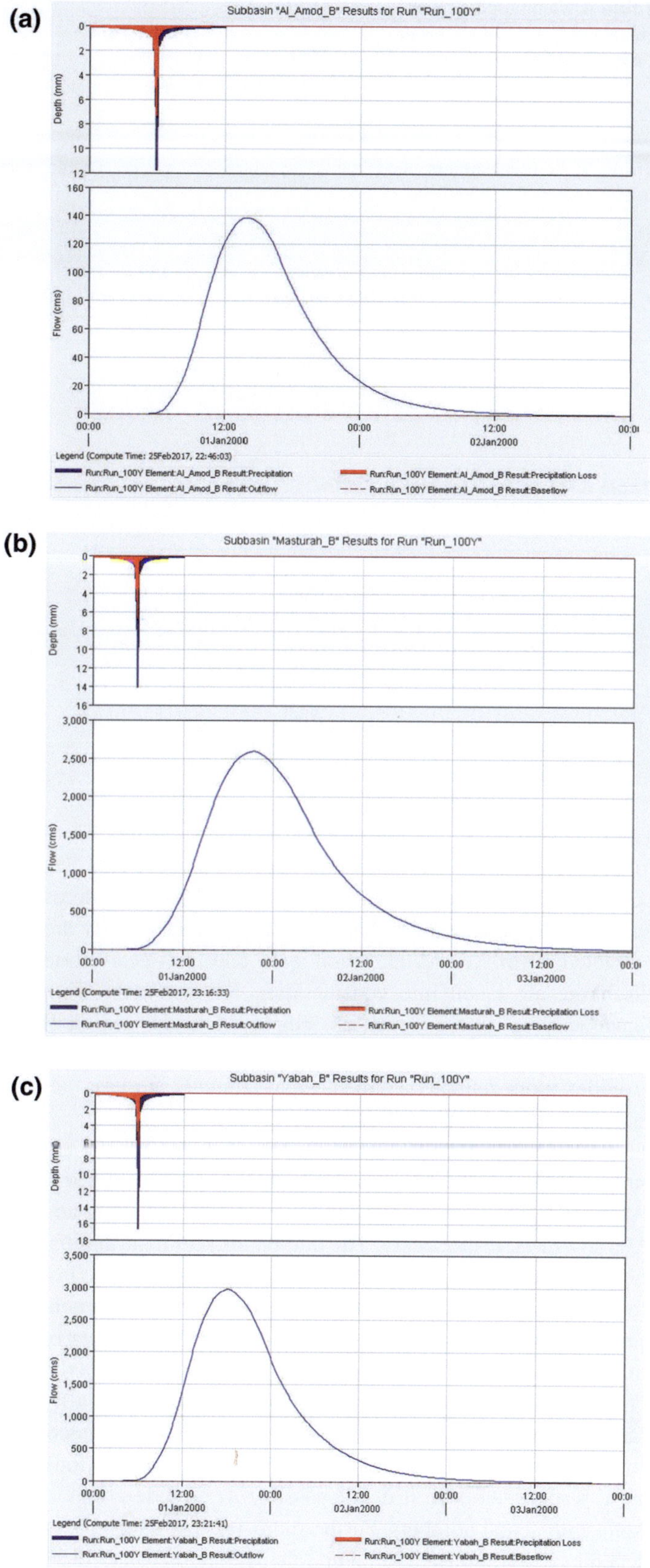

Fig. 4 Hydrographs of wadis: **a** Al-Amud, **b** Masturah, **c** Yabah

Table 4 Geomorphological characteristics of Wadi Jazan (using DEM90 m)

Name	Area (km^2)	MFD slope (m/m)	Drainage density[a] (1/m)
Wadi Jazan dam basin	1430.0	0.0313	0.00063

[a]Drainage density was calculated by selecting the threshold value (1 km^2)

Table 5 Linear sediment yield model results

Basin name	Peak discharge (m^3/s)	Area (km^2)	MFD slope (m/m)	Model-1	
				Sy ($m^3m^{-2}s^{-1}$)	Sy (m^3/s)
Wadi Al-Amud	138.7	214.4	0.0188	3.70E-12	0.00079
Wadi Masturah	2,597	4694	0.0053	8.91E-13	0.00418
Wadi Yabah	2,971	3151	0.0135	3.87E-12	0.01219

Table 6 Non-linear sediment yield model results

Basin name	Peak discharge (m^3/s)	Area (km^2)	MFD slope (m/m)	Drainage density (1/m)	Model-2	
					Sy ($m^3m^{-2}s^{-1}$)	Sy (m^3/s)
Wadi Al-Amud	138.7	214.4	0.0188	0.00077	2.26E-12	0.00048
Wadi Masturah	2,597	4694	0.0053	0.00074	4.61E-12	0.02165
Wadi Yabah	2,971	3151	0.0135	0.00070	9.72E-12	0.03064

yield rates in three drainage basins, Wadis Al-Amud, Masturah and Yabah, in western Saudi Arabia along the Red Sea coast. For the calculations, the geomorphological features such as the drainage area, slope and the drainage density are extracted from SRTM elevation data and 100-year flood discharge are taken into consideration. Two models, with linear and non-linear forms, are employed in the sediment yield calculations. It is notable that the non-linear approach provides more reliable sediment yield amounts per area per time and also volume per time.

According to the non-linear model, the sediment yield amounts for Wadis Al-Amud, Masturah and Yabah for a 100-year return period are 0.00048 m^3/s, 0.02165 m^3/s and 0.03064 m^3/s, respectively. The amount of sediment that the Red Sea takes from each wadi per second is shown in the last column of the tables. For calculation of annual sediment amounts these numbers must be multiplied by the number of similar flood occurrence numbers that are known exactly. These are the amounts that are yielded to the Red Sea by each one of the wadis due to a single flood occurrence. Unfortunately, there are no sediment measurements along these wadis, therefore it is not possible to provide comparisons with actual numbers. Furthermore, it is not advisable to depend on the annual sediment calculations, because there are no floods continuously over each second, but once in 5-year, 10-year, 25-year, 50-year and 100-year discharges. It is, therefore, not possible to calculate the total amount of sediment yield to the Red Sea.

References

Al-Khafif SM (1986) Sedimentation control of wadi Jizan reservoir. FAO, Rome, UTEN/SAU/013/SAU field document, No 22, 58 p

Al-Suba'i KAMG (1991) Erosion-sedimentation and seismic considerations for dam siting in the central Tihamat Asir region. Unpublished PhD thesis, King Abdulaziz University, Faculty of Earth Sciences, Kingdom of Saudi Arabia, 343 p

Anderson HW (1951) Physical characteristics of soils related to erosion. J Soil Water Conserve 6:129–133

Andre JE, Anderson HW (1961) Variation of soil erodibility with geology, geographic zone, elevation and vegetation type in northern California wildlands. J Geophys Res 66:3351–3358

Backer HA (1976) Dimensionless parameters: theory and methodology. Applied Sciences, London, p 128

Barenblatt GI (1979) Similarity, self-similarity and intermediate asymptotics. Consultants Bureau, New York, p 218

Barenblatt GI (1987) Dimensional analysis. Gordon & Breach, Science Publishers Inc., Amsterdam

Bird ECF (2010) Saudi Arabia, Red Sea Coast. In: Bird E (ed) Encyclopedia of the world's coastal landforms. Springer, Heidelberg, pp 1023–1025

Bridgman P (1922) Dimensional analysis. Yale University Press, New Haven

Campbell IA (1977) Stream discharge, suspended sediment and erosion rates in the Red Deer river basin, Alberta, Canada. In: Proceedings of erosion and solid matter transport in Inland waters symposium, Paris, July 1977, IAHS-AISH Publ no 122

Douglas I (1967) Man, vegetation and sediment yield of rivers. Nature 215:925–928

Einstein HA (1968) Deposition and suspended particles in a gravel bed. ASCE J Hydraul Div 94:1197–1205

El-Isa Z, Al-Shanti A (1989) Seismicity and tectonics of the Red Sea and western Arabia. Geophys J Int 97:449–457

Foster GR, Meyer LD (1975) Mathematical simulation of upland erosion by fundamental erosion mechanics in present and perspective technology for predicting sediment yields and sources. USDA Agr Res Ser Publ ARS-S-40:190–207

Hadley RF (1986) Fluvial transport of sediment in arid and semi-arid regions. In: Proceedings of international symposium on erosion and sedimentation in Arab countries, vol 5, pp 335–348 (Iraqi J Water Res)

Hubbert M (1937) Theory of scale models as applied to the study of geological structures. Bull Geol Soc Am 48:1459–1520

Hudson NW (1981) Soil conservation. Batsford Ltd., London

Jansson MB (1982) Land erosion by water in different climates. Uppsala University, Departmental Physical Geography, UNGI Rapport 57, 151 p

Judson S, Ritter DF (1964) Rates of regional denudation in the United States. J Geophys Res 69:3394–3401

Lal R (1976) Soil erosion on alfisols in western Nigeria, III effects of rainfall characteristics. Geoderm 16:377–387

Langbein WB, Schumm SA (1958) Yield of sediment in relation to mean annual precipitation. EOS Trans Am Geophys Un 39:1076–1084

Lee YH, Singh VP (1999) Prediction of sediment yield by coupling Kalman Filter with instantaneous unit sediment graph. IUSG. Hydrol Process 13:2861–2875

Maidment DR (1993) Handbook of hydrology. McGraw-Hill Book Co, pp 12.1–12.61

Meyer LD, Monke EJ (1965) Mechanics of soil erosion by rainfall and overland flow. Trans Am Soc Agr Eng 8:572–577

Middleton HE (1930) Properties of soils which influence soil erosion. US Dept Agr Tech Bull 178:1–16

Morgan RPC (1986) Soil erosion and conservation. Longman, Essex, p 298

Neff KL (1967) Discharge compared to long-term sediment yield. In: Proceedings of CERN symposium, IAHS-AISH Publ no 75

Onstad CA (1986) Current techniques for modelling and predicting erosion and sediment yield. In: Proceedings of international symposium on erosion and sedimentation in Arab countries, vol 5, pp 530–550 (Iraqi J Water Res)

Osborn HB, Lane L (1969) Precipitation-runoff relations for very small semi-arid range-land watersheds. Water Resour Res 5:419–425

Pearce AJ (1986) Geomorphic effectiveness of erosion and sedimentation events. In: Proceedings of international symposium erosion and sedimentation in Arab countries, vol 5, pp 551–569 (Iraqi J Water Res)

Ramberg H (1967) Gravity, deformation and the Earth's crust. Academic Press, London, New York, 214 p

Rasul NMA, Stewart ICF, Nawab ZA (2015) Introduction to the Red Sea: its origin, structure, and environment. In: Rasul NMA, Stewart ICF (eds) The Red Sea. The formation, morphology, oceanography and environment of a young ocean basin. Springer Earth System Sciences, Berlin, pp 1–27

Roose E (1970) Importance relative de l'erosion, du drainage oblique et vertical dans la pédogénèse actuelle d'un sol ferrallitique de Moyenne Côte d'Ivoire, deux années de mesure sur parcelle expérimentale. Cah OSTROM Ser Pédol 8:469–482

Schick AP, Sharon D (1974) Geomorphology and climatology of arid watersheds. Mimeograph report, Department of Geography, Hebrew University, Jerusalem

Scott KM, Williams RP (1978) Erosion and sediment yields in the transverse ranges, southern California. USGS Prof Paper 1030, 38 p

Şen Z (2014) Sediment yield estimation formulations for arid regions. Arabian J Geosci 7(4):1627–1636

Staicu GI (1982) Restricted and general dimensional analysis: treatment of experimental data. Abacus Press, Kent, p 303

Strahler AN (1950) Dimensional analysis applied to fluvially eroded landforms. Bull Geol Soc Am 69:279–300

Thompson DW (1959) On growth of form, vol I and II. Cambridge University Press

USDA-SCS (1985) National Engineering Handbook. Section 4—hydrology. US Department of Agriculture, Soil Conservation Service, Washington, DC

Walling DE (1977) Limitations of the rating curve technique for estimating suspended sediment loads with particular reference to British rivers. In: Proceedings of Paris symposium, IASH-AISH Publ no 122, pp 34–38

Walling DE (1986) Sediment yield and sediment delivery dynamics in Arab countries. In: Proceedings of international symposium erosion and sedimentation in Arab countries, vol 5, pp 775–799 (Iraqi J Water Res)

Walling DE, Webb BW (1986) Solute transport by rivers in arid environments: an overview. In: Proceedings of international symposium erosion and sedimentation in Arab countries, vol 5, pp 800–822 (Iraqi J Water Res)

Whitney JW (1983) Erosional history and surficial geology of western Saudi Arabia. Saudi Arabian Deputy Ministry of Mineral Resources, Technical Record USGS-TR-04-1, 96 p

Willis JC (1971) Erosion by concentrated flow. US Department Agriculture, ARS 41-179, 16 p

Wischmeier WH, Smith DB (1978) Predicting rainfall erosion losses. US Department Agriculture handbook, No. 537, 58 p

Landscape Archaeology, Palaeolithic Survey and Coastal Change Along the Southern Red Sea of Saudi Arabia

Anthony Sinclair, Robyn H. Inglis, Andrew Shuttleworth, Frederick Foulds, and Abdullah Alsharekh

Abstract

Since 2012, a new phase of landscape survey for archaeological remains from the Palaeolithic has been undertaken in the provinces of Jizan and Asir in Southwestern Saudi Arabia. This is the first Palaeolithic landscape survey in this area since the Comprehensive Survey of the Kingdom undertaken between 1977 and 1982. More than 100 Palaeolithic sites have been identified from the Early Stone Age to the Late Stone Age, evidencing a regular association between archaeological remains and the Harrat deposits of basalt. The analysis of two major newly discovered sites, Dhahaban Quarry and Wadi Dabsa, has demonstrated the quality of archaeological and behavioural information that can still be recovered through landscape surveys in this region. At the site of Dhahaban Quarry, the survey has confirmed that Middle Stone Age lithic artefacts can be found in situ in the preserved beach deposits of ancient shorelines suggesting the use of marine resources. At Wadi Dabsa the technological study of a large assemblage of lithic artefacts suggests variations in expertise in lithic technology, and possibilities for understanding the process of learning the skills of lithic technology.

A. Sinclair (✉)
Department of Archaeology, Classics and Egyptology, University of Liverpool, 12–14 Abercromby Square, Liverpool, L69 7WZ, UK
e-mail: a.g.m.sinclair@liverpool.ac.uk

R. H. Inglis
Department of Archaeology, University of York, King's Manor, York, YO1 7EP, UK

R. H. Inglis
Department of Environmental Sciences, Macquarie University, North Ryde, Sydney, NSW, Australia

A. Shuttleworth
Queen's University, Belfast, UK

F. Foulds
Northern Archaeological Associates, Marwood House, 28 Harmire Enterprise Park, Barnard Castle, DL12 8BN, UK

A. Alsharekh
Department of Archaeology, King Saud University, Riyadh, Saudi Arabia

1 Introduction

It is generally agreed that the first hominin species evolved in Central and Eastern Africa sometime after 7 million years ago (Senut et al. 2001; Brunet et al. 2005; Suwa et al. 2009; Brunet 2010). Following a long period of evolution within that continent, fossil evidence from Dmanisi in Georgia then indicates that hominins left Africa sometime before 1.8 million years ago (Garcia et al. 2010). As the closest landmass to Eastern Africa, Arabia must have been one of the first regions into which hominins dispersed when leaving Africa. We should therefore expect to find in Arabia archaeological evidence dating through most of the Pleistocene and indicative of the ways in which the first and later hominin migrations from Africa adapted to new environments. This simple understanding has led to a new wave of Palaeolithic archaeological research in the Arabian Peninsula since 2000 (see papers in Petraglia and Rose 2009 as an early example).

It is along the Red Sea coast where hominins entered the Peninsula either by land migration from the north, or, possibly, via a sea crossing at the southern end at times of lower sea level (Walter et al. 2000; Bailey 2009; Armitage et al. 2011; Lambeck et al. 2011; Groucutt and Petraglia 2012), and it is here that archaeological evidence is likely to be most abundant. Yet, despite the importance of the Red Sea coastal plains to an understanding of hominin dispersal, landscape surveys for evidence of Palaeolithic age have been limited, especially so at its southern end in the provinces of Asir and Jizan. An initial survey in the late 1970s identified archaeological sites with Palaeolithic evidence and, on the basis of typological similarity to the sites in Africa and the Levant, suggested that Arabia might have been colonised from 1.3 million years ago (Whalen et al. 1989; Whalen and Pease 1990). No further survey, however, happened until a more recent programme by

N. M. A. Rasul and I. C. F. Stewart (eds.), *Geological Setting, Palaeoenvironment and Archaeology of the Red Sea*,
https://doi.org/10.1007/978-3-319-99408-6_24

the authors started in 2012. This recent programme has significantly increased the number of Pleistocene age sites in the Red Sea coastal plain and greatly improved our knowledge of hominin landscape settlement, resource use and technological behaviours (Inglis et al. 2014; Bailey et al. 2015).

In this chapter, we shall give a brief overview of the results of landscape surveys conducted from 2012, and compare the evidence from the first archaeological survey to that gathered from recent work to assess the reliability of this earliest research. We shall also discuss in more detail the archaeological and sedimentary information gathered from two specific sites that give an indication of the quality of evidence that still survives. Recent archaeological work confirms some of the typological and anecdotal description recorded by the earliest survey teams, but advances in archaeological and landscape analysis indicate that there still remains a wealth of high-quality archaeological and environmental evidence to be found along the Red Sea that can inform us about a wide range of hominin behaviour, such as the relationships between hominins, changing coastal landscapes and potential marine resources at the coastline, and perhaps about less obvious behaviours such as the transmission of patterns of learning between generations. Even though these deposits and their associated archaeological remains have survived for many thousands of years, it is clear that the archaeological and associated sedimentary evidence, especially that related to the marine deposits preserved along the current Red Sea, is now very fragile and has become fragmented as a result of local development over the last thirty years. It faces imminent threat of destruction from further development. Without a proper appreciation of the evidence preserved and the sedimentary context in which such evidence may be found, including further research in the immediate future to record its distribution and quality, such evidence for the lives of the first human communities of the Arabian Peninsula along the Red Sea may soon be lost forever.

2 Landscape Survey in Archaeology

Interpretations of human behaviour based on evidence from landscape surveys rest on an understanding of the nature of the chronological, behavioural and taphonomic relationships between samples of archaeological material. First coherently described by David Clarke (1973), this may be the relationship between the sample of material that was discarded in the past out of the larger set that was made and used, or between the smaller sample that has been preserved from that which was discarded, or between the preserved sample and a smaller sample that remains visible to the survey team, or between that visible sample and a yet smaller sample that might be collected or recorded. When the focus of interpretation is the Palaeolithic, the interpretive problems are multiplied by the scales of temporal and environmental change involved and the difficulty of identifying discrete episodes of behaviour. Palaeolithic landscape surveys must balance an understanding of the effects of environmental change with the surviving evidence of hunter-gatherer activities distributed across a much wider range of spatial and temporal scales than is the case for the archaeology of post-Pleistocene settled communities. To be effective, investigation requires a knowledge of the taphonomy of archaeological materials—their differential preservation along with the various processes of destruction, burial and exposure in any landscape setting (e.g., Holdaway and Fanning 2014). It also requires an understanding of the ways in which mobile hunter-gatherers might exploit different plant and animal resources as environments change, recognising that the activities undertaken in one place during one environmental setting may be different from those undertaken in that same place at times of different environmental conditions (Binford 1982). In summary, the interpretation of survey data requires the disentangling of potential long-term archaeological palimpsests of discrete episodes of hominin life built up over different time scales and comprising many different activities (Rossignol and Wandsnider 1992; Stern 1993; Bailey 2007). Finally, it must be recognised that the potentially observable behaviour of hominins has changed considerably through the Pleistocene as physical, cognitive, and social capacities of different hominin species evolved.

In the southern Red Sea (Fig. 1), Palaeolithic landscape surveying is particularly difficult for a variety of reasons. A continuous history of changing sea level means that the location of the coast has been subject to considerable movement because of marine transgression and regression. Those coastal deposits which are currently accessible offer evidence from a short period of interglacial time, whilst deposits formed at other times are now submerged and lie many kilometres out from the modern shoreline (Lambeck et al. 2011). The coastal plain in the southern Red Sea has also been subject to episodes of deposition of aeolian, alluvial and colluvial sediments, whilst the lava flows and cinder cone deposits that are so prominent in this area and that have been thought to predate any Palaeolithic archaeology (Dabbagh et al. 1984) might not represent a physical remnant backdrop of ancient volcanic activity, but evidence of episodes of volcanic activity happening throughout the Pleistocene until quite recently (Bailey et al. 2007). Even though physical access to deposits at the coast and in parts of the coastal plain has become easier due to improvements in roads and vehicles, much of the coastal plain, the inner lava flows and the escarpment still remain very difficult to access and survey. Finally, the archaeological materials themselves are also difficult to see; the overwhelming majority are both made from and deposited on local volcanic lava flows.

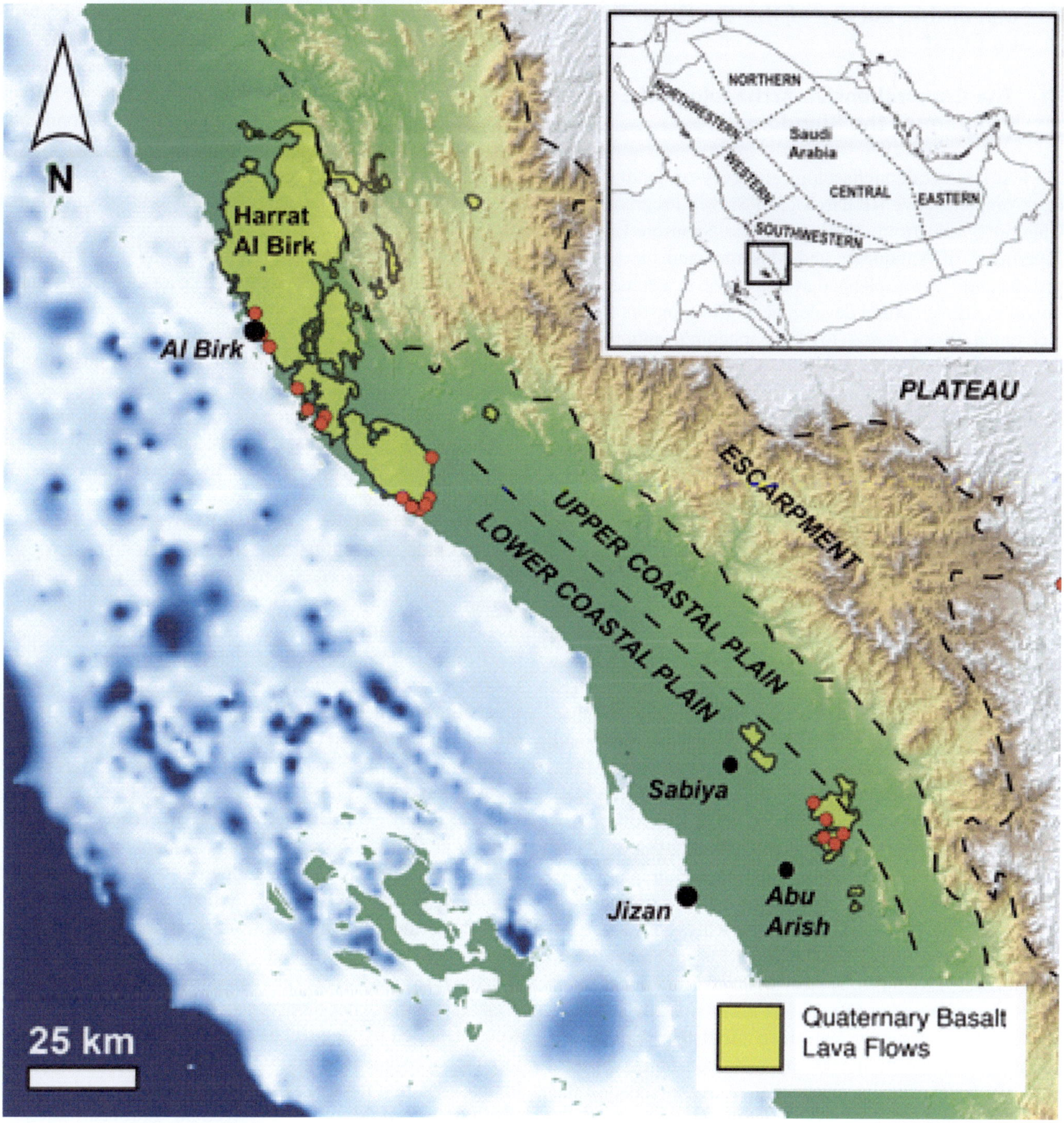

Fig. 1 Map of the location of the study area in Jizan and Asir Provinces, Southwestern Saudi Arabia. Major landscape zones defined by the DISPERSE project (following Devès et al. 2013). Red dots indicate findspots of Palaeolithic artefacts from the CASP survey of the Southwestern Province (Zarins et al. 1981). Elevation data © CGIAR-CSI SRTM 90 m v4.1 database; bathymetric data from GEBCO_08 One Minute Grid

3 Palaeolithic Survey of the Southern Red Sea Coast

3.1 The Comprehensive Archaeological Survey Program of the Kingdom

The first large-scale archaeological survey of Saudi Arabia, the "Comprehensive Archaeological Survey Program of the Kingdom", ran from 1977 to 1981. Sponsored by the Department of Antiquities and Museums and the Ministry for Education, it was planned as a response to a considerable growth in infrastructural development. The sponsors foresaw a potential threat posed to the Kingdom's antiquities if development continued without any prior understanding of the heritage assets that might be destroyed. The timing of the Comprehensive Survey was also significant; it was only from the mid-1970s that there were sufficient Saudi Arabian personnel trained in archaeology and heritage to work with foreign specialists to make such a survey possible (Adams et al. 1977; Potts 1998).

The Comprehensive Survey divided the Kingdom of Saudi Arabia into six large provinces: Northwestern; Northeastern; Western; Central; Eastern; and Southwestern. For each region, a team completed two or three seasons of field survey looking for archaeological sites of all periods. Preliminary reports were published in *Atlal,* then a new journal dedicated to Saudi Arabian archaeology and heritage. Major monographs for each province were planned, but, unfortunately, never came to fruition. Although entitled as a Comprehensive Archaeological Survey of the Kingdom, even at the time, it was recognised that the scale of the task made a total systematic survey impossible. Matters were also not helped by the absence of detailed topographic maps better than 1:500,000 scale (Petraglia 2003) with which to plan and record any work. To compensate, the Survey targeted areas where sites were already known, or areas close to known oases, shorelines, ancient wells and traditional travel routes and where access for survey teams was also possible. In doing so, the survey team recognised that their results would necessarily be affected by these constraints (Adams et al. 1977). The survey teams also clearly understood that any behavioural interpretations of the archaeology of the Pleistocene (the Palaeolithic record) would necessarily be tentative since the environmental constraints that have structured life in Saudi Arabia over the last few thousand years might have been considerably relaxed during earlier episodes with different climate regimes, allowing possible habitation and use of the landscape beyond the targeted survey areas (Adams et al. 1977). Recent research on changing palaeohydrology and landscapes in Arabia shows the wisdom of this observation (Parker 2009; Groucutt and Petraglia 2012; Bailey et al. 2015; Jennings et al. 2015, amongst others), confirmed by new archaeological finds on former wadis and lakes (e.g., Petraglia et al. 2011, 2012, and this volume; Delagnes et al. 2012; Crassard and Hilbert 2013). Despite these limitations, the Comprehensive Survey proposed some clear typological patterns and environmental associations for the Palaeolithic archaeology of Saudi Arabia (Petraglia 2003; Petraglia and Alsharekh 2003) which remain a testament to the observational skill of the original survey teams.

The Red Sea coastal plain in Jizan and Asir provinces lies within the Southwestern province of the Comprehensive Survey. Over six weeks in 1980, survey teams examined an area greater than 21,000 km^2, ranging from the Red Sea coastal plain to beyond the escarpment to the east, as well as including the Farasan Islands in the Red Sea as part of their second season of survey (Zarins et al. 1981, Plate 1A). They identified 20 sites that were typologically assigned to the Palaeolithic, with 4 sites in the coastal plain (Zarins et al. 1981). These sites were identified by stone tool typology as Lower Palaeolithic (including Oldowan and Acheulean), Middle Palaeolithic (including Mousterian) or Upper Palaeolithic by reference to typological or technological similarities to lithic materials discovered elsewhere in Europe or Africa. In Africa, archaeologists refer to these three periods as Early Stone Age (ESA), for Lower Palaeolithic, Middle Stone Age (MSA) for Middle Palaeolithic, and Later Stone Age (LSA) for Upper Palaeolithic. Presuming a colonisation route into Arabia primarily from Africa, these African chronological terms will be used here. This choice of name also indicates a need to look for parallels, where appropriate, in Africa rather than Europe as is often done by other researchers. Palaeolithic sites were located close to lava flows east of Abu Arish and in the Harrat al Birk (Fig. 1). Acheulean sites, within the ESA, were identified by the presence of handaxes, cleavers and a series of worked tools made on flakes that can be described as scrapers, knives and so forth that were always used as simple hand-held tools. Middle Stone Age sites were identified by the presence of prepared core technology in which a core is prepared in such a way so as to facilitate the removal of a flake of pre-determined shape and by tools that conformed to types observed in European sites of Mousterian age (Zarins et al. 1980, 1981). Prepared core technology, originally called "Levallois technique", results in flakes that can be broad and flat, or pointed and flat, that can be worked into a variety of tools of a more standardised shape, and that are also thin enough to haft into handles (see Debenath and Dibble 1994). Along the Red Sea coast, no discrete Palaeolithic localities younger than the MSA were discovered; indeed, it was noted that such materials were in fact difficult to identify in Arabia in general (Zarins et al. 1981).

From the context of sites and finds, the Survey suggested a relationship between hominin settlement and lithic raw material sources, water courses, coastal exploitation of marine resources and environmental or sea level change (Zarins et al. 1981, Plate 5b). Acheulean sites were found on the lava flows close to Al-Birk and Al Shukayk in Harrat al Birk, and these flows appeared to overlie remnant coral terraces approximately 10 m above modern sea level. The location of Acheulean sites was thought to indicate the exploitation of marine foods. For the MSA, the use of marine resources seemed even more certain; survey reports state that they had found tools embedded in remnant coral terraces approximately 2 m above sea level (Zarins et al. 1981). Away from the coast, Palaeolithic materials were found close to other lava flows, for example east of Abu Arish, and along old wadi courses. No absolute dates were produced, but it was suggested that Acheulean sites might date from 1,200,000 years ago, and MSA sites from 100,000 years ago, by comparison to dated sites in Africa, the Levant and Europe with typologically similar artefacts (Zarins et al. 1981). As was typical for Palaeolithic research at the time, the reports are also lacking in a detailed analysis of the techniques of working and the variability of working of stone materials within and between sites that might facilitate interpretations of movement of materials across a landscape.

Despite the real accomplishments of the Comprehensive Survey, the speed and the requirements of the geographical and chronological scale of the Survey meant that many questions remain. We have no idea of the real quantity of Palaeolithic archaeological materials surviving at the time. There are no absolute dates for any of these sites or for the lava flows with which stone tools were associated and which were reported to overlie remnant coral terraces. Aside from the statement that tools were associated with or embedded within coral terraces, no photographs were published to prove this relationship. The Survey reports provide no detail on the archaeological assemblages from the Red Sea coastal margin beyond their basic typological details; only one in the Hima region, east of the escarpment, has such detail. None of the sites found on the coastal plain was revisited for further excavation or analysis, unlike the Acheulean sites of the Wadi Fatimah and Dawadmi in the Western survey province (Whalen et al. 1981, 1983, 1989; Whalen and Pease 1990). As should be expected given the age and broad scope of the Survey, there is no detail on the techniques of working at any one site, how the raw material was exploited and how sites might have fitted into larger patterns of settlement as evidenced by their sequences of production. Such analyses were not common until the mid-1980s. Finally, no samples were collected to provide information on the sediments with which the archaeological materials were associated.

3.2 Palaeolithic Landscape Surveys in Jizan and Asir Provinces Since 2012

Following preliminary reconnaissance in 2004 and 2006 (Alsharekh and Bailey 2013), from 2012, the DISPERSE project has been conducting archaeological fieldwork along the coastal strip of southwestern Saudi Arabia in the provinces of Jizan and Asir as well as in the Farasan Islands, led by Geoff Bailey (University of York, United Kingdom), Geoffrey King (Institut de Physique de Globe de Paris, France) and Abdullah Alsharekh (King Saud University, Saudi Arabia). The broad aim of the Project has been to address the manner of and evidence for the dispersal of hominins, with a focus on dispersal from Africa into Arabia and beyond, recognising that (i) active tectonic and volcanic landscapes offer an attractive set of resources for hominins, and that (ii) much of the archaeological evidence relating to hominin dispersals is currently submerged (Bailey and King 2011; Winder et al. 2015). Fieldwork in Arabia has, therefore, included extensive research on the exploitation of marine resources on the Farasan Islands and on parts of the submerged shorelines, supported by detailed surveys of the topographical features of the sea bed between the mainland and the Farasan Islands to predict places which might have been used by hominins during periods of lower sea level, and where underwater surveys might find preserved archaeological evidence (Bailey et al. 2015, this volume; Momber et al., this volume; Sakellariou et al., this volume). Finally, the DISPERSE Project also initiated a new programme of landscape survey and, where appropriate, detailed site analysis along a 200 km by 100 km strip of the upper and lower coastal plain along the Red Sea from Jizan and Abu Arish in the south to north of the Harrat al Birk (Devès et al. 2013; Inglis et al. 2014). This is the first landscape survey for Palaeolithic archaeology in this region since the original Comprehensive Survey.

In the thirty years and more that have passed since the Comprehensive Survey, archaeological field surveys have been transformed by the availability of satellite imagery, GPS location and GIS mapping to investigate relationships between finds and landscape features. In Arabia, this has an even greater impact due to the continued absence of detailed paper maps. The availability of scalable high-resolution imagery transforms the speed and accuracy of survey planning, whilst accurate GPS location data, available from the late 1990s, allows the rapid recording and representation of survey findings in a graphical manner to facilitate immediate feedback into the planning process. Indeed, the availability of free, up-to-date imagery through Google Earth now makes it possible to observe and interrogate a landscape topographically to find potential types of deposit to be surveyed before entering the survey area, as well as transport routes to plan a series of locations and transects that can be accessed. Online access to Google Earth during the survey season makes it possible to modify plans in the light of daily survey results from an area, and to identify landform or sediment types that may be most profitable. Finally, regular updates to the available satellite imagery permit an examination as to how potential archaeological deposits are being damaged or destroyed by natural processes and human development. Despite these considerable technological developments, surveys are still ultimately dependent on the time-consuming process of walking across a landscape and being able to identify archaeological stone artefacts. This is further complicated in southwestern Arabia by the fact that identifying these pieces is not a case of spotting material anomalies in a landscape (as is often the case in regions where the materials used are not local) but one of looking over many thousands of possible pieces, many of which have been rolled and naturally flaked over time, to find a very much smaller number of real artefacts.

This recent programme of landscape survey initially focussed on re-visiting sites identified by the Comprehensive Survey to check on their current state of survival and the accuracy of their earlier observations. We also examined deposits of a similar type (remnant coral terraces, lava flows) to evaluate the proposed patterns in distributions of archaeological materials. Where possible we have examined naturally or artificially cut sections to look for archaeological materials that are buried by later sediments but exposed in section. We have also explored topographical locations that correspond to known hunter-gatherer patterns of landscape use, specifically examining old water courses and remnant tufa deposits that indicate the presence of ancient bodies of water, as well as locations with potential vantage points from which visual information about local animal and plant resources and other hominin groups might have been gained, and routes between. Recent research suggests that wayfaring and landscape legibility are significant factors in hominin behavioural evolution (Guiducci and Burke 2016; see also Kübler et al., this volume), and taking these factors into account has proven to be successful in landscape surveys elsewhere (Sinclair et al. 2003). Survey localities included small intensively surveyed areas (~200 m by 200 m) associated with specific resources or sediments, through to transects of perhaps 1 km length and 200 m width walked across the landscape. Survey locations were identified and planned from imagery available on Google Earth, and GPS data for all survey locations was recorded, whether archaeological materials were recovered or not. The resulting data was plotted as a layer on Google Earth and on smaller scale GIS-based models of the landscape. Archaeological samples, where found, had their locations recorded using a handheld GPS, were collected, typologically and technologically analysed and put into storage under the care of the regional antiquities service for Jizan and Asir provinces. Samples suitable for sedimentary analysis and for dating have also been collected where possible.

In four seasons of landscape surveys, more than 85 localities out of 110 visited yielded Palaeolithic artefacts (Table 1). From these locations, several relationships seem evident. There are three distinct concentrations of Palaeolithic materials, close to the lava flows at Abu Arish, Sabya and especially in the largest flows of Harrat al Birk (Fig. 2). Typologically, these localities include a few of predominantly ESA (Acheulean) character with handaxes, cleavers, large cutting tools and large flake tools (n = 9), a larger number of MSA character with evidence of prepared-core techniques and their retouched flake products (n = 18), and many localities with both (n = 47). Some localities could not be typologically diagnosed (n = 13), and a number of locations were surveyed without finding archaeological materials (n = 25). There are no localities that are specifically, and only, of LSA character, but there are either a small number, or just isolated pieces, in some places that are likely to be LSA (see below).

In general terms, recent landscape surveying has corroborated earlier findings by the original Comprehensive Survey. It has identified frequent concentrations of Palaeolithic material at Abu Arish and the Harrat al Birk, and supported the proposed association between lava flows and Palaeolithic archaeology by finding new localities close to the lava flows at Sabya. Comparing the quantities and character of evidence, we can now identify patterns with greater confidence than was possible from the original survey data. For example, there is a significant variation in the quantity of Palaeolithic lithics found at different localities; two sites (Dhahaban Quarry and Wadi Dabsa) have produced assemblages of 1000 or more pieces (Fig. 3), though the remainder have much smaller assemblages. Whilst most of the newly surveyed Palaeolithic localities contain lithics of both ESA and MSA typology, there are localities that are of a single typological period. This variability in quantity and quality indicates that it should be possible to examine changes in hominin behaviour through spatial data. There are clear associations between Palaeolithic hominins, water courses and lava flows in Harrat al Birk (Fig. 4), as well as possible evidence for movement of MSA hominins further into the lava fields. A reduced presence of LSA hominins is suggested by the reduced number of localities, and this will be clearer still when comparisons are made of absolute artefact numbers.

Whilst it would be reasonable to have expected the identification of a greater number of Palaeolithic localities in the recent programme of fieldwork when compared to the original survey, given the greater length of time devoted to surveying, the abundance of localities confirms the presence of a rich archaeological record in this region that is still no

Table 1 Survey localities 2012–2017. Latitude/Longitude in WGS1984. ESA/MSA/LSA counts are defined as number of diagnostic techno-typological pieces. Numbers in brackets include possible identifications of broad categories, e.g. ESA/MSA, or MSA/LSA. Undiagnostic artefacts are those lithics that cannot be tentatively assigned to a broad period based on their typological form

Locality	Latitude	Longitude	Province	Description	Lithic artefacts				
					Total	ESA	MSA	LSA	Undiagnostic
L0001	17.187926	42.725637	Jizan	Basalt flow, Jebel Akwah (South)	1	0 (0)	0 (0)	0 (0)	1
L0002	17.288010	42.791759	Jizan	Undulating hills near Al Kadami	2	0 (0)	0 (0)	0 (0)	2
L0003	17.007015	42.926441	Jizan	Basalt flow, Wadi Jizan	20	5 (5)	2 (3)	0 (1)	12
L0004	17.027127	43.039292	Jizan	Escarpment foothills near Al Aridah	0	0 (0)	0 (0)	0 (0)	0
L0005	16.878146	42.939741	Jizan	Cinder cone, Jebel Umm Al Qummam (North)	1	0 (0)	0 (0)	0 (0)	0
L0006	17.055552	42.963629	Jizan	Basalt flow, Wadi Jizan Dam	66	0 (0)	8 (8)	5 (21)	35
L0007	17.113061	42.940373	Jizan	Wadi Jizan Tributary	0	0 (0)	0 (0)	0 (0)	0
L0008	17.083452	42.847372	Jizan	Basalt flow, near Abu Arish (Lower flow)	0	0 (0)	0 (0)	0 (0)	0
L0009	17.076882	42.856067	Jizan	Basalt flow, near Abu Arish (Upper flow)	12	2 (2)	10 (10)	0 (0)	0
L0010	16.869092	42.940231	Jizan	Basalt flow, Jebel Umm Al Qummam (North)	67	31 (33)	17 (26)	1 (8)	10
L0011	16.887337	42.986865	Jizan	Alluvial fan East of Al Wahmah	0	0 (0)	0 (0)	0 (0)	0
L0012	16.874363	42.994776	Jizan	Alluvial fan East of Al Wahmah	0	0 (0)	0 (0)	0 (0)	0
L0013	17.267884	42.683631	Jizan	Basalt flow, Jebel Akwah (North)	8	3 (3)	3 (4)	0 (1)	1
L0014	17.259316	42.708638	Jizan	Wadi draining Jebel Akwah (North)	9	1 (3)	3 (5)	0 (0)	3
L0015	17.241765	42.695073	Jizan	Basalt flow, Jebel Akwah (North)	7	1 (1)	4 (4)	0 (0)	2
L0016	17.248540	42.800845	Jizan	Alluvial deposits, Wadi Nakhlan	2	0 (0)	0 (0)	0 (0)	2
L0017	17.240254	42.804980	Jizan	Hills above Wadi Nakhlan	28	1 (6)	20 (24)	0 (0)	2
L0018	17.193206	42.782904	Jizan	Incised deposits in Wadi Sabiya Quarry	0	(0)	0 (0)	0 (0)	0
L0019	17.219985	42.713273	Jizan	Wadi Nakhlan between Jebel Akwah cinder cones	0	0 (0)	0 (0)	0 (0)	0
L0020	17.187597	42.781249	Jizan	Wadi Sabiya Quarries	13	0 (0)	0 (0)	4 (5)	8
L0021	17.189416	42.800086	Jizan	Wadi Sabiya—Incised deposits	0	0 (0)	0 (0)	0 (0)	0
L0022	17.193590	42.809743	Jizan	Wadi Sabiya Quarries	0	0 (0)	0 (0)	0 (0)	0
L0023	17.188525	42.845222	Jizan	Basalt exposed along Wadi Sabiya	2	0 (2)	0 (2)	0 (0)	0
L0024	17.191863	42.860800	Jizan	Hills above Wadi Sabiya	63	2 (5)	4 (16)	10 (30)	24
L0025	17.254514	42.845795	Jizan	Wadi Nakhlan, escarpment foothills	5	0 (0)	0 (0)	0 (0)	5
L0026	16.981653	42.894690	Jizan	Basalt flow, SW of Abu Arish lava flows	13	1 (6)	3 (5)	0 (1)	3
L0027	17.032474	42.908738	Jizan	Basalt flow, North of Wadi Jizan	11	1 (5)	1 (2)	1 (1)	4

(continued)

Table 1 (continued)

Locality	Latitude	Longitude	Province	Description	Lithic artefacts				
					Total	ESA	MSA	LSA	Undiagnostic
L0028	17.061549	42.914820	Jizan	Centre of Abu Arish lava flows	3	1 (1)	0 (0)	1 (1)	1
L0029	17.056896	42.875615	Jizan	Basalt flow, Western edge of Abu Arish fields	10	0 (0)	9 (9)	0 (0)	1
L0030	17.042681	42.998008	Jizan	Schist hills above Wadi Jizan	3	2 (2)	0 (0)	0 (0)	1
L0031	17.039973	42.981798	Jizan	Alluvial sediments, Wadi Jizan Dam Lake	0	0 (0)	0 (0)	0 (0)	0
L0032	17.858012	41.935777	Asir	Volcanic cone with flat top	91	2 (68)	16 (84)	0 (0)	5
L0033	17.931643	41.978747	Asir	Granite Outcrops E of Harrat Al Birk	80	11 (17)	14 (27)	0 (0)	41
L0034	18.071406	41.623357	Asir	Wadi Dhahaban Quarry	711	14	398	1	138
L0035	18.172324	41.565232	Asir	Coral terrace over basalt, Al Birk coastline	35	6 (6)	8 (13)	0 (0)	16
L0036	18.167809	41.566148	Asir	Coral terrace over basalt, Al Birk coastline	18	14 (14)	3 (3)	0 (0)	1
L0037	18.148964	41.577697	Asir	Coral terrace over basalt, Al Birk coastline	2	0 (0)	1 (1)	0 (0)	1
L0038	18.134898	41.577301	Asir	Coral terrace over basalt, Al Birk coastline (CASP 216-208)	48	11 (30)	8 (27)	0 (0)	10
L0039	18.121848	41.704216	Asir	Hill in headwater basin of Wadi Najla	31	6 (6)	11 (23)	0 (0)	2
L0040	18.104065	41.694961	Asir	Head of gorge, Wadi Najla	23	13 (13)	6 (7)	0 (0)	5
L0041	18.093014	41.698197	Asir	Base of gorge, Wadi Najla	2	0 (0)	0 (1)	0 (0)	1
L0042	18.084579	41.692803	Asir	Gorge base, Wadi Najla	12	3 (3)	3 (3)	0 (0)	5
L0043	18.065674	41.674188	Asir	Hill at coastal mouth of Wadi Najla	0	0 (0)	0 (0)	0 (0)	0
L0044	18.605248	41.480481	Asir	Area of rock art on basalt flow, Al Moalmat	4	0 (0)	2 (2)	0 (0)	2
L0045	18.595227	41.555002	Asir	Basalt flow, Wadi Shafqah	11	0 (0)	2 (11)	0 (0)	0
L0046	18.602847	41.539418	Asir	Alluvium and Basalt, Wadi Shafqah	3	0 (0)	2 (2)	0 (0)	1
L0047	18.632677	41.561697	Asir	Alluvium and Basalt, Wadi Shafqah	5	1 (1)	1 (1)	0 (0)	3
L0048	18.634902	41.544491	Asir	Area of carbonate deposits, Wadi Shafqah	0	0 (0)	0 (0)	0 (0)	0
L0049	18.070333	41.690620	Asir	Young basalt flow, Wadi Najla	20	2 (2)	2 (5)	0 (0)	13
L0050	18.002664	41.858373	Asir	Alluvial fan near Wadi Urayk/Wadi Nahab	2	0 (0)	0 (0)	0 (0)	2
L0051	18.009341	41.776442	Asir	Rockshelters in volcanic Jebel Hashahish	26	0 (0)	0 (14)	0 (0)	12
L0052	17.920381	42.079484	Asir	Linear schist outcrops cut by Wadi Aramram	1	0 (0)	0 (0)	0 (0)	1
L0053	18.041980	42.090410	Asir	Escarpment foothills, Wadi Aramram	21	0 (0)	5 (5)	0 (0)	16
L0054	18.013357	42.074013	Asir	Jebel Bagarah volcanic cone, Wadi Aramram	10	0 (1)	4 (7)	0 (0)	3
L0055	16.862098	42.954549	Jizan	Dune abutting Jebel Umm Al Qumam (North)	0	0 (0)	0 (0)	0 (0)	0
L0060	17.212833	43.031100	Jizan	Escarpment foothills	0	0 (0)	0 (0)	0 (0)	0

(continued)

Table 1 (continued)

Locality	Latitude	Longitude	Province	Description	Lithic artefacts				
					Total	ESA	MSA	LSA	Undiagnostic
L0061	17.249880	42.705320	Jizan	Sediment on basalt flow, Jebel Akwah North)	2	0 (0)	0 (0)	0 (0)	2
L0062	17.249610	42.709800	Jizan	Colluvial slopes of cider cone, Jebel Akwah (North)	11	0 (0)	0 (0)	6 (6)	5
L0063	17.243605	42.704993	Jizan	Basalt surface, Jebel Akwah (North)	13	0 (0)	1 (1)	4 (4)	8
L0064	17.019560	42.946990	Jizan	Quarry adjacent to Wadi Jizan	32	3 (3)	4 (4)	4 (1)	24
L0065	17.035400	42.529430	Jizan	Deflated shell midden, North of Jizan	0	0 (0)	0 (0)	0 (0)	0
L0066	17.184389	42.780317	Jizan	Wadi Sabiya quarry sections	1	0 (0)	0 (0)	0 (0)	1
L0067	17.018126	42.923189	Jizan	Quarries through basalt flows and sediments, Wadi Jizan	14	1 (1)	1 (1)	0 (5)	7
L0068	17.022422	42.929729	Jizan	Quarries through basalt flows and sediments, Wadi Jizan	1	0 (0)	0 (0)	0 (0)	1
L0069	17.041193	43.020466	Jizan	Area of schist bedrock close to Wadi Jizan Dam	1	0 (0)	0 (0)	0 (0)	1
L0070	17.027626	42.974781	Jizan	Area of schist bedrock close to Wadi Jizan Dam	0	0 (0)	0 (0)	0 (0)	0
L0071	17.030012	42.944420	Jizan	Basalt terraces below Wadi Jizan Dam	12	0 (0)	4 (4)	0 (0)	8
L0072	17.162252	42.919970	Jizan	Alluvial terrace, Wadi Dhamad	4	1 (1)	0 (0)	0 (0)	3
L0073	17.153165	42.904202	Jizan	Linear shist hills above Wadi Dhamad	2	1 (1)	0 (0)	0 (0)	1
L0074	17.106895	42.859326	Jizan	Basalt flow, distal end, near As Shugayri	15	1 (2)	3 (4)	6 (6)	4
L0075	17.137263	42.881558	Jizan	Sediment sections beneath basalt	0	0 (0)	0 (0)	0 (0)	0
L0076	17.140631	42.902273	Jizan	Basalt flow and alluvial terrace at base of wadi	2	0 (0)	1 (1)	0 (0)	1
L0077	17.225168	42.717739	Jizan	Sediment and basalt flows between Jebel Akwah North and South cones	3	0 (0)	0 (0)	1 (1)	2
L0078	18.267436	41.515297	Asir	Shell scatter on coral terrace	16	1 (4)	6 (11)	0 (0)	9
L0079	17.100630	42.931879	Jizan	Centre of Abu Arish basalt flows	3	0 (0)	1 (1)	0 (0)	3
L0080	17.092157	42.968446	Jizan	Linear schist hills above Wadi Jizan Dam	11	0 (0)	3 (3)	3 (3)	5
L0081	17.217560	42.965900	Jizan	Escarpment foothills, Al Henaya	0	0 (0)	0 (0)	0 (0)	0
L0082	17.204495	42.990423	Jizan	Escarpment foothills, Al Henaya to N of Wadi Dhamad Dam	0	0 (0)	0 (0)	0 (0)	0
L0083	17.206003	43.019666	Jizan	Escarpment foothills, Al Henaya	0	0 (0)	0 (0)	0 (0)	0
L0084	17.173910	42.734090	Jizan	Basalt flow south of Jebel Akwah (South) cut by road	0	0 (0)	0 (0)	0 (0)	0
L0085	17.165620	42.738590	Jizan	Basalt flow south of Jebel Akwah (South) cut by road	0	0 (0)	0 (0)	0 (0)	0
L0087	18.396523	41.499354	Asir	Basalt flow above wadi confluence, near Markaz as Shurtah	17	1 (2)	5 (6)	0 (1)	9
L0088	18.040502	41.658652	Asir	Volcanic hill on coast at mouth of Wadi Najla	1	0 (0)	0 (0)	0 (0)	1
L0089	18.036455	41.653339	Asir	Coral terraces around northern edge of volcanic cone at mouth of Wadi Najla	14	1 (1)	9 (9)	0 (0)	4
L0090	18.202787	41.524803	Asir	Coral terrace on basalt, Al Birk headland	1	0 (0)	1 (1)	0 (0)	0
L0091	18.066465	41.635032	Asir	Coral and lava terraces on S side of basalt hill adjacent to L0034	2	1 (1)	0 (0)	0 (0)	1

(continued)

Table 1 (continued)

Locality	Latitude	Longitude	Province	Description	Lithic artefacts				
					Total	ESA	MSA	LSA	Undiagnostic
L0092	18.119056	41.597420	Asir	Beachrock exposure (1 km) on basalt	8	0 (0)	6 (6)	0 (0)	2
L0093	18.239446	41.528332	Asir	Coral terrace over basalt	4	0 (0)	1 (4)	0 (0)	0
L0094	17.839722	41.952032	Asir	Hajambar Quarry	7	1 (1)	0 (0)	0 (0)	6
L0095	17.652420	42.688550	Jizan	Terraces above Wadi Bayish Dam Lake	2	0 (0)	0 (0)	0 (0)	1
L0096	18.004440	41.882770	Asir	Dykes rising out of alluvium in tributatries of Wadi Urayk	26	0 (5)	0 (5)	0 (0)	21
L0100	18.465869	41.473108	Asir	Basalt escarpment parallel to coast	8	5 (7)	0 (2)	0 (1)	0
L0101	18.463013	41.492603	Asir	Basalt terrace above Wadi Amq	18	3 (4)	1 (4)	11 (11)	0
L0102	18.476082	41.492370	Asir	Basalt terrace above Wadi Amq	0	0 (0)	0 (0)	0 (0)	0
L0103	18.406560	41.461806	Asir	Beachrock exposure south of Amq (correlated with L0105)	23	6 (10)	3 (10)	5 (7)	2
L0104	18.406078	41.482014	Asir	Basalt ridge 3 km inland from present day coastline	28	20 (21)	2 (4)	3 (4)	1
L0105	18.417158	41.464038	Asir	Beachrock exposure south of Amq (correlated with L0103)	16	4 (5)	2 (6)	5 (8)	1
L0106	18.306375	41.563742	Asir	Wadi Dabsa Tufa Exposure—Main Grids	2847	175 (67)	478	0 (13)	2194
L0107	18.313622	41.559729	Asir	Wadi Dabsa Tufa Exposure—Transect up basalt flow	20	7 (8)	9 (11)	0 (0)	2
L0108	18.305406	41.551689	Asir	Wadi Dabsa—Tufa downstream from main exposure	4	4 (4)	0 (0)	0 (0)	0
L0109	18.301462	41.536807	Asir	Wadi Dabsa—basalt flow	6	1 (4)	1 (5)	0 (0)	0
L0110	18.196402	41.555418	Asir	Wadi terraces next to Al Birk Baladiyah	12	7 (9)	0 (2)	0 (0)	3
L0111	18.212968	41.585135	Asir	Wadi and tufa deposits inland from L0111	3	2 (2)	0 (1)	0 (1)	0
L0112	18.177545	41.596737	Asir	Terraces above Wadi Dahin	4	1 (1)	2 (2)	1 (1)	0
L0113	17.948380	41.749505	Asir	Terrace above Wadi Hamidah	58	1 (1)	0 (2)	55 (57)	0
L0114	17.920450	41.780271	Asir	Wadi draining Jebel Huwwah	30	6 (11)	0 (4)	19 (19)	0
L0115	17.895032	41.799279	Asir	High basalt terrace above wadi draining Jebel Khurmah	6	6 (6)	0 (0)	0 (0)	0
L0116	17.884521	41.795242	Asir	Low undulating basalt c. 1 km to SW of L0115	13	4 (5)	0 (1)	8 (8)	0
L0117	18.369736	41.456772	Asir	Low lying undulating basalt hills close to coast	2	0 (0)	0 (1)	1 (2)	0
L0118	18.132236	41.640305	Asir	Basalt terraces above Wadi Dhaban dam lake	0	0 (0)	0 (0)	0 (0)	0
L0119	18.116808	41.631170	Asir	Isolated basalt jebel in Wadi Dhahaban floodplain	12	0 (0)	3 (3)	7 (7)	2
L0120	18.084993	41.630403	Asir	Low basalt rise 3 km from present-day coastline	6	5 (6)	0 (1)	0 (0)	0
L0121	18.308182	41.485023	Asir	Beachrock exposure overlying low basalt peninsula	40	7 (21)	4 (22)	5 (10)	4
L0122	18.141990	41.585491	Asir	Basalt terrace where Wadi Dahin flows into sea	9	4 (5)	0 (1)	2 (2)	2

(continued)

Table 1 (continued)

Locality	Latitude	Longitude	Province	Description	Lithic artefacts				
					Total	ESA	MSA	LSA	Undiagnostic
L0123	17.851005	41.816456	Asir	Basalt and coral/beachrock terrace next to Moath bin Jebel	8	3 (3)	0 (0)	0 (0)	5
L0124	17.831206	41.834735	Asir	Low lying, undulating basalt hills	0	0 (0)	0 (0)	0 (0)	0
L0125	17.812721	41.845927	Asir	Southern/eastern slopes of Jebel ar Raqabah	4	1 (3)	0 (2)	1 (1)	0
L0126	18.307269	41.568197	Asir	Wadi Dabsa Tufa Exposure—tufa transect	32	5 (21)	3 (20)	5 (8)	0
L0127	18.308949	41.559993	Asir	Wadi Dabsa Tufa Exposure—basalt flow transect	5	3 (5)	0 (2)	0 (0)	0
L0128	17.974918	41.684844	Asir	Raised beach deposits S of Al Qahma	1	1 (1)	0 (0)	0 (0)	0
L0130	18.306375	41.563742	Asir	Wadi Dabsa Tufa Exposure—Main Grids	399	43	121	0	235

more than partially explored. For example, the apparent absence of localities along much of the eastern edges of the lava flows in Harrat al Birk arises from an absence of surveys, resulting from the difficulty of access, and not from a known absence of archaeology. If access were to make surveys possible, there is no reason to think that further Palaeolithic localities would not be found. With one exception, we have also found it very difficult to access and survey potential places going inland toward and into the escarpment. Finally, finds of ESA materials (including a worked shale clast, and a flake from a discoidal core) from artificially-exposed, deep sections below the lava flows (see Bailey et al. 2015) suggest a younger date for some of the lava flows, and the probable presence of many more archaeological remains now covered by overlying sediment.

Beyond these general observations, the two sites found in this survey programme with the largest archaeological assemblages, Dhahaban Quarry and Wadi Dabsa, indicate the extraordinary quality of archaeological and environmental evidence that remains in the Red Sea coastal plain, as well as the potential that exists for addressing more specific questions about marine resource use and the ways of life of past hunter-gatherer communities.

3.3 Dhahaban Quarry and the Evidence for the Use of Coastal Environments

South of the town of Dhahaban, about 1 km inland from the present shoreline, a complex series of coral terraces and marine sediments have been preserved adjacent to the remnants of a heavily-eroded cinder cone and its associated basalt lava flows (Inglis et al. 2014, and this volume; Bailey et al. 2015). From north to south, the deposits, all overlying the basalt flows, include exposures of coral terrace and overlying beachrock, a deep section through shallow marine sediments, potentially formed in a lagoonal environment, and beach sediments, capped by aeolianite at the southernmost extent. Toward the centre of the site, a wadi flowing from east to west has cut through the shallow marine/beachrock sediments (Fig. 5), exposing a unit of rounded cobbles of basalt and coral (for further stratigraphic detail, see Inglis et al., this volume). Absolute dating of the sediments (optically stimulated luminescence), basalt (^{39}Argon/^{40}Argon) and coral (Uranium series) at Wadi Dhahaban is a complex process and still ongoing (Inglis et al., this volume; Sanderson and Kinnaird this volume). However, the elevation of this complex of deposits, about 7 m above current sea level, is broadly consistent with a Last Interglacial high sea stand, suggesting that the artefacts buried below in the cobble unit pre-date MIS 5e (see discussion in Inglis et al., this volume). Examination of the conditions of deposition of the cobble unit is also ongoing,

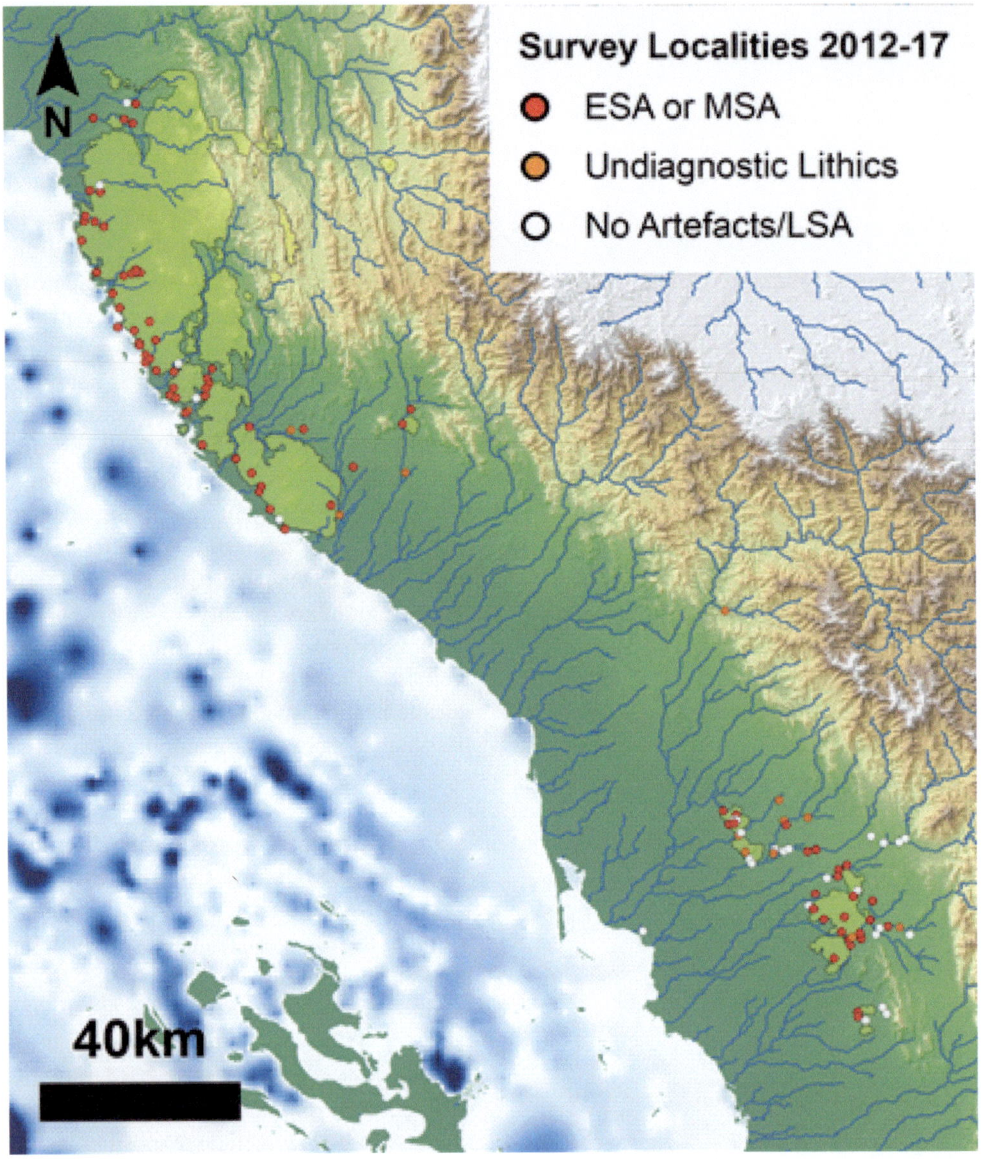

Fig. 2 Palaeolithic localities identified by recent landscape surveys in the southern Red Sea coastal region from 2012 to 2017. Red circles show localities where Lower Palaeolithic/Early Stone Age and/or Middle Palaeolithic/Middle Stone Age artefacts were identified; orange circles show localities where only undiagnostic lithic artefacts were observed; white circles show locations surveyed that yielded no artefacts, or only those of post-Palaeolithic character, i.e., those whose typologies were assigned to the LSA

but the presence of rounded cobbles of coral may suggest reworking of an earlier coral deposit, potentially from an earlier high sea stand, within a terrestrial debris flow, which also incorporated the artefacts.

The deposits at Dhahaban Quarry show the range and complexity of marine deposits that remain in patches along this coast and attest to the complex development and later reworking of marine deposits at times of changing sea level (see Inglis et al., this volume). It also highlights the threats facing these deposits today. This site was first identified because a large area in the central and southern part of the site, possibly equivalent to 50% of the total area of the marine deposits, has been extensively bulldozed, removing the upper deposits of aeolianite and beach rock and forming sections through these deposits. The western edge of the site has also been extensively quarried, removing much more of the same deposit and leaving a large exposed section 6–7 m high. Much of this extraction of sediment presumably occurred to provide aggregate material for building the coastal highway that runs close to the western edge of the site, or perhaps for new housing that has been built in local towns.

Dhahaban Quarry is the source of the second largest assemblage of Palaeolithic materials recovered during this

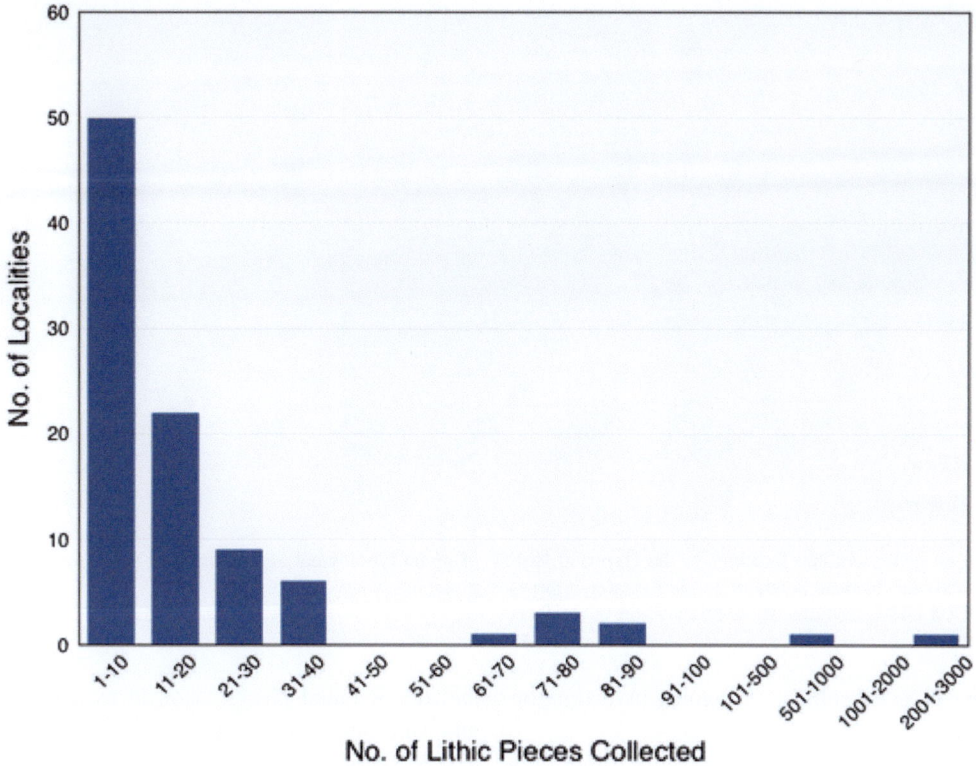

Fig. 3 The variability in numbers of artefacts found at Palaeolithic localities through field surveys from 2012 to 2015

survey. Across the preserved coral and beachrock deposits, more than 900 MSA and ESA basalt artefacts have been collected, containing a range of retouched tool forms (hand axes, picks, scrapers, large cutting tools from the ESA, prepared cores, retouched points, burins and retouched flake blades from the MSA), and other types of prepared and non-prepared cores and unretouched flakes. Most importantly, still embedded within the shallow marine and beachrock deposits that were cut by the wadi, we have found 19 sharp basalt flakes of MSA character with another lithic embedded within a lump of beach rock displaced by the bulldozing activity, suggesting that bulldozing has probably removed many other embedded lithics (Fig. 5c). Spatial analysis is ongoing, but broadly-speaking, the geographical spread of diagnostic and typologically identified stone tools suggests that the ESA materials are to be found at the southern and northern ends of the site, often closest to the lava flows, from where they may have moved downslope onto the surface of the younger marine and aeolian sediments. MSA materials are more common in the central part of the site, and within the marine deposits. Dhahaban Quarry, therefore, confirms the report of the Comprehensive Survey team that MSA period artefacts can be found embedded in marine deposits along this part of the Red Sea coast, as well as on their surface, raising the possibility of finding more embedded stone tools where other deposits of the same type are preserved in other places along the coast. Whilst the lithic evidence indicates that ESA and MSA communities exploited the local lithic resources available here at the coast, the taphonomic complexity of marine deposits, lava flows, and the mixture of depositional and erosional processes at Dhahaban Quarry, make it difficult at present to determine whether they were also here for the specific purpose of exploiting its marine resources. This highlights the necessity of being able to study other similar deposits in the future (Inglis et al., this volume).

3.4 Wadi Dabsa and Technological Evidence for Cultural Transmission

The Wadi Dabsa basin is about 6 km inland from the present-day shoreline. The basin is approximately 1 km^2 in area, and filled by a complex series of tufa carbonate deposits, deposited by the flowing and pooling of carbonate-rich water during periods of increased humidity. Its limits are defined by basalt flows emanating from numerous adjacent cinder cones (Inglis et al. 2015, 2017). The date and sequence of development of these tufa deposits, and that of emplacement of the basalt flows

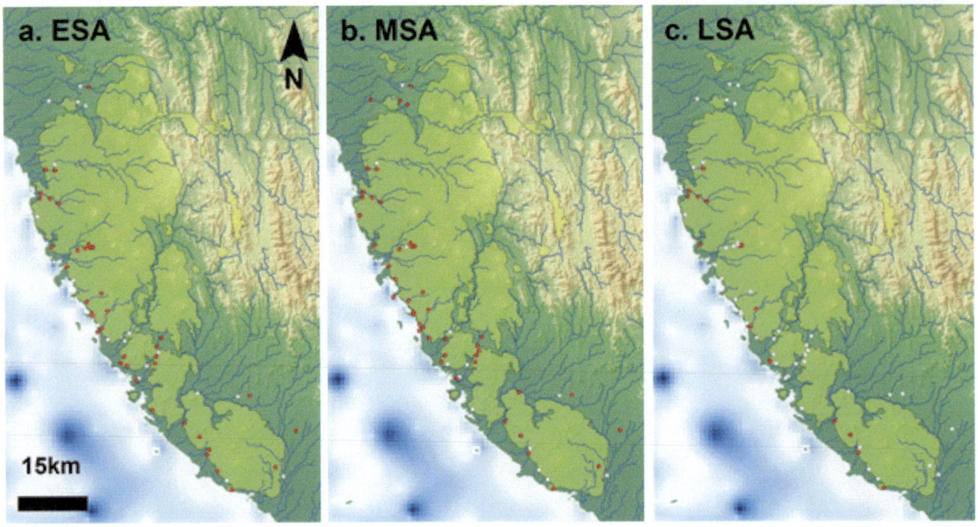

Fig. 4 The distribution of Palaeolithic localities in the Harrat al Birk mapped by typological age against potential modern water courses mapped from the SRTM DEM. Red dots show observation of artefacts of given typological age, white dots show localities where artefacts of given age were not observed

surrounding the basin, is the subject of ongoing investigation (Inglis et al. in prep.).

In 2015, numerous Palaeolithic artefacts were found during an initial survey of the surrounding basalt flows at the limits of the basin, leading to the discovery of a much larger concentration of Palaeolithic artefacts located on the top of a tufa spur overlying a basalt outcrop in the centre of the basin (Inglis et al. 2015; Foulds et al. 2017). The tufa spur is bordered on its southern and western side by a drainage depression filled with sandy-silt sediment. Artefacts of ESA and MSA character were identified and recorded across two 50 × 60 m grids (localities L0106 and L0130) in 2015 and 2017 (Fig. 6). All artefacts identified from L0106 were collected for later technological study, but those from L0130 were typologically recorded, photographed and left in situ. The density of artefacts across the grids (Fig. 7) suggests that their numbers are controlled by geomorphological processes that are deflating an artefact-bearing unit that overlies the tufa (Inglis et al. 2017, in prep.). The artefacts would therefore post-date tufa formation. However, a handaxe encased in tufa (Fig. 8) found on the surface of the drainage depression in 2017 suggests a longer, more complex history for both tufa deposition and hominin activity.

Palaeoenvironmental and dating evidence from the tufa deposits and lava flows will contribute significantly to an understanding of the complexity of local environments inland from the Red Sea coast, but it is the quality of the technological information accessible from this site that is more surprising. Wadi Dabsa has produced an assemblage of more than 2900 artefacts of both ESA and MSA character. ESA artefacts include large cutting tools, scrapers, notched and denticulated pieces, as well as a number of handaxes, and one unusually large handaxe (Foulds et al. 2017). The MSA assemblage contains a variety of retouched flake tools including scrapers, burins, and point forms amongst others. Two small test pits excavated in L0106 in 2017 have produced examples of small shatter flakes that typically result from in situ core working, and demonstrate that the archaeological deposits are not completely deflated and offer potential research gains through archaeological excavation in the future.

The MSA assemblage is particularly well preserved and appears to offer the possibility for refitting pieces making it possible to reconstruct sequences of technological activity (Fig. 9). It contains an extensive sample of prepared cores and prepared-core flakes across all stages of reduction. The production of good flakes through prepared-core technology requires a knowledge of the way in which this technique balances the volume and angles of stone above and below a middle "horizon" (Boëda 1995). It also requires sufficient prior individual experience to be able to adjust the exact placing and weight of any striking of the core to respond to the individual qualities of the nodule being worked and the success or failure of prior blows (see van Peer 1992 for examples from Egypt). At Wadi Dabsa, the lithic evidence for prepared-core reduction has examples of cores that are worked with knowledge and experience, as well as others found in a restricted, but central, part of the site grid (B3-4, C3-4) that show technological mistakes typical of individuals learning to work stone. Specifically, these cores show repeated attempts to strike and remove preparation flakes from the same place, where previous strikes have been

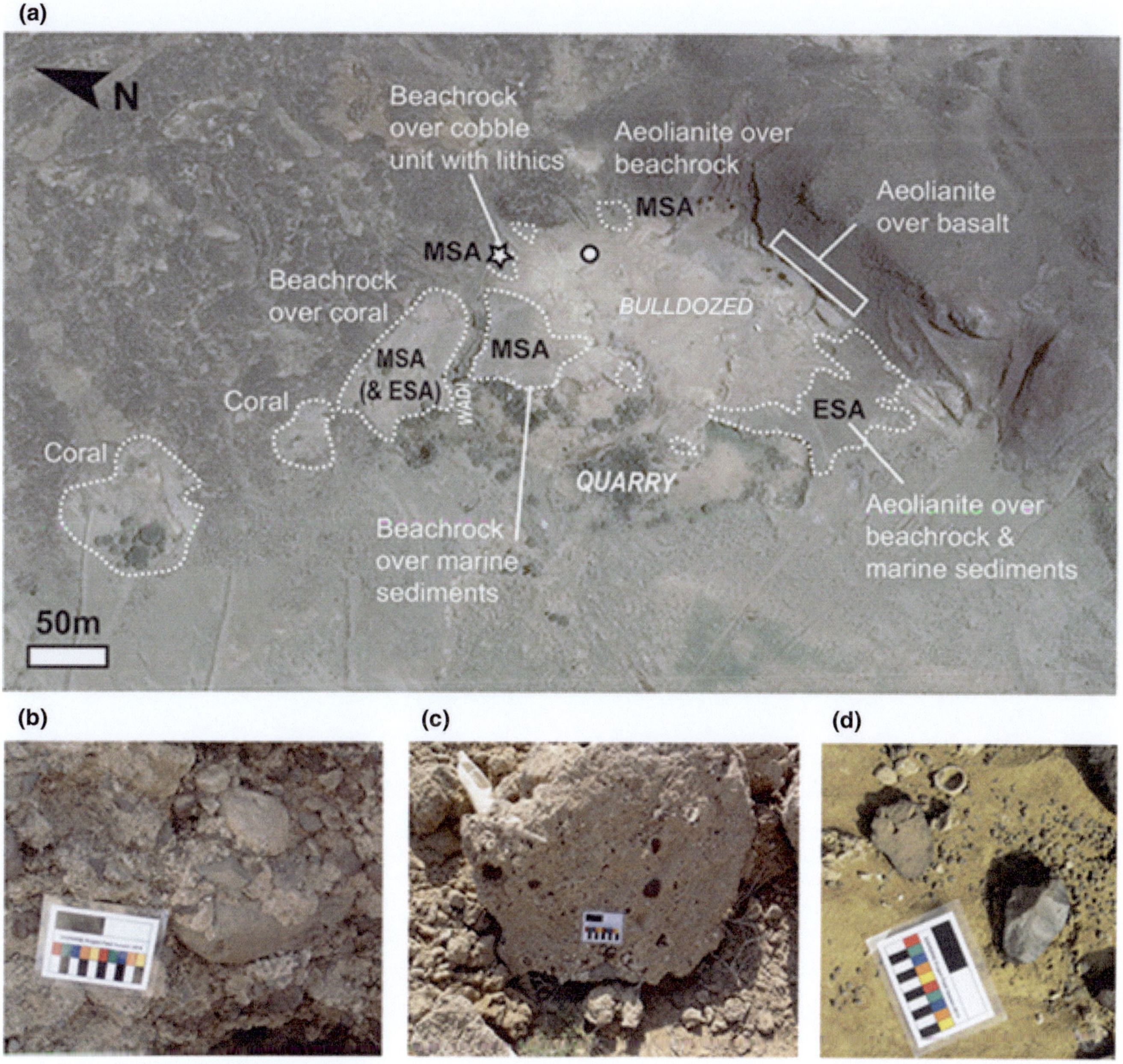

Fig. 5 **a** Overview of geomorphological units at Dhahaban Quarry (after Inglis et al., this volume) and summary of artefact technotypological characteristics found (Imagery © Google Earth, imagery date 19/1/2014); **b** basalt flake embedded within cobble unit, location marked by star on (**a**); **c** basalt flake embedded in bulldozed block of beachrock, location marked by circle on (**a**); **d** example of basalt MSA artefact (prepared core) found on surface of deposits at Dhahaban Quarry. Photos: A. Sinclair

unsuccessful, resulting in cores that must be discarded before the preparation process is complete. These unsuccessful cores demonstrate a knowledge of what needs to be done, the "recipe" for working, but a lack of experience with which to adjust the process when faced by problems. A detailed study of this evidence is ongoing (Sinclair et al. in prep.) with the hope that refitted sequences of reduction will support the technological interpretation derived from the examination of single artefacts.

4 The Research Potential of Archaeological Deposits Along the Southern Red Sea Coast

A new programme of landscape surveying along the southern Red Sea coast has confirmed the basic typological observations of the Comprehensive Survey in this area more than 30 years previously concerning surviving

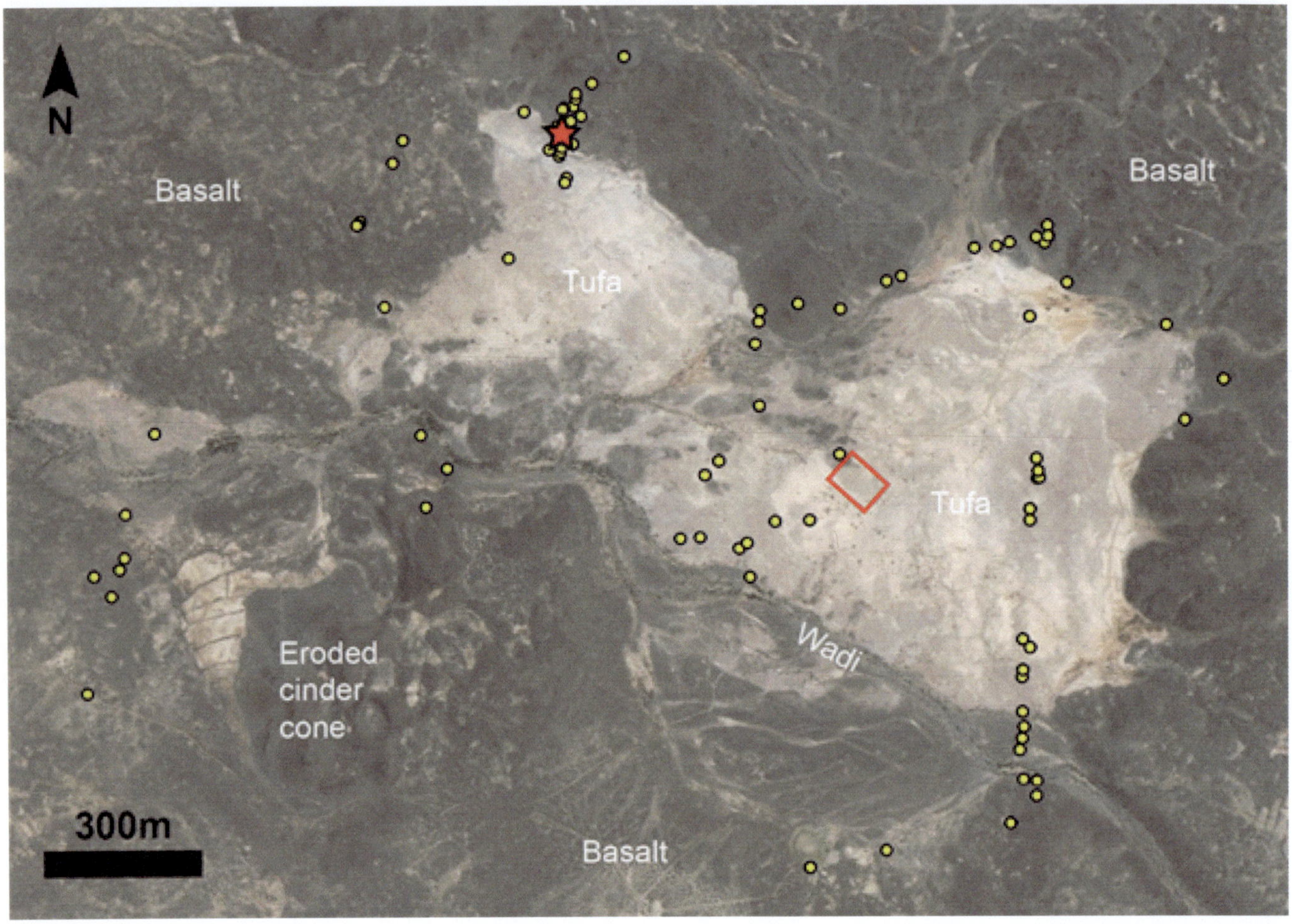

Fig. 6 An aerial view of the Wadi Dabsa showing the main geomorphological units and the locations of observed artefacts in 2015 and 2017. Red box indicates location of L0106/L0130 grids (see Fig. 7), and red star marks find of large handaxe reported in Foulds et al. (2017). Satellite Imagery © CNES/Astrium, imagery date 15/11/2015

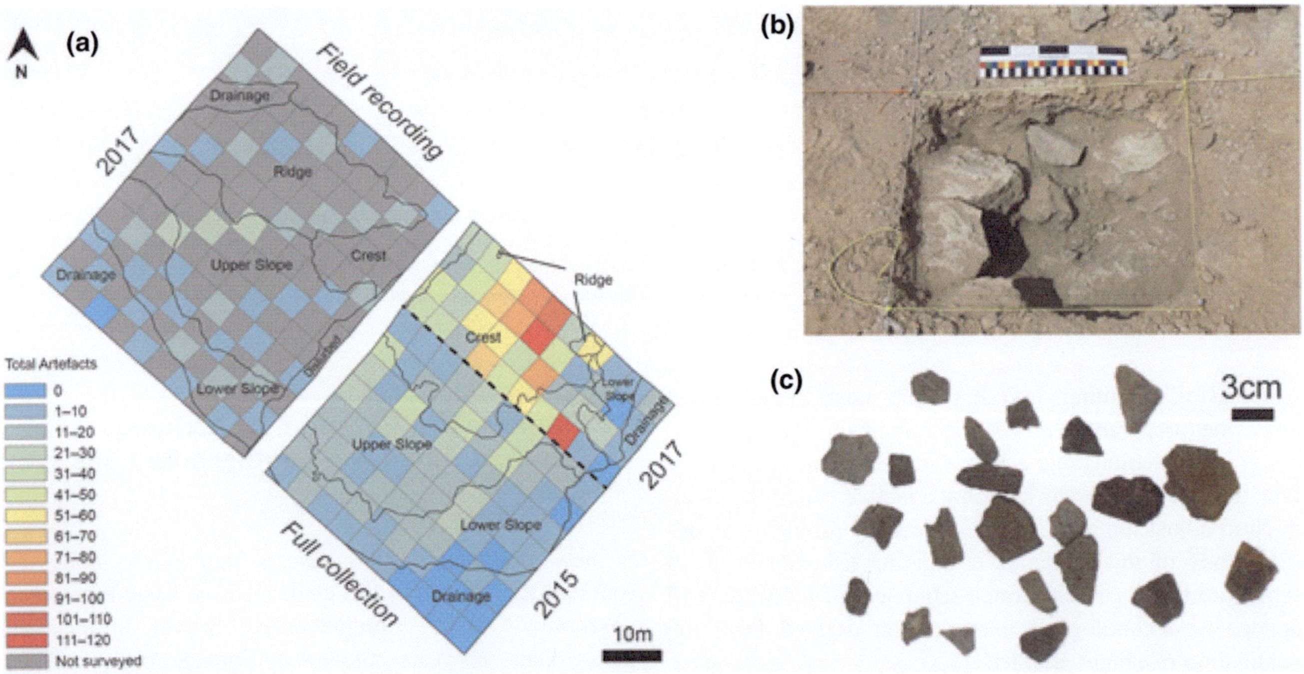

Fig. 7 Artefact survey grid at Wadi Dabsa (L0106/L0130). **a** artefact counts across the grid, showing geomorphological units; **b** basalt artefacts uncovered in situ in Test Pit 1, Photo: G. Bailey; **c** basalt shatter flakes from test pit 2

Fig. 8 ESA handaxe encased in tufa recovered from surface of L0130. Photo: H. Robson

archaeological material of Palaeolithic age. It has identified extensive concentrations of sites close to lava flows at Abu Arish, Sabya and especially in the Harrat al Birk. It has also been able to show, at Dhahaban Quarry, that there are still Palaeolithic artefacts embedded in preserved marine deposits along the coastline, even though the complexity of their depositional history makes it difficult to prove directly the earlier claims for the exploitation of Red Sea marine resources by hominins. Finally, the quality of behavioural information still recoverable from the lithic artefacts of a largely surface deposit at Wadi Dabsa is convincing evidence that considerable future research potential for understanding hominin behaviour remains in the surviving sedimentary deposits along the Red Sea.

Realising the full research potential of other remaining archaeological localities will be challenging. The typological descriptions that have been given to this material are too broad as evidence of date, particularly when more than half of the localities contain palimpsests of different-age materials. A comprehensive programme of dating associated sediments will be necessary. The clear association between Palaeolithic artefacts and lava flows, alongside the

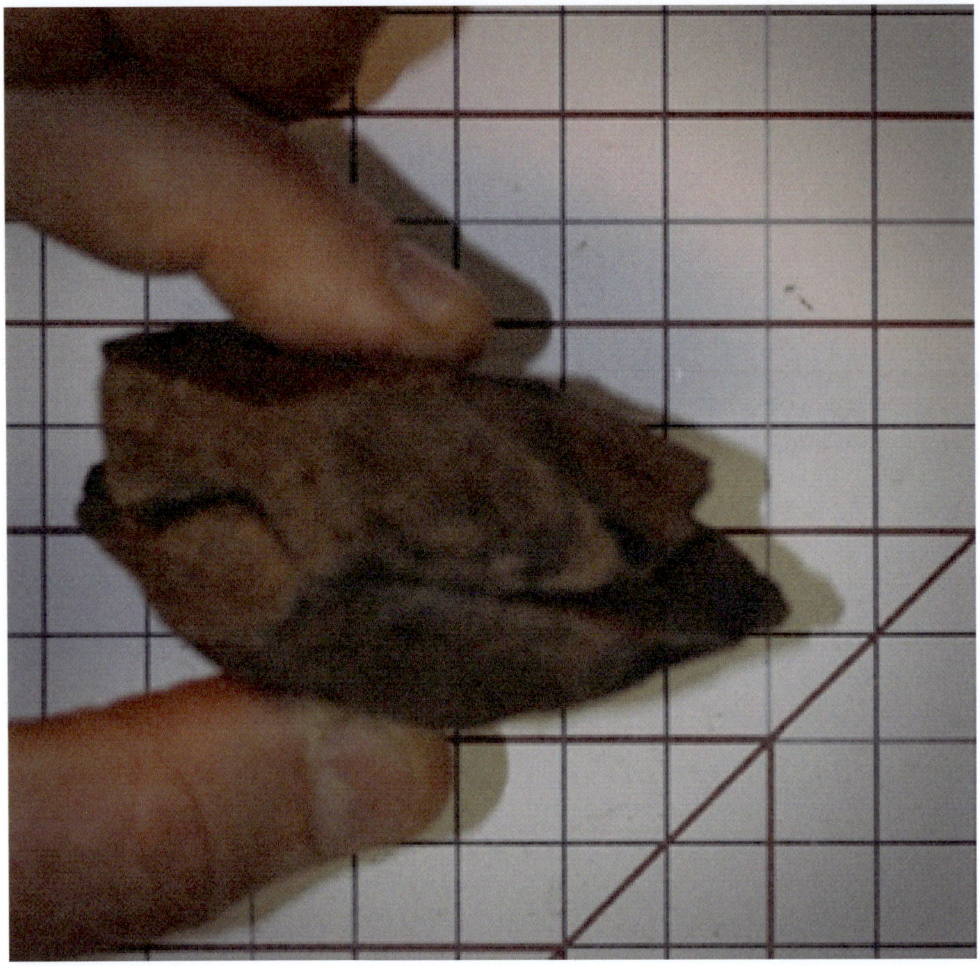

Fig. 9 A prepared core flake (upper) fitting into the negative flake scar of a prepared core (lower) both collected at the Wadi Dabsa indicating the continuity of technological activity at this site. The grid squares behind are 1 × 1 cm

recognition that volcanic activity might have generated new lava flows until quite late in the Pleistocene, indicates that a programme of dating of the lava flows is necessary to provide a series of dates to bracket the upper age of certain localities. Likewise, the presence of tufa deposits and marine deposits with associated archaeological remains that may be dated by OSL (See Sanderson and Kinnaird this volume) also offers great potential for narrowing down the time range of hominin behaviour.

The considerable time depth and geographical scale of hominin life and the spatial distribution of different activities means that we require archaeological assemblages from many different localities if we wish to interpret the dispersal, migration and colonisation of hominins into Arabia. Landscape surveys for Palaeolithic materials have demonstrated that there is genuine potential for this evidence in the southwestern Red Sea. The original Comprehensive Survey began as an essential first step to map the surviving heritage assets across the Kingdom that were faced with potential destruction through development. There are, however, likely to be more sedimentary deposits, as yet undiscovered, with research potential. It is these deposits that need proper mapping and assessment. The recent damage to the deposits at Dhahaban Quarry confirms that the threat from development remains as great as ever. In the case of the Red Sea coast, there is still a (short?) window of opportunity made possible by the availability of aerial imagery and digital mapping to map these deposits, with the potential to inform future development proposals.

Acknowledgements We thank HRH Prince Sultan bin Salman bin Abdul Aziz, President of the Saudi Commission for Tourism and National Heritage (SCTH) and Dr. Ali Al-Ghabban, Vice-President, for granting permission for the archaeological fieldwork reported here (or on which this research is based) and for their ongoing support. We also thank the President of the Saudi Geological Survey, Dr. Zohair Nawab and his staff, in particular Dr. Najeeb Rasul, for additional support and for their invitation to participate in the Jeddah Workshop.This work was supported between 2011 and 2015 by the European Research Council through ERC Advanced Grant 269586 'DISPERSE: Dynamic

Landscapes, Coastal Environments and Human Dispersals'. Fieldwork from 2015 to 2017 was funded by grants from the British Academy (Arthur Reckitt Fund), the Gerald Averay Wainwright Fund for Near Eastern Archaeology at the University of Oxford, and the British Foundation for the Study of Arabia, with additional funding from the European Union's Horizon 2020 research and innovation programme under the Marie Skłodowska-Curie grant agreement No. 660343, "SURFACE: Human-Landscape-Interactions and Global Dispersals: The Surface Record of Palaeolithic Arabia". This is DISPERSE contribution number 41.

References

Adams RMcC, Parr P, Ibrahim M, Al-Mughannum AS (1977) Saudi Arabian archaeological reconnaisance 1976. The preliminary report on the first phase of the comprehensive survey program. Atlal 1:21–40

Alsharekh AM, Bailey GN (eds) (2013) Coastal prehistory in southwest Arabia and the Farasan Islands: 2004–2009 field investigations. Saudi Commission for Tourism and Antiquities, Riyadh

Armitage SJ, Jasim SA, Marks AE, Parker AG, Usik VI, Uerpmann H-P (2011) The southern route "Out of Africa": evidence for an early expansion of modern humans into Arabia. Science 331:453–456

Bailey GN (2007) Time perspectives, palimpsests and the archaeology of time. J Anth Archaeol 26(2):198–223

Bailey GN (2009) The Red Sea, coastal landscapes, and hominin dispersals. In: Petraglia MD, Rose JI (eds) The evolution of human populations in Arabia: paleoenvironments, prehistory and genetics. Springer, Netherlands, pp 15–138

Bailey GN, King GCP (2011) Dynamic landscapes and human dispersal patterns: tectonics, coastlines, and the reconstruction of human habitats. Quatern Sci Rev 30:1533–1553

Bailey GN, Flemming NC, King GCP, Lambeck K, Momber G, Moran LJ, Al-Sharekh AM, Vita-Finzi (2007) Coastlines, submerged landscapes, and human evolution: the Red Sea Basin and the Farasan Islands. J Island Coast Archaeol 2(2):127–160

Bailey GN, Devès MH, Inglis RH, Meredith-Williams MG, Momber G, Sakellariou D, Sinclair AGM, Rousakis G, Al Ghamdi S, Alsharekh AM (2015) Blue Arabia: palaeolithic and underwater survey in SW Saudi Arabia and the role of coasts in Pleistocene dispersals. Quat Int 382:42–57

Bailey GN, Meredith-Williams M, Alsharekh A, Hausmann N (this volume) The archaeology of Pleistocene coastal environments and human dispersals in the Red Sea: insights from the Farasan Islands

Binford LR (1982) The archaeology of place. J Anth Archaeol 1(1):5–31

Boëda E (1995) Levallois: a volumetric construction, methods, a technique. In: Dibble HL, Bar-Yosef O (eds) The definition and interpretation of Levallois technology. Prehistory Press, Madison, Wisconsin, pp 41–68

Brunet M (2010) Two new Mio-Pliocene Chadian hominids enlighten Charles Darwin's 1871 prediction. Phil Trans R Soc B 365:3315–3321

Brunet M, Guy F, Pilbeam D, Lieberman Likius A, Mackaye HT, Ponce de León MS, Zollikofer CP, Vignaud P (2005) New material of the earliest hominid from the Upper Miocene of Chad. Nature 434:752–755

Clarke DL (1973) Archaeology: the loss of innocence. Antiquity 47:6–18

Crassard R, Hilbert YH (2013) A Nubian complex site from central Arabia: implications for Levallois taxonomy and human dispersals during the Upper Pleistocene. PLoS One 8(7):e69221. https://doi.org/10.1371/journal.pone.0069221

Dabbagh A, Emmermann R, Hötzl H, Jado AR, Lippolt HJ, Kollman W, Moser H, Rauert W, Zötl JG (1984) The development of Tihamat Asir during the Quaternary. In: Jado AR, Zötl JG (eds) Quaternary period in Saudi Arabia volume 2: sedimentological, hydrogeological, hydrochemical, geomorphological, geochronological and climatological investigations in Western Saudi Arabia. Springer, Vienna, pp 150–173

Debenath A, Dibble H (1994) Handbook of palaeolithic typology: lower and middle palaeolithic of Europe, vol 1. Philadelphia, University Museum of Archaeology and Anthropology

Delagnes A, Tribolo C, Bertran P, Brenet B, Crassard R, Jaubert J, Khalidi L, Mercier N, Nomade S, Peigne S, Sitzia L, Tournepiche JF, Al-Halibi M, Al-Mosabi A, Macchiarelli R (2012) Inland human settlement in southern Arabia 55,000 years ago. New evidence from the Wadi Surdud Middle Paleolithic site complex, western Yemen. J Hum Evol 63:452–474

Devès MH, Inglis RH, Meredith-Williams MG, Al Ghamdi S, Alsharekh A, Bailey GN (2013) Palaeolithic survey in southwest Saudi Arabia: methodology and preliminary results. Adumatu 27:7–30

Foulds F, Shuttleworth A, Sinclair A, Alsharekh A, Al Ghamdi S, Inglis RH, Bailey G (2017) A large handaxe from Wadi Dabsa and early hominin adaptations within the Arabian Peninsula. Antiquity 91(360):1421–1434

Garcia T, Féraud G, Falguères C, de Lumley H, Perrenoud C, Lordkipanidze D (2010) Earliest human remains in Eurasia: new ^{40}Ar/^{39}Ar dating of the Dmanisi hominid-bearing levels, Georgia. Quat Geochronol 5(4):443–451

Groucutt HS, Petraglia MD (2012) The prehistory of the Arabian Peninsula: deserts, dispersals, and demography. Evol Anth 21 (3):113–125

Guiducci D, Burke A (2016) Reading the landscape: legible environments and hominin dispersals. Evol Anth 25:133–141

Holdaway S, Fanning PC (2014) Geoarchaeology of Aboriginal Landscapea in Semi-arid Australia. CSIRO Publishing, Collingwood

Inglis RH, Sinclair A, Shuttleworth A, Alsharekh A, Al Ghamdi S, Devès M, Meredith-Williams MG, Bailey GN (2014) Investigating the Palaeolithic landscapes and archaeology of the Jizan and Asir regions, southwest Saudi Arabia. Proc Sem for Arab Stud 44:193–212

Inglis RH, Foulds F, Shuttleworth A, Alsharekh AM, Al Ghamdi S, Sinclair AG, Bailey GN (2015) The Palaeolithic occupation of the Harrat Al Birk. Preliminary Report on the 2015 Fieldwork in Asir Province, Southwest Saudi Arabia. University of York/Saudi Commission for Tourism and Antiquities. http://eprints.whiterose.ac.uk/84382/

Inglis RH, Sinclair AG, Alsharekh AM, Barfod DN, Chang HC, Fanning PC, Al Othaibi DT, Robson HK, Shuttleworth A, Stone A, Bailey GN (2017) Preliminary report on UK-Saudi 2017 fieldwork at Wadi Dabsa, Asir Province, Saudi Arabia. University of York/Saudi Commission for Tourism and National Heritage. http://eprints.whiterose.ac.uk/119267/

Inglis RH, Bosworth W, Rasul NMA, Al Saeedi A, Bailey GN (this volume) Investigating the palaeoshorelines and coastal archaeology of the Southern Red Sea

Jennings RP, Singarayer J, Stone EJ, Krebs-Kanzow U, Khon V, Nisancioglu KH, Pfeiffer M, Zhang X, Parker P, Parton A, Groucutt HS, White TS, Drake NA, Petraglia MD (2015) The greening of Arabia: multiple opportunities for human occupation of the Arabian Peninsula during the Late Pleistocene inferred from an ensemble of climate model simulations. Quat Int 382:181–199

Kübler S, Devès MH, Bailey GN, King GCP (this volume) Tectonic geomorphology and soil edaphics as controls on animal migrations and human dispersal patterns

Lambeck K, Purcell A, Flemming NC, Vita-Finzi C, Alsharekh AM, Bailey GN (2011) Sea level and shoreline reconstructions for the Red Sea: isostatic and tectonic considerations and implications for hominin migration out of Africa. Quat Sci Rev 30(25–26):3542–3574

Momber G, Sakellariou D, Bailey GN, Rousakis G (this volume) The multi-disciplinary search for underwater archaeology in the southern Red Sea

Parker AG (2009) Pleistocene climate change in Arabia: developing a framework for hominin dispersal over the last 350 ka. In: Petraglia MD, Rose JI (eds) The evolution of human populations in Arabia. Springer, Dordrecht, Netherlands, pp 39–49

Petraglia MD (2003) The Lower Palaeolithic of the Arabian Peninsula: occupations, adaptations, and dispersals. J World Prehist 17:141–179

Petraglia MD, Alsharekh A (2003) The Middle Palaeolithic of Arabia: implications for modern human origins, behaviour and dispersals. Antiquity 77(298):671–684

Petraglia MD, Rose JI (eds) (2009) The evolution of human populations in Arabia. Springer, Dordrecht, Netherlands

Petraglia MD, Alsharekh AM, Crassard E, Drake NA, Groucutt H, Parker AG, Roberts RG (2011) Middle Paleolithic occupation on a marine isotope stage 5 lakeshore in the Nefud Desert, Saudi Arabia. Quat Sci Rev 30(13–14):1555–1559

Petraglia MD, Alsharekh AM, Breeze P, Clarkson C, Crassard R, Drake NA, Groucutt HS, Jennings R, Parker AG, Parton A, Roberts RG, Shipton C, Matheson C, Al-Omari A, Veall M-A (2012) Hominin dispersal into the Nefud Desert and Middle Palaeolithic settlement along the Jubbah palaeolake, northern Arabia. PLoS One 7(11):e49840

Potts DT (1998) The Gulf Arab states and their archaeology. In: Meskell L (ed) Archaeology under fire: nationalism, politics and heritage in the Eastern Mediterranean and the Middle East. Routledge, London, pp 189–199

Rossignol J, Wandsnider L (eds) (1992) Space, time and archaeological landscapes. Plenum, New York

Sakellariou D, Rousakis G, Panagiotopoulos I, Morfis I, Bailey GN (this volume) Geological structure and Late Quaternary geomorphological evolution of the Farasan Islands continental shelf, south Red Sea, SW Saudi Arabia

Sanderson D, Kinnaird I (this volume) Optically stimulated luminescence dating as a geochronological tool for Late Quaternary sediments in the Red Sea region

Senut B, Pickford M, Gommery D, Mein P, Cheboi K, Coppens Y (2001) First hominid from the Miocene (Lukeino Formation, Kenya). Compt Rend de l'Acad de Sci 332(2):137–144

Sinclair A, McCraith L, Nelson E (2003) Understanding hominid landscapes at Makapansgat, South Africa. In: Mitchell P, Haour A, Hobart J (eds) Researching Africa's past: new contributions from British archaeologists. Oxford University School of Archaeology, Oxford, pp 11–24

Sinclair A, Inglis RH, Shuttleworth A et al (in prep) Technological errors, practical experience and informal learning in the Middle Stone Age at Wadi Dabsa, Saudi Arabia

Stern N (1993) The structure of the Lower Pleistocene archaeological record. Curr Anth 34(3):201–225

Suwa G, Asfaw B, Kono RT, Kubo D, Lovejoy CO, White TD (2009) The Ardipithecus ramidus skull and its implications for hominid origins. Science 326(5949):68e1–7

Van Peer P (1992) The Levallois reduction strategy. Prehistory Press, Madison, Wisconsin

Walter RC, Buffler RT, Bruggemann JJ, Guillaume MMM, Berhe SM, Negassi B, Libsekal Y, Cheng H, Edwards RL, von Gosel R, Neraudeau D, Gagnon M (2000) Early human occupation of the Red Sea coast of Eritrea during the last interglacial. Nature 405:65–69

Whalen N, Pease D (1990) Variability in developed Oldowan and Acheulean bifaces of Saudi Arabia. Atlal 13:43–48

Whalen N, Killick A, James N, Morsi G, Kamal M (1981) Saudi Arabian archaeological reconnaissance: preliminary report on the western province. Atlal 5:43–58

Whalen N, Sindi H, Wahida G, Siraj-Ali JS (1983) Excavation of Acheulean sites near Saffaqah in ad-Dawadmi 1402–1982. Atlal 7:9–19

Whalen N, Davis WP, Pease D (1989) Early Pleistocene migrations into Saudi Arabia. Atlal 12:59–75

Winder I, Devès M, King G, Bailey G, Inglis RH, Meredith-Williams MG (2015) Evolution and dispersal of the genus homo: a landscape approach. J Hum Evol 87:48–65

Zarins J, Whalen N, Ibrahim M, Al-Jawad Mursi A, Khan M (1980) Comprehensive archaeological survey program. Preliminary report on the Central and Southwestern provinces survey. Atlal 4:9–36

Zarins J, Al-Jawad Murad A, Al-Yish KS (1981) The comprehensive archaeological survey program: the second preliminary report on the southwestern province. Atlal 5:5–42

Investigating the Palaeoshorelines and Coastal Archaeology of the Southern Red Sea

Robyn H. Inglis, William Bosworth, Najeeb M. A. Rasul, Ali O. Al-Saeedi, and Geoff N. Bailey

Abstract

Numerous palaeoshoreline features including coral platforms, beachrock and wave-cut notches are present on the Red Sea coastline of SW Saudi Arabia and on the Farasan Islands. Some are associated with prehistoric archaeological material, which has been the focus of ongoing archaeological investigations over the past decade. Dating and interpretation of these features are therefore of considerable interest and relevance to the deep history of human coastal adaptation and colonization in a key zone for the understanding of early human expansion out of Africa, as well as to the study of relative sea-level changes and tectonic movements. This chapter provides details of a field survey carried out in 2014 and presents new information on the location, geological setting, geochronological sampling and archaeological associations of these palaeoshoreline features. The results of dating are still awaited, so that some of our interpretations are still hypotheses in need of further testing. At this stage, it is clear that the most prominent shoreline features on the mainland coast are at elevations similar to those dated elsewhere in the Red Sea as belonging to MIS 5e, and that in at least one exposure Middle Stone Age artefacts can be stratigraphically linked with this period of high sea level. On the Farasan Islands, coral platforms have undergone more variable and localised rates of movement associated with salt tectonics. We set out the field data in support of these interpretations and consider their wider archaeological and tectonic implications.

R. H. Inglis (✉) · G. N. Bailey
Department of Archaeology, University of York, The King's Manor, YO1 7EP York, UK
e-mail: robyn.inglis@york.ac.uk

R. H. Inglis
Department of Environmental Sciences, Macquarie University, North Ryde, Sydney, NSW 2109, Australia

W. Bosworth
Apache Egypt Companies, 11 Street 281, New Maadi, Cairo, Egypt

N. M. A. Rasul
Center for Marine Geology, Saudi Geological Survey, Jeddah, Saudi Arabia

A. O. Al-Saeedi
Survey Department, Saudi Geological Survey, Jeddah, Saudi Arabia

G. N. Bailey
College of Humanities, Arts and Social Sciences, Flinders University, GPO Box 2100 5001 Adelaide, SA, Australia

N. M. A. Rasul and I. C. F. Stewart (eds.), *Geological Setting, Palaeoenvironment and Archaeology of the Red Sea*,
https://doi.org/10.1007/978-3-319-99408-6_25

1 Introduction

The purpose of this chapter is to present the results of a field excursion conducted in November and December 2014 to examine palaeoshoreline features in the southern Red Sea along the mainland coastline of Asir Province in SW Saudi Arabia and on the Farasan Islands in Jizan Province (Fig. 1), and to set out the intellectual and scientific issues that have informed our fieldwork.

Cemented coral platforms and beach deposits indicating the presence of palaeoshorelines are widely distributed above present sea level around the Red Sea, and some have been recorded below sea level (Faure et al. 1980; Dullo 1990; Hoang and Taviani 1991; El Moursi et al. 1994; Gvirtzman 1994; Hoang et al. 1996; Plaziat et al. 1998, 2008; Walter et al. 2000; Lambeck et al. 2011; Manaa et al. 2016; Bosworth et al., this volume; Sakellariou et al., this volume). Those on land mostly represent earlier periods of high sea level during Pleistocene interglacials. They occur at varying elevations above present sea level, up to 100 m in the Gulf of Aqaba, reflecting differences of age, variations in eustatic sea level in different interglacials, and vertical crustal movements associated with tectonic processes of various sorts.

Similar palaeoshoreline features are present in our region. In particular, exposures of cemented coral platforms or terraces and beach deposits at least 4 m above the present sea level, and assumed to relate to higher sea-level stands of the

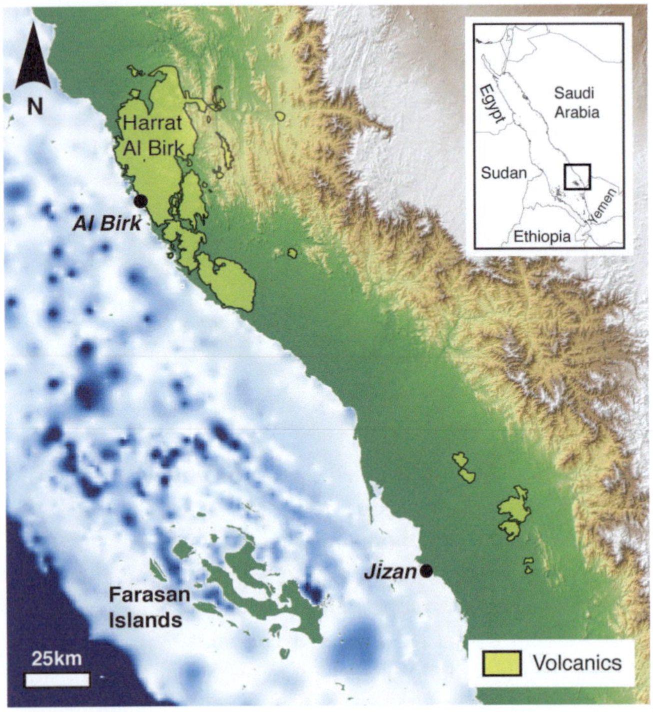

Fig. 1 The southwestern Saudi Arabian Red Sea coastline showing the Harrat al Birk and Farasan Islands. Area of Fig. 4 outlined in black. Pale blue areas show the continental shelf exposed when sea level is lowered to 120 m. The dark blue circular patches on the shelf are deep depressions resulting from solution of Miocene evaporites. Data from GEBCO 08 and SRTM 4.1

Last Interglacial period (MIS 5e) or older, have been noted during earlier archaeological expeditions in the region (Zarins et al. 1980, 1981; Alsharekh and Bailey 2013; Bailey et al. 2007a, b; Inglis et al. 2013, 2014a, b). However, they had not received systematic or extensive geological investigation or geochronological sampling before our 2014 expedition. As such, the region is one of the least well studied in this respect compared to most other coastlines of the Red Sea. It is also the focus of ongoing research on coastal prehistory (Bailey et al. 2012, 2015; Devès et al. 2013; Inglis et al. 2013, 2014a, b, 2015; Meredith-Williams et al. 2013, 2014; Bailey and Alsharekh, in press; Bailey et al., this volume; Hausmann et al., this volume; Sinclair et al., this volume), and this provides an additional incentive for more detailed geological research. Moreover, many coastal features are under threat of destruction by intensifying pressures of development, road-building, other infrastructural projects and tourism, adding urgency to the collection of field data.

For most published studies, the primary interest of these palaeoshoreline features is their contribution to an understanding of sea-level change, and that is often taken to be the main goal of their investigation. Together with evidence from deep sea cores (Siddall et al. 2003; Rohling et al. 2013), this makes the Red Sea an important study region for the wider understanding of global sea level change and its relationship to climate change. However, palaeoshoreline features also have relevance to a wider range of geological, geoarchaeological and archaeological interests, and it is important to be clear about the different aims, data requirements and assumptions that different disciplines bring to shoreline studies. We identify here broadly three different themes:

1. Measurement and modelling of sea-level change in relation to global climate change, often with the ultimate goal of producing better predictions of future sea-level change.
2. Understanding of regional tectonic effects associated with rifting, plate motions and other processes of crustal deformation and their contribution to the wider understanding of Earth deformation processes and geodynamics
3. Charting the role of coastlines in the global expansion of human populations out of Africa during the Pleistocene period and the deeper history of human interest in coastal and marine resources and their exploitation.

These three themes are, of course, interlinked. Shoreline features cannot be interpreted as sea-level indicators without taking into account the interplay between eustatic variations of sea level and vertical crustal movements associated with rifting, plate motions, changes of mass loading and other tectonic effects. This is especially important in the Red Sea, which has a complex geotectonic history. Palaeoshorelines in their turn provide sensitive measures of tectonic movements that can be calibrated against independent evidence of eustatic sea level change. Eustatic changes of sea level also have effects on the mass loading of the Earth's crust in coastal regions and on the erosion or submergence of geological features. In relation to archaeological issues, both tectonic and sea-level changes alter the physical landscape setting within which past human populations made their living; they can alter the nature and accessibility of the plants, animals and water supplies available for subsistence, pathways of movement and communication, and the preservation and visibility of archaeological evidence. Geological studies can also provide chronological control on the archaeology. Geologists and sea-level specialists in their turn are interested in the potential of archaeological data to provide independent age-constraints on geological processes, and are increasingly interested in the relevance of their research to the study of human evolution and the scale of human activities (e.g., Lambeck et al. 2011; Rohling et al. 2013; Benjamin et al. 2017; Kübler et al., this volume).

Shoreline features are viewed differently in each of these types of investigation—as sea-level indicators, as measures of tectonic movement, and as evidence of human exploitation of shorelines, respectively. The measurement accuracy differs accordingly. For sea-level studies, measurements of the highest possible accuracy are required—metre to decimetre scale—with clear specification of measurement errors and other sources of uncertainty, and calibration to a uniform spatial and temporal framework. This is to facilitate differentiation of eustatic sea levels in different interglacials or interglacial sub-stages that may differ by only a small amount, and to facilitate interregional and global comparisons. For tectonic studies, a coarser scale—metres to tens of metres—may be sufficient to distinguish between stable conditions and long-term vertical movements. For archaeological purposes, the absolute height of a shoreline is less important than stratigraphic evidence that the archaeological material can be directly associated with an elevated shoreline feature at the time when sea level was high, rather than representing material deposited on the surface at a later date. For example, stone tools found on the surface of Last Interglacial coral terraces could have been deposited during the subsequent glacial stage, when sea level was far lower than present and the shoreline many tens of kilometres distant.

Despite these inter-relationships and the widely acknowledged virtues of interdisciplinary research, the investigation of these three themes is often pursued by scientific communities working within separate disciplinary or subdisciplinary compartments with poor communication across the boundaries between them or even mutual incomprehension. As a multi-disciplinary group with combined expertise in geoarchaeology, marine science, coastal prehistory, geophysics and tectonic geology, we are interested in all three of these themes and the relationships between them, and our approach to field surveying has been shaped accordingly.

In this chapter, we set out the wider geological and archaeological issues that inform our fieldwork, define our field objectives and methods, present our preliminary results, and discuss their wider implications.

2 Geological and Archaeological Issues

2.1 Geological Context

The Red Sea has a geotectonic setting involving the interaction of several different processes, including: continental rifting involving crustal extension, normal faulting and volcanism; seafloor spreading; plate motions; isostatic changes of mass loading; and more localised tectonic movements resulting from the mobility of Miocene salt deposits, also referred to as evaporites or halites. The basin began to take its present form as a continental rift at about 30 Ma as a northward extension of the East African Rift system, and this whole rift structure was probably triggered at least in part by the eruption of the 4 km-thick flood basalts that occurred over the Ethiopian hot spot—a swelling and thinning of the Earth's crust over a hot plume rising from deep within the Earth's mantle—acting on pre-existing weaknesses in the lithosphere (Hubert-Ferrari et al. 2003; Bonatti et al. 2015; Bosworth 2015). The central Red Sea basin has the typical features and asymmetric cross-section of a rift, with the main faulting running through the centre of the basin. Ongoing extension and normal faulting have resulted in progressive uplift of the rift flanks and deepening of the rift floor. The footwall of the main rift is on the eastern margin, comprising the western Arabian escarpment, which reaches a maximum height of ~3000 m above sea level (asl) in the south; the lower, hanging wall is on the western margin of the Red Sea; and the deepest part of the axial trough in the centre of the basin is ~2800 m below present sea level.

By 19 Ma, seafloor spreading had commenced in the Gulf of Aden. The Arabian Peninsula continued separating from the African Plate, moving to the north, with extension in the Afar region of Ethiopia and the widening of the 'Proto Red Sea'. The Red Sea remained isolated from the world oceans, apart from intermittent incursions of seawater from the Mediterranean, and filled with thick deposits of evaporite because of high rates of evaporation in a closed basin. By 5 Ma, a connection with the Indian Ocean was established, with the onset of seafloor spreading and formation of oceanic crust within the Red Sea Basin. The whole process has been accompanied by episodes of magmatism on land on the Arabian side with extensive areas of volcanic cinder cones and basaltic lava flows, notably during the Miocene (>5 Ma) and the Quaternary (<2 Ma and extending in some regions into the Holocene), although it remains unclear to what extent these originate from shallow sources in the mantle associated with local extension and faulting or from deeper mantle sources associated with the Ethiopian plume.

Many of the details of these processes including their timing and the nature of ongoing deformation are still a matter for debate. An active oceanic spreading centre is present along the basin axis, with the Arabian Plate separating from Africa/Nubia at ~1.7 cm/yr (ArRajehi et al. 2010), and continuing to slide along the large transform faults that define its western and eastern boundaries and to collide with the Eurasian continent to the north. However, there is debate concerning when the transition from continental rifting to seafloor spreading began (Girdler and Styles 1974; Girdler and Whitmarsh 1974; Cochran 1983; Coleman 1993; Cochran and Karner 2007; Mohriak 2015). Plate reconstructions and extrapolation of spreading rates suggest that the tectonic regime has been fairly constant over about the past 11 Myr (McQuarrie et al. 2003; Reilinger et al. 2015).

Despite considerable understanding of the nature of the plate boundary running through the centre of the southern Red Sea, there remain fundamental questions about both the long- and short-term response of the continental margins to this relatively young oceanic rifting. There is no question that the uplifted margins of the southern Red Sea rift underwent a significant period of uplift and denudation during the Early Miocene after the commencement of continental rifting (Bohannon 1986; Bohannon et al. 1989; McGuire and Bohannon 1989; Bosworth and Stockli 2016). However, it remains unclear whether uplift is still occurring and, if so, to what extent it is the result of faulting and extension associated with rifting and ocean spreading or of thermal expansion linked to the Ethiopian plume. The stark relief of the Arabian Escarpment, with peaks of over 3000 m (Spohner and Oleman 1986; Coleman 1993) might suggest ongoing uplift. Data discussed below indicate that some parts of the coastline appear to have been stable during the Late Pleistocene. However, such data need to be interpreted in the light of models of isostatic adjustment (Lambeck et al. 2011). In other regions, the localized effects of plate motions have resulted in significant vertical movement (tens of metres within the Quaternary period), particularly in the north along the Dead Sea transform fault (the Sinai Peninsula and Gulf of Aqaba). Other areas have been subjected to local instability with faulting and uplift resulting from the movement of deposits of Miocene evaporite, notably the Farasan Islands (Bosence et al. 1998; Dullo and Montaggioni 1998).

Providing new constraints on the evidence for vertical movements of the southern Arabian margin will help address these important issues. Better understanding of these tectonic processes also has obvious relevance in correcting the derivation of sea-level position from palaeoshoreline features, and in addition it has an important role to play in the interpretation of archaeological evidence for the early human colonization and occupation of the Arabian Peninsula.

2.2 Archaeological Context

The coastal regions of the Red Sea Basin play a key role in understanding the global dispersal of human populations from Africa during the Pleistocene (Bailey et al. 2007a, b). At least two major episodes of dispersal are recognised, the first of *Homo erectus* or *H. ergaster*, at about 1.8 Ma, using a Lower Palaeolithic/Early Stone Age (ESA) technology, and a later dispersal of *H. sapiens* populations with Middle Palaeolithic/Middle Stone Age (MSA) technology at some time between 190 ka (the earliest dated appearance of anatomically modern *H. sapiens* in Africa) and 60 ka (the arrival of *H. sapiens* in Australia). It has generally been assumed that the main pathway of expansion was via the Nile-Sinai land route in the north, but there is growing interest in the possibility that human populations were able to cross at the southern end in the vicinity of the Hanish Sill and the Bab al Mandab during periods of low sea level. This idea has found particular favour in discussion of *H. sapiens* dispersal, along with the hypothesis that their ability to make sea crossings was facilitated by new adaptations in the exploitation of marine resources and the use of simple sea craft, but use of the southern route by earlier populations cannot be excluded (Beyin 2006; Bailey 2009, 2015; Lambeck et al. 2011; Mellars et al. 2013; Erlandson and Braje 2015). Certainly, the sea channel in the south would have been very narrow for long periods of the Pleistocene and the coastal landscapes on the Arabian side would have benefited from the orographic rainfall attracted by the highest parts of the Arabian Escarpment. These relatively humid conditions would have provided attractive environments for human occupation during all climatic phases of the Pleistocene compared to the more arid interior (Bailey et al. 2015).

These contrasting opinions have recently crystallised into a debate between two contrasting models of population expansion for *H. sapiens* (Groucutt et al. 2015; Petraglia et al., this volume). One, the 'Late Dispersal/Coastal Highway' model, draws on phylogenetic trees constructed from similarities in the mitochondrial DNA of modern populations in Asia and Africa, and on comparisons between the South African and South Asian archaeological records, to propose a rapid dispersal of populations adapted to marine resources and coastal environments around the Indian Ocean rim at 60–50 ka (Macaulay et al. 2005; Mellars et al. 2013). The second, the 'Early Dispersal/Green Arabia' model, highlights ambiguities in the DNA evidence and emphasises the evidence for the extension of grasslands and lakes into the interior of the Arabian Peninsula during a wetter climatic interval in early MIS 5 and the presence of MSA archaeological sites in lake edge settings dating to the same time interval at about 130–120 ka (Parker 2009; Armitage et al. 2011; Petraglia et al. 2011, this volume; Rose et al. 2011; Rosenberg et al. 2013). Of course, these models are not mutually exclusive and both earlier and later dispersals are possible using both interior and coastal pathways of expansion.

The major difficulty with the coastal model is the extreme rarity or absence of archaeological sites in coastal locations. For the Late Dispersal Model, this may be explained by the fact that sea levels were substantially lower at 60 ka and the relevant shorelines and sites are hidden underwater or destroyed. However, even on coastlines with steeply shelving offshore bathymetry and on Last Interglacial shorelines exposed above modern sea level, such sites are very rare, the notable exception being Abdur on the Eritrean shore of the Red Sea dated at 130 ka (Walter et al. 2000).

Absence of evidence is not decisive, and may reflect lack of discovery or a variety of taphonomic factors at the coast edge affecting the visibility and preservation of archaeological material (see Bailey et al., this volume). Evidence of Pleistocene coastal activity is very rare everywhere, and the best-known examples occur in sediments accumulated in coastal caves in South Africa and on various coastlines of the Mediterranean and Atlantic Europe, where the cave setting provides a protective environment for the preservation and easy discovery of dateable and stratified sediments with archaeological evidence (Erlandson 2001; Bailey et al. 2008; Jerardino 2017). In any case, these debates place a high premium on the search for Pleistocene archaeological sites in coastal settings, and beach deposits of Last Interglacial date (MIS 5e) are a prime target for investigation.

Coral terraces associated with earlier periods of high sea level of Last Interglacial date or earlier are especially prominent in the Harrat Al Birk region of the Asir Province, where they are found along many parts of the coastline in close association with an extensive outcrop of Pleistocene volcanic cones and lava fields (Dabbagh et al. 1984), and on the Farasan Islands, where they are a dominant land form and have clearly been subjected to differential warping and uplift by salt doming (McFadyen 1930). In both regions, archaeological material is associated with these coral terraces. In the Harrat Al Birk, MSA stone tools made on basaltic lava are located on the surface of many coral terraces as well as in locations further inland (Fig. 2; see also Sinclair et al., this volume). Major issues here are the chrono-stratigraphic links between the archaeology, the coral terraces and the basaltic lava flows, and more specifically the question of whether the stone tools can be stratigraphically or chronologically linked to the period of high sea level when the corals were forming.

On the Farasan Islands, some 3000 mid-Holocene shell middens radiocarbon-dated between 6.5–4.5 ka, including substantial mounds easily visible on satellite imagery, are located on the edge of coral platforms along the modern coastline. In places, these platforms are elevated some 2–3 m above the present sea level and have been undercut by marine erosion to form a well-defined notch (Fig. 3; Bailey et al. 2013a, this volume; Meredith-Williams et al. 2013, 2014; Alsharekh and Bailey 2013; Bailey and Alsharekh 2018). These are all cultural deposits where mollusc shells have accumulated as food debris. Some are shell scatters, while others have formed impressive mounds as much as 5 m thick resulting from repeated use of the same place over many decades or centuries for the processing of shellfood. Rare stone tools are associated with these shell deposits, including ones made from volcanic stone brought from the mainland, very rare potsherds, hearths, fish bones and occasional bones of gazelle. The cultural material and non-molluscan food remains are sparse, reflecting both the rapid accumulation and large volume of deposits composed mostly of shell debris, and the rarity of suitable raw material on the islands for making stone artefacts. Seasonality studies of the shells show that they were eaten during all periods of the year (Hausmann and Meredith-Williams 2017), but the vast quantities of discarded shells almost certainly

Fig. 2 Examples of Palaeolithic artefacts from the Harrat al Birk collected during January 2015 fieldwork: 1 Potential prepared core (MSA); 2 Small exhausted discoidal core (potentially MSA); 3 Cleaver (ESA); 4 Large flake/scraper (ESA). Photos: Frederick Foulds and Andy Shuttleworth, January 2015

Fig. 3 Shell mound in Janaba Bay, Farasan al Kabir, Farasan Islands. The shell mound sits on the edge of a coral platform, and the notch formed by marine erosion is clearly visible beneath the mound. Photo: Geoff Bailey, December 2014

over-represent the contribution of the molluscs as food, and other resources, particularly fish, were undoubtedly important.

In both regions, the archaeology provides a minimum age for the underlying coral platforms, but the gap in time between the archaeology and the formation of the underlying surface may be very large. Hence, the archaeology cannot be relied on to provide an age constraint on the geology, except at best a minimum date. For example, the coral platform on which the Farasan shell mounds are located is thought to be of Last Interglacial date even though the shell mounds are clearly mid-Holocene in age; MSA stone tools were in use over many tens of thousands or even hundreds of thousands of years. Dating of geological formations is more likely to help the archaeological interpretation rather than the reverse, but in all cases the question of stratigraphic association between geological deposits and archaeological material is critical to understanding the past use of coastal landscapes

3 Field Objectives and Methods

3.1 Field Objectives

Field objectives were as follows, to:

- Study the structural geology of coastal land forms and associated evidence of faulting, fracturing and volcanism.
- Locate fossil marine and beach deposits that represent earlier periods of high sea level.
- Measure the elevation of these deposits in order to track local and regional changes in the Earth's crust resulting from rifting and salt tectonics, and to provide new data on sea-level change.
- Collect samples for dating by Uranium-series (U-series), Amino Acid Racemization (AAR) or Argon-Argon ($^{40}Ar/^{39}Ar$) dating according to local circumstances.
- Relate the geological history of these coastal areas to the archaeological sites which are frequently found at the coast edge in association with marine deposits.

3.2 Definitions and Field Methods

Coral and marine terraces in the Harrat al Birk and the Farasan Islands were surveyed over seven days in November–December 2014. Sites for investigation were selected through examination of satellite imagery (e.g., Google Earth, Landsat GeoCover Mosaics) and knowledge of locations derived from previous field surveys by ourselves and others (Dabbagh et al. 1984; Bantan 1999; Inglis et al. 2014b). Locations visited were traversed by 4-wheel drive vehicles and on foot. Within the Farasan archipelago a small motorboat was also used to access the coastlines of several islands.

3.3 Palaeoshoreline Feature Description and Sampling

In this paper, we use the term 'beachrock' to refer to a lithified coastal deposit consisting of beach sediments that have become cemented through carbonate precipitation on the immediate shoreline (following Mauz et al. 2015). 'Coral terrace' refers to extant deposits of fossil coral that consist primarily of coral skeletons in growth position, with varying amounts of broken coral. Other types of marine deposits are discussed where relevant.

Most of the coral and beach deposits we examined are heavily cemented, and particular attention was devoted to the collection of samples of coral for U-series dating. The degraded nature of the coral material necessitated an intensive search for suitable in situ specimens that had undergone the least alteration. Moreover, the heavily cemented nature of the deposits required heavy-duty hammers and chisels and prolonged effort to extract suitable samples. Most of the coral we examined is unsuitable for dating because of its fibrous open structure and risk of diagenetic alteration and contamination. Many coral samples when broken open showed streaks of mineral staining, a sign of probable contamination. The most suitable material for dating is large coral heads with a dense and uniform structure, and we preferentially selected corals of this type occurring in growth position and containing little detritus or obvious diagenetic alteration. Weathered material was removed mechanically from the samples in laboratories at the Center for Marine Geology at the Saudi Geological Survey in Jeddah, and these were then sent to the Research School of Earth Sciences at the Australian National University for U-series dating.

Shells of large gastropods and the *Tridacna* clam were also collected and submitted to the Department of Chemistry, University of York, for AAR dating. Shell material of this type, as well as coral, is potentially amenable to AAR dating, and can provide reliable age estimates and measures of post-depositional alteration that can aid in assessing the results of U-series dating (Penkman et al. 2008; Hendy et al. 2012).

Basalt samples were removed from key exposures for $^{40}Ar/^{39}Ar$ dating to constrain the geological history and stratigraphic relationships of marine deposits and have been submitted to the USGS in Denver, USA. OSL dating of marine and aeolian sands at Dhahaban Quarry is also underway at the Scottish Universities Environmental Research Centre, East Kilbride (Sanderson and Kinnaird, this volume).

3.4 Feature Measurement and Correction to Elevation Above Mean Sea Level

The heights and locations of the surfaces of beachrock and coral terraces, and of dating samples collected from them, were recorded using a Trimble DGPS. We used the same equipment to make additional traverses and spot elevations of specific features, and to geo-reference the base stations of a fuller total station survey undertaken earlier in 2014 at Dhahaban Quarry (Inglis et al. 2014a). Each locality is designated by a sequential number and the date (e.g., '2014-12-05-5' is the fifth station surveyed on December 5, 2014). In some locations additional elevation data were also obtained by tape measure to sea level. In the Harrat al Birk region, all of these localities had been previously, or were subsequently, surveyed for archaeological artefacts, by the UK-Saudi DISPERSE project and assigned locality numbers (e.g., L0103), numbers that are also referenced here to aid cross-referencing between publications (Inglis et al. 2013, 2014b, 2015; Sinclair et al., this volume).

Each DGPS base station was set up using the sea-level on the nearest beach as the local datum, with the measurements from each station later adjusted to their position above mean sea level through reference to local tide data. According to Bruckner et al. (2011), the annual tidal range in the Red Sea today varies along its length, being highest in the north and the south (2.0 m and 1.4 m respectively), and lowest in the middle (0.5 m), but these are clearly general averages. The nearest sources of tidal data for Al Birk and the Farasan Islands are Al Qunfudah and Jizan, respectively. Here the maximum tidal range derived from daily measurements throughout 2014 is 1.76 m and 1.69 m respectively (data supplied by the General Commission for Survey of Saudi Arabia). We have used the tidal position closest to the time of our DGPS readings to measure the offset between the sea-level surface at the time of the reading and mean sea level (Table 1), and used this to correct our field measurements of palaeoshoreline features to heights above modern mean sea level. For example, a coral terrace measured in the field using Al Birk Station 2 (offset +0.58 m) at an elevation of 3.45 m on the Al Birk coastline represents an elevation of

Table 1 Correction factors for DGPS readings taken at different stations in the study area in order to convert elevations of palaeoshoreline features to elevations above modern mean sea level. The original elevations were measured above a local datum consisting of the sea surface at the time of measurement. The correction factors have been calculated by reference to the nearest available tide tables. See the notes below the table and the text for further explanation

Station Name	Date[a]	Time[a]	Sea-Level Datum (m)	Measurement Error (m)[b]	Tidal Station[c]	Tidal Position (m)[d]	Tidal Error[e]	Annual Max Tidal Range (m)	Mean Sea Level (m)[f]	Correction (m)[g]
Sulayn	30/11/2014	11.45	0	±0.1	Jizan	0.4402	±0.02	−0.10 to +1.59	0.75	−0.31
Farasan Station 2	01/12/2014	9.20	0	±0.03	Jizan	0.566	±0.01	−0.10 to +1.59	0.75	−0.18
Farasan Station 3	01/12/2014	11.10	−1.7	±0.05	Jizan	0.9018	±0.01	−0.10 to +1.59	0.75	+1.55
Dhahaban Station	03/12/2014	13.23	0	±0.03	Qunfudah	0.6092	±0.01	−0.803 to +0.975	0.08	+0.53
Al Birk Station 1	04/12/2014	16.23	0	±0.02	Qunfudah	0.492	±0.01	0.803 to +0.975	0.08	+0.41
Al Birk Station 2	04/12/2014	13.02	0	±0.01	Qunfudah	0.6609	±0.00	−0.803 to +0.975	0.08	+0.58
Al Birk Station 3	05/12/2014	12.51	0	±0	Qunfudah	0.659	±0.01	−0.803 to +0.975	0.08	+0.58
North of 'Amq	05/12/2014	15.04	0	±0.1	Qunfudah	0.5813	±0	−0.803 to +0.975	0.08	+0.50
Qamah	06/12/2014	09.51	0	±0.02	Qunfudah	0.5442	±0.01	−0.803 to +0.975	0.08	+0.46

[a]Time of sea-water surface measurement
[b]Range of variation in measurement of sea-water surface
[c]Location of nearest tidal gauge with hourly records of tidal position
[d]Estimate of water level to the nearest half hour based on 2014 tidal tables
[e]Allows for difference between estimated and actual tidal position at time of measurement
[f]Mid-point of the maximum annual tidal range
[g]Figure to be added to field elevations to give height above modern mean sea level

4.03 m above modern mean sea level. The measurements referred to throughout the text in the results section below are these measurements with the offsets applied, and are therefore metres above mean sea level (amsl).

3.5 Palaeo Sea-Level Estimates

Important in palaeo sea-level estimates are regional measures of the vertical relationships between shoreline features and mean sea-level position, using modern analogues and taking account of such factors as the depth over which a particular indicator is formed ('the indicative range'), tidal range and measurement errors or uncertainties (Mauz et al. 2015; Rovere et al. 2016). In order to estimate the palaeo mean sea-level position in this paper, we make three assumptions: that the tidal range was the same as the modern one; that the surface of the living coral reef approximates the low water mark (0.89 m below mean sea level on the Al Birk coastline); and that the surface of beachrock approximates high water mark (0.89 above mean sea level at Al Birk). In both cases, for ease of reference and in recognition of the difficulties with establishing precise constraints, we round these figures to 1 m.

This gives an offset of +1 m for estimating palaeo mean-sea-level from coral terraces, and −1 m when using elevations of beachrock. Thus, we assign a coral terrace measured at an elevation of 4.03 m amsl on the Al Birk coastline an estimated palaeo mean-sea-level of 5.03 m above modern mean sea level. Similarly, a beachrock elevation of 6.85 m amsl gives an estimated palaeo mean sea level of 5.85 m.

These palaeo sea-level corrections may be too large given that beach deposits may form at different levels across a beach profile in relation to tidal range (Mauz et al. 2015), and that there is some variation in the depth relationship between mean sea level and the surface of the living coral, depending amongst other factors on whether it is a reef flat, a reef crest or a fore reef. We do not have detailed measurements or geomorphic descriptions of beach profiles or living coral reefs in our region to provide suitable analogues. We therefore rely on data from the published literature; in the Red Sea, the living coral surface is usually taken to approximate the low water mark (Lambeck et al. 2011; Manaa et al. 2016). Lambeck et al. (2011) cite figures for the height of mean sea-level above the coral reef surface as +0.5 to +1.0 m and use +0.5 m while Manaa et al. (2016) use an average of +0.3 m based on local measurements of the modern coral surface and the local tidal range. For beach deposits, Lambeck et al. (2011) cite a variety of figures indicating heights of +1.5 to +2.5 m above the coral surface. Our figures differ from those used by Lambeck et al (2011) and Manaa et al. (2016), although not substantially, and we attribute these differences to differences in tidal range (or the data used by other authors to approximate tidal range) in different parts of the Red Sea, and to variations in local beach profiles. Nevertheless, there are clearly uncertainties in our data about the precise position of shoreline and coral features in relation to mean sea level, and we apply a margin of uncertainty of ±0.5 m in our estimates of palaeo sea level from coral terraces, and ±1.5 m when using beachrock.

Additional errors, of course, may arise from the erosion of the original surfaces or other irregularities in their surface morphology, an expectation reinforced by variations shown by our data in the measured elevation of different exposures of the same coral platform or beachrock formation in the same locality, and we refer to these in our discussion of the results.

3.6 Archaeological Survey

Following the 2014 expedition, a renewed season of archaeological fieldwork, in collaboration with the Saudi Commission for Tourism and National Heritage (SCTH), was undertaken to complete archaeological surveying of the newly-identified terraces that had not been examined in previous seasons of fieldwork (Inglis et al. 2015), the findings of which are included in this chapter.

3.7 Geochronology

At the time of writing, the dating samples are still awaiting analysis. We therefore concentrate here on the description of the sample locations, their elevation, and their geomorphological, geological and archaeological implications. We acknowledge that further refinement in estimates of palaeo mean-sea-level would be possible by specifying the species of individual corals selected for dating and their growth position. Since the dating programme is still in progress, however, and given the purposes of this chapter, we do not pursue such refinements here. In any case such refinements need to be weighed against the other sources of uncertainty or error discussed above.

4 Results

4.1 Harrat Al Birk

Thirteen locations were visited along the coastline between the town of 'Amq and just south of the town of Al Qahma (Table 2, Fig. 4). The sedimentary facies present within the terraces varied between locations and included coral, beachrock and finer-grained shell/sand facies consistent with shallow marine deposition.

Table 2 Palaeoshoreline features in the Harrat Al Birk, showing locations visited and samples taken for dating, details of elevations of palaeoshoreline features and estimates of palaeo sea level. The elevations are given in two forms: (a) the raw data as measured in the field by DGPS above the local sea surface at the time of measurement; (b) the estimated height of mean palaeo sea level above modern palaeo sea level, using the correction factors presented in Table 1 for modern mean sea level, and an additional correction and margin of uncertainty to estimate the elevation of the mean palaeo sea level. See the text for further explanation and discussion

Locality	GPS	Description	DGPS Station name	Measured Elevation of Upper Surface (m above local datum)	Corrected elevations (m amsl)	Palaeo sea level (m amsl)	Margin of uncertainty m	Coral samples (U-series)	Basalt Samples ($^{40}Ar/^{39}Ar$)	Shells
L0103 (2014-12-05-5)	N 18.405430° E 41.459820°	2 km terrace of beachrock overlying coral, overlying basalt.	North of ‘Amq	Beachrock 4.55–5.34	Beachrock 5.05–5.84	Beachrock 4.05–4.84	±1.5	–	–	–
L0078 (2014-12-04-1)	N 18.268010° E 41.514970°	Coral terrace adjacent to present-day shore with shell scatter.	Al Birk Station 2	3.45	4.03	5.03	±0.5	HAB-33 to -35	–	HAB-36 (Tridacna) HAB-37 (surface shells)
L0129 (2014-12-04-2)	N 18.260167° E 41.514550	Weathered coral remnant overlying basalt.	Al Birk Station 2	3.44–3.57	4.02–4.15	5.02–5.15	±0.5	HAB-38	–	HAB-39, 40 (Tridacna)
L0093 (2014-12-04 & -03)	N 18.240180° E 41.528170° & N 18.240183° E 41.528100°	Terrace of sandy/shelly beachrock sediments, coral and broken coral. *In situ* flake.	Al Birk Station 2	Beachrock 4.87–6.27	Beachrock 5.45–6.85	4.45–5.85	±1.5	HAB-40 (redeposited)	–	HAB-42 (surface shells)
L0090 (2014-12-04-5)	N 18.204086° E 41.524858°	Coral terrace overlying basalt headland at Al Birk.	Al Birk Station 2	3.45–3.88	4.03–4.46	5.03–5.46	±0.5	–	–	–
L0035 (2014-12-4-7 and -8)	N 18.173730° E 41.565170° and N 18.172830° E 41.564740°	Coral terrace remnant overlying basalt.	Al Birk Station 1	3.20–3.50 and 3.30–3.60	3.61–3.91 and 3.71–4.01 m	4.61–4.91 and 4.71–5.01 m	±0.5	HAB-44	–	HAB-43 (Tridacna)
L0036 (2014-12-04-9)	N 18.168730° E 41.566490°	Coral terrace overlying basalt. Heavily bulldozed.	Al Birk Station 1	4.53–4.85	4.94–5.26	5.94–6.26	±0.5		–	HAB-45 (Tridacna)
L0038 CASP site 216-208 (2014-12-05-1)	N 18.135010° E 41.578940°	Coral terrace overlying basalt.	Al Birk Station 3	2.96–5.27	3.54–5.85	4.54–6.85	±0.5	–	–	–
L0092 (2014-12-05-2)	N 18.122060° E 41.596840°	1 km exposure of beachrock overlying coral and basalt.	Al Birk Station 3	Beachrock 3.57–6.87	Beachrock 4.15–7.45	3.15–6.45	±1.5	–	–	–
L0034 Dhahaban Quarry (2014-12-05-1 to -7)	N 18.072444° E 41.623544°	Complex of shallow marine, coral and beach deposits overlying cobble unit containing rolled coral heads and sharp artefacts.	Dhahaban Station	Coral 5.09–7.52 Beachrock 5.24–7.81	Coral 5.62–8.05 Beachrock 5.77–8.34	Coral 6.62–9.05 Beachrock 4.77–7.34	±1.5	HAB-010, -012, -013, -015, -016, -026.	HAB-001, 007, 008, 017, 019, 030, 031	–

(continued)

Table 2 (continued)

Locality	GPS	Description	DGPS Station name	Measured Elevation of Upper Surface (m above local datum)	Corrected elevations (m amsl)	Palaeo sea level (m amsl)	Margin of uncertainty m	Coral samples (U-series)	Basalt Samples ($^{40}Ar/^{39}Ar$)	Shells
L0091	N 18.066220° E 41.635390°	Sandy conglomerate of shells and coral heads overlain by beachrock and associated with basalt terrace.	Al Birk Station 3	5.93–6.63	6.51–7.21	5.51–6.21	±1.5	–	–	–
L0089 (2014-12-05-4)	N 18.035758° E 41.652261°	Major sequence of partially-bulldozed coral terraces overlain by beachrock and potential aeolianite.	Al Birk Station 3	Coral 3.15–5.81; beachrock 5.81–7.40	Coral 3.73–6.39 Beachrock 6.39–7.98	Coral 4.73–7.39 Beachrock 5.39–6.98	±1.5	–	–	–
L0128	N 17.975367° E 41.684250°	400 m long exposure of beachrock abutting basalt.	Qahmah	5.19–6.32	5.65–6.78	4.65–5.78	±1.5	–	–	–

L0103

At 800 m from the present-day shoreline and 6 km south of ‘Amq, the southern extent of a ~2 km exposure of beachrock facies overlies heavily-weathered coral (Fig. 5a, b). At its southern extent, the top of the beachrock terrace varies between 5.05 and 5.84 m above mean sea level (all elevations given in this form in the text have been corrected to the local mean sea level). No material suitable for U-series dating was identified in the terrace.

During the 2015 archaeological survey, a single lithic artefact was observed and collected from the beachrock surface, a basalt ESA (Early Stone Age)/MSA core fragment. At the foot of the low, undulating, basalt jebels behind the marine deposits, twenty artefacts made on quartz or basalt were collected, including ESA and MSA cores and flakes, as well as small flakes and cores made on quartz of potential LSA (Late Stone Age) date.

L0078

This is 7 km north of Al Birk, where a coral terrace is preserved adjacent to the present-day shore. The top of the coral terrace was measured at 4.03 m (Fig. 5c, d). Three coral samples were removed from the coral deposit, samples HAB-033 to -035, along with a *Tridacna* shell (HAB-36) and a sample of shells from the surface (HAB-37). A thin shell-midden deposit comprising edible marine molluscs collected as food lies on the surface of the terrace and has been radiocarbon dated to 5560 ± 70 BP (Beta–191460); a radiocarbon sample from shell embedded in the underlying coral terrace gave an infinite age (Bailey et al. 2007b). In 2014, six basalt and andesite artefacts were collected from the surface of the terrace, including MSA prepared cores, and flakes struck from prepared cores (Inglis et al. 2014b).

L0129

A very weathered and partially damaged remnant of coral terrace overlies a basalt lava flow adjacent to the present sabkha, with its uppermost (albeit degraded) extent measured at 4.02–4.15 m amsl (Fig. 5e, f). A single sample of fibrous coral was recovered (HAB-38) at 2.62 m amsl. Complete and fragmentary *Tridacna* shells were also collected (HAB-39 and -40). No archaeological material was observed.

L0093

A beachrock terrace overlying a basalt flow consists of sandy sediments containing redeposited cobbles of coral, broken pieces of smaller coral, and shells (Fig. 5g). Spot heights from the top of the terrace varied between 5.45–6.85 m amsl. A single sample of redeposited coral (HAB-40, 4.22 m) was collected from within the terrace, as well as a sample of shells from the terrace surface (HAB-42). An undiagnostic, sharp basalt flake (HAB-41) was recovered from within the terrace at 4.78 m amsl (Fig. 5h). In 2014 a single MSA basalt flake was observed on the surface of the terrace (Inglis et al. 2014b).

L0090

A headland formed by a basalt lava flow extends from the town of Al Birk into the sea. At its tip, and adjacent to the present-day beach, a coral terrace overlies the basalt, but has been heavily damaged by bulldozing. The terrace continues beyond a security fence into a fenced area where it remains intact (Fig. 6a, b). The height of the undamaged terrace was measured at the fence line (closest to its undisturbed height) at 4.03–4.46 m amsl. Suitable material for U-series dating was not identified. In 2014, a single MSA prepared core preform on basalt was found on the surface of the coral terrace (Inglis et al. 2014b).

L0035

Overlying a basalt lava flow, coral terrace remnants were preserved adjacent to the modern day sabkha (Fig. 6c, d). At 2014-12-04-7 the top of the terrace was measured at 3.61–3.91 m amsl and a *Tridacna* shell was collected (HAB-43). Suitable coralline material for U-series dating was not identified. At 2014-12-04-8, close by, and probably contiguous with the previous station, the top of the coral terrace varied from 3.71–4.01 m amsl. A large block of coral had been broken off from the terrace by bulldozing, and a sample of fresh-looking coral was removed from this block (HAB-44). Thirty-five basalt flakes and cores, including artefacts with ESA and MSA affinities, were observed on the lava flows above the raised coral terrace, and on the surface of the terrace itself (Inglis et al. 2014b).

L0036

A remnant of coral terrace less than half a kilometre to the south of L0035 was surveyed (Fig. 6e, f). The top of the terrace was measured at 4.94–5.26 m amsl. The coral was heavily weathered, and no suitable dating material was located, but a *Tridacna* shell was collected from the terrace surface (HAB-45). In 2013, eighteen MSA and ESA basalt flakes and cores were observed on the upper, disturbed parts of coral terraces overlying the lava flow (Inglis et al. 2014b).

L0038, CASP Site 216-208

First recorded by the Comprehensive Archaeological Survey Programme (CASP; Zarins et al. 1981), a coral terrace at this location overlies a lava flow adjacent to the modern sabkha (Fig. 6g, h). It was visited in 2004 (Bailey et al. 2013b) and comparison with photographs in Zarins et al. (1981) showed that considerable damage had already taken place in the interval. Since 2004, much of the immediate locality has been further bulldozed for the construction of a coastguard station, and only part of the coral terrace remained intact as of December 2014. The coral terrace contains two facies, a lower facies of consolidated carbonate and basalt clasts overlain by an upper facies of coral, both in growth position and with a matrix of broken pieces of branching corals. The upper surface of the terrace was measured at 3.54–5.85 m amsl. No material suitable for U-series dating was identified.

Emplacement of the underlying lava flow has been dated by K/Ar to 1.37 ± 0.02 Ma and 125 ± 0.02 Ma (Bailey et al. 2007b). The CASP survey recorded Palaeolithic artefacts with ESA and MSA affinities on the surface of the terrace and underlying basalt. These observations were confirmed by the DISPERSE project, with extensive scatters of ESA and MSA basalt artefacts present on the surface of both sides of the lava flow and coral terrace, including a crude handaxe, radial cores and flakes (Inglis et al. 2014b). Above the coral terrace, an area of SCTH protection encloses boulders engraved with South Arabic script (~9–5th century BC) and shallow tunnels excavated into the volcanic deposits.

L0092

A >1 km-long exposure of weathered coral and beachrock deposits overlies a basalt flow 1.5 km from the present shoreline (Fig. 7a, b). Elevation of the beachrock surface was measured at 4.15–7.45 m amsl. No material suitable for U-series dating was identified. A survey of the exposure yielded eight lithics with MSA affinities, including cores made on basalt and a convergent flake (Inglis et al. 2014b).

L0034 Dhahaban Quarry

Initial DISPERSE reconnaissance in May 2012 identified marine deposits exposed in a modern quarry section close to the coastal road, overlying the basalt flow from an adjacent cinder cone (Devès et al. 2013). Further geoarchaeological investigations were undertaken in 2013 (Inglis et al. 2013). This site is of particular importance because of the number of artefacts that have been recovered here. Many are on the surface of the marine deposits but there are also 19 stone tools in sharp and unrolled condition found stratified within a water-laid cobble unit underlying a complex series of beach, shallow marine and aeolian deposits (Inglis et al. 2014b; Bailey et al. 2015). Because of quarrying activity and erosion by stream action that has cut a wadi bed through part of the deposits, there is no single exposure where the full sequence of deposits can be seen in one section, complicating the task of unravelling their stratigraphic relationships (Fig. 8).

The major depositional units are as follows:

In situ Coral Terrace. These deposits extend to the north of the locality, including large coral heads and facies of broken coral (Fig. 9a, b). Quarrying activity, wadi incision and other disturbances have removed parts of this unit, which is now preserved in several spatially discrete areas (Fig. 8). From the appearance and condition of these coral terrace fragments, their spatial configuration, and their surface elevation, we consider that they are parts of the same unit formed during the same sea-level episode. The upper surface of this coral terrace varies between 5.62 and 8.05 m amsl with large coral heads observed in growth position at 3.78 m amsl in the bulldozed area to the north of the wadi cut, opposite the cobble unit exposure. MSA stone tools are present in places on the surface, especially near the north bank of the present-day stream bed.

Beachrock. This is present mainly in the central and southern area of the locality, but also in fragments overlying coral in the northernmost extent of the exposed deposits (Fig. 9c). The upper surface of the beachrock varies between 5.77 and 8.34 m amsl. Some of the beachrock is located at a

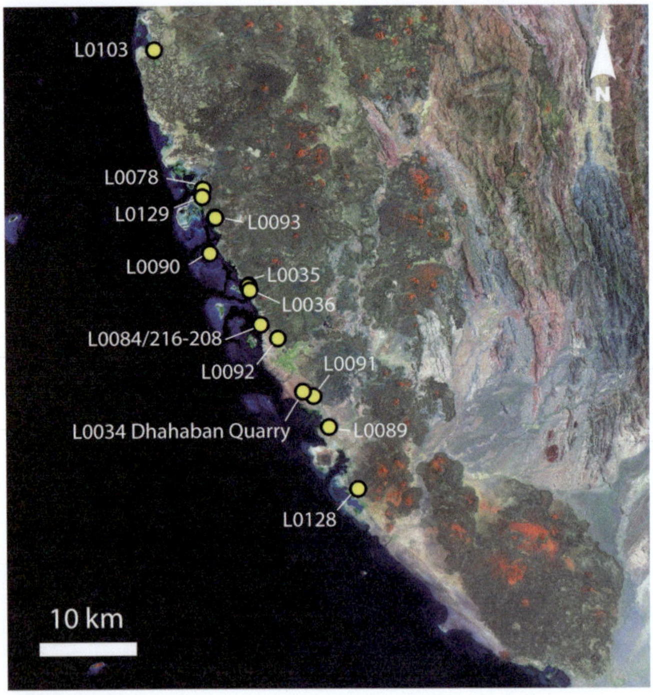

Fig. 4 Locations of marine terraces surveyed and sampled in the Harrat al Birk. Satellite imagery © USGS Landsat ETM + 200 Geocover Mosaics. The orange patches are volcanic cinder cones and the dark green areas are basaltic lava flows

higher elevation than the coral terrace and stratified above it, as would be expected if they belonged to the same sea-level episode. However, some beachrock is lower than the highest points of the coral terrace, suggesting at least two different episodes of high sea level, or a period of variable sea level within an interglacial period, with at least some of the beachrock deposited during marine regression. MSA stone tools are present in places on the beachrock surface.

Cobble Unit. A unit of rounded pebbles, cobbles, re-deposited coral heads and occasional mollusc shells including a large oyster shell. The unit is restricted to a small area where the wadi drains the *harrat* behind the quarry and has exposed the cobble unit in a section extending for about 20 m along the southern margin of the wadi, where it is overlain by lithified shallow marine sands and beachrock (Fig. 9d). The upper limit of the unit, at the contact with shallow marine sands/beachrock, slopes between 7.74 to 7.34 m amsl, with its lowest exposure from 6.02 to 5.2 m amsl at the edge of the present wadi bed; the unit probably extends deeper. The cobbles are highly variable in size ranging from <5 cm up to 30 cm in maximum dimension. They consist of heavily rolled pieces of basalt that have clearly been derived from the basaltic bedrock immediately inland. The deposit has the appearance of a debris flow deposited by torrential stream action. Nineteen sharp basalt flakes were collected from this unit. The fact that they show no evidence of rolling or abrasion, unlike the rest of the clasts, suggests that they have not moved far from their original discard location. Samples from five coral heads were also removed from this unit, including a large coral head oriented in growth position, but not necessarily in primary position.

Marine Sands. Silty sand and shell sand facies exposed in a quarry cut ~6 m deep at the southern end of the site and capped by aeolianite (Fig. 9e); the top of this section is at 8.74 m amsl, with the transition between aeolianite and marine/beach sediments at ~1.25 m below the top of the section (~7.49 m amsl). Samples were removed from this section in 2014 and 2015 for OSL dating at the Scottish Universities Environmental Research Centre, UK (Sanderson and Kinnaird, this volume).

Aeolianite. At the southern end of the site, cemented aeolian deposits abutting the slopes of the cinder cone, and directly overlying the shallow marine sediments in the main quarry section, were observed extending up to at least ~23 m amsl on the volcanic cone. (Fig. 9f).

Analysis of the complex depositional history at Dhahaban Quarry and its implications for reconstructing sea level change is ongoing. The main sequence of events from earliest to latest appears to be as follows: (a) formation of a volcanic cinder cone and associated lava flows at some indeterminate stage in the Pleistocene; (b) marine transgression followed by formation of coral, deposition of shallow marine sands and formation of beachrock deposits during one or more high sea-level stands; (c) accumulation of wind-blown sands derived from the exposed sea bed during the subsequent marine regression, and later cemented through carbonate precipitation into aeolianite.

The most difficult unit to place in this sequence is the cobble unit. We consider three hypotheses:

(a) A storm deposit formed at the back of a beach broadly coeval with the coral terrace and beachrock formations and part of the same high sea-level stand.
(b) A fluvial unit accumulated before the high sea-level stand associated with the coral terrace and beachrock formations, based on the fact that the cobble unit is stratified beneath an overlying beachrock deposit and is at a lower elevation than the adjacent coral terrace on the opposite bank of the wadi.
(c) A fluvial unit formed during an interval of minor sea-level regression and incision of the stream bed between an earlier high sea-level stand associated with the coral terrace and a later high sea-level stand of similar or slightly lower elevation associated with the overlying beachrock.

Fig. 5 Views of marine terraces surveyed in 2014 (1): **a**, **b** L0103 beachrock overlying heavily-weathered coral; **c** L0078 coral terrace overlain by shell scatter, and coral sampled (**d**); **e**, **f** L00129 heavily weathered coral terrace remnant overlying basalt; **g** sandy sediments with broken coral and shell at L0093 with **h** worked basalt flake removed from deposits to left of centre of geological hammer (circle marked). Photos: Robyn Inglis, December 2014

We reject (a) on the grounds of the highly variable clast size, the lack of sorting, the thickness of the deposit, its alignment along the line of the wadi, and the fact that it contains well-embedded within it stone artefacts with sharp edges that show no signs of rolling or water abrasion (Fig. 10).

Hypothesis (b) is possible, but we then have to explain the source of the coral heads embedded in the cobble unit and to suppose that they were eroded from a coral terrace formed during an earlier period of high sea level. There is no evidence for such a coral terrace in the vicinity, nor any evidence that the coral terrace deposits at the site are of

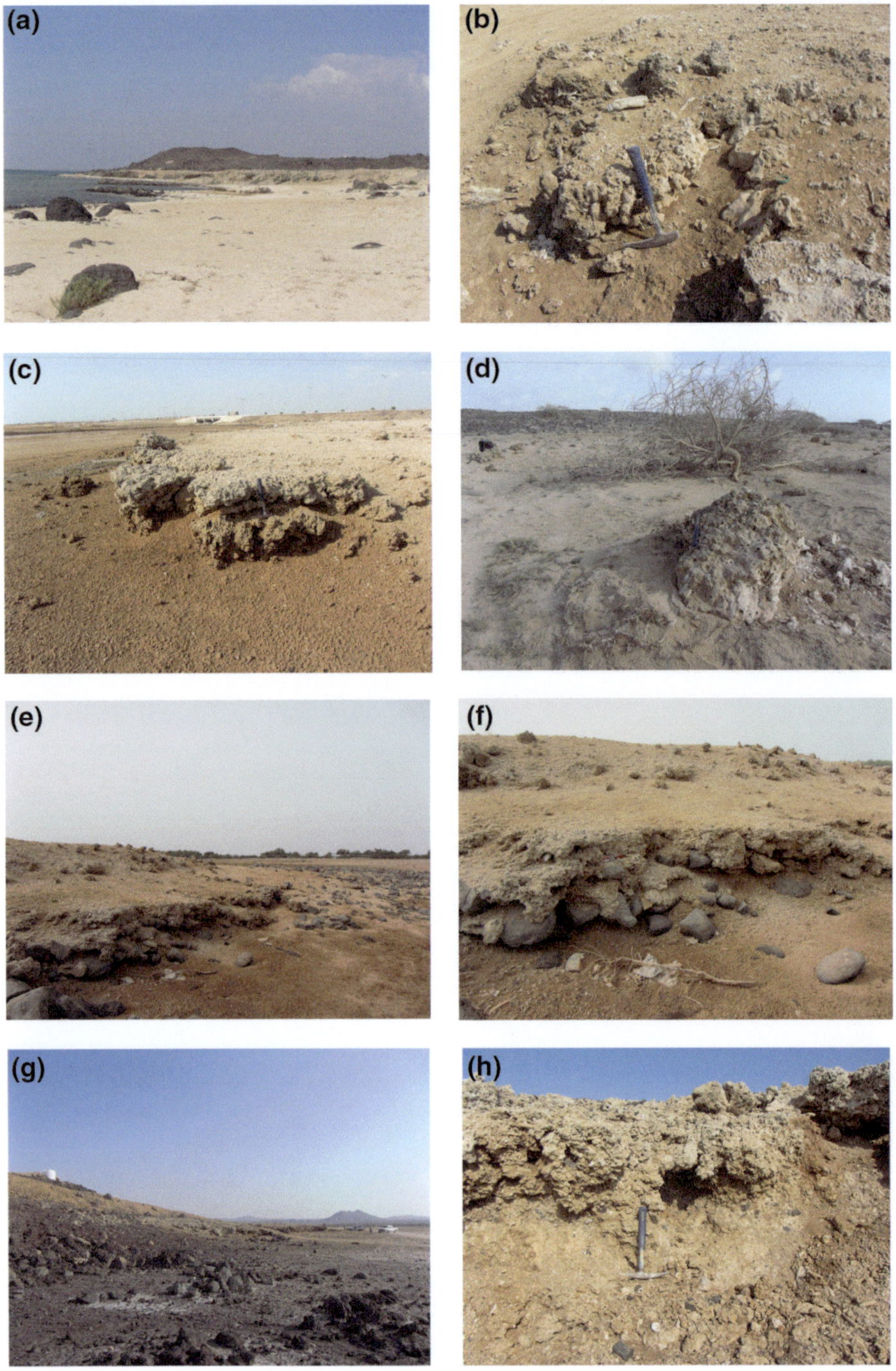

Fig. 6 Views of marine terraces surveyed in 2014 (2): **a**, **b** partially-bulldozed coral terrace overlying basalt at end of Al Birk headland; **c** coral terrace remnants adjacent to sabkha at L0035, and **d** sampled block of coral from terrace; **e**, **f** coral terrace remnant overlying basalt at L0036; **e**, **f** partially bulldozed coral terrace at L0038, CASP site 216-208. Photos: Robyn Inglis, December 2014

composite structure, formed of coral deposited during two periods of sea level separated by $\sim$100 kyr (e.g., evidence of unconformity or differences in the condition and degradation of the coral). But, these observations are not decisive.

Hypothesis (c) is also possible, and would account for the presence of the coral heads and marine shell found within the cobble unit, as well as the corals observed in growth position as low as 3.78 m amsl exposed in the bulldozed area of the wadi. Minor oscillations of eustatic sea-level are recorded within MIS 5e and are consistent with this hypothesis (see Rovere et al. 2016, and references therein). We consider association of any of the observed palaeoshoreline features with the high sea-level stands of the later stages of MIS 5 as unlikely because these were formed at sea levels some 20 m lower than the MIS 5e sea level. For the moment, we consider this the most likely hypothesis, while emphasising that a decisive test in its favour in preference to (b) will be a demonstration by U-series dating of broad contemporaneity between the large coral heads within the cobble unit and the adjacent coral terrace.

From the archaeological point of view, both hypotheses (a) and (c) imply a significant episode of human activity on or very close to shoreline of the contemporaneous sea coast, whereas under hypothesis (b) the cobble unit and its embedded artefacts could have been deposited at any period in the sea level cycle, including a period of marine regression when the coastline was located many tens of kilometres to the west.

L0091

A small coral terrace 1 km to the east of the Dhahaban Quarry, overlying the same basalt flow, was preserved in the mouth of a wadi draining the *harrat* (Fig. 7c, d). The terrace consists of a conglomerate of shells and coral heads, some in growth position. This was overlain by a beachrock facies, the upper surface of which was coincident with a break in slope in the lava terrace that continued to the east. The top of the beachrock was measured to between 6.51–7.21 m amsl. Mangrove whelks and gastropods were observed in and on the sediments. No material suitable for U-series dating was identified in the terrace. In 2014, a potential ESA handaxe roughout on basalt and a rolled basalt clast with retouch were observed on the basalt (Inglis et al. 2014b).

L0089

An eroded cinder cone and its surrounding basalt flows to the west of the coastal road, isolated from the main harrat, are overlain on their northwestern and western flanks by a complex of coral terraces, beachrock and aeolianite (Fig. 11). In places, the coral terraces (which have been subject to some bulldozing and disturbance) contain >1 m coral heads in growth position. Profiles were measured from the top of the largest of these exposures to the sabkha at the foot of the terrace. The surface of the coral, partially bulldozed in places, varies between 3.73 and 6.39 m amsl, with the surface of the beachrock between 6.39 and 7.98 m amsl. No material suitable for U-series dating was identified in the deposits. In 2014, 13 artefacts with MSA affinities, manufactured in basalt, quartz and andesite, were collected from the surface of these deposits, mainly the beach rock surfaces, including prepared cores and points (Inglis et al. 2014b).

L0128

South of Al Qahma, a 400-m long exposure of beachrock, 500 m inland from the present shoreline, overlies a basalt lava flow (Fig. 7e, f). Part of the beachrock has been quarried, leaving large piles of boulders of the material. The intact surface of the exposure was measured to between 5.65–6.78 m amsl. Numerous funerary cairns, some of relatively recent origin, have been built on the beachrock terrace and surrounding basalt hillsides and pottery was found on the surface of the beachrock.

4.2 Farasan Islands

Three days of surveying were spent on the Farasan Islands, targeting three principal areas (Fig. 12):

(1) The north of Farasan Al Kabir, the largest island of the Farasan group, near the village of Sair and the fishing village of Al Hussain. This area shows obvious signs on satellite images of tectonic uplift associated with salt tectonics.
(2) The Sulayn and Abalat group of islands, north of the main port on Farasan al Kabir.
(3) Ras Sheida, near the southern extremity of Farasan al Kabir.

The surface geology of the Farasan Islands is dominated by highly-faulted cemented coral reef terraces, often described as 'limestone', and intervening depressions floored by modern, fine-grained alluvium (Fig. 13). The limestone units have been collectively referred to as the 'Reef Limestone' (Macfadyen 1930) or 'Farasan Island Limestone' (Bantan 1999) and based on limited U-series dating all are of Pleistocene date and in some cases Holocene (Bantan 1999). Bedding dips are generally horizontal to very shallow but locally attain ~30°. Their maximum elevation is now ~75 m asl, but most of the coral terraces lie at 20 m or less. Many of the terraces are cut by pronounced open linear fissures and metre-scale grabens that generally parallel nearby faults (Fig. 14). Both these secondary structures are indicative of predominantly horizontal extension without any evidence of significant strike-slip movement.

Shallow-well data summarized in Bantan (1999) demonstrate that the surficial reef limestones attain a maximum thickness of at least 125 m and are underlain by 10–54 m of interbedded gypsum, claystone and marl. Beneath this interbedded sequence there is at least 155-m thickness of massive evaporite (halite) on Farasan al Kabir. At nearby Zifaf Island (Fig. 12) the halite is more than 350 m thick (Macfadyen 1930). The depth to the base of the halite has not been determined in any published well records. Regional correlations suggest that the massive halite beds are Middle to Late Miocene in age.

In addition to the extensive faulting, the stratigraphy of the island is also locally gently folded. This produces kilometre-scale closed and open basins and domes that are best observed in satellite imagery (Bantan 1999). Large

Fig. 7 Views of marine terraces surveyed in 2014 (3): **a**, **b** weathered beachrock overlying coral terraces at L0092; **c**, **d** coral terrace overlying basalt at L0091; **e**, **f** beachrock overlying basalt at L0128. Photos: Robyn Inglis, December 2014

circular structures are also given expression in much of the shoreline geometry, such as at Al Homrah Bay and Mandhar (Fig. 12). Smaller circular and elliptical depressions with tens to hundreds of metres dimensions occur sporadically across Farasan al Kabir. Examples are seen just east of the hamlet of Al Hussain (Fig. 13). The formation of these circular geomorphological features and the associated gentle folding have been attributed to flow of the underlying Miocene salt (halite) beds (Bantan 1999). This is supported by the fact that the thickness of the post-salt stratigraphic column is highly variable in the drilled wells.

5 The North of Farasan al Kabir

Northwest of the village of Seir at Jabal Sudain, coral deposits have been uplifted to ~15 m asl. Bantan (1999) measured 12 m of dolomitic and highly fossiliferous packstone and wackestone overlain by a cap of cemented coral reef. The promontory is bounded by several faults striking ENE–WSW and dipping ~55–70° to the north and south. This location is significant because several of the faults bear well-defined slickenlines that demonstrate largely dip-slip extensional movement. Most faults cutting the Pleistocene limestones elsewhere do not display kinematic indicators.

On the coast to the southwest of the village of Al Hussain there is a bay used by the local fishermen. Immediately behind the modern shoreline is a narrow coral terrace some 3 m in height with a notch formed by marine erosion (Fig. 14, Table 3). This coral platform is typical of the coastal geomorphology found around many of the other coastlines of the Farasan Islands and often has shell mounds located on it (Fig. 3). Behind this terrace is a partly eroded cliff 25–34 m high forming an impressive sequence of uplifted sediments composed of thick deposits of white

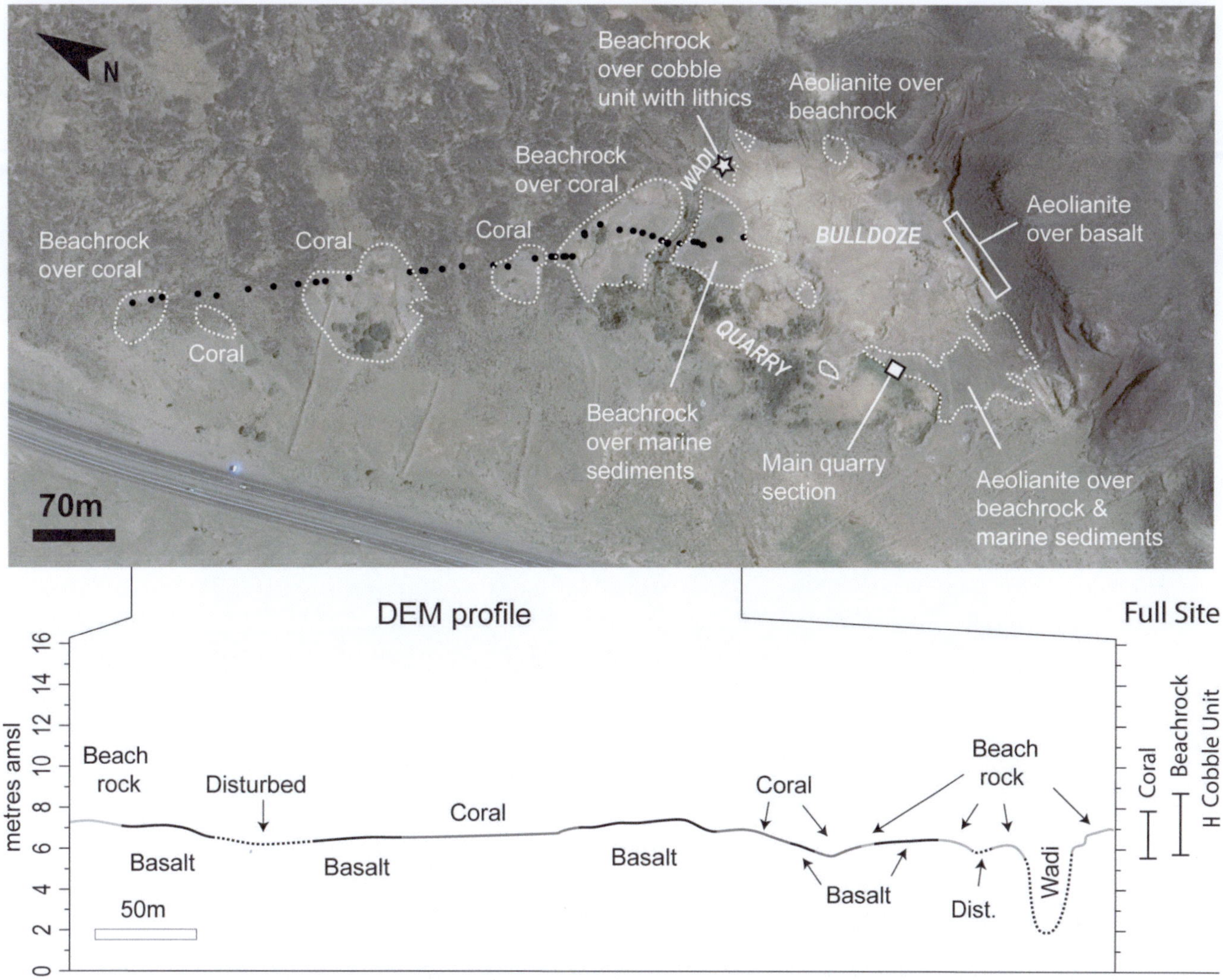

Fig. 8 Aerial view and elevation profile of sediments at Dhahaban Quarry (L0034). Image © Google Earth, dated 19/01/2014. The DEM profile, surveyed in December 2014, has been expanded horizontally for the sake of clarity. The black dots indicate the line of the DEM profile and the individual measurement stations. Black dotted lines in the DEM profile indicate lack of information because of modern disturbance or accumulation of recent sediments that have obscured the original profile of the wadi and the lower sections of Pleistocene deposits that formed the original wadi infill. Alongside the DEM profile is displayed the full range of the upper surface of key stratigraphic units, surveyed using a total station in 2014. See text for further details

biogenic/bioclastic marine sediments with occasional red banding, the whole sequence capped by a coral reef deposit 2 m thick, which we sampled for dating (16.71686°N; 41.90857°E; Fig. 14). The surficial appearance of this uplifted terrace is not significantly different from the lower-lying terraces.

6 Sulayn and Abalat Islands

Most of the smaller islands surrounding Farasan al Kabir display coral terraces at varying, generally relatively low elevations. At Abalat Island a taped spot elevation of the well-developed terrace was 9 m asl (16.80606°N; 42.18356° E). The topography at Am Murabaah is more complex with a terrace at about 8 m asl (16.73954°N; 42.20501°E) and a higher terrace at about 13 m asl (16.74007°N; 42.20469°E). Each of these sites was sampled for U-series dating (Fig. 15).

7 Ras Sheida

Along the present-day coast of Janaba Bay, the modern shoreline is backed by a low coral-reef platform about 2–3 m above present-sea level, which has been undercut by marine erosion, and upon which there are numerous shell mounds of mid-Holocene date.

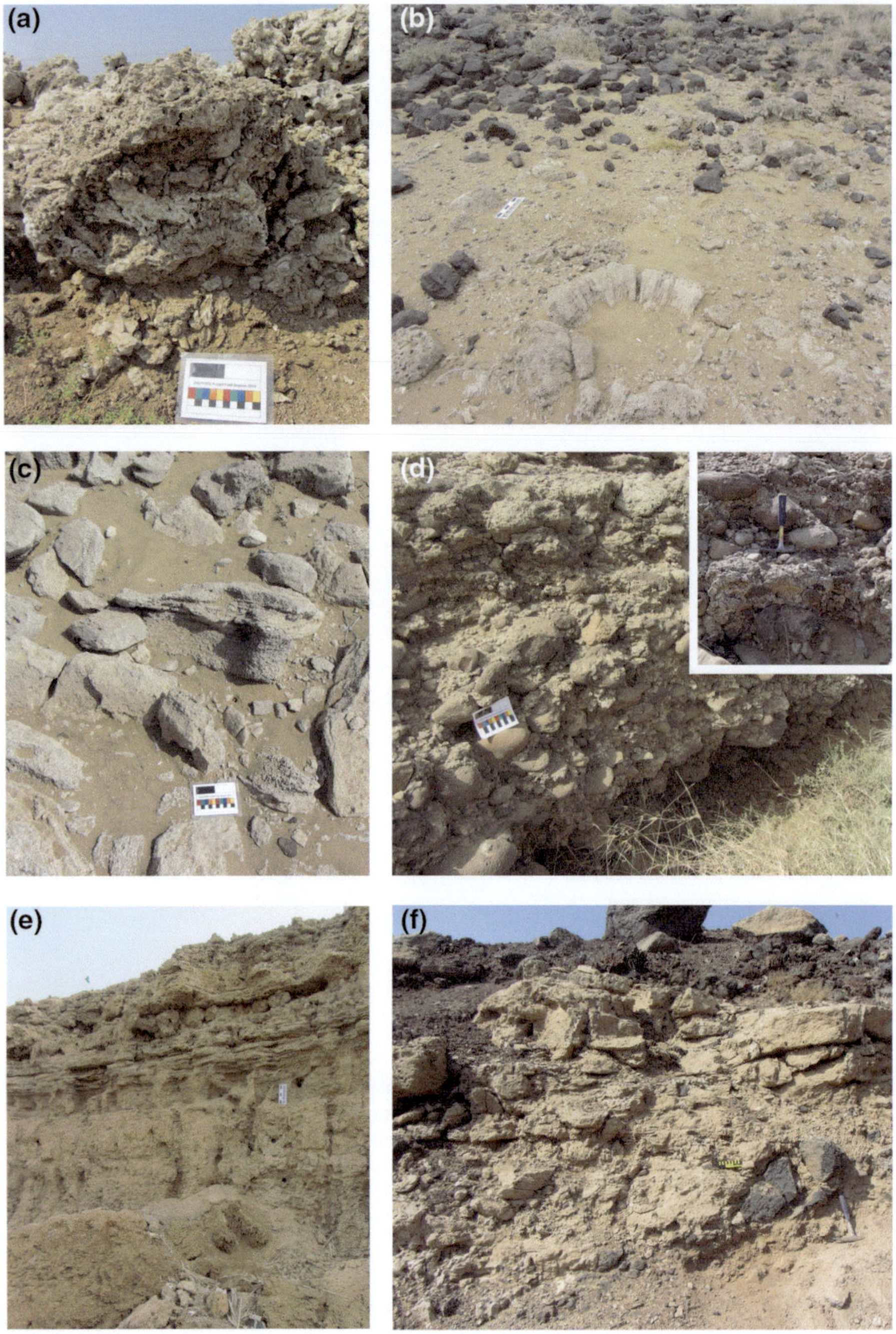

Fig. 9 Sediment and coral facies at Dhahaban Quarry: **a** coral in growth position in terrace of broken coral; **b** large coral in growth position, partially buried by sediment; **c** beachrock weathering and fragmenting in situ; **d** cobble unit beneath marine sediments containing sharp basalt lithic artefacts as well as rounded coral heads (inset); **e** series of shallow marine sediments capped by aeolianite exposed in quarry; **f** aeolianite overlying scoria and basalt jebel. Photos: Robyn Inglis, December 2014

Ras Sheida, at the southern extremity of Farasan al Kabir, contains a series of four uplifted coral platforms (Fig. 16). The highest is Jebel Sheida, the crest of the reef/terrace standing at 26 m (Fig. 17). Several terrace levels below this can be recognised, but their age relationships are complex and require age dates to ascertain (Table 3).

The 3 m terrace is evident at many locations and may be the MIS 5e terrace. The photograph in Fig. 17 is taken from a position where elevations are generally about 7 m—this could be another candidate for MIS 5e or it could be a degraded older terrace. The 'gently dipping terrace' shown in Fig. 17 lies generally at about 10–11 m but does not

Fig. 10 Close-up views of cobble unit at L0034 Dhahaban Quarry: **a** basalt cobbles and coral head showing range of sizes; **b** stone artefact in section (to lower right of scale). Photos: Geoff Bailey (a); Bill Bosworth (b), December 2014

connect physically with the 11-m notch on the southern coast. Rather it projects up higher in the structure to about 18 m. Near the south coast, the average dips in the higher carbonate beds are about 3°. This probably reflects footwall uplift/rotation of a fault running along the south side of the Jebel Sheida. Evidence in support of this is that the beds flatten out to the north away from the fault.

8 Discussion

Whilst the results from the absolute dating have not yet been obtained, the field observations collected have allowed the development of a set of working hypotheses related to the position of past shorelines in the region surveyed, and their implications for understanding tectonic movement and archaeological evidence of coastal activity.

8.1 Geological Discussion

The shallow marine deposits, coral terraces, and beachrock occurrences that we have observed in the Harrat al Birk coastal province all presently lie at elevations of ~3.0–8.5 m although the upper surface of the Dhahaban Quarry beachrock remains slightly undefined (Fig. 18). In general, beachrock sits at elevations slightly higher than nearby coral terraces, and both overlie the shallow marine units when they are exposed; in many cases beach deposits directly overlie coral terrace deposits. Taking account of the corrections for mean sea level discussed earlier, the maximum palaeo sea level estimate from the study is 9.05 m asl ±0.5 m (coral at Dhahaban Quarry), with the rest of the estimates lying between ~5 and 7.5 m, subject to errors of up to ±1.5 m. In light of this, the 9.05 m coral at Dhahaban Quarry may merit closer inspection and confirmation of its 'in situ' setting.

We do not yet have absolute ages for our coral terraces or their associated near-shoreline deposits. However, we can compare our elevations with dated examples from other parts of the Red Sea. Hoang and Taviani (1991) determined U-series dates of 126–138 ka (3 samples) for a coral terrace on Zabargad Island that sits 6–8 m asl. Ages of 125–135 ka (3 samples) were also found for the coral terraces on nearby Rocky Island and North Brother Island in the northern Red Sea (Hoang and Taviani 1991). These terraces have similar elevations of 6–8 m asl. On the Saudi Arabian coastline near Duba, coral terraces dated 119–123 ka (4 samples) are at 4–5.5 m asl (Manaa et al. 2016). On the corresponding Egyptian margin from near Hurghada to Marsa Alam, a coral terrace consistently located at ~6 m asl produced a broader range of ages of ~87–131 ka (10 samples; El Moursi et al. 1994), but this is clearly the MIS 5e highstand. Lambeck et al. (2011), referring to a larger dataset, note that MIS 5e elevations of coral reefs lie in the range of 5–9 m with a mean of 7 ± 1.5 m, except in areas of known tectonic deformation and uplift.

These results give elevations similar to our own data. The simplest interpretation of our dataset, then, is that the beachrock at L0093, L0092, Dhahaban, L0091, L0089, and L0128 represents the geographic position and elevation (~6–7 m above present sea level) of the MIS 5e shoreline. The associated coral deposits were formed near this shoreline, at approximately the same elevation or perhaps 1–2 m lower if they are not the true reef-crest. Further investigation at Dhahaban will elucidate whether the deposits there were deposited by a single high sea stand/cycle, or whether the corals reworked in the cobble unit date from an earlier high sea stand. Our simple and preferred single-high-stand model

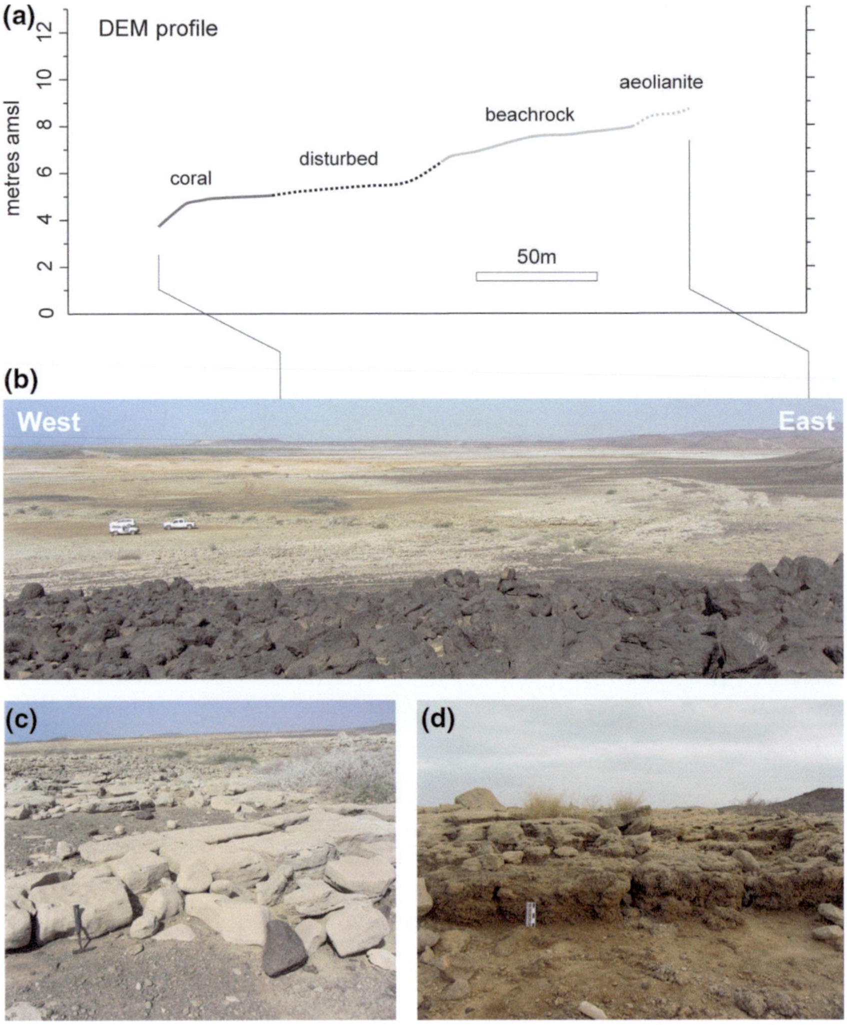

Fig. 11 Profile and photos of coral and beachrock sediments from L0089: **a** DEM profile across centre of coral and beach deposit complex (**b**); **c** beachrock at L0089; **d** coral terrace deposits. Photos: Robyn Inglis, December 2014

for the Harrat al Birk localities is compatible with this limited regional information.

Worldwide, eustatic sea level during MIS 5e is thought to have ranged between +3 and +7 m above the present (Kopp et al. 2009; Benjamin et al. 2017). On the face of it, data from the Red Sea including our results presented here appear to be consistent with this figure, and therefore to indicate either vertical stability or very minor tectonic subsidence in the central and northern Red Sea over the past ~125 kyr.

However, as Lambeck et al. (2011) point out, coastlines are subject to significant changes in elevation as a result of glacio-isostatic and hydro-isostatic effects resulting from the mantle response to re-distribution of loads on the Earth's crust caused by changes in the volume of ice and water, and these effects are likely to show significant geographical variation. These effects can be modelled mathematically and need to be extracted from the elevation of palaeoshoreline indicators before interpretation of the tectonic signal. Detailed modelling by Lambeck et al. (2011) predicts that the position of eustatic sea level in MIS 5e throughout much of the Red Sea should be close to present sea level assuming an absence of vertical tectonic movement. If that is correct, then the presence of MIS 5e palaeoshoreline indicators at 7 m above present sea level is not evidence of tectonic stability, but evidence of uniform tectonic uplift of about that amount over the past 125 kyr. This in its turn raises questions about the nature of tectonic deformation in different parts of the Red Sea Basin. These may require further investigation of regional tectonic deformation or revision of the assumptions underlying the isostatic model, but further discussion of these issues lies beyond the scope of this chapter.

In the Farasan Archipelago the situation is more complex because of the effects of salt tectonics. The influence of movement of the underlying salt bodies is evident from both satellite imagery (Figs. 12 and 13) and field structural observations. Present-day terrace elevations will therefore reflect the complex interaction between eustatic sea-level changes, regional isostatic and tectonic processes, and local halokinesis. Although eventual determinations of

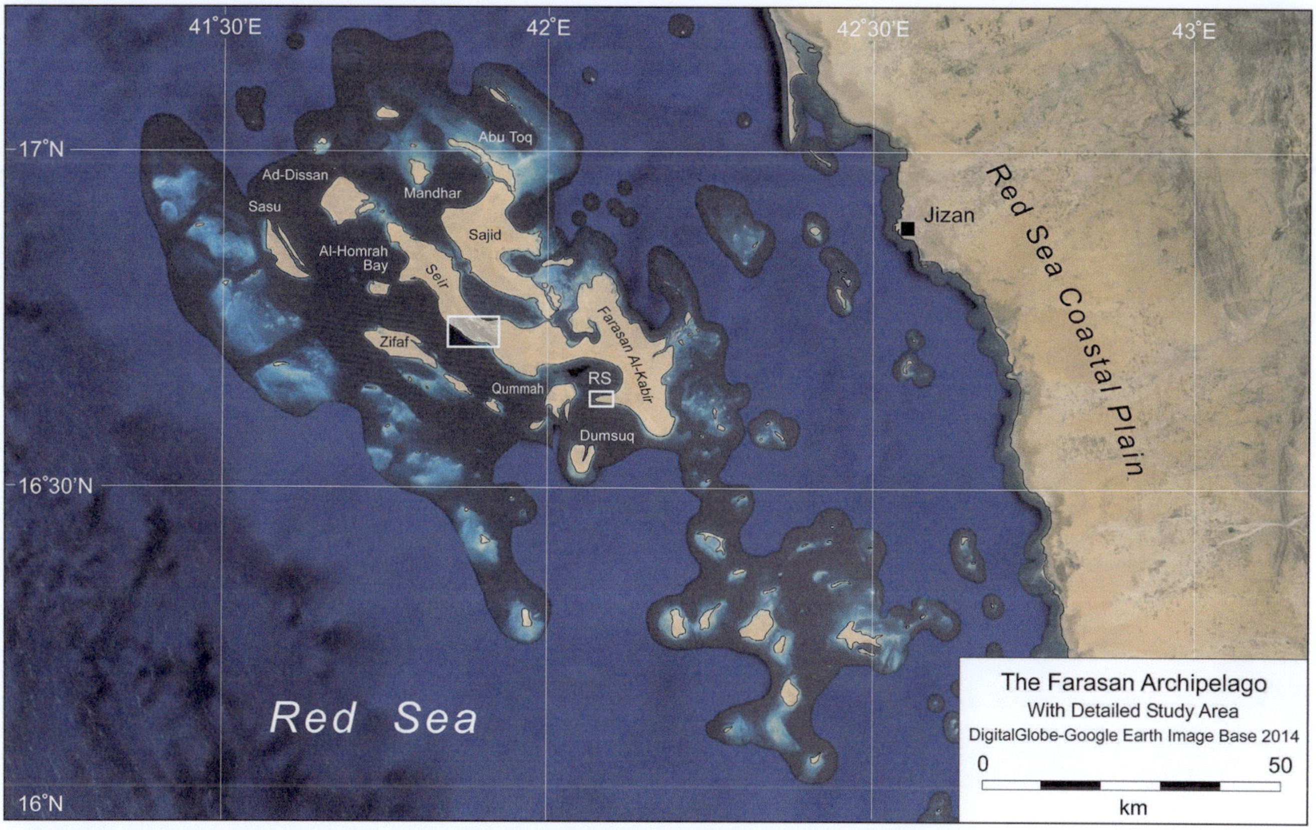

Fig. 12 Farasan Location Map. Figure 13 location is the box below 'Seir' of Farasan al Kabir. RS = Ras Sheida (see Fig. 15)

uplift/subsidence rates will have mostly local significance, they may help to define when salt movement began in this region and whether its movement was constant through time or varied. Detailed interpretation of the vertical tectonic history here must await radiometric dating of the ages of each of the observed coral terraces.

8.2 Archaeological Discussion

Because so much of the archaeological data in our study region comprise stone artefacts lying on the surface with an unknown and uncertain age relationship to their underlying deposits, linking the presence of stone tools to a contemporaneous coastline is problematic. However, two locations in the Harrat al Birk provide the best evidence to date of potential exploitation of the coastline by Palaeolithic populations. At L0093, the presence of a sharp basalt flake embedded within the beachrock terrace indicates that it had been deposited on the beach with little or no geomorphological transport or weathering and surface exposure (either of which would round the flake's edges). It therefore provides a direct link between human activity and beach sediments, indicating the presence of human activity in the immediate coastal zone.

The stratified artefacts at Dhahaban quarry pose a more complex picture of the relationship between artefact deposition and marine environments. This site is of particular interest and importance not only because it is one of the very few sites in the region where MSA stone artefacts are in a stratified and potentially dateable deposit. It has also yielded one of the largest concentrations of stone tools in the wider region, suggesting that the locality had unusual attractions for human exploitation. Most of these stone tools, of course, are on the surface of the coral terrace. Like the artefacts stratified in the cobble unit, they are mostly of MSA type, although we cannot be sure that they all belong to the same time interval. Nevertheless, their unusual abundance is suggestive.

We have outlined the alternative hypotheses about the interpretation of the cobble unit within which the stratified artefacts occur, and we emphasise that a final interpretation must await the outcome of the dating programme currently underway. Our preferred hypothesis is that the cobble unit formed at or close to the time of the high sea-level stand indicated by the elevated coral terrace and beachrock

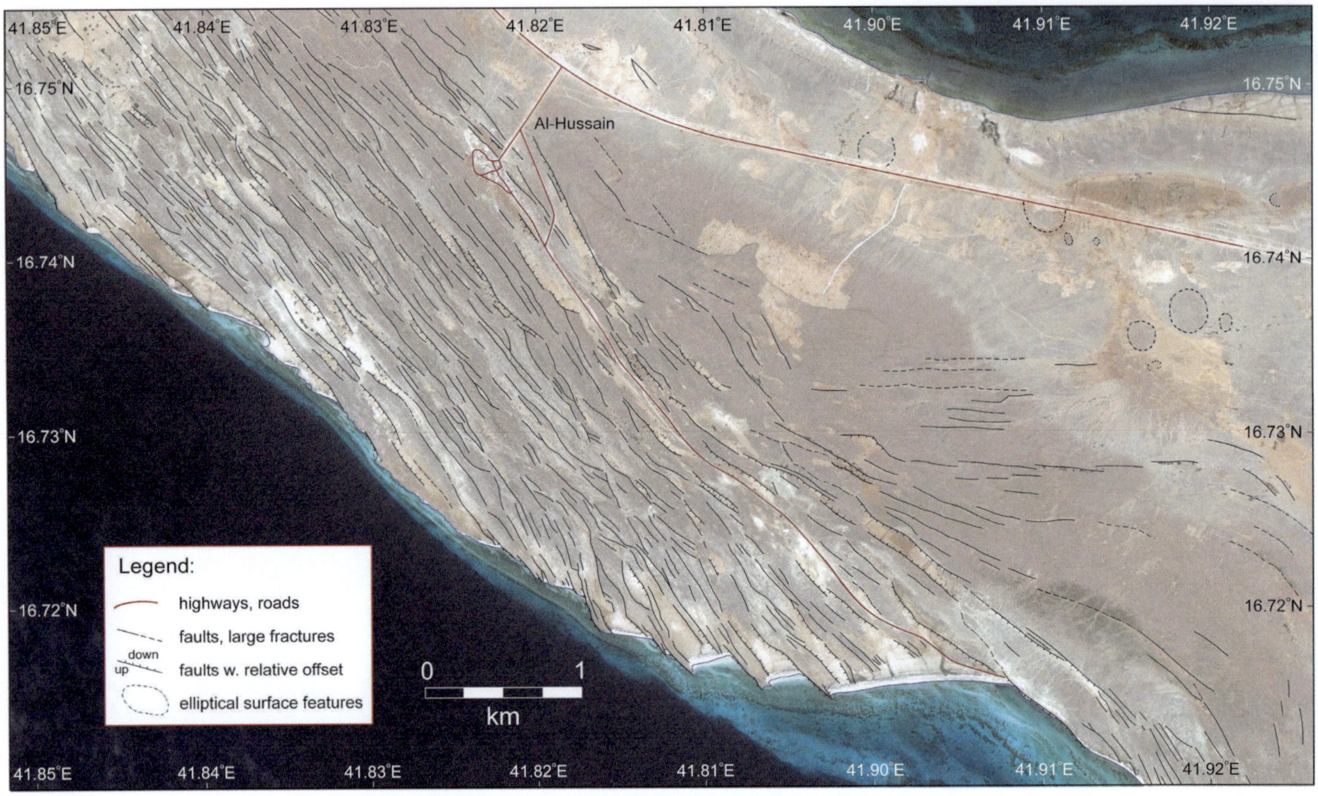

Fig. 13 Integration of field structural observations and satellite image interpretation for a section of Farasan al Kabir around the village of Al Hussain. Location is shown in Fig. 12. (Google Earth; Image DigitalGlobe 2013)

deposits, perhaps during an interval of minor sea-level regression during an interglacial when the coral and beachrock surfaces were exposed as dry ground suitable for human occupation and activity. The artefacts would, then, represent activity carried out on the nearby bank of the wadi at a time of abundant water supply, close to its confluence with the seashore. This is also consistent with palaeoclimatic data, which indicate that the wettest intervals during long-term palaeoclimatic cycles in the Arabian Peninsula, at least during the late Pleistocene and early Holocene, occurred during the earlier part of interglacials, notably during MIS 5e (Rosenberg et al. 2013). Subsequently the stone artefacts were incorporated into the cobble unit by lateral erosion of the stream bank with minimal disturbance and transportation.

Just what activities were carried out there, apart from stone-tool manufacture, and whether they included the exploitation of molluscs or other marine resources, cannot be known from the evidence currently available. However, the combination of a freshwater stream outlet and a marine shoreline represents a well-known combination of ecological advantages and ecotonal variety generally attractive to plant and animal life and therefore to human settlement, with the addition of marine resources at the shore edge. The ready availability of raw materials for making stone tools from the local cinder cone, the basaltic lava flows and the cobbles in the stream bed would have offered an added attraction, though this cannot have been the only reason for concentrating activity in this location, given the widespread distribution of these raw materials more widely in the region. Topographic opportunities for trapping large mammals may have been an additional factor (see Kübler et al., this volume). These possibilities reinforce the importance of obtaining dates, and further evaluation must await the outcome of the dating programme.

In the Farasan Islands, the most important observation from an archaeological point of view is the relative rapidity with which shorelines have been elevated because of salt tectonics. This has implications both for where we might expect to find archaeology and coastlines associated with earlier periods of high sea level, and also for the instability of coastal substrates and molluscan habitats on coastlines subject to rapid changes of elevation and geomorphology (see Bailey et al., this volume).

Since the Farasan Islands would have been connected to the mainland during periods of low sea level, there is no reason, in principle, why we should not expect to find MSA or earlier stone artefacts. However, only two surface finds of stone artefacts that might be of MSA type have so far been found on the Farasan Islands, both on the highest elevations

Fig. 14 Coral terraces and structural features southeast of Al Hussain: **a** ~ 3 m (amsl)-coral platform showing undercutting by marine erosion; **b** ~30-m (amsl) high uplifted terrace, showing marine sediments capped by coral; **c** 120° striking large open fracture cutting high terrace (width is ~60 cm); **d** NW–SE striking small graben on same terrace. Location shown in Fig. 12. Photos: Geoff Bailey (a, b), Bill Bosworth (c, d), November 2014

in the north of Farasan al Kabir, close to the village of Al Hussain (Bailey et al. 2013b). Critical to the evaluation of the potential for habitation in earlier periods of the Pleistocene is the dating of the different coral platforms we have described, and in particular the lowest in the sequence at 2–3 m. This platform, on the shore edge of which many shell mounds are located, forms an extensive land surface in many parts of the Islands, and would have provided important territory for hunting animals and exploitation of other resources as well as opportunities for marine exploitation at the shore edge. Clearly, when this coral platform was forming underwater, the area of land available for terrestrial exploitation would have been much reduced. Age constraints on the dates of formation of these coral terraces will help to evaluate the potentials of the region for human occupation during earlier periods before the establishment of modern sea level and the appearance of the shell mounds at about 7 ka.

Table 3 Palaeoshoreline features in the Farasan Islands showing locations visited and samples taken for dating. Data are presented in the same form as in Table 2. Note that all shorelines have been affected to some degree by salt tectonics. Approximate elevations are not corrected for mean sea level or converted to estimates of palaeo sea level

Locality Name	GPS	Description	DGPS Station name	Measured Elevation of Upper Surface (m)	Corrected elevations (m amsl)	Palaeo sea level (m amsl)	Margin of uncertainty (m)	Coral samples for U-series dating	Basalt Samples for $^{40}Ar/^{39}Ar$ dating	Shells
Farasan Kabir Is. (2014-11-29-7)	N 16.71564° E 41.89341°	Low terrace in central part of island		~4 5.93				FA-001, -002, -003	–	FA-004, -006 (Tridacna) F-005 (Spondylus) F-007 (surface shell)
Farasan Kabir Is. 2014-11-29-8	N 16.71945° E 41.89085°	Large bank of small shells on central fault block terrace		~13				–	–	FA-008 (surface shells)
Am Murabaah Is. 2014-11-30-1	N 16.73954° E 42.20501°	Shells on top of island coral terrace	Sulayn	~8				–	–	FA-012 (surface shells)
Am Murabaah Is. 2014-11-30-2	N 16.73982° E 42.20495°		Sulayn	~10				FA-013	–	–
Am Murabaah Is. 2014-11-30-3	N 16.74007° E 42.20469°	Coral terrace face	Sulayn	~13				FA-014, -015, -015A (~ 40 cm below 15)	–	–
Al Abalat Is. 2014-11-30-5	N 16.80652° E 42.18283°	Top of main coral terrace	Sulayn	~9				FA-016 (several m's lower in terrace face)	–	FA-017 (surface shells)
Farasan Kabir Is. DGPS St. 3 2014-12-01-3	N 16.62941° E 42.08151°	Uppermost of Ras Sheidah multiple terraces	Farasan Station 3	22.95	24.50	25.55	±0.5	FA-018	–	–
Farasan Kabir Is. DGPS St. 3 2014-12-01-4A	N 16.62950° E 42.08172°	Uppermost of Ras Sheidah multiple terraces	Farasan Station 3	19.99	21.54	22.54	±0.5	FA-019	–	–
Farasan Kabir Is. DGPS St. 3 2014-12-01-4B	N 16.63005° E 42.08369°	Uppermost of Ras Sheidah multiple terraces	Farasan Station 3	20.81	22.36	23.36	±0.5	FA-020	–	–
Farasan Kabir Is. DGPS St. 3 2014-12-01-5	N 16.63056° E 42.08177°	Ras Sheidah intermediate terrace w. sinkholes	Farasan Station 3	15.59	17.14	18.14	±0.5	FA-021	–	–
Farasan Kabir Is. DGPS St. 3 2014-12-01-8	N 16.63229° E 42.08458°	Ras Sheidah intermediate terrace	Farasan Station 3	8.52	10.07	11.07	±0.5	FA-022	–	–
Farasan Kabir Is. DGPS St. 3 2014-12-01-9	N 16.63368° E 42.08331°	Ras Sheidah intermediate terrace	Farasan Station 3	9.07	10.62	11.62	±0.5	FA-023A	–	FA-023B (Tridacna)
Farasan Kabir Is. 2014-12-01-11	N 16.64675° E 42.11082°	Wave cut notch below low ~3 m terrace.		~3			±0.5	–	–	FA-024 (Tridacna from ~ 1 m below terrace top)
Farasan Kabir Is. Power Station 2014-12-01-13	N 16.68409° E 42.10396°	Low coral terrace		~3			±0.5	FA-025	–	–

Fig. 15 Coral platforms near Ras Sheida on Farasan al Kabir and on the island of Am Murabaah: **a** Deeply undercut ~3 m coral platform to the east of Ras Sheida looking north towards Farasan Town (below letter a); to the right in the distance a line of shell mounds is visible as a series of white mounds on the edge of the bay; **b** Tridacna shell (sample FA023) embedded in section of ~3-m coral platform at modern shore edge before removal; **c** ~8 m elevated coral platform on Am Murabaah; **d** closer view of the Am Murabaah terrace. Photos: Geoff Bailey, 1 December 2014 (a, b); 30 November 2014 (c, d)

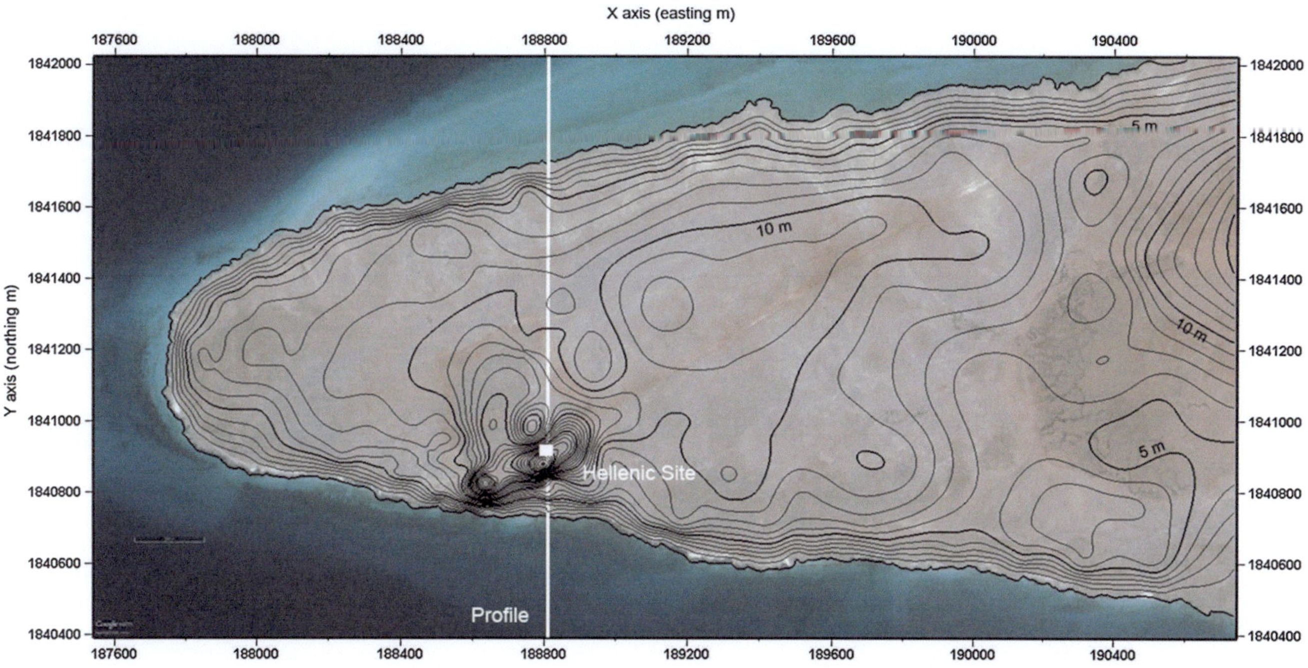

Fig. 16 Farasan Al Kabir Station 3. Digital elevation model for Ras Sheida peninsula. Contour interval is one metre. Data sources and methodology are described in the text

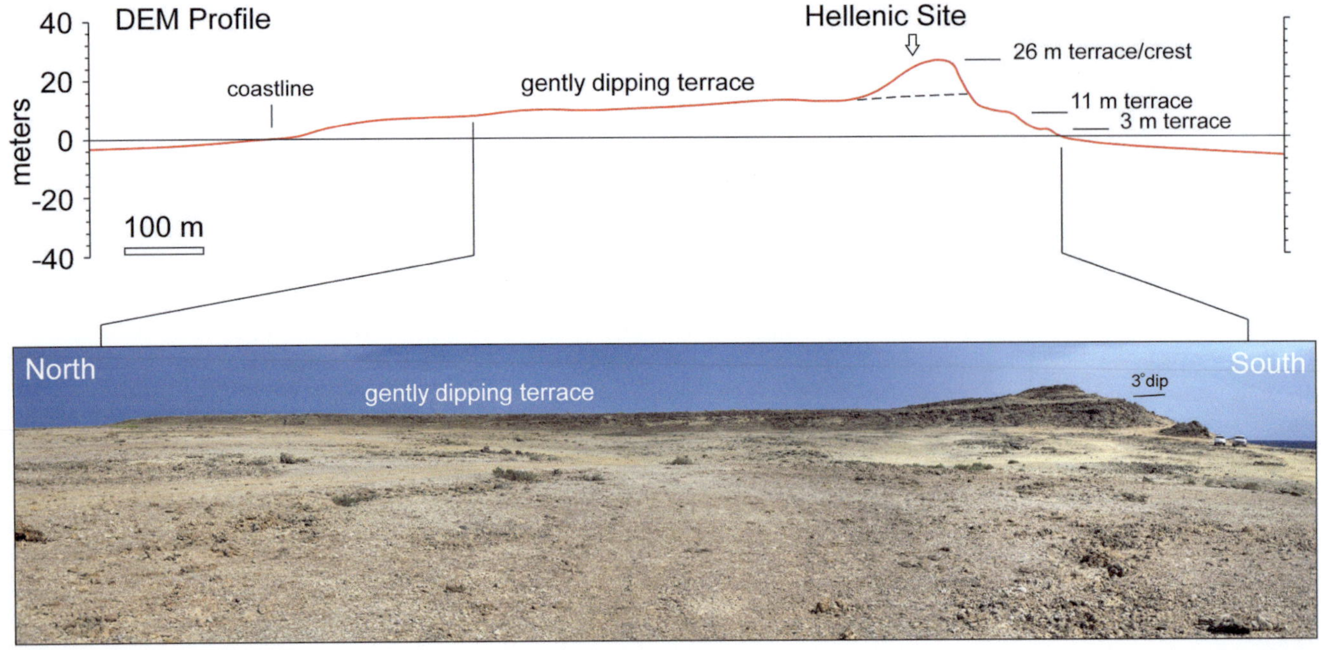

Fig. 17 Topographic profile across Farasan Station 3 and field view facing east; profile location is shown in Fig. 16. Photo: Bill Bosworth, December 2014

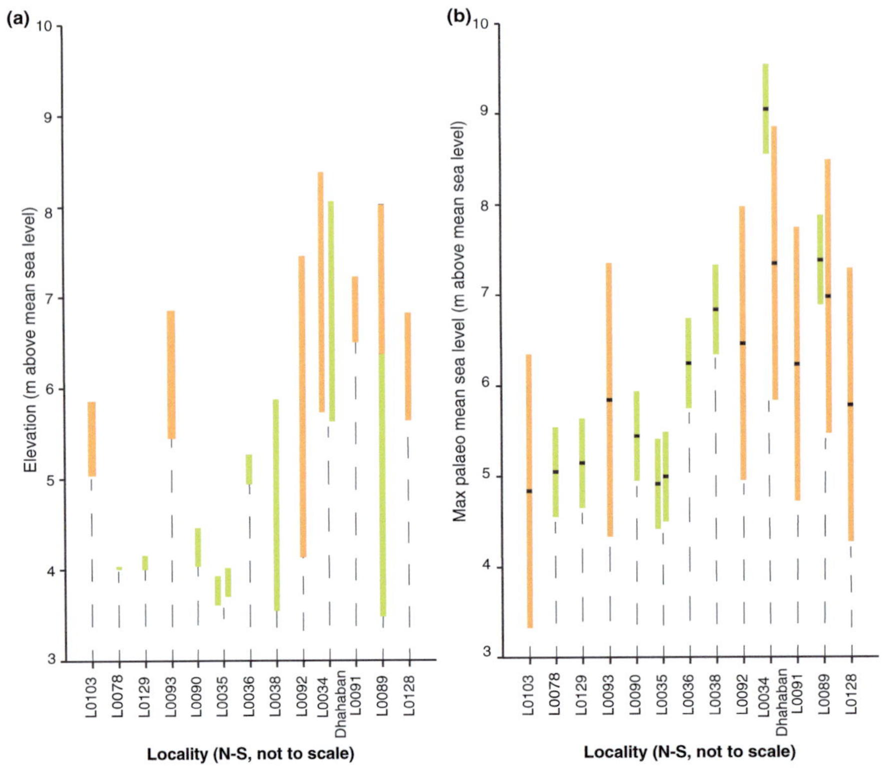

Fig. 18 Schematic of surveyed marine deposit heights in the Harrat al Birk (**a**) and the maximum extrapolated palaeo sea level they represent (**b**). Orange bars for beachrock, green bars for coral. Elevations in (**a**) were measured from the uppermost extant surface of each terrace or unit (in the case of L0034, L0089), and bars in this graph show the maximum variability in elevation of the upper surface of the unit across the area surveyed, not unit depth. In (**b**), the black line represents the maximum elevation of extrapolated palaeo sea level, and the coloured bars the errors associated with this extrapolation

9 Conclusion

This work, whilst in its preliminary stages, has recorded and sampled a key archive of fossil marine deposits in the south-eastern Red Sea, deposits that provide benchmarks for relative sea level change and their specific relationship to the archaeological remains left by the populations that inhabited these regions. At least some of the archaeological material we have studied on the coastline of the Harrat al Birk can be stratigraphically linked to an earlier shoreline, and our preliminary investigations indicate that this belongs to MIS 5e at about 125 ka. Our field data are also bringing to light emerging differences of interpretation in relation to the interactions between eustatic changes in sea level, isostatic adjustment and tectonic movements.

Whilst the absence of chronometric dates means that the hypotheses outlined above remain, at present, in need of further testing, it is clear that the emerging data are highlighting new issues in debates about the interactions between sea level change, tectonic movements and the use of coastlines by human populations in the Red Sea during the Pleistocene. Our field data have helped to sharpen the focus on issues in need of further investigation and to identify specific and competing hypotheses and how they can be resolved by chronometric dating. Above all they highlight the ongoing interest and importance of the Red Sea Basin as a 'laboratory' for investigating the multiple inter-relationships between sea-level change, tectonic processes, changes in coastal geomorphology and palaeogeography, and the archaeology of early human interest in coastlines and marine resources.

Acknowledgements We thank the Saudi Geological Survey for their generous support of this research. We also thank the SCTH for their support of our archaeological research in the region. We also thank two anonymous reviewers for their useful comments. The November–December 2014 fieldwork team comprised Salem Al Nomani, Ali Al Saeedi, Fahad Al Rashidi, Thamer Bakarman, Adel Jurais, Mustafa Khorsheed, Najeeb Rasul, and Nawaf Widinly (Saudi Geological Survey), Geoff Bailey and Robyn Inglis (University of York, U.K.), and William Bosworth (Apache Egypt Companies, Cairo, Egypt). Bailey and Inglis also acknowledge funding from the European Research Council (ERC) under the Ideas Programme of the 7th Framework Programme as Advanced Grant 269586 'DISPERSE: Dynamic Landscapes, Coastal Environments and Human Dispersals'. Inglis also acknowledges funding from the European Union's Horizon 2020 research and innovation programme under the Marie Skłodowska-Curie grant agreement No. 660343, 'SURFACE: Human-Landscape-Interactions and Global Dispersals: The Surface Record of Palaeolithic Arabia'. This is DISPERSE contribution no. 40.

References

AlSharekh AM, Bailey GN (eds) (2013) Coastal prehistory in southwest Arabia and the Farasan Islands: 2004–2009 field investigations. Series of archaeological refereed studies No. 15. Saudi Commission for Tourism and Antiquities, Riyadh

Armitage SJ, Jasim SA, Marks AE, Parker AG, Usik VI, Uerpmann HP (2011) The Southern route "Out of Africa": evidence for an early expansion of modern humans into Arabia. Science 331:453–456

ArRajehi A, McClusky S, Reilinger R, Daoud M, Alchalbi A, Ergintav S, Gomez F, Sholan J, Bou-Rabee F, Ogubazghi G, Haileab B, Fisseha S, Asfaw L, Mahmoud S, Rayan A, Bendik R, Kogan L (2010) Geodetic constraints on present-day motion of the Arabian Plate: Implications for Red Sea and Gulf of Aden rifting. Tectonics 29(3)

Bailey GN (2009) The Red Sea, coastal landscapes, and hominin dispersals. In: Petraglia MD, Rose JI (eds) The evolution of human populations in Arabia. Springer, Dordrecht, pp 15–37

Bailey GN (2015) The evolution of the Red Sea as a human habitat during the Quaternary period. In: Rasul NMA, Stewart ICF (eds) The Red Sea: The formation, morphology, oceanography and environment of a young ocean basin. Springer Earth System Sciences, Berlin Heidelberg, pp 595–610

Bailey GN, Alsharekh AM (eds) (2018) Palaeolithic Archaeology, Coastal Prehistory and Submerged Landscapes in Southwest Saudi Arabia and the Farasan Islands: DISPERSE field reports, 2012–2015. Saudi Commission for Tourism and National Heritage, Riyadh

Bailey G, Carrión JS, Fa DA, Finlayson C, Finlayson G, Rodríguez--Vidal J (2008) The coastal shelf of the Mediterranean and beyond: Corridor and refugium for human populations in the Pleistocene. Quat Sci Rev 27(23/24):2095–2099

Bailey GN, Alsharekh A, Flemming NC, Lambeck K, Momber G, Sinclair A,Vita-Finzi C (2007a) Coastal prehistory in the southern Red Sea basin, underwater archaeology, and the Farasan Islands. In: Proceedings of the Seminar for Arabian Studies, vol 37, pp 1–16

Bailey GN, Flemming NC, King GCP, Lambeck K, Momber G, Moran LJ, Al-Sharekh A, Vita-Finzi C (2007b) Coastlines, submerged landscapes, and human evolution: The Red Sea Basin and the Farasan Islands. J Island Coast Archaeol 2(2):127–160

Bailey GN, Alsharekh A, Flemming NC, Momber G, Moran LJ, Sinclair A, King GCP, Vita Finzi C, Al Ma'Mary A, AlShaikh NY, AlGhamdi S (2012) Coastal archaeology and prehistory in the southwest region of Saudi Arabia and the Farasan Islands: Report on the 2004 and 2006 surveys of the joint Saudi-UK Southern Red Sea Project. Atlal: J Saudi Arabian Archaeol 22:134–157

Bailey GN, Williams MGM, Alsharekh A (2013a) Shell mounds of the Farasan Islands, Saudi Arabia. In: Bailey GN, Hardy K, Camara A (eds) Shell energy: mollusc shells as coastal resources. Oxbow, Oxford, pp 241–254

Bailey GN, Alsharekh AM, Flemming NC, Momber G, Moran LJ, Sinclair A, King GCP, Vita-Finzi C, Al Ma'Mary A, Al Shaikh NY, Al Ghamdi S (2013b) Report on the 2004 and 2006 surveys of the joint Saudi-UK Southern Red Sea Project. In: Alsharekh AM, Bailey GN (eds) Coastal prehistory in southwest Arabia and the Farasan Islands: 2004–2009 field investigations. Saudi Commission for Tourism and Antiquities, Riyadh, pp 15–74

Bailey GN, Devès MH, Inglis RH, Meredith-Williams MG, Momber G, Sakellariou D, Sinclair A, Rousakis G, Al Ghamdi S, Alsharekh A (2015) Blue Arabia: Palaeolithic and underwater survey in SW Saudi Arabia and the role of coasts in Pleistocene dispersals. Quat Int 382:42–57

Bailey GN, Meredith-Williams MW, Alsharekh A (this volume) The archaeology of Pleistocene coastal environments and human dispersals in the Red Sea: Insights from the Farasan Islands

Bantan RA (1999) Geology and sedimentary environments of Farasan Bank (Saudi Arabia) southern Red Sea: a combined remote sensing and field study. PhD Thesis, Royal Holloway, University of London

Benjamin J, Rovere A, Fontana A, Furlani S, Vacchi M, Inglis RH, Galili E, Antonioli F, Sivan D, Miko S, Mourtzas N, Felja I, Meredith-Williams M, Goodman-Tchernov B, Kolaiti E,

Anzidei M, Gehrels R (2017) Late Quaternary sea-level changes and early human societies in the central and eastern Mediterranean Basin: an interdisciplinary review. Quat Int 449:29–57

Beyin A (2006) The Bab al Mandab vs the Nile-Levant: an appraisal of the two dispersal routes for early modern humans out of Africa. Afr Archaeol Rev 23:5–30

Bohannon RG (1986) Tectonic configuration of the Western Arabian continental margin, Southern Red Sea. Tectonics 5:477–499

Bohannon RG, Naeser CW, Schmidt DL, Zimmerman RA (1989) The timing of uplift, volcanism, and rifting peripheral to the Red Sea: a case for passive rifting? J Geophys Res 94:1683–1701

Bonatti E, Cipriani A, Lupi L (2015) The Red Sea: Birth of an ocean. In: Rasul NMA, Stewart ICF (eds) The Red Sea: the formation, morphology, oceanography and environment of a young ocean basin. Springer Earth System Sciences, Berlin Heidelberg, pp 29–44

Bosence DWJ, Al-Aawah MH, Davison I, Rosen BR, Vita-Finzi C, Whitaker E (1998) Salt domes and their control on basin margin sedimentation: a case study from the Tihama Plain, Yemen. In: Purser BH, Bosence DWJ (eds) Sedimentation and tectonics of rift basins: Red Sea-Gulf of Aden. Chapman and Hall, London, pp 448–464

Bosworth W (2015) Geological evolution of the Red Sea: historical background, review, and synthesis. In: Rasul NMA, Stewart ICF (eds) The Red Sea: the formation, morphology, oceanography and environment of a young ocean basin. Springer Earth System Sciences, Berlin Heidelberg, pp 45–78

Bosworth W, Stockli DF (2016) Early magmatism in the greater Red Sea rift: timing and significance. Can J Earth Sci 53:1–19

Bosworth W, Taviani M, Rasul N (this volume) Neotectonics of the Red Sea, Gulf of Suez and Gulf of Aqaba

Bruckner A, Rowlands G, Riegl B, Purkis S, Williams A, Renaud P (2011) Atlas of Saudi Arabian Red Sea marine habitats. Khaled bin Sultan Living Oceans Foundation, Panoramic Press, Phoenix, AZ

Cochran JR (1983) A model for development of the Red Sea. Bull Am Ass Pet Geol 67:41–69

Cochran JR, Karner GD (2007) Constraints on the deformation and rupturing of continental lithosphere of the Red Sea: The transition from rifting to drifting. In: Karner GD, Manatschal G, Pinheiro LM (eds) Imaging, mapping and modelling continental lithosphere extension and breakup, vol 282. Geological Society London Special Publications, pp 265–289

Coleman RG (1993) Geologic evolution of the Red Sea, Oxford monographs on geology and geophysics, vol 24. Oxford University Press, Oxford

Dabbagh A, Emmermann R, Hötzl H, Jado AR, Lippolt HJ, Kollman W, Moser H, Rauert W, Zötl JG (1984) The development of Tihamat Asir during the Quaternary. In: Jado AR, Zötl JG (eds) Quaternary period in Saudi Arabia: Sedimentological, hydrogeological, hydrochemical, geomorphological, geochronological and climatological investigations in western Saudi Arabia, vol 2. Springer, Vienna, pp 150–173

Devès MH, Inglis RH, Meredith-Williams MG, Al Ghamdi S, Alsharekh A, Bailey G (2013) Palaeolithic survey in southwest Saudi Arabia: methodology and preliminary results. Adumatu 27:7–30

Dullo W-C (1990) Facies, fossil record, and age of Pleistocene reefs from the Red Sea–Saudi Arabia. Facies 81:175–192

Dullo W-C, Montaggioni L (1998) Modern Red Sea coral reefs: a review of morphologies and zonation. In: Purser BH, Bosence DWJ (eds) Sedimentation and tectonics of rift basins: Red Sea-Gulf of Aden. Chapman and Hall, London, pp 583–594

El Moursi M, Hoang CT, El Fayoumy IF, Hegab O, Faure H (1994) Pleistocene evolution of the Red Sea coastal plain, Egypt: Evidence from Uranium-series dating of emerged reef terraces. Quat Sci Rev 13:345–359

Erlandson JM (2001) The archaeology of aquatic adaptations: Paradigms for a new millennium. J Archaeol Res 9:287–350

Erlandson JM, Braje TJ (2015) Coasting out of Africa: the potential of mangrove forests and marine habitats to facilitate human coastal expansion via the Southern Dispersal Route. Quat Int 382:31–41

Faure H, Hoang CT, Lalou C (1980) Datations ^{230}Th/^{234}U de calcaires corallines et mouvements verticaux à Djibouti. Bull Societé Géologique de France 22:959–962

Girdler RW, Styles P (1974) Two-stage Red Sea floor spreading. Nature 247:7–11

Girdler RW, Whitmarsh RB (1974) Miocene evaporites in Red Sea cores, their relevance to the problem of the width and age of oceanic crust beneath the Red Sea. In: Whitmarsh RB, Weser OE, Ross DA et al (eds) Initial reports of the deep sea drilling project, vol 23. Government Printing Office, Washington, pp 913–921

Groucutt HS, Petraglia MD, Bailey GN, Scerri EM, Parton A, Clark-Balzan L, Jennings RP, Lewis L, Blinkhorn J, Drake NA, Breeze PS, Inglis RH, Devès MH, Meredith-Williams MG, Boivin N, Thomas MG, Scally A (2015) Rethinking the dispersal of *Homo sapiens* out of Africa. Evol Anth 24:149–164

Gvirtzman G (1994) Fluctuations of sea level during the past 400,000 years: The record of Sinai, Egypt (northern Red Sea). Coral Reefs 13:203–214

Hausmann N, Meredith-Williams MG (2017) Seasonal patterns of coastal exploitation on the Farasan Islands. Saudi Arabia. J Island Coast Archaeol 12(3):360–379

Hausmann N, Kokkinaki O, Leng MJ (this volume) Red Sea palaeoclimate: Stable isotope and element-ratio analysis of marine molluscs

Hendy E, Tomiak P, Collins MJ, Hellstrom J, Tudhope A, Lough J, Penkman KEH (2012) Assessing amino acid racemization variability in coral intra-crystalline protein for geochronological applications. Geochim Cosmochim Acta 86:338–353

Hoang CT, Taviani M (1991) Stratigraphic and tectonic implications of Uranium-series-dated coral reefs from uplifted Red Sea Islands. Quat Res 35:264–273

Hoang CT, Dalongeville R, Sanlaville P (1996) Stratigraphy, tectonics and palaeoclimatic implications of uranium-series-dated coral reefs from the Sudanese coast of the Red Sea. Quat Int 31:47–51

Hubert-Ferrari A, King GCP, Manighetti I, Armijo R, Meyer B, Tapponnier P (2003) Long-term elasticity in the continental lithosphere: modelling the Aden Ridge propagation and the Anatolian extrusion process. Geophys J Int 153:111–132

Inglis RH, Sinclair AGM, Shuttleworth A, Alsharekh AM, Al Ghamdi S (2013) Preliminary report on 2013 fieldwork in southwest Saudi Arabia by the DISPERSE Project: (2) Jizan and Asir Provinces, February-March 2013. http://eprints.whiterose.ac.uk/80964/

Inglis RH, Sinclair A, Shuttleworth A, Alsharekh A, Al Ghamdi A, Devès M, Meredith-Williams M, Bailey GN (2014a) Investigating the Palaeolithic landscapes and archaeology of the Jizan and Asir regions, southwest Saudi Arabia. In: Proceedings of the Seminar for Arabian Studies, vol 44, pp 193–212

Inglis RH, Sinclair AGM, Shuttleworth A, Meredith-Williams MG, Hausmann N, Budd W, Alsharekh A, Al Ghamdi S, Bailey GN (2014b) Preliminary report on 2014 fieldwork in southwest Saudi Arabia by the DISPERSE Project: (1) Jizan and Asir Provinces. http://eprints.whiterose.ac.uk/80965/

Inglis RH, Foulds F, Shuttleworth A, Alsharekh AM, Al Ghamdi S, Sinclair AG, Bailey GN (2015) The Palaeolithic occupation of the Harrat Al Birk: Preliminary report on the 2015 fieldwork in Asir Province, southwest Saudi Arabia. http://eprints.whiterose.ac.uk/84382/

Jerardino A (2017) On the origins and significance of Pleistocene coastal resource use in southern Africa with particular reference to shellfish gathering. J Anthropol Archaeol 41:213–230

Kübler S, Devès MH, Bailey GN King GCP (this volume) Tectonic geomorphology and soil edaphics as controls on animal migrations and human dispersal patterns

Kopp RE, Simons FJ, Mitrovica JX, Maloof AC, Oppenheimer M (2009) Probabilistic assessment of sea level during the last interglacial stage. Nature 462:863–868

Lambeck K, Purcell A, Flemming NC, Vita-Finzi C, Alsharekh AM, Bailey GN (2011) Sea level and shoreline reconstructions for the Red Sea: isostatic and tectonic considerations and implications for hominin migration out of Africa. Quat Sci Rev 30:3542–3574

Macaulay V, Hill C, Achilli A, Rengo C, Clarke D, Meehan W, Blackburn J, Semino O, Scozzari R, Cruciani F, Taha A, Shaari NK, Raja JM, Ismail P, Zainuddin Z, Goodwin W, Bulbeck D, Bandelt HJ, Oppenheimer S, Torroni A, Richards M (2005) Single, rapid coastal settlement of Asia revealed by analysis of complete mitochondrial genomes. Science 308:1034–1036

Macfadyen WA (1930) The geology of the Farasan Islands, Gizan and Kamaran Island, Red Sea. Part 1. General Geology. Geol Mag 67:310–315

Manaa AA, Jones BG, McGregor HV, Zhao J, Price DM (2016) Dating Quaternary raised coral terraces along the Saudi Arabian Red Sea coast. Mar Geol 374:59–72

Mauz B, Vacchi M, Green A, Hoffman G, Cooper A (2015) Beachrock: a tool for reconstructing relative sea level in the far-field. Mar Geol 362:1–16

McGuire AV, Bohannon RG (1989) Timing of mantle upwelling: Evidence for a passive origin for the Red Sea Rift. J Geophys Res 94:1677–1682

McQuarrie N, Stock JM, Verdel C, Wernicke BP (2003) Cenozoic evolution of Neotethys and implications for the causes of plate motions. Geophys Res Lett 30:1–4

Mellars P, Gori KC, Carr M, Soares PA, Richards MB (2013) Genetic and archaeological perspectives on the initial modern human colonization of Southern Asia. Proc Natl Acad Sci USA 110 (26):10699–10704

Meredith-Williams MG, Bailey GN, Hausmann N, Al Ghamdi S, Alsharekh AM, Douka K, Beresford C, Larsen B (2013) Preliminary report on 2013 fieldwork in southwest Saudi Arabia by the DISPERSE Project: (1) The Farasan Islands. http://www.disperse-project.org/sites/disperse-project.org/files/uploads/DISPERSE_2013%281%29_01_Txt%2Bfigs.pdf

Meredith-Williams MG, Hausmann N, Inglis R, Bailey GN (2014) 4200 new shell mound sites in the southern Red Sea. In: Fernandes R, Meadows J (eds) Human exploitation of aquatic landscapes. Special Issue, Internet Archaeol 37. https://doi.org/10.11141/ia.37.2

Mohriak WU (2015) Rift basins in the Red Sea and Gulf of Aden: Analogies with Southern South Atlantic. In: Post PJ, Coleman JL, Rosen NC, Brown DE, Roberts-Ashby T, Kahn P, Rowan M (eds) Petroleum systems in "Rift" basins. 34th annual GCSSEPM foundation perkins-rosen research conference, December 13–16, 2015, Houston, CD-ROM, pp 789–826

Parker AG (2009) Pleistocene climate change in Arabia: developing a framework for hominin dispersal over the last 350 ka. In: Petraglia MD, Rose JI (eds) The evolution of human populations in Arabia. Springer, Amsterdam, pp 39–49

Penkman KEH, Kaufman DS, Maddy D, Collins MJ (2008) Closed-system behaviour of the intra-crystalline fraction of amino acids in mollusc shells. Quat Geochron 3(1–2):225

Petraglia MD, Alsharekh AM, Crassard R, Drake NA, Groucutt H, Parker AG, Roberts RG (2011) Middle Paleolithic occupation on a Marine Isotope Stage 5 lakeshore in the Nefud Desert, Saudi Arabia. Quat Sci Rev 30(13–14):1555–1559

Petraglia MD, Breeze PS, Groucutt HS (this volume) Blue Arabia, green Arabia: examining human colonisation and dispersal models

Plaziat J-C, Baltzer F, Choukri A, Conchon O, Freytet P, Orszag-Sperber F, Raguideau A, Reyss J-L (1998) Quaternary marine and continental sedimentation in the northern Red Sea and Gulf of Suez (Egyptian coast): Influences of rift tectonics, climatic changes and sea-level fluctuations. In: Purser BH, Bosence DWJ (eds) Sedimentation and tectonics of rift basins: Red Sea-Gulf of Aden. Chapman and Hall, London, pp 537–573

Plaziat J-C, Reyss J-L, Choukri A, Cazala C (2008) Diagenetic rejuvenation of raised coral reefs and precision of dating. The contribution of the Red Sea reefs to the question of reliability of the Uranium-series datings of middle to late Pleistocene key reef-terraces of the world. Carnets de Géologie 4:1–40

Reilinger R, McClusky S, ArRajehi A (2015) Geodetic constraints on the geodynamic evolution of the Red Sea. In: Rasul NMA, Stewart ICF (eds) The Red Sea: the formation, morphology, oceanography and environment of a young ocean basin. Springer Earth System Sciences, Berlin Heidelberg, pp 135–149

Rohling EJ, Grant KM, Roberts AP, Larrasoaña J-C (2013) Paleoclimate variability in the Mediterranean and Red Sea regions during the last 500,000 years. Curr Anthr 54(S8):S183–S201

Rose JI, Usik VI, Marks AE, Hilbert Y, Galletti HCS, Parton A, Geiling JM, Morley MW, Roberts RG (2011) The Nubian complex of Dhofar, Oman: An African middle stone age industry in Southern Arabia. PLoS ONE 6(11):1–22

Rovere A, Raymo ME, Vacchi M, Lorschied T, Stocchi P, Gomez-Pujol L, Harris DL, Casella E, O'Leary MJ, Hearty PJ (2016) The analysis of Last Interglacial (MIS5e) relative sea-level indicators: Reconstructing sea-level in a warmer world. Earth-Sci Rev 159:404–427

Rosenberg TM, Preusser F, Risberg J, Plikk A, Kadi KA, Matter A, Fleitmann D (2013) Middle and Late Pleistocene humid periods recorded in palaeolake deposits of the Nafud desert, Saudi Arabia. Quat Sci Rev 70:109–123

Sakellariou D, Rousakis G, Panagiotopoulos I, Morfis I, Bailey G (this volume) Geological structure and Late Quaternary geomorphological evolution of the Farasan Islands continental shelf, South Red Sea, SW Saudi Arabia

Sanderson D, Kinnaird T (this volume) Optically stimulated luminescence dating as a geochronological tool for late quaternary sediments in the Red Sea region

Siddall M, Rohling EJ, Almogi-Labin A, Hemleben C, Meischner D, Schmeltzer I, Smeed DA (2003) Sea-level fluctuations during the last glacial cycle. Nature 423:853–858

Sinclair AG, Inglis RH, Shuttleworth A, Foulds FW, Bailey GN, Alsharekh A (this volume) Landscape archaeology, Palaeolithic survey and coastal change along the Southern Red Sea of Saudi Arabia

Spohner R, Oleman P (1986) Topography, Red Sea Region (1:8,000,000). German Research Foundation, Karlsruhe

Walter RC, Buffler RT, Bruggemann JJ, Guillaume MMM, Berhe SM, Negassi B, Libsekal Y, Cheng H, Edwards RL, von Gosel R, Neraudeau D, Gagnon M (2000) Early human occupation of the Red Sea coast of Eritrea during the last interglacial. Nature 405:65–69

Zarins J, Murad A, Al-Yaish K (1981) The second preliminary report on the southwestern Province. Atlal 5:9–42

Zarins J, Whalen NM, Ibrahim I, Morad A, Khan M (1980) Comprehensive archaeological survey program: preliminary report on the Central southwestern Provinces survey. Atlal 4:9–117

The Archaeology of Pleistocene Coastal Environments and Human Dispersals in the Red Sea: Insights from the Farasan Islands

Geoff N. Bailey, Matthew Meredith-Williams, Abdullah Alsharekh, and Niklas Hausmann

Abstract

This chapter examines the different sources of evidence—phylogenetic, palaeoclimatic and archaeological—that have been used to investigate the hypothesis that early human dispersals from Africa during the late Pleistocene were facilitated by exploitation of marine resources and seafaring abilities and followed a predominantly coastal route including a crossing of the southern end of the Red Sea. We examine critically the current evidence and arguments for and against such a hypothesis and highlight the need for a more sophisticated understanding of the taphonomic factors that determine the formation, preservation and distribution of coastal archaeological deposits such as shell mounds. We present new data on the mid-Holocene shell mounds of the Farasan Islands and examine their spatial and temporal distribution in relation to a coastal environment that has been subject to rapid changes of sea level, geomorphology and ecological potential. We demonstrate that substantial shell mound deposits can accumulate rapidly over a matter of decades, even in a dynamic shoreline environment undergoing changes in relative sea level, that the ecological conditions that provide an abundant supply of marine molluscs as food are highly episodic in time and space, and that the resulting archaeological record is extremely patchy. We highlight the problem of dealing with negative evidence in the archaeological record and the need for a more detailed investigation and understanding of the various factors that determine the survival and visibility of archaeological deposits.

G. N. Bailey (✉) · N. Hausmann
Department of Archaeology, University of York, King's Manor, York, YO1 7EP, UK
e-mail: geoff.bailey@york.ac.uk

G. N. Bailey
College of Humanities, Arts and Social Sciences, Flinders University, GPO Box 2100 Adelaide, SA 5001, Australia

M. Meredith-Williams
Department of Archaeology and History, La Trobe University, Bundoora, VIC 3086, Australia

A. Alsharekh
Department of Archaeology, King Saud University, Riyadh, Saudi Arabia

N. Hausmann
Foundation for Research and Technology–Hellas, Institute of Electronic Structure and Laser (IESL), P.O. Box 1527 711 10 Heraklion, Greece

1 Introduction

Over the past decade, worldwide interest has focussed on the Red Sea region, and especially on the western escarpment and coastal regions of Saudi Arabia, because of the unusually favourable conditions that it offers for the understanding of Quaternary sea-level change and the earliest human dispersals out of Africa over the past 2 million years. A number of different sources of evidence have contributed to this new interest:

1. New archaeological field projects and new finds of early Stone Age archaeology in many parts of the Arabian Peninsula, including the desert interior, the western escarpment of Saudi Arabia and the coastal hinterland of Yemen, Oman and the UAE (Petraglia and Rose 2009; Armitage et al. 2011; Bretzke et al. 2013; Inglis et al. 2014; Bailey et al. 2015; Petraglia et al., this volume; Sinclair et al., this volume)
2. Phylogenetic inference, which has led some authors to argue for a pattern of rapid dispersal of early human populations around the Indian Ocean rim from the Horn of Africa to the shores of Australia, taking in a crossing of the southern Red Sea and the coastlines of southern Arabia (Macaulay et al. 2005; Mellars et al. 2013)
3. Renewed interest in the likely early significance of marine resources such as shellfish and fish in the palaeodiets of early human populations and the possibilities of early sea travel (Erlandson 2001; Bailey and

N. M. A. Rasul and I. C. F. Stewart (eds.), *Geological Setting, Palaeoenvironment and Archaeology of the Red Sea*,
https://doi.org/10.1007/978-3-319-99408-6_26

Milner 2002; Anderson et al. 2010; Marean 2010; Jerardino 2016)

4. A growing realisation that the low sea levels that have persisted throughout much of the Quaternary period have dramatically altered the palaeogeography of the Arabian coastline by (a) creating extensive areas of now-submerged terrestrial landscape attractive to earlier human settlement, particularly in the southern Red Sea, and (b) narrowing sea channels, in particular in the region of the Hanish Sill and Bab al Mandab, which would have made them more easily crossable than today (Siddall et al. 2003; Lambeck et al. 2011; Bailey et al. 2015)
5. New data on climate change (Parker 2009; Drake et al. 2013; Rohling et al. 2013; Rosenberg et al. 2013)

Particular interest in the past decade has focussed on the hypothesis of a major dispersal of modern humans (*H. sapiens sapiens*) out of Africa at about 70 ka (70,000 years ago) or perhaps earlier, accompanied by new adaptations involving seafaring abilities and intensified marine exploitation, with the main pathway of dispersal across the southern Red Sea, around the rim of the Indian Ocean and into New Guinea and Australia.

This idea has taken powerful hold of the scientific and popular imagination, but is largely speculative, deriving support primarily from phylogenetic inference and other indirect clues. Supporting field evidence for or against this hypothesis of an early coastal dispersal is almost non-existent, arguably because the relevant areas and time periods in question are now submerged below modern sea level.

Our aim in this chapter is, firstly, to summarise the arguments for and against the suitability of a dispersal pathway across the southern end of the Red Sea and around the shorelines of the Arabian Peninsula, and secondly, to examine the sorts of archaeological evidence that currently exist for the use of early shorelines and marine resources, or that we need to look for and might expect to find in support of such a proposition. Since so little archaeological evidence is currently available before the establishment of modern sea level at about 6 ka, we focus on the shell mounds of the Farasan Islands, which date from about this time. These are typical of the types of highly visible archaeological deposits that are associated with exploitation of marine foods and sea-travel in a hunter-gatherer setting, and are the type of evidence that we might expect to find in earlier periods and in other coastal regions as markers of a coastal pattern of settlement and dispersal associated with substantial reliance on marine resources. We will present new field evidence relating to the Farasan shell mounds, and use this evidence to examine the ecological and geological circumstances associated with their formation, the ways in which their formation interacts with dynamic changes in shoreline position during the Holocene period, and the likelihood of finding similar deposits at other times and places, in particular during earlier periods of low sea level.

The basis of our approach is that we need to understand the variable processes that determine the differential accumulation, preservation, destruction and visibility of archaeological deposits, before we can draw reliable inferences about past patterns of human demography and dispersal from archaeological site distributions in time and space. Deposits of stone tools, especially in large numbers, usually accumulate very slowly and acquire their archaeological visibility from repeated visits to the same location over long periods (decades to millennia). Once established, the stone tools are resistant to decay but are vulnerable to processes of burial under later sediments, or to displacement and dispersal by erosion. Shell mounds, in contrast, can accumulate very rapidly (days to decades) but only under quite restricted ecological conditions; they are also highly sensitive to burial or removal by small changes in sea level and coastline geomorphology. Both types of deposits are vulnerable to destruction by human impact from modern land use and industrial development. It is these processes that we seek to highlight in the following investigation.

2 Expansion Out of Africa

Both fossil evidence of human remains and phylogenetic data point to Africa as the homeland of our ancient ancestors, with at least two waves of expansion out of Africa: an early one after about 2 Ma by the early members of the genus *Homo* (most likely *H. ergaster* or *H. erectus*), and a later one by anatomically modern humans (*H. sapiens sapiens*). The date of the later dispersal is unclear. The earliest known and dated specimens of anatomically modern *H.s.s.* come from the Kibish formation in Ethiopia dated at 196 ± 2 ka (Fleagle et al. 2008), but the earliest human fossil specimens of *H.s.s.* outside Africa are much younger with dates ranging from, at earliest, about 120 ka in SW Asia, 70 ka or later in SE Asia and Australia, and 40 ka in Europe.

Whether any of these broad dates are representative of the earliest dates of entry of *H.s.s.* populations into their respective continents is a matter of conjecture, subject to all the vagaries of differential preservation and discovery of fossil human remains that are inherently rare. The European dates are quite well constrained thanks to the presence of numerous well-dated remains of Neanderthals (*Homo sapiens neanderthalensis*), who evolved separately in Europe from an earlier common ancestor, and who are reliably associated with distinctive Mousterian tool assemblages

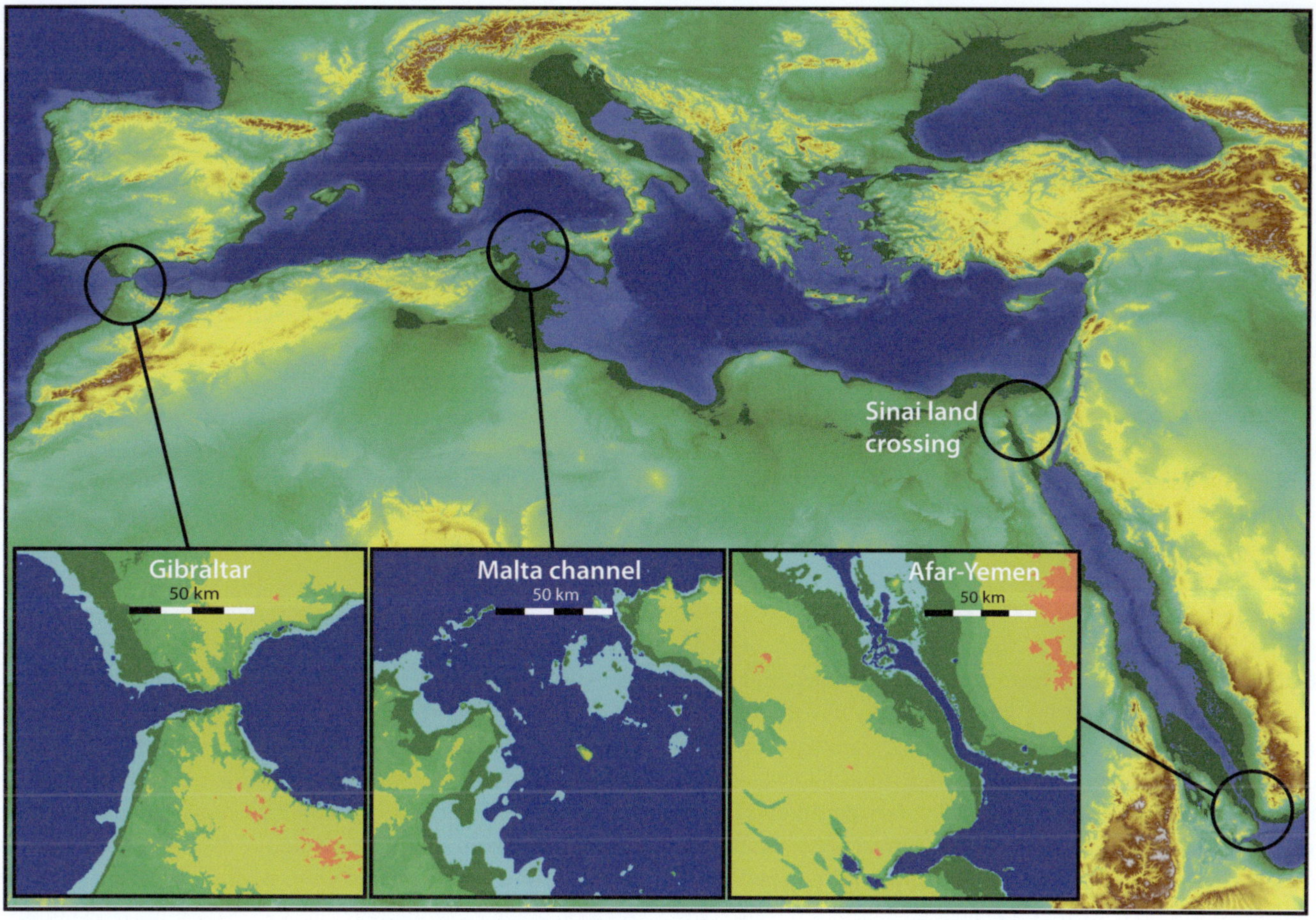

Fig. 1 Map of North Africa and the Mediterranean Basin showing major potential exit points for human dispersal out of Africa. Insets show the distance of sea crossings at the present day and when sea-levels were at −130 m (light blue shading). Courtesy of Maud Devès and the DISPERSE project

quite different from those associated with incoming *H.s.s.* populations. In Asia the picture is much less clear with few human fossil finds, huge gaps in time and space between them, and no characteristic stone-tool assemblages that can be used as reliable proxies for different hominin taxa.

Africa is a very difficult continent from which to escape. All the possible exit routes represent narrow bottlenecks. It has been generally assumed that the most likely exit for all early movements out of Africa is the narrow land corridor of the Sinai Peninsula connecting the Nile Valley to the Levant, on the assumption that land routes would have been the only viable ones for early human populations. The possibility of sea crossings of the Mediterranean during the Palaeolithic era has also been raised. The narrowest crossing at any position of sea level is about 12 km across the Strait of Gibraltar between opposing shores that are clearly inter-visible (Fig. 1). But neither here nor elsewhere in the Mediterranean is there decisive evidence for sea crossings based, for example, on indisputably dated human presence on islands, until a relatively late date from about 13 ka onward, despite claims for earlier evidence (Alimen 1975; Derricourt 2005; Strasser et al. 2010; Ammerman 2013, 2014). In any case, if the land exit via the Sinai is a bottleneck difficult to get through, sea crossings are likely to have been even more so.

2.1 The Southern Dispersal Route

A crossing at the southern end of the Red Sea, with or without a land bridge, has also periodically been proposed, most notably by Lahr and Foley (1994), who first raised this possibility as an alternative pathway for human populations moving out of Africa and into SE Asia. This particular pathway has gained new popularity from a variety of genetic, palaeogeographical and archaeological inferences, often coupled with the idea of new adaptations involving sea travel and use of marine resources as distinctive features of *H.s.s.* that gave them an evolutionary advantage and opened up new environments for colonisation and new migration pathways involving sea crossings (Beyin 2006, 2011; Mellars 2006; Oppenheimer 2012a, b; Mellars et al. 2013).

2.2 Phylogenetic Evidence

Genetic inferences about the dates of human dispersals have been derived from the construction of phylogenetic trees based on genetic variations in present-day human populations and mutation rates of DNA. These give patterns of divergence that suggest a date of 70–60 ka for the *H.s.s.* dispersal out of Africa (Macaulay et al. 2005; Soares et al. 2012; Oppenheimer 2012a, b; Mellars et al. 2013). In some cases, even the geographical routes of dispersal between Africa and SE Asia have been inferred from this evidence.

However, the assumptions on which these conclusions are based and their accuracy have been challenged and remain uncertain (Bailey 2009; Boivin et al. 2013; Field and Lahr 2005; Field et al. 2007; Groucutt et al. 2015). The mutation rates on which the DNA clock is based are uncertain and likely to have varied over time, and some studies indicate that during genetic bottlenecks the rate can actually increase (e.g., Ho et al. 2011). Divergence dates estimated from genetic modelling are subject to very large margins of error, with confidence intervals amounting to many tens of thousands of years. Divergence times also depend on anchoring the extrapolation to independently dated events. By their very nature, these are rare, and comprise a limited number of identifiable and dated migrations of genetically-known modern human populations splitting from their source populations and moving into new territory, for example the expansion of Bantu populations within Africa at 3.5 ka, the entry of anatomically modern humans into Australia supposedly at 50 ka, and the first colonisation of remote Pacific Islands within the past millennium. Other difficulties are that the genetic signature of earlier population expansions may be erased or swamped by later ones, and that nothing in the genetic relationships between modern populations living in Africa and on the Indian subcontinent can possibly specify a preference for a coastal pathway as opposed to a hinterland pathway for the expansion of the common ancestral populations out of Africa. The genetic data are equally compatible with an expansion out of Africa at 130 ka rather than 70 ka or indeed with a yet earlier date, and with a variety of alternative geographical pathways including crossings of the Arabian deserts during periods of wetter climate (Drake et al. 2013; Groucutt et al. 2015).

As often when dealing with investigations that take place in the no-man's-land at the intersection of different disciplinary boundaries, the risk of circular argument is high. For example, geneticists may depend on dates for human fossils and their archaeological proxies in the form of stone-tool assemblages to calibrate the DNA clock; and archaeologists may rely on the inferences of geneticists to provide dates for the fossil and archaeological record; both are inclined to rely on global climatic records of low geographical resolution derived from ice cores, speleothems and marine or lacustrine sediments to reinforce a preferred hypothesis.

Recovery of ancient DNA from fossil bone, though still rare—and likely to remain so because of the limited chronological range and temperature conditions in which ancient DNA is preserved—offers another source of genetic data. If anything, however, this seems likely to complicate rather than simplify the evolutionary narrative, since recent data suggest genetic mixing across taxonomic boundaries, for example between Neanderthals and anatomically modern humans (Green et al. 2010). Similar complications are emerging from DNA analysis of other mammalian taxa, for example mammoths (Enk et al. 2016).

Many genetic interpretations produced by independent analyses of different modern populations appear to be mutually reinforcing, but since they all depend on a similar set of assumptions about rates of genetic change and dates of divergence that are difficult to test independently, they need to be treated with caution and a critical eye when used to examine or reinforce archaeological interpretations about the demography and dispersal patterns of prehistoric human populations. As genetic studies extend more widely amongst modern human populations, so the patterning is becoming more complex, with evidence for two-way migrations and multiple dispersals, some much earlier than the original estimates (Malispinas et al. 2016; Pagani et al. 2016; Bohlender et al. 2016).

Another indirect source of genetic information of relevance to the southern dispersal route is the evidence from DNA analysis of modern baboon populations (*Papio hamadryas*) (Kopp et al. 2014). *Papio* evolved in Africa, splitting from *Theropithecus* according to fossil evidence at about 5 Ma, a date which provides a chronological anchor for phylogenetic reconstruction based on DNA analysis of modern populations. Baboons are present today in Arabia and in northeast Africa with major concentrations in SW Arabia, Eritrea and Ethiopia. DNA analysis shows that these populations form different genetic clades with evidence that the Arabian populations are derived from the African ones, together with some evidence of back migration from Arabia to Eritrea. Kopp et al. (2014) conclude from an analysis of the geographic distribution of genetic diversity and estimates of divergence times and population expansion that the most parsimonious interpretation of these data is that baboons dispersed across the southern end of the Red Sea on at least two occasions: At a mean date of about 150 ka with a range of 222–88 ka; and at 31 ka with a range of 55–11 ka (both ranges at the 95% confidence interval). These results are of considerable interest both in providing a transparent analysis and interpretation of modern genetic data that highlights the margins of error associated with inferred dates, and also in highlighting the likelihood of movement across the southern end of the Red Sea by non-human primates. If baboons were able to make the crossing, seemingly on more than one occasion, why not early humans?

2.3 Palaeogeographical and Palaeoclimatic Variables

The first question to address is the nature of the sea crossing at the southern end of the Red sea. At the present day, the shortest sea crossing is across the Bab al Mandab Strait, a sea-distance of about 30 km. Although it would be possible in principle for an endurance swimmer to cross this distance, a 30 km-wide sea channel is likely to have been an effective barrier to population movements without the use of sea-worthy water craft. We know that Neolithic communities on the Arabian side used artefacts made of obsidian derived from sources in Ethiopia from as early as 8 ka (Khalidi 2009; Khalidi et al. 2010, 2012), so that sea crossings must have been taking place at that time. What sort of sea craft were used is not known for certain, but remains of barnacle-encrusted reeds and bitumen from the Gulf coast of Kuwait dated at about 7 ka suggest that boats made from bundles of reeds were in use at this time, and possibly boats made from sewn wooden planks (Lawler 2002; Carter 2010; Boivin et al. 2009). As with evidence from elsewhere in the region and from more recent periods, boats made from bundles of reeds are the most obvious candidate.

How much earlier such a technology was in use is impossible to say with any certainty. Isolated surface finds of obsidian artefacts have been found in surveys in SW Arabia, which could be from earlier periods (http://www.disperse-project.org/field-reports). However, we do not know their date or the source of the obsidian—there are sources in the Yemen as well as in Ethiopia—nor is the absence of obsidian or lack of evidence for its movement across the Red Sea an argument against the presence of boat technology or sea crossings.

A more relevant consideration, and one for which we do have some evidence is the effect of lowered sea level on the distance across the southern Red Sea (Fig. 2). At the maximum lowering of the Red Sea during the last glacial maximum at about 25–20 ka, global sea levels were 120–130 m lower than present because of the amount of sea water locked up in the continental ice sheets. The southern Red Sea was reduced to a relatively long, narrow and shallow sea channel extending for about 100 km from the Bab al Mandab to the Hanish Sill and no more than about 10 km wide (Siddall et al. 2003; Rohling et al. 2010, 2013). Palaeogeographic reconstructions of palaeoshorelines in the region taking account of global seawater volumes as well as isostatic and tectonic movements of the Earth's crust show that at the shallowest point of this channel, in the vicinity of the Hanish Sill, there was a group of islands in mid channel (Fig. 3). These reduce the sea distances to 4 km or less by island hopping from one side of the channel to the other (Lambeck et al. 2011).

In addition, the geometry of the channel is such that, although it is very wide at present sea-level, over 100 km, below a depth of about 40 m it narrows to a channel with quite steep sides, meaning that the possibility of short crossings would have persisted for any period when sea level was lower than about 50 m below present. Given the chronology of sea level change, this means that short sea crossings would have been available for as much as 40 kyr during the last 100-kyr-sea-level cycle and for some 140 kyr during the past 400 kyr (Lambeck et al. 2011).

Rohling et al. (2013) incorporating climatic proxies have further suggested that the most favourable windows of opportunity for sea crossings would have been narrowed to periods when climatic amelioration and low sea level coincided (see also Fig. 6 in Bailey et al. 2015). Other authors have also used the evidence of broad Pleistocene climatic cycles that periodically opened up the deserts of Arabia and the Sahara to the spread of grasslands and surface water to further constrain the timing of human dispersal (e.g., Drake et al. 2013; Kopp et al. 2014; Breeze et al. 2015, 2016). However, it should be remembered that the key for animal and human populations on the ground is the availability of soil nutrients, vegetation and water, and these are often highly variable at the sub-regional and local scale depending on local factors of topography, tectonic activity and hydrology (see Kübler et al., this volume). Human populations are likely to be attracted to combinations of environmental features that are atypical of the broader environmental or climatic zone within which they occur, and which afford a measure of insensitivity to fluctuations in rainfall or other regional climatic parameters. These especially occur in geological unstable regions where faulting and other tectonic or geomorphological processes rejuvenate soils and soil nutrients, create spring lines, and trap sediments and water in local basins (King and Bailey 2006; Bailey and King 2011; Winder et al. 2015; Kübler et al. this volume).

The southwest region of Arabia is persistently 'green' under most climatic conditions that have been experienced during the late Pleistocene, including those of the present day (Bailey et al. 2015). These attractions are likely to have been further enhanced when sea levels were low by the increased availability of water supplies on the exposed continental shelf because of the presence of fault-bounded basins, and the increased hydraulic head from groundwater

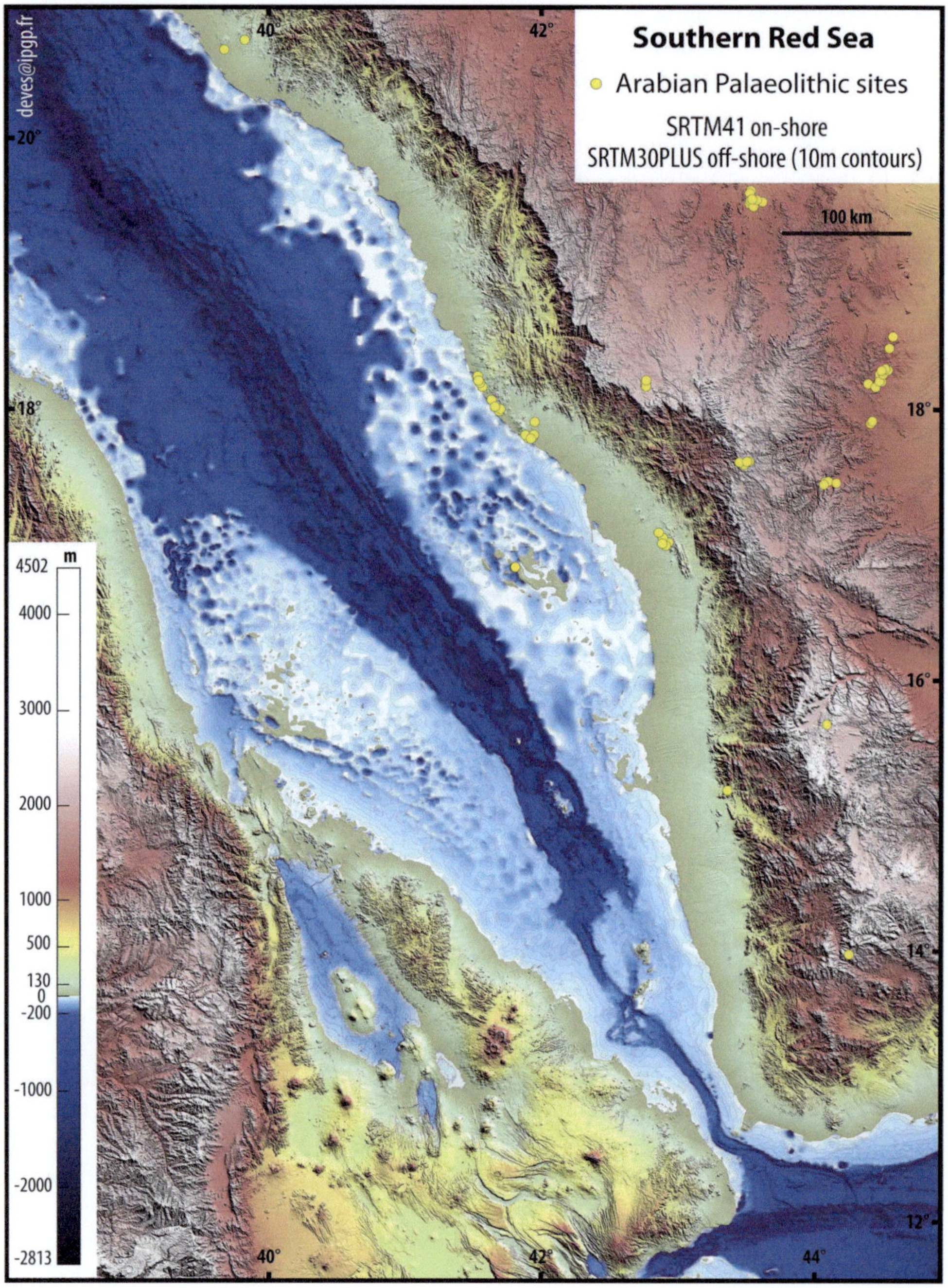

Fig. 2 Enhanced satellite imagery of the southern Red Sea showing the position of the Farasan Islands, and the extent of the submerged landscape at maximum sea-level regression during glacial periods, and the general nature of the seabed topography. ASTER GDEM is a product of METI and NASA. Courtesy of Maud Devès and the DISPERSE project

reservoirs, creating groundwater springs and relatively attractive local conditions regardless of climatic aridity at the continental scale (Faure et al. 2002; Bailey et al. 2015; Sakellariou et al., this volume).

In addition, the isotope data from deep sea cores from within the Red Sea and the evidence of what is known or can be inferred about crustal movements associated with rifting indicate that these palaeogeographical and palaeoclimatic scenarios are likely to have persisted for at least the past half million years. Before that time, the position is less clear because of lack of data from deep-sea sediments extending that far back in time, and because of uncertainties associated with the separation of the Arabian and African plates.

The key point is that opportunities for short sea crossings would have persisted for relatively long periods of thousands of years at a time, if not tens of thousands, over the past half million years. Given some degree of inter-visibility between opposing shores and availability of vegetation and water on

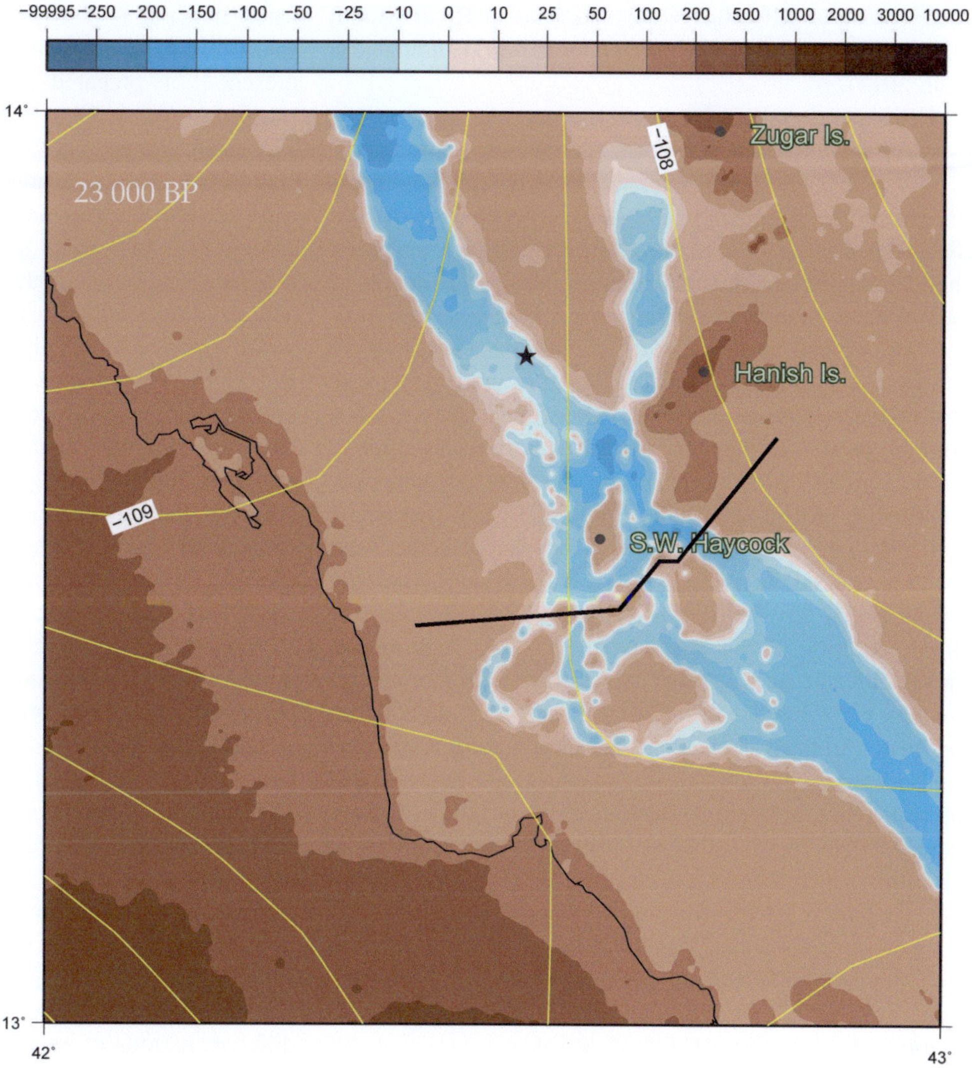

Fig. 3 Palaeoshorelines at the southern end of the Red Sea during the maximum sea-level low-stand at the Last Glacial Maximum. The present-day coastline is indicated by the thin black line. The black star shows the location of the Hanish Sill and the yellow lines represent contours of equal sea level position relative to the present at 23 ka. The shortest sea crossings at this period would have been in the region of the Hanish Sill as shown by the thick black line. After Lambeck et al. 2011

both sides, distances would have been short enough and the windows of opportunity long enough to suggest a high probability of crossings by swimming, floating or rafting on natural mats of vegetation whether by chance or by design. However, the distances are not so large that they require any need to invoke the construction of water craft or seafaring abilities as practised during the Holocene period. The fact that baboons, as noted above, appear to have colonised SW Arabia from NE Africa across the southern channel on at least two occasions between about 220 ka and 11 ka reinforces this conclusion. It should be added that many other terrestrial mammals including hippopotami, elephants, cervids, bovids, equids, ursids and some carnivores are capable of swimming over relatively short distances whether by choice or compulsion. We see no need in any of this evidence to suppose that short sea crossings during favourable windows of opportunity could not have been accomplished by human populations at any time in the past half million years or indeed earlier.

2.4 Archaeological Evidence

Palaeogenetic, palaeoclimatic and palaeogeographical information offer, at best, indirect clues as to the pattern of human settlement and dispersal. The ultimate test must lie in archaeological field data. Here the available evidence for late Pleistocene coastal sites around the southern shores of the Red Sea, especially in shore-edge settings, is meagre. Only one site is unquestionably on the shore edge, and that is the site of Abdur in Eritrea (Walter et al. 2000). Here, artefacts of Middle Stone Age (MSA) type including flakes made on obsidian have been found stratified within beach deposits associated with reef corals which have been dated to about 125 ka. Also present in these deposits are mammal bones and mollusc shells of edible species including oysters and mussels, which the authors have interpreted as evidence of subsistence activity on the beach, the marine foods indicating a novel adaptation comparable to the fish bones and mollusc shells found at a similar date in the caves of South

Africa, such as Blombos Cave and Klasies River Mouth (Henshilwood et al. 2001; Van Niekerk 2011; Langejans et al. 2017).

There is no reason to doubt the age or stratigraphic associations of the Abdur deposit, and the connection to *H.s. s.* is plausible, given the date and the geographical region. The rest of the interpretation, however, is open to challenge. Since the deposit as a whole is a beach deposit including shell and coral-reef material that is clearly the debris of natural death assemblages, including whole oysters described as in growth position, the question of what is food shell and what is natural shell needs to be posed and investigated. Distinguishing between the two types of shell material when both are present in the same deposit is notoriously difficult and has given rise to an extensive literature on diagnostic criteria (e.g., Attenbrow 1992; Rowland 1994; Watters et al. 1992; Rosendahl et al. 2007). The fact that artefacts were found next to edible marine mollusc shells is not necessarily evidence that the shells were eaten as food, nor does the sharp condition of the artefacts rule out post-depositional movement and displacement. A reef flat is not the most obvious place for people to butcher large mammals or for that matter to process and consume locally collected marine foods. Without a thoroughgoing taphonomic analysis that examines the source and condition of all these various materials and the possible agencies of transportation and deposition, including natural as well as cultural ones, the palaeoconomic interpretation must remain in doubt. This evidence remains a very slender basis on which to hypothesise the exploitation of marine resources, coastal environments and the crossing of the southern Red Sea as significant factors in human dispersal out of Africa.

Elevated coral reefs of presumed similar age associated with an MIS 5 high sea-level have been extensively surveyed on the SW Saudi Arabian coastline of the Red Sea (Inglis et al., this volume), but so far with no evidence comparable to Abdur. Middle Stone Age or Middle Palaeolithic artefacts have been found on the surface of some of these coral terraces, but almost none in demonstrable stratigraphic association with them. The exception is Dhahaban Quarry, where stone artefacts are embedded in a water-laid cobble unit beneath a series of shallow marine and beach deposits, and associated with a coral platform ~6.5 m above modern sea level (Inglis et al. 2014, and this volume; Sinclair et al., this volume). Isolated shells of oyster, giant clam and mangrove whelk and pieces of coral are in the same depositional unit, but the shells are not food remains

Otherwise, artefacts of Middle Stone Age type, and in many cases Early Stone Age type, have been recovered in many areas in the Arabian Peninsula. Many are surface finds where dating is impossible except by reference to typological and technological comparisons with better studied and dated material in the neighbouring regions of Africa and southwest Asia. MSA material is typically identified by the presence of prepared core technologies of discoid or Levallois type. At Jebel Faya, material of this type is dated at 125 ka (Armitage et al. 2011; Bretzke et al. 2013), but initial claims that this was evidence for the presence of anatomically modern humans have not been supported by finds of associated human fossil material. Since prepared core technologies have been widely used over at least the past 250 ka and are associated with hominins such as Neanderthals as well as early anatomically modern humans such as the Skhul and Qafzeh remains in Israel, they are clearly unreliable proxies for the presence of anatomically modern humans. Other recently investigated material in the Nefud dates back to about 100 ka (Petraglia et al. 2011), and some of the Arabian material includes bifacially worked flakes of Acheulean type that very likely date back to the Middle Pleistocene, perhaps as early as 0.8 Ma (Petraglia et al. 2009; Bailey et al. 2015).

Most of this material is in the Arabian hinterland and much of it in the Arabian deserts, associated with periods of climate change when hydrological corridors with lakes or wetlands and grasslands were extensively distributed across the deserts of the Rub al ‘Khali and the Nefud (Breeze et al. 2015, 2016; Petraglia et al., this volume). None of this material can be described as coastal except in the very broad sense that some of the sites are in coastal regions broadly defined, such as the recently discovered site of Wadi Dabsa in the Harrat Al Birk region of southwest Saudi Arabia (Foulds et al. 2017; Inglis et al. 2017; Sinclair et al. this volume), with an extensive palimpsest of material of many different periods in an inland basin about 10 km from the present-day coastline.

At first sight the distribution of the archaeological material suggests an overwhelming preference for hinterland locations and little or no interest in the coastline or its marine resources until the mid-Holocene. However, we should beware of taking this at face value.

For example, Usik et al. (2013) have cast doubt on the hypothesis that there was a rapid coastal dispersal of modern humans around the Indian ocean coastline at 60 ka on the grounds that sites of this date are absent on coastlines where we should expect to find them if the hypothesis is correct. On the Oman coastline, for example, MSA (MIS5 or earlier) and Neolithic (mid-Holocene) sites are present but none of Upper Palaeolithic or Late Stone Age type that could be assigned to the intervening period (MIS 4–2). One could argue against this that 60 ka coastal sites would now be on submerged coastlines, but Usik et al. (2013) note that sites of this period are absent even on the uplifted coastline of the Musandam Peninsula. One difficulty here is that sites of this period are rare or absent everywhere in the Arabian

Peninsula. Either there was a prolonged period of human absence, or the archaeological sites of this period are generally of very low archaeological visibility.

How then, are we to deal with negative evidence, that is the absence of archaeological material on or near coastlines, even in areas where the conditions for preservation or visibility on the coastline appear to be good? One place to start investigating this relationship in more detail is to look at more recent coastlines where archaeological material with clear evidence of marine exploitation is well preserved, and to examine the relationship between the formation processes of archaeological deposits, and their spatio-temporal distribution in relation to variations in the geomorphology and ecology of the contemporaneous coastline.

3 Shell Mounds of the Farasan Islands

The Farasan Islands have several advantages for such an examination. In the first place they have one of the largest concentrations of shell mounds known anywhere in the world, comparable to the better known and studied concentrations of shell mounds in Europe, Japan, the Americas, South Africa, West Africa and Australia (Bailey et al. 2013a). This concentration is in part the product of the high ecological fertility of the surrounding inshore waters, and in part due to the remoteness of the islands, the small population size, and the lack of modern development. Shell mounds are notoriously vulnerable to destruction by quarrying activities to exploit the shell for building material or simply as obstacles in the way of road building or other construction activities, often accompanied by the belief that they are natural deposits with no cultural significance. The Farasan Islands have witnessed very little of these developmental impacts until recently, offering a near-pristine distribution of archaeological material along the coastline. Also, the islands are located on a salt dome, with quite rapid uplift of some shorelines because of salt tectonics, and down warping of others (Inglis et al., this volume). In this respect the Islands offer an unusual insight into the impact of changing relative sea levels on the coastal geomorphology of shorelines, molluscan habitats and the occurrence and distribution of shell mounds.

The Farasan shell mounds are often described as 'Neolithic' in the Arabian context, mainly because of their date. Elsewhere shell mounds of similar form in Europe are described as 'Mesolithic', in North America as 'Archaic', or in other parts of the world by some other label according to the local nomenclature. The great majority of the larger shell mounds and concentrations of open air shell middens found all across the world first appear in the mid-Holocene at about 7–6 ka. The implication of these labels is that the shell mounds are a relatively late cultural stage in the broader history of human development, a rung on the ladder of human progress. However, this is also the period when sea level stopped rising from a low sea-level stand of −120 m at the Last Glacial Maximum. There is no reason to suppose that earlier concentrations of shell mounds could not have existed at earlier periods during the Last Glacial on shorelines that are now submerged. Earlier evidence for the exploitation of marine molluscs is certainly present, back to at least 160,000 years ago (Jerardino and Marean 2010), and is especially prominent in deposits dating to the Last Interglacial period (MIS 5) when sea levels stood at heights generally similar to the present day. But the evidence is mostly in caves on the present-day coastline, and none of the deposits is on the scale of the mid-Holocene mound concentrations. This is most likely because these earlier sites are on rocky coastlines that are much less productive of marine molluscs than the bays and estuaries where the mid-Holocene mounds are concentrated.

We make no assumptions here about whether or not the Farasan shell mounds are a cultural phenomenon unique to their particular time and place. That is possible, but we maintain an open mind as to the possibility that similar shell mounds existed on earlier coastlines that are now submerged. Rather our intention here is to use the Farasan example as a high-resolution case study to examine the relationship between the variable preservation and visibility of shell-dominated coastal archaeological deposits and the variable ecology and geomorphology of the coastlines where they occur. In this way, the case study can serve as a benchmark against which to assess the significance of the more sparsely distributed evidence from earlier periods.

3.1 General Features

Over 3000 shell midden deposits have been recorded on the Farasan Islands (Williams 2010; Alsharekh et al. 2013, Bailey et al. 2013a, b; Meredith-Williams et al. 2013, 2014a, b; Hausman and Meredith-Williams 2016a, b). The largest are 5 m tall and roughly oval in plan, forming highly visible features of the landscape (Fig. 4). The largest may contain as much as 1000 m^3 of deposit, representing 1000 tonnes of shell and over 100 million molluscs. Not all deposits are so impressive in size; many are low mounds less than 1 m in thickness and some are surface scatters of shell less than 5 m in diameter with little accumulation of deposit. Nevertheless, the larger mounds are striking and highly visible features in the present-day landscape, sometimes forming a quasi-continuous line of mounds of varying size along the shoreline that are easily identifiable on satellite images (Fig. 5; Meredith-Williams et al. 2014b). Many of these larger mounds form clusters with mounds of varying size and smaller deposits nearby. These are usually lined up

Fig. 4 Line of shell mounds in Janaba Bay West, looking south. The shell mounds are sitting on a coral platform which has been eroded by marine action to form a low undercut cliff at a time when the sea penetrated further inland than today. The height of the cliff is c. 2 m Photo by Abdullah Al Zahrani

along the shoreline or else they are clustered in groups with the larger mounds along the shoreline and smaller mounds and shell scatters extending up to several hundred metres inland from the shoreline.

Nineteen shell mounds of varying size and location have been excavated with systematic exposure of sections, removal of bulk shell samples, collection of faunal and artefactual material where present, and recovery of individual specimens of charcoal and shell from the sections for radiocarbon dating (Table 1; see also Meredith-Williams et al. 2013). The bulk of the mounded shell deposits were formed between about 6.5 and 4.5 ka. Some younger-dated shell deposits are present, some as recent as the Islamic period, but these all have the form of shell scatters, or deposits <50 cm thick.

The dominant mollusc species in all the mounded deposits is the lineated conch, *Conomurex fasciatus*, a small gastropod 4–6 cm in length (Hausmann et al., this volume; Fig. 6). A variety of other species are present in varying proportions. These include the large gastropods, *Chicoreus ramosus* (the branched murex shell), and *Pleuroploca trapezium* (the horse conch). All the gastropod species are found on reef flats, usually on sandy substrates. Bivalve species include *Chama reflexa* (jewel box clam), *Pinctada* cf. *radiata* (pearl oyster), *Arca avellana* (hazelnut ark shell), *Plicatula plicata* (plicate kittens paw), *Spondylus marisrubri* (spiny oyster), *Beguina gubernaculum* (rudder cardita) and *Modiolus auriculatus* (eared horse mussel). These mostly occur on coral reefs and attach themselves to a hard surface with byssus threads or cement-like secretions. All of these species are present today, but the *Conomurex* shells are quite rare, in contrast to their obvious abundance in the prehistoric period. It is not clear whether this is due to temporary and episodic population collapse or long-term habitat degradation. In 2009 we found large quantities of live specimens on one of the smaller offshore islands in a sheltered sandy bay where they could be scooped up by the bucket-full. However, on returning in 2015, there was no trace of them.

The species that is most commonly used as food today is the large strombid gastropod, *Strombus tricornis* (Queen conch). The shells of this species are often found in small scatters near the modern shoreline where they have been left by modern picnickers, along with other typical modern artefacts such as empty drink cans. This species is relatively rare in the prehistoric shell mounds, but is certainly present and occasionally the dominant mollusc in some layers.

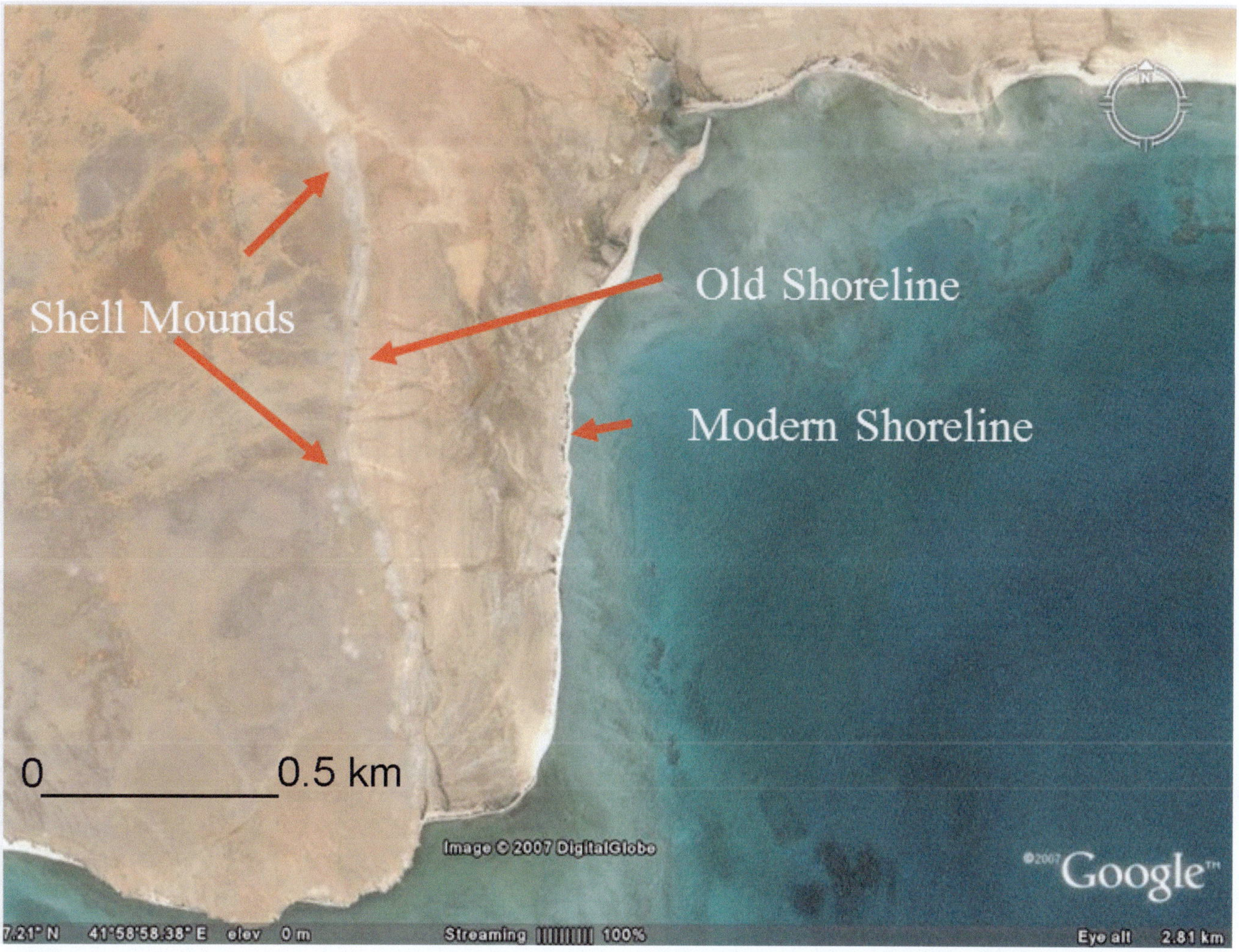

Fig. 5 Google Earth image showing aerial view of Janaba Bay West. Large shell mounds, including the mounds shown in Fig. 4, are clearly visible on a now-abandoned shoreline

Table 1 Radiocarbon dates for shell mounds in Janaba West. See Figs. 9 and 10 for location of sites. Calibration according to Reimer et al. (2013)

Lab Number	Site	Layer	Radiocarbon Age BP	Material	Species	Modelled age 2σ range Cal BP
OxA-31168	JW1705	1	3411 ± 31	marine shell	*C. fasciatus*	3325–2980
OxA-31166	JW1705	6	4842 ± 32	marine shell	*C. fasciatus*	5209–4836
OxA-31167	JW1705	4-basal	6870 ± 38	marine shell	*C. fasciatus*	7413–7172
OxA-27890	JW1727	17	4202 ± 29	charcoal	unknown	4844–4627
OxA-27889	JW1727	23	4287 ± 29	charcoal	unknown	4958–4825
OxA-28617	JW1727	23	4701 ± 28	marine shell	*C. fasciatus*	4970–4630
OxA-28009	JW1727	2	4851 ± 31	marine shell	*C. fasciatus*	5217–4842
OxA-34098	JW1727	8 Top	4759 ± 31	marine shell	*C. fasciatus*	5041–4712
OxA-34099	JW1727	8 Base	4539 ± 33	marine shell	*C. fasciatus*	4786–4433
OxA-31169	JW1727	27-basal	5044 ± 35	marine shell	*Brachidontes variabilis*	5444–5064
OxA-31170	JW5694	3	2767 ± 30	marine shell	*C. fasciatus*	2573–2177
OxA-30870	JW5694	4-basal	2902 ± 29	marine shell	*C. fasciatus*	2700–2365
OxA-31171	JW5697	3-basal	2220 ± 27	marine shell	*C. fasciatus*	1845–1555
OxA-31172	JW5719	basal	2500 ± 29	marine shell	*C. fasciatus*	2199–1870
OxA-31173	JW5719	upper	2554 ± 27	marine shell	*C. fasciatus*	2283–1955

Fig. 6 *Conomurex fasciatus* shell. Photo by Niklas Hausmann

The variable presence in the mounds of these different mollusc species most likely reflects differences in the intertidal and offshore habitats within reach of a given mound or on different parts of the coastline, and to some extent long-term changes in habitat availability, particularly for the *Conomurex* and *Strombus* species of the Strombidae family

Excavations show fairly uniform conditions of shell deposition, with a dominant matrix of whole or broken shells, variable proportions of fine ashy sediment, occasional interleaved ash lenses and scattered small pieces of charcoal (Fig. 7). Fish bones are present, mostly of small specimens most likely caught by net. The most common species are reef fish such as parrotfish (Scaridae), sea bream (Sparidae), emperors (Lethrinidae), trevallies (Carangidae) and groupers (Serranidae) (Beech 2018). There are also occasional bones of gazelle (*Gazella gazella*). Stone artefacts are rare, and include a ground stone axe and a worked flake, both made of volcanic rock only available on the mainland, manuports of coralline limestone of uncertain function, and occasional pieces of worked *Tridacna* (giant clam) shell. Potsherds are very rarely present. A human burial, comprising pits with the remains of two individuals, is present in the top of one of the excavated shell mounds, and is of the same date as the surrounding midden deposit.

The inhabitants of this period were clearly capable of regular sea travel, judging by the presence of shell mounds on smaller offshore islands and the contacts with the mainland, most likely using sea craft made of bundles of reeds as discussed earlier. Seasonality analyses of the mollusc shells also indicate that molluscs were exploited throughout the year, which rules out the hypothesis that the Islands were visited only for seasonal activities (Hausmann and Meredith-Williams 2016a).

3.2 Interpretation of Shell Quantities

It is important at this point to dispel two myths that commonly grow up around the existence of shell mounds on this scale. The first is that the quantities of shell are so vast that they must have been accumulated by natural agencies as natural shell banks or storm deposits rather than by human action. The second is that they represent the remains of people who lived mainly on shellfood.

The first hypothesis is refuted by a number of lines of evidence. First, there is no known natural agency that could dump shells in the form of discrete and mound-like structures at regular intervals along a shoreline or at distances of up to several hundred metres inland. Natural shell banks tend to form long low sub-parallel banks parallel to the modern shoreline and contain layers with large quantities of crushed shell, a very wide-range of molluscan species, beach sand and gravel, evidence of water abrasion and shell specimens of every size and age. The shell mounds we are talking about are dominated by a relatively restricted range of molluscan species, all of which are edible, whole and broken shells of mature-size specimens, ash lenses representing fire places used for cooking, artefacts, and bones of fish and terrestrial mammals.

It is true that artefacts are rare but this is the result of several factors, the scarcity of suitable stone material on the islands for making flaked or ground stone artefacts, the relatively rapid rates of accumulation of the shell deposits due to the bulk of the shell in comparison with other deposited materials, and, most likely, the specialised nature of the shell mounds as locations for the processing and consumption of marine mollusc shells and fish, rather than as settlement sites. The latter may be represented by some of the thinner shell mounds and shell scatters that are located inland from the bigger shell mounds on the immediate shoreline. Some of these have blocks of coral embedded in the shell deposit representing the remains of simple structures, and more abundant surface remains of broken potsherds.

As for the second hypothesis, a simple calculation shows that despite the impressive quantities of shell involved, the amount of food they represent is actually quite small. At a conservative estimate the known shell mounds on the Farasan Islands represent about 500,000 tonnes of shell

Fig. 7 Excavated trench showing section through a large shell mound, JE0087, in Janaba Bay East

material. Taking a shell-weight to meat-weight ratio of 5:1, a calorie content of 50 kcal per 100 g of meat, and a daily human individual requirement of 2000 kcal, this gives a total quantity of 25 million person-days of food or 68,500 person-years. Assuming a total duration of the Farasan shell mounds of 2000 years, this figure in its turn represents enough food to feed 34 people, assuming that they ate nothing but shellfood. Of course a diet of nothing but shellfood would rapidly lead to protein poisoning (Noli and Avery 1988), and these figures are not intended to represent an economic reality. Rather, they indicate parameters that are compatible with a relatively small resident population on the islands—of the order of hundreds rather than thousands of people—and a diet in which shellfood played an important but by no means dominant role. The belief in the dominant role of shellfood results from the great bulk of discarded shell relative to the meat content and the greater resistance of the discarded shells to decay and destruction compared to the material remains of animal and plant food. Nevertheless, here as in other regions of the world (Meehan 1982) the mollusc meat clearly played an important role in the subsistence economy. In particular, seasonality data show an increase of mollusc consumption during the more arid times of the year when other food sources would have been more limited (Hausmann and Meredith-Williams 2016a). Ethnographic observations of living people also show that the converse situation can arise, namely one where people collect large quantities of shellfood but do not necessarily deposit the discarded shells in the form of impressive shell mounds (Hardy et al. 2016).

One other point to comment on is the dominant presence of *Conomurex fasciatus*. This small gastropod at first sight does not look a promising source of food compared to the larger molluscs. It would need some form of mass capture using a bag or basket, and heat to kill the animal and loosen the meat from the shell. Since small gastropods requiring similar techniques of collection and processing are widely used today as food, for example as observed in the markets of China (pers. obs., 2012) and recorded in Australian Aboriginal ethnography (Rowland 1994), there is no reason to consider their small size and processing requirements as an argument against their collection as food.

3.3 Spatial Distribution

The first point to make is that the shell mounds are not uniformly distributed around the shorelines of the islands and the largest mounds and the largest concentrations of mounds are highly restricted in their distribution.

A closer examination shows that the large clusters of shell mounds are limited in their distribution compared to the smaller deposits and shell scatters, which are distributed more extensively around the coastlines of the Islands (Fig. 8). These large clusters are associated with very large shallow bays and this almost certainly reflects the fact that the extensive intertidal and subtidal sand flats hosted by these shallow bays provide the optimal habitat for molluscs, particularly the gastropods, and the most extensive habitat, with vast quantities of living shell. Along the more open shorelines, shell mounds are smaller, exist only as small surface scatters, or are completely absent over long stretches of shoreline, reflecting much less suitable local habitats for large quantities of molluscs, and also on some shorelines greater difficulties of access to the shore because of more irregular terrain and the presence of low cliffs created by marine erosion.

3.4 Chronological Distribution, Rates of Deposition and the Impact of Coastal Change

The chronology of the mounds indicates that the main concentrations, and especially the larger mounds are not only clumped in space, but they are also clumped in time, with a major concentration in the period between 6.5 ka and 4.5 ka. The principal reason for this appears to be the

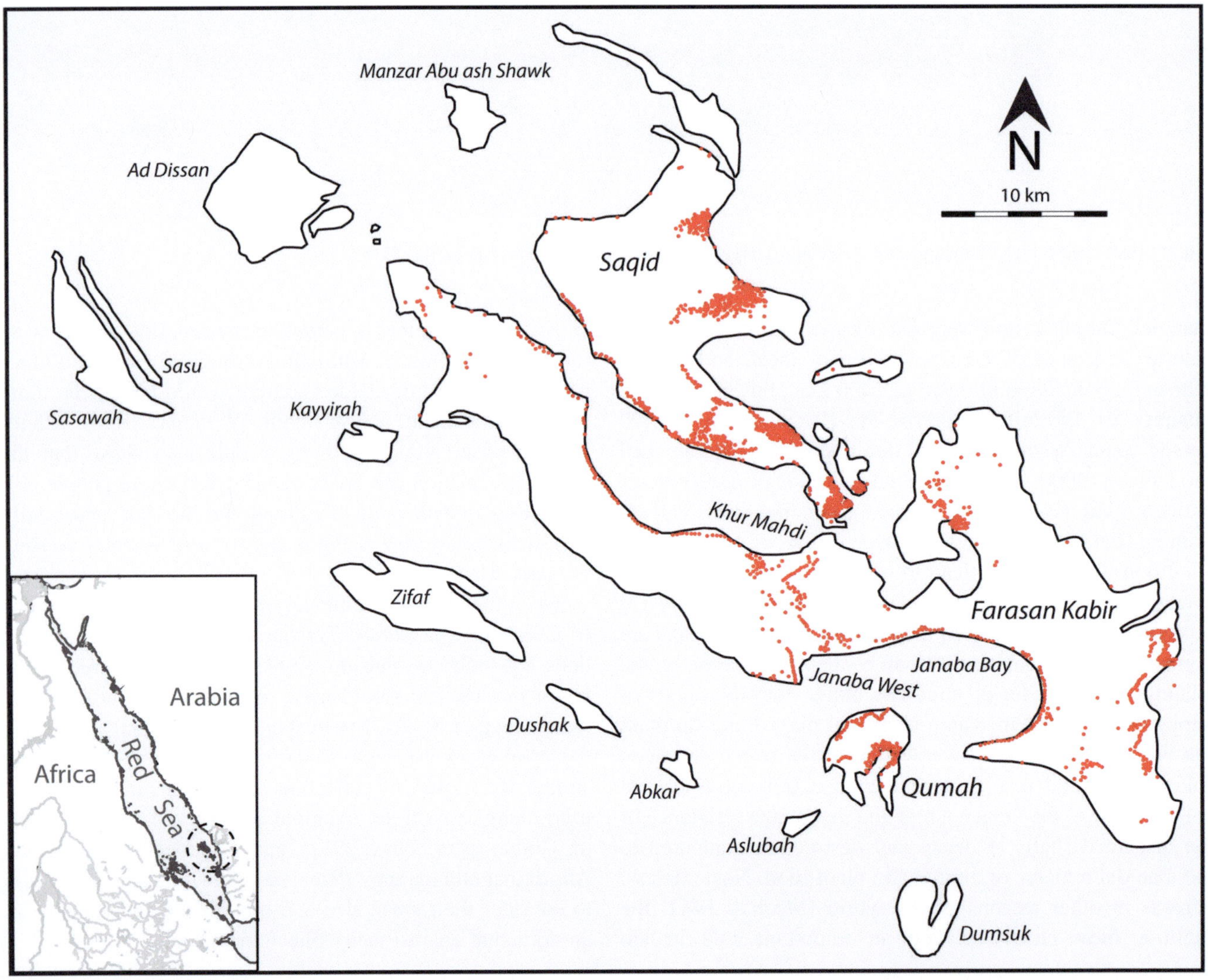

Fig. 8 Distribution of shell mounds (red dots) in the Farasan Islands. Drawn by Niklas Hausmann

relatively short-lived nature of the shallow habitats that generated large supplies of *C. fasciatus*. Before about 6–7 ka, sea level was significantly lower than today, and any shallow sandy intertidal habitats and associated shell mounds must now be under water. After about 4.5 ka, progressive geomorphological and ecological changes removed the extensive areas of *C. fasciatus* habitat.

Today, the shallow bays associated with all the major clusters of large shell mounds are now dry terrestrial surfaces filled with sand. This effect is most dramatically visible at Janaba West, where the original bay and the shorelines and shell mounds that originally lined the inner edge of the bay are clearly visible on satellite imagery (Fig. 5). At some point between the active accumulation of the major shell mounds and the situation visible today, these bays must have begun to dry out, either because of tectonic uplift associated with the Farasan salt dome, or more simply because of progressive accumulation of marine sand and progradation of the beach front, or possibly both processes acting together, amplified by accumulations of wind-blown sand removed from the newly exposed seabed. The result is the transformation of these pre-existing shallow intertidal

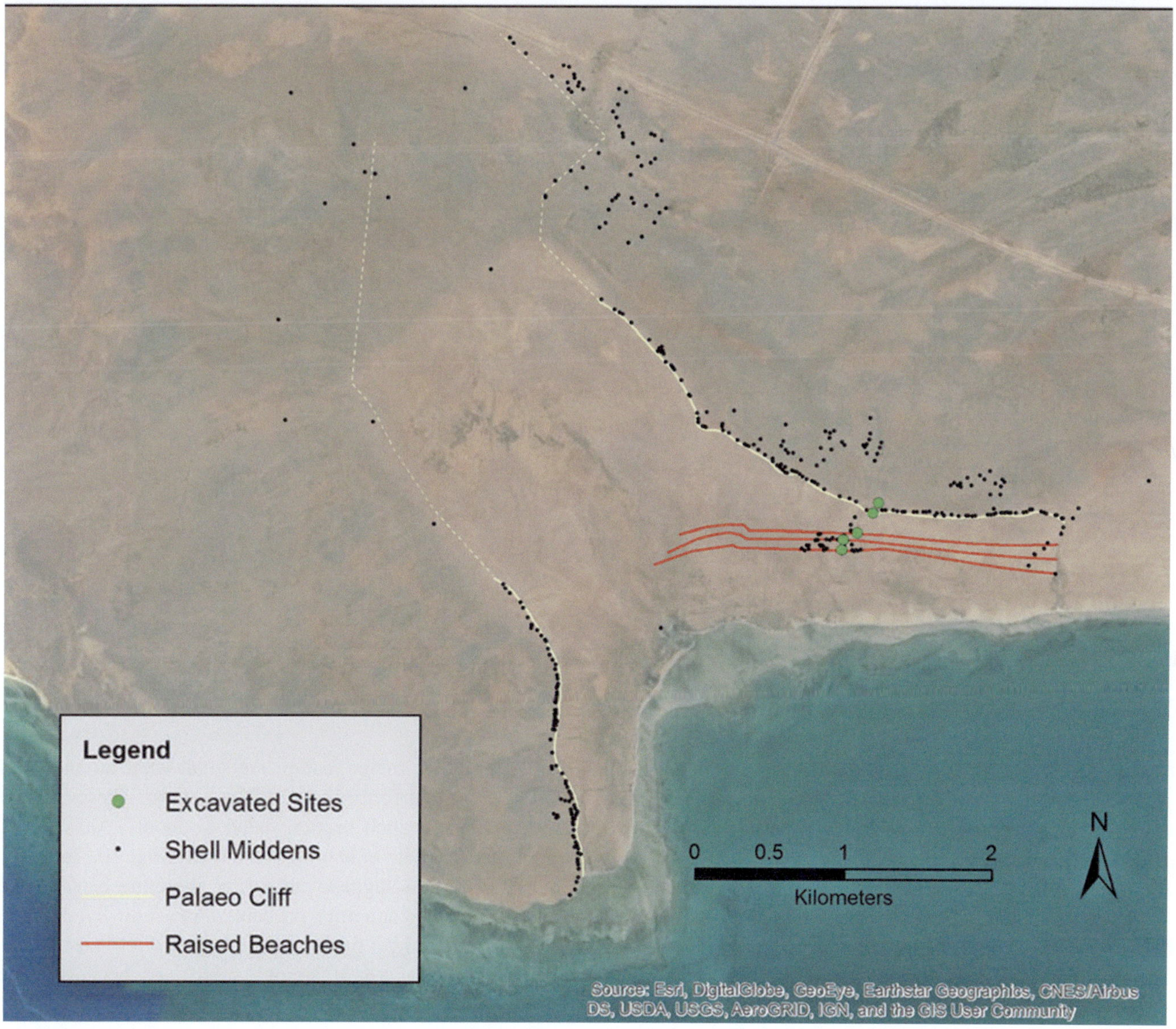

Fig. 9 Satellite image showing detailed distribution of shell middens in Janaba Bay West (see also Figs. 4 and 5). The largest shell mounds are on the original palaeoshoreline (shown in white) which partially circumscribes a now-dry bay. Red lines indicate palaeoshorelines formed during the regression of the shoreline toward the present-day position. For further detail see Fig. 10

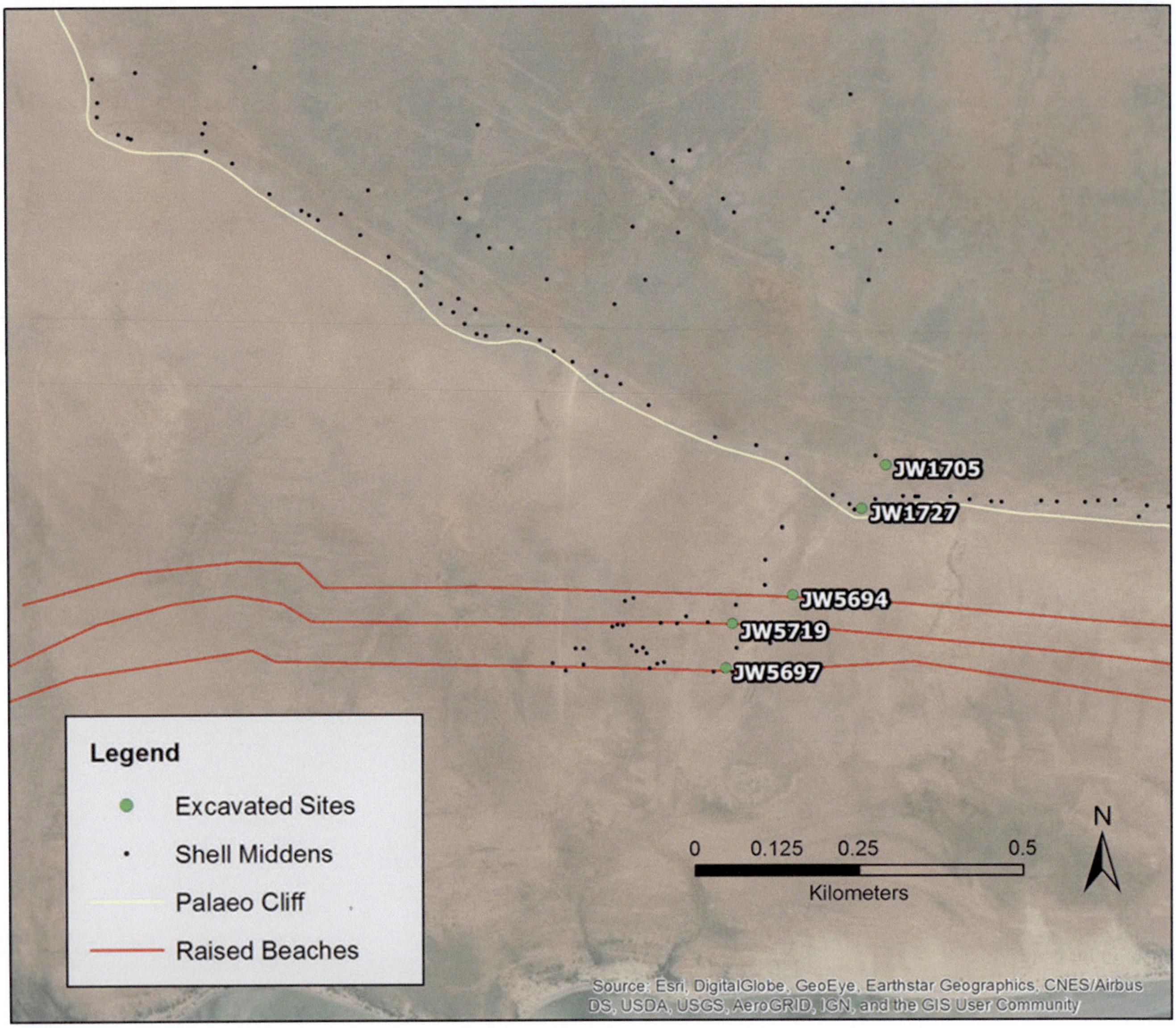

Fig. 10 Satellite image showing detail of shifting shorelines and location of excavated shell mounds in Janaba Bay West

marine embayments into terrestrial surfaces comprising supratidal sand flats, or exposed coral reef surfaces covered with a thin veneer of mobile sand that forms windblown drifts and dunes. These effects are clearly visible in the bays associated with the other major clusters of shell mounds.

A more detailed insight into these processes is afforded by a sequence of shell mounds on the inner margin of the Janaba West Bay. Here there are numerous shell mounds of varying size forming distinctive patterns of distribution. A majority of the sites, including the largest shell mounds in the group, are located along the former shoreline of a major embayment, both clearly visible on satellite images (Figs. 5 and 9). This shoreline was static for long enough to form a low cliff created by marine erosion.

Inland of this major feature are small shell mounds and shell scatters that form a less regular pattern. These include site JW 1705, a shell scatter, dated to about 7300 cal BP, with a younger date of about 5000 cal BP (Fig. 10, Table 1). Their distribution suggests an earlier shoreline or series of shorelines formed at a time, probably quite short-lived, when the sea penetrated yet further inland and then retreated again, either because of slight tectonic uplift, or because of a mid-Holocene high sea level.

The shell mounds formed along the major shoreline are typical deposits dominated by *C. fasciatus*, and are clearly associated with the large embayment when it was an intertidal basin providing a suitable habitat for very extensive beds of shellfish (Fig. 10). Site JW 1727 is typical of the

Fig. 11 Excavated trench of site JW 5694, showing shallow depth of shell deposit

mounds formed along this shoreline, and consists of an oval mound about 30 m long with a maximum thickness of 2 m, a total volume of 163 m^3, and a radiocarbon date of about 4850 cal BP (Hausmann and Meredith-Williams 2016b).

Seaward of this major shoreline are shell deposits of progressively younger date, formed on sandy beach ridges which formed a sequence of shorelines that moved progressively seaward with the retreating shoreline. JW 5694 is a deposit of *Chicoreus* shells and is about 5 m in diameter and 30 cm thick, dated to about 2500 cal BP (Fig. 11, Table 1). Vertebrate remains are also present including bones of fish and gazelle. JW 5719 (Fig. 12) and JW 5697 are scatters of *Chicoreus* shells 5 m in diameter, dated respectively to about 2000 cal BP and about 1700 cal BP (Table 1). These are clearly short-lived deposits associated with a rapidly changing shoreline. Not only are the shell species different, indicating changes in the available molluscan habitat, but the character and function of the deposits appear to be quite different from the larger mounds, suggesting ephemeral camp sites used for a variety of subsistence practices.

One other feature of interest in this group of shell deposits is the evidence that even the larger mounds accumulated very rapidly, perhaps over a matter of decades. Bayesian analysis (Bronk Ramsey 2009) of the radiocarbon dates at JW 1727 demonstrates that this mound accumulated over a period of between 16 and 88 years (65.4% and 95.4% confidence interval respectively), a rate of accumulation further reinforced by detailed analysis of seasonality estimates of the mollusc shells (Hausmann and Meredith-Williams 2016b). This is of particular importance because it refutes the belief that, during a period of rising sea level or a short-lived still-stand, the shoreline would have been moving fast enough that it was never located in one place long enough for accumulations of shell to form archaeologically visible deposits before the shoreline moved to a new position (Fischer 1995, p 382; Bailey 2011, p 322). The above data clearly indicate that this is not a valid argument in the Farasan case, and that substantial archaeological deposits can accumulate over a very short time span. Whether similar arguments can be applied elsewhere will depend on the abundance and stability of the available molluscan supply, and the rate of shoreline displacement. Certainly examples are now coming to light in other parts of the world of equally rapid shell mound growth based both on ethnographic observations (Hardy et al. 2016) and on

Fig. 12 Excavated trench of site JW 5719. The shell matrix is dominated by shells of *Chicoreus ramosus*, clearly visible in the section

multiple radiocarbon dating of archaeological mounds (Holdaway et al. 2017).

4 Discussion

The Farasan example demonstrates that in a dynamically changing shoreline environment subject to changes in relative sea level or other geomorphological changes such as sedimentation and erosion, the habitats and abundance of molluscs may change quite rapidly. Molluscan species such as *C. fasciatus* that provide the very large quantities of shellfood necessary to generate rapid growth of shell mounds are especially vulnerable to these sorts of changes. This is because they depend on very shallow intertidal conditions to thrive in large numbers, and it is precisely these conditions that are most sensitive to minor changes in coastal geomorphology or sea level. Our results demonstrate that shell mounds can grow very quickly to create archaeologically significant and substantial deposits, even when shoreline conditions are changing quite rapidly. But, equally, the conditions that sustain rapid shell-mound growth represent short-lived windows of ecological opportunity, opportunities that may be widely spaced in the geographical dimension as well as the temporal dimension.

The Farasan example therefore has important implications in the search for similar sorts of deposits in earlier periods, whether on submerged shorelines associated with lower sea-levels, or on the emerged shorelines associated with earlier periods of high sea level. Underwater investigation in the Farasan Islands has demonstrated that submerged shorelines are present and have the characteristic undercut notch formed by marine erosion, but so far, the only shell deposits discovered in association with these shorelines are natural death assemblages and not food remains (Bailey et al. 2007, 2013a; Alsharekh et al. 2013; Momber et al., this volume). Destruction of shell mounds by inundation during sea-level rise, subaerial erosion on exposed coastlines, or lack of interest in coastal and marine resources by earlier human populations, are typically invoked to account for the absence of shell mounds in earlier periods. Analysis of the Farasan shell mounds on the present-day coastline suggests another major factor, and that is that the ecological and geomorphological preconditions for rapid shell-mound growth are very limited both spatially and temporally. Much more intensive investigation of taphonomic factors, and in particular more extensive underwater surveys taking advantage of remote-sensing techniques to identify potential diving targets, will be needed before we can be sure that 'absence of evidence' is indeed 'evidence of absence'.

5 Conclusion

Coastlines have always been attractive places for human settlement and the concentration of human populations and human activities at all periods of the past up to and including the modern era, and at all levels of technology, whether Stone Age or Oil Age in character. Just as today a very large proportion of the world's human population is concentrated in coastal lowland regions on or close to the coastline, so this is likely to have been the case in the past. Coastal regions offer diversity of food resources, including marine foods at the shore edge and offshore, often better soils and sediments to support plant and animal life, and better water supplies and more equable climate conditions than hinterland regions, a variety of raw materials for technological production, ranging, in the Arabian case, from basaltic lavas for making stone tools to rare minerals valuable in many modern industrial processes, and generally better opportunities for population movement, and for social contact and interaction. The rim of the Arabian Peninsula as a whole represents a 'fertile crescent' equivalent to the more famous Fertile Crescent to the north comprising the Levantine coastal region and the foothills of the Taurus and Zagros ranges circumscribing the upper reaches of the Rivers Jordan, Tigris and Euphrates. The Arabian Fertile Crescent has the added advantage of abundant marine resources around much of its coastline. Moreover, these advantages are likely to have persisted throughout the climatic fluctuations of the Pleistocene (Bailey et al. 2015).

This concentration of population at the coast edge, also, of course, adds an increasingly powerful anthropogenic imprint to the dynamic changes of the coast edge, modifying many natural processes through modification of land surfaces and pollution, and obscuring or destroying the archaeological evidence of earlier human presence. It is no surprise that some of the most easily accessible and largest concentrations of Stone Age archaeology are to be found in the desert interior, where earlier evidence has suffered relatively little disturbance or destruction by later land use processes and industrial development, whereas areas closer to the coast, where we would expect to find the largest concentrations of Stone Age remains, are also the areas where the risk of burial or destruction of earlier evidence is greatest.

In addition, coastlines are the most dynamic and changeable parts of the wider landscape because of tectonically and climatically induced changes in relative sea level. Also, much of the most attractive terrain, available for long periods of the Pleistocene climatic cycle as noted earlier, is now submerged because of sea-level change, and underwater exploration of this hidden world has scarcely begun (see Sakellariou et al., this volume; Momber et al., this volume, 2018).

The interlinked nature of all these processes is both a distinctive feature of coastal regions and a formidable challenge for the future, not only for archaeologists but for all the scientific disciplines interested in the history of coastal regions. An interest in the 'taphonomy' of landscape change—the processes that have variously buried, obscured, removed, destroyed, modified or preserved different sorts of evidence—is emerging as a new and unifying research focus. These problems are especially obvious in the underwater realm, but they are no less important on dry land. Here an equally powerful array of factors has affected the distribution of archaeological material. These include agricultural development, changes in land-use practices that have variously exposed or obscured archaeological material, the impact of wind and rain, and the growing impact of modern development through large-scale infrastructural developments, including the building of new towns and roads and the quarrying away of large parts of the landscape to provide the raw materials for new developments. Given the many different factors of deposition, site formation, preservation, exposure and destruction that affect archaeological remains, we should be very wary of assuming that the distribution of archaeological remains that have survived and are visible today are in any way an accurate reflection of the distribution of past human populations. These are not simply negative factors that impede archaeological interpretation, but ones that are amenable to investigation, and need to be built into archaeological research design and interpretation at every scale if the discipline is to progress, and such investigations are likely to grow in importance in the coming decades.

Acknowledgements We thank HRH Prince Sultan bin Salman bin Abdul Aziz, President of the Saudi Commission for Tourism and National Heritage (SCTH), Dr. Ali Al-Ghabban, advisor to the President, and Jamal Omar, Vice President, for granting permission for the archaeological fieldwork reported here and their ongoing support and that of their SCTH colleagues in Riyadh, especially Dr. Saad al Rashid, Dr. Abdullah Al Saud, Dr. Hussein Abu Hassan and Dr. Abdullah Al Zahrani. Grateful thanks are also extended to the staff of the SCTH offices in Jizan and Farasan, to the Governor of Farasan, Hussein Aldajani, and to Dr. Faisal al Tamaihi, Jizan University. We also thank the President of the Saudi Geological Survey, Dr. Zohair Nawab, and his staff, in particular Dr. Najeeb Rasul for additional support. We thank Katerina Douka for assistance in the collection of the radiocarbon samples and the interpretation of the results and Eva Laurie for the identification of marine molluscs. The research is funded by the European Research Council (ERC) under the Ideas Programme of the 7th Framework Programme as Advanced Grant 269586 'DISPERSE:

Dynamic Landscapes, Coastal Environments and Human Dispersals'. This is DISPERSE contribution no. 38.

References

Alimen M-H (1975) Les isthmes Hispano-Marocain et Siculo-Tunisien aux temps Acheuléens. L'Anthropologie 79:399–436

Alsharekh AM, Bailey GN, Momber G, Moran LJ, Sinclair A, Williams MGW, Al Shaikh N, Al Ma'Mary A, Alghamdi S, Al Zahrani A, Aqeeli A, Laurie EM, Beech M (2013) Report on the 2008 fieldwork of the joint Saudi-UK Southern Red Sea Project. In: Alsharekh AM, Bailey GN (eds) Coastal prehistory in southwest Arabia and the Farasan Islands: 2004–2009 field investigations. Saudi Commission for Tourism and Antiquities, Riyadh, pp 77–158

Ammerman A (ed) (2013) Island archaeology and the origins of seafaring in the eastern Mediterranean. Eurasian Prehistory 10

Ammerman A (ed) (2014) Island archaeology and the origins of seafaring in the eastern Mediterranean. Eurasian Prehistory 11

Anderson A, Barrett J, Boyle K (eds) (2010) The global origins and development of seafaring. McDonald Institute for Archaeological Research, Cambridge

Armitage SJ, Jasim SA, Marks AE, Parker AG, Usik VI, Uerpmann H-P (2011) The southern route "Out of Africa": evidence for an early expansion of modern humans into Arabia. Science 331:453–456

Attenbrow V (1992) Shell bed or shell midden. Aust Archaeol 34:3–21

Bailey GN (2009) The Red Sea, coastal landscapes and hominin dispersals. In: Petraglia M, Rose J (eds) The evolution of human populations in Arabia. Springer, Dordrecht, Netherlands, pp 15–37

Bailey GN (2011) Continental shelf archaeology: where next? In: Benjamin J, Bonsall C, Pickard K, Fischer A (eds) Submerged prehistory. Oxbow, Oxford, pp 311–331

Bailey GN, Milner NJ (2002) Coastal hunters and gatherers and social evolution: marginal or central? Before Farm Archaeol Old World Hunt-Gatherers 3–4(1):1–15

Bailey GN, King GCP (2011) Dynamic landscapes and human dispersal patterns: tectonics, coastlines and the reconstruction of human habitats. Quat Sci Rev 30(11–12):1533–1553

Bailey GN, AlSharekh A, Flemming N, Lambeck K, Momber G, Sinclair A, Vita-Finzi C (2007) Coastal prehistory in the southern Red Sea Basin: underwater archaeology and the Farasan Islands. Proc Semin Arab Stud 37:1–16

Bailey GN, Alsharekh AM, Momber G, Moran LJ, Gillespie J, Satchell JS, Williams MGW, Reeler C, Al Shaikh N, Robson H, Kamil A (2013a) Report on the 2009 fieldwork of the joint Saudi-UK Southern Red Sea Project. In: Alsharekh AM, Bailey GN (eds) Coastal prehistory in southwest Arabia and the Farasan Islands: 2004–2009 field investigations. Saudi Commission for Tourism and Antiquities, Riyadh, pp 161–215

Bailey GN, Meredith-Williams MGM, Alsharekh A (2013b) Shell mounds of the Farasan Islands, Saudi Arabia. In: Bailey GN, Hardy K, Camara A (eds) Shell energy: mollusc shells as coastal resources. OxBow, Oxford, pp 241–254

Bailey GN, Devès MH, Inglis RH, Meredith-Williams MG, Momber G, Sakellariou D, Sinclair AGM, Rousakis G, Al Ghamdi S, Alsharekh AM (2015) Blue Arabia: Palaeolithic and underwater survey in SW Saudi Arabia and the role of coasts in Pleistocene dispersal. Quat Int 382:42–57. https://doi.org/10.1016/j.quaint.2015.01.002

Beech M (2018) Archaeological fish bones from the Farasan Islands. In: Bailey G, Alsharekh A (eds) Palaeolithic archaeology, coastal prehistory and submerged landscapes in southwest Saudi Arabia and the Farasan Islands: DISPERSE field reports, 2012–2015. Saudi Commission for Tourism and National Heritage, Riyadh, pp 286–302

Beyin A (2006) The Bab-el-Mandab vs the Nile-Levant: an appraisal of the two dispersal routes for early modern humans out of Africa. Afr Archaeol Rev 23:5–30

Beyin A (2011) Upper Pleistocene human dispersals out of Africa: a review of the current state of the debate. Int J Evol Biol 2011:615904. https://doi.org/10.4061/2011/615094

Bohlender R, Yu Y, Huff C, Rogers A (2016) A complex history of archaic admixture in modern humans. American Society of Human Genetics 2016 Annual Meeting, Vancouver, B.C., Canada (abstract)

Boivin N, Blench R, Fuller DQ (2009) Archaeological, linguistic and historical sources on ancient seafaring: a multidisciplinary approach to the study of early maritime contact and exchange in the Arabian Peninsula. In: Petraglia MD, Rose JI (eds) The evolution of human populations in Arabia. Springer, Dordrecht, Netherlands, pp 251–278

Boivin N, Fuller DQ, Dennell R, Allaby R, Petraglia MD (2013) Human dispersal across diverse environments of Asia during the Upper Pleistocene. Quat Int 300:32–47

Breeze PS, Drake NA, Groucutt HS, Parton A, Jennings RP, White TS, Clark-Balzan L, Shipton C, Scerri EML, Stimpson CM, Crassard R, Hilbert Y, Alsharekh A, Al-Omari A, Petraglia MD (2015) Remote sensing and GIS techniques for reconstructing Arabian palaeohydrology and identifying archaeological sites. Quat Int 382:98–119

Breeze PS, Groucutt HS, Drake NA, White TS, Jennings RP, Petraglia MD (2016) Palaeohydrological corridors for hominin dispersal in the Middle East ~250–70,000 years ago. Quat Sci Rev 144:155–185

Bretzke K, Armitage SJ, Parker AG, Walkington H, Uerpmann H-P (2013) The environmental context of Paleolithic settlement at Jebel Faya, Emirate Sharjah, UAE. Quat Int 300:83–93

Bronk Ramsey C (2009) Bayesian analysis of radiocarbon dates. Radiocarbon 51(1):337–360

Carter R (2010) The social and environmental context of Neolithic seafaring in the Persian Gulf. In: Anderson A, Barrett JH, Boyle K (eds) The global origins and development of seafaring. McDonald Institute for Archaeological Research, Cambridge, pp 191–202

Derricourt R (2005) Getting "out of Africa": sea crossings, land crossing and culture in the hominin migrations. J World Prehist 19:119–132

Drake NA, Breeze P, Parker A (2013) Palaeoclimate in the Saharan and Arabian deserts during the Middle Palaeolithic and the potential for hominin dispersals. Quat Int 300:48–61

Enk J, Devault A, Widga C, Saunders J, Szpak P, Southon J, Rouillard J-M, Shapiro B, Golding GB, Zazula G, Froese D, Fisher DC, MacPhee RDE, Poinar H (2016) *Mammuthus* population dynamics in Late Pleistocene North America: divergence, phylogeography, and introgression. Front Ecol Evol 4:42. https://doi.org/10.3389/fevo.2016.00042

Erlandson JM (2001) The archaeology of aquatic adaptations: paradigms for a new millennium. J Archaeol Res 9:287–350

Faure H, Walter RC, Grant DR (2002) The coastal oasis: ice age springs on emerged continental shelves. Glob Planet Change 33:47–56

Field JS, Lahr MM (2005) Assessment of the southern dispersal: GIS-based analyses of potential routes at oxygen isotopic stage 4. J World Prehist 19:1–45

Field JS, Petraglia MD, Lahr MM (2007) The southern dispersal hypothesis and the South Asian archaeological record: examination of dispersal routes through GIS analysis. J Anthropol Archaeol 26 (1):88–108

Fischer A (1995) An entrance to the Mesolithic world below the ocean. Status of ten years' work on the Danish sea floor. In: Fischer A (ed) Man and sea in the Mesolithic: coastal settlement above and below present sea level. Oxbow, Oxford, pp 371–384

Fleagle JG, Zelalem A, Brown FH, Shea JJ (2008) Paleoanthropology of the Kibish Formation, southern Ethiopia: introduction. J Hum Evol 55:360–365

Foulds FWF, Shuttleworth A, Sinclair A, Alsharekh AM, Al Ghamdi S, Inglis RH, Bailey GN (2017) A large handaxe from Wadi Dabsa and early hominin adaptations within the Arabian Peninsula. Antiquity 91(360):1421–1434

Green RE, Krause J, Briggs AW, and 53 other co-authors (2010) A draft sequence of the Neandertal genome. Science 328(5979): 710–722

Groucutt HS, Petraglia MD, Bailey GN, Parton A, Clark-Balzan L, Jennings RP, Lewis L, Blinkhorn J, Scerri EML, Drake NA, Breeze P, Boivin N, Inglis RH, Devès MH, Meredith-Williams MG, Thomas MG, Scally A (2015) Rethinking the dispersal of *Homo sapiens* out of Africa. Evol Anthropol 24:149–164. https://doi.org/10.1002/evan.21455

Hardy K, Camara A, Piqué R, Dioh E, Guèye M, Diadhiou HD, Faye M, Carré M (2016) Shellfishing and shell midden construction in the Saloum Delta, Senegal. J Anthropol Archaeol 41:19–32

Hausmann, N, Meredith-Williams, M (2016a) Seasonal patterns of coastal exploitation on the Farasan Islands, Saudi Arabia. J Isl Coast Archaeol (Online). https://doi.org/10.1080/15564894.2016.1216178

Hausmann N, Meredith-Williams M (2016b) Exploring accumulation rates of shell deposits through seasonality data. J Archaeol Method Theory (Online). https://doi.org/10.1007/s10816-016-9287-x

Hausmann N, Kokkinaki O, Leng MJ (this volume) Red Sea palaeoclimate: stable isotope and element-ratio analysis of marine molluscs

Henshilwood CS, Sealy JC, Yates R, Cruz-Uribe K, Goldberg P, Grine FE, Klein RG, Poggenpoel C, van Niekerk K, Watts I (2001) Blombos Cave, southern Cape, South Africa: preliminary report on the 1992–1999 excavations of the Middle Stone Age levels. J Archaeol Sci 28(4):421–448

Ho SYW, Lanfear R, Bromham L, Phillips MJ, Soubrier J, Rodrigo AG, Cooper A (2011) Time-dependent rates of molecular evolution. Mol Ecol 20(15):3087–3101. https://doi.org/10.1111/j.1365-294X.2011.05178.x

Holdaway SJ, Fanning PC, Petchey F, Allely K, Shiner JI, Bailey GN (2017) Temporal variability in shell mound formation at Albatross Bay, northern Australia. PLOS One 12(8):e0183863. https://doi.org/10.1371/journal.pone.0183863

Inglis RH, Sinclair A, Shuttleworth A, Alsharekh A, Al Ghamdi S, Devès M, Meredith-Williams MG, Bailey GN (2014) Investigating the Palaeolithic landscapes and archaeology of the Jizan and Asir Regions, southwest Saudi Arabia. Proc Semin Arab Stud 44:193–212

Inglis RH, Sinclair AGM, Alsharekh AM, Barfod DN, Chang HC, Fanning PC, Al Othaibi DT, Robson HK, Shuttleworth A, Stone A, Bailey GN (2017) Preliminary report on UK-Saudi 2017 fieldwork at Wadi Dabsa, Asir Province, Saudi Arabia. http://eprints.whiterose.ac.uk/119267/1/JanFeb_2017_Field_Report_small.pdf

Inglis R, Rasul N, Bailey GN, Bosworth W (this volume) Investigating the fossil coral terraces and coastal archaeology of the southern Red Sea

Jerardino A (2016) On the origins and significance of Pleistocene coastal resource use in southern Africa with particular reference to shellfish gathering. J Anthropol Archaeol 41:213–230

Jerardino A, Marean C (2010) Shellfish gathering, marine paleoecology and modern human behavior: perspectives from cave PP13B, Pinnacle Point, South Africa. J Hum Evol 59:412–424

Khalidi L (2009) Holocene obsidian exchange in the Red Sea region. In: Petraglia M, Rose J (eds) The evolution of human populations in Arabia. Springer, Dordrecht, Netherlands, pp 279–291

Khalidi L, Oppenheimer C, Gratuze B, Boucetta S, Sanabani A, Mosabi A (2010) Obsidian sources in highland Yemen and their relevance to archaeological research in the Red Sea region. J Archaeol Sci 37:2332–2345

Khalidi L, Lewis K, Gratuze B (2012) New perspectives on regional and interregional obsidian circulation in prehistoric and early historic Arabia. Proc Semin Arab Stud 42:1–22

King GCP, Bailey GN (2006) Tectonics and human evolution. Antiquity 80:265–286

Kopp GH, Roos C, Butynski TM, Wildman DE, Alagail AN, Groenveld LF, Zinner D (2014) Out of Africa, but how and when? The case of hamadryas baboons (*Papio hamadryas*). J Hum Evol 77:154–164

Kübler S, Devès MH, Bailey GN, King GCP (this volume) Tectonic geomorphology and soil edaphics as controls on animal migrations and human dispersal patterns

Lambeck K, Purcell A, Flemming NC, Vita-Finzi C, Alsharekh AM, Bailey GN (2011) Sea level and shoreline reconstructions for the Red Sea: isostatic and tectonic considerations and implications for hominin migration out of Africa. Quat Sci Rev 30(25–26):3542–3574

Lahr M, Foley R (1994) Multiple dispersals and modern human origins. Evol Anthropol 3(2):48–60

Langejans GHJ, Dusseldorp GL, Thackeray JF (2017) Pleistocene molluscs from Klasies River (South Africa): reconstructing the local coastal environment. Quat Int 427:59–84

Lawler A (2002) Report of oldest boat hints at early trade routes. Science 296(5574):1791–1792

Macaulay V, Hill C, Achilli A, Rengo C, Clarke D, Meehan W, Blackburn J, Semino O, Scozzari R, Cruciani F, Taha A, Shaari NK, Raja JM, Ismail P, Zainuddin Z, Goodwin W, Bulbeck D, Bandelt HJ, Oppenheimer S, Torroni A, Richards M (2005) Single, rapid coastal settlement of Asia revealed by analysis of complete mitochondrial genomes. Science 308:1034–1036

Malispinas A-S, Westaway MC, Muller C, Sousa VC, Lao O et al (2016) A genomic history of Aboriginal Australia. Nature 538:207–214

Marean CW (2010) Pinnacle Point Cave 13B (Western Cape Province, South Africa) in context: the Cape floral kingdom, shellfish, and modern human origins. J Hum Evol 59:425–443

Meehan B (1982) Shell bed to shell midden. Australian Inst Aboriginal Studies, Canberra

Mellars PA (2006) Going east: new genetic and archaeological perspectives on the modern human colonization of Eurasia. Science 313:796–800

Mellars PA, Gori KC, Carr M, Soares PA, Richards MB (2013) Genetic and archaeological perspectives on the initial modern human colonization of southern Asia. Proc Nat Acad Sci 110:10699–10704

Meredith-Williams M, Bailey GN, Hausmann N, Al Ghamdi S, Alsharekh AM, Douka K, Beresford C, Larsen B (2013) Preliminary report on 2013 fieldwork in southwest Saudi Arabia by the DISPERSE Project: (1) The Farasan Islands. http://www.disperse-project.org/sites/disperse-project.org/files/uploads/Wainwright_Jan_%202015_Field_Report.pdf

Meredith-Williams M, Hausmann N, Bailey GN, King GCP, Alsharekh A, Al Ghamdi S (2014a) Mapping, modelling and predicting prehistoric coastal archaeology in the southern Red Sea using new applications of digital imaging techniques. World Archaeol 46(1):10–24. https://doi.org/10.1080/00438243.2014.890913

Meredith-Williams MG, Hausmann N, Bailey GN, Inglis RH (2014b) 4200 new shell mound sites in the southern Red Sea. In: Fernandes R, Meadows J (eds) Human exploitation of aquatic landscapes. Special Issue, Internet Archaeology, vol 37. https://doi.org/10.11141/ia.37.2

Momber G, Gillespie, J, Mason B, Mason C, Tidbury L, Mozayen W, Hamzi F, Al Sadiq J, Al Haiti A, Meredith–Williams M, Hausmann N, Aqeeli A, Mofta M, Bailey GN (2018) Preliminary report on 2014 fieldwork in Southwest Saudi Arabia by the DISPERSE project: (2) underwater research in the Farasan Islands. In: Bailey GN, Alsharekh AM (eds) Palaeolithic archaeology, coastal prehistory and submerged landscapes in southwest Saudi Arabia and the Farasan Islands: DISPERSE Field Reports, 2012–2015. Saudi Commission for Tourism and National Heritage, Riyadh, pp 230–254

Momber G, Sakellariou D, Bailey GN, Rousakis G (this volume) The multi-disciplinary search for underwater archaeology in the southern Red Sea

Noli D, Avery G (1988) Protein poisoning and coastal subsistence. J Archaeol Sci 15(4):395–401

Oppenheimer S (2012a) A single southern exit of modern humans from Africa: before or after Toba? Quat Int 258:88–99

Oppenheimer S (2012b) Out-of-Africa, the peopling of continents and islands: tracing uniparental gene trees across the map. Phil Trans R Soc B 367:770–784

Pagani L, Lawson DJ, Jagoda E, Morseburg A, Eriksson A et al (2016) Genomic analyses inform on migration events during the peopling of Eurasia. Nature 538:238–242

Parker AG (2009) Pleistocene climate change in Arabia: developing a framework for hominin dispersal over the last 350 ka. In: Petraglia M, Rose J (eds) The evolution of human populations in Arabia. Springer, Dordrecht, Netherlands, pp 39–49

Petraglia MD, Rose JI (eds) (2009) The evolution of human populations in Arabia. Springer, Dordrecht, Netherlands

Petraglia MD, Drake NA, Alsharekh AM (2009) Acheulean landscapes and large cutting tools assemblages in the Arabian Peninsula. In: Petraglia MD, Rose JI (eds) The evolution of human populations in Arabia. Springer, Dordrecht, Netherlands, pp 103–116

Petraglia MD, Alsharekh AM, Crassard R, Drake NA, Groucutt H, Parker AG, Roberts (2011) Middle Paleolithic occupation of a marine isotope stage 5 lakeshore in the Nefud desert, Saudi Arabia. Quat Sci Rev 30:1555–1559

Petraglia MD, Breeze PS, Groucutt HS (this volume) Blue Arabia, Green Arabia: examining human colonisation and dispersal models

Reimer PJ, Bard E, Bayliss A, Beck JW, Blackwell PG, Ramsey CB, Buck CE, Cheng H, Edwards RL, Friedrich M, Grootes PM (2013) IntCal13 and Marine13 radiocarbon age calibration curves 0–50,000 years cal BP. Radiocarbon 55(4):1869–1887

Rohling EJ, Grant KM, Roberts AP, Larrasoaña J-C (2013) Paleoclimate variability in the Mediterranean and Red Sea regions during the last 500,000 years: implications for hominin migrations. Curr Anthropol 54(S8):183–201

Rohling EJ, Braun K, Grant K, Kucera M, Roberts AP, Siddall M, Trommer G (2010) Comparison between Holocene and marine isotope stage-11 sea-level histories. Earth Planet Sci Lett 291: 97–105

Rosenberg TM, Preusser F, Risberg J, Plikk A, Kadi KA, Matter A, Fleitmann D (2013) Middle and Late Pleistocene humid periods recorded in palaeolake deposits of the Nafud desert, Saudi Arabia. Quat Sci Rev 70:109–123

Rosendahl D, Ulm S, Weisler MI (2007) Using foraminifera to distinguish between natural and cultural shell deposits in coastal eastern Australia. J Archaeol Sci 34(10):1584–1593

Rowland M (1994) Size isn't everything: shells in mounds, middens and natural deposits. Aust Archaeol 39:118–124

Sakellariou D, Rousakis G, Panagiotopoulos I, Morfis I, Bailey G (this volume) Geological structure and Late Quaternary geomorphological evolution of the Farasan Islands continental shelf, south Red Sea, SW Saudi Arabia

Siddall M, Rohling EJ, Almogi-Labin A, Hemleben C, Meischner D, Schmeltzer I, Smeed DA (2003) Sea-level fluctuations during the last glacial cycle. Nature 423:853–858

Sinclair A, Inglis RH, Shuttleworth A, Foulds F, Bailey GN, Alsharekh A (this volume) Landscape archaeology, Palaeolithic survey and coastal change along the southern Red Sea of Saudi Arabia

Soares P, Alshamali F, Pereira JB, Fernandes V, Silva NM, Afonso C, Costa MD, Musilová E, Macaulay V, Richards MB, Černý V, Pereira L (2012) The expansion of mtDNA haplogroup L3 within and out of Africa. Mol Biol Evol 29(3):915–927

Strasser TF, Panagopoulou E, Runnels CN, Murray PM, Thompson N, Karkanas P, McCoy FW, Wegmann KW (2010) Stone age seafaring in the Mediterranean: evidence from the Plakias region for Lower Palaeolithic and Mesolithic habitation of Crete. Hesperia 79 (2):145–190

Usik VI, Rose JI, Hilbert YH, Van Peer P, Marks AE (2013) Nubian Complex reduction strategies in Dhofar, southern Oman. Quat Int 300:244–266

Van Niekerk KL (2011) Marine fish exploitation during the Middle and Later Stone Age of South Africa. Unpublished Ph.D., University of Cape Town, South Africa

Walter RC, Buffler RT, Bruggemann JJ, Guillaume MMM, Berhe SM, Negassi B, Libsekal Y, Cheng H, Edwards RL, von Gosel R, Neraudeau D, Gagnon M (2000) Early human occupation of the Red Sea coast of Eritrea during the last interglacial. Nature 405: 65–69

Watters DR, Donahue J, Stuckenroth R (1992) Paleoshorelines and the prehistory of Barbuda, West Indies. In: Lewis Johnson L, Stright M (eds) Paleoshorelines and prehistory: an investigation of method. CRC Press, Boca Raton, pp 15–52

Williams MGM (2010) Shell mounds of the Farasan Islands, Saudi Arabia. Proc Semin Arab Stud 40:357–366

Winder IC, Devès MH, King GCP, Bailey GN, Inglis RH, Meredith-Williams M (2015) Dynamic landscapes and complex topography as agents in human evolution: the dispersals of the genus *Homo*. J Hum Evol 87:48–65. https://doi.org/10.1016/j.jhevol.2015.07.002

The Multi-disciplinary Search for Underwater Archaeology in the Southern Red Sea

Garry Momber, Dimitris Sakellariou, Grigoris Rousakis, and Geoff N. Bailey

Abstract

During the height of the last glacial maximum about 20,000 years ago, the sea-level was 120–130 m lower, making movement out of Africa into Arabia relatively easy. The Hanish Sill at the southern end of the Red Sea would only have been a few metres deep, less than 10 km wide and interspersed with small islands. Extensive evidence of archaeological artefacts dating to the Middle Palaeolithic has been found on the southern Arabian Peninsula demonstrating an earlier hominin presence. These movements might well have been facilitated by former periods of low sea level, as for the last million years or so, sea levels have averaged 40–60 m lower than today. These were times when large areas of continental shelf around the Farasan Islands would have been exposed as a terrestrial landscape, providing a coastal environment that would have been attractive for animals and humans. This paper looks at a series of fieldwork projects that have helped characterise the submerged landscape and assess the potential for human occupation of the drowned lands around the Farasan Islands. Significant submerged wave-cut notches, platforms and lacustrine features were recorded, evidence for tectonic realignments was identified and areas with the potential for human occupation were investigated. The fieldwork has provided new information on the nature of the drowned landscape, characterised potential sites of human occupation and identified the challenges that need to be addressed by archaeologists as the investigations continue.

1 Introduction

At the beginning of the millennium, a series of field projects led by the University of York was initiated in southwest Saudi Arabia to develop an understanding of the early prehistory of the region. A Palaeolithic sequence of considerable time depth but consisting mainly of surface finds of artefacts of Lower, Middle and Upper Palaeolithic type had already been established in earlier surveys (Zarins et al. 1979, 1980, 1981). New work was structured around the investigation of the potential role of the southern Arabian route for human dispersal out of Africa and across the southern end of the Red Sea. From an early stage, an underwater component was incorporated into the research strategy, initially under the aegis of the Africa-Arabian Connections Project—AFRAC, or AFRAC MAR referring specifically to the offshore work (2003–2009), and later (2011–2015) as part of the DISPERSE Project (Dynamic Landscapes, Coastal Environments and Human Dispersals) (Bailey et al. 2007a, b; Alsharekh and Bailey 2013; Bailey et al. 2015, 2018; Bailey and Alsharekh in press). The aim of this chapter is to review the different types of underwater work that have informed the evolution of the research strategy over a period of more than ten years, with particular reference to the Farasan Islands and surrounding areas of the submerged continental shelf, to highlight the technical and logistical factors that constrain offshore and underwater research, and to summarise the results.

From high above the Earth, the Farasan archipelago emerges as a patchwork of turquoise and sand coloured islands surrounded by dark blue water. These reefs and irregular rocky outcrops break the tranquillity of the Red Sea

G. Momber (✉)
Maritime Archaeology Trust, Room 014/11, National Oceanography Centre, Southampton, SO14 3ZH, UK
e-mail: garry.momber@maritimearchaeologytrust.org

G. Momber · G. N. Bailey
Department of Archaeology, University of York, King's Manor, York, YO1 7EP, UK

D. Sakellariou · G. Rousakis
Hellenic Centre for Marine Research, Athinon—Souniou Ave (46.7th km), 19013 Anavyssos, Greece

G. N. Bailey
College of Humanities, Arts and Social Sciences, Flinders University, GPO Box 2100 Adelaide, SA 5001, Australia

N. M. A. Rasul and I. C. F. Stewart (eds.), *Geological Setting, Palaeoenvironment and Archaeology of the Red Sea*,
https://doi.org/10.1007/978-3-319-99408-6_27

today, but this was not always the case. Past climatic changes caused fluctuations in the sea level, which dropped to reveal an extensive, flat landscape dotted with deep fault-bounded basins and occasional low hills. When the globe was at its coldest and the ice caps were most extensive around 21,500 years ago the sea-level was 120–130 m lower (Chappell and Shackleton 1986; Lambeck and Chappell 2001; Bailey 2009; Lambeck et al. 2011; Harff et al. 2017). This revealed a coastal plain stretching over almost 1000 km north to south and 50 km east to west, doubling the width of the current coastal plain.

A review of seabed bathymetric data recorded on the Admiralty charts compiled by the UK Hydrographic Office and collected by the Saudi Geological Survey showed that the seabed generally dipped gently toward the west and the edge of the continental shelf. However, the dip is not consistent as it is interrupted by elevated areas of bedrock and submarine pinnacles while being interspersed with deep circular and oval hollows (see Sakellariou et al. 2018). These circular features are particularly interesting as they measure 10–20 km across with many dropping to over 500 m below the surrounding sea floor, making them ideal potential reservoirs of fresh water

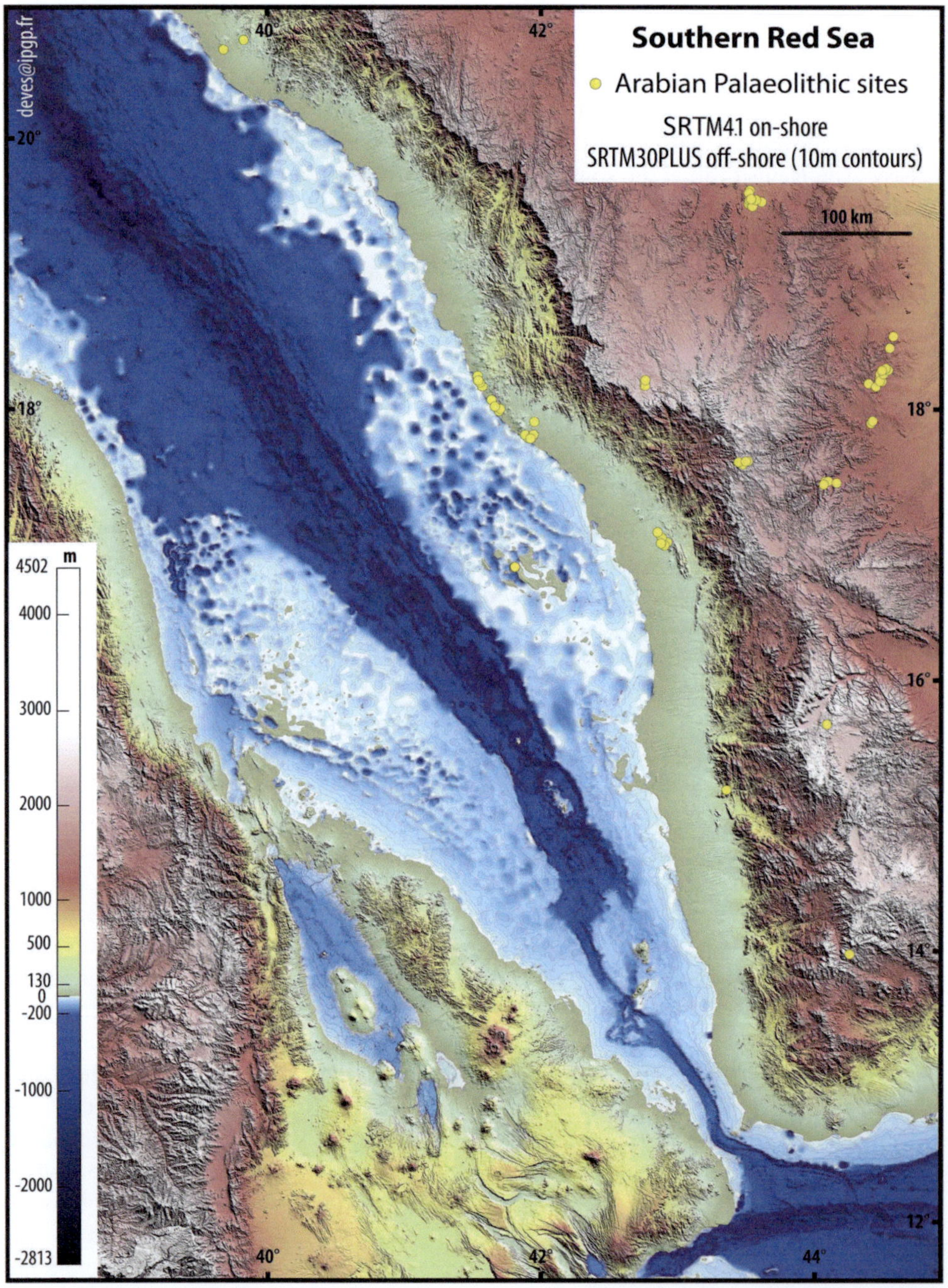

Fig. 1 Location map of Red Sea and Farasan Islands showing exposed land at lower sea level and Palaeolithic sites

when they were free from marine influence. If this was the case and an extensive landscape with lakes and springs existed (see also Faure et al. 2002), this would have been an attractive region for large mammals and human hunters. Furthermore, these environments could have presented ideal conditions for preservation where archaeological and organic materials could survive, and often in better condition than their terrestrial counterparts. In addition, the area is only 500 km north of the Hanish Sill (Fig. 1).

This is the narrowest point of the Red Sea between the Arabian Peninsula and Africa, and it could have been crossed by *Homo sapiens* (or earlier hominins) with relative ease during glacial maxima, when sea levels were at their lowest (Ambrose 1998; Rohling et al. 1998; Lambeck and Chappell 2001; Siddall et al. 2003; Bailey 2009; Lambeck et al. 2011). Clearly, the Farasan Islands, being easily accessible and containing dynamic and complex landscapes with varied ecological opportunities for human exploitation, had the potential to answer significant archaeological questions concerning human dispersal out of Africa (Walter et al. 2000; Oppenheimer 2003; Bailey 2004a, b, 2009; King and Bailey 2006; Armitage et al. 2011; Bailey and King 2011; Mellars et al. 2013; Winder et al. 2013). However, to achieve this, interpretation of the palaeoclimatic, palaeoenvironmental and geomorphological variability of the ancient land surface was necessary (Dullo and Montaggioni 1998; Purser and Bosence 1998; Bantan 1999; Sakellariou et al. 2013, 2018). Evidence for these changes is found in the geology and the sediments. Accordingly, there was a need to visualise the submerged landscape, collect sedimentary samples and physically inspect the seabed. Here we review the multi-disciplinary investigations that have been used to interpret the drowned landscape and the search for archaeological remains.

2 Investigating the Drowned Landscape of the Farasan Region

The Farasan Islands were visited by the project team in 2004. During a season of terrestrial fieldwork, they recorded hundreds of shell middens that demonstrated past coastal exploitation dating back at least 7000 years (Bailey et al. 2007a, b, 2012, 2013, 2018). These discoveries paved the way for a diving project on board Saudi Aramco's 70-m long vessel MV Midyan to search below the water (Fig. 2), followed by three further seasons of underwater diving fieldwork and a two-week marine geophysical survey (Fig. 3). However, before any underwater investigations were conducted information was gathered from the land surveys and assessments of the hydrographic charts helped to identify areas that had the potential to act as occupation sites for early human populations. The location that offered the greatest promise was a 120-m deep depression in the bay to the south of Qumah Island. Here the depression was surrounded by a shallower seabed that is now only 50–60 m below sea level; therefore, when sea levels dropped lower than this during the last glacial cycle there would have been a basin up to 60 m deep. If the basin was filled with fresh water it would have been a very attractive place for wild fauna and human hunters and gatherers.

3 Mixed Gas Diving and Maritime Archaeology

In 2006, the first marine archaeological project was conducted in the area to investigate deeply submerged palaeo-landscapes. The objectives were to characterise geological and geomorphological formations, notably

Fig. 2 MV Midyan demonstrating its fire hoses on site (Photo: Garry Momber)

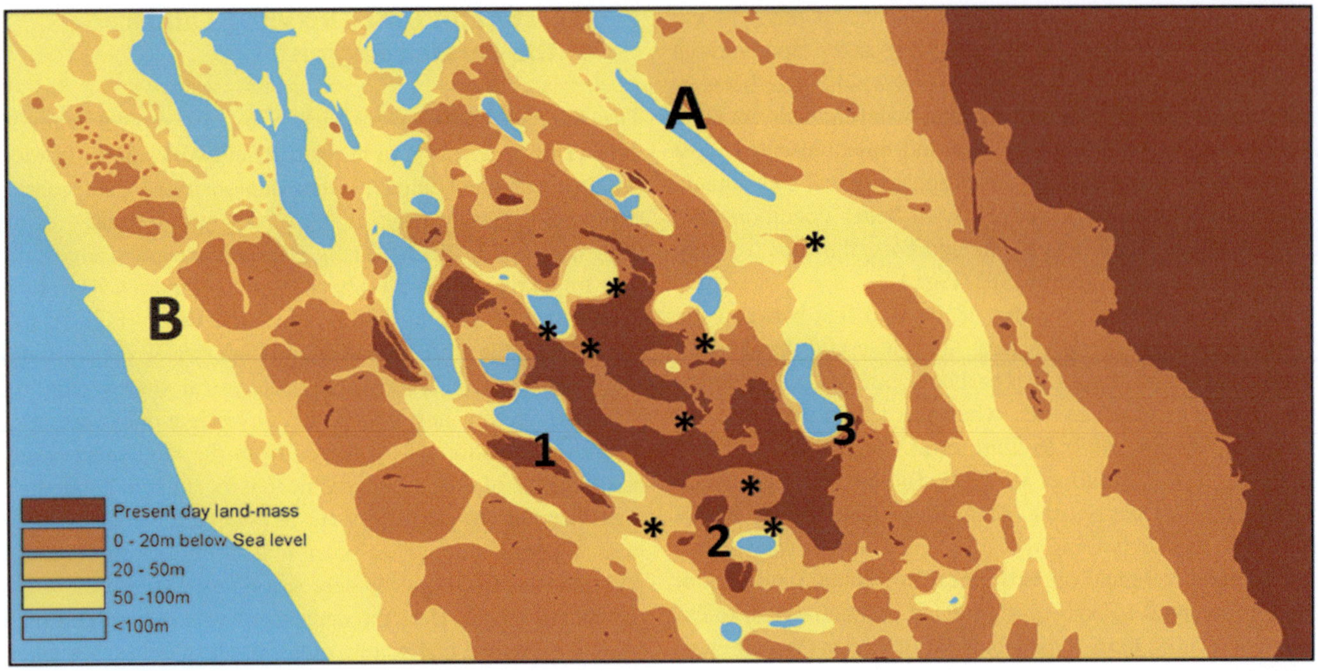

Fig. 3 Map of the Farasan Islands showing the bathymetry coloured at 20, 50 and 100 m depths. When the sea level dropped below 100 m the shelf was largely exposed while the sink holes formed by collapsed diapirs were probably filled with water. The asterisks indicate locations inspected by the dive team in shallow water. The numbered areas are (1) Zufāf Island, (2) Qumah Island and (3) the Sulayan archipelago. A and B indicate the areas of geophysical survey represented in Figs. 18 and 19 respectively (image courtesy R Smith, adapted by J Gillespie)

submerged palaeoshorelines, basaltic formations and possible springs, with the aim of locating areas with the greatest potential for human occupation. This necessitated the use of technical diving. The MV Midyan provided the perfect platform from which to run the diving operation. It was of sufficient size to host an air compressor, a gas blender, dive cylinders to provide mixed gas for the dive team, plus dive equipment, underwater archaeological equipment and all the tools required to maintain a smooth-running project. Different gases had to be mixed and blended for diving purposes that required space for dozens of 50 L J-bottles containing pure oxygen and pure helium, to be managed on the back deck (Fig. 4).

The use of mixed gas is ideal for deep diving but it is technologically and logistically challenging. This imposes constraints on the diver's actions and increases the task load faced by the project manager, dive supervisor and participants. Accordingly, the logistical considerations should be taken into account as a key element of an archaeological project. There are many training manuals and publications describing the detailed methods used for mixed gas diving and books describing the physiological effects of the different gases (Joiner 2001; Brubakk and Neuman 2003; Gerth 2006; Mitchell 2007). However, for this report, a short outline of the practice is given as a backdrop against which this and future operations have to be planned.

All gases used are derivations of air, which is approximately 21% oxygen, 78% nitrogen and 1% other gases. It is the most convenient gas as it can be taken straight from the atmosphere and compressed. It can be breathed underwater for limited periods without harm; however, at depth, high concentrations of compressed gases become toxic. Problems can arise when they enter the cardiovascular system and circulate within the body for extended periods. The longer a diver is under pressure, the greater the amount of gas absorbed. When air is breathed at 10 m underwater it is compressed to half the volume of air at the surface, meaning that the lungs need twice as much gas to fill them. Therefore, the 21% of oxygen per unit of air in the lungs will double, effectively increasing the relative percentage to 42%; this percentage is otherwise referred to as a partial pressure, in this case being 0.42. The pressure increases proportionally with depth; at 20 m it is tripled, at 30 m it is quadrupled, and so on. When oxygen reaches a partial pressure concentration of 1.4, it becomes poisonous and the diver can have a fit or become unconscious. This can occur at a depth of around 65 m.

Nitrogen can also be toxic in high concentrations, causing disorientation and confusion known as nitrogen narcosis or the 'Narcs'. This can affect a diver from around 30 m. In addition, when the diver returns to the surface the pressure is released allowing the gas to expand. If the ascent is too fast,

Fig. 4 Pure oxygen on the back deck of the MV Midyan to be mixed with helium and air to form Nitrox (Photo: Garry Momber)

nitrogen in the body comes out of solution and forms bubbles that can block blood flow, thereby cutting off circulation and starving tissues of oxygen. This is decompression sickness, commonly known as the bends, and is the greatest threat to divers. The most common method used to reduce the amount of nitrogen taken on by a diver is to enrich the gas with oxygen and reduce the amount of nitrogen; this mixture is Nitrox. When Nitrox is breathed by a diver, the amount of nitrogen absorbed by the blood is reduced and the problems of nitrogen narcosis and decompression sickness are lessened. This allows the diver to stay underwater safely for longer. However, the risk of oxygen poisoning at increased partial pressure remains. Due to these constraints, in practice, the value of enriched air Nitrox is limited to depths of no more than around 40 m.

To go deeper without oxygen becoming toxic while limiting nitrogen loading in the blood within the body's tissues, a harmless, inert gas can be added, such as helium. By introducing helium and halving the concentration of oxygen breathed by a diver, the diver can go to twice the depth before oxygen becomes poisonous. Therefore, with 10% oxygen in the mix rather than the normal 21%, an average diver could go to 120 m and beyond without inducing a detrimental 'oxygen hit'. Similar benefits are seen when nitrogen is reduced. When helium is added to air or Nitrox it becomes Trimix. The relatively small proportions of oxygen in the mix make it perfect for breathing at depth; however, if you reduce the amount to less than 17% it will be too weak to breathe at the surface and will cause oxygen starvation. To manage this, the concentrations must be adjusted to suit the best depth and different mixes have to be taken underwater. As a result, a deep diver will normally carry one cylinder with enriched air Nitrox to breath on the way down, two cylinders with Trimix to breathe near and on the bottom, and one with 100% oxygen to breathe while decompressing near the surface (Fig. 5).

Fig. 5 Heavily laden Trimix diver working underwater (Photo: Garry Momber)

The implications of mixed gas diving for archaeological investigations are the limitations on time and mobility when underwater. The need for tight control during a dive operation restricts the actions of the marine archaeologist when compared with their terrestrial counterparts; therefore, diving fieldwork will inevitably take much longer. This is necessary to ensure safety but arguably is worth it when uncovering data that cannot be found elsewhere.

4 Deep Water Investigations off Zufāf and Qumah Islands

Deep diving operations were carried out off the coast of Zufāf Island, on the western side of the archipelago and off the south of Qumah Island, next to the main Farasan Island. Zufāf Island was identified as a suitable dive site by merit of the steep drop-offs plotted on Admiralty Chart "Approaches to Jījān no.15. 2000". These were verified by an echo sounder survey that recorded a steep incline averaging about 33° which levelled off at just over 50 m below modern sea level to form a gently sloping terrace, inclined at 6° for 165 m to the east until it reached 80 m. At this point it dropped almost vertically. The depth in the centre of the channel has been sounded at about 140 m. Investigation of the deep drop-offs and terrace was conducted with mixed gas diving.

The diving operation used Trimix to reach the wave cut terrace at a depth of 56–60 m. The descent down the slope recorded little change in the seabed type although the occasional outcrop of old grey coral rising from under the sand was noted. This suggested that the underlying geology is close to the surface, offering the potential to find areas of the old land surface. At −56 m the seabed levels off and drops continually toward the deep. A blanket of sand covers this terrace making it impossible to see the bedrock below. This continued for over 100 m to the northeast as the seabed dropped to −62 m. A small sondage was excavated 40 cm into the sand by hand fanning but the underlying substrate was not detected as the blanket of sand was too deep at this location.

At the mouth of Qumah Bay, Trimix diving was used to follow a wave cut platform from the Slick Point Peninsula underwater toward the centre of the depression. The divers traced the platform down slope for over 300 m to a depth of 30 m where it runs beneath the sandy sea floor. The divers continued to a depth of 38 m. It was apparent that the rock formation continues below the seabed as occasional exposures of the reef were recorded; however, by −40 m the area became dominated by silty sand. Beyond this point, the flat seabed dips gently to the south. As with the diving off Zufāf Island, the deeper surfaces could not be examined by divers without excavation and it was concluded that a geophysical survey would be necessary to calculate the thickness of the sand. In the shallows, however, reefs, terraces and rock formations were exposed, so the next phase of the fieldwork was to focus on the accessible submerged coastal features at 10–20 m depth.

5 Shallower Water Evidence of Wave Cut Notches and Deformation

The diving confirmed the existence of wave cut and erosional notches at varying depths, and that most of the visible geological features were in the shallower waters; accordingly, this was where the next phase of activity was concentrated. In Qumah Bay, the notches are generally located within cliff sides or steep slopes, particularly toward the entrance in the south. Many of these notches measure around 1 m in height although some measured up to 2 m.

At the eastern entrance to Qumah Bay the area investigated was marked by a west-facing cliff that forms the edge of the wave cut platform as it passes below and adjacent to Slick Point. The cliff is 10 m high with a laterally consistent row of notches running toward its base (Fig. 6). The depth to the back of the largest notch averages around 3–4 m while the entrance is 1.5–2 m high. The platform reef at the base of the notch forms a ledge that protrudes 1–1.5 m beyond the mouth of the cave. Studies of wave cut notches in the Mediterranean show that the height of the average wave cut notch is invariably higher than the mean tidal range but less than the maximum tidal range (Antonioli et al. 2015). The tidal range at Jizan, being the closest mainland port immediately to the east of the Farasan Islands, is 1.69 m (see Inglis et al. 2018). This maximum tidal range is slightly less than the height of the notch suggesting some of them are too large to have been formed by tidally controlled wave action alone. However, it should also be noted that the Farasan Islands are on a relatively shallow shelf and the arrangement of islands within the archipelago can channel water to create a funnel effect that creates a larger tidal range, although long term observations have yet to be collected near the site to confirm this.

Mechanical erosion could be another factor that contributed to the formation of notches at the base of the cliff if this occurred prior to a regression and at a time when this part of the seabed might have been able to retain pebbles. Mechanical erosion would not have had the same impact when the sea level transgressed the site as the platform reef ledge that the notches now sit above is truncated to seaward by a reef step. Beyond this, the seabed drops off steeply into the sand filled bay below. This means that while pebbles could have contributed to erosion when they were being moved by the energy of relatively powerful intertidal waves, once the forces diminished as the sea level rose, the pebbles would be washed off the platform downslope. It is also

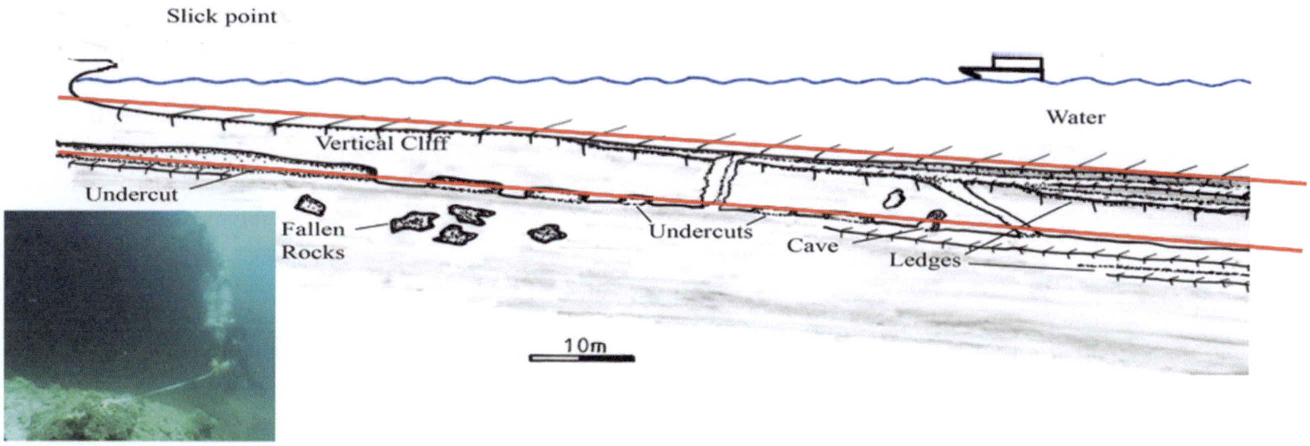

Fig. 6 Plan of submerged cliff, terraces and wave cut notches off Slick Point. The laterally consistent linear features are indicated with red lines. The notches at the foot of the cliff sit above a ledge that is truncated by a vertical drop-off. Sand covers the seabed in front of this rock face, masking much of the underlying seabed. Intermittent, exposures of this smaller cliff were measured at 0.6–0.7 m. The erosion features at the top and within the cliff are parallel and dip toward a collapsed diapir to the south (Drawn by G Momber)

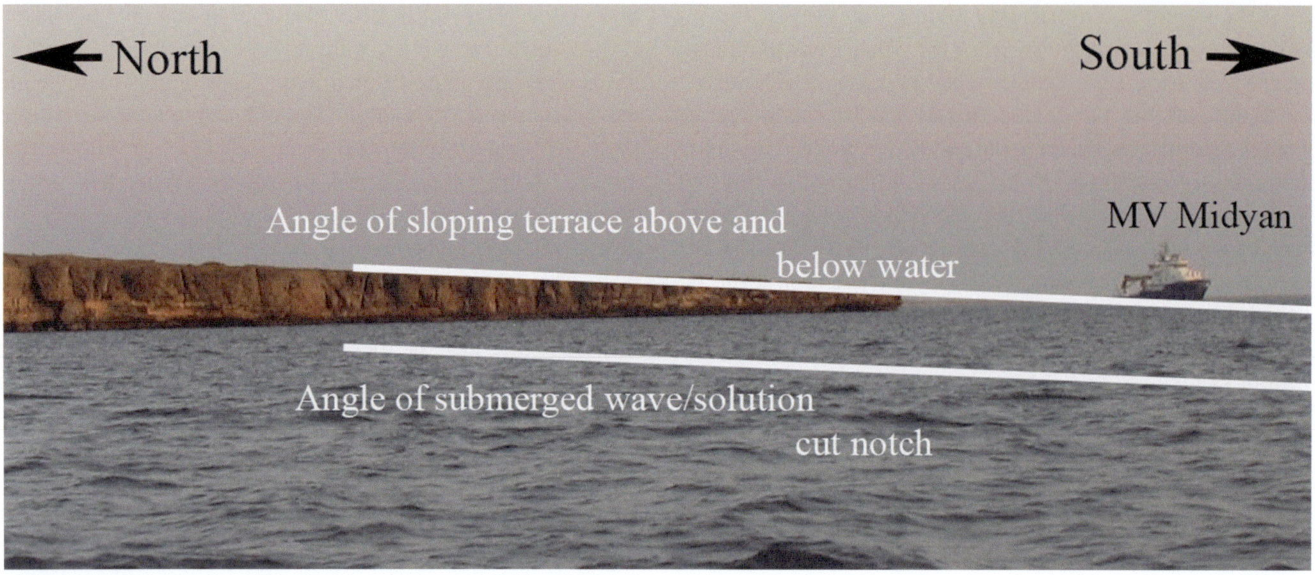

Fig. 7 Slick Point from the west. The white line indicates the approximate position of a stratigraphic break which continues underwater as a wave cut platform and which has been tilted by subsequent tectonic uplift. An equivalent feature has been identified at a lower level, from a depth of 9.6 m below sea level, recorded to a depth of 20 m. It runs in a north–south direction and is believed to relate to a period when sea level was 20–22 m below present (Photo: Garry Momber)

probable that the ledge in front of the cliff was less prominent when the notches were formed as the bay is part of the subsidence associated with salt tectonics. This vertical step in front of the platform reef now protrudes through the sand toward the northern end of the submerged notches where it is 0.6–0.7 m in height.

The system of notches and reefs runs parallel to the platform at the top of the cliff and follows the angle of dip to the south. The submerged notches also continue into the bay to the north where these become shallower as they mirror the strike of the land above water (Fig. 7). Our understanding of the formation processes is essential if we wish to establish a relationship between these features and sea level. However, in doing so we recognise the value of the notches as indicators of a changing land-sea relationship and our attention turned to these submerged exposed reefs where they were free from sediment cover and therefore accessible.

The height of the land above modern sea level across the Farasan Islands is commonly around 2–5 m. The elevation of this landscape may equate to the formation of coral reefs around 125 thousand years ago, during MIS Stage 5e when eustatic sea levels were 5–7 m higher (Rohling et al. 1998; see Inglis et al.2018). However, this can vary across the islands where many areas are subject to localised uplift and downwarping caused by the salt tectonics (Hudec and Jackson 2007). This is particularly the case on Qumah Island where linear scars traversing the land surface indicate faulting and fracturing. Fortunately, the deformation is localised and clearly identifiable. This is the process evident at the entrance to Qumah Bay by Slick Point (see above and Fig. 7). Despite the tilting of the land, the surface is relatively flat, providing a point of reference against which to record submerged notches.

It was noted that the distance between the land surface and the first sequence of notches is consistently in the order of 15–17 m depth. This equates to around 20–22 m below the MIS Stage 5e sea level. Periods when the sea level was stable for long enough to impact the coastline by eroding or dissolving features have been calculated by a number of researchers (Shackleton 1987; Rohling et al. 1998; Lambeck and Chappell 2001; Siddal et al. 2003; Bailey 2009; Lambeck et al. 2011). The research has identified these still stands at 20–25 m depth around 110,000 and 80,000 years ago, at 40–60 m depth 30–45,000 years ago, at 30 m depth 39–44,000 years ago, and 40–50 m depth about 12,000 years ago (Fig. 8).

The still stand at 20–25 m depth is particularly significant for the archaeologist as this would have occurred about 80,000 years ago (when calibrated against the elevated marine terraces in New Guinea as presented in Fig. 8). Notches or undercuts formed at these times, being before a regression, would have dried out and remained above water level as the sea dropped. This would have provided a habitable shelter for over 70,000 and probably 100,000 years while the sea level was lower. Diver surveys along the −10 and −15 m contours confirmed that wave cut notches can be found running along both sides of Qumah Bay and also around the adjacent shoreline of Qumah Island.

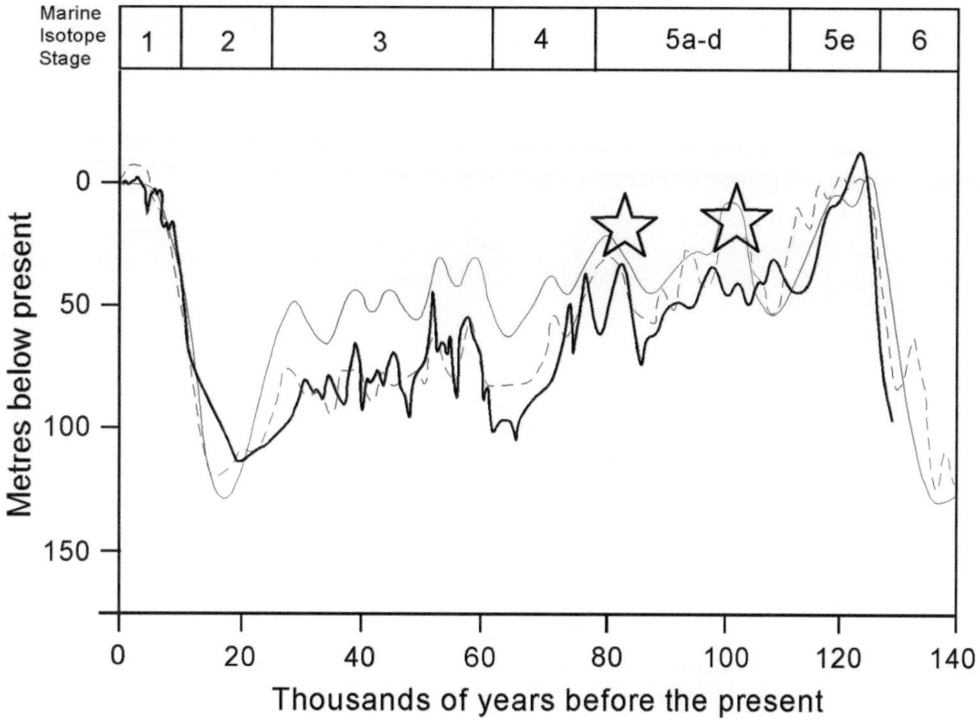

Fig. 8 Global sea-level change over the past 140 kyr. The dashed grey line is based on deep-sea isotope records of planktonic and benthonic fauna, the solid grey line shows the same curve corrected for temperature effects using dated and elevated marine terraces in New Guinea, the dark solid line is based on isotope records of planktonic fauna from the Red Sea. Stars show the probably dates when some of the larger submerged notches were formed around Qumah Island. Modified from Chappell and Shackleton (1986), Shackleton (1987), Lambeck and Chappell (2001), Siddall et al. (2003)

6 Shallow Water Diving to Identify Archaeological Potential Around the Islands

The insights into the submerged landscape features identified in 2006 informed the research questions for 2008, which aimed to gather more data about the submerged terraces around the islands and assess the viability for associated human occupation. Firstly, it was necessary to identify the suitability of a site to provide shelter near possible sources of food and water, and secondly there was a need to qualify the potential for survival of material.

When determining the possibility for preservation of material at a particular location, we should address the implications of morphological changes that occur when the sea level fluctuates. A consequence of rising sea level can be the displacement and dispersal of archaeological material. Alternatively, material can be buried and protected in areas that act as a sedimentary sink. Both scenarios create challenges for the archaeologist, but the biggest threat is physical erosion. Degradation and deflation of deposits will be particularly acute when they are within the surf zone. During this phase, archaeological artefacts are likely to get dispersed. The chances of finding such dispersed sites are minimal unless material remains trapped in crevices. Where the currents and waves are weaker, the possibility exists for artefacts, such as stone axes, to remain in their original position with only finer material removed. This would be good for the archaeologist but it is very rare. A more likely scenario is where sheltered sites are protected from wave energy, enabling sediment to fall from the water column. This can bury large areas of palaeo-landscapes, sealing them for many tens of thousands of years (Andersen 1993; Flemming 2004; Momber 2011, 2014; Peeters and Momber 2014; Momber and Peeters 2017; Flemming et al. 2017). The problem occurs when trying to find such sites as they may be buried beneath many metres of sediment.

It is therefore the case that if areas of sedimentation or erosion are known to be extensive, they do not warrant exploration by divers. Areas that do provide the greatest possibilities for the discovery of archaeological material exist where there is a transition between the two systems, at a point where erosion and sedimentation meet. If cover is thin enough, artefacts are more easily accessible but the challenge is to find these areas. However, the seabed is not static and opportunities can open up following storm events or changes in currents that can disturb the seabed or cause

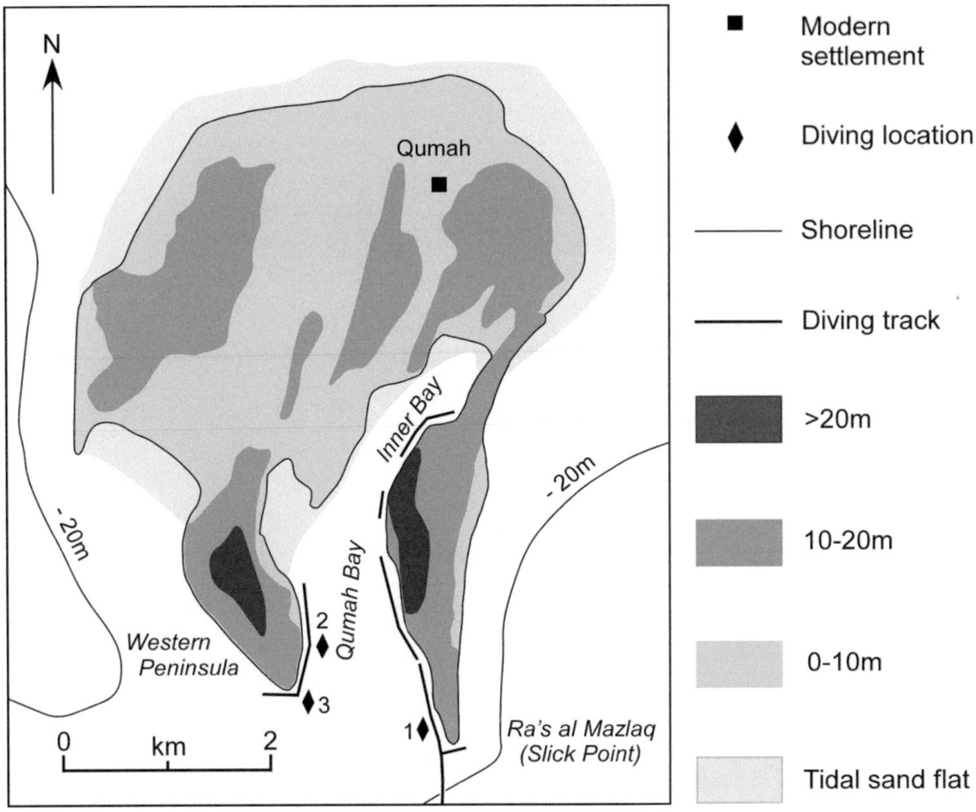

Fig. 9 Map of Qumah Island, showing diving locations and places mentioned in the text. Diving locations are: (1) Slick Point; (2) Western peninsula; (3) Shark Point; the site of RS QB01 is at location 2 (Drawn by G. Bailey)

scouring of sediment. Here, submerged landscapes that have been protected by silt since inundation may become subject to erosion and exposed.

7 Survey Strategy and Methods

With the aid of bathymetric charts and the experience gained during the previous expedition, near-shore palaeo-coastlines were targeted. The archaeological potential for each site was evaluated against a set of criteria that looked at the cultural nature of material (i.e., whether it had characteristics such as a shell midden that suggested longer term occupation or an isolated artefact such as a worked stone tool of local or exotic material), the date of the artefacts or ecofacts and the proximity of the archaeological evidence to the coastal feature of interest. This evidence was considered along with the potential for life sustaining resources nearby, primarily being sources of fresh water. This assessment was coupled with a consideration of coastal geomorphological processes that would have enabled archaeological material to survive the last transgression without burying it below metres of marine sediment (Flemming 2004; Flemming et al. 2017).

During the 2008 season, the dive team inspected 15 new sites around the Farasan Islands including extensive visual searches across large tracts of Qumah Bay (Fig. 9). Notches and wave cut platforms were recorded at all the coastal sites, most of which compare favourably in depth with the 80,000-year still stand. Of the 15 different areas investigated, no archaeological or palaeo-environmental material was uncovered, but two sites were identified as offering the greatest potential for the discovery of archaeological material. One site on the western peninsula of Qumah Bay is associated with overhangs that could offer shelter to early humans, while the other site in the Sulayn Islands compared favourably with the adjacent terrestrial landscape that supports many shell middens.

8 Seabed Excavation and Sampling at Qumah Bay

The submerged notches on the western peninsula of Qumah Bay were selected for excavation and sampling during the 2009 expedition. These possible shelters are opposite Slick Point and are positioned at a favourable vantage point overlooking the depression off the south-east of Qumah Island that could have offered an attractive terrestrial environment when sea levels were lower. The wave cut notches are relatively free from sand cover as the location sits around the interface between deep sediment and scoured exposures.

One particular notch forms a small cave that would have made an ideal rock shelter. The site lies in 10 m of water half way along the inner side of the western peninsula. It has a 2-m-high opening and is 3.5 m deep, with a sloping roof that tapers to the back of the cave (Fig. 10). There are flat areas inside and above the cave that are covered by thin layers of sand and these would have provided suitable platforms for activities such as flint knapping or shell processing. These areas were searched, the sand was cleared and samples were recovered from crevices at the back of cave but no archaeological material was found. The lack of material suggests that if it was present it must have been removed by the waves as the sea level rose during the transgression. We therefore looked below the shelter to see if any objects had found their way into deeper water.

The rock shelter sits above a steep 7-m-high slope that has a small outcropping rocky reef at its base. If artefacts had washed down from the shelter, the more resistant items could have been caught in the reef below. Here, a metre-high notch forms a perfect trap that was full of accumulated sand and shells. It was excavated to record the stratigraphy and recover samples.

Excavations in front of the notch created an 85 cm section (Fig. 11). The top is capped with sand above layers of shell. These form distinct layers with different shell types and become sandier at the bottom. In the entrance to the notch, the remains of a coral reef were uncovered that form a ledge. It appears that the coral did not grow within the small cave-like feature, most probably due to the lack of direct sunlight. Beyond this, toward the back of the notch, was a deeper area that retained material in a well stratified deposit. Five samples were collected vertically from within the section and analysed to look for concentrations of edible shells by comparing them with midden sites on land. The shells were mainly unsuitable for human consumption although

Fig. 10 Garry Momber evaluating south facing entrance to the possible shelter on the western side of Qumah Bay. It offered a view over a low-lying area to the southwest when sea level was lower (Photo: Julie Satchell)

Fig. 11 The small cave filled with layers of shells was revealed following excavations below the wave cut arch in 17 m of water. The relic coral reef can be seen outcropping in the bottom right hand corner (Photo: Garry Momber)

there was one distinct horizon, 45 cm from the top, which was dominated by an assemblage of edible bivalves (Fig. 12). There was also an anomalous piece of stone that contrasted to the lightly coloured calcareous material evident in the rest of the section, but there were no cut marks on the shells and the stone was not sufficiently diagnostic to prove the sample is an artefact. This evidence is inconclusive but the distinct stratigraphic units established that there were areas where material could remain undisturbed following deposition. It is within these shallow sedimentary sinks that we are most likely to find archaeological artefacts (Flemming 2004).

Above water, it is worth noting that the terrestrial cliff over the submerged rock shelter reaches the 10-m contour (Fig. 13). It is slightly domed due to salt tectonic uplift. If the coral platform above the cliff dates to MIS 5e, when it would have been covered by a sea level that was about 2–5 m higher than today, it would suggest the land has risen in the order of 10 m. The depth of the submerged rock shelter from the top of the cliff is 20 m, presenting a likelihood that it could have formed during the last −20-m still stand. This again dates to about 80,000 years ago, so the overhang would have offered welcome shelter for tens of thousands of years before becoming inundated during the last transgression.

9 Seabed Excavation and Sampling at Sulayn al Janub Archipelago

The second site excavated was in a sheltered, shallow water channel south of Sulyan Island. The Sulayn al Janub archipelago is situated on large shallow plateaux that would have been largely dry land 7,000 years ago when sea levels were 5–6 m lower than today (notwithstanding local uplift or submergence). Shell middens visible on the islands demonstrate occupation and the availability of marine resources (Fig. 14).

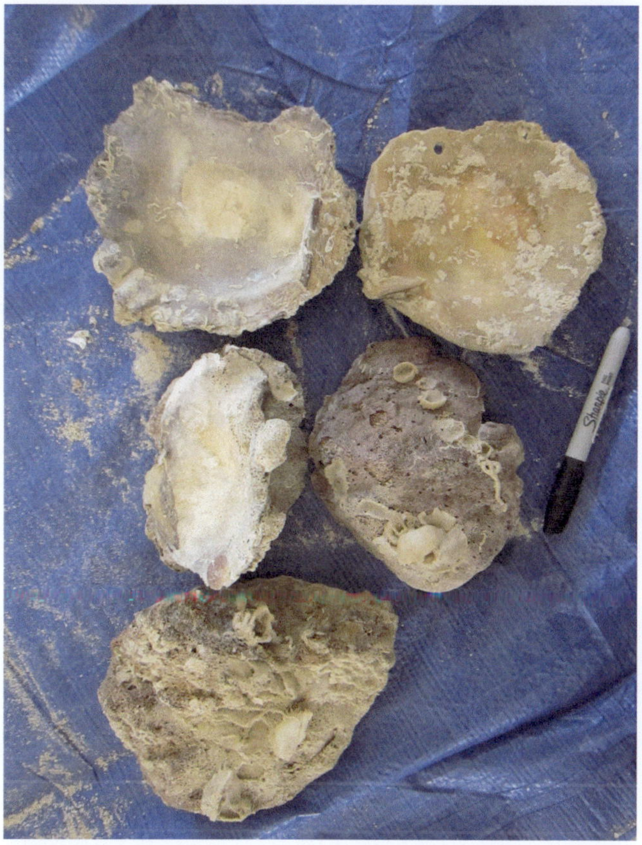

Fig. 12 The sorted finds included an assemblage of edible shells from Qumah that were found alongside an anomalous lithic (Photo: Garry Momber)

The area under investigation is a drowned valley surrounded by three islands with three entrances (Fig. 15). The entrance to the southwest is wide and shallow, only a metre or two deep. The channel to the north is narrow and a little deeper at 2–3 m. In the east, the channel is recorded at 5 m deep and is fairly wide. Interpretation of the modern bathymetry shows that this eastern channel saw the first ingress of water into the basin. A break-through in the north would have followed and finally the wide western channel would have been inundated.

The north and west channels would have been dry land 7,000 years ago. This was a time when the harvesting for molluscs was prolific and several large middens can be seen along the shores today (Bailey and Flemming 2008; Bailey et al. 2013; Meredith-Williams et al. 2014a; Hausmann et al. 2015). Any that were built nearer the water's edge 7,000 years ago would now be underwater. Searching across the centre of the drowned basin proved futile as the seabed is covered with sand; however, where water movement is sufficiently strong the sand has been prevented from settling or has been removed, and the underlying surface is accessible. This occurred when the new channels were opened to the west and north, particularly the northern channel, which is narrower and deeper. Therefore, the northern channel was chosen for excavation as it is near an inshore environment rich in marine molluscs. Also scour caused by the current passing through the channel has resulted in a thinner layer of

Fig. 13 Cliff inside west peninsula and above submerged wave/solution cut features (Photo: Garry Momber)

Fig. 14 Shell midden on South Sulayn Island near the area of underwater excavation (Photo: Garry Momber)

sand toward its centre and a large deposit of shells was recorded at this location during the 2008 fieldwork (Fig. 16).

10 Excavation at Sulayn al Janub

A 10 m long evaluation trench was excavated down-slope from west to east. The working depth was shallow. The west end was recorded at about 2 m below mean high water while the eastern end was just under 3 m. On completion of the trench, sections were recorded and five samples were recovered. Three contexts were noted. From top to bottom these are sand, fine grey organic silty clay and a sandy shell mix. The total thickness of the sediments was recorded as 0.7 m in the west and 0.35 m in the east. The relative concentrations of shells increased to the east as the sand became thinner.

Samples recovered from the trenches were dried and sorted. Mollusc shells were removed from the deposit and the species typical of those found in the on-land middens (see Bailey et al. 2018) were separated. The sub-samples were weighed and the proportional relationship recorded. The shells were densely packed in the lower context but the vast majority were too small to be edible and the wide spectrum of species was too broad to represent a midden deposit. It was concluded that the material from all the contexts was natural rather than archaeological. However, the results have helped to inform our understanding of the depositional process. A layer of fine grey silts below the mobile sandy sediment shows that the underlying deposits are stable in the sheltered waters around the islands. In addition, the presence of the fine silt along the entire length of the trench indicates that protective sediments, which could create anaerobic conditions, accumulate even in the shallower water.

Archaeological items were not recovered but a fragment of charcoal was discovered that suggested human activity. This result warranted further investigation and in 2014 a larger excavation and sampling programme was conducted. During this phase, a new trench running parallel to the initial

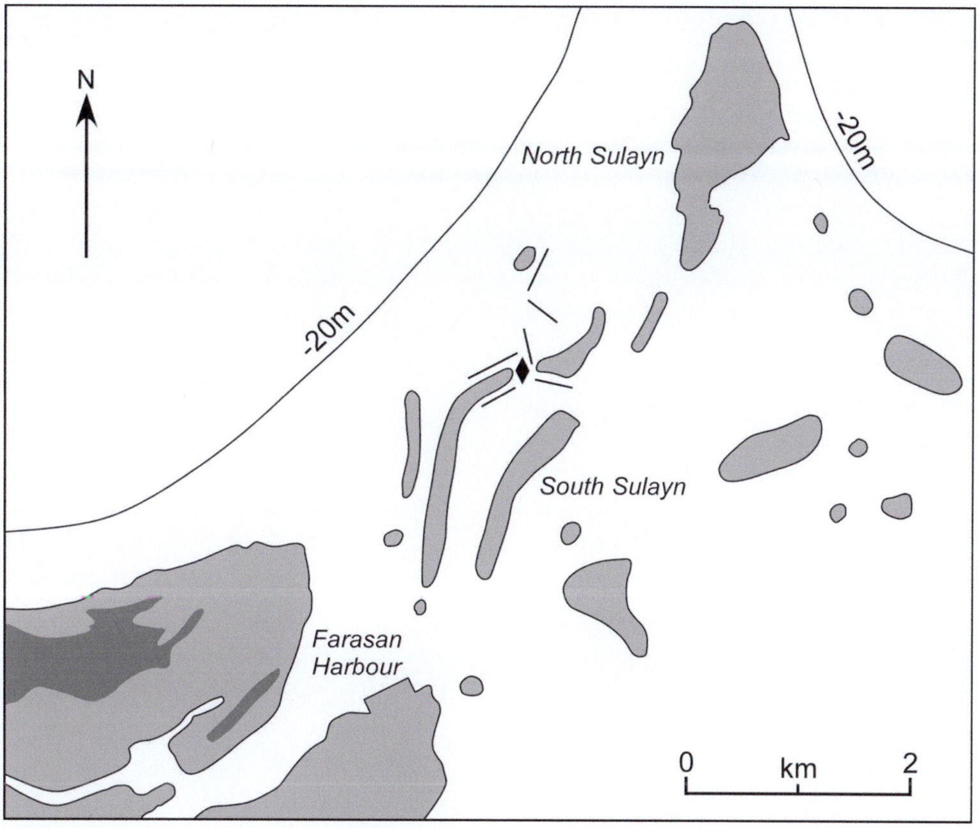

Fig. 15 Chart showing the 20 m contour around the Sulayn al Janub archipelago. The shaded areas are the islands with the darker areas indicating higher ground above 20 m. The location of diver inspections and excavation are indicated by the lines and the diamond respectively (Drawn by G. Bailey)

Fig. 16 The north channel investigated in Sulayn al Janub archipelago. Diving operations were conducted from left to right below the water (Photo: Garry Momber)

evaluation trench was opened. It was widened and extended to form an arm from north to south nearer the centre of the channel. The methodology focussed on increased lateral and horizontal sampling.

Excavations to the underlying bedrock were conducted in selected areas to investigate variations in colouration and with the intent of investigating any taphonomy, to see if any palaeo-soils remained, and to find the source of the charcoal. The new excavations concurred with the results from the first trench, but this time the trench revealed larger areas of bedrock which exposed a layer of buried coral on top of a hard, basal coral terrace. An airlift was used to remove overburden and collect samples from gullies within the coral basement rocks.

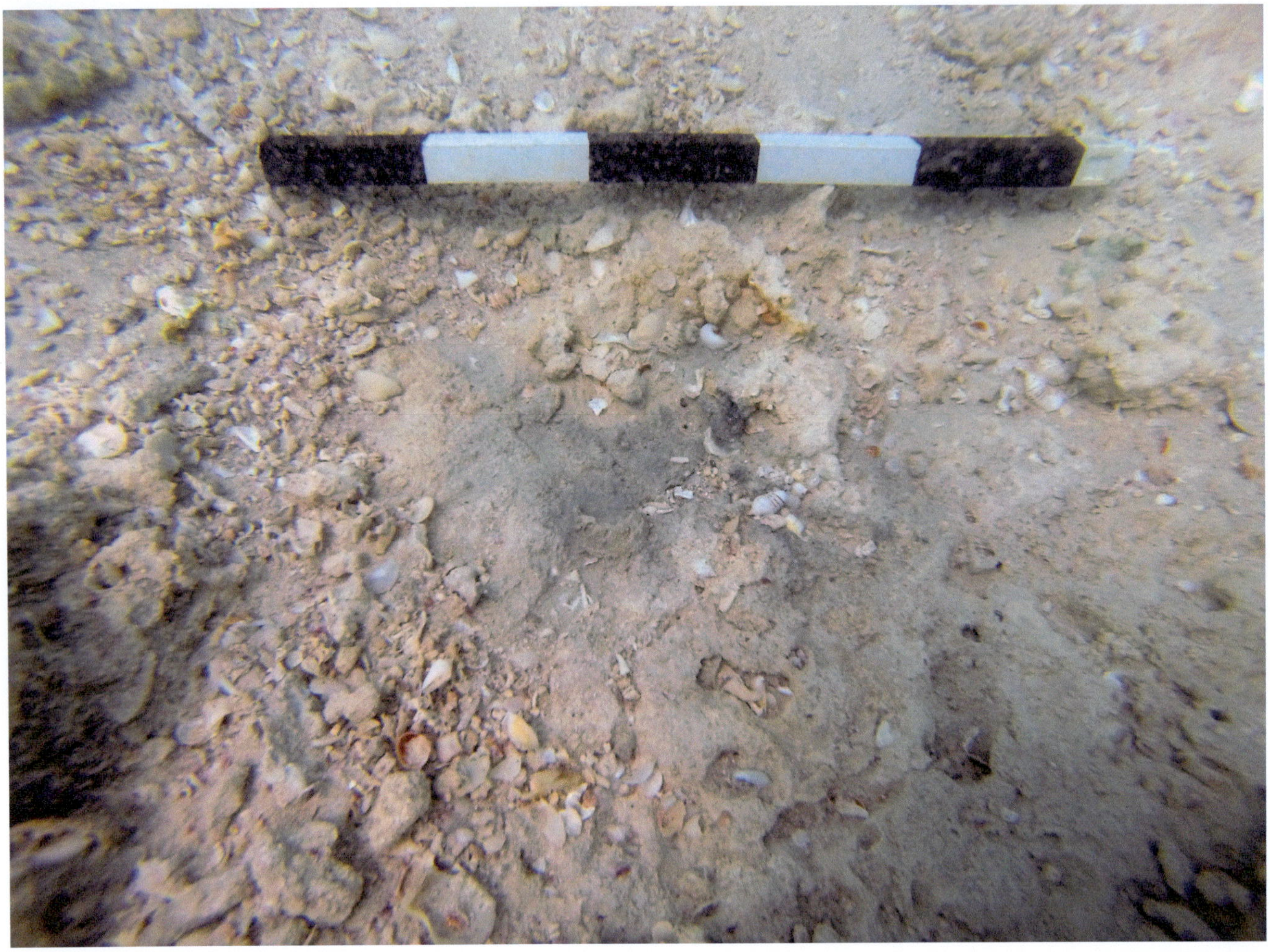

Fig. 17 Area of dark staining being uncovered on the coral bedrock (Photo: Garry Momber)

In the deeper water, to the east of the surveyed area, the sediment cover of sand and shell was thinnest, while the horizon of broken coral below the shell but above the basement/bedrock was thickest. On the northern-most baseline, dark grey stains or patches were recorded that extended below heads of coral that must have grown after the 'stains' were formed. Samples of the seabed surface were collected for analysis, but rather than burn marks, it was concluded they were caused by organic decomposition (Fig. 17).

The 98 samples recovered from the excavation were processed and sampled. Another small piece of charcoal was identified although, yet again, it was not possible to confirm any anthropogenic patterns within the shell collections.

The underwater work conducted by the dive team has provided a great deal of geological and geomorphological information about the coastal waters around the Farasan Islands. We are now much better informed about the elusive nature of the archaeological signature in the area and recognise that middens will deflate as they go through the tidal zone while coral hand tools will degrade when in the sea. However, we have identified submerged shorelines, wave cut platforms and solution notches. By recording these features, we have located sites suitable for habitation by prehistoric peoples and found areas with stable stratified units that could hold material. For the geologist, we have identified markers for relative sea level change and helped interpret the geomorphological effects of salt tectonics as it deforms the landscape. We have also learnt that there are no outcrops of basaltic material around the islands, which is unfortunate as field-walking on the mainland showed these were the main targets of human activity and provided the most distinct indicators of an early human presence.

As the project progressed and understanding increased, it was recognised that there was a growing need to collect more information about the submerged landscape so we could further investigate the hypothesis that the basins formed by salt tectonics held fresh water during periods of lowered sea level. There was also a need to better understand the geomorphological processes across the wider shelf exposed around the Farasan Islands during lower sea level and locate nodes of activity, ideally near basalt outcrops, with a greater potential to find evidence of human activity. To achieve this, we needed to cover a much larger area than we could do with divers, so a multidisciplinary geophysical survey was conducted.

11 Acoustic, Geophysical and Coring Equipment Used in the Farasan Islands

In June 2013, an opportunity arose to conduct a geophysical and acoustic survey to investigate the underlying geological processes and the presence of sediments. These were tools that had not been used around the Farasan Islands for geo-archaeological reasons before but they are becoming increasingly important to the maritime archaeologist when trying to interpret the palaeo-landscape.

Enormous technological progress has been made in recent decades in marine geophysical survey methods and a large spectrum of techniques is available for the systematic mapping of seafloor and sub-seafloor of the continental shelf at various scales, allowing the reconstruction of submerged landscapes and terrestrial palaeo-environments and the evaluation of possible locations for prehistoric archaeological sites. The main advantages of using marine geophysical techniques are: (i) large areas can be surveyed in high resolution (decimetre to metre resolution) and over a relatively short time, (ii) deep seafloor areas can be easily and systematically surveyed, (iii) information on the shallow sub-seafloor structure and stratigraphy can be collected without the need to excavate the seafloor, and (iv) geophysical sub-bottom profiles can be calibrated by coring.

Multibeam echo sounders (MBES) are designed to acquire simultaneously co-registered bathymetric and backscatter (acoustic reflectivity) data covering a swath of the seafloor along the survey track. They are fairly complex systems requiring sophisticated motion sensor units in order to rectify vessel pitch, roll and heave when positioning the data relative to the seafloor (Le Bas and Huvenne 2009). The coverage on the seafloor is proportional to the water depth. Since MBES are usually surface-towed or hull-mounted systems, the resolution of the seafloor decreases as the water-depth increases.

Unlike MBES, side scan sonars (SSS) are deep-towed systems designed to map the acoustic reflectivity of a swath of the seafloor acquired along the survey track. SSS typically consist of an underwater vehicle ("tow-fish") with two transducers mounted on either side (port-starboard). The tow-fish is deployed via a "tow-cable" behind the vessel to maintain constant height (about 20 m) above the seafloor during the survey. Thus, SSS provide high resolution mapping of the seafloor's acoustic reflectivity throughout the survey, both in shallow and deeper depths.

Multibeam echo sounders (MBES) and side scan sonars are powerful tools in surveying, in particular in areas where submerged landscape features are exposed on the seafloor. This is not the case in many areas, where marine sediment deposition has led to the burial of the submerged palaeo-landscape and any remaining prehistoric features below a blanket of marine deposits, the thickness of which may vary between a few decimetres and a few tens of metres. For these cases, a wide range of sub-bottom profilers (SBP) are designed to provide information from below the seafloor, on the structure and stratigraphic pattern of the sub-seafloor. According to the sound source and the frequency (wavelength) and strength of the emitted sound, SBPs are distinguished as "pingers", "chirps", "parametric echosounders", "sparkers", "boomers" and "air guns". In general, higher frequencies provide higher resolution but limited penetration while lower frequencies can penetrate deeper below the seafloor at compromised resolution. Sub-bottom profilers can be either towed behind the survey vessel or be hull- or pole-mounted. Conventional SBPs provide a 2D profile of the stratigraphy along the survey track. Recent technological advances have led to the development of 3D profilers, which can provide three-dimensional imaging of the sub-seafloor but require extensive and complex processing (Missiaen et al. 2017a, b).

Coring can provide physical access to sedimentary layers of the seafloor's substrate and physical constraints for the interpretation of the seismic sub-bottom profiles. Gravity and piston corers can theoretically reach several tens of metres beneath the seabed, but in most cases they rarely exceed 5–10 m penetration. The deployment and recovery of corers require powerful winches and therefore they are used from purpose-built vessels.

12 Deep Seabed Inspection: Methods, Results and Limitations

Manned submersibles (or Human Operated Vehicles, HOVs) and remotely operated vehicles (ROVs) have been used for the inspection of the deep seabed for over 40 years. During recent years, the use of autonomous underwater vehicle (AUV) technology for the survey of the seafloor has become increasingly available and efficient (Missiaen et al. 2017a, b). The most commonly used sensors mounted on ROVs, HOVs and AUVs include navigation sensors for positioning, optical sensors (video, photographic, stereoscopic still cameras), sonar sensors for mapping the seafloor and its features (multibeam, side scan sonar, subbottom profiler) and chemical/environmental sensors for quantifying the oceanographic environment.

Submersibles are best suited for direct-observation mapping and sampling rather than fine-resolution systematic surveys. They are not tethered and thus can move freely and independently from the ship. The bottom time is relatively short, depending on the consumption of power and the need for recharging of the batteries. ROVs are tethered to the survey vessel through an umbilical which provides power and commands to the vehicle and transmits data from the vehicle to the vessel. ROV dives are not constrained on bottom time; however, the tethered configuration restricts their mobility, as the vehicle and the surface ship have to move in concert. AUVs have recently emerged as stable, near-bottom survey platforms. They are not tethered and are designed to run pre-programmed missions by flying along survey lines at constant altitudes from the seabed (Bingham et al. 2010).

13 The Farasan Islands Research Cruise: Survey Methodology and Results

The large-scale imaging of the Farasan continental shelf was carried out by the HCMR survey vessel, R/V Aegaeo. The marine survey was designed to collect as much data as possible and integrate the different data sets in order to map important palaeo-morphological features on the seafloor and its substrate, including palaeo-shorelines and terraces, river-valleys and lakes, and to reconstruct the submerged, terrestrial landscape of the Last Glacial Maximum (Sakellariou et al. 2018).

During the 12 days of the cruise two areas were surveyed systematically, in the inner and the outer continental shelf, northeast and west of the Farasan archipelago respectively. Swath bathymetry and backscatter mapping of the seafloor were performed with the use of two hull-mounted, multi-beam systems (20 and 180 kHz), while certain areas were surveyed in more detail with a deep-towed side scan sonar. A 3.5 kHz pinger was used for the high-resolution survey of the seafloor's shallow substrate while a 10 ci airgun seismic profiler provided information on the deeper geological structure and the seismic stratigraphy (see Sakellariou et al. 2018). Coring with a 5-m-long gravity corer and a box corer was performed on sites selected on the basis of the on-board preliminary interpretation of the acoustic and geophysical recordings. Finally, ROV dives were scheduled to inspect visually specific sites of potential palaeo-environmental interest.

The wide continental shelf off southwestern Saudi Arabia and around the Farasan Islands comprises a 70–80 m deep terrace, dipping very gently toward the Red Sea, which was exposed during the Last Glacial Maximum. Post-LGM marine sedimentation has been negligible on the shelf of the surveyed areas. Marine deposits with a maximum thickness not exceeding a few metres occur only in local depressions of the shelf. Thus, the primarily terrestrial landscape with its main morphological features is preserved and exposed on the seafloor. The survey area in the outer shelf is incised by well-preserved river valleys, developed along fault scarps and outflowing at the LGM shoreline. The latter has been identified at about 120 m below present depth (Fig. 18). A series of flat-topped ridges surrounded by steep, faulted slopes has been mapped at a short distance off the shelf edge and separated from it by deep troughs. The top of the ridges lies at depths comparable to the depth of the shelf; thus, they were exposed during the last low sea-level stage, forming an archipelago at a short distance off the shoreline. The survey area on the inner shelf comprises a 120–130-m-deep valley incised in the 70-m-deep terrace, which connects to a sub-circular 250-m-deep depression through a narrow gorge. Both the valley and the sub-circular depression were probably filled with water and formed lakes during the last glacial maximum, when the surrounding shelf was exposed (Fig. 19). The results of the Farasan Islands research cruise are presented in detail by Sakellariou et al. (2018).

14 Does it Give Clues About Areas with Archaeological Potential?

The seabed surface and sub-surface categorization provided by the geophysical data gives a baseline of anomalies that can be explored in detail by divers. It shows that a great deal of the pre-submergence morphology remains. It also reveals that the thickness of sediment covering the submerged terrestrial surface is relatively thin (see Sakellariou et al. 2018), although it would mask prehistoric archaeological artefacts. This was something that was apparent from the diver and ROV inspections. However, close up human examination by an experienced diving archaeologist would increase understanding and enhance our ability to interpret the character of the seabed. Visual inspection as a method of ground truthing the geophysical data is essential to help determine subtle geo-morphological features, short term erosive or depositional processes, along with indications of palaeo-land surfaces or signatures of human activity. These include any stone artefacts or outcrops of underlying deposits that could indicate buried landforms.

15 Discussion: What Have We Learnt and Where Next?

The AFRAC MAR and DISPERSE projects have provided unique insights into the drowned landscape around the Farasan Islands (Momber et al. 2014). Diving investigation inspected over 20 km of seabed in eighteen distinct areas of coastal waters at different locations over four seasons of fieldwork. The work resulted in a broad understanding of the near-shore submerged landscape around the present-day

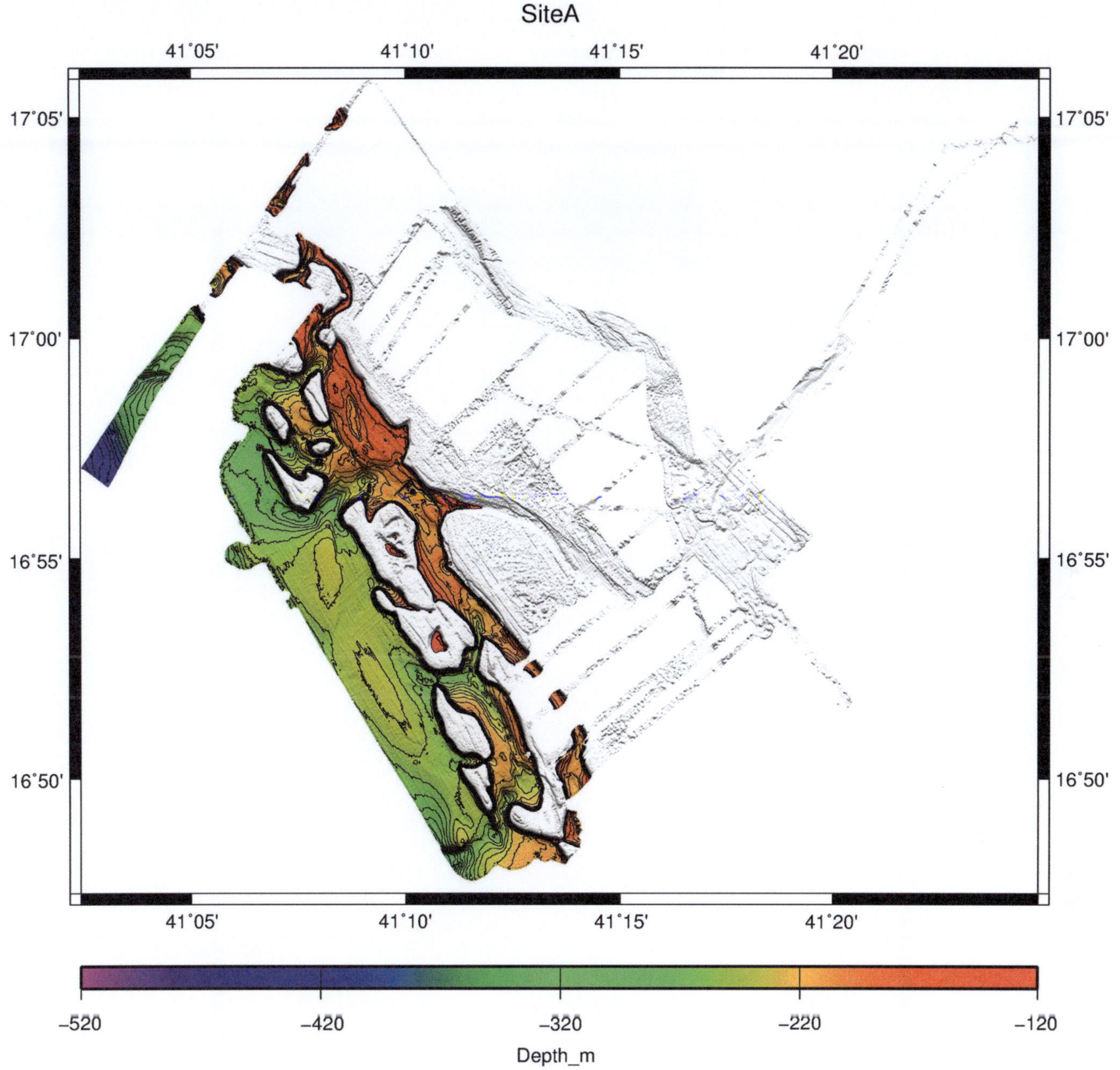

Fig. 18 This series of submerged outcrops off the west end of the continental shelf would have been islands during the LGM. The colours signify contours and the depths below Ordnance Datum (OD) are annotated on the scale. Core samples from within the hollows in the centre of one of the islands identified possible lacustrine deposits (see Sakellariou et al. 2018)

Farasan archipelago where wave cut platforms and notches indicate past shorelines. Attractive sites for human occupation that would have provided shelter and vantage points looking out over resource-rich environments were also identified. Excavation of the sea bed at selected locations revealed stratified units of marine deposits above palaeo-land surfaces, demonstrating that deposits could remain stable following deposition and inundation.

Further offshore, the comprehensive geophysical survey and coring over significant areas of the submerged shelf in 2013 helped us to understand the submerged geology and the geomorphological processes that shaped the wider landscape around the Farasan Islands (See Sakellariou et al. 2018). The data collected is helping to reconstruct a complicated palaeo-landscape interspersed with lakes that would have been fed by wadis or springs when sea levels were

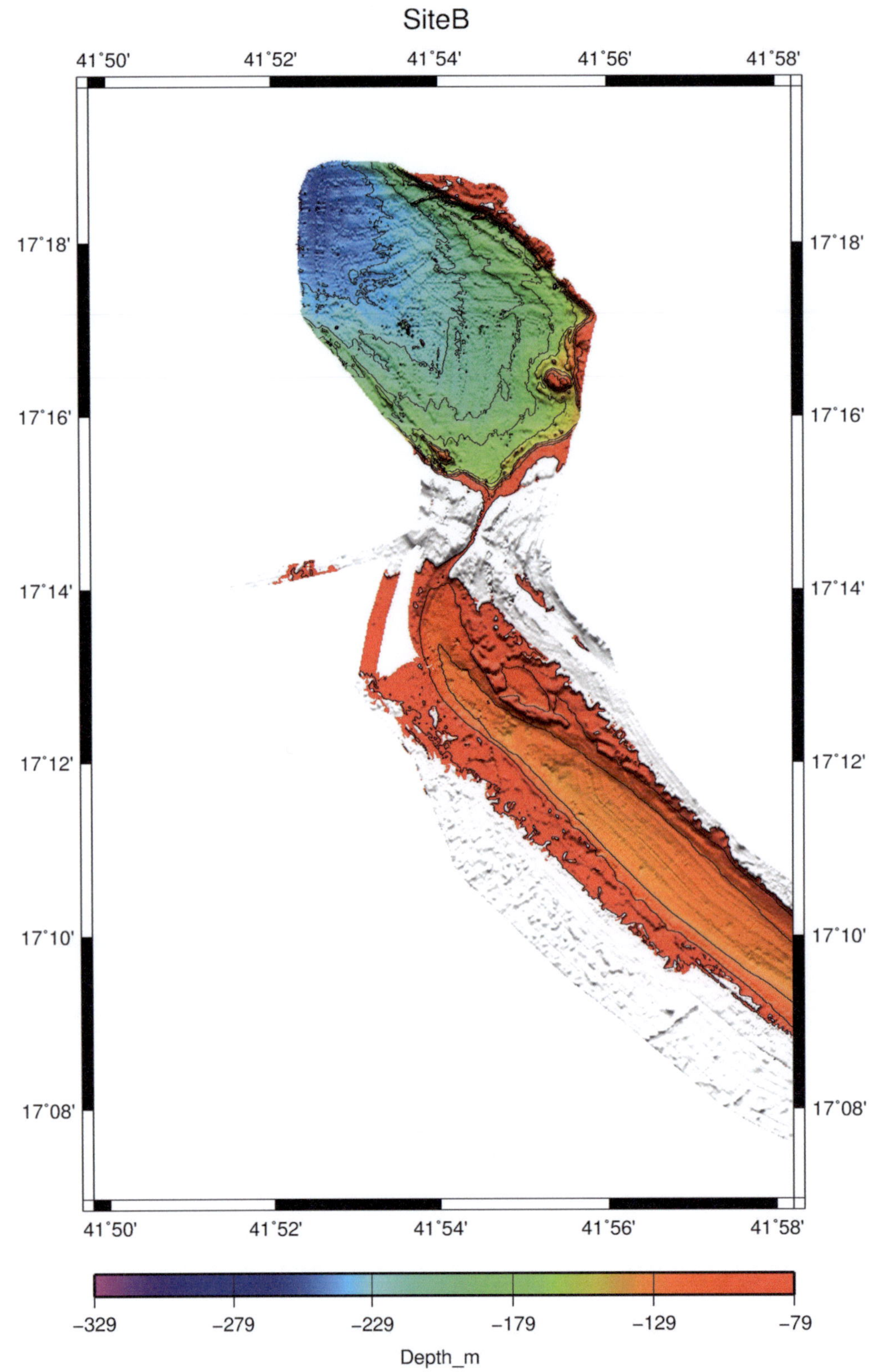

Fig. 19 3D bathymetric relief of interesting submerged landscape features northeast of the Farasan Islands. Survey collected by Hellenic Centre for Marine Research (HCMR) from R/V Aegaeo. The coloured contours begin at −79 m below OD. The large linear feature to the south east was identified as a fault. The orange colour indicates a depth of around 130 m, being approximately 50 m deeper than the surrounding landscape and the narrow gorge to the north. At the last glacial maxima, sea-level was approximately 120–130 m below OD. During this time and until the sea level was high enough to pass through the gorge, the fault was isolated from marine influences and could have contained fresh water. The dark blue area is the southern edge of a circular sinkhole; this formed when a subterranean salt dome or diapir collapsed, and it is over 350 m deep (Courtesy D Sakellariou)

lower. This is of particular importance given the relatively more arid climates that persisted throughout much of the global cycle in this region (see Petraglia et al. 2018). More importantly it is giving us a better understanding of the potential for human exploitation of the now-submerged landscape and environment and its value that is helping to address the issue of human dispersal into and through Arabia.

What we have yet to find underwater is unequivocal archaeological evidence of occupation. Unfortunately, sedimentation, erosion, and the processes of dispersal make archaeological discovery much more challenging under water than on land. The dominant archaeological features surviving today on the present-day land surface that relate to coastal exploitation on the Farasan Islands are the mounded shell middens, of which over 3000 have now been discovered. The midden building began about 7,000 years ago and lasted a couple of thousand years. The excavated structures show accumulations of marine molluscan species that were harvested during different seasons and from different localities (Meredith-William et al. 2014a; Hausmann et al. 2015; see also Bailey et al. 2018, and Hausmann et al. 2018). Some of the shellfish were recovered from a few metres underwater, demonstrating that the islanders had to take to the water and dive, while others came from the intertidal zone. It is most probable that some of the earliest shell middens would have been built on land that is now underwater. Unfortunately, our work has shown that the drowning process would have caused these to deflate and mix with natural assemblages making their signature on the seabed very unclear.

Lithic artefacts recorded during field-walking included small ground stone implements made from fine-grained volcanic material that would have been imported from the mainland as volcanic material has not been identified in or around the Farasan Islands. In contrast, surveys on the mainland between 2013 and 2014 identified rich assemblages of stone tools. Some 119 localities in the Jizan, Asir, and Harrat al Birk basaltic regions resulted in the recording of over 3000 artefacts (Alsharekh and Bailey 2013; Devès et al. 2013; Inglis et al. 2014a, b; Meredith-Williams et al. 2014b; see also Sinclair et al. 2018). On the Farasan Islands, the larger tools discovered were fashioned out of the local coral limestone. These are not as durable as basalt and are prone to weathering, particularly underwater where salt water and marine boring organisms will degrade them beyond recognition.

Although the underwater excavations to date have yet to produce any archaeological material, only a very small section of the submarine landscape has been investigated. A main challenge is the widespread deposition of sand and shell that blankets the seabed. However, a great deal of information about the archaeology and geology of the submerged palaeo-landscape around the Farasan Islands has been generated over the past decade and many lessons have been learnt. The combination of research on land and underwater has helped to identify the nature of stone tools in the region, areas where concentrations are greatest and the conditions that would allow for the greatest chance of preservation. It has demonstrated that the potential to discover stratified archaeological deposits within underwater sites that were protected during the last transgression remains high, but the probability of finding worked lithics would he greatest nearer outcrops of basaltic lava. These outcrops are to be found further north in the vicinity of Harrat Al Birk, where abundant stone tools have been discovered on land, and where volcanic deposits appear to occur also offshore from the present coastline. Here, any stone tools present on the submerged seabed would be more abundant and their darker colour would make them easier to identify against a backdrop of coral terrace or yellow sand.

16 Conclusion

In some respects, purposeful underwater archaeological surveying for prehistoric material is very different from an archaeological survey on dry land and is subject to very different challenges and constraints. Notwithstanding the huge technological and technical advances in recent decades in underwater acoustic surveys, photography and remote sensing and the development of underwater vehicles and robotics, there remain severe limitations on how large an area of the seabed can be explored in a given project.

The financial costs are also much greater than on land, especially where ship time is involved, although these can be mitigated in some cases, as in our work in Saudi Arabia, by collaboration with industrial partners. Diving in shallow water with SCUBA and normal air mixtures is a cheaper option, but issues of health and safety, the provision of logistical support, and the unpredictability of weather conditions and underwater visibility impose their own constraints on the area of seabed that can be examined. In any case, the depths accessible to divers are for the most part in the shallower areas of the continental shelf exposed at low sea level.

Also, features of the original terrestrial land surface and especially archaeological deposits are liable to dispersal and destruction or burial under sediment during and after inundation by sea level rise. We know from the many underwater prehistoric finds that have been discovered that archaeological material can survive and can be exposed and discovered under favourable circumstances. Nevertheless, finding those favourable locations remains a challenge, given the other constraints on underwater surveys.

If the discovery of archaeological sites, often taken to be the primary objective of an archaeological survey, were the sole test of success, our underwater work in the Red Sea would have to be judged, so far, as a failure. However, archaeological surveying should involve much more than the opportunistic discovery of a well stratified and dateable site. What we have learned from our research is the importance of two additional objectives that should inform archaeological surveys over and above the discovery of archaeological sites. The first is the need for a much more detailed understanding of the bigger geological picture of landscape evolution and dynamics within which human occupation took place. This provides the framework for defining the environmental conditions that may have variously attracted or deterred human activity, the geomorphological conditions of erosion or sedimentation that may have destroyed or buried archaeological material, and the basis for predictive models aimed at identifying suitable targets for more careful archaeological investigation.

The second lesson, which follows from the first, is the importance of understanding the taphonomic conditions that determine the preservation, visibility, exposure and destruction of archaeological sites and archaeological material. And that in its turn depends on an understanding of the geological processes of formation and deformation operating in different parts of the landscape. No archaeological survey—and no map of archaeological sites and findspots—can be said to be complete or even in any way meaningful without addressing these two objectives. And this applies as much on land as under water. If anything, the situation on land is as bad if not worse. Archaeological material on land is subject to severe degradation and destruction by subaerial weathering, and is just as liable to burial or exposure and dispersal by erosion or accumulation of sediment as underwater material. To this must be added the increasingly powerful impact of human activity in the form of industrial development, agricultural land-use practices, road-building, quarrying, land fill, and other infrastructural developments.

What is different about underwater archaeological surveying is that the geomorphological evolution and dynamics of the landscape and the complexities involved in determining archaeological site taphonomies are more obvious. However, they are no less present on land. Archaeologists who conduct surveys, whether on land or under water, tend to regard the discovery of a big site or sites as the prime objective, investing much time and resources in excavation of such sites once discovered, and only working outward to the wider landscape at a later stage and then rarely progressing very far away from the primary site. Geologists involved in archaeological surveys, in contrast, tend to start at the other end of the spectrum, wanting to gain an overall understanding of the regional geology and geomorphological processes first, and only later working inward to individual locations. Successful archaeological surveying requires both approaches. Only when sites can be placed in their wider geomorphological and taphonomic setting can the project be judged a success. Surveys that produce sites without that wider understanding are as incomplete as surveys that fail to produce sites. In this respect archaeological surveys on land and underwater have much to learn from each other.

Acknowledgements We thank HRH King Salman bin Abul Aziz Al Saud, formerly Crown Prince and Minister of Defense, and the Hydrographic Department of the Saudi Ministry of Defense, for granting permission to undertake the offshore work, and HRH Prince Sultan bin Salman bin Abdul Aziz, President of the Saudi Commission for Tourism and National Heritage (SCTH) for granting permission and support for our archaeological research in Saudi Arabia from its inception. We also thank Dr. Ali Al-Ghabban, Vice-President of SCTH, other SCTH staff, notably Dr. Saad al Rashid, Jamal Omar, Dr. Abdullah Al Saud, Dr. Hussein Abu Hassan and Dr. Abdullah Al Zahrani for ongoing support, and Dr. Abdullah Alsharekh, Department of Archaeology, King Saud University, for crucial support and participation in all phases of our research. Funding for the underwater work has been supplied by the British Academy, the Natural Environment Research Council (NERC) UK, through its EFCHED programme (Environmental, Factors in Human Evolution and Dispersal), the Leverhulme Trust, the National Geographic, and the European Research Council (ERC) under the Ideas Programme of the 7th Framework Programme as Advanced Grant 269586 'DISPERSE: Dynamic Landscapes, Coastal Environments and Human Dispersals'. For additional support and funding for the offshore work, we thank in particular Saudi Aramco for the provision of their vessel M/V Midyan in 2006, Shell Companies Overseas, the Saudi British Bank (SABB), the Hellenic Centre for Marine Research (HCMR), and the King Abdullah University of Science and Technology (KAUST). This is DISPERSE contribution no. 42.

References

Alsharekh AM, Bailey GN (eds) (2013) Coastal prehistory in southwest Arabia and the Farasan Islands 2004–2009 field investigations. Saudi Commission for Tourism and Antiquities, Riyadh

Ambrose SH (1998) Late Pleistocene human population bottlenecks, volcanic winter and differentiation of modern humans. J Hum Evol 34:623–651

Andersen SH (1993) Coastal adaptations and marine exploitation in Late Mesolithic Denmark—with special emphasis on the Limfjord region. In: Fischer A (ed) Man and Sea in the Mesolithic. Oxbow, Oxford, pp 41–66

Antonioli F, Presti VL, Rovere A, Ferranti L, Anzidei M, Furlani S, Mastronuzzi G, Orru PE, Scicchitano G, Sannino G, Spampinato CR, Pagliarulo R, Deiana G, Sabata E, Sans P, Vacchi M, Vecchio A (2015) Tidal notches in Mediterranean Sea: a comprehensive analysis. Quat Sci Rev 119:66–84

Armitage SJ, Jasim SA, Marks AE, Parker AG, Usik VI, Uerpmann HP (2011) The southern route "Out of Africa": evidence for an early expansion of modern humans into Arabia. Science 331:453–456

Bailey GN (2004a) World prehistory from the margins: the role of coastlines in human evolution. J Interdiscip Stud Hist Archaeol 1 (1):39–50

Bailey GN (2004b) The wider significance of submerged archaeological sites and their relevance to world prehistory. In: Flemming NC (ed) Submarine prehistoric archaeology of the North Sea: research priorities and collaboration with industry. CBA research report 141. Council for British Archaeology, York, pp 3–10

Bailey GN (2009) The Red Sea, coastal landscapes and hominin dispersals. In: Petraglia MD, Rose JI (eds) The evolution of human populations in Arabia. Springer, Dordrecht, Netherlands, pp 15–37

Bailey GN, Alsharekh A (in press) Palaeolithic archaeology, coastal prehistory and submerged landscapes in southwest Saudi Arabia and the Farasan Islands: DISPERSE field reports, 2012–2015. Saudi Commission for Tourism and National Heritage, Riyadh

Bailey GN, Flemming NC (2008) Archaeology of the continental shelf: marine resources, submerged landscapes and underwater archaeology. Quat Sci Rev 27:2153–2165

Bailey GN, King GCP (2011) Dynamic landscapes and human dispersal patterns: tectonics, coastlines and the reconstruction of human habitats. Quat Sci Rev 30:1533–1553

Bailey GN, King GCP, Flemming NC, Lambeck K, Momber G, Moran LJ, Al-Sharekh AM, Vita-Finzi C (2007a) Coastlines, submerged landscapes and human evolution: the Red Sea basin and the Farasan Islands. J Island Coast Archaeol 2:127–160

Bailey GN, Al-Sharekh AM, Flemming NC, Lambeck K, Momber G, Sinclair AGM, Vita-Finzi C (2007b) Coastal prehistory in the southern Red Sea Basin, underwater archaeology and the Farasan Islands. Proc Seminar Arabian Stud 37:1–16

Bailey GN, King GCP, Devès M, Hausmann N, Inglis R, Laurie E, Meredith-Williams M, Momber G, Winder I, Alsharekh A, Sakellariou D (2012) DISPERSE: dynamic landscapes, coastal environments and human dispersals. Antiquity 86(334). http://antiquity.ac.uk/projgall/bailey334/

Bailey GN, Williams MG, Alsharekh AM (2013) Shell mounds of the Farasan Islands, Saudi Arabia. In: Bailey GN, Hardy K, Camara A (eds) Shell energy: mollusc shells as coastal resources. Oxbow, Oxford, pp 241–254

Bailey GN, Devès MH, Inglis RH, Meredith-Williams MG, Momber G, Sakellariou D, Sinclair AGM, Rousakis G, Al Ghamdi S, Alsharekh AM (2015) Blue Arabia: Palaeolithic and underwater survey in SW Saudi Arabia and the role of coasts in Pleistocene dispersal. Quatern Int 382:42–57. https://doi.org/10.1016/j.quaint.2015.01.002

Bailey GN, Meredith-Williams M, Alsharekh A, Hausmann N (2018) The archaeology of Pleistocene coastal environments and human dispersals in the Red Sea: insights from the Farasan Islands

Bantan RA (1999) Geology and sedimentary environments of Farasan Bank (Saudi Arabia) southern Red Sea: a combined remote sensing and field study. PhD thesis, Department of Geology, Royal Holloway, University of London

Bingham B, Foley B, Singh H, Camilli R, Delaporta K, Eustice R, Mallios A, Mindell D, Roman C, Sakellariou D (2010) Robotic tools for deep water archaeology: surveying an ancient shipwreck with an autonomous underwater vehicle. J Field Robot 27(6):702–717. https://doi.org/10.1002/rob.20350

Brubakk AO, Neuman TS (2003) Bennett and Elliott's physiology and medicine of diving, 5th edn. Saunders Ltd., US, 304 p

Chappell J, Shackleton NJ (1986) Oxygen isotopes and sea level. Nature 324:137–140

Devès MH, Inglis RH, Meredith-Williams MG, Al Ghamdi S, Alsharekh AM, Bailey GN (2013) Palaeolithic survey in southwest Saudi Arabia: methodology and preliminary results. Adumatu 27:7–30

Dullo WC, Montaggioni L (1998) Modern Red Sea coral reefs: a review of their morphologies and zonation. In: Purser BH, Bosence DWJ (eds) Sedimentation and tectonics in rift basins: Red Sea-Gulf of Aden. Chapman & Hall, London, pp 583–594

Faure H, Walter RC, Grant DR (2002) The coastal oasis: ice age springs on emerged continental shelves. Global Planet Change 33:47–56

Flemming N (2004) The prehistory of the North Sea floor in the context of continental shelf archaeology from the Mediterranean to the Nova Zemlya. In: Flemming NC (ed) Submarine prehistoric archaeology of the North Sea. CBA research report 141, York, pp 11–20

Flemming NC, Harff J, Moura D (2017) Non-cultural processes of site formation, preservation and destruction. In: Flemming NC, Harff J, Moura D, Burgess A, Bailey GN (eds) Submerged landscapes of the European continental shelf. Wiley Blackwell, pp 51–82

Gerth WA (2006) Decompression sickness and oxygen toxicity in US surface-supplied He-O_2 diving. In: Lang MA, Smith NE (eds) Proceedings of advanced scientific diving workshop. Smithsonian Institution, Washington, DC, pp 24–35

Hausmann N, Meredith-Williams M, Bailey GN (2015) Results of recent excavations on the Farasan Islands and studies of large-scale prehistoric shellfish gathering in the Red Sea. In: Bicho N, Detry C, Price TD, Cunha E (eds) Muge 150th anniversary conference, vol 2. Cambridge Scholars Press, Cambridge, pp 301–313

Hausmann N, Kokkinaki O, Leng MJ (2018) Red Sea palaeoclimate: stable isotope and element-ratio analysis of marine mollusc shells

Harff J, Flemming NC, Groh A, Hunicke B, Lericolais G, Meschede M, Rosentau A, Sakellariou D, Uscinowicz, S, Zhang W, Zorita E (2017) Sea level and climate. In: Flemming NC, Harff J, Moura D, Burgess A, Bailey GN (eds) Submerged landscapes of the European continental shelf. Wiley Blackwell, pp 11–50

Hudec MR, Jackson MPA (2007) Terra infirma: understanding salt tectonics. Earth-Sci Rev 82:1–28

Inglis RH, Sinclair AG, Shuttleworth A, Alsharekh A, Devès MH, Meredith-Williams M, Al Ghamdi S, Bailey G (2014a) Investigating the Palaeolithic landscapes and archaeology of the Jizan and Qunfudah regions, southwestern Saudi Arabia. Proc Seminar Arabian Stud 44:193–212

Inglis RH, Sinclair AGM, Shuttleworth A, Al Maamary A, Budd W, Hausmann N, Meredith-Williams MG, Alsharekh AM, Al Ghamdi S, Bailey GN (2014b) Preliminary report on 2014 fieldwork in southwest Saudi Arabia by the Disperse Project: (1) Jizan and Asir Provinces. http://eprints.whiterose.ac.uk/80965/

Inglis RH, Rasul N, Bailey GN, Bosworth B (2018) Investigating the fossil coral terraces and coastal archaeology of the southern Red Sea

Joiner JT (2001) NOAA diving manual: diving for science and technology, 4th edn. Best Publishing, US, 660 p

King GCP, Bailey GN (2006) Tectonics and human evolution. Antiquity 80:265–286

Lambeck K, Chappell J (2001) Sea level change through the last glacial cycle. Science 292:679–686

Lambeck K, Purcell A, Flemming N, Vita-Finzi C, Alsharekh A, Bailey GN (2011) Sea level and shoreline reconstructions for the Red Sea: isostatic and tectonic considerations and implications for hominin migration out of Africa. Quat Sci Rev 30:3542–3574

Le Bas TP, Huvenne VAI (2009) Acquisition and processing of backscatter data for habitat mapping. Comparison of multibeam and sidescan systems. Appl Acoustics 70:1248–1257

Mellars PA, Gori KC, Carr M, Soares PA, Richards MB (2013) Genetic and archaeological perspectives on the initial modern human colonization of southern Asia. Proc Natl Acad Sci 110:10699–10704

Meredith-Williams MG, Hausmann N, Bailey GN, Inglis RH (2014a) 4200 new shell mound sites in the southern Red Sea. Internet Archaeol 37. https://doi.org/10.11141/ia.37.2

Meredith-Williams MG, Hausmann N, Bailey GN, King GCP, Alsharekh A, Al Ghamdi S (2014b) Mapping, modelling and predicting prehistoric coastal archaeology in the southern Red Sea using new applications of digital imaging techniques. World Archaeol 46(1):10–24

Missiaen T, Sakellariou D, Flemming N (2017a) Survey strategies and techniques in underwater geoarchaeological research: an overview with emphasis on prehistoric sites. In: Bailey G, Harff J, Sakellariou D (eds) Under the Sea: archaeology and palaeolandscapes of the continental shelf. Springer, Dordrecht, pp 21–38

Missiaen, T Pieters M, Maes F, Kruiver P, De Maeyer P, Seys J (2017b) The SeArch project: towards an assessment methodology and sustainable management policy for the archaeological heritage of the North Sea in Belgium. In: Bailey G, Harff J, Sakellariou D (eds) Under the sea: archaeology and palaeolandscapes of the continental shelf. Springer, Dordrecht, pp 415–424

Mitchell SJ (2007) Technical diving. In: Moon RE, Piantadosi CA, Camporesi EM (eds) Dr. Peter Bennett symposium Proceedings, Held 1 May 2004, Durham, NC

Momber G, Peeters JHM (2017) Postglacial human dispersal and submerged landscapes in north-west Europe. In: Harff J, Sakellariou D (eds) Bailey G. Under the Sea: archaeology and palaeolandscapes of the continental shelf, Springer, pp 321–334

Momber G (2011) Submerged landscape excavations in the Solent, southern Britain: climate change and cultural development. In: Benjamin J, Bonsall C, Pickard C, Fischer A (eds) Submerged prehistory. Oxbow Books, Oxford, pp 85–98

Momber G, Gillespie J, Mason B, Mason C, Tidbury L, Mozayen W, Hamzi F, Al Sadiq J, Al Haiti A, Meredith–Williams M, Hausmann N, Aqeeli A, Mofta M, Bailey G (2014) Preliminary report on 2014 fieldwork in southwest Saudi Arabia by the DISPERSE Project: (2) Underwater research in the Farasan Islands. http://www.disperse-project.org/field-reports

Momber G (2014) Submerged archaeology and cultural responses to climatic amelioration. In: Evans A, Flatman J, Flemming NC (eds) Prehistoric archaeology on the continental shelf: a global review. Springer, New York, pp 193–212

Oppenheimer S (2003) Out of Eden: the peopling of the world. Constable, London

Peeters JHM, Momber G (2014) The southern North Sea and the human occupation of northwest Europe after the Last Glacial maximum. Neth J Geosci 93(1–2):55–70

Petraglia MD, Breeze PS, Groucutt HS (2018) Blue Arabia, green Arabia: examining human colonisation and dispersal models

Purser BH, Bosence DWJ (eds) (1998) Sedimentation and tectonics in rift basins: Red Sea-Gulf of Aden. Chapman & Hall, London

Rohling EJ, Fenton M, Jorissen FJ, Bertrand P, Ganssen G, Caulet JP (1998) Magnitudes of sea level stillstands in the past 500,000 years. Nature 394:162–165

Sakellariou D, Bailey GN, Momber G, Meredith-Williams MG, Alsharekh A, Rousakis G, Panagiotopoulos I, Morfis I, Stavrakakis S, Pampidis I, Renieris P, Georgiou P, Kalogirou S, Mantopoulos P, Stasinos V, Kallergis M, Manousakis L, Al Nomani SM, Devès M (2013) Preliminary report on underwater survey in the Farasan Islands by the R/V Aegaeo, May–June 2013. http://www.disperse-project.org/field-reports

Sakellariou D, Rousakis G, Panagiotopoulos I, Morfis I, Bailey G (2018) Geological structure and late quaternary geomorphological evolution of the Farasan Islands continental shelf, south Red Sea, SW Saudi Arabia

Shackleton NJ (1987) Oxygen isotopes, ice volume and sea level. Quat Sci Rev 6:183–190

Siddall M, Rohling EJ, Almogi-Labin A, Hemleben C, Meischner D, Schmeltzer I, Smeed DA (2003) Sea level fluctuations during the last glacial cycle. Nature 423:853–858

Sinclair A, Inglis RH, Shuttleworth A, Foulds F, Bailey GN, Alsharekh A (2018) Landscape archaeology, Palaeolithic survey and coastal change along the southern Red Sea of Saudi Arabia

Walter RC, Buffler RT, Bruggemann JJ, Guillaume MMM, Berhe SM, Negassi B, Libsekal Y, Cheng H, Edwards RL, von Gosel R, Neraudeau D, Gagnon M (2000) Early human occupation of the Red Sea coast of Eritrea during the last interglacial. Nature 405: 65–69

Winder I, King GCP, Devès M, Bailey GN (2013) Complex topography and human evolution: the missing link. Antiquity 87:333–349

Zarins J, Ibrahim M, Potts D, Edens C (1979) Saudi Arabian archaeological reconnaissance 1978. The preliminary report on the third phase of the comprehensive archaeological survey program—the coastal province. Atlal J Saudi Arabian Archaeol 3:9–42

Zarins J, Whalen N, Ibrahim M, Mursi AJ, Khan M (1980) The comprehensive archaeological survey program. Preliminary report on the Central and Southwestern provinces. Atlal J Saudi Arabian Archaeol 4:9–36

Zarins J, Al-Jawad Murad A, Al-Yish KS (1981) The comprehensive archaeological survey program. The second preliminary report on the Southwestern province. Atlal J Saudi Arabian Archaeol 5:9–42

Geological Structure and Late Quaternary Geomorphological Evolution of the Farasan Islands Continental Shelf, South Red Sea, SW Saudi Arabia

Dimitris Sakellariou, Grigoris Rousakis, Ioannis Panagiotopoulos, Ioannis Morfis, and Geoff N. Bailey

Abstract

A marine geological-geophysical survey of selected areas on the Saudi Arabian continental shelf of the Farasan Archipelago in the southeastern Red Sea has been conducted in the framework of the ERC-funded DISPERSE project. The aim was to explore systematically the geological, tectonic and sedimentary structure of the shelf and provide insights into the submerged prehistoric landscapes. The survey targeted features believed to be of significance in relation to the archaeological potential of the submerged landscape, including geological structure, palaeoenvironment, and sea-level change. Swath bathymetry, seismic and subbottom profiling data, side scan sonar imaging and sediment coring indicate that the flat areas belonging to the 70–90 m deep shelf may have developed as part of an erosive marine terrace during the latter part of MIS 3, between 30 and 45 ka BP. A second terrace lying at 115–120 m water-depth is associated with MIS 2, whereas a third one, mapped at about 40 m water-depth, may correspond to the MIS 5.1 period, 80–85 ka BP. Extensional tectonics, possibly driven by basin-ward flow of underlying Miocene evaporites below the shelf, is responsible for the rupturing of the latter along NW–SE trending faults. The resulting fault-bounded blocks, which are composed of Plio-Quaternary rocks, were dragged and drifted away from the shelf edge to create isolated flat-topped ridges surrounded by steep slopes and troughs. The largest part of the shelf along with the 90 m deep, flat tops of the ridges were exposed when the sea-level was at 115–120 m bpsl (below present sea level) during the Last Glacial Maximum (LGM). This geomorphological configuration may also be valid for the previous low sea-level period, 140 ka BP (MIS 6). Shallow and deep depressions and valleys on the main terrace of the shelf, which formed by solution of evaporite diapirs or domes, were permanent or ephemeral lakes when the shelf was exposed during MIS 2. Similar lakes possibly formed in the many deep sinkholes which occur on the available hydrographic charts along the 120 km wide and several-hundred-km long Farasan shelf. Finally, the presence of valleys and canyons on the seafloor of the survey areas indicates erosion of the shelf under subaerial conditions due to surface water-flow.

D. Sakellariou (✉) · G. Rousakis · I. Panagiotopoulos · I. Morfis
Hellenic Centre for Marine Research, 19013 Anavyssos, Greece
e-mail: sakell@hcmr.gr

G. N. Bailey
Department of Archaeology, University of York, King's Manor, York, YO1 7EP, UK

G. N. Bailey
Arts and Social Sciences, Flinders University, College of Humanities, GPO Box 2100 Adelaide, SA 500, Australia

1 Introduction

In this paper we present the results of a marine geological-geophysical survey of the Farasan Islands continental shelf, in the southern Red Sea, SW Saudi Arabia. This is a rare example of a marine geological-geophysical survey informed by joint marine-geoscientific and archaeological thinking with the aim of exploring systematically submerged landscapes in deeper areas of the continental shelf and targeting features of potential archaeological significance in relation to geological structure, palaeoenvironment, and sea-level change (for other examples see Dixon and Monteleone 2014). The Red Sea region is of wide interest and significance both to geoscientists and archaeologists: to geoscientists because it represents an unusual 'laboratory' for investigating Pleistocene sea-level change (Lambeck et al. 2011) superimposed on a dynamically changing landscape controlled by rift-tectonics and salt-tectonics (Bosworth et al. 2005); and to archaeologists because it is regarded as one of the primary pathways of dispersal for early human populations expanding out of Africa during the Pleistocene, in which the now-submerged landscape of the extensive continental shelf is believed to have played a key role.

N. M. A. Rasul and I. C. F. Stewart (eds.), *Geological Setting, Palaeoenvironment and Archaeology of the Red Sea*,
https://doi.org/10.1007/978-3-319-99408-6_28

This study has developed as an international and interdisciplinary collaboration arising directly out of the SPLASHCOS COST Action (www.splashcos.org; Bailey and Sakellariou 2012) and within the ERC-funded project DISPERSE (Bailey et al. 2012). The main objectives were (i) to understand the underlying Quaternary geology and the role of active geological processes in changing the coastal/submerged landscape; (ii) to reconstruct the broad outlines of the now-submerged landscape and (iii) to identify specific locations where archaeological evidence of past human settlement when sea level was lower than present might have been preserved. This work builds on two strands of earlier investigation, one concerned with the impact of active tectonics on the early landscapes of human evolution (King and Bailey 2006; Bailey and King 2011; Bailey et al. 2011), the other with the impact of sea level change and submerged landscapes on the potential connections between Africa and Arabia and the dispersal of early humans expanding out of Africa during the Pleistocene (Bailey et al. 2007; Lambeck et al. 2011; Bailey et al. 2015).

2 Regional Setting

The Red Sea is the youngest oceanic basin on earth, developed as part of an extensive rift system between the Nubian (African) plate to the west and the Arabian plate to the east. Continental rifting in the southern Red Sea and the Gulf of Aden began 30–24 Ma ago (Purser and Bosence 1999, and references therein; Bosworth et al. 2005, and references therein; Bosworth 2015) (Fig. 1). The principal phase of rift shoulder uplift and rapid syn-rift subsidence along the Red Sea followed shortly after 24 Ma. Extension evolved initially perpendicular to the Red Sea rift axis until approximately 14 Ma, when the Aqaba-Levant Transform Fault cut through Sinai and the Levant margin (Bayer et al. 1988) and linked the Red Sea rift with the Bitlis-Zagros convergence zone. Since then, extension across the Red Sea has changed from rift-normal (N60E) to highly oblique (N15E), parallel to the Aqaba-Levant transform fault. During that period, the marine connection with the Mediterranean Sea became restricted and consequently sedimentation in the Red Sea changed from open marine to evaporitic. It was by then, during the Middle to Late Miocene, when massive, up to 2500 m thick evaporites (mainly halite) accumulated on the Red Sea floor (Bosence 1998). The top of the Miocene salt marks a basin-wide unconformity (Hughes and Beydoun 1992; Mitchell et al. 1992), recognized throughout the Red Sea Basin and coincides with the onset of oceanic seafloor spreading in the south-central Red Sea, at latitudes between 15°N and 20°N (Girdler and Styles 1974; Roeser 1975; Searle and Ross 1975; Cochran 1983; LaBreque and Zitellini 1985).

The Red Sea rift is markedly asymmetric. The western (African) margin is narrow and steep, whereas volcanism occurs almost exclusively on the eastern (Arabian) side together with the most significant uplift (Bosence 1998). This asymmetric uplift of the rift shoulders led to the formation of the Great Escarpment, the mountain-front that runs parallel to the Arabian shoreline (Fig. 1). According to Schmidt et al. (1982) the first stage of uplift of the Red Sea Great Escarpment began during Middle Miocene time.

Increased marine water influx through Bab al Mandab in the Early Pliocene (Stoffers and Kuhn 1974; Stoffers and Ross 1977) led to the re-establishment of open marine conditions in the Red Sea (Bosworth et al. 2005). Marine oozes and marginal clastic sediments together with carbonates of Plio-Quaternary age were deposited on top of the Miocene evaporites. Coral reef limestone deposits occur all along the present shorelines and on the shelf and may reach a thickness of up to 300 m, for example, in the Farasan Bank (Mideast Industries Ltd. 1966; Bosence 1998). On the southeastern Red Sea margin, along the littoral zone near Jizan, these deposits are buried under clastic debris derived from erosion of the high rift escarpment.

Salt diapirism began in the Red Sea soon after the widespread deposition of massive halite in the Middle to early Late Miocene. Salt domes reached the surface in many parts of the basin by the end of the Miocene, and most were subsequently buried. A few domes are presently at or very near the seafloor along the Egyptian, Saudi Arabian, and Yemeni margins (Bosworth et al. 2005). Rejuvenation of pre-evaporite extensional fault-systems has stimulated salt mobilization and diapirism in the Red Sea (Purser and Bosence 1999; Hudec and Jackson 2007).

Two shelves have developed along both margins of the Red Sea axial trough, which extend from the shoreline seaward for distances of 30 to more than 100 km. In the southern Red Sea, south of latitude 21°N, where the Red Sea reaches its maximum width of 360 km, both shelves extend for more than 120 km each and are capped by predominantly carbonate sediments (Abou Ouf and El-Shater 1992; Carbone et al. 1998). Two main groups of islands occur in this part of the Red Sea, the Dahlak Islands on the Eritrean shelf and the Farasan Islands on the Arabian shelf. The water depth around the two archipelagos is less than 100 m while the semicircular geometry of the islands and reefs suggests ascending salt domes (Dullo and Montaggioni 1998).

Pleistocene shallow marine reefal limestones, deformed by salt diapirism, form the bedrock of the Farasan Islands (Bantan 1999). The uplift of the Farasan Islands was probably initiated by differential loading of Miocene salt, triggered by tectonic stretching of the overburden during the regional extension of the Red Sea rift. The islands themselves comprise more or less a reef-flat with faulting and uplift strongly influenced by salt diapirism (Dullo and

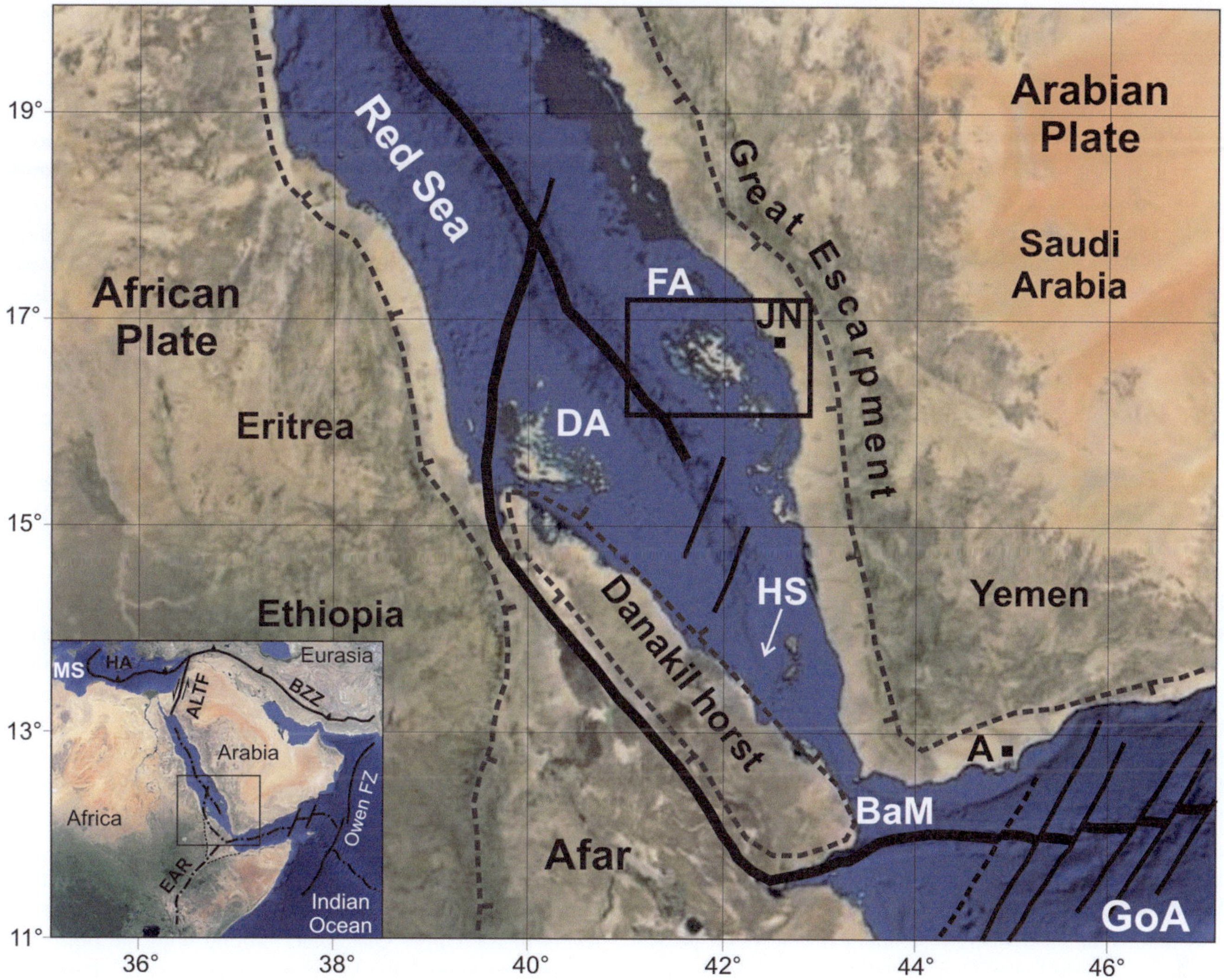

Fig. 1 Google Earth image of the southern Red Sea-west Gulf of Aden with major geodynamic elements modified after Bosworth et al. (2005) showing locations mentioned in the text. A: Aden; BaM: Bab al Mandab Strait; DA: Dahlak Islands; FA: Farasan Islands; GoA: Gulf of Aden; HS: Hanish Sill; JN: Jizan. The black box marks the area shown in Fig. 2

Montaggioni 1998). The tectonic pattern exhibits graben and horst structures developed parallel to the Red Sea trough. Therefore, salt diapirism has created not only dome-like structures but also NW–SE salt walls. The exposed reef limestones are mainly of Eemian age (MIS 5.5, 125 ka BP) and occur at an elevation not higher than 15 m above present sea-level (see Inglis et al., this volume). They are underlain by marly limestones onlapping gypsum and anhydrite domes (Dabbagh et al. 1984; Bantan 1999).

3 Sea-Levels, Palaeogeography and Submerged Landscapes

The MIS 5.5 reef-terrace marks a very prominent and widespread palaeoshoreline at elevations of up to 20 m along the present Red Sea shoreline (Dullo and Montaggioni 1998). Older high stand terraces, particularly the MIS 7.1 (206 ka BP) and 9.3 (310 ka BP) ones are around 15 m and 30 m above present sea-level (Dullo 1990). On the submerged shelf, three prominent submarine terraces at −20 m, −60 m and −90 m have been recognized in various areas off the Red Sea coasts. The −20 m terrace has been attributed to MIS 5.2 (90 ka BP) and the −60 m terrace formed during MIS 4.2 (64 ka BP) (Gvirtzman 1994). For the most prominent, −90 m deep, terrace, Dullo and Montaggioni (1998) suggested a MIS 3 age, 28–37 ka BP, in accordance with observations in the Indian Ocean by Dullo et al. (1997).

Analyses of sediment cores from the Red Sea floor have shown that it has been open to the Indian Ocean throughout the last 500,000 years at least (Rohling et al. 1998, 2009; Siddall et al. 2004). Palaeosea-level modelling and shoreline reconstructions by Lambeck et al. (2011) confirm that the Red Sea has remained connected with the Gulf of Aden throughout the last 400,000 years, albeit the cross-sectional areas at times of glacial maxima were about 2% of that today. The minimum channel widths (less than 4 km) connecting the Red Sea to the Gulf of Aden during lowstands occur south of the Hanish Sill (Fig. 1) and remain narrow for as long as local sea levels are below −50 m.

Phylogenetic analyses and interpretations are in favour of the hypothesis that Anatomically Modern Humans originating in East Africa dispersed rapidly about 60,000 years ago across the southern end of the Red Sea to the Arabian Peninsula (Walter et al. 2000; Macaulay et al. 2005; Mellars 2006; but see Bailey et al., this volume). New dates for Stone Age sites in the Arabian Peninsula indicate the presence of humans already since at least 130,000 BP (Petraglia and Rose 2009; Armitage et al. 2011; Petraglia et al. 2011; Rose et al. 2011; Delagnes et al. 2012). Whether the archaeological material at these sites was deposited by Anatomically Modern Humans or not is unclear, but in any case, there is now known to be much earlier archaeological material in the Arabian Peninsula most probably associated with archaic hominin species (Bailey et al., this volume; Petraglia et al., this volume). At any rate, the available sea-level and shoreline reconstructions for the Upper Pleistocene and the Holocene show that the sea channel between the Bab al Mandab Strait and the Hanish Sill remained open throughout this period. Hence, migration from Africa to Arabia across the southern Red Sea would have required sea crossings. However, according to Lambeck et al. (2011), suitable periods for crossing occurred for long periods at times of sea level lowstands, involving sea crossings of no more than 4 km.

Continental shelves, now submerged below the present sea level, should have offered relatively mild climate conditions compared to cold or arid hinterlands during low sea level periods (Bailey and Flemming 2008). In addition, the coastal oasis hypothesis, proposed by Faure et al. (2002), suggests that as sea level dropped, the water from underground springs would have found an easier exit onto the adjacent coastal areas, creating a well-watered coastal lowland—or at any rate coastal oases— potentially attractive to plant and animal life during glacial periods precisely over the time range when overall climatic conditions were becoming drier in the adjacent hinterland.

Within the context described above, the aim of this paper is to present new geological (tectonic, sedimentological) and geomorphological data from the continental shelf of the Farasan Islands, to produce palaeogeographic reconstructions of Pleistocene lowstands and to highlight specific palaeomorphological features which might have been exploited by prehistoric humans.

4 Survey Strategy, Materials and Methods

The continental shelf around the Farasan Archipelago is several hundred kilometres long and over 120 km wide. During the 14 days of the DISPERSE-Farasan cruise aboard R/V Aegaeo, it was necessary to focus on relatively small areas from which to draw conclusions applicable to wider areas of the continental shelf. Two areas were surveyed.

Farasan 1 area (Figs. 2 and 3) is located on the outer edge of the shelf and offers the opportunity to (i) map precisely the seafloor relief, (ii) study the shallow geological structure and the role of faulting and diapirism in the shaping of the shelf edge and the continental slope, (iii) study sedimentary environments and retrieve sediment cores from the outer shelf and the slopes, (iv) map low sea-level palaeoshorelines, particularly the LGM shoreline, and (v) assess hydrogeological conditions and predict possible locations of water points on the outer shelf, particularly springs or spring lines, often located at the foot of low cliffs and fault scarps.

Farasan 2 area (Figs. 2 and 4) is located on the inner shelf, between the Farasan Islands and the Arabian shoreline. The available bathymetric data (GEBCO, International Chart Series Sheet 157, UK Hydrographic Office) show a prominent, NW–SE trending valley, parallel to the main rift axis and a complex topography that appears to drain into a >200 m deep, circular basin, possibly formed by solution or withdrawal of underlying evaporites. The latter could have been freshwater-filled when sea level was low, and may contain a sediment sequence showing the transition from marine to lacustrine conditions as related to changing sea levels in the Upper Quaternary. Along with the objectives already mentioned for the previous study area, Farasan 2 area offers the opportunity to understand the role, character and sedimentation of the deep, circular, sinkholes and reconstruct the submerged landscape of the inner Farasan continental shelf during low sea-level periods.

In addition to the two areas above (Farasan 1 and Farasan 2), two seismic transects (Transect 1 and Transect 2) were shot by using the airgun seismic profiler (Fig. 2).

The marine survey in both areas included a wide variety of geological and geophysical techniques (Figs. 3 and 4): (1) Multi-beam bathymetry with two hull-mounted systems (20 kHz and 180 kHz), (2) Airgun 10ci single channel seismic reflection profiling, (3) High-resolution sub-bottom profiling with a 3.5 kHz pinger, (4) Deep-towed, 110/410 kHz frequency, digital side scan sonar, (5) 3–5 m long gravity coring, (6) 40 × 40 × 60 cm box coring, and (7) ROV dives to visually inspect sites selected from bathymetric, side scan sonar and seismic data.

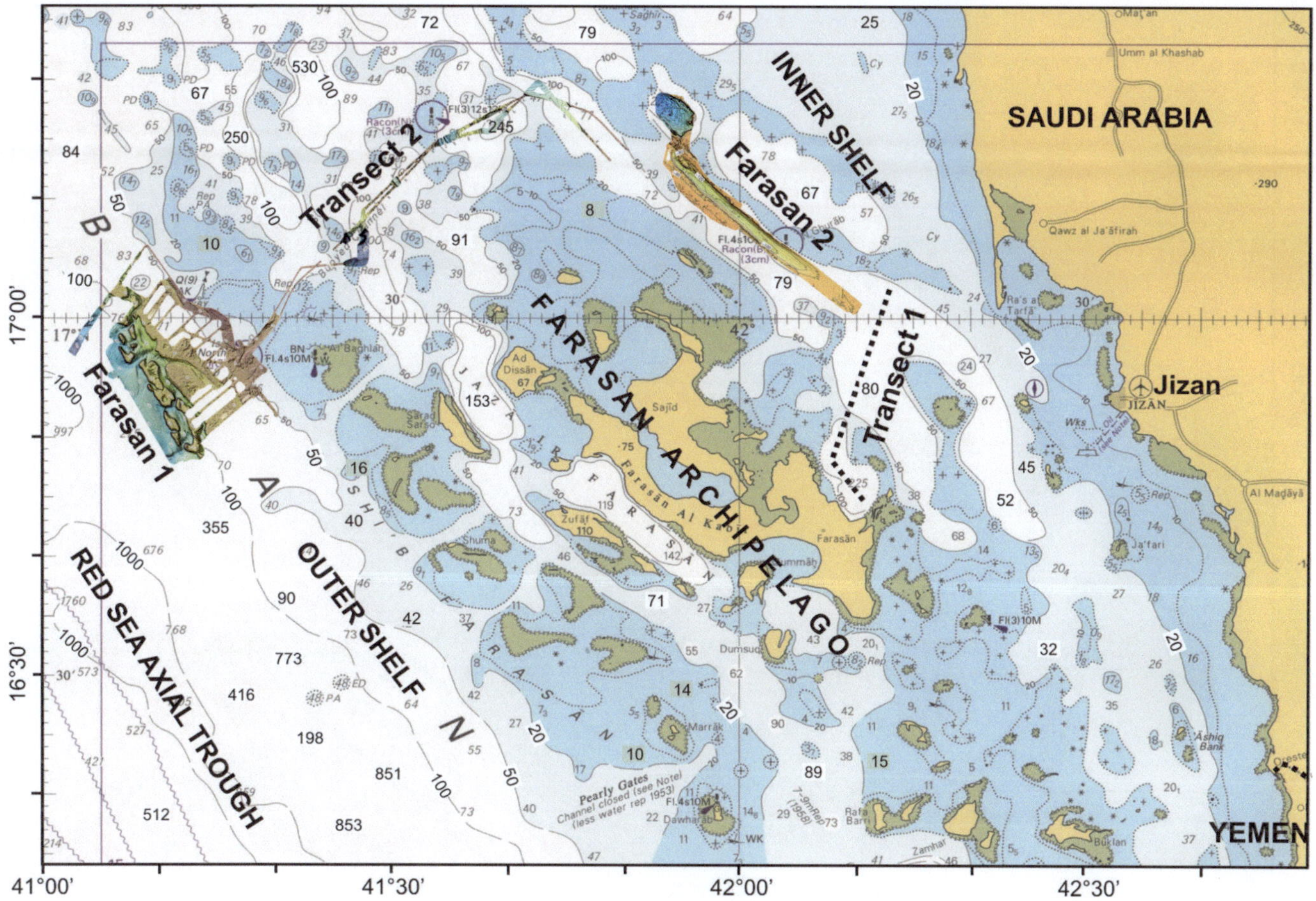

Fig. 2 Extract from the International Chart Series, UK Hydrographic Office Sheet 157, showing the bathymetry of the Farasan Islands continental shelf and the surveyed areas FARASAN 1 and FARASAN 2 during the DISPERSE-Farasan cruise aboard R/V Aegaeo

In total, about 500 km^2 of the seafloor was mapped with the multi-beam echosounder systems, 170 nautical miles of airgun seismic reflection profiles, 250 nautical miles of 3.5 kHz sub-bottom profiles and 140 nautical miles (260 km) of side-scan sonar lines were acquired. Finally, 18 gravity cores and 2 box-cores were recovered and 5 ROV dives were accomplished.

5 Outer Shelf

5.1 Seafloor Morphology

The most prominent morphological feature of the surveyed outer Farasan shelf is a flat terrace dipping gently toward the axis of the Red Sea, at depths ranging between 70 m landward and 85–90 m toward the shelf edge (Figs. 5 and 9). It corresponds to the 90 m deep terrace widely observed in the Red Sea, which, in accordance with observations in the Indian Ocean (Dullo et al. 1997) has been correlated by Dullo and Montaggioni (1998) with MIS 3, at 28–37 ka BP. A second terrace, lying at about 40 m depth, has been mapped along the landward edge of the survey area (Fig. 5a). It is separated from the main, deeper terrace by steep slopes locally 30–40 m high (Figs. 5 and 9). A third, narrow terrace occurs off the shelf edge, at about 115–120 m on the steep slopes of the shelf edge and around the flat-topped highs (Figs. 5 and 9). Its width does not exceed a few hundred metres.

Numerous, up to about 20 m high, mostly circular mounds of different sizes occur on the main terrace. They represent coral reefs that have developed following the sea level rise after the LGM (Dullo and Montaggioni 1998). Two valleys incise the flat, 70–90 m terrace (Fig. 3). The first one, at the northern end of the survey area, has developed along the foot of the slope in front of the −40 m terrace (Fig. 5a). Its floor is at about 80 m depth and becomes deeper and wider toward the NW. The second, more prominent valley runs from southeast to northwest in the central part of the survey area (Fig. 5b). The depth of the valley floor increases gradually toward the northwest from about 80–85 m to over 140 m at the outflow to the deep trough off the shelf. Shallow

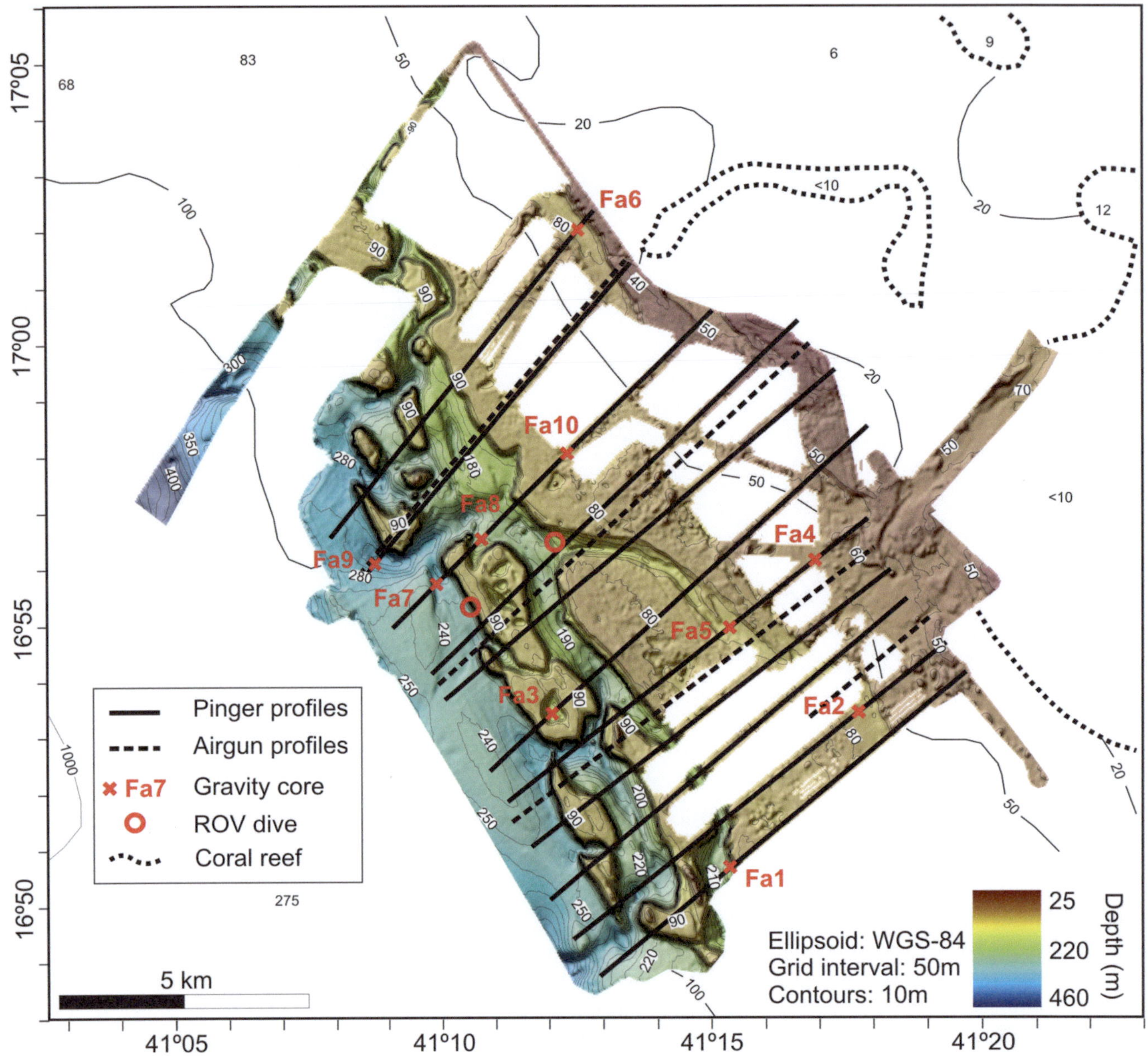

Fig. 3 Survey area FARASAN 1 on the outer shelf of the Farasan Islands with location of seismic and pinger profiles, side scan sonar tracklines (same as pinger profiles), sediment coring sites and ROV dives. Background: International Chart Series, UK Hydrographic Office Sheet 157 "Masamirit to Bab el Mandab", scale 1:750.000 at 20°00'N latitude, depths in metres

isolated depressions occur randomly on the main terrace. They display circular, elongate and irregular shapes and are 10–40 m deeper than the surrounding shelf floor, reaching maximum depths of 140 m water depth.

A 200 m deep, narrow trough aligned SE–NW and bounded by steep margins runs parallel to and off the shelf edge and separates it from a series of shallow, flat-topped ridges or shoals, similar to underwater mesas or drowned, flat-topped islands. The depth of their flat tops below present sea level is about 90 m, very similar to the depth of the shelf edge. The slopes all around them are very steep, with slope values exceeding 45°. Towards the SW, the steep slopes pass into the continental slope of the Red Sea which displays smooth morphology. The overall impression from the bathymetry is as if these shallow, flat-topped blocks were initially parts of the shelf, then were cut off from it and somehow drifted laterally toward the SW, giving birth to the narrow troughs behind and in between them. More evidence on the nature and origin of the shallow blocks is provided in the following section.

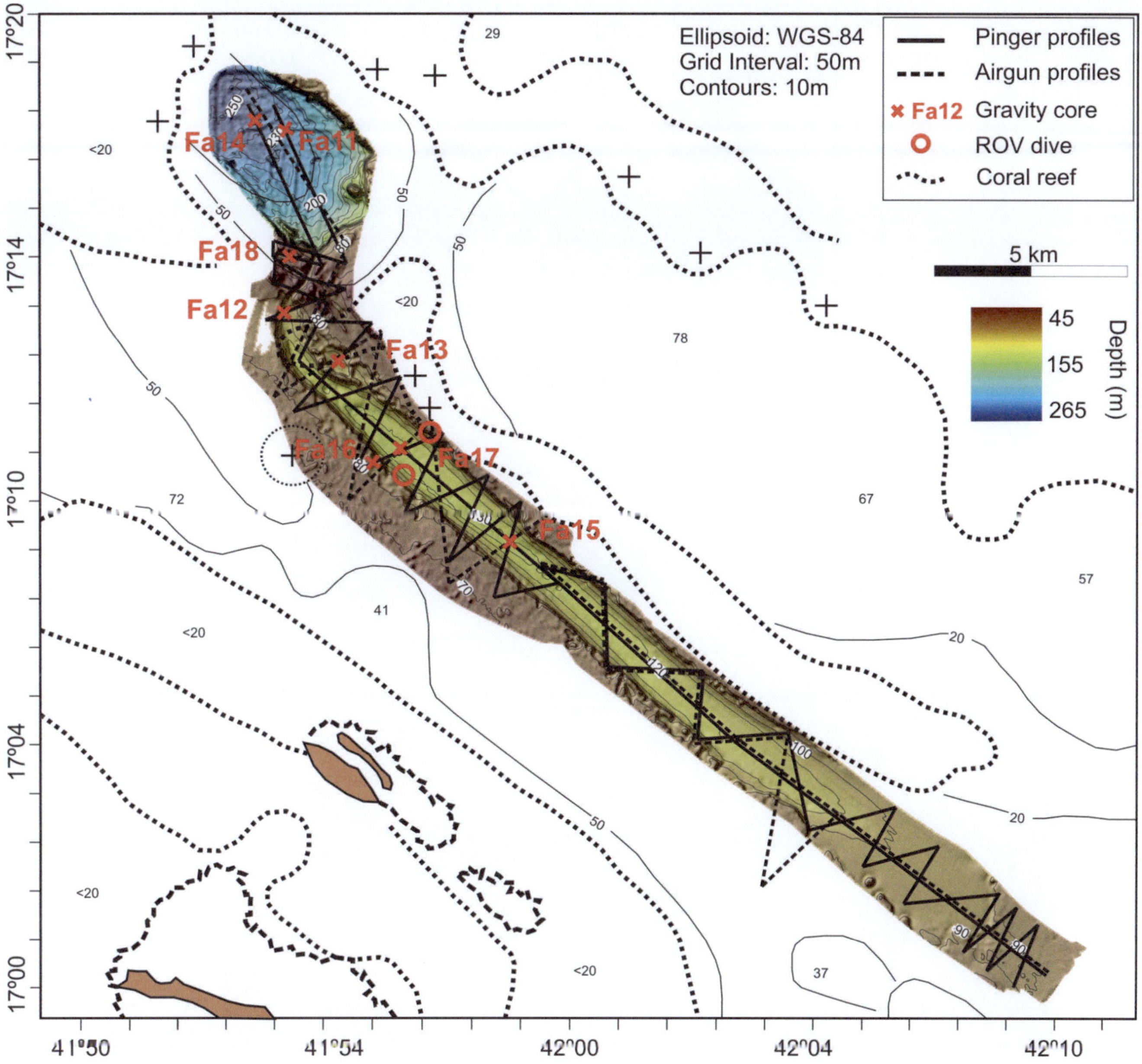

Fig. 4 Survey area FARASAN 2 on the inner shelf of the Farasan Islands with location of seismic and pinger profiles, side scan sonar tracklines (same as pinger profiles), sediment coring sites and ROV dives. Background: International Chart Series UK Hydrographic Office Sheet 157 “Masamirit to Bab el Mandab”, scale 1:750.000 at 20°00’ North latitude, depths in metres

5.2 Geological Structure

The interpretation of the seismic profiles shows that the main geomorphological features of the FARASAN 1 survey area have developed under the control of SE–NW running normal faults and fault zones (Figs. 6 and 9). The observed faults do not exceed 10–12 km in length and most of them dip toward the SW. Traces of faults on the seafloor of the main terrace are recognized as morphological scarps and are responsible for the development of valleys incised along the faults on their hanging walls, as is the case for the two valleys described above and shown in Fig. 5. Total offset produced by the F2 fault is about 35–40 m while the F1 offset is less. Subsidence of the hanging wall of the F2 fault led to the formation of the valley.

The seismic stratigraphic configuration of the shelf shows horizontal packages offset by sub-parallel normal faults (Fig. 6). Strong reflectors mark the boundaries between the six seismostratigraphic packages identified. No angular unconformity can be observed between the older packages 5 and 6. The internal reflectors of package 4 onlap unconformably the top reflector of package 5. Similarly, package 3

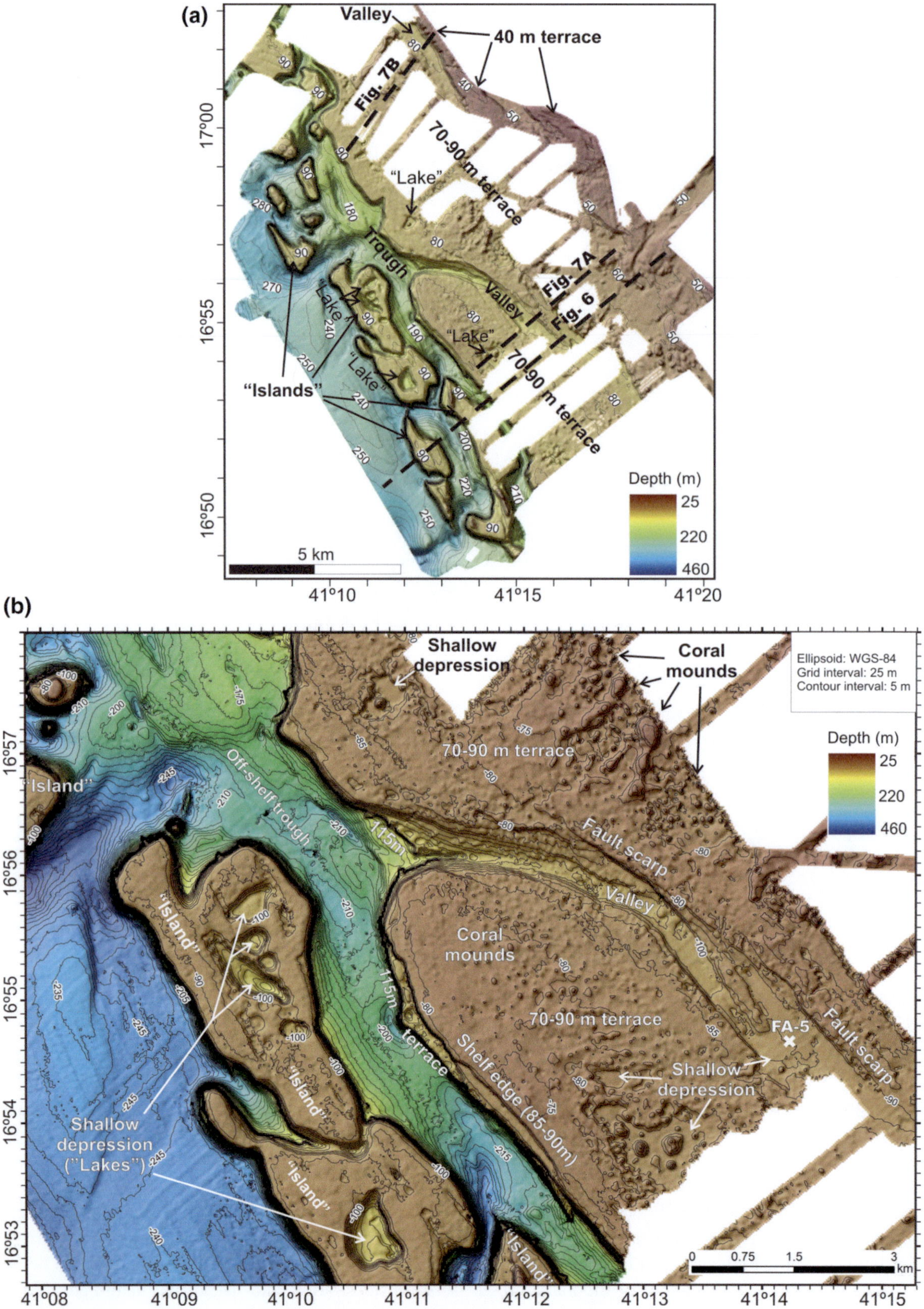

Fig. 5 **a** Geomorphological elements of FARASAN 1 survey area on the outer shelf. **b** Enlargement of the central part of the survey area with detailed morphological features and location of core FA-5 shown in Fig. 8

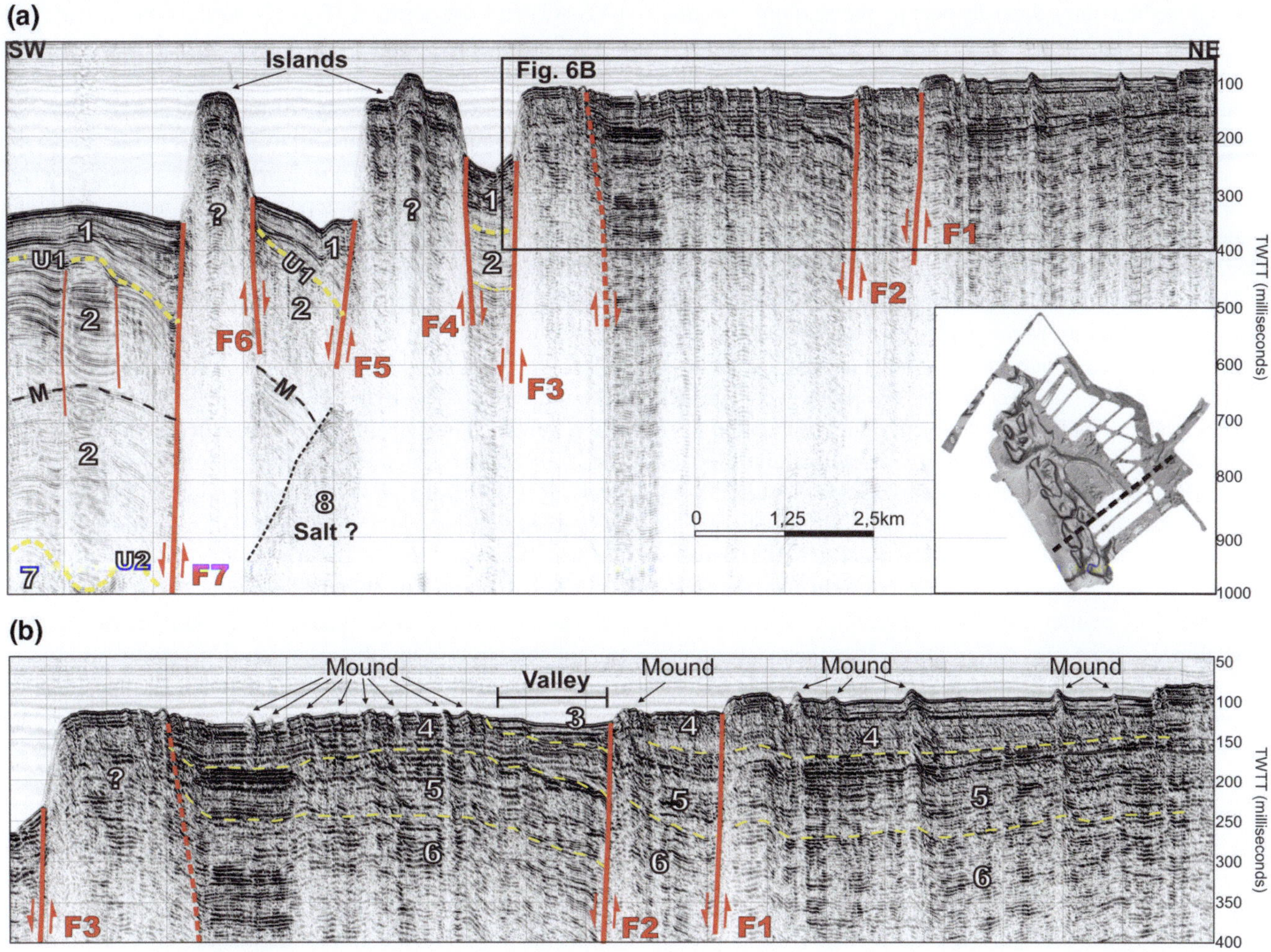

Fig. 6 **a** Airgun 10ci seismic profile across the outer shelf of Farasan Islands. **b** Magnification of the profile on the main shelf. White numbers with black outline correspond to seismostratigraphic packages. U1 and U2 are unconformities. Red numbers with black outline correspond to individual faults. See text for further explanation. M: multiple. Location of the profile in inset and in Fig. 5a. Vertical exaggeration × 10

lies unconformably over the top reflector of package 4. The seismostratigraphic package 3 occurs only locally, in the small depression in the hanging wall of one of the SW-facing normal faults. This depression coincides with the SE–NW oriented valley that runs through the central part of the survey area (Fig. 5b).

The seismostratigraphic package 3, with a maximum thickness of about 30 ms (roughly 22–23 m), occurs only locally, in the shallow depressions or valleys of the shelf. It represents the youngest deposit on the shelf, probably of Upper Quaternary (Holocene and Uppermost Pleistocene) age. Packages 4, 5 and 6 are characterized by parallel, continuous, locally undulating reflectors and may correspond to the Pleistocene shallow marine reefal limestone, which forms the bedrock of the Farasan Islands (Bantan 1999).

The shelf terminates sharply toward the SW and is followed by two shallow, flat-topped blocks (Islands on Fig. 6a) which are separated from each other and from the shelf by deep, narrow, fault-bounded troughs (Fig. 6a). Although the seismic stratigraphy of the ridges is poorly imaged due to the very irregular topography of their tops, internal sub-parallel reflectors are recognized and it is supposed that the sedimentary sequence of their substrate is comparable to the one described below the main shelf. Despite the faults separating them, no vertical offset can be measured between the main shelf and the shallow blocks. The narrow troughs formed between the shallow blocks and the shelf's edge host sedimentary deposits exceeding 150 m in thickness. The seismic units 1 and 2, which fill the troughs, are characterized by parallel and continuous reflectors and are separated from each other by an angular unconformity U1.

Thick sedimentary deposits occupy the seafloor substrate southwest of the steep shallow ridges. Seismic unit 1 displays a maximum thickness of about 200 ms (>150 m) next to the marginal fault F7 and covers unconformably unit 2. The latter is >500 ms (>400 m) thick and is separated from the underlying unit 7 by the unconformity U2. Finally, one more seismic unit, unit 8, is observed at depth below the islands. Its chaotic to transparent seismic character may indicate the presence of mobilized salt deposits.

The seismostratigraphic description of the seismic profile and the interpretation of the geological structure of the shelf edge, as described above, strongly indicate that the shallow, flat-topped blocks (islands) off the shelf-edge were initially belonging to the shelf. They were cut off and separated from the shelf due to predominantly NW–SE trending faulting and the subsequent formation of the deep troughs between them. The irregular shape of the blocks, along with the absence of any vertical offset between them and the shelf-edge indicate that the breakup process may not be of purely tectonic origin. There is neither evidence of compression nor any indication of strike slip deformation. Extensional tectonics with normal faulting might be able to offer a solution to their formation but the fact that there is no vertical offset between the ridges and the shelf is against it. A possible explanation can be found in deformation processes related to the underlying Miocene evaporites. Extensive basin-ward salt flow in the form of salt glaciers or namakiers has been observed and documented in the central Red Sea (Augustin et al. 2014; Feldens and Mitchell 2015; Augustin et al., this volume). Alongslope and downslope ridge and trough morphologies have developed on the flow surfaces, parallel to the local seafloor gradient, presumably due to extension of the sedimentary cover or strike-slip movement within the evaporites (Feldens and Mitchell 2015). The morphology of the deep troughs and the flat-topped ridges and shoals mapped off the Farasan islands shelf-edge is very similar to one of the ridges and troughs observed in the central Red Sea. Therefore, it is reasonable to assume that basinward flow of the Miocene evaporites has occurred below the Farasan shelf too. Dragging of the overlying sedimentary cover due to the salt flow underneath has led to breakup and drifting of sedimentary blocks toward the axis of the Red Sea and away from the shelf, leaving space behind them for the formation of the deep troughs. The absence of vertical offset between the blocks and the shelf supports this drifting hypothesis.

5.3 Sedimentation

High resolution subbottom profiles across the outer shelf (Fig. 3) show that the recent Holocene sediment deposition has been very limited. Two profiles from the central (A1–A2) and the northern (B1–B2) parts of the survey area (Fig. 7) show representative cross sections of the outer shelf. The main terrace on the outer shelf is flat and dips gently southwestward from roughly 70 m landward to 85–90 m seaward. Numerous mounds, up to 20 m high, rise above the flat shelf-floor and give it a unique geomorphological character. The 40 m deep terrace occurs at the northeastern end of A1 profile in Fig. 7.

The largest part of the shelf is free of Holocene sedimentary deposits, and the Quaternary shallow marine sedimentary rocks which outcrop on the Farasan Islands (Bantan 1999) are exposed on the shelf floor. Recent sediments have been observed only in the depressions and valleys on the main terrace. They display a transparent seismic character and their thickness depends on the morphology of the substrate. It ranges from a few decimetres in the shallowest parts of the depressions and valleys to 15–20 m maximum in the deepest parts.

Detailed sedimentological, geochemical and radiometric analyses on the cores retrieved from the shelf are in progress, in collaboration with the Scottish Universities Environmental Research Centre (see Sanderson et al., this volume) and the Saudi Geological Survey. Here we present some basic observations indicating that homogeneous, marine, up to 3 m thick, fine-grained deposits pass downward into coarser sediments. The latter display evidence of lacustrine depositional environments.

Ten cores have been collected from the outer shelf survey area with the objective to unravel the nature of the recent sedimentary deposits and the sedimentation environment. Core FA-5 (Fig. 8) has been taken from the southeastern part of the valley that runs through the central part of the survey area (Fig. 5) and provides insight into the sediments deposited in the depressions of the shelf. The coring site, at the centre of the valley, constitutes a local deep surrounded by shallower depths. The upper part of the core, from 0 cm down to 286 cm below the seafloor, consists of fine, olivine green sand. The physical properties of the sediment remain stable, apart from a decrease in the magnetic susceptibility at 150 cm bsf (below seabed surface). After a 3 cm-thick layer with shell debris at 286–289 cm, all properties show marked changes, including a sharp decrease of the magnetic susceptibility. A gradual change in colour from olivine green to grey occurs between 300 cm and 330 cm bsf while the grain size becomes coarser. Large bivalves and shell fragments occur between 330 cm and 360 cm, while the grain size decreases again below 360 cm until the bottom of the core at 363 cm.

The sedimentological description of FA-5 (see Fig. 5 for location) along with the down-core physical properties indicate a significant change in the depositional environment. The upper part of the core represents homogenous marine sedimentation during the present high-stand period.

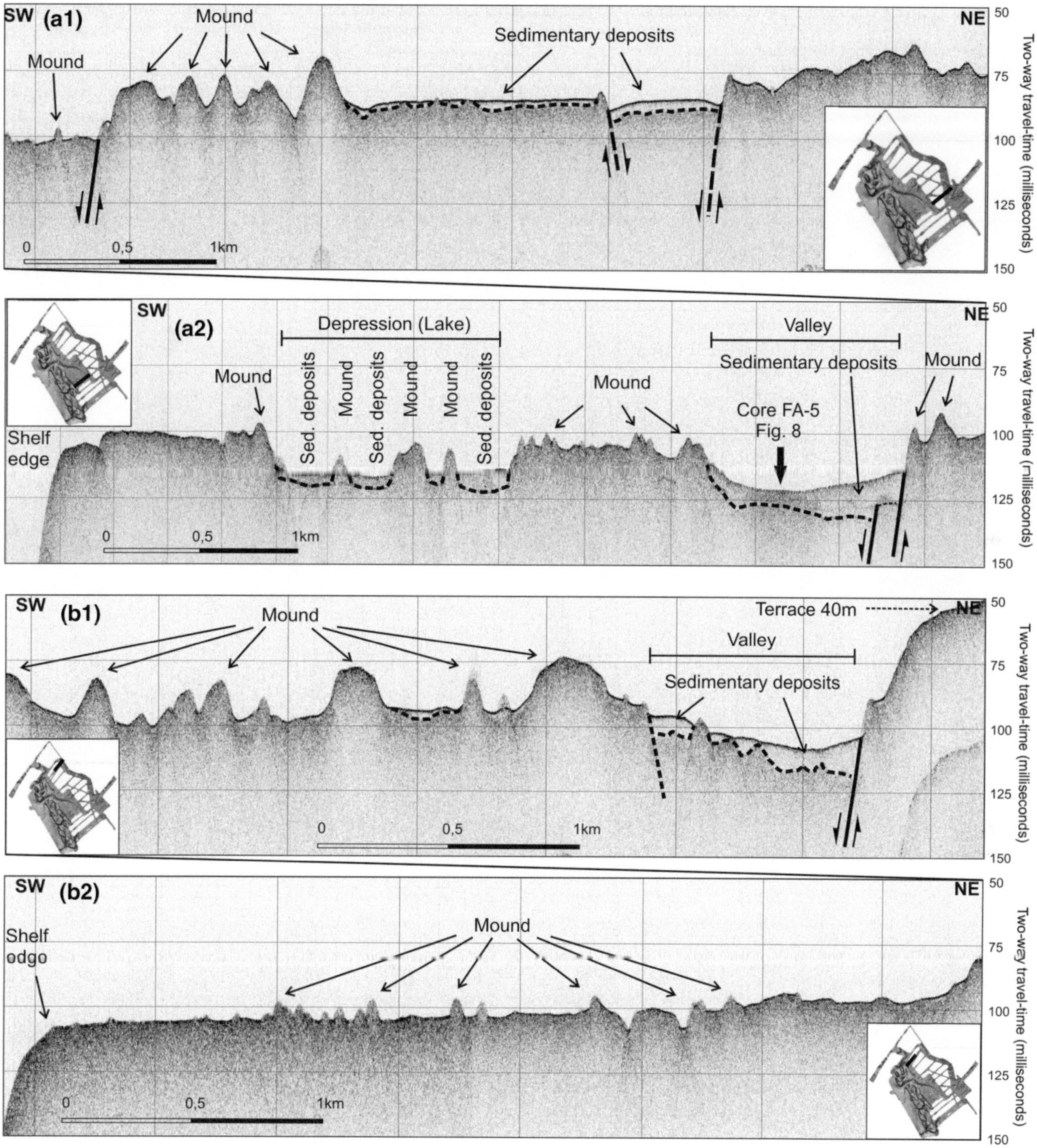

Fig. 7 Two subbottom profiles (3.5 kHz) across the outer shelf of Farasan Islands. A1 and A2 are the two halves of one profile in the central part of the survey area. B1 and B2 are the two halves of a second profile in the northern part of the survey area. Holocene sediments occur only in the depressions of the shelf. Note the numerous mounds rising from the shelf's seafloor. Location of the profiles in Fig. 5a. Vertical exaggeration × 20

The lower part of the core, below 3 m bsf, displays sedimentological evidence of lacustrine environments prevailing in the valley during the last low sea-level stand, when the shelf was exposed. Significant change of the environmental conditions at the end of the LGM has also been recorded in the sediments deposited off the shelf. Core FA-9 (see

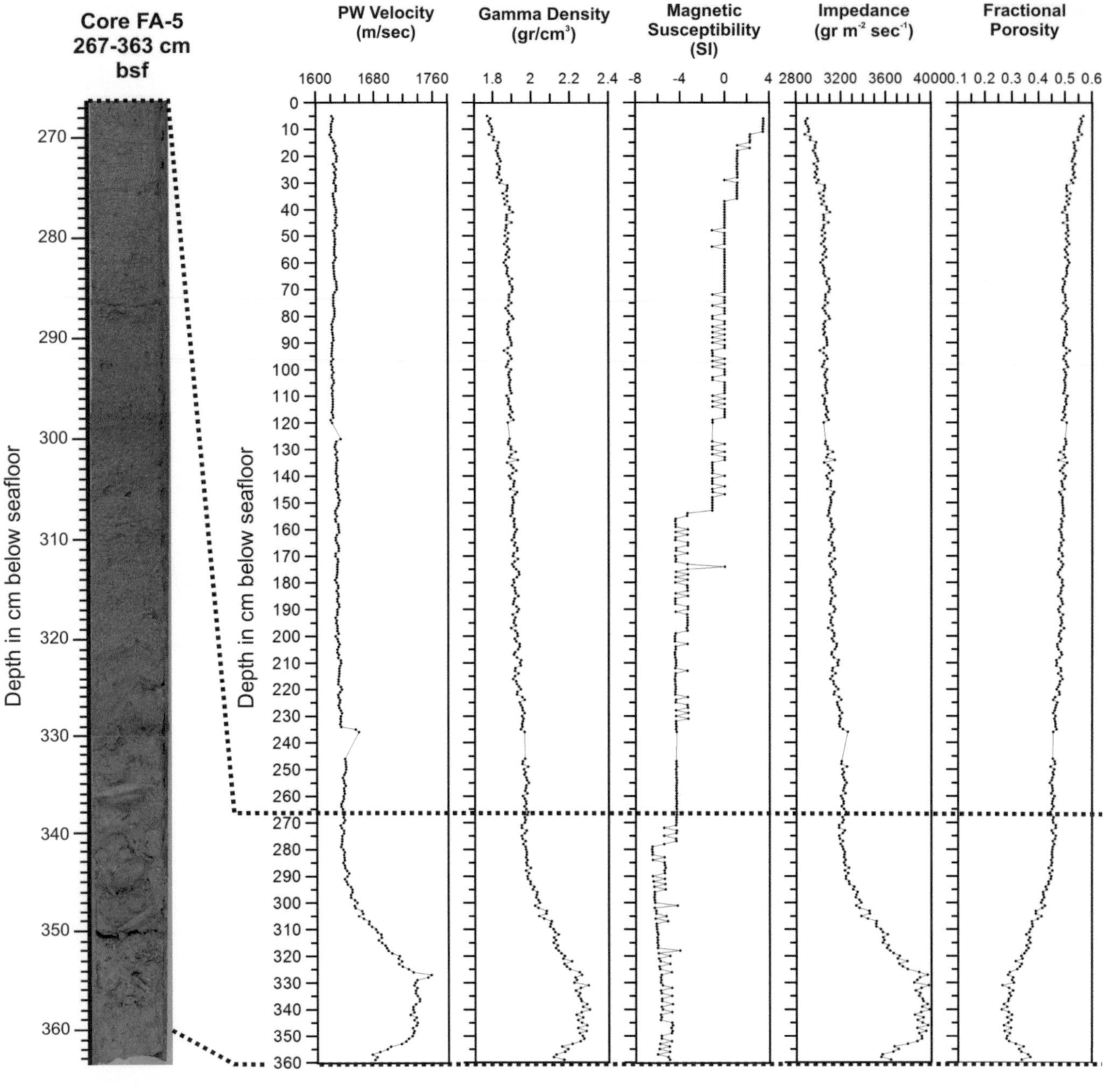

Fig. 8 Physical properties of core FA-5 and photo of the lower part of the core, between 267 and 363 cm below the seafloor. The location of the core (lat: 16°54.647', long: 41°14.237', depth 92 m) is shown on the map of Fig. 5b and on the subbottom profile A2 of Fig. 7. Note the gradual change in sediment colour from olive green to grey between 300 cm and 330 cm and the change in the properties below 270 cm

location in Fig. 3) has been retrieved from the margin of the Red Sea basin at 302 m depth, close to the foot of the basin-ward facing slope of one of the "Islands" (Fig. 3). Micropaleontological analyses on the sediments of the core confirm the presence of a major change in the productivity of the Red Sea waters at 130 cm bsf, which is associated with the transition from the Late Glacial to the Holocene periods (see Geraga et al., this volume, for details).

5.4 Outer Shelf Submerged Landscapes and Palaeoshorelines

Integration of the swath bathymetry, seismic and subbottom profiling data and sediment cores presented above reveals the palaeo-geomorphological evolution of the Farasan Islands outer shelf with respect to Upper Quaternary fluctuating sea-level.

Seismic profiling and swath bathymetry data show that the area has been affected by extensive faulting, presumably associated with salt flow in the deeper stratigraphic levels. Dragging and deformation of the overlying Plio-Quaternary deposits due to the flow of the underlying Miocene salt deposits toward the axis of the Red Sea has been well documented in the central Red Sea (Augustin et al. 2014; Feldens and Mitchell 2015; Augustin et al., this volume). Apparently, salt flow occurs below the Farasan Island shelf too and results in the separation of blocks from the edge of the shelf and their drift towards the SW. No significant vertical offset has been observed between the drifted blocks and the shelf. The faults running along the outer shelf display rather limited vertical throws while the shelf remains flat. These are good evidence that vertical tectonics during the Quaternary has been very low.

In the absence of direct chronological constraints, we attempt to estimate the age of the observed underwater terraces indirectly, by comparing their depths with the available sea-level curves for the Red Sea. This comparison can provide realistic hypotheses on the ages of the terraces in particular because vertical tectonics in the area of the Farasan outer shelf is negligible as discussed above. Predicted sea levels in the Red Sea during the LGM, 20,000 years BP, show considerable geographic variability, with relatively low values for the northern (105 m bpsl [below present sea level]) and southern (115 m bpsl) ends and deeper values (130 m bpsl) for the central part (Lambeck et al. 2011). The 115–120 m terrace observed locally along the steep slopes off the shelf edge and around the isolated ridges ("islands") must have been formed during the LGM and is thus indicative for the position of the LGM shoreline at the latitude of the Farasan Islands (Fig. 9). In that case, the outer shelf, the depth of which does not exceed 90 m, as well as the tops of the isolated ridges off the shelf, were subaerially exposed during the LGM. The shelf was part of a wide, flat coastal plain attached to the Arabian landmass while the tops of the isolated ridges were flat-topped islands separated from each other and from the mainland by narrow and deep channels or troughs. Note that the shape of the LGM islands as outlined by the palaeoshoreline is very similar to the shape of the present islands which form the Farasan Archipelago.

The 70–90 m deep terrace occupies the largest part of the outer shelf. Assuming negligible vertical tectonics in the area, the time of the formation of the terrace can be estimated by comparing its depth with the predicted sea-level curve. The inset diagram in Fig. 9 shows part of the sea-level curve of Rohling et al. (2013) for the Red Sea for the last 150 kyr. The depth of the main terrace coincides with the predicted depth on the curve between 30 ka and 45 ka BP, which is the upper half of MIS 3. According to this curve, the sea-level has fluctuated between 75 m and 85 m bpsl during this period. Similarly, the 40 m deep terrace can be correlated with the period 80–85 ka BP (MIS 5.1) when the sea-level was between 35 m to 45 m bpsl.

Based on the subbottom profiling and the sediment coring data, the small depressions on the main shelf display a dual character: firstly, they constitute the depocentres of the marine sedimentation on the shelf during the post-LGM sea-level rise, and secondly, they hosted the deposition of lacustrine sediments when the shelf was subaerially exposed during the LGM. Consequently, the small and shallow depressions and parts of the valleys on the shelf were lake basins during the LGM (Fig. 9).

6 Inner Shelf

6.1 Seafloor Morphology

The survey area in the inner shelf of the Farasan Islands is characterized by two main morphological features (Fig. 10): a 35 km long, 2–2.5 km wide, up to 130 m deep, SE–NW trending, straight valley and a 250 m deep, ellipsoidal depression in the northern end. The southeastern part of the valley is wider and 90–100 m deep. Toward the NW it becomes narrower and reaches 130 m depth close to its northwestern end, where it turns northward and terminates against a shallower area with irregular morphology. The bottom of the valley is rather flat and is bounded by steep, up to 60 m high slopes (Fig. 10c). On both sides of the valley the seafloor is rather flat, with mean depth around 70 m and with numerous mounds. It corresponds to the 70–90 m terrace on the outer shelf.

A 2 km long, 200 m wide and 85 m deep canyon with steep slopes connects the northernmost tip of the SE–NW trending valley with the southernmost tip of the deep, ellipsoidal depression. The latter, the Deep "Lake" in Fig. 10a, b, is 8.5 km long by 4.5 km wide and 250 m deep with the maximum depths occurring close to its northwestern edge. It is surrounded by very steep slopes, locally steeper than 50–60%, which pass upward to very shallow reefs, not suitable at all for navigation. The shape of this deep and steep depression along with the widespread presence of evaporites and diapirs in the subsurface indicates that it has been formed by the solution of a salt dome or diapir.

The backscatter image on Fig. 10B2 depicts very nicely the differences in the acoustic character of the seafloor. The deep depression displays very low reflectivity, which points to very fine sedimentary deposits and a smooth seafloor. The irregular morphology of the shallow parts coincides with the high backscatter (dark) area where the substrate is expected to be exposed on the seafloor and eroded. Finally, the seafloor in the valley and in the small depression on the

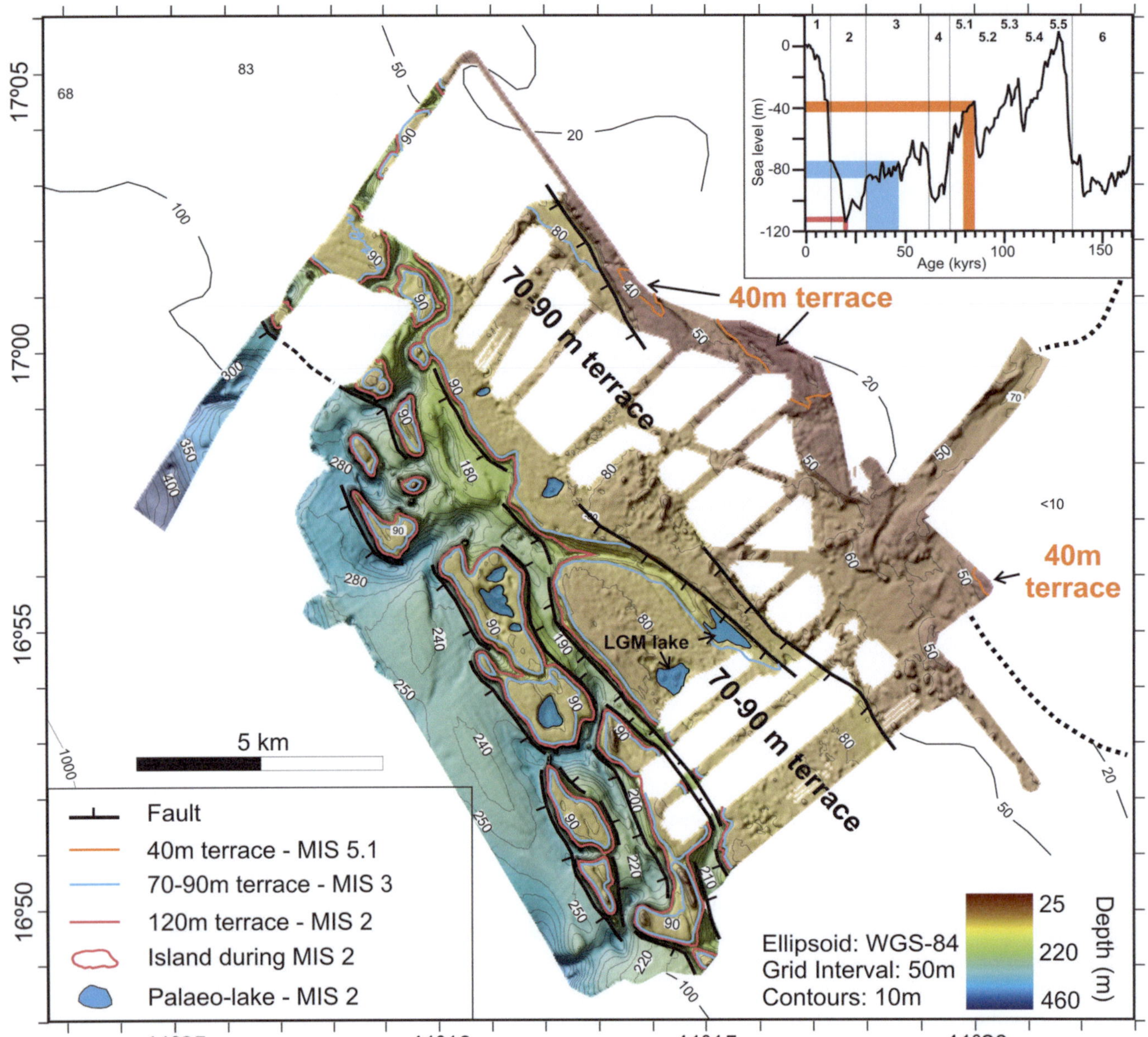

Fig. 9 Shaded bathymetry with tectonic and geomorphological features of the outer shelf. The inset diagram shows the correlation of the three observed underwater terraces (40 m, 70–90 m, 115–120 m) with the sea-level curve of Rohling et al. (2013) for the Red Sea

northern side of it display moderate to low reflectivity, which is indicative for the deposition of relatively coarser sediments with respect to the ones in the deep depression. The slopes along the two sides of the valley are in general linear (Fig. 10c). In a closer view, they display high morphological irregularity, with many invaginations and embayments, which are probably evidence for subaerial erosion on the flanks of the valley. The lower parts of the flanks, deeper than 100 m, dip at an angle of roughly 5% toward the axis of the valley. The upper parts, between 100 m and 70 m, are significantly steeper and display slope values of 25–30%.

6.2 Geological Structure and Sedimentation

The seismic and subbottom profiles shown on Figs. 11 and 13 provide insight into the geological structure and sediment accumulation on the Farasan Islands inner shelf, in particular in the long, straight valley and deep depression.

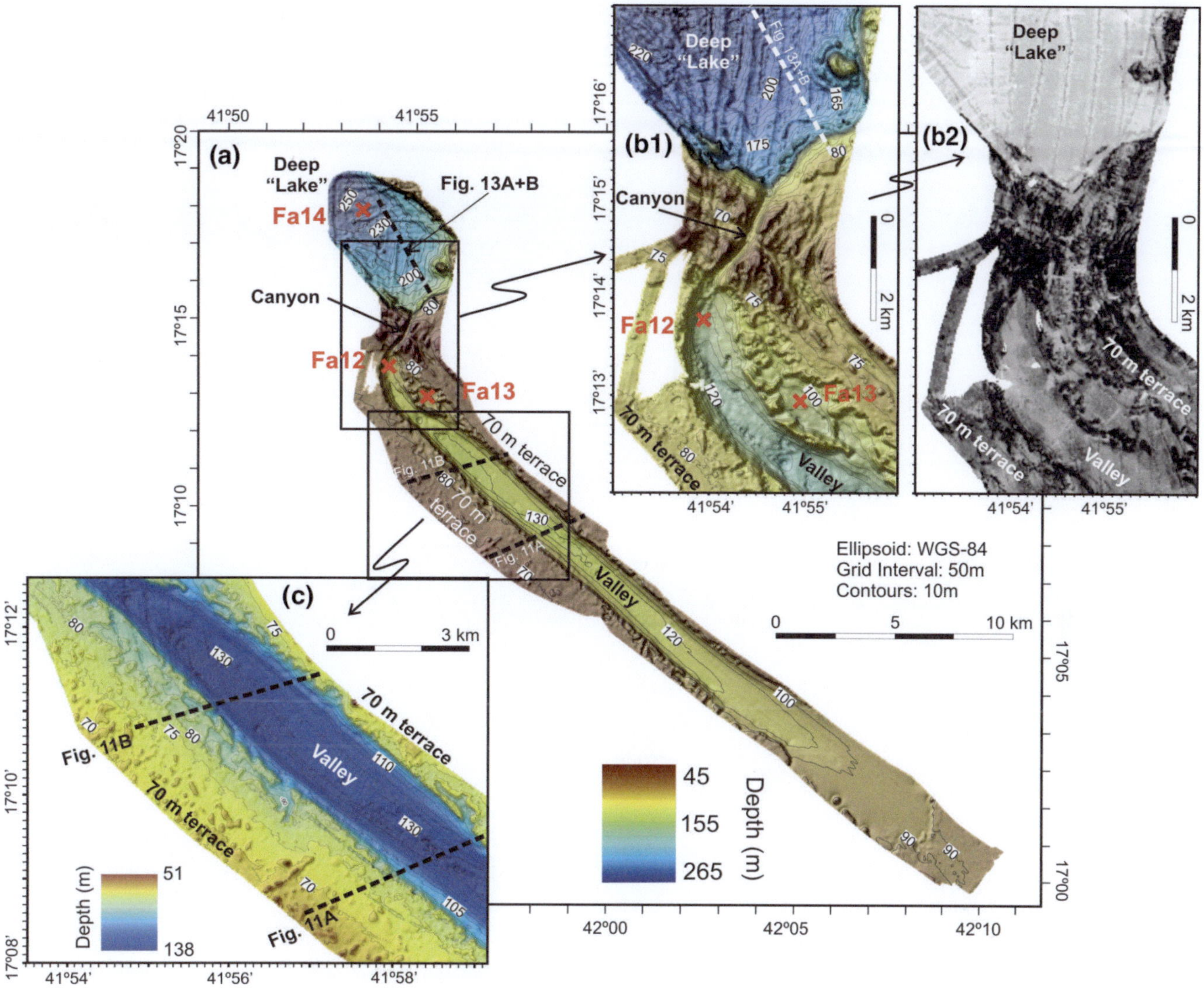

Fig. 10 **a** Shaded bathymetry and main morphological elements of Farasan 2 survey area on the inner continental shelf. **b1** Detailed bathymetry of the northern part of the Farasan 2 area. Note the canyon connecting the valley with the deep depression. **b2** Backscatter image of the same area as in B. Note the large variability in seafloor reflectivity. **c** Detailed bathymetry of the northern part of the valley. Note the irregular morphological configuration of the two slopes of the valley. Location of the profiles of Fig. 11 are shown on C. Location of the profiles in Figs. 11 and 13 are shown on **a** and **c**. Locations of cores FA-12 (Fig. 12) and FA-14 (Fig. 14) are shown on A

The substrate of the inner shelf (Fig. 11a) displays a stratigraphic succession that is very similar to the one observed below the outer shelf (Fig. 6b). The seismostratigraphic package 3 is the youngest deposit on the inner shelf, and occurs only in the valley with a maximum thickness of 20 ms (roughly 15 m) and must have been deposited during the Upper Quaternary (Holocene and Uppermost Pleistocene). The packages 4, 5 and 6 occur below the valley and the shelf south of it; they are characterized by parallel, continuous, locally undulating reflectors and may correspond to the Pleistocene shallow marine reefal limestone, which forms the bedrock on the Farasan Islands (Bantan 1999). Two opposite, normal faults run along the walls of the valley and are responsible for its formation and subsidence. The total vertical offset of the southern fault is 70 m, as inferred from the depth difference of the bottom of the seismic package 4 at both sides of the fault. The seismic stratigraphy on the foot wall of the northern fault is different from the one below the valley. No stratigraphic correlation can be made between seismic reflectors on both sides of the fault. Thus, the offset produced by this fault cannot be measured in the same way as for the southern fault. Therefore, using the seafloor as a marker, we suggest that the vertical offset of the northern fault is comparable to that of the southern fault, and that the valley has been formed within a symmetric graben.

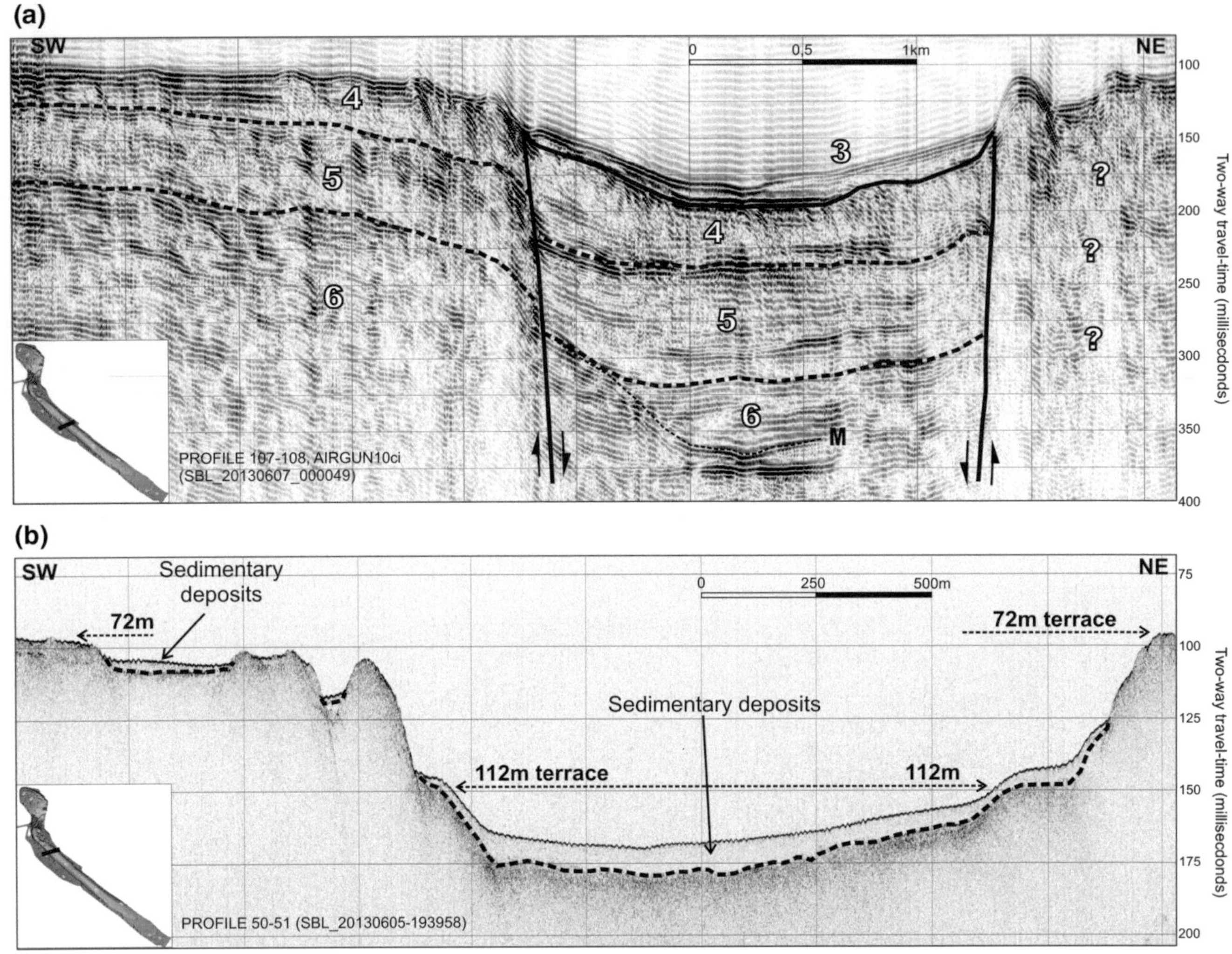

Fig. 11 **a** Airgun 10ci seismic reflection profile across the valley in the inner shelf of Farasan 2 survey area. Vertical exaggeration × 10. **b** High resolution subbottom profile across the same valley. Vertical exaggeration × 20. The locations of both profiles are shown on Fig. 10

The subbotom profile in Fig. 11b displays a higher resolution image of the stratigraphy across the valley. Recent sediments with transparent acoustic character have been deposited in the valley and cover the acoustic basement of the area. The thickness of the acoustically transparent deposits does not exceed 10 ms (7–8 m roughly) which is about half of the thickness of the seismostratigraphic package 3, that is, the topmost package observed on the airgun profile (Fig. 11a) in the valley. The top of the acoustic basement, below the transparent deposits, displays an erosional terrace at 112 m on both sides of the valley (Fig. 11 b). This terrace has formed due to the activity of water when this area was exposed and before the deposition of the transparent sediments. The valley has been incised on a flat plain which now forms the shelf seafloor at about 70 m depth. There is no evidence of any morphological element connecting the valley with the LGM sea, the shoreline of which has been identified at 115–120 m in the outer shelf survey area. Consequently, the 112 m terrace in the valley could not be formed as a marine one. Instead, it may constitute good evidence that the valley was a lake with water level at this position during the LGM.

The sediment core FA-12 (Fig. 12) retrieved from the northern end of the valley at 105 m depth bpsl confirms a major change of the depositional environment downcore. The upper part of the core, from 0 cm down to 90 cm bsf, consists of medium to coarse, olivine green sand with bivalves and shell fragments. The sediment colour changes at 90 cm bsf from olive green above to bluish gray below along with a gradual fining of the grain size to mud with some thin layers of sand. Similarly, a major change in the physical properties of the sediment, including a significant increase in magnetic susceptibility, occurs at 90 cm bsf. The sedimentological description of FA-12 along with the down-core physical properties indicate

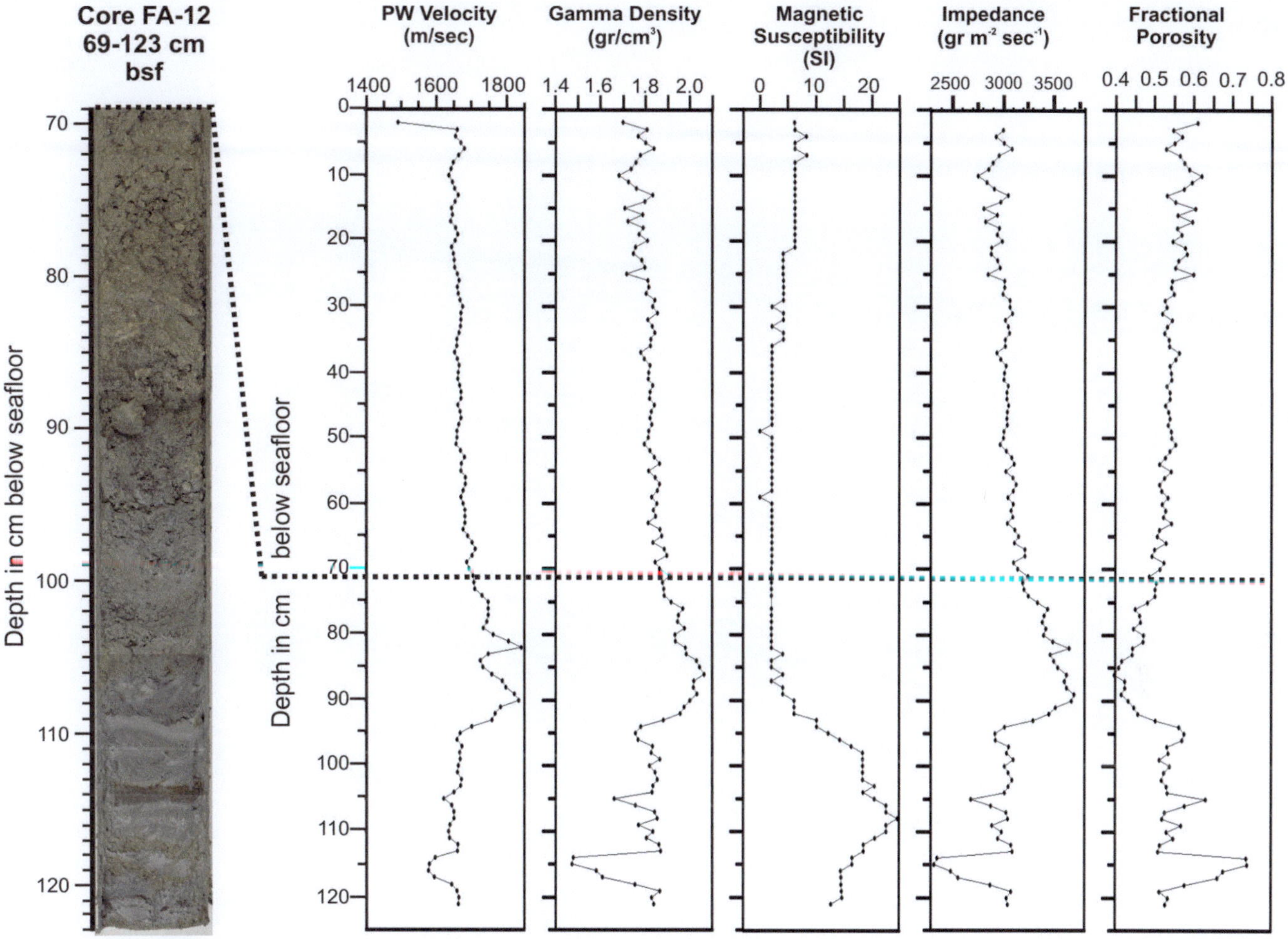

Fig. 12 Physical properties of core FA-12 and photograph of the lower part of the core, between 69 and 123 cm below the seafloor. The location of the core (lat: 17°13.719', long: 41°54.070', depth 105 m) is shown on Fig. 10. Note the change in sediment colour from olive green to grey at 90 cm bsf which coincides with the change in the properties at the same depth

a significant change in the depositional environment at 90 cm bsf. As with core FA-5 from the outer shelf (Fig. 8), the upper part of core FA-12 represents homogeneous marine sedimentation during the present high-stand period. The lower part, below 90 cm bsf, displays evidence of a lacustrine sedimentary environment prevailing in the valley from which the core has been recovered, during the last low sea-level stand, when the shelf was exposed. The sediment core FA-13 was retrieved from 102 m depth within a small depression at the northern margin of the valley (Fig. 10a, b). Sedimentological, geochemical and radiometric data from this core display a stratigraphic pattern very similar to the one observed in core FA-12 and confirm that the change in the sedimentary environment is associated with the post-LGM sea-level rise (see Sanderson et al., this volume, for details).

The deep depression at the northern part of the Farasan 2 survey area is the second major feature of the inner shelf (Fig. 10a, b). This 250 m deep depression hosts sedimentary deposits more than 600 ms (roughly 500 m) thick (Fig. 13a). The uppermost, less than 20 ms thin, seismostratigraphic package A drapes the irregular top of package B. The high-resolution profile of Fig. 13b provides a precise image of the stratigraphic architecture of the uppermost sedimentary deposits. The seismostratigraphic package A displays a rather transparent character with one layered horizon in the middle. It has been deposited on top of package B, on a fairly irregular reflector that marks the interface between packages A (transparent) and B (acoustic basement on the sub-bottom profile). The latter, with a maximum thickness of 100 ms (75–80 m) at the north-western edge of the depression (Fig. 13a), displays a rather chaotic seismic character with few discontinuous, strong, parallel, internal reflectors. Two seismostratigraphic packages, C and D, are separated from each other by an angular

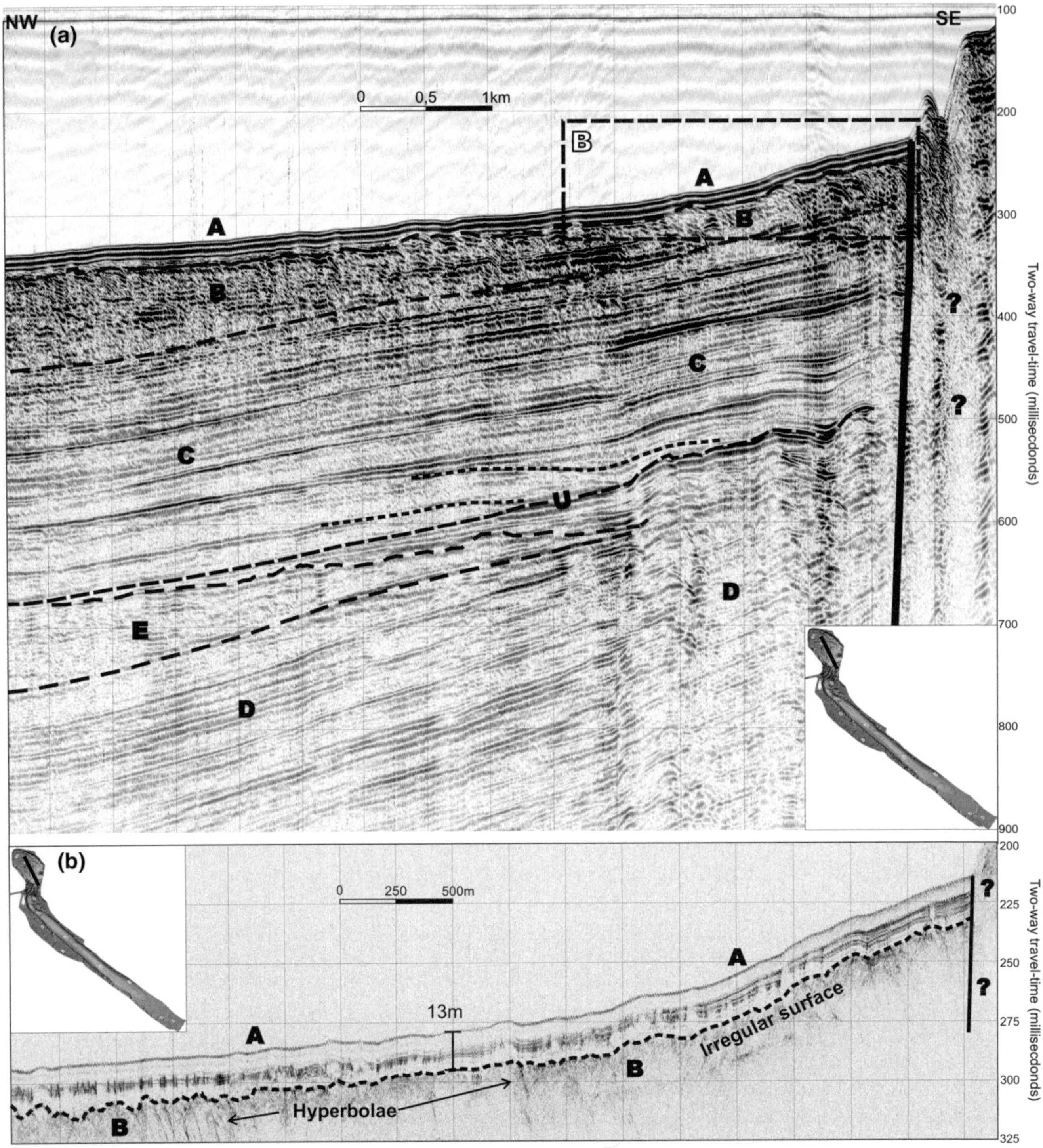

Fig. 13 **a** Airgun 10ci seismic reflection profile from the deep depression in the inner shelf of Farasan Islands. Vertical exaggeration × 10. **b** High resolution subbottom profile from the same depression. Vertical exaggeration × 20. Both profiles have been shot along the same track (see location on Fig. 10). The dashed box in A marks the position of the subbottom profile shown on B

unconformity U. Both packages are characterized by continuous, parallel reflectors, indicating uninterrupted sediment deposition. The seismostratigraphic package E, with its chaotic to transparent character, may represent a mass failure deposit derived from a possible landslide on the slopes of the depression.

The seismic reflectors of all packages in the seismic profile (Fig. 13a) terminate abruptly against the acoustic basement of the inner shelf. The interface between the sedimentary infill of the depression and the acoustic basement is steep and displays no evidence of active vertical tectonic movement. The age of the sediments deposited in the depression is not known. From their thickness, which exceeds 500 m, it can be assumed that the formation of the depression and the onset of sedimentary deposition may well be of Middle to Lower Pleistocene or even Pliocene age. In the absence of any age constrains, it is not possible to assess the chronological relationship between the sediments in the depression and the sedimentary formations of the shelf's substrate. The overall shape and structure of the depression strongly suggest that it may represent a sinkhole created due to the solution of a former salt diapir or dome. In addition to the deep depression studied here, a careful examination of the available bathymetric charts (Fig. 2) reveals the existence of several depressions-sinkholes distributed on the Farasan Islands shelf, with similar size, shape and depth.

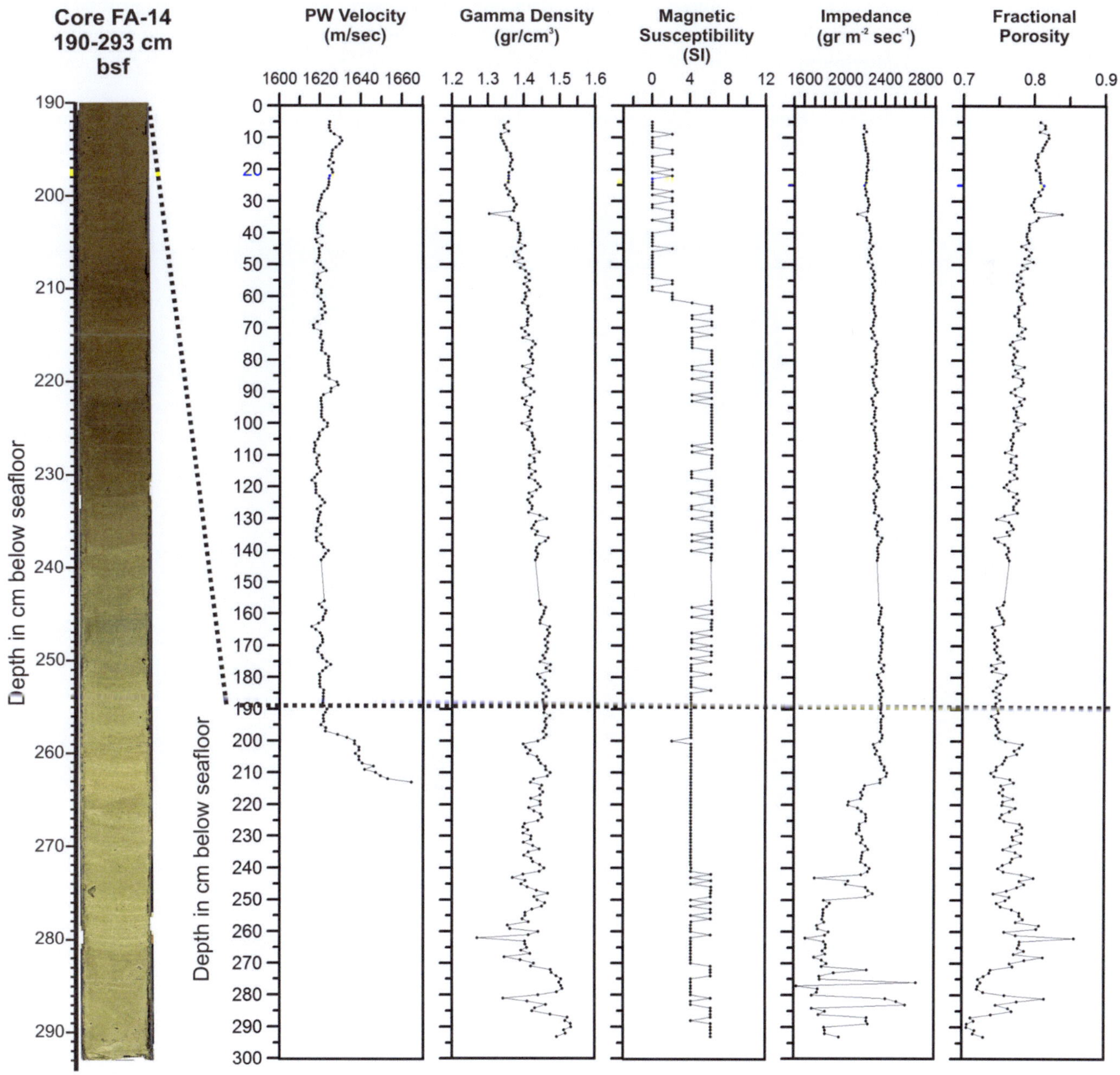

Fig. 14 Physical properties of core FA-14 and photograph of the lower part of the core, between 190 and 293 cm below the seafloor. The location of the core (lat: 17°18.281', long: 41°53.043', depth 245 m) is shown on the map of Fig. 10a. Note the gradual change in sediment colour from olive grey to light greenish grey below 220 cm bsf, which coincides with the change in the other properties at the same depth

Core FA-14 (Fig. 14) confirms a change in the sedimentary environment within the sinkhole, similar to the change observed in the valley of the inner shelf (Fig. 12) as well as in the valley of the outer shelf (Fig. 8). The upper part of the core, between 0 and 230 cm bsf is composed of homogeneous, olive grey silt. Silty sediments persist to the lower end of the core at 293 cm bsf but their colour changes gradually to light greenish grey and remains like that. The physical properties of the sediment, with the exception of the magnetic susceptibility, change down-core too. The gamma density and the fractional porosity show opposite patterns with many variations below 200 cm. The impedance also displays strong variability below 200 cm.

The sedimentological description of the core FA-14 along with the down-core physical properties indicate a gradual change in the depositional environment below 220–230 cm bsf. As with cores FA-12 (Fig. 12) from the valley and FA-5 from the outer shelf (Fig. 8), the upper part of core FA-14 represents homogeneous marine sedimentation during the present high-stand period. The lower part may have been deposited in a different, very probably lacustrine environment, like the one prevailing in the nearby valley.

6.3 Inner Shelf Submerged Landscapes and Palaeoshorelines

The analyses and interpretations of the swath bathymetry, the seismic and subbottom profiling and the sediment coring data presented above lead to interesting results on the palaeo-geomorphological evolution of the inner shelf survey area with respect to Upper Quaternary sea-level fluctuations (Fig. 15).

The seafloor of the Farasan 2 inner shelf survey area lies at about 70–75 m depth and is in general flat, with numerous mounds developed on it. This depth is close to the depth of

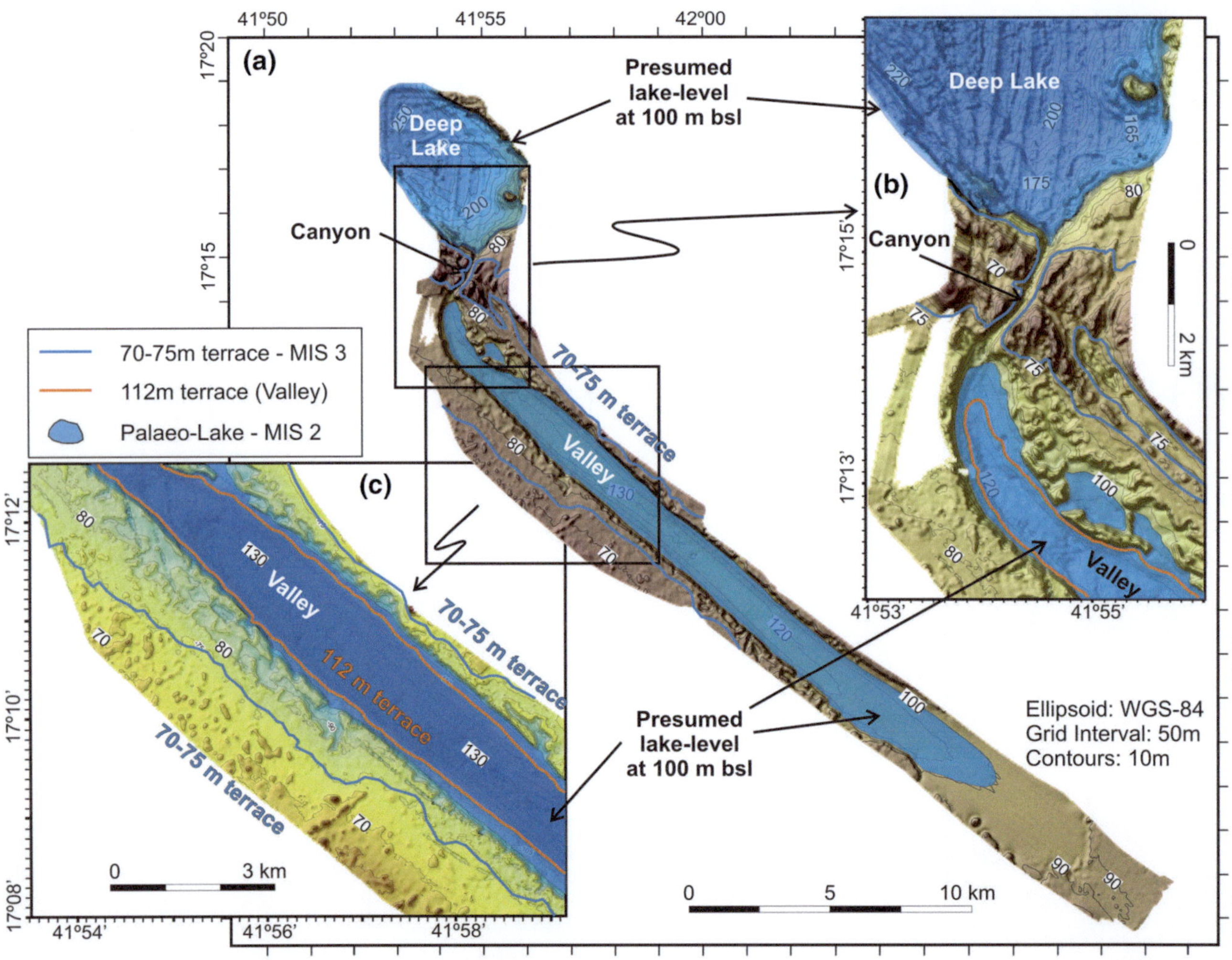

Fig. 15 **a** Interpretative shaded bathymetry maps of the inner shelf of the Farasan Islands area. **b** Detailed map and interpretation of the northern part of the survey area. **c** Detailed map and interpretation of part of the valley

the main terrace mapped on the outer shelf between 70 m and 90 m (Fig. 9) which has been assigned to MIS 3. This is good evidence to suggest that the MIS 3 marine terrace is the one that forms the largest part of the present continental shelf around the Farasan Islands.

Both the long, straight valley and the deep circular to ellipsoidal sinkhole display evidence for a major environmental change at the transition from the LGM to the Holocene high sea-level stand. A morphological terrace at 112 m depth on both sides of the valley, covered by a drape of transparent, recent deposits (Fig. 11), indicates that the valley was filled with water up to this depth in a certain period, possibly during parts of the last low sea-level period. The deposition of lacustrine sediments in the valley has been confirmed with cores FA-12 (Fig. 12) and FA-13. Still, the lacustrine sediments in these cores occur at 106 m and 103 m bsl respectively, which is higher than the observed terrace. This may mean that the lake-level within the valley may have fluctuated by several metres before the area was drowned by the rising sea level. A second possible lake level may be at about 100 m depth, which is the depth of the foot of the steep parts of the slopes of the valley (Fig. 15).

The deep depression to the north of the valley is interpreted as a sinkhole, based on the steepness of the surrounding walls and its shape. It also hosted a lake, as indicated by the bathymetry of the surrounding seafloor and confirmed by the analyses of the sediment in core FA-14 (Fig. 14). Unlike the valley, there is no data to estimate or infer the position of the water-level in the sinkhole. By analogy with the assumed lake-level in the valley, the shoreline of the lake in the sinkhole has been drawn at about 100 m depth on Fig. 15.

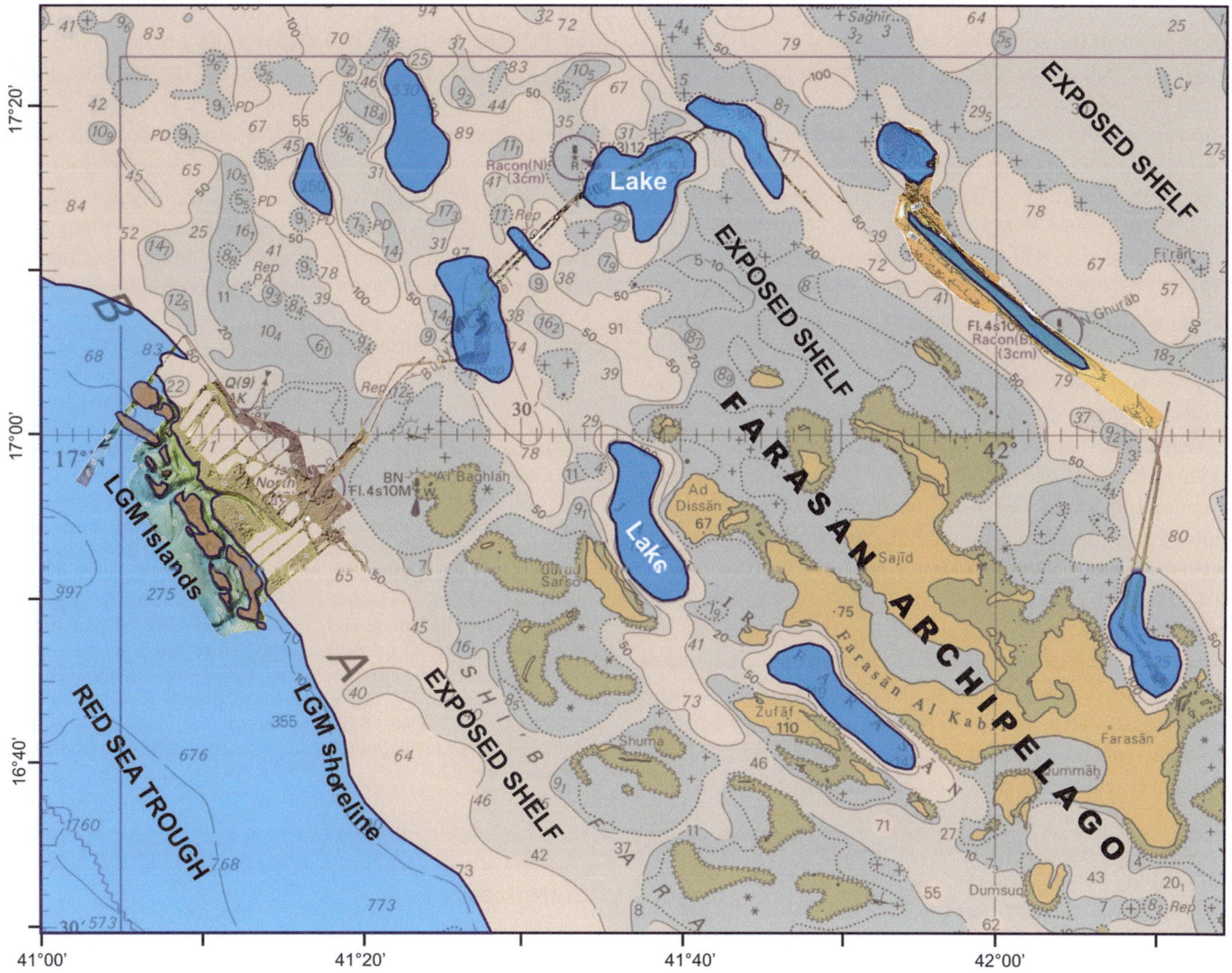

Fig. 16 Palaeogeographic reconstruction of the Farasan Islands shelf during the Last Glacial Maximum. Note the inferred lakes where deep sinkholes occur on the shelf. The LGM shoreline has been drawn at 115–120 m depth. Note the islands (brown colour) at short distance from the LGM shoreline

No matter what was the exact water-level in the valley and the sinkhole, the fact is that during the LGM, when the shelf was subaerially exposed, two lakes, a long and shallow one and a deep and ellipsoidal one, existed on this part of the inner shelf. The analyses of the sediment cores retrieved from both are in progress and so far we have no evidence to indicate whether they were filled with freshwater or not. The narrow, 80 m deep canyon, which connects the northernmost tip of the valley with the southernmost tip of the sinkhole, is fairly clear evidence for erosion due to surface flowing water when the shelf was exposed.

7 Synthesis—Discussion

The observations and results on the geological and tectonic structure, the seismic stratigraphy, the sedimentation and the geomorphological evolution of the two survey areas in the outer and the inner shelf of the Farasan Islands, as presented above, lead to the following conclusions.

The flat part of the Farasan Islands inner and outer shelf, which has been mapped at depths between 70 m and 90 m, dips very gently toward the southwest and may have developed as an erosive marine terrace during the upper half of MIS 3, between 30 and 45 ka BP. This hypothesis is supported by the sea-level curve of Rohling et al. (2013) for the Red Sea. The deeper terrace that has been mapped on the outer shelf at 115–120 m depth corresponds to the sea-level of MIS 2. Thus, the entire shelf may have been subaerially exposed during the LGM (Fig. 16). A shallower terrace, at about 40 m depth, may be correlated with MIS 5.1, 80–85 ka BP, when the sea-level fluctuated between 35 m and 45 m bpsl.

Extensional tectonics, possibly driven by the inferred, basin-ward flow of Miocene evaporite deposits in deeper stratigraphic levels below the shelf, is responsible for the rupturing of the latter with predominantly NW–SE trending normal faults. Dragging of the overlying, faulted, Plio-Quaternary rocks due to the underlying salt flow led to the drifting away of blocks off the shelf edge and the creation of isolated ridges and shoals surrounded by steep slopes and troughs. The 90 m deep flat tops of the ridges and shoals were also subaerially exposed when the sea-level was at 115–120 m bpsl during the Last Glacial Maximum and were islands located at short distances from the shelf edge and the LGM shoreline.

The troughs separating the islands from the shelf host sedimentary sequences, the thickness of which exceeds 150 m. There are not enough constraints to estimate the mean sedimentation rate in the troughs, but it is reasonable to assume that the troughs existed for at least several hundreds of thousands of years. In that case, and if vertical movements can be considered as negligible in the area, then it seems reasonable to assume that the geomorphological configuration suggested here for MIS 2, with the exposed shelf and the flat-topped palaeo-islands, may also be valid for the previous low sea-level period at 140 ka BP (MIS 6).

Shallow or deep, circular, irregular or elongate depressions and valleys have been mapped on the main terrace of the Farasan Islands shelf. Seismic stratigraphic and sedimentological data confirm that they were filled permanently or ephemerally with water, thus forming lakes, when the shelf was exposed during MIS 2. Geochemical, micropaleontological and radiometric analyses on the sediments retrieved from below the seafloor of these depressions are in progress, aiming at unravelling the environmental conditions prevailing in the LGM lakes, and in particular at resolving whether the lakes were filled with fresh water.

The hydrographic charts of the Farasan Islands shelf show many deep depressions with depths in excess of 200 m, very similar to the one interpreted as a sinkhole lake studied here, in the inner shelf Farasan 2 survey area. If they were lakes too, the 120 km wide and several-hundred-km-long Farasan shelf may have been a place where numerous lakes existed when it was exposed during the LGM (Fig. 16). In addition, the presence of valleys and canyons indicates that surface water run-off eroded the shelf under subaerial conditions. Finally, the seismic stratigraphy observed below the shelf, with horizontal or gently dipping sedimentary layers, may have favoured the development of groundwater aquifers and the occurrence of springs along the slopes of the valleys when subaerially exposed.

Acknowledgements We thank the Saudi Commission for Tourism and National Heritage (SCTH) for permission to undertake fieldwork, and the President, HRH Prince Sultan bin Salman bin AbdulAziz al Saud, the Vice-President, Professor Ali Al-Ghassan, and the Director General, Jamal al Omar, for their support of our research. We also thank HRH King Salman bin Abul Aziz Al Saud, formerly Crown Prince and Minister of Defense, and the Hydrographic Department of the Saudi Ministry of Defense, for permission to undertake the cruise of R/V AEGAEO. We also thank Lt. Fahad Al Shwish, Observer from the Hydrographic Department of the Saudi Ministry of Defense, for his support and valuable assistance in overcoming unexpected logistical difficulties during the cruise, and Captain Theodoros Kanakaris and the crew of R/V AEGAEO for their untiring efforts to ensure the smooth running of the scientific operation during the offshore survey work. The research is funded by the European Research Council through Advanced Grant 269586 DISPERSE under the 'Ideas' Specific Programme of the Seventh Framework Programme. This is DISPERSE contribution no. 43.

References

Abou Ouf MA, El-Shater A (1992) Sedimentology and mineralogy of Jizan shelf sediments, Red Sea, and Saudi Arabia. J King Abdulaziz Univ Marine Sci Jeddah 3:39–54

Armitage SJ, Jasim SA, Marks AE, Parker AG, Usik VI, Uerpmann H-P (2011) The southern route "Out of Africa": evidence for an early expansion of modern humans into Arabia. Science 331:453–456

Augustin N, Devey CW, van der Zwan FM, Feldens P, Tominaga M, Bantan RA, Kwasnitschka T (2014) The rifting to spreading transition in the Red Sea. Earth Planet Sci Lett 395:217–230

Augustin N, Devey N, van der Zwan F (this volume) A modern view on the Red Sea Rift: tectonics, volcanism and salt blankets

Bailey GN, Flemming NC (2008) Archaeology of the continental shelf: Marine resources, submerged landscapes and underwater archaeology. Quat Sci Rev 27:2153–2165

Bailey GN, King GCP (2011) Dynamic landscapes and human dispersal patterns: tectonics, coastlines and the reconstruction of human habitats. Quatern Sci Rev 30:1533–1553

Bailey GN, Sakellariou D (2012) SPLASHCOS: Submerged Prehistoric Archaeology and Landscapes of the Continental Shelf. Antiquity 86(334). http://antiquity.ac.uk/projgall/sakellariou334/

Bailey GN, King GCP, Flemming NC, Lambeck K, Momber G, Moran LJ, Al-Sharekh AM, Vita-Finzi C (2007) Coastlines, submerged landscapes and human evolution: The Red Sea Basin and the Farasan Islands. J Island Coastal Archaeology 2:127–160

Bailey GN, King GCP, Devès M, Hausmann N, Inglis R, Laurie E, Meredith-Williams M, Momber G, Winder I, Alsharekh A, Sakellariou D (2012) DISPERSE: dynamic landscapes, coastal environments and human dispersals. Antiquity 86(334). http://antiquity.ac.uk/projgall/bailey334/

Bailey GN, Reynolds SC, King GCP (2011) Landscapes of human evolution: models and methods of tectonic geomorphology and the reconstruction of hominin landscapes. J Human Evolution 60 (3):257–280

Bailey GN, Devès MH, Inglis RH, Meredith-Williams MG, Momber G, Sakellariou D, Sinclair AGM, Rousakis G, Al Ghamdi S, Alsharekh AM (2015) Blue Arabia: Palaeolithic and underwater survey in SW Saudi Arabia and the role of coasts in Pleistocene dispersal. Quatern Int 382:42–57. https://doi.org/10.1016/j.quaint.2015.01.002

Bailey G, Meredith-Williams M, Alsharekh A, Hausmann N (this volume) The archaeology of Pleistocene coastal environments and human dispersals in the Red Sea: Insights from the Farasan Islands

Bantan RA (1999) Geology and sedimentary environments of Farasan Bank (Saudi Arabia) southern Red Sea: A combined remote sensing and field study. Ph.D. Thesis, Department of Geology, Royal Holloway, University of London, January 1999

Bayer HJ, Hotzl H, Jado AR, Roscher B, Voggenreiter W (1988) Sedimentary and structural evolution of the northwest Arabian Red Sea Margin. Tectonophysics 153:136–151

Bosence DWJ (1998) Stratigraphic and sedimentological models of rift basins. In: Purser BH, Bosence DWJ (eds) Sedimentation and tectonics in Rift basins: Red Sea-Gulf of Aden. Chapman & Hall, London, pp 9–25

Bosworth W, Huchon P, McClay K (2005) The Red Sea and Gulf of Aden Basins. J African Earth Sci 43:334–378

Bosworth W (2015) Geological evolution of the Red Sea: Historical background, review, and synthesis. In: Rasul NMA, Stewart ICF (eds) The Red Sea. Springer Earth System Sciences, Springer-Verlag Berlin Heidelberg, pp 45–78. https://doi.org/10.1007/978-3-662-45201-1_1

Carbone F, Matteucci R, Angelucci A (1998) Present day sedimentation of the carbonate platform of the Dahlak Islands, Eritera. In: Purser BH, Bosence DWJ (eds) Sedimentation and tectonics in Rift basins, Red Sea-Gulf of Aden. Chapman & Hall, London, pp 523–536

Cochran JR (1983) A model for development of the Red Sea. Bull Am Assoc Petrol Geologists 67:41–69

Dabbagh A, Hoetzl H, Schnier H (1984) Farasan Islands. In: Jado AR, Zöttl (eds) Quaternary Period in Saudi Arabia. Springer, Wien, pp 212–220

Delagnes A, Tribolo C, Bertran P, Brenet M, Rémy C, Laubert J, Khalidi L, Mercier N, Nomade S, Peigne S, Sitzia L, Tourne-Piche J-F, Al-Halibi M, Al-Mosabi A, Macchiarelli R (2012) Inland human settlement in southern Arabia 55,000 years ago. New evidence from the Wadi Surdud Middle Paleolithic site complex, western Yemen. J Human Evolution 63(3):452–474

Dixon JE, Monteleone K (2014) Gateway to the Americas: Underwater archaeological survey in Beringia and the North Pacific. In: Evans A, Flemming N, Flatman J (eds) Prehistoric Archaeology of the Continental Shelf: A Global Review. Springer, New York, pp 95–114

Dullo WC (1990) Facies, fossil record and age of Pleistocene reefs from the Red Sea (Saudi Arabia). Facies 22:1–46

Dullo WC, Blomeier D, Camoin GF, Casanova J, Colonna M et al (1997) Morphological evolution and sedimentary facies on the fore-slopes of Mayotte, Comoro Island: Direct observations from submarine. In: Camsin G, Bergerson D (eds) Carbonate Platforms of the Indian Ocean and the Pacific. IAS Special Publication

Dullo WC, Montaggioni L (1998) Modern Red Sea coral reefs: A review of their morphologies and zonation. In: Purser BH, Bosence DWJ (eds) Sedimentation and Tectonics in Rift Basins: Red Sea-Gulf of Aden. Chapman & Hall, London, pp 583–594

Faure H, Walter RC, Grant DR (2002) The coastal oasis: Ice age springs on emerged continental shelves. Global Planetary Change 33:47–56

Feldens P, Mitchell NC (2015) Salt flows in the central Red Sea. In: Rasul NMA, Stewart ICF (eds) The Red Sea: the formation, morphology, oceanography and environment of a young ocean basin. Springer Earth System Sciences, Berlin Heidelberg, pp 205–218. https://doi.org/10.1007/978-3-662-45201-1_1

Geraga M, Sergiou S, Sakellariou D (this volume) Preliminary results of micropaleontological analyses on sediment core FA09 from the southern Red Sea continental shelf

Girdler RW, Styles P (1974) Two stage sea-floor spreading. Nature 247:7–11

Gvirtzman G (1994) Fluctuations of sea level during the past 400,000 years: the record of Sinai, Egypt (Northern Red Sea). Coral Reefs 13:203–214

Hudec MR, Jackson MPA (2007) Terra infirma: Understanding salt tectonics. Earth-Sci Rev 82:1–28

Hughes GW, Beydoun ZR (1992) The Red Sea-Gulf of Aden: Biostratigraphy, lithostratigraphy and palaeoenvironments. J Petrol Geol 15:135–156

Inglis RH, Bosworth W, Rasul N, Al Saeedi A, Bailey G (this volume) Investigating the palaeoshorelines and coastal archaeology of the southern Red Sea

King GCP, Bailey GN (2006) Tectonics and human evolution. Antiquity 80:265–286

LaBreque JL, Zitellini N (1985) Continuous sea-floor spreading in Red Sea, an alternative interpretation of magnetic anomaly pattern. Bull Am Assoc Petrol Geol 69:513–524

Lambeck K, Purcell A, Flemming N, Vita-Finzi C, Alsharekh A, Bailey GN (2011) Sea level and shoreline reconstructions for the Red Sea: Isostatic and tectonic considerations and implications for hominin migration out of Africa. Quatern Sci Rev 30:3542–3574

Macaulay V, Hill C, Achilli A, Rengo C, Clarke D, Mcchan W, Blackburn J, Semino O, Scozzari R, Cruciani F, Taha A, Shaari NK, Raja JM, Ismail P, Zainuddin Z, Goodwin W, Bulbeck D, Bandelt HJ, Oppenheimer S, Torroni A, Richards M (2005) Single, rapid coastal settlement of Asia revealed by analysis of complete mitochondrial genomes. Science 308:1034–1036

Mellars P (2006) Why did modern populations disperse from Africa ca. 60,000 years ago? A new model. Proc Nat Acad Sci USA 103:9381–9386

Mideast Industries Ltd. (1966) Salt production possibilities of the Jizan and Farasan Island area. Saudi Arabian Directorate General for Mineral Resource Report DGMR-279, 13 pp

Mitchell DJW, Allen RB, Salama W, Abduzakm A (1992) Tectonostratigraphic framework and hydrocarbon potential of the Red Sea. J Petrol Geol 15(2):187–210

Petraglia MD, Rose JI (eds) (2009) The evolution of human populations in Arabia. Springer, Dordrecht

Petraglia MD, Alsharekh AM, Crassard E, Drake NA, Groucutt H, Parker AG, Roberts RG (2011) Middle Paleolithic occupation on a Marine Isotope Stage 5 lakeshore in the Nefud Desert. Saudi Arabia. Quatern Sci Rev 30(13–14):1555–1559

Petraglia MD, Breeze PS, Groucutt HS (this volume) Blue Arabia, Green Arabia: Examining human colonisation and dispersal models

Purser BH, Bosence DWJ (1999) Sedimentation and tectonics in rift basins: Red Sea-Gulf of Aden. Chapman & Hall, London

Roeser HA (1975) A detailed magnetic survey of the southern Red Sea. Geol. Jb. D13:131–153

Rohling EJ, Fenton M, Jorissen FJ, Bertrand P, Ganssen G, Caulet JP (1998) Magnitudes of sea-level lowstands of the past 500,000 years. Nature 394:162–165

Rohling EJ, Grant K, Bolshaw M, Roberts AP, Siddall M, Hemleben C, Kucera M (2009) Antarctic temperature and global sea level closely coupled over the past five glacial cycles. Nat Geosci 2:500–504

Rohling EJ, Grant KM, Roberts AP, Larrasoaña JC (2013) Paleoclimate variability in the Mediterranean and Red Sea regions during the last 500,000 years: Implications for hominin migrations. Curr Anthropol 54(S8):S183–S201. https://doi.org/10.1086/673882

Rose JI, Usik V, Marks A, Hilbert Y, Galleti C, Parton A, Geiling JM, V. Cerné V, Morley M, Roberts R (2011) The Nubian complex of Dhofar, Oman: an African Middle Stone Age industry in southern Arabia. PLoS ONE 6(11):e28239. https://doi.org/10.1371/journal.pone.0028239

Sanderson D, Kinnaird T (this volume) Optically Stimulated Luminescence dating as a geochronological tool for Late Quaternary sediments in the Red Sea region

Schmidt DL, Hadley DG, Brown GF (1982) Middle Tertiary continental rift and evolution of the Red Sea in south-western Saudi Arabia. Ministry of Petroleum and Mineral Resources, Deputy Ministry for Mineral Resources, Jeddah, USGS-OF-03-6, 56 pp

Searle RC, Ross DA (1975) A geophysical study of the Red Sea axial trough between 20.5 and 22 N. Geophys J Roy Astron Soc 43:555–572

Siddall M, Smeed DA, Hemleben C, Rohling EJ, Schmelzer I, Peltier WR (2004) Understanding the Red Sea response to sea level. Earth Planet Sci Lett 225:421–434

Stoffers P, Kuhn R (1974) Red Sea evaporites - A petrographic and geochemical study. In: Whitmarsh RB, Weser OE, Ross DA (eds) Initial reports of the Deep Sea Drilling Project, vol. 23. U. S. Government. Printing Office, Washington, pp 821–847

Stoffers P, Ross DA (1977) Sedimentary history of the Red Sea. In: Hilpert LS (ed) Red Sea Research 1970–1975. Saudi Arabian Ministry of Petroleum and Mineral Resources, Directorate General of Mineral Resources, Jeddah, Bulletin 22, pp i–iv and H1–H19

Walter RC, Buffler RT, Bruggemann JJ, Guillaume MMM, Berhe SM, Negassi B, Libsekal Y, Cheng H, Edwards RL, Von Gosel R, Neradeau D, Gagnon M (2000) Early human occupation of the Red Sea coast of Eritrea during the Last Interglacial. Nature 405:65–69

Tectonic Geomorphology and Soil Edaphics as Controls on Animal Migrations and Human Dispersal Patterns

Simon Kübler, Geoffrey C. P. King, Maud H. Devès, Robyn H. Inglis, and Geoff N. Bailey

Abstract

This chapter examines the relationship between the changing geomorphology of physical land forms in tectonically and volcanically active regions, topography, soil nutrients, movements of large mammals, and patterns of human subsistence and dispersal in the early stages of human evolution. We place particular emphasis on the ways in which minor topographic barriers—for example, river gorges, fault scarps and basaltic lava flows—constrain the movements of large mammals during their seasonal migrations and offer opportunities for early humans to take advantage of predictable natural constrictions to ambush animals. We also emphasise the importance of soil edaphics—the mineral composition of soils as a source of trace elements essential for animal growth and health—as another key variable in determining the distribution and movements of animals and their human hunters. Soil edaphics are closely related to the nature of the underlying regolith or bedrock, and are consequently highly variable in their distribution, providing additional constraints on animal movements. We show how the combination of topographic and soil-edaphic mapping in conjunction with the observed locations of stone-tool or fossil assemblages can highlight patterns of early human behaviour, using examples from the East African and Jordanian Rifts and the Arabian margin of the Red Sea. Finally, we note that these methods have the potential to be applied more widely in other regions of the world and to problems of animal and human health at the present-day.

1 Introduction

The physical landscape forms the basis for the development of animal and human habitats. Studying the interaction between the fauna and the environment therefore requires an understanding of the character and evolution of landscapes. Here we provide examples demonstrating that the knowledge and techniques derived from tectonic geomorphology can provide this information and be adapted to research in palaeoanthropology, archaeology and ancient and modern land use. Combining evidence in the fossil and archaeological record with an understanding of landscape processes allows an additional insight into past human and animal behaviour, and is an important complement to the data derived from excavation and the study of artefacts and fossils alone. The approach we have developed combines extensive fieldwork with satellite-derived data to produce maps of the landscape as it would have existed at an earlier time, taking account of the animal species that would have been the primary food supply and their habitat preferences, and incorporating evidence of changes in land forms brought about by tectonic activity and other geomorphological processes.

This work was originally inspired by the need to place archaeological sites and other evidence of past human behaviour into their wider landscape setting in order to

S. Kübler (✉)
Department of Earth and Environmental Sciences, Ludwig Maximilians University, Luisenstrasse 37, 80333 Munich, Germany
e-mail: s.kuebler@lmu.de

G. C. P. King · M. H. Devès
Institut de Physique du Globe de Paris, 1 rue Jussieu, 75238 Paris Cedex 5, France

M. H. Devès
Centre de Recherche Psychanalyse Médecine et Société, CNRS EA 3522—Université Paris Diderot, Paris, France

R. H. Inglis · G. N. Bailey
Department of Archaeology, University of York, King's Manor, York, YO1 7EP, UK

R. H. Inglis
Department of Environmental Sciences, Macquarie University, North Ryde, Sydney, NSW 2109, Australia

G. N. Bailey
Arts and Social Sciences, College of Humanities, Flinders University, GPO Box 2100 Adelaide, SA 5001, Australia

N. M. A. Rasul and I. C. F. Stewart (eds.), *Geological Setting, Palaeoenvironment and Archaeology of the Red Sea*,
https://doi.org/10.1007/978-3-319-99408-6_29

illuminate the patterns and causes of early human dispersal and colonisation of new territory in the very earliest stages of human development in Africa and adjacent regions. As the work has progressed, it has become clear that the results of our approach have considerable relevance not only to the past but also to a better understanding of environmental factors that affect animal and human health at the present day (Winder et al. 2013; Devès et al. 2014, 2015; Bailey et al. 2015; Kübler et al. 2015, 2016).

In this chapter we consider selected aspects of reconstructing the ancient use of the landscape, with particular emphasis on the role of topography and soil edaphics in constraining the distributions and seasonal movements of the large herbivores on which human subsistence largely depended for long periods of the past (Owen-Smith 1988). By soil edaphics we mean the ability of the soil to provide essential nutrients in the form of trace elements necessary for the growth and survival of large herbivores. We summarise the relationship between tectonic topography and the early stages of human evolution, and set out in more detail the variables we consider as significant in palaeolandscape reconstruction and the methods we use to reconstruct ancient topography and soil edaphics. We show how soil edaphics are closely linked to the nature of the underlying rock substrate, and how the constraints imposed on animal movements by a combination of topography and soil edaphics would have facilitated the capture of large mammals by early human hunters. We do not enter here into the debates about whether early hominins obtained meat by scavenging or by hunting (see debates in Blumenschine et al. 1994; Domínguez-Rodrigo 2002; Pickering 2013), but note that our reconstructions are consistent with ambush hunting of large and dangerous animals by early members of the genus *Homo* with no need for equipment other than hand-thrown missiles or thrusting spears.

For examples, we draw on case studies from the East African Rift in Kenya, from the Dead Sea Rift in the southern Levant and from SW Arabia (Fig. 1). Although these are widely disparate regions, they are united by several features in common. First, they are all regions with evidence of human occupation in the earliest (Lower Palaeolithic) periods of the Stone Age; more specifically they are associated with similar stone-tool assemblages of Acheulean type, made by some of the earliest members of the genus *Homo* (*H. ergaster* or *H. erectus*), the earliest of our examples extending back in time to 1.4 Ma. We examine selected archaeological sites from each region, chosen because they are substantial sites indicating repeated use of favoured locations in a wider landscape and because they provide good illustrations of landscape reconstruction. Secondly, they occur at different points along one of the major axes of earliest dispersal of the genus *Homo*, namely the East African Rift and the Red Sea and Levantine corridors, and illustrate features that we consider to be significant in promoting or permitting population dispersal. Thirdly, they are associated with different geomorphological expressions of the Afro-Arabian tectonic system.

The dominant feature of this tectonic system is the East African Rift. Rifting is thought to have been initiated with the massive flood-basalt eruptions that occurred in Ethiopia at 30 Ma, with propagation southwards for over 1000 km as far as Mozambique and the borders of South Africa, exploiting weaknesses in the Earth's crust. Throughout, the Rift displays typical features of rifting, with progressive uplift of the valley margins by normal faulting, creation of lake basins on the rift floor formed by internal drainage, and substantial volcanic activity. North of Ethiopia, the 'Proto Red Sea' was a northward extension of the East African Rift, but subsequently widened into a young ocean as the Arabian Plate began to move north away from Africa, creating the Gulf of Aden in the south and colliding with Europe to the north. North of the Red Sea, the transform faults along the western margin of the Arabian plate as it moved northward created the so called Syrio-Jordanian Rift. This is not, strictly speaking, a rift but is the result of strike-slip motion. The cross-sectional profile of the resulting valleys is narrower than in a true rift, but in other respects there are many similar tectonic features including high levels of earthquake activity, significant episodes of magmatism and internally draining lake basins.

We begin with examples from Kenya in the classic landscape setting of the East African Rift. This is one of the earliest and best studied centres of early human origins and development, and the region which offers the most detailed application and illustration of our methods. We then move to the southern Levant, another region with a long history of geological and archaeological investigation that provides a solid platform for palaeolandscape reconstruction. Finally, we turn to the intermediate region of the western Arabian escarpment. Here, there is a shorter history of investigation, and discovery of new archaeological sites and palaeoenvironmental investigations are ongoing (see the other chapters in this volume).

2 Landscape and the Evolution of *Homo*

It is widely accepted that a progressive drying of the climate resulted in the gradual disappearance of tropical forest and thus affected the range of forest species in substantial parts of Africa, and specifically in East Africa (Bobe 2006; Cerling et al. 1997; deMenocal 2004). As forest cover reduced after 5 Ma, fauna began to exploit more open savannah conditions characterised by grasslands, small trees and shrub vegetation. Today, although forest contains a larger diversity of species, savannah supports a very much larger faunal

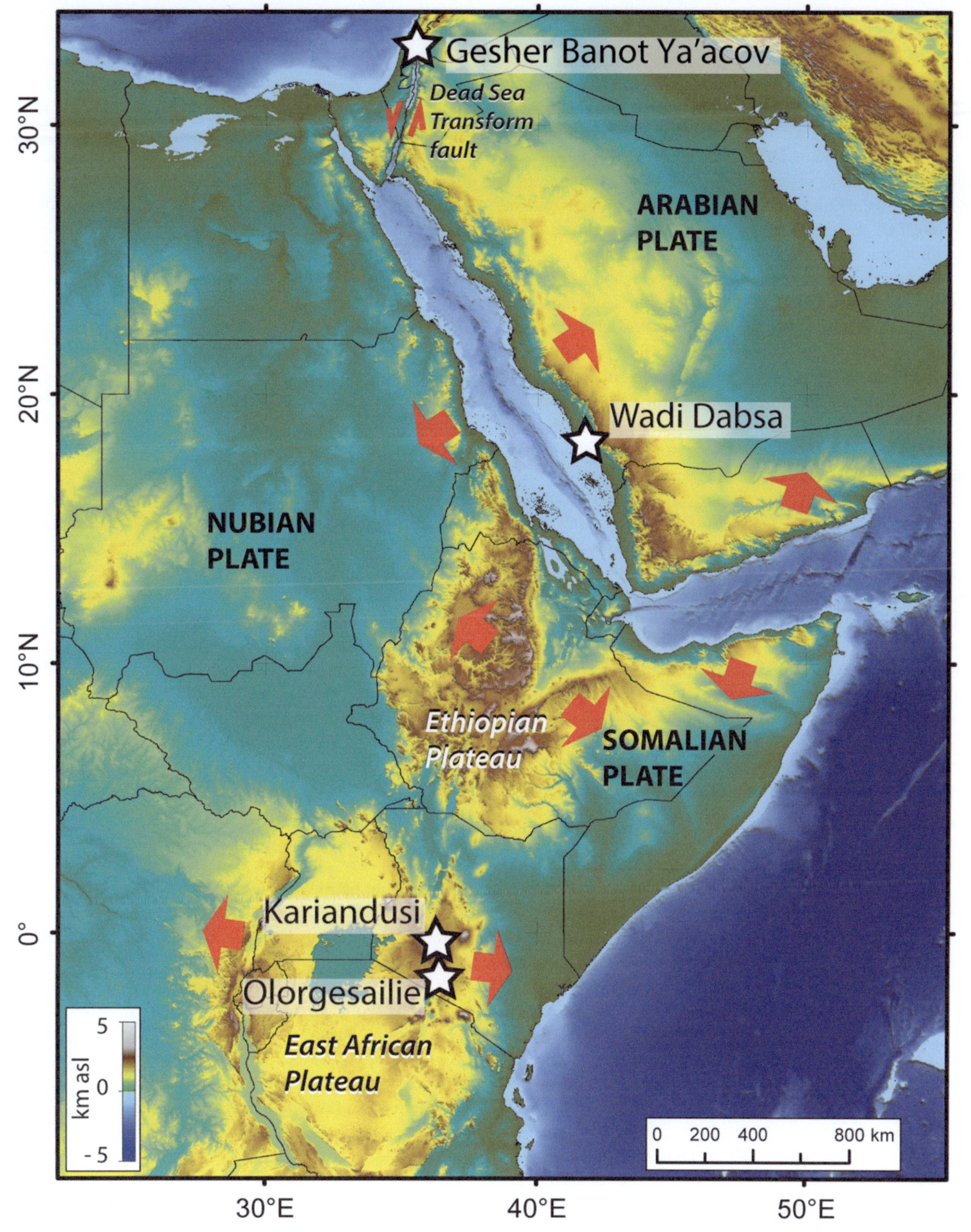

Fig. 1 Location map of case studies and archaeological sites plotted on a map of the Afro-Arabian tectonic system showing general relief and major tectonic features. Red arrows indicate separation of the African Rift and the relative motions of the Arabian and African plates. Uplift of the Ethiopian and East African Plateaus is the result of thinning and swelling of the Earth's crust over hot plumes rising from deep within the Earth's core. The Ethiopian plume broke through the crust with massive eruption of flood basalts at ~ 30 Ma. This is thought to have triggered the initial separation of the African Rift to the south and the proto Red Sea to the north, with subsequent separation and northward motion of the Arabian Plate by ocean spreading in the Gulf of Aden and the Red Sea, creating the Dead Sea transform fault in the north-west and collision with Plates to the north (not shown)

biomass per square kilometre than any other zone of the planet (Maglio and Cook 1978). Several faunal lineages show radiations as species began to exploit the rich new ecological niches associated with the increasing grassland (Bobe and Eck 2001; Bobe and Behrensmeyer 2004). This is also true for hominins; the fossil record indicates several early hominin genera (*Sahelanthropus, Orrorin, Kenyanthropus, Ardipithecus, Australopithecus*), which may reflect different evolutionary strategies to living in new habitat types. Of these genera, only *Homo* has remained, becoming with time the top predator (e.g., Wood 2005).

The key to how hominins may have adapted to living in novel environments may well lie in the types of landscapes they exploited. Here we refer to the physical landscape that Earth scientists are used to exploring, that is, landforms, rocks, sediments and soils, and we use the term environment to mean the combination of vegetation and the physical landscape. Within this context, savannah corresponds more to a vegetation type as part of the environment, rather than to the physical landscape. While vegetation types are quite sensitive to climate change, features of the physical landscape are not affected to the same extent. Landscapes evolve in response to geodynamical processes such as active tectonics, volcanism, erosion and deposition. These processes shape landscape geomorphology and hydrology, and act on the geology and on soil composition, which can in turn affect the edaphic properties of the soils. The simple picture one might have of savannah becomes more complex when one starts considering the features of the underlying physical landscape. This can be made of various types of rocks and soils. It can be rough or smooth, covered by volcanic lavas or deformed by active tectonics. As a result, it can vary in its sensitivity to climatic oscillations. Habitats are the suitable environmental (vegetation and landscape) conditions that pertain to a particular species (Vrba 1982). Suitable regions for hominins need to meet key habitat requirements, namely a range of forage (C3 browse and C4 grass food types), refuge from predation and a source of drinking water (e.g., Reynolds et al. 2011). There are specific geomorphological contexts that can provide these habitat requirements, notably those affected by volcanic or tectonic activity.

We have shown that the locations of Palaeolithic sites in many regions are closely associated with tectonically active and topographically complex landscapes (King and Bailey 2006; Bailey et al. 2011). In almost all cases, tectonic and/or volcanic activity was ongoing when Palaeolithic people were present in these regions. Continuing activity rejuvenates features and prevents them from being smoothed by erosion. Such areas are associated with a wide variety of landforms, for example, rocky outcrops, cliffs, gorges, ridges, offering plenty of look-out and observation points, opportunities for shelter and protection from most predators and key strategic advantages for hunting (e.g., King et al. 2010; Bailey and King 2011).

Early humans, being slow moving, would have been vulnerable to fast-moving cursorial predators in flat, featureless landscapes. For the same reason, throughout history, persecuted people have always moved to mountainous regions to escape cavalry, trucks or tanks. Tectonic activity is particularly efficient at creating complex landscapes including: (1) cliffs and river gorges, which hominins and other primates could exploit for safety and sleeping roosts; (2) catchment of drinking water; and (3) heterogeneous environments, potentially associated with a range of food types. A key advantage of tectonically active areas is that faulting or folding activity can disrupt water tables and create lakes and swamps that are constantly renewed as long as the activity persists. Faulting can thus provide a range of uplifted, dry areas where grazing animals would predominate, with down-dropped, wetter areas that would provide ecological heterogeneity within a relatively small spatial range and sources of drinking water and other resources. This pattern fits well with existing inferences about hominin habitats from many sites in Africa (see Reynolds et al. 2011 for a review).

Early hominins, such as *Australopithecus*, are considered to have exploited trees for security and adapted to the savannah by means of anatomical adaptations, progressively acquiring effective bipedalism. The energetically (relatively) efficient *Homo* body form appeared at about 2 Ma (Wood 2005). The most widely accepted view is that this adaptation allowed more rapid movement on flat ground, and freed the arms for carrying. An extreme version of this is the 'Endurance Running Hypothesis' (Bramble and Lieberman 2004), which proposes an ability to run prey animals to exhaustion as a primary hunting technique (but see Pickering 2013, p. 99–102 for a critique). Other possibilities are thermoregulatory efficiency that allowed exploitation of a day-time hunting niche (Wheeler 1993). Recently, an alternative hypothesis that considers the role of topography in the evolution of the early *Homo* body form has been proposed (Reynolds et al. 2011; Winder et al. 2013, 2014; Devès et al. 2015). This 'Scrambler Man' hypothesis argues that rough topography would have selected for a progressive transition to an upright posture associated with bipedal locomotion through climbing and scrambling activities. This view contrasts with previous models based on adaptations to forest or flat savannah in favour of physical incentives presented by steep, rugged terrain. Using rugged topography to monitor and trap migratory mammals could have selected for greater speed and agility in hominin populations over time. This hypothesis explains well the key anatomical changes observed over the course of human evolution (e.g., Winder et al. 2013).

3 Key Variables in the Reconstruction of Landscapes and Animal Movements

We argue that complex topography has not only played a role in human evolution and land use but is also a critical factor in defining the habitats and seasonal movements of large herbivorous mammals, imposing constraints that facilitate human exploitation. We consider two physical attributes of a landscape that can affect animal movements: (1) the roughness of the topography itself; and (2) soil characteristics that indicate areas of good and poor pasture quality.

By roughness we mean irregularities in surface morphology (Bailey et al. 2011, p. 4–5). A rough topography may be contrasted with a smooth topography—one with surfaces that are smooth and level. But roughness can occur at different scales, ranging from steep mountain slopes and vertical cliffs to surfaces that appear flat but are difficult to cross because of jagged erosional features, such as some limestone karsts and lava flows. In assessing roughness, we use measurements of slope angles derived from Digital Elevation Models (DEM) (Bailey et al. 2011, p. 277), supplemented by field observations. Our interest is particularly in identifying topographic features that are sufficient to block or deter the movements of large herbivores, channel their movements along predictable pathways, and make them more vulnerable to human predation.

The majority of herbivores are thought to migrate in response to changes in available water, grazing and seasonal temperature variations. In modern landscapes, seasonal movements of wild animals are limited by humans fencing off large areas, but historical documents record these types of migrations. For example, early European explorers to southern Africa recorded mass migrations of springbok (*Antidorcas marsupialis*) and other animals to areas of fresh grazing (Skinner and Louw 1996). The seasonal migrations of wildebeest (*Connochaetes taurinus*) and other grazing animals in the Serengeti continue today. Seasonal resource variations are still exploited by transhumant shepherds in many parts of Europe, who move their animals in the same way that wild species such as the red deer (*Cervus elaphus*) would have moved in the past without human intervention (Sturdy et al. 1997).

With regard to soil attributes, we pay particular attention to edaphic factors as a guide to pasture quality, though we also note the importance of water retentiveness, especially in arid or semi-arid regions. Edaphic factors, or the shorthand term 'edaphics', concern the ability of the regolith (i.e., soils and subsoils) to supply, by plant take-up, the nutrients necessary for herbivore growth, health and reproduction. These edaphic factors are critical to the growth of young animals and the continued health of younger and older individuals. They can shape animal movements within a landscape as a result of the presence or absence of vital nutritional components in the vegetation of certain areas. The availability of soluble phosphate is especially important, as this supports bone development. Its limited availability plays an important role in Serengeti wildebeest migrations today (Murray 1995).

Simple phosphorus or phosphate levels in soils and subsoils are not, on their own, guides to the edaphic quality of the soils. Tricalcium di-orthophosphate, the main constituent of animal bones, is only soluble in ionised acidic water, and its release into soils in a form which can be taken up by plants is normally very slow. So, for high edaphic quality in relation to phosphates, not only is an adequate source of the minerals required, but the conditions must be such that the minerals can actually be taken up by the fodder which the herbivores eat (Henkin et al. 1995). Even where the regolith provides abundant sources of the main minerals and nutrients in a form which can be taken up by plants, specific trace elements may be missing, for example, elements, such as selenium, cobalt, copper or potassium (Burrows et al. 1979; Corah 1996; Formigoni et al. 2011; Kadim et al. 2003; Ruess 1984). In such cases, animals may need to make periodic movements to areas that supply the missing elements. Nowadays, fodder supplements and fertilisers often provide this requirement for domestic animals without the need for extensive movements.

We distinguish soil edaphics from soil fertility. Soils may be fertile in the sense that they can support lush vegetation, but may nevertheless be lacking in the nutrients essential for animal health, especially for the growth of young animals. A given soil might support abundant vegetation, but this may be of poor quality when viewed as a source of food and trace elements for herbivores. Over wide areas of the Mediterranean, for example, until the recent substantial fall in grazing pressures resulting from changes in modern human economies, the edaphically poorer areas often carried more abundant plant vegetation precisely because they were less exploited by animals, while the vegetation in the edaphically richer areas was bitten down hard, reflecting their differential attractiveness to herbivores (Sturdy and Webley 1988; Sturdy et al. 1997).

4 Methods of Palaeolandscape Reconstruction

The aim of palaeolandscape reconstruction is to derive a digital landscape model (palaeoDEM) at the time of hominin occupation for the desired study region. The starting point for reconstructing ancient landscapes is a present-day DEM, usually derived from freely available Shuttle Radar Topographic Mission (SRTM) data. This is then corrected by subtracting a correction displacement field (CDF) across the DEM to restore earlier elevations. Creating a CDF requires information on (1) fault motions, (2) erosion and deposition of sediments, (3) volcanic activity, (4) other sources of surface motion such as subsidence from groundwater disposal. Each of these variables needs to have a temporal component (absolute or relative) to allow the CDF to be determined for a specific time interval. This can be judged by analysing the relationships of geological units and structures, for example, tilted lake beds or lava flows overlying tectonically faulted structures, and combining them with geochronological data, which are available with varying degrees of accuracy in our individual study regions. Tectonic motion can be restored with computer models and codes using dislocations in a 2D half-space based on analytical expressions (Okada 1982). Vertical motion from non-tectonic processes such as groundwater disposal or collapse of magma chambers can be reconstructed by creating palaeo-contour lines from a large number of topographic profiles, which are then corrected for subsidence and erosion using Geoinformation Software packages (GIS). A more detailed description of this methodology can be found in the supplementary material of Kübler et al. (2015), and has been applied in greatest detail to the Olorgesailie case study discussed below, where tectonic changes have had the most significant effects on landscape morphology.

5 Edaphic Analysis

The objective of chemical analysis of soils and plants in the context of understanding edaphic properties is to gather information on nutrient variability in the individual rock units and sediments in the study region. Usually three to five soil and plant-tissue samples per rock unit are taken to allow for nutrient variability. Composite soil samples are taken from five sampling points in a 5 × 5 m square from the uppermost 25 cm of a soil profile. For deep soil profiles in unconsolidated sediments, additional samples along a depth profile (core or soil pit) are taken. Plant sampling usually focuses on collecting grass and other shallow-rooted plants at the soil sampling site. Soil and plant samples are then tested for the concentration of the most important macronutrients and trace elements such as calcium, copper, iron, manganese, magnesium, nitrogen, phosphorus, potassium, sodium, and zinc. Further soil samples are tested for pH-value, electrical conductivity and total organic carbon.

6 The Kenyan Rift

Olorgesailie in the southern Kenya Rift lies in the centre of a 60 km-wide rift floor (Fig. 2) and is famous for its abundance of Acheulean artefacts, fossil mammals and palaeoenvironmental indicators together with a cranial specimen of *Homo* preserved in sediments spanning ~1.2 to <0.5 Ma (Isaac 1977; Potts 1989; Behrensmeyer et al. 2002; Potts et al. 2004). The site is adjacent to a palaeolake. Prey animals include elephant, hippopotamus, giant baboon, equids and bovids, with evidence of carcase butchery.

The geology of the region comprises extensive trachyte flows laid down between 0.7 and 1.4 Ma, overlapping the period when the site was in use (Fig. 3). These are the dominant rock type on the rift floor. Also present are Plio-Pleistocene basalts and tuffs dated to 1.4–2.7 Ma associated with the volcanoes of Mt. Olorgesailie and Mt Esayeti. There are many sub-parallel fault scarps, and these are nearly vertical in places, especially on the trachyte, which is particularly resistant to erosion, and these constrain east–west movements of large animals and humans (Kübler et al. 2015). Sediments carried from the north and the rift flanks are present in small patches, but uplift and back tilting prevent the entry of sediments from outside the main rift.

The landscape around Olorgesailie today is significantly different from when the hominins were present, resulting from ongoing faulting and the partial collapse of the Olorgesailie volcanic caldera (Fig. 4a). Tectonic motion on two north–south trending normal faults has resulted in tilting of the Legemunge lake beds that contain the archaeological material. These deposits were originally laid down horizontally, so that the fault motions must postdate the archaeological occupation. The caldera collapse of the northwestern Olorgesailie edifice resulted in draining of the palaeolake. The past and present geometry of the landscape is defined by its tectonic structure, allowing reconstruction of the palaeo-morphology by modelling fault displacements. By removing the effects of fault motion and making corrections for erosion and deposition of sediment (Fig. 4c, d), a palaeo-DEM of the landscape can be created as it would have appeared during the period when hominins were present (Fig. 4b).

Modern analyses of macronutrients and trace elements from soils on a representative sample of lithological and sedimentary units demonstrate a consistent relationship between edaphic quality and the underlying regolith (Kübler et al. 2015, 2016). Trachytes with poor-quality soils dominate the region (Fig. 3), with richer soils developing only on some other volcanic rocks and on sediments brought by rivers from the north and east. This is supported by

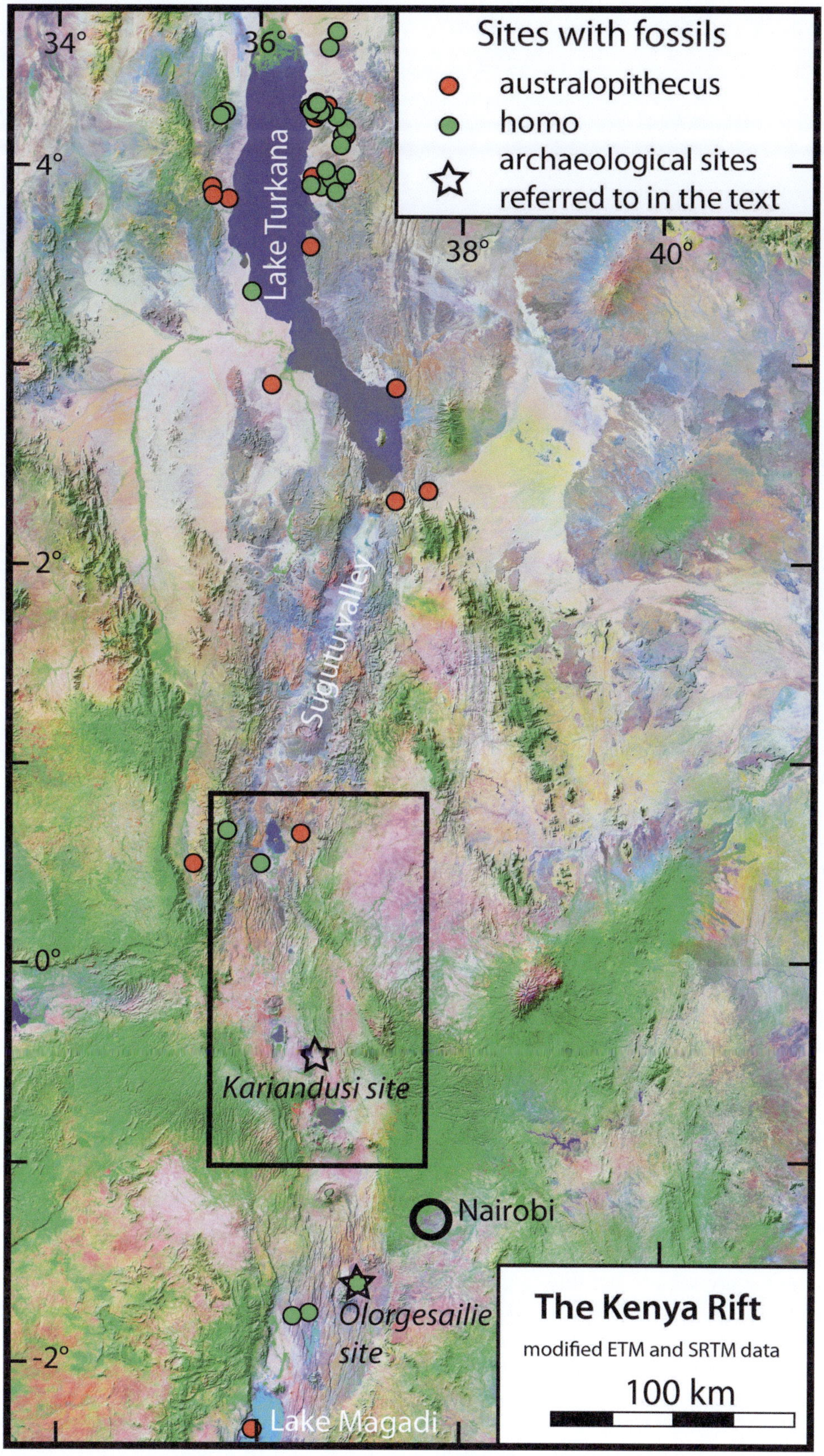

Fig. 2 General location map of sites in the central and southern Kenyan Rift in relation to topography, showing position of Olorgesailie and Kariandusi. The locations of other sites with hominin fossils are also shown. Data from the Paleobiology database (https://paleobiodb.org). After Kübler et al. (2016). The image is based on ETM + legacy data. Topography from SRTM v4.1. Maps created by SK and GK using Adobe Illustrator CS 5.1, Adobe Photoshop CS 5.1, MAPublisher 9.4.0, and Global Mapper

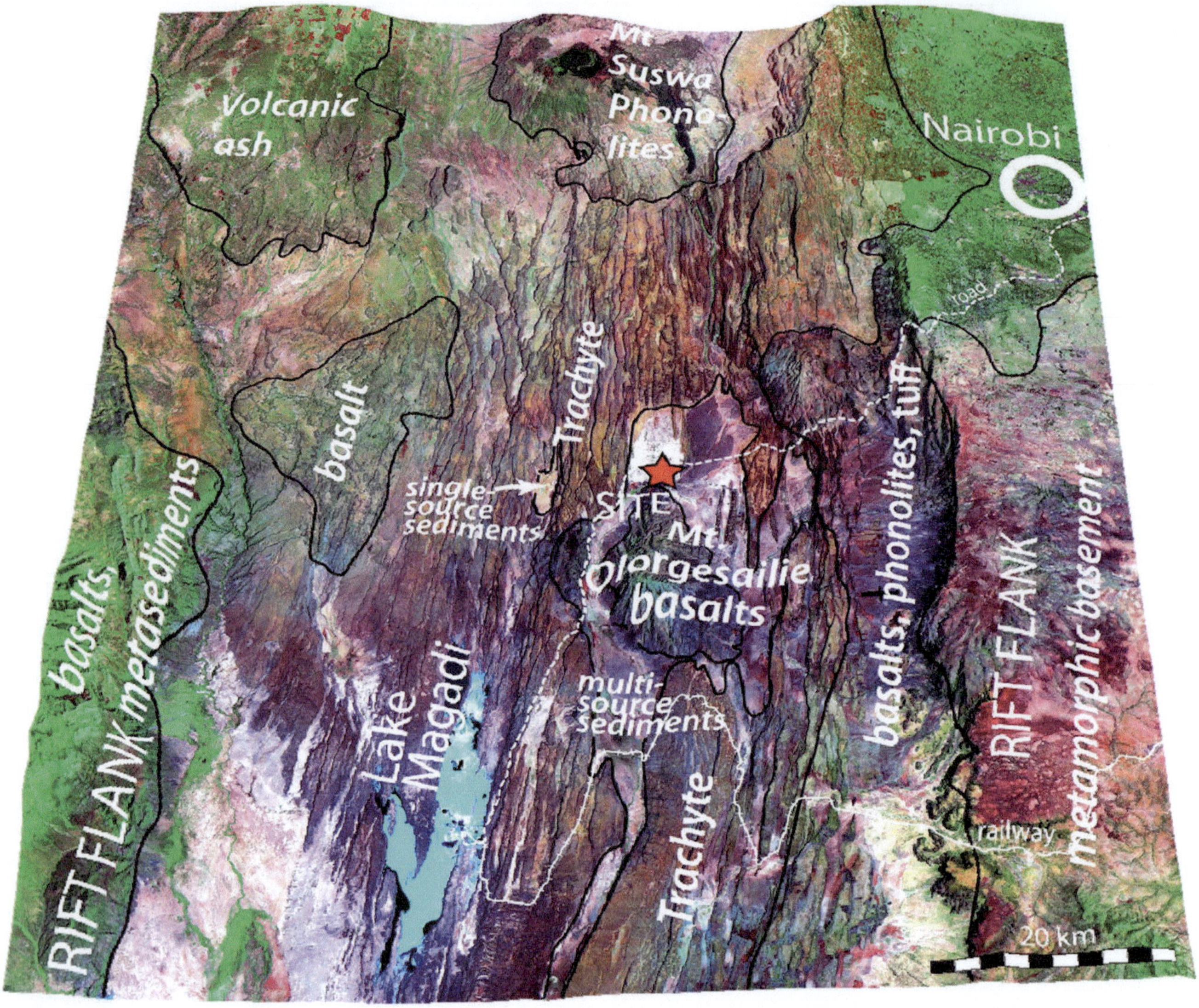

Fig. 3 Oblique 3D view of the South Kenya Rift centred on the Olorgesailie hominin site, showing faults and other geological features. The region is heterogeneous with some areas well-vegetated and others with thin vegetation. Thin vegetation can result from soils that do not favour plant growth, but can also result from heavy grazing and browsing of favoured vegetation. The image is based on ETM + legacy data. Topography from SRTM v4.1 data with a vertical exaggeration of ~8. A red star indicates the Olorgesailie site. Maps created by SK and GK using Adobe Illustrator CS 5.1, MAPublisher 9.4.0, Global Mapper 16, and ENVI 5.1

interviews with Masai shepherd families who are well aware where animals must graze and browse to remain healthy.

In the past, soils close to the palaeolake would have provided an attractive focus for animal grazing and browsing, especially during the dry season. But this area is comparatively small, so that large herds would need to move to the more distant rift flanks 30 km away, where more extensive grazing would have been available particularly during the wet season. Routes of animal migration to or from the flanks were greatly constrained by the numerous north–south fault scarps and would need to pass along a predictable route between the lake and the volcanic edifice (Fig. 4b). It is on this route that the archaeological site is located.

We conclude that a key reason why the Olorgesailie site was attractive to hominins and repeatedly used over a long period is its proximity to a nearby area that has excellent edaphics in a wider region that is highly deficient, and that it controlled the only route allowing movement of large animals between east and west, facilitating trapping by ambush hunting. Proximity to a reliable source of drinking water, suitable volcanic stone for making artefacts, and good look-out points for monitoring animal movements are additional advantages of the Olorgesailie location, but are not sufficient to explain why the site remained a repeated focus of human activity over such a long period. When the caldera collapsed and the lake was drained, the topographic

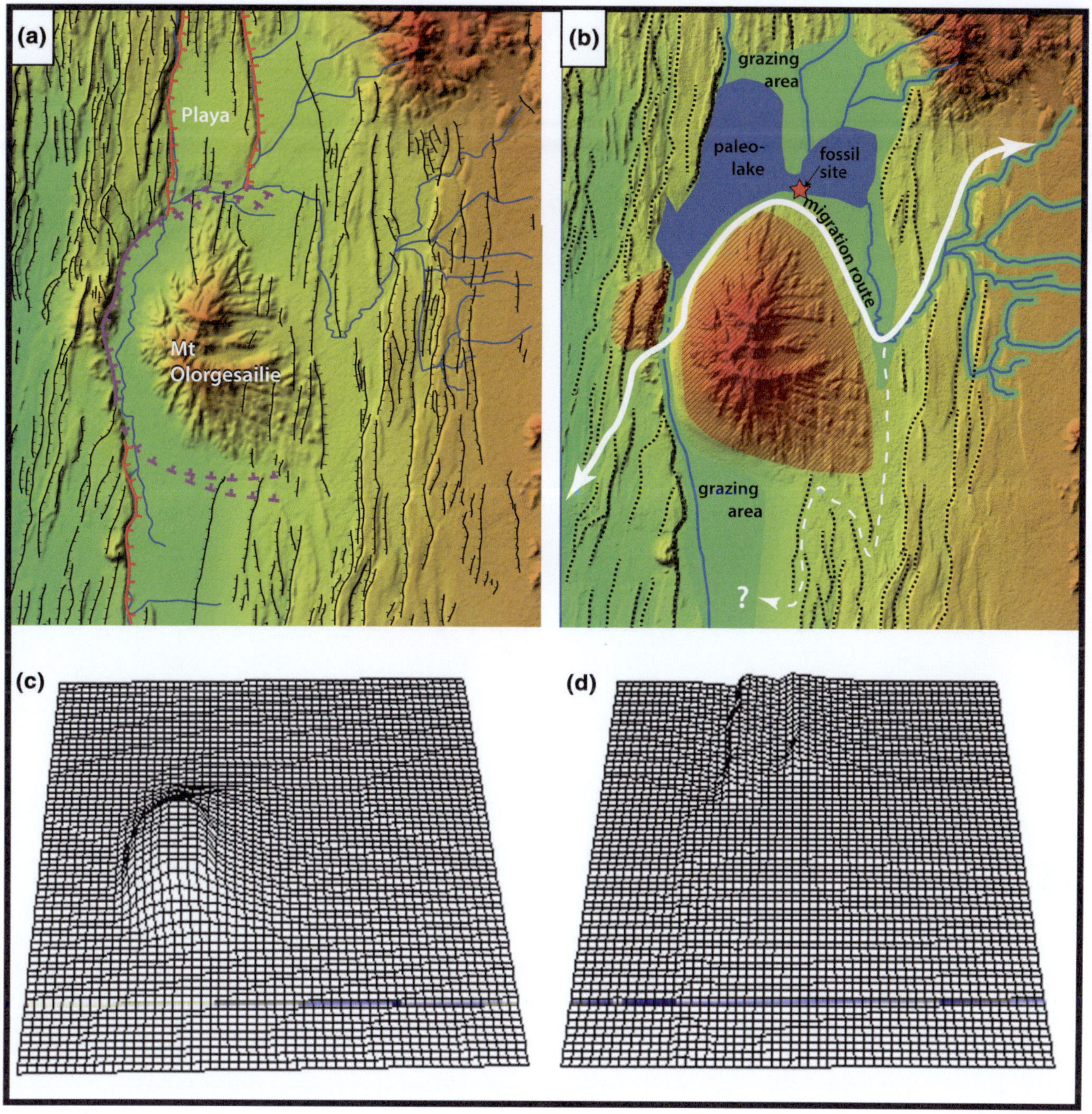

Fig. 4 Digital elevation models (DEMs) of the Olorgesailie region. **a** Present day topography. Prominent fault scarps are indicated in black and those indicated in yellow and white are used in the modelling of earlier topography in (**b**). The yellow faults are young faults, and post-date the lake and the period of hominin activity at Olorgesailie. The white faults result from caldera collapse. **b** PalaeoDEM of the Olorgesailie region. White circle indicates the position of the Olorgesailie site. The volcanic edifice of Mt. Olorgesailie was already in place and impeded drainage to the south, resulting in the formation of the lake. A possible late lower lake level is indicated in dark blue. Some faults in the trachytes do not cut basalts of the Olorgesailie edifice and therefore clearly pre-date it. Dotted lines show faults that formed barriers to animal movement and are thought to have existed when the site was used by hominins. The drainage system was limited by a barrier in the same place as hypothesized by Behrensmeyer et al. (2002), so the barrier could have been higher than the lake without fully blocking drainage from the lake. Likely grazing areas are indicated. Routes to the flanks of the Rift negotiable by large animals are shown. Fault scarps and the volcanic edifice would only be accessible to smaller and more agile animals. **c** Correction displacement field (CDF) for caldera collapse. (d) CDF for faulting. Only the largest effects are visible. Images are based on SRTM v4.1 data. Maps created by GK using Adobe Illustrator CS 5.1, and Global Mapper 9.4.0. Landscape reconstruction and fault mechanisms are calculated with Almond 7.05 software (www.ipgp.jussieu.fr/ king) based on the program developed by Okada (1982)

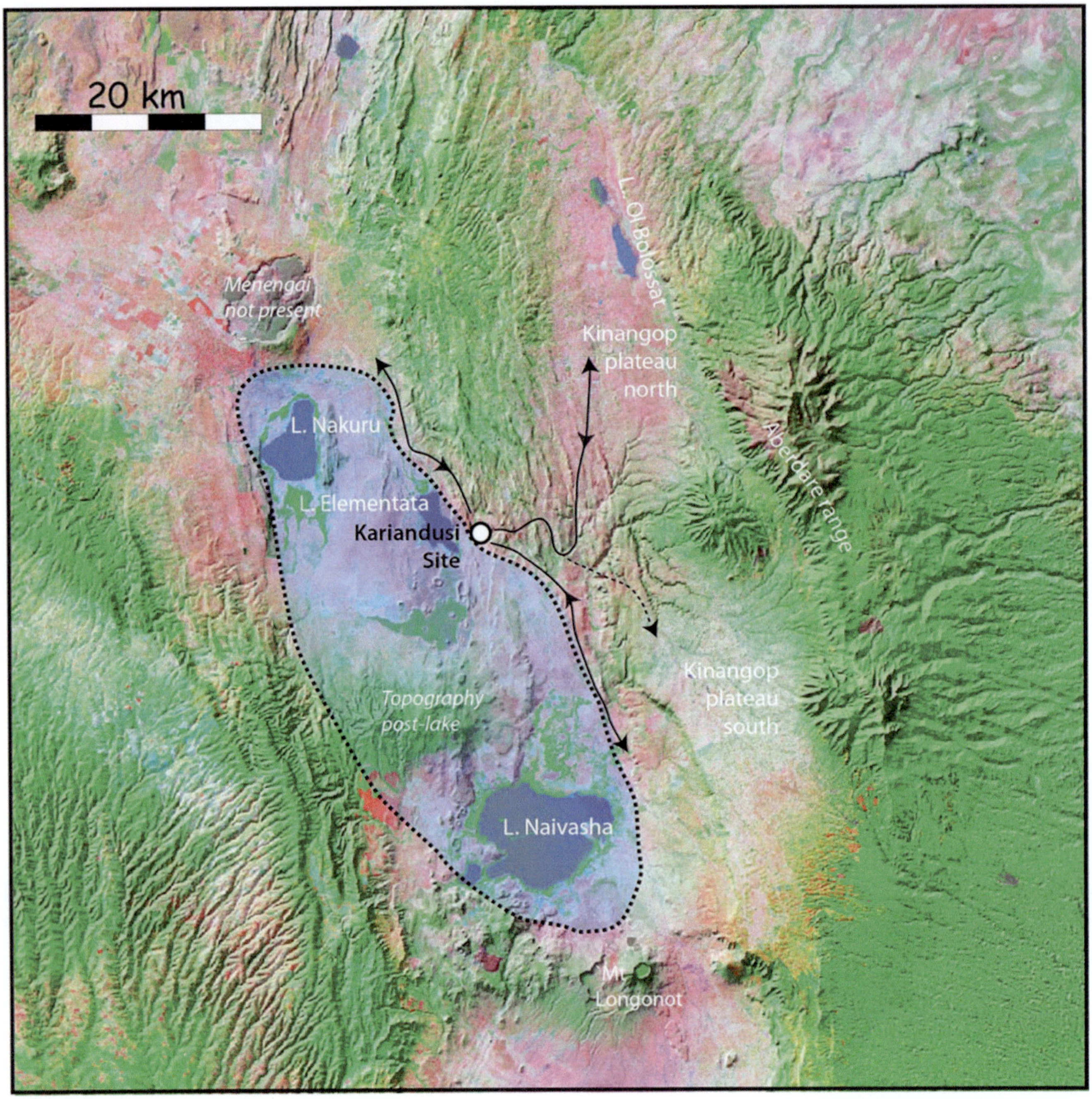

Fig. 5 Map of the Kariandusi region, showing the location of the site, major topographic features, proposed movements of large mammals and the outline of the expanded lake at the time when the site was in use. Maps created by SK and GK using Adobe Illustrator CS 5.1, Adobe Photoshop CS 5.1, MA Publisher 9.4.0, and Global Mapper 16

constraints that favoured ambush hunting disappeared and occupation ceased.

The site of Kariandusi in the central Kenyan Rift provides an informative comparison, showing similar features to Olorgesailie, except that it was used over a much shorter period. The material was originally deposited on a tributary of the Kariandusi River draining into the adjacent Lake Elmentaita, and comprises numerous Acheulean bifaces and other stone artefacts, with dates bracketed dated between 0.79 and 0.98 Ma (Leakey 1931; Gowlett and Crompton 1994; Shipton 2011). There are few faunal remains, consisting only of some equid teeth because of poor conditions for preservation of organic materials.

The distribution of edaphically-rich soils in the wider region is patchy, with the best concentrations on the rift flank to the east of the site. Here, an extensive upland basin is

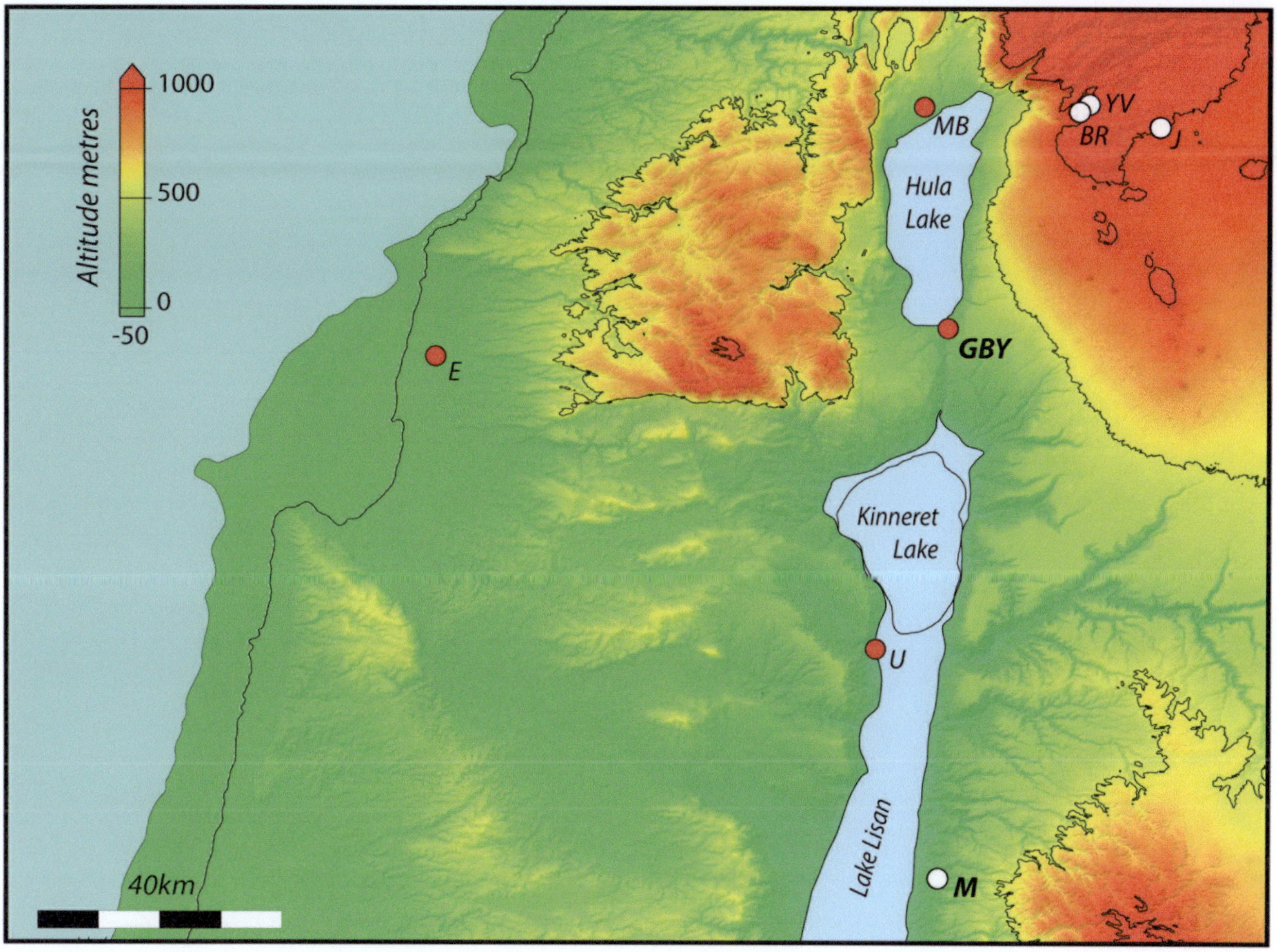

Fig. 6 The central Dead Sea Rift and adjacent regions showing relief, the coastline at lowered sea level and Lake Lisan at its maximum extent. A selection of major Lower Palaeolithic sites is shown. Red circles are Acheulean sites with faunal remains including elephant, white circles are Acheulean sites without fauna. After Devès et al. (2014). Sites referred to are as follows: J. Joubbata; YV. Yaafouri Valley; BR. Berekhat Ram; MB. Maayan Baruch; E, Evron; GBY, Gesher Banat Ya'aqov; U. Ubeidiya; M. Mashari'a (Ronen et al. 1980; Goren 1981; Goren-Inbar 1985; Tchernov 1988; Bar-Yosef and Goren-Inbar 1993; Tchernov et al. 1994; Macumber and Edwards 1997; Goren-Inbar et al. 2000; Ronen 2003). After Devès et al. (2014)

enclosed by steep faults and easily accessible to herds of large animals only from the rift floor to the west and northwest (Fig. 5). At the time when the site was occupied, much larger areas of the rift floor were submerged under an enlarged freshwater lake, and the lakeshore was at a higher level, forming a barrier to animal movements from west to east except through a narrow corridor close to Kariandusi (Kübler et al. 2015). As at Olorgesailie, freshwater, raw material for making stone artefacts and good viewpoints were all locally available throughout the Pleistocene. However, the Kariandusi site appears to have been used only during the period when high lake levels gave the location strategic advantage in relation to topographic barriers to ambush large mammals during their seasonal migrations. This reinforces the importance of complex faulted terrain and topographic barriers as a determinant of Acheulean site location and use.

7 Southern Levant

The Jordanian or Dead Sea Rift, comprising the system of valleys and lakes that extend northward from the Gulf of Aqaba on the Red Sea through to the Lebanon and Syria, is not strictly speaking a rift in the same sense as the East African Rift. It is, rather, a strike-slip structure resulting from the northward rotation of the Arabian Plate away from Africa. Nevertheless, it has many similar geological and topographic features, including high levels of earthquake activity and faulting, volcanic activity with extensive basaltic lava flows, elongated valleys with lake-filled basins subject to variations in lake level, marked changes of elevation over relatively short distances, and generally-speaking a complex topography (Enzel and Bar-Yosef 2017).

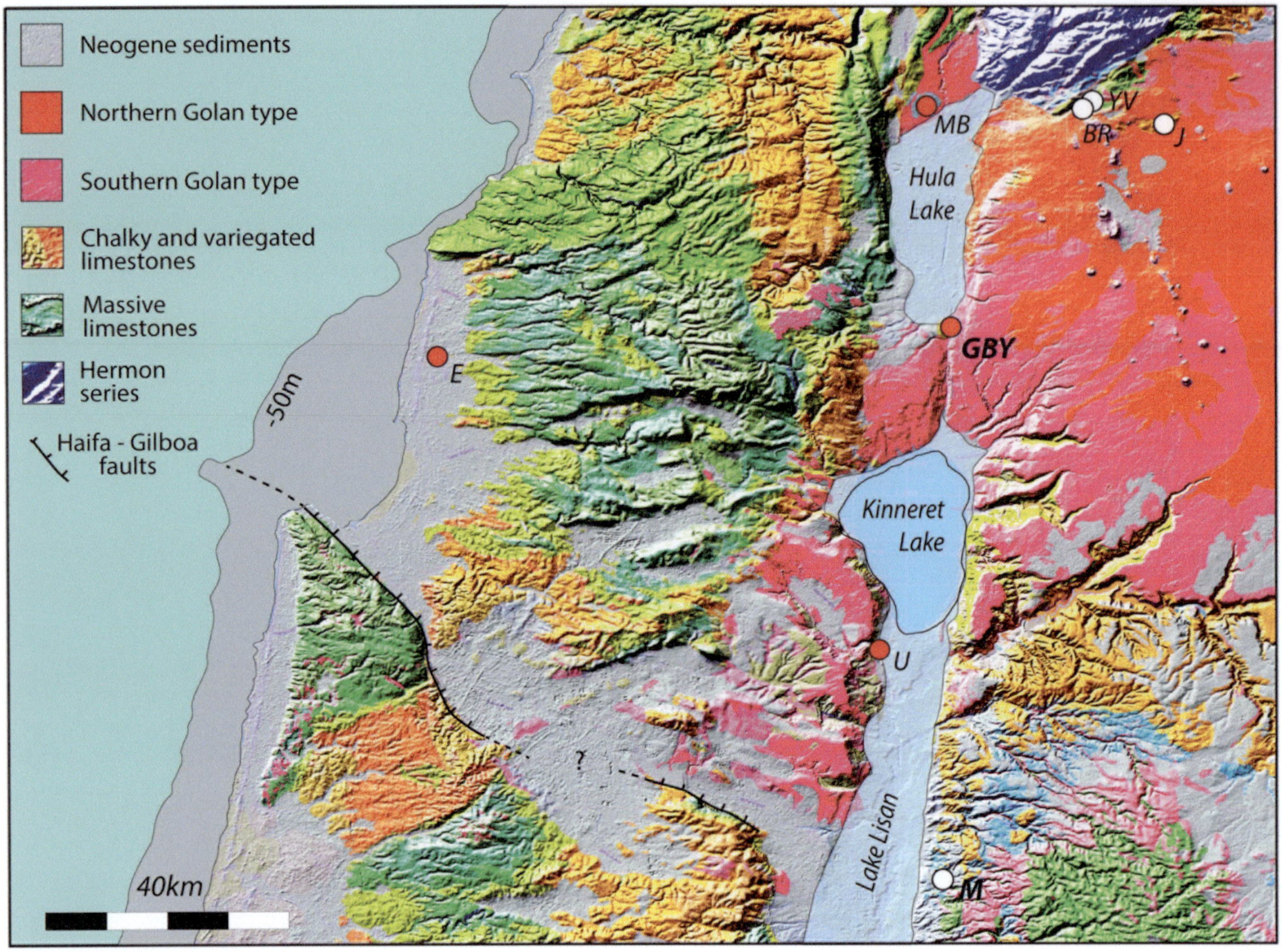

Fig. 7 Simplified geology plotted on a relief map of the area shown in Fig. 6 and used to identify areas of differing soil edaphic quality. Of particular importance are the basalt regions, which develop soils with high levels of soluble phosphate. Site information as in Fig. 6. After Devès et al. (2014)

The long history of archaeological investigation has resulted in many sites of Lower and Middle Palaeolithic age (Fig. 6; Enzel and Bar-Yosef 2017). The Galilee region of Israel, centred around and west of the Sea of Galilee, is particularly rich in sites. Possible reasons for this are the complexity of the landscape with numerous cliffs and deep, steep-sided valleys that provided human groups with security and tactical access to large mammals. The geology of the region is dominated by limestone and basalt, which can offer attractive environments for herbivores including good edaphic properties. These bedrock-types can also erode to form surfaces with small-scale roughness (known as angry karst on limestone). And these rougher surfaces provide significant impediments to animal movement and opportunities for human hunters.

Devès et al. (2014) considered topographic and edaphic constraints on access and how the region could have been exploited by large animals (e.g., elephant, rhinoceros, aurochs) in the Lower Palaeolithic (Fig. 6). The altitude can reach 1000 m in the Golan, and other regions are above 500 m, resulting in substantial differences in temperature between summer and winter which promote seasonal movements of large herds of animals between upland and lowland pastures. Many areas would have been inaccessible to large mammals such as elephants because of steep slopes and complex topography. These features provide major constraints on seasonal animal movements. The geology includes sedimentary rocks typical of many regions around the Mediterranean plus basalt flows, some of which have been active in historic times (Fig. 7). An earlier study in Greece (Sturdy and Webley 1988) has characterized soils on similar geology and provides a guide to edaphics and water retentiveness in this region (Fig. 8). While some of the sedimentary rocks provide edaphically adequate soils, water retention can be poor. The best soils, edaphically speaking, are associated with basalts on the Golan. The superiority of

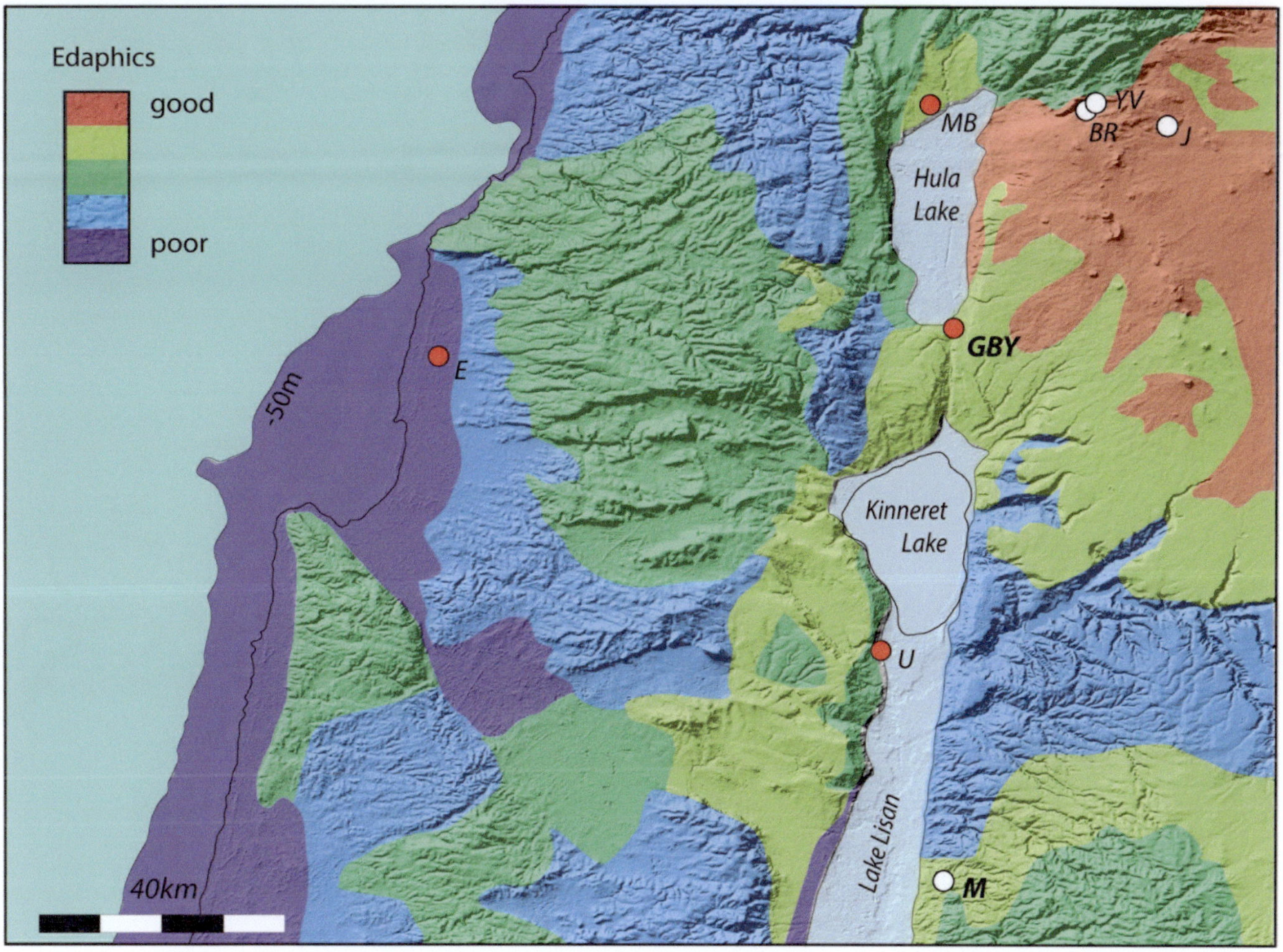

Fig. 8 Interpreted edaphic categories based on geology and plotted on a relief map. Site information as in Fig. 6. After Devès et al. (2014)

animals raised on these soils has been known since biblical times and results from high levels of soluble phosphate (Devès et al. 2014).

Mapping of these different features makes it possible to reconstruct the likely annual movements of large animals (Fig. 9). Winter graze and browse could be found near the coast but in general would lack important trace nutrients. The Golan would have attracted animals in spring and summer. Here, they they would have had their young in a region with the best soils to provide important nutrients for growth. In late summer and autumn, they would have moved to lower altitudes to take advantage of a region of high water retentiveness.

One of the major sites of the Lower Palaeolithic in the region is the Acheulean site of Gesher Banot Ya'acov (GBY) with thousands of stone artefacts including Acheulean bifaces, smaller flakes designed for hafting, and evidence for the use of fire (Goren-Inbar et al. 2000; Rabinovich et al. 2012; Alperson-Afil and Goren-Inbar 2016). The site extends over several kilometres with a deep sequence of sediments and a long series of archaeological layers indicating repeated used over a period of about 150,000 years between about 0.7 and 0.85 Ma. The stone-tool assemblage shows similarities with African sites, suggesting a new wave of human expansion out of Africa at about this time. The site is located near the edge of the palaeo-Hula Lake, and part of it remains waterlogged, with preservation of organic remains including wood, bark, fruits and seeds. Animals exploited include elephants, hippo, rhino, gazelle, horse and bovids, with evidence of carcase butchery. Subsistence also included plant foods and fish (Rabinovich et al. 2012; Melamed et al. 2016).

At the time of its occupation, the site was strategically located to intercept large mammals moving through one of the few available corridors for east–west movement across the Jordan valley (Fig. 10). Immediately to the north were the margins of the expanded Hula Lake. To the south was the palaeo-Lake Lisan, with an intervening area dissected by

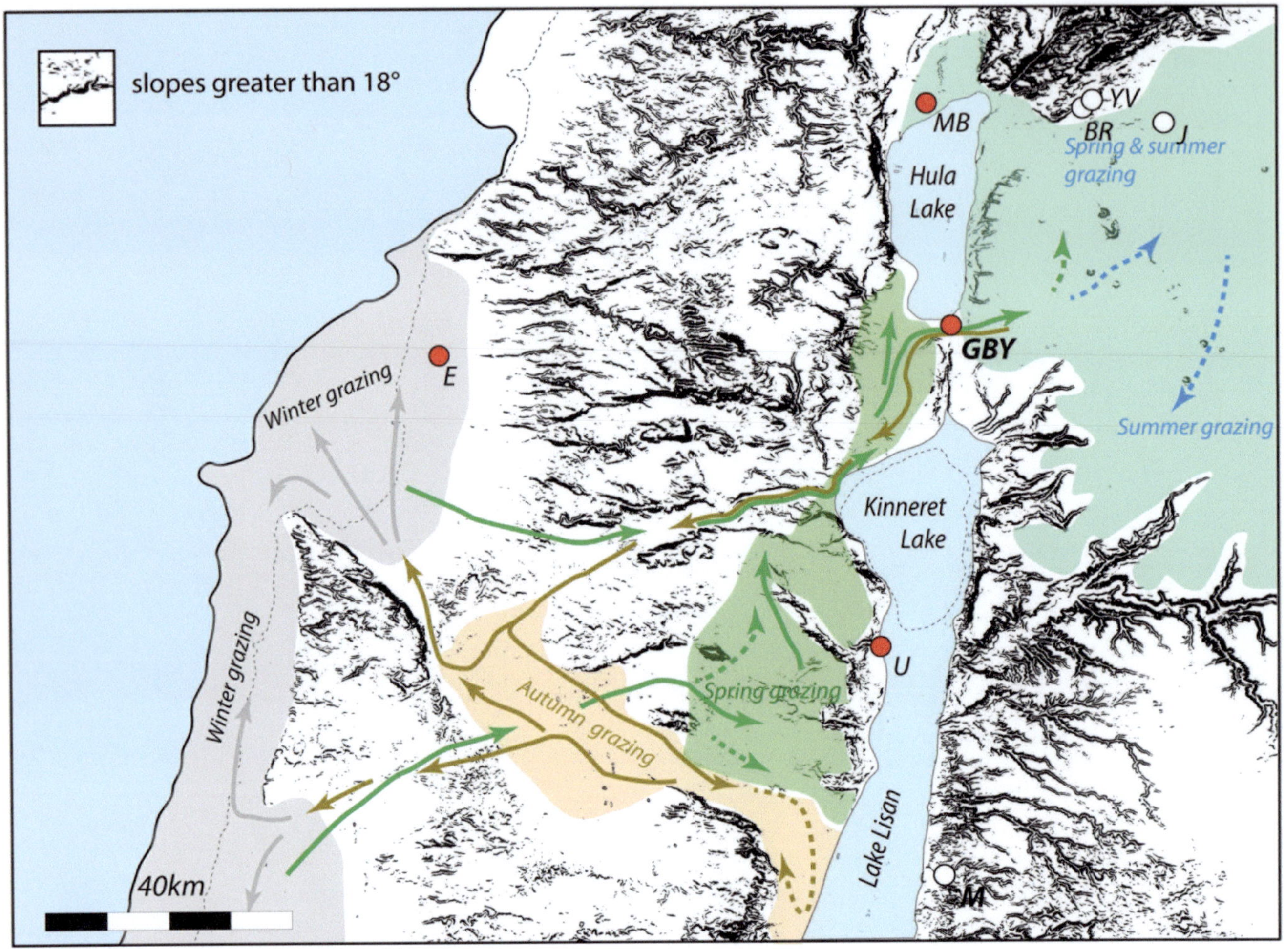

Fig. 9 Reconstructed seasonal movements and grazing areas of large mammal herbivores. The map also shows slopes greater than 18°, which constrain areas accessible to large mammals and routes for seasonal movement. Water retentive soils play an important role on the inner edge of the coastal plain and contribute to the identification of favoured spring, summer and autumn areas. Site information as in Fig. 6. After Devès et al. (2014)

rough topography and steep slopes. Like the African sites already discussed, the location of the site has many advantageous features including proximity to freshwater, volcanic material for artefact manufacture, and, in the GBY case, additional food supplies of non-migratory animals such as wild boar, plants and fish. However, given that the large herbivores, particularly elephant and fallow deer, were the major sources of the subsistence economy, the location of the site on one of the few crossing points from west to east, combined with the tactical opportunities for bringing down large prey, including miring along lake edges, must be a prime factor in the importance of this site (Devès et al. 2014, p. 152–3). By the end of the Lower Palaeolithic, large animals such as elephants had disappeared, the distribution of archaeological sites changed, and evidence of human occupation in regions inaccessible to large animals becomes important.

8 Southwest Saudi Arabia

The western escarpment of the Arabian Peninsula is the result of uplift associated with rifting and seafloor spreading of the Red Sea since at least the Miocene (Jado and Zötl 1984; Coleman et al. 1983; Purser and Bosence 1998; Bonatti et al. 2015; Bosworth and Stockli 2016). In contrast to the other areas already discussed, there is little earthquake activity or evidence of fault scarps on the present-day land surface, apart from localised earthquakes in the vicinity of volcanoes. This is because the active rift margin where major earthquake activity and faulting take place is in the central axial trough of the Red Sea over 2000 m below present sea level. The present terrestrial landscape is on the uplifted rift flank and the major features of relevance to the present discussion are the numerous volcanic cones and extensive

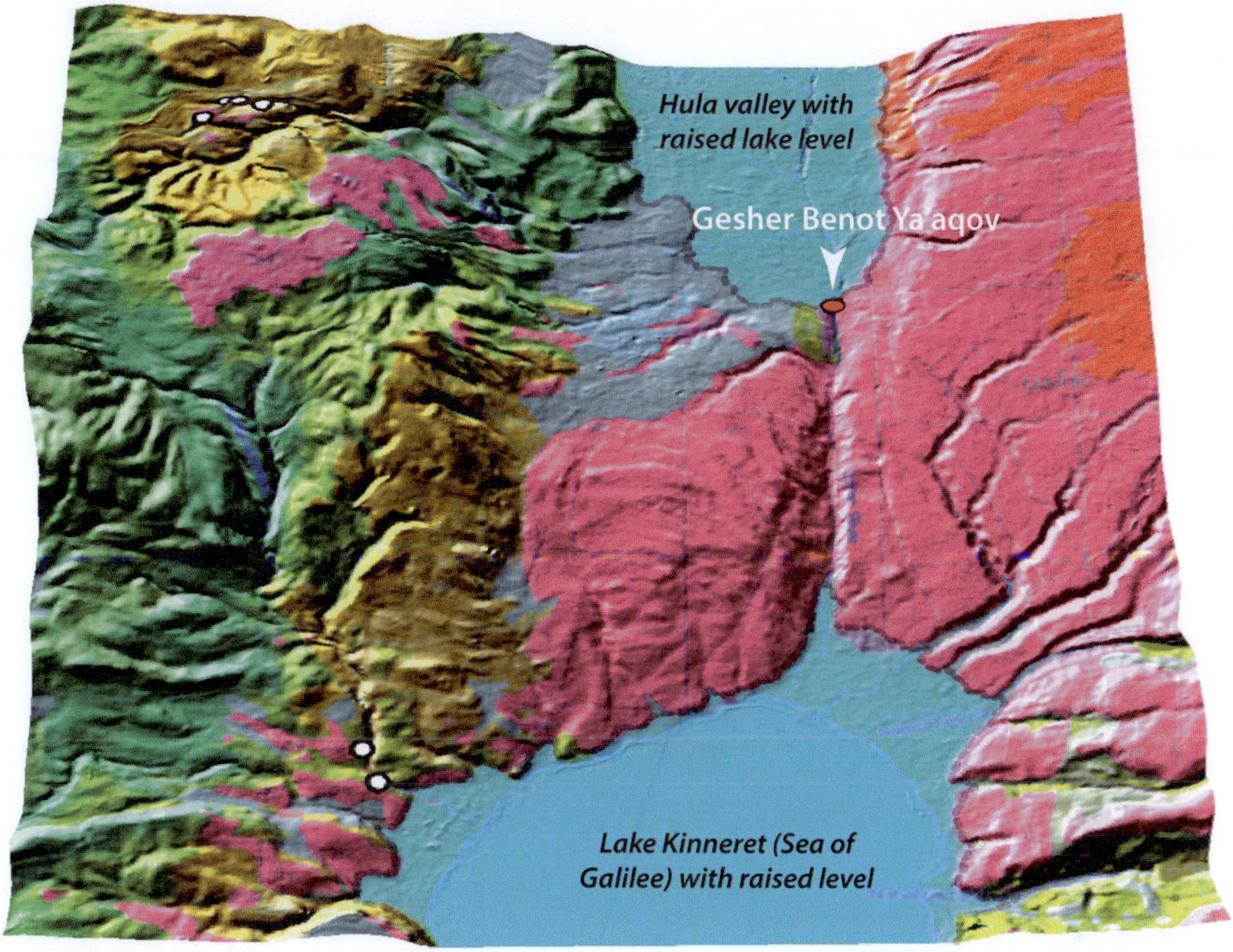

Fig. 10 Oblique 3D view of the region around Gesher Banat Ya'aqov in relation to geology and topography at the time of its occupation. The image is based on etm + legacy data. Topography from SRTM v4.1 data with a vertical exaggeration of ~8. Colour coding of geology as in Fig. 7. Note the restricted corridor between the expanded Lake Hula to the north and the steep-sided Upper Jordan valley to the south. This forms a natural ambush for trapping large mammals moving between the grazing lands west of the Rift valley and the Golan Heights. After Devès et al. (2014)

lava flows that form a belt of volcanic activity extending in a broadly north–south orientation. Dating and interpretation of these volcanic deposits are ongoing but substantial volcanic activity took place during the Pleistocene, extending in some regions into the Holocene period.

Also, in contrast to the other case studies, the history of archaeological investigation is much shorter. Major surveys during the 1980s identified numerous surface finds of Palaeolithic material over large areas, but intensive investigation using modern survey and excavation methods has taken place only very recently (see Petraglia et al., this volume; Sinclair et al., this volume). Very few sites have so far been found with stratified remains that can be independently dated or with preservation of animal bone. An added complication in coastal regions is that thousands of square kilometres of new territory were made available during periods of lower sea level. Investigation of this now-submerged landscape has only just begun, but it is clear that extensive tectonic activity associated with salt deposits has created a rapidly evolving landscape of complex topography with hills, fault-bounded depressions, springs and lake basins likely to have been attractive to the earliest human populations of the region (Bailey and King 2011; Dabbagh et al. 1984; Devès et al. 2011, 2013; Bailey et al. 2015, and this volume; Sakellariou et al., this volume; Momber et al., this volume).

Surveys in the Jizan and Asir provinces over the past 5 years have identified over 100 locations with Palaeolithic artefacts (Fig. 11). These have yielded some 6000 stone artefacts, mostly of Lower or Middle Palaeolithic type (Devès et al. 2012; Inglis et al. 2013, 2014a, 2014b, 2015, 2017; Sinclair et al., this volume). Most are surface finds with small numbers of artefacts, reflecting the heavily deflated nature of

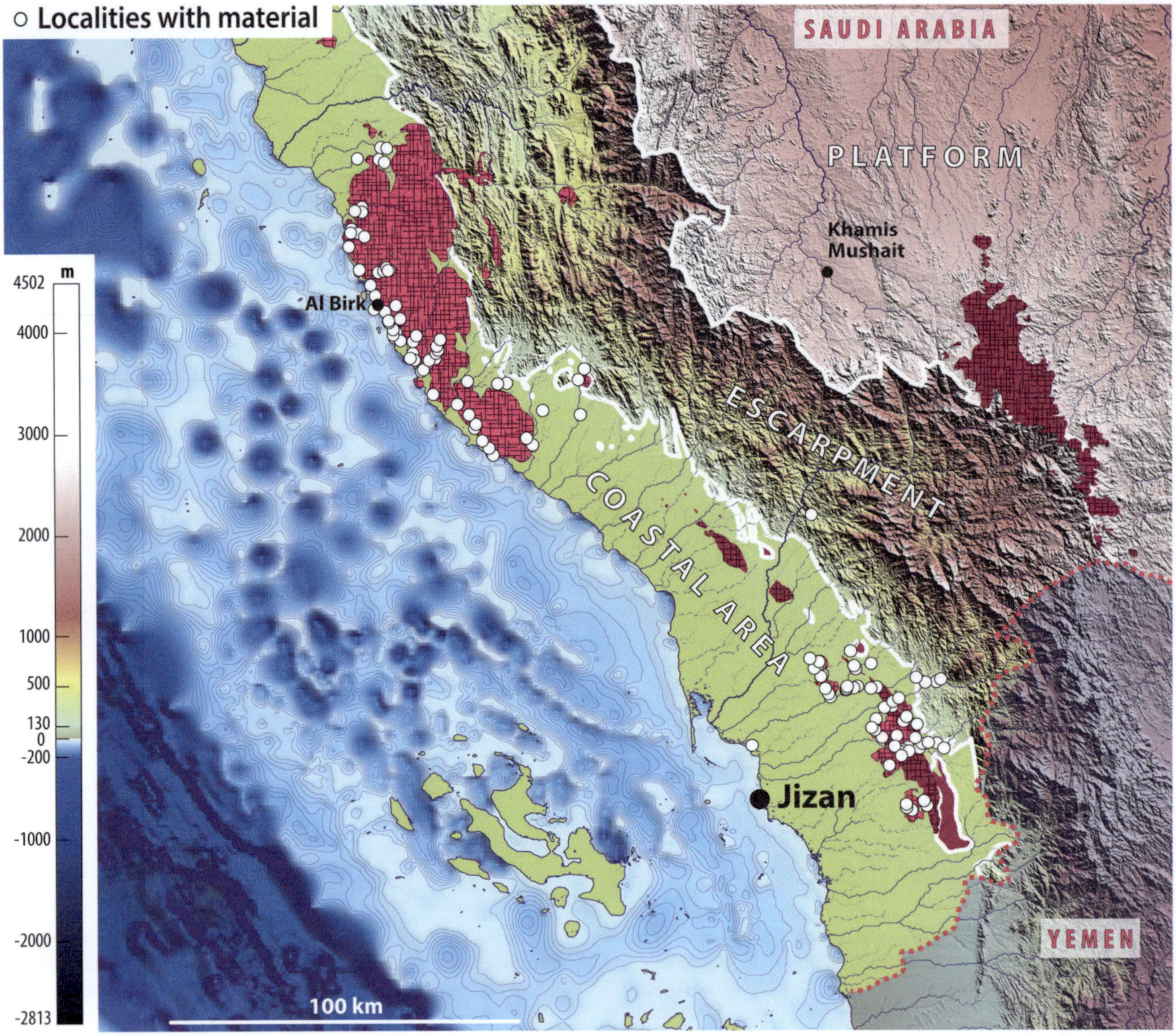

Fig. 11 The Asir and Jizan provinces of Southwest Saudi Arabia showing major features of topography and geology, and the distribution of localities with Palaeolithic artefacts. Note the amount of new land exposed at lowest sea level (light blue). The dark blue circles are deep depressions formed formed by withdrawal of evaporites. Topographic data from SRTM41 and bathymetry from SRTM30PLUS. Image prepared using Adobe Illustrator and MaPublisher by Maud Devès

the land surface, but some have been found in exposures of stratified deposits. Preserved faunal remains have not yet been discovered, and known faunal assemblages of Pleistocene age are rare in the Arabian Peninsula. By analogy with discoveries elsewhere, large mammals present likely included elephant, equids, and bovids, particularly *Oryx* (Thomas et al. 1998; Stimpson et al. 2016). Surveyed localities are mostly distributed on the inner edge of the coastal plain and many are associated with Quaternary volcanic deposits. The distribution is likely biased by the fact that sites that might once have existed on the coastal plain have been buried under more recent alluvial and aeolian deposits, and those in the more mountainous hinterland have been removed by erosion (Devès et al. 2013). The surviving artefacts may also owe their discovery in part to the easy identifiability, and general durability, of stone artefacts made from nearby basaltic lavas. However, this association is by no means the only factor, and other widely available raw materials were used for making artefacts including metamorphic and sedimentary rocks. Despite these cautions, the available artefact distribution provides some insights into the relative attractiveness of different landscape settings.

By far the largest concentration of artefacts occurs at Wadi Dabsa, where a concentration of over 3000 artefacts,

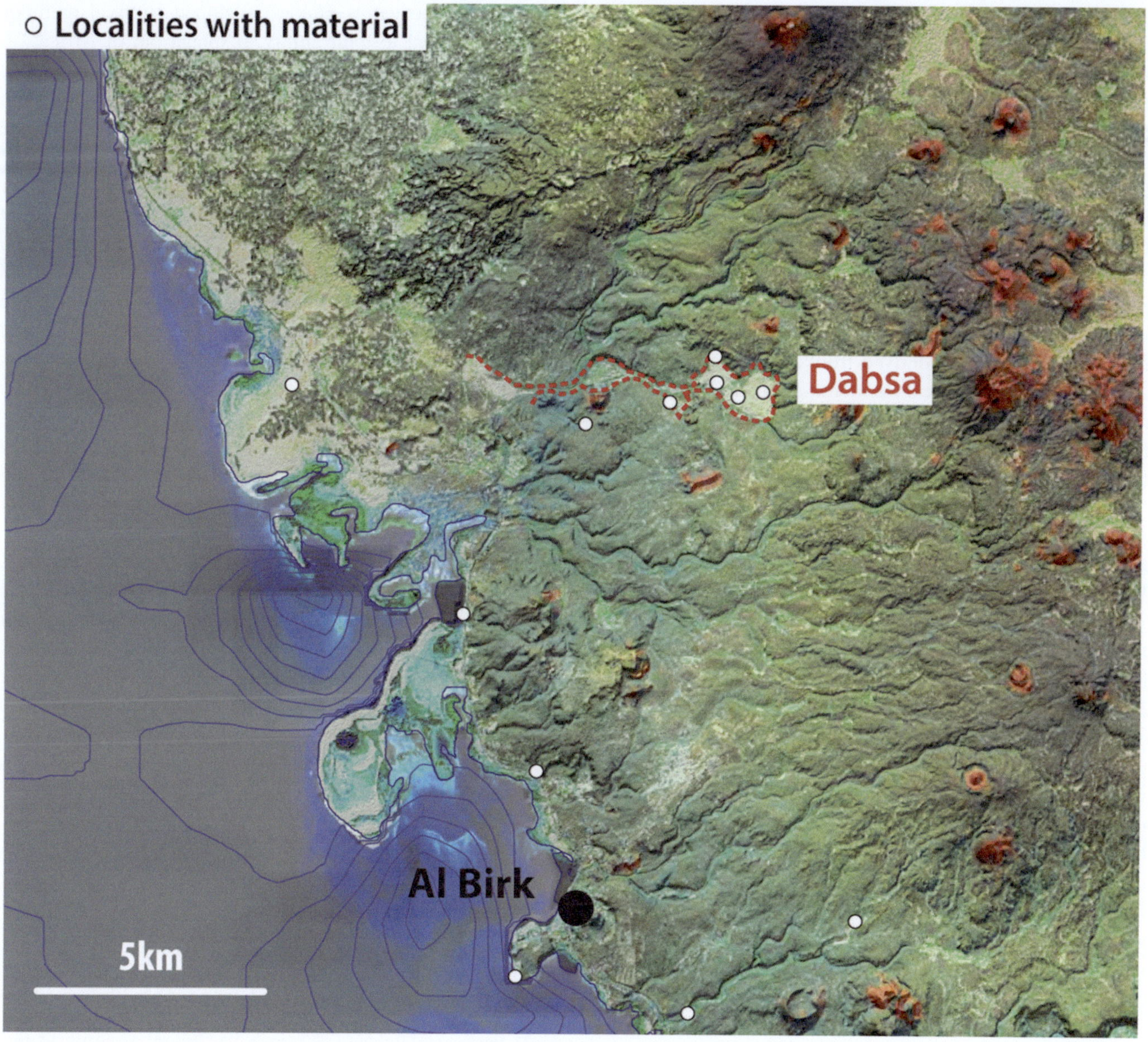

Fig. 12 The region around the site of Wadi Dabsa, showing its relationship to major topographical features and geology. Topographic data are from TanDEM-X. Bathymetry is from SRTM30PLUS with bathymetric contours at 10 m intervals. Topographic information combines slope angles superimposed on elevation with a vertical exaggeration of 4:1. This in its turn is superimposed on a LANDSAT image. Features in orange are volcanic cones and the geology deeply incised by rivers is predominantly basaltic lavas. The tufa in the Wadi Dabsa basin shows up as a light-coloured patch. The location of Palaeolithic localities are shown as white circles. The red dotted line shows the approximate line of barriers formed by basaltic lava flows that would have provided significant constraints on the movement of large mammals, and enclose a series of basins that open out to the west onto the flatter terrain of the coastal plain, which would have extended further out from the present coastline during periods of lower sea level (see also Fig. 11). Image prepared by Maud Devès

mostly of Lower and Middle Palaeolithic type, has been found in an enclosed basin (Inglis et al. 2015, 2017; Foulds et al. 2017; Sinclair et al., this volume). The site occurs in the Harrat al Birk volcanic region, where volcanic cinder cones and extensive basalt flows dissected by stream channels create an extensive and characteristically rough and complex topography (Fig. 12). The most extensive areas of grazing for large animals and large herds would have been on the coastal plain. Access to alternative seasonal grazing inland and at higher altitude is heavily constrained by volcanic lava flows and boulder fields.

The Dabsa site is unusual in its location in the centre of a large area of tufa extending for over 1 square kilometre and almost completely enclosed by lava flows that are difficult to cross. It is clear from stratigraphic observations that the tufa postdates the surrounding lava flows and that at least some of the artefacts were deposited at or before the time of tufa formation. The tufa indicates extensive conditions of water flow and water retention, creating conditions that are likely to have been especially attractive to grazing animals, and particularly so during dry seasons or during semi-arid climate episodes. This forms one of a series of smaller, partially enclosed basins that could have channelled animal movements inland toward attractive conditions, culminating in a virtual cul-de-sac at Dabsa.

Stratigraphic, geochronological, archaeological and palaeoenvironmental analysis is ongoing, and edaphic information is not presently available, but we hypothesise that alluvial sediments derived from the surrounding basalt are likely to have concentrated the best available mineral nutrients. As in our other examples, raw materials for artefact manufacture, water supplies and topographic constraints that facilitated easy access to large mammals are key factors in the choice of site location.

9 Discussion and Conclusions

By common consensus Africa is considered to be the original evolutionary centre of the genus *Homo* (*Homo habilis, H. ergaster, H. erectus*), and members of this genus first expanded out of Africa by about 1.8 Ma. We suggest that the use by hominins of complex and dynamic landscapes was critical to allowing our ancestors to develop and expand their range. We further propose that climbing and scrambling to exploit the tectonically active and topographically complex environments of Africa promoted the evolution of the *Homo* body form. Today we are better than most other animals at exploiting complex topography to our strategic advantage. Historically, minority groups have commonly retreated to the safety of mountainous regions. The Druze and Christians of Mount Lebanon and the Shia in Iraq retreated from the reach of Ottoman cavalry, and the English found the Welsh in their mountain retreats difficult to subdue. In these cases, escaping from cavalry was a motivation, but in more recent conflicts these advantages apply equally well to escaping from tanks and other mechanical adjuncts to war.

There is agreement that the large brain of *Homo* required a reliable supply of fatty protein. It is commonly thought that this must have been supplied by scavenging the kills of other predators and that systematic hunting did not appear until around 400 ka, as documented by the discovery of wooden spears at the Schöningen site in Germany (Thieme 1997). The analysis of the Olorgesailie site in southern Kenya suggests that at 1 Ma or earlier, *Homo* was using ambush hunting to trap large and dangerous animals. Analysis of sites of similar age in other regions provides additional support for this hypothesis and for the view that our earliest ancestors of the genus *Homo* developed survival methods that were critical to their ability to spread out of Africa.

Finally, we emphasise that the analysis of soil edaphics in the Kenyan Rift, based on large numbers of measurements in the field (Kübler et al. 2015, 2016) has demonstrated a close relationship between the nutrient properties of the soils and the mineral constituents of the underlying regolith, whether that is the underlying bedrock, or underlying sediments that have acquired their mineral properties by erosion from bedrocks elsewhere. Moreover, this work has highlighted areas that are deficient in critical nutrients today. In the Kenya Rift, nutrient related problems of human and animal health are common. Examples are severe cobalt deficiency in the Lake Nakuru region leading to starvation of ruminants (locally known as Nakuruitis) or calcium deficiency combined with excessive fluoride levels causing dental and skeletal deformations in humans and wildlife (Maskall and Thornton 1996). Our analysis shows that in a region of variable bedrock geology, the edaphic properties of the soils are highly variable. Extrapolation from bedrock geology, supported by mineralogical analysis of soil samples in the field, represents a method with considerable potential for application elsewhere, and not only to studies of ancient land use but also to the present-day, with major implications for animal and human health.

Acknowledgements The research on which this chapter is based was funded by the European Research Council through ERC Advanced Grant 269586 DISPERSE, under the 'Ideas' Specific Programme of FP7 to GB and GCPK in 2011. Fieldwork in East Africa was supported by the National Museums of Kenya, the Kenya Agricultural and Livestock Organisation, the Kenyan National Commission for Science, Technology and Innovation, and the British Institute in Eastern Africa. In Saudi Arabia, we thank HRH Prince Sultan bin Salman bin Abdul Aziz, President of the Saudi Commission for Tourism and Antiquities (SCTH), other members of SCTH, particularly Professor Ali Al-Ghabban and Dr. Abdullah Al Saud, and Dr. Abdullah Alsharekh of King Saud University, for their support. We also acknowledge additional funding to Robyn Inglis from the British Academy (Arthur Reckitt Fund), the Gerald Averay Wainwright Fund for Near Eastern Archaeology at the University of Oxford, the British Foundation for the Study of Arabia, and the EU Horizon 2020 research and innovation programme under the Marie Skłodowska-Curie grant agreement No. 660343, "SURFACE: Human-Landscape-Interactions and Global Dispersals: The Surface Record of Palaeolithic Arabia". The text has also benefited from the comments of three anonymous reviewers. This is DISPERSE contribution number 46 and IPGP Publication No. 3922.

References

Alperson-Fil N, Goren-Inbar N (2016) Acheulian hafting: Proximal modification of small flint flakes at Gesher Benot Ya'aqov, Israel. Quatern Int 411(Part B):34–43

Bailey GN, King GCP (2011) Dynamic landscapes and human dispersal patterns: Tectonics, coastlines, and the reconstruction of human habitats. Quatern Sci Rev 30:1533–1553

Bailey GN, Reynolds SC, King GCP (2011) Landscapes of human evolution: Models and methods of tectonic geomorphology and the reconstruction of hominin landscapes. J Human Evolution 60:257–280. https://doi.org/10.1016/jjhevol.2010.01.004

Bailey GN, Devès MH, Inglis RH, Meredith-Williams MG, Momber G, Sakellariou D, Sinclair A, Rousakis G, Al Ghamdi S, Alsharekh A (2015) Blue Arabia: Palaeolithic and underwater survey in SW Saudi Arabia and the role of coasts in Pleistocene dispersals. Quatern Int 382:42–57. https://doi.org/10.1016/j.quaint.2015.01.002

Bailey GN, Meredith-Williams M, Alsharekh A, Hausmann N (this volume) The Archaeology of Pleistocene Coastal Environments and Human Dispersals in the Red Sea: Insights from the Farasan Islands

Bar-Yosef O, Goren-Inbar N (1993) The lithic assemblages of 'Ubeidiya: A Lower Palaeolithic site in the Jordan Valley. Qedem. Monograph 14. Institute of Archaeology, Hebrew University of Jerusalem, Israel

Behrensmeyer AK, Potts R, Deino A, Ditchfield P (2002) Olorgesailie, Kenya: A million years in the life of a rift basin. Soc Sediment Geol Spec Publ No. 73:97–106

Blumenschine RJ, Cavallo JA, Caplado SD (1994) Competition for carcasses and early hominid behavioural ecology: A case study and conceptual framework. J Human Evolution 27:197–213

Bobe R (2006) The evolution of arid ecosystems in Eastern Africa. J Arid Environ 66:564–584

Bobe R, Behrensmeyer AK (2004) The expansion of grassland ecosystems in Africa in relation to mammalian evolution and the origin of the genus Homo. Palaeogeog Palaeoclimat Palaeoecol 207:399–420

Bobe R, Eck GG (2001) Responses of African bovids to Pliocene climatic change. Paleobiol Mem 27(2):1–48

Bonatti E, Cipriani A, Lupi L (2015) The Red Sea: birth of an ocean. In: Rasul NMA, Stewart ICF (eds) The Red Sea: The Formation, Morphology, Oceanography and Environment of a Young Ocean Basin. Springer Earth System Sciences, Berlin Heidelberg, pp 29–44

Bosworth W, Stockli D (2016) Early magmatism in the greater Red Sea rift: Timing and significance. Can J Earth Sci 53(11):1158–1176

Bramble DM, Lieberman DE (2004) Endurance running and the evolution of Homo. Nature 432:345–352

Burrows GE, Brodie B, Thedford TR (1979) Selenium supplementation for beef cattle in Kenya. Bull Animal Health Production in Africa 27:139–143

Cerling TE, Harris JM, MacFadden BJ, Leakey MG, Quade J, Eisenmann V, Ehleringer JR (1997) Global vegetation change through the Miocene/ Pliocene boundary. Nature 389:153–158

Coleman RG, Gregory RT, Brown GF (1983) Cenozoic volcanic rocks of Saudi Arabia. US Geol Survey Open-File Report 83–788. https://doi.org/10.3133/ofr83788

Corah L (1996) Trace mineral requirements of grazing cattle. Animal Feed Sci Technol 59(1–3):61–70

Dabbagh A, Emmermann R, Hötzl H, Jado AR, Lippolt HJ, Kollman W, Moser H, Rauert W, Zötl JG (1984) The development of Tihamat Asir during the Quaternary. In: Jado AR, Zötl JG (eds) Quaternary Period in Saudi Arabia, vol 2. Sedimentological, hydrogeological, hydrochemical, geomorphological, geochronological and climatological investigations in Western Saudi Arabia. Springer-Verlag, Vienna, pp 150–173

DeMenocal PB (2004) African climate change and faunal evolution during the Pliocene-Pleistocene. Earth Planet Sci Lett 220:3–24

Devès M, King GCP, Klinger Y, Amotz A (2011) Localised and distributed deformation in the lithosphere: The example of the Dead Sea valleys. Earth Planet Sci Lett 308:172–184. https://doi.org/10.1016/j.epsl.2011.05.044

Devès M, Inglis R, Meredith-Williams M, Alsharekh A, Al Ghamdi S, Bailey G (2012) Preliminary report of reconnaissance fieldwork in southwest Saudi Arabia, May–June, 2012. http://www.disperse-project.org/field-reports

Devès M, Inglis R, Meredith-Williams M, Al Ghamdi S, Alsharekh AM, Bailey GN (2013) Palaeolithic survey in Southwest Saudi Arabia: methodology and preliminary results. Adumatu 27: 7–30

Devès M, Sturdy D, King GCP, Godet N, Bailey GN (2014) Hominin reactions to herbivore distribution in the Lower Palaeolithic of the southern Levant. Quatern Sci Rev 96:140–160. https://doi.org/10.1016/j.quascirev.2014.04.017.15

Devès MH, Reynolds S, King GCP, Kübler S, Sturdy D, Godet N (2015) Insights from earth sciences into human evolution studies: The examples of prehistoric landscape use in Africa and the Levant. Comptes Rendus Geosci 347(4):201–211

Domínguez-Rodrigo (2002) Hunting and scavenging by early humans: The state of the debate. J World Prehistory 16:1–54

Enzel Y, Bar-Yosef O (eds) (2017) Quaternary of the Levant: Environments, climate change, and humans. Cambridge University Press, Cambridge

Formigoni A, Fustini M, Archetti L, Emanuele S, Sniffen C, Biagi G (2011) Effects of an organic source of copper, manganese and zinc on dairy cattle productive performance, health status and fertility. Animal Feed Sci Technol 164:191–198

Foulds F, Shuttleworth A, Sinclair A, Inglis R, Alsharekh A, Al Ghamdi S, Bailey GN (2017) A large handaxe from Wadi Dabsa and early hominin adaptations within the Arabian Peninsula. Antiquity 91(359):1421–1434

Goren N (1981) The Lower Palaeolithic in Israel and adjacent countries. In: Cauvin J, Sanlaville P (eds) Préhistoire du Levant: Chronologie et organisation de l'espace depuis les origines jusqu'au VIe millénaire. Colloques internationaux du Centre national de la recherche scientifique, 598. Éditions du CNRS, Lyon, pp 193–205

Goren-Inbar N (1985) The lithic assemblages of Berekhat Ram Acheulian site. Golan Heights. Paléorient 11(1):7–28

Goren-Inbar N, Feibel CS, Verosub KL, Melamed Y, Kislev ME, Tchernov E, Saragusti I (2000) Pleistocene milestones on the Out-of-Africa corridor at Gesher Benot Ya'aqov. Israel. Science 289(5481):944–947

Gowlett JAJ, Crompton RH (1994) Kariandusi: Acheulean morphology and the question of allometry. African Archaeol Rev 12:3–42

Henkin Z, Noy-Meir I, Gutman M, Seligman NG (1995) The effect of phosphorus nutrition, shrub control and fire on the dynamics of Mediterranean batha vegetation in Galilee. Israel, FAO/ CIHEAM Network on Pastures and Fodder Crops, Avignon, France

Inglis RH, Sinclair AGM, Shuttleworth A, Alsharekh AM, Al Ghamdi S (2013) Preliminary report on 2013 fieldwork in southwest Saudi Arabia by the DISPERSE project: (2) Jizan and Asir provinces, February–March 2013. Report to the Saudi Commission for Tourism and Antiquities. http://www.disperse-project.org/field-reports; http://eprints.whiterose.ac.uk/80964/

Inglis RH, Sinclair A, Shuttleworth A, Alsharekh A, Al Ghamdi S, Devès M, Meredith-Williams MG, Bailey GN (2014a) Investigating the Palaeolithic Landscapes and Archaeology of the Jizan and Asir Regions, Southwest Saudi Arabia. Proc Seminar for Arabian Studies 44:193–212

Inglis RH, Sinclair AGM, Shuttleworth A, Meredith-Williams MG, Hausmann N, Budd W, Alsharekh A, Al Ghamdi S, Bailey GN (2014b) Preliminary Report on 2014 Fieldwork in Southwest Saudi Arabia by the DISPERSE project: (1) Jizan and Asir Provinces. Report to the Saudi Commission for Tourism and Antiquities. http://www.disperse-project.org/field-reports; http://eprints.whiterose.ac.uk/80965/

Inglis RH, Foulds F, Shuttleworth A, Alsharekh AM, Al Ghamdi S, Sinclair AG Bailey GN (2015) The Palaeolithic Occupation of the Harrat Al Birk: Preliminary Report on the 2015 Fieldwork in Asir

Province, Southwest Saudi Arabia. Report to the Saudi Commission for Tourism and Antiquities. http://www.disperse-project.org/field-reports; http://eprints.whiterose.ac.uk/84382/

Inglis RH Sinclair AGM, Alsharekh AM, Barfod DN, Chang HC, Fanning PC, Al Othaibi DT, Robson HK, Shuttleworth A, Stone A, Bailey GN (2017) Preliminary report on UK-Saudi 2017 fieldwork at Wadi Dabsa, Asir Province, Saudi Arabia. http://eprints.whiterose.ac.uk/119267/

Isaac GL (1977) Olorgesailie: Archaeological studies of a Middle Pleistocene lake basin in Kenya. University of Chicago Press, Chicago

Jado AR, Zötl JG (eds) (1984) Quaternary Period in Saudi Arabia Volume 2: Sedimentological, hydrogeological, hydrochemical, geomorphological, geochronological and climatological investigations in Western Saudi Arabia. Springer-Verlag, Vienna

Kadim IT, Mahgoub O, Al-Ajmi DS, Al-Maqbaly RS, Al-Saqri NM, Ritchie A (2003) An evaluation of the growth, carcass and meat quality characteristics of Omani goat breeds. Meat Sci 66(1):203–210

King GCP, Bailey GN (2006) Tectonics and human evolution. Antiquity 80:1–22

King GCP, Bailey GN, Reynolds SC (2010) Dynamic landscapes and human evolution. In: Sintubin M, Stewart IS, Niemi T, Altunel E (eds) Ancient earthquakes: Archaeoseismology along the Alpine-Himalayan seismic zone. Geol Soc America Special Paper 471, pp 1–19. https://doi.org/10.1130/2010.2471(01)

Kübler S, Owenga P, Reynolds SC, Rucina SM, King GCP (2015) Animal movements in the Kenya Rift and evidence for the earliest ambush hunting by hominins. Nature Sci Reports 5(14011):1–7. https://doi.org/10.1038/srep14011

Kübler S, Owenga P, Rucina S, Reynolds SJ, Bailey GN, King GCP (2016) Edaphic and topographic constraints on exploitation of the Central Kenya Rift by large mammals and early hominins. Open Quatern 2(5):1–18. https://doi.org/10.5334/oq.21

Leakey LSB (1931) The Stone Age cultures of Kenya Colony. Cambridge University Press, Cambridge

Macumber PG, Edwards PC (1997) Preliminary results from the Acheulean site of Mashari'a 1, and a new stratigraphc framework for the Lower Paleolithic of the east Jordan Valley. In: Kebel HGK, Kafafi Z, Rollefson GO (eds) The prehistory of Jordan II: perspectives from 1997. Studies in early Near Eastern production, subsistence and environment, vol 4. Ex Oriente, Berlin, pp 23–44

Maglio VJ, Cook HBS (eds) (1978) The evolution of African mammals. Harvard University Press, Cambridge, MA

Maskall J, Thornton I (1996) The distribution of trace and major elements in Kenyan soil profiles and implications for wildlife nutrition. In: Appleton JD, Fuge R, McCall GJH (eds) Environmental geochemistry and health. Geol Soc, London, Special Publ 113, pp 47–62. Ddoi:https://doi.org/10.1144/gsl.sp.1996.113.01.05

Melamed Y, Kislev ME, Geffen A, Lev-Yadun S, Goren-Inbar N (2016) The plant component of an Acheulian diet at Gesher Benot Ya'aqov. Israel. Proc National Acad Sci 113(51):14674–14679. https://doi.org/10.1073/pnas.1607872113

Momber G, Sakellariou D, Rousakis G, Bailey G (this volume) The multi-disciplinary search for underwater archaeology in the southern Red Sea

Murray MG (1995) Specific nutrient requirements and the migration of the wildebeest. In: Sinclair ARE, Arcese P (eds) Serengeti II—dynamics, management and conservation of an ecosystem. University of Chicago Press, Chicago, pp 231–256

Okada Y (1982) Internal deformation due to shear and tensile fault in a half-space. Bull Seismol Soc America 82:1018–1040

Owen-Smith N (1988) Ecological links between African savanna environments, climate change and early hominid evolution. In: Bromage TG, Schrenk F (eds) African biogeography, climate change, and human evolution. Oxford University Press, New York, pp 138–149

Petraglia MD, Breeze PS, Groucutt HS (this volume) Blue Arabia, green Arabia: examining human colonisation and dispersal models

Pickering TR (2013) Rough and tumble: Aggression, hunting, and human evolution. University of California Press, Berkeley and Los Angeles, CA

Potts R (1989) Olorgesailie: New excavations and findings in Early and Middle Pleistocene contexts, southern Kenya rift valley. J Human Evolution 18:477–484

Potts R, Behrensmeyer AK, Deino A, Ditchfield P, Clark J (2004) Small Mid-Pleistocene hominin associated with East African Acheulean technology. Science 305:75–78

Purser BH, Bosence DWJ (eds) (1998) Sedimentation and tectonics of rift basins: Red Sea—Gulf of Aden. Chapman and Hall, London

Rabinovich R, Gaudzinski-Windheuser S, Kindler L, Goren-Inbar N (2012) The Acheulian site of Gesher Benot Ya'aqov Volume III. Mammalian taphonomy. The assemblages of Layers V-5 and V-6. Springer, Dordrecht

Reynolds SC, Bailey GN, King GCP (2011) Landscapes and their relation to hominin habitats: Case studies from Australopithecus sites in eastern and southern Africa. J Human Evolution 60: 257–280

Ronen A (2003) Evron Quarry, western Galilee, Israel. In: Burdukiewicz JM, Ronen A (eds) Lower Palaeolithic small tools in Europe and the Levant. British Archaeological Reports International Series 1115. Archaeopress, Oxford, pp 113–120

Ronen A, Ohel MY, Lamdan M, Assaf A (1980) Acheulian artifacts from two trenches at Maayan Barukh, Upper Galilee, Israel. Isr Explor J 30:17–33

Ruess RW (1984) Nutrient movement and grazing: Experimental effects of clipping and nitrogen source on nutrient uptake in Kyllinga nervosa. Oikos 43(2):183–188

Sakellariou D, Rousakis G, Panagiotopoulos I, Morfis I, Bailey G (this volume) Geological structure and Late Quaternary geomorphological evolution of the Farasan Islands continental shelf, southern Red Sea, SW Saudi Arabia

Shipton C (2011) Taphonomy and behaviour at the Acheulean site of Kariandusi, Kenya. African Archaeol Rev 28:141–155. https://doi.org/10.1007/s10437-011-9089-1

Sinclair A, Inglis RH, Shuttleworth A, Foulds F, Alsharekh A (this volume) Landscape archaeology, Palaeolithic survey and coastal change along the southern Red Sea of Saudi Arabia

Skinner JD, Louw GN (1996) The Springbok Antidorcas marsupialis (Zimmermann, 1780). Transvaal Museum Monograph No. 10:1–49. Transvaal Museum, Pretoria

Stimpson CM, Lister A, Parton A, Clark-Balzan L, Breeze PS, Drake NA, Groucutt HS, Jennings R, Scerri EML, White TS, Zahir M, Duval M, Grün R, Al-Omari A, Al Murayyi KSM, Zalmout IS, Mufarreh YA, Memesh AM, Petraglia MD (2016) Middle Pleistocene vertebrate fossils from the Nefud Desert, Saudi Arabia: Implications for biogeography and palaeoecology. Quatern Sci Rev 143:13–36

Sturdy DA, Webley DP (1998) Palaeolithic geography: Or where are the deer? World Archaeology 19:262–280

Sturdy DA, Webley DP, Bailey GN (1997) The Palaeolithic geography of Epirus. In: Bailey GN (ed) Klithi: Palaeolithic settlement and Quaternary landscapes in Northwest Greece, vol 2. Klithi in its local and regional setting. McDonald Institute for Archaeological Research, Cambridge, pp 541–559

Tchernov E (1988) Les mammifères du Pleistocene inferieur de la Vallée du Jordain à Oubediyeh. Memoires et Travaux du Centre de Recherche Français de Jerusalem 5. Association Paléorient, Paris, pp 191–212

Tchernov E, Horwitz LK, Ronen A, Lister A (1994) The faunal remains from Evron Quarry in relation to other Lower Paleolithic hominid sites in the Southern Levant. Quatern Res 42(3):328–339

Thieme H (1997) Lower Palaeolithic hunting spears from Germany. Nature 385:807–810. https://doi.org/10.1038/385807a0

Thomas H, Geraads D, Janjou D, Vaslet D, Memesh A, Billiou D, Bocherens H, Dobigny G, Eisenmann V, Gayet M, de Lapparent de Broin F, Petter G, Halawani M (1998) First Pleistocene faunas from the Arabian Peninsula: An Nafud Desert, Saudi Arabia. Comptes Rendues de l'Acad des Sci Paris 326:145–152. https://doi.org/10.1016/s1251-8050(97)87459-9

Vrba ES (1982) Biostratigraphy and chronology, based particularly on Bovidae, of southern hominid-associated assemblages: Makapansgat, Sterkfontein, Taung, Kromdraai, Swartkrans; also Elandsfontein (Saldanha), Broken Hill (now Kabwe) and Cave of Hearths. In: de Lumey H, de Lumley MA (eds) Premier Congrès International de Paléontologie Humaine, Tome 2. CNRS, Nice, pp 707–752

Wheeler PE (1993) The influence of stature and body form on hominid energy and water budgets: A comparison of Australopithecus and early Homo physiques. J Human Evolution 24:13–28

Winder IC, King GCP, Devés M, Bailey GN (2013) Complex topography and human evolution: The missing link. Antiquity 87:333–349

Winder IC, King GCP, Devès MH, Bailey GN (2014) Human bipedalism and the importance of terrestriality: A reply to Thorpe et al. (2014). Antiquity 88:915–916

Wood B (2005) Human evolution. Oxford University Press, Oxford

Blue Arabia, Green Arabia: Examining Human Colonisation and Dispersal Models

Michael D. Petraglia, Paul S. Breeze, and Huw S. Groucutt

Abstract

The dispersal of hominins (our species and our closely related bipedal ancestors) out of Africa is a major topic in human evolutionary studies. As a geographic cross-road between continents, the Arabian Peninsula has a significant role to play in understanding the movement of hominin populations and the effect of climate change in shaping demographic history throughout the Pleistocene. However, the Palaeolithic evidence in Arabia has often been marginalized in prominent Out of Africa models, with the assumption that hominins would have avoided the hyper-arid desert belt, and utilized coastlines for their movement, especially along the Indian Ocean rim. Two interdisciplinary archaeological projects, named DISPERSE and PALAEODESERTS, have been conducted in Saudi Arabia in recent years in order to address dispersal models, and to improve our understanding of the hominin occupation history of the Red Sea coastline ("Blue Arabia") and the interior of the peninsula, especially during wet periods ("Green Arabia"). While acknowledging the importance of the Red Sea region as a potential zone for hominin occupation in the Pleistocene, we emphasize the crucial significance of the terrestrial environments for repeated hominin expansions during ameliorated periods in the Pleistocene, when a mosaic of ecosystems and plentiful rivers, wetlands and lakes were present.

M. D. Petraglia (✉) · H. S. Groucutt
Department of Archaeology, Max Planck Institute for the Science of Human History, 07745 Jena, Germany
e-mail: petraglia@shh.mpg.de

P. S. Breeze
Department of Geography, King's College, London, UK

H. S. Groucutt
School of Archaeology, University of Oxford, Oxford, UK

N. M. A. Rasul and I. C. F. Stewart (eds.), *Geological Setting, Palaeoenvironment and Archaeology of the Red Sea*,
https://doi.org/10.1007/978-3-319-99408-6_30

1 Introduction

One of the hottest topics in human evolutionary studies is the dispersal of hominins out of Africa (Templeton 2002; Klein 2008; Stewart and Stringer 2012; Groucutt et al. 2015a). Although traditionally marginalized, or even ignored in the palaeoanthropological literature, the Arabian Peninsula is a critical geographic landmass for understanding hominin dispersals through time, as the region constitutes half of southwest Asia, the key interface with the rest of Eurasia (Petraglia et al. 2015) (Fig. 1).

Generalised sketches about the pathways of hominin movements out of Africa have centred on the argument that humans moved along the coasts as they expanded across southern Asia. With respect to *Homo sapiens*, one prominent view is that littoral margins were used as coastally-adapted peoples rapidly expanded toward Sahul (e.g., Stringer 2000; Oppenheimer 2009; Mellars et al. 2013). In this view, terrestrial Arabia was avoided owing to its hyper-arid state some 60,000 years ago, when such movements were hypothesized to have occurred. Yet, at the same time, with respect to Arabia itself, archaeological field surveys had long shown that Palaeolithic sites were abundant, with most sites situated inland, and with some located along the Red Sea margin (see Groucutt and Petraglia 2012 for a review). The presence of Palaeolithic sites across Arabia was viewed as a consequence of pluvial phases, which have been demonstrated by speleothems and lake records, and are evident in climate models (Rosenberg et al. 2011, 2013; Jennings et al. 2015a; Timmermann and Friedrich 2016). Based on Palaeolithic archaeological information in Arabia, two corridors were envisioned for out of Africa expansions, that is, the Bab al Mandab crossing, linking the Horn of Africa with the southwestern portion of Arabia, and the Sinai, which linked northeast Africa with the Levant and northern Arabia (see Petraglia 2003; Petraglia and Alsharekh 2003) (Fig. 1).

In the absence of verifiable Upper Palaeolithic assemblages in Arabia, an argument was made that *Homo sapiens*

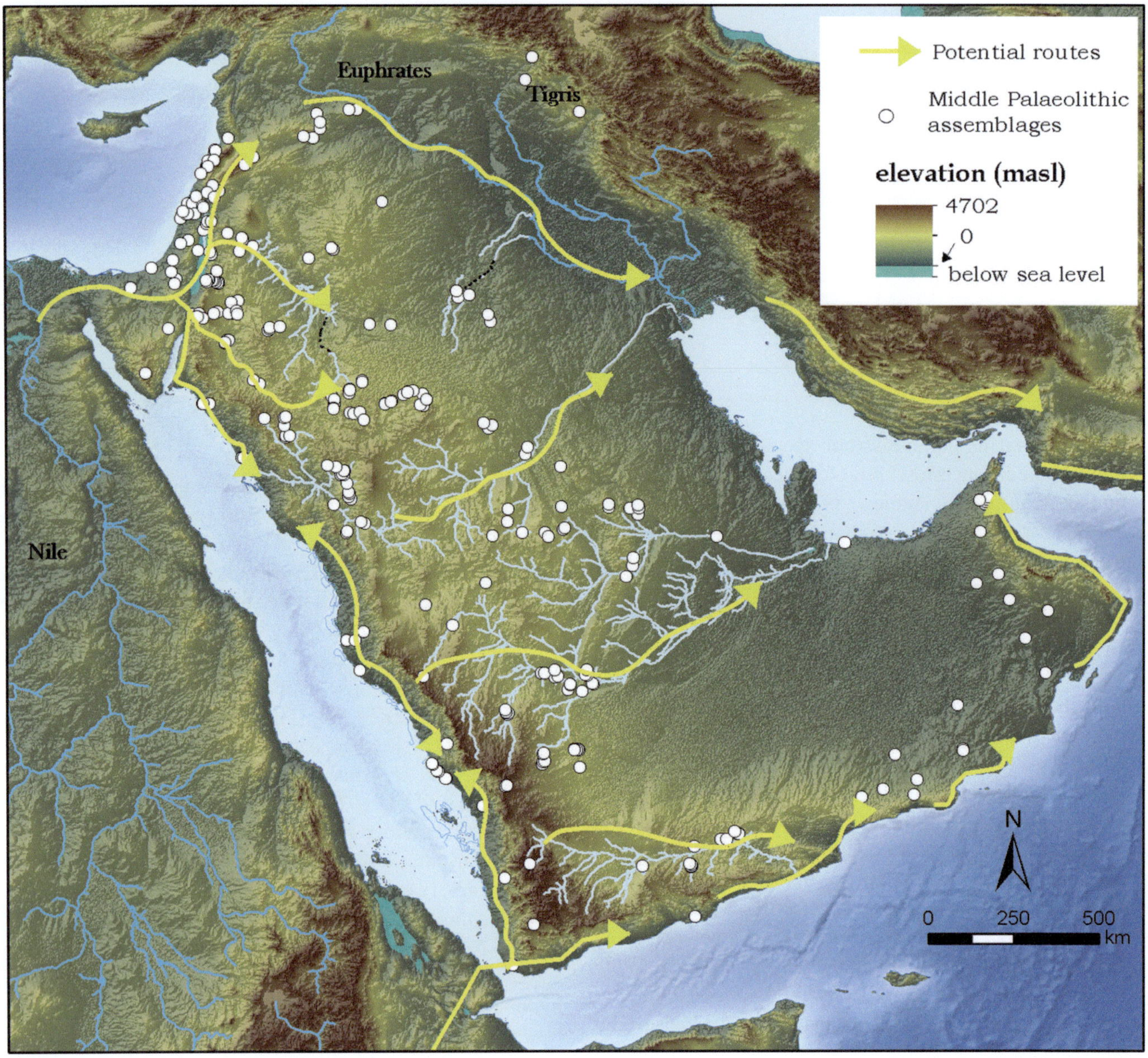

Fig. 1 Potential routes out of Africa and through Arabia during the Pleistocene (yellow arrows) in relation to Middle Palaeolithic archaeological sites (Groucutt and Petraglia 2012). Major allogenic drainage (likely active throughout the Pleistocene) is shown in darker blue, and major drainage in the Arabian Peninsula (potentially active during Pleistocene humid phases) is shown in light blue (Breeze et al. 2015). Notably, the Bab al Mandab route is usually depicted in most literature on out of Africa, although no landbridge was present in the Pleistocene. Terrestrial dispersals were possible at many times in the Pleistocene, allowing populations to disperse northward, eastward and southward into Arabia. Depicted here are Middle Palaeolithic archaeological sites, illustrating that groups may have used the Red Sea as a route, although occupations would have been far more common in the interior

moved out of Africa during wet periods, using Middle Palaeolithic technologies (Petraglia and Alsharekh 2003; Rose 2004). Other than initial survey work in the 1970s, often covering dozens of kilometres per day, few detailed archaeological studies have been implemented in Arabia during recent decades, with a few notable exceptions (e.g., Crassard 2010; Rose et al. 2011). This picture changed dramatically in recent years, as buried and stratified Palaeolithic sites have been found across Arabia (Armitage et al. 2011; Petraglia et al. 2011, 2012; Rose et al. 2011; Delagnes et al. 2012). Additionally, major interdisciplinary archaeology projects have been initiated in Saudi Arabia in the past few years, in collaboration with the Saudi Commission for Tourism and National Heritage (SCTH), and supported by the European Research Council (ERC). These archaeological projects are called DISPERSE, which concentrates its efforts on the Red Sea, and PALAEODESERTS, which conducts its work in various inland areas, as will be described below.

Both projects have considered a mixed model, with hominins utilising any favourable setting for dispersals, both interior continental and coastal fringe, to be the likely pattern for Pleistocene dispersals into Arabia. To investigate the dynamics of these dispersals, each project has focused upon one element of this model. The DISPERSE project emphasizes the need to understand movements along coastlines and accordingly has conducted active research along terrestrial and submerged environments along the Red Sea coast (e.g., Bailey et al. 2015). The PALAEODESERTS project has placed emphasis on understanding inland settings and interior movements during times of improved climate, and has therefore concentrated its efforts on palaeoriver and palaeolake settings in a number of locations across Saudi Arabia (e.g., see papers in Petraglia et al. 2015).

Here we will briefly examine coastal and terrestrial dispersal models, situating Palaeolithic archaeological sites into an over-arching framework in order to better understand the human population history of the Arabian Peninsula.

2 Blue Arabia and the Red Sea Zone

The DISPERSE project, first initiated in 2011, aims to understand how Pleistocene hominins utilized the Red Sea coastal margins, and how geological instabilities resulting from active tectonics, volcanism and sea-level change, both conditioned human populations and differentially preserved the archaeological record (e.g., Bailey 2015). Survey work has been primarily concentrated in southern and southwestern Saudi Arabia, in an area the investigators have referred to as the "Southern Crescent", denoting a region with plentiful rainfall and diverse ecological settings, consisting of mountain escarpment and foothills, coastal zones with alluvial settings, and fertile marine settings, all of which are argued to have been ideal for supporting human populations over the long term (Bailey et al. 2015).

Terrestrial surveys in southwest Saudi Arabia have been conducted, resulting in the identification of 67 localities with lithic artefacts, most commonly attributed to the Middle Stone Age (the Africanist terminology for Middle Palaeolithic). Most of the localities were associated with volcanics and lava flows, with two inland localities producing buried finds (Bailey et al. 2015). On the coastline at Harrat al Birk, 10 localities with Early Stone Age (Lower Palaeolithic) and Middle Stone Age (Middle Palaeolithic) artefacts were identified, consistent with earlier survey results (e.g., Zarins et al. 1981). At Wadi Dhabsa, an Acheulean locality was identified with cores, handaxes, cleavers and other large cutting tools (Foulds et al. 2017). The Dhabban Quarry was perhaps one of the most significant locations, as artefacts occur in association with a fossil beach and coral terrace complex along the Red Sea margin. The raised beach complex is consistent with other stratigraphic evidence to suggest it is associated with Marine Isotope Stage (MIS) 5e (Lambeck et al. 2011; Bailey et al. 2015). In one area, 19 lithic artefacts with possible Middle Palaeolithic affinities were identified in a cobble unit, and overlain by a beach shell deposit grading into a lithified sediment deposited by aeolian processes (aeolianite). The interpretation here is that the Middle Palaeolithic occupations occurred in an alluvial setting, and artefacts were subsequently deposited in or near the wadi and later covered by sea level rise. The Dhabban Quarry is of significance for indicating near-coastal occupations, as noted by Bailey et al. (2015, p 49) "This site therefore potentially acts as evidence for exploitation of the immediate coastal environment, even if it stops short of providing evidence for the use of marine resources."

Tremendous effort by the DISPERSE team has been exerted to understand the submerged landscape of the southern Red Sea, including diver surveys, acoustic surveys and sediment coring (e.g., Meredith-Williams et al. 2014; Bailey 2015). Mapping of the submerged landscape has revealed the presence of stream channels and basins with possible palaeolakes. As demonstrated by this work, the former near-shore and the earlier coastal fringe would have been a potentially productive environment for Palaeolithic populations. Although this is the case, this underwater research has not yet resulted in the identification of any Palaeolithic sites or artefacts.

Taken as a whole, the findings emerging out of a number of years of research along the Red Sea margin suggest that a model of dispersal that is inclusive of near-coastal use is plausible. The DISPERSE team has, in fact, successfully identified a number of Palaeolithic sites near the coast, though these have not provided any clear evidence for the exclusive use of coastal dispersal corridors, or coastally-oriented populations. Hence, at present, there is no direct evidence for the exploitation of marine resources by Palaeolithic populations. Field research continues in submerged landscapes, in areas with terraces and former shorelines, in the hopes of recovering Palaeolithic coastal sites and evidence for marine resource use.

3 Green Arabia and Inland Zones

The PALAEODESERTS project, formerly initiated in 2012, aims to understand the relationship between climate change and hominin population history across the mainland of the Arabian Peninsula. Previous surveys across Saudi Arabia, and initial archaeological research by some team members in the Jubbah Oasis (Nefud Desert), indicated that Palaeolithic

Fig. 2 Detailed view of northern Arabia showing palaeohydrological data (Breeze et al. 2016) and Middle Palaeolithic sites (Groucutt and Petraglia 2012) in relation to potential routes. Regions with palaeoenvironmental evidence for locally enhanced precipitation during MIS 5, and palaeodrainage and palaeolakes (diachronic) mapped through remote sensing are displayed, including the "Tabuk Corridor", that shows evidence for a spatio-temporally coherent route into northern Arabia in relative proximity to water Sources during MIS 5 (see Breeze et al. 2016 for details). Middle Palaeolithic populations dispersing into Arabia may have moved from the Sinai, travelling along the Gulf of Aqaba and the Red Sea. Terrestrial routes during humid periods would have also been utilized owing to plentiful rivers and lakes

archaeological sites were abundant and they were associated with palaeorivers and palaeolakes (Petraglia et al. 2011, 2012; see Groucutt and Petraglia 2012 for background-). The interdisciplinary archaeological research programme included the building of climate models, which showed that Arabia was significantly wetter than today based on an examination of rainfall and monsoon patterns (Jennings et al. 2015a). At the same time, remote sensing and GIS techniques were applied to reconstruct the palaeohydrology of the Arabian Peninsula, demonstrating the ubiquity of river drainages, palaeolakes and wetlands (Breeze et al. 2015, 2016) (Figs. 1 and 2). Compilations of palaeoclimate records

(Parton et al. 2015a) and palaeoenvironmental records relative to archaeological sites (Parton et al. 2015b), have demonstrated dramatic changes in rainfall patterns and fluctuations in lakes, rivers and wetlands in Arabia. After five years of field investigation in seven inland locations across Saudi Arabia, a large number of Palaeolithic sites, ranging from the Lower Palaeolithic (Acheulean) to the Neolithic, were identified.

The PALAEODESERTS survey in the Nefud Desert revealed the presence of Acheulean sites in association with raw material outcrops, alluvial settings and palaeolakes (Shipton et al. 2014), as has been noted from previous surveys across Arabia (Petraglia 2003). Although the Acheulean sites have not been dated, palaeolakes in the Nefud are present in the Middle Pleistocene, dating to humid intervals at about 410 ka (MIS 11) and 320 ka (MIS 9) (Rosenberg et al. 2013), although earlier ages are expected for these Lower Palaeolithic occurrences. Among the most impressive Acheulean sites and landscapes were those in central Arabia, at Dawadimi, where Acheulean sites were closely linked with igneous dykes and palaeoriverine settings extending over an area of at least 100 × 50 km (Jennings et al. 2015b), although the large majority of these technologically distinctive sites remain undated. One of the earliest identified Middle Pleistocene localities in the Nefud Desert is primarily a palaeontological site known as Ti's al Ghadah (Stimpson et al. 2015, 2016). Combined U-series and ESR analyses of *Oryx* teeth indicate that the fossil assemblage dates to about 500 ka, or MIS 13, although OSL estimates indicate that the site may be around 325 ka (Rosenberg et al. 2013). The fauna at Ti's al Ghadah is a biogeographical admixture that consists of likely endemics and taxa of African and Eurasian affinity and includes mammals (*Palaeoloxodon* cf. *recki*, *Panthera* cf. *gombaszogenis*, *Equus hemionus*, cf. *Crocuta crocuta*, *Vulpes* sp., *Canis anthus*, *Oryx* sp.), birds (*Struthio* sp., *Neophron percnopterus*, *Milvus* cf. *migrans*, *Tachybaptus* sp., *Anas* sp., *Pterocles orientalis*, *Motacilla* cf. *alba*) and reptiles (Varanidae/*Uromastyx* sp.) (Stimpson et al. 2016).

A range of Middle Palaeolithic localities were discovered by PALAEODESERTS survey teams across Saudi Arabia, including stratified sites producing evidence of occupations near palaeolakes (Petraglia et al. 2012; Groucutt et al. 2015b; Jennings et al. 2016) (Fig. 3). In addition, colleagues working in southern Arabia have identified sites in rock shelters (Armitage et al. 2011) and in open-air settings along the margins of rivers and streams (Rose et al. 2011; Delagnes et al. 2012). Among the earliest, Jebel Qattar (JQ1), along the Jubbah Palaeolake, dates to about 211 ka (Petraglia et al. 2011; Groucutt et al. 2016). Other Middle Palaeolithic sites are dated to episodes within MIS 5 and MIS 3, such as at Jebel Faya C and A, Mundafan, Jebel Qattar 1, Al Marrat (ALM3), and Shi'bat Dihya 1 (SD1). Although MIS 5 to 3 dates can be supported for other Middle Palaeolithic sites, in some cases dating can be problematic owing to complex stratigraphic settings (e.g., Aybut) and/or the wide range of standard deviations of age estimates (e.g., JSM1, JKF1).

In sum, despite some dating limitations, Palaeolithic archaeological sites and fossil localities in terrestrial settings of Arabia largely associate with humid periods, during MIS 13, 11, 9, 7, 5 and 3. Hominins and mammal communities likely spread across the peninsula during favourable times, when rivers and lakes were present. With respect to routes, mapping of palaeohydrology and Palaeolithic archeological sites has indicated the likelihood of movements across the Sinai Peninsula into the Levant and then northern Arabia, and subsequently into the heart of Arabia. Movements likely followed inland dispersal routes along palaeohydrological networks, such as the "Tabuk Corridor" during ameliorated periods (Breeze et al. 2016) (Fig. 2).

In summary, the PALAEODESERTS research has demonstrated an intimate connection between humid periods, the presence of palaeolakes and rivers, and the expansion of hominins over multiple periods in the Pleistocene. Although these interdisciplinary investigations have highlighted the importance of interior corridors, there is, of course, the possibility that hominins may have used the Gulf of Aqaba and the northern Red Sea margin to disperse southward into Arabia, although only limited evidence exists for near-coastal sites in this vicinity (Ingraham et al. 1981) (Fig. 2). Future work in the northern Red Sea zone will hopefully indicate if this was one of many terrestrial corridors connecting Arabia to the Sinai and the Levant.

4 Discussion

Major strides have been made in understanding the occupation history of hominin populations in Arabia in recent years. The DISPERSE and PALAEODESERTS projects have documented a range of Palaeolithic archaeological sites in Saudi Arabia, both along the Red Sea coastal fringe and across the interior of the peninsula. The results of the PALAEODESERTS project have been especially suggestive of the need to revise Out of Africa dispersal models. Rapid movements and exclusive coastal pathways, particularly those that employ a route along the Indian Ocean rim, are difficult to sustain in light of the overwhelming evidence for hominin presence along palaeorivers, palaeolakes and wetlands across the peninsula (e.g., Petraglia et al. 2010; Boivin et al. 2013).

In assessing the nature of the Palaeolithic evidence along the coastal fringes and in the interior of Arabia, Bailey et al. (2015, p 55) made this perceptive comment: "We suspect

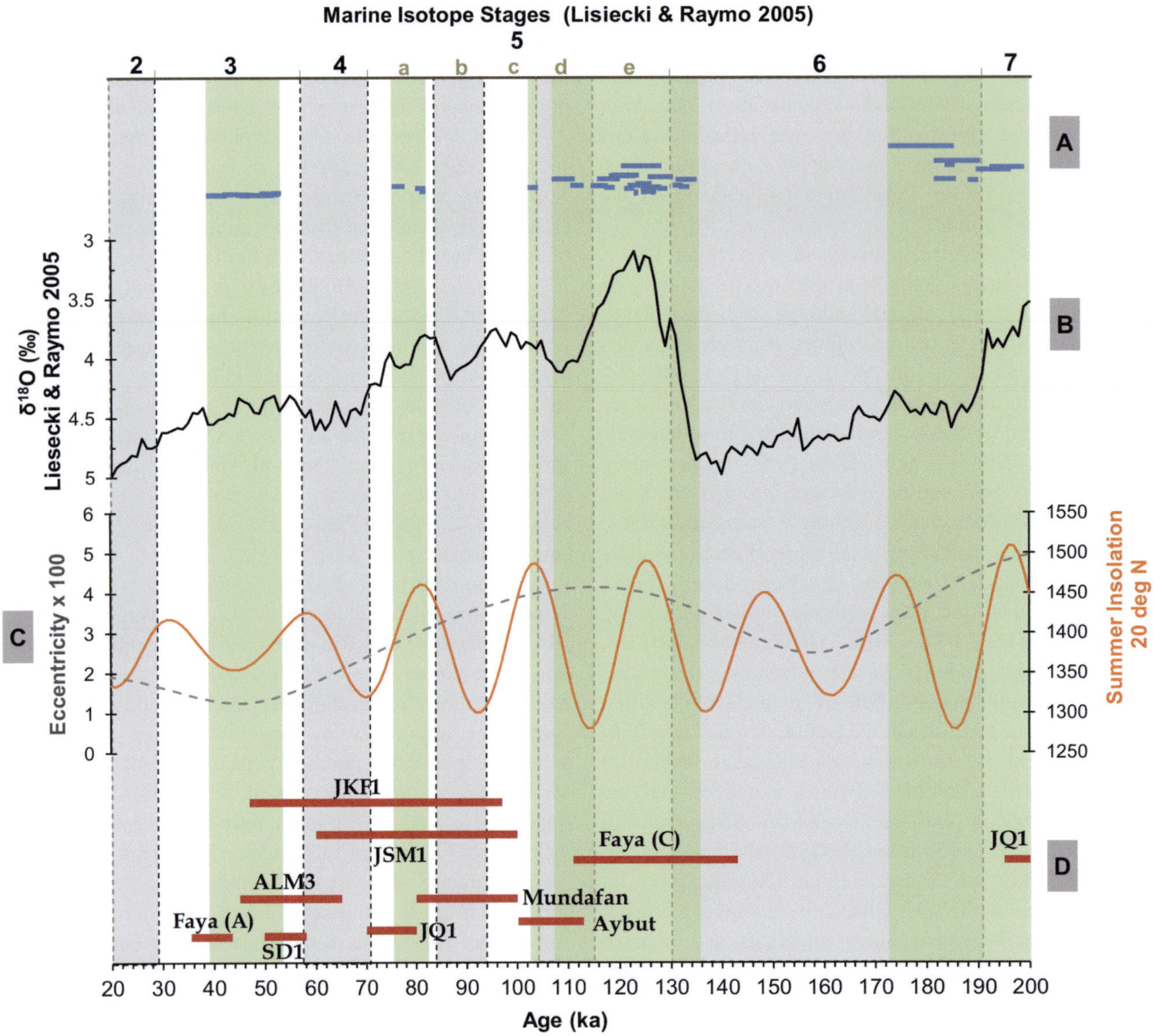

Fig. 3 Palaeoclimate proxies (200-20 ka) in relation to dated Arabian archaeological assemblages. **a** Blue horizontal bars denote dated Arabian speleothem growth phases (one sigma errors) (Burns et al. 2003; Fleitmann et al. 2003, 2011), reflecting regionally increased moisture. Green vertical bars denote potential humid phases indicated by these speleothem records. It should be noted that these records may be a conservative estimate of humidity within Arabia and are spatially constrained to the southern portions of the peninsula. **b** Global stacked benthic oxygen isotope record (inverted) for the period from 200 to 20 ka (Lisiecki and Raymo 2005). **c** Summer insolation values and eccentricity data (× 100) at 20 degrees N (Berger and Loutre 1991). **d** Buried and dated archaeological assemblages from Arabia between 200 to 20 ka (Jennings et al. 2016; Groucutt et al. 2015b; Petraglia et al. 2011, 2012; Delagnes et al. 2012; Rose et al. 2011; Armitage et al. 2011). Bars denote published date ranges (one sigma) for assemblages, labelled with site and assemblage name. Abbreviations: Jebel Qattar-1 (JQ1), Jebel Sanman-1 (JSM1), Jebel Katefeh-1 (JKF1), Al Marrat (ALM3), Shi'bat Dihya-1 (SD1)

that patterns of dispersal in, through and around the Arabian Peninsula will turn out to be neither solely "green"—following well-watered environments wherever they may occur—nor solely "blue" in following the coastline, but a combination of the two, the emphasis varying according to a highly dynamic set of ever-changing palaeogeographical and climatic circumstances." Erlandson and Braje (2015) echoed this astute argument, noting that while ancient coastlines were highly productive habitats for subsistence, they were not universally productive in their resource bases, and that when productive terrestrial and riverine settings were present, foraging populations would have likely exploited these productive food biomes.

Here we second the view that it is logical to expect that hominins utilized both coastal waterways and interior zones as corridors. The joint effort by the DISPERSE and

PALAEODESERTS teams in fact indicates that dispersal models should not be characterized as exclusively "coastal" or "interior". Instead, these ecosystems should be viewed as part of the potential range of habitats that hominins, particularly *Homo sapiens*, may have used and exploited (Groucutt et al. 2015a).

Although some convincing arguments have been put forward about coastal environments and their potential for human adaptations in the Pleistocene, we would like to make the following points, which we believe reinforces the overwhelming significance and evidence for the use of interior environments as routes of hominin movements and behavioural adaptations.

1. We support the view that coastlines are not homogeneous, but rather extremely variable along the fringes of Africa, Arabia (including the Red Sea) and the Indian Ocean rim (see shoreline evidence in Erlandson and Braje (2015) and Groucutt et al. (2015a)). Accordingly, we would contend that Pleistocene human populations were never exclusively adapted to coastal environments and were not dependent on a marine diet all across the African and southern Asian coastlines.
2. There is a paucity of Palaeolithic archaeological sites in coastal settings, including along the Indian Ocean rim, where out of Africa dispersals are assumed to have occurred. We certainly agree that submerged Palaeolithic archaeological sites may be present on or near former coastlines. This potential has been illustrated by the DISPERSE team on the Red Sea, but this is not the case for all areas on the Red Sea, nor along the south Arabian coast, especially along the steep coastal margins of the Indian Ocean.
3. Even if coastal zone sites are identified in submerged zones in the future, we anticipate that it will be difficult to demonstrate that Pleistocene hominin populations were exclusively, or even largely, marine oriented, rather, we would expect that Pleistocene populations would have incorporated marine resources into a flexible diet that included significant terrestrial resources.
4. All available archaeological evidence, including data collected along the Red Sea by the DISPERSE team, shows the primacy of terrestrial adaptations. A large number of palaeorivers and palaeolakes, associated with numerous Pleistocene archaeological sites and fossil localities, have been found all across the interior of Arabia. This provides strong and convincing support to indicate that hominins expanded during multiple wet periods ("Green Arabia"), occupying and adapting to favourable ecological conditions for their subsistence needs.
5. Coastal dispersal routes have been popular in the public imagination and common in the scientific literature. We believe that the significance of inland routes, especially in the context of the Lower and Middle Palaeolithic, have been severely underplayed, probably owing to apriori assumptions that the Arabian Peninsula was arid and not conducive to sustained hominin presence. The PALAEODESERTS project will continue to aim to rectify this situation, examining dispersal routes across many areas of the Arabian Peninsula, but also illustrating how climate variability shaped the demographic history of hominin populations over long periods of the Pleistocene.

Acknowledgements We thank HRH Prince Sultan bin Salman bin Abdulaziz Al-Saud, President of the Saudi Commission for Tourism and National Heritage (SCTH), and Professor Ali Ghabban, Vice President of the SCTH, for permission to carry out this study. Jamal Omar, Abdulaziz Al-Omari and Abdullah Al-Zahrani of the SCTH provided critical support for our work in Saudi Arabia. We thank Dr. Z. A. Nawab, President of the Saudi Geological Survey (SGS), for his support of interdisciplinary archaeological research in Saudi Arabia, and Dr. Najeeb Rasul of the SGS for the invitation to participate in the Red Sea conference. We would especially like to acknowledge Abdullah Alsharekh of King Saud University for his long term support of our archaeological research in Saudi Arabia. Our research was funded by the SCTH and the European Research Council (grant no. 295719, to MDP). HSG acknowledges funding from the British Academy. PSB acknowledges funding from the Leverhulme Trust. We are grateful to Geoff Bailey and two anonymous reviewers for comments on an earlier draft of this paper.

References

Armitage SJ, Jasim SA, Marks AE, Parker AG, Usik VI, Uerpmann H-P (2011) The southern route "Out of Africa": evidence for an early expansion of modern humans into Arabia. Science 331:453–456

Bailey G (2015) The evolution of the Red Sea as a human habitat during the Quaternary period. In: Rasul NMA, Stewart ICF (eds) The Red Sea: The Formation, Morphology, Oceanography and Environment of a Young Ocean Basin. Springer Earth System Sciences, Berlin Heidelberg, pp 599–614

Bailey GN, Devès MH, Inglis RH, Meredith-Williams MG, Momber G, Sakellariou D, Sinclair AGM, Rousakis G, Al Ghamdi S, Alsharekh AM (2015) Blue Arabia: Palaeolithic and underwater survey in SW Saudi Arabia and the role of coasts in Pleistocene dispersals. Quatern Int 382:42–57

Berger A, Loutre MF (1991) Insolation values for the climate of the last 10 million years. Quatern Sci Rev 10:297–317

Boivin N, Fuller DQ, Dennell R, Allaby R, Petraglia MD (2013) Human dispersal across diverse environments of Asia during the Upper Pleistocene. Quatern Int 300:32–47

Breeze PS, Drake NA, Groucutt HS, Parton A, Jennings RP, White TS, Clark-Balzan L, Shipton C, Scerri EML, Stimpson C, Crassard R, Hilbert Y, Alsharekh A, Al-Omari A, Petraglia MD (2015) Remote sensing and GIS techniques for reconstructing Arabian palaeohydrology and identifying archaeological sites. Quatern Int 382:98–119

Breeze PS, Groucutt HS, Drake NA, White TS, Jennings RP, Petraglia MD (2016) Palaeohydrological corridors for hominin dispersals in the Middle East ~ 250–70,000 years ago. Quatern Sci Rev 144:155–185

Burns SJ, Fleitmann D, Matter A, Kramers J, Al-Subbary A (2003) Indian Ocean climate and an absolute chronology over Dansgaard/Oeschger events 9 to 13. Science 301:1365–1367

Crassard R (2010) The Middle Paleolithic of Arabia: The view from the Hadramawt region, Yemen. In: Petraglia MD, Rose JI (eds) The evolution of human populations in Arabia. Springer, Netherlands, pp 151–168

Delagnes A, Tribolo C, Bertran P, Brenet M, Crassard R, Jaubert J, Khalidi L, Mercier N, Nomade S, Peigne S, Sitzia L, Tournepiche JF, Al-Halibi M, Al-Mosabi A, Macchiarelli R (2012) Inland human settlement in southern Arabia 55,000 years ago: New evidence from the Wadi Surdud Middle Paleolithic site complex, western Yemen. J Human Evolution 63:452–474

Erlandson JM, Braje TJ (2015) Coasting out of Africa: The potential of mangrove forests and marine habitats to facilitate human coastal expansion via the southern dispersal Route. Quatern Int 382:31–41

Fleitmann D, Burns SJ, Neff U, Mangini A, Matter A (2003) Changing moisture sources over the last 330,000 years in northern Oman from fluid-inclusion evidence in speleothems. Quatern Res 60:223–232

Fleitmann D, Burns SJ, Pekala M, Mangini A, Al-Subbary AA, Al-Aowah M, Kramers J, Matter A (2011) Holocene and Pleistocene pluvial periods in Yemen, southern Arabia. Quatern Sci Rev 30:783–787

Foulds FWF, Shuttleworth A, Sinclair A, Alsharekh AM, Al Ghamdi S, Inglis RH, Bailey GN (2017) The recent discovery of a Palaeolithic site at Wadi Dabsa. Implications for hominin adaptations in the Arabian Peninsula. Antiquity, Saudi Arabia in press

Groucutt HS, Petraglia MD (2012) The prehistory of the Arabian Peninsula: Deserts, dispersals, and demography. Evol Anthropol 21:113–125

Groucutt HS, Petraglia MD, Bailey G, Scerri EML, Parton A, Clark-Balzan L, Jennings RP, Inglis RH, Devès MH, Meredith-Williams M, Boivin N, Thomas MG, Scally A (2015a) Rethinking the dispersal of *Homo sapiens* out of Africa. Evol Anthropol 24:149–164

Groucutt HS, White TS, Clark-Balzan L, Parton A, Crassard R, Shipton C, Jennnings RP, Parker AG, Breeze PS, Scerri EML, Alsharekh A, Petraglia MD (2015b) Human occupation of the Arabian Empty Quarter during MIS 5: Evidence from Mundafan Al-Buhayrah, Saudi Arabia. Quatern Sci Rev 119:116–135

Groucutt HS, Breeze P, Drake NA, Jennings R, Parton A, White T, Shipton C, Clark-Balzan L, Al-Omari A, Cuthbertson P, Wedage OM (2016) The Middle Palaeolithic of the Nejd, Saudi Arabia. J Field Archaeology 41:131–147

Ingraham M, Johnson T, Rihani B, Shatla I (1981) Saudi Arabian comprehensive survey program: c) Preliminary report on a reconnaissance survey of the northwestern province (with a note on a brief survey of the northern province). Atlal 5:59–84

Jennings RP, Singarayer J, Stone EJ, Krebs-Kanzow U, Khon V, Nisancioglu KH, Pfeiffer M, Zhang X, Parker A, Parton A, Groucutt HS, White TS, Drake NA, Petraglia MD (2015a) The greening of Arabia: Multiple opportunities for human occupation of the Arabian Peninsula during the Late Pleistocene inferred from an ensemble of climate model simulations. Quatern Int 382:181–199

Jennings RP, Shipton C, Breeze C, Cuthbertson P, Bernal MA, Wedage WMCO, Drake NA, White TS, Groucutt HS, Parton A, Clark-Balzan L, Stimpson C, al Omari A, Alsharekh A, Petraglia MD (2015b) Multi-scale Acheulean landscape survey in the Arabian Desert. Quatern Int 382:58–81

Jennings RP, Parton A, Clark-Balzan L, White TS, Groucutt HS, Breeze PS, Parker AG, Drake NA, Petraglia MD (2016) Human occupation of the northern Arabian interior during early Marine Isotope Stage 3. J Quaternary Sci 31:953–966

Klein RG (2008) Out of Africa and the evolution of human behavior. Evol Anthropol 17:267–281

Lambeck K, Purcell A, Flemming NC, Vita-Finzi C, Alsharekh AM, Bailey GN (2011) Sea level and shoreline reconstructions for the Red Sea: Isostatic and tectonic considerations and implications for hominin migration out of Africa. Quatern Sci Rev 30:3542–3574

Lisiecki LE, Raymo ME (2005) A Pliocene-Pleistocene stack of 57 globally distributed benthic δ 18 O records. Paleoceanography 20:1–17

Mellars P, Gori KC, Carr M, Soares PA, Richards MB (2013) Genetic and archaeological perspectives on the initial modern human colonization of southern Asia. Proc Nat Acad Sci 110:10699–10704

Meredith-Williams MG, Hausmann N, Bailey GN, King GCP, Alsharekh A, Al Ghamdi S, Inglis RH (2014) Mapping, modelling and predicting prehistoric coastal archaeology in the southern Red Sea using new applications of digital-imaging techniques. World Archaeology 46:10–24

Oppenheimer S (2009) The great arc of dispersal of modern humans: Africa to Australia. Quatern Int 202:2–13

Parton A, White TS, Parker AG, Breeze PS, Jennings R, Groucutt HS, Petraglia MD (2015a) Orbital-scale climate variability in Arabia as a potential motor for human dispersals. Quatern Int 382:82–97

Parton A, Farrant AR, Leng MJ, Telfer MW, Groucutt HS, Petraglia MD, Parker AG (2015b) Alluvial fan records from southeast Arabia reveal multiple windows for human dispersal. Geology 43:295–298

Petraglia MD (2003) The Lower Paleolithic of the Arabian Peninsula: Occupations, adaptations, and dispersals. J World Prehistory 17:141–179

Petraglia MD, Alsharekh A (2003) The Middle Palaeolithic of Arabia: Implications for modern human origins, behaviour and dispersals. Antiquity 77:671–684

Petraglia MD, Haslam M, Fuller DQ, Boivin N, Clarkson C (2010) Out of Africa: New hypotheses and evidence for the dispersal of *Homo sapiens* along the Indian Ocean rim. Annals Human Biology 37:288–311

Petraglia MD, Alsharekh AM, Crassard R, Drake NA, Groucutt H, Parker AG, Roberts RG (2011) Middle Paleolithic occupation on a marine isotope stage 5 lakeshore in the Nefud Desert, Saudi Arabia. Quatern Sci Rev 30:1555–1559

Petraglia MD, Alsharekh A, Breeze P, Clarkson C, Crassard R, Drake NA, Groucutt NA, Jennings R, Parker AG, Parton A, Roberts RG, Shipton C, Matheson C, al-Omari A, Veall M-A (2012) Hominin dispersal into the Nefud desert and Middle Palaeolithic settlement along the Jubbah palaeolake, northern Arabia. PLoS ONE:e49840

Petraglia MD, Parton A, Groucutt HS, Alsharekh A (2015) Green Arabia: Human prehistory at the crossroads of continents. Quatern Int 382:1–7

Rose JI (2004) The question of Upper Pleistocene connections between east Africa and south Arabia. Curr Anthropol 45:551–555

Rose JI, Usik VI, Marks AE, Hilbert YH, Galletti CS, Parton A, Geiling, JA, Cerny V, Morley MW, Roberts RG (2011) The Nubian complex of Dhofar, Oman: An African Middle Stone Age industry in southern Arabia. PLoS One:e28239

Rosenberg TM, Preusser F, Fleitmann D, Schwalb A, Penkman K, Schmid TW, Al-Shanti MA, Kadi K, Matter A (2011) Humid periods in southern Arabia: Windows of opportunity for modern human dispersal. Geology 39:1115–1118

Rosenberg TM, Preusser F, Risberg J, Plikk A, Kadi KA, Matter A, Fleitmann D (2013) Middle and Late Pleistocene humid periods recorded in palaeolake deposits of the Nafud desert, Saudi Arabia. Quatern Sci Rev 70:109–123

Shipton C, Parton A, Breeze P, Jennings R, Groucutt HS, White TS, Drake N, Crassard R, Alsharekh A, Petraglia MD. (2014) Large flake Acheulean in the Nefud Desert of northern Arabia. PaleoAnthropology, 446–62

Stewart JR, Stringer CB (2012) Human evolution out of Africa: The role of refugia and climate change. Science 335:1317–1321

Stimpson CM, Breeze PS, Clark-Balzan L, Groucutt HS, Jennings R, Parton A, Scerri E, White TS, Petraglia MD (2015) Stratified Pleistocene vertebrates with a new record of a jaguar-sized pantherine (*Panthera* cf. *gombaszogensis*) from northern Saudi Arabia. Quatern Int 382:168–180

Stimpson CM, Lister A, Parton A, Clark-Balzan L, Breeze PS, Drake NA, Groucutt HS, Jennings R, Scerri EML, White TS, Zahir M, Duval M, Grun R, Al-Omari A, Al Murayyi KSM, Zalmout IS, Mufarreh YA, Memesh A, Petraglia MD (2016) Middle Pleistocene vertebrate fossils from the Nefud Desert, Saudi Arabia: Implications for biogeography and palaeoecology. Quatern Sci Rev 143:13–36

Stringer C (2000) Palaeoanthropology: Coasting out of Africa. Nature 405:24–27

Templeton A (2002) Out of Africa again and again. Nature 416:45–51

Timmermann A, Friedrich T (2016) Late Pleistocene climate drivers of early human migration. Nature 538:92–95

Zarins J, Murad AAJ, Al-Yish KS (1981) The comprehensive archaeological survey program. a. The second preliminary report on the southwestern province. Atlal 5:9–42

Optically Stimulated Luminescence Dating as a Geochronological Tool for Late Quaternary Sediments in the Red Sea Region

David C. W. Sanderson and Timothy C. Kinnaird

Abstract

This chapter concerns the use of luminescence methods as geochronological tools for dating Late Quaternary sediments in the Red Sea region. The dating methods all use stimulated luminescence to register signals developed in mineral systems in response to long term exposure to ionising radiation in the environment. The principles of luminescence dating are outlined followed by discussion of its application to the Arabian Peninsula, where, particularly in SE Arabia and parts of the interior, a growing corpus of work is emerging, which is helping to define past arid or humid periods of importance to palaeoclimatology and to archaeology. Turning to the Red Sea, studies conducted within the DISPERSE project are presented both in marine and terrestrial settings. The motivation for much of this work concerns definition of the environmental conditions and chronologies for hominin and human dispersion through Arabia. Data are presented which identify, for the first time, late Pleistocene evidence on the inner continental shelf near the Farasan Islands, using material from the 2013 cruise of RV AEGAEO. Results are also presented from the littoral fringe of southwest Saudi Arabia, identifying units associated with MIS5 which have palaeo-environmental and archaeological significance. It is to be hoped that further research in coming decades will continue to extend the regional chronology for the littoral fringe of the Red Sea. In this respect, luminescence dating has the potential to help define the environmental history of this important area, to assist with assigning marine and terrestrial features into unique stages of Quaternary climate cycles, and to promote better understanding of human-environment interactions in this dynamic area.

D. C. W. Sanderson (✉)
Scottish Universities Environmental Research Centre (SUERC), East Kilbride, UK
e-mail: david.sanderson@glasgow.ac.uk

T. C. Kinnaird
School of Earth and Environmental Sciences, University of St. Andrews, St Andrews, UK

T. C. Kinnaird
School of Biological and Environmental Sciences, University of Stirling, Stirling, UK

N. M. A. Rasul and I. C. F. Stewart (eds.), *Geological Setting, Palaeoenvironment and Archaeology of the Red Sea*,
https://doi.org/10.1007/978-3-319-99408-6_31

1 Introduction

Optically stimulated luminescence, and its related variants, have emerged in recent decades as important geochronometric tools for constraining sediment formation ages and post-depositional processes in many Quaternary palaeo-environments, including those of the Red Sea region. In this chapter, the underlying principles and age limits of luminescence dating are outlined. Examples are discussed of work which has defined the age limits of relic dune systems in the Arabian Peninsula, and studies conducted within the DISPERSE project on the littoral fringe of southwest Saudi Arabia and from marine cores off the coast of the Farasan Islands. The results confirm the utility of luminescence dating to palaeo-environmental work in the Red Sea region, and testify to varied environmental conditions in the past, which make this area critical for studies of human dispersal and occupation during Palaeolithic periods.

2 Luminescence Dating of Sediments

Luminescence dating is part of a class of radiation damage dating methods, which include electron spin resonance (ESR) dating and methods based on stimulated luminescence of minerals, by heat, as in thermoluminescence (TL), or light, as in photostimulated luminescence (PSL) or optically stimulated luminescence (OSL). In these methods, prolonged exposure to ionising radiation in the physical environment deposits energy, releasing electrons and holes from their normal bound states in insulating minerals. Charge-trapping at meta-stable defect centres within the mineral lattice can store energy and, through evolution of populations of trapped charge carriers, can store quantum information concerning radiation exposure, optical exposure and thermal history. By stimulated release of trapped charge, using heat or light, and subsequent relaxation of the system, low level light emission occurs, which can be detected and quantified using photon counting methods.

The intensity of luminescence signals depends on the extent of prior exposure to natural radiation since an earlier zero condition, and on the sensitivity of the individual minerals, which has a complex dependence on formation and post formation conditions. In dating techniques, a series of calibration experiments are undertaken, with laboratory radiation sources, to quantify the natural luminescence signal in terms of "equivalent dose". This corresponds to the radiation dose (in units of Grays, which are a measure of energy deposition per unit mass) which would account for the natural signals observed from the mineral system under study. In parallel with the equivalent dose determination, a combination of low-level radiometric measurements, radionuclide concentration measurements, and dosimetric modelling is used to define palaeo-dose rates from the sample and its environment. The luminescence age is then determined as the quotient of equivalent dose to the estimated palaeo-dose rate.

Figure 1 provides a schematic illustration of the evolution of luminescence signals in a hypothetical mineral from geological formation, to deposition within a sediment, and final laboratory processing. Since primary mineral formation processes, whether from high or low temperature crystallization, should not introduce initial trapped charge populations to the structural defects, the initial post formation state corresponds with no potential for generating stimulated luminescence. The complex geological, thermal and radiation histories experienced by minerals which can include multiple cycles of diagenetic alteration, irradiation, thermal exposure, weathering and sedimentation over long periods, have the potential both to affect the defect configuration within mineral lattices, and their state of electronic and hole population. Early research exploring the potential for using TL to directly date the formation of geological materials (e.g., McDougall 1968) demonstrated that defect filling by ionising radiation (for example the alpha, beta and gamma radiation from the natural decay series, primordial nuclides and cosmic radiation) tends to proceed significantly faster than defect formation (for example by recoil damage during internal alpha decay or spontaneous fission, or by alterations). Therefore, geological-cycle luminescence is generally expected to saturate at levels whereby the thermal stabilities of trapped charge carriers are in a pseudo-equilibrium with their environmental conditions. This view has prevailed as a paradigm, which leads luminescence dating applications toward more recent periods whereby target dating events follow a "zeroing event", such as heating (in synthetic materials like ceramics, during firing, or in the case of heated lithics, by natural or artificial processes), or significant light exposure which "bleaches" the potential signals during terminal cycles of weathering and sedimentary deposition. It is of course easy to see that there are some geological history pathways which can lead to multiple cycles of signal evolution and diminution, as well as to changes in defect populations and configurations prior to such a "zeroing event". Some of these are illustrated schematically by the multiple pathways in the geological part of Fig. 1.

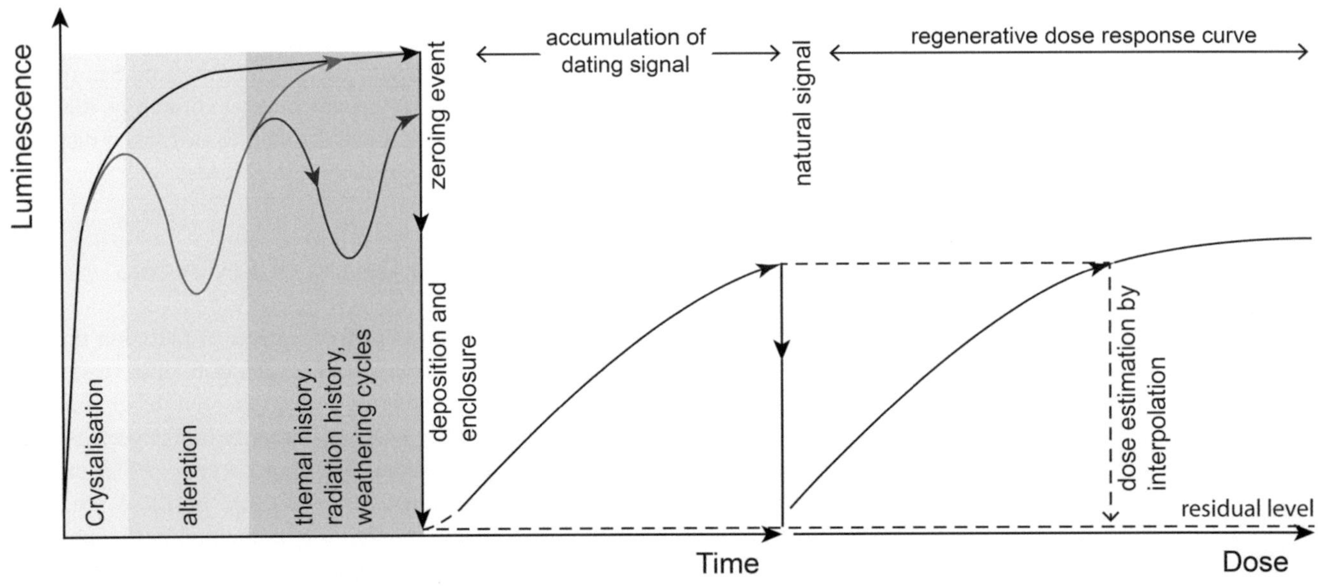

Fig. 1 Illustration of the luminescence signal evolution in a hypothetical mineral during its geological history, the period of natural accumulation following sedimentary deposition, and during laboratory measurement cycles

Zeroing by optical exposure can be envisaged as a consequence of energy capture of photons by trapped electron or hole populations, resulting in promotion from localized states in point defects to conduction and valance bands, from where charge transport and eventual relaxation can take place. The optical transitions associated with de-trapping can be direct or indirect, whereby photon and thermal energies are both involved. The relaxation processes are potentially complex since optical bleaching at environmental temperatures can also be associated with re-trapping. Nevertheless, the end result of bleaching is to reduce the latent luminescence signals to residual levels, close to zero. Enclosure following sedimentary deposition and protection from further light exposure leads to a new period of natural accumulation, in response to further exposure to ionising radiation in the environment. Thus, the potential luminescence signal increases again, during the accumulation period. The aim of luminescence dating, as indicated above, and in general accounts of the method (e.g., Aitken 1985, 1998), is to measure and quantify the radiation dose equivalent to the accumulation of the dating signal, and to assess the elapsed time by comparison with the integrated environmental dose rate required to generate the equivalent dose.

Figure 1 also illustrates the signal trajectory following sampling, and mineral separation in the laboratory, both of which are conducted under subdued lighting conditions, and during dose quantification. Measurement of the natural luminescence signal by stimulation in the laboratory once again depletes the signal. Quantification using regenerative methods involves the use of calibrated radiation sources to develop dose response curves, which can be interpolated to determine the equivalent dose. This approach is illustrated schematically in Fig. 1. Another approach that could be considered would be to prepare multiple portions of the samples and apply the method of standard additions, as illustrated in Fig. 2, to estimate equivalent doses, in which case an appropriate dose response function must be fitted to the data set and extrapolated. This method can also be combined with regenerative approaches.

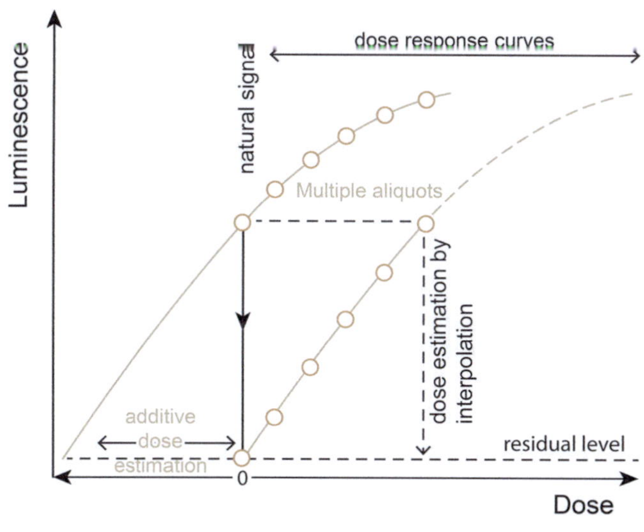

Fig. 2 Illustration of the additive dose estimation approach, and combined additive-regenerative approaches

With these considerations in mind, for sediments which have (i) been effectively zeroed prior to last deposition (for example by light exposure, or by heat), (ii) which have not been subject to post-depositional alterations, (iii) whose luminescence signals are retained with stability, and (iv) whose dose response characteristics can be reconstructed in the laboratory without bias, luminescence ages correspond to the physical time interval since deposition. For systems with more complex behaviour or origins, the system can deliver information on depositional and post-depositional conditions which may be of value in combination with other information to help constrain understanding of sedimentary records. The system is also capable of yielding information about past thermal histories, and of registering the effects of prolonged light exposure to lithic surfaces.

While the luminescence phenomena used for TL, OSL and related dating systems can be observed from most translucent minerals, the majority of geochronometric studies have been applied to quartz or feldspar systems. Age limits depend on defect configuration and luminescence properties of individual quartzes and feldspars. Luminescence sensitivity and the completeness of pre-depositional zeroing set the lower age limit, which can be less than a few years for the feldspar system, and below 100 years for sufficiently bright quartz. Upper age limits are set by a combination of signal stability and by the saturation limits associated with finite defect populations within target minerals.

For quartz OSL dating, the upper limit of saturation is usually found in the 100–200 Gy dose range, corresponding to ages of approximately 2×10^5 years depending on dose rates. As the system approaches saturation, there is some degradation of precision and resolution, typically observed above approximately 100 ka. Notwithstanding this, the quartz OSL dating system is well established now within the 10^2–10^5 year age range as a geochronometer with increasing applications. There are also deep trap signals from quartz with potential for age extension, up to 10^6 years, on a research basis, of which phototransferred thermoluminescence (PTTL) (Sanderson et al. 2001), thermally transferred OSL (TT-OSL) (Wang et al. 2006; Adamiec et al. 2008, 2010), and high temperature phosphorescence (Kinnaird et al. 2010) are perhaps the more promising approaches. Quartz OSL ages are mostly measured using regenerative dose response techniques, such as the single-aliquot regenerative (SAR) method. Here many aliquots, or even single quartz grains are studied, and dose distributional analysis used as a means of identifying mixed-age, or incompletely bleached samples.

The feldspar systems have greater sensitivity and uniformity than quartz-based systems, but are associated with an unenviable reputation for instability, which must be examined explicitly within dating procedures. Feldspar luminescence can be stimulated in many wavebands from UV to near IR, and is associated with complex distributions of trapping systems with a wide variety of kinetic behaviour. Dose response curves continue to grow into the 1–10 kGy region, which corresponds to age limits of approximately 1 Ma, providing stability can be assured. Thermally stabilized single aliquot regenerative-additive dose (SARA) procedures (Kinnaird et al. 2015a), the use of high temperature post-IR infrared stimulated luminescence (pIR-IRSL) (Thomsen et al. 2008; Buylaert et al. 2009), and the use of near UV stimulation bands are some of the approaches under consideration for applying feldspar based luminescence methods to early Quaternary systems.

Effective sampling is of critical importance to successful sediment dating, and there has been significant recent progress in the development of simple luminescence profiling methods (Sanderson and Murphy 2010), which can be applied both in the field, using portable luminescence readers, and in the laboratory, to characterise complex sediment sequences, and identify discontinuities (e.g., the Pleistocene/Holocene discontinuity illustrated below in core FA13), inversions and potentially mixed units. When used in combination with quantitative laboratory dating procedures these approaches seem to be useful for defining the context of dating materials.

3 An Overview of Luminescence Dating Studies in the Red Sea Region and Arabian Peninsula

Given the spatial scale of the Arabian Peninsula (about 3.24 million km^2) and the Red Sea (which encloses an area of 0.438 million km^2, with a linear coastal dimension of some 2250 km) there are, so far, comparatively few luminescence dates from the coastal region. However, the opportunities can readily be perceived, given the nature of land cover and the relatively recent, on a geological timescale, rifting which formed the Red Sea.

One strand of enquiry concerns aridity history. Luminescence dating has been associated with drylands research since the inception of the first thermoluminescence dates for sediments in the late 1970's. Following the initial demonstration that marine cores showed a stratigraphic progression of thermoluminescence ages, and the recognition of the role of optical bleaching in resetting the system (Wintle and Huntley 1979), Singhvi et al. (1982, 1983) reported dating of sand dunes in drylands in Rajastan. The TL methods used at that time had residual signals, and early work explored means of estimating their magnitude. The development of optical stimulation approaches, initially using filtered green lasers (Huntley et al. 1985), and later with diverse stimulation sources (Bøtter-Jensen et al. 2000) has greatly facilitated the analysis of sediments.

In respect of drylands research on the Arabian Peninsula, McClure (1976) identified phases of lake development in the Mundafan basin (Rub al Khali) between 36–21 ka and 9–6 ka (uncalibrated radiocarbon years), indirectly constraining the associated dune formations to the intervening and succeeding arid periods. Glennie and Singhvi (2002) have outlined interactions between glacially forced sea level in the Arabian Gulf, exposing the floor during low-stands, and the palaeoclimatic systems. Given the modern interplay between dry "Shamal" winds from the north, and the moist NE and SW monsoon cycles, an analogous framework was proposed for earlier cycles to account for the transfer of carbonate rich sediments to SE Arabia, during dry cold periods of the Quaternary, and subsequent alteration and redistribution in more humid periods. Issues concerning radiocarbon dating, particularly of carbonate fractions, were discussed and some 30 luminescence dates from dunes sands were presented. When classified by sedimentary setting into Shamal, and Monsoon linked categories, associations between the generalised sea-level curve of Boulton (1993) and the ages of the two types of sediments were notable. These data broadly supported the framework outlined above, linking arid periods to the glacial cycle. On this basis arid conditions have been experienced back to some 350 ka with more humid periods, during the interstadials, when weaker Shamal winds and a stronger SW Monsoon may have been experienced. This framework has been extended further in SE Arabia. Bray and Stokes (2003, 2004) examined a one metre core from Liwa in the UAE, producing 12 OSL dates of which the upper 55 cm ranged from 2.5–3.2 ka, grading upward to 6–7 ka at the base of the core. They suggested that the transition between humid early Holocene conditions (about 10 ka–6 ka) and the arid conditions which followed was accompanied by a local lag time, as older dune sands in the Rub al Khali were reworked. Speleothem records from northern Oman (Burns et al. 1998) provide another record which is broadly consistent with this framework. Preusser et al. (2002) discussed a suite of 74 IRSL dates obtained from cores in the Wahiba sands in Oman, which ranged from MIS1 back to MIS6. From these data, periods of sand movement were inferred in MIS1, the earlier part of MIS2, MIS4 leading into MIS5a, MIS5d leading into 5e, and MIS6.

Additional studies in SE Arabia include the UAE (Glennie et al. 2011; Farrant et al. 2015; Preston et al. 2015), Oman (Blechschmidt et al. 2009; Preusser et al. 2002;

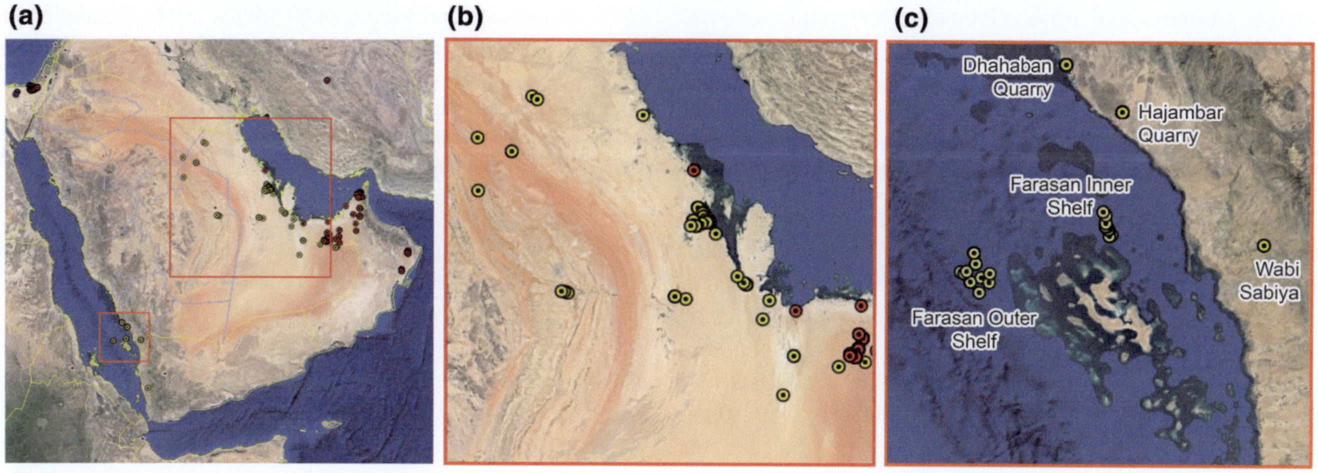

Fig. 3 Location map showing the distribution of luminescence dating samples referred to in the text. Red symbols (left and middle maps) are from the INQUA dune atlas database. Yellow symbols are from SUERC data, combining Saudi Aramco commissioned work in eastern Saudi Arabia, and Munro's samples in Yemen (left and middle maps) with samples from the recent DISPERSE project (right map)

Preusser 2009; McClaren et al. 2009; Matter et al. 2015; Hilbert et al. 2015; Hoffman et al. 2015), and parts of Yemen and Saudi Arabia (Munro and Wilkinson 2007; Munro et al. 2013). Duller (2016) reviewed some 385 luminescence dates from Arabia, associated with the INQUA dune Atlas (Lancaster et al. 2016), the majority of which are from the UAE or Oman (Fig. 3a). He draws attention to the different luminescence dating approaches used in these studies, noting that within the current uncertainties on luminescence dating the task of disentangling individual MIS stages remains challenging. While the majority of determinations reviewed fall in the last 100 ka, some 31% are older, which include IRSL feldspar data, where stability limits may be in question, and quartz OSL data where signal saturation may be a limiting factor. The emergence of age extension methods, such as TT-OSL, pIR-IRSL, and other techniques with potential to respond to stable deep-trap signals with high saturation dose responses, and related techniques are clearly important to extending luminescence chronologies to the early Quaternary.

While Munro's earlier investigations (Munro and Wilkinson 2007) also were concerned with aridity/humidity cycles in Yemen and Saudi Arabia, his later summary (Munro et al. 2013) also had access to more than 70 additional OSL dates from Late Quaternary sediments in Saudi Arabia commissioned by Stephen Franks on behalf of Saudi Aramco and conducted at SUERC (Burbidge et al. 2007a; Cresswell and Sanderson 2009). These were collected (Fig. 3b) with the primary aim of assessing the conditions and durations of formation of clay coatings on quartz sands during cyclic aridity phases, as a means of validating porosity models applied to resource reservoirs. However, the data from these studies, which included a series of samples covering wadi and sabkha systems, also placed constraints on the aridity history of the recent environment, confirming that cemented and altered palaeo-dune surfaces imply a diverse range of environmental and climatic conditions in the past, including humid and arid cycles that reach back to before 275 ka. It was noted in this work that uranium anomalies, with atypical U/Th ratios were observed in some cases, in association with cemented sands. Since cementation itself is the product of at least one cycle of hydration and aridity and can, under some circumstances, re-deposit uranium and radium in series-disequilibrium, it is important that dose rate determination methods with the potential to detect such behaviour are used for such material.

A second strand of enquiry, where the coupling between Quaternary climatic cycles and the environment is critical, concerns hominin and human dispersal from Africa and particularly the southern migration routes to India, SE Asia and beyond. Whether a northern corridor from Africa through the Levant, or a southern crossing of the Red Sea are considered, the Arabian Peninsula presents both topographic and aridity barriers which are sensitive to timing. Dispersal may have been initiated in MIS5 or earlier, with potential routes through the interior of Arabia critically affected by climate, hydrology and seasonality. Potential southern migration routes along coastal fringes, and the possibility of southern crossings of the Red Sea, are also dependent on the effect on coastal regions of changing sea levels during glacial cycles. Petit-Maire et al. (2010) have presented U/Th dates for a palaeolake in Jordan which shows clear evidence of shells from MIS5 and earlier. Recent work conducted in currently arid parts of Arabia (Dennell and Petraglia 2012; Breeze et al. 2015; Groucott et al. 2015a, b, c; Jennings et al. 2015a, b; Parton et al. 2015a, b; Petraglia 2011; Petraglia et al. 2011, 2015; Scerri et al. 2015; Stimpson et al. 2015) has broadened the range of recognized environments in the interior. Further south, others (Preusser 2009; Armitage et al. 2011; Rosenberg et al. 2011) have suggested that palaeo-environments associated with early anatomically-modern human (AMH) activity were present in Oman and the UAE at MIS5e. The role of movements of the inter-tropical convergence zone (ITCZ) in influencing palaeo-climates in Arabia has been noted by these authors. It remains an important research objective to link comprehensive and accurate chronologies for the aridity history of the region to the archaeological and faunal records and to the glacial cycle since access involving potential crossings of the southern Red Sea area and the Gulf would have been facilitated by low global sea levels, for example in MIS6, or in the later stages of MIS 5 or MIS4. From the above it will be seen that a great deal of information has been accumulated in recent years to confirm that current aridity of the region is not representative of the Quaternary past, and that humid periods have occurred, in the early Holocene and at several stages of the Pleistocene where conditions for human occupation and migration were favourable. While the direct chronological evidence linking climates and palaeo-environments still lacks the precision and detail needed, for high resolution modelling reconstruction, larger numbers of luminescence dates from sediments together with radiocarbon dating of the more recent materials, and U/Th dating of speleothems and other carbonates, are now available from Arabian contexts, particularly in the SE.

Turning to the Red Sea there is at present a far smaller body of luminescence dates, and a need for systematic work on the sediments of both the terrestrial margins and marine shelves of the littoral fringe in order to enhance understanding of the interplay between tectonics, climate, sea level and landscapes. Lambeck et al. (2011) have reviewed data and models for shoreline reconstruction in the recent Quaternary, taking account of isostatic and tectonic considerations as well as the available stratigraphies and surviving evidence of former coastlines with temporal constraints. It is clear that a more comprehensive set of chronological constraints is needed. In considering southern corridors for past crossing of the Red Sea during glacial low stands, it is significant to note that the Bab al Mandab is considered to have maintained a connection to the Indian Ocean at all stages, but that current models point to periods where short sea crossings may have been episodically available at times of minimum sea level (e.g., for short periods at 55 ka, 75 ka, and 100 ka; also within the time intervals 12–32 ka, 61–68 ka, and 94–96 ka). Bailey and coauthors (Bailey et al. 2007; Bailey and Flemming 2008; Bailey and King 2011) have made the case for systematic exploration of both marine and contiguous terrestrial areas in order to identify sites associated with Palaeolithic human activities and dispersion. The recent DISPERSE project has explored both parts of this on the Arabian side of the Red Sea (Bailey et al. 2015; Hausmann and Meredith-Williams 2016; Winder et al. 2015; Sakellariou et al., this volume). As part of this work, a series of luminescence profiling and dating analyses have been conducted at SUERC, from marine cores in the vicinity of the Farasan Islands, and from three terrestrial sites shown on the right-hand map of Fig. 3c.

4 Luminescence Dating and DISPERSE

The focus of DISPERSE research is on the western Arabian escarpment and the now-submerged territory of the southern Red Sea and has included field surveys on land and underwater and marine surveys offshore in the Farasan Islands. It also draws on comparative research in adjacent regions in the Near East and southern Levant. The work reported here focuses on two strands of this research: Firstly, the ongoing search for submerged, surviving fragments of Late Pleistocene landscapes (Sakellariou et al., this volume), with archaeological potential, on the Saudi Arabian continental shelf and secondly, archaeological and palaeo-geographical investigations on terrestrial coastal environments in SW Arabia (Sinclair et al., this volume; Inglis et al. 2014a). The following sections summarise the luminescence dating work conducted so far in support of these activities. Technical reports have further details (Kinnaird et al. 2015b, c).

5 Dating Sediment Cores from the Farasan Continental Shelf

As part of the search for surviving fragments of past littoral landscapes that formed during Pleistocene periods of lowered sea-level, a marine survey was conducted using the Hellenic Centre for Marine Research (HCMR) vessel AEGAEO in the Farasan region in May-June 2013. Multiple geophysical techniques were used to characterise the bathymetry of the sea-floor and the nature of the sub-surface (Sakellariou et al., this volume), and gravity cores were collected for later analysis. The aim of this study was to assess the potential of luminescence methods to classify the stratigraphy within such cores to date key samples from the material, to define sedimentation rates, and if possible to identify Pleistocene units. This survey targeted two regions on the outer and inner Farasan shelves respectively, as shown in Fig. 3c. These areas are characterised by different bathymetry; (i) the outer shelf has prominent terraces at 70–80 m depth and 38–40 m depth on the continental shelf, and 120 m depth on the continental slope, whereas (ii) the inner shelf comprises a 120 m deep elongate basin bounded by NW-SE normal faults, with bathymetric terraces at 70–75 m and 122 m depth. Eighteen gravity cores and 2 box cores were recovered to variable depths up to 4.5 m. Of these cores, 10 cores were collected on the outer shelf, and 8 cores were collected on the inner shelf (Fig. 3c).

Sakellariou et al. (2013) summarized the cruise, locating samples and making initial interpretations of the geophysical and sedimentological data available from the gravity cores prior to further laboratory investigations. A subset of the cores was retained un-opened, to avoid potential light exposure, in order to facilitate luminescence profiling and dating.

6 Core Samples and Initial Luminescence Profiles

The available geophysical and sedimentological data for all the cores was reviewed, and on this basis, two cores were selected for initial luminescence investigation. Core FA6 (2.90 m long) was gathered (on 4/6/2013 at 17°02.35'N, 41° 11.23'E, at 83 m depth) from the outer shelf and core FA13 from the inner shelf (on 6/5/2013 at 17°12.838'N, 41° 55.131'E at a depth of 102 m). Both cores show magnetic susceptibility anomalies (Sakellariou et al. 2013); in FA6 at 205 cm depth (with Susceptibility Indices (SI) of 0 to 6–7), and in FA13 at 145 cm depth (from SI 0 to 3–4). A preliminary interpretation of the magnetic data was to raise the conjecture that it may be associated with the boundary between marine and terrestrial sediments (thus the Holocene and Late Pleistocene boundaries) (Sakellariou et al., this volume). Following the selection of the cores for luminescence characterisation and dating, the two cores were split under subdued light conditions at the HCMR and subsampled for screening using an SUERC portable OSL reader (Fig. 4). Sediment was extracted at regular intervals through both cores (about 10 cm), with tighter resolution sampling at the prominent breaks in the geophysical data, including the shift in magnetic susceptibility at 205 cm depth in FA6 and 145 cm in FA13. While the luminescence screening was being conducted under low light conditions as before, the other halves of the split cores were photographed and used to log the sedimentary sequences for the first time.

Luminescence profiles for IRSL, OSL, the IRSL/OSL ratio, and respective depletion indices were established using the methods described in Sanderson and Murphy (2010). These in turn were used to define a series of 'units' representing different episodes in the sedimentary and luminescence records, characterised by different trends and progressions in luminescence intensities with depth. This, in combination with the interpretation of the geophysical data, was used to select samples for further laboratory characterisation and quantitative luminescence dating. Material from the most promising zones for dating was identified, for example, where the initial profiles showed smooth downward progressions in luminescence signals. In addition, samples were taken for dating at prominent lithological breaks, and from above and beneath the magnetic susceptibility change.

Both cores showed increasing signal levels with depth. For core FA6 both IRSL and OSL showed similar depth progressions, spanning a dynamic range of one order of magnitude. Figure 5 shows the OSL profile alongside the original geophysical records. The changing OSL signal gradients around 150 cm may imply changing sedimentation rates, or changes in luminescence sensitivity gradients. To further assess these, and to gain an insight into the apparent age range implied by these signals, it is necessary to calibrate the material. This can be done either using an external radiation source on the bulk profiling samples or in the laboratory using prepared material (cf. Burbidge et al. 2007b). The pronounced magnetic susceptibility anomaly at 205 cm depth did not correspond with a step change in luminescence intensity, nor did the accompanying sedimentary logs show a fabric change, suggesting revision of the initial interpretation prior to opening the core.

The situation with core FA13 from the inner shelf is, however, different as shown in Fig. 6. Here OSL intensities

Fig. 4 Luminescence screening of cores FA6 and FA13 at HCMR using the SUERC portable OSL system (Sanderson and Murphy 2010)

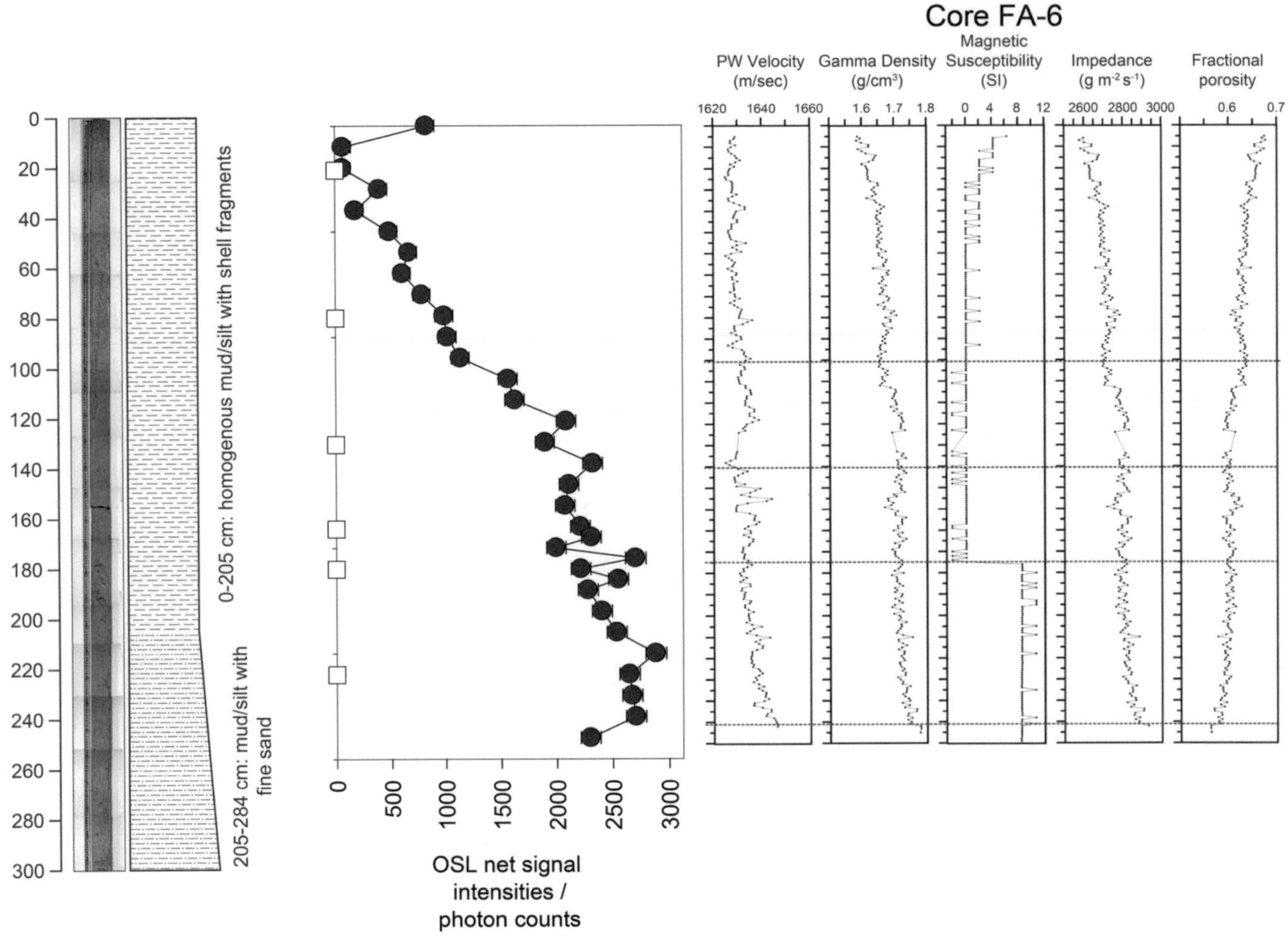

Fig. 5 OSL profile for core FA6 from the outer shelf, together with geophysical data

vary over 3 orders of magnitude, with similar signal levels to those observed in core FA6 for the upper 145 cm but with abrupt increases in signal intensities below this transition. This is accompanied by changes in sedimentary character and a series of magnetic susceptibility anomalies. At this stage of the analysis the evidence pointed toward the basal section of core FA13 as potentially exhibiting an age step corresponding to the Holocene/Pleistocene boundary, with FA6 most probably coming from younger material deposited at higher sedimentation rates in the outer shelf. The laboratory characterization and dating stages therefore aimed to assess these observations and interpretations.

7 Laboratory Profiling Measurements from Cores FA6 and FA13

Having thus established a relative framework, laboratory analysis progressed, first to calibrated luminescence screening and characterization of the profiling samples, then to quantitative dating runs from the samples selected above. Paired aliquots of 150–250 μm "quartz" and "polymineral" concentrates, and polymineral fines (4–11 μm) were prepared (cf., Burbidge et al. 2007b) and subjected to a simplified dose determination using a Risø DA15 OSL/TL reader. The reader was fitted with a combination of Schott BG39 + Corning 7/59 + Schott GG400 filters to detect in the region 390–450 nm, and isolate the main emission of the feldspar fraction. Although it might have been hoped that long range aeolian sands, potentially transported from Saharan or Arabian sources in different palaeoclimatic systems might have been present in the cores, quartz yields in 90–250 μm fractions were extremely low, particularly from the finer sediments in the upper parts of the cores. The quartz fractions recovered also exhibited low OSL sensitivities. Similarly, sand sized feldspar yields were limited with the exception of the coarser units at the base of core FA13. At this stage, it is unclear whether the absence of fine sands in the upper layers of the samples relates to the hydrodynamic processes associated with deposition and reworking of the marine sediments, or the limited quantities of Aeolian quartz being deposited at critical times in the development of the

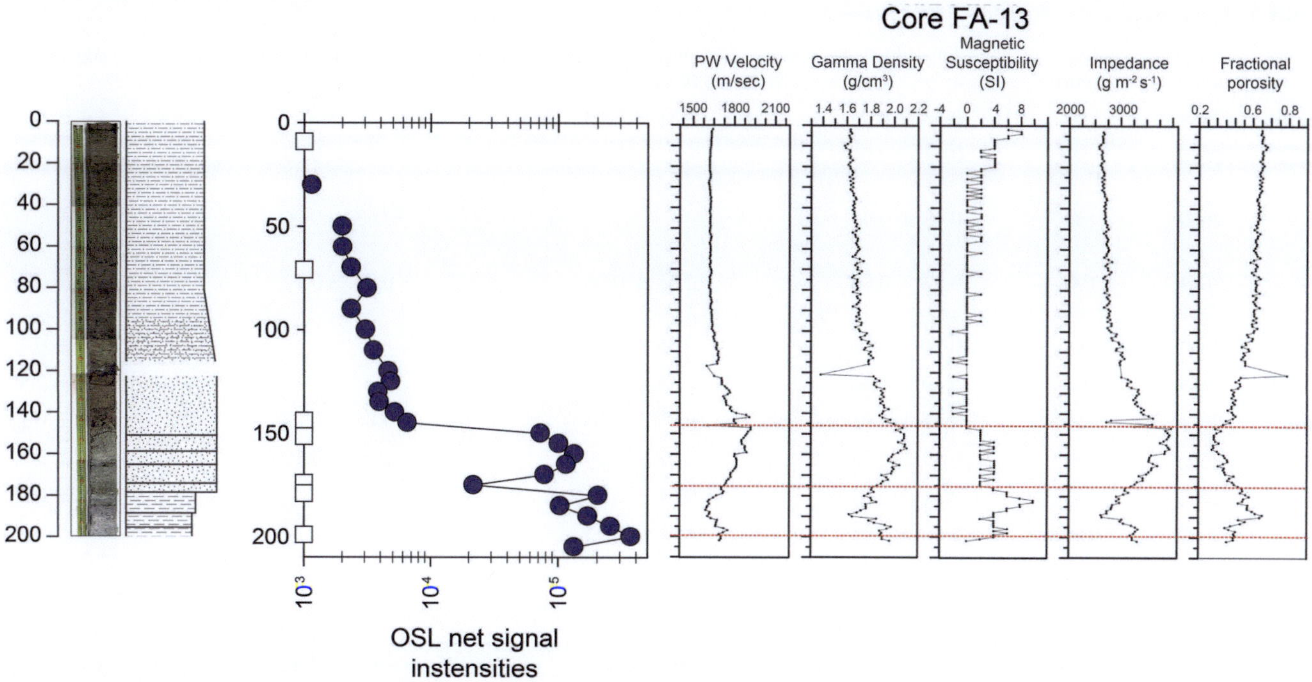

Fig. 6 OSL profiles from core FA13 from the inner shelf, together with geophysical data

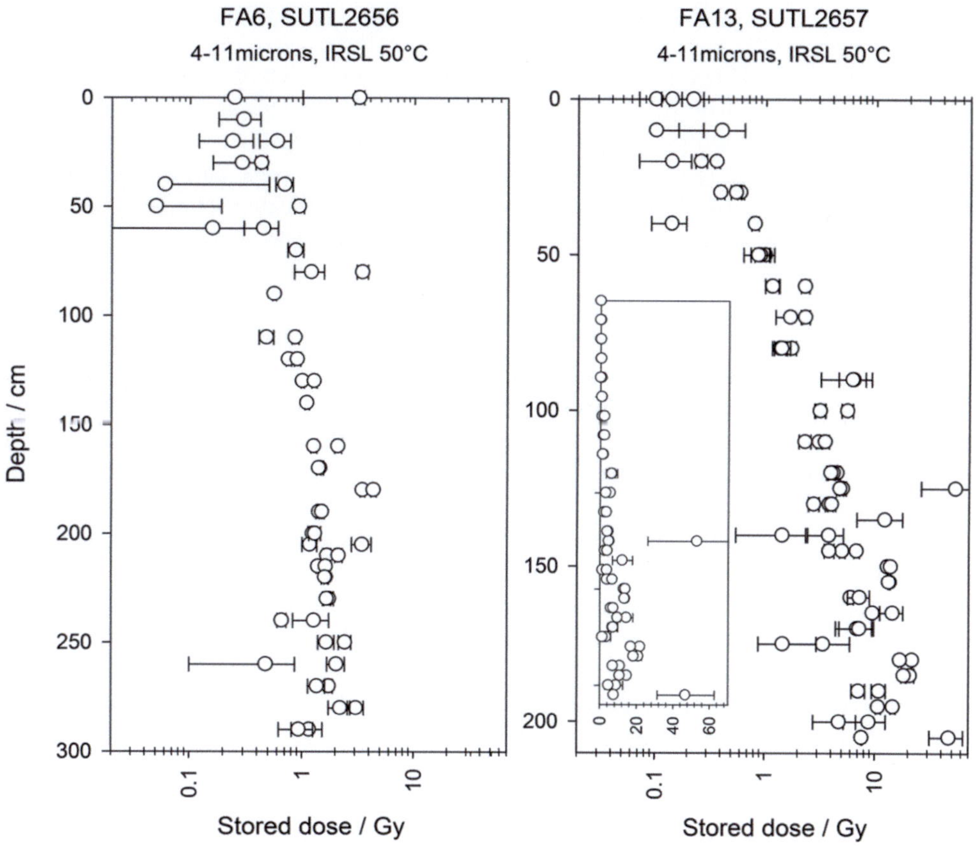

Fig. 7 Laboratory profiles from cores FA6 and FA13 showing the sedimentary logs, range of screening signals and stored dose values from each core

Table 1 Luminescence ages for cores FA6 and FA13

Core	SUTL no.	Depth (cm)	Total dose rate (mGy a^{-1})	IRSL SARA stored dose (Gy)	IRSL SAR stored dose (Gy)	Age (ka)
FA6	2658	20	0.38 ± 0.05	0.51 ± 0.03	–	1.4 ± 0.2
	2659	90	0.34 ± 0.04	–	0.36 ± 0.23	1.1 ± 0.7*
	2660	150	0.48 ± 0.09	–	0.63 ± 0.09	1.3 ± 0.3*
	2661	190	0.53 ± 0.06	2.05 ± 0.06	0.82 ± 0.13	1.6 ± 0.3*
	2662	210	0.51 ± 0.06	0.76 ± 0.20	–	1.5 ± 0.4
	2663	260	0.49 ± 0.06	2.85 ± 1.16	–	5.9 ± 2.5
FA13	2664	10	0.47 ± 0.05	0.95 ± 0.04	0.88 ± 0.22	2.1 ± 0.2
	2665	70	0.49 ± 0.05	2.73 ± 0.09	1.06 ± 0.19	5.6 ± 0.7
	2666	145	0.51 ± 0.05	0.51 ± 0.14	–	1.0 ± 0.3
	2667	150	0.68 ± 0.06	10.1 ± 0.55	19.9 ± 0.7	15.0 ± 1.5
	2668	175	0.64 ± 0.06	14.1 ± 1.5	13.5 ± 0.4	21.9 ± 3.2
	2669	180	1.70 ± 0.11	28.9 ± 9.0	24.2 ± 0.8	17.0 ± 5.4
	2670	200	1.10 ± 0.08	20.1 ± 4.4	25.2 ± 0.6	18.3 ± 4.2

cored sequences. However, fine grained (4–11 μm) polymineral fractions with good infra-red stimulated luminescence (IRSL) sensitivities were available from all samples producing depth-progressive stored dose estimates. This phase was therefore selected for dating both cores. Figure 7 shows the resulting dose profiles from both cores, reproducing the trends and maxima in the luminescence screening data. The stored dose values through the entirety of core FA6 and the upper part of FA13 were low, suggesting that these layers were likely to give later Holocene dates. Larger stored dose estimates, which in a low dose rate environment would correspond with substantially older dates, potentially of Late Pleistocene age, were only found at depth in core FA13.

Following the investigations outlined above, it was decided to exploit the fine (4–11 μm) polymineral phase in dating. Initially, equivalent dose determinations were made on 8 aliquots per sample using an IRSL SAR procedure (cf., Wallinga et al. 2000), with the objective of establishing whether reproducible data could be obtained, and to develop a dose response for the regenerative cycles (Table 1). Thereafter, dose estimates were obtained using an adapted SARA protocol, which incorporated long overnight preheats before first measurement to mitigate short-term fading effects (cf., Kinnaird et al. 2015a). The results of fading tests conducted between 6×10^5 s and 7×10^7 s indicate mean fading rates of 1.7 ± 0.014% per decade of time, suggesting that this worked. No further corrections for fading have therefore been made in the data which follow. In the original SARA method, groups of aliquots were formed with added doses, and a linear dose estimate made for each aliquot by scaling luminescence signals between the first (natural plus added dose) and a regenerated signal. Dose estimates are formed by regression of the added dose response curves to zero signal. In the modification suggested by Alexander (2007), long overnight preheating was introduced before the first readout, so that both natural and added doses are stabilised in the manner suggested by Sanderson (1988a). Automated readout with regenerative dose pre-heated in the instrument completes the sequence. This was implemented here with sets of 8 aliquots, sub-divided into 4 groups of 2 aliquots (natural, natural + β1, natural + β2, natural + β3). Irradiations were performed using an automated ELSEC irradiator equipped with a 1.85 GBq ^{90}Sr source (dose rate 1.76 Gy/min at the time of the experiment). The experimental conditions were as follows: IRSL from the natural and natural + β irradiated aliquots was recorded in a Risø DA15 reader, following 16 h preheating at 120°C. The samples were then irradiated within the reader, and their test dose responses were recorded after 30 s preheating at 220°C within the instrument. All IRSL measurements were conducted at 50°C for 60 s, with background signals recorded before and after stimulation. The data were reduced by integration and late light subtraction, and normalized IRSL

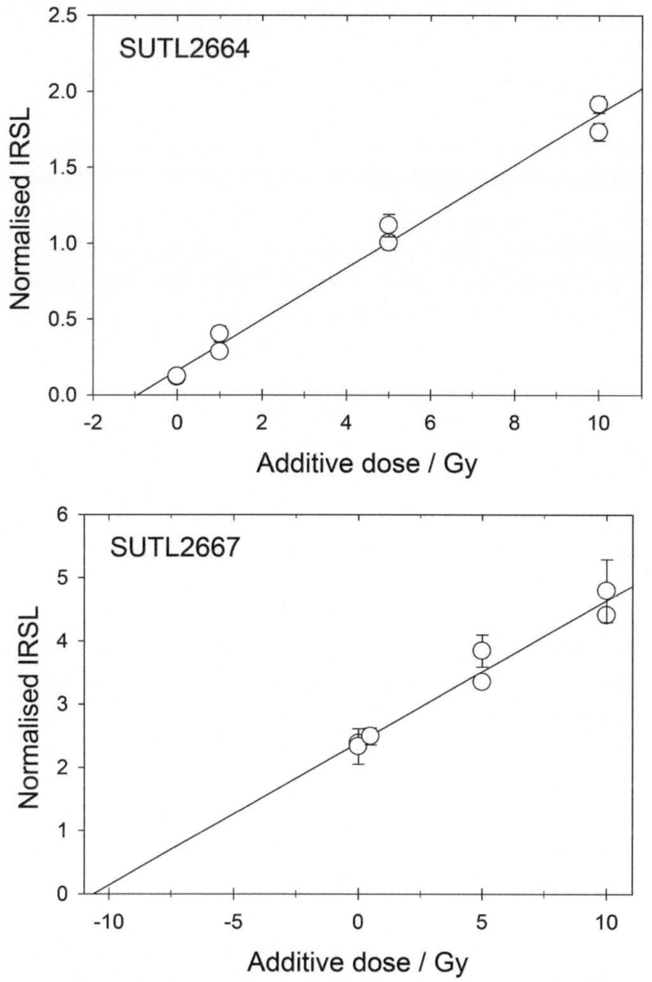

Fig. 8 IRSL SARA additive plots for the uppermost sample in FA13 (SUTL2664) and the first of the coarser units toward the base of the core (SUL2667)

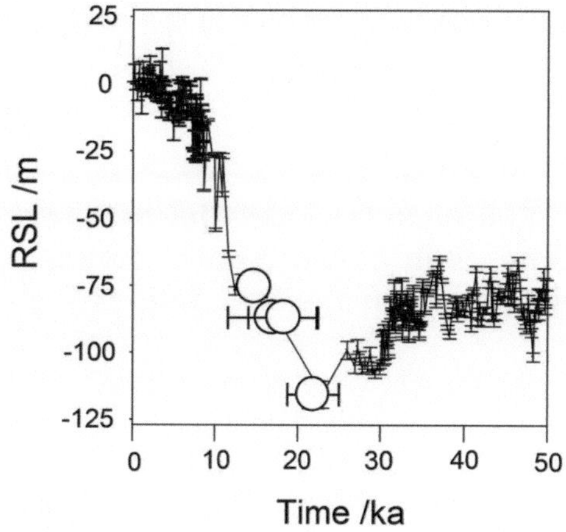

Fig. 9 IRSL SARA ages obtained for the basal unit of core FA13, plotted relative to a sea-level curve for the Red Sea (Grant et al. 2012)

ratios were used to construct additive dose response curves which were fitted by linear regression and extrapolated (using the methods of Scott and Sanderson 1988) to determine the stored doses. Figure 8 illustrates the process.

Dose rates to the dating materials were evaluated using analyses of the uranium, thorium and potassium concentrations obtained by HRGS coupled with beta dose rate measurement using TSBC (Sanderson et al. 1988b). As the sample quantity was limited, TSBC and HRGS measurements were performed on the same 20 g sub-sample, protected from light, and later used for mineral separations. To accommodate the low dose rates of these materials, duplicate 80 ks HRGS measurements were made. HRGS data, converted to dry infinite-matrix dose rates, were combined with beta dose rates, estimated water contents and alpha efficiency as inputs for the effective dose rate estimates. The contributions from the cosmic dose were modeled (Prescott and Stephan 1982; Prescott and Hutton 1988) by combining latitude and altitude specific dose rates (0.17 ± 0.01 mGy a^{-1}), with time-dependent corrections for water depth and overburden (for the period the terrestrial sediments accumulated). For the silts sampled in FA6 (0–300 cm depth) and FA13 (0–150 cm depth), the depth of water above the core attenuated the contribution of the cosmic dose rate to a few percent of the total dose. However, for the coarser-grained units sampled at the base of FA13 (from 150 cm depth), consideration was given to the palaeo-environment(s) at the time of deposition; from reconstructions of Red Sea relative sea-levels, it is apparent that the sea flooded rapidly in after deposition of the coarse-grained units, indicating that the cosmic dose contribution needs only to be modeled for a few percent of the total time it could have potentially been accumulating. The net effect of this is that the cosmic dose contribution is very minor relative to the total environmental dose rate at each of the dating positions. Subsequent investigations have shown evidence of excess uranium revealed through analysis of the low energy gamma ray lines from ^{234}Th in comparison with those from further down the decay series. Evidently this is not accompanied by the full decay series, and will be investigated further in future work.

Table 1 lists the IRSL SARA dating results based on the method described above. For core FA6 all data fall into the later part of the Holocene as expected from the profiling results discussed in the previous section. These samples are consistent with sedimentation rates between 0.4 and 1.4 mm a^{-1} with the lowest sample being consistent with a 6 ka marine highstand. For core FA13 the upper part (above 150 cm) covers a similar time interval in the Mid to Late Holocene. At 150 cm, there is a temporal discontinuity, beneath which Late Pleistocene material is encountered. In Fig. 9 the ages obtained from these lower samples have been superimposed on the sea level curve of Grant et al. (2012). Given the water depth of 102 m at time of sampling, the positioning of the earliest samples on this curve is consistent with a terrestrial Pleistocene environment at the time of deposition.

8 First Evidence for Late Pleistocene Sediments in a Near-Shore Marine Environment in the Red Sea

The cores examined here are the first two appraised out of a collection of 20 recovered offshore at the Farasan Islands by the research cruise of R/V AEGAEO in 2013. The IRSL SARA dates from one of them, core FA13, provides the first evidence so far which confirms that there are surviving fragments of formerly terrestrial palaeo-landscapes preserved under the Red Sea, confirming the hypotheses raised by Bailey et al. in the EFCHED and DISPERSE projects. Furthermore, it has highlighted the value of initial luminescence screening made using portable OSL equipment as an important tool in appraising sediment stratigraphies within cores, offering information on the range and magnitude of the ages recorded in each, and a means to map out and correlate discrete depositional events. The direct evidence from the HCMR cruise and the dating work confirms that such environments are conserved under some marine silts of 1.5 m depth in the position of core FA13. Having conducted both profiling and quantitative methods there appear to be potential for using profiling of similar core materials to generalize the chronology and extend the search for further examples of such environments. This is an important part of the longer-term search for Palaeolithic materials and sites in the nearshore marine environment. While there is a need for further work to replicate these results from other cores to extend the geographic range and to resolve U series behaviour as indicate above, this represents a significant step forward.

9 Analysis and Dating of Sediments Associated with Landscape Evolution and Archaeological Work

The aim of the terrestrial OSL investigations in SW Saudi Arabia was to construct chronologies for sediment stratigraphies in the Jizan and Asir regions and to produce the temporal framework to reconstruct palaeo-environments and -geographies through the Mid Pleistocene and Holocene relative to Palaeolithic migrations and occupations. Three areas were identified for detailed investigations: (1) Wadi Sabiya (Jizan), (2) Dhahaban Quarry (Asir), and (3) Hajambar Quarry (Asir) (Fig. 3c). The significance of sites 1 and 2 is that at each, lithic tools with Middle Stone Age (MSA) affinities were recovered in sediments indirectly dated by association to the Late to Middle Pleistocene. The significance of the third site is that it contains an environmental record, with evidence for a humid phase in the early Quaternary, and thus it is a valuable palaeo-environmental marker for this region.

Robyn Inglis visited the sites in January-February between 2012 and 2015 to collect samples from key units within the sediment stratigraphies for luminescence dating. In the absence of field spectrometry, bulk sediment samples were collected from around each dating position in different geometric schemes to permit reconstruction of the external gamma dose rate from adjacent stratigraphic units (full details are provided for each case study below). In this work, dose rates for the bulk sediment were quantified using HRGS and TSBC in the laboratory coupled with water content analysis. The gamma dose rates for each dating position were determined from the reconciliation of the estimates from the dating position and the adjacent bulk sediment samples. Equivalent dose determinations were obtained, in the first instance, by a quartz OSL SAR method (Murray and Wintle 2000). In some cases, the results go beyond the age limits of quartz saturation. In other cases, issues concerning the interpretation of mixed age dose distributions need to be considered relative to the context of samples, the age limits of the methods, and the association with external age constraints by direct or indirect reference to landscape classification or external information.

Full technical descriptions are provided for each case study in Kinnaird et al. (2015c). In brief, standard sample preparation procedures were used to extract and concentrate sand sized 150–250 μm quartz from each dating sample, the purity of which was checked by scanning electron microscopy, and when impurities or contaminants were noted, these separates were subjected to repeated HCl and HF washes. All measurements were made on the same Risø DA15 OSL/TL readers as described above, but with U340 detection filter packs to detect in the region 270–380 nm. OSL was measured at 125°C for 60 s, integrated over the first 2.4 s of measurement and subtracting an equivalent signal taken from the last 9.6 s. Dose response curves were constructed from 5 to 6 regenerative doses extending up to 200 Gy, or to 500 Gy for those samples whose natural cycle signals exceeded 200 Gy. SAR acceptance checks included recycling and internal dose recovery tests on the dose response curves, zero level checks and IRSL response tests. Aliquots were rejected from further analysis if they failed sensitivity checks (based on test dose response), SAR acceptance checks or where significant IRSL response was associated with anomalous equivalent doses.

Here the outcomes from three sites, Wadi Sabiya, Hajambar Quarry, and Dhahaban Quarry are discussed. The outcomes provide chronological constraints for Late Quaternary environments, including results with MIS3 and MIS5 associations. There are, however, unresolved challenges concerning the extent to which sampling methods can overcome post-depositional influences associated with weathered or reworked elements, and the ability, particularly in coastal marginal settings, of the data themselves to resolve unique parts of the sea level curves associated with the different parts of the glacial cycle. These challenges are discussed at the end of the section.

10 Wadi Sabiya

To the south of Jebel Akwah, 4 km distant, Wadi Sabiya and its tributaries incise wadi floodplain.

Sediments up to 10 m below the surrounding landscape. These deposits are capped by a volcanic tuff which extends for tens of kilometres over the surrounding area. It has been related to the eruption of Jebel Akwah dated by the K-Ar method to 0.44 ± 0.26 Ma (K-Ar; Dabbagh et al. 1984). The Wadi Sabiya has been extensively quarried, revealing thick sedimentary sequences consisting of basal wadi gravel deposits, overlain and cut into by subsequent wadi channel and associated floodplain deposits. Loose blocks of tuff were observed on the very top of the section that had been isolated by quarrying, consistent with but laterally offset to in-situ tuff deposits. On this basis it has been assumed that the sedimentary sequence is overlain by the aforementioned volcanic tuff. The site was visited in January 2014 by Robyn Inglis (Inglis et al. 2014a, b). In the sediment stratigraphy investigated here, the channel deposits contain a worked shale clast, with three long blade removals (potentially the result of soft-hammer percussion, indicative of a MSA technology; Bailey et al. 2015). If this typological assessment is correct, and it is found that the wadi and channel deposits are older than the tuff, then this artefact becomes a contender for the earliest MSA evidence in Arabia given the existing K-Ar dates. To address this, samples were collected for OSL dating from throughout this quarry section, from (i) the silty, fine- to medium-grained, semi-indurated sands which form the fill of the lower channel deposit (SUTL2673), (ii) the finer, sandy silt of the overlying channel deposit (SUTL2672), and (iii) the coarser, sandy silt matrix of the uppermost channel deposit (SUTL2671). The samples were collected with no field spectrometry or portable OSL profiling to characterise the luminescence stratigraphies. However, in addition to the formal dating samples, bulk sediment samples were collected around each of the dating positions, for use in reconstructing the external gamma dose rate. For each position between 18 and 24 dosimetry samples were collected to represent stratigraphic units within 25–30 cm of the dating tube positions.

Dose rates to the dating materials were obtained by HRGS, coupled with beta dose rate measurement using TSBC. To estimate environmental dose rates for the 3 dating tubes examined, composite samples were constructed from the bulk sediments and measured by HRGS, giving effective gamma dose rates between 0.4 and 0.5 mGy a^{-1}. These are comparable with the measured values recorded from dating tubes (0.4–0.5 mGy a^{-1}). Both data sets were combined to estimate the total dose rates for the samples, which ranged from 1.4 and 1.5 mGy a^{-1}. Equivalent doses were determined, initially, by a quartz SAR OSL protocol on 32 aliquots per sample. All samples showed heterogeneity in equivalent dose estimates, with common central tendencies (~35–50 Gy), but with low dose (~10–20 Gy) and high dose (>100 Gy) outliers, leading to some ambiguity in assigning luminescence ages (e.g., Figure 10 (right, a)). Subsequently, 32 additional aliquots per sample were subjected to a simplified two-step SAR protocol to increase the statistical power (e.g., Figure 10 (right, b, c)). For each sample, linear and weighted means and central and common age model estimates of stored dose were indistinguishable between the 32 aliquot and 64 aliquot sets (Fig. 10 (right)). Luminescence ages (Table 2) were calculated using standard microdosimetric models, with uncertainties that combined measurement and fitting errors from the SAR analysis, all dose rate evaluation uncertainties, and allowance for the calibration uncertainties of the sources and reference materials.

The quartz OSL SAR sediment ages obtained from the strata at Wabi Sabiya range between 15 and 29 ka, when weighting the stored dose value toward the low dose component, and 22 and 30 ka, when weighting the stored dose value toward a central tendency. It is clear from the equivalent dose distributions that the majority of quartz aliquots from these sediments are incompatible with the geological expectations, that is, deposition within MIS5 (unless there are other mechanisms for reworking or partially resetting the OSL signals in these sections after deposition). This raises questions about the assumed stratigraphic relationship between these wadi deposits and the potentially overlying tuffs, as well as the post-depositional history of the sediments. It would take further fieldwork and investigations to resolve these questions.

11 Hajambar Quarry

A quarry 10 km northwest of Hajambar (Fig. 3c), adjacent to the road connecting Al Haridhah and Al Edati, was identified as containing wadi deposits overlying a lava flow. The quarry is approximately 15 m deep, and 100 m wide. The top of the sequence is at ~30 m asl. A reconnaissance visit in June 2012 identified laminated silts and grits at the base of the quarry, interpreted as wadi overbank deposits, laid down during a humid phase and potentially providing a

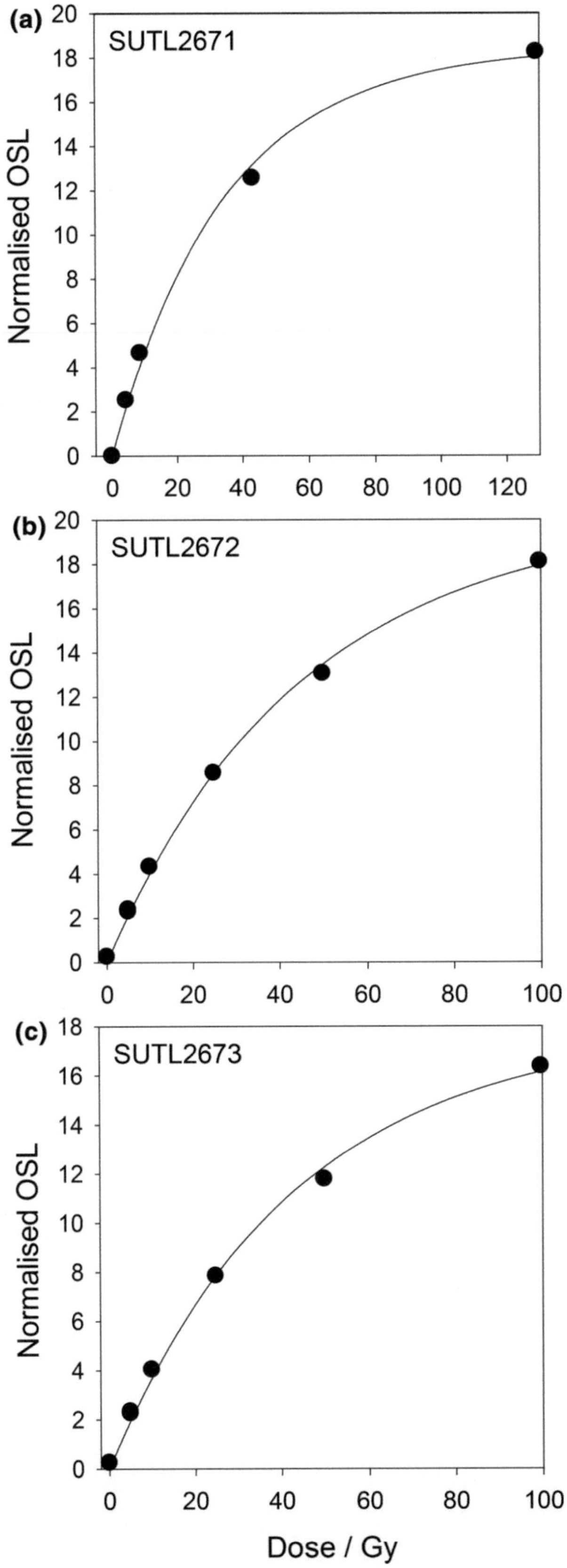

Fig. 10 Illustrative dose response curves (left hand figures) from quartz SAR-OSL determinations from Wadi Sabiya, and (right hand side) dose distributional plots. It is notable that the dose distributions observed for the **a** 32 aliquot and **b** and **c** 64 aliquot sets are indistinguishable within the limits of error; although over-dispersion is emphasised in the latter

Table 2 Luminescence ages for the sediments sampled at Wadi Sabiya

Sutl no.		Weighted mean		Central dose model	
	Dose rate (mGy a^{-1})	Stored dose (Gy)	Age (ka)	Stored dose (Gy)	Age (ka)
2671	1.40 ± 0.08	21.8 ± 1.3	15.5 ± 1.3	37.3 ± 2.8	26.6 ± 2.4
2672	1.50 ± 0.08	22.3 ± 0.7	14.9 ± 0.9	32.1 ± 2.4	21.5 ± 1.9
2673	1.35 ± 0.07	39.1 ± 1.1	29.0 ± 1.8	40.7 ± 2.4	30.0 ± 2.4

rare palaeoenvironmental marker from SW Arabia. These deposits are overlain by a silty sand unit dominated by rhyzoliths and carbonate concretions. The quarry section is cut by a number of faults and fractures, which offset the sediments, and have also acted as conduits for fluid movement and precipitation of carbonates. These were logged and sampled in November 2012 (Bailey et al. 2012) for micromorphological and sedimentological analysis, and three tubes were taken for OSL dating (SUTL2805-07). The site was revisited in February 2014 to collect further OSL samples (Inglis et al. 2014b). This later visit was complicated by the partial flooding of the quarry, with water covering the base of the sediments sampled in 2012 and the trench dug below. It is possible that the water had, at some point in the very recent past, covered the laminated sediments. Two OSL samples and background samples were taken from the upper part of the laminated silts (SUTL2808-09). In stark contrast to conditions in 2012 when the sections were completely dry, the sediments were wet in 2014 at time of sampling. In addition to the formal dating samples, single composite bulk sediment samples were collected around each of the 2012 samples, with sets of 18 bulk samples from within 0–30 cm of the 2014 samples, to reconstruct the external gamma dose rate.

These sediments were dated using the combination of equivalent dose determinations by quartz SAR OSL, coupled with dose rate determinations using the combination of HRGS and TSBC. Composite sediment samples were constructed using the bulk surrounding each dating position at spacings of 0–8 cm, 8–15 cm, and 15–25 cm. To estimate environmental dose rates for the 2014 dating tubes examined, composite samples were constructed from the bulk sediments and measured by HRGS, giving effective gamma dose rates between 0.3 and 0.5 mGy a^{-1}. These are comparable with the measured values recorded from dating tubes (0.4–0.6 mGy a^{-1}). Both data sets were combined to estimate the total dose rates for the samples, which ranged from 0.9–1.7 mGy a^{-1}. 16 aliquots of quartz were prepared for each sample and analysed by the quartz OSL SAR method. The data derived from the SAR dose determinations were analysed at two scales; composite data sets were explored using Excel spreadsheets and Jandel Sigmaplot software, while individual dose response curves were analysed using the Risø 'Analyst' program. Individual decay curves were scrutinised for shape and signal consistency, and compared relative to other aliquots in their 16 aliquot and preheat set. Individual (aliquot-specific) dose response curves were determined using an exponential fit. Only one produced a finite age (SUTL2805; Fig. 11), although with considerable scatter in its equivalent dose distribution, indicating some mixing of materials with residual ages. The remaining four samples showed natural luminescence signals in excess of the saturation signals for the regenerative dose curves. In an attempt to extract age data from the two remaining samples, additional regenerative dose points were added up to a nominal 500 Gy; for one of these samples (SUTL2809), a slight increase in individual normalised dose response was noted; however, for the second of these (SUTL2808), the individual normalised dose responses showed no additional growth.

The uppermost sample in the quarry (SUTL 2805; Fig. 11) yielded an age estimate of approximately 110 ka (Table 3), with weighted standard deviation across the equivalent dose distribution (30 ka) and the weighted standard error (10 ka) indicated. The lower samples in the quarry face did not yield finite ages. For two of these samples, a number of individual aliquots did yield natural normalised signals within the limits of their respective regenerative dose curves, suggesting sediment ages in excess of >90 ka (SUTL2806) and in excess of >55 ka (SUTL2807). Minimum age estimates for the lower strata are in excess of >120 ka. Notwithstanding the limitations of the Hajambar quartz, which saturates early, the quartz OSL SAR data do provide a temporal constraint to the end of the humid phase based on the upper two samples on the environmental section of some 90–100 ka. Since the quartz OSL signals from the majority of samples are in saturation there would be scope for exploring age extension methods in the future work if finite age estimates for these deposits are required.

12 Dhahaban Quarry

A quarry 3 km south of Dhahaban (Fig. 3c), adjacent to the coastal road connecting Hajambar and Dhahaban, was selected for study following the discovery of abundant surface lithics (>700 in number), as well as lithics embedded with the sediments (Bailey et al. 2015; Inglis et al. 2014a).

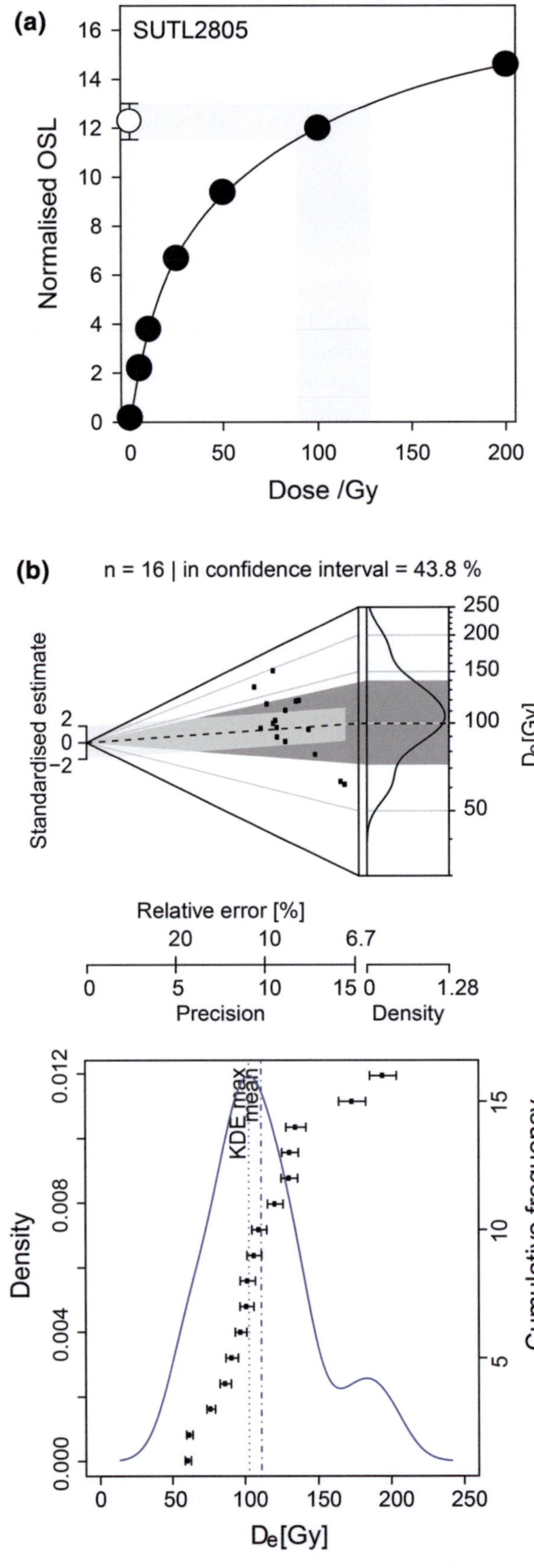

Fig. 11 Dose response curve and dose distributions from sample SUTL2805 from Hajambar, leading to a finite quartz SAR OSL age within MIS5

Table 3 Luminescence ages for the sediments sampled at Hajambar quarry

Sutl no.	Dose rate (mGy a^{-1})	Stored dose (Gy)	Age (ka)
2805	0.94 ± 0.04	100 ± 25 (7)	110 ±30 (10)
2806	1.01 ± 0.04	>90 Gy	>90
2807	1.58 ± 0.05	>50–55 Gy	>30
2808	1.06 ± 0.04	>200–250 Gy	>120–150
2809	1.15 ± 0.04	>150 Gy	>130

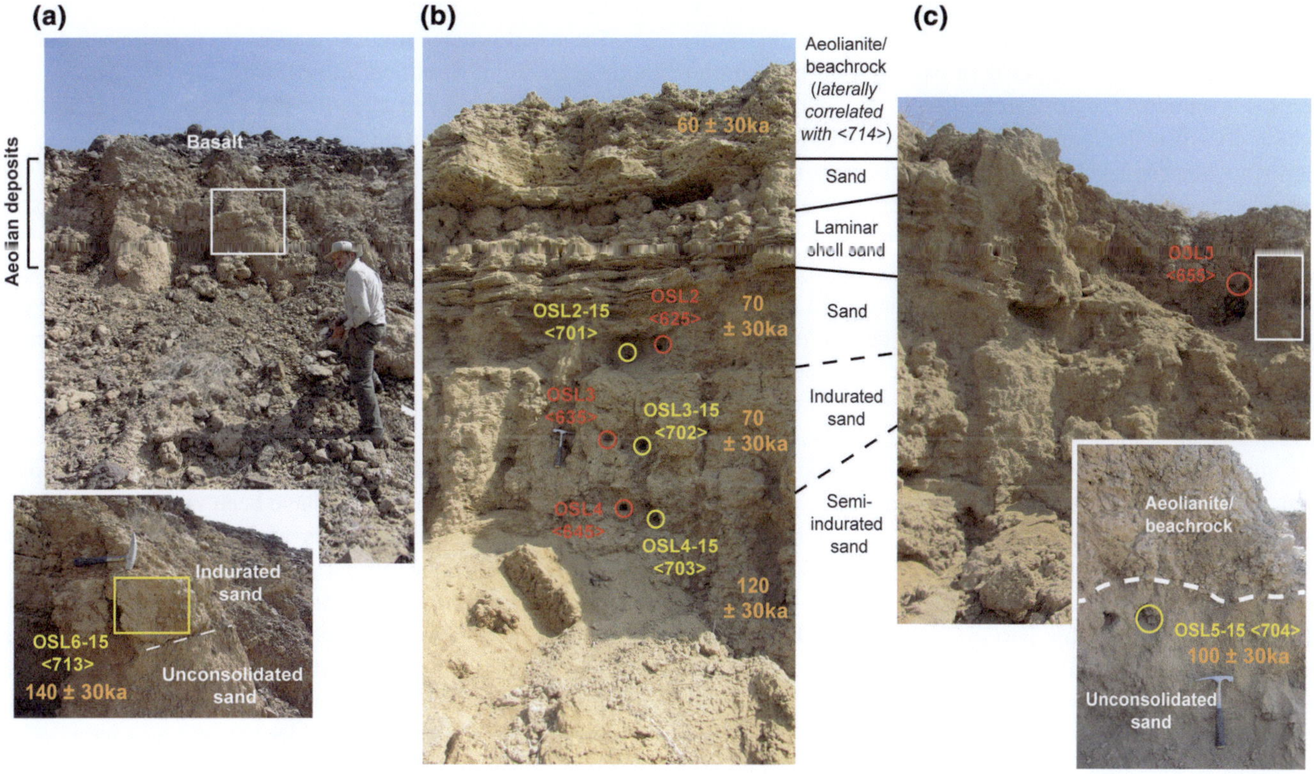

Fig. 12 The stratigraphy at Dhahaban quarry and positions of samples gathered in 2014 (red symbols) and 2015 (yellow symbols) together with the age determinations from the later series of analyses. Samples were collected in three parts of the quarry labelled a, b and c below, with additional cutting back between the 2014 and 2015 sampling

The deposits at the Dhahaban Quarry site comprise a basal wadi cobble unit (restricted to one area and containing 19 andesite artefacts with MSA affinities), overlain by coral terraces and shallow marine sediments in different parts of the locality, the latter capped by beachrock and aeolianites (Inglis et al., this volume). The upper surfaces of the marine/beach deposits are ~6 m asl, and the complex overlies a degraded, westward dipping, basalt slope. A ~6 m deep section in the main quarry area sampled for OSL analysis comprised (from the base up): (1) a 50 cm-thick, semi-indurated fine- to medium-grained sand, with laminae of coarser sands; (2) approximately 70 cm thick, indurated medium-grained sands, containing marine shells; (3) about 50 cm of unconsolidated sands; (4) about 170 cm of laminated consolidated, shell-dominated sands grading into more poorly consolidated sands; and (5) the capping aeolianite. The site was visited in reconnaissance in 2012, then again in 2014, for more systematic geomorphological mapping and archaeological surveying (Inglis et al. 2014b). The section was first sampled for OSL in 2014, with 4 samples taken from two locations within the quarry; units in the main marine quarry sedimentary stratigraphy were sampled three times across in two locations ~40 m laterally offset along the main cut (Fig. 12a, b); the fourth sample was removed from 100 m E of the quarry section, the aeolianite overlying the basalt jebel (Fig. 12c).

Following preliminary OSL analyses, outlined below, which produced age estimates which were systematically younger than expected, the sections were re-sampled in 2015 with further cleaning and cutback (Inglis et al. 2015). Samples were taken to date deposition of the shallow marine deposits within the raised beach complex (SUTL2674-76 and SUTL2810-13) to provide age estimates of the high sea stand in this part of the Red Sea and terminus ante quem (TAQ) estimates for deposition of the unit containing stratified MSA lithics. Additional samples were collected from the aeolianites capping the sequence (SUTL2677 and SUTL2814-15).

The samples collected in 2014 were subjected to OSL analyses in 2015 (prior to the re-visit of the site that year). The outcomes were younger than expected on geomorphological grounds, ranging from 25–40 ka, depending on which dose distributional model was favoured, but all from periods when sea level was expected to be very much lower than contemporary settings or former Quaternary high stands. This raised questions concerning sample integrity and the extent to which potential weathering, bleaching or thermal signal erosion prior to collection might have led to underestimation of the original depositional ages. The subsequent set of samples were cut back further, although for the indurated sands there were limits to what was practical in the field, and it may be necessary to return for further sampling with mechanical cut-backs to ensure that surface alteration has been overcome. Ages were obtained using separated quartz SAR OSL analysis, coupled with dose rate analysis by HRGS and TSBC. Given the close association between the 2014 and 2015 samples, the gamma dose rates measured by HRGS for the paired sets were reconciled in a dosimetric model. This yielded effective gamma dose rates between 0.4 and 0.5 mGy a^{-1}, indistinguishable from those measured directly on the dating materials (0.4–0.5 mGy a^{-1}). Total effective dose rates for the 2015 samples ranged between 1.0–1.5 mGy a^{-1}, comparable to those obtained in 2014, which varied over the same range. Equivalent dose determinations were made using 16 small aliquots per sample. All samples showed heterogeneous dose distributions, with common central tendencies (>60 Gy), but with both low dose (30–80 Gy) and high dose (100–180 Gy) components (outliers at 2σ). Also, for some samples, individual aliquots returned normalised natural OSL signals above the saturation limits of their dose response curve. For each sample, different dose components were modelled with respect to the elevation of the raised beach complex, the expected age of the marine terrace, and the required uplift rates (calculated with respect to the present elevation and the potential age of the terrace). In comparison with the results from the 2014 samples, quartz SAR OSL ages for the marine sediments fall in the range 120 ± 40 ka to 50 ± 20 ka, for SUTL2812 and 2811 respectively, and for the aeolianites, 90 ± 30 ka to 60 ± 30 ka, for SUTL2813 and 2815 respectively (Table 4).

The fossil beach complex overlies (in part) a wadi cobble unit containing stratified lithics with Middle Stone Age affinities; the sediment ages for the terrace provide a TAQ for cultural activity in this area from 120 ka.

While these data have extended the estimated ages of the Dhahaban units back to early parts of the glacial cycle including MIS5 and provide a general link to the Middle Palaeolithic artefacts obtained from the area, there are still difficulties in reconciling the sample stratigraphy, the heterogeneous dose distributions and sedimentary sequence within reconstructed sea level histories which need to be given further attention in future work. The difficulties are illustrated in Figs. 13 and 14. Taking the sea level curves of Grant et al. (2012), Fig. 13 identifies the times in the past 150 ka when sea levels would have been within range, under a series of uplift scenarios (as average rates), to position coastal material in the current location (Wilson et al. 2014). This indicates that three parts of MIS5 might have been associated with near-coastline positions at (a) 80–88 ka (MIS5a–b), (b) 100–110 ka (MIS5 c–d), and (c) 120–135 ka (MIS5e and before), with mean uplift rates of 0.5, 0.2 and 0 mm/a. It is instructive to consider how the data fall relative to these potential periods for possible deposition.

This is shown in Fig. 14 for the three marine samples at the base of the section. The coloured bands on the figure indicate the three uplift scenarios within MIS5 which could potentially provide plausible periods for deposition of sediments in this location. Clearly there are aliquots within each sample which fall both above and below the shaded areas, and at this stage the data sets do not provide an indication of a unique assignment to one of the three uplift scenarios. The

Table 4 Luminescence ages for the sediments sampled at Dhahaban quarry

Sutl no.	Dose rate (mGy a^{-1})	Stored dose (Gy)	Age (ka)
2810	1.50 ± 0.07	109 ± 45 (14)	70 ± 30 (10)
2811	1.26 ± 0.06	86 ± 33 (9)	70 ± 30 (10)
2812	1.33 ± 0.06	153 ± 45 (17)	120 ±30 (15)
2813	1.49 ± 0.05	143 ± 38 (17)	100 ±30 (10)
2814	1.04 ± 0.06	145 ± 68 (26)	140 ±70 (30)
2815	1.09 ± 0.07	64 ± 30 (9)	60 ± 30 (10)

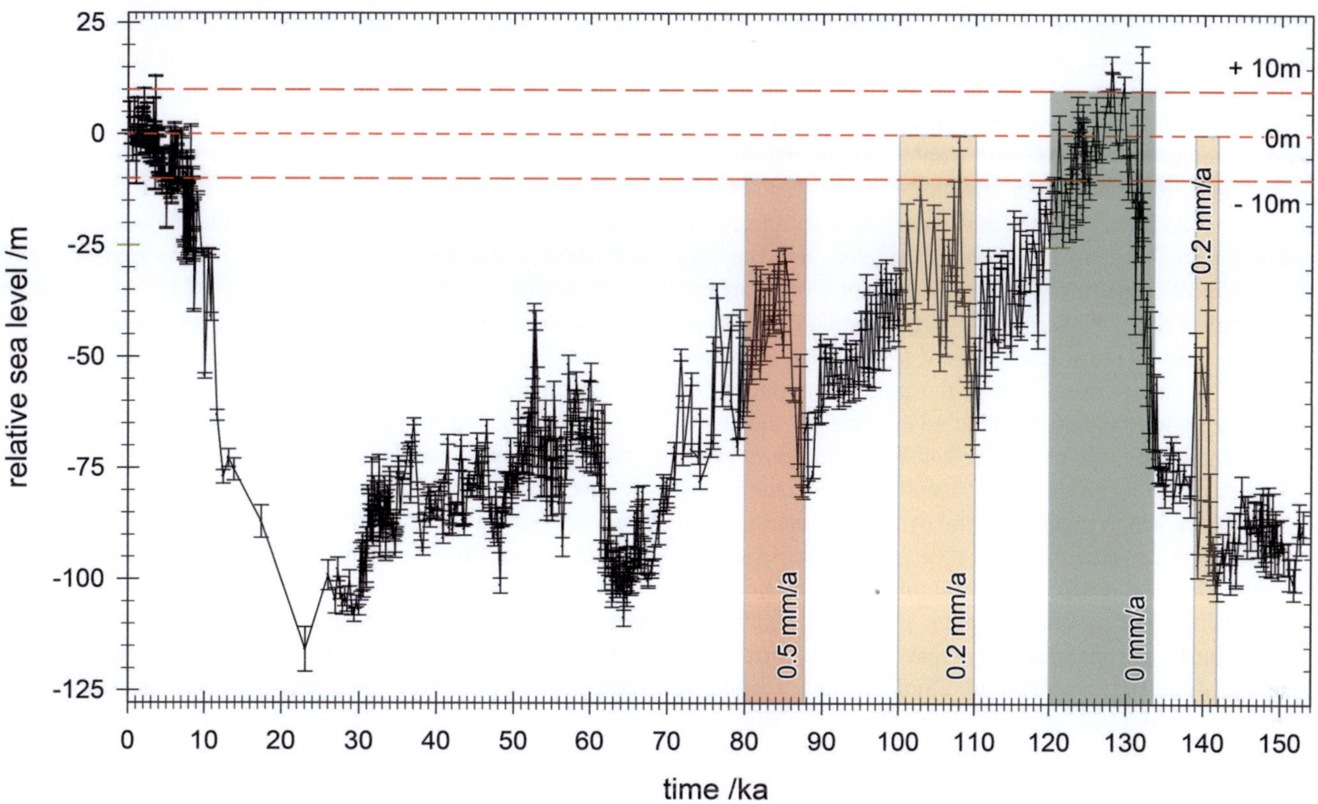

Fig. 13 Relative sea level curve for the Red Sea Region (Grant et al. 2012), annotated with the approximate uplift rates necessary to place the Dhahaban fossil beach complex at its present position assuming different luminescence depositional ages

Fig. 14 KPE plots (after Dietze et al. 2013) for SUTL2810–2812, the marine sediments at the base of the section superimposed on the uplift scenarios of Fig. 13

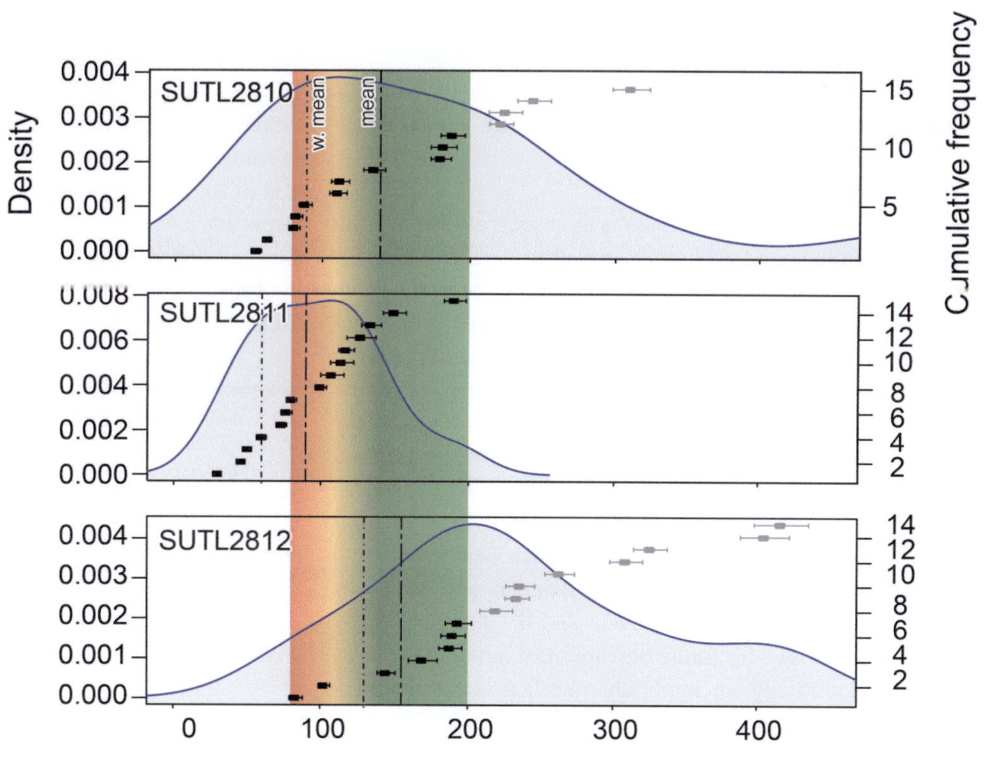

central dose estimates for SUTL2810 span the youngest and oldest plausible MIS stage 5 possibilities, depending on whether weighted or non-weighted analyses are preferred. For sample SUTL2811 only the unweighted estimate falls within a plausible period, and would require the highest uplift rate considered (0.5 mm a^{-1}). Only for sample SUTL2810 do both weighted and unweighted estimates fall within MIS5e, which might be considered the scenario which is most consistent with current estimates of uplift rate in the region (cf., Lambeck et al. 2011; Pedoja et al. 2014, and references therein). Even there it is clear that some aliquots tail to higher doses raising the possibility that earlier Quaternary high stand cycles, beyond the range of the quartz SAR method, may also need to be considered in future work. This site, which has important Palaeolithic associations and a stratigraphy which implies several cycles of induration and sedimentation, appears therefore at this stage to raise several possible scenarios relative to coastal processes and humidity cycles. While the second sampling campaign resulted in earlier central age estimates and makes MIS5 attributions plausible, challenges remain to eliminate any remaining limitations associated with sampling and to achieve a chronology which can separate the specific MIS stage from the plausible candidates listed above, and perhaps from earlier Quaternary highstands. To our knowledge this is challenging too for other dating methods, and it would seem that additional field and laboratory studies are needed to resolve the limitations of current data sets in these important landscapes.

13 Discussion and Implications for Future Research

As indicated in this chapter there is emerging evidence from sediment dating studies in the terrestrial environment for cycles of humidity and aridity going back to at least 200–300 ka. This sets an important series of challenges to ongoing research in the region. Current methods have the capacity to work within the timeframe up to 200–300 ka and, with new developments, to perhaps extend it to earlier Quaternary periods. However, new work will be needed both on method development and validation and on identification, detailed sampling and analysis of such earlier Quaternary units in order to refine chronologies for landscape and palaeoenvironmental evolution studies. Coupling of the Arabian and Red Sea environments to the monsoon cycle and the African aridity cycle is also an area for future development, and the Red Sea region is potentially an important interface zone for such studies. In this chapter, work associated with the DISPERSE project has also been presented in both near-shore marine settings and in terrestrial settings. Linking the results of direct sediment analyses and chronologies to current knowledge of sea level history during the last glacial cycles is both important to setting the palaeoenvironmental context for human dispersal and current archaeological exploration.

In the marine application presented here the utility of luminescence profiling has been clearly demonstrated as a new and critical part of the search for Late Pleistocene sediments with terrestrial associations in today's marine environments. The case illustrated of the Farasan core FA13, which has an erosional discontinuity below which Late Pleistocene material survives from the end of the last glacial cycle in a contemporary sea water depth of 80 m, shows the combined utility of both profiling and luminescence dating. It provides the first concrete verification of the long-searched-for evidence that this part of the Red Sea margin contains the predicted environments where the search for Palaeolithic material in former littoral fringes can be conducted. The other case studies presented here, in association with DISPERSE, are related to the ongoing search for landscapes and sites associated with Palaeolithic archaeology and associated palaeo-landscapes in terrestrial settings. The examples presented include former water-lain sediments at Wadi Sabiya, Hajambar and Dhahaban, with palaeoenviromental and archaeological significance. Results obtained so far at these sites span a period from 20 ka up to >150 ka, based on quartz SAR dating. At this stage, there are outstanding questions concerning the extent to which sampling cut-backs of the older units are needed to overcome post-depositional alteration and potential thermal erosion of signals from earlier Quaternary periods. In the case of the older units such as those at Hajambar, and potentially at Dhahaban, age extension work may be needed to push the age limits back to and beyond MIS5e. There is more work needed to refine chronologies sufficiently to provide unique separation of some of the subdivisions of MIS5, which are of critical importance to materials in the littoral fringes of the Red Sea. But notwithstanding those future challenges, it is quite clear the luminescence dating is making important contributions to understanding the environments of the Red Sea.

Acknowledgements The work presented in the section on luminescence dating and DISPERSE was supported by ERC Advanced Grant 269586 DISPERSE (Dynamic Landscapes, Coastal Environments and Hominin Dispersals) and by the Saudi Commission for Tourism and National Heritage (SCTH) through its President, Prince Sultan bin Salman bin Abdul Aziz, and his staff. The support, involvement and contributions from Prof. Geoff Bailey, Dr. Dimitris Sakellariou and Dr. Robyn Inglis are gratefully acknowledged. Dr. Alan Cresswell, Dr. Lorna Carmichael and Mr. Simon Murphy are also acknowledged for their help in the laboratory. The work presented here formed an invited presentation at the Jeddah workshop on the Geological Setting, Oceanography and Environment of the Red Sea, held in February 2016. We would like to thank the Saudi Geological Survey, and particularly Dr. Najeeb Rasul, for hosting this workshop and supporting the publication. This is DISPERSE contribution no. 44.

References

Adamiec G, Bailey RM, Wang XL, Wintle AG (2008) The mechanism of thermally transferred optically stimulated luminescence in quartz. J Phys D: Appl Phys 41:135503–135517

Adamiec G, Duller GAT, Roberts HM, Wintle AG (2010) Improving the TT-OSL SAR protocol through source trap characterization. Radiat Meas 45:768–777

Aitken MJ (1985) Thermoluminescence dating. Academic Press, London

Aitken MJ (1998) An introduction to optical dating: The dating of Quaternary sediments by the use of photon-stimulated luminescence. Oxford Science Publications, Oxford

Alexander SA (2007) The stability of the remnant luminescence emissions of alkali feldspar. Ph.D. Thesis, University of Glasgow. http://theses.gla.ac.uk/1001/

Armitage SJ, Jasim SA, Marks AE, Parker AG, Usik VI, Uerpmann H-P (2011) The southern route "Out of Africa": Evidence for an early expansion of modern humans into Arabia. Science 331:453–456

Bailey GN, Flemming NC, King GCP, Lambeck K, Momber G, Moran LJ, Al-Sharekh A, Vita-Finzi C (2007) Coastlines, submerged landscapes and human evolution. The Red Sea Basin and the Farasan Islands. J Isl Coast Archaeol 2:127–160

Bailey GN, Flemming NC (2008) Archaeology of the continental shelf: Marine resources, submerged landscapes and underwater archaeology. Quat Sci Rev 27:2153–2165

Bailey GN, King GCP (2011) Dynamic landscapes and human dispersal patterns: Tectonics, coastlines and reconstruction of human habitats. Quaternary Sci Rev 30:1533–1553

Bailey GN, Inglis RH, Meredith-Williams MG, Hausmann NBMJ, Alsharekh AM, Al Ghamdi S (2012) Preliminary Report on Fieldwork in the Farasan Islands and Jizan Province by the DISPERSE Project. November-December 2012. University of York/Saudi Commission for Tourism and Antiquities. http://eprints.whiterose.ac.uk/80966/

Bailey GN, Devès MH, Inglis RH, Meredith-Williams MG, Momber G, Sakellariou D, Sinclair AGM, Rousakis G, Al Ghamdi S, Alsharekh AM (2015) Blue Arabia: Palaeolithic and underwater survey in SW Saudi Arabia and the role of coasts in Pleistocene dispersals. Quat Int 382:42–57

Blechschmidt I, Matter A, Preusser F, Rieke-Zapp D (2009) Monsoon triggered formation of Quaternary alluvial megafans in the interior of Oman. Geomorphology 110:128–139

Bøtter-Jensen L, Bulur E, Duller GAT, Murray AS (2000) Advances in luminescence instrument systems. Radiat Meas 32:523–528

Boulton GS (1993) Ice ages and climate. In: Duff PMcLD (ed) Holmes Principles of physical geology, 4th edn. Chapman and Hall, London, pp 439–469

Bray HE, Stokes S (2003) Chronologies for Late Quaternary barchan dune reactivation in the southeastern Arabian Peninsula. Quaternary Sci Rev 22:1027–1033

Bray HE, Stokes S (2004) Temporal patterns of arid-humid transitions in the south-eastern Arabian Peninsula based on optical dating. Geomorphology 59:271–280

Breeze PS, Drake NA, Groucutt HS, Parton A, Jennings RP, White TS, Clark-Balzan L, Shipton C, Scerri EML, Stimpson CM, Crassard R, Hilbert Y, Alsharekh A, Al-Omari A, Petraglia MD (2015) Remote sensing and GIS techniques for reconstructing Arabian palaeohydrology and identifying archaeological sites. Quatern Int 382:98–119

Burbidge CI, Sanderson DCW, Fülöp R (2007a) Luminescence dating of dune sand, wadi, sabkha and playa sediments, Saudi Arabia. Technical Report, SUERC, 120 pp. http://eprints.gla.ac.uk/76538/

Burbidge CI, Sanderson DCW Housley RA, Jones P (2007b) Survey of Palaeolithic sites by luminescence profiling; a case study in eastern Europe. Quaternary Geochronology 2:296–302

Burns SJ, Matter A, Frank N, Mangani A (1998) Speleothem based paleoclimatic record from northern Oman. Geology 26:499–502

Buylaert JP, Murray AS, Thomsen KJ, Jain M (2009) Testing the potential of an elevated temperature IRSL signal from K-feldspar. Radiat Meas 44:560–565

Cresswell AC, Sanderson DCW (2009) Luminescence dating of dune sands and sabkha sediments in Saudi Arabia. SUERC dating report, SUERC, p 37

Dabbagh AR, Emmermann H, Hötzl AR, Jado HJ, Lippolt W, Kollman H, Moser W, Rauert ZJ (1984) The development of Tihamat Asir during the Quaternary. In: Jado AR, Zötl JG (eds) Quaternary Period in Saudi Arabia, vol 2. Sedimentological, Hydrogeological, Hydrochemical, Geomorphological. Geochronological and Climatological Investigations in Western Saudi Arabia. Springer-Verlag, Vienna, pp 150–173

Dennell R, Petraglia MD (2012) The dispersal of Homo sapiens across southern Asia: How early, how often, how complex? Quaternary Sci Rev 47:15–22

Dietze M, Kreutzer S, Fuchs MC, Burow C, Fischer M, Schmidt C (2013) A practical guide to the R package Luminescence. Ancient TL 32:11–18

Duller GAT (2016) Challenges involved in obtaining luminescence ages for long records of aridity: Examples from the Arabian Peninsula. Quatern Int 410:69–74

Farrant AR, Duller GAT, Parker AG, Roberts HM, Parton A, Knox RWO, Bide T (2015) Developing a framework of Quaternary dune accumulation in the northern Rub' al-Khali, Arabia. Quatern Int 382:132–144

Glennie KW, Singhvi AK (2002) Event stratigraphy, paleoenvironment and chronology of SE Arabian deserts. Quaternary Sci Rev 21:853–869

Glennie K, Fryberger S, Hern C, Lancaster N, Teller J, Pandey V, Singhvi A (2011) Geological importance of luminescence dates in Oman and the Emirates: An overview. Geochronometria 38:259–271

Grant KM, Rohling EJ, Bar-Matthews M, Ayalon A, Medina-Elizalde M, Ramsey CB, Satow C, Roberts AP (2012) Rapid coupling between ice volume and polar temperature over the past 150,000 years. Nature 491:744–747

Groucutt HS, Scerri EML, Lewis L, Clark-Balzan L, Blinkhorn J, Jennings RP, Parton A, Petraglia MD (2015a) Stone tool assemblages and models for the dispersal of Homo sapiens out of Africa. Quaternary International 382:8–30

Groucutt HS, Shipton C, Alsharekh A, Jennings R, Scerri EML, Petraglia MD (2015b) Late Pleistocene lakeshore settlement in northern Arabia: Middle Palaeolithic technology from Jebel Katefeh, Jubbah. Quaternary International 382:215–236

Groucutt HS, White TS, Clark-Balzan L, Parton A, Crassard R, Jennings RP, Parker AG, Breeze PS, Scerri EML, Alsharekh A, Petraglia MD (2015c) Human occupation of the Arabian Empty Quarter during MIS 5: Evidence from Mundafan Al-Buhayrah, Saudi Arabia. Quaternary Sci Rev 119:116–135

Hausmann N, Meredith-Williams M (2016) Seasonal patterns of coastal exploitation on the Farasan Islands. J Island Coastal Archaeology, Saudi Arabia. https://doi.org/10.1080/15564894.2016.1216478

Hilbert YH, Parton A, Morley MW, Linnenlucke LP, Jacobs Z, Clark-Balzan L, Roberts RG, Galletti CS, Schwenninger JL, Rose JI (2015) Terminal Pleistocene and Early Holocene archaeology and stratigraphy of the southern Nejd, Oman. Quatern Int 382:250–263

Hoffmann G, Rupprechter M, Rahn M, Preusser F (2015) Fluvio-lacustrine deposits reveal precipitation pattern in SE Arabia during early MIS 3. Quatern Int 382:145–153

Huntley DJ, Godfrey-Smith DI, Thewalt MLW (1985) Optical dating of sediments. Nature 313:105–107

Inglis RH, Sinclair A, Shuttleworth A, Alsharekh A, Al Ghamdi S, Devès M (2014a) Investigating the Palaeolithic landscapes and archaeology of the Jizan and Asir regions, southwest Arabia. Proc Seminar for Arabian Studies 44:193–212

Inglis RH, Sinclair AGM, Shuttleworth A, Meredith-Williams MG, Hausmann N, Budd W, Alsharekh A, Al Ghamdi S, Bailey GN (2014b) Preliminary Report on 2014 Fieldwork in Southwest Saudi Arabia by the DISPERSE project: (1) Jizan and Asir Provinces. University of York/Saudi Commission for Tourism and Antiquities. http://eprints.whiterose.ac.uk/80965/

Inglis RH, Foulds F, Shuttleworth A, Alsharekh AM, Al Ghamdi S, Sinclair AG, Bailey GN (2015) The Palaeolithic Occupation of the Harrat Al Birk: Preliminary Report on the 2015 Fieldwork in Asir Province, Southwest Saudi Arabia. University of York/Saudi Commission for Tourism and Antiquities. http://eprints.whiterose.ac.uk/84382/

Jennings RP, Shipton C, Breeze P, Cuthbertson P, Bernal MA, Wedage WMCO, Drake NA, White TS, Groucutt HS, Parton A, Clark-Balzan L, Stimpson C, al Omari A-A, Alsharekh A, Petraglia MD (2015a) Multi-scale Acheulean landscape survey in the Arabian Desert. Quaternary International 382:58–81

Jennings RP, Singarayer J, Stone EJ, Krebs-Kanzow U, Khon V, Nisancioglu KH, Pfeiffer M, Zhang X, Parker A, Parton A, Groucutt HS, White TS, Drake NA, Petraglia MD (2015b) The greening of Arabia: Multiple opportunities for human occupation of the Arabian Peninsula during the Late Pleistocene inferred from an ensemble of climate model simulations. Quatern Int 382:181–199

Kinnaird TC, Cresswell AJ, Bishop P, Sanderson DCW (2010) Luminescence Dating of Samples from Raised Beaches in Tanzania. Technical Report SUERC. http://eprints.gla.ac.uk/76481/

Kinnaird TC, Sanderson DCW, Bigelow GF (2015a) Feldspar SARA IRSL dating of very low dose rate aeolian sands from Sandwick South, Unst, Shetland. Quat Geochronol 30:168–174

Kinnaird TC, Sanderson DCW, Sakellariou D, Bailey G (2015b) Palaeographic reconstructions of the submerged prehistoric landscapes of the Farasan continental shelf, South Red Sea: Chronological constraints. Technical Report, SUERC, 49 pp. http://eprints.gla.ac.uk/110542/

Kinnaird TC, Sanderson DCW, Inglis R, Bailey G (2015c) OSL dating of terrestrial sediments from SW Saudi Arabia: Wadi Sabiya (Jirzan region) and Harrat Al Birk (Asir region). Technical Report, SUERC, 56 pp. http://eprints.gla.ac.uk/110550/

Lambeck K, Purcell A, Flemming NC, Vita-Finzi C, Alsharekh AM, Bailey GN (2011) Sea level and shoreline reconstructions for the Red Sea: Isostatic and tectonic considerations and implications for hominin migration out of Africa. Quaternary Sci Rev 30:3542–3574

Lancaster N, Wolfe S, Thomas D, Bristow C, Bubenzer O, Burrough S, Duller G, Halfen A, Hesse P, Roskin J, Singhvi A, Tsoar H, Tripaldi A, Yang X, Zárate M (2016) The INQUA Dunes Atlas chronologic database. Quatern Int 410:3–10

Matter A, Neubert E, Preusser F, Rosenberg T, Al-Wagdani K (2015) Palaeo-environmental implications derived from lake and sabkha deposits of the southern Rub' al-Khali, Saudi Arabia and Oman. Quatern Int 382:120–131

McClure HA (1976) Radiocarbon chronology of Late Quaternary lakes in the Arabian Desert. Nature 263:755–756

McDougall DJ (1968) (ed) Thermoluminescence of geological materials. Academic, New York

McLaren SJ, Al-Juaidi F, Bateman MD, Millington AC (2009) First evidence for episodic flooding events in the arid interior of central Saudi Arabia over the last 60 ka. J Quaternary Sci 24:198–207

Munro RN, Wilkinson TJ (2007) Environment, landscapes and archaeology of the Yemeni Tihamah. In: Starkey J, Starkey P, Wilkinson T (eds) Natural Resources and Cultural Connections of the Red Sea. British Archaeological Reports International Series 1661. Archaeo Press, Oxford, pp 13–33

Munro RN, Walkington H, Franks S, Wilkinson TJ, Sanderson DCW (2013) Aspects of Late Cainozoic aeolian landscapes in Arabia: Implications for early man. In: Al Ansary AR, Al-Muaikei KI, Alsharekh AM (eds) Man and Environment in the Arab World in Light of Archaeological Discoveries. Abdul Rahman Al-Sudairy Foundation, Riyadh, pp 7–46. ISBN 978-603-90326 5-6

Murray AS, Wintle AG (2000) Luminescence dating of quartz using an improved single-aliquot regenerative-dose protocol. Radiat Meas 32:57–73

Parton A, Farrant AR, Leng MJ, Telfer MW, Groucutt HS, Petraglia MD, Parker AG (2015a) Alluvial fan records from southeast Arabia reveal multiple windows for human dispersal. Geology 43:295–298

Parton A, White TS, Parker AG, Breeze PS, Jennings R, Groucutt HS, Petraglia MD (2015b) Orbital-scale climate variability in Arabia as a potential motor for human dispersals. Quatern Int 382:82–97

Pedoja K, Husson L, Johnson ME, Melnick D, Witt C, Pochat S, Nexer M, Delcaillau B, Pinegina T, Poprawski Y, Authemayou C, Elliot M, Regard V, Garestier F (2014) Coastal staircase sequences reflecting sea-level oscillations and tectonic uplift during the Quaternary and Neogene. Earth-Sci Rev 132:13–38

Petit-Maire N, Carbonel P, Reyss JL, Sanlaville P, Abed A, Bourrouillh R, Fontugne M, Yasin S (2010) A vast Eemian palaeolake in southern Jordan (29 N). Global Planet Change 72:368–373

Petraglia MD (2011) Archaeology: Trailblazers across Arabia. Nature 470:50–51

Petraglia MD, Alsharekh AM, Crassard R, Drake NA, Groucutt H, Parker AG, Roberts RG (2011) Middle Paleolithic occupation on a Marine Isotope Stage 5 lakeshore in the Nefud Desert, Saudi Arabia. Quaternary Sci Rev 30:1555–1559

Petraglia MD, Parton A, Groucutt HS, Alsharekh A (2015) Green Arabia: Human prehistory at the crossroads of continents. Quatern Int 382:1–7

Prescott JR, Stephan LG (1982) The contribution of cosmic radiation to the environmental dose for thermoluminescent dating. Latitude, altitude and depth dependencies. PACT 6:17–25

Prescott JR, Hutton JT (1988) Cosmic-ray and gamma-ray dosimetry for TL and electron-spin- resonance. Nuclear Tracks Radiation Measurements 14:223–227

Preston GW, Thomas DSG, Goudie AS, Atkinson OAC, Leng MJ, Hodson MJ, Walkington H, Charpentier V, Méry S, Borgi F, Parker AG (2015) A multi-proxy analysis of the Holocene humid phase from the United Arab Emirates and its implications for southeast Arabia's Neolithic populations. Quatern Int 382:277–292

Preusser F, Radies D, Matter A (2002) A 160,000-year record of dune development and atmospheric circulation in southern Arabia. Science 296:2018–2020

Preusser F (2009) Chronology of the impact of Quaternary climate change on continental environments in the Arabian Peninsula. Comptes Rendus Geosci 341:621–632

Rosenberg TM, Preusser F, Fleitmann D, Schwalb A, PenkmanK ST, Al-Shanti MA, Kadi K, Matter A (2011) Humid periods in southern Arabia: Windows of opportunity for modern human dispersal. Geology 39:1115–1118

Sakellariou D, Bailey GN, Momber G, Meredith-Williams M, Alsharekh A, Rousakis G, Panagiotopoulos I, Morfis I, Stavrakakis S, Pampidis I, Renieris P, Georgiou P, Kalogirou S, Mantopolos P, Stasinos V, Kallergis M, Manousakis L, Al Nomani SM, Devès M (2013) Preliminary report on underwater survey in the Farasan Islands by the R/V Aegaeo, May-June 2013. Technical Report, Hellenic Centre for Marine Research (HCMR), Athens, Greece

Sakellariou D, Rousakis G, Panagiotopoulos I, Morfis I, Bailey G (this volume) Geological structure and Late Quaternary geomorphological evolution of the Farasan Islands continental shelf, south Red Sea, SW Saudi Arabia

Sanderson DCW (1988a) Fading of thermoluminescence of feldspars—characteristics and corrections. Nuclear Tracks Radiation Measurements 14:155–161

Sanderson DCW (1988b) Thick Source Beta-Counting (TSBC)—a rapid method for measuring beta-dose-rates. Nuclear Tracks Radiation Measurements 14:203–207

Sanderson DCW, Bishop P, Houston I, Boonsener M (2001) Luminescence characterisation of quartz-rich cover sands from NE Thailand. Quaternary Sci Rev 20:893–900

Sanderson DCW, Murphy S (2010) Using simple portable OSL measurements and laboratory characterization to help understand complex and heterogeneous sediment sequences for luminescence dating. Quat Geochronol 5:299–305

Scerri EML, Breeze PS, Parton A, Groucutt HS, White TS, Stimpson C, Clark-Balzan L, Jennings R, Alsharekh A, Petraglia MD (2015) Middle to Late Pleistocene human habitation in the western Nefud Desert, Saudi Arabia. Quatern Int 382:200–214

Scott EM, Sanderson DCW (1988) Statistics and the additive dose method in TL dating. Nuclear Tracks Radiation Measurements 14 (1/2):345–354

Sinclair A, Inglis RH, Shuttleworth, A, Foulds, FW, Alsharekh, AM (this volume) Landscape archaeology, Palaeolithic survey and coastal change along the southern Red Sea of Saudi Arabia

Singhvi AK, Sharma YP, Agrawal DP (1982) Thermoluminescence dating of sand dunes in Rajastan, India. Nature 295:313–315

Singhvi AK, Sharma YP, Agrawal DP, Dhir RP (1983) Thermoluminescence dating dune sands—some refinements. PACT 9(II):499–504

Stimpson CM, Breeze PS, Clark-Balzan L, Groucutt HS, Jennings R, Parton A, Scerri E, White TS, Petraglia MD (2015) Stratified Pleistocene vertebrates with a new record of a jaguar-sized pantherine (Panthera cf. gombaszogensis) from northern Saudi Arabia. Quatern Int 382:168–180

Thomsen KJ, Murray AS, Jain M, Bøtter-Jensen L (2008) Laboratory fading rates of various luminescence signals from feldspar-rich sediment extracts. Radiat Meas 43:1474–1486

Wang XL, Wintle AG, Lu YC (2006) Thermally transferred luminescence in fine grained quartz from Chinese loess, basic observations. Radiat Meas 41:649–658

Wallinga J, Murray A, Wintle A (2000) The single-aliquot regenerative-dose (SAR) protocol applied to coarse-grain feldspar. Radiation Measurements 32:529–533

Wilson JWP, Roberts GG, Hoggard MJ, White NJ (2014) Cenozoic epeirogeny of the Arabian Peninsula from drainage modeling. Geochem Geophys Geosyst 15:3723–3761

Winder IC, Deves MH, King GCP, Bailey GN, Inglis RH, Merridith-Williams M (2015) Evolution and dispersion of the genus *homo*: A landscape approach. J Human Evolution 87:48–65

Wintle AG, Huntley DJ (1979) Thermoluminescence dating of a deep-sea sediment core. Nature 279:710–712

Results of Micropalaeontological Analyses on Sediment Core FA09 from the Southern Red Sea Continental Shelf

Maria Geraga, Spyros Sergiou, Dimitris Sakellariou, and Eelco Rohling

Abstract

This study presents results of an examination of planktonic and benthic foraminiferal assemblages from the upper half of a 2.64 m-long sediment gravity core retrieved from the southern Red Sea continental shelf. The examined interval corresponds to the time period of the last 16 kyr. The microfaunal associations show concurrent and concomitant variations at long and short time scales. The examined deglacial interval suggests that the Strait of Bab al Mandab most likely remained open, connecting the Red Sea with the Indian Ocean, although this connection was extremely limited. Productive waters associated with inflow from the Gulf of Aden into the Red Sea prevailed during the Late Glacial and Early Holocene periods ($\sim$10 to $\sim$6 ka BP), a phase of intensified summer monsoons in the Arabian Sea. The Late Holocene period shows a reduction of productivity and sea floor oxygenation during which time the winter monsoon was stronger. Short-term variations in the abundances of planktonic and benthic foraminiferal assemblages have been linked to events of increased aridity within the Late Glacial and Holocene intervals.

1 Introduction

The Red Sea is an interesting basin from various perspectives, including geology, oceanography, biodiversity, and archaeology. It is a narrow but deep oceanic trough which is connected to the Indian Ocean through a narrow and shallow channel with a sill depth of 137 m (Biton et al. 2008) extending from the Strait of Bab al Mandab to the Hanish Sill. Today, the climate of the Red Sea is affected mostly by the Indian Ocean monsoon system (Trommer et al. 2010). These topographic and climatic signatures make the Red Sea sensitive to sea level and climatic changes, and the sediments of its seafloor are important archives preserving information regarding these changes. The glacial-interglacial signal is dominant because the full drop in sea-level during glacial maxima is of the same order as the total sill depth (Rohling et al. 1998; Siddall et al. 2003, 2004; Lambeck et al. 2011). The glacial periods with their drop in sea level are marked by the disappearance of planktonic species driven by hypersalinity, and complex reorganizations of hydrography, productivity and subsurface oxygenation in the Red Sea (Fenton et al. 2000). During interglacials, when the connection with the open sea was restored, changes in the intensity and duration of the monsoonal system affected the oceanographic patterns of the Red Sea (Arz et al. 2003; Gupta et al. 2003; Siccha et al. 2009; Trommer et al. 2010).

Topography and climate also make the Red Sea region interesting from an archaeological point of view. The southern Red Sea especially has become a focus of attention, being one of the more fertile regions of the Arabian Peninsula (Bailey et al. 2007, 2015) with a relatively narrow crossing from Africa at low sea-level stands (e.g., Rohling et al. 2013), and extensive areas of territory that are now submerged (Bailey et al. 2015). This interest underpins the DISPERSE project (Bailey et al. 2015). A critical component of this project concerns the investigation of the palaeoceanographic, palaeogeographic and palaeoclimatic evolution of the southern Red Sea, in particular in the region

M. Geraga (✉) · S. Sergiou
Laboratory of Marine Geology and Physical Oceanography, Department of Geology, University of Patras, Patras, Greece
e-mail: mgeraga@upatras.gr

D. Sakellariou
Hellenic Centre for Marine Research, 19013 Anavyssos, Athens, Greece

E. Rohling
Research School of Earth Sciences, Australian National University, Canberra, ACT 2601, Australia

E. Rohling
Ocean and Earth Science, National Oceanography Centre, University of Southampton, Southampton, SO14 3ZH, UK

N. M. A. Rasul and I. C. F. Stewart (eds.), *Geological Setting, Palaeoenvironment and Archaeology of the Red Sea*,
https://doi.org/10.1007/978-3-319-99408-6_32

of the Farasan Islands (Bailey et al. 2015, and this volume; Inglis et al., this volume; Momber et al., this volume).

Within the framework of the DISPERSE project, gravity core FA09 was retrieved from the southern Red Sea continental shelf near the Farasan Islands. Study of this undisturbed hemipelagic sediment core aims to provide new insights into the climatic and oceanographic variability in the southern Red Sea during the last glacial-interglacial cycle.

In this chapter, we present results for the upper part of the core, which spans the late glacial to Holocene interval. The core sediments were examined for variations in the abundances of planktonic and benthic foraminifera. The well-preserved record of these planktonic and benthic foraminifera in Red Sea sediments and their sensitivity to oceanographic changes make them a useful indicator of past oceanographic and climatic changes (Edelman-Fürstenberg et al. 2001; Badawi et al. 2005; Abu-Zied et al. 2011).

Geological and oceanographic setting

The Farasan Islands are located in the southern Red Sea, on the southwestern side of the Arabian Peninsula (Fig. 1). They were formed by progressive uplift due to salt diapirism, which resulted in a relative sea-level fall (Bantan 1999; Bailey et al. 2007; Demarchi et al. 2011). Ages determined from strontium and carbon isotopes suggest that this uplift is ongoing (Bantan 1999). The same regime is thought to have prevailed during the period of deposition of FA09. The low sea-level stand during the LGM (Last Glacial Maximum) resulted in exposure of the Farasan continental shelf and Pleistocene reefs around the Farasan Islands to weathering and erosion. The following sea-level rise flooded the area and left the Farasan Islands exposed. Because of local (salt-diapirism) tectonics, the relative sea-level record for the Farasan region is a complicated combination of ice-volume related eustasy (e.g., Grant et al. 2012, 2014; Lambeck et al.

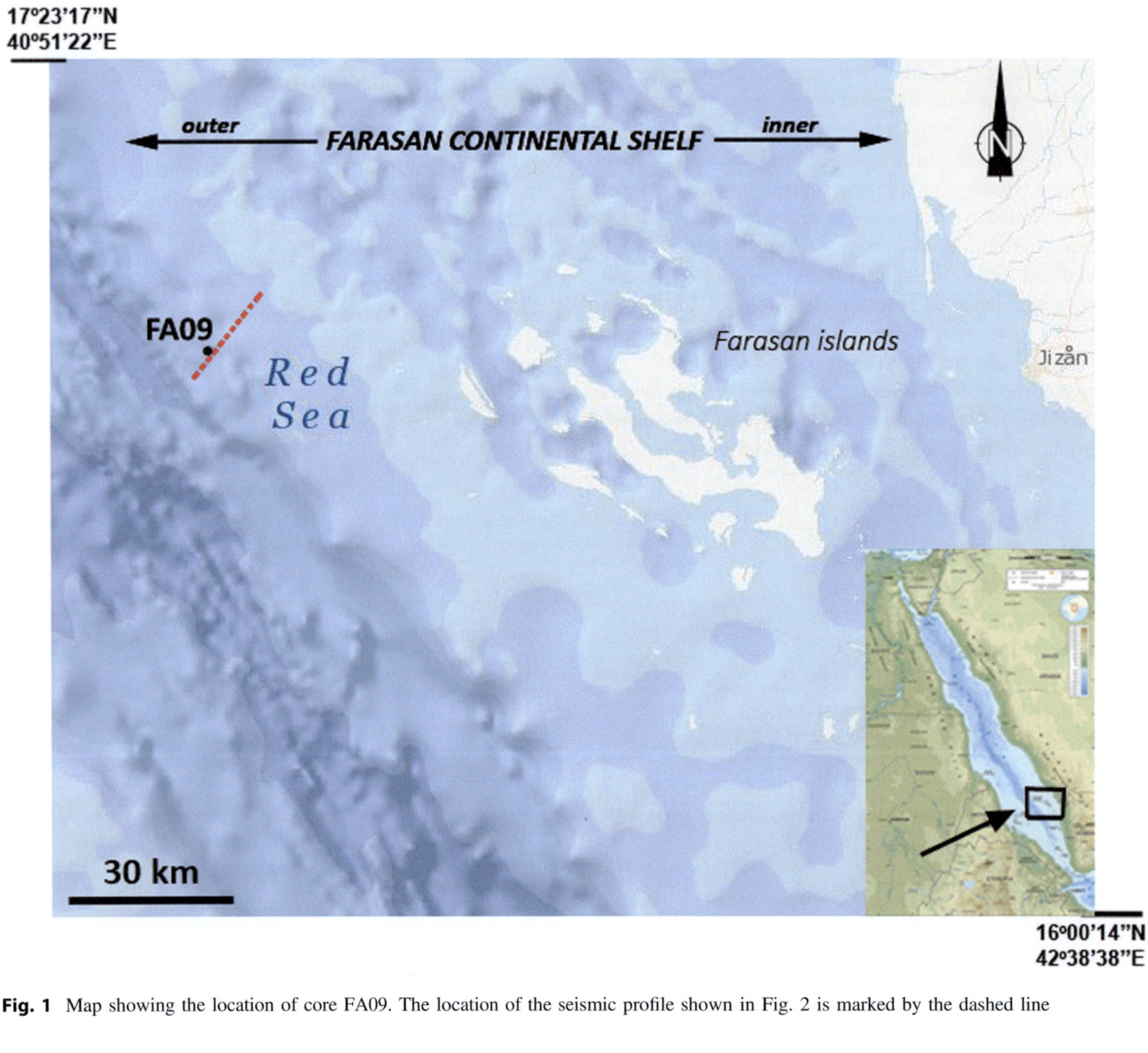

Fig. 1 Map showing the location of core FA09. The location of the seismic profile shown in Fig. 2 is marked by the dashed line

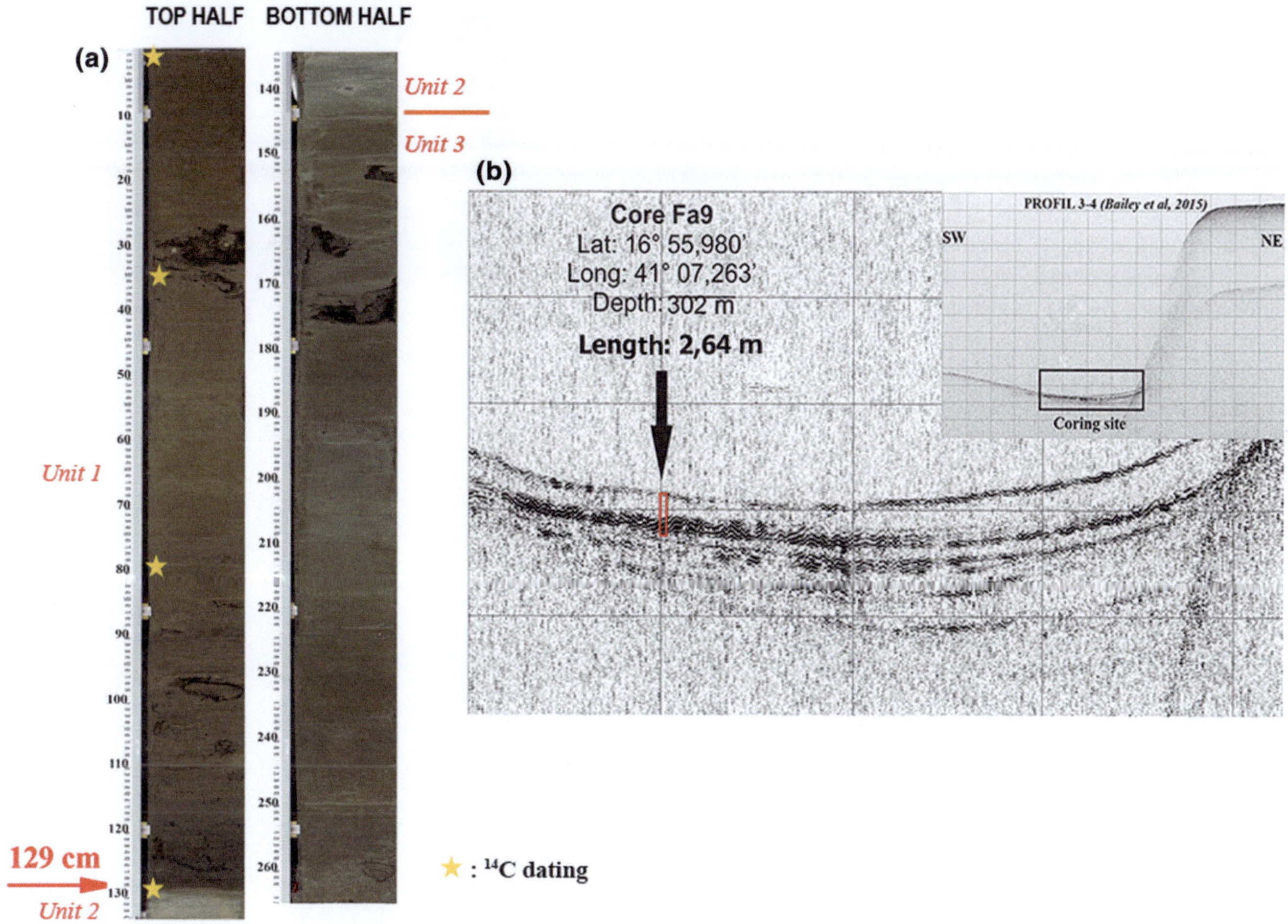

Fig. 2 **a** Photograph presenting the lithology of core FA09 (see text for lithological units 1–3). The position of radiocarbon-dated samples and the lithological change at 129 cm are marked. This work was focused on the top half; **b** Seismic profile from the coring site

2014), local uplift resulting from salt-diapirism as mentioned above, and glacio-isostatic changes related to water-loading on the wider regional shelf (e.g., Lambeck et al. 2011). In general, the (sub-) tropical climate and shallow waters lead to high production of carbonates (Abu Zied et al. 2011), and colonization as well as proliferation of reefs.

Our core FA09 was retrieved from the outer edge of the Farasan continental shelf, which is characterized by complex topography mostly due to extensive diapirism of massive Middle to early Late Miocene evaporites, along with extensional, rift-related faulting (Bosence et al. 1998). Bailey et al. (2015) proposed that a large part of the Farasan continental shelf is occupied by a prominent, gently seaward-dipping morphological terrace developed on Pliocene-Quaternary marine limestones (Dabbagh et al. 1984; Bantan 1999), intruded by salt diapirs which derive from the afore-mentioned underlying Miocene evaporites. The same authors indicate a large number of fault-bounded basins that were potentially water-filled lakes (i.e., potentially valuable water holes) during the Pleistocene. These basins are now covered by Holocene marine sediments. The core was collected from within one of those basins, adjoining the morphological terrace (Fig. 2).

The southern Red Sea lies under the influence of the South Asian monsoon climatic regime. Northwesterly winds dominate the area during summer and turn to southeasterly winds during winter (Morcos 1970). Oceanographic studies have shown that water exchange between the Red Sea and the Gulf of Aden (Indian Ocean) through the Bab al Mandab Strait (at the southern end of the Red Sea) is controlled by the dominant seasonal wind system (Wyrtki 1974; Woelk and Quadfasel 1996; Siddall et al. 2002; Sofianos and Johns 2007). During winter, surface water from the Gulf of Aden (GASW) enters the Red Sea moving northward, while dense, saline Red Sea water flows out into the Gulf of Aden. During summer, Gulf of Aden Intermediate Water (GAIW) enters the basin as a subsurface layer from the Gulf of Aden (Siddall et al. 2002; Sofianos and Johns 2007). Above this layer, the Red Sea Surface Water (RSSW), which is warmer and saltier than GAIW, is moving southward. Underneath the GAIW, two Red Sea water masses are found that both originate in the northern Red Sea: Red Sea Outflow Water (RSOW) with very high salinities (around 40 psu) and Red Sea Deep Water (RSDW) which fills the basin from about 200 m to the bottom and is characterized by very low oxygen concentrations. The sea surface temperature in the area is high (>30 °C). The salinity is also high, ranging between 36 and 41 psu throughout the water column, showing lower values in the layer associated with the GAIW mass. Oxygen concentrations vary with water depth and present two zones of minimal values, one centred at around 75 m depth ($\sim$10 mmol/kg) associated with GAIW, and the second at around 400 m depth associated with the "old" waters of the RSDW mass. The highest oxygen concentrations occur in the surficial waters (>100 mmol/kg).

The intrusion of more nutrient-enriched waters from the Indian Ocean fuels the Red Sea ecosystem (Triantafyllou et al. 2014). During winter, the influx of nutrients and high influx of organic matter into the Red Sea from the Gulf of Aden is limited, and occurs in the surface layer. During summer, more nutrient (phosphate) and to a lesser extent TOC-enriched inflow occurs by means of the GAIW.

2 Materials and Methods

Core FA09 was collected by *R/V Aegaeo* (Hellenic Centre of Marine Research, HCMR) near the Farasan Islands (Fig. 1), within the framework of ERC-funded project DISPERSE (Bailey et al. 2015). It was obtained from 302 m water depth at 16°55.980'N, 41°07.263'E and is 264 cm long (Fig. 2a).

FA09 shows a distinct change in sedimentary characteristics (Fig. 2a). The upper unit (unit 1, 0–129 cm) consists of brownish mud with a small proportion of sand. This overlies an intermediate unit (unit 2, 129–144 cm), which consists of light grey mud and which is separated from the upper unit by a sharp contact. Then there is a lower unit (unit 3, 144–264 cm), which consists of dominant brownish mud that is sparsely interbedded with discontinuous light grey muddy laminates (Fig. 2a). We performed a number of radiocarbon assays (see below), and found that the upper 134 cm of FA09 corresponds in age to the last $\sim$16 kyr. These radiocarbon dates indicate a steady accumulation of marine sediments since that time to the present (Fig. 3). Because of the presence of the above-mentioned abrupt transition between unit 1 and unit 2, and uncertainty about the chronology of the intermediate and lower units, we here focus only on the upper-(marine) section of the core, that is, the latter half of the last deglaciation and the Holocene interglacial.

We took a total of fifty (50) sediment samples from the upper 134 cm, which we examined for foraminiferal populations. For the analysis, samples of wet sediment of 1 cm thickness and of specific volume were collected throughout the core. After weighing, the wet samples were dried at 50–55 °C for 24–48 h and weighed again. The samples were then washed over 63, 150 and 500 micron sieves with distilled water. An ultrasonic cleaning procedure was necessary for some samples in order to avoid coatings. Each sample from these three sieved fractions was dried at 50–55 °C for 24–48 h and weighed. In this chapter we present the results of analyses of the fraction between 150 and 500 μm.

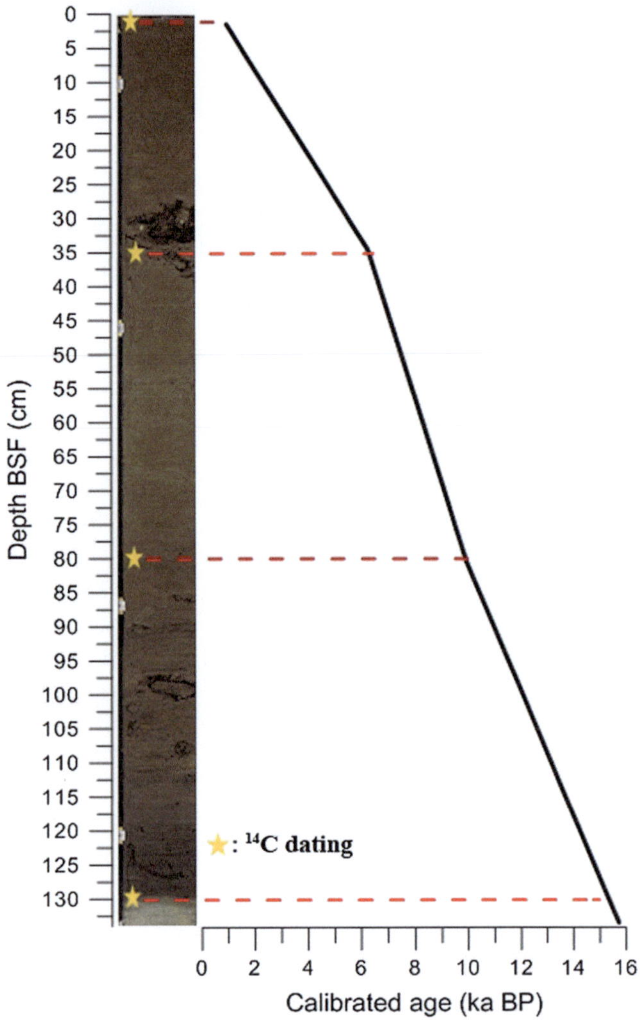

Fig. 3 The sedimentation rate as deduced from the radiocarbon dates

For each sample, an aliquot containing at least 300 specimens of planktonic foraminifera was selected. Planktonic foraminifera were counted and identified up to species level. Samples with an extremely high number of foraminiferal specimens were split in appropriate aliquots by use of an Otto splitter. Benthic foraminifera were also counted and identified at least to the genus level for each sample. Species identification was mainly based on the generic classification of Loeblich and Tappan (1987) and the studies of Hottinger et al. (1993).

Each planktonic taxon is expressed as a percentage of the total assemblage and as the number of specimens per gramme of dry sediment. Each benthic taxon is expressed only as the number of specimens per gramme of dry sediment, due to low abundances of the benthic foraminifera in the aliquots (less than 200 specimens per sample). The numbers of planktonic and benthic specimens per weight of dry sediments >150 µm were estimated as indices of planktonic (PFN) and benthic (BFN) foraminiferal production, respectively. The downcore variations in planktonic percentage of the total foraminiferal assemblage (P%: (Np 100)/(Np + Nb), where Np and Nb are the number of planktonic and benthic specimens, respectively) was estimated as a quantitative index to approximate water depth (Van der Zwaan et al. 1999).

Multivariate statistical analysis (R-mode Principal Component Analysis, PCA) was carried out on the foraminiferal data in order to discriminate groups of foraminiferal species and taxa that reveal common trends and, consequently, indicate changes on the palaeoceanographic evolution. We conducted the analysis using the SPSS program (version 24). In the data set the planktic and benthic foraminifera are expressed as numbers of specimens per gram of dry sediment. Prior to multivariate statistics, rare species with extremely low contributions were omitted from the raw data set.

Four samples of about 10 mg of *Globigerinoides ruber* individuals from selected depth intervals were submitted for ^{14}C AMS analyses (Figs. 2a and 3). The analyses were carried out at the laboratories of the Scottish Universities Environmental Research Centre (SUERC). The radiocarbon results were calibrated using the Calib v.7.10 program (Table 1). The MARINE 13 curve was used and the nearest available ΔR value was selected (ΔR: 110 ± 38; Southon et al. 2002).

3 Results

3.1 Sediments

The three aforementioned lithological units of FA09 were identified on the basis of colour, grain size estimates, and sedimentary structures. The finding of fine grained sediments in the core is in agreement with the coring site's acoustic signal in geophysical surveys (Bailey et al. 2015) (Fig. 2b).

The available dates indicate that the examined part of the core (0–134 cm) corresponds to the last ~16 kyr (Table 1), which spans the final part of Marine Isotope Stage2 (MIS2, the latter half of the deglaciation that followed the Last Glacial Maximum), and the Holocene. The mean sedimentation rate for this interval is 7.34 cm/kyr (Fig. 3). Based on this rate it is suggested that the intermediate unit may correspond to MIS2, and the lower unit to MIS3, but we do not have a robust age model for these parts, which is a concern because they are topped by a very sharp transition (discontinuity?). Hence, we focus only on the upper half of FA09.

3.2 Planktonic Foraminifera

Planktonic foraminifera are present throughout the examined interval . They show large variations in abundance, in generally low-diversity assemblages. *Globigerinoides ruber*, *Globigerinoides sacculifer*, *Globigerinella calida*, *Globigerinella siphonifera*, *Hastigerina pelagica*, *Globoturborotalita rubescens*, *Globoturborotalita. tenella*, *Orbulina universa*, *Globigerina bulloides*, *Globigerina falconensis*, *Globigerinita glutinata, Turborotalita clarkei, Turborotalita quinqueloba, Globorotalia menardii, Neogloboquadrina incompta, Neogloboquadrina pachyderma*, and *Neogloboquadrina dutertrei* are the main species. *Globigerinoides ruber* is the most abundant species and alone accounts for 33% on average of the total population. This species together with *G. glutinata*, *Gs. sacculifer* and *Globigerinella* spp. dominate the upper 80 cm of the core and the interval between 133 and 129 cm (at the base of the examined section) with total percentages up to 90%. At the same sediment depths, *G. bulloides* and Neogloboquadrinids contribute to the microfaunal assemblages but their numbers are increased between 80 and 129 cm, where together with *Gr. menardii* they dominate the planktonic assemblages (up to 60%) (Fig. 4).

A strong contribution of *Gs. ruber*, *G. glutinata* and *Gs. sacculifer* is common in extant planktonic foraminiferal assemblages of the region (Auras-Schudnagies et al. 1989), in recently deposited marine sediments of the Holocene in the Red Sea (e.g., Auras-Schudnagies et al. 1989; Locke and Thunell 1988; Siccha et al. 2009; Abu-Zied 2013), and in late Holocene sediments (Siccha et al. 2009; Trommer et al. 2010). This is attributed mainly to the tolerance of these species of the high sea-surface temperatures and salinities

Table 1 Radiocarbon ages and calibrated dates for the FA09 core

Sample depth BSF (cm)	Radiocarbon age BP	Calibrated date ka BP
1.5	1515 ± 32	0.95
34.5	5916 ± 32	6.23
80.5	9252 ± 32	9.91
129.5	13366 ± 56	15.33

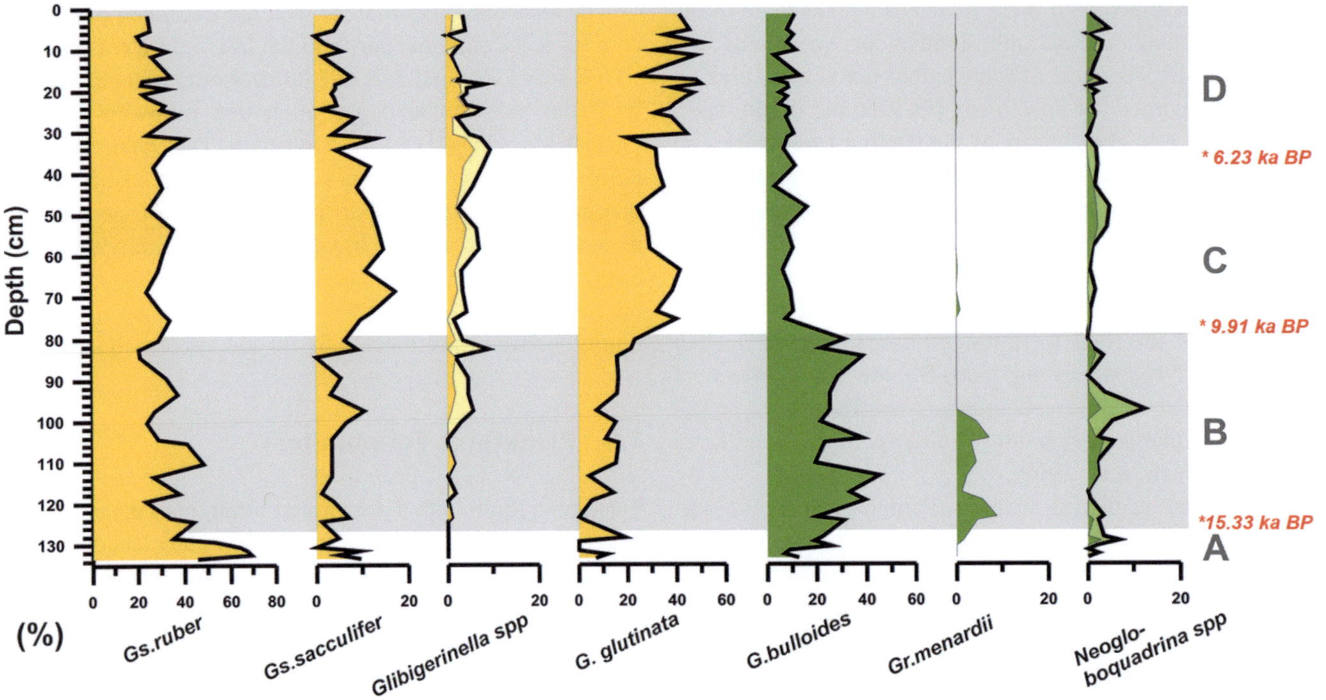

Fig. 4 Downcore variations in the abundance of planktonic foraminifera (given in percentages of the total assemblage) through the examined interval of core FA09. The intervals A–D and the calibrated radiocarbon dates are indicated. In the diagrams for *Globigerinella* and *Neogloboquadrina*, the light-coloured areas show the abundance of *G. calida* and *N. incompta*, respectively

Table 2 Rotated factor loadings for the 5-factor model (Rotation method: Varimax with Kaiser Normalization)

Varimax rotated factor loadings					
	Component				
	1	2	3	4	5
Gs. ruber	**0.762**	0.431	0.000	–0.126	–0.201
Gs. sacculifer	**0.898**	0.009	0.077	0.077	0.080
G. glutinata	**0.848**	–0.222	–0.198	–0.095	–0.164
Globigerinella	**0.828**	–0.242	–0.212	–0.098	–0.015
G. bulloides	0.082	**0.920**	–0.020	0.017	–0.097
Gr. menardii	–0.234	**0.773**	–0.039	–0.165	–0.049
Neogloboquadrina	0.040	**0.772**	–0.155	0.065	0.095
M. millettii	–0.122	**0.826**	0.077	–0.155	0.062
Milliolids	–0.424	0.107	**0.765**	–0.121	0.109
Cibicides	–0.127	–0.254	**0.850**	0.021	0.038
Gyroidinoides soldanii	–0.069	–0.209	**0.836**	0.046	0.007
Discorbinella	–0.008	–0.034	0.098	–0.112	**0.890**
EpifaunaSum	0.021	0.325	**0.647**	0.132	0.053
Bolivina	–0.036	–0.071	0.096	**0.938**	0.102
Bulimina	–0.057	–0.104	0.037	**0.944**	0.025
InfaunaSum	0.427	–0.025	**0.560**	0.116	0.025
Agglutinants	–0.146	0.054	0.029	0.343	**0.767**

that characterize the waters of the Red Sea, and secondly to the preference of these species for low nutrient availability. Among the three species, the proportional distribution of *Gs. sacculifer* is not clear (Fenton et al. 2000; Siccha et al. 2009).

Today, the spatial distributions of *G. bulloides*, *Gr. menardii* and neogloboquadrinids present increasing trends toward the southern Red Sea; their abundances have been attributed to advection of water masses that enter from the Gulf of Aden (Auras-Schudnagies et al. 1989; Siccha et al. 2009).

3.3 Benthic Foraminifera

Benthic foraminifera are present in almost all examined samples, except for the intervals between 34 and 38 cm and at 6 cm in the core (Fig. 5). Abundances are low between 110 and 90 cm and in the upper 34 cm of the core. Although less abundant than planktonic microfauna, the benthic faunas comprise highly diverse assemblages. In total 76 benthic foraminiferal species have been identified, with *Quinqueloculina*, *Spiroloculina*, *Triloculina*, *Miliolinella*, *Bolivina*, *Bulimina*, *Cibicides*, *Discorbinella*, *Rosalina*, *Textularia*, and *Reophax* as the dominant genera.

The downcore distribution of each benthic foraminiferal taxon shows frequent variations, including intervals of disappearance of the species. Miliolids show elevated concentrations in the lower sediments between 90 and 134 cm. Above this interval, miliolids are rare or absent. High abundances of *Quinqueloculina* spp. and of general miliolids are found in oligotrophic environments and bottom waters where oxygen concentrations are sufficient (Blackwelder et al. 1996; Platon et al. 2005; Murray 2006). In addition, their abundances seem to be commonly elevated during high-salinity aplanktonic intervals in Red Sea cores (Rohling et al. 1998; Murray 1991). Epifaunal species belonging to *Amphistegina*, *Cibicides*, *Discorbis*, *Discorbinella*, *Gavelinopsis*, *Glabratella*, *Gyroidinoides*, *Millettiana*, *Neoconorbina*, *Oridosalis*, *Planorbulina* and *Rosalina* show almost concurrent appearances and abundance variations throughout the sequence. Among them *Amphistegina*, *Cibicides*, *Discorbis*, *Gavelinopsis*, *Millettiana*, *Neoconorbina*, *Planorbulina* and *Rosalina* are known as epiphytic genera and/or genera that thrive on hard substrates such as coral banks (Mateu-Vicens et al. 2010; Murray 2006). Furthermore, species belonging to *Discorbinella* and *Oridosalis* have been associated with suboxic (Kaiho 1994; Edelman-Fürstenberg et al. 2001; Murray 2006) and hypersaline bottom environments (Murray 2006; Abu-Zied et al. 2011). However, Gupta et al. (2011) consider *Oridosalis* as a genus that prefers high-oxygen conditions.

Species belonging to *Bolivina* and *Bulimina* show high abundances between 130 and 80 cm, peaking between 90 and 75 cm. Buliminid and bolivinid specimens occur in benthic assemblages of oxygen-poor and organic-rich environments (Sen Gupta and Machain-Castillo 1993; Mendes et al. 2004; Murray 2006). Specimens belonging to the infaunal genera of *Ammonia*, *Cassidulina*, *Elphidium*, *Nonion*, *Nonionella*, *Neouvigerina* and *Uvigerina* were also present in the examined samples. Their abundance in the benthic microfauna was higher between 80 and 50 cm and at the base of the core (130–134 cm).

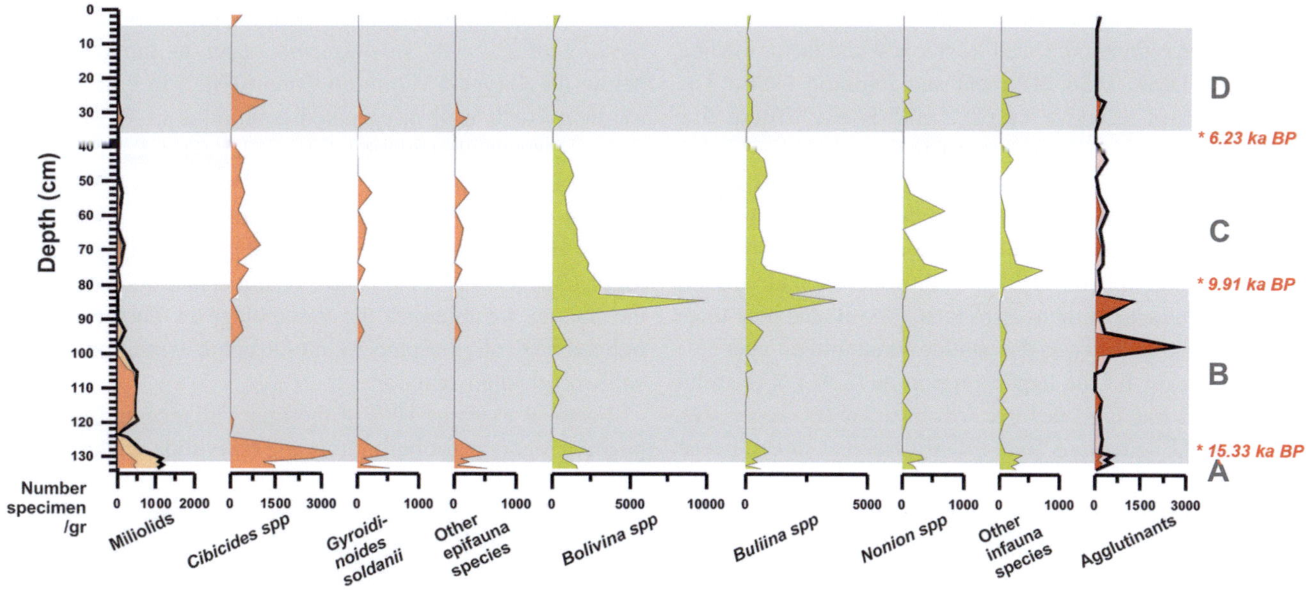

Fig. 5 Downcore variations in the abundance of benthic foraminifera (given in number of specimens per gramme of dry sediment) through the examined interval of core FA09. The intervals A–D and the calibrated radiocarbon dates are indicated. In the diagrams of miliolids and agglutinants the light-coloured areas show the abundance of *Quinqueloculina* and *Textularia*, respectively

Agglutinated benthic foraminifera belonging mostly to *Textularia* and *Reophax* showed an increased abundance at the base of the core (130–134 cm) and between 100 and 80 cm. Agglutinants are common assemblages in high-salinity environments (Abu-Zied et al. 2011; Murray 2006).

3.4 Statistical Treatment

R-mode factor analysis (PCA; Davis 1986; Reyment and Joreskog 1996) was performed using a dataset of variables representing the abundances of planktonic and benthic foraminiferal species/genera. The purpose of this analysis is to illustrate possible links and relationships between the observed planktonic and benthic foraminiferal species and genera, reducing the overall complexity of the original data and examining concurrent changes throughout the water column, thus improving our evaluation of palaeoceanographic variations. For the R-mode factor analysis, we constructed a database of seventeen variables that represent the abundances of planktonic and benthic foraminiferal species, expressed as number of specimens per unit weight of dry sediment (>150 μm). The variable *Globigerinella* includes the sum of *G. calida* and *G. siphonifera* (Weiner et al. 2015) and the variable *Neogloboquadrina* represents the sum of all species belonging to *Neogloboquadrina* genera (*N. incompta*, *N. pachyderma*, *N. duertrei*, *N. humerosa*). In the benthic assemblages, the group labeled *Epifauna Sum* includes the sum of *Discorbis*, *Neoconorbina*, *Planorbulina*, and *Rosalina*, while the group labeled *Infauna Sum* includes the sum of *Ammonia*, *Cassidulina*, *Elphidium*, *Neouvigerina* and *Nonion*. In order to select the most suitable factor model, a combination of criteria was followed. These criteria include total variance estimation, Catell's scree test and factor loading values (see Davis 1986; Reyment and Joreskog 1996). The final number of selective factors found is five. These five factors explain 76% of the total variance and each variable shows communalities higher than 0.5 (Table 2). This means that the 5-factor model expresses sufficiently the analyzed variables. The first two factors represent relations between planktonic foraminiferal taxa (including the benthic species *Millettiana millettii* in the second factor), while the next three indicate relations between the benthic foraminiferal taxa.

Factor 1 explains the largest proportion (23%) of the total variance and has high positive loadings for *Gs. ruber*, *Gs. sacculifer*, *G. glutinata* and *Globigerinella*. Open ocean observations show that *Globigerinoides ruber*, *Gs. sacculifer* and *Globigerinella* thrive in the upper water column, up to 200 m water depth, where the temperature ranges between 11 and 32 °C and the salinity ranges between 22 and 49 psu (Fenton et al. 2000, and references therein). *Globigerinita glutinata* shows a more cosmopolitan character (Fenton et al. 2000, and references therein) and is an indicator of SW monsoon variations in the open-ocean upwelling area of the Arabian Sea (Anderson and Prell 1991). Based on the above, we interpret the assemblage of Factor 1 as an indicator of warm surface waters of high salinity, while the oxycline (if any) is at least below the depth of reproduction of these species (80–100 m; Fenton et al. 2000, and references therein).

Factor 2 explains a significant proportion (20%) of the total variance and has high positive loadings for *G. bulloides*, *Neogloboquadrina*, *Gr. menardii* and the Cymbaloporidae *M. millettii*. *Globigerina bulloides* is typical of regions of increased productivity (Thiede 1975; Overpeck et al. 1996; Siccha et al. 2009). Today this species is found only in the southern Red Sea and the Gulf of Aden (Auras-Schudnagies et al. 1989; Siccha et al. 2009) and is thought to reflect an influence of upwelling of nutrient-rich deep water toward the surface layer (Kroon and Ganssen 1989) in response to strong monsoon activity (Gupta et al. 2003). As this upwelling progresses, neogloboquadrinids and *Gr. menardii*—which are usually abundant at a deeper level in the water column—appear in abundance in the early stages, while *G. bulloides* appears to thrive during the final stage (Kroon and Ganssen 1989). In addition, Gupta et al. (2010) documented that, in the Indian Ocean, benthic species belonging to Cymbaloporidae are found as indicators of productive surface waters following the trends of *G. bulloides*. Thus, we interpret the assemblage of Factor 2 as an indicator of productive waters associated with upwelling of nutrient-enriched deep waters to the surface.

Factor 3 explains 14% of the total variance and has high positive loadings for miliolids, the groups of *Epifauna Sum* and *Infauna Sum* and *G. soldanii*. In the present central Red Sea, species belonging to genera such as *Cibicides*, *Gyroidinoides*, and *Neouvigerina* occur at the margin or below the Oxygen Minimum Zone (OMZ) in bottom environments with well oxygenated conditions (1–2.5 ml O_2/l) and moderate to low organic fluxes (2.5–0.8 g C/m^2/yr), and moderate to low Total Organic Carbon (TOC) content ranging from 0.3 to 0.1 wt% (Edelman-Fürstenberg et al. 2001). Elevated oxygen levels are also implied by the miliolids (Blackwelder et al. 1996; Platon et al. 2005). Based on the above, we interpret the assemblage of Factor 3 as an indicator of oligotrophic to mesotrophic conditions and a well oxygenated seafloor.

Factor 4 explains 12% of the total variance and has high positive loadings for buliminids and bolivinids. In the present central Red Sea, species belonging to *Bolivina* and *Bulimina* genera occur in the centre of the OMZ (Oxygen Minimum Zone) where the oxygen concentrations are low (<0.5 ml O_2/l) and the organic carbon flux and TOC content are high (~3.5 g C/m^2/y and 0.7 wt%, respectively; Edelman-Fürstenberg et al. 2001). In addition, species belonging to these genera are recorded in the Arabian Sea where enhanced surface primary production leads to high organic flux on the seafloor with low to intermediate oxygen conditions (De and Gupta 2010). Based on the above, we interpret the assemblage of Factor 4 as an indicator of eutrophic seafloor levels and moderate to low oxygen conditions.

Factor 5 explains 8% of the total variance and has high positive loadings for *Discorbinella* and agglutinants. Today agglutinants (together with miliolids) dominate hypersaline lagoons in the Red Sea (Abu-Zied et al. 2011). Furthermore, taxa such as *Textularia* show an affinity to dysoxic and hypoxic conditions and to increases in salinity (Moodley et al. 1998; Almogi-Labin et al. 1996). In addition, *Textularia agglutinans* is found in the OMZ in the Arabian Sea (De and Gupta 2010). *Discorbinella* is a major component in the centre of the OMZ in the central Red Sea (together with *Bolivina* and *Bulimina*; Edelman-Fürstenberg et al. 2001). Thus, we interpret the assemblage of Factor 5 as an indicator of low oxygen and likely high salinity bottom conditions.

Figure 6 shows downcore variations of the factor scores for each factor.

4 Discussion

Our results for core FA09 allow us to recognize four environmentally distinct intervals in the southern Red Sea within the last 16 ka:

Interval A (15.7–15.3 ka BP, base of the examined core interval). This interval is characterized by extremely low numbers of planktonic foraminifera (PFN) and high scores on factors 3 and 5 (Fig. 6).

Intervals where the planktonic foraminifera are almost absent ("aplanktonic" zones) have been previously recorded in sediment cores from the Red Sea, and they have been attributed mainly to the prevalence of extremely high sea surface salinity (about 50 psu or higher) due to near-isolation of the Red Sea at low full-glacial sea levels (Almogi-Labin et al. 1991; Fenton et al. 2000, and references therein). The youngest of these intervals occurred at the Last Glacial Maximum (LGM, or MIS2). Interval A corresponds to the late stages of this period. The presence of planktonic foraminiferal species in FA09 suggests that salinity in the study area did not rise to levels beyond the tolerance of these microorganisms. This is in agreement with observations from other cores from the southernmost Red Sea, which do not show "truly" aplanktonic layers (Fenton et al. 2000, and references therein). These microfaunal associations within the "aplanktonic" interval are different relative to the central and northern Red Sea in terms of foraminiferal population and diversity (Fenton et al. 2000). This can be ascribed to advection of specimens from the Gulf of Aden through the Bab el Mandab Strait. This layer in FA09 therefore supports the notion that water-exchange through the strait remained active during at least the late phase of the LGM, as was also inferred by Rohling et al. (1998), (Fenton et al. 2000); Siddall et al. (2003); and Fernandes et al. (2006). Any separation from the Indian Ocean in this interval would not only have resulted in immediate aplanktonic conditions, but also in drawdown of the Red Sea water level below the sea-floor depth of FA09 within a century or two, with associated evaporite deposition. Instead, we see continuous low-abundance planktonic foraminiferal faunas.

Although high salinity may have had an influence as well (see above), the low P% values (Fig. 6) suggest that the core location was much shallower than it is today and was probably part of the inner shelf (Van der Zwaan et al. 1999; Murray 1991). The co-occurrence of high scores of factors 3 and 5 indicates that miliolids and agglutinants were dominant components in the benthic microfauna (Figs. 5 and 6). Similar associations occur in hypersaline lagoons in the present Red Sea close to the core site (Abu-Zied et al. 2011). Furthermore, similar assemblages have been obtained during glacial intervals in marine cores in previous studies (Locke and Thunell 1988; Badawi et al. 2005). This suggests the establishment of low oxygen and high salinity conditions in waters over the seafloor during the late phase of the LGM in the southern Red Sea (Locke and Thunell 1988; Fenton et al. 2000, and references therein). It also confirms that, while the Strait of Bab al Mandab may have remained open, it did become seriously restricted, causing high salinities in the basin (see also Locke and Thunell 1988; Rohling et al. 1998; Fenton et al. 2000; Fernandes et al. 2006).

This interval ends with an increase in the BFN numbers associated with high scores of factor 3 (Fig. 6), suggesting an improvement of sea-floor oxygenation and reduction in salinity in association with post-glacial sea-level rise. Similar trends have been seen in benthic microfaunas at the end of the LGM interval in other cores from the southern and central Red Sea (Badawi et al. 2005).

Interval B (15.3–10 ka BP). This interval corresponds to the latest glacial period and onset of the Holocene. It is characterized by high PFN values and high scores of factors 2 and 3 (Fig. 6).

The high abundance of planktonic foraminifera may be explained by improvements in the water exchange over the Hanish Sill, from the Arabian Sea into the Red Sea, as sea level rose (Rohling et al. 1998; Siddall et al. 2003). Today the intrusion of relatively fresh, cold and nutrient-enriched waters from the Gulf of Aden constitutes the major source of nutrients in the southern Red Sea (Raitsos et al. 2015). It is likely that the sea level of this time-interval, which was still low and therefore created a shallow strait in the south, permitted only a two-layered water exchange between inflow

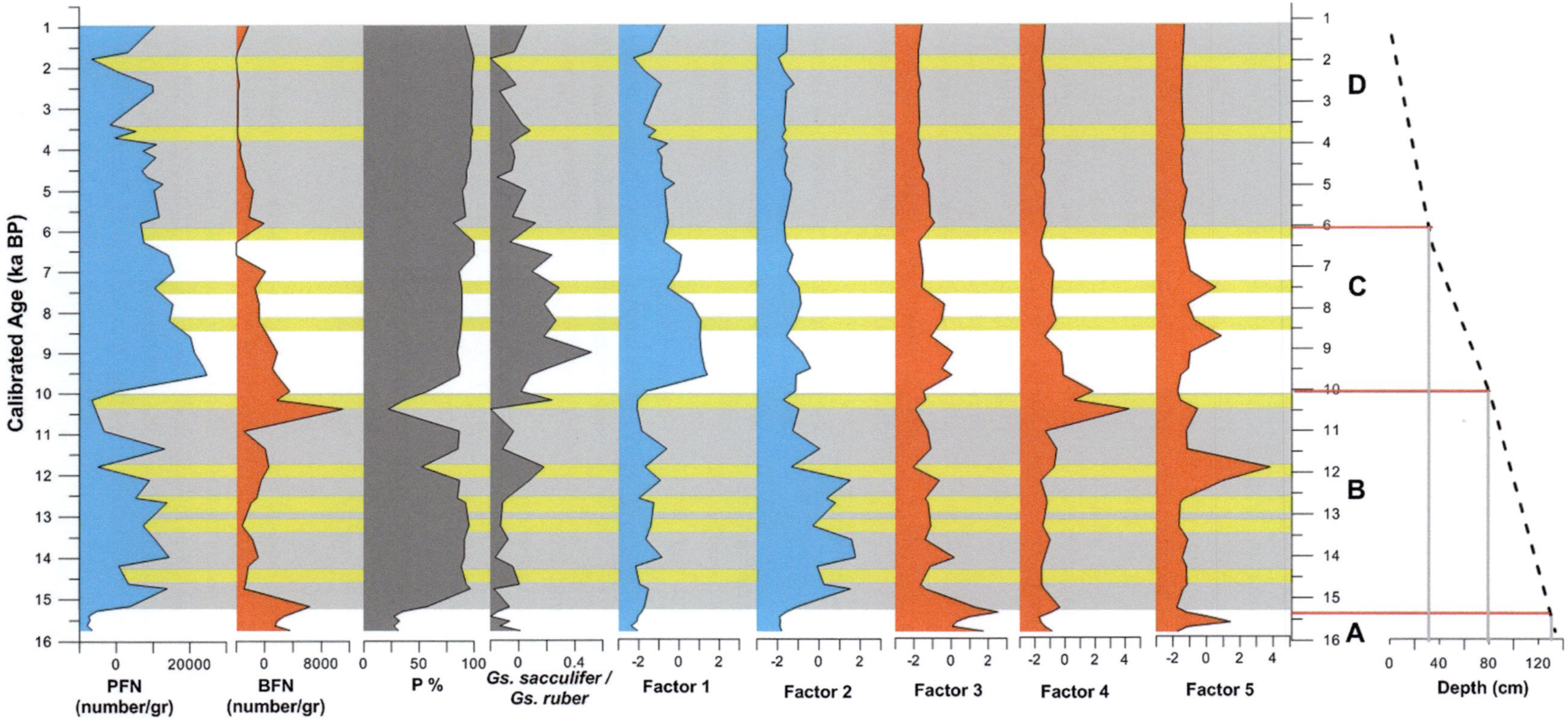

Fig. 6 Chronostratigraphy and downcore variations in the abundance of benthic and planktonic foraminifera (given in number of specimens per gramme of dry sediment) together with the percentage participation of planktonic forams (P%) and *Gs. sacculifer* versus *Gs. ruber* abundances, and the downcore variations of the scores of factors 1–5. The age-depth correlation is depicted. The intervals A–D and intermediate time periods discussed in the text are indicated

from the Gulf of Aden and outflow from the Red Sea (Siddall et al. 2002). The core site was situated in shallower waters than today, where coastal upwelling could have developed. Today coastal upwelling—which induces local productivity—is observed in the shallow zone of the Farasan Archipelago (Sofianos et al. 2016; Dreano et al. 2016). The relatively shallow inflow of Gulf of Aden waters of time-interval B, along with this potential development of local coastal upwelling, would have increased productivity in the southern Red Sea, leading to an increase of the planktonic assemblages of factor 2 (high factor 2 scores; Figs. 4 and 6). At the same time, the abundance of deeper-dwelling planktonics together with high factor-3 scores and moderate factor-4 values (Figs. 4 and 6) suggest that the water column above the site of FA09 was sufficiently oxygenated throughout and that mesotrophic conditions had been established on the seafloor.

Within this interval, short-term perturbations are observed in the abundance of the planktonic population, which is dominated by the factor-2 assemblage. The most pronounced changes are observed at around 14.5 ka and 11.5 ka BP, as brief reductions in the planktonic populations. Today, the ecosystem and the oceanographic conditions in the southern Red Sea are more prone to impacts from the Indian Ocean than other parts of the Red Sea (Sofianos and Johns 2007; Triantafyllou et al. 2014). Therefore, we propose that the short-term changes in FA09 may be linked to changes in the Arabian Sea. Marine palaeoclimatic, palaeo-productivity and denitrification records from the Arabian Sea suggest that the intensity of the summer SW monsoon was stronger during interstadials and weaker during stadials, within the last glacial cycle (e.g., Schulz et al. 1998; Ivanochkoa et al. 2005). The stronger SW monsoon caused stronger upwelling in the Arabian Sea during both the Holocene and interstadials, enhancing fluxes of buried organic matter in the area and reducing the water-column oxygenation. On the contrary, the LGM and the stadials were times of reduced organic burial fluxes and improved water-column oxygenation in the Arabian Sea (Ivanochkoa et al. 2005). The intermittent Interval-B reductions in planktonic populations in core FA09, and in particular in species related to eutrophic waters, likely reflect weakened inflow of subsurface Gulf of Aden water due to reduced SW-monsoon-driven upwelling in the Gulf of Aden. This reduced the nutrient availability at the site of core FA09. Almogi-Labin et al. (1991) have also documented temporal changes in the planktonic fauna of the central Red Sea (foraminifera and pteropods) during this time-interval (Late Glacial period). Almogi-Labin et al. (1991) attributed these changes to oscillations between humid and arid conditions, where arid conditions produced unfavourable conditions for survival of the organisms. From the present study, it seems that oceanographic conditions in the southernmost Red Sea never reached sufficiently severe conditions to cause total absences of planktonic populations, as seen in more northern parts of the Red Sea at this time. However, it is noticeable that the event, at around 11.5 ka coincides with an increase of the factor 5 scores (Fig. 6), implying a reduction in seafloor oxygen concentration and an increase in salinity that coincided with the harsher conditions that prevailed in more northern parts of the Red Sea. A comprehensive explanation for all combined signals remains elusive in the absence of detailed dating, but it is evident that some coherent pattern existed across the basin.

The end of interval B is marked by an increase of factor-4 scores (Figs. 5 and 6). The abundances of buliminids and bolivinids increase sharply at around 10.5 ka and remain high in the subsequent interval, which implies an increase in the organic flux on the sea floor and/or a reduction of seafloor oxygenation at the core site. This may be explained by an onset of strong GAIW intrusion into the Red Sea. At around 10.5 ka BP, sea level had risen sufficiently (to $\sim$50 m below the present level) to allow the intrusion of the GAIW over the Hanish sill (Siddall et al. 2002, 2003). A three-layer water exchange was established and remained active until the present. It is likely that at 10.5 ka, the onset of GAIW inflow with significant quantities of nutrients and organic matter increased productivity and thus the organic fluxes on the seafloor in the southern Red Sea, and so increased the abundances of buliminids and bolivinids at FA09.

Interval C (10–6 ka BP). The main characteristics of this interval are increased planktonic abundances and high factor-1 scores (Fig. 6). Moderate factor-3 scores suggest that, within this interval, the seafloor was mesotrophic and sufficiently oxygenated (Fig. 6).

A sharp increase of planktonic populations related to warm and salty surface waters (factor 1) reflects the establishment of warm and salty surface conditions at the site of FA09 in the early Holocene. Planktonic assemblages reflecting warm, high salinity and oligotrophic waters similarly developed throughout the Red Sea in the early Holocene (Siccha et al. 2009; Trommer et al. 2010). We propose that flooding of the vast Red Sea shelves (doubling the sea's surface area) when sea-level rose higher than about –50 m may have caused some of this shift to predominance of species indicative of warm, salty, and oligotrophic conditions. In these assemblages, variations in the relative abundances between individual species have been attributed to changes in salinity and productivity along the Red Sea. These variations were not simultaneous between the northern and central Red Sea and they are attributed to differentiations in the water circulation mode of the Red Sea (Siccha et al. 2009; Trommer et al. 2010), and relatively larger shelf areas in the south than in the north may also have played a role.

The onset of the PFN-increase and of increases in the abundances of *Gs. ruber*, *Gs. sacculifer*, *G. glutinata*

(Figs. 5 and 6) at ~10 ka in FA09 closely coincide with similar variations in the central Red Sea (core KL09; Siccha et al. 2009; Trommer et al. 2010), suggesting a synchronicity between the central Red Sea and the region of FA09. However, in detail the planktonic assemblages show some dissimilarities between these regions. The most pronounced is the (low-abundance) contribution of productivity-indicator species (factor 2) in FA09, and their absence in the central Red Sea. We propose that this is related to GAIW influences that were present at this time only in the southern Red Sea region of FA09.

The early Holocene is a known period of increased monsoon intensity and northward migration of the Intertropical Convergence Zone (ITCZ; Fleitmann et al. 2007, and references therein). The strong summer monsoons of that time caused intensified upwelling of nutrient-enriched waters in the Arabian Sea (Ivanochkoa et al. 2005; Gupta et al. 2011), which included increased upwelling of nutrient-enriched GAIW in the Gulf of Aden. This facilitated enhanced GAIW (and nutrient) penetration into the Red Sea. Today, in summer, the GAIW is observed in the southern Red Sea at depths between 50 and 100 m (Sofianos and Johns 2007). Although its properties are strongly diluted northward, the main part of the GAIW intrusion can reach up to ~16°N (only ~100 km south of FA09), where it is detected mainly as a mass of low salinity (37–38 psu), elevated nutrient concentrations (higher than 10 μmol/l) and a dissolved oxygen content of 22.5 mmol/kg, at ~75 m depth (Sofianos and Johns 2007 Dreano et al. 2016). With intensified monsoon circulation and increased GAIW penetration during the early Holocene, we suggest that our observed planktonic foraminiferal changes indicate that the site of FA09 came under the direct influence of the enhanced GAIW intrusion. The resultant productivity increase is further reflected by increases of buliminids and bolivinids in the benthic assemblage, which are related to increased organic fluxes to the seafloor (factor 4; Figs. 5 and 6). Indeed, this assemblage peaks between 10.5 and 8 ka, almost in phase with the most intense monsoon period (Fleitmann et al. 2007). Given that these specific planktonic and benthic assemblages are seen in FA09 and not in the central Red Sea (Siccha et al. 2009), we infer that the location of FA09 was close to the northernmost edge of the early Holocene GAIW intrusion.

Comparison between the planktonic assemblages of FA09 and those of the central Red Sea reveals some further interesting differences. Notably, *Gs. Sacculifer* (Fig. 6), is the dominant species in the central and northern Red Sea, where its abundance exceeds that of *Gs. ruber* during the early Holocene (Siccha et al. 2009; Trommer et al. 2010). In the southern Red Sea (core FA09), however, *Gs. sacculifer* shows elevated numbers (Fig. 4) but does not exceed *Gs. ruber* (see ratio *Gs. sacculifer* versus *Gs. ruber*; Fig. 6). Although open-ocean observations show that both *Gs. ruber* and *Gs. sacculifer* dominate surficial waters of high temperature and high salinity (Fenton et al. 2000, and references therein), there are studies which show *Gs. sacculifer* as a relatively stenohaline species and *Gs. ruber* as a euryhaline species (Hemleben et al. 1987; Bijma et al. 1990). In addition, *Gs. sacculifer* seems to dominate oligotrophic areas but *Gs. ruber* dominates water masses of higher productivity (Halicz and Reiss 1981; Bijma et al. 1990). If temperature and salinity control the proportional distribution of *Gs. sacculifer*, then this difference should be explained by the contribution of the water from the Gulf of Aden in the southern Red Sea. Today, the inflow of water from the Gulf of Aden at the surface in the winter and subsurface in the summer affects the southern Red Sea (FA09) and provokes surface waters of lower salinity and higher productivity in comparison to the surface water of the northern parts of the Red Sea. If this oceanographic pattern was the same during the early Holocene, then the differences in the relative abundance of *Gs. sacculifer* observed between the southern Red Sea and the northern parts of the Red Sea could reflect the prevalence of less saline and more fertile waters above the site of core FA09.

Both the typical planktonic and benthic populations show progressive reductions toward the end of interval C at 6 ka (Fig. 6), suggesting a gradual increase in salinity and food-availability stress in surface waters, and a resultant decrease in organic flux to the seafloor.

Interval D (6–1 ka BP). The main characteristics of this interval, which corresponds to the late Holocene, consist of a reduction in planktonic populations (low PFN numbers, low scores of factors 2 and 3), and low benthic foraminiferal abundances (low BFN).

Interval D contains similar planktonic foraminiferal species to interval C, but at lower abundances for almost all species (Figs. 4 and 6). Among them, *Gs. sacculifer* shows the largest reduction. This decline of this species coincides with a near-absence of the factor-4 benthic assemblage (Figs. 5 and 6). In addition, for a time-interval between 4.0 and 2.0 ka, benthic populations almost completely disappear, which suggests a complete lack of food availability (or, alternatively, anoxic bottom waters, for which we have no further indications). Together, these observations suggest a prevalence of less eutrophic waters, relative to the preceding interval (C). The late Holocene is marked by shorter and weaker summer monsoon circulation and southward migration of the ITCZ (Fleitmann et al. 2007, and references therein). Trommer et al. (2010) also observed changes in the microfauna associations between 3.7 and 1.7 ka in the central and northern Red Sea, which they attributed to weakening summer monsoons along with intensification of winter monsoons. In addition, an intensification of the OMZ (Oxygen Minimum Zone) and weakening of upwelling in the Arabian Sea during the Late Holocene also reflect weakening summer monsoons (Das et al. 2017). The above suggests that the inferred reduction in nutrient availability in this interval (D) of FA09 may be related to weakening of the GAIW inflow into the Red Sea.

5 Conclusions

Four intervals were distinguished, based mainly on variations in the planktonic foraminiferal assemblages and benthic foraminiferal abundances in the sequence of core FA09. These intervals correspond to the upper part of MIS2, the Late Glacial and the early to late Holocene. The examined interval of MIS2 suggests that the Strait of Bab al Mandab most likely remained open during the LGM, albeit in a very restricted form. During the Late Glacial, initiation of coastal upwelling and of inflow from the Gulf of Aden into the Red Sea resulted in more productive conditions in the southern Red Sea. It is likely that significant inflow of GAIW started at around 10.5 ka. From that time onward, the planktonic and benthic microfauna at the site of FA09 reflect changes in the southern Red Sea that followed changes in the intensity of (summer) monsoon winds. During the Late Glacial and early Holocene, the water column was in general well oxygenated throughout, but during the late Holocene a reduction of productivity and oxygen levels occurred during which time the winter monsoon was stronger. Temporal and simultaneous variations between the various proxies investigated seem linked to a combination of sea-level and monsoon influences on the Red Sea in general, and in its southernmost sector in particular (where FA09 is located) through modulation of GAIW inflow through the Strait of Bab al Mandab.

Acknowledgements This work was supported by the European Research Council through ERC Advanced Grant269586 'DISPERSE: Dynamic Landscapes, Coastal Environments and Human Dispersals' to Geoff Bailey and Geoffrey King, 2011–2016. We thank HRH Crown Prince Salman bin Abul Aziz Al Saud and the Department of General Survey of the Ministry of Defense and HRH Prince Sultan bin Salman bin Abdul Aziz, President of the Saudi Commission for Tourism and National Heritage (SCTH) for permits and general support, and the President of the Saudi Geological Survey, Dr. Zohair Nawab and his staff, in particular Dr. Najeeb Rasul, for additional support and for their invitation to participate in the Jeddah Workshop. In addition, we thank Geoff Bailey and the personnel at Scottish Universities Environmental Research Centre (SUERC, Glasgow, Scotland) for arranging the radiocarbon analysis. Finally, we give special thanks to the crew of the R/V Aegaeo for their great and responsible work during the survey. This is DISPERSE contribution no. 45.

References

Abu-Zied RH (2013) Effect of the Red Sea brine-filled deeps (Shaban and Kebrit) on the composition and abundance of benthic and planktonic foraminifera. Arab J Geosci 6:3809–3826

Abu-Zied RH, Bantan RA, Basaham AS, El Mamoney MH, Al-Washmi HA (2011) Composition, distribution, and taphonomy of nearshore benthic foraminifera of the Farasan Islands, southern Red Sea, Saudi Arabia. J Foramin Res 41:349–362

Almogi-Labin A, Hemleben C, Meischner D, Erlenkeuser H (1991) Paleoenvironmental events during the last 13,000 years in the central Red Sea as recorded by pteropoda. Paleoceanography 6:83–98

Almogi-Labin A, Hemleben C, Meischner D, Erlenkeuser H (1996) Response of Red Sea deep-water agglutinated foraminifera to water mass changes during the Late Quaternary. Mar Micropaleontol 28:283–297

Anderson DM, Prell WL (1991) Coastal upwelling gradient during the Late Pleistocene. In: Proceedings of the ODP Scientific Results, vol 117, pp 265–276

Arz HW, Lamy F, Pätzold J, Müller PJ, Prins M (2003) Mediterranean moisture source for an Early Holocene humid period in the northern Red Sea. Science 300:118–121

Auras-Schudnagies A, Kroon D, Ganssen G, Hemleben C, van Hinte JE (1989) Distributional pattern of planktonic foraminifers and pteropods in surface waters and top core sediments of the Red Sea, and adjacent areas controlled by the monsoonal regime and other ecological factors. Deep Sea Res Part A 36:1515–1533

Badawi A, Schmiedl G, Hemleben C (2005) Impact of late Quaternary environmental changes on deep-sea benthic foraminiferal faunas of the Red Sea. Mar Micropaleontol 58(1):13–30

Bailey G, Al-Sharekh A, Flemming N, Lambeck K, Momber G, Sinclair A, Vita-Finzi C (2007) Coastal prehistory in the southern Red Sea Basin, underwater archaeology, and the Farasan Islands. In: Proceedings of the Seminar for Arabian Studies, vol 37, pp 1–16

Bailey G, Deves MH, Inglis RH, Meredith-Williams MG, Momber G, Sakellariou D, Sinclair AGM, Rousakis G, Al Ghamdi S, Alsharekh AM (2015) Blue Arabia: Palaeolithic and underwater survey in SW Saudi Arabia and the role of coasts in Pleistocene dispersals. Quat Int 382:42–57

Bailey G, Meredith-Williams M, Alsharekh A, Hausmann N (this volume) The archaeology of Pleistocene coastal environments and human dispersals in the Red Sea: Insights from the Farasan Islands

Bantan RA (1999) Geology and sedimentary environments of Farasan Bank (Saudi Arabia) southern Red Sea: A combined remote sensing and field study. Unpublished PhD dissertation, University of London, U.K., 296 pp

Bijma J, Faber J, Hemleben C (1990) Temperature and salinity limits for growth and survival of some planktonic foraminifers in laboratory cultures. J Foraminiferal Res 20(2):95–116

Biton E, Gildor H, Peltier WR (2008) The Red Sea during the Last Glacial Maximum: Implications for sea level reconstruction. Paleoceanography 23, PA1214. https://doi.org/10.1029/2007pa001431

Blackwelder P, Hood T, Alvarez Zarikian C, Nelsen TA, McKee B (1996) Benthic foraminifera from the NECOP study area impacted by the Mississippi River plume and seasonal hypoxia. Quat Int 31:19–36

Bosence DWJ, Al-Awah MH, Davison I, Rosen BR, Vita-Finzi C, Whittaker E (1998) Salt domes and their control on basin margin sedimentation: a case study from the Tihama plain, Yemen. In: Purser BH, Bosence DWJ (eds) Sedimentation and tectonics in Rift Basins: Red Sea-Gulf of Aden. Chapman and Hall, London, pp 448–466

Dabbagh A, Emmermann R, Hötzl H, Jado AR, Lippolt HJ, Kollmann W, Moser H, Rauert W, Zötl JG (1984) The development of Tihamat Asir during the Quaternary. In: Jado AR, Zotl JG (eds) Quaternary Period in Saudi Arabia, vol 2. Sedimentological, Hydrogeological, Hydrochemical, Geomorphological, Geochronological and Climatological Investigations in Western Saudi Arabia. Springer, Vienna, pp 150–173

Das M, Singh RK, Gupta AK, Bhaumik AK (2017) Holocene strengthening of the Oxygen Minimum Zone in the northwestern Arabian Sea linked to changes in intermediate water circulation or Indian monsoon intensity? Palaeogeogr Palaeoclimatol Palaeoecol 483:125–135

Davis JC (1986) Statistics and data analysis in geology. Wiley, New York, p 647

De S, Gupta AK (2010) Deep-sea faunal provinces and their inferred environments in the Indian Ocean based on distribution of Recent benthic foraminifera. Palaeogeog Palaeoclimat Palaeoecol 291:429–442

Demarchi B, Williams MG, Milner N, Russell N, Bailey G, Penkman K (2011) Amino acid racemization dating of marine shells: amound of possibilities. Quat Int 239:114–124

Dreano D, Raitsos DE, Gittings J, Krokos G, Hoteit I (2016) The Gulf of Aden Intermediate Water intrusion regulates the southern Red Sea summer phytoplankton blooms. PLoS ONE 11(12):e0168440. https://doi.org/10.1371/journal.pone.0168440

Edelman-Fürstenberg Y, Scherbacher M, Hemleben C, Almogi-Labin A (2001) Deep-sea benthic foraminifera from the central Red Sea. J Foramin Res 31:48–59

Fenton M, Geiselhart S, Rohling EJ, Hemleben C (2000) Aplanktic zones in the Red Sea. Mar Micropaleontol 40:277–294

Fernandes C, Rohling EJ, Siddall M (2006) Absence of Quaternary Red Sea land bridges: biogeographic implications. J Biogeogr 33:961–966

Fleitmann D, Burns S, Mangini A, Mudelsee M, Kramers J, Villa I, Neff U, Al-Subbary A, Buettner A, Hippler D, Matter A (2007) Holocene ITCZ and Indian monsoon dynamics recorded in stalagmites from Oman and Yemen (Socotra). Quat Sci Rev 26:170–188

Grant KM, Rohling EJ, Bar-Matthews M, Ayalon A, Medina-Elizalde M, Bronk Ramsey C, Satow C, Roberts AP (2012) Rapid coupling between ice volume and polar temperature over the past 150 kyr. Nature 491:744–747

Grant KM, Rohling EJ, Bronk Ramsey C, Cheng H, Edwards RL, Florindo F, Heslop D, Marra F, Roberts AP, Tamisiea ME, Williams F (2014) Sea-level variability over five glacial cycles. Nat Commun 5:5076. https://doi.org/10.1038/ncomms6076

Gupta AK, Anderson DM, Overpeck JT (2003) Abrupt changes in the Asian southwest monsoon during the Holocene and their links to the North Atlantic Ocean. Nature 421:354–357

Gupta AK, Sarkar S, De S, Clemens SC, Velu A (2010) Mid-Brunhes strengthening of the Indian Ocean Dipole caused increased equatorial East African and decreased Australasian rainfall. Geophys Res Lett 37:L06706, 6

Gupta AK, Mohan K, Sarkar S, Clemens SC, Ravindra R, Uttam RK (2011) East-west similarities and differences in the surface and deep northern Arabian Sea records during the past 21 Ka. Palaeogeog Palaeoclimat Palaeoecol 301:75–85

Halicz E, Reiss Z (1981) Palaeoecological relations of foraminifera in a desert enclosed sea: the Gulf of Aqaba. Mar Ecol 2:15–34

Hemleben C, Spindler M, Breitinger I, Ott R (1987) Morphological and physiological responses of Globigerinoides sacculifer (Brady) under varying laboratory conditions. Mar Micropaleontol 12:305–324

Hottinger L, Halicz E, Reiss Z (1993) Recent Foraminifera from the Gulf of Aqaba, Red Sea. Slovenska Akademija Znanosti in Umetnosti, Ljubljana, p 179

Ivanochkoa TS, Ganeshrama RS, Brummerb GJ, Ganssenc G, Jungc S, Moretond S, Kroon D (2005) Variations in tropical convection as an amplifier of global climate change at the millennial scale. Earth Planet Sci Lett 235(1):302–314. https://doi.org/10.1016/j.epsl.2005.04.002

Kaiho K (1994) Benthic foraminiferal dissolved-oxygen index and dissolved oxygen levels in the modern ocean. Geology 22:719–722

Kroon D, Ganssen G (1989) Northern Indian Ocean upwelling cells and the stable isotope composition of living planktic foraminifers. Deep-Sea Res 36:1219–1236

Lambeck K, Purcell A, Fleming NC, Vita-Finzi C, Alsharekh AM, Bailey G (2011) Sea level and shoreline reconstructions for the Red Sea: Isostatic and tectonic considerations and implications for hominin migration out of Africa. Quat Sci Rev 30:3542–3574

Lambeck K, Rouby H, Purcell A, Sun Y, Sambridge M (2014) Sea level and global ice volumes from the Last Glacial maximum to the Holocene. Proc Natl Acad Sci USA 111(43):15296–15303

Locke S, Thunell RC (1988) Paleoceanographic record of the last glacial/interglacial cycle in the Red Sea and Gulf of Aden. Palaeogeog Palaeoclimat Palaeoecol 64:163–187

Loeblich AR, Tappan H (1987) Foraminiferal genera and their classification, vol 2. Van Nostrand Reinhold, New York

Mateu-Vicens G, Box A, Deudero S, Rodriguez B (2010) Comparative analysis of epiphytic foraminifera in sediments colonized by seagrass Posidoniaoceanica and invasive macroalgae *Caulerpa spp*. J Foramin Res 40:134–147

Mendes I, Gonzalez R, Dias JMA, Lobo F, Martins V (2004) Factors influencing recent benthic foraminifera distribution on the Guadiana shelf (southwestern Iberia). Mar Micropaleontol 51:171–192

Momber G, Sakellariou D, Bailey G, Rousakis G (this volume) The multi-disciplinary search for underwater archaeology in the southern Red Sea

Moodley L, Schaub BEM, Van der Zwaan GJ, Herman PMJ (1998) Tolerance of benthic foraminifera (Protista: Sarcodina) to hydrogen sulphide. Mar Ecol Prog Ser 169:77–86

Morcos SA (1970) Physical and chemical oceanography of the Red Sea. Oceanogr Mar Biol Ann Rev 8:73–202

Murray JW (1991) Ecology and paleoecology of benthic foraminifera. John Wiley & Sons, New York, p 397

Murray JW (2006) Ecology and applications of benthic foraminifera. Cambridge University Press, pp 426

Overpeck J, Rind D, Lacis A, Healy R (1996) Possible role of dust-induced regional warming in abrupt climate change during the last glacial period. Nature 384:447–449. https://doi.org/10.1038/384447a0

Platon E, Sen Gupta BK, Rabalais NN, Turner RE (2005) Effect of seasonal hypoxia on the benthic foraminiferal community of the Louisiana inner continental shelf: the 20th century record. Mar Micropaleontol 54:263–283

Raitsos DE, Yi X, Platt T, Racault M, Brewin RJW, Pradhan Y, Papadopoulos VP, Sathyendranath S, Hoteit I (2015) Monsoon oscillations regulate fertility of the Red Sea. Geophys Res Lett 42:855–862

Reyment RA, Joreskog KG (1996) Applied factor analysis in the natural sciences. Cambridge University Press, Cambridge, p 371

Rohling EJ, Fenton M, Jorissen FJ, Bertrand P, Ganssen G, Caulet JP (1998) Magnitudes of sea-level lowstands of the past 500,000 years. Nature 394:162–165

Rohling EJ, Grant KM, Roberts AP, Larrasoaña JC (2013) Palaeoclimate variability in the Mediterranean and Red Sea regions during the last 500,000 years; implications for hominin migrations. Curr Anthropol 54(No. S8, Alternative Pathways to Complexity: Evolutionary Trajectories in the Middle Paleolithic and Middle Stone Age):S183–S201

Schulz H, Rad U, Erlenkeuser H (1998) Correlation between Arabian Sea and Greenland climate oscillations of the past 110,000 years. Nature 393:54–57

Sen Gupta BK, Machain-Castillo ML (1993) Benthic foraminifera in oxygen-poor habitats. Mar Micropaleontol 20:183–201

Siccha M, Trommer G, Schulz H, Hemleben C, Kucera M (2009) Factors controlling the distribution of planktonic foraminifera in the Red Sea and implications for the development of transfer functions. Mar Micropaleontol 72:146–156

Siddall M, Smeed DA, Matthieen S, Rohling EJ (2002) Modeling the seasonal cycle of the exchange flow. Deep-Sea Res 49:1551–1569

Siddall M, Rohling EJ, Almogi-Labin A, Hemleben C, Meischner D, Schmelzer I, Smeed DA (2003) Sea-level fluctuations during the last glacial cycle. Nature 423:853–858

Siddall M, Smeed DA, Hemleben C, Rohling EJ, Schmelzer I, Peltier WR (2004) Understanding the Red Sea response to sea level. Earth Planet Sci Lett 225:421–434

Sofianos SS, Johns WE (2007) Observations of the summer Red Sea circulation. J Geophys Res 112:C06025. https://doi.org/10.1029/2006JC003886

Sofianos S, Papadopoulos V, Abulnaja Y, Giouroukou D, Bolonaki E, Hoteit I (2016) Summer-time monsoon-driven variability in the Red Sea and the effects of the exchanges with the Indian Ocean. In: International conference on the marine environment of the Red Sea. King Abdullah University of Science and Technology

Southon J, Kashgarian M, Fontugne M, Metivier B, Yim WW-S (2002) Marine reservoir corrections for the Indian Ocean and Southeast Asia. Radiocarbon 44:167–180

Thiede J (1975) Distribution of foraminifera in coastal waters of an upwelling area. Nature 253:712–714

Triantafyllou G, Yao F, Petihakis G, Tsiaras KP, Raitsos DE, Hoteit I (2014) Exploring the Red Sea seasonal ecosystem functioning using a three-dimensional biophysical model. J Geophys Res Oceans 119:1791–1811

Trommer G, Siccha M, Rohling EJ, Grant K, van der Meer MTJ, Schouten S, Hemleben C, Kucera M (2010) Millennial-scale variability in Red Sea circulation in response to Holocene insolation forcing. Paleoceanography 25. https://doi.org/10.1029/2009PA001826

Van der Zwaan GJ, Duijnstee IAP, den Dulk M, Ernst SR, Jannink NT, Kouwenhoven TJ (1999) Benthic foraminifers: proxies or problems? a review of paleoecological concepts. Earth Sci Rev 46:213–236

Weiner A, Weinkauf M, Kurasawa A, Darling K, Kucera M (2015) Genetic and morphometric evidence for parallel evolution of the *Globigerinella calida* morphotype. Mar Micropaleontol 114:19–35

Woelk S, Quadfasel D (1996) Renewal of deep water in the Red Sea during 1982–1987. J Geophys Res 101(C8):18155–18165

Wyrtki K (1974) On the deep circulation of the Red Sea. L'oceanographie physique de la Mer Rouge. Cent Natl pour l'Exploitation des Oceans, Paris, pp 135–163

Red Sea Palaeoclimate: Stable Isotope and Element-Ratio Analysis of Marine Mollusc Shells

Niklas Hausmann, Olga Kokkinaki, and Melanie J. Leng

Abstract

The southern Red Sea coast is the location of more than 4,200 archaeological shell midden sites. These shell middens preserve archaeological and climatic archives of unprecedented resolution and scale. By using shells from these contexts, it is possible to link past environmental information with episodes of human occupation and resource processing. This chapter summarises current knowledge about the marine gastropod *Conomurex fasciatus* (Born 1778) and discusses its use in environmental and climatic reconstruction using stable isotope and elemental ratio analysis. It offers a review of the most recent studies of shell midden sites on the Farasan Islands, their regional importance during the mid-Holocene, theories about seasonal use of the coastal landscape, and preliminary results from new methods to acquire large climatic datasets from *C. fasciatus* shells.

1 Introduction

On most of the world's coastlines, the exploitation of marine resources in both modern and past societies was and continues to be a common practice with a longstanding tradition (Erlandson 2001). Evidence is found in countless archaeological sites that date far into prehistory and are partly or completely composed of mollusc shells, the food leftovers of the human inhabitants (Cortés-Sánchez et al. 2011; Marean 2014; Steele and Klein 2008). These shell middens feature several characteristics from which archaeological and climatic research benefits. Firstly, the hard shell matrix protects the remains that they cover, such as bones and artefacts (Thomas 2015a). Secondly, the shell carbonate itself allows for ample material suitable for radiocarbon dating at a high stratigraphic resolution (Stein et al. 2003). And lastly, each individual mollusc shell contains information about the environment in which it grew, because the chemical compositions of shell growth increments are controlled by several environmental factors (e.g., water temperature and salinity). Depending on the growth rate of the shell species, these records can provide daily to centennial information on past environmental conditions (Ridgway et al. 2011). Archaeological shell midden deposits therefore provide dateable, well preserved, high-resolution climate records that are directly related to human activity.

In this chapter, we build on this potential and present studies on the Farasan Island shell midden complex dating to the mid-Holocene (6–4,000 cal BP, or calibrated radiocarbon years before the present) and the middens' main constituent, the marine gastropod *Conomurex fasciatus* (Born 1778). First, we provide a description of the Arabian Neolithic and the sites on the Farasan Islands (Fig. 1). This is followed by an outline of how C. *fasciatus* shells were used in the past and are preserved today, as well as a description of how it is possible to use shells for environmental and climatic analysis. The section concludes with a modern reference study that focuses on modern *C. fasciatus* and how its shell geochemistry is influenced by the local environment. Second, we describe three applications of the climatic data derived from shell stable isotope and trace element analysis. The section is introduced by a study that compares modern and ancient shells, to assess climatic changes preceding the main occupation period of the Farasan shell middens, which could be linked to the reason why people

N. Hausmann (✉)
Foundation for Research and Technology—Hellas, Institute of Electronic Structure and Laser (IESL), P.O. Box 1527 711 10 Heraklion, Greece
e-mail: niklas@palaeo.eu

N. Hausmann · O. Kokkinaki
Department of Archaeology, University of York, King's Manor, York, YO1 7EP, UK

M. J. Leng
NERC Isotope Geosciences Facilities, British Geological Survey, Nottingham, NG12 5GG, UK

M. J. Leng
Centre for Environmental Geochemistry, University of Nottingham, Nottingham, NG7 2RD, UK

N. M. A. Rasul and I. C. F. Stewart (eds.), *Geological Setting, Palaeoenvironment and Archaeology of the Red Sea*,
https://doi.org/10.1007/978-3-319-99408-6_33

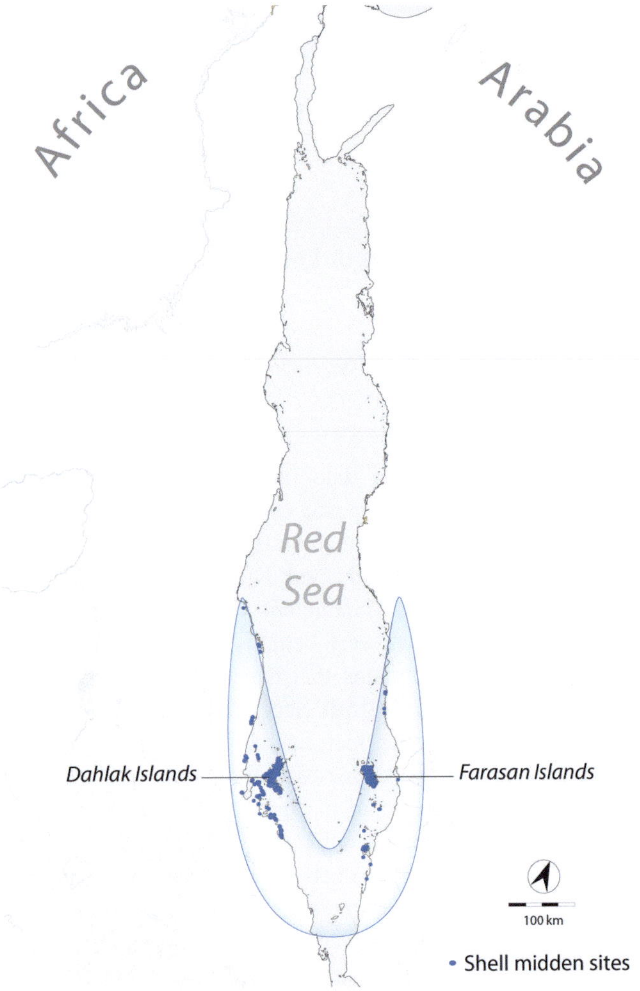

Fig. 1 Locations of shell midden sites in the southern Red Sea. Highlighted area indicates region of intense marine exploitation

settled on the islands. Lastly, we also show how short-term seasonal records within individual shells can be used to reconstruct seasonal eating habits of the prehistoric population. We also give a perspective on ongoing work involving the use of rapid measurements of elemental ratios in shell carbonate to reconstruct seasonal patterns.

2 Archaeology

2.1 Regional Overview

The Arabian Neolithic is broadly defined as the prehistoric period linked to specific archaeological sites and lithic artefacts on and from the Arabian Peninsula within the first half of the Holocene. In Arabia, this period is far from being as comprehensively researched as is the case in Europe or the Levant (Edens and Wilkinson 1998; Harrower 2008; Martin et al. 2009; McCorriston 2013). Generally, it is assumed that the Neolithic population of Arabia was much more mobile than its Neolithic counterparts in the north (McCorriston 2013). McCorriston et al. (2002, 2005, 2012; also Crassard et al. 2006) researched the domestication and early herding of cattle in the Wadi Sana (Yemen), and suggested a divide of Neolithic ways of life into people that practised hunting and herding, and people that almost solely relied on herding practices.

In addition to these food procurement strategies, it is now also apparent that a significant portion of the population made use of marine resources (Zazzo et al. 2014). Over 4,000 shell midden sites have been found in the last 15 years (Meredith-Williams et al. 2014b), suggesting that the consumption of marine resources was a way of living that was just as important as herding and hunting. Along the southern Red Sea coastline and on its multitude of islands, people throughout the Holocene took advantage of the rich marine wildlife and collected shellfish in large amounts. The majority of shell remains were found as large middens on the Dahlak Islands, Eritrea, and the Farasan Islands, Saudi Arabia. Many of those sites are large enough to be seen on modern satellite imagery (Meredith-Williams et al. 2014a). In other places, especially the Arabian mainland, middens are less commonly found from satellite imagery and therefore are more likely to be covered by dunes or to be eroded by wadis (Khalidi 2005).

Only a small fraction of middens in the southern Red Sea have been dated. The earliest of these are found on the Eritrean coastline and start at 8,000 cal BP with intermittent occupations continuing to 5,000 cal BP. The earliest sites on the eastern side of the Red Sea date to 7,500 cal BP on the Farasan Islands with almost constant shell accumulation spread over multiple sites between 6,500 and 4,500 cal BP and thereafter intermittent depositions into historical times (Hausmann and Meredith-Williams 2016b; Bailey et al., this volume).

In addition to Holocene sites, it is assumed that shell middens have been used globally by multiple coastal populations and reach back into periods of (pre-Holocene) lower sea level (Bailey and Craighead 2003). This coastal activity might be, albeit below current sea level, archaeologically very rich (Bicho and Haws 2008; Finlayson 2008; Erlandson et al. 2008; Jerardino 2013; Woodman 2013). Dispersal out of Africa, during the times of lower sea level, makes it likely that there is a plethora, so far undiscovered, of archaeological sites in the underwater areas in the Red Sea (Bailey 2004; Bailey and King 2011; Bailey et al. 2007, and this volume; Flemming et al. 2003; Momber 2000, 2004, and this volume).

2.2 Farasan Island Shell Midden Complex

Earlier research studies on the Farasan Islands tried to find suitable Palaeolithic (Pleistocene-Holocene) sites to identify the possible migration out of Africa through a southern pathway (Bailey et al. 2014a, b, and this volume). The location of the Farasan Islands on the eastern shore of the Red Sea, and possible connections to other islands, makes them a favourable destination for hominids crossing from Africa to Arabia.

Investigations on land provided several unstratified localities that are likely to be related to the Palaeolithic, but no dateable material from this period was identified. Nearly 3,000 shell midden sites were found, ranging from the mid-Holocene to Islamic periods. The structural preservation of thousands of sites is likely mainly due to the lack of urban development in the area until recently. Generally, shell middens are destroyed by construction workers as shells provide an easily accessible construction material, despite the archaeological remains within them, which often include human bones.

The unexpected richness of the archaeological material, as well as the increasing urban development on Farasan potentially leading to their destruction, has shifted research focus to these sites and triggered multiple excavations in 2004, 2006, 2008, 2009, 2012 and 2013 (Alsharekh and Bailey 2014) (Fig. 2). More than 20 shell middens have been excavated as well as 71 test pits along major palaeo-shorelines on the large islands of the archipelago: Farasan Kabir, Qumah and Saqid (Bailey et al. 2014a, b, and this volume).

Shell middens occur along modern shorelines and palaeo-shorelines, which have been tectonically uplifted in some areas. The association with specific shorelines, and hence specific sea-levels, allows a tentative assignment of the same date to sites sharing the same shoreline (Meredith-Williams et al. 2014b).

One particular site has been analysed in significant detail. Site JW1727 is located in the uplifted western part of Janaba Bay (Fig. 3). It is part of a large group of shell middens along one palaeoshoreline, and is part of a line of middens that follow the direction of sea level change. While JW1727 itself is approximately 0.75 km away from the modern shoreline, there are dozens of smaller sites in between, dating to progressively younger periods, the closer they get

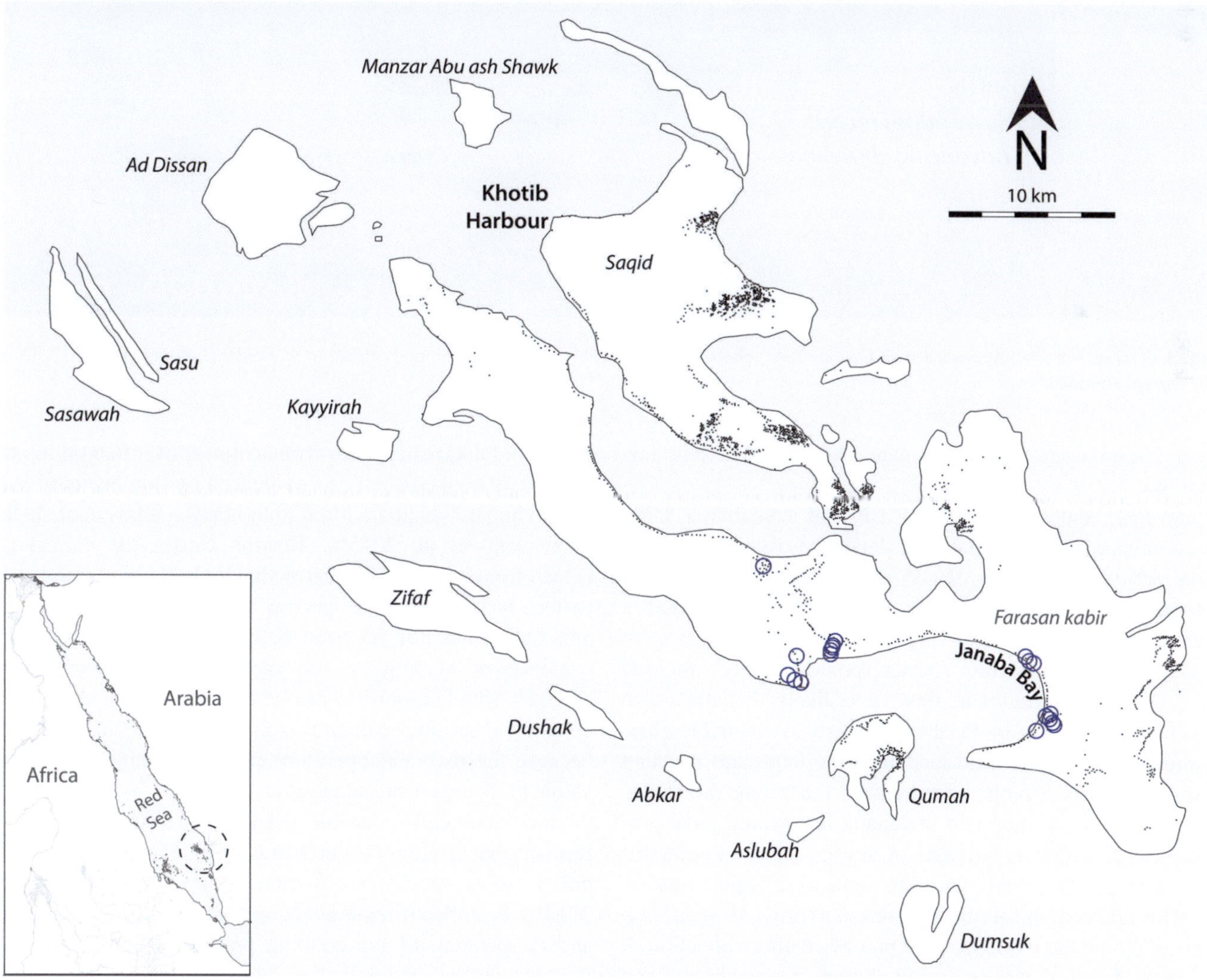

Fig. 2 Locations of excavated (blue circles) and unexcavated (black dots) shell middens on the Farasan Islands

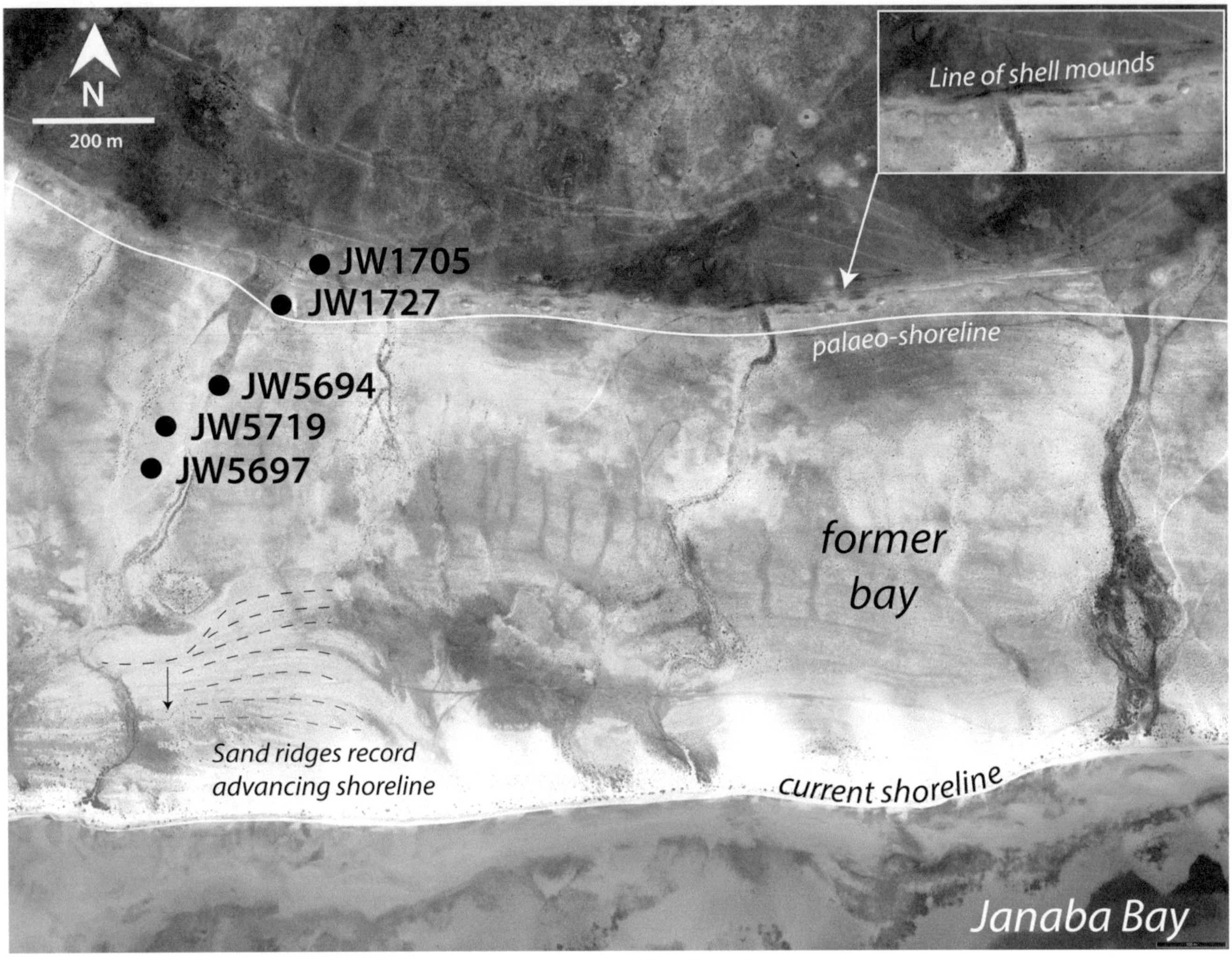

Fig. 3 Transect of shell mounds across Janabay Bay palaeo-shoreline, with younger shell mounds closer to and older shell mounds farther from the current shoreline

to the current sea level (Hausmann and Meredith-Williams 2017b).

JW1727 dates to 4,800 cal BP and contains a large accumulation of shells that were deposited over a very short time period (0–88 years, 95.4% probability, Hausmann and Meredith-Williams 2016a). Over the course of less than a century, people collected living specimens of *C. fasciatus* in addition to other mollusc species (predominantly *Pinctada* sp. and *Arca avellana*) as food. It is likely that the warm shallow water areas of Janaba Bay were an amenable environment for molluscs and facilitated the large and constant supply of marine protein for humans inhabiting the shoreline. The now uplifted part of Janaba Bay would have provided shell-gathering grounds of several square kilometres that were easily accessible to the prehistoric populations.

The archaeological record does not provide clear evidence of how the molluscs were processed after collection. It is likely that they were gathered close to the waterfront and heated in bulk, using non-permanent hearths, often on top of previously cooked shells, as evidenced by thin layers of ash and charcoal, which are often found between layers of shells (Hausmann et al. 2015b). Heating causes the animal to detach from its shell, thus extracting the meat becomes much easier. It is likely that people made use of this effect, although it has not yet been possible to confirm this from evidence of heating on the shell surface. Mineralogical changes from aragonite to calcite can occur when shells that contain aragonitic calcium carbonate are being heated, because the rise in temperature causes the mineral composition to change from aragonitic to calcitic, as aragonite is thermodynamically unstable (Sand et al. 2012). However, Raman spectroscopy on archaeological *C. fasciatus* shells did not observe changes in mineralogy (Hausmann et al. 2015b). Significant mineral changes depend on the duration and temperature of the cooking process. Small shells like *C. fasciatus* can be cooked within a few minutes in boiling water, which is unlikely to be enough to cause any measurable changes in the mineral or isotopic composition (Milano et al. 2016).

There is little evidence of processing tools or cooking equipment in the archaeological remains at JW1727. Some pottery material, in the form of a small sherd, has been found (Hausmann et al. 2015b), but it is unclear how pottery was used or if it was part of the shell processing activities. Since preservation conditions are generally good, it is possible that shellfish processing happened at a different location and JW1727 was only a waste dump for shells. It is also possible that processing was undertaken on site but artefacts involved were deposited elsewhere and are perhaps more likely to be found in connection with habitational sites. Habitational sites are not preserved on Farasan, possibly as they would have lacked protection from soil erosion. As such, middens are the primary source of archaeological information on the Farasan Islands. Furthermore, since few middens are found outside Farasan in Arabia, they are also some of the most important archaeological sites in the region. There is evidence of similar patterns of shell midden sites on the western side of the Red Sea, suggesting that the Farasan islands are not simply an anomaly found in Arabia, but are a result of the richness of the Red Sea environment, the coastal dynamics of the tectonic landscape, and the efficiency of the coastal exploitation patterns that led to the construction of the shell mounds.

3 Shell Midden Calcium Carbonate as an Environmental Proxy

3.1 C. fasciatus Stable Isotope Geochemistry

Here, we combine sclerochronology, the study of physical variations in the accretionary hard parts of marine organisms (Jones et al. 2007), with the measurement of stable isotopes of this same material. Specifically, we measured the ratios $^{18}O/^{16}O$ and $^{13}C/^{12}C$. These ratios may vary according to external environmental or internal metabolic factors (Cusack and Freer 2008; Wanamaker et al. 2007), and are used as a well-constrained method in archaeological science (Andrus 2011; Leng and Lewis 2016; Schöne and Gillikin 2013).

Growth increment formation in molluscan shells follows a biochemically complex process that mainly includes the formation of calcium carbonate and organic material during aerobic and anaerobic states of the metabolism. In the state of aerobic metabolism, molluscs deposit calcium carbonate in the form of aragonite or calcite in combination with organic material enclosed within the shell structure (Lutz and Rhoads 1977). The resultant mineralogy is specific for each mollusc species and can both occur on its own (i.e., all aragonite or all calcite), mixed together (i.e., alternating aragonite and calcite bands) or separated into different parts of shells (Leng and Lewis 2016).

Molluscs precipitate their shell layers ($CaCO_3$) and acid (H^+) by the following chemical reaction:

$$Ca^{2+} + HCO_3^- \rightleftarrows CaCO3 + H^+ \quad (1)$$

The aerobic metabolism is associated with periods when molluscs are exposed to well-oxygenated waters, such as at high tide, and when the shell is open. In contrast, in times of anaerobic state (i.e., shell closure) the amount of acidic components in the shell-building fluids rises and is gradually neutralised by the calcium carbonate of the shell, resulting in the dissolution of calcium carbonate and a larger amount of organic material content in the layer deposits (Crenshaw 1980; Lutz and Rhoads 1977).

Many other factors can play a role in the changing precipitation patterns of calcium carbonate. Extreme events like spring tides, storms, or high temperatures can affect a mollusc's metabolism and can cause it to rapidly convert into an anaerobic state and result in hiatuses in the isotopic record. Changes linked to the physiology of a mollusc are equally important. Stress such as spawning events (Kennish 1980) and the availability of food can have a significant influence on gastropod shell growth (Schöne 2008).

There is a theoretical and empirical relationship between $^{18}O/^{16}O$ (expressed in reference to established standards as $\delta^{18}O$) of the shell carbonate and temperature due to equilibrium fractionation. The higher/lower the ratio the colder/warmer the temperature, so long as the oxygen isotope composition of the seawater remains constant (Grossman and Ku 1986). Seawater is isotopically heavier than freshwater so rivers and streams entering the coastal environment can affect shell $\delta^{18}O$. The oxygen isotope ratio of shells is a record of the environment over a mollusc's lifetime, which can then be analysed to produce a local environmental record. Shell $\delta^{18}O$ can thus allow for the calculation of temperature or salinity (by way of empirical $\delta^{18}O$-salinity relationships) of surrounding waters, the seasonal variations of either of these, and long-term changes related to climatic or oceanographic changes.

Here we use $\delta^{18}O$ values from aragonitic shells of *C. fasciatus* (hereafter called $\delta^{18}O_S$), and the empirically derived temperature equation of Grossman and Ku (1986), adjusted for the $\delta^{18}O$ of collected modern sea water (hereafter called $\delta^{18}O_W$) (Dettman et al. 1999), to estimate the Sea Surface Temperature (SST) at the Farasan Islands for the moment of shell growth.

$$SST\ (^\circ C) = 20.6 - 4.34\left(\delta^{18}O_S - \left(\delta^{18}O_W - 0.27\right)\right) \quad (2)$$

3.2 Modern Seawater Isotopic Composition and Temperature Measurements

In order to calibrate Eq. 2 for *C. fasciatus* we compared modern measured temperature and $\delta^{18}O$ of seawater with modern shell $\delta^{18}O$. Local measurements of temperature and salinity were taken over a 1.5-year period using two data loggers (Star Oddi DST CTD) to investigate the local environment around Farasan, and to confirm that the archipelago follows the pattern of the regional climate or experiences any local anomalies (Fig. 4). These were then combined with the Sea Surface Temperature (SST) and Sea Surface Salinity (SSS) data that were accessible via the National Oceanic and Atmospheric Administration (NOAA). The data were measured hourly by a GLOBAL-HYCOM remote sensor (available for the Farasan Islands (FARS1) at http://ecoforecast.coral.noaa.gov/index/0/FARS1/station-home). Temperatures are reported in °C and salinity in practical salinity units (psu), which is equivalent to parts per thousand.

Results show a significant change in SST throughout the year in an overall range of about 7–8 °C, which translates into a $\sim 2‰$ change in estimated $\delta^{18}O$ ($\delta^{18}O_{Est.}$). The change is regular and occurs at similar times of the year, with little variation between the different years or measurement locations (Fig. 4). The seasonal salinity changes are similarly consistent from one year to another. However, they have a range of only 2 psu. This nearly constant salinity is likely to be due to the generally low precipitation throughout the year and thus low amounts of fresh water that enter the coastal bays. However, fresh water inflow from ocean currents can have a strong impact on the $\delta^{18}O_W$ values of the local marine water and needs to be accounted for. Water samples from Janaba Bay and eight other sites around Farasan were analysed in order to assess spatial differences and to explore the water exchange in different parts of the research area.

Because the salinity values from the NOAA record seem to coincide with the inflow of surface water from the Indian Ocean for the autumn and winter months, which is less saline than Red Sea surface water (Raitsos et al. 2013), it is possible that this inflow has an influence on the $\delta^{18}O_W$ of the marine water around Farasan. This issue requires further investigation since it may substantially affect the $\delta^{18}O_S$ record of *C. fasciatus* shells—both modern and archaeological.

The comparison of $\delta^{18}O_W$ from different study areas around Farasan showed how evaporation fractionation can vary spatially, and helped to single out areas with little variability to produce more precise estimates of past climate changes. $\delta^{18}O_W$ values from spring (May) and winter (December) were used to compare the way different locations are being affected by evaporation (Fig. 5). Spatial analysis was performed by interpolation of stable $\delta^{18}O_W$ isotope data from different locations. Subsequently, they

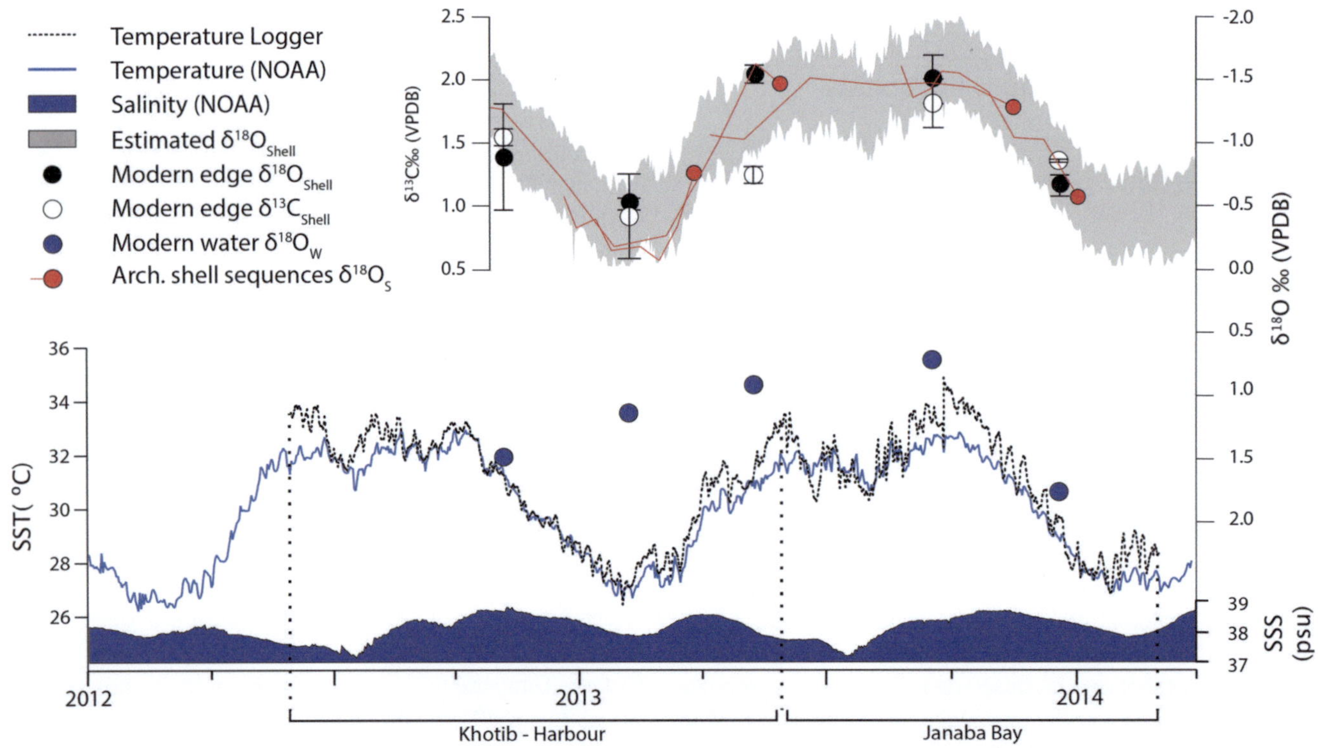

Fig. 4 Seasonal change of temperature and salinity with corresponding stable isotope values ($\delta^{18}O$ in black and $\delta^{13}C$ in white) measured in shell carbonate of *C. fasciatus* and in $\delta^{18}O$ water samples (blue) as well as the estimated $\delta^{18}O$ for each season (grey area). Sequential $\delta^{18}O$ values from archaeological shells as examples for determining season of death (red line, circle for terminal value)

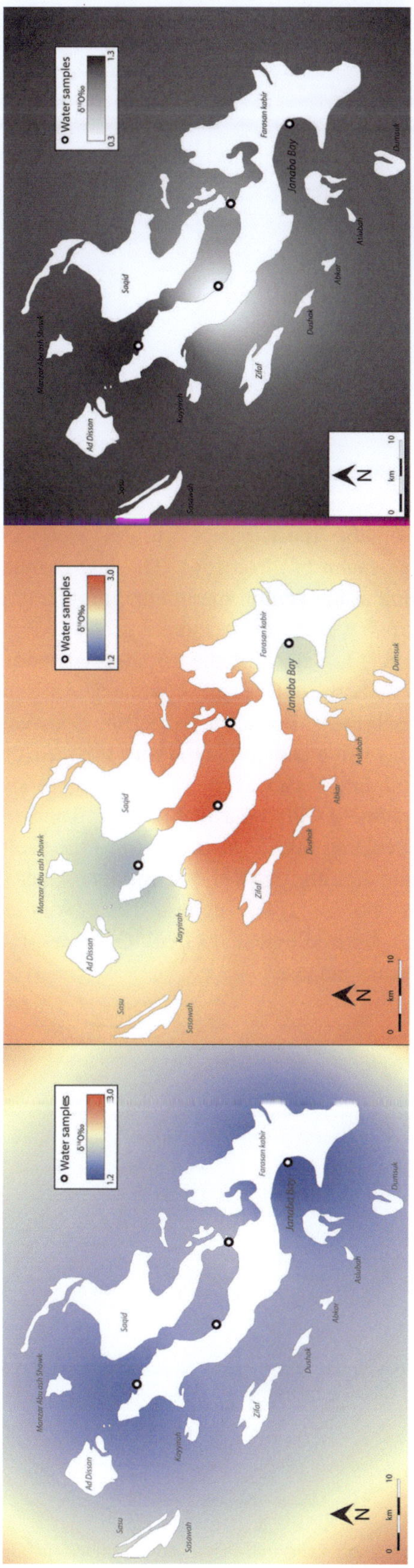

Fig. 5 Spatial distribution of $\delta^{18}O_W$ values in spring (left) and autumn (centre). Difference between seasons (right) is most visible in the sample areas between the two large islands (Farasan Kabir and Saqid). Their sheltered location causes little water exchange with the Red Sea and makes the impact of evaporation fractionation during summer more apparent

were used to calculate a raster through an inverse-distance-to-power-method (IDW). This method estimates values between sample locations in accordance with the values of nearby samples. The colouring of areas is arbitrary and the range is based on the minimum and maximum of the isotopic values that were used for the calculations. This isotope distribution-map is only preliminary and it can be significantly improved by adding more data from other areas in order to achieve a more detailed analysis.

The isotope distributions corresponding to May and December differ significantly. While the May samples are more homogenous, the December samples show distinct differences in isotopic values between sites that are directly exposed to the Red Sea marine water and sites that are more enclosed. The enclosed sites experience less water exchange and hence are more susceptible to fractionation due to evaporation. Evaporation is increased during the summer with higher temperatures and no precipitation. This aspect of the spatial differences in isotopic composition around Farasan is an especially important consideration for comparisons of shell carbonate from archaeological sites that are not from the same areas within the islands. It is likely that the shells growing in more enclosed waters experienced a relative increase of $\delta^{18}O$ during summer, similar to the water samples above.

Overall, the stable isotope data from the marine water samples strengthens the assumption that no major influences from other water sources are present and that the change in temperature is the dominant factor that affects the $\delta^{18}O$ recorded in the shell carbonate. The overall isotopic landscape of Farasan reflects evaporation and surface water exchange with the surrounding Red Sea. However, the modern marine environment of Janaba Bay, the main location of the archaeological studies, is not significantly affected by evaporation, and exhibits sufficient exchange with the open water to avoid anomalies. As a result of these findings, the $\delta^{18}O_{est}$ dataset, based on the local environment of the *C. fasciatus* shell, provided a simple baseline for the examination of the shell carbonate as a proxy for palaeotemperature.

4 Biology of *C. fasciatus*

Comprehensive studies of *C. fasciatus* (Fig. 6) are scarce and much information has been drawn from other species of the *Conomurex* and *Strombus* genera under the assumption that they share a similar shell anatomy and ontogenesis. A common characteristic of *Strombus* shells is the characteristic lip, which grows on the aperture when they reach adulthood. The presence of the lip is used as a way of distinguishing juvenile and adult parts of the shell, the main body (=juvenile) and the lip on the aperture of the shell (=adult).

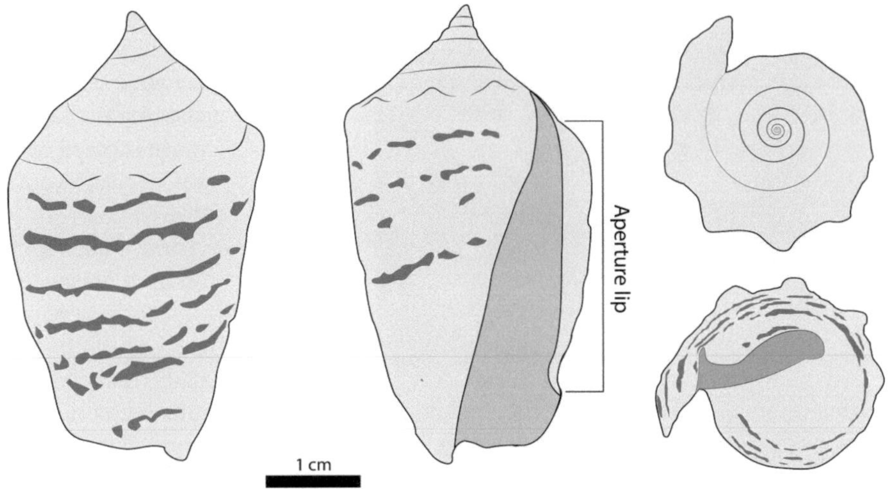

Fig. 6 Adult specimen of *C. fasciatus* shell with location of aperture lip

C. fasciatus is a herbivore that feeds on detritus and algae. It is adapted to the warm climate and the highly salinity of the Red Sea. Adult specimens range in sizes between 25 mm and 50 mm in length. It is generally thought that this species prefers to live on fine sediment in calm waters (Sharabati 1984), although this does not coincide with our experience during collection. Around the Farasan Islands we found that live specimens of *C. fasciatus* were most abundant in areas with little vegetation and sand, and were subjected to heavy wave action.

Modern specimens of *C. fasciatus* were collected during 2012, 2013 and 2014 (Hausmann 2015). From earlier research (Bailey et al. 2013) it was known that the animal can be present in large numbers and groups of 50 have been seen on Soulayn Island, northeast of Farasan Kabir. In comparison, the coastline of Janaba Bay has smaller colonies of *C. fasciatus*. Live specimens were found in an area near the JE0004 shell mound. Many fruitless expeditions were carried out all around the islands, looking at different types of shorelines and micro-environments, despite help from local fishermen and members of the coast guard. Apart from the location near JE0004 only a beach north of Seir village in the northwest of Farasan Kabir had *C. fasciatus* specimens present. Both sites were exposed to heavy wave action and had little to no sand cover over a rough coral bedrock. Underwater vegetation was present but not in large quantities, although this varied with the season. Summer months had very little vegetation, while winter months had more extensive vegetation.

Over the course of several seasons, an experiment was carried out to observe the stable isotope composition of shells from molluscs that died at different times of the year. By solving Eq. 2 for the $\delta^{18}O_S$ value using known values for temperature and $\delta^{18}O_W$, we can estimate the $\delta^{18}O$ values in shells (hereafter $\delta^{18}O_{est}$) and create a baseline (Fig. 4, grey area). We can then compare the $\delta^{18}O_S$-value of the collected modern shell to the $\delta^{18}O_{est}$, and assess the use of the shell carbonate of *C. fasciatus* as a proxy for palaeotemperature. In total, we collected 16 live specimens with at least two specimens per season. Their shells were cleaned and the $\delta^{18}O_S$-value of the most recent growth increment was measured (Hausmann et al. 2015a). The last growth increments are found at the ventral margin and are likely to have grown in the 1–2 weeks before sampling.

Shell-edge $\delta^{18}O_S$ values of the 16 shells ranged from −0.5‰ to −1.7‰ between February (winter) and September (autumn), respectively. For each collection date, we calculated the $\delta^{18}O_{est}$ value and compared it to the measurements of the corresponding season. The estimated values showed a very good agreement with the actual measurements of $\delta^{18}O_S$ (R^2 = 0.88) revealing that *C. fasciatus* form the shells close to or in isotopic equilibrium with the ambient water (Fig. 7) according to Eq. 2. This suggests that we can use well-preserved archaeological shell $\delta^{18}O$ values to reconstruct temperature ranges.

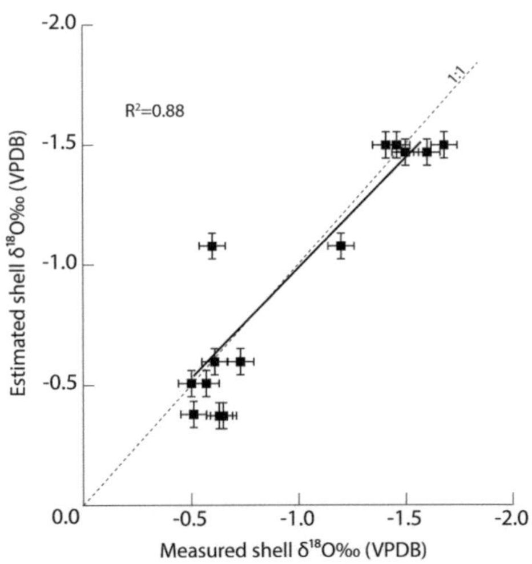

Fig. 7 Correlation of measured shell edge values $\delta^{18}O_S$ versus estimated values $\delta^{18}O_{Est}$, indicating that all major factors controlling the isotopic composition of C. fasciatus are accounted for

5 Applications for *C. fasciatus* as a Climate Proxy

5.1 Introducing Three Case Studies

Building on the archaeological background and the study of $\delta^{18}O$ records in modern specimens of *C. fasciatus*, we briefly describe three recent case studies. The first study compares the isotope composition of shells from two periods of occupation on Farasan to investigate whether they reflect the climatic changes found in other palaeoclimate proxies from Arabia (Hausmann et al. 2015b). This approach combines data from different temporal resolutions and integrates short-term archaeological 'snapshots' with the long-term records from marine cores or stalagmites. The second study employs the sub-annual records of individual shells to determine the time of year the shells were being exploited (Hausmann and Meredith-Williams 2017b). This technique can be applied to the study of the mobility of human populations as well as to reconstruct accumulation rates and seasonal structure of the shell matrix. The third study presents preliminary elemental ratio data in *C. fasciatus* shells and discusses this type of data as a means of assessing environmental change.

1. *Comparison of modern and mid-Holocene climates*

The majority of the Farasan shell middens accumulated during a period from 6,500 to around 4,500 cal BP (Bailey et al. 2013, and this volume). This period is on the border of a climatic change from a humid to a more arid climate between 8,000 and 6,000 cal BP, covering the north of Africa, Arabia and parts of India (Adamson et al. 1980; Arz et al. 2003; Fleitmann et al. 2007; Gasse 2000; Hoelzmann et al. 1998; Pachur and Kröpelin 1987; Rossignol-Strick 1983, 1985; Van Campo et al. 1982). For human populations, this change in climate seemed to cause a shift in population densities and subsistence strategies, which were inherently linked to the annual summer monsoon transporting water from the Indian Ocean to the inland areas, precipitating regularly and intensively and providing life sustaining conditions for central Arabia (Lézine et al. 2010). In detail, the humid conditions in the early Holocene are associated with a northward displacement of the summer monsoon due to a shift in the intertropical convergence zone (ITCZ) to 23°N (Biton et al. 2010). The influence can be seen in the large number of environmental records available from wetlands and lakes from all over Arabia. From this climate optimum to 5,000 cal BP, a climatic deterioration is seen in lacustrine and spring deposits (Lézine et al. 2014). The marine records of aeolian dust transported from arid areas and deposited in the Arabian Sea (KL74; Sirocko et al. 1993) also show an increase of sand deposition around 7,000 cal BP after a 3,000-year period of less sand deposition. Van Rampelbergh et al. (2013) compared different proxies from the southern Arabian Peninsula and found a gradual change toward arid landscapes at around 7,000 cal BP for the Yemeni lowlands and for the Yemeni highlands at 5,000 cal BP. This is contrasted by an abrupt change in salinity from 7,000 to 6,500 cal BP found in $\delta^{18}O$ records of planktonic foraminifera *Globigerinoides ruber* from the northern and the central Red Sea (Arz et al. 2003).

Archaeological shells of *C. fasciatus* on Farasan were used to provide a local marine record for this long-term climate change seen elsewhere in Arabia. Two shell middens sites, JE0087 and JW1727, dated to 6,500 cal BP and 4,800 cal BP, respectively were used. Comparing the $\delta^{18}O$ of shells from both sites provides insight into when climate change occurred around Farasan. The above climatic data also shows a change from humid to a more arid environment over this period and is perhaps the cause of the large-scale shellfish exploitation that happened during the occupation of JW1727.

For the isotope analysis, sequential carbonate samples were drilled from the shell edge to the protoconch (the very first increments of shell growth). The range of $\delta^{18}O_S$ values from shells collected from JE0087, JW1727, and the modern environment were compared. Shells from around 6,500 cal BP showed $\delta^{18}O_S$ ranging from –0.1‰ to –1.9‰, shells from 4,800 cal BP showed $\delta^{18}O_S$ ranging from –0.2‰ to –2.2‰, while the modern shell $\delta^{18}O_S$ ranged from –0.3‰ to –2.8‰. These ranges in $\delta^{18}O$ are fairly similar, although the modern values have a slightly wider range and have slightly lower average values. This difference might imply a lower seasonal contrast, as well as cooler summers in archaeological times compared to today, assuming the isotope composition of the seawater was constant.

Other records suggest a decline in relative humidity connected to the southward displacement of the ITCZ between 7,000 and 6,000 cal BP (Fleitmann et al. 2007). Accordingly, Arz et al. (2003) found an increase of $\delta^{18}O$ values from –0.6‰ to –1.0‰ in the Gulf of Aqaba and from –0.6‰ to –1.4‰ outside the Gulf of Aqaba in the north of the Red Sea, by measuring planktonic foraminifera (*Globigerinoides ruber*) from sediment cores. But on Farasan the average $\delta^{18}O_S$ change is insignificant between 6.5 kyr cal BP (–1.1‰) and 4.8 kyr cal BP (–1.2‰). Equally puzzling, the average $\delta^{18}O_S$ found in Farasan shells decreases by 0.3‰ between the mid-Holocene and now, indicating an increase in humidity or warmer temperatures. It should be noted that our $\delta^{18}O$ data are from the coastal environment and the foraminifera from the open ocean. In recent times, Raitsos et al. (2013) have shown that the Farasan Islands are in an area where temperatures have increased dramatically since 1994 (+0.8 °C when comparing with two time frames of 1985–1993 and 1994–2007). A temperature increase of +0.8 °C would decrease $\delta^{18}O_S$ by ~0.2‰ and would be enough to explain the lower values found in the modern shells.

In summary, the shell carbonate from the two archaeological periods does not indicate large climatic variations, implying that the amount of precipitation and the overall aridity of the environment of both periods was likely to have been very similar. There is no indication of a less rich marine environment and it was likely similarly productive throughout the Holocene. Based on the stable isotope data, it is likely that there has been no dramatic environmental change between the small-scale exploitation that is represented in the first layers of JE0087 and the large-scale exploitation represented by JW1727. The gradual change to an arid environment over several thousand years that was found in some studies (Fleitmann et al. 2007; van Rampelbergh et al. 2013) was not as evident as the abrupt change around 7,000 cal BP that was found in others (Arz et al. 2003; Sirocko et al. 1993). Following this it is more likely that the Tihama lowlands and also Farasan became arid much earlier and more rapidly than the highlands.

2. *Seasonality study of archaeological shells*

The isotope record in *C. fasciatus* can also be used to analyse seasonality. For this we analysed a sequence of shells from the shell mound JW1727. Individual shell records indicate seasonal minima and maxima in the water temperature (assuming no change in the oxygen isotope composition of the water), and the terminal value of the shell (at the aperture) may provide an indication of the season when the mollusc was captured (Thomas 2015b). From this we are able to draw general conclusions about how the ancient Farasan people carried out shellfish exploitation. Shells from five layers in the midden were selected and, where possible, grouped in stratigraphic sequence (top, centre, base). It was only possible to determine the season of death of 66 of the 86 analysed specimens. In some cases, the exact season of death (winter, spring, summer, autumn) was not distinct, in which case we grouped them in an intermediate seasonal group (winter-spring, spring-summer, etc.).

The results from all stratigraphic levels show that the shellfish were collected throughout the year (Fig. 8), although more were collected in the arid summer/autumn months. It is perhaps not surprising that the southern Red Sea with its subtropical environment and general richness in marine wildlife provided year-round coastal resources. In contrast, the summer aridity of this region and the seasonally negative effect on the terrestrial vegetation is perhaps why there was an increase in shellfishing during the warmer seasons.

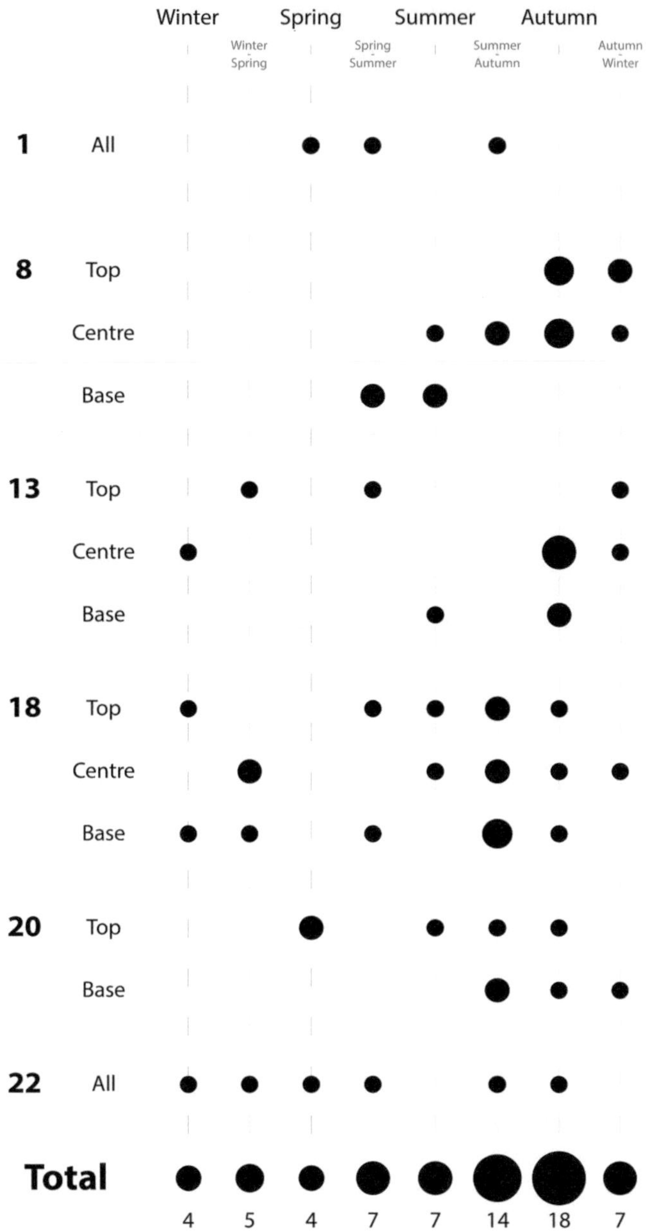

Fig. 8 Seasonality results of analysed layers in JW1727 by layer and location within layer (top, centre, base). Note the gradual change in season with a change in stratigraphic sequence in layer 8

In layer 8, there is a change in the season of collection (spring to autumn) from the base to the top (Hausmann and Meredith-Williams 2017a) suggesting that the layer is likely to have accumulated over a year or so as the season of collection is progressive, rather than the layers being formed from several unconnected seasons in different years. The length of the period that is covered by the sequential seasons, combined with the size of the layer, results in an accumulation rate that suggests a temporal scale of weeks and months.

Furthermore, we can use the fact that the analysed shells in layer 8 lived at the same time and that their records cover the same years. Hence, any variation from one specimen to another is due to the variation of how individual mollusc specimens record temperature and how the temperature varies within their shared habitat. This provides a good estimate of how accurate temperature estimates are that are based on this species in this area. Without the shells' stratigraphic and seasonal relation, any differences in their records could be due to variations between the different years in which the molluscs grew.

Lastly, by stratigraphically grouping shells of the same season of death and connecting other archaeological remains to them, it is possible to give more meaningful interpretations to the archaeological deposit and artefacts within it. One of the conditions for providing a temporal scaffolding like this is a large number of specimens (dozens to hundreds) to be analysed from a small area within the deposit. This directly translates into research costs and is difficult to carry out using the methods applied above.

3. *Preliminary data on Mg/Ca and Sr/Ca ratios as palaeothermometer proxies for* ***C. fasciatus***

Here we provide preliminary data on elemental ratios in *C. fasciatus* shells and discuss their potential as a climate proxy. Generally, the incorporation of Mg^{2+}- or Sr^{2+}-ions in relation to Ca^{2+}-ions into calcium carbonate has been proposed as a proxy for palaeotemperature (Elderfield and Ganssen 2000; Watanabe et al. 2001). However, specifically in mollusc shells elemental ratios have been analysed with mixed success and were largely shown to not be as reliable as stable isotope values (i.e., $\delta^{18}O$) for temperature reconstruction (Cusack and Freer 2008). Most mechanisms of shell precipitation are not universally applicable to all shell species and seldom apply without variation between species or specimens of the same species (e.g., Graniero et al. 2017). Nevertheless, elemental ratios have the advantage that they can be assessed very quickly (Cobo et al. 2017; Durham et al. 2017; Giakoumaki et al. 2007) and can easily be analysed in the form of spatially detailed maps (Cáceres et al. 2017; Hausmann et al. 2017). Additionally, fast analyses have the benefit of lower costs for individual samples, which can result in larger archaeological datasets of seasonal coastal exploitation. The Farasan Island shell middens with over 3,000 sites and thousands of cubic metres of shell are a good example of a large shell assemblage and the immense amount of time and money that can be required for its study. As shell seasonality analysis becomes more established in archaeology, the weaknesses of studies that are too small are becoming more apparent (Thomas 2015b). For instance, Jew et al. (2013, 2014) showed that 35% of seasonal signals were falsely interpreted due to small sample numbers of analyses performed on individual shells. Also, Hausmann and Meredith-Williams (2017a) showed the loss of information when analysing small numbers of shells per stratigraphic unit.

To work toward larger shell seasonality datasets, we use the ability of rapid elemental mapping through Laser Induced Breakdown Spectroscopy (LIBS) to assess Mg/Ca and Sr/Ca ratios in *C. fasciatus* as potential palaeoclimate proxies for the Farasan Islands. To achieve this, we selected archaeological specimens that had previously been analysed using stable oxygen isotopes, to provide a comparative framework. LIBS is a simple, straightforward and versatile technique for rapid analysis of the chemical composition of a huge variety of materials. It is based on the atomic emission spectroscopy of plasma, generated by focusing a high intensity (in the order of GW/cm^2) laser beam onto the material. The plasma emission is collected and is spectrally resolved, thus providing qualitative and quantitative results by monitoring the emission line intensities of the chemical constituents of the material. LIBS has been shown to have significant potential in environmental studies (Harmon et al. 2005; Marín-Roldán et al. 2014; Qiao et al. 2015). The method presents many advantages over conventional analytical techniques, such as the ability to perform in situ measurements without any sample preparation, nearly instantaneous measurement time and easy-to-use equipment of low-maintenance cost. Hence, it is possible to carry out a multitude of measurements with a high spatial resolution and within a small period of time (one measurement takes a few seconds). For all these reasons, LIBS appears to be the most promising technique for the rapid analysis of the vast number of shell middens.

Sample preparation for LIBS is minimal and because the analysed shell specimens were used in earlier studies (Hausmann and Meredith-Williams 2017a, b), cleaning and sectioning of the shells was already finished. However, LIBS analysis benefits from flat sample surfaces, which is why it was necessary to regrind the samples using metallographic grit paper (P800) and to remove any holes left by previous sampling steps. The experiments were carried out using the LIBS setup described in Hausmann et al. (2017), which can be used to provide maps of elemental ratios with a resolution of ~ 90 µm. However, sample locations were measured using single pulses instead of 10 to achieve a faster acquisition speed. Single pulse analysis has been shown to provide sufficiently accurate information on Mg/Ca ratios in elemental maps of carbonates (Cáceres et al. 2017). To assess the ratios of Mg and Sr towards Ca, we measured the peak intensity ratios of the respective emission lines with the wavelengths of 279.55 nm (Mg II), 407.77 nm (Sr II), and 317.93 nm (Ca II). Intensity values of Mg and Sr were then

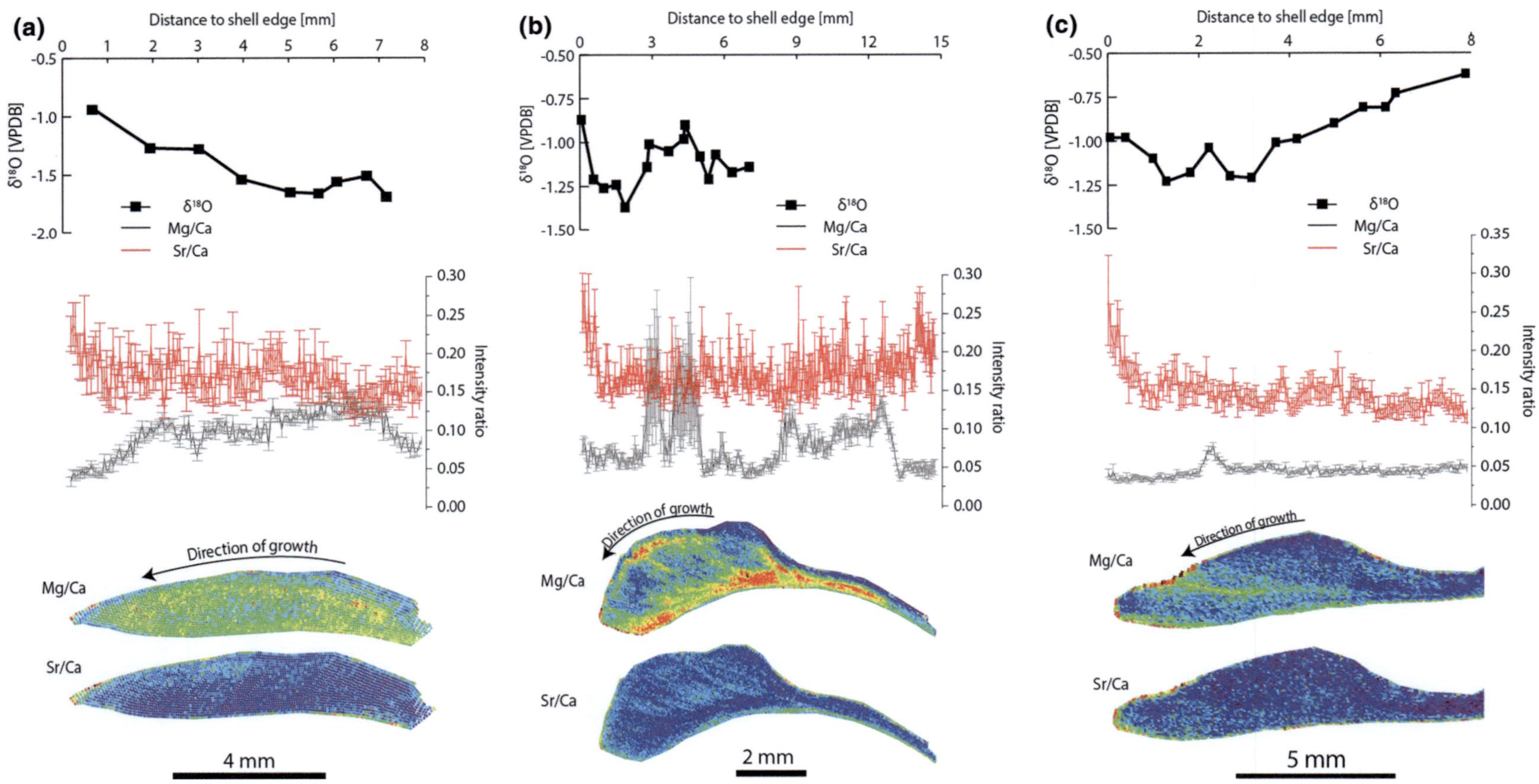

Fig. 9 Comparison of sequential $\delta^{18}O$ values from archaeological shells with line scans and maps of Mg/Ca and Sr/Ca intensity ratios

divided by the intensity value of Ca, to provide an intensity ratio that is proportional to the absolute concentration of Mg/Ca and Sr/Ca.

Four shells from the shell midden JW1727 were analysed (Sample ID: JW1727-8-T-2, JW1727-8-M-12, JW1727-8-M-13, for clarity now samples A, B, C, respectively).

All specimens previously provided a $\delta^{18}O$ record of seasonal temperature change (Fig. 9). The $\delta^{18}O$ values produced records of temperature change of varying length, covering different seasons of the year. Due to the short time period covered by the individual shell records, a more detailed annual overview of elemental composition was not possible. Shell A covers a period of summer until late autumn or early winter based on the $\delta^{18}O$ values. A Mg/Ca line scan following the direction of growth shows an increase of Mg/Ca, a plateau phase and a subsequent drop of Mg/Ca toward the shell edge. This is not reflected in all parts of the analysed shell, as is shown by the mapped values. The variation within coeval parts of the shell increments is as high as the variation covered by the line scan, suggesting that Mg/Ca intensity ratios are not controlled by a time-resolved factor (i.e., environmental change). The same results apply to Sr/Ca line scans and elemental maps. More specifically, exterior parts of the shell aperture seem to be more enriched in Sr than coeval interior parts of the same growth increment. Additionally, the error of individual sample locations is almost as large as the overall variation of Sr/Ca intensity ratios in the whole, making it difficult to assess any seasonal change in the record.

Shell B covers a longer time span of over one year, based on the $\delta^{18}O$ values. While the Mg/Ca line scan shows several Mg enriched areas that far exceed the range of intensity ratios measured in shell A, they were not in agreement with $\delta^{18}O$ values. Moreover, the elemental map of shell B shows the spatial variability of Mg/Ca intensity ratios and the inconsistency of enriched areas in regards to following the structure of growth increments. This is most prominent in the thin part of the whorl grown prior to the shell lip, where high and low Mg/Ca intensity ratios grow in parallel. Sr/Ca intensity ratios show a larger variation than in shell A, however, they are not corresponding with $\delta^{18}O$ values. The mapped Sr/Ca data shows a general enrichment in the central part of the shell aperture, however, this enrichment does not persist to the interior or exterior parts of the shell.

Shell C produced similar results to shells A and B, with little correlation of stable isotope values and elemental intensity ratios. The Mg/Ca line scan shows almost no variation, with a slight decrease toward the shell edge and a single enriched area at about 2 mm distance from the shell edge. This increase is tentatively mirrored by the $\delta^{18}O$ values. The map of Mg/Ca intensity ratios shows that the enriched area intensifies towards the shell edge but largely follows the constraints of the growth increment. Considering the lack of responses of Mg/Ca intensity ratios in *C. fasciatus* to environmental change, it is likely that this time-resolved increase reflects a stop in shell growth. This results in organically rich growth increments, that have an increased concentration of Mg (Schöne et al. 2010). However, this enrichment is not reflected in Sr/Ca intensity ratios, which do not show any change despite a first gentle and then sharp increase toward the shell edge.

The elemental analyses presented here are preliminary, but largely indicate that Mg/Ca and Sr/Ca ratios in *C. fasciatus* do not reflect the environmental changes experienced by the animal and cannot be used as palaeothermometer. Responses of elemental ratios to environmental changes are various but ideally show regularly repeating patterns (Warter and Müller 2017). These patterns are entirely missing from our samples. Moreover, in all shells the spatial patterns of elemental ratios largely did not follow the structure of growth increments and thus were not consistently connected to the increments' times of growth or connected to environmental changes or events happening at specific times or over certain periods.

Equally, most growth increments produced values of high and of low elemental ratios, reflecting larger spatial patterns that were not restricted by the extents of the growth increments of which they are part. These spatial patterns were not as predictable as has been shown in other shell species, that had 'preferred' sampling zones, which predictably provided more consistent results (Lazareth et al. 2013).

Stable oxygen isotope values indicate that the analysed areas are records of one year or less, which would explain the lack of patterns that occur annually. However, the missing correlation between the stable oxygen isotope values and elemental ratios shows that even sub-annual temperature changes are not reflected in the elemental ratios and that determining the season of harvest is not possible.

Elemental ratios and the degree to which they reflect changes in ambient temperature can change throughout the lifetime of the mollusc (Ferguson et al. 2011; Freitas et al. 2005). In this study, we solely sampled adult sections of *C. fasciatus* shells, as they contained the sampling locations for the stable isotopes analysis, which aimed to determine the mollusc's season of death. Thus, we cannot claim that elemental ratios consistently do not reflect changes in temperature. However, in the areas of the shell that contain information important to seasonality studies (i.e., the aperture lip), it is evident that elemental ratios cannot be used as a temperature proxy. LIBS analysis potentially covers all elements and future analyses should aim to also measure other elemental concentrations (e.g., Li, Ba, Mn) and further expand on our preliminary data.

6 Conclusion

This chapter shows that archaeological shell midden sites, in particular sites in the southern Red Sea coast, have considerable potential for interdisciplinary research, involving archaeology, biology, and palaeoclimatic reconstruction. Specifically, they would have provided a focus point for African-Arabian coastal populations. The stable isotope composition of *C. fasciatus* shells found within the shell midden sites provides a high-resolution record of the local marine environment from modern times until the mid-Holocene. Elemental shell records provided by LIBS showed no obvious environmental control on Mg/Ca and Sr/Ca ratios. Future analyses should aim to employ more comprehensive modern data as well as a larger variety of elemental ratios. Archaeologically, the Farasan shell middens are a prime example of intensive shellfish exploitation, where there are shell mounds of up to several metres in height that accumulated within only a few centuries. The shells are evidence of the richness of marine landscapes in general and the southern Red Sea.

At present the growing industry and population of the Farasan Islands pose a threat to the preservation of these shell mounds. To further exploit these rich archives of prehistoric life in the southern Red Sea region it is becoming increasingly important to protect these archaeological sites.

Acknowledgements We greatly thank Najeeb Rasul for organising the meeting in Jeddah and for his help with the publication of this chapter. Research on the Farasan Islands has been made possible through the support and collaboration of the Saudi Commission for Tourism and National Heritage (SCTH), in particular its President, HRH Prince Sultan bin Salman bin AbdulAziz al Saud, Professor Ali Al-Ghassan, and Jamal Omar. Underwater work on modern shells has been supported by permits from the Department of General Survey of the Ministry of Defense. The work has been supported by grants from the European Research Council (Advanced Grant 269586 DISPERSE under the 'Ideas' Specific Programme of FP7), and EU Marie-Curie Action, "Intra-European Fellowships for Career Development" (Archaeological and climatic data from elemental ratios using rapid analysis of shell carbonate—ACCELERATE) "Grant Agreement No. 703625". The research has greatly benefited from comments and discussions with Geoff Bailey, Robyn Inglis, Matthew Meredith-Williams, Harry Robson, Andre Colonese, Demetrios Anglos, Panayiotis Siozos and Costas Fotakis. This is DISPERSE contribution no. 38.

References

Adamson DA, Gasse F, Street FA, Williams MAJ (1980) Late quaternary history of the Nile. Nature 288:50–55

Alsharekh A, Bailey G (2014) Coastal prehistory in southwest Arabia and the Farasan Islands: 2004–2009 field investigations. Saudi Commission for Tourism and Antiquities, Riyadh

Andrus CFT (2011) Shell midden sclerochronology. Quatern Sci Rev 30(21):2892–2905

Arz HW, Lamy F, Pätzold J, Muller PJ, Prins M (2003) Mediterranean moisture source for an early-Holocene humid period in the northern Red Sea. Science 300:118–121

Bailey GN (2004) World prehistory from the margins: the role of coastlines in human evolution. J Interdiscip Stud Hist Archaeol 1:39–50

Bailey GN and Craighead AS (2003) Late Pleistocene and Holocene coastal palaeoeconomies: a reconsideration of the molluscan evidence from northern Spain. Geoarchaeology 18:175–204

Bailey GN and King GCP (2011) Dynamic landscapes and human dispersal patterns: tectonics, coastlines, and the reconstruction of human habitats. Quatern Sci Rev 30:1533–1553

Bailey GN, Flemming NC, King GCP, Lambeck K, Momber G, Moran LJ, Al-Sharekh A, Vita-Finzi C (2007) Coastlines, submerged landscapes, and human evolution: the Red Sea Basin and the Farasan Islands. J Island Coastal Archaeology 2:127–160

Bailey GN, Meredith-Williams M, Alsharekh AM (2013) Shell mounds of the Farasan Islands, Saudi Arabia. In: Bailey GN, Hardy K, Camara A (eds) Shell energy, pp 241–254

Bailey GN, Alsharekh A, Flemming NC, Momber G, Moran LJ, Sinclair A, King GCP, Vita-Finzi C, Al-Ma'Mary A, AlShaikh NY, AlGhamdi S (2014a) Coastal archaeology and prehistory in the southwest region of Saudi Arabia and the Farasan Islands: report on the 2004 and 2006 surveys of the joint Saudi-UK southern Red Sea Project. In: Alsharekh A, Bailey GN (eds) Coastal Prehistory in Southwest Arabia and the Farasan Islands—2004–2009 Field investigations, pp 16–74

Bailey GN, Alsharekh A, Momber G, Moran LJ, Williams MGW, Satchell JS, Gillespie J, Reeler C, AlShaikh N, Robson H (2014b) Coastal archaeology in the Farasan Islands: report on the 2009 fieldwork of the joint Saudi-UK Southern Red Sea project. In: Alsharekh A, Bailey GN (eds) Coastal prehistory in Southwest Arabia and the Farasan Islands—2004–2009 field investigations, pp 160–215

Bailey GN, Meredith-Williams M, Alsharekh A, Hausmann N (this volume) The archaeology of pleistocene coastal environments and human dispersals in the Red Sea: insights from the Farasan Islands

Bicho N, Haws J (2008) At the land's end: marine resources and the importance of fluctuations in the coastline in the prehistoric hunter–gatherer economy of Portugal. Quatern Sci Rev 27:2166–2175

Biton E, Gildor H, Trommer G, Siccha M, Kucera M, van der Meer MTJ, Schouten S (2010) Sensitivity of Red Sea circulation to monsoonal variability during the holocene: an integrated data and modeling study. Paleoceanography 25:PA4209

Cáceres JO, Pelascini F, Motto-Ros V, Moncayo S, Trichard F, Panczer G, Marín-Roldán A, Cruz JA, Coronado I, Martín-Chivelet J (2017) Megapixel multi-elemental imaging by laser-induced breakdown spectroscopy, a technology with considerable potential for paleoclimate studies. Scientific Reports 7(1):5080

Cobo A, García-Escárzaga A, Gutiérrez-Zugasti I, Setién J, González-Morales MR, López-Higuera JM (2017) Automated measurement of magnesium/calcium ratios in gastropod shells using laser-induced breakdown spectroscopy for paleoclimatic applications. Appl Spectrosc 71(4):591–599

Cortés-Sánchez M, Morales-Muñiz A, Simón-Vallejo MD, Lozano-Francisco MC, Vera-Peláez JL, Finlayson C, Rodríguez--Vidal J, Delgado-Huertas A, Jiménez-Espejo FJ, Martínez-Ruiz F, Martínez-Aguirre MA (2011) Earliest known use of marine resources by Neanderthals. PLoS ONE 6(9):e24026

Crassard R, McCorriston J, Oches E, ʿabd Al-Aziz Bin A, Espagne J, Sinnah M (2006) Manayzah, early to mid-Holocene occupations in Wādī Ṣanā (Ḥaḍramawt, Yemen). In: Proceedings of seminar for Arabian studies 36:151–173

Crenshaw MA (1980) Mechanisms of shell formation and dissolution. In: Rhoads DC, Lutz RA (eds) Skeletal growth of aquatic organisms. Plenum Publishing Corp, New York, pp 115–132

Cusack M, Freer A (2008) Biomineralization: elemental and organic influence in carbonate systems. Chemical Rev 108(11):4433–4454

Dettman DL, Reische AK, Lohmann KC (1999) Controls on the stable isotope composition of seasonal growth bands in aragonitic fresh-water bivalves (*Unionidae*). Geochim Cosmochim Acta 63:1049–1057

Durham SR, Gillikin DP, Goodwin DH, Dietl GP (2017) Rapid determination of oyster lifespans and growth rates using LA–ICP–MS line scans of shell Mg/Ca ratios. Palaeogeog Palaeoclimat Palaeoecol. https://doi.org/10.1016/j.palaeo.2017.06.013

Edens C, Wilkinson TJ (1998) Southwest Arabia during the holocene: recent archaeological developments. J World Prehistory 12:55–119

Elderfield H, Ganssen G (2000) Past temperature and $\delta^{18}O$ of surface ocean waters inferred from foraminiferal Mg/Ca ratios. Nature 405 (6785):442

Erlandson JM (2001) The archaeology of aquatic adaptations: paradigms for a new millennium. J Archaeol Res 9:287–350

Erlandson JM, Moss ML, Des Lauriers M (2008) Life on the edge: early maritime cultures of the pacific coast of North America. Quatern Sci Rev 27:2232–2245

Ferguson JE, Henderson GM, Fa DA, Finlayson JC, Charnley NR (2011) Increased seasonality in the western mediterranean during the last glacial from limpet shell geochemistry. Earth Planet Sci Lett 308:325–333

Finlayson C (2008) On the importance of coastal areas in the survival of Neanderthal populations during the Late Pleistocene. Quatern Sci Rev 27:2246–2252

Fleitmann D, Burns SJ, Mangini A, Mudelsee M, Kramers J, Villa I, Neff U, Al-Subbary AA, Buettner A, Hippler D, Matter A (2007) Holocene ITCZ and Indian monsoon dynamics recorded in stalagmites from Oman and Yemen (Socotra). Quatern Sci Rev 26:170–188

Flemming NC, Bailey GN, Courtillot V, King G, Lambeck K, Ryerson F, Vita-Finzi C (2003) Coastal and marine palaeo-environments and human dispersal points across the Africa-Eurasia boundary. In: Brebbia CA, Gambin T (eds) The maritime and underwater heritage. Wessex Institute of Technology, Southampton, pp 61–74

Freitas P, Clarke LJ, Kennedy H, Richardson C, Abrantes F (2005) Mg/Ca, Sr/Ca, and stable-isotope ($\delta^{18}O$ and $\delta^{13}C$) ratio profiles from the fan mussel *Pinna nobilis:* Seasonal records and temperature relationships. Geochem Geophys Geosyst 6(4). https://doi.org/10.1029/2004gc000872

Gasse F (2000) Hydrological changes in the African tropics since the last glacial maximum. Quatern Sci Rev 19:189–211

Giakoumaki A, Melessanaki K, Anglos D (2007) Laser-induced breakdown spectroscopy (LIBS) in archaeological science—applications and prospects. Anal Bioanaly Chem 387(3):749–760

Graniero LE, Surge D, Gillikin DP, Briz i Godino I, Álvarez M (2017) Assessing elemental ratios as a paleotemperature proxy in the calcite shells of patelloid limpets. Palaeogeog Palaeoclimat Palaeoecol 465:376–385

Grossman EL, Ku TL (1986) Oxygen and carbon isotope fractionation in biogenic aragonite: temperature effects. Chem Geol 59:59–74

Harmon RS, De Lucia FC, Miziolek AW, McNesby KL, Walters RA, French PD (2005) Laser-induced breakdown spectroscopy (LIBS) —an emerging field-portable sensor technology for real-time, in-situ geochemical and environmental analysis. Geochem Explor Environ Analy 5:21–28

Harrower MJ (2008) Hydrology, ideology, and the origins of irrigation in ancient southwest Arabia. Curr Anthropol 49:497–510

Hausmann N (2015) The shell mounds of the Farasan Islands—an isotopic study of seasonality and coastal exploitation. Doctoral Dissertation, University of York

Hausmann, N. and Meredith-Williams M.(2017a) Exploring accumulation rates of shell deposits through seasonality data. Journal of Archaeological Method and Theory Vol. 24(3): 776–795.

Hausmann, N. and Meredith-Williams M. (2017b). Seasonal patterns of coastal exploitation on the Farasan Islands, Saudi Arabia. Journal of Island and Coastal Archaeology, Vol. 12(3):360–379.

Hausmann N, Colonese AC, de Lima Ponzoni A, Hancock Y, Meredith-Williams M, Leng MJ, Bailey GN (2015a) Isotopic composition of *Conomurex fasciatus* shells as an environmental proxy for the Red Sea. Quatern Int. https://doi.org/10.1016/j.quaint.2015.08.051

Hausmann N, Meredith-Williams M, Bailey G (2015b) Results of recent excavations on the Farasan Islands and studies of large-scale prehistoric shellfish gathering in the Red Sea. In: Bicho N, Detry C, Price TD, Cunha E (eds) Muge 150th: the 150th anniversary of the discovery of mesolithic shell middens. Cambridge Scholars Publishing, pp 301–314

Hausmann N, Siozos P, Lemonis A, Colonese AC, Robson HK, Anglos D (2017) Elemental mapping of Mg/Ca intensity ratios in marine mollusc shells using laser induced breakdown spectroscopy. J Analyt Atom Spectrom 32:1467–1472

Hoelzmann P, Jolly D, Harrison SP, Laarif F, Bonnefille R, Pachur HJ (1998) Mid-holocene land-surface conditions in northern Africa and the Arabian Peninsula: a data set for the analysis of biogeophysical feedbacks in the climate system. Global Biogeochem Cycles 12 (1):35–51

Jerardino A (2013) Mega shell middens and hunter-gatherer resource intensification along the west coast of South Africa. In: Bailey G, Hardy K, Camara A (eds) Shell energy, 207–228

Jew NP, Erlandson JM, Watts J, White FJ (2013) Shellfish, seasonality, and stable isotope sampling: $\delta^{18}O$ analysis of mussel shells from an 8,800-year-old shell midden on California's Channel Islands. J Island Coast Archaeol 8(2):170–189

Jew NP, Erlandson JM, Rick TC, Reeder-Myers L (2014) Oxygen isotope analysis of California mussel shells: seasonality and human sedentism at an 8,200-year-old shell midden on Santa Rosa Island California. Archaeological Anthropol Sci 6(3):293–303

Jones DS, Arnold B, Quitmyer I, Schöne BR, Surge D (2007) First international sclerochronology conference. http://conference.ifas.ufl.edu/sclerochronology. Accessed 21 August 2017

Kennish M (1980) Shell microgrowth analysis, mercenaria mercenaria as a type example for research in population dynamics. In: Rhoads D, Lutz R (eds) Skeletal growth of aquatic organisms. Plenum Publishing Corporation, New York, pp 255–294

Khalidi L (2005) The prehistoric and early historic settlement patterns on the Tihāmah coastal plain (Yemen): preliminary findings of the Tihāmah Coastal Survey 2003. Proc Semin Arab Stud 35:115–127

Leng MJ, Lewis JP (2016) Oxygen isotopes in molluscan shell: applications in environmental archaeology. Environ Archaeol 21 (3):295–306

Lazareth CE, Le Cornec F, Candaudap F, Freydier R (2013) Trace element heterogeneity along isochronous growth layers in bivalve shell: consequences for environmental reconstruction. Palaeogeog Palaeoclimat Palaeoecol 373:39–49

Lézine AM, Robert C, Cleuziou S, Inizan ML, Braemer F, Saliège JF, Sylvestre F, Tiercelin JJ, Crassard R, Méry S, Charpentier V (2010) Climate change and human occupation in the southern Arabian lowlands during the last deglaciation and the Holocene. Global Planet Change 72(4):412–428

Lézine AM, Bassinot F, Peterschmitt JY (2014) Orbitally-induced changes of the Atlantic and Indian monsoons over the past

20,000 years: New insights based on the comparison of continental and marine records. Bull Soc Geol Fr 185:3–12
Lutz RA, Rhoads DC (1977) Anaerobiosis and a theory of growth line formation. Science 198:1222–1227
Marean CW (2014) The origins and significance of coastal resource use in Africa and Western Eurasia. J Hum Evol 77:17–40
Marín-Roldán A, Cruz JA, Martín-Chivelet J, Turrero MJ, Ortega AI, Cáceres J (2014) Evaluation of laser induced breakdown spectroscopy (LIBS) for detection of trace element variation through stalagmites: potential for paleoclimate series reconstruction. J Appl Laser Spectrosc 1:7–12
Martin L, McCorriston J, Crassard R (2009) Early Arabian pastoralism at Manayzah in Wādī Ṣanā, Ḥaḍramawt. Proc Semin Arab Stud 39:271–282
McCorriston J (2013) Pastoralism and pilgrimage. Curr Anthropol 54:607–641
McCorriston J, Harrower M, Martin L, Oches E (2012) Cattle cults of the Arabian Neolithic and early territorial societies. Am Anthropol 114:45–63
McCorriston J, Harrower M, Oches E, ʿAqil AB, Heyne C, Noman A, Crassard R, BaDhofary K, Anderson J (2005) Foraging economies and population in the Middle Holocene highlands of southern Yemen. Proc Semin Arab Stud 35:143–154
McCorriston J, Oches EA, Walter DE, Kl Cole (2002) Holocene paleoecology and prehistory in highland southern Arabia. Paleo 28:61–88
Meredith-Williams M, Hausmann N, Bailey GN, King GCP, Alsharekh A, Ghamdi S, Inglis RH (2014a) Mapping, modelling and predicting prehistoric coastal archaeology in the southern Red Sea using new applications of digital-imaging techniques. World Archaeol 46:10–24
Meredith-Williams M, Hausmann N, Inglis R, Bailey G (2014b) 4200 new shell mound sites in the southern Red Sea. In: Fernandes R, Meadows J (eds) Human exploitation of aquatic landscapes. Special Issue, Internet Archaeology. https://doi.org/10.11141/ia.37.2
Milano S, Prendergast AL, Schöne BR (2016) Effects of cooking on mollusk shell structure and chemistry: Implications for archeology and paleoenvironmental reconstruction. J Archaeological Sci Reports 7:14–26
Momber G (2000) Drowned and deserted: a submerged prehistoric landscape in the Solent, England. Internat J Naut Archaeol 29:86–99
Momber G (2004) The inundated landscapes of the western Solent. In: Flemming NC (ed) Submarine prehistoric archaeology of the North Sea: research priorities and collaboration with industry. CBA Research Report 141, London, pp 37–42
Momber G, Sakellariou D, Bailey GN, Rousakis G (this volume) The multi-disciplinary search for underwater archaeology in the southern Red Sea
Pachur HJ, Kröpelin S (1987) Wadi Howar: paleoclimatic evidence from an extinct river system in the southeastern Sahara. Science 237:298–300
Qiao S, Ding Y, Tian D, Yao L, Yang G (2015) A review of laser-induced breakdown spectroscopy for analysis of geological materials. Appl Spectrosc Rev 50:1–26
Raitsos DE, Pradhan Y, Brewin RJW, Stenchikov G, Hoteit I (2013) Remote sensing the phytoplankton seasonal succession of the Red Sea. PLoS ONE 8:e64909
Ridgway ID, Richardson CA, Austad SN (2011) Maximum shell size, growth rate, and maturation age correlate with longevity in bivalve molluscs. J Gerontol Series A Biomed Sci Med Sci 66(2):183–190
Rossignol-Strick M (1985) Mediterranean quaternary sapropels, an immediate response of the African monsoon to variation of insolation. Palaeogeogr Palaeoclimatol Palaeoecol 49:237–263
Rossignol-Strick M (1983) African monsoons, an immediate climate response to orbital insolation. Nature 304:46–49
Sand KK, Rodriguez-Blanco JD, Makovicky E, Benning LG, Stipp SLS (2012) Crystallization of $CaCO_3$ in water–alcohol mixtures: spherulitic growth, polymorph stabilization, and morphology change. Cryst Growth Des 12:842–853
Schöne BR (2008) The curse of physiology—challenges and opportunities in the interpretation of geochemical data from mollusk shells. Geo-Marine Lett 28:269–285
Schöne BR, Gillikin DP (2013) Unraveling environmental histories from skeletal diaries—advances in sclerochronology. Palaeogeog Palaeoclimat Palaeoecol 373:1–5
Schöne BR, Zhang Z, Jacob D, Gillikin DP, Tütken T, Garbe-Schönberg D, Soldati A (2010) Effect of organic matrices on the determination of the trace element chemistry (Mg, Sr, Mg/Ca, Sr/Ca) of aragonitic bivalve shells (Arctica islandica)—comparison of ICP-OES and LA-ICPMS data. Geochem J 44:23–37
Sharabati D (1984) Red Sea shells. Routledge and Kegan Paul Plc, London
Sirocko F, Sarnthein M, Erlenkeuser H, Lange H, Arnold M, Duplessy JC (1993) Century-scale events in monsoonal climate over the past 24,000 years. Nature 364:322–324
Steele TE, Klein RG (2008) Intertidal shellfish use during the middle and later stone age of south Africa. Archaeofauna 17:63–76
Stein JK, Deo JN, Phillips LS (2003) Big sites—short time: accumulation rates in archaeological sites. J Archaeol Sci 30(3):297–316
Thomas KD (2015a) Molluscs emergent, Part I: themes and trends in the scientific investigation of mollusc shells as resources for archaeological research. J Archaeol Sci 56:133–140
Thomas KD (2015b) Molluscs emergent, Part II: themes and trends in the scientific investigation of molluscs and their shells as past human resources. J Archaeol Sci 56:159–167
Van Campo E, Duplessy JC, Rossignol-Strick M (1982) Climatic conditions deduced from a 150-kyr oxygen isotope–pollen record from the Arabian Sea. Nature 296:56–59
Van Rampelbergh M, Fleitmann D, Verheyden S, Cheng H, Edwards L, De Geest P, De Vleeschouwer D, Burns SJ, Matter A, Claeys P, Keppens E (2013) Mid- to Late Holocene Indian Ocean Monsoon variability recorded in four speleothems from Socotra Island, Yemen. Quat Sci Rev 65:129–142
Wanamaker AD, Kreutz KJ, Borns HW, Introne DS, Feindel S, Funder S, Rawson PD, Barber BJ (2007) Experimental determination of salinity, temperature, growth, and metabolic effects on shell isotope chemistry of Mytilus edulis collected from Maine and Greenland. Paleoceanography 22(2):PA2217
Warter V, Müller W (2017) Daily growth and tidal rhythms in Miocene and modern giant clams revealed via ultra-high resolution LA-ICPMS analysis—a novel methodological approach towards improved sclerochemistry. Palaeogeog Palaeoclimat Palaeoecol 465:362–375
Watanabe T, Winter A, Oba T (2001) Seasonal changes in sea surface temperature and salinity during the little ice age in the Caribbean Sea deduced from Mg/Ca and $^{18}O/^{16}O$ ratios in corals. Marine Geol 173(1):21–35
Woodman P (2013) Shell energy: an overview. In: Bailey G, Hardy K, Camara A (eds) Shell energy, pp 299–308
Zazzo A, Munoz O, Saliège JF (2014) Diet and mobility in a late neolithic population of coastal Oman inferred from radiocarbon dating and stable isotope analysis. Am J Phys Anthropol 153:353–364

Ancient Ports of Trade on the Red Sea Coasts —The 'Parameters of Attractiveness' of Site Locations and Human Adaptations to Fluctuating Land- and Sea-Scapes. Case Study *Berenike Troglodytica*, Southeastern Egypt

Anna M. Kotarba-Morley

Abstract

The Red Sea region is unfavourable for long-shore nautical activity as it lacks natural topographic features that could be used as harbours; there are only a few suitable bays for landing along its coasts, where wadi mouths allow for a break in the reef. However, experiencing seasonally variable winds and currents, parts of the Red Sea constituted favourable marine environments for sea voyaging, contact and trade for millennia. This paper focuses on the influence that the local environmental and climatic context (including land- and sea-scape), had on the location, development, and ultimate success or decline of key Classical (Greco-Roman) ports of trade on the Red Sea coast, most pertinently those involved in exchange on the Spice, Incense and Maritime Silk Routes. The importance of changes in geomorphological, climatic, landscape and sea level configurations that led to the alternation of these human-adapted landscapes will be discussed within the new theoretical framework of 'Parameters of Attractiveness' developed whilst focusing on a case study, the Greco-Roman port town of *Berenike Troglodytica* on the southern Red Sea coast of Egypt. These parameters—grouped into 4 main categories: Sea, Land, Resources, and Socio-Economic and Political—were designed in order to statistically quantify the attractiveness of particular sites along the rims of the Red Sea for use as trade ports.

A. M. Kotarba-Morley (✉)
Department of Ancient History, Faculty of Arts, Macquarie University, Sydney, NSW, Australia
e-mail: ania.kotarba@mq.edu.au

A. M. Kotarba-Morley
School of Archaeology, Oxford Centre for Maritime Archaeology, University of Qxford, Oxford, UK

A. M. Kotarba-Morley
Centre for Archaeological Science, School of Earth, Athmospheric and Life Sciences, University of Wollongong, Wollongong, Australia

N. M. A. Rasul and I. C. F. Stewart (eds.), *Geological Setting, Palaeoenvironment and Archaeology of the Red Sea*,
https://doi.org/10.1007/978-3-319-99408-6_34

1 Introduction

Ports and harbours can be located in extremely diverse political, landscape and geomorphological settings, forming hubs that not only connect different environments and cultures, but also act as centres of social and cultural diversity and the hybridisation of ideas. Owing to their favoured positions in the coastal, riverine or lacustrine landscape, and the functional and cultural sub-divisions within them, ports constitute excellent study areas for understanding past landscapes 'of the shores', geomorphological dynamics, and —in the case of sea ports—sea level changes, as well as ancient marine technologies, cultural interactions and large-scale geographic networks. It is therefore important to understand what drives the choice of location of these sites since cursory examination appears to reveal an unpredictable, non-linear pattern for the distribution of these important infrastructural nodes. It is also crucial to develop a methodology through which we can attain a deeper understanding of the land- and sea-scape settings, the contemporary environmental context, and the availability of natural resources (to build and maintain them as well as for trade) for these ancient commercial harbours. Defining and understanding their physical configuration and the level of human adaptation required to utilise and maintain the ports over long periods of time also remain key research issues.

Due to the location of ports and harbours on the dynamic intersection between the waterscape (sea, lake or river) and the terrestrial landscape, prograding and regressing coastlines provide excellent areas for the study of temporal changes in settlement patterns and the degree of adaptation and modification required from the people using and occupying them. It is therefore necessary to develop a deeper understanding of the environmental context and the landscape setting of ancient ports in order to fully appreciate the reasoning behind the choice of their particular locale on the Red Sea coast (our study region) as well as the level of geomorphological and environmental change within these littoral 'coast-scapes'.

The near constant adaptation to geomorphologically dynamic shorelines in ports such as *Berenike Troglodytica* on the southern Egyptian coast emphasises the impact that natural factors such as sea level change, coastal erosion and sedimentation, and climate change had on the viability of these settlements and towns (e.g., Kotarba-Morley 2015, 2018).

The theoretical concept including a suite of variables, called the 'Parameters of Attractiveness' (PoA), developed for the purposes of this research, will allow the investigation of the rationale behind the location of ports, and the evaluation of whether they were related to landscape, cultural, economic or socio-political factors, or a combination thereof. These variables were primarily derived from a comprehensive review of both modern and ancient literature pertaining to the physical and socio-political characteristics of port sites and supplemented by available data from the case study site, the Greco-Roman port of trade *Berenike Troglodytica* on the southern Red Sea coast of Egypt. Four primary categories of variables have been identified as playing a role in the siting of new ports. These can be broadly divided into: (i) Sea, (ii) Land, (iii) Resources, and (iv) Socio-political and economic factors (hereinafter SPE), and were identified in order to determine the extent to which the landscape and physical environment dictated locations of Hellenistic or Roman provincial ports of trade (or their trading counterparts) on the Red Sea, and whether other variables (e.g., geopolitical location, proximity to existing trade routes) were equally, or more, important.

The Berenike case study (Kotarba-Morley 2015, 2018) demonstrated the utility of the PoA approach for quantifying and evaluating many competing factors that were likely to have influenced the choice of ancient port location. This pilot study determined the extent to which the landscape and physical environment, and their changes through time, dictated initial site selection and the further development of a new port of trade at Berenike. It also sought to understand, more objectively, how ports, as hubs of socio-political interactions between different cultural groups, were designed to generate profit and prosperity. The results from the pilot study also suggest that, not only did these parameters influence the initial sitting of a port, but also, once founded, they affected the degree to which a harbour and the associated town may have developed into an important port or a regional hub and how it changed through time.

This is particularly important since very little is known to date about Classical (Greco-Roman) ports on the Red Sea beyond brief references in classical texts (see de Romanis 2009; Wilson 2015; Kotarba-Morley 2015). Since over 100 years ago, when Couyat (1910) calculated the potential locations of five important Red Sea ports, primarily using Ptolemy's *Geography* (Berggren and Jones 2000), very little progress has been made to locate any 'missing' sites. Aside from Myos Hormos, Adulis and Berenike, no other exact and unequivocally confirmed location of an actual harbour within a Classical port town is known. Although ancient authors mention 49 ports situated on the Red Sea, only 9 of them have ever been successfully located (see Table 1 and Fig. 1). The remaining 39 are divided between 34 port towns provisionally identified or with a probable location, and 5 that have not been identified at all. Additionally, six locations have been identified that have not yet been correlated with any harbours mentioned in the ancient texts, but which, according to the 'Parameters of Attractiveness', discussed further in this paper, have great potential for accommodating a significant port and/or have yielded archaeological material relevant to the study period. A similar situation can be observed in the Gulf of Aden, the Arabian Gulf and the Indian Ocean, where most site locations from this period are still yet to be confirmed.

Even though, as we will see below, the paucity of archaeological and geoarchaeological research concerning Red Sea ports can look, at first glance, rather alarming, it is important to consider the contemporary social, political and economic factors that might have exacerbated this situation. The coastlines of the Red Sea are generally sparsely inhabited and often rather inhospitable in terms of climate and the shortage of modern infrastructure. In many regions, there is a lack of facilities and resources meaning that scientists hoping to work there find that the logistics are too difficult to overcome. Moreover, the relatively unstable political situation and harsh climate can similarly hamper research in the region. It is clear that there is still a great deal of work to be done to begin to fill in the gaps in our knowledge of ports and harbours in the Red Sea region and it is certainly time that greater research effort is expended to bring the level of research in line with that undertaken in the Mediterranean.

2 A Brief Introduction to Greco-Roman Ports of the Red Sea

As mentioned above, very little research has been undertaken in the Red Sea region, with studies mainly concentrated on the Pharaonic ports of the Egyptian Red Sea coast (e.g., Tallet 2009, 2012a; Tallet and Mahfouz 2012), including Ayn Soukhna (Abdel-Raziq et al. 2012; Tallet 2010, 2012b, c, 2013), Wadi Gawasis (Sayed 1977; Bard and Fattovich 2003; Fattovich 2005; Bard et al. 2007; Hein et al. 2008, 2011; Fattovich and Bard 2012) and Wadi El-Jarf (Tallet 2012c; Tallet and Marouard 2012, 2014; Tallet et al. 2012). However, some interesting research has been undertaken on a number of Greco-Roman sites, mostly in Egypt and some recent excavations in Saudi Arabia (see below).

Although the northernmost Late Roman and Islamic Red Sea port of Aila (Aela/Aelana/modern Aqaba) is never

Table 1 Locations of ancient ports mentioned in PME, Ptolemy's and Strabo's Geographies and in other ancient texts (confirmed or probable in bold and unidentified in regular font) Adapted from de Graauw's (2014) on-line database and supplemented by the author

	Ancient port	Modern location	Country
1	Aelana, Aila, Elaea, Berenice, Ezion Geber	Aqaba	Gulf of Aqaba
2	Iotabe	Jezirat Fara'un, Coral Island 14 km southwest of Eilat, on the Sinai coast; or Tiran island at the mouth of the Gulf of Aqaba?	Gulf of Aqaba
3	Saba	Near Eilat (?)	Gulf of Aqaba
4	Ankale	Near Haql	Gulf of Aqaba
1.1	*Potential port location*	Nuweiba	Gulf of Aqaba
1.2	*Potential port location*	Tayyib al Ism	Gulf of Aqaba
1.3	*Potential port location*	Tell el Mashraba, near Dahab	Gulf of Aqaba
5	Makna	Near Magna	Gulf of Aqaba
6	Isle of sea-calves (dugongs?)	Tiran and Sanafir Islands, at the entrance of the Gulf of Aqaba (?)	Gulf of Aqaba
7	Leuke Kome, Leukos Limen, Albus Portus, Onne (?)	Port site on the Arabian side of the Red Sea where the road to Petra starts. Potentially located in the Bay of Aynunah near Al Khuraybah at 28°03'N(?). Ptolemy locates Leukos at 2°10' N of Berenike, but that is on the other side of the Red Sea. Others locate the port at Sharm Yanbu, also known as Charmuthas (opposite Berenike, 15 km north of Yanbu). However, Ali al-Ghabban (pers. comm. 2010) believes it to be Umm Useila/Useyla in the northern Red Sea on the entrance of the Gulf of Aqaba. Others believe it to be the site 47 km up the coast at Al Wajh (e.g., Nappo 2010)	Saudi Arabia
8	Modiana (?)	Al Muwalih	Saudi Arabia
9	Dabba, Modiana (?)	Dhuba at 27°21'N, 35°41'E	Saudi Arabia
10	Hippos Kome	Unidentified	Saudi Arabia
11	Egra and Phoinikon Kome	Al Wajh/Wedjh? Recognised as Egra since at least the 19th c. However, no associated archaeological evidence has ever been recovered from the site	Saudi Arabia
12	Ampelone, Akra (?)	At the outlet of Wadi al Hamd South of Al Wajh (?)	Saudi Arabia
13	Raunathou Kome	Near Khurayyim Said	Saudi Arabia
1.4	*Potential port locations*	Al Haura and Umm Lajj	Saudi Arabia
14	Charmothas, Charmute	Sharm Yanbu 15 km north of Yanbu? However, if Sharm Yanbu should be Leuke Kome then Charmothas could be located further south at the lagoon of Khor al-Kharar near Rabigh	Saudi Arabia
15	Farasan Islands	Roman naval base in front of Jizan (Van Hecke et al. 1861, p 747; Phillips et al. 2004, pp 244–245; Adams 2007, p 35; Cooper and Zazzaro 2012, p 408)	Saudi Arabia
16	Arsinoe, Cleopatris, Port Daneon, Klysma, Clysma, Clysina	Suez	Egypt
17	Phoinikon	Unidentified	Egypt
18	Marah	Near Ras Matarma	Egypt
19	Rhaithou	El Tor	Egypt
1.5	*Potential port location*	Tell el Raya	Egypt
20	Abu Shar	Roman fort at the end of the Via Hadriana, between El Gouna and Hurghada	Egypt
1.6	*Potential port location*	Wadi Safaga	Egypt

(continued)

Table 1 (continued)

	Ancient port	Modern location	Country
21	Philoteras Portus, Philotere, port of Aennus	Marsa Gawasis, 23 km south of Safaga. Ptolemy locates Phîloteras at 30' S of Myos Hormos, which would be near the airport of Marsa Alam and Ras Toronbi, where some creeks like that of Coraya beach and of Port Ghalib may have been used as ancient shelters. Pliny speaks of Aennus. However, some scholars seem to agree to locate this port at Marsa Gawasis near Safaga	Egypt
22	Arsinoe Troglodytika	Unidentified but somewhere between Philotera and Myos Hormos, perhaps near Kalawy Imperial Resort or Hamrawein (?)	Egypt
23	Myos Hormos, Port d'Aphrodite	Quseir al-Qadim, at the Mövenpick hotel site, 8 km north of Quseir. Ptolemy locates Myos Hormos at 3°25' N of Berenike. *PME 1* indicates that this site is at 1800 stadia (approximately 330 km) from Berenike, which would lie near Safaga. However, the location of the site is agreed upon 8 km north of Quseir	Egypt
24	Nechesia	Potentially Marsa Nakari, 18 km south of Marsa Alam	Egypt
25	Berenike Troglodytika, Berenike with ancient lighthouse	Pliny indicates that there is no shadow at noon on the day of the summer solstice, which is the definition of tropic (located at 23° 26'N). The present latitude of Berenike is 23°56'N. Large modern naval base—Baranis—is located some 8 km to the north of the site	Egypt
26	Ophiôdès, Agathonis, Tytis, Topasus Island	Isle of Zabargat/d, St John's Island, the island with topaz, some 80 km south of Berenike. The isle of Ophiôdès is well located as it seems to be the only one producing olivine stone in this area	Egypt
27	Bathus Profondus Portus	Dungunab (?) Ptolemy locates the 'deep port' of Bathus at 2°50' S of Berenike. This situates it in the large bay of Dungunab, which is sheltered from the northern waves prevailing in this area	Sudan
28	Dioscuror	Dungunab (?) Ptolemy locates Dioscuror at the same place as Bathus, perhaps on the west side of Mukawwar Island	Sudan
29	Theon Soterum, Deorum Salutarium, Sotira	Suakin (?) Ptolemy situates Theon Soterum at 1° N of Ptolemais, potentially corresponding with Suakin at 19°08'N	Sudan
30	Evangeliorum, Evangelon	Protected bay of the island Saqir, near Trinkitat (?) Ptolemy situates Evangelon at 30' N latitude of Ptolemais, at 18°44'N	Sudan
31	Port Elaea	Ptolemy's Evangelon (?)	Sudan
32	Ptolemais (Epi) Theron, Ptolemais of Hunters, Epitherias	Probably Aqiq Kebir or Adobona village and the nearby island of Badhur	Sudan
33	Port Melinus	Unidentified	Sudan
34	Port of Colobônalsos, Cape Colobon	Ptolemy locates Colobon at 2° N of Adulis, which does not correspond with any prominent cape or promontory	Sudan
35	Berenice of Saba, Epidire, Berenice Panchrysos?	Massawa (?) Ptolemy locates Saba at 50' N of Adulis, which corresponds with Massawa at 15°38'N. According to Pliny (NH, 6.170) Berenike Panchrysos might be located at the same place	Eritrea
36	Ery	Large bay of the Isle of Dahlak, Isle of Disset, Isle of Deses (?)	Eritrea
37	Gabaza, port of Adulis	Massawa (?) or Zula. PME locates Adulis at 3000 stadia (550 km) south of Ptolemais inside a south-facing bay, which corresponds to Zula at 15°15'N. Pliny locates it at 5 navigation days from Ptolemais, which leads to more or less the same location. It may be noted that Ptolemy is widely mistaken when locating it at 40' N of Diré; it lies near Assab or Beilul, 400 km further south. It may be noted also that Assab still is a port nowadays, while Beilul does not show any nautical interest (today) except for a cape located about 15 km further east	Eritrea
38	Isle of Diodore	Island located in the bay of Zula and now joined with land as 'Galala Hills' 6 km south of Zula	Eritrea
39	Port of Eumene	Port of Eumène (?)	Eritrea

(continued)

Table 1 (continued)

	Ancient port	Modern location	Country
40	Arsinoe	Ras Dumeira at Rahayta (?). Ptolemy locates Arsinoe at 20' S of Diré, which corresponds with the lagoon of Godoria on the north coast of Djibouti near 12°09'N. However he mentions it north of Diré on his list, which makes some authors think the site is at Ras Dumeira	Eritrea
41	Cape Dire	Ras Siyan, on the west side of Bab el Mandeb. Cape Diré is located in front of Cape Acila (Murat in Yemen) as Strabo indicates. It provides good shelter against the eastern waves prevailing in this area. The place called Fagal is at 12°27'N and seems to be an attractive location for ports mentioned by Ptolemy	Djibouti
42	Port Antiphile	Tadjoura (?). Ptolemy locates Antiphile at 45' S of Diré, which corresponds with Tadjoura located at 11°47'N on the coast of the Gulf of Djibouti. Strabo locates it 'leaning against the territory of creophags', which probably lies to the south of the Bab el Mandeb Strait	Djibouti
43	Port of Isis	Djibouti (?). The Port of Isis is at ten days rowing (500–1000 km) from Adulis according to Pliny, which lies near the coast of Djibouti (at 660 km) with its two islands located just in front of the port (as indicated by Pliny)	Djibouti
44	Avalitae, Avalites, Abalites, Aualis	Zeila in Somalia, 50 km east of Djibouti. Avalites seems to be located with some certainty at Zeila, where just a sand spit remains today or at the promontory of Djibouti	Somalia
45	Burnt Island	Volcanic island Jabal al-Tair northwest of Hodeida (?)	Yemen
46	Bolicas	Mokha (?). Port of the Omerits, in front of Adulis, on the Arabian side.	Yemen
47	Masala, port of Mouza, Muza	Mokha (?)	Yemen
48	Sosippi	Dhubab seems to be the only natural shelter between Murad/Océlis and Mokha/Mouza	Yemen
49	Acila, Ocelis, Artemidore	Murad (?), south of Mokha and in front of Ras Siyan	Yemen

actually referred to as a harbour (Ward 2007, p 163), Eusebius (*Onomasticon* 6.17–20 and 8.1) mentions ships sailing between Aila, Egypt and India, attesting to its maritime character. A number of geotechnical, geoarchaeological, archaeoseismical and GIS studies have been conducted over the years on the site and in its surroundings (Russell 1980; Niemi and Smith 1999; Niemi et al. 2006; Mansoor 2002; Mansoor et al. 2004; Al-Tarazi and Korjenov 2007) but more work will hopefully build on this in the future. Also in the north of the Red Sea some early surveys and excavations (e.g., Bourdon 1925) were undertaken at the important port of Clysma (Cleopatris/Arsinoë/Islamic Qolzoum) in the 1920s and 1930s (Bruyère 1966), and renewed fieldwork at the site is likely to start soon (Blackman, pers. comm.). The early studies suggested that the location of the ancient harbour was evidenced by concrete quays, a feature more common in the Mediterranean than the Red Sea. Recent work on the Nile–Red Sea canal (Cooper 2009, 2014) has brought into focus the importance of this port.

The shortest route between the River Nile and the Red Sea in the southern part of Egypt is marked by Wadi Quseir, which forms a direct link between the important port of Myos Hormos and the site of Coptos (Quft) in the Nile Valley (Zitterkopf and Sidebotham 1989). Myos Hormos has been studied in great detail (Peacock and Blue 2006, 2011), ascertaining the location and depth of the harbour in the Roman and Islamic periods. Blue (2011) also modelled the silting up of the harbour,[1] with the migration of a sandbar, originally in the inner reaches of the lagoon, gradually moving toward the entrance of the lagoon, closing off the channel, and ultimately disabling the harbour. The discovery of waterfronts constructed from empty amphorae proved extremely valuable in developing our understanding not only of Myos Hormos and its social context, but also of technological connectivity across the Empire (Blue 2011). This

[1]Excavations in Trench 15 suggest that the harbour experienced siltation problems from at least the 1st c. CE.

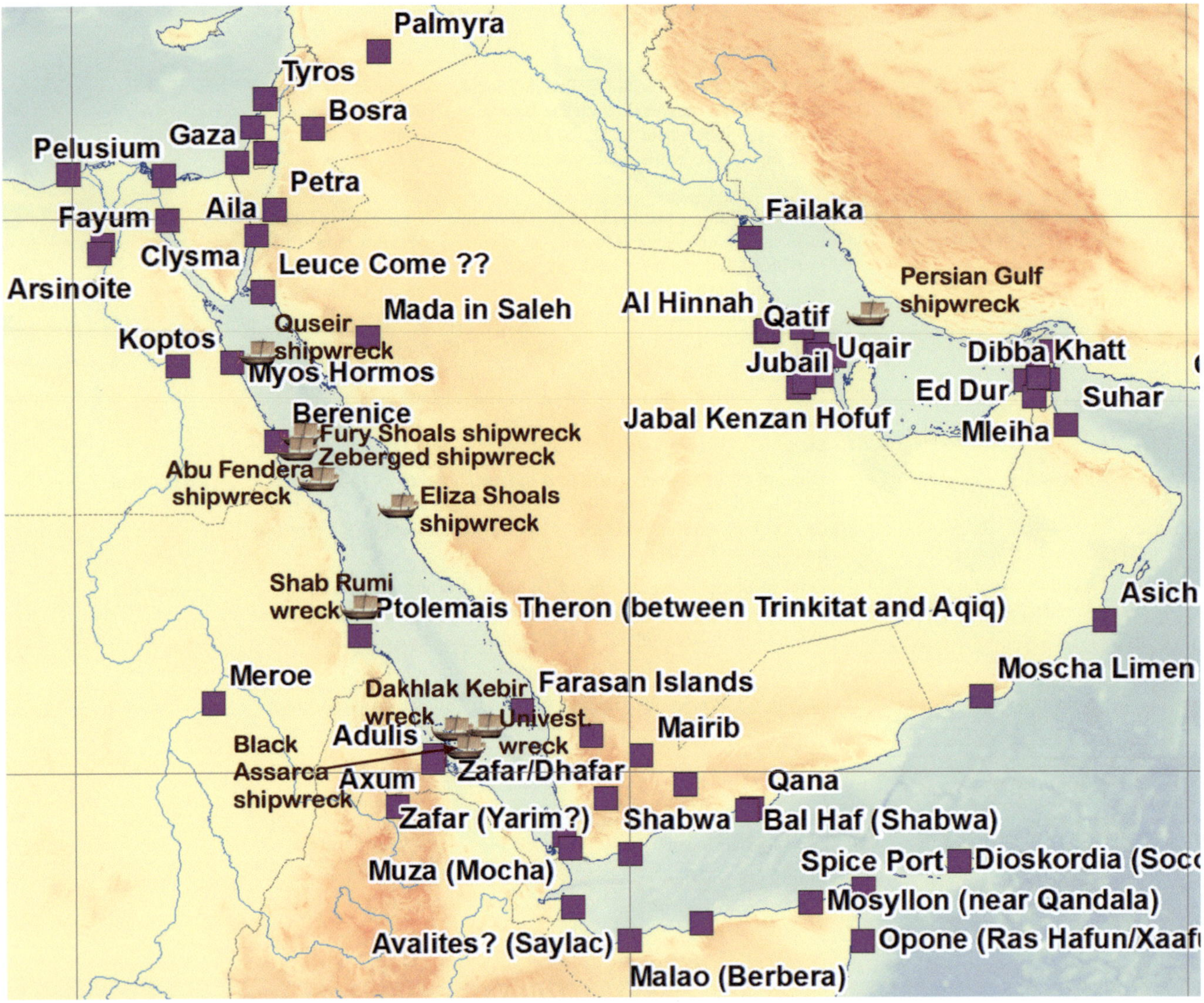

Fig. 1 Topographic and bathymetric view of the Red Sea with locations of Greco-Roman ports mentioned in Table 1 as well as known Roman-era shipwrecks in the area (drawn by AM Kotarba-Morley)

type of construction was used across the Mediterranean, for example, in Naukrais (Thomas et al. 2016), Cadiz on the Atlantic coast of Spain (Bernal et al. 2005) or in sites in northern Italy and southern France (Pessavento Mattioli 1998).

Berenike Troglodytica was one of the most important port cities on the Egyptian Red Sea coast during the Ptolemaic and Roman periods (Sidebotham 2011). The city intermittently prospered for eight centuries, from about 275 BCE, when it was established as a small outpost, initially involved in the trade of African elephants, to the mid-6th century CE, when it finally ceased to operate (e.g., Sidebotham 2011). The port of Berenike was sited just south of the large peninsula of Ras Benas, which offered the site protection from the elements (Kotarba-Morley 2015, 2018). The location of Berenike Troglodytica benefits from at least two basins, 'the Lagoon' and 'the Embayment', both of which are highly conducive to accommodating significant mooring areas with little need for human modification of the existing natural landscape. However, given that these basins are also very efficient sediment traps, with the potential to infill with coastal, fluvial and windblown sediments, high levels of maintenance must have been necessary to prevent them from silting up and becoming unnavigable (Kotarba-Morley 2015, 2018).

Recent excavations on the site of Aynuna by a Polish team from the University of Warsaw unearthed a significant Nabatean coastal city with an abundance of Roman pottery, coins and other evidence for trade in the Red Sea. Located some 2 km from the modern fishing village in the Aynuna Bay, which the excavators believe to be the location of an ancient port, the extensive ruins at Aynuna have only been cursorily surveyed in the 1980s. Whilst it is still debatable whether the site is indeed the port of Petra, Leuke Kome (Albus Portus/Leukos Limen), the findings themselves seem to be of importance for understanding the trade connections of the region (e.g., Gawlikowski and Juchniewicz 2016).

To date, we do not have a clear understanding of the whereabouts of the harbour of ancient Nechesia, the smallest Ptolemaic–Roman Red Sea emporium, although Marsa Nakris has been put forward as a potential location (Sidebotham 2011, p 186). Nor do we have any definite answers regarding the locations of Philoteras/Aenum (Sidebotham 2011, p 184), or Ptolemais (Epi)Theron (Seeger et al. 2006; Sidebotham 2011, p 186).

Further to the south, the city of Adulis on the Eritrean Red Sea coast was a key port of the Late Antiquity Axumite Empire, connected with the capital, Axum, by a meandering mountain passage, and has been subject to significant research effort (Peacock and Blue 2007). The two harbours of Adulis—Diodorus Island and Orienê—have been located and geoarchaeological research, including coring and a geomorphological survey, was initiated in 2004 in different areas of the site. The current location of the site, 7 km from the present shoreline, indicates considerable coastal change over the intervening years. While the results of coring were deemed inconclusive in terms of direct evidence for a marine environment (Peacock and Blue 2007, p 47), they did provide indications as to the location of the ancient shoreline. This work also helped to identify the seastack bisecting the Galala Hills in the southern part of the site as Diodorus Island, one of the harbours mentioned in the *Periplus Maris Erythraei* (PME4, Casson 1989), in whose lee ships could safely moor (Blue and Peacock 2007, p 47). The second port at the site—Orienê—had a magnificent lagoonal harbour and an adjacent settlement in the central valley of Dese (Peacock and Blue 2007, p 137). Renewed excavation at the site by the Italian team (e.g., Zazzaro 2013; Zazzaro et al. 2014) might yield further detail regarding this port town.

The port towns and harbours described briefly here form only a part of the picture of coastal exchanges and commercial dynamics on the Red Sea in Greco-Roman times. As mentioned above many of these hugely important nodes of communication are yet to be discovered and studied, with some of them probably remaining under the desert sands for many years to come. What is important to keep in mind at this stage is that locations of those sites are not spread at random along the Red Sea rim and have been chosen with particular parameters in mind, due to their attractiveness to the settlers and sailors and with a clear understanding of the constraint and limitations of contemporary engineering, shipbuilding technologies, supply chains and political needs. With ancient landscape, climatic and environmental reconstructions available for this period we would be able to better imagine the relationships between the sites and the level of adaptation required from the settlers to tame their chosen locations.

3 Methodology

In order to assess the relative contribution of a series of social-political and environmental parameters to the choice of location of Greco-Roman ports of trade located on the Red Sea coast a case study site, *Berenike Troglodytica*, was selected. Previous research (Kotarba-Morley 2015, 2018) reconstructed the changing environment at the site over key stages of its development using geological and geoarchaeological datasets. This environmental dataset is additionally supplemented with existing historical data regarding political, social and economic factors that are of importance to produce a list of potential variables and links and relationships between them.

During the process of building up this model four main categories of so-called 'Parameters of Attractiveness' (PoA) have been distinguished, as mentioned above: (1) Sea, (2) Land, (3) Resources, and (4) Socio-Economic and Political. Each of these categories consists of between 3 and 5 parameters valued at 1–5. To carry out a semi-quantified assessment of the Parameters of Attractiveness a series of hierarchical sliding scales (1–5) is employed specific to each parameter in the four categories identified. In this model the higher the value, the more favourable the location. As can be seen, the highest score in each of the parameter groups represents the most favourable site from the point of view of its sea or land position, access to resources, or its position in the socio-political-economic landscape.

Given that each of the parameter groups has a different number of variables within them, the data were normalised using a simple formula: *total score/total possible score*. The final score of each category, as well as the final score for the whole site (calculated by: *sum of all total scores for each category/number of categories*) will then fall under a simple denomination of:

- 0–0.25—Not Attractive
- 0.25–0.50—Poorly to Reasonably Attractive given the circumstances
- 0.50–0.75—Moderately Attractive
- 0.75–1—Very Attractive

The pilot study site of Berenike has been subject to systematic analyses of the contemporary environmental characteristics of its port's/ports' location/s and in turn the assessment and quantification of social, political and economic aspects that also had an impact on the attractiveness of its location. Such 'port attractiveness' patterns emerging from the assessment of the contemporary natural and political landscape will be presented as the main focus of this study. In the future, more advanced statistical analyses could be performed in order to assess the past landscapes and environments of known Greco-Roman and contemporary ports and harbours on both sides of the Red Sea and set them within an analytical and interpretative framework of 'Parameters of Attractiveness'.

4 Results: Parameters of Attractiveness

The concept of this theoretical framework of PoA was partly based on the 'landscape learning idea' (Rockman and Steele 2003), the 'common sense geography' approach (Geus and Thierling 2014), as well as the 'transaction cost theory' (Williamson 1989). Some of the questions listed below, originally part of a list posed by Rickman (1985) and discussed also by Karmon (1985) in a study of Mediterranean harbours, were also highly relevant to the development of the PoA framework and as such drove forward the initial strategy to quantify the variables that may be included in a list of Parameters of Attractiveness.

i. What factors influenced decisions to position ports and harbours in particular areas of the coastal landscape?
ii. Was the choice purely pragmatic (relating to landscape and resources for instance) or potentially also linked into the politics of the region?
iii. How did port towns integrate and communicate with hinterland routes and other systems of communication, such as rivers, canals and lagoons, and with each other?
iv. What was the relationship between the ports and areas producing raw materials (agricultural goods, quarries, etc.), and were the ports also centres of manufacture?
v. Which of these ports were the centres of population and consumption?
vi. What factors influenced the economic and financial success of a port—natural factors or human need?
vii. What factors triggered the decline of a port city (e.g., geographical, political, social or economic changes)?
viii. Who maintained the harbours and at what cost?
ix. How would individual municipalities have financed the harbour maintenance? Was it from harbour taxes, loans, benefactions from local magnates, or merchant guilds, or was it always an external source such as a state treasury?
x. Can ancient harbour sediments provide information about human-environment interactions?

These have proven to be particularly important in the assessment of the pilot study site. The city of Berenike was founded in a harsh, marginal environment where relatively small environmental changes (e.g., fluctuations in sea level, changes in coastal geomorphology, an increase in seasonal floods) might have had a dramatic effect on the short- and long-term viability of the site. Whilst the current coastal situation of the site (see Fig. 2) does not seem propitious for hosting a large volume of maritime traffic, the design and construction techniques of ancient merchantmen, the available navigational and mooring techniques of the time (e.g., using oars for entering and exiting ports), the location of the Red Sea at the crossroads of Indian Ocean and Mediterranean commerce, and the configuration of natural features of the coastal landscape, all meant that Berenike must have possessed a sheltered harbour with sufficient water depth to accommodate high volumes of incoming and outgoing merchant vessels that was at least reasonably attractive (see Table 2) to settlers and sailors alike.

Below I present the results of the analyses conducted on the site set against the general Red Sea coast context. Each of the categories is briefly introduced and then followed by a description of a particular parameter and the results from the pilot study at Berenike. All of those results are also shown in Table 2 for the convenience of the reader.

5 Sea Parameters

Sea parameters are those variables that influence marine environments, such as winds, currents, waves, reefs and maritime resources. The achievements of ancient sailors and navigators reveal a deep knowledge of the marine environment and a near-symbiotic relationship with these variables (e.g., Facey 2004). Similarly, designs of ancient ships prove that ancient shipwrights had a deep understanding of the materials used to construct vessels that could operate in a diverse range of weather conditions (e.g. Kotarba-Morley 2017). The parameters used here were selected following a review of literature (ancient and modern) regarding the environmental conditions on the Red Sea and a detailed appraisal of ancient navigational techniques and ship designs on the ancient Red Sea and beyond (e.g., Kotarba-Morley 2017). It is hoped that they will allow for a better appreciation of the natural, physical and geographical characteristics of Hellenistic and Roman provincial harbours in this region. In turn, this will permit an establishment of possible navigation patterns on the Red Sea in this period and draw out the major factors that could have been influential in types of vessel steering in the region.

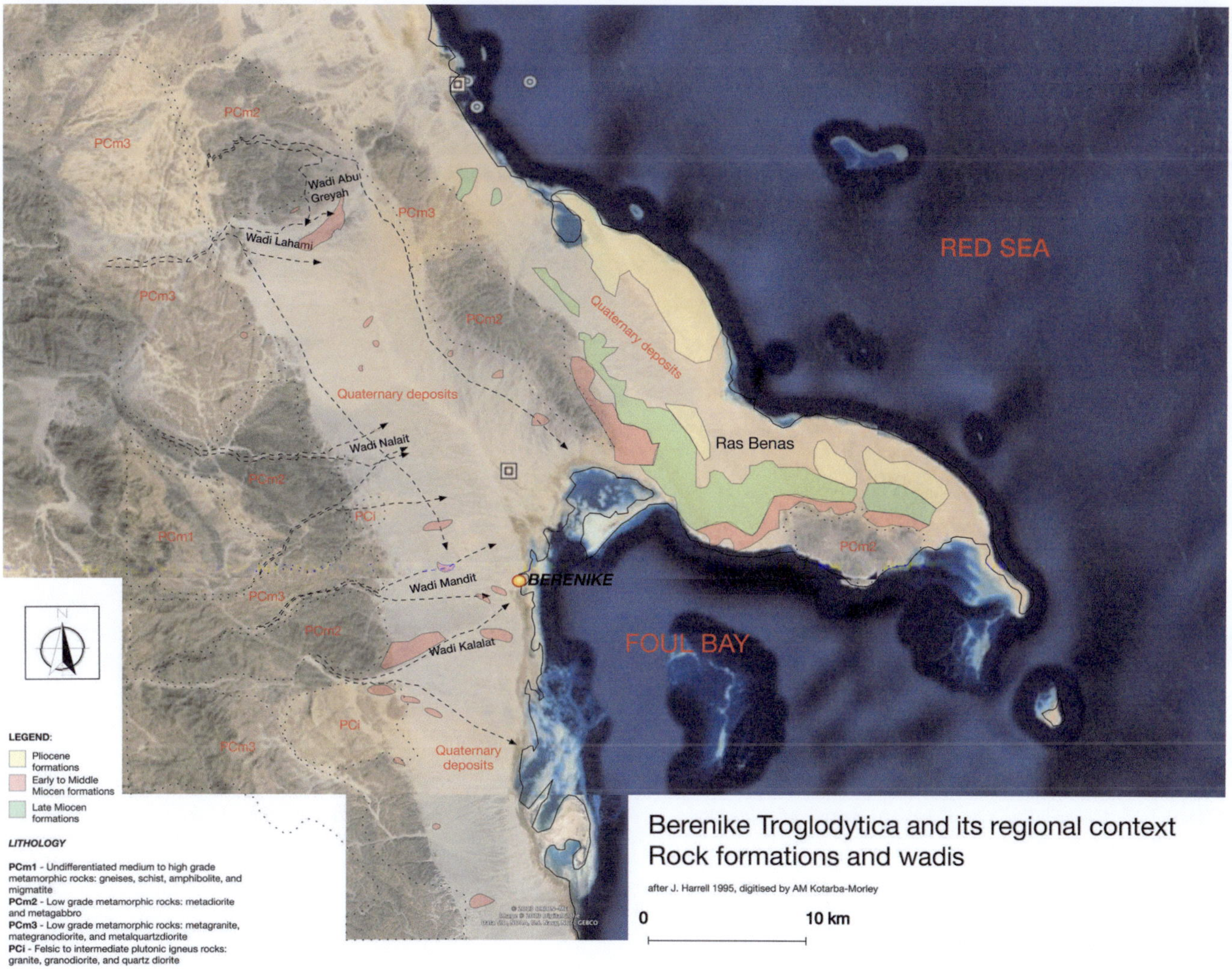

Fig. 2 Berenike Troglodytica and its regional context. Rock formations and wadis (drawn by AM Kotarba-Morley, based on Harrel 1996)

5.1 Winds: Seasonality, Strength and Direction (Also Ship Technology)

The prevailing forces propelling ancient sailing vessels along trade routes were atmospheric winds, and merchant vessels plying the Red Sea and the Indian Ocean in antiquity, and using ports such as Berenike, were usually propelled by sails rather than the oars (although some might have also carried oars on board). The Red Sea circulation is dominated by winds, evaporation and the monsoon system, with northerly–northeasterly winds dominating (see Fig. 3). The prevailing northerlies in the northern part of the Red Sea (where Berenike and Myos Hormos are located) blow almost all year round, with the central part of the basin experiencing more mixed conditions, with prevailing northerlies supplemented by some westerly blows (as a result of trough pattern anomalies promoting Sub-Tropical Westerly Jets (STWJ) over the Red Sea (Williams 2011, p 36). In the southern part of the Red Sea the northerlies still prevail in the summer, with the reverse pattern in the winter when southerlies prevail approximately 55–70% of the time (Davies and Morgan 1995, pp 29–30). The strong seasonal currents (0.5 knot) follow the directions of winds aiding the northward journeys (Whitewright 2007, p 85). During the northern hemisphere summer, intertropical fronts move northward bringing strong winds and thunderstorms, thus making a passage in the peak summer months of June and July almost impossible for smaller vessels (Curtin 1984, p 99). In the winter months, when this region is under the influence of the Mediterranean cyclones, strong but short-lived westerly to southwesterly storm winds occur (e.g., Edwards and Head 1987; Davies and Morgan 1995, p 29). The border between the northerly and southerly winds is approximately 20°N, representing the most northerly position of the ITCZ during the summer time (Patzert 1972a, b, 1974; Edwards and Head 1987; Stenchikov 2011).

Table 2 Parameters of attractiveness of the site of Berenike Troglodytica (Kotarba-Morley 2015)

Category: land parameters			
Parameter	*Parameter characteristics*	*Berenike score*	*Berenike justification*
Natural lagoon	*Level of availability of natural bay or lagoon*	5	Large sheltered bay
1	No bay or natural lagoon		
2	N/A		
3	Half natural, improved by humans		
4	N/A		
5	Very suitable natural lagoon or bay, potentially with an inlet		
Siltation rate	*Speed and volume of siltation, available dredging technology*	2	High siltation rate
1	Very high unmanageable siltation rate or high siltation rate with no technology to rectify it		
2	Fast and high siltation rate with little technology to rectify it but with efforts made by local population to deal with a problem		
3	Fast and/or high and/or seasonable siltation rate with technology in place to rectify it but only partially and at a slow pace		
4	Slow and relatively low siltation rate but technology in place		
5	No siltation or very slow siltation rate, which can be easily dealt with using locally available technology		
Rivers, creeks etc.	*Accessibility, size of estuary, degree of protection, depth, current*	2	Powerful wadis; providing source of fresh water only from deep wells; no irrigational properties but fresh water influx allowed for an opening of the channel at the entrance of the harbour; also used as corridors of connectivity with the hinterland
1	Difficult to access, few natural food and water resources, hindered access to the hinterland via river transport		
2	Difficult approach to the river or a riverine port, difficult or hindered access to hinterland and its resources through the river		
3	Decent access to the river or a riverine port, some resources available easily through the riverine transportation routes, others not		
4	Good access to the riverine port and good availability of resources		
5	Good and safe mooring facilities, easy access to the site and into the hinterland		
	Berenike Sea parameter total score	8/15	0.53

Category: resource parameters			
Parameter	*Parameter characteristics*	*Berenike score*	*Berenike justification*
Fresh water	*Fresh water availability, accessibility, proximity, quantity, and quality*	1	Fresh water available, but the closest well is located 7.6 km away on the edge of the mountains; caravan of donkeys had to ferry water to town at all times
1	Very low availability and/or low accessibility of freshwater; sources located far from the site and/or having only low water reserves and/or poor quality		

(continued)

Table 2 (continued)

Category: resource parameters			
Parameter	*Parameter characteristics*	*Berenike score*	*Berenike justification*
2	Low availability and accessibility; poor quality water, not in sustainable quantities, has to be transported to the site		
3	Water available in the vicinity of the site, however it needs to be imported on pack animals; water quantity and quality varies from fair to good or is seasonally dependent		
4	Good water available on site or in its proximity, easy or easy enough to access; sustainable quantity producing limited amount of surplus		
5	Good quality water available on site in sustainable quantities for humans, animals and agriculture leaving enough surplus for merchants and their crews		
Marine resources & game	*Demand for fish, availability of fish, available technology, possibility of trade in seafood and fishproducts (if surplus)*	4	Plentiful fish, large quantities of fish bone from archaeological record; also hunting for game in the mountains and access to donkey and camel meat and to birds
1	Very little fish supply, no available technology and little demand (subsistence strategies lie in other areas of food resources)		
2	Demand in place, but not much fish available in the area that could be sustainably fished, or there is no technology in place to fish it		
3	Demand in place with a good supply of fish and good enough technology or/and fishing conditions but not producing surplus		
4	Demand in place producing enough surplus for some trade		
5	Sustained demand, producing plenty of surplus, which when traded provides the community with a good income source		
Agriculture & pastures	*Quantity and distribution of cultivable land, crop production, animal power and transport. Quality, quantity, sustainability, seasonality of pastures*	1	Potentially, brackish water could have been used for horticulture or light agriculture. Macrobotanic evidence shows that there could have been a small-scale agriculture in Berenike (Jarosław Zieliński, pers. comm. 2011). Pastures probably only suitable for camels and donkeys
1	Very poor conditions, including no arable land, lack of soils and insufficient water to sustain irrigation. No pastures, and fodder has to be imported		
2	Poor soils, marginal irrigation potential, small-scale agriculture and horticulture possible with great effort. Seasonable small pastures with fodder imported from hinterland		
3	Fair quality soils, however irrigation cost possibly high. Some moderate quality pastures but small and seasonal, abundance of seaweed for collection (but high effort)		
4	Good quality soils, moderate irrigation required. Good pastures, either in the vicinity of the site or in nearby satellite villages, sustainable fodder acquisition		
5	Optimal conditions for agriculture, large tracts of arable land, good quality fertile soils and abundance of water. Excellent pastures and plenty of fodder for animals		

(continued)

Table 2 (continued)

Category: resource parameters			
Parameter	*Parameter characteristics*	*Berenike score*	*Berenike justification*
Bldg. Mats.	*Level of availability of biotic (e.g., forests) and abiotic (e.g., building materials, ore) resources*	4	Abundance of gypsum anhydrite and coral, however of a very low quality
1	No building material available in the area; all materials have to be imported		
2	No suitable, good quality building material available in the immediate area		
3	Limited supply of decent building materials; need to transport		
4	Good source of building materials but either a limited supply and/or no maintenance facilities and/or skilled craftsman or some issues with transport/access		
5	Very good quality of local quarries for stone and ore; easily available good quality building materials		
Wood	*Level of availability of biotic (e.g., forests) and abiotic (e.g., building materials, ore) resources for ships and buildings*	1	Only acacia and mangrove wood and the rest of the wood has to be brought through the caravan route and Nile route from the Mediterranean or from the south/Indian Ocean
1	No decent quality wood available in the area. Only shrubs		
2	No suitable wood available for ship repairs; wood can be imported or collected (such as mangrove) for urgent ship repairs		
3	Limited supply of decent wood. Need to transport		
4	Good source of wood but either a limited supply and/or no maintenance facilities and/or skilled craftsman or some issues with transport/access		
5	Very good wood freely available, skilled craftsman, shipyard with dry docks		
Raw materials & trade goods	*Availability, portability, quality*	4	Raw materials such as gold, emeralds, topaz, etc
1	Very little or an almost total lack of raw materials and/or trade goods in the area		
2	Some raw materials and natural resources are available near the site but they are either of a poor quality or are difficult to obtain/hard and distant to transport		
3	Good quality or easily accessible type or raw resources or trade goods that justifies an effort expended to obtain it		
4	Sustainable quantity of decent quality resources within a manageable distance		
5	Abundant good quality natural resources that are easily accessible and in the vicinity of the site		
	Berenike Sea parameter total score	15/30	0.5

Category: socio-political-economic parameters			
Parameter	*Parameter characteristics*	*Berenike score*	*Berenike justification*
Demand	*Scale of demand*	5	Need to situate a port on the Red Sea coast at the southernmost edge of the Empire and within the reach of monsoon winds

(continued)

Table 2 (continued)

Category: socio-political-economic parameters			
Parameter	*Parameter characteristics*	*Berenike score*	*Berenike justification*
1	No demand		
2	Low demand		
3	Sufficient demand		
4	Significant demand		
5	Strong demand		
Strategic position	*Connection with hinterland and provision of safety for the inhabitants in case of an attack*	4	Fort, defensive walls, island locale; Caravan road through the desert, hydreumata stations, forts and praesidia on the way; Closer than Myos Hormos for the sea journey and no need to fight the northerly winds; but longer route for the caravan through the desert
1	Difficult and dangerous, seasonally impassable road, no watchtowers, difficult or impossible to obtain supplies		
2	Difficult or dangerous route, possibility of attacks, could be fully or partially seasonally impassable, no or very little access to supplies		
3	Connection with hinterland by good route, with reasonable safety, however one which is either seasonally impassable or has little access to supplies		
4	Good connection with hinterland through safe and sound road or river which becomes more difficult to pass seasonally, some supplies available en route		
5	Good connection with hinterland through a network of safe and sound roads or rivers that are not particularly affected by seasonal changes, facilities and supplies available en route		
Stability & safety	*Political situation in a region; civil unrest; war; predictability of conflict*	4	Nomadic tribes in the hinterland, however the routes are well protected by praesidia and forts
1	Permanent state of war, dangerous place		
2	Unpredictable situation in the region		
3	Generally safe port but with a limited safety in its surrounding and tensions that may turn violent		
4	Safe and stable port with some neighbouring tribes susceptible to occasional revolts		
5	Safe harbour under stable rule		
Port facilities & infrastructure	*Accessibility of human expertise, administrative facilities, hospitality service and built infrastructure*	4	Tetarte (25% tax) charged in Alexandria or duties levied in Coptos
1	Lack of facilities		
2	Poor facilities		
3	Decent facilities exist on site, however in a limited supply		
4	Good facilities and decent administration		
5	Very good administrative and hospitality services lower taxes, no harbour duties, no queues to enter the harbour; good workshops and shipwrights; easy and cheap supplies; a bar; a brothel		
	Berenike Sea parameter total score	17/20	0.85

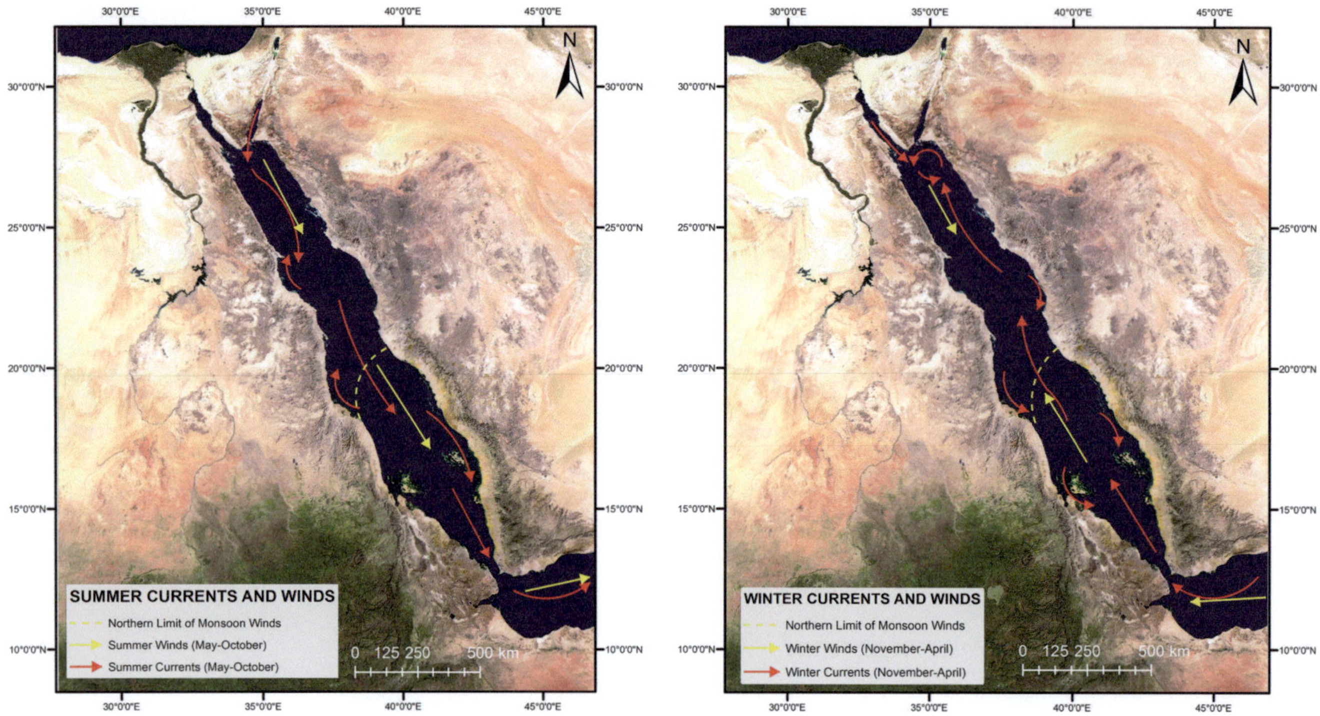

Fig. 3 Summer and winter currents and winds on the Red Sea (drawn by AM Kotarba-Morley)

In weak winds the course would be north-east by north and due west (Whitewright 2008, p 139), whilst in a strong wind, as Sulaimān al-Mahrī tells us, sailors would allow for an east by north or due east course for the Arabian coast and west-southwest or southwest by west for the African coast (Tibbetts 1961, p 327).

Traditionally, it is understood that vessels which left Egypt early enough (around August) arrived in India some time in September (Curtin 1984, p 99), and then set out for the return journey from India after three months of trading and ship maintenance, any time after the onset of the northeasterly monsoon around December and no later than mid-January (Pliny, *NH* 6.106; Whitewright 2007, p 78), bringing the vessels to the southern part of the Red Sea by the end of March at the latest. This was in time for southeast winds prevailing from January to March that are reliable up to Berenike's latitude (Facey 2004, pp 9–11; most northerly position of the ITCZ). This would enable the crews to exploit any 'decent' southerly and offshore (diurnal) wind that was experienced on the Red Sea (Whitewright 2007, p 78). After that, from April to December, northerly winds prevailed over the southern Red Sea making the passage very difficult if not impossible (de Romanis 2009; Seland 2011, p 401). The journey from India to Arabia could have been performed any time between late October and early March (Tibbetts 1971, p 231) with no 'closing' date of arrival in the Persian Gulf. However, according to the speed calculated by Seland (2011, p 401) if a vessel left the northern Red Sea at the beginning of the season and took two months from India to the Gulf the merchants would have arrived there around mid-December. The so-called southern passage to Africa was also available from July to August (Horsburgh 1841, p 484).

5.2 The Winds at Berenike

Berenike's location is extremely favourable with respect to winds, being situated in a part of the Red Sea/Indian Ocean that receives seasonal, strong winds with northerly seasonality, and as such scores highly—5—in this PoA category. The wind patterns over the Red Sea meant that vessels arriving from the south would often be sailing into Berenike's harbour between January and March, and from April/May onward if arriving from the north (Facey 2004, pp 9–11). The passage north was viable owing to the southeast monsoonal winds that attained Berenike's latitude, supplemented by the southerly and offshore (diurnal) winds. The northerly winds that prevail on the Red Sea for most of the year allowed vessels from the northern Red Sea to sail to Berenike but ensured that the passage north was much more laborious. This also had an important implication for port sites such as Myos Hormos or Clysma, which would score poorly on this parameter (e.g., 2) owing to the presence of seasonally weaker winds. The favourability, or not, of wind also depended upon the technology of the time and its ability to deal with wind conditions, making this parameter culturally specific and dependent upon equipment and expertise.

5.3 Currents: Strength, Direction, Seasonality and Predictability

Ocean currents are movements of water in response to prevailing wind patterns, variations in salinity, water temperature and water density that are important for seafaring and harbour location. Wind-driven currents are strongest when wind drives surface water to the gulfs, through narrow straits, or into and out of estuaries and lagoon straits (Thurman et al. 1999, p 208), as is the case at the southern end of the Red Sea (Fig. 3) and, on a smaller scale, at the entrance to Berenike's harbour. Tidal currents in the open ocean rarely exceed 2 knots, but strengthen where flow is channelled through gulfs, island straits, or estuary and lagoon entrances, where they can attain speeds of up to 9 knots (Admiralty 1995), creating extremely challenging navigational conditions and making mooring extremely dangerous. Currents are to a large extent predictable, and so their position and speed would have been well known to experienced navigators. Information about currents would have been passed on via oral tradition and via specialist texts such as the *Periplus Maris Erythraei* (e.g., *PME* 13, 25, 43, 46; Casson 1989). Their strength, direction, seasonality and predictability would have been important in prospecting for the site of a new port.

5.4 Sea Currents at Berenike

The currents at Berenike are strong but only favourable in a southerly direction, thus the site attracts only a moderate PoA score of 3 in this category. The other part of the justification for such a score is that strong currents that might have existed at the entrance to the lagoon at the time when it was much deeper could have, potentially, hindered smooth operation of entering and exiting the port (bringing in a requirement for tug boats and availability of oarsmen on trading vessels).

5.5 Wave Action: Strength, Size, Length, Seasonality

Waves are extremely efficient energy transporters and are important in navigation, with implications for sailing conditions and harbour location. Upon reaching the coast, waves change height and direction due to shoaling and the effects of refraction and diffraction (de Graauw 1986) and can be potentially devastating to vessels (see Casey 2011 on the subject of rogue waves). Assessing wave strength, size, length, and seasonality would have to therefore been taken into consideration by ancient port prospectors. The hydrological conditions on the Arabian Sea indicate that almost 50% of swells for the months of August are classified as 'heavy', whilst around the Arabian Peninsula the ratio is 56% (Beresford 2013, p 222); however, the Red Sea waves are recorded as being much lower. The ability to withstand large swells would have been one of the considerations of ship designers and even though capsizing a sailing vessel is difficult even in 'big seas' (due to the trigonometry of the behaviour of mast and keel and their length ratio) examples of large, breaking and barrelling swells violently breaking masts and then toppling vessels are well-known. Sailing on top of long waves, if performed proficiently, could however increase a vessel's speed as long waves move quickly and lose relatively little energy, therefore propelling the boat, whilst short waves move more slowly, consequently losing more energy and slowing the craft (Pethick 1995, p 13).

5.6 The Effect of Waves at Berenike

The coastal configuration at Berenike, when the harbour was operational, would have meant that waves broke on the southern promontory, which sheltered the lagoon (see Fig. 4). The approach to the harbour mouth would have been treacherous due to the extent of the underwater coral reef but at the same time this reef acted as a first wave break before much lowered and reduced in strength waves that reached the southern promontory. This would have been obvious to the ancient prospectors and the configuration of Berenike's immediate coastline would have been highly attractive owing to this sheltering feature in topography. Berenike scores very highly in Wave PoA (score = 5) due to the fact that the harbour is well-sheltered and that the wave range at Berenike is generally low in this part of the Red Sea and does not disrupt the incoming or outgoing traffic on site.

5.7 Near-Shore Coastal Shelf Geomorphology and Bathymetry

The geomorphology includes shoals, reefs, rocky coast, as well as space to manoeuvre and to set out beacons at the entrance of the port. The configuration of the near-shore environment would have been an extremely influential coastscape parameter in the locating of a new port, most likely only second in importance to the presence of a natural sheltered harbour or a lagoon. When considering the final approach to a port, a number of factors would have been taken into consideration beyond those already mentioned such as

Fig. 4 The site of Berenike Troglodytica with marked extents of Ptolemaic and Roman areas and supposed extent of port in different periods (drawn by AM Kotarba-Morley using Google Earth satellite image)

prevailing winds, currents, and tidal cycle, including submerged obstacles such as rocks, reefs and shoals, incline and depth of the coastal shelf, and space to manoeuvre (Figs. 2 and 4). However, the way in which they were dealt with by ancient mariners was culturally and temporally specific and depended upon the level of technology available to a particular human group at a particular time in history. A technical innovation to help mitigate for near-shore, submerged obstacles was the introduction of sounding weights, used in both navigation and mooring (Oleson 2000, 2006, 2008; Kirkland 2010, p 3). The technique of the sounding line was first reported by Poseidonius who calculated that the Mediterranean was a thousand fathoms deep (Strabo, *Geogr.* 1.1.37). This method was also used to assess the distance from the coast, the bathymetry of the sea floor, and to identify a safe passage to (or from) the port (Morton 2001, p 207).

5.8 Berenike's Coastal Geomorphology

Berenike scores 3 in the coastal geomorphology category for a number of reasons. Although taking into consideration each of the above-described variables separately, it is clear that they would not have generated such a low score. However, when combined they show that Berenike's near-shore geomorphological setting was only partially satisfactory. Firstly, the underwater reefs and shoals around the site are clearly visible on satellite imagery and empirical water depth testing by the author recorded a depth of between 1–1.5 m depending on the tide level. Even with the ~0.85 m higher sea level in antiquity (Kotarba-Morley 2015, 2018) they would still have represented hazards for ancient sailors. With an over half a metre increase in water depth the shallow coastal shelf surrounding the site would have remained risky for ancient navigators. Additionally, as Pedersen (2015, p 128) points out, the ubiquitous Red Sea reefs blocked access to most of the coast, preventing vessels from reaching the shore. At Berenike, however, a break in the reef located at the mouth of a wadi or a sharm[2] (Murray and Warmington 1967, p 25) was present due to the flow of fresh water precluding excessive reef growth. Whilst size and depth data for the lagoon entrance remain unclear, the deepest part of the access channel is likely to have been almost as narrow as

[2]Passage, usually between mountains.

it is today, whilst the overall width would have been in a range of 100 m (Kotarba-Morley 2015, 2018). It seems that increased coral growth within the channel in the last 2000 years could have been a result of decreased fresh water output from the wadis, most likely due to sediment accumulation on the sandbanks of the lagoon entrance in the later stages of site occupation (and further creation of sabkha fields). Even if the entrance was much wider, the reefs surrounding the area ensured a potentially perilous entrance to the port, raising our expectations for finding shipwrecks in the vicinity of the roadstead.

5.9 Tidal Range: Range and Size

The tidal cycles dictated the rhythm of life in the port city, with ships arriving and offloading, crews resting on shore, whilst taking care of resupplying of the ships and undertaking essential repairs, and finally re-loading and timing to leave the harbour on the rising tide. These cycles could have also affected the port administration, as it does today in smaller natural havens. The seasonal tides (neap and spring), related to the earth-sun orbital cycles, would have in turn affected the efficiency of trade, in particular in the period of the year when the tides were especially high, and could have been connected with an intensified period of fishing, trading and sailing activities additionally associated with favourable winds (such as monsoons in the Indian Ocean). The tidal range of different basins, seas and oceans can be very diverse depending on a number of variables. Given that the Red Sea is a semi-enclosed basin (similar in fact to the Mediterranean) the tidal range is very small, in the region of 0.6 m. The local tidal cycles and their connection with the phases of the moon would have been well known in antiquity (e.g., Flemming 1995), and the dates of major culminations could have been calculated and used in planning shipping voyages.[3] Tides could have also been used to the advantage of skilled seafarers who were able to sail out of the harbour using the tide as a mean of propulsion, hence incorporating the occurrence of this natural phenomenon with the techno-culturally driven factors—vessel design and sailors' expertise.

5.10 Tides at Berenike

Entry into the harbours of the Red Sea would have been totally dependent upon understanding the tides and the ocean's bathymetry, especially important at Berenike owing to the danger of grounding at low tide on the shallow (1–1.5 m) reefs or sandbanks. This is why Berenike scores only 2 in this category—although the 0.5 m tidal range is low the shallowness of the coastal shelf means that a large area is exposed at low tide making for a dire possibility of grounding.

6 Land Parameters

The Land parameters are those physical aspects of the coastal landscape that are likely to have been attractive to the prospectors of new port sites, most probably because they already serve a purpose in its role as a port. The littoral environment represents a third of all ecological and economic resources in the world at present, and has been widely explored by humans throughout history (Benedetti 2007). Coastal landscapes can comprise a wide diversity of physical environments, from wetlands, lagoons, estuaries, bays, mangroves and high cliffs, as well as a very broad range of biological ecosystems. The process of prospecting for new port sites in the Classical world in marginal areas of the Hellenistic and Roman Empires and their counterparts is essentially unknown. It is likely that already existing local harbours would have been utilised in the first instance, either modified to suit current needs or retained as already suitable. However, it is equally likely that there were so many of these small moorings along any given stretch of the coast that a sort of crude PoA would still be used by the prospectors to assess which of these would have been the best for larger-scale development.

The most obvious of these parameters would be the presence of a sheltered bay that would protect a number of ships from the destructive powers of stormy seas and high winds, coastal geomorphology, as well as the speed and volume of siltation and the proximity of inland waterways such as rivers (including seasonal wadis), creeks and lakes. The physical configuration of the landscape in which the new port is to be situated is of the utmost importance as in most instances it will dictate a city's layout and connections with the hinterland. The exception to this rule is when the political or strategic position of a proposed port, or a market demand, is so critical that the port is ultimately founded in an otherwise unattractive location regardless of the configuration of the landscape.

6.1 Presence and Physical Configuration of a Natural Bay, Lagoon or Inlet

The spatial configuration of the site determines the level of availability of a natural bay or lagoon. One of the major prerequisites for the siting of the new facility, once the demand for a port serving a particular region was identified,

[3]For example, Julius Agricola is said to have gained the island of Mona (Ynys Môn or Anglesey Island off the north-west coast of Wales) only because of his exploitation of an extremely low tide (Tacitus, *Agricola* 14–15).

would have been the existence of a natural sheltered bay, lagoon or inlet that could be used to safely moor the required quantity of vessels. Exploitable basins could be either entirely natural or artificially manipulated landscape features, and ancient port prospectors and builders would have certainly favoured locations with reasonably deep natural bays or lagoons. The exception to the rule is where harbour basins would, for whatever reason, be constructed entirely artificially, such as the port of Caesarea Maritima on the coast of Israel, Portus in Italy, or Carthage in Tunisia.[4] It should be added, that given a high enough demand, even the most inhospitable coastline without bays or inlets could have hosted an artificial harbour. However, in the contemporary context of the Red Sea and Indian Ocean littoral, it is unlikely that such costly endeavours would have been approved in areas where a large expense could not be justifiably explained. Specifically, even though we know of shipments of *opus signinum/pulvis puteolanus*[5] over 2,000 km distance from Italy for the purpose of building the port of Sebastos at Caesarea (Oleson and Branton 1992; Oleson et al. 2014), the availability of this material in marginal areas of the Empire, such as Berenike of even Myos Hormos (where stacked-up amphorae are used instead to build waterfronts), is unlikely.

Sometimes, the size and depth of the harbour basin could have been dictated by the type of facility needed for trade and other associated activities (e.g., large international trade hub or smaller regional stopping point) or additional military outpost. Therefore, it can be assumed that when a suitable natural location was identified, ancient engineers carried out a basic survey to establish the depth as well as the maximum and minimum extents of the basin, determined in part by the tidal range and coastal geomorphology, to ascertain whether they were suitable to accommodate their certain needs as found or whether they needed to be expanded.

6.2 Natural Bay or Lagoon at Berenike

Berenike scores very highly (5) in the 'Natural lagoon' category because of the large sheltered bay with an adjacent smaller embayment that could both support mooring of a significant number of vessels (approximately 85–105 medium to large merchantmen, Kotarba-Morley 2018) in two separate basins. This would have safely sheltered a large number of boats and ships from strong northerly and offshore winds (especially in the winter) as well as the associated waves and storm surges.

6.3 Siltation Rate and Regime

The siltation regime depends on the speed and volume of siltation and available dredging technology. Siltation, or sediment infilling, from terrestrial (rivers, wadis), marine (tides and currents), and atmospheric (windblown sand) sources, was, and is, one of the most significant problems faced by engineers, port builders and their maintenance teams (e.g., Blackman 1982; Yin et al. 2000). Evidence for primitive harbour dredging used to maintain basins has been recorded in the Mediterranean as dating to the Bronze Age (e.g., Rabān 1990, 2009; Pydyn 2011), and reached a peak of advancement in Roman times. The gaps in the reef along the Red Sea coast, where vessels could enter bays and harbours, were usually located opposite wadi or river mouths (Pedersen 2015, p 25), emphasising the importance of understanding siltation rates for the engineers planning new Red Sea ports. Whilst most known Red Sea and Indian Ocean harbours show no signs of major dredging activities to date, such a practice could have been undertaken using simple and relatively low-technology equipment when undertaken on a small scale (e.g., Hoyle 1989). Although Roman engineers may have had the technological capability to estimate the future degree of sediment infilling of a harbour basin, we do not know if they carried out such calculations and, if so, how accurate these would have been. Such estimates would allow for the evaluation of the time and resources needed to keep dredging operational. There is evidence, however, that rates of siltation were ignored if the location seemed particularly suitable in other ways (e.g., strategic location). This was the case at Ostia Antica, the harbour city of Ancient Rome, where fluvial influx from the Tiber caused significant long-term siltation issues (e.g., Giraudi et al. 2006, 2009; Bellotti et al. 2011; Goiran et al. 2012; 2014a, b; Salomon et al. 2012; Millet et al. 2014; Vittori et al. 2015).

6.4 Siltation at Berenike

Although harbour dredging was a routine activity in the Roman Mediterranean (e.g., Bini et al. 2009; Morhange and Marriner 2010, p 29; Mazzini et al. 2011), the level of maintenance and associated costs are little known for provincial cities such as Berenike, which, as mentioned above, would have necessitated at least occasional dredging to keep the harbour operational and accessible for large ships. Berenike's siltation rate is very high and so the site scores

[4]Dug harbour-basins opening onto the sea through a channel, such as in Carthage, are referred to as cothons (Carayon 2005).

[5]Hydraulic concrete was arguably used in harbour installations from at least the mid-1st century BCE (when it was used at the entrance channel to Portus Iulius in the Gulf of Pozzuoli; Oleson et al. 2004, 2014, Chap. 4) to build underwater structures as it sets subaqueously (Vitruvius *De Archit*. 2.6.1, 5.12.2-3; Gazda 2001, pp 171–177; Lancaster 2005, p 58; Brandon et al. 2008).

low (2) in this category. Seasonal flash floods flowing through the wadis resulted in a large flux of fluvial sediments to the coast and harbour basin area each year. Additionally, the marine sediment accumulation also contributed to the siltation of the lagoon and the embayment. These rates of sediment accumulation may have ultimately become unmanageable by the local maintenance teams, and the harbour was eventually abandoned.

6.5 Rivers, Creeks and Backwaters

Factors affected by the existence of rivers, creaks or backwaters on the site include accessibility, size of an estuary, the degree of protection, depth and currents. Ports and harbours can be effectively and strategically located in the estuaries and backbeach areas of rivers, which in turn provide crucial conduits for the inland movement of cargo using riverine craft (Salway 2004). These backbeach or backwater environments represented excellent locations for founding a port town and viable fishing, farming and foraging communities, and were widely exploited in Europe (most Roman cities were located on floodplains), and also in South and Southeast Asia, at sites such as Pattanam (Kerala, southern India), believed to be the ancient port of Muziris (Shajan et al. 2004; Cherian et al. 2009; Selvakumar et al. 2009). Rivers are also sought-after locations for settlement and trade as the estuarine areas encompass many different littoral environments and represent one end of a riparian corridor that might penetrate many hundreds (or seasonally thousands) of kilometres inland, connecting markets located up the navigable river. Propitious locations are often available at the mouth of a creek sheltering the harbour from the winds and sea currents, and allowing for an extensive roadstead without the need for significant harbour installations. Saltwater fish are also readily available and the connection with the hinterland is more accessible. It should be noted, however, that sailing in the tidal reaches presents a number of unique navigational challenges, and river estuaries could pose great hazards for sailing (Strabo, *Geogr.* 1.3, 6–7; Seneca, *Nat. Quest.* 6.26,1; Pausanias, *Perieg.* 5.5, 7; 7.2, 11). Shifting sandbars, shaped in or near estuaries due to alluvial sedimentation debouching from the river and exposed at low tide, would create obstacles for incoming and outgoing shipping and affect siltation. These could be increased by strong onshore and offshore winds, vortices or swell, creating additional hazards for those attempting to enter a river harbour.

6.6 River and Creek Access at Berenike

The location of Berenike on the semi-arid Red Sea coastline with only ephemeral wadi-systems means that the site scores only 2 in this category. Whilst there are no perennial rivers in the region, the wadis replenished the water table, providing sources of fresh water from deep wells. Fresh water output to the sea, generated by precipitation in the mountains, created a break in the coral reef range that in turn provided access to the lagoon. Wadis served as corridors of connectivity between the people of the Nile Valley and those of the Red Sea—usually separated by the vast and inhospitable Eastern Desert.

7 Resource Parameters

Natural resources are understood here to be materials and components that are derived from the regional or local environment, such as water from aquifers or rivers, stone from quarries, and wood from forests or plantations. These natural assets can be divided into biotic resources, such as animals and animal products, plants and plant products, including timber for shipbuilding and building construction, as well as fossil fuels such as coal, and inorganic materials such as metal ores, rocks or sediments. They can be used as raw materials or as trade commodities. This section explores the major Resource parameters, including availability, quantity and quality of fresh water, marine and terrestrial food products, building materials, pasture and arable land, and evaluates their attractiveness in the Red Sea Classical ports. The accessibility of raw materials that represent trade commodities, including good quality wood, rock (e.g., marble), metal ore and luxury goods such as spices, frankincense, tortoise shell, emeralds and topaz are also taken into consideration. Ancient prospectors would likely have assessed the availability and proximity of their sources as well as the effort required to procure and safeguard them, their value on the open market, and 'portability' (i.e., different methods were required and different costs incurred to transport elephants or marble compared to spices or tortoise shell).

7.1 Availability and Proximity of Fresh Water

This includes fresh water availability, accessibility, proximity, quantity and quality. Access to potable water is arguably the most important consideration when prospecting for a new port city, although we might be surprised how counter-intuitive the siting of ports (and presumably other facilities) is with respect to this resource when political machinations have to be taken into account. Where fresh water is not easily available, it immediately becomes apparent how difficult and expensive it is to purify and transport it (either through 'manual or animal carriage' or via aqueducts or pipelines). Naturally, water is impossible to compromise on and essential to food production, economic development, and life itself, and access to it would have been a major factor of attractiveness of any given location to ancient settlers. Even with limited bathing and cooking, a human being comfortably needs about 15–20 L of water a day during a hot summer, especially in the desert, and a little less in a winter (Zitterkopf and Sidebotham 1989, p 164; Sidebotham 2011, p 89). Although slightly brackish water can be imbibed if necessary—potentially leading to a water hierarchy in which the sweetest water is consumed by the higher ranks or used to re-victual the ships, for example—it was nonetheless a top priority in any administration agenda to access, acquire, store, protect and distribute sufficient quantities of clean, fresh water (e.g., Sidebotham 2011, p 87).

The growth of ports imposed responsibility not only for supplying water to the town but also for both incoming and outgoing ships and inland merchants (caravans) (e.g., Hansen 1983; Crouch 1993; Angelakis et al. 2012). Water was required not only to re-stock ship supplies, but also for drinking, bathing and laundry whilst in port. Estimates of water requirements for ship crews have recently been re-evaluated in light of a series of experiments and sea trials with a requirement of around 8 L per person per day (Morrison and Coates 1986; Shaw 1993; Hattendorf and Unger 2003).[6] Moreover, both incoming and outgoing caravans would have to take water for themselves as well as their pack animals. During summer months, and with a heavy load, a donkey requires around 10 L per day, a camel about 20 L per day (Schmidt-Nielsen 1964, pp 85–92, 1965),[7] similar to a horse. These animals would have to water well before, and after, a 2-week (or longer) journey across the deserts surrounding the Red Sea rim. Whilst desert animals could go for long periods without water, this was only possible in cooler seasons and with light loads. In all, it seems that supplying ships, crews and terrestrial merchants would account for an additional two or even three times the ordinary supply of the port's permanent population during the trading season. An estimate like this would have to have been kept in mind by the port prospectors when setting up a port in the area with scarce water supply.

7.2 Berenike's Fresh Water Access

The freshwater issue at Berenike was clearly significant. Not only was the site without a direct water source, but a number of wells, or *hydreumata*, located at the foothills of the mountains that were used, were quite distant, approximately 7.5 km away. A very rough estimate, assuming that 100 large ships visited Berenike throughout the seasons (Kotarba-Morley 2018) with crews of 6–10, with each crew member requiring 6.5–8 L of freshwater a day (average between the amount recorded for quarry workers and a galley crew in an Olympias trial) for a period of 2 weeks (approximate time required for unloading the cargo and re-provisioning) would amount to 150,000 L (to sustain crews visiting a port per season only). Additionally, ships would have to re-supply with water to last through their journey (depending on the direction, sometimes for the whole length of it). A rough calculation shows that a similar amount of water to that needed by ship crews would be required by a local population of 2,000 people (Steven Sidebotham, pers. comm. 2011) with each person consuming a ration of over 6.5 L (1 *keramion*) of freshwater per day. Similar amounts of water required for both incoming and outgoing caravans should also be added to this equation.

Berenike scores only 1 in this category, because the distance and effort of supplying large qualities of water to town—potentially up to 35,000 L a day[8] in the peak season—was a colossal undertaking and one that would require a constant flow of water ferried on the backs of pack animals such as donkeys. This would then need to be stored (although no cisterns have been found in Berenike to date) and distributed around town and into the harbour. It perhaps reveals the strategic importance of Berenike as a port when we consider that it is most likely that it did not have a direct fresh water source on site and had to rely on the transport of water from around 7–8 km away.

7.3 Availability of Food Resources

The food resources category includes here marine and terrestrial procurement (hunting and gathering), with the demand for marine species and game, availability of marine

[6]The Olympias and other trials have shown, for example, that a member of a galley crew would need a minimum of 8 L per day (approximately half of the 'comfortable' amount), including their main meal (e.g., Morrison and Coates 1986; Shaw 1993; Hattendorf and Unger 2003), substantially more than earlier estimates as low as 2.25 L (Sleeswyk and Meijer 1998). However, these approximations are for oarsmen who would have needed to drink more water than a regular sailor.

[7]However, a large and thirsty camel, which did not drink for an extended period of time (i.e., travelling in the desert) can drink up to 200 L a day (pers. observ.).

[8]Calculated using rough estimates of water use for the local population (of approximately 2,000 inhabitants consuming a minimum of 6.5 L a day and comfortably 15–20 L), moored vessel crews for a period of approximately 1 month at a time (approximately 100 ships × 6–10 crew plus passengers), ships re-supplying for a lengthy journey to the Indian Ocean and the incoming, ongoing and resting caravans (humans and pack animals).

resources and game, available procurement technology, and the possibility of trade in seafood, fish and terrestrial animal products (if in surplus). The availability of marine resources, access to terrestrial game, and the availability of food that can be foraged must have hugely influenced the potential choice of a port location. Fishing and procurement of other marine resources (e.g., fish, molluscs, shellfish, seaweed, coral, marine mammals, birds, reptiles) provided a relatively comfortable subsistence strategy for coastal populations and people who worked on the sea. Seafood represented not only easily harvestable food, but also a trade commodity for local populations, especially in some of the arid regions of the Red Sea or North Africa (e.g., Wilson 2002). The Red Sea is particularly renowned for the diversity of its coral reefs and its fish resources (Morgan 2004), and much is known about the available fishing techniques in antiquity (Thomas 2010). While it is not one of the 'big game' fishing zones, the reef environments are active perennially, providing a wealth of resources, making them more attractive to human exploitation than seasonally propitious basins (e.g., Mannino and Thomas 2002; Dulvy and Polunin 2004). Most of the Red Sea coasts are fringed with 1–3 km wide reefs, with abundant fish and marine mammals, adjacent to an outer barrier of several hundred metres depth, allowing for the fishing of deep marine species. Notably, recent zooarchaeological data confirms consumption of large quantities of fish (and even shark) endemic to Red Sea (only one fish bone belonging to the Nile riverine fish species) in the Eastern Desert forts such as *Mons Claudianus* and *Mons Porphyrites*, located 2-days camel ride away from the sea (Hamilton-Dyer 1990, pp 76–77, 2001; Marzano 2013). It is worth mentioning that the collection and 'cultivation' of seaweed and algae could have also been an important part of a subsistence strategy as well as a tradable commodity (e.g., Woodward 1951; Fleurence 1999) although not very likely to be performed along the Red Sea.

Hunting wild game would have been an important means of supporting or supplementing the seafood diet of the Red Sea inhabitants. Eastern Desert-dwelling animals such as the Dorcas gazelle, the fennec (a small, desert-dwelling fox), the Nubian ibex, the Egyptian/Cape hare, and two kinds of jerboa (a mouse-like rodent with long hind legs for jumping), as well as feral camels and goats could have been hunted for food, with the diet possibly supplemented by smaller animals such as turtles, frogs, lizards and birds (Hughes 2014, p 93) and maybe even by beekeeping (Crane 1999, pp 203–211). Hunting with dogs was particularly popular, and indeed there were specially bred animals that were regularly exported to the Empire—maybe even ones that have been discovered in the unusual animal cemetery at Berenike (Osypińska 2016).

7.4 Availability of Food Resources Around Berenike

Given the coastal location, populations of port cities would undoubtedly have relied predominantly on fish for their sustenance. However, despite the propensity for a seafood-based diet (e.g., in Berenike an estimated 30% of the archaeozoological assemblage is fish bones—Marta Osypińska, pers. comm. 2013), cultivated plants and domestic animals would have also constituted part of the diet (Cappers 2006; Zieliński 2011), as well as supplying materials for clothing and other secondary products. The availability of different strands of subsistence strategies means that Berenike scores 4 in the 'Food' category. Plentiful fish, attested by large quantities of fish bones in the archaeological record, especially in the 'port area', show the high level of exploitation of marine resources. However, it is possible that many of them could be just natural waste making it unambiguous whether all of the fish bone material was a part of post-consumption remains. On another hand, the rate of recovery of small fishbones is always a widely discussed issue even on excavations that employ strict sieving. Additionally, some evidence for hunting for game in the mountains is available, with domestic animal meat—such as of donkey, pig or a camel—also being often utilised for food. Bird and rodent bones are also present in the archaeozoological record, however, mostly in the trash dump deposits suggesting an unlikely food source (Osypińska 2011, 2017). Unusually for this period, carefully interred dog burials found in Berenike (e.g., in trench BE10-63/65) might indicate an individual's attachment to their hunting dogs (Osypińska 2016).

7.5 Agriculture—Cultivable Lands and Pastures

The agriculture aspects include here the quantity and distribution of cultivable land, crop production, animal power and transport, and quality, quantity, sustainability and seasonality of pastures. Ancient assessments of agricultural value must have considered the availability of cultivable land and the fertility of soils. Although access to local supplies would have been an important consideration, the ability for the port to at least partly sustain its population, or even create an agricultural surplus in order to feed visiting ship crews and land-based merchants, or for trade, is likely to have been important in the creation of a new harbour town. Port city populations would have also been dependent to some extent upon animals domesticated for power (plough oxen, pack animals such as donkeys, camels and horses), meat, and by-products (hides, dairy). Given the coastal locations of new ports it was likely that the need for primary meat products would have been offset by the use of marine resources, such as was the case at Berenike. However, if secondary products such as dairy and hides were sought then they may have required pastureland to allow grazing of fodder, pack and work animals.

Although many attempts have been made to quantify the agrarian economy in the Ptolemaic and Roman Empires (cf., Kron 2012; Bowman and Wilson 2013), scholars admit that the results have been largely inconclusive (cf., Bowman 2013). At the peak of its development (Principate, 27 BCE–284 CE), the Roman agrarian system was diversified and well-developed, not only benefiting from a prosperous network of local and urban markets, and "the wide-spread use of convertible husbandry, crop rotations, heavy manuring, drainage and irrigation, pruning and grafting, and improved fodder crops producing remarkably high yields and large livestock", but also the general security and excellent overland and maritime transport infrastructure (Kron 2012). This implies that good quality agricultural products could have been delivered to port cities such as Berenike, and technological advances in farming would have allowed for more efficient use of available soils and local microclimates making a development of small-scale agriculture or horticulture viable.

The significance of particular parameters for sites located in specific climatic and environmental zones is worth a comment. There is evidence from Myos Hormos, for example, that caravans would have supplied hay brought all the way from the Nile Valley for animals (Van der Veen 2011). This must have been a complex logistical issue and implies a lack of viable pasturelands. Even though some grasslands occur in warmer climates (30 °C) and with lower rainfalls (around 500–600 mm) (Woodward et al. 2004), with a lack of good pastureland directly adjacent to the sites and suitable for grazing stocks of animals, the existence of shrubs or small grasslands on the coastal plain or in the wadis could be seen as an added benefit. The most popular desert-crossing pack animal in the Eastern Desert and along the Red Sea coasts, the camel, requires approximately 10–20 kg of fresh food a day (or 5–10 kg of dry fodder) depending upon the size of the animal, but up to 30–50 kg when working. As camels eat mostly grasses, herbs, bushes, trees such as acacia, and salt-tolerant plants they are well-adapted to travelling around the Red Sea region. It should be noted, nevertheless, that a density of 2–3 kg of fresh food per hectare is considered favourable for camel grazing, whilst 1–1.5 kg as poor (Sultan al-Dowaish, pers. comm. 2009). Additionally, some animals (such as pigs that were kept on sites in the hinterland of the Eastern Desert) could have been partially fed on human waste, this solving two problems with one solution.

7.6 Agriculture or Horticulture at Berenike

In terms of this PoA, Berenike scores a very low 1, chiefly because of the absence of sustainable agricultural land or suitable pastureland. It is possible that brackish groundwater could sustain some types of salt-tolerant pastureland, horticulture or even light agriculture or oasis-like plantation (of, e.g., date palms), but to date there is no unequivocal evidence for any. Macrobotanical evidence appears to show, however, that there could have been small-scale farming at Berenike (Jarosław Zieliński, pers. comm. 2011). Evidence for potential pastures, dominated by C_4 grasses, have been identified by Zieliński who claims that, if indeed in existence, it is possible that they could only be suitable for camels and donkeys.

7.7 Availability of Local Building Materials

Building materials under consideration are represented by the biotic (e.g., forests) and abiotic (e.g., building materials, ore) resources. The availability of construction and repair materials (for boats, infrastructure and buildings) would have been important for setting up and maintaining a port town. The building materials would not have been used exclusively for port infrastructure, but also for the various storage, industrial and domestic buildings and installations. Whilst materials, in some cases, could be imported from 'outside' areas (such as *pozzolana* imported from up to 2,000 km away within the Mediterranean), it is likely that the availability of decent basic building materials would have been an attractive asset in the siting of a new port. A wide variety of materials could be used for building, including easily available sediments such as clay and sand, which could be fired or sun-dried to produce mud bricks, lime plaster, rubble, cement, gravel, as well as quarried rocks of various types ranging from perishable gypsum anhydrite, poor quality limestone or coral, to hard and extremely durable granites, marbles or basalts. Access to quarries, from which ships can transport acquired material to other locations (Russell 2012), could have been one of the reasons that prompted the prospection for ports, or could have at least been a serious consideration when prospecting for new ports. The development of roads, aqueducts and other types of infrastructure, the foundations of new ports and cities, and the expansion of old ones, also necessitated extensive quarrying (Russell 2013, p 16).

Organic materials could also have been widely used for the construction of houses, administrative buildings, fences and enclosures, as well as jetties and wharfs. These could include, depending on environmental conditions, cob (a mixture of clay and grass/hay), wood, reed mats, thatch, sod/peat, bitumen, coral, animal skins, leather, bone, and textile. An interesting example of the use of amphorae as the base for a wharf is also known from a couple of Roman port sites in Egypt including Myos Hormos on the Red Sea (Peacock and Blue 2006, pp 68–74) and Naukratis—a river port in the delta of the Nile next to its Kanopic branch (Thomas et al. 2016).

7.8 Building Materials Around Berenike

Three different types of stone were sourced in the area around Berenike: (i) ordinary stone for building and infrastructure such as gypsum anhydrite as well as hard coral, (ii) high-quality stones such as granite, marble, and basalt used for decorative purposes (Crouch 2004; Russell 2013) and exported to the Mediterranean via the Nile Valley, and (iii) gemstones such as olivine, topaz or emerald, also sourced for export. Since the landscapes around Berenike support an abundance of gypsum anhydrite and coral that can be used as low-quality building materials these were widely utilised on site for domestic edifices but also for more prestigious buildings such as temples. Large quantities of high-quality stone are also available from the mines exploited in Roman times in the area surrounding Berenike; however, there is no evidence, to date, for their use on site and it is envisaged that most of quarried high-quality material was exported to the Mediterranean. Berenike would therefore score 4 in this category.

7.9 Availability of Wood

The level of availability of forests and quality of wood for ship repairs, ship building and other production purposes would have been a very significant aspect in the siting of an ancient port. A growing population and rapid economic development would have resulted in an increased demand for fuel to produce light and heat in both domestic and industrial contexts, as well as in public buildings (such as baths) (Veal 2013). In an ancient urban context wood was not only used as fuel but also for industrial processes such as mining, smelting, and pottery making, as well as for military and domestic purposes. Most importantly for a port city, though, it was used for shipbuilding and for repair material.

Wood such as Indian teak, African blackwood or Lebanese cedar was ideal for ship construction, but many other species could have been used for a diverse range of repairs. Whilst top-quality wood was often exported over large distances and therefore was sometimes readily available in transhipping ports such as Berenike, access to moderate quality wood for repairs would have been an important consideration. Typical features of the Red Sea coasts are lagoons and sheltered bays (*mersas*) fringed by mangrove trees (Khalil 1994, p 126), which could sometimes be good locations for harbours (if not too shallow). An ethnohistorical study undertaken by Pierre Schneider (2017) on the coasts of the Red Sea confirms that mangrove wood could have been used as a freely available ship repair material. Theophrastus (*Hist. Plant.* 4.7.1–2) also mentioned using mangrove wood for building boats, and specifically *Avicennia maritima* or *Avicennia officinalis* (white mangrove) to repair them. Dionisius Agius (pers. comm. 2012) also remarks about sources from Djibouti corroborating the use of mangrove for shipbuilding, most likely for repairs. This information shows the need to re-evaluate our understanding of the availability of and the demand for wood on the Red Sea and other deforested coasts.

7.10 Wood at Berenike

In terms of wood, only acacia and mangrove are locally available. Other wood species, recorded in the archaeological material, such as cedar, pine or oak were imported via the caravan route from the Nile valley and the Mediterranean. Those originating in Africa or India such as blackwood and teakwood, were imported through the maritime route. The availability and use of low-quality building materials local to Berenike and the great scarcity of wood scores the site only 1 in this category.

7.11 Accessibility of Raw Materials and Trade Commodities

The category of raw materials and trade commodities needs to be considered in light of their availability, portability, quality and types. As already discussed above the accessibility of natural resources is likely to be a key factor in positioning any production centre that could have been associated with a port of trade. Apart from wood and building materials (see categories above), an economic boom during the early Principate and increased demand for tools, weapons, ornaments and jewellery (Craddock 1995) also amplified the need by the Roman Empire, and ore-deprived Central Italy, for heavy metals (gold, silver, copper, tin, lead, zinc, iron, mercury, arsenic, and antimony; Healy 1978). As such, heavy minerals mined on the fringes of the Empire were usually carried by sea—departing and approaching via ports of trade. The importance of metal ores to the State was so huge that the army oversaw mining undertaken by locals and convicts condemned to the *metalla*,[9] some of them located in Roman Egypt. Some scholars even suggest an existence of a centralised bureau in Rome that was responsible for mining (Hirt 2010). Specific sediments such as clays were also required for pottery manufacture, and sand, with a high quartz (silica) content, was needed for glass production. Other trade goods such as gems and turtle shell, as well as luxury organic materials were available along the Red Sea coast and exploited by coastal port-dwelling communities for exchange with visiting merchant vessels.

7.12 Raw Materials and Trade Goods at and Around Berenike

Berenike scores highly (4) in this category as it is located in an area with plentiful raw materials. Gold, emeralds, olivine, topaz, and basalt, exploited in the area in the Ptolemaic and Roman times, were excellent trade commodities representing valuable assets when located nearby, boosting the economy of a port, not by direct trade but by connection with the quarries. Although sand is certainly plentiful at Berenike and there is also some clay available in the area, there have been no production centres of glass or pottery production identified to date, and the high quantity of decent wood required for such manufacture would mean that it was not viable, on a large scale, at the site where wood resources were scarce (see previous parameter). Other raw materials such as metal ore, as well as non-essential foodstuffs and goods for elite consumers including spices, frankincense, tortoise shell, or ivory, were not immediately available at Berenike but were traded from this port. For example, the Alexandrian Tariff mentions 54 commodities (including 20 plant products) (Miller 1969, pp 278–280) subjected to import tax in Alexandria (Cappers 2006, p 3), whilst *Periplus Maris Erythraei* mentions all together 34 plant products, 18 of which were traded from Berenike (Casson 1989; Cappers 2006). Such luxury items could easily be obtained in some areas of the African Red Sea coast and therefore convenient access to their sources and a method of re-distributing them could have been a valid consideration when setting up a port such as Berenike.

8 Socio-Political-Economic Parameters

The social, political and economic parameters presented in this section, in contrast with the environmental (sea, land, resources) parameters set out in the sections above, relate to human decision-making and are used as a counter-balance for the factors such as the availability of raw materials and trade goods as well as the attractiveness of the local sea- and land-scape setting. Admittedly, these variables are not only 'archaeologically invisible' but also very difficult to quantify and if not approached rigorously, with enough historical data available to seek them, could create a circular argument. Since they could have also changed throughout the lifetime of the port, or have been quite short-lived, they should be used with the highest level of caution and with appropriate levels of analyses of historical sources. However, they should not be entirely excluded from consideration since not only would they affect the choice of location made by those who were setting up the new port, but they also mattered to those who were to inhabit and use the facility. Demand aside, strategic location, ease of access to the hinterland, and political stability, including the safety of the inhabitants and visiting ships' crews and cargo, would play an important role and would ultimately determine the popularity of the ports in respect of the expectations of sailors and merchants. Nevertheless, it is worth noting at this point that many or most of the parameters and criteria can be overridden if economic circumstances increase potential profit beyond a certain threshold.

8.1 Demand: Scale of Demand

Probably the most important of all parameters, and one that was in the hands of the higher political powers rather than the prospectors and engineers, was the actual demand for the new port (including both market demand for a particular commodity or a range of them, as well as a potential local or regional need for a new port facility). Although this parameter is difficult to assess quantitatively and highly contextual and historically contingent it would have only been after the demand for a new port was identified, communicated and made official that the environmental variables would then come into play to decide its exact location (although the location of the new port would have been constrained to some greater or lesser extent at this time).

The 'Linder Hypothesis' regarding international trade patterns (Linder 1961) is very applicable in describing the demand network created between the Roman Empire and its eastern counterparts. The increasingly complex and aspirational societies, kingdoms and empires of India, Southeast Asia, East Asia and the Arabian Peninsula (Casson 1974; Manguin et al. 2011; Ray and Salles 2012) increased the demand for trade goods. Empires and states such as the Axumite Kingdom in Ethiopia, the Himyarite, Qataban, Sabaean and Ma'in Kingdoms of Southern Arabia, the Parthian and Sassanian Kingdoms of Iran, along with the empires of the East including the Ruhuna Kingdom in Sri Lanka, and the areas ruled by, amongst others, the

[9] *Metalla* were the imperially owned mining and quarrying districts of the Roman and Byzantine Empires (Friedman 2008).

Sātavāhana, Chola and Guptas dynasties in India, were highly developed civilisations that had similar luxurious demands to those expressed by Roman élites. For example, in exchange for gold *denarii* and wine Indians sent spices to the Roman markets. Similarly, the Funan in Vietnam and Cambodia, the Champa kingdoms in Vietnam, the Langasuka and Pan Pan kingdoms in Malaya, the Arakan and Pyu monarchies in Myanmar/Burma, along with the empires of Han, Three Kingdoms, Jin and Northern and Southern dynasties in China had very advanced societies in a similar need of luxury items. They were interested in taking part in the emerging global economy and pursuing its benefits.

Whilst in some contradiction to Wallerstein's World-system theory (Hopkins and Wallerstein 1982; Wallerstein 2009), Linder proposes (see discussions in: Frankel 1997, p 60, pp 133–134; Choi 2002) that the closer the preference structure between two countries, the larger the trade volume becomes (whilst creating a specialisation within the production of differentiated goods between the two nations). For example, the growing demand in Rome for luxury items was similar to that observed in Chinese markets where the demand for goods from the borderlands of South Asia (such as lapis lazuli from northeastern Afghanistan) and the western regions (such as jade from Khotam) was similarly high (Srinivasan 2007, p 91). Analogously, the notion of 'opposite shores' developed through observation of the popularity of ports situated on the western shores of 19th-century Britain during the increased Atlantic trade (e.g. Harvie 2008) might be successfully applied to the Indo-Roman trade, with the Mediterranean representing Western Europe and the Indian Ocean region, especially India, representing North America and the Red Sea the transitional area in-between—that is, the Atlantic. It is an obvious point that the ends of the supply chain have a major impact on the development and growth of particular modes of transhipment—that is, ports during periods of intensified trade activity.

8.2 Demand for Berenike

The highest score of 5 in this category for Berenike may indicate that 'Demand' played a key role in siting of the site. Both the Ptolemies and the Romans required a port located as far south on the Egyptian Red Sea coast as possible in order to connect with the southernmost edge of the Empire, and within easy reach of the monsoon winds. Whilst the initial demand differed, with Ptolemies interested mainly in the transportation of African elephants and Romans concerned with the efficient transhipment of Indian Ocean goods, the chosen location for a port was required to, similarly, remain easily connected with the hinterland via wadi corridors and within easy distance of one of the large Nile riverine ports and commercial centres.

8.3 Strategic Position

This factor includes the defendability of the site, and security of its connection with the hinterland. The strategic position of the proposed site would have been an important factor in the setting up of a port city or town. Depending on the needs of the port community, the coastal landscape, weather conditions, and relations with local tribes and surrounding political entities would have to be taken into consideration in evaluating the attractiveness of a particular location. The key factors that would have been taken into consideration by ancient port builders when thinking about good strategic positions would have therefore included:

i. Safety of the position in connection to the hinterland routes.
ii. Provision of safety for the inhabitants in case of an attack:
    - Location on a hill, uplifted terrace or other easily defendable elevated position (good visibility and good observation points).
    - Readily defended approach from the sea, possibly blocked by harbour chains or watched over by a fortified building.

The contact with the hinterland would have been a major concern when setting out a maritime trading emporium. The supply and distribution of goods by land (or river) would have to occur by pack animal, hauled by slaves (or by river-boat). This would often have to be supplemented with supplies of water and food for both human and animal consumption, as well as building materials and ship parts. The quality, (seasonal) accessibility, and safety of roads and rivers toward the hinterland, including rest areas where porters and pack animals could rest and water, would have been critical for merchants distributing goods inland. The assurance of safe passage inland would have determined the price of goods being distributed and therefore might have swayed the choice of port from which they would set sail. It was therefore in the interest of the port community, or even, at an earlier stage, of port planners, to be within easy distance of a good road inland and a corridor to connect with the hinterland. Such a connection would allow for the procurement of goods and supplies from inland, and for the distribution of sea-trade commodities enabling merchants to charge higher prices on their goods to the middleman.

The Romans were great tacticians who understood the importance of strategic positions in the landscape and are known to have taken advantage of them wherever possible, usually exploiting higher ground and natural defences (e.g., promontories and headlands such as, for example, for setting up a Roman lighthouse of the *Portus Dubris* at Dover in England, in Alexandria on the Mediterranean or in Berenike on the Red Sea). Such strategies were used for creating an advantage over a potential enemy and for defending their interests in marginal areas of the Empire (e.g., Luttwak 1976, pp 173–180). Some ports and landing places on the Red Sea were so strategically positioned that they also served, as in the Mediterranean, as naval bases (such as Jizan on the Farasan Islands; Phillips et al. 2004, pp 244–245; Adams 2007, p 35). These were usually located in a separate harbour basin from the commercial harbour (e.g., Carthage) but in smaller ports, such as Berenike, they might have only had a special delineated (or not) area.

8.4 Berenike's Strategic Position

The recently discovered Ptolemaic fort and substantial defensive city walls indicate conscious improvements in the defensibility of the site of Berenike in its early stages of existence. Berenike is naturally strategically located on top of an uplifted promontory with good all-round views, including the entire Foul Bay and the entrance to a number of surrounding wadis, and this position would have enabled the sending of signals to forts and stations located in the foothills of the mountains. A score of 4 is achieved by Berenike not only due to its defensive position and viewshed location, but also because it benefitted from good connections with the hinterland via a number of caravan routes supported by well-positioned and well-defendable stations, forts and *praesidia* (Bagnall et al. 2001; Sidebotham and Zitterkopf 1995).

8.5 Political Stability and General Safety

Coastal sites are threatened from both the sea and hinterland, as Polanyi (1963, pp 30–45) notes in his seminal *Ports of Trade in Early Societies*. These nodes of activity are, accordingly to his 'port of trade' concept, the neutral zones, administered by the powerful and centralised state in the hinterland and potentially located within hostile surroundings. They are situated outside a riparian village or coastal city, where foreign and native merchants interacted and the 'royal' officers dictated prices and maintained the neutrality zone. The political stability of a region must have therefore played a crucial role in deciding which port to trade in. This included factors such as the political situation in a region, as well as a potential for civil strife, war and predictability of conflict (internal due to, for example, high taxes, social inequality or through slave revolt or external aggression). An unstable political situation would be a serious impediment to trade, potentially driving up prices and creating an unreliable trade situation. Conversely, a stable situation and safe harbour would have allowed for the development of unhindered trade. Similarly to the case of 21st-century Horn of Africa, the presence of pirates, competing tribes, marauders and other outlaws would have made some areas of the coast and inland routes difficult to pass or stopover at, affecting sea traffic, trade prices and the general wellbeing of sea merchants and their agents.

8.6 Political Stability Around Berenike

Ten *praesidia*, part of a Roman early-warning system, were situated within a distance of 7–35 km from Berenike (Bagnall et al. 2001; Sidebotham and Zitterkopf 1995). They not only supplied the town with water but also protected valuable trade routes through the wadis, with many of them housing garrisons consisting of soldiers from different regions of the Empire (i.e., Roman auxiliary troops). For example, inscriptions and votive dedications provide evidence for Palmyrene archers patrolling the Koptos–Berenike desert road in the time of Marcus Aurelius. Although 'unruly' nomadic tribes populating the Eastern Desert could have occasionally caused strife, the well-protected caravan routes, with their *praesidia* and forts, meant that Berenike scores 4 in this category.

8.7 Port Infrastructure and Facilities

The attractiveness of any given port to the merchants includes the accessibility of human expertise, administrative facilities, appropriate port infrastructure and hospitality services. It is also based on existing networks of kinship and faith as well as practicalities and journey times (e.g., Seland 2013), and is therefore often difficult to assess quantitatively. Meeting the expectations of the sailors and merchants who would use the facility would also represent an important part of the planning of a port. Information about the particular types of human expertise in the area, such as specialist workshops, talented craftsman or artists, or a renowned and capable shipwright, a good brothel (see the Coptos Tariff for a tax paid for the transport of prostitutes from the Nile valley to Red Sea ports; Young 2001, p 44), bar, theatre, fighting arena or other form of entertainment, could potentially result in the modification of a route as the crew of the vessel chose a more attractive port for their needs, if one was available. Therefore, access to these 'human resources' and their 'willingness' to settle in a new port town would be a crucial consideration when assessing the attractiveness of a port. Whilst such facilities, as mentioned above, were unavailable prior to the setting of the harbour, the availability of human resources, the capacity of the port to develop such amenities, the availability of particular types of goods and a particular type of landscape and environmental setting, as well as the location on a particular node of a trade network (e.g., Leidwanger et al. 2014; Seland 2016) with, for example, a pre-set taxes zone, would have been a crucial consideration.

Another influential factor would be the potential risk of queues to enter the harbour, and the level of taxes and custom duties,[10] which could have convinced the merchants to travel to other locations to trade. Customs duties were high in some areas of the ancient world and in many cases predestined the success or otherwise of a particular port, therefore this could, in the future, justify making custom duties and taxes a parameter on their own. For example, the *tetarte*, as mentioned by the Muziris Papyrus (Rathbone 2001), was a 25% customs fee on eastern trade imports from Red Sea harbours such as Myos Hormos or Berenike, levied in Alexandria that had to be paid on all commodities (Temin 2004; Fitzpatrick 2011; Wilson 2015). Nabatean ports such as Leuke Kome would have had their own rates of taxes. Usually the imported goods would be raw materials that were then sold in the cities like Alexandria itself (Cottier 2010), so subsequent taxes of a value close to 2–2.5% would have been less significant. Only a limited number of bureaucratic centres had the power to levy taxes and so the queues could be long and frustrating, potentially lasting for days whilst tax officers were highly officious (Cottier 2005). Whilst a similar level of tax would be levied in all Red Sea ports under Roman political rule, ports located outside the Empire could have competed for popularity by adjusting their customs duties. Under such circumstances the attractiveness of the port would depend on its level of connectedness, its size (and ability to accommodate vessels waiting to pay the levy on the roadstead) and how quickly merchants would be able to move the delivered goods making highest possible profit.

8.8 Berenike's Port Infrastructure and Facilities

The closest major port from Berenike was Myos Hormos but a number of other smaller anchorages were available en route. Although some of them, such as Nechesia, Ophiodes, Bathus Profondus Portus, and Dioscuror, are mentioned in texts as receiving moderate traffic, a lack of archaeological data regarding their location and character does not allow us to assess how competitive they could have been in comparison with Berenike or Myos Hormos. Berenike attains 4 in this category based on the evidence from ostraca and papyri (Bagnall et al. 2000, 2005) and ancient texts such as *Periplus Maris Erythraei* attesting to the importation of variety of foodstuffs and wine to the town (Cappers 2006). There are both references to prostitutes and brothels (i.e., Muziris Papyrus), as well as temples and religious and cultic worship, serving a diverse array of creeds and interests, from explorer and merchant to pious traveller. Based on this written evidence it seems clear that the city provided good quality services to merchants and sailors, comparably much better than anything they could obtain in the immediate area.

9 Summary of Berenike's 'Parameters of Attractiveness'

The Parameters of Attractiveness (PoA) scored by Berenike for each of the categories can be seen in Table 2 and we can draw a number of conclusions from this semi-quantitative dataset. As can be seen from the tables the Sea parameters that score 0.72 (Moderately Attractive) and Socio-political-economic parameters that score a very high 0.85 (Very Attractive) put the site in the Moderately Attractive or Very Attractive. At the same time, values for both Land parameters and Resource parameters are low at 0.53 and 0.50 respectively and located at the border of Poorly to Reasonably Attractive and Moderately Attractive. Berenike scored a total of 2.60 out of 4 (0.65) overall making it a practical and a rather attractive location.

As we can see above, these data suggest that the decision behind the siting of Berenike was not entirely due to environmental parameters. Whilst Berenike scores highly in terms of the excellent marine and political context of the site, it falls well short in terms of the physical landscape setting, and is especially poor with regard to water supply. This seems counter-intuitive, although, in light of other very strong parameters (i.e., natural bay, winds, demand), can be explained. Despite this rather obvious hindrance, Berenike flourished in that challenging landscape for almost a millennium. This shows that political will, economic demand and marine setting were often—especially in the Roman Empire—much more influential than fundamental considerations such as the scarcity of fresh water available at the site. The lack of other ports in the region, and indeed the lack of demand for more meant that, due to its specific location, Berenike was able to flourish, with barely any competition, intermittently for 800 years.

So, why was there such a strong political or economic impetus to locate Berenike in this particular location? There must have been a great deal of thought that went into the weighing up of the pros and cons of this location when plans were being made and advisors were petitioning for this old

[10] Compare with establishing Delos as a freeport in 167 BCE to counterbalance Rhodes 2% harbour tax (Duncan-Jones 1990, p 37; Adams 2011). Within a year, trade immediately bypassed Rhodes in favour of Delos (Adams 2001, p 90). Similarly, the growth in importance of the 18th-century CE port of Az-Zubarah in Qatar, which created a duty-free zone led to the quick decline of the otherwise popular and successful port of Muharraq in Bahrain (Rob Carter, pers. comm. 2010).

Ptolemaic port to be used. Strategically it must have filled a gap along the young but flourishing trade network, but could there have been other locations nearby? Whilst the site of the modern naval base of Baranis seems extremely propitious for locating a maritime port the inaccessibility of the site makes it difficult to assess whether there are any ancient remains and why the southern location, a potentially less attractive one, has been chosen instead.

The prevailing wind patterns reaching Berenike from the south, along with the natural harbour and lagoon system available at the current site must have nevertheless represented a landscape configuration so suitable that it could not be ignored. The sheltered location just to the south of the large peninsula of Ras Benas, with the wadi corridors (logistical transport super-highways) connecting the site with the hinterland, must have been crucial factors in the decision-making process. Although there are other sheltered bays along this stretch of coast within 100 km north and south of the site, Berenike's Parameters of Attractiveness—potentially those connected with strategic location or access to hinterland routes (i.e., the presence and proximity of a number of wadi systems)—must have tipped the balance.

We cannot know whether the initial decision to exploit the coast at Berenike in the Hellenistic period was made with haste, but the sudden need for war elephants in the time of Ptolemy's II struggles with the Seleucids may have added an element of urgency to the decision-making process. Later, in the Roman period, situating a port in a 'known' Ptolemaic location may have made it the most obvious strategic choice, since it already had some developed infrastructure to connect it with the Nile and was a 'known' quantity rather than a totally new area to explore. It would also be the fastest port to utilise with the necessity to provide for the steadfast 'picking-up' of the trade networks.

As mentioned above, the ancient port of *Berenike Troglodytica* supported its citizens and transient visitors for almost a millennium. This was primarily due to the urgent demand for commodities feeding the needs of the Empire, enabling its survival for a sustained period of time in this rather hostile, marginal environment. However, as the above study shows, its location was ultimately attractive enough to successfully provide for the port, its inhabitants and its visitors.

10 Conclusions and Moving into the Future

The Berenike case study presented here demonstrates the utility of the Parameters of Attractiveness approach for quantifying and evaluating the many competing factors that were likely to have influenced the choice of ancient port locations. It is envisaged that future work will be carried out to enlarge and enhance this dataset. To this end a number of port sites on the Indian Ocean and its associated basins that the author has worked on, or visited, would be examined in more detail and the 'Parameters' already employed at Berenike will be used to assess different aspects of their 'attractiveness'.

The possibility of performing such analyses on an enlarged dataset, discussed in brief in this paper, would allow for a more objective and hence robust evaluation of these parameters and a better confidence-scoring scheme of each variable with a greatly refined and enhanced system ready to observe patterns on a larger sample size. A more diverse dataset would also help to create a buffer against the misinterpretations of secondary data, particularly when assessing sites that cannot be visited or examined first hand. The enlarged dataset will include 'potential locations' as well as localities which do not host successful ports, or indeed settlements of any kind. Those randomly selected locations on the coast will allow for testing of the robustness of this approach and methodology.

Should larger sample sizes become available (more definite locations within Table 1 will be assessed during the upcoming fieldwork, led by the author and conducted in Saudi Arabia by the University of Wollongong and Saudi Geological Survey team) this simple scheme of semi-quantitative sliding scales would also allow for cross-checking of the viability of the models themselves (and links and relationships between them). An enlarged dataset will also allow for a more objective and hence more robust evaluation of each parameter and for setting up relevant connections between different groups and individual parameters. It would similarly allow for lists of relevant, irrelevant and neutral variables and the assessment of their relationships to be created (e.g., *presence of an active wadi* versus *availability of a good bay with a decent entrance*, but also *presence of an active wadi* versus *dormant wadi; a good bay with a narrow, shallow channel* versus a *good bay with a wide, deep channel*). It is anticipated that the evaluation and scoring system will be greatly refined and enhanced after patterns have been observed on a larger sample size. For example, if 'Demand' is observed repeatedly to be one of the key groups of parameters, it may need to be valued between 1–10, whilst 'Tidal range', which might turn out to not necessarily play such a huge role in the siting of a port, might be valued between 1–3. More advanced normalisation can then be employed along with the introduction of a more robust confidence-scoring system.

The hypothesis that the attractiveness of the sites of harbours on the Red Sea was dependent primarily on landscape and environmental factors, with a strong influence of political will from Ptolemaic rulers in Alexandria, from Rome, Petra or Axum, or other dominating powers, can then be tested more robustly.

Acknowledgements The author would like to thank Dr. Damian Robinson and Prof. Andrew Wilson for unfailing support during this research, and Dr. Lucy Blue and Prof. Chris Gosden for invaluable comments. Thanks are also extended to Prof. Tom Hillard and two other anonymous reviewers for further insights and comments relating to Red Sea ports and ancient Roman port technologies, as well as very valuable comments regarding socio-economic and political parameters, especially in relation to demand and port infrastructure and facilities. This research has been so far carried out with generous support of The Griffith Egyptological Fund, a Polish Centre for Mediterranean Archaeology scholarship and Meyerstein Award from the University of Oxford.The deepest gratitude is also given to Saudi Geological Survey and Dr. Najeeb Rasul for the invitation to the Red Sea workshop in Jeddah in 2016 and SGS support for a 'Landscapes and Environments of late Holocene Red Sea ports of trade: Geomorphological and Geoarchaeological survey' and a planned fieldwork season of the 'Coastal Cultural Landscapes of the Saudi Red Sea', led by the author of this paper.

References

Abdel-Raziq MA, Castel G, Tallet P, Marouard G (2012) The pharaonic site of ayn soukhna in the Gulf of Suez 2001–2009 progress report. In: Tallet P, Mahfouz E-S (eds) The Red Sea in pharaonic times: recent discoveries along the Red Sea coast. Proceedings of the colloquium held in Cairo/Ayn Soukhna, 11–12th January 2009, Cairo. L'Institut Français d'Archéologie Orientale, pp 3–20

Adams C (2001) For good and evil: the impact of taxes on the course of civilization, 2nd edn. The Madison Books, Lanham, Maryland and Oxford

Adams C (2007) Land transport in Roman Egypt: a study of economics and administration in a Roman province. Oxford University Press, Oxford

Adams C (2011) Tax-haven tales: kooks, crooks, and con men in the offshore world. Laissez Faire Books, Baltimore, Maryland

Admirality, Great Britain Hydrographic (1995) Admiralty tide tables: Atlantic and Indian Oceans, vol 2. The Hydrographer of the Navy, Taunton, Somerset

Al-Tarazi EA, Korjenkov AM (2007) Archaeoseismological investigation of the ancient Ayla site in the city of Aqaba. Jordan Nat Hazards 42(1):47–66

Angelakis AN, Mays LW, Koutsoyiannis D, Mamassis N (2012) Evolution of water supply throughout the millennia. IWA Publishing, London

Bagnall RS, Bülow-Jacobsen A, Cuvigny H (2001) Security and water on the eastern desert roads: the prefect Iulius Ursus and the construction of praesidia under Vespasian. J Roman Archaeol 14:325–333

Bagnall RS, Helms C, Verhoogt AMFW (2000) Documents from Berenike: Greek ostraka from the 1996–1998 seasons, vol 1. Fondation Egyptologique Reine Elisabeth, Papyrologica Bruxellensia 31, Bruxelles

Bagnall RS, Helms C, Verhoogt AMFW (2005) Documents from Berenike: texts from the 1999–2001 seasons, vol 2. Association Egyptologique Reine Elisabeth, Papyrologica Bruxellensia, p 33

Bard KA, Fattovich R (2003) Mersa Gawasis: a pharaonic coastal site on the Red Sea. Bull Am Res Center Egypt 184:30–31

Bard KA, Fattovich R, Ward CA (2007) Sea port to Punt: New evidence from Marsa Gawasis, Red Sea (Egypt). In: Starkey JCM, Starkey P, Wilkinson TJ (eds) Natural resources and cultural connections of the Red Sea. Soc Arabian Studies Monographs No. 5. Archaeopress, Oxford pp 143–148

Bellotti P, Calderoni G, Di Rita F, D'Orefice M, D'Amico C, Esu D, Magri D, Preite Martinez M, Tortora P, Valeri P (2011) The Tiber river delta plain (central Italy): coastal evolution and implications for the ancient Ostia Roman settlement. The Holocene 21:1105–1116

Benedetti A (2007) The Littoral: an essential space for humanity. Special Issue: Endangered coastlines. CNRS International magazine 8. Editorial. http://www2.cnrs.fr/en/981.htm. Accessed 6 June 2014

Beresford J (2013) The ancient sailing season. History and archaeology of classical antiquity, vol 351. Brill, Leiden and Boston

Berggren JL, Jones A (2000) Ptolemy's geography: an annotated translation of the theoretical chapters. Princeton University Press

Bernal D, Sáez AM, Montero R, Díaz JJ, Sáez A, Moreno D, Toboso E (2005) Instalaciones fluvio-marítimas de drenaje con ánforas romanas: a propósito del embarcadero flavio del caño de Sancti Petri (San Fernando, Cádiz). Spal 14:179–230

Bini M, Chelli A, Durante AM, Gervasini L, Pappalardo M (2009) Geoarchaeological sea-level proxies from a silted up harbour: a case study of the Roman colony of Luni (northern Tyrrhenian Sea, Italy). Quatern Int 206(1/2):147–157

Blackman DJ (1982) Ancient harbours in the mediterranean. Part 2. Int J Naut Archaeol 11(3):185–211

Blue LK (2011) The Amphora Wharf: the context and construction of Roman amphora installations. In: Peacock DPS, Blue LK (eds) Myos Hormos—Quseir al-Qadim. Roman and Islamic ports on the Red Sea. Finds from excavations 1999–2003, vol 2. Archaeopress, Oxford, pp 35–42

Blue LK, Peacock DPS (2007) The ancient shoreline and maritime landscape. In: Peacock DPS, Blue LK (eds) The ancient Red Sea port of Adulis, Eritrea: results of the Erito-British expedition, 2004–2005. Oxbow Books, Oxford, pp 39–56

Bourdon C (1925) Anciens canaux, anciens sites et ports de Suez. Mémoires de la Société Royale de Géographie D'Égypte, vol 7. Imprimerie de L'Institut Français D'Archéologie Orientale, Cairo

Bowman A (2013) Quantifying Egyptian agriculture. In: Bowman A, Wilson AI (eds) Quantifying the Roman economy: methods and problems. Oxford Studies on the Roman Economy. Oxford University Press, Oxford, pp 177–204

Bowman A, Wilson AI (2013) Quantifying the Roman economy: integration, growth, decline? In: Bowman A, Wilson AI (eds) Quantifying the Roman economy: methods and problems. Oxford Studies on the Roman Economy. Oxford University Press, Oxford, pp 3–84

Brandon CJ, Hohlfelder RL, Oleson JP (2008) The concrete construction of the Roman harbours of Baiae and Portus Iulius, Italy: the ROMACONS 2006 field season. Int J Naut Archaeol 37(2):374–392

Bruyère B (1966) Fouilles de Clysma-Qolzoum (Suez) 1930–1932. Imprimerie de L'Institut Français d'Archéologie Orientale, Cairo

Cappers RTJ (2006) Roman food prints at Berenike. Monograph, vol 55. Cotsen Institute of Archaeology, Los Angeles

Carayon N (2005) Le cothon ou port artificiel creusé. Essai de définition. Méditerranée 104:5–13

Casey S (2011) The wave: in pursuit of the rogues, freaks, and giants of the ocean. Anchor

Casson L (1974) Travel in the ancient world. Hakkert, Amsterdam

Casson L (1989) The Periplus Maris Erythraei: text with introduction, translation, and commentary. Princeton University Press, Princeton

Cherian JP, Prasad GVR, Dutta K, Ray DK, Selvakumar V, Shajan KP (2009) Chronology of Pattanam: a multi-cultural port site on the Malabar coast. Current Sci 97(2):236–240

Choi C (2002) Linder hypothesis revisited. Appl Econ Lett 9(9):601–605

Cooper JP (2009) Egypt's Nile-Red Sea canals: chronology, location, seasonality and function. In: Blue LK, Cooper JP, Thomas R, Whitewright J (eds) Connected hinterlands. In: Proceedings of Red

Sea project IV held at the University of Southampton September 2008. Soc Arabian Studies Monographs, vol 8. Archaeopress, Oxford, pp 195–210

Cooper JP (2014) The Medieval Nile: route, navigation, and landscape in Islamic Egypt. The American University in Cairo Press, 392 p

Cooper JP, Zazzaro C (2012) A stone anchor from the Farasan Islands Saudi Arabia. Int J Naut Archaeol 41(2):407–411

Cottier M (2005) The organisation of customs duties in Ptolemaic and Roman Egypt (circa 332 BC–AD 284). Unpublished PhD thesis, University of Oxford

Cottier M (2010) The customs districts of Roman Egypt. In: Gagos Y (ed) Proceedings twenty-fifth international congress of papyrology, Ann Arbor 2007. American Studies in Papyrology. Scholarly Publishing Office, University of Michigan Press, Ann Arbor, pp 141–148

Couyat J (1910) Ports gréco-romains de la Mer Rouge et grandes routes du Désert Arabique. CRAI 54(6):525–542

Craddock PT (1995) Early metal mining and production. Edinburgh University Press, Edinburgh

Crane EE (1999) The world history of beekeeping and honey hunting. Routledge, New York

Crouch DP (1993) Water management in ancient Greek cities. Oxford University Press, New York

Crouch DP (2004) Geology and settlement: Greco-Roman patterns. Oxford University Press, New York

Curtin PD (1984) Cross cultural trade in world history. Cambridge University Press, New York

Davies S, Morgan E (1995) Red Sea pilot. Imray Laurie Norie & Wilson Ltd, London

de Graauw A (1986) Wave statistics based on ship's observations. Coastal Eng 10:105–118

de Graauw A (2014) Ancient ports and harbours, the catalogue, 4th ed., Port Revel. Potential ancient harbours. Ancientportsantiques.com. https://www.academia.edu/7628768/Potential_Ancient_Harbours. Accessed 4 August 2014

de Romanis F (2009) Patterns of trade in the Red Sea during the age of the Periplus Maris Erythraei. In: Blue LK, Cooper JP, Thomas RI, Whitewright J (eds) Connected hinterlands. In: Proceedings of Red Sea Project IV held at the Unversity of Southampton September 2008. Soc Arabian Studies Monographs. Archaeopress, Oxford, pp 31–36

Dulvy NK, Polunin NV (2004) Using informal knowledge to infer human-induced rarity of a conspicuous reef fish. Animal Conserv 7 (4):365–374

Duncan-Jones RD (1990) Structure and scale in the Roman economy. Cambridge University Press, Cambridge

Edwards AMC, Head SM (1987) Red Sea (key environments). Pergamon Press, Oxford

Facey W (2004) The Red Sea: the wind regime and location of ports. In: Lunde P, Porter A (eds) Trade and travel in the Red Sea region. In: Proceedings of the Red Sea project I held in the British Museum October 2002, vol 2. Soc Arabian Studies Monographs. Archaeopress, Oxford, pp 7–17

Fattovich R (2005) Marsa Gawasis: a Pharaonic coastal settlement by the Red Sea in Egypt. In: Starkey JCM (ed) People of the Red Sea. Proceedings of the Red Sea project II held in the British museum October 2004. Archaeopress, Oxford

Fattovich R, Bard KA (2012) Archaeological investigations at Wadi/Mersa Gawasis, Egypt: 2006–2007, 2007–2008 and 2009 field seasons. In: Tallet P, Mahfouz E-S (eds) The Red Sea in pharaonic times: recent discoveries along the Red Sea coast. Proceedings of the colloquium held in Cairo/Ayn Soukhna, 11–12th January 2009. L'Institut Français d'Archéologie Orientale, Bibliotheque d'Étude, Cairo, 155, pp 21–26

Fitzpatrick MP (2011) Provincializing Rome: the Indian Ocean trade network and Roman imperialism. J World Hist 22(1):27–54

Flemming BW (1995) Tidal signatures in modern and ancient sediments. Spec Publ Int Assoc Sedimentologists. Blackwell Science Ltd, Oxford

Fleurence J (1999) Seaweed proteins: biochemical, nutritional aspects and potential uses. Trends Food Sci Technol 10(1):25–28

Frankel J (1997) Regional trading blocs in the world economic system. Institute for International Economics, Washington, DC

Friedman HA (2008) Industry and empire: administration of the Roman and Byzantine Faynan. Unpublished PhD thesis, University of Leicester

Gawlikowski M, Juchniewicz K (2016) Aynuna on the Red Sea. Soc Arabian Studies Conference 2016 poster

Gazda EK (2001) Cosa's contribution to the study of Roman hydraulic concrete: an historiographic commentary. In: Goldman NW (ed) New light from ancient cosa: classical mediterranean studies in Honor of Cleo Rickman Fitch. Hermeutics of art, vol 10. University of Michigan Press, New York, pp 145–177

Geus K, Thiering M (2014) Features of common sense geography: implicit knowledge structures in ancient geographical texts, vol 16. LIT Verlag, Münster, Antike Kultur und Geschichte

Giraudi C, Paroli L, Ricci G, Tata C (2006) Portus (Fiumicino—Roma) Il colmamento sedimentario dei bacini del Porto di Claudio e Traiano nell'ambito dell'evoluzione ambientale tardo-antica e medievale del delta del Tevere. Archeologia Medievale 33:49–60

Giraudi C, Tata C, Paroli L (2009) Late Holocene evolution of Tiber river delta and geoarchaeology of Claudius and Trajan Harbor, Rome. Geoarchaeology—an International J 24(3):371–382

Goiran J-P, Salomon F, Pleuger E, Vittori C, Mazzini I, Boetto G, Arnaud P, Pellegrino A (2012) Port antique d'Ostie. Résultats préliminaires de la première campagne de carottages. Chronique Des Activités Archéologiques De l'École Française De Rome

Goiran J-P, Salomon F, Mazzini I, Bravard J-P, Pleuger E, Vittori C et al (2014a) Géoarchéologie du delta du Tibre: évolution géomorphologique holocène et contraintes hydrosédimentaires dans le système Ostie-Portus (Italie). J Archaeol Sci 41(8):389–398

Goiran J-P, Salomon F, Mazzini I, Bravard J-P, Pleuger E, Vittori C, Boetto G, Christiansen J, Arnaud P, Pellegrino A, Pepe C, Sadori L (2014b) Geoarchaeology confirms location of the ancient harbour basin of Ostia (Italy). J Archaeol Sci 41:389–398

Hamilton-Dyer S (1990) The animal remains. In: Bingen J (ed) Quatrième Campagne de Fouille au Mons Claudianus, Rapport préliminaire. Bulletin de l'Institut Français d'Archéologie Orientale 90. Cairo

Hamilton-Dyer S (2001) The faunal remains. In: Peacock DPS, Maxfield VA (eds) Survey and excavation Mons Claudianus, vol 2, pp 251–312

Hansen RD (1983) Water and wastewater systems in imperial Rome. J Am Water Res Assoc 19(2):263–269

Harrell JA (1996) Geology. In: Sidebotham SE, Wendrich WZ (eds) Berenike'95, Preliminary report of the excavations at Berenike (Egyptian Red Sea Coast) and the survey of the eastern desert. Special Series, vol 2. Research School of Asian, African and Amerindian Studies (CNWS), Leiden, pp 99–126

Harvie C (2008) A floating Commonwealth: Politics, culture, and technology on Britain's Atlantic coast, 1860–1930. Oxford University Press, New York

Hattendorf JB, Unger RW (eds) (2003) War at sea in the Middle Ages and the Renaissance. Boydell Press, Woodbridge

Healy JF (1978) Mining and metallurgy in the Greek and Roman world. Thames and Hudson, London

Hein CJ, Fitzgerald DM, Bard KA, Fattovich R, Zazzaro C (2008) Geological investigations of a Middle Kingdom harbor: Wadi Gawasis, Egypt. In: 2008 joint meeting of the Geological Society of America, Soil Science society of America, American Society of Agronomy, Crop Science Society of America, Gulf Coast

Association of Geological Societies with the Gulf Coast Section of SEPM, Session No. 255, Archaeological Geology, Houston. Geological Society of America

Hein CJ, FitzGerald DM, Milne GA, Bard K, Fattovich R (2011) Evolution of a Pharaonic harbor on the Red Sea: implications for coastal response to changes in sea level and climate. Geology 39 (7):687–690

Hirt AM (2010) Imperial mines and quarries in the Roman world: organizational aspects 27 BC–AD 235. Oxford Classical Monographs. Oxford University Press, New York

Hopkins TK, Wallerstein IM (1982) World-systems analysis: theory and methodology, vol 1. Sage Publications Inc

Horsburgh J (1841) The India directory, or directions for sailing to and from the East Indies, China, Australia, and the adjacent ports of Africa and South America: Comp. chiefly from original journals of the honourable company's ships, and from observations and remarks, resulting from the experience of twenty-one years in the navigation of those seas, 5th ed. W. H. Allen and Co, London

Hoyle BS (1989) The port–city interface: trends, problems and examples. Geoforum 20(4):429–435

Hughes JD (2014) Environmental problems of the Greeks and Romans: ecology in the ancient mediterranean. Johns Hopkins University Press, Baltimore

Karmon Y (1985) Geographical components in the study of Ancient Mediterranean ports. In: Rabān A (ed) Harbour Archaeology. In: Proceedings of first international workshop on Ancient Mediterranean harbours, Caesarea Maritima. Archaeopress, Oxford, pp 1–6

Khalil ASM (1994) Fishes in the mangrove ecosystem of Egypt and the Suakin area, Sudanese Red Sea: Population ecology, community structure and species diversity. Unpublished M.Sc. thesis, University of Khartoum

Kirkland K (2010) Marine sciences: notable research and discoveries. Infobase Publishing, New York

Kotarba-Morley AM (2015) The port of Berenike Troglodytica on the Red Sea: a landscape-based approach to the study of its harbour and its role in Indo-Mediterranean trade. PhD thesis, University of Oxford

Kotarba-Morley AM (2017) The Maritime context of the trans-Mediterranean–Indian Ocean trade: Critical review of the Roman era vessels of the Red Sea. In: Agius DA, Khalil E, Scerri E, Williams A (eds) Proceedings of Red Sea VI conference, Tabuk, Saudi Arabia. University of Exeter, Brill, pp 171–206

Kotarba-Morley AM (2018) Port town and its harbours: sedimentary proxies for landscape and seascape reconstruction of the Greco-Roman site of Berenike Troglodytica, Red Sea coast of Egypt. Polish Archaeology in the mediterranean (PAM 25) Special Issue: 'Studies' from the Red Sea to the Gulf

Kron G (2012) Roman agriculture. In: Bagnall RS, Broderson K, Erskine A, Heubner S (eds) The encyclopedia of ancient history. Blackwell, Oxford, pp 217–222

Lancaster LC (2005) Concrete vaulted construction in imperial Rome: innovations in context. Cambridge University Press, New York

Leidwanger J, Knappett C, Arnaud P, Arthur P, Blake E, Broodbank C, Brughmans T, Evans T, Graham S, Greene ES, Kowalzig B, Mills B, Rivers R, Tartaron TF, Van de Noort R (2014) A manifesto for the study of ancient Mediterranean maritime networks. Antiquity 342 online, pp 1–5

Linder SB (1961) An essay in trade and transformation. Wiley, New York

Luttwak E (1976) The grand strategy of the Roman empire: from the first century A.D. to the third. John Hopkins University Press, Baltimore, Maryland

Manguin P-Y, Mani A, Wade G (2011) Early interactions between south and southeast Asia: reflections on cross-cultural exchange. Institute of Southeast Asian Studies

Mannino MA, Thomas KD (2002) Depletion of a resource? The impact of prehistoric human foraging on intertidal mollusc communities and its significance for human settlement, mobility and dispersal. World Archaeol 33(3):452–474

Mansoor NM (2002) A GIS-based assessment of active faults and liquefaction potential of the city of Aqaba, Jordan. M.S. Thesis, Geosciences Department, UMKC

Mansoor NM, Niemi TM, Misra A (2004) A GIS-based assessment of liquefaction potential of the city of Aqaba. Jordan Environ Eng Geosci 10(4):297–320

Marzano A (2013) Harvesting the sea: the exploitation of marine resources in the Roman Mediterranean. Oxford Studies on the Roman Economy. Oxford University Press, Oxford

Mazzini I, Faranda C, Giardini M, Giraudi C, Sadori L (2011) Late holocene palaeoenvironmental evolution of the Roman harbour of Portus. Italy J Paleolimnol 46(2):243–256

Miller JI (1969) The spice trade of the Roman Empire 29 B.C.–A.D. 641. Clarendon Press, Oxford

Millet B, Tronchère H, Goiran J-P (2014) Hydrodynamic modeling of the Roman harbor of Portus in the Tiber Delta: the impact of the north-eastern channel on current and sediment dynamics. Geoarchaeol Int J 29(5):357–370

Morgan G (2004) Review of the state of world marine capture fisheries management: Indian Ocean. Country review: Egypt (Red Sea coast). Fisheries Aquaculture Dept. http://www.fao.org/docrep/009/a0477e/a0477e0i.htm. Accessed 9 June 2014

Morhange C, Marriner N (2010) Mind the (stratigraphic) gap, Roman dredging in ancient Mediterranean harbours. Portus, Ostia and the ports of the Roman Mediterranean. Contributions from archaeology and history. Bolletino Di Archeologia on Line. Edizione Speciale. In: Roma 2008—International Conference of Classical Archaeology Meetings Between Cultures in the Ancient Mediterranean, pp 23–32

Morrison JS, Coates JF (1986) The Athenian trireme: the history and reconstruction of an ancient Greek warship. Cambridge University Press, Cambridge

Morton J (2001) The role of the physical environment in ancient Greek seafaring. Brill, Leiden, Boston and Koln

Murray GW, Warmington EH (1967) Troglodytica: the Red Sea littoral in Ptolemaic times. Geograph J 133(1):24–33

Nappo D (2010) On the location of Leuke Kome. J Roman Archaeol 23:335–348

Niemi TM, Smith AM (1999) Initial results of the southeastern Wadi Araba, Jordan geoarchaeological study: implications for shifts in Late Quaternary aridity. Geoarchaeol Int J 14(8):791–820

Niemi TM, Thomas R, Parker ST (2006) A 2nd century earthquake at the time of the Roman annexation of Nabataea? Geol Soc America. In: 2006 Philadelphia annual meeting (22–25 October 2006). Abstracts with Programs 38(7):178

Oleson JP (2000) Ancient sounding-weights: a contribution to the history of Mediterranean navigation. J Roman Archaeol 13:293–310

Oleson JP (2006) Herodotus, Aristotle, and sounding weights: the deep sea as a frontier in the classical world. In: Tassios TP, Palyvou C (eds) Ancient Greek technology: proceedings 2nd international congress. Technical Chamber of Greece, Athens, pp 32–48

Oleson JP (2008) Testing the waters: the role of sounding-weights in ancient Mediterranean navigation. In: Hohlfelder RL (ed) The maritime world of Ancient Rome. Memoirs American Academy in Rome, Suppl. 6. University of Michigan Press, Ann Arbor, pp 119–176

Oleson JP, Branton G (1992) The technology of King Herod's harbour. In: Vann RL (ed) Caesarea papers: Straton's tower, Herod's harbour, and Roman and Byzantine Caesarea. JRA Supplementary Series, vol 5. Ann Arbor, pp 49–67

Oleson JP, Brandon C, Cramer SM, Cucitore R, Gotti E, Hohlfelder RL (2004) The ROMACONS project: a contribution to the historical

and engineering analysis of hydraulic concrete in Roman maritime structures. Int J Naut Archaeol 33(2):199–229
Oleson JP, Brandon CJ, Hohlfelder RL, Jackson MD (2014) Building for eternity. The history and technology of Roman concrete engineering in the sea. Oxbow Books, Oxford and Philadelphia
Osypińska M (2011) Archaeozoological remains. In: Sidebotham SE, Zych I (eds) Berenike 2008–2009. Report on the excavations at Berenike, including a survey in the eastern desert. Polish Centre of Mediterranean Archaeology, University of Warsaw, pp 67–76
Osypińska M (2016) Pet cats at the early Roman Red Sea port of Berenike, Egypt. Antiquity 90(354)
Osypińska M (2017) Archaeozoological remains. In: Sidebotham SE, Zych I (eds) Berenike 2010–2011: report on the excavations at Berenike, including a survey in the eastern desert. PCMA Excavation Series. Polish Centre for Mediterranean Archaeology, University of Warsaw
Patzert WC (1972a) Seasonal reversal in Red Sea circulation. L'Oceanographie physique de la mar Rouge: symposium de l'Association Internationale des Sciences Physique de l'Ocean, Paris, 9–10 Octobre 1972. Centre National pour l'Exploitation des Océans, Paris, pp 55–89
Patzert WC (1972b) Seasonal variations in structure and circulation in the Red Sea. Hawaii Inst Geophys Reports. Hawaii Institute of Geophysics, University of Hawaii, Honolulu
Patzert WC (1974) Wind-induced reversal in Red Sea circulation. Deep Sea Res Oceanogr Abstr 21(2):109–121
Peacock DPS, Blue LK (2006) Myos Hormos—Quseir al-Qadim, Roman and Islamic ports on the Red Sea, Volume 1: the survey and report on the excavations 1999–2003. Oxbow Books, Oxford
Peacock DPS, Blue LK (2007) The ancient Red Sea port of Adulis, Eritrea. Report of the Eritro-British expedition, 2004–5. Oxbow Books, Oxford
Peacock DPS, Blue LK (2011) Myos Hormos—Quseir al-Qadim Roman and Islamic ports on the Red Sea, Volume 2: finds from the excavations 1999–2003. University of Southampton Monograph Series in Archaeology, vol 6. Archaeopress, Oxford
Pedersen RK (2015) A preliminary report on a coastal and underwater survey in the area of Jeddah. Saudi Arabia Am J Archaeol 119 (1):125–136
Pessavento Mattioli S (ed) (1998) Bonifiche e drenaggi con anfore in epoca romana: aspetti tecnici e topografici. Materiali d'archeologia, 3. Módena
Pethick J (1995) An introduction to coastal geomorphology. Wiley, Chichester
Phillips C, Villeneuve F, Facey W (2004) A Latin inscription from South Arabia. In: Kennet D (ed) Papers from the thirty-seventh meeting of the seminar for Arabian studies held in London, 17–19 July 2003. Proceedings of seminar for Arabian studies, vol 34. Archaeopress, Oxford, pp 85–100
Polanyi K (1963) Ports of trade in early societies. J Econ Hist 23(1):30–45
Pydyn A (2011) Argonauci epoki kamienia, Wczesna aktywność morska od pierwszych migracji z Afryki do końca neolitu. Wydawnictwo Naukowe Uniwersytetu Mikołaja Kopernika, Toruń
Rabān A (1990) Medieval anchors from the Red Sea. Int J Naut Archaeol 19(4):299–306
Rabān A (2009) The harbour of Sebastos (Caesarea Maritima) in its Roman Mediterranean context. In: Artzy M, Goodman B (eds) Gal Z. Archaeopress, Oxford
Rathbone D (2001) The 'Muziris' papyrus (SB XVIII 13167): financing Roman trade with India. Alexandrian Studies II in Honor of Mostafa el Abbadi. Bulletin de la Societe Archeologique d'Alexandrie, Alexandria 46:39–50
Ray HP, Salles J-F (eds) (2012) Tradition and archaeology. Early maritime contacts in the Indian Ocean. The Nalanda-Sriwijaya Series. Institute of Southeast Asian Studies, New Delhi
Rickman GE (1985) Towards a study of Roman ports. In: Rabān A (ed) Harbour archaeology. proceedings of first international workshop on ancient mediterranean harbours, Caesarea Maritima. Archaeopress, Oxford, pp 105–114
Rockman M, Steele J (2003) Colonization of unfamiliar landscapes: the archaeology of adaptation. Psychology Press
Russell B (2012) Shipwrecks and stone cargoes: some observations. In: Gutierrez A, Lapuente P, Roda I (eds) Interdisciplinary studies on ancient stone. ICAC, Tarragona, pp 533–539
Russell B (2013) The economics of the Roman stone trade. Oxford Studies on the Roman Economy. Oxford University Press, Oxford
Russell KW (1980) The earthquake of May 19, A.D. 363. Bull Am School Oriental Res 238:47–64
Salomon F, Delile H, Goiran J-P, Bravard J-P, Keay SJ (2012) The Canale di Comunicazione Traverso in Portus: The Roman sea harbour under river influence (Tiber delta, Italy). Géomorphologie: Relief, Processus, Environnement 1:75–90
Salway RWB (2004) Sea and river travel in the Roman itinerary literature. In: Brodersen K, Talbert RJA (eds) Space in the Roman world: its perception and presentation. Lit Verlag, Münster, pp 43–96
Sayed AM (1977) Discovery of the site of the 12th Dynasty port at Wadi Gawasis on the Red Sea shore. Revue d'Égyptologie 29:140–178
Schmidt-Nielsen K (1964) Terrestrial animals in dry heat: desert rodents. Clarendon Press, Oxford
Schmidt-Nielsen K (1965) Desert animals. Physiological problems of heat and water. Oxford University Press, New York
Schneider P (2017) 'On the Red Sea (…) the trees are of a remarkable nature' (Pliny the Elder): The Red Sea mangroves from the Greco-Roman perspective. In: Agius DA, Khalil E, Scerri E, Williams A (eds) Proceedings of Red Sea VI conference, Tabuk, Saudi Arabia. Brill, University of Exeter
Seeger JA, Sidebotham SE, Harrell JA, Pons M (2006) A brief archaeological survey of the Aqiq region (Red Sea Coast) Sudan. Sahara 17:7–18
Seland EH (2011) The Persian Gulf or the Red Sea? Two axes in ancient Indian ocean trade, where to go and why. World Archaeology 43(3):398–409
Seland EH (2013) Networks and social cohesion in ancient Indian Ocean trade: geography, ethnicity, religion. J Global Hist 8(3):373–390
Seland EH (2016) The Periplus of the Erythraean Sea: a network approach. Asian Rev World Hist 4(2):191–205
Selvakumar V, Shajan KP, Tomber R (2009) Archaeological investigations at Pattanam, Kerala: new evidence for the location of ancient Muziris. In: Tomber R, Blue L, Abraham S (eds) Migration, trade and peoples. British Academy, London, pp 29–41
Shajan KP, Tomber R, Selvakumar V, Cherian PJ (2004) Locating the ancient port of Muziris: fresh findings from Pattanam. J Roman Archaeol 17:312–320
Shaw T (1993) The trireme project: operational experience 1987–90, lessons learnt. Oxbow Books, Oxford
Sidebotham SE (2011) Berenike and the ancient maritime spice route. University of California Press, Berkeley and Los Angeles
Sidebotham SE, Zitterkopf RE (1995) Routes through the eastern desert of Egypt. Bull University Museum University of Pennsylvania 37 (2):39–52
Sleeswyk AW, Meijer F (1998) The water supply of the ARGO and other oared ships. Mariner's Mirror 84(2):131–138
Srinivasan DM (ed) (2007) On the cusp of an era: art in the Pre-Kuṣāṇa world. Brill's Inner Asian Library. Brill, Leiden
Stenchikov G (2011) Modeling of regional climate over Red Sea and Arabian Peninsula. In: Red Sea Research Center Symposium April 2011. King Abdullah University of Science and Technology (KAUST), Jeddah

Tallet P (2009) Les Égyptiens et le littoral de la mer Rouge à l'époque pharaonique. CRAI 153(2):687–719

Tallet P (2010) Prendre la mer à Ayn Soukhna au temps du roi Isési. BSFE 177–178, pp 18–22

Tallet P (2012a) The Red Sea in Pharaonic Egypt. Assessment and prospects. In: Tallet P, Mahfouz E-S (eds) The Red Sea in Pharaonic times: recent discoveries along the Red Sea coast. Proceedings of colloquium held in Cairo/Ayn Soukhna 11–12th January 2009. L'Institut Français d'Archéologie Orientale, Cairo, pp 171–174

Tallet P (2012b) A new Pharaonic harbour in Ayn Sokhna (Gulf of Suez). In: Agius DA, Cooper JP, Trakadas A, Zazzaro C (eds) Navigated spaces, connected places: proceedings of Red Sea project V held at the University of Exeter, 16–19 September 2010. British Foundation for the Study of Arabia Monographs, vol 12. Archaeopress, Oxford

Tallet P (2012b) Ayn Sukhna and Wadi el-Jarf: two newly discovered Pharaonic harbours on the Suez Gulf. Br Mus Stud Anc Egypt Sudan 18:147–168

Tallet P (2013) Deux notes sur les expéditions au Pays de Pount à la lumière de nouvelles données archéologiques. Revue d'Ègyptologie 64:189–209

Tallet P, Mahfouz E-S (2012) The Red Sea in Pharaonic times: recent discoveries along the Red Sea coast. In: Proceeding of the colloquium held in Cairo/Ayn Soukhna 11th–12th January 2009. L'Institut Français d'Archéologie Orientale, Bibliotheque d'Étude, Cairo 155

Tallet P, Marouard G (2012) Wadi al-Jarf—an early Pharaonic harbour on the Red Sea coast. Egypt Archaeol 40:40–43

Tallet P, Marouard G (2014) The harbor of Khufu on the Red Sea coast at Wadi al-Jarf. Egypt Near East Archaeol 77(1):4–14

Tallet P, Marouard G, Laisney D (2012) Un port de la IVe dynastie au Ouadi al-Jarf (mer Rouge). Bull De L'institut Français D'archéologie Orientale 112:399–446

Temin P (2004) Financial Intermediation in the Early Roman Empire. J Econ Hist 64(3):705–733

Thomas R (2010) Fishing equipment from Myos Hormos and fishing techniques on the Red Sea in the Roman period. In: Bekker-Nielsen T, Casasola DB (eds) Ancient nets and fishing gear. Proc international workshop on "nets and fishing gear in classical antiquity: a first approach" Cadiz, 15–17 November 2007. Monographs of the SAGENA Project, vol 2. Universidad de Cádiz, Servicio de Publicaciones, Cadiz, Aarhus University Press, pp 139–159

Thomas R, Pennington B, Villing A (2016) Naukratis in its riverine setting. In: IKUWA 6 conference presentation

Thurman HV, Trujillo AP, Abel DC, McConnell R (1999) Essentials of oceanography. Prentice Hall

Tibbetts GR (1961) Arab navigation in the Red Sea. Geograph J 127 (3):322–334

Tibbetts GR (1971) Arab navigation in the Indian Ocean before the coming of the Portuguese: Being a translation of Kitab al-Fawa'id fi usul al-bahr wa'lqawa'id of Ahmad b. Majid Al-Najdi, together with an introduction on the history of Arab navigation, notes on the navigational techniques and the topography of Indian Ocean, and a glossary of navigational terms. Oriental Translation Fund Series, vol 42. Routledge, Royal Asiatic Society of Great Britain and Ireland, London

Van der Veen M (2011) Consumption, trade and innovation: Exploring the botanical remains from the Roman and Islamic ports at Quseir al-Qadim, Egypt. Africa Magna Verlag, Frankfurt

Van Hecke J, Bossue B, De Buck V, Carpentier E (eds) (1861) Acta Sanctorum Octobris, ex Latinis & Græcis, aliarumque gentium Monumentis, servata primigenia veterum Scriptorum phrasi, Collecta, Digesta, Commentariisque & Observationibus Illustrata a J. van Hecke, B. Bossue, V. de Buck et E. Carpentier, e Societate Jesu Presbyteris Theologis. Tomus X, quo dies vigesimus tertius et quartus continentur. Typis Henrici Goemaere, Brussels

Veal RJ (2013) Fuelling ancient Mediterranean cities: a framework for charcoal research. In: Harris WV (ed) The mediterranean environment in ancient history. Brill, Leiden and Boston, pp 37–58

Vittori C, Mazzini I, Salomon F, Goiran J-P, Pannuzi S, Rosa C, Pellegrino A (2015) Palaeoenvironmental evolution of the ancient lagoon of Ostia Antica (Tiber delta, Italy). J Archaeol Sci 54:374–384

Wallerstein I (2009) World system analysis. In: Modelski G, Denemark RA (eds) World system history (Vol. 1 of Encyclopaedia of life support systems). EOLSS Publications, Oxford

Ward W (2007) Aila and Clysma: The rise of northern ports in the Red Sea in late antiquity. In: Starkey J, Starkey P, Wilkinson T (eds) Natural resources and cultural connections of the Red Sea. Archaeopress, Oxford, pp 161–171

Whitewright J (2007) How fast is fast? Technology, trade and speed under sail in the Roman Red Sea. In: Starkey JCM, Starkey P, Wilkinson TJ (eds) Natural resources and cultural connections of the Red Sea. Soc Arabian Studies Monographs, vol 12. Archaeopress, Oxford, pp 77–88

Whitewright J (2008) Maritime technological change in the ancient mediterranean: the invention of the lateen sail. Unpublished PhD thesis, University of Southampton

Williams CJR (2011) African climate and climate change: physical, social and political perspectives. Springer, London and New York

Williamson OE (1989) Transaction cost economics. Handbook of industrial organization, vol 1, pp 135–182

Wilson AI (2002) Marine resource exploitation in the cities of coastal Tripolitania. In: Khanoussi M, Ruggeri P, Vismara C (eds) L'Africa Romana XIV. Carocci di Roma, Roma volume 1. Rome, pp 429–436

Wilson AI (2015) Red Sea trade and the state. In: de Romanis F, Maiuro M (eds) Across the ocean: nine essays on Indo-Mediterranean trade. Brill, Leiden and Boston, pp 13–32

Woodward FI, Lomas MR, Kelly CK (2004) Global climate and the distribution of plant biomes. Phil Trans Roy Soc B 359:1465–1476

Woodward N (1951) Seaweeds as a source of chemicals and stock feed. J Sci Food Agric 2(11):477–487

Yin J, Falconer RA, Chen Y, Probert SD (2000) Water and sediment movements in harbours. Appl Energy 67(3):341–352

Young GK (2001) Rome's Eastern Trade: International commerce and imperial policy 31 BC–AD 305. Routledge, London and New York

Zazzaro C (2013) The ancient Red Sea port of Adulis and the Eritrean coastal region. Previous investigations and museum collections. Cambridge Monographs in African Archaeology, vol 85. Archaeopress, Oxford

Zazzaro C, Manzo A, Cocca E (2014) Towards a chronology of the Eritrean Red Sea port of Adulis (1st–Early 7th Century AD). J Afr Archaeol 12(1):43–73

Zieliński J (2011) Archaeobotanical remains. In: Sidebotham S, Zych I (eds) Berenike 2008–2009. report on the excavations at Berenike, including a survey in the eastern desert. PCMA Excavation Series, vol 1. Polish Centre for Mediterranean Archaeology, University of Warsaw, Warsaw, pp 59–66

Zitterkopf RE, Sidebotham SE (1989) Stations and towers on the Quseir Nile road. J Egyptian Archaeoly 75:155–189

Ancient Sources

Eusebius of Casearea (*Onomasticon* 2005) Translated by R.S. Notley and Z. Safari. *Onomasticon*. A Triglott Edition with Notes and Commentary. Brill, Leiden

Pausanias (*Perieg* 1918) Translated by W.H.S. Jones and H.A. Ormerod. *Periegesis Hellados*—Description of Greece. Loeb Classical Library, 188. Harvard University Press, Cambridge

Pliny (Plinius Secundus, Gaius, NH 1942) Translated by H. Rackham. Pliny. Natural History. Loeb Classical Library. Harvard University Press, Cambridge, MA

Seneca, Lucius Annaeus (*Nat. Quest.* 1972) Translated by T.H. Corcoran. *Naturales quaestiones*. University of Chicago Press, Chicago

Strabo from Amaseia (*Geogr*. 1917) Translated by H.L. Jones. The geography of Strabo, Vol. I. Books 1–2. William Heinemann Ltd, London

Tacitus, Publius Cornelius (*Agricola* 1914) Translated by M. Hutton and W. Peterson. Agricola. Germania. Dialogue on Oratory. Tacitus Vol. 1. Loeb Classical Library, 35. Harvard University Press, Cambridge

Theophrastus from Lesbos (*Hist. Plant*. 1916) Translated by A.F. Hort. Enquiry into plants (*Historia Plantarum*), vol I. Books 1–5. Loeb Classical Library, 70. Harvard University Press, Cambridge

Vitruvius, Marcus Pollio (*De Archit*. 1931) On architecture (*De Architectura*), vol I. Loeb Classical Library, 251. Harvard University Press, Cambridge

Authors' Biography

Najeeb M.A. Rasul
Technical Adviser
Center for Marine Geology, Saudi Geological Survey, Jeddah, Saudi Arabia
rasul.nm@sgs.org.sa;
najeeb_rasul@hotmail.com

Dr. Najeeb M.A. Rasul holds a Master's degree in Marine Geology, and has a Ph.D. in Geological Oceanography from the University of Wales, U.K. He is presently a Technical Adviser at the Saudi Geological Survey (SGS) in Jeddah. For the past 15 years he has been associated with several organizations and has held key research positions at Geological and Geophysical Research Systems in Canada, the Challenger Division for Seafloor Processes at the National Oceanography Centre, Southampton, U.K., Environment Canada in Burlington, Canada, and the Center for Marine Geology at the SGS, where he was the head of the centre from 2003 to 2010. He has conducted research cruises with European Union Research centres, including the Institute of Marine Sciences (ISMAR), Italy and the IOC-UNESCO (TTR programme) and has been a EUROMARGINS and Saudi Scientific Coordinator. He has been involved in several research projects and cruises as chief scientist and principal investigator in the Pacific, Indian and Atlantic Oceans, including the Mid-Atlantic Ridge, Strait of Gibraltar, Gulf of Cadiz, Mediterranean Sea, Tyrrhenian Sea, Alboran Sea, Henderson Lake, Lake Erie, Arabian Gulf, Strait of Hormuz, Arabian Sea, and currently the Red Sea for the last 15 years. His research interests have concentrated on the fields of both shallow and deep water sedimentary processes, depositional mechanisms and environmental aspects. These include research on modern underwater sedimentary features and processes involved in their formation, deep sea turbidite systems, sediment dispersal and pathways, sediment plumes, fluxes and deposition, contaminants and chemical pathways for pollution and their environmental impact in both time and space, as well as marine mineral resources. He has also worked on the formation and exploration of oil and gas in deltas and offshore regions.

Ian C.F. Stewart
Consultant
Stewart Geophysical Consultants Pty. Ltd., Adelaide, South Australia 5069, Australia
stewgeop@senet.com.au

Dr. Ian C.F. Stewart received a Ph.D. in seismology from the University of Adelaide, South Australia. From 1974 until 1982 he was with the Department of Earth Sciences at Memorial University of Newfoundland, Canada, where he taught geophysics. He then carried out research and worked on potential fields for Saudi Aramco in the Exploration Department in Dhahran, Saudi Arabia from 1982 until 1996, and later for Normandy Exploration in Australia for 2 years in gold exploration. He was the Geophysical Technical Adviser to the Saudi Geological Survey, Jeddah, between 2003 and 2013. He is an independent geophysical consultant based in Adelaide, applying proprietary methods for processing gravity and magnetic data for numerous mining and oil companies in Australia, the Middle East, Africa and elsewhere.

Abdullah Alsharekh
Professor
King Saud University, Department of Archaeology, Riyadh, Saudi Arabia
asharekh@hotmail.com

Dr. Abdullah Alsharekh has been a Professor in the Department of Archaeology at King Saud University (KSU) since 1996 undertaking teaching and research. He obtained a BA degree in archaeology from KSU and then in 1987 received a scholarship to Cambridge University to undertake his M. Phil. and Ph.D. degrees. He has participated in a number of research and consultation projects, including being a member of the scientific committee and the supervisor of the Man and Nature gallery in the National Museum, Riyadh. He was the first archaeologist in Saudi Arabia to win a grant from the prestigious grant programme at King Abdulaziz City for Science and Technology to start a research project at the site of Thumamah, northeast of Riyadh. In 2005, he won the

N. M. A. Rasul and I. C. F. Stewart (eds.), *Geological Setting, Palaeoenvironment and Archaeology of the Red Sea*,
https://doi.org/10.1007/978-3-319-99408-6

Prince Salman (now King) for the study of the history of the Arabian Peninsula for best published paper. In 2003, he started a collaborative research project with Prof. Geoff Bailey from the University of York to look at the prehistory and palaeoenvironment of the Red Sea region, and in 2011 another major project was started in collaboration with Prof. Michael Petraglia from Oxford University to look at early human migrations into Central Arabia. He has been an advisor to the Deputy Minister for Antiquities and Museums, and currently is an advisor to the Ministry of Higher Education. Since 1999 he has been an editor of the academic archaeological journal Adumatu. His research interests include the study of the archaeology and palaeoenvironments of the Arabian Peninsula, rock art and pastoral archaeology. In 2015, he was nominated as the best researcher at the King Saud University prestigious award ceremony for the quality of publishing in the Humanities and Social Sciences.

Abdulnasser S. Al-Qutub
Research Associate
Center for Marine Geology, Saudi Geological Survey, Jeddah, Saudi Arabia
qutub.as@sgs.org.sa

Abdulnasser Al-Qutub is with the Center for Marine Geology, Saudi Geological Survey. He has an M.Sc. from King Abdulaziz University, Jeddah, specializing in coral reef sedimentology. He has been involved in several research projects dealing with bio-geological aspects of various sub-environments of the Red Sea, focusing on water quality and anthropogenic input from different sources that affect coral reefs. He participated in a research cruise on RV Poseidon, jointly conducted by KAU and IFM-GEOMAR, in the Hatiba Deep and Port Sudan Deep in the Red Sea to understand the processes of hydrothermal venting and volcanism. He has also been on a research cruise with KAUST and WHOI dealing with the health of coral reefs, and has studied seabed morphology during a cruise between Jeddah and Port Sudan aboard RV Certaman.

Alessio Sanfilippo
Researcher in Petrology and Geochemistry,
Dipartimento di Scienze della Terra, Università di Pavia, Pavia, Italy
alessio.sanfilippo@unipv.it

Dr. Alessio Sanfilippo has recently earned a researcher position in petrology and geochemistry at the University of Pavia, Italy. During his Ph.D. and Postdoctoral studies he also worked at the Woods Hole Oceanographic Institution (USA) and at Kanazawa University (Japan), taking part in two international oceanographic expeditions. His studies are mainly devoted to understanding the magmatic processes concurring into the accretion of the oceanic lithosphere, exploring the geochemical relationships between residual mantle, igneous crust and melts erupted to the surface.

Ali O. Al-Saeedi
Senior Surveyor
Survey Department, Saudi Geological Survey
Saeedi.AO1@sgs.org.sa

Ali O. Al-Saeedi received his M.Sc. in Surveying from the University of Tabuk, Saudi Arabia. He has been the Head of the Survey Department, Saudi Geological Survey, since 2002. From 1990 to 1994 he was with the Bin Laden Contracting Company, and then with the Ministry of Mineral Resources in Jeddah until 1998. He then worked on projects with the American Mission investigating submerged tunnels in Britain and Makkah. From 2008 to 2012 he was in private business, with projects including processing satellite imagery and laser scanning at sites in Makkah and Madinah. His interests include geodesy and point cloud equipment and processing, and recent work at SGS has involved using UAV technology to study the movements of sand dunes in the province of Al Qunfudhah.

Ana Paula Gouveia Jácome
Research Associate
Universidade Federal de Ouro Preto
Ouro Preto, Minas Gerais, Brazil
anapaulagjacome@gmail.com

Ana Paula Gouveia Jácome a Research Associate in Geological Engineering at Federal University of Ouro Preto (UFOP). In 2013, she worked on Quartzite mapping in the Itacolomi Mountain as a recipient of the Young Talent for science scholarship, a research program sponsored by the Brazilian government. She studied at State University of New York, Plattsburgh U.S.A. from 2014 to 2015 as a recipient of the Brazilian Science Without Borders program. She worked on Ar-Ar geochronology and isotope geochemistry of volcanic rocks from the Red Sea region at Lamont-Doherty Earth Observatory of Columbia University in 2015.

Andy Shuttleworth
Lecturer in Human Palaeoecology
Department of Archaeology and Palaeoecology
Queen's University Belfast,
U.K.
A.Shuttleworth@qub.ac.uk

Dr. Andy Shuttleworth is a Lecturer in Human Palaeoecology at Queen's University Belfast and a Honorary Research Fellow at the Department of Anthropology, Durham University. He took his B.Sc. (Hons) in forensic and investigative science at the University of Central Lancashire in 2007 with a focus on human osteology and biological and forensic anthropology. He undertook his M.Sc. in Palaeoanthropology at the University of Liverpool in 2008 and stayed in Liverpool to complete his Ph.D. *('An Anthropological Assessment of Neanderthal Behavioural Energetics')* in 2013. Following the completion of his Ph.D., he undertook a post-doctoral curatorial position at NML World Museum, Liverpool, cataloguing the Museum's Palaeolithic and prehistoric lithic collection. Between 2014 and 2017 he was an Assistant Professor (Teaching) at the Department of Anthropology, Durham University before moving the Queen's University Belfast. He has archaeological field experience in Palaeolithic archaeology in its landscape and human behavioural context in the UK, Near East, East Africa, Iraq and Saudi Arabia. His current research interests include understanding hominin behavioural adaptations to new landscapes, Neanderthal energetics and their influence on behaviour, and communication of prehistoric heritage to the wider community. He is a Fellow of the Higher Education Academy (HEA).

Anna M. (Ania) Kotarba-Morley
Lecturer
Department of Ancient History Australian Hearing Hub, South Wing, Macquarie University, NSW 2109, Australia
and
Centre for Archaeological Science, School of Earth and Environmental Sciences, University of Wollongong, Australia
and
Oxford Centre for Maritime Archaeology, School of Archaeology, University of Oxford, U.K.
ania.kotarba@mq.edu.au

Dr. Anna Kotarba-Morley is a Lecturer in Archaeology and Ancient History at the Macquarie University in Sydney. She holds a PhD/DPhil in Archaeology from the University of Oxford, UK. In order to elucidate mechanisms of far-distance maritime trade in the Early Historic (Pre-Islamic) period Ania's research focuses on maritime archaeology of ancient ports of trade around the coasts of the Red Sea and the Indian Ocean and better understanding of human adaptations to—and interactions with—the natural environment and changing land- and sea-scape. Ania's extensive archaeological and heritage experience have been gained on sites in the Near and Middle East: Egypt, Syria, Lebanon, Bahrain, Kuwait, Qatar, U.A.E., Oman, Turkey and Saud Arabia; in South Asia: India and Bangladesh; in Africa: Sudan, Kenya, mainland Tanzania, Mafia Archipelago, Zanzibar archipelago, Madagascar, Lesotho; in the Mediterranean and the Black Sea: Greece, Italy, Montenegro, Ukraine; on numerous sites in the Northern and Western Europe including Belgium, UK and Poland; and recently in Australia and the Pacific region including in NSW and the Republic of Kiribati.After a few years of teaching and course development at the University of Wollongong Ania has moved to Macquarie University, Sydney where she currently teaches Landscape and Environmental Archaeology and contributes to other teaching in Archaeology and Ancient History. She is actively involved in research projects in Egypt, Lebanon and East Africa and she is developing research projects and student exchange programmes with a number of Mediterranean countries. Ania is very active in the field of Cultural Heritage protection, preservation and management and acts an expert member of ICAHM (The International Scientific Committee on Archaeological Heritage Management), ICOMOS Australia (International Council on Monuments and Sites) and Australia-New Zealand Joint Working Group on Risk Preparedness for Cultural Heritage under patronage of Blue Shield, as well as a reviewer for World Heritage Committee (on WH nominations). She is an Australia's representative on ICOMOS Emerging Professionals Working Group as well as key heritage advisor on a World Bank Wadi Bisri Dam project in Lebanon under the auspices of Council for Development and Reconstruction.

Anthony Sinclair
Professor
Department of Archaeology, Classics and Egyptology, University of Liverpool, 12–14 Abercromby Square, University of Liverpool, L69 7WZ, U.K.
a.g.m.sinclair@liverpool.ac.uk

Dr. Anthony Sinclair is a specialist in Palaeolithic archaeology and is currently Professor of Archaeology in the Department of Archaeology, Classics and Egyptology at the University of Liverpool. His research interests are the manufacture and use of lithic technology and its social and cognitive analysis, the human perception of landscapes and conspecific communities, and the archaeology of the 18th century. He has fieldwork experience in Greece, Japan, Saudi Arabia, South Africa and the U.K. In DISPERSE he is a member of the survey team in southwest Arabia, and is responsible for lithic analysis. He has had considerable experience of survey work in southwest Saudi Arabia and the Farasan Islands over the past seven years.

Antonio Langone
Research Scientist
CNR-IGG U.O.S. Pavia, Via Ferrata 1, 27100 Pavia, Italy
langone@crystal.unipv.it

Dr. Antonio Langone is a researcher at the Institute of Geosciences and Earth Resources, National Research Council of Italy, in Pavia and since 2015 he has been the manager of the LA-ICP-MS laboratory. In 2008 he obtained his Ph.D. in Earth Sciences from Bologna University. From the start of his research activity he has focused on the petrology and tectono-metamorphic reconstructions of different sectors of the Variscan chain, the Italian Alps and the Himalayan belt. A large part of his work has been dedicated to the reconstruction of the metamorphic history of basement rocks with particular attention to the link between U-Pb ages of accessory minerals and microstructures. He has also been involved in several projects dealing with the petrology and geochronology of magmatic rocks. He is the author and or co-author of more than 20 publications in international journals.

Antonio Schettino
Aggregate Professor
University of Camerino, School of Science and Technology–Geology Division, Via Gentile III da Varano, 62032 Camerino (MC), Italy
antonio.schettino@unicam.it

Dr. Antonio Schettino is the Aggregate Professor of Geophysics at the University of Camerino, Italy. From 1982 to 2002 he was a software developer and consultant in Milan, Italy, with scientific interests in the application of advanced algorithms to the geosciences. During this period he collaborated with academic institutions in Italy (Universities of Milan and Camerino) and the USA (University of Texas at Arlington) to the design and implement software for plate tectonic modelling. His academic career started in 2002, when he commenced a Ph.D. in Earth Sciences at the University of Camerino. From 2002 onward his scientific interests focussed on global tectonics, plate kinematics, and geodynamics, with a special emphasis on the tectonic evolution of the Red Sea, central Atlantic, and Mediterranean regions. He participated in six financed research projects, for one of which he was the scientific coordinator. He led four scientific expeditions in Italy, Saudi Arabia, and Albania. He has produced 25 scientific papers (5 before his academic career) and one book. From 2006 onward he taught courses of solid earth geophysics and theoretical principles of geophysical prospecting (potential fields), and supervised two graduate students and one post-doctoral researcher in geophysics. He is a Fellow of the Royal Astronomical Society and regular member of the American Geophysical Union.

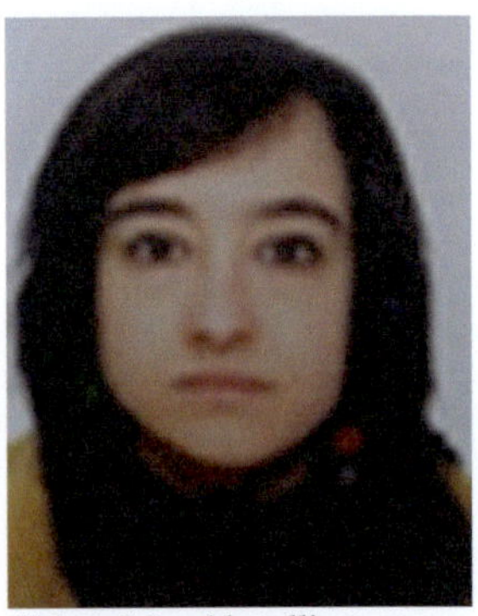

Chiara Macchiavelli
Post Doctoral Researcher
Institute of Earth Sciences Jaume Almera, ICTJA-CSIC, Lluis Sole i Sabaris s/n, 08028 Barcelona, Spain chiara.macchiavelli@unicam.it

Dr. Chiara Macchiavelli is a post–doctoral researcher in Marine Geophysics at the CSIC Jaume Almera of Barcelona, Spain. She received her Master's degree in Geophysics and Applied Geophysics with first class honours from the University of Naples and her Ph.D. in Earth Sciences from the University of Camerino, where she also spent two years as a post-doc research fellow. Her scientific interests are focused on marine geophysics and plate kinematics of the Atlantic, Red Sea, and Mediterranean regions. Currently the subject of her research is the analysis of marine magnetic anomalies from the central Atlantic and the kinematics of the Iberian plate. She published seven scientific papers on international peer-reviewed journals.

Colin W. Devey
Professor
GEOMAR Helmholtz Centre for Ocean Research Kiel, Wischhofstraße 1-3, 24148 Kiel, Germany
cdevey@geomar.de

Dr. Colin W. Devey is Professor at the University of Kiel and the GEOMAR Helmholtz Centre for Ocean Research Kiel, Germany. He holds a doctoral degree from the University of Oxford (UK) for his work on the stratigraphy and geochemistry of the Indian Deccan Trap lava series and habilitated 1996 in Geology/Paleontology at the University of Kiel (Germany). After some years as senior research scientist in Kiel he became Professor for "Petrology of the Ocean Crust" at the University of Bremen where he also was Dean of the Faculty of Geosciences. In 2004 he moved back to Kiel and became Head of the research division "Dynamics of the Seafloor" and later Deputy Director of the former Leibniz Institute of Marine Sciences IFM-GEOMAR. He also was Chairman of InterRidge, Member of the IODP Science Steering and Evaluation Panel as well as German representative of the Scientific Committee for Ocean Research (SCOR). Colin Devey is an extensively experienced scientist in the fields of volcanic geochemistry, submarine volcanology and hydrothermal systems and was scientific leader of many marine research cruises to e.g., the Atlantic, Arctic, Red Sea and Pacific Ocean. He has written more than 150 scientific publications and was scientific advisor and moderator for TV-shows, exhibitions and books.

Daniel Stockli
Professor
Department of Geological Sciences, Jackson School of Geosciences, The University of Texas at Austin, 1 University, Station C9000, Austin, TX 78712-0254, U.S.A.
stockli@jsg.utexas.edu

Dr. Daniel Stockli is an expert in rift tectonics and geo- and thermochronology. He received his M.S. from ETH Zurich in Switzerland and his Ph.D. from Stanford University. He was an Assistant and Associate Professor at the University of Kansas from 2001–2011 and has been a Full Professor at the University of Texas at Austin since 2011. He is currently the Chevron (Gulf) Centennial Professor in Geological Sciences at the University of Texas and the Director of the UTChron Geo- and Thermochronometry Laboratory. His research focuses on the integrated application of thermochronology and geochronology to tectonic and geological problems to better understand the temporal and thermal aspects of continental rifting and break-up. His expertise is in combining structural geology with low-temperature thermochronology to elucidate the spatial and temporal distribution of deformation in intra-continental rifting, orogen-parallel extension, and continental rupture leading to seafloor spreading. In addition he works on geo- and thermochronometric technique development, calibration, and bench marking, with special emphasis on the development of new thermochronmeters (e.g. monazite, rutile, and magnetite) and novel applications to constrain the timing and thermal evolution of tectonic and structural processes. He has extensive experience working in the Red Sea and the Gulf of Suez and has collaborated with the Saudi Geological Survey, industry, and academia on the on-shore extensional record of Red Sea/Gulf of Suez rifting in Saudi Arabia and Egypt.

David Sanderson
Professor of Environmental Physics
Scottish Universities Environmental Research Centre (SUERC), East Kilbride, U.K.
David.Sanderson@glasgow.ac.uk

Dr. David Sanderson is Professor of Environmental Physics at the University of Glasgow and the Scottish Universities Environmental Research Centre (SUERC). His research interests are in environmental and archaeological physics and chronology, measurement of environmental radioactivity using airborne, in-situ and laboratory methods to assess environmental dosimetry and track environmental processes, luminescence dating, and the physical analysis of food for authenticity purposes. He has worked on the application of luminescence methods to the dating of late Pleistocene terrestrial sediments in southwest Saudi Arabia and to the dating of sediments from marine cores on the continental shelf in the vicinity of the Farasan Islands as part of the DISPERSE project.

Davide Zanoni
Researcher in Structural Geology
Università degli Studi di Milano, Dipartimento di Scienze della Terra "A. Desio", Via Mangiagalli 34, 20133 Milano, Italy
davide.zanoni@unimi.it

Dr. Davide Zanoni earned M.Sc. and Ph.D. degrees from Milano University (Italy) in 2002 and 2007, respectively. From 2007 to 2010 and from 2011 to 2012 he was a post-doctoral fellow at the Department of Geology of the University of New Brunswick (Canada) and at the Department of Earth and Environmental Sciences of Pavia University (Italy), respectively. Since 2012 he has been a researcher in structural geology at the Earth Science Department of Milano University, teaching Introduction to Geology and field schools (first year B.Sc.) and Regional Geology and Crystalline Rock Microstructure (M.Sc. level). His research interests are focused on the tectono-metamorphic evolution of crystalline basement units to reconstruct the geodynamic evolution of active margins. In particular he is interested in the late- to post-orogenic exhumation of continental tectono-metamorphic units and in recent years he has focused also on subduction and exhumation tectonics of ophiolites embedded in collisional belts. Since 2005 he has supervised seven B.Sc. and fifteen M.Sc. students. He has published 25 papers and 5 geological maps. His research is focused on the western Alps and the southeastern Canadian Cordillera, cooperating with the University of Milano, University of Pavia, University of Nice (France), University of New Brunswick, and Colorado School of Mines (USA). Recently he started cooperating on Red Sea rifting studies with the University of Camerino (Italy).

Derek Keir
Associate Professor
National Oceanography Centre Southampton, University of Southampton, U.K.
and
Dipartimento di Scienze della Terra, Universita degli Studi di Firenze, Florence, Italy
D.Keir@soton.ac.uk

Dr. Derek Keir is Associate Professor in Geophysics at the National Oceanography Centre Southampton, University of Southampton. He completed a 4-year MSci in Geology in Geophysics at Imperial College London in 2002, and his Ph.D. on seismicity of the Ethiopian Rift in 2006. His research interests are mainly using seismicity to understand plate boundary and volcano deformation, and has devoted his entire research career to studying the East African, Red Sea, the Gulf of Aden, and Cameroon Volcanic Line. During the last decade, he has contributed to all three major UK projects in Ethiopia, from being a PhD student on project EAGLE, a NERC research fellow on the Afar Consortium, and now a PI on RiftVOLC. In 2011, he was awarded the Jason Morgan award from the Tectonophysics section of AGU, and in 2013 was the BGA Bullerwell lecturer.

Dimitris Sakellariou
Geologist, Research Director
Hellenic Centre for Marine Research, Athinon - Souniou Ave (46.7th km), Anavyssos 19013, Greece
sakell@hcmr.gr

Dr. Dimitris Sakellariou is a Structural/Marine Geologist and Research Director at the Institute of Oceanography of the Hellenic Centre for Marine Research, Greece. He holds a Bachelor's degree in Geology from the Department of Geology of the Kapodistrian University of Athens, Greece. He took postgraduate courses on tectonics and sedimentology at the Ruhr University, Bochum, Germany in 1984-1985. He completed his doctoral thesis in 1989 in the Institute of Geology of Johannes-Gutenberg University, Mainz, Germany. He was Head of the Marine Geology-Geophysics Department of the Institute of Oceanography from 2004 to 2009. Since 2009 he has held the position of the Scientific Coordinator of the Underwater Activities Department (Underwater Vehicles) of HCMR where he managed and coordinated the use and maintenance of the manned submersible Thetis, four remotely operating underwater vehicles (ROVs) and the scientific diving team of HCMR. He has coordinated or has participated in more than 40 national, international and European research projects with over 150 papers in peer-reviewed journals, congress proceedings, contributions and invited lectures. His research interests fall in the field of Structural/Marine Geology with particular focus on neotectonics and active faulting, evolution and sequence stratigraphy of sedimentary basins, tectonics–sedimentation–sea level changes relationships, Holocene geology, submarine landslides and slope stability, submarine volcanism, cold seeps and mud volcanoes. He is active in marine geoarchaeological research, particularly in the search for, surveying and mapping of submerged archaeological remains in shallow and deep water and the reconstruction of submerged prehistoric landscapes.

Eelco J. Rohling
Professor
Research School of Earth Sciences, Australian National University, Canberra, ACT 2601, Australia
and
Ocean and Earth Science, University of Southampton, National Oceanography Centre, Southampton SO14 3ZH, U.K.
eelco.rohling@anu.edu.au

Dr. Eelco J. Rohling started a Lectureship at Southampton in 1994, becoming Professor in 2002. The award of an Australian Laureate Fellowship in 2012 took him in March 2013 to ANU in Canberra, Australia. He completed a Ph.D. in 1991 at Utrecht University, The Netherlands, and undertook postdoctoral research at Utrecht and the Woods Hole Oceanographic Institution, U.S.A. He has been a member of the Royal Netherlands Academy of Arts and Science since 2008, and received a Royal Society Wolfson Research Merit Award in 2011. He was vice-chairman and chairman of the 26-nation International Marine Global Changes Study Programme (2003-2008), and editor of Paleoceanography (2006-2009) and Reviews of Geophysics (2010-present). He has authored or co-authored about 160 peer-reviewed publications and book chapters, of which 18 are in Nature and Science group journals. His research interests span: (1) High-resolution investigation of ocean/climate changes during the Neogene, and in particular the Quaternary, to determine the nature, timing and magnitude of natural climate variability; (2) Theoretical and applied (integrated with proxy records) modelling of present-day and past states of circulation and property distribution; (3) Theoretical and practical/analytical research on the use of conservative properties and δ18O to trace deep-water formation, advection and mixing processes in the modern ocean. This mainly concerns theoretical arguments and new interpretations of existing data sets; (4) Assessment of changes in deep-sea ventilation states and nutrient distribution mechanisms, the interactions with the organic and inorganic carbon cycle, and the impact of ecological responses on proxy records; and (5) Integration of Holocene palaeoclimate research with archaeological records.

Enrico Bonatti
Professor
Istituto di Scienze Marine, CNR, Bologna, Italy
and
Lamont Doherty Earth Observatory, Columbia University, Palisades, New York, U.S.A.
enrico.bonatti@bo.ismar.cnr.it

Dr. Enrico Bonatti holds degrees in Geology from the University of Pisa, Italy. In 1960 he was at Yale University with a Fullbright Fellowship. He then moved to Scripps Institution of Oceanography at the University of California as an Assistant Research Geologist. In 1965 he moved to the University of Miami School of Marine Science where in 1971 he became a Full Professor of Marine Geology. In 1975 he moved to Lamont Doherty Earth Observatory of Columbia University, U.S., as Doherty Senior Scientist. In 2002 he became Professor of Geodynamics at the University of Rome La Sapienza, and Director of the Institute of Marine Sciences of the Italian Research Council (CNR). He is presently an Associate at the Institute of Marine Sciences of the CNR and a Special Scientist at Columbia University Lamont Doherty Earth Observatory. He is a member of the Russian Academy of Science, of the European Academy of Science and a Fellow of the American Geophysical Union. He was awarded the Patterson Medal in Oceanography by the Royal Swedish Academy of Science and the Feltrinelli Prize by the Italian Accademia dei Lincei. He has led over 30 oceanographic expeditions in the Atlantic, Pacific, Indian and Antarctic Oceans, in the Red Sea and the Mediterranean Sea. He has participated in DSDP-IODP drilling legs. He dived several times to the floor of the Atlantic with the Alvin and Nautile submersibles. He has done field work in the Afar Rift (Ethiopia), Polar Urals (Russia), Tunguska (Siberia), Zabargad Island (Red Sea) and elsewhere. He has done research on various aspects of the geology of the oceans and has written about 200 scientific papers, including many in journals such as Nature and Science. He has also published many more popular articles in journals such as Scientific American.

Frederick Foulds
Northern Archaeological Associates
28 Harmire Enterprise Park
Barnard Castle
DL12 8BN
U.K.
frederick.foulds@googlemail.com

Dr. Frederick W.F. Foulds is an independent researcher and lithic specialist. He obtained his B.A. in Archaeology at Durham University in 2006, followed by a M.A. in Archaeology in 2007 and a Ph.D. in Palaeolithic Archaeology in 2013 at the same institution. In 2014, he worked as part of the AHRC funded Songs of the Caves research project, which assessed the association of acoustics with Palaeolithic cave art using state of the art analytical techniques, and he acted as a lithic analyst for the 2015 DISPERSE field season. His research interests lie in lithic technology and what it can reveal about the individuals behind the creation of stone tools, and he has used metrical analyses, three-dimensional scanning and explored aspects such as symmetry in order to try and understand the processes behind lithic reduction. His main focus is the Acheulian of Britain and Europe and the Middle to Upper Palaeolithic transition, and he has carried out fieldwork within Britain, Spain and Saudi Arabia. He has been the named lithic specialist for a wide range of research and commercial projects and has written multiple specialist reports and academic papers for periods ranging from the Palaeolithic to the Bronze Age. In 2017, he became a committee member of the Lithic Studies Society and the editor of its journal, Lithics.

Froukje M. van der Zwan
Postdoctoral Research Fellow
Geomar Helmholtz Centre for Ocean Research, Kiel, Germany
and
Christian Albrechts University, Kiel, Germany
fzwan@geomar.de

Dr. Froukje M. van der Zwan received her M.Sc. in geosciences with a specialisation in geochemistry and petrology from the Vrije Universiteit Amsterdam, the Netherlands. Her research topics at this time were xenoliths from the Kaapvaal lithospheric mantle found in kimberlites (South Africa) and crystal size distribution studies of lavas from Merapi volcano (Indonesia). After her studies she moved to Kiel, Germany, where she performed her Ph.D. at GEOMAR Helmholtz Centre for Ocean Research Kiel, within the 'Jeddah Transect Project'. The topic was Hydrothermal Activity at Slow-Spreading Mid-Ocean Ridges for which she developed a new method to measure basaltic chlorine with high precision on electron microprobes. She applied this method together with petrology and additional geochemical data to basalts from mid-ocean ridges with a focus on the Red Sea. The aim was to understand multiple processes at varying scales, from magmatism to hydrothermal circulation, prospection of submarine deposits and volcano morphology. During this work she took part in several seagoing expeditions to the Red Sea and the Greenland Sea and she was involved in developing new geological models of the Red Sea. The current focus of her post-doctoral project at the University of Kiel, for which she received a grant from the German Science Foundation is on intraplate volcanism. She studies intraplate volcanism both on land, performing a project on volatiles as a potential cause for magmatism of the Cameroon Volcanic Line (West Africa) and in the ocean at e.g. the Bathymetrists Seamounts. She further continues her work on continental breakup in the Red Sea and in the South China Sea.

Garry Momber
Director
Maritime Archaeology Trust, Room 014/11, National Oceanography Centre, Southampton SO14 3ZH, U.K.
garry.momber@maritimearchaeologytrust.org

Garry Momber is the Director of the Maritime Archaeology Trust, Director of Maritime Archaeology Ltd. And Research Associate for the University of York. He has been a practicing maritime archaeologist for over 25 years during which time he has assisted in the development of the discipline in the UK and overseas. He has led many pioneering projects that have been concerned with developing underwater archaeological practice, training, education, dissemination and capacity building. His area of research expertise has been in relation to archaeological remains within drowned landscapes, working in British waters, particularly the Solent, the Mediterranean, the Red Sea and the Arabian Gulf. He has over 60 publications, has given hundreds of presentations and directed a dozen national and international, high level archaeological projects. He has represented UNESCO as an expert in maritime archaeology in Europe, North Africa and the Middle East.

Geoff N. Bailey
Anniversary Professor of Archaeology Emeritus
University of York, Department of Archaeology, King's Manor, York YO1 7EP, U.K.
and
Flinders University, College of Humanities, Arts and Social Sciences, GPO Box 2100, Adelaide, SA 5001
Australia
geoff.bailey@york.ac.uk

Dr. Geoff N. Bailey is Anniversary Professor of Archaeology Emeritus at the University of York and Visiting Professor at Flinders University, Australia. He took his BA in Archaeology at the University of Cambridge in 1970 and his PhD in 1976, and stayed in Cambridge as Postdoctoral Research Fellow, University Lecturer, and Senior Tutor of Clare Hall. In 1996, he moved to the Chair of Archaeology at the University of Newcastle upon Tyne, and in 2004 to a newly-founded Anniversary Chair at York. He has world-wide interests in the evolution of terrestrial landscapes and the ways in which geological instabilities resulting from sea-level change and active tectonics at plate margins and in rifts have shaped human lives, livelihoods and long-term evolutionary trajectories. He has particular research interests in Palaeolithic cave sites, in mounded shell middens as the most visible archaeological expression of past coastal settlement, in the analysis of marine mollusc shells for information on palaeodiet and palaeoclimate, and in the relationship between coastal archaeology and changes in sea-level and coastal geomorphology. He has field experience in many parts of the world, and has led major multi-national and multi-disciplinary projects in Australia, Greece and Saudi Arabia. He is Chairman of the EU Trans-domain COST network SPLASHCOS (Submerged Prehistoric Archaeology and Landscapes of the Continental Shelf) and Principal Investigator (with Geoffrey King) of the European Research Council DISPERSE Project (Dynamic Landscapes, Coastal Environments and Human Dispersals). He has published 20 books and over 200 papers, and is a Fellow of the Society of Antiquaries of London, Member of the Chartered Institute for Archaeologists, Corresponding Member of the German Archaeological Institute, Corresponding Fellow of the Australian Academy of the Humanities, Member of the Academia Europaea, and winner (with Abdullah Alsharekh) of the 2017 prize of the King Abdulaziz Foundation for a book published on the history of the Arabian Peninsula in a foreign language.

Geoffrey C.P. King
Senior Research Professor
Institut de Physique du Globe de Paris, 1 rue Jussieu, 75238 Paris cedex 05, France
king@ipgp.fr

Dr. Geoffrey C.P. King is Senior Research Professor (Classe Exceptionel) in the Laboratoire Tectonique of the Institut de Physique du Globe Paris, and Honorary Professor at the University of York, U.K. His research interests are in earthquake hazard, modelling of tectonic deformation, and the role of landforms in hominin evolution. His fieldwork experience includes China, Ethiopia, Greece, India, Italy, Kenya, the Levant, Nepal, New Zealand, Pakistan, South Africa, Tibet, Turkey, and the U.S. A. He is Co-Investigator of the DISPERSE project, and has participated in fieldwork in Saudi Arabia.

Grigoris Rousakis
Associate Researcher
Hellenic Centre for Marine Research, Athinon-Souniou Ave (46.7th km), Anavyssos 19013, Greece
rousakis@hcmr.gr

Dr. Grigoris Rousakis graduated from the University of Patras in 1984 with a degree in geology, and completed his M.Sc. in marine geology and his Ph.D. in marine sedimentology (geotechnical properties) at the Technical University of Athens in 1999 and 2005 respectively. Currently he is an Associate Researcher at the Institute of Oceanography of Hellenic Centre for Marine Research. His main research fields, in which he has coordinated 15 projects, are marine geology, sedimentology and geophysics, Holocene geology, sea level changes and sediment sequences/facies, paleoceanography, underwater geoarchaeology, geotechnical properties of sediments, slope stability, cold seeps and mud volcanoes, neotectonics-active faulting-marine paleoseismology, submarine volcanism, deep sea habitat mapping, submarine surveys for manufactures (cable and pipeline route surveys, artificial reefs etc). During the last 29 years he has been involved in a variety of research projects, most of which have been funded by the EU (CINCS, MATER, METROMED ASSEM, ANAXIMANDER, 3HAZ CORINTH, KM3NeT, SEAHELLARC, MEDECOS, CORAL FISH, EMSO, AEGEAN EXPLORATIONS, TERRASUMERSA, DISPERSE etc.). He has been chief scientist in more than 20 national and international research cruises onboard oceanographic vessels. Dr. Rousakis has published more than 23 papers in peer-reviewed journals, with more than 85 original contributions in international and Greek conferences.

Håkon Rueslåtten
Senior Research Scientist
N-7000
Trondheim
Norway
hg.rues@gmail.com

Dr. Håkon Rueslåtten graduated in geology at the University of Oslo in 1976. From 1979 to 1984 he held an Associate Professor position in Engineering Geology, University of Trondheim. In 1984 he joined the Department of Petroleum Technology in SINTEF, Trondheim, and from 1986 to 2005 he was a Research Advisor at STATOIL ASA's R&D facility in Trondheim. Joining Numerical Rocks (later Lithicon) in 2005, Håkon held the position of Chief Geoscientist. Since 2012, Håkon has worked as an independent consultant. His special field of competence is within reservoir characterization, mineralogy and geochemistry.

Hans Konrad Johnsen
R&D Manager
N-7000
Trondheim
Norway
Hans.Konrad.Johnsen@ntebb.no

Dr. Hans Johnsen obtained a M.Sc. in 1976 and a Ph.D. in 1982 from the Norwegian Institute of Technology in Trondheim. From 1983 to 1992 he was the managing director of petroleum consultant PETRECO As. He was the manager of the Centre for Applied Gas Technology in Statoil from 1992 to 1995, and later the project manager for SOFC fuel cell development in Statoil from 1995 to1996. From 1996 to 1999 he was a member of the R&D Strategic Advisory Board in Statoil, and then the R&D manager, New Ideas, Radical Research Program for Statoil until 2008. Until end of 2015 he was R&D manager at Det norske oljeselskap ASA. He is now working as an independent consultant.

Huw Groucutt
Postdoctoral Fellow
Department of Archaeology, Max Planck Institute for the Science of Human History, Jena 07745, Germany
and

School of Archaeology, University of Oxford, U.K.
huw.groucutt@googlemail.com

Dr. Huw Groucutt is a British Academy Postdoctoral Fellow at the University of Oxford. Dr Groucutt previously obtained degrees in archaeology (BA) and palaeoanthropology (MSc) from the University of Sheffield, before completing his doctoral studies at the University of Oxford (archaeological science). He was subsequently employed as a postdoctoral researcher on the ERC funded Palaeodeserts project. His research focuses on hominin demography and behaviour in the Saharo-Arabian arid belt and its surroundings, between 500-5 thousand years ago, with a particular emphasis on the period between 130 and 50 thousand years ago. Dr Groucutt has led publications in a variety of journals including Quaternary Science Reviews and Evolutionary Anthropology. He has co-edited several volumes of Quaternary International including The Middle Palaeolithic in the Desert, Lithics of the Late Middle Palaeolithic, and Green Arabia: Human Prehistory at the Cross-Roads of Continents.

Ioannis Morfis
Electronic and Electrical Engineer
Hellenic Centre for Marine Research, 19013 Anavyssos, Greece
gianmor@hcmr.gr

Dr. Ioannis Morfis has been an electronic and electrical engineer at the Institute of Oceanography of the Hellenic Centre for Marine Research (HCMR) since 2004. He holds a B.Eng. in electronic and electrical engineering from the University of Leeds, a M. Sc. in mobile and satellite communications from the University of Surrey, and a M.B.A. in engineering systems management from the National Technical University of Athens (NTUA). He is completing a Ph.D. in seafloor characterization by remote sensing techniques using sparse signal reconstruction methods, in the Department of Geology at the University of Patras. At HCMR, he has participated in research projects with a wide range of applications including route surveys for underwater pipelines and cables, dredging, coastal works, marine geoarchaeology, environmental monitoring, and mineral resource exploration. He is currently working on the acquisition and processing of bathymetric data as well as on the operation, maintenance and technical support of the multi-beam systems of the research vessels Aegaeo (SeaBeam 2120 and 1180) and Alkyon (RESON 7125). He has participated in oceanographic cruises in the Aegean Sea, Mediterranean Sea and Red Sea, widening his expertise in the use of the state of the art seafloor and sub-seafloor geophysical techniques (side-scan sonar, sub-bottom profilers). His scientific work has been published in peer-reviewed international journals and has also been presented in Greek and international conferences.

Ioannis Panagiotopoulos
Research Fellow
Hellenic Centre for Marine Research, 19013 Anavyssos, Greece
jpanagiot@hcmr.gr

Dr. Ioannis Panagiotopoulos has been working in the Hellenic Centre for Marine Research (HCMR) as a Research Fellow for the last 9 years. He has a degree in geology from the University of Patras (Greece) and a M.Sc. in geological oceanography from the National Kapodistrian University of Athens, Dr. Panagiotopoulos completed his doctoral studies in the Oceanography Department of Southampton University with the main subject of his research being the erosion threshold of sand-mud admixtures. This research was funded by the European Commission (Marine Science and Technology) from 1992 to 1995. His research interests include sediment dynamics: physical mechanisms of sediment transport and physico-chemical controls of sedimentation; measurement and estimation of sediment transport rates; geotechnical and hydraulic interpretation of sediment textures and structures; origin and nature of bed forms (ripples, dunes, bars); wave boundary layers; sediment transport in steady and oscillatory flows; initiation of motion; Shields criterion; bed load motion; sediment entrainment and suspension; bed forms in steady and oscillatory flow; nearshore wave-current processes, wave-induced currents; long-shore sediment transport and cross-shore sediment transport. He also studies coastal geomorphological formations (e.g., river deltas) and the dynamic evolution of coasts, under the action of currents and waves, in relation to their geological identity (rock type, resistance to erosion). He deals with coastal zone management issues associated with human intervention (e.g. diversion of river mouths, construction of ports and dams) and the resulting adverse effects on the operation of associated ecosystems and various end-users. Currently, Dr Panagiotopoulos operates the multibeam systems of the HCMR research vessels *Aegaeo* (SeaBeam 2120 & 1180) and *Alkyon* (Reson SeaBat 7125) and also works on the processing and analysis of bathymetric data and the production of high-resolution bathymetric charts and maps of seabed reflectivity and rugosity. He has participated in numerous shallow- and deep-water hydrographic cruises in the Aegean Sea, Mediterranean Sea and Red Sea. His scientific work has been published in peer-reviewed international journals and has been presented in Greek and international conferences as well.

Jordi Julià
Faculty of the Departamento de Geofísica, Universidade Federal do Rio Grande do Norte, Brasil
jordi@geofisica.ufrn.br

Dr. Jordi Julià is a Professor of the Departamento de Geofísica and a member of the Programa de Pós-Graduação em Geodinâmica e Geofísica at the Universidade Federal do Rio Grande do Norte (UFRN) in Brazil. He graduated in Physics from the Universitat de Barcelona, Spain, in 1999. Prior to his affiliation with UFRN that started in August 2011, he undertook research at several universities in the United States, including Penn State, the University of South Carolina, and Duke University. Since 2011,

Prof. Julià has supervised 1 Ph.D. and 4 M. Sc. students, as well as a number of undergraduate research projects. He is an active member of the American Geophysical Union, the Socidedade Brasileira de Geofísica, and the Societat Catalana de Física. He has given numerous invited talks, lectures, conference presentations, and webinars, and published over 35 peer-reviewed papers in international journals.

Lorenzo Angeletti
Research Scientist
Istituto di Scienze Marine (ISMAR-CNR), Via Gobetti 101, 40129 Bologna, Italy
lorenzo.angeletti@bo.ismar.cnr.it

Dr. Lorenzo Angeletti has a degree in Earth Sciences and a Ph.D. in Palaeontology and Palaeoecology. For the past 10 years he has been undertaking research in marine bio-sedimentology at the Marine Science Institute of the Italian National Research Council. He is interested in the sedimentology and geology of past and present deep water habitats, with a focus on the evolution of bioconstructions in aphotic environments (cold-water corals), and benthic analysis of hard and soft substrata from shelf to bathyal plain and hydrocarbon habitats (cold-seeps). He also has an extensive background in the systematics of mega- and macro-benthic organisms, from the shallow to the deep water environment, especially from the Mediterranean basin, but also in the Red Sea, Ponto-Caspian and Antarctic habitats, as well as experience in the systematics of macro-benthic organisms from hydrocarbon habitats from the Plio-Pleistocene deposits in the Apennines, to recent submarine environments. His current research is focussed mainly on habitat mapping of deep water environments, with a particular emphasis on cold-water coral ecosystems and cold seep benthic communities of the Mediterranean Basin. At ISMAR he is involved with the on-going revision and upgrading of the classification scheme of Mediterranean and the Black Sea marine habitats in the framework of the EU project CoCoNet. As a marine bio-sedimentologist he has been on more than 25 multidisciplinary oceanographic cruises, on 5 of which he was Chief-Scientist.

Luigi Vigliotti
Senior Scientist
Istituto di Scienze Marine, CNR Bologna, Italy
luigi.vigliotti@bo.ismar.cnr.it

Dr. Luigi Vigliotti is a senior scientist at the Institute for Marine Sciences (ISMAR) CNR, Bologna (Italy). Since 1985 has been responsible for the Paleomagnetic Laboratory at ISMAR-BO and is a project leader in this field. He has held fellowships at the University of Budapest (Hungary), Columbia University (New York, USA), the University of California at Berkeley/Stanford and Davis (USA), the University of Liverpool (UK), the University of Newcastle (Australia), and the University of Florida (Gainesville, USA). He was the paleomagnetist on board the Joides Resolution for the Leg 127 of the Ocean Drilling Program (ODP) in the Japan Sea and a participant in the Tenth Italian Expedition in Antarctica. He is an expert in the acquisition, processing and interpretation of paleomagnetic and rock-magnetic data of volcanic rocks as well as of marine and lacustrine sediments, and his research includes environmental magnetism, geodynamics, structural geology and magnetostratigraphy applied to plates and microplate rotation. He is a national vice-delegate and member of the National Committee for the International Association of Geomagnetism and Aeronomy (IAGA) and coordinator of the Division I (Internal Magnetic Field) of IAGA-Italy.

Marco Ligi
Director of Research
Istituto di Scienze Marine, CNR, Bologna, Italy
marco.ligi@bo.ismar.cnr.it

Dr. Marco Ligi is a Director of Research at the Institute of Marine Sciences, National Research Council (CNR) in Bologna, Italy. His scientific career has included being a consultant in numerical cartography and database management systems at ENEA in La Spezia, a one-year Associate Researcher on multichannel seismic processing at Rice University, Houston, and more than 20 years at the Institute of Marine Sciences. His research started with numerical cartography, but soon shifted to geophysics and structural aspects of the origin and evolution of ocean basins, ranging from geophysical modelling to the tectonics of oceanic transform plate boundaries, to the transition from continental to oceanic rifting in the Red Sea. In particular, his present scientific activity is focused on the geophysics of basic processes at accretionary plate boundaries such as mantle flow and melting modelling, multichannel seismic data acquisition, processing and interpretation, multibeam acquisition and processing, and inversion and modelling of gravity data along mid ocean ridges. He has led several oceanographic expeditions in the Mediterranean, Atlantic and Indian Oceans and has written over 128 scientific papers of which 60 are in JCR journals and 4 are in Nature and Science. He has supervised several graduate students from the Universities of Bologna and Rome.

Marco Taviani
Research Director
Istituto di Scienze Marine (ISMAR-CNR), Via Gobetti 101, 40100 Bologna, Italy
Biology Department, Woods Hole Oceanographic Institution, 266 Woods Hole Road, Woods Hole,
Massachusetts 02543, U.S.A.
Stazione Zoologica Anton Dohrn, Villa Comunale, 80121 Napoli, Italy
marco.taviani@bo.ismar.cnr.it

Dr. Marco Taviani is Research Director at the Italian National Research Council Institute of Marine Sciences (ISMAR-CNR) in Bologna since 2001. After graduating from the University of Bologna in 1976 he held various research positions and fellowships until 1989 in the Laboratory of Marine Geology at Bologna, at Lamont-Doherty Geological Observatory, Columbia University, New York, and also at Louisiana State University and the University of Houston. Research lecturer at various USA, China and European Universities. He became a tenured researcher at the CNR in Italy in 1982 and received a Ph.D. in palaeontology from the University of Roma in 1987. His recent research has focussed upon marine ecosystems (especially deep water corals), Cenozoic-Recent marine paleontology and paleoecology (Antarctica, Mediterranean, Red Sea, Authors' Biographies 631 Western Indian Ocean), extreme environments (polar, cold seeps), bio-sedimentology (biogenic carbonate factories), paleoclimatology and paleoceanography based upon sediment cores (Mediterranean, Red Sea), drillholes (Antarctica, Mediterranean), corals (Mediterranean, Red Sea, Atlantic Ocean, Antarctica) and other natural archives, including multiple geochemical approaches. He has participated in over 50 oceanographic missions in the Mediterranean, Red Sea, Atlantic Ocean and Antarctica), often as chief scientist. He was the National Coordinator in 1991–1993 for the E.U. RED SED (The Red Sea and Gulf of Aden Sedimentology), and in 1994–1997 for the E.U. TESTREEF programme (Temporal and SpaTial variability of western Indian Ocean REEFs). He has been the coordinator of other national and international projects (such as the Euromargins Moundforce: Forcing of Carbonate Mounds and Deep Water Coral Reefs along the NW European Continental Margin and WP of the E.U. HERMES, HERMIONE, COCONET programmes). He has published over 300 scientific papers, as well as book chapters and various popular science articles in Italian and international magazines. He is in the UNEP list of coral experts and has been member of the CITES Scientific Commission (Italy).

Maria Geraga
Assistant Professor
Laboratory of Marine Geology and Physical Oceanography, Department of Geology, University of Patras, Greece
mgeraga@upatras.gr

Dr. Maria Geraga is an Assistant Professor in the Geology Department of the University of Patras, Greece. She has more than 20 years' experience in oceanography. Her research activities include the study of the evolution of coastal palaeogeography and of the palaeoclimate based on the examination of microfauna, surveying with marine remote sensing techniques for the detection and mapping of submerged cultural heritage sites and the mapping of marine habitats and monitoring of marine and coastal pollution. She has participated in many national and EU research projects (ASSEM, HYPOX, APREH and SASMAP) and has been a member of the organizing committees of national and international conferences (IAS, HCMR, EGE, ICGC12). She has 29 publications in international refereed journals and has also contributed to over 50 full papers in international conferences and special publications in the field of marine sciences.

Martin Hovland
Senior Research Scientist
N-4055 Sola, Norway
mthovland@gmail.com

Dr. Martin Hovland FGS is an Emeritus Professor who worked for the Norwegian energy company Statoil from 1976 to 2012 as a senior engineer, marine geology specialist and researcher. He is associated with the Universities of Tromsø and Bergen (Norway), as a lecturer. He has been adjunct professor at the Centre for Geobiology, University of Bergen. Since 1987 he has also been member of the expert Environmental Protection and Safety Panel (EPSP) for the Ocean Drilling Program (ODP) and the Integrated ODP (since 2003), and participated as an organic geochemist on ODP Leg 146, to sample gas hydrates on the Cascadia Accretionary Wedge ('Hydrate Ridge'). During his long marine survey experience and research on features and processes in the seafloor he has published four books and about 150 peer reviewed papers. They range in topics from deep-sea geohazards, including pockmarks, seepage, gas hydrates, and mud volcanoes, to the enigmatic Norwegian deep-water coral reefs (*Lophelia* reefs). He continues to work on theories for the formation of salt on Earth and Mars, which include the studies of high-temperature/high-pressure (HTHP) systems in the ocean, including hydrothermal processes, serpentinization, and abiotic hydrocarbon formation. He is currently working as a senior consultant and advisor for the Norwegian company Tech Team Solutions, Stavanger.

Matthew Meredith-Williams
Lecturer
La Trobe University, Department of Archaeology and History, Bundoora, VIC 3086, Australia
M.Meredith-Williams@latrobe.edu.au

Dr. Matthew Meredith-Williams took his BSc at the University of Leeds where he was mentored by Jamie Woodward in geoarchaeology, and his MPhil at Cambridge where he read Quaternary Science, specialising in geoarchaeology and prehistory. Following this he became a Senior Geoarchaeologist at the Museum of London, before taking his PhD at the University of York under the guidance of Prof Geoff Bailey. This focused on the provenance of prehistoric shell mounds on the Farasan Islands, which his first post doctorate (on the DISPERSE Project) continued, and expanded to take in the submerged landscape of the region. A second postdoc under the Marie Curie International Outgoing Fellowship scheme has seen him travel to the University of Auckland to undertake comparative research of the Farasan Island shell mounds with others around the globe. Dr. Meredith-Williams is primarily a field archaeologist and geoarchaeologist, and is building experience around the world. He utilises a number of techniques in research, from landscape archaeology using remote sensing, integrating this down to the site level through GIS and digital recording techniques. He is currently Lecturer in the Department of Archaeology and History at La Trobe University, Australia.

Maud H. Devès
Associate Professor
Institut de Physique du Globe de Paris, 1 rue Jussieu, 75238 Paris cedex 05, France
and
Centre de Recherche Psychanalyse Médecine et Société, CNRS EA 3522–Université Paris Diderot
deves@ipgp.fr

Dr. Maud H. Devès is Associate Professor at the Université Paris Diderot (France). She obtained her MSc in Material Sciences at the Institut National des Sciences Appliquées (INSA-Lyon) and an MSc in Earth Sciences at the École Normale Supérieure (ENS-Lyon) in 2006. She took her PhD in Geophysics at the Institut de Physique du Globe de Paris (IPGP) in 2010, followed by a Postdoctoral Research Fellowship (2010–2014) on the ERC-funded DISPERSE project (Dynamic Landscapes, Coastal Environments and Human Dispersals) as a joint appointment between IPGP and the Department of Archaeology, University of York, U.K., and further developed her trans-disciplinary approach to environmental issues with an MSc in Psychology in the Université Paris Diderot (2013). She worked as a Research Fellow in Sciences Po Paris on the issue of disasters (from threat to risk, expertise and media coverage) and was appointed Associate Professor in the Université Paris Diderot in 2016, jointly affiliated to IPGP and the Centre de Recherche Psychanalyse, Médecine et Société. She has world-wide interests in the ways in which physical instabilities shape human physical, psychological and social lives, on short-term as well as on long-term evolutionary trajectories, and in combining qualitative and quantitative approaches that draw on humanities and physical sciences perspectives. She has field experience in many parts of the world, notably in Middle East and the Caribbean, and has coordinated interdisciplinary and inter-institutional research projects: Earth Politics 2015–2017; and ANR RAVEX 2016–2020 funded by the French National Agency for Research. Since 2016, she has been the Chairman of the Scientific Council of the French Platform for Disaster Risk Reduction (AFPCN). She has published more than thirty papers in different disciplines (earth science, archaeology, political and environmental sciences, and psychology) and languages (French, English and Portuguese).

Mazen Abuabdullah
Technical Adviser
Department of Applied Geology, Saudi Geological Survey, Jeddah, Saudi Arabia
Abuabdullah.mm@sgs.org.sa

Mazen Abuabdullah has been the GIS and Databases Adviser to the Saudi Geological Survey, Jeddah, since 2005. After receiving a B.Sc. in Mathematical Science (IT Section) from the University of Damascus, Syria, in 1993, he joined the Arab Center for the Studies of Arid Zones and Dry Lands (ACSAD), where he worked in the water resources division on the development of software for analyzing pumping tests and to link GIS with mathematical models of groundwater and surface water, and the development of water resources database management systems. As well as contributing to water resource studies in various parts of the Arab world and training technicians on the use of GIS and mathematical models, he then worked with other divisions in ACSAD to develop systems to document and analyze environmental data, such as vegetation and to estimate the efficiency of the use of saline water in irrigation for crops in arid and semi-arid zones. In 2011 he obtained a diploma in Integrated Water Resources Management from the UN University in cooperation with Arabian Gulf University, Manama, Bahrain. Since joining the SGS he has participated in hydrological and hydrogeological studies in addition to engineering geology projects.

Melanie Leng
Professor
NERC Isotope Geosciences Facilities, British Geological Survey, Nottingham NG12 5GG, U.K.
and
Centre for Environmental Geochemistry, University of Nottingham, Nottingham NG7 2RD, U.K.
mjl@bgs.ac.uk

Dr. Melanie Leng is an isotope scientist with a first degree in earth sciences and a doctorate in sedimentology and basin dynamics from the University of Wales, Aberystwyth, UK. Melanie is currently the Director of the Centre for Environmental Geochemistry and a Science Director at the British Geological Survey where she also manages the Stable Isotope Facility. She is also the Professor of Isotope Geosciences at the University of Nottingham. Her research focus includes using isotopes in understanding: (1) Current and past freshwater contributions into the Antarctic Ocean and effects on ocean circulation; (2) Climate influences on northern Europe and effects on migration and endemism; (3) Climate-driven human evolution, innovation, and dispersal through Africa; (4) Major climatic boundary thresholds through geological time; (5) Understanding biogeochemical cycling using isotope techniques; (6) Development of stable isotope techniques in relation to current environmental issues; (7) Developing research within the International Continental scientific Drilling Program.

(Merry) Yue Cai
Associate Research Scientist
Lamont Doherty Earth Observatory, Columbia University, Palisades, New York, U.S.A.
merrycai@gmail.com; cai@ldeo.columbia.edu

Dr. (Merry) Yue Cai received her Ph.D. in Isotope Geology from Columbia University, U.S.A. in 2009. For her thesis she applied traditional geochemical tracers and long-lived radiogenic isotope systems (Pb-Sr-Nd-Hf isotopes) to study the nature and dynamics of the Earth's mantle using magmas from the Arctic Gakkel ridge and the Mexican Volcanic arc. She was a Columbia Science Fellow from 2008 to 2012, during which time she led seminars on a variety of current science topics, including Earth science, Neuroscience and Astronomy, for the core curriculum course *Frontiers of Science* at Columbia University. Since 2013, she has been an Associate Research Scientist at Lamont-Doherty Earth Observatory where she and her collaborators apply geochemical and geochronological tools on a wide range of research topics, including mantle geochemistry, subduction zone processes, continental crust formation, sediment provenance and isotope stratigraphy.

Michael John Roobol
Consultant
The Anchorage, Sandy Haven, St. Ishmaels, Haverfordwest, Pembrokeshire SA62 3DN, U.K.
roobolinc2000@hotmail.com

Dr. M. John Roobol obtained his Ph.D. at Imperial College 1970 for work on the petrology of the Vesturhorn Peninsula in Iceland. In 1968-69 he participated in two Royal Society expeditions to Deception Island volcano in the Antarctic and in 1969-70 in another to the Andes volcanoes. At the University of the West Indies from 1970 to 1974 he started work on the Lesser Antillean volcanoes and in 1975 worked at the University of Montreal on chemistry of the Deep Drill Project Leg 37 and Caribbean volcanoes. From 1975 to 1976 he carried out yacht-based field studies in the Caribbean with laboratory work at Puerto Rico. From 1977 to 1983 was at King Abdulaziz University in Jeddah, and became Professor of Mineralogy and Petrology, also continuing Caribbean studies. He was with the Directorate General of Mineral Resources in Jeddah from 1983 to 1999 as a petrologist, then formed and became the head of the Geohazard Section, then Chief Geologist, while continuing studies of the Caribbean. In 1999-2000 he spent a year exploring the Pacific by yacht. From 2000 to 2007 he joined the newly-formed Saudi Geological Survey as an adviser. Since 2007 he has been based in the U.K., continuing Caribbean work, consulting with the Saudi Geological Survey, and in Australia is working on the prototype of new type of drill for mining placer gold that does not destroy the environment.

Michael Petraglia
Professor of Human Evolution and Prehistory School of Archaeology, Department of Archaeology, Max Planck Institute for the Science of Human History, Jena 07745, Germany
petraglia@shh.mpg.de

Dr. Michael Petraglia is Professor of Human Evolution and Prehistory at the Max Planck Institute for the Science of Human History, Jena, Germany. He is a member of the Human Origins Program, Smithsonian Institution, Washington, DC, USA. Professor Petraglia has conducted archaeological research in the Arabian Peninsula since 2001, examining the role of Arabia in global prehistory. Since 2011, he has led several large-scale, interdisciplinary field projects in the Kingdom of Saudi Arabia, examining the relationships between climate change and population history, in collaboration with the Saudi Commission for Tourism and National Heritage. Hehas authored and co-edited 9 books and over 150 journal articles and book chapters. He is the co-editor of *The Evolution of Human Populations in Arabia: Palaeoenvironments, Prehistory and Genetics* (Springer, 2009) and *Green Arabia: Human Prehistory at the Cross-Roads of Continents* (2015, *Quaternary International*, no. 382).

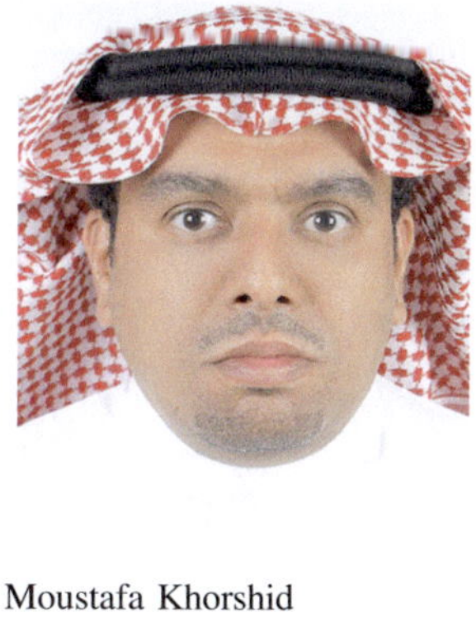

Moustafa Khorshid
Research Associate
Center for Marine Geology, Saudi Geological Survey, Jeddah, Saudi Arabia
Khorsheed.mm@sgs.org.sa

Moustafa Khorshid is a marine sedimentologist in charge of the Marine Geological Laboratory at the Saudi Geological Survey. He holds a Master's degree in Information Technology (IT) from Univeristy of Malaya, Malaysia in 2010. His research interests involve the coastal and marine sediments of the Red Sea, specializing in sediments impacting the mangroves in hyper-saline lagoons. He is presently a leader of the Umm-Lujj research project that deals with the distribution of sediments and their environmental impact on the coastal area including the coral reef and its flora and fauna.

Nasser L. El Agami
Professor
Applied Geology Division, Department of Environmental Geology, Saudi Geological Survey, Jeddah, Saudi Arabia
ellagami.nl@sgs.org.sa

Dr. Nasser L. El Agami has been a full Professor of radioactive geology and environmental geochemistry, at the Nuclear Materials Authority, Egypt, since 2006, and is currently an expert on radioactive geology in the Environmental Geology Section of the Saudi Geological Survey. His Ph.D. (1996) was based on studies of the geology and radioactivity of the Palaeozoic rock units in the Sinai Peninsula, Egypt. His research has concentrated on the geology of Egypt and Saudi Arabia, with over 32 years experience in the fields of sedimentology, radioactivity, geochemistry, environmental studies, uranium exploration and mineral deposits in Egypt, Saudi Arabia and elsewhere. He has participated in several projects on groundwater contamination with radionuclides and heavy metals and has used facies concepts and depositional environments to determine the source of radioactive contamination. He is member of several international and national societies and has published 44 papers.

Neil Mitchell
Reader
School of Earth and Environmental Sciences, University of Manchester, Williamson Building, Oxford Road, Manchester M13 9PL, U.K.
neil.mitchell@manchester.ac.uk

Dr. Neil Mitchell is a marine earth scientist who holds a readership in the University of Manchester. He gained his D.Phil. from Oxford University in 1989 for his work on the tectonic evolution of the Indian Ocean Triple Junction based on GLORIA long-range sidescan sonar. Since then, he has broadened his interests to address a variety of problems concerning the creation of oceanic crust at ridges and subsequent modification by faulting, sedimentation, erosion and growth of volcanic structures. The work is based largely on multibeam and sidescan sonar and is supported by technical work on sonar data processing and analysis. This research has been funded by research fellowships from the Royal Commission for the Exhibition of 1851 (held at Lamont- Doherty Earth Observatory, New York), the NSERC (the University of New Brunswick, Canada), the NERC (the University of Durham) and the Royal Society (the University of Oxford and subsequently Cardiff University, where he also worked as a university lecturer). In Manchester, he is pursuing a range of research on quantifying and modelling seabed morphology in a variety of settings including continental margins and around volcanic islands. He has served on the Site Survey Panel of the IODP and Steering Committee of NERC/IODP. He currently teaches geophysical techniques, submarine geology and earthquake hazards at undergraduate level, and aspects of geophysics and diffusion at masters level.

Nicholas Harmon
Lecturer in Marine Geophysics and Seismology,
Dipartimento di Scienze della Terra, Universita degli Studi di Firenze, Florence, Italy
N.Harmon@noc.soton.ac.uk

Dr. Nicholas Harmon is a marine geophysicist and seismologist, focusing on oceanic lithosphere and mantle dynamics. He was born and raised in New York. He attended Boston University, graduating Summa Cum Laude majoring in History and Geology in 2000. He did his Ph.D. at Brown University working with Prof. Donald Forsyth, graduating in 2007. After an internship with ExxonMobil in Houston, Texas, he became a RIDGE2000 postdoctoral fellow working with Donna Blackman at Scripps Institution of Oceanography in 2007. In 2009, he became a lecturer at the National Oceanography Centre at the University of Southampton, which is his current institution.

Nico Augustin
Senior Research Scientist
Geomar Helmholtz Centre for Ocean Research, Kiel, Germany
naugustin@geomar.de

Dr. Nico Augustin is a senior research scientist at the GEOMAR Helmholtz Centre for Ocean Research Kiel, Germany. He received his Ph.D. in geosciences from the University of Kiel in the field of geochemistry and petrology of submarine mantle rocks in the vicinity of hydrothermal activity. During his PostDoc time he performed two-years of scientific excurses in isotope geochemistry of hydrothermal fluids and on the petrology and genesis of submarine authigenic carbonates. With his capability to look at the problems from a different perspective, he was able to develop some new, but controversial ideas. Since 2010 he moved his research focus from geochemistry to bathymetric mapping and geomorphology of mid-ocean ridges. At this time he started his research in the Red Sea Rift and was part of two major research expeditions there. He developed a model that brings the Red Sea Rift, that was geologically thought to be special, back in line with the known mid-ocean ridges worldwide. Even if his main research interest still lies in the Red Sea rift and research there is ongoing, he was further involved in numerous other seagoing expeditions, often working in internationally and interdisciplinary projects together with e.g. biologist, geophysicists and oceanographers. The aim of his current work is to understand volcanology and hydrothermalism in all oceans, but particularly in rift zones, by means of seafloor imaging and morphological analyses combined with information from other disciplines.

Niklas Hausmann,
Research Associate
University of York, Department of Archaeology, King's Manor, York YO1 7EP, U.K.
and
Foundation for Research and Technology–Hellas, Institute of Electronic Structure and Laser (IESL), P.O. Box 1527, GR-711 10 Heraklion, Greece
niklas@palaeo.eu

Dr. Niklas Hausmann received a B.Sc. in archaeology and geosciences from the Christian-Albrechts-Universität zu Kiel in Germany and an M.A. in archaeology from the University of York. For both degrees he carried out analyses of mid-Holocene sites (Satrup and Duvensee) in southern Scandinavia. Of main interest were the subsistence of people inhabiting the sites as well as the environmental characteristics of the regional area. During his M.A. he also studied temperature changes and seasonality along the coastline of northern Spain during the Mesolithic and upper Palaeolithic using stable oxygen and carbon isotopes. Following this he started working on the Farasan Islands shell mounds. He composed a methodological framework to analyse seasonal and long-term climatic changes in the southern Red Sea using marine gastropods from archaeological contexts, which is the subject of his Ph.D., completed in 2015.

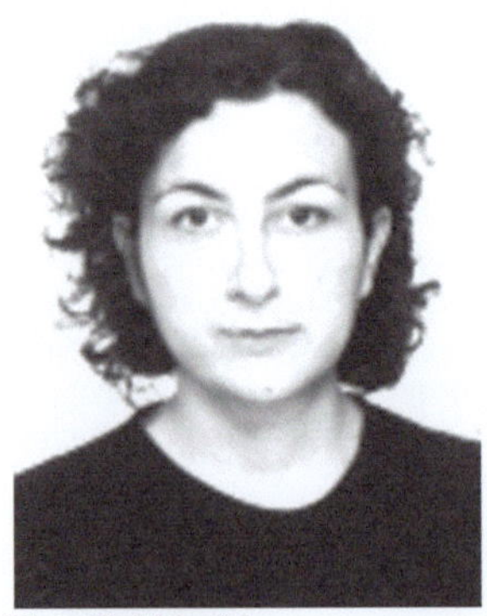

Olga Kokkinaki

Foundation for Research and Technology–Hellas, Institute of Electronic Structure and Laser (IESL), P.O. Box 1527, GR-711 10 Heraklion, Greece

olga@iesl.forth.gr

Dr. Olga Kokkinaki received a B.Sc. in physics, with specialty in atomic and molecular physics, from the Physics Department at the University of Crete. In her diploma thesis she studied the non linear response of the third order of silicon and carbon nanocrystals. She continued her studies at the same Department from which she received a Master degree in Optoelectronics-Microelectronics and a PhD, which was focused on the dynamics of phase transitions in the UV laser irradiation of molecular solids. She is a post-doctoral fellow at the Institute of Electronic Structure and Laser, part of the Foundation for Research and Technology-Hellas (IESL-FORTH). Currently, her research interests include the application of various spectroscopic techniques such as LIBS, TOF-MS and LIF in order to investigate: (a) the chemical and structural analysis of cultural heritage materials (b) the interaction of pneumatically nebulized aerosols of metalloproteins with laser pulses and (c) the functionality and degradation of high voltage field insulators.

P. Martin Mai

Professor of Earth Science and Engineering,

King Abdullah University of Science and Technology, Division of Physical Sciences and Engineering, Thuwal 23955-6900, Saudi Arabia

martin.mai@kaust.edu.sa

Dr. P. Martin Mai is Professor in the Physical Sciences and Engineering Division at KAUST. He joined KAUST in June 2009 as founding faculty member. Prior to his affiliation with KAUST, Prof. Mai worked as senior research scientist at ETH Zurich, Switzerland. Since 2003, Prof. Mai has supervised 15 Ph.D. and six M.Sc. students, guided eight postdoctoral scientists, and served on numerous student evaluation committees. Prof. Mai is an active member of the American Geophysical Union, the Seismological Society of America, the European Geophysical Union (serving as its Division President Seismology, 2015–2019), and the German Geophysical Society. He has given numerous invited talks, lectures, and conference presentations, and has published over 70 peer-reviewed articles in high-ranking international journals. Prof. Mai has (co-)organized several international workshops and conferences, and chaired many special sessions at international conferences.

Paolo Montagna

Research Scientist

Istituto di Scienze Marine (ISMAR-CNR), Via Gobetti 101, 40129 Bologna, Italy

and

Lamont-Doherty Earth Observatory, Columbia University, 61 Route 9W, Palisades, NY 10964, U.S.A.

paolo.montagna@bo.ismar.cnr.it

Dr. Paolo Montagna is a researcher at the Italian National Research Council Institute of Marine Sciences (ISMAR-CNR) in Bologna with interests in the geochemistry of biogenic carbonates for palaeoclimate reconstructions and biomineralization studies. He obtained his Ph.D. in Earth Sciences from the University of Padova (Italy) in collaboration with the Australian National University. He was awarded a three year post-doctoral Marie Curie International Outgoing Fellowship, for which he spent two years at Lamont-Doherty Earth Observatory at Columbia University and one year at the Laboratoire des Sciences du Climat et de l'Environnement in Gif-sur-Yvette. His research focuses on the development and application of geochemical proxies to address fundamental problems in palaeoceanography and palaeoclimatology. This includes the analysis of minor and trace elements, as well as stable ($^{11}B/^{10}B$) and radiogenic ($^{143}Nd/^{144}Nd$, $^{87}Sr/^{86}Sr$, $^{230}Th/U$) isotopes in shallow and deep-water coral skeletons. He has participated in 15 oceanographic missions (one as Chief Scientist and one as Co-chief Scientist) in the Mediterranean Sea, Atlantic Ocean, and the Ross Sea off Antarctica. He has authored more than 40 peer-reviewed scientific papers, 3 book chapters, more than 80 abstracts and 15 invited talks. He has served as a reviewer for the NSF (USA, 2006-2011, 2014-2015), CNRS/INSU (France, 2014), ANR (France, 2014), DFG (Germany, 2015), CNR cruise proposal (2012) and FONDECYT (Chile, 2014) as well as for numerous scientific journals.

Paul S. Breeze

Research Associate

Department of Geography, King's College, London, U.K.

paul.breeze@kcl.ac.uk;

paulstephenbreeze@gmail.com

Dr. Paul Breeze is a post-doctoral Research Associate at King's College London (KCL). Dr Breeze's research primarily focuses upon hominin dispersals in the Saharo-Arabian deserts and how climatic fluctuations affected past population dynamics in these areas. His speciality is using GIS and Remote Sensing to explore environmental change and human responses and identify and map palaeogeographic features and archaeological sites. He was formerly a landscape archaeologist in the Visual and Spatial Technologies Centre of the University of Birmingham (UK), after starting his career working as a commercial archaeological excavator. Dr Breeze received his doctorate from KCL, and MA (Practical Archaeology) and BSc (Geology and Archaeology) from the University of Birmingham. His current projects include the Palaeodeserts, Peopling the Green Sahara, and WIDER-SOMA projects (respectively ERC, Leverhulme and NERC/GCF-funded). To date he has authored or co-authored more than 20 publications.

Peter Johnson

Consultant

6016 SW Haines Street, Portland, Oregon 97219, U.S.A.

petergeo@earthlink.net

Dr. Peter Johnson has been a student of Saudi Arabian geology for over 35 years. He began work in the Arabian Shield in 1977 as an exploration geologist with Riofinex Ltd, helping to map the geology of, and explore for, VMS deposits in the southeastern Asir terrane. His later projects with Riofinex involved compilations of shield geology at scales of 1:250,000 and 1:1 million, exploration for Sn-W, and preliminary tectonic modeling of the shield concurrent with the initial introduction of the terrane concept to the region. From 1982 to 1990, he consulted for Riofinex Ltd, the U.S. Geological Survey Saudi Arabian Mission, and the Saudi Arabian Deputy Ministry for Mineral Resources, making geologic compilations and completing explanatory notes to accompany 1:250,000-scale GM maps, preparing 5-year reports of work completed in the shield, and making assessments of the mineral resource potential of parts of the shield. For the next 10 years, he was a Technical Advisor with the USGS Mission, collaborating in exploration projects for orogenic gold and VMS deposits, helping with campaigns of SHRIMP dating, and continuing a project to digitize the geology of the Arabian Shield. These activities continued between 2000 and 2007, when he was a Technical Expert with the Saudi Geological Survey. Since 2007, he has been a geological consultant based in Portland, Oregon, USA. He is the author of many internal reports on Saudi Arabian geology, more than 25 published papers, and a recent book on the geology of the Saudi Arabian Shield. He is currently a member of an IGCP project to compile a new map of the Gondwana supercontinent and is contributing to a volume for the 2016 IGC meeting in Cape Town on the mineral fields of Africa.

Robert J. Stern
Professor
Geosciences Department, University of Texas at Dallas, Richardson, TX 75080-0688, U.S.A.
rjstern@utdallas.edu,
bobdadbob@yahoo.com

Dr. Robert J. Stern is Professor of Geosciences at the University of Texas at Dallas. He and his students and colleagues have studied the uplifted continental crust around the Red Sea known as the Arabian-Nubian Shield for 40 years. These rocks formed ~880-580 Ma, and Stern's studies of them include geochemistry, geochronology and isotopic compositions of plutonic, volcanic, and metamorphic rocks; ophiolite studies; remote sensing; studies of banded iron formations, diamictites, and lower crustal xenoliths. Structural studies focus on the Najd strike-slip system in Saudi Arabia and in Egypt, which provide powerful constraints for reconstructing the Red Sea. These studies involved several MS and PhD students from the region and include aspects of the geology of Egypt, Ethiopia, Jordan, Saudi Arabia, and Sudan. He and his students and colleagues have written important regional syntheses, include the East African Orogen, Nd isotopes, ophiolites, and the lithosphere of the Arabian Plate. He has visited the SGS several times and worked with them on various projects. He is Editor-in-Chief of International Geology Review and a Fellow of the Geological Society of America and the American Geophysical Union.

Robyn Inglis
Marie Skłodowska-Curie Global Fellow, University of York, Department of Archaeology, King's Manor, York YO1 7EP, U.K.
and
Department of Environmental Sciences, Macquarie University, North Ryde, Sydney, NSW, Australia
robyn.inglis@york.ac.uk

Dr. Robyn Inglis is a Marie Skłodowska-Curie Global Fellow in the Department of Archaeology, University of York, U. K., and a Visiting Associate of the Department of Environmental Sciences, Macquarie University, Australia. Trained in geoarchaeology at the University of Cambridge and Reading, she utilises geoarchaeological techniques from the site to landscape scale (soil micromorphology, sedimentology, geomorphology) to examine human-environment interactions during the Late Pleistocene, with a particular emphasis on the emergence and global spread of modern humans in presently arid and semi-arid environments. Her research focuses mainly on the Middle East and North Africa, and most recently on the relationship between landscape evolution, lithic artefact distribution and hominin behaviour in Southwestern Saudi Arabia, and she was a Postdoctoral Research Associate on the DISPERSE project from 2011-2015. She also has strong research interests in reconstructing Quaternary sea level change, and its implications for examining the utilisation of coastlines by humans throughout history.

Ryan Gallacher
Postdoctoral Researcher in Seismology
Department of Earth and Environmental Sciences, Tulane University, New Orleans, LA, U.S.A.
rgallacher@tulane.edu

Dr. Ryan Gallacher is currently a Postdoctoral Researcher at Tulane University in New Orleans. He completed his Ph.D. from 2012–2017 at the National Oceanography Centre Southampton, University of Southampton. He used various passive seismology techniques including surface wave tomography, SKS-splitting, and receiver functions to image the mantle beneath the southern Red Sea region. The surface wave tomography study is published in Nature Communications. Prior to the Ph.D., he completed an MRes in Science of Natural Hazards at the University of Bristol. His MRes thesis on seismic structure beneath the Cameroon Volcanic Line is published in Tectonics. In 2013, he was awarded an honorable mention by GeoPrisms for his AGU presentation on mantle structure beneath Afar and the Red Sea. He completed his undergraduate degree in Geophysics with Meteorology at the University of Edinburgh in 2010.

Salem M.S. Al-Nomani

Research Associate
Center for Marine Geology, Saudi Geological Survey, Jeddah, Saudi Arabia
Nomani.sm@sgs.org.sa

Salem M. S. Al-Nomani carries out research at the Center for Marine Geology, Saudi Geological Survey, Jeddah. He obtained a B.Sc. in Marine Geology from the Faculty of Marine Science, King Abdulaziz University, Saudi Arabia in 2004, where he studied sediment transport processes in Wadi Al-Kura and the sediment dispersal pattern in Sharm Obhur. He is also a PADI's Open Water SCUBA Instrucor (OWSI). He has undertaken numerous surveys of coastal waters and the seabed along the Red Sea coast of Saudi Arabia. He has also participated in several research cruises in the Red Sea aboard the RV Urania, RV Pelagia and RV Aegaeo and at present is working on numerous projects related to the Red Sea.

Sara Ronca
Research Scientist
Dipartimento di Scienze della Terra, Sapienza Università di Roma, Italy
sara.ronca@uniroma1.it; sara.ronca@gmail.com

Dr. Sara Ronca is currently a researcher in petrology at the Department of Earth Sciences, Sapienza University, in Rome (Italy) where she is also a lecturer in petrology for the Geological Sciences B.S. degree. She graduated in Geological Sciences (B.S. and M.S.) from Perugia University in 1990, and obtained a Ph.D. in Earth Sciences in 1996. She has experience in field work, petrographic and microstructural analysis, inorganic analytical research, and petrological modelling. Her research activities have mainly addressed the mineralogy, petrology and isotope geochemistry of igneous rocks, including alkaline-carbonatitic magmatism in Brazil, late to post-Hercynian dyke magmatism of the Sardinia-Corsica domain, alkaline ring complex magmatism of Egypt, and calcalkaline magmatism of the Mediterranean area. Other fields of research focus on the sedimentology of mixed carbonate-siliciclastic deposits from the western Mediterranean area and the mineralogy and petrography of cultural heritage stone materials. She is presently involved in research projects dealing with the Cenozoic basaltic magmatism of Egypt and the Red Sea area.

Sarantis Sofianos
Assistant Professor
Department of Physics, University of Athens, University Campus Buildings Phys-5, Athens 15784, Greece
sofianos@oc.phys.uoa.gr

Dr. Sarantis Sofianos is an Assistant Professor at the University of Athens, Faculty of Physics, and coordinator of the Ocean Physics and Modeling group. After graduating from the University of Athens he obtained his Ph.D. in Meteorology and Physical Oceanography from the University of Miami in 2000. His research is focused on the dynamics of semi-enclosed seas, coastal processes and climate variability using seagoing and numerical modelling techniques, including operational oceanography. He participates in the board of the Mediterranean Oceanography Network for the Global Ocean Observing System.

Simon Kübler
Postdoctoral Researcher
Department of Earth and Environmental Sciences, Ludwig Maximilians University, Luisenstrasse 37, 80333 Munich, Germany
s.kuebler@lmu.de

Dr. Simon Kübler received a M.Sc degree in Geosciences in the Leibniz University of Hannover in 2008. He carried out his doctoral studies in geology at the Ludwig Maximilians University of Munich (LMU Munich) and received his Ph.D. in 2013. Since then he has been working as postdoctoral researcher and lecturer in geology at the Department of Earth and Environmental Sciences at LMU Munich. His research focuses on analyzing the surface expression, landscape response and environmental implications of active faults in the continental regions of East Africa, Europe and North America. He is furthermore interested in understanding the physical constraints of dynamic landscapes, earthquake processes and the interplay between tectonics, soil edaphics and human and animal nutrition by combining conventional geological mapping approaches with remote sensing analysis of high resolution terrain and optical satellite data and lab analysis of rocks, soils and vegetation. He has published over twenty papers in international peer-reviewed journals covering the fields of palaeoseismology, tectonic geomorphology, soil science and palaeoanthropology.

Spyros Sergiou
Researsch Associate
Laboratory of Marine Geology and Physical Oceanography, Department of Geology, University of Patras, Greece
sergiou@upatras.gr

Spyros Sergiou is a marine geologist/sedimentologist with a M.Sc. who is currently studying the Late Quaternary paleogeographic and paleoceanographic evolution of the southern Red Sea toward his Ph.D. His main research interests include marine sedimentology, paleoceanography, sea level changes, coastal geology and sedimentary geochemistry-mineralogy. He has participated in many research programs, with experience in sediment sampling, and sedimentological, micropaleontological and geochemical analyses. He has been involved in the "IODP Expedition 381- Corinth Active Rift Development" program as a sedimentologist, participating in both offshore and onshore operations.

Timothy Kinnaird

Research Associate
School of Earth and Environmental Sciences, University of St. Andrews, U.K.
and
School of Biological and Environmental Sciences, University of Stirling, U.K.
tk17@st-andrews.ac.uk

Dr. Timothy Kinnaird directs the luminescence laboratories at the School of Earth and Environmental Sciences at the University of St. Andrews. He completed a Ph.D. in Geology at the University of Edinburgh in 2007, where his doctoral work focused on the tectonic and sedimentary responses to incipient continental collision in the easternmost Mediterranean (Cyprus) as archived in the Mesozoic to Recent cover sediments of the island. He is an Honorary Research Fellow at the University of St. Andrews and at the School of Biological and Environmental Sciences at the University of Stirling. Previously he was a Research Associate in Environmental Physics at the Scottish Universities Environmental Research Centre (SUERC), when he contributed to the monograph 'Optically Stimulated Luminescence dating as a geochronological tool for the dating of Late Quaternary sediments in the Red Sea region'. Prior to this, he worked at CASP (University of Cambridge), as part of the East Greenland Team, with a research focus on the Cenozoic evolution of the East Greenland margin. His research interests span the disciplines of Earth Science and Archaeology.

Webster Mohriak
Associate Professor
UERJ–State University of Rio de Janeiro, Rio de Janeiro, Brazil
webmohr@gmail.com

Dr. Webster Mohriak has been working in the oil and gas industry for the last 30 years, conducting several basin analysis projects in the eastern Brazilian and West African continental margins. After obtaining a Ph.D. in geology from Oxford University, studying the tectonic development of the Campos Basin, offshore Brazil, he coordinated several regional basin analysis projects at Petrobras. After retiring in 2011 he became an Associate Professor at the Rio de Janeiro State University and is also a consultant, focusing on basin analysis, petroleum geology and petrophysics of pre-salt carbonates.

He has published more than 100 scientific papers and was the main editor for the book "Atlantic Rifts and Continental Margins" published by American Geophysical Union in 2000. He is one the editors of the Geological Society of London book "Conjugate Divergent Margins" published in 2013. In 2014 he was invited to participate in the AAPG Distinguished Lecture program, presenting the "Birth and development of continental margin basins: analogies from the South Atlantic, North Atlantic and the Red Sea".

William (Bill) Bosworth
Senior Geological Advisor
Apache Egypt Companies, 11 Street 281, New Maadi, Cairo, Egypt
bill.bosworth@apachecorp.com;
bill.bosworth@apacheegypt.com

Dr. William (Bill) Bosworth is a Senior Geological Advisor for Apache Egypt Companies based in Cairo, Egypt. He has over 30 years petroleum exploration experience, mostly in North and East Africa and the Near and Middle East. His education included Bachelor and Master of Science degrees at Rensselaer Polytechnic Institute and a Ph.D. in structural geology at the State University of New York at Albany. Prior to joining the oil industry, he taught at Colgate University and worked on research projects in the northeastern United States and Canada. Later research interests focused on continental rift systems, intra-plate stress fields, and the geodynamics of North Africa. He is an Associate Editor of the Journal of African Earth Sciences and of Marine and Petroleum Geology. Bill has been an active member of the American Association of Petroleum Geologists and is presently president-elect for its Africa Region. In 2014 he received the AAPG Distinguished Service Award for his work in support of Egyptian university students. He is a Board Member of the Egyptian Petroleum Exploration Society and frequently leads field trips to the Red Sea and Gulf of Suez.

Zekâi Şen
Department of Applied Geology, Saudi Geological Survey, Jeddah, Saudi Arabia
and
Turkish Water Foundation, Libadiye Caddesi Doğanay Sokak No:10 Kat: 4, Üsküdar, İstanbul, Turkey
zekaisensu@gmail.com

Dr. Zekâi Şen obtained B.Sc. and M.Sc. degrees from the Department of Reinforced Concrete, Civil Engineering Faculty, Technical University of İstanbul, in 1971. Further post-graduate studies were carried out at the Imperial College of Science and Technology, University of London, where he was granted a Diploma of Imperial College (D.I.C.), an M.Sc. in engineering hydrology in 1972 and a Ph.D. in stochastic hydrology in 1974. He has worked in England, Norway, Saudi Arabia and Turkey in a number of different faculties. His main interests are hydrology, water resources, hydrogeology, hydrometeorology, hydraulics, science philosophy and history. He has published numerous papers in international journals on various topics and has written many books in Turkish, one book in Arabic and several in English. He has supervised many students for M.Sc. and Ph.D. degrees. He has received several national awards (Science Encouragement Prize for Young Scientists in 1978, and Science Prize in 1993, both from the Scientific and Technological Research Council of Turkey), the administration and Science Prize from the King Abdulaziz University, Jeddah, and international science prizes, the most recent one given for his contribution as part of the IPCC team from 2002-2007 that was concerned with Climate Change Impact on Fresh Water Resources, resulting in the award of the Nobel Peace Prize. He also has a chair at Prince Sultan Research Center, King Saud University, Riyadh. He worked with the Saudi Geological Survey for three years as a full time consultant on various water, climate change and groundwater issues in Saudi Arabia. He was a consultant with ARAMCO for hydrogeology and groundwater training program development and has had numerous other national and international consulting projects on water resources, climate change, renewable energy alternatives, earthquake engineering, and the earth sciences. He is currently working at the Civil Engineering Faculty, Technical University of Istanbul and is also the president of Turkish Water Foundation.

Zheng Tang
Ph.D. student of Earth Science and Engineering,
King Abdullah University of Science and Technology, Division of Physical Sciences and Engineering, Thuwal 23955-6900, Saudi Arabia
Zheng.Tang@kaust.edu.sa

Dr. Zheng Tang is a Ph.D. from the Department of Earth Sciences and Engineering at KAUST. Before joining KAUST in 2012, he obtained his master's degree in geology (2012) from the University of Illinois at Urbana-Champaign in the U.S. He received his bachelor's degree in geophysics in 2010 from the University of Science and Technology of China. His research interests involve exploring the crustal and upper mantle structure using geophysical datasets, and his current main research project is studying the crustal and uppermost mantle shear-wave velocity structure beneath Saudi Arabia utilizing the joint inversion of P-wave receiver functions and Rayleigh wave group velocities.

Zohair A. Nawab
President
Saudi Geological Survey
Jeddah, Saudi Arabia
nawab.za@sgs.org.sa

Dr. Zohair A. Nawab is a former President of the Saudi Geological Survey, Kingdom of Saudi Arabia. He has a Ph.D. from the University of Western Ontario in London, Ontario, Canada as well as a Diploma in Applied Geology from the South Dakota School of Mines and Technology in Rapid City, South Dakota, USA. From 1978 to 1985 he was the Assistant Secretary General for the Saudi-Sudanese Red Sea Commission, and was a member of the Saudi Arabian Delegation to the UN Law of the Sea Conference and to the UN Sea-Bed Authority for many years. In 2001 he became the Deputy Minister for Mineral Resources, and became a Senior Advisor in the Ministry of Petroleum and Mineral Resources in 2003. He has also served as a board member of the Saudi Company for Precious Metals and the Saudi Arabian Mining Company (Ma'aden), the Saudi Geological Survey, as well as the Steering Committee for Railway Expansion Projects. He is the co-author of three books and has published many papers.

Reviewers' List

Abbas Mansour
Faculty of Science, Geology Department, South Valley University, Qena, Egypt
aman@svu.edu.eg
Keywords: sedimentology, environmental geology, marine geology, Quaternary geology

Abu Mohamed El-Rus
Geology Department,
Assiut University, Egypt
aliabuelrus@yahoo.com
Keywords: geochemistry, igneous petrology, metamorphic rocks

Ahmed Hassan Ahmed
Faculty of Earth Sciences,
King Abdulaziz University,
Jeddah, Saudi Arabia
ahmh2jp@yahoo.com
Keywords: economic geology, mantle petrology, geochemistry, mineralogy

Ahmed Rushdi
ETAL, 2951 SE Midvale Dr.
Corvallis, OR 97333, U.S.A.
and
Department of Earth and Environmental Sciences, Faculty of Sciences, Sana'a University, Sana'a, Yemen
aimrushdi@gmail.com
Keywords: marine chemistry, oceanography, environmental chemistry, biogeochemistry

Albert Matter
Institute of Geological Sciences, University of Bern, Bern, Switzerland
albert.matter@geo.unibe.ch
Keywords: environment, sedimentology, diagenesis, geochemistry

Alessio Rovere
MARUM, University of Bremen and Leibniz ZMT, Germany
arovere@marum.de
Keywords: sea level changes, paleoclimate, coastal geomorphology

Alim Sidiqqui
Saudi Geological Survey, Jeddah, Saudi Arabia
Sidiqqui.AA@sgs.org.sa
Keywords: petrology, ore microscopy, Scanning Electron Microscopy (SEM)

Amy Prendergast
School of Geography, University of Melbourne, Australia
amy.prendergast@cantab.net
Keywords: molluscs, geochemistry, palaeoenvironments, human-environment interaction

Andre C. Colonese
Department of Archaeology, BioArCh, University of York, U.K
andre.colonese@york.ac.uk
Keywords: coastal archaeology, environmental archaeology, stable isotopes

Andrea Argnani
ISMAR-CNR,
Institute of Marine Sciences,
via Gobetti 101, 40129, Bologna, Italy
andrea.argnani@ismar.cnr.it
Keywords: active tectonics, kinematic reconstructions, marine geology, submarine landslides

Anil K. Gupta
Department of Geology and Geophysics
Indian Institute of Technology
Kharagpur, India
anilg@gg.iitkgp.ernet.in
Keywords: Indian Ocean, Indian monsoon, Quaternary, extreme events

Antonio Caruso
Dipartimento di Scienze della Terra e del Mare,
Università degli Studi
Italy

N. M. A. Rasul and I. C. F. Stewart (eds.), *Geological Setting, Palaeoenvironment and Archaeology of the Red Sea*,
https://doi.org/10.1007/978-3-319-99408-6

antonio.caruso@unipa.it
Keywords: foraminifera, paleoclimatology, paleoecology, paleoceanography

Anthony Sinclair
Department of Archaeology, Classics and Egyptology, University of Liverpool, 12–14 Abercromby Square, University of Liverpool, L69 7WZ, U.K.
a.g.m.sinclair@liverpool.ac.uk
Keywords: palaeolithic archaeology, lithic technology, landscape archaeology, archaeological theory

Axel Ehrhardt
Marine Seismic Group, Bundesanstalt für Geowissenschaften und Rohstoffe (BGR), Federal Institute for Geosciences and Natural Resources, Germany
Axel.Ehrhardt@bgr.de
Keywords: multichannel seismics, northern Red Sea, eastern Mediterranean, Arctic

Ayman G. Awadallah
Faculty of Engineering, Fayoum University, Fayoum, Egypt
aga00@fayoum.edu.eg
Keywords: Statistical hydrology, water resources

Bernard LeGall
UMR/CNRS Géosciences Océan European Institute for Marine Studies, Plouzané, France
blegall@univ-brest.fr
Keywords: structural geology, strain/magmatism field investigations, East African rift

Christian Timm
Department of Marine Geoscience, GNS Science, Lower Hutt, New Zealand
c.timm@gns.cri.nz
Keywords: marine geology, petrology and volcanology

Christel Tiberi
CNRS researcher, Geosciences Montpellier, Montpellier, France
christel.tiberi@umontpellier.fr
Keywords: seismology, gravimetry, lithospheric rifting processes

Christophe Hémond
Laboratoire Géosciences Océan Université de Brest et CNRS Plouzané, France
chhemond@univ-brest.fr
Keywords: mantle, geochemistry, ocean ridge, hot spot, isotopes

Christopher I. Burbidge
Office of Radiation Protection & Environmental Monitoring, Environmental Protection Agency, Dublin, Ireland
c.burbidge@epa.ie
Keywords: radiation dosimetry, environmental radioactivity, geochronology, archaeometry

Chuck DeMets
Department of Geoscience, University of Wisconsin-Madison, Madison, Wisconsin, U.S.A.
chuck@geology.wisc.edu
Keywords: plate tectonics, earthquake cycle

Claudio Natali
Department of Physics and Earth Sciences, University of Ferrara, Italy
ntlcld@unife.it
Keywords: geochemistry, petrology, geodynamics

Cynthia Ebinger
Marshall-Heape Chair and Basin Research Editor-in-Chief
Earth and Environmental Sciences
Tulane University
New Orleans, LA,U.S.A.
cebinger@tulane.edu
Keywords: crustal dynamics, rifting, fault-magma interactions, seismicity

Edoardo Dallanave
Ludwig Maximilians University, Muenchen, Germany
dallanave@geophysik.uni-muenchen.de
Keywords: paleomagnetism

Emmanuel Masini
Total EP, R&D department, Frontier Exploration program U.K.
emmanuel.masini@total.com
Keywords: geodynamics, structural geology, petroleum geology

Eric Hellebrand
Department of Geology and Geophysics
University of Hawaii, Honolulu, U.S.A.
ericwgh@hawaii.edu
Keywords: mantle petrology and geochemistry

Eugenio Fazio
University of Catania, Department of Biological, Geological and Environmental Sciences - Earth Sciences Section, Catania, Italy
efazio@unict.it
Keywords: metamorphic petrology, structural geology, microtectonics

Fabio Manuella
Department of Biological, Geological and Environmental Sciences - Earth Sciences Section, University of Catania, Italy
fabiocarmelomanuella@gmail.com
Keywords: serpentinization, hydrocarbons, diamonds, xenoliths

Fabrizio Antonioli
ENEA - Casaccia, Laboratorio Modellistica Climatica e Impatti, Rome, Italy
fabrizio.antonioli@enea.it
Keywords: Sea level changes, coastal geomorphology and sea flooding risk, coastal tectonics, coastal karst evolution

Geoff N. Bailey
University of York, Department of Archaeology, King's Manor, York YO1 7EP, U.K.
and
Flinders University, College of Humanities, Arts and Social Sciences, GPO Box 2100, Adelaide, SA 5001, Australia
geoff.bailey@york.ac.uk
Keywords: coastal prehistory, bioarchaeology, Submerged landscapes, Underwater archaeology

Geraint W. Hughes
Natural History Museum; King Fahd University of Petroleum and Minerals, Saudi Arabia - Applied Microfacies Limited
wynapgwilym@gmail.com
Keywords: microfacies, palaeoenvironments, foraminifera

Geraint Owen
Department of Geography
College of Science
Swansea University, U.K .
g.owen@swansea.ac.uk
Keywords: clastic sedimentology, soft-sediment deformation, paleoenvironments, geology of South Wales

Giacomo Corti
National Research Council of Italy, Institute of Geosciences and Earth Resources, Italy
giacomo.corti@igg.cnr.it
Keywords: tectonics, continental rifting, modelling

Gianreto Manatschal
IPGS_EOST
Université de Strasbourg/CNRS
Strasbourg, France
manat@unistra.fr
Keywords: formation and reactivation of rifted margins, Alps, fossil and present-day distal rifted margins

Giovanna Rizzo
Department of Sciences
Director of University Center for Students' Orientation
University of Basilicata
Campus Macchia Romana, Potenza, Italy
giovanna.rizzo@unibas.it
Keywords: ophiolitic sequences, environment, zircons, petrology

Gregory Henkes
Department of Geosciences, Stony Brook University, Stony Brook, NY, U.S.A.
gregory.henkes@stonybrook.edu
Keywords: stable isotope geochemistry, carbonate geology, paleoclimatology, biogeochemistry

Hamdy Hamed Abd El Naby
Nuclear Materials Authority, Cairo, Egypt
hhabdel@yahoo.com
Keywords: geochemistry, mineralogy, igneous and metamorphic petrology

Hassan A. Eliwa
Geology Department, Faculty of Science
Menoufia University, Shebien El Kom, Egypt
eliwa98@yahoo.com
Keywords: Geochemistry, geochronology, regional geology

Ian Davison
Earthmoves Ltd.
Camberley, Surrey, U.K.
and
Visiting Professor, Department of Earth Sciences, Royal Holloway, University of London, Egham, Surrey
i.davison@earthmoves.co.uk
Keywords: frontier exploration, salt tectonics

Ikram Osmani
Faarnad Geological Consulting Inc., Burnaby, BC, Canada
ikeosmani50@gmail.com
Keywords: mineral exploration, base and precious metals, rare earth/rare metals, geophysics

Ingrid Ward
Archaeology, Flinders University, Adelaide, Australia
and
Adjunct in the School of Social Sciences at the University of Western Australia, Australia
ingrid.ward@uwa.edu.au
Keywords: geoarchaeology, prehistory, maritime archaeology

James R. Cochran
Lamont Research Professor
Lamont-Doherty Earth Observatory of Columbia University, Palisades, NY, U.S.A.
Palisades NY 10964
jrc@ldeo.columbia.edu
Keywords: marine geophysics, gravity modeling, continental rifting

Janis Thal
University of Bremen/MARUM, Bremen, Germany
ja_th@uni-bremen.de
Keywords: marine geology, geological seafloor mapping, subaqueous volcanism, GIS

Javier Escartín
Equipe de Géosciences Marines. Institut de Physique du Globe de Paris, France
escartin@ipgp.fr
Keywords: marine geology, marine geophysics
tectonics, rock mechanics

Jean Malan
Senior Geoscientist,
Getech, Leeds, U.K.
jean.malan@getech.com
Keywords: petroleum geology, sedimentology, exploration

Jelmer W. Eerkens
Department of Anthropology
University of California at Davis
Davis, C.A. U.S.A.
jweerkens@ucdavis.edu
Keywords: stable isotope analysis, archaeometry,

subsistence studies, seasonality studies

Joan Gardner
Naval Research Laboratory, U.S.A.
joan.gardner@nrl.navy.mil
Keywords: geology, geophysics, oceanography, acoustics

Joël Ruch
Stanislas Torrents, Marseille, France
ruchjo.ch@gmail.com
Keywords: volcano-tectonics, rift zone, structural geology, satellite imagery, analogue experiments

Joel Q.G. Spencer
Department of Geology,
Thompson Hall,
Kansas State University,
Manhattan,KS,U.S.A.
joelspen@ksu.edu
Keywords: optically stimulated luminescence and thermoluminescence dating, Late Quaternary and archaeological science

John P. Coakley
Hamilton, ON
and
National Water Research Institute (NWRI), Environment Canada (Retired), Ontario, Canada
jcoakley@cogeco.ca
Keywords: sediment distribution, coastal and nearshore zone, contaminant dynamics, sediment tracers

Kamal A. Ali
Faculty of Earth Sciences,
King Abdulaziz University,
Jeddah, Saudi Arabia
alik6588@yahoo.com
Keywords: geochemistry, igneous petrology, isotope geology

Karen Hardy
ICREA, Pg. Lluís Companys
Barcelona, Catalonia, Spain
and
Departament de Prehistòria,
Facultat de Filosofia i Lletres,
Universitat Autònoma de
Barcelona, Bellaterra,Spain
karhardy2@gmail.com
Keywords: palaeolithic, Mesolithic, shell middens, plants, diet

Knut Bretzke
University of Tübingen,
Department of Early Prehistory and Quaternary Ecology, Germany
knut.bretzke@uni-tuebingen.de
Keywords: Stone Age archaeology, Middle East, field archaeologist, archaeoinformatics

Luc Lavier
Department of Geosciences,
Jackson School of Geosciences,
University of Texas at Austin,
TX, U.S.A.
and
Institute for Geophysics, Jackson School of Geosciences,
University of Texas at Austin,
TX, U.S.A.
luc@jsg.utexas.edu
Keywords: geodynamics, plate boundaries, computational geosciences, lithospheric deformation

Lucien Montaggioni
CEREGE (Centre Européen de Recherche et d'EnseignementenGéosciences de l'Environnement), UMR (Research Unit) - CNRS, Aix-Marseille University, Marseilles, France
montaggioni@cerege.fr
Keywords: reefs, carbonates, Quaternary, Neogene

Luis M. Pinheiro
Department of Geosciences and Associated Laboratory CESAM
University of Aveiro, 3810-193-Aveiro, Portugal
lmp@ua.pt
Keywords: marine geophysics, seismic methods, tectonics, cold seeps

Marco Oliverio
Department of Biology and Biotechnologies "Charles Darwin", Sapienza University of Rome, Italy
marco.oliverio@uniroma1.it
Keywords: mollusca, evolutionary, biology, systematics, ecology

Martin Zuschin
University of Vienna, Department of Palaeontology, Austria
martin.zuschin@univie.ac.at
Keywords: conservation paleobiology, ecology, taphonomy, coral reefs

Michael Hudec
Bureau of Economic Geology,
The University of Texas at Austin, U.S.A.
michael.hudec@beg.utexas.edu
Keywords: salt tectonics

Michele Morsilli
Department of Physics and Earth Science, Ferrara University, Italy
mrh@unife.it
Keywords: carbonate sedimentology, facies analysis, geological mapping, GIS

Mick O'Leary
The School of Molecular and Life Sciences, Curtin University, Kent Street, Bentley, WA 6102, Australia
mick.oleary@curtin.edu.au
Keywords: paleosea level and climate reconstructions
coastal and coral reef geomorphology
submerged landscapes,
marine geoscience

Miquel Canals
CRG Marine Geosciences
Department of Earth and Ocean Dynamics
Faculty of Earth Sciences
University of Barcelona, Spain
miquelcanals@ub.edu
Keywords: seafloor mapping, marine sedimentary processes and environments, global change and the ocean, marine anthropogenic impacts

Mohamed Abdel Khalek El Sayed
National Institute of Oceanography and Fisheries, Academy of Scientific Research and Technology,
Alexandria, Egypt
mohelsayed@hotmail.com
Keywords: biogeochemical cycle, natural elements, trace elements, marine environment

Mohamed El Tokhi
Geology Department,
United Arab Emirates University, U.A.E.
meltokhi@uaeu.ac.ae
Keywords: mineral chemistry, geochemistry of basic and ultrabasic rocks, fluid inclusion and marine pollution

Naila M.O. Babiker
Seismological Research Unit, Remote Sensing and Seismology Authority, National Center for Research, Khartoum Sudan
nailamoh@yahoo.com
Keywords: geophysics, seismology, seismic hazard studies, earthquake data processing

Natalie Accardo
Postdoctoral Researcher,
Department of Geosciences,
Pennsylvania state University, PA, U.S.A.
nza216@psu.edu
Keywords: seismology, tectonophysics

Neil Munro
Institute of Climate and Society, Department of Land Resources Management - Environmental Protection (LaRMEP), Mekelle University, Tigray, Ethiopia
neil.munro@dial.pipex.com
Keywords: land resources, geo-archaeology, sand dune stabilization

Nick Raterman
Chevron Europe Eurasia and Middle East Exploration and Production Company
nsraterman@chevron.com
Keywords: structural geologist, geomorphology, geophysics, computational geoscience

Nuno Bicho

ICArEHB - Interdisciplinary Center for Archaeology and the Evolution of Human Behaviour
Universidade do Algarve
Campus de Gambelas
Portugal
nbicho@ualg.pt,
nunofbicho@gmail.com
Keywords: archaeology, African Stone Age, hunter-gatherers, coastal archaeology

Nuzhat H. Hashimi
NIO, Goa, India and King Abdulaziz University, Saudi Arabia
hashimi_nh@hotmail.com
Keywords: marine sedimentology, geophysics, paleoclimatology, mineralogy

Pablo Arias
Universidad de Cantabria (Cantabria International Institute for Prehistoric Research), Spain
pablo.arias@unican.es
Keywords: Neolithization, Late Paleolithic, prehistoric ritual and symbolism, archaeology of death

Paraskevi Nomikou
Department of Geology and Geoenvironment,
National and Kapodistrian University of Athens, Greece
evinom@geol.uoa.gr
Keywords: submarine volcanism, seabed morphology, tectonics

Parth Chauhan
Department of Humanities and Social Sciences, Indian Institute of Science Education and Research Mohali, India
parth73@gmail.com
Keywords: Old World, paleoanthropology, paleolithic archaeology

Peter Betts
School of Earth,
Atmosphere and Environment,
Monash University,
Clayton, Victoria 3800,
Australia
peter.betts@monash.edu
Keywords: tectonics, aeromagnetics, geodynamics, structural geology

Peter Feldens
Leibniz Institute for Baltic Sea Research Warnemünde, Germany
peter.feldens@io-warnemuende.de
Keywords: geomorphology, habitat mapping

Philip J. Ball
Saudi Aramco, Saudi Arabia, and
Honorary Senior Research Fellow, Keele University, U.K.
philipjball@hotmail.com
Keywords: tectonics, structural geology, geodynamics, basin analysis

Philippe Huchon
Sorbonne Université,
Institut of Earth Sciences, Paris, France
philippe.huchon@upmc.fr
Keywords: tectonics, plate kinematics, marine geology and geophysics

Ramadan H. Abu-Zied
Faculty of Marine Sciences, King Abdulaziz University, Jeddah, Saudi Arabia
rabuzied@kau.edu.sa
Keywords: paleoceanography, micropaleontology, isotopic geochemistry

Remy Crassard
CNRS - CEFAS
French Center for Archaeology and Social Sciences
Kuwait City, Kuwait
remy.crassard@cnrs.fr
Keywords: Paleolithic, Neolithic, Arabian Peninsula, lithic technology

Robert Carter
Arabian and Middle Eastern Archaeology
UCL Qatar

robert.carter@ucl.ac.uk
Keywords: Arabian archaeology, maritime cultures, urbanism, maritime exchange

Robert Duncan
College of Earth, Ocean, and Atmospheric Sciences
Oregon State University
U.S.A.
rduncan@coas.oregonstate.edu
Keywords: plate tectonic controls of volcanism, hotspots and mantle plumes, mantle geochemistry

Salah Y. El Hadidy
Technical Advisor, Saudi Geological Survey, Jeddah, Saudi Arabia
el_hadidy.sy@sgs.org.sa
Keywords: seismology, seismicity and seismotectonic, phase identification

Sally C. Reynolds
Institute for Studies in Landscape and Human Evolution and Department of Archaeology, Anthropology and Forensic Science, Faculty of Science and Technology, Bournemouth University, U.K.
sreynolds@bournemouth.ac.uk
Keywords: African palaeoecology, mammalian evolution, hominin landscape use, climate change and human evolution

Samantha Hansen
Department of Geological Sciences, University of Alabama, U.S.A.
shansen@ua.edu
Keywords: seismology, tectonophysics

Saowanee Wijitkosum
Environmental Research Institute, Chulalongkorn University, Thailand
w.m.saowanee@gmail.com
Keywords: area-based environmental management, environmental impact assessment of land use, spatial development and planning, biochar for food security and carbon sequestration

Sara Lafuerza
Institut de Sciences de la Terre de Paris, ISTeP, Université Pierre et Marie Curie, Sorbonne Université, Paris, France
sara.lafuerza@upmc.fr
Keywords: submarine mass movements, geohazards, geotechnics

Silvia Ceramicola
IstitutoNazionale di Oceanografia e di Geofisica Sperimentale, OGS, Trieste, Italy
sceramicola@inogs.it
Keywords: marine geology, geophysics, geohazards assessment, Mediterranean Sea

Simon Holdaway
School of Social Sciences, University of Auckland, Auckland, New Zealand
sj.holdaway@auckland.ac.nz
Keywords: coastal archaeology, shell mounds, geoarchaeology

Sönke Reiche
Institute for Applied Geophysics and Geothermal Energy, RWTH Aachen University, Aachen, Germany
sreiche@eonerc.rwth-aachen.de
Keywords: marine geophysics, seismic reflection, salt tectonics

Srikumar Roy
Irish Centre for Research in Applied Geosciences, School of Earth Sciences, University College Dublin, Ireland
srikumar.iitkgp@gmail.com
Keywords: seabed-fluid-flow, gas-hydrates, CCS, igneous sills

Stefan Feuser
Institute of Classics, Christian-Albrechts-University Kiel, Germany
feuser@klassarch.uni-kiel.de
Keywords: classical archaeology, geoarchaeology, port cities, Mediterranean Sea

Stephen Franks
RockFluid Systems, Inc.
Stonebridge Geological Associates
McKinney, TX, U.S.A.
franks_steve@hotmail.com
Keywords: petrography, geochemistry, sedimentology, field geology

Sung-Joon Chang
Division of Geology and Geophysics, Kangwon National University, Chuncheon, South Korea
sjchang@kangwon.ac.kr
Keywords: seismology, tectonics

Sylvie Leroy
Director Istep, Sorbonne Université, Paris, France
sylvie.leroy@upmc.fr
Keywords: marine geosciences, continental margins, oceanic spreading initiation

Thomas Hillard
Department of Ancient History University of Sydney, Australia
thomas.hillard@mq.edu.au
Keywords: harbours, landscape archaeology, Roman politics

Victoria Pease
Department of Geological Sciences, Stockholm University, Sweden
vicky.pease@geo.su.se
Keywords: tectonics, magmatism

Vittorio Scribano
University of Catania, Department of Biological, Geological and Environmental Sciences, Italy
scribano@unict.it
Keywords: petrology, volcanology, mantle xenoliths, oceanic crust

Walter Mooney
Earthquake Science Center, U.S. Geological Survey, Menlo Park, California, U.S.A.
mooney@usgs.gov
Keywords: seismology, geophysics, tectonics, earthquakes and tsunamis

Wendy Nelson
Department of Physics, Astronomy, and Geosciences, Towson University, Towson, MD, U.S.A
wrnelson@towson.edu
Keywords: igneous petrology, high-T geochemistry, mantle dynamics

William Bosworth
Senior Geological Advisor
Apache Egypt Companies, 11 Street 281, New Maadi, Cairo, Egypt
bill.bosworth@apachecorp.com, bill.bosworth@apacheegypt.com
Keywords: petroleum geology, tectonics structural geology, basin analysis

Yamandú H. Hilbert
CNRS, Archéorient, Environnements et sociétés de l'Orient ancient, CNRS/Université Lumière Lyon 2
Maison de l'Orient et de la Méditerranée
7 Rue Raulin, France
yamandu.hilbert@CNRS
Keywords: human behavioural evolution, landscape archaeology, stone tool technology, field archaeology

YavuzSelim Güçlü
Istanbul Technical University, Istanbul, Turkey
gucluya@itu.edu.tr
Keywords: hydrology, fuzzy logic, solar radiation, modeling

Ziying Li
President, Beijing Research Institute of Uranium Geology, China
zyli9818@126.com
Keywords: radioactive geology, mineral deposits, geochemistry

MIX
Papier aus verantwortungsvollen Quellen
Paper from responsible sources
FSC® C105338

If you have any concerns about our products,
you can contact us on
ProductSafety@springernature.com

In case Publisher is established outside the EU,
the EU authorized representative is:
Springer Nature Customer Service Center GmbH
Europaplatz 3, 69115 Heidelberg, Germany

Printed by Libri Plureos GmbH
in Hamburg, Germany